Handbook of Vehicle Design Analysis

To Paul, Peter and Denise

Handbook of Vehicle Design Analysis

by

John Fenton

Mechanical Engineering Publications Limited
London and Bury St Edmunds, UK

First published 1996

ISBN 0 85298 963 6

A CIP catalogue record for this book is available from the British Library.

Printed in Great Britain by Antony Rowe Ltd, Chippenham, Wiltshire

CONTENTS

PART 4: SPECIALIST VEHICLE DESIGN

PART 5: VEHICLE SYSTEMS

Foreword

The publication of this volume is coincident with the normal retirement date of the author, after a working lifetime in automotive engineering communication. Following early industrial appointments with Vauxhall Motors and ERA, as a draughtsman and development engineer in car suspensions, he developed an interest for vehicle structures at Unilever Ltd where he was involved in the design of special purpose commercial vehicles in the early days of refrigerated transport and bulk carriage of food products. He was sponsored by the company in the first year of the postgraduate automotive engineering course at Cranfield where his thesis involved the application of aeronautical structural techniques in sandwich panels and matrix structural analysis to an integral-construction van.

This early taste for applying analysis techniques in vehicle design was to be followed by a 33 year span with the technical magazines, *Automotive Design Engineering* and *Automotive Engineer*, the latter an MEP/IMechE title which he edited until his retirement date. In a five year interval between his working for each title he helped to set up the design teaching at the then newly formed vehicle engineering degree course at Hatfield Polytechnic as well as working in technical publicity for Leyland Truck and Bus. Both the above magazines have been centrally involved with the dissemination of developing analytical techniques, alongside reporting on new automotive system developments for the vehicle industry.

This volume covers both system developments and analytical techniques, much of the latter being condensed extracts from particularly significant published works during this time-span. Some of the earlier works involved analysis techniques developed from empirical experimentation and predate the widespread introduction of SI units. In such cases conversion to SI units, and reconstruction of the related graphical illustrations, was felt to be inappropriate. Much of the analytical work is centred around impending legislation and, where this is quoted in the text, it is for illustration only and it is, of course, important to examine the latest statutes in the locality concerned. The list of references given at the end of each specialist chapter is a key element of the publication, providing where possible a link to the original publication source. Where the original publication is no longer in print many of the sources can be found in such libraries as those of the Institution of Mechanical Engineers, London, or the Motor Industry Research Association, in Nuneaton, as well as the British Library

PREFACE

The widespread availability of computing power to designers and engineers has created a renewed interest in predictive techniques in automotive design analysis. Analytical methods that were once considered too laborious and time-consuming can now be handled with considerable speed by programming in design calculations for solution over a wide range of input variables.

Weight reduction continues to drive many areas of automotive design and development. However efficient the means of converting energy of the fuel into tractive force at the wheel-to-ground contact area, the fuel itself has a limited energy capacity. So the 'potential' energy of the fuel, if equated to the kinetic energy (again proportional to weight) of a vehicle, moving at a given speed, reveals a finite energy value which no fuel economizing device can ever exceed.

The case for reducing fuel consumption by reducing mechanical component and equipment weight, in turn reducing translation and rotational inertial drag on acceleration, as well as rolling resistance and gradient resistance, has been well argued. The configuration and assembly of components and structural elements is also of importance from a weight reduction standpoint.

As excess weight at one part of a vehicle has to be supported by extra weight added elsewhere, conversely a weight reduction in one part may result in a meaningful saving in others. Also the fitting of a lighter system, such as a smaller engine, in one place could result in a mechanical change in another, smaller brakes, for example, so there is a chain reaction. A lighter body structure may result in smaller wheels, and so on. The overall system design is thus of major importance.

A quantitative measure of the effect of weight saving on fuel economy is indicative in studies carried out to investigate automatic gearboxes by a UK manufacturer. These showed that the proportions of fuel used to accelerate the mass of a typical European car, when driven according to the USA Environmental Protection Agency 'Highway', 'Urban', and 'EEC 15' cycles, were 18, 38, and 25 per cent, respectively. Thus it was deduced that a one-third reduction in vehicle weight may be expected to yield some 6-12 per cent fuel economy. This comparatively modest yield meant that new component development for lighter weight was normally accompanied by design for lower cost in material and labour.

Vehicle weight also has a close relationship with body size so compact vehicles reap rich rewards in fuel savings, not only in weight saving but with proportionally reduced air drag as cross-section area is reduced. Now engineers are looking at the body structures themselves to see if the weight in them is efficiently distributed to react applied loads with the minimum of redundancy, to reduce 'skin' drag by smoothing surfaces and to provide passenger protection in accidents.

In this book the analytical techniques are summarized which are necessary to design the structure, as a preliminary stage, prior to finite-element analysis. Succeeding chapters also look at the fundamentals of vehicle systems design as well as the more analytical techniques of performance calculation. The analytical processes within the text will be of particular value to students in course levels between Technical T-5 and second year degree in vehicle engineering. Parts of the book will further be of value in the final year degree and post graduate studies in vehicle engineering.

Until recent years students and apprentices in automotive engineering still entered the industry with a background of mechanical engineering training in such subjects as thermodynamics, theory of machines, hydraulics, mathematics, mechanical design, and strength of materials. The teaching of theory of structures had been more common in courses of aeronautical engineering, civil (structural) engineering, and naval architecture. Within the vehicle manufacturing industry other trends are throwing a new emphasis on body structural design: these intensify the need for wider training in structural theory for body designers. At the same time there are strong commercial pressures to shorten the lead times of new model introductions and to meet the projected standards of structural safety, emission control, ride/handling refinement and reduced fuel consumption, all without resort to extensive road proving programmes.

The increasing application of analytical techniques to the solution of these problems demands a greater knowledge of automotive-engineering theory from those about to enter vehicle design/construction, and from those already employed in vehicle body design and manufacture. The third category of reader, the specialist vehicle builder, will also find much to attract him in the techniques applied both in his own and other related specialities.

The book is an introduction to vehicle design analysis in the sense of providing readers with an insight into analytical methods given in a wide variety of published sources such as technical journals, conference papers, and proceedings of engineering institutions, for which a comprehensive list of references is provided. The analyses are therefore not necessarily fully developed or rigorously evaluated; recourse to the original references is necessary particularly in order to understand the limiting assumptions on which the analyses are based.

JOHN FENTON

PART 1:
STRUCTURAL DESIGN

Chapter 1: Applying classical structural design methods

Constraints on design
Structural design basics
Framework analysis
Beam elements
Stiff-jointed frames
Torsion of box-beams
Column and sheet stability
Designing sheet steel structures

While advanced finite element techniques of computerised structural analysis are widely used for predicting crashworthiness of new body shell designs, it is also useful to apply traditional structural techniques in order to get a better feel for the design in the early concept stages.

Applying classical structural design methods

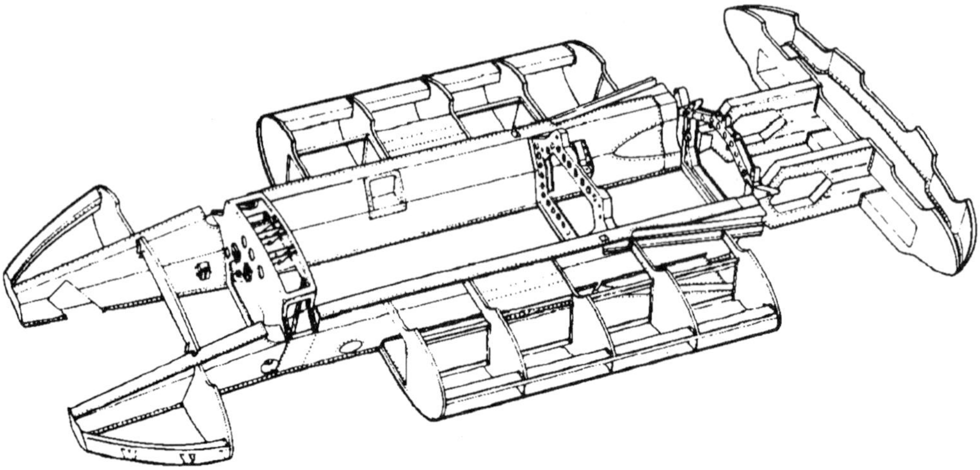

Fig 1: Sigma racing car structure

Constraints on design

Expensive capital equipment used in forming and fabricating vehicle body panels has always to be in the mind of the vehicle body designer; legal and safety constraints on design must be understood, too. He or she must also be aware of how the structural concept for a new vehicle affects the load paths through its principal elements. In addition, vehicle design is becoming increasingly an exercise in packaging; trucks, vans, buses, cars and specialist vehicles all have to be designed within strict spatial envelopes. Power units, transmissions, suspensions, and support systems of all kinds have to be found a place among the occupants and payload goods. These have to be transported in comfort, safety, and often at very high speeds, and be allowed wide apertures for entering/dismounting or loading/unloading.

Rather than the structural requirements of the vehicle dictating the layout of the propulsion, transmission, and suspension systems, according to the optimum reaction of the load inputs involved, the tendency in recent years has been to leave structural considerations to later in the design process. In the early days of horseless traction, of course, it was different. When one looked at a vehicle one would nearly always see a stout chassis frame, not just wonder whether or not there might be one somewhere within. But then, of course, even the car chassis, with its sub-systems, was manufactured in one plant and sent to another for its coachbuilt superstructure, so clearly the technique and the design method were different. With the increasing importance of crashworthiness considerations, however, there is now a much earlier attention paid once more to the structure.

The other main purposes of the vehicle structure are, of course, to link up the mounting points for the vehicle's rear suspension and final drive, front suspension and steering, engine and gearbox, tank for fuel, and seats for occupants. It requires a rigidity to maintain accurate handling, lightness to reduce inertia and rolling drag, and toughness to sustain punishing fatigue loads from road, power unit, and driver. Most trucks still retain chassis frames; in buses they are being 'developed out' and in cars they have virtually disappeared. Even in racing and sports cars unitary construction has become the rule rather than the exception.

In this chapter we shall look at the basic scantlings of the structure and the purpose of the various elements, the way structures are developing, and application of the classical analytical methods.

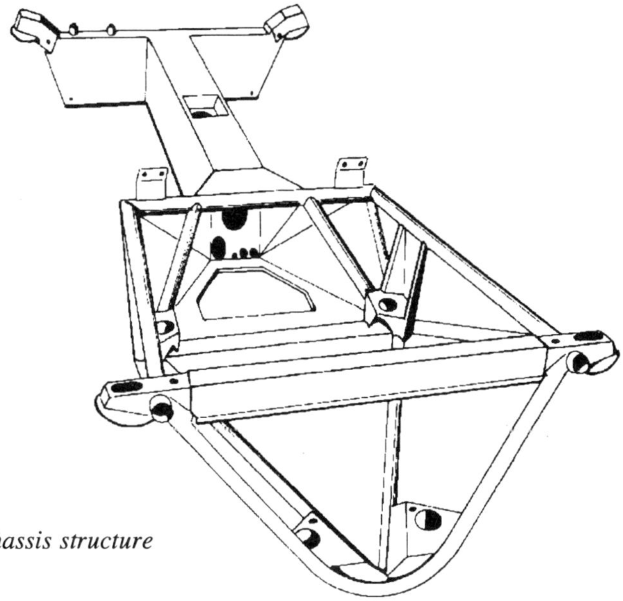

Fig 2: Lotus backbone chassis structure

Approach for the specialist builder

For the specialist builder the initial approach to 'shell' structure design can commence with a drawing layout aimed at optimum positioning of the main structural elements. The designer will need a minimum of four drawings to one fifth scale to depict front, side, and rear elevations, and a plan view. Since thin-walled shell construction involves large hollow closed members, additional cross sections might also be useful.

The first step is to draw in front and rear roll centre heights, from a suspension analysis, and the hub support members at either end. This enables the suspension linkage to be laid out and the vertical wheel travel determined. Next draw in front/rear axle centre lines, ground line, and undertray line to give adequate ground clearance at full bump. Reach diagrams for the drivers and occupants will help to fix pedal and hand control positions, from ergonomic theory, and the space left for packaging engine, transmission, radiator, and fuel tank can then be analysed.

Structure design starts with the major transverse members or bulkheads, generally one each at front and rear axles and two in between for fascia and seat back, positioned according to whether the car is front or rear-engined. On a front-engined sports car the front bay is normally occupied by the power unit, the central bay by the occupants' legs, and the rear bay by the seats; on a rear-engined car, feet and legs are in the front bay, seats are in the centre, and rear bay is occupied by the engine/transmission.

Bulkheads through which driver's legs and engine/transmission protrude must have large cutouts; these therefore need careful design to ensure that they are structurally stable. All the transverse bulkheads must be part of a structure which is rigid enough for minimum bending and torsional deflection. Large regular sectioned torque boxes are required in forming the longitudinals which must be deep enough for adequate bending rigidity. In the design shown in Fig 1 for the Pininfarina Sigma racing car structure, the two rearmost bulkheads are joined by deep box beams, themselves cut out in the webs to allow for movement of the drive shafts. The front three bulkheads are joined by more widely spaced box beams, in turn stabilised by outrigged half oval section members forming the fuel tanks and the streamline fairings for the wheels. In the critical area of the engine/transmission mountings additional members link the two support bulkheads, seen tapering in plan view. Beside it is a contemporary road car understructure in the 'high performance' category, Fig 2.The detail design of mountings on the bulkheads and longerons for engine/transmission, and all the others involving concentrated loads, is of crucial importance. Thin walled shell structure design is only successful if the full section proportions are exploited in resisting the loads, so that design of web stiffeners, stabilising rings, and spreaders need attention.

Deflection as a design criterion

Colin Campbell[1] describes the limitations of the traditional chassis, likening it to a good bridge structure. He points out that whereas a good bridge is designed to sit on solid piers, a good chassis has to sit on a road that flutters and wobbles underneath—like the waves of a choppy sea. When diagonally opposite wheels lift simultaneously on an uneven surface severe racking strains are imposed and a considerable twist can take place. Torsional deflections of the somewhat flexible coach-built bodies and disturbance of the steering geometry could result in a rough ride. Further advances in vehicle development were to be an even greater threat to the traditional design of chassis frame. As car buyers, and later bus riders, demanded more comfort, softer suspensions had to be considered. These required much larger vertical wheel deflections than were possible with the beam-axled front (steering) suspensions of the early vehicles. Large vertical wheel motions with beam-axled suspensions would have meant severe changes in wheel inclination and the resultant development of high gyroscopic torques. Only by providing independent suspension of the front (steered) wheels could the high wheel displacements be achieved, which, with minimal camber and steer angle change, were the prerequisites of a soft ride. But with the discarding of the beam axle and single piece track rod of the coupled front suspension a new rigidity was required of the vehicle structure to which the independent front suspension linkages were anchored.

It is usual to express torsional stiffness in lbf ft/deg (Nm/deg) of twist between front and rear 'axles'. In a typical family saloon 4500 lbf ft/deg (6100 Nm/deg) should be a minimum with 5000-5500 (6500-7500 Nm/deg) preferred. It is thus apparent that criterion for vehicle structures is normally for stiffness rather than strength, the structural analyses being of deflections rather than stresses. Mid span bending deflections for a car should not exceed 0 05 in (1.27 mm) and door aperture deformation should not exceed 0.05 in (1.27 mm) for a 1500 lb (680 kg) mid span load.

Car structure classification

The main categories of structure are:

Integral: in which the main loads are carried by the members making up the structure. In practice this means that all the bodywork, except doors and hatches,

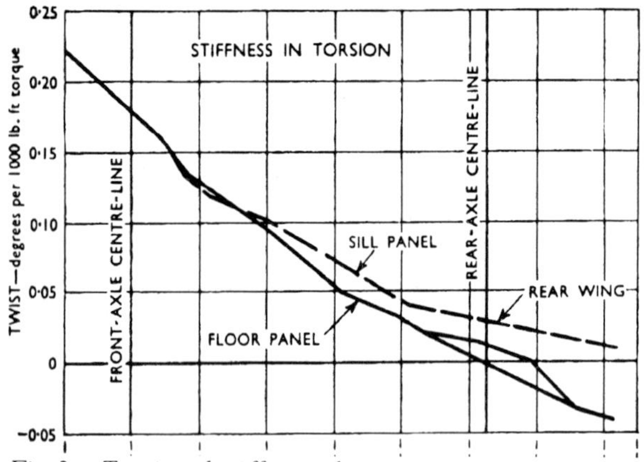

Fig 3a: Torsional stiffness plots

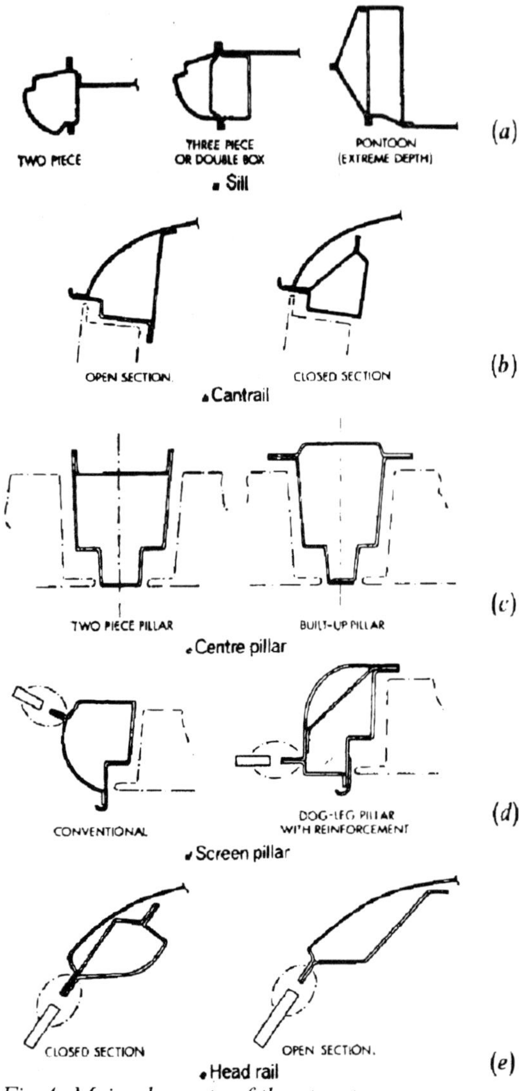

Fig 4: Main elements of the structure

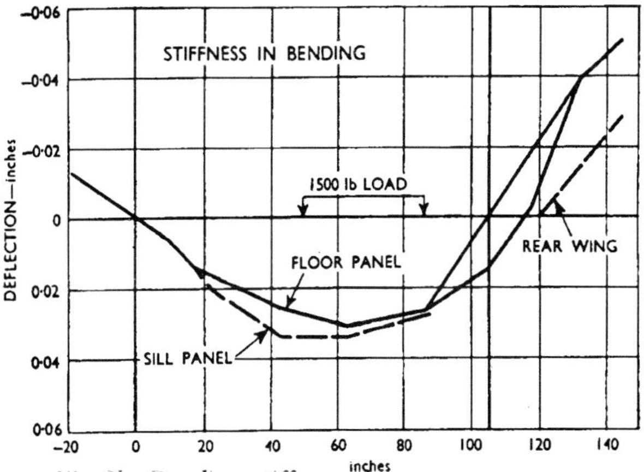

Fig 3b: Bending stiffness plots

is load carrying.

Semi-integral: in which there is a separate chassis frame but the remainder of the structure contributes substantially to the overall stiffness.

'Punt' structures: in which the floor is a box section with adequate torsional stiffness so that open-sports or lightweight GT bodywork can be fitted.

For the small family saloon, Platt[2] suggested that the ascendancy of the unitary (integral) body in Europe was due to the following.

1. More effective use of material resulting in weight saving or higher strength and rigidity for given weight.
2. Some saving in factory cost at the penalty of higher tooling cost.
3. 'All of a piece' construction giving freedom from destructive relative movements between chassis structure and body shell.
4. Well suited to the design of cars with low floors.

Platt points out that tests of torsional stiffness of complete shells do not merely provide absolute values of stiffness. By placing numerous dial gauges along the length of the structure, graphs can be obtained such as that shown in Fig 3 which would reveal any undesirable discontinuities in stiffness along the body.

The discontinuities would reveal possible failure points in road service under the action of cyclical racking strains as the vehicle negotiates uneven surfaces. Clearly the reinforcing at such points and repeating the stiffness test in search of negligible discontinuity would lead to a more durable structure.

The main objectives in structural design are to make the best use of material by arranging for each member to support as near as possible its maximum load potential and to make the structure direct and continuous by providing an unbroken path from point of application to point of reaction. For the specialist builder the advantages of mild pressing steel enjoyed by the mass producers are insignificant on a one-off special or short series build. By careful examination of the specific strength and specific stiffness of different materials, an optimum choice can be made, governed as well by the fabrication facilities available. In considering material in sheet form the flexural stiffness is a sounder criterion than the direct strain.

Typical metal thicknesses used in private car construction are 0.036 in (0.915 mm) for large area panels such as roofs, quarters, doors, and floors; these can reduce down to between 0.030 and 0.026 in (0.765 and 0.66 mm) where curved contours add stiffness and 'tightness'. In major structural parts such as cross members, sills, rails, and pillars, thicknesses of 0.039 and 0.049 in (1-1.25 mm) are common with 0.064 in (1.625 mm) used for local reinforcements.

Panels are normally spot welded together, typical strengths for given panel thickness being shown in Table 1. Pitch for spot welds normally varies between 1 and 22 in (25.4 and 63.6 mm) according to strength and sealing requirements; minimum pitch is

Table 1: Spot-weld strengths for given panel thicknesses

SWG	*Thickness (in) (mm)*	*Spot weld dia. (in) (mm)*	*Average shear failing load (ton) (tonne)*
20	0·036 (0·915)	0·18 (4·575)	0·63 (0·64)
18	0·048 (1·22)	0·22 (5·6)	0·80 (0·813)
16	0·064 (1·62)	0·25 (6·35)	0·88 (0·895)

three times the spot-weld diameter.

In looking at structural configurations in integral structured cars Platt summarises the main elements of the structure (Fig 4) as an assembly of rails, sills, and pillars with practical sections as shown. He sites the GM Holden, Fig 5, to demonstrate the triangulation in elevation and plan designed to provide rigid connection between front suspension and bulkhead. This design had proved successful even on gruelling roads in the Australian outback.

Attachment of mechanical units to an integral body needs special attention to the spreading of load in avoiding local distortion and failure, also in limiting undesirable relative movement. The rear spring front hanger bracket is an important area in conventional leaf sprung front-engine/rear-drive layouts. This mounting has to transmit braking and drive loads horizontally as well as vertical loads due to weight and braking/driving torque reaction, and side loads due to cornering. The example illustrated in Fig. 6 shows that the boxed (longitudinal) body member has an internal stiffener and side plates with through bolts.

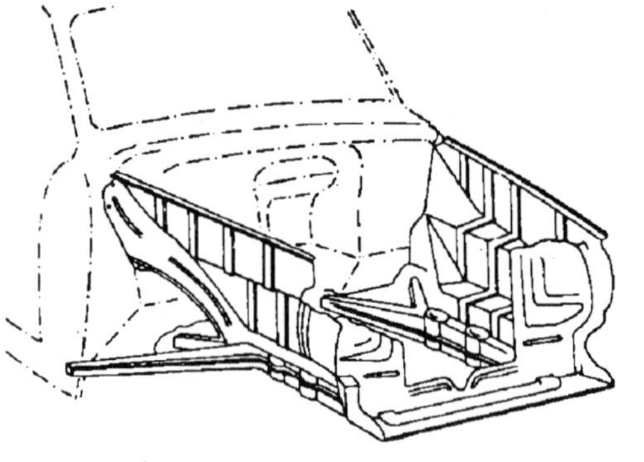

Fig 5: GM Holden front-end

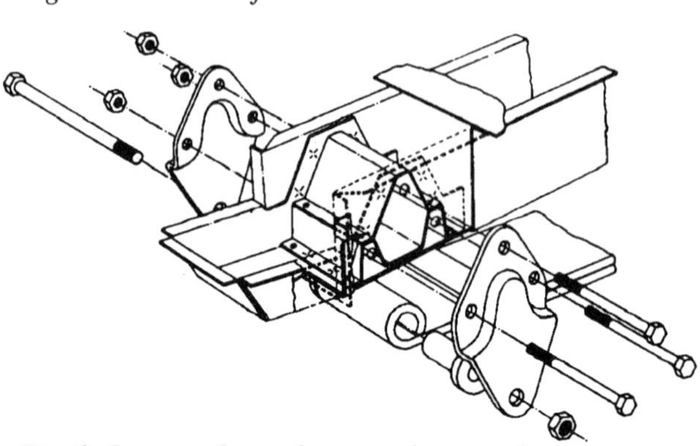

Fig 6: Structural attachment to box-member

Structural design basics

Solid elastic materials such as steel and, to a degree, aluminium alloy or even plastics heavily reinforced with elastic materials, can be readily examined in respect of internal loads and deflections. This is on the assumption that stress s (internal load per unit area) is proportional to strain e (change in size per original size).

It is useful to think in terms of SI units because the gravity force on one kilogram load is 9.91 (approximately ten) and most units in the system are expressed in multiples of tens, hundreds, thousands and so on. Internal load is assumed distributed equally across the section of the solid body under strain in simple, linear, stress analysis and the ratio stress/strain is the elastic modulus, E, 70 kN/mm^2 for aluminium.

As well as direct stress, tensile and compressive, shear stresses often need to be considered. Here a shear modulus, G = stress/strain, applies. Note that any element under shear stress τ, Fig 7, is subject to a complementary shear stress, τ' — arising from the force balance on the element, to maintain equilibrium, as:

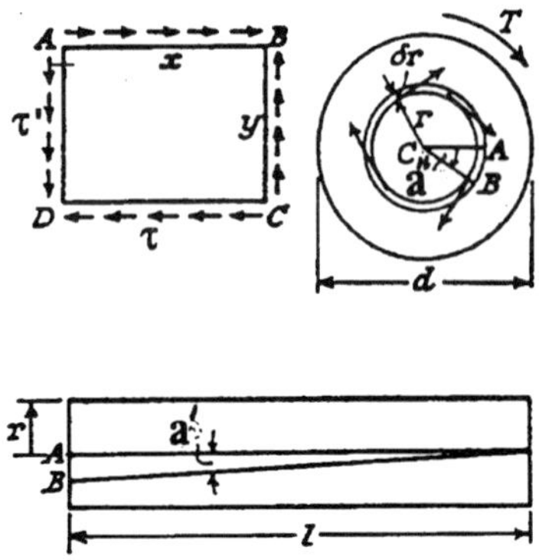

Fig 7: Shear in flat panel and circular section

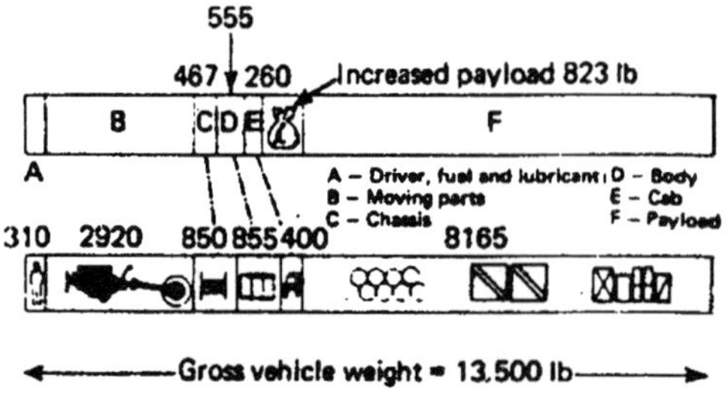

Fig 8: Economic rationale for structural efficiency

$$\tau_{xz} y = \tau'_{yz} x$$

This is important in materials like timber which are weak in shear along the grain.

A structural element under twist is subject to shear strain. As well as strain on the transverse sections of the element, there will be complementary shear straining of the longitudinal planes. In Fig 7b, twist α over length l can be equated to the shear angle α'' and radius r of the symmetrical transverse section, and the strain expressed as stress in:

$$\tau / r = G\alpha$$

The torque T on the element can be equated to the sum of those of each tangential stress, and the expression for the polar moment of inertia, J, substituted, to obtain $T/J = \tau/r$ showing that, for a given torque, shear stress is proportional to the radius and the torsional stiffness is:

$$T/\alpha'' = GJ/l$$

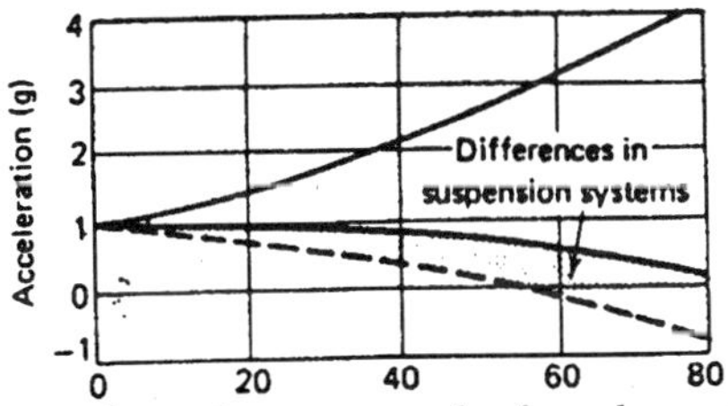

Fig 9: Load envelopes per mph of road speed

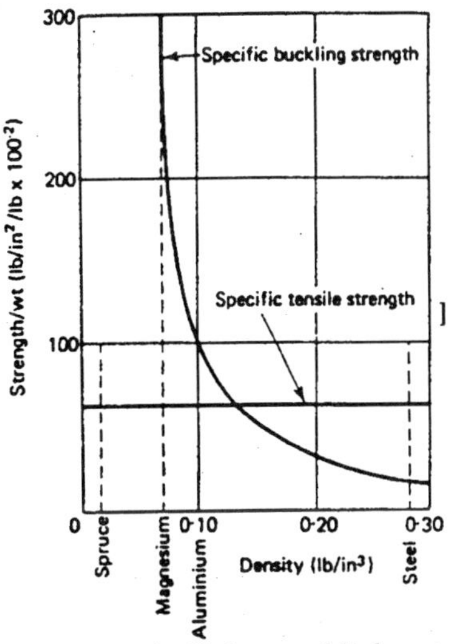

Fig 10: Structural efficiency of lightweight materials

Structural efficiency

Immediately after World War 2, it was realised in the USA that lessons learned from aircraft design could be applied to ground vehicles. In a 1945 SAE paper presented by Mac Short[3], a particularly far-sighted view was presented. Fig 8 reproduced here demonstrated potential savings made possible by using structural design techniques. The example applied to a truck described in US units of measurement. He showed that by careful use of aluminium alloy, 30% weight saving could be achieved in the cab and 35% in the body and that 15 hours of stress analysis would be required per pound weight of structure saved, a figure that has been subsequently slashed by the emergence of computers.

Part of the suggested approach is to draw up load envelopes as shown in Fig 9 for torsion, bending and shock loads. Another key factor is use of careful design at the points of load application, to avoid structural discontinuity, by spreading concentrated loads as much as possible. Once stress intensities have been kept low by these means, the other main design objective, with monocoque structures, is to avoid buckling of the thin sheet materials. The immediate attraction of the lower density materials, for such purposes, is shown in Fig 10. This illustration merely reflects the effect of increased thickness on a plane panel under end load; even greater buckling resistance is, of course, obtainable with framework-stiffened panels and sandwich construction.

It is a good discipline to consider the purpose of the vehicle structure on starting a new design: to link up the mounting points for suspension and drive systems with the support platform for passengers and/or payload. Rigidity of the structure is required to maintain accurate handling; lightness to reduce inertia and rolling drag; toughness to provide fatigue resistance against vibration and shock loads from the ground surface.

When diagonally opposite wheels lift simultaneously considerable torsional loads can be applied to the vehicle structure. With the trend to softer suspensions, involving wheel displacements that could cause steer fight if careful geometric control is not maintained, the rigid disposition of suspension anchorages is paramount. In Fig 11, for spring stiffness c (N/mm) and road displacement d, load transmitted to the structure is cd. If torsional stiffness is C, torsional deflection of the chassis $D = cd/C$ (in mm).

For spring rate 15 N/mm, a torsional stiffness of over 2000 N/mm will be required to limit structure deflection at the suspension anchorage to 1 mm when negotiating 100 mm road bumps.

The use of hollow closed section members for withstanding torsional loads can be visualised by considering the effect of slitting the tube along its length and twisting it. Some two thirds loss of stiffness results from the ability of the free edges to slide with respect to one another. For the rectangular section closed tube shown in Fig 12, the angle of twist in radians per unit length is given by

$$T(b' + d') / 2Gb'^2 d'^2 t$$

and average shear stress by

$$T/2tb'd'$$

There will be a local increase in the stress levels at the internal corners according to the graph shown if Fig 13. In many cases the structural member is loaded by a combination of stresses: direct stresses, direct stresses due to bending and shear stresses. Normal practice is to add direct to bending stresses; add bending stresses in planes at right angles to each other and then add shear stresses to the above. This results in principal stresses, in tension/compression and shear respectively:

$$(\sigma/2) +/- [(\sigma/2)^2 + \tau^2]^{1/2}$$

$$[(\sigma/2)^2 + \tau^2]^{1/2}$$

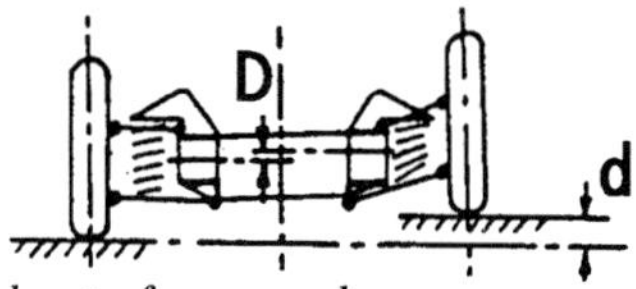

Fig 11: Twist due to frame and suspension deflection

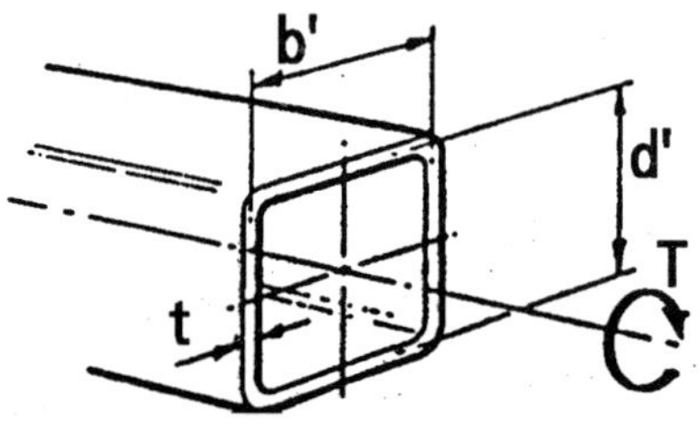

Fig 12: Twist in rectangular section tube

Finite element method

Division of the structure into small parts (finite elements) allows analysis by the FE method. The SSS technique, introduced in the next chapter, involves a form of FE idealisation but, as explained there, the relatively 'course mesh' division into large parts allows a good appreciation of the behaviour of the main elements of the structure so important in design. Fine mesh division into small parts, either planar or solid, allows a more accurate prediction of the structural stress or deflection pattern.

The method is an extension of matrix structural design techniques also to be considered in a later section. In its simplest form, plane triangular elements are used and stress strain relationships, described above, used to obtain equations which are solved by a computer for structural displacements and stresses. Fig 14 shows some panel elements used and Fig 15, an early representation of a car body made from both panel and beam elements. The numbers show the node points of the model where the elements interconnect. Typically six equations, corresponding to three perpendicular force and torque balances at these points would define the internal loading of the structure.

For 'solid' shapes and multi-curved surfaces, more complicated elements, including three-dimensional ones, can be used. These involve a greater

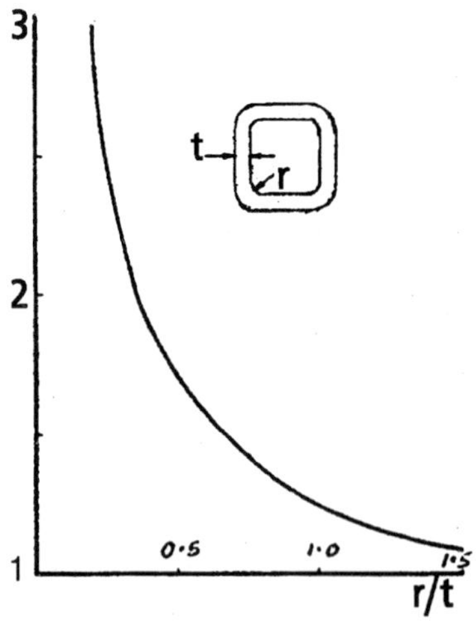

Fig 13: Stress concentration factor vs r/t ratio

number of equations to be solved simultaneously. The use of matrices to obtain computer solutions is really seen to advantage in such cases. Proprietary computer programs for such purposes are, of course available. One such, under the acronym of PAFEC (Program for Automatic Finite Element Calculation) is available from the well known company of that name founded at the University Science Park, Nottingham.

Thin-wall structural analysis

The many simplifying assumptions that can apply to panelled structures make them surprisingly easy to calculate for loads and deflections. In applying the basic rules of structural mechanics, the designer soon develops 'structural feel' which is invaluable in visualising new concepts. Structural engineers have developed the design technique of 'thin-walled-structures' from the early theories of aircraft stress analysis and achieved surprisingly accurate results.

The first section of the chapter introduced one of the approaches to analysing thin-wall structures, the representation of one surface of a load bearing structure by a rectangular panel enclosed by end-load carrying bars at its edges. There are a number of other simplifications which can help turn real life structures into ones which are readily solved by uncomplicated calculation techniques. As well as assuming panels only carry shear and tensile forces in their own plane, it can also be taken that all loads can be ignored

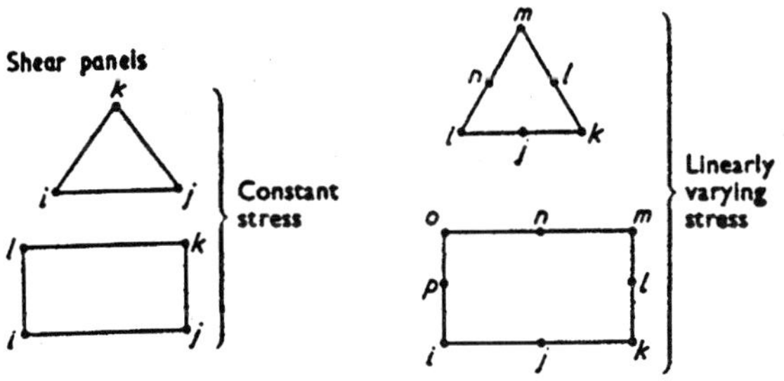

Fig 14: Finite elements for panels

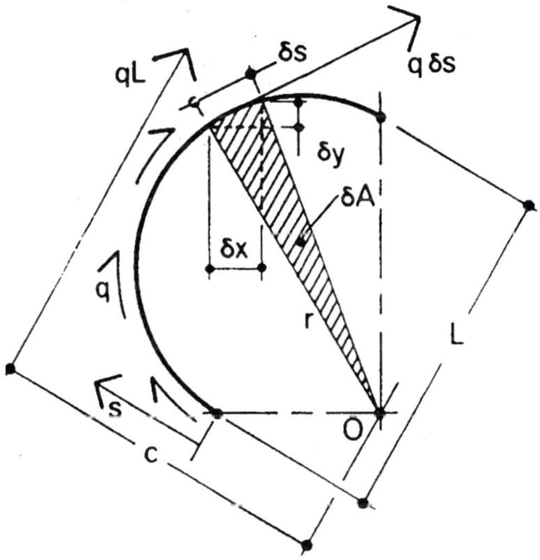

Fig 17: Shear flow in generalised beam section

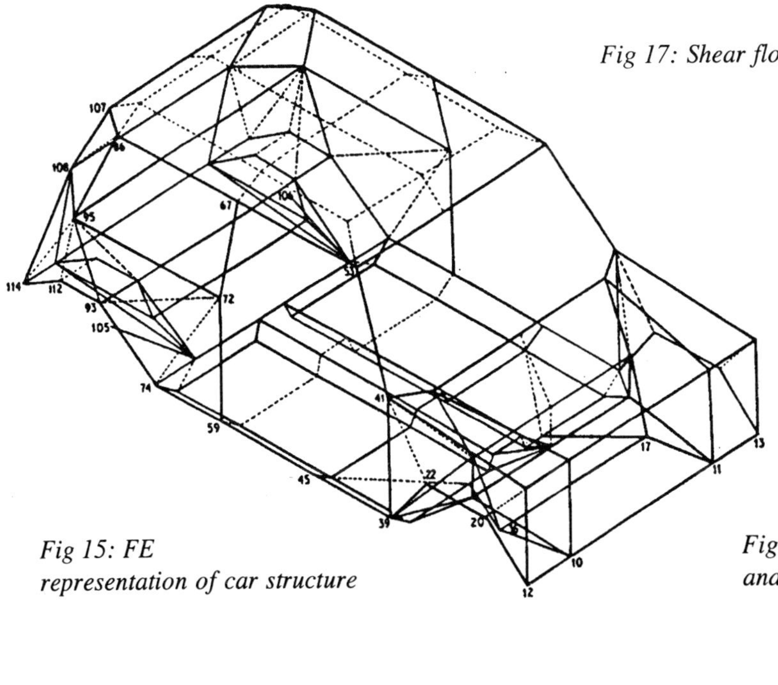

Fig 15: FE representation of car structure

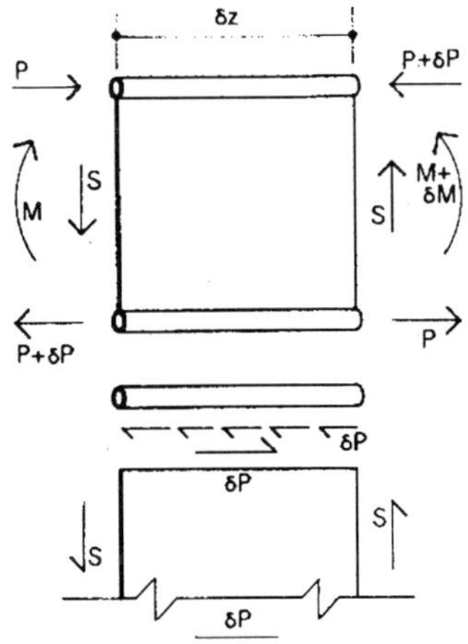

Fig 16: Idealisation into shear panels and end-load bars

to which the thin wall is flexible or would collapse by buckling. Flanged edges of panels, and intermediate stiffening swages, can be represented by end-load carrying area added to the cross-sections of those used to idealise the basic surface panel.

While the application of simple stress/strain relationships to any structural design problem in a vehicle body might suggest the use of very fine gauges of metal, the practical limitations of sheet handling and accidental damage caused by highly concentrated loads normal to the surface set minimum gauge thicknesses. In the case of volume-built cars and vans, the general body material employed is deep-drawing mild steel to EN 2A/1 with an ultimate tensile stress of 45,000 lbf/in^2 (310 x 10^6 N/mm^2) and elastic modulus of 30 x 10^6 lbf/in^2 (210 x 10^9 N/mm^2), 20 gauge being the common minimum thickness for general purpose. Many specialist vehicles use special materials and therefore thickness is much more variable; so too are the structural design criteria for the wide variety of possible special-purpose bodies.

Beam structural elements and bending theory are to be considered in a later chapter; that theory is not required for short thin walled beams as the principal panel loads are shear ones. Thus a small element of a thin-walled beam can be considered as in Fig 16 for design purposes. The magnitude of 'pure' shear stress in the panel, $t = S/ht$ for thickness t and shear force S. It is useful for calculation to express the shear in 'flow' terms: shear flow = force per unit length, so that the shear flow, $q = S/h = tt$. Under the action of the bending moment M longitudinal stress in the bar $s = P/A$ where A is the cross-section area of the edge bar or 'boom' in thin-wall parlance. Therefore $s = M/DA$ where D is the distance between booms. The moment of inertia, or second moment of area, of the whole beam cross section $I = sAy^2$ where y is the distance of the boom area from the bending axis (the centre, for a symmetrical section beam). At any distance L along the beam, from the point of application of S, as $M = SL$, boom end load $= qL$ and it is assumed, for short (elemental) lengths of the beam, elemental force $dP = dM/D = Sdz/D$ so that $dP/dz = S/D = q$. This implies that the shear flow in the 'web' of the beam is equal to the rate of change of end load in the boom.

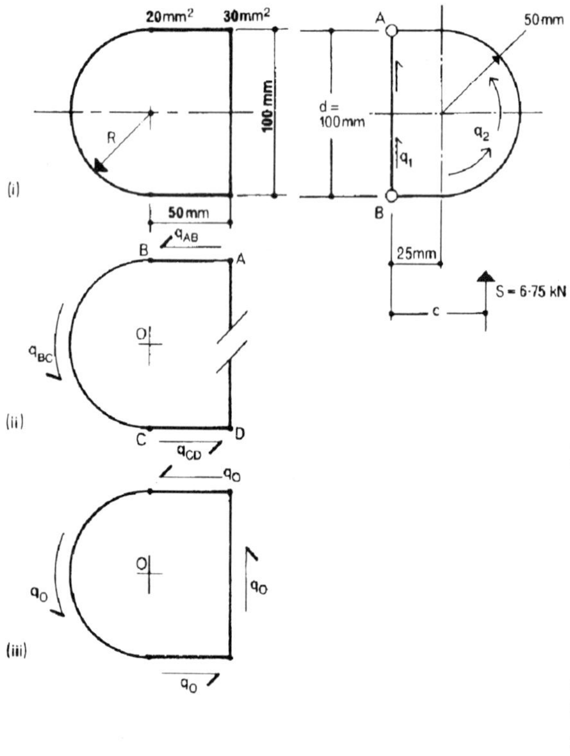

Fig 18: Car or van sill section idealisation

The assumption of constant shear flow in the web is also made for a curved web and an important theorem by Peery[4] showed that for a beam section such as that shown in Fig 17, twice the area enclosed by the curved web $[A]$ divided by 'chord' of the curve (100mm in this case) gives the distance c -- moment arm of the resultant of the shear flow in the curved web. A car sill beam could well be idealised into the section shown in Fig 18. Generally, in closed section thin wall beams, the idealisation is made to enclose the shear panels between booms -- the shear flows in the panels being in equilibrium with the externally applied load. This example is solved by using the expression $S/D = q$, derived above; with the two shear panels, in this case, $q_1 D + q_2 D = S$ so that $q_1 = (S/D) - q_2$ and in taking moments about boom B, $Sc = 2q_2 [A]$ giving

$$q_1 = (S/D) - Sc/2[A] \text{ and } q_2 = Sc/2[A]$$

Then by substituting the numerical value of $[A]$, and choosing thicknesses of 3 mm and 1 mm for straight and curved webs, these two equations can be solved simultaneously for the condition of equal shear stress in the webs to find the value $c = 33.33$ mm which meets this condition. Thus, if the beam was to be loaded within its span, this would be the most efficient position across the section to fix the load attachment.

Another example to consider is a rectangular section beam having, say, four angle corner flanges of area A_f, depth d, breadth b and corresponding panel thicknesses t_d and t_b. This would be idealised by a four-boom box beam with boom section areas

$$A_f + (bt_b/2) + (dt_d/6)$$

Calculation technique starts by making an imaginary cut to make the closed section momentarily into an open one by removing an element of the section -- then to find 'open-section' shear flows. Next an opposite sense shear flow q_o is introduced on all elements, equal to the 'resultant shear flow' q^*, on the cut element. Final stage is to balance moments of the combined shear flows against the moment arm of the externally applied load. The q_o flows are assumed to set up torque $T = q_o 2[A]$.

Here it is useful to introduce an extended form of the expressions for torque and twist, given in the later section on torsion, for solid and thick walled sections. Akin to the Peery expression above, it can be shown for a closed thin-wall section member that twist per unit length

$$\alpha' \{q/2[A]G\}\sigma(s/t)$$

and torque, $T = q2[A]$, for angle of twist a'', shear modulus G and s/t summed for elements around the section such that in the generalised box tube of Fig 19,

$$a''/l = [T/4(bd_2)G](b/t_1 + d/t_2 + b/t_3 + d/t_4)$$

Reverting to the box beam of the preceding paragraph, it is now possible to use these expressions to obtain $\sigma q^*(s/t) = 0$ when equating the twist per unit length to zero (as this symmetrically sectioned beam is only in bending and not torsion).

A triangular section beam can be evaluated in the same way; Fig 20, with equal boom areas 1,2,3 = a, is an example for which shear panels have width h and thicknesses t_1 and t for vertical and angled panels respectively. To find the position across the section at which an external load will cause no twist of the beam (known as the 'shear centre'), the right hand panel is removed to first obtain open-section shear flows; but by taking moments about (2), q_{12} and q_{23} are isolated and q_o is balanced by the moment of the external force S at the shear centre, with offset e. By putting $S = I$,

$$(\alpha h^2/2)e = q_o h(3)^{1/2}/2.$$

Then by putting $a''l = 0$, as open-section shear flows $q_{12} = q_{23} = ah/2$, the value of e becomes

$$(3)^{1/2}ht_1/(2t_1 + t).$$

At a joint between box members such as the important rear-quarter to sill junction of a car, Fig 21, consideration should be given to load transfer at the corner. In figures taken from the design of the Austin A30 structure by TK Garrett[5], bending moment of 51000 lbf in (5.76 kNm) induced end loads of 33000 lbf (146.7 kN) in the top and bottom flanges of the 5 in (127mm) sill. The 18.5 MN/m² safe working stress steel used in this application dictated an area of 0.123 in² (0.795 cm²) edge material -- a 3.4 in (86.5mm) wide flange of 20 gauge sheet. The webs of sill and quarter panels overlap as shown by the dotted lines to provide the necessary transfer of the 1900 lbf (8.45 kN) shear force applying in this example. It is also important to maintain the cross-section shape of thin-walled box-beams, both along their length and, crucially, at the corner joint where a diagonal diaphragm (bulkhead) might be required.

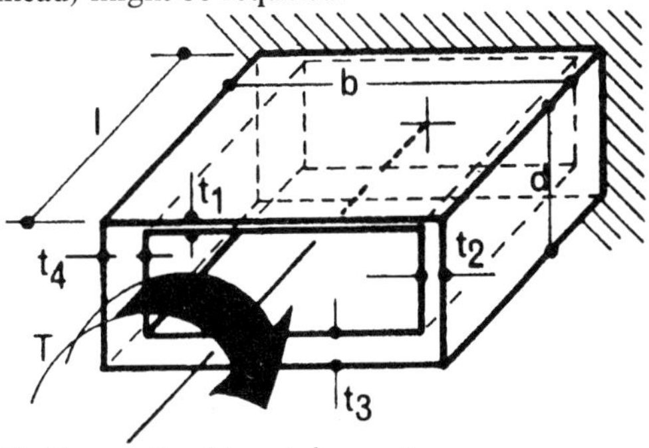

Fig 19: Generalised box-tube section

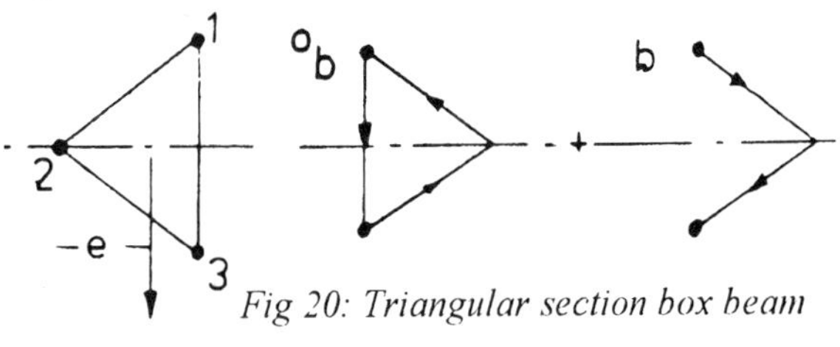

Fig 20: Triangular section box beam

In the case of a curved corner joint between box beams, Fig 21, the web is subject to direct stress M/hRt where R is the radius of curvature and t the web thickness. For the lightly angled joint shown in the same figure, a 'kink-strut' may be required at the junction to maintain the cross section against collapse. Load in the strut will be $M/d(\cos a + \cos b)$. If the beam has particularly deep webs, it may be necessary to provide vertical stiffeners which help stabilise shear panels against buckling. One technique is to swage the panel rather than fasten on additional parts. General rules for swaging are that all swages should be straight and not intersect; that they should run along the shortest distance between supported edges of the panel. In calculation they can be represented by end-load carrying bars.

A useful description of the twist of open-section thin walled members, for use with the expression $T/\alpha'' = GJ/l$, mentioned above, refers again to the generalised section in Fig 17. Polar moment of inertia $J = \sigma s t^3/3$ leads to the values for particular shapes shown in Fig 22 alongside some J values for representative closed sections.

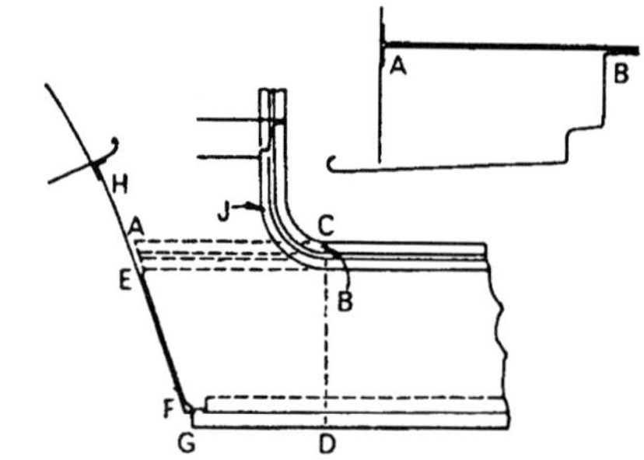

Fig 20: Sill to quarter panel junction

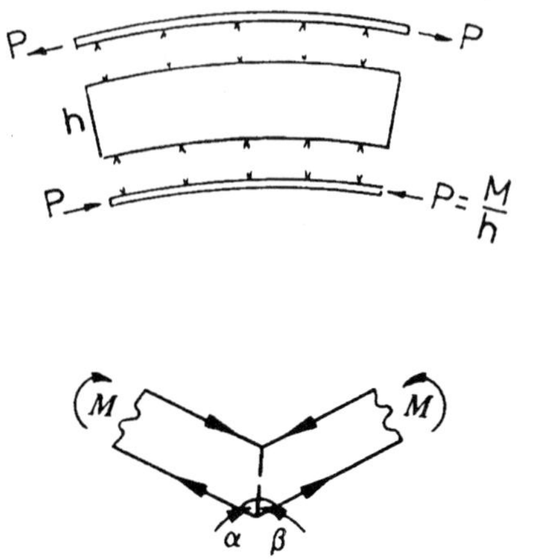

Fig 21: Curved and angled corner joints

OPEN SECTIONS

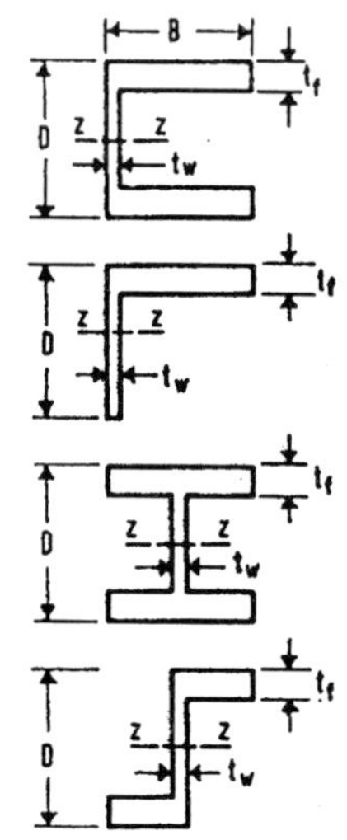

'J' $= \dfrac{2\ Bt_f^3 + (D - 2t_f)t_w^3}{3}$

'Izz' $= \dfrac{BD^3 - (B - t_w)(D - 2t_f)^3}{12}$

'J' $= \dfrac{B.t_f^3 + (D - t_f)t_w^3}{3}$

'Izz' $= \dfrac{B.D^3 - (B - t_w)(D - t_f)^3}{12}$

'J' =
'Izz' = } as for channel above

'J' =
'Izz' = } as for channel above

CLOSED SECTIONS

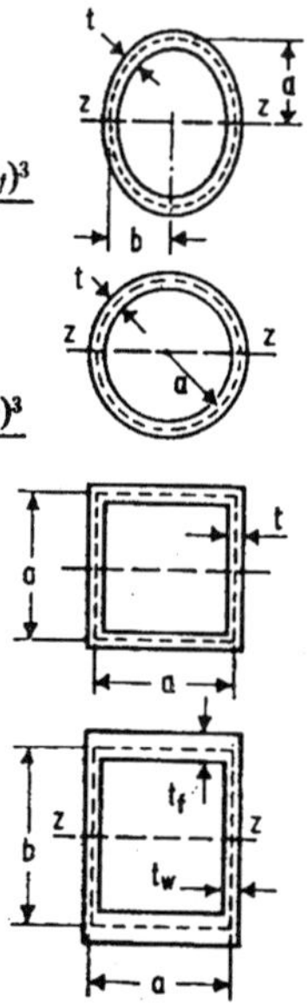

'J' $= \dfrac{4\pi a^2 b^2 t}{\sqrt{2}\ (a^2 + b^2)}$

'Izz' $= \dfrac{\pi}{4}\left\{\left(a+\dfrac{t}{2}\right)^3\left(b+\dfrac{t}{2}\right)-\left(a-\dfrac{t}{2}\right)^3\left(b-\dfrac{t}{2}\right)\right\}$

'J' $= 2\pi a^3 t$

'Izz' $= \dfrac{\pi}{64}\left\{(2a + t)^4 - (2a - t)^4\right\}$

'J' $= a^3 t$

'Izz' $= \dfrac{(a + t)(b + t)^3 - (a - t)(b - t)^3}{12}$

'J' $= \dfrac{2a^2 b^2 t_w t_f}{b t_w + a t_f}$

'Izz' $= \dfrac{(a + t_w)(b + t_f)^3 - (a - t_w)(b - t_f)^3}{12}$

Fig 22: Section characteristics

Framework analysis

Both triangulated pinjointed frames and rectangular stiff-jointed frameworks are used in real and idealised body structures. The pillars and rails of car and van bodies are the key frame elements of private and commercial vehicles. In many specialist sports vehicles tubular space frames, which can be analysed by framework methods, form the structures which support the GRP shells. Framework methods derived from civil and aeronautical engineering are valuable analytical tools which solve these structures and form the basis of other methods used in solving integral and monocoque ones.

Earliest attempts to predict stresses and deflections, within the first all-welded steel car bodies, were based on the assumption that the body sides behaved like truss frames. The action of vertical loading from engine and occupant seat mounts was examined against the reaction at the axle spring hangers. These analyses were subsequently refined to take in the effects of stiff joint conditions at the A,B and C post connections with sole bar and cant rail. This last approach, however, requires the application of bending theory to be considered in a later section.

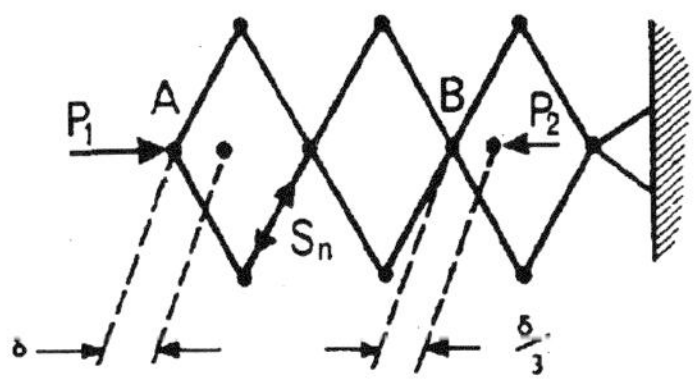

Fig 23: Linkage mechanism

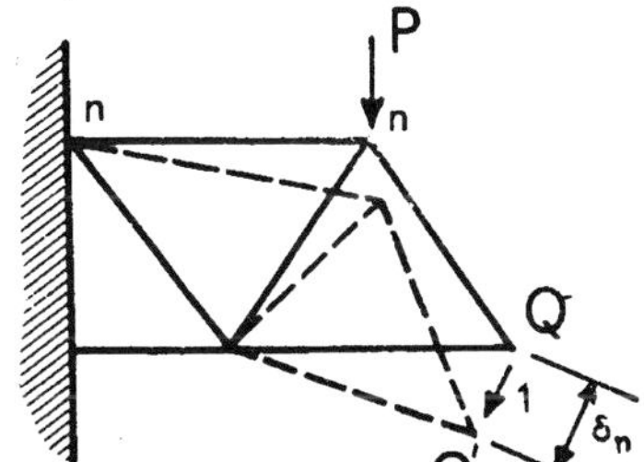

Fig 24: Stable braced frame

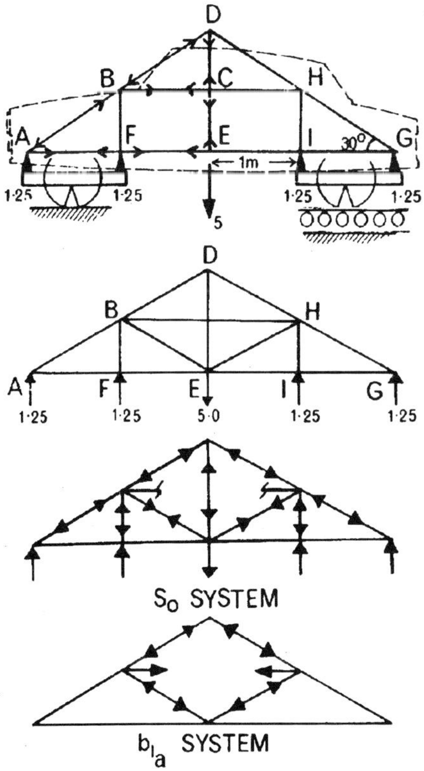

Fig 25: Car sideframe representation

	S_0	b_1	l	$S_0 b_1$ l/AE
(2X) AF	$+1{\cdot}25\sqrt{3}$ (X2)	$+{\cdot}25\sqrt{3}$ (X2)	1	·94 = 1·84
(2X) AB	− 2·5 (X2)	− ·5 (X2)	$\frac{2}{\sqrt{3}}$	1·44 = 2·88
(2X) BF	− 1·25 (X2)	− ·25 (X2)	$\frac{1}{\sqrt{3}}$	0·18 = 0·36
(2X) BC	$+2{\cdot}5\frac{\sqrt{3}}{2}$(X2)	$+\frac{{\cdot}5\sqrt{3}}{2}$(X2)	1	0·94 = 1·84
(1X) DC	+ 5	+ 1	$\frac{1}{\sqrt{3}}$	2·88 = 2·88
(1X) CE	+ 5	+ 1	$\frac{1}{\sqrt{3}}$	2·88 = 2·88
(2X) BD	− 5 (X2)	− 1 (X2)	$\frac{2}{\sqrt{3}}$	5·78 = 11·56
(2X) FE	$+1{\cdot}25\sqrt{3}$ (X2)	$+{\cdot}25\sqrt{3}$ (X2)	1	0·74 = 1·46
				26·54 /AE = S_E

Member	S_0	b_{1a}	l	$S_0 b_{1a}$l	b_{1a}l	$S=S_0+b_{1a}X$	b_{1b}	$S_0 b_{1b}$l
(2X) AB	− 2·5	0	$\frac{2}{\sqrt{3}}$	0	0	− 2·5	− 0·5	1·46
(2X) AF	$+1{\cdot}25\sqrt{3}$	0	1	0	0	$+1{\cdot}25\sqrt{3}$	$+\frac{\sqrt{3}}{4}$	0·94
(2X) BF	− 1·25	0	$\frac{1}{\sqrt{3}}$	0	0	− 1·25	$-\frac{1}{4}$	0·18
(2X) FE	$+1{\cdot}25\sqrt{3}$	0	1	0	0	$+1{\cdot}25\sqrt{3}$	$+\frac{\sqrt{3}}{4}$	0·94
(2X) BD	− 3·75	$-\sqrt{\frac{2}{3}}$	$\frac{2}{\sqrt{3}}$	10	1·54	−2·975	−0·595	2·05
(2X) BE	+ 1·25	$-\sqrt{\frac{2}{3}}$	$\frac{2}{\sqrt{3}}$	−3·33	1·54	+ 2·025	+ 0·505	1·18
BH	0	+ 1	2	0	4	− 1·485	− 0·297	0·43
DE	3·75	$+\sqrt{\frac{3}{3}}$	$\frac{2}{\sqrt{3}}$	5	0·77	+ 2·975	+ 0·595	1·03

$$X_{BH} = -\frac{\Sigma S_0 b_1\, l/AE}{\Sigma b_{1a}^2\, l/AE} = -1{\cdot}458 \text{ kN} \qquad S_E = \Sigma S_0 b_{1b}\, l/AE = \frac{16{\cdot}4}{AE}$$

Fig 26: Member force tabulation

Frameworks of discrete links can be analysed by methods first derived for framed bridges, roof trusses and like structures. Such frames are often hinged linkage mechanisms, cross-braced (triangulated) — to form rigid structures. The method is thus best understood by an initial understanding of an unbraced mechanism, Fig 23. The concept of 'work', taken from basic mechanics theory, is used to make the rule that in a system in equilibrium (force balance) under a set of loads, the work done by the various applied loads is equal to the work done by their resultant. Internal (virtual) work is also done by the strained links of the mechanism and those who have operated a 'lazytong' riveting tool will know of the load magnification seen in the figure where

$$P_1 s = P_2 s/3$$

which on applying the rule above gives $P_2 = 3P_1$. Internal forces S do virtual work U_i which can be equated to the external work $U_e = Ps$.

For a stable framework that has been braced, Fig 24, a two-stage visualisation of the load system is helpful. The first is the equilibrium of bars and joints, ignoring the effects of deformation, while the second is from the forces developed when compatibility of strain is considered at the joints. When trying to find displacement of any point in a frame, stage-one forces can be considered as fictitious (virtual) and an imaginary unit load be assumed to act (for convenience of calculation) at the point where the displacement is to be found. In stage two, the real event gives rise to displacements obtained from the work equation:

$$U_e = U_i \text{ or } Sn_1.dn(o)=\Sigma\, \sigma.1.\varepsilon_0$$

for member force S, deformation d, stress s and strain e in the nth member under virtual unit 1 and real original o loads. When member cross section A and length l are considered, then

$$1.s = \Sigma(n)\ \sigma_1 .\varepsilon_o/Al$$

and for the framework shown, if only member n is extended:

$$1.S_n = b_{1n} dl_o$$

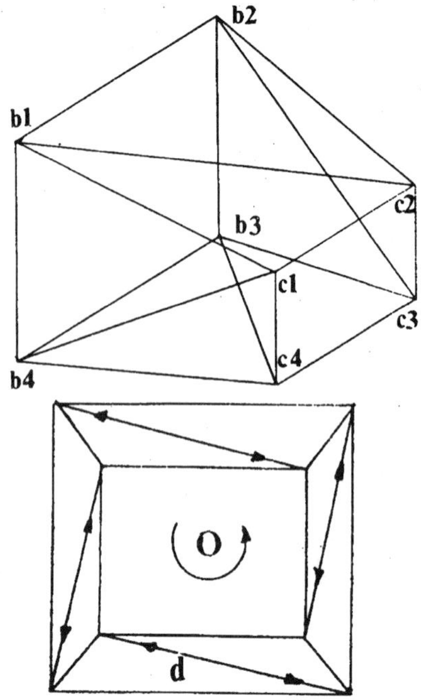

Fig 27: Single bay of space truss

where b_l is member force due to the unit load. And when all members deform,

$$S = \Sigma S_o.b_1/(l/AE)$$

since $e_o.l = s_o.l/E = S_o.l/AE$ where E is the elastic modulus.

To show how the method is applied, a somewhat oversimplified car side-frame is considered in Fig 25. A table is made up, Fig 26, for the member forces with unit load being applied at E where the vertical displacement downwards is to be found. Note that the frame is symmetrical so that only forces in half the frame members need be calculated and doubled up as appropriate.

The procedure is to start with the external reaction points A and F and use trigonometry to resolve the proportion of the reaction force which passes through to the member. As all joints are assumed to be pinned, the structure gains its rigidity by end-loading of the members only with a geometry giving a triangulated bracing effect. Members at right angles to the external forces at the joint receive no load and those in line receive equal and opposite load at each side of the joint. Load directions at the joint (see arrows) pair up with those at opposite ends of the member to define whether a tensile or compressive force exists. Away from the external load points, at B, say, resolving vertically to AD gives

$$BF\cos 30 = BC\sin 30$$

therefore internal force BC (tensile, positive) is $1.25(3)^{1/2}$.

Space frames

The last example was for a plane frame (two dimensional frame); for a three dimensional frame, resolving is a little more difficult and a method known as tension coefficients is applied. This is based on the fact that proportionality exists between both resolved components of length and of force. Tension coefficient

$$t = S/L = F_x/X = F_y/Y = F_z/Z$$

for member force S, in a length l, projecting force vectors and length components, F_x, F_y, F_z and X,Y,Z on perpendicular coordinate axes.

Fig 27 shows one bay of a space truss, that might be an extension used to support an engine/gearbox unit behind a monocoque bodyshell structure. For analysis, the frame is assumed to have rigid plates at $b_{1,2,3,4}$ and $c_{1,2,3,4}$ which offer no force reaction perpendicular to their own planes (zero axial warping constraint). When considering the torsional load case, for the bay, b_1c_1 and b_4c_4 member forces are zero by resolving at b_1 and c_4, observing zero axial constraint of the truss. Remaining members (the envelope) have equal force components F_2 at the bulkheads and tension coefficients for all of them are F_z/Z. This common value can be determined by projecting envelope member forces on to the bulkhead planes, as shown. By taking moments about any axis O, for any one member (c_3b_4, say, with projected length d), contribution to torque reaction is = rtd (half the area of the triangle formed by joining O to c_3 and b_4). Thus total torque is $2tA$, by summation.

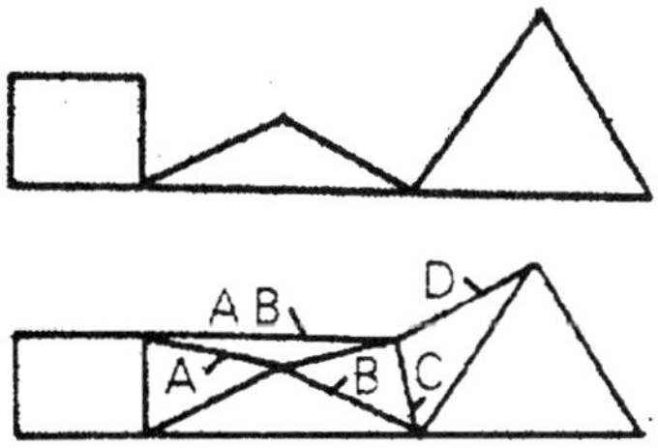

Fig 29a: Cockpit bay

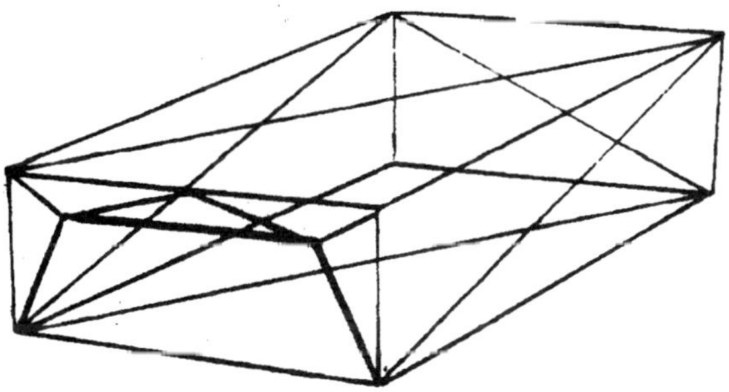

Fig 29:b Front-engine bay

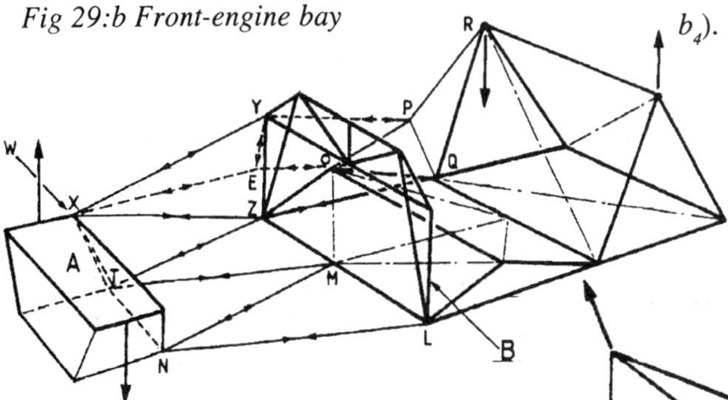

Fig 29: Triangulated frame

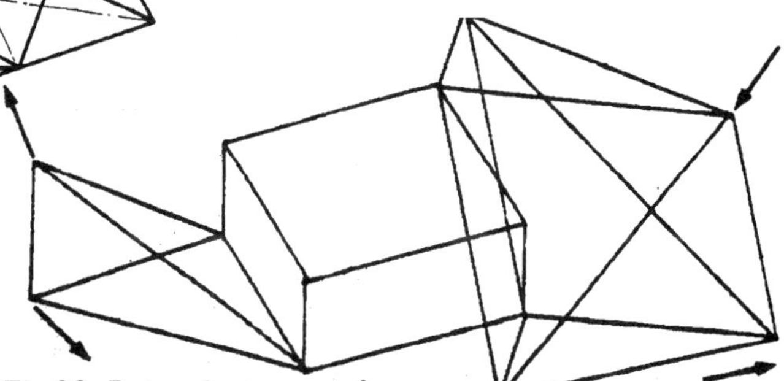

Fig 28: Determinate space frame

Costin and Phipps[6] show how loads in a complete carspace frame are calculated similarly. Frank Costin, too, is a notable pioneer of space frame analysis; he conjectured that any two torsionally stable systems pin-jointed together, at any two node points, would result in torsional continuity between bays, which like that in Fig 28, is statically determinate. This led Costin to approach the design of a cockpit bay as in Fig 29a; tubes A,B,C,D are added to provide bending continuity, also. The front-engine bay is represented by a simple box with triangulated bracing, Fig 29b, and the complete vehicle frame emerges in Fig 30 (the Costin Amiga).

Statically indeterminate frames

The great majority of vehicle structures are indeterminate, due, for example, to stiffness at member joints assumed for the simplified analysis, above, to be pinned. Such frames need to be idealised as determinate structures or analysed by an extension to the above method. Real-life frames usually involve additional members which are redundant from the standpoint of statical determinacy but might still perform some useful additional purpose.

Slightly better idealisation of the car sideframe shown in the top view of Fig 25 would be that of the frame in the view beneath it in which the stiff-jointed door openings are represented by cross-braced pinned frames rather than linkage mechanisms. While the crossbraces are redundant from the standpoint of the frame retaining its shape under moderate load, and from the standpoint of calculation, they are more truly representative of the anti-lozenging stiffness of the openings.

The technique of analysing single-redundancy frames is to cut any redundant member and then calculate member loads for the remaining, determinate, frame. Here BH is cut to give the S_o and b_1 force systems shown below and a third force system b_{1b} determined for finding the required central deflection after a force equal to that in the cut redundant member is inserted. This is done again by applying unit load vertically downwards at E then obtaining b_{1b} forces by dividing S forces (from stage-one) by the real external load of 5 kN at E, giving member forces for unit external load. The table in the lower part of Fig 26 shows the central deflection to be reduced by a factor of 26.54/16.9 = 1.5 by the bracing which simulates the stiffness of the aperture frame.

Finding the degree of redundancy of a framework can be difficult when only considering the number of possible extra load paths in the structure. Another approach is to consider that equilibrium equations alone are insufficient to find the forces at any member connection, or external load input, point. This has led to the expression for number of redundancies = $3f + 2h + r + 3m - 3j$ where f is the number of fixed supports, h the number of hinged supports, r number of roller supports, m the number of members and j the number of joints. This reduces to $m + r - 2j$ for a two-dimensional (plane) frame. Of the three equilibrium conditions of horizontal and vertical force, and moment, balance, only the first two apply at a pinned joint, though pinned joints can be given extra degrees of freedom as shown in Fig 31.

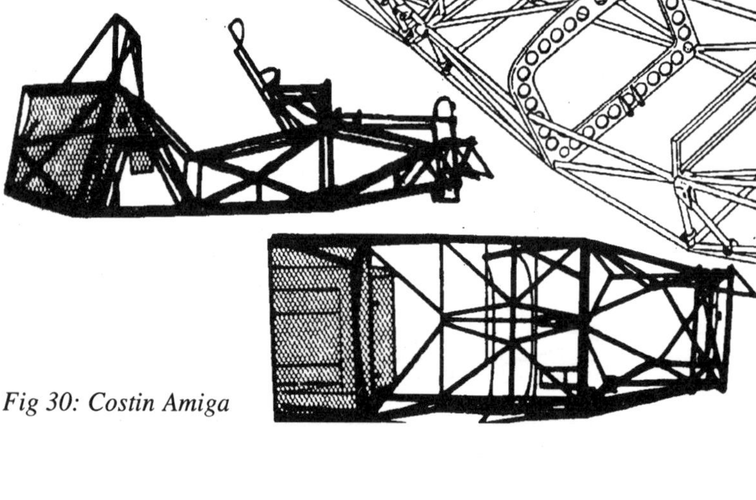

Fig 30: Costin Amiga

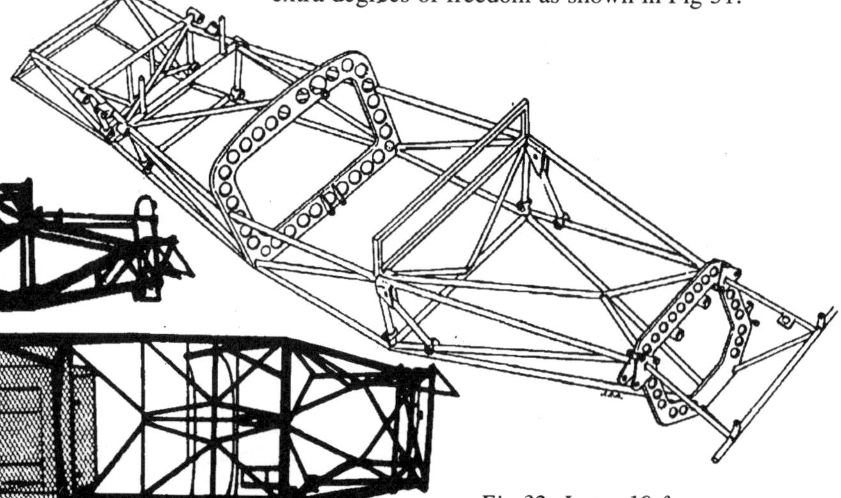

Fig 32: Lotus 19 frame

Stable determinate frames are made by adding triangles formed by two extra members and one joint as shown. Addition of more or fewer members involves either a mechanism or a redundant frame in which some, or all, members can only deflect with accompanying deformation of the others. Note that a determinate frame has members which can deform independently. Determinate space frames are made by adding pyramids of three extra members. The Lotus 19 is an example, Fig 32 — before the adoption of the later backbone tube and space frame powertrain support.

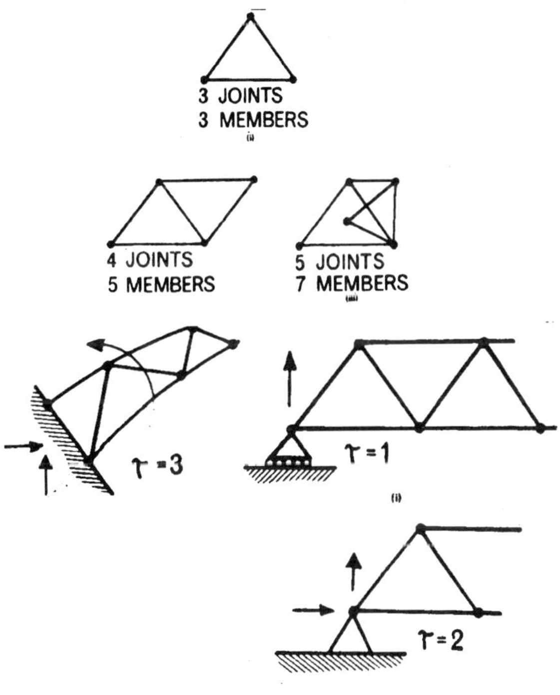

Fig 31: Number of redundancies

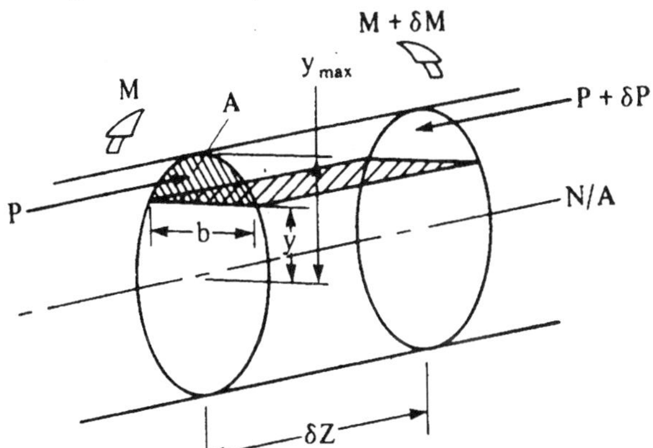

Fig 34: Beam section notation

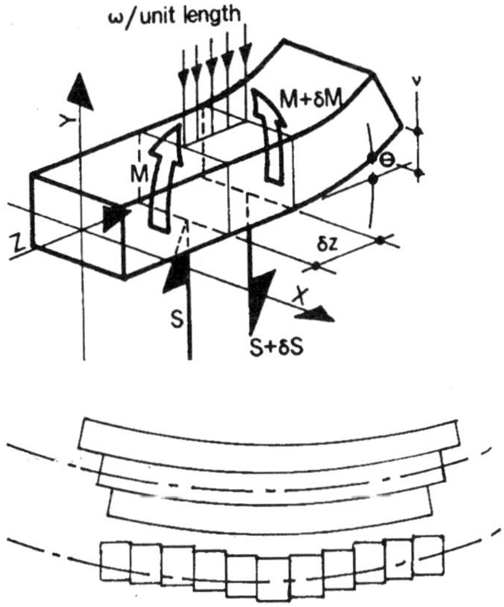

Fig 33: Beam calculation notation

Beam elements

Consideration of thin-walled structures in an earlier section introduced beam-elements under the combined action of shear forces and bending torques (or moments). For the relatively short and deep beams there considered, external shear forces caused the predominant internal loading of the beam. Here we consider the relatively longer and shallower beam for which the predominant internal loading is bending stress caused by external bending moments.

Familiar structural elements on vehicles, subject to bending rather than shear stresses, in flexure, are chassis sidemembers on trucks and integral body crossbearers on cars — which carry seating loads reacted at their end-connections with the body side-panels and rails.

Fig 33 is useful in establishing a sign convention for the analysis of beams so that positive and negative values of forces follow a constant rule for different load cases. Shear force S is taken as positive clockwise; bending-moment M is positive upwards towards the compressed top fibres of the beam (clockwise left, anticlockwise right); load is negative applied downwards on the compressed edge of the beam and changes in slope θ and translational-deflection v of the beam are both positive in the positive direction of upward loading on the beam. The balance of forces over a short slice of the beam, length dz, along with basic stress/strain relationships established earlier, leads to the equation of flexure, for a beam bent to radius of curvature R, used in calculation: $\sigma/y = M/I = E/R$ for bending stress σ at y distance over the cross-section of the beam from the 'neutral-axis' and I section second moment of area — where, at the centre of symmetry of the section, bending stresses are zero between increasing compression, above, and increasing tension, below, for the sort of loading depicted in Fig 34. Inset in Fig 33 is an imaginary picture of the beam element to help visualise the concept of 'complementary shear' by considering the beam to be comprised of vertical and horizontal slipplanes. To calculate the value of these complementary vertical and horizontal shear stresses an integration needs to be performed, for irregular shaped sections, based on further relationships between external and internal forces. Moving from left to right along the elemental length dz in Fig 34, the external load w per unit length equals the rate of change of shear force (first derivative, dS/dz) which in turn equals the second derivative of bending moment (and in turn the fourth derivative of distance y times the flexural stiffness, EI) where the elastic modulus is E. This leads to the formula for shear stress in beams:

$$\tau = SAY/Ib$$

where Y is the distance from the neutral axis to the centroid of the shaded area of the cross section seen in Fig 34.

A use for this formula is in determining the required pitch of rivets (or spot welds) on the closing plate which might be boxing in a beam of otherwise open cross section (Fig 35). Table 2 shows a logical way to set out the calculation and the distance of the neutral axis from the lower face of the bottom reinforcing plate can be found from $\sigma[Ay]/\sigma[A]$ in this case = 2.64/2.15 = 1.23 cm. The section second moment of area for the total section is found from $I = I_g + Ah^2$ where h is the distance of the centroid of each element of the section from the neutral axis and I_g the element's inertia about its own axis. Here I = 3.877 cm^4 and the distance of the neutral axis from the farthest stressed layer of the beam = 3.3-1.23 = 2.07 cm. As the centroid distance of the reinforcing plate from the neutral axis is 1.18 cm, then for a given rivet shear load capacity, required pitch can be determined. For a bending moment of 300 Nm, this would work out to be a value of 2.05 cm in this particular worked example.

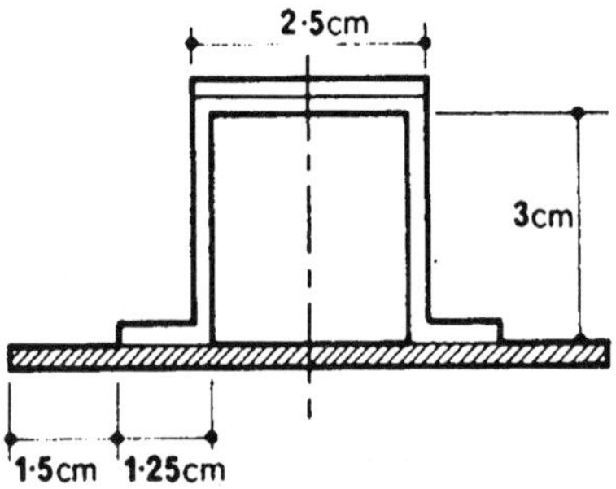

Fig 35: Fabricated box-beam

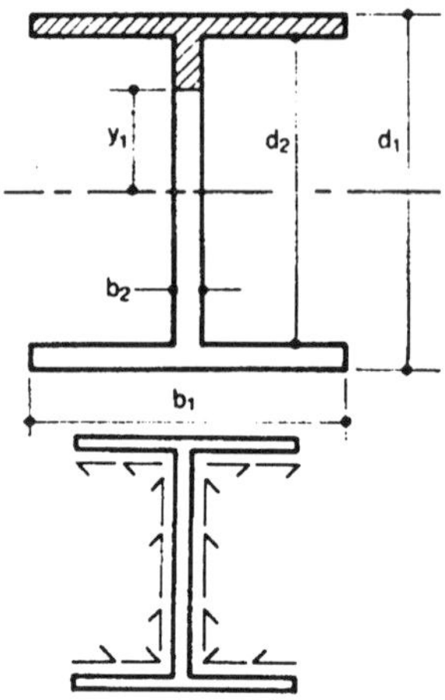

Fig 36: Thin-wall I-beam

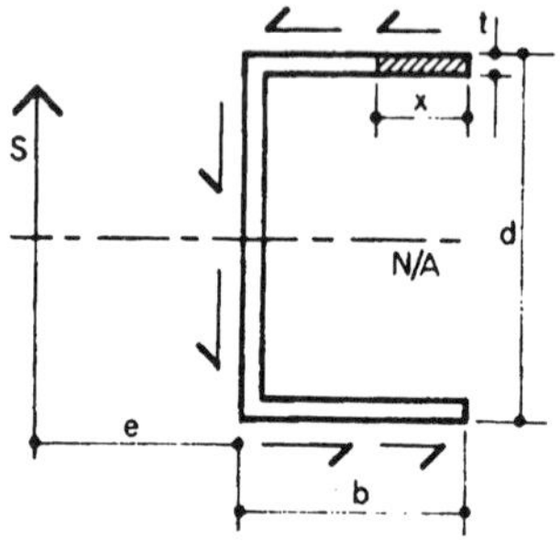

Fig 37: Thin-wall channel

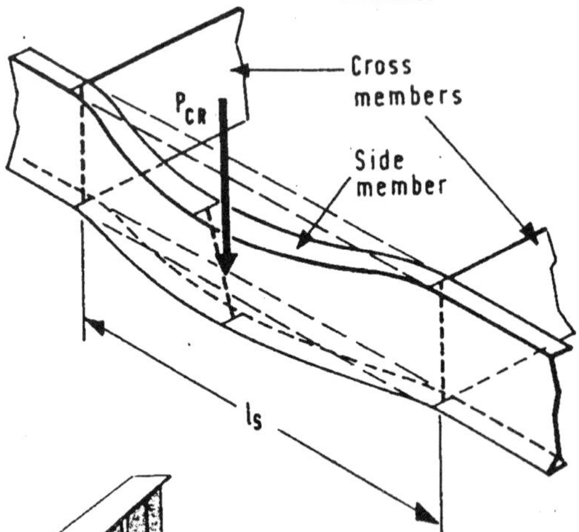

Fig 38: Lateral buckling

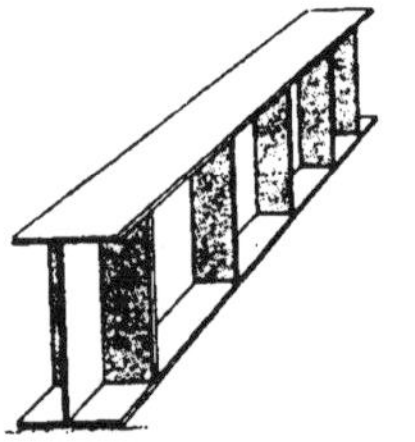
Fig 39: Web stiffeners

Table 2: Section modulus calculation

	Area A (cm^2)	Centroid distance y (cm)	Ay (cm^3)	I_g (cm^4)	distance to N/A, h (cm)	Ah^2 (cm^4)
Top 'flanges'	0·5	3·2	1·60	$(2{\cdot}5 \times 0{\cdot}2^3)/12 = 0{\cdot}00167$	1·97	1·94
Webs	0·6	1·6	0·96	$(0{\cdot}2 \times 3^3)/12 = 0{\cdot}45$	0·37	0·082
Bottom flanges	0·25	0·15	0·04	$(2{\cdot}5 \times 0{\cdot}1^3)/12 = 0{\cdot}00021$	1·08	0·293
Reinf. plate	0·8	0·05	0·04	$(8 \times 0{\cdot}1^3)/12 = 0{\cdot}00067$	1·10	1·11
	$\sum = 2{\cdot}15$		$\sum = 2{\cdot}64$	$\sum = 0{\cdot}4525$		$\sum = 3{\cdot}425$

Table 3: Values of gamma

$I_s^2 . GJ/EI_w$	0·4	4	8	16	32	64	96	128	200	400	
for P_{CR} concentrated at mid-span	466	154	114	86·4	69·2	—	54·5	52·4	49·8	47·4	*Load applied at section centroid*
for P_{CR} distributed evenly between lateral supports	673	221	164	126	—	101	79·5	76·4	72·8	—	*Load applied at section centroid*
	587	194	145	112	—	91·5	73·9	71·6	69·0	—	*Load applied at top flange*

Table 4: Permissible stress for main alloys — tsi (N/mm^2)

Alloy	Condition	Form	Thickness in (mm) from	Thickness in (mm) to	Axial p_t p_c	Bending p_{bt} p_{bc}	Shear p_q	Bearing p_b	λ_s
H30	WP	Extrusions, solid and hollow	¼ (6·4)	3(c) (76)	9·5 (147)	10·5 (162)	5·7 (88)	14·4 (222)	59
		Sheet	—	¼ (6·4)	9·0 (139)	9·9 (153)	5·4 (83)	13·7 (212)	61
		Drawn tube	—	1/16 (1·6)	9·1 (141)	10·1 (156)	5·5 (85)	14·4 (222)	
			1/16 (1·6)	—	8·6 (133)	9·6 (148)	5·2 (80)		64
		Plate	¼ (6·4)	1 (25)	8·5 (131)	9·5 (147)	5·1 (79)	13·7 (212)	
N8	M	Extrusions, solid and hollow	—	—	5·4 (83)	6·3 (97)	3·2 (49)	13·0 (201)	96
		Plate	¼ (6·4)	1 (25)					
	O	Extrusions, solid	—	—	5·3 (82) 5·0 (77)	6·1 (94) 5·9 (91)		12·2 (188)	101
		Plate	¼ (6·4)	1 (25)	5·4 (83) 5·1 (79)	6·3 (97) 6·0 (93)		13·0 (201)	
H9	WP	Extrusions, solid and hollow	—	6(e) (152)	5·7 (88)	6·3 (97)	3·4 (53)	9·0 (139)	81
	P	Extrusions, solid and hollow	—	⅛ (3·2)	5·2 (80)	5·7 (88)	3·1 (48)	8·2 (128)	86
			⅛ (3·2)	½ (13)	4·2 (65)	4·7 (73)	2·5 (39)	7·6 (117)	102

Table 5: C1 and C2 values for webs

Alloy	C_1		C_2	
H30-WP sheet	65		120	
plate	67		124	
N8-O, N8-M plate	84		156	
Others	$150/\sqrt{p_q}$	p_q in tonf/in²	$280/\sqrt{p_q}$	p_q in tonf/in²
	$590/\sqrt{p_q}$	p_q in N/mm²	$1100/\sqrt{p_q}$	p_q in N/mm²

The analysis of the beam section in separate elements is particularly valuable in calculating thin walled beam performance. Such sections have shear stresses varying widely across the section — in contrast to a slender solid-section beam, the shear stress across which can often be assumed constant in calculation. For the I-section thin-wall beam of Fig 36, complementary shear flow acts horizontally in the flanges, as well as vertically in the web, which runs out to zero at the edges. Thus for thin-wall section beams, the assumption of constant shear stress is made only for the 'vertical webs' normal to the neutral plane of bending. For sections which are non-symmetrical such as the channel of Fig 37 the sum of the shear stresses in individual elements has a resultant at distance e from the web which passes through the shear-centre of the beam.

To avoid twisting of the beam, induced by bending loads, the external load should be applied through the shear centre, the value of e being obtained from $e = td^2b^2/4I$. To ensure best performance from thin-wall section beams, means must be provided to ensure that cross-sectional shape is maintained — if necessary by employing lateral stiffeners in the form of bulkheads or ribs. There is otherwise a tendency for certain sections to buckle in bending,

Fig 38. Another problem to avoid is that of fatigue cracking of compression flanges on open section beams; this can be caused by local buckling of flanges which are too wide to be stable in compression. This tendency can be reduced by the use of web stiffeners as shown in Fig 39. There is also a possibility of overall lateral buckling of flanges if too narrow a flange is chosen. A formula due to Timoshenko can be used to calculate optimum spacing of crossmembers to avoid such problems in which 'warping rigidity', rather than bending rigidity, is expressed as the product of elastic modulus and warping constant, $I_w = t_f h^2 b^3$ for use in the formula $P_{cr} = g\ (EI_w GJ)^{1/2}$ where γ takes values from Table 3; *b* and *h* are section height and width and t_f is flange thickness.

A common condition in vehicles is the support of deck flooring by cross-bearers. If the decking is made up of a number of discrete beams, or planks, and an elastic crossbearer is supporting them at mid-span, the deflection of the planks is given by:

$$(5wl^4/384EI) - (Rl^3/48EI)$$

from which can be found the radius of curvature of the deflected combination $R = (5wl/8) - (48EIy/l3)$ and the centre deflection of the crossbearer is $(w/a)[1 - 2(\cos bl/2)(\cosh bl/2)/(\cos bl + \cosh bl)]$ where $a = 48EI/l^3$ and $b = (a/4EI)^{1/4}$

Beam design methods

The design of beam systems in aluminium alloy is covered in British Standards[7]. This states that the direct flange stress in bending shall not exceed the permissible value given in Table 3. The average shear stress in the webs should not exceed that in Fig 40 for *d/t* not exceeding *C1* in Table 5. It is also necessary to guard against failure in lateral buckling in those cases where the external loads might give rise to this situation — that is a tendency to twisting and sideways deflection. Thin-walled beams should not be stressed beyond compressive limits shown in Fig 41 where the lamda value (λ) of the y-axis is calculated from $k(t_f/t_2)^{1/2}$ for I-sections and channels or $k(t_f/b)^{1/2}$ for solid and hollow rectangular sections where *k* is found from Fig 42, t_2 is flange thickness and *b* is section width.

Intermediate transverse web-stiffeners must be provided throughout the length of a beam where the *d/t* ratio exceeds the value of *C1* in Table 4 where *t* is web thickness and *d* is web depth. Stiffeners must be designed so that second moment of area of each, about the face of the web, is not less than $1.3d^3 t^3 /s^2$ where *s* is stiffener spacing. Normally it will not be required to have stiffeners at spacings greater than *1.5d*. Where the ratio *d/t* exceeds *C2* in Table 5, longitudinal stiffeners, Fig 43, will be required as well as transverse ones. Fig 44 gives values of *h/d* such that panels above and below the stiffener are of equal strength. Stiffeners should be so designed that second moment of area about the face of the web is not less than $4dt^3$. Outstanding leg-widths of both transverse and longitudinal stiffeners must not exceed 12 times plate thickness.

Curved and cranked beams

The Unit Load method of structural analysis is particularly suitable for more complex shape real-life beams under a variety of load conditions. The method

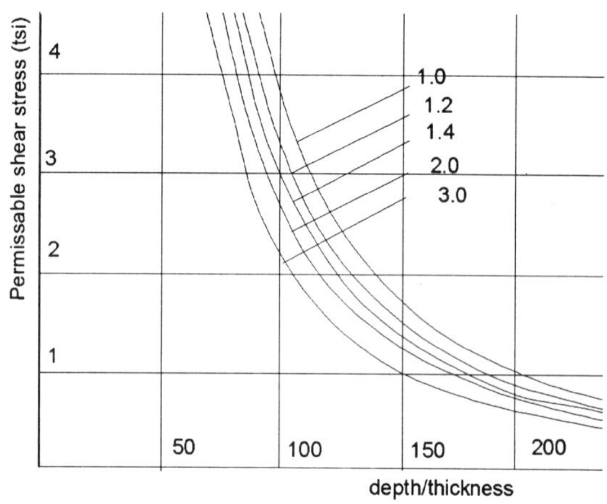

Fig 40: Permissable shear stress in webs -- tsi

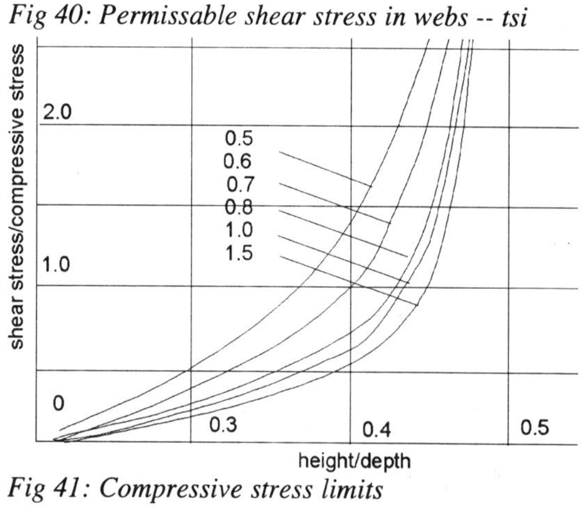

Fig 41: Compressive stress limits

can be understood by imagining a small elastic element in a curved beam which is otherwise considered to be rigid. An imaginary load *w*, at *P* in Fig 45, is then considered to cause vertical displacement *D*; the beam deflects Δ under the influence of the bending moment and it can be shown that $\Delta = \int Mm/EI.ds$ where *m* is bending moment due to imaginary unit load at P and *M* due to the real external load system. An example might be a car floor transverse crossbearer having a 'tunnel' portion for exhaust-system and prop-shaft clearance, Fig 46. To find the downward deflection at B, the imaginary unit load is applied at that point and it develops reactions of 2/3 at A and 1/3 at D. The bending moments for the real and imaginary loads are shown in Table 6 which also illustrates the subsequent calculation to obtain the deflection due to the driver and seat represented by distributed load *w* over length *l*.

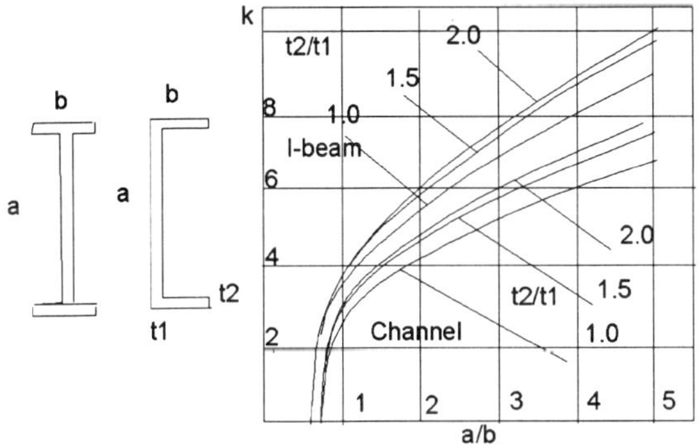

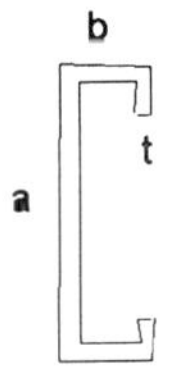

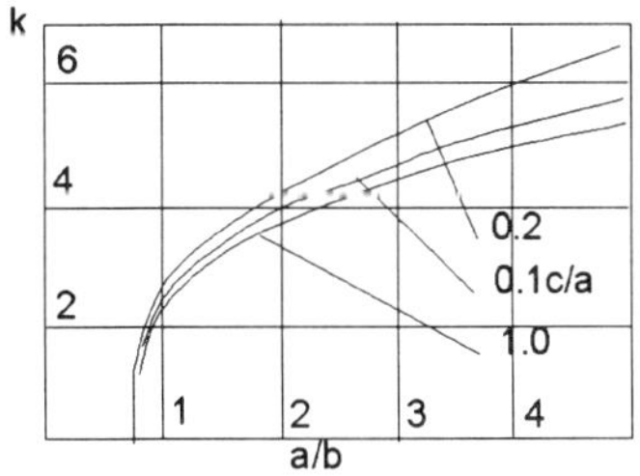

Fig 42: K values

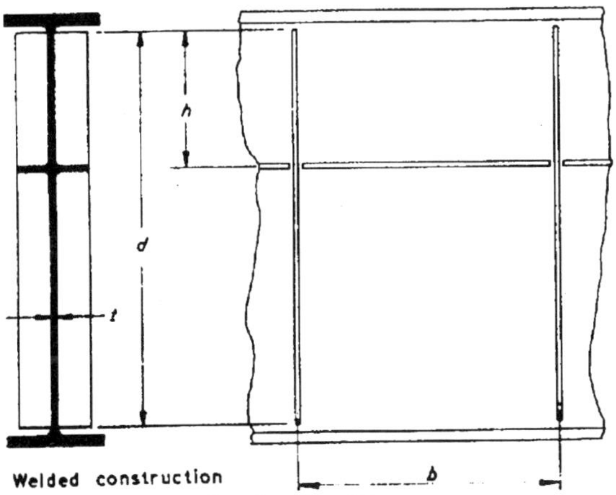

Fig 43: Longitudinal web stiffeners

Continuous beams

Where a beam passes over one or more mid-span supports, a 'redundant' structure exists which is 'statically indeterminate'. A more refined analysis therefore has to be undertaken known as the Moment Distribution method. This uses a hypothetical cut to determine distribution of bending moment along the beam. For the beam and load shown in Fig 47, firstly each support is hypothetically locked with a fixing which makes slope of the beam equal to zero at these points. The so-called fixing endmoments (FEMs) are then calculated at these points and their difference in value for adjacent bays noted (the sign convention states that all FEMs are assumed positive). These moments are then progressively relaxed by balancing moments which revert the 'fixed' supports back to pinned ones. The relaxing moments are apportioned according to the relative stiffness of the 'bays' each side of the support, distribution factors for this purpose being proportional to elastic modulus and section second moment of area but inversely proportional to bay length:

$$K_1 = (EI/l)_1/[(EI/l))_1 + (EI/l)_2]$$

and $K_2 = 1 - K_1$

As the supports are 'relaxed', a carry-over moment is transferred to the next support, either side,

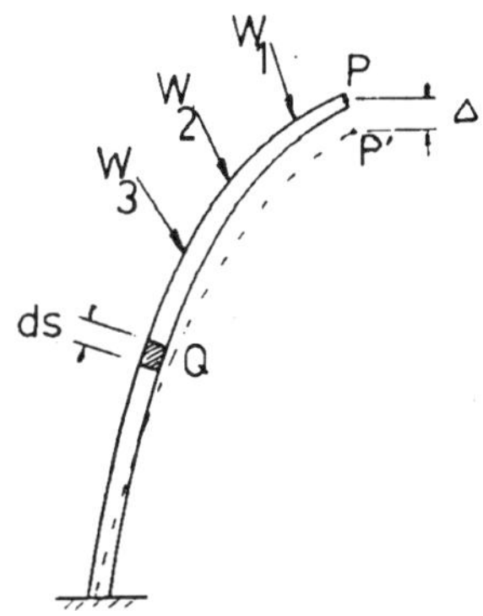

Fig 45: Unit Load method

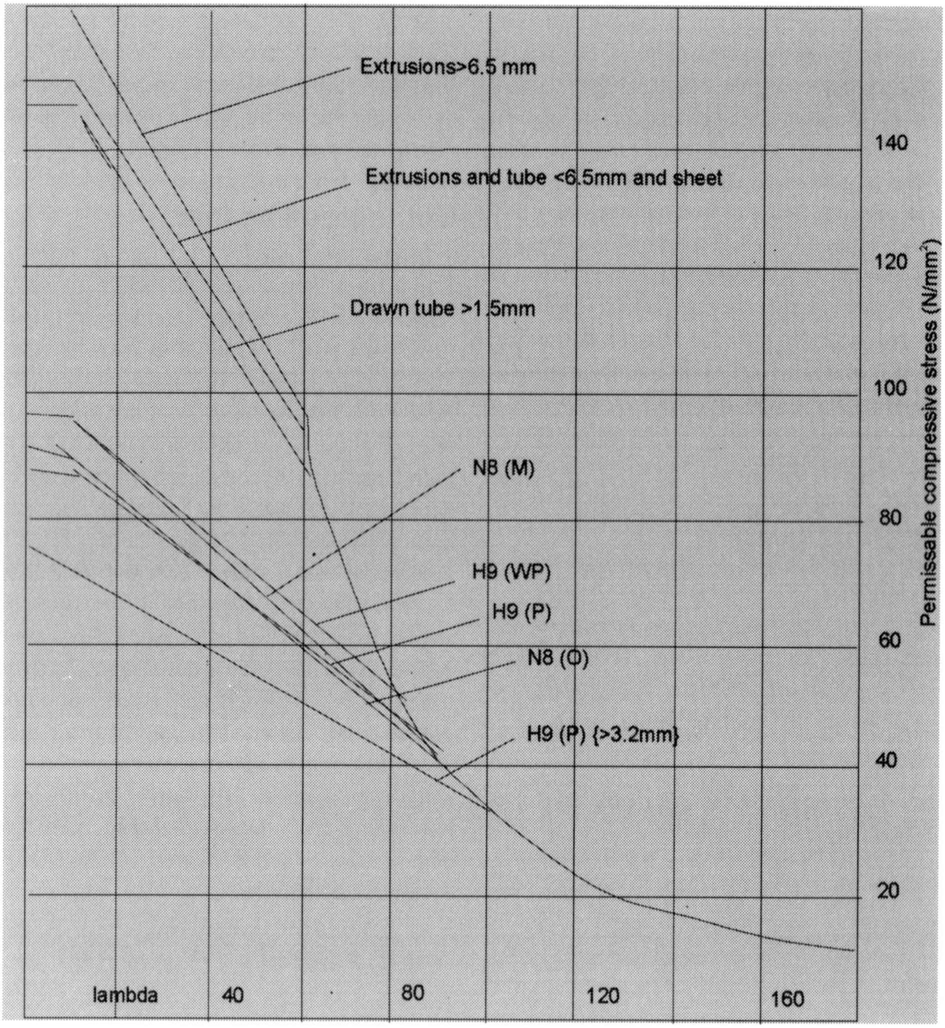

Fig 44: Values of h/d

according to the relation $Ma = (1/2)Mb$. The sign convention makes the moment 'differences' into an algebraic sum so that the negative sign becomes positive. In the figure, the bending moment diagrams for the two bays in the example, after 'fixing' the intermediate support, are shown beneath the loading diagram; these indicate the FEM imbalance. The relaxing operation in stage-3 is equivalent to the successive imposition of BM diagrams such as in (iv) — with the final BM diagram appearing in (v).

An automotive example of the use of the method is seen in Fig 48 to represent the chassis sidemember of a medium-weight truck having an applied load system representing front and rear engine/transmission mounts and payload acting through bodymount points. Stepping of the frame approxi-

Table 7: Sidemember calculation stages, from + 0.98 at B

B		C		D		E
1	0.555	0.445	0.378	0.622	1	D.F
-0.35 -0.63	+0.35 +2.69	-5.2 +2.15	+5.2 -1.48	1.3 2.43	+1.3 +0.5	-1.8 F.E.M
+1.34 -1.34	-0.32 +0.59	0.74 +0.47	+1.07 -0.50	+0.26 0.82	-1.22 +1.22	
+0.30 0.30	-0.67 +0.51	0.25 +0.41	+0.24 0.32	+0.61 0.63	-0.41 +0.41	
+0.25 0.26	-0.15 +0.17	0.16 +0.14	+0.21 0.16	+0.21 0.26	-0.27 +0.27	
+0.09 0.09	0.12 +0.11	0.08 +0.08	+0.07 0.07	+0.13 0.12	0.13 +0.13	
0.98	+3.18	-3.18	+4.26	-4.26	+1.8	1.8 kNm

mates to the change in section modulus due to tapering and the distributed load pattern in (ii) approximates the real one with additional conversion to SI units. Pinned supports represent the ends of the laminated suspension springs and section second moment of area for the beam in mm^4 x 10^6 are for each bay as follows: AB 8.5, BC 12.5, CD 20, DE 16.5 and EF 12.5. Handbook values of the FEMs for the two bays are shown at (iii) and the analysis is carried out as in Table 7, from which the following distribution factors are obtained:

$$K_{cl}=(1.25/1.25)/(1.25/1.25+2.0/1.25)=0.555$$
$$K_{dl}=(2.0/1.25)/(2.0/2.5+1.65/1.25)=0.378$$

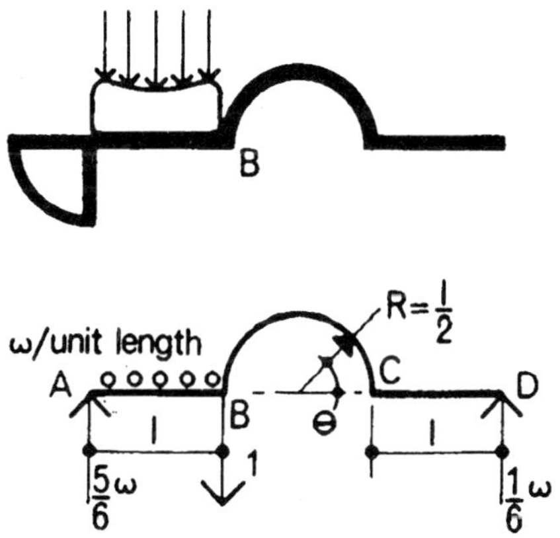

	M_0	m_1
AB	$\frac{5wl}{6}s - \frac{ws^2}{2}$	$\frac{2}{3}s$

BC $\quad \frac{wl}{6}\left(l+\frac{l}{2}-\frac{l}{2}\cos\theta\right) \quad \frac{1}{3}\left(l+\frac{l}{2}-\frac{l}{2}\cos\theta\right) = \frac{l}{6}(3-\cos\theta)$

$$= \frac{wl^2}{12}(3-\cos\theta)$$

CD $\quad wls/6 \quad s/3$

So

$$\delta_B = w\int_0^l \left(\frac{5}{9}ls^2 - \frac{1}{3}s^3\right) ds +$$

$$\frac{wl^4}{144}\int_0^{\pi} \left(3^2 - 6\cos\theta + \cos^2\theta\right) d\theta + \int_0^l \frac{wl}{18}s^2\, ds$$

Fig 46: Seat support bearer

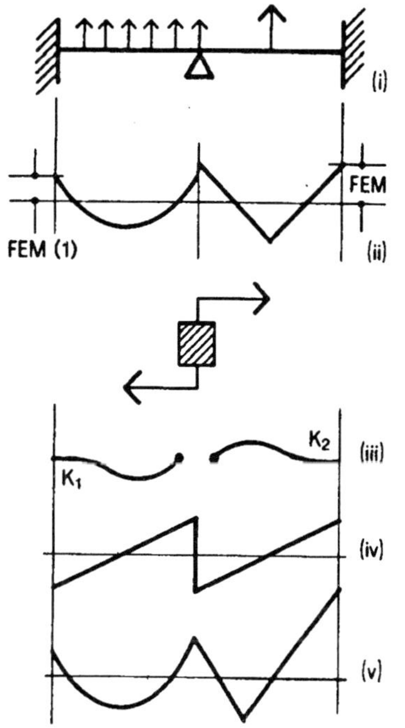

Fig 47: Moment distribution method

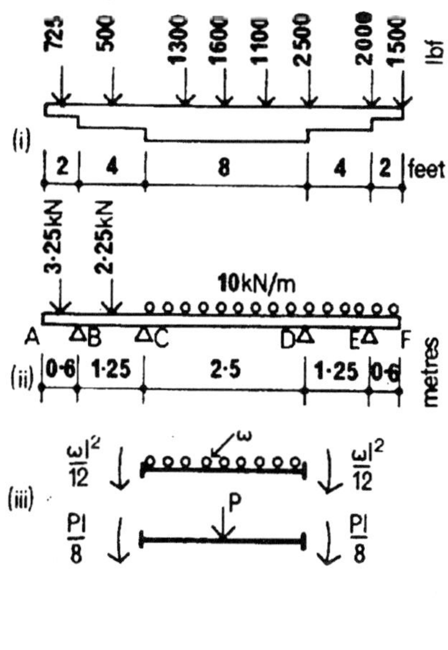

Fig 48: Chassis sidemember calculation

Stiff-jointed frames

A variety of structural elements of the vehicle body and chassis comprise stiff-jointed frames. These may be either simple rectangles or made up of arched and cranked members. All can be readily analysed by the Unit Load technique introduced in the previous sections.

The methods of calculation used for curved and cranked beams were also described above and can be adapted for the calculation of deflections in stiff-jointed frames. When the frame is stiff-jointed as opposed to pinned, bending moments are transferred from one element to another across the joint.

In the part of the frame shown as Fig 49 subject to twisting and bending, deflection due to twist = $\int T dx / GJ$. This deflection would then be added to that due to bending, obtained from the formula given in the earlier section. Usually the deflection due to axial straining of the elements is so small that it can be discounted.

A good example of a frame subject to twisting and bending is the portal shape frame of Fig 50 having loading and support, in the horizontal plane, indicated. If it is required to find the vertical deflection at D for the load P applied at B, equating vertical forces and moments is first carried out to find the reactions at the supports. Next step is to break the frame up into its elements and determine the bending/moments in each — ensuring compatibility of these and that associated loads are 'transferred' across the artificially broken joints.

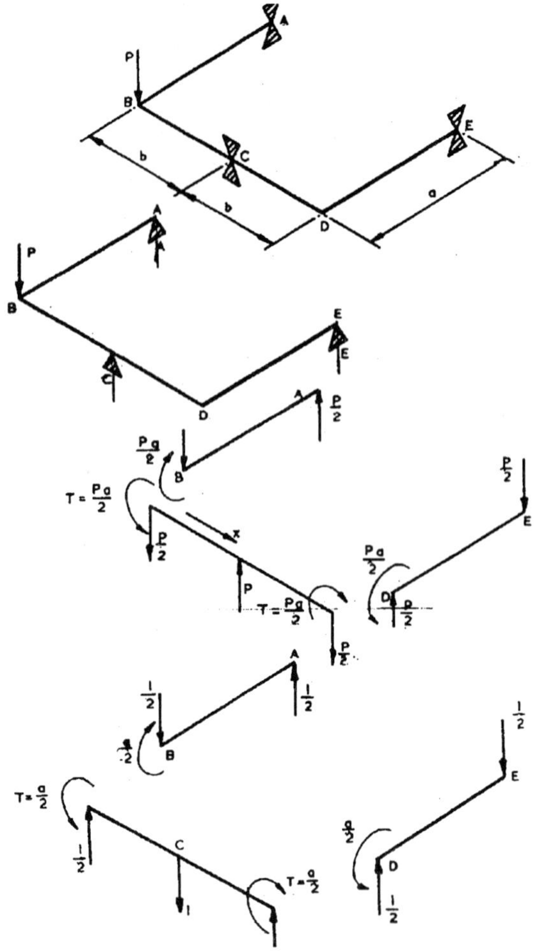

Fig 50: Portal frame under combined loading

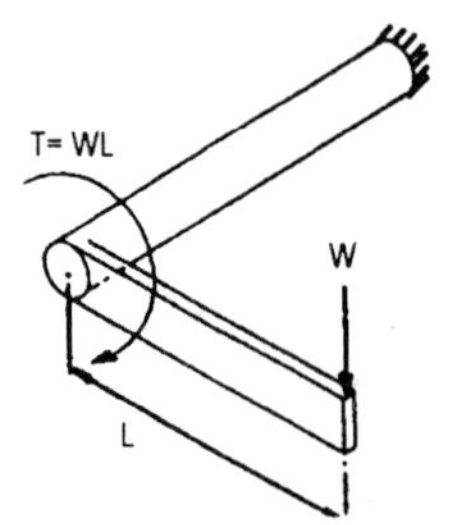

Fig 49: Element subject to twisting and bending

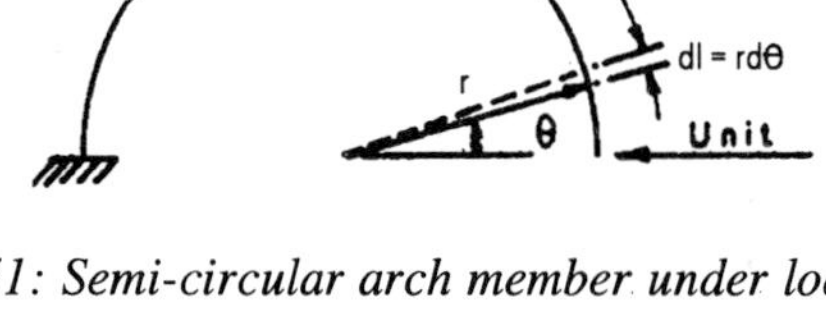

Fig 51: Semi-circular arch member under load

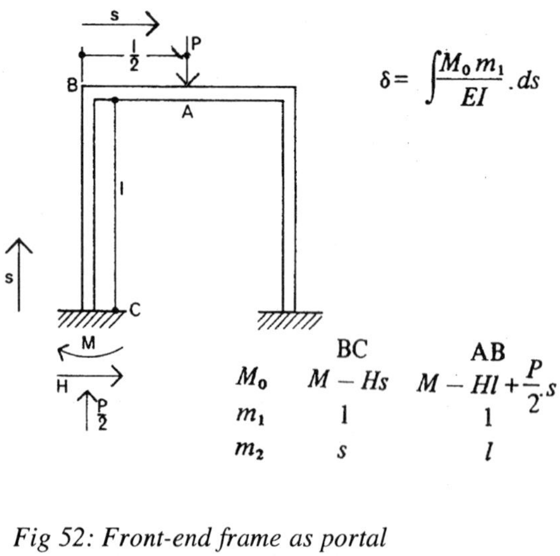

Fig 52: Front-end frame as portal

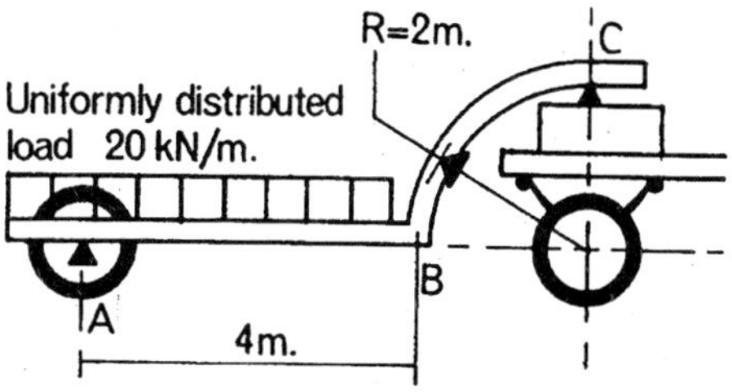

Fig 53: Goose-neck semi-trailer frame

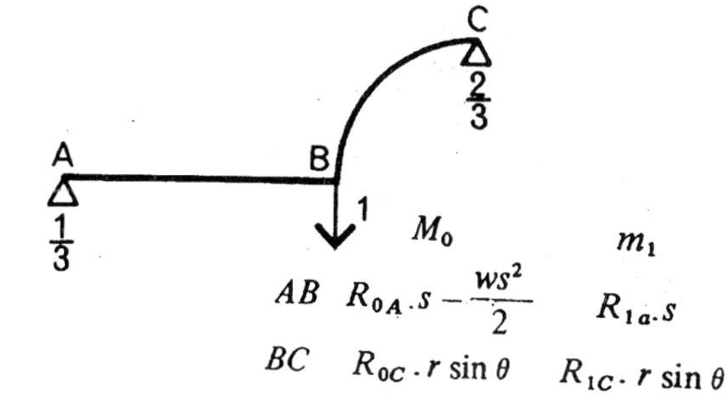

Fig 54: Bending moments for trailer frame

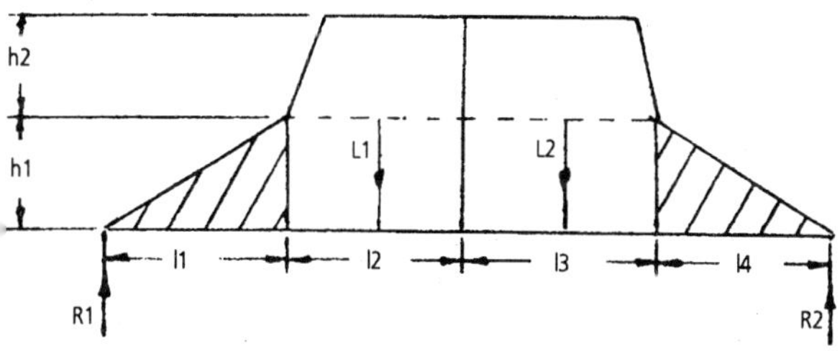

Fig 55: Simplified sideframe idealisation

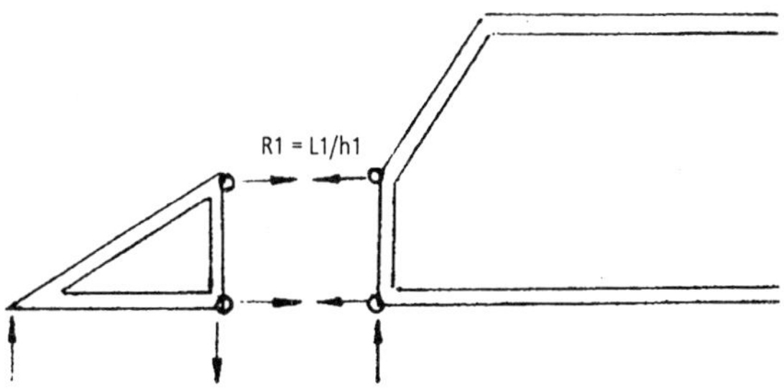

Fig 56: Reaction load referred into sideframe

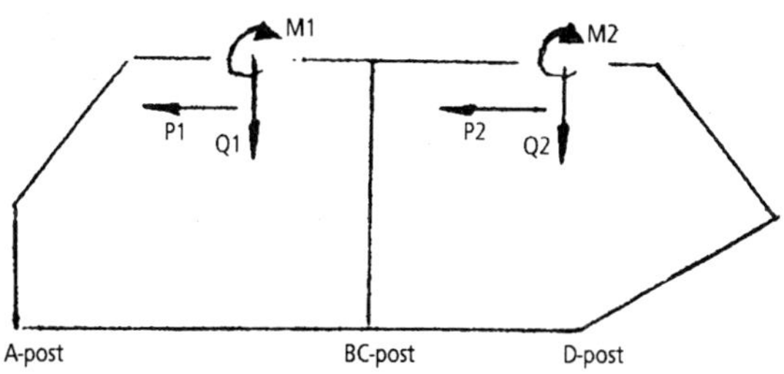

Fig 57:Alternative idealisation for sideframe

Integrations have to be carried out for the three deflection modes, as follows

$$\int Sb.ds/AE + \int Mm.ds/EI + \int Ff.ds/AG$$

to obtain the combined deflection. Both vertical and horizontal components can be obtained by applying vertical and horizontal imaginary unit loads, in turn, at the point where the deflection is to be determined. For the example shown in the figure, elastic and shear moduli, E and G, are 200 and 80 GN/m^2 respectively.. S is axial load and F the shear load, the latter being negative in compression. The values of these, with the bending moments, are as follows:

	S	b	M	m	F	f
Vert.						
AB	0	0	Ps	1s	P	1
BC	-P	-1	1P	1.1	0	0
Hor'l.						
AB	0	1	Ps	0	P	0
BC	-P	0	1P	1s	0	1

These can readily be substituted in the expression above to determine deflections as a factor of P.

The method can also be used for 'continuum' frames incorporating large curved elements provided curvature is high enough for the engineer's simple theory of bending to be applicable. In the semi-circular arch of Fig 51, at any point along it defined by variable angle θ, bending moment is given by $Wr \sin\theta$ due to the external load and $r \sin\theta$ due to the imaginary unit load at the position and in the direction of the deflection which is required. In this case it is the same as that of the external load and the deflection due to bending is equal to the integral of the product of these between $q = 0$ and p, divided by the flexural rigidity of the section EI, product of elastic modulus and section second moment of area.

A common example of a portal frame in a horizontal plane is the front crossmember and front ends of the chassis sidemembers of a vehicle imagined 'rooted' at the scuttle. The load case of a central load on the crossmember, perhaps simulating striking a pole, can be visualised in Fig 52.

An example of a frame having both curved and straight elements is the idealised goose-neck semi-trailer shown in Fig 53. Again a table of bending moments can be written, as shown, and the vertical deflection at B calculated by using the Unit Load formula. Given the value of I for the sidemember of 0.25 x 10^4 metre^4 and E of 210 GN/m^2, the the deflection works out to be 53 mm.

It is generally best to integrate along the perimetric co-ordinate for curved elements and, as above; when requiring a translational deflection, apply unit force, and for a rotational displacement, apply unit couple. In the case of redundancies, remove the redundant constraint by making a cut then equate the deflection of the determinate cut structure with that due to application of the redundant constraint.

Idealised car-body side-frames

Saloon-car sideframes can be idealised into stiff-jointed frames with a relatively small number of redundancies — for analysis. Fig 55 suggests an idealisation with a considerable degree of symmetry and Fig 56 the feeding in of reaction loads. Normally a four door vehicle side frame would have six degrees of redundancy as shown in Fig 57. Various simplifications can be made to obtain an approximate solution of the structure. Pin-joints assumed at the ends of the B-C post would reduce the redundancy to 4. A strong D-post top-joint might also justify the assumption of the A-post top joint being pinned, reducing the structure to 3 redundancies.The overall bending moment on the frame can be replaced by compression force in the cant rail and tension in the sill. If the maximum moment is in front of the B-C post, BM at the base of the post is Ph^2. The balancing compressive force will be shared by the B-C and D-post in proportion to their relative stiffnesses. Other vehicle sideframes can be simulated by 'L' and 'U' shaped members as shown in Fig 58 and again Unit Load equations can be set up in terms of the main elements. Flexibilities can be added in series to obtain a picture of the whole structure including shear panels whose contribution will be At^2/Gt where G is rigidity modulus, A the panel area, t its thickness and t the shear flow due to unit torque. Method of assessing the degree of redundancy of stiff jointed frames such as the door closure ring shown at Fig 59 is to cut the ring and consider the forces necessary to reclose it (*centre*). Here two linear displacements and a rotation are required giving an order of redundancy of three. For a three dimensional ring structure this rises to six, Fig 60. Generally for a stiff-jointed structure of M members and N nodes, the number of rings will be $M - N + 1$. Fig 61 shows a worked example using algebraic symbols for a pinned parabolic arch to demonstrate the Unit Load method.

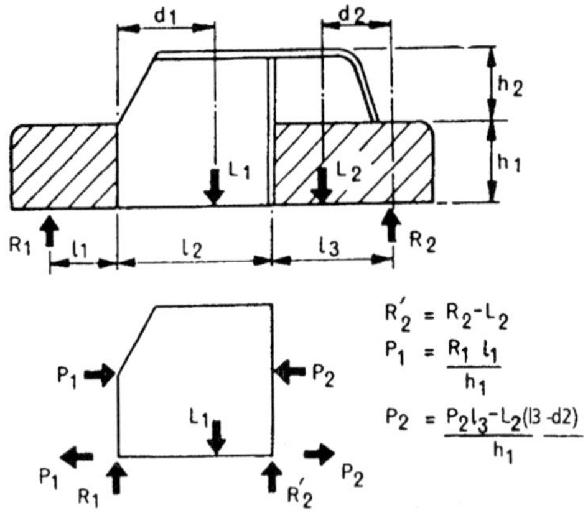

Fig 58: Redundant sideframe idealisation

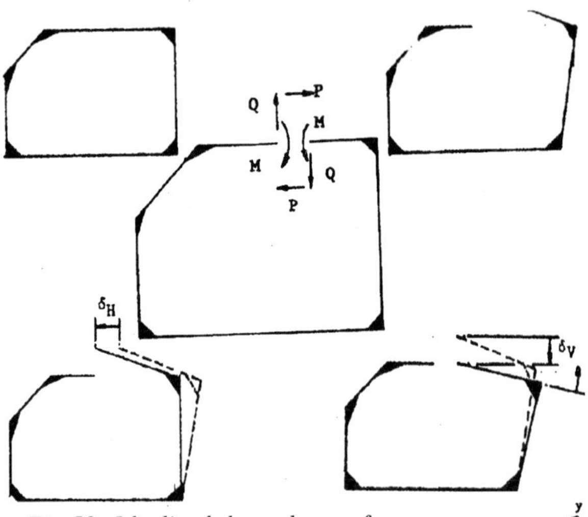

Fig 59: Idealised door-closure frame

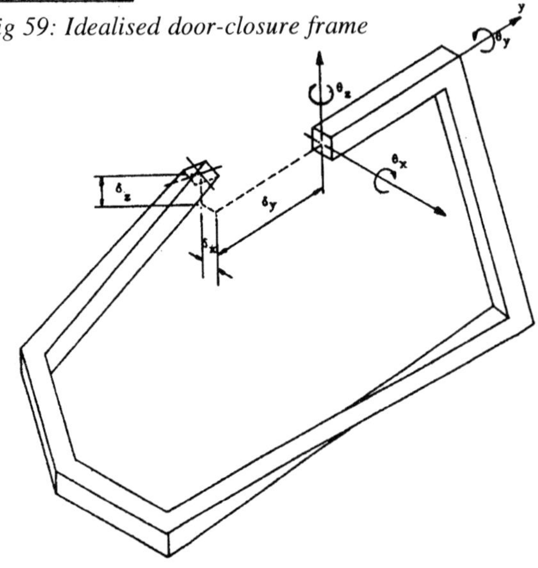

Fig 60: Three-dimensional ring-frame

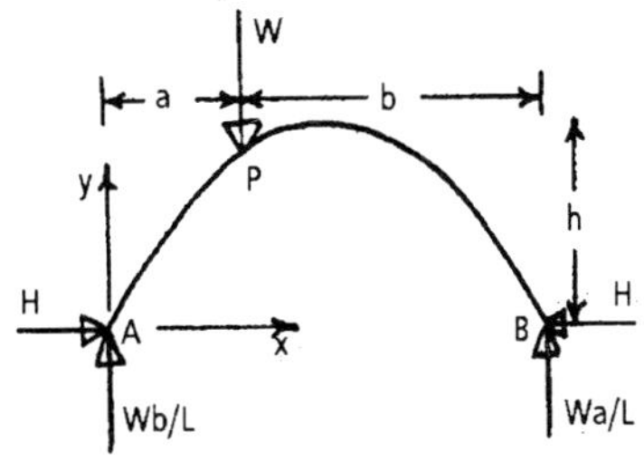

Part AP origin of x and y at A

$$M = \frac{Wb}{L}x - Hy, \quad m = -y \quad \therefore \int_A^P Mm\,ds \simeq H\int_0^a y^2dx - \frac{Wb}{L}\int_0^a xy\,dx$$

Part PB (origin of x and y at B).

$$M = \frac{Wa}{L}x - Hy, \quad m = -y \quad \therefore \int_B^P Mm\,ds \simeq H\int_0^b y^2dx - \frac{Wa}{L}\int_0^b xy\,dx$$

Adding the integrals for the two parts and equating to zero gives

$$0 = H\int_0^L y^2dx - \frac{Wb}{L}\int_0^a xy\,dx - \frac{Wa}{L}\int_0^b xy\,dx$$

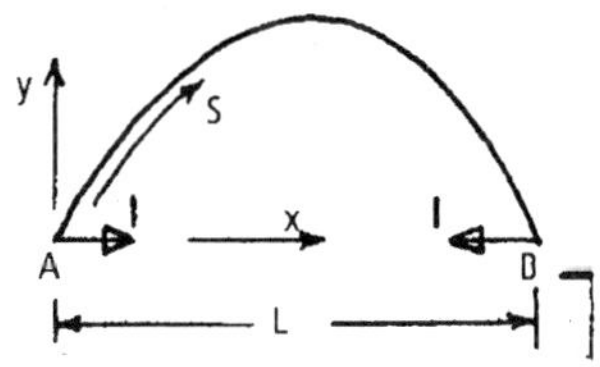

$$i.e., \; 0 = 16h^2H\int_0^L \left(\frac{x^2}{L^2} - \frac{2x^3}{L^3} + \frac{x^4}{L^4}\right)dx$$

$$-\frac{4Wbh}{L}\int_0^a \left(\frac{x^2}{L} - \frac{x^3}{L^2}\right)dx - \frac{4Wah}{L}\int_0^b \left(\frac{x^2}{L} - \frac{x^3}{L^2}\right)dx$$

$$i.e., \; 0 = 16h^2HL\left(\tfrac{1}{3} - \tfrac{1}{2} + \tfrac{1}{5}\right) - \frac{4Wbh}{L}\left(\frac{a^3}{3L} - \frac{a^4}{4L^2}\right)$$

$$-\frac{4Wah}{L}\left(\frac{b^3}{3L} - \frac{b^4}{4L^2}\right)$$

Which reduces to: $H = \dfrac{5\,Wab}{8\,L^3h}\left(4a^2 + 4b^2 - \dfrac{3a^3}{L} - \dfrac{3b^3}{L}\right)$

Fig 61: Worked example of pinned parabolic arch

Torsion in box beams

As many vehicle structures can be idealised as box-beams, the behaviour of these structural elements in torsion is fundamental in body stressing and deflection evaluation. The theory is a valuable teacher in design prior to the more accurate prediction of loads and displacements by finite element techniques.

Torsion of box beams was introduced at the end of the earlier section on thin-walled structural analysis. It is useful to examine the change from a fully closed section box beam to an open section one, in Fig 62, which is really a limiting case of the closed tube as it collapses. The shear stress distribution has a basic influence on the torsional rigidity of the cross-section. Closed tubes resist torsion by the area of the cross section bounded by the mean thickness of the wall. Shear flows around an elemental section of the wall are equal and constant (Fig 63). Also by integrating along the circumference of the section, the sum of the horizontal and vertical components of tangential forces dP can be summed to zero and their tangential components equated with the externally applied torque T to give shear flow, $q = T/2[A]$ where $[A]$ is the area of the closed tube bounded by the mean thickness line of the wall.

The twist per unit length of a closed tube is found by equating the shear energy per unit volume with the work *done* on the twisted cross section. This gives twist/unitlength:

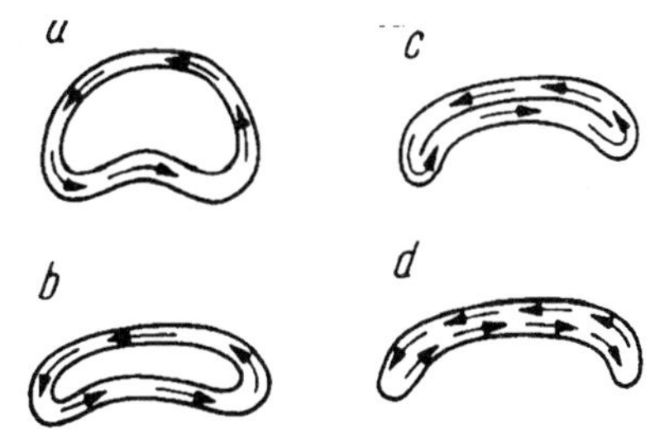

Fig 62: Closed to open section beam

$T/4GA^2\ \Sigma ds/t$

for shear modulus G and wall thickness t.

These formulae can also be used to examine the torsion in closed tubes with stringers (using the aircraft terms to describe the box beam reinforced by longitudinal bars, Fig 64). A saloon car body in torsion (Fig 65) is seen in contrast to the van type structure in Fig 66. The first illustrates the significant factor governing passenger compartment torsional stiffness is equal and opposite transverse shear forces in the roof and floor influenced substantially by the

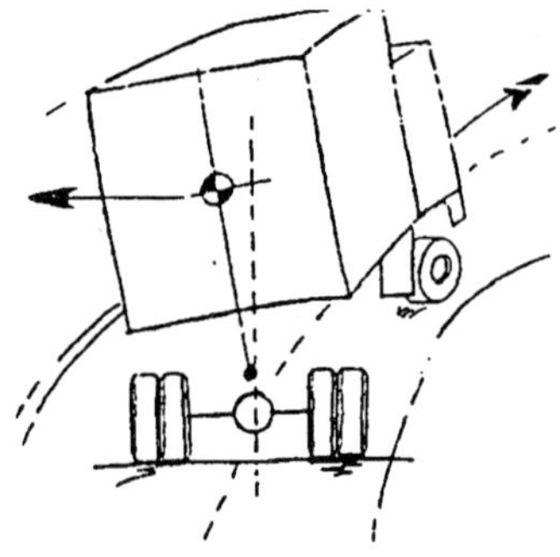

Fig 66: Van as box tube

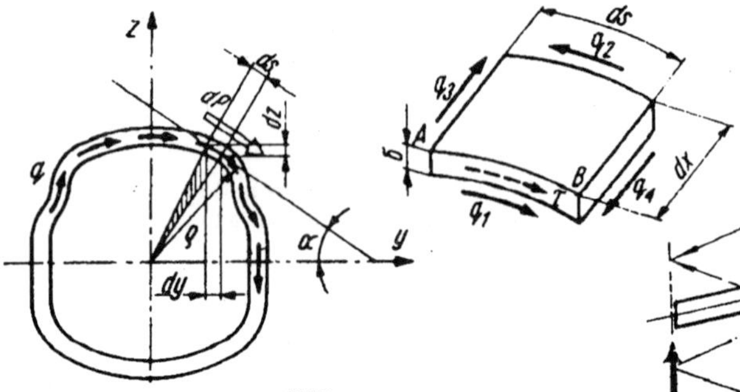

Fig 63: Shear flows in beam element

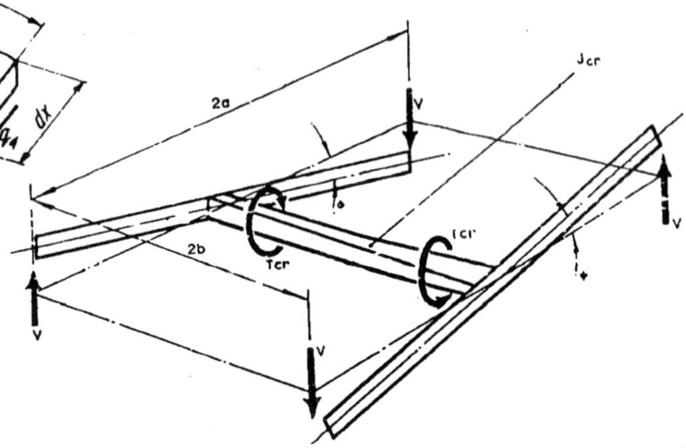

Fig 67: Underbody cross-member in torsion

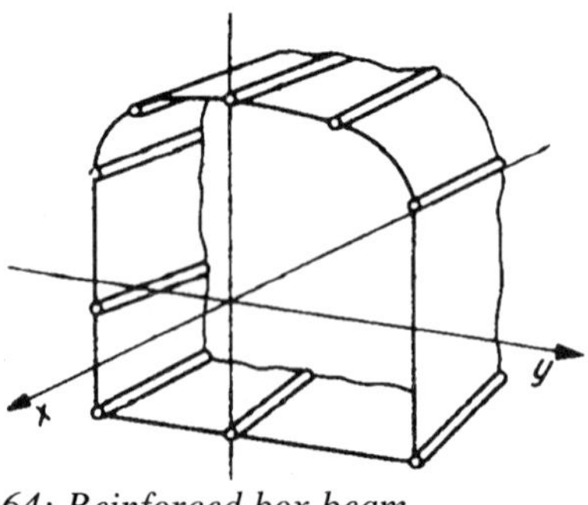

Fig 64: Reinforced box beam

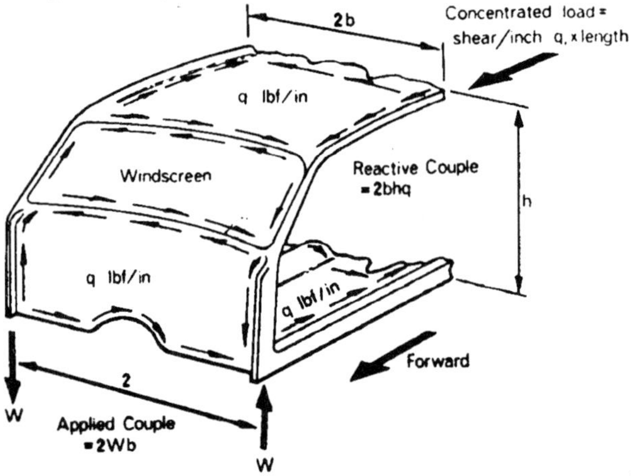

Fig 65: Saloon body as box tube

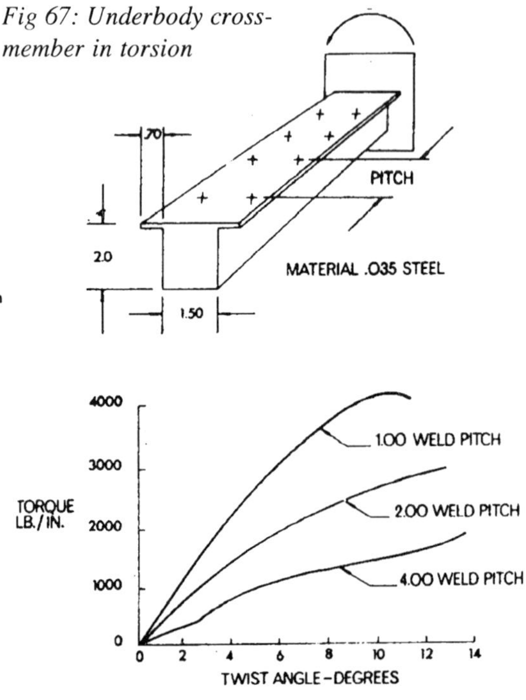

Fig 68: Spot-weld pitch effect on twisting

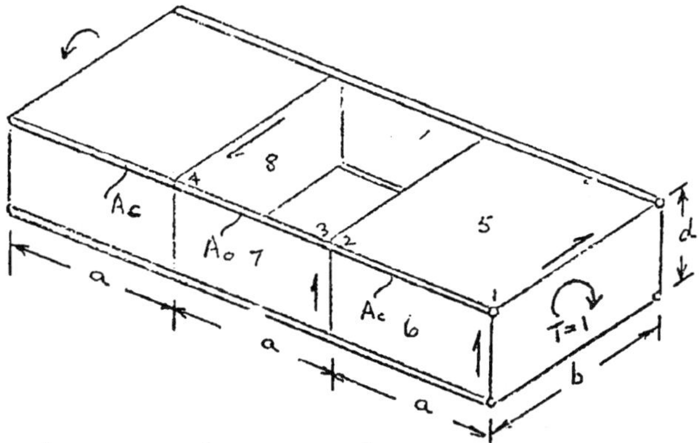

	Internal force due to T = 1 b	Element flexibility matrix	Number of identical elements	Contribution to structural flexibility $b^T f b$
1 2	0 $a/2bd$	$\frac{a}{6EAc}\begin{bmatrix}2 & 1\\1 & 2\end{bmatrix}$	8	$8a^3/12b^2d^2EAc$
3 4	$a/2bd$ $-a/2bd$	$\frac{a}{6EAo}\begin{bmatrix}2 & 1\\1 & 2\end{bmatrix}$	4	$4a^3/12b^2d^2EAo$
5	$3/4bd$	ab/Gt	4	$9ab/4b^2d^2Gt$
6	$1/4bd$	ad/Gt	4	$ad/4b^2d^2Gt$
7	$1/bd$	ad/Gt	2	$2ad/b^2d^2Gt$
8	$3/4bd$	db/Gt	2	$9db/8b^2d^2Gt$
				$\frac{9(b+d)a}{4b^2d^2Gt}$

Fig 69: Effect of panel removal on box tube

windscreen which closes a rectangular framework otherwise subject to lozenging deformation. There is also a contribution to torsional stiffness made by the two sills acting as a pair of torque tubes and of these acting with tranverse members to form a floor frame. Fig 67 shows torsional resistance set up in the underbody cross member due to the difference in slopes of the sills at each end, for which

$$V = T_{cr}/2a = \phi\,(J_{cr}G/2ab)$$

where J_{cr} is the polar moment of inertia of the crossmember and 2f = pb/360a. It is important to note that spot-weld or rivet spacing has an effect on torsional rigidity as indicated by Fig 68.

Torsion is often the overall structural design criterion for vehicle bodies. Open sections of the body are torsionally much more flexible than closed ones unless the axial deflection of the ends of the open section can be restrained. In this case the torsion can be resisted by differential bending. This type of restraint might be achieved in an open car body if the front bulkhead, A-posts and dash panel form a closed torsion box. In the idealised structure of Fig 69, the effect on torsional stiffness of removing the panels from torsion boxes can be seen in the accompanying

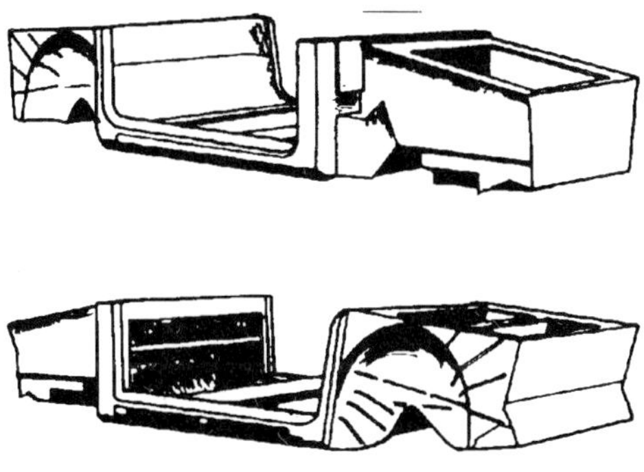

Fig 70: Car structure with enlarged sills and bulkhead

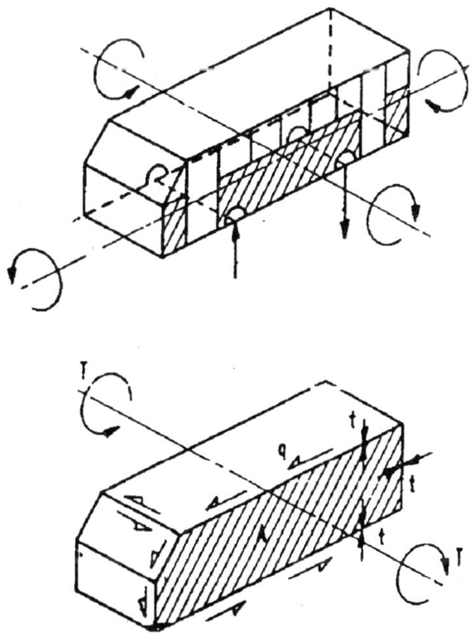

Fig 71: Coach as box tube

table, due to Dr J M Howe of Hertfordshire University. Torsional flexibility is 50% greater than that of the closed tube (or open tube with a rigid jointed frame of similar shear stiffness to the removed panel surrounding the cut-out) if the contributions of flanges and ribs are neglected.

Enlargement of the sill sections is another way to achieve the overall torsional stiffness of a vehicle body as well as partially boxing in the luggage and engine compartments in the case of the car body, Fig 70. For the example of the passenger coach structure the kind of idealisation shown in Fig 71 is useful in estimating torsional resistance. A yet more generalised idealisation for several types of semi-closed body structure is due to C J Cooke[8] of Rolls Royce Motors (Fig 72). Torsional stiffness is then given by

$$[(bh/6a)(A+B)]/[1+(e_1+e_2)/2a-(h/a)(k_A+k_D)]$$

due to roof and floor acting in differential shear and

$$b^2/688F[(ab^2/J_a G)+(b^3/EI_b)]$$

due to the sills dash and rear squab acting as a rectangular frame. The terms A,B,k_A,k_D,F may be evaluated for unit angle of twist by methods given in Cooke's original paper.

The effect of joint flexibility on vehicle body torsional resistance must also be taken into account. Experimental work carried out by P W Sharman[9] at Loughborough University have shown how some joint configurations behave. The importance of adding diaphragms at intersections of box beams was demonstrated, Fig 73. Without such stiffening, the diagram shows the vertical webs of the continuous member are not effective in transmitting forces normal to their plane so that horizontal flanges must provide all the resistance. The distortions shown inset were then found to take place if no diaphragms were provided.

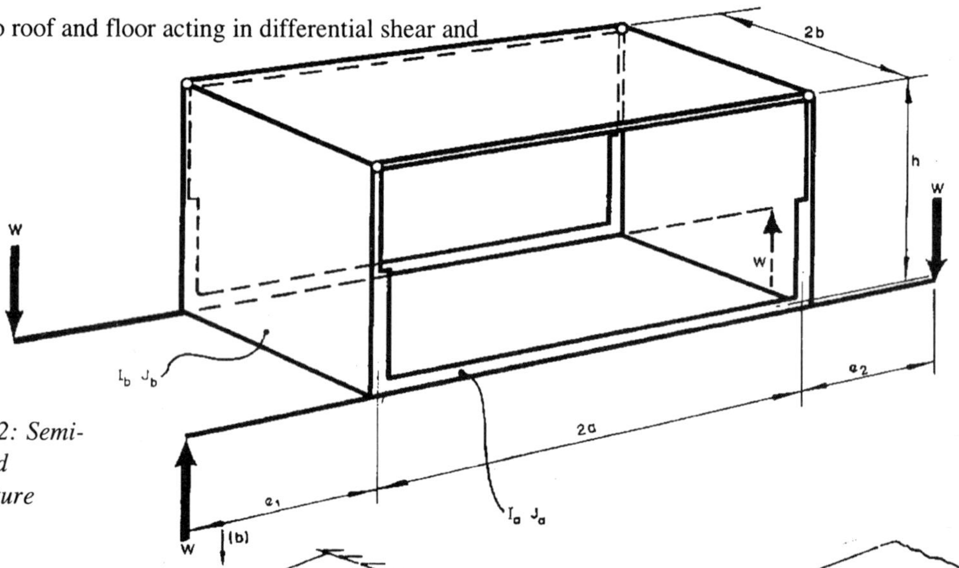

Fig 72: Semi-closed structure

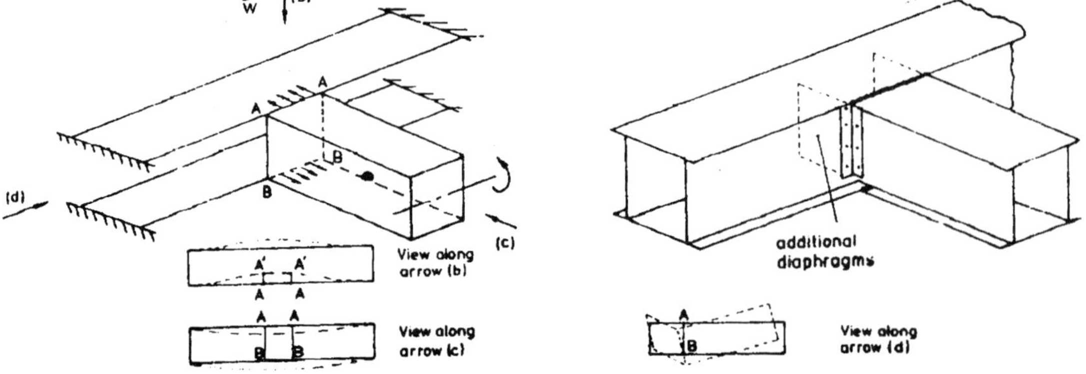

Fig 73: Use of diaphragms at beam intersections

Stability of columns and plates

A key factor in the successful structural design of thin sheet panels is examination for structural instability. Too often great efforts are made to compute the predicted stresses in a vehicle body due to bending, end-load or shear without realising that in many cases the structure will suffer an instability collapse at much less than the theoretical maximum 'direct' stress due to external loading.

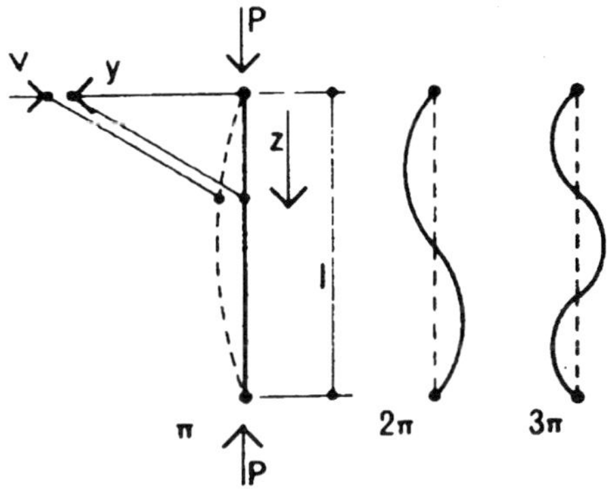

Fig 74: Strut under end-load

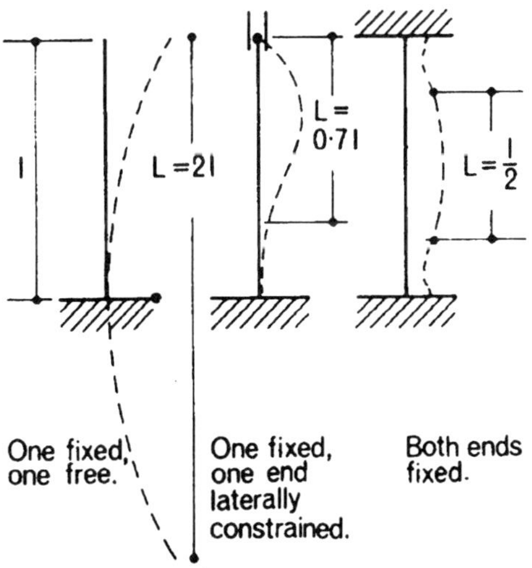

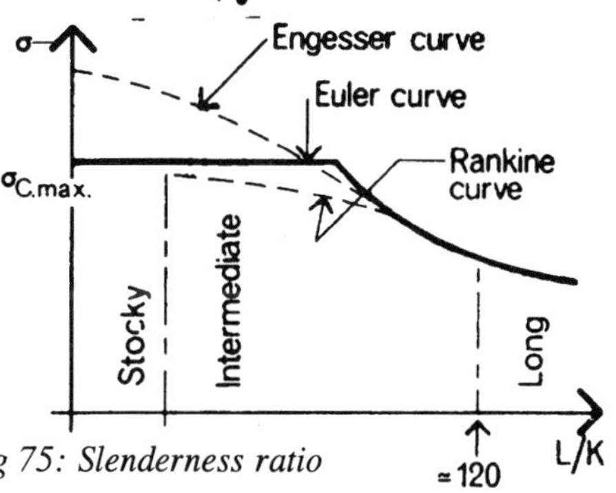

Fig 75: Slenderness ratio

The theory of panel instability is merely an extension of that applying to a narrow strut so that is an obvious starting point in the analysis. Fig 74 shows an end-loaded strut, diagrammatically, with some possible buckling modes alongside. If the strut is short enough then buckling instability will not apply; it is thus necessary to determine a critical slenderness-ratio above which the member becomes susceptible to instability failure. As the strut bows, in the single curvature mode, it is clear from the diagram that a bending moment develops in the strut due to the displacement of the deflected centre line from the line of action of the end loads. The classic analysis due to Euler involved equating this with the second derivative of lateral deflection of the strut y at any distance z along its length from one end. This is an extension of the equation of flexure considered in the section on beams earlier in this chapter. This is written as

$$(d^2y/dz^2) = -Py/EI$$

where flexural rigidity EI is product of elastic modulus and section second moment of area of the strut. The standard solution for this differential equation, putting $a = (P/EI)^{1/2}$ is:

$$v = A\sin az + B\cos az$$

whcrc A and B arc constants which arc climinatcd by substituting the values of the variables at the ends of the strut. Thus at $z = 0$, $v = 0$ and so $B = 0$ while at $z = L$, again $v = 0$ so that $A\sin aL = 0$ and hence aL = p, 2p, 3p, successive solutions which show the theoretical possible mode shape of the strut if it were perfectly straight and the end loads perfectly in line.

Due to inevitable slight eccentricities, only the single curvature mode occurs in most practical cases, for which $a^2 = \pi / l^2 = P/EI$, a relationship leading to the formula for Euler critical load (onset of buckling):

$$P_e = \pi^2 EI/l^2$$

An 'effective length' can be substituted for l, Fig 75, when other than pinned-end conditions apply to the end-fixing of the strut. L is determined from the points of contraflexure (zero bending moment and directional change in slope) arising from the bending modes induced by the end fixing moments. Experiments show that the Euler theory is inaccurate below a slenderness ratio of $L/K = 120$ (K is the least radius of gyration in the expression for section second moment of area $I = AK^2$ where A is the section area. If slenderness ratio is plotted against critical buckling stress, three broad categories of strut show themselves. The shortest, 'stocky' struts fail by direct stress without buckling while those above 120 S-R are Euler struts; between the two an 'intermediate' series exists which requires empirical formulae to allow for the fact of initial curvature or load eccentricity affecting 'Euler' critical values. Rankine's formula states:

$$1/\sigma_{crit} = 1/\sigma_{c.max} + 1/\sigma_c$$

from which the critical stress

$$\sigma_{crit} = \sigma_{c.max}/[1 + (\sigma_{c.max}/\pi^2 E)(1/K)^2]$$

Classic examples of strut members in vehicle bodywork include the windscreen pillars and B-posts of saloon cars and CV cabs in the roll-over accident situation. Such a B-post section, idealised for analysis is shown as Fig 76. To determine its critical end load for buckling in the roll-over situation, its neutral axis of bending has first to be found — using a method such as the tabular one in Fig 77. Assuming the roof end of the pillar to impose 'pinned' end fixing conditions (so that $L = 2l$) and that the pillar is one metre in length, then critical load is

Element	A	Y	Ay	Ig	h	Ah²	Ig + Ah²
1.	100	½	50	100/2 = 8·3	18·5	34 200	342 083
2.	2 X 15	1½	45	30/12 = 2·5	17·5	9250	9252·5
3.	2 X 51	26½	2750	2·50³/12 = 1040	7·5	5620	6660
4.	50	51	2550	50/12 = 4·14	32	34 000	34 004·17
						Σ = 83 070	Σ = 84 124·91

Fig 77: Neutral axis determination

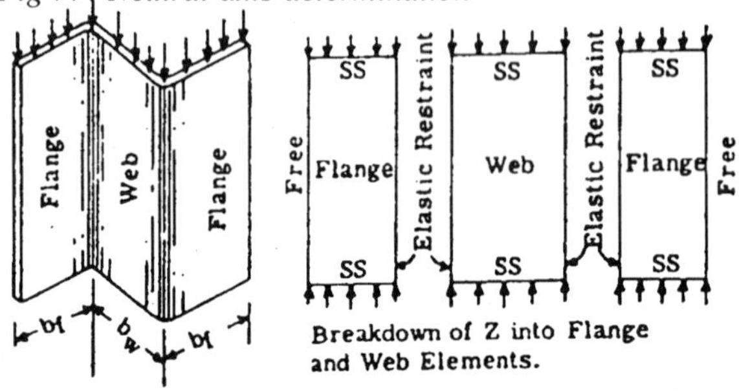

Fig 79: End-loaded zed or channel-section

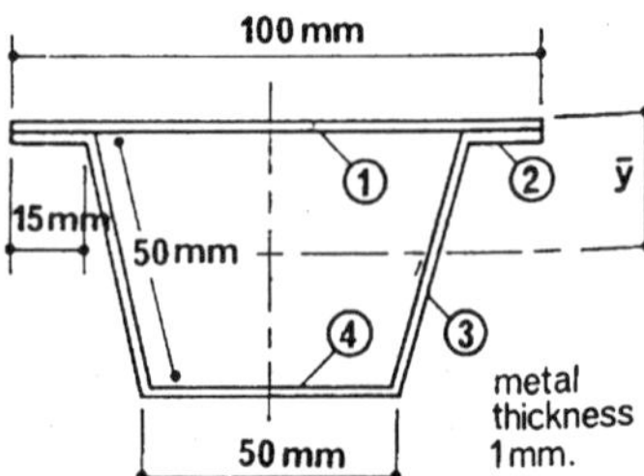

Fig 76:B-post section

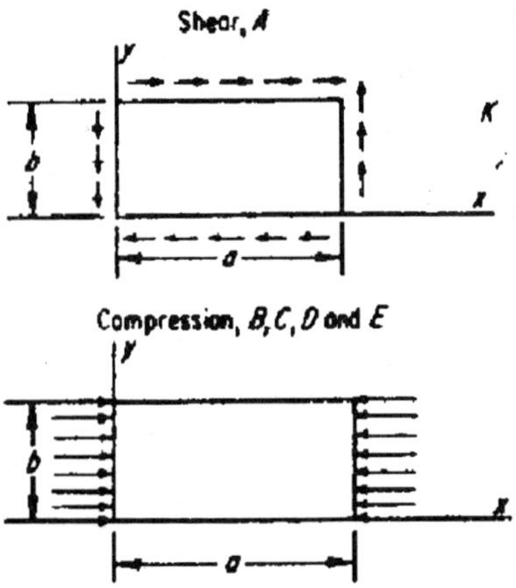

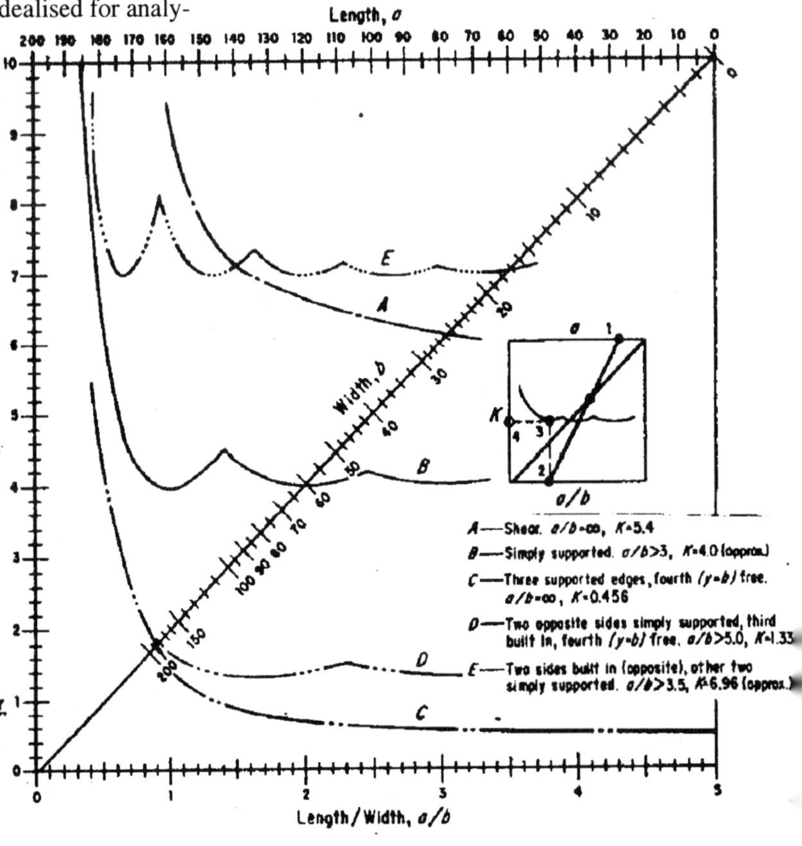

Fig 78: K-factor curves

$10.210.10^9.8.4.10^{-8}/22$. Taking E for steel as 210 x 10^3 N/mm^2, the stress at this load is $44.10^3/280.10^6$ = 157 MN/m^2 since $A = 250$ mm^2 — which is above the critical buckling stress. If, however, the cant rail is assumed to provide lateral support at the top end of the pillar then $L = 0.7l$ and critical buckling stress is 1280 MN/m^2 and the strut would fail at the direct yield stress of 300 MN/m^2. Other formulae, such as those due to Southwell and Perry-Robertson, will allow for estimation of buckling load in struts with initial curvature.

Plate buckling

An extension of the Euler theory can also be used to predict the buckling resistance of flat rectangular plates under end load. For a relatively narrow rectangular 'strip' it is sufficient merely to substitute strip width b in the formula for second movement of area for a rectangular cross-section of $I = bt^3/12$ for strip thickness t and then substitute this in the Euler formula. However, as the strip gets wider — to assume the proportions of a panel — then the effects of anticlastic bending (that is in the plane at right angles to the one of primary concern) have to be considered. The buckling load is then factored by $1/(1 - \nu^2)$ where ν is the Poisson's Ratio of the material. This same factoring is necessary when the panel is bending under the effect of lateral rather than end loading.

The formula for critical buckling load of a plate under end load thus becomes

$$K\sigma 2Et^2/12b(1 - \nu^2)$$

with values of K, depending on edge fixing conditions taking values from the curves in Fig 78 which also gives values for buckling under shear loading, to be considered below. For a plate under end load, Fig 79, it has been found that the stress distribution across the plate is of the order of that shown in Fig 80. This has given rise to the concept of 'effective width' whereby load is assumed to be carried only by the edge strips of the plate.

Under shear loading, the critical buckling load for a flat rectangular plate is $K\pi^2 EI/b^2$. A common method of treating shear buckling is to represent the panel by a framework of end-load

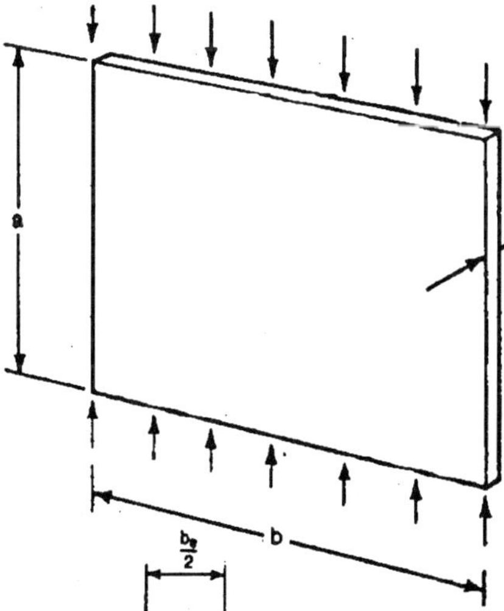

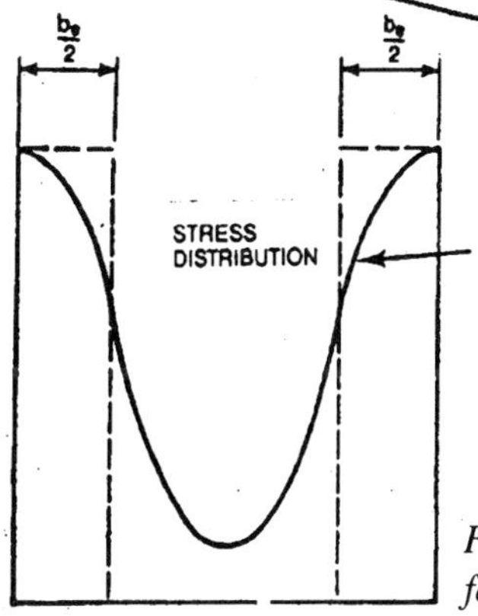

Fig 80: Stress distribution for end-loaded plate

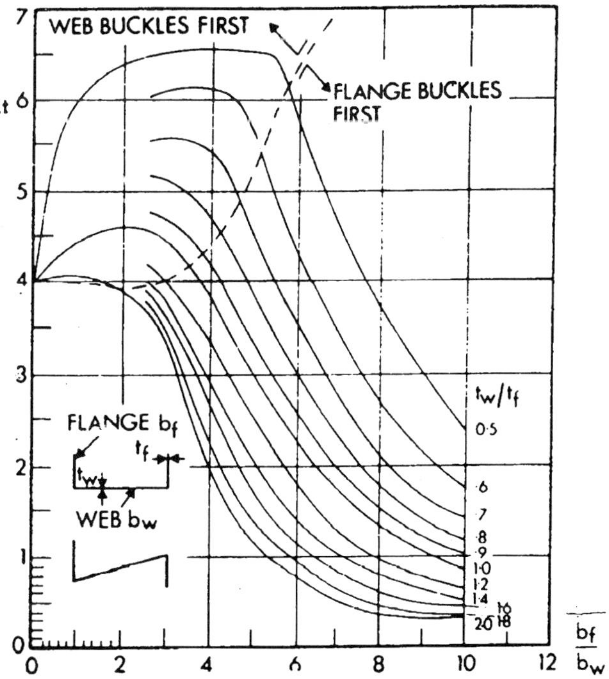

Fig 81: K-values for zed and channel sections

carrying bars along each edge and stabilised by two diagonals, the whole being pin-jointed at the corners. This was illustrated in an earlier section. Buckling of the compression diagonal under shear loading of the whole frame is analogous to shear buckling of an equivalent panel. The tension in the opposing diagonal is analogous to the tension fields which exist in the equivalent panel and give rise to its shear resistance. The zed or channel section under end load can also be considered to be an assembly of 'plates' subject to buckling deformation, Fig 79. The plates supply mutual elastic support for each other and curves due to Bruhn have been constructed, Fig 80, to obtain the K values for use in the buckling formulae. As well as the overall column buckling mode the local instability of elements of such sections should also be examined. The approach is first to determine limiting stress level based on overall instability then to evaluate local instability at this level of stress.

Steel-section framed and skinned structures

High strength steels as well as mild steels can be used in structural sections but it is important to observe the so-called tension-to-yield ratio as a measure of the material's elongation. A minimum ratio of 1.08 between the yield to maximum tensile stress has been recommended for materials with limited elongation. Yield strength can of course be increased by cold forming particular elements of the section and is a funtion of the sharpness of bend between section elements.

Presence of corners in a section also permits it to carry load after buckling-collapse of the flat elements of the section. End load carrying capacity of comparatively long members is first considered in terms of overall buckling by the methods given in the previous section. The critical stress for overall column buckling is then used to evaluate the local instability of flat elements of the section. For this

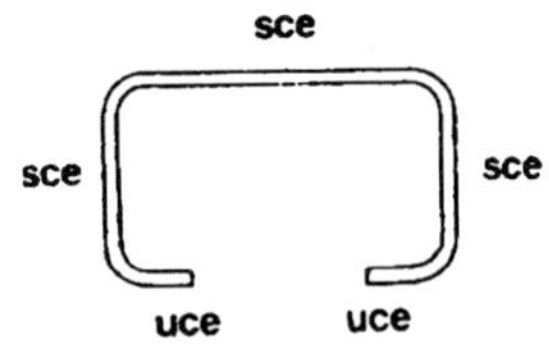

Fig 81: Stiffened (sce) and unstiffened (uce) compression elements

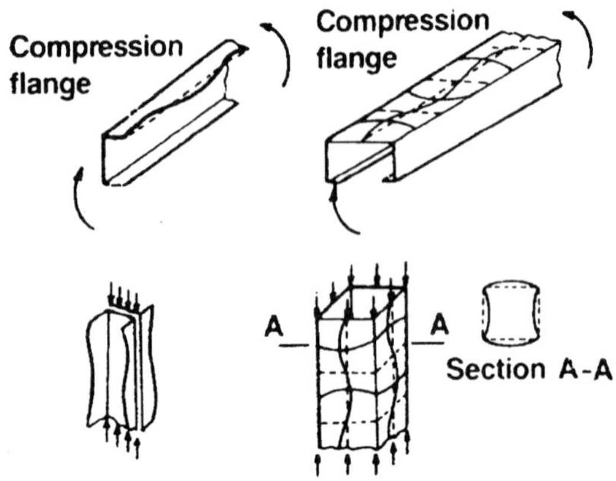

Fig 82: Local buckling of stiffened (left) and unstiffened (right) elements

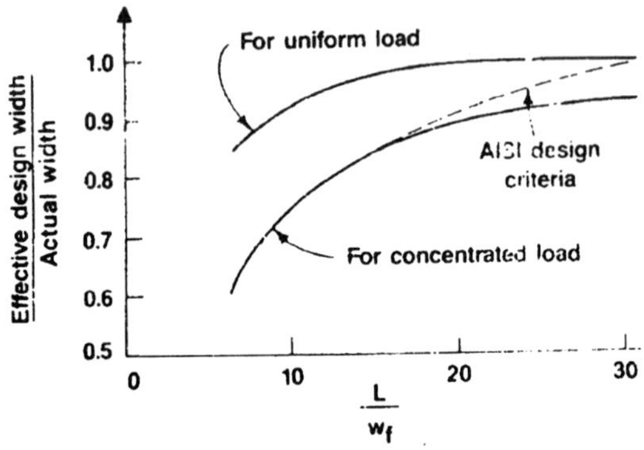

Fig 84: Shear lag effect

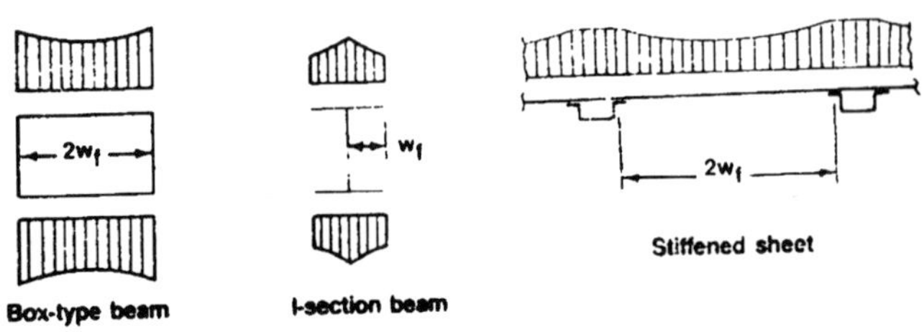

Fig 83: Effective flange width

purpose the elements are classified as stiffened and unstiffened as shown in Fig 81 in an analytical method due to the American Institute of Steel Industries (AISI)[10].

For unstiffened elements, the overall buckling stress is reduced to account for loss of full section capacity under end load due to local buckling. The latter behaviour is assumed to be as depicted in Fig 82 as governed by the slenderness ratio of the element expressed as

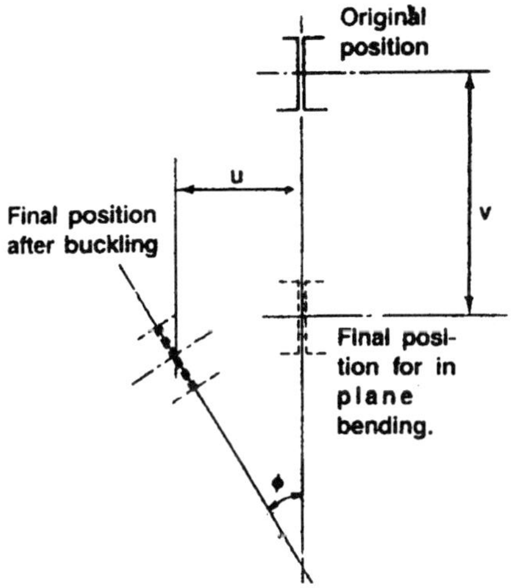

Fig 85: Lateral buckling

Cross Section	Shear Stress	Torsional Stiffness Constant J; $\phi = \frac{T}{JG}$
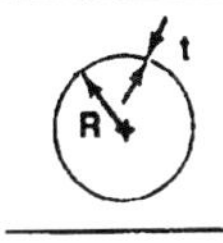	$v_t = \frac{T}{2\pi R^2 t}$	$J = 2\pi R^3 t$
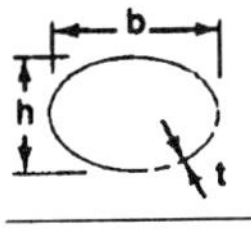	$v_t = \frac{T}{\frac{\pi}{2} bht}$	$J = \frac{\pi b^2 h^2 t}{2\sqrt{2(b^2+h^2)}}$
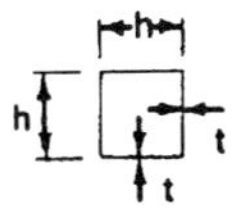	$v_t = \frac{T}{2h^2 t}$	$J = h^3 t$
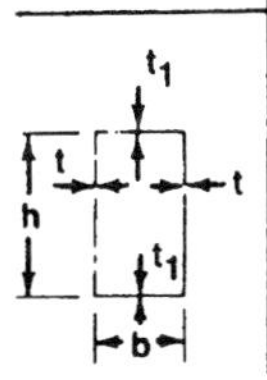	$v_t = \frac{T}{2bht}$ $v_{t_1} = \frac{T}{2bht_1}$	$J = \frac{2tt_1 b^2 h^2}{bt + ht_1}$ $t_1 = t;\ J = \frac{2b^2h^2t}{b+h}$

Fig 86: Torsional shear stress and stiffness

$$1.052(w/t)(\sigma/E)^{1/2}(k)^{-1/2}$$

where s is stress, *E* the elastic modulus, *k* the flat-plate buckling coefficient and *w/t* the width to thickness ratio. For unstiffened elements a maximum *w/t* value of 60 is recommended by AISI.

The bending capacity of thin wall section profiles is governed by yield strain in compression, local instability and overall lateral instability as well as the effect of local stress build-ups due to combined loading. In short beams, shear stress becomes a significant factor, as shown earlier, and the effective flange-width (Fig 83) reduces due to the effects of shear lag, Fig 84.

While it is usual to calculate web shear stress as a uniform value of *F/ht*, where *F* is shear force, the maximum shear stress at the neutral axis may be 25% larger. With web stiffeners which divide the web into rectangular panels, an increase in shear capacity is effected. The capacity is similarly increased if the web is curved to a maximum force carrying capacity of

$$4.895Et^3/h + 0.10Et^2h/R$$

where *R* is the web radius of curvature.

Lateral instability has to be considered where section second moment area is considerably less in the transverse plane to that in the vertical plane of bending. Lateral buckling involves both bending and twisting, Fig 85, the latter due to the compression flange wanting to buckle while the tension one stays straight. As a result, torsionally stiff tubular members have little tendency to lateral buckling in bending. Rectangular section tubes have little strength reduction from this effect even down to depth to width ratios of 10. Torsional shear stress and stiffness formulae for closed sections are shown in Fig 86.

Sections with curved elements which are open rather than closed have different behaviour. These are not torsionally weaker than similar closed sections but have resistance made up of both shear due to lozenging tendency of the developed 'wall' (St Venant, so-called; see earlier section) and that due to axial warping. To examine the second effect, of 'differential bending', common on beams with widely spaced flanges likely to bend as well as 'lozenge' the critical factors of beam end restraint and span have to be considered.

Another consideration is torsional buckling and singly symmetric sections such as those in Fig 87 can buckle about the y-axis when the axial load is to the left of the shear centre. With loads to the right of the shear centre, failure can also occur in a torsional/ flexural mode (flexurally about the x-axis). In the case of curved members, inner fibres in bending have higher stresses than determined by conventional theory and radial stresses are developed in the webs due to bending moments. Those parts of the flanges distant from the webs lose their effectiveness and tranverse bending occurs. When the ratio a/c (Fig 88) is less than 10, stress increase at the inner fibres should be considered. Stress in a curved member can be calculated from

$$(M/Aa)[1 + y/Z(a + y)]$$

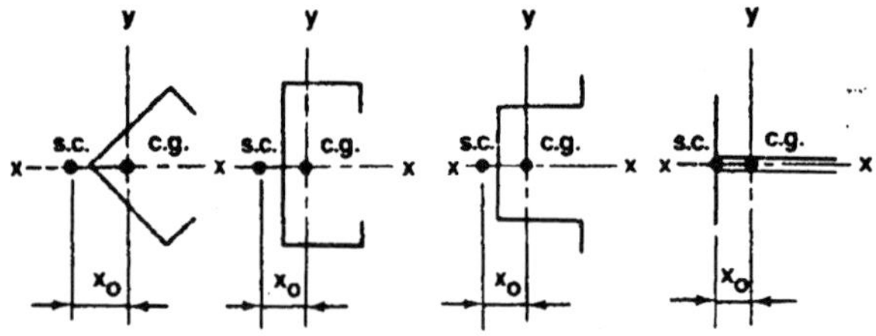

Fig 81: Sections subject to torsional buckling

where A is section area, y the distance from the centroidal axis and Z takes a value from Fig 89, for some example sections. This is due to the neutral axis moving towards the centre of curvature as seen in Fig 88. The result is distortion of the flanges and an effective flange width can be calculated as $0.7(a_f t_f)$.

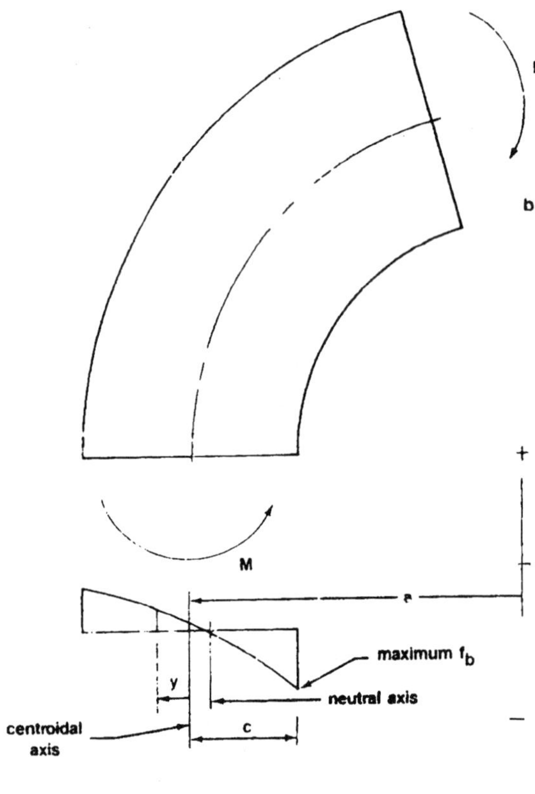

Fig 88: a/c ratio

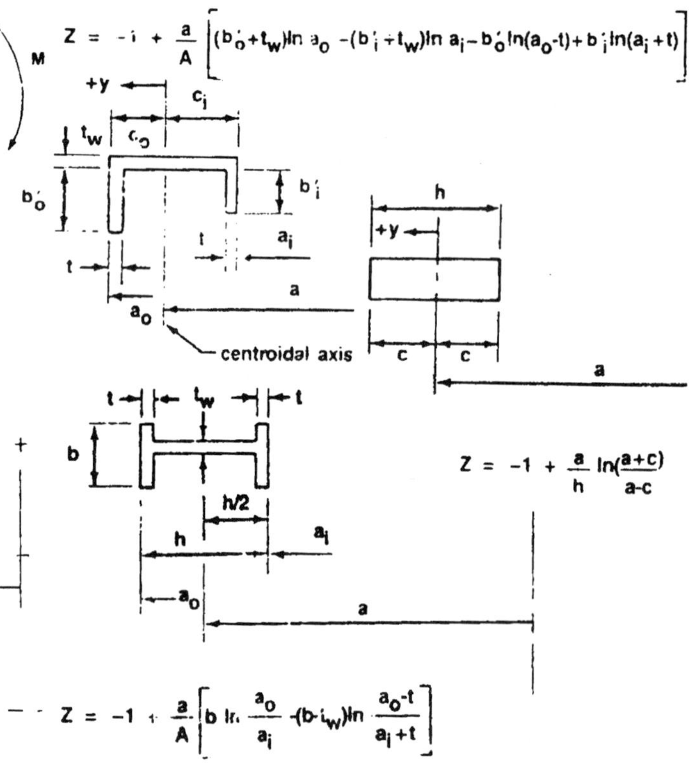

Fig 89: Zed values for three sections

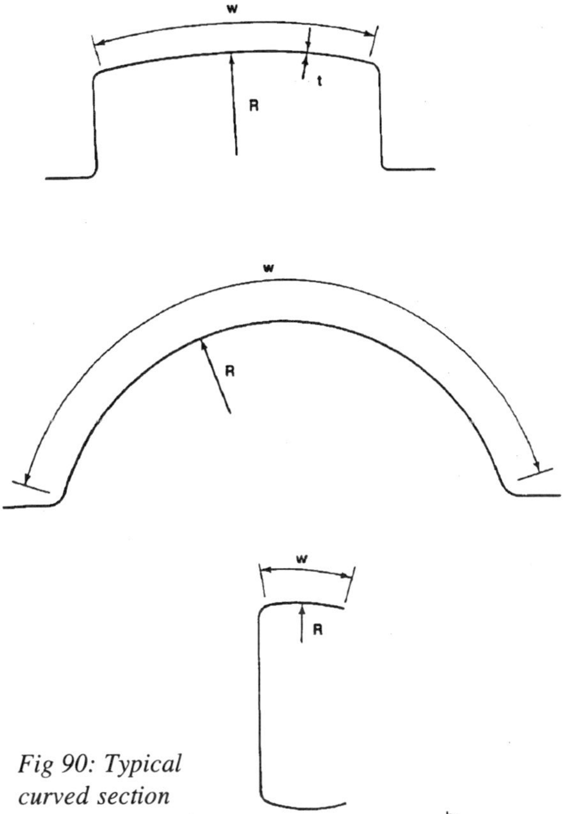

Fig 90: Typical curved section elements

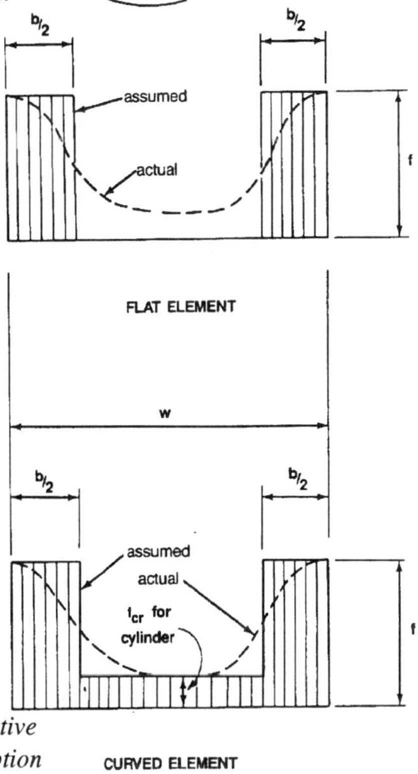

Fig 91: Effective width assumption

Buckling under end load

In the case of curved section elements such as those in Fig 90, both stiffened and unstiffened, but typically braced along one or both sides by a corner bend, an extension of buckling theory for flat plate elements can be applied. For flat elements, the buckling stress is given by:

$$f_{crf}=180,760(t/L)^2$$

But for most elements, which are straight over their length, the length dimension is such that it is beyond that where buckling, evaluated by the above analysis, would occur. Local buckling will be similar to that derived for a cylinder where critical buckling stress in MPa is given by:

$$f_{crc}=0.605Et/R$$

where t is thickness, L the length and R the sheet curvature radius. Elastic buckling of curved plate elements is between these two values and given by:

$$f_{crp}=[f_{crc}^{\ 2} + (f_{crf}/2)^2 + f_{crf}/2]$$

Inelastic buckling can also be obtained from the above by substituting the secant modulus for E.

There is also post buckling resistance offered by curved elements such as the ones illustrated, due to their perpendicular flanges. An effective width of plate to represent the flanges is considered, Fig 91, analogous to that assumed for a flat element. However, unlike the latter, the ineffective portion of a curved element is assumed to carry the critical buckling stress as a circular cylinder having the same thickness and radius. Instead of neglecting the centre portion, this is assumed to have an effective thickness:

$$t_e=(A_o/A)(F_y/f)t$$

where A/A_o is taken from Fig 92.

When the curved segment is tangential to the straight segment, rather than separated by a sharp bend, the evaluation is more complex. According to AISI[9] a curved segment with R/t </= 7 is normally used to stiffen a flat element and in this case the flat element is measured from the point of tangency, Fig 93. For a curved element with radius thickness ratio

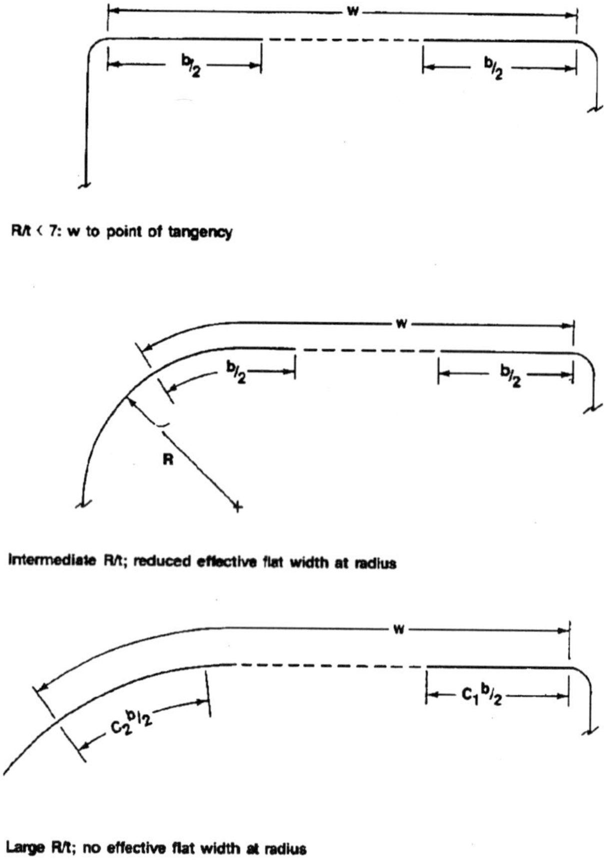

Fig 93: Combined flat and curved elements - a,b and c from top to bottom

</= 33 (assuming 345 MPa yield steel), the complete yield strength of the element can be realised before buckling. But measuring *w* from the point of tangency may not be safe, nor the assumption that the curved element provides the stiffened edge. At some dividing point the flat element is taken as half the stiffened element or even as an unstiffened element, as illustrated.

For the case of element (b), in the figure, an approach to evaluating this problem is seen by example. Here *(R/t)>7* and the radiused portion will thus reach yield. If the flat portion has width 90mm, thickness 1.8mm and radius 36mm, a new *w* is established, arbitrarily, which is larger than 90mm, by estimating the appropriate additional amount of arc to add from *(p/4)t[(R/t)-7]* such that *w* =108.4mm. Equivalent width *b* is then calculated as 80 mm so that 40mm would be consider effective adjacent to the sharp bend and 21.6mm of the flat portion adjacent to the large radius corner would be effective. For the element at (c) where *R/t* is large enough for the equivalent area A_o of the radius is less than the actual area. The effective region near the sharp radius can thus achieve a higher stress than the large radius portion. So by changing, as before, the radius to 84mm and using the same relationship for amount of arc, then *w*=146.1mm. Fig 91 can be used to find *b* and then the stress on the section will be shown to be

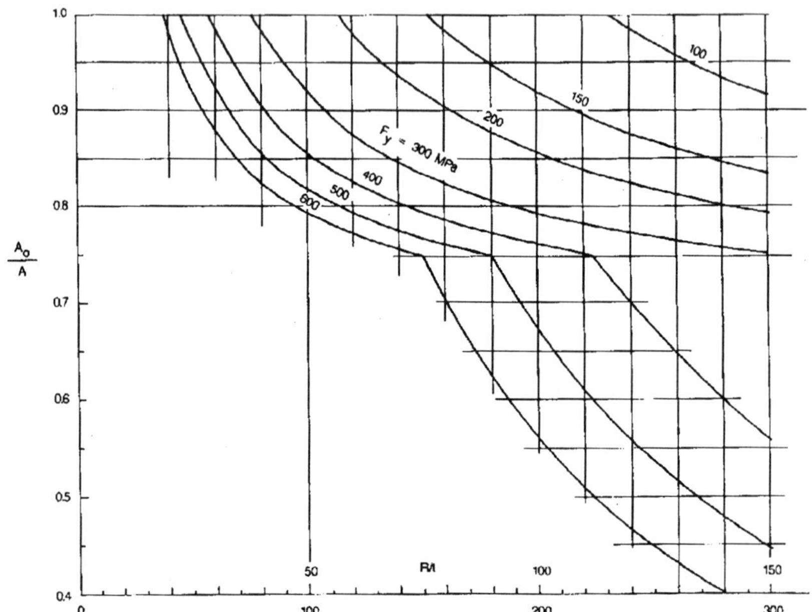

Fig 92: Curves for A/Ao

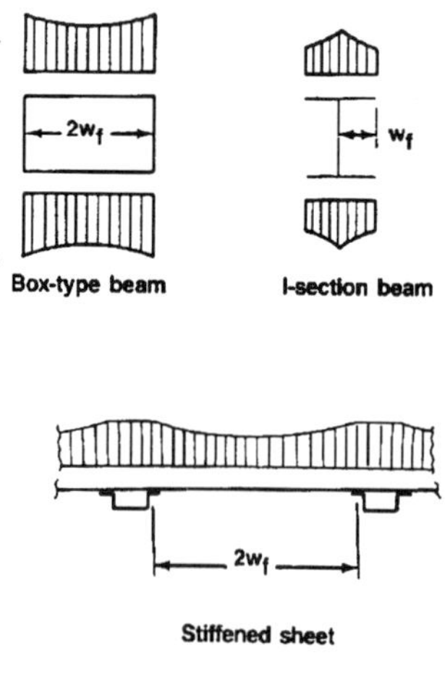

Fig 94: Method for obtaining w_f

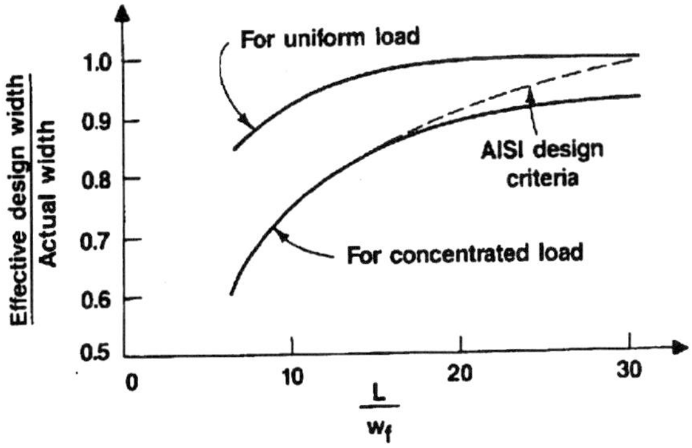

Fig 95: Curves for w_f

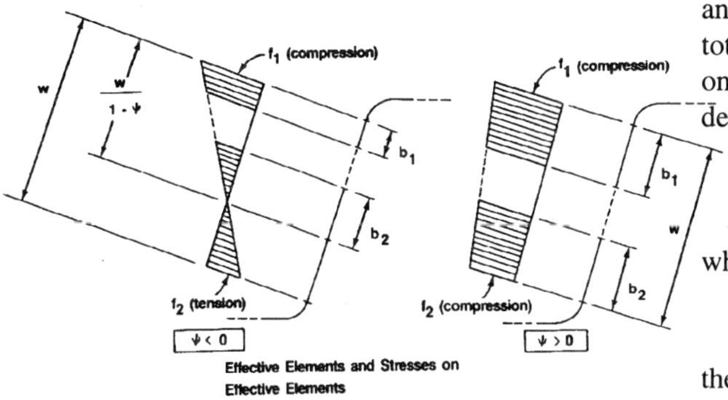

Fig 96: Stress distribution on web

governed by the limiting stress on the radius portion.

Sections comprising a number of flat and curved elements pose greater difficulty and requires an approach which considers all components in the section and their relationship to one another. If such a section has a sharp corner with $(R/t)<0.057E/F_y$ then post-buckling strength can be expected; if not the radius portion, or flat portion tangential to the radius, will limit capacity. If the sharp corners are not evenly spaced over the section, post-buckling strength may be useless as the section properties in that condition may produce a load less than the elastic buckling load.

Overall stability of structural members has been calculated using the total cross-section area A and a factor Q representing the ratio of effective-to-total area at yield. Now the approach is to calculate on the effective area calculated at the ultimate stress determined by lateral buckling of the member:

$$\text{For } f_e<F_y/2,\ f_{cu}=f_e;\ \text{for } f_e>f_y/2,\ f_{cu}=f_y[1-f_y/4f_e]$$

where critical Euler stress

$$f_e=p^2E/(KL/r)^2$$

the above applying to any sections not subject to torsional-flexural buckling. Axial load capacity:

$$P_u = A_e f_{cu}$$

where f_{cu} is the ultimate compressive stress and K an effective length factor related to the slenderness ratio.

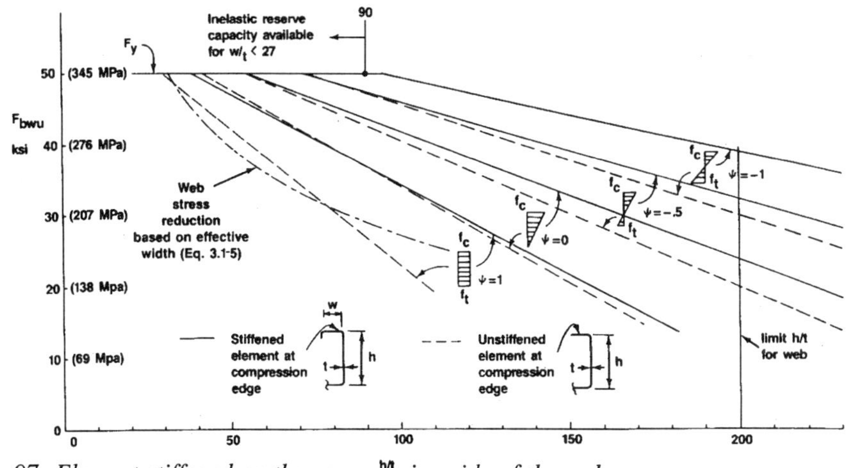

Fig 97: Element stiffened on the compression side of the web

Bending of thin-walled beams

For short span thin walled beams the effect of shear lag is to cause an effective flange width which is less than the distance between the webs. The width w_f is defined for different beam sections in Fig 94. Since the load enters the flange from the webs, and develops gradually with bending moment, other factors than compressive stress and element slenderness can affect effective width which is determined from the curves in Fig 95. If the *w/t* ratio of a stiffened compression element is lesss than $(E/f_y)^{1/2}$ it may be posssible to permit inelastic stress distribution in the web so as to increase ultimate bending moment capacity beyond the yield moment.

The flat elements are the webs of the flexeral member section and these have a linear stress variation across them as in Fig 96, according to the symmetry of the section about its neutral axis of bending. Generally a web height is limited to *h/t=200* if no transverse or longitudinal stiffening is provided. If the section assymetry makes tension the larger stress, the likelihood of compression-stress control is small if *h/t<200*. When compression is the larger stress the limiting stress can be found from the expressions which follow. A stiffened element at the compression side of the web exerts a rotational constraint leading to a higher buckling load than can be achieved by an unstiffened one, Fig 97.

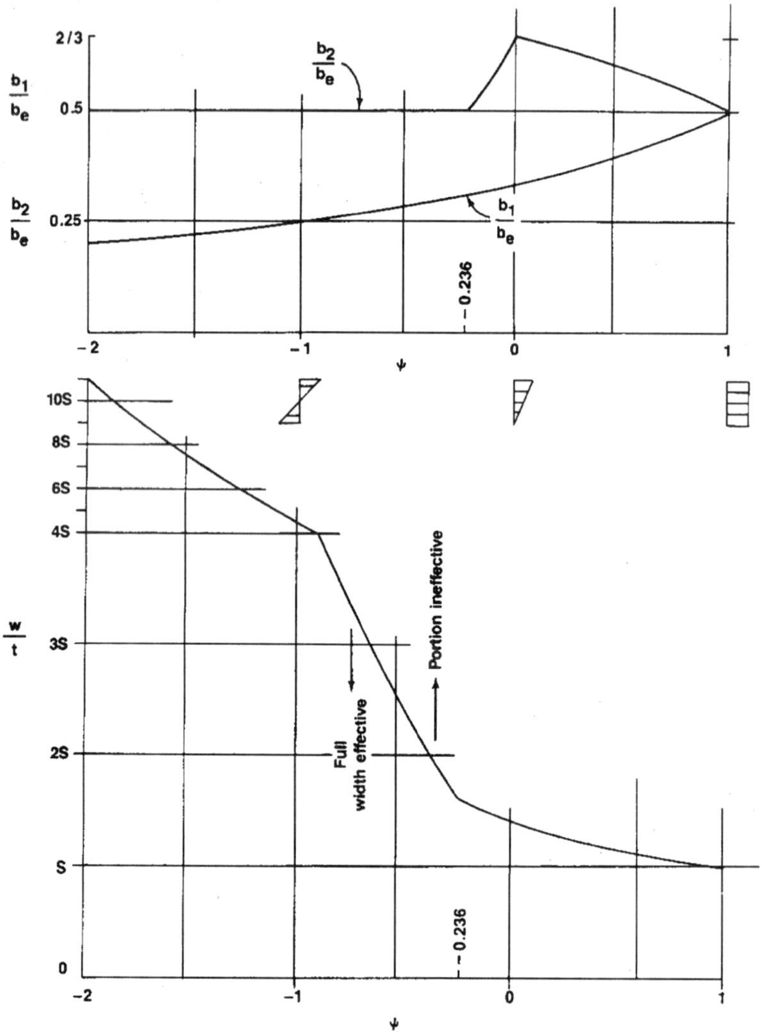

Figs 98 and 99: Effective flange widths

For the stiffened element:

$$f_{bwu}=[1.197 - 0.247(h/t)(f_y/kE)^{1/2}]f_y$$

and for the unstiffened one:

$$f_{bwu}=[1.229 - 0.374(h/t)(f_y/kE)]f_y$$

where $k=4 + 2(1 - j)^3 + 2(1 - j)$ for $j=f_t/f_c$ (Fig 98 and 99). k is the buckling factor as a function of variation in tensile to compressive stress ratio as plotted in the figure. When $j = 1$ (uniform stress) an h/t between 30 and 39 is where buckling begins to reduce stress limit. As compression on the element decreases and pure bending approaches, the element can achieve stress levels shown for $j = -1$. The effective width of the compression portion of the flat element is shown as b_1 and b_2 in Fig 98 with $b_1 = b_e/(3 - j)$. For mainly bending stress variation $(j=/<-0.236)$ width $b_2=b_e/2$ and for $j>-0.236$ $b_2=b_e-b_1$ w.

These are plotted in Fig 99 over the practical range of j, the w/t at which the flat width becomes effective being shown in the lower part of the figure.

Inelastic stress redistribution allows the section to strain beyond yield in some cases. If w/t for the flange under compression is less than $1.11/(f_y/E)^{1/2}$ and the depth to thickness ratio of the compression portion of the web is less than the same amount, the section can be strained beyond yield, provided there is no instability, shear stress is below $0.577F_y$ and web inclination is less than 30 degrees. The compressive strain is limited to three times the yield strain but there is no limit to tensile strain. The flange must also be stiffened by two webs so generally only tubular or top-hat sections can achieve this extra capacity. For preliminary design it is best to limit the compression strain to 0.005. Fig 100 shows stress distribution which can occur in bending; (a) and (b) are governed by the equations in (c).

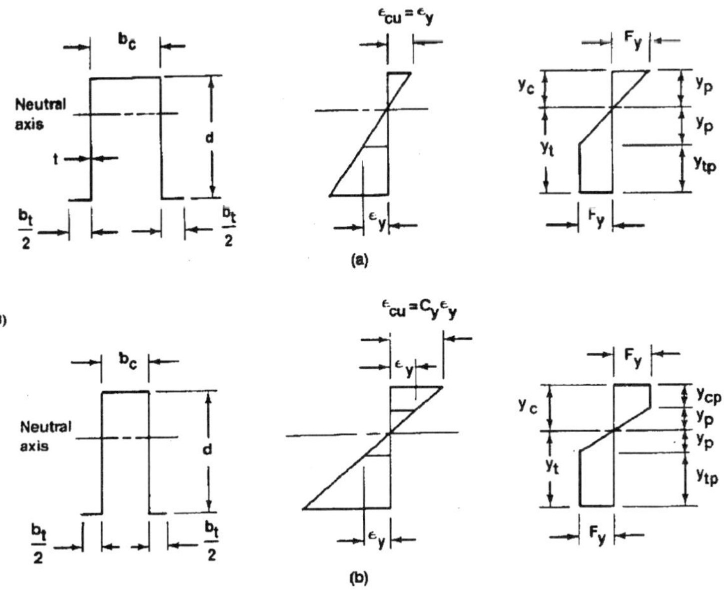

$$y_c = \frac{b_t - b_c + 2d}{4} \qquad y_t = d - y_c$$

$$y_p = \frac{y_c}{\epsilon_{cu}/\epsilon_y} \qquad (y_p = \frac{y_c}{3} \text{ when } C_y = 3)$$

$$y_{cp} = y_c - y_p \qquad y_{tp} = y_t - y_p$$

$$M_u = F_y t\left[b_c y_c + 2y_{cp}\left(y_p + \frac{y_{cp}}{2}\right) + \frac{4}{3}(y_p)^2 + 2y_{tp}\left(y_p + \frac{y_{tp}}{2}\right) + b_t y_t\right]$$

$$M_u = F_y t\left[b_c y_c + \frac{52}{27} y_c^2 + d(d-2y_c) + b_t(d-y_c)\right]; \; C_y=3$$

$$M_u = F_y t\left[b_c y_c + \frac{4}{3} y_c^2 + d(d-2y_c) + b_t(d-y_c)\right]; \; C_y=1$$

(c)

Fig 100: Stress distribution in bending

Chapter 2: Open/closed structures

Open platform structures
Closed body structures
Chassis-analysis
Sub-assembly analysis
Body shell analysis; shell structure loading, mounting and stability
Stress concentration in cut-outs
Finite-element analysis and FEA applied to car bodies

Flat-platform structures

Since the beginnings of the motor vehicle the prediction, at the design stage, of how the body structure behaves in terms of internal stresses and deflections, has challenged body engineers. Historically the presence of a chassis structure strong enough to react the high loads occurring at the mountings of the engine, gearbox and suspension springs ensured the adequate support of even the flimsiest of body structures. In recent times the integration of chassis/ body structures in vehicles like saloon cars and road tankers has required sophisticated predictions of structural strength — aided by the arrival of computer programs to make the handling of calculations less tedious.

Even where a separate chassis frame exists, changes in the severity of vehicle operation, and the need to make bodies lighter, means that ways of assessing strength before the body is committed to metal are likely to appeal to engineers seeking to improve vehicle efficiency. Approaches to analysis vary with the category of body structure involved and in the first part of this chapter the open-top body will be considered. This applies to both cars and commercial vehicles — open, or coupé type, performance cars and platform, or dropside, bodied trucks.

Such flat, or punt-like, structures are characterised by having all torsion and bending transferred through the 'floor' (Fig 1) which requires a rigidity in both these respects. In the case of a car, the floor panel is bounded by edge-members (sills) and beams (crossmembers) used to mount power-train and suspension. A useful approach to the analysis of truck body structures is due to a former Cranfield Institute of Technology visiting lecturer, Dr Ing Pawlowski[1]. It is an admirably simple and flexible technique known as the simple structural surfaces (SSS) method. It is a systematic procedure for applying simple idealisation and analysis procedures developed during the early days of monocoque aircraft structures. For the purpose of analysis, the vehicle is represented by mostly rectangular sheet panels and associated rails (or bars). These do not exactly correspond to the panels, rails and pillars of the actual final structure for they are 'representative' in the sense only of the principal internal structural load-carrying capacity of the members — shear, in the case of the panels, and end-load, in the case of the bars. This gives a very simple, yet quite accurately representative, 'idealisation' of the body structure.

Bending forces are represented indirectly in the sense that a beam element of the structure is ide-

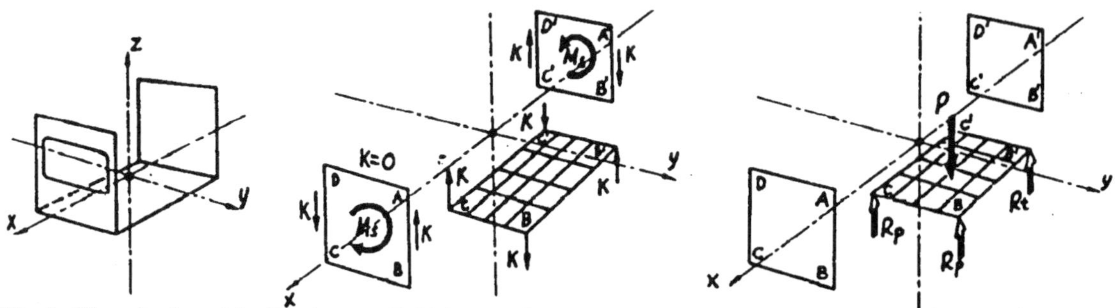

Fig 1: Flat-platform idealisation: a. (left) layout, b. (centre) torsion, c. (right) bending

alised by a panel representing the web of the beam and by bars representing the flanges (very similar to the way civil engineers analyse the behaviour of deep 'composite' I-beams). In the structure as a whole, which is divided into 'structural surfaces for analysis', all forces are assumed to act in the plane of the surface and, conversely, that all forces perpendicular to the surface (except where another surface joins it at right angles) are zero.

The simplest structural surface idealisation is a single shear panel with one bar at each edge to take the end loads, Fig 2. To help understand the validity of the idealisation, first consider a flexible hinged framework of the edge bars. This flexible frame will become rigid with respect to the load *P* by adding the shear panel. The next part of the diagram shows the actual stress distribution and the following part, the one assumed for the idealisation (in a very similar way the early monocoque airframe designers analysed their sheet/stringer constructions by assuming the stringer, for the purposes of analysis, had not only its own cross-section area but also an additional one equivalent to the thickness of the adjacent sheet for an assumed distance from the edge of the panel. As can be seen, the panels are assumed to act in 'pure' shear, that is with equal and opposite forces on their opposing edges, and that the shear force is transferred to the end-load carrying members which form the main skeleton of the structure, as seen in the final part of the diagram. The co-operation of the adjacent skin panels with the bars is accounted for in the same way as in early aircraft-structural-analysis, as mentioned above. Typically a strip of skin 20-60 times its thickness is added to the 'bar' cross-section area depending on the real section shape and the method of connection to the skin, also the curvature of the skin close to the 'bar' member, Fig 3. It is convenient to consider shear on the edge of the panel as 'flow', or force per unit length.

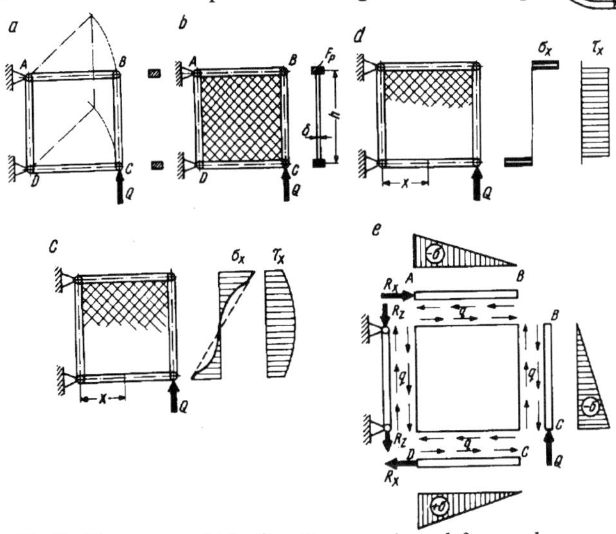

Fig 2: Shear panel idealisation: a. pinned-frame, b. panel added, c. stress pattern (actual), d. stress assumed, e. equilibrium

Truck body loading

The punishing treatment received by truck bodies in service, together with the great variety of use, and abuse, means that combinations of many load cases have to be considered in finding the worst one, for realistic use in the analysis. Considering first the downward force of the payload, in the uniformly distributed case, the following surface loadings typically apply for these five ranges of payload capacity corresponding to average platform sizes for each range:

Load capacity	*Surface loading*
Tonnes	kg/m
0.8-2.0	190-330
2.0-4.0	300-450
4.0-7.0	450-650
7.0-12.0	450-650
12.0 & more	600-1000

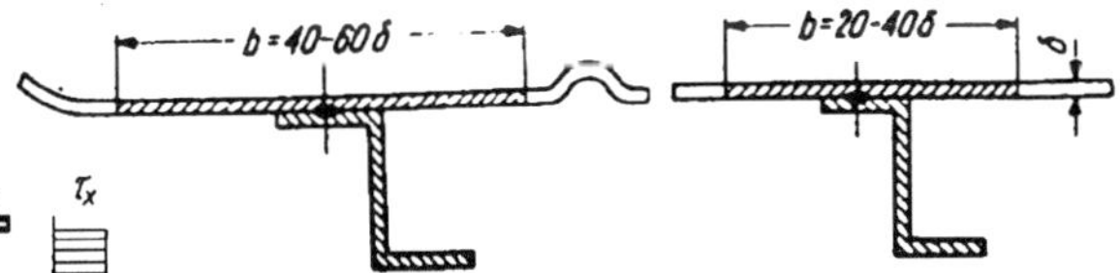

Fig 3: Width of skin added to end-load carrying member

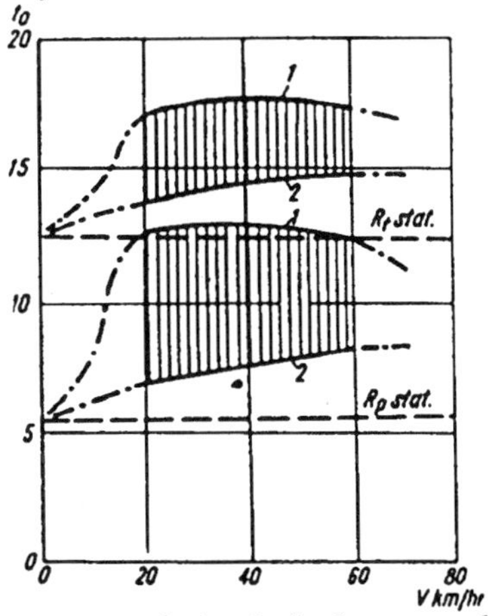

Fig 4: Service axle-loads for heavy truck

These 'best-case' stress distributions can then be modified, in a general-haulage vehicle, to allow for highly concentrated loading that might occur from the legs of stillages or perhaps the mounting pads of heavy machinery. Dynamic factors should be built in to allow for travelling over rough roads, at speed, on harsh suspensions. Combinations of inertia loads due to acceleration, braking, cornering and kerbing should also be considered. One approach might be to detach a platform body of the type mounted on horizontal runners and remount it on four cornerwise load cells; by taking readings before and after static loading of the vehicle — and again, by means of suitable recording equipment, during testing over a severe trial-course. Then a good idea of the factors needed for severity of operation could be obtained. Large scale truck manufacturers will normally have large banks of such 'road-load' data.

Since it is argued that a vehicle structure which is strong enough to endure the maximum dynamic loads, which occur sparodically, will also have sufficient resistance against repeated dynamic forces at less than their average value (fatigue loading), then key factor in design is the determination of dynamic coefficients which can be used in static analysis. Fig 4 shows the results of an investigation of the vertical forces on the front and rear axles of a heavy truck with unladen weight 8 tonnes, gross weight 18 tonnes; static axle loads 5.4 tonnes front and 12.6 tonnes rear; wheelbase 4.9 metres. Curves (1) apply to driving into a cylindrical segment, radius 200mm, height 35mm, and curves (2) apply to driving over a concrete surface with joints every 10 metres. Horizontal, dashed, lines are static axle loads — front, lower, and rear, upper. Typically a co-efficient of 3 is used in design to factor up the static vertical loads, for trucks, in the absence of more detailed knowledge.

As well as the above loads which apply to two wheels riding a vertical obstacle, the case of the single wheel bump, which causes twist of the structure, must be considered. The torque applied to the structure is assumed to be 1.5 x the static wheel load x half the track of the axle. Depending on the height of the bump, the individual static wheel load may itself vary up to the value of the total axle load. If, as in Fig 5, it is assumed the body is torsionally rigid compared with the chassis of the truck (as might apply to a tipper), and similar bumps are encountered by opposite sides of front and rear axles simultaneously, the bump height is equal to the tyre deflection + spring deflection x track/spring-base, an expression which can be converted into forces by substituting tyre and spring stiffnesses. Bump heights of +/- 300 mm are reported to have been used in design.

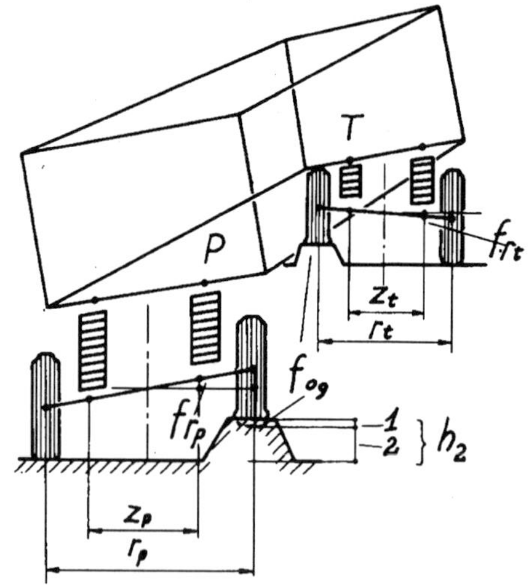

Fig 5: Twist from single wheel bump

For longitudinal forces, largest braking deceleration could be considered as 7 metres/second for trucks. Higher forces can be generated by the driving of the vehicle into a kerb and higher still if the vehicle has side forces due to cornering superimposed as well. For safety in design, these combined forces would then be assumed to act simultaneously with vertical 'bending' and 'torsion' loads arising from the 'double' or 'single' wheel bump cases above. The well-known technique of constructing bending-moment and shear-force diagrams for

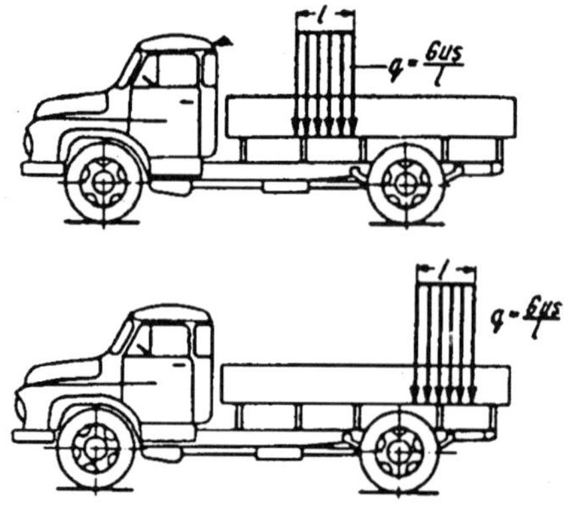

Fig 6: Determining worst case payload position

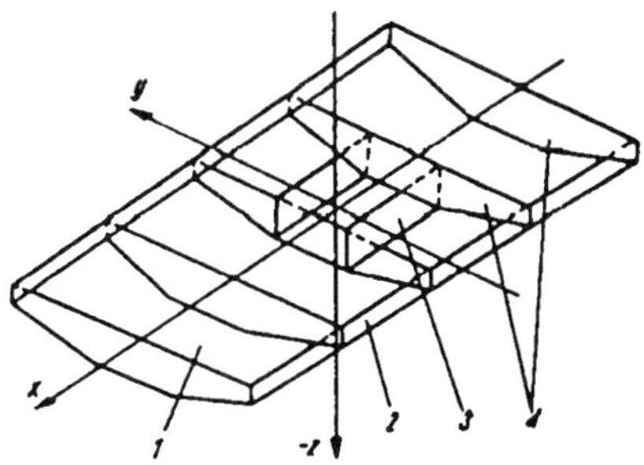

Fig 7: Platform body elements: 1. floor plate, 2. side beams, 3. longitudinals, 4 crossmember

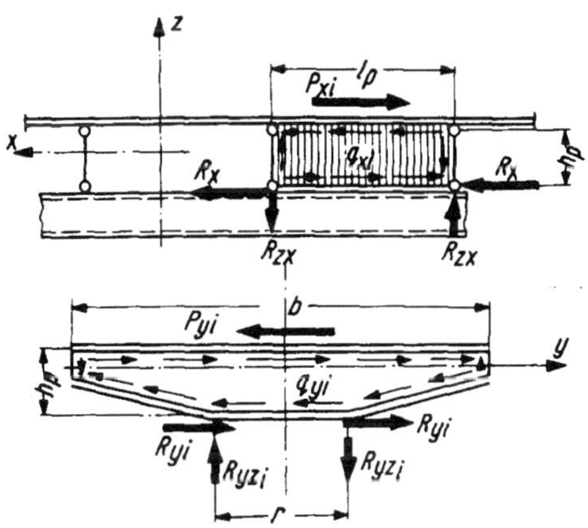

Fig 8: Floor-bearer rigid in shear

the vehicle-structure as a horizontal beam supported by its axles can then be used to determine worst-case positions of the payload for different parts of the structure, Fig 6.

Platform body stressing

Taking this particular body structure to comprise floor-plate, edge beams and longitudinal cross-members, its behaviour will depend on whether the floor-plate comprises a homogeneous sheet with bending stiffness in all directions in the horizontal plane — or corrugated/planked in which case only 'lengthwise' bending stiffness will apply. Longitudinal loads require the presence of an anti-lozenge bay as shown in the idealisation of the structure, Fig 7. This is made rigid by the addition of additional longitudinal members or cross-ties. Side loads are transferred to the chassis mounting points because the cross-member is rigid with respect to shear, Fig 8, while under vertical loading the structure reacts with internal forces as shown in Fig 9.

Without factoring for dynamic inertial effects, the reactions between body and chassis are:

$$R_l = pbl_l/2$$

$$R_x = q_x\, l_p/2$$

$$R_{xz} = q_x\, h_p$$

$$R_{yi} = q_{yi}\, b/2$$

$$R_{zi} = q_{yi}\, h_p\, b/r$$

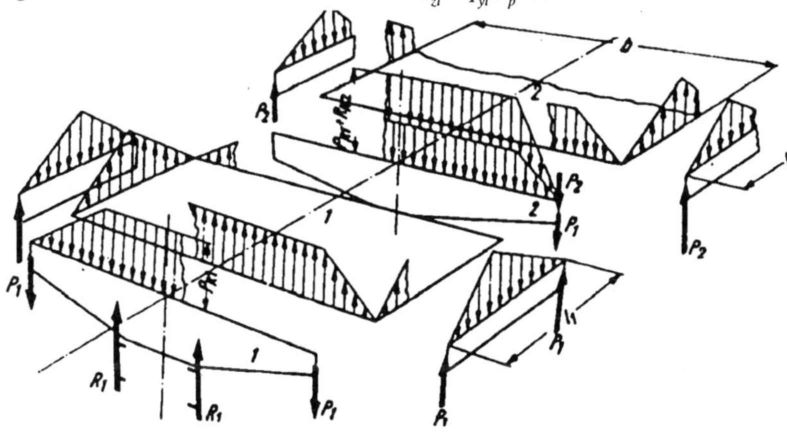

Fig 9: Internal forces, due to bending, in truck

where q = shear stress for the respective directional axes shown.

In the case of cars with self supporting underbody structures, or the ladder-frame chassis of the truck, the idealisation of the 'chassis' is shown in Fig 10. The co-operation of the remainder of the body is ensured through the joint of the sill with the sidewall — and the front and rear bulkheads with the crossmembers of the flat structure. The degree of contribution will of course depend on the number and flexibility of joints to the remainder of the body assembly.

Under symmetrical loading, overall bending, the structure comprises members in shear bending and torsion as in the case of assymmetrical loading, or torsion, when the pattern of loading shown in Fig 11 exists. If in the layout shown, and with the particular design constraints applying to the rest of the vehicle, the sidemembers, or sills, cannot be made sufficiently rigid, then an alternative layout as shown in Fig 12 is typical. The floor panel, in the case of the car, takes no part in carrying the overall bending and torsion loads, its function being to transfer vertical loads placed on it to the adjacent structural elements.

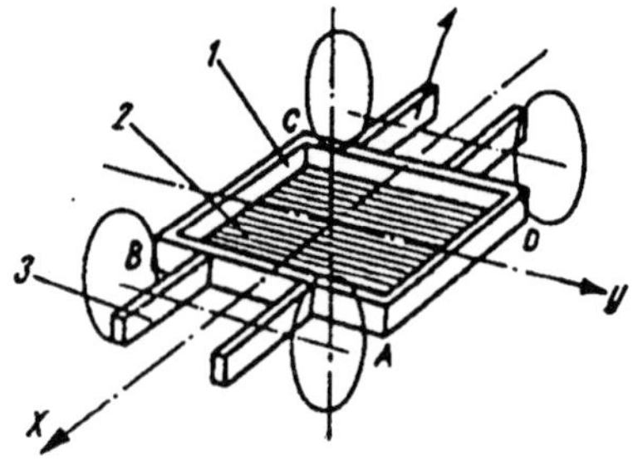

Fig 10: Car underbody elements: 1. sill, 2. floorpan, 3. front longitudinals, 4. rear longitudinals

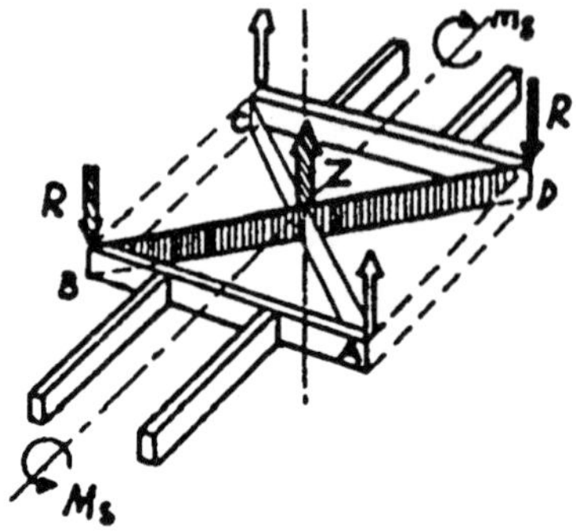

Fig 12: Cruciform-braced frame

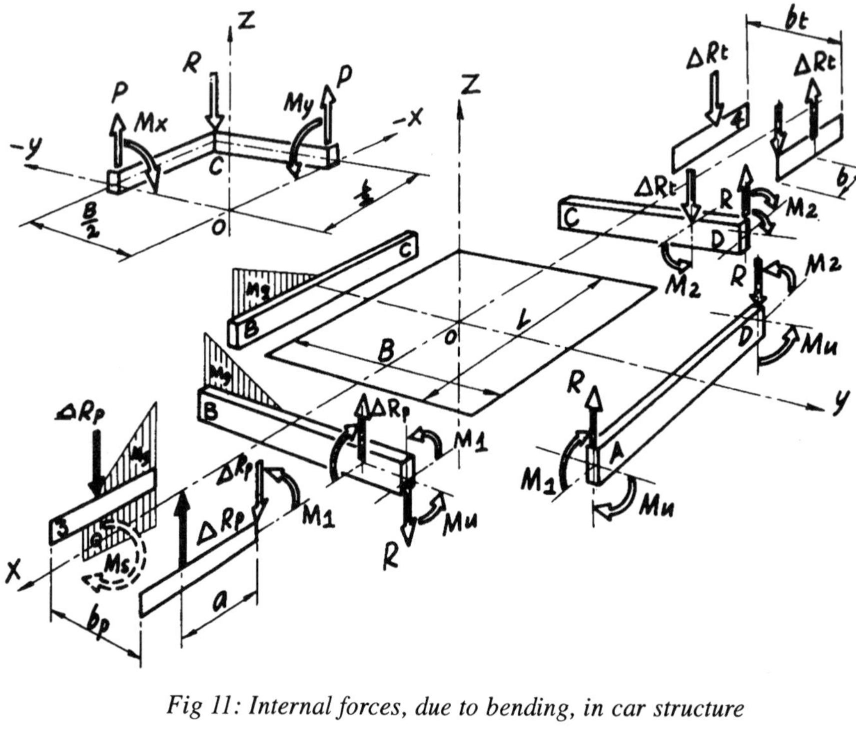

Fig 11: Internal forces, due to bending, in car structure

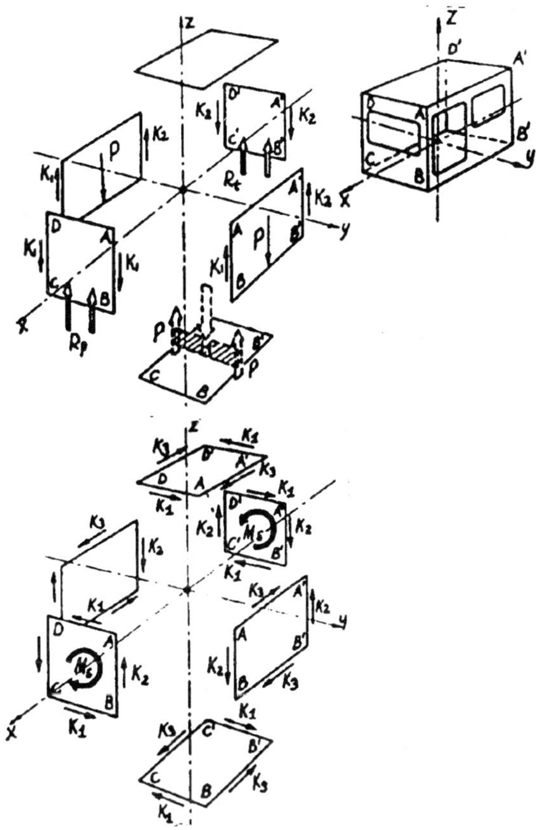

Fig 13: Closed body idealisation

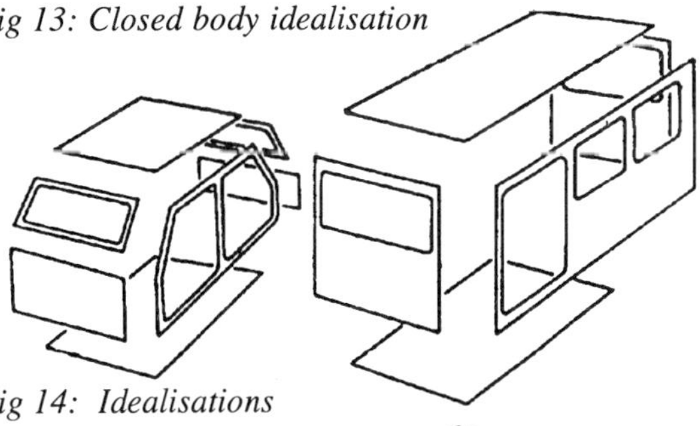

Fig 14: Idealisations for car and bus structures

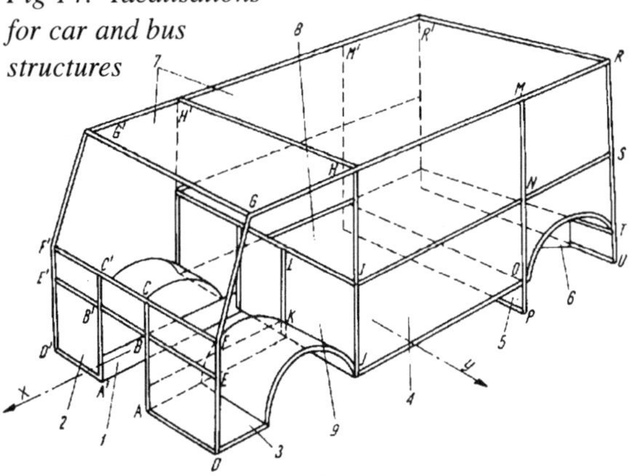

Fig 15: Van structure idealisation

Closed body structures

The first part of this chapter examined the design-analysis of flat-platform structures; here we apply the same techniques to consider closed, 'box-like', structures typical of saloon-cars, vans and buses. Shown in simple, idealised, form in Fig 13, this illustration gives the general distribution of forces in the alternate torsion and bending modes. External loading is reacted by all sides, of the 'box', behaving as shear panels. If either the roof panels, or front and rear end-walls, are non-structural then the floor panel must again be rigid with respect to forces at right-angles to its plane — as in the case of the flat-platform structure considered in the first part. Fig 14 shows how typical car-shell and bus structures are idealised into 'Simple Structural Surfaces' as previously defined.

In the case of an integral box van, the idealisation of Fig 15 can be considered. This assumes the frames enclosing windscreen, cab doors and rear doors are able to resist shear deformation — otherwise 'open-structure' theory has to be applied. When the force balance for all the structural surfaces is equated, Fig 16, edge forces caused by the torque Ms amount to

$$R_p = M_s / S_p$$
$$R'_p = R_p \, a_2 / (a_1 + a_2)$$
$$R''_p = R_p \, a_1 / (a_1 + a_2)$$

The separate elements of the structure, such as the sidewall in Fig 17, can be examined by analysing them as panel and bar systems — as discussed in the previous section. The problem of van design is that the most heavily loaded structural surfaces have the largest openings cut in them. This results in high bending moments in the corners of the surrounding frameworks given by

$$M_n = q_2 \, H \, l_2 / 4$$

for opening height, H, length.

For saloon-cars, a parallel breakdown would give the force distribution shown in Fig 18 — in this case for torsional loading. Structural surface idealisations, such as for the sideframes, require stiff-jointed beam representation. Such frames are normally redundant structures which imply the presence of more than one (statically determinate) load path between external load(s) and reactions. A redundant (or statically indeterminate) structure requires knowledge of the elastic properties of the members to establish the load distribution paths. Since real structures contain an infinite number of alternative load paths they are infinitely redundant so will always require idealisation for analysis into a workable number of redundancies. The sideframe for a four door saloon, as shown in the figure, has six degrees of redundancy visualised by cutting the two ringframes and counting the fixing reactions required at the cuts. To satisfy fixing against vertical, horizontal and rotational displacement of the members would require three reactions at each cut — giving a total of six redundancies for the sideframe.

Van as box beam

A semitrailer box-van body enjoys a degree of symmetry and structural uniformity that can allow it to be considered as a box-beam in structural design — particularly so if a stiff sandwich construction is employed for its main surfaces. With such 'fully stabilised' walls, and other outer panels, shear loads can be transferred across the panels in the same way as assumed for the much smaller car and panel-van structures. In a large trailer van, loading on the horizontal floor planks is transferred by the cross-bearers to the sidepanels. The latter act as beams which pass the

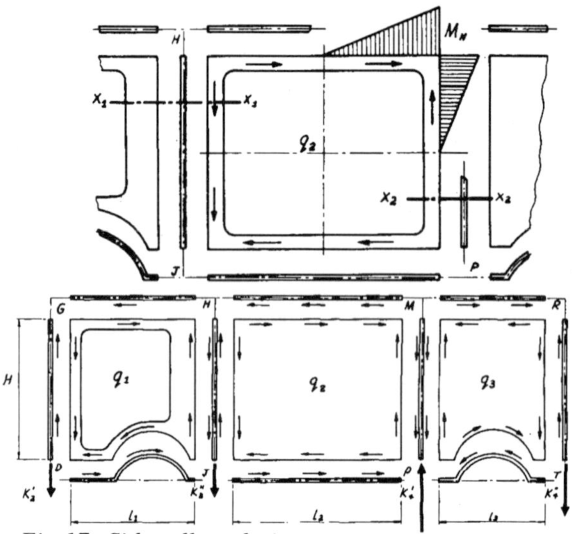

Fig 17: Sidewall analysis

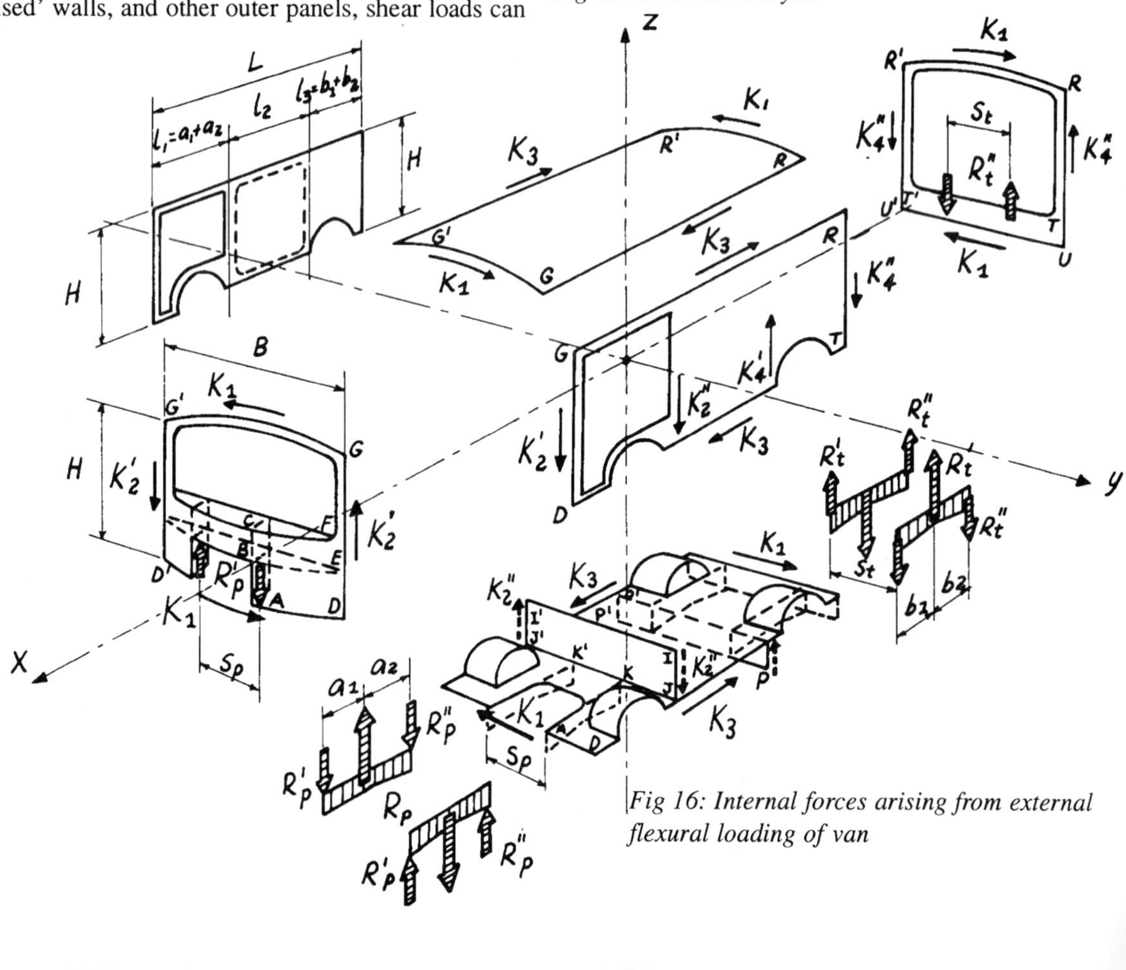

Fig 16: Internal forces arising from external flexural loading of van

load through to the coupling-gear and wheel-suspension respectively. Twisting is resisted by the entire body acting as a tube.

The bending and torsion of the box tube can be analysed directly if certain conditions are met for the application of simple beam theory. In bending, plane cross-sections must remain plane to validate the theory (there must be no axial warping) and cross-sections should also retain their planar outline (no swelling or bowing-in); there is also the proviso that the load should be applied uniformly across the cross-section. Departure from these ideals can be understood for practical box beams by considering the deformation mechanism in the box cantilever of Fig 19. Loaded as shown, there is a tendency for the 'essing' deformation of the sidewalls to take place, arising from the parabolic distribution of shear stress across the vertical panels. There is also axial warping of the horizontal panels caused by shear lag. This is due to prevention of bending forces applied at the panel edges, by shear flexibility, from being able to maintain even value across the panel width.

When fully stabilised sandwich panels are not employed, systems of pillars and rails are the rule for reinforcing side-panels. Part of the objective is to divide the sidewall into panels which are small enough to avoid the possibility of buckling. This buckling has traditionally been analysed using 'tension-field' beam theory which assumes, as the panel is sheared, a diagonal tension field, only, applies in the panel. Fig 20 nicely describes this concept, in which a single bay truss framework, top, is compared with a plane sheet panel, bottom. Under load *P* the tendency for the diagonal brace in the frame to react tension load is paralleled by the tension field assumed to exist in the panel. Bowing of the other brace is

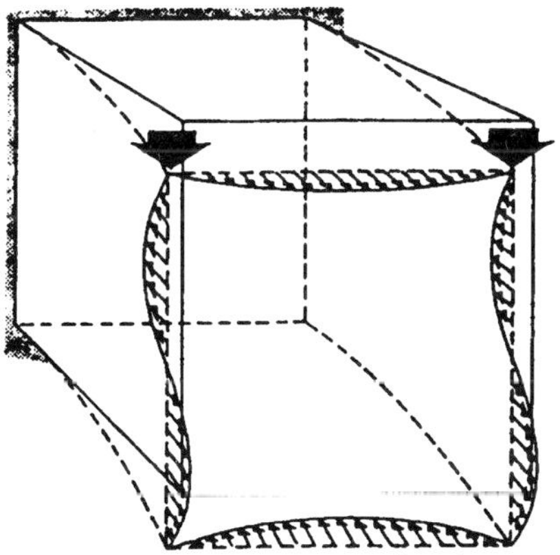

Fig 19: Cantilevered box beam

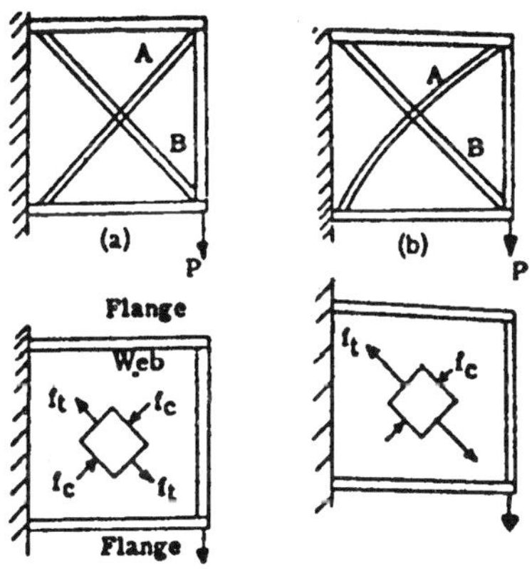

Fig 20: Tension-field beam concept

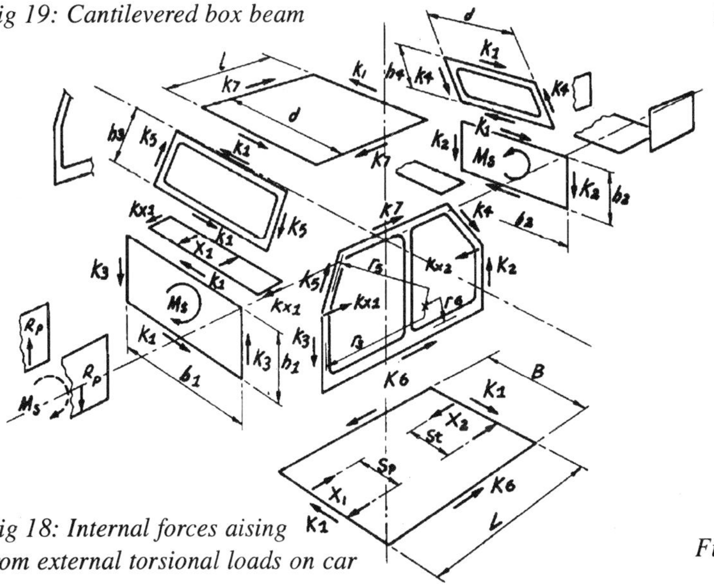

Fig 18: Internal forces aising from external torsional loads on car

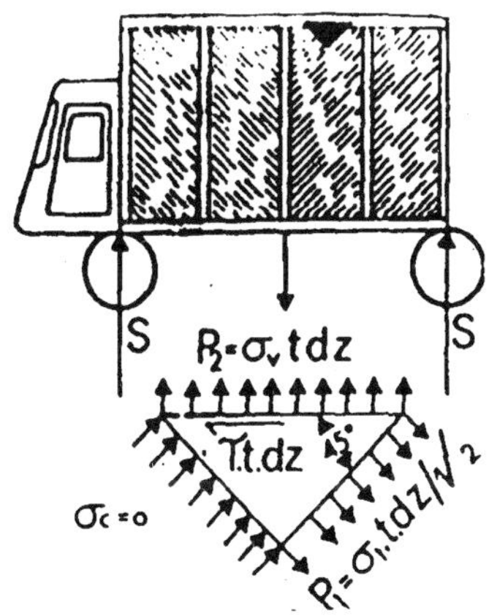

Fig 21: Tension field in van sidewall

paralleled by compressive wrinkling of the panel.

In the sidewall depicted in Fig 21 assumed to be under the action of a central concentrated load 2S then tension bands, corresponding to the broken lines in the figure, can be considered. If the pillars and rails provide a condition in which pure shear of the panels takes place, then the forces acting on a triangular element of the panel can be used to obtain the relationship:

$$\sigma_v\, t\, dz - \sigma t(tdz/2^{1/2})\; 1/2^{1/2} = 0$$

for direct stress sigma and a similar one for shear stress τ — the two of which can be equated to give

$$\sigma t = 2\tau \text{ and } \sigma_v = \tau$$

If h is the height between upper and lower rows of rivets securing the panel to its top and bottom rails then $\tau = S / ht$ and rivets are subjected to loading τtdz, horizontally and vertically, with resultant:

$$2^{1/2}\tau tdz$$

or rivet load per unit length of $1.41\ S/h$. With rails held apart by pillars, compressive load in them is $P_l sin\ 45 = Sd/h$ for pillar spacing d.

Pillars and rails in body stressing

The effect of horizontal intermediate rails (stringers) related to the corner rails (booms) can be analysed by early aircraft structural techniques. For the reinforced box beam shown in Fig 22 the effects of shear lag on the departure (full lines) from simple bending theory is shown in Fig 23 from which it can be seen: the longer the beam the more accurately can simple beam theory be applied.

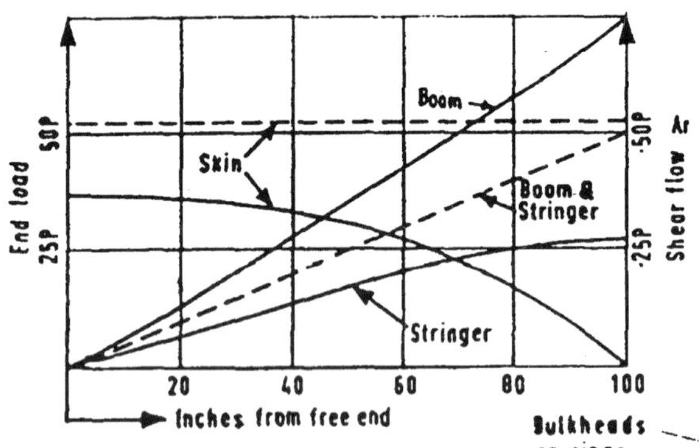

Fig 23: Skin, stringer and boom load distribution

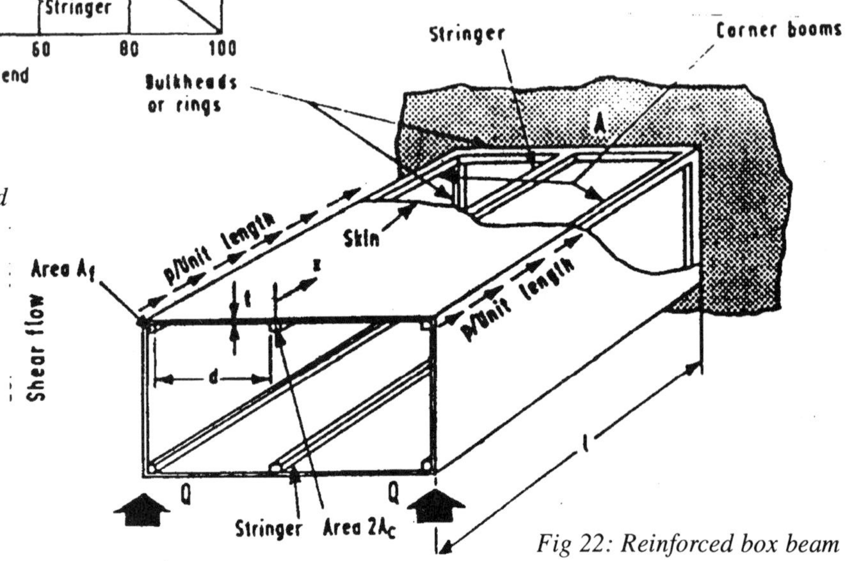

Fig 22: Reinforced box beam

Chassis frames

The conventional commercial vehicle, and traditional coach-built car chassis is a shallow semi-stiff-jointed frame subject to bending and torsional loading and often substantial deflection. The compromise structures used on volume built vehicles in this category, to match a whole range of body types on one chassis configuration, can often be improved upon for specific specialist vehicle categories.

The goods and passenger vehicle chassis frame, in forming a mounting for a great many different body configurations, is inevitably a compromise in the torsional stiffness it provides for many body types. Because flexibility has to be provided for flat-platform trucks involving cross country travel in military operation there is little relative stiffness available to relieve a stiff box, or tanker-shell, body from wracking loads. Codes of practice stipulated in readiness for World War 2 required a typical 3-ton payload platform truck to be capable of having one wheel lifted from the ground without affecting the others. Bodybuilding manuals provided by truck manufacturers thus lay down elaborate systems of reinforcement for such special applications as tankers, tippers and mixers.

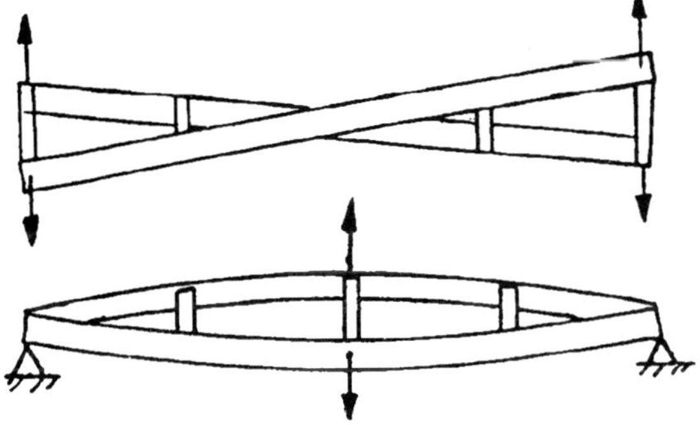

Fig 24: Twist modes for different loadings

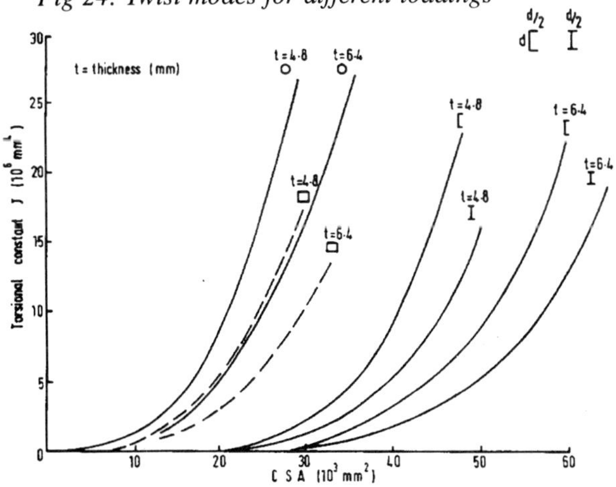

Fig 25: Stiffnesses of various sections

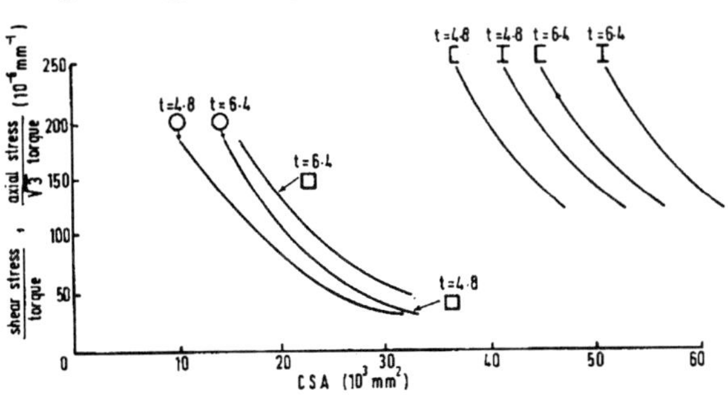

Fig 26: Related stresses in twisting

Specialist chassis builders, including some trailer makers, have the opportunity to tailor chassis stiffness to body type and, of course, to provide, in some instances, chassis structures integrated with the bodies. This creates the need for being able to predict stiffness at the design stage in order to avoid the high degree of structural-redundancy arising from gross differences in stiffness between chassis and body. The different ways a chassis can be loaded in torsion include an asymmetric payload on the load platform; uneven ground surface; lateral accelerations acting on high centre-of-gravity loads and severe cornering manoeuvres.

The different types of torsional load result in extremes of torsional deflection mode as shown in Fig 24. On balance it is best to have the torsionally stiffest crossmembers towards the ends of the chassis frame. Closed rather than open section members are obviously an advantage for this purpose. However, deep open sections can also exert a considerable influence on overall torsional stiffness of the chassis frame if the ends of the crossmembers are rigidly attached to the sidemembers. Remember, though, that high stresses will be built up at the interface — due to constraining the axial warping of the crossmember sections. Figs 25 and 26, due to PW Sharman of Loughborough University[1] compare stiffnesses and associated stresses for various sections.

The former Swiss company Saurer[2] built an all-welded chassis on an experimental basis to establish the desirability of greater torsional stiffness. Sufficiently soft springs had to be provided to replace the frame's traditional supplementary role of maintaining wheel/ground contact. The separate effects of frame and suspension stiffnesses can be understood by referring to Fig 27 and using the expression for wheel deflection that will just lift the front of the vehicle given by

$$h = F_l f = W_v f l_g / l$$

which equals sum of flexibilities due to spring, tyre and frame stiffnesses given by

$$(1/k_{fs})(t_{fw}/t_{fs})^2 + 1/k_{ft}(t_{fw}/t_{rw})^2$$
$$x[(1/k_{rs})(t_{rw}/t_{rs})^2 + (1/k_{rt})(t_{fw}^{\,2}/2k_{ch})(\pi/180)]$$

Frame stiffnesses measured by Saurer were 2175 lb ft (2948 Nm)/deg for a conventional frame and 10,875 lb ft (14745 Nm)/deg for the experimental welded frame with tubular crossmembers. These resulted in values of h around 6.5 ft (1.98 m) for the conventional frame and 5 ft (1.52 m) for the new frame. The theoretical results obtained are reduced by the effect of helper springs, the company point out. Fig 28 shows a comparison of frame stresses with the two frames, each twisted around 0.5 deg about the longitudinal axis.

An approximation for the torsional stiffness of a complete chassis frame can be obtained by summing individual sidemember and crossmember stiffnesses as seen in the representation of Fig 29 — using the reciprocals of stiffness as appropriate (inset). However there is an incompatibility of strain between assumed bending and twisting modes which makes the following expression for torsional flex-

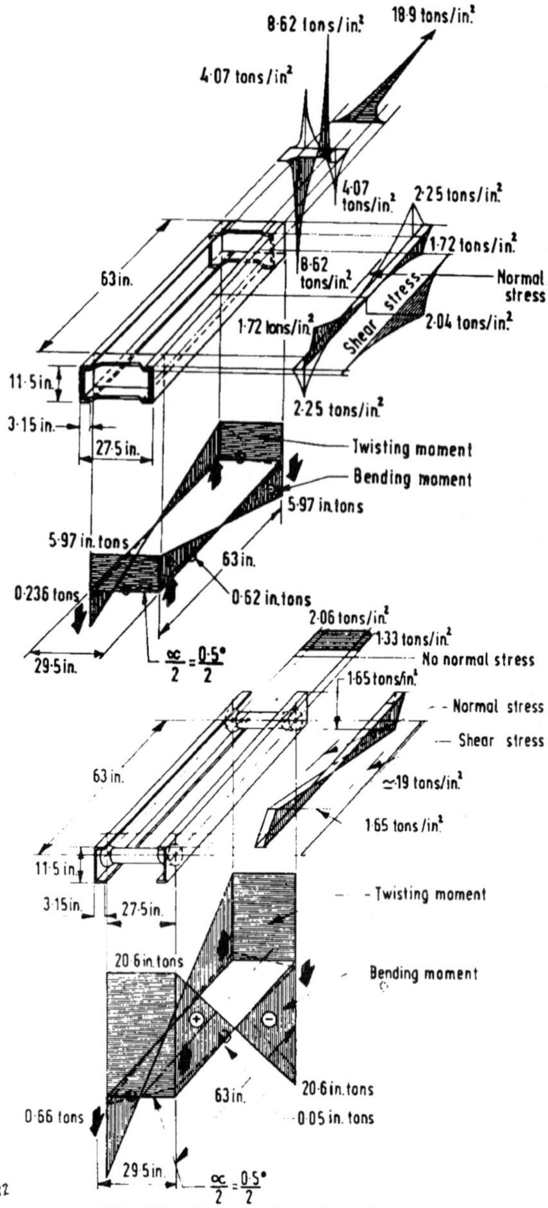

Fig 28: Conventional and welded frames compared

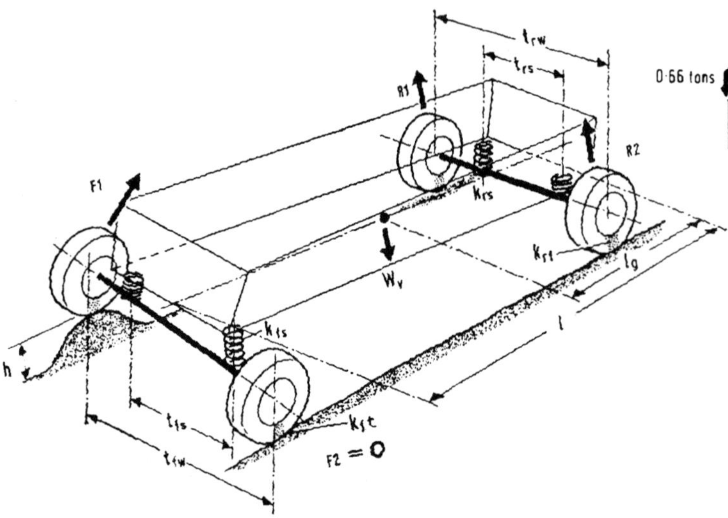

Fig 27: Torsional parameters

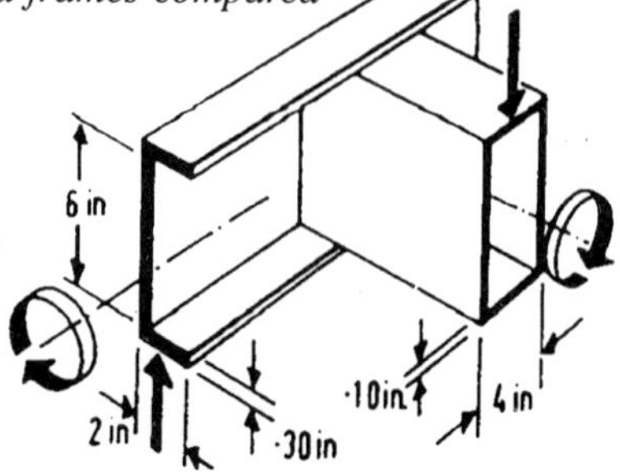

Fig 30: Example frame joint dimensions

ibility α/T only an approximate one:

$$[5l-1/G(J_1+J_2)] + 10/[3l_2E\{(I_1/l_1^3)+(I_2/l_2^3)\}+(3G/2l_1)(J_1+J_2)]$$

and *E*, *G* are section second moments of area and moduli in bending and twisting respectively. For the example sections of Fig 30, a figure of 1320 lb ft/deg is obtained for torsional stiffness which compares with a figure of 1594 lb ft/deg using a more exact (but more laborious) strain-energy method[3] involving more accurate compatibility of strains shown in the deflection modes of Fig 31.

Conventional designs usually involve ladder frames of channel section members as shown in Fig 32 with variations of crossmember section and attachment as shown in Fig 33. There are other loadings on a typical crossmember due to such effects as offset fuel tank mounted to the frame; also lozenging of the frame in the horizontal plane due to effects such as asymmetric braking. These can add up to the kind of crossmember loading combinations shown in Fig 34.

Another aspect of frame torsional stiffness is the effect it has on roll moment distribution between front and rear suspensions. The ability of the suspension to transfer roll-moment on to the wheels depends then on spring rate and spring base. Rear wheels usually take a greater proportion, on goods vehicles, due to the sprung mass centre being nearer to the rear axle — resulting in both a smaller torsional compliance of the frame and a greater rear spring stiffness. There is also greater roll moment transfer at the rear of a conventional truck chassis due to the higher roll centre of the rear suspension. The combined effect can lead to a limiting oversteer condition in cornering unless the frame is stiffened, and front suspension roll stiffness increased so that a more equable roll moment distribution is obtained.

There is need too for ensuring sufficient stiffness of the chassis frame against bending in the horizontal as well as the vertical plane, to avoid differential steer effects on front and rear axles. The ride

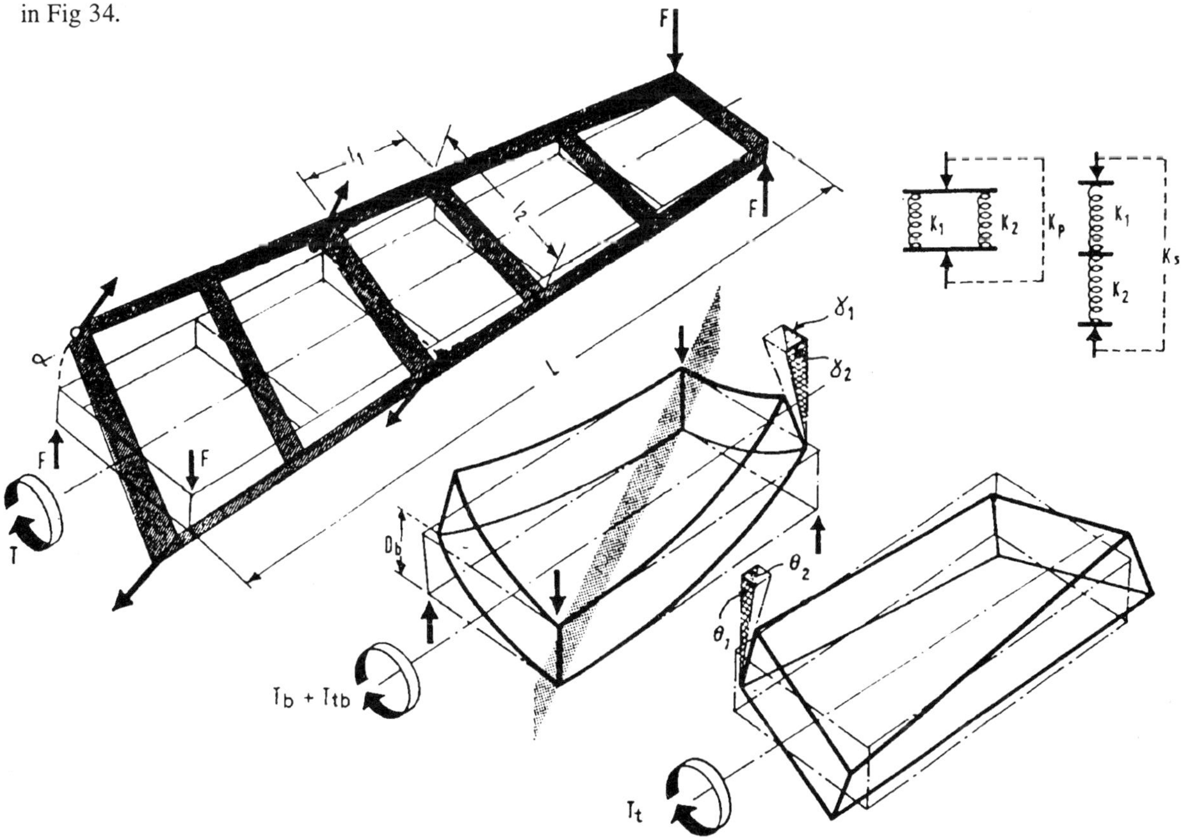

Fig 29: Simplified chassis-representation

of the vehicle can also be affected by frame stiffness according to the bending and twisting mode frequencies excited by dynamic road inputs. Using thin walled sidemembers, very high natural frequencies can be obtained with deep webs and these can be made more economical in weight by circular or diamond shape cut-outs along the length of the web. Weight can also be saved by profiling the sidemember section so its depth of section more nearly corresponds to the bending-moment distribution.

Sidemember to crossmember attachment methods vary with vehicle type and conventional practice has seen lighter trucks with flange attachment and heavier ones attached through the webs. Spacing of crossmembers affects both lozenging resistance and the resistance to buckling of the compression flange of the sidemember under bending load. Diagonal bracing can also be employed to advantage. The frame as well as being strong enough in bending and torsion, must also have sufficient resilience (strain energy) to allow the designed deflection without over-stressing — an important criterion in crossmember to sidemember joint design. The 'alligator' type crossmember is one of the design solutions to this problem.

Frame dimensions on special purpose commercial vehicles are often set by vehicle legal dimensional limits, axle loads and spacing. SAE practice is to space sidemembers at 34 in (0.86 m) overall width and their height is often governed by suspension movement requirements. Sidemember depth follows from bending moment calculations, the allowable bending stress being conventionally taken as one third of the yield stress. The crossmember corresponding to the rear suspension anchorage sets an important design criterion. Fig 35, due to Sherman[4], shows a flange connected alligator crossmember under load due to 'single wheel bump' of the suspension. Because chassis twist is accompanied by lateral displacement of the axle, the latter exerts a high bending moment, via the spring, on the hanger bracket which can cause diaphragming of the web unless properly restrained by the crossmember.

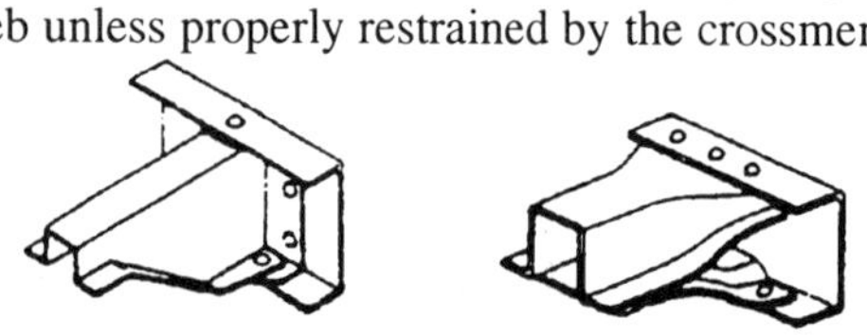

Fig 33: Typical crossmember/sidemember joints

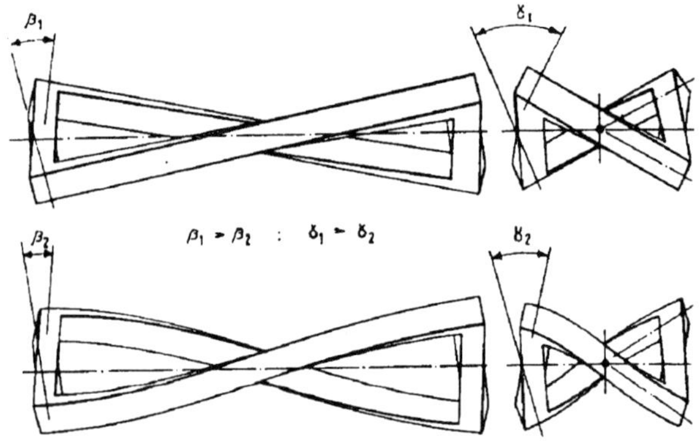

Fig 31: Correct twisting modes for strain compatibility

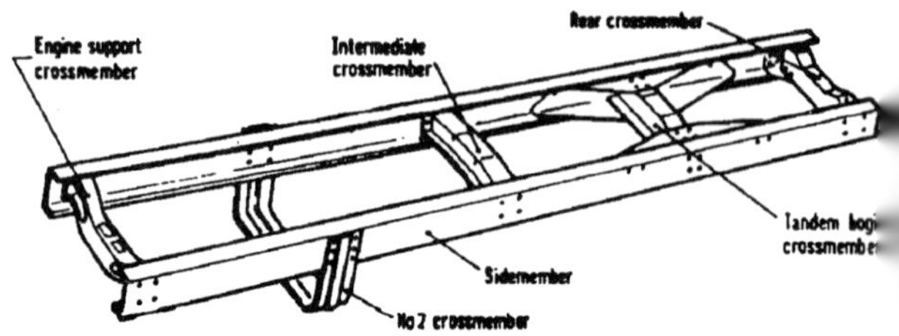

Fig 32: Conventional channel-sectioned ladder chassis

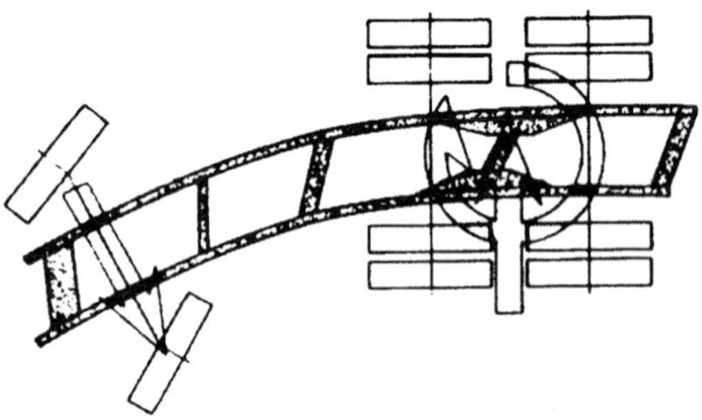

Fig 34: Lateral loading of frame

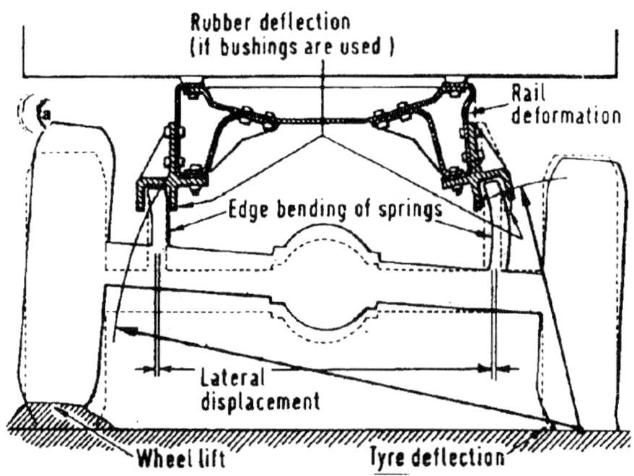

Fig 35: Diaphragming of the sidemember web

Body sub-assembly analysis

Structural analysis of individual elements such as beams, can be combined to examine the behaviour of body sub-assemblies prior to analysis of the complete vehicle body shell. This adds to the overall 'feel' of the structure during the design process.

Examination of sections of the body as a separate entity can help the visualising of deflections influenced by remoter sections to which the first might be anchored. An example of this is the flexing of the forward extensions from the bulkhead of a saloon car. The overall distortion in Fig 36 results from the fact that the principal axis of the beams formed by the wing and valence assemblies are inclined to the vertical. The result is that a component of the applied force bending the beams about their

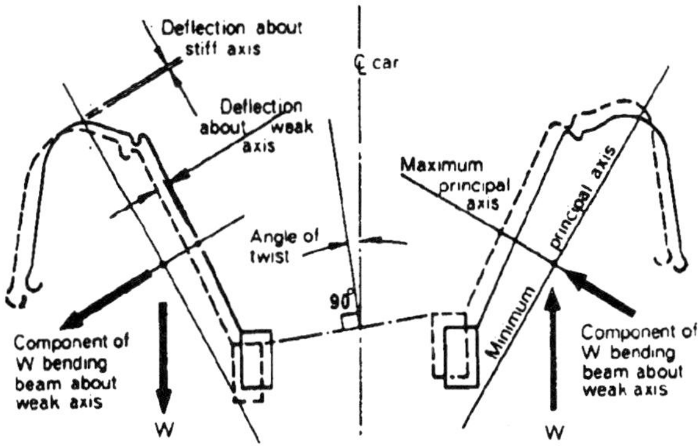

Fig 36: Body front-end twist

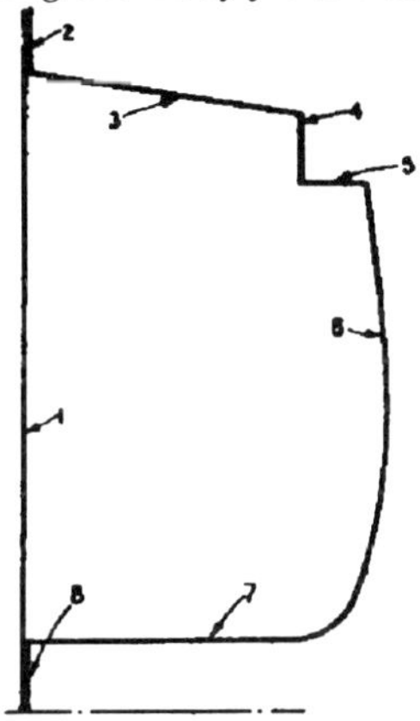

Fig 38a: Sill section

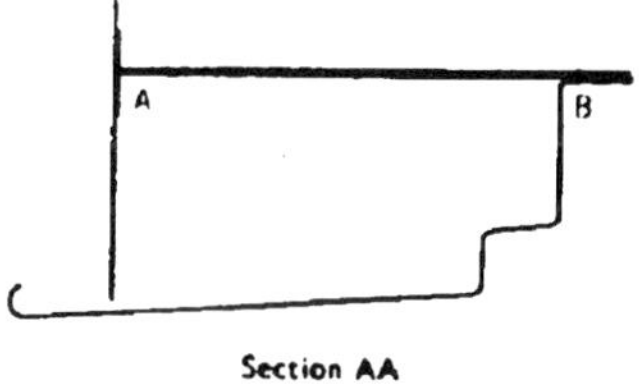

Fig 39: Joint in sill

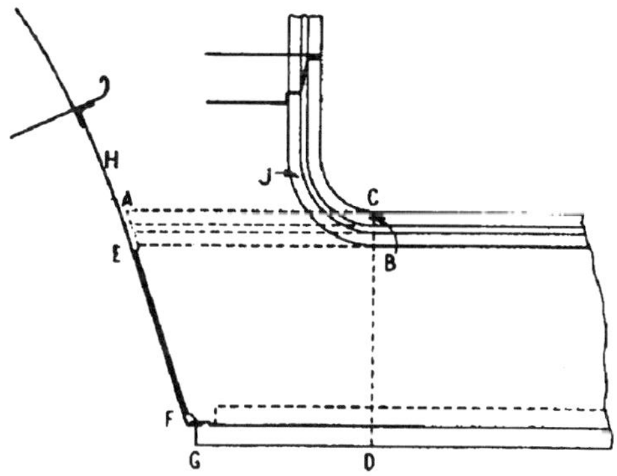

Fig 40:Elevation of sill joint

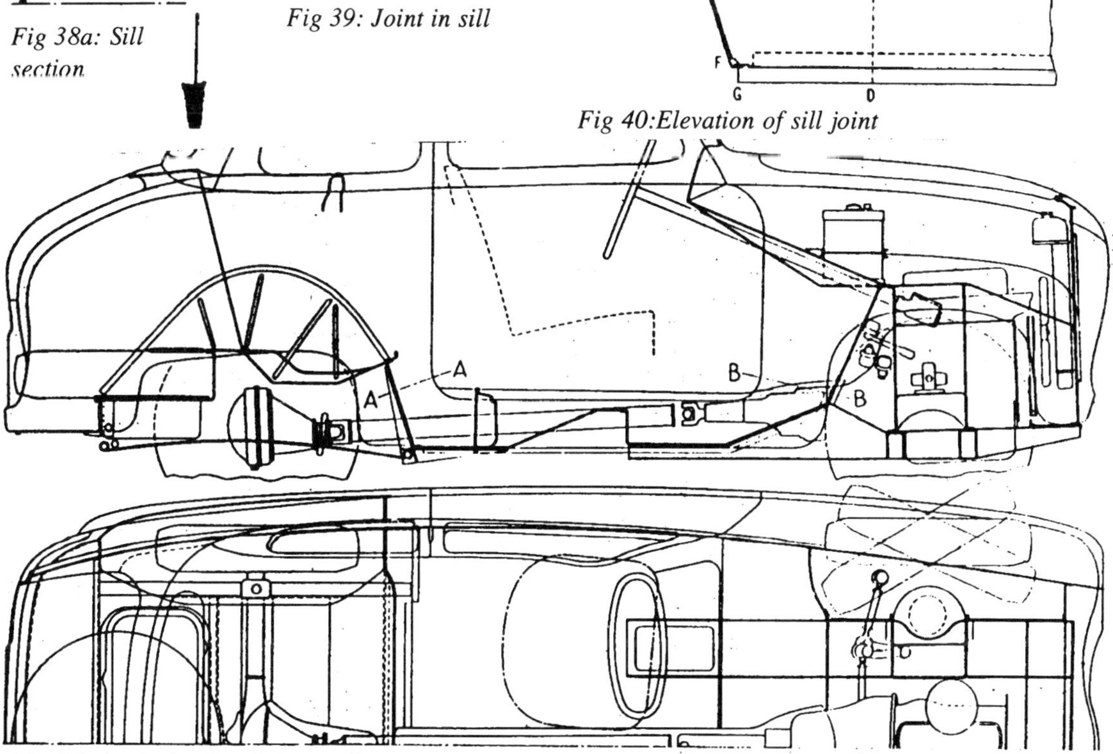

Fig 37: Analysis of early integral structure

Item	Size	Area	y	y^2	Ay	Ay^2	Local I
1	0·036 × 5·0	0·180	2·5	6·25	0·4500	1·1250	0·374
2	0·036 × 0·5	0·018	4·75	22·5	0·0855	0·4050	0·0004
3	0·036 × 2·0	0·072	4·4	19·36	0·3170	1·3950	—
4	0·036 × 0·5	0·018	4·0	16·00	0·0720	0·2880	0·0004
5	0·036 × 0·5	0·018	3·75	14·05	0·0675	0·2530	—
6	0·036 × 3·25	0·117	1·875	3·52	0·2190	0·4120	0·103
7	0·036 × 2·0	0·072	0·5	0·25	0·0360	0·0180	0·024
8	0·036 × 0·5	0·018	0·25	0·0625	0·0045	0·0001	0·0004
		ΣA = 0·513		Σ =	1·252	3·896	0·502

Fig 38b: Sill secton properties

weak axes to the extent that the radiator frame is unable to stabilise the assembly. Only good continuity between wing and underbody side rail extension, via the bulkhead, ensures that the lozenging displacement of the lower A-post has a positive stiffening effect on the assembly.

The approach of analysing individual assemblies was taken by the engineer who carried out the structural evaluation of one of the UK's first integral car body structures.

The Austin A30 body shell was analysed by a former aeronautical stressman, Ken Garrett, who was one of the first to apply aircraft thin-wall structure techniques in road vehicle bodies. The complete structure is shown as Fig 37. The sill section of Fig 38a was considered as a beam with the help of the table of elemental properties shown in Fig 38b. Study of Fig 37 shows that to the rear of the sill, bending loads are taken by the outer quarter panel — also by an inner quarter panel attached to the sill inner plate and rear pillar. This inner panel extends over the rear wheel arch, which joins it and its upper edge, is stabilised by both its shape and its attachment to the outer panel; to the rear it is stabilised by joining the side of the parcel shelf and to the drip channel in the boot opening.

The junction of the inner quarter panel with the heelboard on Section AA of Fig 37 is similar to A in Fig 39. If there were only single thickness between A and B, it may be necessary to extend the inner panel of the sill upwards and the inner quarter panel downwards so that they overlap as shown, to double the strength of the web portion. At the rear of the two panels one is flanged inwards and the other outwards so as to connect both with the heelboard.

The importance of structural continuity, in order to transfer loads from one part of a structure to another is clear from the example shown in Fig 40. Here it is necessary to transfer both shear and bending moment between the sill and the rear end of the vehicle. The sill outer panel is joggled under the rear quarter panel and there is a vertical joint line CD. At the rear end the sill is closed by flanging outwards of the outer panel and its connection to the heel board at EF. End loads to be transferred from the top and bottom of the section to the rear quarter panel are found by dividing the section depth into the bending moment. These loads will govern the number of spot welds required at the connection. The direction of the loads will determine whether a 'kink strut' is required at the joint, its dimensions being found by resolving the forces along the line of the strut.

The rear seat and heelboard structure of a saloon car can also be considered as a separate assembly for analysis if the assumption is made that its edges are rigidly rooted in the vehicle sidewalls, Fig 41. In this example, interior vehicle width is assumed to be 1.5 metres and the four constituent panels are rectangular and set at right angles to one another. The shorter sides measure 0.2 (parcel-shelf), 1.0 (squab), 0.5 (seat) and 0.2 metres (heelboard). The latter has a cut-out of 0.15 metre radius and the resulting necked portion of the heelboard is given an

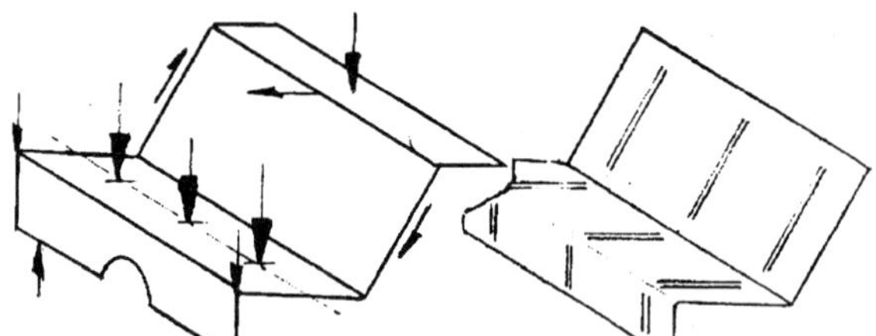

Fig 41: Rear seat pan

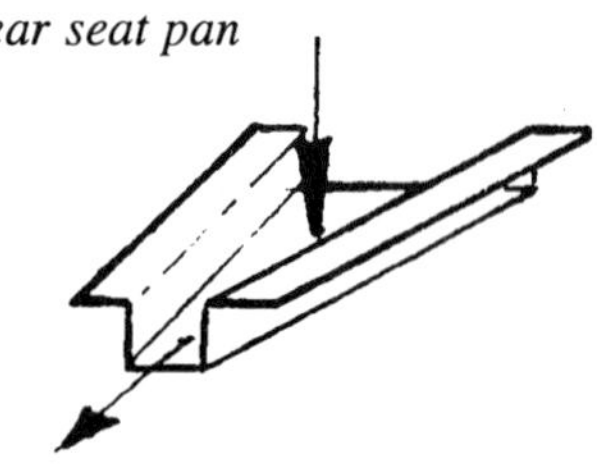

Fig 42: Parcel shelf stiffener

inward folded lower flange. This is boxed into the the seat panel with a zed section reinforcing member.

A load system can be used to simulate a spare-wheel carrier suspended from the parcel-shelf; shear loads of 2.5 kN on the edges of the squab to represent reaction to overall twist of the body shell; horizontal forward pull at the centre of the leading edge of the parcel shelf to simulate forward inertia loads on a seat-belt anchorage at that point; passenger loads vertically downwards on the seats assumed to act on the seat centre line at three equi-spaced intervals of 0.375 metres; finally loads of 2 kN applied in the plane of the heelboard panel to represent upward forces from the spring hanger brackets which cause four-point bending of the panel arising from the vertical downward reactions at the sidewalls.

These loads are factored for inertia effects: 3g bump assumed on the spare-wheel carrier; 1.5g bump assumed on the squab side-loads; 10g impact load assumed for the forward pull of the parcel-shelf. This order of load could be used in selecting panel gauge, joint configuration, swaging patterns or additional reinforcement as necessary. A maximum working stress of 100 MN/m² could be assumed for the mild steel panels and a 0.5 kN working load for each spot weld. A reinforcement for the parcel shelf could be in the form of Fig 42 where a common member both diffuses the restraint force across the width of the panel and avoids bending of the panel in its plane by transferring vertical support loads to the edges. Since there will be diagonal tension fields set up in the squab due to the alternating shear reactions in body twist, either diagonal or vertical swaging is probably appropriate for this panel. Vertical ones could be used, in conjunction with parallel ones on the seat, to give both panels continuity of end load and flexure resistance within their own planes.

A critical area of the assembly will be the necked section of the heelboard with its associated box reinforcement. Here analysis of an idealised

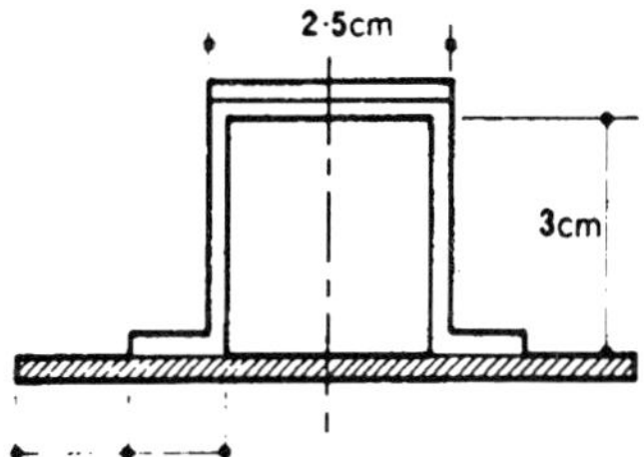

Reinforced top-hat beam section

$$y_{n.a.} = \frac{\sum A \cdot y}{\sum A} = \frac{2{\cdot}64}{2{\cdot}15} = 1{\cdot}23 \text{ cm};$$

$$I = \sum (I_g + A \cdot h^2) = 3{\cdot}877 \text{ cm}^2$$

Fig 43: Heel-board section property calculation

	Area A (cm²)	Centroid distance y (cm)	Ay (cm³)	I_g (cm⁴)	Centroid distance to N.A. h (cm)	Ah^2 (cm⁴)
Top 'flanges'	0·5	3·2	1·60	$(2{\cdot}5 \times 0{\cdot}2^3)/12 = 0{\cdot}00167$	1·97	1·94
Webs	0·6	1·6	0·96	$(0{\cdot}2 \times 3^3)/12 = 0{\cdot}45$	0·37	0·082
Bottom flanges	0·25	0·15	0·04	$(2{\cdot}5 \times 0{\cdot}1^3)/12 = 0{\cdot}00021$	1·08	0·293
Reinf plate	0·8	0·05	0·04	$(8 \times 0{\cdot}1^3)/12 = 0{\cdot}00067$	1·10	1·11
	$\sum = 2{\cdot}15$		$\sum = 2{\cdot}64$	$\sum = 0{\cdot}4525$		$\sum = 3{\cdot}425$

Fig 44: Second-moment calculation

Distance from neutral axis to extreme fibre $y = 3{\cdot}3 - 1{\cdot}23 = 2{\cdot}07$ cm.
Given a bending moment $M = 300$ Nm, bending stress

$$\frac{M \cdot y}{I} = 300 \times 2{\cdot}07 \times 10^{-2}/3{\cdot}877 \times 10^{-8} = 160 \text{ MN/m}^2$$

Distance from neutral axis to centroid of reinforcing plate

$$\bar{y} = 1{\cdot}18 \times 10^{-2} \text{ m}$$

Given $R = 500$ N for each spot weld and assuming the flanges are held by two rows of spots

$$p = \frac{R \cdot I}{S \cdot A \cdot \bar{y}} = \frac{2 \cdot 500 \times 3{\cdot}878 \times 10^{-8} \times 10^3}{2{\cdot}0 \times 10^3 \times 0{\cdot}8 \times 10^{-4} \times 1{\cdot}18 \times 10^{-2}} = 20{\cdot}5 \text{ mm}$$

Fig 45: Bending deflection calculation

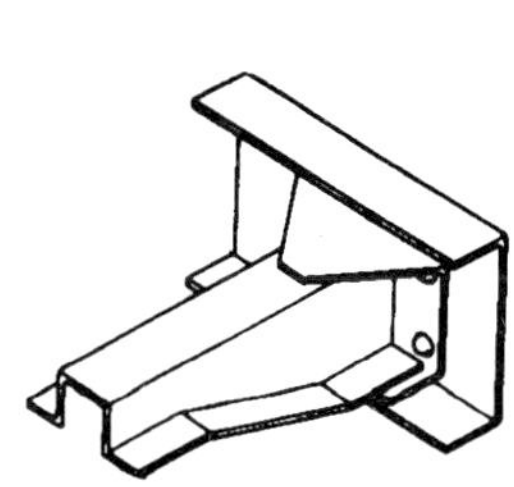

Fig 47: Crossmember/ sidemember joint for CV

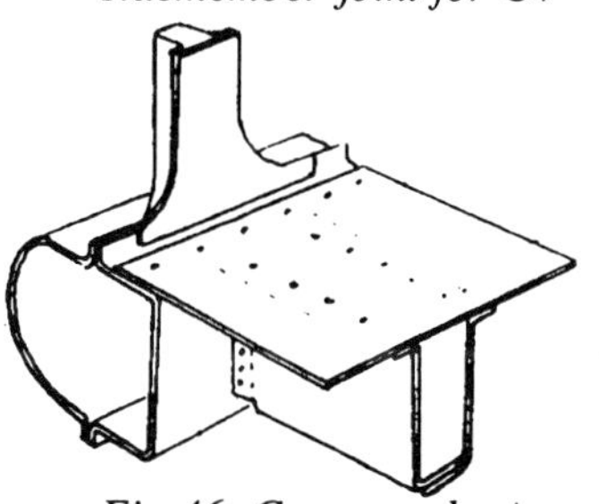

Fig 46: Crossmember/ sidemember joint for car

cross section, Fig 44, would be useful. Section second moment of area could be found from the tabular method in Fig 45 and the vertical deflection of this reinforcing box beam from the calculation shown in Fig 46.

Joints of sidemembers with crossmembers are critical to both car, Fig 47, and commercial vehicle, Fig 48, structures. According to Tidbury *et al*[5] the axial warping of the crossmember beam section is an important criterion in considering the intersection with the sidemember. If the axial warping of the crossmember is constrained by the intersection with the sidemember, end-loading is induced in the crossmember. When loads do not act at the centroid of the section, these cause bending moments as, of course, the loads which do not pass through the section shear centre cause torques. The axial load is conveniently expressed as a bi-moment for the sake of the analysis. This is a pair of opposite moments separated by known distance, Fig 48. Also if a bending moment is applied in a plane parallel to the longitudinal axis, it can be resolved into a moment in a parallel plane through the shear centre and a bimoment. In Fig 49, a moment is introduced to a channel member by forces on the flange as shown — resulting in a bending moment *P.d* and a bi-moment *M.h* as well as a shear force *P* and torque *P.h* over the distance d between the forces.

It is necessary to evaluate the bi-moments caused by direct loads in the crossmember which do not pass through points of zero warping. In Fig 50 this applies to forces *F3* and *F4*; the exception to this is shown of the two equal and opposite forces *F1* and *F2* whose effects cancel each other out as they are equidistant from the point of zero warping. Forces *F5* and *F6*, however, do cause both a bi-moment and a bending moment. It is necessary to add these effects to the normal sets of forces and moments which are passed across the joint when setting up finite element models of the overall structure under bending and torsion.

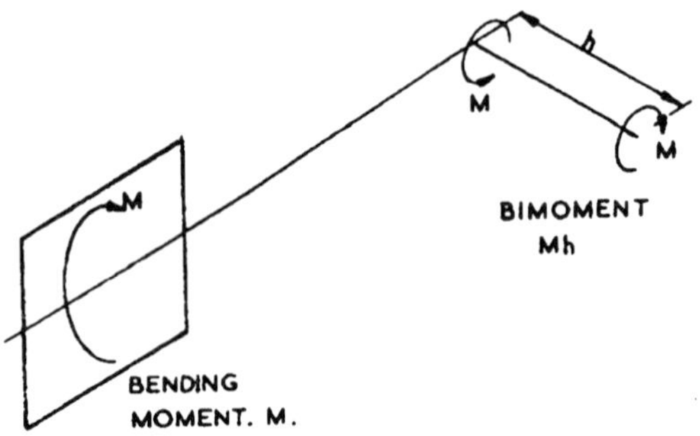

Fig 48: Bi-moment

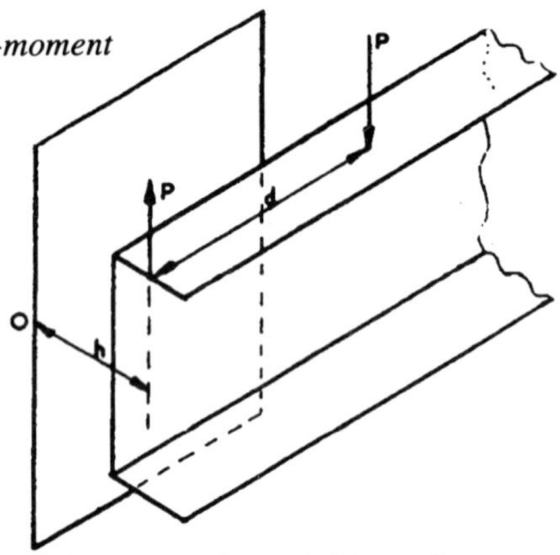

Fig 49: Force system on channel sidemember

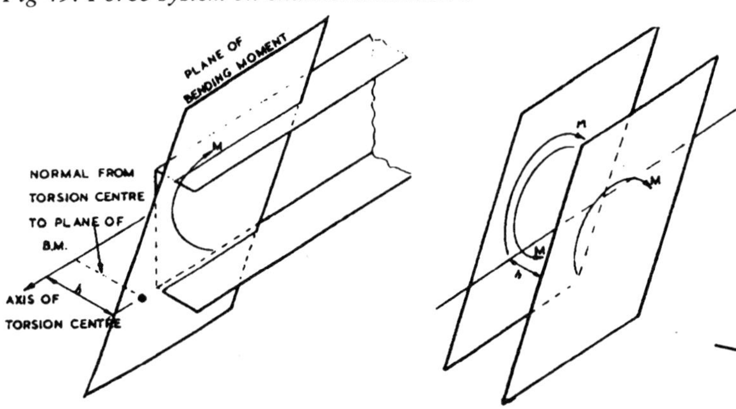

Fig 50: Shear force plus bi-moment

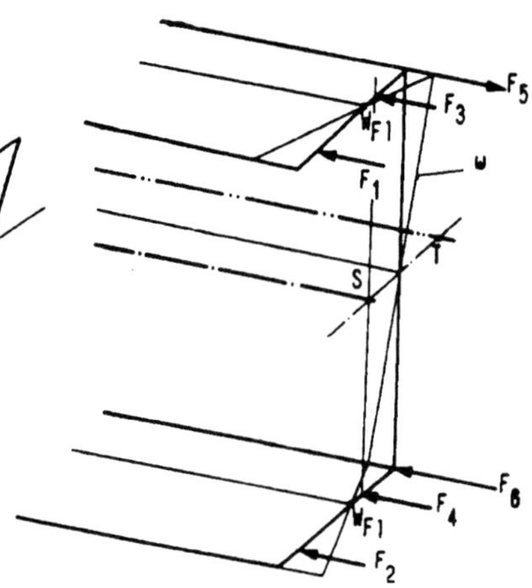

Fig 51: Bi-moment evaluation

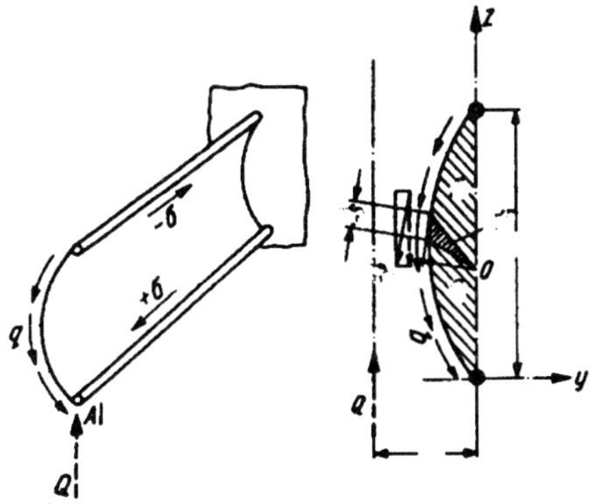

Fig 52. Idealisation of curved panel

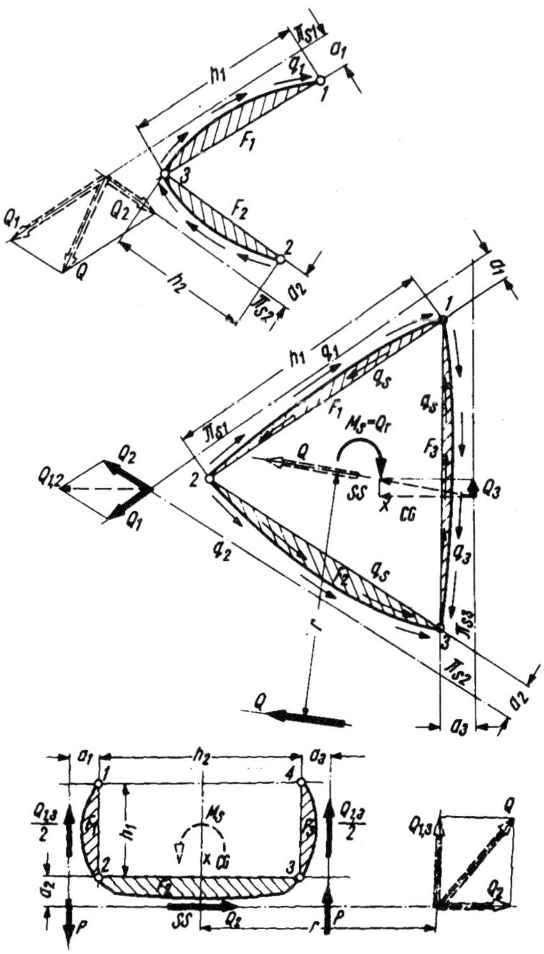

Fig 53: Shell made from curved panels

Body shell analysis

The bending and torsion of idealised box beams has been covered in earlier sections. Here the body shell is considered in a less idealised form to see, by examining the likely load paths, how elements of the structure are best represented for analysis.

Janusk Pawlowski argued that his Simple Structural Surfaces method of analysis, in which structures are represented by surfaces comprising shear panels with end-load carrying bars at their periphery, depended on classifying parts of the structure into the appropriate combinations of bars and panels. The simplest combination comprises two bars and a panel as shown in Fig 52. Shear flow q in the panel is found from the sum of the projections on the related reference axis. Thus on the x and y axis respectively it is

$$\int_0^h q\,ds.\cos a \text{ and } \int_0^h q\,ds.\sin a$$

Shear flow does not depend on the shape of the wall but only on height of the element h which is effectively the depth of the beam which the element represents. Location of the resultant shear, through the shear centre, is found by taking moments, which transform to give $a = 2A/h$ where A is the area bound by the mean width of the panel

Cumulative combinations of this basic beam element can be made to form open 'vee' section beams, closed tubes and open channels as depicted in Fig 53. In closed shell form the multi-stringer arrangement of Fig 54 is rigid to a force Q which causes shear, bending and twisting of the structure. The bending is statically determinate but the shear not so; thus the methods are used which were introduced in the earlier section on thin-walled-structures to evaluate the flows — a figure then shown depicting the shear flow diagram for the 'cut' section alongside a diagram illustrating calculation of the unknown shear flow q_0. Full exploitation of the strength and rigidity of shells built up in this way depend upon maintenance of the cross section shape by the use of bulkheads.

The tendency for axial distortion of the shell in twist should also be accounted for. With the box-tube shell of Fig 55, subject to torque T, there is axial defection of the corner:

$$(T/16abG)[(b/t_1) - (a/t_2)]$$

There is no axial distortion (warpless tube) when $b/t_1 = a/t_2$.
Bending of a multi-stringer tubular shell can be analysed relatively easily with the aid of some simplifying assumptions. For the arbitrary section of Fig 56, stringer cross-section-areas *A* are enlarged by allowing for the contribution made by adjacent panelling. For width of skin panel *b*, thickness *t* and distance of effective stringer area from neutral axis y, direct stress is notated as σ and buckling stress as σ_b. When there is no skin buckling, effective stringer area on the tension side of the shell

$$A = A_s + (1/2)b_u t_u + (1/2)b_l t_l = A_s + b_m t$$

when $t_u = t_l = t_s$ meaning stringer; *u* is upper and *l* lower. After the onset of buckling, the effective area of the panel is assumed to be

$$(1/3)bt + (2/3)bt(\sigma_b/\sigma)$$

and the effective stringer area is approximately

$$A_l = A_s + (1/3)b_m t = (2/3)b_m t(\sigma_{bm}/\sigma_s)$$

where *m* is mean value between upper and lower panels. The method involves tabulating *A* and *Y* from a chosen datum and finding its distance from the neutral axis as sAY/sA and the section second moment of area about the neutral axis as

$$I = \Sigma AY^2 - y\Sigma A$$

If some panels are assumed to buckle then their effectiveness will have been over-estimated and a reduction of area is necessary of $dA = A - A_l$. A new neutral axis can then be found from $y_1 = -\Sigma dAy/\Sigma A_l$ and a new value of area moment $I_1 = I - \Sigma dAy^2 - y_1\Sigma A1$ and the stress becomes $\sigma_1 = M(y - y_1)/I_1$. Successive further approximations can also be made following the same procedure.

To simplify the determination of degree of redundancy of a shell structure, shear panels can be simulated as pin jointed frames for this purpose, in a method due to G B Bolland of Loughborough University. Fig 57 shows representations for simple beams and torque tubes. The formulae given earlier in the section on frameworks can then be used to add up the

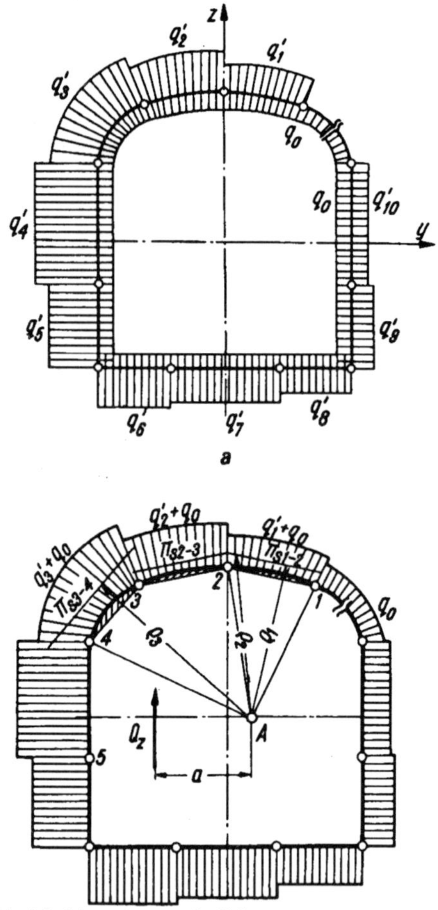

Fig 54: Multi-stringer shell beam

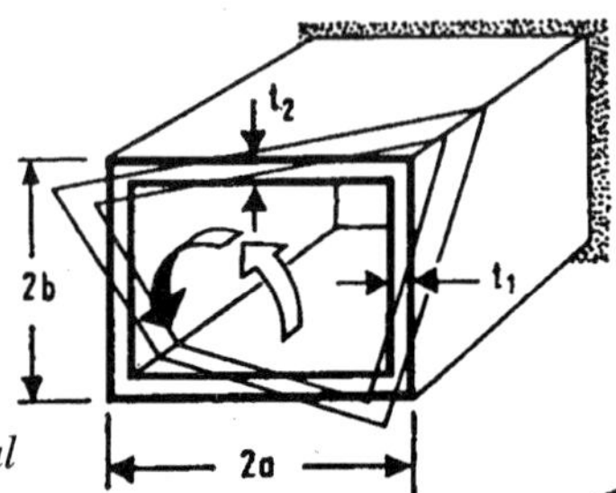

Fig 55: Axial warping of box tube in torsion

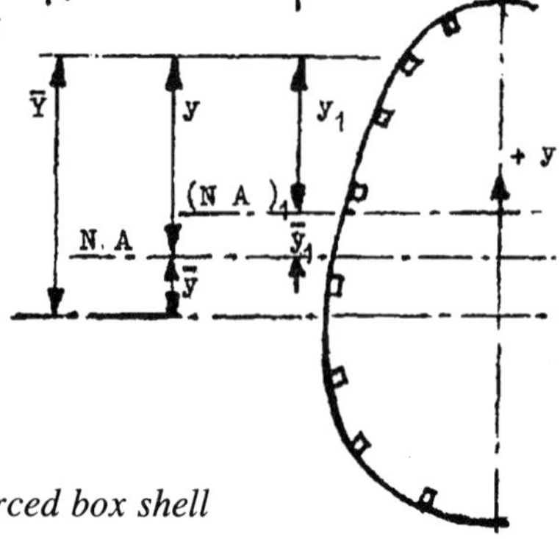

Fig 56: Reinforced box shell

redundancies for the whole shell. Fig 58 shows a curved panel represented in this way and Fig 59 a box with framed cut-outs and longitudinal torsion members. In the latter case, the structure is regarded as three-dimensional with two-dimensional elements hung on and a general form of the equation for counting redundancies becomes

$$R = m - (3j_3 + 2j_2 - 6)$$

where m is the number of members and j_2, j_3 the numbers of two and three dimensional joints respectively. In this case, number of redundancies $R = 8$ since $m = 74$, $j_3 = 16$ and $j_2 = 12$.

Often vehicle structures are idealised as combinations of pin- and stiff- jointed frames as in the case of the saloon car treated as a symmetrical side frame in Fig 60. Members BI and DH are pin-jointed but all others are rigid and the system has five redundancies, as a plane frame. Integral bus structures have been idealised as shells such as that shown as Fig 61 and box vans as shown, due to G H Tidbury of Cranfield University, as Fig 62.

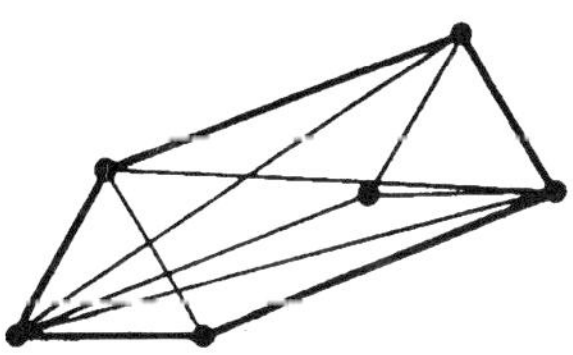

Fig 57: Frame representaton of shell

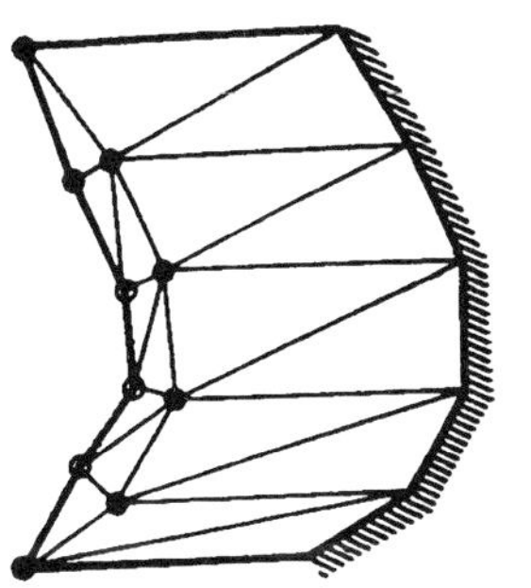

Fig 58: Frame representation of curved panel

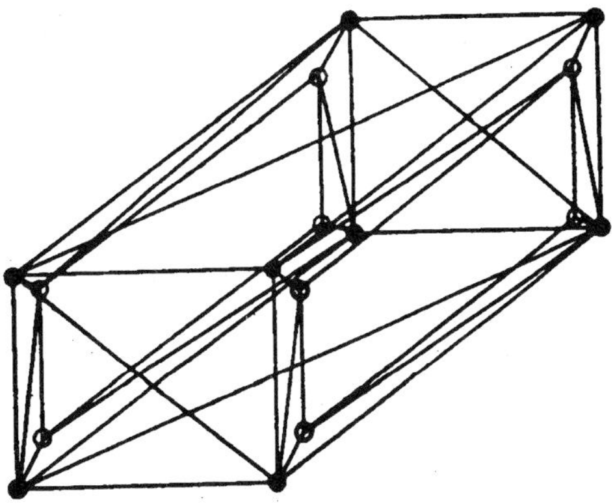

Fig 59: Frame representation of box with cut-outs

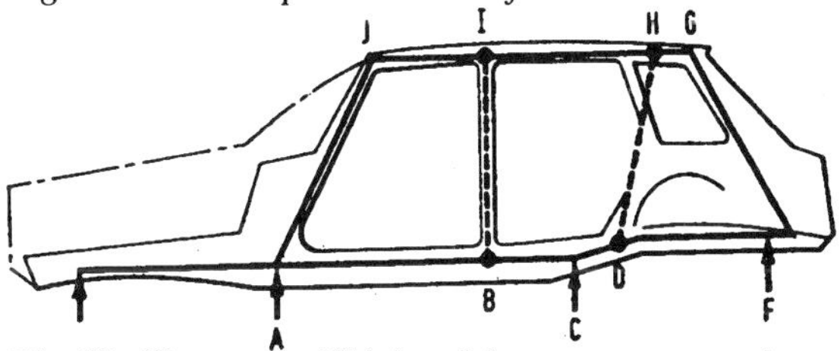

Fig 60: Pin- and stiff-jointed frame representation

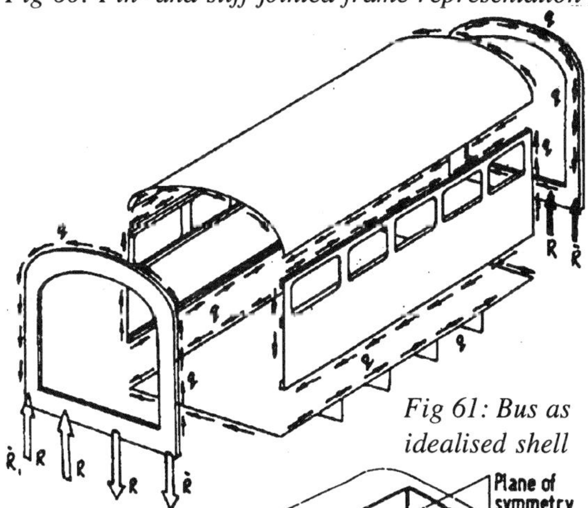

Fig 61: Bus as idealised shell

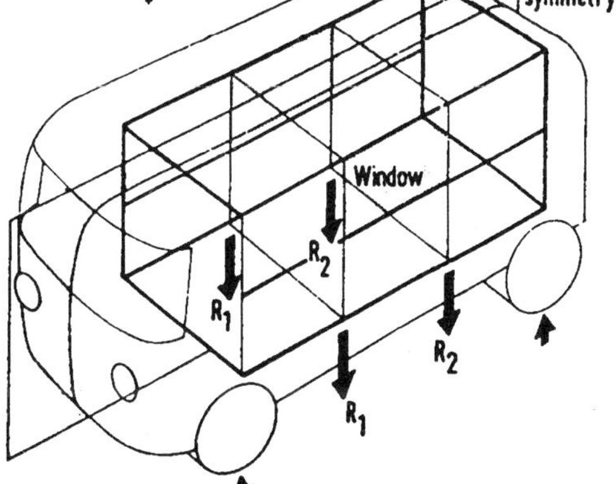

Fig 62: Van shell idealisation

Shell structure loading

In the section on finite element analysis of body structures , the construction of the mesh is shown to require a knowledge of the behaviour of the structural elements of the body shell. The effectiveness of the finite-element analysis and the economy of the meshing procedure depends on an understanding of the load paths and likely deflection modes, a number of which are discussed in this section.

Fig 63 shows the results of a matrix force analysis carried out by Tidbury[7] on the idealised van body of Fig 62. As the structure is divided into relatively few panels, a good visualisation of the the structural behaviour, Fig 64, can be obtained — prior to a more comprehensive analysis using finite-element techniques. As opposed to the simpler structure analysed in an earlier section, this one required computer solution of the matrices.

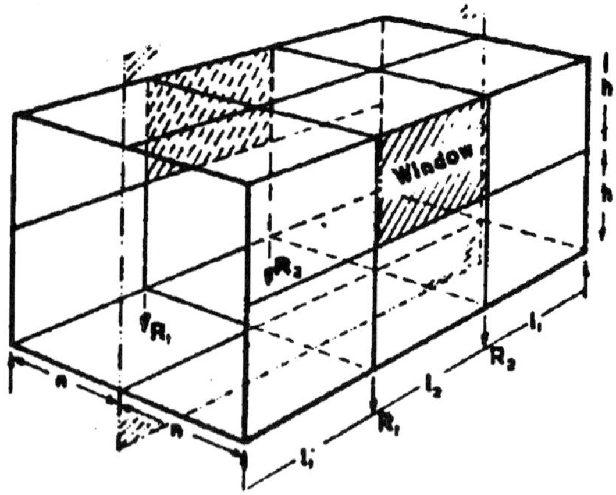

Fig 63: Van body as 3-bay box tube

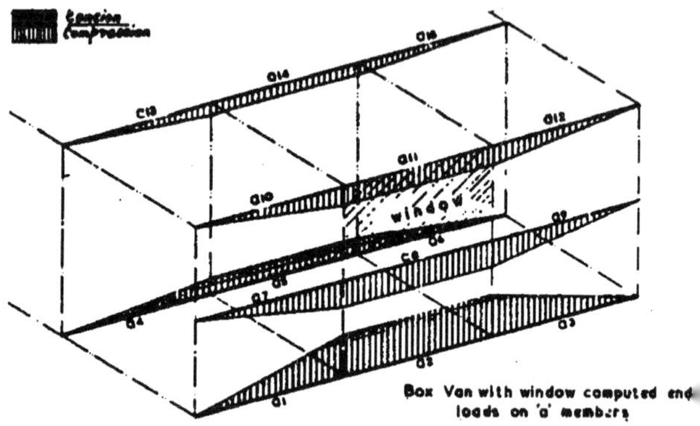

Fig 64: Stressing a 3-bay box-tube

Curved shell sections

For curved sections of a shell, the treatment of cylindrical plates under compression is a useful starting point. Critical stress for local buckling depends on ratio of radius-of-curvature to wall thickness. An empirical formula gives critical stress as

$$=kE/(R/\delta)$$

where k is the experimental bucking co-efficient obtained from the graph in the lower part of Fig 65. E is the elastic modulus of the shell material and the increase in buckling strength depends upon dimension R, decrease of which substantially increases critical stress.

In considering those parts of the shell which are flat plates, the shear deformation is an important criterion. The formula discussed in an earlier section, relating shear stress and direct stress at an angle 45 degrees. In Fig 66, $\sigma_1 = \sigma_2 = \tau$ and when σ_2 exceeds a given value the plate will begin to buckle, waves appearing at right angles to the compressive stress. Critical shear stress is given by

$$KE(B/\delta)^2$$

where K is another experimental co-efficient depending on the edge fitting as shown in the lower part of the figure. Decreasing dimension B is the most

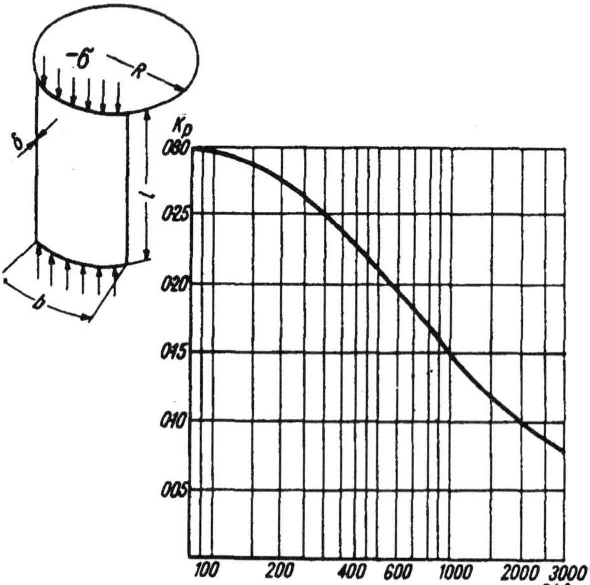

Fig 65: Buckling of cylindrical plates

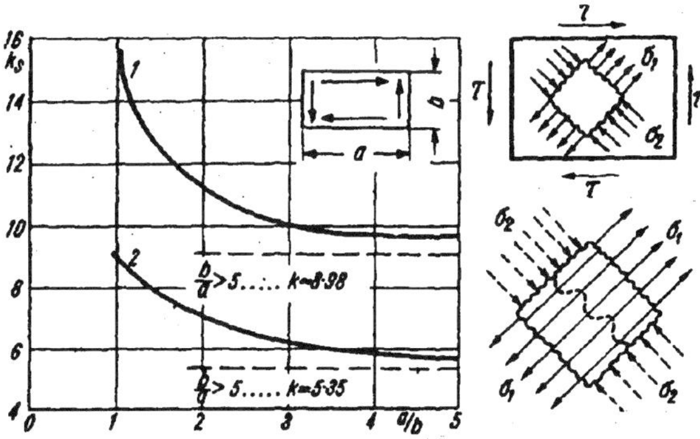

Fig 66: Shear and direct stress relationship

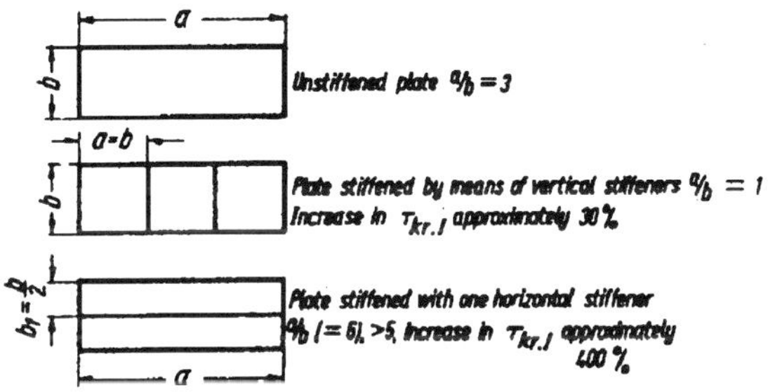

Fig 67: Stiffening of a shear panel

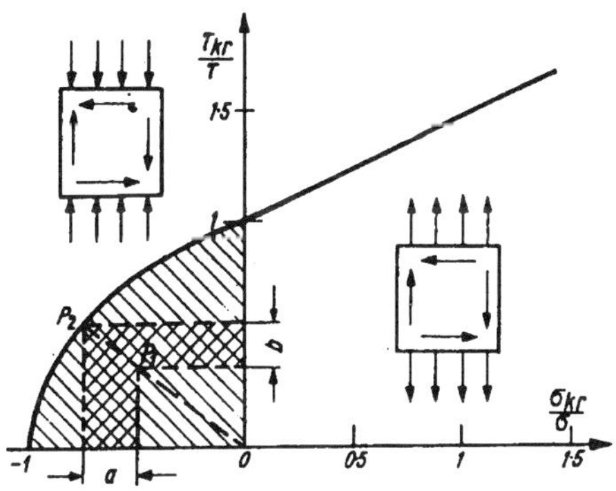

Fig 68: Critical stress diagram

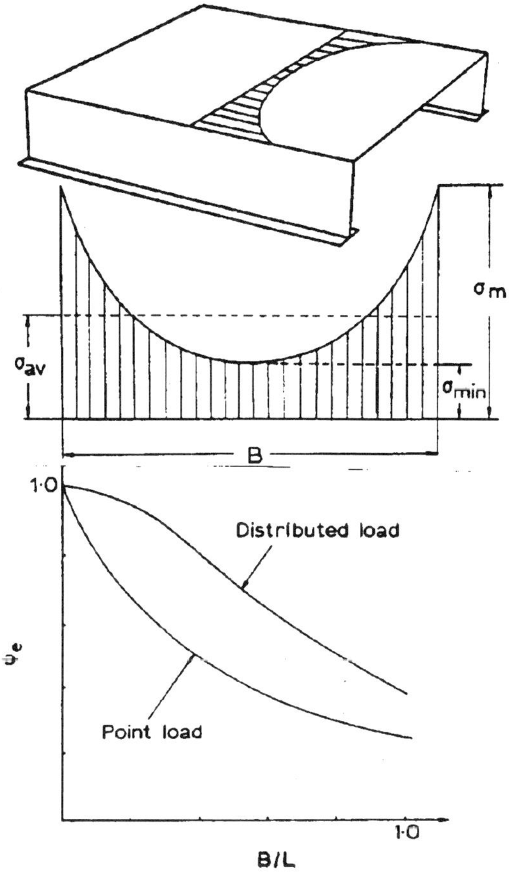

Fig 69: Shear lag effect

(Uniform load over length of each web)

α	$\frac{B}{L}$	Stress effective breadth			Deflection effective breadth		
		Mid-span	*Quarter-span*	*Support*	*Mid-span*	*Quarter-span*	*Support*
0·0	0	1·00	1·00	1·00	1·00	1·00	1·00
	0·05	0·98	0·98	0·84	0·98	0·98	0·98
	0·10	0·95	0·93	0·70	0·94	0·94	0·93
	0·20	0·81	0·77	0·52	0·79	0·79	0·77
	0·40	0·50	0·46	0·32	0·48	0·47	0·47
	0·60	0·29	0·28	0·22	0·31	0·30	0·30
	0·80	0·20	0·19	0·16	0·21	0·20	0·20
	1·00	0·16	0·15	0·12	0·17	0·16	0·16
1·0	0	1·00	1·00	1·00	1·00	1·00	1·00
	0·05	0·97	0·96	0·77	0·93	0·92	0·92
	0·10	0·89	0·86	0·60	0·84	0·84	0·83
	0·20	0·67	0·62	0·38	0·62	0·62	0·60
	0·40	0·35	0·32	0·22	0·33	0·33	0·32
	0·60	0·22	0·20	0·15	0·21	0·20	0·20
	0·80	0·16	0·15	0·11	0·16	0·16	0·16
	1·00	0·12	0·11	0·09	0·12	0·12	0·12

Fig 70: Effective width ratios for siply supported box girders

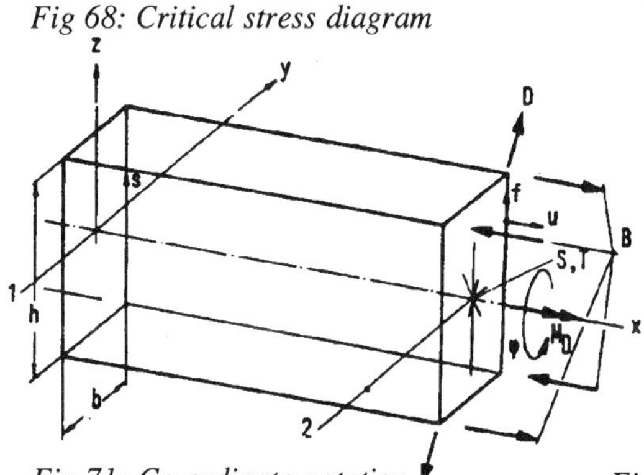

Fig 71: Co-ordinate notation

effective way of increasing critical shear stress. Another technique is by suitable choice of edge dimensions, Fig 67.

Simultaneous compression and shear stresses are given by the equation

$$\sigma/\sigma_k = -1+(\tau/\tau_\kappa)^2$$

attributed to Pawlowski; also that between tensile and tangential stress as

$$\tau/\tau_k = 1+(\sigma/2\sigma_k)$$

These are shown graphically as Fig 68 in which the region of stability is limited by a parabola which becomes a straight line near the ordinate and the abcissa. Points with values of t/t_k and s/s_k within this region do not buckle.

Shear lag in bending

Distribution of elastic shear lag in the wide flange of a box beam is introduced in a separate section. The effect of this is to increase the overall deflection of a box-beam under load — also an increase of the longitudinal stresses at the web/flange intersection and a decrease at mid-section. In design it is convenient to replace breadth b by an equivalent breadth b_e such that the application of simple bending theory gives correct stress and deflection values, in a method described by Donald[8]. This has become standard practice as set out in BS 5400 (Part 3). Shear lag is shown to be significantly dependent on plan dimensions of the flange *B/L* as seen in Fig 69, bottom half. The effective breadth of the flange in a continuous box beam can be estimated by treating each portion of the beam, between points of contraflexure, as an equivalent simply-supported span. The beam is considered to consist of an assembly of simply-supported L-section beams each having a web plate and a flange plate. Effective breadths of these are determined from the table in Fig 70 which applies to the case of a uniform load over the length of each web.

Axial warping in torsion

Another important effect in box-beams is the out-of-plane distortion of cross-sections of a box-beam known as axial warping; this is described for open-section beams in a separate section. It has also been described by Beerman[9] who points out that while

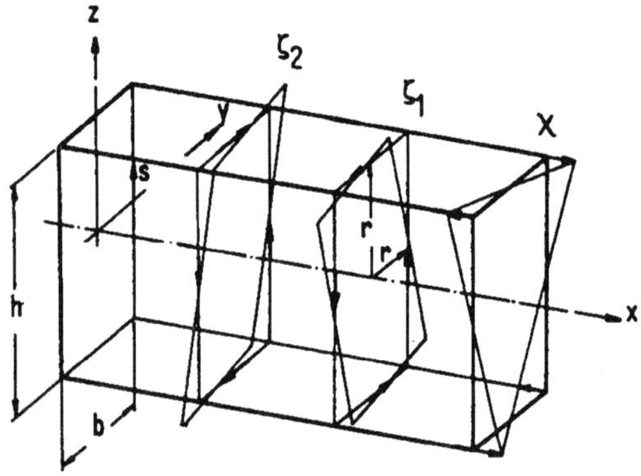

Fig 72: The three deformations

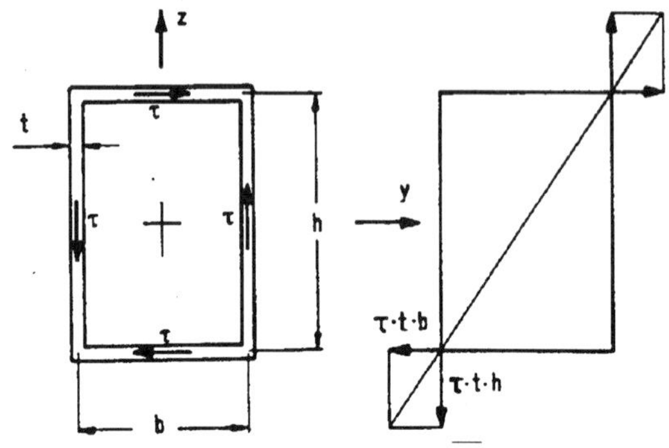

Fig 73: Force system

inhibition of warping does not produce large torques in closed section beams, the compatibility of any warping displacement at the end connections of the beams must be taken into account. Fig 71 shows a co-ordinate notation used in analysis; *u* displacements represent warping while *f* displacements are both rotation and lozenging of the cross-section. The total deformation can be separated into three parts as shown in Fig 72.

The lozenging displacement is associated with a lateral bi-moment *D*, a set of self-equilibriating shear forces, seen in Fig 73 to be the product of shear stress τ, length of the sides and material thickness t; this is not to be confused with the longitudinal bi-moment *B*. Fig 74 shows the distribution of these bi-moments as well as warping and lozenging functions for two rectangular and two square sections with equal total length of sides and different thicknesses — subjected to to equal bi-moments at each end and with lozenging inhibited at these ends. Note that the equal bi-moments cause opposite internal loads at the two ends so that the bi-moment will be zero at the centre of the beam.

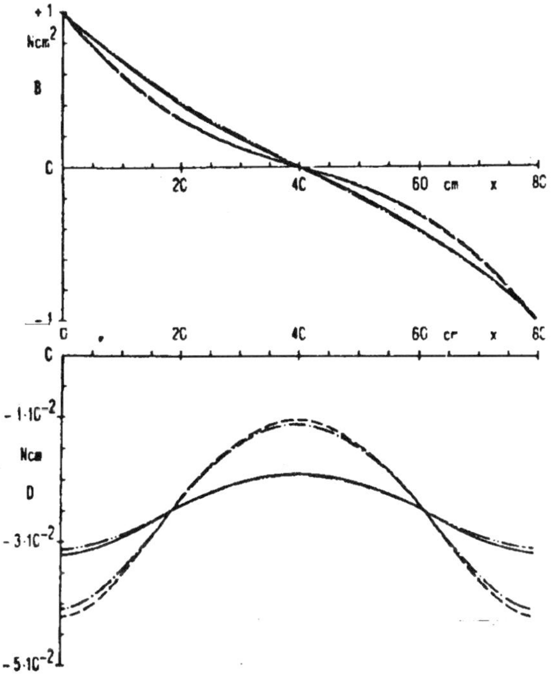

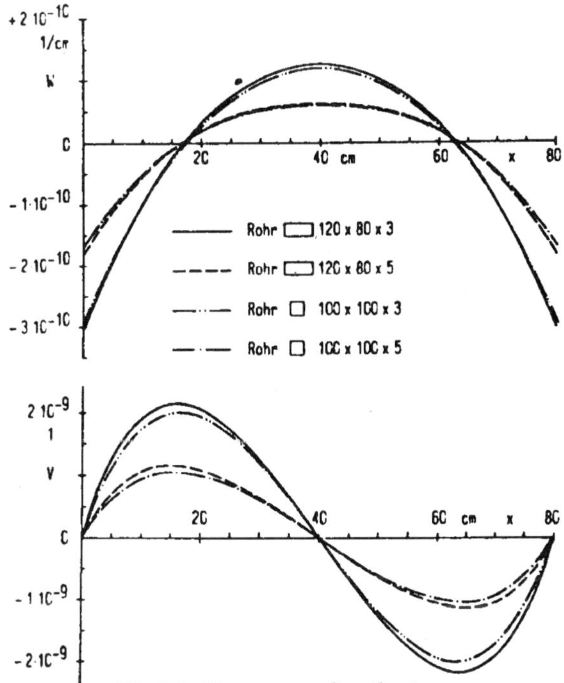

Fig 74: Bi-moment distribution

Mounting shell structures

The mounting of structures such as CV cabs and tanker shells needs particular care in design — to diffuse the concentrated loads at the mount over as wide an area as possible of the supported structure. Some of the analytical approaches are discussed here alongside an examination of some of the accepted mounting arrangements for cabs and tanks.

The loading of the shell through the mounts can also be controlled by supporting the mounts on pivoting subframes, for example, to relieve the shell of torsional loads. The subframe can also sometimes be in the form of a cradle which itself has a load-spreading effect over the shell. The cradle can then be trunnion mounted to a further cross-tube, Fig 75, which would be resiliently-mounted to the chassis-

Fig 76: Cab-support system (left)

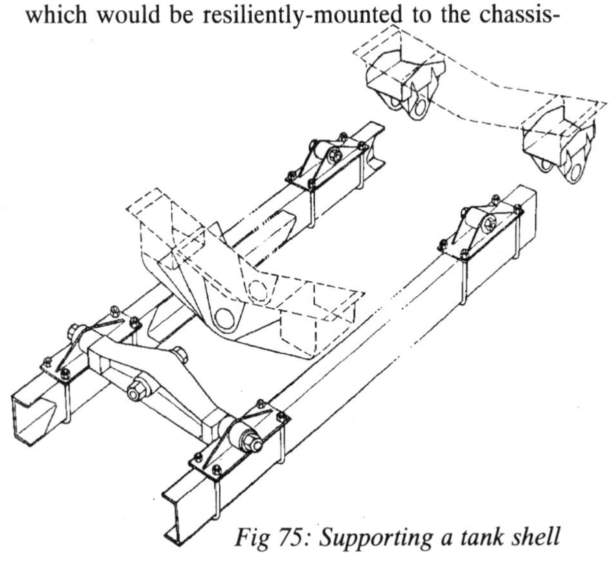

Fig 75: Supporting a tank shell

frame. In the case of cabs, traditional practice was to ensure relative stability for the occupants in the transverse plane, by having at least one stiff mount either side of the cab, while including further flexible mounts at the rear to allow for torsional weaving of the chassis frame.

Beyond these generalities, a much more detailed analysis is now usual in considering load-spreading, mounting spring-stiffness and a controlled degree of damping the relative motion between chassis and shell. A sophisticated damping system for a contemporary truck cab is seen in Fig 76. Four air-springs are used for the Renault AE cab — one of the tallest in production — the fronts at 4.5 bar and the rears at 3.5 bar. Longitudinal stability is obtained by Panhard rods; anti-roll and rear lateral location bars are also fitted. Three levelling valves are employed and a ride frequency of about 1 Hz was the design target.

In tankers, the approach is often one of using the bulkhead of the tank shell as the position for mounting the tank on to a saddle-bracket, Fig 77. The weight of the cargo is carried first by circumferential tension in the shell; the tension in turn creates shear stresses, resulting in an overall beaming action. The shear is a maximum at the supports, of course, where it feeds down from the shell wall via the interior bulkheads, Fig 78. By augmenting the boot at the base of the bulkhead stiffeners, it is possible to make the stiffener carry all the load and dispense with the saddle. The support reaction is then transferred directly from the longitudinal to the boot, Fig 79. If the boot is replaced by an internal longitudinal between two bulkeads, see inset, then the shear load at each end of the internal longitudinal will be transmitted to the stiffener by an appropriate welding bead.

Cab mounting considerations

In cab suspension the theoretical ideal, two mounts at the front and one at the rear, to allow chassis-frame weaving is often made impractical by the requirements for cab tipping. The tilting mechanism may be torsion bar or hydraulic, Fig 80, the latter being favoured for heavier cabs. One way of incorporating

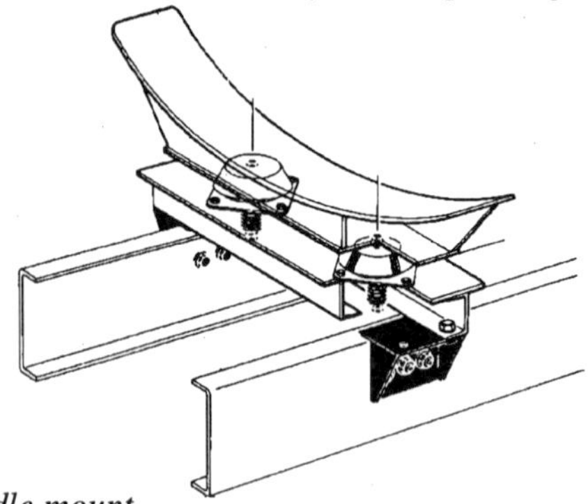

Fig 77: Saddle mount

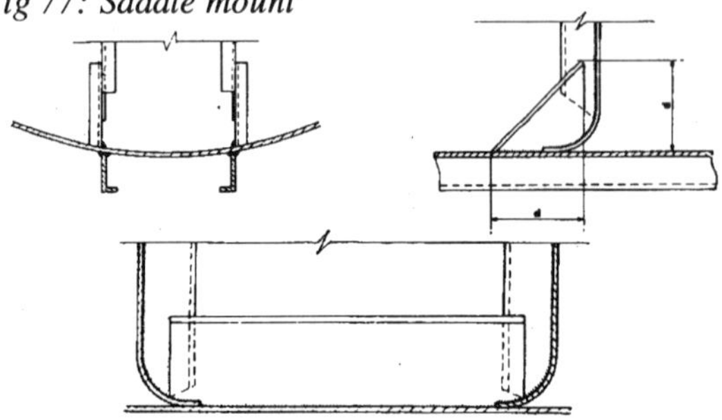

Fig 79: Loading trough boot

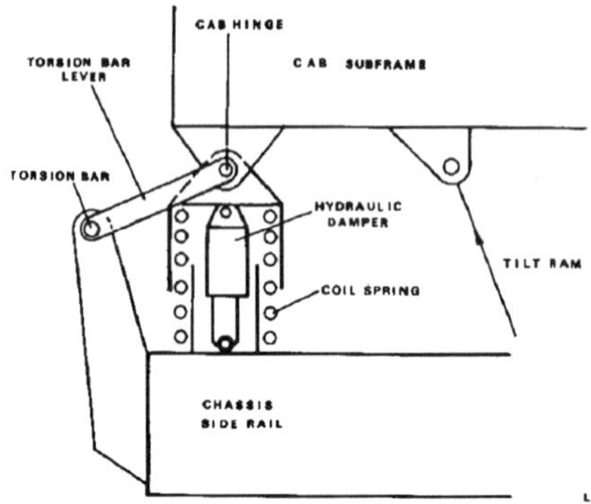

Fig 81a: Mount with hinge

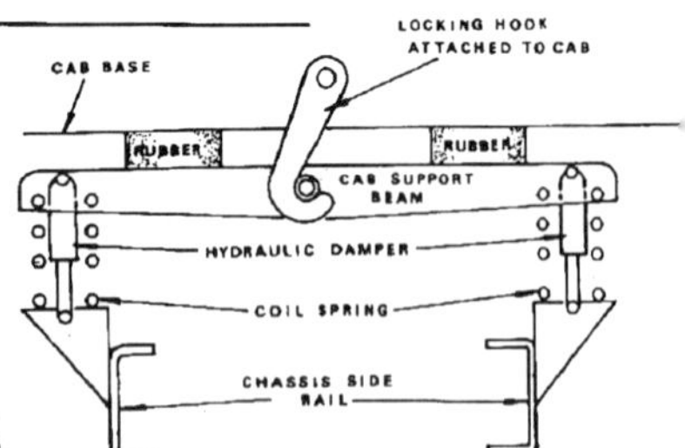

Fig 81b: Rear-mount with tilt-lock

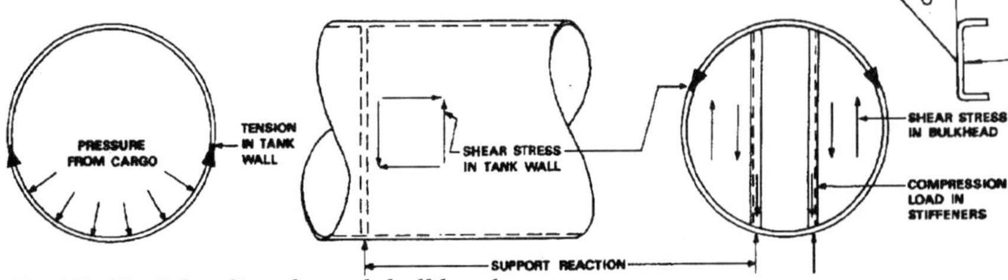

Fig 78: Tank loading through bulkhead

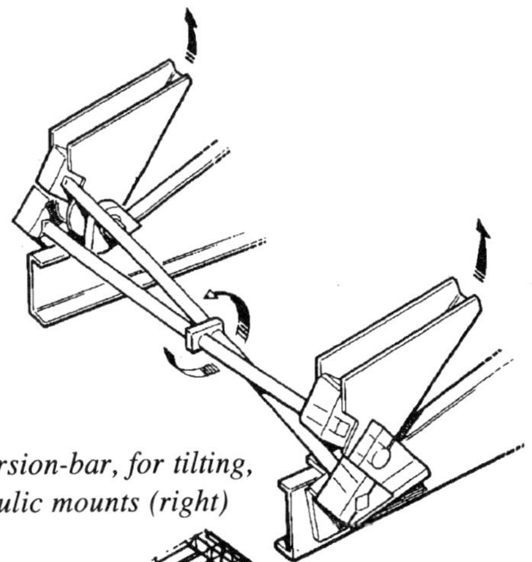
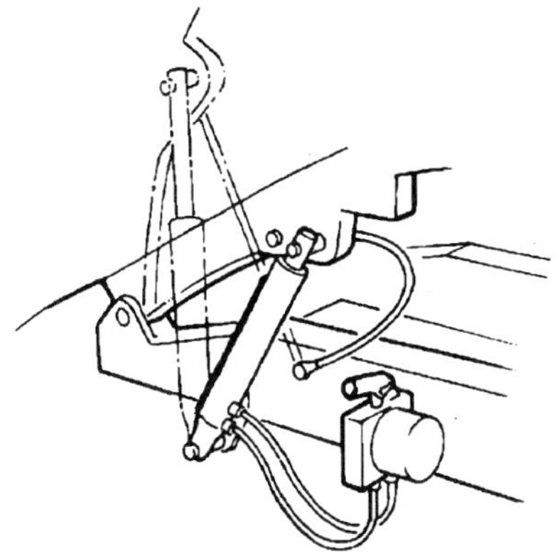

Fig 80: Torsion-bar, for tilting, and hydraulic mounts (right)

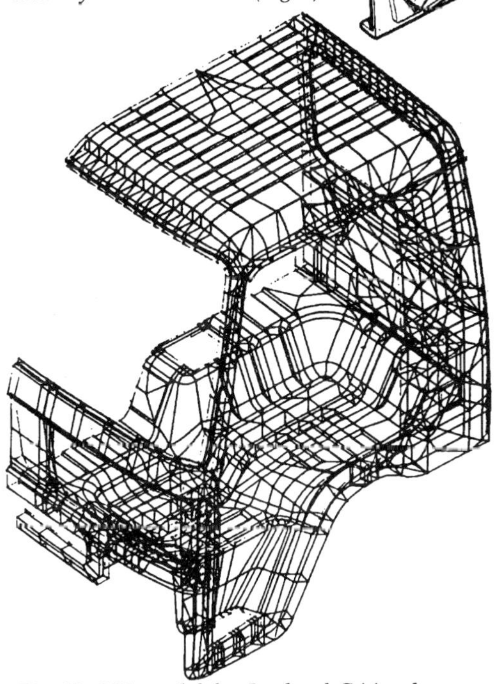

Fig 82: FE model for Leyland C44 cab

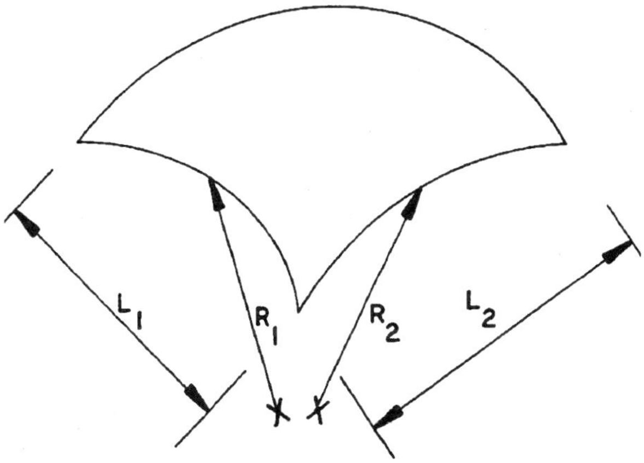

Fig 83:Double-curved panels

the front mount with the tilt hinge is shown diagrammatically in Fig 81. This arrangement minimises stresses imparted into the cab structure. A rear mount for a tilt cab is usually incorporated with a cab tilt-lock. Conventionally two rubber springs are mounted close together to simulate single-mount characteristics with respect to chassis twist.

The pattern of stress distribution into the cab arising from the load inputs at the mounts, is an ideal candidate for analysis by finite-element methods. For the Leyland C44 cab introduced on its Roadtrain truck range, the FE model was as shown in Fig 82.

Analyzing simple shells

If some estimate is required of a shell's ability to sustain external load, without FEA, the classical spherical case provides some indication. A uniform pressure q produces a membrane stress of qa/rt where a is the spherical radius and t is the thickness of the sheet. The value of q can be increased up to a point where the shell buckles when its critical value is approximately $(E/2)(t/a)^{1/2}$ where E is the elastic modulus. If the surface is stiffened by ribs, the last equation can be factored by $(t_B / t_m)^{3/2}$ where t_m is substituted for t, being the ratio of stiffener area to stiffener spacing S_s and t_B the equivalent bending thickness evaluated from the stiffener moment of inertia I_s as $(12I_s / S_s)^{1/3}$. In the case of a concentrated load P on a spherical shell, the theoretical stiffness is given by:

$$P/d = CEt^2/a(1-u^2)^{1/2}$$

where d = deflection, C is a constant. The expression is generalised for double-curved panels which are not necessarily spherical by replacing the curvature $1/a$ (Fig 83) by:

$[(L_1^2/R_1) + (L_2^2/R_2)]/2L_1L_2$

The constants and a fuller description of the technique can be obtained from an SSRC publication[10].Simplified design charts are also provided by Marshall[11].

The case of box-tubes intersecting, as shown in Fig 83, is a useful aid to design of brackets which connect mountings to shells. Here the tendency for the end-on box-tube to punch through the wall of the shell is expressed as a punching shear efficiency E_v given by maximum allowable punching shear stress divided by allowable tensile stress in the shell. E_v is shown in Fig 84 as a function of non-dimensional parameters given in the text. The first of these figure shows the yield pattern. The first localised failure is punching at the material yield stress thereafter sidewall yielding of the main shell occurs. For compression loads this is in the form of web crippling with capacities shown on the right hand side of the second figure.

Tanker shells use a variety of sandwich rubber mounts for isolation from chassis motion. A four point mounting system may have either simple plane sandwich units at each corner or vee-configurations to give additional transverse or longitudinal stiffness, as required. Conical sandwiches can be used to obtain stiffness in all directions within the horizontal plane. Another interesting approach taken by a French company for a 12,000 litre tanker was the arrangement shown in Fig 85. On each side of the rear, a tee shaped fitting was used with rubber bushes on each arm of the tee which are parallel to the longitudinal axis of the vehicle. A circular rubber mount secures the leg of the tee to the chassis frame while the two arm bushes are carried by a tank shell bracket. At the front end, the tank is connected by a single rubber-bushed pivot, also with its axis parallel to that of the vehicle, to the centre of a yoke. Each end of this yoke is connected by a rubber-bushed shackle to a bracket beneath it on the frame. Axes of the bushes in the shackles are arranged transversely in the vehicle so as to take length changes due to frame deflections.

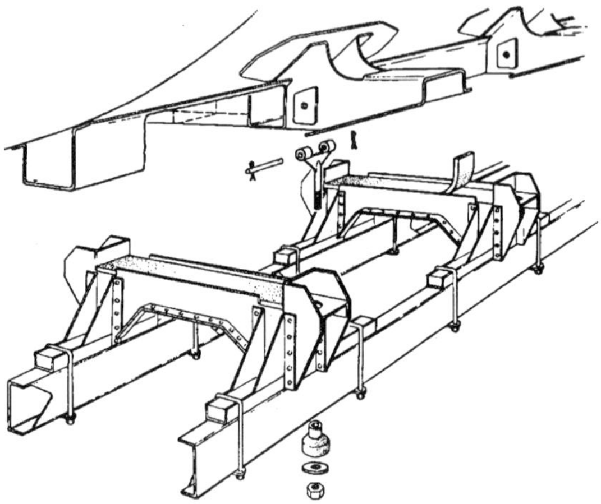

Fig 85: Special tank mounting

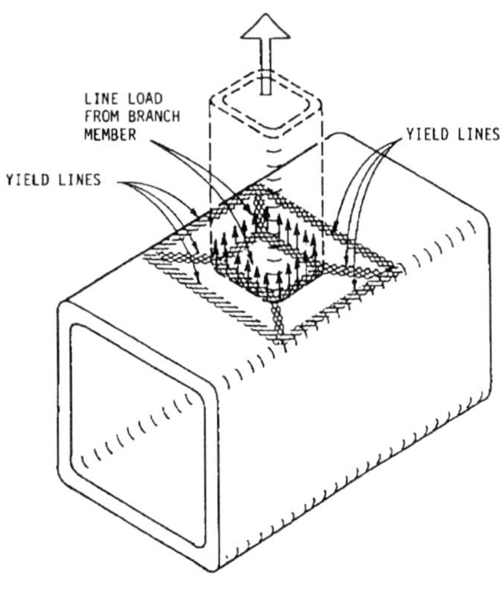

Fig 83: Intersecting box tubes

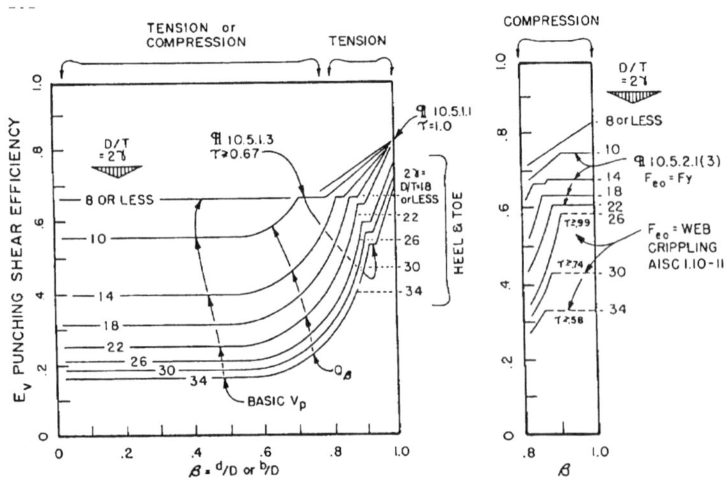

Fig 84: Allowable 'punching' stress

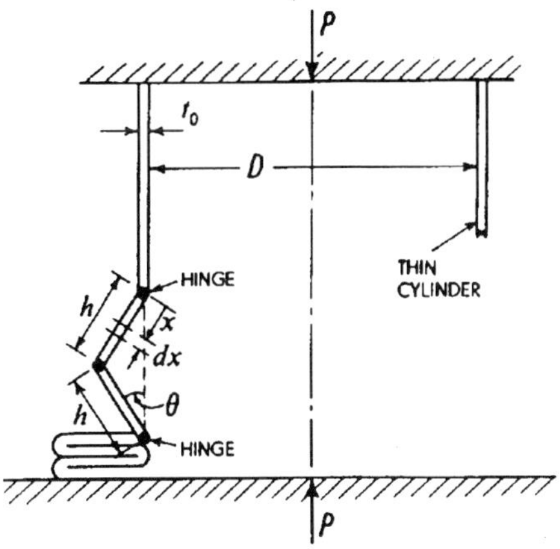

Fig 86: Collapse idealisation

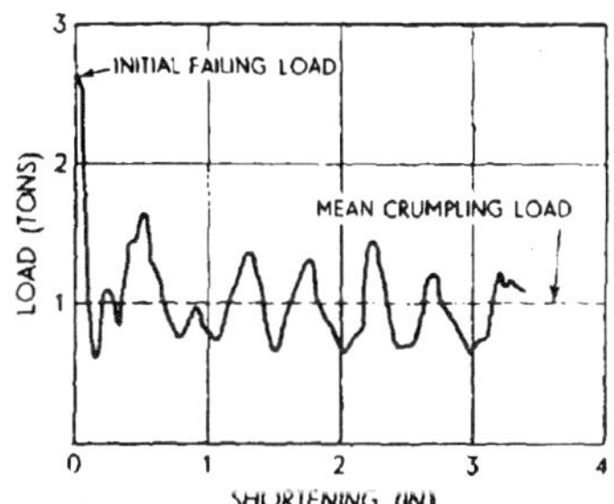

Fig 87: Static crush load trace

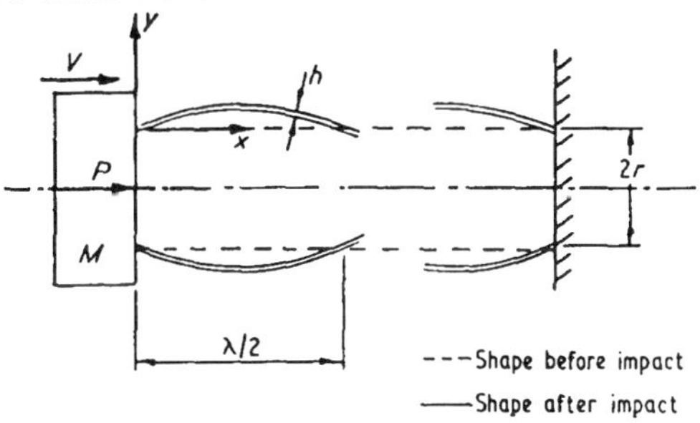

Fig 88: Pre-failure phase

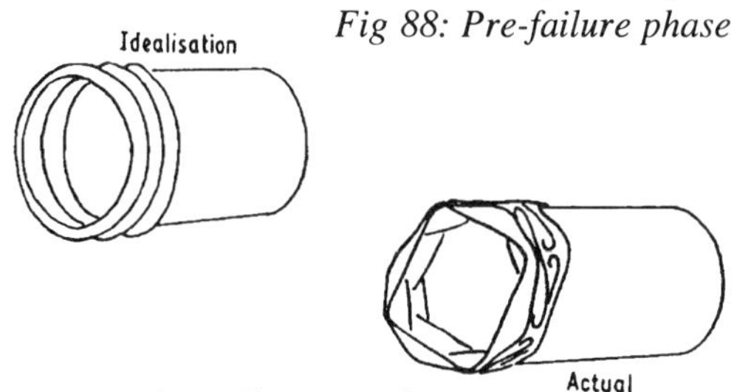

Fig 89: Cylinder collapse mode

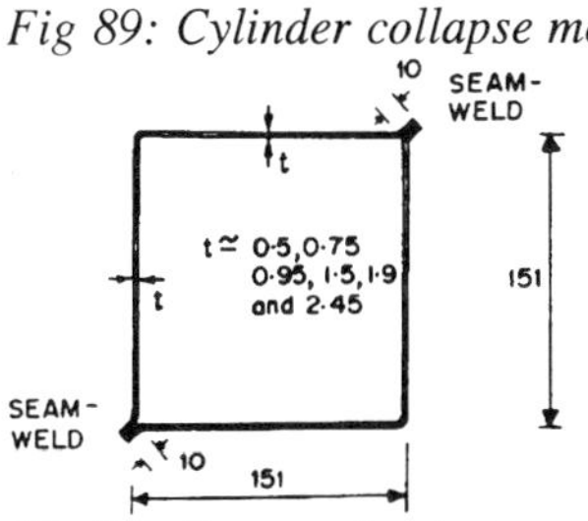

Fig 90: Box-section test specimen

Stability of shell structures

As well as considering the stability of struts and plates under end loading, shell stability is also of importance in the design of thin-walled structures under many conditions of loading and is a pre-requisite to the study cab structures, for example. A thin tube or box-beam under end loading can suffer stability failure as a buckling shell, the controlled behaviour of which can be used in the production of impact-absorbing structures for impact protection.

Another section deals with impact collapse of structures, within which plastic-hinges are introduced as an analytical tool. This theory can be extended to the consideration of thin cylinders and box-tubes under axial loading. On the assumption of a concertina-type collapse mode, the collapsing member can be idealised for analysis as shown in Fig 86. The energy dissipated in plastic bending of the four circular hinges can be summed to arrive at an overall figure for the axially loaded shell. Also to be added is the energy dissipated in stretching, under a substantially uniform tensile hoop stress, in the metal between the hinges. This will depend on whether the buckling is entirely external as shown or entirely internal, at the other extreme. For calculation purposes a mean value of energy is chosen between these extremes and P/Y becomes $= 6t_o(Dt_o)^{1/2}$

In crushing a tube under static load conditions, as each new row of buckles forms, the axial load rises to a maximum and falls to a minimum as the buckles fold flat, the load varying about a mean value as shown in Fig 87. When the tube is subject to dynamic impact, the ratio of the dynamic to the static crumpling load increases with speed. The ratio reflects changes in buckling mode due to inertia — also as a consequence of the rate-of-strain effect on the material of the tube.

Experiments with mild steel tubes preliminary to the design of railway coaches, for example, have shown a linear increase in the ratio from 1 to 2.5 with impact speed increase from 0 to 125 ft/sec. A moving tubular structure colliding with a flat stationary body will experience a resisting force $F = \sigma A$ where σ is the mean crushing stress for the structure and A its cross section area. Mass of the structure is ρAL, for density ρ and length L. Uniform retardation is therefore $\sigma/\rho L$, by dividing the two expressions above. Retardation is thus inversely proportional to length in this simplified analysis.

Only the front part of the structure, closest to the impact area, is subject to crushing. This length x can be estimated from the assumption that all the kinetic energy is dissipated in the plastic deformation, such that $Px = Mv^2/2$ and

$$x/L = (\pi/12)[(D/t_o)(\rho v^2 / Y)]^{1/2}$$

where Y is the yield stress of the material.

Workers at Birmingham University showed the pre-failure phase of a thin cylinder being impacted as Fig 88 and devised an expression for the failure load as $P_e = n^2\pi^2/EI/L^2$.

They showed in Fig 89 how a cylinder actually collapses compared with the idealised mode which is assumed in a simplified analysis and suggested an expression for the mean load during crumpling as $P = KYh^{1.5}(2r)^{1/2}$ where K is approximately 6 — to give reasonable accuracy for the post-failure stage of a dynamic test. It is suggested that this is valid even for cylinders with cut-outs in them provided the cut-out does not exceed 50 % of the cylinder circumference.

Civil engineering techniques of analysing box columns are also of potential value to automotive designers in calculating the instability of shell structures. Work at the Australian Monash University[10] has involved testing box specimens as shown in Fig 90. Others involved b/t ratios ranging from 60 to 158 and exceptionally as high as 200 and 300.

Those with low b/t developed pent-roof shaped buckles while the higher ones had so-called flip-disc

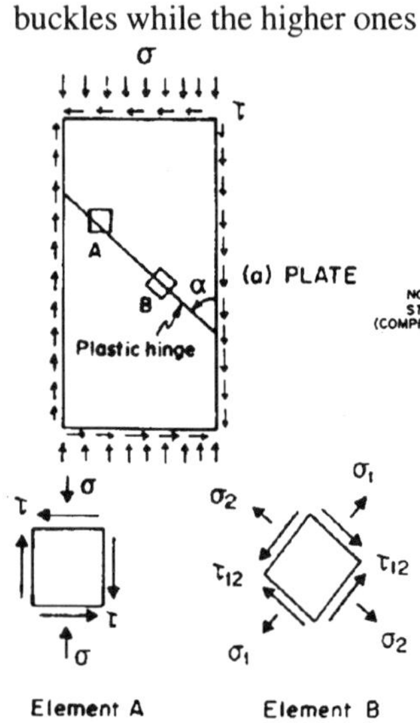

Fig 91: Analysis of symmetrical side

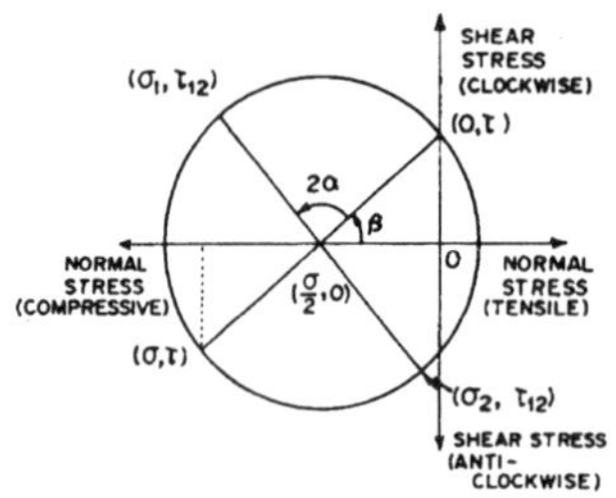

Fig 92: Mohr's circle

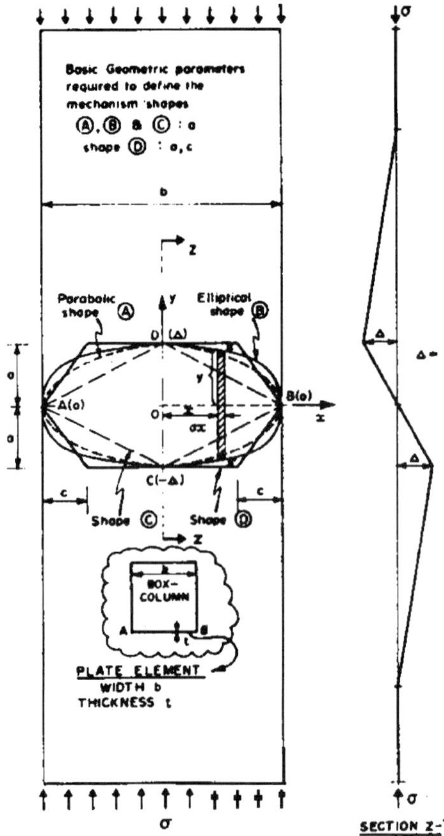

Fig 93: Roof type failure

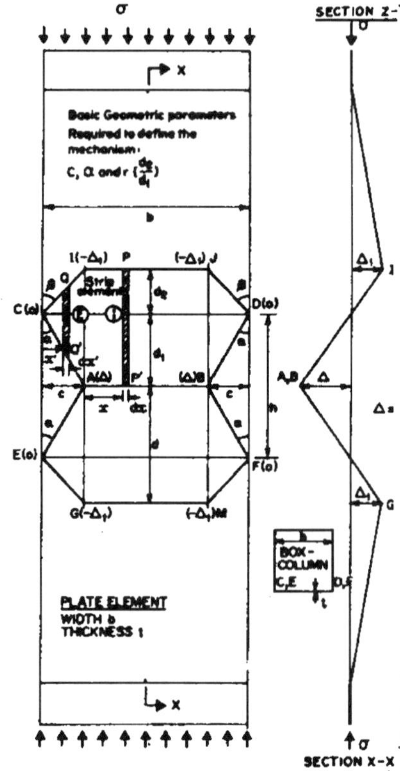

Fig 94: Flip-disc failure

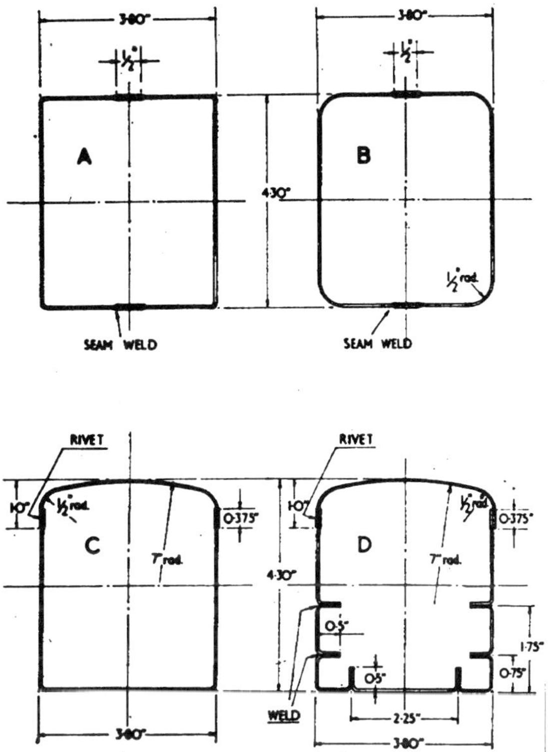

Fig 96: Bus structure test models

shapes with general corrugations having curved tops. On increasing the applied load, specimens pass through three deformation phases: elastic, elasto plastic and plastic. The last stage is analysed using rigid plastic theory based on observation of the test specimen failure modes.

The assumption of similar modes applying to each side of the box is made and the consideration of one side, only, in the analysis, Fig 91. The plate is considered to carry average compressive stress σ and shear stress τ, subscripts 1 and 2 applying respectively to normal-to and in the direction of the plastic hinge.

Average shear stress τ_{12} is found using the Mohr's circle technique, Fig 92. Plastic moment per unit length of hinge is: $M_p R[1 - (2d/t)^2]$ where R depends on the stress distribution, established in sample testing, and d is the distance from the neutral axis of the plastic hinge to the mid-plane of the plate, again found empirically. Fig 93 shows the idealised 'roof' mechanism of failure and Fig 94 the flip-disc.

In work aimed at finding the crush resistance of monocoque bus structures, Leyland and UMIST researchers[12] have examined the theory of crumpling in square tubes and suggest the Macauley formula:

$$P_v/P = n(1 + nKV/\lambda)$$

relating static to dynamic crumpling load P_v where n is a factor depending on the variation of strength with impact velocity V and l is a geometric scaling factor. The researchers simplified the 8 tonne bus , 9 ft high x 8 ft wide x 34 ft (2.74x2.44 x10.36 m) long to a plain rectangular tube of mild steel (to use in the empirical relationship:

$$P/P_o = 3/p\ (A/A_o)^{1/2}$$

with A_o, the enclosed area of tube and the tube cross-section area, giving a so-called solidity ratio of 0.0069. $P_o = YA = 1432$ tonnes. The results shown in the table of Fig 95 compare the two equations. Because of the discrepancies, the researchers constructed test models as in Fig 96 for testing under a drop hammer — with the results shown in Fig 97.

P/P_o	Static Mean Crumpling Load P (tons)	Dynamic Mean Crumpling Load P_v (tons)	Shortening x (inches)	Deceleration a (ft/sec²)	k (g's)
0.0552	79.1	158.2+	8.1 (1.99%)	640	19.8
0.0794	113.8	113.8	11.25(2.75%)	457	14.2

Fig 95:Results from two formulae

COACH MODELS

MODEL NUMBER	TUBE SECTION	SIDE CUT-OUTS RIGHT HAND	SIDE CUT-OUTS LEFT HAND
1	A	PLAIN TUBE	
2	B	—‖—	—‖—
3	C	—‖—	—‖—
4	D	—‖—	—‖—
5	A	A	A
6	A	B	NONE
7	A	C	C
8	A	D	A+C
9	C	D	A+C
10	D	D	A+C

The models were manufactured from a mild steel EN 2c and its yield stress was taken to be 20 ton/in² throughout the calculations. The hammer weight was 77lb. Kinetic energy = ½ (mass of hammer + mass of model) V_0^2.

						Static Tests		Dynamic Tests				
Model No.	Weight lbs.	Net C.S.A. 'A' in²	Enclosed C.S.A. 'A'$_0$ in²	Solidity Ratio 'A/A$_0$'	$P_0 = YA$ ton	Approx Mean Crumpling Load 'P' (tons)	$\frac{P}{P_0}$	Velocity 'V_0' ft/sec	Kinetic Energy lb–ft	Shortening 'x' ins	Mean Crumpling Load 'P_v' (tons)	$\frac{P_v}{P_0}$
1	3.75	0.829	16.32	0.0507	16.55	2.8	0.169	34.1	1455	1.5	5.0	0.302
2	3.65	0.806	16.16	0.050	16.12	3	0.186	38.5	1855	1.5	6.625	0.41
3	3.53	0.78	15.9	0.0491	15.6	2.8	0.18	34.5	1485	1.5	5.32	0.341
4	4.96	1.098	15.9	0.069	22	7.5	0.341	35.0	1560	9/16	14.85	0.675
5	3.28	0.829	16.32	0.0507	16.55	2	0.121	35.4	1560	1¾	4.78	0.239
6	3.6	0.829	16.32	0.0507	16.55	2.1	0.127	35.2	1550	1.15/16	4.29	0.214
7	3.63	0.829	16.32	0.0507	16.55	1.8	0.109	35.2	1550	2⅜	3.5	0.175
8	3.15	0.829	16.32	0.0507	16.55	1.7	0.103	36.2	1630	2⅝	3.32	0.166
9	3.07	0.78	15.9	0.0491	15.6	1.2	0.077	35.2	1540	3.5/16	2.08	0.104
10	4.04	1.098	15.9	0.069	22	3	0.136	35.0	1540	1.11/16	4.9	0.245

Fig 97: Results from model tests

Stress concentration in cut-outs

In fully-monocoque body shell structures, panel cut-outs give rise to stress concentrations which must be allowed for in design. Techniques originally developed for aerospace structures can be used to determine the reinforcements required to alleviate the effects of such stress raisers.

Structural analysis used for determining the behaviour of bars or tubes under direct or shear stress; beams, plates and shells under bending or internal pressure; all these are assumed, for simplicity of analysis, to have distribution of stress following some simple law. For direct stress in a tension member, for example, it is assumed that stress is uniformly distributed and that shear stress varies proportionally to the distance from the centre, under torsional loading. The effect of a hole in a tension member, however, is to produce the stress distribution shown in Fig 98. The high concentration of stress at the edges of the hole show that weakening effect is due to more than just loss of area.

The hole is termed a stress raiser and the ratio of the true maximum stress to that calculated by elementary theory, assuming uniform stress distribution, is defined as the stress concentration factor. Such factors for a whole variety of stress raisers such as notches; fillets, grooves and holes of many different shapes are provided by Young's classic work[13].

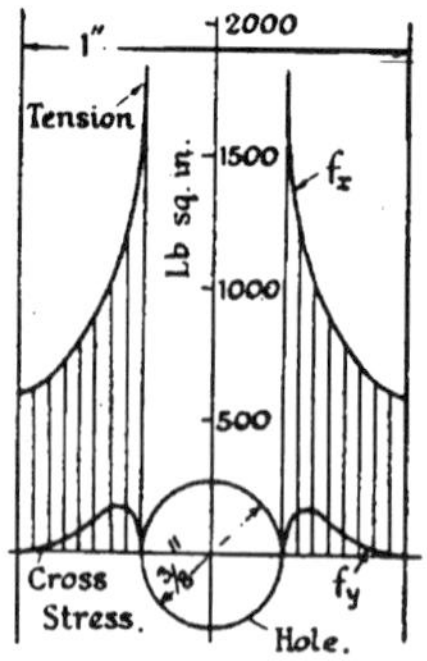

Fig 98: Stress distribution around hole

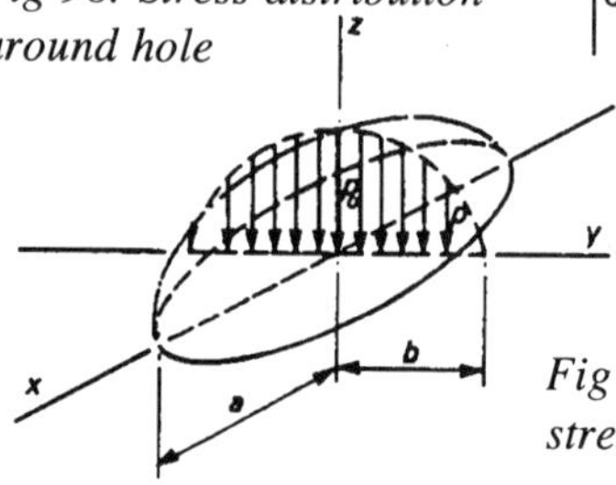

Fig 100: Hertzian stress concentration

Stress condition and manner of loading	Factor of stress concentration k for various dimensions
Elastic stress, in-plane normal stress 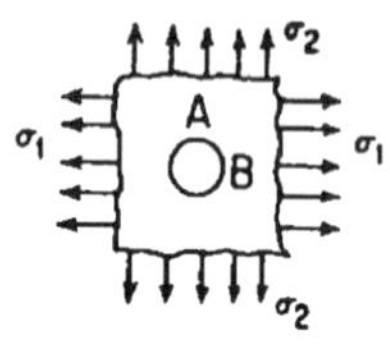	Uniaxial stress, $\sigma_2 = 0$ $\sigma_A = 3\sigma_1 \quad \sigma_B = -\sigma_1$ Biaxial stress, $\sigma_2 = \sigma_1$ $\sigma_A = \sigma_B = 2\sigma_1$ Biaxial stress, $\sigma_2 = -\sigma_1$ (pure shear) $\sigma_A = -\sigma_B = 4\sigma_1$
Elastic stress, out-of-plane bending 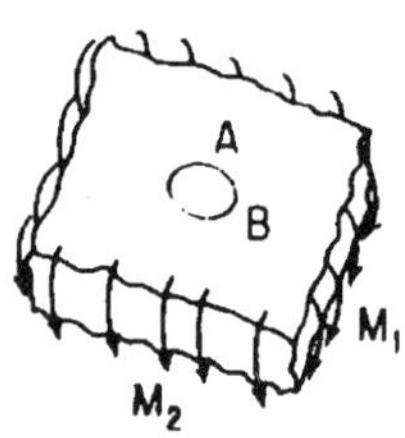	Simple bending, $M_2 = 0$ $\sigma_A = k\frac{6M_1}{t^2}$ where $k = 1.79 + \frac{0.25}{0.39 + (2r/t)} + \frac{0.81}{1 + (2r/t)^2} - \frac{0.26}{1 + (2r/t)^3}$ Cylindrical bending, $M_2 = \nu M_1$ $\sigma_A = k\frac{6M_1}{t^2}$ where $k = 1.85 + \frac{0.509}{0.70 + (2r/t)} - \frac{0.214}{1 + (2r/t)^2} + \frac{0.335}{1 + (2r/t)^3}$ for $\nu = 0.3$ Isotropic bending, $M_2 = M_1$ $\sigma_A = k\frac{6M_1}{t^2}$ where $k = 2$ (independent of r/t)

Fig 99: Stress concentration factors (Young)

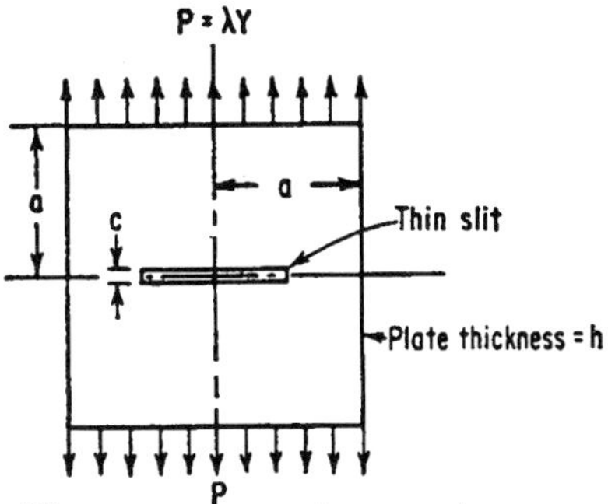

Fig 101: Slit cut-out in tension panel

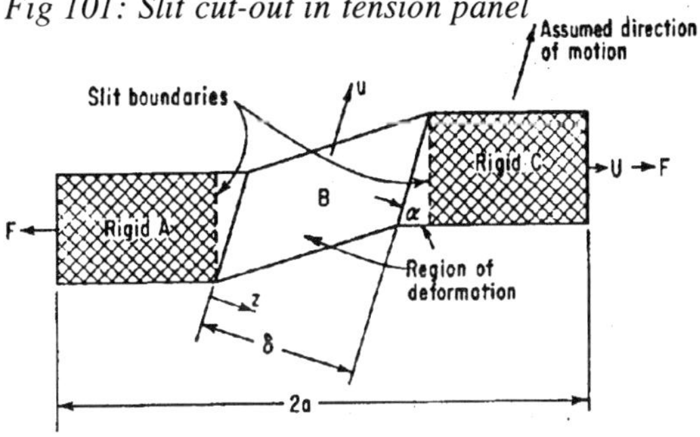

Fig 102: Assumed deformation of slit panel

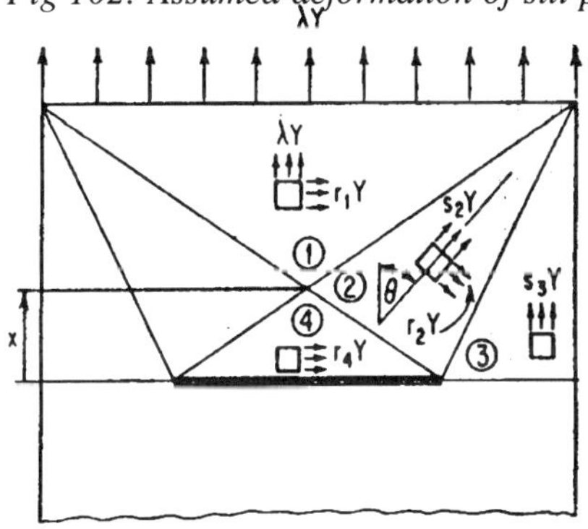

Fig 103 : Analysis of segments

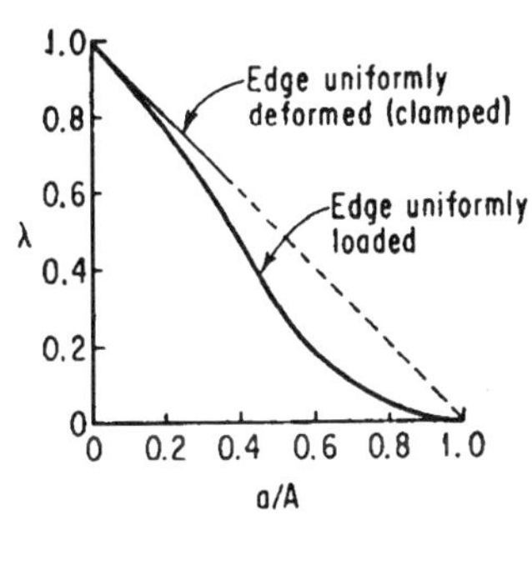

Fig 104: Cut-out factors for square panel

[14,15], Fig 99.

These usually assume elastic behaviour and are 'safe' limits for design in that in reality plastic yielding which occurs on over-stressing mitigates the concentration effect. This is why the stress concentration at rivet holes in ductile materials under static loading can safely be ignored. Under fatigue loading, however, this may not be the case and the familiar term of 'notch sensitivity' is used to describe the effect. A separate stress concentration factor in fatigue is thus defined, being the ratio of the endurance limit of a plain member to the nominal stress at the endurance limit of the member containing a stress raiser. An expression due to Neuber defines this factor as:

$$k_f = 1 + (k-1)/[1+\pi(\rho/r)^{1/2}/(\pi-\theta)]$$

where q is the flank of the notch, or stress raiser, *r* the radius of curvature at the root of the notch and r a dimension related to the grain size tof the material (= 0.0189 inch for steel).

Stress concentration arises too from contact of a 'sharp' pointed body with an otherwise uniform member. Hertzian contact stresses can be estimated for the general case of two curved surfaces being pressed together, for which the pressure distribution over the contact patch is given by the ordinates of a semi-ellipsoid, Fig 100, contained by the surface of contact, such that: $p = p_o[1 \quad (x/a)^2 + (y/b)^2]^{1/2}$

The total load *F* is represented by the volume of the semi-ellipsoid: $F = 2pabp_o/3$ such that the peak

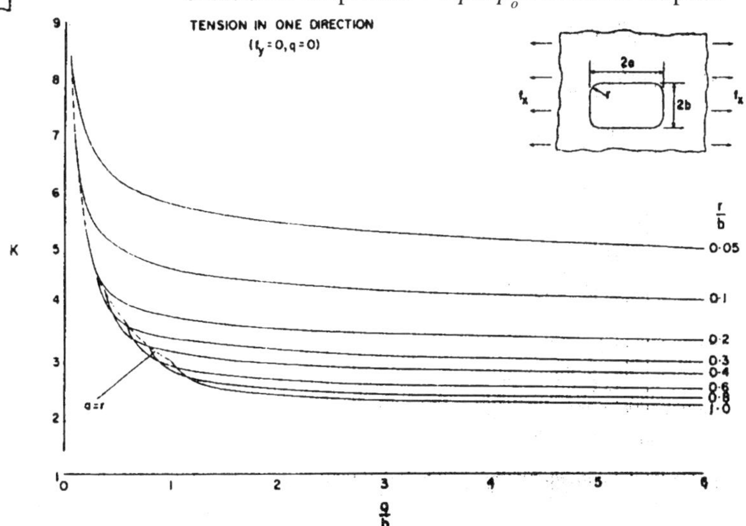

Fig 105: ESDU stress concentration factors

pressure p_o is 1.5 times the average pressure *($F=p\ ab$)*

This theory is useful for determining stress concentrations in cut-outs of stressed panels in such structures as vehicle bodies. For a large panel containing an elliptical cut-out the stress concentration factor for the points at the end of the a-axis is:

$$k_t = 1 + 2a/b$$

thus for a circular hole, $a=b$, where the local stress is three times the average stress, dropping off rapidly as distance is increased from the discontinuity. The magnitude of k_t depends on the sharpness of the discontinuity and, in the limiting case of cracks, the theory of fracture mechanics needs to be applied.

For larger cut-outs in panels of significant thickness a cut-out factor l is defined to determine the maximum edge load that can be carried by, say, a slit panel as in Fig 101. The objective is to find the weakening of the panel so that it can be properly reinforced. The assumed deformation model is seen in Fig 102 as the panel under tension kinks at the slit. Technique for analysis is to divide the panel up into segments as in Fig 103 and stresses in each determined. For a square panel of side $2A$, with a circular cut-out of radius a, for example, the approximate cut-out factor is as shown in Fig 104.

Engineering Science Data Unit (ESDU) Data Sheet 65004, available from 27 Corsham Street, London N1 6UA, UK, provides stress concentration factors for both reinforced and unreinforced rectangular holes with rounded corners in flat panels. The data is presented in the form of curves as shown in Fig 105.

Plastics and composites

Stress concentration effects vary between linear-elastic and plastic materials. Fig 106 shows the case of a circular hole in an infinite flat plate for the two materials. Whereas the stress concentration factor is taken as 3 for both cases the different stress/strain curves will result in different peak stresses for the different materials.

Polymer suppliers Du Pont point out that the effect of strain rate is also important, Fig 107, for the shape of the stress-strain curve. The company's Computer-Aided Technical Service group in Geneva recommends that the allowable stress should be the smallest value of s_{ult} or SEe/S_c where S is a safety factor taken from the table in Fig 108 and S_c is the stress concentration factor discussed above.

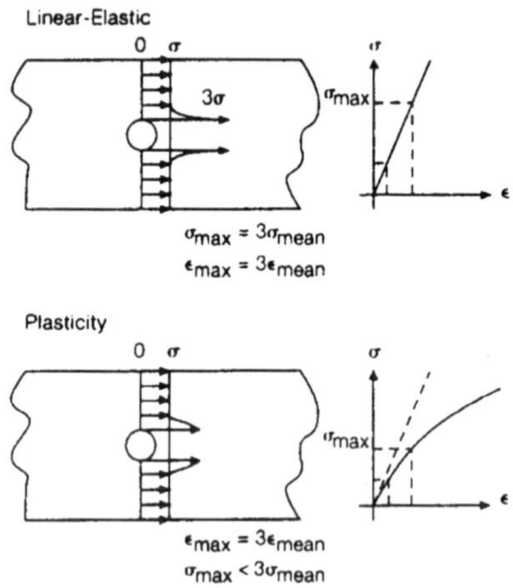

Fig 106: Effect of plasticity

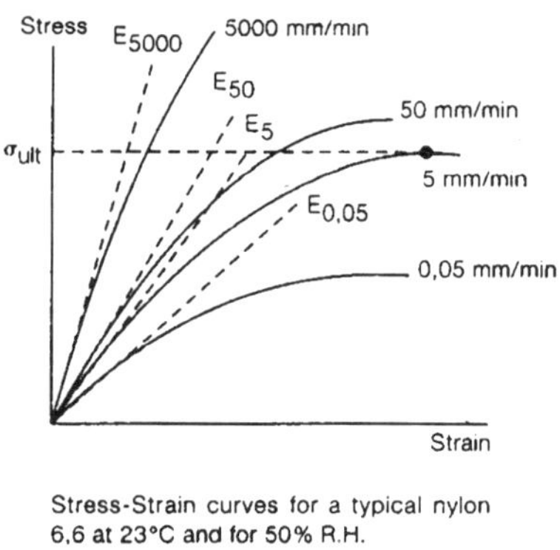

Fig 107: Effect of strain rate

s1 = Risk of Injury to People in Case of Failure
no risk = 1,0
possible injury = 0,7
probable injury = 0,5
(*Example: If an accelerator pedal fails, injury is possible; if a brake pedal fails, it is probable.*

s2 = Processing Factor *†
For materials that are not reinforced, processing factor s2 = 1,0. For most glass-reinforced materials, where processing conditions are not known, assume fibres are not optimally oriented and calculated stresses – to be on the safe side – are: S = 0,5 for glass-reinforced "Zytel" and 0,35 for "Rynite".

s3 = Stress Calculation Accuracy
Accurate model, conservative approach = 1,0
Use of analytical formulae = 0,75
Estimates = 0,5

s4 = Material Degradation
Natural material or carbon black = 1,0

Fig 108: Safety factors

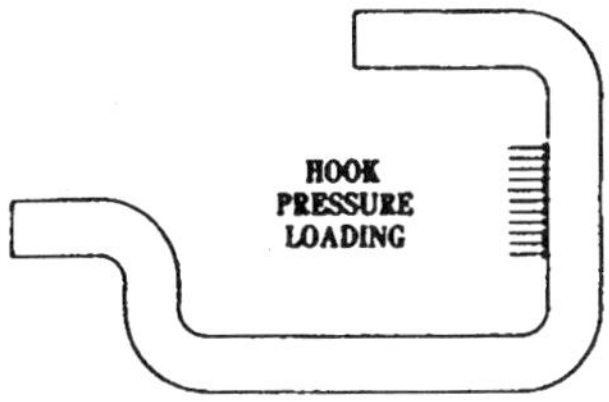

Fig 109: Towing hook as FEM structure example

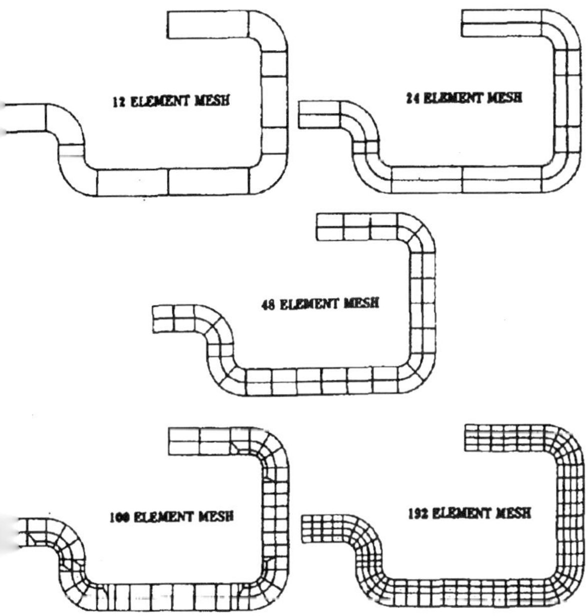

Fig 110: Various mesh densities

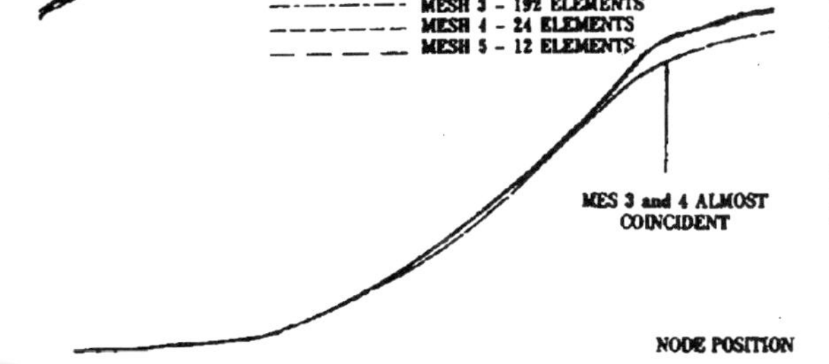

Fig 111: FEM vs elasticity theory

Finite element analysis

This computerised structural analysis technique has become the key link between structural design and computer-aided drafting. As well as fine-mesh analysis which gives an accurate stress and deflection prediction, course mesh analysis can give the structural feel vital in conceptual design.

One of the longest standing and largest FE software houses is PAFEC who have recommended a logical approach to the analysis of stuctures. This is seen in the example of a constant-sectioned towing hook shown in Fig 109. As the loading acts in the plane of the section the elements chosen can be plane. Choosing the optimum mesh density (size and distribution) of elements is a skill which is gradually learned with experience. Five meshes are chosen in Fig 110 to show how different levels of accuracy can be obtained.

Next step is to calculate several values at various key points — using basic bending theory as a check. In this example nearly all the meshes give good displacement match with simple theory but the stress line up is another story as shown in Fig 111. The lesson is: where stresses vary rapidly in a region, more densely concentrated smaller elements are required; over-refinement could of course, strain computer resources.

Each element is connected to its neighbour at a number of discrete points, or nodes, rather than continuously joined along the boundaries. The method involves setting up relationships for nodal forces and displacements involving a finite number of simultaneous linear equations. Simplest plane elements are rectangles and triangles and the relationships must ensure continuity of strain across the nodal boundaries. Fig 112 shows a force system for the nodes of a triangular element along with the dimensions for the nodes in the one plane. The figure shows how a matrix can be used to represent the co-efficients of the terms of the simultaneous equations.

Another matrix can be made up to represent the stiffness of all the elements *[K]* for use in the general equation of the so-called 'displacement method' of structural analysis:

$$[R] = [K].[r]$$

where *[R]* and *[r]* are matrices of external nodal

forces and nodal displacements; the solution of this equation for the deflection of the overall structure involves the inversion of the stiffness matrix to obtain $[K]^{-1}$. Computer manipulation is ideal for this sort of calculation.

A parallel structural method is the 'force', rather than displacement method, which uses forces rather than displacements as the unknowns in the equation. Prior to FEA, this was used with simple structural surfaces, discussed earlier, to divide the structure into idealised sections and similarly write down stress-strain relationships for matrix analysis. This approach is sometimes preferred to FEA, in the earlier stages of design, so as to obtain a better visualisation of the force patterns in the structure — in relation to the external loads.

An example might be the idealised van body shell of Fig 113 in which concentrated loads are considered to act at the centres of the lower edges of the sidewall panels and reacted at the ends of these edges. The structure comprises three bulkhead frames having pin-jointed members connected by eight sandwich panels — assumed for the analysis to be hinge-jointed at their edges. The structure is then divided into stress systems in the same manner as the redundant frameworks considered in a previous section. The determinate *[R]* system conveniently comprises the plate and four bars making up the lower half of the sidewalls while the three redundant *[X]* systems are 'self-equilibriating', Fig 114, *X1*, *X2* and *X3*.

Akin to the equation describing the stress/strain relationship for redundant frameworks, a similar one applies to the panelled structure as follows:

$$[S] = \{[b_0] - ([b_1][b_1]'[f][b_1])^{-1}\}\ ([b_1]'[f][b_0])[R]$$

used to evaluate matrices when unit loads are applied to the stress systems. The b_0 matrix is multiplied by $1/2l$, since for simplicity in algebraic expression of the matrix elements, the unit load due to the *[R]* system is multiplied by $2l$ and the member forces are then quickly determined by simple statics.

The Unit Load theorem is used to evaluate sub-matrices for the bars and plates in the overall flexibility matrix, Fig 115. For this example, modulus G is taken as $E/2$ and half the section area of the sandwich panel on each side of the equivalent 'a' bar considered concentrated at the bar for end-load resistance. Section area of the intermediate bulkhead

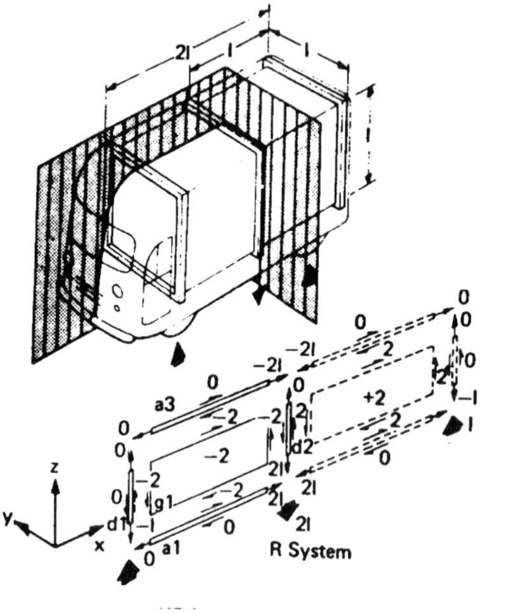

Fig 113: Box van idealised for analysis

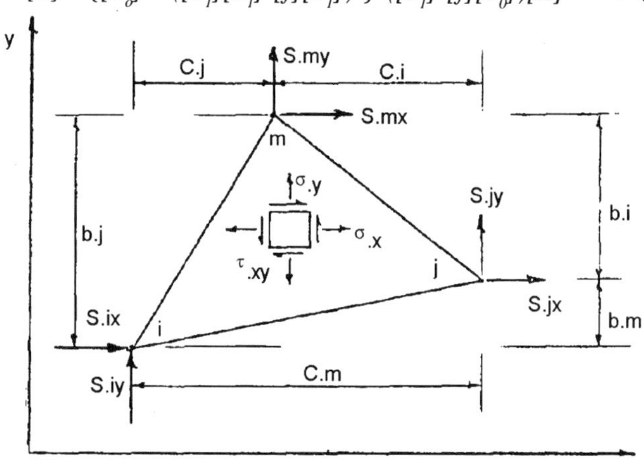

$$\begin{Bmatrix} S_{ix} \\ S_{jx} \\ S_{mx} \\ S_{iy} \\ S_{jy} \\ S_{my} \end{Bmatrix} = -\frac{1}{2} \begin{bmatrix} b_i & 0 & c_i \\ -b_j & 0 & c_j \\ b_m & 0 & -c_m \\ 0 & c_i & b_i \\ 0 & c_j & -b_j \\ 0 & -c_m & b_m \end{bmatrix} \begin{Bmatrix} \sigma_x \\ \sigma_y \\ \tau_{xy} \end{Bmatrix}$$

$$\underline{S} = \underline{C\sigma} = \underline{B^T\sigma}$$

Fig 112: Node equations in matrix form

bars is taken as a quarter of that of the adjacent panel while those at the end take halves — resulting in sub-matrices for the bulkheads as in Fig 116. This leads to the matrix formulation and manipulation of Fig 117 which is about the largest one would want to undertake without a computer. The final matrix of member forces results from taking a central side load of 10 tonnes and these forces are plotted on the

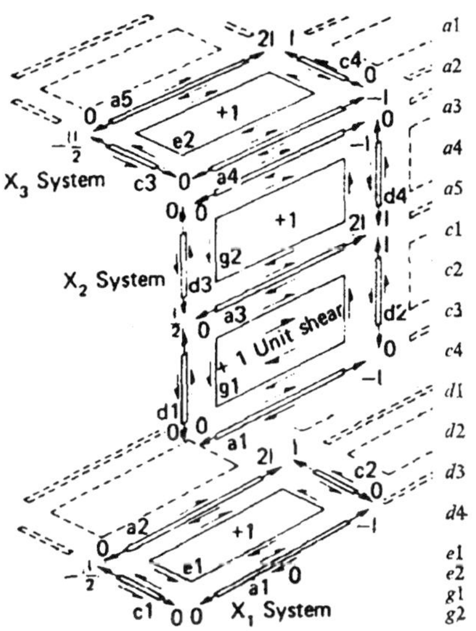

Fig 114: Self-equilibriating force system

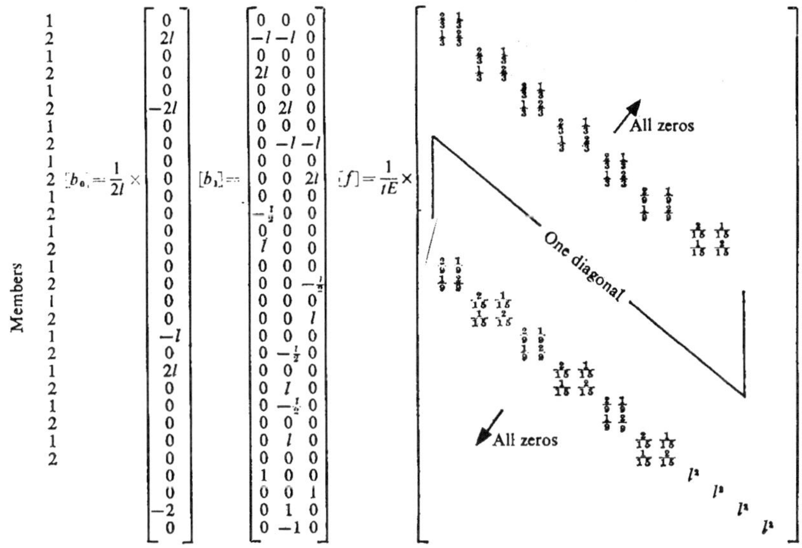

Fig 115: Matrix relationship

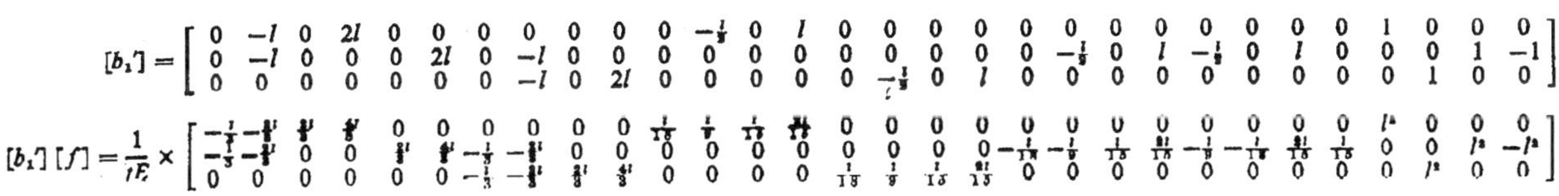

Fig 116: Sub-matrices

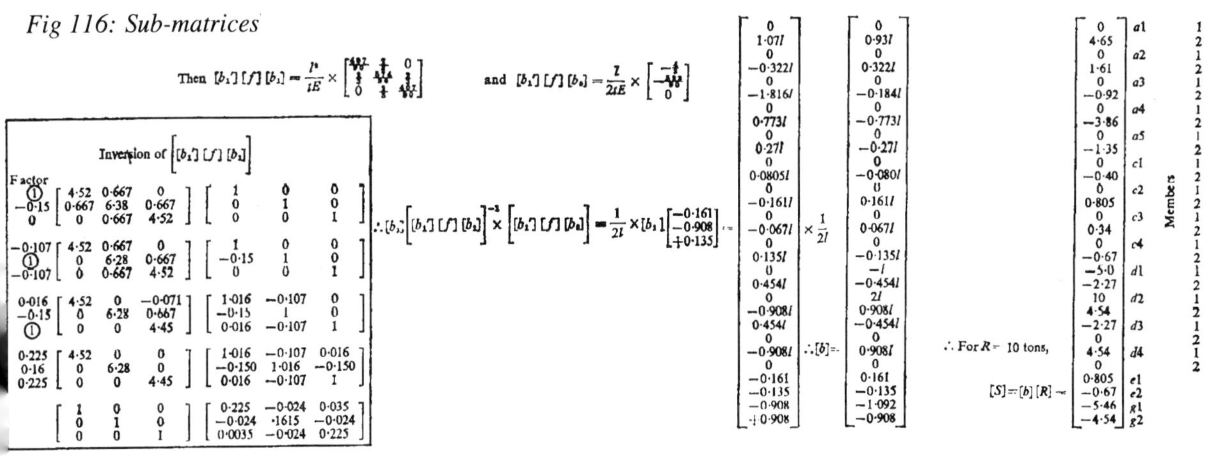

Fig 117: Matrix manipulation

distribution diagram for the van structure in Fig 118. This gives a good 'feel' for the structure in that, by applying the external load at the bottom of — rather than uniformly over — the cross section of the box beam formed by the van structure, it has resulted in uneven compressive and tensile forces in the top and bottom corner booms. Also a reduced loading of the intermediate top and bottom longitudinal stringers shows the action of shear lag.

By applying the matrix displacement analysis to a vehicle structure, Fig 119, in a very similar manner shows how course-mesh finite element analysis can similarly be used to obtain structural feel. The structure which represented a Land Rover-like, but monocoque, field car was evaluated some years ago by workers at Cranfield and Birmingham University, being idealised for analysis as shown. A uniformly distributed loading over the deck produced shear loading in the panel idealisations as shown — the substantial load distribution throughout the structure justifying the choice of monocoque construction.

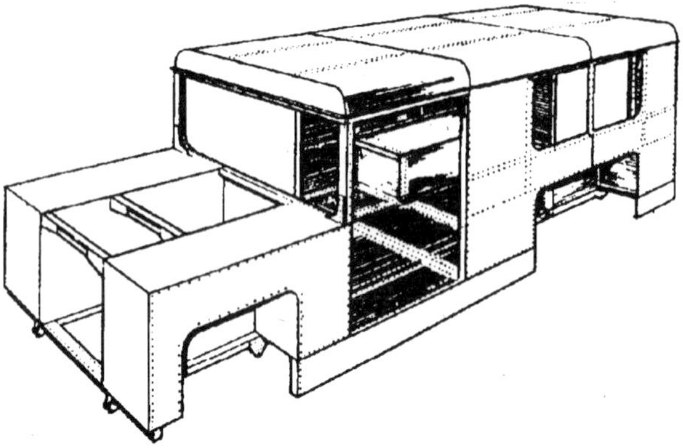

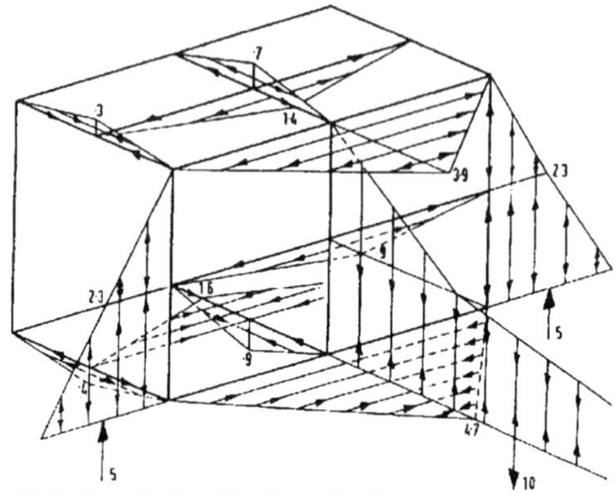

Fig 118: Load distribution diagram

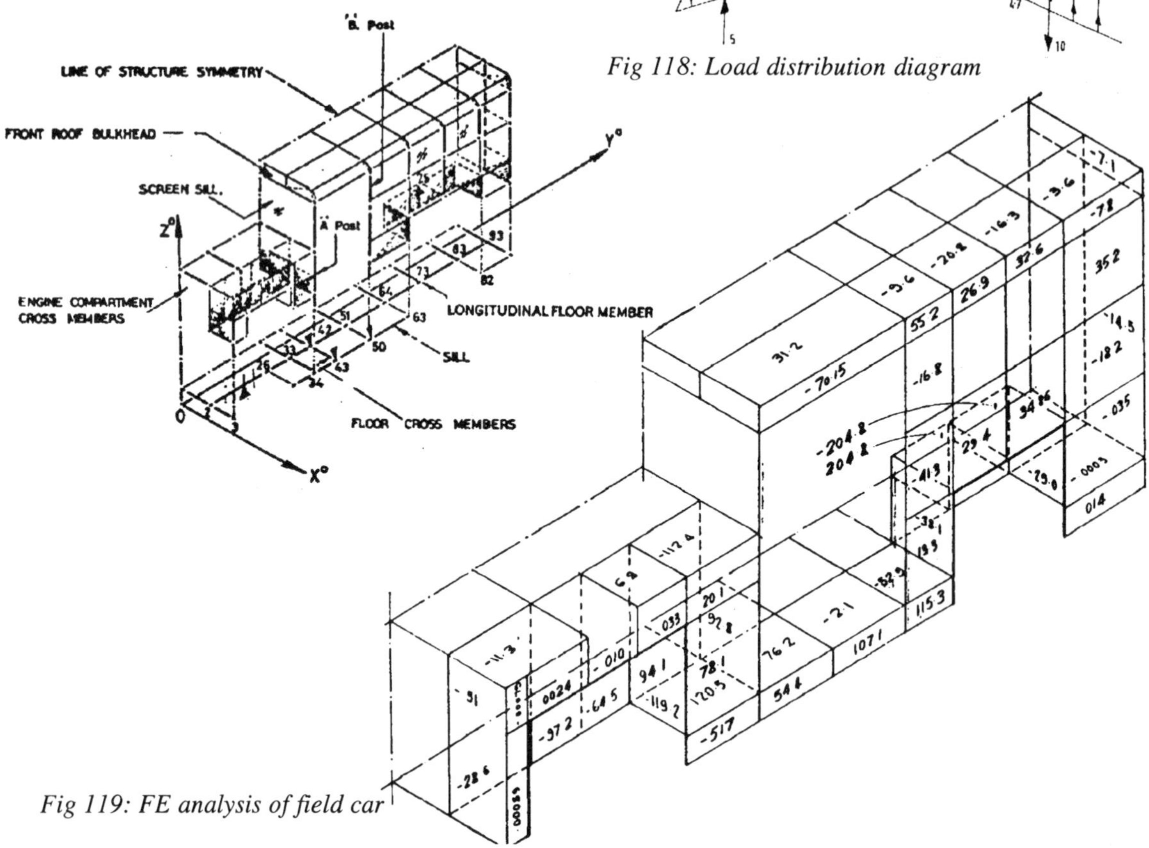

Fig 119: FE analysis of field car

FEA for car bodies

The Finite Element method of structural analysis was introduced in the previous section which described its evolution from matrix methods of solution for unit-load (force) and displacement techniques of analysis. Here some of the accepted procedures are discussed for applying to body structural analysis.

As well as for loads and displacements, FE techniques, of course, cover temperature fields and many other variables and the structure, or medium, is divided up into elements connected at their nodes between which the element characteristics are described by equations. The discretisation of the structure into elements is made such that the distribution of the field variable is adequately approximated by the chosen element-breakdown.

Equations for each element are assembled in matrix form to describe the behaviour of the whole system. Computer programs are available for both the generation of the meshes and the solution of the matrix equations such that use of the method is now much simpler than it was during its formative years.

Economies can be made in the discretisation by taking advantage of any symmetry in the structure to restrict the analysis to only one half or even quarter — depending on degree. As well as planar symmetry, that due to axial, cyclic and repetitive configuration, Fig 120, should be considered. The latter can occur in a bus body, for example where the structure is composed of identical bays corresponding to the side windows and corresponding ring-frame.

Element shapes are tabulated in Fig 121 — straight-sided plane elements being preferred for the economy of analysis in thin wall structures. Element behaviour can be described in terms of 'membrane' (only in-plane loads represented); in bending only or as a combination entiled 'plate/shell'. The stage of element selection is the time for exploiting an understanding of basic structural principles; parts of the structure should be examined to see whether they would typically behave as a truss-frame. beam or in plate-bending, for example. Avoid the temptation to over-model a particular example; however, because number and size of elements are inversely related, as accuracy increases with increased number of elements.

Different sized elements should be used in a model — with high mesh densities in regions where a rapid change in the field variable is expected. Different ways of varying mesh density are shown in Fig 122, in the case of square elements. All nodes must be inter-connected and therefore the fifth option shown would be incorrect because of the

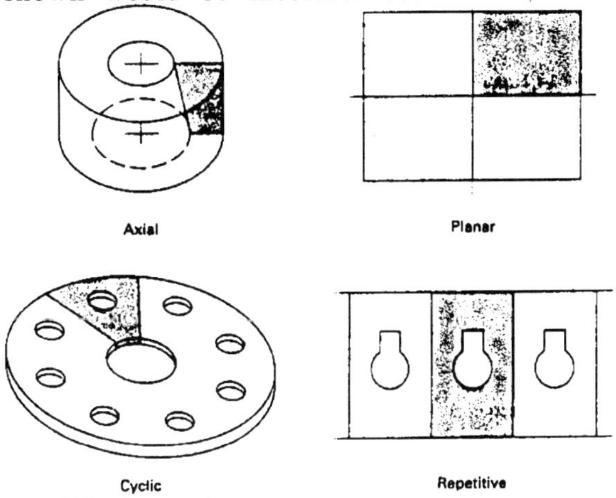

Fig 120: Types of symmetry

Shape	Type	Geometry
Point	Mass	
Line	Spring, beam, spar, gap	
Area	2D solid, axisymmetric solid, plate	
Curved area	Shell	
Volume	3D solid	

Fig 121: Element shapes

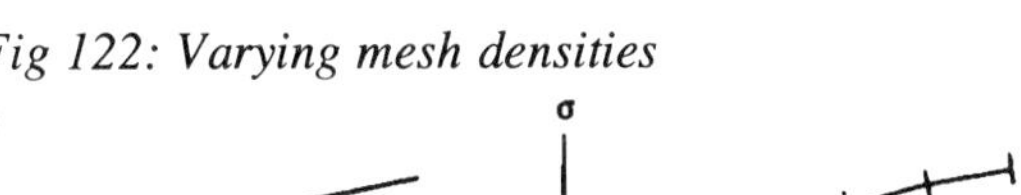

Fig 122: Varying mesh densities

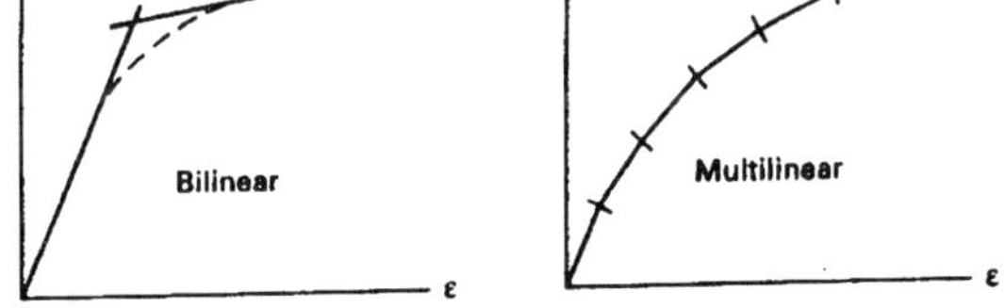

Fig 123: Stress-strain curve representation

discontinuities.

As element distortion increases under load, so the likelihood of errors increases, depending on the change in magitude of the field variable in a particular region . Elements should thus be as regular as possible — with triangular ones tending to equilateral and rectangular ones tending to square. Some FE packages will perform distortion checks by measuring the skewness of the elements when distorted under load. In structural loading beyond the elastic limit of the constituent material an idealised stress-strain curve must be supplied to the FE program — usually involving a multilinear representation, Fig 123.

When the structural displacements become so large that the stiffness matrix is no longer representational then a 'large-displacement' analysis is required. Programs can include the option of defining 'follower' nodal loads whereby these are automatically re-orientated during the analysis to maintain their relative position. The program can also recalculate the stiffness matrices of the elements after adjusting the nodal co-ordinates with the calculated

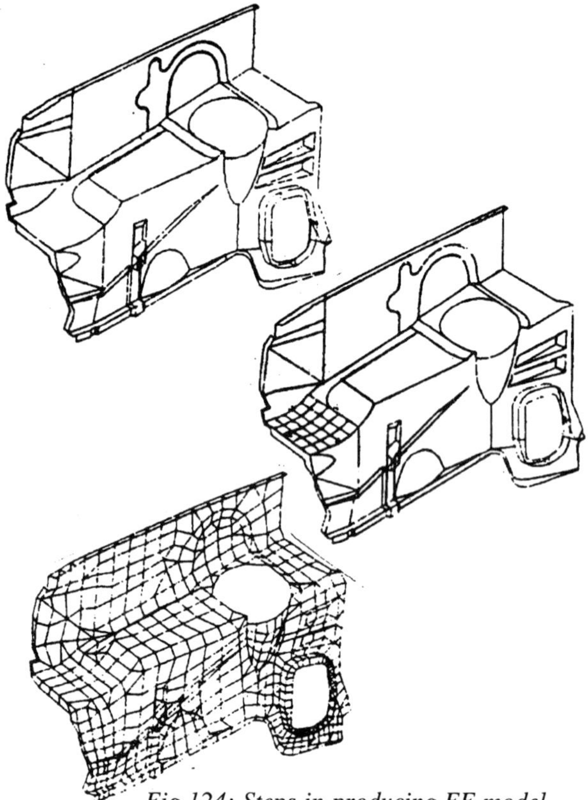

Fig 124: Steps in producing FE model

Structural Analysis.

NASTRAN	linear static analysis
ABAQUS	non-linear static and dynamic analysis
MOTRAN	body loadings from road surface
OPUS	optimisation of body
WRECKER-F	crash analysis
MENTOR	crash analysis
ANSYS	small strains on impact components
FATSUP	fatigue analysis

Pre- and Post- Processors.

TIDE	inter-face between NASTRAN and MOVIE.
MOVIE	graphics display post-processor of F.E. results.
FELS	pre processor for ABAQUS
NAVIE	pre/post processor for NASTRAN
PATRAN-G	pre/post processor for F.E.packages generally
NASTECK	NASTRAN display package for Tektronix terminals.
NASPLOT	NASTRAN plotting package.
Lundy/Prime	generation of F.E. models and display of NASTRAN stresses and deformations.

Applications.

BELTFIT	fit of seat belts
GRASP	liftgate opening geometry and efforts (gas strut assistance)
HOODLUM	hoodtop opening geometry and efforts
CALSPAN CVSP	kinematic dummy predictions (Crash Victim Simulation Program)
TMIRROR	external mirror field of view checks using Tektronix display
TENSP/TORSP	various spring programs
DOORLIFT	door swing as function of hinge inclinations
PROFTEK/PROFTAB	cross-section properties with digitised input
IMIRROR	inside mirror field of view
WIPER	wind screen wiper layout

Fig 125: FE programs

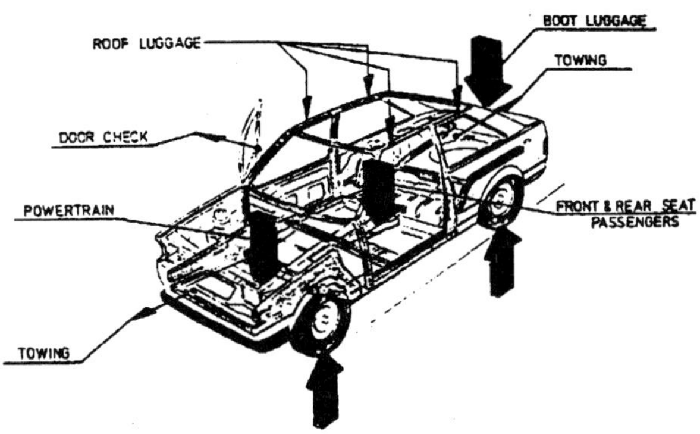

Fig 126: Load inputs

displacements. Instability and dynamic behaviour can also be simulated with the more complex programs.

The principal steps in the FEA process are:

* idealisation of the structure (discretisation)
* evaluation of stiffness matrices for element groups
*assembly of these matrices into a super-matrix
* application of constraints and loads
* solving equations for nodal displacements
* finding member loading.

For vehicle body design, programs are available which automate these steps, the input of the design engineer being, in programming, the analysis with respect to a new model introduction. First stage is usually the obtaining static and dynamic stiffness of the shell, followed by crash performance based on the first estimate of body member configurations. From then on it is normally a question of structural refinement and optimisation based on load inputs generated in earlier model durability cycle testing. These will be conducted on relatively course mesh FE-models and allow section properties of pillars and rails to be optimised and panels thicknesses to be established.

In the next stage, projected torsional and bending stiffnesses are input as well as the dynamic frequencies in these modes. More sophisticated programs will generate new section and panel properties to meet these criteria. The inertias of mechanical running units, seating and trim can also be programmed-in and the resulting model examined under special load cases such as pot-hole road obstacles. As structural data is refined and updated, a fine-mesh FE simulation is prepared which takes in such detail as joint design and spot-weld configuration. With this model a so-called sensitivity analysis can be carried out to gauge the effect of each panel and rail on the overall behaviour of the structural shell.

Joint stiffness is a key factor in vehicle body analysis and modelling them normally involves modifying the local properties of the main beam elements of a structural shell. Because joints are line connections between panels, spot-welded together, they are difficult to represent by local FE models. Combined FE and EMA (experimental modal analysis) techniques have thus been proposed to 'update' shell models relating to joint configurations. Vibrating mode shapes in theory and practice can thus be compared. Measurement plots on physical models excited by vibrators are made to correspond with the node points of the FE model and automatic techniques in the computer program can be used to update the key parameters for obtaining a convergency of mode shape and natural frequency.

An example car body FE analysis at Ford was described at one of the recent Boditek conferences, Fig 125, outlining the steps in production of the FE model. An extension of the PDGS computer package used in body engineering by the company — called FAST (finite element analysis system) — can use the geometry of the design concept existing on the computer system for fixing of nodal points and definition of elements. It can check the occurrence of such errors as duplicated nodes or missing elements and even when element corners are numbered in the wrong order. The program also checks for misshapen elements and generally and substantially compresses the time to create the FE model.

The authors considered that upwards of 20,000 nodes are required to predict the overall behaviour of the body-in-white. After the first FE analysis was carried out, the deflections and stresses derived were fed back to PDGS-FAST for post-processing. This allowed the mode of deformation to be viewed from any angle — with adjustable magnification of the deflections — and the facility to switch rapidly between stressed and unstressed states. This was useful in studying how best to reinforce part of a structure which deforms in a complex fashion. Average stress values for each element can also be displayed numerically or by graduated shades of colour.

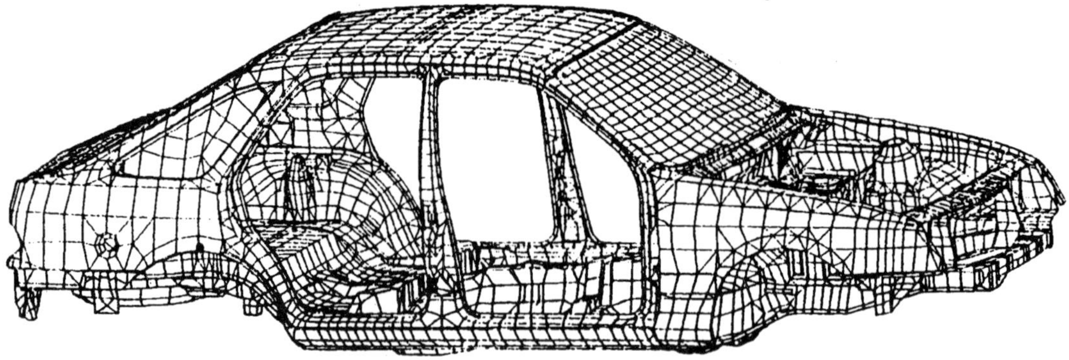

Fig 127:Global model for body-in-white

The authors released a list of the programs used, as Fig 125, by the company about a decade ago. Load inputs were as shown in Fig 126 and the FE model for the BIW, in Fig 127.

Chapter 3. Safety under impact

Impact protection basics
Design for crashworthiness
Occupant & cargo restraints
Passive restraint systems
Side-impact analysis
Bumper systems
Energy-absorbent foams
Laws of mechanics applied to safety
Crashworthiness of road tankers

While the need to obtain advantageous trade-off between fuel-ecomomy and exhaust emission control may be the prime mover in structural design for reduced weight, probably the main stimulus for body structural analysis has been secondary-safety design, in recent years.

Impact protection

Design for impact-protection has now become one of the key factors in the structural engineering of volume cars and the same regulations often apply to specialist cars and commercial/industrial vehicles when used on public roads. Much of the safety protection can be built in at the concept stage if predictive techniques are used.

Impact protection for vehicle occupants has been in focus since the pioneering work of Ralph Nader in the 1960s recognised, in particular, the poor protection offered by standard production cars of the day. This work has now, of course, spread to most categories of vehicle including trucks, buses and racing cars. UK investigations at that time[1] showed saloon cars typically offering up to 3 ft (0.91 m) crush

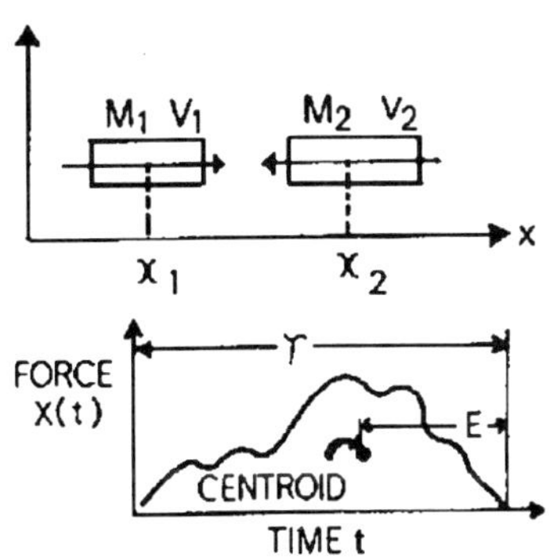

Fig 1: Collision parameters

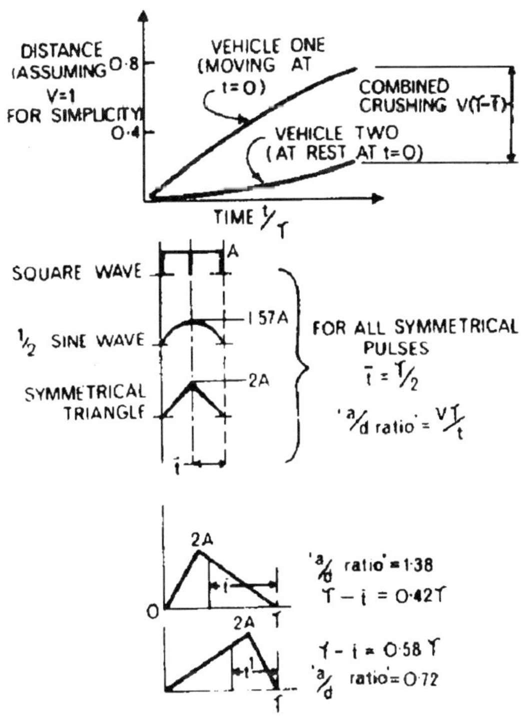

Fig 2: Idealised impact pulses

resistance, with 0.1 second deceleration, in front-end barrier impacts at about 35-40 mph (55-65 kph). This was for average sized cars of typically 13.75 ft (4 m) long by 5.25 ft (1.6 m) wide and weighing around 2350 lb (10.5 tonnes).

When vehicles collide with one another, head-on, as represented in Fig 1a, an impact pulse might be typically as shown in Fig 1b. Each vehicle has momentum (mass x velocity) and there will be a common velocity after impact:

$$V_3 = (M_1V_1 + M_2V_2)/(M_1 + M_2)$$

and it can be shown that the combined crushing of the two vehicles is $(V_1 - V_2)(T - t)$ the shape of the impact pulse determining time t, Fig 2 showing the effect of different idealised standard pulse shapes.

Plastic collapse

Crush resistance depends upon front-end structural layout and for many vehicles, the bending resistance of the front crossmember is crucial. Bending beyond the elastic range is the case to examine. Consideration of the stress/strain curve for mild steel, for example, Fig 3a, shows that initial yield occurs at a strain of 0.001 while the plastic range continues to a strain of 0.010 to 0.020 when strain hardening commences. Because of the material's ductility, and long plastic range, increasing the bending moment beyond that at which yield is just reached results in a spread of yielding across the depth of the beam section. Distribution of bending stress across the section of a beam undergoes the transformation shown in Fig 1b, the attainment of the upper yield stress is followed by a falling off to a lower yield stress, as shown, until failure at the lower level applies across the whole of section and a 'plastic hinge' is said to develop. The corresponding plastic moment:

$$M_o = (\sigma Lbd/2)d/2 = \sigma Lbd^2/4$$

against $M_b = sBbd^2/6$ for bending in the elastic range; the ratio of the two bending moments M_o/M_b is known as the 'form factor' for the beam. It is a measure of the additional load which can be carried under part plastic strain. Typical values are 1.15 for an I-beam and 1.7 for a circular section tube.

In considering again the front crossmember of a vehicle, one legal requirement for cars in certain

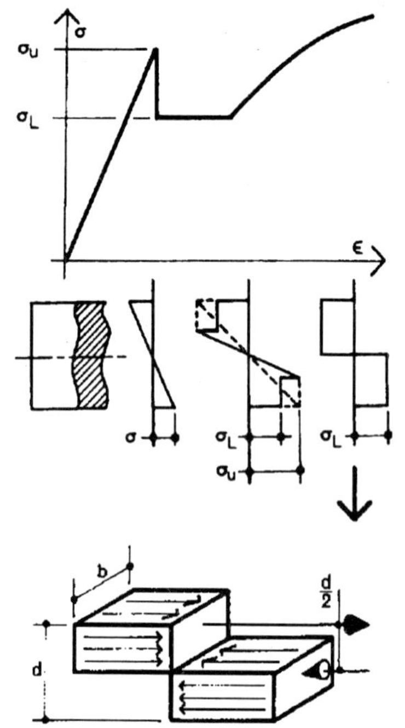

Fig 3: Plastic bending

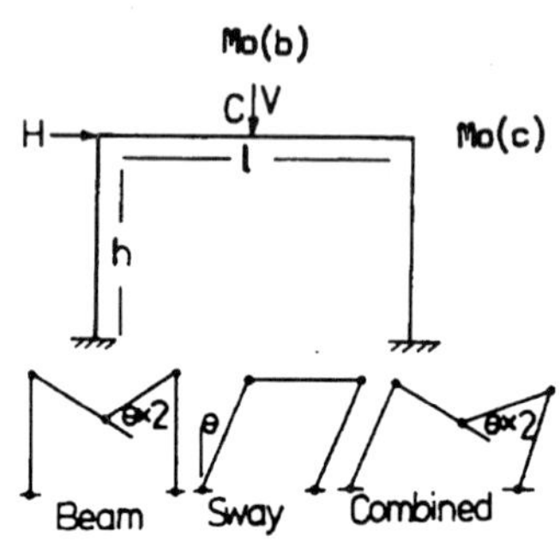

Fig 4: Front-end as portal frame

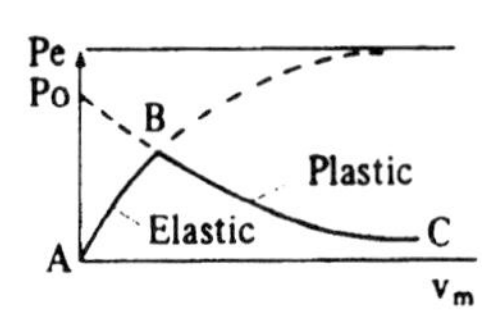

Fig 6: Buckling load vs lateral displacement

countries is for structural integrity to be maintained up to a certain impact velocity when striking a post-like object at mid-span. Collapse of the beam occurs in this situation when the central load reaches either $4M_o/L$ or $8M_o/L$ according to whether the crossmember 'beam', length L, is 'simply-supported' (unrestrained) at its end or embedded in fixed supports, for calculation purposes. It can also be shown that when the central load rises to $45.6M_o/L$, plastic hinges also form at points $0.202L$ either side of mid-span, the hinges moving inwards towards mid-span for even greater loads.

Technique in designing for impact-resistance is to use as much permanent deformation of the structure as possible without encroaching on the occupants' 'clearance zone'. Rails and pillars can be arranged so that plastic hinges occur at their intersections and subsequent deformation will be as a hinged 'mechanism'.

Collapse mode can be modified by gusseting of joints, to alter the effective linkage lengths of the mechanism and the addition of shear panels further affects behaviour. Fig 4 suggests that the front-end of a conventional vehicle, the structure around the engine bay in plan view, can be likened to a portal frame — again under the action of the central impact load. Mode of collapse is seen to depend on the geometry and relative stiffness of the structural elements, as well as the load pattern. In the beam-collapse case, work done by the central load, $VL/2$, is equal to the energy dissipated in the plastic hinges $M_o(b)2\theta + 2M_o(c)\theta$ giving $VL/4 = M_o(b) + M_o(c)$. Similarly, for sway collapse, $Hh/2 = M_o(c)$ and for combined collapse, $VL/4 + Hh/2 = M_o(b) + M_o(c)$, assuming the portal has pinned feet. If fixed feet are assumed,

$VL/4 + Hh/2 = M_o(b) + M_o(c) + M_o(p)$.

A typical vehicle might have a front-end portal 1.5 metres wide x 1.5 high and be subject to 10 kN central impact load with a lateral component of 2.5 kN. Thus $Hh/2 = 1.875$ kN and $Vl/2 = 3.75$ kN and if pinned feet are assumed, a section modulus of $M_o/Ss(b)$ is required for $M_o(c) = M_o(b) = M_o$ where S is shape factor.

US researchers[2] have shown that the collapse rate of specially designed front-end structures can readily be predicted. For the arrangement in Fig 5, curved beam theory can be used to show controlled-collapse can be predicted for the S-shaped longitudinals. For moment arm lm of the frontal impact force the radius defining the neutral axis of the beam can be calculated from:

$$A/[(W_1 l_m r_2/r_1)+(W_2 l_m r_4/r_2)+(W_3 l_m r_5/r_3)+(W_4 l_m r_6/r_5)]$$

For l_m = 4.7 inches, the 50,000 lb/in² steel of 0.132 inch thickness can be shown to have R = 5.64 inches and a collapse load per longitudinal of 21.700 lbf. Even practically straight members subject to end load tend to bend because of unavoidable load-

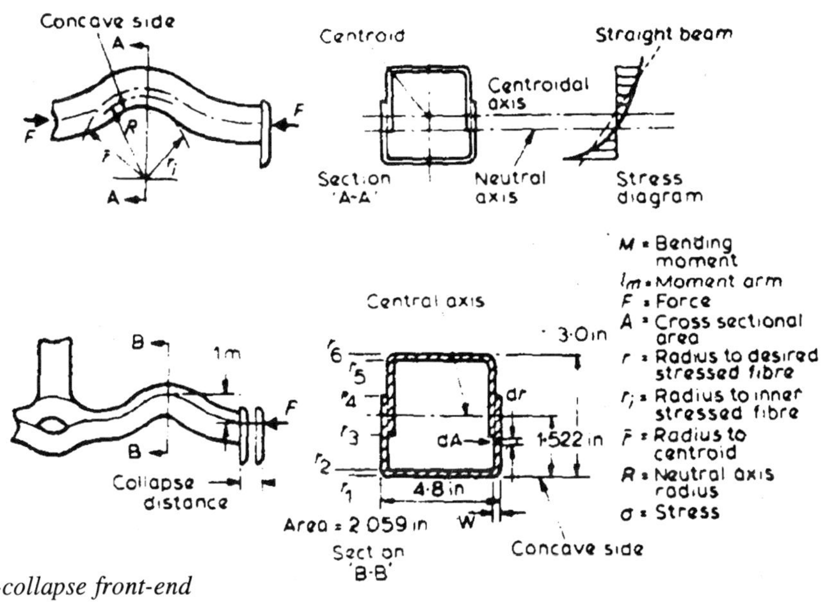

Fig 5: Controlled-collapse front-end

eccentricity creating a bending moment. In a plot of axial buckling load against lateral displacement, Fig 6, the latter grows at an increasing rate, with end load, until at B, a plastic hinge forms. The collapse load does increase under impact conditions, however, because high lateral inertia forces usually restrain the longitudinal. When thin-walled closed section members with substantial relative width are employed, crumpling under axial impact usually only occurs in the front part of the member because of longitudinal inertia effects and the formula for static crumpling has been found to describe adequately the dynamic impact condition, for which crush load is:
$6s_y t(Dt)^{1/2}$ and crush length $MV^2/(12s_y t^{1.5}D^{0.5})$
for a circular tube of diameter D and thickness t hit by mass M at closing speed V, where s_y = yield stress of the tube material.

When the whole vehicle is modelled as a rectangular section tube (akin to a bus structure without doors and windows) an approximation can be made at predicting the crush length of the vehicle shell as it impacts a completely rigid object as
$P/(V^2/2g)(1 + c)dAL$;
for $P/P_o = [(6A/A_o) + 0.04][1 + (v/50)]$
for cross section area A and A_0 the area enclosed by A (A/A_o is the solidity ratio); P_o is As_y.

Safety frames

Industrial vehicles such as farm tractors, and increasingly CV cabs, have safety frames affording a substantial degree of protection to the driver in both overturning and impact situations. Early work reported by Massey Ferguson on farm tractor safety frames was based on exploiting the considerable increase in the yield strength of mild steel when it is subject to rapidly applied loads, Fig 7. For calculation purposes the frames were modelled as links connected by plastic hinges and after applying the test load in the position shown, observe how far the point of application of the load can move before any of the links encroaches on the driver's zone of clearance. For prescribed test energy levels given by British Standards for the weights of tractor concerned, a critical 'test' load is established.

This load is then applied to the same shaped frame, but under elastic rather than plastic deformation conditions, to determine member proportions and deflections.

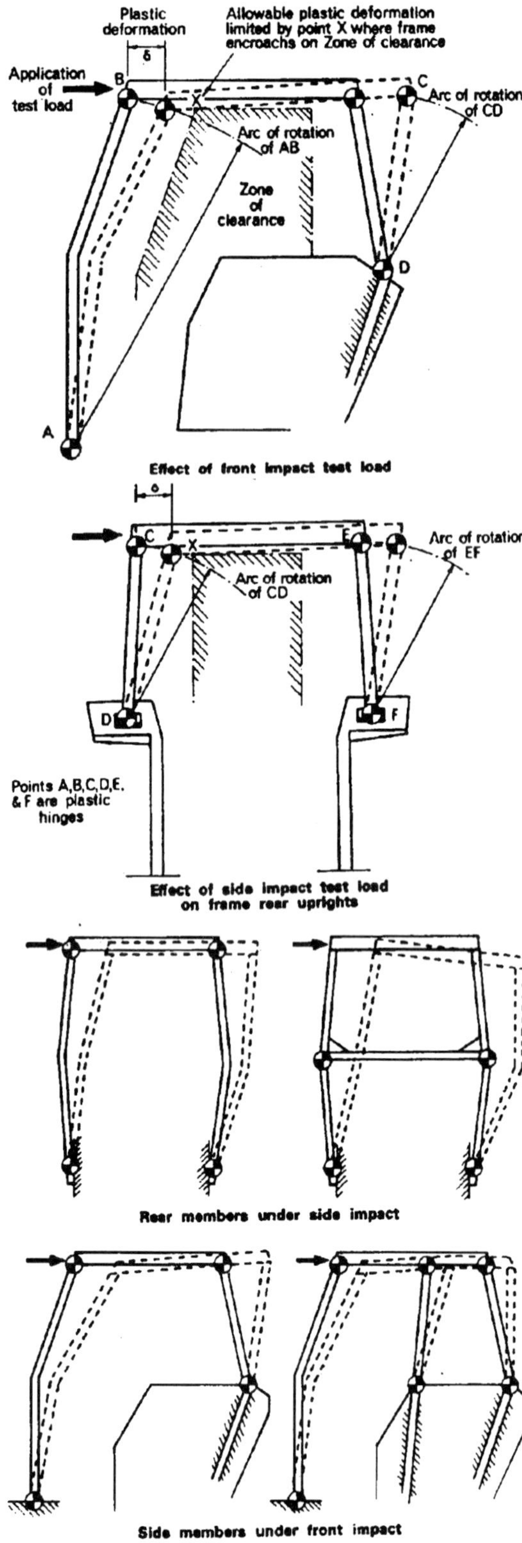

Fig 7: Tractor cab safety frame

Design for crashworthiness

The behaviour of structural elements is described in a way that can lead to a better understanding of the frontal impact resistance of vehicle structures. The previous section showed how the shape of the crash time history pulse can be altered by building in particular collapse rates into the impacting elements of the body structure; here the design of such elements is considered.

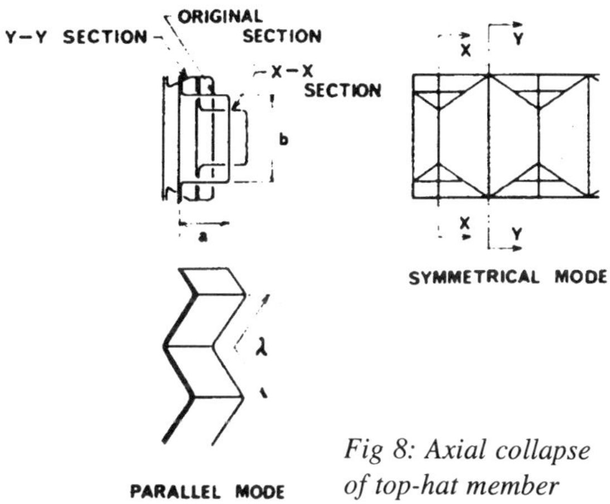

Fig 8: Axial collapse of top-hat member

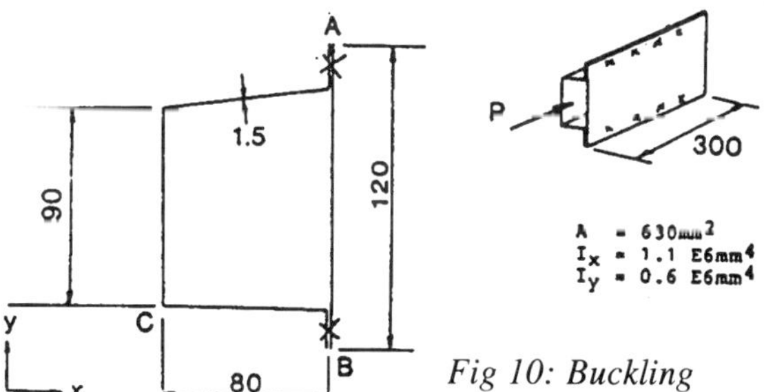

Fig 10: Buckling stress calculation

It is estimated that some 100 kJ of energy is required to be absorbed by the individual front-end structures of two car-weight vehicles impacting head on at a combined speed of 30 mph. Such values need to be equated with absorption capacities of members in either bending or end-on crush mode.

Accurate predictions of crashworthiness can only be obtained by building inertia effects and strain rates into the calculations. However, prior to the linear dynamic modelling necessary to simulate these effects, some useful preliminary calculations can help understand the mechanisms of failure and provide a first guide to the required structural behaviour.

In the case of heavier commercial vehicles, very little of the crash energy is absorbed into the structure when impacting a car or lighter vehicle, therefore crashworthiness has to be built in by means of such devices as front underrun guards. These must be positioned with ground clearances no greater than 300 mm so that impact is made with the strong part of the car's front-end. The truck bumper must also be a yielding structure and the stroke of the underunn guard should be about 300-400 mm for compatibility with car front-end collapse characteristics. It should be designed to collapse at a constant horizontal force equivalent to that which a car can withstand without producing serious intrusion or too violent a deceleration — such that the car front and underrun struts collapse together.

Tests have shown that a force of 230 kN is developed when car crush typically reaches 300 mm, representing a 60 kJ energy absorption — equivalent to a light car moving at 40 km/h. In practice, progressive rates are used so that the 'bumpers' are tolerant

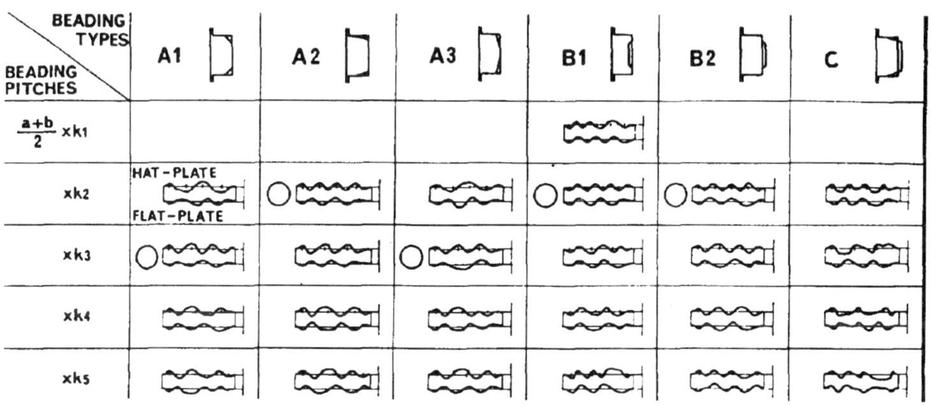

Fig 9: Folding controlled by beading

of lower speed impacts and initial yielding is typically arranged at 100 kN or less. Struts are often made of Invertube cartridges; each of a pair would be about 450 mm long and made to compress, under a lever system by 200 mm.

In a design reportedly due to the Transport and Road Research Laboratory, the Invertubes comprised steel tubes of 100 and 80 mm diameter, the smaller of which was turned back on itself at one end and then peripherally welded to the open end of the large tube. In impact the smaller tube progressively turns itself inside out. Tubes of 16 swg each had a collapse load of about 50 kN to give a total energy absorbing capacity of the bumper of around 20 kJ.

One technique of stabilising the axial collapse of lighter vehicle front-end side rails is to make them deform into a series of folds. This is considered by some to be more favourable than allowing collapse by overall bending deformation of the siderails — as energy can be absorbed more efficiently. Fig 8 shows axial collapse of a top-hat sectioned member in which stability is achieved by initial folding near the impact end with peaks and troughs of waves perpendicular to the axis — subsequent folding progressing towards the root end. Beading is used to initiate folds and so ensure a stable collapse.

In early work carried out by Toyota, Fig 9 shows how different beading arrangements can produce different folding effects and below it the resulting folding modes. A method of calculating the absorption characteristics of front rail sections has been reported by Hawtal Whiting based on the critical buckling stress of the constituent panels. In the rail section of Fig 10, when considering the axial stability of side AB, buckling stress can be calculated from:

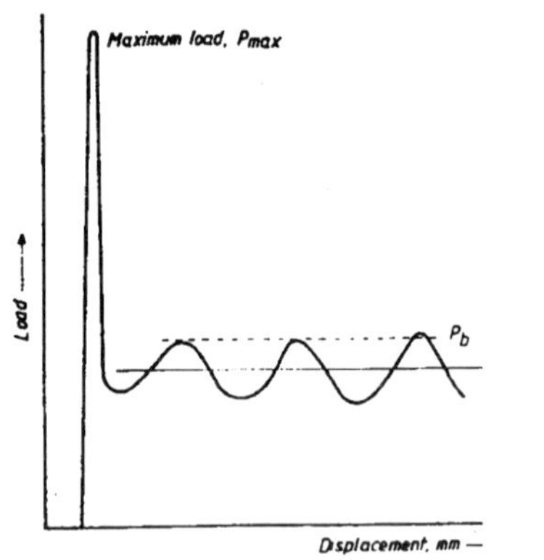

Fig 11: Box-tube crush performance

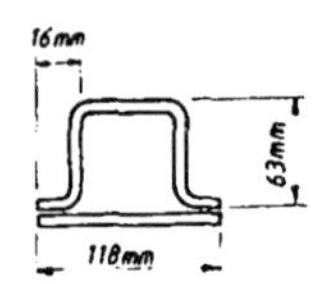

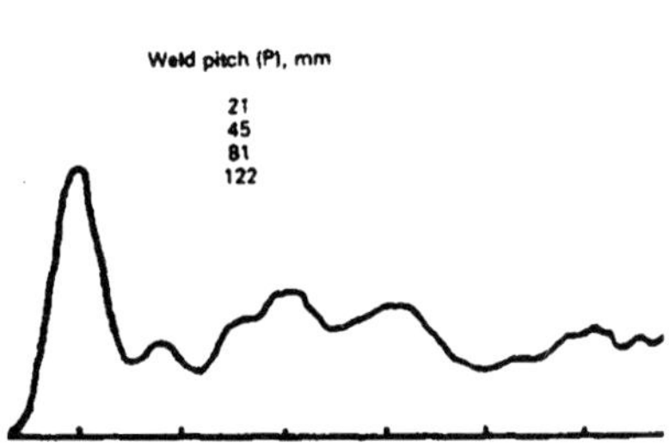

Fig 12: Actual dynamic impact pulse

Effect of weld pitch on collapse of spot welded box sections tested under 30 mile/hr impact loading (6000J energy)

Batch number	Type	Thickness, mm	Yield strength, N/mm²	Weld pitch, mm	Reduction in sample height, mm	Energy to deform sample 10mm*, J	Mode of collapse
146	Low carbon	1.2	186	21	161, 169, 149, 170, 151, 161	375	6A
				45	184, 177, 183	330	2A, 1B
				81	226, 199, 216, 212, 201, 211	285	6A
				122	236, 208, 222	270	2A, 1B
151	HSLA	1.2	340	21	132, 137, 139	440	[illegible]
				45	139, 150, 168	395	3A
				81	168, 174, 172	350	3A
				122	198, 188	310	2A
192	Dual phase	1.2	370	21	123, 136, 121	475	3A
				45	130, 135	485	2A
				81	147, 148, 144	410	3A
				122	161, 183, 141	370	3A
193	Dual phase	1.2	405	21	100, 107, 102	580	3A
				45	107, 111, 122	530	3A
				81	120, 111, 120	510	3A
				122	137, 158, 148	405	3A
212	5182 Al	1.6	131	21	182, 186, 186, 188	325	4A
				45	201, 219, 199	290	3A
				81	258	235	1B
				122	367	165	1B

Effect of yield strength on collapse of box sections tested under 30 mile/hr impact loading (6000J energy)

Batch number	Type	Thickness, mm	Yield strength, N/mm²	Weld pitch, mm	Reduction in sample height, mm	Energy to deform sample 10mm*, J	Mode of collapse*
146	Low carbon	1.2	186	21	170, 151, 161, 169, 161, 149	375	6A
				81	212, 201, 226, 199, 211, 216	285	6A
151	HSLA	1.2	340	21	132, 137, 139	440	3A
				81	168, 174, 172	350	3A
192	Dual phase	1.2	370	21	123, 136, 121	475	3A
				81	147, 148, 144	410	3A
193	Dual phase	1.2	405	21	100, 107, 102	580	3A
				81	120, 111, 120	515	3A
508	Dual phase	1.2	425	21	101, 106	580	2A
				81	122, 115	505	2A
212	5182 Al	1.6	131	21	182, 186, 186, 188	325	4A
				81	258	235	1B

$$\sigma_b = k.E/(1 - 0.32)^2(t/b)^2$$

where E = 208000 N/mm^2; t = 1.5 mm; b = 120 mm and k = 3.29, giving s_b = 117 N/mm^2 and P = 73 kN. A swage would be incorporated into the closing plate AB designed to make it stable until yield of the material is reached. Peak axial load capacity for the section is given by

$$1112t_1 86b_0 14.\sigma_y 0.57.B_0.43$$

where B = 1.0 for t/b < 0.016 so that peak capacity is about 105 kN and mean load capacity is 33.5 kN — given fold initiators to trigger the crush mode. For a circular section thin tube the classic formula due to Macauley gives the collapse load as

$$2.\sigma_y.A[10(t/r) - 0.03]$$

where R is tube radius, t the wall thickness and A the enclosed area of the cross section.

Spot welding patterns affect the crush performance of fabricated box tubes and this has been studied by TWI using a drop test rig. At crushing rates up to 25 mm/min, quasi-static loading, load/compression characteristics are typically as shown in Fig 11 with permanent deformation after peak load Pmax is reached. Subsequent ‘waves’ represent the generation of plastic hinges along the section, fluctuating about a mean load P_m. Area under the load/deflection curve gives the energy absorbed. For this form of loading, tests showed effect of spot weld pitch, below a value of 50 mm, to be insignificant. The shape of a typical dynamic impact curve is theoretically different from the quasi-static case, as seen in Fig 12.

Dynamic impact loading was found to give substantially different results — and these depended

. Effect of spot weld pitch on collapse of weld bonded box sections tested under 30 mile/hr impact loading (6000J energy)

Batch number	Steel type	Thickness, mm	Yield strength, N/mm^2	Weld pitch, mm	Reduction in sample height, mm	Energy to deform sample 10mm*, J	Energy to deform equivalent box section 10mm without adhesive*, J	% increase in energy absorption due to presence of adhesive
146	Low carbon	1.2	186	21	148, 137, 132, 140	430	375	15
				45	148, 148, 139	415	330	26
				81	154, 161, 182	360	285	26
				122	174, 199, 200, 166	325	270	20
151	HSLA	1.2	340	21	111, 109, 122	525	440	19
				45	125, 121, 133	475	395	20
				81	156, 130, 142	420	350	20
				122	156, 150, 145	400	310	29
192	Dual phase	1.2	370	21	106, 96, 99	600	475	26
				45	98, 109, 108	570	485	17
				81	144, 140, 110	455	410	11
				122	120, 126, 125, 132	480	370	30

Effect of adhesive type on the collapse of weldbonded box sections tested under 30 mile/hr impact loading (6000J energy)

Batch number	Steel type	Thickness, mm	Yield strength, N/mm^2	Adhesive type	Weld pitch, mm	Reduction in sample height, mm	Energy to deform sample 10mm*, J	% increase to energy absorption due to presence of adhesive
146	Low carbon	1.2	186	No adhesive	21	161, 169, 149, 170, 151, 161	375	–
					122	236, 208, 222	270	–
				A15	21	148, 137, 132, 140	430	15
					122	179, 199, 200, 166	325	20
				A16	21	133, 119, 130	470	25
					122	205, 207, 203	290	8
				A17	21	139, 128, 136	445	19
					122	191, 178, 184, 164	335	24
151	HSLA	1.2	340	No adhesive	21	132, 137, 139	440	–
					122	198, 188	310	–
				A15	21	111, 109, 122	525	19
					122	156, 150, 145	400	29
				A16	21	110, 103, 120	540	23
					122	161, 182, 156	360	16
				A17	21	136, 120, 118	480	9
					122	179, 154, 146	410	32
192	Dual phase	1.2	370	No adhesive	21	123, 136, 121	475	–
					122	161, 183, 141	370	–
				A15	21	106, 96, 99	600	26
					122	120, 126, 125, 132	480	30
				A16	21	101, 95, 89	630	33
					122	138, 137, 134	440	19
				A17	21	89, 101, 98	625	32
					122	142, 118, 137	420	14

Fig 13: Effect of weld pitch and yield strength

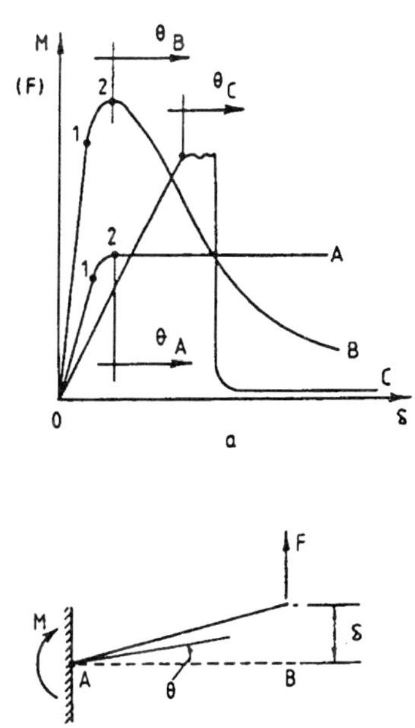

Fig 14: Constant moment curve

considerably on different jointing methods for the fabrication, such as spot-welding, weld-bonding and adhesive bonding. TWI therefore impact-tested specimens with these different jointing methods, as well as different materials and thicknesses. Tests were carried out at 30 mph impact with a drop mass of 66.8 kg giving impact energy of 6 kJ. Tables in Fig 13 show the effect of weld pitch and yield strength on the spot-welded box sections compared with weld- and adhesively- bonded types. Spot-welds were found to act as sites for fold initiation with deformation increasing for longer weld pitch. With weld-bonding, presence of adhesive increased the energy required to deform the section. With pure bonded joints a quite different failure mode was experienced.

The Cranfield Impact Centre has been prominent in the prediction of structural crash behaviour and has developed computer programs for the analysis of both structural elements and complete systems. These can deal with the plastic hinges formed in gross bending deformation of the box-beam elements. Unlike the compact section members referred to in the previous section, hinges in large thin-walled sections develop in a different way. Instead of a constant moment, curve 'A' in Fig 14, a variety of curves can be obtained depending on the location of beam and joint — also on the nature of local failure. Local buckling usually causes a moment drop-off, curve 'B', or if the material separates, the kind of behaviour in curve 'C'. A typical box section hinge is shown in Fig 15, in which the plastic deformations are concentrated along the yield lines of bending and rolling deformation. The figure represents the kinematics of the joint between about 5 degrees and the angle of 'jamming' when the two opposite sides of the compression flanges come into contact. Experimental (dotted) and theoretical (solid) load/deflection curves are compared in Fig 16, alongside their respective collapse modes, for the quasi-static (left) and quasi-dynamic, right, load cases.

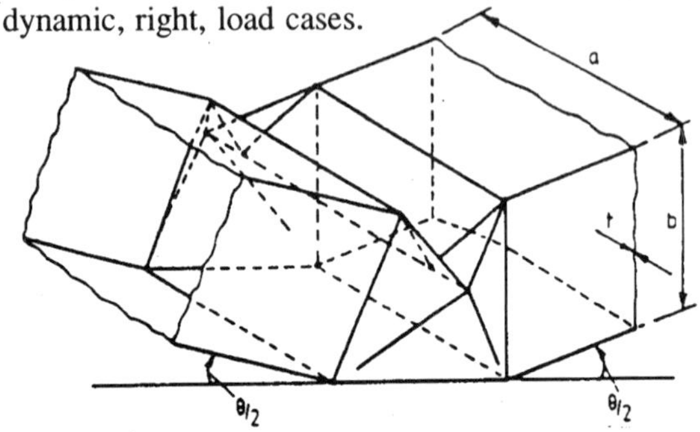

Fig 15: Box-member plastic-hinge

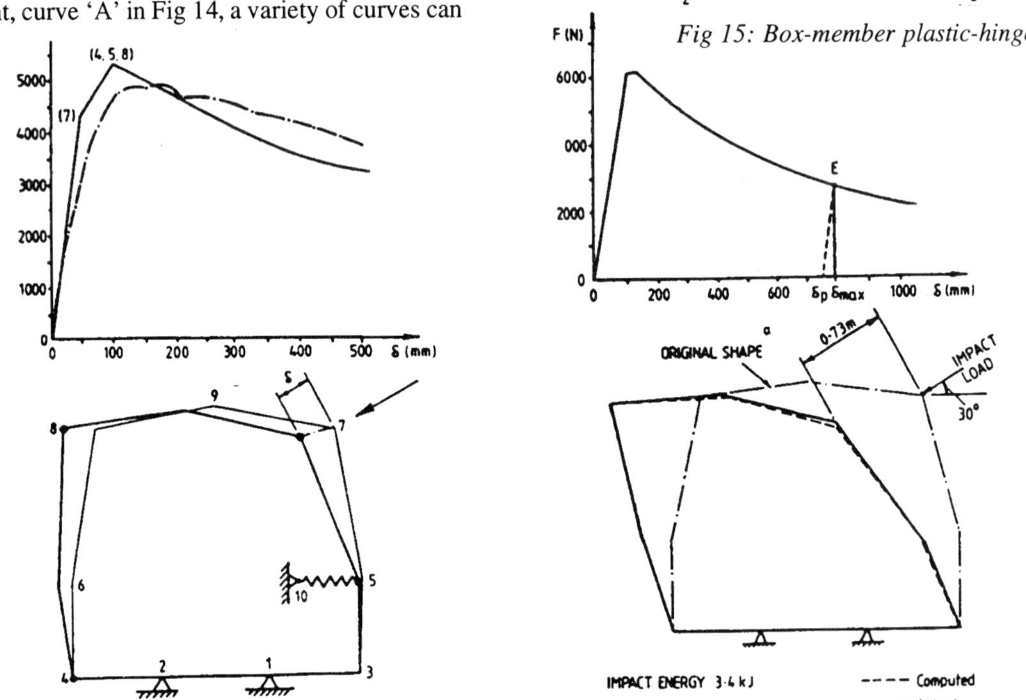

- Experimental (dotted) and theoretical (solid), quasi-static load-deflection curves of a bus ring (a) and the overall collapse mode (b)

- The theoretical 'quasi-dynamic' load-deflection curve of an impacted bus ring (a) and its final deflections (b).

Fig 16: Quasi-static and quasi-dynamic performances

Occupant and cargo restraints

The physical restraint of occupants and cargoes on road vehicles is vital in the accident situation and, in the case of cargoes, is increasingly important in normal running as traffic densities increase. It is a case where the 'village-blacksmith' approach to vehicle design of yesteryear has less and less relevance.

Hunch assumptions about inertia loads on passengers and cargoes during violent braking nearly always tend to be wrong and the efforts spent in a few simple calculations will always be rewarded. Many times have truck operators forgotten the huge inertia effects of a moving load and have made the assumption that the the heavy static weight of a payload can keep it in place on the open deck of a lorry-body. Equally remiss was the assumption, before the age of safety belts, that it was only necessary for an occupant to brace him/herself against a car bulkhead for protection in head-on collision.

Even at modest town speeds, of 30 mph, striking a rigid object such as a concrete lamp standard will cause the car to be brought to rest in a little more than one tenth of a second over a front-end crush distance of perhaps two feet. An unrestrained occupant will strike the car interior, in secondary impact, at virtually the initial speed of the car — with minimal protection even given from a padded dash.

The effect of the seat belt is not just to restrain the occupant away from projections which can imposed high localised forces but also to reduce the forces which act on the whole body. Stretching of the belt in impact allows the body to move forward about eight inches and because the body is tied to the car some of the crushing distance is added to the stretch of the belt in what is known as a 'ride-down' effect. A resulting stopping distance of some 12 inches for the body could typically be assumed in the 30 mph 'rigid' impact situation. Belts have, of course, to comply with standards applying to the country of

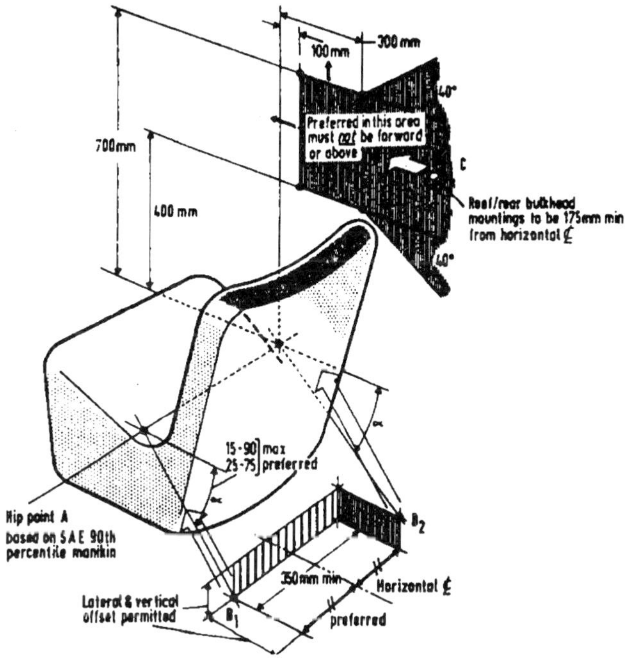

Fig 17: BS seat belt anchorage points

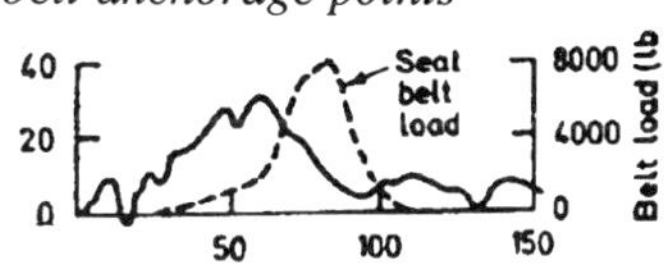

Fig 18: Decceleration pulse

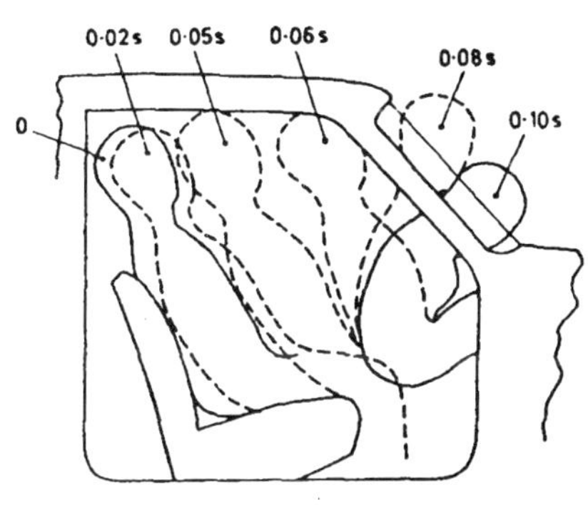

Fig 19: Motion of unrestrained dummy

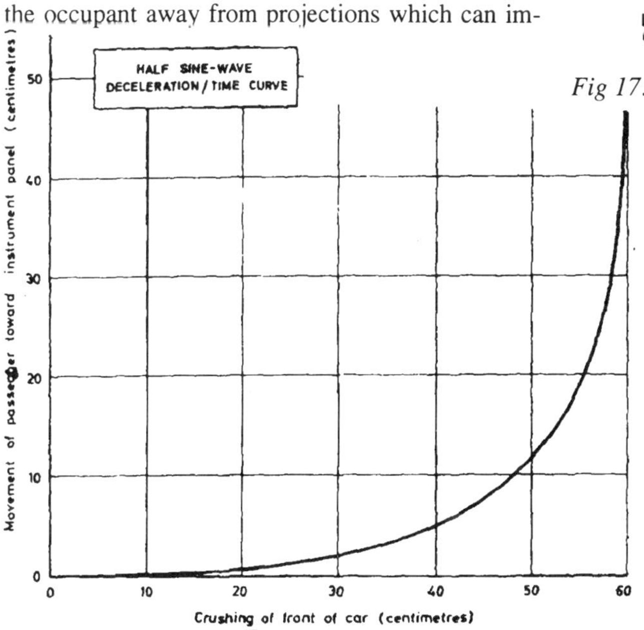

Fig 20: Passenger movement in frontal impact

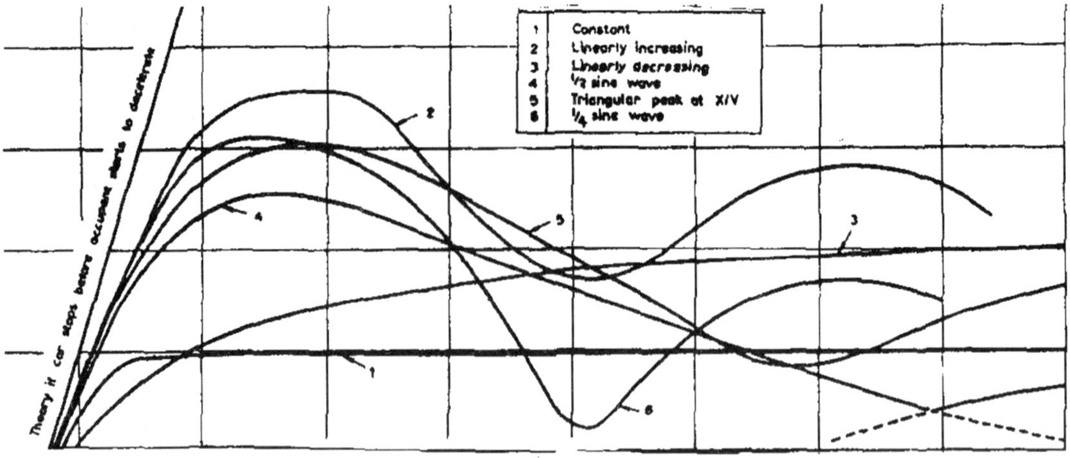

Fig 21: Belt stretch for different crash pulses

operation. In the UK British Standards define the belt and its attachments as well as prescribing positions of the anchorage points as shown in Fig 17.

The three-point belt with retractor reel has become almost the accepted norm. The retractor mechanism is such that when a certain vehicle deceleration value is reached a fast-response interlock inhibits the webbing roller. Extended protection can be obtained using belt pre-tensioner devices and even air bags sensitive to different levels of deceleration. Very quick responses are necessary for successful operation and typically a bag must be inflated by the time the occupant has moved forward eight inches — say, 30 milliseconds. The pre-tensioner would need to be fired after about 10 milliseconds. Sensors may be electro-magnetic or electronic.

Pioneering work by the Transport and Road Research Laboratory was important in setting the standards for passenger restraint. Fig 18 shows measurements taken of head-on car impacts which included traces of seat belt loading during the accident

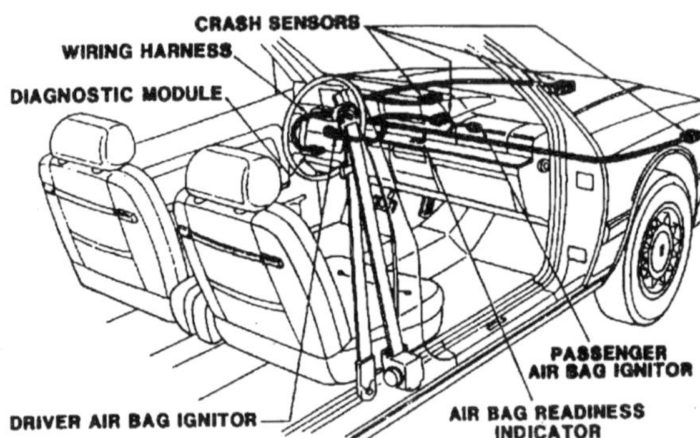

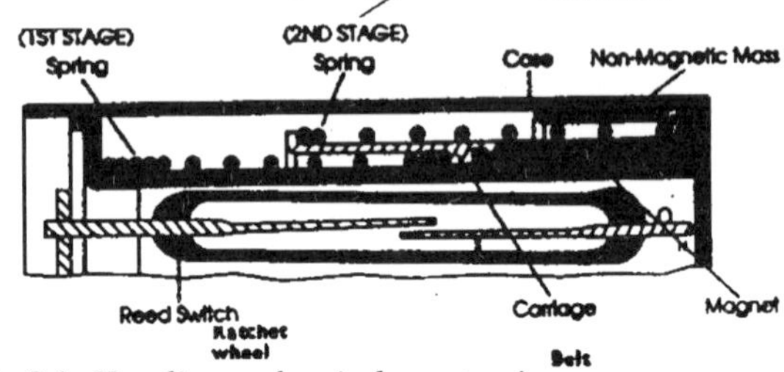

Fig 24: Hamlin reed-switch pretensioner

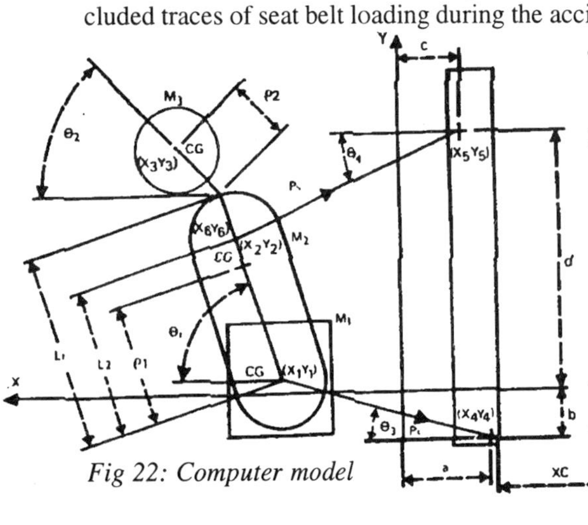

Fig 22: Computer model

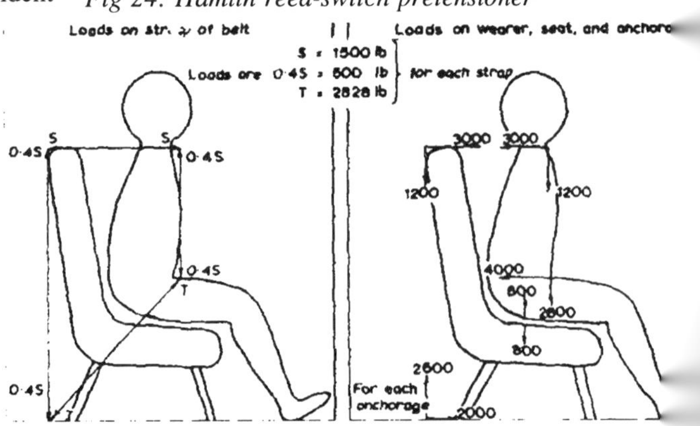

Fig 23: Peak loads on occupants

situation. The 0.1 second or so duration of impact is sufficient to use the energy absorption of the structures to the benefit of occupant survival. An earlier section suggested approximations for the shape of crash impact pulses. A deceleration time curve of a half-sine wave is found to be a reasonable approximation in calculating the movements of a restrained occupant in an impact situation.

Fig 19 shows calculation results for the motion of an unrestrained dummy. The relation of occupant movement to front end crushing of the vehicle is shown in Fig 20 based on such calculations. In the case of a seat belt with 1 in (25.4 mm) slack, at this amount of forward movement of the occupant, the car has crushed by about 13 in (330 mm) and 11 in (279.4 mm) remains — which is added to 14 in (356 mm) stretching of the seat belt. Body deceleration was found to be about 17 g — only about one quarter of that imposed by even a well-designed padded instrument panel. Even a collapsible steering column on its own was found to impose an acceleration of some 35 g after 10 in (254 mm) deformation.

There is always scope for improved design of the vehicle interior, with reference to the projected movement of the occupants on impact, to allow the belt as much stretch as possible. There is also scope for retractor mechanism designs which would result in a constant belt force during impact rather than a rising one. At the time of the research it was suggested that such work could result in belt protection for occupants with impact speed raised from a limit of 35 to 50 mph. The calculations were based on assumed occupant deceleration of 30 g and a 17 inch survival space in front of the occupant — together with a vehicle front-end crush distance of 40 inches. There is also great scope for improved comfort and convenience for the vehicle occupant and recent trends by such makers as Mercedes Benz, to incorporate belt anchorages into the seat structure, for open-roadster applications, could also have important general vehicle usage spin-offs.

The TRRL caried out work to establish whether belt systems designed to BS 3254 could be improved upon. By simplifying assumptions to make vehicle crush and belt stretch stiffnesses both linear a convenient analysis can be carried out. Such was the basis for the characteristics obtained in Fig 21 for different assumed impact pulses shown. Fig 22 shows the kind of computer model used to predict impact levels between occupant and vehicle interior when the survival space is insufficient or to demonstrate the reduced impact loads arising from the use of the seat belt. Subsequent analysis has been carried out which allow for the non-linearities in the various stiffnesses. Fig 23 shows peak loads on occupants calculated on the basis of a 50 g occupant deceleration impact. A recent example of a system introduced for pre-tensioning of seat belts or actuating an air bag uses the Hamlin 59500 series sensor based on a reed-switch, Fig 24. It incorporates a coil spring and torroidal shaped ferrite core. Under a shock force, its magnet deflects against the spring until positioned over the reed switch. Magnetic force then actuates the switch to deliver a signal when a shock force of a particular magnitude is reached. Operate time and

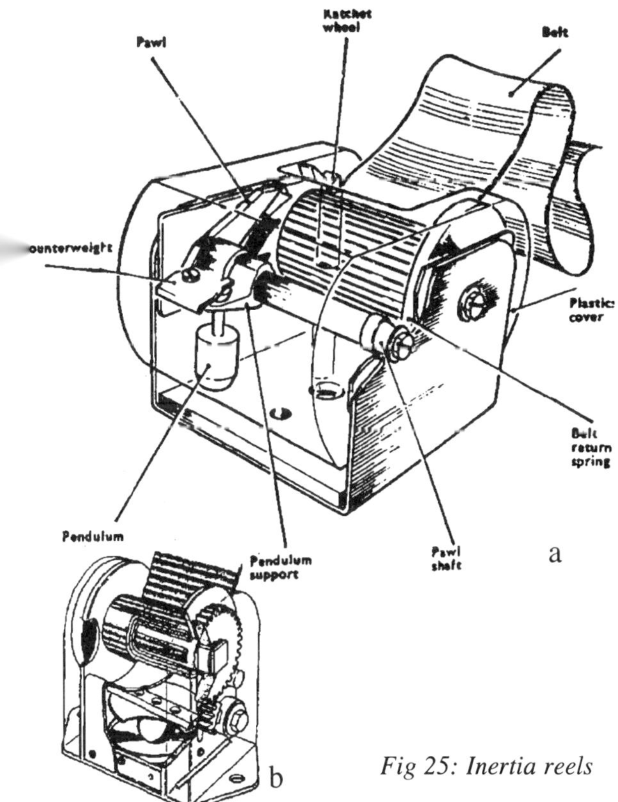

Fig 25: Inertia reels

Fig 31: Roll retention

threshold sensitivity can be preset. The development of inertia reels is seen in the examples of Fig 25. At (a) is an early Kangol system showing the separate belt and vehicle inertia mechanisms acting in the webbing shaft cog. A more recent design is the Britax Kolb system seen at (b) which has spindle locking built into both sides of the frame yet uses the minimum number of moving parts, to improve reliability. Performance data is summarised in Fig 26.

Recent design of air-bag restraint is exemplified by the Honda system of Fig 27. This is of the so-called progressive-power type and comprises four sensors to detect impact in frontal collision; a compact propellent-gas generator in the steering wheel hub with the folded air-bag. There is also a subsidiary sensor, Fig 28, in the passenger compartment which fires at 2 g rather than 12 g for the frontal sensors. Relative settings are based on the crush resistance of the vehicle to allow for time delay between front-end and interior response to impact. The inflater comprises an igniter (zirconium potasium nitrate plus a heater) and sodium nitride to generate hydrogen. The 60-litre air bag is of nylon cloth with a rubber-coated inside and it inflates in 0.2 seconds.

Cargo restraints

The high masses associated with many goods cargoes result in very high inertia loads when sudden braking takes place and in special-purpose vehicles particularly, with greater design freedom to match the

Locking distances
car sensitivity up to 0.3 g — no locking
at 0.45 g — locking within 25 mm
at 0.7 g — (same)
webbing sensitivity up to 1.0 g — no locking
at 1.5 g — locking within 25 mm
angle sensitivity up to 17 deg — no locking } **in all directions**
at 21 deg and more — locking
Outer dimensions
height: 85 mm (with 2.1 m stored webbing: 86.5 mm)
width: 86 mm
depth: 62mm (with 2.1 m stored webbing: 67 mm)
***Webbing capacity* 2.1 m (webbing thickness 1.2 mm)**

Fig 26: Reel performance data

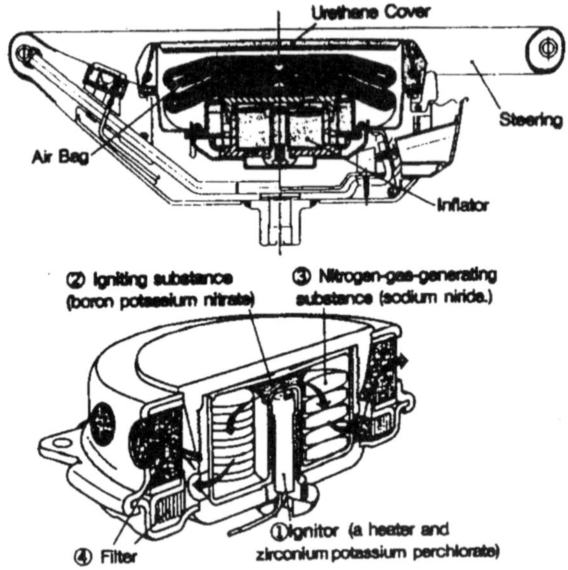

Fig 28: Sensors for air-bag firing

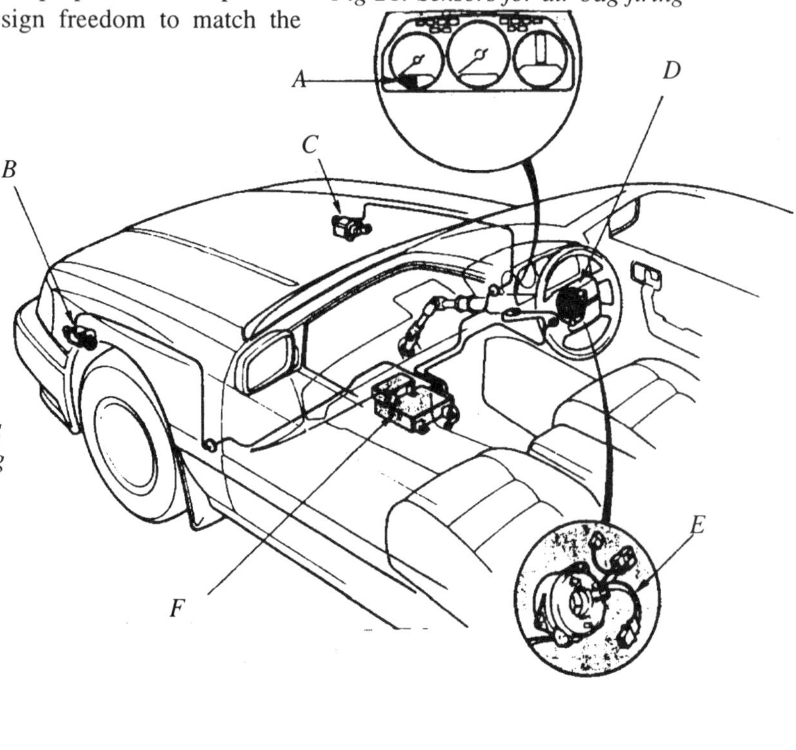

Fig 27: Honda air-bag restraint — A. warning light, B. front service (left), C. front service (right), D. air bag inflator, E. cable unit, F. diagnostics unit

vehicle to the payload, there is considerable scope to make loads totally secure, even in the collision situation. The importance of lateral acceleration on turns and even horizontal acceleration in sudden 'take-offs' should also be accounted for. Any shifting of the load, en route, can also create serious stability problems due to the alteration of inter-axle load distribution. In the UK alone there are said to be 4500 successful prosecutions for unsafe loads per year — a figure that might rise to 10,000 when undetected cases are included.

The scale of the problem is such that legislation could well force the design of special purpose vehicles for such potentially dangerous loads as paper rolls and even steel girders. Specialist coil carriers are already widely used for heavier rolls of steel strip and horrifying reports have been heard of items like steel bar piercing the back of driver's cabs on open platform lorry-bodies, under emergency decelerations.

Construction and Use Regulations are clear in the legal requirement to ensure safety and stability of payloads. One of its clauses insists on suitability of use for purpose. While the interpretation of this is clear for such loads as liquids and bulk 'solids', it is less obvious in the case of awkwardly shaped loads — other than abnormally large ones requiring transport under Special Types orders.

Of considerable help is the Department of Transport publication 'Safety of loads on vehicles' which provides a code of practice for construction requirements of load retention equipment. The requirement is to withstand inertia forces equal to the load weight, forwards and to half the load weight sideways and backwards. Reliance on the headboard of an open-platform truck may not be sufficient and it is extremely difficult to have a standard design to meet a wide variety of payload requirements.

The code recommends varying the number of anchorage points for different types of payload. Some guidance is given in Fig 29 for reasonably symmetrically shaped cargoes according to overall dimension. Useful advice is also given on sheeting, ropes, chains, strapping and nets for general cargoes. More specific practices apply to rolls and coils (Figs 30 and 31), sacks and bricks.

A specific British Standard (BS 5073) covers the special case of loads inside containers while the Health and Safety Commision have published codes, available from HMSO, for dangerous substances. While many of these are specifically operator requirements, some fall squarely in the court of the specialist vehicle designer. An important case in point are low-loading machinery transporters. The particular case of load restraint in light vans has been covered in MIRA research reports made public in recent years. These include a code of practice covering bulkheads, load anchorages, storage bins and restraint systems.

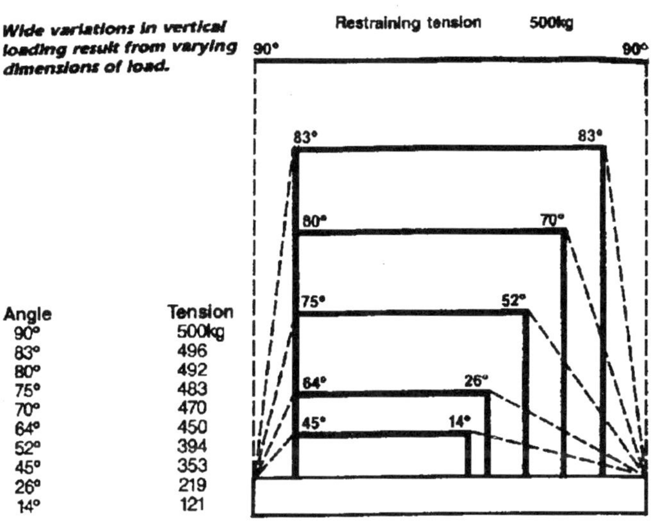

Fig 29: Anchor points for cargo restrait

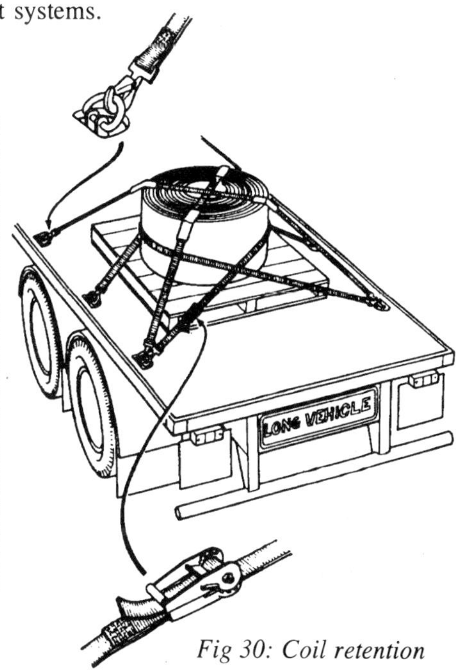

Fig 30: Coil retention

Passive restraint systems

Long experience with passive restraint systems for vehicle occupants in the USA has led automotive engineers to be wary of public reaction to a variety of restraint systems which either cause inconvenience or add considerable extra cost to the vehicle. Legislation is likely to increase for all categories of vehicle as the years roll by so it is prudent to examine the rival systems in some detail.

The requirement for passive restraint systems on 1987 model-year cars in the USA stimulated the considerable development of different solutions and brought into focus the rivalry between air-bags and passive seat-belt systems. At the time of the impending legislation, GM in the USA examined a variety of seat-belt systems which were 'passive' in the sense of not requiring a conscious effort by the user when entering or leaving his or her vehicle. These are described below after consideration of different air-bag solutions.

A representative air-bag system is that developed by Honda and known as the SRS in which a

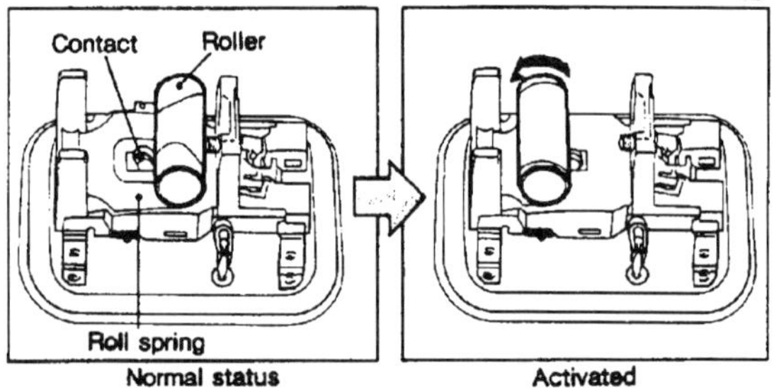

Fig 32: Honda SRS sensors

sensor unit is attached to each side of the body front-end. These comprise a roll-spring, roller and base, enclosed in a sealed casing, Fig 32. With impacts over 12g (equivalent to a frontal- barrier impact of 16 kph), the roller actuates the switch. Two cowl sensors, of similar design, are enclosed in a diagnostics unit on the tunnel structure between front-seat occupants. These sensors switch at 2g typical of a deceleration of this part of the structure after the front-end has collapsed.

The steering-wheel hub inflator was introduced and illustrated in the previous section; it

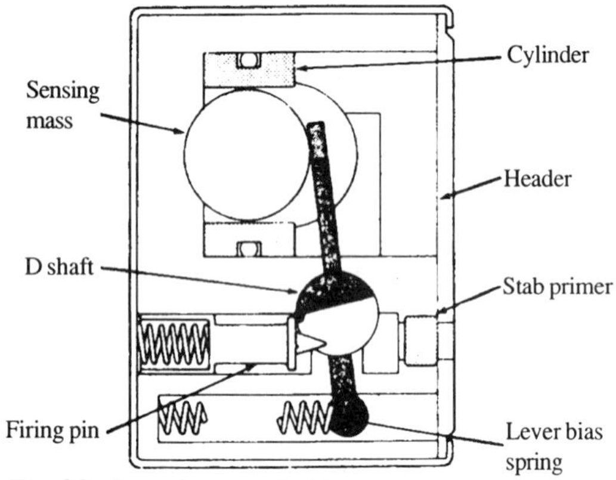

Fig 33: Breed sensor/initiator

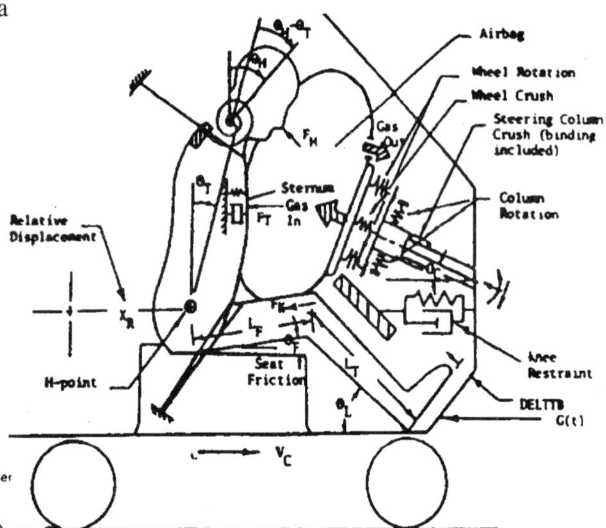

Fig 35: Computer model

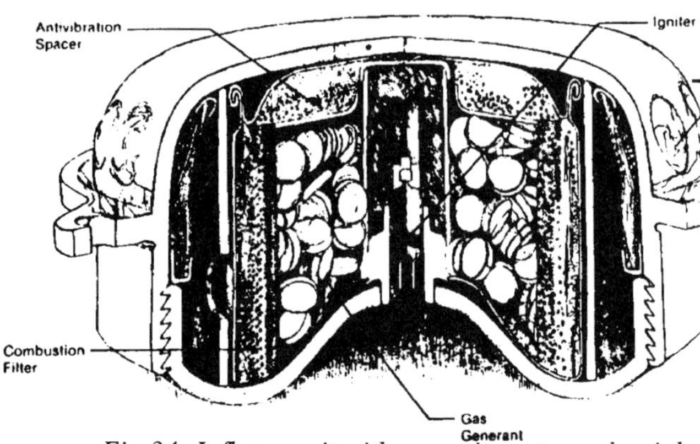

Fig 34: Inflator unit with gas exit ports on the right

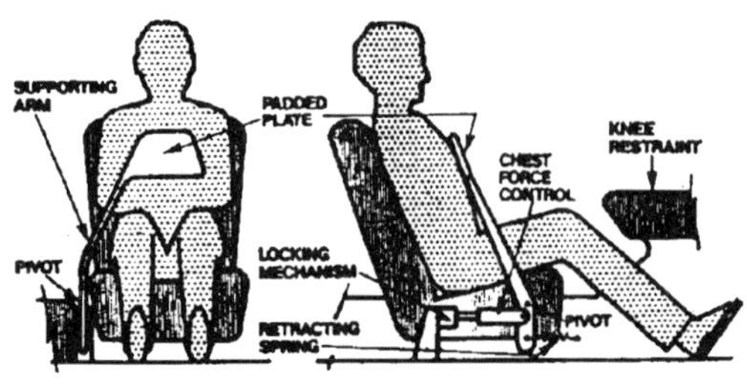

Fig 39 (below): Cushion restraint

consists of an igniter (heater and zirconium potassium perchlorate) and a nitrogen generator (sodium nitrate). When both front-end and cowl sensors record impact, the inflator activates and the chemicals 'burn' to inflate the air-bag. The bag is of nylon cloth with an inside rubber coating. The nitrogen fills the 60 litre bag, two vent holes releasing gas to prevent too high an impact on the driver's face.

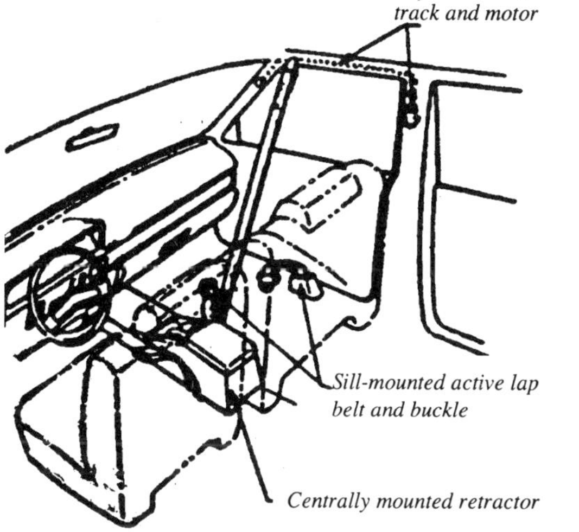

Fig 36: 2-point passive belt

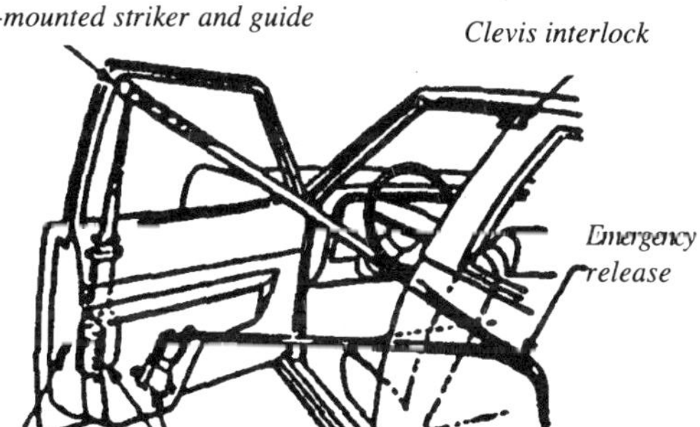

Fig 37: Door-mounted 2-point belt

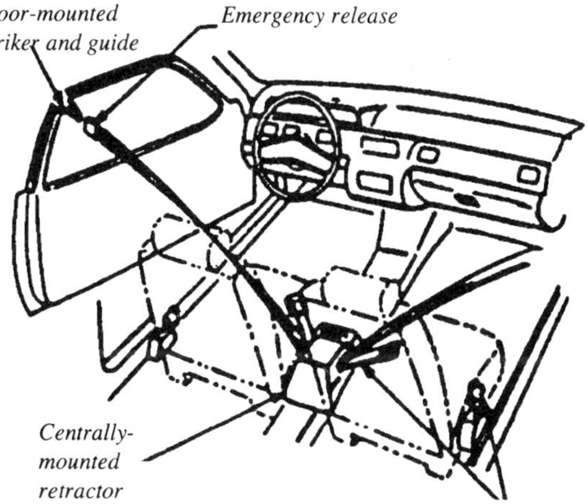

Fig 38: 3-point passive belt

Simplicity in design and operation is obviously the key to guaranteed working of air-bag systems. equipment is only effective if the timing of the bag inflation is appropriate to the events of the very fast impact cycle which occurs in collisions. This has led to the use of computer simulation of both occupant movement and dynamic structural deformation of the vehicle in a wide variety of accident situations so that the air-bag effectiveness can be optimised.

This procedure was used in the development of the US Breed Corporation air-bag system whose crash sensor initiator is shown as Fig 33. If a deceleration of over 4g occurs the ball-sensing mass moves forwards against the spring. The ball will push the bias lever far enough to release a cocked firing pin to initiate the gas generating material for inflating the air-bag. Two such sensor-initiators are built in to each system to provide high statistical reliability. The system differs from conventional ones in that, at the time of its introduction, most of the latter used electrical switches in the crush zone of the body, with associated electrical-circuits and diagnostics, to signal inflation of the bag.

Pyrotechnic sensors for air-bags have been developed for US systems by Morton Thiokol Inc. Through a percussion mechanism a thin-layer explosive (TLX) transfer line is initiated to transfer a pyrotechnic shock wave to the inflator at a speed of some 4500 ft (1372 m)/sec. The sensor is mounted to the air-bag module housing next to the inflator, and on the opposite side is a cavity containing the energy storage capacitors. Power from the horn circuit is used to charge the capacitors and when the sensor responds to a crash impact it closes a set of electrical contacts to let current pass to the squib which fires the inflator, Fig 34.

Considerable debate still takes place about the choice of electronic or pyrotechnic firing, also about the best position for the sensor in the steering wheel. A number of companies have developed computer models which help in the selection of such design criteria. Fig 35 shows the Volvo VDRACR model which is a relatively simple 2-dimensional representation using lumped masses, the air-bag being modelled as an ellipsoid.

Non air-bag solutions

The body-mounted 2-point passive seat-belt system, Fig 36, uses a single retractor mounted to the centre of the vehicle with the upper, shoulder, belt anchorage located on thee upper frame or B-pillar. Knee bolsters are required with the system and the upper anchorage requires a means of powering itself along a guide track to facilitate entrance into the vehicle. The door-mounted 2-point system, Fig 37, again uses a single retractor mounted on or near the central tunnel — and has an emergency release latch mounted in the door area. Knee bolsters are required, too, and the upper anchorage is door-mounted to clear the entrance area as the door is opened. A considerable challenge presents itself with both 2-point systems in providing protection for a wide variety of occupant sizes.

The alternative is a 3-point door-mounted system having two door-mounted retractors for both the shoulder and lap belts, Fig 38. This arrangement allows a third, central, seating position with just the two outer occupants having the passive belt facility. Knee restraints are usually not required and dynamic performance is similar to an 'active', manually fastened, belt system given proper location of the anchorages.

A third major category of passive restraint is the automatic cushion restraint comprising an arm hinged from the central tunnel or console. This supports a chest pad held by light spring pressure against the occupant and used in connection with a knee bolster. The pad allows free movement in normal travel but locks up on impact. Fig 39 shows an early system developed by Accles Britax in which an energy-absorbing collapsing tube device is used to control the movement of the arm and resulting chest loading when locked-up above 6g deceleration.

An interesting derivative of this idea is the integrated child safety seat devised by Volkswagen which the company designed into one of its concept cars. The right hand rear seat of the Futura car has a leading cushion-edge which can be lifted upwards and positioned in front of the child to act as an adjustable safety restraint, Fig 40. This transverse bolster can be moved forward and backward and is unlocked with a lever underneath the seat cushion to spring back to its original position.

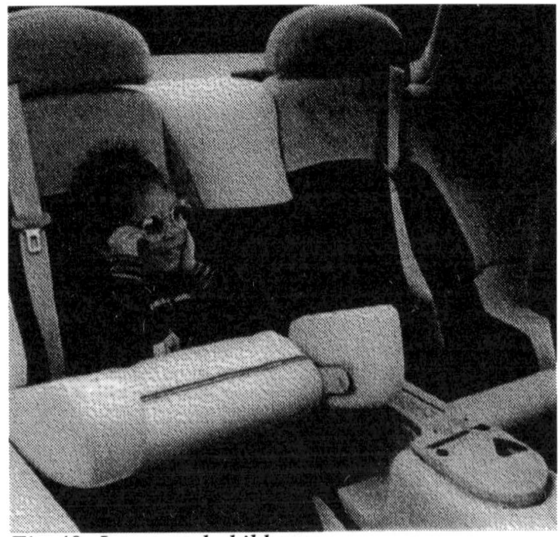

Fig 40: Integrated child-seat

Side-impact protection

Differences in available survival space mark a major difference between frontal- and side-impact analysis. Compute simulation techniques are described which are tending to simulate structural deformation and occupant movement simultaneously.

In 1990 a federal standard for side-impact protection was introduced in the USA (FMVSS 214) and a proposed standard is now being discussed in the European market for 30 mph impacts. Experts consider that the restraint system cannot be considered separately from the crash pulse, as in front-impact analysis, because of the limited space available to absorb impact energy. A single analytical technique is thus favoured to simulate both structural collapse and occupant kinematics.

In a car or cab structure key parameters are door interior stiffness and topography, door to occupant initial separation, door intrusion profile after impact and door to bodyshell interaction. In determining the door intrusion profile the structural analysis requires the collapse velocity of the vehicle's side structure and the compliance of the vehicle structure adjacent to the occupant's body. It is also necessary to relate occupant contact time and distribution of contact with inner structure compliance.

Test work at the Cranfield Impact Centre has resulted in simplifying the shape of the velocity pulse at the occupant's chest area, for analysis purposes, to that shown in the straight lines in Fig 41. Before point A, rate of change of velocity with time is -35g;

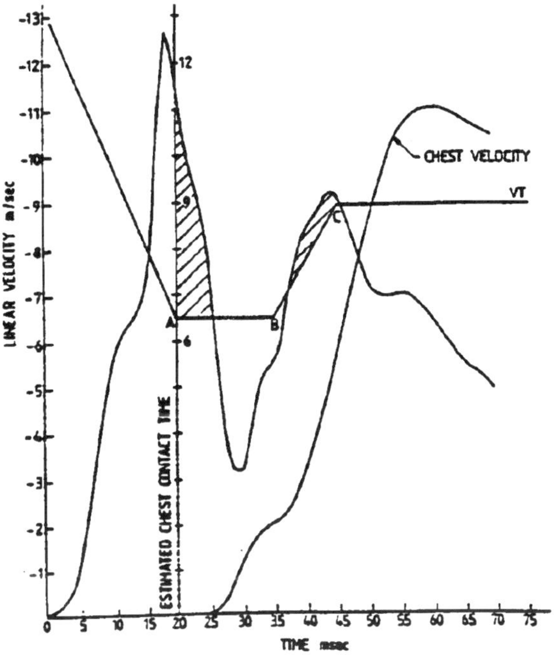

Fig 41: CIC standard impact pulse

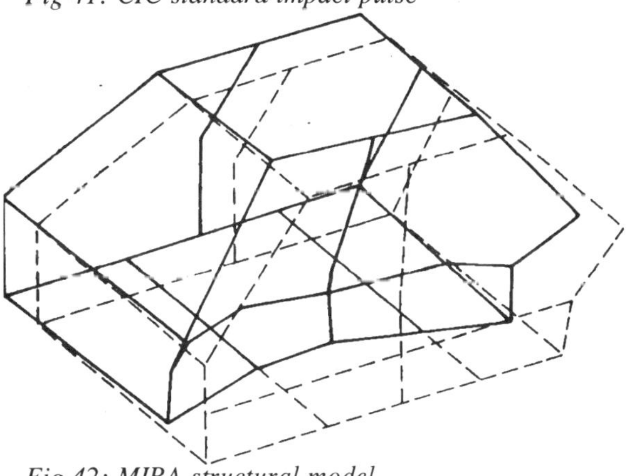
Fig 42: MIRA structural model

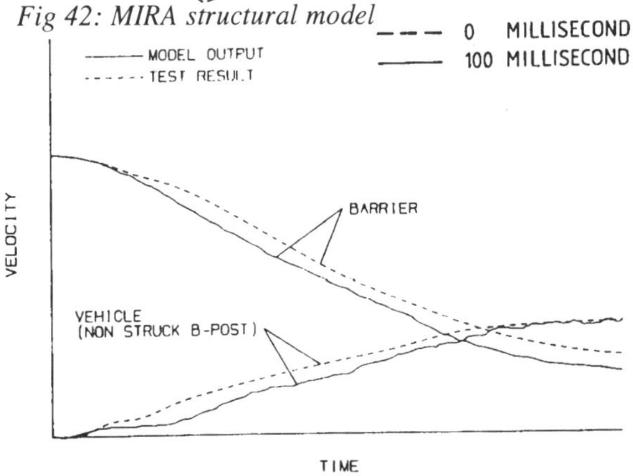

Fig 43: MIRA model verification

change in time in ms is 15 and 10, and change in velocity in m/s is 0 and 2.5, for AB and BC respectively. The resulting velocity template, obtained after testing, proved valuable in identifying the type of collapse velocity pulse likely to cause serious injuries. For a given geometry and inner wall compliance, if the side structure velocity remains below that on the template then injury level is likely to be low. Initial peak velocity influenced chest acceleration while the second peak influenced rib deformation. Tests showed that by placing an occupant next to a door, scope for improved protection through better door design and use of energy-absorbent material will be significantly enhanced.

Modelling of side-impact has been carried out by MIRA using the DYNA3D package involving distributed masses to represent the inertia of the structure. Fig 42 shows the simulated structure before, and 100 ms after, impact. The model included a simple representation of the Eurosid dummy. Good correlation with crash test results is seen in Fig 43. A separate detail analysis of occupant kinematics was used for a detail study of injury levels using the MADYMO package, with the time history of structural deformation and vehicle acceleration taken from the first, structural, simulation, Fig 44.

MIRA investigated a number of structural elements in detail as shown in Fig 45. The B-post has a tendency to fail at waist rail height and this was a prime candidate for reinforcement to prevent premature buckling failure. Non-linear properties of these structural elements were determined by quasi-static tests on the individual elements. Comparisons of door velocity and intrusion before and after modification are shown in Figs 46 and 47 taken from full-scale dynamic sled-testing which takes account of strain-rate effects. Note that a 30% reduction in average door velocity was achieved while the dummy was in contact with the door. Because peak injury levels have been found to take place about 15 ms after initial dummy/door contact, it is very important to reduce velocities during contact and not just at the initial instant of contact.

Earlier work by the Cranfield Impact Centre, with Ford showed how crashworthiness criteria could be included in concept design. The study started with only the interior package data for a future 3 or 5 door hatch-back car for which structural behaviour of member joints and B-pillar deformation, from tests on an earlier vehicle, were used. A simulation program called SIMSIM1 was used to obtain the collapse mechanism of the structure when impacted into a standard lateral test barrier. Figs 48 and 49 show the respective 3 and 5 door vehicles and their respective structural models.

The work was able to quantify the structural contribution of each failed beam element by calculating the crash energy absorbed, Fig 50 . This showed that the bottom of the B-pillar (A) in the 3 and 5-door designs absorbed up to 25% of the crash energy, resulting in extensive deformation of the B-pillar. Next stage was to limit the side structural collapse by distributing more even loading by the impacting vehicle into as many beam elements as possible, to obtain the optimised results shown in these figures. This included improving properties of the B-pillar (A), rear-seat pan (B) and adding a door intrusion beam (C). The figures show the percentage contribution to crash energy absorbed of each member throughout the structure, the reduced B-pillar contribution ensuring reduced deformation of the B-pillar and hence greater survival space. In fact the mid-span deflection was reduced 147 to 70 and 175 to 67 for 3 and 5 door vehicles respectively. The table in Fig 51 shows the resulting reduced injury levels to the occupants.

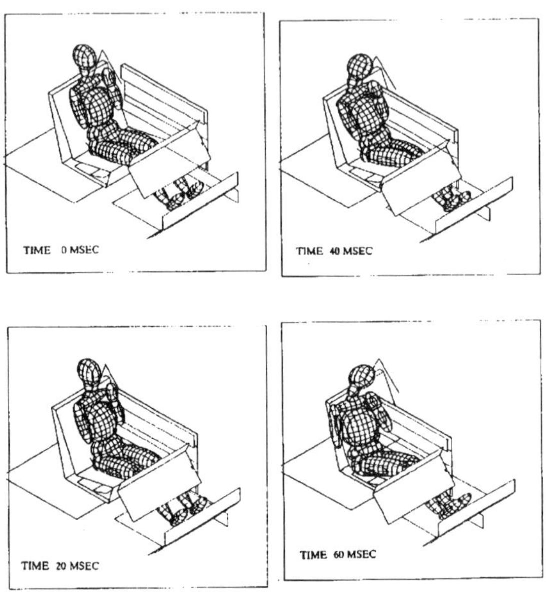

Fig 44: MADYMO simulation of occupant motion

The SIMSIM1 program is a coarse-mesh beam-element non-linear finite-element code suited to side collisions which essentially involve beams in bending. Compression and torsion can also be simulated and the program is hybrid in the sense of being able to analyse elastic behaviour in traditional beam theory as well as non-linear material behaviour from test-derived experiments. It can simulate large deformations plus local and global collapse mechanisms. Simulation of non-linear behaviour includes both

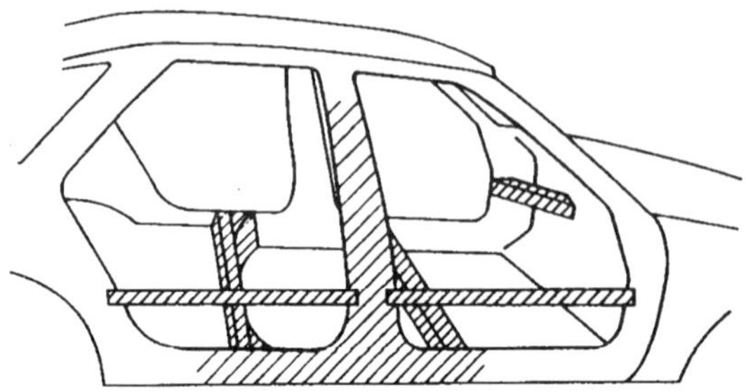

Fig 45: Elements investigated

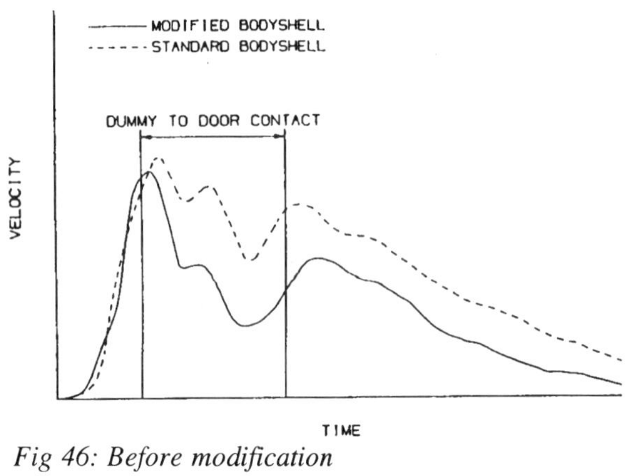

Fig 46: Before modification

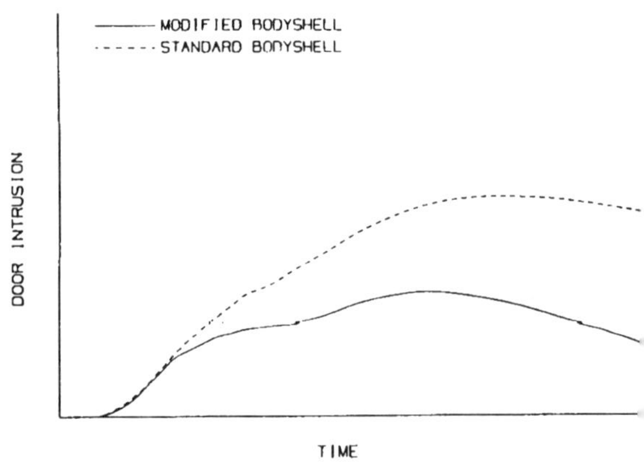

Fig 47: After modification

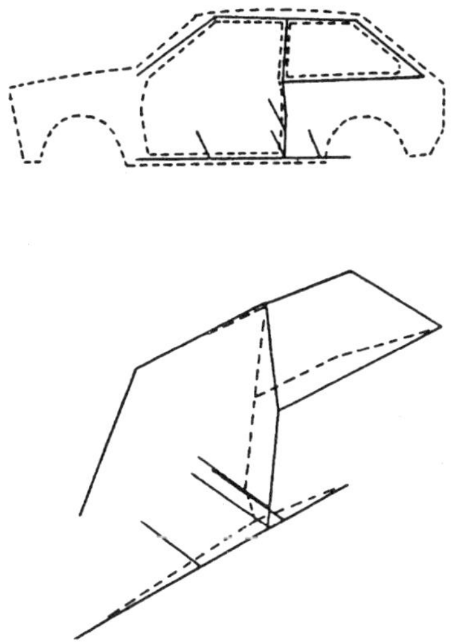

Fig 48: 3-door model

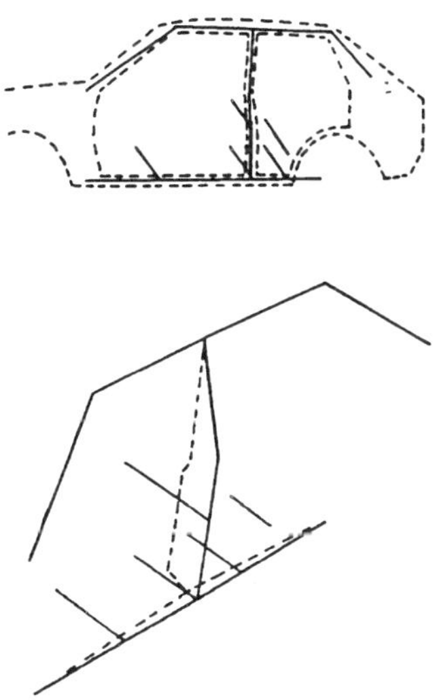

Fig 49: 5-door model

`perfectly-plastic' (beam can maintain load-bearing capacity after failure) and `moment drop-off' (beam's supported load falls off rapidly after failure). The finite stiffness of the joints can also be taken into account; these can otherwise significantly degrade the analysis results.

Another program employed by Cranfield Impact Centre is SIMSIM2 which can predict the rigid body motions and vehicle side deformations in a 90 degree side collision. This is achieved with a model comprising a series of unidirectional spring-mass systems representing both vehicles as they contact one another. General equations of motion, together with impacted vehicle tyre and suspension data, are used to predict the whole vehicle motions after impact.

CVS Analysis	Initial 3 door Vehicle	Optimized 3 door Vehicle	Initial 5 door Vehicle	Optimized 5 door Vehicle
Max Velocity of Side Structure (m/s)	12.5	10.7	15.2	12.7
Velocity of Side Structure on Contact with Dummy (m/s)	12.5	10.2	15.1	12.2
Velocity of Side Structure at Chest Velocity of 0.1 m/s (m/s)	7.9	4.7	9.9	7.2
Head Injury Criteria (HIC)	138.0	67.0	224.0	94.0
Max Lateral Head Acceleration (g)	29.5	21.2	31.8	22.7
Chest Severity Index (CSI)	375.0	154.0	497.0	229.0
Max Lateral Chest Acceleration (g)	67.3	49.5	76.2	51.8
Max Lateral Chest Velocity (m/s)	10.5	7.8	12.5	8.9
Max Lateral Pelvis Acceleration (g)	110.0	80.6	117.5	83.9
Max Lateral Pelvis Velocity (m/s)	12.0	8.5	14.4	9.8

Fig 51: Reduced injury levels

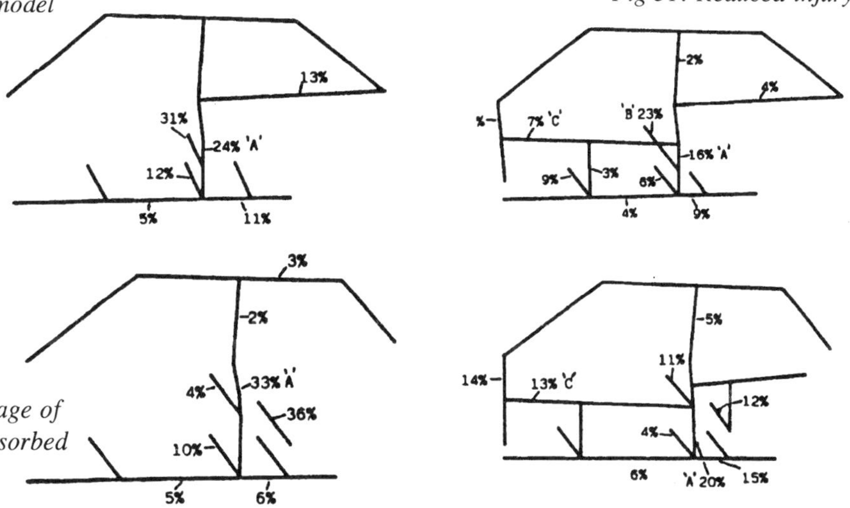

Fig 50: Percentage of crash energy absorbed in each element

Bumper systems

Large frontal impacts account for two-thirds of occupant injuries in motor vehicles and any contribution the bumper system can make to lessening this incidence is clearly worth striving for. Increasingly plastic-foam systems are being used to fulfil the major energy-absorbing role. Here a technique of optimising foam performance is suggested and an interesting way of combining foam with metal structures to deal with different parts of the impact speed spectrum.

While direct energy absorption capability of the bumpers is limited, careful design can ensure the protection in minor impact of those more important controlled-collapse structures, both at the front and rear of the vehicle, which do make a major contribution. They can also offer protection to other safety systems such as lamps, locks and latches for the various cover panels whose correct working could otherwise be jeopardised by minor impact. The prescribed dimensional envelopes for the positioning of lights on vehicles, plus the requirements of the engine cooling air inlets and rear hatch openings restrict the space available to fulfil these functions and of shielding such items as fuel filler pipes and lighting harnesses. Another factor which affects design is the incidence of minor damage to the sides of the vehicle as streets become more congested and parking more hazardous. This is leading to increased consideration of all-round bumper systems. Front and rear fascia panels are a further trend in which the bumper is integrated with 'soft' end structures for the vehicle. Design of such front/rear ends requires particular attention to the minimum legal impact absorption requirements. The 5 mph capability set initially by the US market, in FMVSS part 81, has been important in this respect.

Usually shock absorption may be accomplished by such materials as moulded foam, or honeycomb cellular matrices, attached to the back-up beam — historically made from steel. Hydraulic dampers in the place of these materials are another solution. Attempts have, of course, been made in volume-produced cars to use moulded composites in place of the steel bumper beam, Fig 52. Such a system is designed to absorb energy by the foam interposed between beam and fascia, the required characteristics for the beam being shown in Fig 53 and a typical reinforcement scheme for the composite being seen

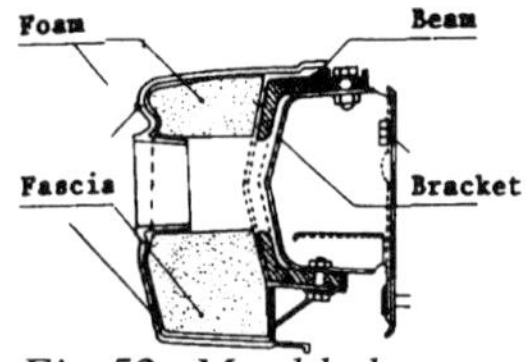

Fig 52: Moulded composite bumper

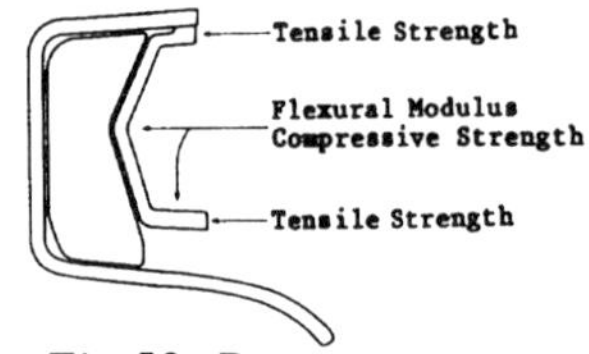

Fig 53: Beam characteristics required

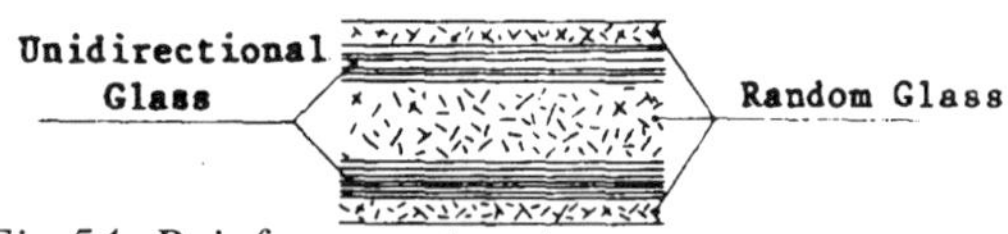

Fig 54: Reinforcement system

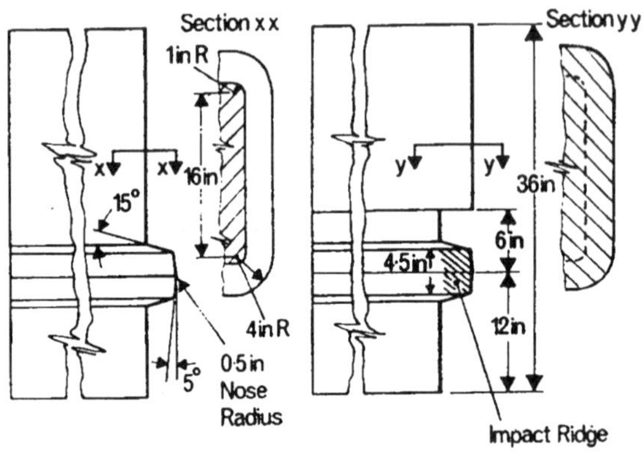

Fig 55: Test pendulum

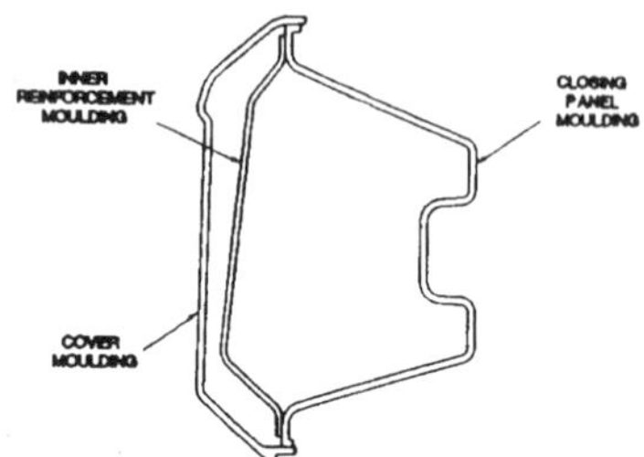

Fig 56: Cross section

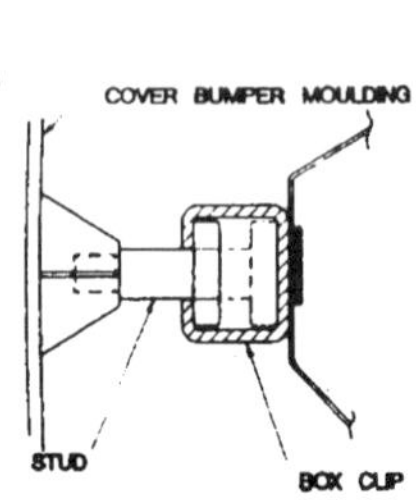

Fig 57: Side fastening

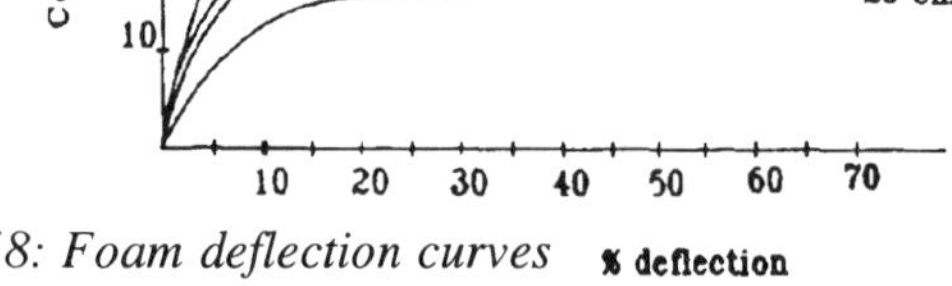

Fig 58: Foam deflection curves

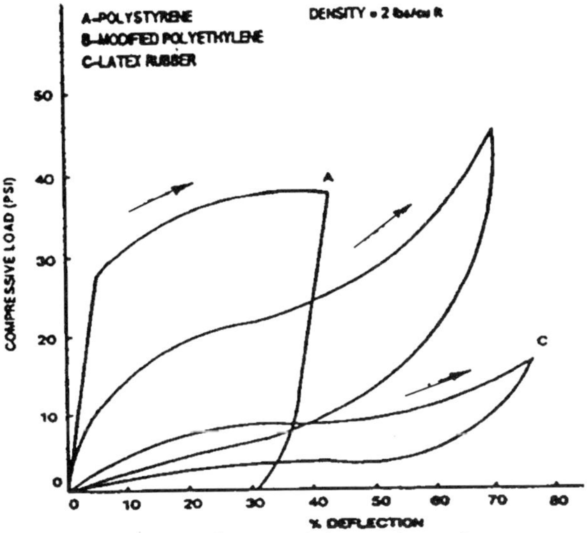

Fig 59: Foam characteristics compared

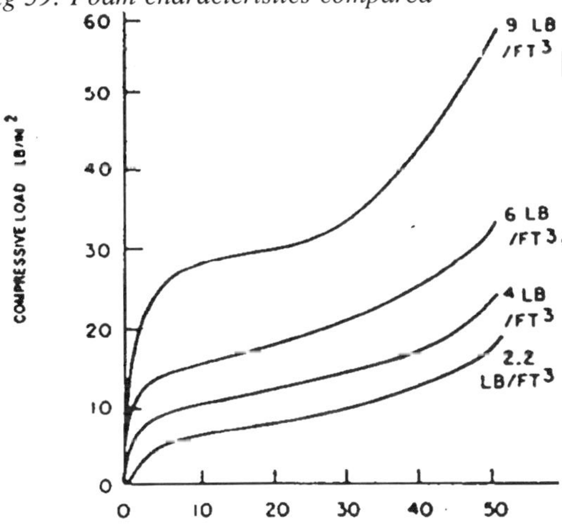

Fig 60: Density effect

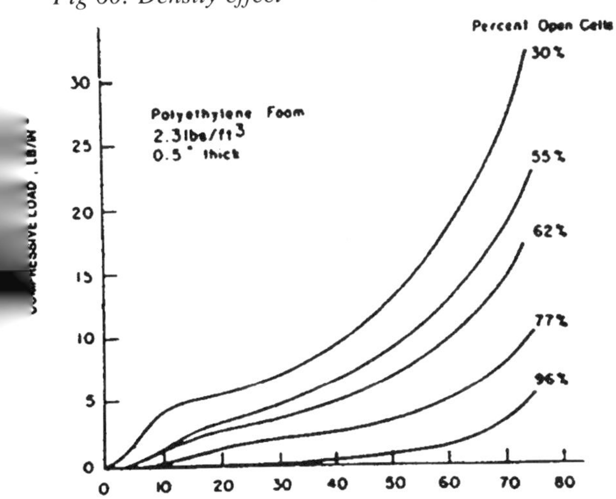

Fig 61: Grain effect

in Fig 54. An earlier US standard, FMVSS 215, gave guidance on the dimensioning of bumpers to restrict damage to lights, and prevent over-riding of bumpers, causing vehicles to become interlocked in an accident situation and leading to punctured coolant radiators. Associated with this standard was the introduction of pendulum tests. Fig 55 shows the type of pendulum head shape required for simulating an impacting vehicle. The test involved three 5 mph impacts at 20 in (508 mm) above the ground with type-b and three with type-a at 16-20 in (406-508 mm) from the ground. Corner impacts at 3 mph were also mooted at the time of the code's introduction. Pendulum mass is equal to the unladen weight of the vehicle.

In Europe, ECE 42 sets the legal requirements for bumper design which specifies 4 kph pendulum impact tests. An example of a system designed to that standard is the Rover Montego unit which broke new ground in being integrated with a spoiler and being finished in body colours. Weight requirements limited the thickness of the combined cover moulding to 3 mm and the part had to withstand 2 hours at 130 C in a stoving oven; material chosen was Bayer Pocan modified polybutylene-teraphthalate. The centreline cross section is shown in Fig 56, the structure comprising a box section simply supported at its longitudinal mountings corresponding to the main longitudinal members of the body. The elements of the box beam were assembled together using a copper/polyester sacrificial tape heated to form the weld. The wrap-around portions of the bumper are fixed to the sides of the vehicle by means of the fastenings seen in Fig 57, involving a boxed clip, supporting a long shanked stud on the bumper. This allows an expansion of 25 mm on the full width of the bumper. For the longitudinal fixings at the front of the vehicle, these are pre-loaded by a double coil spring washer and a shouldered bolt. A nylon washer plate is sandwiched between the steel mounting bracket and the PBT mounting surface of the bumper closing panel, allowing the surfaces to move over one another during expansion. In developing polypropylene foams for bumper cores, Montedison-Bollate, Via Principe Eugenio 1, 1-20155 Milan, have established a calculation technique for the isotropic cellular structure of the foam. The elastic modulus E relates to the foam density ρ, with m a constant, according to $E = K\rho^m$.

Fig 58 shows load deflection curves for different plastic foam thicknesses; the threshold line represents the requirements of a particular bumper system showing the compromise that has to be made between choosing thickness to obtain constant load for a given (plastic) deflection and sufficient percentage defection (and recovery).

Fig 59 plots plastic foams and latex foams at the extremes of these requirements against a good compromise modified polyolefine foam. Obtaining the compromise behaviour is a function of foam density, open-cell content and polymer phase composition as seen in Figs 60 and 61. The objective is to obtain a rectangular cellular structure of gas-filled closed cells. Fig 62 gives a mechanical model of the foam and the effect of cell size while Fig 63 shows the wide variety of characteristic curves which can be obtained. Compressive stress in a foam, stress-strain rate for which is nearly independent of the rate of deformation, is a function only of strain-energy required W_d to deform a sample, being given by the area under the loading curve, Fig 64: $W_d=\int_0^{\varepsilon} \sigma(\varepsilon)d\varepsilon$ for stress σ and strain ε. Velocity of the impacting masses is given by: $mv^2/2=mv_o^2/2-\int_o^{\varepsilon} \sigma(\varepsilon)d\varepsilon$ for impacting object mass m and v_o,v velocities before and after impact. Total energy lost during impact is predicted from the hysteresis W_h (area between loading and unloading curves):

$$W_h = m(v_o^2 - v^2)/2 = \int_o^{\varepsilon} [\sigma(\varepsilon)-h(\varepsilon)]d\varepsilon$$

Montedison have used these techniques to develop a new-concept bumper for high speed collisions (16 kph impacts to give no damage). Fig 65 shows how energy absorption increases rapidly with speed. The design was based on a 2-stage energy management system: flexible foam to provide absorption at impacts below 2.5 mph and a bow-structure replacing the conventional steel bumper beam, for higher speed impacts.

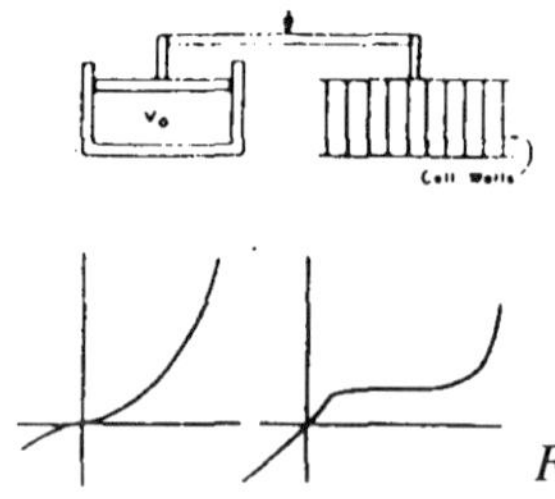

Fig 62: Foam model

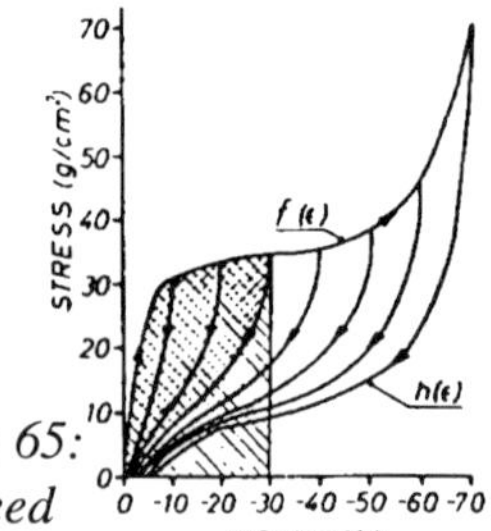

Fig 65: Speed effect

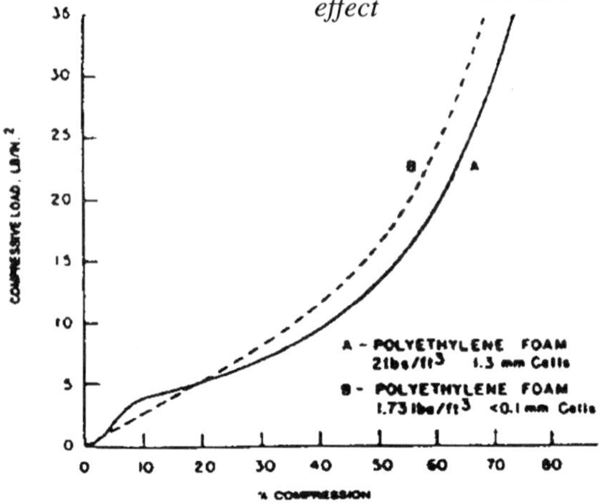

Fig 63: Different foam performances, kilojoules

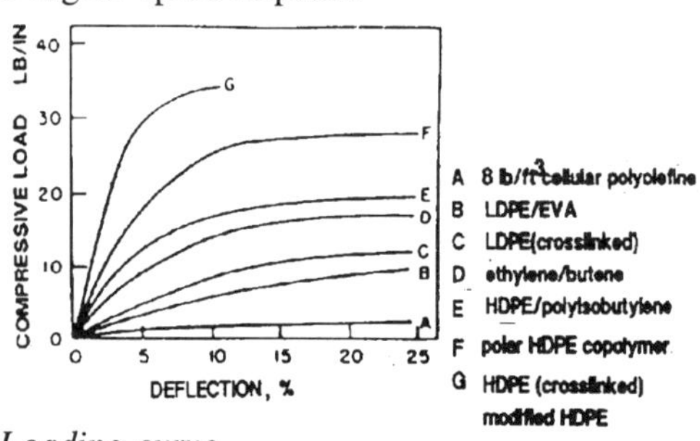

Fig 64: Loading curve

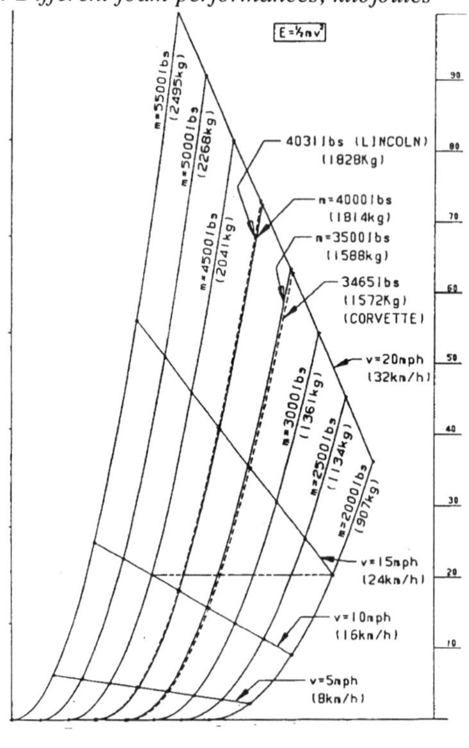

Fig 66: New departure bumper structure

Energy-absorbent foams

An understanding of the different types of expanded plastic foams is necessary for their successful use in energy-management. Rigid, semi-rigid and elastomeric PUR foams are compared here alongside bead foams from expanded polystyrene

ICI's USA polyurethanes operation is addressing the NHTSA regulations for occupant safety and the company argue that PU systems are increasingly used for impact energy management. A new range of EM foams has been developed to meet FMVSS 202 for head-impact and FMVSS 208 for knee impact. The phasing in of side impact requirements (FMVSS 214/216) over the next few years is also involving development work for applications which can involve impact energies of 150 to 450 ft lbs (203-610 J).

Foams classified as rigid, semi-rigid and viscoelastic are each being developed. The first absorbs energy by a non-recoverable crushing of the foam-cells whereas the second and third create a viscous dissipation of energy followed by a slow recovery. Typical processing parameters for all three types are shown in Fig 67.

Dynamic impact testing of rigid EA foams produces a force-deflection curve such as that in Fig 68, which reveals much about energy absorption characteristics. The area within the rectangle 1+2+3 is the maximum which the foam could possibly absorb — a measure of the foam's ability to minimise high peak loads. Area 1 is unrealised energy, area 2 that absorbed by the sample and area 3 the total rebound energy. The ideal efficiency is percentage of maximum energy and input efficiency the ratio of energy absorbed to the total energy input, the latter giving a measure of the energy absorbed by the foam *vs* the energy rebounded.

Typical impact curves are seen as Fig 69 a,b,c and d. As impact energy increases, penetration into the foam, as well as the peak load, also increases. While the foam exhibits fairly uneven deceleration of the impacting body, as the impact energy increases

Component Temperatures (°F)	75-95
Component Mix Pressures (psi)	1500-2000
Demold Time (min)	2-3
Machine Throughput (g/sec)	100-350
Mold Temperature (°F)	120-140
Mold Release Agent	Wax/Water-based

Fig 67: Processing parameters

penetration and peak load also increase. Effect of surface skin is seen as a sudden increase in initial load followed by a fall-off — a useful characteristic. The figure shows that the foam crushes evenly until it

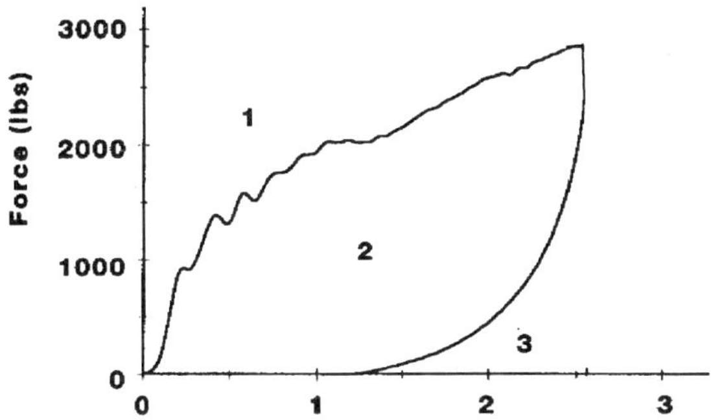

Fig 68: Force/deflection characteristics

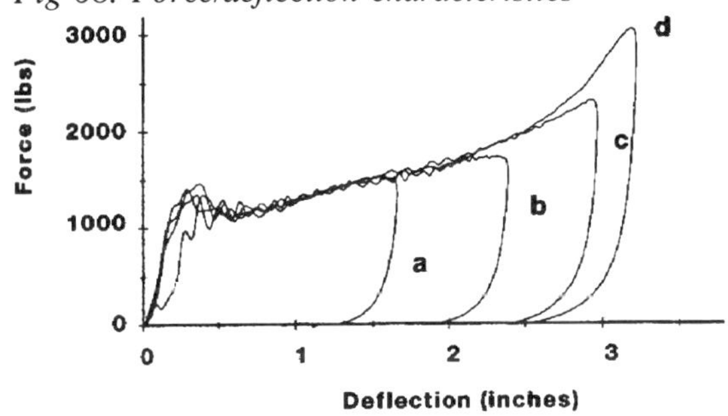

Fig 69: Typical impact curves

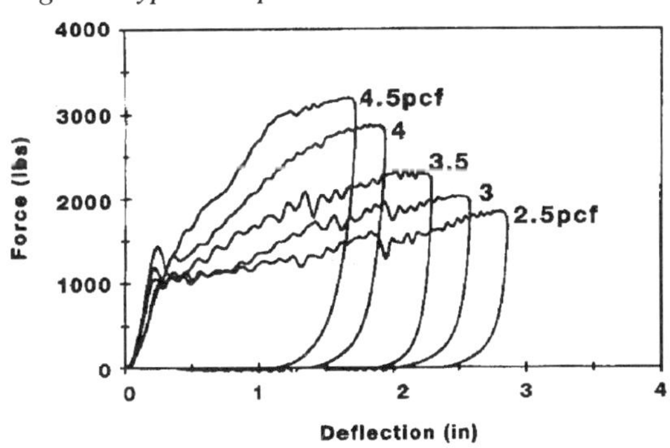

Fig 70: Effect of density increase

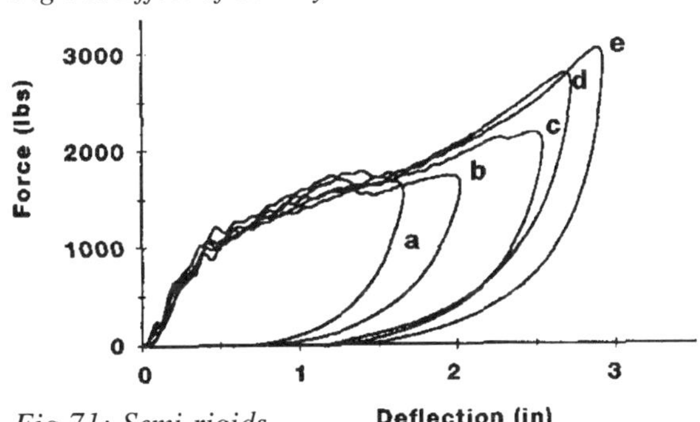

Fig 71: Semi-rigids

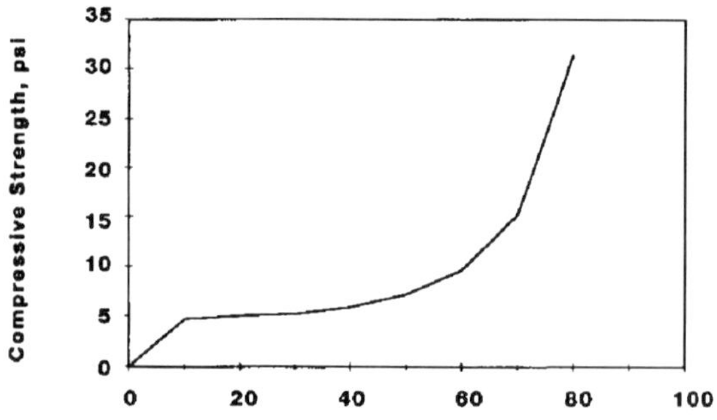

Fig 72: Visco-elastics

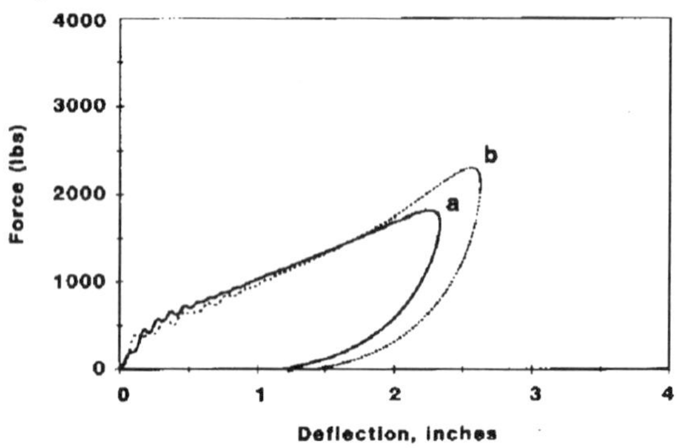

Fig 73: Dynamic impact curves

begins to stack up under the impacting body. Analysis of test results has shown the efficiency of the foam ranges from around 50-70% for 2.5 lb/ft^3 (0.04 g/m^3) material. Fig 70 shows the effect of increasing foam density, keeping the impact energy constant. Both density and formulation of the foam can be altered to tailor the foam for a particular application.

Semi-rigid foams are moulded at higher densities than rigids, generally. Characteristics for a 4.5 lb/ft^3 (0.072 g/m^3) system are seen in Fig 71. The increase in peak loads at higher energies is more pronounced because a flat impacting surface was used to obtain these particular results. If the striker has a rounded surface, representative of a head or a knee, the force tends to ramp up more evenly with increase in area during impact.

In the case of visco-elastic foams, a recovery is registered but it is not immediate. Static compressive strength properties are seen in Fig 72. Compressive strength is much lower than rigids and semi-rigids but it remains stable up to 60% strain. The dynamic impact curves of Fig 73 show that the foam absorbs up to 85% of the input energy with a relatively low

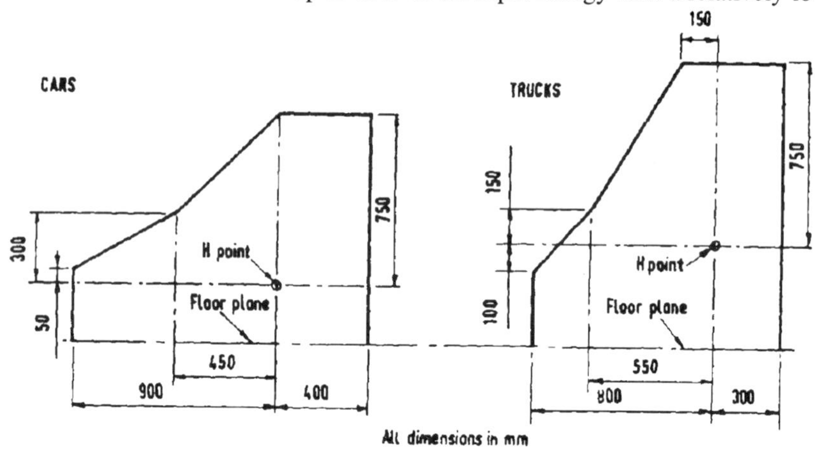

Fig 74: Survival spaces

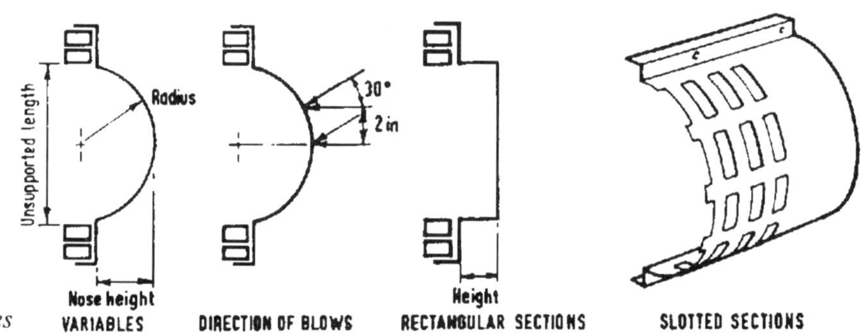

Fig 75: Armature shapes

level of rebound but does not achieve the `plateau' situation of the rigids. At lower levels of impact energy it does show 45-50% efficiency and its unique properties make it useful for such applications as anti-submarining ramps.

The design of crash pads involves the use of survival space envelopes, typical ones for cars and trucks being shown as Fig 74 for belted occupants. Armature test shapes are shown in Fig 75 and these can be designed to absorb energy in a way that relieves the foam of all the impact duty. Thinner layers of foam can then be used and the purpose of the foam becomes that of efficiently spreading the load over the surface of the armature. Human tolerance of impact loads is crucial to design studies and the maximum energy which can be absorbed is the product of that load and the total deflection of the padding.

Thermoplastics which are foamed at low densities while in the form of small particles can produce expanded bead foams which have been successfully applied to automotive interior components. One of the best known is expanded polystyrene (EPS), used in packaging, which is of course below the strength requirements for crash pads. Expanded high-heat copolymers such as styrene-maleic anhydride from ARCO Chemical are suitable candidates, however. The material is sold as solid plastic particles having a bulk density of 40 lbs/ft^3 (6.4 g/m^3). These contain the blowing agent pentane and the material transforms to a foam, of between 2 and 10 lbs/ft^3 (0.032-0.16 g/m^3) on heating. This foam is introduced into a mould and exposed to pressurized steam, the secondary heating causing further expansion and fusing together of the foam particles.

Knee bolsters are a common application in US cars where FMVSS 208 requirements call for a 30 mph vehicle crash into a fixed barrier with a maximum occupant femur or upper leg compressive load of 2250 lbs (1 tonne). At such vehicle impact speeds the knee typically impacts the bolster at about 12-15 mph with about 3000 in lbs (339 J) of energy per knee being dissipated by the bolster. Tests have shown S-MA foams to have 63% efficiency under these conditions. The Eagle Premier was the first full vehicle-width knee bolster made from the material, Fig 76. It consists of a steel back-up rail, the moulded foam section, and a PUR foam and vinyl skin.

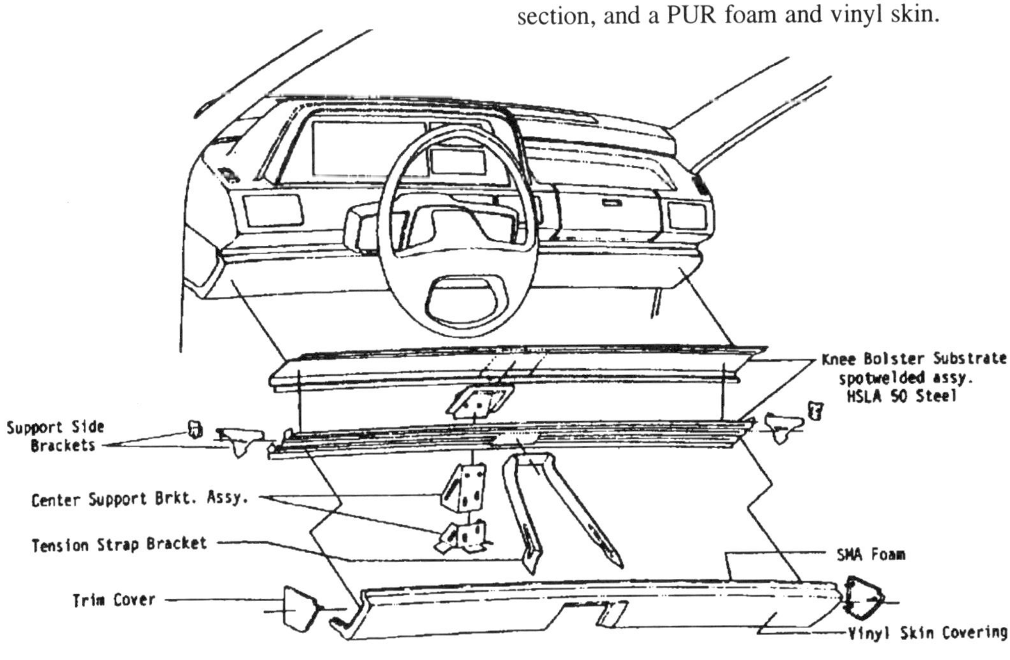

Fig 76: Knee bolster assembly

Laws of mechanics applied to vehicle safety

While the previous sections consider most aspects of secondary safety, it is also necesseary to examine the external force systems, and related rigid-body dynamic analysis of the accident situation. According to Professor Noel Murray[7] of Monash University in Australia (noted for its civil engineering structural research) the application of the basic laws of mechanics to the crash situation has revealed serious omissions and inconsistencies in current practice.

The author contrasts the situation of the racing driver who, more often then not, walks away unharmed, from relatively high speed crashes, with the public-roads environment where huge causality tolls are now all too familiar worldwide. Although the industry spends large sums on body structure design and manufacture to meet existing legal requirements, commensurate with the light weight structures required for low fuel consumption, perhaps not enough is done for the safety of the vehicle buying public in studying the crash behaviour of vehicles as a whole.

Murray likens the occupant cell of a typical saloon car with an unbraced Vierendeel truss structure where no triangulated bracing is supplied for structural integrity. Such unbraced frames rely on the integrity of the joints between pillars and rails, where very high bending moments can occur in certain crash situations — which lead to lozzenging deformation of the door and window openings. Joints have thus to be as least as strong as the members radiating from it. Where the joint relies entirely on spot-welding, the spacing of welds is obviously crucial and any load case which can cause unzipping of a line of spots is clearly to be avoided.

A car structure is cited in which the B-pillar is fixed to the rocker panel by only four spots and where the spot-weld pitch on the A-pillar is so wide as to induce local buckling in the flanges of the pillar between spots. The quality of production spot welds is also shown to be far below the quality obtained on laboratory test specimens in a number of cases.

The author provides an interesting analysis of a crash pulse, Fig 77, to show the effects of different parts of the vehicle (stiff and flexible) reacting to the impact. It helps make the point that in applying basic dynamic laws the average force for a particular crush

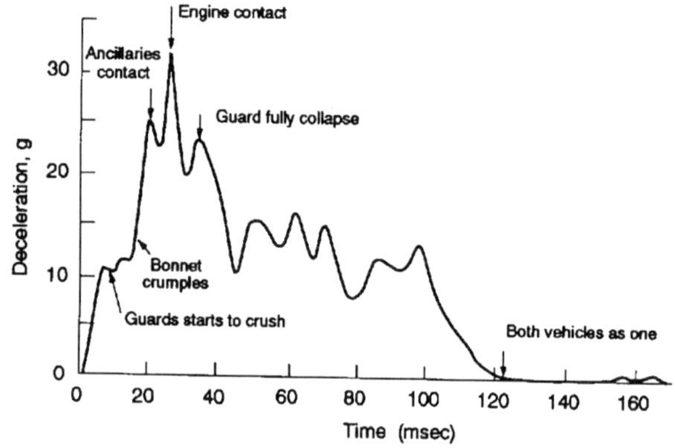

Fig 77: Representative crash pulse

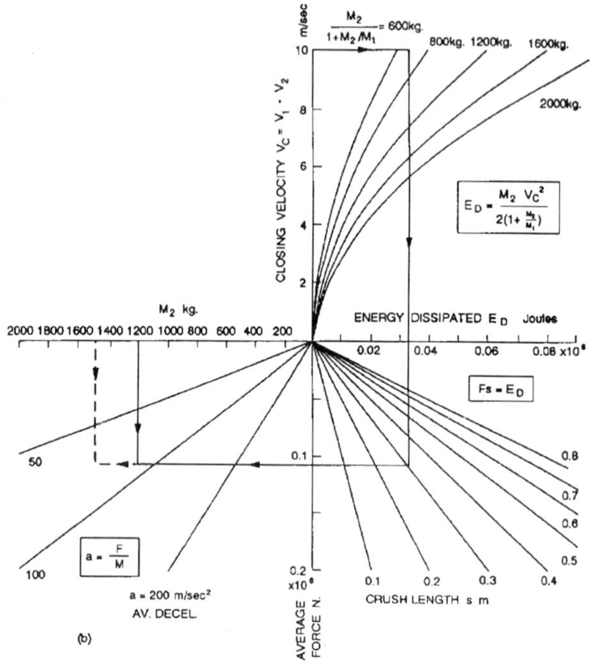

Fig 78: Relating closing velocity with energy to be dissipated

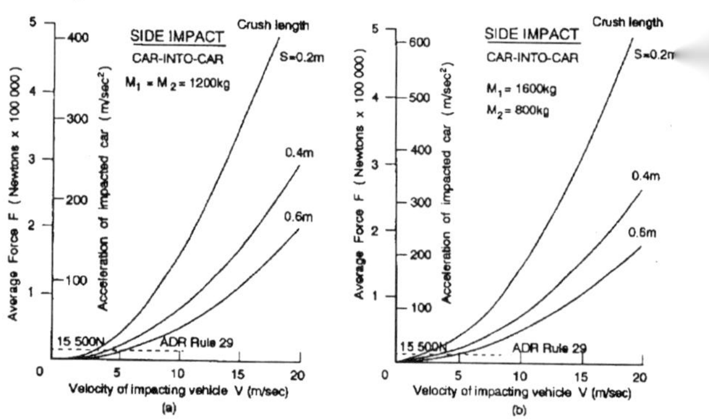

Fig 79: Average force/acceleration for various crush lengths

length has to be considered in working out the energy to be dissipated by a particular part of the structure.

For a car impacting a stationary heavy truck with mass of order of magnitude higher than the car, and assumed infinitely rigid, the author suggests that it is reasonable to assume the work done in crushing the structure is equal to the loss in kinetic energy of the car as it is brought to rest by the truck

$$Fs = MV^2/2$$

for car mass M and speed V, crushing force F and crush distance s. A 1200 kg car impacting at 60 kph would involve 166,667 joules of energy and if crushed by 0.4 metre, average force is 417 kN or 42 tonnes force. This would, of course, double if the crush length was only 0.2 metre. The force also is seen to rise as the square of impact speed. The average deceleration of the car can be estimated from

$$a = Fs/Ms$$

as 347 metres/sec^2 or 35 g. Where two comparable vehicles impact these expressions are modified by putting $s1 + s2$ for s in allowing for the deformation of each structure and the author has constructed the graph in Fig 78 to solve these expressions without calculation.

In the side impact situation, when two 1200 kg cars impact so that the front of the moving one contacts the side of a stationary one, the author shows the energy dissipated to be $300V^2$ joules for a closing velocity in metres/sec. At 50 kph this amounts to 58,800 joules. If just half this is assumed to crush the side of the stationary car, then a crush distance of 0.2 metre would involve an average force of 15 tonnes and acceleration of 12.5g. A generalised graphical interpretation is given by the author in Fig 79. The adequacy of side impact protection legislation is questioned by the author in the light of these results.

In the case of frontal impact of a car against the rear of a stationary platform truck, test results obtained by Monash University laboratory are shown in Fig 80 to reveal the effects of various elements in the crushing process and leads to a plea for appropriate underrun protection legislation. A problem is to decide how much energy should be absorbed by an underrun barrier. While UK researchers are cited as recommending a figure of 20 000 joules, Fig 81 shows the effect of a 1200 kg car impacting at 50 kph; it has some 117 600 joules to dissipate. By subtracting the suggested barrier absorption figure, the car is left with 97 600 joules which would reduce impact velocity to 46 kph, probably an insufficient margin in the author's opinion.

Besides the casualty aspects of accidents, the

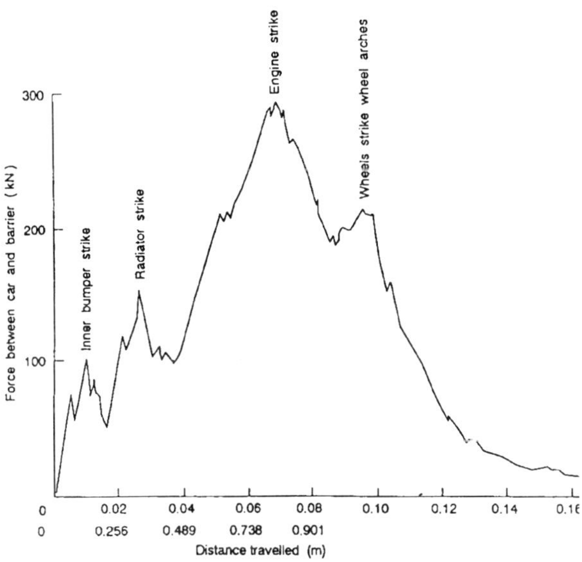

Fig 80: Underrun barrier test at Monash

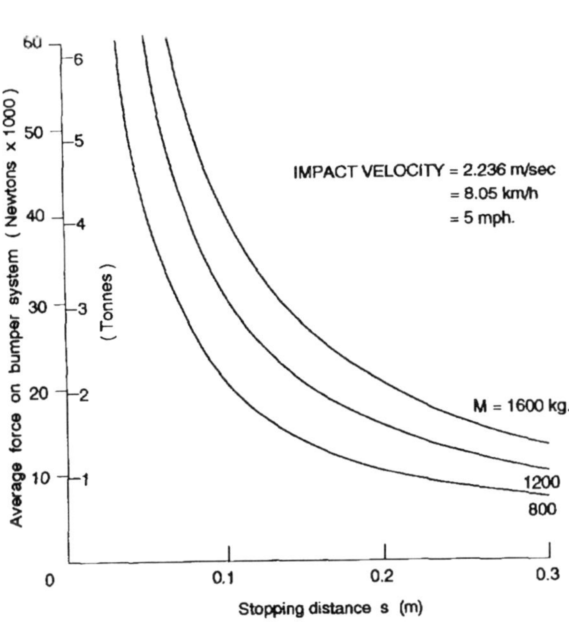

Fig 81: Bumper forces at low velocity impact

question of physical damage in low speed impacts is considered in the book. In particular the US requirements for bumper bars are examined. For the 5 mph impact requirement, the energy to be dissipated for a car mass M striking a stationary object is 2.5M joules. On equating this with an average force times stopping distance, the results shown in Fig 81 are obtained for three different car masses. The most striking point is how rapidly the average force increases as the stopping distance decreases. In the case of the peak force, it is argued that this is at least 50% greater than the average — an important factor in design. Current bumper designs, it is argued, allow typically only 0.05 metre deflection before expensive parts of the bodywork are damaged and in fact the figure of 60 kN shown by the graph could not be sustained by typical modern bumpers. A solution might be to use an energy-absorbing bumper of the type seen in Fig 82, says the author.

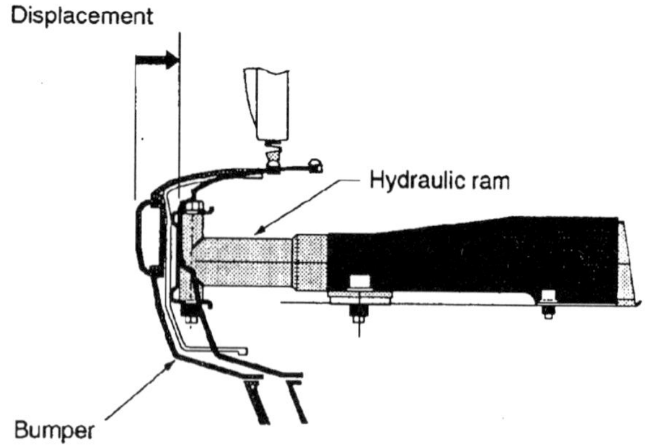

Fig 82: Hydraulic energy-absorbing device

A better feel for studying real-world accidents is obtained by using graphical techniques, it is argued. By expressing the time taken to bring an impacting vehicle to rest as

$$t = 2s/V$$

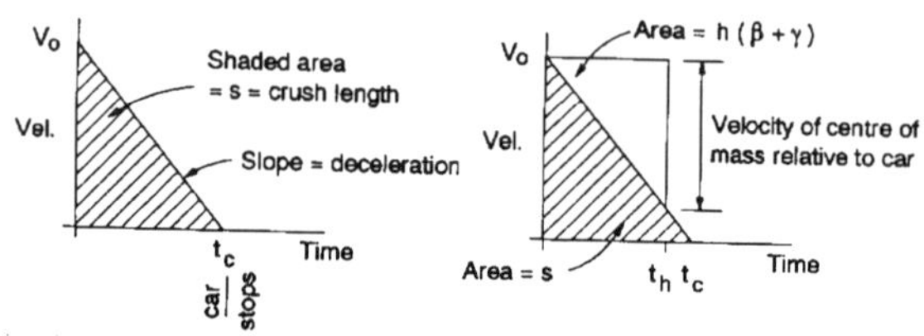

Fig 83: Crush length representation

This can be expressed graphically as in Fig 83a. When the driver is represented as in the diagram of Fig 84, the length of the path of its centre of mass is *h(b + g)*and over this length it is assumed the original speed of the mass is maintained. In Fig 83b this information is added to that of Fig 83a. The unshaded are is equal to the distance travelled by the centre of mass up to the instant of head strike. In Fig 85 an example is given in which the deceleration is relatively low and parameter values are h = 0.6 metre, g = 0.524 radian, b = 0.785 radian, V = 8.33 metre/sec and d = 30 metre/sec^2 (3.06g). Having established line AB, a horizontal line is drawn from A to represent uniform speed of the centre of mass. End point t_h can be found by measuring the appropriate area and the speed of the centre of mass at impact is given by length CD. The speed which the head strikes the steering wheel is thus 13.72 metres/sec.

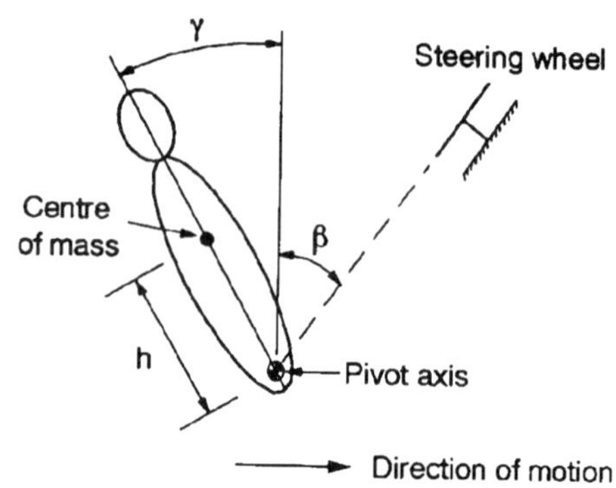

Fig 84: Occupant representation

In his book, some estimate is made by the author of the non-uniform deceleration which occurs as the real vehicle crushes at different rates during the impact and a similar analysis then shows the head striking the wheel at around 25.8 metre/sec (93 kph!).

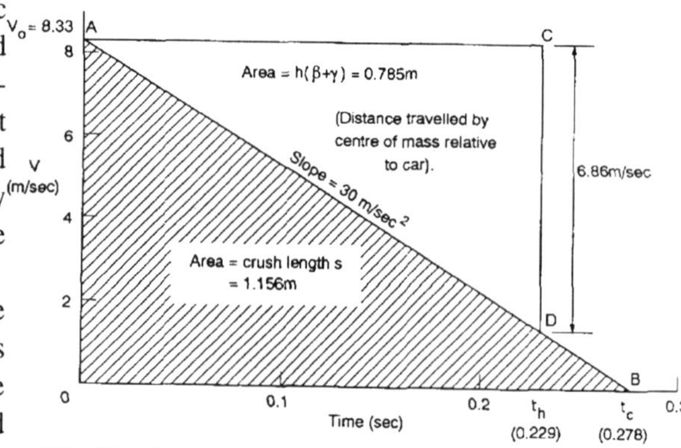

Fig 85: Construction for driver striking wheel

Case study: crashworthiness of tanker shells

In applying crashworthiness theory to tankers, assessing the several failure modes and crash scenarios, involved in tanker accidents lends itself to computer simulation as real-life tests would prove to costly. This is the conclusion of AR Duffin[8] of Frazer-Nash Consultancy in work on the impact resistance of road tankers based carried out for the Health and Safety Executive.

The finite element code DYNA3D was shown to be used for simulation of four scenarios, rear-impact, side-impact, roll on to side and rolled tank sliding impact with kerb. The models used in the simulations were arranged to investigate the effect of changing material type, tank thickness, baffle spacing and side protection within the accident scenarios depicted in Fig 86. Failure criterion was onset of rupture of the tank. The effect of side protection in the form of lengths of motorway guard rail attached to the tank sides was also examined.

Materials and thicknesses employed for the different cases are listed in the table of Fig 87. The representative tanker was a MAX-section design, of 2 + 3 axle articulated configuration with a trailer length of 11.4 metres. Flat tank baffles were assumed and a full liquid load with 5% ullage. Fig 88 shows the finite-element mesh, the 5 mm thick aluminium alloy being assumed to have yield stress of 190 MN/m^2, UTS of 250 MN/m^2 and plastic strain at failure of

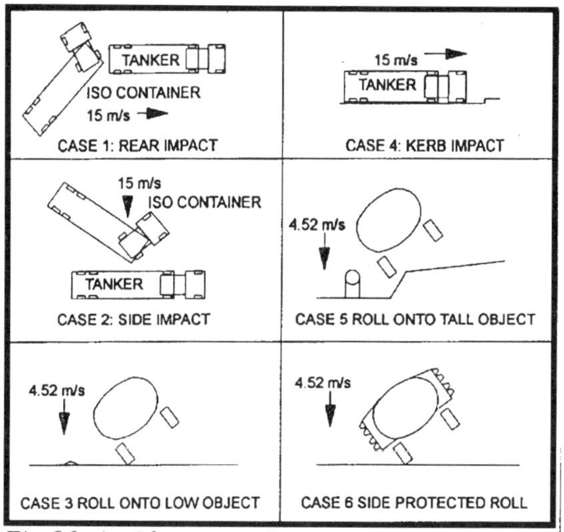

Fig 86: Accident scenarios

Case No.	Impact Scenario	Initial Velocity	Final KE	Failed Y/N
1	Rear impact	15 m/s	99.9%	Y
2	Side impact	15 m/s	99.3%	Y
3	Roll onto low object	4.52 m/s	N/A[1]	Y
4	Rolled, sliding into kerb	15 m/s	99.8%	Y
5	Roll onto tall object	4.52 m/s	88%[1]	Y

Fig 89: Cases examined in detail

Case Number	Impact Scenario	Tank	
		Material[1]	Thickness
Case 1	Rear impact by ISO container	Al	5 mm
Case 2.i	Side impact by ISO container	Al	5 mm
Case 2.ii	Side impact by ISO container. Tanker with side protection	Al	5 mm
Case 3	Roll onto low roadside object & ground	Al	5 mm
Case 4	Tanker rolled & sliding into kerb	Al	5 mm
Case 5.i	Roll onto tall roadside object	Al	5 mm
Case 5.ii	"	Al	7.5 mm
Case 5.iii	"	St	3 mm
Case 5.iv	"	St	5 mm
Case 5.v	"	St	10 mm
Case 5.vi	"	SSt	2.5 mm
Case 5.vii	Tanker with narrow baffles rolling onto tall object	St	3 mm
Case 6	Tanker with side protection roll onto flat ground	Al	5 mm

Fig 87: Materials and thicknesses used

5%. The FE model was refined to suit each impact case by increasing density of the struck area. Rupture was considered to have occurred when plastic strain at failure was exceeded in over four elements in the mesh. The kinetic energy of the impacting body (the tanker itself in the roll-over cases) at the point of failure was compared with the initial value to obtain an index of resistance.

Five cases examined in detail are shown in the table of Fig 89. In Case 1 only part of the trailer length was modelled (the two rear compartments) the remainder of the trailer and load being represented by distributed masses at the non-struck end. The resulting greatest depth of the deformed profile, due to the impacting corner of the ISO container, was 110 mm. The container was taken to weigh 38 tonnes and impact at a speed of 15 metres/sec assumed. This velocity was found to have dropped by 0.012 m/s (0.15 % drop in KE) and indicated serious rupture under this localised high energy impact. For Case 2, the results of the simulation were presented in the form of von-Mises stress plots on the deformed profile as seen in Fig 90. Greatest stresses can be detected in the area of folding which occurs in the upper part of the tank. Location of initial failure was predicted to be at the fold closest to impact. Again the impacting container retained most of its kinetic energy (99.3%), the velocity dropping by only 0.05 m/s and maximum depth of deformed profile was 0.35 metres. With Case 3 the roadside object was assumed to be a 150 mm high spherical hump and the tanker rolled on to it at a velocity of 4.52 m/s. In this case failure did not occur at the primary impact location which just wrapped around the object until the rest of the tank struck the ground. At that point baffles began to fold.

For Case 4 a rolled tanker was assumed to be sliding at 45 degrees into a 150 mm high kerb to impact, without rotation, at 15 m/s. A greatly refined mesh was required at the highly localised impact area of the tank top corner striking the top corner of the kerb. Failure occurred soon after impact with little tank distortion and negligible change in tanker velocity. Case 5 differentiated from Case 3 in that tank failure was caused at the local impact site prior to the tank main body hitting the ground. The object has height of 1.4 m and diameter of 0.4 m to simulate this condition. Fig 91 shows the considerable tank side deformation which resulted in 0.28 m/s velocity

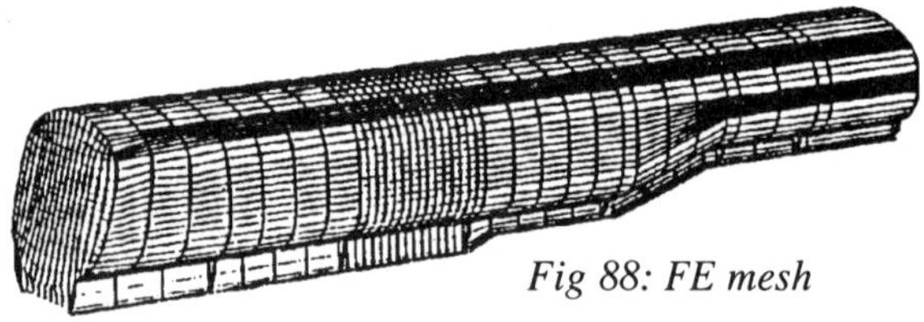

Fig 88: FE mesh

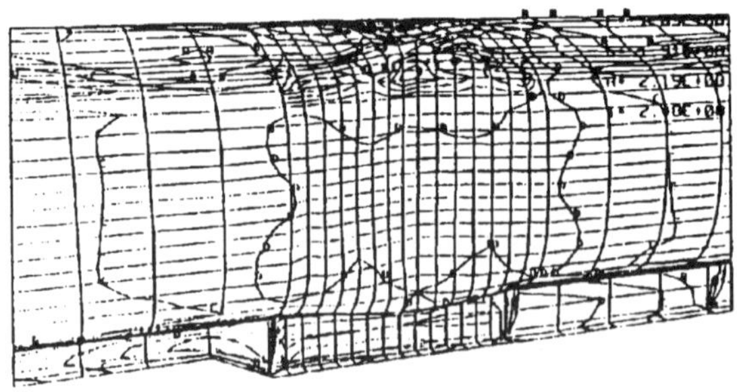

Fig 90: Case 2 impact

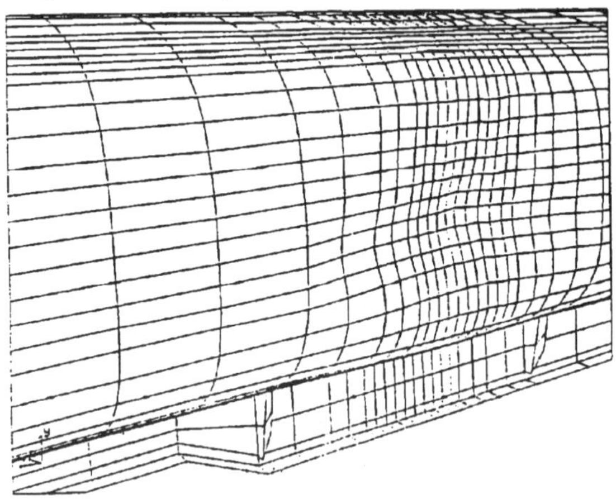

Fig 91: Case 5 impact

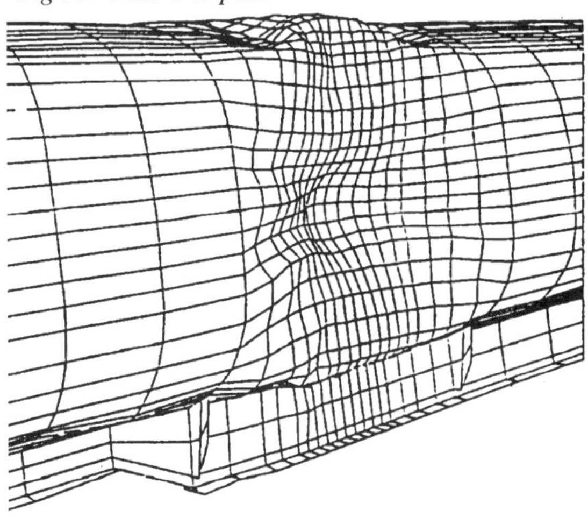

Fig 92: Case 5 repeated with parametric changes

change, involving 88% of KE remaining in the trailer. The Case 5 simulation was rerun to test the effect of parametric changes, the results being tabulated as Fig 92. Because low ductility of the baseline aluminium alloy leads to low energy absorption, mild steel to BS 1501-225 490 and stainless steel to 316 S62 were examined for comparison, the increased reduction in the kinetic energy being seen in the table. With the stainless steel model wall thickness could be halved and Fig 93 shows the considerably greater deformation available to absorb impact energy. A modified mesh was used to model the effect of additional side impact protection but this showed little value against localised high energy impacts, the sideguards merely flattening and bending. The mesh was further modified and refined to test sideguard effectiveness in pure rollover.

Case No	Material[1]	Thickness (mm)	Failed (Y/N)	Velocity Change[2] (m/s)	Remaining Kinetic Energy
5.i	Al	5	Y	0.28	88 %
5.ii	Al	7.5	Y	0.39	83 %
5.iii	St	3	Y	1.24	53 %
5.iv	St	5	Y	2.52	20 %
5.v	St	10	N	4.52	0
5.vi	SSt	2.5	N	4.52	0
5.vii[3]	St	3	Y	0.77	69 %

Fig 93: Effects of parametric changes

Chapter 4. Noise, Vibration, Harshness

Noise & vibration basics
Body structural vibration
Chassis beaming vibration
Designing against fatigue
Rubber as an isolator
CV-body mounting
Anti-noise enclosures
Sandwich-panels
Structural dynamics applied

Noise and vibration

Audible noise and uncomfortable high frequency vibrations are close to one another on the frequency spectrum and can often be reduced by similar design treatments. The classification of noise sources is dealt with here alongside measures for isolation and/or insulation /damping.

Those making attempts to design against noise inside vehicles must recognise the parallel paths of air-borne and structure-borne noise when making predictions of likely noise levels. The orders of sound level attenuation need also to be appreciated before embarking on relative design solutions. Experience has shown the possibility of lowering engine noise transmission, to the vehicle chassis/body structure, by 15-25 dB, using well designed soft-mount systems. The power-train is of course the main noise source — with transmission gearbox and final drive being up to 20 dB lower than engine noise level. Lower still are levels of noise transmitted by wind and road. However the level of all these sources added together can be amplified by 20-30 dB due to the effect of structural and cavity resonances within the car-body, truck-cab or bus saloon.

In general terms acoustic power radiated by vibrating panels is proportional to the square of their area. Fig 1 shows typical sound pressure levels inside a vehicle body relative to the excitation frequency of different noise sources. The 'components' of sound need to be extracted from such results in order to determine the relative annoyance levels of different sources. As well as weighted sound level, dB(A), measurable by portable meters, 'loudness', in sones, and 'articulation' are further important criteria. Sones provide an approximate measure of auditory sensation while articulation index measures the ease with

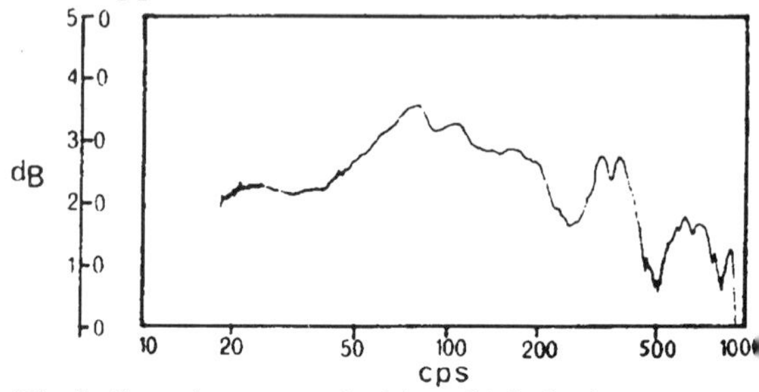

Fig 1: Sound pressure inside vehicle body

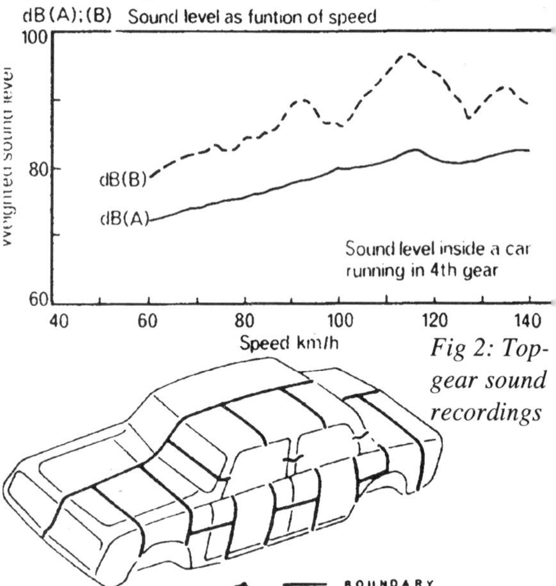

Fig 2: Top-gear sound recordings

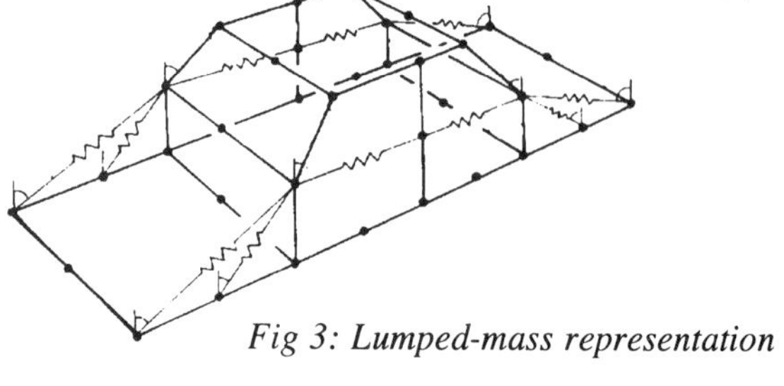

Fig 3: Lumped-mass representation

which two people at close quarters can converse.

Also helpful in appreciation of the problem are sound level *vs* top-gear vehicle speed records such as that in Fig 2. Here dB(A) is the normal 'comfort' sound level while dB(B) gives a good picture of behaviour at low frequencies. Panel vibrations are generally responsible for high frequency annoyance while low frequency effects come from overall torsion and flexure of the structure in the so-called 'shake' mode. One approach to analysing the latter effects is to represent the structure, as in Fig 3, by 'lumped' masses of different parts of the structure concentrated at points connected by springs representing stiffness of panels, rails and pillars.

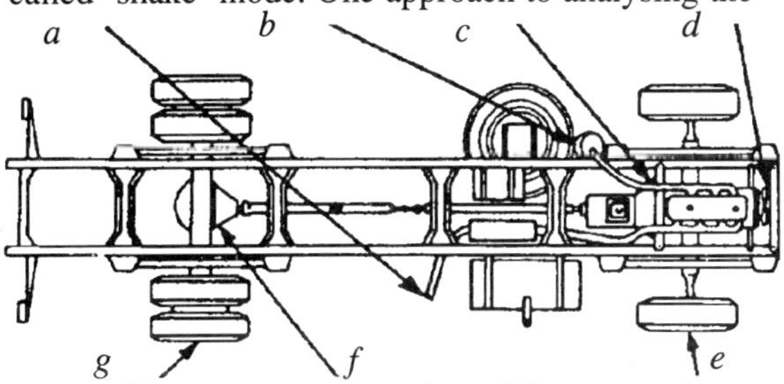

Fig 4: CV noise component intensities — a. exhaust, b. air intake, c. engine/gearbox, d. cooling fan (all $N^{4.5}$); e. front tyres (V^3W^3), f. rear axle (V^2), g. rear tyres (V^3W^3)

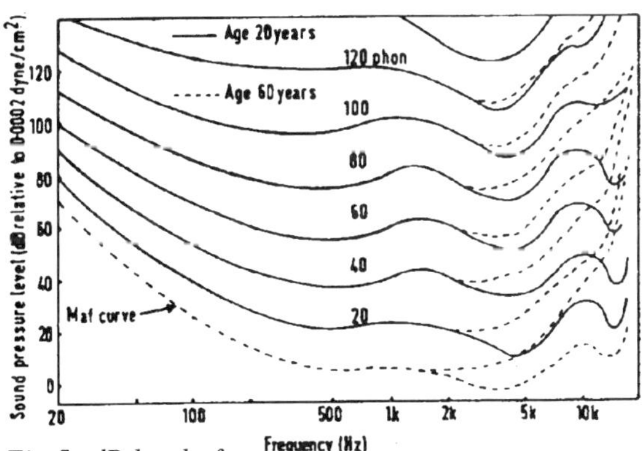

Fig 5: dB level of pure tones at different frequencies

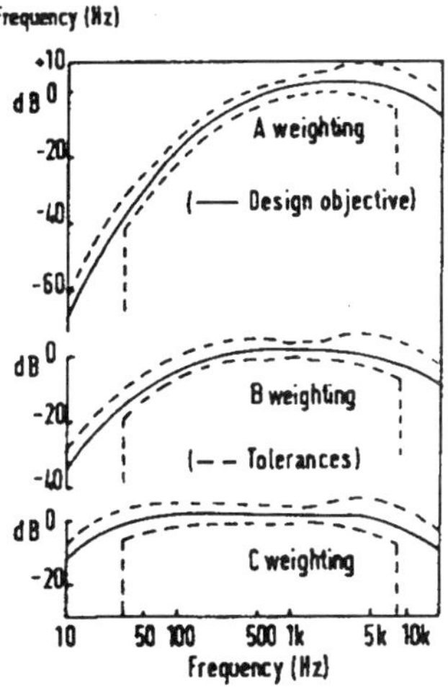

Fig 6: Weightings for different dB levels

The decibel measure of sound pressure level is $20\ log_{10}(P/P_{ref})$ where P_{ref} is a reference pressure of 0.0002 microbars. When making a measurement it is usual to insure the background noise level is 10 dB below that of the object being measured. When considering noise levels inside a body it must be remembered that there is also a 'cavity resonance' effect akin to a stationary sound wave inside a tube. A body has a natural 'boom' frequency at which this resonance is excited at a particular road speed. A 'stationary' wave is one in which outgoing and returning waves add to produce a wave of zero intensity while a 'standing' wave is one where power is absorbed at one end and therefore the intensity is not zero. Approximations for the noise levels of various 'components' of the power train are shown in relation to vehicle weight *W*, speed *V* and engine speed *N* — useful as a starting point in designing sound attenuation and shielding systems in commercial vehicles, Fig 4. Sound pressure levels, being logarithmnic, cannot be added arithmetically and, for example, two pressure levels of 80 dB would combine to give an output of 83 dB. In general the total sound pressure level is given by

$$10log_{10}[(N_1/10^{10})+(N_2/10^{10})+(N_3/10^{10})+\ etc]$$

where N_1, N_2 etc are the levels being added.

Loudness is not just a function of sound pressure as the ear is not equally sensitive to all frequencies. One measure of loudness is the phon which is based on comparing the sound source being measured with another pure sine wave of 1000 Hz coming from directly in front of the observer. When the two are judged equally loud the pressure level of the reference tone is the loudness in phons. Fig 5 shows the decibel level of pure tones at various frequencies required to produce equal sensations of loudness — based on the average of a number of observers.

This work was carried out at the National Physical Laboratory in the course of establishing British Standard 3383. Weightings for the different decibel levels are shown in Fig 6. As the phon scale gives no indication of the change in noise necessary to, say, double the loudness of a sound, another measure, the sone, had to be defined, Fig 7. Loudness level P in phons relates to loudness S in sones by $P = 40 + 10\log_2 S$

Airborne noise enters the vehicle interior through inadequately sealed holes and closures. Parallel noise transmission through the structure may be difficult to detect if it is about at equal level with the airborne noise. Decoupling the noise generating source from the structural path is the prime method of structural noise reduction while a combination of hole sealing and panel lining can be used to prevent ingress of exterior airborne noise as well as that generated by vibrating panels on the inside of the body. A designers-guide published by the US Lord Corporation is available from Metzeler and Lord Gimetall of The Brook Trading Estate, Deadbrook Lane, Aldershot GU12 4XB. This provides the basic theory of decoupling using spring/damper mounting systems. Fig 8 shows the conditions for controlling the free vibration by providing a predictable spring rate and damping co-efficient in the mounting. This determines the natural frequency of the system and the amplitudes of successive vibrations (x_1 and x_2) of the mass. Transmissibility curves for damped mounting-springs can be obtained as in Fig 9 to define the ratio of input to output amplitude. For an undamped spring, transmissibility is $1/[(f_d/f_n)^2 - 1]$ for values of f_d/f_n less than or equal to $2^{1/2}$ where f_d is the disturbing frequency and f_n is the spring natural frequency. The figure reflects this formula and its modification for different damping ratios is given by $(C/C_c) = (1/2)K''/K'$. In considering overall noise and vibration levels in a vehicle a distinction should be made between the periodic excitations originating from the rotating parts and the random nature of road input vibrations. Successful evaluation of road input effects is usually based on the examination of in-service input data collected on suitably instrumented vehicles. As well as for ride sensitivity, the vibrations of the vehicle can also be classified in three groups corresponding to known human sensitivities.

In the 0-3 Hz frequency range, motions are evoked by the suspension springing and levels of discomfort worsened by pitching and rocking motions overlaid on the basic bounce motions. In the 3-30 Hz range, vibrations of the sprung masses come into effect while from 300 to several hundred Hz drive train and audible vibrations from the tyres and wind come into play. In the 40-70 Hz sector, unpleasant boom resonances occur just on the audible threshold.

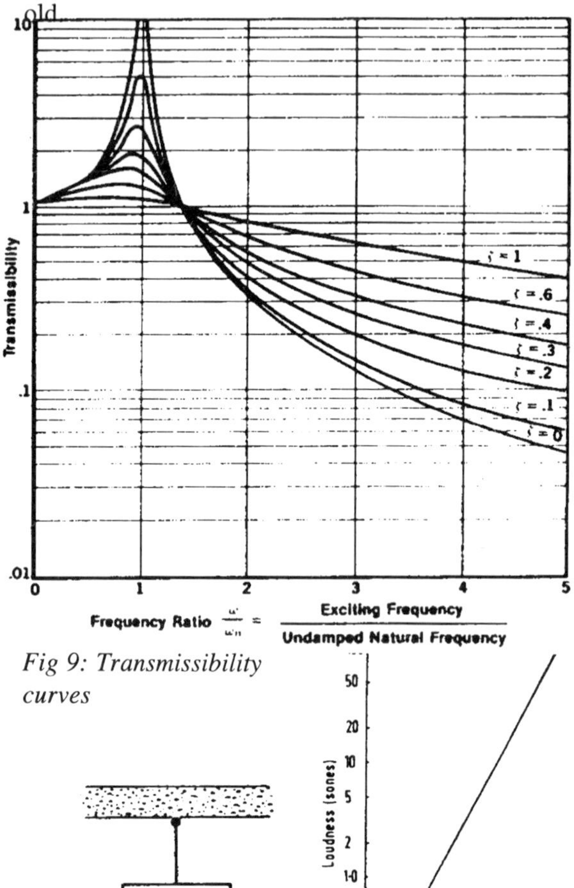

Fig 9: Transmissibility curves

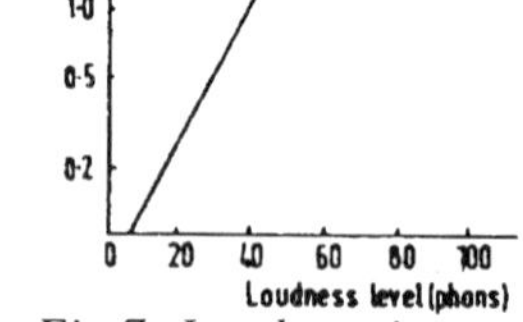

Fig 7: Loudness in sone

$$f_n(\text{Hz}) = \frac{1}{2\pi}\sqrt{\frac{K'}{M}} \approx 3.13\sqrt{\frac{1}{d_{stat}}}$$

$$\tau\,(\text{sec}) = \frac{1}{f_n}$$

$$\frac{K''}{K'} \approx 2\frac{C}{C_c} = \frac{2}{\omega_n \tau}\ln\frac{X_1}{X_2}$$

where τ = Period of Free Oscillation
C = Damping Coefficient (lbs.-sec./in.)
C_c = Critical Damping Coefficient
$\frac{C}{C_c}$ = Viscous Damping Ratio
ω_n = Undamped Circular Natural Frequency in rad/sec. ($2\pi f_n$)

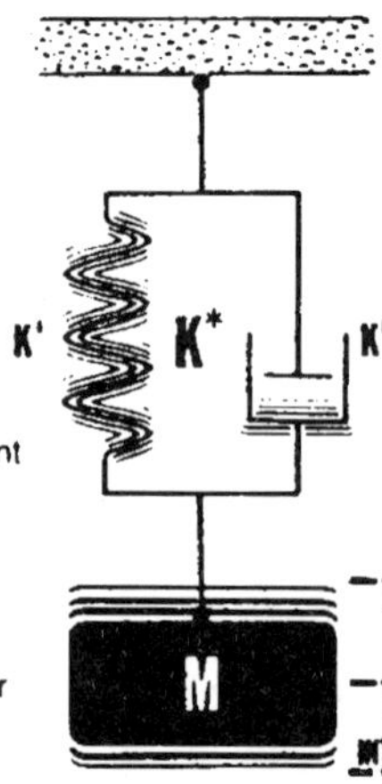

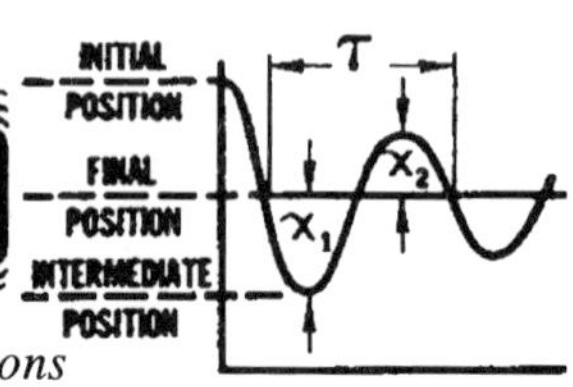

Fig 8: Decoupling conditions for free vibrations

Body structural vibration

Avoidance of discomfort caused by the 'harshness' and 'shake' vibration frequencies can be achieved at the design-stage if the shape of the vehicle is virtually finalised, a complex FE analysis carried out and appropriate remedial measures taken during build such as the addition of vibration absorbers.

Without the luxury of these facilities it is possible to avoid serious vibration effects by having a thorough understanding of the possible causes as the design is built up and arranging for the structure to be outside the range of likely exciting resonances. Because road input vibrations are generally of a random nature, the variable frequency and amplitude content can excite different body panels into vibration, or cause overall bending and/or twisting motions of the structure, according to road surface and handling inputs. Generally it is the aim to design enough bending and torsional stiffness into the structure so that natural frequencies fall in a range of 20-30 Hz — to avoid the danger of excitation by suspension frequencies and the consequent loss of handling control due to movement of the anchorages. The higher road input frequencies, in the 'harshness' region of the overall spectrum, can cause difficulties which are sometimes overcome by the use of soft-mounted subframes which maintain 'chassis' and 'body' in a different vibrational environment.

Pioneer work at Birmingham University identified the vibration modes of a typical volume-built panel-van shell structure and found the associated resonant frequencies. By suspending the shell on flexible elastic cords and exciting the suspension anchorages with vibrators sweeping across a wide frequency range, localised natural frequency responses were determined. Ten measured frequencies and mode shapes are listed in Fig 10. In fact the symmetric modes were not just bending ones because of the disparity between the stiffnesses of different parts of the shell and the nature of the design which involved large cut-outs and some flexible connections. Motions involved a combination of lozenging of the load-carrying area with torsion or bending of the floor, roof and sides, Fig 11.

Subsequent studies on integral car structures have yielded vibration modes typified by those in Fig 12 for one phase condition at 140 Hz. This is obtained by affixing transducers to the surface and filtering signals to the frequency being studied. Single point harmonic forcing is also used to to obtain point responses in terms of driving force and resultant motion. The variation of point response against forcing frequency is known as 'mobility' — the inverse of mechanical impedance. The instrument used comprises a piezo-electric transducer incorporating a force and acceleration sensor on the one probe head. A sweep oscillator is then tuned to a particular resonant frequency and a roving accelerometer placed on the structure to read off RMS amplitude and phase at each grid point. The resulting quadrature amplitude response is plotted on a plan of the structure as shown in Fig 13.

Random vibration can be analysed to find its frequency and amplitude content by statistical tech-

Measured symmetric modes

Mode no.	Natural frequency (Hz)	Figure number	Description
	24·90	15.11	Breathing mode, floor and roof in antiphase, sides in antiphase moving together while roof and floor move apart. Longitudinal bending of the floor.
	32·56	15.12	Same as mode 1 but with greater motion of sides and front of roof.
	38·03	15.13	Combination of longitudinal and transverse bending in both floor and roof with their motion in phase.
	46·87	15.14	Same as mode 3 but the longitudinal and transverse motions are combined in a different sense.
	47·55	15.15	Transverse bending of the roof and rear floor together with longitudinal bending of the front.
	52·16	15.16	Same as mode 5 but with a reversed phase relationship between the transverse bending of the rear floor and the longitudinal bending of the front.

Measured anti-symmetric modes

Mode no.	Natural frequency (Hz)	Figure number	Description
1	20·54	15.17	Primary lozenging of the open-ended load-carrying area with the magnitude increasing towards the rear. Floor and roof moving laterally in antiphase, sides moving vertically in antiphase.
2	27·12	15.18	Torsion of the floor and roof with secondary lozenging of the load-carrying area so that the front and rear lozenge in antiphase. Lateral bending about the door opening was also present.
3	39·60	15.19	Secondary torsion of the floor but not the roof together with tertiary lozenging of the load-carrying area so the centre of the van was lozenging in antiphase with the front and rear.
4	49·36	15.20	Secondary torsion of both floor and roof with vertical bending of the sides giving a form of secondary lozenging across the load-carrying area.

Fig 10: Typical mode frequencies

niques. A relatively simple way of describing a random surface profile of a road or track is to plot the vertical and horizontal displacement from the origin of co-ordinate axes as its horizontal projection traverses at constant linear velocity, Fig 14. The key statistical 'means' are shown in Fig 15. The time in which a point lies in any narrow band about any height *x*, when divided by sample time *T*, gives the so called probability density. This can be plotted on a Gaussian distribution curve to give information on the amplitude and frequency of bumps — based on the number of times the vertical displacement variable crossed a given level. The concept of 'density' is best understood by considering a bar of constant cross-section having variable density along its length. The mass of such a bar between distances of *X1* and *X2* from end could be shown by the expression in Fig 16a and by considering distance along a frequency scale w in lieu of a length scale *X*, the spectral density *S(w)* is analogous to physical density *p(x)* and the mean square amplitude to the mass as shown in Fig 16b. Panel vibrations are usually excited by engine or suspension motions transferred through their respective mounts, causing referred movement of panel edge supports. The panels are relatively flexible in bending so are easily excited into vibration if they are flat. Rectangular plates have different vibration harmonics as shown in Fig 17, the natural frequency, in the case of simply-supported plates, being:

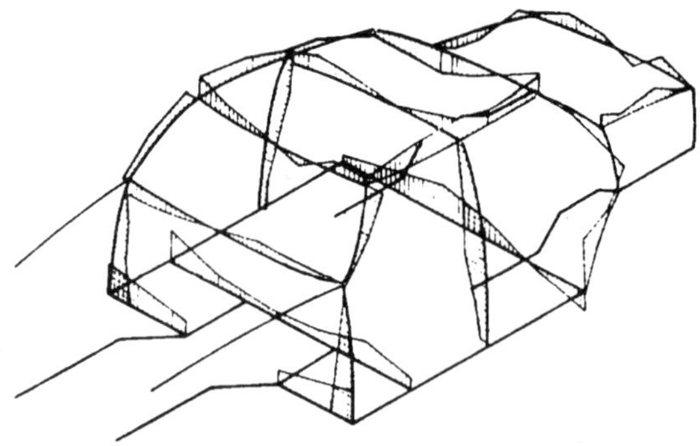

Fig 12: Carbody shell vibration

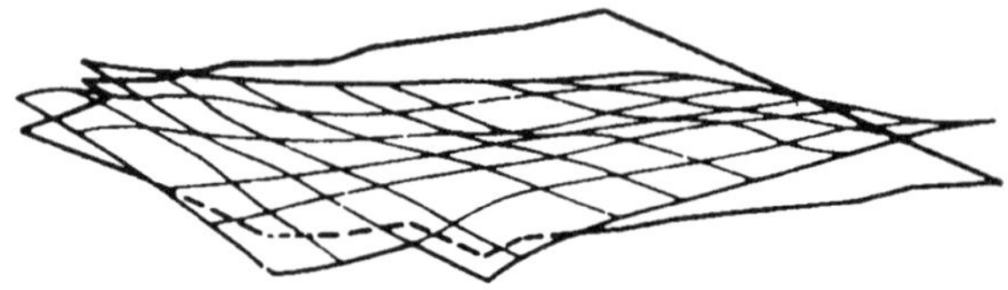

Fig 13:Response of floor structure

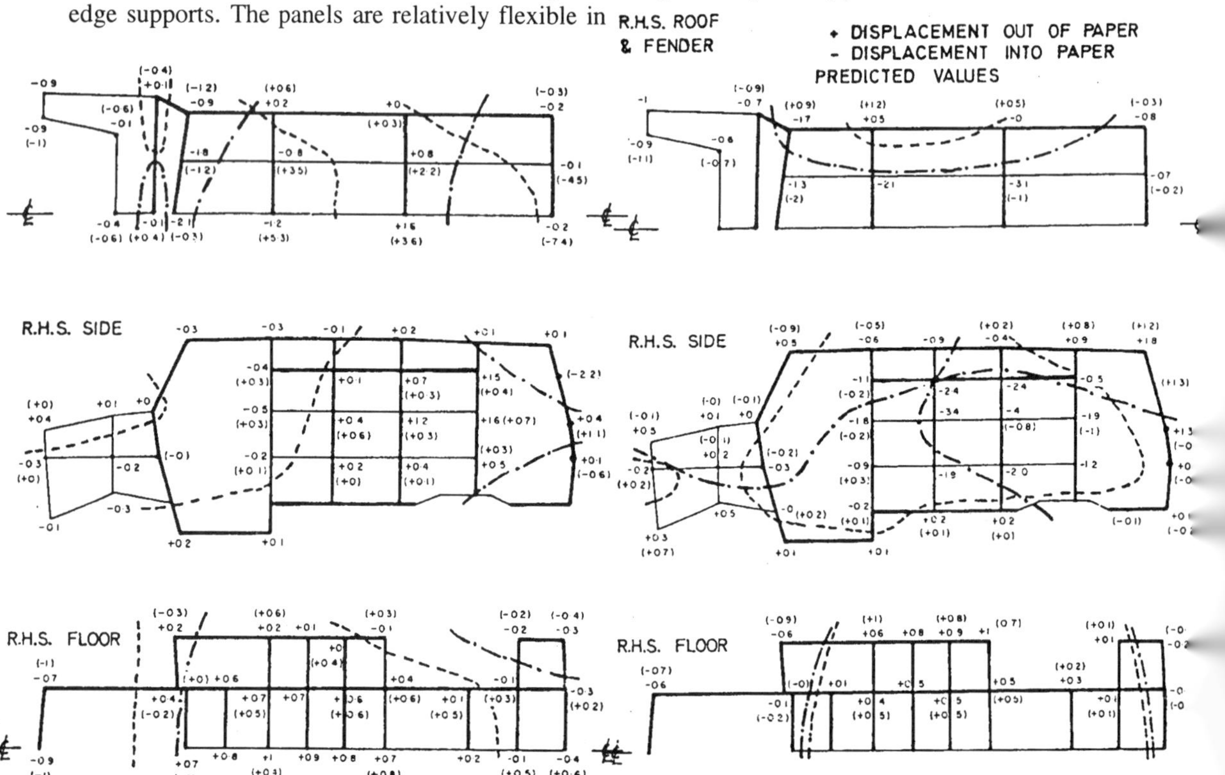

Fig 11: Combined bending and twisting motions

$$\pi^2(gD/dt)^{1/2} / [(m^2/a^2) + (n^2/b^2)]$$

where *m* and *n* are the harmonics, *a* and *b* the lengths of the sides and *g* the acceleration due to gravity.

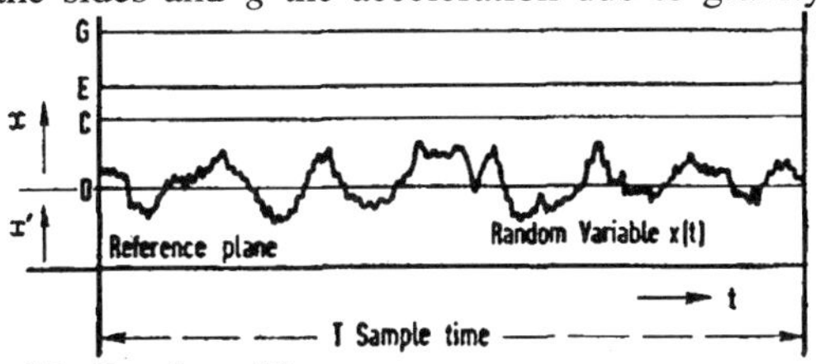

Fig 15: Road profile

Mean value,

$$\text{line } OO = \frac{1}{T}\int_0^T x(t)\,dt$$

Mean amplitude,

$$\text{line } CD = \frac{1}{T}\int_0^T |x(t)|\,dt \qquad \text{signified } \langle |x(t)| \rangle$$

Mean square amplitude,

$$\text{line } GH = \frac{1}{T}\int_0^T x^2(t)\,dt \qquad \text{signified } \langle x^2(t) \rangle$$

Root mean square amplitude,

$$\text{line } EF = \sqrt{\frac{1}{T}\int_0^T x^2(t)\,dt} \qquad \text{signified } \sqrt{\langle x^2(t) \rangle}$$

Fig 16a: Statistical mean values

$$m_{X_1 \to X_2} = A\int_{X_1}^{X_2} p(X).dX \qquad \langle x^2(t) \rangle_{\omega_1 \to \omega_2} = \int_{\omega_1}^{\omega_2} S(\omega)\,d\omega$$

Fig 16b: Spectral density concept

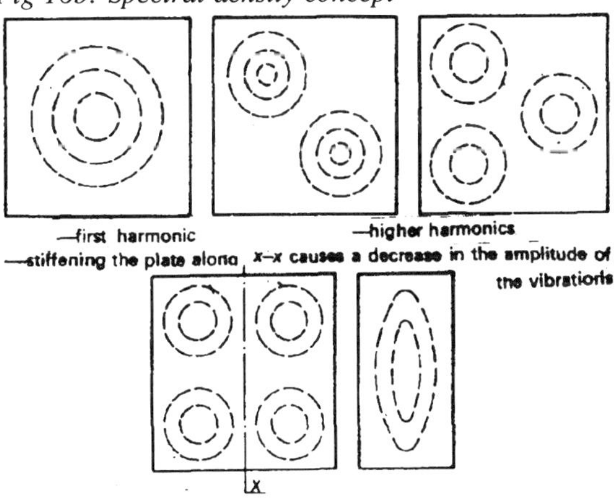

Fig 17: Flat plate harmonics

Material	Young's Modulus E (GN/m²)	Density ρ (Mg/m³)	E/ρ	$\sqrt{(E/\rho)}$	Ratio of Nat. Freqs (Plates of Equal Thickness)	Ratio of Nat. Freqs (Plates of Equal Weight)	Ratio of Thickness (Plates of Equal Weight)
iron	120	7	17	4·12	1	1	1
	210	7·9	25	5	1·2	1·35	0·9
inium	70	2·7	26·5	5·15	1·25	3·24	2·6
esium	43	1·7	26·5	5·15	1·25	5·15	4·1
reinforced :							
ndom glass	5·6–10·3	1·3–1·8	5·1	2·26	0·55	2·48	4·5
directional ass	16–24·2	1·3–1·8	13	3·6	0·88	3·95	4·5
ex	6 2	1·15	5·4	2·32	0·56	1·94	6·1

Fig 18: Panel natural frequencies

Panel density and thickness are *d* and *t* respectively. That of the first harmonic, which produces most noise, is obtained by putting *m* and *n* equal to 1. Fig 18 shows a comparison of panel natural frequencies.

No simplified method exists for estimating mode shape and natural frequencies of structures like integral saloon car bodies. A not too complex approach was developed some years ago by General Motors before finite-element methods became widespread. It has the advantage that the designer can visualise vibration effects at a very early design stage. The local masses of the structure are lumped at nodes and the properties of the weightless beams assumed between the nodes can be adjusted as experience is built up in physical model testing. As a very first estimate of calculating natural frequency, the Unit Load method can be employed assuming the body to behave as a uniform beam with the same flexibility as calculated for the front passenger load point. Then the value of *f1*, displacement due to unit load, found by applying the method to a structure such as Fig 19a can be considered within the uniform beam at 19b:

An equivalent uniform torsional stiffness *GJ* can be found from the natural frequency in torsion which can be estimated from: $(1/2L)(GJ/I)^{1/2}$ where *I* is the mean rolling moment of inertia per unit length of the body which can be assumed to roll about an axis through the mid-points of front and rear suspension mounts.

Over AB, $M1 = (b/l)x1$ and $m1 = (b/l)x1$ Over CB, $M2 = (a/l)x2$ and $m2 = (a/l)x2$ thus $f2 = (1/EI).\int[(b/l)^2x1^2.dx1 + (a/l)^2x2^2dx^2]$ from which the equivalent beam stiffness can be obtained: $EI = a^2b^2/3lf1$ since $f1 = f2$ Assuming a uniform mass distribution *m* per unit length of the body, the first natural frequency is: $(22.4/2\pi L^2)(EI/m)^{1/2}$

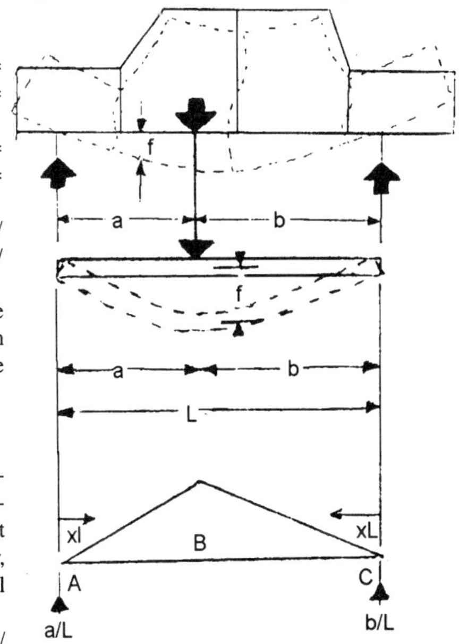

Fig19: Car shell simplified as a beam

Chassis beaming vibration

It is possible to predict the dynamic behaviour of a chassised vehicle if the body is inherently flexible or very soft mountings are used between it and the chassis. A graphical technique can be employed to determine the modes of beaming deflection of the chassis frame and its natural frequency in bending.

While a number of analytical methods for predicting stuctural vibration behaviour depend on the assumption that mass can be concentrated at discrete points, this is usually done for complex structures, including shells, with frequent changes in section. For more regular structural members such as chassis longitudinals the distribution of the mass can be taken into account — as well of the elasticity of

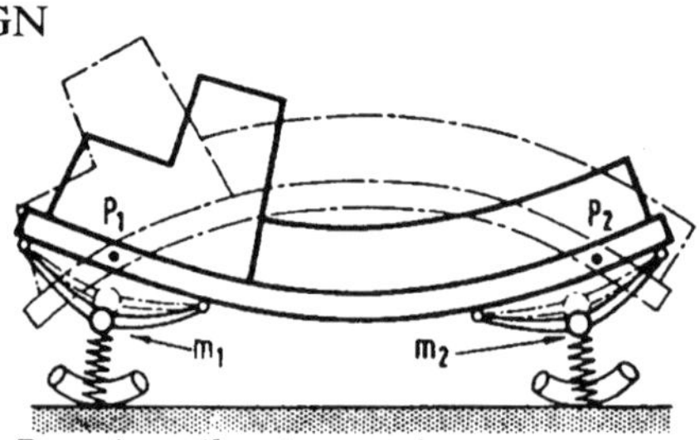

Fig 20: Beaming vibration mode

the member in bending. If a truck chassis, for example, is set into vibration under laboratory conditions and its vibration modes measured, the vehicle has one prominent mode as demonstrated in an exaggerated sense in Fig 20. By making the assumption that the longitudinals take up a half sine-wave shape in beaming vibration, the classical Rayleigh's method can be used to obtain a formula relating natural frequency to section dimensions and the elastic properties of the material. The strain energy of the beam

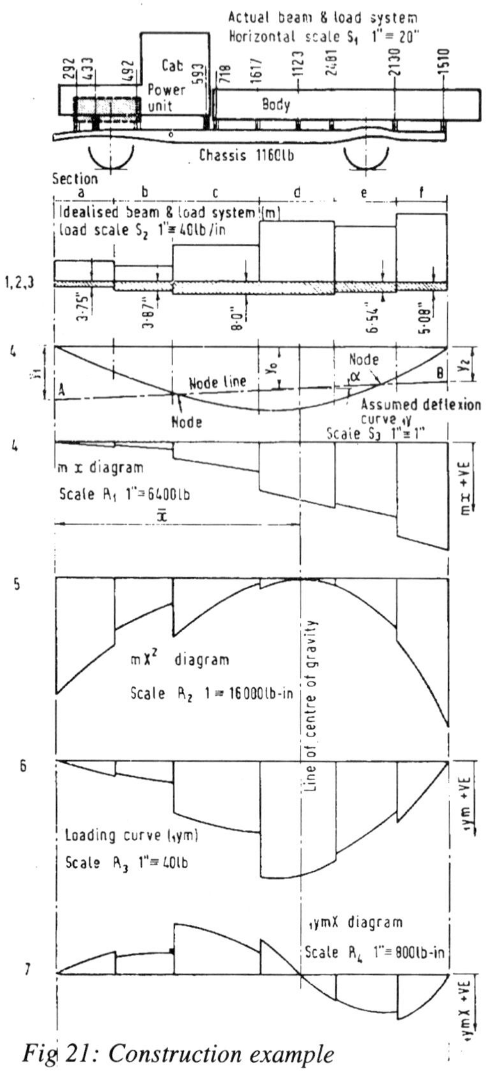

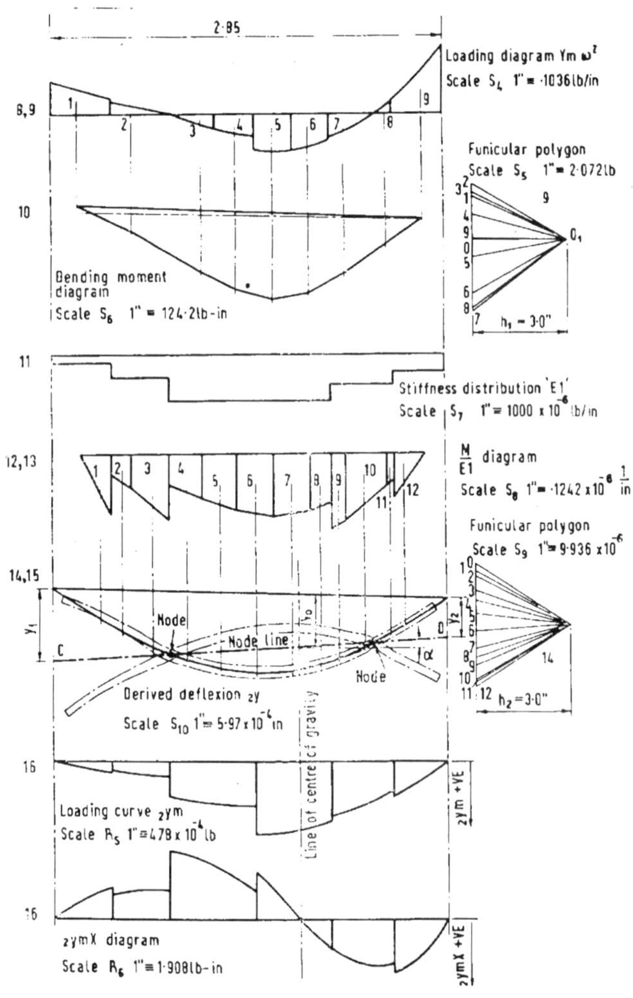

Fig 21: Construction example

at maximum deflection, and zero velocity, is then equated to the kinetic energy of the beam elements at zero deflection, and maximum velocity, as it passes through the central, non-deflected shape, within the vibration cycle. An imaginary small element of the beam is considered and its behaviour is simulated by the presence of an effective centralising spring, stiffness *k*, exerting a restoring force *ky*. The 'potential' strain energy is obtained by integrating these instantaneous forces over the full amplitude of vertical displacement. By assuming simple harmonic motion of the beam the kinetic energy of each element expressed as mv^2, can be integrated along the length of the beam and equated with the potential energy to obtain an approximate fundamental frequency of

$$[(EI/m)\pi/4l^2\{(1/4)-(2/\pi^2)\}^{1/2}]^{1/2}$$

The avoidance of 'cab nod' is a key stimulus to the better understanding of chassis frame beaming vibration. So as to obtain more accurate vibration mode profiles a graphical method was proposed by Cranfield researchers which allowed changes in section property along the length of the beam to be taken into account. It is based on a method due to Stodola which assumes an initial deflection curve, as in the Rayleigh method above, but progressively refines it to more nearly predict the actual one. The initial curve gives an inertia loading profile based on actual variable load distribution along the chassis. The real beam is drawn out to a scale of *S1* and divided into equi-spaced ordinates. Beneath it a mass distribution curve is drawn, which will vary by stepped increments; mass per unit length *m* will have units such as lb sec^2/in^2. Below that the assumed deflection *1y* curve is plotted at a scale of *S2*. At this stage the inertia loading curve can be drawn, putting *w2 = 1* in the load expression *1y.m.S2*. Areas between the ordinates of this curve can be replotted as a diagram of distinct inertia loads as in Fig 21, respective areas being given by an integral (*from 0 to S1*) of the above expression, to a scale *S3*. From that point the familiar technique of drawing a polygon of forces, to the same horizontal scale, is carried out. A vertical scale *h1* is chosen for convenience of drawing construction and the result leads to a bending moment diagram with ordinates *M* in scale *h1.S1.S2* in lbs per unit of measure on the drawing. The key factor in the execution of the method is carefully adhering to the scales chosen and understanding their effect at each stage. The stiffness distribution is next constructed *EI* with *E* in lb/in^2 and *I* in in^4, to a scale *S4*, such that *S4* = lb/in^2 per unit of measure on the drawing. Dividing the ordinates obtained from the last two curves will allow a further one for *M/EI* to be constructed. A similar technique to the one used on the inertia load curve can then be used, of integrating the areas between the ordinates, to produce a new set of ordinates which in turn will provide a more accurate beaming deflection curve for the vibrating chassis sidemembers.

The Cranfield workers essentially developed this Stodola method, covering beams simply sup-

---

1 Draw out chassis frame to tlength scale S_1, with point loads at intervals, from Fig 21.
2 Idealize the non-uniform chassis into lengths of uniform section. Draw out idealised 'distributed' load diagram m to vertical scale S_2.
3 Write down assumed deflexion curve $_1y$ to scale S_3.
4 Construct moment diagram *mx* from ordinates *m* in (2) and distance *x* from left-hand end, to scale R_1. Obtain *x* by dividing area of the resulting diagram by that of the *m* diagram (factoring for S_2 and R_1)
5 Using the c.g. as the new datum, plot the mx^2 diagram (Fig21a) to scale R_2 (area here is *I*).
6 Derive loading curve $_1ym$ as product of ordinates (2) and (3) to scale $R_3 = S_3S_2$ and find its area, *A*
7 Plot the $_1ymx$ diagram (Fig 21b) to scale R_4 and carefully find resultant area as large positive and negative parts.
8 Calculate y_1 and y_2 (Fig 21(b), and draw node line AB on curve $_1y$, measure the ordinates *Y* and multiply these by the *m* ordinates (2) to give new load $Ym\omega^2$ to scale S_4.

Fig 22a: Drawing procedure

9 Divide resulting load diagram into small areas (loads) and plot them as the vertical line of the first funicular polygon to scale S_5, then complete the polygon.
10 Calculate y_1 and y_2 (Fig 21b), and draw bending oment diagram by normal methods, and if necessary ensure that it closes, thus overriding any small errors in calculating areas; scale is $S_6=S_5h_1S_1$.
11 Draw the stiffness distribution *EI* to scale S_7.
12 Divide major ordinates of BM diagram (10) bythose of diagram (11), and plot *M/EI* diagram to scale $S_8 = S_6/S_7$; divide conveniently into 11, say, areas and measure these and their centroids.
13 Plot these as the vertical line of the second funicular polygon to scale S_9 and complete it.
14 Draw derived deflexion curve $_2y$ with lines parallel to 'rays' of the polygon as in a BM diagram (lines 00_2 and $12\ 0_2$ are produced back to the ends of the beam); scale is $S_{10} = S_9h_2S_1$.
15 Ordinate $_1y_{max}/_2y$ measured at centre of beam gives the natural frequency of the loaded beam.
16 Draw $_2ym$ and $_2ymX$ diagrams to scales $R_5$and R_6 so that y_1 and y_2 can be calculated and CD drawn in to give the node points.

ported at their ends, to one for free-free beams vibration mode. Fig 22a is a drawing procedure and Fig 22b an example construction.

SECTION ALONG BEAM:	LH	a		b		c		d		e		f	RH
1. LOAD DIST'N, lb/in		24·05		18·45		44·35		72·35		66·25		83·95	
2. ORD, m (S_2, 1 in = 40 lb/in)		0·60		0·46		1·11		1·81		1·66		2·10	
3. ORD, $_1y$ (S_3, 1 in = 1 in)	0	0·38	0·75	1·11	1·38	1·73	1·88	1·85	1·61	1·30	0·87	0·45	0
4. DIST, x (S_1, 1 in = 20 in)	0		1·87		3·67		6·35		8·70		10·62		12·25
ORD, mx (in = 6400 lb)	0		1·12,0·86		1·69,4·07		7·04,11·48		15·73,14·40		17·59,22·3		25·7
AREA (1 in² = 128,000 lb in)		0·13		0·29		1·86		4·0		3·84		4·89	
$\Sigma A = \Sigma mx$ = 15·01 in² = 1 921 000 lb in, Σm = 12,550 lb, $\bar{x}$ $\Sigma mx/\Sigma m$ = 153·1 in													
5. DIST, X (1 in = 20 in)	−7·65	−6·70	−5·78	−4·85	−3·98	−2·67	−1·30	0	1·05	2·0	2·97	3·79	4·60
ORD, mX² (1 unit = 16,000 lb in)	36·2	27·0	20·05,15·40	10·84	7·30,17·57	7·32	1·87,3·06	0	1·99,1·82	6·62	14·60,18·51	30·15	44·40
AREA (1 in² = 3·2 x 10⁶ lb in²)		5·10		2·01		2·28	0·13		0·07	1·40		4·96	
1 = Σmx^2 = 15·96 in² = 51·0 x 10⁶ lb in²													
6. ORD, $_1ym$ (ω^2) (1 in = 0·104 lb/in)	0	0·23	0·45,0·35	0·51	0·64,1·53	1·92	2·08,3·40	3·45	2·91,2·67	2·15	1·44,1·83	0·94	0
AREA (1 in² = 800 lb in)		0·42		0·89		5·00		7·71		4·05		1·52	
ΣB = 19·60 in² = 15,670 lb in, $y_0 = \Sigma BIM$ = 1·25 in													
7. ORD, $_1ymx$ (1 in = 3200 lb in)	0	−1·53	−2·61,−2·00	−2·48	−2·53,−6·09	−4·93	−2·71,−4·42	0	3·06,2·80	4·30	4·28,5·42	3·58	0
AREA (1 in² = 64,000 lb in²)		0·65		−1·07		−3·20	−0·08		0·42	1·93	0·38		0·95
8. D = −1·99 in² = − 126,400 lb in², $\alpha = -D/I$ = 0·0025; y = 1·25 + 153·1 (0·0025) = 1·63 in, y_2 = 1·25 − 91·9 (0·0025) = 1·02 in													
ORD, Y (1 in = 1 in)	−1·63	−1·21	−0·80	−0·40	−0·06	0·37	0·57	0·60	0·41	0·15	−0·23	−0·60	−1·02
ORD, mY (ω^2) (1 in = 0·104 lb/in)	−0·98	−0·73	−0·48,−0·37	−0·19	−0·03,−0·07	0·41	0·63,1·03	1·08	0·74,0·68	0·25	−0·38,−0·48	−1·26	−2·14
9. AREA = (S_9 1 in² = 2·07 lbf)		−1·36		−0·36	0·28		0·69,1·27		1·09,0·52		−0·10	−2·08	
10. ORD, M (S_6, 1 in = 124·22 lb in)		0	0·44	0·71	1·33	1·87	2·36,2·62		2·39,2·02	1·76	0·81,0·68	0	
11. ORD, EI (S_7, 1 in = 1000 x 10⁶)		0·24		0·68		1·40		1·40		0·88		0·50	
12. ORD, M/EI (S_8, 1 in = 0·124 in⁻¹)		0	1·83,0·64	1·04	1·95,0·95	1·34	1·69,1·87	1·71	1·44,2·29	1·99,0·92	0·77,1·35	0	
13. (1 in² = 2·49 x 10⁻⁶) (S_9, 1 in = 9·94 x 10⁻⁶)		0·96	0·55		1·72,1·18	1·63	2·08	1·97	1·05,0·98	1·85	0·16	0·66	
14. ORD, $_2y$ (S_{10}, 1 in = 5·97 x 10⁻⁴)	0·38		0·90,1·09,1·25		1·70,1·86,2·04		2·27,2·36,2·28		2·06,1·94,1·87		1·55,1·01,0·83		0·51
15. $\omega^2 = {}_1y/{}_2y$ (max) = 1335, ω = 36·8 rad/s, f = 5·85 c/s													
16. ORD, $_2ym$ (1 in = 478 x 10⁻⁴ lb) AREA 1 in² = 0·96 lb in)		0,0·54, 0·29,0·66, 0·32		0·50,0·78, 0·58,0·86, 0·65		2·06,2·52, 2·26,2·62, 3·26		4·27,3·73, 4·12,3·51, 4·70		3·21,2·57, 3·10,1·67, 2·42		2·12,1·07, 1·74,0 0·87	
ΣB = 12·22 in² = 11·69 lb in, $_2y_0$ = 11·69/12,550 = 9·31 x 10⁻⁴/(1·56) in													

DIST. X	−7·65 −7·0 −6·15 −5·78 −5·46 −4·52 −3·98 −3·47 −2·42 −1·30 −0·20 0·69 1·05 1·26 2·05 2·97 3·28 3·88 4·60
ORD, $_2ymX$ (1 in = 1·91 lb in)	0. −3·33, −2·91, −3·54, −8·20, −6·09, −5·55, +2·57, +3·37, +5·26, +6·30, +4·06
	−1·60, −3·78, −3·15, −3·41, −7·84, −3·41, −0·82, +3·68, +3·90, +4·96, +5·71, 0
AREA (1 in² = 38·16 lb in²)	−0·98 −1·51 −2·89 −1·33, −0·91 0·49,1·12 1·18,1·12 0·41
D = −3·31 = −126·0 lb in², $\alpha = -D/I$ = 126/51·0 x 10⁶ = 2·47 x 10⁻⁶	
y_1 = 0·00093 + 153·1 (2·47 x 10⁻⁶) = 2·195 in, y_2 = 0·00093 −91·9 (2·47 x 10⁻⁶) = 1·18 in	

Fig 22b: Example construction

Designing against fatigue

As well as either steady or impact loading, repetitive cyclic loading has to be considered in relation to the effective life of a structure. The aerospace industry has been central to the study of fatigue problems and is fully familiar with the concept of structural life — in, for example, the need to satisfy a 'Certificate of Airworthiness'.

Fatigue failures, in contrast to those due to steady load, can of course occur at stresses much lower than the elastic limit of the structural materials. Failure normally commences at a discontinuity or surface imperfection such as a crack which propagates under cyclic loading until it spreads across the section and leads to rupture. Even with ductile materials failure occurs without generally revealing plastic deformation. Fig 23 shows the terminology for describing stress level and the loading may be either complete cyclic reversal or fluctuation around a mean constant value. Fatigue life is defined as the number of cycles of stress the structure suffers up until failure. The plot of number of cycles is referred to as an *S-N* diagram, Fig 24, and is available for different materials based on laboratory controlled endurance testing. Often they define an endurance range of limiting stress on a 10 million life cycle basis. A log-log scale is used to show the exponential relationship $S = C.Nx$ which usually exists for C and x as constants depending on the material and type of test, respectively. The graph shows a change in slope to zero at a given stress for ferrous materials — describing an absolute limit for an indefinitely large number of cycles. No such limit exists for non-ferrous metals and typically, for aluminium alloy, a 'fatigue limit' of 5×10^8 is defined. It has also become practice to obtain strain/life (Fig 25) and dynamic stress/strain (Fig 26) for materials under sinusoidal stroking in test machines. Total strain is derived from a combination of plastic and elastic strains and in design it is usual to use a stress:strain product from these curves rather than a handbook modulus figure. Stress concentration factors must also be used in design.

When designing with load histories collected from instrumented past vehicle designs of comparable specification, signal analysis using rainflow counting techniques is employed to identify number of occurrences in each load range. In service testing of axle beam loads it has been shown that cyclic loading has also occasional peaks due to combined braking

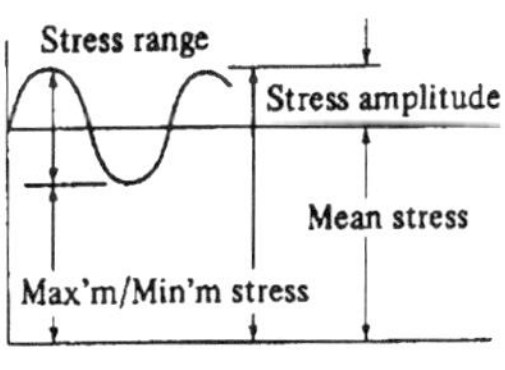

Fig 23: Terminology for cyclic stress

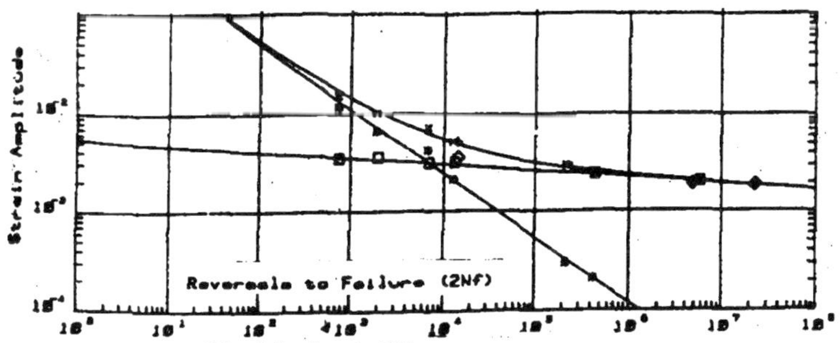

Fig 25: Strain/life curves

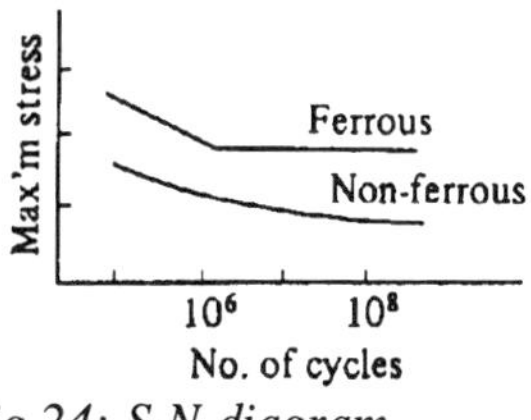

Fig 24: S-N diagram

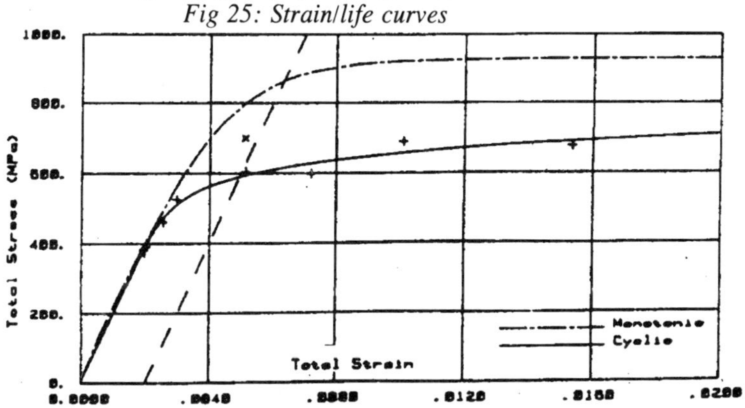

Fig 26: Dynamic stress/strain curves

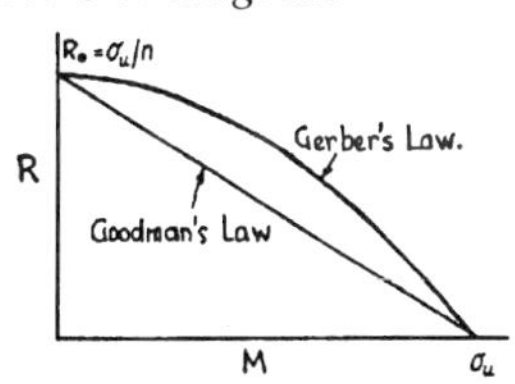

Fig 27: Fatigue limit diagrams

and kerbing equivalent to four times the static wheel load. Predicted life based on specimen test data could be twice that obtained from service load data. Calculation of the damage contribution of the individual events counted in the rainflow analysis can be compared with conventional cyclic fatigue data to obtain the necessary factoring. In cases where complete load reversal does not take place and the load alternates between two stress values, a different (lower) limiting stress is valid. The largest stress amplitude which alternates about a given mean stress, which can be withstood 'infinitely', is called the fatigue limit. The greatest endurable stress amplitude can be determined from a fatigue limit diagram, Fig 27, for any minimum or mean stress. Stress range (R) is the algebraic difference between the maximum and minimum values of the stress. Mean stress (M) is defined such that limiting stresses are M +/- R/2.

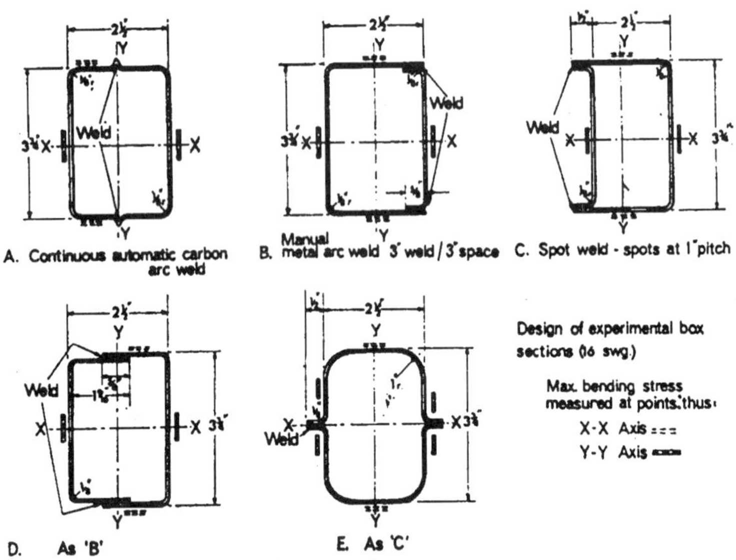

Fig 28: Box-sections used in TWI fatigue tests

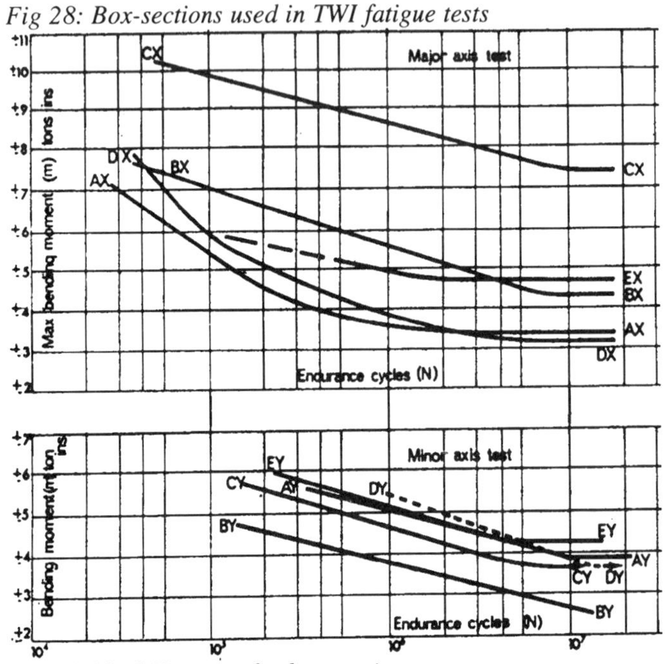

Fig 29/30: S-N curves for box sections

Section moments of inertia, in⁴

Section	I_{XX}	I_{YY}	J
A	1·55	0·79	1·57
B	1·67	0·81	1·31
C	1·98	1·09	1·43
D	1·75	0·75	1·24
E	1·07	0·94	1·30

Fig 31: Section properties

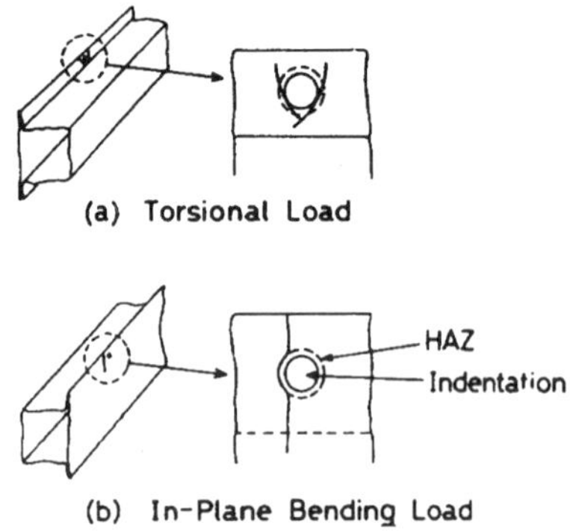

Fig 32: Crack comparison

Strength reduction factors in bending

Specimens	Fatigue strength at 10^7 cycles ($\pm$ tonf/in^2)	Strength reduction factor
EX	8	1·0
CX	7	1·1
BX	4·5	1·8
AX	4	2·0
DX	3	2·7
EY	5·5	1·4
BY	4	2·0
CY	3·5	2·3

Strength reduction factors in torsion

Specimens	Applied torque corresponding to 10^7 cycles (tonf/in)	Strength reduction factor
A	1·8	1
B	0·75	2·38
C	1·7	1·05
D	0·75	2·38
E	1·3	1·3

Fig 33: Box-member test results

Fatigue limit in reverse bending is generally about 25% lower than in reversed tension and compression, due, it is said, to the stress gradient — and in reverse torsion it is about 0.55 times the tensile fatigue limit. Frequency of stress reversal also influences fatigue limit — becoming higher with increased frequency. An empirical formula due to Gerber can be used in the case of steels to estimate the maximum stress during each cycle at the fatigue limit as $R/2 + (\sigma_u 2 - nR\sigma_u)^{1/2}$ where σ_u is the ultimate tensile stress and n is a material constant = 1.5 for mild and 2.0 for high-tensile steel. This formula can be used to show the maximum cyclic stress σ for mild steel increasing from one third ultimate stress under reversed loading to 0.61 for repeated loading. A rearrangement and simplification of the formula by Goodman results in the linear relation $R = (\sigma_u/n)[1 - M/\sigma_u]$ where $M = \sigma - R/2$. Fig 27 shows the relative curves in either a Goodman or Gerber diagram frequently used in fatigue analysis. If values of R and σ_u are found by fatigue tests then the fatigue limits under other conditions can be found from these diagrams.

Where a structural element is loaded for a series of cycles $n1$, $n2$ .... at different stress levels, with corresponding fatigue life at each level $N1$, $N2$ ... cycles, failure can be expected at $\Sigma n/N = 1$ according to Miner's Law. Experiments have shown this factor to vary from 0.6 to 1.5 with higher values obtained for sequences of increasing loads.

Fatigue of structural box sections

Spot welded box section members typical of vehicle side-rail elements have been tested by the Welding Institute and different joint configurations evaluated as in Fig 28. These were made in mild steel to EN 24 and subject to both free-free bending and torsional oscillations about both xx and yy axes at 53 to 83 Hz. S-N curves in terms of bending moment are shown in Fig 29 for the sections whose inertia properties are shown in Fig 30. Section E proved to be the best performer in bending and the deduced reduction factor relative to that section, for the others, is shown in Fig 31. In the torsional tests, a vibrator was used to oscillate the sections at their resonant frequency given by $(60/2\pi)[GJ.12g/I_oL]^{1/2}$.

For shear modulus $G = 11.7 \times 10^6$ lbf/in^2 (80 x 10^6 kN/m^2), length $L = 36$ in (0.91 m), vibrator inertia $I_o = 3260$ lbf/in^2 (22 MN/m^2) and torsional constant J = 1.82, the resonant frequency is 40Hz for gravity acceleration g = 32.2 ft/sec^2 (981 cm/sec^2). In the torsional case A proved to be the best performer with reduction factors for the others shown in Fig 31. Tests carried out by Nihon University for the Japan SAE used single and double closed top hat sections. Cracks were found to occur on spot welds close to centre span of the torsional specimens where effect of end constraint on developed stress is minimal. Cracks occurring differ from those in bending tests as shown in Fig 32 while Fig 33 shows how increased metal gauge favourably affects torsional fatigue life. Useful advice on the effects of arc welding on fatigue life is given in British Standard BS 5400 Part 10.

Service load data

SAAB as aircraft maker as well as a car producer were early into the use of service load data for design life calculation. Early work on suspension parts used strain gauges to measure surface stresses from which bending moments were derived and a frequency distribution obtained for repesentative road and track routes for 10^6 cycles. The load spectrum was typically divided into half this number to obtain a mixture of high and low loads representative of operational conditions — and programmed rig-testing of the part carried out commencing with a medium load step and running with rising and descending steps until failure occurred. It was practice to scale up all stages of the test load spectrum to reduce test time — and repeat the tests at different severities. Plotting results on log-log paper produces a 'working life line' from which it is possible to read off life associated with level of bending moment experience.

Another approach, Fig 34, is to obtain random load histories from service tests, by plotting load against occurrences. Several individual histograms can be combined to give a total effect and the resultant extended to 106 cycles by extrapolation. For test purposes the curve is approximated by a step program. Cycles with an amplitude less than 10% of the maximum were omitted to achieve a 30% test-time saving. Fig 35 is an example of the eight steps into which a curve is typically divided. Then, as before, a series of accelerated tests with different scale factors can be run to obtain the straight life line on the log-log plot which essentially becomes a 'mixed-cycle' S - N curve for life prediction .

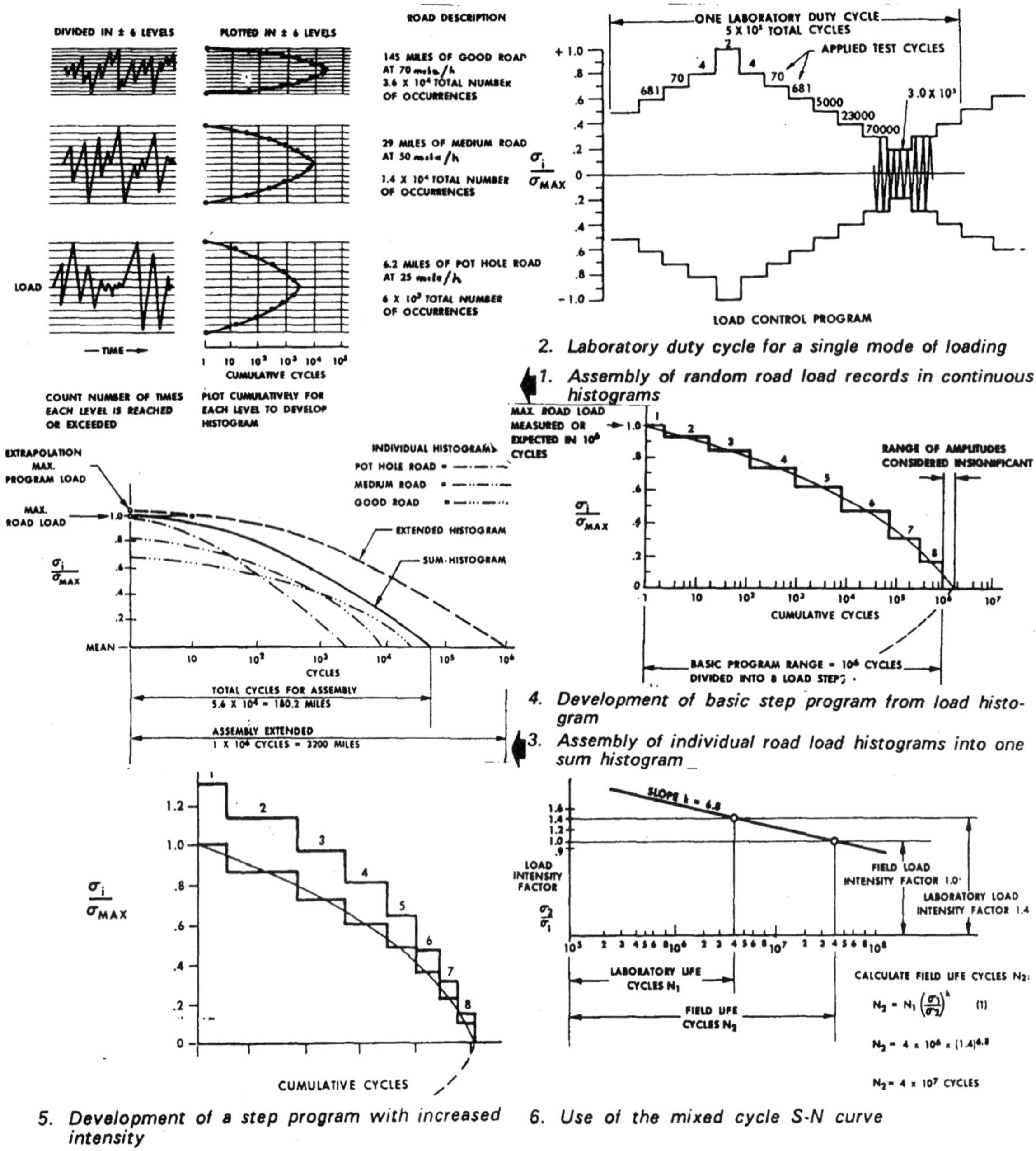

2. *Laboratory duty cycle for a single mode of loading*

1. *Assembly of random road load records in continuous histograms*

4. *Development of basic step program from load histogram*

3. *Assembly of individual road load histograms into one sum histogram*

5. *Development of a step program with increased intensity*

6. *Use of the mixed cycle S-N curve*

7. *Principle of the mixed-cycle" S-N curve*

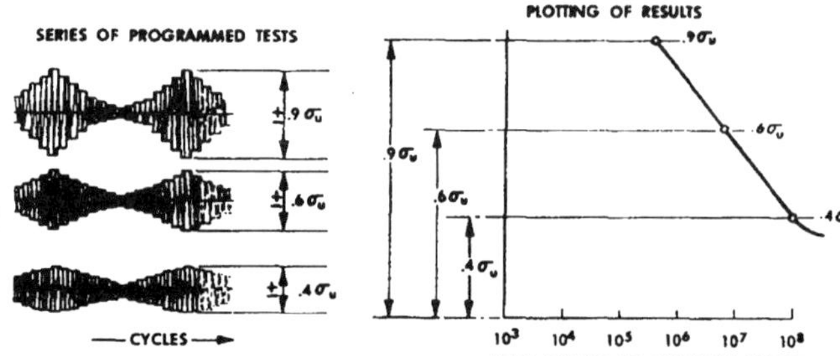

Step	σ_i/σ_{max}	Number of cycles	Cumulative total cycles
1	1	1	1
2	0·95	4	5
3	0·85	70	75
4	0·725	681	756
5	0·575	5,000	5,756
6	0·425	23,000	29,000
7	0·275	70,000	99,000
8	0·125	1·5 x 10⁵	2·5 x 10⁵

Fig 34: Method based on cyclic load histograms

Rubber as an isolator

For alleviation of many vibration problems an understanding is needed of available rubber isolation-systems; here some of the basic physical characteristics of the material are examined and a guide given to information sources on special mixes and anchorage configurations.

Both natural and synthetic rubbers are used in anchorages and compliant bushes for vibration isolation. The key mechanical properties of vulcanised natural rubber are shown in Fig 36. This data is contained in a useful guide 'Engineering design with natural rubber' obtainable from the Natural Rubber Producers Association, 19 Buckingham Street, London WC2. The very high bulk modulus compared with the elastic (Young's) modulus means that rubber hardly changes in volume even under high loads and therefore space must be available for it to expand into on deformation. By restricting the expansion, effective increase in stiffness of a spring unit can be obtained.

Breaking strength of rubber in tension is in the range 14-28 MN/m^2 based on the unstretched cross section area. This can rise to 200 for the cross section at break — compared with compression strength at failure of 160 MN/m^2. Some typical stress/strain curves for natural rubber, depending on hardness, are shown in Fig 37. Creep and stress relaxation must also be allowed for in design. These vary linearly with the logarithmn of time as shown in Fig 38. When, conversely, rapid rates of strain are applied, rubber exhibits a substantial degree of damping, hysteresis curves being shown in Fig 39. This property varies substantially with the amount of filler (usually carbon black); left curves are rubbers with 50 parts reinforcing black and right curves for 50 parts non-reinforcing black. Dotted curves are after one cycle and full curves after ten.

		Gum rubber	Filled rubber
Hardness	IRHD	45	65
Tensile strength (T.S.)	MN/m^2	28	21
Elongation at break (E.B.)	%	680	420
Young's modulus (E_0)	MN/m^2	1·9	5·9
Shear modulus (G)	MN/m^2	0·54	1·37
Bulk modulus (E_∞)	MN/m^2	1000	1200
Poisson's ratio		0·4997	0·4997
Resilience	%	80	60
Velocity of sound	m/s	37	37
Specific gravity		0·93	1·16
Specific heat		0·45	0·41
Thermal conductivity relative to water		0·25	0·31
Coefficient of cubical expansion/deg C		67×10^{-5}	56×10^{-5}
Electrical resistivity	ohms/cm cube	$1{\cdot}7 \times 10^{16}$	3×10^{10}
Dielectric constant		3	15
Power factor		0·002	0·1

Fig 36: Vulcanised rubber properties

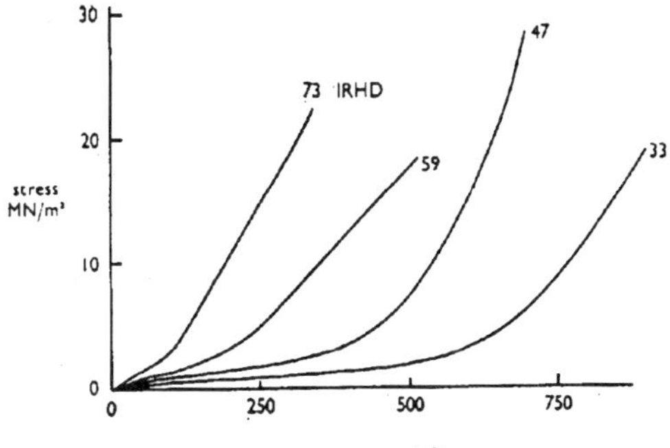

Fig 37: Stress/strain curves for natural rubber

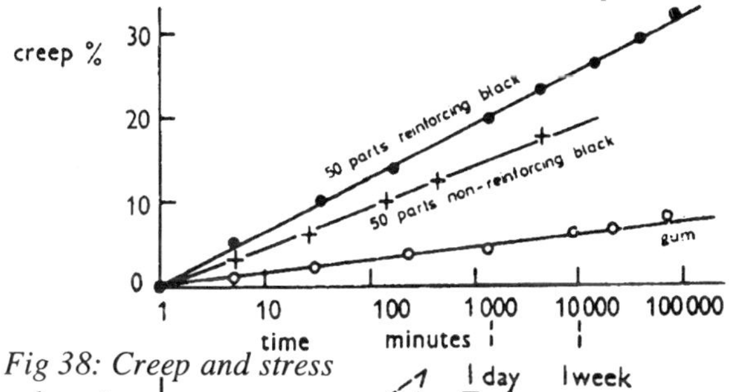

Fig 38: Creep and stress relaxation

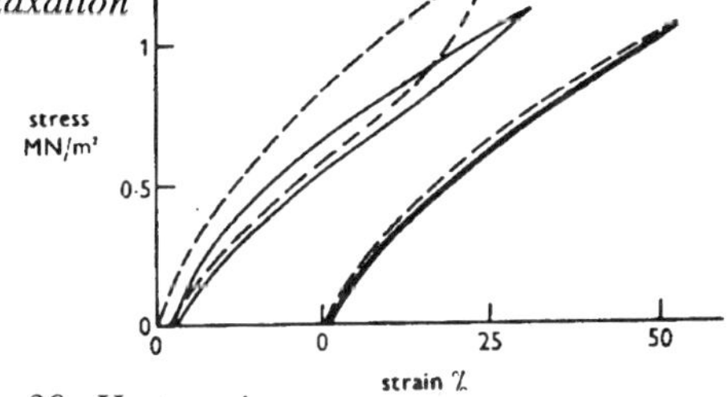

Fig 39: Hysteresis curves

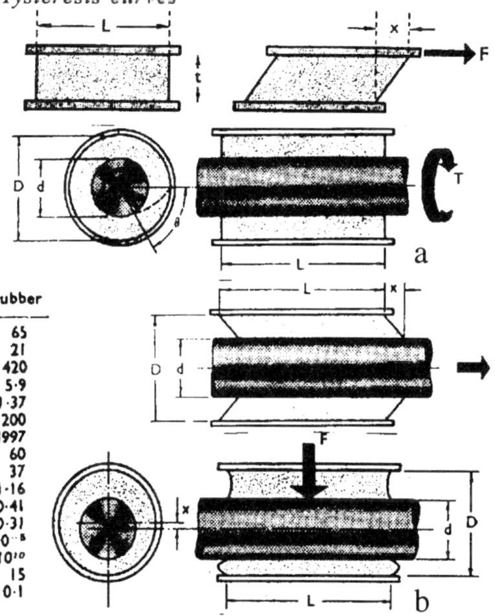

Fig 40: Bonded rubber mountings

In many applications, rubber is bonded to metal; by preventing slipping at the load carrying surfaces, more predictable and repeatable performances can be obtained. In a bonded rubber shear mounting as in Fig 40a (or in circular form, 40b), the normal elasticity formulae apply. When ratio of thickness to length exceeds about 0.25, deflection due to bending should be allowed for as

$$Ft^3 / 36AGk_r^2$$

where k_r is radius of gyration about the neutral axis $= L^2/12$ for a rectangular section. Stiffness of rubber in compression is calculated with the help of a shape factor. Provided no slipping occurs at the load surfaces, this is given by

$$S = LB / 2t(L + B)$$

for breadth B, length L and thickness t. Variation of compression modulus with shape factor is shown in Fig 41. Synthetic rubbers include polyisoprene, styrene-butadiene, butadiene, ethylene-propylene, butyl, chloroprene, nitrile, polyurethane, chlorosulphonated polyethylene, polyacrylic, ethylene-acrylic, silicone, fluorsilicone, fluorocarbon, polysulphide, epichlorohydrin, polypropylene oxide and polynorborene. There is also a range of thermoplastic rubbers. Properties and types of most of all these are listed in British standard BS 6716.

One manufacturer[1] listsas proprietary mounts and couplings, handling loads up to 14 000 lb and powers up to 2000 bhp; also, engineering theory is included in an associated Designers Guide. Different configurations can provide the required combination of stiffness and damping, Fig 42.

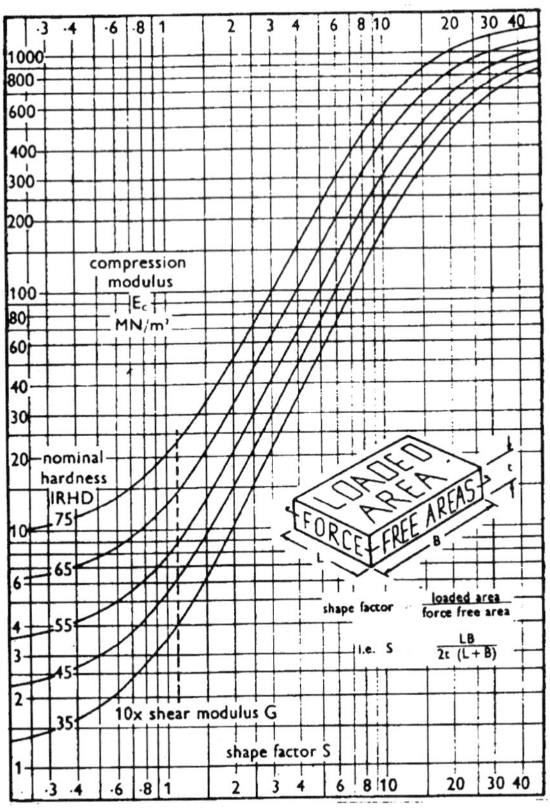

Fig 41: Compression modulus vs shape factor

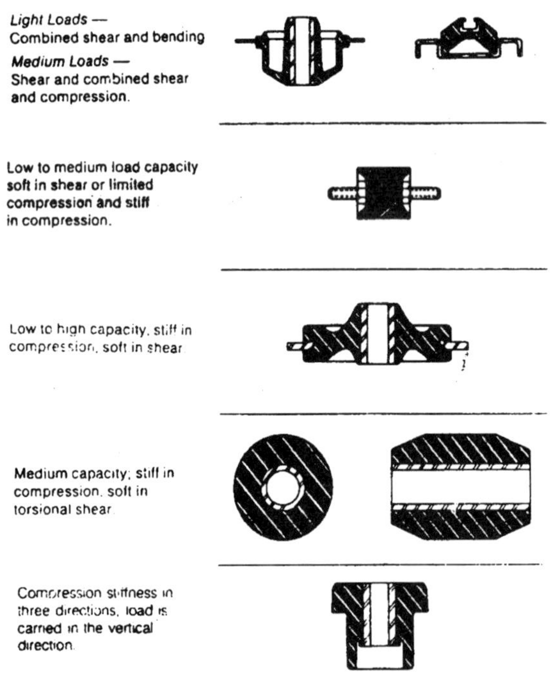

Fig 42: Isolator configurations; from top to bottom: Soft in all directions; soft in two directions, stiff in one; stiffin two directions, soft in one; stiff in two directions translationally, soft torsionally,; equal stiffness in three directions

CV body mounting

The prospect of product liability legislation has brought a considerably increased application of engineering-science into the study of chassis-to-body interaction and the publications of bodybuilder codes of practice, compliance with which will make both parties less vulnerable to product-liability claims. Vehicles such as road tankers, of the non-integral type, have historically required greater attention to the chassis/body interface in order to protect the shell structures against fatigue-failure at the mounting locations. Some important lessons were learned from this experience which now benefit the mounting of box-van shells, and related configurations, which have themselves increased in torsional rigidity as construction methods have developed.

The considerable disparity between the stiffness of CV chassis and body is likely to continue as long as any one chassis design has to serve the requirements of a proliferation of body types and modes of operation. With the exception of purpose built military, and related, all-terrain vehicles, having long-travel 'soft' suspensions, truck chassis have relatively 'hard' short-travel ones. These suspensions require a torsionally flexible frame, Fig 43, to permit sufficient cross articulation of the axles for retaining wheel ground contact on rough terrains.

Chassis frame design was considered in an earlier section; besides the constraints described in it, there is also thought to be unnecessary unladen weight penalty in making chassis frames too rigid by conventional structural techniques. Truck makers

Fig 43: Flexible chassis frame

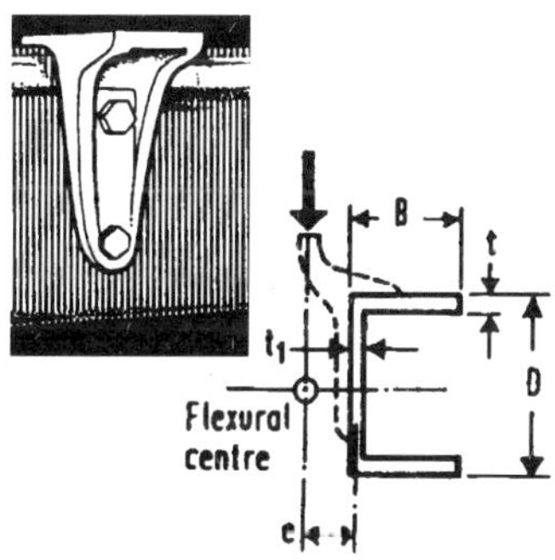

Fig 44: Mounting at the shear centre

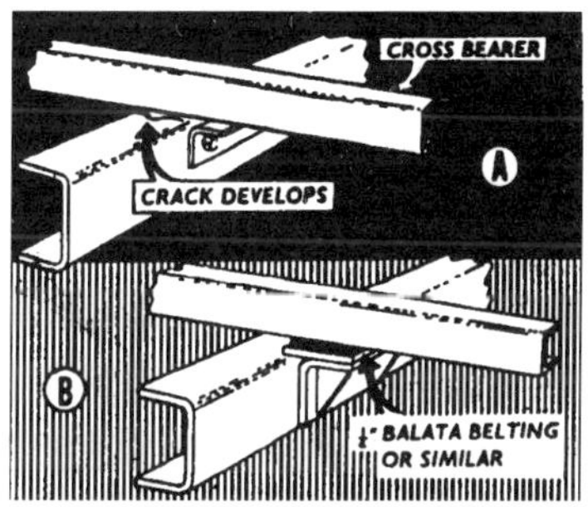

Fig 45: Protective bracket

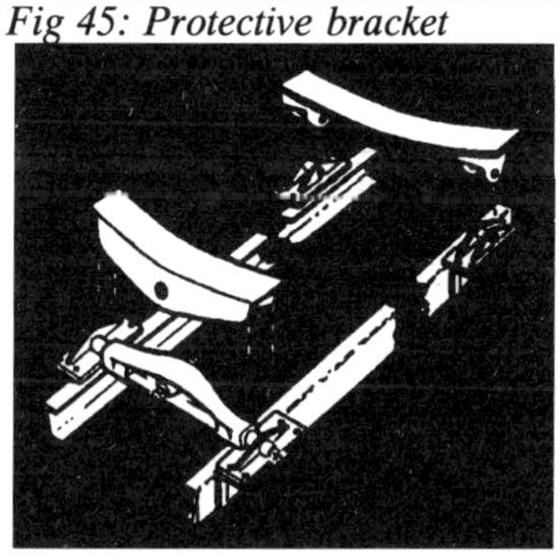

Fig 46 : tanker mounts

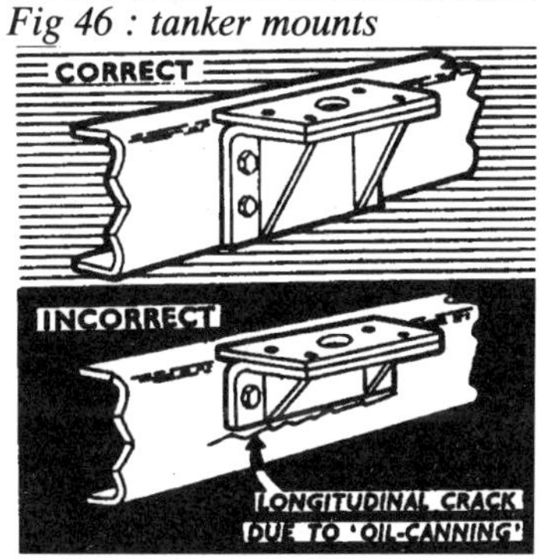

Fig 47 : Avoiding oil-canning

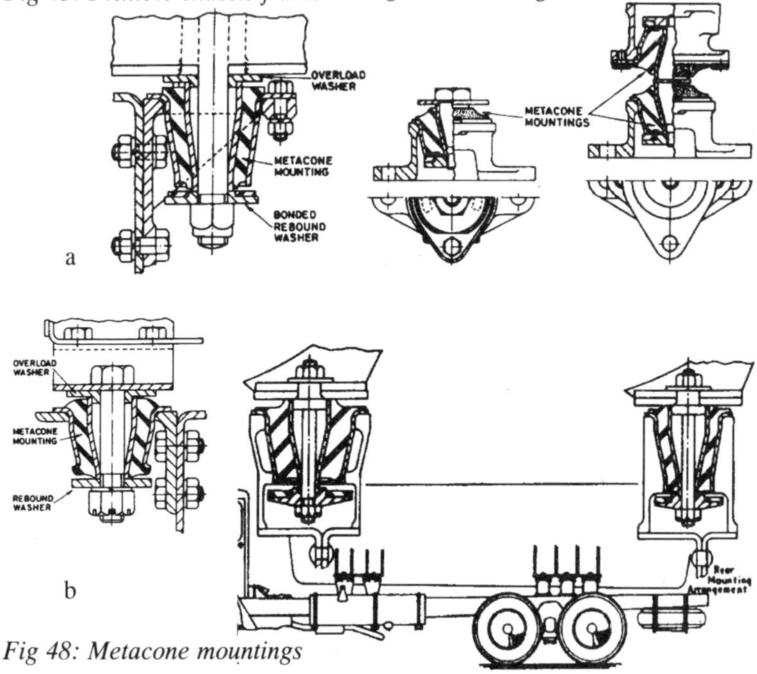

Fig 48: Metacone mountings

such as GM Bedford had considerable early experience with four-wheel-drive trucks for military application and were among the first to recommend that bodies were not direct-mounted to upper flanges of the chassis sidemembers, offering factory-supplied brackets which involved clearance between sidemembers and body cross-bearers.

This bracket design, Fig 44, was said to avoid frame cracking arising from fidgeting motions, between chassis and body, causing wear and looseness between the two, Fig 45. The distance e, defining the flexural centre, at which the line of action of vertical load gives no twist to the channel section is given by: (Bt/4I)(B-t1)(D-t)(D-2t) where I is section second moment of area for the channel. Lips of these mounting brackets were arranged so they did not overlap the top flange of the sidemember by more than one inch; this allowed vertical movement of the flanges during weaving caused by torsional flexure on rough ground. The object of the lip was to relieve the bracket fixing bolts from shear loading; as well as maintaining chassis/body clearance as discussed above.

The company also advised against longitudinal body runners direct-mounted to the sidemember top flange on the grounds that it is not designed for supporting the entire weight of body and payload. The increase in section depth is also said to stiffen the frame in the area of body support such that any flexing takes place between the foremost edge of the body and the front-spring mounts — leading to possible fatigue cracking at the stress concentration point at the leading edge of the body bearer. Hence the need for brackets connecting direct to the sidemember webs, and, the company maintained, at the shear centre of the channel section, in order to avoid localised twisting moments. The company also used a U-shaped pressing between the two brackets to the rear of the back-axle, thus making any horizontal lozenging motions impossible.

For tankers, or even torsionally-stiff insulated box vans, the company recommended a layout which allows the stiff body to float on the chassis, unaffected by the latter's weaving. Three or five-point resilient mountings, Fig 46, were recommended for tanks, with a centrally located trunnion mount used to support the front end. As fewer outrigger brackets are used, in such cases, than for conventional bodies, greater strength must be built into them and the company recommended their vertical faces extending to the full depth of the sidemember web in order to avoid 'oil-canning', Fig 47.

Proprietary resilient mounts are obtainable from such companies as Metalastik. The company recom-

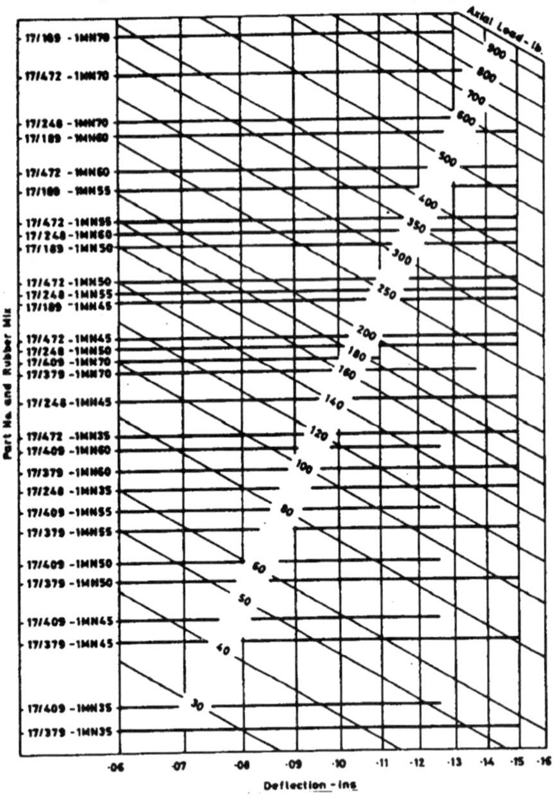

Fig 49: Metacone characteristics

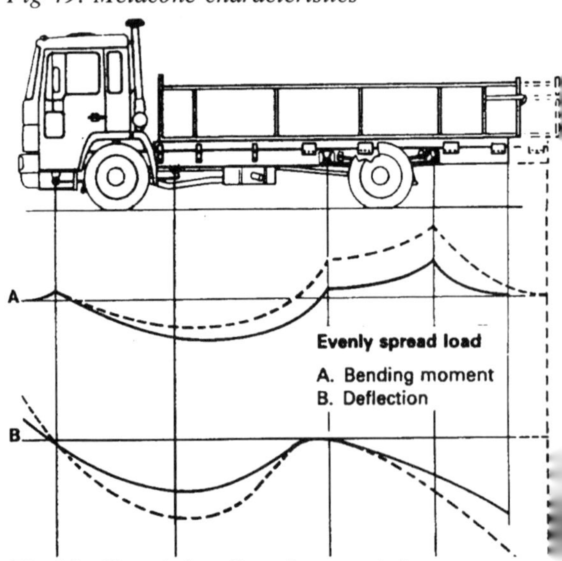

Fig 50: Chassis bending characteristics

mend that mountings should be designed to give ample flexibilty over a strictly limited range of movement — beyond which stiffness of the resilient elastomer should increase rapidly and progressively. Positioning along the chassis has, of course, to be calculated so that the deflection of such units is fully utilised without overstraining. The Metacone mounting, Fig 48, is available in four types; one has integral buffers and and metal washers for bump and rebound (a). A second has integral buffer and metal washer at one end plus separate bonded rebound washer at the other (b). A third comprises mounting plus bump and rebound washers, assembled into a robust housing with two-hole fixing at the base while a fourth includes the SL derivative mounting having cut-out portions in the 'rubber' — with integral buffers and metal washers for bump and rebound. Typical characteristics are shown in Fig 49.

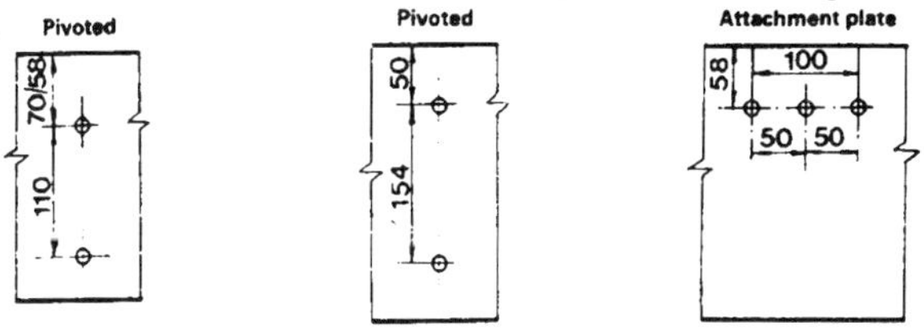

Fig 51: Pre-punched hole arrays

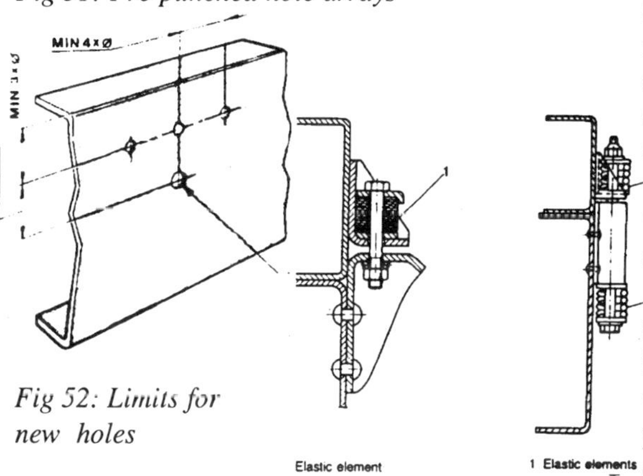

Fig 52: Limits for new holes

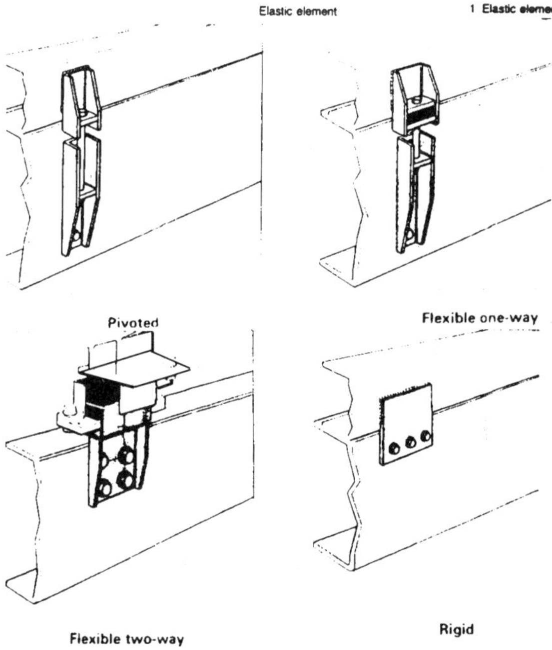

Fig 53: Standardised attachments

Guidance figures for chassis frame deflections, constrained by mounted bodies, have been provided in pioneering quantitative analyses by Erz who quotes frame twist of +/- 25 deg and wheel deflections of 53 cm from ground level with maximum obstacle heights ranging from +/- 30 cm, on-road, to +/- 50 cm on road. Softer vehicle suspensions are of course continuously reducing frame deflection, with the passage of time, relative to spring deflection. One of Metalastik's approaches was to recommend stiffest mounting in the transverse plane of thc payload centre of-gravity then softening progressively towards the end of the body.

The main CV chassis/cab manufacturers have now issued comprehensive recommendations applying to their own chassis-frames. Volvo have provided frame bending diagrams for mounting bracket positioning; recommended bracket types for bodies of varying torsional stiffness and related bolting recommendations. Superstructure Instructions give specific advice for such variants as tippers, tankers and vans. Fig 50 is a typical bending moment and deflection plot for the FL6 'rigid' model with platform body. Reference holes are provided on the delivered chassis to facilitate measurements from the axle centrelines. Ready-punched holes (Fig 51) are also now provided for different attachment (mounting) type. Controlling dimensions are also given (Fig 52) for extra holes in the web to suit specialist bodies. Four types of regular attachments have been standardised (Fig 53) for pivoted (a), flexible one-way (b), Flexible two-way (c) and rigid (d) according to body type (Fig 54). The Leyland DAF Superstructures manual uses the concept of matrices for giving specific recommendations according to wheelbase, axle configuration and body type. These work around the basic mounting systems shown in Fig 55 of single tie-rod attachment (a) and retainer plate (b). Iveco Ford issue the Bodybuilders

Instructions manual which recommends a number of 'auxiliary frames' and mounting instructions for distributing loads from different body types uniformly on the chassis, Fig 56. These frames may be flexible or rigid according to recommendation for a particular body type and its connection to the chassis (Fig 57) varies accordingly. For a box-van body the minimal dimensions for the longitudinal members of the auxiliary frame are shown in Fig 58. Where the box body uses crossbearer supports brackets rising above the auxiliary frame, these need triangulated anchorages to contain lengthwise thrusts, Fig 59.

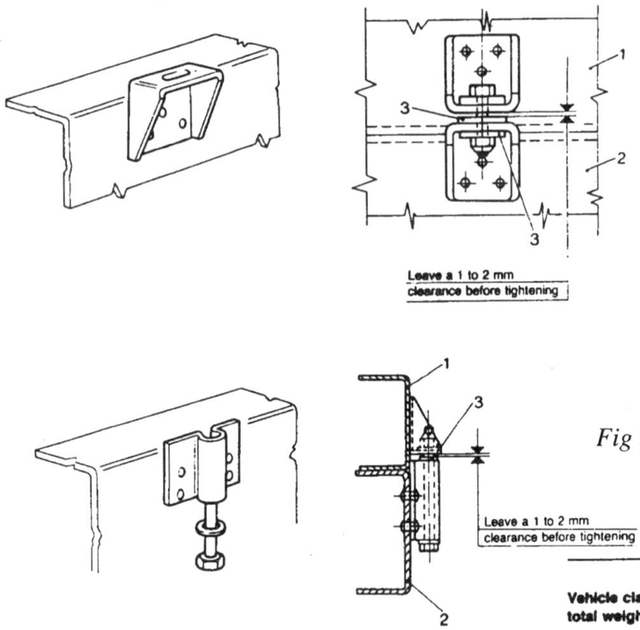

1 Auxiliary frame
2 Frame
3 Shims

Fig 54 : Flexible attachments

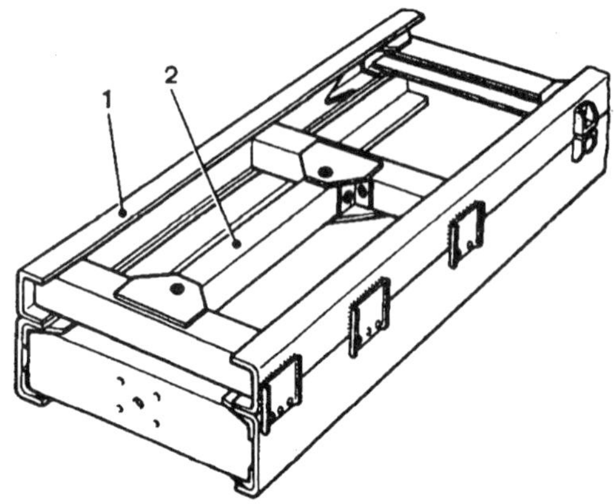

Fig 56: Auxiliary frame

Vehicle class total weight	MINIMUM SECTIONAL REINFORCING IRON	
	Moment of resistance for each section iron Wx (cm³)	Dimensions (mm)
3 ÷ 7 ton	16	80x50x4
7,9 ÷ 10 ton	26	100x50x5
11 ÷ 33 ton	46	120x60x6

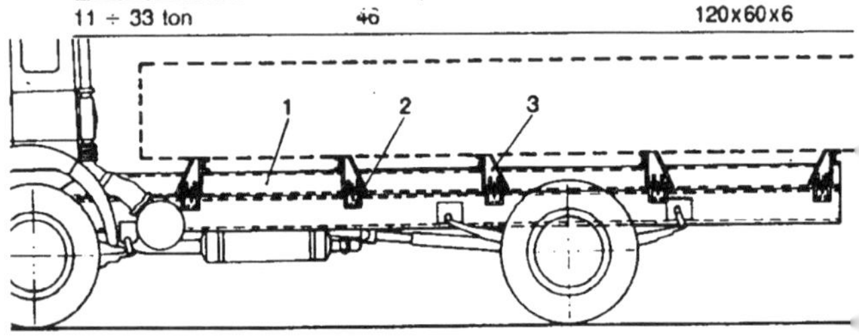

Auxiliary frame
Brackets
Securing anchorages

Fig 58: Auxiliary frame sizes

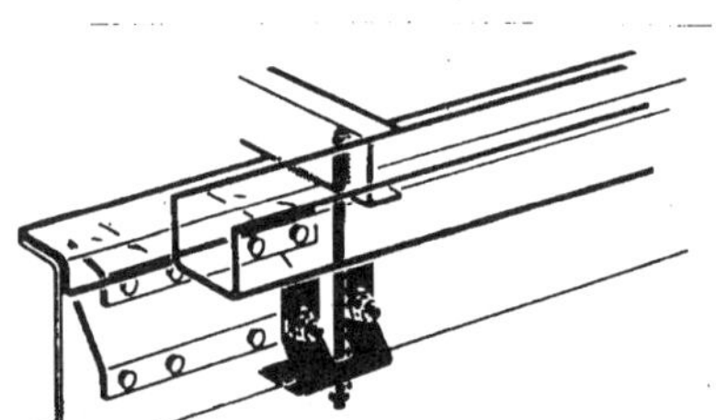

Fig 59: Triangulated anchorages

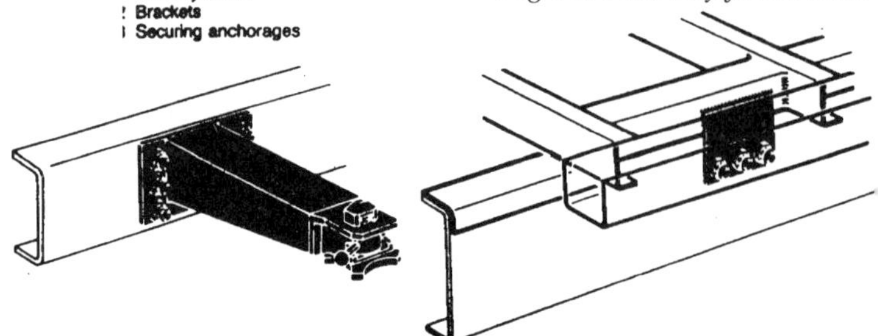

Fig 55: Attachment variants

Anti-noise enclosures

Here the method of isolating noise by means of 'structural enclosures' is considered whereby the noise source is enclosed in a box which insulates the emanation of airborne noise as much as possible. This can be the quickest and cheapest method of reducing perceived noise, which of course is airborne to the human ear, especially so if the design and prototype build is already well advanced.

The full enclosure is an extension to the use of noise shields which are sometimes preferred if the noise source is an engine which requires frequent maintenance access to it and needs airflow over its surface. Low stiffness and high damping are the requirements of such shields and it is necessary to isolate them structurally from the vibrating source so that they do not act as vibration amplifiers, typical methods of mounting being shown in Fig 60. Distance between shields and surface of the engine, say, must also be minimised in order to prevent noise build up in the intervening space. Obviously the shields do not seal in the airborne noise like an enclosure but merely interfere with the radiation of sound from the adjacent vibration surface.

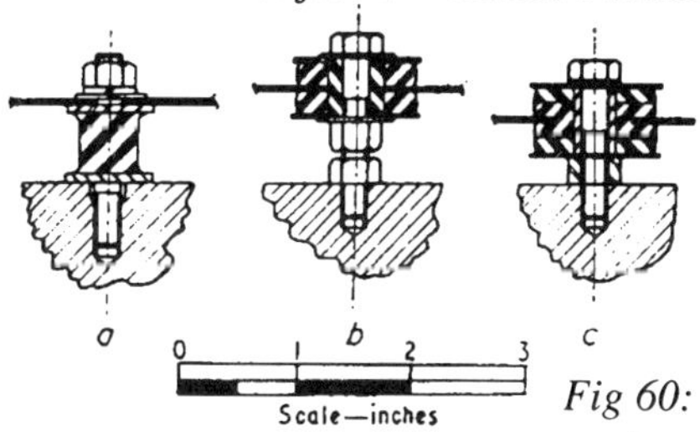

Fig 60: Types of enclosure mounting

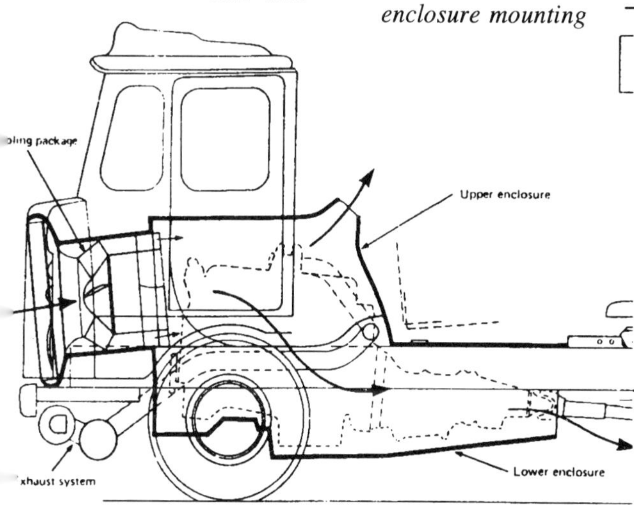

Fig 61: QHV enclosure

It is also possible to use a semi-sealed enclosure where the maintenance of airflow is vital but maintenance access less so. Fig 61 shows such an enclosure on the TRRL QHV 90 quiet heavy vehicle project where a so-called tunnel enclosure was used. This comprised an upper enclosure covering the top and back of the engine which lifted with the cab; separate panels fixed to the chassis covered the top of the gearbox and bottom of the power-unit; the complete set formed a duct through which cooling air passed.

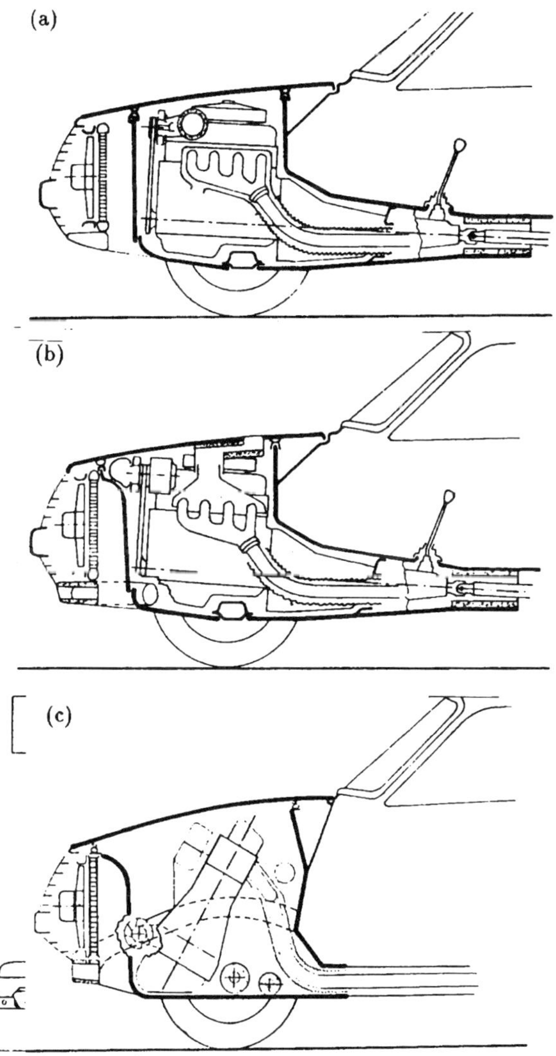

Fig 62: Car configurations to suit different engine/radiator installations

Similar enclosures have been reported by AVL for use in cars but with the cooling airflow kept out of the enclosure by arranging for separate coolant-radiator ducting. Fig 62 shows alternative solutions for different engine and radiator configurations. At (a) the radiator cooling air is forced downwards by the bonnet and leading bulkhead assisted by additional air passages through the wheelarch and bonnet as necessary. A supplementary fan on the water pump pulley provides a limited airflow within the enclosure, air entering it passing through a silencer and exiting both around the exhaust pipe and the transmission shaft. At (b) the leading bulkhead is replaced by a duct behind the radiator to achieve cost saving. Airflow in the enclosure is also simplified by replacing the encapsulation of the exhaust system by a chimney and air-silencer just beneath the bonnet air-outlets. The chimney creates a thermal draft and the exhaust system passes through the enclosure within an acoustic sleeve. This same philosophy applies to the transverse engined layout at (c) but without the chimney element, airflow inside the enclosure being assisted by the dynamic pressure head in front of the vehicle. Fig 63 shows the noise reductions obtained. In commercial vehicle noise research, reported by Ricardo, it is considered possible to reduce an engine noise source by 10dB(a) using full enclosures. Using ones which give acceptable maintenance access can also give reasonable attenuation. Fig 64 shows simple side shields having a 25 mm thick PUR coating — used together with an unlined sump undertray. The existing engine cover was used but a lining added. Shields were made from 1.6 mm thick damped steel laminate and weighed a total of 30 kg to achieve 6dB(a) sound reduction. Antiphon Ltd of 170 Park road , Peterborough, PE1 2UF are prominent in the supply of highly damped panels suitable for noise shielding. Their SA material is produced for the reduction of airborne noise and is a self-supporting absorber comprising a rigid, resinated felt-like carrier, moulded in a checked pattern with cavities and ridges. A 1 mm polyester film is heat-sealed to the panel and bonds to the ridges so that an air gap back exists between it and the base of the carrier. Noise causes the film to vibrate freely into the cavities between the ridges allowing sound waves to enter the porous carrier, the makers claim. The material weighs 0.25 lb/ft² (1.22 kg/m²) and is about 1/4 in (6.35 mm) thick, Fig 65. Absorbtion co-efficients are in Fig 66.

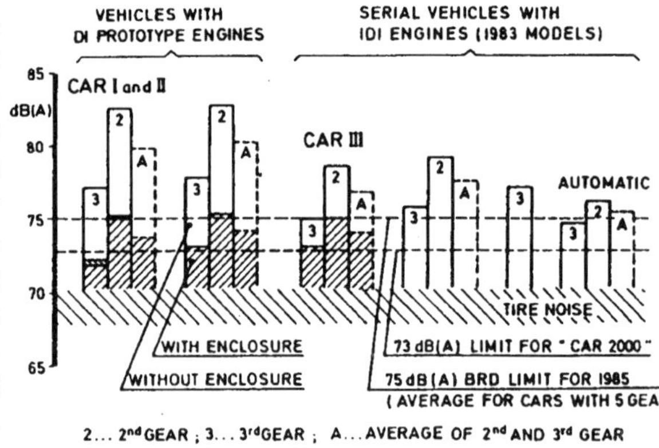

Fig 63: Noise reduction for each configuration

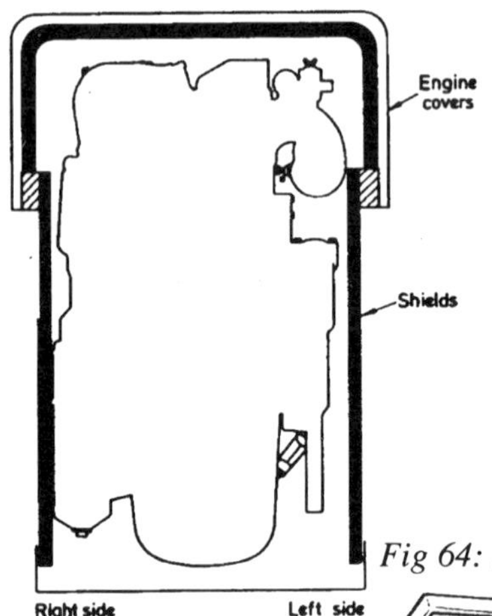

Fig 64: Side shields

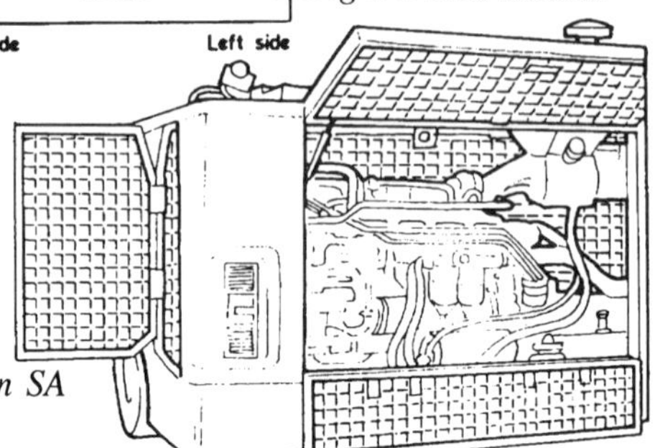

Fig 65: Antiphon SA system

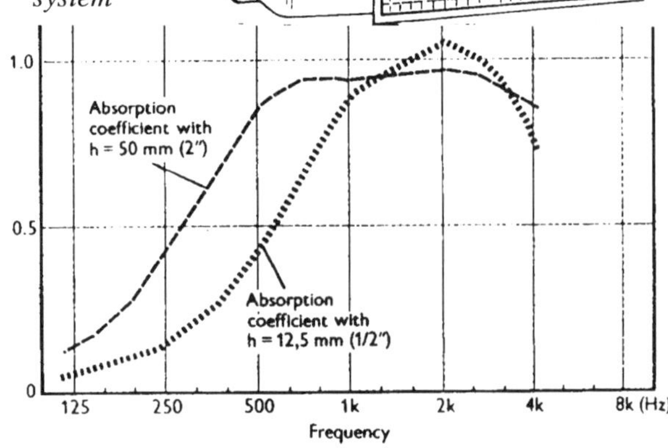

Fig 66: Absorption co-efficients

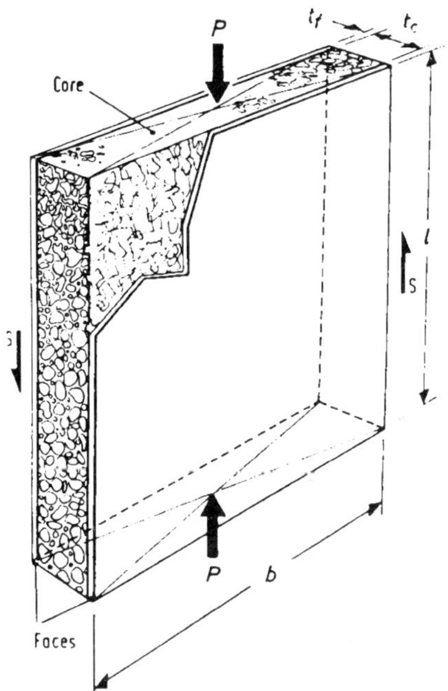

Fig 67: Sandwich panel parameters

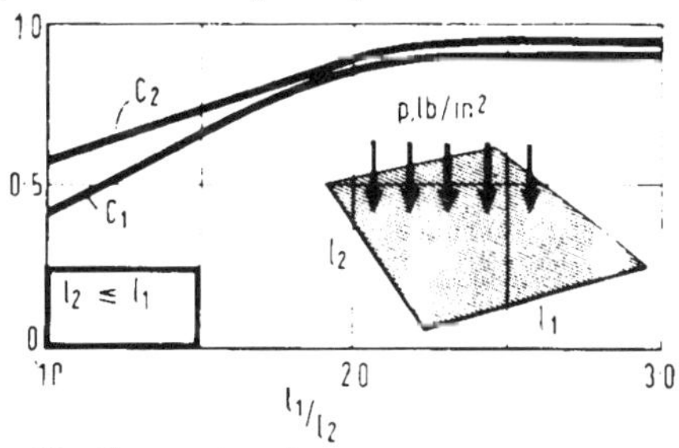

Fig 68: Correction factors

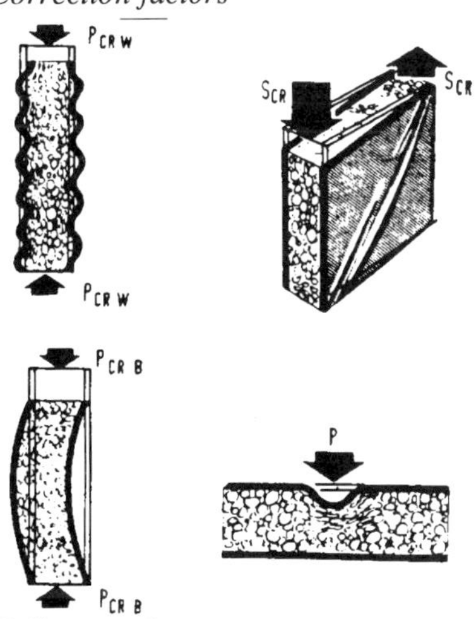

Fig 69: Failure modes

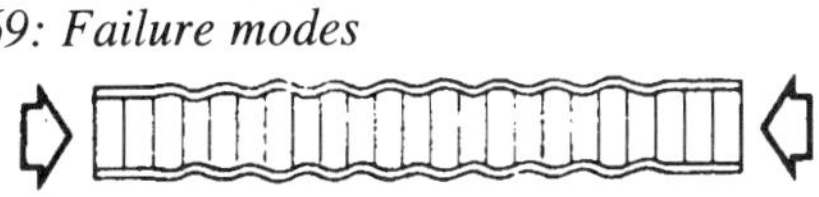
Fig 71: Honeycomb-cored panel

Sandwich panel analysis

Analysis of forces and deflections in faces and cores as a means of optimising the main parameters is important for structural performance. Choice of skin and core modulus, density and sustainable working stress can be made so that the lightest possible panel is stable with respect to the external loading and that acoustic properties are optimised.

Principal dimensions of sandwich panels and a representation of the external forces for analysis is shown in Fig 67. Direct compressive and shear loads shown, rather than lateral bending or local indentation forces, are the ones for which the resistance of the panel to buckling, or local wrinkling, must be considered. The concept of shear buckling in unsupported skin panels has been introduced in an earlier section alongside buckling of unsupported skins under direct end load. At this stage formuli will be provided, but not fully derived, to show the critical buckling loads.

The objective of a structural sandwich panel is to fully exploit the strength of the face material without premature instability of the panel. Even the most elaborately reinforced 'sheet/stringer' arrangement of panel and framework is said to be liable to failure under end load — with induced stress below 50% of the maximum stress sustainable by the panel. This is because of a lack of bending stiffness in the panel and its frame. In the case of the sandwich panel, the core material, being capable of transmitting shear force between the sandwiching skins, effectively acts as the web between two 'flanges' represented by the skins. The efficiency of the sandwich panel is thus usefully defined as the ratio: maximum load per given width of panel by the weight of the given area of panel — divided by the same ratio for an equi-sized panel which has faces fully stabilised by a weightless core. This points to the key properties of the core material as specific stiffness and specific strength and the failure modes under end load being local instability of the faces or overall column buckling of the panel (or, if over stabilised, direct overstress of facings or core). Sandwich panels were first seriously analysed in the 1940s co-incident with their use in aircraft structures such as the Mosquito of World War 2 fame. Researchers at the Royal Aircraft Establishment (Farnborough) used a strain-energy method to derive the critical end-load for wrinkling failure as

$$0.63 E_f^{1/3} E_c^{2/3}$$

where E is the elastic modulus for face and core. For column buckling of the panel, the parameters of panel length and edge support must also be considered — together with thicknesses of face and core plus the bending stiffness of the combination.

To calculate sandwich panel bending stiffness it is generally assumed that shear stress is uniformly distributed across the core. Central deflection for a uniformly loaded sandwich panel of moderate width, simply supported at its ends, is given by

$$(5wa^4/384E_fI) + (wa^2/8G_cA)$$

where a is the length between supports, A is the core cross section area, I the section second moment of area and G_c the shear modulus of the core material. To allow simply supported panels to be designed without resort to complex calculation, correction factors shown in Fig 68 have been derived which, if used with simple beam theory, give good approximations in calculation. The deflection, or bending moment, is first calculated by simple beam theory, above, and then multiplied by the appropriate factor, C_1 in the case of central deflection and C_2 in the case of bending moment. The critical end load P_{cr} for unit width of panel, in overall column buckling, can then be found from the formula

$$(1/P_{cr}) = (2a^2/\pi^2E_ft_ft_c^2) + 3/2Gt_c$$

factored by $(1 + a^2/b^2)^2$ for panel length a and width b where dimensions shown in Figs 67 and 69 apply.

For shear instability of the panel depicted in Fig 69c, critical shear stress in the panel is given by

$$(2.75K_sE_c\text{-}f\ t_c^2/b^2)/(1+ 5.5E_ft_ft_c/[b^2k_{sp}G_c])$$

where K_s and K_{sp} are factors relating to single faces and sandwich panels respectively as obtained from Fig 70. The above calculation formulae apply to isotropic core materials such as rigid foams for which it can be assumed that physical and mechanical properties are uniform in all directions. A modified approach is necessary for honeycomb cores such as those used for racing car construction. Formulae due to Ciba Geigy which apply to that company's Aeroweb honeycomb core material used in sandwich panels give critical buckling stress for the face skins as: $(d/b)^2EK$ where d is defined in Fig 71 and K is obtained

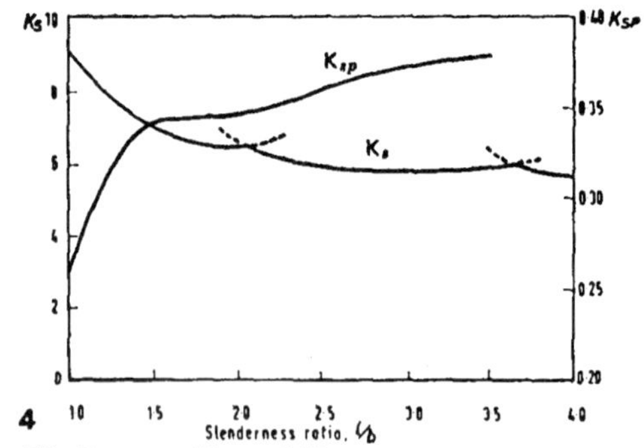

Fig 70: Factors Ks and Ksp

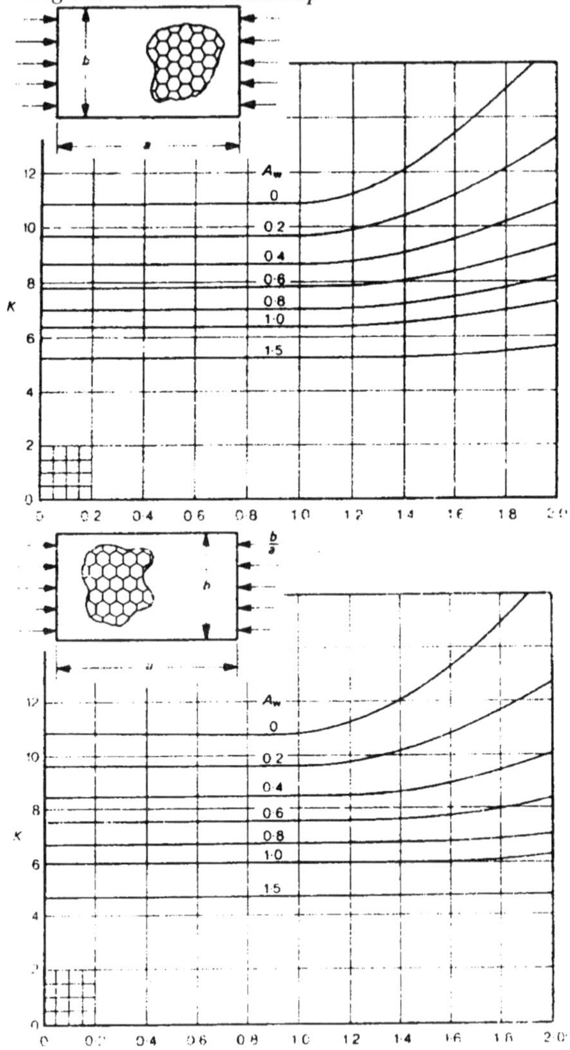

Fig 72: k-factors for honeycomb cores

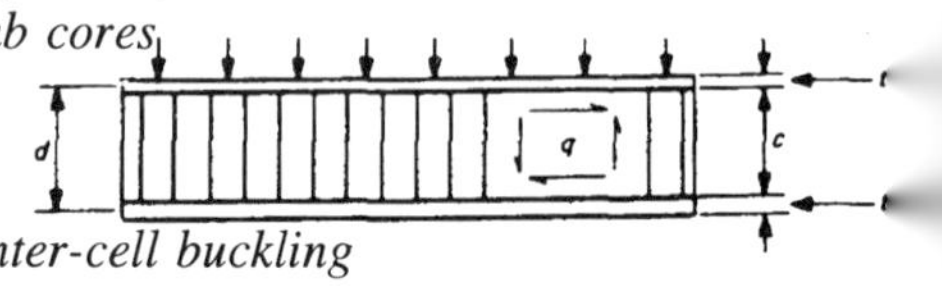

Fig 73: Inter-cell buckling

from Fig 72a or b according to the core ribbon direction. The various failure modes for honeycomb core panels are described in the company's Construction Trends publication series (July 1989) and the critical wrinkling stress for end-loaded panels becomes $0.50(G_c E_c E_f)^{1/3}$. A further possible failure mode is inter-cell buckling shown in Fig 73. Here the critical buckling stress in the skins is given by $3E_f(t/d)^2$ where d is now the diameter of a circle inscribed within that cell.

For a rectangular honeycomb-cored sandwich-panel subject to uniform normal pressure, the maximum value of stress in the direction of the long l or short s side of the panel is given by $\sigma_{l,s} = s^2pC_2,C_3/dt_1$ where C_2 and C_3 are obtained from either Fig 74a or b depending on the core ribbon direction of the honeycomb. For the lower facing skins, the stresses are obtained by substituting t_2 for t_1 in the formula above. Shear stresses in the core under this load condition can be found from $q_{1,2} = spC_4,C_5/d$ where $C_{4,5}$ are obtained from the appropriate graph in Fig 75a and b. The central-deflection of panels under uniform side loading p can be calculated from

$$s^4p\ C_1\ /EJ \text{ where } J = t_1t_2d_2/(t_1 + t_2)$$

and C_1 takes values from the graphs in Fig 76a and b.

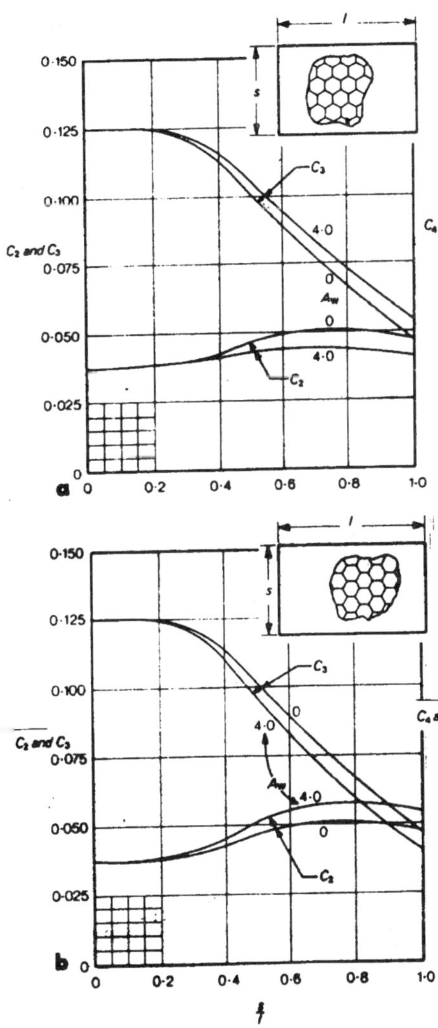

Fig 74: Factors C2 and C3

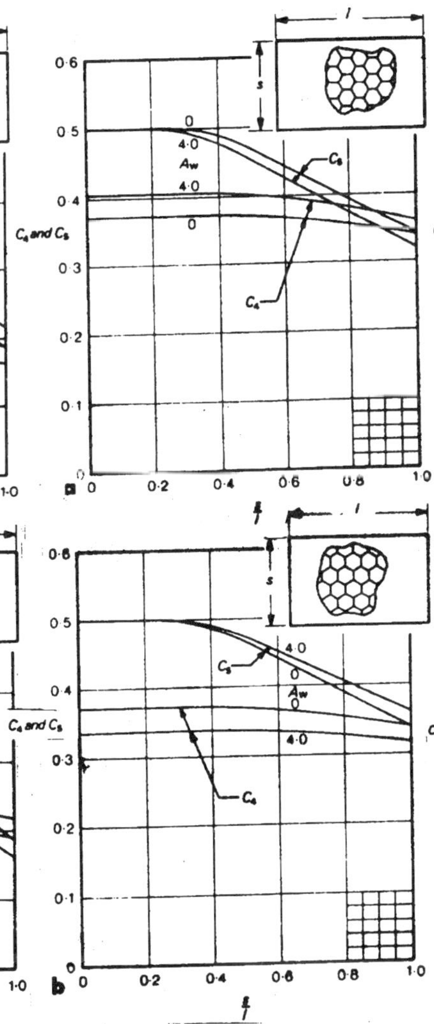

Fig 75: Factors C4 and C5

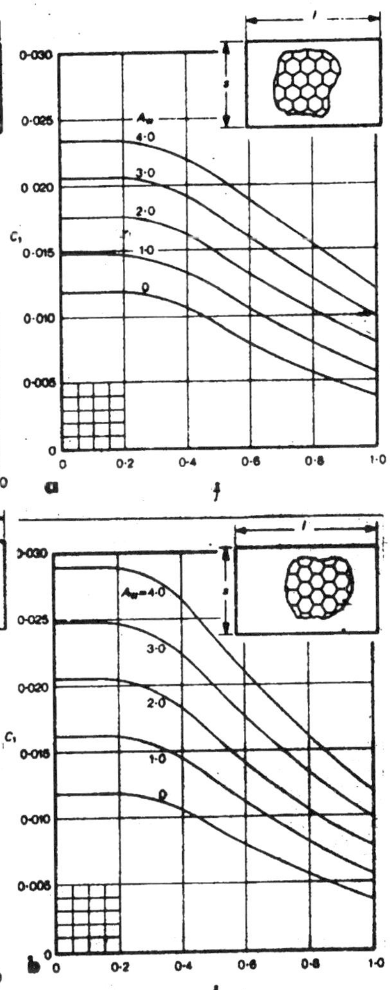

Fig 76: Factor C1

Structural dynamics applied

Recent computer analysis packages have the potential to automate complex structural dynamics calculations. Several recent papers are advancing the approach of integrating dynamic structural and fatigue analyses.

Optimal design changes to solve vibration induced fatigue failures can only be derived by including structural dynamics into fatigue life calculations, according to Liefooghie *et al* of LMS International[2]. This was particularly so when desire for weight saving precludes the time-worn technique of beefing up the failed component in the area of the fatigue crack. The integration is effected by a third input to the analysis: strain histories from service records of the component.

The first step is to derive forces by representing the operating data, accelerations and strains, in the frequency domain in relation to frequency response functions (FRFs). The influence of a vibration mode can then be ascertained by synthesizing a strain signal from these forces and FRFs. Also, by applying structural modifications to calculate modified FRFs, the effect of design changes on fatigue life can be predicted.

Within the Esprit Project package 2486 DYNAMO, Fig 77, such a procedure has been demonstrated for the resonance fatigue of a car body. Design modifications can be built in to the software to test the effect of 'what if' changes in material properties and stress concentration factors, say. Resonance fatigue problems have been studied in this way by Porsche and the results confirmed in track and laboratory fatigue tests. The approach is to drive a first test car over an endurance track until a fatigue failure occurs; then a second car is instrumented to measure strain and acceleration signals of relevance. In one case, spectral analysis of these signals identified vertical acceleration of the gearbox mounting and front differential as the most important ones which caused a fatigue crack.

These are shown in Fig 78(a), along with the critical strains close to where the crack occurred, as a function of track distance covered calculated from an integrated wheel speed signal — which indicate the four sections of the track with bad surface. Power spectra of these signals over these bad sections show a peak in the same frequency range, Fig (78b). Corre

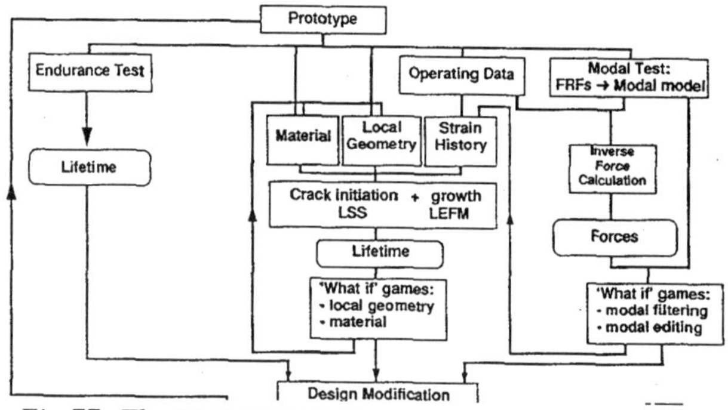

Fig 77: The LMS DYNAMO package

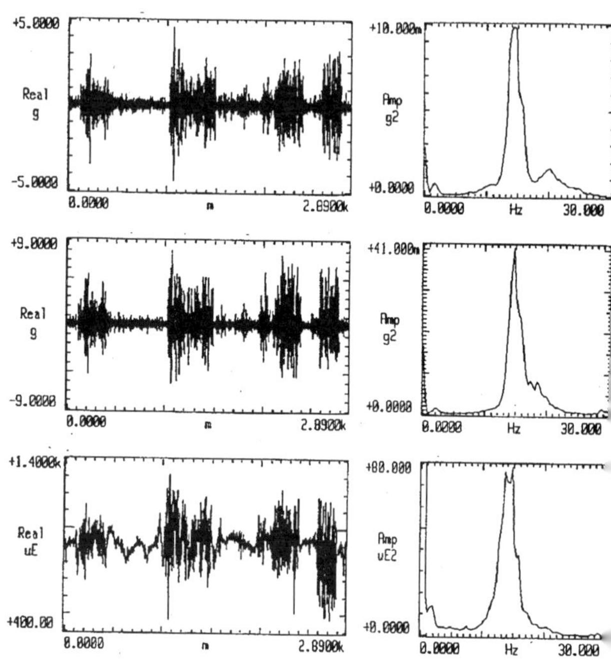

Fig 78: (a) Test recordings, left; (b) Power spectra, right

Mode No.	Reference (Hz)	Initial (Hz)	Initial Diff. (%)	Final Diff. (%)	MAC(%
7	6.77	6.77	0.01	0.01	100.
8	11.32	11.32	0.02	0.00	100.
9	17.34	17.48	0.77	0.03	99.4
10	27.21	29.95	8.76	1.22	96.0
11	33.13	33.74	1.82	0.57	70.9
12	33.53	35.39	5.53	0.98	90.4
13	41.21	43.85	6.41	0.48	97.2
14	43.21	45.71	4.13	0.31	85.4
15	47.35	49.75	5.06	1.46	53.3
16	49.17	51.01	3.73	0.84	57.0
17	53.95	54.33	0.70	0.23	51.5
18	54.24	55.61	2.52	0.37	88.8

Fig 80: Model updating procedure

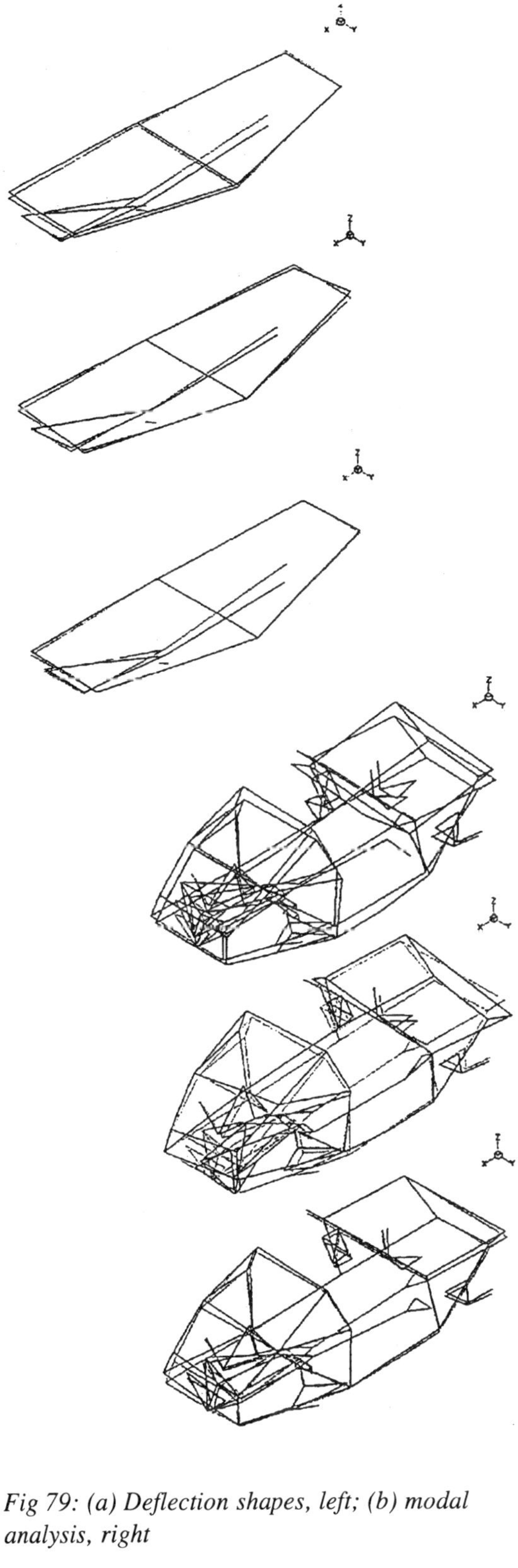

Fig 79: *(a) Deflection shapes, left; (b) modal analysis, right*

-sponding operational deflection shapes in a multi-source environment (signals from transducers throughout the structure) can be calculated by a principal-component analysis, the authors explain.

Traditional techniques for determining such shapes assume the whole structure behaves coherently with respect to one reference transducer. The coherent motion can, of course, be made up of several components when closely coupled modes are excited by driving over a test track. Principal Component analysis extracts more information from the data so as to introduce dimensionality of the operating phenomena. It results in a statement of the (independent virtual) signals called 'principal components'. By calculating cross-powers between the measurements and each of these, deflection shapes coherent to the three principal components can be displayed as a function of frequency. Fig 79(a) shows these at the frequency of the spectra peaks for the accelerations and critical strain. These correspond well to the three of the modes identified in modal analysis of the second test car, Fig 79(b):

* 1st critical component at The critical frequency cor responding to a vertical powertrain mode
* 2nd corresponding to powertrain rolling coupled to body torsion
* 3rd corresponding to a vertical front differential mode

The three modes play a key role in the structural dynamics mechanism causing the fatigue crack and the authors demonstrate good correspondence between experimentally observed and analytically predicted initiation with total life.

Sensitivity analysis

The way in which FE models can be validated and updated, using sensitivity techniques and weighted updating of physical parameters, is explained by McCulloch and Dascotte[3] of Dynamic Engineering. The technique helps the analyst to assess why a simulation model differs from a real structure. Modelling of stiffness and mass can be inaccurate, for example, because of an insufficient number of elements. The modal behaviour of the model may be taken as a basis for assessment but because of limited degrees of freedom results will only be valid in a limited frequency range.

Sensitivity-based updating is a technique for validating the FE solution automatically. It involves the calculation of a correlation criterion such as the Modal Assurance Criterion (MAC) to identify correlation between analytical modal results and those derived by Experimental Modal Analysis (EMA). Identifiable errors are those due to the FE formulation used, those due to geometrical simplification (mesh density) and those due to the input values of mass or stiffness.

SYSTUNE is the name of the program which has been developed for interfacing databases from FE and EMA sources. It covers all possible situations related to the purpose of the FE model, information available on modelling assumptions and the number/type of analytical/experimental results available. Fig 80 shows the model updating procedure applied to the example of a car body; essentially a sequence of updating runs is performed in which mass, static- and then dynamic-behaviour is corrected.

Substitution of polymer-composites

Obviously the material properties need to be accurately known to obtain an accurate FE analysis. Where properties are uncertain then approximated values can be updated in the technique, reflecting the real material as built into actual components. When several sub-structures are assembled together updating is best carried out in steps: first the components themselves and then the joints between them. Also coarse FE models can be derived from fine models and forced to adopt the same dynamic behaviour, thus saving on computer time. Such detailed effects as ribs, flanges and solid transitions can be merged into the coarse models in a dynamically correct way, to give accurate resonant frequencies and mode shapes.

Fig 81: Car model with modified joint conditions

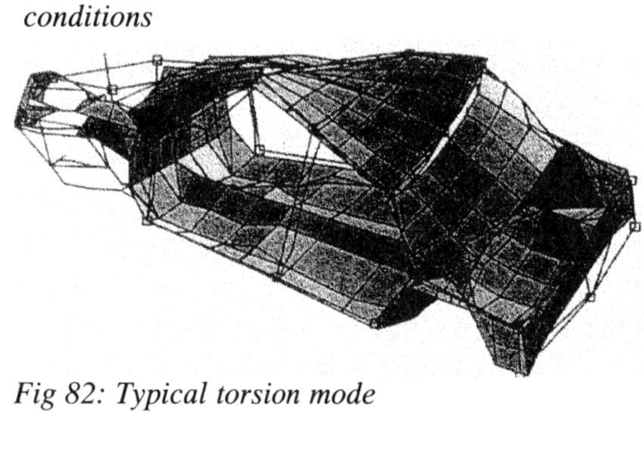

Fig 82: Typical torsion mode

Fig 81 is an example of a car model for which the joints were updated by changing their rotational spring stiffnesses to obtain proper correlation, comparing actual dynamic behaviour with a reference one. No boundary conditions were considered since the structure was freely suspended. Fig 82 shows a typical torsion mode, for which the table in the earlier Fig 80 lists all modes before and after updating. Errors on resonance frequencies were reduced to an average of less than 1% and some rotational stiffnesses were reduced to zero, showing that the joints were effectively 'pinned'.

Effect of trim on dynamic performance

According to Cheng and Olatunbosun of Birmingham University[4], a vibrating structure can be considered analagous to a large number of simple sources distributed geometrically on the enclosure panels. Knowing the contribution of each of these, and relevant phase information, sound pressure level can be predicted at any point within the cavity. The sound pressure level also yields the vibro-acoustic relationship which leads to damping energy dissipation due to trim coverings.

When FE analysis is otherwise used to model the modal behaviour, with assumed structural damping, there is the possibility of errors, the authors maintain. With the 'simple source' technique the structure can be assessed in an absolute sense over a given bandwidth in terms of performance figures which account for effects of both resonant and non-resonant modes. Once structural problem areas are identified, the affects of damping trim pads can be evaluated.

The authors define 'point mobility' as the input point velocity response normalised by the driving force. This parameter therefore describes the ability of a point in a structure to admit vibration inputs and hence is a measure of performance at an attachment point where vibration energy is introduced. Transfer mobility is used to describe the velocity response of one point with respect to a vibration input at another. Modal mobility is the average transfer mobility over part of, or the whole of, the vehicle structure, as a single parameter. It allows elemental phase cancellation due to structural damping and is determined by summing the transfer mobility vectors for elemental areas dividing the structure. Thus it can be extended to provide a measure of vibro-acoustic performance in respect of structure-borne radiation.

Structural dynamic performance standards determined from a large sample of monocoque-structured vehicles over a considerable time-span are -10 and -15 for point mobility and -28 and -36 for modal mobility. The figures refer to 10Hz and 10-100Hz bandwidths, in turn — units being dB, ref 10^{-3}m/Ns. A comparison of these parameters is obtained between various stages of trim. In a test programme on five trim phases, three of them are presented in the paper, fully trimmed, partially trimmed (damping pads) and untrimmed body-in-white. The structure is hung from A-frames on soft springs. Fig 83 shows the layout of the test structure which is a luxury car monocoque carrying 156 transfer mobility measurement stations distributed over nine panels enclosing the interior cavity. Some 55 tests were carried out at the 22 input stations shown by the table of Fig 84, the right hand columns showing the results. These show that the sample structure had poor broadband point mobility performance with a number of attachment points failing to meet the -15dB standard. The results are discussed in detail within the original paper.

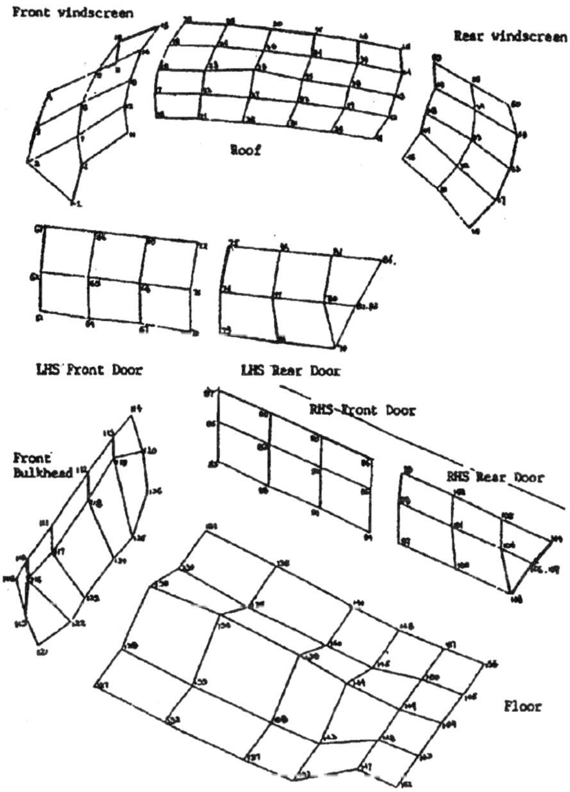

Fig 83: Layout of test luxury-car structure

Excitation Station & Station Code	Direction	Point Mobility dB, ref 10^{-3} m/Ns 10-250 Hz	Modal Mobility dB, ref 10^{-3} m/Ns 10-250 Hz
LHS Front Suspension Lower Arm Mount (R1)	Vertical	-7.78	-29.51
Front Subframe Engine Mount (R2)	Vertical	-3.71	-29.87
Rear Subframe Engine Mount (R3)	Vertical	-10.72	-29.97
RHS Engine Mount (R4)	Vertical	-19.83	-36.30
LHS Engine Mount (R5)	Vertical	4.55	-29.25
LHS Front Suspension Upper Arm Mount (R6)	Vertical	-23.47	-38.61
Bulkhead Engine Mount (R7)	Fore/Aft	-11.46	-31.66
LHS Front Suspension Tie Rod Mount (R8)	Fore/Aft	-15.83	-35.80
LHS Front Suspension Tower (R9)	Vertical	-18.98	-37.04
LHS Front Anti-roll Bar Mount (R10)	Vertical	-5.34	-29.10
LHS Rear Suspension Shock Absorber Tower (R11)	Vertical	-17.84	-39.36
LHS Rear Suspension Coil Spring Tower (R12)	Vertical	-15.55	-37.97
LHS Rear Suspension Lower Arm Mount (R13)	Vertical	-15.59	-36.46
LHS Rear Suspension Trailing Arm Mount (R14)	Fore/Aft	-22.54	-42.93
LHS Rear Anti-roll Bar Mount (R15)	Vertical	-10.15	-38.17
RHS Engine Mount (R16)	Fore/Aft	1.92	-33.00
LHS Engine Mount (R17)	Fore/Aft		
LHS Front Suspension Upper Arm Mount (R18)	Lateral	-16.40	-36.36
Front Subframe Engine Mount (R19)	Fore/Aft	-8.49	-32.50
Rear Subframe Engine Mount (R20)	Fore/Aft	-7.01	-32.67
LHS Front Suspension Lower Arm Mount (R21)	Lateral	-11.59	-34.72
LHS Rear Suspension Lower Arm Mount (R22)	Lateral	-22.45	-40.35

Fig 8: Test details and results

Chapter 5: Structural applications

Bus structural design basics
Case study in bus structural design
SPV idealisation
Designing in structural-composites
Analysis of composite structures
Hydroformed tubular structures
Bonded aluminium alloy structures

Bus structural design

The development of design techniques is traced from the introduction of the first integral structures, through the application of regular stressing procedures for subassemblies to the overall finite element analysis of the shell. First, the approximate methods, based on considering cut sections of the shell, are examined as a means of obtaining a feel for contributions made by different elements to the overall structural performance.

Much structural analysis of bus bodies dates back to work done in East European countries who pioneered the application of aeronautical techniques to early forms of integral construction. Bus production at the Hungarian Ikarus company ran at some 12,000 units per year in the 1960s and 70s and many of the techniques were developed at the Autokut Research Institute in that country. These techniques were developed into structural design procedures at the Cranfield Structural Design Group[1]. An approximation for stresses in an integral bus structure can be obtained by assuming the bending loads are carried by the side frames. The side door apertures are critical areas as they represent large cut-outs in the structural surfaces. Shear force at the centre of each opening is assumed to be shared equally between the cant rail and sole bar according to their respective stiffnesses. These in turn cause root bending moments at the corners of the opening which can be calculated as in Fig 1. Redundant-frame analysis of this ring-frame can be avoided by making the assumption of the frames being formed of a series of cantilevers, subjected to a shear force 50% greater than the value obtained from the side-frame shear force diagram.

Also, the bending moment at the centre of the door will give rise to equal and opposite end loads in the cant rail and sole bar which are readily reacted by the sidewall structure up to the waist-rail. Force P on the cant rail, however, will be balanced by a sum of forces Qs developed by bending of the window pillars. These can be calculated according to the stiffnesses of the respective pillars from the formulae shown in Fig 2/3. To allow for the fact that the joints at the top of the pillars are gussetted rather than pin-jointed the bending moment in each pillar is assumed to be reduced by one third. Torsional loading of the body, Fig 4/5/6, gives rise to a torque which can be considered to act across the structure which acts as a short tube of length equal to the bus width.

The shear flow q is the n given by $q = T/2A$ from the torsion formula developed in an earlier section. This shear flow is carried down from the roof by the pillars in the same way as the bending case but causing bending moments at right angles to the bending-case ones. Fig 7 shows the bending moments in the windscreen pillars for example. To compensate for the poor resistance of the side pillars to shear loading on the side frame, an anti-lozenge panel, Fig 8, can be provided.

Semi-integral construction vehicles with negligible stiffness in the pillars can be analysed as trough structures, Fig 9, with the side wall up to the waist rail carrying the substantial part of the bending loads. These will be carried by both chassis and body, in proportion to their relative stiffnesses. Whether the body is timber, aluminium-alloy or steel-framed makes a dramatic difference to the contribution made towards stiffness if it is assumed that the frame reacts only bending and the skin-panels only shear. The body contribution might be higher in those coaches which have no central door openings on the sideframes and which have ring frames made up of the chassis/body floor cross-bearers, side pillars and roof sticks.

Fig 10 shows the 'trough' representation and

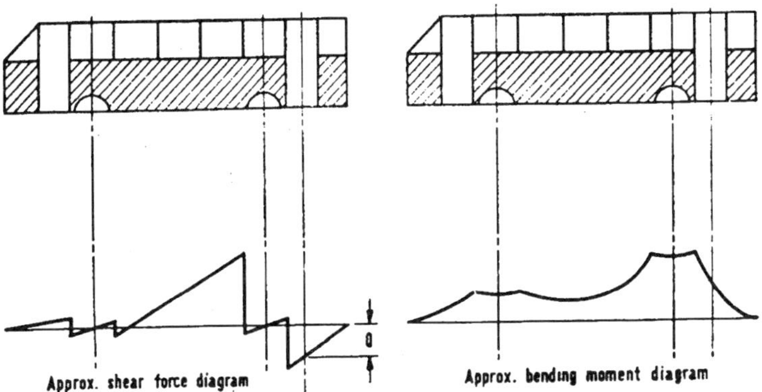

Fig 1: Bending moments in door aperture frames

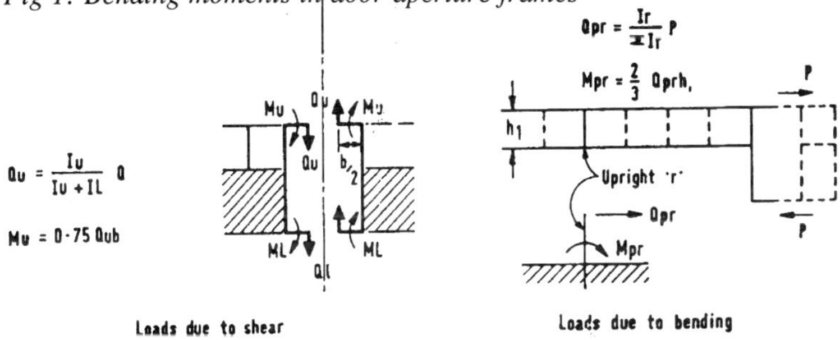

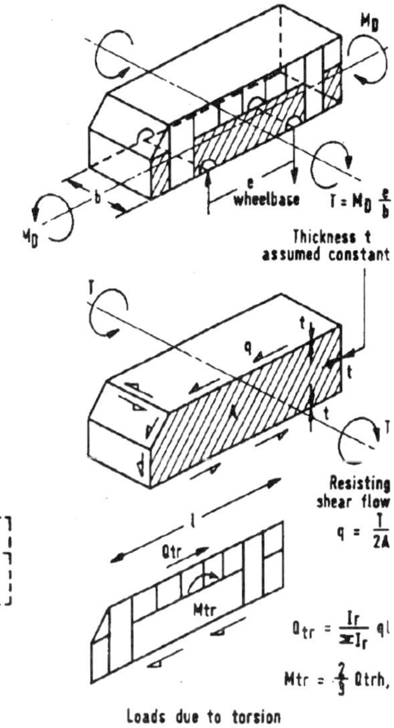

Fig 4/5/6: Torsional loading

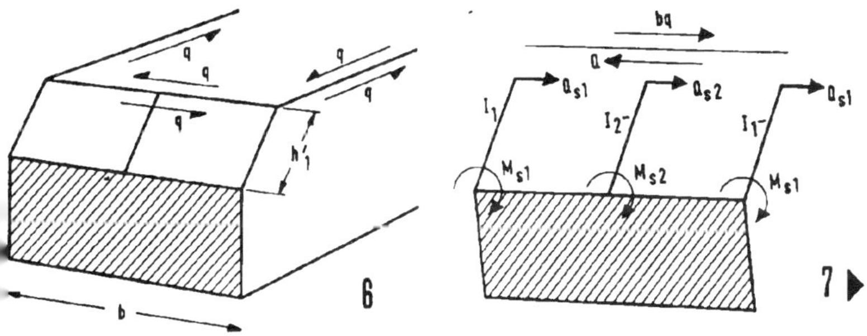

Fig 7: Windscreen pillar loading

$2Q_{s1} + Q_{s2} = Q = bq$
$Q_{s1} = I_1 bq/(2I_1 + I_2)$
$M_{s1} = \frac{2}{3} Q_{s1} b_1$

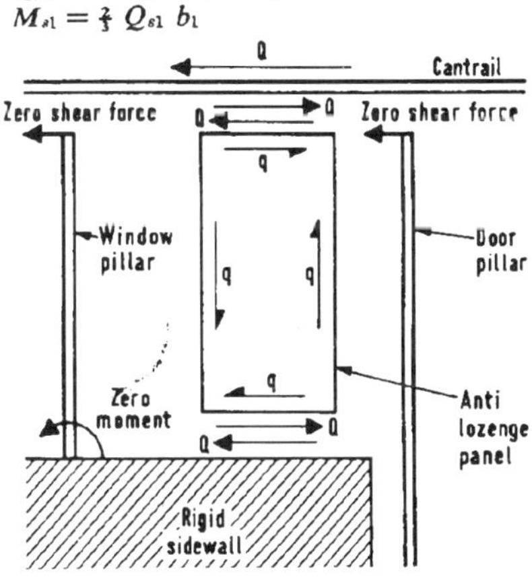

Fig 8: Anti-lozzenge panel

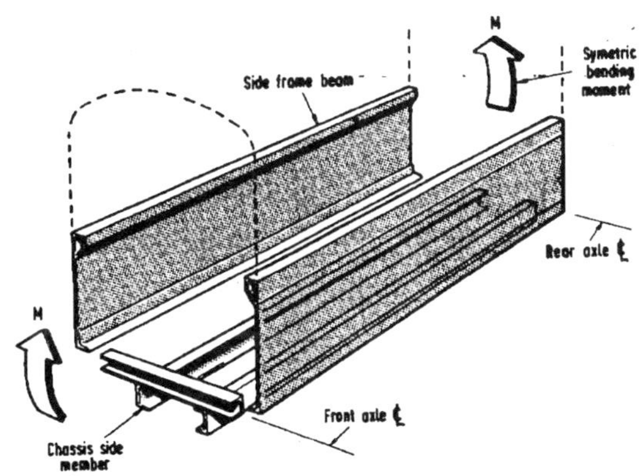

Fig 9:Trough structure

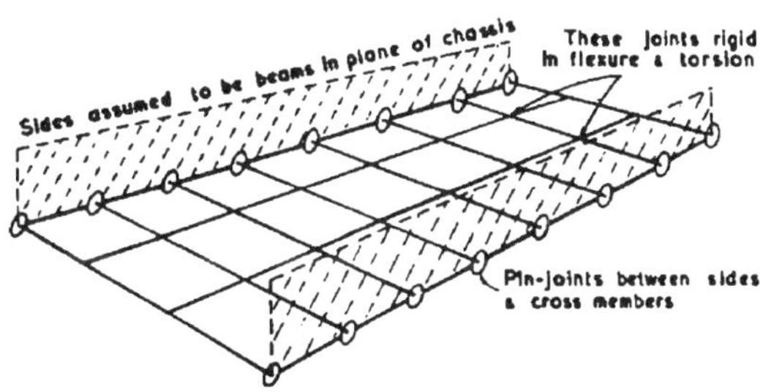

Fig 10: trough idealisation

Fig 11 a model test that was carried out at Cranfield to compare the stiffnesses of sideframes with and without door 'cut-outs'. The deflections of the two frames are compared in Fig 12. When the structure is made assymmetric by having a door cut-out(s) only on one side the two sides of different flexibilities are joined by a series of rings whose shear flexibility will affect the overall structural flexibility. If the sides of the rings are very stiff, shear will be carried by the roof-sticks and floor-crossbearers, as by the cant-rails and door-sills at the cut-out.The effect of one such ring, at mid-span, is calculated in Fig 13. Its flexibility, (f3) can vary widely with the assumed stiffness of the sides. Applying this analysis to a composite structure has suggested ratios of body to chassis bending stiffness of about 3.5:1.

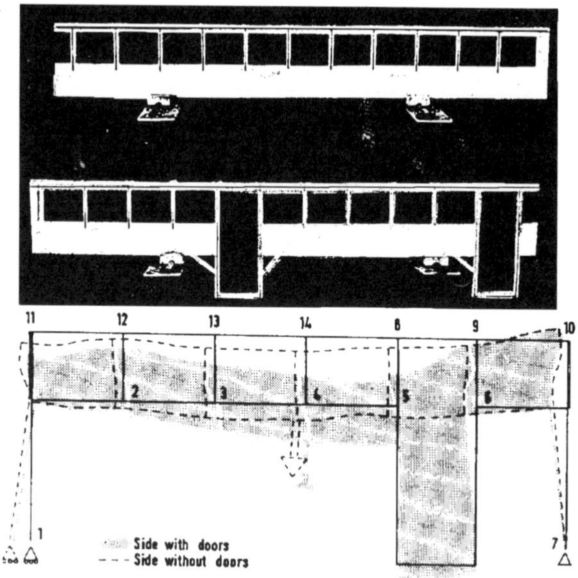

Fig 11: Model sideframe tests and computer simulation (below)

In the pioneering Leyland National bus structure (Fig 14), galvanised steel coated with 18 microns of zinc was used in thicknesses up to 5 mm on the underframe. Epoxy powder coating from 50-75 microns was used to overcoat the zinc. Main assemblies were Avdel-riveted together, these comprising floor, roof, sideframes, front and rear assemblies of spot-welded steel pressings. Target life of 15 years was the design criterion and normal working stress was made to be one third the material proof stress. The structure was given twice the torsional and beam stiffness of traditional built structures and had also to meet rollover requirements of the American Federal School Bus Roof Load Test. Overall loading criteria included 3g vertical intensity generally, with 4g in some areas; 2g vertical plus horizontal braking loads; 2g vertical plus lateral 1g 'radial' loading and torsional loading of 50,000 ft lbs (67.8 kNm) at the axles.

Individual elements and joints such as pillars, transverse roof sticks, underbody crossmembers, wheelarch risers and underframe joints were first analysed. Next stage was the evaluation of large structural systems such as wheelarch groups, doorframes,step and engine beam assemblies.To prove these techniques a structural modulebay was constructed and tested physically.

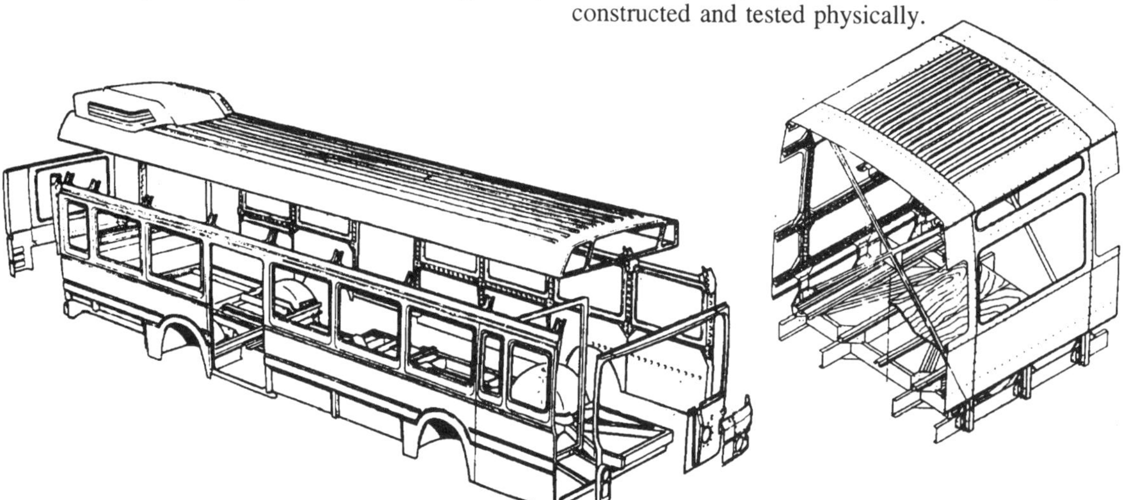

Fig 14: Leyland National structure and bay model

CV structures case study

The Pawlowski Simple Structural Surface method for open-platform and closed structures, introduced in earlier sections, is here applied to bus structures. The closed integral-bus structure is considered as one of the most suitable for this method and the advantage of providing a structural feel for concept designs is apparent.

In such structures use is made of sidewalls and roof; Fig 15 shows the classic idealisation in which the following basic elements play the major role: cross-members, sidewall pillars and roof arches form the ring-frames which are connected by floor plates, side and roof panels. The shear deformation of the window area is crucial to the analysis. Since shearing occurs of the window cut-out frames, which depends on the bending stiffness of the window pillars, anti-lozenge panels are needed to stabilise the openings, as discussed in the earlier section.

Edge forces for the structural surfaces in the bending mode reduce the loading of the side-wall, Fig 16. Payloads P_{zi} are transferred to the lower parts of the ring frames by the floor-plates while the cross-members transfer load to the wall pillars and therefore to the side panels. There is also a portion of the roof panel, width b_w, which is included in the cant

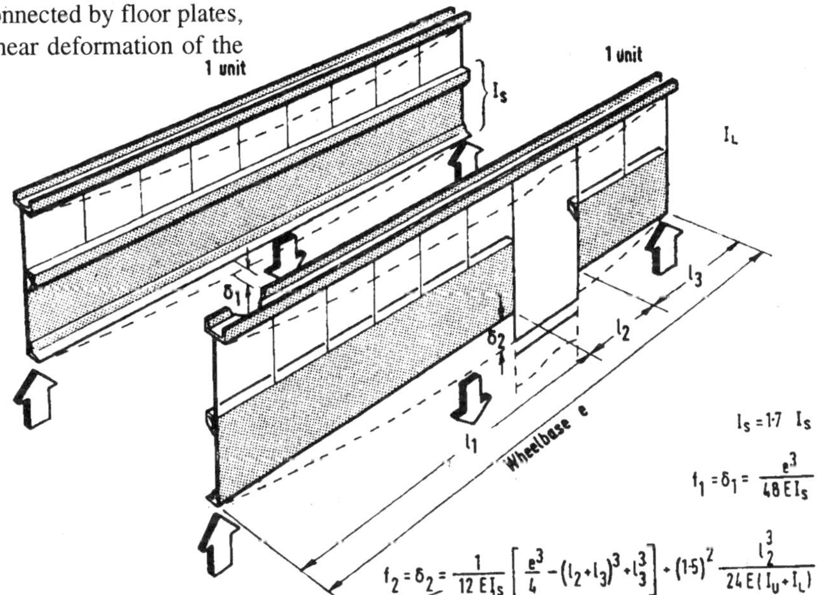

Fig 12: Side-frame deflections

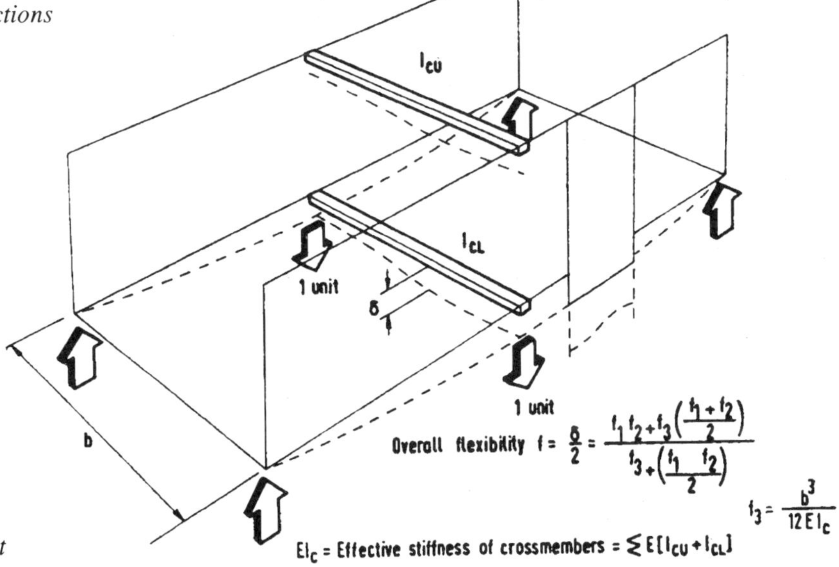

Fig 13: Ring-frame effect

rail. Two reaction forces $0.5R_t$ correspond with the spring hanger beams. Structural surfaces 'O' make no contribution in the bending mode.

Analysis of these forces, Fig 17, in the separate structural surfaces would normally involve a statically indeterminate problem. By treating the sidewall, in a shear panel design, as a beam, sufficient accuracy can be obtained for the purpose of a qualitative estimation. The doorless wall in the top half of the figure is the simpler one and where the pillar F_0-D_0 is loaded with force P_{z0}, this is balanced by panels E_0-D_0-D_1-E_1 and E_0-F_0-F_1-E_1 exerting shear flow:

$$q_0 = P_{z0}/2(h_0 + h_s)$$

Pillar F_1-D_1 is in equilibrium with this shear flow, together with force $P_{z1}/2$, , reaction $R_{pl}/2$ and shear flow:

$$q_1 = [(R_{pl}/2 - P_{z1}/2)/(h_0 + h_s)] - q_0$$

Shear flows $q_2 ... q_6$ are obtained similarly from the equilibrium equations and all cause bending moments in the walls which are weakened by window cut-outs. These reach a magnitude, at the window

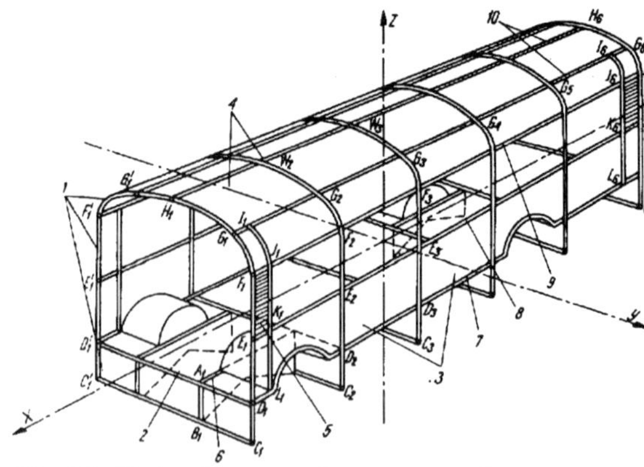

Fig 15: Main elements of monocoque bus

corners, of:

$$M_g = q_i l_i h_0/4$$

In the end-load carrying members, F_0- F_6 and D_0-D_6, direct forces P_n arise from the shear flows in the panels and have a magnitude at any point x in a

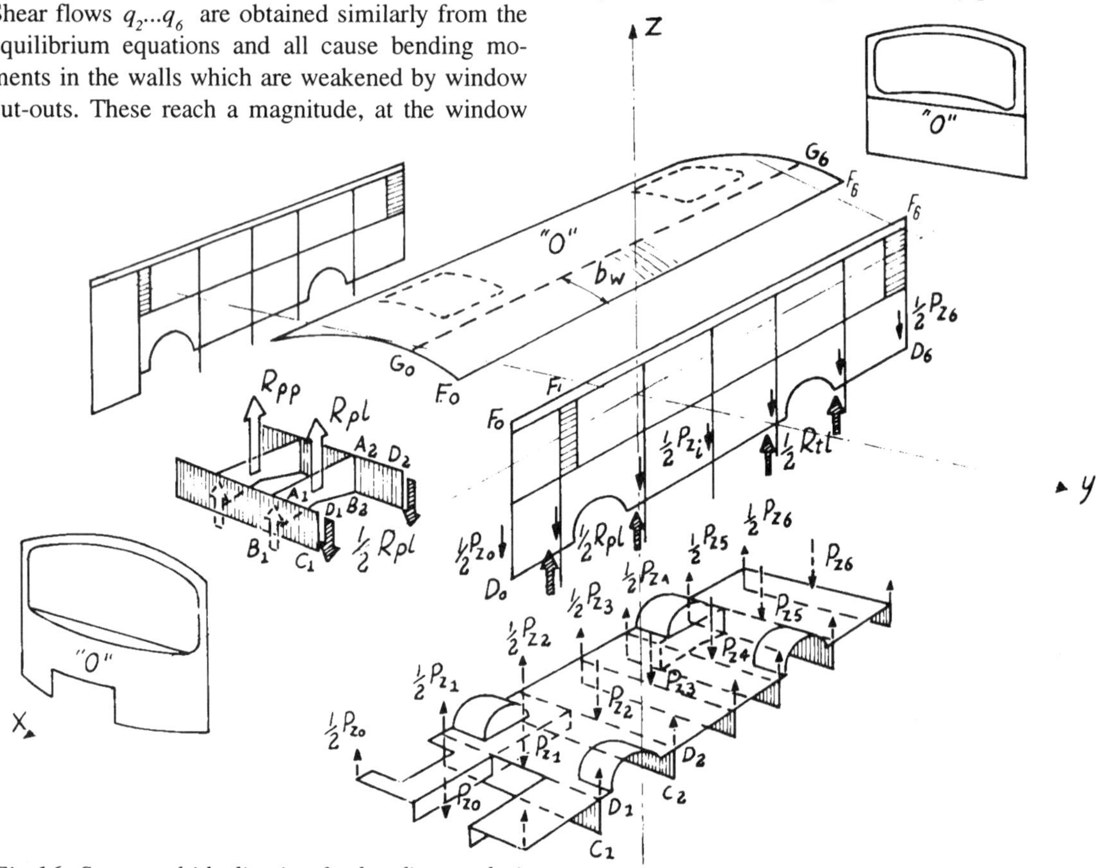

Fig 16: Structural idealisation for bending analysis

cross-section of:

$$P_{nx} = \Sigma^{\xi}_{0}\theta_{i}\lambda_{i}$$

In the case of a sidewall with doors, reinforcements are required to make up for the loss in bending rigidity. In the lower half of the figure, the effect of part of the roof acting as cant rail is examined. Force $P_{z0}/2$ loads the pillar G_0-D_0 and is balanced by shear flow q_1 in the panel G_0-F_0-F_1-G_1 where:

$$q_0 = 0.5P_{z0}/h_p$$

Since this pillar is heavily loaded by the reaction force it should be given a large cross-section so that even a high shear flow q_1 will not seriously bend the corners. This flow is calculated from the equilibrium equation of the pillar G_1-D_1:

$$q_1 = (R_{pl}/2 - P_{z1}/2 - q_0h_p)/(h_p + h_0 + h_s),$$

further shear flows $q_2 ... q_6$ being similarly obtained from the equilibrium equations of the pillars. Use of part of the roof as a cant rail has also reduced bending at the corners, by the ratio: $-\Delta M_g = h_p/h_0$. It is possible thus to reduce pillar bending moment by some 30%, with appreciable effect on the strength of the body.

Torsional loading of van body

Twisting of the bus body by asymmetric vertical loading of the wheels produces edge forces shown in Fig 18. Moment M_s is transferred from the spring hanger beams A_a-B_1-B_2-A_2 etc. to the bottom of the main ring and hence, as the reactions R_1 and R_2, via side pillars D_1-C_1, D_2-C_2, D_4-C_4 and D_5-C_5 to the side

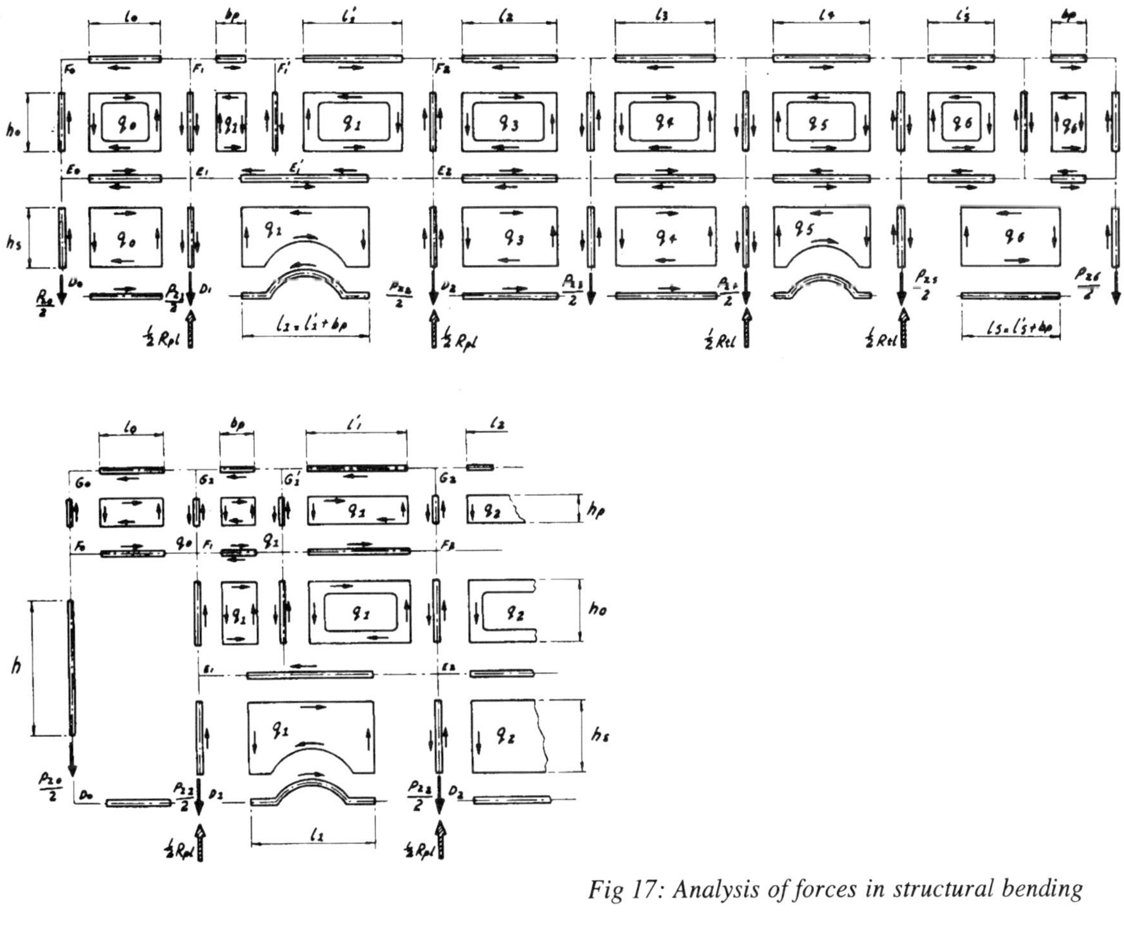

Fig 17: Analysis of forces in structural bending

wall, producing edge forces K_1, K_2 and K_3. Again the magnitude of the edge forces may be determined from a system of equilibrium equations from the separate structural surfaces:

$$K_1L + K_2h + R_1(2l_1 + l_2) - R_2(2l_1 + 2l_2 + 2l_3 + l_4) = 0$$

$$K_1B - K_3H = 0; K_2B - K_3L = 0$$

During twisting of the structure the most heavily loaded elements are the window pillars. These are in bending, under force K_2, and deflecting according to their individual rigidity, the force bending the ith column being:

$$P_i = K_2I_i/\Sigma_1^nI_i$$

where I_i is the pillar second moment of area and K_2 the edge force. Moment at the base of the column is:

$$M_{gi} = P_iZ[_{ho/2}^{ho}$$

This formula underlines the importance of the anti-lozenge areas in a side-wall under torsion; typically two anti-lozenge panels are worth considerably more than 5-6 normally dimensioned pillars. The moment is also dependent upon the kind of fixing to the roof. If there is little bending resistance, Fig 19, the moment-arm of force P_i is the full height h_0 of the opening. If on the other hand the cant rail can transfer the fixing moment to the same extent as the panel below (c), then the moment arm length is reduced by half. Corner pillars are particularly heavily loaded since force in two planes, caused by K_2 and K_3, occurs in them; this is dangerous when the tendency is to reduce cross-section for the sake of greater visibility. Total bending moment for the pillar F_0-D_0 is:

$$M_{gn} = K_2(I_0/\Sigma_1^nI_i)(h_0/2) + K_3h_0/4$$

When the direct load on the pillar due to the force P_z, and the weakening due to the lack of panel in the door opening takes effect, large stresses are involved, given by:

$$\sigma = (MK_2/W_y) + (MK_3/W_x) + (P_{zo}/F)$$

where MK_2 is the moment caused by the force P_0 as part of the edge load, MK_3 the moment caused by the force P, P_{zo} the vertical force in the column, W_y and W_x the stiffnesses with respect to y and x axes and F the cross section area of the column. The corresponding load diagrams for bending and torsion for a low-floor integral coach are seen as Fig 20. Here the payload and tare load F_{zi}, together with the reactions of the suspension R_f and R_r are contained in the area bounded by crossmembers, sidewalls, longitudinals and front/rear end frames. The effects of step loadings in both modes are shown inset.

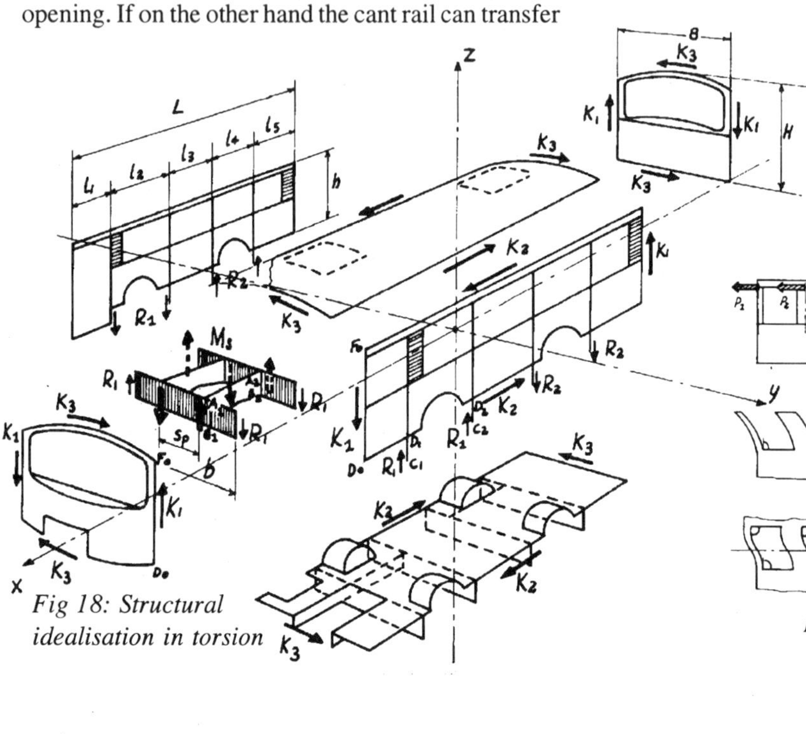

Fig 18: Structural idealisation in torsion

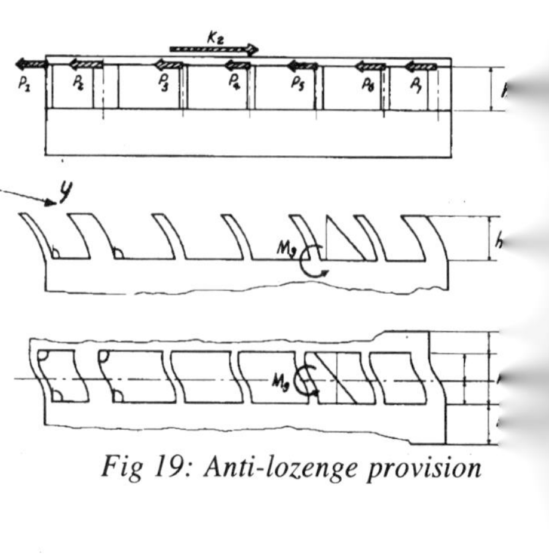

Fig 19: Anti-lozenge provision

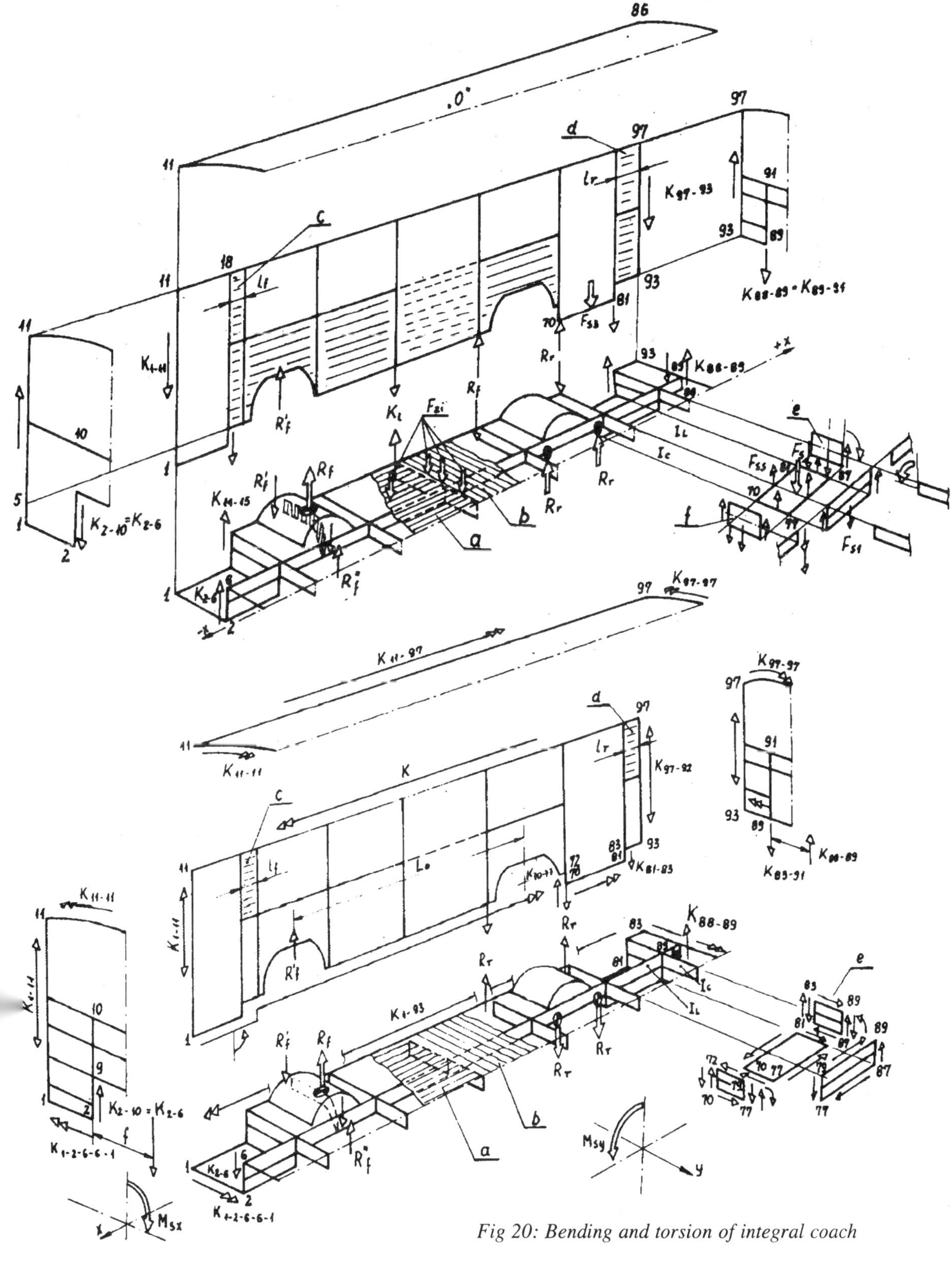

Fig 20: Bending and torsion of integral coach

Idealising van/LCV structures

An earlier section dealt with the idealisation of regular car and CV structures by the Pawlowski Simple Structural Surface method. It introduced the former aeronautical structural technique of representing a real panel by an idealised one comprising a flat rectangular panel taking only shear load bounded by uniform section end-load carrying bars. Here more irregular shaped structures are considered typifying a van and LCV.

The vehicle structural analysis pioneer Janusz Pawlowski considered that the finite element techniques now widely used in structural design were less suitable at the initial concept stage because they depended on fairly precise modelling of the shapes and related mass centres which were unlikely to be fixed at this stage. By using the SSS method, the designer could progress from a relatively simple beam model, step by step, into one which represented the final structure more closely — and in so doing arrive at a more efficient structural solution. By contrast, he considered, the FEM modelling route was 'passive' rather than interactive.

The SSS approach was one in which local stiffnesses could be arranged to meet the requirements of external load inputs; the general elasticity depends on the characteristics of the whole body under the action of service loads and partial plasticity;

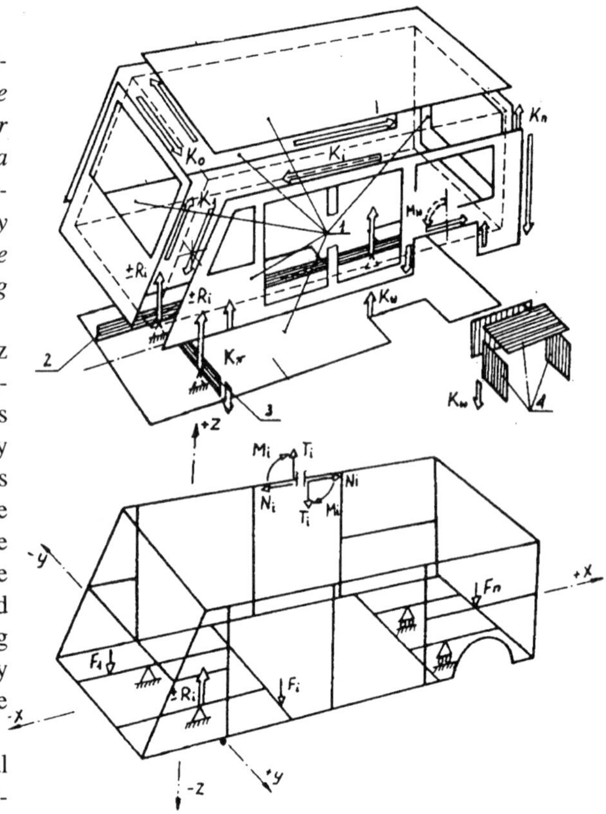

Fig 22: Beam representation for panel van

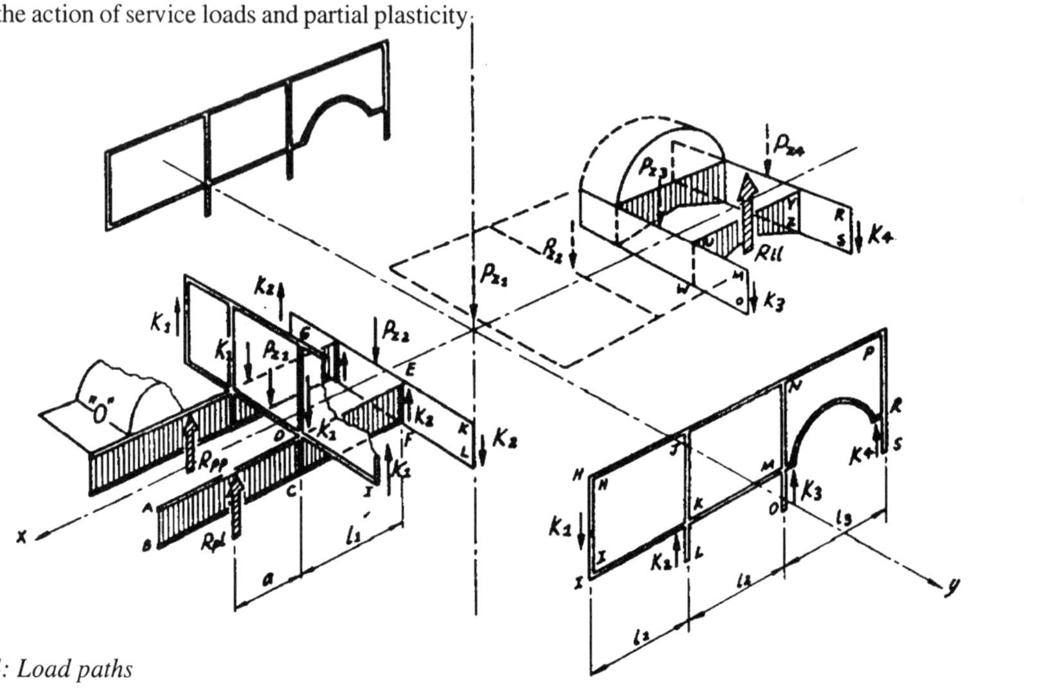

Fig 24: Load paths

of the body meets the energy absorption requirements of crash impact. This leads the designer, at an early stage, to direct loads into the thin wall structure in the planes of the thin walled elements. The preliminary structural model in Fig 22 starts as a beam representation with 15 levels of redundancy and then progresses to an SSS model comprising six main structural surfaces plus an adequate number in the underbody group depending on the support conditions. Edge loads in each group of SSSs are obtained from the equilibrium conditions applying to each one. Interaction between SSSs occurs as edge loads such as *K1*. On top of these is superimposed loading cases due to bending, tension and planar shear for the whole structure to obtain a set of linear equations which can easily be solved.

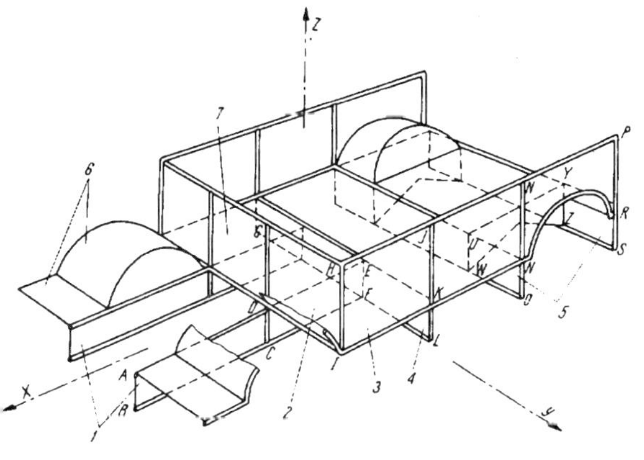

Fig 23: Non-participating cab

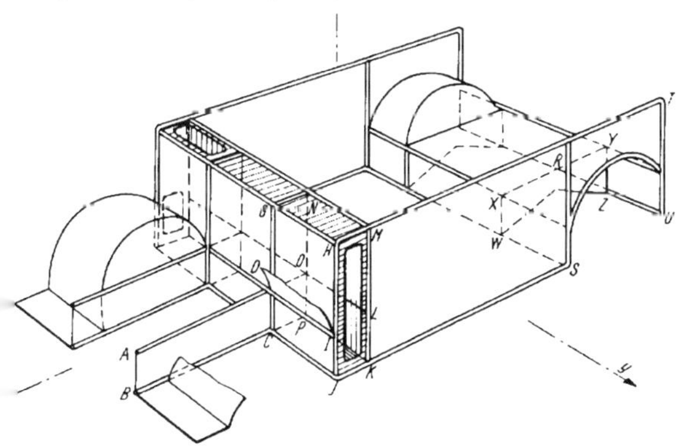

Fig 25: Pick-up with box bulkhead

SSSs occurring in one vehicle body can also be included in a generalised description of others in a family so that the same structural efficiency can be translated to derivative models. Such a vehicle might be a pick-up truck designed for ultimate structural efficiency with an integral load-bearing cab. In the case where the cab does not participate in reacting structural loads, the structure would be reduced to that shown in Fig 23; this is reduced to front longitudinal members, wheel and cover plates, bulkhead, rear crossmembers, sidewalls and floor plate as the main load bearing elements.

Such a structure is sufficiently rigid for vertical symmetrical loads P_z. The front reactions R_{pp} and R_{pl} are introduced into the corner post H-L by the first crossmember E-F-K-L, Fig 24, and the bulkhead is in equilibrium under the edge force *K2* of the sidewall. Forces P_{z1} and P_{z2} are also introduced into these elements and from equilibrium:

$$K_2 = (R_{pl}a/l_1) + P_{z2}/2$$
$$K_1 = R_{pl} + K_2 - (P_{z1}/2) - (P_{z2}/2)$$

Note that the large load *K1* on the bulkhead, which is transferred through the corner post H-I to the sidewall; force *K2* is also introduced into the sidewall through the post K-L, which is common for the crossmember and the wall.

In order for the sidewall to be in equilibrium, edge forces *K3* and *K4* must exist in the pillars M-O and R-S:

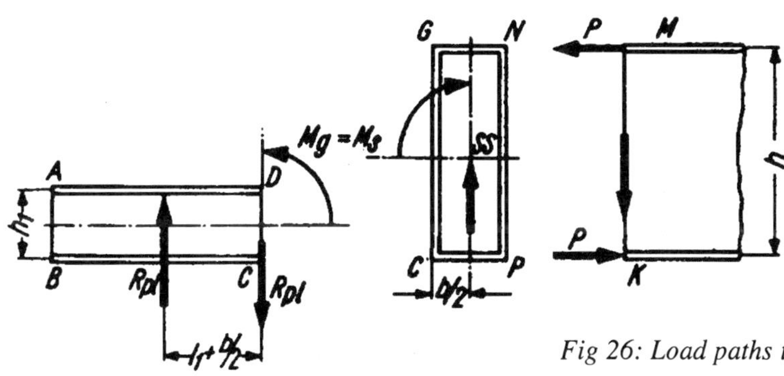

Fig 26: Load paths through bulkhead

$$K_3 = [K_1(l_1 + l_2 + l_3) - K_2(l_2 + l_3)]/l_3$$

$$K_4 = K_1 - K_2 - K_3$$

Force *K3* is usually larger than *K4* and both act on crossmembers M-O-W-U and R-S-Y-Z respectively. From these the forces are transferred to the spring beams U-W-Y-Z and the spring supports.

Normally a pick-up structure is not sufficiently rigid with respect to torsion since the torque M_s can only be absorbed through the lower members in bending — leading to large deformations. This disadvantage, along with the impossibility of lowering the floor level of the load platform, because of the height of the auxiliary longitudinal members, can be overcome by use of a box such as H-M-K-J formed from the bulkhead parts of the sidewalls and a closing wall; it can also be used to house the fuel tank and spare wheel. The analysis of such an arrangement, Figs 25 and 26, is made on the assumption of the front support beam A-B-C-D being a cantilever fixed in the box — transferring the root moment to the sidewall as a torque.

With addition of the integral forward-control cab, Fig 27, the pick-up becomes a structure with good resistance to bending and torsion, with its rear part fixed in the box formed by the crew cab, Fig 28. The twisting moment M_s, introduced from the wheels into the suspension by the couple $R_t S_t$, is balanced with the edge forces *K1* and *K2*; the latter arise from the action of the sidewalls as a result of their being fixed in the cab with the forces *X* and *(K1+K2)*:

$$K_1 + K_2 = M_s/b$$

$$X = [K_1 l_2 + K_2(l_2 + l_3)]/h_s$$

The cab is supported by the two front spring beams, separated by a width S_p, and joined to the rear wall of the cab at point K and to the front wall at A. The force *(K1 + K2)* is transferred via the rear bulkhead to the points K and K' in the form of the reaction P on the front wall. The latter, which is fixed to the front spring beams at A, acts on the springs with a couple $4R_p''$. Assuming that the lower force *X* acts at the level of the cab mounts, these forces amount to:

$$R'_p = (K_1 + K_2)B/S_p$$

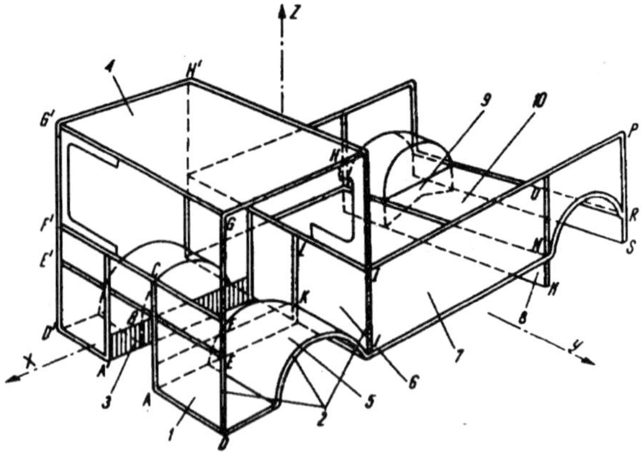

Fig 27: Integral cab structure

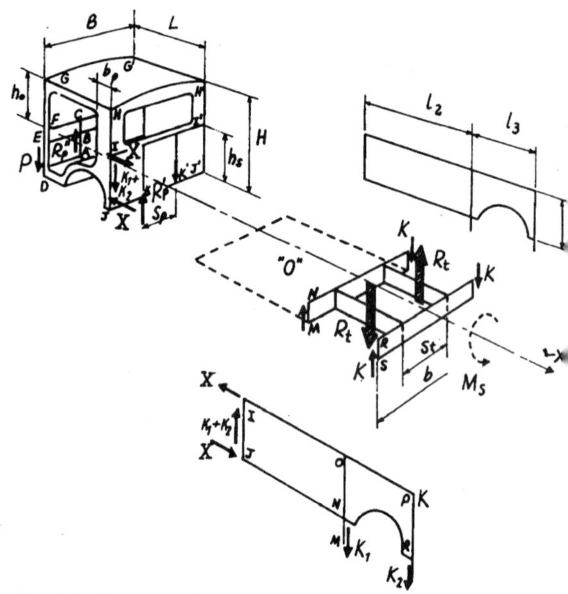

Fig 28: Rear-end section

$$P = Xh_s/L$$

$$R_p'' = PB/S_p$$

The forces R'_p and R''_p are transferred from the front spring beam to the spring supports as the couple R_p, which gives a moment M_s on a lever arm S_p.

All six SSSs which form the cab have forces in them and if the sidewall of the cab has the form of a door frame, then the beam H-J should be substantial in order to absorb force *X*.

Structural composites

While racing and sports cars are currently the principal users of structural composites, it is likely that other categories will soon follow as polymer to metal price ratios change. Special techniques of analysis are required to allow for directional properties which follow the reinforcement configurations.

Structural composites can be based on a variety of resin and reinforcement systems, the properties of some of the main ones being given by Phillips[2] in Fig 29. The effectiveness of fibre reinforcement depends on the fibre aspect ratio of length/diameter which should be a minimum of 100. Most commercial fibres are less than 20 microns in diameter but the area of each available for adhesion to the resin is relatively high compared with the compressed bulk of the individual fibres. Resins, in general, are an order of magnitude lower in strength/stiffness than fibre reinforcements.

Fig 30 provides a useful relationship between fibre and resin matrix to show percentage strength against ratio of moduli. Fig 31, however, provides a more detailed examination from which basic calculation formulae can be derived. Between the two extremes of submerged depth shown, for the fibre inside the resin, there is a critical length L_c where fibre failure may be either pull-out or tensile fracture and the failing load is balanced between shear and tension. These can be equated to give: $\tau = \sigma_f r/2l_c$

At the end of the fibre tensile stress is zero and rises to a constant maximum once l_c is exceeded.

In the case of carbon-fibre reinforcement, low molecular weight epoxy resins have become the favoured matrix. Properties of commercial grades are shown in column 3 of the table in Fig 32, the first two

Fibre/material	Specific gravity density (g/cm³)	Tensile strength (MPa)	Tensile strength (× 10³ lbf/in²)	Young's modulus (GPa)	Young's modulus (× 10⁶ lbf/in²)
E-glass	2.54	2410	349	69	10
S-glass	2.49	2620	380	87	12.6
carbon type 2	1.75	2410	349	241	35
aramid (high modulus)	1.44	3450	500	124	18
boron	2.63	2760	400	379	55
alumina	3.30	3000	435	297	43
asbestos (chrysotile)	2.40	1490	216	183	26.5
wood	0.50	69	10	7	0.95
light alloy	2.69	476	69	72	10.50
steel (structural)	7.85	413	60	207	30
titanium alloy	4.52	711	103	117	17

Fig 29: Reinforcement properties compared

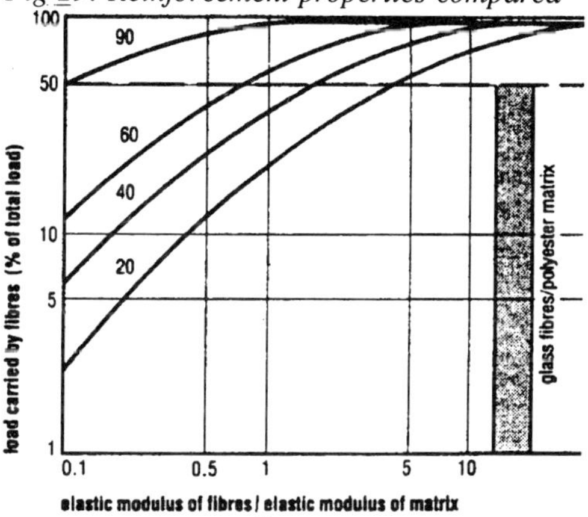

Fig 30: Percentage strength vs modulus ratio

	Tensile strength (MPa)	Tensile strength (× 10³ lbf/in²)	Young's modulus (GPa)	Young's modulus (× 10⁶ lbf/in²)	Specific gravity	Elongation of break (%)
lower-modulus fibre	2,650	384	59	8.6	1.44	4
higher-modulus fibre	2,650	384	127	18.4	1.45	2.4

Fig 33: Aramid fibre properties compared

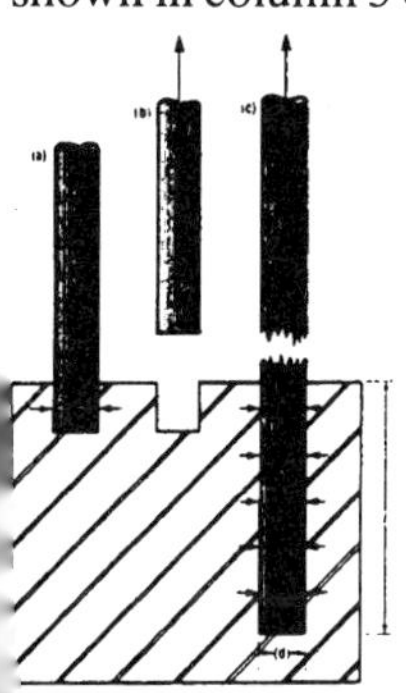

Fig 31: Embedded fibre analysis

Fibre value \ Fibre type	1		2		3	
Young's modulus, lbf/in² × 10⁶ (GPa)	50	(345)	35	(241)	29	(200)
ultimate tensile strength, lbf/in² (GPa)	290,000	(2.00)	350,000	(2.41)	320,000	(2.21)
specific gravity	1.92	—	1.75	—	1.70	—
Unidirectional composite (V_f = 60%)						
Young's modulus, lbf/in² × 10⁶ (GPa)	26	(179)	17.5	(121)	14.5	(100)
ultimate tensile strength, lbf/in² (GPa)	145,000	(1.00)	200,000	(1.38)	175,000	(1.21)
specific gravity*	1.72	—	1.55	—	1.52	—

* Specific gravity of epoxy resin taken as 1.25

Fig 32: Carbon fibre grades compared

columns showing potential properties available when more exotic types 1 and 2 carbon-fibres are used. The table of Fig 33 shows the properties available with the available forms of aramid fibre. Large panels and shells are now commonly moulded in autoclaves — cylindrical pressure vessels which can generate several atmospheres as well as a means of producing vacuum at one side of an air-tight membrane. Fig 34 shows a typical construction.

Tensile modulus of a composite laminate is:

$$E_k = E_f V_f B + E_m V_m$$

where E is modulus, k composite, f fibre and m the matrix, an expression due to Hollaway[3]. V_m is the volume fraction of the matrix and the lay-up factor B = 1, 1/2, 3/8 for unidirectional, bidirectional or random (in-plane) orientation respectively. Stress/strain curves for glass and aramid fibres are shown in Fig 35, with effect of fibre geometry on stress/strain curves in Fig 36. Jointing configurations are, of course, crucial in design with the critical effect of adherend thickness being shown in Fig 37. Weakest joints are those where failure is limited by the interlaminar failure of the adherend or peel of the adhesive. Next strongest are those in which load is limited by adhesive shear-strength and strongest of all are those which fail outside the joint area at a load equivalent to the strength of the adherend.

The compressive strength and stiffness of laminates is difficult to predict and analysis is usually based on the buckling of slender columns on an elastic foundation. The two idealised buckling modes which have been suggested are given in Fig 38; these are such that adjacent fibres buckle either in or out of phase with each other. The critical compressive stresses are given by:

$$\sigma_c = G_m/(1 - V_f)$$

for shear mode in phase buckling and

$$\sigma_c = 2V_f [E_f E_m V_f/3(1 - V_f)]^{1/2}$$

for extension-mode out-of-phase buckling.

Stress/strain relationships for uniaxially aligned discontinuous fibres stressed in parallel to the fibre direction can also be estimated by calculation, Fig 39. There is a portion at the end of each finite length which is stressed to less than the maximum stress of a continuous fibre. This is known as the transfer length and depends on the maximum fibre stress and the effective shear strength of the fibre-matrix interface σ_{xy} and is given by:

$$l_l = \sigma_f d_f / 4\,\sigma_{xy}$$

where d_f is fibre diameter and σ_f the maximum stress in fibre length l_l. Critical transfer length in which fibre stress drops from σ_{fmax} to zero is l_c. A quantity α

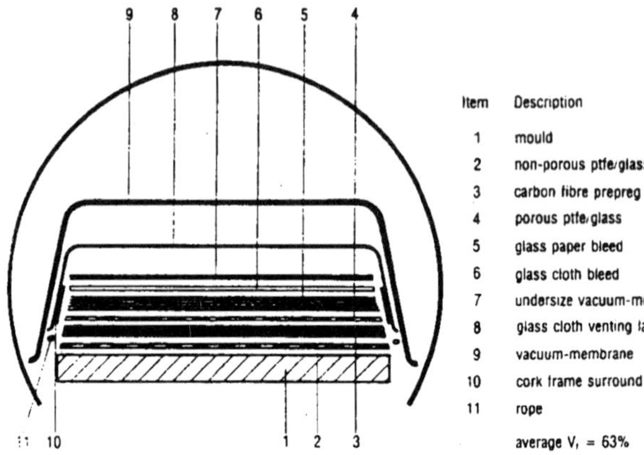

Fig 34: Autoclave construction

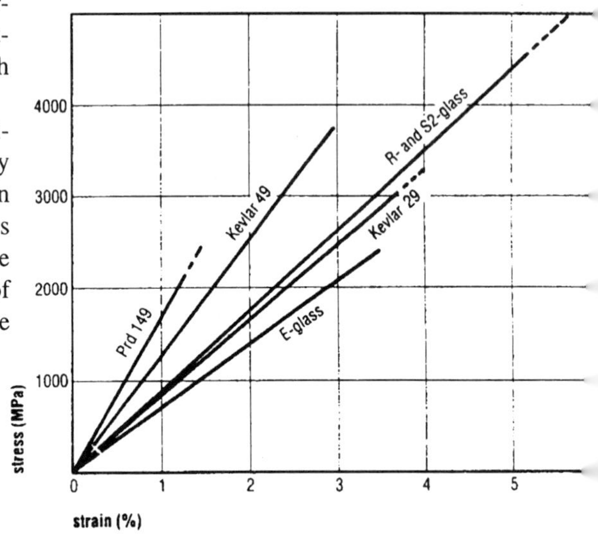

Fig 35: Stress/strain curves for aramids

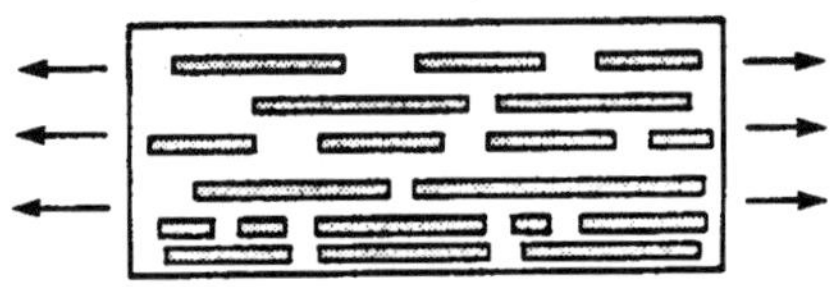

Fig 36: Fibre geometry effect

is defined as the area under the stress distribution curve over the length $l_c/2$; then the fibre end of the length l_c may be considered as supporting a reduced average stress $\alpha\sigma_{fmax}$ or it may be reduced to length $\alpha l_c/2$ and subjected to σ_{fmax}. Referring to the figure, average axial stress $\sigma_{f.av}$ for fibre lengths not less than l_c can be deduced from:

$$\sigma_{fmax}(1 - l_c/2l)$$

and for those less than the critical length as

$$\sigma_{fmax}\, l/2l_c$$

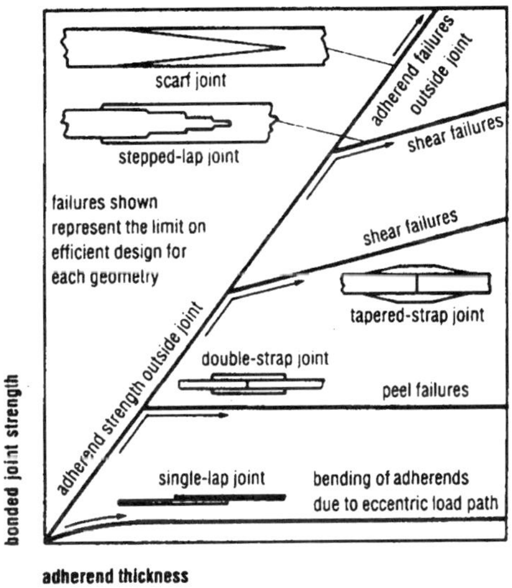

Fig 37: Joint configurations

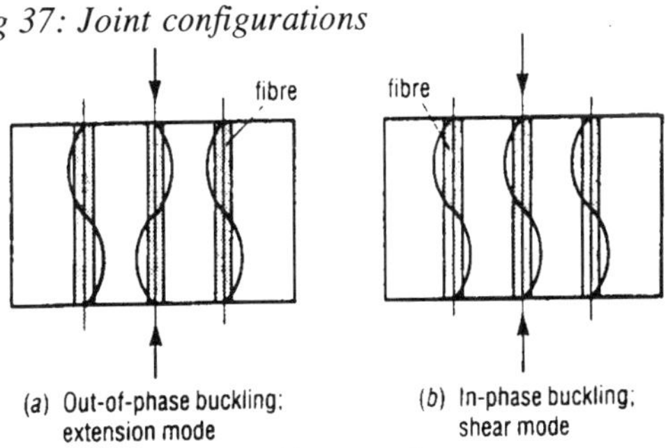

Fig 38: Idealised buckling modes

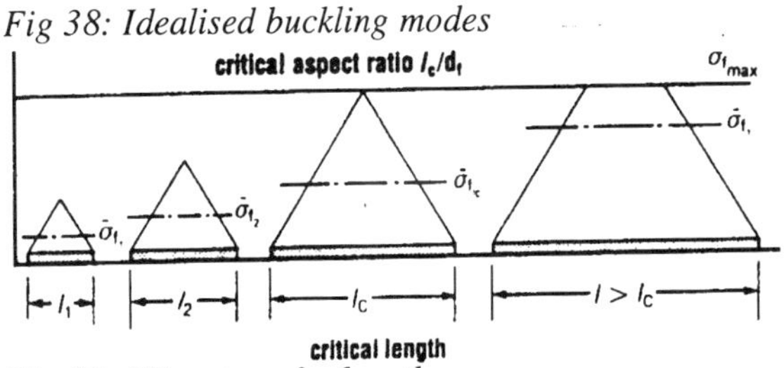

Fig 39: Fibre transfer length

Composite onstruction techniques

The structural evaluation of the Panther Solo car structure has been described by Parry-Williams[4]. Fig 40 shows the body shell components, notable for the two-piece roof structure. Major loadings on the shell are seen in Fig 41. Uni-directional fibres are used in strip form around all the apertures in the primary structure with aluminium inserts positioned where threaded anchorage holes are needed. A torsion test was carried out between front and rear suspension anchorages. Deflections were measured at ten places, at stations a, b, c, d and e, and torque applied at the three levels shown by the table in Fig 42. Dial gauge readings, in mm, taken at 750 mm from the vehicle centreline are shown.

Application of advanced composite materials in racing cars has been described by Dr HA Potts[5] of Advanced Composites Group Ltd. At ACT two in-house tooling systems are used for mould making — gel-faced HT5C Precom or LTM Prepreg form of the same material. The former is a room-temperature

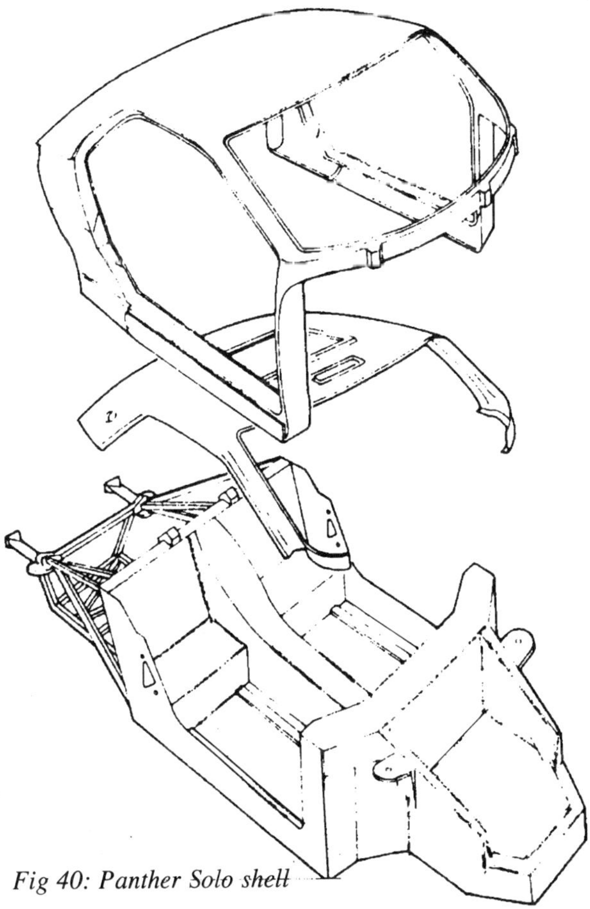

Fig 40: Panther Solo shell

cure system which is subsequently post-cured to around 150 C under a vacuum bag at 0.5 bar while the latter requires full autoclave pressure at 20-50 C. Near zero shrinkage carbon fibre mould tools are utilised to ensure absolute accuracy of anchorage dimensions. Warping and twist of the moulds is prevented by a rigid backing structure also in carbon-fibre laminate.

Following polishing of the moulds, and the application of release agents, the first skin is laminated in stages, with intermediate debulks using either vacuum or autoclave pressure. On completion, the lay-up is cured under full autoclave pressure of 90-100 psi. When cured, the bag, breather and peel ply are removed and the laminate surface lightly abraded and degreased. A resin film ply is then fitted to the skin and preshaped honeycomb pieces added. Next the many anchorage inserts are fitted and the second stage of the lamination process takes place, usually at full vacuum in an oven. After this cure cycle, core and insert detail is carefully checked for any movement prior to the third stage in the lamination, the inner skin; here less than full autoclave pressure is used in order to avoid crushing of the core splices.

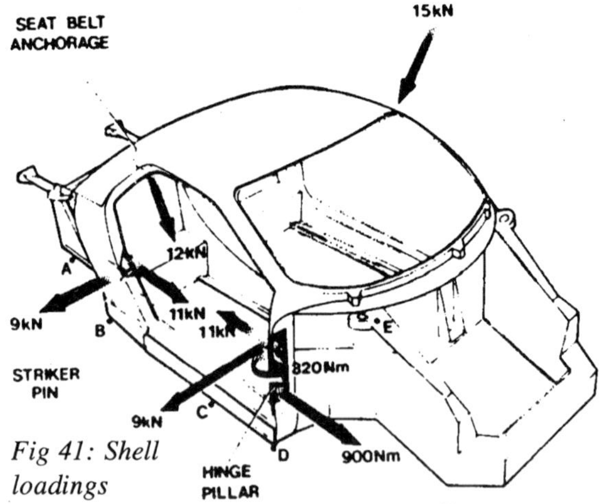

Fig 41: Shell loadings

POSITION	APPLIED TORQUE (Nm)							
	0		500		1052		1552	
	RH	LH	RH	LH	RH	LH	RH	LH
a	0	0	+0,13	-0,03	+0,26	-0,06	+0,38	-0,10
b	0	0	+1,58	-0,42	+2,12	-0,85	+2,71	-1,30
c	0	0	+1,81	-0,61	+2,56	-1,19	+3,38	-1,83
d	0	0	+0,95	-0,75	+1,83	-1,15	+2,79	-1,24
e	0	0	+1,07	-0,75	+1,98	-1,42	+3,05	-2,13

Fig 42: Deflections measured during tests

		Abbreviation	Cost £tonne^{-1}	Density kgm^{-3}	UTS MNm^{-2}	Modulus GNm^{-2}	Impact Strength Jm^{-1}	Fatigue Strength MNm^{-2}	Friction Coefficient	Maximum Service Temperature °C	Expansion Coefficient 10^{-6} C^{-1}	Thermal Conductivity Wm^{-1} °C^{-1}	Specific Heat	Volume Resistivity ohm cm	Dielectric Strength kVmm^{-1}	Dielectric Constant	Power Factor	Refractive Index	
Alkyd	– Asbestos Filled		1500	1600	45	16	15			230	30	0.6		10^{14}	14	4	0.02		A3
Epoxy	– Cast		2500	1200	60	3.5	20			150	50	0.2	0.4	10^{14}	16	4	0.02	1.57	A7
	– Glass Filled		3000	1800	100	25	600			200	20	0.3		10^{14}	14	4	0.01		A3
	– Glass Laminate			1820	270	19	370			200									A5
	– Glass Wound(60% unidirectional)			2000	1500	40		300		260	6	0.3	0.24		16	5	.025		A6
	– Carbon Fibre(")			1600	1400	120		840		70									
	– Aluminium Whisker (14% vol)				750	40													
Ester	– Cast	UPR	1000	1300	60	3	15		0.35	120	70	0.2	0.5	10^{15}	18	3.5	0.02		A7
	– Dough Moulding Compound	DMC	1000	2000	50	7-14	600			170	30	0.25		10^{15}	15	5	0.01		A3
	– Sheet " "	SMC	1000	2000	65	7-16	750			170	25	0.25	0.22	10^{15}	15	5	0.02		A3
	– Glass Matt Laminate	GRP		1600	150	12	800	100		175	12	0.25		10^{14}	24	5	.015		A5
	– Glass Wound(60% unidirectional)			1800	1000	30		280											A6
Imides	– Cast			1400	120	3	35			260	26	0.32	0.3	10^{17}	22			1.70	A3
	– Glass Filled			1400	190	22	800			260	15	0.48	0.27	10^{16}	20	4.7	.005		A3
Melamine	– Cellulose Filled	MF		1500	50	9	15		0.2	100	40	0.6	0.4	10^{12}	15	7.5	0.03		A
	– Glass Laminate			1900	300	18	500			140	100	0.5		5×10^{11}	8	14	0.1		A5
Methacrylate	– Glass Matt Laminate				150	12				200									A
Phenolic	– Asbestos Filled	PF		1800	40	11	15		0.3	250	20	0.6	0.3	10^{10}	8	10	0.1		A
	– Glass Filled		200	1800	40	18	500			250	12	0.4	0.3	10^{13}	10	5	0.02		A
	– Asbestos Laminate			1720	80	16	190			250					4	15	0.3		A
	– Glass Laminate			1650	160	10	340			270	180				24	5.5	0.03		A
Urea	– Asbestos Filled	UF		1500	60	10	15		0.2	75	30	0.3	0.4	10^{12}	15	8	0.03		
	– Cellulose Filled			1520	65		13			75				10^{12}	9	0.3	0.18		
Urethanes	– Reaction Moulded	RIM		1100	22	0.3				120	130								
	– " " (reinforced)	RRIM			25	1					45								
	– " " (foam)			500	15					90	72	0.06		3×10^{14}	10	2.4	0.007		
Amides	– " "	RIM		1130	40	1.1	50			180									

Fig 43: Basic mechanical properties

Analysis of composite structures

Design calculation techniques for composite structures have generally been based on pseudo-elastic methods. These have increased in value as users build up life-time confidence in engineering light-weight polymer composites, the advanced end of which is, of course, the racing car structure.

West[6] has provided basic data for thermosets in the table shown as Fig 43. He maintains that pseudo-elastic design is still the most commonly used method, in which classic elasticity theory is applied to polymer composites but that time-dependent material elastic properties are used in design formulae. The procedure is thus to specify service conditions such as life-duration and operating temperature, and select the worst combination in design, then substitute properties appropriate to the conditions.

Fig 44 shows the example of a ribbed plate subject to a uniform pressure *p* of 5 kPa, required to sustain a minimum life of 10^5 hours. Sheet thickness *t* is 2 in (50.8 mm) and 0.32 in (8 mm) width ribs are employed. Here, grid spacing is required for two materials, unreinforced PVC and polyester dough moulding compound reinforced with chopped glass. The unsupported area *a x a* is considered as a plate with built-in edges for which maximum stress is given by 0.308 $a^2 p/t^2$. This gives values of a as 100 for UPVC and 200 for DMC.

A thin-walled rectangular box-beam made from long-fibre reinforced polymers, and having diaphragm stabilising webs along its length, would be designed against the criterion of buckling collapse. In the example shown, it is required to determine the maximum centre-point loading for a 2-cell beam, Fig 45, having uni-directional fibres on the top and bottom flanges with randomly-oriented fibre for webs and diaphragms at 50% by volume and a polyester type resin.

Composite elastic modulus in the fibre direction is :

$$E_{C1} = E_R V_R + E(1-V_R) \text{ and } E_{C2}=E_R E/[EV_R(1-V_R)]$$

for the transverse directon. Corresponding Poissons ratios are:

$$n_{C1}= n_R V_R + n(1-V_R) \text{ and } n_{C2}= n_{C1}E_{C2}/E_{C1}$$

Composite shear modulus in the fibre direction is:

$$G_{C1}=G_R G/[GV_R + G_R(1-V_R)]$$

where G_R and G are for fibre and matrix respectively and

$$G = E/[2(1 + n)]$$

Using values from Fig 1, results are seen in the table of Fig 4 and critical buckling stress is:

$$s_B = (4p^2/b^2t)[(D_L D_T)^{1/2} + H)$$

where D_L , D_T and H are the section second moments of area corrected for Poissons ratio effect.

Du Pont have applied pseudo-elastic design methods for such structures as rectangular section seat backs, Fig 46 using the expression for breadth:

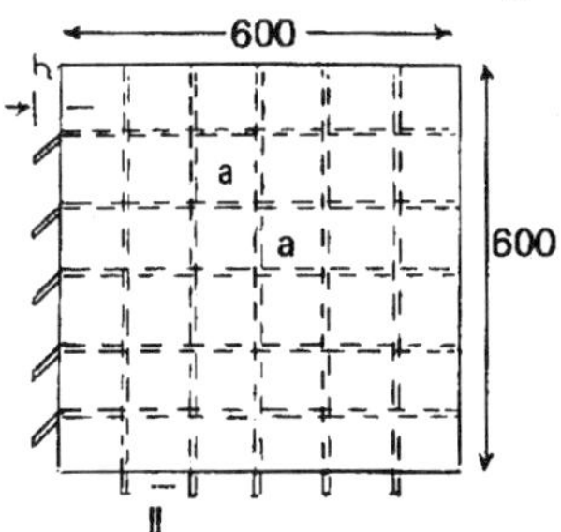

Fig 44: Ribbed-plate example

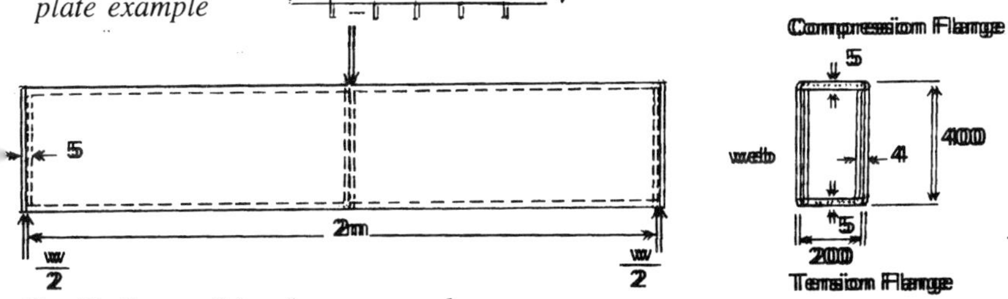

Fig 45: Two-cell box beam example

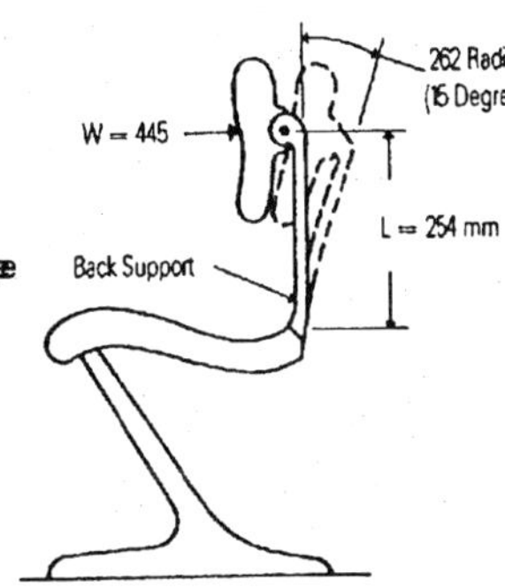

Fig 46: Seat-back example

$b = 6WL^2/Eqh^3$

Putting various thicknesses and the relevant moduli from Fig 47 for different materials into the equation gives a series of widths; from these the amounts of material used and an idea of relative part costs is obtained. Thinner sections will have shorter moulding cycles, important in determining part cost.

For the stiffest resin, Rynite 555, the flexural modulus at a thickness of 3.2 mm is = 17.9 x 10^3 MPa, giving breadth of:

$$\frac{[16 \times 445 \times (254)^2]}{[17.9 \times 0.262 \times (3.2)^3]}$$

= 1121 mm and the c/s area — proportional to the mass of material used — is 3.2 x 1121 = 3587 mm^2.

This width is obviously too great, so Du Pont repeat the exercise with some thicker sections/smaller widths which look more acceptable: for h = 4.8 mm, E= 13.8 x 10^3 MPa then b = 431 and c/s area is 2069 mm^2, and so on for h = 6.4, 8.0, 9.5 mm, and by extrapolating the experimental data slightly, for h = ll.1 mm. All the results are :

h mm	E MPa x10^3	b mm	area mm^2
3.2	1 7.9	1121	3587
4.8	1 3.8	431	2068
64	10.3	244	1559
8.0	8.1	159	1269
9.5	59	130	1235
11 1	4.5	107	1186

Property	Compression flange	Tension flange	Web and diaphragm
Modulus			
– Longitudinal E_{C1} GNm^{-2}	36.5	36.6	19.7[illegible]
– Transverse E_{C2} "	5.75	5.75	19.7[illegible]
– Shear G "	2.14		7.53
Poissons ratio			
– Longitudinal ν_{C1}	0.275	0.275	0.31
– Transverse ν_{C2}	0.043	0.043	0.31
Strength			
– Longitudinal GNm^{-2}	0.42(X_{C1})	1.26 (X_{T1})	0.3[illegible]
– Transverse "	0.050(X_{C2})	0.025(X_{T2})	0.3[illegible]

Fig 47: Derived "elastic" properties

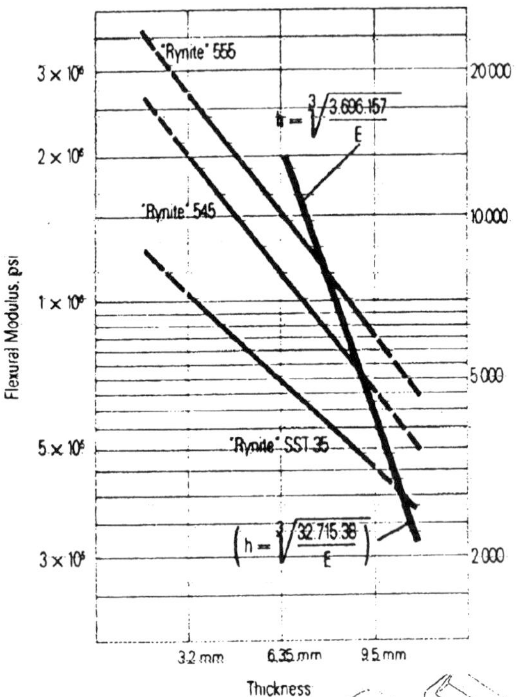

Fig 48: Material performance comparison

To find the thickness to satisfy the 15 degree (0,262 radian) deflection criterion, using the same equation, but this time solving for h, to determine the two unknowns h and E we insert the known values for load, deflection, length and width then plot this on Fig 48, — for example, by using two values of E and the corresponding calculated values of h (E = 13800 MPa, h = 6.4 mm; E = 4140 MPa, H = 9,7 mm) and connecting the points by a straight line. This intersects the modulus/thickness line for Rynite 555 to give the solution for h as 7.8 mm.

The lines for Rynite 545 and SST 35 intersect the equation plotted at greater thicknesses, and since

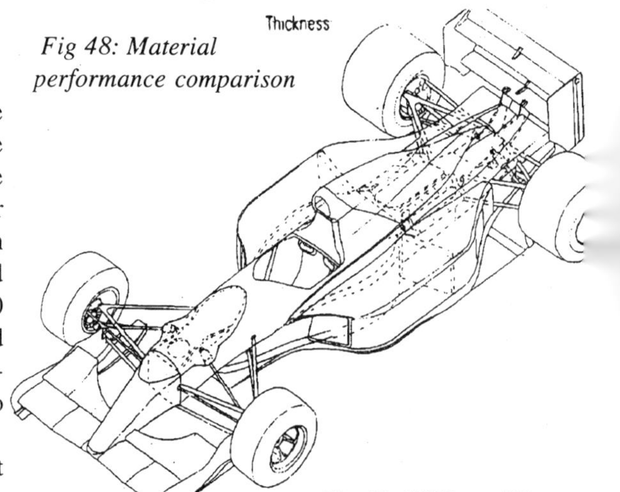

Fig 49: Williams F1 car

these resins can take more impact than Rynite 555, one has the choice of playing safe — at the cost of more resin in the back support. But the stress in the thicker section would be lower, and this could be especially important. Keeping stresses low in the back support will help to extend fatigue life, reduce creep, and ensure that the part can safely take overload several times the design load.

Racing car structures

Advanced composite materials are of course widely used in race-car structures such as the Williams FW14B seen in Fig 49. O'Rourke[6] explains that FISA load cases apply to specific portions of the structure. The structure also has to sustain inertia loads of 4.5*g* laterally and 4*g* under braking as well as 10*g* instantaneous bump load. Downforces may be as much as 5kN and a combined load envelope would be as in Fig 50. The main 'chassis' comprises five principal components, Fig 51, the outer shell being reduced to two, Fig 52.

The structural composite materials used at Willliams comprise a range of carbon and aramid fibre-reinforced epoxy resin prepregs, aluminium and Nomex honeycomb core materials, and epoxy adhesives and fillers in film and paste forms. The majority of components are made from a 120 C-curing, modified epoxy prepreg but where mechanical property retention at temperature is important a 175 C-curing resin system is employed. Carbon fibre is used in both woven and uni-directional tape forms, intermediate modulus type carbon fibre being chosen where components are of a fully structural nature and high strength type for the remainder of applications. Ablative materials are incorporated around the exhaust outlet where temperatures of 500 C are typical.

Prepreg lay-up is accomplished by hand and plies are laminated to the orientations specified in an appropriately configured mould; a vacuum bag is applied and curing takes place in an autoclave to the recommended cycle. Total factory floor space given over to composites activities amounts to 21500 ft^2 (2000 m^2) whilst curing facilities include 3 autoclaves, the largest of which is 2.5 metres in diameter and 5 in length. Structural performance is such that the levels of chassis torsional stiffness that are now attained are in excess of 250% of those measured on the last generation of metal structures, for no increase in the mass of the component.

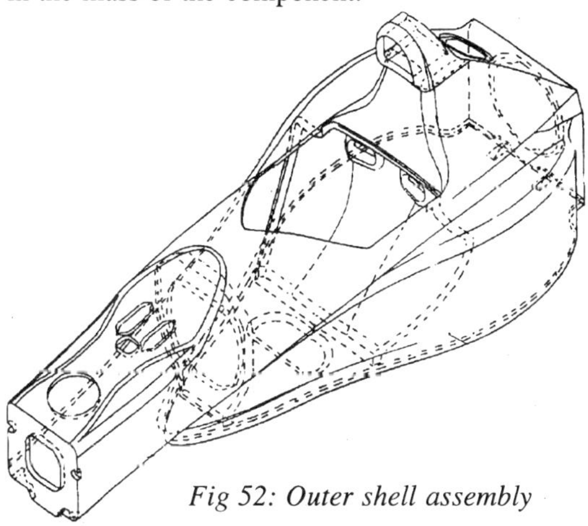

Fig 52: Outer shell assembly

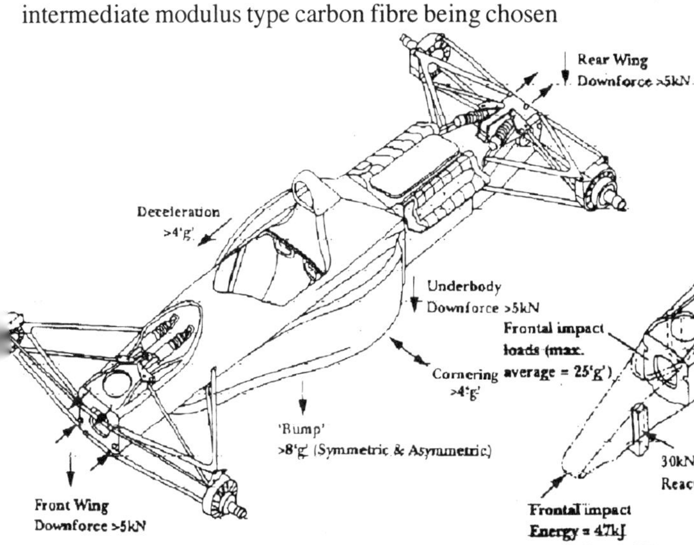

Fig 50: Load envelope

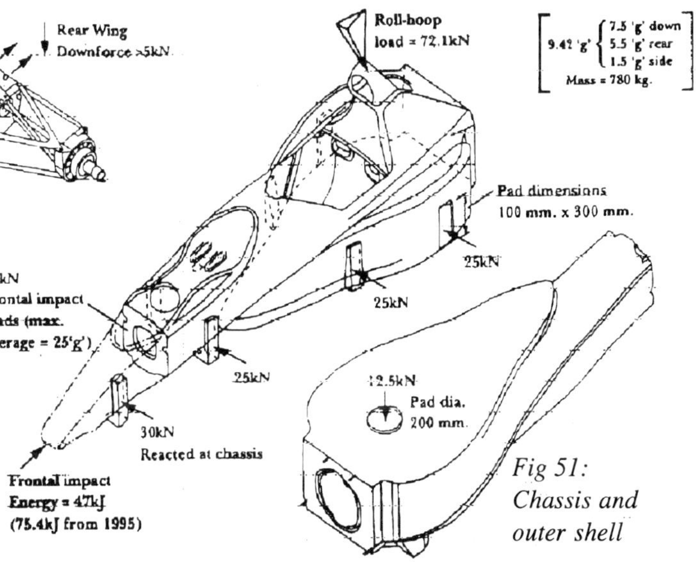

Fig 51: Chassis and outer shell

Hydroformed tubes for vehicle structures

As well as new construction techniques offered by composites for automotive structures advanced forming methods are realising new metal structures. Lotus Engineering have carried out a case study using hydroformed tube where the ability to develop complex cross-section shapes and curved axial paths is used to advantage in a function-based design approach.

The case study applies to the design of a saloon-car B-post as typically represented by Fig 53. The function-based design approach involved defining the member as part of the overall body structure and then as a member for supporting hinges locks and side impact forces.

The overall structure is designed rationally from first concept in this function-based approach even from the point of choosing between triangulated space-frame, monocoque, platform or ladder frame construction for the structure. With regard to a more detailed study of the B-post, alternative sideframe options are considered, as in Fig 54, and a rational choice made. Joint and beam section properties are next identified for primary load path directions. The following step takes in manufacturing considerations in which tests can establish the advantages of different processes for the key elements as in Fig 55. Best design options are next assessed for predicted performance based on the criteria of structure performance, mass, manufacturability, cost and robustness of design. This structural scheme then allows detail design and packaging to proceed.

Key structural requirements for the B-post involved the load cases shown in Fig 56. Overall structure studies identified a requirement for high shear stiffness in the plane of the sideframe while 'sensitivity' studies showed stiffness of B-post to roof-rail joint about a transverse axis Ry to have the largest effect on overall torsional stiffness and the third largest effect on bending stiffness. Design philosophies for the post are summarised in Fig 53. The first gives particularly high robustness to end joint rotational stiffness while this parameter has very low value in the second. The third places emphasis on the packaging allowance above the waistline (Fig 57) while the fourth is a combination of the first and second and is well suited to the packaging requirements. Bending loads are limited to the short free-beam length above the waistline and the waistline-joint can be enhanced by making the upper beam and one lower leg in one piece. There is also a smooth transition of properties at the joint. Tumblehome of the upper part involves the vertical lower half changing into a curved profile to meet the cant rail, the

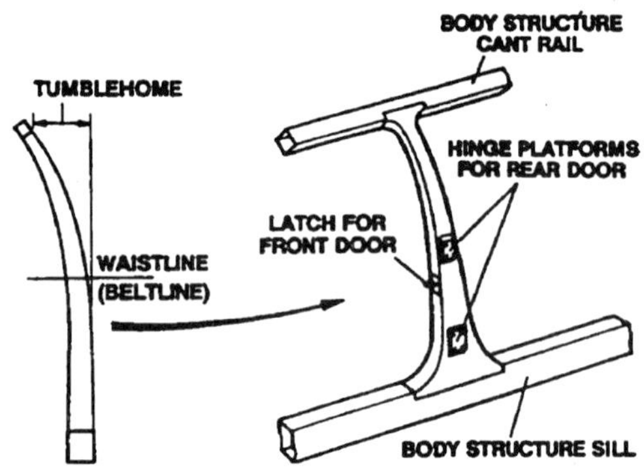

Fig 53: B-post under investigation

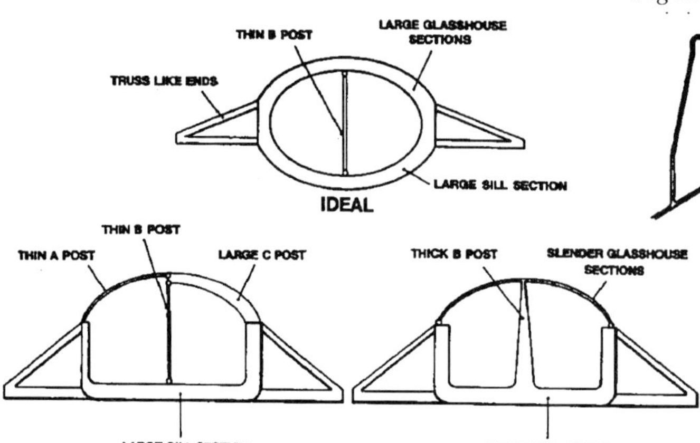

Fig 54: Alternative sideframe options

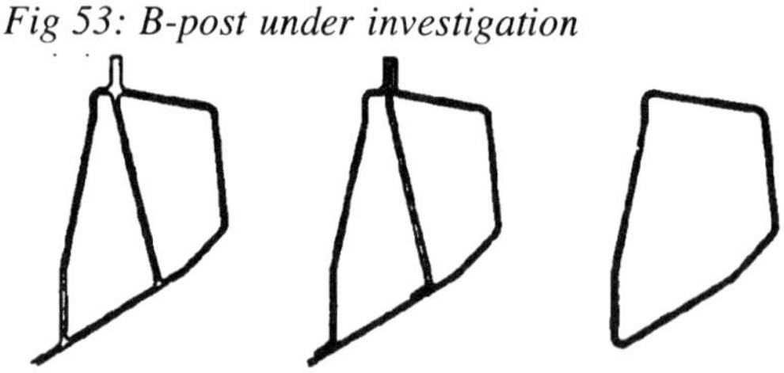

Fig 55: Aternative manufacturing processes for sill: from left, a extruded, b pressed or rolled, c hydroformed

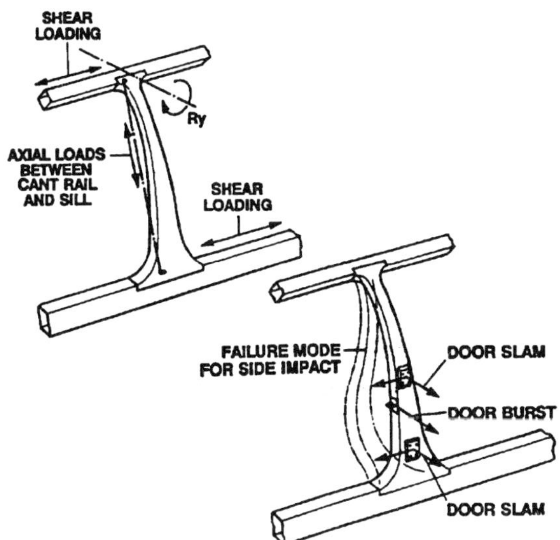

Fig 56: B-post load cases

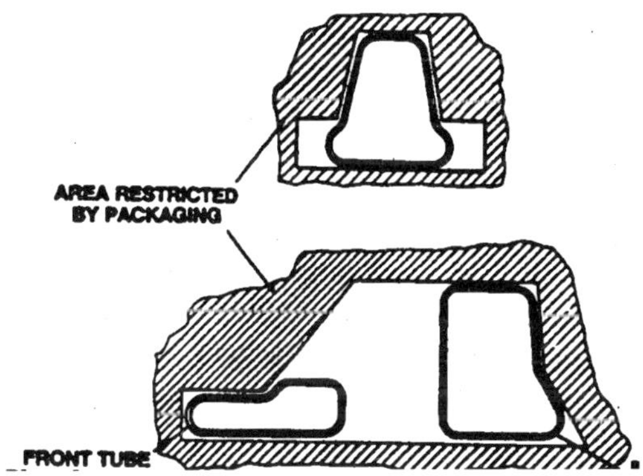

Fig 57: Package allowance above waistline: top, above waist-line, bottom, below waist-line; front tube is 70 mm and rear tube 90 mm diameter.

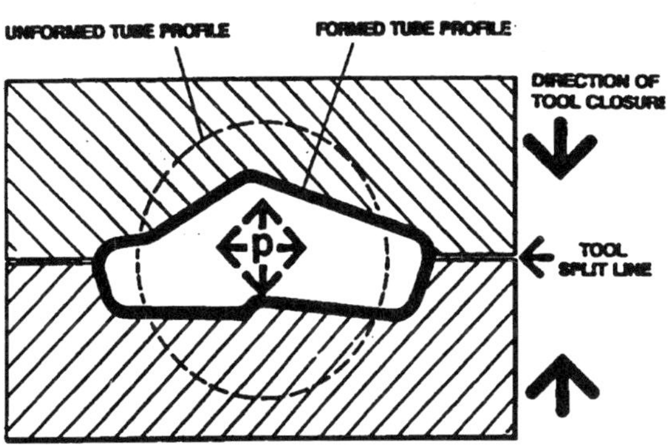

Fig 58: Priniciples of hydroforming

structural offset resulting in torsion and bending loads as well as shear.

Packaging space defined the structural cross-sections for an analysis to determine section properties. These predicted 1.2 mm thick mild steel would be satisfactory for maximum stress in the static load cases. This is a key dimension for hydroformed tube as the fillet radii for structural packaging are dependent on it. Another analysis, of the manufacturing processes had already revealed the ability to form complex sections in curved paths to be crucial in maximising the structural package, particularly the ability to alter section shapes at the ends of the post to suit optimised joining with the upper and lower rails.

Fig 58 shows the principles of hydroforming, in which a circular tube is forced simultaneously into cross section and curved path shapes. Fig 59 shows the resulting structural solution under finite element analysis which demonstrated good continuity at the waistline joint but some manufacturing complexity which had to be reassessed and solved by laser-cutting of the formed tube to ease blending of the formed tubes.

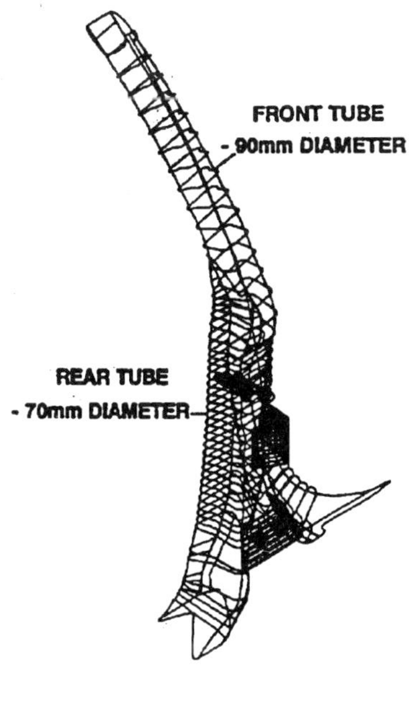

Fig 59: FE analysis

Bonded aluminium structures

Recent work at Alcan[7] has shown how adhesives could be the key to producing aluminium sheet monocoque body structures in modest volumes. The paper also refers to earlier work, reported here, in which a formal procedure is proposed for ensuring the integrity of joints.

Fig 60 shows the production process used to produce six Metro cars over a decade ago. Earlier work on Alcan's Aluminium Vehicle Technology (AVT) project had determined the necessary sheet pre-treatment for making the surface compatible with adhesives and more amenable to resistance spot welding. This was important for the weld-bonding process which the company believed to be the best way to fabricate body structures.

A suitable drawing lubricant had also to be developed which allowed panels to be pressed and adhesive applied subsequently without the need to clean down the panels after pressing. Specialised automotive sheet materials were also developed, these being typically formed in AA5754 AlMg alloy, a medium strength material offering good formability and corrosion resistance. For surface skins the AA6111 material is preferred, an AlMgSi alloy which combines good corrosion resistance with high strength after paint bake.

Fig 61 shows commercially available structural adhesives ranked according to cure temperature, an important consideration being that the adhesive must not cure during the assembly stages. Single part epoxies become the main contenders on this basis; also their peel strengths are high above ambient temperatures. A modified form of epoxy is needed, however, to provide an adequate impact strength throughout the range of vehicle operating temperatures, even at the critical sub-zero level. An advanced formulation developed with Ciba-Geigy solved this and the problem of high-temperature creep performance so as to eliminate joint failure at the hot spots on the vehicle.

For structural optimization on monocoque vehicles, Alcan developed a method for joint design of weld-bonded structures which allowed for geometric variations, manufacturing variability and complex joint loading. This involves finite-element modelling with bond line generation software involving joint-line elements, modelled joint stiffness and joint design envelopes, for the 50% of the car's structural

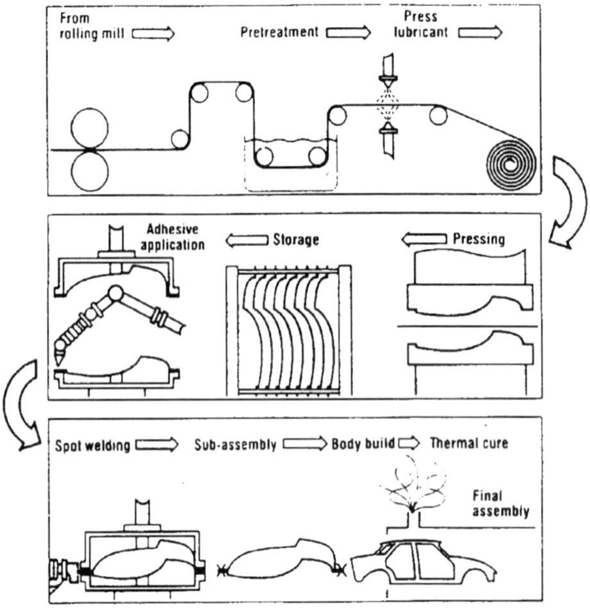

Fig 60: Production process for aluminium car bodies

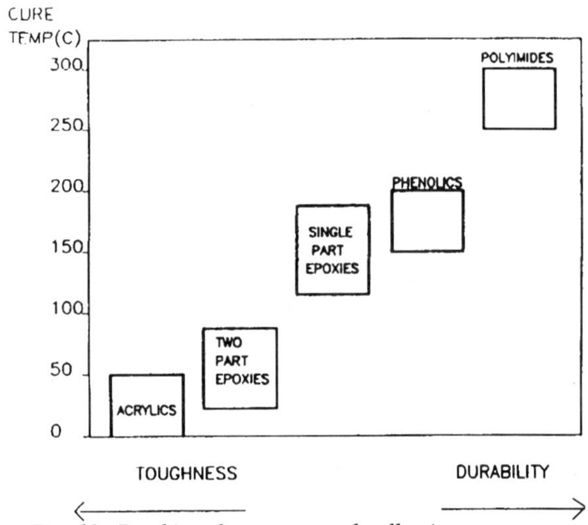

Fig 61: Ranking for structural adhesives

Joint Type	Static Strength (N/mm)	Fatigue Strength** (N/mm)
Bonded Aluminium (25% Fillet), 2mm 5754-O	68.4	24.4
Spot-Welded Steel (1.16mm CR2) 25mm pitch	28.95 (723.8)*	4.6 (115.0)*
Spot-Welded Aluminium, 2mm 5754-O 25mm pitch	30.4 (760.0)*	2.0 (50.0)*

* Values in Brackets are Load per Spot-Weld (N).
** Maximum Load for 10^7 Cycles (Stress Ratio R=0.1)

Fig 62: Static and fatigue strengths compared

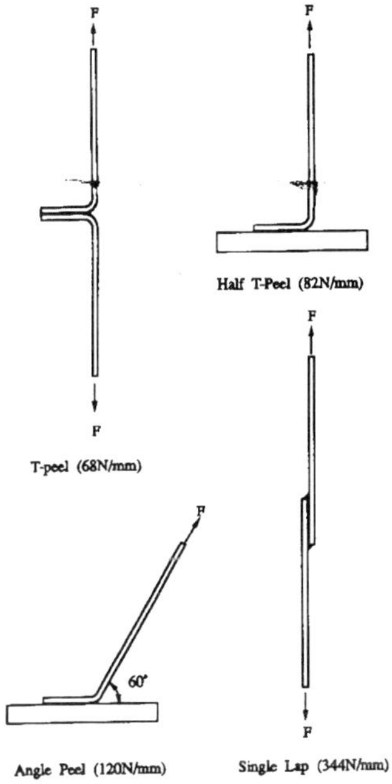

Fig 63: Different joint configurations

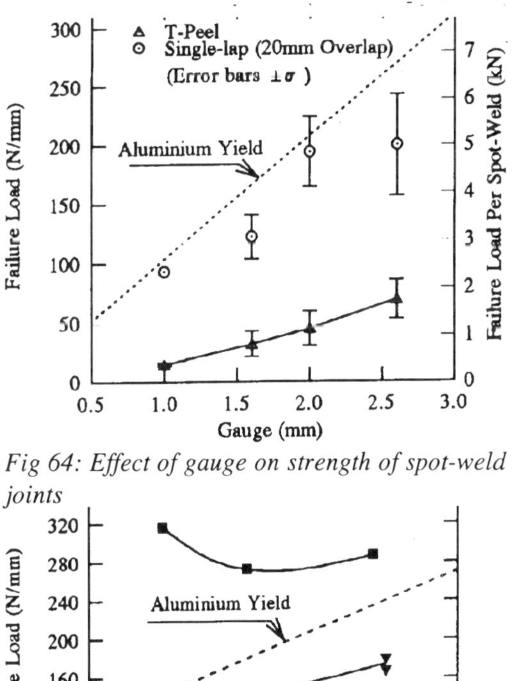

Fig 64: Effect of gauge on strength of spot-weld joints

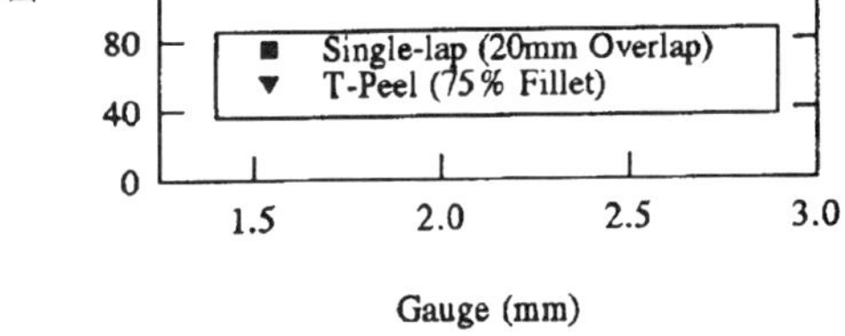

Fig 65: Gauge effect on bonded joint strength

joints in which adhesives formed a crucial element. First car to benefit from this treatment was the Jaguar XJ220.

McGregor *et al*[8] had earlier described the joint-design approach in detail, in which the table of Fig 62 compares static and fatigue strength of various aluminium joining techniques which demonstrated the superiority of the bonded joint. Fig 63 gave the static strength of various bonded joints with the special epoxy adhesive, a trend which is similar for spot-welded joints. In Fig 64, the aluminium yield line is the load per unit width for a given gauge at which the whole of the cross-section of the specimen is at yield (90N/mm^2). For the spot-welded joint it is seen that the aluminium adherend has not fully yielded before the spot-weld fails.

For bonded joints, however, Fig 65 shows that single lap joints fail at a load much greater than that required to yield the aluminium. Comparing the figures also shows there is much greater influence of the gauge on the strength of the spot-welded joint than on the bonded joint. The table of Fig 66 gives the influence of bondline (adhesive layer) thickness on the performance of T-peel joints. The most critical parameter has been found to be the fillet size, Fig 67. Its influence is shown in Fig 68 where fillet size is increased from 20 to 100% to give a joint strength increase of four times. By examining the stress distribution at the bondline the reason for this becomes clear, Fig 69. Most of the load transfer is seen to take place in the fillet, with almost no load in the flange of the specimen.

The manufacturing process must therefore be designed around this important criterion. The three stages in the Alcan procedure for joint design are:

Bondline Thickness (mm)	Failure Load/Width (N/mm)
0.22	141.1
0.30	140.0
0.95	145.0

2mm AA5754-O Aluminium, Fillet Size: 75%

Fig 66: Joint failure loads

* modelling of the joint-line areas within the full-scale FE structural model of the vehicle
* extraction of the important loads acting on the joint from this model
* comparison of the loads on the joints with the joint failure envelopes, which provide crite ria based on detailed modelling and tests for each configuration.

By representing the joints within the structure in a simplified way, complexity of the overall model is reduced. Such a model normally involves shell elements representing its thin-walled parts and the joint design approach has a row of special shell elements along the joint lines known as Joint Line Elements. These have the special feature that all local co-ordinate systems are oriented so that the y-axis always points along the joint line, the element tension force per unit length in the y-direction *Ty* representing joint-line tension. The *Ty* value is an element output value is one which can be obtained from most shell elements. These joint line elements are also arranged to contain modified physical and geometrical properties, representational of real joints, such as tension and bending stiffnesses. Software has been developed to produce these automatically.

The Joint Line Generator accepts as input the FE mesh data including nodes and elements of a structure then creates new elements at the boundaries. These are oriented in relation to the joint line for post-processing of jointline forces. It can incorporate several of the most common joint modelling methods

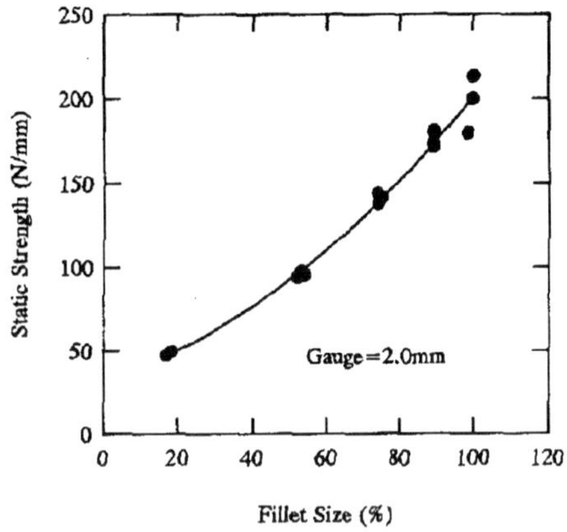

Fig 68: Effect of fillet size on static strength

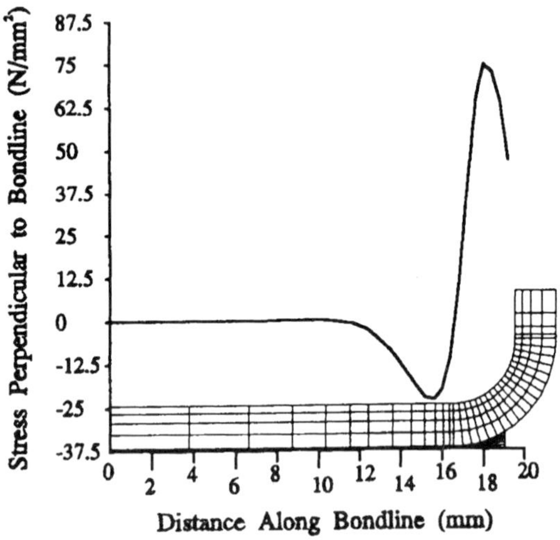

Fig 69: Stress distribution in the bondline

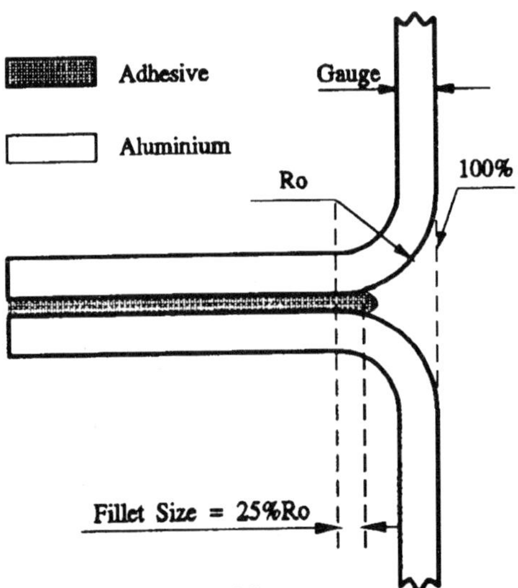

Fig 67: Defining the joint fillet size

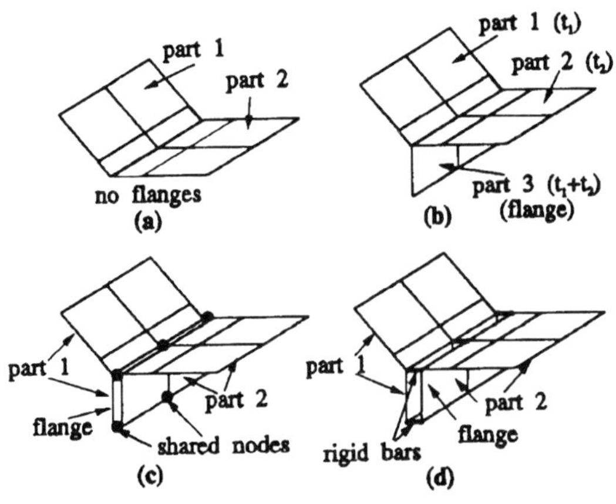

Fig 70: Joint modelling methods

including, Fig 70, no flanges, flanges modelled as one set of elements with combined thicknesses and those represented by two sets of elements but sharing the same set of nodes, also flanges modelled as two sets of elements and connected by rigid bars. Fig 71 shows an example of the use of the Generator in a four part structure.

Correct prediction of tension force and bending moment distribution depends on stiffness modelling. The table in Fig 72 compares predicted stiffnesses of weld-bonded and spot-welded T-peel joints, under tension and bending, with aluminium sheet stiffness, based on detailed FEA. This shows the bending of the joint to be much closer to the sheet performance than the tension.

In determining the critical loads on joints, FE modelling has come up with the stress distribution shown in Fig 73 based on the analysis of a 3-D model of a half-peel joint using 20-node solid elements. The figure shows the tension force produces the largest stresses in the joint with stress concentrated in the adhesive fillet while shear loading produces a concentration further into the main bondline. A critical

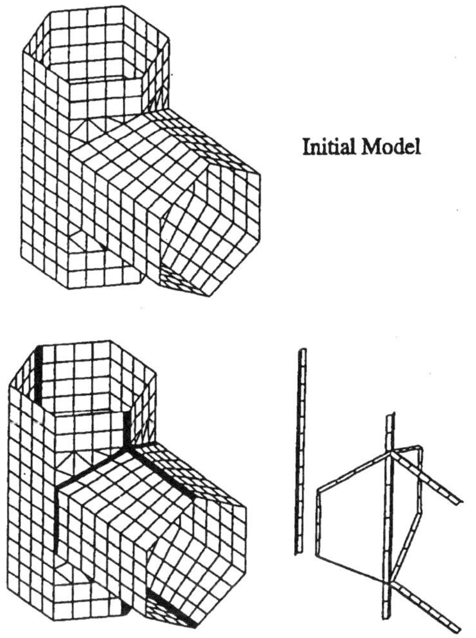

Fig 71: Use of joint line generator

TENSION STIFFNESS per UNIT LENGTH [(N/mm)/mm]			
Gauge (t) (mm)	Sheet	Weld-Bonded	Spot-Welded
1.2	8400 (100%)	954.6 (11.4%)	29.2 (0.35%)
1.6	11200 (100%)	1134 (10.1%)	57.5 (0.51%)
2.0	14000 (100%)	1337.0 (9.5%)	94.9 (0.68%)
BENDING STIFFNESS per UNIT LENGTH [(Nmm/rad)/mm]			
Gauge (t) (mm)	Sheet	Weld-Bonded	Spot-Welded
1.2	1131 (100%)	1051 (92.9%)	551 (48.7%)
1.6	2681 (100%)	2430 (90.6%)	1282 (47.8%)
2.0	5237 (100%)	4628 (88.4%)	2447 (46.7%)

Fig 72: Predicted stiffness of wel-bonded and spot-welded joints

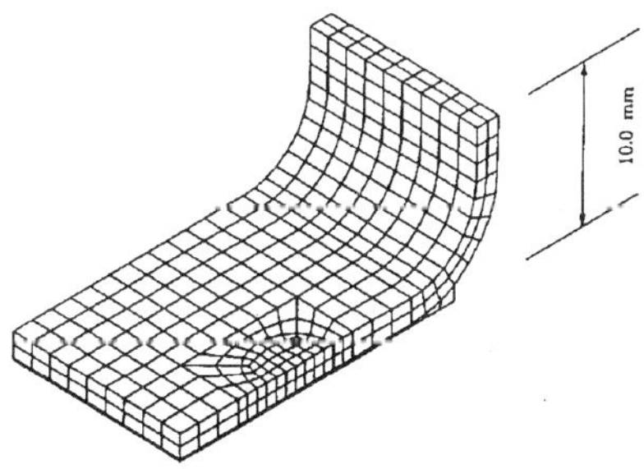

Fig 73: FE model of weld-bonded joint

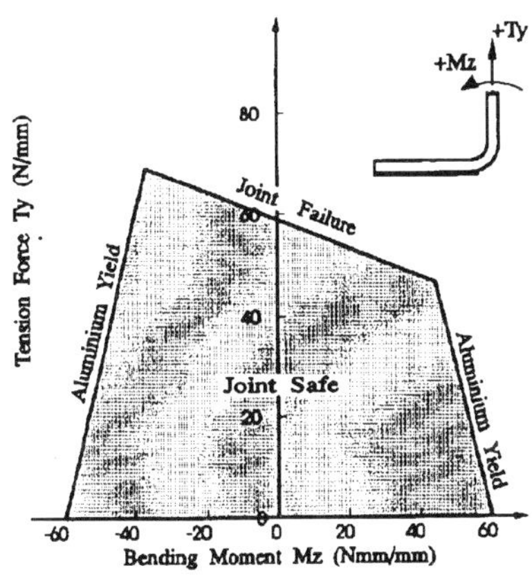

Fig 74: Typical design envelope

combination of tension and moment loading was therefore chosen as the joint failure criterion in the design method.

Joint design envelopes were developed as a logical part of the design approach, the forces obtained from the full scale FE model being checked against them. The envelope, Fig 74, defines a safe region and every point represents a tension and moment combination. In this case the top boundary, representing adhesive failure, was obtained by a combination of FE modelling, tensile tests of peel joints and FE simulation of the test to obtain the exact combination of tension and moment forces.

The joint design procedure has been used on a number of full vehicle structures. In that shown modelled for analysis in Fig 75, the half-structure contains 16 500 elements and 14 700 nodes, involving some 71 different pressed parts. The Joint Line Generator was used to introduce 4 mm high joint line elements which led to 3280 further ones and 3382 more nodes, increasing the model size by 20%, Fig 76. Loads used for the analysis were fatigue ones measured at the suspension mounts, the structure being of aluminium gauges ranging from 1.0 to 3.0 mm weighing 132 kg. The assumption that all the joints were peel ones gave a highly conservative estimate of performance and normally this would be followed by the assumption of a more sensible joint configuration. On the basis of tension loads only, no moments, the table in Fig 77 summarises the static and fatigue strengths of various techniques of T-peel jointing which compares the modelled torsional stiffness of the structure.

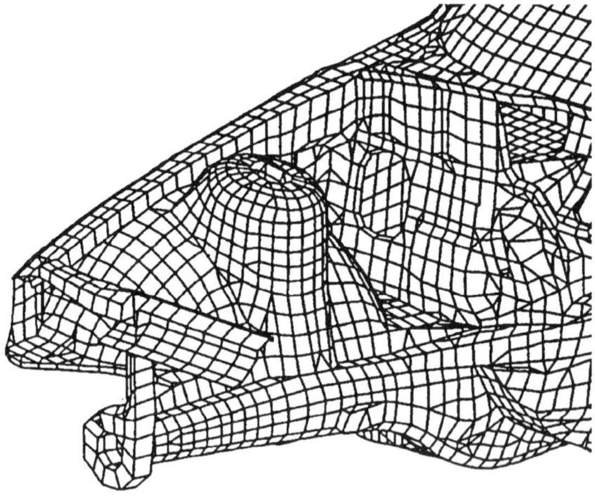

Fig 75: FE model of body front-end

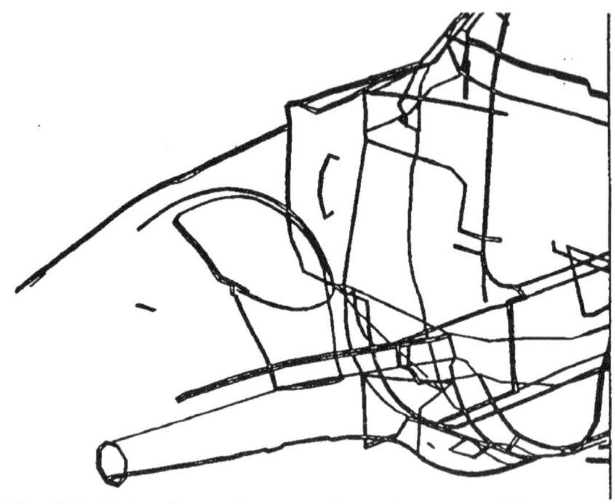

Fig 76: Joint line elements for above model

Configuration	Stiffness (kNm/deg)
Bonded Joints (fillet size = 75%)	11.04
Spot-Welded Joints (30mm pitch)	9.44
Continuous Joints	10.71

Fig 77: Torsional stiffness of aluminium vehicles

PART 2:
BODY CONSTRUCTION

Chapter 6: Fabrication methods

Snap-fit and spring-clip fasteners; special springs;
specialised fastening techniques; bolting and riveting systems;
arc and gas welding techniques; spot, seam and projection welding;
adhesive bonding, structural adhesives and welding of plastics;
design for automated and robotic assembly
Shearing, bending and profiling; flame and jet cutting;
press-working basics and advanced developments;
deep-drawn press-work and superplastic forming;
roll-forming, flex-forming and spinning/flow-forming
Anti-corrosion pre-treatment and coatings;
finish painting; body sealants;
coated steels, plastic coatings and pre-coated panels;
glazing techniques; trim pad construction

While most vehicle makers have divested themselves of machining operations, with the passing down of mechanical system construction to the tier-one systems suppliers and their sub-contractors, forming and fabrication of metals and composites used in body construction is often an in-house activity. Ever developing construction techniques mean that both product and production engineers must be ever vigilant of changing methods in order to maintain high quality at economical price.

Fastening techniques

Fabrication of vehicle and body systems can be considerably eased by the correct choice of fastening technique, of which threaded, riveted, tabbed and spring-clip mechanical fasteners are among the alternatives. There are also, of course, a wide range of welding/brazing alternatives. Sometimes the design of the part to be joined can be arranged to give it a fastening ability for mating with the adjacent part.

Snap fits

It is possible by careful design of plastic mouldings to avoid the use of separate mechanical fastenings or adhesives — by using the snap-fit potential of the materials. According to Du Pont, all snap-fits are variations on two basic designs, the cylindrical (Fig 1) or cantilevered lug (Fig 2). They work by using an undercut in one piece and on the other, a lip or hook. Unlike press fits they do not involve permanent deformation since the parts return to an unstressed state after fixing. Their main possible disadvantage is the additional tooling cost for making the constituent parts. If the cylinder of the joint is slotted then effectively the joint can be considered as a series of lugs.

The lugs can be made easy to separate, or not, by varying the return angle. Undercuts of 90 degrees can if necessary be moulded by using the side cores of corresponding slots in the mating part. Lugs should be designed with thickness or width tapering from the root to give uniform distribution of stress — and an adequate bending length must always be allowed. The basic strength and deflection equations are given in Fig 3.

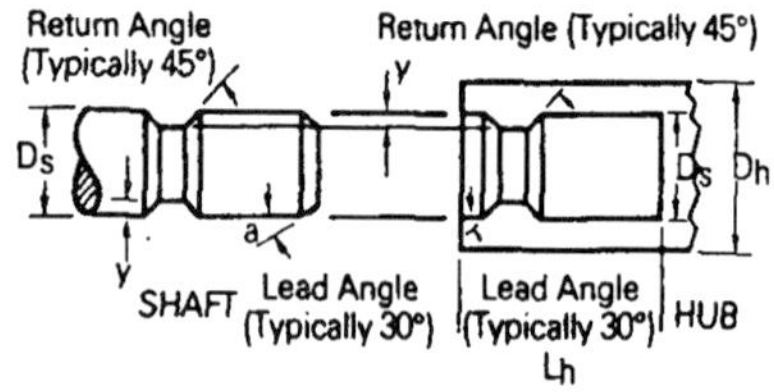

Fig 1: Cylindrical snap-fit

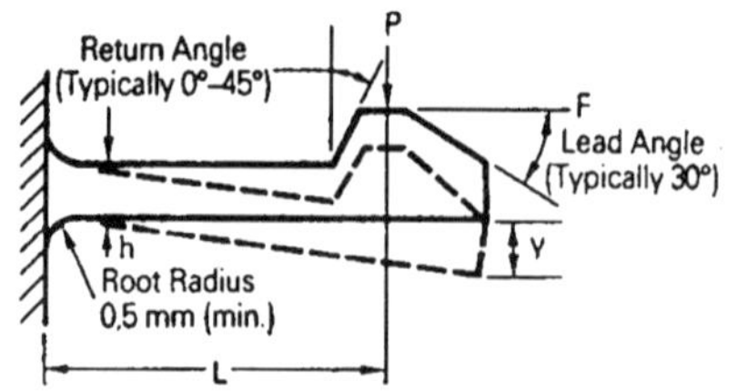

Fig 2: Lug-type snap-fit

h = Root thickness
b = Root width (not illustrated)
L = Length of arm
a = Lead angle (°)
r = Return angle (°)
μ = Coefficient of friction
P = Deflection force
E = Flexural modulus
I = Cross-section moment of inertia

Stress = $S = \frac{PLh}{2I}$ Deflection = $y = \frac{PL^3}{3EI}$

Assembly force = $F = P\left(\frac{\mu + \tan a}{1 - \mu \tan a}\right)$

$$F = \frac{(1+f)\ \mathrm{Tan}\ (a) S_d D_s L_h}{W}$$

Where:
F = Assembly force
f = Coefficient of friction
a = Angle of beveled surfaces
S_d = Stress due to interference
D_s = Shaft diameter
W = Geometry factor (see below)
Lh = Length of hub expanded*
y = Undercut

$$W = \frac{1 + (D_s/D_h)2}{1 - (D_s/D_h)2}$$

Fig 3: Assembly force calculation

Spring clips

Spring clip fasteners can simplify and speed the assembly of body and vehicle systems in a way that is difficult to match by threaded fasteners. By careful attention to design rules, reliable performance can be obtained for the assembled system.

Retaining rings are possibly the best known form of spring clip fastener and these may be used to replace machined shoulders, set collars, rivets, cotter-pins, nuts and other bulky fastening devices — as well as the associated machining operations. Rings divide between internal and external locking types, Fig 4, and bevelled edge types can be used to perform

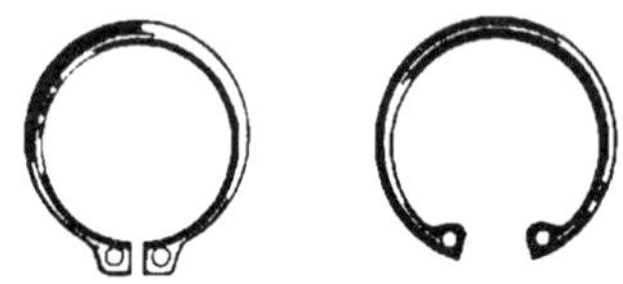

Fig 4: Internal/external retaining rings

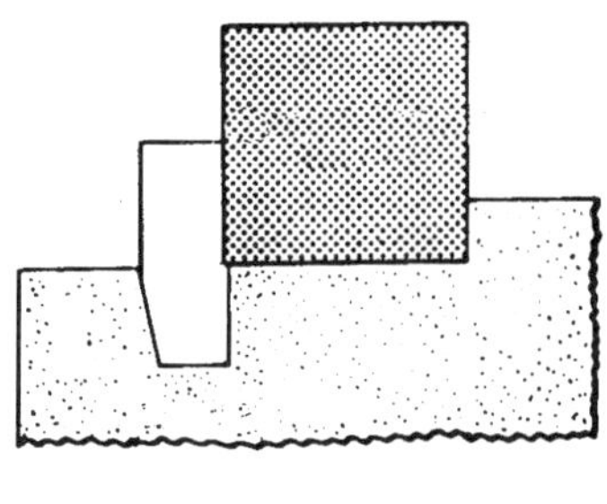

Fig 5: Bevel-edge retaining ring

Fig 6: Retaining clips

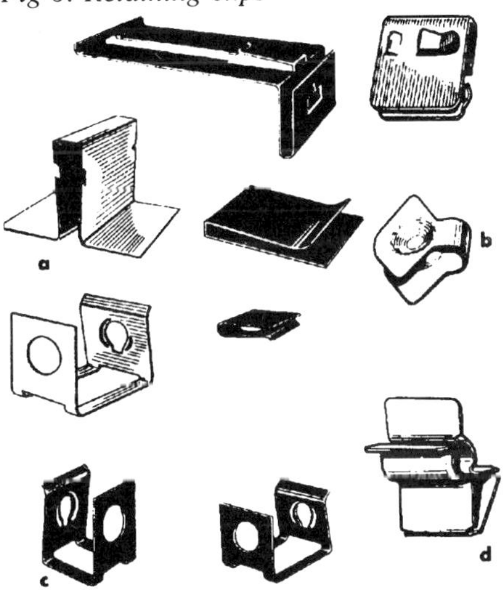

Fig 7: Assembly clips

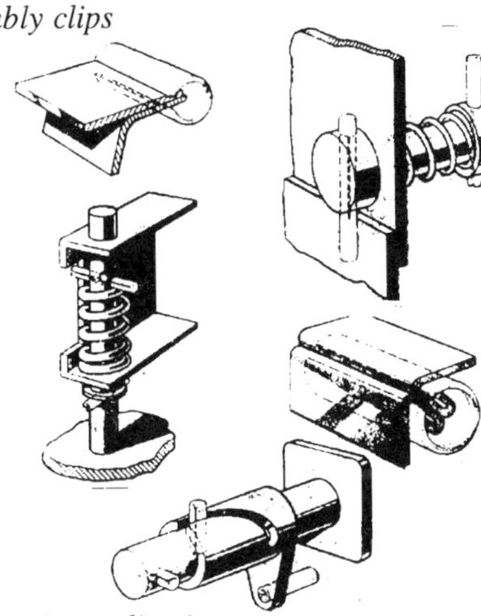

Fig 8: Spring-pin applications

precision locking jobs such as the retention of bearings, as shown in Fig 5 .

Other clips are of the push-in or nut retainer type and a large number of hybrids have been developed incorporating barbs, tags, riveting or other facilities. A selection are shown in Fig 6 from the designs developed by Salter. At (a) is a fastener which is attached to a body by a single rivet and used to retain a circular moulded emblem which is snapped in place, without tools, and conceals the fixing. At (b), another variant snaps into a rectangular hole and studs engage the fastener, on the part to be fixed. Triangular, arched-shape clips at (c) are intended for metal assemblies requiring a high retaining force. At (d) large clips used in pairs, to this design, are employed in retaining a loudspeaker on a trim panel, typically. The clip at (e) is designed to mate with a circular stud and can withstand an axial load on the stud of some 175 lb; it can be quickly assembled with a push-on tool or arbor press. That shown at (f) is for securing flexible materials, the upturned ends preventing digging into the soft surfaces. Like (e) in function, (g) is used to retain against forces on a larger diameter shank but with minimal radial operating clearances and might typically be used to secure an indicator lamp in a dash panel. Type (h) is used for retention within a cavity ranging in size from 3/8 to 2 inch diameter.

Metal-to-metal assemblies have been known to use a variety of spring clips, some of which are typified in Fig 7. They exert a firm grip on the edge of panels and can be positively anchored by piercing a small hole in one member to receive a tongue. One of this last type is also shown (left) at (b) developed for attaching weatherstrip. The complementary part, (right), snaps in place over the abutting corners of the strip. Captive types are illustrated at (c) which snap into place and provide some latitude for engagement with slightly off-position fixing screws. A special captive clip is seen at (d) which was developed as a squab adjuster for the rear seat of a car; the appropriate mating bolt is merely 'zipped' into place to achieve the required positional adjustment.

Spring-pin fasteners are another resourceful assembly aid — some applications devised by Firth Cleveland Fastenings being shown in Fig 8. That company's Rollpins are essentially slotted, chamfered, cylindrical spring-pins which are heat-treated and tempered to achieve optimum toughness, resilience and shear strength. These also can be used to make inexpensive handles and latches as shown in the figure as well as securing knobs, joining shafts or guiding control cables. Other examples shown are for spring retention and cam driving.

High-torque blind cage nuts are another application for the spring clip, that shown in Fig 9a being produced by Lintite. A variant shown alongside has a torque arm projecting from the cage to increase torque capacity. Projections at the end of the arm spring into location holes. Lintite also make the Rapid series of spring clip fasteners shown at 9b.

Clips used in conjunction with rivets, studs and bolts are shown from the Spire range in Fig 10 while an interesting variant in Fig 11 is produced by Dzus. It is called Pilot and comprises a sprung receptacle for a mushroom-headed stud. The clip is inserted in a panel while the stud is collared to hold a grommet which serves to retain the stud in the closure panel, when open, as well as providing a cushion between the mating parts. The end of the stud is laterally grooved to allow the receptacle to grip it in the closed position.

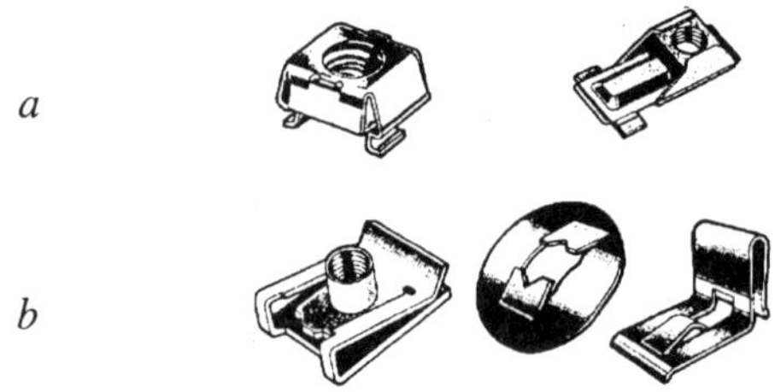

Fig 9: Lintite fasteners

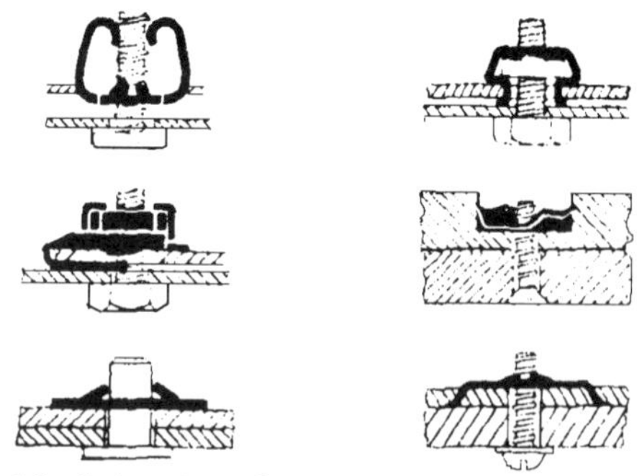
Fig 10: Spire rivet clips

Fig 11: Dzus rivet clips

Specialized fastening system

An interesting mechanical joining technique which is a possible alternative to spot-welding, riveting or bolted connection is the Eckold press-joining process. The joint is made with relatively simple equipment and heat-treatment properties of the substrate sheets are unaffected.

The method has been called H-forming and the tooling consists of a punch and shearing/forming die, Fig 12. The centre of the die has a convex anvil which has cutting inserts set on either side. The punch forces the top sheet, and in turn the bottom sheet, on the cutting edges to form a step. As the material is pushed down it is forced out sideways by the anvil. The inner edges of the cutting blades are thus forced to spring out slightly whilst constraining the material so that it flows sideways under the bottom sheets and the joint is wider than the cut-step.

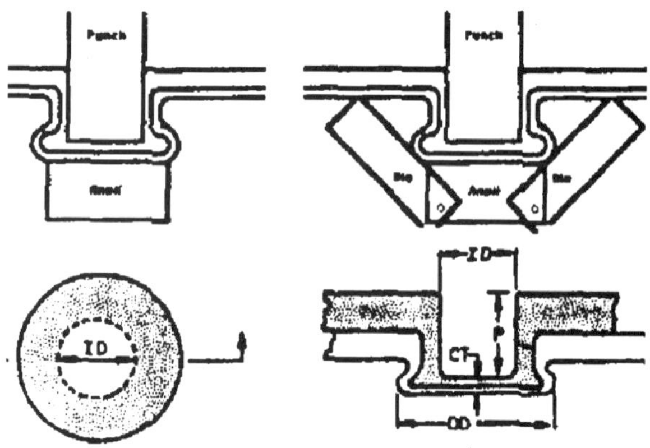

Fig 12: Punch, die and anvil

The form of the top part of the joint is determined by the contours of the punch but in any case is leak-proof. Fig 13 shows the characteristic dimensions of the joint: degree of penetration and thickness of the cold-formed web beneath the joint. Penetration depth DT is determined by distance between the top of the shearing blades and the bottom of the die while web thickness is a function of the sheet characteristics. Fig 14 shows the final form of the joint from which it can be seen that the shape varies in the X and Y planes as the die is open-ended in the X-direction.

Thickness in the Y-direction depends on punch width and separation of the cutting leaves. Forming of the inner joint element is crucial to the strength. Fig 15 relates the stages of forming the joint with the force/penetration plot The H-joint has the advantage over spot-welding that non-conductive

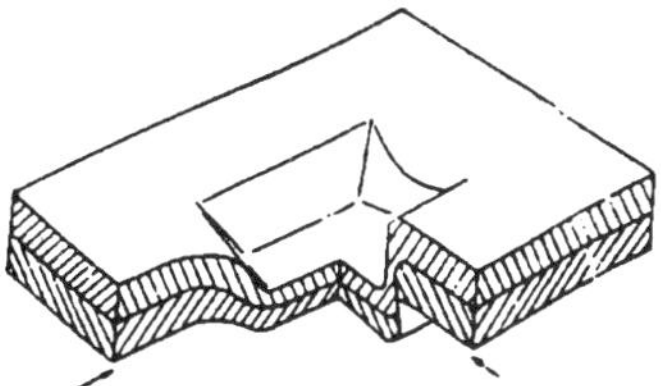

Fig 14: Final form of joint

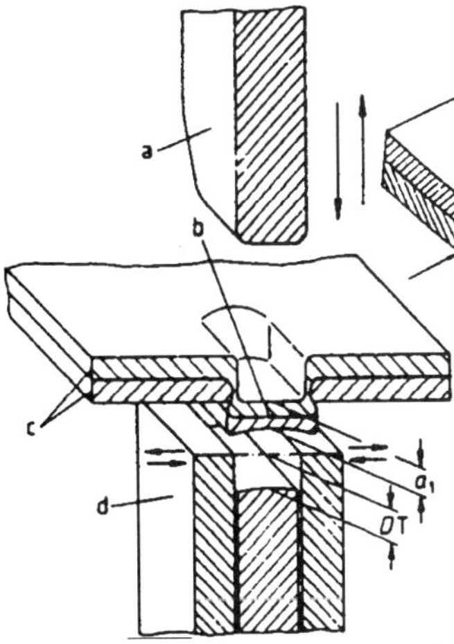

Fig 13: Characteristic dimensions

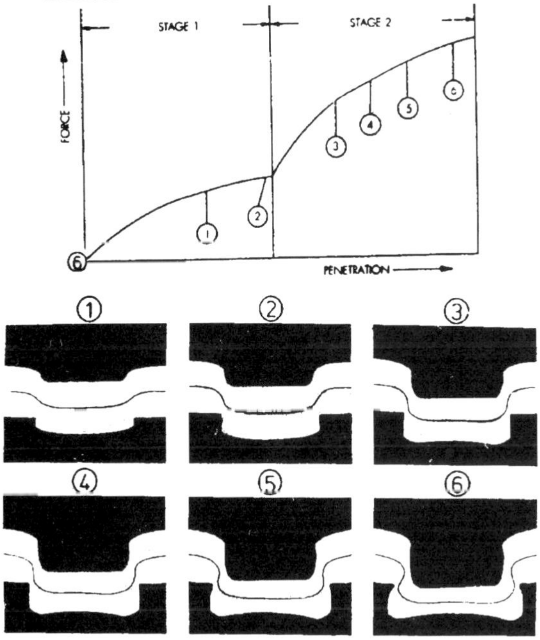

Fig 15: Stages of forming

surfaces can be applied to the sheets and power requirements are considerably less to form each joint. Sometimes the joint can be made within the forming operation of the sheet parts. The higher cost of the punch-and-die set than spot-welding electrodes is offset by their life which is usually 200,000 joints before re-machining. This is 20 times greater than electrode life and 100 times greater in the case of welding galvanised steels.

The process had its origins in the Tog-l-loc joint developed by the US BTM Corporation, and was referred to as a button-type joint. During its development it was found that to insure the punch reaches a positive stop, and will not stall during the process, a press of at least 2 ton capacity was required for joining mild-steels up to 0.030 inch thickness. The extent the button protrudes from the sheets should be such that anvil depth - thickness = 0.005 inch for sheets up to 0.05 inch thickness. Less formable steels may be joined with a smaller anvil depth, to eliminate tearing at the internal surface of the joint. Strength of the joint was found mainly to be a function of the cap thickness of the button.

Threaded fasteners

Critical bolted joints on vehicles require a knowledge of the theory of threaded fasteners so the exact capacity of the joint can be calculated. This is crucial to braking, steering and suspension components, for example, where the safety of road users can easily depend on the clamping effectiveness of the joint.

The forces acting on a threaded fastener are shown in Fig 16 for the thread interface. The torque applied to a threaded fastener not only has to overcome friction at the thread interface but also has to overcome bolt head to clamping surface friction and to do work in inducing the clamping tension into the bolt. It is usually assumed, for calculation purposes, that annular friction torque acts at the mean radius of the bearing faces and that the thrust N between two threads is on a helix. The total tightening torque can then be approximated to

$$FD[\mu_a(R_i + R_o)/2D + \mu_t(R_t sec\beta/D) + (P/2\pi D)]$$

where D is bolt diameter, R thread helix radius (inner/outer) and F bolt tension. The three bracketed terms are represented by K_a, K_{tf} and K_u in the table of Fig 17 which gives numerical values for head, thread and induced friction constants. In practice some 90% of the total applied torque does no more than overcome friction (which can vary by up to 20% between fasteners in practical systems). Use of molybdenum disulphide lubricants can reduce the mean K value to about 0.15. It is also advisable to make the bearing face of the nut and mating surface flat to within +/- 0.001 inch.

Choice of threaded fastener should resolve itself into determination of expected applied tension load and the selection of bolt which will withstand at least the same load in the pre-stressed condition. A safety margin of 80% of the yield strength is further added as a protection against over-stretching in the plastic range. By measuring the length change in the bolt during the tightening process, the stress, and hence the induced load, can be calculated from the elastic modulus.

Critical bolts should be tightened with torque spanners to prescribed values, remembering that only 10% will be inducing clamp load. The table in Fig 18 gives tightening torques and approximate tensions for a range of bolt sizes.

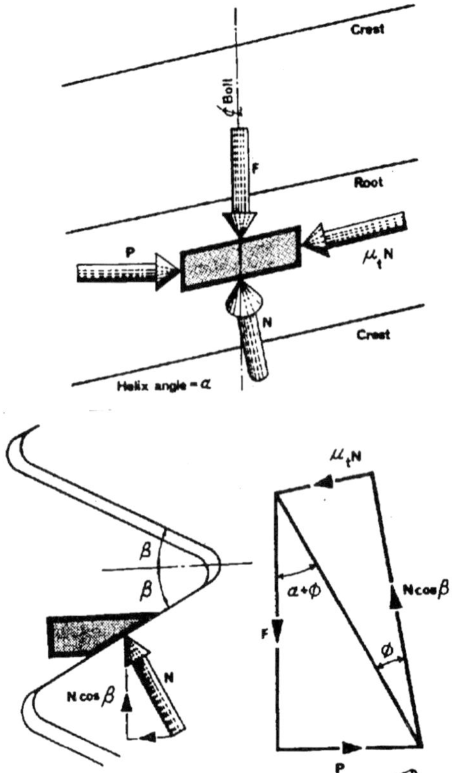

Fig 16: Forces at thread interface

Size	A QUALITY (28 tonf/in²) Torque (lbf.ft)	A QUALITY (28 tonf/in²) Tension (lbf)	R QUALITY (45-55 tonf/in²) Torque (lbf.ft)	R QUALITY (45-55 tonf/in²) Tension (lbf)
THREADS				
10	0·104	946	0·168	151
9	0·146	118	0·233	187
8	0·227	158	0·368	256
7	0·339	209	0·543	335
6	0·074	261	0·768	422
5	0·730	346	1·165	560
4	1·045	441	1·680	703
3	1·545	579	2·490	935
2	2·360	771	3·800	1240
1	3·420	985	5·470	1580
0	4·760	1273	8·000	2100
$\frac{1}{4}$ in	6·05	1460	9·55	2300
$\frac{5}{16}$ in	11·70	2260	19·00	3680
$\frac{3}{8}$ in	2·10	3375	33·80	5440
$\frac{7}{16}$ in	33·75	4650	54·70	7520
$\frac{1}{2}$ in	50·60	6110	81·20	9860
$\frac{5}{8}$ in	101·00	9720	163·00	15 780
$\frac{3}{4}$ in	181·00	14 550	283·00	22 800
$\frac{7}{8}$ in	284·00	19 600	458·00	31 600
1 in	430·00	25 900	688·00	41 600

Fig 18: (left) Tightening torques and approximate tensions

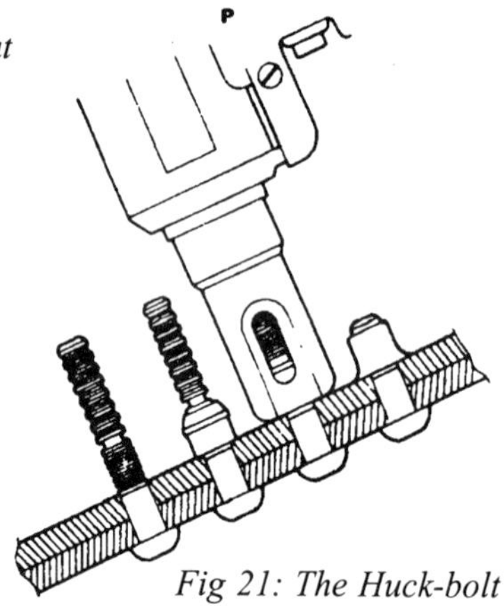

Fig 21: The Huck-bolt

NOMINAL DIA. (in)	K_a		K_{tf}				K_u			
	BSW and BSF	UNC and UNF	BSW	BSF	UNC	UNF	BSW	BSF	UNC	UNF
$\frac{1}{4}$	0·696	0·688	0·488	0·509	0·504	0·522	0·032	0·025	0·032	0·023
$\frac{5}{16}$	0·669	0·649	0·497	0·512	0·510	0·528	0·028	0·023	0·029	0·021
$\frac{3}{8}$	0·651	0·624	0·505	0·517	0·515	0·536	0·027	0·021	0·026	0·018
$\frac{7}{16}$	0·657	0·643	0·505	0·516	0·515	0·534	0·026	0·020	0·026	0·018
$\frac{1}{2}$	0·660	0·624	0·503	0·518	0·519	0·540	0·027	0·020	0·025	0·016
$\frac{5}{8}$	0·654	0·626	0·510	0·523	0·523	0·545	0·023	0·018	0·023	0·014
$\frac{3}{4}$	0·651	0·626	0·516	0·523	0·527	0·547	0·021	0·018	0·021	0·013
$\frac{7}{8}$	0·622	0·625	0·518	0·527	0·529	0·547	0·020	0·017	0·020	0·013
1	0·620	0·625	0·519	0·528	0·523	0·546	0·020	0·016	0·020	0·013

Fig 17: Values of Ka, Ktf and Ku

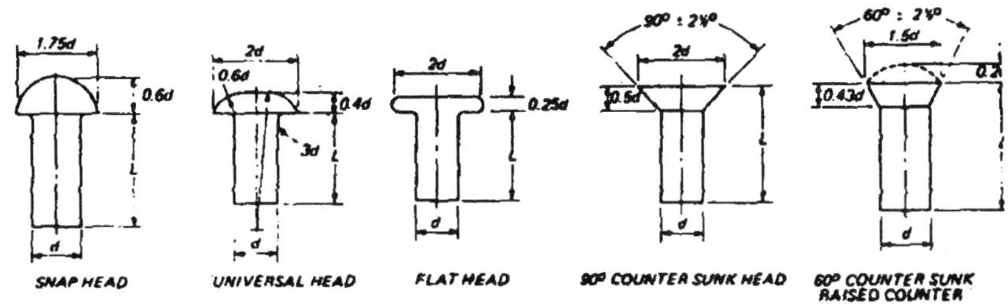

Fig 19: Types of common rivet

Riveting systems

As possibly the most common body panel jointing technique in specialist vehicle construction, riveting is governed by standard codes of practice which can lead to improved efficiency. A number of variations on the rivet are available for special purposes and for configurations where substantial loads are being transferred, stressing procedures can be adapted to desktop computer programs which allow optimum layouts to be predicted with the minimum of fuss.

British Standard BS 4620 covers solid rivets for general purpose and shows the standard heads as in Fig 19. (a) refers to cold-forged and (b) to hot-forged rivets, usually applying to larger sizes. The sizes involved are from 1mm to 39mm diameter; small sizes only are covered in a further standard, BS 641:1951, heads for these being snap, pan, mushroom, flat and countersunk to 60, 90 and 120 degrees. Semi-tubular and bifurcated designs are covered in BS 4895:1981 and 4894:1973 respectively.

Pop-rivets widely used in body construction fall into the tubular category and are classified as shown in Fig 20. It is recommended that rivets for joining aluminium alloy be made from NR6 or HR30 material — alloys suitable for cold driving. Accepted practice is to use rivets with a diameter of one to three times the thickness of the thinnest member through which they pass — and pitched at three to six times their diameter. Pitch is usually increased by an amount depending on the thickness of the materials being joined and the need to avoid separation of the sheets between rivets.

Falling between the categories of 'rivet' and 'bolt', the Huck-bolt (Fig 21) has a positive-lock collar which results in a vibration resistant fastening. The clamp force applied during the fixing results in a predictable uniform preload and a combination of the high shear resistance of the rivet is combined with the tensile resistance of the bolt. Installation rate is as high as 25/minute; the joint is waterproof and can be undone with a special tool. A high-tensile version, the HMP suits special applications like the mounting of concrete pumps on heavy-duty chassis. Greater

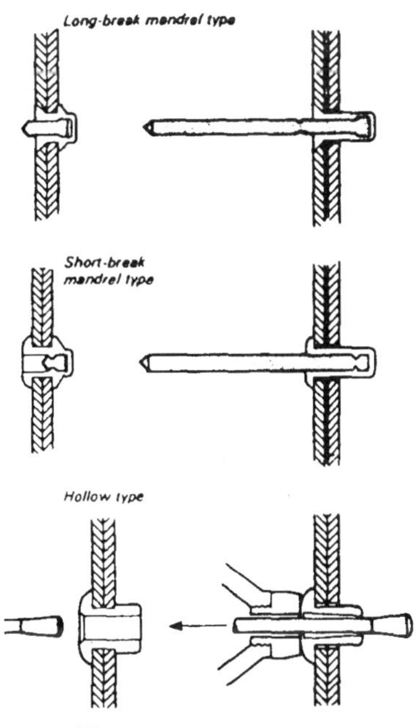

Fig 20: Types of pop-rivet

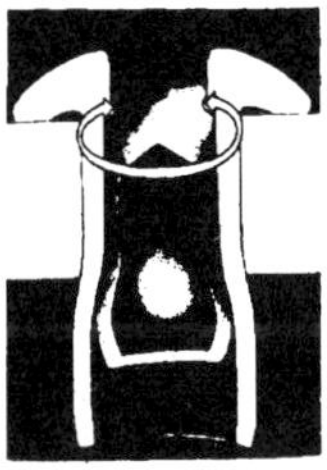

Fig 22: The Magna-Lok

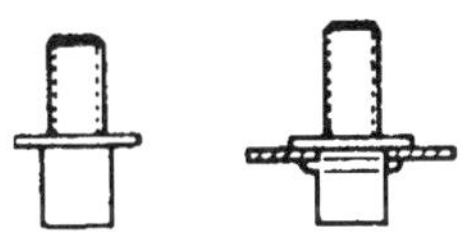

Fig 23: Riveting captive nut

fatigue resistance is claimed against comparable bolted systems due to the relative smoothness of the contoured grooves.

Another variation on the rivet is the Furma self-piercing fastener which can be used to join even pre-painted panels. A variant on the pop-rivet is the Unitac Multigrip blind riveting system. Here the shank of the rivet has peripheral grooves which collapse in a manner such as to expand the shank in the hole as the mandrel is pulled through them. The mandrel ultimately breaks to form a solid head on the blind side of the workpiece, thus sealing it. Huck's Magna-Lok fastener (Fig 22) is claimed to achieve a similar result. The riveting principle can also be used to fix a captive nut or threaded stud as in the Fastener Techniques and Delta Metal Products systems shown in Figs 23 and 24. For softer panel materials, provision for high pull-through resistance, and avoidance of 'volcano' effect, can be obtained with the systems shown in Fig 25.

Riveted joint load distribution

When calculating the stresses and strains on riveted joints, it is usually assumed the load is transmitted entirely by the rivets — friction between the connecting plates being ignored. When the centre of gravity of the rivets is on the line of action of the load, all rivets are assumed to carry equal parts of the load. If the line of action of the load is eccentric to the centre of gravity of the rivet group then the torque due to eccentricity must be added to the direct load in computing individual rivet shear forces.

Lap joints such as that in Fig 26 can be analysed by considering just a pitch width strip of the joint and noting that the minimum cross sectional area of the sheet occurs between the rivet holes. Three main failure criteria for the joint are tearing of the sheet, shear of the rivets or crushing of the sheet/rivet at the interface. For the forces causing these three failure modes to be equal

$$(p - d)t\sigma = \pi n \tau d^2/4 = ndtb$$

where p is rivet pitch, d diameter, t sheet thickness, n number of rivets, s limiting tensile stress in sheet, t limiting shear stress in rivet and b limiting compressive stress in sheet/rivet. Usually the third term is ignored as for most materials too large diameter of rivet would theoretically be required if the clamping action of the rivet is not taken into account. The diameter of the rivet is therefore normally chosen empirically and the pitch determined from the first two terms. A corner or flange joint as in Fig 27 often involves torsional forces as the load would not pass through the centroid of the joint. In the case shown

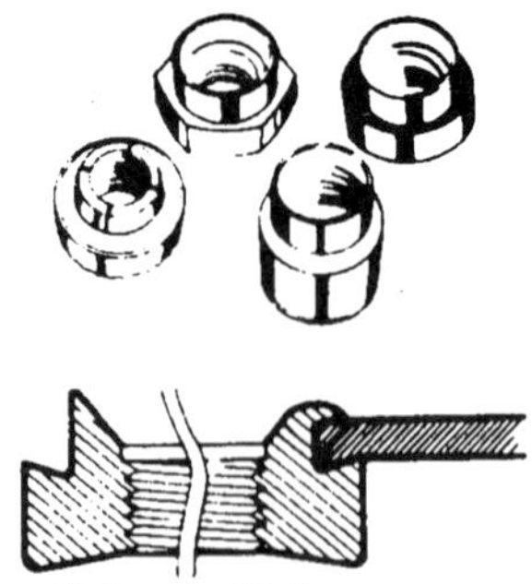

Fig 24: Alternative nut fixing by Delta

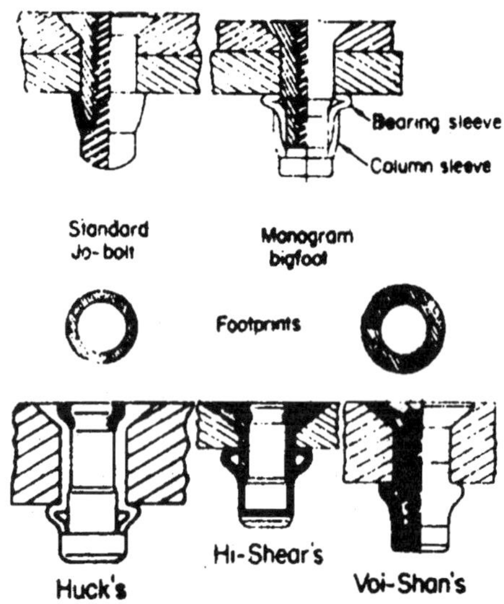

Fig 25: Systems to avoid pull-through

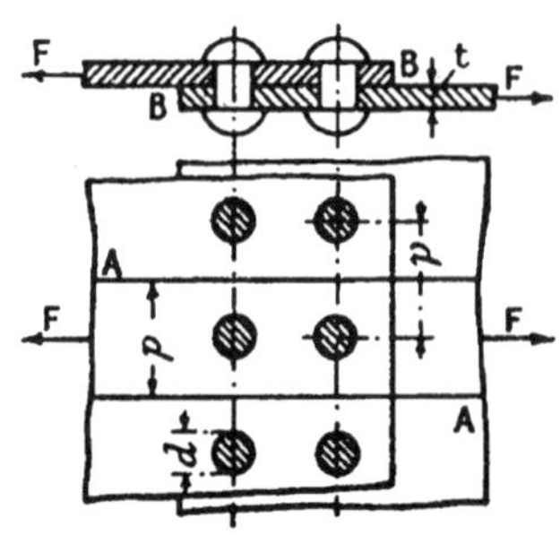

Fig 26: Lap-joint analysis

each rivet would be subject to load Qd^2/sds in the direction of loading together with a force QLr_id^2/sr_id^2 perpendicular to the line between the rivet andthe joint centroid, due to the torque of the applied load about the centroid.

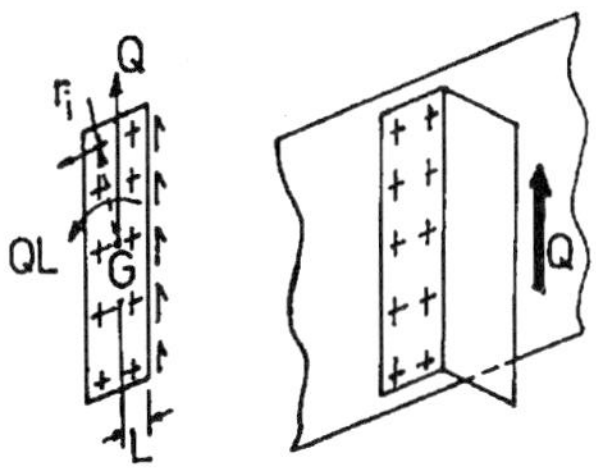

Fig 27: Flange-joint analysis

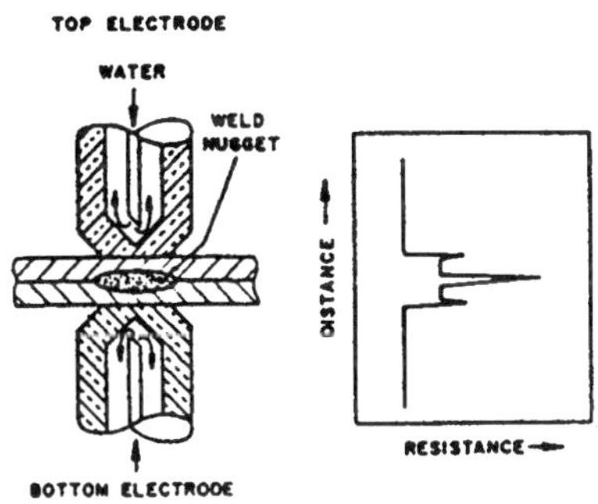

Fig 28: Spot-welding

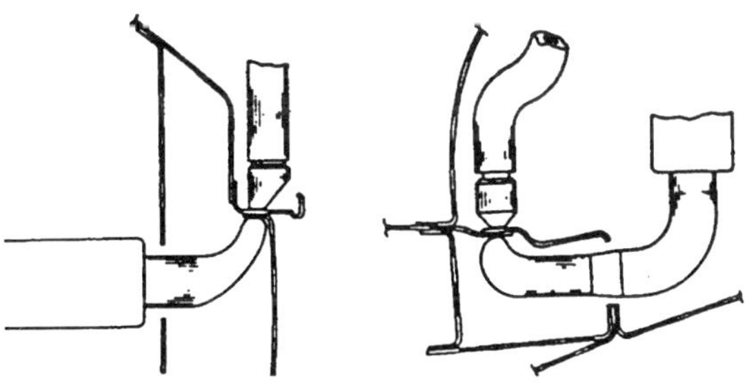
Fig 29: Reaching inaccessible flanges

a

b

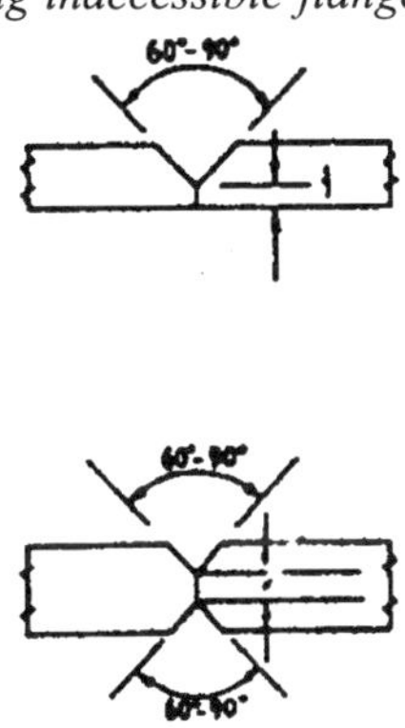

Fig 30: Single and double vee fillet welds

Welding technology and systems

For bodywork welding, the SMMT Bodybuilders Code of Practice requires adherence to British Standard BS 5750:Part 2 in which it is pointed out that welders should be provided with laid-down procedures and that finish welds should be inspected in accordance with BS 5289 in the case of fusion welds. There is thus a new focus on weld quality and the need to understand the basic processes involved.

Resistance welding, by spot or seam techniques, is appropriate to a number of relatively thin sheet materials — while for plate, tube and sections, fusion welding is often more appropriate, with melting temperatures obtained either by oxy-acetylene or electric-arc. Different conditions are required for welding steel, rather than aluminium alloys, and for welding heat-treated, extra high strength, materials.

Spot-welding (Fig 28) has advantages of high application speed and relatively low cost (after capital costs are defrayed) but very careful control of welding parameters is important to the achievement of satisfactory quality. There is also the problem of surface deformation on aesthetically critical areas. Part fit-up is important — particularly when using very high strength alloys. In the case of coated materials, such as zinc/steel, longer weld times are involved in 'breaking through' the coatings. Accessibility of relatively hidden flanges is possible with careful electrode design, Fig 29

Fusion welding of steel can involve a variety of processes. Arc welding may be metal inert gas (MIG) or tungsten (TIG) shielded. Gas metal arc welds (GMAW) can be made in all positions given control over arc current and voltage; electrode wire and shielding gas feed, also torch, speed. Advantages are high productivity and reliability coupled with low cost. Disadvantages include the bulkiness and unwieldiness of the equipment. A similar process is flux cored arc welding (FCAW), with shielding provided by gas formed in burning the flux. This uses simpler plant but can involve welding slag. Another fusion process is electron beam welding in which the weld is created by focusing a high energy stream of electrons on to the workpiece — with the advantage of providing a deep narrow weld zone, offset by the cost and difficulty of providing vacuum conditions.

Aluminium alloys can be welded using oxy-acetylene gas welding up to thicknesses of 0.125 inches when nearly pure aluminium is involved. Generally TIG or MIG arc welding is used in bodybuilding — where structural alloys are the rule — but it is important to note that any strength of the parent material obtained by heat treatment, in the joint region, will be lost. Minimum thickness for welding by MIG is usually 1/8 to 3/16 th inch but there is practically no upper limit. A fine-wire process extends the lower limit down to 14 SWG and even lower if support is provided by jigging incorporating backing chills.

TIG is not quite so economic as MIG for aluminium alloys. As the arc is struck between an argon-shielded non-consumable electrode, and the material to be welded, filler wire is added separately as in gas welding. It can be used for thicknesses from 20SWG to 3/16 th inch. From 3/16 to 1/4 inch either TIG or MIG can be employed but above 1/4 inch, MIG is usually more economical and effective. TIG is a slower process than MIG, thus degree of control over weld shape is high. This is useful in welding extruded body members.

Butt welds in aluminium alloy require a temporary backing strip with MIG but not with TIG. Material up to 3/16 th inch can be welded without a bevel, in one pass, by both MIG and TIG. From 3/16 to 3/8 inch, a single 70 degree included angle vee bevel is required with root face equal to 1/3 rd of the plate thickness, Fig 30a. Above 3/8 th inch, a double-vee, Fig 30b, is required. Fillet welds do not require bevelling of abutting edges unless severe dynamic loading is required of the finish part. However, a leg length must be deposited equivalent to the member thickness. Where dynamic loading is envisaged, excessive convexity of the fillet should be avoided. It is possible to compensate for some of the strength loss, mentioned above with heat treatable alloys, by using a permanent fillet welded backing to a butt joint. With non heat treatable alloys a weld strength over 90 percent of the annealed strength can be expected.

Mechanised welding, in which the welding is done by machine, but not as in automatic or robot welding where the operator makes little input, involves constant supervision and manual correction of the welding parameters and joint tracking. According to ESAB, the opportunity could arise of replacing a highly trained welder with a machine operator. Key to success is planning of the production project, it is claimed. Choice of joint, Fig 31, should depend upon process, welding position and thickness. Another proposal by the company is for improving the welding of stainless steel components, using the pulsed-arc process. Here a low background current is used just enough to keep the arc alive without migration of metal filler. Pulses of higher current then cause successive deposition of the metal — but without the short-circuiting that takes place in conventional short-arc welding. Advantages are minimum spatter, more uniform welding and a lower heat input than with the alternative spray arc process, Fig 32.

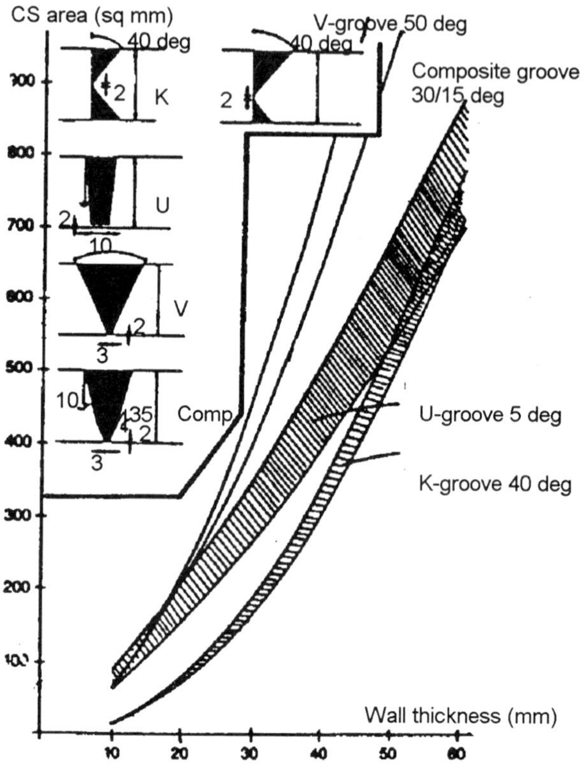

Fig 31:Choice of joint criteria

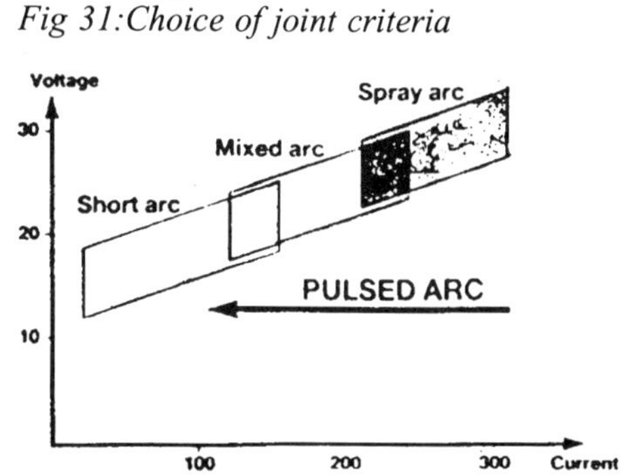

Fig 32: Pulsed arc welding alternatives

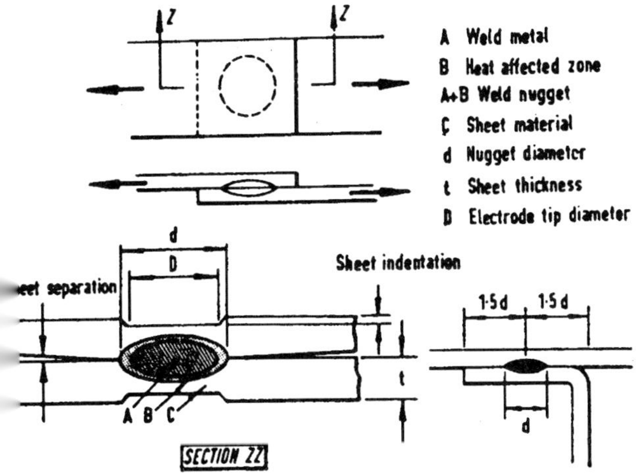

Fig 33: Spot-welding parameters

a

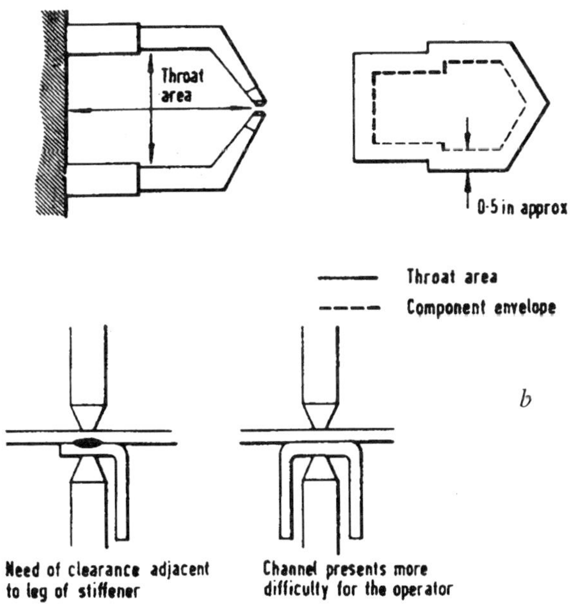

Fig 34: Spot-weld access and clearance

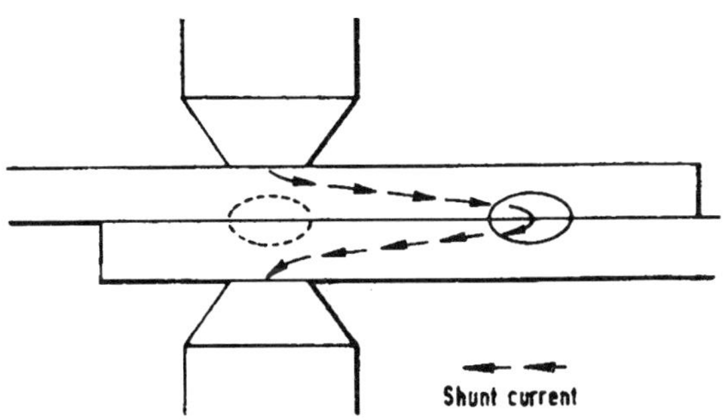

Fig 35: Shunting effect

Spot-welding

The basic advantages of spot welding have been listed in the previous section. Here the techniques and limitations of-spot welding are considered in detail and guidelines suggested for design of flanges and other joint configurations.

The heat and pressure required for the spot weld both need careful control to obtain optimum joint properties. Fig 33 is useful for identifying the important parameters affecting joint performance. In the case of welding mild steel sheet, electrode tip diameter is usually chosen to equal the square root of the sheet thickness in contact. Thus, where different thicknesses are being joined, different electrode sizes are required either side of the joint. The resulting nugget diameter is closely related to the electrode tip diameter and determines the strength of the joint. Sheet indentation and separation should be minimised for best performance of the joint.

Distribution of the spots on the workpiece may be limited by the throat area of the welding gun and only in the case of very high volume production is it economical to use different arm lengths to obtain different 'reaches' on, say, one particular vehicle body fabrication. Local difficulties of electrode access should also be considered, Fig 34a, as should sufficient edge distance to contain the weld nugget, Fig 34b. The two principal failure modes for a spot-welded joint are shearing through the nugget and pull-out of the nugget. While load values corresponding to these two modes can be found from tests on single joints, care is needed in extrapolating loadings for multiple-spot joints due to the so-called 'shunt' effect, Fig 35, in which it is seen that a portion of the current can pass through an adjacent, previously made, weld. Thus load per spot can also reduce with decreasing spot pitch.

To optimise the appearance of spot welded structures, it is normal to conceal or partially shield the joints. A relatively inexpensive way to join panels is to use the coach joint of Fig 36a with flanges turned towards the inside of the body. The visible line of the joint is preferably made to correspond with a styling feature; a PVC plastisol sealer can be used between the flanges to obtain weather proofing. If, by contrast, a lap joint is necessary the spot-weld can be made less visible by using a flat area of copper to replace the electrode on the show side of the joint, Fig 36b.

Fatigue testing of spot-welded joints is gaining in acceptance as a design proving technique, Fig 37, due to AISI, showing a comparison in performance between carbon steels and high-strength-low-alloy steels. Such tests have shown that there is a critical weld diameter above which failures behave in a normal fatigue crack mode. The crack grows around the nugget and pull-out failure occurs. Below the critical diameter, however, rather than pull-out, an inter-facial failure occurs. Fig 38 gives a table showing weld capacity in terms of ultimate shear strength per spot, V_{su}.

Special set-ups can be designed for spot-welding pre-coated steels which are painted on one side only. Galvanised steels can also be welded, but with higher current and longer weld time required to overcome the interface resistance and shunting effects due to current flow through the higher conductivity surface coating. Special procedures are also available for welding aluminium and lead/tin coated steels. In situations where only one side of a jointed sheet can be accessed, the technique of argon-arc fusion spot welding can sometimes be employed to advantage. In so-called puddle welding the heat source is an arc struck across a pre-determined gap between a non-consumable tungsten electrode (negative) and the parts to be welded (positive). A shielded argon atmosphere is provided and the upper component is melted through to the lower one. Stud-welding is another related process which uses short duration of arc(38 milliseconds) to produce stud attachments to panels without causing any back-marking. A small projection on the weld face, Fig 39, first contacts the mating part and a very high current passes which disintegates the projection. Current continues to flow across the resulting gap, in the form of an arc, while the stud is driven home by spring or pneumatic force.

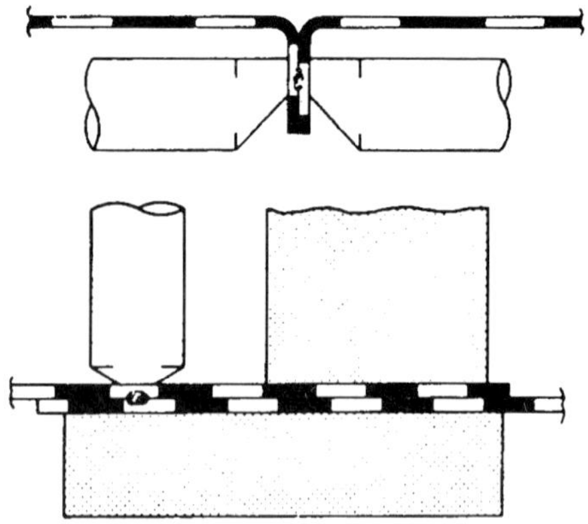

Fig 36: Concealed flange and lap joints

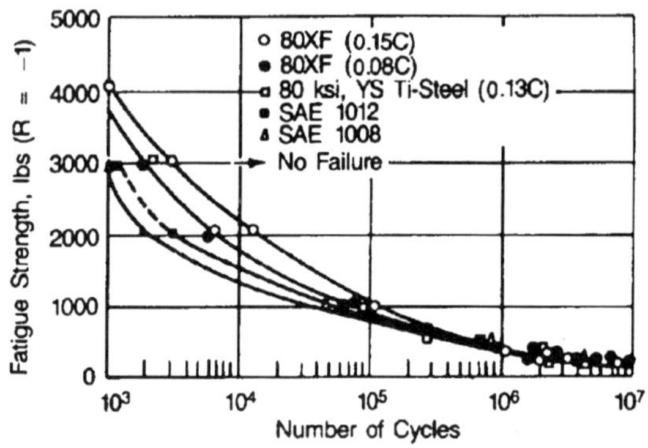

Fig 37: Fatigue strength of spot welds

Thickness of thinnest outside sheet (inches)	V_{su} (kips)	Thickness of thinnest outside sheet (mm)	V_{su} (kN)
0.010	0.130*	0.25	0.58
0.020	0.437	0.51	1.94
0.030	1.000	0.76	4.45
0.040	1.423	1.02	6.33
0.050	1.654	1.27	7.36
0.060	2.282	1.52	10.15
0.070	2.827	1.78	12.57
0.080	3.333	2.03	14.82
0.094	4.313	2.39	19.18
0.109	5.987	2.77	26.63
0.125	7.200	3.18	32.03
0.188	10.000*	4.77	44.48
0.250	15.000*	6.35	66.72

Fig 38: Weld capacity

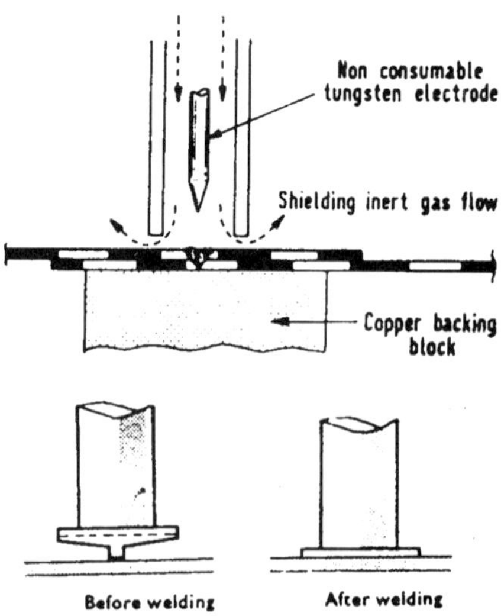

Fig 39: Fusion and stud welding

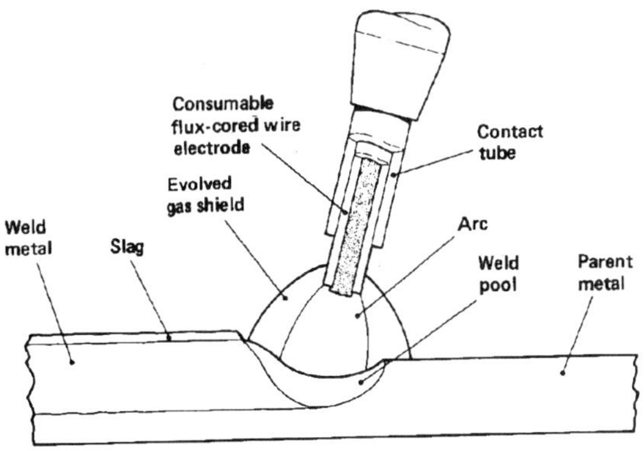

Fig 40: Flux-cored arc-welding

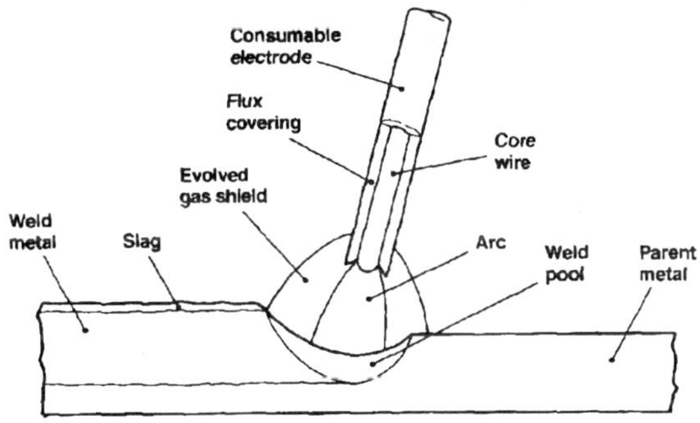

Fig 41: Manual metal arc -welding

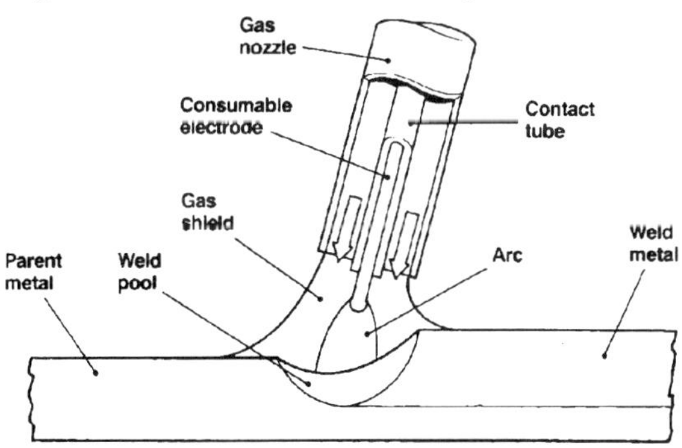

Fig 42: MIG welding

Electrode size (mm)	(SWG)	*Typical current in amps*	*Length of electrode (mm)*
1.6	16	25	250
2.0	14	45	350
2.5	12	65	350
3.5	10	115	450
4.0	8	145	450
5.0	6	215	450
6.0	4	265	450
6.3	(¼ in.)	285	450
7.0	2	320	450
8.0	(5/16 in.)	360	450

Fig 43: Typical welding current data

Arc-welding

While arc-welding is a manual skill which must be learnt by practice, a number of rules should be observed to optimise the performance of the resulting joint. There are also various factors which affect the particular technique of arc-welding which are discussed here.

In flux-cored arc-welding, a hollow electrode filled with flux provides shielding gases, deoxidisers, alloy additions and slag formers. Provided there are enough materials in the core there may not be need for external shielding gas provision and such wires allow welding which is resistant to draughts and is said to be self-shielding; however, a higher capacity power source will be needed for the heavier currents needed. High deposition rates and deep penetration is attainable but greater fuming can be a penalty. Heavier work such as might be met in construction vehicles is the usual application. The principle of the technique is seen in Fig 40, due to Houldcroft[1] of TWI, compared with the manual metal arc process in Fig 41 and metal inert gas (MIG) welding in Fig 42.

In the MIG process a continuous solid electrode is used, rather than a fixed-length rod with MMA, thus intermittent 'pools' of weld, corresponding to the electrode length, are avoided. A consistent arc length is maintained by self-adjustment such that the constant voltage power source causes current changes which oppose arc-length changes. Shielding gas and other parameters can be set to give desired metal transfer rate from the electrode.

Temperatures in MMA welding reach up to 6000C, the heat at the upper end of the arc melting the electrode while that at the lower melts the parent metal being welded. The arc is struck by tapping down the electrode to make contact then quickly raising it to the required arc gap — usually equal to the electrode diameter — then steadily feeding down the electrode as it burns away. As heating is more concentrated, higher welding speeds are possible compared with gas welding. Fig 43, due to Gibson and Smith[2], shows typical welding current data for a mild steel electrode. When a wide bead of weld is needed this can be obtained by weaving the electrode from side to side as it is advanced forward, Fig 44 showing some typical weave patterns.

The basic terminology for arc-welding is seen in Fig 45, due to Hicks[3]. This also differentiates between butt and fillet welding. Only if the metal is thin, less than 2 mm, is it possible to make a full penetration weld between abutting edges. Otherwise a gap between edges has to be left which can be minimised by bevelling the edges almost to full depth. With thicker metal the metal can be bevelled on both sides. To minimise distortion it is good practice to distribute welds equally about the neutral axis of a section and intermittent welding can also be helpful in this respect.

Strength of welds can be calculated on the throat thickness, Fig 46a, or in the case of a fillet weld, Fig 46b. In the latter case the stress on the weld is obtained by dividing the load by the throat area. Normally it is preferred to have a chart such as that in the figure giving strength of fillet welds in load per unit length.

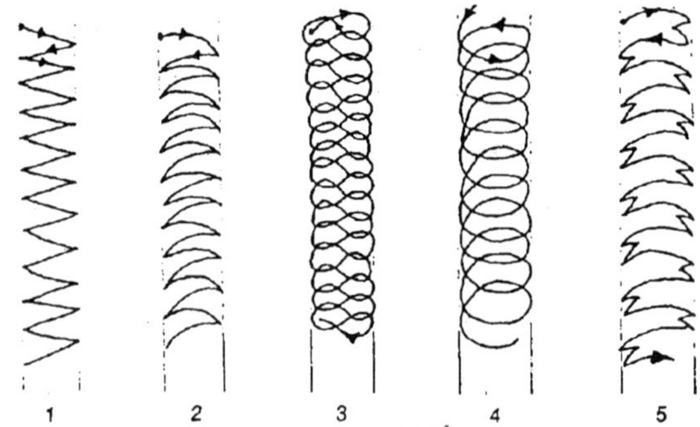

Fig 44: Weave patterns

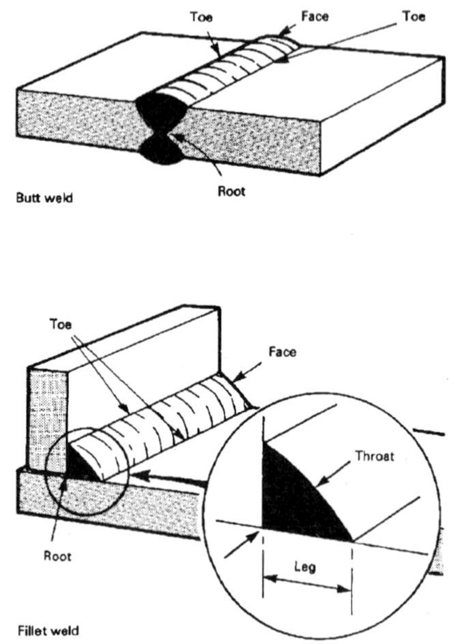

Fig 45: Nomenclature

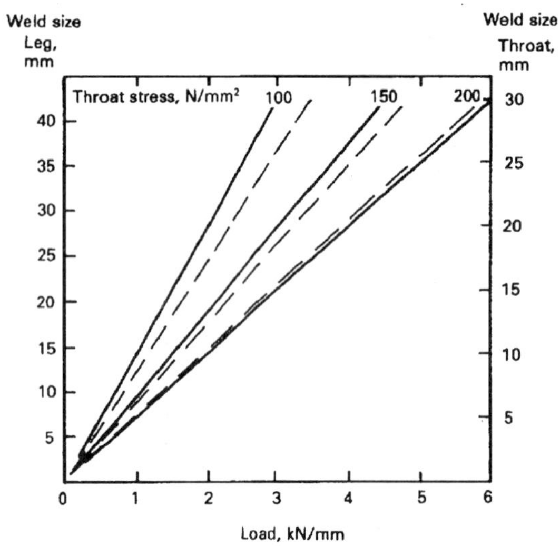

Fig 46b: Fillet weld design chart

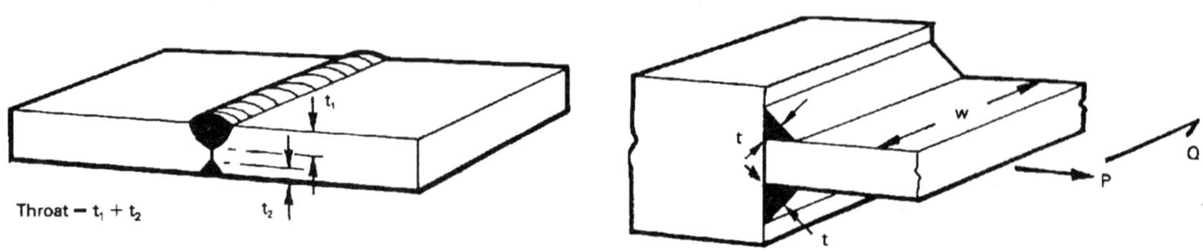

Fig 46a: Stress calculation on butt and fillet weld;tw = weld throat area; P/2bw = weld stress for load P and Q/2tw weld stress for load Q

Projection, stud and seam welding

While spot-welding is probably the most important category of resistance welding, seam-, stud- and projection welding are very useful alternatives for particular applications. One of the most famous automotive applications of stud welding was the US Ford wire wheel which had spokes welded at either end to the rim and hub of the wheel.

In projection welding, current and pressure are localised at the weld section by means of an embossed upset on one or both sides of the work. The operation is carried out on a press-welder with the workpiece placed between copper platens which are within the single winding of a large transformer. It is particularly suited for fastening attachments such as brackets, spigots and weld-nuts to sheet metal which can only be accessed from one side.

Electrodes are flat and so can contact the parts over a wide area allowing more accurate aligning of parts. Use of several projections within the electrode area also allows several welds to be performed simultaneously and individual welds can be closer together than with spot-welding as the shunting problem does not apply. Even with a single weld the conditions of the weld guarantee a higher quality than that obtained from spot-welding and it is sometimes possible to weld in an area inaccessible to spot-welding electrodes. This is particularly true of parts too small to spot-weld with a narrow electrode without burning.

In joining sheet material a round form of projection is ideal, Fig 47, having radius 1-2 times sheet thickness. For curved surfaces elongated projections are recommended, Fig 48. Low projections give minimum marking but high ones (not exceeding 90% metal thickness) give greater strength. Where different metals are being welded the one with the higher conductivity should carry the projections. Aluminum alloys need special attention. Controlling slope to suppress the first few cycles of welding current is the secret — so as to reduce current density while the projection is collapsing and so avoiding metal expulsion.

A special case of projection welding is stud-welding, Fig 49, which is used extensively for attaching fastenings for such applications as soft trim attachments. Sheets down to 26 gauge in steel and 22 in aluminium alloy can have studs attached without back-marking.When the firing current is closed, a current of up to 3500 A is discharged through the stud thus disintegrating the stud. Current continues to flow across the gap between stud and parent metal while the stud is driven home by spring or pneumatic pressure.

In seam welding, Fig 50, the electrodes are replaced by wheels whose peripheral form controls the nugget shape. Either a continuous seam can be produced, for such applications as the flanges of fuel tanks, by the roll-spot technique, or the method can be used to produce intermittent welds. The continuous seam is generated by overlapping spots. Welding current is usually supplied in pulses, with wheel rotation being continuous. In some cases the work is driven through by the wheels while in others the wheels idle and the work is fed through by an external force. Material thicknesses for seam welding are usually in the range 0.5 to 3mm and most common metals and alloys can be handled. It is a fast method of welding and readily controllable for quality. A higher current than that used for spot-welding is required as the previously made nugget provides a shunt path for the current flow.

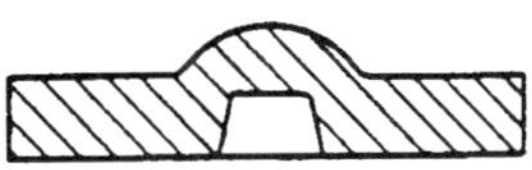

Fig 47: Projection for flat sheet

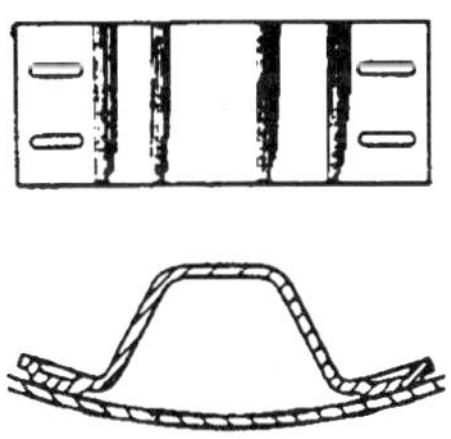

Fig 48: Projection for curved sheet

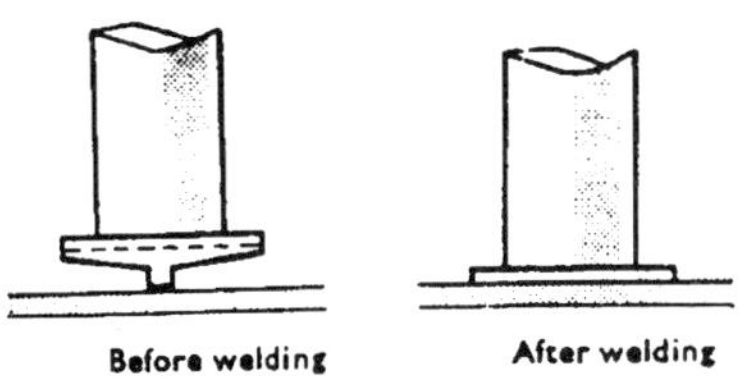

Fig 49: Stud welding

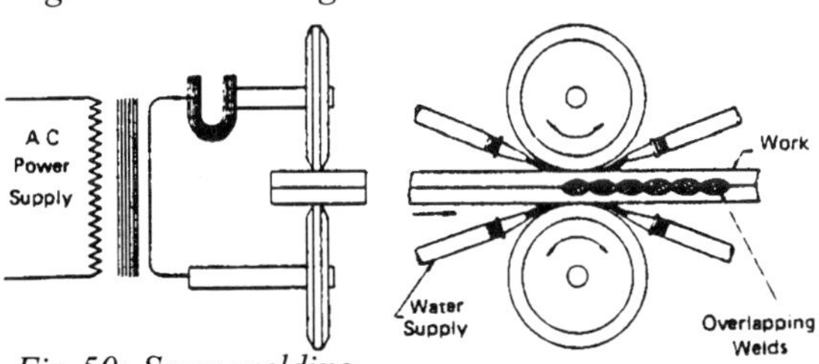

Fig 50: Seam welding

Adhesive bonding

The ability to bond a whole surface rather than a few points is one of the top attractions of adhesives. Advances in materials and applicator technology are also making bonding more price and time competitive than bolting riveting and spot welding. And as load bearing structures become more refined the need to maintain structural continuity at the joint line will further the preference for adhesives.

Provided the ground rules of designing for bonded structures are obeyed, high load capacities are available from modern adhesives. It must be remembered that bonded joints perform best under pure tension, compression or shear, less well under cleavage and poorly under peel loading, Fig 51. Of the many available joint configurations, a number of which are shown in Fig 52, the tapered and double-lap joints are better than the single lap but, of course, more costly to produce. Fig 53 shows a number of ways to reduce peeling tendency which might otherwise destroy a joint. These are some of the ideas put out to support the sale of Araldite epoxy adhesives by Ciba Geigy

Very broadly, adhesives divide between the 'natural' animal (casein), vegetable (gum), mineral (bitumen) and the 'synthetic', elastomers, thermoplastics and thermosets. These divisions are by chemical source and a more useful classification follows the main uses which have arisen from the commercially developed forms. In that respect ten major families have been defined by Permabond Adhesives as follows:

Anaerobic	Plastisol
Cyanoacrylate	Polyurethane
Epoxy	Solvent rubber
Hot melt	Tape
Phenolic	Toughened types

The last category sub-divides again into anaeorobic, acrylic and epoxy adhesives. Application varies according to viscosity, obviously, with liquids applied by brush or spray, at a price, but a little ingenuity can often save money in this area. Solvent based adhesives normally involve a time penalty in waiting for the solvent to dry either between application and jointing, or with porous surfaces, after jointing. Film adhesives are often used particularly when bonding the skins of sandwich panels to honeycomb cores. This usually involves a mixed adhesive and requires an elevated temperature to cure. As well as the question of joint configuration and load capacity, key factors also to be considered are type of material and surface condition; relative expansion coefficients and joint tolerances. For working loads above 3.5 MN/m^2 (500 lb/in^2), some experts recommend that only structural adhesives of the thermoset type be used.

For an epoxy, such as Araldite, recommendations are to clean, degrease and abrade surfaces;

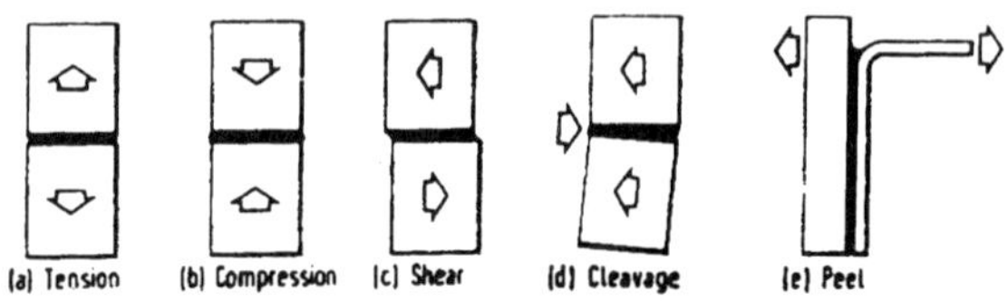

Fig 51: Bonded joint failure modes

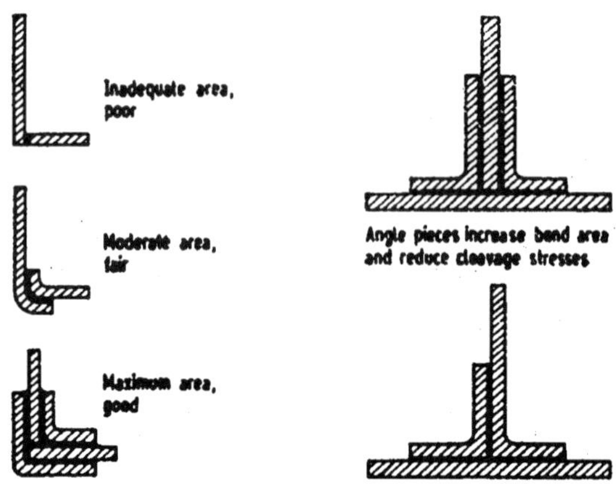

Fig 53: Ways to prevent peeling

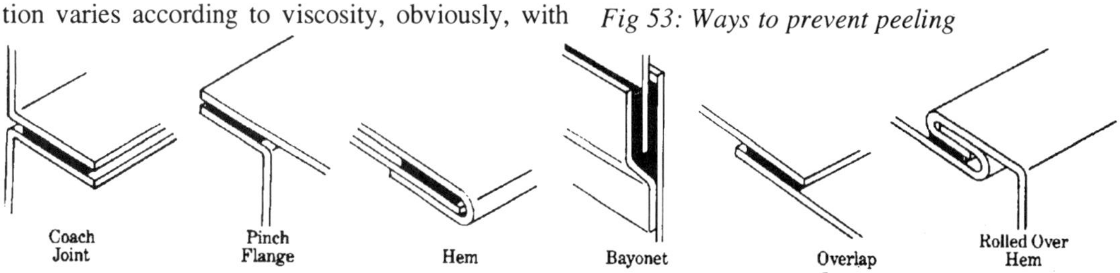

Fig 52: Joint configurations

observe curing time specified and apply clamping pressure; avoid operation above 100 C. Select viscosity according to joint condition thin for close-fitting, thick for gap-filling; seek special advice when jointing materials which have grossly differing expansion coefficients and choose a low modulus grade where

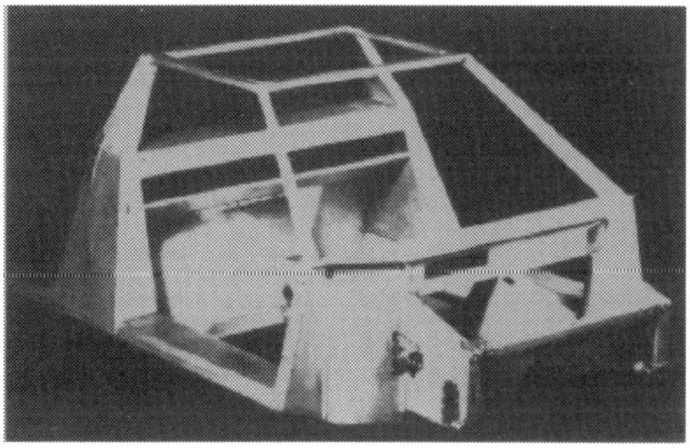

Fig 55: Part-bonded centre section

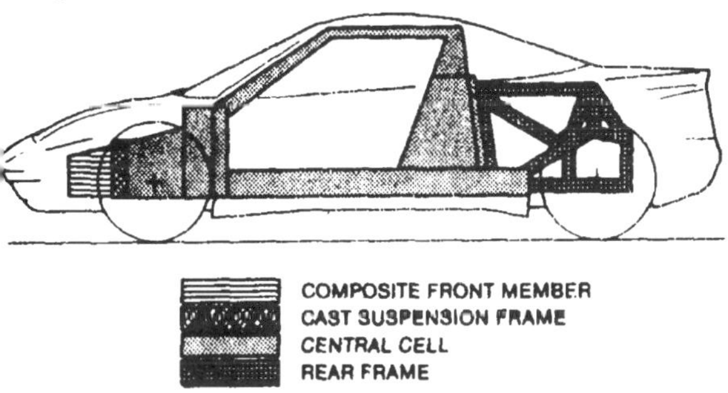

Fig 56: Assembly of sections

high vibration or impact loads are expected; observe 5 percent weight tolerance in the mix components of the adhesive and for large quantities, mix in shallow trays to avoid heat build up which can shorten life at this stage. Heat applied after jointing, during the cure period, can enhance joint strength. The table in Fig 54 gives joint strength ratings, key for which is (1) moderate, (2) less strong than jointing materials, (3) bond strength of 2000 lb/in^2 on standard test specimen.

Alcan International recently completed a bonded structure project on the Ferrari 408 research vehicle using a toughened epoxy adhesive heat cured to 180 C. The part-riveted / part-bonded centre-section (Fig 55) connects the front and rear end-frames (Fig 56) and has torsional stiffness of 10.4 kNm/deg. This part of the structure is made up of parts which are folded on simple press brakes. Assembly technique is to put the parts together with self-tapping screws in the rivet holes to check for fit and then bond them with the adhesive which is specially formulated to have good balance between slump-resistance and extrudability. Ordinary paint stoving ovens can be used for the cure.

Structural adhesives

The work at Alcan associated with the above example has led to some manufacturing recommendations formulated by the company. Adequate peel resistance of the adhesive is, of course, the prime requirement and this, say the company, must stand up over

JOINTING MATERIAL	LIN EXPAN COEFF X 10^{-6}/°C	OPTIMUM JOINT STRENGTH (a)	ACID ETCH TREATMENT	JOINT STRENGTH RATINGS (see chart key)						
Aluminium	24	4750	sulphuric and chromic	3	3	3	3	2	2	3
Brass	19	1000–1500	—	2	2	1	1	1	1	2
Ceramics	2·5–4	ceramic breaks	—	3	3	3	3	3	3	3
Chromium plate	—	1500–2000	hot dilute hydrochloric	3	3	3	3	1	1	3
Copper	17	1000–1500	ferric chloride	2	2	1	1	1	1	2
Glass	8·5–10	glass breaks	—	3	3	3	3	2	2	3
Glasscloth laminates—epoxy	12–25	laminate breaks	—	3	3	3	3	3	3	3
polyester	12–35	laminate breaks	—	3	3	3	3	3	3	3
Graphite and carbon	6–7	carbon breaks	—	3	3	3	3	3	3	3
Lead, tin and solder	22–27	500–1000	—	2	2	2	2	2	2	2
Magnesium	27	2000	chromate	3	3	3	3	3	3	3
Nickel	13	2000	—	2	2	2	2	1	1	2
Nylon	90	—	—	2	2	1	1	1	2	2
Paper laminates—melamine and phenolic	15–35	laminate breaks	—	3	3	3	3	3	3	3
'Perspex'	90	800	—	1	1	1	1	2	2	1
Polystyrene	60–80	—	—	1	1	1	1	2	2	1
Polyethylene and polypropylene	150–300	1000	chromic acid or flame	1	1	1	1	1	1	1
P.t.f.e.	60–70	1000	(b)	2	2	2	2	2	2	2
P.v.c.	50–60	1000	trichloroethylene	2	2	2	2	2	2	2
Rubber	130	rubber breaks	conc. sulphuric or nitric for 5 min	3	3	3	3	3	3	3
Mild steel and iron	11	2000–3000	dil. phosphoric	3	3	3	3	2	2	3
Stainless steel	18	2000–3000	oxalic	2	2	2	2	1	1	2
Wood (c)	3–60	wood breaks	—	3	2	2	3	3	3	3
Zinc	26	500–700	—	2	2	3	2	1	1	2

Fig 54: Joint strength ratings

a temperature range from -40 to +120 C. As part of the Aluminium Structured Vehicle Technology (ASVT) programme, the next most important requirement is precision of placement to avoid contamination of jigs and fixtures. Application speeds from 10 mm/sec, for complex joints, to 500 mm/sec must be sustainable, with bead diameters from 1 to 5 mm on inclined surfaces from horizontal through to vertical. Robotic application of adhesive is seen as a way of obtaining predictable control of quality — with a careful integraton of adhesive, robot and dispensing system being required.

As well as the structural performance requirement of the adhesive, viscosity, thixotropy, filler abrasivity and slump resistance have each to be controlled. A 10 kg 6-axis electric-drive robot is suggested but there is a special requirement for the dispenser to adjust flow exactly in proportion to the speed of the nozzle over the workpiece.

Weld bonding has become a familiar way of using structural adhesives in which just a few spot-welds help to locate the joint elements and hold the structure together prior to cure of the adhesive. The cost configurations for the ASV and conventional vehicle, comparing weld-bonded and plain spot-welded lap joints is shown in Fig 57. There is also the advantage of improved corrosion resistance. The weld bonded structure is also said to be typically 25% stiffer than its spot-welded counterpart.

The Ferrari 408 structure exemplified in the previous section has become a mobile laboratory for the assessment of bonded structures in aluminium alloy, Fig 58. It is now recognised that the advantage

ASV	ASV TO STEEL COST VARIANCE	COATED STEEL
Aluminum Materials for Base Unit and Door Structures	+ +	Coated Steel Materials for Monocoque Body and Bolt-on components
Pressing of Aluminum Components		Pressing of 25% more coated steel components
Structural adhesives and epoxy pads	+	Interweld and seam sealers plus metal to metal adhesives
Adhesive application and spot-welding and adhesive curing for BIW assembly		Sealant and adhesive application and 3 times as many spot welds
Mill and drill operations	+	BIW finishing operations
No body structure painting or protection	-	Full systems paint finish and underseal, waxes
Automated trim and final assembly	–	Difficult access for automated assembly
Fitting of finish painted plastic panels	+ -	Rectification of painted body following assembly
Downsized power train and chassis components for equivalent performance	-	Normal power train and chassis components

Fig 57: Weld bonding and spot welding compared

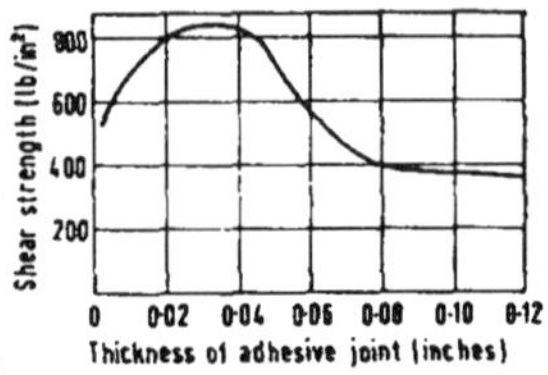

Fig 59: Strength variation with thickness

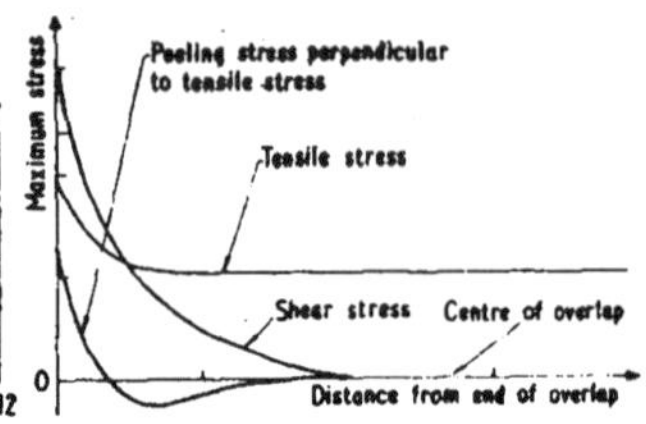

Fig 60: Stresses in single lap joints

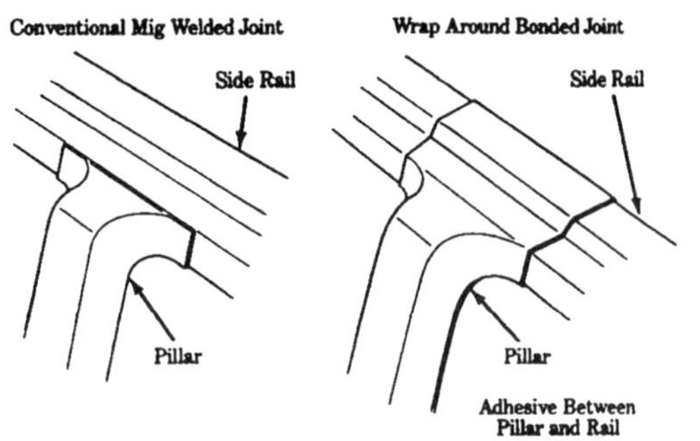

Fig 62: Joint modification for bonding

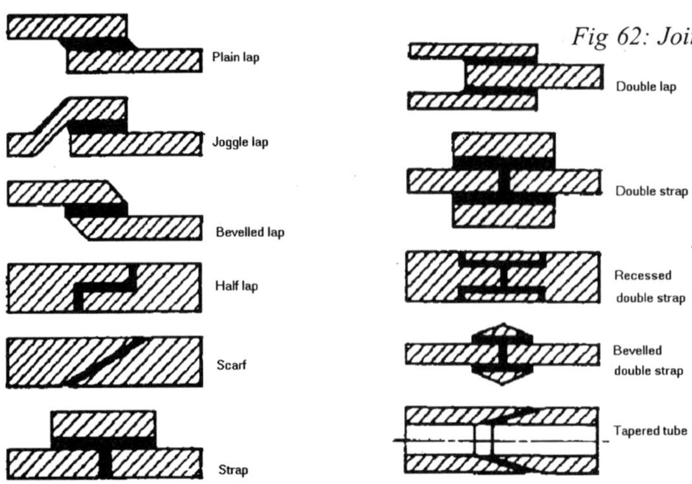

Fig 61: Joint configurations

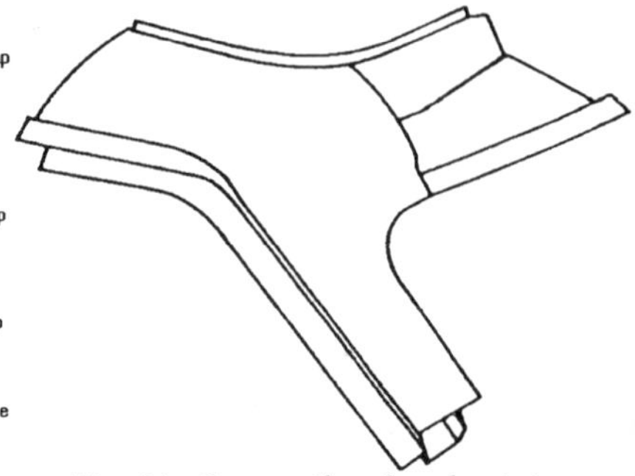

Fig 64: Cant rail to header joint

Fig 58: Ferrari 408 bonded structure

of bonding can only be realised if the early concept design of the vehicle allows for this construction technique. Design should be based on the need to exploit the ability of the adhesives to exploit their highest strengths in tension and shear; this implies the minimisation of peel and cleavage forces. Gap thickness is also critical and Fig 59 shows the variation in shear strength of a PVC plastisol adhesive with thickness of the jointing material. Conventional weld-type flanges should be avoided and it should be noted that in the common over-'lap' joint, while bond strength increases linearly with joint width, this is not the case with overlap length. A joint factor has been proposed, to allow for this effect in calculation, of $(\text{thickness})^{1/2}$ / lap-length. Stresses in a single lap joint are shown as Fig 60. Of the joints shown in Fig 61, tapered and double-lap joints are better than the single lap.

A suggested body joint modification for bonding steel panels, due to AISI, is shown in Fig 62. To prevent undue squeeze out of adhesive, either formed dimples or recessed flanges can be used, Fig 63, for prevention of distortion by intermittent spot-welds. In the case of aluminium alloy, finite-element analysis work by ASVT on the typical A-post/cant-rail/header joint of Fig 64 showed that a bonded aluminium alloy joint could equal the spot-welded steel one in terms of structural stiffness.

A single-part heat-cured toughened epoxy material was used in the ASVT work plus occasional spot-welds, blind-rivets or self-drilling drive screws for tacking. Fig 65 shows typical strength values for the lap and tee geometries of Fig 66 to compare 1.6 mm thick bonded aluminium with 1 mm thick spot-welded steel. This does not take into account the discontinuous nature of the spot-welded joint and with a 40 mm pitch spot joint the aluminium/steel figures for lap and tee configurations were 238/200 and 86/25 N/mm.

FAILURE LOADS FOR LAP AND T-JOINTS LOADED IN TENSION FOR BONDED ALUMINUM AND SPOT-WELDED STEEL

	LAP JOINT		T-JOINT	
	FAILURE LOAD (N)	NOMINAL METAL STRESS (MPa)	FAILURE LOAD (N)	NOMINAL METAL STRESS (MPa)
ALUMINUM 1.6mm 5251-0 BONDED	5960	149	2150	54
SINGLE SPOT-WELD 1.0mm (SOFT) SPOT-WELD	8030	320	984	39

Fig 65: Lap and tee geometry characteristics compared

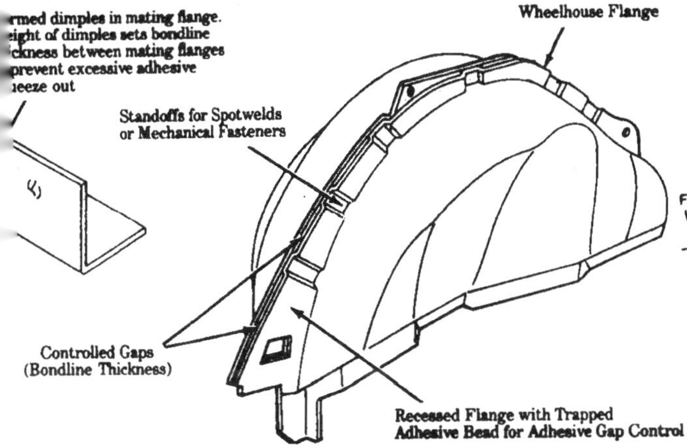

Fig 63: Use of recessed flanges

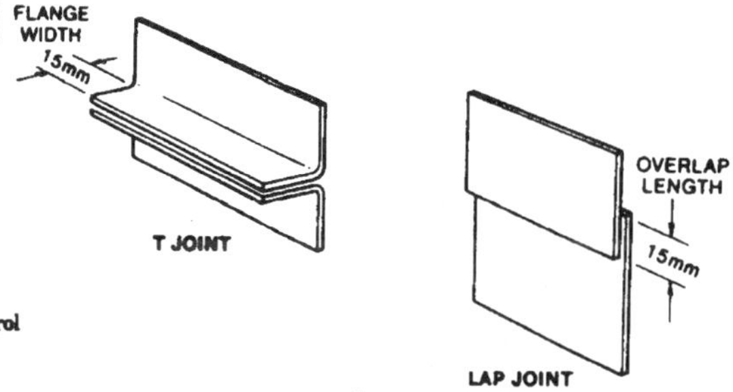

Fig 66: Lap and tee geometries

Mechanical bonding

Hook and loop fasteners are providing a valuable alternative 'bonding' technique which is of particular value to interior trim applications. The pioneering Velcro brand can be woven, knitted or moulded into plastics and composites.

Peel strengths range from 0.35 to 3.5 lb/in^2 (246-2460 kg/m^2), tensile strengths from 2 to 24 (14-168) and shear strengths from 6.5 to as high as 120 lb/in^2 (45.5-840 N/m^2). Higher performances can be obtained when the system is used in other than planar overlap configuration. Examples are curved surfaces and interference shear traps, both of which combine the various modes of separation in order to enhance joint strength. There is a tendency in shear separation for the mating surfaces to move laterally and so an interference shear set-up as in Fig 67 substantially increases the pull-out force. Another technique is the so-called manipulative shear trap, in which the configuration works as a hinge which closes to secure the joint for high shear force though relatively low peel force is required to break the joint, Fig 68.

In general the hook is a monofilament and the loop a multifilament fibre. To form the hook the fibre goes through a cutting process in which the hook eye is positioned between a set of comb teeth with cutting blades at every other tooth. Separation of hooks and loops occurs over a single line of engagement and tension forces are slightly higher than peeling ones because a greater area of hook engagement is involved. In shear the hooks and loops are being raked over each other, thus increasing the separating force yet further.

A special development used to close two panels on a flush surface without any gap is called anti-

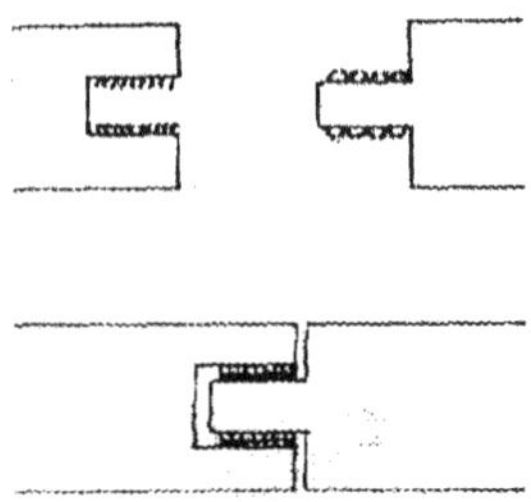

Fig 67: Interference shear

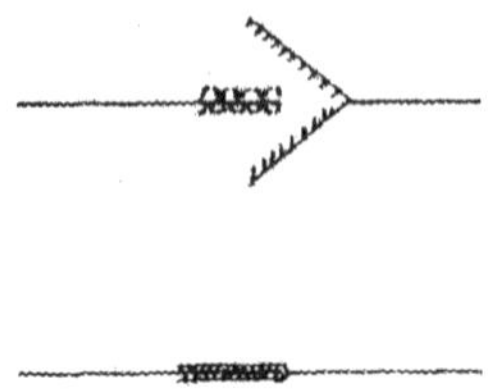

Fig 68: Manipulative shear

TAPE PRODUCTS	TYPE/BRAND	DESIGNED FOR
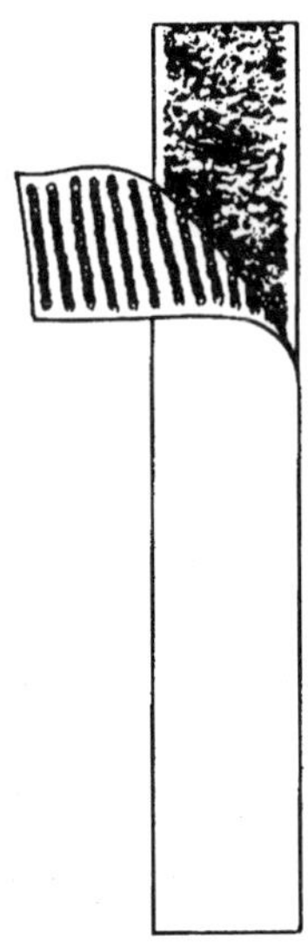	Nylon	General Purpose
	Polyester	Water and Chemical Resistant
	HI-AIR® (NOMEX*)	Flame Retardant
	HI-GARDE® (Stainless Steel)	Very High Heat Resistance
	MID-TEMP®	Moderate Heat Resistance
	HI-MEG®	Electrically Conductive or for R.F.I. Shielding
	ASTRO-VELCRO®	"Non-Burning" "Non-Igniting"
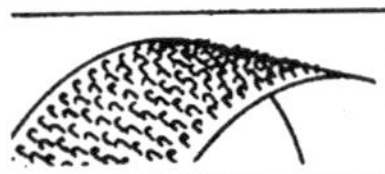	Molded Nylon Hook MVH Series	Extra Strength
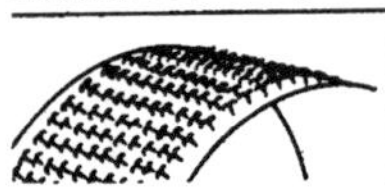	Molded Nylon Arrowhead MVA Series	Specialty Fastening

Fig 69: Velcro tapes

SPECIALTY PRODUCTS	TYPE/BRAND	INSTALLATION
	VELSTRAP® (VST)	Reuseable, for High Strength Cinching (with Ring)
	BACK TO BACK STRAP	Reuseable Self-Wrapping Tape
	FACESTRAP (VFS®)	Adjustable Reuseable Cinching Strap (without Ring)
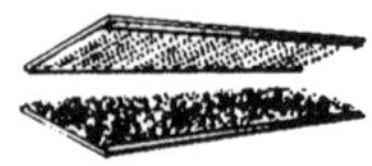	VELSTICK®	Semi-Rigid Screw, Nail or Rivet
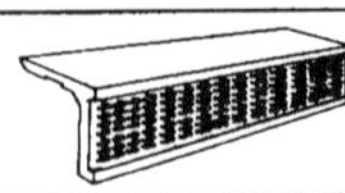	INSTANT VALANCE	Screw, Nail or Rivet
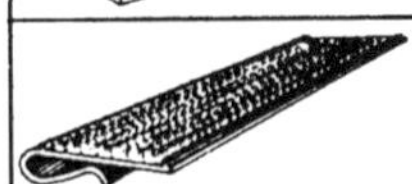	EDGECLIP®	For Slipping on Open Edges
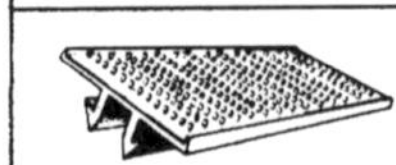	VELTRACK® BRAND	Slotted Insertion
	VELWASHER	Surface Mounted Screw, Nail or Rivet

Fig 70: Speciality products

backlash closure. This involves a flexible component, to create flush attachment, which collapses under closing pressure. Another new departure is moulding of the fastener as an integral part of the closure panels. This eliminates secondary attachment operations, of course, and avoids piercing the substrate with atttachment holes.

A benefit of hook-and-loop fastening which is not so obvious is the ability to absorb impact over a jointing surface. This has already been put to use by one motor-manufacturer, in the fixing of a lightweight bumper requiring less than normal energy-absorbing properties to meet the statutary 5 mph impact test. Compared with discrete metallic fasteners, moulded-in Velcro also overcomes the problem of pull-out caused by differential thermal expansion rates between a composite substrate and its fasteners.

Specialist Velcro fasteners, in the range supplied by Selectus Ltd are available as tapes shown in Fig 69 and speciality products in the table of Fig 70. Row 1 includes types for securing convertible hoods, truck cab headlings and floor carpeting. Row 2 types would be used for jobs like speaker grille attachment. Row 3 is for 'spot' fastening applications where a double faced foam pressure-sensitive backing is required. Row 4 applications include boot carpeting, car headliners and door panel assemblies. Row 5 has uses in secondary dashboard attachments and those for truck cab interiors. Row 6 covers removeable valances and Row 7 for securing acoustic materials. Fig 71 shows technical data for automotive industry products and Fig 72 shows key application areas.

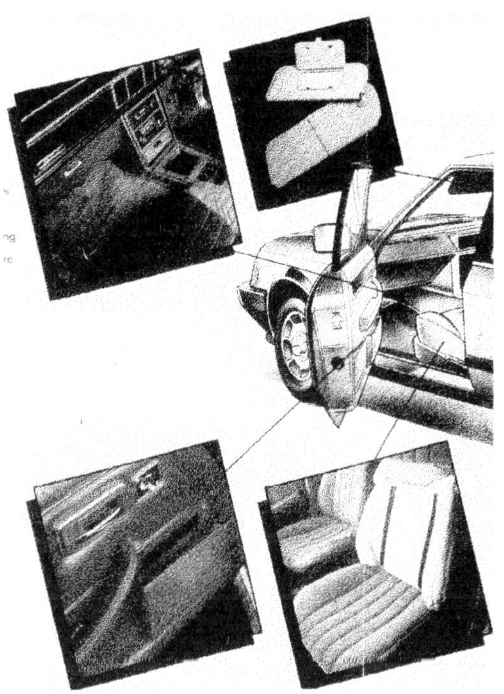

Fig 72: Velcro application areas

	STANDARD VELCRO	TYPE 19P	MVA 8 HOOK	MVA 10 HOOK
ERIAL	Nylon 6.6.	Polyester	Injection moulded nylon with standard loop	Injection moulded nylon with standard loop
AR STRENGTH GTHWISE) M²	0.84	1.70	1.30	1.30
. STRENGTH GTHWISE) M²	0.21	0.16	0.45	0.45
ION/ HING	0.42	0.84	0.67	0.91
F OPENINGS/ INGS	In excess of 5,000	3,500 max	350	1,000
ERATURE E HIGH –	200°F – 70°F 93°C – 56°C	200°F – 70°F 93°C – 56°C	200°F – 70°F 93°C – 56°C	200°F – 70°F 93°C – 56°C
ABLE HS MM	10, 15, 20, 25, 30, 50, 100	15, 20, 30	8, 12.5, 25	8, 12.5, 25
ABLE URS	20	natural, black	black, white	black, white
HES ABLE	Pressure sensitive adhesive, Welding (U.F. & R.F.) Solvent reactivated	Pressure sensitive adhesive, Welding (U.F. & R.F.) Solvent reactivated	Pressure sensitive adhesive, Welding (U.F. & R.F.)	Pressure sensitive adhesive, Welding (U.F. & R.F.)

Fig 71:Technical data for Velcro tapes

COMPONENTS AND COLOURS

SOFT HARDWARE® COMPONENT	PLASTIC SUBSTRATE COLOUR	VELCRO FASTENING TAPE	TAPE COLOUR(S)
VELSTICK FASTENERS	Black White Black	Woven nylon hook & loop Woven polyester hook and loop Moulded nylon hook	Black White Black
VELSTICK VELWASHER FASTENERS	Black White Black	Woven nylon hook & loop Woven polyester hook & loop Moulded nylon hook	Black White Black
VELSTICK VELCOIN FASTENERS	Black White Black	Woven nylon hook & loop Woven polyester hook & loop Moulded nylon hook	Black White Black
VELTRACK FASTENERS	Light grey Light grey Light grey	Woven nylon hook & loop Woven polyester hook & loop Moulded nylon hook	Light Grey White Grey
EDGECLIP FASTENERS	Clear Clear Clear Clear Clear Clear	Moulded nylon hook Moulded nylon hook Moulded nylon hook Woven nylon hook & loop Woven nylon hook & loop Woven nylon hook & loop	Black White Beige Black White Beige
INSTANT VALANCE FASTENERS	White White	Woven nylon hook & loop Woven polyester hook & loop	White White
VELSTUD FASTENERS	Black Black Black	Moulded nylon hook 8	Black

Welding ceramics and composites

Future automotive parts, and some already used in racing cars, are likely to be combinations of ceramics, metals and polymer composites. Techniques for joining them may not strictly be welding but they have a good deal in common with early pressure-welding techniques.

The Welding Institute (TWI) have showed that the successful exploitation of material properties can often be achieved by using newly-developing joining techniques to combine certain properties of one material, or composite, with the complementary properties of those of another. Thus, it is argued, whereas materials like ceramics, with attractive thermal and wear performance, have often been compromised by a brittleness of the material, this might be overcome by joining it to a tougher substrate. TWI have shown already that even a ceramic faced engine tappet is feasible with such techniques.

Hot isostatic pressing is one such method; it can also be used to make ceramic parts from silicon carbide and nitride. It differs from conventional hot-pressing in that pressure is applied on the three perpendicular axes, rather than in one direction. The pressing involves a closing up of pores and consequent slight loss in volume, Fig 73. The process can be carried out with or without a secondary encapsulating material of, say, glass or other ceramic. Diffusion-bonding of ceramics to metals would involve the sort of pressure distribution shown in Fig 74. The TWI's facility offers pressure up to 2000 bar and temperatures to 2200C and component size limits are 100 mm diameter by 150 mm high.

Ceramic materials can also be sprayed on to substrates using plasma or high velocity oxygen fuel spraying (HVOF), Fig 75. The plasma process uses a DC arc struck between non-consumable electrodes, an inert gas being fed into the arc to form a high-temperature plasma (15,000C) which is compressed and accelerated by the torch nozzle. The coating material is fed into the plasma in powder form, the whole being ejected at velocities up to 600 metres/sec. Oxyfuel spraying, already used for ceramic coating of piston crowns, involves combustion of a fuel gas with oxygen at high pressure and/or high flow rates. A high velocity gas flame thus created, propels powdered coating materials on to a substrate, Fig 76.

Interestingly, the welding together of dissimi-

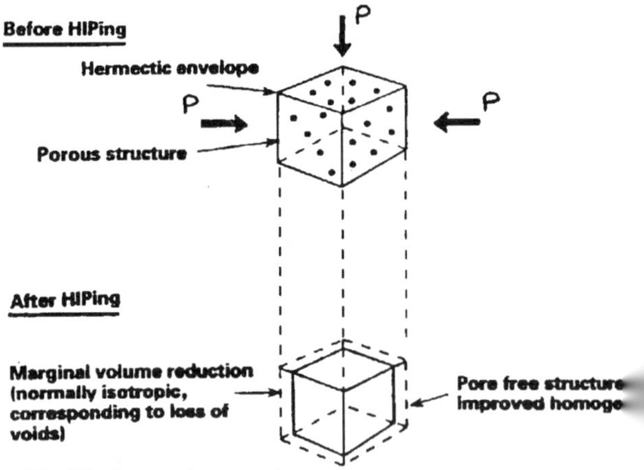

Fig 73: Isostatic pressing

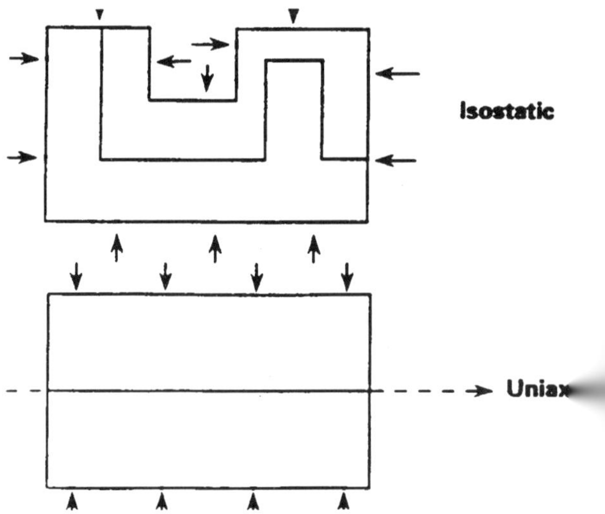

Fig 74: Pressure distribution

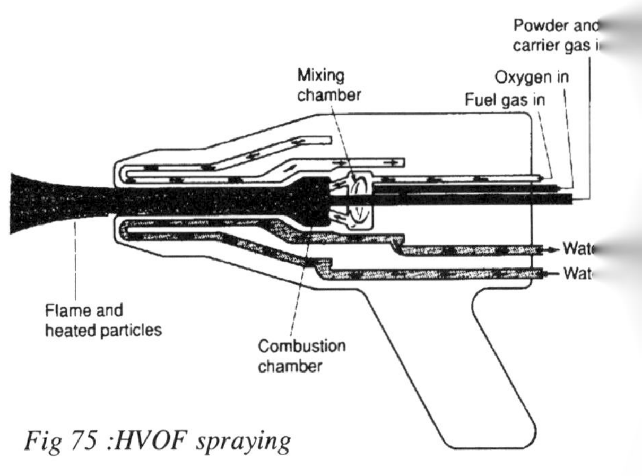

Fig 75 :HVOF spraying

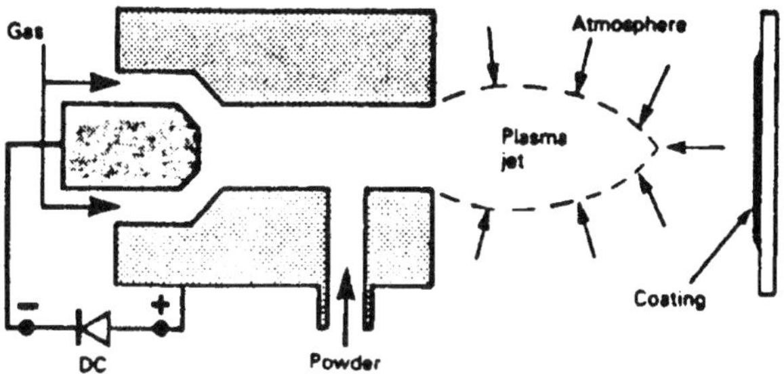

Fig 76: Plasma spraying

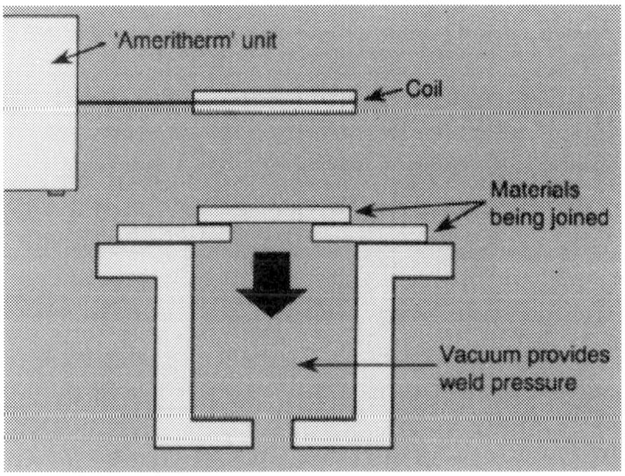

Fig 77: PCM welding

Fig 78: Focused infrared welding

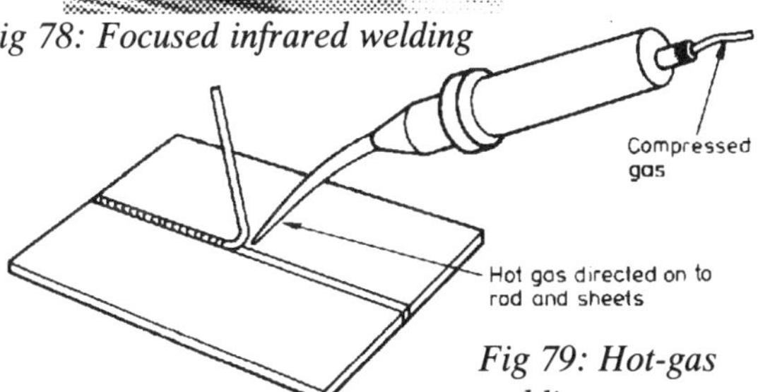

Fig 79: Hot-gas welding

lar materials can be aided by coating techniques, too. In so-called Polymer Coated Material (PCM) welding, developed by TWI, carbon-fibre thermoplastic composites can be joined to aluminium alloy, for example. One component is pre-treated with a thermoplastic coating, after which a proven plastic welding technique such as induction welding, Fig 77, can be used to join the parts.

Welding composites

The basic plastic welding processes, which also include hot-plate and ultrasonic techniques, are now being supplemented by others which may one day make the welding together of structural composite vehicle body panels of the future as easy as joining steel ones. One such process is focused infrared welding, Fig 78, which TWI has demonstrated to have potential for body panels. It is an extension of hot-plate welding in which the hot-plate is replaced by an infrared lamp and water-cooled elliptical reflector. The angled copper mirror splits the focused single heat source and directs it to each welding surface of the two parts to be joined. A typical lamp would be powered at 2400 W, sufficient to provide 16W per mm of weld length. The surfaces are pushed together, on reaching the correct temperature, and the lamp/mirror simultaneously withdrawn. The surfaces heat in about 1 second and welding is complete after a further 3 seconds. Tests have already been carried out on filled and unfilled polypropylene — also two polycarbonate alloys.

Some time ago, TWI carried out trials to show that welding techniques applied to join unreinforced thermoplastics could also be applied to composites, the results being contained in Bulletin Volume 28 for hot-plate, ultrasonic, vibration and induction-heating techniques. Comparable work on hot-gas welding, Fig 79, is not mentioned. While the technique is used mainly for on-site work inaccessible for heavy equipment, the technique might find future application for tankers— already it is widely used in the automotive repair trade for rectifying damaged bumpers and the like. In recent years the TWI have stated that the process suffers from a lack of standardisation and a widely accepted procedure for fabricators. They believe most thermoplastics can be hot-gas welded provided care is taken to optimise welding conditions. Long term creep behaviour should be careful watched, when structural parts are considered.

Automated assembly

This important new discipline has added benefits to plastics and composites moulding for it can avoid many of the common fastening techniques by consolidation of individual parts into a composite whole. Bayer's Technical Applications Division has shown that by consolidating as many parts as possible into the individual mouldings, substantial economies should be possible.

The case study considered here uses an instrument panel, Fig 80, the commercial success of which, as a design, depends on the consideration, at the early concept stage, of such factors as: flowability, shrinkage, tolerances and wall thickness to rib ratios; also of surface quality and visibility of flow lines; finally, aspects relating to the mould such as position of the gate, number of hot-runner nozzles, demouldability and general cycle feasibility must be considered.

Computer programs are now available for simulating the mould filling process without actually constructing the mould. With knowledge of material properties, flow lengths can even be predicted for a particular wall thickness. A general relationship is shown in Fig 81 for different injection times — and for wall thickness of 2.6 mm. When making corrections to the mould, shrinkage of the material must be determined both during and after the moulding process. This relates to pressure drop in the cavity according to material characteristics — as shown for typical polycarbonate/ABS blends in Fig 82.

If bosses to take screws are required on the backs of visible surfaces, the standard boss (left) should be replaced by the design (right), in Fig 83, to provide a wall-thickness ratio of 1.2 to 1.3, required for best visual results. The same requirements apply to ribs as well as bosses, in these situations. Another surface effect is particularly noticeable with grained surfaces (obtained by photo-etching the inside of the

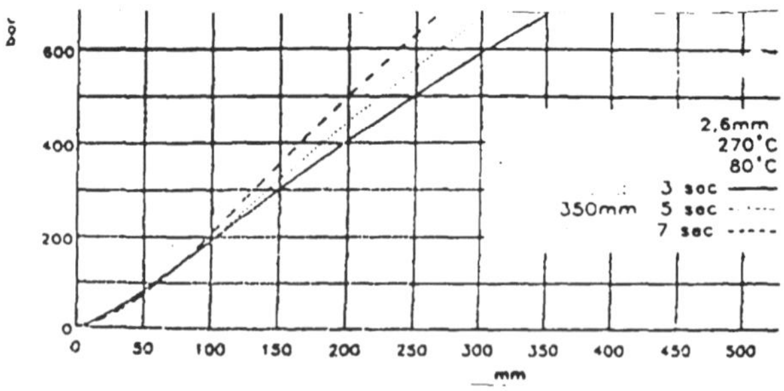

Fig 81: Flow-length/pressure relationship

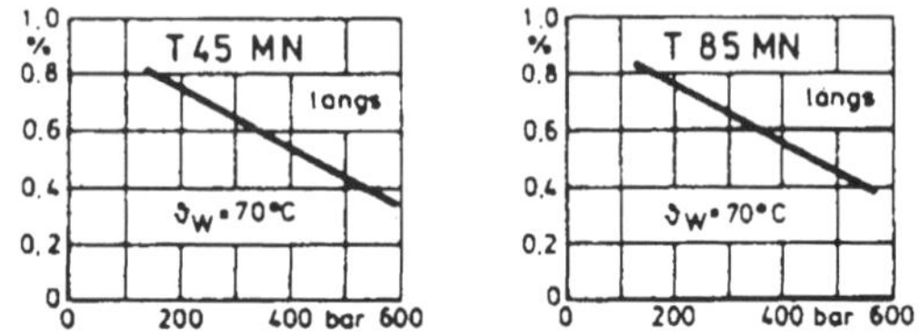

Fig 82: Cavity pressure drop for ABS mould

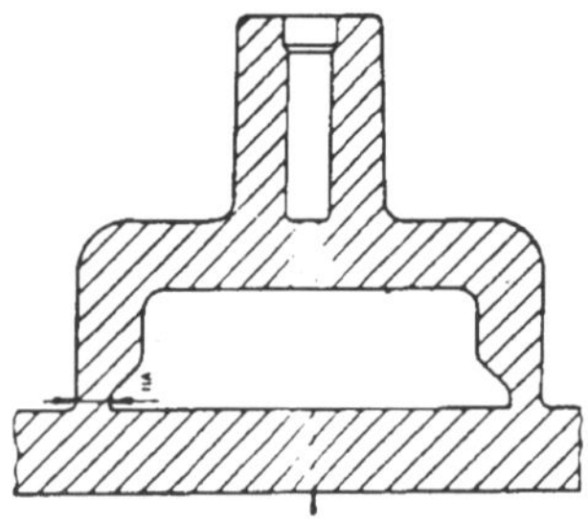

Fig 83: Optimising boss design

Fig 80: Dash panel case study

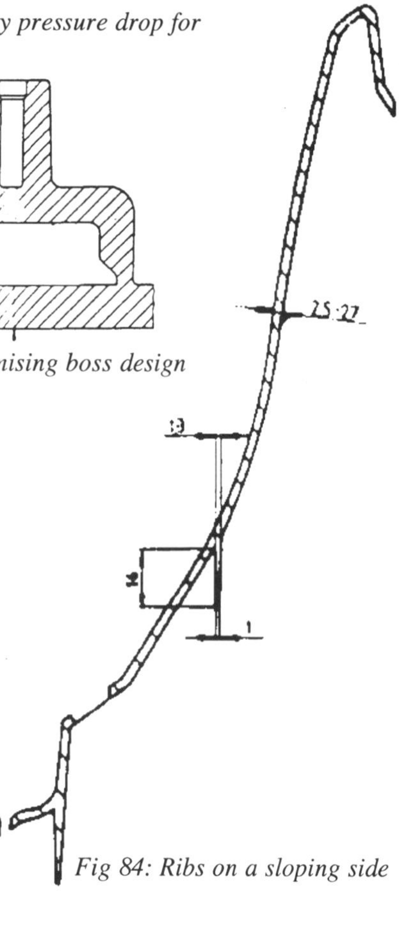

Fig 84: Ribs on a sloping side

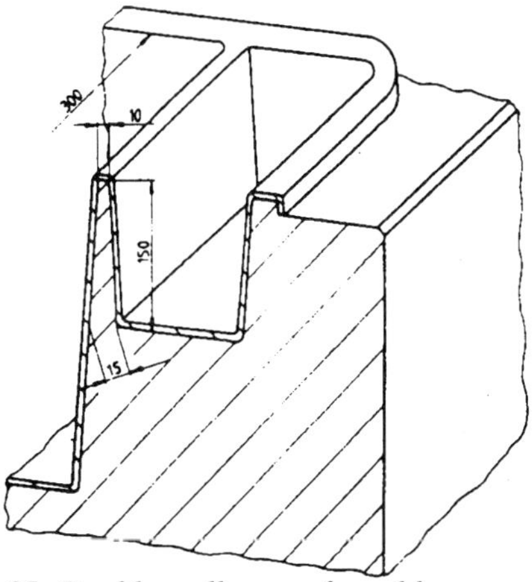

Fig 85: Double-wall area of mould

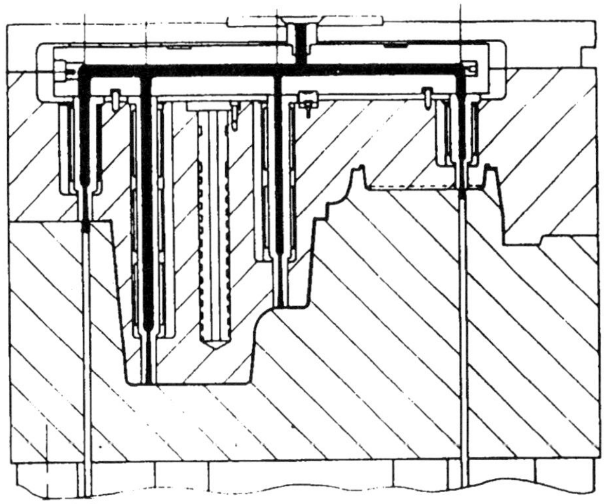

Fig 86: Hot-runner system

mould's cavity); this is the weld line, the position of which can be varied, to find a less visible area, by repositioning the gate. To obtain an evenness of gloss, moulded parts which are located in one and the same visible area should all be made of the same material.

The grained surface can also be disfigured if inadequate draft is provided. Generally there should be one degree per 20 microns mean roughness in addition to the normal one degree of draft. Another problem area is ribs positioned on surfaces of the moulded parts that slope inside the mould (Fig 84). This is because the tip of the core that forms in the mould can be deformed by the injection pressure. This, in turn, can exert pressure on the grained surface after cooling. Double-walled areas of mouldings are also critical (Fig 85). Cores of this type are readily deflected through pressures acting on one side during the mould filling process. This leads to problems in demoulding; also it is no longer possible to cool the cores — which themselves determine the length of the cycle. In tall moulds, the importance of core centring, with respect to the direction of mould opening, must also be stressed.

Plastics mould parameters

As a more detailed part of the case study, the glove locker section of the overall panel is considered as a separate moulding, shown in Fig 86 with its hot runner system designed by computer simulation. The principal visible surface makes it necessary for the core to be located on the fixed half of the mould — because of the injector pin marks. Despite central gating, with which the melt is injected from the inside, outwards, elastic deflection could still be expected with this layout which would influence the filling process. Thus the core must be centred.

The computer drawn filling pattern predictions in Fig 87 help to determine where flow lines occur and if any trapped air is likely. Provision was made for two types of gating in this example — both on visual grounds and to suit the mould. In the compartment area there are short sprue gates and in the parting surface region for the openings there are rectangular plates. The simulation also helps to see whether all parts near the sprue are filled uniformly and simultaneously. Inspection shows the paths taken by the fill are very different. To even these out further calculations are necessary to balance pressure according to temperature profile. In this example runner diameters vary from 11 to 24mm depending on their length and the melt throughput.

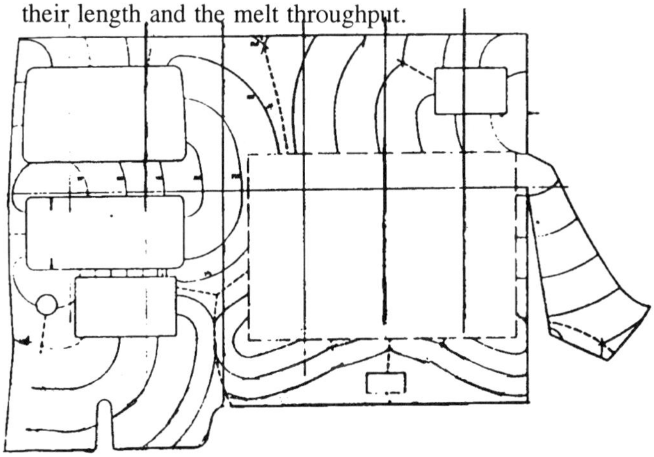

Fig 87: Computerised filling-prediction

Design for robotic assembly

While basic considerations of automated assembly were considered in the last section with the particular case study of a large plastic moulded instrument panel showing the benefits of integrating separate parts into a single moulding, here techniques are examined for easing the assembly of those parts which cannot be integrated.

When considering the wider implications of robotic assembly benefits are available even to the small-series builder if a sufficiently flexible robot is programmed for a wide variety of assembly operations. Considerable design changes may be needed to the vehicle in order to exploit the technique fully, however. The volume sector teaches us, for example, that the basic body shell may need to have a detachable front-end in order to place more easily the major mechanical assemblies into the vehicle.

At a more elemental level careful planning of the stations in an assembly system can reap rewards in the automated feeding of small parts for a larger assembly, Fig 88. In the illustration, a spindle assembly is considered and Fig 89 is the associated process model used in planning the procedure.

Minimising part count is an equally valuable discipline in simplifying the assembly process. With sheet metal parts it is obviously an advantage to create features from bending the parts rather than welding on stand-offs, inserts or snap features, Fig 90. The deformable properties of plastic can often be used to take up tolerances within an assembly, Fig 91, and integral 'springs' can be created to eliminate parts, Fig 92, or tabs in the case of sheet metal.

By designing a product so that subsequent parts are located on previously assembled ones, reorientation can often be avoided by use of features on the earlier part to locate the subsequent one. Self-locating features can be used in this context, Fig 93, such as the appropriate creation of a chamfer. In restricted areas features can be used to positively locate a part before it is released, Fig 94, or even adding a dimple can facilitate the alignment of blind features, Fig 95.

Parts should also be made symmetrical where possible to reduce orientation and if this is not possible it may be feasible to design obvious asymmetrical features into the part to facilitate orientation, Fig 96. Obviously it is wise to avoid parts that might nest or tangle and adjustments can often be reduced by using a long positioning hole on one side of a two-screw fixing, for example, Fig 97.

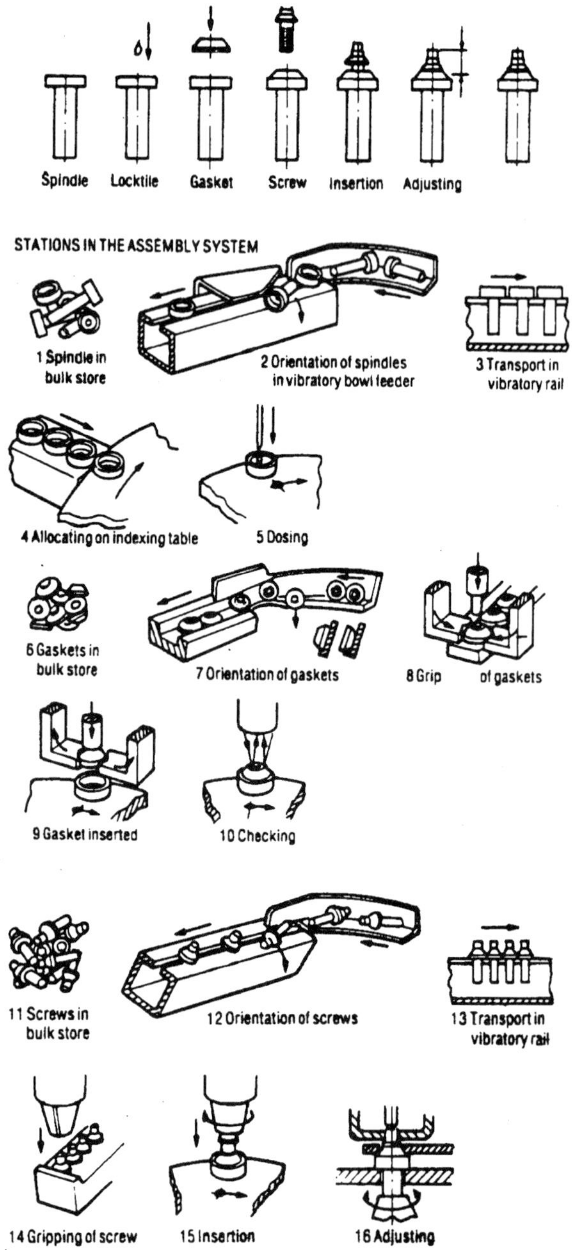

Fig 88: Logical build sequence for spindle assembly

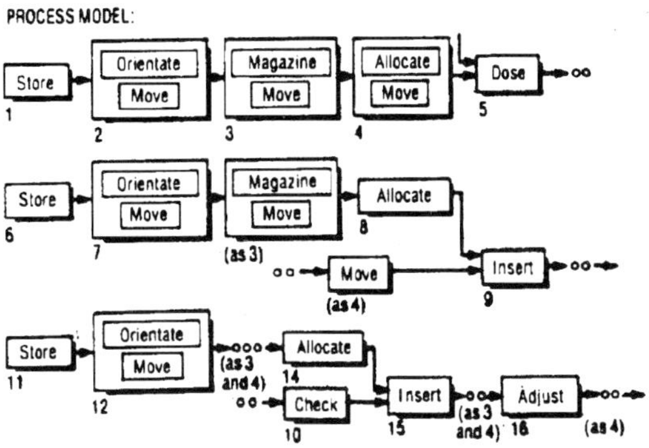

Fig 89: Assembly operation plan for Fig 88

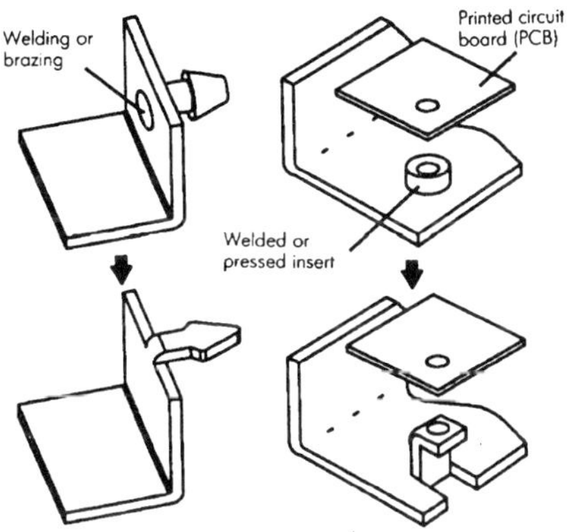

Fig 90: Use of bending instead of welding

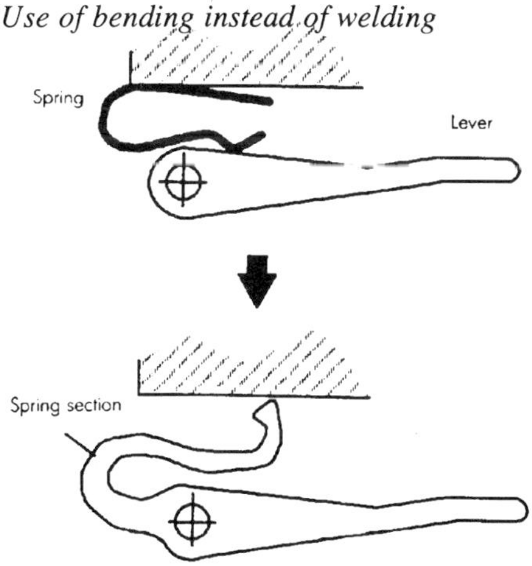

Fig 91: Eliminating spring-clip by plastics integration

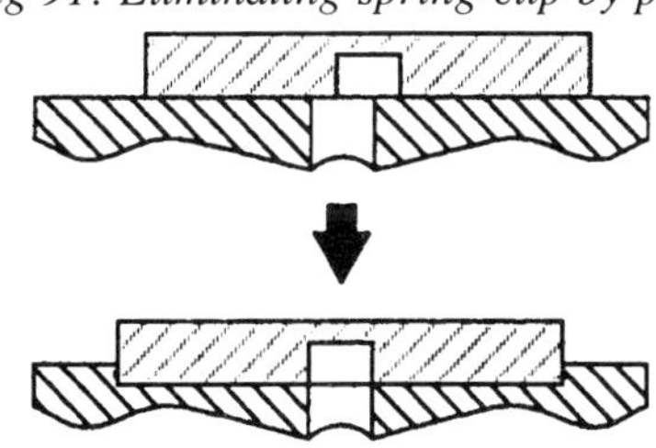

Fig 92: Dimpled cut-out to ensure line-up of bores

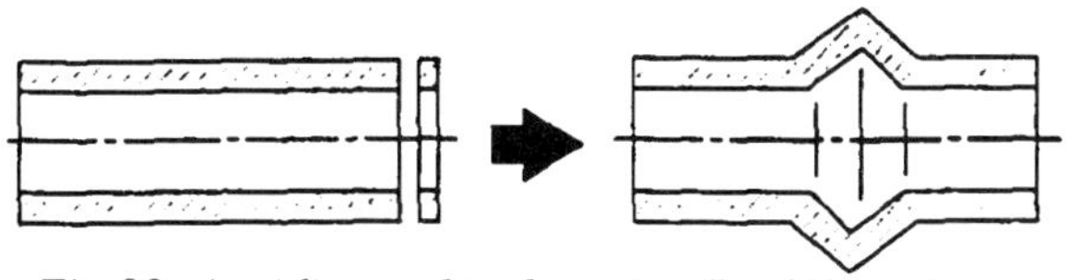

Fig 93: Avoiding a shim by using flexibility of part

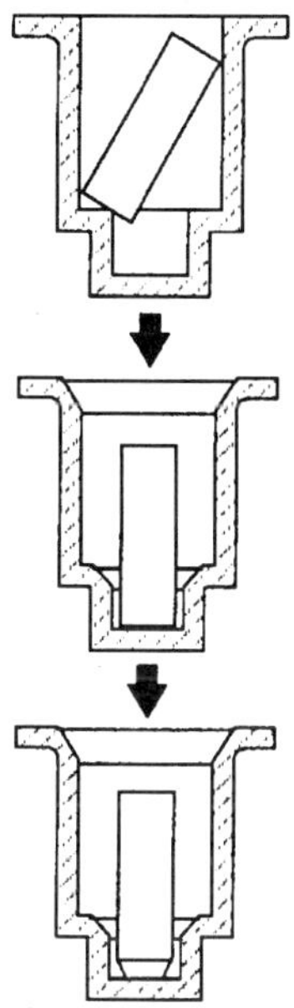

Fig 94: Use of chamfer to assist location

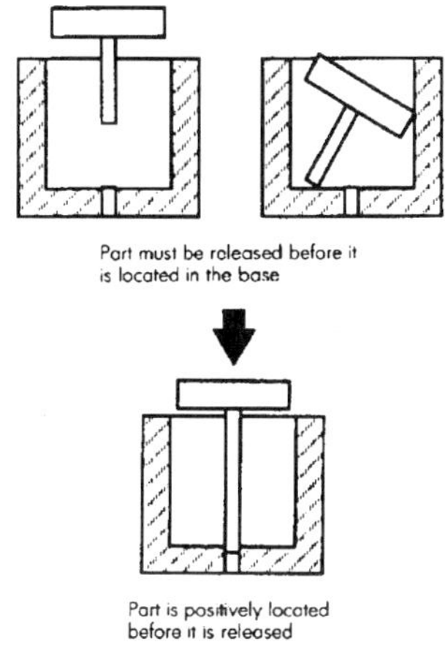

Fig 95: Ensuring a bore fit by means of a secondary location

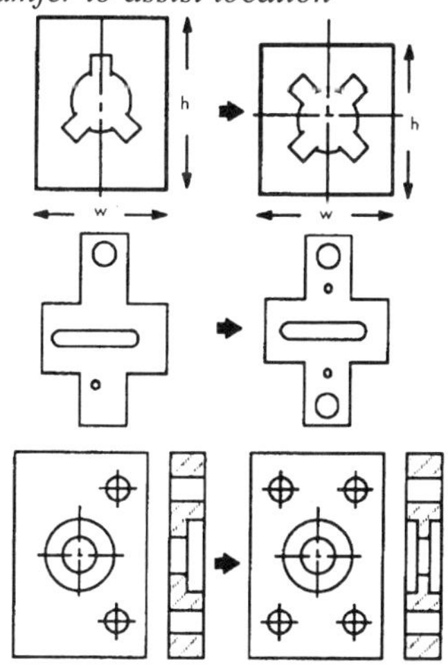

Fig 96: Building-in symmetry

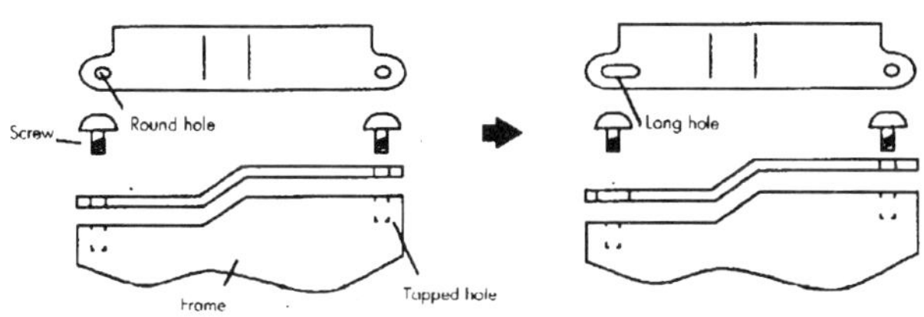

Fig 97: Use of slotted hole to avoid adjustment

Cutting and forming

The profiling, and cutting, of sheet metal has been markedly changed by the introduction of programmable computer-numerical-control techniques. It has affected most of the various ways of cutting sheet-metal including blanking, shearing, gas-cutting and laser/water-jet profiling —providing greater accuracy, repeatability and the ability to set up quickly from drawing.

The latest CNC punch presses are permitting the ready inclusion of simple forming operations into the metal-blanking cycle. Stiffening ribs and louvres are among the possibilities, without removing the workpiece from the press. Turret-mounted tooling allows quick change between operations. Such devices as quick set-up tool-length adjustment are also valuable — in changing between one job and another. Tool height is obviously critical in operations like swage-forming and such devices mean that by the twist of a knurled collar, the height can be stepped by small discrete increments.

Another useful device is the positive stop, shown as Fig 98, which is the product of USA tooling specialists, CE Tooling. This is useful for such operations as coining and making push-out blanked-holes to make subsequent wiring apertures in parts. The stop causes the the tool to lightly crimp the part near the end of the stroke. It can also be used to speed the set-up of such operations as swage-forming, by progressively bringing the tool-height up to tolerance while carrying out test-strikes on the part. This is important because the higher cycle speeds possible with CNC blanking mean that changes in design for swage-forming might be necessary.

The cost of blanking tools can often be reduced by thoughtful dimensioning of the part-design. In Fig 99, for example, none of the radii shown may be necessary for the performance of the part. Although the tool for producing blank B would cost more than that for blank A, the tool for producing blank C would cost less than that for Blank D — the reason being that far fewer punch replacements would be necessary during the production run.

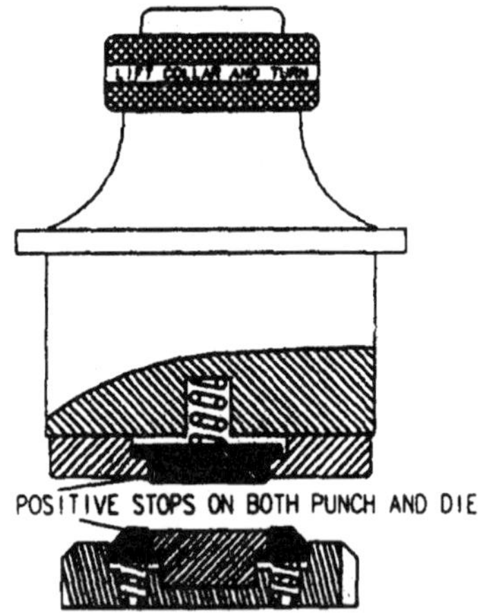

Fig 98: Positive stop by CE Tooling

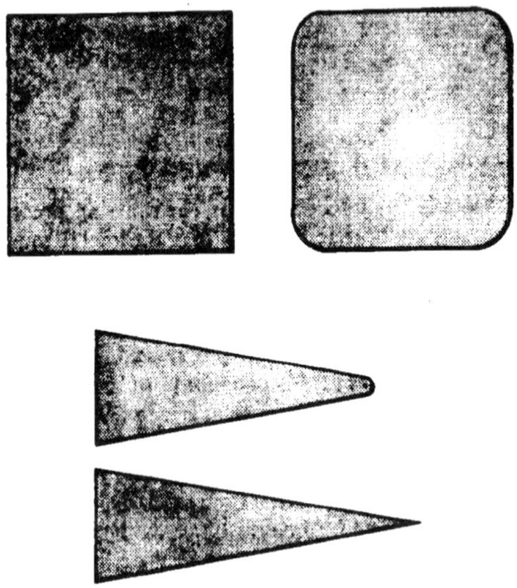

Fig 99: Attention to detail design

Fig 100: Hydraulic press-brake

Shearing and folding

The use of CNC front-gauging can make considerable improvements in part accuracy. Mazak Nissho Iwai make a typical state-of-the-art machine, the 50/

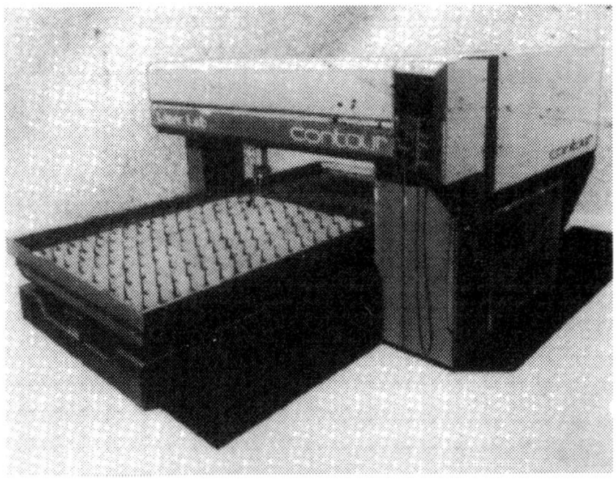

Fig 101: Laser cutter

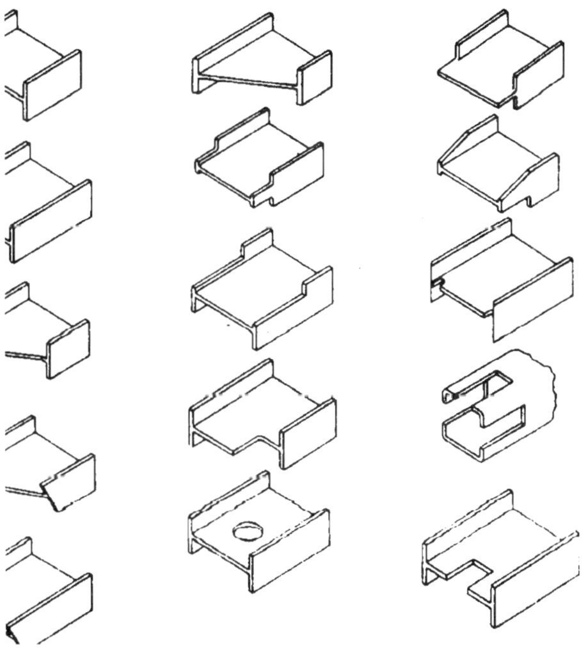

Fig 102: Flame-cut sections

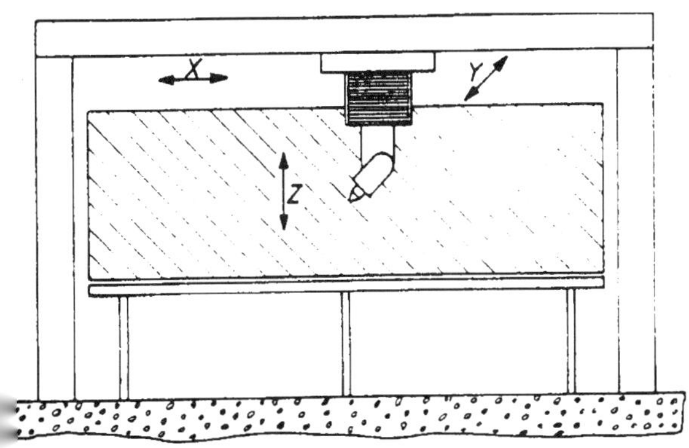

Fig 103: Gantry laser system

100/150 tonne Apex CNC-controlled hydraulic press-brake, shown in Fig 100, which has ram positioning effected by a linear scale feed-back system giving accuracy to 20 microns. CNC back-gauging, and automatic die-clamping are also available. Fold lengths of 1.25-2.0, 2.5-3.1 and 3.1-4.0 are attainable, according to the machine specified. A 20 mm/sec fold rate is possible in an opening height of 420 mm.

Gas, plasma, laser and water-jet

CNC profile cutting is also speeding the cutting out of more complex shapes. Among recent machine introductions is the Technicut Por-Tech 20 from Electro-Craft. This is a twenty foot capacity unit using plasma-flame to cut sheet from 16 to 20 swg. It uses DC servo motor/tachometers to cut at 10 m/min with 0.1mm accuracy on receipt of instructions from a CAD workstation.

The introduction of lasers with power in excess of 1kW has made possible the high speed cutting of thick plate and both articulated arm and gantry machines have been built to exploit the process. A recent introduction by Laser Lab is said to have made an important advance in the reduction of cost of such systems. The Contour, Fig 101, is an integrated design such that the operator controls the machine from one working position, the control taking care of both the motion system and the laser. A dual-pallet table means that plate loading down-time can be minimised. Power ramping and two or three axis motion are also optionally available.

Flame and jet cutting

Gas cutting is applicable to ferrous metals for providing rough cuts in plate adequate for subsequent welding in fabrication work. The oxy-acetylene torch uses a central jet of oxygen which is under the control of the operator, the cutting action being a combination of chemical action and erosion by the narrow, high velocity, jet. Material is locally heated to ignition temperature by the oxy-acetylene flame, and the metal subsequently oxidised by the flame. The width of cut is about 3/16th of an inch and, with mechanically guided torches, a dimensional tolerance of about 1/32nd of an inch can be held for material thicknesses up to two inches.

In plasma cutting, a gas-stabilised electric arc passes through an orifice to generate intensely high temperature, high velocity, ionised flow of gas — the plasma stream which melts and blows the metal away. A higher quality of cut, and faster rates, than oxy-fuel cutting can thus be achieved. A wider range of materials can be cut, too, when mixed-gas techniques are used.

According to ESAB Automation, robotised cutting is gradually coming on stream for cutting steel structural sections. Fig 102 shows some sections involved in a recent collaborative contract. The robot uses a 3-D sensor to measure the actual dimensions of the profile and these are matched with stored design data in order to download the guidance program to the controller on set-up. Tenfold productivity increase in comparison with manual cutting is claimed. Fig 103 shows a typical overhead gantry laser system used for cutting. The company have also reported progress in cutting-nozzle development for oxy- and plasma cutting which have resulted in cutting speeds being raised by 20%. For oxy-cutting a Laval socket has been incorporated into the oxygen cutting chan- this results in a mach 2.5 flow velocity and the cutting of thin sheets has been particularly improved.

In abrasive water-jet cutting, involving the addition of abrasive to the water jet stream, Fig 104, the kinds of performance occur as listed in the table in Fig 105, according to ESAB. Advantages claimed for the process include no heat distortion, a narrow cut (kerf) and no mechanical deformation. Pure water-jet cutting, without abrasives, is useful for cutting carpets, plastics and grp composites.

For gas-cutting , the ESAB-Hancock Ultralex UXD-P portal-type cutting machine is capable of speeds of 12 metres/minute. Shown in Fig 106, it is made in two sizes for cutting widths of 1500 and 2000 mm. It is numerically controlled and can be programmed quite easily for fabrication work.

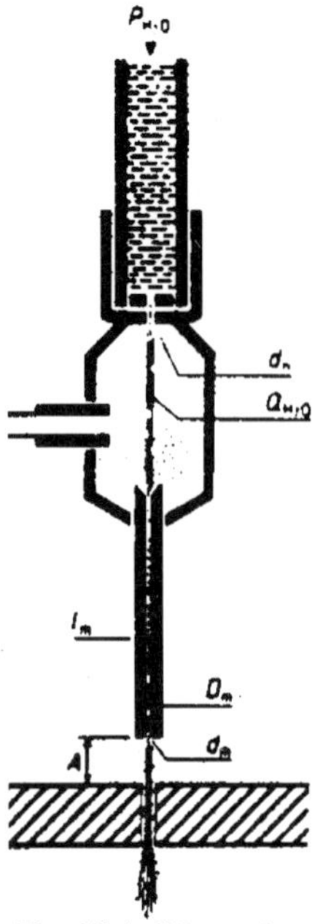

Fig 104: Water-jet cutting

Thickness (mm)	Cutting speed (mm/min)
1– 70	3300– 20
1– 50	2500– 25
1– 60	3500– 30
2– 30	2500– 60
1– 30	5000– 70
1– 60	3000– 25
10–150	200– 20
4– 50	5000–200
5–100	8000–200
15–100	2000–120
3– 60	9000–200
1– 40	20000–300

Fig 105: Cutting performances

Fig 106: Portal-type cutter

Presswork techniques

A number of conflicting requirements between the optimum design, on the one hand, and production on the other, apply to pressed parts. It is important that form design meets the problems posed by both the toolmaker and pressworker before new shapes are committed for sheet metal working.

A major panel can be affected by something as basic as strip-mill rolling widths: these might govern the depth of draw for a panel-van roof panel, for example. Despite this, the need to use as large a

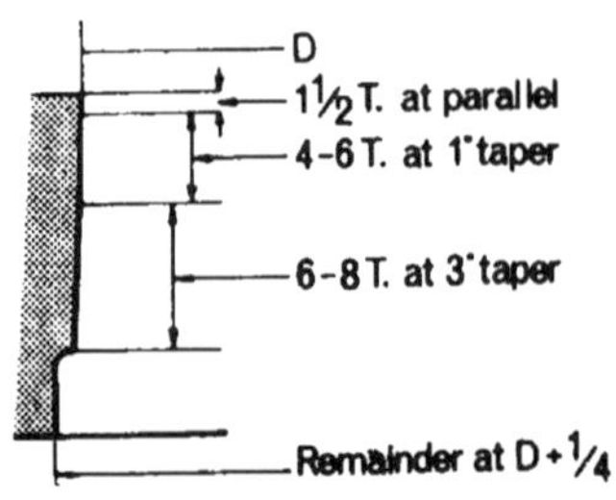

Fig 107: Die clearance considerations

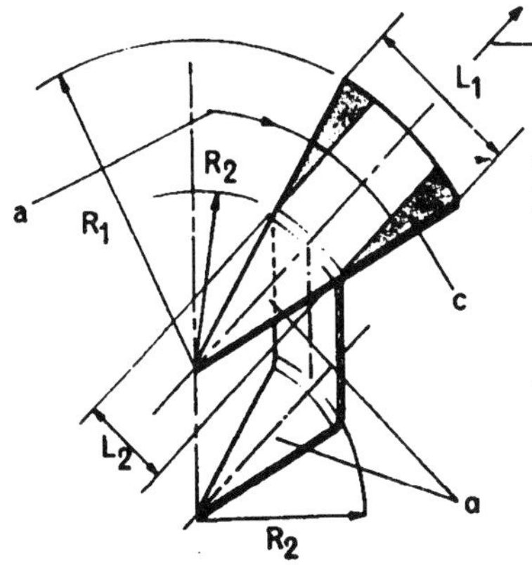

At (a), hoop compressive stress set up by annulus between radii R1/R2 so arc L1 shortens to L2 when passing across the side face. At (b) material ungoes elongation in radial direction and at (c) is displaced relative to area of sector then at (d) this material remains unchanged.

Fig 108: Effect of deep drawing

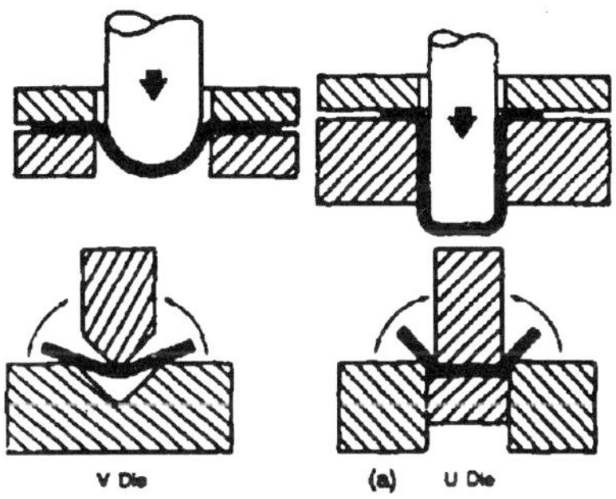

Fig 109: Round and flat-bottom punches

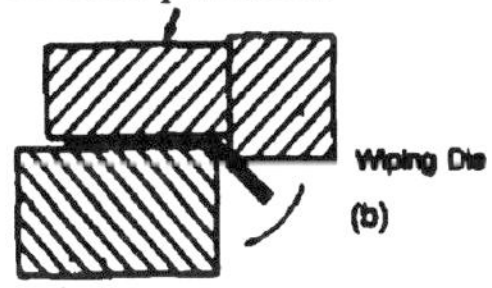

Fig 110: Bending deformation

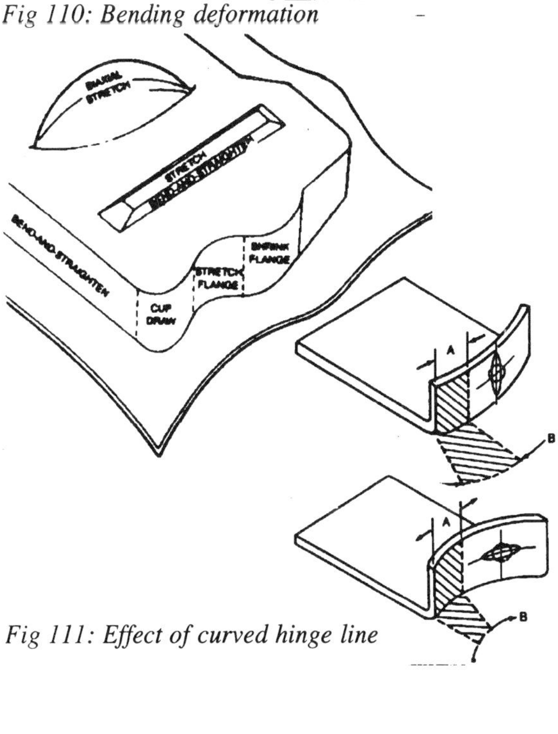

Fig 111: Effect of curved hinge line

single-piece panel as possible applies to almost every pressing situation. To produce a typical panel it may be necessary to use a blank and pierce tool; a crop or notch tool; a preform and then a flange tool — finally a setting tool. The need to ensure that a thickness reduction beyond 10 percent does not occur is likely to require more than one 'hit' on all but the simplest deep drawn parts. In view of the number of operations involved, it is of course desirable that the minimum number of panels be used in all cases.

Cutting clearance between punch and die is important in avoiding blank dishing and high loading requirement on the one hand, or, on the other, a break in the blank edge that might induce crack propagation. Die clearance, Fig 107, is also important — with 'backing off' chosen to avoid either distorted blanks or a burst die. In these connections it is necessary for the product engineer to be aware of metal gauge variation due to 'mill' tolerances; also to any variation in temper and grain run in relation to bend direction. Elastic recovery to the flat state will vary with temper and the appropriate overset to compensate for this effect must be determined. Fig 108 shows the reason for using a pressure-plate to stop wrinkle formation during deep drawing.

The illustration shows a sector of a flat circular blank before and after deep drawing. During deformation R1 must reduce to R2 and L1 to L2. Displacement of the material shown shaded sets up hoop stresses in the annular region outside the die aperture, buckling and direct stresses from which influence the amount the blank diameter can be reduced in one draw.

Press-forming modes

A complex part may involve the use of several forming operations. In the stretching operation the blank is clamped at the so-called die ring either by hold-down pressure or lock beads and a contoured punch pushed through the die opening. All deformation takes place in the metal over the area of the die opening only, the bi-axial tension resulting in a thinning over the total 'dome'. In the deep draw situation, the blank is drawn into a die, circular as shown in Fig 109a, by a flat bottom punch, b, which pulls the flange radially towards the die opening and the deformation as shown in each case occurs. More common, though, is plane bending as shown for 'V' and 'U' dies alongside a 'wiping' bend, Fig 110, in which one edge is clamped while the punch wipes the free end down. In all cases one or more edges swing through space and the swing must be calculated to allow sufficient room in the tooling; the metal outside the 'plastic hinge' is, of course, unstrained.

When the hinge line is changed from straight to curved, the metal outside the hinge no longer remains unstrained and either shrink or stretch flanging takes place as shown in Fig 111.

To achieve the same shape as that generated by the above bending operation, a 'bend-and-straighten' can be used which will give the finish product different characteristics. Here the swing of the metal is prevented by the blank-holder and the bottom radius formed round the punch by a bend operation. A simultaneous radius is formed around the die and thereafter each individual element in the final wall begins in the flange and is pulled towards the radius zone. As the die radius line is straight, no compression is created along it and upon entering the die radius zone, the element is bent to form the radius contour. On leaving it, the element is likewise unbent or straightened to conform again to the wall, hence the name of the operation. The main difference between it and 'bending' is that in the latter case, the final wall remains in the unworked state whereas in the former, the bending and unbending sequence hardens the metal and reduces subsequent formability.

Fig 112 shows the combination of forming modes which could go into one part — indicating the normal function breakdown required before analysis of a pressing. One analysis method is the length-of-line technique, illustrated, in which the salient stages are listed. Line 1 is generated by a radial or cup drawing mode obtained by examining a plan view of the pressing, Fig 113, where it can be seen that the four corners can be combined to form a cup created by deep drawing. Line 2 is generated by bending and straightening, in contrast. In the secondary forming operations, trimming and reverse flanging may present problems if thickness remains constant over Line 2 and increases, to 40 percent say, over Line 1 in the corner cup draw. One solution may be to increase blank width to provide an additional flange, after trimming, to meet subsequent welding requirements. This will restrict metal flow from the hold-down area and increase the depth of the pressing required by stretching over the punch. If such action created a failure over the punch, such a trend could be reversed by increasing the die radius to allow easier flow of the extended blank into the die cavity. If, however, the die radius is now too large then a restrike operation may be required to sharpen the radius.

Metalworking terminology

Shearing is a straight cut across the coil width of the delivered metal to form a square or rectangular

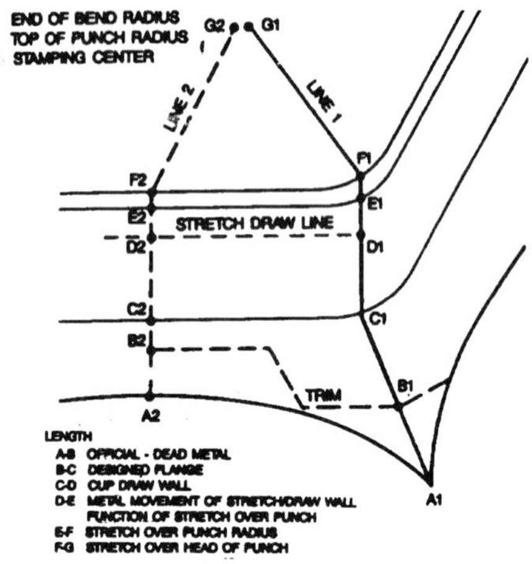

Fig 112: Line length analysis technique involving blank edge, trim line, die radius, stretch-draw line, end-of-bend radius, top-of-punch radius and stamping centre

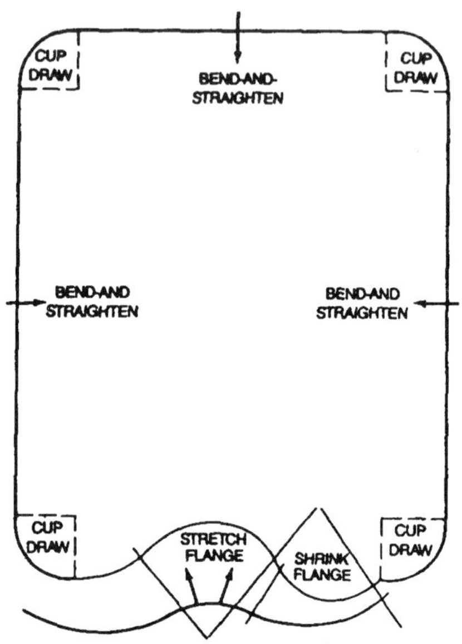

Fig 113: Plan view of pressing

blank. More often, 'blanking' will be required where the contour is both straight and curved. Metal can be saved in blanking by the best outline for 'nesting' within the strip or coil, Fig 114, in which sharp corners are seen sometimes to be preferred to more aesthetically curved shapes. Contouring should also be arranged to match as near as possible the die opening. This will encourage more uniform metal flow into the die cavity preventing buckling of the flange and reducing drag of extra flange metal behind critical zones. Finally, contouring should be arranged to minimise any post-forming trim operations.

Preforming is an additional stage for gathering metal into a zone for later use before the edge metal is restrained. This allows the necessary length of line to be generated without excessive tension that might cause breakage due to metal thinning over

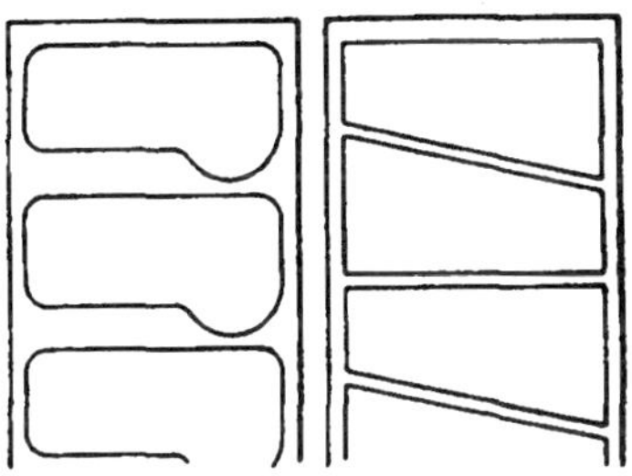

Fig 114: Blank nesting

Fig 115: Developed length

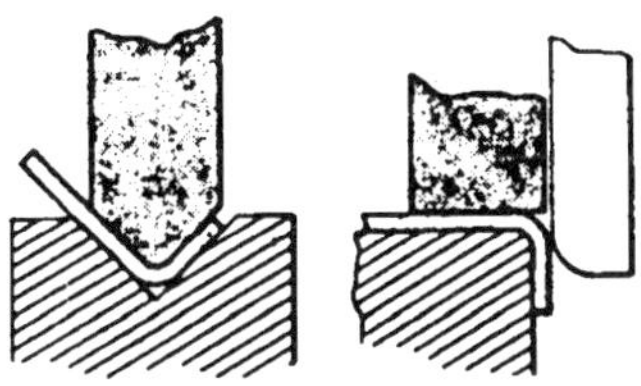

Fig 116: Bend radii minima

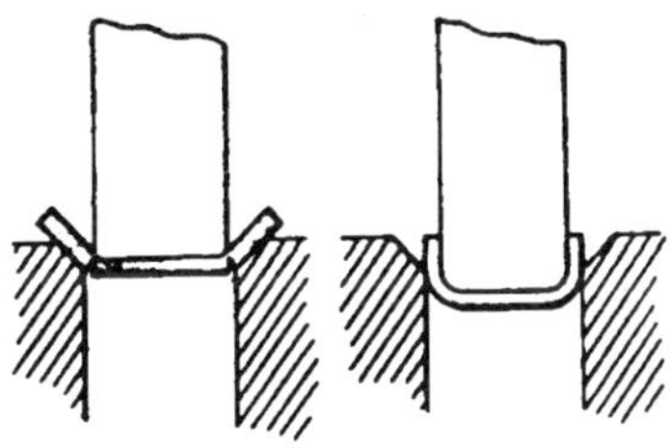

Fig 117: Draw radii minima

tight radii. Thereafter, blank insertion, positioning and closing the hold-down ring are the next operations. The ring may be formed in one or two planes, in its function of preforming the blank closer to the contours of the punch. This avoids metal being trapped under the punch and balances the forces on it in the interest of avoiding skidding of the character line. Punch action and retraction takes place in the main forming operation. Ejection and transfer of the part then usually complete the operation.

Simple bending is affected considerably by relation of thickness, t, to bend radius. The latter dimension can range from $0.5t$ for soft materials up to about $7t$ for full temper steels. If bending is with the metal grain the radii can increase by up to a factor of two. In calculating the developed length, Fig 115,

$$l = b + c + 1.5708(t/3 + r)$$

this value could alter with ductility and grain direction, however. It would only be correct for the open-vee tool but not the side-drawing tool illustrated in Fig 116. Draw dimensional rules are also much variable with conditions. About 2 degrees of draft is helpful on most work to avoid wrinkling and in general it is helpful if depth of draw does not exceed 1.5 x punch diameter for each strike. Radii at the bottom of the part should be as large as possible to avoid the situation of Fig 117. A 'square' component can be drawn to a depth at least equal to its width providing the ratio of depth to corner radius is no more than five. Die aperture radii should generally be within 4 to 10 times metal thickness and the punch nose radius should not fall below twice metal thickness.

Pressworking do's and dont's

In Fig 118, raising of a flange where the blank contour passes acutely across the bend line should be avoided. Otherwise the metal diminishes in width towards the bend line and where it fails to meet the die wall an unsightly bulge results,(a). When beads, stiffening ribs or depressions are required in a panel their root radius should be as generous as possible consistent with the strengthening effect required; allow ample clearance around swages to allow possible easing of the radii (b).

For a 'clean' finish panel, work must be carried out at all points or the remaining, unworked, metal will remain soft and easily distort; for a straight drawn shell an adequately strong bottom is only possible by stiffening in the form of beads or a radiused die line at the closed end (c). Incorporate flanges where possible in long shells to avoid distortion of the panel by releasing stresses during the trimming operation; ideally wall curvature, in plan, should be provided to counteract the effect introduced by the die radius (d). Use symmetrical cups where possible so that the area of the final pressing is the same as that for the blank; cylindrical portions are then equal in area to annular ones, on the blank (e).

Pressworking steel behaviour

Advanced methods are considered below for avoiding the various pitfalls which can occur during the set-up process in preparing blank and die for a production run. The determination of correct retaining pressure on the side-flanges of a boxed-shaped deep-drawn part can be estimated by predictive techniques, Figs 119 and 120, without the expense of building prototype dies. Insufficient retaining pressure allows excessive material to be drawn in from the blank, which results in wrinkling. Conversely, too high a retaining pressure, together with too much friction on the upper die surface, can lead to excessive straining of the centre of the panel, resulting in spring-back. At British Steel's Welsh Laboratories the effect of surface texture on formability is being studied: surface texture provides the main control over friction in the press. A compromise has to be found between rough surfaces which retain lubricant, and avoids pressure welding of the strip to the tools, and providing a texture which is fine enough to present a smooth surface after painting. Rolls of strip are subject to electro-discharge and laser texturing techniques, seen as the best means of controlling strip texture to the optimum levels. A double-acting press has been installed for these experiments; it has two moving platens, one inside the other. Forces up to 500 tonnes on the inner platen are used to form the part while the outer platen can exert forces up to 300 tonnes for restraining the edges of the blank which can be of up to 2.74 x 1.52 metres projected area. Press loading is computer-controlled and time histories of loadings can be printed out for recording optimum load cycles. The experiments are also being used to determine degree and mode of strain required

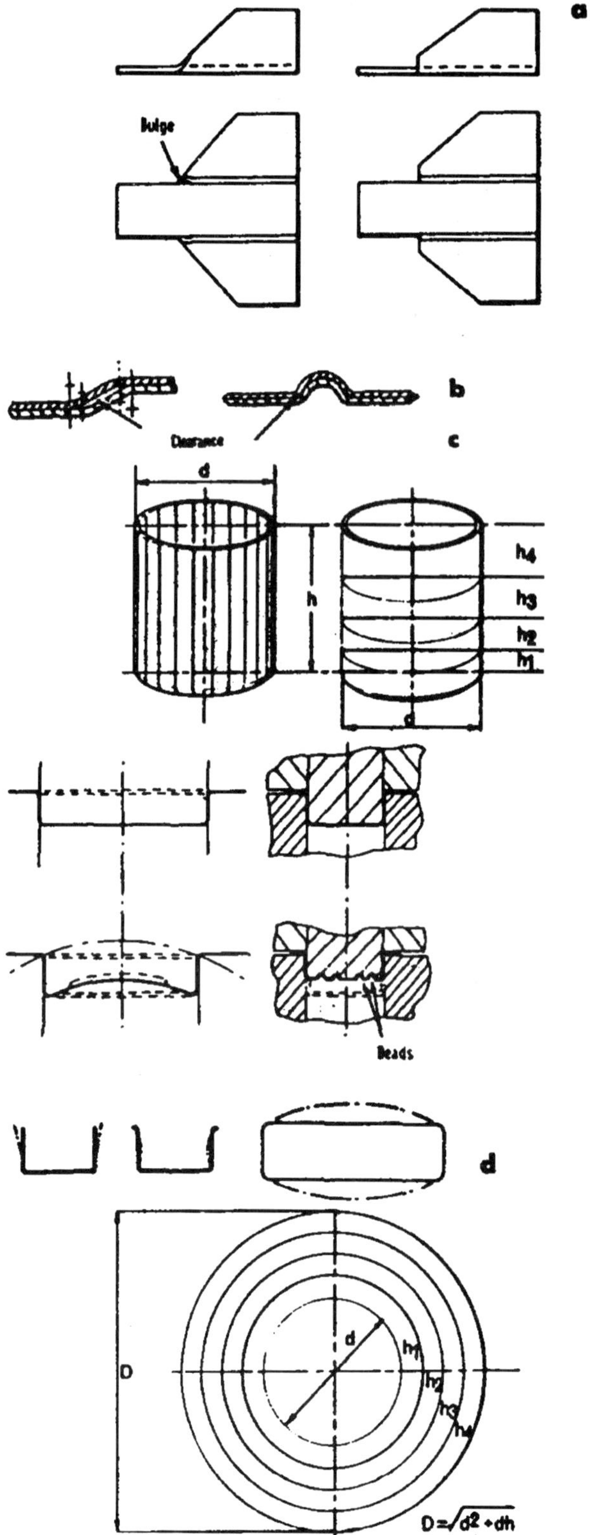

Fig 118: Do's and dont's

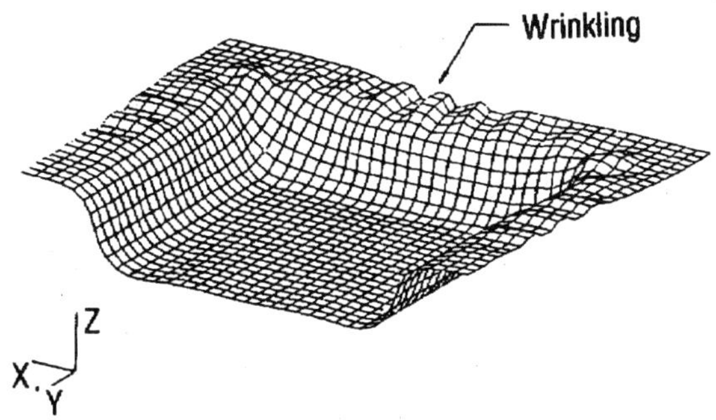

Fig 119: DYNA3D code used by HW Structures can predict the onset of wrinkling in an incorrect die set-up

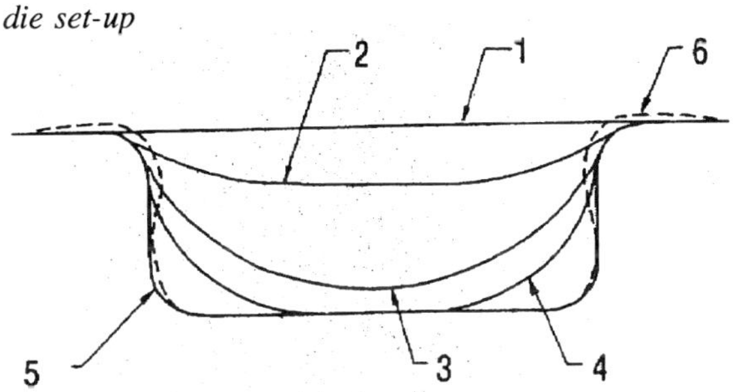

Fig 120: Spring-back prediction: 1 blank; 2-4 deformation under increasing pressure; 5 spring

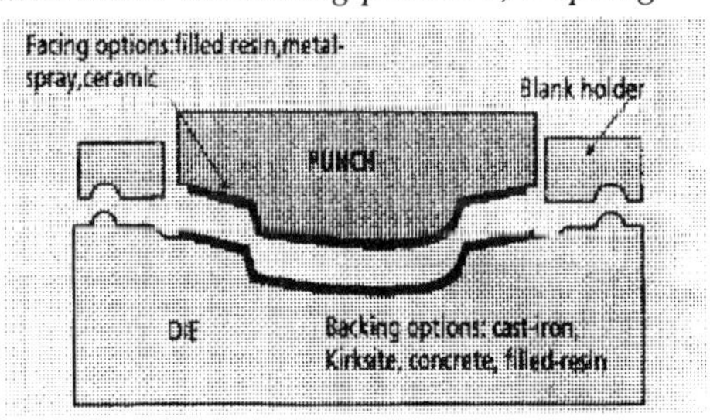

Fig 121: Low-cost tooling (Warwick University)

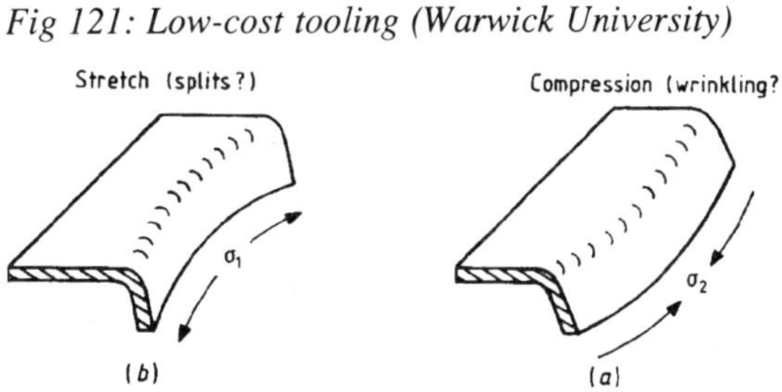

Fig 122: Bending along a curved line

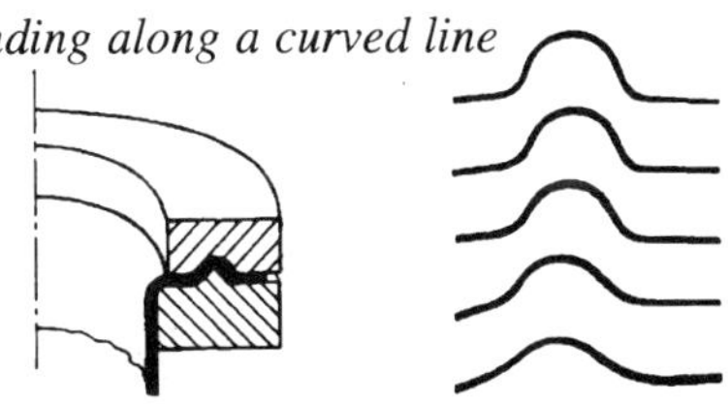

Fig 123: Effect of bead penetration on sheet contour

for imbuing curved panels with maximum dent resistance. The work is also showing how suitable positioned holes in the edges of a blank can be used to stress-relieve the centre section for avoiding splitting tendency after deep draw.

At the University of Warwick studies are being made on low-cost press-tooling which could have value for the production of panels for low-volume niche-market vehicles, Fig 121. Alternatives to cast iron dies under examination include zinc alloys, resins, ceramics and sprayed metals. Production runs of up to a million pressings are being mooted for zinc dies from alloys with much higher durability than that used in the past for prototyping dies. Resin systems under investigation include epoxy and polyurethane combinations with wear resistant fillers such as slate dust incorporated. The resin forms the surface layer of the die while the backing may be concrete, zinc alloy or cast iron. Even the more conventional cast iron dies can be reduced in price substantially by using cheaper grey iron in place of alloyed nodular iron and then chromium plating to producc surface hardnesses up to 900 VPN. These are particularly suited to coated steels as the chromium layer prevents the build up of zinc particles. In recent work reported by Roger Pearce[4], spring-back is examined in some detail. In the uniaxial bending case practical solutions include successive overbending until correct shape is obtained; bending with applied tension so that the neutral axis is moved out of the section depth of the sheet. The latter can be achieved by setting negative clearance between punch and die so that both surfaces are put in tension. When bending over a curved line, shrink or stretch flanging can occur, Fig 122, according to whether a concave or convex curve is involved.

Draw-beads are one of the ways of controlling the flow of metal between a die and blank-holder. They are positioned around the die-opening such that by subjecting the moving sheet to a series of bending and unbending cycles, sensitive control over metal flow is obtained. The bead acts like a brake which can give high restraining forces with relatively low blank-holding pressures. In Fig 123, the resulting strip sections are shown for various bead penetrations while Fig 124 shows increase in hardness of the sheet due to the bending and unbending process at an average bead penetration.

Computer programs are now available for simulating the effect of pressworking process variables. Matra Datavision have recently announced OPTRIS which can accept data from a variety of 3-D CAD systems. A fine finite-element mesh is used to describe the sheet metal blank and its deformation can be observed on-screen. It can predict thickness distribution and forecast wrinkles and tears. Volume vehicle builders have found that proving-out time for dies has been reduced by half, using the program. It can also analyse multi-stage drawing and thus determine the minimum number of redrawing operations required. Following one operation, the residual strain at each mesh node is stored for the simulation of the subsequent operation.

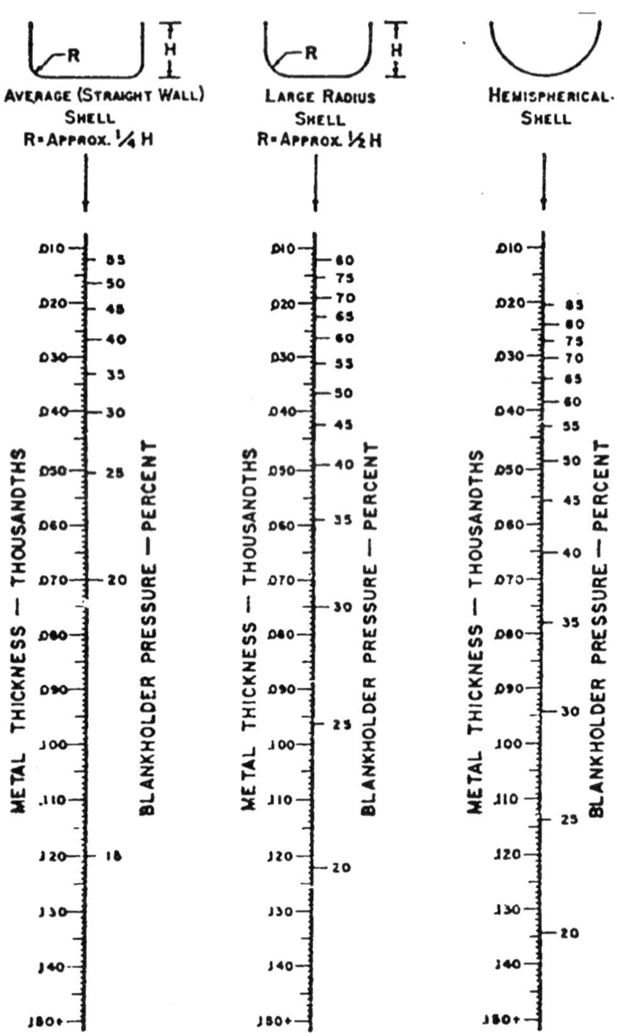

Fig 127: Parameter optimisation

Deep-drawn presswork

While general engineering knowledge may suffice for designing components and panels for shallow presswork, deep drawing requires a much closer understanding of the production process if costly mistakes are to be avoided. Some of the ground rules are provided below.

British Standard BS 4342 sees it as a process in which blank or workpiece (usually controlled by a pressure-plate or blankholder), is forced into and/or

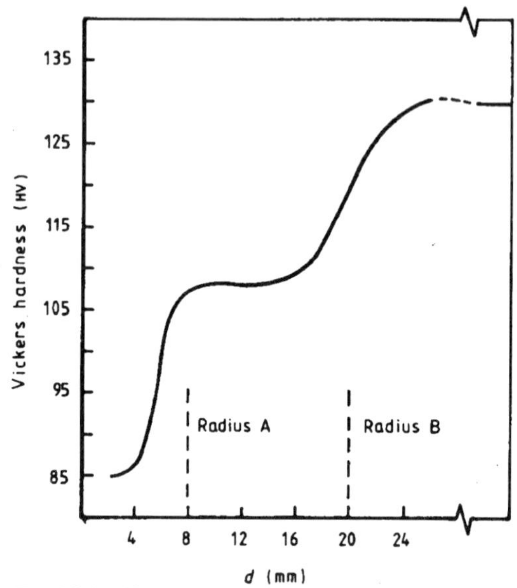

Fig 124: Change in properties as strip passes through draw-bead

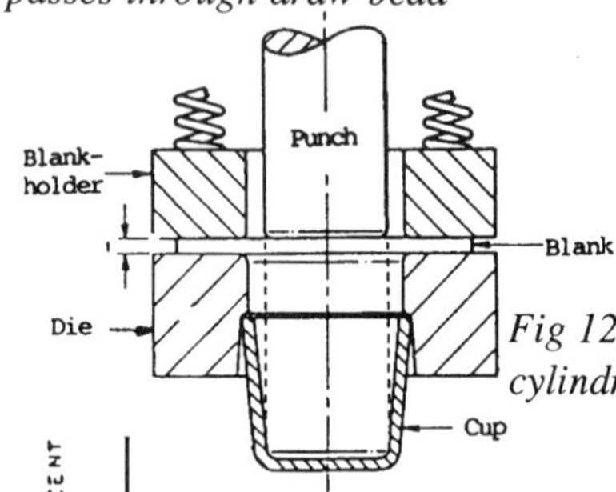

Fig 125: Producing a cylindrical shell

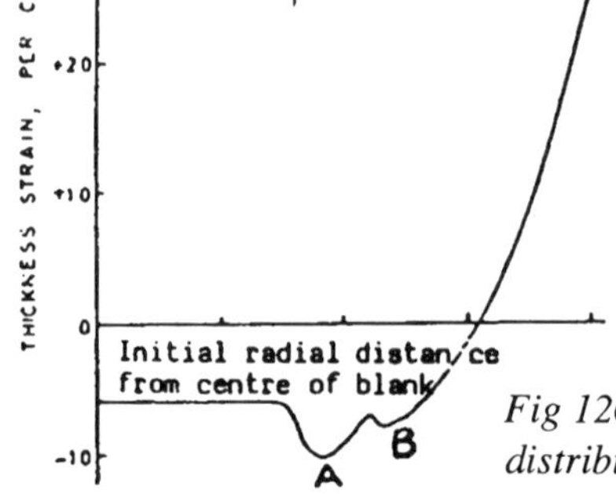

Fig 126: Thickness distribution

through a die by means of a punch to form a hollow component in which thickness is substantially the same as that of the original material. The process by which a component has its thickness reduced and length increased is known as ironing.

Deep drawn shells can be further processed by additional drawing operations (redrawing) or by reverse drawing, in which the punch travels in the opposite direction to the component with that used for the previous draw. The material undergoes compression in the direction of the periphery of the shell — so that it tends to flow along its length and sheet thickness, virtually, does not change.

Fig 125 shows the set up for producing a cylindrical shell from a circular blank using a flat headed punch. The advancing punch causes an elemental ring of sheet to be bent round the punch head profile. Simultaneously the material in the flange is dragged radially inwards to the die opening where it is bent and unbent as it passes over the die edge radius. The circumference of each elemental ring shortens and sustains a compressive stress — as illustrated in the earlier article. Also a tensile radial stress is imposed as the flange is dragged inwards.

Fig 126 shows the typical thickness distribution which results across the sheet. In a correctly designed tooling set, failure caused by an oversized blank, or faulty material/lubrication properties, should occur by fracture at neck (A). Generally the limited drawing ratio (LDR) of maximum blank diameter to punch diameter, is equal to 2 — producing a cup with height slightly less than the punch diameter. Amount of draw can be expressed as percentage: *D - d/D x 100* where *D* is blank diameter and *d* punch diameter.

Blankholder force should be kept as low as possible commensurate with avoidance of wrinkles caused by the radial friction force between the blank and the tool. Avoidance of puckering is also necessary — caused by the circumferential compressive stress in unsupported material between punch and die where radial drawing is taking place. It is controlled by restricting the size of the profile radii of punch and die to decrease volume of unsupported material —

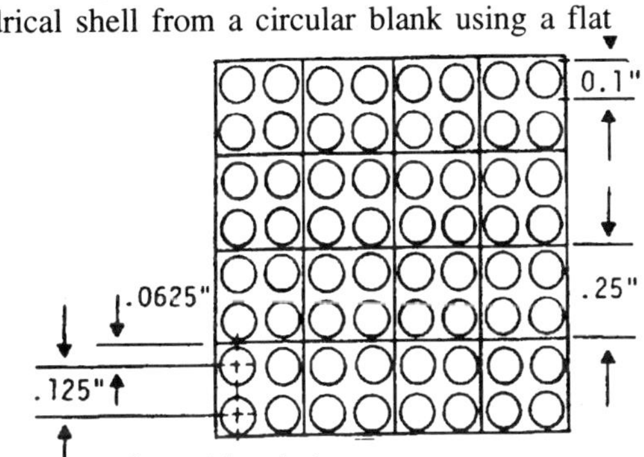

Fig 128; Circular grid technique

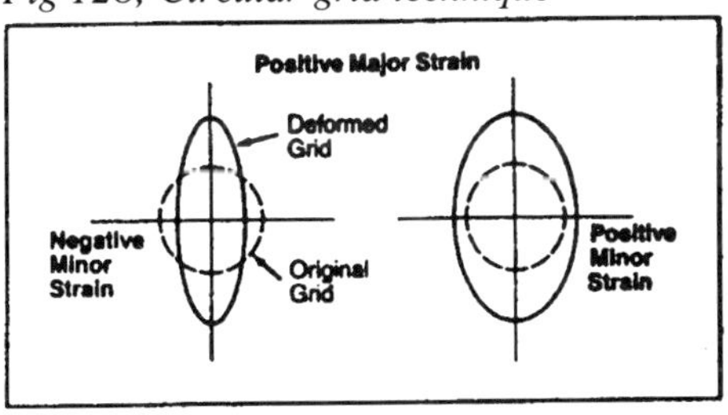

Fig 129: Major and minor axes

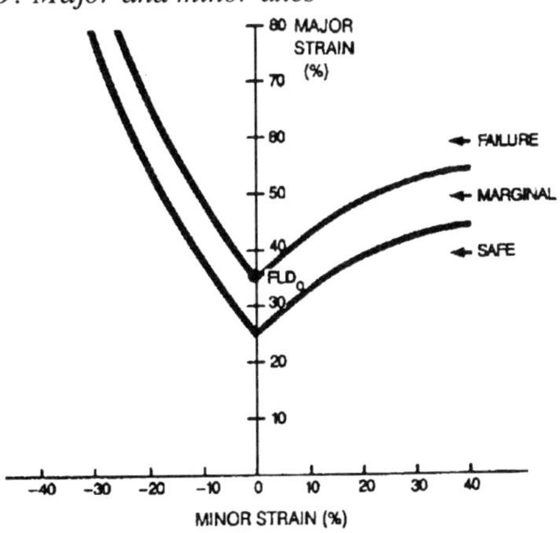

Fig 130: Forming limit diagram

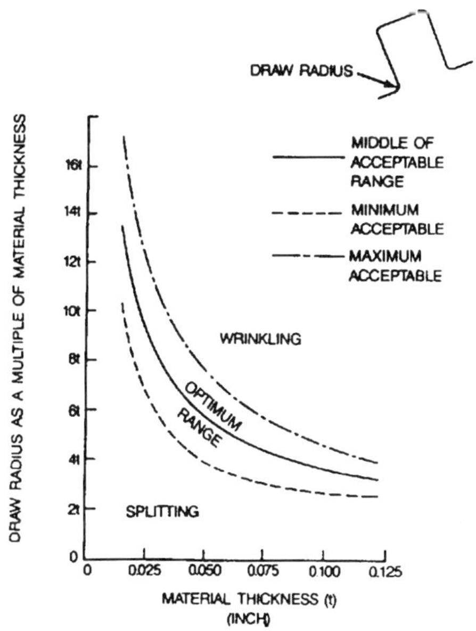

Fig 131: Draw radius vs thickness

also by increasing blankholding force. Fig 127 shows how, by expressing blankholder pressure as a percentage of draw pressure, that tables can be drawn up to optimise the different parameters for various punch shapes and material thicknesses.

There are also basic parameters of the pressing steel of interest to formability. These are work-hardening exponent, plastic strain ratio and strain rate hardening exponent. Circle grid strain analysis is a useful technique for seeing how the material deforms. The sheet to be formed is etched with a circle grid, Fig 128. The amount of plastic strain at each surface can be measured at different stages in forming by observing major and minor axes of the resulting deformed ellipses, Fig 129. This is used to derive a forming limit diagram, Fig 130.

By intelligent use of such diagrams, variables like blankholder pressure, die clearance , lubrication, steel grade and thickness can be altered to optimise the forming operation. Draw radius relates to material thickness as shown in Fig 131. If a deeper cup is required, relative to the diameter of the cup, one or more redraw operations are necessary, Fig 132.

Superplastic forming

A number of metal alloys can be produced in a superplastic state such that their strength drops to zero and elongations of some 1000% are possible. The products of two suppliers are reviewed here to show the potential for producing contoured panels economically in comparatively low volumes.

To achieve superplasticity the alloy must have a very fine grain size which must remain stable up to at least half the melting temperature of the material. The alloy is worked above this temperature threshold and carefully chosen strain rates are required in

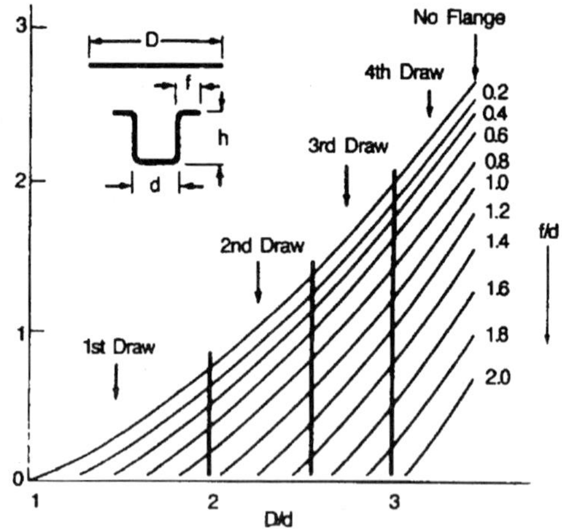

Fig 132: Multiple redraw

Supral 100	0	130	220	12
Supral 100*	T6	300	420	9
Supral 150*	0	120	195	9
Supral 150*	T6	270	390	9
5083 SPF	0	150	300	20
7475 SPF	T5	340	420	11
7475 SPF	T5**	360	450	8
7475 SPF	T6**	525†	574†	10
8090 SPF	T6**	350	450	4

Fig 133: Alloy properties compared
Alloy Condition 0.2% Proof Stress UTS Elongation

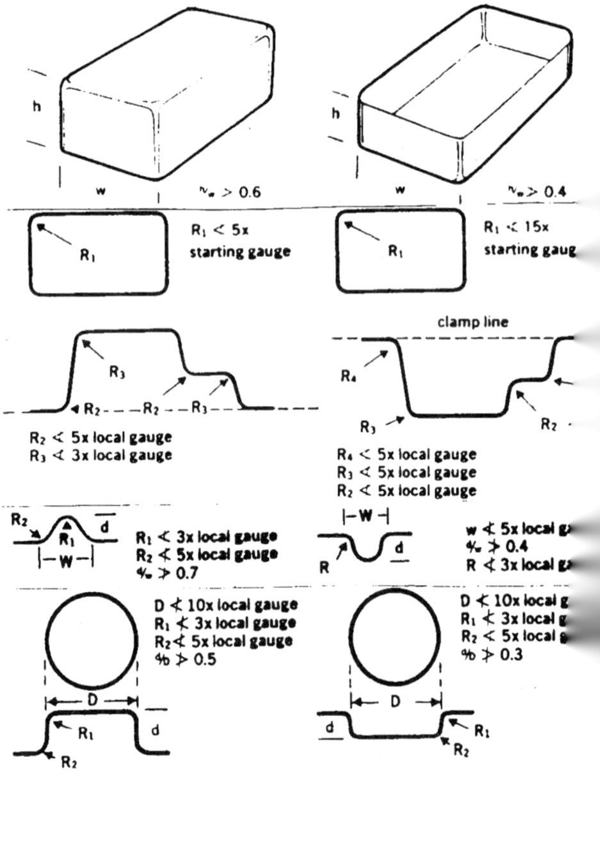

Fig 134: Corner radius recommendations

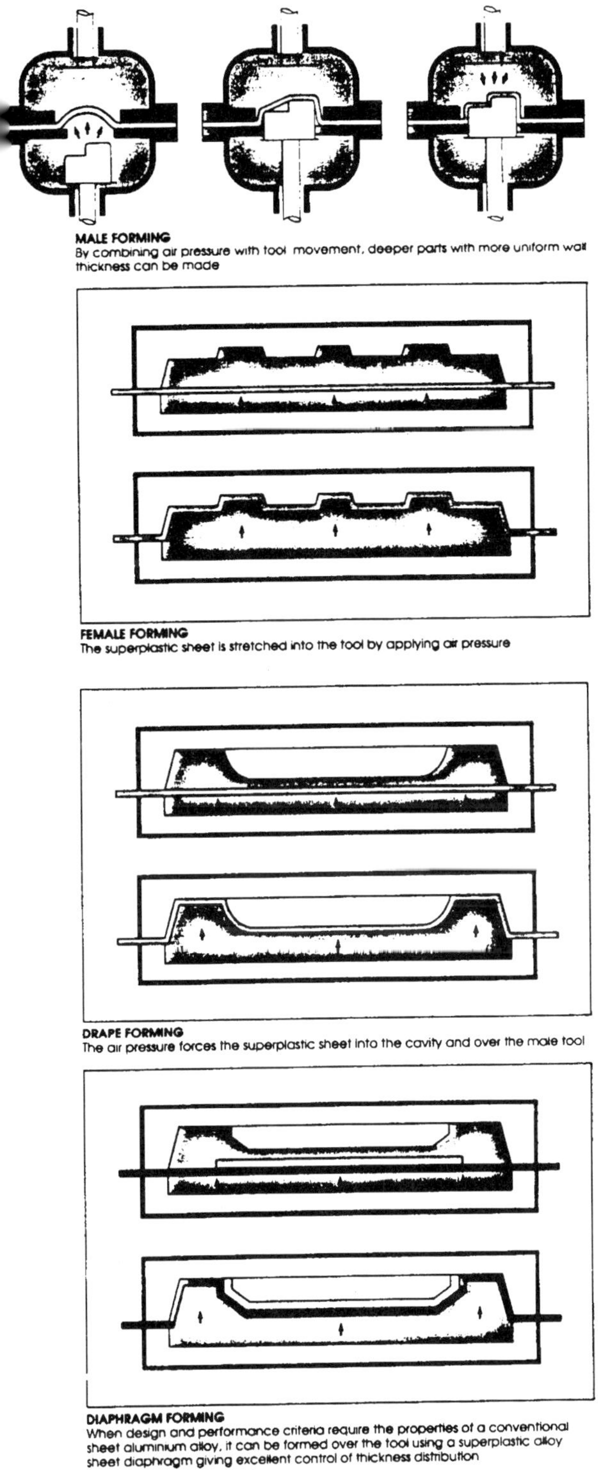

Fig 135: Alcan forming methods compared

forming. A popular alloy for superplastic forming is 78% zinc and 22% aluminium which has a room-temperature tensile strength comparable with low-carbon steel. When raised to 250C this alloy will become superplastic at a strain rate of 20 in (508 mm)/minute.

With properties in this state similar to thermoplastics the material can be formed by such methods as vacuum-forming — using relatively low-cost tooling. Alongside the ability to undergo large elongations in tension is the ability to conform well in compression such that very fine detail can be reproduced in a forging process, for example.

One approach to achieving the superplastic state is closed-die forming in which slugs of alloy are forced through conical dies to reduce the grain size prior to forming. Forgings thus produced will not fracture in a brittle manner like die-castings and properties shown in Fig 133 are typical. The resulting alloy can be fabricated by conventional techniques such as spot-welding, brazing and soldering while special techniques are available for fusion-welding the material to steel.

Aston Martin was a pioneer user in the industry, using the Supral alloy for the Lagonda luxury car from TI Superform. With this material both male and female forming can be carried out over a single tool using 10 bar air pressure at around 500C. Cycle time can be from 8 to 50 minutes depending on sheet thickness; sheets from 4mm down to 0.8 mm can be handled.

Spring-back problems are virtually eliminated with the process and panels are reproducible to a tolerance of +/- 1mm over a length of 2.5 metres. There is a limit on depth of drawer, however, of 60\% relative to panel width — with an overall limit of 400 mm using male forming and a draft angle of 2 degrees. Corner radii are critical, recommended values being given in Fig 134.

Superform Metals describe four superplastic forming techniques as in Fig 135. All operate at temperatures between 470 and 520 C. The company can provide a trimming service for formed panels, as well as the incorporation of machined cut-outs. Some interesting automotive applications include the sliding door for a Volvo ambulance shown in Fig 136 and a 4mm thick racing car bulkhead for March Engineering, Fig 137.

Another supplier is Custom Metalforms whose

product is branded Super Plastic Zinc (SPZ), with properties as shown in Fig 138. Their thermoforming process is seen in Figs 139-140 . The four hydraulic rams give 100 tonnes clamp load capacity and will form sheets up to 1.2 x 0.65 metres. Dies or moulds are usually precision sand-cast in aluminium alloy or iron, then finish-machined and polished.

Fig 136: Ambulance door

Fig 137: Racing car bulkhead

	Imperial units	SI units
Tensile strength (a)	64,000 psi	441 MPa
Elongation	10%	10%
Design strength (b)	7000 psi	50 MPa
Density	0.187 lb/in^3	5.3 gm/cm^3
Thermal conductivity	85 BTUft/hr/ft^2/°F	147 W/M/K
Specific heat	0.13 BTU/lb/°F	544 J/Kg/°C
Electrical conductivity	34%IACS	34%IACS
Coeff of Therm Exp	13.5 × 10^{-6}in/in/°F	25 × 10^{-6}°C
Melting range	790-930°F	420-500 °C
Electro magnetic shielding	Zinc offers protection against magnetic fields	-
Corrosion resistance	Zinc alloys have good corrosion resistance	-
Finishing	Can be plated, painted, buffed or left in the as-formed state	-

Fig 138: SP zinc properties

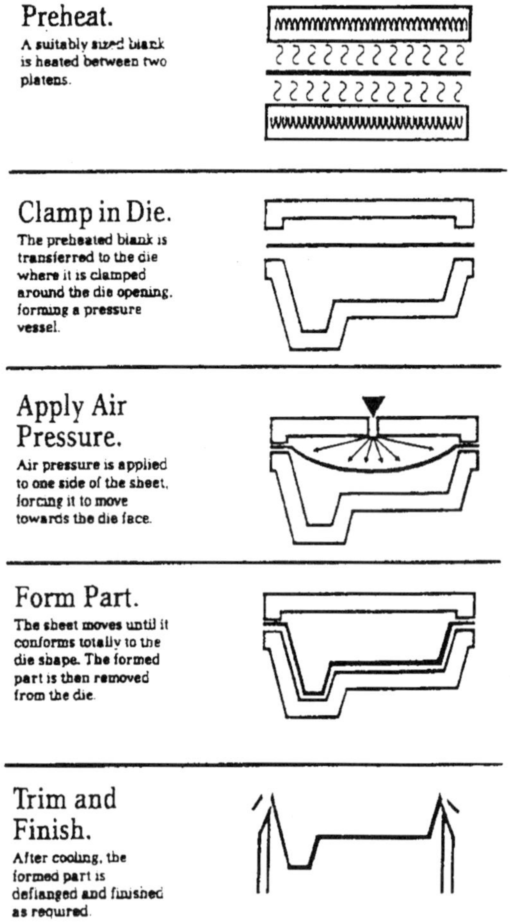

Figs 139-140: SPZ thermoforming

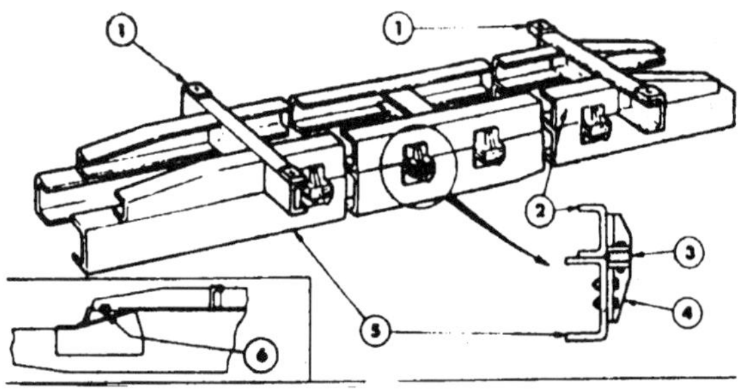

Fig 141:Chrysler truck fabricated by press brake and six press-braked parts of a Ford chassis (above)

Press-braking/roll-forming

Bending of metal sheet in one plane is carried out either by press-brake or roll-form in most chassis and bodybuilding situations. A number of successful specialist vehicles (Fig 141) have been designed for production in these techniques, saving the capital cost of draw-forming on conventional presses. Some guidelines are given here to optimise the use of the former equipment.

The bending of metal involves changing shape without foreshortening length or width so that the 'blank' area remains intact and no stretching takes place. Between the plastically deformed extreme fibres and the undeformed neutral axis are elastically deformed fibres which cause spring-back. This requires a degree of over-bend — necessary to reach an exact angle. There is also a shift of the neutral axis at the bend towards the compression side, of 0.3 to 0.5 times the metal thickness. An allowance has to be made on the blank to allow for the bend radius. For bend angle a and radius r, length allowance (normal to the bend) is $(a/180)\pi(r + 0.4t)$ when r is about 3 times the thickness t. For bending along a straight line the press-brake can be set to either 'air-bend' or 'bottom' the workpiece. The first case involves three point bending by the tip of the punch and the edges of the die while the second involves full contact by the total working surface of punch and die with the workpiece, Fig 142. Reasonably complex sections can be built up by progressive bend operations. Special tooling can also be used on press-brakes to obtain edge seaming and joggling — also tube forming. Load required to make a simple right-angle bend is a function of the die ratio (bottom tool vee opening/ metal thickness) which is optimum at 8. Above this value, load is reduced; when high tensile materials are employed, however, the ratio itself is increased. The value of the load (tonnes) is obtained from $8L_f t/D_w M_f$ where L_f is load factor; D_w is die width (mm) and M_f the material factor (which is 1 for mild steel, 2 for

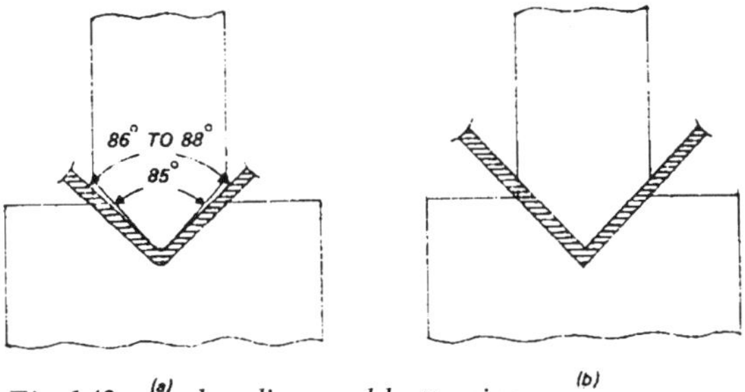

Fig 142: Air bending and bottoming

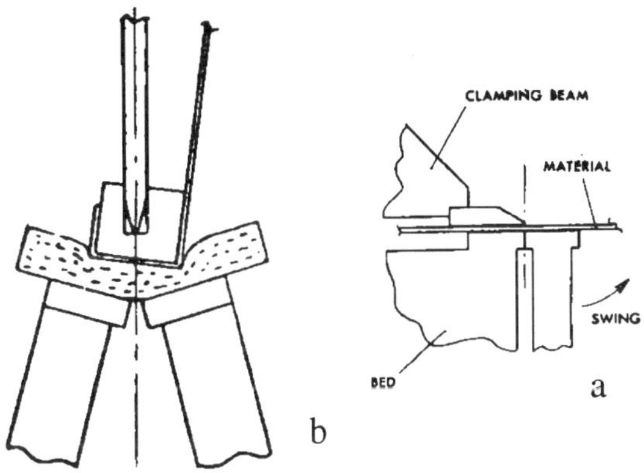

Fig 146: Folding machine and hydro-form compared

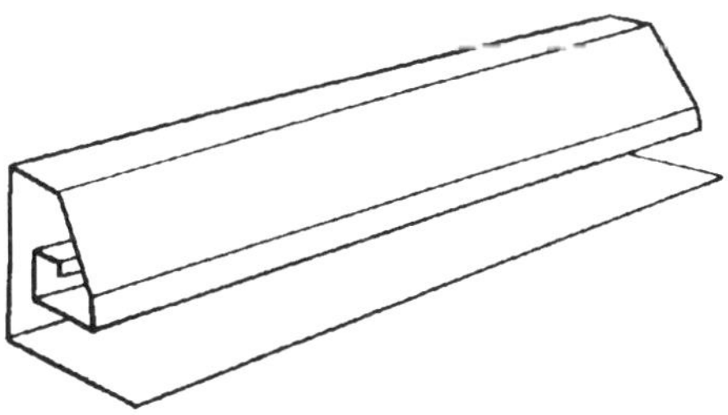

Fig 144; Part with five internal bends

Metal Thickness	0·91	1·22	1·63	2·03	2·64	3·25	4·76	6·35
Load Factor. F	2	2·5	3·5	4·25	5·5	6·75	10	14
Metal Thickness	7·94	9·53	12·70	15·88	19·05	22·23	25·40	--
Load Factor. F	16·75	20	27·75	33·5	40·25	46·75	53·5	—

Fig 143: Load factors

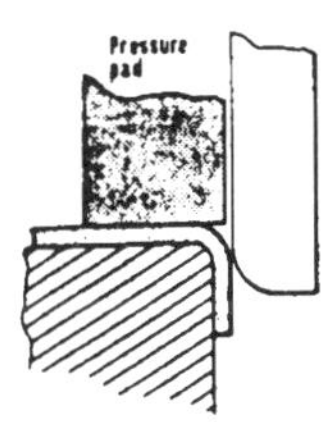

Fig 145: Simple right angle bend

aluminium and 0.67 for stainless steel). Load factors are shown in Fig 143 for different metal thicknesses.

A number of refinements have been developed for press-brakes which widen the scope of metalforming. The Belgium LVD company are reported to have supplied a machine to the US Caterpillar concern which has a 6 metre bed length. It has automatic thickness compensation built in and linear encoders each side of the ram to ensure parallelity and bending accuracy. Tool changing is also automatic and a plate support system is provided to prevent deformation during forming operations. LVD have also developed a CADCAM system which automatically calculates bending process parameters, after a part has been designed, and generates an NC program. Available on Promecam hydraulic pressbrakes is a CNC unit which provides synchronised back gauge and bending depth control. A typical product which can be made at one handling, with 8 internal bends, is shown in Fig 144.

Trial-and-error is often the best procedure for obtaining optimum set-up in press-braking. Even on the moderately simple shape of Fig 145 any 'rule' for producing the right angle bends would be subject to exceptions. Minimum bend radius, for example, can vary from $t/2$ for soft material up to $7t$ for full temper steels. This applies, however, only if the bend is across the grain; should the bend be with the grain, these radii would cause cracking and would need raising by 50 to 100%. According to whether an open-vee or 'sidedrawing' tool is used, the length allowance for the bend could vary considerably.

An alternative to press-braking is hydro-forming which involves lower thrust forces to achieve bends. While a press-brake overcomes the restrictions of a simple folding machine (Fig 146a), the hydro-form can operate within the same space freedom by with lower loads, Fig 146b. This is because the load is applied to the swinging wings, or folding beams, of the machine. The beams are located in trunnions which pivot on a fixed fulcrum pin, giving a sweeping action on the material, making it ideal for coated and polished workpieces.

Roll-forming

This technique is of course used to obtain higher speeds of bending of strip stock — but will usually involve longer set-up times in positioning rollers for a particular profile. The strip is pulled through and

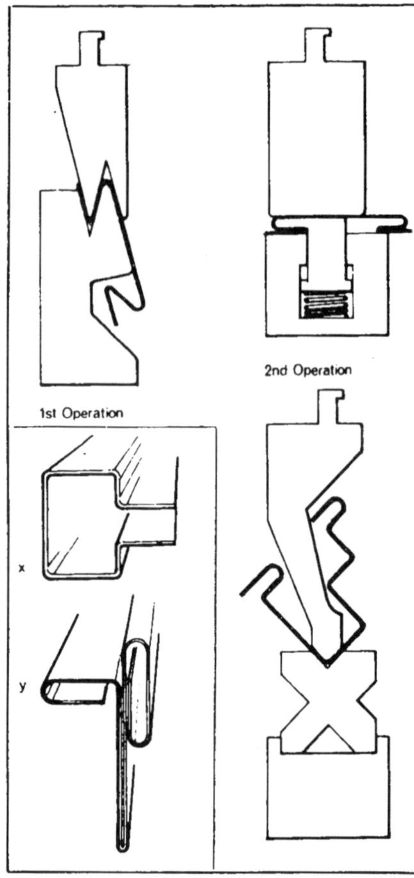

Fig 147: Roll-forming

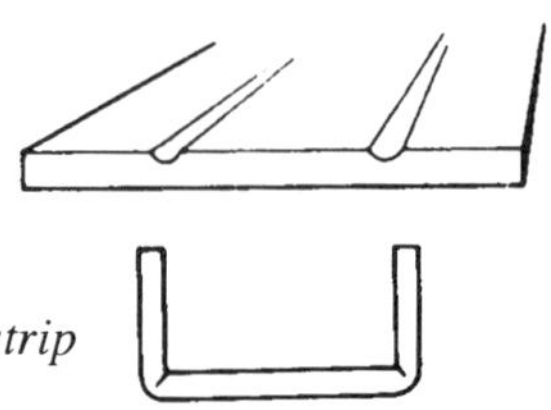

Fig 148: Pre-grooved strip

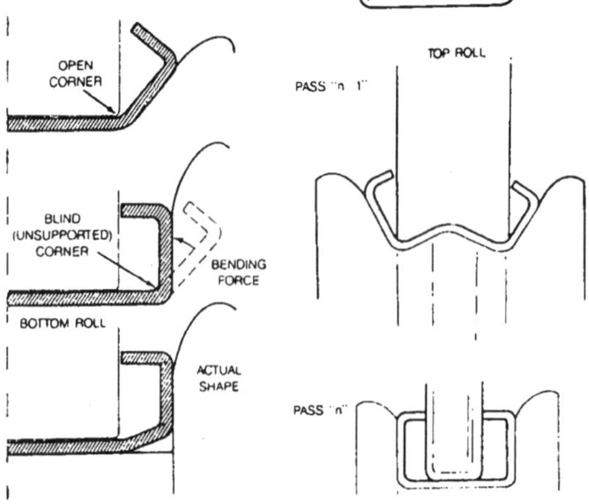

Fig 149: Secondary rolling

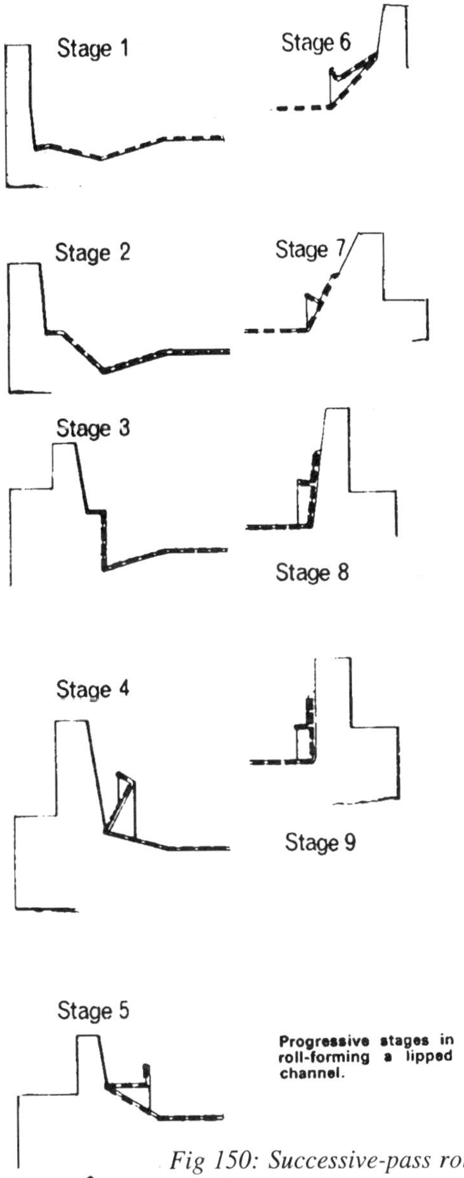

Fig 150: Successive-pass rolling

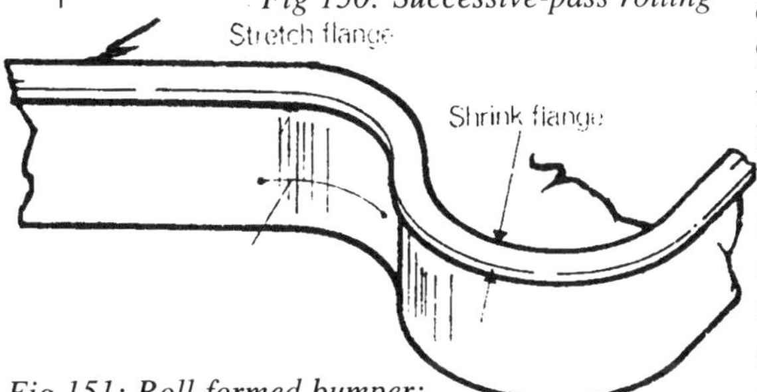

Fig 151: Roll-formed bumper:
Stress puckers will form in the bottom at the point arrowed unless control is exercised to make the metal exceed the plastic limit

rolled into the desired section shape simultaneously. Roll speeds of about 20-70 metre /min are typical for mild steel of 0.15 to 10 mm thickness. Forming lines may incorporate pre-notching and/or pre-punching presses to achieve profiles with shaped holes and cut-outs; alternatively the rolls can be fed with pre-cut blanks. Beyond the rolls, cut-off presses can also be adapted for extra operations such as embossing. Fig 147, due to AISI, shows the step-by-step formation of a typical profile. Note that while the corner line travels in a straight line, the edge is constrained to a helical path which involves some stretching and recompressing. Care must be taken in design to avoid any deformation of the finish part duc to locked-up strains which result.

As well as pre-judging the flow of material as it passes through the mill, the tool designer must also consider the shape of the finish section with respect to the lengths of unstiffened projections — also the orientation of the workpiece as it emerges from the mill. Spring-back effects must also be allowed for and minimum radii with respect to different materials have to be observed. It is sometimes possible to roll below these minima if the strip is pre-grooved as shown in Fig 148. When the shape of the section is such as to preclude access of the rolls to certain corners, techniques of secondary rolling, as shown by Fig 149, may be adopted to straighten back the section.

Complex sections will be rolled in successive passes as shown in Fig 150. Working tolerances quoted by experienced formers are +/- 0.010 in (0.25 mm) on the profile and a length L to straightness factor of $L/300$. Bend radii down to half sheet thickness are possible without grooving on hot-rolled grades of steel and zero radius for cold-rolled grades of deep-drawing quality to BS 1449. Steels up to BS 968 grade have been successfully cold-rolled when very high specific strength ratios have been required. Bumper bars are often roll-formed but a preform may be produced by press-forming, for wrap-around designs — the preform being subsequently rolled in a machine having one roll replaced by a form-block slipper. The straight preform is clamped to the slipper on its front profiled edge and a matching form roller applies reaction pressure against the form-block which is itself rotated to form, simultaneously, the bumper in cross-section/contour, Fig 151.

Flex-forming

Sheet metal forming with flexible tools can offer potential benefits to short-run work, with the possibility of reducing die-cutting cost as only one half the die need be made. Recent developments of the flex-forming process now allow relatively complex shapes to be formed and high accuracies required of prototype production can be realised.

Either the male block or the cavity can be made from relatively inexpensive materials, the other tool half being a flexible rubber diaphragm. In the case of the system now being sold by ABB, pressurised hydraulic fluid actually forms the metal at the other side of the diaphragm and presses are available for forming pressures up to 2000 bar. The industrial realisation of the process has depended on the successful reaction of the very high thrust forces these pressures inflict on the press. This has been achieved by enshrouding structural parts in pre-stressed wire, so that they react with the forces without change in shape which would impair accuracy of forming. Fig 152 describes the process in which the blank (2) is placed on the rigid tool-half (3); the pair are then run into the press cylinder where the diaphragm (1) is located. Oil pumped into the fluid cell forms the blank by wrapping it around the rigid tool half, even into undercuts. Presses with trays up to 2.0 x 4.0 metres have been built.

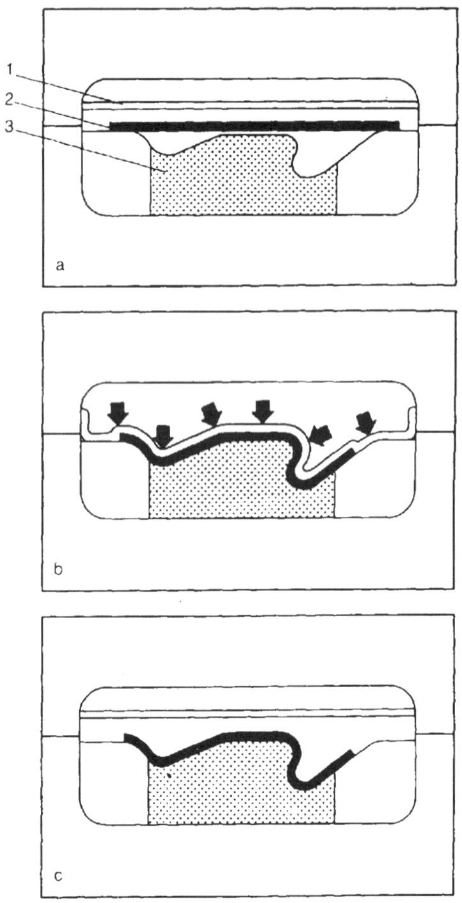

Fig 152: Stages in the flex-forming process

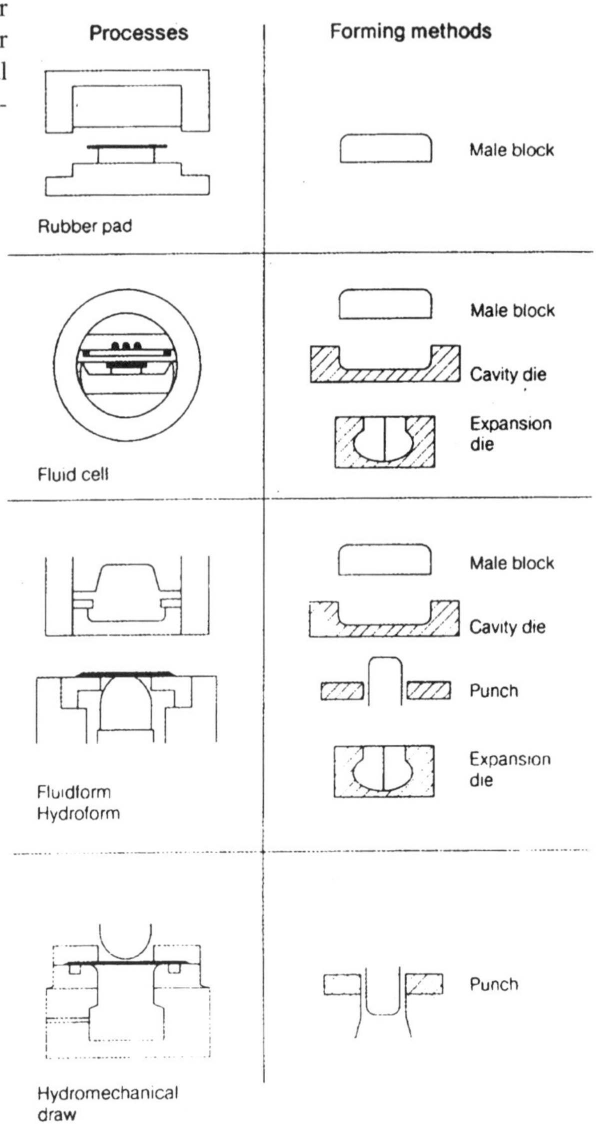

Fig 153: Fluid-forming comparisons

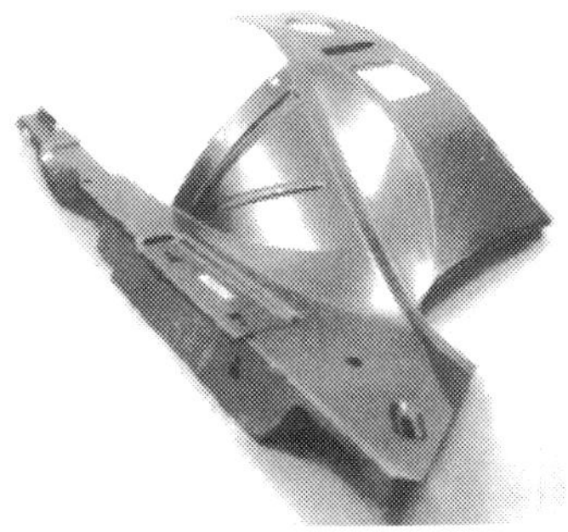

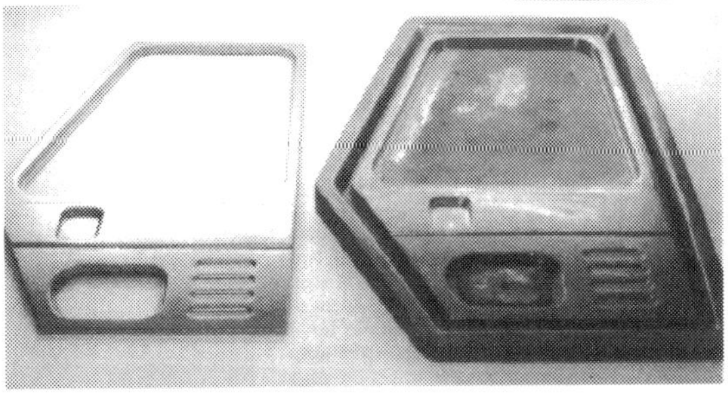

Fig 154: Rear fender and door pressings

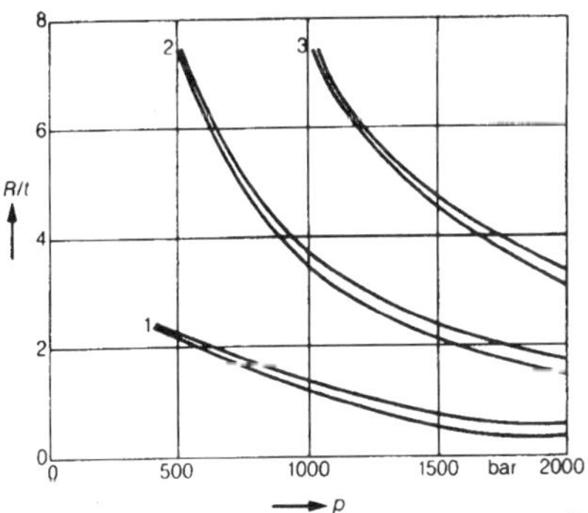

Ratio of radius *R* to the sheet metal gauge *t* for different materials as a function of the forming pressure *p*

a Cavity die
b Punch
1 Pure aluminium
2 Deep drawing steel
3 Stainless steel

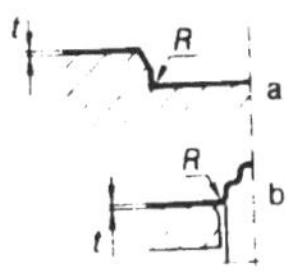

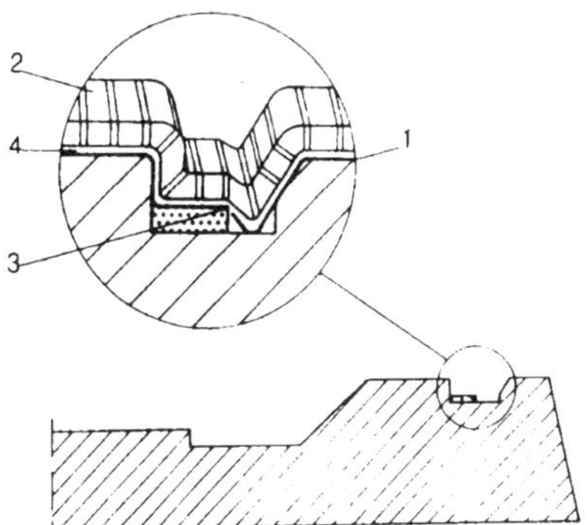

Fig 155: Integration of trimming

How trimming can be integrated in the flexforming process. The blank (1) is stretched over a sharp cutting edge (3) by the diaphragm (2) until the excess material is torn off. The diaphragm is protected by a layer of rubber (4).

In prototyping work, sheets of different blank thicknesses can easily be tried out. There is also the advantage that in-built stresses arising from hand-forming are largely avoided and 90% cost reductions have been claimed as well as halving of lead times. Series production runs of several thousand units are also feasible if complexity of shape is not too great and provided a tougher tooling material for the rigid half is used. Choice of the tool material is also governed by the blank material and minimum forming radii.

Considerably greater depth of draw is achievable with the rubber diaphragm process than earlier fluid forming processes which used soft rubber pads. Sometimes more than one hit may be necessary for deep draw and, exceptionally, use of the Fluidform rather than Fluidcell process may be necessary. The Fluidform press has a telescopic ram system carrying a moving punch, the diaphragm being carried in the circular pressure unit. A similar process but without the diaphragm is hydromechanical drawing; Fig 153 shows the processes compared diagrammatically. Two different forming radii are involved in flex-forming, that by the rigid tool half and that by the diaphragm. The latter is known as the free-form radius and governs the pressure required. Mild pressing steels generally need pressures below 1400 bar. Pressure intensifiers are used to achieve these higher pressures over the 200 bar 'system' pressure. Cycle time of the operation is normally 1-3 minutes, depending on forming pressure — but productivity can be increased by using pallets pre-loaded with tools and blanks for simultaneous processing.

Fig 154 shows the rear fender of the BMW Z1 roadster made in galvanised 0.8 mm steel. It is formed in several steps, with intermediate trimming, and uses a pressure of 700 bar.The rigid tool half is made from Kirksite low melting point zinc alloy. The latest technique is to integrate the trimming of the sheet component. Here the rigid tool half has a cutting edge which trims the metal under the pressure of the diaphragm, a protective layer of rubber being inserted between diaphragm and blank, Fig 155.

Other potential benefits of fluid-forming processes, generally, are freedom from elongation of the blank when the part is open-ended and avoidance of surface scuffing. The latter is of importance in the pressing of coated steels or pre-printed panels.

Spinning and flow-forming

Very large circular forms and those requiring heavy metal gauges can be expensive to produce on press-tooling but relatively cheap by spinning. Here the basic parameters of the method are considered alongside the slightly more sophisticated process of flow-forming.

Both spinning, and the related flow-forming process, are used in the production of many other vehicle parts, from wheels to air-tanks. The advent of computer numerical control has enabled both low and high volume manufacture of spun parts. Tooling costs are relatively low compared with presswork and prototype production is a particularly attractive proposition, Fig 156; the range of sizes available from Metal Spinners (Newcastle) are seen in Fig 157.

Spinning is basically a technique of kneading sheet metal about a rotating form into a desired circular shape. The process can be carried out on a lathe with suitably large swing clearance and the form can be made of wood for low production quantities. The form is fastened to the headstock and the blank pressed against it by the tailstock. Blank and form are rotated at high speed and the operator, beginning at the centre, works the metal against the form with blunt ended hardwood or steel rods. Tolerances down to 1/8 th in (3.2 mm) can be held on work over 4 ft (1.22 m) diameter. Thicknesses are limited by the strength and work-hardening of the material.

In flow-forming the spinning blank is formed by shaped rolls. Wheel discs for drop centre wheels have been made for some time by the method and more recently discs for taper bead seat and tractor wheels are being produced. Rotational cold forming is another name given to the process and continuous reduction in wall thickness can be obtained from base to edge. Wheel discs thus produced have accurate diameter and section profile, good surface finish and low out-of-balance. Blank tolerance has no effect on finish-part tolerance.

The blanks are centred in front of the mandrel by an automatic feeder and clamped by a tailstock pressure pad. Two or three work rollers flow-form the blank, Fig 158. Rollers are either controlled by copying attachments which follow templates — or more recently by CNC. When finish thickness is reached, rollers are retracted, the work saddle drives to its limit position, the main spindle is braked and the disc stripped from the former. Saddle feed rates are particularly important. Since feed rate in a radial direction is governed by the axial feed rate and the copying profile, forming relationships change in the course of one pass. Radial feed must be matched to the degree of deformation and thus machines are provided with as many as four axial rates, switchable from a control panel.

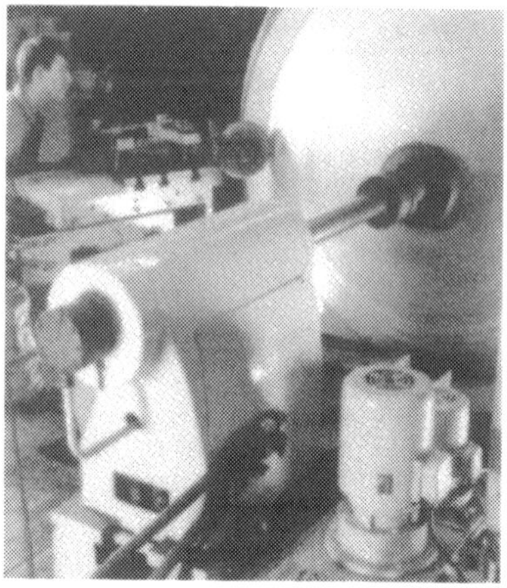

Fig 156 (left): Tanker dished-end production

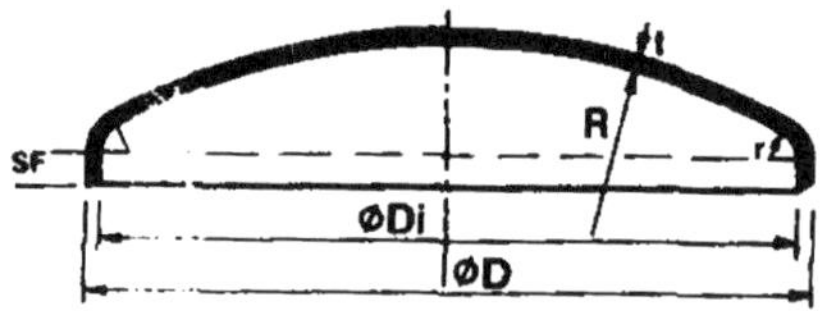

MAXIMUM BLANK DIAMETER: 4500mm
MAXIMUM THICKNESS MILD STEEL: 16mm
MAXIMUM THICKNESS STAINLESS STEEL: 12mm
MAXIMUM THICKNESS ALUMINIUM: 40mm

Spherical Radius	Corner Radius	Corner Radius
1700mm	35mm	0.5/8"
1900mm	38mm	0.3/4"
2100mm	40mm	1 1/2"
2300mm	45mm	4"
2600mm	50mm	6"
2900mm	115mm	8"
3200mm	150mm	
3600mm	200mm	
4000mm	225mm	
	250mm	
	300mm	
	325mm	
	350mm	
	400mm	
	450mm	

Fig 157: Dished end dimen-sions for a main supplier to the tanker market

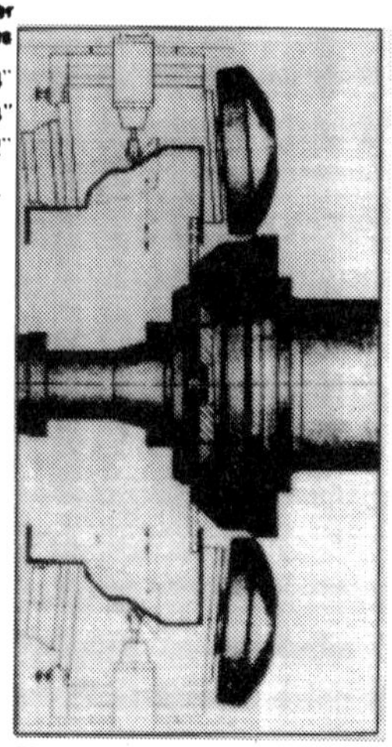

Fig 158: Flow-forming

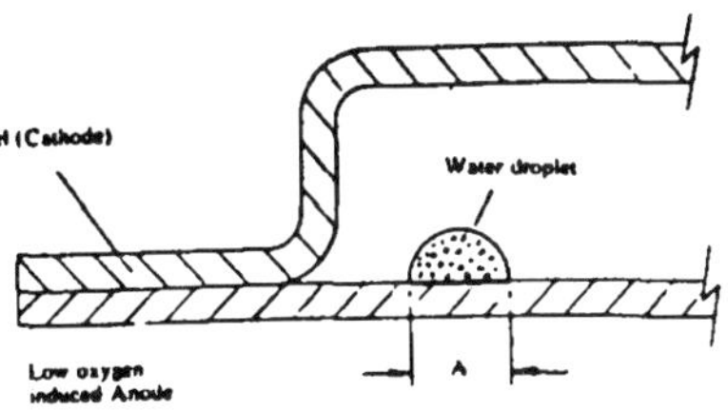

Fig 159: Cathodic corrosion

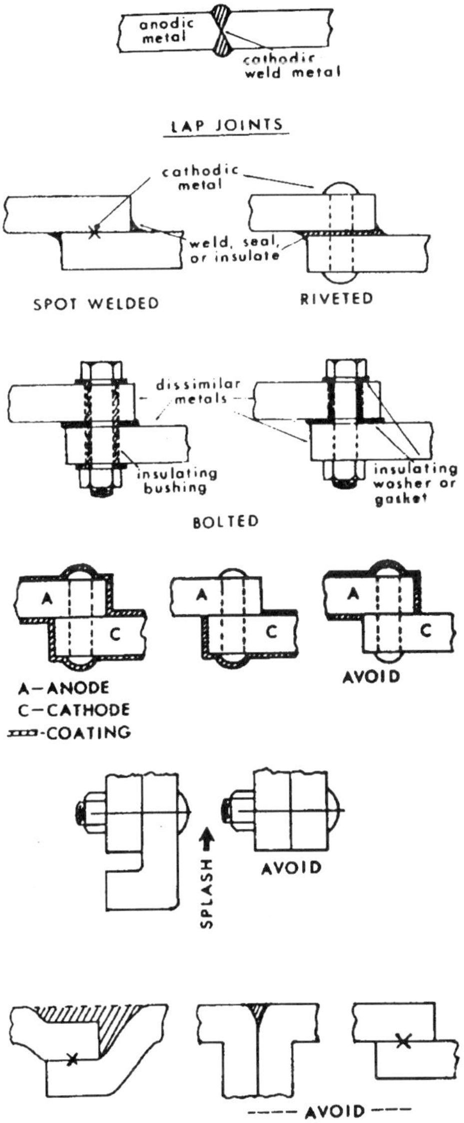

Fig 160: Do's and don'ts of joint design

Body finishing techniques for corrosion prevention

An understanding of the basic mechanisms of corrosion is useful for the selection of the appropriate treatment. Not only steel, also aluminium alloys and even stainless steel need appropriate protection. Careful attention to joint design and 'ventilation' of box members reap valuable rewards in subsequent service of the vehicle.

Corrosion is generally understood to be a nature-cycle phenomenon of a metal slowly reverting to the state in which it occurs as a natural ore. The corrosion process is electro-chemical as in a battery — with the dilute acid arising from road salt and atmospheric carbon/sulphur dioxides dissolved in water. As in a battery, the anode dissolves and the cathode collects positive ions from the electrolyte (acid). A corrosion 'cell' of this sort can be within a single drop of impure water (dilute acid) on the surface of the vehicle body. The area beneath the drop is starved of oxygen and becomes anodic while the surrounding area becomes a cathode, Fig 159.

Serious body corrosion is generally accepted to start from the inside on unprotected areas of metal and then work outwards to the paint film, which blisters. Two techniques used to protect the interior surfaces are 'wax' spraying and sacrificial protection. The latter involves using a coating such as zinc (more negative in the electro-chemical series than iron), on the surface of the steel, thus promoting cathodic action and protecting the steel by sacrificing the zinc.

Spraying with petroleum- or waxed-based inhibitors is a valuable technique — especially when bodywork has also received an electrocoat primer. The last process involves making the body an anode of a dc electric circuit and the paint tank the cathode. The specially formulated pigment blend is electro-conductive and tends to deposit evenly on all exposed surfaces.

Besides low-oxygen corrosion, the problem of crevice corrosion is prominent in vehicles with a multiplicity of spot welded flanges. Paint molecules can be simply too large to enter joint crevices and subsequent 'working' of the mating surfaces in service can promote the sucking in of moisture by capillary action. Another means of attack is by pitting corrosion caused by rupture of the paint film by stone chipping. Lastly, galvanic corrosion caused by metal-to-metal contact of metals wide apart in the electrochemical series, is another common enemy.

While aluminium alloys have good resistance to corrosive environments, due to the protective surface oxide film, this can be breached by acids formed from de-icing salts and subsequent pitting corrosion occurs. The alloys are also anodic to most metals and therefore good protection is required against galvanic corrosion. Stainless steels offer even better protection — again due to formation of an oxide film which itself requires some protection against local rupture.

Joint design for corrosion prevention

Avoidance of both crevice and galvanic corrosion is necessary in joint design. Crevices should either be eliminated or shielded and insulation should be provided to separate dissimilar metals, with preferably the surface area of the cathodic metal being less than that of the anodic one. Fig 160 gives some 'avoid' and 'adopt' situations while Fig 161 gives pictorial advice on paint access and drainage conditions.

Surface treatment to improve paint finish

Work reported by Nissan[5] has shown the manufacturer to be exploiting laser-textured dull-steel in improving paint finish on its vehicles. Working with Kawasaki Steel, the company has overcome the problem of shot-blasted dull-steel, used to obtain suitable surface roughness for improving formability, by rolling the steel strip between laser-processed rollers.

The criteria shown in Fig 162 were used for assessing painted surfaces, the company having noted that glossiness and smoothness accounted for 70% of the subjective judgement of paint appearance. A fine dull surface finish is traditionally recognised as providing better image clarity but too fine a surface gives rise to galling and handling problems during forming. Frequency analysis of several samples of auto-

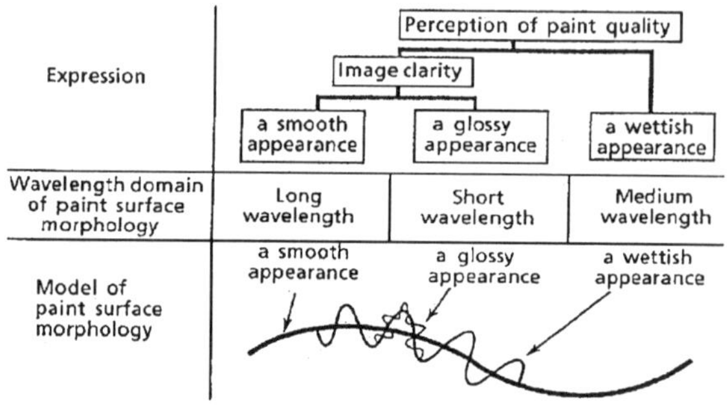

Fig 162: Elements forming perception of quality

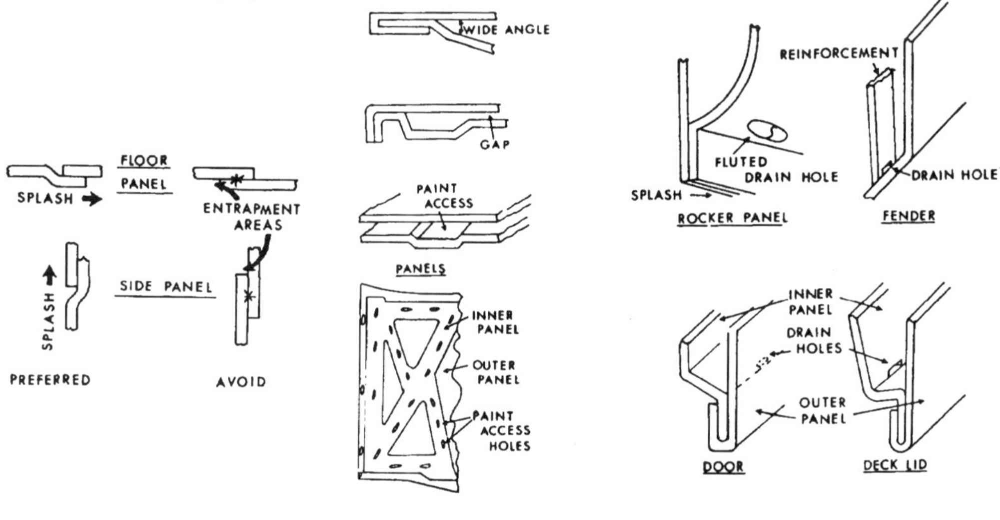

Fig 161: Draining paint from closed sections

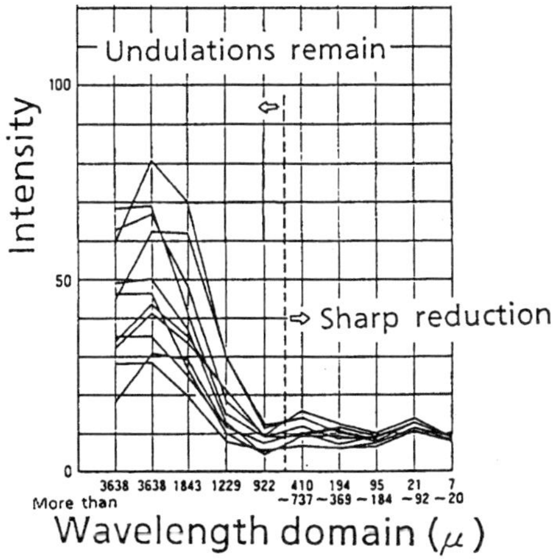

Fig 163: Frequency analysis of finish-painted steel sheet

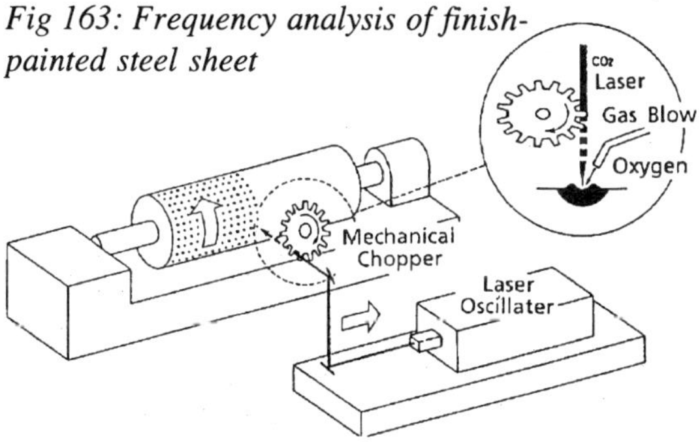

Fig 164: Roll-engraving by laser beam

Specimens	YP (N/mm^2)	TS (N/mm^2)	El (%)	n	r	SRa (μm)
Low Type of Laser-Texture						0.72
High Type of Laser-Texture	171	298	47	0.25	1.80	2.36
Shot-blast						1.17

Fig 165: Sheet specimens tested

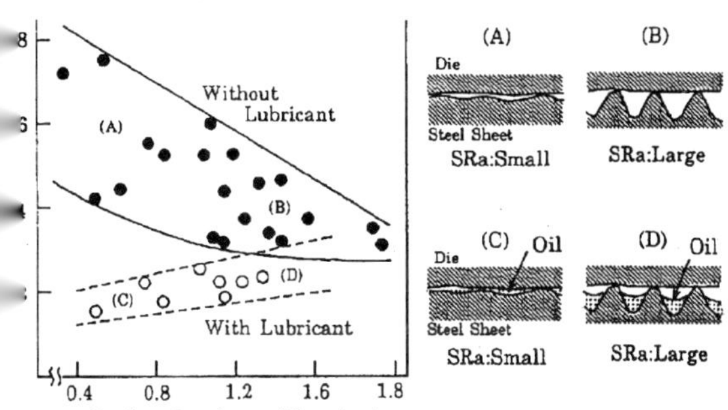

Fig 166: Friction and roughness relationship

motive sheet steel showed that sheet with surface undulations component were retained in the paint film surface as seen by Fig 163.

When conventional sand-blasted rollers are used, these undulations were difficult to eliminate because of the random pattern of craters produced in the roller surface by the process resulting in wave-lengths over 800 microns. By using a laser beam to engrave the surface, using the process shown in Fig 164, a regular pattern of craters can be obtained, which resolves the problem by producing only short wave-length undulations.

To test the effect on formability the company used the specimens shown in the table of Fig 165 where *SRa* is the arithmetic mean deviation of the surface profile accounting for the area, Fig 166 showing its relationship to friction co-efficient. Trials with and without press lubricant showed that without it, galling is less likely when the area of contact between the die and the steel sheet is smaller; with lubricant, the latter is more effective when the peaks and valleys of the sheet surface are smaller. Fig 167 showed the results of deep-drawing tests in which the low-roughness type of laser-textured steel showed best formability, attributable to its lower friction. The practical case of an inner wheelbox panel, Fig 168, was also used to examine both formability and galling. Similar conclusions were reached. The effect of surface roughness on fracture behaviour was also examined on the basis of the reduction in panel thickness of the side wall corners where risk was greatest. The results in Fig 169a indicated that high-roughness type of laser-textured steel sheet allowed a large decrease in panel thickness even at a low blank-holding force, suggesting that this steel would be susceptible to fracture.

Galling tests were based on stamping 100 consecutive panels using a medium roughness laser-textured sheet and two other types, having rejected the high roughness one on grounds of fracture proneness. All three samples passed successfully. The effect of varying the amount of metal flow was also investigated, the results being shown in Fig 169b. Here the vertical axis indicates the length of the flange end *L1* along the long side of the panel. This length increased with increased number of stamping, making metal flow more difficult. Results indicated that steel with a higher *SRa* value are somewhat superior in respect of anti-galling performance.

Overall, the summary in Fig 170 showed that to improve paint image clarity a certain degree of surface roughness and crater density is required, better formability being obtained in the region of high crater density and low surface roughness. Anti-galling property is improved in the region of relatively rougher surface finish.

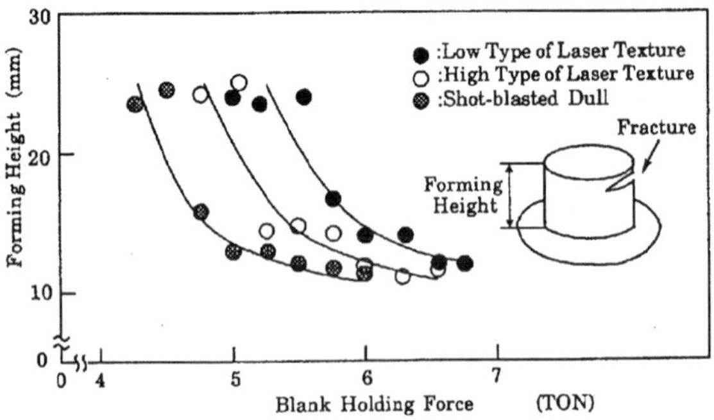

Fig 167: Deep-drawing test results

Fig 168: Inner wheelbox panel

Finish paint techniques and equipment

Painting is an area in which the volume vehicle builder is likely to lead the specialist builder in new techniques and processes. Here we examine some recent developments which could apply to both 'sides' of the industry and apply respectively to metal and plastic/GRP substrates. In recent years, emphasis has been on making metal bodies resist rusting out rather than making top coats more durable. This has led to phosphating plants, dipping tanks and electropaint.

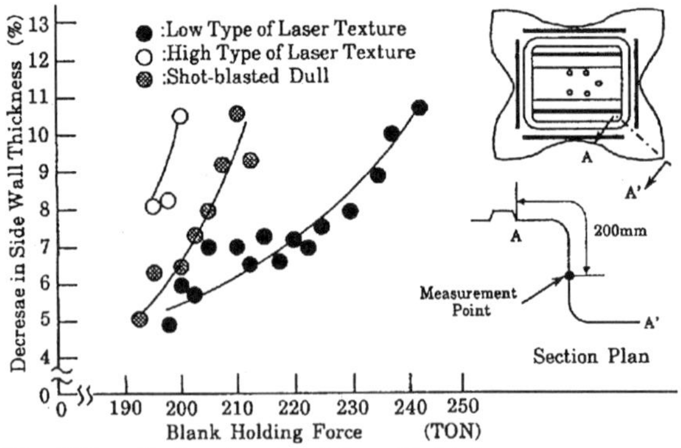

Fig 169a: Reduction in sidewall thickness

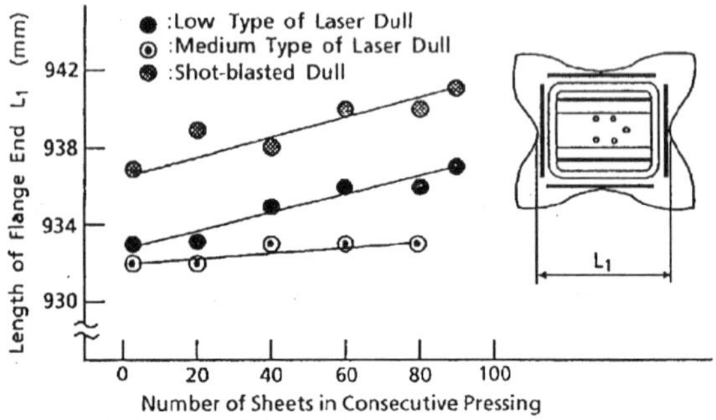

Fig 169b: Metal flow change in successive stamping

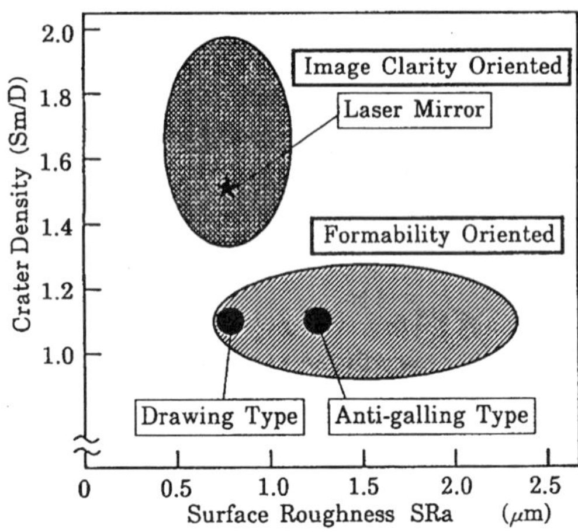

Fig 170: Application of laser-textured steel

Basic components of a paint system pigment, binder and solvent are worth considering separately for a proper understanding of comparative systems Pigment provides both colour and opacity (hiding or covering power) but can also contribute to durability of both finish and priming coats. In the priming coa the pigment's main task is to resist corrosion and promote adhesion. Binders are the film formers which also contribute to the substrate bond as well a binding the pigment together. The solvent makes the mixture of pigment and binder thinner for ease of application and evaporates by 'air flash' at room temperatures or by heating up to 80 C panel temperatures for quick dry.

Broadly, paint types divide between cellulose synthetics, oil-based, low-bakes, two-pack materials and thermosetting acrylic enamels. Celluloses require multiple coats because of the low thickness of each film — also solvent retention can cause softness for some time after application. Oil and synthetic based paints rely on final film hardening (arising from oxidation of the paint) and re-coat periods can be critical for subsequent rectification procedures; however their high solid content gives them good covering power. Low-bakes involve a full cure of up to 60 minutes followed by a cooling down period during which oxidation makes the materials non-thermoplastic. The 'twin packs' require the paint, and a separate hardener (usually a toxic isocyanate), to mix together for the cure. An unusually hard film results which can be force-dried to allow the vehicle to enter service quickly.

The most common system in use by the volume builders is thermosetting acrylic enamel based on the chemistry of methylol acrylamide and hydroxylated acrylic monomers. The latter vary in chemistry according to their four main contributions of either providing: hardness, flexibility, cross-linkages or accelerated cure. The systems are suitable for the formulation of metallic finishes — due to the high order of gloss, polishing and weathering qualities. Melamine resins are now commonly used to accelerate the cure down to 30 minutes at a stoving temperature of 120-130 C. A key advantage of the system is the ability to reflow during 'repair in process' rectification procedures on the assembly lines of the major OEMs. Two such procedures are bake-sand-bake and bake-and-respray, both involving initial bake at 70-100 C during which solvents evaporate but the material remains thermoplastic though hard enough to be rubbed down. After tak-wiping, in the

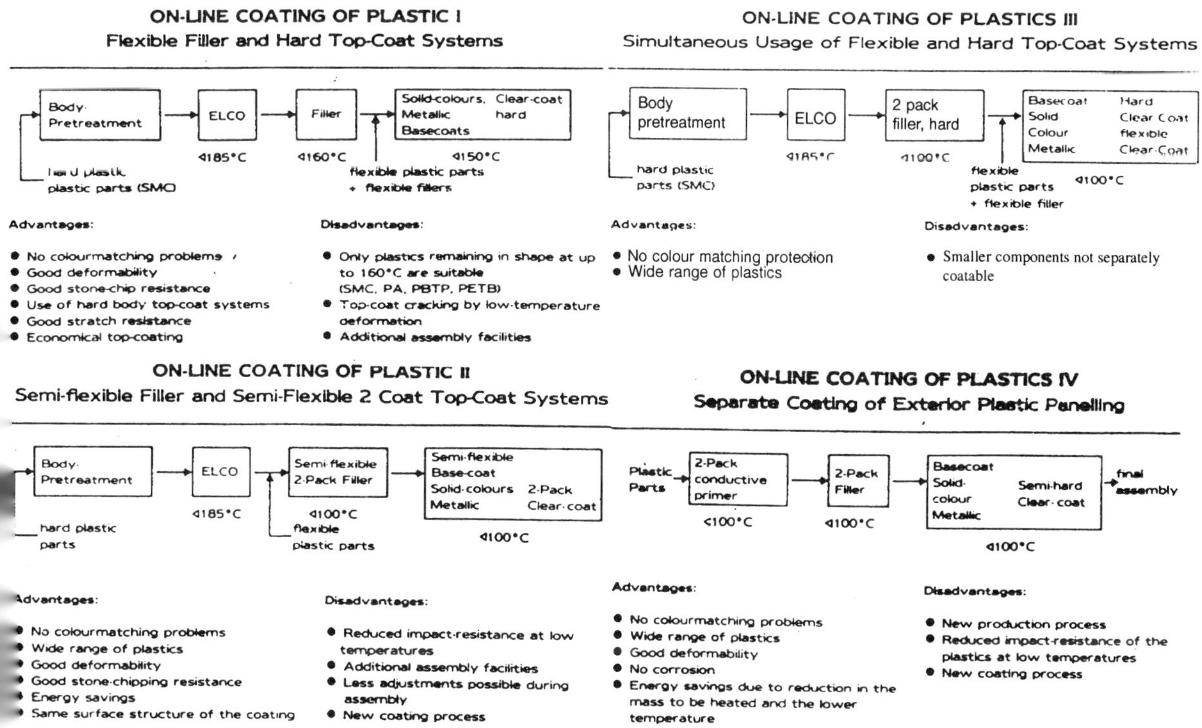

Fig 171: Azco plastic painting process

first procedure, restoving takes place at 140 C during which reflow hides the sanding marks — while in the second these are resprayed over without the need for masking.

Ford spec ESB.M32J.102A released in 1967 is widely accepted as a standard for reflow performance. Hydroxylated acrylic finishes provide an ideal carrier for the aluminium flakes used in metallic finishes and are superior to the alternative alkyds. Key to good reflow performance with metallics is the insurance that flakes are evenly dispersed initially.

Powder painting systems

The first materials used in powder coatings were the epoxy resins, applied at high air pressures to achieve film thicknesses of 100-125 microns after 30 minutes of stoving at 200 C. Now a wide range of materials can be sprayed with film thicknesses down to 35 microns, involving a 10 minute cure at 160 C. Polyesters are now the common coatings and usually entail only a single coat system. Generally, mechanical properties are superior to wet painted coatings and a variety of machining and forming processes can be applied to the finish panel. Resin, pigment, hardener and additives are pre-mixed as solids and fed to an extruder, the barrel of which is heated to a temperature at which the mix melts. Cooled extrudate then passes through a kibbler (crushing machine) from which pellets are subsequently ground to a powder prior to spraying. Total cleanliness of the substrate is necessary, including the removal of oxide films in the case of aluminium alloys.

Future painting developments?

There is obviously considerable impetus to reduce the number of stages in any painting process for vehicle bodies. Increasing the thickness of electrocoat primer films from the current 17 to around 35 microns is one way of reducing the number of subsequent coats. A single coat E-coat primer is already in view — for application after phosphate pretreatment. Next stage will be to eliminate primer-surfacer films. Other possible developments include a twin-coat metallic finish involving flashing-off between base and clear coat as well as paints which are sympathetic to both metallic and plastic substrates.

The Netherlands company Akzo Coatings have been prominent in the development of paints for plastic substrates. Problems to be overcome in painting of plastic include surface etching by solvents, mechanical strength reduction, reduced low-temperature flexibility, differential thermal expansion and blister formation. Successful twin-pack flexible PUR coatings are available for plastic panels of nylon, polycarbonate, PUR-RIM, PBTP and polypropylene; there are also special systems for GRP and ABS. Akzo point out that flexible paints can never quite rival the finish of hard coatings used on metal substrates.

For a part like the ZMC tailgate of the Citroen BX car, a post-coating of 2-pack epoxy based conductive primer is used which is dried for 20 minutes at 140 C. The part is subsequently coated with a conventional filler and top-coat then baked at 140 C. Next stage in development will be an in-mould coating which is targeted to harden within 60 seconds. Fig 171 shows the process procedures under development by the company. A variation of the fourth is used for the BMW Z-1 Roadster-special. This involved the so-called Varioflex system which suited the different types of plastic used on different panels of the body.

Pre-coated metal sheet subsequently formed into body panels is a further line of development. Another is the use of paint-film 'prepregs' in lieu of sprayed paint. According to Bruce Johnson of 3M's Automotive Systems Laboratory, the system shown diagrammatically in Fig 172 has been produced with a conformable paint coating, reinforced with a flex-

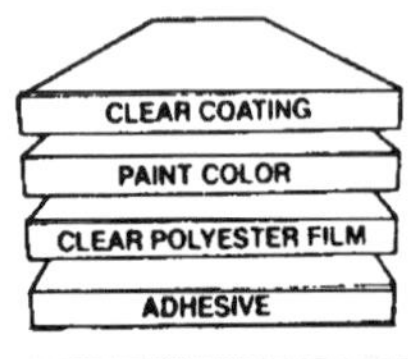

Fig 172: Paint film prepreg

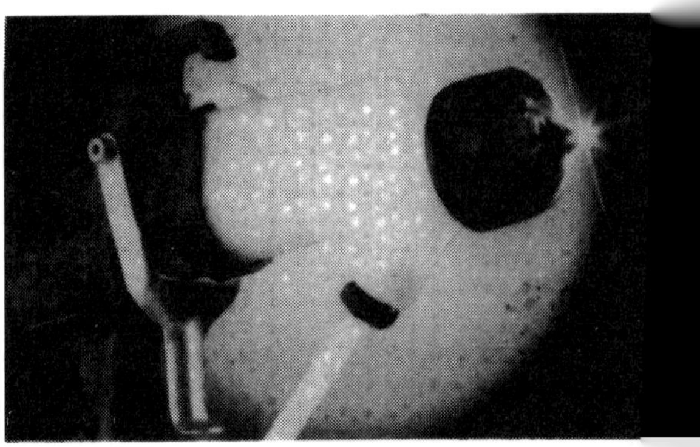

Fig 173: Electrostatic spray gun

ible layer and bonded with a hot-melt adhesive. Tests carried out by the company have shown that such dry films actually perform better than sprayed paint for certain attributes such as resistance to impact and road chippings. Application methods for dry films include vacuum thermoforming, for porous parts, and compression moulding for other 'solid' parts. For E-coated steel panels, bonding is carried out prior to forming — with the carrier film removed after the operation.

The DeVilbiss Company are pioneering the use of electrostatic paint application systems in the medium volume automotive industry and recently introduced the Aerobell rotary atomiser which operates on air-bearings at 60,000 rpm. The company have also introduced a manual spray gun with the benefit of electrostatic operation, Fig 173. The EFX-100 receives a low-voltage input which is converted to 20-80 kV for maximum transfer efficiency. The guns will apply solvent-based coatings using either air atomisation or air-assisted 'airless' systems and water-borne coatings using air atomisation.

Becker Industrial Coatings have introduced a new application process for polyurethane paint known as Vapour Injection Curing. It is suitable for coating three dimensional objects in either steel, plastic, aluminium or wood substrate. The atomising air carries a tertiary amine gas catalyst which mixes with the paint on release from the gun-head — to give accelerated curing speeds, Fig 174. A new family of PUR coatings has also been developed for the process including primers, pigmented and base/clearcoat finishes.

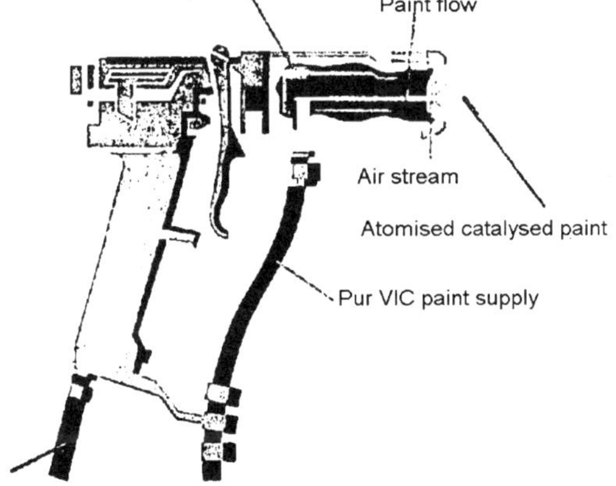

Fig 174: Vapour injection curing gun

Body sealants

The demarcation between gap-filling adhesives and sealants proper is blurred. Compared to adhesives, sealants have less cohesive strength but unlike adhesives may be required to have such properties as non-sagging and ability to adhere to oiled substrates. The need to match formulation to specific duty is considered here

In volume car and CV manufacture many sealants are putty mastics which can be pumped to different parts of the production lines for application by extrusion-guns, either manually or robot-operated. Lower volume specialist vehicles can benefit from particular sealants such as those in uncured tape form which can be cured either by heat or chemical activators.

Most sealants are, of course, of mastic consistency and are based on filled or unfilled elastomers — with thixotropic agents added to prevent sagging. The elastomeric binder is an important constituent as it largely determines mechanical and physical properties. Pigments are added for colourising, obviously, but their presence also increases density to reduce flow tendency. Conversely, solvents may be added to achieve a more pourable consistency.

Two basic groups of elastomer-sealants are the high and low molecular-weight categories. The low weight materials are generally one-part gun-sealants. The higher weight materials are used to compound ribbons and tapes or are dispensed with two-part guns in which the elastomer changes after mixing from a syrup to a rubber-like solid.

The other major categorisation is between interior or exterior vehicle application. The latter, of course, often require compatibility with paint finishes while the former may be used for spot-welded joints on the floor pan and require special expansion characteristics to obtain a water-proofing function. Such expandable systems can be applied to the joint prior to welding, the effect of the weld-through being to expand the sealant material.

For windscreens, different systems are used for the bonded and non-bonded systems, necessarily; with the latter, the screen is mounted in a moulded weatherstrip and the sealant is required to provide a watertight joint, reduce glass movement and reform ruptured bonds when movement takes place. Polyisobutylene systems are used for this purpose whereas the bonded-in type employs a fast heat-

Chemical Type	Base	Pretreatment	Remarks
Oleo-Resins	Drying and non-drying oils, resins and fillers.	Require primers to give adhesion to porous surfaces.	Best suited to lap joints, but can be tamped into butt joints. Available in strip form, or for knife or gun application
Rubber-Bitumen	Natural or synthetic rubber compounded with bitumen or pitch.	Movement accommodation of the order of 10%	Cold poured (set by solvent action) or hot poured.
Butyl	Butyl rubbers or degraded butyl rubbers, used alone or in combination with solvents, oils, extenders, etc.	Can be skin-forming, non-skinning, non-drying or curing (curing types cure by solvent evaporation).	Very wide range of compositions from soft, sticky compounds to hard rubber strips.
Polysulphide	Single or two-part mixtures based on polysulphide polymers. Two-part mixtures cure by chemical reaction fast.	Primers required on most porous and friable surfaces.	Wide range of compositions from soft elastic materials to hard rubbers. Available in a wide range of colours. Usual base of caulking compounds.
Silicone	Single-part chemically curing silicone elastomer.	Require primers on most porous and friable surfaces.	Remain fully elastic. Available in a range of colours
Polyurethane	Two-part self-curing mixtures based on polyurethane polymers.	Primers required on most surfaces for adequate adhesion.	Remain fully elastic.
Acrylic	Mixtures of acrylic polymers.		Plastic in nature.

curing polysulphide. An alternative to this is a preformed butyl self-adhesive tape applied to the glass before insertion. There are also systems using PUR sealants and an interesting alternative using a neoprene tape with embedded resistance wire. Exterior floor-pan seam sealants are required to be non-flow types which are impermeable and immune from cracking and blistering. For under-vehicle waterproofing, the sealant can be applied as a liquid and cured to an expanded solid in paint stoving ovens. For other exterior situations non-flow, anti-corrosive and self-levelling systems are available. Welded seams with small gap irregularities such as drip-rails can be sealed with vinyl plastisols prior to finish paint. Self-levelling ensures a smooth finish on which to apply the paint. The anti-corrosive systems are used on areas inaccessible for painting. *Tables above give sealants, applications and costs.*

Anticorrosion coatings

The interaction between coating systems and the basic mechanisms of corrosion are discussed here alongside precoated zinc and plastic systems as well as traditional pretreatments.While the best protection against corrosion lies in good body design , nevertheless there are difficulties inherent in such areas as interweld crevices (Fig 175) and the remoter regions of box section interiors which require supplementary coatings. Good protection is obtained by ventilation of the enclosed area, using wax injection and where possible specifying pre-coated metals.

Capillary penetration of wax into interweld crevices is crucial to overcome the lack of oxygen replenishment in these regions which promotes the development of corrosion cells. Electrocoat primers also may not reach such areas and there could be a tendency for road de-icing salts to remain in trapped liquid left behind after winter slush has dried by evaporation. Waxes are effective in repelling water which forms into globules on the surface of the wax.

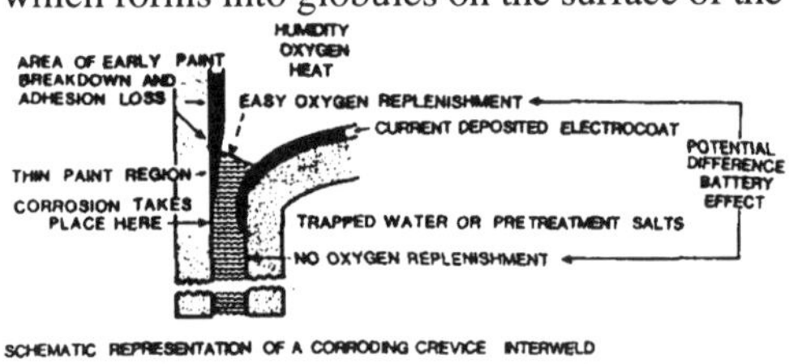

Fig 175: Crevice corrosion

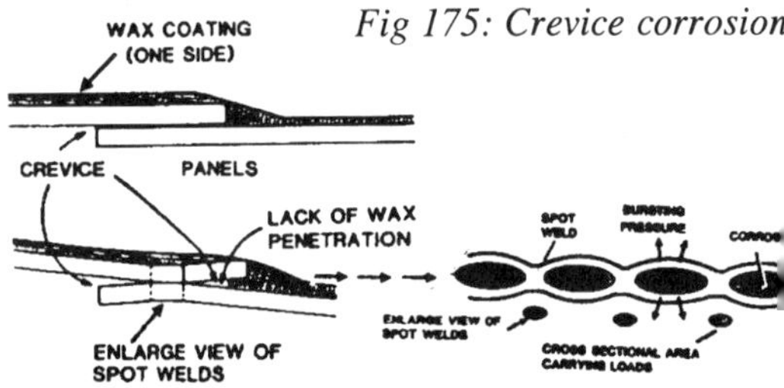

Fig 176: Danger of bridging

	Material	Application(s)	Relative Cost
NON-HARDENING	Bituminous	Caulking, general purpose sealant	1 – 4
	Butyl	Metal-to-metal, metal-to-glass, caulking	5 – 10
	Acrylic	Caulking metalwork, fibre joints, gaskets	8 – 10
	Polybutane	General purpose construction caulking and sealers	3 – 7
HARDENING FLEXIBLE	Neoprene (solvent-release)	General purpose sealant; also for dissimilar metals	8 – 12
	Butyl (solvent-release)	Sealant for metal, glass, masonry, etc	5 – 9
	Acrylic	General purpose construction sealant, caulking and glazing	8 – 10
	Silicone (one part)	General purpose construction sealant, caulking and glazing	20 – 40
	Silicone (two part)	Sealant for heat shields, etc; porting and encapsulation	40 – 100
	Polysulphide (one part)	General purpose sealant, caulking, glazing	17 – 25
	Polysulphide (two part)	Caulking, porting and encapsulation	12 – 15
	Polysulphine/Bitumen	Flame and fuel resistant caulking medium, sealant for expansion joints	4 – 6
	Polyurethane (one part)	General purpose sealant, caulking, glazing	12 – 15
	Polyurethane (two part)	Caulking, porting and encapsulation	12 – 15
	Epoxy (two-part modified)	Caulking, porting and encapsulation	7 – 15
	Fluorinated hydrocarbon	Oil and heat resistant sealant	80 –120
HARDENING RIGID	Oleo-Resin	Caulking, general purpose sealant	1 – 6
	Epoxy	Pipe sealing, porting, encapsulation of electrical compounds	7 – 14
	Polyester	Gaskets, pipe thread sealant, porting, moulding, encapsulation	6 – 12

Its viscosity must be chosen between the requirements of ability to penetrate the crevice on the one hand and provide a suitable thick surface film to prevent wicking of the water up the crevice on the other. Relatively high viscosities are necessary to maintain a thick film on a vertical surface but it must not be too high as to cause bridging of the crevices as shown by Fig 176. It is also important that the crevice does not draw up only the solvent suspending the wax and not the wax itself.

Electrolytic zinc-coat steels used in protection against corrosion are increasingly being applied. Fig 177 shows possible applications for British Steel's Nizec zinc-coated steel for one side and both-side coated cases. The CR code in the designation, signifying cold rolling, is usually followed by a number depicting the thickness of the zinc layer in microns. Zinc itself corrodes to sacrificially protect the steel and in so doing forms a colourless salt which itself protects the zinc. Since the zinc coating can be

Key: Exploded panels are those considered by British Steel to be suitable for Nizec: White panels — one side only: black panels — both sides

Fig 177: Nizec panels on car body

ALKALINE CLEANERS / **COLD EMULSION**

	ALKALINE CLEANERS	COLD EMULSION
FERROUS	No. 3, 4, 5, 7, 8, 14, 16, 17, 18, 19, 20, 22, 23, 24, 26, 28, 30, 62	EMULSION CLEANER 'R'
ALUMINIUM & ALLOYS	7, 10, 11B, 19, 22 26, 28, 62,	EMULSION CLEANER
ZINC & ALLOYS MIXED METAL	6, 7, 14, 18, 19, 22, 26, 28	CONCENTRATE NO.1

ACID CLEANERS

DERAN, 'B' D & P.S.	Warm, derust & passivate
DERAN 'C' & WALTERGEL P.5	Brush/swab on derust and passivate.
D.P.1	Cold Combined degrease/pickle
D.P.2 & D.P.3	Warm Combined degrease/Pickle
P.S.4 & P.S.5	Heavy duty acid pickles
P.S.7	Cold – Cuprous metals

Complete degreasing, cleaning, rust & scale removing processes. Pickling & passivating. Immersion soak, spray or brush/swab. Power washer cleaners.

PHOSPHATING PROCESSES approved to D.E.F 29A & BS 3189

'C.1'	Iron Class 1. Immersion	
'MC'	Manganese Class 1. Immersion	
'E' SR & S.R.H	Zinc – Class II	Immersion / Low Sludge & Scale
Fasbond	Zinc	Paint Bonding
A & F		Class II Immersion
'J'		Class III Immersion
Cold		Class III Immersion ambient

Fasbond A.S.	Zinc Class III Spray	
D.W. & I.L.W.	Zinc Wire drawing – Immersion	
DXL & DXH	Zinc Cold Extrusion – Immersion	
A.P.	Class IV Lightweight Iron – paint bonding. Tube drawing	Spray
A.P.3		Mixed –metal
A.P.4. A.P.5 A.P.6		Spray or Immersion

Phosphating processes for Corrosion protection Paint bonding Rubber bonding Tube & wire drawing Deep pressing & extrusion Chemical blacking Paint strippers.

CHEMICAL BLACKING PROCESSES · Walterblak & Walterblak 4

CHROMATING PROCESSES

Cromcote	AL	Gold coatings for aluminium	Paint bonding & protection
	GL	Blue/Green coatings for aluminium	
	ZC	Zinc & Galvanised	
	ZH	Galvanised	
	ZF	Galvanised	

D.D.3 – Sealant additive for hot rinses

CHEMICAL OXIDISING PROCESSES

Walterisation 'L'	Aluminium, Paint bonding & corrosion resistance
SATIN ETCH PROCESS for ALUMINIUM	
WALTERBRYTE C.H. Aluminium & alloys	
COLD BRUSH ON TREATMENTS	
FOSCOTE R.S. – AL – GA For Steel – Aluminium – Galvanise	

Chromate conversion processes for:– Corrosion protection and paint bonding on:– aluminium, zinc, copper and their alloys.

Chemical oxidising & satin etches for aluminium & alloys.

DEWATERING FLUIDS

D.F.F.	Film free – limited protection
D.F.F. 2	Thin film – good protection
D.H.	Black variant of above
BLACK D.H.	Dry resin film – good protection
D.M.	Thin mineral oil – good protection

SEALANTS

V.S. 1 & 2	Temporary – for storage
SD6 & 9	Black – spirit stains
GM OIL	Mineral – corrosion inhibited
G.C. OIL	Mineral – exceptional protection
EMULSIFIABLE OIL No.2	Water/oil emulsion – dry finish excellent protective

Dewatering fluids – filming & non filming: lanolised, mineral or resinous. Corrosion inhibited oils and protectives. Shellac dyes & water stain One pack metal conditioners etching primers & lacquers.

Fig 178: Traditional pre-treatments

vapourised locally around spot welds there is a case for further wax injecting to protect the weld crevices, as before. The objective with zinc coating is to make metallic contact with the steel to promote a cathodic reaction in which both metals are equally exposed to the liquid environment. The mechanism is one of the electrons produced by the dissolving zinc travelling through the metallic contact and being changed into negative ions at the steel surface. These go back through the liquid environment to form a new zinc compound with the dissolved zinc ions.

In vehicles a problem occurs in that the high resistance of a thin film of water absorbed into the paint film reduces the flow of ionic current. However, a compromise is reached in that the very reduction of this current is the basis of the separate barrier mechanism of the paint for reducing corrosion. It is also necessary to overcome the difficulty of a less good adhesion of paint with galvanised steels.Finish painting systems are being underlaid with increasingly sophisticated primer coatings. Fig 178 shows the range of products and processes offered by Walterisation. Much of these treatment processes are for ensuring good adhesion of the finish coats — and indeed the primers and intermediates. Procedure for building up a good substrate is to abrade the surface with 180 'wet & dry'; wash down and dry then apply the phosphoric acid etch. After subsequent hot-water wash and blow dry a single coat of self-etch primer is applied. Next, three coats of filler are applied and baked for half an hour and this layer rubbed down with 600 'wet & dry'. Finally the surface is inspected and stopped up with stopper as necessary and a further filler/sealer coat applied as necessary after blocking down.

An interesting alternative to plastic coating of metals is the use of pre-painted metals in specialist body assembly. Several factors need to be considered in the choice of coating in the ability of such panels to withstand subsequent forming and fabrication. The key coating parameter is dry heat resistance being the resistance to film fracture after the film has been stretched and exposed to heat. Generally this varies inversely with the molecular weight of the coating polymer and amount of three dimensional cross-linking undergone in the cure cycle. Preferred primers for the coatings are the higher molecular weight epoxides, urethanes and linear polyesters. Among the special techniques used in welding precoated sheets are the use of weldable primers on the reverse side, indirect projection welding and the use of short bursts at very high energy levels.

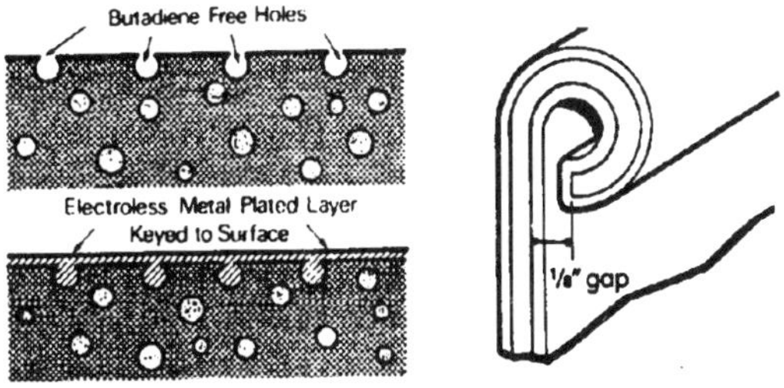

Fig 179a (left): Etching and coating; b(right): Rolled edges;

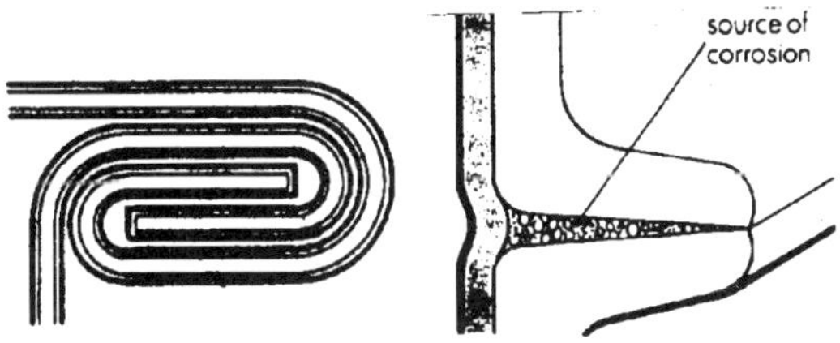

Fig 180a (left): Roll joint; b(right): Flange joint

Plastic coating

Plastic coatings and coatings for plastics generally involve the metal to polymer bond which needs care in design to avoid either peeling or corrosion. Careful adherence to rules of form and application will ensure the proper bond which is key to satisfactory in-service life. Most of the bright metal coating is applied to ABS plastics but higher temperature-tolerant plastics such as PPO can also give good results. In general, if a part has to flex, it should not be plated. However, selective plating may be employed and if, say, it is just a question of avoiding snap-fit lugs, then selective plating for the remainder of the part is quite feasible. The approach is normally to electroless coat the part with nickel and then regard it as metal for subsequent chromium plating. Uniform wall thickness in the order of 3 mm is preferred by platers; 5mm is considered to be a maximum and where possible, abrupt change in thickness should be avoided. Where ribs are used, these too should be restricted in depth to around 50-60% of wall thickness. Because plating exaggerates the very smallest surface imperfection in the plastic, it is necessary to avoid any chance of sink-marks. Ribs should thus be well radiused and subject to normal draft of about 1 deg. In the part as a whole, external radii should be not less than 1mm and internal ones 0.5mm minimum. A groove of some 1.5 mm deep should be used at the boundary of plated and unplated surfaces.

Plating thicknesses will of course depend on part application. Tests reported on ABS have shown that below 0.4 micron, microporsity can occur but above 0.8 micron, only crazed chromium films are considered to produce good quality. Adhesion strengths on ABS of up to 4-6 kgf/mm^2 can be expected but it is best to retain a temperature spread within 40 - 90 C. After cleaning the substrate prior to plating it is usual to etch the surface with acids which remove the butadiene, Fig 179a. Following the electroless nickel coating stage, applied by dipping into a catalytic reduction solution, very low current density is used in the electroplating so as to maintain the optimum thin coating of copper prior to the finish chromium layer. To avoid surface blistering at high temperatures, higher grade plastic substrates like PPO are chosen. A brand such as Noryl PN235 has heat deflection temperature of 112 C and water absorption at 23 C of about 0.08% in 24 hours.

Coating metals with plastic

A number of dip-moulding and similar processes have been available for the plastic coating of metals but particularly strong coating with materials like polyamides are possible with plastic powder coating. ATO Chemicals' Rilsan technique is one such and interestingly the nylon is developed from caster oil rather than the mineral oil which is the source of most plastics. Parts to be be coated must, however, be capable of sustaining the fusion temperatures involved. Coating film thickness is commonly 75 to 200 microns and fusion temperature between 200 and 225 C. An electrostatic projection coating process is used for parts too unwieldy for dip-coating — normally large thin-sheet fabrications. Powders are passed through a nozzle carrying a high electrical potential; particles become charged and attract to the metal substrate at zero potential.

In the alternative fluidised bed coating process, the part is oven-heated and then dipped into a fluidized bed of powder, akin to dip coating, for thicknesses from 325 to 375 microns. Really large items can be coated by a third process known as flame spraying which causes the powder to melt and flow. Again, radiused edges are the rule for good coating. Rolled edges, Fig 179b, should be left open with a gap of about 5mm. Roll or crush joints, Fig 180a, can be coated before rolling. By contrast, edge-welded flange joints, Fig 180b, cannot be used as it is virtually impossible for the powder to penetrate the entire joint.

Pre-coated panels

With heightening awareness of ecological matters, pre-coated (painted) steel and aluminium sheet is now of increasing interest to the motor industry as painting pollution problems can be dealt with earlier and perhaps more effectively by the coil-coater.

Coaters say because the bond between the metal substrate and the coating system is continuous, coil-coated metal can also withstand subsequent fabrication without damage to the constituent elements. Tests by the American Institute of the Steel Industry have shown that body panels can be formed from top-coated pre-painted sheet in existing metal forming presses using existing dies. Because the pre-coaters apply paints to both sides of the metal, corrosion problems during storage and pre-fabrication are minimised and panel galling is reduced by the absence of metallic contact with the dies. Dissimilar metal substrates can also be used without the usual risk levels of bi-metallic corrosion.

Prime candidates for the process are instrument panels and bumpers also cover-panels such as bonnet, boot and door. In general, roll-forming rather than press-braking is preferred for folding pre-coated sheets since there is less chance of coating damage. Then, in subsequent drawing stages, the main concern is to ensure that press-tools are designed to allow for the coating thickness. Edges can be protected with edge-trims or touch-up paints; rolled edges or returned flanges can also be incorporated, Fig 182.

A number of fastening systems have been developed for the material, and recommended by the European Coil Coating Association, which avoid the need for welding. Self-piercing rivets, self clinching nuts and self-drilling/self-tapping screws are the three principal contenders, Figs 183/4. Since they do not require pre-drilled holes they eliminate possible damage to the coating in machining. A variety of spring tension fasteners are also suitable — and, for

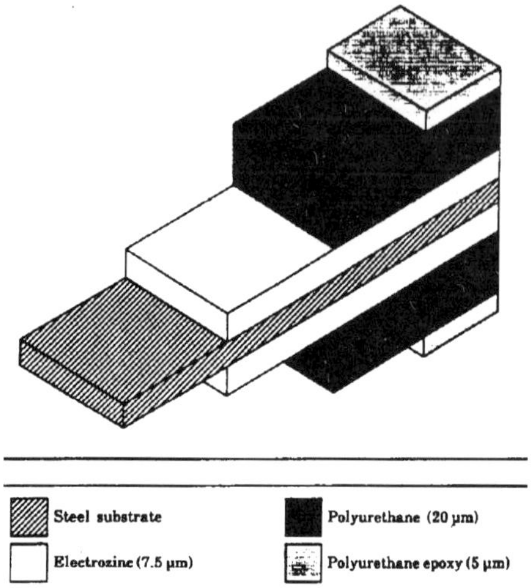

Fig 188: Flexgard T pre-coated sheet

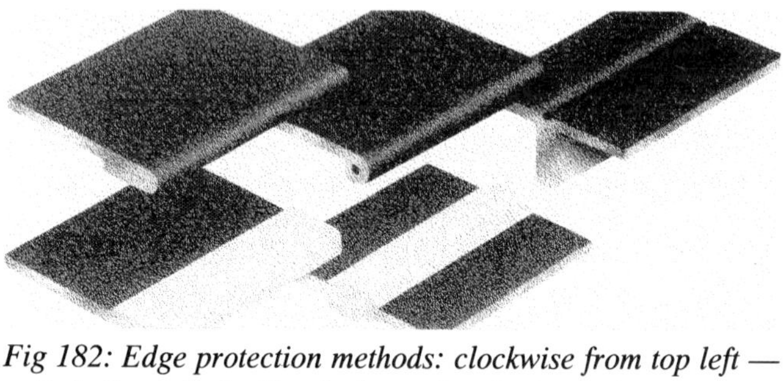

Fig 182: Edge protection methods: clockwise from top left — folded; rolled; flanged; double closing strip; single closing strip

Fig 185: Clinched open section

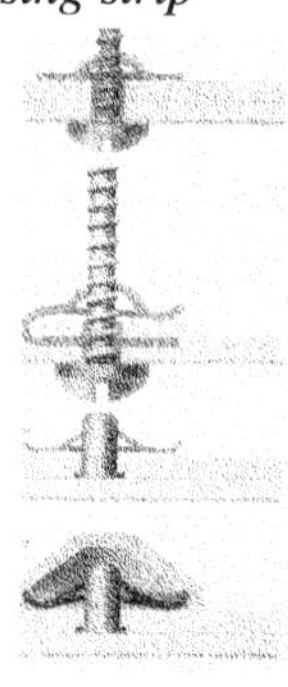

Fig 183: (far left) Insert fastening methods

Fig 184: (left) Clip fastening methods

Fig 186: Clinched closed sections

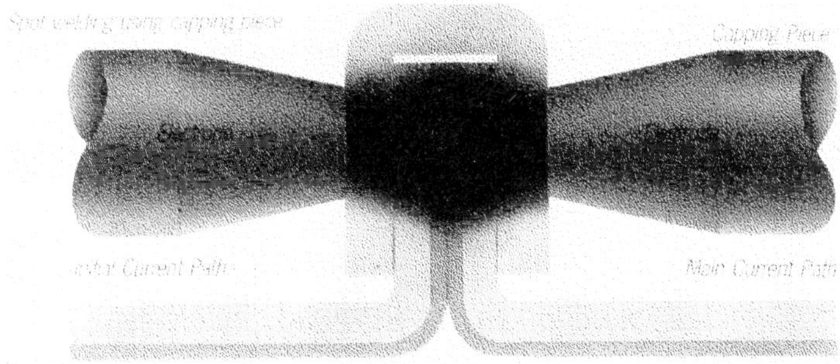

Fig 187: Bridge-conductor for spot-welding

PROPERTY	SPECIFICATION
% Gloss	25 - 45
Hardness	F - 2H Eagle Turquoise
Film Thickness Dry Wet	0.15 - 0.25 0.35 - 0.55
Cure Test	50+ Double MEK Rubs
Reverse Impac t	1.5 x Metal Thickness No pickoff
Flexibility	OT Bend : No pickoff

Fig 189: Flexgard T specifications

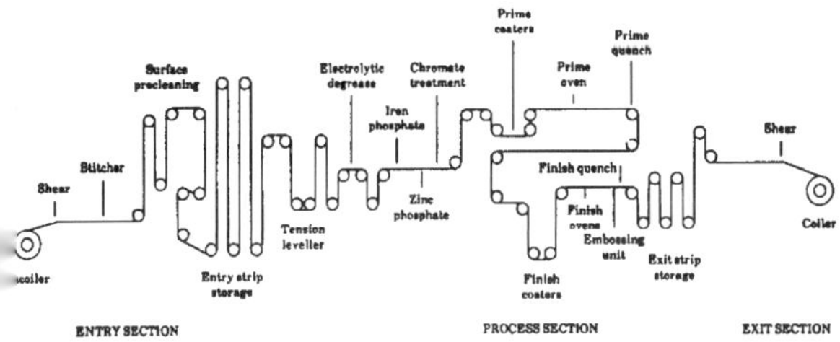

Fig 190: Production of Colorcoat

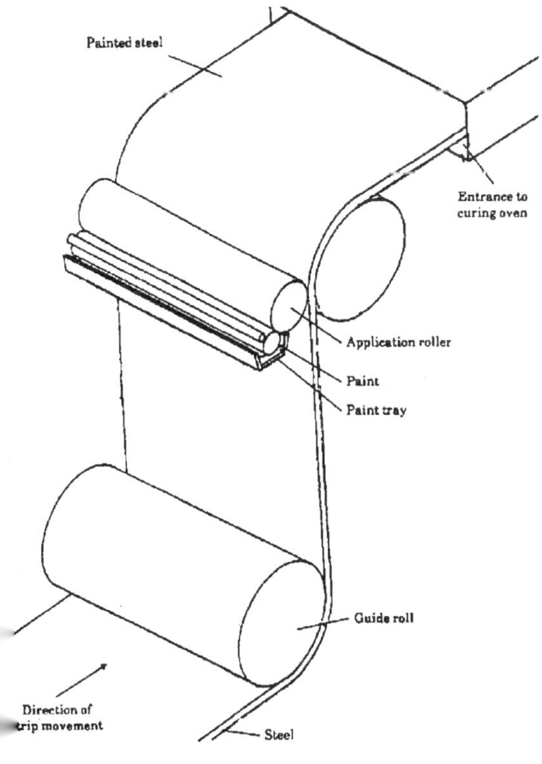

Fig 191: Rolling of Colorcoat

non-structural flanged joints, spring clips are a convenient fastener.

Clinched sections have also been developed to provide interlocking joints suitable for box-sections, Figs 185/6. Portable stitch-folding guns are also available which punch out and fold two tabs over one another in a single operation. While resistance-welding by conventional methods is not possible, due to the electrical properties of the coating, spot-welding of in-turned flanges can be carried out by placing a channel capping piece over the flanges to provide a low resistance contact, Fig 187. Capacitor-discharge stud welding can also be used to fix a stud to a hidden area of a component. A combination of short welding time and high current meets the requirements for welding without marking the components. Adhesive-bonding is feasible provided a correct choice of adhesive is made according to the particular organic coating.

British Steel have recently performed industry-standard testing of pre-primed sheet steel systems and have found superior crevice corrosion performance over an equivalent electropainted panel. The pre-prime system tested was Flexgard T developed by PPG, Fig 188. Its specification is seen in the table of Fig 189. Shown in Fig 190 is the Colorcoat line of British Steel at Shotton where both PUR-based primer and epoxy-PUR top-coat are applied wet to the pre-treated zinc-coated steel by roller coaters spanning the width of the sheet strip, Fig 191. After the primer is applied, the painted strip enters an oven to cure the organic film and expel excess solvent also to produce cross-linking of the polymers.

Glazing techniques

Glazing of vehicles has reached the interesting stage where glass is potentially a substantial contributor to the structural integrity of the body, Fig 192. Since the direct bonding of glass into the body apertures the material has made a contribution to stiffening the 'closure' frames by the addition of a diaphragm member.

Now the flexural stiffness of the material is being exploited — thus far mainly on concept vehicles — in the construction of frameless glass canopies or even structural roof systems. Traditionally the key application has been the windscreen — in which there have been recent trends to single and double curvatures — and for which Motor Vehicle Construction and Use Regulation 1017 requires the use of safety glass. This is usually laminated plate of 4.4 to 8 mm thickness and must provide distortion-free visibility even in areas of high curvature. General specification is provided in British Standard BS 857 and the latest type encompasses the best features of both toughened and laminated glass. It comprises high stressed inner glass ply bonded to a low-stressed outer ply by a highly penetration-resistant interlayer.

Variations are available from major suppliers Triplex — with clear or tinted glass, silk screened obscuration bands, graduated shade bands and fine wire heating elements. Toughened 'body' glass for side and rear windows (and to a limited extent for windscreens) is available in 3 to 6 mm thicknesses — again with a variety of silk-screened patterns also in Sundym and bronzed varieties. Another product is Hotscreen, the windscreen glass with a grid of very fine tungsten wires embedded in the laminate — and, similarly, Defogger backlights.

The Triplex Hotscreen product typically involves 500 yards of 20 micron tungsten wire — dissipating 600 watts per square metre on a 12 volt system enabling a uniform defrost within two minutes. If higher voltages can be made available, gold and silver films can be used for demisting and defrosting which are more efficient again — with the potential to dissipate 1500 watts per square metre.

The trend in volume cars towards higher glazed areas has been accompanied by a parallel one of reducing glass thickness — so as to avoid nett increases in body weight. This trend is likely to apply

Fig 192:Glass contributing to structural integrity

Sound attenuation values

Typical car body shell	26 dB
5 mm toughened glass	26 dB
3 mm toughened glass	23 dB
5 mm laminated glass	28 dB

Resonance of glass plate, 1500 mm × 500 mm

Shape	Thickness (mm)	Edge support	Resonant frequency (Hz)
Flat	4	Rigid	45
Simple curve	4	Rigid	146
Simple curve	3	Rigid	143
Complex curve	4	Loose	15

Typical heat transmission factors (ΔT = 20°C)

Road speed (km/h)	U for toughened glass ($W\ m^{-2}\ {}^{\circ}C^{-1}$)	Glass thickness (mm)
7	5·6	5
	5·6	3
47	6·7	5
	6·8	3

Fig 193: Properties of glass

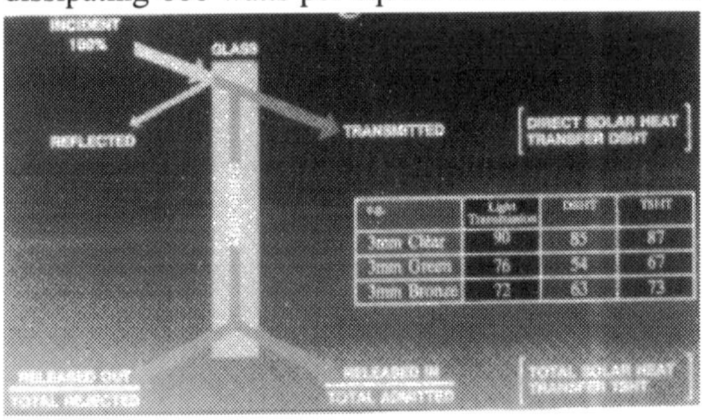

Fig 194: Optical effects of double glazing

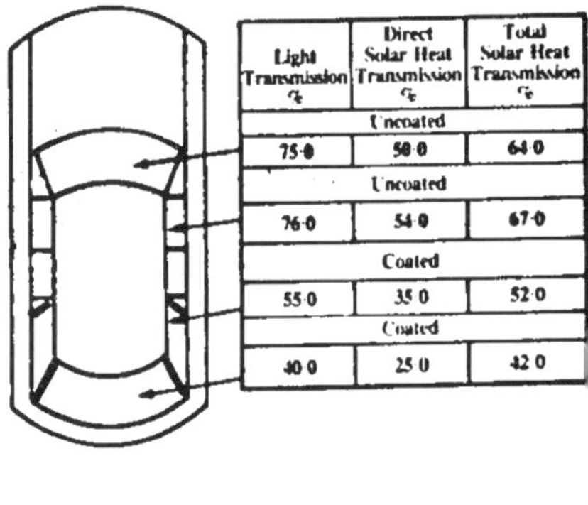

Fig 195: Tinted glass effect — combined with coati

to many other vehicle sectors and already 3mm toughened glasses and 5 mm laminates are being specified. There can be a cost penalty with thinner glass arising from its greater fragility at the manufacturing stage and the greater difficulty in controlling any curvature required. On the other hand thin glass can be used advantageously if curvature, coupled with rigid mounting, is present — as suggested in Fig 193 which also shows thickness effect on sound attenuation and heat transmission. It is even possible to bend and toughen glass, on a pilot scale, down to 2mm. Solar control systems under development include the use of photochromic and heat-rejecting glass — also special forms of double glazing, Fig 194.

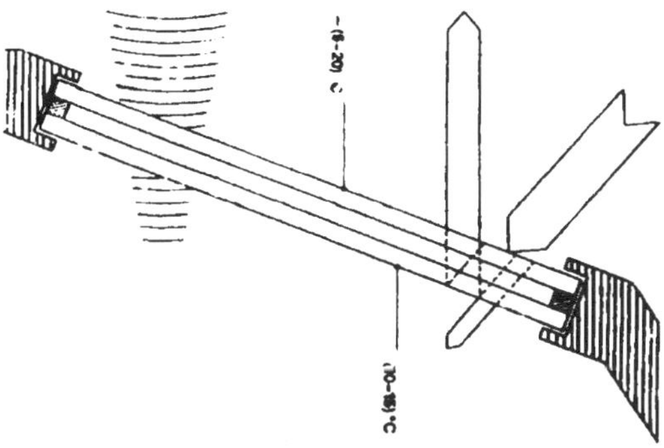

Fig 195:Reflectance effects in double glazing

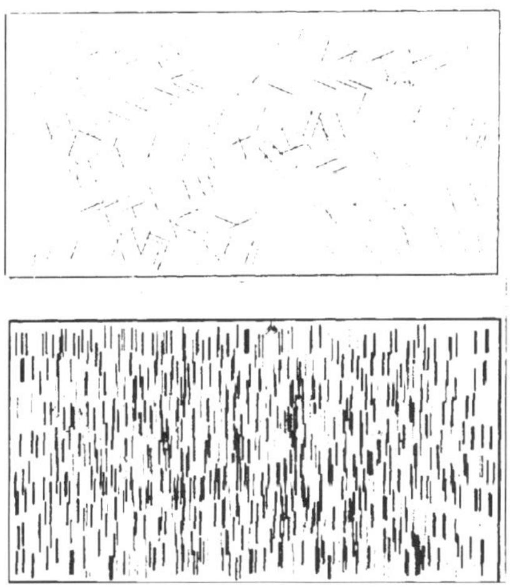

Fig 196: Switching of liquid crystals: with (above) and without (below) electric field

European Safety Regulations limit the amount by which 'tinted' glasses can be used — in the interests of maintaining optical clarity, Fig 195a. On front screens light intensity may not be reduced by more than 25%. On side windows through which rear-view mirrors are viewed this figure becomes 30% and for the vehicle as a whole 45% with the rear window alone having an allowance of 60%. Even 'clear' glass can reduce intensity by as much as 10% and it has to be remembered that regulations also impose limits on maximum image distortion. However, while clear glass transmits about 84% of solar energy, it is possible to procure optically acceptable tinted glass which will reduce heat build-up in stationary vehicles by 'between a quarter and a half', say the makers.

When glass becomes part of the vehicle structure, and subject to diaphragm stresses, there is risk of noise due to 'drumming'. As glass cannot easily be swaged without detriment to its optical performance alternative means have to be found to reduce this effect. One such is a form of double glazing comprising two sheets of optical glass, precisely matched in curvature, with an air gap between them, Fig 195b. In the case of roof lights, liquid crystal systems can be used with which light transmissibility can be changed subtantially by the operation of a switch, Fig 196. An

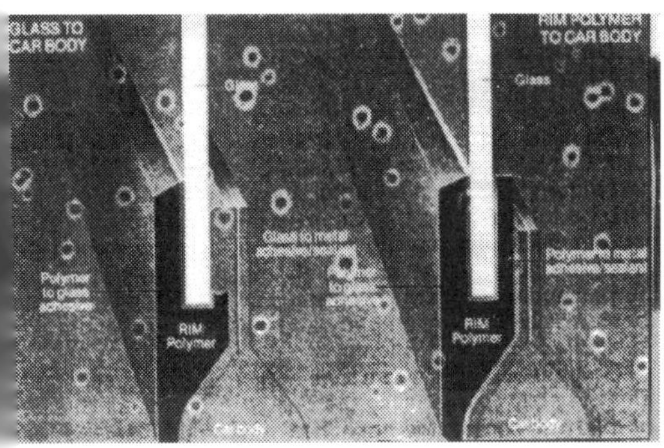

Fig 197: RIM encapsulation of glass

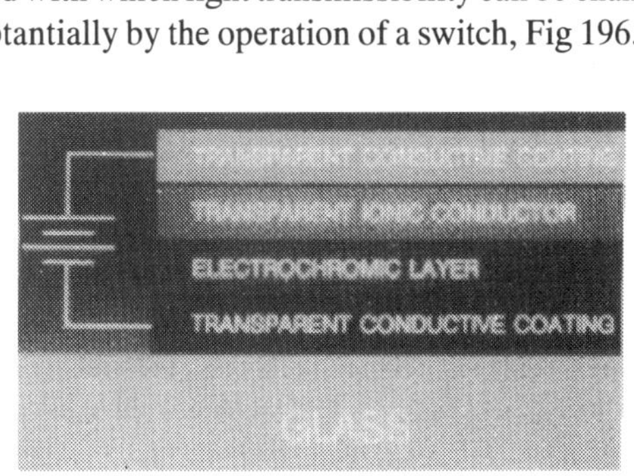

Fig 198: Electrochromic glass

alternative is an electrochromic system, Fig 198, in which an electric current generates lithium and hydrogen ions from the ionic conductor layer. These then flow to the electrochromic tungsten dioxide layer which changes its colour to a blue tint.

Encapsulation of glass is another trend, in which the glass, seals, trims and clips are pre-assembled as a module which is relatively easy to install. Encapsulation can also encompass flush glazing systems and indeed provide easier achievement of the tight tolerances involved. Dow have developed a technique using reaction injection moulding (RIM), Fig 199, involving relatively low pressures and temperatures that thus keep costs to a reasonable level. The process is even said to be capable of producing complete sliding-window units.

The two reactive components are blended by high pressure impingement and then injected at low pressure into a closed mould. Reactive edge adhesive is applied to the glass which is then placed in the mould for a curing time of about 30 seconds, Fig 200.

Trim pad construction

For pouring foam into thermoformed skins, when making up such parts as crashpads and spoilers, polyether-based PUR is usually employed. Either TDI or MDI foaming can be used, the latter allowing a one-shot process. Density can be varied by altering the amount of blowing agent used; the higher the density the easier it is to obtain a well finished moulding. Resilience, too, can be altered by compounding techniques. Fig 201 shows tables relating four typical formulations for semi-rigid foam and indicating how water and triethanolamine content affect density and hardness.

Cycle time involves several phases: the cream time takes from 2-60 seconds and allows the unreacted chemicals to reach the extremities of the mould; then rise time is that for the liquid of 1.15 gm/cm^3 to foam into a density of around 0.07 gm/cm^3; cure time is the period in which polymerisation takes place. Catylists are used to vary the relative timings.

Self-skinning PUR foams now form a substantial part of those used for automotive trim panels. These high-modulus foams can be produced by either the prepolymer or one-shot processes. The former involve reacting excess TDI with polyol then completing the foam-reaction with addition of water and catalysts. With the one-shot system, the individual

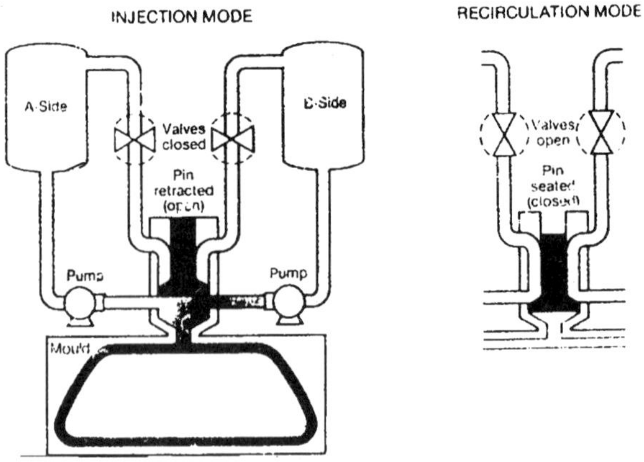

Fig 200: RIM encapsulation equipment

	A	B	C	D
Propylan 555	100	100	100	100
Triethanolamine	5	5	10	10
Water	2·5	3·5	2·5	3·5
Propamine A	0·8	0·7	0·8	0·7
Crude MDI	61·0	76·8	75·3	90·5

	A	B	C	D
Core Density lb/ft³	4·9	4·0	6·0	5·2
Compressive Strength (Load in lb/in² to 50% compression)	3·1	2·7	6·6	4·3
Compression Set (50% compression for 22 hrs @ 70°C)	8·1	15·7	10·1	17·4
Tensile Strength lb/in²	18·9	15·6	25·2	16·7
Elongation	55	50	43	40

Fig 201: Semi-rigid foam formulations

Property	Value
Density kg/m³	48–160
Tensile Strength kN/m² (BS3379)	138–414
Tear Strength N/m ASTM-D 1564	157–525
Compression Set 75%	25–35
(BS3379) 90%	80–90
Elongation at Break % (BS3379)	50–90

Fig 202: Crash-pad foam characteristics

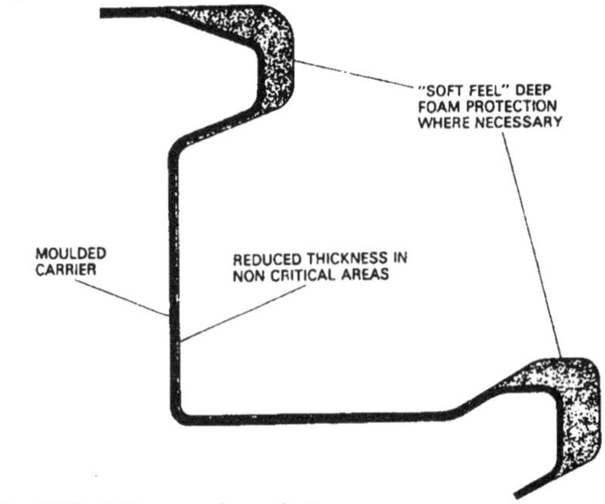

Fig 203: Microsorb technique

chemicals are mixed together at the time of moulding. While such processes can effect cure without heat, cycle times can be shortened by passing through an oven at 100-160 F prior to demoulding. Moulds of glass or epoxy reinforced resin can be used, or for higher volumes, cast aluminium alloy.

Typical characteristics for a crash pad foam are shown in Fig 202. Covers may be vac-formed ABS or PVC, blow-moulded or blow-formed PVC — also rotationally cast as slush- moulded PVC. A variety of support structures or inserts can be used including metals, plastics or moulded woodfibre. Particular care is needed with deep drawn parts to avoid excessive thinning of the core or loss of surface grain on the skins.

Flooring systems from PUR foam are made by spray forming a PVC skin into an electro-form, the skin, after curing, being filled with semi-rigid foam then backed with a thin PVC sheet. In the case of door trim panels, a typically blow-formed PVC sheet of about 0.3 mm might be used; the pressure developed during foaming then forms the PVC blank against a contoured mould surface to eliminate the expense of the vac-forming operation and permitting the use of a less-expensive cover material. Fibre-board or similar inserts are required to ensure thermal stability of the moulded pad.

Self skinning foams involve a cellular core whose structure gradually changes towards the outside of the moulding to form a closed integral outer skin. Hardness can vary from that of a flexible cushion to a rigid form — in densities from 100 to 800 kg/m^3. Such foams cure without external heat normally but post-curing can be applied to obtain a particular physical characteristic. Again a wide variety of mould materials can be used, with or without grained surface. An important consideration is the need to completely vent the air, to avoid surface imperfections. A vac-formed polystyrene sheet liner can be used for this purpose, which has the additional attraction of dispensing with a release agent. Acrylic and PUR mould coatings permit colour keying with other body parts. A comparatively recent advance by Marleyfoam is the Microsorb technique of obtaining soft-touch surfaces for such parts as crashpads, Fig 203.

Chapter 7: Metal and polymer systems

Designing in GRP and high-strength composites;
Thermoplastics; designing in polypropylene;
ABS and the styrenes; thermo-forming of plastics
Load-bearing plastics; reinforced thermoplastics and engineering plastic alloys;
Semi-rigid PUR foams and sandwich panel construction;
Reaction injection moulding; resin transfer moulding
Intelligent steel selection; HS steels;
Aluminium alloy sheet, extrusions and castings;
Austenitic and ferritic stainless steels; magnesium alloys; copper alloys
Hot and cold forging; ferrous casting
Metal-matrix composites; structural timbers; metal injection moulding

Designing in GRP and high-strength composites

A specialist technology is building up around GRP and allied compositesand is beginning to take over as a standard construction technique for specialist sports cars and low-volume production truck cabs.

In the broad classification of particle, laminate and fibre reinforced composites, GRP falls mainly into the last category. Reinforcement can, however, range between almost particle-shaped chopped-strands to woven mats akin to those used in laminates. The resin generally used is polyester but phenolics and epoxies are two of the alternatives. The glass reinforcement can be extremely strong, depending on conditions; for example, a single filament drawn to about 12 microns has an ultimate tensile stress of

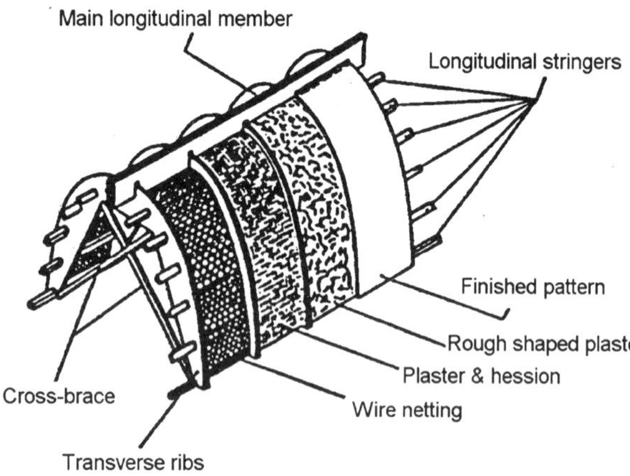

Fig 1: Typical mould construction

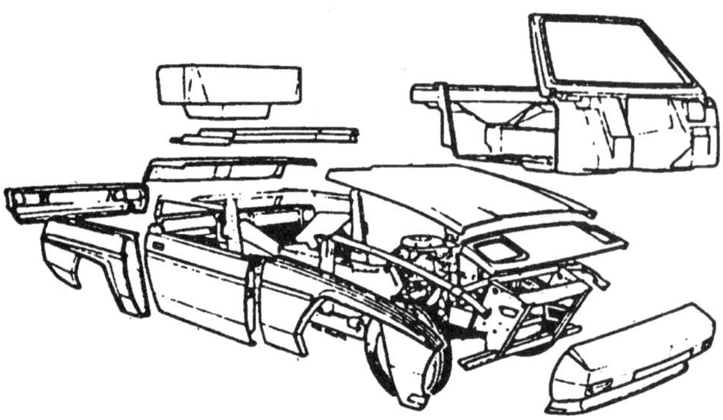

Fig 2: Reliant Scimitar mouldings

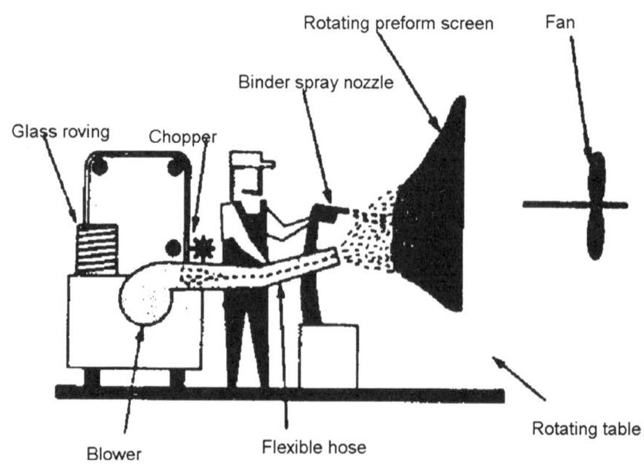

Fig 3: Making a rotating preform

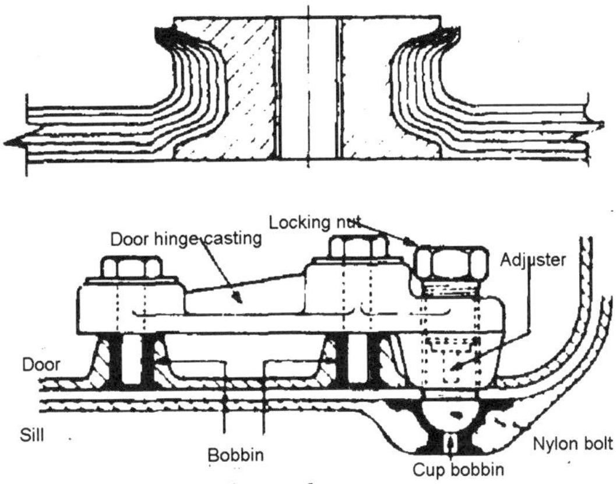

Fig 5: Mechanical attachments

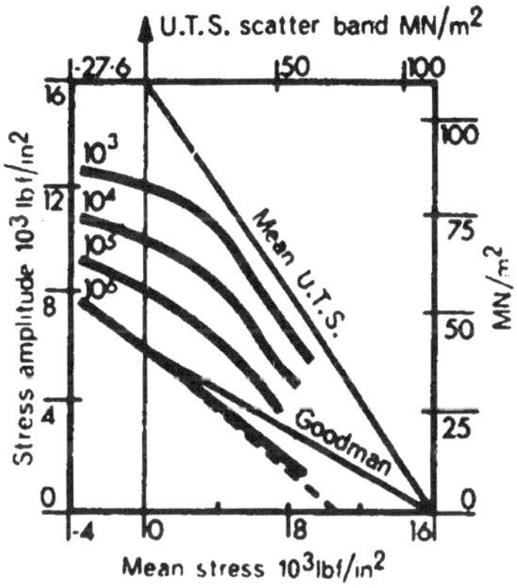

Fig 6: Goodman fatigue diagram

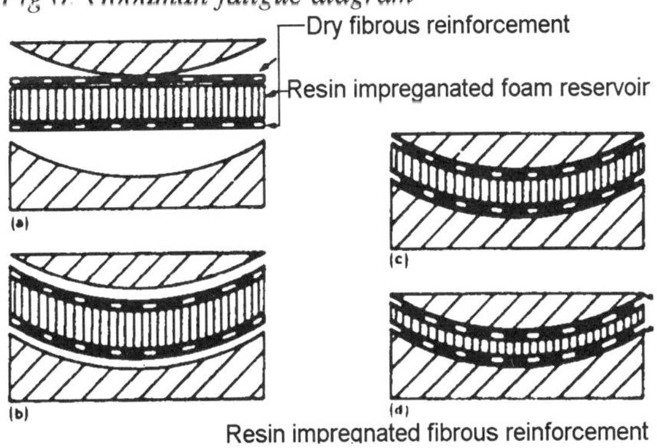

Fig 7: Reservoir moulding

Material	Glass fibre by weight %	Flexural strength lbf/in² x 10^3	Flexural strength kN/m²	Flexural modulus lbf/in² x 10^5	Flexural modulus GN/m²	Tensile at Yield lbf/in² x 10^3	Tensile at Yield kN/m²	Tensile modulus lbf/in² x 10^5	Tensile modulus GN/m²	Ult. Tensile elongation %	Izod Impact ft. lbf/in notch	Flammability	Continuous heat resistance °F	Continuous heat resistance °C
SMC	15-30	18-30	124-207	14-20	9.6-13.7	8-20	55-138	16-25	11-17	0.3-1.5	8-22	Slow to non-burning	300-400	150-200
DMC (BMC)	15-35	10-20	69-137	14-20	9.6-13.7	4-10	28-69	16-25	11-17	0.3-0.5	2-10	Slow to non-burning	300-400	150-200
Preform/mat	25-50	10-40	69-276	13-18	8.9-12.4	26-30	172-207	9-20	6-14	1-2	10-20	Slow to non-burning	150-400	65-200
Cold press moulding polyester	20-30	22-37	152-255	13-19	8.9-13.1	12-20	82-138	—	—	1-2	9-12	Non-burning	150-400	65-200
Spray-up polyester	30-50	16-28	110-193	10-20	6.9-13.7	9-18	62-124	8-18	5.5-12.4	1.0-1.2	4-12	Burn to SE	150-350	65-176
Moulding compound (phenolic)	5-25	18-24	124-165	30	20.6	7-14	48-117	26-29	17.9-20	0.25-0.6	1-6	SE	325-350	165-176
Nylon	6-60	7-50	48-345	2-26	1.3-18	13-33	90-228	2-20	1.3-13.7	2-10	0.8-4.5	Slow to SE	300-400	150-200
ABS	20-40	23-26	158-179	9.2-15	6.3-10.3	11-16	76-110	6-10	4.1- 6.9	3-3.4	1-2.4	Slow	200-230	93-110
Polypropylene	20-40	7-11	48-76	3.5-8.2	2.4- 5.6	5.5-10.5	38-72	4.5-9	3.1- 6.2	1-3	1-4	0.85-0.75 in/min	300-320	150-160
Thermoplastic polyester	20-35	19-29	131-200	8.7-15	6.0-10.3	14-19	96-131	13-15.5	9-10.7	1-5	1.0-2.7	Slow	275-375	135-190

Fig 4: GRP properties compared

3.5 GN/m².

The unsaturated (crystic) polyesters can be cured from liquid to solid state either with heat plus catalyst or, at room temperature, using an accelerator plus catalyst. A wide variety of techniques can be used to speed up the curing time but the present state of the art precludes GRP from very high volume manufacture (above 200,000 parts per annum). The main attraction, to low volume manufacturers is the relatively low tooling cost. A typical mould construction is shown in Fig 1, from Scott Bader's "Crystic Polyester Handbook".

A variety of methods are used for producing the composite. Traditional hand layup (bucket and brush) is still used where large complex shapes are involved; Reliant are keen exponents and use the method on the inner body moulding of their 2-seater Scimitar car, Fig 2. Its bonnet skin is made by the VARI process, perfected by Lotus, while the boot skin is a cold pressing using matched dies. Matched dies are also used to produce the body shells of the Chevrolet Corvette, at a rate of 90 units per day, working at pressures of 125,400 lbf/in² (850 kN/m²) but with dies heated to 215-300 F (102-149 C). Curing of the resins under these conditions takes 25 minutes. The glass reinforcement is preformed by spraying chopped fibres and resin binders on to rotating preform screens, Fig 3. The preform is stoved for 23 minutes before placing on the mould; a total of 89 mouldings is used in each body.

GRP material properties

A typical GRP mix for heavy duty automotive use is glass fibre 40.0, resin 40.0, monomer 0.41, filler 16.5, lubricant 0.08 and preform binder 2.0 percent. Mechanical properties for such a mix are tensile strength 24,000 lbf/in² (165 kN/m²), elastic modulus

1.4 x 10^6 lbf/in^2 (9.6 MN/m^2) and shear strength 16,000 lbf/in^2 (110 kN/m^2). Fig 4, due to Owens/ Corning Fibreglass, provides a more complete description of properties giving an indication of the effects of different resin systems. Failure in a plastic material occurs usually at the flexural, or cross-breaking, strength limit of the material with no definable yield point. Experience shows that minor impacts cause only a spring-back deflection while limiting stresses cause fracture but no permanent set. With hollow sections, however, tension rather than compression failure is likely to be the criterion.

Thin walls are generally desirable in press moulding for reducing cure times. Thickness should be in the range $0.3 < t < 0.25$ in ($7.6 < t < 6.35$ mm) for sheet. Design approach is to integrate parts into as small a number of individual mouldings as possible so as to save fabrication cost and avoid stress concentration. Double curvature should be used, where possible, to increase stiffness with well-radiused curves preferred to assist fibre distribution and air removal. As much taper as possible should be allowed in deep-draw mouldings and undercuts avoided to assist removal from mould. Local strength and stiffness can be enhanced by moulding in ribs, hollow sections and lightweight cores; as stress concentration is not relieved by yielding, it is best to use large area washers and metal inserts to spread shear and bending loads at attachment points. Fig 5 shows mechanical attachment systems used on the Lotus Elan body. Diecast oval bobbins set in prior to moulding have high resistance to twist and tension. The door hinge attachment comprises a diecast plate into which is screwed a bolt with a nylon ball at the end. The balls engage cup bobbins laminated into the adjacent bodywork. Recommended safety factors in design are:

Short term static	2
Long term static	2
Variable unidirectional	4
Repeated	5
Reversed alternating	5
Impact loads	10

Fatigue strength has been found to lie between 0.20 and 0.35 times the mean static strength under fully reversed stress conditions at one million cycles. Fig 6 shows a Goodman diagram in which it is seen that experimental mean stress amplitude falls below the Goodman line for structural metals.

GRP special construction techniques

Reservoir moulding is a method of producing GRP sandwich panels. Here a reservoir of flexible open-cell foam is impregnated with a resin and sandwiched between two layers of dry fibre reinforcement. The three layer sandwich is placed between dies under a

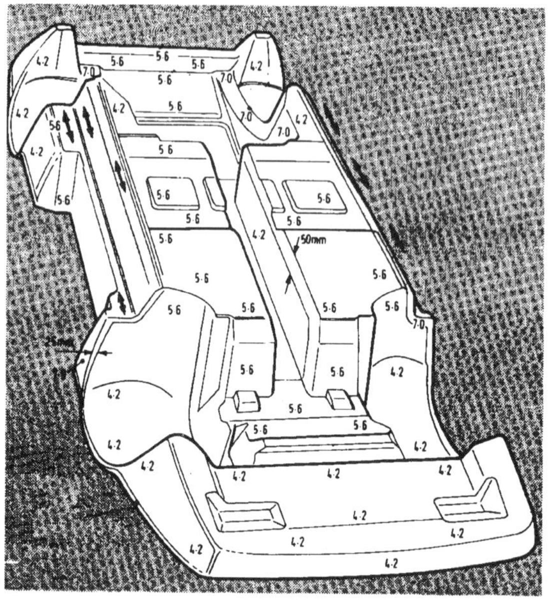

Fig 8: Weight of glass in moulding

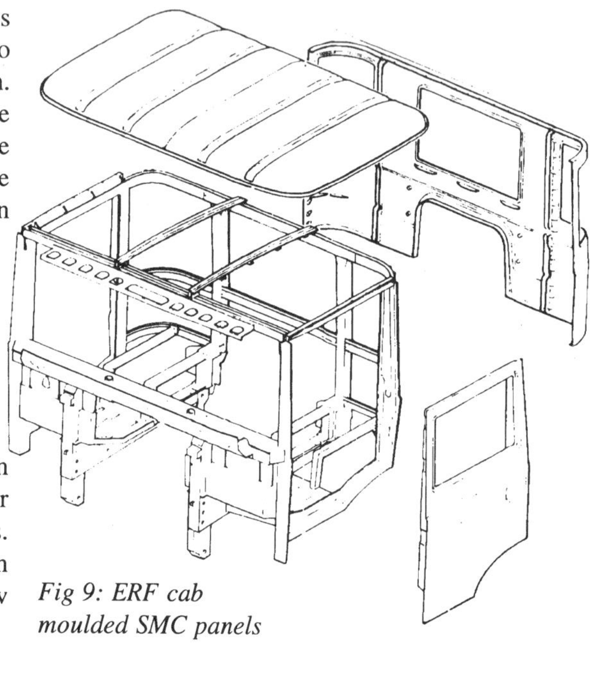

Fig 9: ERF cab moulded SMC panels

pressure of just 12 kg/cm^2. The foam acts as a sponge which, during moulding, squeezes the resin into the fibre, Fig 7. Sandwich stiffness can be altered by varying compression pressure and the tooling can be simplified by using one rigid and one flexible die-face such as a liquid-filled bag. The VARI (vacuum assisted resin injection) process was pioneered by Lotus for car body shells but is now available for a variety of licensed manufacture applications, including truck cabs. Here the part is made between matched mould surfaces after laying up of the reinforcement by hand. Vacuum is then applied to the space between the mould faces and resin automatically drawn in to the cavity. Preformed foam cores can also be placed in the mould to achieve localised box sections within the main part. It is also now possible to preform the glass reinforcement to speed the lay-up process. Another recent development is a technique for making metal faced moulds which can be heated to further shorten the cure cycle. Fig 8 shows the bottom half of a Lotus car structure, made by VARI, with numbered panels indicating the weight of glass (in lb/ft^2) used in each area of the moulding.

Sheet moulding compound (SMC)

The ability to produce sheets of glass fibre reinforcement impregnated with a catalysed layer of polyester, within protective layers of polyethylene film has led to sheet moulding compounds (SMC). These simplify the process of matched die moulding to give relatively high production rates. Low profile resin systems can be employed for exterior body skin panels requiring high surface smoothness. Four types of SMC reinforcement are common: short chopped glass rovings in a random pattern for equal strength in all directions; endless rovings arranged parallel to chopped fibres for unidirectional strength; medium length rovings arranged parallel but staggered to obtain better mould flow than in the last case; finally a wound formation of crossed roving tapes laid at 20 deg again to effect directional strength. Fig 9 shows the larger panels of ERF's truck cab which are fitted over a steel frame.

Non-glass composites

Structural performance comparable with metals can be had with the addition of materials such as para-aramids and carbon fibres. These materials are already commonplace in racing car structures and offer possibilities for such parts as slim-line windscreen pillars on road cars and trucks. The Kevlar brand of para-aramid, by Du Pont, exists in two states: firstly as a 0.12 mm thick endless fibre with cut length varying from 6 to 100 mm, secondly in the form of 'pulp'. It is said to be 30 percent stronger than glass. The company also make a so-called meta-aramid, Nomex, which is a 'paper' that can be formed into lightweight honeycomb cores of sandwich panels. Courtaulds Carbon Fibre Division have tabulated the properties of high strength composites, compared with metals, as in Fig 10, which includes S and E-glass composites as a datum.

High-strength composites

Thermosetting composites are now being specified for 'mechanical' as well as 'structural' parts. The former require close attention to form design for properly exploiting the special properties of both the reinforcing materials and the resin matrix.

The National Engineering Laboratory has been prominent in the UK development of advanced composites, co-ordinating research work of a consortium of materials, component and vehicle manufacturing companies. An important outcome has been the

	Tensile strength (GPa)	Tensile modulus (GPa)	Specific gravity	Specific strength	Specific modulus
A-S/epoxy	1.59	113	1.5	1.06	75
XA-S/epoxy	1.90	128	1.5	1.27	85
HM-S/epoxy	1.65	190	1.6	1.03	119
S-glass/epoxy	1.79	55	7.0	0.90	27
E-glass/epoxy	1.0	82	2.0	0.50	21
Aramid/epoxy	1.29	83	1.39	1.00	60
Steel	1.0	210	7.8	0.13	27
Aluminium L65	0.47	76	2.8	0.17	26
Titanium DTD 5173	0.96	110	4,5	0.21	25

Fig 10: High-strength composite properties

development of a composite suspension system for light vehicles. It is based on a hot press moulded arm incorporating continuous glass reinforcement. It has been made using filament winging and resin transfer moulding techniques and follows on NEL work to co-ordinate the production of GRP springs and suspension arms.

Volvo researchers on the team came up with the twin-leaf design, Fig 11, which integrated suspension arms and springs in one piece. This comprised a crossmember connecting the twin-leaf trailing arm/springs. Additional constraints on the 'springs' were located at the forward end where the trailing arms attach to the vehicle and at an intermediate point restricting the arm's vertical motion. Epoxy resins were used as the matrix for the glass. Reaction blocks between the leaves were given shear flexibility by layers of composite laminated with elastomeric material; the blocks are bonded in position such that there are no sliding surfaces in the whole design.

NEL have listed the different fabrication processes for composites as shown in Fig 12. Reinforcing materials now include glass, aramid and carbon fibres — also a number of other mineral fibres and mats including cloth and paper. A number of different weaves are used with the woven fibre reinforcement configurations. These include plain-weave, with warp and weft woven alternately; satin-weave in which the weft passes beneath three warp yarns and over one — and vice-versa; twill-weave in which groups of fibres are plain woven to produce a diagonal pattern; finally uni-directional weave which signifies fibres lain together in parallel strands for maximum directional effect in physical properties.

The common thermosetting-resin systems are polyester, vinylester, epoxy, phenolic and bismalyamide. Because thermosets will not revert to their liquid state when heated, their temperature tolerance is generally higher than for thermoplastics. Compressing resin and fibres together sufficiently to get a reliable bond between them has led to systems of curing mouldings under pressure such as vacuum-bag and autoclave. Designing in press-moulded GRP involves a similar approach to form-design in unreinforced plastics. Maximum parts integration in to a single moulding; increase stiffness by using

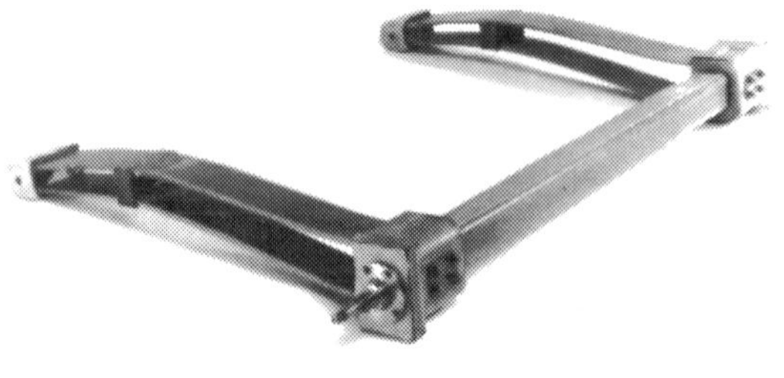

Fig 11: Volvo twin-leaf composite spring

Hand lay-up	**Layers of reinforcing mat or fabric are cut to fit, laid in the mould, and saturated with resin mix. The product has only one finished side. Boat hulls are often manufactured with this process.**
Spray-up	**The resin/reinforcement mixture is sprayed into a mould of relatively complex shape. Used for objects like tubs and showers.**
Pultrusion	**As rods and tubes are pulled through a moulding machine, resin is applied. This process is used when high strength along the length of the finished product is desired. Fishing poles and construction beams are fabricated with this process.**
Compression moulding	**If volumejustifies tooling and set-up cost, this is the most efficient way to produce large, contoured, high-strength components. The resin/reinforcement mixture is applied between matched male and female parts and subjected to high heat and pressure.**
Injection moulding	**Resin and reinforcement are forced by a screw or plunger into a matched metal mould where it takes shape and cools. This process is suitable for manufacturing a large quantity of small parts quickly. Decorative knobs and handles, pulley wheels, and electrical junction boxes are examples of injection moulded products.**
Filament winding	**Fibre is would around a suitably shaped mandrel in a predetermined pattern. Resin impregnation can occur during or after winding. This process produces products with exceptional strength. Underground storage tanks and pipes are the most common filament-wound products.**

Fig 12: Fabrication processes

Material	Glass fibre by weight %	Flexural strength lbf/in² x 10³	Flexural strength kN/m²	Flexural modulus lbf/in² x 10⁵	Flexural modulus GN/m²	Tensile at Yield lbf/in² x 10³	Tensile at Yield kN/m²	Tensile modulus lbf/in² x 10⁵	Tensile modulus GN/m²	Ult. Tensile elongation %	Izod impact ft. lbf/in notch	Flammability	Continuous heat resistance °F	Continuous heat resistance °C
SMC	15-30	18-30	124-207	14-20	9.6-13.7	8-20	55-138	16-25	11-17	0.3-1.5	8-22	Slow	300-400	150-200
DMC (BMC)	15-35	10-20	69-137	14-20	9.6-13.7	4-10	28-69	16-25	11-17	0.3-0.5	2-10	to	300-400	150-200
Preform/mat	25-50	10-40	69-276	13-18	8.9-12.4	25-30	172-207	9-20	6-14	1-2	10-20	non-burning	150-400	65-200
Cold press moulding polyester	20-30	22-37	152-255	13-19	8.9-13.1	12-20	82-138	—	—	1-2	9-12	Non-burning	150-400	65-200
Spray-up polyester	30-50	16-28	110-193	10-20	6.9-13.7	9-18	62-124	8-18	5.5-12.4	1.0-1.2	4-12	Burn to SE	150-350	65-176
Moulding compound (phenolic)	5-25	18-24	124-165	30	20.6	7-14	48-117	26-29	17.9-20	0.25-0.8	1-6	SE	325-350	165-176
Nylon	6-60	7-50	48-345	2-26	1.3-18	13-33	90-228	2-20	1.3-13.7	2-10	0.8-4.5	Slow to SE	300-400	150-200
ABS	20-40	23-26	158-179	9.2-15	6.3-10.3	11-16	76-110	6-10	4.1- 6.9	3-3.4	1-2.4	Slow	200-230	93-110
Polypropylene	20-40	7-11	48-76	3.5-8.2	2.4- 5.6	5.5-10.5	38-72	4.5-9	3.1- 6.2	1-3	1-4	0.85-0.75 in/min	300-320	150-160
Thermoplastic polyester	20-35	19-29	131-200	8.7-15	6.0-10.3	14-19	96-131	13-15.5	9-10.7	1-5	1.0-2.7	Slow	275-375	135-190

Fig 13: Properties for a heavy-duty mix material

double curvature and well radiused curves (this also assists fibre distribution and air removal); as much taper as possible to facilitate mould removal; use of ribs, hollow sections and lightweight cores for localised stiffness; use of large area washers and metal inserts to spread shear and bearing loads at attachment points (as stress concentrations are not relieved by yielding as with ductile metals): these are the main ground rules. Fig 13 shows mechanical properties for the heavy-duty mix shown and Fig 14 gives design rules in pictorial form.

Thermoplastic materials/composites

Separate, related, features deal with grp composites and moulding design for ease of production; here we examine the properties of unreinforced thermoplastics and some of the application opportunities. These materials span end-uses from soft trim through to mechanical linkages and the choice of polymer system is crucial to a proper exploitation of properties.

Wall thickness variation	None recommended (maximum possible: 1·5:1 for air plenum preform 2·1 for directed fibre preform)	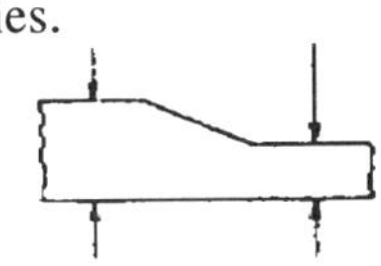	Changes must be tapered (running out from the middle to the edges)
Local stiffness	Combine with metal reinforcement, properly moulded in	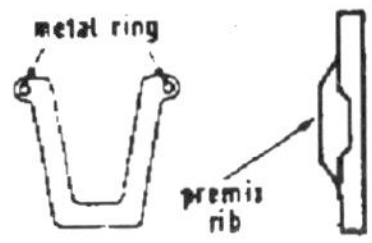	Continuous ribs are preferred to individual bosses or lugs
Tolerances	$t \pm$ 0·005 in $l \pm$ 0·003–0·005 in/in (out-of-round 0·003 in/in)	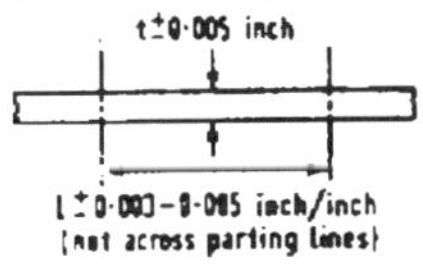	Uneven distribution of glass is the principal cause of high tolerance level
Draft	Minimum is 1°; for parts over 18 in deep use 2–3°	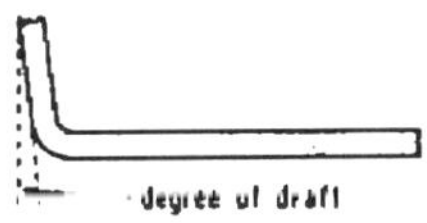	Amount depends on depth of draw; it can be reduced with highly polished moulds. Greater draft is required on interior than exterior, to protect against shrinkage effects
Corner radii	Minimum ¼ in	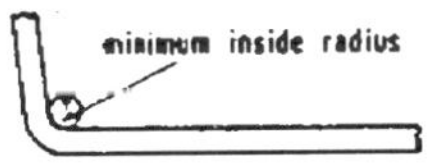	Maximize reduced stress concentration and better glass distribution
Undercuts	None recommended	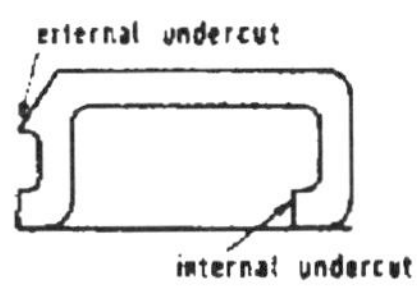	Mould can often be tilted, and drafts used, to overcome an *apparent* undercut; otherwise, split moulds are necessary
Flash lines	Ensure vertical	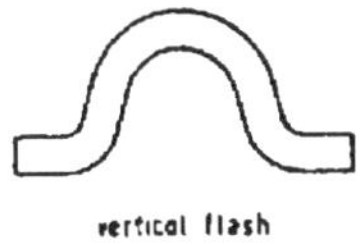	Horizontal flash lines can be avoided by designing with extra material for cropping after moulding
Warpage	Use rounded in preference to flat forms		Adopt narrow lips and flanges on edges on any flat parts
Holes	Adopt moulded-in holes if $A_1 > 5°$		For moulding, holes must be on a surface perpendicular to press travel; otherwise adopt machined holes

Fig 14: Design rules

Classification of thermoplastics is best rationalised by going back to the beginnings of synthetic materials and tracing the family heritage of some of the present key contenders amongst the existing family of useful polymers. The 1860s saw some of the early successes with synthetic materials based on the co-valent bond with carbon of such elements as hydrogen, oxygen, nitrogen and chlorine. The large molecule 'polymers' (meaning 'many bricks'), having extended chains of elements, gave rise to solid substances with unusually strong forces holding the elements together. These forces were modified with plasticisers to achieve varying degrees of flexibility and, with those materials known as thermoplastics, by applied heat which causes melting rather than decomposition. Operating extremes of rigid thermoplastics are defined by low and high temperatures leading to either 'glass-like' or 'rubber-like' states. As well as plasticisers, fillers are added which reduce cost for a given bulk; also stabilisers to improve weathering, colourisers and, to improve mould flow, lubricants.

PVC exists in many forms according to the amount of plasticiser added; cracking after long use usually signifies loss of plasticiser. Polythene exists in both crystalline and amorphous states representing two markedly different densities; as the material is difficult to keep stable under ultraviolet light, it is used mainly in hidden situations such as cable insulation. Polypropylene and acetal resins are similar in derivation but very much tougher. Polyamides, like nylon, come from the separate phenol resin derivative, with the ability to align crystals in the direction of a common fibre giving rise to relatively high strengths though when untreated, water absorption is also high. Another resin derivative is polyester leading to the styrene polymers such as ABS. Higher strength polymers come from the epoxy resin branch. Non crystaline plastics, with lightly cross-linked polymer chains, are the synthetic 'rubbers'; as well as the flexible forms, certain polyurathane rubbers can be cast as solids.

Engineering thermoplastics

As well as injection-moulding, these can be thermoformed, extruded and blow-moulded and are often reinforced. Projected price reduction, relative to hot-rolled steel and aluminium, suggests economic substitution by the early 1990s with likely exploitation in all classes of special vehicle bodywork. In Europe the main, multi-application, contenders in engineering plastics are considered by the British Plastics Federation to be ABS, polyamides, polyacetals, polycarbonates, polyphenylene oxides and reinforced polypropylenes.

Polypropylenes and polyamides can be modified in many ways to improve tensile strength, creep resistance, thermal expansion characteristic, low-temperature impact strength and high-temperature stiffness. Chemical modification includes addition of ethylene as a co-monomer to polypropylene and rubber to polyamides — both to obtain greater impact strength. Particulate fillers, such as glass and mineral beads, can be used to improve heat distortion temperature, rigidity, dimensional stability and creep resistance. Thirdly, the use of glass and other fibres (usually shorter than those used with thermosetting resins in GRP composites) can improve heat ageing as well as most of the characteristics listed above — but with the disadvantage of causing some

Material	Flexural modulus GPa	Notched Izod impact strength J/m	Heat distortion temperature °C
Polypropylene homopolymer	1.7	45	65
Polypropylene + 40% calcium carbonate	2.8	53	75
Polypropylene + 40% talc	4.2	35	97
Polypropylene + 30% coupled glass fibre	6.5	100	148
Development grades PXC 51370	3.3-5.3	300-230	144-152

Fig 15: Polypropylene co-polymer properties

To meet motor industry requirements for adhesion, wqter-soak and stone-chip resistance:
1 Apply adhesion-promoting primer and dry in hot air from 70-90 C.
2. Apply surfacer coat for exterior parts, typically an alkyd stoved at 120-140 C.
3. Apply solid colour top-coat, typically a polyester system stoved at 135 C.

Fig 16: Painting procedure for polypropylene

directionality of strength and rigidity.

Generally the amorphous thermoplastics — PC, PPO, and ABS — are valued for their low shrinkage and good impact strength but fatigue and chemical resistance characteristics may not suit certain operating environments also reinforcement is said to do little for these polymers. By contrast, crystalline thermoplastics such as PA and PP have a lower temperature tolerance — but this can be corrected by reinforcement. PP has been used as a fascia panel material, with both mineral and rubber additions — in a polymer termed olefinic thermoplastic elastomer (OTE). Glass reinforced versions can be used for such applications as bumpers and spoilers. Further specialist polypropylene copolymers have properties with potential for exterior body panel application, Fig 15. A suitable painting procedure for such panels is shown in Fig 16. A stampable sheet form of the material is particularly attractive for general structural and enclosure work — with properties as shown in Fig 17. Here a continuous glass mat is used for the reinforcement, giving good strength but not particularly smooth surface finish. Complicated ribs and bosses can readily be formed into finish structures and moulding times are from 20-50 seconds using cheap cast metal tooling.

Stampable sheet for bodywork

Understanding of plastics fabrication as a 'flow' process, in which there is only one step to a completely finished part, is crucial to successful design. As the material flows into the heated mould cavity, it permits varied section thicknesses, bosses, ribbing and complex contours. As the finish plastic has an 'elastic' modulus about 1/30th that of steel it will normally be required to have a thickness three times that which is effective for a design in steel. The ability to use ribbing to obtain flexural rigidity does, however, mean that reduced thicknesses are possible. Fig 18, due to the US company Allied Fibres & Plastics, shows the effectiveness of different rib types. The technique can also be used to stiffen brackets by the addition of a flange as shown in Fig 19. Also corrugated walls can be a useful form of stiffening, Fig 20.

At points of attachment, either spreader-washers, for through bolts, or metal inserts, for tapped holes, should be used. Technique for calculating stresses for a press-fit insert are shown in Fig 21.

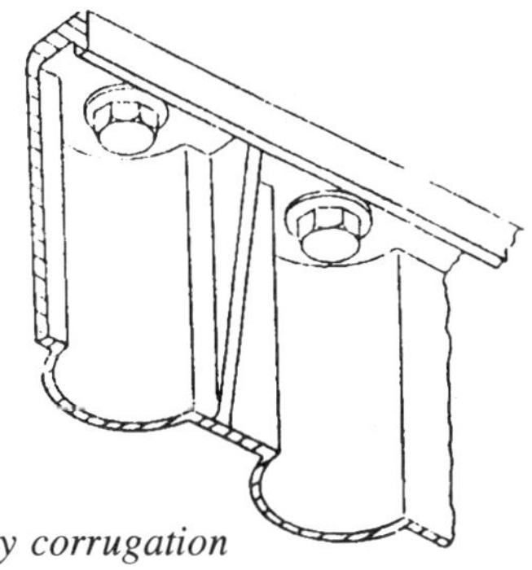

Fig 20: Stiffening by corrugation

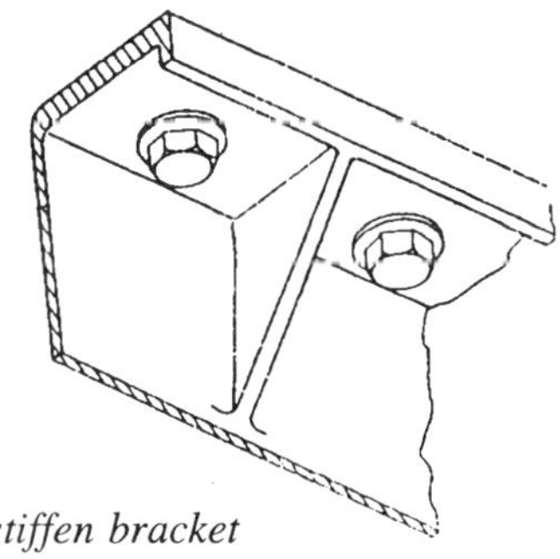

Fig 19: Rib to stiffen bracket

Property	HW60GR30	'Azdel'
Density/(g/cm³)	1.12	1.19
Flexural modulus/GPa	6.5	5.5
Tensile yield stress/MPa	86.2	75
Heat distortion temperature/°C (18.6 kg/cm²)	148	155
Notched Izod impact strength/(J/m)	100	535
Coefficient of thermal expansion/(10^{-5} °C^{-1})	3	2.7

Fig 17: Properties of Azdel stampable pp sheet

Case	Shape	Change	Moment of inertia	Increase in I ΔI	Increase in weight ΔW	Ratio I/wt $\Delta I/\Delta w$
1		Base 2" x ¼"	0.0026			
2		Double height	0.0208	700%	100%	7
3		Add ⅛" wx ¼" H Rib	0.0048	85%	6.25%	14
4		Add ¼" wx ¼" H Rib	0.0064	146%	12.5%	12
5		Add ⅛" wx ½" H Rib	0.0118	354%	12.5%	28
6		Add ¼" wx ½" H Rib	0.0194	646%	25%	26

Fig 18: Ribbing configurations

A common stampable thermoplastic is Radlite GMT which is 30-40 percent glass-reinforced polypropylene; this is suitable for non-painted semi-structural applications such as aerodynamic undertrays. Typical properties are shown in Fig 22 and modulus distribution in a flow-formed part in Fig 23. Effect of mould flow is to produce glass orientation which can give directionality of strength and stiffness. Usually the 'flats' receive the greatest moulding pressure and thus have the best material properties. Side walls are moulded at lower pressure which results in lower strength. If the flow path is restricted by a long thin area, the resin can 'outrun' the glass resulting in further loss of strength. The moulding process involves cutting sheet into specific blank sizes, preheating the blanks in an infrared oven and placing them in a presstool. Pressure is applied for 30-45 seconds.

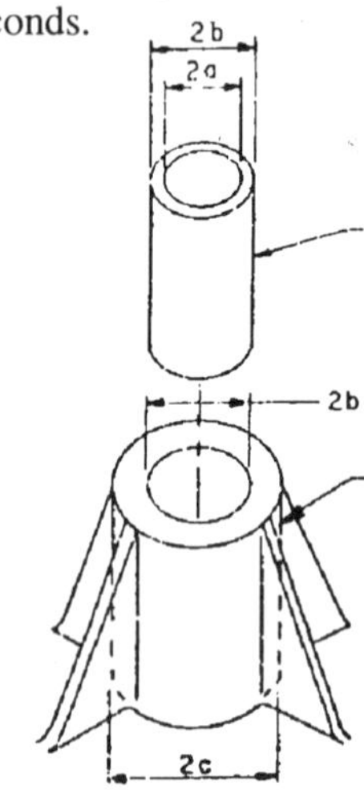

$$\Delta = pb\left[\frac{1}{E_1}\left(\frac{b^2+c^2}{c^2-b^2}+\nu\right)+\frac{1}{E_2}\left(\frac{b^2+a^2}{b^2-a^2}-\nu\right)\right] = pbK$$

$$K = \frac{1}{E_1}\left(\frac{b^2+c^2}{c^2-b^2}+\nu\right)+\frac{1}{E_2}\left(\frac{b^2+a^2}{b^2-a^2}-\nu\right)$$

$$P = \frac{\Delta}{bK}$$

Δ = Sum of radial deformations

$E_{1,2}$ = Moduli of Elasticity of Outside & Inside Cylinders

P = Shrink Fit Pressure

S_D = Allowable Design Stress

ν = Poisson's Ratio

$$\max S_{hoop} = P\frac{b^2+c^2}{c^2-b^2}$$

$$S_D > S_{hoop}$$

Fig 21: Press-fit insert stress calculation

ABS and the styrene plastics

For semi-structural applications at relatively low cost ABS is a useful plastic. Related styrene plastics such as SAN also have significant automotive application as does the sheet and laminate forms of ABS

The styrene plastics involve four main products of commercial interest, styrene homopolymer, butadiene-styrene (BS) copolymer, styrene-acrylonitrile (SAN) copolymer and the familiar ABS copolymers. Styrene homopolymer has limited use at ambient, or above, temperatures but toughened polystyrene has structural uses, particularly in expanded form, as a self-skinning foam. SAN is best known for such applications as household cups but has a structural application in an 'alloyed' form to be discussed below. ABS is one such alloy, different grades of which depend on relative proportions of nitrile rubber, butadiene, styrene and SAN. Alloying with PVC and polycarbonate also yields useful properties. The material is, of course, widely used for interior body

Composition by weight			
polypropylene:	75	60	per c
glass fibre, 13mm	25	40	per c
Density (consolidated)	1.08	1.22	g/cc
Flexural Properties			
modulus	5500	6500	MPa
peak strength	130	165	MPa
Heat Distortion Temperature (1.8 MPa)	154	159	°C
Impact Strength			
(notched) Izod	400	650	J/m
(Notched) Charpy	45	70	KJ/m²
Falling Dart (3.5mm thickness)	17	20	J

Fig 22: Properties of Radlite

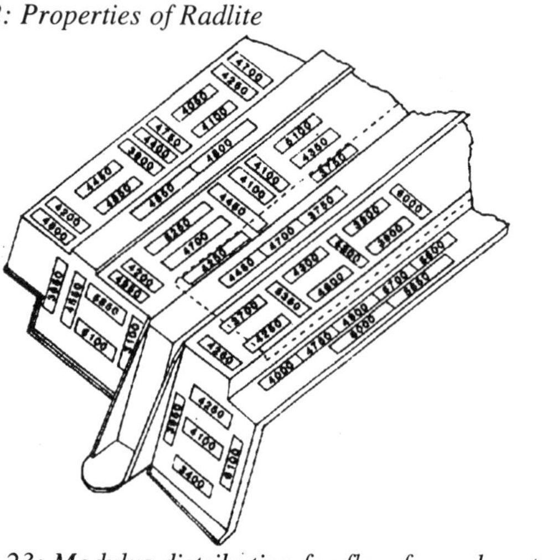

Fig 23: Modulus distribution for flow-formed part

As the butadiene content is reduced, these properties diminish	these properties increase
Impact resistance	Tensile strength
Elongation	Creep resistance
Deflexion under load	Rigidity
Flex life	Flexural strength
Opacity	Gloss
	Abrasion resistance
	Chemical resistance
	Surface hardness
	Translucency

Fig 24: Varying butadiene content

	styrene acrylonitrile copolymer	Lowest Type 1	Intermediate Type 2	High Type 3	Highest Type 4	Heat resistant
Impact, Izod, notched $\frac{1}{8}$ in × $\frac{1}{2}$ in bar (ft lb/in)						
@ 73°F	0·40	1·1	2·0	3·2	4·3	2·0
@ 0°F	—	0·8	1·4	1·7	2·0	—
@ − 40°F	0·40	0·6	0·8	1·0	1·4	0·5–1·0
Elongation (%)	2·5	10–15	15–20	20–25	25–40	5–10
Tensile strength at yield (lb/in²)	11 500	8000	7000	6500	6000	8500
Flexural strength at yield (lb/in²)	15 500	13 000	11 500	10 000	9000	12 000
Flexural modulus (lb/in² × 10^{-5})	5·1	4·3	3·8	3·5	3·2	3·5–4·0
Rockwell hardness (R scale)	122	113	110	95	90	112
Heat distortion temp. (°F), 264 lb/in² load	205	190–196	186–193	185–191	183–190	225

Fig 25: Physical properties: ABS derivatives of SAN

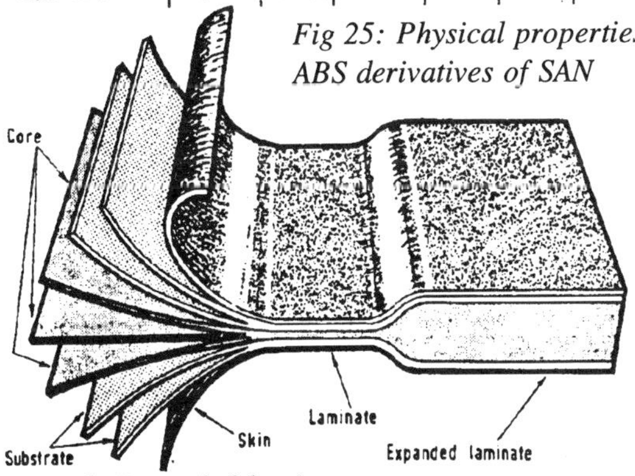

Fig 26: Expanded laminate

hardware and is ideal for ducting and utility trim-panels, under the general heading of 'light-engineering' plastics. General grades suit applications where heat requirements are not excessive. ABS alloys can range from low-impact high-rigidity grades to very high impact ones having considerable flexibility. The effect of varying the butadiene content is seen in the table of Fig 24; its complete removal from ABS leaves SAN. This last material possesses unusually high rigidity and is transparent so has found automotive application in cover panels for dashboard instruments. Typical physical properties in quantitative terms are in Fig 25. The distinguishing properties corresponding to most of the ABS copolymers are easy fabrication, good chemical resistance and outstanding dimensional stability within temperature limitations. Durability and creep resistance under load are both good by plastics standards.

The temperature-dependence of mechanical properties requires that special steps need to be taken in design calculation. Sheet forms of ABS can be calendered to obtain a range of surface textures, including simulated leather frequently used in automotive trim panels. Transparent SAN can also be used in conjunction with vacuum-metalizing to obtain pleasing display panels. Electro-plating of ABS is also a valuable surface treatment property — unsurpassed by most other thermoplastics. Details of plating ABS were given in the earlier section on coatings for plastics in Chapter 6, as was the thermoforming of ABS. An interesting variant on the sheet form is expanded laminate, Fig 26, which can also be thermoformed and is rigid enough for vehicle body shell construction. The performance of Royalex grade, from US Rubber, is seen in Fig 27; this brand has been used in cab production for the White Truck Corporation and in a variant of the Ford Cobra as a body shell material.

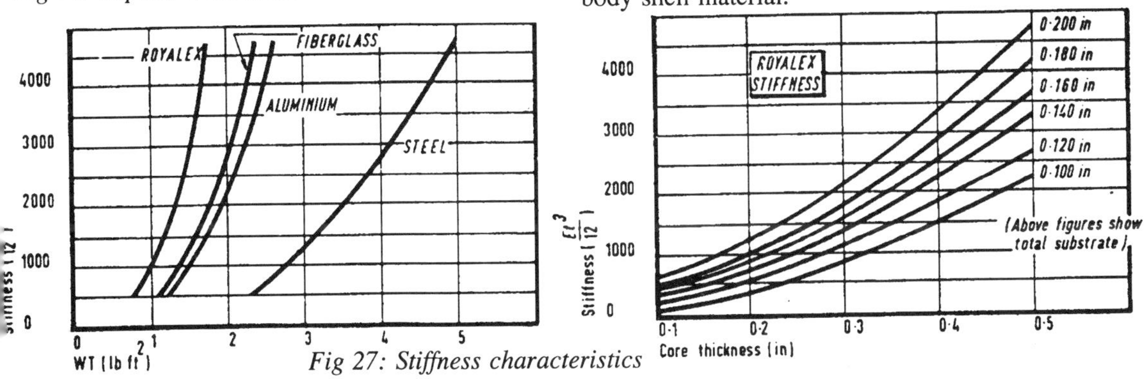

Fig 27: Stiffness characteristics

Thermoforming plastic sheet

Vacuum and pressure forming are used for shaping thermoplastic sheet around a former and as low-cost processes, are attractive for small runs of relatively low stress-bearing materials. Electric car bodies have been formed in materials such as polypropylene copolymer and acrylic. More common automotive uses, however, are for cab and body, Fig 28, interior parts in ABS and its copolymers.

ABS is particularly suitable for vacuum-forming, in which a heated and softened sheet is drawn into a mould by creating a vacuum between it and the mould surface. Moulds can be of wood, sheet-metal or filled epoxy resins. The limitation of vacuum-forming is the excessive drawing down at sheet corners which can lead to undue thinning of the material. This has led to drape-forming which involves insertion of the mould into the softened sheet before the application of vacuum. Two further variations are air-assist and plug-assist forming. The former employs partial inflation of the heated sheet by air pressure and the latter has a plug in the form of a mould lowered into the softened sheet prior to application of the final vacuum.

Large formings beyond the scope of injection mouldings can be produced. In pressure, rather than vacuum, forming, considerable accuracy of detail can also be obtained. However, it is not generally possible to obtain abrupt changes in wall thickness as is the case with injection moulding. Since internal stresses in the material are consequently low, with thermoforming, a stronger thin-walled item can generally be obtained. Generally, a minimum draft angle of 2-3 degrees applies to male moulds but female ones are more tolerant because of the tendency of the material to shrink away from the mould; and draft angles down to one degree are possible.

Shallow female moulds can even tolerate parallel sides. Generous radii are required, however, in order to minimise internal stresses. Absolute values depend upon wall thickness; see Fig 29a. In Fig 29b is seen how the inside and outside surfaces of thermoformed parts must essentially be parallel in contrast to those of injection mouldings. It can also be difficult to obtain sharp detail on the side opposite the mould when thick walls are being formed, Fig 30. Accurate consistency of wall thickness across the part certainly cannot be relied upon although this can to some extent be controlled by zoned heating and cooling. As a guide, parts formed on male moulds have thick bases and thin sides while those formed on female ones have the opposite.

Fig 31a, due to West[1], shows the air-assist technique used to improve wall thickness distribution. Vents are best drilled in areas of the mould where the material is last to form. Materials suppliers should be consulted on the appropriate hole size for

Fig 28: BMW Z-13 with thermoformed plastic body

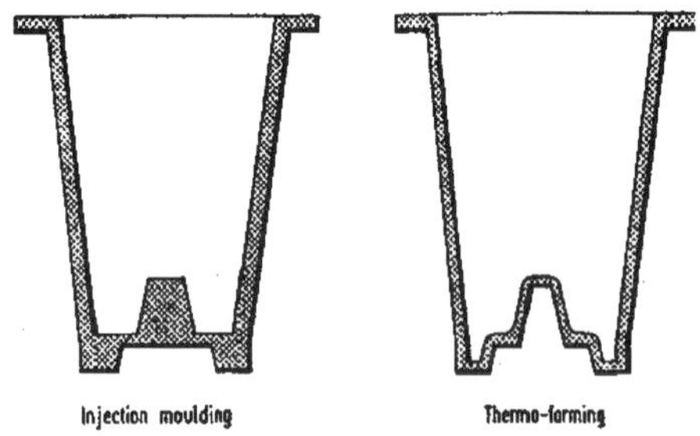

Fig 29a: Radius dependence on wall thickness

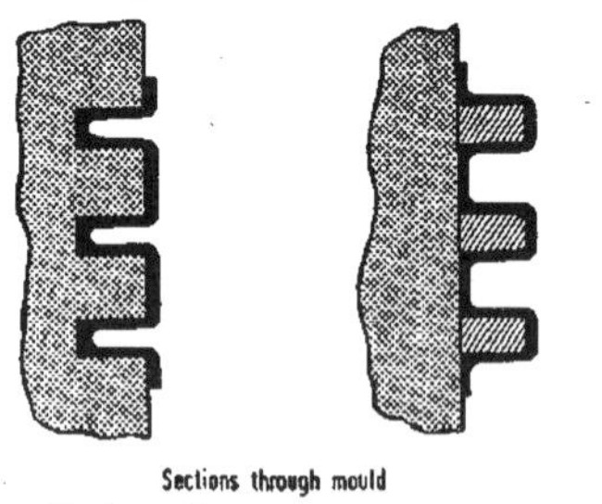

Fig 29b: Parallel wall requirement

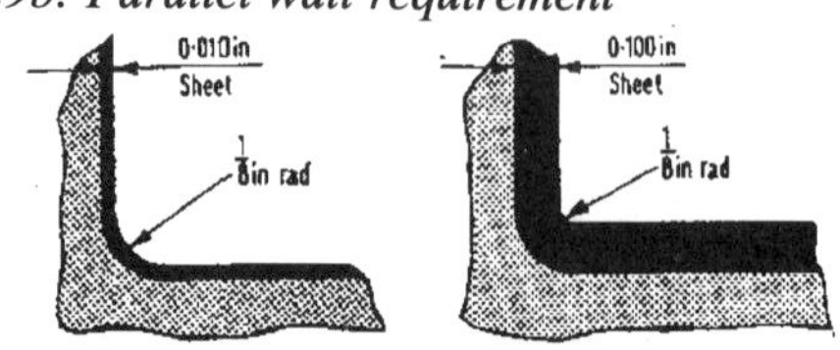

Fig 30: Sharp wall detail

their material, which will minimise witness marks. Sometimes it is possible to use a slot as seen in Fig 31b, in lieu of holes, depending on the shape of the part. Borg Warner have produced a useful guide to vac-forming ABS sheet to help determine draw ratio (Ratio of surface area S to area of blank within the clamp frame s). For a rectangular shaped part, length l, width w, and height h the ratio is obtained from:

$$S/s = [lw + 2lh + 2wh + 2(1.15l \times 0.1w) + 2w \times 0.75l]/1.15l \times 1.2w]$$

Sheet blank starting thickness T can be found from $T = CRp$ where $C>1$ for $R>3$ and is $=1.3$ for negative forming, 1.2 for positive forming and 1.1 for air-assist forming. Tolerances on finish-parts should allow for 0.7-0.8, for negative, and 0.3-0.5% shrinkages for positive moulds respectively. Heat cycle time varies with sheet gauge, surface finish, colour and heat source. In the case of ABS sheet with double-side heating to 165 C surface temperature, gas fired machines typically require from 20 to 310 seconds for sheet gauges from 1 to 10 mm. These should be factored by 1.4 for gauges up to 7 mm with single-sided heating systems. Cooling channels should be provided in the moulds to reduce component temperature down to 65 C prior to withdrawal from the mould. Surface finish on the mould should be within 5-10 microns for optimum air evacuation and recommended vacuum line cross-sectional area is 2000 mm^2/m^2 of mould area and total area of vacuum holes is equal to 1.2 times that of the vacuum line.

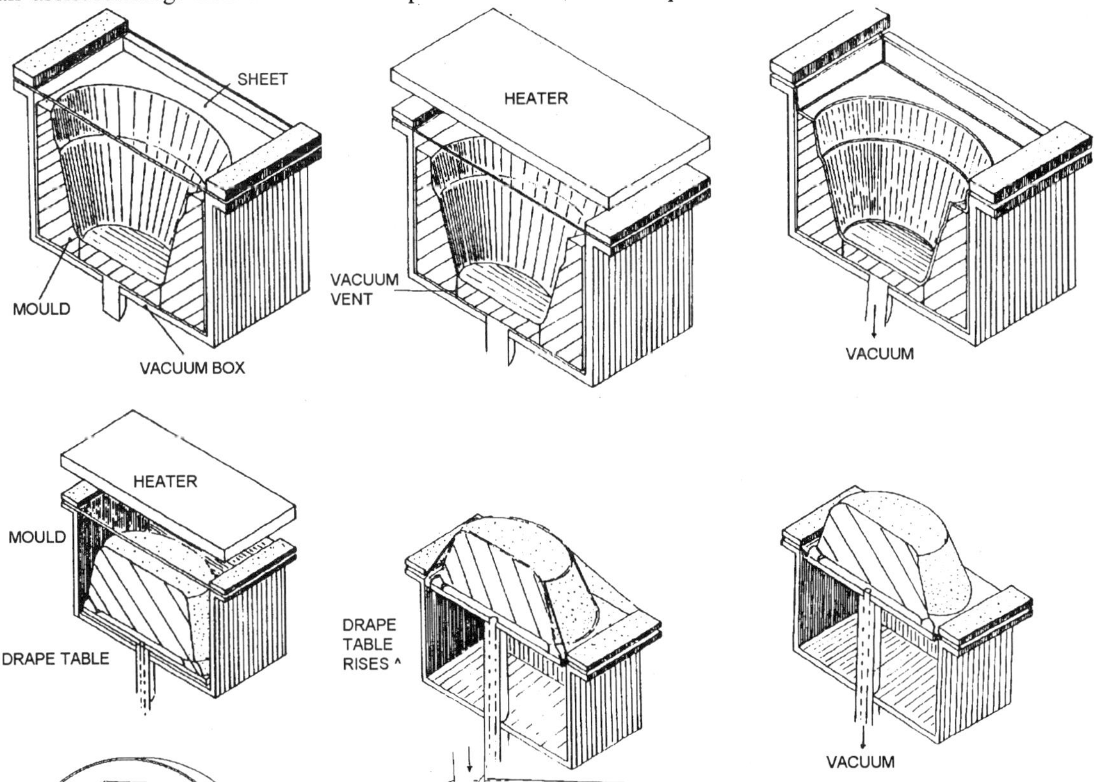

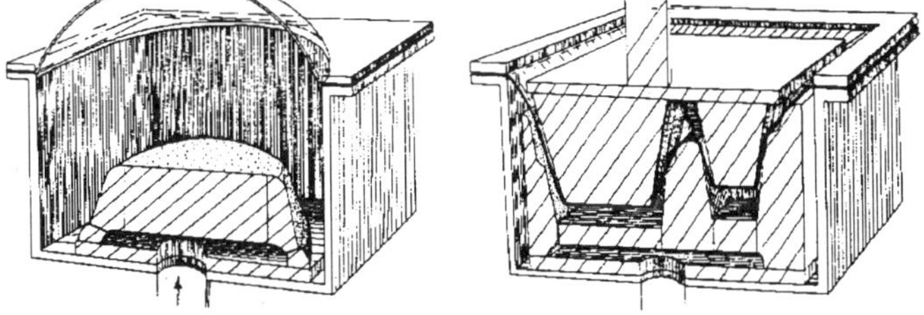

Fig 31a: Vacuum-forming (top) and drape forming (centre) compared with air-assist technique (bottom)

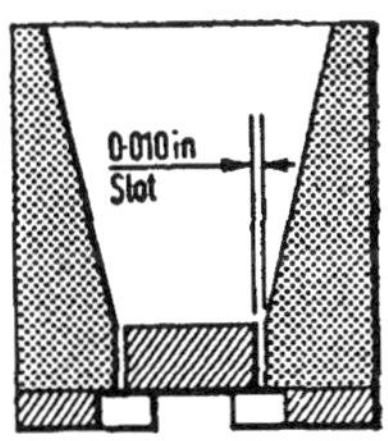

Fig 31b: Use of slot

Load-bearing plastics

Design techniques for load-bearing plastics inevitably differ from those applying to elastic materials and continuous attention must be paid to creep properties when predicting stresses and deflections. This section gives some rules for predicting stresses and deflections of loaded parts.

Plastic material specifications must state service temperatures and duration of loads when using the so-called pseudo-elastic design method. The value of 'elastic' modulus used must be relevant to these service conditions — and because, for most plastics, modulus value is unlikely to be above a few GN/m^2 at the very best, high inertia cross sections are the rule for minimising deflections.

Creep is defined as the strain, which is time dependent, resulting from applied load — and associated stress relaxation is the stress, which is time dependent, resulting from the applied deformation. Given a defined duration of loading, a (secant) creep modulus may be calculated as the ratio of the applied stress to the applied strain at the loading time and temperature of interest. Modulus is generally strain-dependent and in carrying out material tests it is useful to represent creep data by cross-plotting to obtain isochronous stress-strain curves for different times under load.

Design approach is to first identify maximum service temperature and duration of load; next to calculate maximum stress applying to the particular component; then to read the strain from the appropriate creep data and calculate the value of creep modulus; finally to use this value in elasticity calculations to predict deformations as stress relaxation in the component — arriving at the final dimensions by an iterative process.

Considerable increases in bending stiffness per unit weight can be obtained by exploiting the section-inertia increases which are so easily achieved with plastics parts. Deep thin webs or wide flanges must however must be examined against the possibility of premature buckling failure.

The following worked example applies to an extruded T-section strip in acetal co-polymer, 1 in wide x 1 in deep (2.5x2.5 mm), with general wall thickness of 0.2 in (5 mm). It forms a beam fixed at both ends and is assumed to have to support its own weight at a temperature of 20 C for 10 years. Simple calculations show the section inertia (second mo-

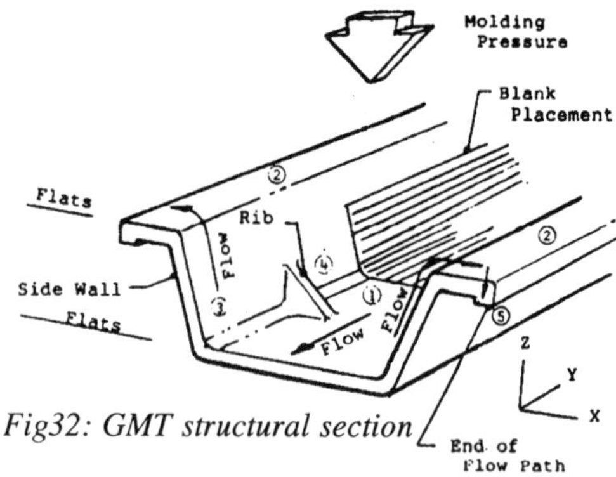

Fig32: GMT structural section

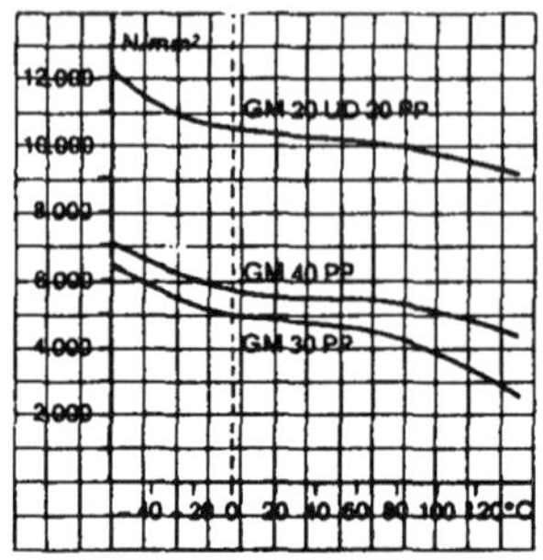

Fig 33: Flexural modulus vs temperature

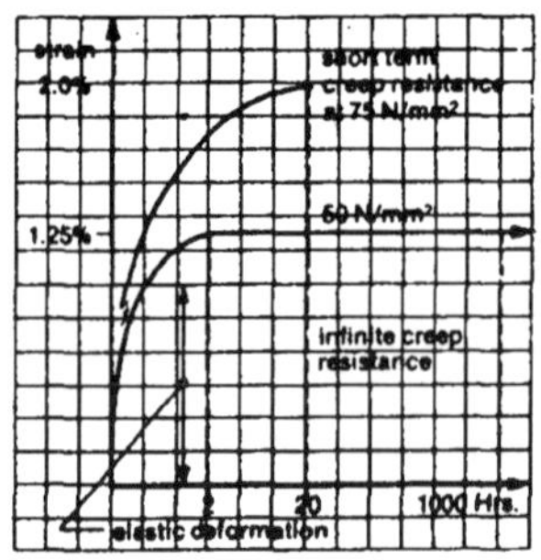

Fig 34: Creep resistance of GM 40 PP

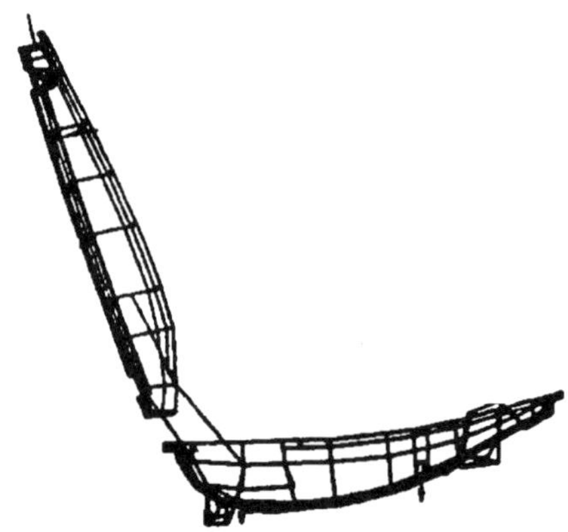

Fig 35: Chrysler seat structure

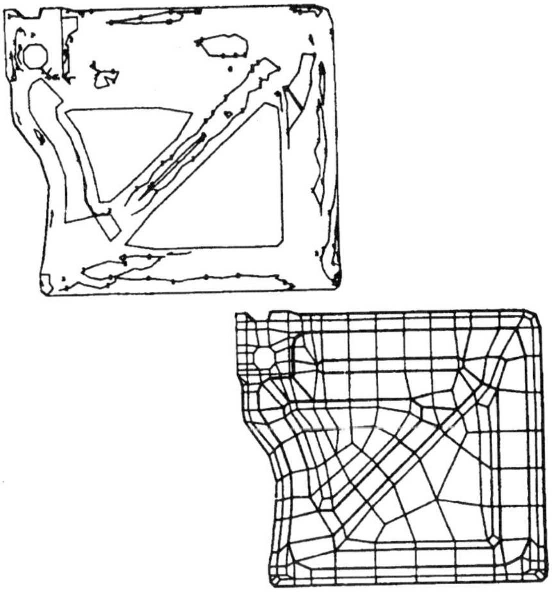

Fig 36: Von Mises stress plot

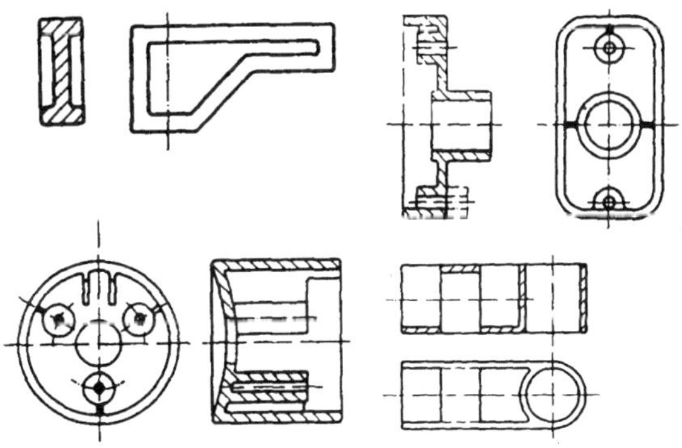

Fig 37: Rib design

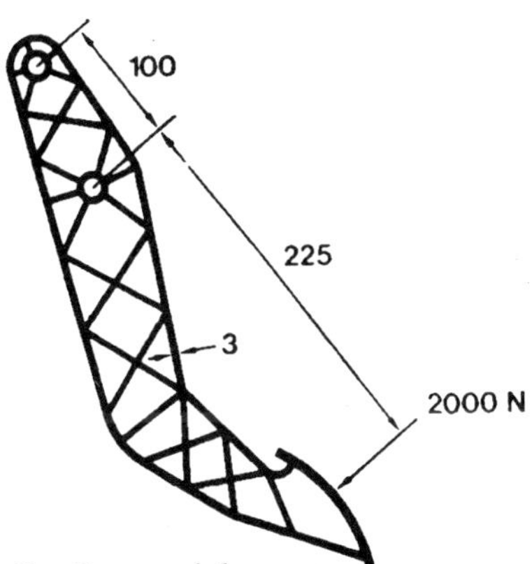

Fig 38: Du Pont pedal structure

ment of inertia) to be 0.0313 in^4 (from imperial tables), cross-section area 0.36 in^2 (2.32 cm^2) and density 0.51 lb/in^3 (14 g/cm^3). Assuming a strain of less than one per cent; estimated tensile creep modulus, from tables, is 1.1 x 10^5 lbf/in^2 (758.5 MN/m^2) at 1% strain, 10 yr, 20 C. The appropriate bending deflection formula then gives a maximum central deflection of 0.18 in after 10 years.

It is important in design to note that notch-sensitivity to impact failure is particularly prone with some plastics. Different grades of a generic class of materials can also have quite different impact properties. Stress concentrations must be reduced by the use of generous blend radii at corners and elsewhere. When load is removed from a thermoplastic, strain is recoverable provided the material has not yielded. Initially the recovery rate is high but it decreases with time. So for periodic loading conditions, the strain at the end of each loading period increases with time at a lower rate than if the load were maintained constant.

For a component like a seat structure, the load imposed by the occupant is rarely constant and effects such as braking and acceleration must be taken into account. The latter increases deflection response and vice versa. It has been found by experiment that strain response to increase in load decreases with number of times the event has occurred.

The form of the component affects its mechanical properties. In a member such as that in Fig 32, the 'flats' receive the greatest moulding pressure and have the best material properties; the sidewalls are moulded at lower pressure and so have lower strength. Typical modulus and strength figures for various elements of the GMT section shown in the diagram should be ascertained in its assessment. Fig 33 shows the material properties vs temperature for GMT while Fig 34 shows creep resistance of the unreinforced polypropylene which is the base material.

In seat design, deflection is the usual limiting criterion and tests by Chrysler have shown, for example, that a seat may reach limiting US Federal Motor Vehicle Safety Standards deflections while experiencing only 30 MPa stress levels. However, fatigue stress levels are important, it is argued; under a 800 N rearward load, the seat should not experience more than 50 MPa to obtain 'infinite' life. If the seat design can be modelled for FE analysis, a Von Mises stress plot should be predicted (this is a combination of 3-dimensional tensile and shear stresses that represent a failure criterion) — as shown in Fig 36 for the split back-folding rear car seat of Fig 35.

An example of rib design to minimise deflections is shown in Fig 37. In the pedal member designed by Du Pont in Fig 38, test deflection criterion for the part was taken to be a maximum of 10 mm at full load which is considered as 200 N for a throttle pedal, 500 N for a clutch and 2000 N for a brake pedal. The way different elastic moduli effect the result, for four of Du Pont's thermoplastics, is shown in Fig 39. For its Delrin brand co-polymer the company have released time-dependent stress-strain curves as shown in Fig 40. Some designers favour calculating long-term creep modulus from secant modulus E_c for a given deformation h; a value is suggested of $(1/2)E_c/0.1h$ for 1% strain. This is factored by 0.7 per 70 F to allow for thermal loading. Additionally, safety factors are used to approximate the static load condition as 1.5-1.9 for amorphous and 1.2-1.8 for semi-crystaline thermoplastics. VW designers suggested the approximate load factors of Fig 41 for moderately stressed shaped plastic parts. This shows how little remains of the original elastic modulus after long-term loading at 212 F.

For their Delrin resins, Du Pont recommend designing where possible with uniform wall thickness to avoid internal stresses caused by moulding. The transition from one wall thickness to another should also be gradual. Ribs should be specified with caution otherwise they will distort the part after moulding. Ribs and strengthening members should be 1/2 - 3/4 as thick as the walls they reinforce and deep ribs may require 1/4 - 1/2 degrees of taper for easy ejection from the mould. By drawing a circle at the intersection of the rib and wall, section thicknesses can be compared, as shown in Fig 42. Rib thickness combined with blend radius could easily overstep the wall thickness limit criterion; see Fig 43 for creep resistance.

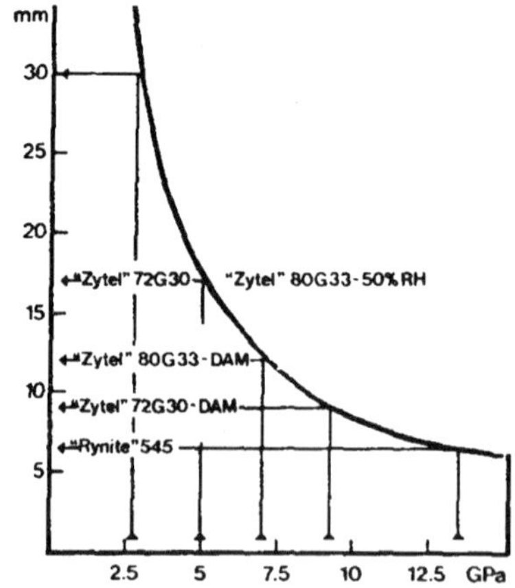

Fig 39: Modulus variation with grade

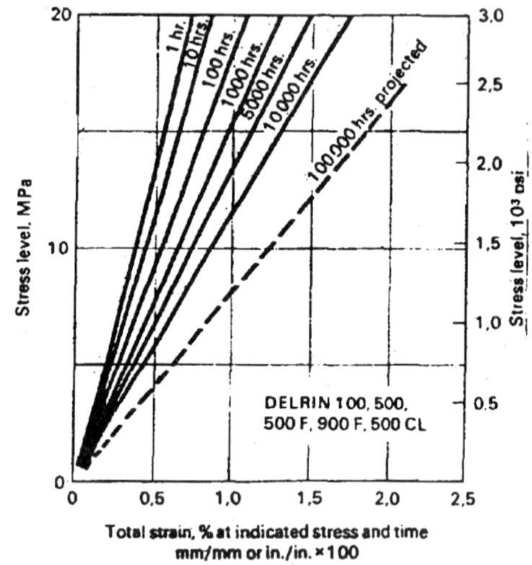

Fig 40: Time-dependent stress-strain curves

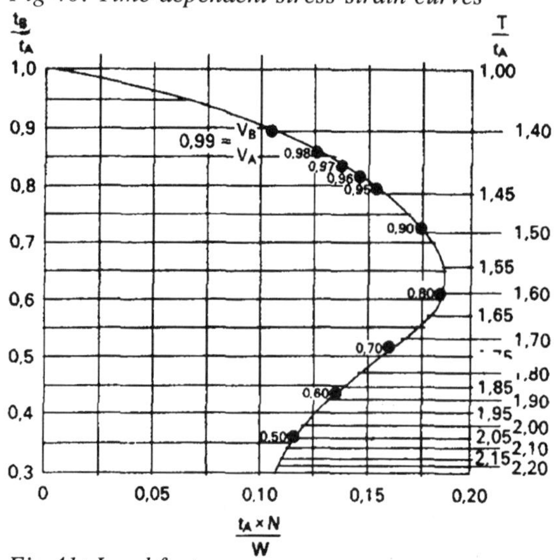

Fig 41: Load factors

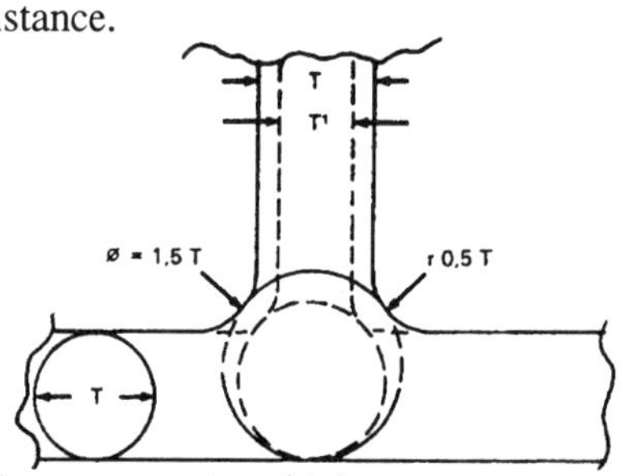

Fig 42: Finding true section thickness

Designing in polypropylene

This important semi-structural thermoplastic is widely favoured for the balance between its mechanical properties and ease of moulding. ICI were early in the introduction of the material into automotive applications and have built up a useful bank of design and application data, some of which is reproduced here.

ICI produce some 40 homopolymers and co-polymers of polypropylene under the Propathene brand — including filled, reinforced and grades which can be plated. Common to all are the properties of good temperature and chemical resistance also suitable flow properties for easy processing. Mechanically they are mostly semi-structural with low weight for a given strength; they also possess good fatigue and environmental stress cracking resistance.

The relatively low cost of the material has also ensured polypropylene's widespread use for automotive applications. Homopolymers are suitable for less structurally demanding interior trim applications but for areas like foot-well kick panels, requiring greater toughness at low temperatures, general-purpose co-polymers are appropriate. For fascia crash-pad work, tougher high-grade co-polymers are needed.

Grade GYM 202 of 10,000 kP/cm^2 flexural modulus is used, for example, in the full-width one-piece fascia panel for a cross-country vehicle. This is a large thin-section panel requiring the combination of easy-flow mould filling and toughness in service. GWM 101 of 14,000 kP/cm^2 flexural modulus has been used for 'organ-type' accelerator pedals with integral hinge of the flexing plastic. CV and tractor cab sound-insulating panels are often made of foamed polypropylene. Reinforced grades are also used in foamed form for such applications as tractor seat pans and rear parcel shelves in cars.

Commonly used for radiator fan shrouds, heater casings and ducts are the talc-filled grades such as 22T20/40H, of 35,000 kP/cm^2 flexural odulus, because of their high rigidity and heat distortion temperature. In chassis and exterior applications generally, 20% glass filled grades are popular for such applications as headlamp bodies while greater toughness is available in GWM 203, with 350 J/m Izod impact strength at 23 C, for bumper inserts and whccl-arch liners. PXC 4717 is a grade specially developed for plating, the pre-treatment process achieving a plating peel strength of 4.5 - 7 kg/cm.

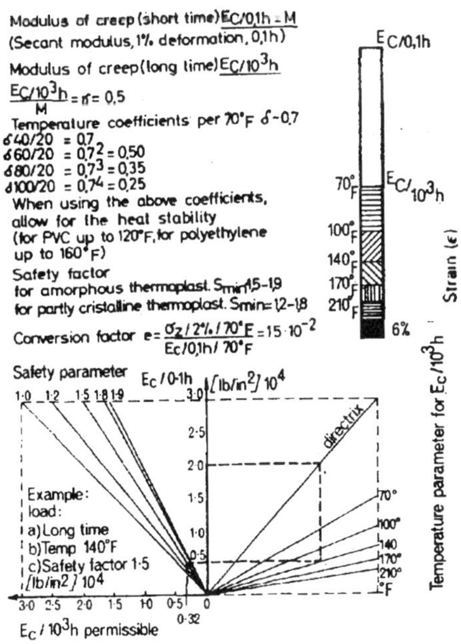

Fig 43: Nomogram for creep resistance

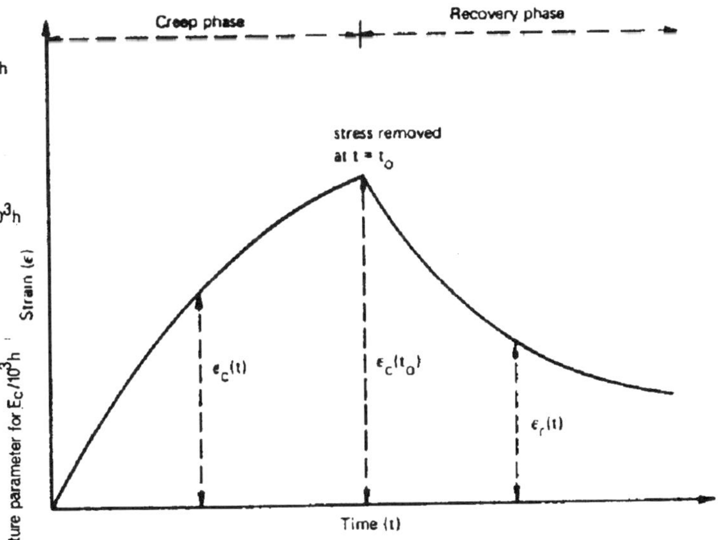

$$\text{Fractional recovered strain} \equiv \frac{\text{strain recovered}}{\text{creep strain when load is removed}} = \frac{\epsilon_c(t_0) - \epsilon_r(t)}{\epsilon_c(t_0)}$$

$$\text{Reduced time} \equiv \frac{\text{recovery time}}{\text{duration of creep period}} = \frac{t - t_0}{t_0}$$

Fig 44: Creep and relaxation of a thermoplastic

Mechanical design of plastics

Deformation, fracture after dynamic fatigue and resistance to impact are the important mechanical properties. ICI supply data based on both creep, Fig 44, and stress-relaxation tests. The first involves observation of strain as a function of time when the specimen is held at a constant load while the second involves measuring stress as a function of time while maintaining constant strain.

Creep data may be represented as strain vs log time, Fig 45. These can be presented in different ways, to suit particular requirements, by drawing sections at constant times or constant strains. A further derivation of possible use is tensile creep modulus vs log time.

ICI point out that Propathene will recover after long periods of creep and this property can be exploited in components subject to intermittent loading. However, the recovery, like creep itself, is time-dependent and the data an be presented most concisely if new co-ordinates are defined, Fig 46. General recommendation by the company, for design, is to use the criterion of maximum strain to signify 'failure'. Limits are 1% where welded joints are used, 2% for injection-moulded components and 3% as an absolute upper limit.

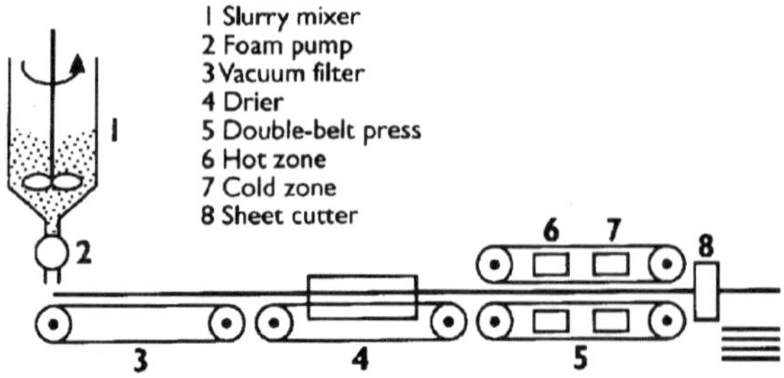

Fig 48: Slurry process for GMT

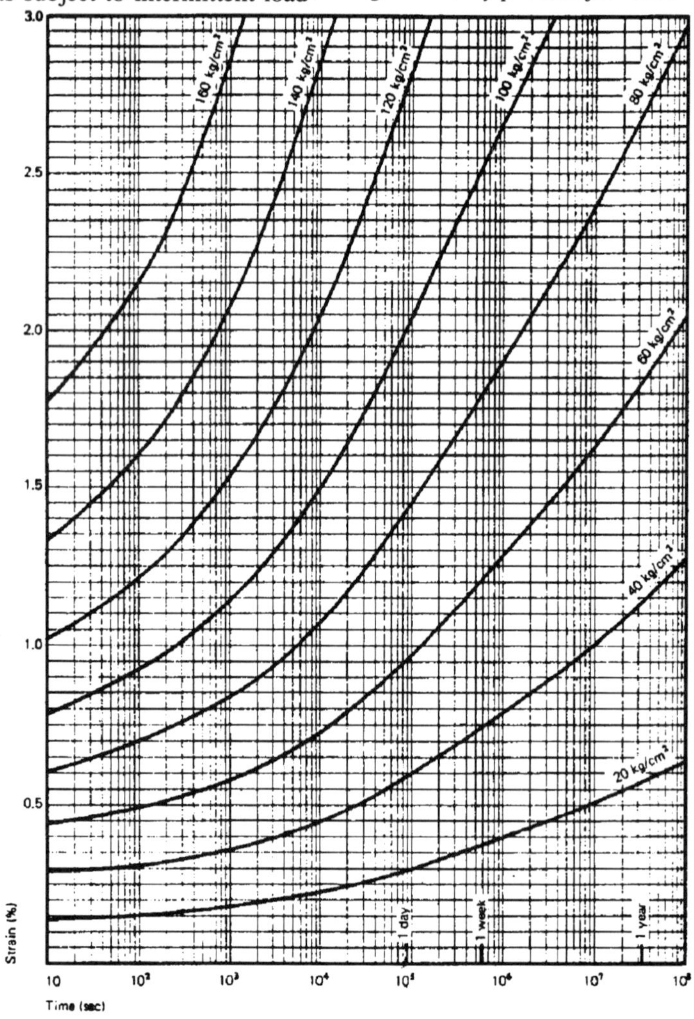

Fig 45: Strain vs log time

Reinforced thermoplastics

Apart from structural RIM, a number of other reinforced thermoplastics, including those based on conveneniently handlable prepregs, are available as a ready means of providing structural panels for medium volume levels of production.

An early entrant into the market for reinforced plastic structural materials was ICI whose Sandwich-Moulding process of the early 1970s was successfully used to produce structures of relatively high specific stiffness. It involved a new injection moulding technique, for the time, which allowed the production of 'solid' skins and foamed cores in one operation — but skins were not exclusively glass-reinforced. As well as the greater flexural rigidity of thermoplastic panels produced by the process, surface quality, too, was improved over 'solid' panels by virtue of absence of sink marks at the intersection of the panel with reinforcing ribs or wall normal to the surface. Also the skin materials did not have to be the same as that used for the core material, with foaming agent. An early prototype part made in the process was a bonnet panel which weighed just 10 lb (4.5 kg) and had a thickness of 0.4 in (10 mm). A conventional steel equivalent at that time was three times less stiff in bending and weighed 22 lb (10 kg)while a solid plastic one, of the same material as the sandwich skins, required a thickness of 0.21 in (5 mm) yet still did not have sufficient torsional stiffnesss.

Relatively high volumes had to be used because tool costs were even 10% higher than conventional injection moulds. Two polymer formulations

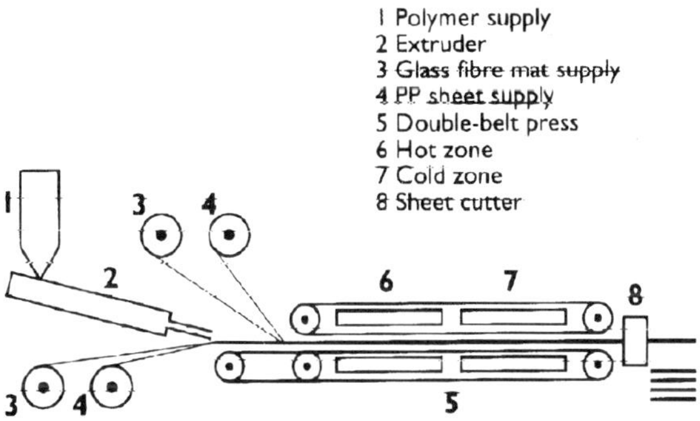

Fig 49: Separate process for random fibre mats

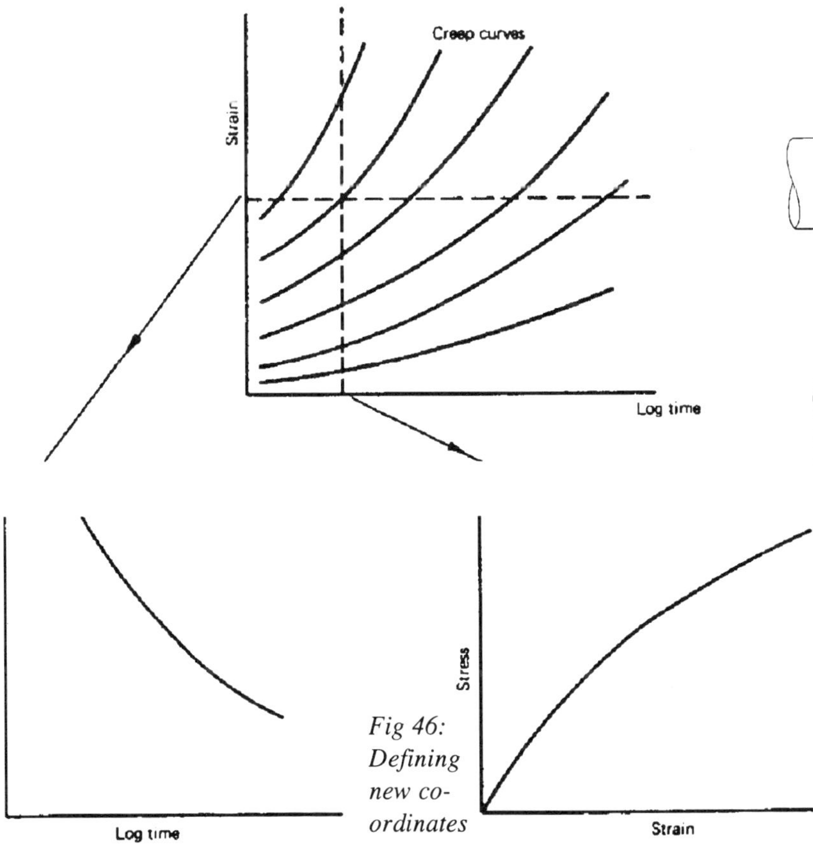

Fig 46: Defining new co-ordinates

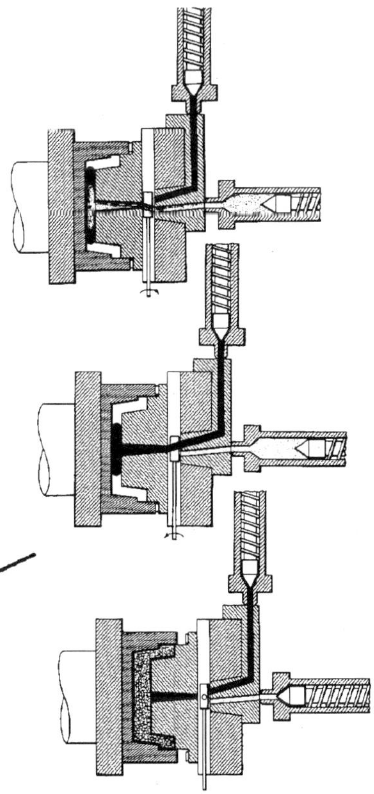

Fig 47: Stages in ICI sandwich moulding process

were injected from separate injection units, one after another, into a mould via a common sprue. A short time after filling, the mould was opened to allow the foamed sandwich to expand, Fig 47.

In designing structures made in the process ICI found that calculations used for solid plastics could be employed but with flexural modulus divided by the expansion ratio in the moulding process, given by final thickness of moulding/ unexpanded thickness. Where different skin and core polymers were used, the two moduli were summed and then divided by twice the expansion ratio to arrive at an equivalent modulus figure. For a 25 % coupled-glass filled polypropylene skinned panel of 0.5 m^2 area at 20 C subject to short-period loading, approximate mid-point deflection is 0.1mm per kg uniformly distributed. At 1.8:1 expansion rato thickness required would be 10.4 mm and weight 2.8 kg assuming an expanded polypropylene core and an equivalent elastic modulus of 1.58 GN/m^2 .

GMT systems

The general case for thermoplastic matrix composites has been well made. A DTI pamphlet[2] argues that the limited shelf-life of GRP thermoset prepregs, due to chemical instability, advanced the use of glass mat thermoplastics (GMTs) and advanced thermoplastic composites (ATCs). The latter resulted from a need, initially by aerospace users, for tougher materials with fibre volume fractions above 50% using a greater proportion of aligned filaments. While polypropylene was once the predominant material, usage has widened to include nylons, PBT/PET, polycarbonate and polyphenylene sulphide.

Generally chopped fibres are processed in a wet slurry process, akin to paper-making during which chopped fibre, polymer powder and processing aids cause fibre bundles which form filaments. The slurry is pumped on to a vacuum filter belt and the water removed, the resulting non-woven web of mixed fibres and polymers is then dried, Fig 48. This process is particularly well suited to high fibre loadings (60-70% by weight). For random fibre mats, these are impregnated with molten polymer by sandwiching the extrudate between two layers of mat as in Fig 49.

GMT is usually itself processed either by stamping or high speed compression moulding, the former being used for simpler-shaped parts. Stamping involves placing a near net shape pre-heated blank in a

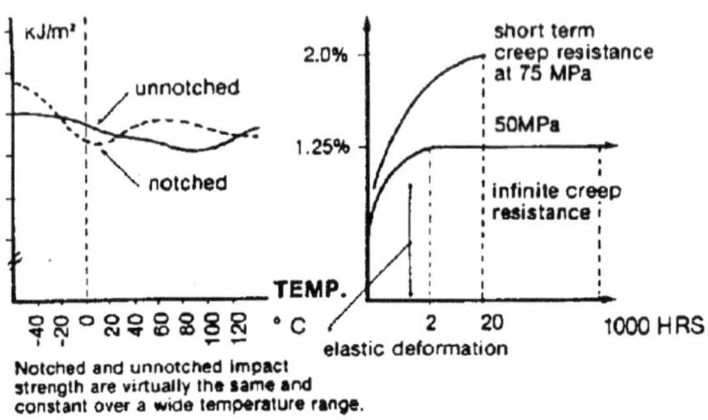

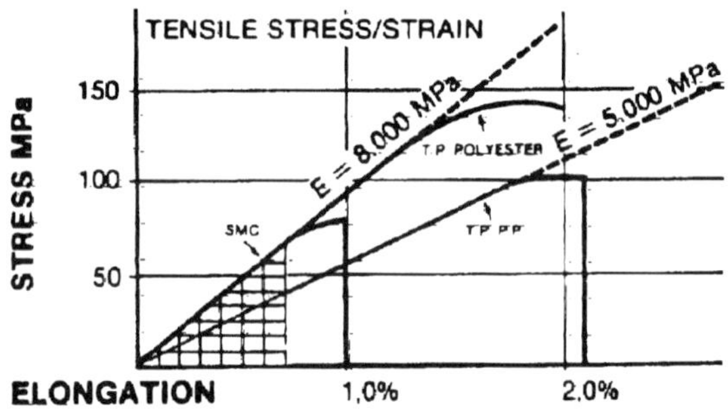

Figs 51 and 52: Performance comparisons with polyesters (top left: energy; top-right: strain)

	ISO standard	PP-GMT*	PET-GMT**	PA-GMT**
Glass content	1172	30	30	30
Tensile strength	527	60	100	120
Elongation at break	527	--	2	2.5
Modulus of elasticity	527	--	9000	6500
Charpy impact 23 °C	179/2D	45	60	70
HDT-A	75	140	245	215
Relative density	R1183	1.13	1.60	1.37

Fig 50: Akzo GMT properties

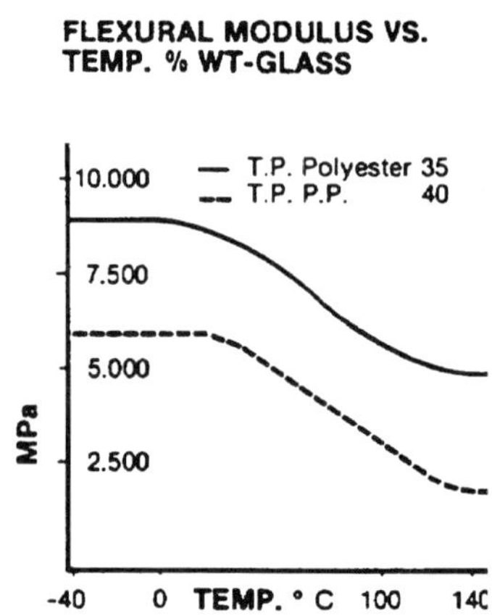

Fig 53: Akzo GMT properties

lower temperature moulding press and pressing, in a cycle time from 15-45 seconds, sufficiently high pressure producing good surface finish. High speed flow-moulding of larger, more complex parts can allow variable wall thicknesses to be obtained and flow can provide the part with directional properties.

The table in Fig 50 summarises selected properties of three GMT types available from Akzo. With stamping of panels in these materials surface finish would be inappropriate for exterior applications, so main uses are in such areas as boot floors and hidden seat supports. Because the product cools in the mould, wall thickness is limited to a maximum of 5 mm and to prevent stress concentrations minimum fillet radius of 0.1 mm should apply. To ensure good filling, ribs of no more than 3-5 times wall thickness should be used. Forms of reinforcement possible include short fibres, 0.2-2 mm, random mat with fibres up to 10 mm, uni-directional tapes and woven or knitted fabrics when compression moulding is used. Figs 51 and 52 show performance comparisons with polyester systems.

Engineering plastics alloys

As well as the familiar plastic monomer categories, a series of co-polymers has been developed by the material suppliers; these can provide very useful combinations of properties . Here the products of one company, which specialises in this area, are examined.

A plastics materials supplier which has been particularly involved with plastics alloys is G E Plastics. Among this supplier's principal engineering copolymers are Lexan polycarbonate, Noryl polyphenylene oxide (PPO), the Valox terephthalates, The Xenoy blends of semi-crystalline plastics, Noryl GTX PPO alloy and Ultem polyetherimides.

The alloy blends are a series of products rather than one particular percentage mix of polymers. The Valox family, for example, involves polybutylene, polyethylene and polycyclohexylene in three separate copolymers PBT, PET, and PCT. All combine good mechanical, thermal and di-electric properties and so find application in electrical applications such

NORYL* GTX Typical properties*	SI UNITS	TEST METHODS DIN	ASTM	900	901	910	810	820
Mechanical								
Tensile strength at yield	N/mm²	53455	D 638	60	54	60	—	—
at break	N/mm²	53455	D 638	57	56	57	90	120
Elongation at yield	%	53455	D 638	5-7	5-7	5-7	—	—
at break	%	53455	D 638	100	50	100	15	10
Tensile modulus	N/mm²	53457	D 638	2150	2150	2150	4500	6000
Flexural strength	N/mm²	53452	D 790	100	90	100	130	170
Flexural modulus	N/mm²	53457	D 790	2100	2000	2100	3000	4000
Charpy notched impact								
23°C	kJ/m²	53453	—	13	12	13	6	7
−20°C	kJ/m²	—	—	9	8	9	4	5
Izod notched impact								
23°C	J/m		D 256	240	175	250	80	80
−20°C	J/m			190	125	190	60	70
Hardness, H 358/30	N/mm²	53456	—	115	85	115	124	132
Thermal								
Oxygen Index**	%	—	D 2863	< 20	< 20	< 20	26	26
Flame retardancy*** (1.6 mm)		—	UL Stand 94	94 HB	94 HB	94 HB	94 HB	94 HB
Heat Resistance:								
Vicat, method B	°C	—	D 1525	> 220	215	> 220	> 240	> 240
Vicat, B/120	°C	53460	—	195	170	195	210	225
DTUL, 1,82 N/mm²	°C	—	D 648	—	—	—	175	205
Mould Shrinkage	%	—	D 1299	1.2-1.6	1.2-1.6	1.2-1.6	0.5-0.7	0.4-0.6
Thermal Conductivity	W/m°C	VDE 0304/1	C 177	0.23	0.23	0.23	0.24	0.25
Coefficient of linear thermal expansion	m/m°C	VDE 0304/1	D 696	9×10^{-5}	9×10^{-5}	9×10^{-5}	$4\text{-}5 \times 10^{-5}$	$3\text{-}4 \times 10^{-5}$
Physical								
Specific gravity	—	53479	D 792	1.10	1.10	1.10	1.16	1.23
Water 23°C/50%RH/24h	mg	53495	—	29	28	29	31	30
absorption 23°C/50%RH/24h	%	—	D 570	0.4	0.3	0.4	0.5	0.4
Equilibrium, 23°C/50%RH	%	—	D 570	0.9	0.9	0.9	1.1	1.0
Saturation, in water 23°C	%	—	D 570	3.5	3.4	3.5	3.6	3.5

Fig 52: Noryl GTX alloys

as heavy-duty connectors. There are a variety of glass-filled grades, too, some of which suit electric motor frames.

Ultem polyetheimide is inherently flame retardant and combines the physical properties of the more exotic plastics with the processing ability of normal engineering plastics. Supec polyphenylene sulphide (PPS)-based resins are high-performance crystalline polymers which combine short-term high heat resistance with long-term thermal stability, plus chemical resistance, which allows under-the-bonnet application. The Zenoy series was specifically developed for the auto-industry and is generally noted for mechanical strength, chemical resistance and dimensional stability combinations. Particular examples have high resistance to petrol and lube-oils and have high impact strength down to -40C; some grades (CL200 series) are off-line paintable, too.

Noryl GTX has both the inherent advantages of the PPE monomer as well as the chemical resistance of a semi-crystalline material. The result is a material with heat performance of over 190 C Vicat B which makes on-line painting possible. Its low-density also gives it a weight advantage over a number of glass-filled thermoplastics on a strength-for-strength basis. Noryl GTX is however, itself available in glass-reinforced form (GTX 800) and a 20% glass-filled form (GTX 820) has a flexural modulus of 4000 MPa while GTX 830 goes up to 7000 MPa. The chart in Fig 52 covers the properties for whole range. The unfilled GTX 901 is particularly notable for being able to withstand braking temperatures in the wheel trim application. The material requires a three-zone screw for plasticising in injection moulding, Fig 53. The unfilled GTX 901 is notable particularly for being able to withstand braking temperatures — useful in the wheel trim application.

Case-study in plastics alloys

G E Plastics' Scope concept car had its tailgate moulded from this material to demonstrate a structural application for a part which was to be painted on-line. Both interior and exterior skins, Fig 54, are moulded in the material. The complete part was designed with 25 degrees of rake for the rear window so as to preserve lamina air-flow across it So as to avoid loss in rear vision with this arrangement the vertical section of the tailgate above the rear bumper is fitted with a transparent panel of Lexan. The part is also notable for having the rear-lamp bodies, hinge and strut anchorages integral with the inner shell.

After assembly, the part is primed with a 2-component conductive primer prior to mounting on the body shell for painting on the electrostatically applied top-coat paint line, the paint being baked at 150 C. The assembly operation involves both structural adhesive bonding and heat-staking, the pins for the latter being positioned through holes in the window flange and under the license plate —so that the glass sealant and lifting grip cover the heatstake points. The stakes serve the purpose of locating the two shells in place until the adhesive has cured. The electrical connections are also heat-staked, being metal strips which attach to the ribs of the interior shell. A connector block combined with the lock-pin transmits electrical power into the door. These pins are also heat- staked and a soft buffer moulding hides the joints from view.

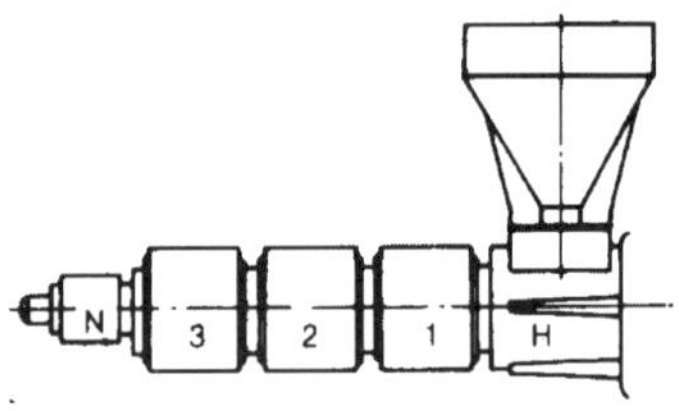

Fig 53: Three-zone screw

Fig 54: Scope tailgate in GTX

Semi-rigid polyurethane foam

With this polymeric material, which is in a separate category from most others used in automotive work, a variety of techniques can be used to obtain compounds of different physical properties. Here the semi-rigid foam variant is considered in some detail.

Urethane polymers are strictly speaking 'rubbers' and are difficult to classify in 'thermoplastic' and 'thermosetting' terms. They are made from combining polyesters or polyethers with di-isocyanates and are unusual in that their physical properties do not depend on compounding materials. In fact they cross-link and undergo chain extension to produce a wide variety of useful materials including hard and soft foams, elastomers and rigid-panels, when suitably reinforced. These reinforced reaction injection moulded panels are to be considered separately in a future article as are the flexible foams. The semi rigid foams considered here are of interest in that they can be foamed in situ or cast in slabs — even with a so-called 'self-skinning' high-density outer surface. Urethane polymers have good tensile strength, load-bearing capacity, elongation potential and both good hardness and abrasion resistance.

Energy absorption is a key property for the material's application in bumpers and interior crash-padding. The material is also used as an acoustic insulator in combination with solid PVC or EPDM sheet, the latter providing a dense layer to reduce sound transmission while the PUR foam absorbs the sound energy. The material is also combined with carpeting to form shaped floor mats which are convenient for assembly purposes and avoid joints and gaps in the floor-covering. Ability to use low-cost moulds and relatively simple production equipment makes the material valuable in its more structural role in specialist vehicles. Complete vehicle front-ends can be made — also such aerodynamic aids as spoilers and air-dams moulded at modest cost. For items such as fascias and crash pads, PUR may be foamed into a skin of PVS or ABS, previously shaped by vacuum-forming, drape-forming or slush-moulding — exploiting the material's valuable adhesive properties. Self-supporting liner panels can be produced in the so-called Press Foam process with which fabric skins can be incorporated. The PUR is foamed in a flat tray then clamped in a press before curing is

	Material type	Density kg/m³	K	r	Temperature variation
1	Urethane	31.5	312	0.02	None
2	Urethane	75.5	730	0.06	- 20%
3	Isocyanurate	50	385	0.06	- 20%
4	Isocyanurate	53.5	374	0.05	- 20%
5	Phenolic	43.5	67	0.08	None
6	Phenolic	50.5	327	0.03	None

(Units for K are such that $K\dot{\varepsilon}^r$ is in kN/m^2.)

	Material type	Density kg/m³	r	Temperature variation
7	Urethane	95	0.11	- 70%
8	Urethane	85 - 89	0.16	- 90%
9	Urethane	48,66	0.18	?
10	Urethane	120	0.09	?
11	Urethane	90	0.15	?
12	PVC nitrile	161	0.18	- 90%
13	Ethylene	154	0.09	?
14	Ethylene	37	0.04	?
15	Scotfoam	52	0.30	?

Fig 56: Strength increase with density

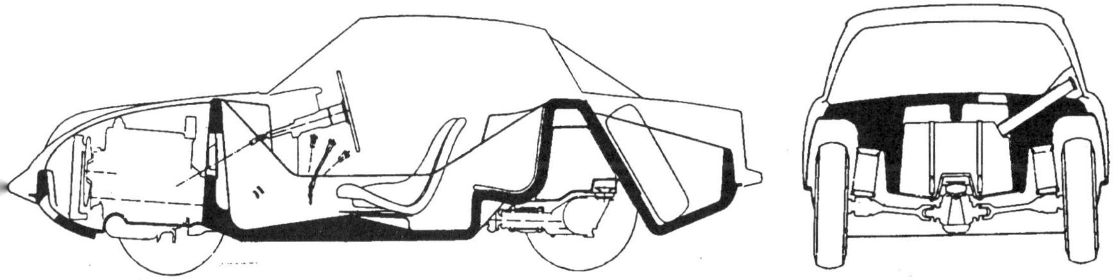

Fig 55: Bayer experimental sandwich construction car

completed. Differential compression of such partly cured foams enables its density to be varied so that strength and hardness can be varied across a panel.

One of the most interesting structural uses for PUR was proposed by Bayer who built a complete car structure using a rigid foam within GRP sandwich skins, Fig 55. Another important application is in the core of refrigerated vehicle bodies where its closed cellular structure and low thermal conductivity play crucial roles. Again structural strength can be varied according to density as seen in Fig 56. Three in-situ foaming techniques are available, pour-in-place, spray-in-place and froth-in-place. The first allows bodies to be built with minimum capital expenditure. All that is required for a typical hand mixing kit is protection equipment, contact thermometer, heater unit, power-drill and propeller-type stirrer. The foam typically exerts a pressure of 2-3 lb/in^2 (0.14-2.0 bar) as it rises, after mixing and pouring, so the cavity must be adequately jigged to avoid any panel distortions.

Where jigging is too difficult to achieve, spray-in-place permits panel insulation on one side of a panel to be applied and it is of course possible to trim the expanded surface and bond on the second panel to obtain a sandwich structure. For spraying, a system based on methylene diphenyl di-isocyanate (MDI) is preferred to the longer established toluene di-isocynate (TDI). Frothing-in-place involves the addition of a frothing agent, under pressure, into the mix to give rapid expansion prior to the foaming action. Jigging costs can be reduced by this method and greater heights can be filled from a single 'pour'.

The assessment of physical properties of foams for crash-padding effectiveness can be carried out on test machines which apply a fixed rate of strain to the foam specimens. The compressive stress on the foam is related to strain by *Ker* where *K* and *r* are material constants, Fig 57. Typical behaviour for semi-rigid foams is shown in Fig 58. For rigid foams, yield occurs at approximately 10 percent strain, beyond which the cell structure collapses. Semi-rigid foams, however, show good recovery at these strains, by contrast, and can survive up to 60 percent strain levels. The semi-rigid foams are seen to be more strain-rate dependent than the rigid ones. The above apply to flat test-loading surfaces and the effect of a cylindrical indenter is seen in Fig 59 superimposed on the 'flat' case.

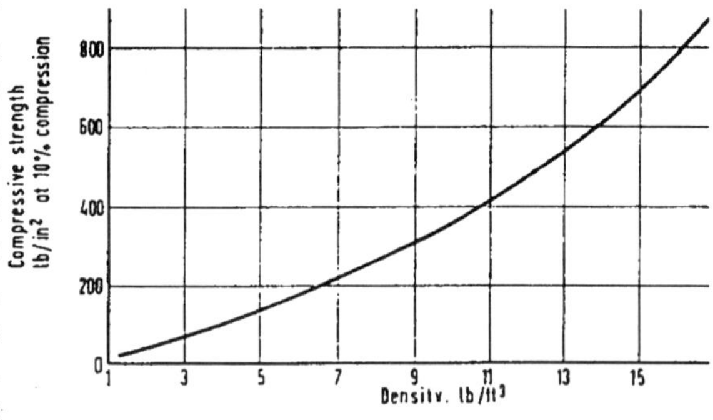

Fig 57: Stress/strain relationship

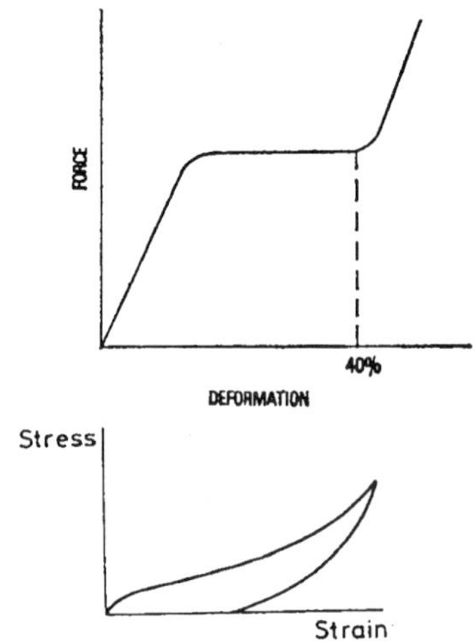

Fig 58: Semi-rigid foam characteristics

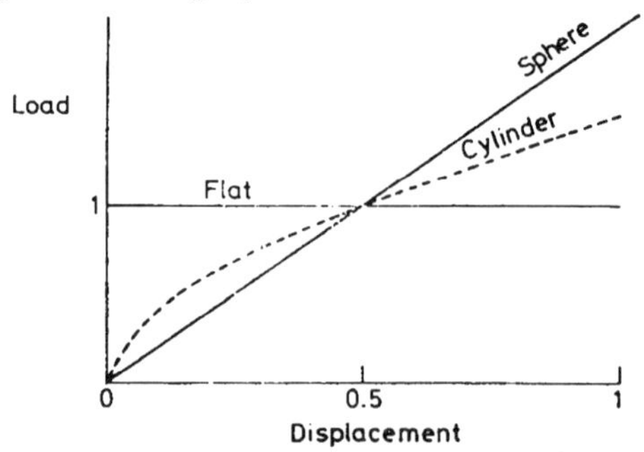

Fig 59: Semi-rigid foam testing

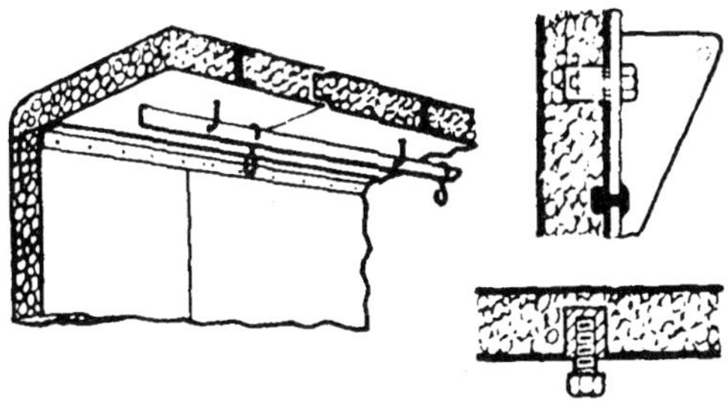

Fig 60: Plugs and inserts

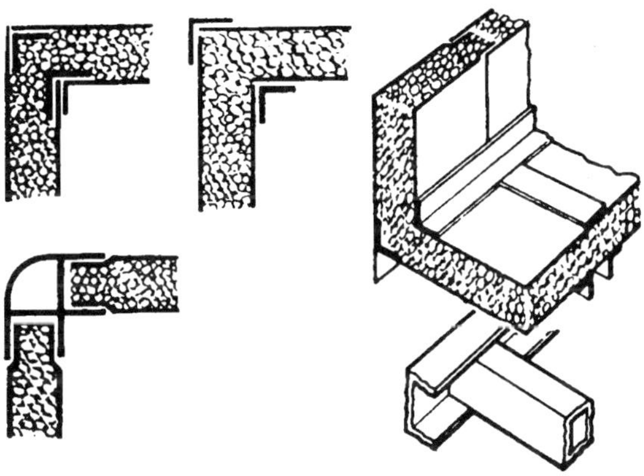

Fig 61: Corner joint techniques

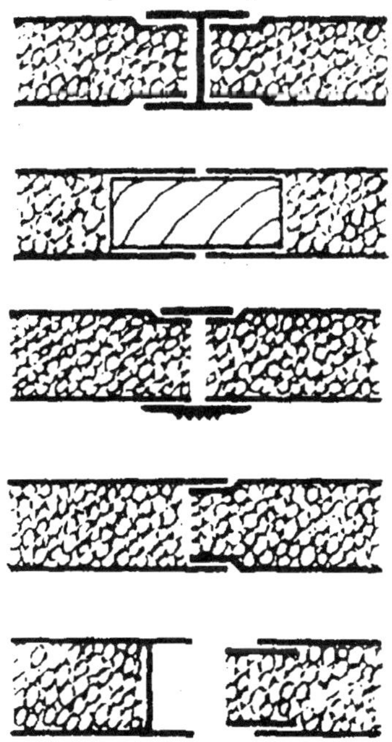

Fig 62: Reinforcement techniques

Sandwich panels

The need to stabilise thin sheet structures against buckling and wrinkling, under either compressive or shear loading, has led to the use of sandwich construction. Cost and complication in assemblies of pillars and rails, to support large area panels, is overcome by using light-weight core materials sandwiched between thin skins of higher strength facings.

Besides the clear structural advantages of sandwich panels, those of heat and sound insulation, also the ability to achieve good surface smoothness over wide areas, add to the appeal of this form of construction in many special vehicle categories besides refrigerated vehicles, considered in Chapter 14. Recent automated build techniques have further widened their scope and developments of plastic foam technology has enabled extremely light-weight core materials to be used of high structural efficiency.

The key parameters of the panel which have to be related to the external loads and the maximum stress capabilities of the face and core materials, as discussed in a later chapter. Any lateral loading of the panels (at right-angles to the faces) must also be considered in relation to the reaction and support points of the structure using bending and stability theory to be discussed later.

Bonding of the core to the skins is a critical factor of structural integrity. Foamed in-situ core systems are usually self adhesive to the skins; otherwise complete coverage of the surface must be ensured if the continuity of shear load path between core and skin, on which the strength and rigidity of the panel relies, is to be guaranteed. A common method of applying pressure between skins and core, during bonding, is the vacuum bag which allows atmospheric pressure to be evenly applied across the surfaces.

Concentrated loads, at attachment points on the panel, must be fed through to the core without inducing peeling loads between it and the skin. A variety of plugs and inserts can be bought for this purpose, Fig 60. Corner joints will normally require reinforcement of the type shown in Fig 61, again to protect against peeling loads. Butt-jointing of panels needs to be as strong as the main area of the panel and should involve the use of gap-filling adhesives. Some reinforcing techniques are shown as Fig 62.

Foam and honeycomb cored panels

One technique of automating the application of bonding resin between core and faces was perfected by Duramin Engineering in the 1960s. Shown in Fig 63, the system depends on cutting transverse irrigation grooves into the surface of the slab-stock expanded polyurethane core, (A), as well as inlet and outlet manifolds, (B) and (C). The core is sandwiched by glass-cloth laminates, (E), held between aluminium alloy mould faces (F) backed by timber stiffeners (Z). A colourised gelcoat is shown at (G) and (H) shows extrusions used to close the open ends of the mould. An encapsulating vacuum bag (I) is sealed at (J). Air is evacuated from the mould by an exhauster (K) and resin drawn in through container (L). The Lotus VARI process, described in a separate chapter, has also some similarities as well as the ability to handle curved panels and incorporate a wide variety of core shapes.

Proprietary foam-cored sandwich panels can also be purchased from a number of sources. Some properties of Schmitz Ferroplast panels are shown in Fig 64. Another popular core material is honeycomb cell aluminium or plastic available in Aeroweb sandwich panels from CibaGeigy of Duxford in Cambridgeshire. Such cores are favoured for demanding structural duties such as racing car body construction. Typical fastening, edging and cornering methods for such panels are shown in Fig 65. Aeroweb arimid honeycombs are also available in hexagonal or 'overexpanded' rectangular format, the latter (Fig 66) particularly suiting curved panels. A variety of cell sizes and 'paper' thicknesses can be specified to give differing strengths and densities. The material must also be designed against the various possible failure modes shown in Fig 67.

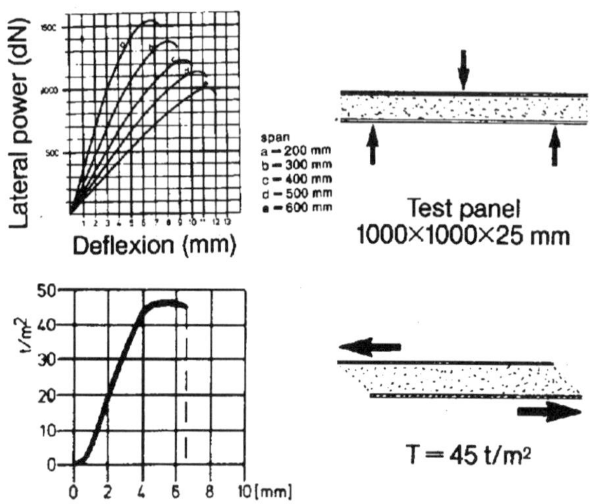

Fig 64: Ferroplast panel properties

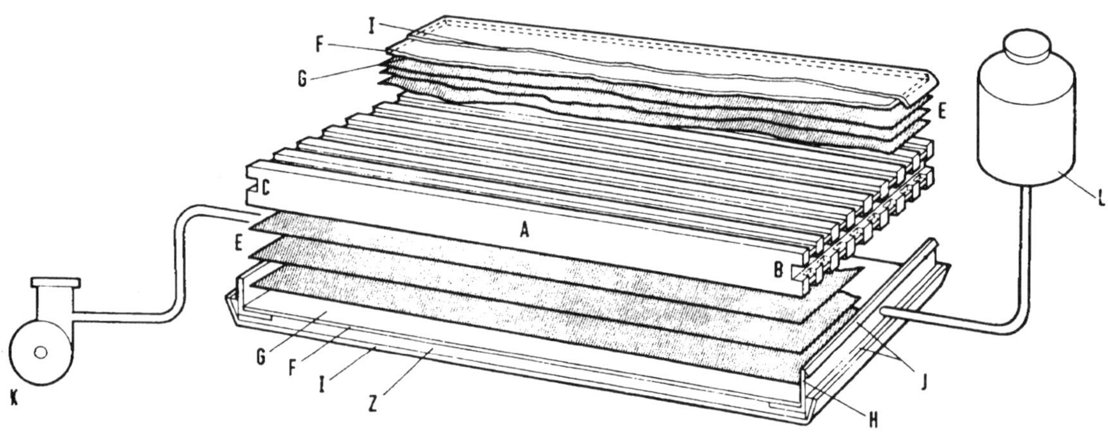

Fig 63: Automated resin-injection

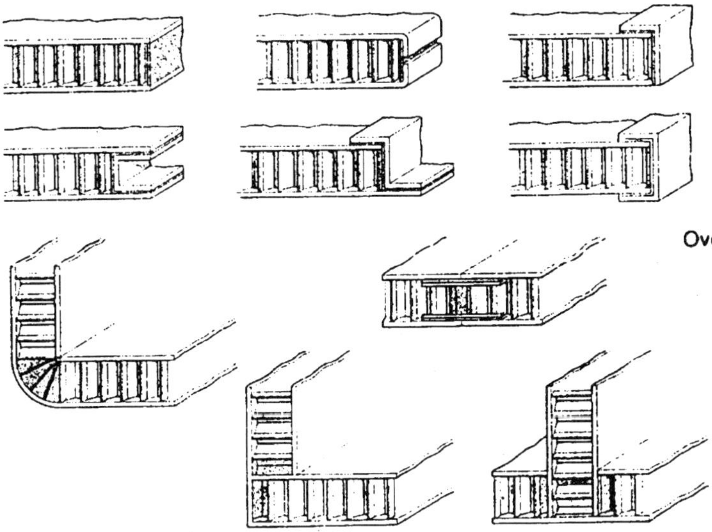

Fig 65: Aeroweb jointing techniques

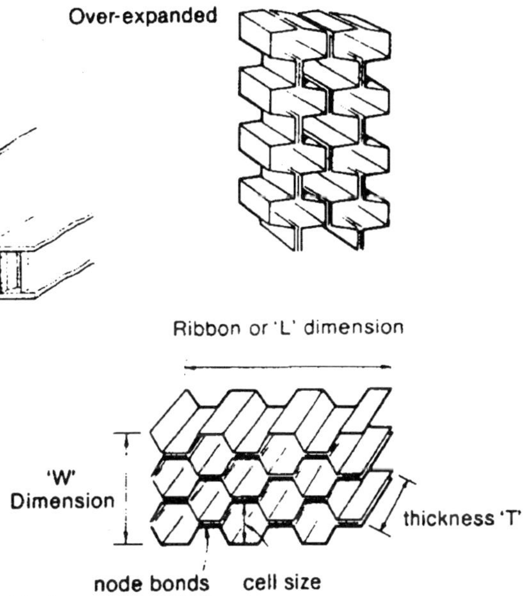

Fig 66: Aeroweb variations

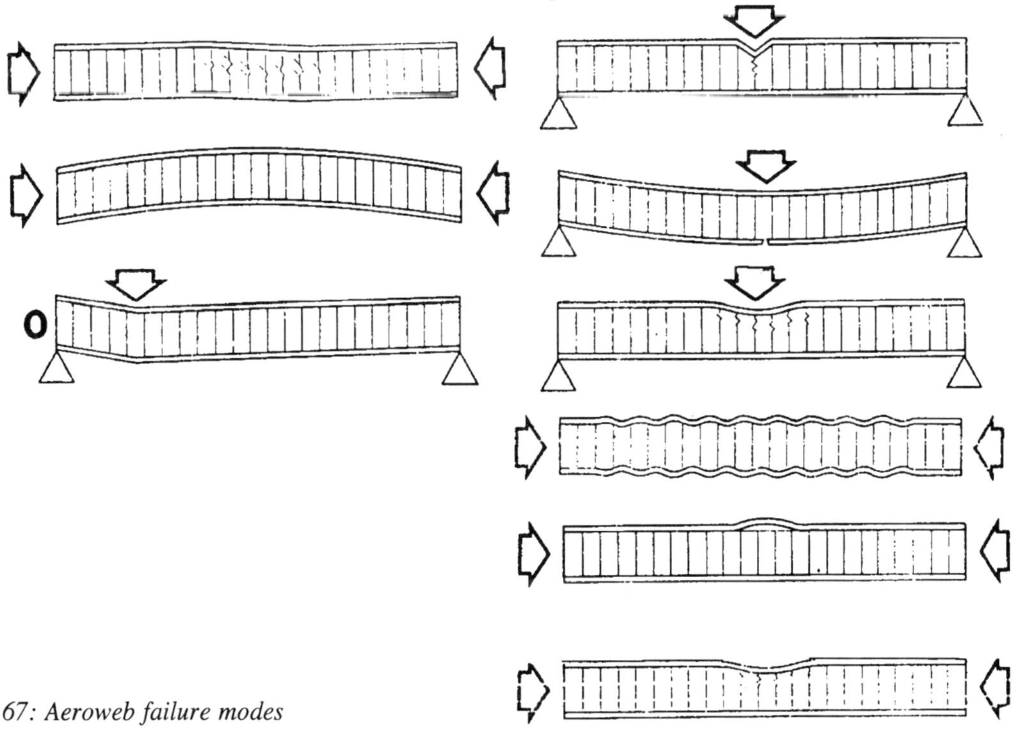

Fig 67: Aeroweb failure modes

Reaction injection moulding

Rapidly becoming to thermoplastics and urethanes what GRP-moulding is to thermosets, RIM is now challenging GRP for composite body panels in relatively low volumes. The process is modest in tool costs and produces panels of high stiffness when the recently developed polyurea systems are employed.

By using polymerisation in the mould, the RIM technique is quite different from other plastic moulding methods and can be used for producing quite complex parts and panels without unduly high tooling investment —since mould pressure is low. Two or more components flow into the mixing chamber, Fig 68, at relatively high pressure (100-200 bar) and then are expanded into the mould at much lower pressure. The streams impinge at high speed to obtain thorough mixing and intiate polymerisation as they flow into the mould cavity at a pressure of about 100 bar. Low viscosity during mould filling is one of the key attractions of the process as a relatively small metering machine can make large parts. The low viscosity also simplifies reinforcement with, for example, the possibility of using continuous-fibre mat placed in the mould.

Some 90% of RIM production is in polyurethanes and urea-urethanes, the latter being uniquely suited to the process as they do not melt-flow like normal thermoplastics and therefore conventional injecton-moulding is not possible. ICI have developed a family of polyureas for body panel applications with unusually good processability and physical properties. Gel times of 2 seconds are possible and mould temperatures are less than 200 F. Overall cycle time is about 1.5 minutes and further development promises less than one minute, Fig 69. Filler packages are also becoming available which allow part surface finish comparable with steel; moisture stability is high compared with competing thermoplastics and the materials can tolerate temperatures of 375 F.

Figs 70 and 71 show show heat sag properties, of the so-called 9002 material, which are seen to be dependent on post-curing and mould temperature. Fig 72 shows typical properties for a formulation that would suit body panels but others are available which raise the elastic modulus as high as 200,000 psi. The company also produces medium-modulus systems for fascias and bumpers, properties of two such being shown in Fig 73.

As well as increasing strength and rigidity of panels, the addition of glass-fibre or other reinforcements considerably improves the compatibility of

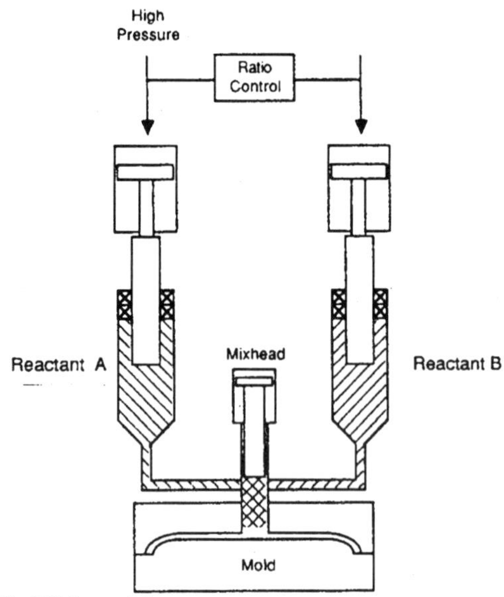

Fig 68: RIM machine

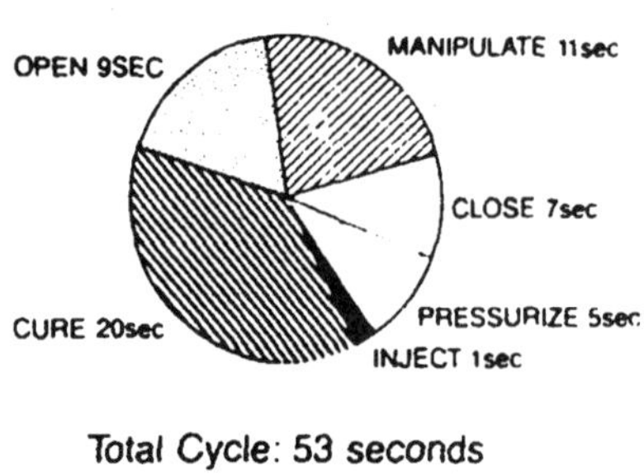

Fig 69: High modulus urea cycle

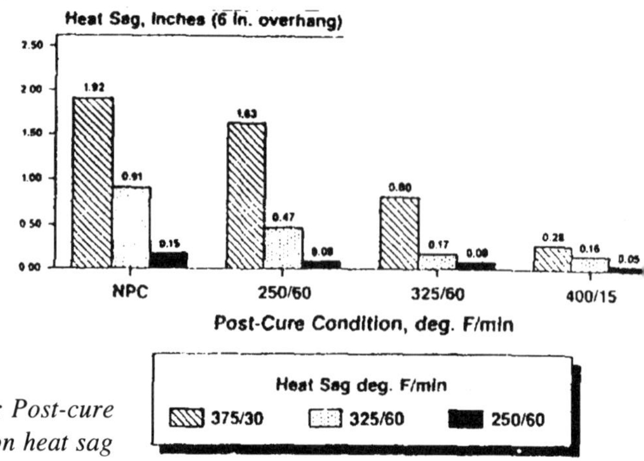

Fig 70: Post-cure effect on heat sag

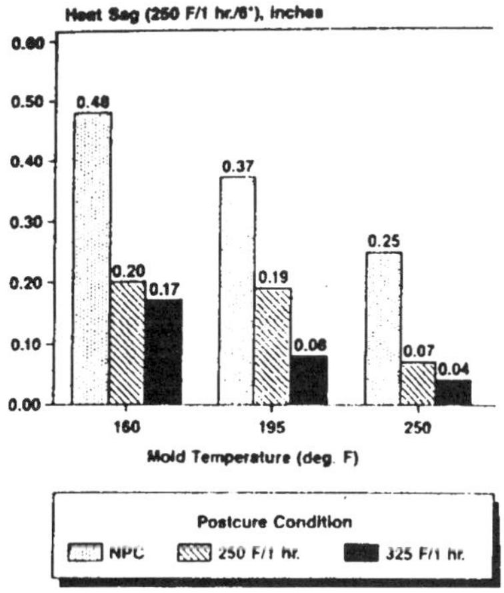

Fig 71: Mould temperature effect on heat sag

Typical properties	
Specific gravity	1.10
Flexural modulus, psi	
at 73°F	144 000
at −20°F	220 000
at 158°F	95 000
Modulus ratio −20°F/158°F	2.31
Tensile strength, psi	5100
Elongation at break, %	95
Gardner impact, ft-lb at −20°F	10.1
Mould temperature, °F	160
Component temperature, °F	110

Fig 72: Body-panel RIM formulation

Milled Glass	0	15
Specific gravity	1.02	1.12
Flexural modulus, psi		
at 73°F	35 000	56 000
at −20°F	80 200	125 000
at 158°F	23 500	45 000
Modulus ratio, −20F/150°F	3.40	2.78
Tensile strength, psi	3250	2400
Elongation at break, %	180	100
Tear strength, Die C, lbs/in	430	480
Heat sag, 6" overhand, inches		
250°F/1hr	0.28	0.18
Mould temperature, °F/1hr	250	250

Fig 73: Fascia/bumper formulation

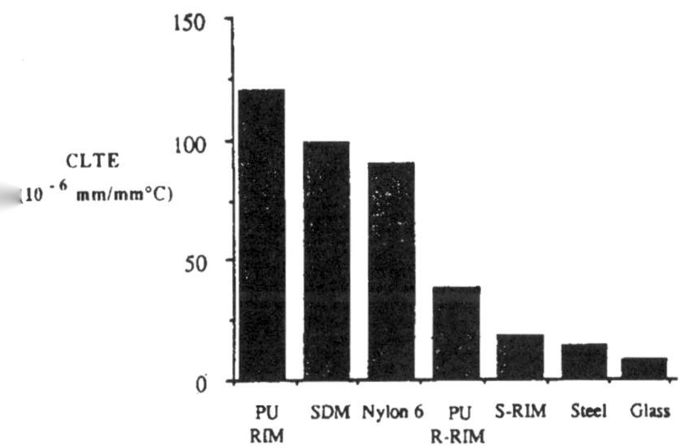

Fig 74: Comparison of processes

thermal expansion co-efficient with such materials as steel and aluminium. Since polymers typically have co-efficients some ten times greater than steel, a metre-long part hung on to a steel body could change in length by 1 cm between summer and winter temperatures. The way in which this can be modified by reinforcement is clear from the comparisons of Fig 74. In this figure CLTE is co-efficient of linear thermal expansion; SDM is styrene dimethyacrylate; RRIM is reinforced RIM which in this case is 10% by volume 1.6 mm chopped glass fibres; S-RIM involves a styrene dimethacrylate injected into 10% by volume continuous fibre-glass mat.

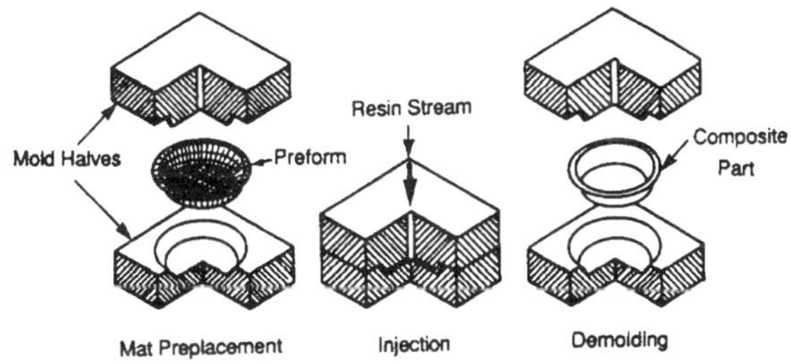

Fig 75: Steps in the SRIM process

	RTM	S-RIM
Equipment Cost	$30,000	$500,000
Flow Rate (kg/min)	2.3	55
Mixing	static mixers	impingement
Mold Pressure (MPa)	0.3	2.4
Void Content (vol%)	0.1 - 0.5	0.5 - 2.0
Mold Materials	epoxy	steel
Mold Temperature [(b)] (°C)	25 - 40	95
Component Viscosities (MPa·s)	100-550	< 200
Cycle Time (min)	10-60	2-6

Fig 76: SRIM and RTM compared

	Isocyanurate [(a)]	Urethane [(b)]	Acrylamate [(c)]	Epoxy [(d)]
Random Glass Mat (wt%) (3 mm thick part)	38	44.8	40	40
Specific Gravity	1.54	1.53	1.46	--
Void (vol%)	1.5	1.5	--	--
E_f (MPa at 25°C)	8,100	9,600	8,700	9,200
Tensile Strength (MPa)	150	150	125	160
Elongation (%)	7.3	2.0	2.1	1.2
Izod Impact (J/m)	510	660	790	~800
Heat Distortion (°C)	184	189	240	>200
Thermal Expansion (m/m°C) x 10^{-6}	~20		27	18

Fig 77: SRIM-composite properties

S-RIM is the process in which long-fire mat is placed in the mould and reactive monomers injected on to it, Fig 75, the process being akin to resin-transfer-moulding but with high-pressure impingement mixing to accelerate the reaction, Fig 76. Typical properties of S-RIM composites are shown in Fig 77. A recent development is so-called low-density S-RIM which is particularly attractive for trim-panel applications; its comparative properties are seen in Fig 78.

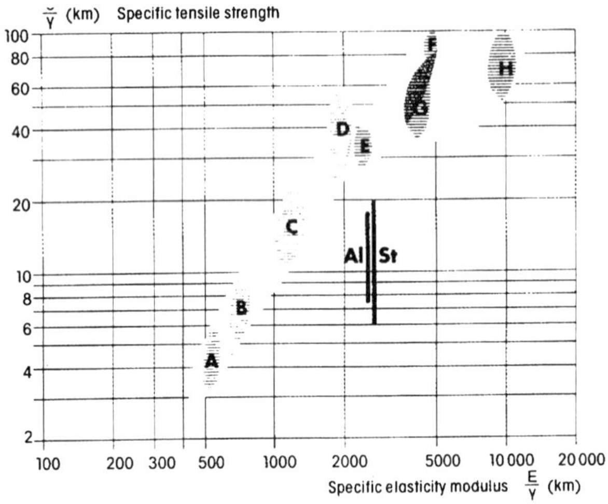

A · SMC
B · Random Fibre (RTM)
C · Glass Cloth (RTM)
D · UD Glass (RTM)
E · Aramid Cloth (RTM)
F - UD Aramid (RTM)
G - Carbon Fibre Cloth (RTM)
H - UD Carbon Fibre (RTM)
Al - Aluminium
St - Steel

Fig 81: Properties of RTM reinforcement systems

Resin transfer moulding

In resin transfer moulding (RTM) a low viscosity fluid is injected into a mould containing pre-placed fibre reinforcement. RTM has been used to mould large non-automotive FRP parts as an alternative to open moulding because it can provide two good surfaces and make parts with sandwich cores.

For relatively short model runs (10-20 000), RTM is a lower capital cost process compared with SMC compression moulding. The stages of the process are shown schematically in Fig 79. First the glass reinforcement or preform is placed in the mould. Once the mould is closed the resin is injected with no or little movement of the glass. After mould-filling the part is left curing in the mould until it is dimensionally stable so that it can be demoulded without losing its shape. The fact that the reinforcement is pre-placed in the mould gives the process the potential for making parts with better surface and mechanical properties than SMC where the fibre orientation is usually less controlled and favourable because of its flow.

During the filling stage, the resin being injected usually contains filler and can exhibit a viscosity with some shear thinning behaviour. However, for typical RTM compositions this effect is small and as a first approximation can be neglected.

Property	Units	LD-SRIM	Woodfiber	ABS	PP
Specific gravity	g/cc	0.55	0.85	1.05	1.04
Filler content	%	20 (glass mat)	-	-	20 (talc)
Flexural modulus	MPa	2370	3790	2000	2090
Tensile strength	MPa	41	13	36	25
Dart impact	J	5.7	3.0	11	13
HDT @ 1.8 Mpa	°C	95	90	83	130

Property	Requirement	Result
Thickness, mm	per drawing	2.72
Specific gravity	per drawing	0.55
Warpage, %	4 max.	0.40
Dimensional stability, % Δ	0.6 max.	0.40
Tensile modulus, MPa	12 min.	41
Flexural modulus, MPa	1750 min.	2370
Flexural strength, MPa	25 min.	67
Impact resistance, J	0.9 min.	no cracking
Flammibility, mm/min	101.6 min.	self extinguishing
Odor	no objectional	none
Stain resistance	no staining	non-staining

Fig 78: LD-SRIM properties

Epoxy Resin		100 P.B.W.
Hardener System		20 P.B.W.
Mixed Viscosity at 25°C		7.0 - 10.0 Poise
Tensile Strength	MPa	87.1
Tensile Modulus	GPa	2.92
Elongation at Break	%	6.9
Flexural Strength	MPa	140.8
Flexural Modulus	GPa	2.97
Compression Strength (Yield)	MPa	121.7
Heat Distortion Temperature		153°C
Tg		164°C

Fig 82: Properties of resin systems

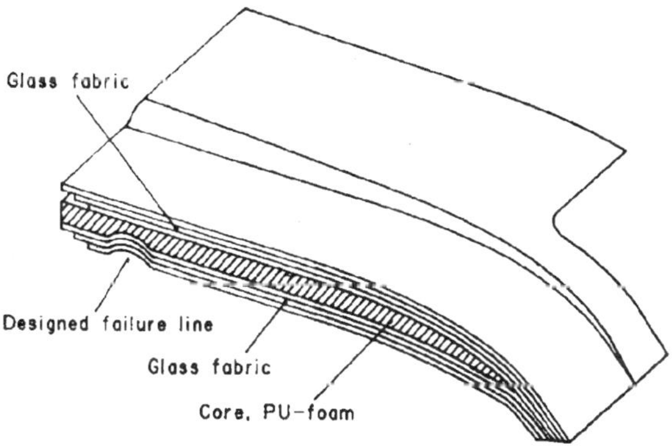

Fig 83: Designed failure line in bonnet panel

According to the Seger & Hoffmann subsidiary of Dow Chemical, the areas in which development effort has been concentrated are preform technology, essential for industrial processing; obtaining a class 'A' surface from the mould; modifications of resin systems to improve flow and fibre wet-out; resin metering, mixing and injection procedures; optimisation of the 'fibre package' for flow, mechanical and visual properties. To increase the efficiency of the moulding stage it is best to perform as many operations as convenient outside the RTM mould; the optimal placement of fibre into the RTM mould can be time-consuming particularly if the shape is relatively complex.

The preform procedure should satisfy the following requirements: produce a precisely formed shape that can be easily placed in the RTM tool; be capable of handling fibres which may need accurate orientation; produce a preform having sufficient integrity to withstand industrial handling. Methods available for preforming complex shapes include spraying fibres and binders on to a perforated mould or the use of mats and/or fabrics pre-treated with thermoplastic binder which can be shaped by pressure when heated. The former process tends to be slow and labour-intensive but has better control of the preform shape and generally has low material wastage. A disadvantage is, however, the inability to accurately orient fibres. A particular problem when preforming cloths, which have to be drape-formed, is that fibre orientation and density change at sharply changing shapes in the tool. This means that the properties of the composite at these points can be difficult to predict.

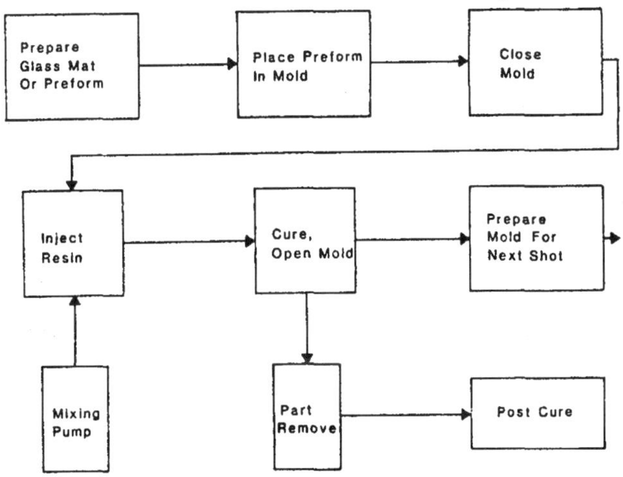

Fig 79: Stages in RTM process

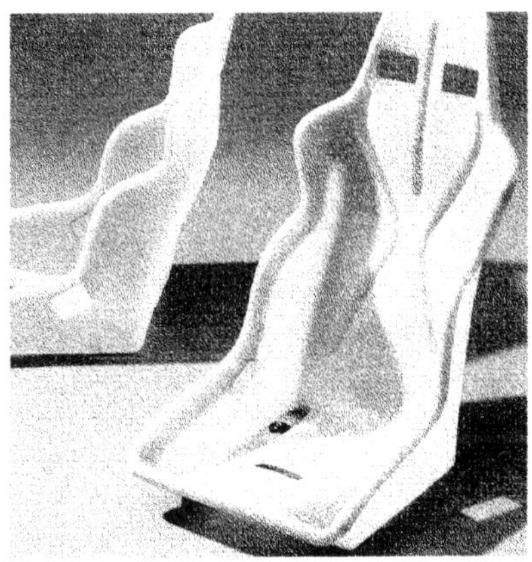

Fig 80: Seat shell made by Dow

The seat shell of a sports seat is one example successfully produced by the company. The part had previously been produced by hand lay-up, producing shells of inconsistent quality and surface finish. The part is complex with a deep draw, multi-curvature shape and multi plane shut-off. The method of preforming chosen used continuous filament random glass fibre mats (Fig 80). The first-attempt preforms and mouldings had been made with a single plane split-line with the associated problems of waste. Optimisation of the preform blank, preform procedures and RTM mould modifications resulted in a 30 — 40% saving in material during preforming compared to the first stage mouldings.

The preforming of fibre reinforcement for flat mouldings such as a bonnet may not be absolutely necessary. The handling of the fibre mats, however, can be difficult as they are rarely self-supporting. A bonnet can be designed as a slim sandwich structure instead of inner and outer mouldings. The rigid foam core thus provides a convenient method of supporting the fibre reinforcement during transportation of the fibre to the RTM mould.

An RTM bonnet moulding has to satisfy both the normal horizontal body panel mechanical requirements and have a class 'A' surface finish from the mould. Many fibre types and cloth constructions can be used in the RTM process as seen in the property diagram (Fig 81). Because of its overall price versus performance, E-glass is expected to be the most widely used reinforcement material for automotive applications.

Resin injection

Epoxy resin systems, with their low volume shrinkage of between 1-3% together with their good mechanical properties, are ideally suited to class 'A' surface body panel applications. Painting at temperatures up to 150 C can also be met with specifically formulated epoxy resin systems. The table in Fig 82 gives the most important mechanical properties of a typical resin system developed for RTM. Faster curing systems based on vinyl ester resins can be used for RTM structural moulding where surface finish is less critical. Vinyl ester resin-based systems are capable of giving mould closed times of under 3.5 mins. High speed mixing and dispensing technologies developed for the polyurethane industry are also likely to be adapted for the RTM process.

Metal inserts to spread loads at high stress areas, such as hinge attachment points, can be incorporatcd into the RTM moulding. The shape of the polyurethane core moulding can provide a method of locating and transporting these inserts during the preform process. A particular design feature of the bonnet is the designed failure line (Fig 83). During high speed impacts the bonnet fails along this line.

As with most composite processes, consolidation of parts is the key to successful and economic exploitation; Dow cite the apron structure in Fig 84 as a good example.

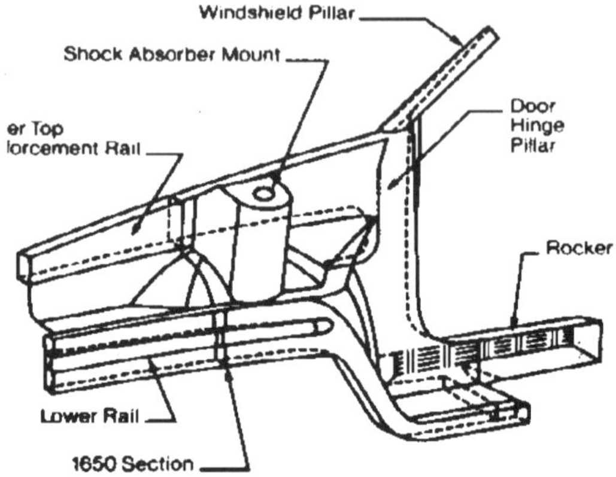

Fig 84: Parts consolidation in apron structure

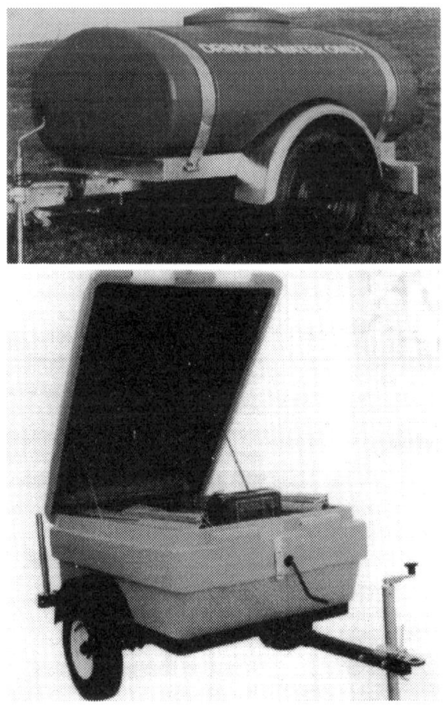

Fig 85a: Rotational moulding applications

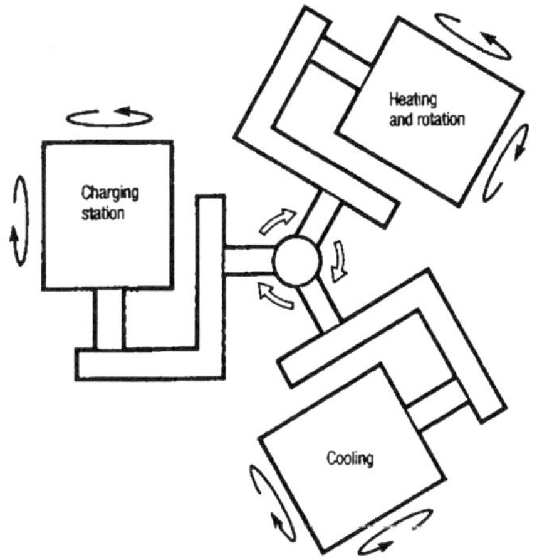

Fig 85b: Process configuration

	INSIDE RADIUS	OUTSIDE RADIUS
	(in inches)	
POLYETHYLENE		
Ideal	.500	.250
Commercial	.250	.125
Minimum	.125	.060
POLYVINYLCHLORIDE		
Ideal	.375	.250
Commercial	.250	.125
Minimum	.125	.080
NYLON		
Ideal	.760	.500
Commercial	.375	.375
Minimum	.187	.187
POLYCARBONATE		
Ideal	500	.750
Commercial	.375	.375
Minimum	.125	.250

Fig 88:Recommended draft angles.

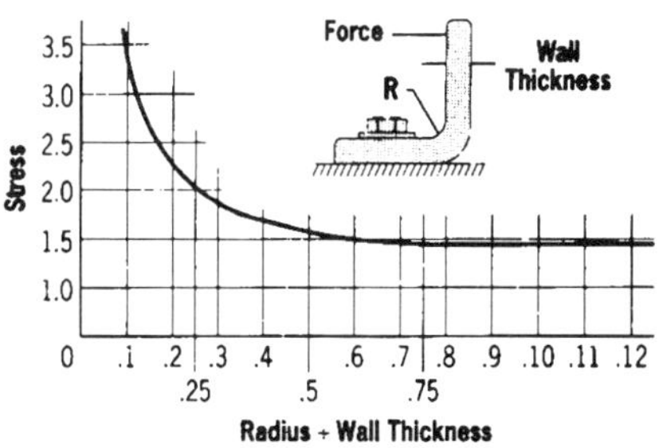

Fig 89: Corner radii

	OUTSIDE SURFACES		INSIDE SURFACES	
	Minimum	Better	Minimum	Better
POLYETHYLENE	0°	1°	1°	2°
PVC	0°	1.5°	1°	3°
NYLON	1°	1.5°	1.5°	3°
POLYCARBONATE	1.5°	2°	2°	4°
POLYCARBONATE	1°	1.5°	1.5°	3°

Fig 90: Recommended radii

Rotational moulding

Hollow sectioned end-products ranging from on-board fuel tanks to tanker bodies up to 9000 litre capacity, for small trailers, are among the potential applications of rotationally-moulded plastics, Fig 85a, with the basic process configuration being seen in Fig 85b.

T&D Rotomoulding[3] have shown that advances in the process have made it a cost-competitive alternative to blow-moulding, vacuum-forming and GRP moulding. The ability of the thermoplastic material to free-flow within the cavity of the mould as opposed to being forced, as in blow- or injection-moulding, is said to be an important benefit. The ideal shape for the process is a part which has smooth blends between adjacent contours of its surface.

The inner wall of the part will be free-formed, as the outside is defined by the mould, and chosen wall thickness is critical to the the design. A size of 3 mm is said to provide good compromise between cycle time, ease of processing, strength and cost but ranges form 1-20mm are not uncommon. Tolerance of 20% on this dimension would be typical. By arranging for differential temperarures over the surface of the mould, a degree of differential wall thickness can be obtained.

The steps in Fig 87 are a guide to optimum design. Flatness tolerance on typical PE or PVC parts would be around 0.5 mm per 50 mm, reducible to 0.3 mm for nylon and polycarbonate. Generally it is best to minimise the use of flat surfaces on the part. Adequate distance must be allowed between parallel walls, *A*, to allow powdered plastic to contact all cavity surfaces. A minimum of five times wall thickness is recommended. Likewise corner angles must be of adequate size, *B*. PE, PVC and nylon parts can be moulded with angles down to 45 degrees but harder flow materials require a higher minimum. PE, PC and PVC can go down to 30 degrees and nylon has been taken down to 20 degrees. There must also be an adequate radius where two walls meet.

Stiffening ribs are a valuable feature of roto-moulded parts and generally small multiple ribs are preferred to large single ones. The ribs must themselves be hollow and normally depth *Y*, and width *X*, *C*, should be 4 and 5 times wall thickness respectively. Rounded ribs,*D*, can also be used but sidewalls *Z* need to have draft sufficient for easy mould release; individual material specifications should be referred to. Use of `kiss-off' ribbing, *E*, is also a possibility where inner and outer walls can be joined at intervals. The contact area is normally 1.75 times wall thickness. It is also a useful place to introduce holes into the part.

Liberal use of draft angle throughout the design will reduce the mal-effect of high mould-release forces. As the part cools it shrinks and the external surfaces draw away from the cavity, making mould removal easy; however, the internal surfaces shrink on to the mould and removal is difficult without generous draft, *F*. Different materials have different draft-angle requirements, Fig 88. Where textured surfaces are used, higher draft angles will be required.

Inreasing the size of corner radii distributes corner stress over a wider area and improves the moulding of corners. Stress raisers occur when the inside corner radius is less than 25% wall thickness and corner strength enhancement occurs when the value is over 75%, *G*, Fig 89. Inside and outside of corners should be radiused to minimise thickness reduction effects at the corners, *H*. Recommended radiuses for different plastics are seen in Fig 90.

A vent hole is required somewhere on the part, to equalise pressure inside and ouside during moulding, unless the part is almost spherical in shape. An inward or outward projection parallel to the parting line of the mould is an `undercut' which must be deformed in withdrawing the part from the mould. Most moulds are of two or three piece construction so as to accommodate undercuts, side-cored holes or inserts, *I*.

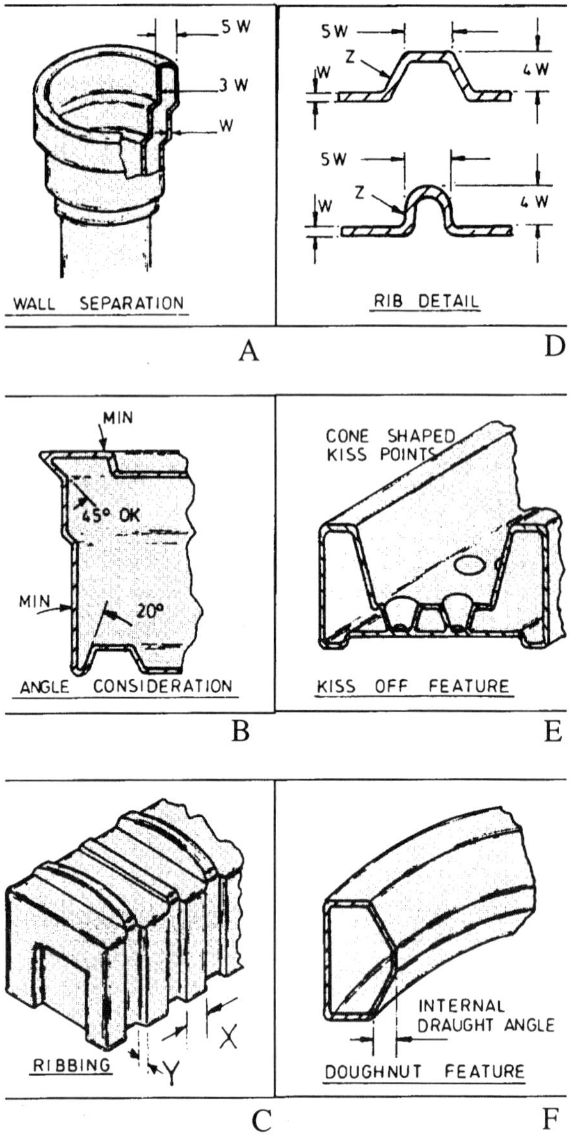

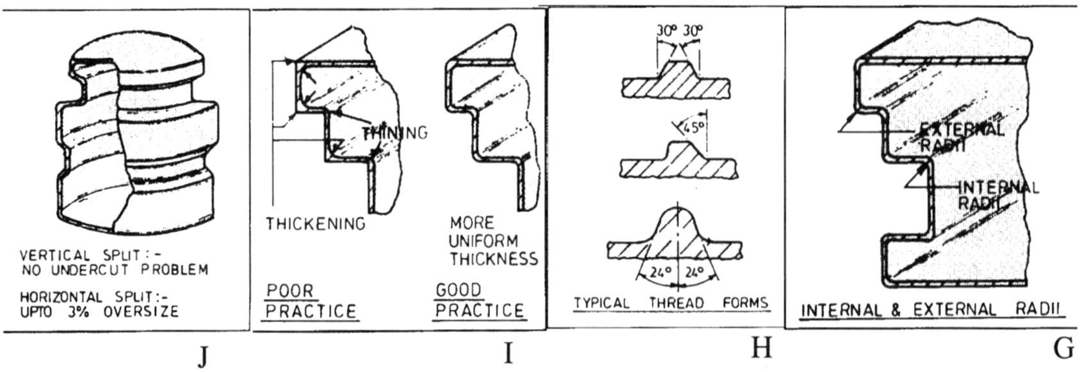

Fig 85: Optimising the design

Intelligent steel selection

Steel has been proven to be one of the most successful structural materials and its high resilience has proved it be an effective competitor to the so-called 'light-alloys' now that reliable anti-corrosion treatments have been developed. Because there are so many varieties of the material, according to alloying elements and heat-treatment, the selection of the appropriate grade is often a daunting task to the non material-scientist.

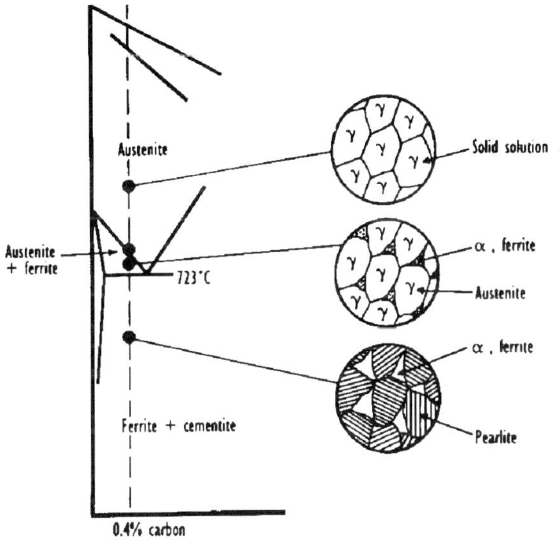

Fig 91: Slow cooling of a 0.4% carbon steel

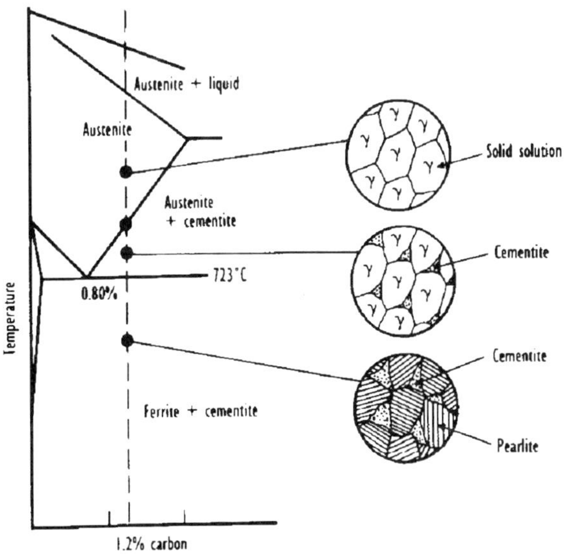

Fig 92: Slow cooling of a 1.2% carbon steel

Steel is classified according to a variety of crystal structures, dependent on the concentration of each alloying element; it is also affected by the way in which the material is cooled from furnace conditions and upon any cold-work carried out. The Materials Information Service of the UK Dept of Trade and Industry list the principle ones as ferrite, austenite, cementite, pearlite, martensite and bainite. Ferrite is the crystal arrangement for pure iron, which is found as part of the structure of most steels, and can absorb carbides by diffusion in the solid-state. Austenite is the single phase solid solution in which all steels exist at sufficiently high temperatures and with some alloys it is present at room temperature.

The crystal arrangement for these two principle forms is body-centred cubic and face-centred cubic respectively. Figs 91 and 92 show the effect of slow-cooling of 0.4 and 1.2% carbon steels respectively while Fig 93 shows the way in which chill rate can be varied to produce the other forms; Fig 94 shows the iron to iron-carbide systems available with different heat-treatments.

Work hardening is the resistance to continuing plastic flow which deforms the crystal structure — usually by microscopic shearing along definite planes. The extent of work-hardening depends on the number of intersections between crystal orientations at which slipping can take place. Most alloy steels have a complex structure with many such intersections and so are particularly susceptible to cold-work — which can usually improve strength and hardness at the expense of ductility — and the effects can only be cancelled by annealing.

Carbon steels

With up to 0.05% carbon, steels are ductile and similar in property to iron; they cannot be heat-treated. This is also the case for 0.05-0.2% steels but their hardness can be increased by surface carburising and there are no problems with heat-affected-zones in welding. Because ductility is high, machinability is poor unless high cutting speeds are employed or elements added to form free-machining inclusions.

Medium-carbon steels, of 0.2-0.5% C can be modified by heat-treatment and work-hardening; then the hardening effects in the heat-affected-zones of welds must be accounted for. Such steels are plentifully produced so are relatively cheap and their medium-strengths are useful in some structural applications. An intermediate range of 0.5-0.8% C are even more susceptible to heat-treatment and work-hardening; they can also be flame-hardened -- and can be cold-worked to give yield strengths of up to 2000MPa — but impact strength is poor.

With the high-carbon steels, over 0.8% C, cold-working is not possible and these steels fracture at very low elongation. Machinability is good when normalised to reduce surface hardness.

Alloy steels

Key features of the effects of alloying elements are as follows. Chromium improves high-temperature properties and corrosion resistance. Nickel lowers critical heat treatment temperatures and toughens steel; it also imbues good low-temperature performance without detriment to impact strength; vanadium generally improves mechanical properties due to its grain- refining effect. Molybdenum also gives a fine-grained structure with consequent improvement in strength. Tungsten is valued in steels requiring hardness and high-temperature stability. Boron can considerably increase hardness of low to medium carbon steels, while maintaining good weldability. Copper enhances corrosion resistance while aluminium is useful for nitriding and grain refining. Manganese lowers heat-treatment temperature and can provide high-carbon steels that are tough and workable. High chromium steels also offer good corrosion resistance and when a steel has more than 18% chromium it becomes a stainless steel whose properties are discussed later.

High Strength Steels

As with aluminium alloys, so with steel: appropriate alloy selection for a given set of design and loading conditions can result in substantially enhanced strength without altering structural rigidity. With the recent development of High Strength Steels (HSS) in sheet form, the opportunities for higher specific strength panels and pressings present themselves.

British Steel supply the motor industry with high strength steel in narrow and wide strip form. For volume cars in the UK something like 10% (200 kg per car) of the steel is HyPres High Strength Low Alloy (HSLA). The material first made its name for CV chassis stampings and a variety of grades correspond to the available fatigue lives shown in Fig 95.

Somewhat higher priced rephophorised steels are also available where even higher strengths and thinner gauges are required. In the case of HSLA, relatively low carbon content ensures reasonable ductility. According to grade, HSLA yield strengths

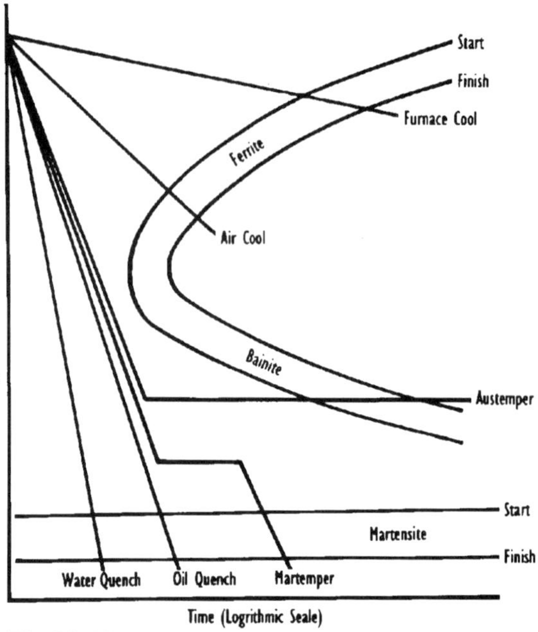

Fig 93: Time (x-axis)-temperature (y-axis) transformation diagram

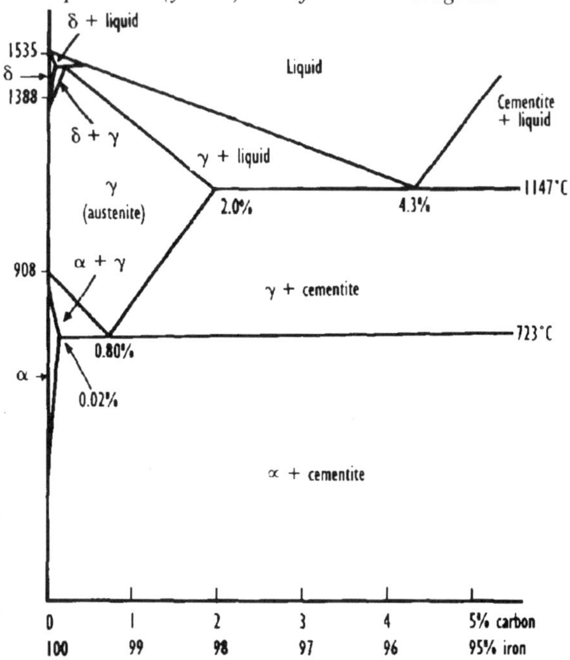

Fig 94: Iron carbide system (y-axis: temperature C)

can range from 275 to 550 MN/m^2 and elongation from 12 to 24%. Elastic modulus is virtually the same as mild steel and therefore strength-critical rather than deflection-critical panels are the most promising candidates. On a direct-substitution basis it has been suggested that the materials will offer sheet thickness reductions of 15-35%, for similar panel configuration and loading, at a cost penalty of about 20%. Some prefer the niobium rather than titanium grades, for ease of welding, correct choice of alloy allowing the use of the same spot welding plant as for mild steel. There was the further advantage, in HS steels other than HSLA, of bake-hardening which can take place in paint stoving ovens, with a potential yield strength enhancement of 80 N/mm^2 or so. In the UK, most truck chassis members were made from hot-rolled HSLA in gauges ranging from 3 mm on crossmembers to 8 mm on sidemembers. Fig 96 provides a comparison of performance for different steel grades.

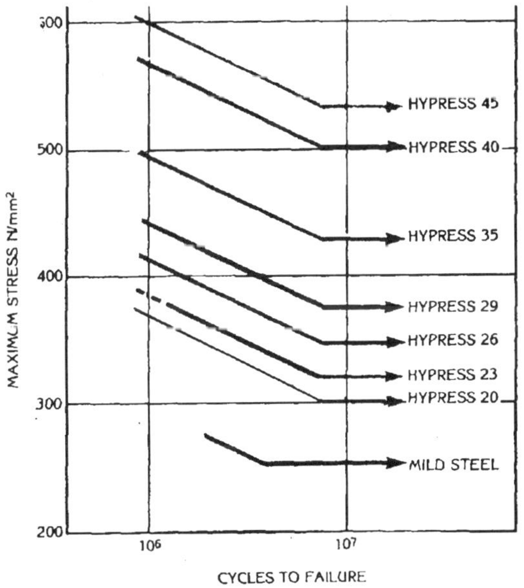

Fig 95: Fatigue curves for Hypress

Grade	Minimum Yield Strength ksi (MPa)	Minimum Tensile Strength* ksi (MPa)	Elongation (percent) Minimum in 2 in. (50 mm) Hot Rolled	Cold Rolled	Galvanized
35	35 (241)	50 (345)	23	23	23
40	40 (276)	55 (379)	21	20	20
45	45 (310)	60 (414)	19	18	18
50	50 (345)	65 (448)	17	16	16
55	55 (379)	70 (483)	15	N/A	N/A
60	60 (414)	75 (517)	N/A	10	10
70	70 (483)	80 (552)	N/A	8	8
80	80 (552)	-	N/A	4	4
120	120 (827)	-	N/A	2	N/A
140	140 (965)	-	N/A	-	N/A

Fig 96: HSS performance comparisons

In so-called Dual-Phase steels, strength is enhanced by a transformation hardening due to successive heating and then quenching. The resulting microstructure exhibits two phases, ferrite and martensite. The former is relatively soft while the latter exhibits strengths up to 1250 N/mm^2. Proportion of the two phases is controlled by the temperature from which the steel is quenched — and by the rate of quenching. Continuous annealing lines now used in the most modern steel plants have water quenching systems which allow rates up to 1000 C per second to be achieved.

Pressworking techniques

The degree of panel thickness reduction is somewhat limited by the stretch marks which can occur during pressing. Compared with the press settings for mild steel, there should be modification of the gap between the mating dies to avoid either wrinkling due to uneven strain or buckling due to radial compression. In the case of parts requiring deep drawing, some modification of detail design will probably be necessary compared with that used for mild steel and the number of stages in the forming process often has to be increased.

Work hardening is also an important element in strength enhancement for some HS steels. Standard tension tests on specimens can be used to determine work-hardening exponent n and plastic strain ratio r as the strain rate hardening exponent m. Typical values, due to the American Iron and Steel Institute are given in Fig 97. The n value is related to the steepness of the stress/strain curve in the plastic deformation region and co-relates with elongation

	Yield Strength ksi (MPa)	Tensile Strength ksi (MPa)	Percent Elongation in 2 in. (50 mm)	n value	r_m value	HRB hardness
Commercial Quality	40 (276)	51 (351)	34	0.18	1.1	60
Drawing Quality	37 (255)	50 (345)	37	0.19	1.1	47
Drawing Quality Aluminum Killed	35 (241)	49 (338)	39	0.19	1.1	49
Deep Drawing Quality Aluminum Killed	30 (207)	46 (317)	41	0.20	1.5	45
Interstitial Free	25 (172)	43 (296)	42	0.21	1.8	40

Fig 97: Work hardening exponents

and ratio of tensile strength to yield-strength. If the metal strain hardens during the forming process, initial straining will cause the deformation zone to increase in strength — until deformation stops in that region and spreads to another. The higher the strain hardening the more uniformly will deformation be spread throughout the pressing.

The *r* value is a measure of the resistance to thinning as the metal is drawn. The weakest zone of a typical 'cup' pressing is at its base; in order for the base to deform and elongate, the thickness must be small. By specifying a material that resists thinning, a more uniform stress distribution will result. The *m* value works in the same way as the *n* value in delaying the start of local 'necking'. While high *n* extends the range of uniform deformation, high *m* delays the onset of the local neck.

Series	Principal Alloying Elements
1000	Aluminium of 99.0% minimum purity
2000	Copper
3000	Manganese
4000	Silicon
5000	Magnesium
6000	Magnesium + Silicon
7000	Zinc
8000	Lithium and others
9000	Unallocated

Fig 98: BS numbering system

Melting point	660°C
Boiling point	2480°C
Thermal conductivity (0-100°C)	0.57 cal/cm/s/°C
Temperature coefficient of linear expansion (0-100°C)	23.5×10^{-6} per °C
Electrical resistivity at 20°C	2.69μ Ω cm.
Temperature coefficient of resistance (0-100°C)	4.2×10^{-3} per °C
Density at 20°C	2.69 g/cm³
Modulus of elasticity	68.3 GPa
Modulus of Torsion	25.5 GPa
Poisson's ratio	0.34

Fig 99: Physical properties of 'pure' aluminium

Description	U.K. Designation	Min. tensile strength MPa 1200 alloy*
Annealed, fully soft	O	70
As manufactured	M	—
Quarter-hard	H2	95
Half-hard	H4	110
Three-quarters hard	H6	125
Fully-hard	H8	140

Fig 100: Strength enhancement

Aluminium alloy

Recent increases in the price of aluminium alloys prompts the intelligent use of the material to fully optimise its value. Rather than making substitution of materials in existing designs, a redesign for the new material is always worthwhile if time permits.

In many areas of special vehicle building, like container van bodies, aluminium alloy is of course the rule rather than the exception. In other areas, such as luxury and high-performance cars, aluminium alloys were used historically, were next displaced by steel now compete for varying shares of the market with steels and composites. Small cars, of course, were first produced with aluminium alloy and then, in the volume sector particularly, made in steel. With

E - Excellent G - Good
V - Very Good. N - Not recommended

Alloy	TIG MIG	Resistance	Brazing	Soldering
1080A	V	G	V	V
1050	V	V	V	V
1200	V	V	V	V
2014A	N	E	N	N
3003	V	E	V	V
3103	V	E	V	V
3105	V	V	G	G
5005	E	E	G	G
5083	E	E	N	N
5154A	E	E	N	N
5251	V	E	N	N
5454	E	E	N	N
6061	V	V	V	G
6082	V	V	G	G
7010	V	V	N	N
7020	V	V	N	N
7075	N	V	N	N

Fig 101: Jointing data

	Specific gravity	Density, kg/m³	Melting point °C	Linear coefficient expansion, 10^{-4}/°C	Specific heat joules/kg/°C	Modulus of elasticity G.P.A.	Torsional modulus G.P.A.	Poisson's ratio
Aluminium	2.69	2699	580-660	24.0	0.96	69	25	34
Mild Steel	7.85	8097	1900	12.6	0.42	207	84	30
Copper	8.93	9200	1083	16.5	0.38	120	45	33
Magnesium	1.75	1803	565	29.0	1.02	44	17	33
Timber (Ash)	0.60	618	—	4.5	—	10	—	—
Glass	2.64	2520	1500	8.4	—	69	—	24
A.B.S.	1.05	1082	125	90.0	—	1.4-2.7	—	—
P.V.C.	1.40	1442	95	70.0	1.40	2.8	—	—
Nylon	1.14	1174	125	98.0	1.67	3.6	—	—

Fig 102: Material properties

the increasing use of plastics skin panels on small specialist sports cars there is some sign of aluminium alloys competing with steel for the construction of the base structures.

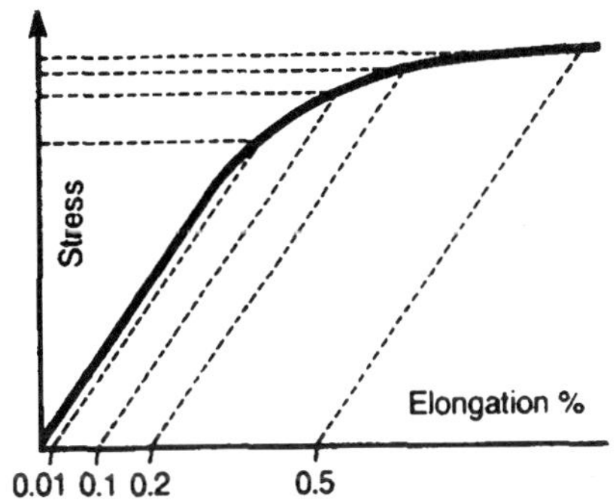

Fig 103: Typical stress/strain curve

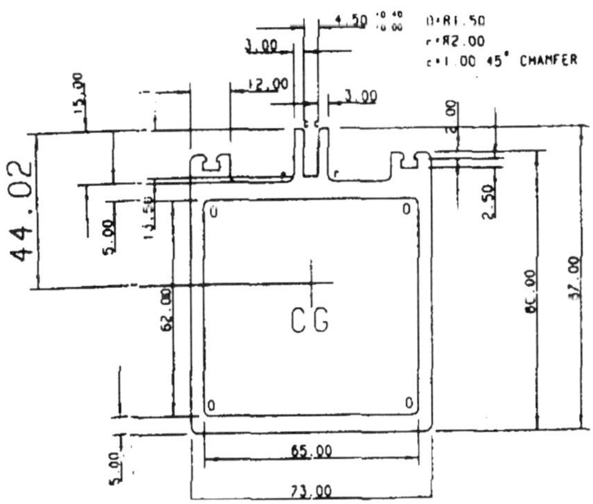

Fig 104: Calculating section properties

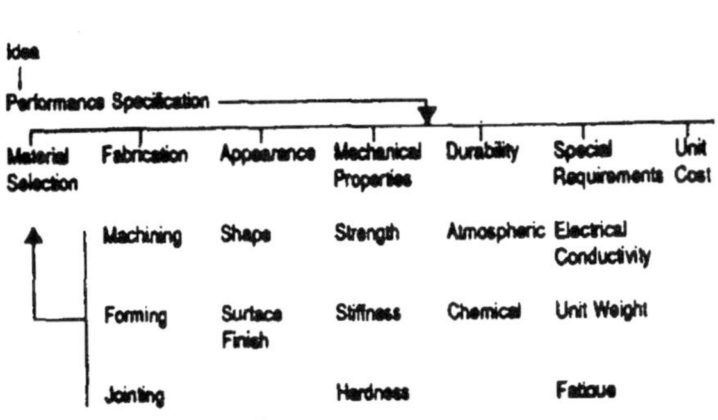

Fig 105: Design plan for aluminium alloy build

Fig 98 shows the British Standard numbering system for designating aluminium alloys while Fig 99 gives some of the physical properties of 99.99% pure aluminium. Most of the alloys have elastic modulus one third that of steel and weight about three times that of steel — hence the danger of direct substitution for a particular design. The strength of aluminium alloys, depending on composition and heat-treatment, can virtually match a number of steel alloys; however, selection must be made with care so as not to unduly compromise some other property or pay too high a price.

Two distinct groups of alloys are those that can and those that cannot be heat-treated. Those which cannot can have their strength enhanced by varying degrees of cold-working. This is expressed in the alloy by the letter H and a number from 1 to 8 indicating strength increase. There will normally be progressive loss of ductility with cold working and if this quality is at a premium then the O-designated alloy should be selected — signifying the fully annealed condition which removes any cold-working effect. The non heat-treatable series are 1000, 3000, 4000 and 5000 while Fig 100 shows strength enhancement of the heat-treatable ones. Heat-treatment has two basic stages; heating at prescribed high temperature followed by rapid cooling is solution heat treatment; heating at a prescribed low temperature for a prescribed time is known as ageing. The former increases the metal's ductility and the latter causes hardening; forming of the metal is thus sometimes carried out between the two stages. Alloys suitable for these processes are 2000, 6000, 7000 and 8000; while designations TB, TE and TF refer to solution treated and naturally aged; artificially aged and the combination, respectively.

The Aluminium Federation provide much data, such as that in Fig 101, on the joining of aluminium alloys. Because all of the alloys have a tenacious oxide film covering, provision has to be made for breaking through this prior to welding. Because the metal has much higher conductivity than steel, higher heat inputs for fusion welding are required; its high reflectivity also means that a different appearance from steel is exhibited in the solid to molten transition phase. Because heat applied in soldering, brazing or welding has a softening effect in the region of the join, work-hardened alloys will be locally annealed — an aspect which must be allowed for in design. Material properties for aluminium alloy appear in Fig 102 alongside those for rival materials. Recently developed aluminium-lithium alloys have a lower than normal density and strengths and stiffnesses compare well with more expensive copper and zinc alloys with aluminium. It is important in design to realise that aluminium alloys have about twice the expansion co-efficient of steel — also that, unlike steel, the material does not have a clearly defined yield point, leading to the 0.2% strain proof stress being used in design, Fig 103. For applications like truck bodies the natural 'mill' finish of aluminium has been found acceptable. Upgrades on this are provided by anodic oxidation (anodising), liquid or powder organic coatings — for which BS 1615, BS4842 and BS 6496 apply respectively.

A recent change from riveting to adhesive bonding of aluminium alloy panels in vehicle construction has been reported for mini-bus production. Quieter operation of the vehicle is said to result and the continuous join line has resulted in a lighter gauge of framing. This and other design and fabrication problems is covered in the Shapemakers Extrusion Design Package, available from the Aluminium Federation. It comprises a comprehensive manual and a number of computer diskettes containing CAD programs for many design applications. Accent is on extrusions which are of course widely used in specialist vehicle body building; the software allows rapid calculation of section properties for complex extrusions — with considerable benefit to structural analysis, Fig 104. The 2000 and 6000 alloy series are described for the extrusion application and a flow chart for assessing design ideas is suggested as Fig 105. Extrusion allows the convenient addition of bulbs and root buttresses, Fig 106, to improve section properties and can provide even for integral hinges, Fig 107, for specialist application. Built-in mechanical fasteners, Fig 108, are another possibility. Alloys 2014 and 2024 have been used for truck chassis frames; one of the key factors in choosing high-performance alloys is the fatigue to ultimate tensile strength ratio depicted in Fig 109.

In this illustration, Al-Zn-Mg and Al-Mg-Si alloys are compared for fatigue property which like toughness and stress corrosion resistance reduces with increased ultimate strength. Code of practice CP118 from the BS I is particularly valuable for the structural use of the material.

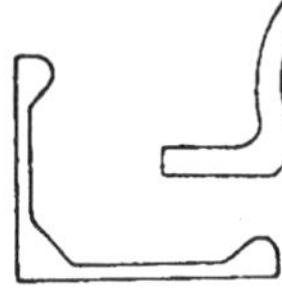

Fig 106: Section bulbs

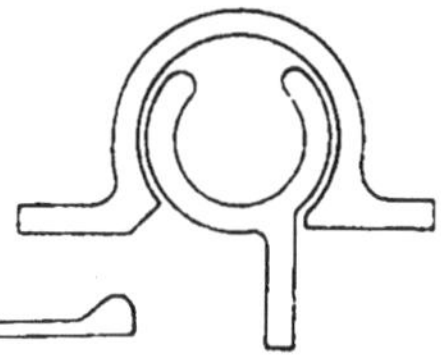

Fig 107: Extruded hinges

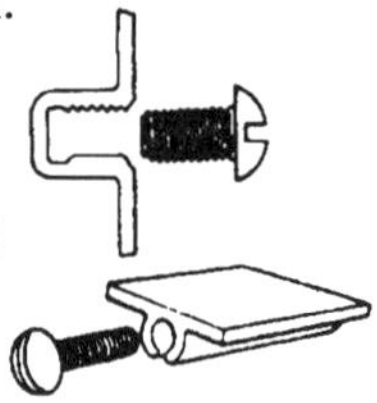

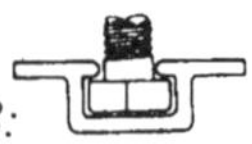

Fig 108: Fastener fixtures

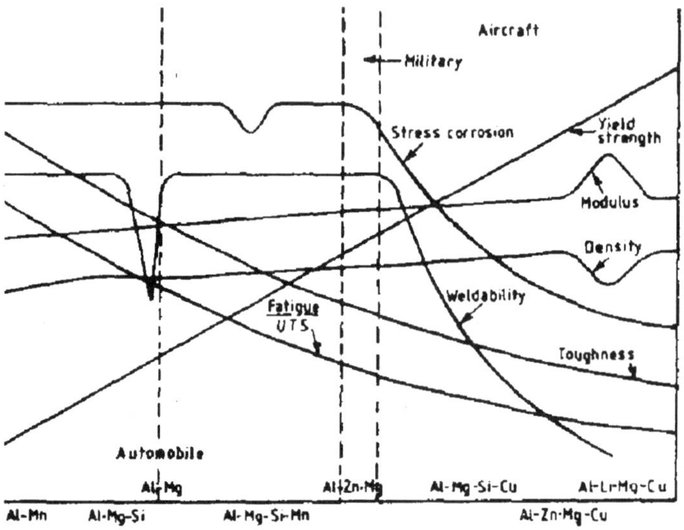

Fig 109: Fatigue to ultimate strength comparisons — far left band, windows, doors, car bodies, containers and electrical; near left band, road transport structures and tank ends

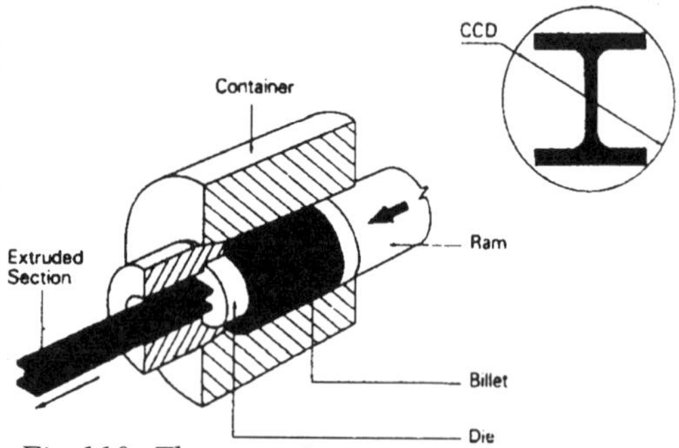

Fig 110: The extrusion process

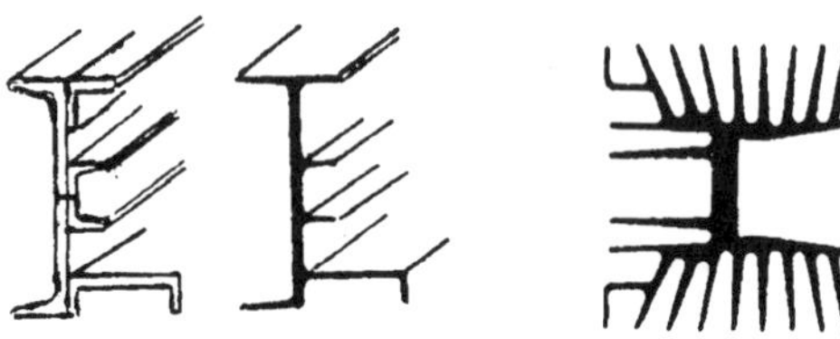

Fig 111: Section intricacy with extrusion

Aluminium-alloy extrusions

Careful fabrication control together with imaginative structural design can bring about pleasing body shapes, for a variety of end uses, with the help of aluminium alloy sections. Given here are some design guidelines on how to exploit the imaginative sections available from the aluminium suppliers.

Use of aluminium extrusions for vehicle body structures —particularly in commercial vehicles — has revolutionised design in recent decades and brought to the smallest of workshops the ability to fabricate clean-lined vehicles with accuracy. Lengthy fitting and finishing operations are largely avoided by exploiting imaginative sections created by the aluminium companies for solving common assembly problems. To produce a complex extruded cross-section, a pre-heated ingot in its plastic state is forced through a steel die cut out to the section shape, Fig 110. For hollow sections, a mandrel is attached to a 'bridge', the extruded stream being split at the bridge and re-united before emerging from the die. Section advantages include intricacy, fine-tolerance and the ability to replace several parts, Fig 111. The ability to produce sections previously only achievable by roll-forming and machining are also great benefits.

Section shapes should be chosen where possible to have their various elements as near as possible of equal thickness; this is to ensure consistency of metal flow and hence resultant physical properties. Sections of high overall dimension, or circumscribing circle diameter (CCD), are more expensive to produce with high accuracy because of the difficulty of maintaining tolerance — arising from the tendency for slower metal flow at the periphery. Accuracy in compact scctions is normally good enough to provide successful mating components — such as those in the designs of British Alcan in Fig 112. These also show how flush fitting surfaces can be obtained as well as sliding, 'drawer', effects. The elasticity of aluminium alloy also allows for the use of clip-fits, Fig 113, either permanent or releasable. Tolerances for straightness, twist and concavity/convexity are shown in Fig 114 alongside those for angular warpage of section elements. Tables are also available from aluminium suppliers for tolerances of extruded round tubing in terms of eccentricity caused by variation of mean wall thickness.

Aluminium alloy extrusions in platform and box van bodies are now familiar technology. Platform bodies have traditionally been at the more expensive end of the chassis/body combinations. Fig 115 shows typical crossbearer and longitudinal runner sections. The rear bearer may incorporate a shroud plate, as shown; usually bearers are solid-

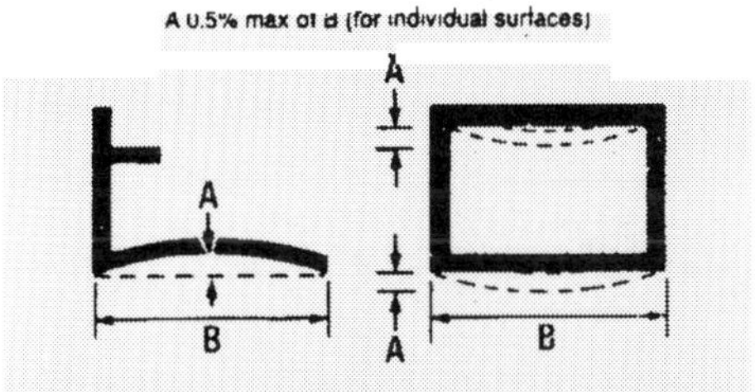

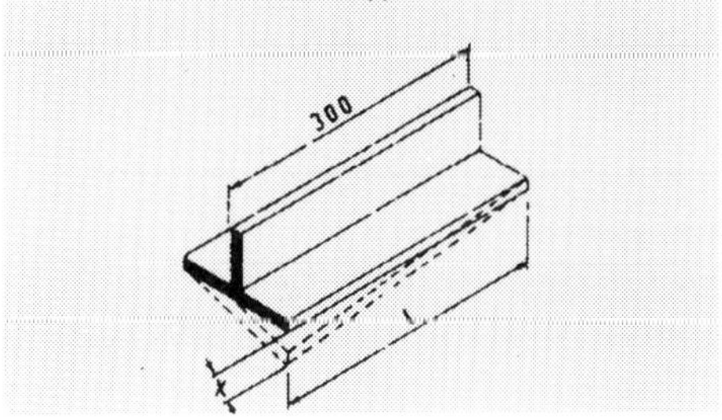

Circumscribing circle mm	Angle of twist per 300 mm run x°	Total angle of twist per length L maximum
Under 40	1°	5°
40 up to and inc 80	½°	3°
Over 80	¼°	2°

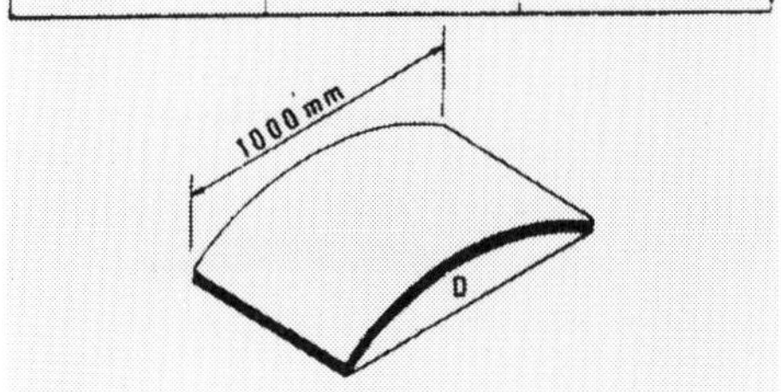

For diameter, or sections within circumscribing circle mm	Conditions	Departure from straightness over any length of 1000 mm D mm
Up to 100	All	1.5
Over 100	M	2.0
	All others	2.5

Fig 112: Secton variation with extrusion: bow, top: twist, centre; straightness, bottom

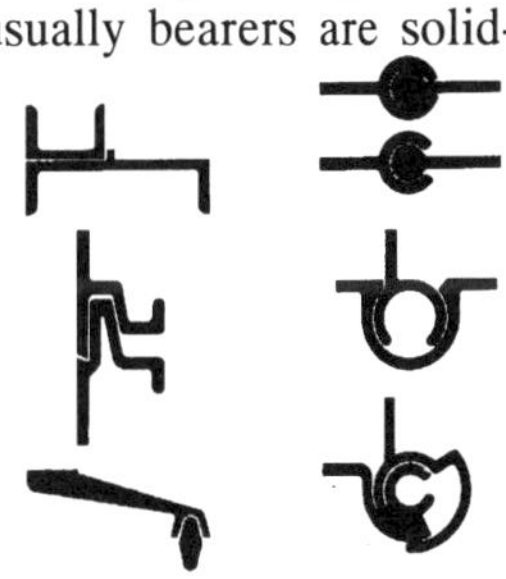

Fig 113: Clip-fit designs

riveted or bolted to longitudinals using angle brackets, as seen in Fig 116a. Side boards are increasingly made from extrusions with plank sections shown in Fig 116b; the conventional box van body uses a variety of sections including the sill and corner pillars, as in Fig 117.

Extrusions are also used with timber in so-called composite construction bodywork as seen in the cant rail section of Fig 118 used to connect plywood sidewall and roof panels. Usually the top corners of box vans are castings which enclose the cant rail extrusions. The kind of sill section used will depend on the requirement for a rub-rail. Extruded floor panel sections are also now commonplace and Fig 117 shows recommended spacing and carrying capacities. It is also necessary to check the carrying capacity of the 'planks' in respect of the wheel loadings from hand- and power- operated pallet trucks. Individual planks may have to withstand concentrated loads of 1650 lb (750 kg) under these conditions. Timber inserts are also used where there is a requirement for securing loads by wood battening. Fig 119 shows a side rave section used to form the edge of platform floor. A wide variety of jointing recommendations for vehicle body extrusions is in the aluminium supplier's handbooks. It is important to realise, however, that many of these are not applicable to integral structures — requiring greater attention to the load paths between the bulkhead rings and the longitudinal 'stringers' of the structure.

Straightness tolerances for extruded bar, regular sections and extruded round tube

Sections within circumscribing circle	Condition	Departure from straightness (S) over any selected length of 1000mm
Up to 100mm	All	1·5mm
Over 100mm	As extruded (M)	2·0mm
	All other	2·5mm

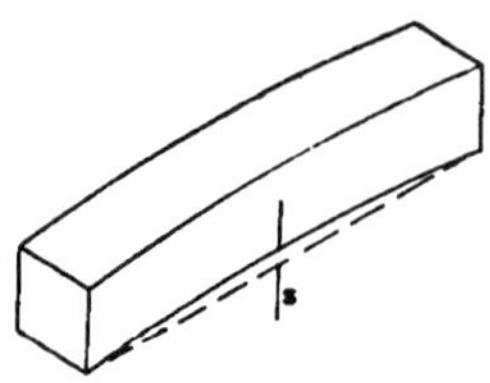

Twist tolerances for extruded solid and hollow sections

Diameter of circumscribing circle	Angle of twist (t) per 300mm run	Total angle of twist per length
Under 40mm	1°	5°
40 up to and inc. 80mm	½°	3°
For diameters over 80mm:		
Lengths up to 8000mm.	¼°	2°
Lengths over 8000mm.	¼°	3°

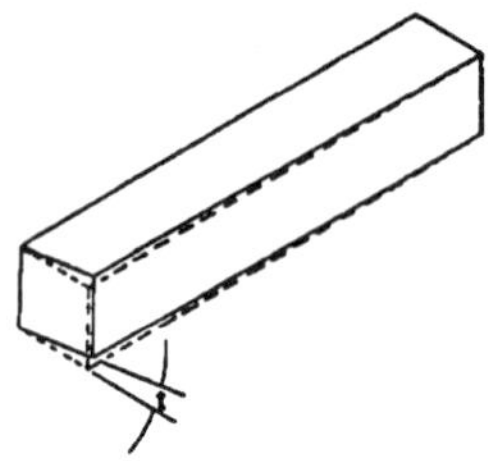

Concavity and convexity tolerances

Over the width (A) of the section the maximum tolerance on concavity and convexity (B) shall be 0·05mm per 10mm of width

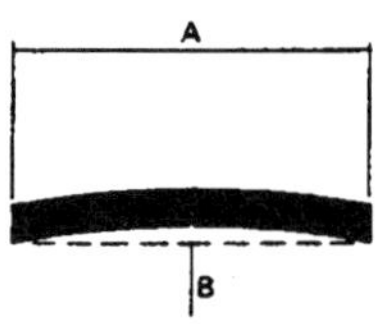

Normal angular tolerances

Nominal thickness of thinnest leg	Allowable deviation from angle specified (θ)
Up to and inc. 1·6mm	±2°
Over 1·6mm, up to and inc. 5·0mm	±1½°
Over 5·0mm	±1°

Angles are measured at the extremities of the section.

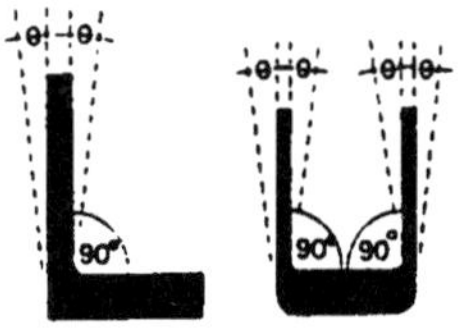

Fig 114: Tolerance requirements

Recently revealed properties for a range of structural extrusions recently introduced by Alcoa are shown in Fig 120. These high strength thin walled sections have predictable energy absorption characteristics and therefore can be used in car spaceframes such as that from the company shown in Fig 121. This kind of construction is said to result in 30-40% weight savings compared with steel unibody structures.

The Shapemakers organisation set up by the extrusion suppliers have published a comprehensive handbook, and associated software disks, of 2D computer-aided-design procedures, known as the Extrusion Design Manual. On a desk-top computer, section characteristics can be quickly obtained in terms of area, periphery, weight per metre, also section second moment of area. Among the sample calculation procedures provided is the following one for unloading ramps for a vehicle, Fig 122. The requirement was for the ramps to have maximum weight of

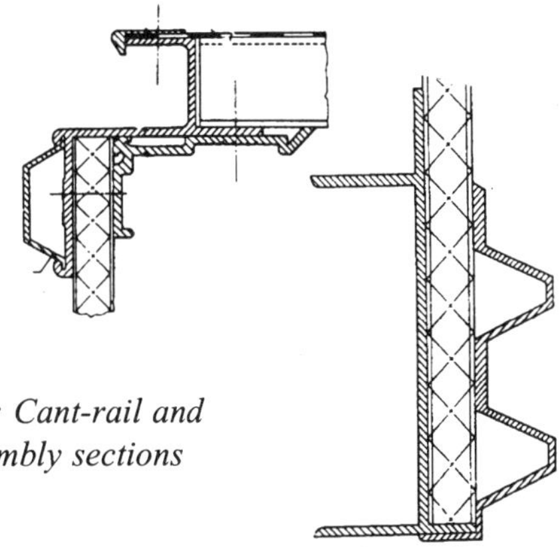

Fig 118: Cant-rail and sill assembly sections

Fig 116a: Joining bearer to longitudinal

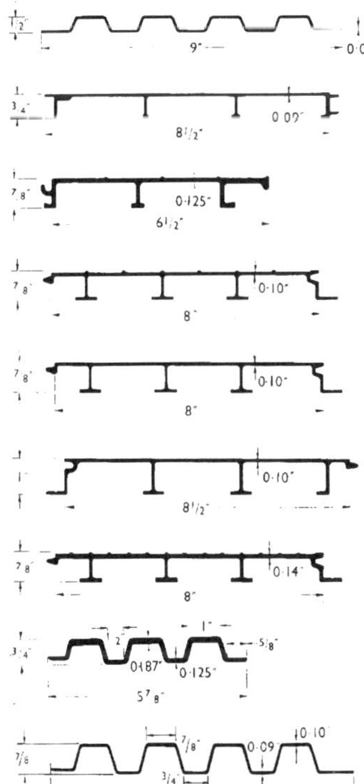

Weight as laid (lb. per sq. ft.)	Maximum Distributed (lb. per sq. ft.) Cross-bearer Spac 15"	18"	21"	24"
1·60	670	470	340	260
1·86	650	450	330	250
2·99	1,330	920	680	520
2·52	1,420	980	720	550
2·48	1,520	1,050	770	590
2·42	1,580	1,100	800	610
3·36	1,750	1,210	890	680
3·27	1,830	1,270	930	710
2·02	1,900	1,320	970	740

Fig 115: Crossbearer and runner sections

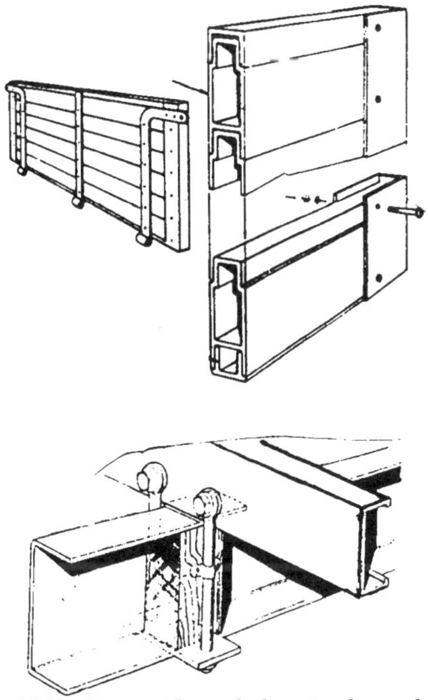

Fig 116b: Drop-side and chassis clamp details

50 kg each, span of 2.5 metres, operating angle up to 30 degrees and maximum wheel loading from pallet truck of 2 tonnes shared by four tyres of maximum width 200 mm. Specified channel section is 254 x 88(web) x 14(flange) mm.

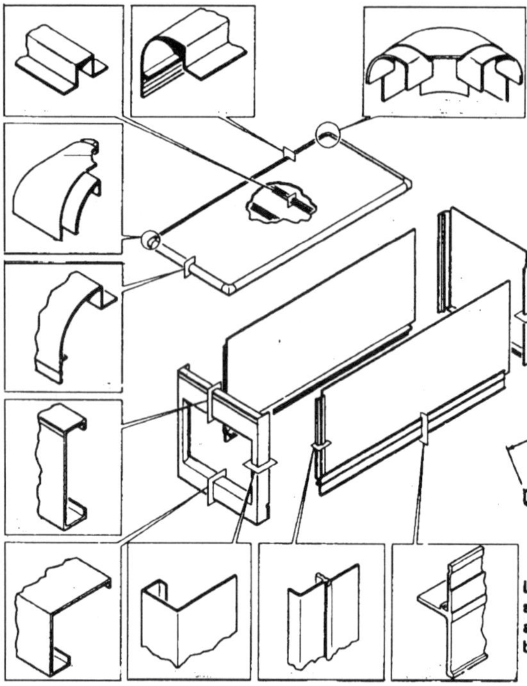

Fig 117: *British Alcan CV body extrusion examples*

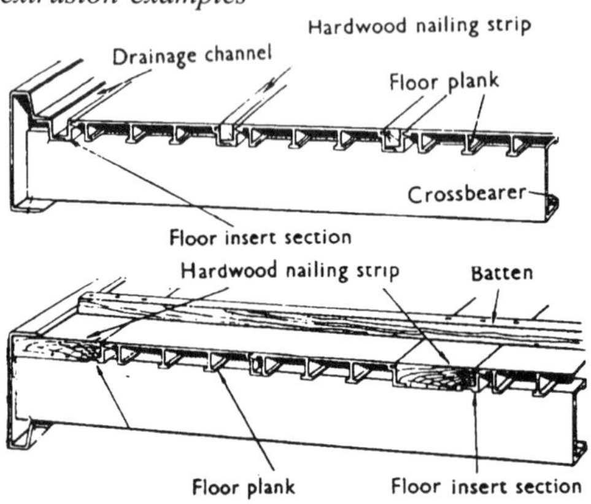

Fig 119: Floor-plank and side rave sections

Yield Strength, T4, ksi	16	(110 MPa)
Tensile Strength, T4 ksi	34	(235 MPa)
Percent Elongation, T4	21	
Yield Strength, T6, ksi	39	(269 MPa)
Tensile Strength, T6, ksi	44	(304 MPa)
Percent Elongation, T6	11	

Fig 120: ALCOA extrusion properties

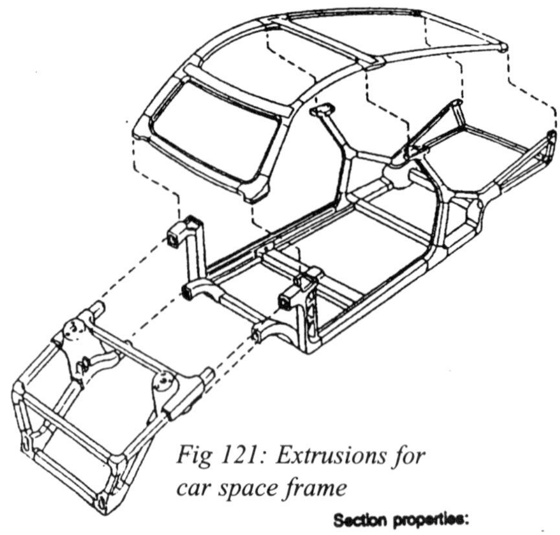

Fig 121: Extrusions for car space frame

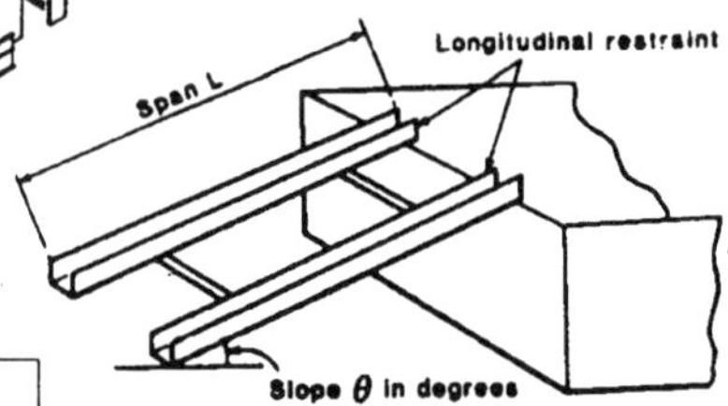

Section properties:

Area		5030 mm²
Modulus	Zxx	54620 mm³
Inertia	Ixx	3459100 mm⁴
Radius of Gyration	r	26.2 mm
Weight/metre		13.39 kg/m
Alloy		6082 T6

Loading. As the vehicle is unloaded it moves out of the horizontal with a considerable shift in its neutral axis and the loading on the first set of wheels increasing. This will be a feature of the individual vehicle. For the purposes of this calculation it is assumed to be 10%, hence -

$$\text{Maximum individual wheel load} = \frac{19640\text{N (2 tonnes)} \times 110}{4 \quad 100} = 5400\text{N}$$

Bending Stresses. The ramp acts as a simply supported beam and with normal wheelbased vehicles will have a central load as the worst condition. (Load Case 2.)

$$M = \frac{WL}{4} = \frac{5400\text{N} \times 2500\text{mm}}{4}$$

Maximum bending moment = 3375000 Nmm

$$\text{Maximum Stress} = \frac{3375000}{54620} = f_{bc} = 61.8\text{N/mm}^2$$

Allowable Stress Levels. See Table 3.2 (From British Standards CP118)

6082 T6 alloy

Bending $P_{bc} = 154\text{N/mm}^2$

Deflection

$$\delta = \frac{WL^3}{48EI} \qquad \text{For 6082 } E = 68{,}900\text{N/mm}^2$$

$$\delta = \frac{5400 \times 2500^3}{48 \times 68900 \times 3459100}$$

$$\delta = 7.45\text{mm}$$

The deflection/span factor = $\frac{\text{Span}}{336}$
which is well inside the recommended value of $\frac{\text{Span}}{200}$

Fig 122: Sample calculation for loading ramps

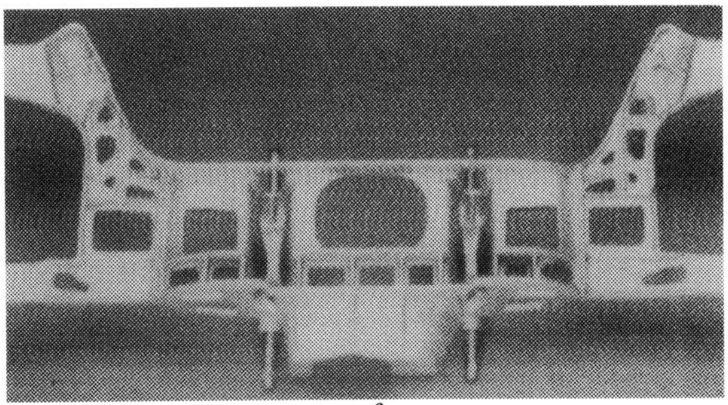

Fig 123: rear suspension frame

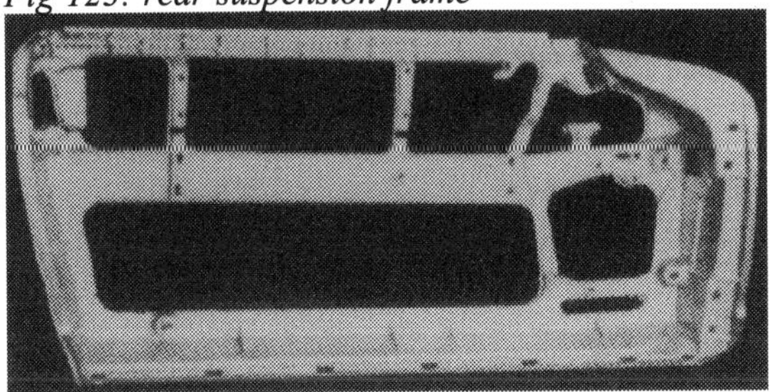

Fig 124: Door framework

Aluminium alloy castings

The volume vehicle sector is demonstrating that aluminium alloy is increasingly being used for structural members such as engine support crossmembers and rear suspension frames, Fig 123. In the luxury and sports sectors of the specialist vehicle market there are such components as door frameworks (Mercedes Benz), Fig 124, and suspension wishbones/king-posts (Honda NSX). Low-pressure die-casting was discussed in an earlier section and is particularly suitable for thin sections. The process has been used for special road wheels and involves alloys such as BS LM 25 having elongation from 7-13%. Sand castings might better suit low volume work and some do's and dont's are given in Fig 125.

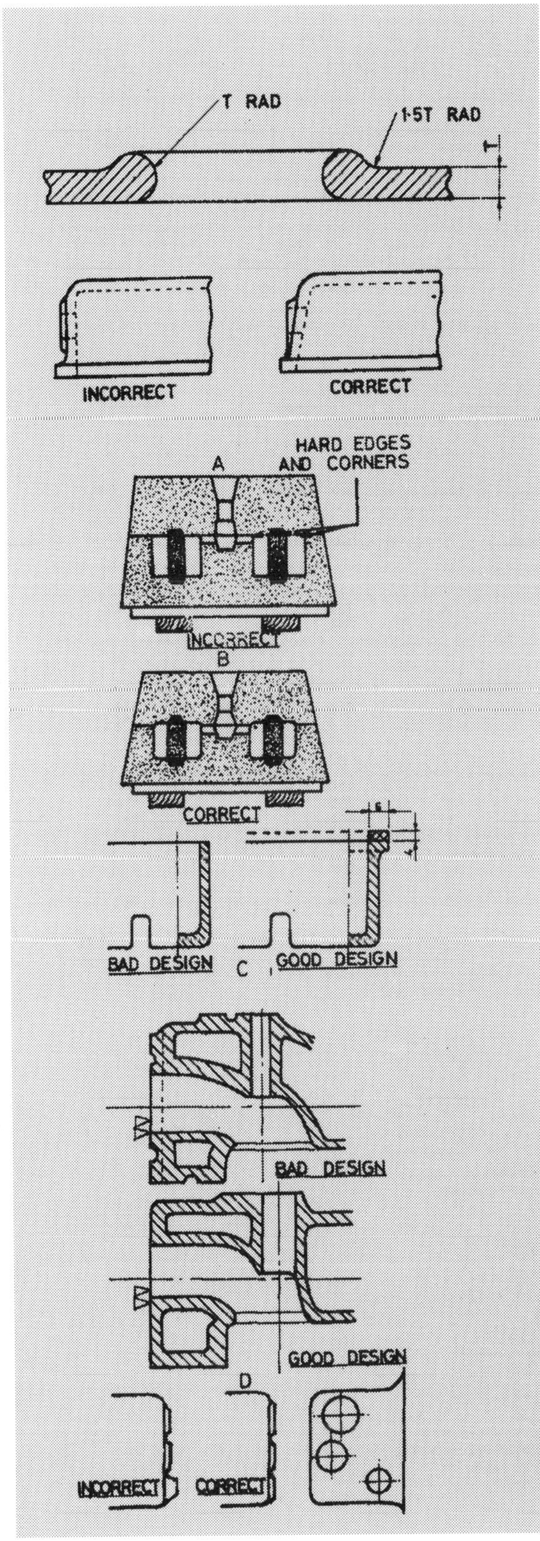

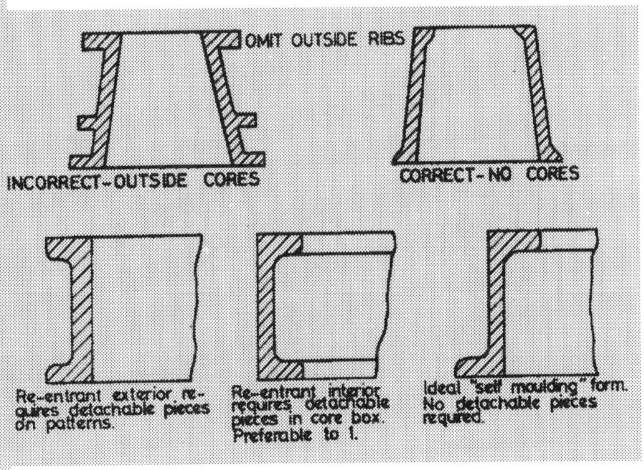

Fig 125: Do's and dont's of sand-casting

BS1490 Reference No.	Minimum Ultimate		Typical Brinnel Hardness Number	Thermal Conductivity W/mK	Castability (Gravity Die)	Pressure-Tightness	Atmospheric Corrosion Resistance	Machinability	Anodic Film	Fusion Weldability	
	Bearing Stress MN/m²	Shear Stress MN/m²								Gas	Argon Arc
General purposes alloys											
LM4	320 560	150 —	85 115	133	Good	V. Good	Good	Good	Cloudy grey	Fair	Good
LM6	370	130	60	162	V. Good	V. Good	V. Good	Fair	Dark grey	Good	V. Good
LM8	320 370 430 560	130 150 170 220	60 70 80 110	179 171	Good	Good	Good	V. Good	Light grey	Good	V. Good
LM25	320 370 430 560	130 150 170 220	60 70 80 110	159	V. Good	Good	Good	V. Good	Light grey	Good	V. Good
LM16	560 650 610	210 220 —	95 115 100	166	Good	Good	Good	V. Good	Light grey	Fair	V. Good

BS1490 Reference	Composition (Nominal) Per cent	Temper or Condition	Alloy (Alcan Ref)	Principal Specification and Requirements[1]			Other Related Spec'ns	Typical 0·2% Proof Stress MN/m²
				U.T.S. MN/m²	Minimum% Elongation on $5{\cdot}65\sqrt{So}$	U.T.S. Ton f/in²		
General Purpose Alloys								
LM4	Al-Si5Cu3	M TF	117	160 280	2	10·0 18·0	L79	108 —
LM6	Al-Si12	M	160	185	7	12·0	4L33	90
LM8	Al-Si5Mg	M TE TB TF	B116	160 185 230 280	3 2 5 2	10·5 12·0 15·0 18·0	716A 722A 727A 735A	100 160 120 270
LM25	Al-Si7Mg	M TE TB7 TF	D135	160 190 230 280	3 2 5 2	10·4 12·3 14·9 18·0		100 160 120 270
LM16	Al-Si5Cu1-5Mg	TB TF TB7	C125[2]	280 325 310	12 4 10	18·0 21·0 20·0		150 300 201

Fig 127: Casting alloys

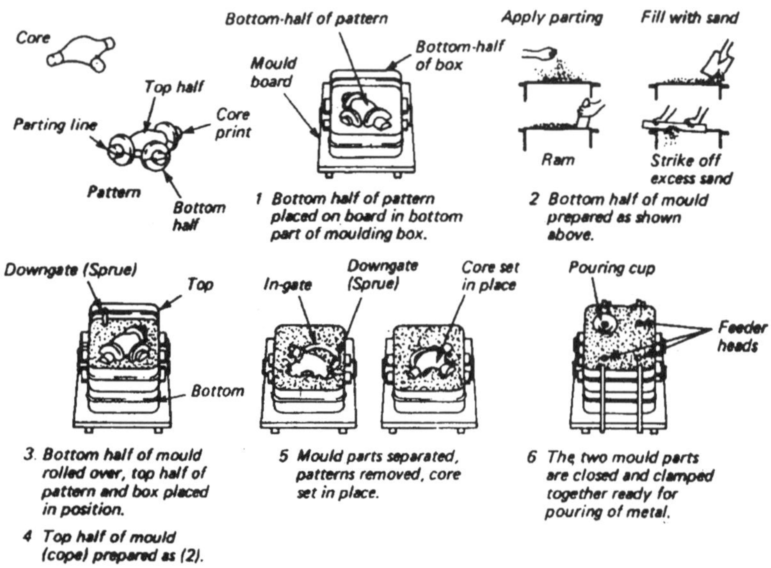

Fig 128: Sand moulding techniques

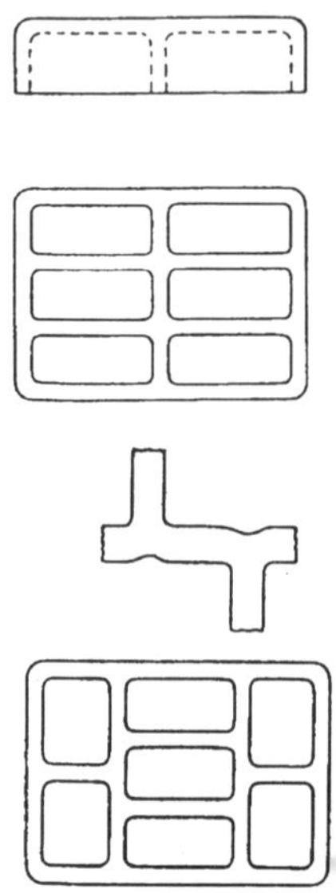

Fig 129: Staggering the ribs

An aluminium casting process overview is iven in Fig 126, due to MBB. The alloys for casting Fig 127) are chosen for good mould-filling capabilies. A357 is the standard casting alloy which conains silicone (G-AlSi7Mg0.6) is particularly faoured — with strength of 330 N/mm². For higher trength requirements, A201 is preferred which has a igure of 414 N/mm². The basic techniques of sand noulding are shown in Fig 128. The core, of course, s made separately by compacting bonded sand in a ore box. Extensions shown on the core represent orints which accurately locate the core in the mould. A variety of additions convert the hand process hown to one of machine control in most cases. In the ase of shell moulding, the sand grains are coated vith resin which makes green sand flow as easily as lry sand. It is blown on to a heated metal pattern to orm a resin bonded shell. Machines thus equipped are capable of higher outputs and the castings proluced characterised by a smoother than normal sandast surface.

When finish casting is to be laid out for mahining the drawing should show reference points. These should not be influenced by core-shift or novement of the drag. In general they should be as far ipart as possible and on the same side of the parting inc. The latter should be on one plane to facilitate oarting. Pattern dimensions must differ from those on he finish-part drawing, to allow for shrinkage. All surfaces of the pattern also need a slight draft angle.

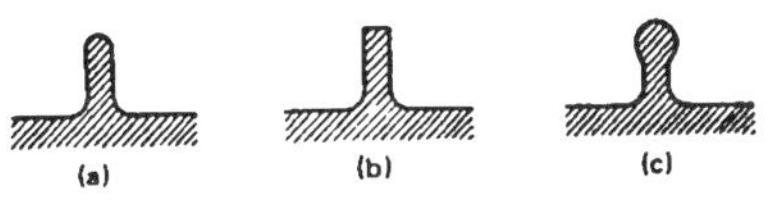

Fig 130: Rib configurations

In designing ribbed sections, relationship of the rib section to the main section should be such as to permit a uniformly graded metal section. Fig 129 is an example where the design is altered by staggering the ribs to eliminate X-junctions and obtain the rib detail, top right. Three types of rib design are compared in Fig 130. The section at (b) is satisfactory but that at (a) poor because of high stress at the extreme edge; most preferred is that in (c) provided moulding procedure does not become unduly complicated.

	Investment casting	*Premium sand casting*	*Conventional sand casting*
Advantages	• Small tolerances • Small wall thickness	• High mechanical properties for all wall thickness • Small wall thickness • Small tolerances	• Large dimensions • Low costs • Local high mechanical properties by local cooling
Disadvantages	• Lower mechanical properties	• Higher costs for castings and tooling	• Larger tolerances • Larger wall thickness
Part size (maximum overall dimensions)	1000 × 800 × 500 mm	3600 × 2000 × 1000 mm	5000 × 1500 × 1500 mm
Wall thickness, depends on: • outer dimensions • part geometry • alloy	T_{min} = 1.5 ± 0.15 mm	T_{min} = 1.8 ± 0.4 mm 0.2 (Local: 1.5 ± 0.2 mm)	T_{min} = 2.5 ± 0.5 mm (Local: 2 ± 0.5 mm)
Tolerances	According to countries standard	T_{min} = ± (0.4 + 1.500/ 1000 mm	According to countries standard
(D = considered dimension)	For example: VDG-P690 for Germany		For example: DIN 1688 for Germany
Surface roughness	R_a = 3.2 µm	R_a = 3.2–6.4 µm	R_a = 6.4–12 µm

Fig 126: Aluminium casting overview

Condition	Tensile strength (tonf/in²)	0·2% Proof stress (tonf/in²)	Elongation (%)	Hardness (H.V.)	Bend test 180°
Softened	35	14	35	200	—
Quarter hard	55	33	25	220	1t
Half hard	65	47	10	270	2t
Three quarter hard	75	58	7	320	4t
Hard	80	60	5	360	6t

Fig 131: 301 stainless steel

Grade Warm Worked	Tensile Strength	1·0% Proof Stress	Elongation
304 (18/10 chromium-nickel)	40 tonf/in²	24	30%
316 (18/10/3 chromium-nickel-molybdenum)	40 tonf/in²	28	20%
Nitrogen Bearing Grade			
304 (18/10 chromium-nickel)	38/51 tonf/in²	20-5 tonf/in²	35%
316 (18/10/3 chromium-nickel-molybdenum)	41/51 tonf/in²	22 tonf/in²	35%

Fig 132: Effect of warm working

Austenitic stainless steels

High-alloy corrosion resistant steels (otherwise known as 'stainless') comprise the ferritic, martensitic, precipitation- hardening and austenitic groups, the last of which has a potential for automotive structures depending on the degree of refinement of the design as well as the prevailing relative price of nickel.

Ferritic stainless steels, or chrome-irons as they are sometimes called, are considered in a later section. Here the higher grade austenitic stainless steels are examined — which can be successfully exploited if used in very thin sections and stressed to higher predicted levels when weight saving is crucially important. While strengths in the annealed condition are not very high, good ductility and work hardening ability means that it can be used in the cold-worked condition and still retain a good deal of formability. Fig 131 shows properties of 301 in various standard work-hardened conditions. Even the half hard state offers a proof stress of 50 tons/in^2 combined with a very useful elongation of 10%.

When used in the form of plate, for road-tanker shells, say, 0.2% nitrogen can be added to the steel melt to beneficial effect. Up to 50% increase in proof stress can be obtained above the normal annealed state — without any strength reduction in subsequent heat treatment of the plate. Another possibility is to 'warm-work' the otherwise untreated material — with benefits as shown in Fig 132.

Typical of tanker construction in stainless steel is the example shown in Fig 133 which also uses a special lightweight tandem bogie suspension. An example of stainless steel cladding section is seen in Fig 134, developed by a vehicle builder in conjunction with Firth Vickers from their 301 17/7 material of 28 gauge. Section properties are shown in Fig 135.

Types 304 and 321 are the normal British Standard austenitic stainless steels. The 321 grade is stabilised with titanium to provide complete immunity from intergranular corrosion and improve high temperature oxidation resistance in such applications as exhaust systems. An 18% Cr 9%Ni austenitic alloy is also made for components requiring even higher corrosion and oxidation resistance; austenitic grades are non-magnetic. The British Standard designation for stainless steels is found in BSI publication PD6290 — the three digits of the grades mentioned above are used for definition. All 300 steels are austenitic and

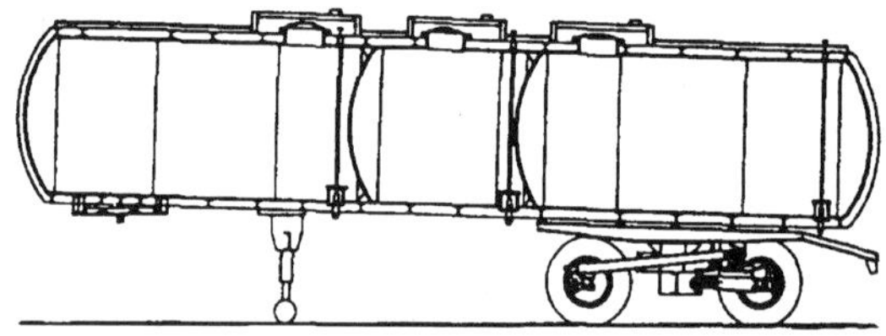

Fig 133: Frameless tanker in stainless steel

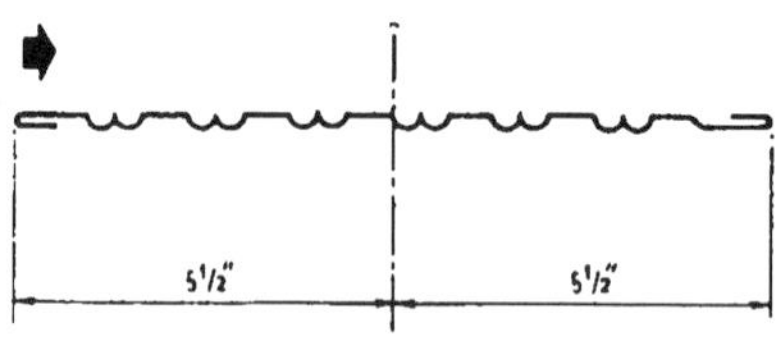

Fig 134: Cladding section

Properties of cladding section

Weight	0·63 lb/ft²
Section area	0·1776 in²
Inertia—about NA 'X-X'	1·958 in⁴
Section modulus—about 'X-X'	0·356 in³
Position of NA	Central in 11 in widths
Yield stress	30 ton/in²
Uniform load capacity per span:	
48·4 tons	1 ft
5·3 tons	3 ft
1·3 tons	6 ft

Fig 135: Section properties

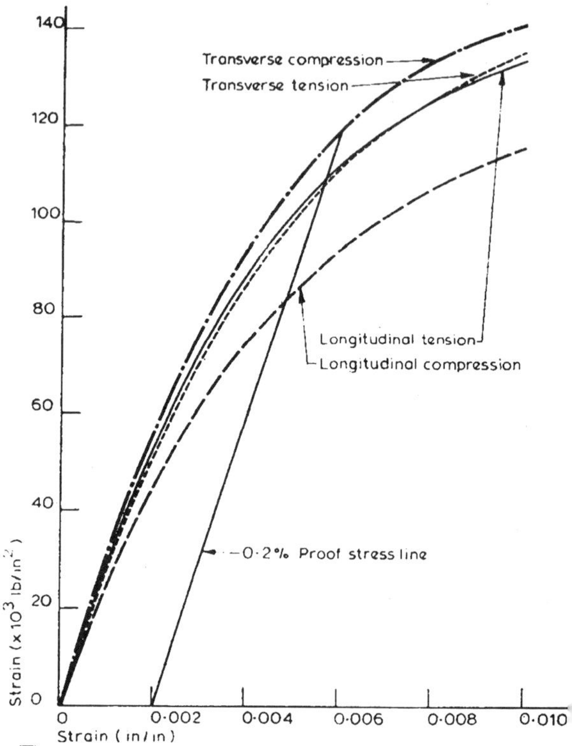

Fig 137: Stress-sttrain curves

the full range with suffixes is shown in Fig 136. Plate, sheet and strip are covered in BS 1449, billets and forgings in BS 970 and wire in BS 1554.

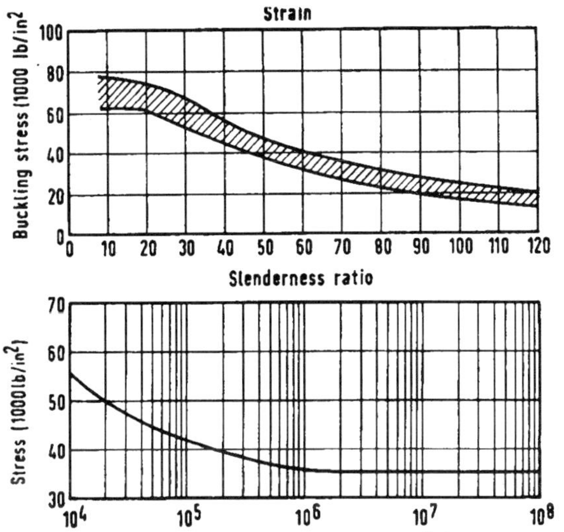

Fig 139: Critical buckling stress for 301 (above) and fatigue curve for 301 (below)

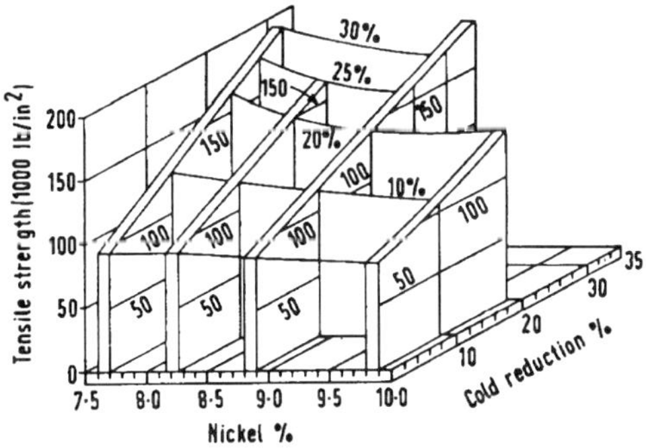

Fig 138: Effect of nickel concentration

The stress/strain relationship of austenitic steels is unlike those of ordinary steels, Fig 137, there being no yield point. Working stress is usually taken as two-thirds of the 0.2% proof stress. It should also be noted that some anisotropy is found after cold working, the compressive properties in the direction of rolling being not strengthened as effectively as the longitudinal tensile or transverse tensile and compressive properties. Another peculiarity of stainless is that increase in compresssive stress level is accompanied by a falling off in elastic modulus. Struts should therefore be designed on the basis of tangent modulus obtained from the stress/strain curve at the operational stress. Fig 136 lists the wide range of surface finishes that sheet and plate forms of the material are made available from the rolling mills. Effect of nickel concentration is seen in Fig 138, due to Sandvik

According to International Nickel, the austenitic group is the most extensively used of all stainless steels because of good corrosion and heat resisting properties and its excellent weldability. If sufficient nickel is added to the chromium steels, the austenite is stabilised so that it is retained at room temperature, particularly advantageous as far as welding is concerned because it eliminates the need for any pre or post heat treatment, which ensures weld joints of good strength and ductility.

The material also has advantages under buckling, shear and fatigue loading. Fig 139a shows the behaviour of annealed 301 as a function of slenderness ratio while Fig 139b shows a typical *SN* curve. Unlike the ferritic steels, these materials are not particularly sensitive to notches.

BS Designation	En Steel replaced	Class*	Principal constituents (percentages)	Forms available†	Remarks
409S17	New grade	F/M	Cr 12, C 0.09 + Ti	1, 2	Low cost weldable type
420S37	56C	M	Cr 13, C 0.24	3	Not suitable for welding, hardenable.
430S15	60	F	Cr 17, C 0.10	1, 3	Not suitable for welding
301S21	New grade	A	Cr 17, Ni 7	1, 2	Work hardens most rapidly of austenitic grades
302S25	58A	A	Cr 18, Ni 9, C 0.12	1, 2, 3	Less suitable for welding than 304S16
303S21	58M	A	Cr 18, Ni 9 + S	3	'Free machining' grade
304S12	New grade	A	Cr 18, Ni 10, C 0.03	1, 2, 3	Low carbon variety more suitable for welding than 304S16
304S16	58E	A	Cr 18, Ni 9, C 0.07	1, 2, 3	
305S19	New grade	A	Cr 18, Ni 11, C 0.10	1	Work hardens least rapidly of austenitic grades
316S16	58J	A	Cr 17, Ni 11, Mo 2.5	1, 2, 3	Resists most common chemical loads
321S20	58B & C	A	Cr 18, Ni 9 + Ti	1, 2, 3	Titanium stabilized for welding in thicker section and high temperature service.

Fig 136: Range of 300 grade steels

Ferritic stainless and galvanised steels

While nickel-rich austenitic stainless steels used in such applications as frameless tank shell construction are important in the context of both high structural strength and high resistance to corrosion, less costly ferritic types, such as Cromweld 3CR12, are now finding increasing application in special vehicle building. Meanwhile an increasingly used alternative to hot-dip galvanised steel is seen in the electrolytically zinc-coated cold-rolled steels such as British Steel's Zintec. The properties and potential of the two materials are compared here.

Weldability is one of the key attributes of Cromweld 3CR12, the relatively low cost stainless steel now used in a variety of structural monocoques which might once have been the province of the higher grade alloys of aluminium. By removing much of the anti-corrosion pre-treatment costs associated with mild pressing steel, its use becomes economically viable — particularly if effort is made to use the material in a structurally efficient way such that its high modulus, relative to aluminium alloy, can be exploited. In this context, the term 'stainless' is seen by its suppliers as applying to steels containing at least 10.5% chromium.

Best known in cutlery applications, early stainless steels had 12% chromium content — but many other compositions have since developed. Main classification is between non-magnetic chrome-nickel steels and the magnetic straight chrome steels. However, there is a main sub-division of the latter category between martensitic and ferritic types. Of the last two, the former are heat-treatable but the latter not and, between the two, comes the comparitively recent addition of 3CR12 whose price, as unity, is compared with other materials in Fig 140.

Because the so-called chrome irons (ferritics) suffered from poor weldability, they became domi-

Material	Sheet (£/Kg)
Mild steel	0.4
3CR12	1.0
Ferritic stainless steel	
430 (17Cr)	1.5
Austenitic stainless steel	
304 (18Cr-10Ni)	1.8
316 (17Cr-12Ni-2.5Mo)	2.2
Duplex stainless steel	
2205 (22Cr-5.5Ni-3Mo-0.2N)	2.6
High alloy steel	
310 (25Cr-20Ni)	3.3
Ti-clad steel	7.5
Alloy 400 (Ni-32Cu)	15
Titanium (pure grade 2)	17
Alloy C276 (Ni-16Mo-15Cr-6Fe-4W)	25
Titanium (0,2Pd grade 7)	33
PTFE	38
Alloy 625 (Ni-21Cr-9Mo-2.5Fe)	40
Zirconium	45
Tantalum	250

Fig 140: Metal price comparison

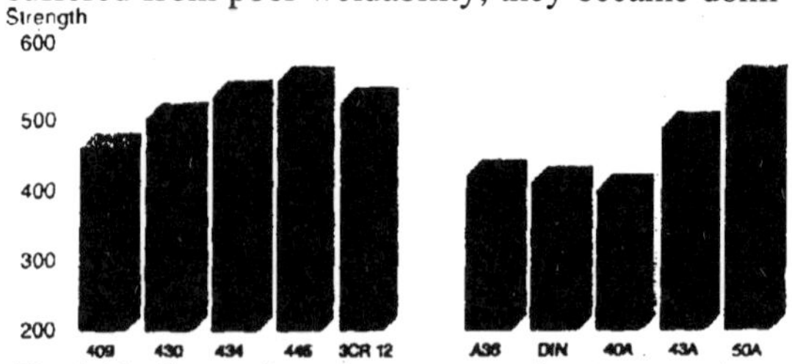

Fig 141: Strength comparison — ferritic stainless, left; structural steels, right

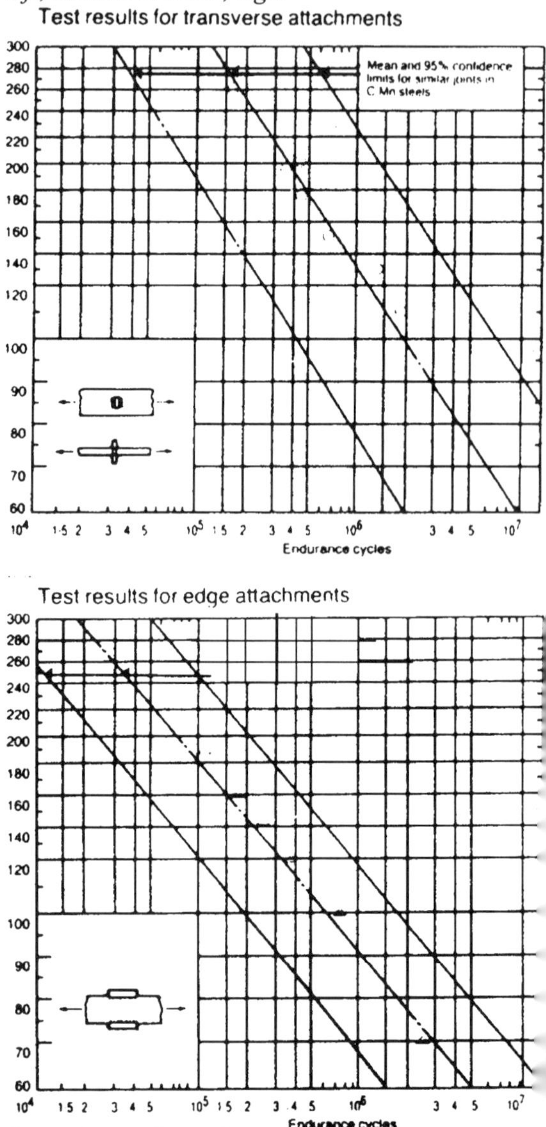

Fig 142: Notch toughness comparison

nant in very thin gauge applications such as specialist exhaust systems and trim — while the cutlery industry continued with martensitics. 'Utility stainless' is seen as the category for 3CR12, whose strength properties are shown as Fig141 alongside competitors. Specialist vehicles incorporating these materials have shown that the ferritics represent about 10 to 20% of the unladen weight; so that total vehicle price is only partially geared to price of the 'stainless'. The material can be welded in thicknesses up to 30 mm and it is available in eleven thicknesses of plate down to 3 mm. Lengths from 2 to 6 metres are available in widths from 1 to 1.5 m. Available sheet thicknesses are 4.0 down to 0.8 mm.

Notch toughness is as in Fig 142 while fatigue properties are revealed in the two sets of S-N curves shown in Fig 143 for different weld geometries. Fig 144 shows corrosion resistance performance compared with other materials — giving rate per year in microns for 2, 5 and 10 year periods. Unlike the austenitics, 2CR12 is said to be immune from stress corrosion.

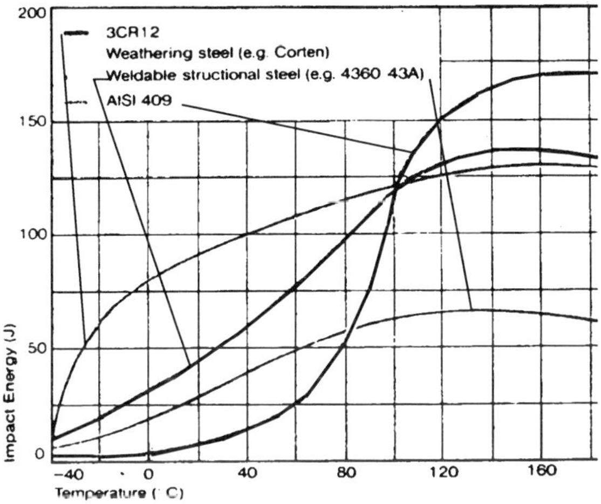

Fig 143: S-N curves for different weld geometries

Corrosion rate (in micrometres/year) after 2 years, 5 years and 10 years at various sites.

Period	2	5	10
Metal			
Mild steel	10,9	6,84	4,32
'Cor-Ten' A	6,10	3,56	2,29
Zinc	0,203	0,330	0,330
Copper	0,838	0,813	0,559
Aluminium			
3S	0,038	0,028	0,028
M57S	0,028	0,038	0,033
50S	0,051	0,056	0,028
B51S	0,046	0,028	0,033
Stainless steel			
430	<0,0025	<0,0025	<0,0025
304	<0,0025	<0,0025	<0,0025
316	<0,0025	<0,0025	<0,0025
3CR12 (Pickled)	0,007		

Fig 144: Corrosion resistance

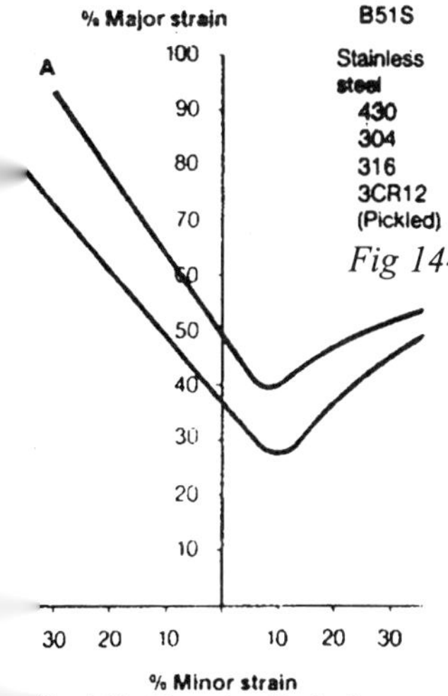

Fig 145: Forming limit diagrams

Electrolytic zinc-coated steel

The substrate for BSC's Zintec material conforms to BS 1449: Part1:1983 and is available in various qualities for different depths of draw — as well as a high strength variant called Tenform. After coating, the material can be formed and fabricated in the same way as uncoated steel of equivalent properties. The coating, of course, offers 'sacrificial' protection yet does not impair welding and soldering. As well as Tenform high strength steel the material is also available with substrates of rephospherised steel. So-called Tenform PK has a forming limit diagram as shown at Fig 145 for a 0.7 mm sheet of 275 N/mm^2 yield stress material (B) compared with 180 N/mm^2 yield mild steel (A). Increases in press loads can be calculated by assuming a drawing force proportional to 0.5 x (yield + tensile strengths). Recommended resistance spot welding conditions for two Zintec sheet of 0.8 and 0.4 mm are shown in Fig 146a alongside typical spot-weld shear strengths in b.

Single Sheet Thickness mm	Weld Diameter mm	Shear Strength kN
0.7	4.2	4.0
0.9	4.7	5.5
1.2	5.5	8.0
1.6	6.3	11.0

Single Sheet Thickness		Electrode tip dia.	Welding Conditions			
Over	Up to and including		Force	Weld Time		Current
mm	mm	mm	kN	cycles	s	kA
0.4	0.6	4	1.5	8	0.16	6.0
0.6	0.8	4	1.5	10	0.20	7.0
0.8	1.0	5	2.4	12	0.24	8.0
1.0	1.2	5	2.4	15	0.30	9.0
1.2	1.6	6	3.4	18	0.36	10.5
1.6	2.0	7	4.6	22	0.44	12.0

Fig 146: Spot-welding conditions and shear values

Magnesium alloys

While magnesium alloys have been used extensively in automotive applications, the historical impetus has come from the need for engine/gearbox weight reduction in such rear-engine vehicles as the VW Beetle car, AEC buses and a number of current Formula One racers. With more sophisticated alloying techniques and careful design analysis the material is finding important new structural roles.

It is still not widely realised that the cost per cubic inch of magnesium alloys is lower than for aluminium and steel; also it is highly competitive with zinc and plastics composites when specific strength is taken into account. As the material is abundant in sea water it seems likely that magnesium's future prospects are good, particularly when used as a metal matrix for fibre-reinforced composites. The material has also been used in CV body and cab construction in the USA, historically, and there are good prospects again in these areas for wrought products provided the properties are properly exploited in design. In the specialist vehicle sector, where specific performance requirements are available for the end-use, particularly good benefits are promised.

Supplying companies like Magnesium Electron, Norsk Hydro and Dow have been good providers of design and specificational data for optimising end-uses of the material. Considerable expertise has been gained in the demanding application of road wheels which have been a successful application. In high performance vehicles the 1.8 gm/cm^3 figure for

Fig 147: Section profiles

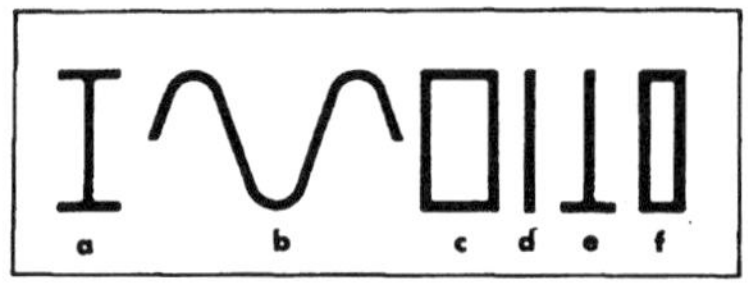

Fig 148: Stiffening webs

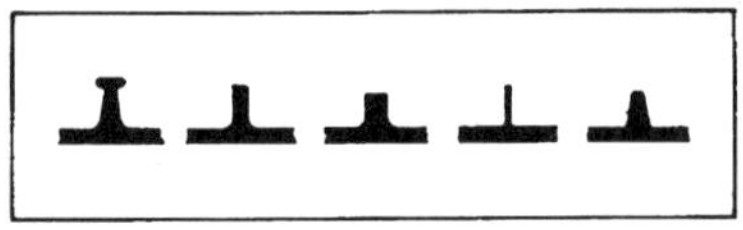

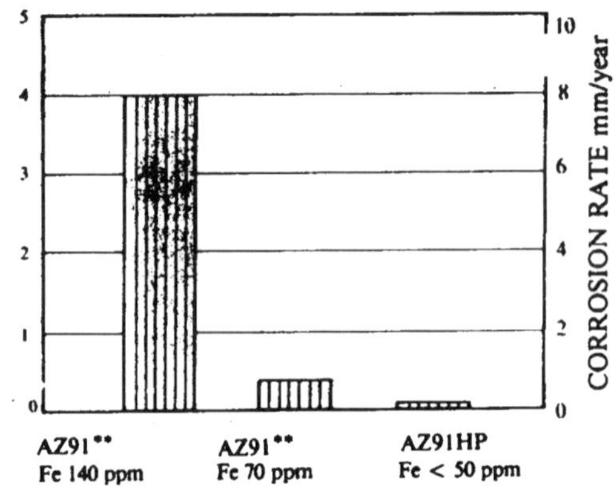

Fig 150: Corrosion rates compared

Alloy	Paint system*	Exposure time hrs	Scribe corrosion Creepage mm	in	Stone chip resistance Rating No.**
AZ91	A	240	7	0.25	—
AZ91	B	240	3	0.12	4
AZ91HP	A	240	0	0	—
AZ91HP	B	1000	0.5	0.02	1-2
AZ91HP	C	1000	0	0	1

Fig 151: Performance improvement with coating

	MAGNESIUM		ALUMINIUM		ZINC			REINFORCED THERMOPLASTICS		
	AZ91D Die Cast	ZE63A-T6 Sand Cast	LM24 Die Cast	LM25TF Chill Cast	BS1004A Die Cast	ZA8 Chill Cast	ZA27 Sand Cast	Lexan 3413	Torion 4203L	Zytel 70G33
Yield Strength (MPa)	160	190	150	220	–	208	255	–	–	–
Tensile Strength (MPa)	230	300	320	280	283	240	320	130	191	124
Compressive Yield Strength (MPa)	160	195	–	185	414	210	255	120	172	414
Elongation (%)	3	10	1-3	2-5	10	1-2	8-11	3-5	15	14
Hardness (H_B)	63	60-85	80-85	90-110	82	85-90	90-110	–	–	–
Shear Strength (MPa)	140	–	185	205	214	241	225	72	–	86
Impact Strength (J)	2.0	0.5	3	8	58	–	58	100	–	133
Fatigue Strength (MPa)	97	115-125	138	95	47.6	51.8	103.5	40	–	–
Elastic Modulus (GPa)	45	45	71	71	–	85.5	80	8.6	0.7	–
Density (g/cc)	1.81	1.87	2.79	2.68	6.60	6.3	5.0	1.43	1.38	1.38
Specific Heat (J/kg°k)	1050	960	963	963	419	435	525	0.27	–	–
Thermal Expansion Coefficient ($\times 10^{-6}$/°k)	26	26.5	21.8	22.0	27.4	23.2	26.0	22	30	23
Thermal Conductivity (W/M°k)	72	109	96.2	151	113	115	122.5	0.21	3.1	–
Electrical Conductivity (% IACS)	11.2	30.9	24	39	27.0	27.7	29.7	–	-	–

Fig 149: Alloy properties

magnesium is a clear advantage over the 2.7-2.8 figure for aluminium alloys. The elastic modulus is 6.4 x 10^6 lb/in^2 (44GN/m^2) compared with 10 x 10^6 (69) for aluminium alloy so that deep, high modulus, sections are the rule in structural design. Resilience of the material is high and some alloys have particularly good damping capacity. Both these factors make the material a promising choice for impact-absorbing structures.

Typically, a magnesium alloy body panel would be twice as thick as a steel one but less than half its weight. The thickness will give benefits both in vibration reduction and resistance to denting by minor impacts. Fig 147 shows somc cxtruded sections. Those at (a), (b) and (c) are preferred while (d), (e) and (f) should be avoided because of high edge stresses and difficulty in production. When using stiffening webs built in to the walls of castings, Fig 148, those with smooth blend radii are preferred; thin webs and sharp corners should be avoided.

Tungsten arc welding can be used for all magnesium alloys; gas welding fluxes tend to be corrosive to the material. When spot-welding is used, spots should be placed at intervals of 8-16 times sheet thickness and be a distance 4-10 times thickness from the edge of the sheet. While normally the surface carries a protective oxide film, contact with other metals should be avoided in damp conditions to avoid electrolytic action. Fire hazard only applies when the material is in very thin sections or finely divided chips. Even then they cannot usually be ignited below their melting point of 600 C.

In the design of flanged joints drawn up by bolts, flange faces should be generous in area and bolts should be spaced closely. Where bolts will be removed frequently, ferrous metal inserts are recommended. Rivets used with the material should be aluminium alloy with 5-7 % magnesium and diameters should be three times sheet thickness, at most. Minimum pitching of rivets should be three times the rivet diameter and minimum edge distance twice the diameter.

Fig 149 shows a comparison of mechanical and physical properties of magnesium alloys with aluminium and zinc based casting alloys and reinforced thermoplastics — due to Magnesium Services. For three popular alloy grades, corrosion rates are compared in Fig 150 based on the ASTM B117-73 salt spray test. When finished with organic coatings, the performance increases to that of Fig 151. The company point out that alloy purity has no significant influence on galvanic corrosion and give the advice in Fig 152 for avoiding problems.

Trapping liquids at places where dissimilar metals are joined should be avoided and aluminium alloy washers from the 5000 or 6000 series should be used with thickness or diameter large enough to provide the clearances shown. When joining magnesium to other metals, fasteners should enter from the non-magnesium side, Fig 153, and aluminium shims should be used at the joint face. The use of shielding, as shown, should also be adopted.

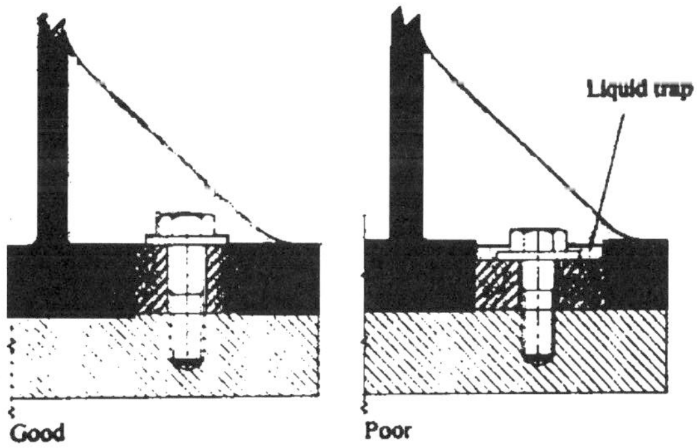

Fig 152: Avoiding trapped liquids

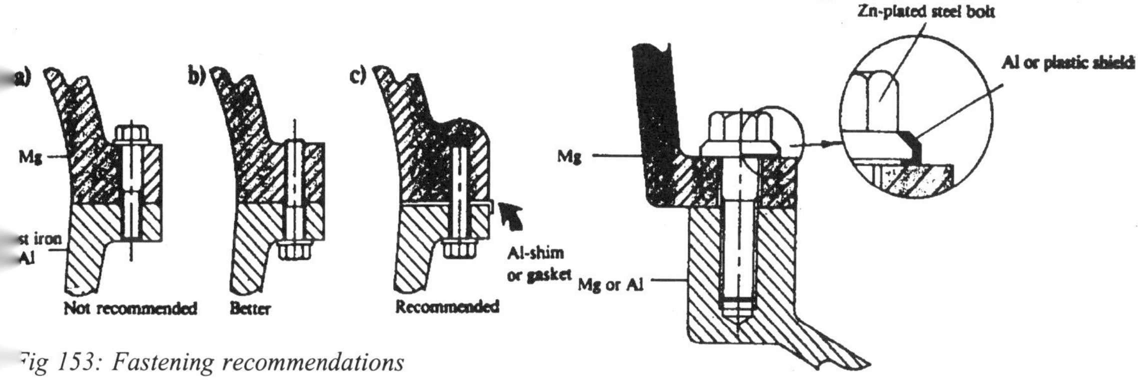

Fig 153: Fastening recommendations

Copper alloys

Corrosion resistance, ductility, and, when suitably alloyed, fast machinability, characterise copper and its derivatives. The material is also important for its high electrical and thermal conductivity and even has valuable spring properties when alloyed with berylium. Brasses are, of course, one of the commonest copper alloys and in earlier times the material was important decoratively in automotive application.

Copper's usage in some notable architectural applications such as the window framing on Coventry Cathedral's famous curtain wall of glass prompts the question whether designers should reconsider the material for prestige vehicles such as luxury coaches and limousines. Widely used in the heat-exchanger cores of radiators, the material could have this role usefully extended to the walls of vehicle air-intakes if the long-heralded concept of surface-radiators comes about.

The material is already making a come-back for brake piping, particularly in passenger service vehicles, where very long service lives are required for this safety-critical application. There is serious consideration being given, too, on top-market cars as longer warranty periods are demanded — and, as quality expectations rise, even on more modest vehi-

Metal	BS No	Specification Designation	Metal Removal Rate cm³/min
Brass	2874	CZ121 Pb4	133
Aluminium	1474	FMA (2011)	80
		HE30 (6082)	44
Mild Steel	970	EN1A	36
Stainless Steel	970	304	6

Fig 154: Metal removal rates

Brass		Availability							
Designation	Description	2870 Sheet Strip Foil	2871 Tube Pt 1	2871 Tube Pt 2	2871 Tube Pt 3	2872 Forgings & Forging Stock	2873 Wire	2874 Rods & Sections	2875 Plate
CZ 101	90/10 Brass	•		(x)			(•)		
CZ 102	85/15 Brass	•		(x)			•		
CZ 103	80/20 Brass	•		(x)			•	(x)	
CZ 104	Leaded 80/20 Brass					(•)		•	
CZ 105	70/30 Arsenical Brass	(x)					(x)		•
CZ 106	70/30 Brass	•					•	(x)	•
CZ 107	2/1 Brass	•					•		
CZ 108	Common Brass	•		•			•		
CZ 109	Lead Free 60/40 Brass	(x)				•	(•)	•	(x)
CZ 110	Aluminium Brass	•		•	•				•
CZ 111	Admiralty Brass				•				
CZ 112	Naval Brass	•				•		•	•
CZ 113	Naval Brass, Special Mixture	Deleted—see CZ 133							
CZ 114	High Tensile Brass (low aluminium)					•		•	
CZ 115	High Tensile Brass (soldering quality)					•		•	
CZ 116	High Tensile Brass					•		•	
CZ 118	Leaded Brass 64% Copper 1% Lead	•							
CZ 119	Leaded Brass 62% Copper 2% Lead	•		•			•		
CZ 120	Leaded Brass 59% Copper 2% Lead	•							
CZ 121 3 Pb	Leaded Brass 57% Copper 3% Lead					•	(•)	•	
CZ 121 4 Pb	Leaded Brass 57% Copper 4% Lead					•	(•)	•	
CZ 122	Leaded Brass 57% Copper 2% Lead					•	(•)	•	
CZ 123	60/40 Brass (Low Lead)	•				(•)			•
CZ 124	Leaded Brass 62% Copper 3% Lead							•	
CZ 125	Cap Copper	•							
CZ 126	Special 70/30 Arsenical Brass	(x)		•	•				
CZ 127	Aluminium/Nickel/Silicon Brass			•		(x)		(x)	
CZ 128	Leaded Brass 60% Copper 2% Lead					•		•	
CZ 129	Leaded Brass 60% Copper 1% Lead					•		•	
CZ 130	Leaded Brass for Sections							•	
CZ 131	Leaded Brass 62% Copper 2% Lead							•	
CZ 132	Dezincification Resistant Brass					•		•	
CZ 133	Naval Brass (uninhibited)							•	
CZ 134	Naval Brass (high leaded)							•	
CZ 135	High Tensile Brass with Silicon					(•)		•	
CZ 136	Manganese Brass					(•)		•	
CZ 137	Leaded Brass 60% Copper, 0.5% Lead					(•)		•	
CZ 138							(•)		

Fig 155 (right and opposite): Copper alloys and properties

les. The brake-pipe application is more likely to be copper-nickel alloy than a copper-zinc one and its rotagonists argue that the long reign of tin/lead oated mild steel may be nearing its end due its mited life expectancy in severe operating conditions. Brasses containing around 37% zinc are the so-alled duplex alloys with excellent ductility at hot orking temperatures used in forging and extrusion. 'hose most commonly cold-worked as sheet are the 0% zinc brasses. Free machining alloys are those which additionally contain 3% lead to aid the chip-raking of the swarf which makes the high speed roduction of such items as pipe fittings so profitable. 'ig 154 shows the remarkably high metal removal ate compared with competitive materials while Fig 55 provides the main categories and properties.

Such specialist vehicles as fire-tenders and ully-emptiers are also big users of brass, arising mainly from its corrosion resistant properties. The availability of brass in precise preformed shapes such as extrusions, hot-stampings and die-castings eliminates much of the machining required by some of its less costly competitors. The alloying of copper with tin, to produce bronze, is important; when containing around 5% tin and 0.3% phosphorous the so-called phosphor-bronze becomes useful as a spring material — particularly when electrical conductivity (while being non-magnetic) is required at the same time. With higher than 0.3% phosphorous, a different phase alloy is formed which has good bearing properties and finds uses in the bed-plates of turntables, for example. With 2% beryllium added to copper an even tougher alternative to phosphor-bronze occurs with better wear resistance. Heavily loaded bushes or springs with almost complete freedom from hysteresis can be made from the metal.

Composition															
Copper %		Tin %		Lead %		Aluminium %		Arsenic %		Iron %		Others %	Zinc %	Total Impurities %	Nearest ISO Designation
Min	Max	Min	Max	Min	Max	Min	Max	Min	Max	Min	Max			Max	
89.0	91.0				0.05						0.10		REM	0.40	CuZn10
84.0	86.0				0.05						0.10		REM	0.40	CuZn15
79.0	81.0				0.05						0.10		REM	0.40	CuZn20
79.0	81.0			0.1	1.0								REM	0.60	CuZn20Fe
70.0	73.0				0.075			0.02	0.06		0.06		REM	0.60	CuZn30As
68.5	71.5				0.05						0.05		REM	0.30	CuZn30
64.0	67.0				0.10						0.10		REM	0.40	CuZn33
62.0	65.0				0.30						0.20		REM	0.50 (Excl Pb)	CuZn37
59.0	62.0				0.10								REM	0.30 (Excl Pb)	CuZn40
76.0	78.0				0.07	1.80	2.30	0.02	0.06		0.06		REM	0.30	CuZn20Al2
70.0	73.0	1.0	1.5		0.07			0.02	0.06		0.06		REM	0.30	CuZn28Sn1
61.0	63.5	1.0	1.4										REM	0.7	CuZn38Sn1
57.5	60.5	0.60	1.25										REM	0.75	
56.0	59.0	0.2	1.0	0.5	1.5		1.5			0.5	1.2	0.3-2.0Mn 0.02 Max Sb	REM	0.50 (Excl Al)	CuZn39AlFeMn
56.0	59.0	0.6	1.1	0.5	1.5		0.2			0.5	1.2	0.3-2.0 Mn	REM	0.50	CuZn39AlFeMn
64.0	68.0					4.0	5.0			0.5	1.2	0.3-2.0 Mn	REM	0.50	
63.0	66.0			0.75	1.5								REM	0.30	CuZn35Pb1
61.0	64.0			1.0	2.5								REM	0.30	CuZn37Pb2
58.0	60.0			1.5	2.5								REM	0.30	CuZn39Pb1
56.5	58.5			2.5	3.5								REM	0.7	CuZn39Pb3
56.5	58.5			3.5	4.5								REM	0.7	CuZn38Pb4
56.5	58.5			1.5	2.5								REM	0.7	CuZn40Pb2
59.0	62.0			0.3	0.8								REM	0.30	CuZn40Pb
60.0	63.0			2.5	3.7						0.35		REM	0.50 (Excl Fe)	CuZn36Pb3
95.0	98.0				0.02						0.05		REM	0.25	CuZn5
69.0	71.0				0.07			0.02	0.06		0.06		REM	0.30	CuZn30As
81.0	86.0		0.10		0.05						0.25	0.80-1.40 Ni 0.80-1.30 Si	REM	0.50 (Excl Sn, Pb, Fe & Mn)	CuZn14AlNiSi
58.5	61.0			1.5	2.5								REM	0.50	CuZn38Pb2
58.5	61.0			0.8	1.5								REM	0.50	CuZn39Pb1
55.5	57.5			2.5	3.5		0.5						REM	0.75 (Excl Al)	CuZn39Pb3
61.0	63.0			1.0	2.5								REM	0.50	CuZn37Pb2
REM			0.2	1.7	2.8			0.08	0.15				35.0/37.0	0.5	CuZn36Pb2As
59.0	62.0	0.50	1.0		0.20						0.10		REM	0.4	CuZn39Sn
59.0	62.0	0.50	1.0	1.3	2.2						0.10		REM	0.2	CuZn37Pb2Sn
57.0	60.0		0.3		0.8	1.0	2.0				0.5	1.5-3.5 Mn 0.3-1.3 Si 0.2 Ni	REM	0.5 (Excl Sn Pb Mn and Ni)	CuZn37Mn3Pb2Sn
56.0	59.0				3.0							0.5-1.5 Mn	REM	0.7 (Excl Pb)	CuZn40Pb2Mn
58.5	61.0			0.3	0.8						0.2		REM	0.5	CuZn40Pb
80		0.5											REM		CuZn20Sn

Hot-forging

While forging is a process for reducing a slug of metal as close as possible to its finish-dimensions, with minimal machining, thus being of considerable value in volume manufacture, it as also valuable in improving mechanical properties by altering grain-flow. This can have important consequences in specialist vehicles such as racing cars where extra manufacturing cost in low volume can be justified for increased performance.

The forging process can be used in the refinement of grain structure, directional control of 'flow-lines' as well as breaking up and redistributing unavoidable inclusions. In hot-forging, working of metal is carried out above recrystallization temperature but below the melting point, to exploit the reduction in reshaping force required compared with cold-working. Finishing temperature is usually arranged close to the upper limit of the cold-working range so that the finest possible grain size is ensured.

Sometimes cold-working is carried out as a post-process to finish closer to final size and enhance strength by work-hardening.

The ability to make the flow of metal, in forging, follow the contour of the part is the important advantage of the process as the grains and inclusions are altered at the same time. The part can also be given anisotropic properties by suitable control of the forging process — such that structural efficiency is considerably increased and a lighter and tougher part results. The grain-flow in a sample part can be checked by slicing the part, in the plane of interest, then polishing and etching the exposed surface. The resulting 'fibrous' appearance are due to the elongated inclusions which define the grain flow.

The US Forging Industry Handbook provides relationships between mechanical properties and grain flow, summarised in Fig 156. The benefits to a part such as a gear-wheel are particularly graphic when compared with machining the part out of solid-bar as might be the temptation for a one-off component. Even the basic up-setting of a slug of metal, from which to machine the part, would be helpful as the 'fibres' are bulged radially such that strong grain orientation exists at the base line of the teeth, where maximum bending stresses occurs. It would be better to form the blank as a section as shown in Fig 157; best of all would be to rough form the teeth as shown by the section in Fig 158. The degree of reduction in

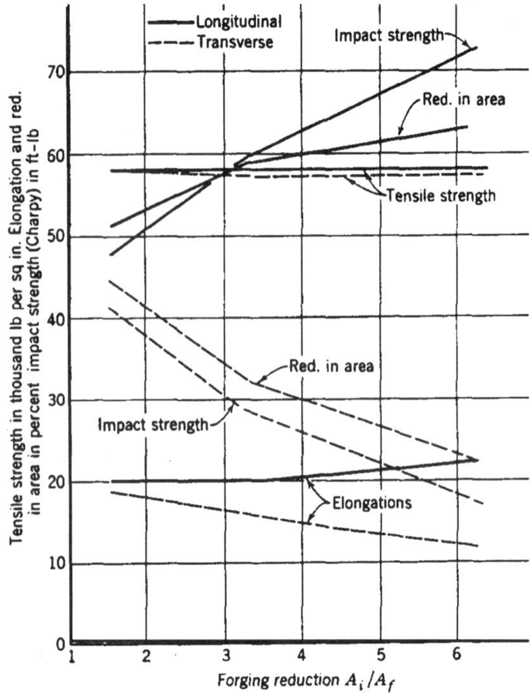

Fig 156: Grain flow effect on properties

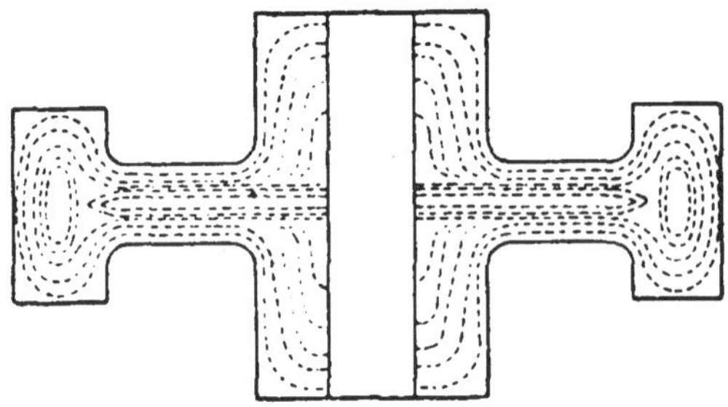

Fig 157: Blank form for better grain flow

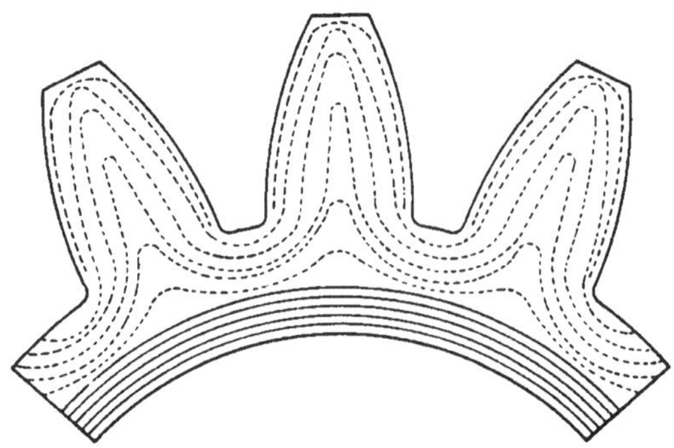

Fig 158: blank form for best grain flow

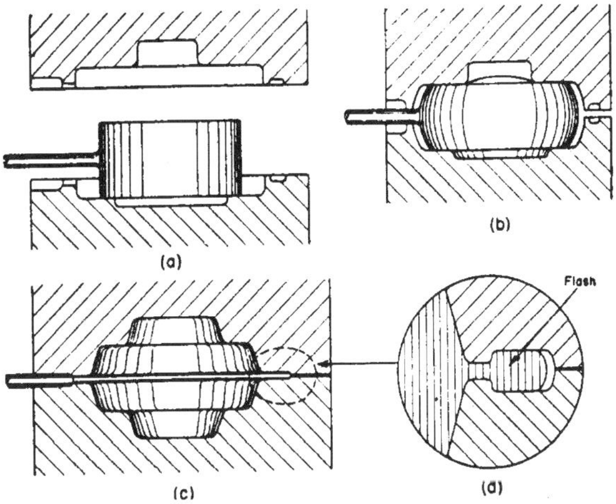

Fig 159: Blank reduction during forming

the forging of the blank from the billet is also significant as can be seen in Fig 156.

The same principles apply to a variety of stressed parts, the individual shape and performance requirement needing consideration by the designer. The flow of metal in a forging die is of considerable importance. It is essential, of course, to consult the forger on the exact metal flow pattern with respect to the surfaces of the die and the chilling action they produce. It may be necessary to carry out preforming operations to distribute the metal more gradually towards its desired shape. British Standard 4114 for forging was compiled on the basis of many international inputs and is a good guide for detail design. A useful paper on more general aspects of design has been produced by GKN[4]

Cold forging and extrusion

Both these are processes introduced to produce parts in high volume with very little loss of metal in swarf thus reducing costs by exploiting economies of scale and minimising use of material. However, the working of the material can also produce parts of better structural performance because grain flow is improved and surface hardness increased.

Drive shafts throughout the transmission line of the vehicle can all benefit from these processes and, weight-for-weight, produce parts with greater resilience and fatigue resistance, Fig 160. With more refined developments such as orbital cold forging, gear wheels, too, of superior performance can be manufactured. Even a low-volume vehicle could benefit by selecting shafting produced by cold-forging from the parts inventories of the high volume builders. Given that the right performance specification can be found it is more than likely that saving in weight and increase in durability will be the result.

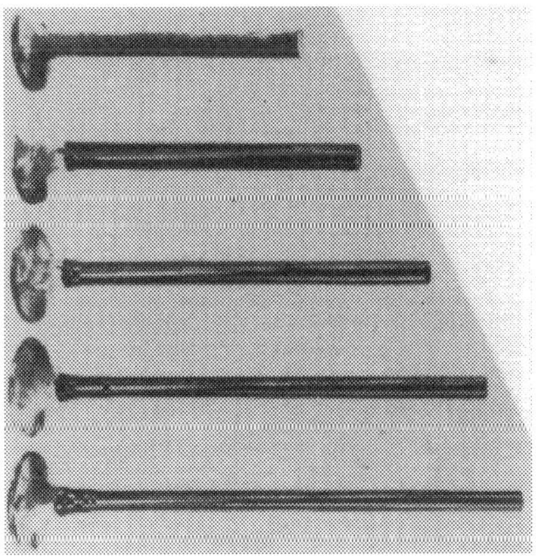

Fig 160: Stages in production of an axle shaft which starts as an upset forging left and goes through four stages of cold-extrusion.

Steel AISI	Condition	Reduction, %	Hardness	Ultimate Tensile Strength, ksi (MPa)	Ultimate Yield Strength, ksi (MPa)
1016	Before extrusion		R_B 67	61.8 (426)	45.3 (312)
	After extrusion	50	R_B 93	109 (752)	105 (724)
		70	R_B 96	114 (786)	111 (765)
1045	Before extrusion		R_B 81	85.8 (592)	43.2 (298)
	After extrusion	50	R_B 96	146 (1007)	140 (965)
		65	R_B 100	154 (1062)	146 (1007)
1340	Before extrusion		R_B 80	79.6 (549)	46.9 (323)
	After extrusion	50	R_B 98	140 (965)	135 (931)
		65	R_C 28	148 (1020)	142 (979)
8620	Before extrusion		R_B 70	67 (462)	48.1 (332)
	After extrusion	50	R_B 98	122 (841)	118 (814)
		65	R_B 99	130 (896)	124 (856)

Fig 161: Steel properties with cold working

Cold extrusion

The increase in hardness and ultimate strength of steels before and after cold extrusion is shown in Fig 161 in which very substantial percentages can be noted. Work hardening during the working of the material is such as to avoid the need for heat treatment after the part has been formed. Depending on the amount of deformation in the process, yield strengths can rise by up to 300%. Extruded parts are limited to the forming of symmetrical parts and a drive shaft can be cold-headed during processing to provide an integral flange; even internal splines can be formed during working of the shaft. With this combined upsetting of the ends, shafts up to 2 metres in length are possible, Fig 162.

Methods employed for cold extrusion include backward, forward, radial, combination, impact and continuous extrusion, the first two being shown in Fig 163. With impact extrusion particularly, large reductions are possible because of the magnitude of the impact force. The generally accepted limits for maximum upset diameters are shown in Fig 164.

Advanced forging techniques

Orbital cold forging is carried out by the Floform company of Welshpool using machines from rhe Swiss Schmid company. In this process the cold metal slug is forced by hydraulic pressure between upper and lower dies, with the upper one orbiting around its vertical axis, Fig 165. The forming forces are thus concentrated on a small area of the part which is being rolled over its complete surface area. This reduces the load required to cold-forge a typical part by a factor of six and leads to the use of smaller machines. Tooling, too, is simpler and cheaper. More accurate form filling is also claimed. A 200 tonne machine at Flowform produces parts up to 85 mm in diameter. Tolerances in the die and punch capacities are quoted as +/- 0.03 and 0.06 repectively — with a length tolerance od +/- 0.2 mm. Finish formed bevel gear wheels can even be formed on the ends of shafts and relatively low runs of 200 parts per week are viable.

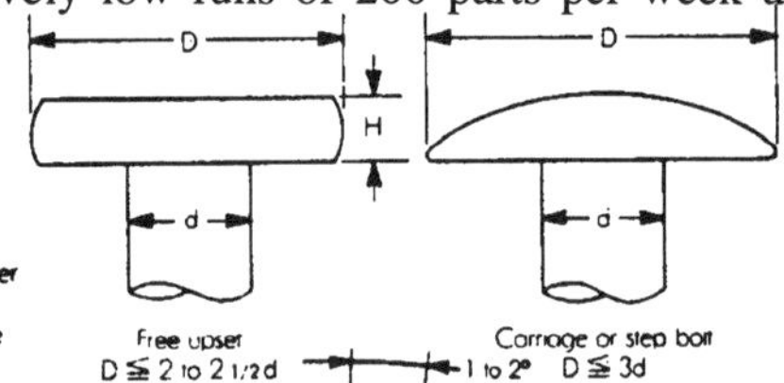

Fig 164: Upset diameter limits

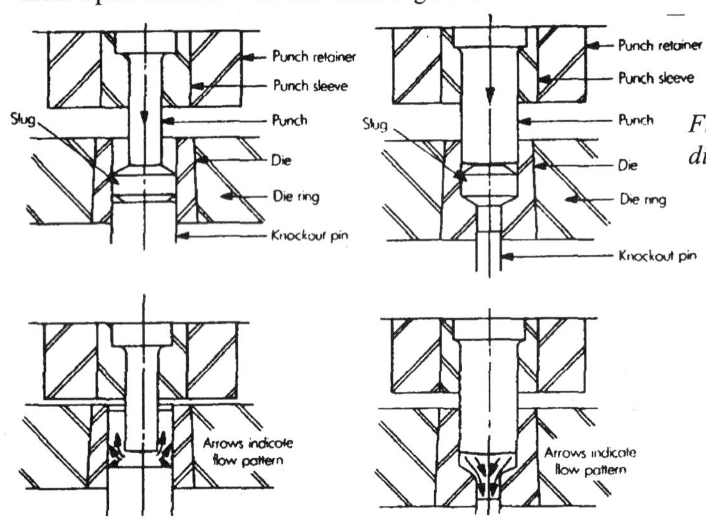

Fig 163: Forward (left) and backward (right) extrusion

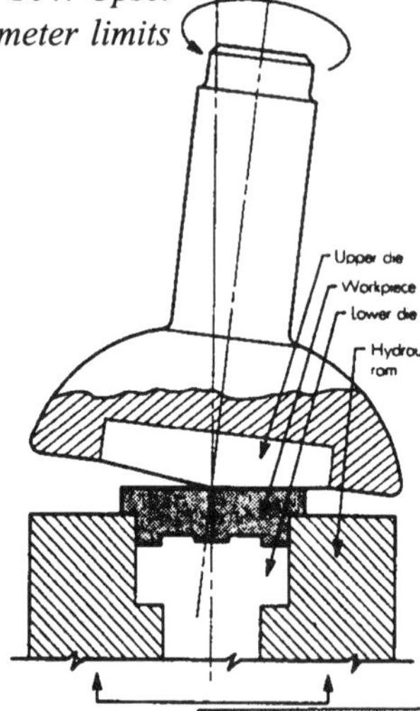

Fig 165: Orbital forging

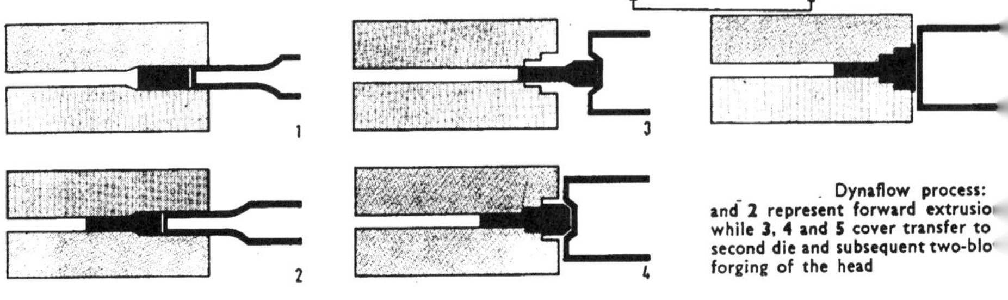

Fig 162: Stages in the Dynaflow process of combined extrusion and forging:

Front suspension system Part name	Source: Alusuisse Buhler Weight in grams			
	Steel	Aluminum	Savings	
Upper control arm, front	750	250	500	67%
Upper control arm, rear	800	300	500	62%
Suspension arm	1850	700	1150	55%
Steering control arm	2100	1100	1000	48%
Support	190	120	70	37%
Suspension arm bearing	300	140	160	53%
Shock absorber bearing	185	130	55	30%
Bearing for steering control arm	370	280	90	25%
Knuckle	6950	3900	3050	44%
Total per side	13495	6920	6575	49%
Total per vehicle	26990	13840	13150	

Fig 166: Weight saving potential of aluminium alloy

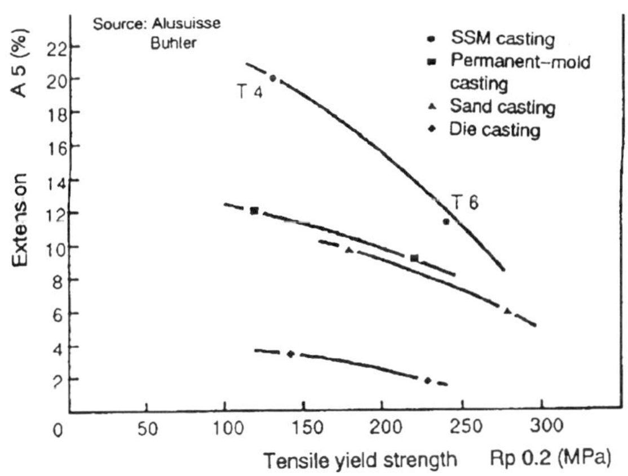

Fig 167: Properties of SSM castings compared with rivals

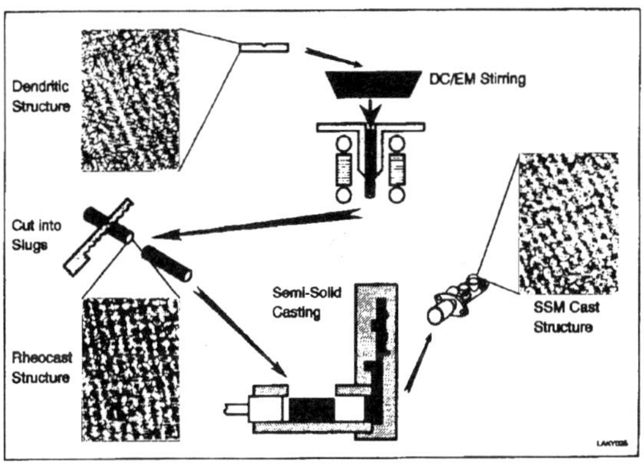

Fig 168: The SSM process

Advanced casting and die-casting techniques

Semi-solid metal (SSM) and squeeze casting processes introduced by Buhler offer potential for weight saving in automotive chassis, steering and body parts. Weight saving is also considerably enhanced in ferrous metals, using the process from Sintercast

. Heat-treatable lightweight components are seen as the promise for the new generation of casting machines from Buhler, some of which can handle composite materials. Although conventional die-casting of light alloys for automotive parts has been widespread, the lack of heat-treatability and inadequate mechanical strength has precluded the process from a number of applications.

The table in Fig 166 shows the potential weight saving in a front suspension system made possible by substitution of aluminium alloy for steel. These new casting processes make the economical production of such parts an increasingly realisable proposition, exploiting shot-control techniques. High injection power and precise velocity control also means that thin-walled weldable components are made possible with application in space-frame vehicle structures and the processes are threatening more expensive forgings for such applications.

In squeeze-casting, the handling of the liquid metal ensures there is non-turbulent, laminar die-filling followed by controlled freezing under high pressure. It is mainly used for thick-walled aluminium alloy parts. With SSM casting a special raw material is used having a microstructure which allows casting when the material is only 40% liquid. Very close to net shape casting is possible, reducing finish-machining to a minimum. Fig 167 shows mechanical properties of SSM castings compared with parts cast by more traditional rival processes.

SSM in detail

Sometimes called Thixocasting, the process has three stages, commencing with the cutting of specially processed alloy billets into slugs. These are heated to a precise temperature at which they have both solid and liquid portions; finally the heat-softened slugs are placed into the shot sleeve of the die-casting machine. The special alloys are produced using direct-chill billet casting in combination with powerful electro-magnetic stirrers. This breaks up dendrites and transforms the fragments into globules which gives the alloy its special properties, Fig 168. The resulting SSM slugs retain their shape when heated but can flow like liquid, under pressure. These are currently available in type 356/357 AlSi7Mg.

Three to four metre long billets are bought from the SSM supplier and cut to length with sufficient accuracy to obtain correct shot weight. Since they retain shape during heating they can be mechanically handled into the shot sleeve to automate the production process. As the plunger compresses the metal its viscosity drops and the material flows like a plastic to fill the cavity with near perfect shape conformity and as the material is only 40% melted solidification, shrinkage is minimised and cycle times are reduced by up to 25%.

Typically a T5 heat-treatment is carried out, involving quenching straight from the die, followed by low temperature ageing (4 hours at 180 C). Fig 169 shows how considerably less machining is required on the final part compared with an equivalent permanent mould casting and the table in Fig 170 shows some typical properties obtained from the centre of a 20 mm cast shaft.

Sintercasting

Compacted graphite iron is an iron matrix containing graphite particles, in a vermicular shape which provides a coral-like network, quite distinctive from grey flake iron or nodular iron, as seen in Fig 171. The continuity of graphite gives good thermal conductivity, vibration damping and machinability while the rounded edges give the high strength, stiffness and fatigue behaviour characteristic of nodular iron. Thus the material combines the benefits of grey and ductile irons and permits a highly castable tough material.

At elevated temperatures its fatigue strength is five times that of aluminium alloy which makes the material a significant contender for lightweight engine blocks and cylinder heads. Its good machinabity also makes it an important contender for axle housings and the material is considerably cheaper than aluminium alloy.

Exploitation of this useful material has been made possible by the Sintercast process. Extremely accurate process control is the key and resort to titanium inclusions, used by earlier methods, is not necessary. Good control is based on Sintercast's claimed unique 'know-how' of the solidification

Fig 169: Potential for net-shape casting

	Ultimate Tensile Strength		Yield Strength (estimated)		Elongation
	MPa	ksi	MPa	ksi	%
T5 heat-treated 4 hours at 180°C	265	38	190	28	10
T6 heat-treated 4 hours at 520°C 4 hours at 160°C	309	45	245	36	11

Fig 170: Properties of an SSM cast shaft

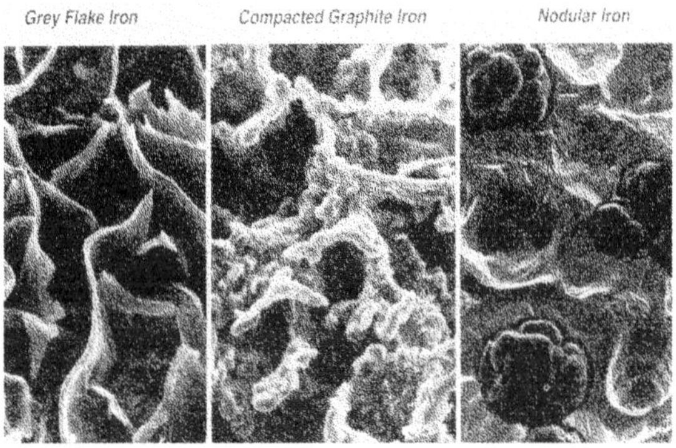

Fig 171: Microstructures compared

Fig 172: Sintercast cylinder block

behaviour of molten iron, a high-resolution measuring device, computer analysis techniques, which do not disturb production flow, and an automated correction of the melt composition just before casting.

Between 10-25% weight reductions have been achieved in engine blocks and it is claimed that CGI cylinder bores expand 20-30% less than with grey iron. Future 80 mpg eco-cars driven by direct injection diesels will be very strong contenders for the material as the company claim that light alloys will be unable to withstand 25:1 compression ratios and associated combustion temperatures of 630 C. Fig 172 shows a Sintercast block.

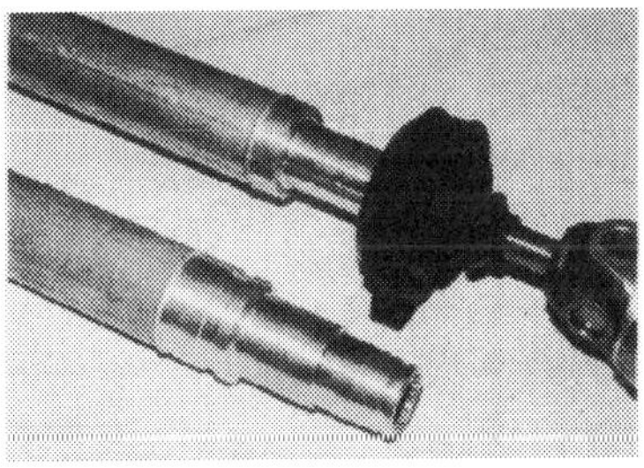

Fig 173: MMC drive-shaft

Metal-matrix composites

Fibrous or particulate reinforcements in a matrix of metal can be used to enhance physical and mechanical properties of the parent material — to the extent that Honda, for example, employed a selectively reinforced aluminium cylinder block in one of the recently introduced Prelude models. This eliminated the need for cast-iron cylinder liners and raised power-output from 135 to 140 bhp without increasing either engine size or weight.

With proprietary systems such as Duralcan, tubular components like propeller shafts, Fig 173, are being produced which promise radical improvements in specific strength and rigidity. Most MMC applications have centred around aluminium and usually involves the impregnation of a preform of ceramic fibre and the squeeze-casting process. Applications have included fibre-reinforced pistons, connecting rods and valve-train parts for engines. The fairly costly material is usually best exploited in rotating or reciprocating parts for reducing inertia as well as dead-weight of the vehicle.

Volume Percent Al_2O_3	Ultimate Strength (ksi)	Yield Strength (ksi)	Tensile Elongation (%)	Elastic Modulus (Msi)
W6A.xxA				
0	45 (38)	40 (35)	20	10.0
10	51 (47)	43 (38)	10	11.8
15	53 (49)	47 (42)	6	12.9
20	54 (50)	51 (46)	4	14.1
W2A.xxA				
0	76 (68)	69 (60)	13	10.6
10	77 (72)	72 (66)	3	12.2
15	77 (72)	73 (67)	2	13.6
20	75 (70)	73 (67)	1	14.7

Fig 174: Tensile properties of two grades of Duralcan (maxima and minima)

Material	W6A.20A (6061/20 vol% Al_2O_3)
Condition	T8511
Nominal Tube Size	4.000" diam x 0.085" wall thickness
Diameter	±0.005"
Wall Thickness	±0.004"
Ovality	±0.006"
Eccentricity	±0.004"
Straightness	0.025" TIR maximum on a 6-ft length

Fig 175:Tolerances achieved by Duralcan drive-shaft

Infiltration of ceramic fibre preforms by the metal has been achieved with the Liquid Metal Forming (LMF) process. This involves liquid metal being displaced upwardly into a die, by gas pressure. The die is pre-evacuated and after filling is complete high pressures are applied to ensure pre-impregnation of the preform. Alternative approaches involve spray deposition of metal and reinforcement to produce semi-finished material which is then rolled or extruded to final size.

Production of near net shape MMCs has also been achieved by casting a slurry of particles in a metal alloy matrix. Brake disc rotors have been made by this process in Alcan's Duralcan material, an Al-SiC or Al-AL_2O_3 particulate MMC. Duralcan tube has now reached the stage of readiness for commercial exploitation; cylinder-liners and propeller shafts are seen as the other main automotive applications initially. This is now seen as a cost-effective means of achieving higher modulus, wear resistance and strength. The commonest materials, to date, contain some 20% alumina reinforcement.

In the 6061 aluminium alloy, addition of alumina increases yield and ultimate strengths as shown by the table in Fig 174, but decreases the elongation. The most noticeable property enhancement is in elastic modulus, however, which rises by 40% for 20% alumina addition. Particle distribution remains basically isotropic after extrusion and little directionality of properties with extrusion has been noted. There is some loss in toughness and an increase in density.

Because the critical speed for whirling of a propeller shaft can be expressed as:

$$(15\pi/L^2)[(E/\rho)g(R_o^2 + R_i^2)]^{1/2}$$

where *L* is length, *R* radius, *r* density and *E* elastic modulus; the only material characteristic affecting critical speed is *E*/r. For steel and aluminium alloy this ratio is similar but for Duralcan the specific stiffness is 32% greater, resulting in a 15% increase in critical speed. Alternatively, for the same critical speed, a 7% increase in shaft length or 13% reduction in shaft diameter is possible.

Achievable tolerances with Duralcan are shown in the table of Fig 175 and seen to be similar to those attainable with the matrix alloy. The MMC most suitable for prop-shaft tubes has the properties shown

Yield Strength	45.7 ksi
Ultimate Tensile Strength	50.8 ksi
Tensile Elongation	2.1%
Elastic Modulus	14.2 Msi
Specific Stiffness	1.35 x 10^8 in.^{2}/s^2
Torque at Yield	
0.1% Offset	4700 ft·lb
JAEL*	4100 ft·lb
Maximum Torque	5100 ft·lb

Fig 176: Properties of a large diameter Duralcan drive-shaft material

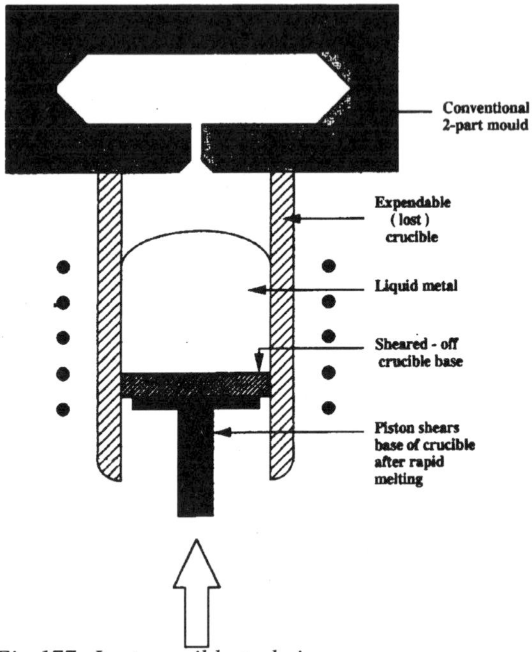

Fig 177: Lost-crucible technique

in the table of Fig 176. In making up prop-shafts the tubes had forged aluminium alloy yokes welded to each end using 5356 alloy welding wire.

In cast rather than extruded forms greater variety of shapes are, of course, attainable. One promising casting technique has been developed by Alcan at their Kingston, Ontario, laboratories. The 'lost-crucible' technique involves melting MMC as a slug in a single shot melting process then, when the material is at the correct temperature, a ram is pushed up the base of the crucible. This shears it off from the walls, causing the crucible base to act as a piston, with the walls acting as a cylinder, so as to displace the MC into the mould. This avoids the problems of settling of the solid phase and reduces the chances of picking up hydrogen or oxide films, Fig 177.

Structural timbers

The use of timber in structural and semi-structural roles was common in cars, trucks and buses — traditionally (Fig 178) — but lost ground with the increasing wage rates of skilled wood-workers and the relative cost reductions of metalworking processes. This is not to mention the increasing price of timber. There are now signs of a fresh interest in 'renewable' raw materials following official initiatives.

The European Commission has a 12 million ECU budget available over the next three years aimed at increasing availability of forest resources. There are also advances in resin-impregnated wood-fibre products which blur the boundaries with plastics-composite technology. The very high specific strength of timbers such as spruce was shown to advantage as skins of sandwich panels for the World War 2 Mosquito aircraft.

Notable automotive applications for other timbers include Frank Costin's historic Marcos structure (Fig 179). This famed vehicle structural engineer was later to use a timber monocoque for the structure of his own Costin Amigo car (Fig 180). The central portion of the Amigo structure comprised two wide and deep box-section side-booms and two transverse bulkheads, also of box section. They are connected underneath by a flat floor panel with a built-up non structural transmission tunnel along the middle. Inboard extensions of the side-booms project forward to form bearers for the engine and frontal space frame, of steel-tubular construction, which also carries the front suspension crossmember to close the torsion loop. Behind the rear bulkhead is an extension of stepped-box form with full depth side-webs to provide rear suspension anchorage points as well as enclosing luggage and spare-wheel housing.

Gaboon plywood of top-grade non-marine specification was employed having 0.8, 1.5, or 3 mm thickness. Parana pine of 12.5 and 25 mm square section was used for the joints — at the time cheaper than the Sitka spruce used in the original Marcos structure. Local reinforcement at anchorage points is particularly easily achieved by glueing on backing blocks. Adhesive used throughout was Ciba Geigy's Aerolite 300. Bonded-in diaphragms helped to stabilise the side booms and the triple-panel bulkheads are stabilised by longitudinal diaphragms. Transverse diaphragms are used in the bearer portions to take

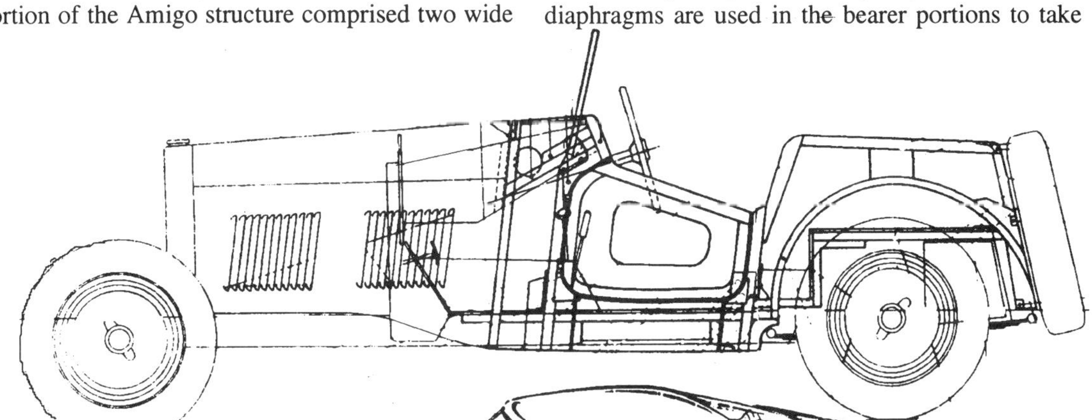

Fig 178: Classic sports car with 46 timber frame members

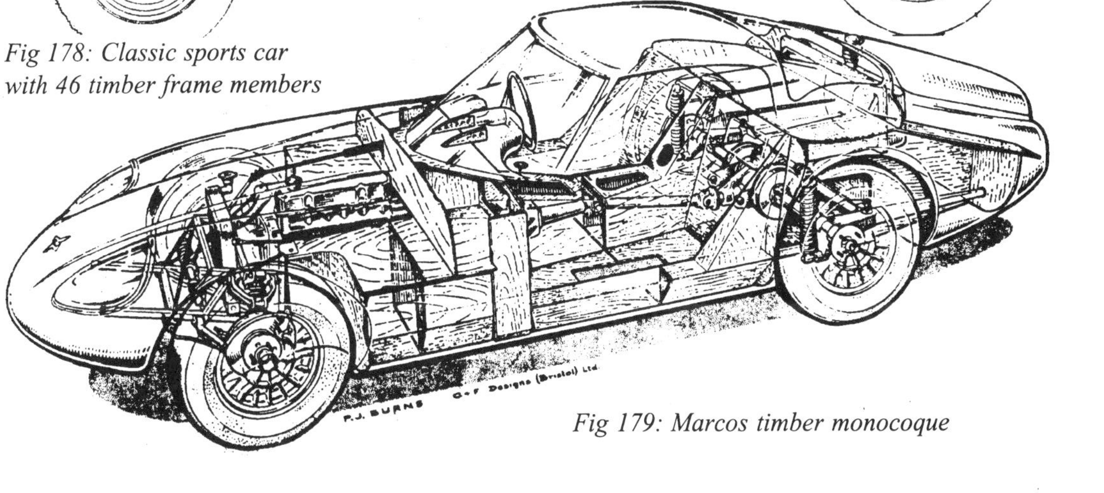

Fig 179: Marcos timber monocoque

shear loads from the engine mountings. For protection, the entire structure was painted with a two-part polyurethane primer and topcoat; also the top surfaces of the plywood was given a nylon sheathing to prevent abrasion and the engine bay treated with fire-resistant paint. Kerb weight of the car was 1468 lb (670 kg) and torsional stiffness over 3500 lbf ft (4745 Nm)/deg. In commercial vehicles, timber has of course played a structural role in decking and dropsides of open-platform trucks as well as in the frameworks of bus and box bodies. It also has some appeal for closure frameworks on refrigerated vehicles because of its relatively low thermal conductivity compared with metal equivalents. Structural plywoods are also widely used for decking — 12.5 mm thick material supported on 355 mm spaced crossbearers typically replacing planked decking of 35mm thick solid timber. Trim panels for cab doors are an application for moulded woodfibre and finisher panels for cab rear walls can involve 3D mouldings made by preforming rather than profiling hardboard sheet. Usually a mixture of 75% mechanical woodpulp and 25% long-fibre kraft pulp is blended in water with small quantities of resin and other additives.

The mixture is preformed on a screen to the shape of the finish product and then pressed between matching dies at 260 C with a pressure of 200 lb/in^2 (1.4 MN/m^2). One die is slotted and covered with a perforated screen to allow the water and steam to escape — thus only one side is smooth. Density varies according to resin impregnation and wall thicknesses usually vary between 0.06 to 0.12 inch (1.5-3 mm). Lateral pressure resistance for a 2.5 mm thick panel is typically 350 lbf/in^2 (2.4 MN/m^2).

Structural properties of timbers

Woods are classified as soft or hard according to whether they come from cone-bearing or broad-

Fig 180: Costin Amigo monocoque detail

Standard name	Botanical species	Density	Maximum bending strength: Equivalent fibre stress at maximum load	Stiffness: Apparent modulus of elasticity	Energy consumed to total fracture: Total work	Resistance to suddenly applied loads: Maximum height of drop of 50 lb hammer	Resistance to suddenly applied loads: Toughness. Energy consumed	Resistance to suddenly applied loads: Izod notched bar test. Energy consumed	Maximum compressive strength parallel to grain
Softwoods		lb/ft^3	lb/in^2	1000 lb/in^2	in lb/in^3	in	in lb	ft lb	lb/in^2
Fir, Douglas	*Pseudotsuga taxifolia (douglasii)*	37	14 800	2260	28·2	34	—	—	8480
Hemlock, Western	*Tsuga heterophylla*	30	11 100	1690	17·6	26	—	—	6390
Pine, Parana	*Arancarla Augustifolia*	33	14 200	1510	15·3	26	—	—	7970
Redwood (Baltic)	*Pinus sylvestris*	30	11 200	1420	—	24	70	—	6360
Spruce, European (Whitewood)	*Picea abies*	23	9700	1140	12·4	19	—	—	5110
Spruce, Sitka	*Picea sitchensis*	27	10 900	1750	22·8	26	90	6·4	5800
Hardwoods		lb/ft^3	lb/in^2	1000 lb/in^2	in lb/in^3	in	in lb	ft lb	lb/in^2
Agba	*Gossweilerodendron balsamiferum*	30	10 200	1070	12·9	21	74	6·5	5310
Ash	*Fraxinus excelsior*	42	15 100	1860	41·8	43	180	—	6990
Balsa	*Ochroma spp. Heartwood*	12	2800	470	—	—	—	1·5	1680
Balsa	*Ochroma spp. Sapwood*	17	5300	710	—	—	—	4·0	3730
Beech	*Fagus sylvatica*	43	16 200	1950	27·4	45	151	10·6	7870
Birch, Canadian yellow	*Betula lutea*	43	17 200	2270	44·6	61	—	—	8880
Dahoma	*Pipta denlasdrum Africanum*	43	15 800	1620	37·6	36	—	—	8520
Danta	*Cistanthera papaverifera*	47	16 100	1790	26·3	39	—	—	9080
Elm	*Ulmus procera*	32	9300	1090	14·6	23	68	4·5	4740
Gurjun	*Dipterocarpus spp.*	46	15 900	2260	28·0	37	—	—	8330
Idigbo	*Terminalia ivorensis*	29	12 000	1400	10·8	19	78	5·8	6320
Keruing	*Dipterocarpus cornutus*	50	18 800	3110	24·2	39	154	13·3	10 580
Makoré	*Mimusops heckelii*	44	12 900	1420	—	—	—	—	6220
Mansonia	*Mansonia altissima*	40	16 800	1690	30·1	45	169	10·2	8190
Oak	*Quercus robur, Q.petraea*	43	13 300	1560	19·1	33	131	8·7	7210
Obeche	*Triplochiton scleroxylon*	24	8400	900	11·5	21	59	5·0	4190
Opepe	*Sarcocephalus diderrichii*	47	16 500	2070	21·3	28	104	8·3	10 020
Red Muranti	*Shorea dasyphylla*	30	12 100	1620	19·9	30	—	—	6990
Sapele	*Entandrophragma cylindricum*	40	9500	1450	—	23	102	7·4	8320

Fig 181: Properties of soft and hard woods

leafed trees and does not always mean hard or soft in the physical sense. The values in Tables 1 and 2 of Fig 181 refer to air-dried timbers (12% moisture content) as for some green timbers (50% moisture content) strengths are markedly lower. Those shown are ultimate strengths which should be appropriately factored for their working design values. Grain effects are obviously most important in predicting design performances. This is to some extent less in recently introduced plywoods having 45 deg angle between grain directions of adjacent laminations. Modulus of elasticity of wood parallel to the grain is 15 to 20 times higher than at right angles to it and in plywoods, bending stress is by no means proportional to distance from the neutral axis (Fig 182).

Bending moment is related to stress by $M = KfI/y$ where f is ultimate stress, I the second moment of area of ply having grain parallel to the span and y the distance from the neutral axis to the outermost face of the included plies. K is normally taken as 1.5 for 3-plies (grain of outer plies parallel to span) and 0.85 for others. A nomograph for solution of this expression is shown as Fig 183.

Scales

M_1, moment of resistance (strip 1 in wide), $K = 0{\cdot}85$ (lb in)
M_2, " " " " " " $K = 1{\cdot}5$ (lb in)
f, stress (lb/in²)
I, second moment of area of plies having their grain parallel to the span (in⁴)
$2y$, distance between outermost faces of included plies (in, mm)

Fix turning point on 'Z' line using known 'y' and 'I'.
Use 'Z' and known 'f' to find 'M'.

*The approximate second moment of area is derived from $I = (BD^3/12)\,(Bd_1^3/12) + (Bd_2^3/12)$ which equals $[D^3 - d_1^3 + d_2^3]$ per unit width. Safety factors of 1.33 have been suggested with an additional factor of 2 for long term loading.

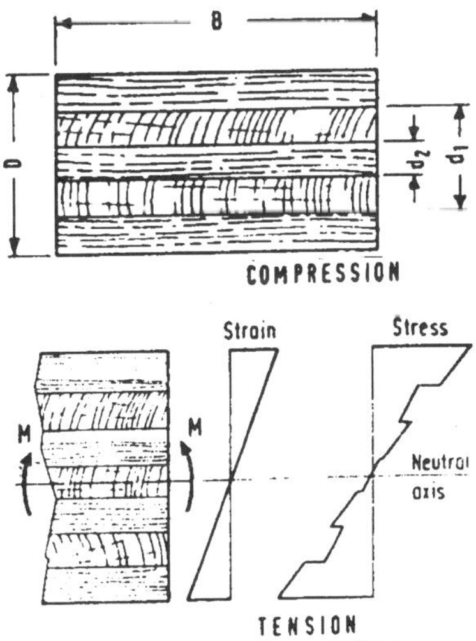

A. $\frac{3}{16}$ in 3-ply equal-thickness veneers
B. $\frac{1}{4}$ in 5-ply with face and back veneers half thickness of core veneers
C. Parallel to face grain
D. Perpendicular to face grain

	A		B	
	C	D	C	D
Mean	9600	1840	6890	6020
Standard deviation	417	87	562	302
Minimum	8783	1669	5789	5428
Basic stress	3710	700	2440	2290

Fig 182: Assymetric neutral axis of ply section and ultimate strengths

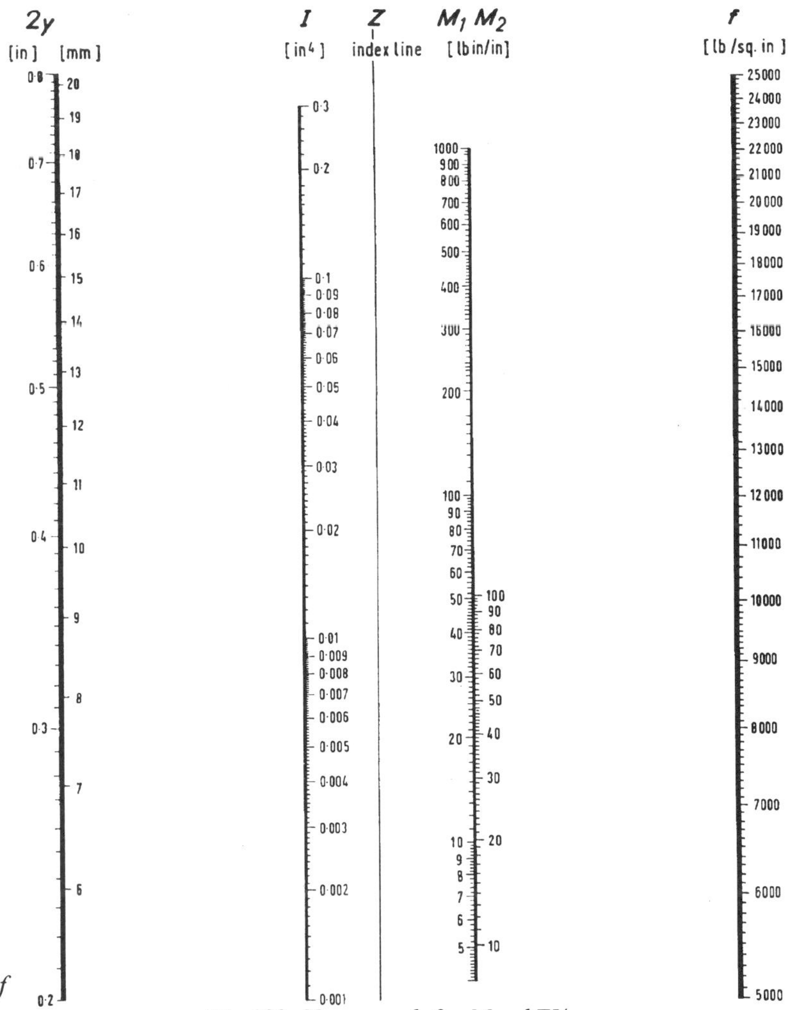

Fig 183: Nomograph for M = kFI/y

Metal injection moulding

For this process of forming metals as though they were plastics, the Israeli company Metalor Ltd has developed the capability to produce parts by MIM with high surface finish and dimensional accuracy of 0.5% which has alluded the technique since its introduction in the mid 1980s.

As well as the higher accuracies the process can now achieve densities of up to 97% of the theoretical solid density of the base material and stronger components than are possible with sintering, are now said to be available. Generally, any shape that can be plastics injection moulded can now be produced in metal and production runs of less than 2000 are claimed to be economically feasible. Better control of the moulding, and subsequent drying, than is available with sintering is the secret. With shrinkage of some 18%, between initial green moulding and finished product, tooling design is paramount and currently the technique is best suited to parts of 13-150 grammes, with a size envelope of 50 x 50 x 50 mm.

Part design follows the recommendations for plastics injection moulding, with uniform wall thickness being critical to the preservation of accuracy during shrinkage. Coring can be used to attain this uniform thickness, as seen in Fig 184. This can also reduce cost by reducing material volume and cutting processing times. Often cores can be achieved by adding holes formed by pins protruding into the mould cavity. Blind holes formed by pins supported only at one end should be limited in depth to twice the pin diameter otherwise deflection might result from flow of feedstock.

In the special case, Fig 185, of holes perpendicular to one another, making one of the holes in a D-section, will allow the tooling to function better, be stronger and minimise flashing. Fig 186 shows the use of ribs to make uniform wall thickness easier to attain, but warpage, sink marks and stress concentrations must be guarded against. Where wall uniformity is impossible a gradual transition between thicknesses is desirable as seen in Fig 187. The mould should be gated at the heavier section to ensure proper packing of the feedstock. Draft angle for mould ejection can be as low as 0.5 to 2 degrees while fillers and corner radii should be from 0.4 to 0.8 mm.

Threads can be moulded into a part and the least expensive way is to locate the parting line on the centre of the thread as in Fig 5. Dimensional tolerance of +/- 3 % can normally be achieved and one third this value for very small parts. Surface finish is 0.8 microns which is better than most investment cast parts.

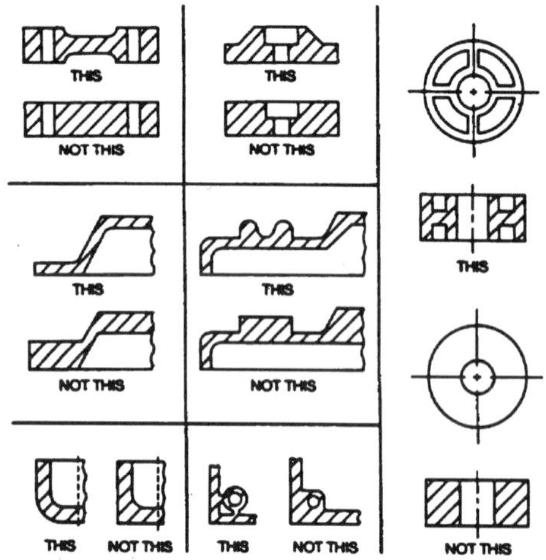

Fig 184: Uniform wall thickness and coring

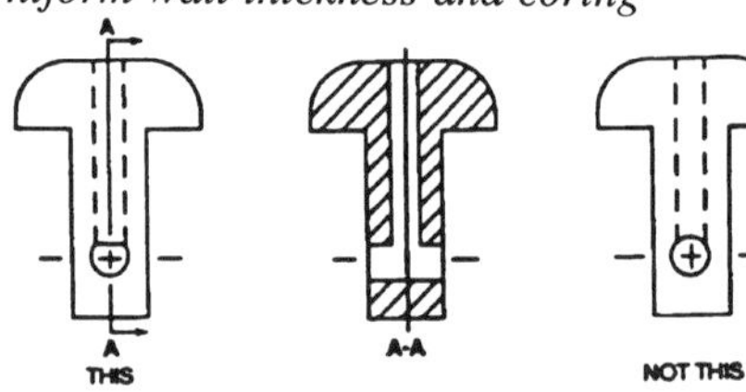

Fig 185: Holes perpendicular to one another

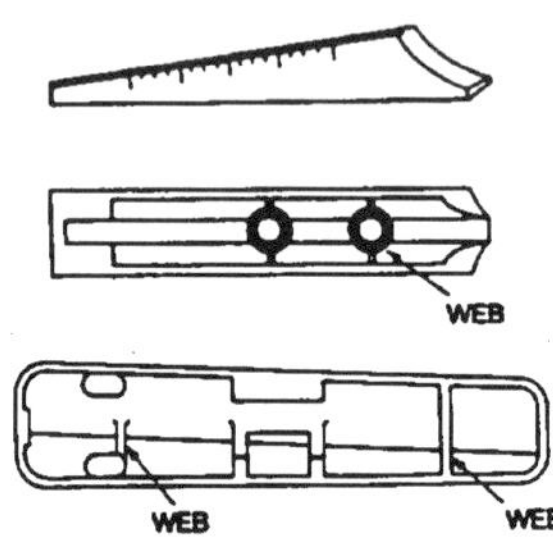

Fig 186: Webbing

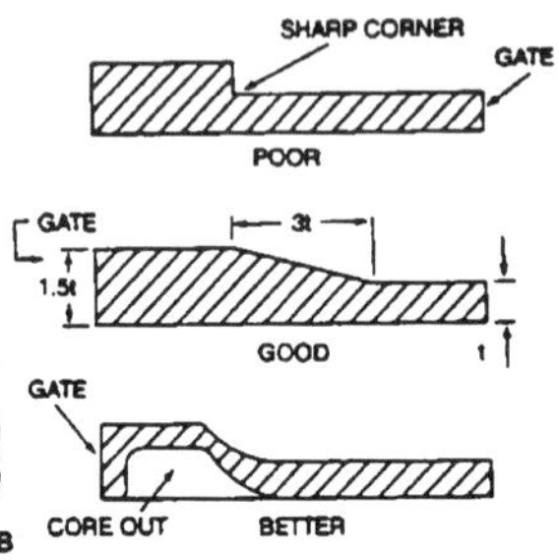

Fig 194: Wall thickness transition

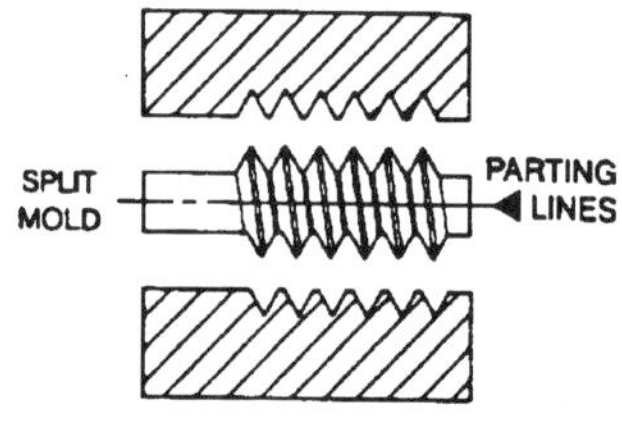

Fig 187: Moulding external threads without side core

PART 3:
BODY SYSTEMS

Chapter 8: Shaping and packaging

A merging of engineering and design disciplines is necessary for successful execution of vehicle body shaping and packaging. This can involve severe clashes in traditional societies and even the most progressive manufacturers have sharp divides between product engineering and styling departments. Body fitting and interior trim also need a multi-cultural approach so as to come up with aesthetically pleasing effects which are also durable and appropriate to the performance behaviour of the vehicle.

Aesthetics and industrial design

As successful aesthetic design comes from practice rather than conventional teaching, much depends on the commitment to good industrial design by the practioner before he or she puts pencil to paper. Fig 1 reveals some of the results of good industrial design practice while the text below argues that the overall design of vehicles is a difficult mix of engineering-science, constructional skills and aesthetics which can only be described by the non-English word 'technik'. This concept is considered here with other factors which ultimately affect appearance — in an introductory article intended to find the right frame of mind for embarking on successful design.

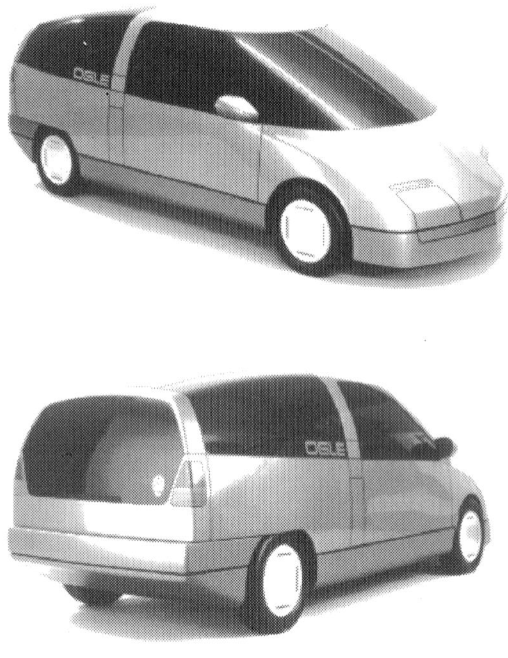

Fig 1: Design studies by Ogle

The market opportunities for well designed vehicles are one of the positive aspects of the worsening market response to the design of conventional volume-built vehicles. A uniform product for all, it is often argued, seems to be the result of aerodynamics and 'clinic' design which boils down a wide sample of individual preferences into a 'something for all but not much for any'. Another prevailing factor could be the high percentage of volume-built vehicles which are purchased by companies either for commercial use in fleets, in the case of CVs, or as 'job' or 'status' cars. This inevitably presents a different set of criteria to the product planners and designers — as the fleet sector will have different market demands from those of the individual-purchase sector.

In the case of cars, the end-user in the company will usually have a 'say' in the choice within a band of offerings made to himor her by senior management or the transport excutive concerned with fleet purchase. The decision to buy might be influenced by something seemingly as trivial as the status of the 'L' pack over the 'GT' pack or how the neighbours will rate the car in the pecking order of consumer-hardware purchases. Strong in the mind of the fleet user will be the knowledge that he or she will not be paying for fuel, service and repairs — and that replacement will occur in as little as one year's time.

In the case of the large CV fleet, the purchasing decision for the volume vehicle is more likely to be made on perceived reliability; running cost based on servicing by commercal garages (it has even been known to make the selection on service-interval multiplied by recommended charge — times projected life) and above all by life-end price in the resale market. Fitness for purpose is likely to be relatively low down on the priority list and the scope for good design very limited indeed.

Product planners in the volume sector seem to have switched from talking about function versus appearance and are now addressing themselves to whether the perceived market demands within a particular known sector should take preference over opening a new niche to widen the sector by appealing to a greater range of customers. In the specialist vehicle sector, by contrast, there seems to be scope for aiming at 'good design' which attracts new customers for its own sake.

Recognition of design

The sympathetic use of materials in design and construction, it is often suggested, can favourably affect the design of the end-product and materials, a real consideration for the end-user's enjoyment in driving or even just sitting inside the vehicle. The way the vehicle responds to the controls — and the sound engineering of its functional systems — can also influence a wide range of buyers.

Design has happily entered the political arena in recent years and is no longer left to the design-

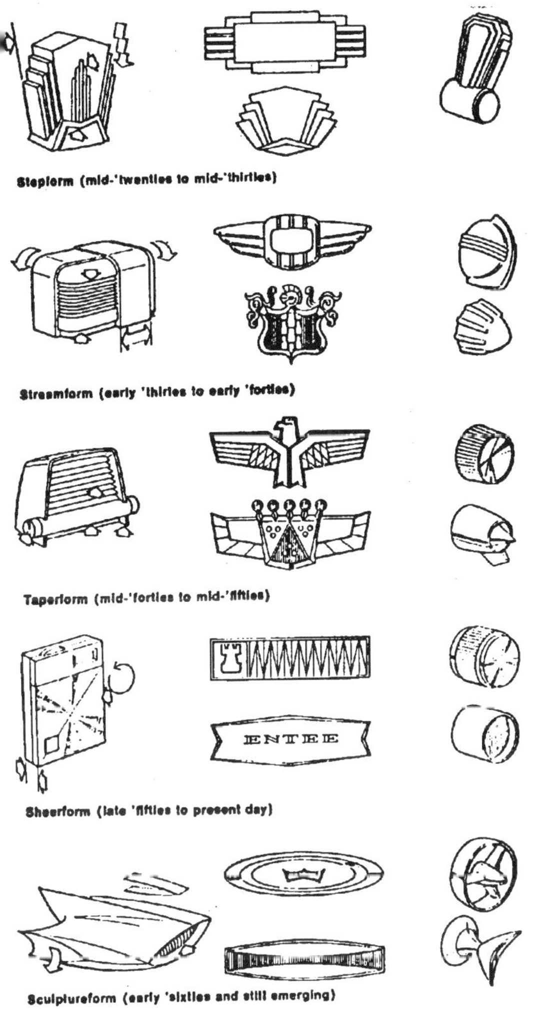

Fig 2: Styling cliches

conscious few who can so often be out of touch with day-to-day marketing and manufacturing. 'Design or decline' is the catchy slogan of the new thinking. Recent reports show that the UK design consultancy business has been growing at a rate of 30 % per year and is now beginning to get somewhere near in total spend to those leading marketing services — sales-promotion and advertising; it has already overtaken market reearch and public relations in overall turnover terms. While this is encouraging in one sense, it hides the fact that 20% of companies have no internal design resource. Nor does it reveal that in many smaller companies, good design no longer necessarily flows from good craft skills of the workforce directed by enlightened management who can properly exploit such a resource. The integrated approach to design is now being advocated — in place of merely 'making things look good'. The tendency to use dedicated stylists is losing ground to sharing many job functions, at all levels, in the key aspects of industrial design and engineering for quality.

Market experience shows that specialist commercial and industrial vehicles, unlike their 'volume' counterparts, can often be operated by those who are very much concerned wih fitness for purpose; vehicles are as likely to be serviced in the operator's own workshops as in 'high-street' service stations and will probably be driven by experts who have a real appreciation of the engineering of their functional systems. Car buyers are likely to be enthusiasts who are concerned not only about the performance but also by sculptural shape of the product.

Stylists, how they took control

The American-originated styling-for-sales strategy historically grew out of the Madison Avenue-based advertising industry and many of its participants had their offices in that neighbourhood — a world away from the manufacturing base. In the UK, it was argued, there was an additional factor to this geographical divide; according to 1960s design consultant Hugh Clausen, organisations like the Institution of Mechanical Engineers separated the traditonal drawing office based designer from the so-called 'professional (chartered) engineers; more recently such institutions have been accused of overloading the contact hours in engineering teaching courses so that students miss the wider education necessary for development into good working designers.

The need for good industrial design is an inevitable result of subordinating design to styling, which consciously makes a visual addition. The development of styling cliches corresponding to the changes in fashion over the years, Fig 2, is completely divorced from function or beauty in the classical sense. A prominent design teacher, John Grant, who advocated this view, argued that industrial design had the ultimate aim of producing the formal properties which related the function and structure of the product to what the buyer sees and uses. The designer, in this context, synthesised what the engineer analysed, Fig 3. He lists objectives as function, utility (permanence of usage), simplicity (ease of manufacture and service) and elegance (in both appearance and the scientific sense). Whereas the styled product had to

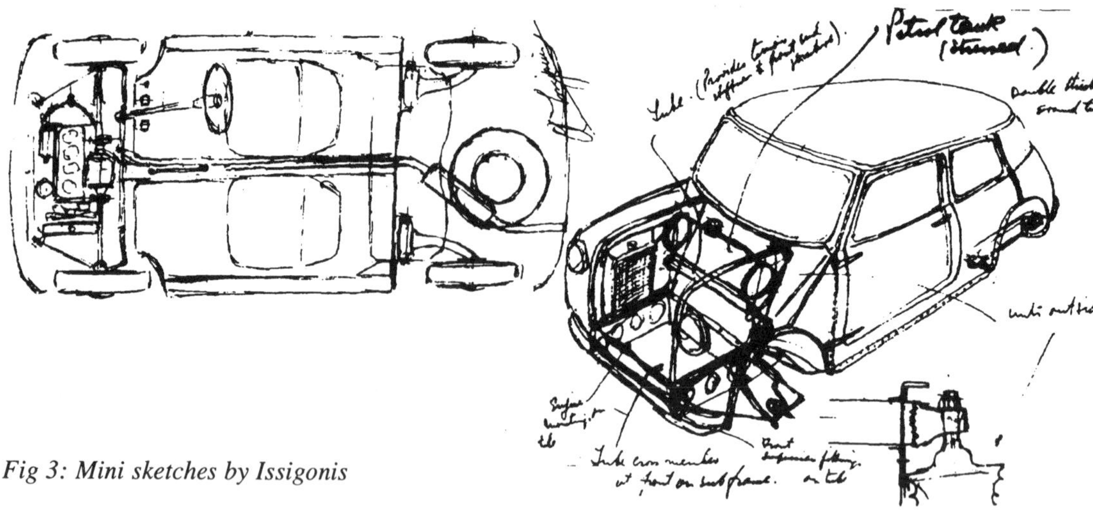

Fig 3: Mini sketches by Issigonis

be changed every time that fashion changed, good design could establish market leadership over a number of years; this was the key message of good industrial design.

In the same way as Prince Charles encourages reverence for the British architectural heyday — the regency era when designers drew on the classical aesthetic proportions of ancient Greece — so also were vehicle design teachers like Carl Olsen reported to have advocated that in car aesthetiscs, one should pay regard to the heyday of UK body design — when sleek knife-edged bodies were built on such chassis as Frazer Nash, Aston Martin, Bentley and Alvis.

Another teacher, Nigel Chapman, was reported as saying 'it is pointless trying to pander to what appears to be public taste — the industrial designer must work with engineers to evolve the most appropriate concept'. Another prominent consultant, Douglas Scott, who had the London Transport Routemaster bus among his design portfolio, stressed the need for the designer to have an intimate knowledge of the construction processes; he is reported as saying, for example, that 'rather than the presswork engineer adding his opinion a the early concept stage, it is far better for the designer to have a real feeling for presstool work and sheet metal assembly in the same way as the sculptor in clay has a feel for his material`.

Fit and finish

Attention to detail is another important element in the visual impressions made on vehicle users. A designer's overall control is necessary to avoid details being irretrievably changed between design sign-off and production-build as has so commonly occurred at the 'volume' manufacturers. Another pitfall is the tendency for a CAD draughtsman to select and arbitrarily blend between surfaces, during development of a shape, rather than visualise the sort of effect achieved by the sculpturer with a few deft strokes of the scalpel. Expert care is necessary, too, in the choice of surface texture, colour and contour, when specifying interior finishes in the immediate vicinity of driver and passenger. And it is vital to understand that detail finishes applying to one material rarely read across to another.

Inter-disciplinary design approach

Technik is an interesting concept taken from the non-English language which seems to have no direct interpretation in modern Britain. The word is used in Germany and Sweden and has different interpretation from 'technology' for which equivalent words also exist in those languages. It is a more realistic way of expressing what in Britain is mistakenly referred to as 'applied science' because it relates much more to engineering technique than the manufactured product. Thus engineers in Sweden and Germany are seen not so much as technologists, closer to science on the one hand, or technicians close to the craft skills, but as 'contrivers' who devise, make and improve artifacts. The lesson is being learned that it is not enough just to employ a scientist and craftsman, in combination, to achieve a well designed vehicle. There must be a master designer in control who has a properly integrated training in the arts and sciences.

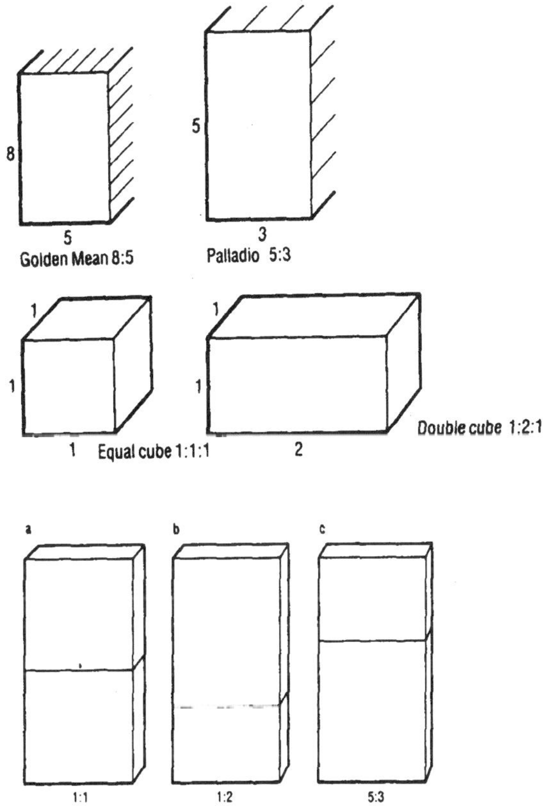

Fig 4: Classic proportions (above) and stability in form (below)

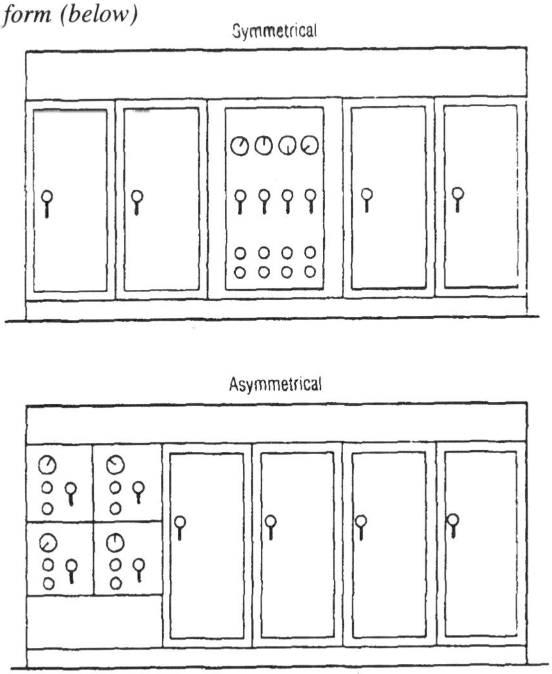

Fig 5: Introducing asymmetry

Formal aesthetics in design

The post-war foundation of UK industrial design was based on the Scandinavian model with an aesthetic based on simple lines, use of natural materials and an overall feeling of clarity and light in interiors. The moving apart of the two design cultures between then and now is attributed to the relative isolation of Scandinavia on the one hand and the heavy influences on UK design of such factors as the 'throw-away' design culture of North America. By the 1970s Scandinavia had lost its earlier design supremacy to countries such as Italy. However, the beginnings of Scandinavian design, much earlier in the century, are said to have been inspired by the German Werkbund example, a derivative of which we examine below.

Balance in form design

According to Flurscheim[1], form has been defined as the image presented by the outer surface of an object — but it is also the manner in which it is able to be perceived. It cannot therefore be treated purely as a functional problem and must be viewed from the perspective of the potential user and to its environment — unlike a work of art which is often viewed in isolation. Form is able to convey certain messages about the object through historical association and other learnt attributes.

'Balance' in form is based on historical perceptions of proportion. Euclid's 'golden mean' calls for a ratio in rectangular dimensions of 5 to 8. Palladio is associated with 5:3 and Adam with the double cube, Fig 4. Stability is another aspect of balance, in considering shapes that generate an impression of equilibrium. Instability can be suggested even on simple rectangular shapes, Fig 4b, according to the positioning of their horizontal dividing lines with respect to the mid-height level. Tall free-standing structures are given stability in another way — the introduction of a very slight taper. Symmetry is also a factor in balance and whereas symmetry might suggest a state of order, Fig 5, asymmetry can often create a greater sense of interest. Repetition, too, can be used to enhance balance; the repeating of individually similar shapes can help to create an impression of unified order.

Functional design will be achieved when all the visual components support a tidy and co-ordinated impression of function without the need for any imposed style. Because of their complexity many products could present an untidy and unco-ordinated impression and a styled co-ordinating medium is then necessary. One method is to base the design on straight lines with flat and rectangular forms. Another is to employ curved lines with two or three dimensional curved forms.

German influence in design

The conscious use of formal aesthetics in contemporary design dates from the foundation of the German Bauhaus design school in the 1920s. The history of the school is described by Whitford[2] who speaks of its revolutionary effect on art education, one of the aims of which was to elevate the status of 'crafts' to that of 'fine art.' The Great Exhibition of 1850, held in Paxton's Crystal Palace was earlier to have cast doubts in the minds of some of its thoughtful visitors who contrasted the wonder of the building with the paucity of design in the objects it contained. While this had led to the end-of-century craft revolution with its art-nouveau style, the later Bauhaus movement was to spawn the more design-conscious art-deco style. During its life-time, however, the Bauhaus spurned any sort of styling cliche at all.

One of the most important of the school's industrial design influences was the work done by its prominent teacher, Peter Behrens, in an earlier appointment, when he started to revolutionise the design of AEG products, including the logo of Fig 6 designed as long ago as 1907. Two fundamentals of the Bauhaus approach were: (a) forbidding the students to specialise and (b) that all art education be based on craft training; however, first director Gropius argued, at the end of World War 1, against 'the dangerous worship of might and the machine which led us over the spiritual to the economic abyss' — not surprising therefore that the Nazis closed the school in the 1930s.

A significant milestone in aesthetic design teaching came in 1925 at the Bauhaus when the modern artist Kandinsky introduced a design course which aimed at an art which was 'intellectually

Fig 6: Gropius logo for AEG

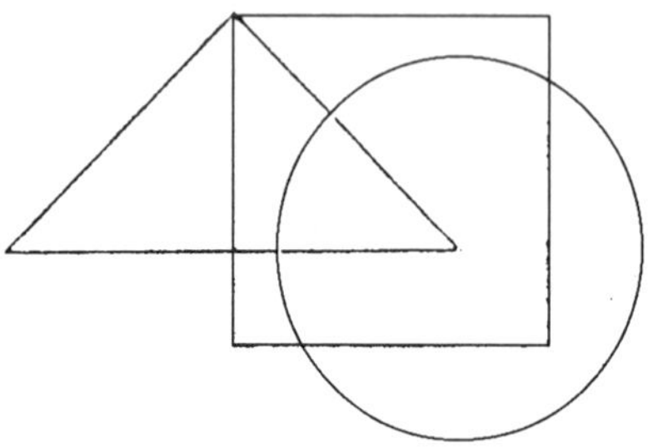

Fig 7: Basic elements of form

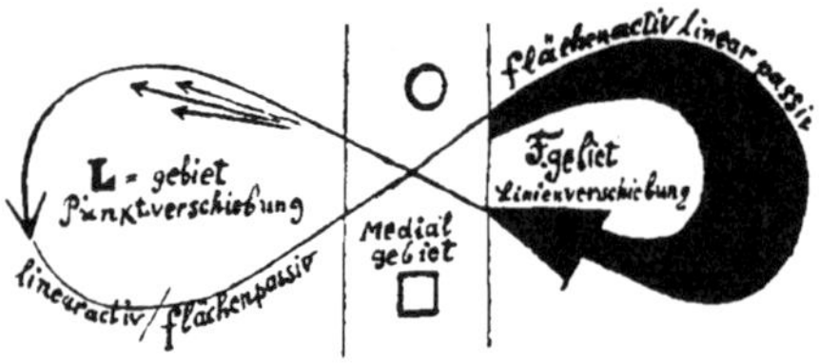

Fig 8: Klee's active and passive lines

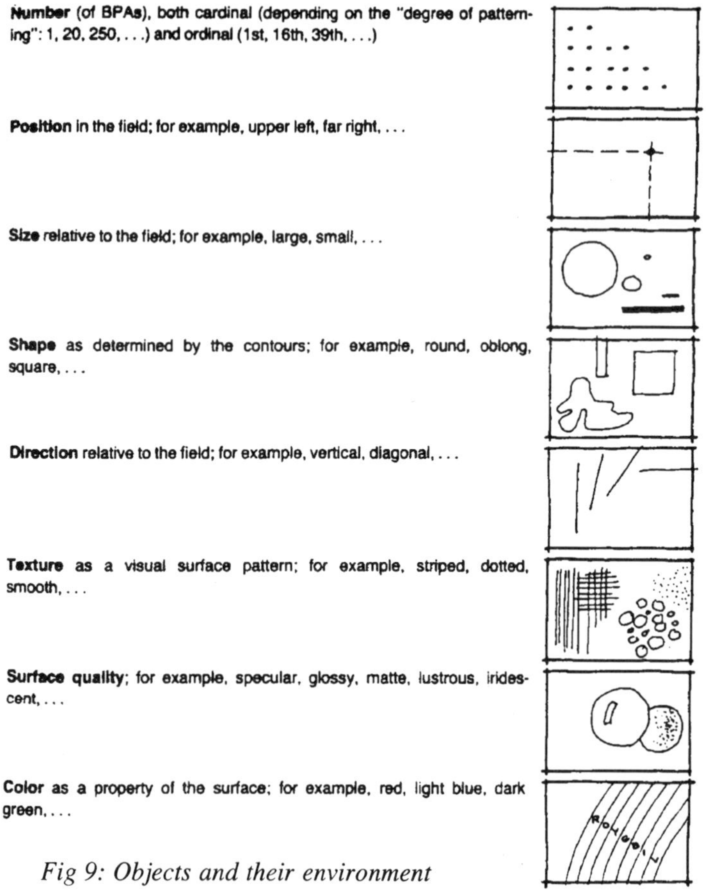

Fig 9: Objects and their environment

controlled as well as emotionally expressive'. A fellow teacher, Itten, argued that the clear geometric form is the one most easily comprehended, its basic elements being the circle, square and the triangle, Fig 7. Form was also colour, he asserted, those of the spectrum being the ones most easily comprehended.

His colour classes were based on the foundation that there were seven distinct types of colour contrast: a simple contrast between one colour and another, a contrast between a colour of a light and a colour of a dark tone, between a warm colour and a cold, and between complementary colours. In place of the familiar 'colour wheel', with the colours of the spectrum around its periphery, he designed a colour sphere to describe all the relationships.

Kandinsky tried to rationalise the phenomenon of synthaesthesia, where as one sense is stimulated another reacts, and postulated the four major 'sounds' of colour contrasts as: warm and bright; warm and dark; cold and bright; cold and dark. 'Temperature' is the tendency of any colour towards yellow (absolute warmth) or blue (absolute coldness). Colours were also considered to have 'meanings' — yellow was earthly and blue heavenly, the former symbolising opposite feelings to the latter. Since green is made by mixing yellow and blue it creates balance and harmony, in his opinion. Kandinsky also believed in the basic elements of geometric form being the circle, square and triangle. The square he considered symbolised peace and the triangle vehemence.

The creation of a line from a point was thought to have its strength dependent on the number of forces applied to produce an end result, a single regular force producing a straight line while two or more forces produce angular or wavy lines. Vertical lines are 'warm' while horizontal lines are 'cold'. Another teacher, the prominent artist Paul Klee, tried to rationalise the 'movement' and direction of line. He produced the diagram in Fig 8 to compare active and 'passive' lines, between which limits it was said to be 'medial' — when defining a coherent form. He maintained that medial lines define either 'structural' or 'individual, forms. Structure is achieved if the same visual elements can be repeated infinitely whereas an individual form cannot have anything added or subtracted.

Self-training in form design

A valuable course of instruction for students of formalised design is provided by Peter Thiel[3] in an admirable American work of cross-disciplinary teaching. Of the many contributions within it from prominent design teachers, one which puts design into focus will be valuable to those who, for example, approach the subject from an engineering background: 'a design may be very beautiful, but it is not art ... the artist wants to make a visual statement ... the designer needs a client to present him a problem'.

The course sees five clear stages in the design process; preliminary design, detail design, approval, implementation and evaluation. The problem of perception is one which is treated in some depth on the course, starting from the understanding that the'material' of visual perception which enters our eyes is a relatively small spectrum of electro-magnetic radiation, emanating either from light sources, direct, or reflected by illuminated sources in the environment. The light energy is modulated, patterned in both space and time dimension, by interaction of the light sources in reflection, refraction and diffraction with respect to the atmosphere and environmental surfaces, Fig 9.

Without such modulation we would see 'nothing' and visual perception is thus the de-coding of the modulated information. This is illustrated by forms of camouflage, Fig 10, when the units of the patterned field have such a strong visual relationship with one another that we are unable to organise them into an alternative arrangement. The four types shown are termed somatolysis (body dissolution), disruptive colouration (dazzle painting), countershading and the suppression of form-revealing cast shadows.

Another useful concept is the bounded visual field which the human eye can perceive. Generally all the eye perceives is an arrangement of patches of colour, variously shaded. Perception of solid form is an experienced-learning process; we 'see' nothing but flat contours and only by a series of 'experiments' do we find, for example, that a stain of grey indicates the dark side of a solid substance or that a faint hue means that the object is far off. There is also the realisation that the eye's field is roughly oval in shape, extending some 150 degrees vertically and 180 degrees laterally, Fig 11. The construction for an equidistant hemispherical projection to simulate the eye's field of vision is shown in Fig 12. Main points to realise are that although the visual world is unbounded and uniformly clear, the actual visible field is bounded and has a centre-to-periphery gradient of clarity. The hemispherical projection is actually a fish-eye lens photograph, with a 180 degree vertical subtense and a uniformly clear field. Fig 13 shows a fish-eye photograph with a hemispherical projection alongside showing the basic pattern areas (BPAs). In this projection, the BPAs can be described as having different shapes and sizes; being in different positions; having different colours and textures; being vertical or horizontal and being either glossy or matte. This leads to the identification of the primary static visual attributes .

One of the Bauhaus' greatest teachers was Laslo Moholy-Nagy[4] who considered the cameraless photograph to be a valuable tool in design teaching. Photo-sensitive paper exposed to light will record the varied intensity of light sources in black, white and scales of grey. Laying solid objects on the paper will project shadows in various grey scales. The concept of 'figure' and back-'ground' was also crucial to understanding visual perception. When we attempt to penetrate a camouflage we try to regroup the basic pattern areas in terms of alternate figures on new grounds.

Fig 10: Forms of camouflage

Fig 11: Field of vision

Organic form

Many express the desirability of natural form so it is useful for the designer to rationalise the attributes of such forms. A good exercise is to consider the tideline jetsum which one encounters during a beach walk. The man-made items washed ashore from ship's waste will stand out a mile unless they are the product of an outstanding designer who can make good claim to the natural origin. The difference between the two classes of object, it is considered, is one of growth. As organic forms develop the effects of time on development are clear — the form is defined by the change in form. Environmental forces to which an object, or creature, is subjected in its lifetime eventually reach an equilibrium in the final form. Man-made objects which reflect natural forms are thus seen in designs for traditional ships, Fig 14.

An interesting exercise in design training is the use of grid-lines for the purpose of scaling up arbitrary shapes. The exercise is valuable not for the means of scaling up, which of course can be done

Given point P, a stationary point at eye level; line PF, the horizontal line of sight; and point 0, any point in the forward field of view; to find 0′, the hemispherical projection of point O, in a circle of any convenient radius H; with horizontal axis L′R′, vertical axis Z′N′, and center P′:

1. Establish plane PSOF through point O and line PF. This intersects vertical plane LR-ZN in line PS.

2. Find β, angle APS in right triangle PAS, using true lengths of AP and AS.

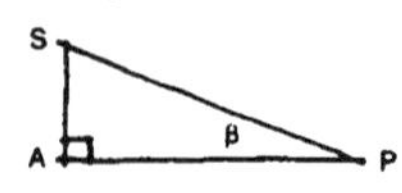

3. Find α, angle FPO in right triangle PFO, lying in plane PSOF; using true lengths of PF and FO. Note that FO = PS.

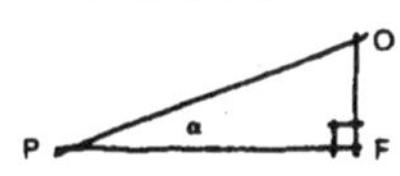

4. Draw radius H from center P′ at angle β.

5. Lay out distance r from P′ on H, where r = (α/90) H. This locates 0′.

Note that when α = 90°, r = H, and that when α = 0°, r = 0. Note also that lines and edges parallel to PF will lie along a radius in the hemispherical projection.

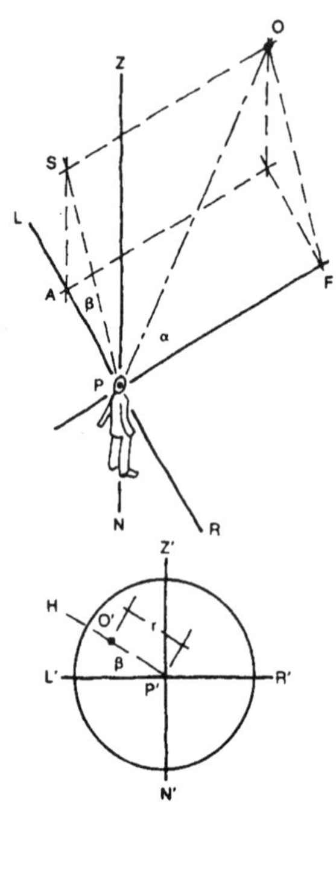

Fig 12: Construction for hemispherical projection

optically, but in helping to develop skills in estimating distances. Freehand 'fairing-in' of the enlarged contour is particularly instructive; reproducing points of inflexion where two curves of opposite slope join together, is also demanding of effort. Where curves join tangentially with straight lines the effect of the curve radius gradually increasing to infinity has to be reproduced and it is helpful for the student to think of the path of an aircraft coming into land as being a good analogy to the shape required. Such flowing transition curves are also those which approximate the ideal for vehicles moving through fluids, traditionally boats through water, but with increasing road speeds, also cars moving through air.

Conveying expression is another task for the designer and a visually aware person has to recognise the key visual features of, let's say, a steep rock, a willow tree or the colours of a sunset and the effects each has in putting over its expressive message. The columns of a temple, Fig 15, are expressive — owing to careful proportioning by their architects in classical times. Location, proportion and shape of the columns are all important in making their image contain the desired impression. The illustration is, in fact, a stored image supplied by Autodesk to purchasers of their Autosketch bottom-of-the-range CAD program and also gives the proportions of the 'golden' rectangle upon which so many Georgian reproductions of classical buildings and styles were based.

Gestalt psychologists are the originators of concepts that govern the organisation of objects within a visual field. Four laws are espoused; that of 'proximity' stating that the relative closeness of some units, compared with others at a greater distance, cause the closer elements to be seen together as a separate entity. The law of similarity concerns the tendencies of commonalities in the attributes of a number of separate elements to be related as a group. The law of continuance refers to the relating tendency of similarities in the changes in the attributes of a series of adjacent basic pattern areas. Finally, the law of closure deals with our perceptual tendency to group certain visual elements and, by 'filling-in' the gaps between them, to establish one simpler larger form. The four rules are illustrated in Fig 16.

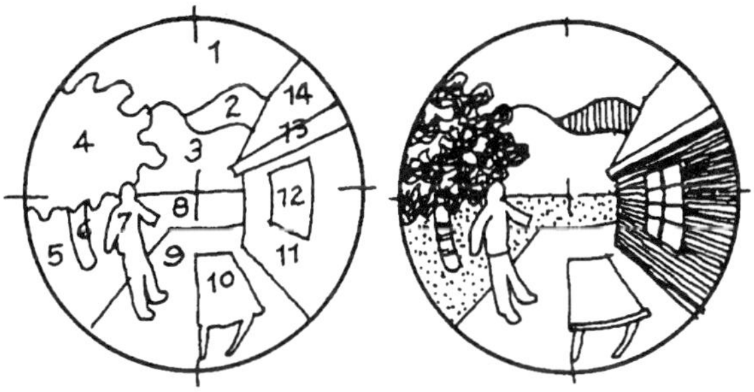

Fig 13: BPAs in fish-eye photograph

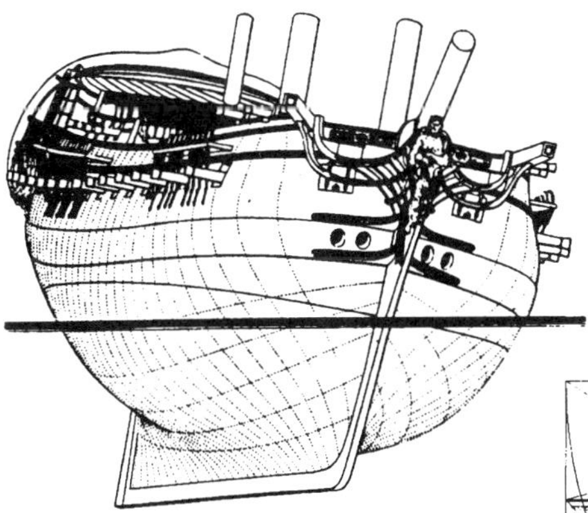

Fig 14: Natural form in classic ship

Colour attributes

Colour is said to be perceived in five different ways; the illuminant mode in which it is perceived as belonging to a source of light; the illumination mode based on illuminated objects casting shadows; the

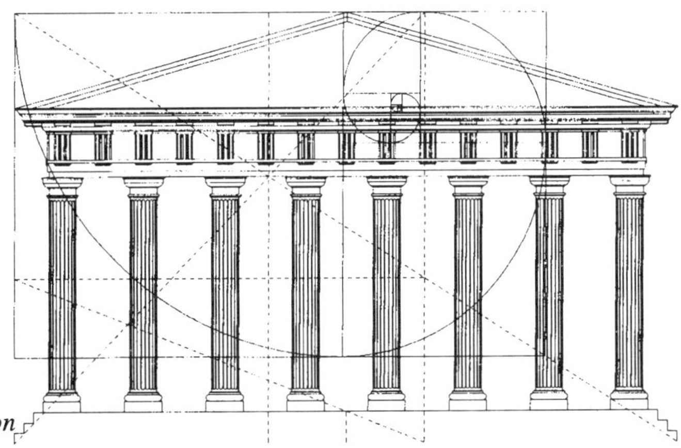

Fig 15: Classic proportion

surface mode in which colour is perceived as belonging to a surface; the volume mode involving passage of light through a semi-opaque mist such as smoke and finally, the film or aperture mode in which a colour is seen through an aperture, as in a mask, isolated from and unrelated to other colours. One of the best known systems of colour notation is due to Albert Munsell whose familiar hue-circle and hue, value and chroma scales are shown as Fig 19. This allows for 100 hues of the five predominant colours of the rainbow spectrum of white light. Extending vertically through the centre of the hue circle is the scale of reflectances, known as 'values'. Radiating out from the central core of colour solid are steps of saturation, known as 'chroma'. The scales are important for the correct specification of colour under standard conditions of illumination and viewing. A contemporary journal [5] gives a step-by-step process for choosing interior colours based on these concepts and taking in those of warmth, spaciousness and excitement level discussed earlier in this chapter. The question of surface texture is also crucial to the perception of colour. The directional pattern of light reflected from a surface depends on the degree of roughness of the surface and on the pattern of that roughness. In the extremes, 'diffuse' or 'specular' light will be reflected from a surface of random micro-facets or of a mirror-like finish. Three dimensional objects illuminated by sunlight, say, will also show a different pattern of illumination at a scale above that of surface facets — due to the chiaroscuro effect. This is because of the overall form of the object and its relationship with the light source. Fig 20 shows typical conditions of highlights and shades for different shaped objects.

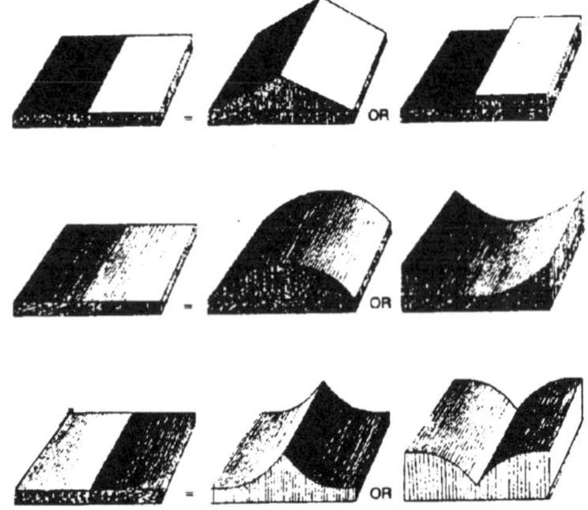

Fig 20: Highlights and shades

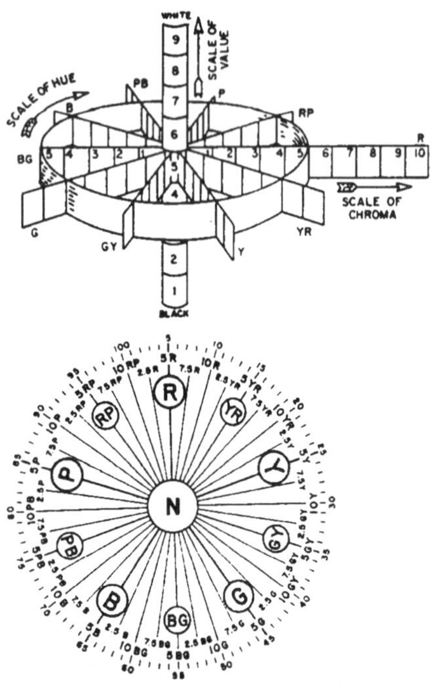

Fig 19: Munsell hue-circle and colour scales

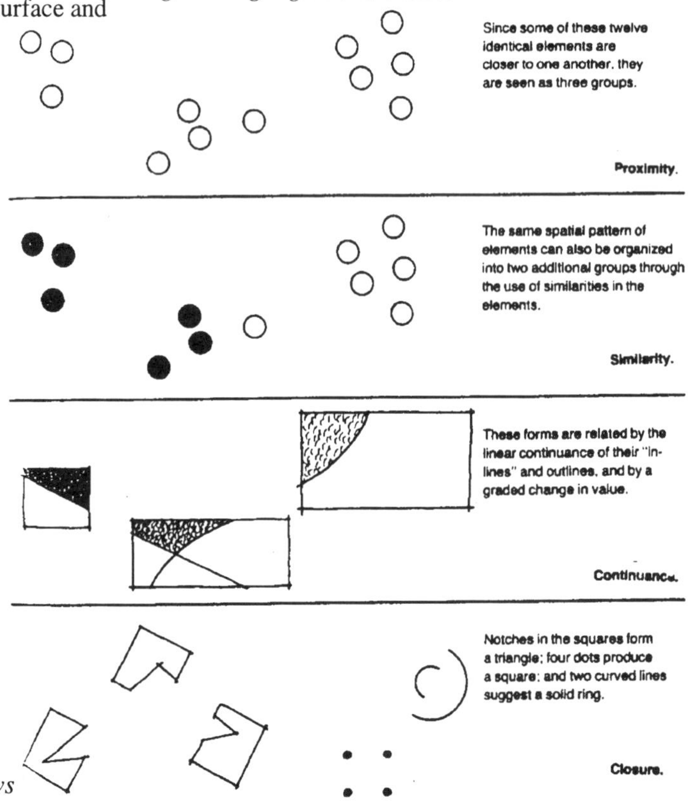

Fig 18: Gestalt laws

CAD and surface development

A complete CAD system involves several layers of software between the electronics and the user starting with, at the computing 'machine' itself, the operating system. This does basic 'housework' such as looking after the disc drives, which hold the next layer up, accepting keyed input and arranging screen display. In most micro-computers (PCs) the operating system is incorporated into the basic 'interpreter' circuitry which converts keystroke data into machine code understandable by the central processor unit (CPU). Other layers of software are the 'programming language' and the 'application program'. Levels of programming language vary according to similarity with the language of the user (high, such as BASIC) or the machine (low, an assembler language). A good programmer is thus a person good at language — not necessarily a conventional engineer. In some cases a programmer may use a 'compiler' to turn the program language into binary instructions for direct access to the operating system.

Graphics are fundamental to both CAD and surface development; they are described in quality terms by the 'resolution' of either screen display or printed image — referring to the numbers of lines in the display times the number of dots per line. A bank of random access memory (RAM) is often reserved for graphics. Normally the graphics display is controlled by a dedicated microprocessor called the 'video controller'; each bit of the RAM corresponds to one dot on the screen which is switched on or off (lit up) by a 1 or an 0 from the CPU — the technique of bit mapping. A more accurate representation, used in 'drawing' rather than 'painting' is vector graphics in which lines are described mathematically and can be accurately scaled.

Only by 1980 had even the mini-computer-based CAD systems acquired the ability to successfully model a part in 3-D. Two of the most capable were considered to be DUCT and Unigraphics. By 1982 the serious commercialisations of desk-top microcomputers, plus the advent of networked engineering workstations, changed the face of CAD. The AutoCAD package has become an important contender with good drafting facilities backed by after-sales support. The Autodesk company have fully exploited the PC's potential and realised a system which offers both 3-D surface and full solid-modelling. Memory limitations in the future for many systems are in prospect of being overcome by new technology introductions such as writeable CD-ROM disks.

Key elements of a CAD system are devices for input, display, storage and output. A 'mouse', joystick controller or light pen might typically be used for input; display screens are usually of the raster-scan or the refresh-tube type; storage is either by machine memory (RAM) or by magnetic (hard or floppy) disc; output may to an impact (dot-matrix), non-impact (bubble-jet, laser) printer or a plotter. Basic elements of a drawing are points, straight lines, arcs and curves. Positioning is achieved either by using a cursor, in conjunction with an accurate background reference grid to which a point is 'snapped to' — or numerically, using cartesian or polar co-ordinates, Fig 21. Some CAD facilities available with a typical package are shown in Fig 22. It is also possible to construct associated databases of useful specificational information, also libraries of symbols and frequently used detail components. Layering is another useful facility — comparable with a number of transparent printed sheets in conventional drawing — and editing functions are a dramatic improvement over rubber erasers.

With an adept knowledge of co-ordinate geometry, it is possible to produce clear drawings with

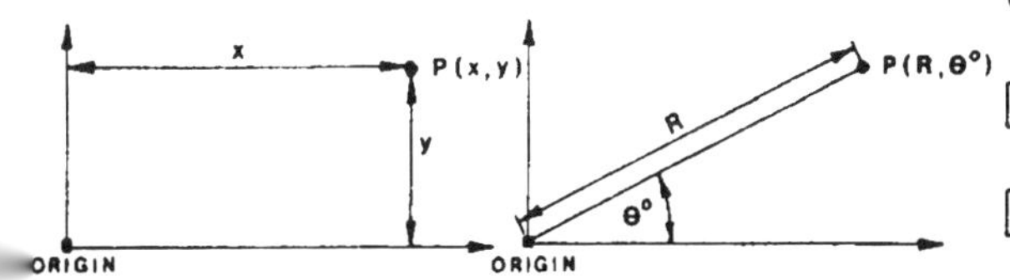

Fig 21:Cartesian and polar co-ordinates

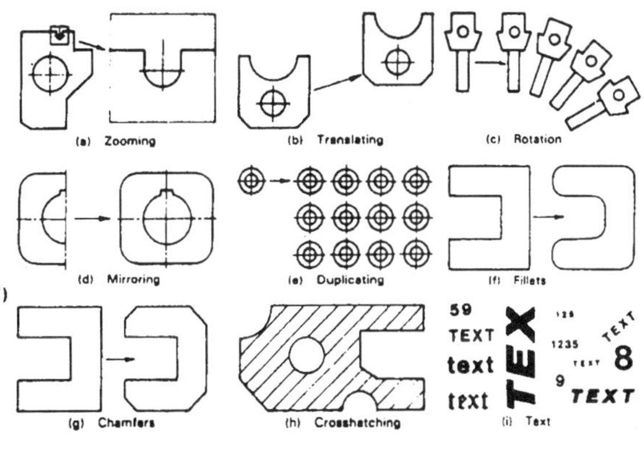

Fig 22: Some facilities available with CAD

a very modest desk-top computer provided it is properly interfaced with a suitable plotter. Working would inevitably be slow, however, and happily the computer-aided-drafting programs carry out the co-ordinate geometry automatically so that spatial points can be defined accurately and joined with lines or pre-defined mathematical curves at great speed. The computer performs graphical geometry calculations for tangent lines to curves, optimised fillet radii and all manner of constructions. It is thus only a short step to plot the motion of linkage mechanisms and check large assembies of parts for foul conditions.

CAD has, of course, only a limited use in the creative side of design and so before making a system purchase, it should be realised that the system is going to be of most benefit after the creative work has been carried out. Creative design is seen as a process in which ideas are generated, sorted and then tested — followed by a process of modification until the desired result is achieved. It is estimated that the designer spends 20% of his or her time searching for information, 65 % generating and sketching possible solutions and 5 % carrying out calculations before the

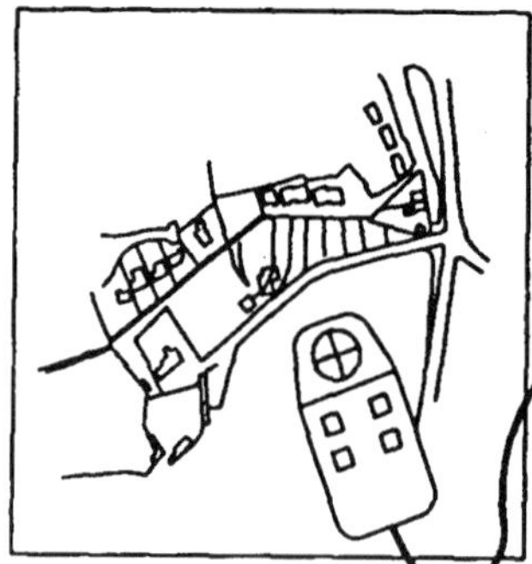

Fig 24: Digitising tablet

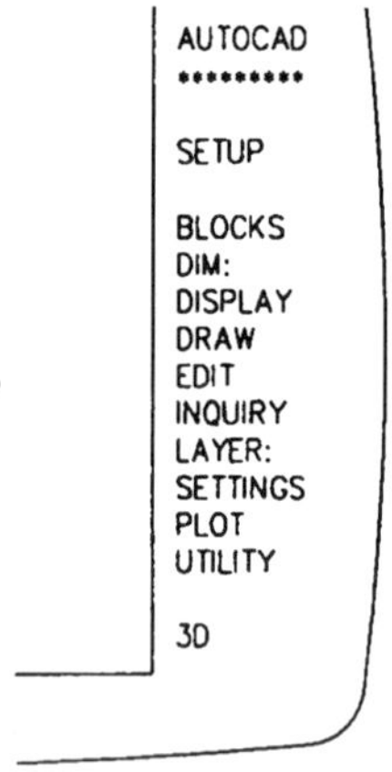

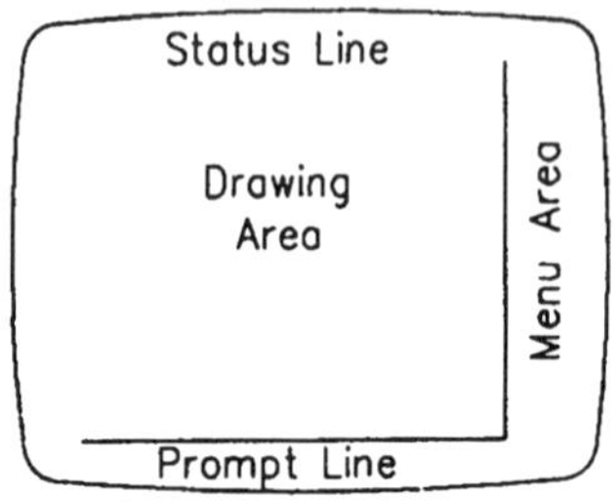

Fig 25: Screen areas

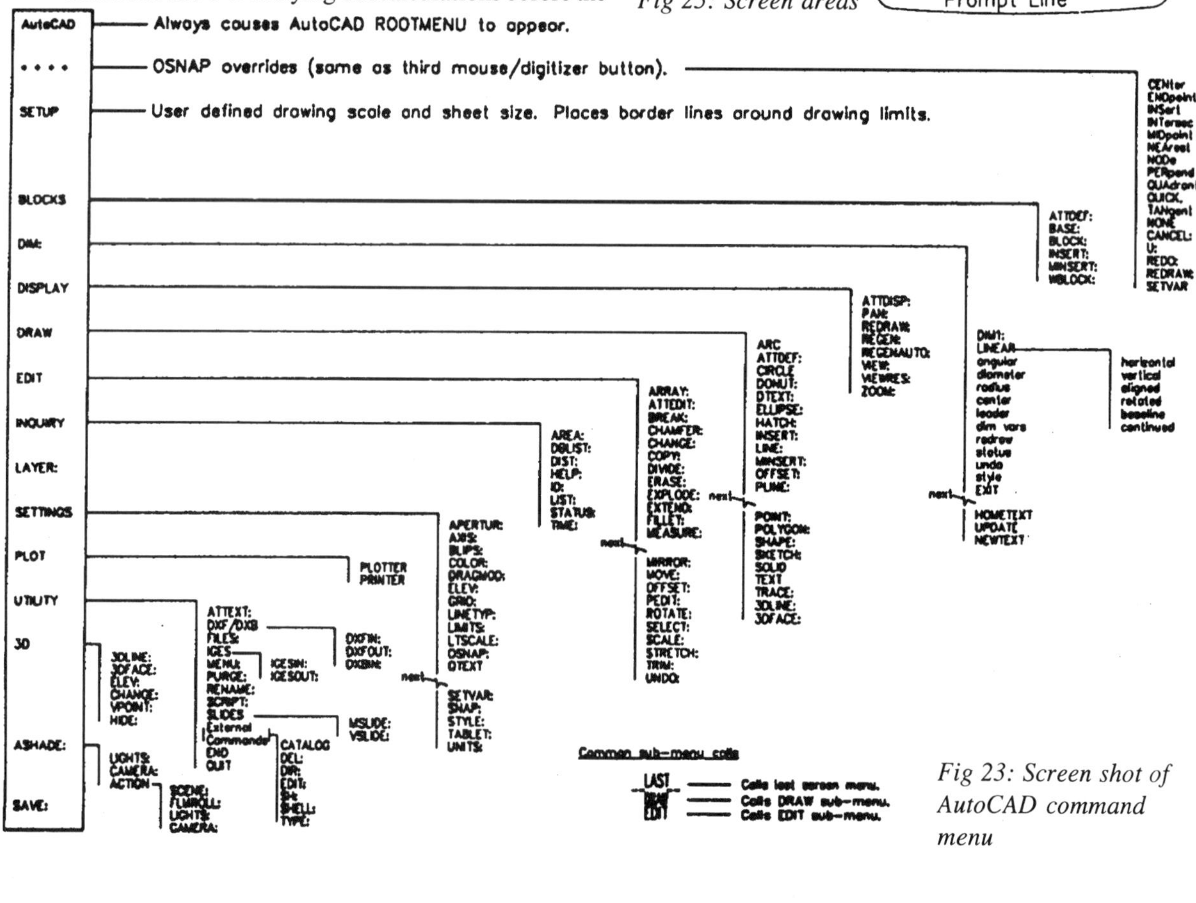

Fig 23: Screen shot of AutoCAD command menu

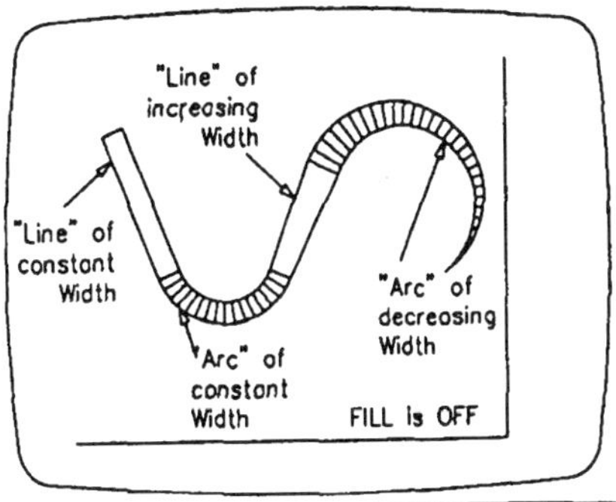

Fig 26: Polylines

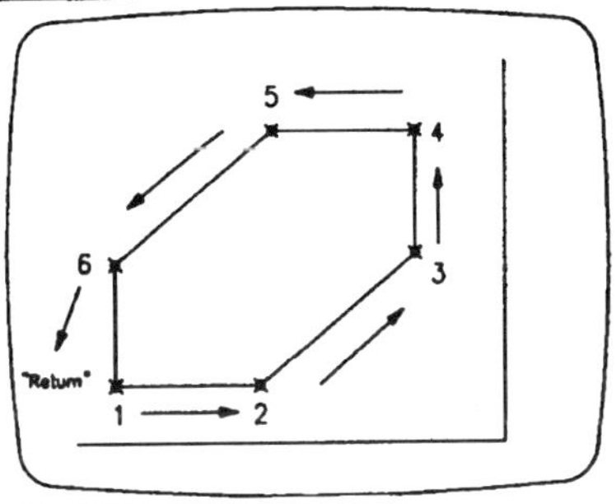

Fig 27: Area measurement

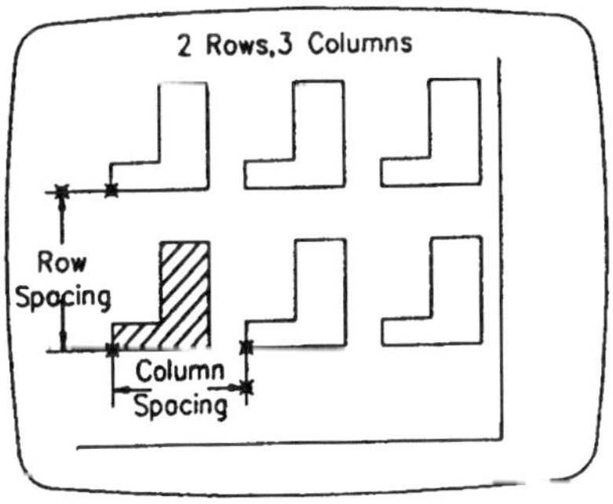

Fig 28: Ordering arrays

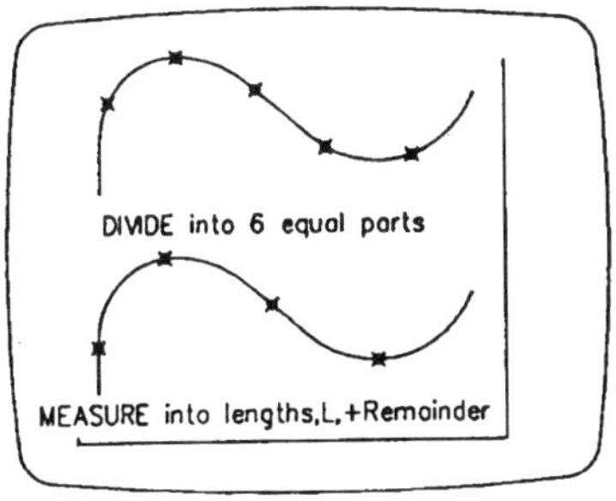

Fig 29: Divide and measure

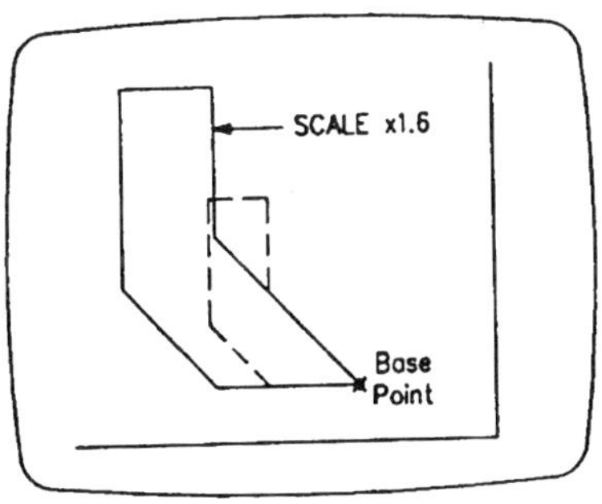

Fig 30: Scaling

stage of carrying out computer-aided-drafting. There is a further 10% of time writing reports to convince others of the viability of the design. The multi-tasking ability of the better desk-top computers does at least allow the more or less simultaneous use of programs for word-processing, calculation and drafting as well as calling on databases of information. Computer-aided-design usually implies carrying out other functions as well as drafting.

A typical drafting-only package uses a rollerball 3-button mouse with 'pick' button to fix a point on the drawing screen, for starting a line. A 'return' button for repeating commands and a third for accessing the command menu completes the trio; Fig 23 shows the AutoCAD comand menu. A digitising tablet, akin to a 2-D scanner, can also be used where points on a 'hard-copy' drawing need to be read into the picture, Fig 24.

The AutoCAD screen has four main areas, Fig 25, the menu being based on a tree structure, root-menu, which is shown inset. To join a new drawing element on to a specific feature of the current drawing, an 'Object SNAP' facility exists whereby a connection can be made to the closest line on a grid underlying the drawing. A feature which is special to the program is Polylines, Fig 26, which enables the drawing of a series of interconnected lines and/or arcs. A further option allows specific widths to be specified. After construction, the combination can be treated as a single object and even a smooth curve fitted to the constructed shape if required. In a similar way to word-processing, 'blocks' of drawings, as with blocks of text, can be separately saved and introduced as required elsewhere.

Objects can be measured to obtain areas for use in calculation — merely by going through the kind of sequence shown in Fig 27. There are also facilities for ordering arrays, Fig 28, and for divide, measure, Fig 29 and scaling, Fig 30. Dimensioning, of course, is automatic and the addition of textural annotation can be added in the same way as a word-processing or desk-top publishing system.

A technique under discussion for CAD is the so-called 'plastic modeller', in which relevant dimensions of objects can be specified as variables rather than numerical values. Thus design work can proceed with loosely designed shapes whose parameters would be specified as a result of calculation or 'fit' to another object — necessitating interfacing with another graphical database.

The fact that it is actually more difficult to operate a sophisticated CAD system than to draw comparatively simple shapes on a home-computer — with the aid of a well produced manual — begs the important question as to at what level does a company, of a particular size and complexity of design involvement, enter the CAD market. This is a question almost impossible to answer by the vendors themselves and for which independent advice is very hard to come by. If an office computer exists at the workplace, it could be very worthwhile experimenting with simple programs if the drafting to be done does not extend into the more complex area of surface-development.

A program such as MLD2 CAD has a 'parametric' facility whereby a drawing's standardised elements can be stored and used at different scales and even different key dimensions for different proportions. There is facility, too, for automatic removal or presentation of hidden detail — facilitating construction of assembly drawings from detail ones. There are also window commands which allow magnification or reduction of separate sections in four areas of the screen, Fig 31. It has links to other proprietary software for drawing and data transfer permitting either finite element structural analysis or computer-aided manufacture by the generation of defined cutter paths.

CAD packages differ from drawing ones in several respects and have a clear concept of the end-product in their make up. Working speed is, of course, increased by having the potential to produce libraries of standard components and symbols. However, there is the potential problem that 'screen refresh' can take a long time in view of the time taken for the program to redraw the complex geometric constructions on each occasion. In advancing from 2- to 3-D facility, the simpler programs have only a wireframe facility and no 'solid' visualisation. Often a rotation feature is provided whereby the current viewing position may be rotated about the azimuth and elevation directions of the drawing in any combination. In the absence of mouse control, various key combinations move the cursor along the three co-ordinate axes. Points can usually also be defined by direct keyboard input.

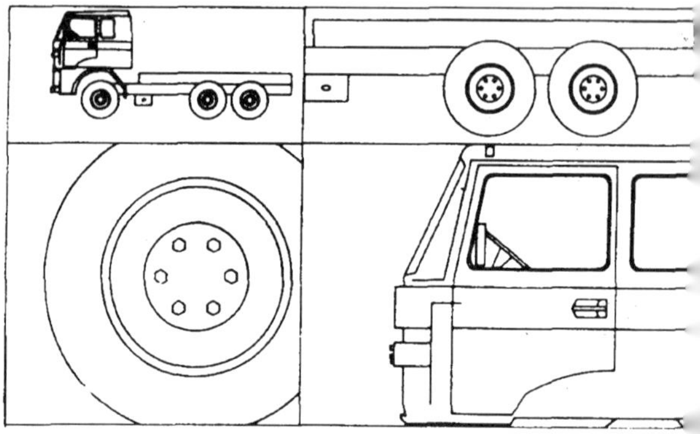

Fig 31: MLD2 windows and parametric design

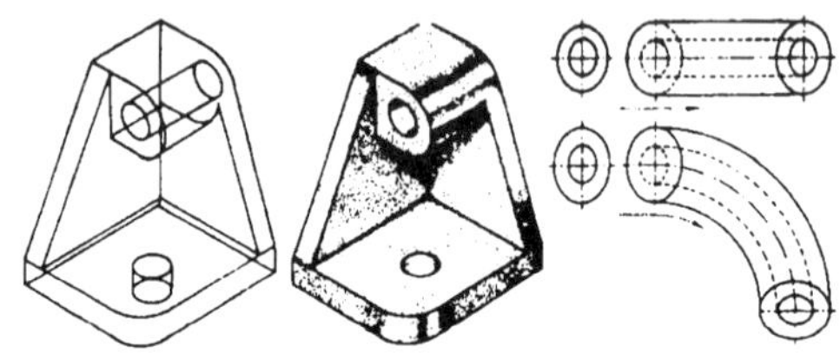

Fig 32: Wire-frame, solid and 2.5D models

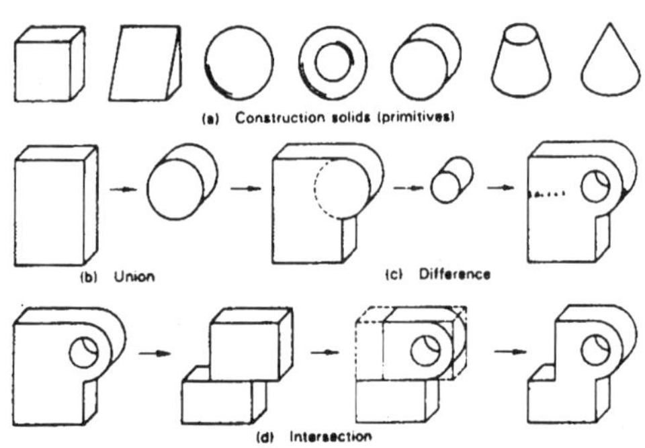

Fig 33: 3D modelling

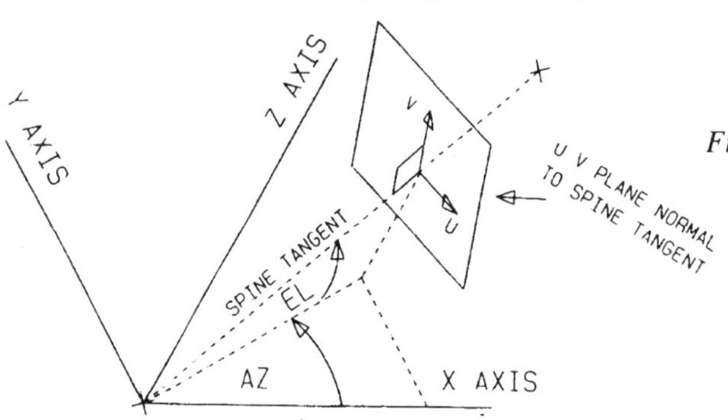

The relationship between the UV and XYZ axes. When looking forwards along the spine, the U axis is to the right of the parallel to the XY plane. The V axis is orthogonal to the U axis and the spine tangent while being as upright as possible.

Fig 34: Spine concept

3-D modelling and surface development

Three-dimensional modelling involves the spatial image being drawn in a pictorial projection using x-y-z co-ordinate geometry — and stored in memory. It can be recalled as either pictorial or orthogonal projection based on several 2-D views. It can also be represented by only its edges (wire-frame) to minimise computer memory as well as in solid form. Further a so-called 2.5 D presentation, Fig 32, can be used in which a wire-frame 2-D shape can be translated and rotated into a 3-D shape. Solid modelling is carried out with the processes of 'union', 'difference' and 'intersection', Fig 33.

Surface modelling lies between solid and wire-frame. The initial wire-frame has its gaps filled by flat or rounded surfaces which can then be shaded. Complex surfaces cannot be simply represented by geometrically simple solids and so a surface therefore has to be developed by defining and joining curves. The 'complexity' of the surface refers to its underlying mathematical description.

For computerising this process, Delcam's DUCT is a valuable package which was developed at Cambridge University in the late 1960s. One of its major claims is its ability to direct the cutter-path of a die machining tool with a reliability sometimes not found in rival CADCAM systems.

Basically it is a modeller working in terms of centre lines and sections. A straight or curved reference line, the Spine, has sections normal to it, Fig 34, and is defined at least by any two points in x,y,z co-ordinates and additionally by an angle if required. Successive sections are described in u,v co-ordinates with axes parallel to x,y. Between the data points on the sections, the program uses the Bezier principle whereby the curve is defined in terms of points and associated control points. So-called 'strength factors', Fig 35, are used in conjunction with angles which can be specified in two directions, at the data points, to describe the surface. Open ducts, Fig 36, singly or in combination, allow body-panel type surfaces to be modelled.

The latest version is DUCT 5 which is a direct surface modeller said to be faster than wire-frame based methods. It incorporates automatic blending, trimming and filleting of intersecting surfaces and a relational database is available which allows surfaces to be assembled together. Areas, volumes and masses of completed panels can also be calculated.

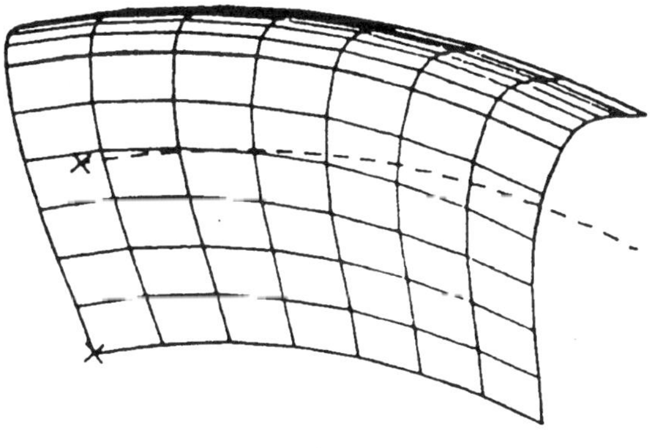

Fig 36: Open Duct panel representation

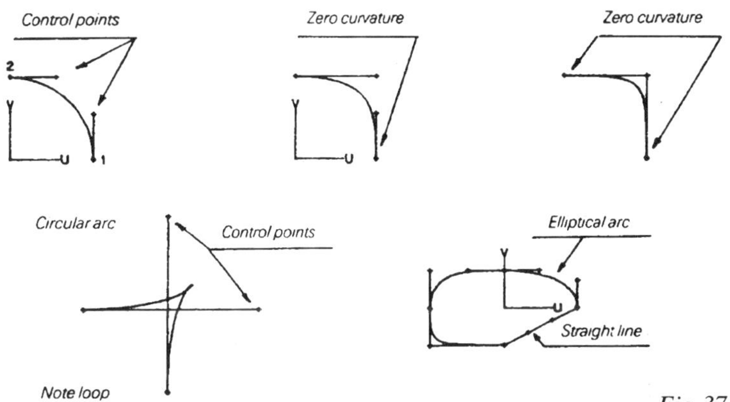

Fig 35: Strength factors

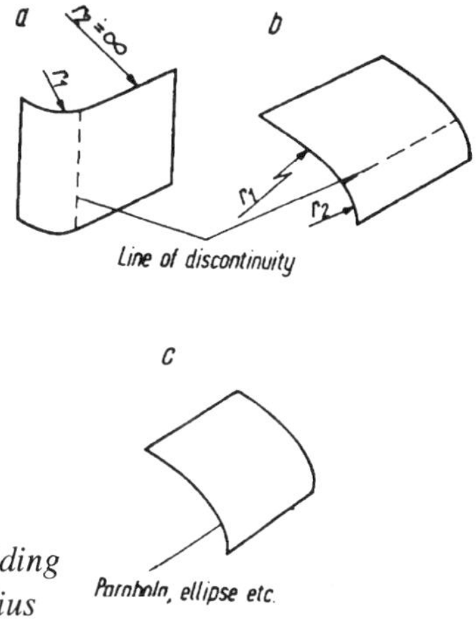

Fig 37: Avoiding constant radius

Surface geometry

Curved body panels are, of course, preferred to flat ones to avoid distorted light reflections and obtain greater panel stability against vibration. There is also an 'aesthetics discord' in joining a flat panel to a curved one. Even in single curved surfaces it is aesthetically preferable to avoid curves with constant radius since the line joining surfaces generated by different radii is also problematical, Fig 37. Parabolic or elliptical curves are therefore preferred. Better still, double curvature surfaces ensure overall fluidity of shape.

In three-dimensional curvature, equality, similarity and affinity (collineation) are three key parameters (Fig 38). In the first, the generators are identical; in the second they generate curves whose slopes are equal at different positions; in the third, curves I to IV are related by the curves at V and W. The varying degrees of affinity are shown in Fig 39; this shows the effect of different initial curves used for interpolating the shape on subsequent drawings between the boundary curves. Different surfaces are thus obtained: a continuous edge from the corner; b a fading edge; c a fading corner.

After the shape has been chosen by clay or computer modelling, the surface is split up into sections corresponding to the major boundaries, prior to collineation, Fig 40. Intermediate curves between the boundaries are collineated using trapezoidal, triangular, radial or rectangular divisors, Fig 41. In the first case, boundary curve K1 is represented in the subsidiary diagram by the base line JE drawn to any convenient length. Normal boundary lines AE and FJ are represented by vertical lines (with the same lettering) drawn to a suitable scale. Straight connecting lines like AF and BG are drawn to complete the diagram. Section lines I and II are represented on the diagram by drawing normals on the base line through K and P which are themselves chosen so that AP:AK:AF is a constant relation in the two figures. The intersections L,M,N and R,S,T can then be transferred back to the first figure and these, when connected, give the lines K3, K4 and K5.

With triangular divisors, only three boundary curves are given and in both methods linear proportionality of the intermediate curves is the simplest form of dependence. Higher forms of collineation, such as radial and rectangular divisors, are used to obtain a more refined aesthetic.

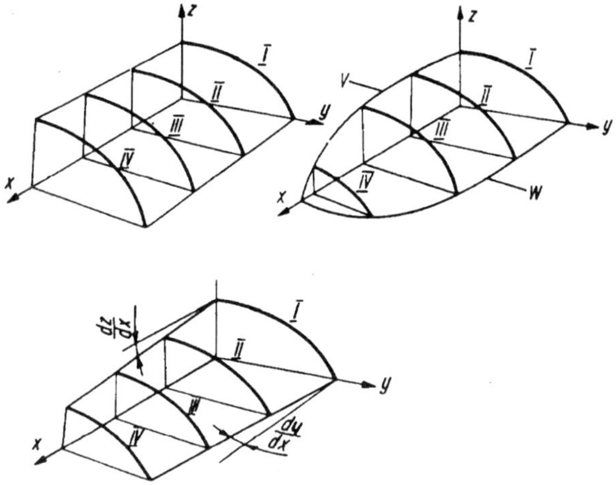

Fig 38: Equality (top-left), similarity (top-right)and affinity

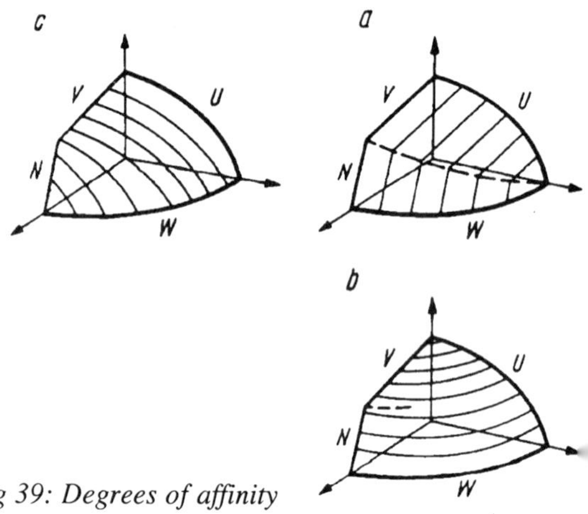

Fig 39: Degrees of affinity

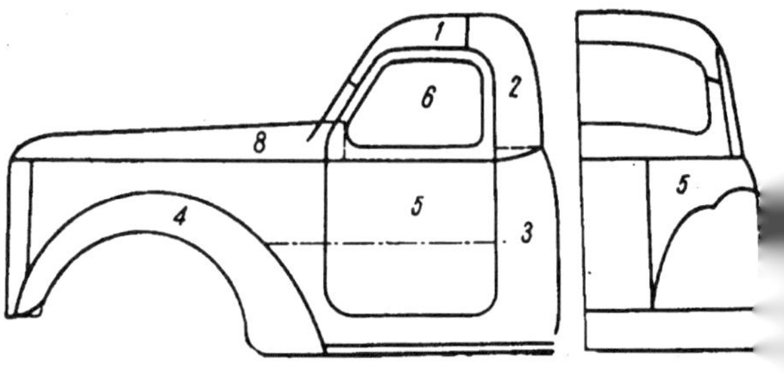

Fig 40: Sections by major boundary

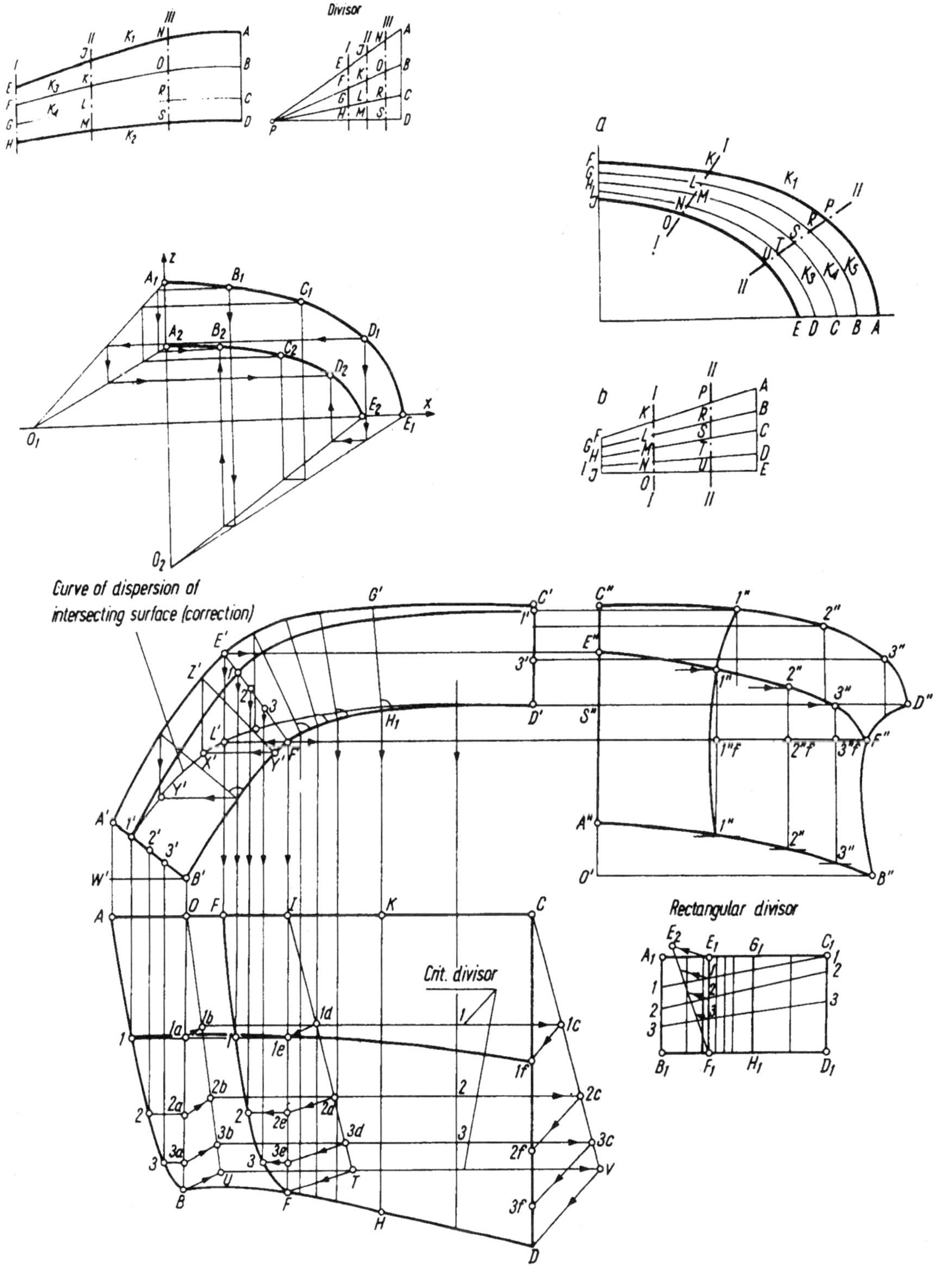

Fig 41: Surface development by rectangular (bottom), trapezoidal (top right), triangular (top left) and radial (top left — lower) divisors

Interior ergonomics (vehicle controls)

Driver and passenger comfort; ease of entry and exit; driver relationship with the primary controls of the vehicle — all these affect the interior package dimensions of passenger-cars and public service vehicles: also the cab layout of goods and industrial vehicles. While the high volume builders invest heavily in studies to optimise ergonomics of particular designs, the specialist builder often needs to make recourse to more generalised published data to optimise driver and passenger spatial envelopes for a relatively low volume production vehicle.

Ergonomics is strictly the study of matching man to machine and is a surprisingly recent science. Much dimensional data used in the UK for matching drivers to the controls of their vehicles came from studies of World War Two fighter plane cockpits.

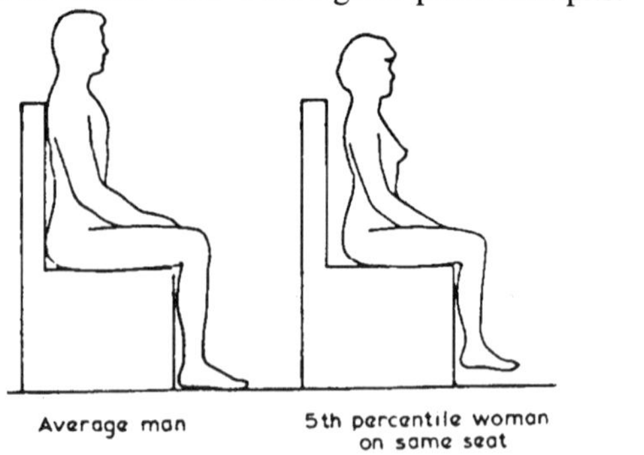

Fig 43: Seat measurements as affected by percentiles

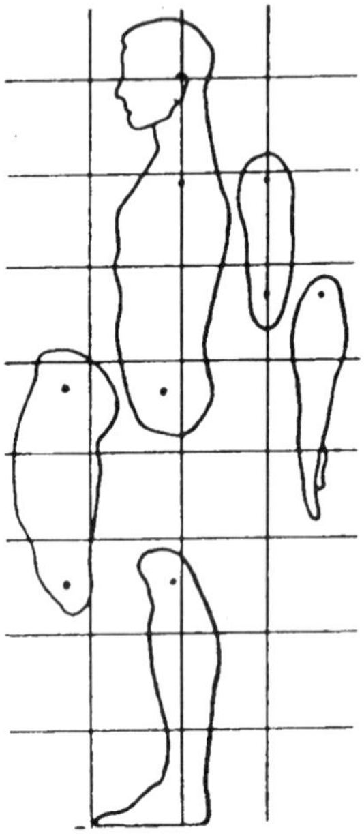

Fig 42: Dimensions of average male on 10 inch grid

1

Dimension	Percentile male			Percentile female		
	5th	50th	95th	5th	50th	95th
1 SRP to knee hinge	20	22	23½	18½	20½	22½
2 Knee hinge to sole of foot	18½	20	22	15½	18	20
3 Buttock to patella	21½	23½	25	20½	22	24
4 Buttock to back of knee	17	19	20½	16	18	19½
5 Knee height (no shoes)	20	21½	23½	17	19½	21½

Dimension	*5th %ile female*	*95th %ile male*
6 Shoulder point above seat (sitting shoulder height less 2 in)	17 in	23 in
7 Shoulder point to elbow hinge (shoulder to elbow less 1½ in)	10½ in	14½ in
8 Elbow hinge to closed hand (elbow to fingertip less 1 in for hinge and 4 in for closed hand)	11 in	15 in
9 Sitting eye level above seat	25½ in	33½ in

2

SEAT PARAMETERS

Optimum E = 105° (90% acceptable, 102°–108°)
C = 6° (90% acceptable, 5°–8°)
B ≤ 16 in for small women
= 17 in as compromise (provided knee pivot of small woman > 2½ in above seat front edge, to allow full pedal depression)
A = 18–24 in
D < 15–16 in

* Failure to meet this dimension causes accident risk due to reduced pedal force arising from popliteal pressure. Seat front edge must be rounded and softly padded.

3

Normal and maximum (emergency) pedal efforts given by shortest women and tallest men in 90% range, for three seat heights; M = Males, F = Female.

Seat height (in)		Knee angle		Norm. force (lb)		Max. force (lb)	
SRP	Front edge	M	F	M	F	M	F
11	13	133°	126°	75	45	135	80
9	11	140°	142°	80	50	155	95
7	9	145°	135°	90	55	175	120

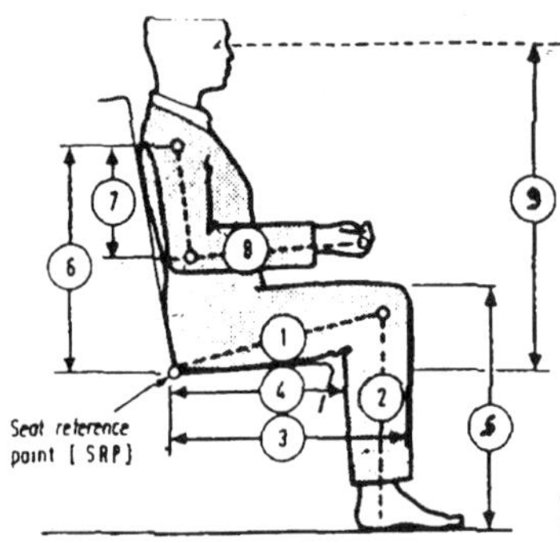

Fig 45: Seated occupant with tables of seat dimensional data

Hence earlier data sources need treating with care in respect of the special levels of physique in the subjects concerned — and often their above average stature. Buses and taxis, for example, need designing for a user-population of much less fit, and often smaller stature, people. The use of templates in drawing work therefore needs some caution. Fig 42 shows the dimensions of an 'average' man; since it is usual to design within percentile extremes, this could be scaled up and down to represent these extremes

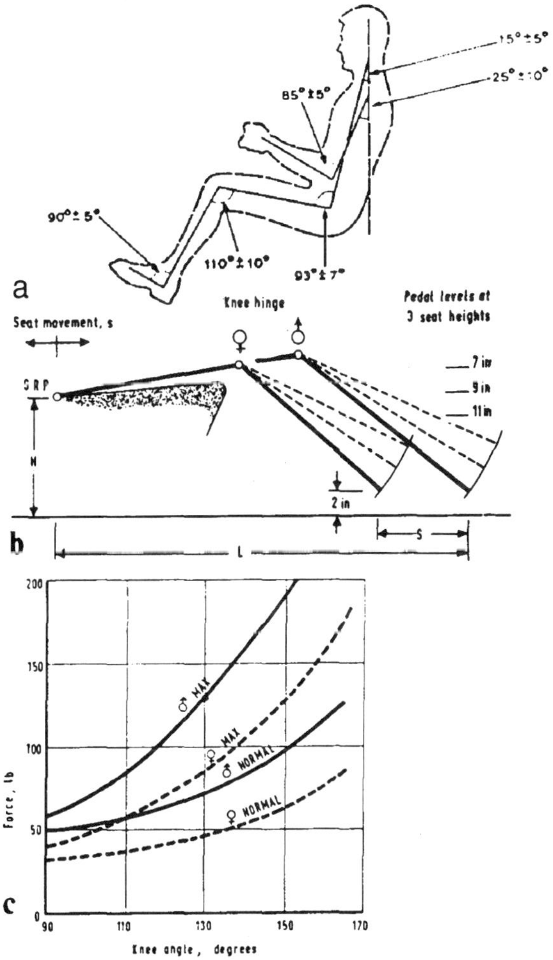

Fig 44: Recommended comfort angles

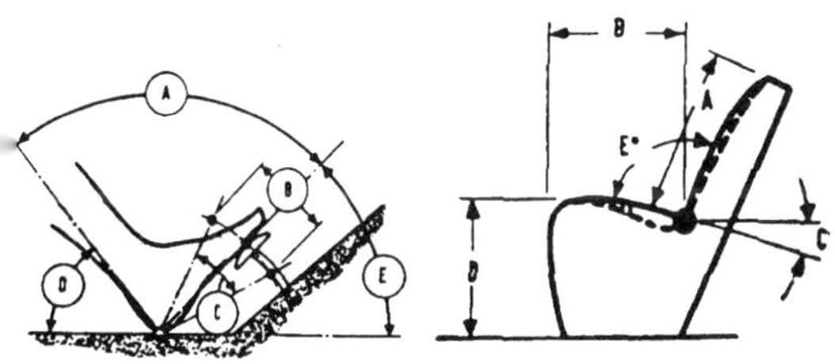

Fig 46: Foot control data

and two jointed models made for fit testing on scale drawings of interiors.

Percentile correction

As there are few women, say, smaller than the 5th percentile and few men larger than the 95th, these correspond to statures of 58 and 72 inches. By definition, percentiles relate to a single dimension for a group of people and it should thus be realised that the 'average man' concept could be invalid for different demographic groups of people, even when factored for stature as described above. Fig 43 shows an unfactored 'average' man alongside a 5th percentile woman sitting on the same seat. Minimum fatigue over long periods in the seated position has led to comfort angles as shown in Fig 44 but another very important consideration is the provision of space for a person to change position during long sustained periods of driving. By defining a seat reference point (SRP) as the intersection of the seat and squab surface, Table 1 in Fig 45 shows the necessary factoring of data to accommodate percentile variations.

Foot controls

Early work by H Murrell of Cardiff University focused on the different ergonomic requirements for the different controls in a vehicle. He pointed out that foot controls are divided between those like brakes, needing considerable force, for which free leg movement is required, and the 'precise' control of throttle, say, where use of ankle articulation is better suited. For braking and declutching a situation of progressive force with virtually zero movement is ideal. Critical factors for pedal force and movement are: knee angle in leg operation optimum 160-165 degrees (cf. most comfortable relaxed range 120-130 degrees) and ankle angle in leg and ankle operation, within 90-125 degrees absolute minimum. Both angles are obtained from a spatial relationship of seat and control; see Fig 46. Optimum lateral offset of pedals from body midline is 25 in (635 mm); at 6.5 in (165 mm) offset, leg force potential is reduced by 75%. Alternative approaches in calculation are either to determine seating/pedal position, given the force required hence knee angle from Fig 44a; or, given seat/pedal position, to determine maximum allowable force. Optimum seat parameters are shown in Fig 44b and Table 2 of Fig 45. A worked example to find leg operation parameters for 95th percentile man to 5th percentile woman is as follows, using the data from Fig 44b: procedure for calculation is to design for adequate vision of the shortest woman and then use available pedal pressure for this position from Table 3 of Fig 45. Pedal movement is then arranged to be in line with the lower leg. The greatest force can be applied by the instep; it is reduced by 50% if toes are used. Hence it is best to site the pedal 6 in (15.25 mm) above floor level with further 2in (7.6 mm) clearance for toes; pedal travel should be restricted to 2in (7.6 mm) maximum before pressure is required.

Ankle controls, such as throttle, are best with pedals hinged at their base. Optimum and maximum angles for barefoot are given in Fig 46 and Table 3 of Fig 45; when pedal is hinged at heel-to-floor contact point, angle E should be increased, to allow for heels, by 5 degrees for men and 10 degrees for women compared with the no-shoes condition. For example, in Fig 46, E = 45 degrees (55 degrees with shoes) for women and E = 55 degrees (58 degrees with shoes) for men (also E = 71 degrees and 73 degrees for men and women, with 7in seat height) so an acceptable pedal can be arranged.

Designing controls for ankle operation starts with the determination, from Table 1 of Fig 45, of seat height for population extremes and finding corresponding D in Fig 46. A pedal angle is then selected to lie within allowable range. Throttle and brake pedal surfaces are then aligned for quick foot interchange, so the throttle pedal is mounted closer to the SRP than the depressed brake pedal (using smaller knee angles than for the brake pedal when calculating

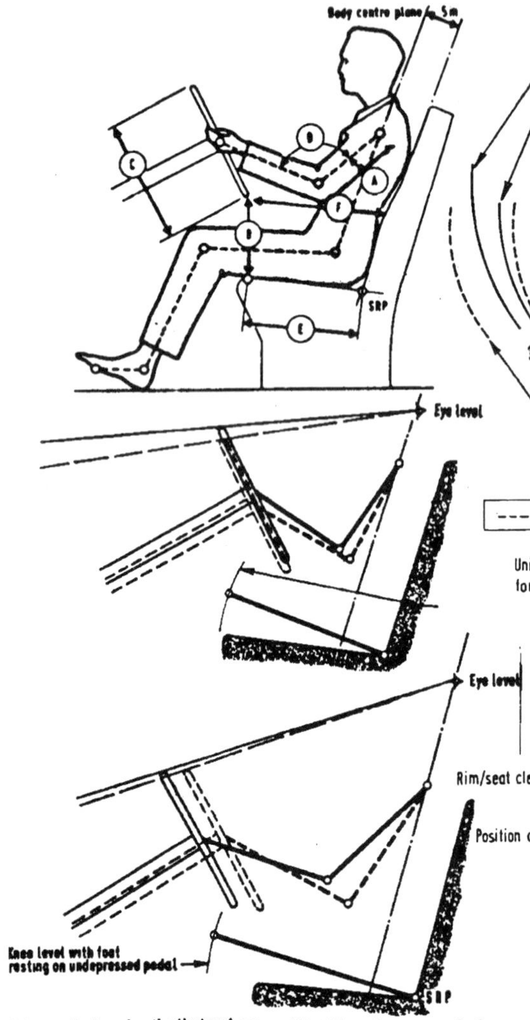

Fig 47: Hand control data

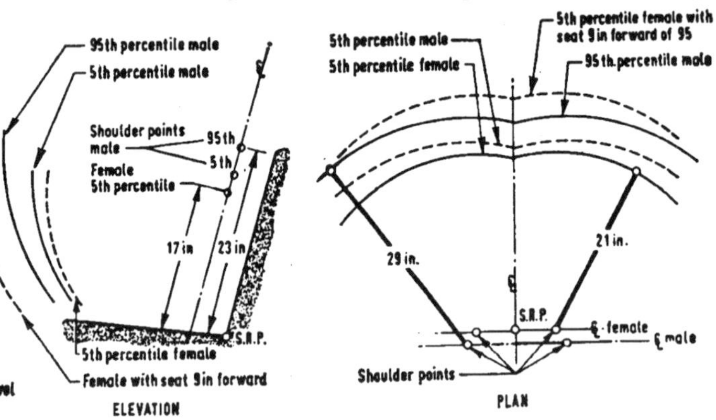

Fig 48: Reach envelopes

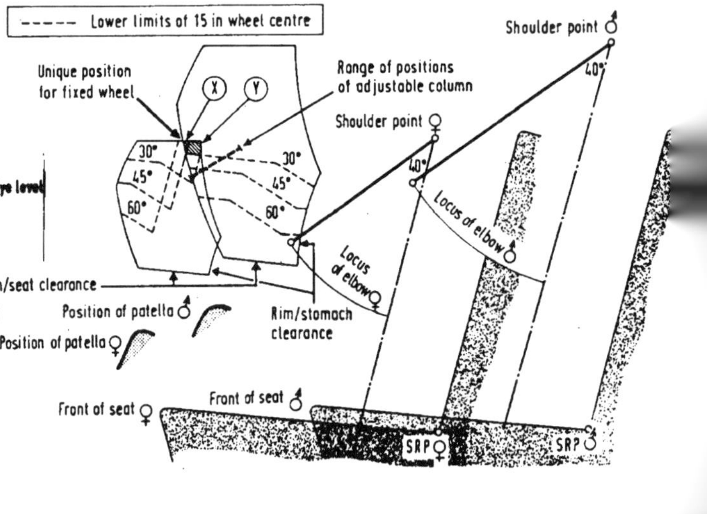

Fig 49: Improved vision with adjustable colum

the position). Sufficient throttle spring loading is required to support the foot (48 lb or 22kg). As an example, when E = 55 degrees to suit the smallest woman on an 11 in (280 mm) seat, A is 115 degrees for the largest man on a 7 in (177.5 mm) seat. If depression is 15 degrees, this gives an unacceptable maximum for A of 130 degrees. The remedy is to increase E to 60 degrees and reduce the depression.

Hand controls

Murrell relates the areas of hand reach to the SRP as in Fig 47 — applying to the seat-belt restrained driver. Note that the large stature person seems unusually badly off in relative terms — but this is due to fact that he/she must sit far back to ensure proper use of foot controls.

Several factors affect steering wheel position; applied hand force depends on column rake, perpendicular distance from SRP to wheel centre and seat back inclination. Maximum force is available when pulling longitudinally with fore-arm and shoulder acting together — while faster, but low efficiency operation, motion is obtained with movement across the body. A wheel tilted 30 degrees to the vertical could have a rim force 1.7 times that with a vertical wheel but the rate of turning is halved and the energy expended reduced considerably. A competition car thus has a near vertical wheel and an extended arm position gives maximum speed for a short-term high stress activity. Heavy industrial vehicles, by contrast, favour the flatter wheel.

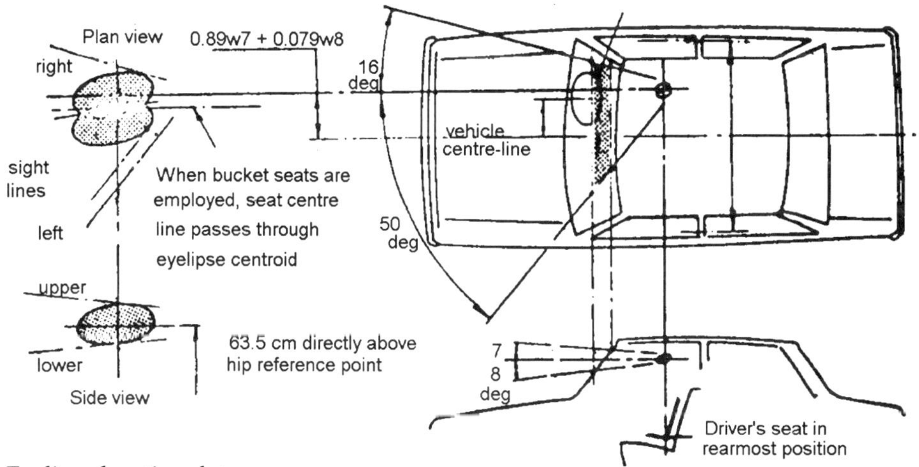

Fig 50: Eyelipse location data

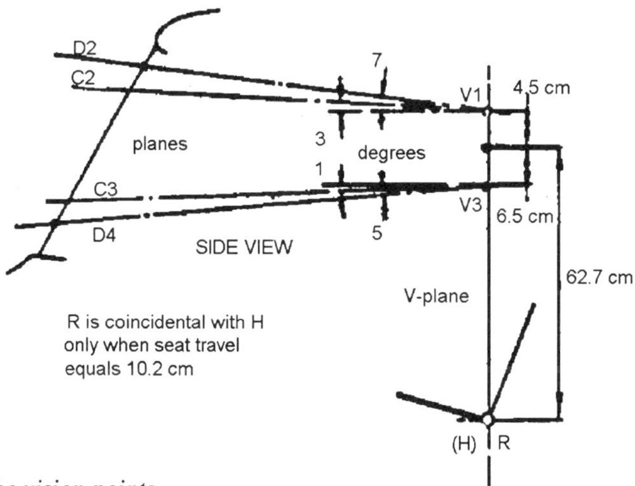

Fig 51: Position of the vision points

Design procedure is to locate wheel centre and inclination for population extremes then determine reach envelopes using dimensions in Table 1 of Fig 45 and Figs 44a and 46. Wheel centre must then fall in the common area shown in Fig 48. If the column length is nonadjustable, no position meets population extremes; for a 45 degree wheel the small area X applies and for a 30 degree wheel the larger area Y is common. If column length is adjustable by 3.5 inches, for this example, then a satisfactory condition exists for both driver extremes — anywhere with a 30 degree wheel and for a small range within the 45 degree inclination. Fig 49 shows the resulting improved woman's vision when an adjustable column is used.

Forward vision

The SAE Eyelipse is a common standard used for optimising forward vision; this is a contour of eye position boundaries for a range of driver statures. Tangents drawn from the contours are constructed to obtain vision areas. Fig 50 shows the method for locating the eyelipse and European practice is to locate the plan-view eyelipses on a longitudinal line passing through the centre of the driving seat. Since the maximum sideways viewing angle of the eye is 30 degrees, some geometric modification to the planview contour is required above this angle. Fig 51 indicates the position of the vision points (V) within the vehicle and it can be seen that a plane in which the V-points are located is required to be positioned 0.4in forward of the hip reference point (joint articulation centre) for each 0.5 inches of longitudinal seat travel in excess of 4 inches, in this example.

Car and CV drivers and passegers

Another USA standard, the AMA measuring system, has been used widely to determine the relative benefits of vehicle space envelopes. This is shown in Fig 52 with standard manikins used in the system and the critical dimensions defining space and comfort for occupants. Comparisons between a sports saloon, bus and limousine, Fig 53, are interesting in this context. In passenger service vehicles, the journey

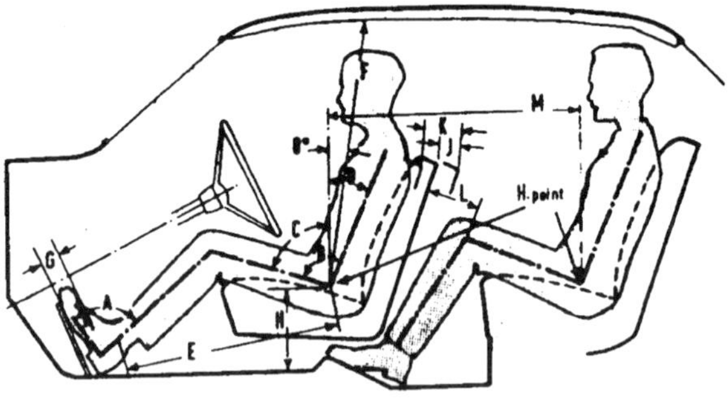

Fig 52: AWA measuring system

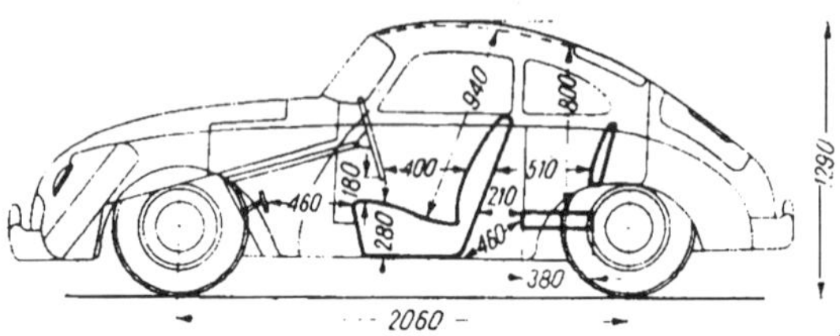

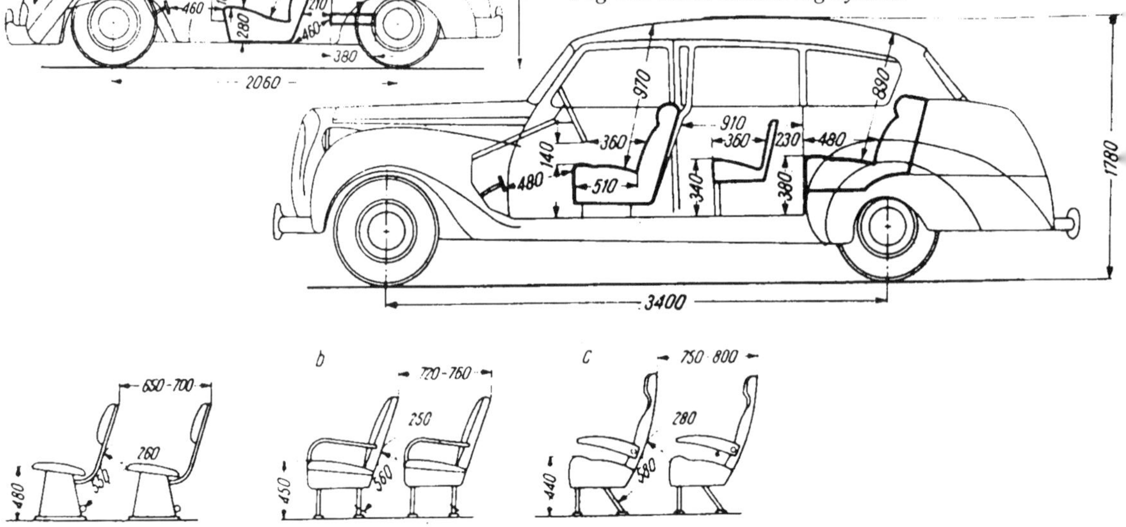

Fig 53: Sports saloon, limousine and bus dimensional comparison

time exerts considerable influence on comfort and thus seating dimension requirement varies for city-bus, inter-city bus and touring coach. The driver's cabin for a PSV also requires different criteria from a passenger car as shown in the Leyland National bus example of Fig 54. This vehicle was also studied extensively from the interior movement and entry/exit condition for passenger comfort, Fig 55.

Computerised approaches to ergonomic system design

The advance of computer simulation into ergonomic package and driver-control layout has revolutionised the interior design process. Standard codes are now available which enable the package to be readily related to the vehicle structure and the layout of mechanical systems.

The Systems Engineering Dept at GM has been prominent in the use of ergonomics to optimise occupant-packaging design of vehicles. The techniques rely on a human factors database which has been built up over many years. Studies on hundreds of human subjects have resulted in standardized drafting templates, leading to analysis of the occupant's space envelope within the vehicle, such as that in Fig 56. As well as the static anthoprometric measurements, such as those described in the previous section, so-called functional, task-oriented, measurements are taken with the human body at work, in motion, in driver- and passenger- space attitudes. These are expressed in three-dimensional co-ordinates around the major reference points, Fig 57a and b. Key points are H (hip), BOF (ball of foot) and AHP (accelerator heel point); such procedures are now standardised under SAE codings such as J1100.

Selection of design limits is carried out to determine what percentage of the population is to be accommodated and to arrive at the amount of seat adjustment required. When seat position is plotted against percentile, an almost linear portion comprises the central 90% of the sample, Fig 58, and in

Fig 55: Entry conditions for Leyland National bus

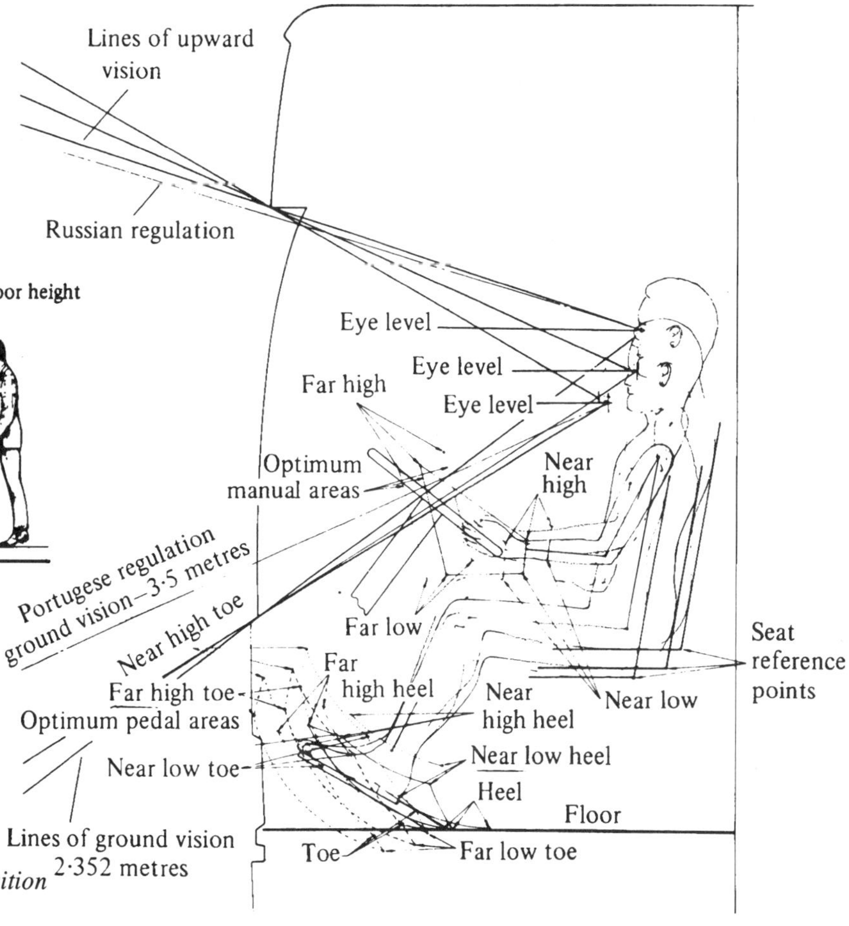

Fig 54: Bus driver seating position

this example horizontal adjustment of 168 mm, if properly positioned, will accommodate this range. The difficulty of simply placing a fifth percentile female manikin on a vehicle drawing, in order to obtain the fifth percentile reach diagram, say, is that the interaction of anthoprometrical variables with the workspace cannot always be predicted.

This has led the SAE to a task-oriented statistical approach involving the development of functional anthoprometric models, including computer-aided-design tools. These describe 'probable' location of the reference points, expressing design limits in percentile levels of accommodation, the table in Fig 59 providing an interesting comparison between task-oriented models and manikin ones. Such techniques were used to draw up SAE J287 Driver Hand Controls Reach (1990) — and a typical envelope is seen in Fig 60. As well as statistical models of the main reference points, other tools used include H-point machines such as that in Fig 61 and vehicle dimensioning procedures that provide standard measurements as a database for analysis and comparison of vehicle interior space.

Usually the packaging engineer develops the vehicle from the ground up, the seat height and inclination being variable to vehicle category between, say, sports car (low) and van (high). Ground clearance and underframe structure define the floor level — which locates accelerator heel point and so on. Finally, sight lines are drawn from the eyelipse and roof clearances measured to the head-position

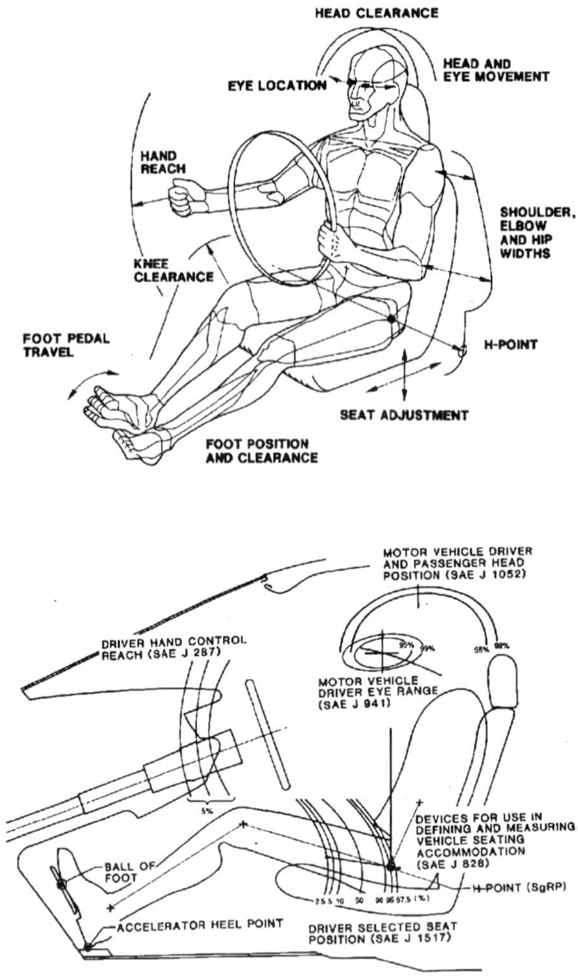

Fig 57: Major reference points

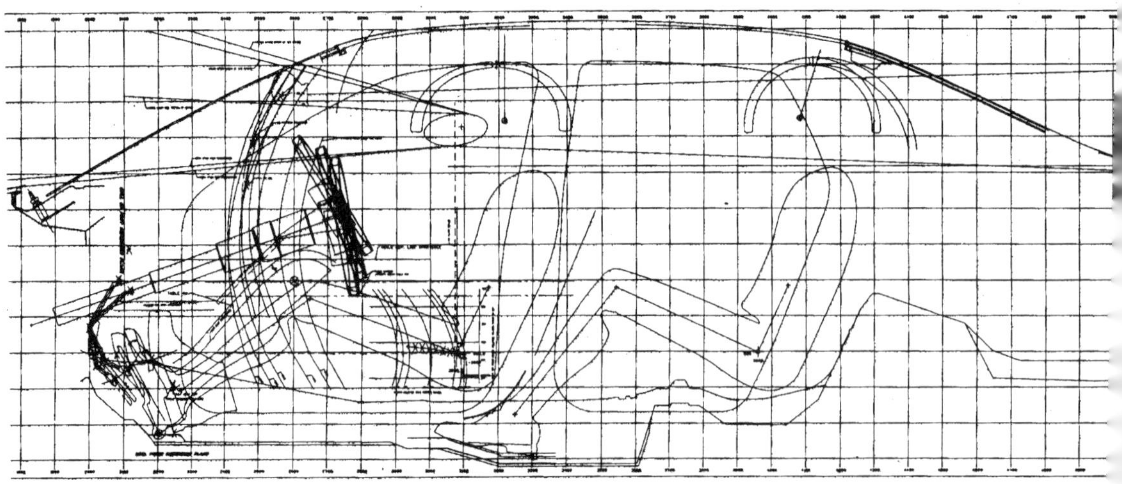

Fig 56: Interior space envelope

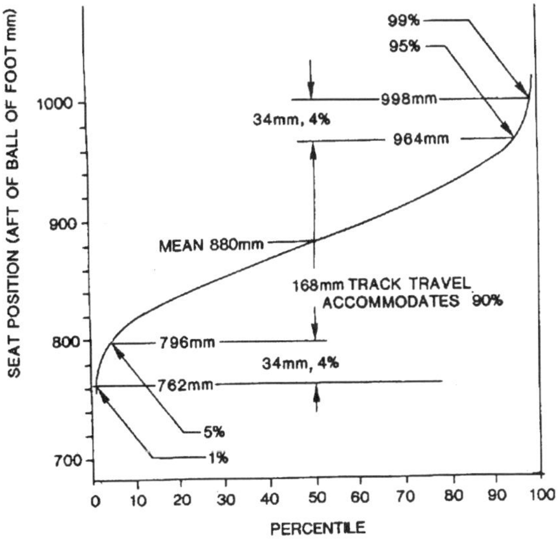

Fig 58: Percentile plots to find seat adjustment

Task oriented percentile models	Manikin oriented models
Uses anthropometric data to define a user population stratified across stature, weight and age.	Uses anthropometric data to define 95 per cent male and fifth percentile female dimensions.
Conducts tests for specific tasks (reach, eye location) in vehicles to develop statistical models defining spatial requirements.	Uses manikins and/or selected large (95 per cent) males and small (5 per cent) females to define spatial requirements.
Assumes that sample populations define user requirements.	Assumes that specified large males and small females comprehend all user requirements.
Assumes user needs are expressed by a central tendency (usually a normal Gaussian distribution) with exclusions at both ends, for example 95 per cent accommodation with 2.5 per cent excluded at each end.	Assumes that 2-D (or 3-D) manikins can predict or model human requirements
	Assumes that a given percentile person is definable from the sum of the parts.
Many levels of accommodation are described, but only for the task studied.	Many tasks are measured and evaluated, but only for a defined small and large user.
Results in well defined *statistical* model that defines accommodation levels for *specific tasks.*	Results in questionably defined *geometric* manikin models that pedict accommodation for only *two extreme percentile people.*

Fig 59: Task-oriented vs manikin models

contours, and the previously discussed reach-clearances included.

One of the most sophisticated computer-aided-ergonomics packages is SAMMIE-CAD, now being developed by Loughborough University. Models are constructed from primitive shapes such as cuboids, spheres, cylinders, meshes and polyprisms which can be specified interactively, by name and key dimension. Irregular solids can also be modelled and curves are represented by multi-facets. A reflection facility allows mirror-image copies to be made, which simplifies defining the second side of a vehicle, say. The program's workplace modeller is strong in its ability to specify logical or functional relationships between components, important for the operation of doors, pedals and adjusters — by the man models of Fig 62. The man model is constructed from 17 pin-joints and 21 rigid links encased in solid modules of 'flesh'; link lengths and joint constraints are data-driven and can easily be controlled by the user. The designer interacts with the system via a menu-driven interface, some 30 menus each having up to 20 commands. The vehicle model is built up from data as previously described and can be complex or relatively simple, as in Fig 63.

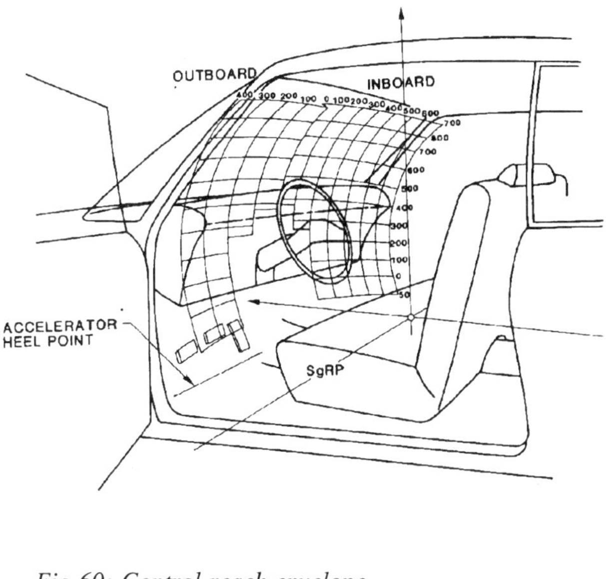

Fig 60: Control reach envelope

Measuring seat comfort

Researchers at Ford have constructed sensitive manikins for measuring seat comfort which are based on a distribution of small load cells over the contacting surface. This leads to pressure distribution 'contour maps' such as that in Fig 64. Such an apparatus is used in conjunction with special seats whose cushion-springing can be adjusted by means of helical springs distributed across the surface which can be adjusted to different rate settings. A computer analysis procedure then co-relates the two distributions for a variety of human test subjects. An optimum comfort distribution, as shown in Fig 65, is then reproduced as closely as practically possible, on the manikin. The manikin can then be used to rate different seating systems for comfort level.

Displacement rather than pressure is also used as a measurement criterion for seat comfort in vehicles such as AFVs and racing cars where the driver needs to be held still by a fairly elaborate support mechanism. In work carried out by Southampton University, a lateral backrest support was constructed from foam blocks for this purpose, Fig 66, and the motion of the driver's head measured by an ultrasonic displacement transducer. Subjects were subjected to 240 seconds of Gaussian random lateral (y-axis) vibration at 1.0 metre/sec/sec rms with a frequency bandwidth of 0-3.15 Hz, in stages, while backrest height was altered in 100 mm increments between 0 and 700 mm. Probability density functions of head displacements were calculated from the time histories given by the ultrasonic transducers and the mean distribution across the 16 subjects for the eight seating conditions is shown in Fig 67.

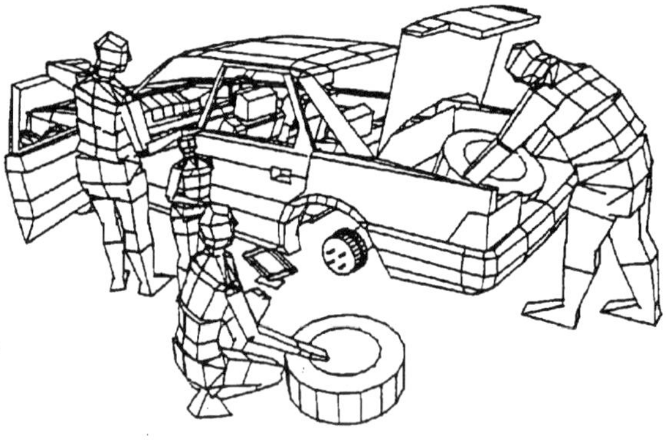

Fig 62: SAMMIE man models

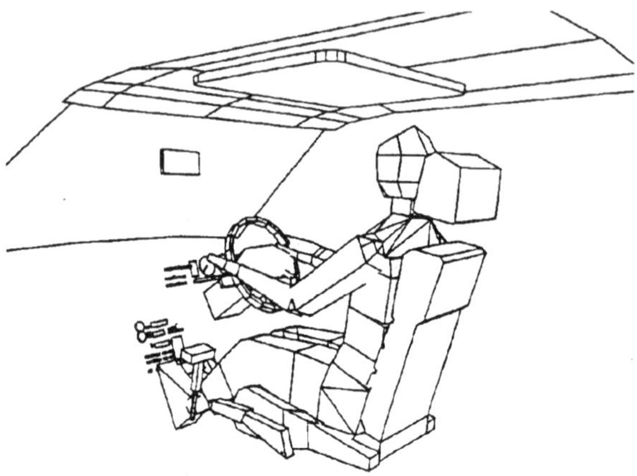

Fig 63: Simple vehicle model

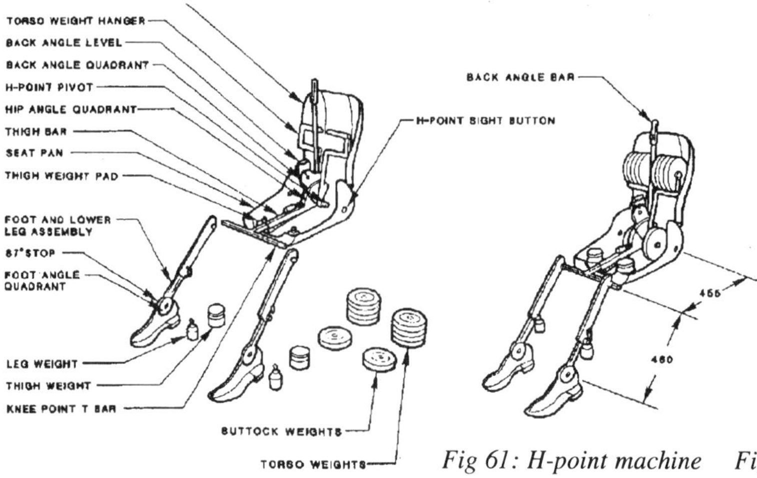

Fig 61: H-point machine

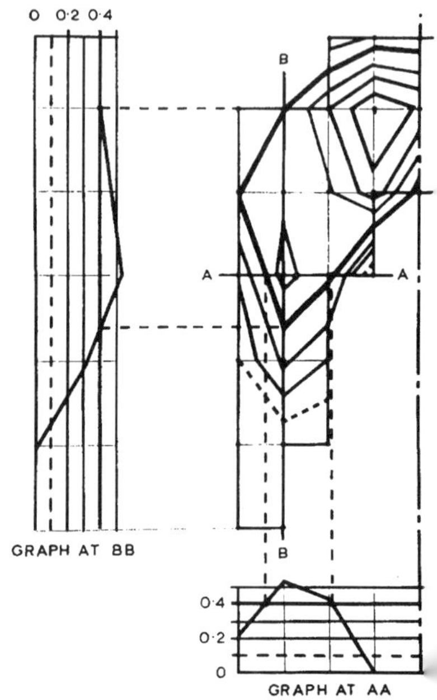

Fig 64: Pressure distribution contours

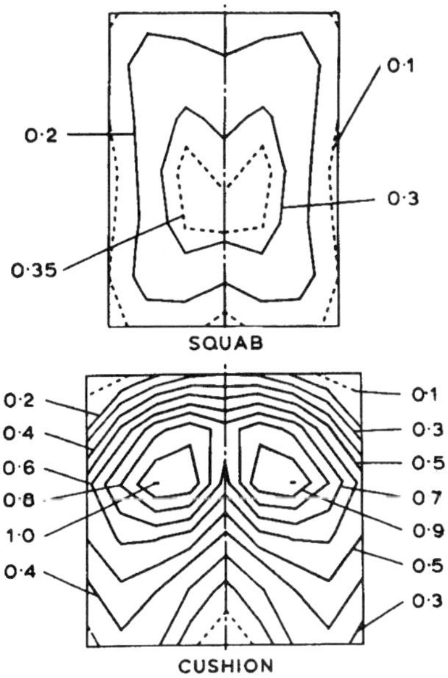

Fig 65: Optimum comfort distribution

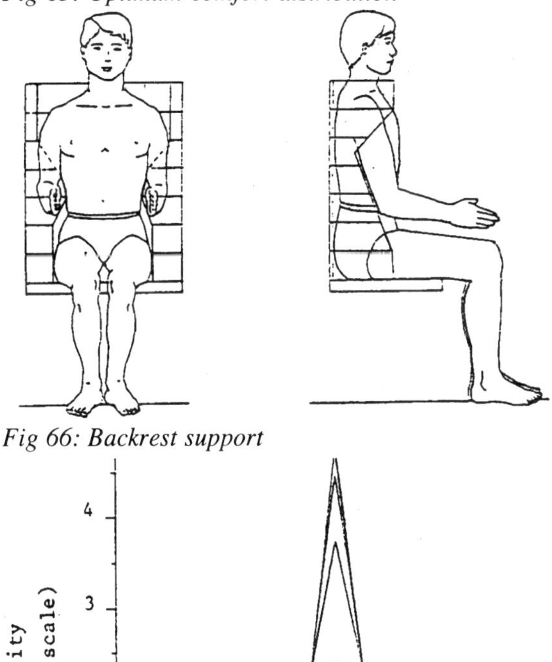

Fig 66: Backrest support

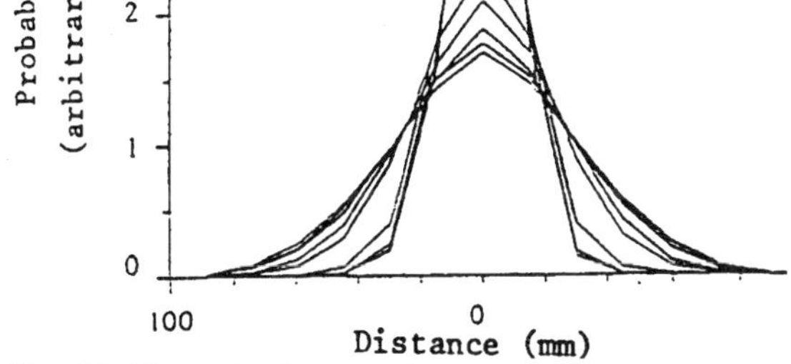

Fig 67: Head displacements measured

Dashboard instruments

The controversy continues as to whether car-customers and CV-operators prefer traditional or 'high-tech', fully or minimally informative, dashboard instrumentation. The answer which the volume vehicle builders continually fail to address is that different customers/operators have different preferences.

Conventional wisdom for the volume builders is provided by theoretical ergonomists and industrial psychologists who seek to generalise human behaviour and present some sort of mythical 'mean' person for product planners to try and satisfy. While the specialist vehicle marketing man will quickly recognise the limitations of such an approach, such research should not be dismissed out of hand because most systematic investigation is likely to yield data which can be selectively exploited in design.

Early ergonomic work was carried out for military purposes in order to optimise the information systems available to those who controlled aircraft and armoured fighting vehicles and this is exemplified by the list of requirements produced by University of Wales researchers in the early 1960s, reproduced as Fig 68. This helps to classify, and rate for performance, all the different information displays on the dashboard. While digital displays are excellent for accurate visualisation of information, their ability to display comparatative values is severely limited.

Speed recording is really best presented by a circular instrument with the pointer angle aiding assimilation. Interestingly, expert studies have shown black figures on a white dial are preferable to the reverse because dark backgrounds tend to impair vision. Otherwise very bold markings are needed and where colours are used, strong contrast is necessary.

'Experts' will argue that oil-pressure, oil-temperature and engine-speed are of value only to those who can properly interpret them. While in itself this is an acceptable view, many would argue that these instruments have immense educational value to the uninitiated — one only has to suffer the expense of engine failure once on a high performance vehicle to learn the importance of maintaining oil pressure.

One of the benefits of re-configurable solid state displays is that they give the opportunity to call up a subsidiary panel that could give more definite information as to whether temperature is too hot. On rear-engine water-cooled engine vehicles, in particular, it is important to signal this information as drivers are likely to be unaware of escaping steam in this situation. The argument could be extended, too, to 'battery charging' indication as many specialists might prefer to know, as well, the rate of charge — and even a switch-out system that would register the rate of discharge (current consumption) and perhaps even the state of charge of the battery, on a selectable basis.

Makers of ammeters have pointed out the very wide range of information that can be learned from that instrument for the discerning driver. The ability to check unduly long initial charge rates (indicating wrongly set output controllers) or inadequate running charge rates (indicating impending electrical failure due to slipping fan belt) could be valued by some drivers. It is also sometimes forgotten that the ammeter is also a good check of ignition function — as slow cranking of the engine will register a fluctuating discharge rate to indicate satisfactory contact breaker functioning.

Experts rightly advise of the safety aspects of providing too much information on the main displays. As the single-channel mechanism of the brain has a strictly limited capacity for data transmission, it is wise to relieve it of any checking function — leaving this to either subsidiary or selectable displays. It takes about 1.25 seconds to read a value and 0.75 to make a check depending on the number of excursions the eyes must make. Remember to locate instruments which have to be read, therefore, in the centre of vision and limit peripheral vision of warning signals to within 60 degrees on each side of the visual centre line. Another valuable lesson is to arrange instruments in a logical order of sequence and if all are of the analogue, pointer, type — having their pointers parallel and upright for the most usual operating condition.

Fig 69: Back-lit LCD

Fig 70: VF/LC hybrid display

Solid state displays

Computer graphic technology has great potential for dashboard display systems if it can be used in sympathy with customer needs — and has particularly revolutionised the availability of selectable information. Liquid crystal displays of the type pioneered for pocket calculators are widening their appeal with the advent of back-lighting which has considerably improved their visibility, Fig 69.

Other developments include the integration of

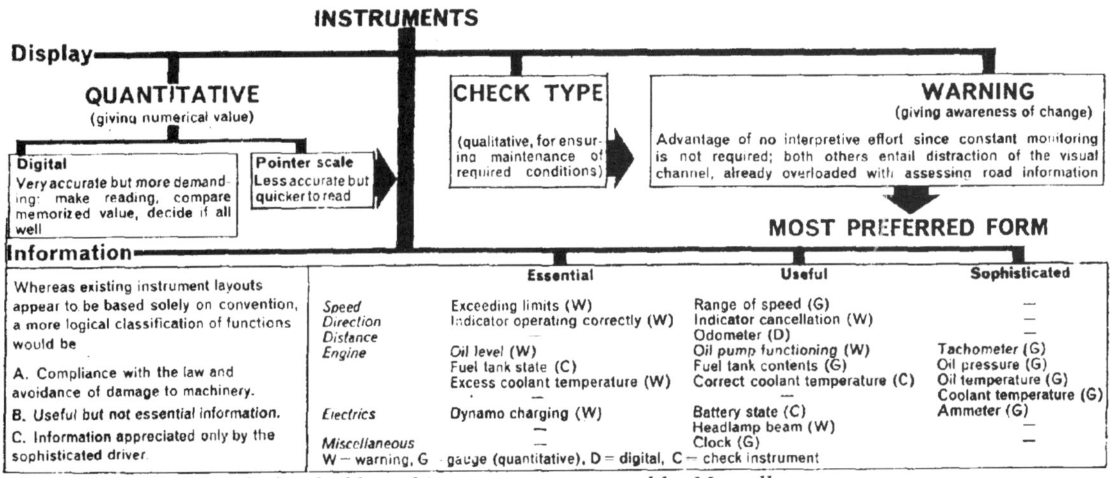

Fig 68: Design rationale for dashboard instruments suggested by Murrell

Fig 71:Test display

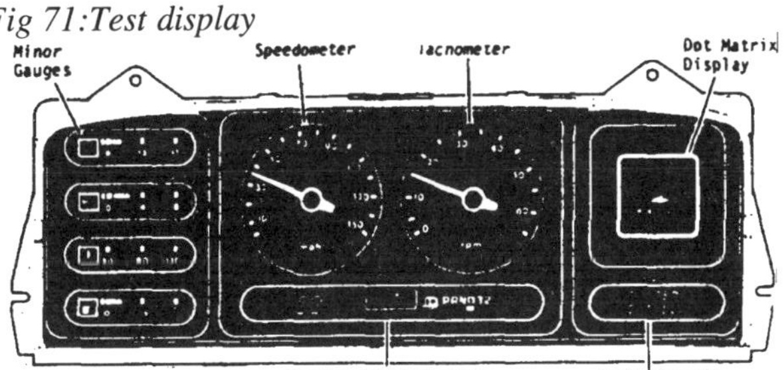

Fig 72: XJ40 display

drive circuits on to the displays so that extensive interconnections between modules will not be required. This technology is now preferred to vacuum fluorescent displays of the type pioneered for the Mark 1 Austin Maestro — though hybrid systems of both technologies are said to be a possibility, Fig 70. Cathode ray tube technology is another important contender and benefits from the considerable development work that has gone into computer monitors. Compact size commensurate with low-cost is one of the problems yet to be solved commercially — and, like LCDs, the advantage of being able to program in the occasional display of less essential information.

The problem of visibility changing on LCDs

MOST PREFERRED

	Men Age (years)			Women Age (years)			Colour	Bifocal
	17-30	31-50	50+	17-30	31-50	50+	Defect	Wearers
Most Attractive	G	G BG	O Y	G	BG Y	BG Y	G	Y O
Easiest to Read	BG G	BG Y	O Y	BG G	BG Y	BG Y	O	Y
Least Distracting	G BG	G BG	Y	G	BG	BG Y	R	BG Y

LEAST PREFERRED

	Men Age (years)			Women Age (years)			Colour	Bifocal
	17-30	31-50	50+	17-30	31-50	50+	Defect	Wearers
Least Attractive	-	R O	R G	O	R	R	R Y	R G
Most Difficult to Read	R	R O	R G	R O	R	R G	R	O R
Most Distracting	O Y	O Y	R	O Y	R	R O	O Y	-

Fig 73: Display preferences

has been overcome by so called 'transflective' designs using a combination of front and backlighting. Background discolouration with temperature change can be developed out and colour displays can be used without fear of predominant colours changing with variation of ambient conditions. In tests carried out by the Institute for Consumer Ergonomics some idea of subject colour preference has been scientifically evaluated. Using the display in Fig 71 preferences were recorded as shown in Fig 72.

Vacuum fluorescence has been successfully used on the Jaguar XJ40 for the warning signals which are seen on a 32x32 dot-matrix display, Fig 73. Below these a 20 character caption or ISO symbol is used as an aid. The trip computer is operated by nine keys to give nine displays and has facility for keying in intended journey length, Fig 74. Bar graphs are used to present fuel-in tank, oil pressure, coolant temperature and battery voltage. The odometer display data is held and updated in an EPROM close to the computer.

Advanced instrument displays

Commonly used forms of twisted nematic LCD have dark figures appearing on a light background. Polarisers on the front and back of the display sandwich long rod-like molecules of an organic fluid which are twisted through 90 degrces from one polarizer to another. When no electric power is applied, light will pass from front to rear polariser plate by a light reflection process. On reaching the polariser plate it is reflected back towards the front of the cell via a multi-purpose surface which acts as a reflector during daylight conditions. In passage the light is twisted again and emerges on the front polariser as a light spot on a bright back-ground, with no image visible. On energising the panel the molecules are altered such that the light is no longer twisted with the result that the rear polariser absorbs the light and a dark spot appears on the light background of the rear polariser.

Contemporary technology employs colour segments with backlighting added for large displays. Fig 75 shows a chip-in-glass fluorescent indicator panel em-bodying such techniques and Fig 76, a suitable operational block diagram. Methods employed to produce a three-dimensional appearance for the alpha-numeric may involve colours (phosphor materials) and filters. When current is applied to the CIG FIP grid the phosphor coated characters emit light which can be filtered to any colours. White light can be generated by a bar segment anode printed with closely spaced blue green and red phosphors.

Multi-colour cells are an important advance which makes use of separately controlled active elemcnts, within each viewed active cell, in a technique known as 'combe patterning'. Each element has a unique fixcd filter which, with alternative or combined activation, will give a perceived colour change. A display can thus be made to adopt a secondary colour when critical levels are reached, Fig 77. According to which basic LCD technology is used, twisted nematic (TN) or super-twisted-nematic (STN), the colour of the base monochrome cells can be created by fixed filters located in a set area to the rear of the cell, Fig 78.

Such a colour ccll may comprise, typically, three grouped pixels, each individually filtered, being individually or simultaneously activatcd, and producing up to seven colours. TN cells, in static drive, are said to offer good contrast, vicwing angle, and colour realisation but for larger areas STN cells are specified where multiplex rates in excess of those controllable by TN are required. Molecules may be twisted in the range 180-270 degrees, some 90 degrees more than in a TN cell. STN cclls are however limited by colour variance and operating temperature extremes. For driver-controlled switching applications, with status information monitoring, the LCD can be overlaid with a membrane touch-sensitive cell. Local pressure on the membrane would activate the function shown on thc relevant part of the display while a single main polling switch would remap the display and the membrane activatcd zones.

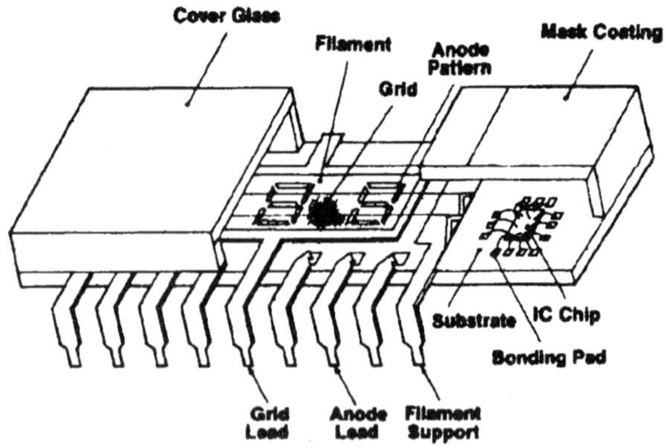

Fig 75: Chip-in-glass fluorescent indicator

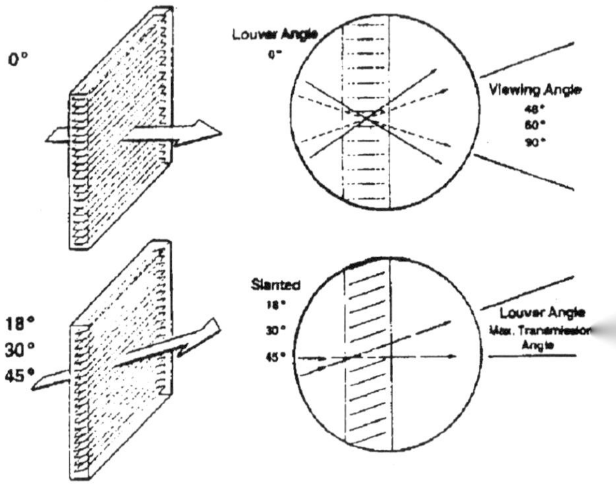

Fig 77: Louvre technique for reducing glare

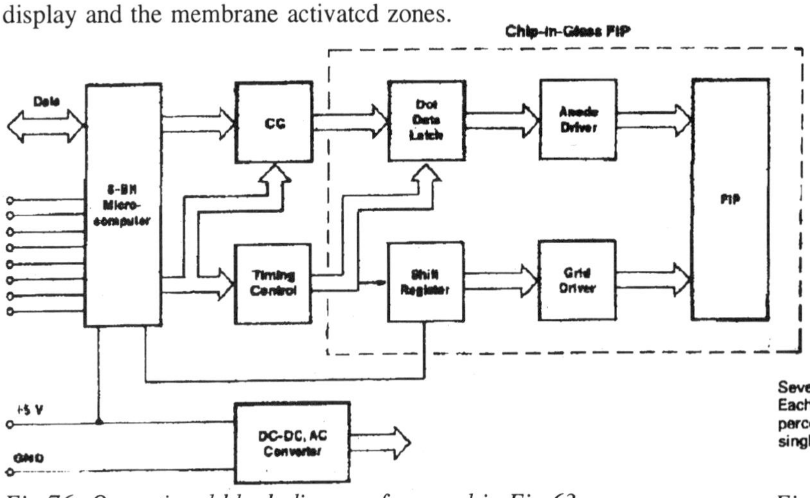

Fig 76: Operational block diagram for panel in Fig 63

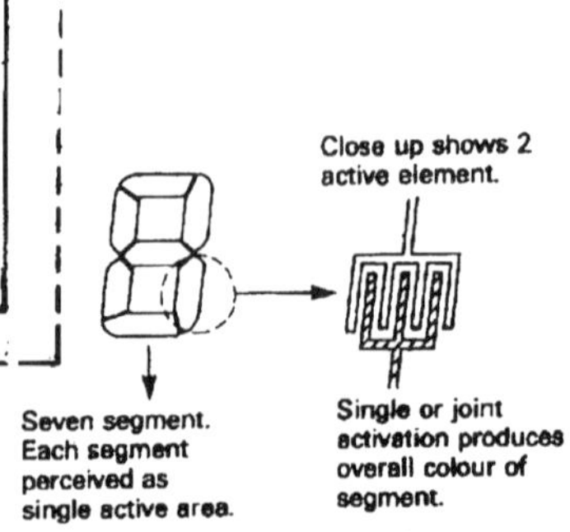

Fig 78: Colour-effect technique

CRTand VF displays

Cathode ray tube displays rcquire primary driver information in a high intensity, high resolution multi-colour image for direct rather than reflected or projected viewing. A suitablc device would be a single-gun CRT with set colour phosphor zones, the primary display driver being a mono graphics controller with a multi-pixel based display driving capability and an integrated serial data link for translating vehicle sensor information. A typical system would involve a display screen energised by a door opening, causing a 'wake-up' signal to display a menu page allowing the selection of all subsequent displays.

Touch technologies available include conductive-membrane, capacitive and scanning infrared beam types. In the resistive membrane system of Fig 79 a resistive coating is applicd to a glass implosion shield in front of the display which has a sheet or metalliscd embossed mylar over its glass. An air gap between thc two layer prevents contact. A grid of electrodes is etched on to respcctivc surfaces in X and Y directions respectivcly. On application of local pressure a short is caused between X and Y electrodes at the contact point and the computer signalled appropriately.

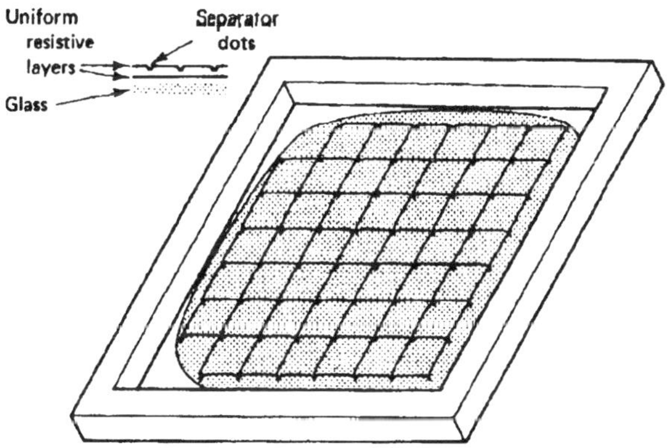

Fig 79: Touch-screen technology

Vacuum fluorescent displays involve a triode, with cathode grid and anode, the filament cathode being a suitably coated tungsten wire. When heated to 600 C, thermo-electrons are emitted towards thc cathode at a rate determined by thc voltage applied to the grid stretched over the anode. The electrons collide with a phosphor layer on the thick film anode to emit light. The technique of 'self-standing grids' allows location of the grid in a position to suit the phosphor pattern. Segments of an analogue bar graph can be progressively illuminatcd by variation of the drive method. When large area displays are required, location and connection of display ele-mcnts can prove difficult and a solution is to display information on to thc unused portions of the instrument front face by means of a mirrored reflection area in the front and dark toned areas at thc rear of the display lens. Transparent symbol slides may be uscd to projcct appropriate images. These can also be the basis of head-up displays which allow the driver to focus on the road-line. Here a VFD may be hidden within the fascia and information projected on the windscreen, suitably coatcd with a transparcnt film

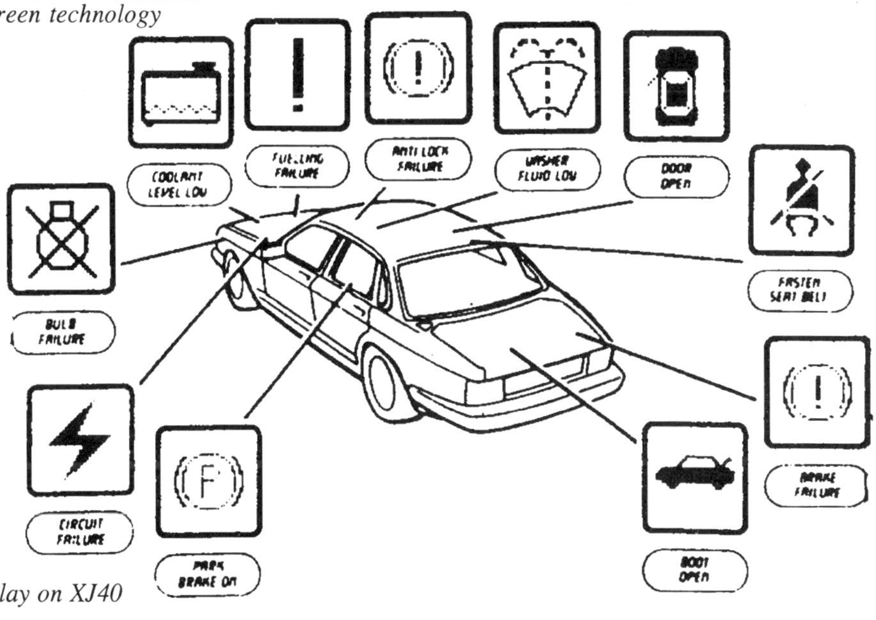

Fig 74: Icon display on XJ40

acting as a half-mirror.

Illumination of displays in future could well utilise new light-emitting plastics, exploiting the technology of electro-lumincsccnce. This permits the emittancc of light when an applied AC voltage excites chemical compounds placed betwcen metal electrodes. Reducing glare from electronic instruments has been carried out by light control film (LCF) techniques. This is an optically transparent film having uniformly-spaced micro-louvres, Fig 80. Louvres shield the unwanted light from the driver's eyes. VDO are probably now the main manufacturer of automotivc instrument displays. The company has been active in liquid crystal displays with negative contrast — bright segments on a dark background — with light provided by transflective design in which a combination of front and back lighting is involved. The company have developed a technique whereby the cell appcars neutral in colour, under transmitted back light, due to a normally grey background coloration. Contrast of the display is almost indepcndent of the indicator colour chosen. This makes it an attractivc proposition for multi-colour displays without the necessity for black-masking outside the active driven areas. As seen in Fig 80, the display can also be supplied in non-rectangular formats to improve viewing through the steering wheel.

In providing secondary displays such as warning messages, the company can meet the requirement for alpha-numeric characters. When symbols are required, a dot-matrix presentation is usually necessary. Fig 81 shows examplcs from the available range which may include arbitrary graphics as well as symbols. Pulsating displays can also be used to animate the action of a windscreen wash. Long texts may appear running vertically in a roll-mode or horizontally as a moving screen. Warning symbols can also be made to appear alternatcly in orange or red, say, to signify the urgcncy of the warning. Dot-matrix formats also open up thc possibility of providing further displays in thc periphcral view of the driver in a different form than is possible on the central speedometer and associated display, Fig 82.

CRTs, EL and VF displays are luminous (active) and so need a sufficient supply of electrical energy. While EL and VF displays typically rcquire an installation depth less than 40 mm, CRTs require a depth equivalent to their screen diagonal and contrast can be impaired by bright ambient light. However, they offer the greatest information density due to their high resolution. By comparison LCDs are passive displays and thc liquid crystal really acts like an electrically controllable shutter for external light, thus dictating the need for backlighting. Power consumption is comparatively low and involves 1 to 10 microwatts per square centimetre. Negative displays, dark segments against a light background, while providing a more attractive optical effect , cost more to manufacture than positive ones. Contrast and range of viewing angles of an LCD abatc with increasing multiplex rates, required for increasing reso-lution.Displays driven at multiplexing rates up to l:16 are currently used in vehicle applications, providing 80x200 image dots for adequate readability. Siemens now offer the TTD display which offers three levels of information prcscntation. Two outer zones display time and outside temperature while the date is shown in the central zone, with an 8-figurc LCD screen; when the radio is switched on, this is over-laid with the transmitter frequency.The size of the display is 132 x 26 mm, with 180 individual segments, to provide clear legibility. Thc central zone display allows the radio to bc remote mountcd and at night its brightncss is automatically dimmed to reducc brightness, Fig 83.

Fig 81: Neutral-background display

Fig 82: Symbols available from VDO

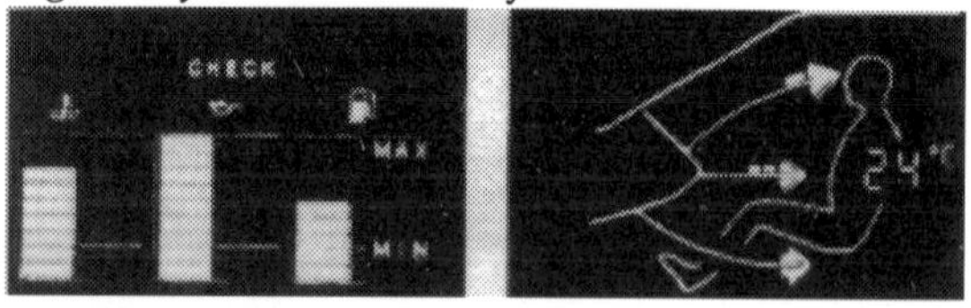

Fig 83: Peripheral-view displays

Fig 84: Siemens TID system

Legal weights and dimensions for goods vehicles

Trucks differ from cars and buses in that their size and shape is very much dictated by commercial and legal considerations. Construction and Use regulations define the UK legal dimensions of goods vehicles. While this section may help the interpretation of statutory instruments published by HMSO, it is important to refer to the original instruments as frequently as possible to record the frequent changes that take place.

The UK and Ireland were provided with a derogation by the EEC which exempts them from EC directives for heavy trucks until the end of 1998. However, weight increases have already been allowed on such categories as four-axled articulated vehicles and drawbar units.

L = 12 metres B = 2.5 metres (with 2.75 concession for heavy tractors and 2.6 for fridge vehicles whose side walls are at least 45 mm thick)
R = 0.6wb (for refuse vehicles, works trucks and tippers R = 1.15 maximum, regardless of wb)
W_A = 4.6 tonnes for one wheeled axles (or 5.09 with wide/twin tyres)
W_A = 9.2 tonnes for two wheeled axles (or 7.12 for steered axles without load compensation)
W_A = 10.17 tonnes for twin-tyred two wheeled axles (or 10.5 for single wide tyres)
W_G = 14.23 tonnes for vehicles less than 2.65 metres wb
W_G = 16.26 tonnes for vehicles at least 2.65 metres wb
W_G = 17.00 tonnes for vehicles at least 3.0 metres wb
For three and four axled rigid vehicles:
W_G = 16.26 tonnes for 3-axled vehicles (increasing with wb as in Table 1)
W_G = 18.29 tonnes for 4-axled vehicles (increasing with wb as in Table 2)
$R_{6/8}$ = 0.6 $wb_{6/8}$ (overhang measured from 110 mm rear of bogie mid-span)

Fig 85b: Rigid vehicle dimensions

Tractor/trailer combinations
For trailers generally:
B = 2.5 metres (or 2.3 for tractor GVW < 3.5 tonnes)
B = 2.6 metres for fridge trailers with walls over 45 mm thick
For drawbar trailers:
L = 12 metres for W_G>/= 3.5 tonnes (or 14.04 metres for composite — dolly + semi-trailer)
L_C = 18 metres W_G = 18 tonnes for wheelbase above 3 metres
W_{GT} = 24.39 tonnes (or 32.52 tonnes with power-assisted brakes)
W_{GT} = 40.65 tonnes when drawn by 'locomotive'
For articulated combinations:
L = 15.5 metres (or 16.5 if unit can turn within concentric circles of 12.5 and 5.3 metres radii)
L = minimum defined by Table 3 for artics with W_{GT}>/= 32.5 tonnes
W_G = 14.23 tonnes for wb>/=2.0 metres (20.33 for wb>/=3.0 with 3 axles)
W_G = 16.26 tonnes for wb>/=2.4 metres (22.50 for wb>/=4.0 with 3 axles)
W_G = 17.00 tonnes for wb>/=2.7 metres (24.39 for wb>/=4.9 with 3 axles)
W_{GT} = 20.33 tonnes (or upwards according to Tables 3 and 4 for three or more axled tractive units).
H = 4.2 metres for W_{GT}>/=32.52 tonnes

Fig 85c: Tractor/trailer combinations

Referring to the diagrams in Fig 85a, for rigid vehicles generally, dimansions in Fig 85b apply and for tractor/trailer combinations, those of Fig 85c:

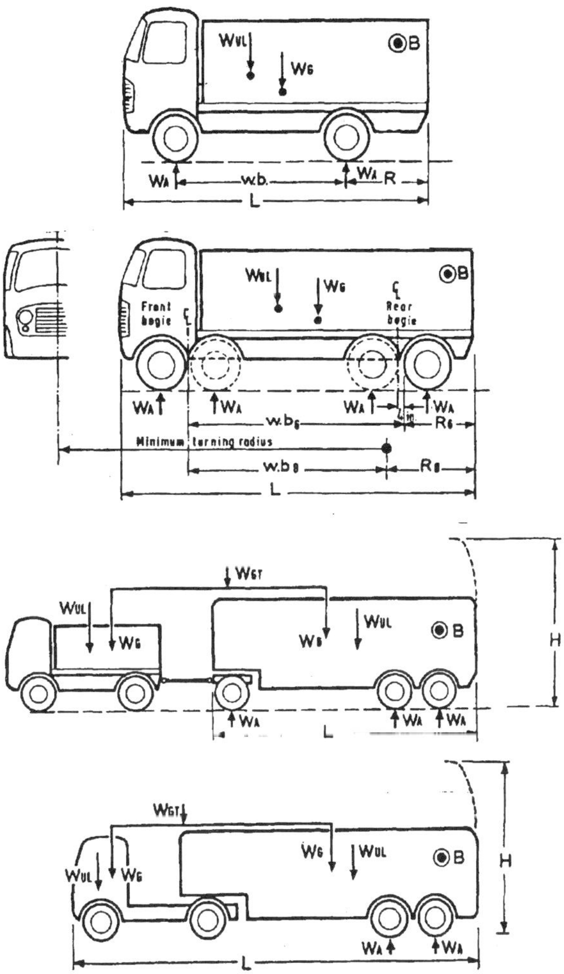

Fig 85a: Truck dimensions

Min. distance between axles (m)	Max weight: - when plated weight of neither axle exceeds half of the specified weight (kg)	- in cases not within previous column when plated weight of neither axle exceeds 10170 kg	- in cases not within either column on left (kg)
1.02	16,260	12,200	10,500
1.05	17,280	15,260	10,500
1.20	18,300	16,270	15,260
1.35	18,800	17,280	16,500
1.50	19,320	18,300	18,000
1.80	20,000	19,000	19,000
1.85	20,340	19,320	19,320

Fig 86: Maximum weights for two closely spaced axles

Axle weights alter for closely spaced tandem axle arrangements according to the data in Fig 86 and the tables of Fig 87.

2 ***Vehicles with four or more axles:***

Must not exceed 18,290kg except as shown below:

distance between front & rear axle at least 3.7m and max. axle weight not more than 8,640kg	**20,330**
distance between front & rear axle at least 4.6m and max. axle weight not more than 8,640kg	**22,360**
distance between front & rear axle at least 4.7m and max. axle weight not more than 8,640kg	**24,390**
distance between front & rear axle is at least 5m and max. axle weight not more than 9,150kg	**24,390**
distance between front & rear axle at least 5.6m and max. axle weight not more than 9,150kg	**26,420**
distance between front & rear axle at least 6m and max. axle weight not more than 9,660kg	**26,420**
distance between front & rear axle at least 5.9m and max. axle weight not more than 9,150kg	**28,450**
distance between front & rear axle at least 6.3m and max. axle weight not more than 9,660kg	**28,450**
distance between front & rear axle at least 6.3m and max. axle weight not more than 9,400kg	**30,490**
distance between front & rear axle at least 6.5m and max. axle weight not more than 9,660kg	**30,490**

1 ***Three-axled vehicles:***

Must not exceed 16,260kg except as shown below:

	kg
distance between front & rear axle at least 3m	**18,290**
distance between front & rear axle at least 3.2m and max. axle weight not more than 8,130kg	**20,330**
distance between front & rear axle at least 3.9m and max. axle weight more than 10,170kg	**20,330**
distance between front & rear axle at least 3.9m and max. axle weight not more than 8,640kg	**22,360**
distance between front & rear axle at least 4.6m and max. axle weight more than 10,170kg	**22,360**
distance between front & rear axle at least 4.9m and max. axle weight not more than 9,400kg	**24,390**
distance between front & rear axle at least 5.1m and max. axle weight more than 10,170kg	**24,390**

5

Relevant axle spacing (m): where motor vehicle has 2 axles	*where motor vehicle has at least 3 axles*	*Max. weight (kg)*	*Min. overall length (m)*
less than 2.2	less than 2.2	20,330	-
at least 2.2	at least 2.2	22,360	-
at least 2.6	at least 2.6	23,370	-
at least 2.9	at least 2.9	24,390	-
at least 3.2	at least 3.2	25,410	-
at least 3.5	at least 3.5	26,420	-
at least 3.8	at least 3.8	27,440	-
at least 4.1	at least 4.1	28,450	-
at least 4.4	at least 4.4	29,470	-
at least 4.7	at least 4.7	30,490	-
at least 5.0	at least 5.0	31,500	-
at least 5.3	at least 5.3	32,520	-
at least 5.5	at least 5.4	33,000	10.0
at least 5.8	at least 5.6	34,000	10.3
at least 6.2	at least 5.8	35,000	10.5
at least 6.5	at least 6.0	36,000	11.0
at least 6.7	at least 6.2	37,000	11.5
at least 6.9	at least 6.3	38,000	12.0

3

10.0m = 33,000kg pmw
10.3m = 34,000kg pmw
10.5m = 35,000kg pmw
11.0m = 36,000kg pmw
11.5m = 37,000kg pmw
12.0m = 38,000kg pmw

4

Number of axles of motor vehicle	*Intermediate axle weight (kg)*	*Spacing between rear-most axle and foremost axle (m)*	*Max weight transmitted to road by all wheels of vehicle (kg)*
2	-	at least 2.0	14,230
2	-	at least 2.4	16,260
2	-	at least 2.7	17,000*
3 or more	not over 8,390	at least 3.0	20,330
3 or more	not over 8,640	at least 3.8	22,360
3 or more	not over 10,500	at least 4.0	22,500
3 or more	not over 9,150	at least 4.3	24,390
3 or more	not over 10,500	at least 4.9	24,390

Fig 87. Dimension tables

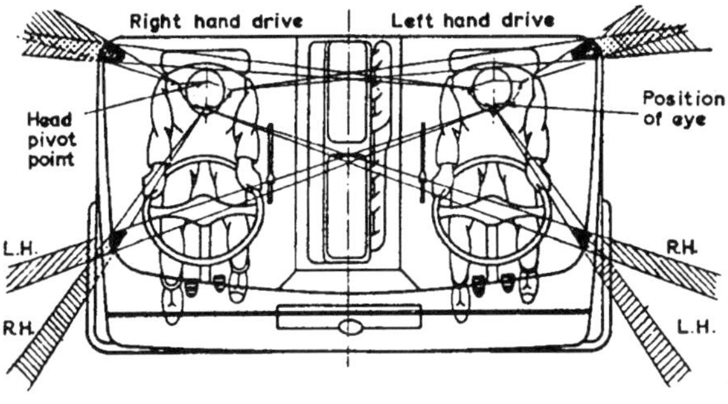

Fig 88: Handing of cab

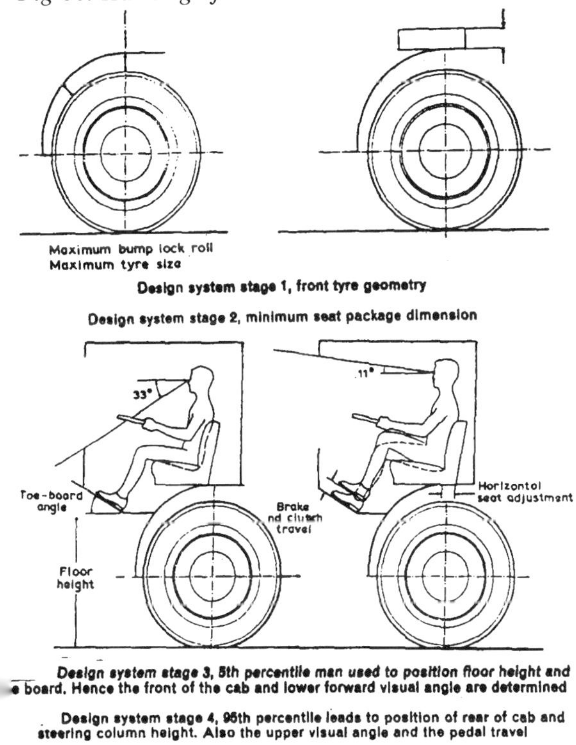

Fig 89: Design scheme for basic layout

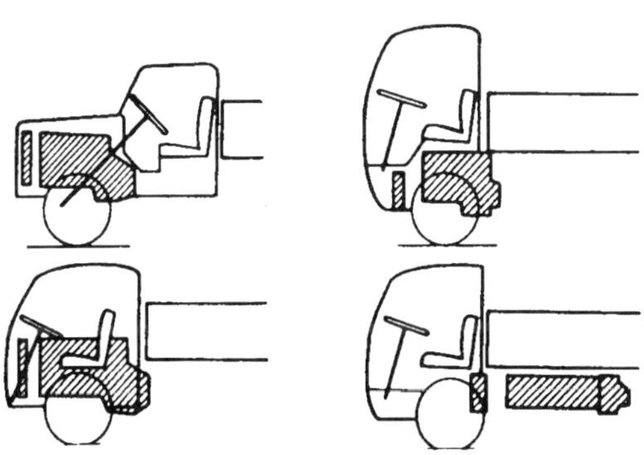

Fig 90: Effect of engine position

Truck cab ergonomics

Given the constraints of legal dimensions on cab layout discussed in the previous section, the approach to ergonomics is similar to that on cars. However, while very necessary improvements have been made to cab comfort and convenience, in recent years, to provide the standards of a private car, it is important to realise that there are some fundamental differences in ergonomic design required of the two workplaces. For the long-distance CV driver in particular it is imperative that proper dimensional relationship between seat and controls are obtained to minimise fatigue and make the working environment as pleasant as possible.

According to Loughborough's HUSAT Research Institute, advances in driver information systems are likely soon to revolutionise the ergonomic design of vehicle cabs. The ability to travel at higher speeds in an increasingly more complex road and traffic network is also an important engine of change. Systems for route navigation, guidance, anti-collision, driver-status (drowsiness) monitoring, parking aids and auto-policing systems are all on the horizon. Research into head-up and head-down displays is well underway in helping to alleviate these problems. In considering driver information systems, it is essential to configure these so that they are a real help to drivers and not not overload them with redundant information. Tests on design concepts should of course be made with real drivers and not information scientists! It is also vital not to test the systems in isolation but as an integral part of the driving task.

The earlier section defined the basic criteria for posture in the driver environment — with the accent on car layouts. The use of servo-assistance for steering and braking has, of course, meant that the CV cab environment is trending towards the car one as the control efforts between the categories tend towards one another. This has been reflected in the steepening of steering wheel rake and postural relationship with pedals based on improved positional control rather than the application of heavy pedal efforts. However the fundamental differences in construction between cars and heavy commercials often means the CV driver is often placed much higher with relation to the road and yet has to make frequent dismounts in the case of delivery operations. Much larger planform areas, bulkier bodies and restricted rear vision also differentiate the CV cab — not to mention the intrusion of a large engine and the need to operate a multi-speed gearbox.

The need to meet 'handing' requirements of international operation (Fig 88) affects both categories of course but the physical size of the engine can make the distance between LH and RH driving positions greater. However, the picture shows how careful siting of the pillars can maximise the field of view. A design procedure established by Leyland engineers for fixing the basic dimensions of a cab is shown in Fig 89. The procedure would, of course, have to be modified according to the engine position (Fig 90). Sometimes a modular construction, Fig 91, is used to build different versions of a basic design without serious cost penalty. The dimensions relating driver to his/her seat and the vehicle controls have been recommended as in Fig 92.

A new approach to cab vision problems is in the use of three-dimensional graphics computer programs such as SAMMIE. This allows considerable evaluation resource, Fig 93, prior to committing to metal. It can involve constructing a grid on up to all six sides of a cube around the cab design in 3-D so as to plot out exactly what can be seen. Legislation for mirror performance, Fig 94, is defined in terms of the areas to be seen on the ground and these can also be modelled by the computer in evaluating mirror sizes and positions. As speeds and traffic densities increase, greater concern is being shown for cab safety design. One suggestion has been to make the cab as effectively a detachable capsule from the chassis frame which itself would take the impact of a collision while the cab and its occupants would be partially cushioned by its support springs, Fig 95. There were safety considerations, too, in mounting and dismounting from cabs and appropriate dimensions needed to apply to widely ranging cab sizes and types, Figs 96 and 97. The real challenge in cab

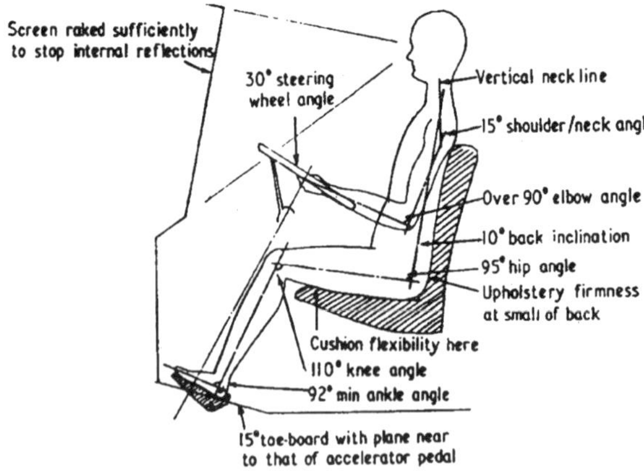

Fig 92: Driver/seat/control relationship

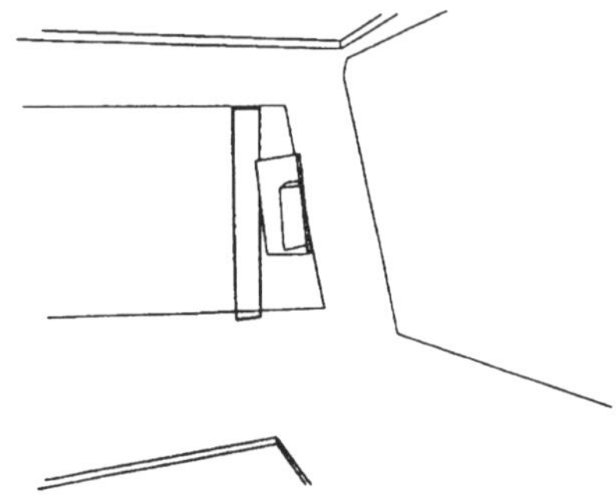
Fig 93: Computer-modelling the interior

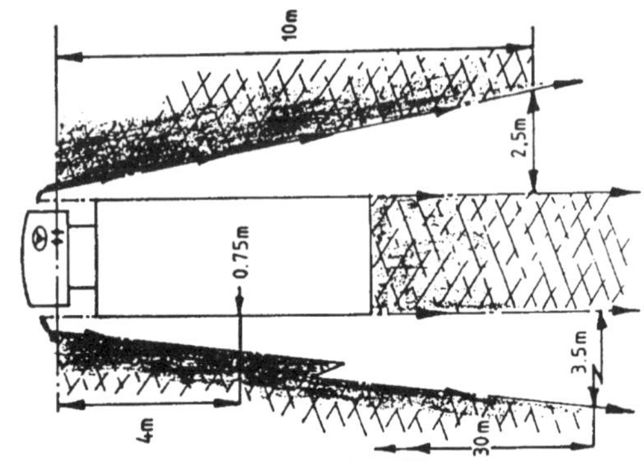

Fig 94: Simulating mirror performance

Segments of the module 1—segments common for all cabs; 2—segments of normal control cab only; 3—segment of forward control; 4—segments of forward control cab; 5—segment of long-distance forward control cab

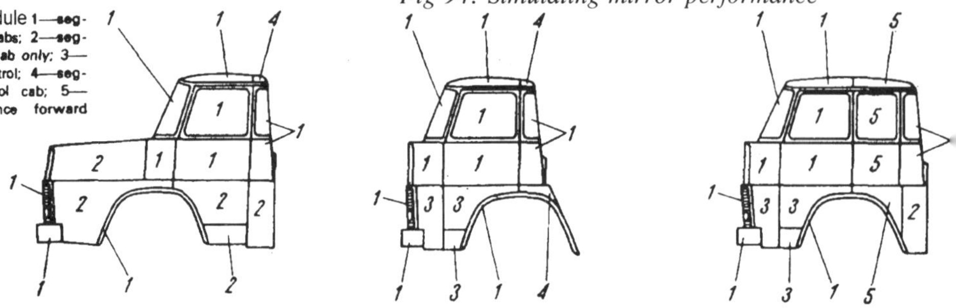
Fig 91: Modular layout

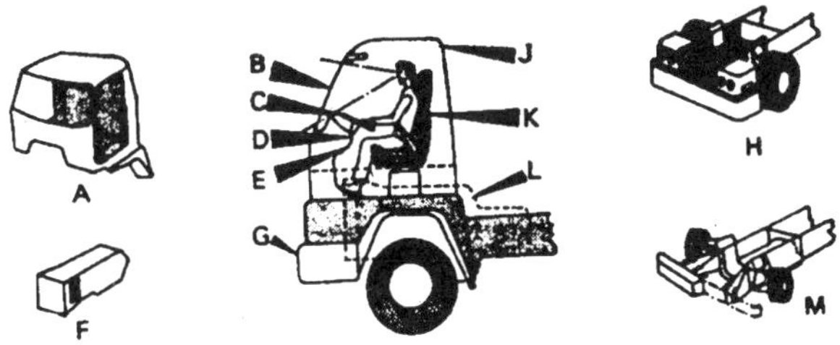

Fig 95: Safety considerations

ergonomics is to provide a workspace that suits a wide range of body statures. For this there is no substitute for a properly executed package drawing layout using largest and smallest percentile anthropometrical data. Pioneering work by Rebiffe portrays the body as a bio-mechanical model consisting of eight links as in Fig 97. To establish control comfort areas the hip reference point (H) for a small person is positioned directly above the contact point between buttock and seat. Hip segment HE is drawn at an angle corresponding to the seat back inclination (25 deg to the vertical), followed by the upper limb links with angles A5 and A6 at 45 and 80 deg respectively. By rotating links EC and CM in turn, the shape Es, Fs, Gs,Hs is traced out in which the hand controls can be positioned. A similar procedure is adopted for the foot controls and the process repeated for the other population extreme, using similar constructions as those given earlier for cars.

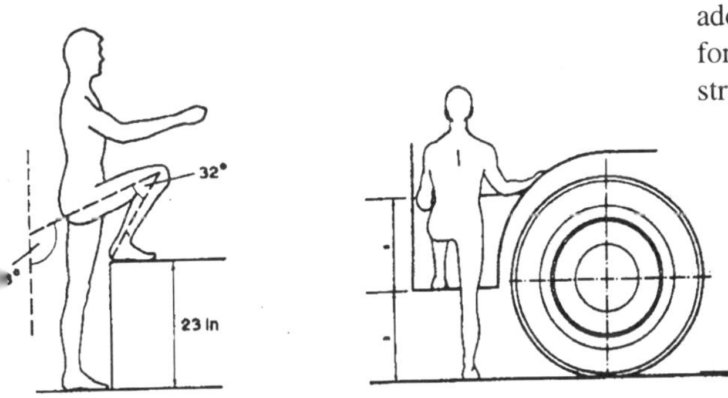

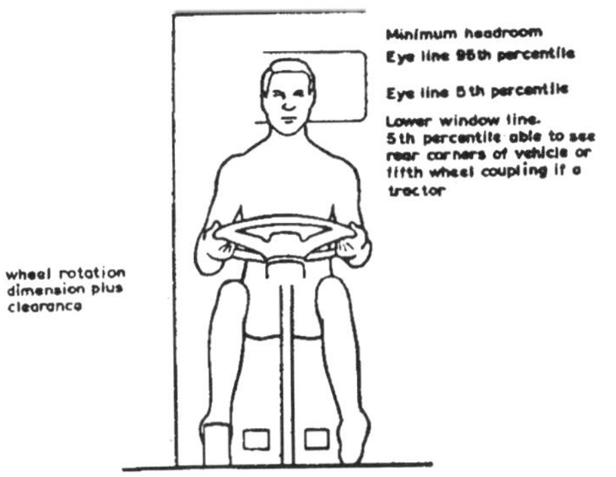

Fig 96: Dismounting dimensions

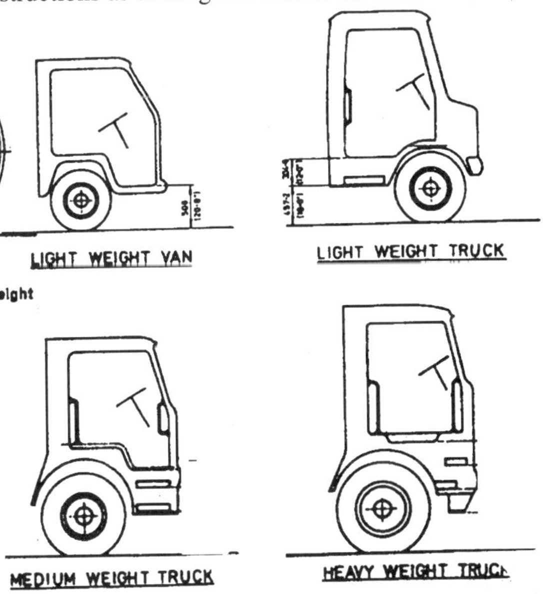

Fig 97: Effect of vehicle type

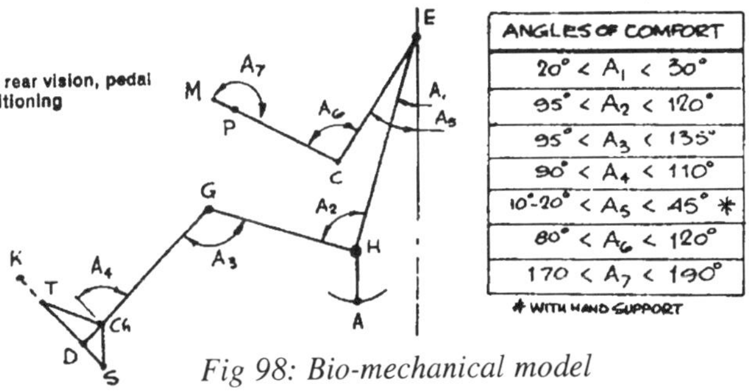

Fig 98: Bio-mechanical model

Mechanical package layout

Important to the body designer is the packaging of the mechanical components in relation to the main operational function of the vehicle. Some of the factors affecting the choice of power train and running gear layout are considered here in relation to the different categories of special vehicle.

An early breakthrough in sports car occupant and mechanical packaging was by the Lotus Europa, shown in Fig 99, whose imaginative package helped to achieve remarkably low frontal area and drag coefficient at the time of introduction. Vehicle weight of around 1350 lb was split 45/55 % front/rear. The central box backbone chassis was a key feature of the design — together with front suspension beam and splayed chassis rear-extensions supporting the power train. The rearward 'mid' engine configuration does, of course, compromise the space required for rear occupants in the conventional 'passenger' car. One of the main debates of mechanical packaging is therefore that of the desirability or otherwise of grouped engine and drives, front or rear, against the use of front and rear drive. Some argue that because of the large clearances swept out by the front wheels in steering, space between them is of little use except for the engine. Basic stability in handling also comes from a concentration of weight at the front — as well as easier engine access, in the case of 'normal control' vehicles. Easier engine cooling and inherently better gear shift linkage is also argued in favour of front-engine/rear-drive. However, front 'grouping' is preferred for proximity of driver to controls as well as better rear-end structural integrity for most body types. Supporters of rear grouping of engine/drive argue that there is no restriction on turning circle as front wheels are free of the engine package between them. Better traction on gradients is also claimed — arising from the greater weight on the driving wheels — and there is said to be a minimum change in centre-of-gravity of the vehicle with change in payload and or passengers. Noise, fumes, heat and exhaust are also more easily removed and there is no problem of having to steer driven wheels. Any grouped system is claimed to be more efficient, arising from the proximity of engine and drive; also the lack of any intrusion by propeller shafting between the front and rear 'axle' positions obviously eases pressure on the passenger space. Fig 100 compares front and rear grouped engines of two former passenger cars, the Morris 1100 and NSU Prinz. The transverse engine layouts are also seen to advantage against the Alpine Renault in-line engine, seen in Fig 101, which encroaches to some extent into boot space. In the Ford Techna concept car of the late 1960s, a variant on the front-engine/rear-drive theme shows that a driveline angled in plan view can alleviate passenger-to-mechanicals clearance problems, Fig 102. In panel vans and LCVs, the conventional layouts are sometimes challenged by new concepts such as those in Figs 103 and 104, proposed respectively by aluminium suppliers ALCOA and steelmakers US Steel.

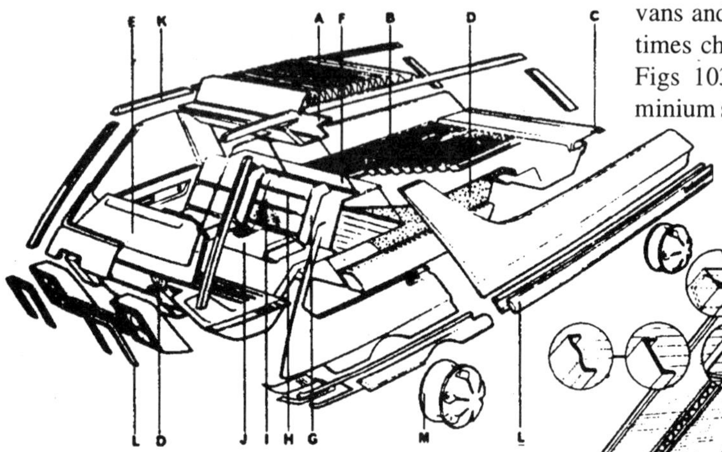

Fig 103: ALCOA concept van

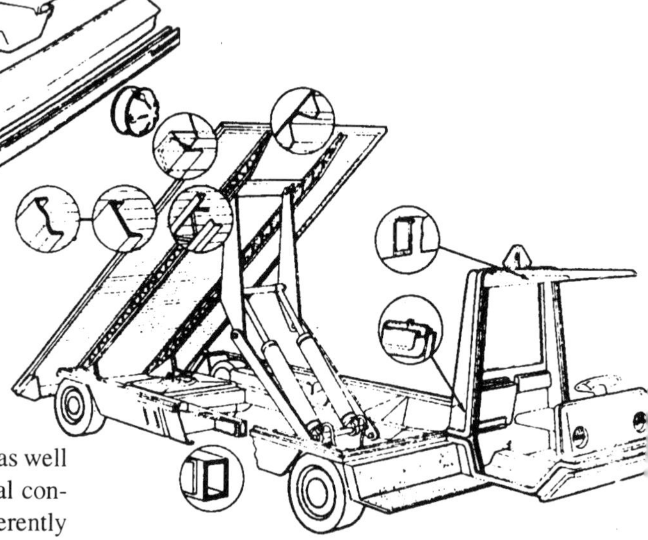

Fig 104: US Steel concept LCV

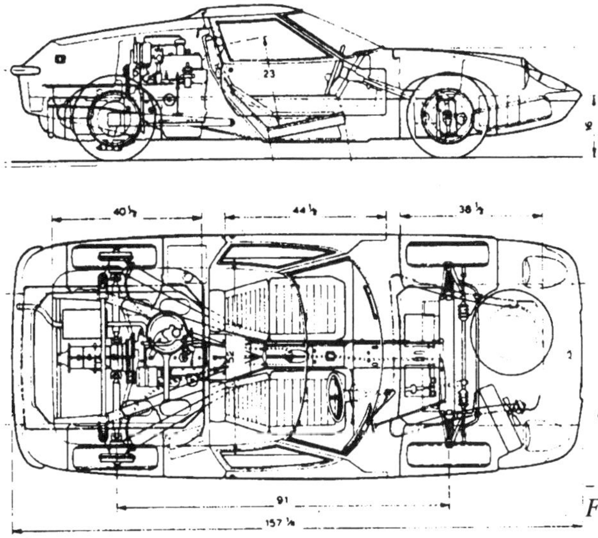

Fig 99: Lotus Europa

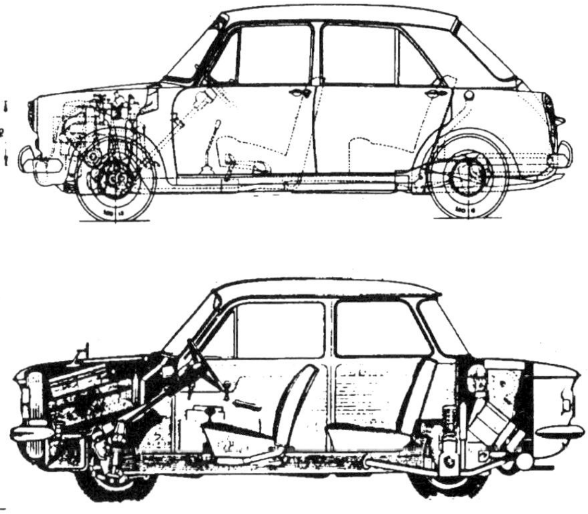

Fig 100: BMC 1100/ NSU Prinz comparison

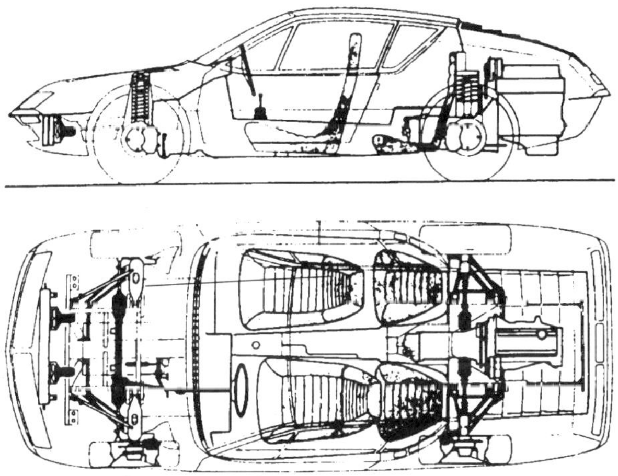

Fig 101: Alpine Renault

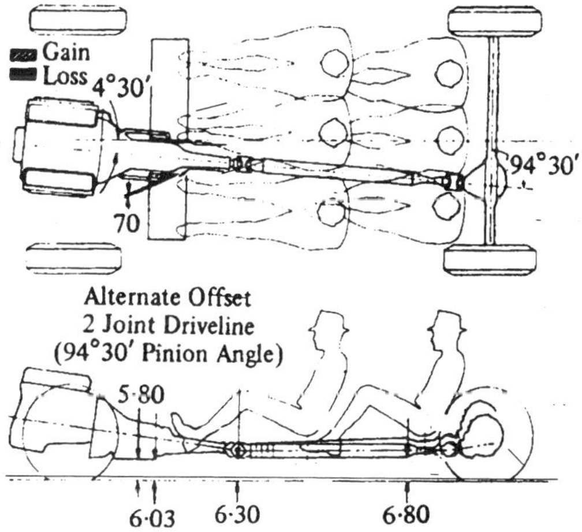

Fig 102: Ford Techna concept

In the first, the chassis (D) is of foam-filled roll-form construction and an extruded tailboard (C) folds flush with the deck for easy loading. (A) is storage area for a sliding tilt while (B) is the deck floor made from interlocking extruded planks. A tubular seat structure (H) abuts a cast frame (G) and supports an aluminium screen mesh (I) around which foam and fabric is bonded. (F) is a storage space, (E) a stamped instrument panel and (K) an extruded windscreen frame. Forged wheels are at (M) and rubber inserts in critical contact areas at (L). The US Steel Venture Wagon uses a flat-four engine in unit with the rear drive to achieve its pleasing low profile. The offset cab mounted on an outrigger permits the carriage of long and awkward shapes.

Goods vehicle layout

Concept commercial vehicles have also been used to show the possibilities for improving the packaging of larger goods vehicles such as the light/medium distribution van and the inter-urban supply vehicle of Fig 105. The Leyland TX450 has unusually low slung power and drive trains to give exceptionally low deck height. It is notable also for the inclusion of a compactly packaged system of independent wheel suspension, front and rear, using low rate air springs. An automatic transmission is also accommodated within the depth of the new-departure fabricated aluminium alloy chassis frame. In general, however, the mechanical packaging of the heavier on-road goods vehicles in current production is unlikely to alter because of the major OEMs' need to keep up volumes to maintain economy and hence the need to retain standardised layouts so as to appeal to as wide a range of end-users as possible.

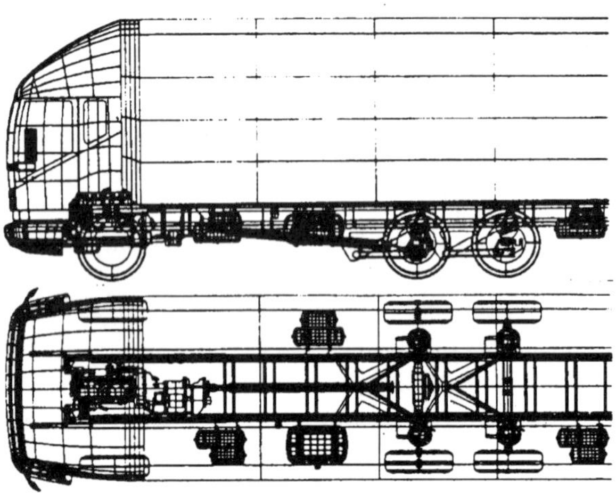

Fig 105: Leyland concept truck

Scope for specialist delivery vehicles

Considerable opportunities therefore exist for special-purpose-vehicle builders to develop designs for specific operations. The large-capacity delivery vehicle is a case in point where, already, imaginative cab designs by the major OEMs, as well ingeniously thought out box-van body designs by bodybuilders have achieved good compromise solutions. The optimum solution, however, might await the commitment of that special purpose vehicle builder who is able to configure the mechanicals, and integrate body and cab, in a way that maximises the efficiency of the overall delivery operation.

Fig 106: Citroen H-van

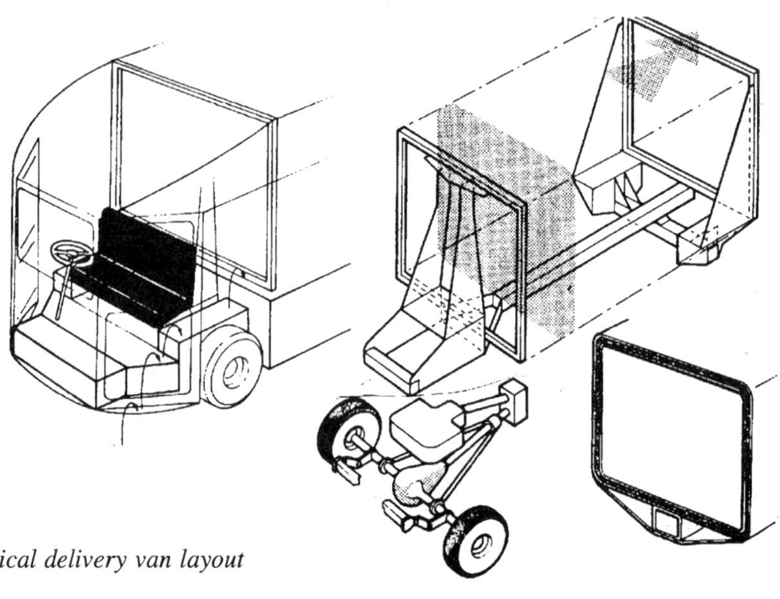

Fig 107: Radical delivery van layout

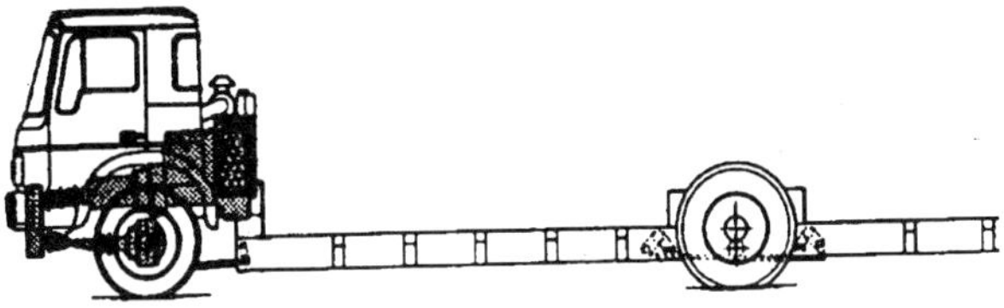

Fig 108:Bruce delivery truck conversion

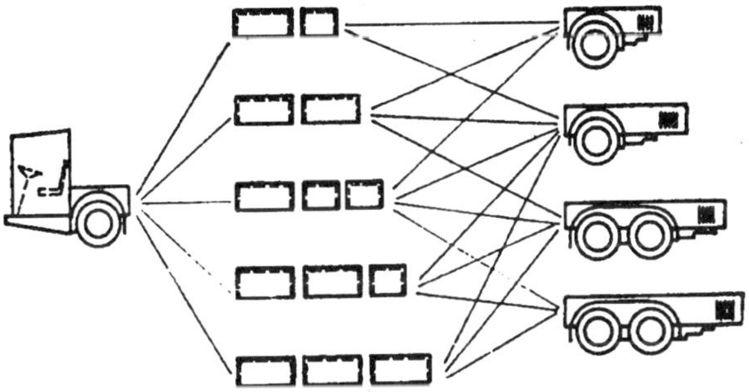

Fig 109: Pawlowski modular system

Among the needs of the delivery vehicle operator are an across-cab and front-rear (cab to load-space) walk-through facilities. Car standards of comfort and convenience in the driving controls and minimum vehicle overall dimensions, for ease of parking, must come with best use of interior space for payload and access compatible with ISO pallet dimensions. A level-control on the suspension system would also be a valuable attribute for street deliveries, as would the minimum possible rear deck dimension. A move in this direction was made by the Citroen H-van, Fig 106, but presumably volumes were not at that time high enough to amortise the tooling costs of high volume production.

A more radical proposal is shown in Fig 107 based on the silhouette of the engine used in an earlier delivery truck, the horizontal flat-four of the Albion Claymore. The main body is made from a sandwich panel tube structure, using structural panels of the type described in the section on sandwich panels. The well-aft disposition of both axles gives acceptable weight distribution for front-drive. The expedient of using a 4WD transfer gearbox, for an easily made-available front-drive arrangement, has already been demonstrated by the Pitt Trailers conversion of the Bedford Rtype truck for the CEGB in the 1960s. It could possibly give a more compact layout than the Bruce Leyland conversion for Scottish and Newcastle Breweries shown in Fig 108 for which the overriding factor was the achievement of a 20 in (508 mm) platform deck height, to ease the unloading of drinks crates from an open platform body. A yet more radical proposal for specialised goods vehicles is due to the late Dr Ing Janusz Pawlowski from Poland's Institute of Industrial Design. Here the inter-axle space occupied by the chassis frame is replaced by structural compartments which slightly increase deck height but result in extra lower-level accommodation for freight. The whole leads to a modular system of construction in which different overall lengths and drive systems can be achieved with minimal modification, Fig 109.

Aerodynamic basics

Widening realisation of the importance of aerodynamics, prompted by the fuel crises of the 1970s, has focused interest by designers in most of the different vehicle sectors. The subject is one in which lessons learned for one vehicle category, in one generation, can very often be applied elsewhere in an another generation. As well as aerodynamic drag reduction, improvements in handling stability and noise prevention can follow careful design in this area.

Key to drag reduction is in the basic expression relating the contributing factors given in the section on performance prediction — whereby air resistance:

$$= C_d A \rho V^2 / 2g$$

where A is vehicle frontal area, r the air density and V the velocity through the air. This expression defines the much discussed drag coefficient C_d which is mostly a function of the shape of the vehicle. With the massive attention taken by the volume builders in recent years, textbook figures for 'typical' vehicle shapes quickly become outdated. Figures for the most basic shapes are useful and some of these are given in Fig 110. A 'teardrop' shape, adopted by water as it falls through the air, is as low as 0.1. The value is much affected by objects which project into the streamlines around a moving vehicle and by rear end shape affecting the turbulence in the wake of the vehicle. Air density varies with atmospheric temperature t, and pressure p, according to the law $\rho = 144p/Rt$ where R is the gas constant, equal to 53.3 for air. In standard atmospheric conditions, $D = 0.0763$ lbf/ft^3. Total drag comprises air resistance, defined above, plus skin friction (depending on surface roughness; this could amount to 10% of air resistance) plus resistance to airflow through cooling and interior ventilation systems.

Product of drag coefficient and frontal area is a term frequently used to compare vehicles aerodynamically. This product can actually decrease with a rise in frontal area if, for example, a protruding part is covered with a streamline fairing. Values for A in square feet (and square metres) are typically 14-20 (1.3-1.85) for small cars, 22-30 (2.04-2.70) for large cars and 60-80 (5.6-7.4) for buses and trucks. Areas can be estimated as 0.9 x body height x wheel track. Engine power required to overcome air resistance is $0.0026C_dAV_r{}^2V/375$ where V_r is the geared road speed. This shows that in still air conditions the power required to overcome air drag is proportional to the cube of vehicle speed — indicating the importance of aerodynamics at higher vehicle speeds.

Fig 111 illustrates air-flow around a car body to show how the flow separates and rejoins as a body passes through the air. As the air approaches the body it has to divide, the point of diversion being termed the stagnation point — at which the dynamic pressure head is $\rho V_r^2/2$. As air flows around the body a pressure distribution is built up as shown; over the roof, increase in air speed causes a suction zone to develop due to the pressure drop. Conversely, at the base of the windscreen, the streamlines separate and a pressure zone results from the low velocity. The air resistance, overall, is the summation of the horizontal components of the pressure distribution vectors. This shows why nose shape is also critical to drag.

Reducing the wake size, by maintaining unseparated air flow further back along the body, is also

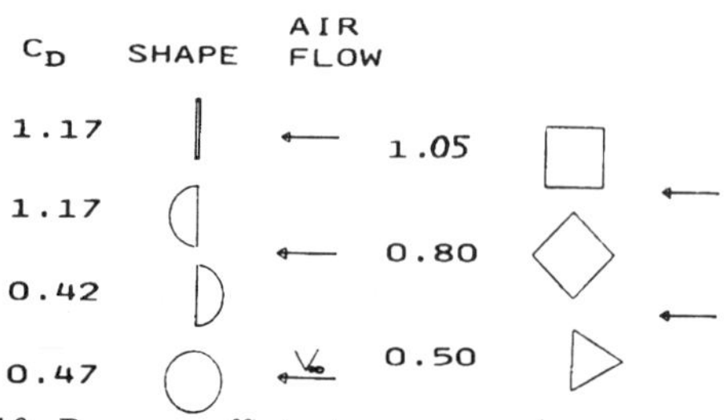

Fig 110: Drag co-efficients

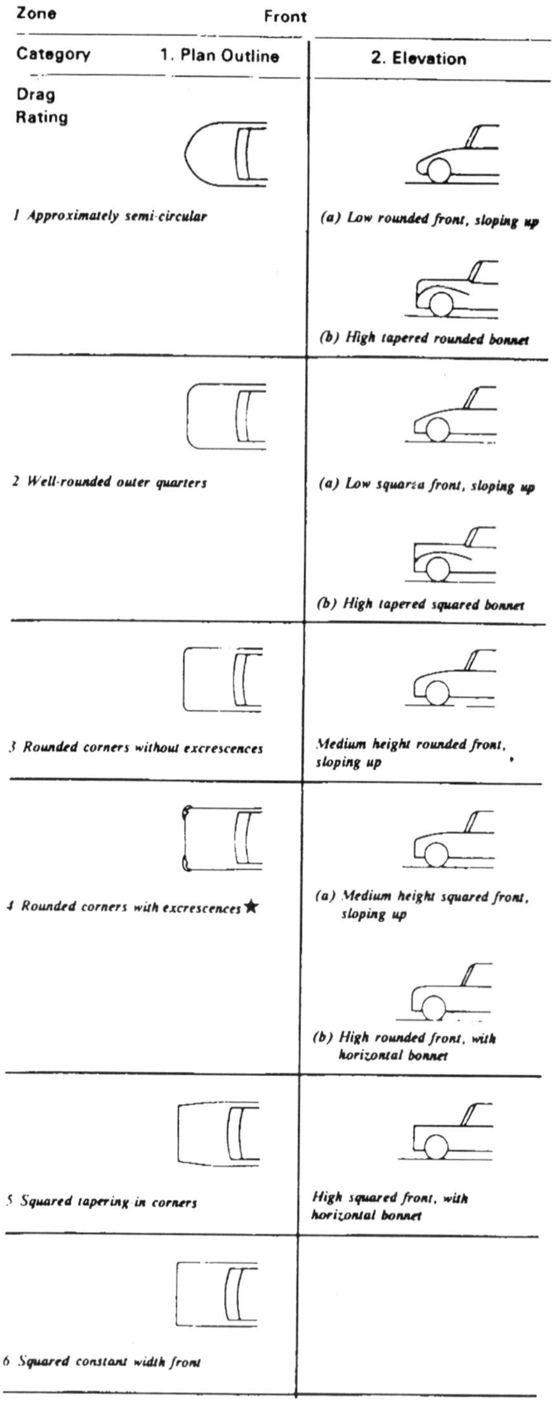

Fig 112: Components of drag

Zone	Windscreen/Canopy junction	
Category	3. Scuttle and wing cross-section	4. Windscreen plan
Drag Rating	*1 Flush bonnet and wings, well-rounded body sides*	*Full wrap-round (approximately semi circular)*
	2 High scuttle, low wings	*Wrapped round ends*
	3 (a) Bonnet flush with rounded top wings *(b) High scuttle, with rounded-top wings*	*Bowed*
	4 Bonnet flush with squared-edge wings	*Flat*
	5 Depressed bonnet, with high squared edged wings	

Zone	Rear canopy/boot	Lower rear quarters	Underbody
Category	7	8	9
Drag Rating	*1 Fast-back, i.e., roof line continuous to tail*	*Well-or medium-tapered to rear*	*Integral, flush floor, little projecting mechanism*
	2 Semi-fast back i.e., with discontinuity in line to tail	*Small taper to rear or constant width*	*Intermediate*
	3 Squared canopy with boot rear edge squared	*Outward taper (or flared-out fins)*	*Integral, projecting structure and mechanism*
	4 (a) Rounded canopy with rounded boot *(b) Squared canopy with short or no boot*		*Intermediate*
	5 Rounded canopy with short or no boot		*Deep chassis*

Zone	Canopy	
Category	5. Windscreen peak	6. Canopy plan
Drag Rating	*1 Rounded*	*Well- or medium-tapered to rear*
	2 Squared (including flanges or gutters)	*Tapering to front and rear (max. width at BC post) or approximately constant width*
	3 Forward-projecting peak	*Tapering to front (max. width at rear)*

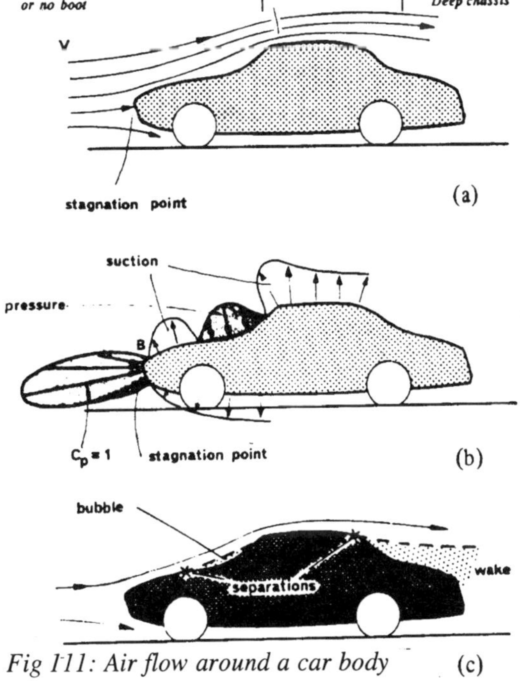

Fig 1.11: Air flow around a car body

helpful — as in an aerofoil wing section. While the flow remains streamline, air adheres to the vehicle over a thin boundary layer whereafter, moving away from the surface, it is sheared until the air velocity builds up to that of the main stream. Surface roughness gives rise to tangential viscous forces, skin friction, while severe surface discontinuities cause separation of the flow which is a starting point for the turbulent wake. At vertical corners, like the windscreen edge, flow slows down causing the pressure to rise and the possibility of a separation. Flow would probably re-attach itself further down stream along the side of the vehicle. The design aim is thus to maintain the boundary layer as far back along the body as possible by using very gentle curvatures. The advantage of sloping sides becomes clear, in this respect, as the blending in of the windscreen will be easier — and air will make its way around the sides to sweep the separation point further towards the tail of the vehicle.

Much valuable work has been carried out by the Motor Industry Research Association at Nuneaton. A number of their earlier reports are now publicly available. Fig112 taken from Report 1967/9 provides a method of adding up the effects of the shapes of different areas of a car body to predict drag coefficient at the conceptual stage in design. Ratings shown for each configuration indicated are summed to a total drag coefficient being obtained from:

$$C_d = 0.16 + 0.0095 \times \textit{drag rating}$$

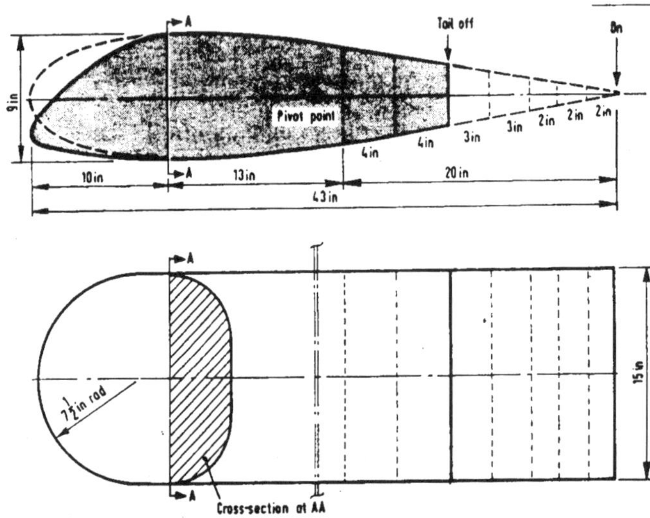

Fig 113: Near tear-drop shape

Lift and pitch

The near tear drop shape shown in Fig 113 was tested some years ago at Imperial College in London. It provided a drag coefficient of just 0.17 with ground clearance at 1/5th the depth of the body. Such an idealised shape would require modification to reduce the effects of changing angle of incidence to the oncoming airstream in a practical road vehicle. With constant incidence and decreasing ground clearance, negative lift begins to appear as a result of the venturi effect beneath the body. Tests showed the pitching moment increased with decreasing ground clearance while drag increased due to negative camber effects. The effect on drag co-efficient of varying incidence angle is shown in Fig 114. Blunting the trailing edge increases both drag coefficient and the lift curve slope.

When using aerofoil wings, such as that sectioned in Fig 115, to induce negative lift, it is usual to define the wing plan area *S* and chord length *C* in calculation: *Pitching moment* = $\rho V^2 SC/2 \times$ *pitching moment co-efficient*. The angle of incidence of the

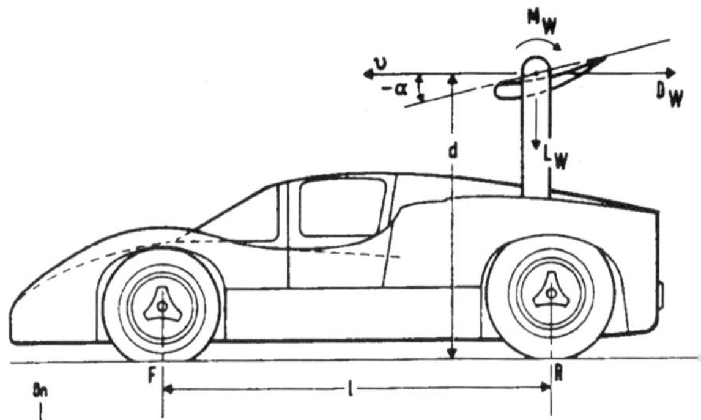

Fig 115: Aerofoil wings

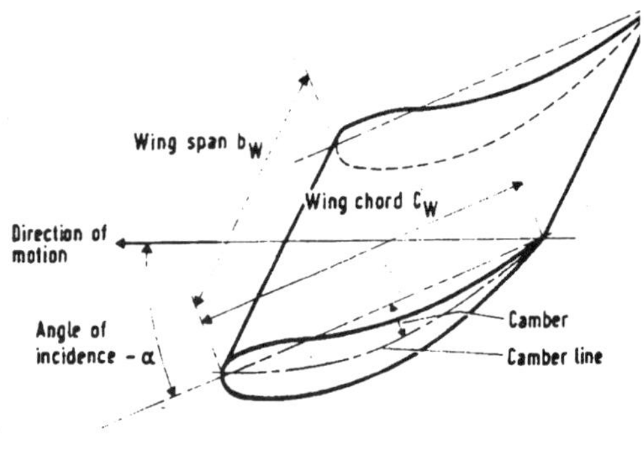

Fig 116: Cambering an aerofoil

wing is measured from the zero-lift attitude of the wing as a datum. Direction of resultant force on a wing is usually downwards and slightly backwards with angle to the vertical called down-wash. Lift per unit span = $\rho V^2 C L_c/2$ where L_c is lift co-efficient. The characteristics of the aerofoil can also be changed by cambering as shown in Fig 116.

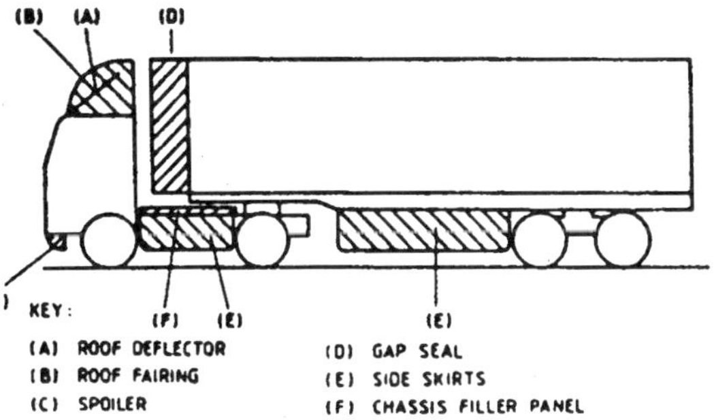

Fig 117: Aero-devices on artic container

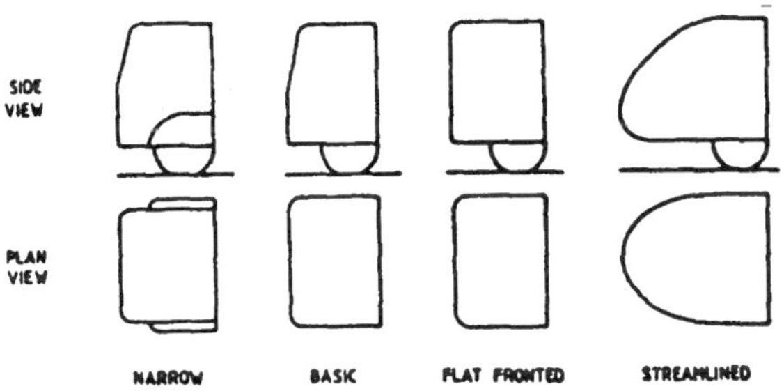

Fig 119: Cab-shape effect

Bluff vehicles

The effect of prevailing wind on drag has led to the use of wind-averaged drag coefficients. For a large box-shaped goods vehicle, side winds can substantially increase the drag coefficient from its 'head-wind' value. Tests at MIRA on an articulated container van model, Fig117, have shown that a range of add-on devices can combine to effect drag reductions up to nearly 40%. Cab leading edge design is critical but streamlining the remainder of the cab is found to provide no benefit unless attached flow is maintained. Of the add-on devices, apart from roof-deflectors for high bodied vehicles, side-skirts were shown as particularly beneficial. Fig 118 shows proportion of drag due to aerodynamic effects — the dominant element at 90 km/h and above for this shape of vehicle. Fig 119 shows the effect of cab shape for different trailer body configurations.

From the US, engineers at the Kenworth truck company have compiled a table, from different sources, of the effects of individual changes on drag reduction; see the table in Fig 120.

Aerodynamic stability

Roll, pitch and yaw modes need to be considered in respect of vehicle stability. Roll stability can be achieved by ensuring that centre of pressure (point of application of wind sideforce) is close to the roll axis of the car; small deviations are tolerable if the vehicle does not have roll-steer characteristics. Distribution

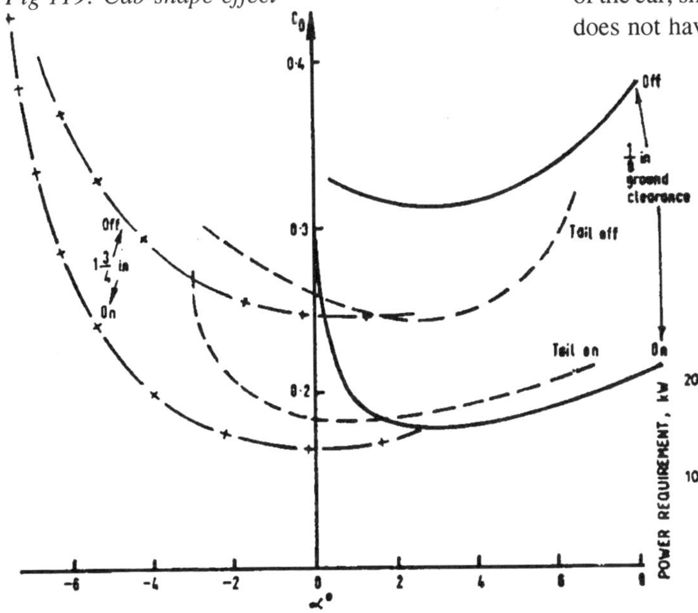

Fig 114:Incidence angle affecting drag

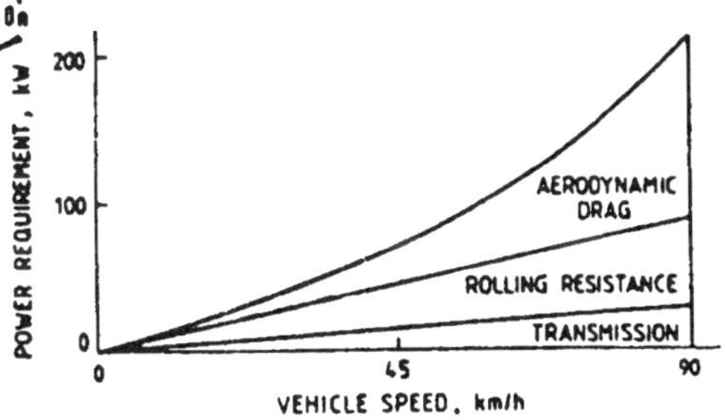

Fig 118: Contributors to motion resistance

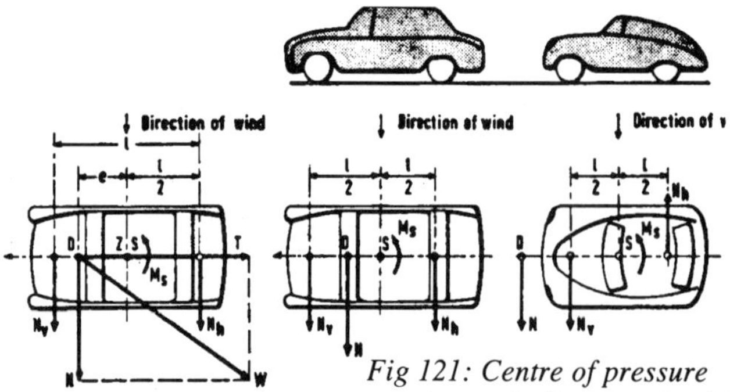

of lift forces acting on a vehicle determines the magnitude of the pitching moment — which tends to vary the wheel loading with speed. Tail-end spoilers used to form the top edge of a drag-reducing cut-off tail can also induce downward stabilising forces on the road-wheels. Light-weight vehicles with centre-of-gravity well aft are less easy to control and are susceptible to zero or small negative lift at the front. Drooping the nose of the vehicle can have a stabilising effect in this respect — equivalent to reducing the

Modification	Tractor-Trailer Low	Tractor-Trailer High	Straight-Truck Low	Straight-Truck High
1. Tractor or Cab				
a) Deflectors	-0.01	0.15	-0.02	0.12
b) Roof fairings	-0.10	0.24	N/A	
c) Front edge rounding	0.00	0.06	N/A	
d) Aero Bumper	-0.05	0.05	N/A	
e) Rear cab extenders	0.04	0.08	N/A	
f) Sunvisor	-0.01	0.005	N/A	
2. Van Trailer or Body				
a) Front face fairing	0.09	0.16	0.14	0.22
b) Edge rounding (vs. square)				
- front posts	0.08	0.10	0.12	
- top front rail	0.05	0.08	0.10	
- posts and rail	0.13	0.18	0.22	
- top longitudinal	0.00	0.02	0.00	
- rear posts and header	0.02	0.10	N/A	
c) Edge beveling				
- 45 deg. front post & rail	0.04	0.05	0.10	
- shallow bevel on rear posts and header	0.03	0.09	0.03	0.05
d) Refrigeration unit	0.03	0.10	0.04	
e) Edge fairings	0.0	0.06	0.12	
f) Front turning vanes	0.0	0.08	0.20	
g) Smooth sides (vs. ext. rib)	0.03	0.09	N/A	
h) Skirts	0.02	0.07	0.02	
i) Bogie fairings	-0.02	0.02	N/A	
j) Streamlined tail (with streamlined front)	0.0	0.13	N/A	
3. Miscellaneous				
a) Reduced gap (1.5m to 1.0m)	0.02	0.10	N/A	
b) Gap seal	0.04	0.10	0.03	
c) Fill gap	0.06	0.12	N/A	
d) Deflector & body fairing	0.0	0.15	N/A	
e) Streamlined truck	0.45	0.55	0.40	

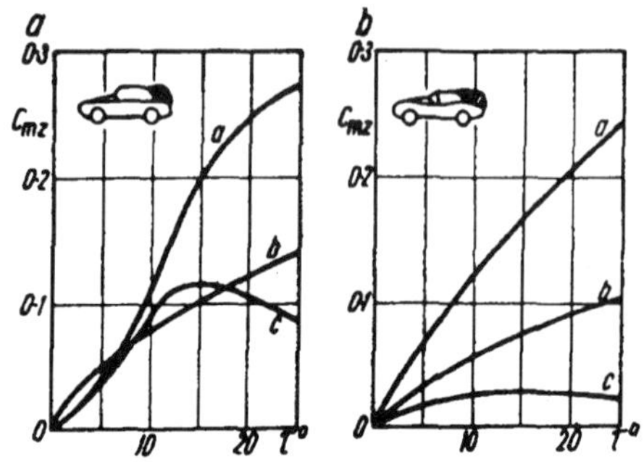

Fig 122: Effect of stabilisers

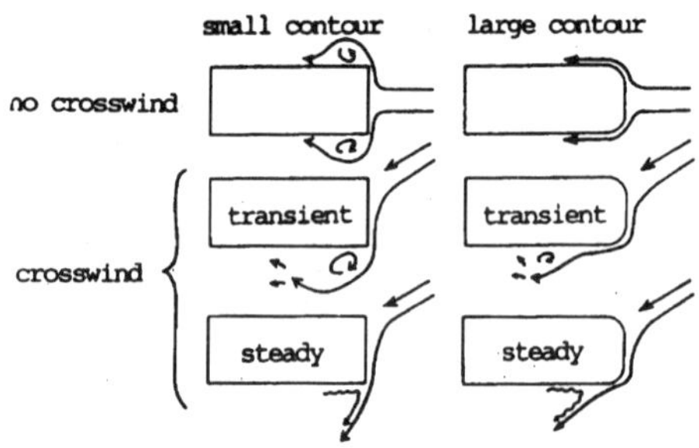

Fig 123: Cross-wind effect on yaw

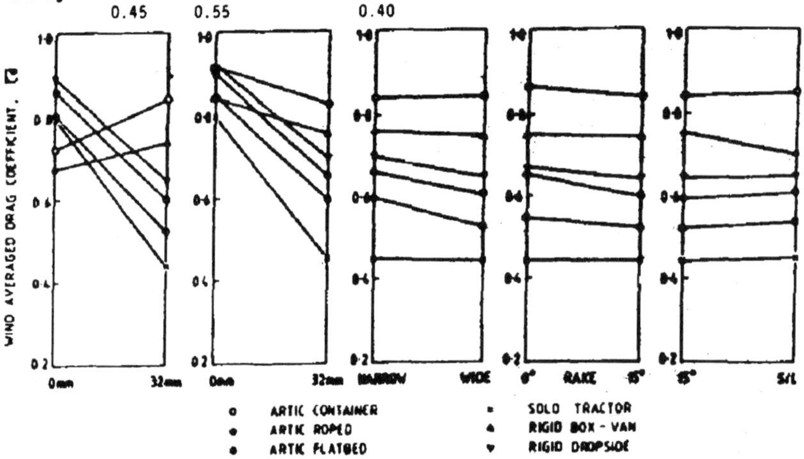

Fig 120: Constituent effect on drag reduction: from base of C_D = 55%

angle of incidence of an aerofoil. The top surface of wedge-shaped bodies also provides a negative angle of incidence resulting in a downforce, as does an underside venturi. Care must, of course, be exercised in reducing ground clearance — as the venturi will stall in the limit with the result that reduced mean velocities can cause destabilising lift forces.

Stability in the yawing plane governs the amount of steering correction required in cross winds. Only when the centre of pressure corresponds with the neutral steer point will the vehicle maintain its heading and merely drift sideways. Ahead of this point instability occurs and it is preferable to arrange the centre-of-pressure just behind the neutral steer point. Sharp corners on a body should be avoided in the interests of reducing sharp pressure changes when the vehicle is subject to side gusts — with consequent changes in yawing moment. Fig 121 shows the force pattern, with centre of pressure at *D*, at which air resistance *T* and normal wind force *N* act; a torque $M_s = Ne$ develops around the vertical axis Z through the centre of gravity. Streamline vehicles for which $Ne = (N_v + N_h)/2$ have a greater turning moment than bluff ones for which $Ne = (N_v - N_h)/2$. Use of stabilisers, or fins, at the rear of the vehicle can give good results. Fig 122 shows measured yawing moment co-efficients of production cars modified by the addition of stabilisers: a saloon car; b coupe; c twin stabilisers.

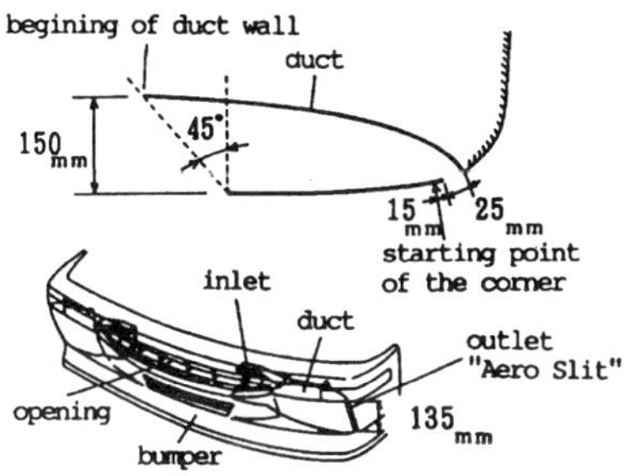

Fig 124: Use of aero slits

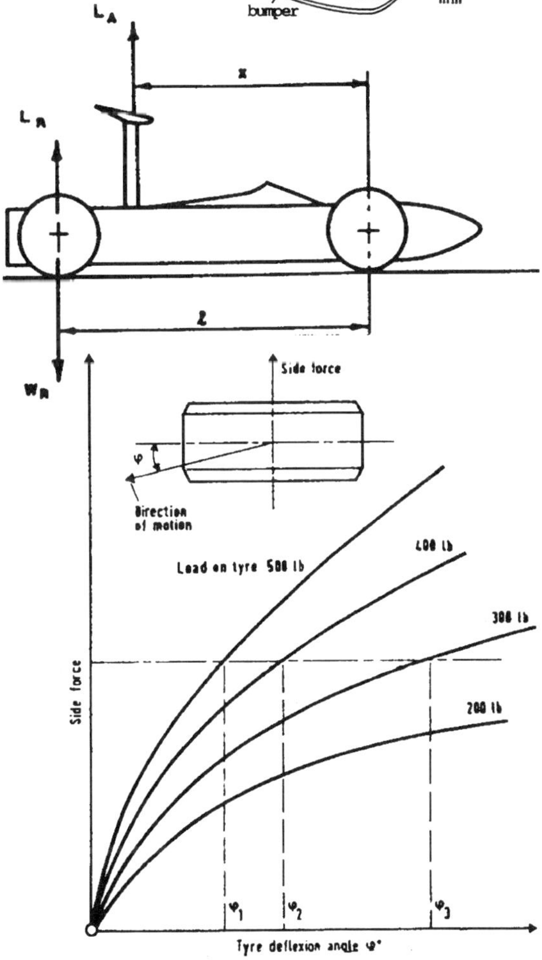

Fig 125: Effect of down-force on handling

Care must be taken in relating results measured in wind tunnels with those expected on the open road. To account for viscous and inertia effects in flows around models in tunnels and full sized vehicles on the road, the Reynolds Number: $R_e = rVW_n n/m$ where W_n is nominal model size and n is viscosity coefficient. The expression needs to be evaluated for both conditions. For equal Reynolds numbers, flow conditions are thought to be identical for model and vehicle. Thus forces can be scaled up in proportion to scale ratio. Normally the nominal model size is defined by the width dimension but if the height is very different from the width, the square root of the frontal area should be used. A fullsize truck at 55 mile/h has a Reynolds number, based on width, of 4.1×10^6. The SAE standard (J1252) states that a minimum of 0.7×10^6 is an acceptable test condition provided wind speed is less than 300 ft (91.5 m)/sec to avoid air compressibility effects.

An interesting use of wind tunnel testing has been carried out by Toyota to study vehicle motion in a cross-wind. Effect on yawing moment co-efficient was investigated — of altering front end corner contours using 1/10 th scale models. Fig 123 shows flow pattern variation. The company devised a method of inducing flow separation in order — to more nearly equalise flow in steady and transient conditions. Ducting is used to connect high and low pressure areas on the vehicle in order to blow air out of slits in the front corners, Fig 124, in order to induce the desired separation.

Race-car aerodynamics

The aerodynamic design of racing car bodies varies with the rules allowable by different race formulae as well as periodic changes made by the authorities in search of better and safer sport. About the only common theme between models is the basic subject discipline which yields the essential principles for making design improvements whether by add-on devices or changes in body shape.

Serious work on racing car aerodynamics commenced, some would say, with the introduction of negative lift devices in the 1960s. Such devices put very high loads on suspension systems and can radically affect cornering characteristics of the tyres as shown in Fig 125. Choice of wing aerofoil sections is thus crucial to overall handling of the racing car. While the aerospace industry has much data available on the performance of aerofoils, it has to be understood that much of this is based on wind tunnel measurements inappropriate to the race-car application. Dynamic similarity is crucial to the extraction of meaningful data in design. This is the consideration of air flows around two similar bodies, but of different size, to ensure that inertia and viscous forces at equivalent points in the flow path are in the same ratio. This is ensured by calculating Reynolds Number for each flow case, as defined in the earlier article. As theoretical relation between lift/drag co-efficients and Reynolds number is mathematically complex, it is usual to determine this experimentally for all but

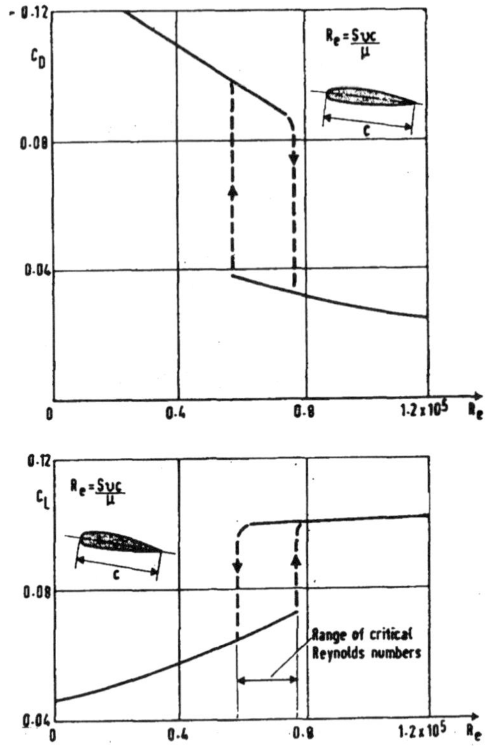

Fig 127: Characteristic curves for aerofoils

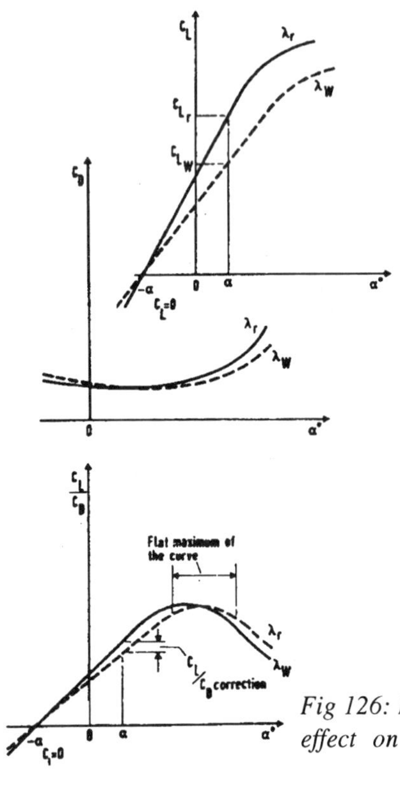

Fig 126: Reynolds number effect on lift/drag co-efficients

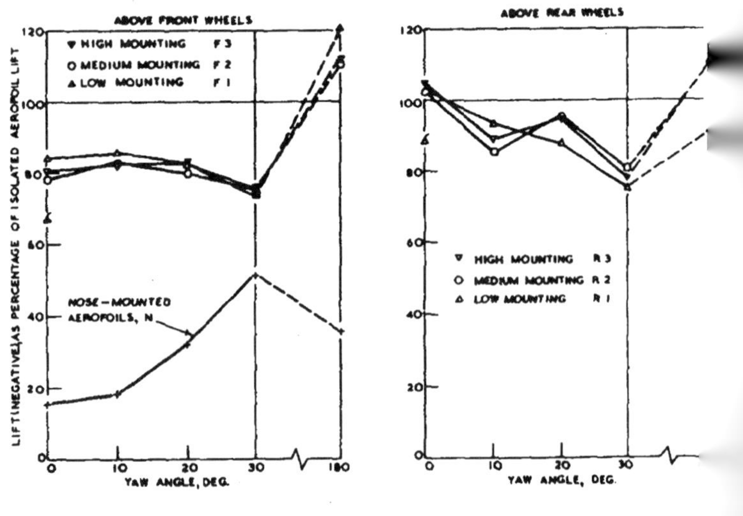

Fig 131: Negative lists obtained in tes

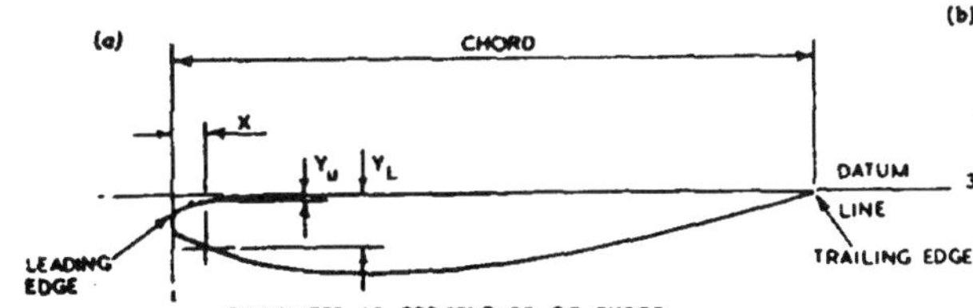

ORDINATES AS PERCENTAGE OF CHORD

X	0	1·25	2·5	5	10	15	20	30	40	50	60	70	80	90	100
Y_U	3·50	1·93	1·47	0·93	0·42	0·15	0·03	0	0	0	0	0	0	0	0
Y_L	3·50	5·45	6·50	7·90	9·60	10·68	11·36	11·70	11·40	10·52	9·15	7·35	5·22	2·80	0·12

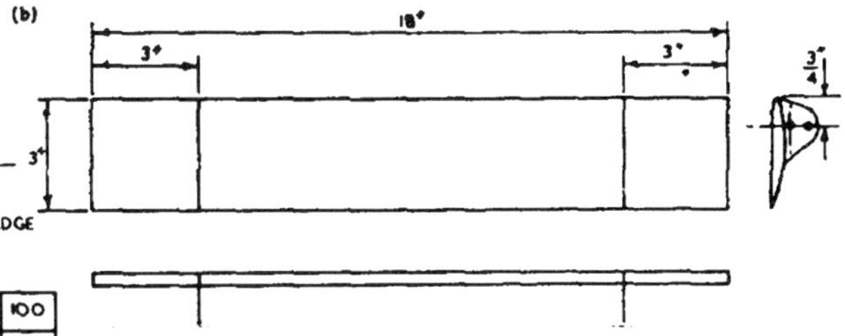

Fig 128: Test section for aerofoil evaluation

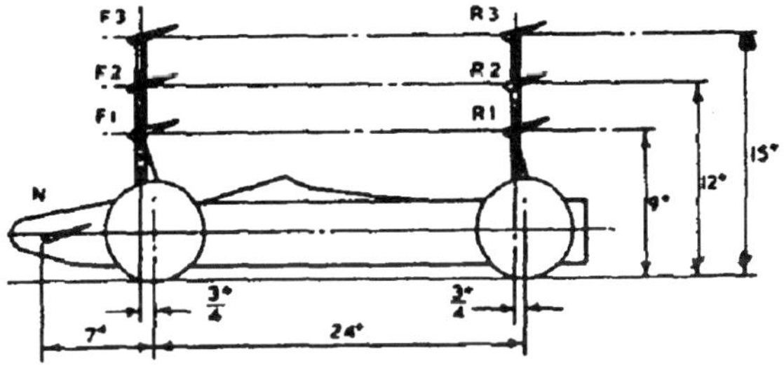

Fig 130: Wing mounting positions

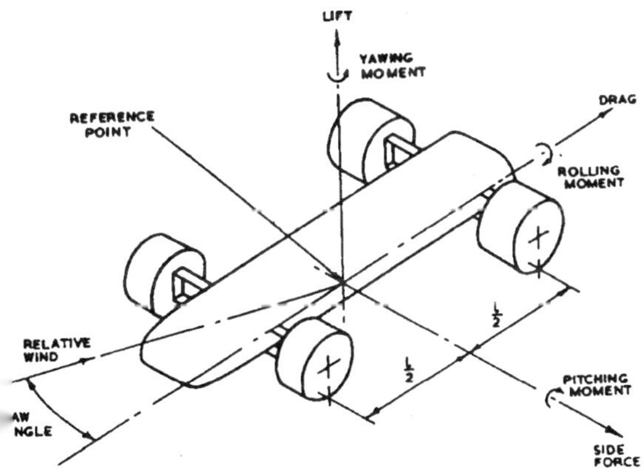

Fig 129: Force components measured

the simplest of shapes.

Fig 126 shows that for aerofoil shapes, only at low *Re* flows is there a significant jump in *Cl* and *Cd* values. The ‘characteristic dimension’ for the *Re* expression is in this case the chord of the aerofoil. Flows corresponding to higher speeds of the vehicle aerofoil would behave predictably; see the data in Fig 126. Design procedure is thus to find the range of angles of incidence for the wing at the design operational speed of the vehicle. Selection of the aerofoil section should be on the basis of highest possible lift force with the minimum of drag at the angles of incidence under consideration. Fig 127 shows characteristic curves for a selection of different aerofoils based on work carried out at the City University.

A comparatively flat maximum for the *Cl*/*Cd* characteristic shows of course that drag is minimum over a wide range of incidence angles. Because the rectangular planforms of the wing used on the car may be different from those of the wind-tunnel

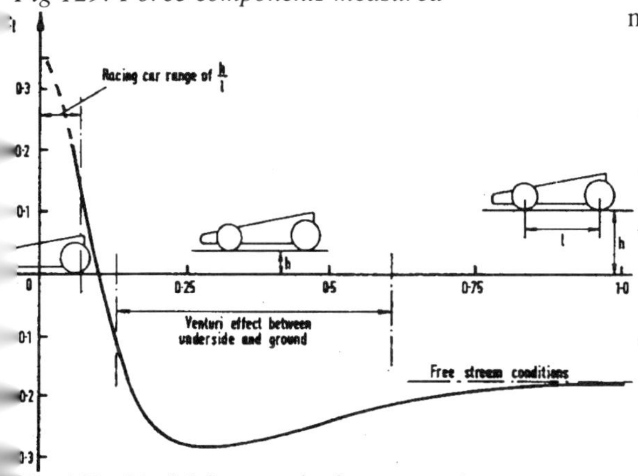

Fig 132: Model for race body aero testing

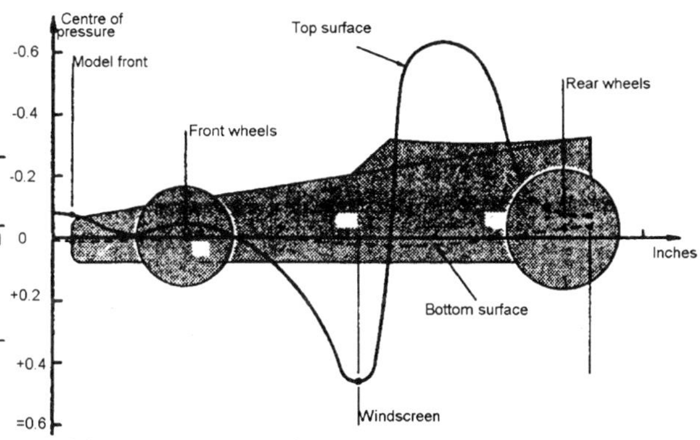

Fig 133: Pressure distribution on test model

testpiece, aspect ratio $\lambda = b2/S$ should be noted in calculation: the ratio of wingspan to constant chord in the case of a rectangular planform. Compared with a wing of infinite span, a typical short wing would suffer additional induced drag (as seen in Fig 126 when comparing lamda values for wind-tunnel (w) and real (r) devices); conversion tables must therefore be used. The change in lift and drag co-efficients can also be calculated from

$$dC_l = (C^2\omega t57.3/a\pi)[(1/L\omega t) - (1/L)]$$
$$dC_d = (C^2\omega t/\pi)[\{1 + 2dC_l/C_lt/L\omega t\} - 1/L]$$

for lift and drag change. Wing positioning must be such as to operate high enough above the car to be in a clean airstream while avoiding an excessive negative pitching moment from the drag of the wing.

Early work sponsored by the Jim Clark foundation was valuable in optimising race-car aerofoils, based on the section shown in Fig 128. Test work involved measuring the force components shown in Fig 129 at a wind tunnel speed of 100 mph. For the wing mounting positions shown in Fig 130, negative lifts produced are shown in Fig 131 expressed as a

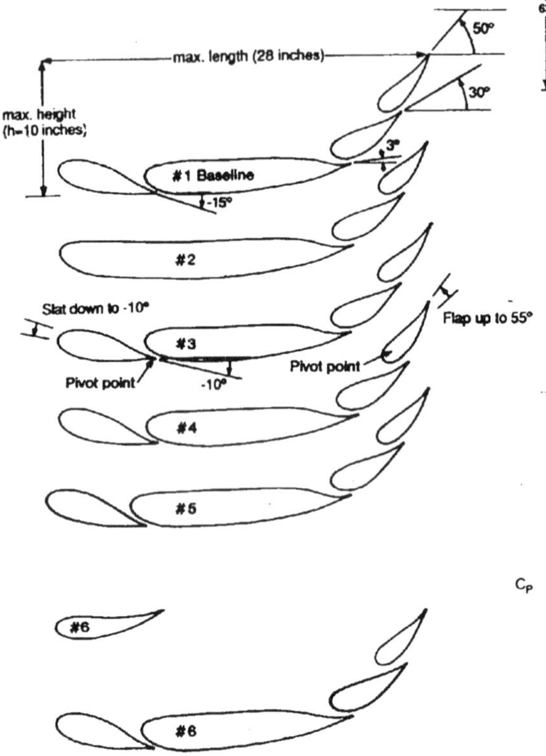

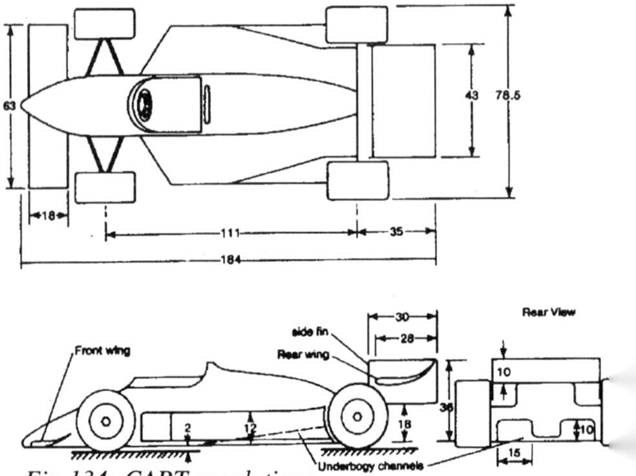

Fig 134: CART regulations

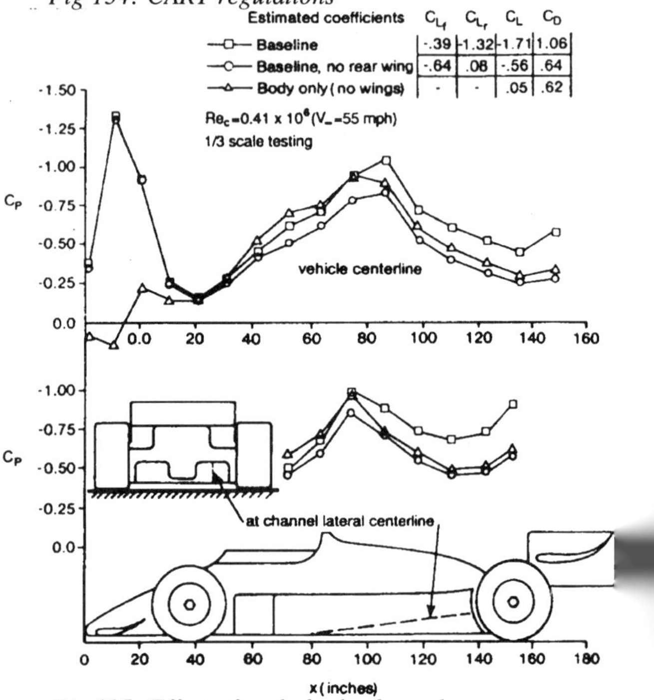

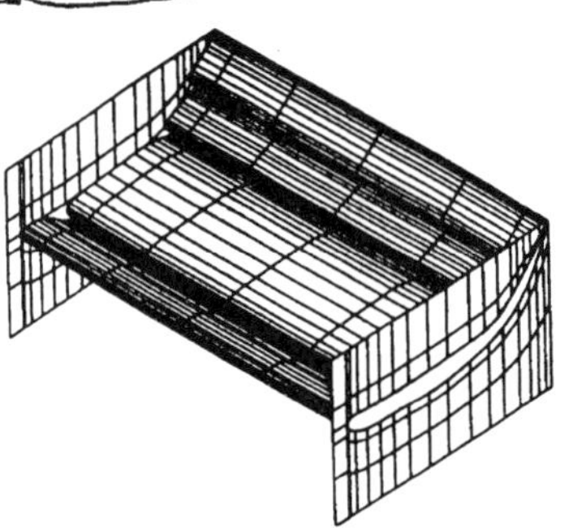

Fig 136: Computer-optimisation of wing

Fig 135: Effect of underbody channels

percentage of the force produced by an isolated aerofoil without the car. Note that when mounted above the front wheels the aerofoil lost 20 % effectiveness at all yaw angles up to 30 degrees. When placed above the rear wheels it only fell to this level at 30 degrees. Also shown is the variation of lift of the stub aerofoils (N) mounted on the nose of the car. Aerodynamic characteristics of the typical wedge shape of racing car bodies was also the subject of pioneering work at City University. The basic form of such vehicles also has a substantial effect on reducing lift. One eighth scale model tests were used to demonstrate the ground effect in Fig 132. When the underside is parallel to the ground there is effectively a negative angle of incidence with consequent downforce — in the high, 'free-flight' condition. With reduction in h there is an increasing venturi effect beneath the car creating a suction which adds to the negative lift. At very low clearances an upward lift force occurs when the airflow stalls beneath the body. In the pressure distribution plot of Fig 133 the nett upward lift resultant can be seen — with a lift co-efficient in this case of 0.23. More recent work at San Diego University in the USA showed that cars designed to CART regulations, Fig 134, could have rear wings developing -2.2 negative lift co-efficient. The small front wing positioned far-forward near the ground operates in the undisturbed air stream and works at fairly low loading to trim downforce distribution on the vehicle. Underbody channels are also permitted in the body configuration, by the regulations, to enable further increase in downforce. The interacting effect of these with the rear wing is shown in Fig 135 found in one third scale model tests. Optimising a high lift rear wing of very modest aspect ratio was achieved by computer modelling of multi-element aerofoil systems, Fig 136. Using a finite-element model of the optimum section computation was carried out and checked against an experimental model made from 664 panels. Test results are shown in Fig 137[5].

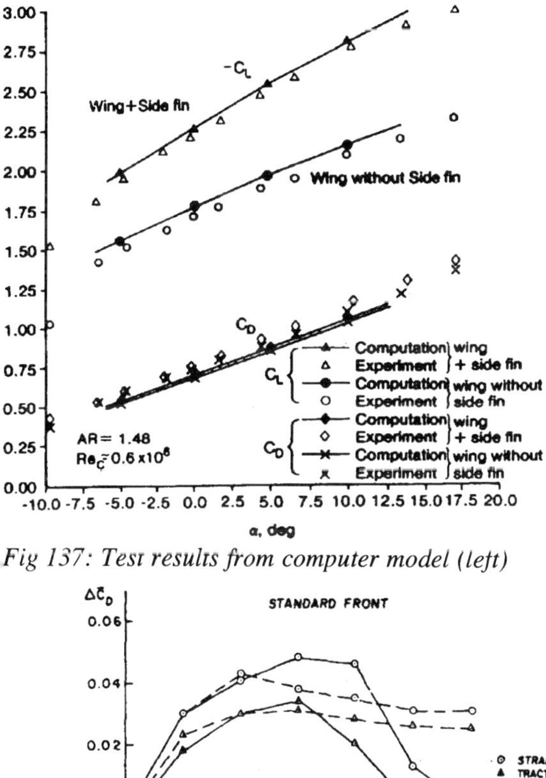

Fig 137: Test results from computer model (left)

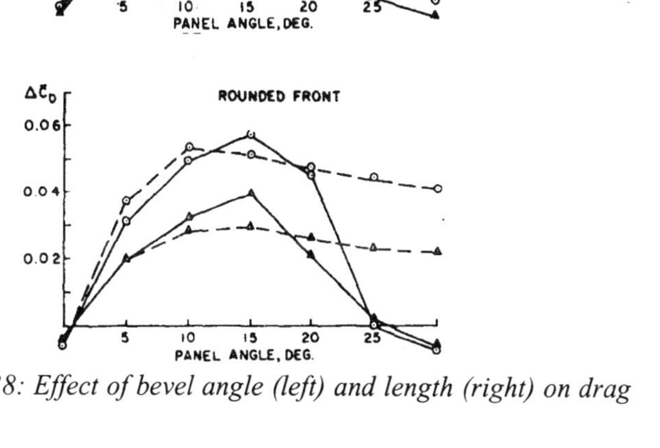

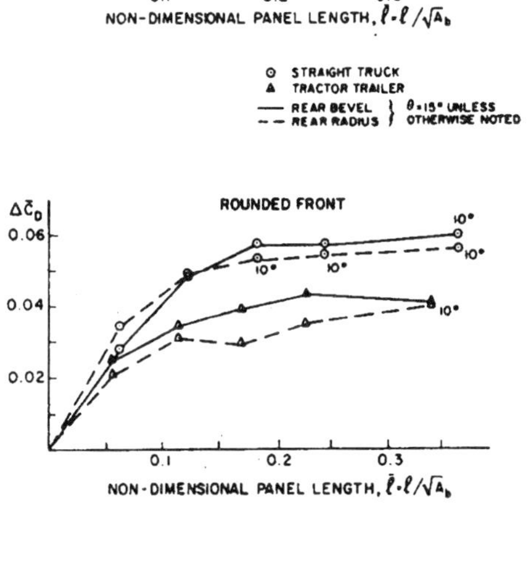

ig 138: Effect of bevel angle (left) and length (right) on drag

Truck aerodynamics

The bluff rather than streamline bodies which apply to many special vehicles require a different aerodynamic analysis from that applying to the more streamline shapes. This section examines both the basic principals involved and the effects of some add-on devices which can be used, given a proper understanding, to effect quite substantial fuel-economy savings.

The previous section was based generally on the assumption of streamline air flow over the body surfaces. Here the effect of turbulence caused by bluff body shapes is considered so that the flow over typical box-cube shapes can be understood.

The viscosity and compressibility of the air becomes significant when air is flowing over a bluff body which grossly disturbs the air stream and drag co-efficient depends upon Reynolds Numbers:

$$= \rho VL/\nu$$

where r is air density, V relative velocity, n viscosity and L is a major dimension of the bluff body with respect to the air stream.

The objective in aerodynamic design of bluff bodies is to try and avoid separation of airflow by using devices which attach the flow to the surface as far as possible along the vehicle towards its wake. SAE Recommended Practice J1252 suggests using the square root of the frontal area for the 'significant dimension' L in the Reynolds Number equation — if vehicle height is very different from vehicle width. A large truck might have a Reynolds Number of 4.1 x 10^6 and generally speaking the high drag of truck container-bodies arises from pressure reduction over the rear face of the vehicle due to the entrainment of air by the shear layers separating from the rear edges.

Either bevelling or radiusing the rear edges help to reduce this effect. Tests made by the National Research Council of Canada show the effect of rear-panel bevel angle and length on drag for both square and radius fronted vehicles in Fig 138. The danger of applying streamline flow theory directly to a bluff body is shown in Fig 139 which compares calculation with experimental result for flow around a circular cylinder. Only around the stagnation point where the flow separates at the front of the body is the flow similar to the theoretical 'streamline' case.

The shearing of fluids can be visualised with the help of the diagram in Fig 140. Here flow is considered to be in the x-direction and its velocity

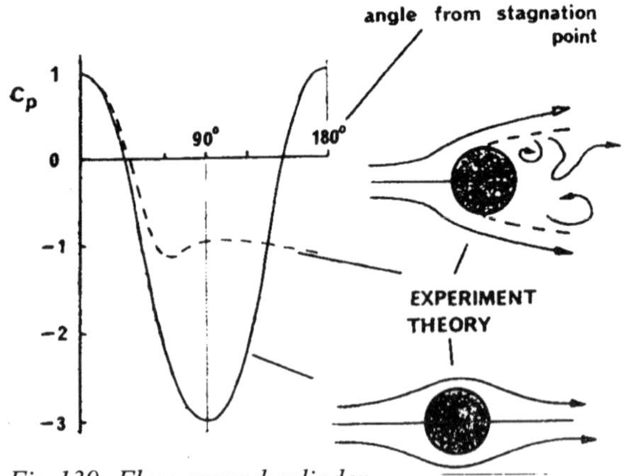

Fig 139: Flow around cylinder

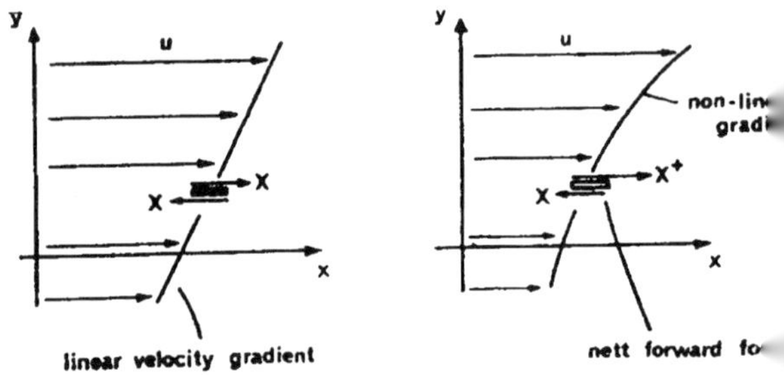

Fig 140: Shearing of a fluid

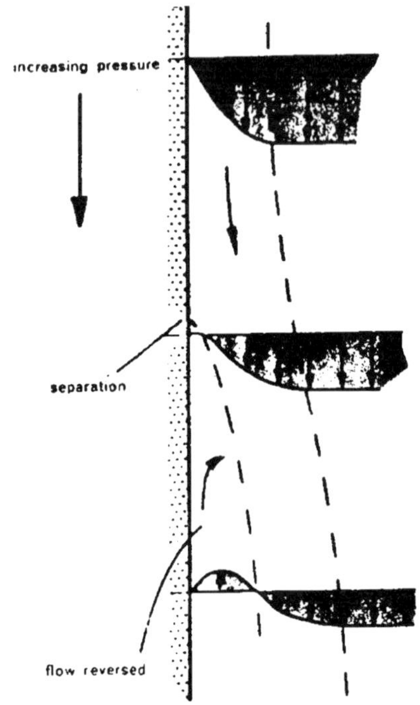

Fig 141: Flow reversal

increasing with the profile shown in the y-direction. For any element a shear stress X exists because the

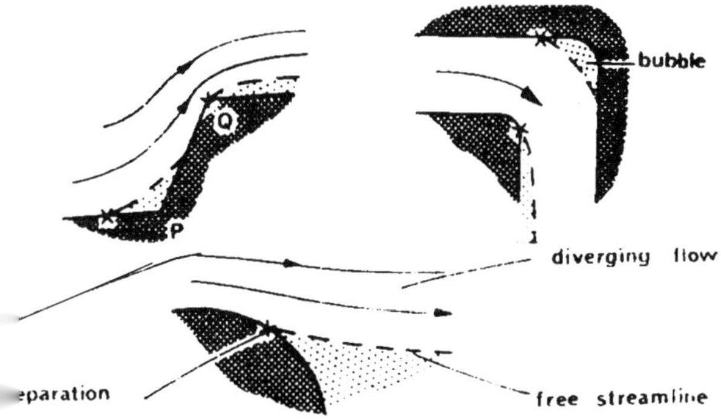

Fig 142: Body shape effects

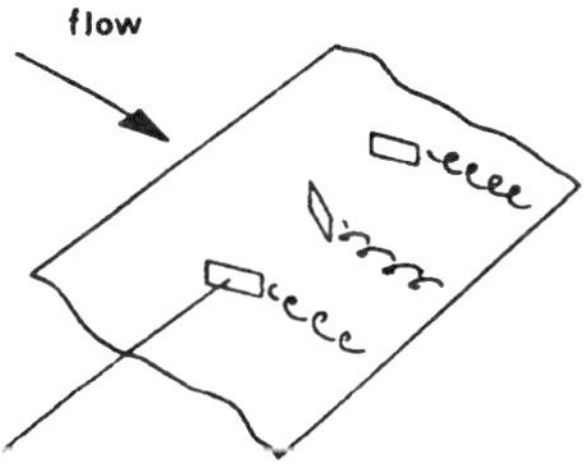

Fig 143: Vortex generators

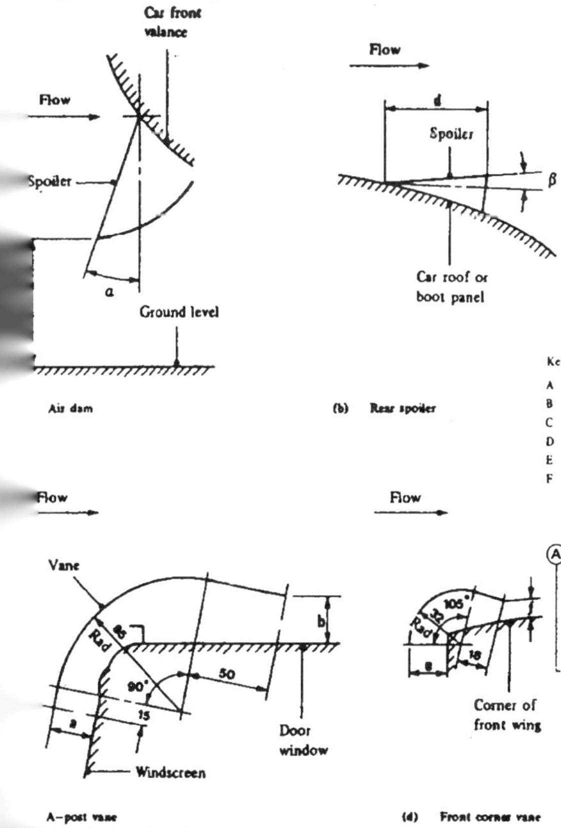

Fig 144: Turning vanes

fluid tries to reduce the relative motions of adjacent layers. When the velocity profile is linear, tangential force each side of the element is zero; otherwise there is a nett forward force due to pressure gradient in addition to the viscous force. This can be represented mathematically and simplified to produce the Reynolds Number equation stated above.

If, due to the shape of the vehicle body, the pressure outside the boundary layer, between air and body-surface, increases in the flow direction, the gradient effect will also be felt within the layer. However, as within the layer, the velocity has already been reduced because of the viscous stresses, there is possibility of the fluid being brought to rest or reversed, Fig 141. In such cases the layer separates from the surface and is no longer under direct control of the body. This explains the formation of the wake. The occurrence of separation depends on several factors including the position of the boundary layer upstream of the area being considered. Various flow patterns can arise from separations caused by different body shape effects, Fig 142. Boundary layer control is principally a matter of moving separation points as far aft of the body as possible, either by arranging for very gradual adverse pressure gradients, using curved front-ends, or using layer control devices. Such devices either speed up or remove stagnant air near the surface; sometimes this can be done by sucking air through slots in the surface which might double as air inlet ducts for another vehicle system. Small inclined slats, or vortex generators, are also used — the small vortices produced having the effect of breaking up the stagnant layer, Fig 143.

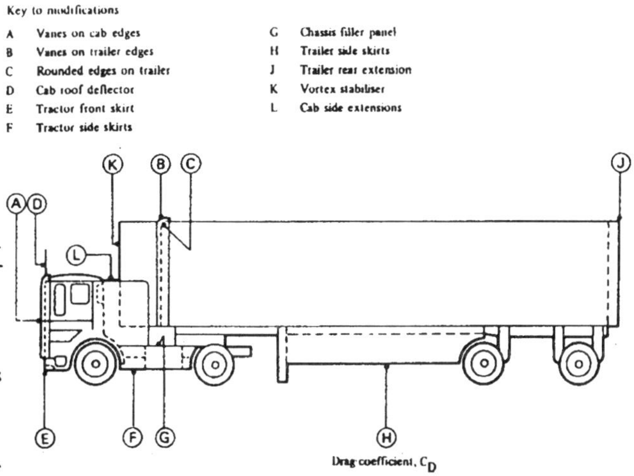

Fig 145: Drag reduction devices

Experimental work carried out at MIRA has provided a useful evaluation of add-on devices. Fig 144 shows turning vanes used for attaching flow around relatively sharp corners such as windscreen pillars with which some 3% drag reduction was measured. Their use on the leading edges of box van bodies, however, is reported to have produced reductions up to 30%. The degree of reduction depends very much on the shape of the cab ahead of the box container and in this context it is important to note that tests have shown vanes mounted to the windscreen corners of a cab have actually increased the overall drag of the vehicle. Rounded leading edges for box bodies have resulted in 12-13% reductions rising to 15% in conjunction with cab-roof deflectors. Effects of tractor and trailer skirts, on articulated

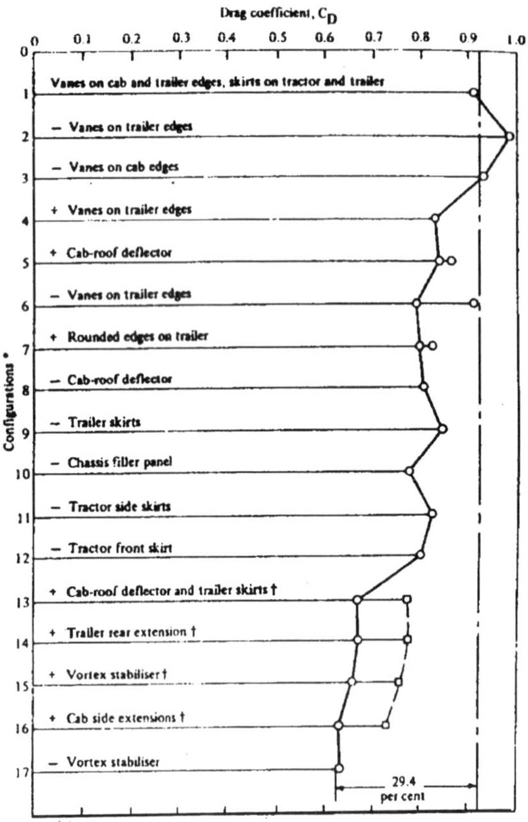

Fig 146: Incremental drag reductions

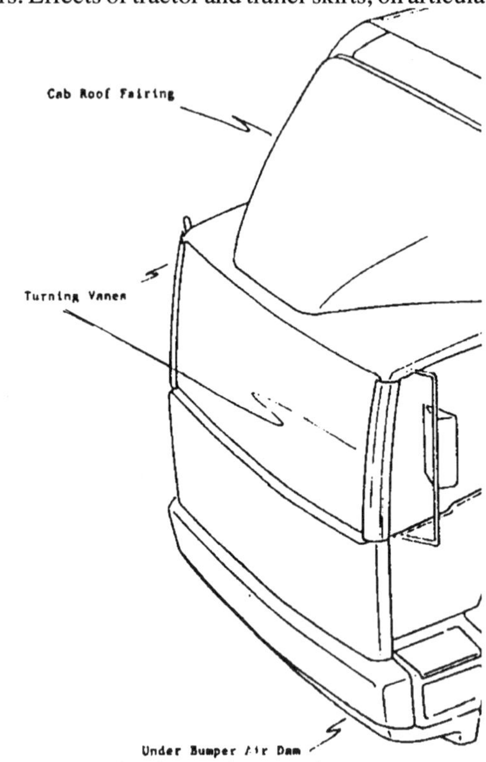

Fig 147: Leyland Freighter mods

Roof deflector) with)high)load/	10 to 12%
Roof fairing)box body	10 to 25%
Air dam (rigid)	up to 4%
(articulated)	up to 2%
Cab side deflectors	up to 2%
Chassis skirts (rigid)	5 to 12%
(articulated on tractor only)	up to 4%
Leading edge turning vanes	up to 1%

Typical wind averaged drag coefficient reduction,

Fuel Consumption (mile/gallon)

Test Speed and Gear	Standard at 10 000kg	Prototype at 10 273kg	Improvement
Test Track			
40 mile/h, 5th gear	14.69	17.86	22%
50 mile/h, 6th gear	12.61	16.49	31%
60 mile/h, 6th gear	9.70	13.56	40%
Motorway (maximum gradient 1:1000)			
40 mile/h, 5th gear	15.60	19.85	27%
50 mile/h, 6th gear	13.01	17.83	37%
60 mile/h, 6th gear	10.18	14.38	41%
Leyland to Kendal (M6 and A6, return) Test Route			
M6	11.42	14.64	28%
A6	13.98	16.35	17%
Overall	12.63	15.46	22%

Fig 148: Fuel consumption reductions for different add-on devices

vehicles, have also been recorded, Fig 145. In combination, a 30% reduction in drag of articulated vehicles is achievable with such modifications, Fig 146. The Leyland Freighter vehicle is shown in detail in Fig 147 with the modifications and their contribution highlighted. Fig 148 shows some of the fuel consumption benefits in tabular form. Model testing carried out at the Cranfield Institute of Technology has also thrown light on the effect of skirts and cab-to-body seals. The cab to body gap can be particularly detrimental in cross winds when there is a tendency for crosswinds within it to separate fore-aft flow prematurely on the leeward side of the vehicle. The same problem applies with crosswinds blowing through the underside of the vehicle. The advantages of skirts in this context are shown in Fig 149.

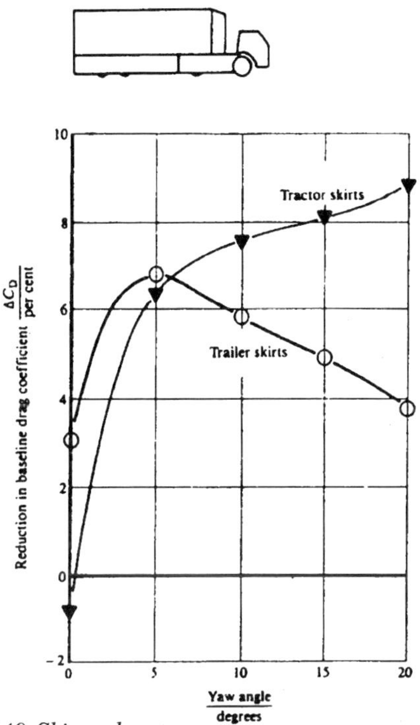

Fig 149:Skirt advantages

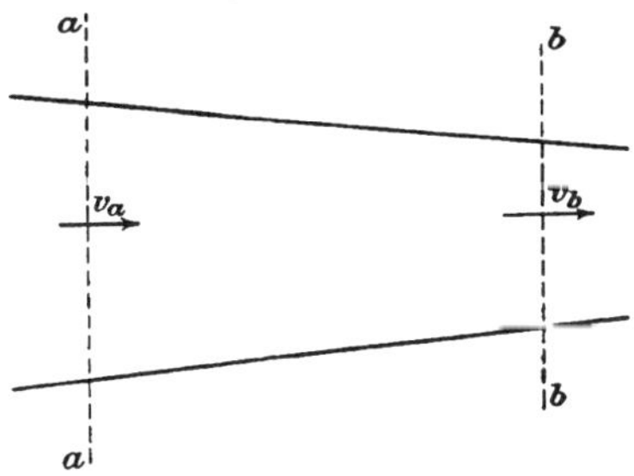

Fig 150: Flow in a duct

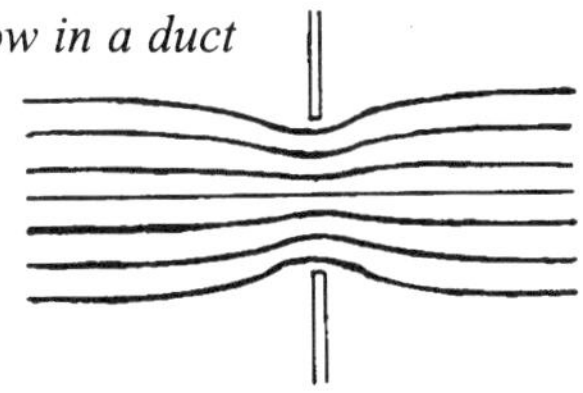

Fig 151: Constriction in a duct

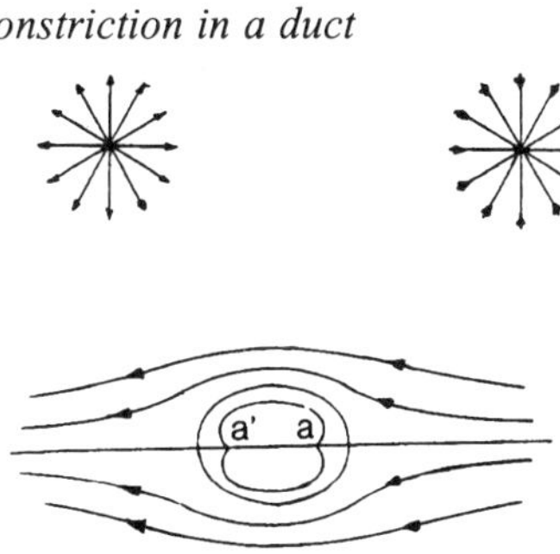

Fig 152: Flow around oval section

Local air-flows

Local airflows in the vicinity of the engine cooling system are examined, which are of interest both in determining the contribution of the system to overall vehicle drag and in determining the air-flow available for cooling heat transfer.

Detailed studies of local airflows require a knowledge of some of the basics of fluid mechanics, starting with the definition of a fluid as a material substance which cannot sustain a shearing stress when at rest. In fact, real fluids are not completely inviscid but often the viscosity effects are negligible, as is the fluid's incompressibility under normal conditions. Crucial to the understanding of forces imposed on bodies in a fluid stream is the significance of relative body between body and stream — as opposed simply to the absolute velocity of the stream. Forces on any point on the inclined surface of the body can be resolved into a pressure normal to the surface and a shearing stress tangential to it. The 'total pressure force' is the summation of all the elemental point forces normal to the surface and the 'skin-friction' is the corresponding summation of tangential forces.

In flow through ducts such as that defined by the air-inlet system, an equation describing the continuity of flow can be written which is useful in calculation. Between two stations (1,2) of the duct, Fig 150, with the assumption of continuity of flow (no fluid added or removed at any point), and with the fluid assumed to be incompressible:

$$s_1 v_1 = s_2 v_2$$

where s is the spacing of the streamlines.

Since v is inversely proportional to s, the effects of constrictions in the ducts are shown as Fig 151. These show the classic increase in velocity at the throat of a venturi, as the streamlines come close together, and on the outer edges of a cylinder crossing a wide, rectangular-section, duct. Corresponding reductions in velocity occur upstream and downstream of the cylinder.

The Bernoulli equation relates fluid velocity to pressure, p such that

$$p + \rho v^2/2 = B$$

where B is the Bernoulli constant and the second term is the dynamic pressure, or the increment above the static pressure (at a stagnation point in the flow) that describes the freestream pressure at any point in the flow. In air at sea level its effect is fairly minimal, and drag forces are the result of pressure changes rather than air-density changes.

A rotating fluid is termed a vortex, and a free vortex if no external forces are applied to it. For tangential flow velocities v_1 and v_2 at radii r_1 and r_2 from its centre, a centrifugal head of pressure is imposed:

$$= (v_2^2/2g) - (v_1^2/2g)$$

the basis of the power developed by a centrifugal blower. Applying Bernoulli's thereom, with constant pressure head, the static pressure decreases rapidly towards the centre of the vortex. When the fluid flow is filled with vortices, a so-called shearing flow occurs as in the case of the shallow layer (boundary layer) close to the surface of a body in the fluid stream or in its wake. If the shearing flow takes place in a very thin layer it is called a vortex sheet.

Sources and sinks are useful concepts in visualizing fluid flow, the former being a point in the fluid from which flow issues symmetrically in radial direction — and the latter is its converse. An oval shaped section body in a flow could be represented by a source and sink pair, Fig 152, the two being analogous to the poles in a bar magnet and their associated magnetic field; a and a' are two stagnation points where the uniform flow velocity is equal and opposite to the velocity caused by the source-sink pair. A circular section body in a flow can be represented by so-called doublet in which the distance between source and sink shrinks to nothing. When a vortex and doublet combine in a uniform stream, Fig 153, the stagnation points are displaced from the vortex field away from the axis of symmetry. The effect of the vortex is thus said to give the flow a camber or curvature as a symmetrical body in the stream which

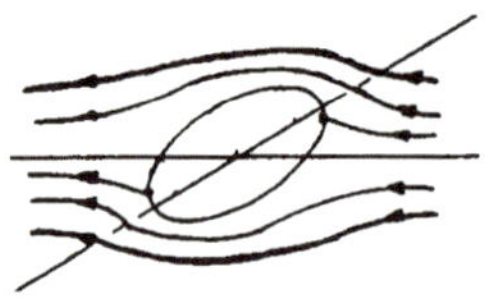

Fig 154: Flow around elongate section

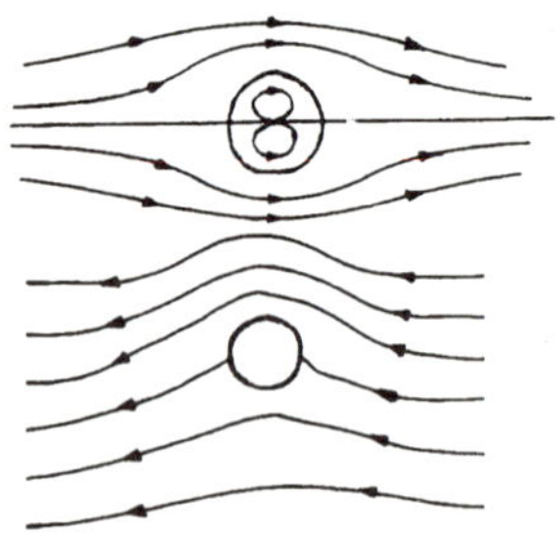

Fig 153: Vortex and doublet

Fig 155: Flow around flat plate

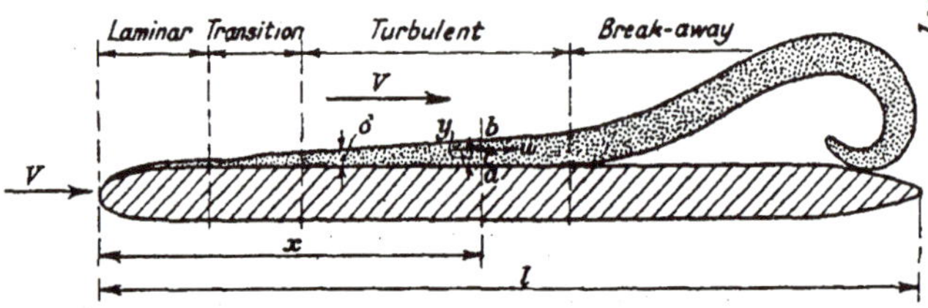

Fig 156: Velocity variation in boundary layer

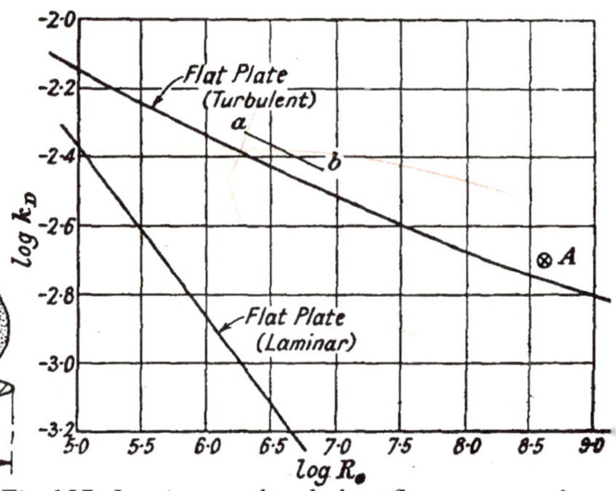

Fig 157: Laminar and turbulent flow compared

was cambered at an an angle to the stream would give a similar flow field.

If there is no circulation around a body of elongated section, Fig 154, inclined to the free-flow the streamline pattern is symmetrical with upstream and downstream stagnation points — and for a flat plate as in Fig 155. Variation in velocity within the boundary layer, thickness δ, can be visualised with the help of Fig 156 representing a surface past which fluid is flowing with velocity V. The body immersed in the fluid will have a Reynolds number of

$$R_e = Vl\nu$$

where l is its linear dimension and n the fluid viscosity. For low R_e, say less than 500,000, the flow within the layer is streamline and for high values, above 1,000,000, turbulent. In practice different combinations as shown apply. In the laminar flow region, thickness of the boundary layer can be approximated as

$$=5.83(lx)^{1/2}/R_e^{\ 1/2}$$

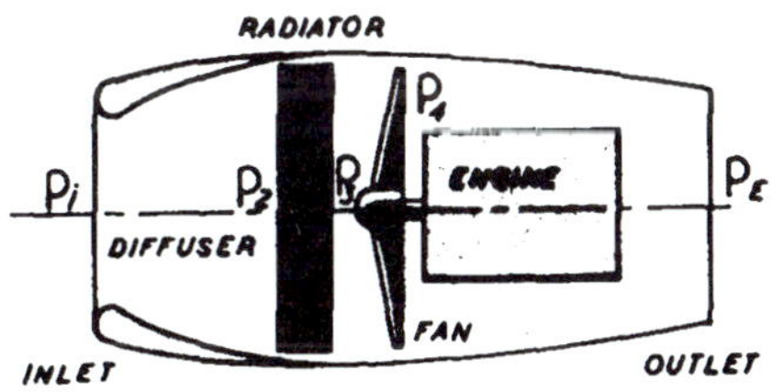

Fig 158: Ideal cooling system

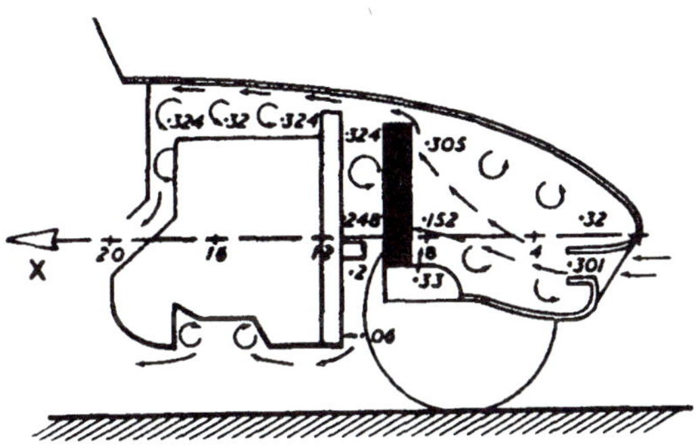

Fig 159: Actual cooling system

Drag co-efficient for a flat plate can be found from the Karman equation to be

$= 1.327(1/R_e)^{1/2}$ for laminar flow and

$=0.072(1/R_e)^{1/5}$ for turbulent flow, Fig 157.

Airflow for engine cooling

Estimating the quantity of excess heat to be dissipated for a given vehicle speed is a useful first step. This can be obtained from

$$K_1A(T_w - T_a)V^{0.8}$$

where A is radiator area, K_1 the radiator constant, T_w the radiator wall temperature and T_a the air temperature; V is airflow velocity at the entrance to the radiator. Rate of air mass flow required is $=\rho AV^2$.

The ideal cooling system, Fig 158, comprises a straight sealed duct, the inlet of which leads to the diffuser and then through to the radiator, followed by an outlet at the rear. In a less idealised system such as Fig 159, with typical internal pressure distribution shown, the aerodynamic efficiency of the radiator can be estimated from a formula due to Skibor-Rylski as

$$2K(T_w - T_a)/\rho V^{1/2}C_d$$

where C_d is its drag co-efficient. As the efficiency increases with airflow velocity, a long duct in front of the radiator is to be preferred with a gradually increasing cross-section. The diffuser should regulate the airflow to obtain the optimum speed. Efficiency is maximised by having the airflow sufficiently turbulent for the air particles to spend the most time collecting heat from the walls. The front grille can be designed to produce the right conditions of turbulence.

Experiments have shown that the diffuser duct should not decline more than 7 degrees to its centreline at any point, to avoid flow separation from its surface. Total drag due to the internal airflow can be expressed as the loss in momentum of the air between inlet and outlet, the product of air mass flow and velocity differential between inlet and outlet. This needs to be added to the pressure drop over the same path times the exit area.

Chapter 9: Body fittings/controls

Wiper/winding/locking mechanisms
Other interior mechanisms
Electric wiring & electronic controls
Body electrical systems and electronic control systems
Advanced body electronics;
Networking of body system controls

Screen-wiper and door-latching mechanisms

The fitment of body mechanisms to vehicles has sometimes been regarded almost as an afterthought within the design process — and understandably there has been a desire to utilise where possible proven products from past applications. Often these units are designed for a restrictive specification imposed by a particular vehicle and their perfomance may be a poor compromise for the application. As computer resources have made possible a simplification of hitherto lengthy design procedures, it is well worth while considering a wide choice of drive linkages and using a mechanism design package to optimise the choice for a particular application.

International safety standards are having a profound effect on the configuration of window and door mechanisms. Among others, FMVSS and ECE legislators lay down rules for wiped areas and wiper speeds, in the case of screen-wipers, also fully and secondary latch positions, with integrity under 10/12 kN transverse/longitudinal loads, for door-latches.

Established design for a latching device is shown in Fig 1. Rotating bolt latches such as the Wilmot Breeden example shown in Fig 2 overcome the problem with sliding latches, of overcoming the frictional force on the slider created by the door seals. The effort needed to withdraw the pawl of this claw latch is

$$(dsa/b)m_1 + m_2r/R)$$

where $m_{1/2}$ are the friction co-efficients at the positions shown. Sometimes a mechanical stop, or detent is applied to the rotating element — in this case a

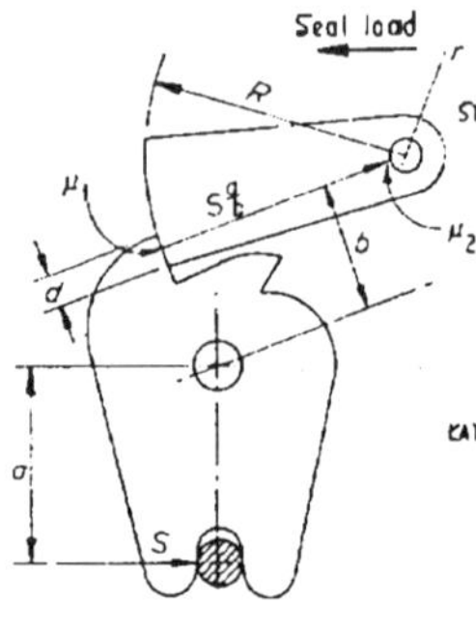

Fig 1: Latching mechanism

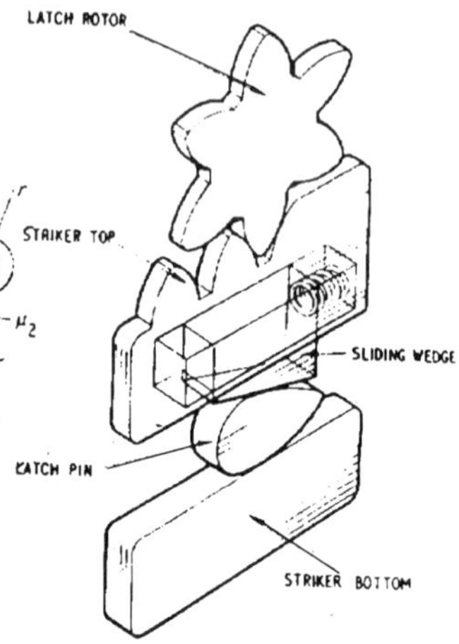

Fig 2: Rotating bolt latch

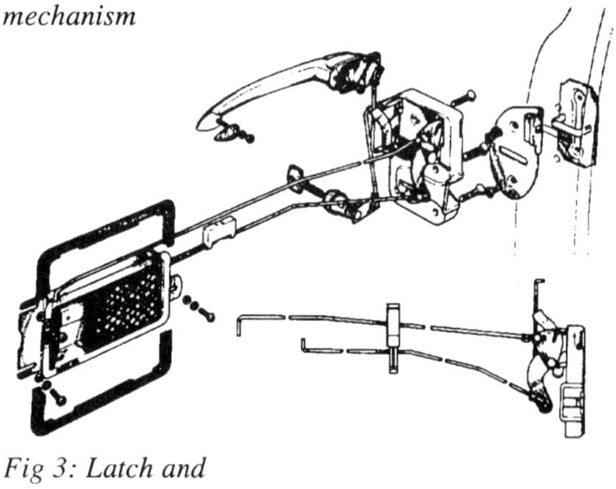

Fig 3: Latch and lock assembly

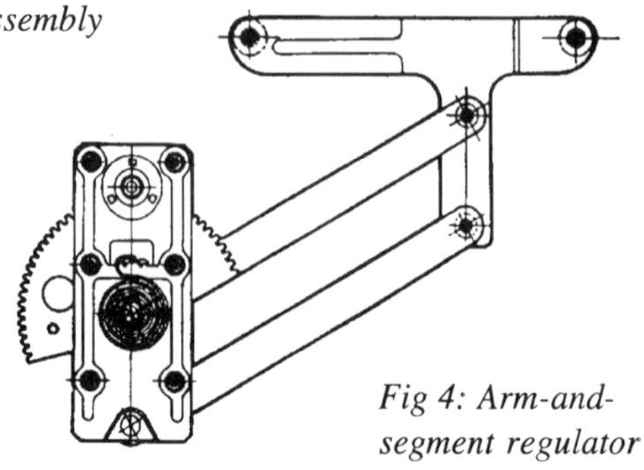

Fig 4: Arm-and-segment regulator

striker pin is engaged with the arm of the claw in the closed position. Further complication of the basic linkage results from the need to hold the door rattle-free over rough terrain where forces of about 50 lbf (1.1 kN) can occur between door and pillar due to body torsional deflection. Means for resisting this separation range between moving-wedge devices to elastically deforming members.

The first type is illustrated in Fig 2 which a sprung wedge moves in to take up any gap between door and pillar members. Shown in Fig 3 is the total latch and lock assembly in plant-on form, with staple which forms the striker.

Window regulators and screen-wipers

In window regulators, the arm and segment type (Fig 4) was the traditional pattern for volume cars and trucks; in recent years there is evidence of a switch to the use of flexible plastic perforated tapes which act as racks to engage pinion gears on the winder. With the first type, a number of multi-arm variants are available for extra large and curved glasses.

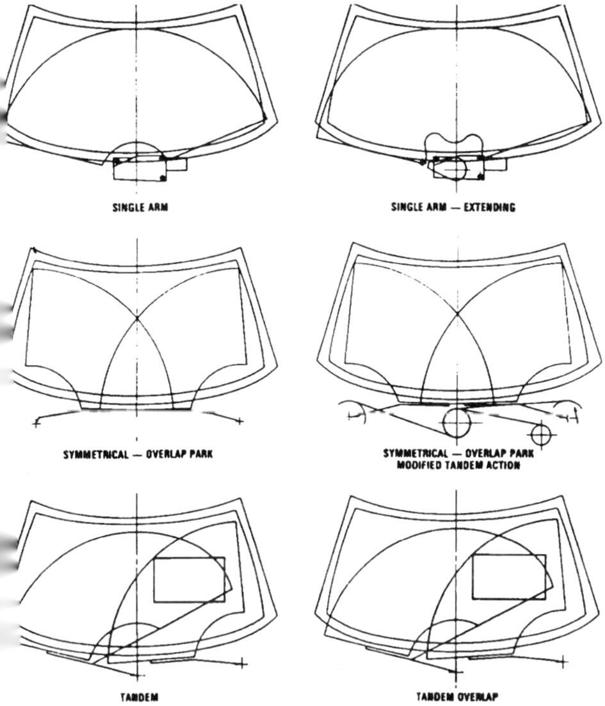

Fig 5: Screenwiper linkage variations

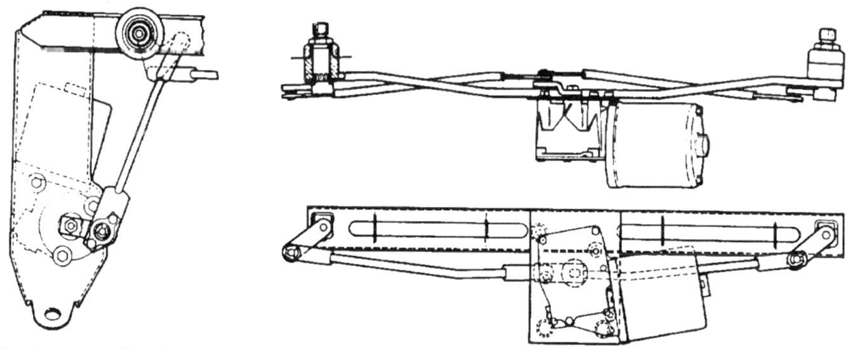

Fig 6: Typical wiper mechanism

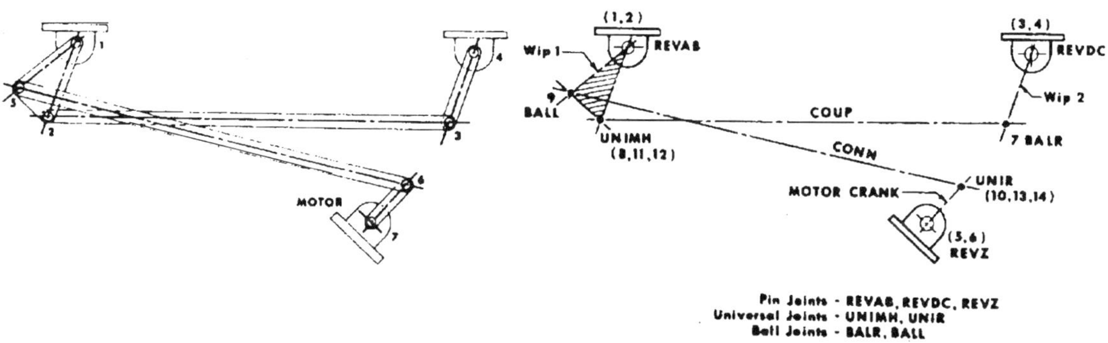

Fig 7: Mechanism modelling

Equivalent KAM model

Windscreen wiper mechanisms are governed by wiper-area patterns quoted in such standards as SAE J903A and J941. Some wiper patterns, available from Delco Dynamics linkages, are shown in Fig 5. Linkage types, like that in Fig 6 from Wilmot Breeden, are said to be more efficient than the flexible rack type but the latter is less demanding in space. A number of software packages on mechanism linkage design particularly suit the screenwiper. One due to Singh of Ford (USA) used the model in Fig 7 to represent the two basic four-bar linkages of the mechanism, for obtaining the characteristics shown in Fig 8 — which can be analysed in respect of proneness to stall, unduly high velocity/ acceleration peaks or poor load distribution on the motor between forward and return strokes.

Synthesis of mechanisms

The rational rather than intuitive design of mechanisms involves both finding the best type of mechanism and the dimensions of the mechanism. Elements of a linkage which connect together are called kinematic pairs, as exemplified by those in Fig 9. So-called lower and higher pairs are those with less or more than two degrees of freedom while 'mobility' of a mechanism is the residual number of degrees of freedom which can be found from:

$$3(n\text{-}1) - j_h - 2j_i$$

where j_h and j_i are the numbers of higher and lower pairs respectively and n the number of links. This will govern the way a number of links can be put together to form different linkages.

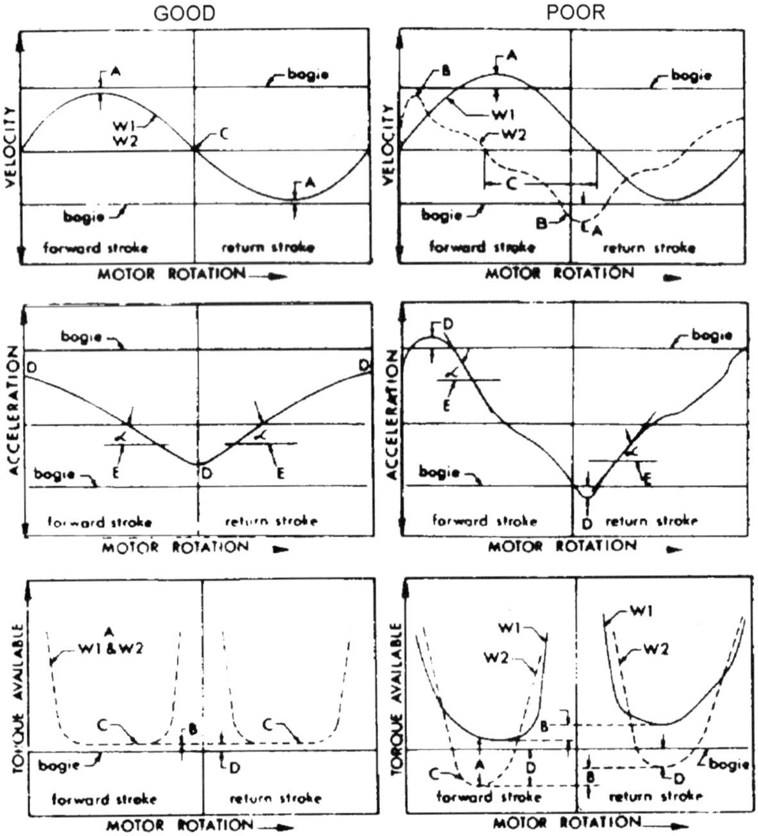

Fig 8: Simulated wiper performances

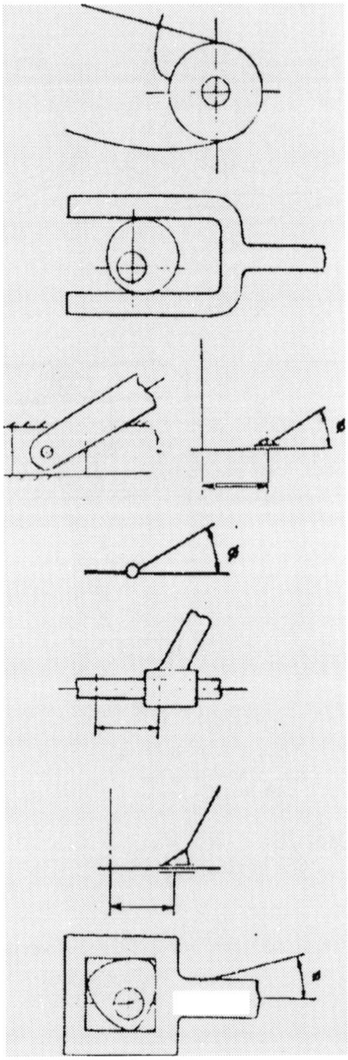

Fig 9: Kinematic pairs

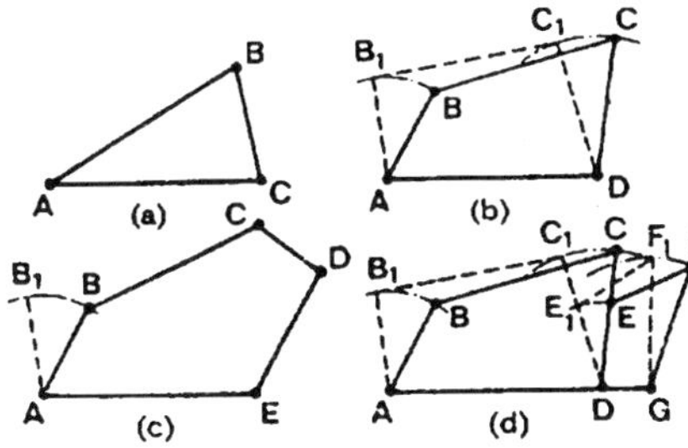

Fig 10: Kinematic chains

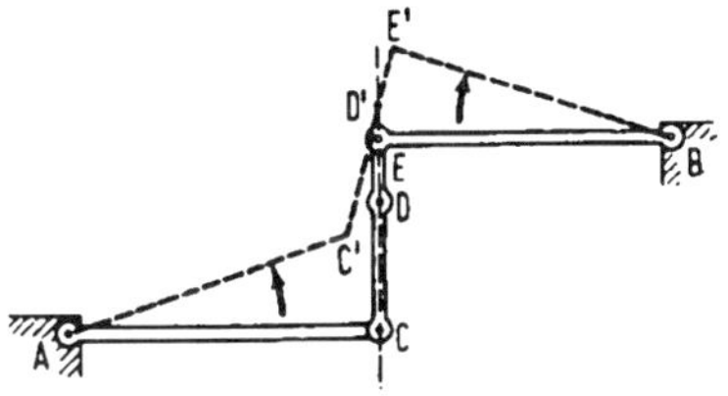

Fig 11: Watts linkage

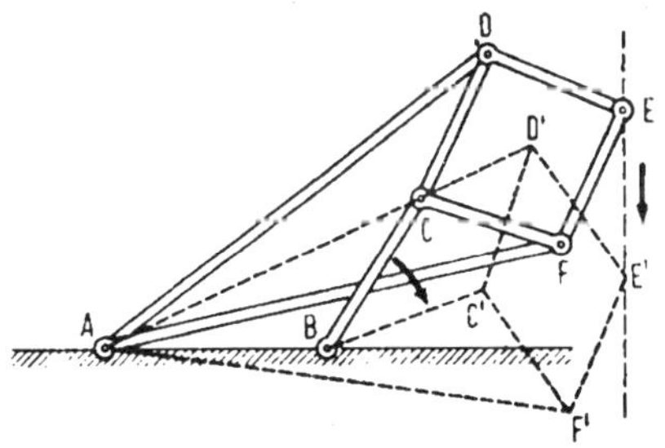

Fig 12: Peaucellier's mechanism

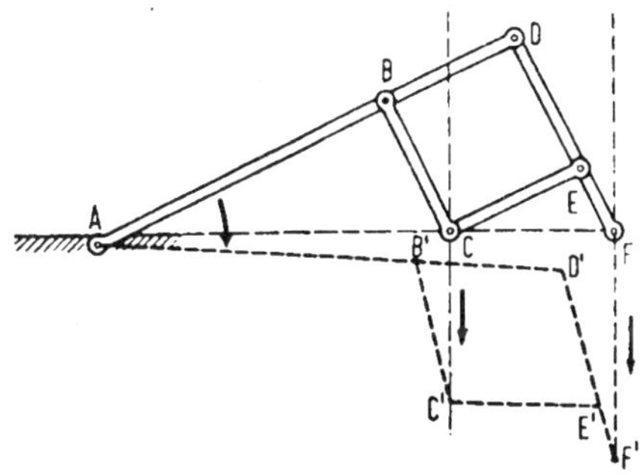

Fig 13: Pantograph mechanism

A kinematic chain is a combination of kinematic pairs in which each link forms part of two pairs and in which relative motion is completely constrained. Fig 10 shows three arrangements in which each link forms part of two turning pairs, so that the first part of the definition of a kinematic chain is satisfied by each arrangement. At (a), a rigid frame is formed, with no relative motion between links, whereas in (b) relative motion is completely constrained. However, if five bars are connected together by pin-joints, as at (c), unpredictable motion takes place. To build a more complex kinematic chain from that shown in (b), two more links may be added provided these form pairs with two existing ones — as well as themselves forming a pair, (d). Proceeding in this way, complex linkages can be constructed, for which the following general relationship applies:

$$n = 2j - 4 \quad or \quad n = (2/3)[j + 2]$$

provided only lower pairs are involved.

Strictly the word 'mechanism' should only be used when one link of a kinematic chain is fixed and the arrangement is used for transmitting or transferring motion. If the number of links is n, the same number of mechanisms can be made by fixing each of the links in turn, each being an 'inversion' of the original chain. When the mechanism has links strong enough to do work as well as just transmit motion, it becomes a 'machine'.

Straight line mechanisms

It is often a requirement to design a mechanism to produce straight-line motion and those without sliding pairs are usually the most useful for automotive purposes. The familiar Watts linkage of Fig 11 has the smallest number of links and pivots of any with only turning pairs. It has the disadvantage of a limited range of motion and an alternative is Peaucellier's mechanism of Fig 12 which has eight links but produces greater relative output. It would also be theoretically possible to combine a Watts linkage with a pantograph mechanism, Fig 13, which has the effect of magnifying the primary output to an enlarged secondary output. Point (C) in the figure could be constrained to move in a straight line by the primary linkage. While such a system could operate satisfactorily at high speeds, wearing of the pivots, especially (A) and (B), affects accuracy quite quickly.

Body interior mechanisms

The application of servo-assistance to body interior mechanisms is radically changing their configuration. There is also a trend for the larger suppliers to offer total system sub-assemblies, such as the door shell, which could lead, if not to integration of the mechanism, at least commonisation of the servo - power source.

The theory of mechanisms related to screen-wiper and door-latching systems was introduced in the last section. Two further important groups are seat-adjustment and window-winder systems, both of which are increasingly power-assisted as lower-cost electrical motors become more readily available. A typical manually adjustable seat mechanism for commercial vehicles uses a parallelogram linkage. At rest, this is locked by a crank which drops into slots in the frame. A tension-spring biasses the mechanism against the weight of the driver and a torsion spring holds the crank in the chosen slot. For cars, a conventional arrangement, the traditional slide adjuster, has been improved by the use of nylon-rollers and friction pads and a further refinement has the lever pivoted directly on the moving member. Now complex servo-adjust systems are employed, Fig 14.

For screen-wiping, the scissor linkage discussed in the earlier article is largely superseded by flexible rack and pinion designs, even on commercial vehicles. A commonly adopted standard for design is the SAE J 903 regulation shown in Figs 15/16, which includes requirements for 2-speed wiping. Fig 17 shows the classic flexible-rack system which gains over the scissor-linkage in compactness. Drive from the permanent magnet motor is stepped down by about 40:1 through a plastic gear-set. The connecting rod big-end embraces a loose eccentric, frictionally connected to the crank and permitted to turn through 180 degrees; see inset. The motor is then made reversible so that in normal operation the pin travel corresponds to symmetrical blade angles. Reversal of the motor causes the eccentric to rotate and displace the travel of the pin so that the wipe is asymmetrical, and brings the blade to the edge of the screen at the limit of travel. At the same time a crosshead causes a plunger to actuate a limit switch and stop the motor, which is quickly brought to rest by the braking effect of the overrun voltage as the armature brushes are shorted out.

Bonnet hinging and locking mechanisms are

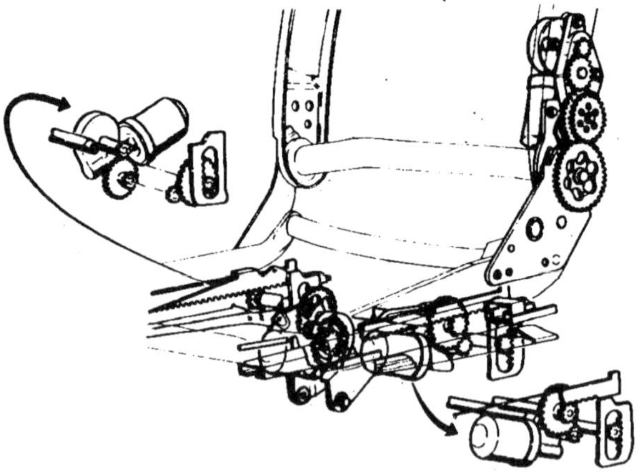

Fig 14: Servo-adjusted car seat slide

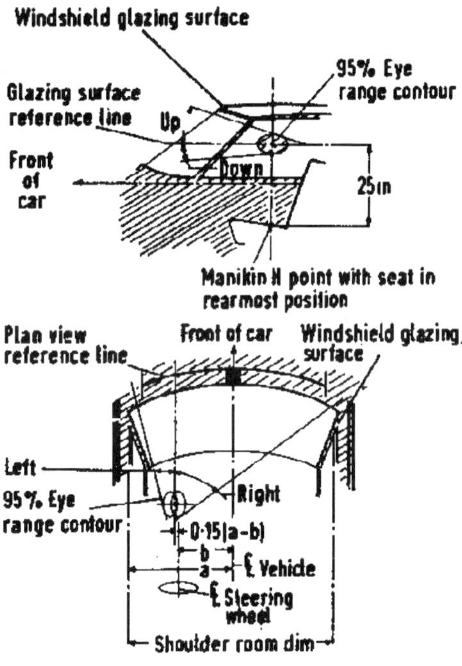

Fig 15: Notation for SAE specification

Area	Area wiped min, %	Eye range contour, %	Angles, deg			
			Left (W)	Right (X)	Up (Y)	Down (Z)
A	80	95	18	56	10	5
B	95	95	14	53	5	3
C	100	95	10	16	5	1

Fig 16: SAE specification

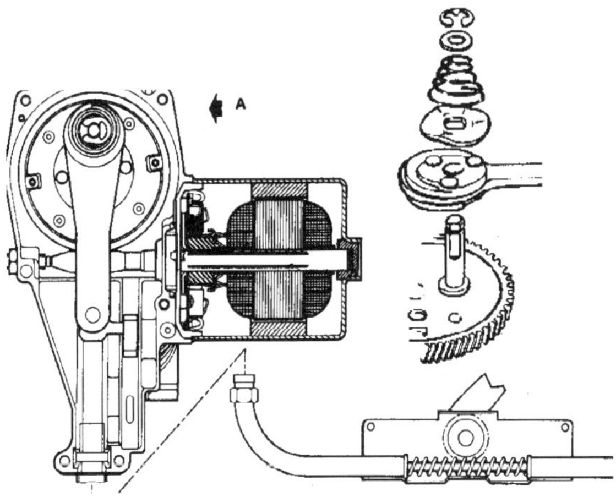

Fig 17: Flexible rack wiper drive

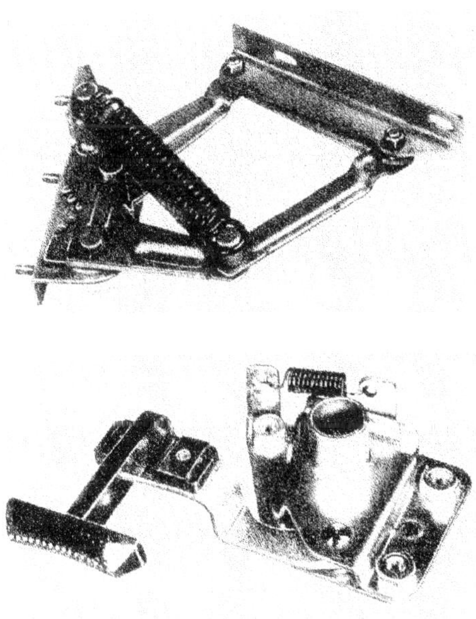

Fig 18: Bonnet hingeing and locking

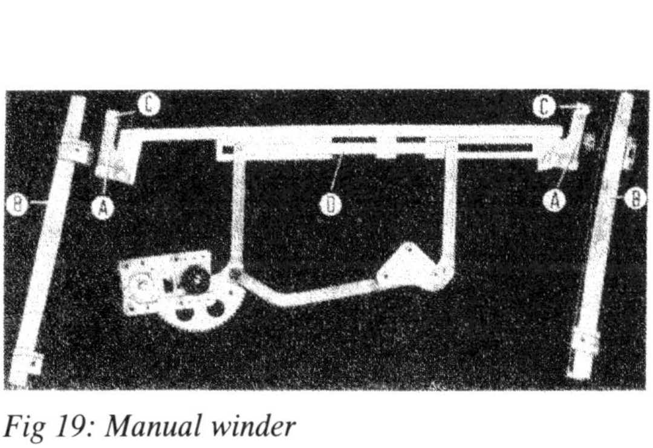

Fig 19: Manual winder

typified by the systems shown in Fig 18, an assisted-lift geared hinge and a secondary latching arrangement. The former uses a 5-bar linkage to bias the bonnet into the open position, two of the links being geared together. A pawl with torsion spring is seen in the latter, access to which is usually through the radiator grille. The traditional window-regulator mechanism, exemplified by the system shown in Fig 19 for open cars is operated by a double arm geared regulator. Moulded nylon guides at A have upper and lower bearing surfaces which run in the window drop channels, B. Upper bearing surfaces C contact specially shaped upper sections of the channels, causing lateral pressure to tilt the glass inwards in the raised position. The lower bearing surfaces A raise the glass while the strip D, connecting the regulator arms, maintains the lower edge of the glass horizontal. In an electric window regulator, the mechanism consists of a lifting screw and nut with a scissor-type linkage attached to the latter. The screw is driven by a reversible electric motor through a lost-motion coupling. A protective device is usually specified which prevents the motor windings from overheating. Fig 20 shows an alternative hydraulic drive system.

Fluid-power servo assistance can also be applied to door-hinging mechanisms, as seen by the Dunlop system shown as Fig 21. In this case pneumatic power is used in a system of inflatable sacs which act against vanes. Nylon pads shown at A, inset, prevent wear due to sliding between vane and sac. A zed linkage has also been considered which will increase door-opening from 90 to 180 degrees. Measuring 8 x 3 x 3 in (203x76x76 mm), air displacement for the powered hinge is 19 in^3 (31.14x10^{-5} m^3).

Seat-reclining mechanisms are the speciality of the PL Porter Company. The Sedgelok system comprises an incremental recliner, Fig 22, having an upper hinge arm which incorporates a gear segment. In the locked position this mates with a slide which prevents the rotation of the upper arm about its pivot. The lever which actuates the locking slide is spring-returned so as to have a cam action when the slide is loaded, thereby increasing locking force in response to seat back loading. Another neat way of achieving locking at different positions of angular displacement is available from the company in the form of the Mechanical Lock system, Fig 23. Here, a steel rod is gripped by two clutch coils which bear against specially configured holding bushings swaged into a surrounding tubular housing. These either have an angled coil-contact face or a flat which allows slipping action at a predictable force input. Inboard ends of the clutch coils are positioned against notches in the actuation lever so that rotation against the coil helix expands their diameter so as to allow infinite angular positioning for subsequent relocking. Another variant of it is the Electro-Hingelok of Fig 24 which can be used to power-drive a recliner mechanism, Fig 25.

Delco Products were pioneers in the development of a number of powered body-interior mechanisms for Europe following their long experience in the US market. Central locking systems based on the compact lock-actuator of Fig 26 can be used with single and multiple control points as seen in Fig 27. Direct drive actuators are available for doors, tailgate, boot and fuel-filler flap. An input at the door 'switch' causes the relay timer to be triggered. Fail-safe features built-in ensure the system does not respond to power supply interruption, open circuiting of the input switch or high voltage transients. It is also proof against internal operation in the event of a break-in through the windows.

The company can also supply a modular door system containing the central locking facility — as well as power window lift and air-flow valve, all on the mounting panel shown as Fig 28. Even window channels, glass, seals, latch mechanisms and audio speakers can be supplied as a total pre-assembled door package.

Adjustable mirrors are themselves giving rise to some interesting mechanism designs. The one in Fig 29 by Magnetex pioneered the compact through-door configurations. Its range of adjustment is 20 degrees vertical and 35 horizontal, allowing the same mechanism to be used on either side of the vehicle. The smaller of the two adjustment knobs controls horizontal inclination, effected by rotation of a captive spigot — the internally threaded bore of which engages a corresponding thread on the acetal actuating shaft which tilts the mirror. Vertical inclination by the larger knob, held captive between a static spacer and the smaller one, rotates to move a screw-mounted plastic sleeve. This motion is further controlled by a keyway in the spacer to tilt the mirror vertically.

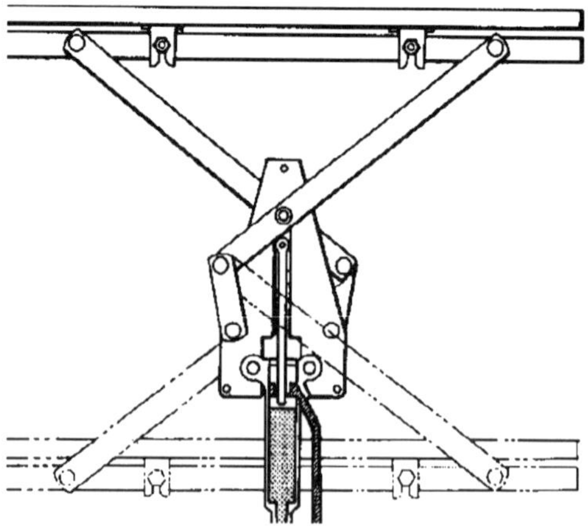

Fig 20: Powered winder

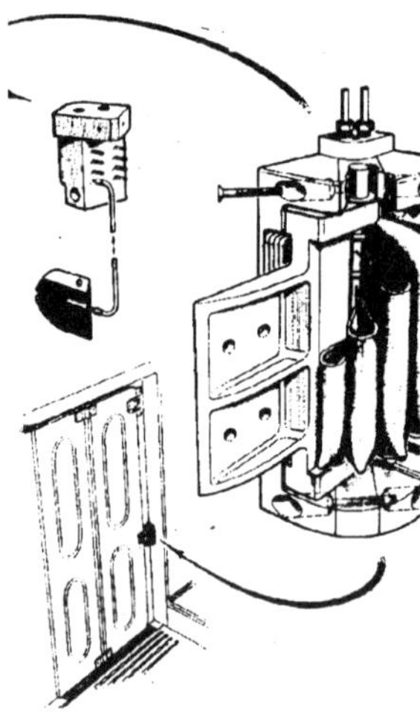

Fig 21: Powered door hinge

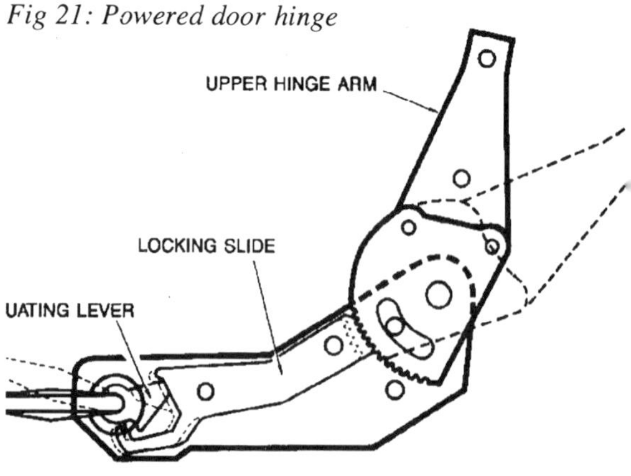

Fig 22: Porter Sedgelok

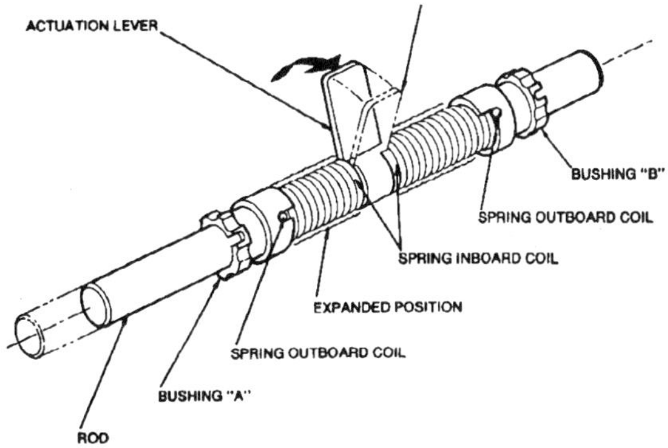

Fig 23: Mechanical lock

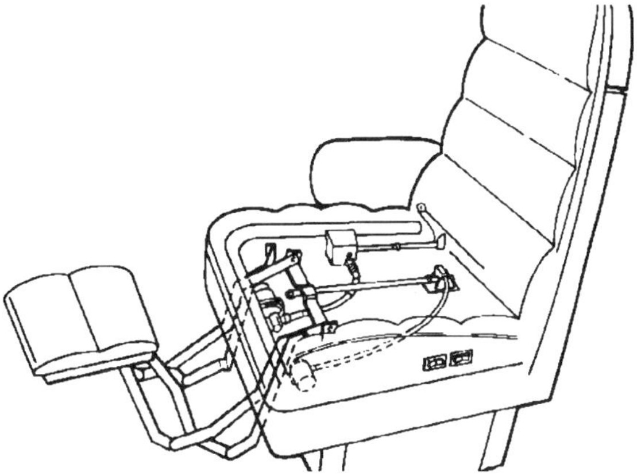

Fig 25: Dual-control of back and foot rests

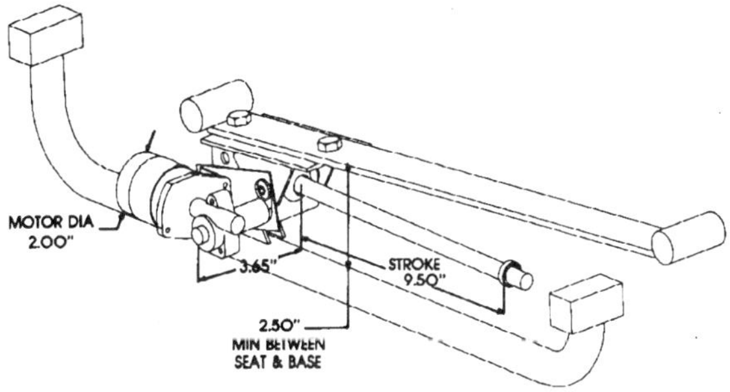

Fig 24: Electro-Hingelok

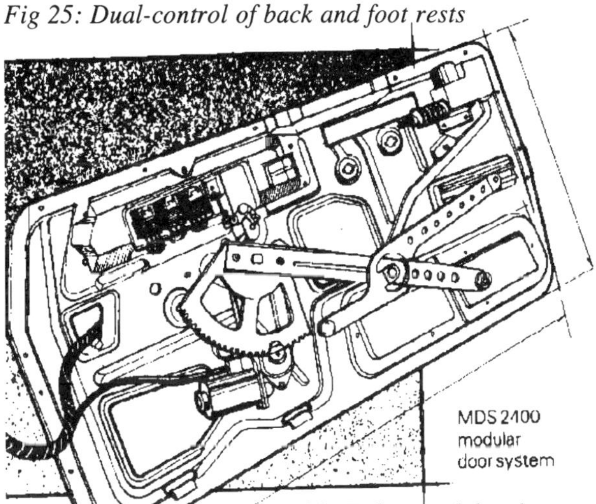

Fig 28: Delco modular door

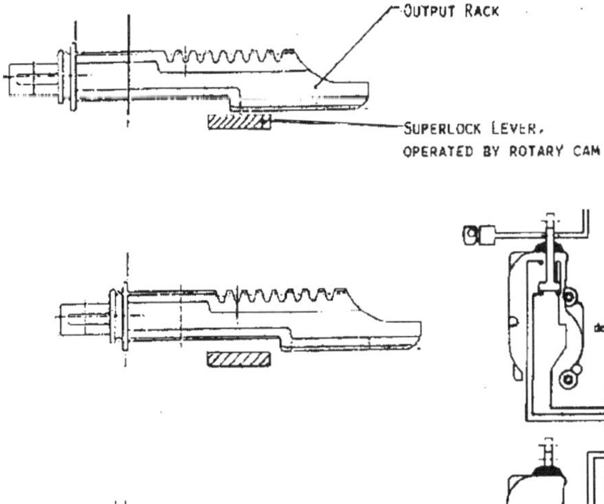

Fig 26: Lock-actuator: top, unlock; middle, lock; bottom, super-lock

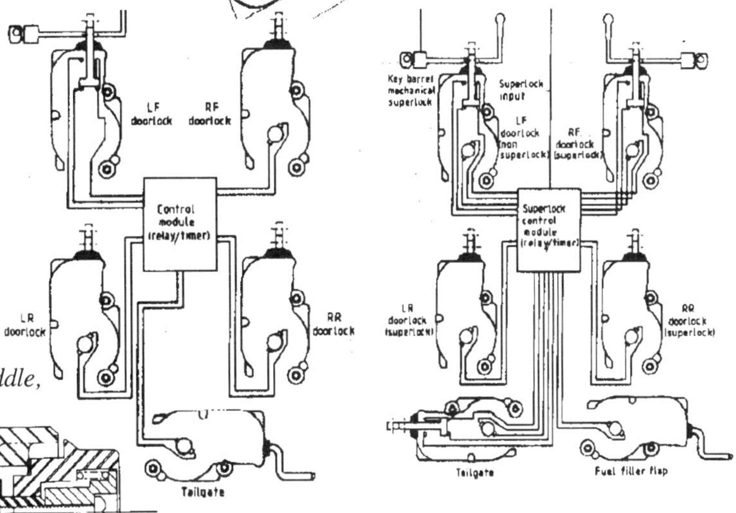

Fig 27: Delco central locking

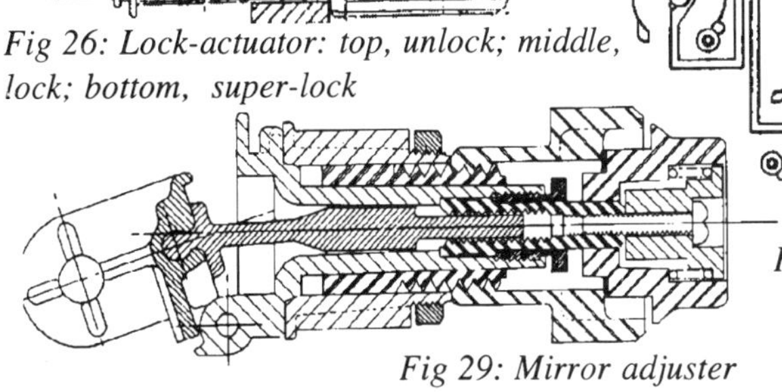

Fig 29: Mirror adjuster

Electric wiring and electronic controls

As vehicle electrical systems multiply there is need to simplify the wiring systems which connect them. This article introduces the reader to standardised layouts and coding conventions. Control systems are next examined and a sound introduction to on-board computer control is given with a description of multiplex networking.

A useful way to approach the design of an electrical system is to consider the balance of energy for the vehicle — from the small proportion of chemical energy burnt in the fuel to feed the electrical equipment to its division and conversion into sound, heat, light and mechanical energy. The need to

Type of Conductor	Permissible voltage drop in positive conductor U_{vl}	Permissible voltage drop in entire circuit U_{vg}	Notes
Lighting conductors from terminal 30 of light switch to lamps < 15 W to trailer socket from trailer socket to lamps	0.1 V	0.6 V	Current at nominal voltage and nominal power
from terminal 30 of light switch to lamps > 15 W to trailer socket	0.5 V	0.9 V	
from terminal 30 of light switch to headlamps	0.3 V	0.6 V	
Charging cable from terminal B+ of alternator to battery	0.4 V at 12 V 0.8 V at 24 V	– –	Current at nominal voltage and nominal power
Control leads from alternator to regulator (terminals D+, D–, DF)	0.1 V at 12 V 0.2 V at 24 V	–	At maximum excitation current (note 1)
Main starter cable	0.5 V at 12 V 1.0 V at 24 V	– –	Starter short-circuit current at +20°C (notes 2 and 3)
Starter control lead From starter switch to terminal 50 of starter Solenoid switch with single winding Solenoid switch with pull-in and hold-in windings	1.4 V at 12 V 2.0 V at 24 V 2.4 V at 12 V 2.8 V at 24 V	1.7 V at 12 V 2.5 V at 24 V 2.8 V at 12 V 3.5 V at 24 V	Maximum control current (notes 4 and 5)
Other control leads from switch to relay, horn, etc.	0.5 V at 12 V 1.0 V at 24 V	1.5 V at 12 V 2.0 V at 24 V	Current at nominal voltage

Nominal cross section mm²	Resistance per meter at 20°C mΩ/m	Conductor diameter Max. dimension mm	Cable diameter Max. dimension mm	Permissible continuous current (approximate values) at +30°C A	at +50°C A	Permissible current density, continuous operation A/mm²
0.5	37.1	1.0	2.3	11	7.8	10
0.75	24.7	1.2	2.5	15	10.6	10
1	18.5	1.4	2.7	19	13.5	10
1.5	12.7	1.6	3.0	24	17.0	10
2.5	7.6	2.1	3.7	32	22.7	10
4	4.71	2.7	4.5	42	29.8	10
6	3.14	3.4	5.2	54	38.3	6
10	1.82	4.3	6.6	73	51.8	6
16	1.16	6.0	8.1	98	69.6	6
25	0.743	7.5	10.2	129	91.6	4
35	0.527	8.8	11.5	158	112	4
50	0.368	10.3	13.2	198	140	4
70	0.259	12.0	15.5	245	174	3
95	0.196	14.7	18.0	292	207	3
120	0.153	16.5	19.8	344	244	3

Figs 30/31 (left and above): Copper conducting cables and their permissable voltage losses

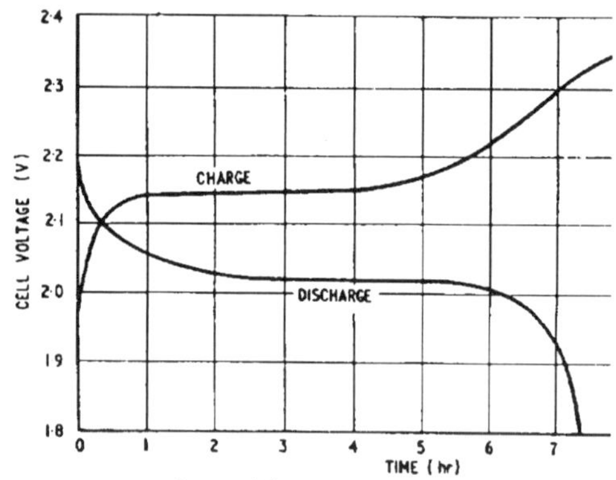

Fig 32: Charge/discharge cycles

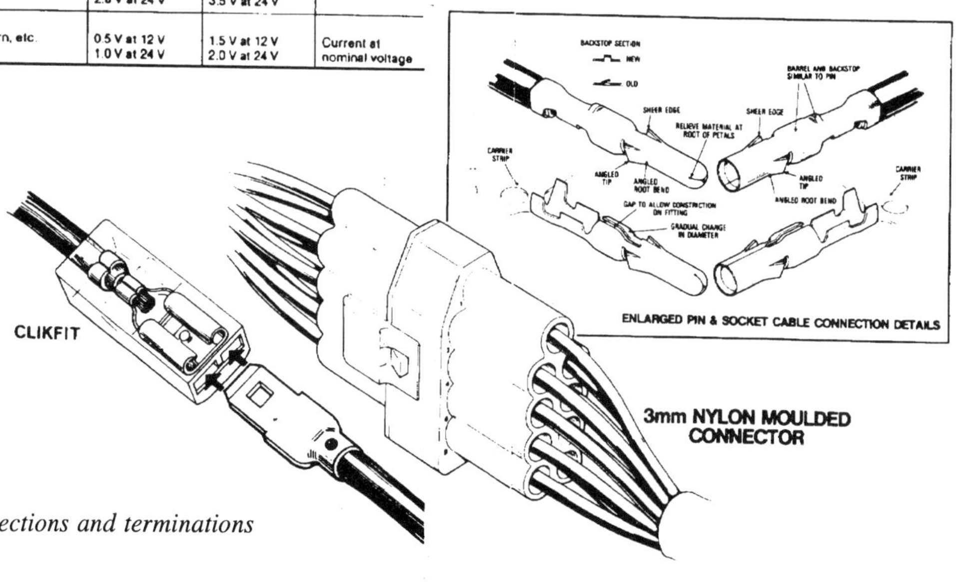

Fig 33: Connections and terminations

supply energy when the engine is at rest in turn creates the need for a storage battery and charging system. This way of thinking leads to the consideration of electrical loads and the respective cable capacity to feed the loads.

Traditionally, the so-called 'main' feeds to control box via ammeter (or charge indicator), from the battery, used 44 strands of 0.012 in wire while from the control box to lighting switch, a cable of 28 similar strands ran to the lighting switch. Proportionally 'lighter' cables then branched out to the various auxiliaries. Where flexibility is required in the cable, to accommodate relative motion, it is usual to increase the number of strands and decrease individual diameter.

There is also a traditional terminal notation in control boxes for auxiliary equipment — with 'A' signifying a battery feed reading on the ammeter. A1 is unfused and passing around load winding of voltage regulator. A2 is fused and passing through the load winding. A3 is unfused, through the load winding and controlled by the ignition switch. A4 is the same but fused.

Standardising wiring layouts

Statutory Instruments set out within Vehicle Construction And Use Regulations define the wattage and positioning of lighting equipment and cables, Figs 30/31. As alternators control output current automatically, only voltage regulation is necessary. Voltage curves for lead-acid batteries vary according to the the charge or discharge cycle as shown in Fig 32. One of the highest current consumers in the car auxiliary circuit (apart from the combined lighting load) is the windscreen wiper with consumptions from 2 to 3 amps in the case of car systems. Wiring terminations may be of the spade type for single cable interconnection or component termination — and of the pin and socket type for most multi-cable interconnections, Fig 33.

A circuit diagram for a petrol engined car using Bosch standard symbols is shown in Fig 34. Here, the traditional naming of parts has given way to a more recent standard of coding for device and terminal designation. Obviously trucks, buses and special purpose commercial/industrial vehicles are likely to have more complex layouts; use of section identifica-

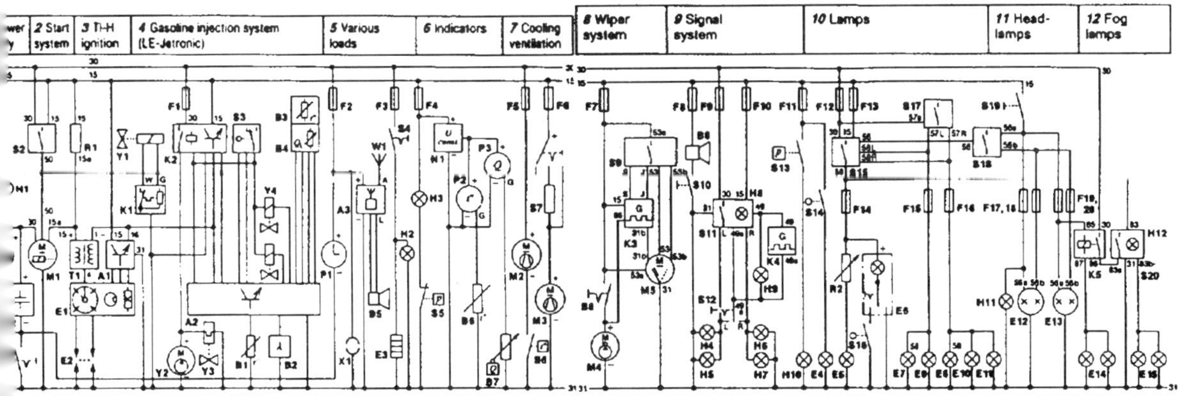

Device desig	Device	Section
	TI-H trigger box	3
	LE-Jetronic control unit (ECU)	4
	Car radio	5
	Engine temperature sensor	4
	Lambda sensor (only for USA)	4
	Air-temperature sensor	4
	Air-flow sensor	4
	Speaker	5
	Cooling-water temp. sensor	6
	Fuel-level sensor	6
	Horn	9
	Ignition distributor	3
	Spark plugs	3
	Rear-window defroster	5
	Back-up lamp	10
	Instrument-panel lamp	10
	Interior lamp with switch	10
	Side-marker lamp, FL	10
	Side-marker lamp, FR	10
	Tail lamp, BL	10
	Tail lamp, BR	10
	License-plate lamp	10

Device desig	Device	Section
E12	High/low beam headlamp, L	11
E13	High/low beam headlamp, R	11
E14	Fog lamp	12
E15	Fog-warning lamp	12
F1...F20	Fuses	
G1	Alternator with regulator	1
G2	Battery	1
H1	Alternator charge-indicator lamp	1
H2	Rear-window defroster	5
H3	Oil-pressure indicator	6
H4	Turn-signal lamp, FL	9
H5	Turn-signal lamp, BL	9
H6	Turn-signal lamp, FR	9
H7	Turn-signal lamp, BR	9
H8	Hazard-warning-signal indicator lamp	9
H9	Turn-signal indicator lamp	9
H10	Stop lamp	9
H11	High-beam indicator lamp	11
H12	Fog-warning-lamp indicator	12
K1	Thermo-time switch	4
K2	Control solenoid, Jetronic	4

Device desig.	Device	Section
K3	Wiper interval relay	8
K4	Hazard-warning-signal flasher	9
K5	Fog-lamp relay	12
M1	Starter motor	2
M2	Radiator fan motor	7
M3	Fan motor	7
M4	Windshield washer motor	8
M5	Wiper motor	8
N1	Voltage stabilizer	6
P1	Clock	5
P2	Cooling-water temperature indicator	6
P3	Fuel-level indicator	6
R1	Ignition-coil dropping resistor	3
R2	Variable resistor for E5	10
S1	Battery switch	1
S2	Ignition/starting switch	2
S3	Throttle-valve switch	4
S4	Switch for rear-window defroster	5
S5	Oil-pressure warning switch	6
S6	Temperature switch (radiator fan)	7

Device desig.	Device	Section
S7	2-speed fan switch	7
S8	Windshield washer switch	8
S9	Wiper switch	8
S10	Horn button	9
S11	Hazard-warning-signal switch	9
S12	Turn-signal switch	9
S13	Stop-lamp switch	9
S14	Back-up lamp switch	10
S15	Headlamp switch	10
S16	Door contact switch for E6	10
S17	Parking-lamp switch	10
S18	Low-beam switch	11
S19	Headlamp flasher button	11
S20	Fog-lamp switch	12
T1	Ignition coil	3
W1	Car antenna	5
X1	Socket	5
Y1	Cold-start valve	4
Y2	Electric fuel pump	4
Y3	Auxiliary-air device	4
Y4	Injection valve	4

Fig 34: Bosch circuit diagram for petrol-engined car

tion according to the DIN standard shown helps to make for relatively easy identification. Tables 1 and 2 show cable sizes and permissible voltage losses, suggested by Bosch, as an aid to harness design.

Electronic control systems

As traditional electrical equipment gives way to that involving electronic control of vehicle services the question of networking the controls together often arises. This then leads to the possibility of replacing the wiring harness by a multiplex system involving power and data 'buses'. The bus in this sense is a wire, or cable of wires, capable of simultaneous transmission of many power lines and control signals. Control systems which are commonplace now on luxury cars are almost certainly destined for a variety of tomorrow's specialist vehicles. On its XJ6 model Jaguar has reduced the signal wire cross-sections from 1.0 to 0.5 mm^2 and uses multi-way connectors of just 1.5 mm diameter and a signal wire earth switching technology is employed. A central microprocessor is used with 30 outputs to control systems for engine management, comfort control, driver information, suspension levelling and cruise-control. The lighting is also on a logic circuit which prevents the illegal selection of certain combinations of lamps. Outputs from the circuit energise relays located in modules at the four corners of the vehicle, the relays feeding power current to adjacent lamps. The modules also contain electronic circuits which can detect bulb failures.

Electronic systems desribed

It is useful to be able to describe how an automotive system works, in general, either 'qualitatively' by block diagrams or 'quantitatively' by the use of models. Subsequent systems analysis can lead to determining the function and performance of the system. The historic work of Joseph Fourier allowed any signal entering a system to be described by a combination of simple sine waves and the response of the system to be measured similarly. An equation can be written to describe the function of the system and the equation, whose plot (graphical representation) is the same as the frequency response of the system, is known as the transfer function. This can be used to show the changes in amplitude (gain) or phase (time delay) of the signal.

A control system is used to direct the operation of other systems and must be accurate, responsive, stable and have 'noise' immunity. Block diagrams for open and closed loop control systems are shown in Fig 35 for a piece of 'plant', the closed-loop system allowing the control to better meet the design criteria. The control shown is 'proportional' because the control signal is proportional to the error signal. Another familiar automotive control mode is called 'limit-cycle' in which the system output is monitored and then feedback control responds only when the output goes beyond preset limits.

Transistor technology

Solid state semiconductors are most commonly used to effect system response. The conductivity of the base silicon is altered by diffusing boron or phosphorous into it when it becomes, respectively a p-type or n-type semiconductor. Many combinations of types

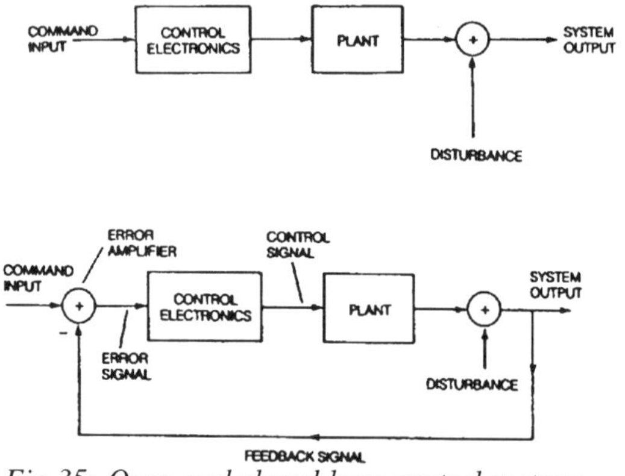

Fig 35: Open and closed loop control systems

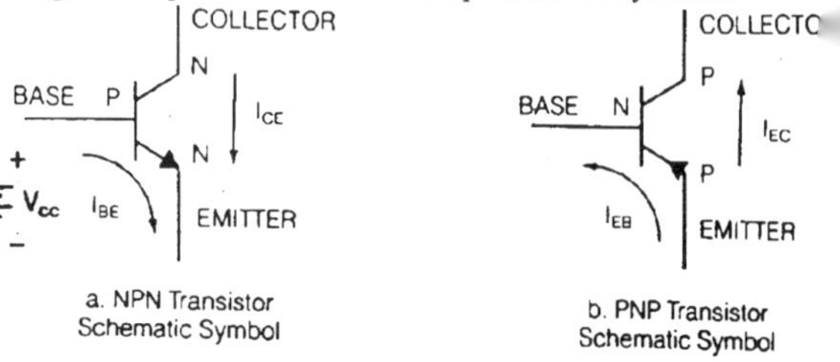

Fig 36: Common bi-polar transistors

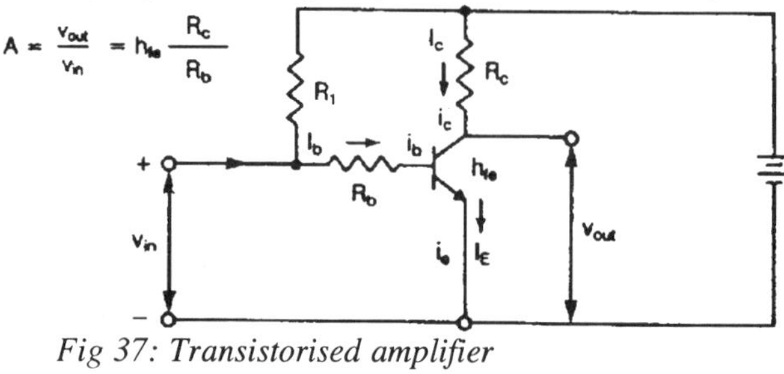

Fig 37: Transistorised amplifier

are used to make diodes or transistors (similar to traditional amplifying valves). A diode rectifier comprising p and n type junctions, anode and cathode, would provide a 'forward' resistance of about 510 ohms in the forward biased condition — of a positive voltage on the anode. In the opposite, reverse biased direction, the device is effectively an insulator with resistance of more than a million ohms. In contrast to diodes, transistors are active devices which either have gain or store energy.

The two common bipolar types, shown in Fig 36, are the NPN and PNP transistors which differ only in respect of current flow direction and otherwise act as amplifiers or switches. The transistor is effectively a combination of base-emitter and collector-base junction diodes as illustrated for an NPN type. This

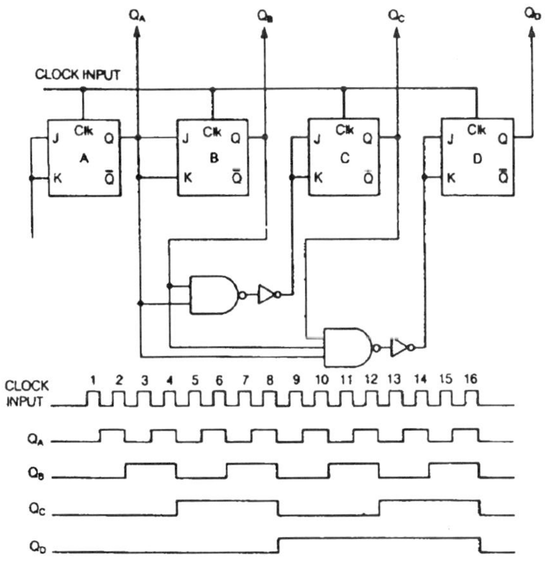

Fig 38: Four stage synchronous counter

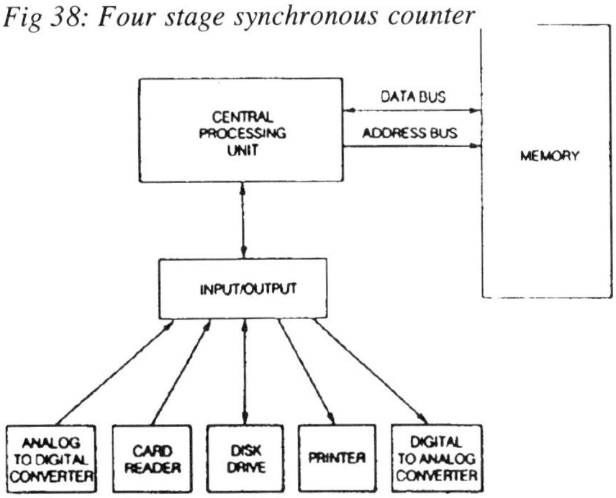

Fig 39: Block diagram for basic computer

would be shown as an amplifier as in Fig 37 with gain

$$= V_{out} / V_{in} = h_{fe} R_c / R_b$$

where h_{fe} is known as the small signal current gain. With suitable choice of resistances, very high voltage gains can be obtained in the case of the operational amplifier — the basic building block of the analogue computer, used to model a variety of automotive systems.

Digital circuits

Binary digital circuits are characterised by having an output which can only be one of two different states, represented by a voltage or current level. In such circuits the transistor is used as a switch between its 'cut-off' and 'saturation' conditions. The latter is beyond the linear amplifying range of the transistor — in which a large increase in base current produce little rise in collector current and voltage drop across the collector-emitter is very small as the collector-based diode becomes forward biased.

The on-off states of the digital circuits are represented 1 or 0 in the binary, or base 2, numbering system. While the decimal system has numbers grouped from right to left, in representing ones, tens, hundreds and so on, the binary system numbers are grouped from left to right on a power base 2 instead of 10. The binary number 0010 is read as 'zero,zero,one,zero' and not 'ten' while decimal 256 is binary 100000000. Digital computers perform binary digit ('bit' for short) manipulations using three basic logic gates: the NOT, AND and OR gates. The first is an inverter for which an input of 1 becomes an output of 0 and vice versa. The second has at least two inputs and one output, the latter being high (1) when both inputs are high (1) and low (0) if both inputs are low (0). The third, like the second, has two inputs and one output which is 'high' when one OR two inputs are 'high'.

Other common gates are NAND, NOR, XOR and 'adders'. These can be combined to build simple 'combinational' circuits or more complex sequential or memory circuits in which the sequence of past input values, together with the logic state at the time, determine the present output state which may be 'latched' in one way or another. Such flip-flop circuits may be synchronised to change state at a particular time determined by a clock pulse. Fig 38 shows a four stage synchronous counter formed from such circuits which can be used to count 2 to the power 4, or 16, pulses before returning to a starting state, resulting in the timed wave-forms shown at the outputs. Such stages can be lined up to store the digits of a binary number as a temporary register or permanent memory. Because digital circuits can be made from combinations of many identical gates they are ideally made in integrated circuit form, the 1958 invention made by Jack Kilby of Texas Instruments which sparked the micro-electronics revolution.

Computer control

A block diagram for a basic computer is shown in Fig 39. When integrated circuits (ICs) are used the central processing unit (CPU) becomes a microprocessor in which all arithmetic and logical decisions are made. The memory can be 'written' into or 'read' from — and is termed a random access memory (RAM) to distinguish it from read-only-memory (ROM) circuits which can be added to introduce special facilities. This is by means of so-called 'firmware' programs — because unlike software they are unchangeable.

A typical engine control computer does arithmetic using only 8 bits, allowing decimal numbers up to 127 to be represented (plus one bit to show +ve or -ve). The machine's internal memory has several thousand data latches (flip-flops), one for each bit stored, each with a location number which can be 'addressed'. Both the instruction 'program' and data can be stored in the internal memory, once loaded from the appropriate peripheral device. The address of bits is stored as a binary number within temporary latches called registers. Fig 40 shows how the address bus is wired relative to the data bus. A 16 bit address register can access 65,536 memory locations containing 8 bits (one byte). The data bus, usually only 8 bits wide, is a two-way highway compared with the single-way highway of the address bus. A clock signal tells the memory when it can take and release control of the data bus. By combining bytes and operating on them one after another (serial operation), in time sequence, very large numbers can be handled. To start the computer, a small start-up (boot) program in memory sets all the CPU's registers to the correct values, thereby 'initialising' the system. The programmer uses an 'assembly' language of mnemonics (abbreviated command words) to compile the operating program. An assembler program is then run to convert the mnemonics into binary numbers which sets the computer into operation. The computer is connected to vehicle systems via interfaces, a parallel one implying, say, 8 lines of data being sampled at one time. Digital to analogue converters are used to control the system with an analogue voltage. A similar communication system to that inside the computer could be used to connect the service systems and their controllers throughout the whole vehicle. The pressure for such a system comes from the need to reduce the number of cables in the wiring harness as more service systems are added. Multiplex (MUX) wiring allows power and data control buses to replace many cable links. Fig 41 shows how such a system, proposed by Motorola, has the advantage of low switching currents and the ability to build in a self-diagnostic facility. Cost of intelligent power switching is the main deterrent which is likely to reduce in the course of future microelectronic development. At this stage a clock rate of 2400 Hz allows about 4 milliseconds to determine the address and store the control signal data. A serial data network uses an 8 bit word, in the system shown. Intelligent load switches being developed use unipolar Field Effect Transistors.

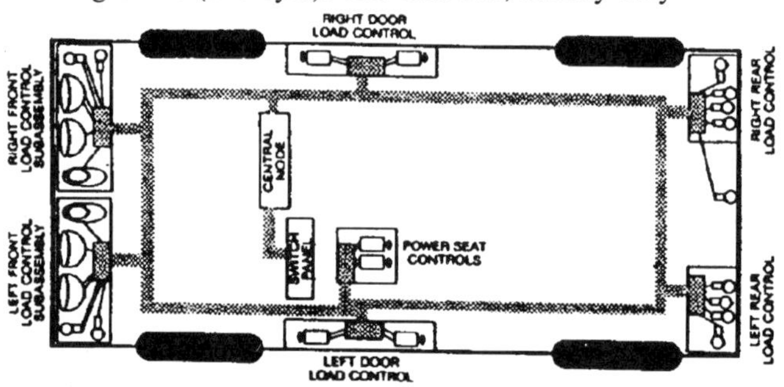

Fig 41: Vehicle multiplex system

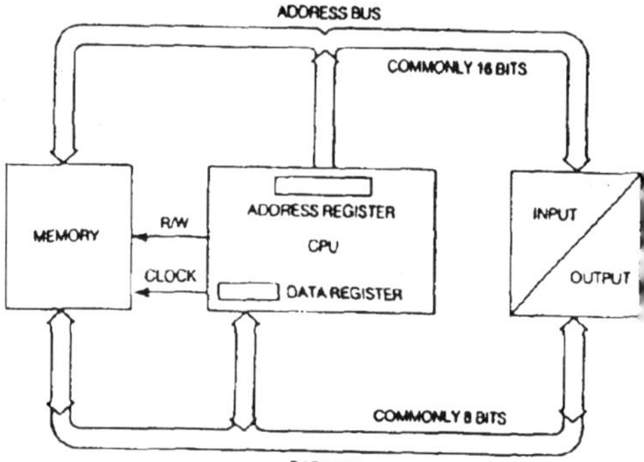

Fig 40: Address and data buses

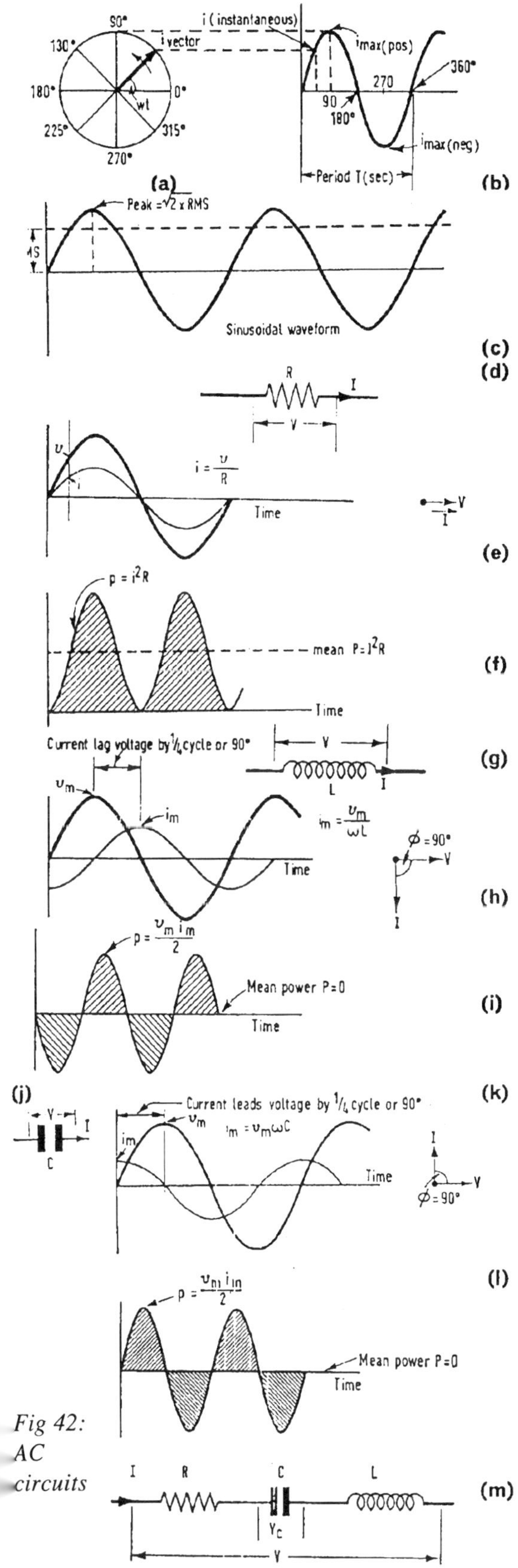

Fig 42: AC circuits

Electronic controls applied

While the previous section introduced electronic control in conceptual terms, here the solid-state electronic elements used as building blocks for electronic control circuits are examined. Some appreciation is also offered as to how these can be applied in digital logic devices for on-board microcomputer control.

Materials which can be classified as conductors, semi-conductors or insulators vary in the amount of free electrons present. The movement of the electrons can be arranged, in suitable primary circuits, to control the operation of secondary power circuits incorporating clectro-mechanical actuators contained in the various systems of the vehicle. Some 6288 x 1015 electrons represents a charge of one coulomb, the passage of which each second is equivalent to a current flow of one amp. The electric force, or potential, of one volt gives each coulomb of flow an energy charge of one joule — defining the basic units for any quantitative analysis to be carried out.

Charge can be stored on a condenser, to the extent that one coulomb charge taken up when one volt is applied represents a capacity of one farad. The material's ability to store charge as the 'plate' of a condenser is expressed as a ratio to air (value unity) as its di-electric constant while the time taken for the condenser to reach 63.2% of its full charge defines its time constant.

The use of alternating rather than direct currents in control circuits is convenient in the ability of the voltage to be easily transformed into a lower or higher one — also in the absence of commutator sparking in any motors used in the associated power circuit, Fig 42. Part (a) shows how vector notation is used to express the sine-wave representation of alternating current, at (b); the root mean square quantification of voltage and current values is seen at (c) while (d,e,f), (g,h,i) and (k,l) show current and voltage relationships for purely resistive, inductive and capacitative elements of a circuit respectively.

In a resistive *R* and inductive *L* series circuit, there is a phase change such that the voltage leads the current by the 'phase angle' (between vectors) and an impedance is set up, to current flow: *Z* is voltage/current *(E/I)*— analogous to ohmic resistance in a DC circuit. With capacitance *C* substituted for *L*, the resulting *C,R* series circuit is such that the voltage lags the current by an equivalent phase angle. When all three exist together in a series circuit, (m), the capacitative and inductive impedances tend to cancel out so that when $\omega = 1/\omega C$ occurs, then in the impedance formula:

$$Z = [R^2 + (\omega L - 1/\omega C)^2]^{1/2}$$

$Z = R$ and resonance occurs so that a magnification condition can result when the voltage across the capacitance could be higher than the applied voltage. Resonant frequency of the alternating current, $w = (1/LC)^{1/2}$, the circular frequency of the rotating vector. The voltage magnification factor = $(1/R)(L/C)^{1/2}$. At the resonant frequency condition of minimum impedance, the arrangement is termed an 'acceptor circuit'. The higher the magnification factor, the better is the 'selectivity' — the ability of the circuit to select one frequency to the exclusion of all others.

In the case of parallel circuits of capacitance and inductance, some interesting characteristics are possible. In practice there is always some resistance in the circuit but in theory, without resistance, a circulating current should build up in the parallel loop. Because of the inevitable resistance present a so-called 'dynamic impedance' exists at resonance. which is the maximum condition, and defines a 'rejector circuit'.

Solid-state 'active-devices'

Voltage gain and direction control in a circuit is usually achieved by transistors made from semiconductor materials doped with impurities to alter their characteristic. Silicon is the usual base material with boron and phophorous used respectively to achieve so-called p-type and n-type semi- conductors respectively. As forward-conducting rectifiers, transistor diodes have a positive voltage on the p-type material (the anode) to obtain the necessary bias. With a negative voltage on the n-type material, a reverse non-conducting bias is obtained. Fig 43a shows the symbol for the device alongside the current/voltage (I/E) transfer characteristic for a silicon diode. The rectifier circuit in Fig 43b is of the 'half-wave' type, the diode being reverse- biassed to block current flow in the reverse direction. Current flow is smoothed from half-wave to near steady value by the condenser which stores charge and opposes voltage changes by discharging to the load resistance.

Active transistors (transfer resistors) have terminal connections to the circuit which can cause

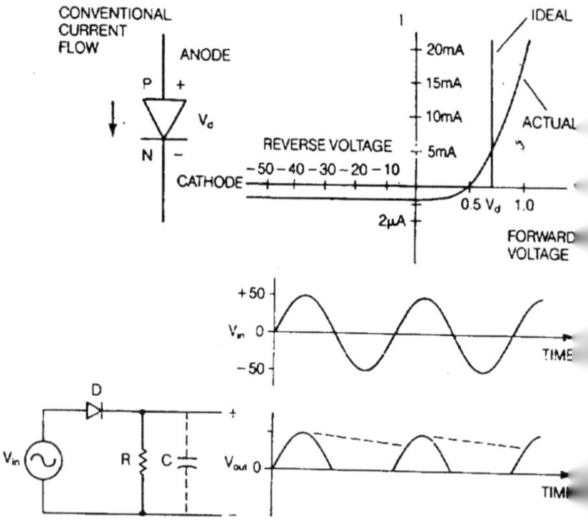

Fig 43: Transistor and e#rectifier circuit

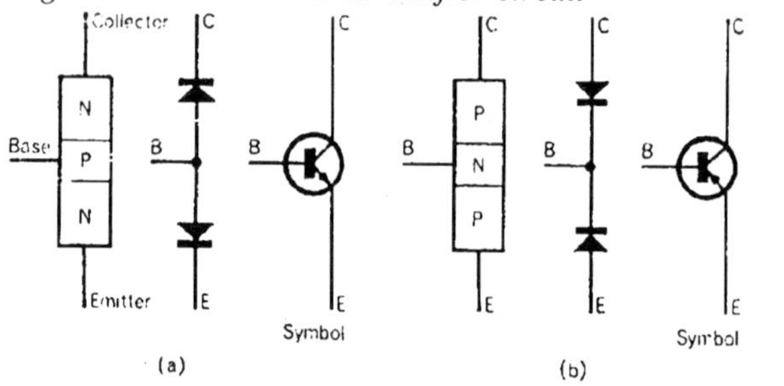

Fig 44: Back-to-back diodes

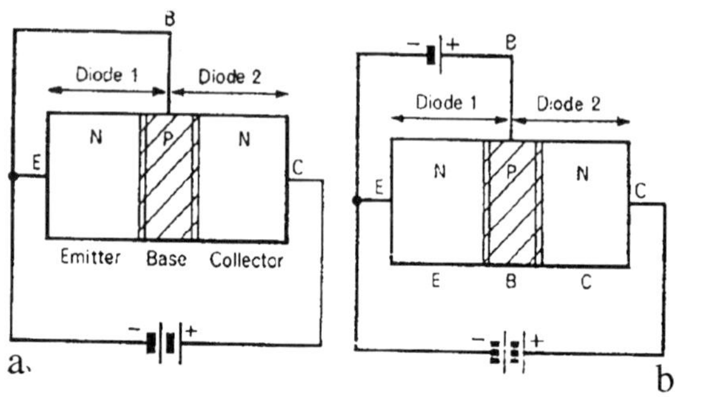

Fig 45: Transistor biassing

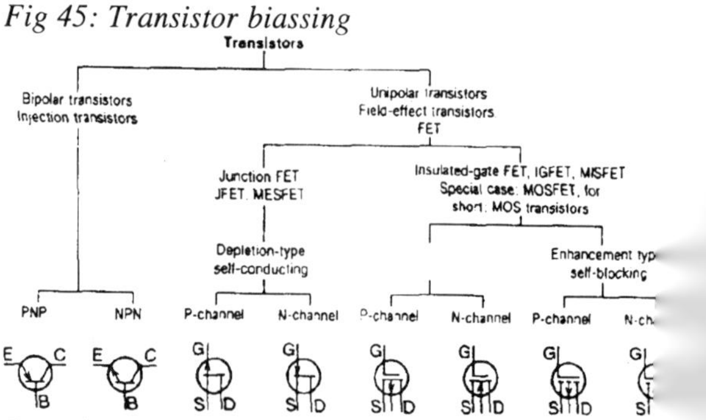

Fig 46a: Transistor types

voltage to be amplified. They can be regarded as back-to-back diodes (Fig 44) and may be of the NPN or PNP type for so-called 'bi-polar' transistors, introduced in the earlier section, which also showed the respective current flows across the PN junction. Both can operate as amplifiers and switches, the collector/base junction becoming forward-biassed in the 'on' condition — in the switching mode. Referring to Fig 45a, the NPN transistor is shown with the base at the same potential as the emitter; no current will flow since diode 1 has the same potential at anode and cathode and diode 2 is reverse-biassed. In Fig 45b, there is a forward bias on diode 1 and when this exceeds the junction barrier potential, electrons will move from the the base (P region) towards the emitter (N region) attracted to the higher potential at the collector; small current change, base to emitter, causing large current change, collector to emitter. An amplifier circuit for a transistor was shown in the introductory section.

In uni-polar, field-effect, transistors (FET), current is controlled by an electric field generated by a voltage applied at a control electrode called the gate. The types and physical representations of different transistors are shown in Fig 46 a and b, due to Bosch, and the resistive capacative and inductive components of a series circuit in Fig 46c. The word uni-polar comes from the FET only employing one type of charge carrier, against the bi-polar's two. MOS stands for metal-oxide-semiconductor — referring to the sequence of layering if the silicon. The P

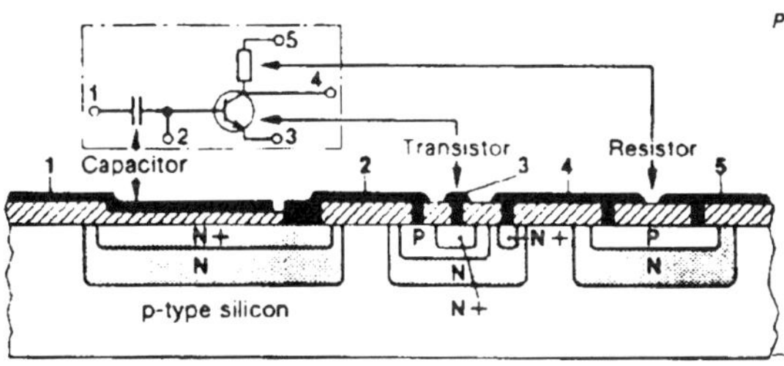

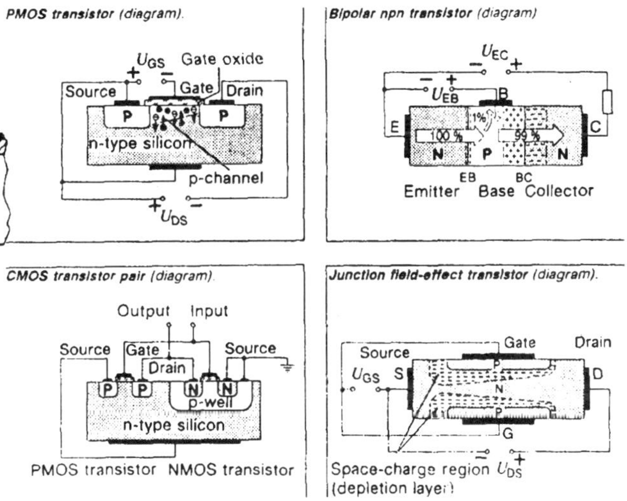

Fig 46b: Transistor types and (above) installation in an integrated circuit

	R	C	L	R & C	R & L
Circuit					
Vector diagram	V, I	I, $\phi = 90°$, V	V, $\phi = 90°$, I	ϕ, I, V	ϕ, V, I
Power $P =$	VI or I^2R	0	0	$VI \cos\phi$ or I^2R	$VI \cos\phi$ or I^2R
Impedance $Z =$	R	$\frac{1}{\omega C}$	ωL	$\sqrt{[R^2+(1/\omega C)^2]}$	$\sqrt{[R^2+(\omega L)^2]}$
Phase angle =	0	90°	90°	$\tan^{-1}(\frac{1}{\omega CR})$	$\tan^{-1}(\frac{\omega L}{R})$
Impedance triangle	R	90°, $\frac{1}{\omega C}$	90°, ωL	ϕ, R, Z, $\frac{1}{\omega C}$	ϕ, Z, R, ωL
Units for calculations	Ohms (Ω)	* Farads (F)	Henries (H)	Ohms & Farads (Ω & F)	Ohms & Henries (Ω & H)

Fig 46c: Table of R, C and L components in series circuit

and N suffixes refer to the doping by either a donor or acceptor impurity. Suffix C stands for complementary and involves the fabrication of NMOS and PMOS transistors on the same silicon chip. One further type is the thyristor, familiar to those involved in speed controllers for electric vehicles. These are semi-conductors with at least three PN junctions, switchable from the forward to the reverse-block state.

An introduction to digital circuits and computer control was given in the previous section so it is now timely to consider the basic concepts of control necessary for designing control-circuits. Fig 47, due to Bosch, compares closed and open loop systems. In the former, the directly controlled variable *x* is continuously measured and compared with a reference variable *w*. It is then changed continuously to reduce variation from the reference variable — within a closed loop. This is also called proportional control because the control signal is proportional to the error signal. In the latter case of open-loop control, one or more input variables acts on the output variable through the inherent characteristics of the system. In open loop control, a series of elements act upon one another as links in a chain. Only disturbance *z1* can be addressed as the open loop has no effect on disturbance *z2*.

Fig 48 shows a control circuit for controlling temperature within a closed space, using the closed-loop concept, This, however, uses a technique known as limit cycle control, which monitors the system's output and responds only when the output goes beyond preset limits. More sophisticated controls use logic circuits in which a certain operation takes place only when a certain set of conditions occur. Transistors in this instance are used only as switches and all data is in terms of 'signal' (1) or 'no signal' (0). According to Trevor Jones[1], the combination of logic circuits and micro-miniaturisation on integrated circuits has most transformed the prospects for vehicle control systems. Now that power devices, too, are being reduced to chip size, the way is open for new technologies to achieve control strategies without the quality problems of a multitude of soldered joints. He shows in Fig 49 the voltage and current ranges achievable with power integrated circuits. As their price reduces, he points out, the need for conventional solenoid relays for power actuation may eventually disappear.

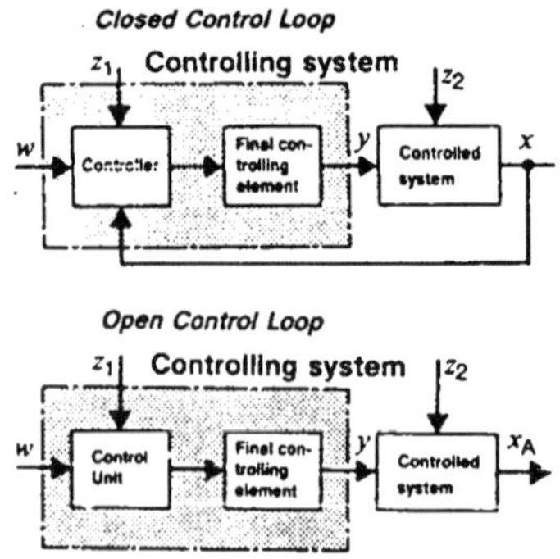

Fig 47: Closed and open loop systems

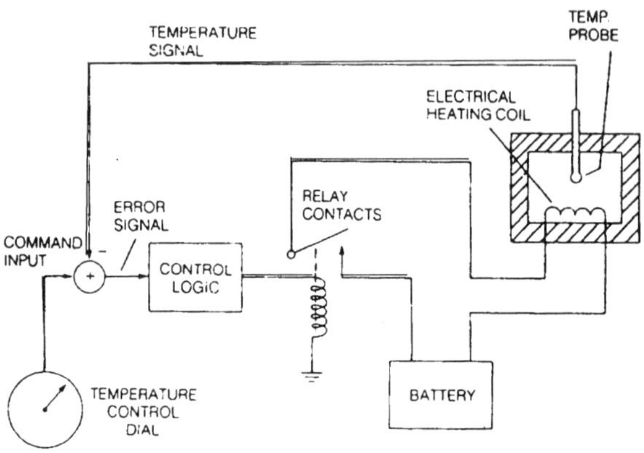

Fig 48: Temperature control circuit

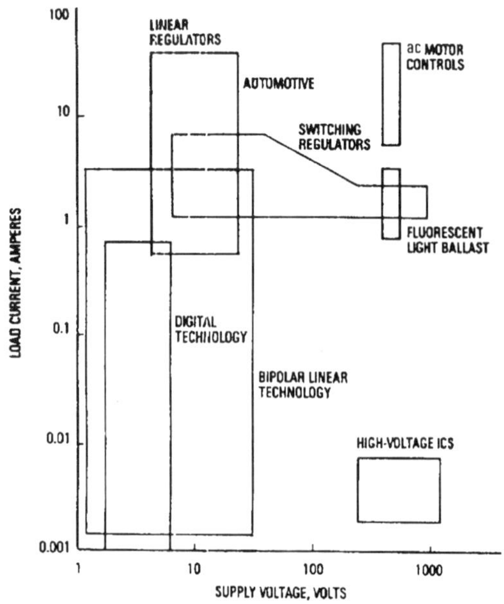

Fig 49: Achievable voltage and current ranges

Body electronics

In the previous section, 'active' devices were introduced. Here the benefits and possible benefits are discussed in applying multiplexing to body system electronics.

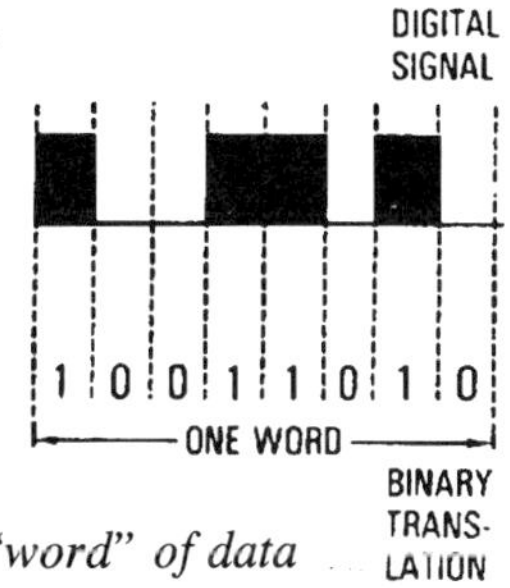

Fig 50: A "word" of data

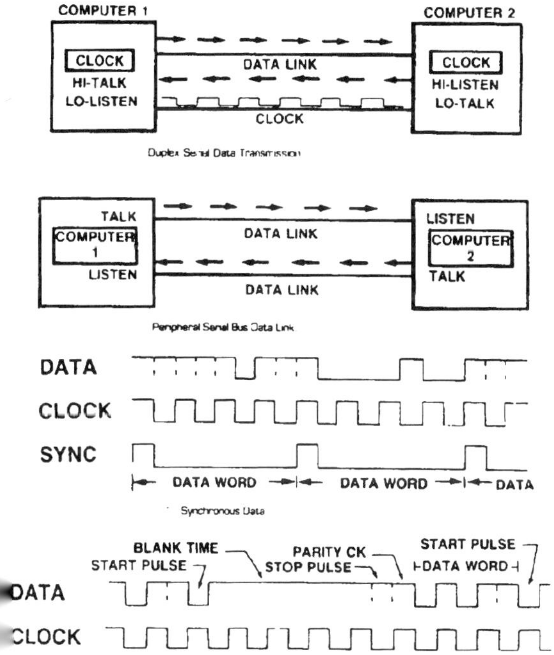

Figs 51 and 52: Parity check and clock pulse

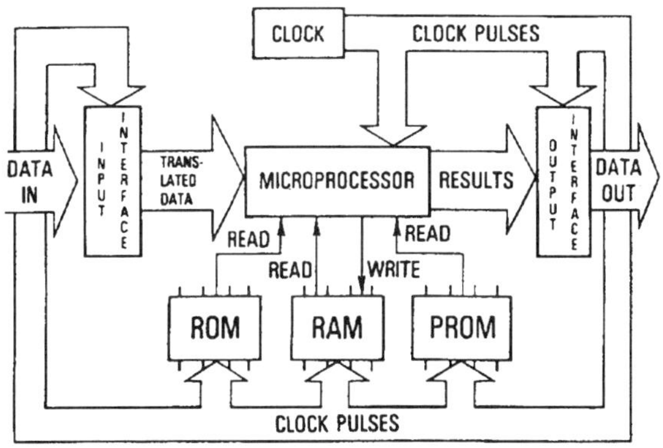

Fig 53: Parallel data transmission

The increasing data processing requirements, above those required of the electronic systems required for engine/transmission management and antlock braking, may lead to ways of increasing the calculation rate of microprocesssors. The three main approaches are to use high speed semi-conducting materials such as gallium arsenide, to revise the architecture of the central processing unit or increase the bit length for processing the data. The expense of providing separate processors around the vehicle for controlling the various body systems would otherwise probably preclude their use to very particular applications.

Suitable microprocessors are now being volume-produced at low price and at last multiplex wiring systems for connecting the systems together on a cable ring with a modest number of wires are appearing at the luxury end of the vehicle market. Another possibility is that so called application specific integrated circuits (ASICs or 'custom chips') will be configurable by consultants to the particular requirements of vehicle builders.

A key function of the on-board computer is the clock generator, discussed in an earlier section, used for producing single bit length pulses. The memories watch these pulses during data processing (sending and receiving information). By knowing how long each bit is supposed to last the computer can distinguish between the different binary signals, Fig 50. Important to vehicle computers are the various interfaces which have built in to the converters for turning the many analogue sensor signals to digital ones for the microprocessor.

Similarly output interfaces need circuits for converting binary code into digital output signals which suit the devices they are controlling. In cases where more than one computer is used per vehicle, data links will be needed between them. When so-called 'serial' data transmission is used, 'words' of data are sent one after the other. The computers are then arranged to be clocked together in synchronisation. Duplex serial data transmission implies a single two-way link between computers and serial data can be sent or received by either. When clock pulse is high, the first is programmed to send while the second listens and the reverse situation occurs for low clock pulse. Alternatively, in a peripheral serial bus data link, one always sends data in one direction while the other always receives it in the opposite direction.

The data link systems are said to be synchronous when the data flow is constant and the clock pulses regular. In some systems, however, data is transmitted only when it is required rather than being sent continuously — asynchronous data transmission, Fig 51. A start pulse signals the computer prior to transmission and a stop pulse signals completion while a parity check may be used to ensure that that the total words have been sent, Fig 52. Parallel data transmission, Fig 53, by contrast, involves several separate data lines with corresponding switches turned on and off simultaneously to send specific data signals. In the case of multiplexing, a central transmitter (microprocessor) acts as a master controller for the several receivers which control the principal loads of the body electrical system. A bidirectional bus data link connects them; this comprises two wires, 'signal' and 'ground'. Switches on the transmitter each have a unique signal and the microprocessor provides bus data link control and sequencing logic. So called 'state sequencing' is used by the control software that cycles through various tables to ensure proper conditions exist to advance from one state to another. Switch input requests may be ignored, passed on as a command to the receiver or interpreted as a specific command to be issued — according to whether, say, the ignition switch is on or off, or whatever. Receivers either operate electrical loads directly or through mechanical/electronic relays. Intelligence level of the receivers is restricted to that required to signal error state (bulb-failure, say) signals back to the transmitter. The Lucas Body Systems arrangement uses the SAE J1850 (Carlink) protocol set up to standardise the datalink procedures for multiplex systems. The company have applied multiplexing to controlling electrical systems in two doors. Standard protocols open up the market so that equipment from on e supplier can be used confidently with that of another operating on the same protocol.

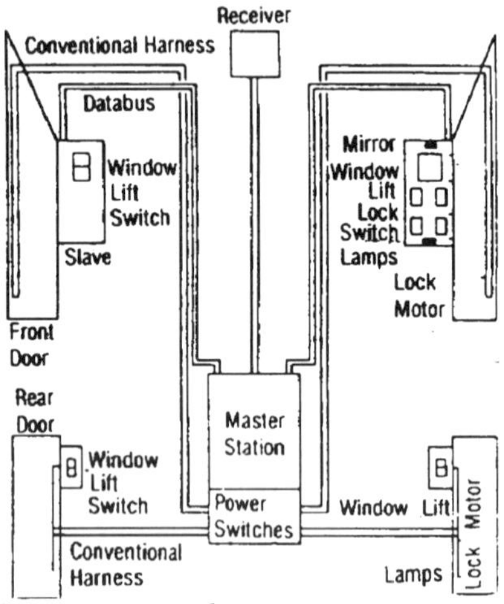

Fig 54: Lucas network system

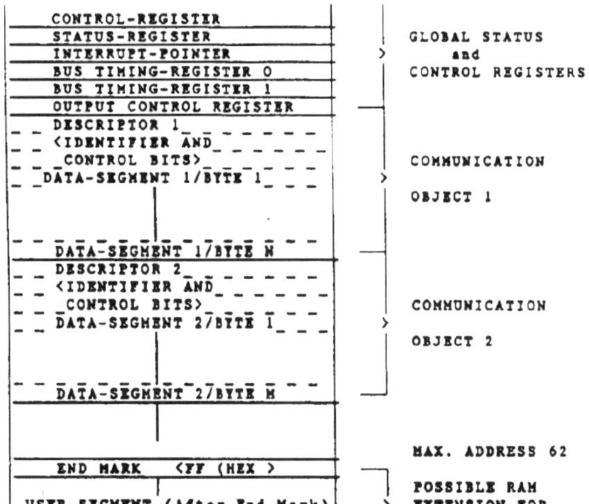

Fig 55: Bosch/Intel system

The following features are controlled: electric mirror, electric window lift, keyless entry, central door locking, courtesy light delay, lock illumination, lock barrel heating and rear door windowlift isolation, Fig 54. An alternative approach is that devised by Bosch and Intel which is known as CAN, Fig 55. Philips Semiconductors have recently introduced an intelligent power switch on this protocol. It comprises a power MOSFET linked with logic circuitry and promises a solution for multiplex connection of the 30 or so motors which can now feature on top-range vehicles, Fig 56. The units are said to have lower drive power requirements than bi-polar switches. Fig 57 shows the four most common motors used in vehicles while Fig 58 shows a bridge of semi-conductors used to control a sun-roof where directional change of the motor is required. Source-drain diodes in the MOSFETs eliminate the need for external switching diodes. PM brushless DC motors typically require six switches to generate the rotating field, Fig 59. Change of direction or speed are controlled by timing of the field pulses; these can be derived from a dedicated IC such as the Philips NE5570. Rotor position sensing is required to determine which windings should be energised. One switch supplied by the company is capable of driving motors up to 25A, Fig 60.

Typical motor and switch requirements in a top of the range car

motor applications	typical power (W)	nominal current (A)	typical number of such motors	type of drive	typical number of switches per motor	proposed MOSFET: standard BUK-	L² FET BUK-	comments
air-conditioning	300	25	1	unidirectional, variable speed	1	456	556	Active suspension may also require such a high power motor
radiator fan	120–240	10–20	1	unidirectional, variable speed	1	455	555	These motors may go brushless, requiring 3 to 6 lower rated switches
fuel pump	100	8	1	unidirectional	1	453	553	
wipers: front			1–2			452/ 453	552/ 553	Reversing action is at present mechanical. This could be done electronically using 2 or 4 switches
rear	60–100	5–8	1	unidirectional, variable speed	1			
headlamp			2					
washers: front			1–2					
	30–60	2.5–5		unidirectional	1	452	552	
rear			1–2					
window lifter	25–20	2–10	2–4	reversible	4	452/ 455	552/ 555	
sun-roof	40–100	3.5–8	1	reversible	4	452/ 453	552/ 553	
seat adjustment (slide, recline, lift, lumbar)	50	4	4–16	reversible	4	453	553	
seat belt	50	4	2–4	reversible	4	453	553	
pop-up headlamp	50	4	2	reversible	4	453	553	
radio aerial	25	2	1	reversible	4	452	552	
door lock	12–36	1–3	6–9	reversible	4	451/ 452	551/ 552	
mirror adjustment	12	1	2	reversible	4	451	551	

Fig 56: Motors on top-range car

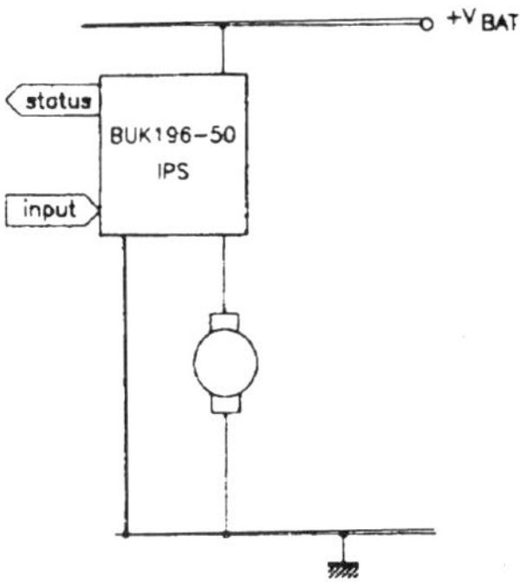

Fig 60: Driver for 25A motor

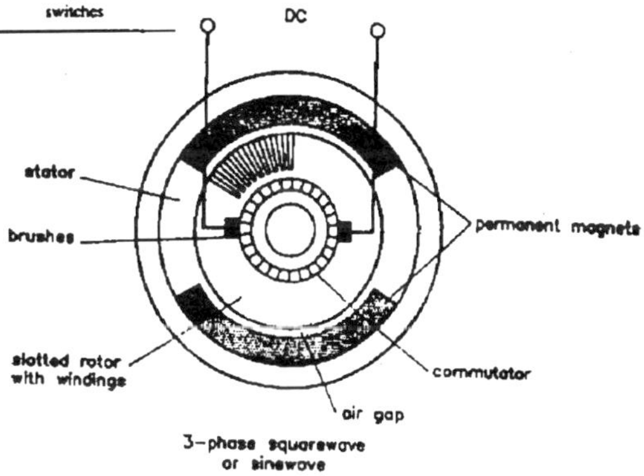

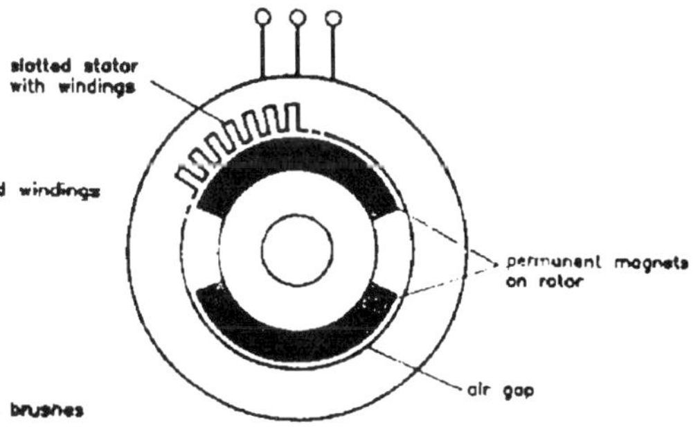

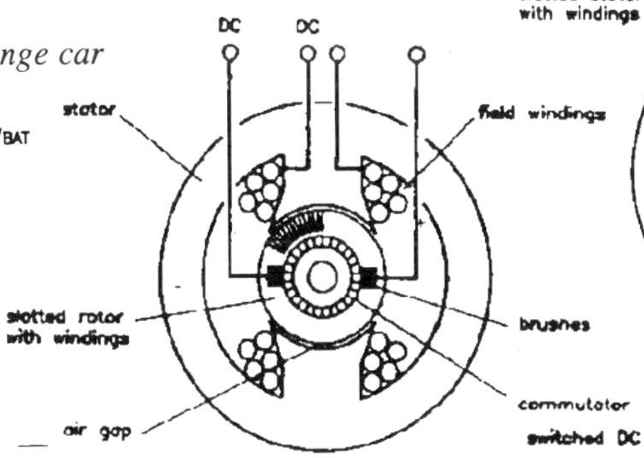

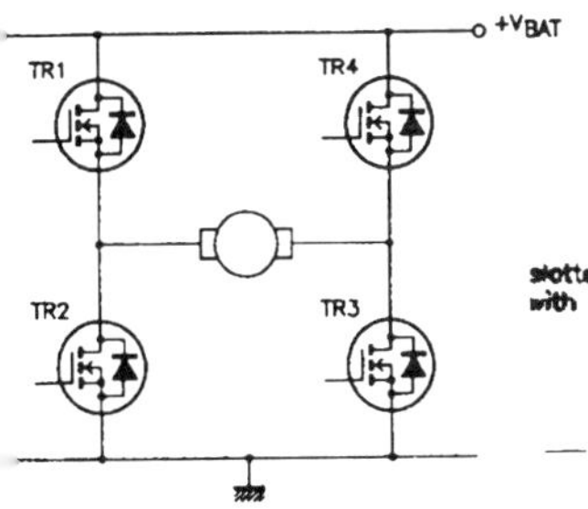

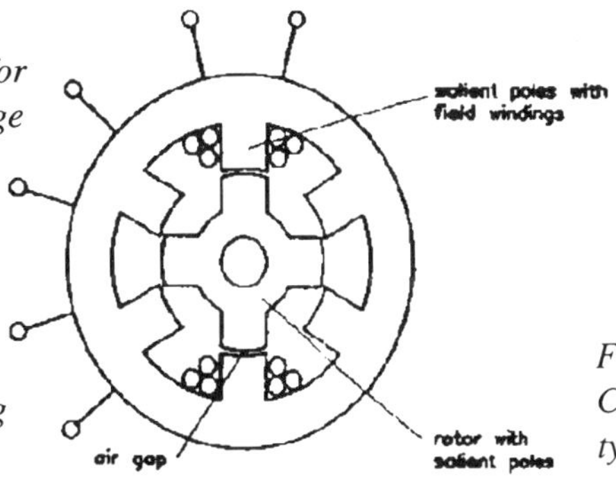

Fig 58: Bridge circuit for motor directional change

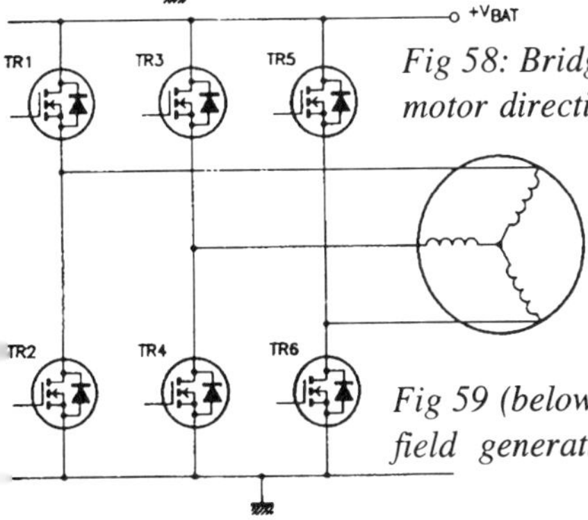

Fig 59 (below): Rotating field generation

Fig 57 : Common motor types

Body electronic systems

Before multiplex wiring becomes the norm on vehicles, to handle the increasing number of on-board electronic systems, datalink connections between sub-harnesses may be the solution. This is discussed here with descriptions of some advances in the electronic systems themselves.

The practicalities of multiplexing the electrical supply to body systems seems to have put something of a brake on its application. Development of junction boxes between main and sub-sections of wiring harnesses by Lucas Rists might encourage the process when active versions are employed to establish serial links. Junction boxes are attractive, too, in conventional wiring harnesses to avoid splicing and doubling up of cables into terminals. The sub-sections of the harnesses can also involve fewer cables by automating some of the functions. Suppression components and routing diodes can also be conveniently located in the boxes — also fuses and relays when power demand is high. Making the boxes 'active' involves the addition of timing functions for electronic systems and possible inclusion of transceivers for serial datalinks. Depending on power requirements either metal stamped interconnections or printed circuit boards can be used to link through the boxes. The company describe an application to a vehicle with some 700 cables in the facia area, 70 of which are routed to the driver's door and 25 to the passenger door, Fig 61. The redesign involved breaking down the facia harness into smaller units which can have automated operation and reducing the dirver's door cable number to 25. The solution was to combine active and passive junction boxes with a serial datalink. As most control signals originate within the driver's door (central-locking, mirror-adjust and window regulation) the serial data link was used between it and the A-post, Fig 62.

The datalink uses pulse width modulation for bit encoding, Fig 63, and is made electrically secure

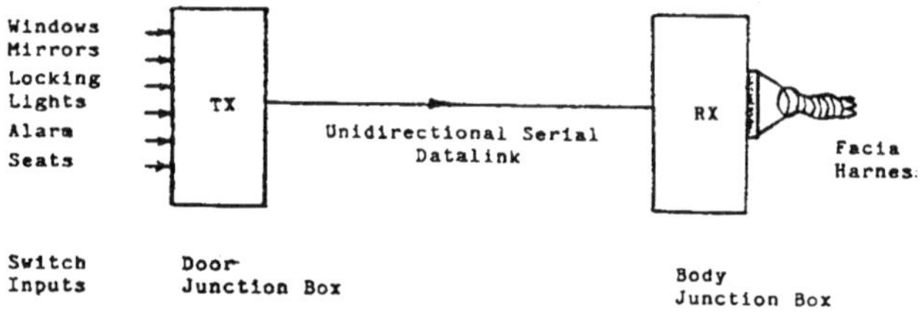

Fig 61: Serial datalink

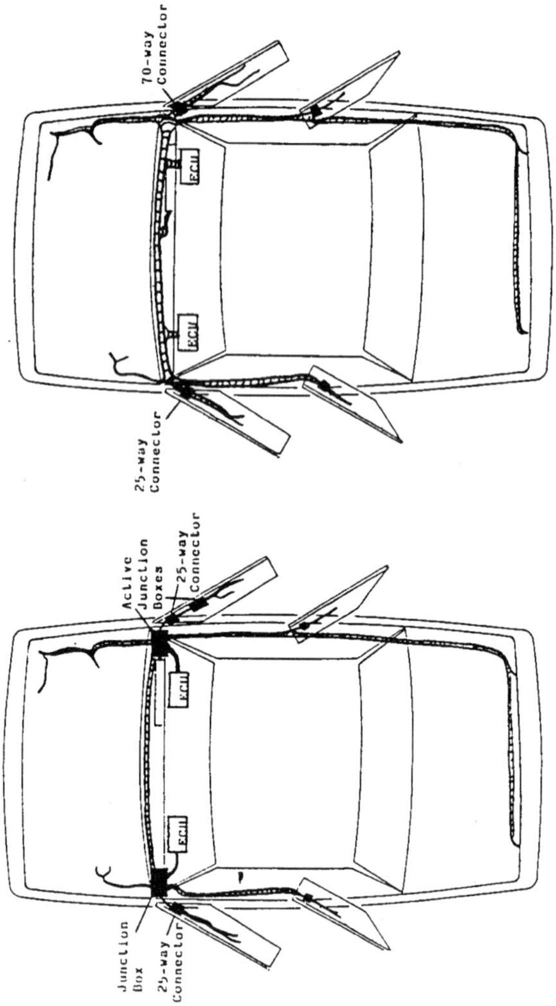

Fig 62: Door system application — above, serial data link; below, layout with junction boxes

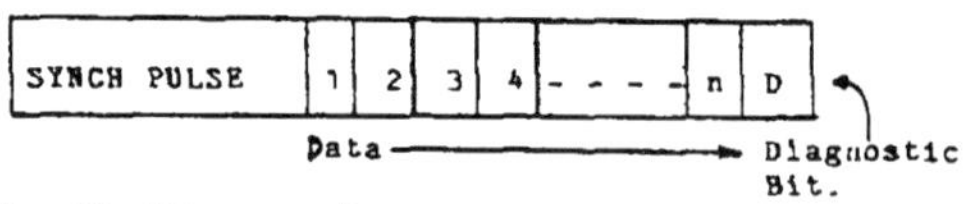

Fig 63: Bit encoding

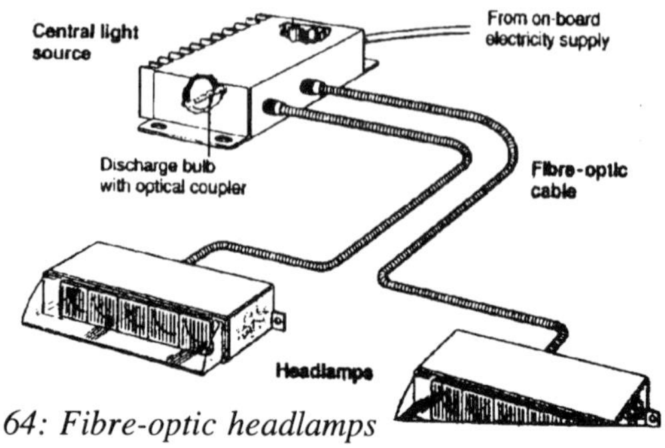

Fig 64: Fibre-optic headlamps

in being immune to short circuits either to 0 or 12V. A diagnostic databit is transmitted in every message frame and on its receipt a light-emitting diode confirms function of the databus. As a further safety measure all systems remain operable from their local switches. In the example application the number of cable splices was reduced from 82 to 15 and the 6 subsections of the facia harness involves one quarter fewer cables.

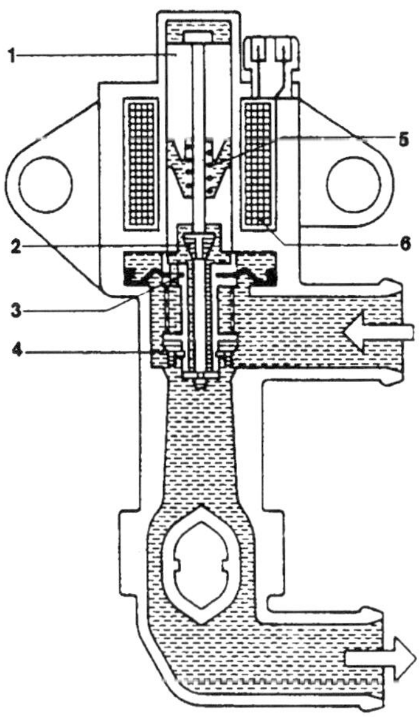

When current flows through the coil, the armature moves downward and closes both sealing cones. When no current is flowing, the valve opens.

1 Armature
2 Sealing cone
3 Bore
4 Main sealing cone
5 Spring
6 Coil

Coolant

Fig 65: Heater solenoid-control

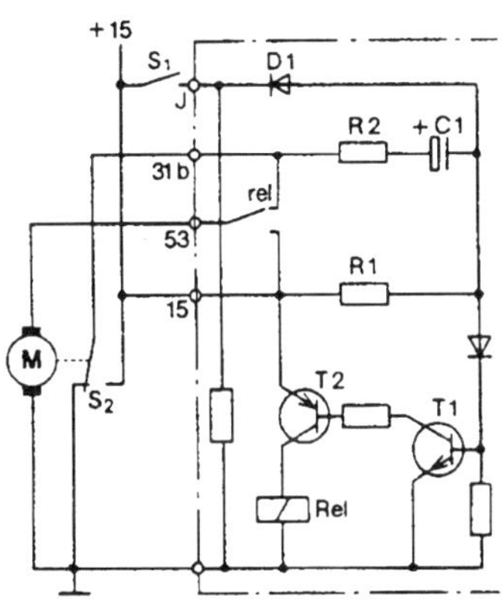

Fig 66: Wiper delay circuit

On-board system developments

Among recent advances by Bosch, a fibre-optic cabled headlamp system promises very low profile lenses. The remote light source is housed in a safe area of the body and transmits its light via cables to the conventional headlamp locations. Feasibility of the system arises from the recent availability of gas-discharge lamps for the central source. These are some 2.5 times the intensity of halogen lamps and produce virtually a daylight colour spectrum. The G-D lamp is enclosed in a double-ellipsoid reflector to maximise the light yield. At the headlamp location there are four projection lamps with associated lens systems, each occupying a 40 x 40 mm area. Each juxtaposed 'lamp' has a specific function. One presents light/dark cut-off and the other three produce correspondingly differing light modulation. The beams overlap so that the total light emitted is homogeneously distributed in a pattern required by EC regulation, Fig 64.

For heating and air-conditioning, the company has also developed new motors — of the elecronically commutated type — which dispense with brushes and mechanical contacts. Another interesting product for heaters is an automatic temperature control system based on negative-temperature-coefficient sensors and a solenoid valve for water control, Fig 65. A potentiometer is used to select the desired temperature and an electronic controller continuously adjusts the system according to signals from temperature sensors close-to and remote-from the heater. A pulse is output by the ECU to the solenoid valve at a four second interval.

For windscreen wipers, the company has introduced an optoelectronic sensor which detects rain or screen-mist and operates the wiper as necessary. Wiper operation is quieter, too, by the recent availability of variable speed motors which avoid the annoying rising and falling whir of conventional systems. A two-motor wiper system is also on offer which will make the normal wiper linkage redundant. As well as appropriately synchronising motor speeds the controllers also improve functional performance of the wipers.

Electronics also comes to the aid of automatic windscreen wash-wipe systems by Bosch. The windscreen is sprayed with washer fluid; the wipers then distribute it, clean the screen and dry it. An intermittent-wiper switch comprises a pulse-generator with a variable pulse frequency. Each pulse drives the wiper via a relay and causes one wipe cycle, the delay being set between one and five seconds, Fig 66. The unit also uses an astable multivibrator as the clock generator, Fig 67, the delay of which can be set within specified limits. Selection of the RC element in the circuit, comprising resistors R5 to R (and capacitor C1), achieves this. C1 has capacity of some 500 μF and the five resistors are connected to it in steps, their resistance varying between 30 and 300 kΩ, corresponding to the timing delays. Pulse period is set by the time constant R2.C2, at about 0.5 second. The relay switches on the motor for this period, via terminal 88a. The wash-and-wipe system overrides the intermittent wiper circuit. This circuit has an RC element which not only controls the lead time but also time for wipe-and-dry.

The system can also be purchased with a monostable, externally triggered (by the motor reverse switch), multi-vibrator. Here the RC element is as shown in Fig 68 with R1 and R2 as charge and discharge resistors, respectively determining lead and wiping times.

The company also use multivibrators as an improved alternative to hot-wire turn-signal flasher units. These have better durability and are less sensitive to prevailing temperatures. The multivibrator is the signal generator and is combined in one unit with monitoring stage, turn-signal relay with check lamp and hazard warning. Since the filament wire is a resistor with positive temperature co-efficient characteristics, high current flows initially, decreasing rapidly with increased temperature.

The multivibrator acts as controller — having two amplifier sections, one switching stage and a switch with a differential amplifier. It is an integrated circuit, R1 and C2 forming the RC element which controls the timing of the flasher-frequency, Fig 69. C1 and C3 control voltage spikes and so serve as noise suppressors. Diode D protects the IC against incorrect polarity should the battery leads become reversed. A switching transistor matches multivibrator power to the turn-signal relay. and the IC has a voltage regulator stage which maintains constant voltage despite battery condition. If a lamp fails the voltage drop across Rk changes so that the transistor conducts and the frequency of the vibrator changes as required by law.

Networking of body system controls

According to Siemens, electrical connections which were part of traditional wiring harnesses are increasingly concentrated in special junction boxes. Rather than having the one central one, future vehicles are expected to have several around the vehicle as the trend to modularisation of harness sections continues. Increasingly the boxes are also expected to contain the electronic controllers for body systems to save on wiring and in the future electronic control of the connections within the boxes will make them programmable from the end of the production line.

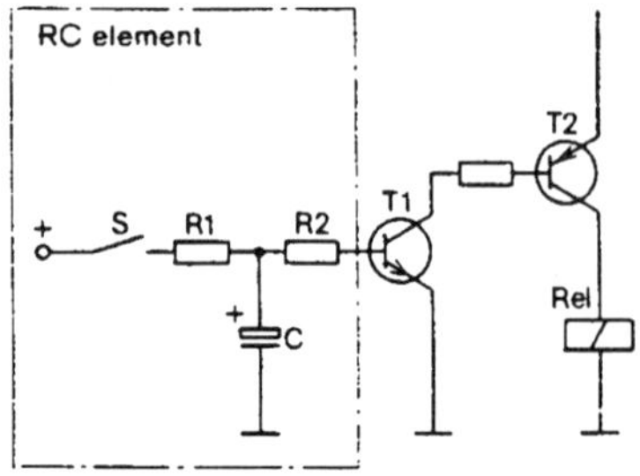

Fig 67: Clock generator

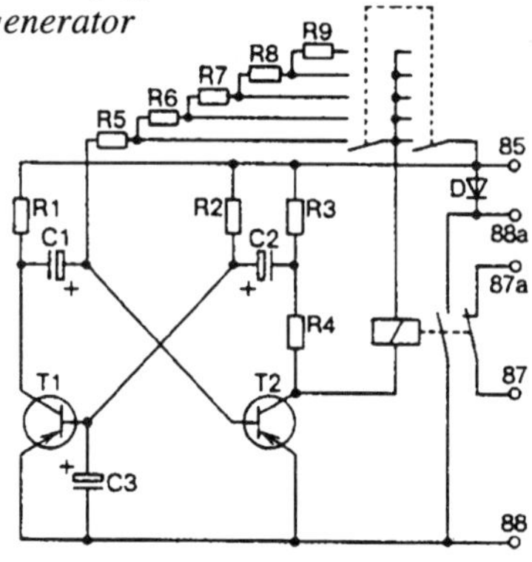

Fig 68: Externally triggered multi-vibrator

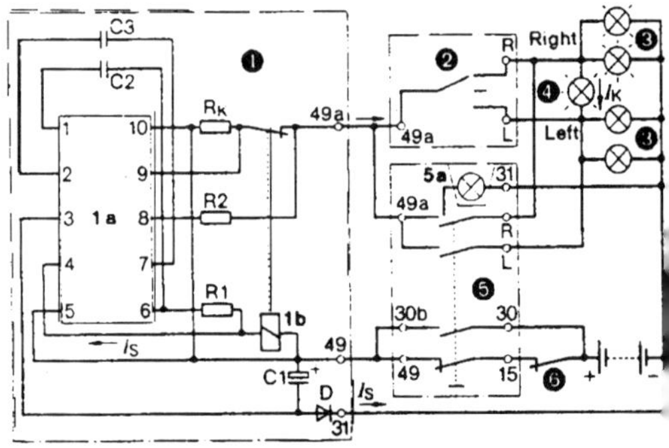

Fig 69: Flasher integrated-circuit

Serial data transfer links are still uncertain, according to the company, with respect to their widespread fitment and what their data transfer rates should be. While very low rates are adequate for doors in which only power windows and central locking were involved, the eventual coming of production vehicle guidance systems will mean fast enough transfer rates to download the contents of a CD-ROM into a display system. The way in which the power supply network is separated from the information network in such systems is shown in Fig 70. According to Bosch, with modular subsystems, the limited capacity of the signal and power supply lines can be taken into account. Other advantages offered by subsystems include greater system safety, decoupling development and flexibility. The partial multiplex systems communicate with one another by means of a hierarchically superior bus. This bus can, for example, relay all the subsystems' diagnostic messages. One disadvantage of segmentation is that it involves considerably more work on the wiring. A car with multiplex only requires a total of around 50 metres of four-conductor multiplex cable. A car in the top bracket with conventional wiring requires about 2.5 kilometres of cable. The CAN protocol with a data transmission speed of up to 100 k baud (= 100,000 bits per second) is especially suitable for communication between multiplex systems in the field of car body electronics. The basic CAN module is the section of the interface which is important where dealing with the protocol is concerned. This module is integrated into various microprocessors.

A self-contained multiplex system also involved a change of wiring. Whereas the point-to-point connection prevails in conventional wiring, a multiplex wiring system basically consists of four connections linking the components with one another: earth line, battery voltage line and two signal lines.

Traditional plug connections are only suitable to a certain extent. The so-called IDC method of termination is ideal for straight-through connections. By simply cutting through the cable insulation with blade contacts, a connection is achieved between the otherwise undamaged cables and the plug contacts. The cable can be prefabricated by way of a flat strip with four parallel lines.

The plug is hermetically sealed using three gaskets. This also ensures that there will not be any corrosion at critical spots. In gold plating all the four 2.5 mm flat contacts, this provides extra reliability for low contact resistances of the signal contacts and also greatly increases the current carrying capacity of the current conducting contacts, Fig 71.

A multiplex wiring system in conjunction with 'intelligent' components can largely replace conventional control units. An example of this is a partial multiplex system in a vehicle door for which there are two solutions: in the first, a local electronic module is connected with actuators, operational elements and sensors of the subsystem in the conventional manner, in other words, by means of a point-to-point wiring system. This certainly solves the design problem of routing from the vehicle body to the door. However, a large area is required to accommodate the system which is relatively inflexible. There are more plug contacts and this reduces reliability. In the second solution, the control electronics (intelligence) system is integrated into the actuators and the operating element accommodates the electronic functions which are necessary to code instructions and relay them. The actuators with their built-in control electronics perform the operating functions (such as adjusting the position of the mirror, opening/closing the windows). The intelligence is integrated into all the basic components in the form of a microprocessor with a basic CAN interface which enables detailed self-diagnosis. A complete check at the end of the assembly line on the vehicle manufacturer's premises and rapid fault diagnosis in the workshop are both possible thanks to this intelligence, Fig 72.

Window actuation with anti-pinch

The extension of a multiplex system for doors to include automatic window actuators with an anti-pinching feature is possible without using any extra cabling and control units. Extra costs are merely incurred with the connection of an intelligent actuator to the multiplex bus. Such convenience circuits are of course subject to strict safety requirements. If the driver does not have to close each window one by one it is easy to overlook the fact that something or other has got trapped in the process. For this reason the electronics system must instantly recognise any pinching and also ensure that the closing procedure is interrupted and reversed.

With this in mind the electronics system monitors the speeds of the individual motors. If the motors are slowing down but their closing power is increasing then an evaluation electronics system recognises in a fraction of a second whether something has got trapped and immediately reverses the motor's rotational direction. Of course, the drive does not switch off until the last few millimetres when the top edge of the glass is securely positioned in the ridge. It requires a lot of power to ensure excellent sealing at this point. The motors only switch off when the windows are properly closed.

Bosch is working to develop intelligent actuators which can also be employed with seat adjustment motors, steering column adjustment, mirror adjustment and actuator for the parking brake.

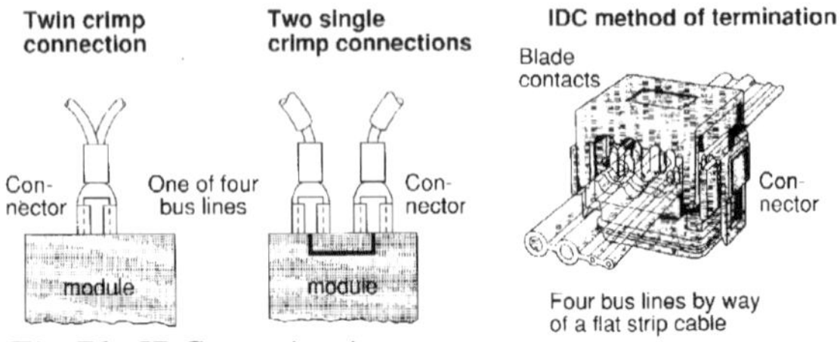

Fig 71: IDC termination system

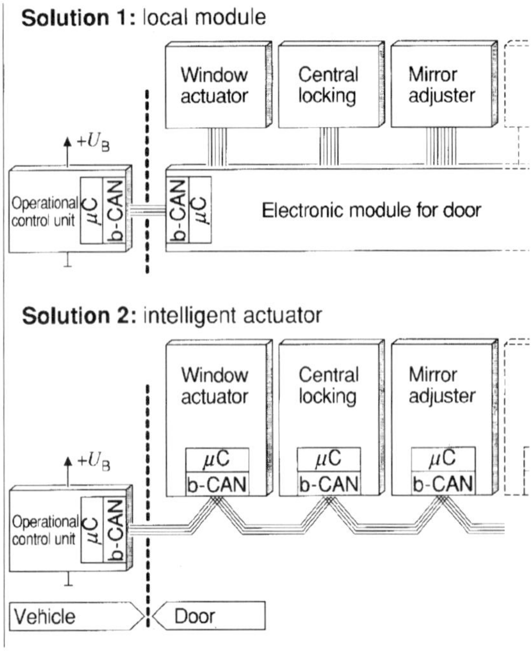

Fig 72: Local module and intelligent actuator solutions

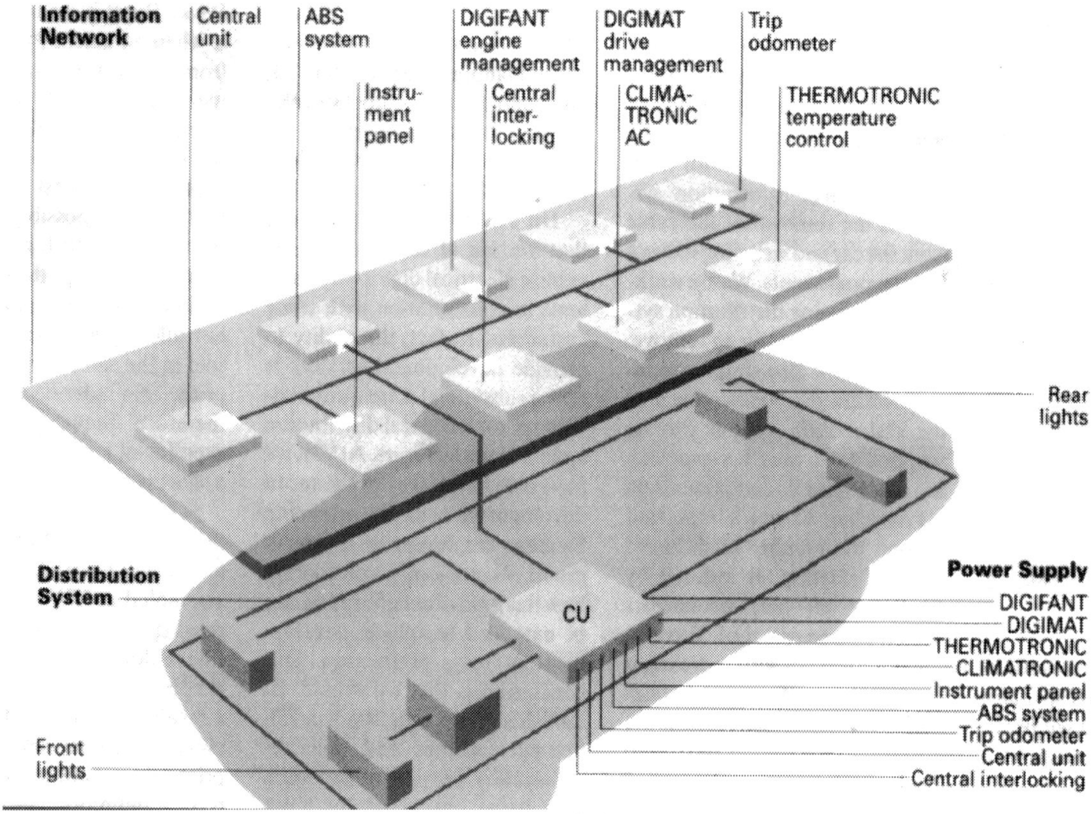

Fig 70: Separation of power supply network and data-link

Chapter 10: Body environment

Body heat transfer; air-conditioning systems
Interior-trim basics
Seating systems
Process advances in body trim
Smart-glass windows
Lighting design/development
Bus body equipment
Electro-magnetic compatibility

Body heat transfer

Good insulation of a car saloon or CV cab is possible by careful design. This requires an understanding of the heat flow mechanisms through and from the skins as well as consideration of the properties of the different alternative insulants. Overall heat transfer co-efficients can then be built up so that materials and layouts can be chosen at the concept stage and related to their predicted heat losses.

Creating a comfortable vehicle environment requires the same rules as used for the achievement of comfortable buildings. In this respect human comfort has been found to be optimised when a person's natural heat loss equals the heat required to be dissipated. The liberation of heat from the body has been found to be 30% by convection, 25% by evaporation and 45% by radiation and this should be the aim to optimise comfort. Different internal temperatures suit different levels of human activity — sedentary work such as driving requires a temperature of 9C.

Heat is lost through the 'fabric' of the vehicle, by conduction and subsequently being dissipated by radiation and convection on the outside. Heat is also lost by 'ventilation' due to intake air from the vehicle's heater air intakes and around the weatherstrips of the doors and windows. The diurnal cycle of daytime and nighttime temperatures obviously affects the 'fabric', and 'ventilation', heat losses. The fabric loss is calculated by estimating the heat loss for each surface enclosing the space and then summating the total:

$$Q=UA(T1 - T2)$$

where Q is rate of heat transfer in watts; A is area of element losing heat in sqare metres; $T1$ is internal environment temperature in C: $T2$ is the outside temperature in C and U is the overall thermal transmittance co-efficient in W/m^{2}C and

$$Qf = Qa + Qb + Qc \text{ etc.}$$

Typical U values are 5.6 for glass and 1.5 for well trimmed and insulated panels. U values can also be calculated by summing the conductivities and resistances of the respective elements of the panels:

$$U = L1/k1 + L2/k2 + Ra + Rb \text{ etc.}$$

with heat path lengths L in mm; thermal conductivity k in W/mC and elemental resistances Rn in m^{2}C/W. Air infiltration losses can be calculated from the general formula:

$$Qv = 0.33NV(T1 - T2)$$

where Qv is the rate of heat transfer to ventilating air; $0.33N$ the ventilation allowance based on the assumption of one complete air change per hour; V is the vehicle internal volume in m^3.

In practice the surface area of the vehicle saloon or cab acts as both reflector and emitter — by re-radiating some of the heat absorbed. The basic mechanisms of heat transfer are described in a later section on refrigerated bodies where it is noted that the conduction mechanism is similar to the conduction of electricity. With heat, however, there are no specific insulators except vacuum since the energy moves not by electrons but by atomic vibration. There are thus differences in temperature profile along a cooling fin to that expected from an electrical analogue. Similarly heat radiation should not be confused with atomic radiation generally where electrons are also involved. In heat radiation, electromagnetic radiation is involved at various wavelengths.

Radiant heat transfer is not stopped by vacuum and is generally transmitted in relation to the fourth power of the surface temperature of a body and its emissivity. Radiation is also received by the body from heat flows in the reverse direction — depending on the fourth power of the surrounding temperatures, which may be far from constant, and on the absorbivity of the body. It is important to realise, in calculation, that absorbivity only equals emissivity at one particular wavelength.

In natural convection, flow is generated by the buoyancy of warm fluid relative to cold fluid and if speeds are low, flow can be laminar rather than turbulent. Factors affecting fluid motion are thermal expansion co-efficient and density of the surrounding fluid — and the relative position of the heat transfer surface and its temperature difference with the fluid. There is no exact proportionality with temperature difference so a heat transfer constant co-efficient cannot be defined; generally flow is related to between the 1.25 and 1.33 power of the temperature difference resulting in co-efficients between 5 and 20 W/m^2. In the case of forced convection there is a proportional relationship but the temperature gradient depends on the thickness of the sluggish 'boundary' layer of fluid between the heat-transmitting surface and the forced draught of fluid. This thickness varies with the inverse square root of the draught velocity, approximately.

The method of calculating heat loss by the combined effect of different mechanisms is best understood by example. A metal cab might be assumed to have an internal volume of 1000 litres and weigh 100 kg, its steel panels having specific heat of 2000 J/kg K and their surface covered by insulation 30 mm thick having conductivity of 1.1 W/mC and

MATERIAL	THERMAL CONDUCTIVITY (Btu. in/ft². °F.h)
METALS	
Copper	2670
Aluminium	1460
Steel	330
MINERALS	
Granite	20·3
Limestone	10·6
Glass	7·3
PLASTICS (UNEXPANDED)	
Polythene	2·5
PVC	1·2
Polystyrene	0·9
GASES	
Hydrogen	1·18
Air	0·18
Trichlorofluoromethane	0·057

Fig 1: Thermal conductivities

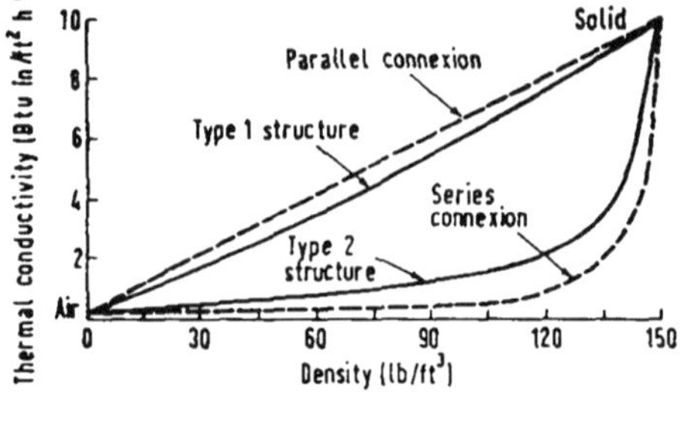

Fig 2: Effect of structure

surface area 15 square metres. A reasonable assumption would be for the convective and radiative 'coefficients' to average 12W/m^{2}C. Overall resistance of the insulation = 0.03/(1.1 x 15) = 0.0018 K/W. Surface resistance (convection plus radiation) = 1/(12 x 15) = 0.0055 K/W. Overall resistance = 0.0018 + 0.0055 = 0.0073 K/W. Heat loss per θ temperature difference with surroundings: = θ/0.0073 W and heat capacity = 100 x 2000 = 200,000 J/K for the metal cab. So for every watt, the temperature falls at 1/200,000 K/s — so the rate of temperature loss with time *t* is = $-\theta/(0.0073 \times 200{,}000)$ so that:

$$\log_e \theta = -t(0.0073 \times 200{,}000).$$

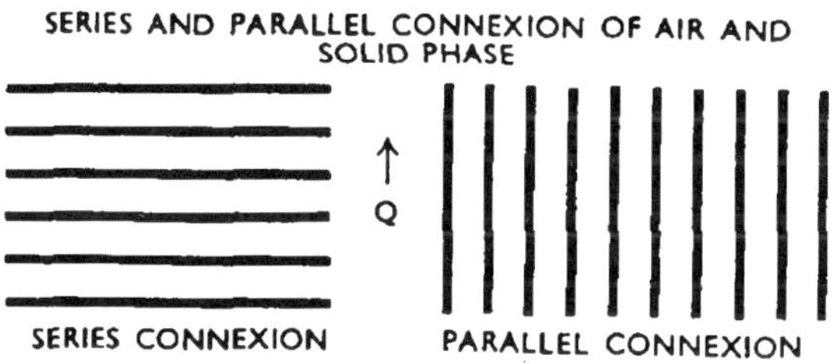

$$k = \frac{k_a}{(1-e) + e \, . \, k_a/k_s} \qquad k = k_s(1-e) + e \, . \, k_a$$

k is thermal conductivity of the structure
k_s is thermal conductivity of the solid phase
k_a is thermal conductivity of the air phase

e is porosity defined by $\dfrac{\text{volume of voids}}{\text{total volume}}$

Fig 3: Calculating conductivity

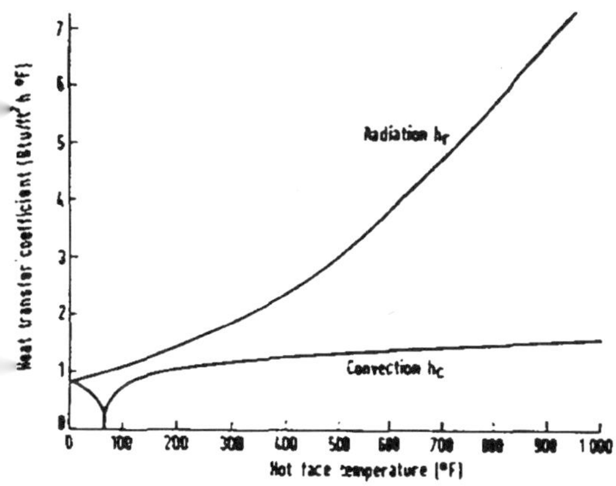

Fig 4: Heat loss to atmosphere

Thus for a surrounding temperature of 10 C, cab temperature will drop from 80 to 40 C in time:

= $\log_e$(80 - 10)/(40 - 10) x 200,000 x 0.0073 x 1/3600
= 9.03 hours

Fig 1 shows the thermal conductivities of a number of media at ambient temperature. In the case of gases the conduction mechanism involves the loss of energy by molecular collision, rather than vibration as is the case with solids. The lower conductivity of gases is an important factor in the choice of insulating materials and underlies the reason for gas expanded, or 'blown', plastics insulants of various types. In blown insulants, assumed to have a regular cubic lattice, two extremes can be considered, type-1 having solid walls with cubical pores and type-2 comprising cubes of solid material surrounded by air passages. Values given in Fig 2 are based on a pore-free solid phase having density of 150 lb/ft^3 (2430 kg/m^3) and thermal conductivity of 10 BTU (2.52 kgcals) in/ft^{2}F. Fibrous insulants tend towards one extreme and cellular materials towards the other. Solid phase and gas phase are essentially in parallel in the type-1 structure and in series for the type-2 structure, the formulae for calculating these two extremes being shown in Fig 3. Fig 4 shows the heat loss to atmosphere from a plane vertical surface by convection and radiation where:

$$Qc = 0.3(T1 - T2)^{5/4}$$
$$Qr = 1.73 \times 10^{-9} E(T1^4 T2^4)$$

E is the surface emissivity taken as 1.0 for the example in the figure which applies to an ambient temperature of 65 F (18.3 C). At low temperatures, thermal radiation wave lengths are in the infra-red region; at elevated temperatures it is the predominant form of heat transmission as shown by the figure.

If a material absorbs all the radiation falling upon it, and reflects and transmits none, it is known as a 'black body'. Such a body also has maximum radiation when it is heated and is thus taken as the standard against which the emission from other bodies is compared. Surface emissivity is thus the ratio of its radiation to that emitted from a black body at the same temperature. A black body is defined as a surface with absorbivity of unity. While non-metallics might have a value of 0.9, that of aluminium is only 0.05 which accounts for the use of foil forms of the material in insulation. The radiation flux per unit area of any body is called the emissive power = 5.67 x 10^8T^4 where T is the thermodynamic temperature of the body, this radiation being spread over a wide spectrum of wavelengths (Fig 5).

Real surfaces are not 'black' and emit less radiation and their monochromatic emissivity is the ratio of the emissive power of a black body at the same wavelength and temperature. It varies considerably with wavelength and such surfaces are called 'selective emitters'. A 'grey body' is thus defined as one having constant emissivity at all temperatures and wavelengths — in order to obtain solutions to real problems by treating selective emitters in small increments of wavelength of constant emissivity. Considering a grey body in black surroundings is a convenient way to reach a solution, Fig 6, for which heat transfer rate:

$$= \varepsilon A_1 \sigma (T_2^4 - T_1^4)$$

suffix 1 referring to the grey body and suffix 2 the black surroundings. This is a useful approximation for real bodies for which a realistic estimate of constant emissivity can be made in 'large' surroundings, implying that the reflected energy is unlikely to be re-incident on the relatively small grey body — and will by successive reflections to the surroundings become less and less. ig 7 shows the heat transfer by solid conduction, air conduction and radiation, as a

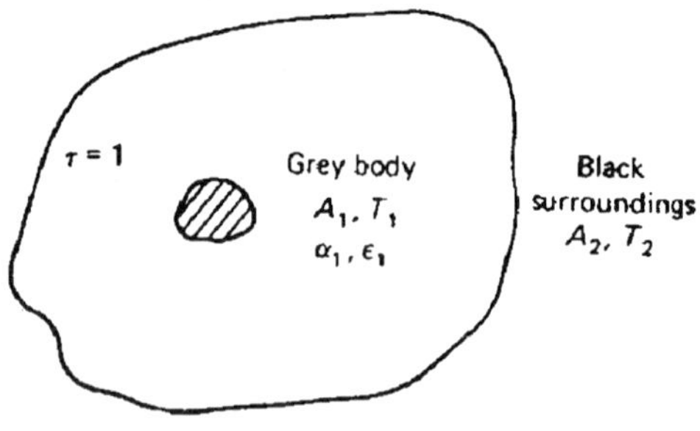

Fig 6: Grey body in black

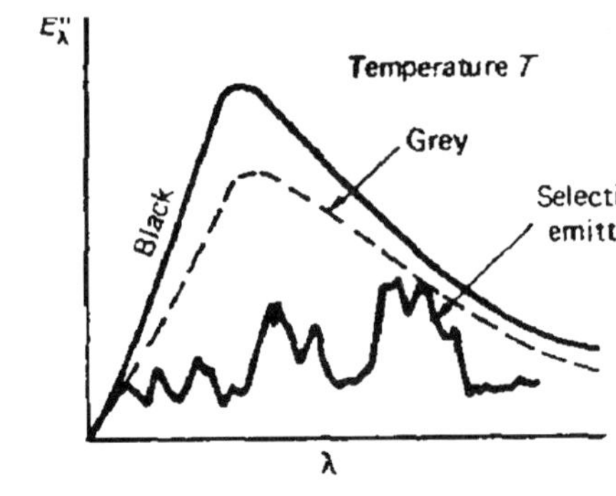

Fig 5: Radiation vs wavelength

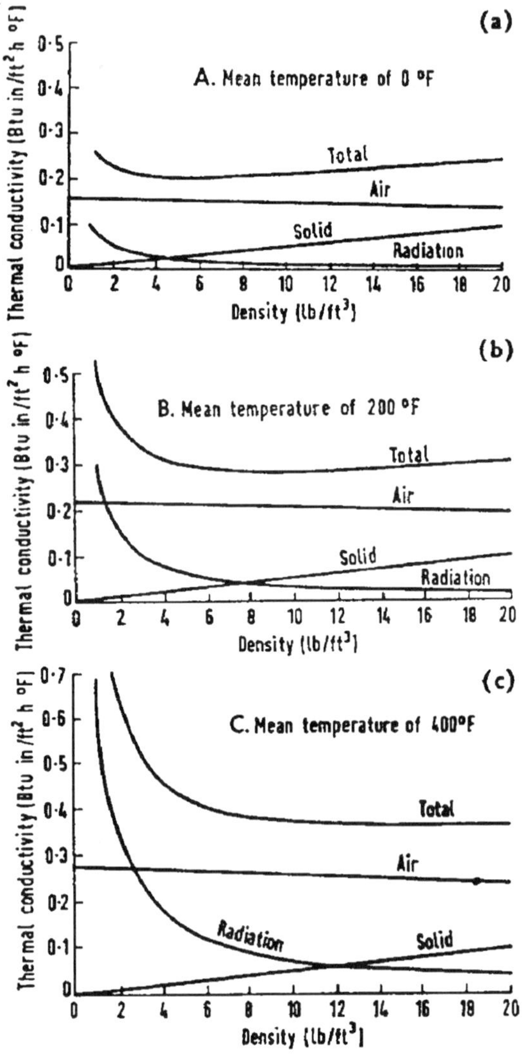

Fig 7: Heat transfer at different mean temperatur

function of density, for an insulating material at mean temperatures of 0, 200 and 400 F, values being representative of a fine-fibred mineral wool material for which solid conduction has been taken as *0.005D* and $kr = 1.08 \times 10^{-9} T^3 m/D$; convection effects are neglected. In solids with a greater degree of parallel connection than mineral wool, conduction increases more rapidly with density — with consequent effect on the overall conductivity. In Fig 8 conductivities of two mineral wool samples are compared, at a density of 7 lb/ft^3, with mean fibre diameters of 5 and 15 microns respectively. From this it is seen that by reducing fibre diameter it is possible to obtain thermal conductivity values not much more than still air.

Hence the trend to reduce this limiting value by replacing air in expanded plastics insulants by such gases as trichlorofluoromethane having thermal conductivity of only 0.057 BTU in/ft^2 h F. Substitutes which are kinder to the ozone layer are, of course, now being sought. In the case of a vehicle body which can be likened to two two fluids being separated by a solid wall, Fig 9, the resistances for conduction and convection in series may be added and the heat transfer rate expressed in terms of the fluid temperatures at either side. Putting $(Tf1 - Tf2) = 0$, the overall temperature difference, then for a plane wall, an overall heat transfer co-efficient can be defined by:

$$1/U = 1/h1 + \sigma(\delta x/k) + 1/h2$$

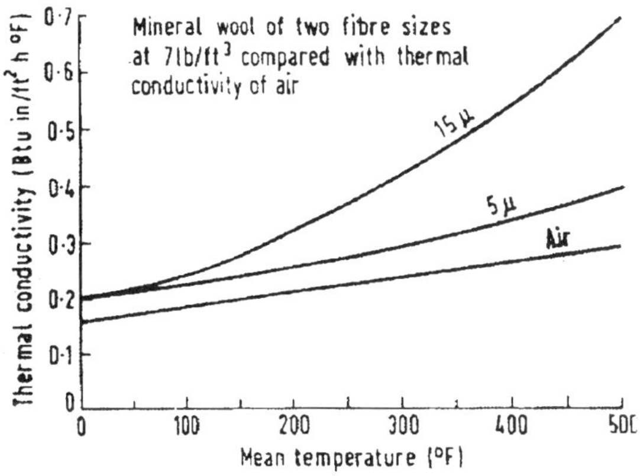

Fig 8: Effect of fibre size

Air-conditioning systems

The combination of heating, cooling, airchange and dehumidification that constitutes air-conditioning can be successfully and economically applied to vehicles only if performance requirement is carefully calculated.

The requirements for full air conditioning of a confined space cover control of air heating, humidity, air movement, air changes odour preventatives and air distribution — not just the cooling for which the load is calculated below. The cooling is usually achieved by passing air over the evaporator plates of a refrigeration system. To avoid a maze of control knobs that would be required for all these facilities, integrated automatic controls are essential in the vehicle application.

Research done on studying the discomfort of close, crowded rooms within buildings has shown it to be due to the physical rather than chemical properties of air — its temperature, humidity and stillness. The presence of a variable quantity, normally around 1%, of aqueous vapour in air is of especial importance and is generally defined by either the absolute or relative humidity. Absolute humidity denotes the weight of moisture in a given volume of moist air. Amount of moisture which air can hold is temperature dependent and is 'saturated' at any temperature according to the maximum amount of moisture it can hold — i.e. 100% humidity. Relative humidity is the proportion to that 100 expressed as a percentage. On cooling moist air, a temperature is reached (the dew point) below which condensation occurs. A guide to preferred air change can be gained from building, and even marine, practice. It has been stated that a seaman's shipboard accommodation should provide a minimum of 120 ft^3 (3.36 m^3) per occupant and traditional building standards suggest that each person requires 1000 to 1200 ft^3 (28-33.6 m^3) of fresh air per hour for satisfactory comfort. A rather larger fresh air requirement has been stated for vehicle drivers — of 1500-1800 ft^3 (42-50.4 m^3)/hour — typically 125 ft^3 (3.5 m^3)/minute for a five-seater car. This allows an increment for reducing inward draughts. Dehumidification can be obtained by cooling the incoming air.

Successful application of vehicle interior air-conditioning depends on a reasonably accurate cooling load calculation for the vehicle interior. Three principal constituents of the cooling load are solar radiation on the vehicle, heat input due to passengers and heat input from the fresh air supply into the vehicle.

Up to 300 btus (75.6 kgcals)/hour can be transmitted into a typical car by the sun's rays and it useful therefore to assess the different transmissilities of glass and innertrimmed steel surfaces. An example may be 6 x 4 ft (1.83 x1.22 m) for the roof area of a saloon car and twice 2.5 x 6 ft (0.76x1.83 m) for the side panels (less windows) also 2.5 x 4ft (0.76x1.22 m) for the rear panel — giving a total of 64 ft^2 (2.42 m^2). Glass area is computed in the same way and in this case is taken as 5ft^2 (0.465 m^2) for the windscreen, 11 ft^2 (1.02 m^2) for the side windows and 3ft^2 (0.28 m^2) for the rear-light — totalling 19 ft^2 (1.77 m^2). The conductivity of trimmed steel panels is quoted as 14 btu (3.53 kgcal)/ft^2/h so that in this example the heat flow through them due to solar radiation on the outside is 895 btu (225.5 kgcal)/hour. The equivalent figures for the glass area (whose quoted heat-transmissibility is 160 btu (40.3 kgcal)/ ft^2/h) are 3040 btu (766 kgcal)/h for the glass alone and a combined panel-and-glass total of 3935 btu (992 kgcal)/h.

The figure above could be substantially reduced by the use of tinted glass. The reduction could be of the order of the quoted heat transmission per person of 400 btu (100 kgcal)/h — which for a 5 seater makes the total of 2000 btu (504 kgcal)/h. Such figures give a clue to the temperature difference on a full and empty bus! Total heat to be removed from the fresh supply can be calculated from the assumption of

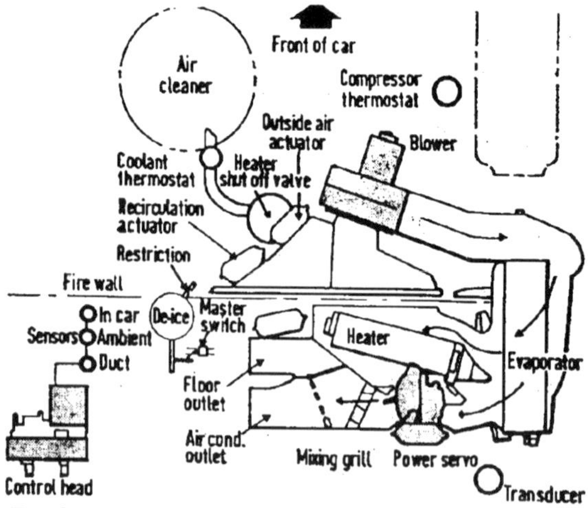

Fig 9: Early Cadillac system

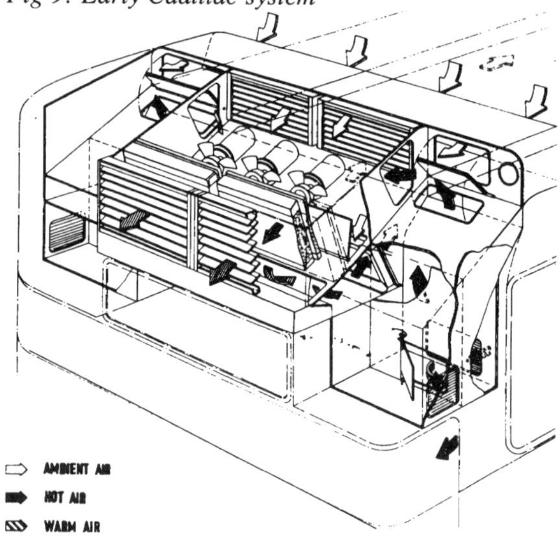

Fig 11: Patented system for buses

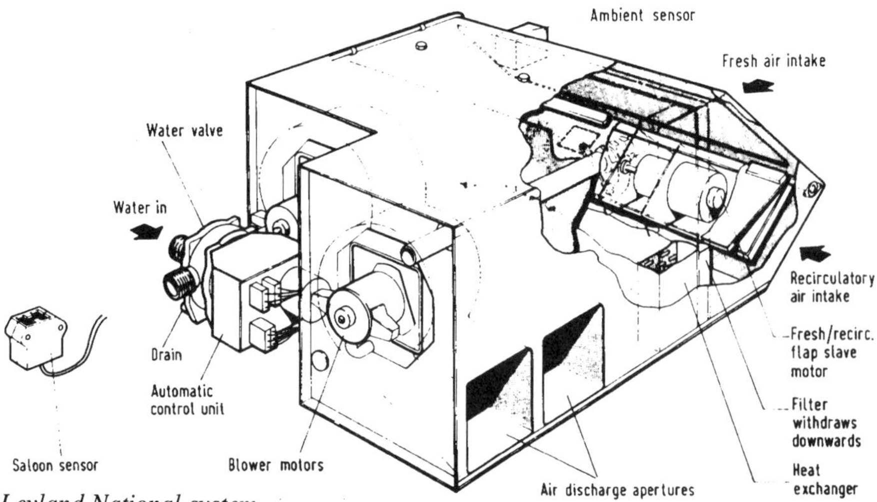

Fig 10: Leyland National system

given climatic conditions which in this case are taken as 100 F (37.8 C) dry bulb temperature, 79 F (26.1 C) dry bulb and outside-air relative humidity of 40%. Given also the interior requirements, of say 80 F ((26.7 C) dry, 57F (13.9 C) wet and 50% RH, sensible heat to be removed from 1 lb of fresh air is 0.238 x 20 = 4.76 btu (1.2 kgcal)/lb of dry air. Heat to be removed from moisture per pound of ambient air for the given reduction in wet bulb temperature (reducing vapour pressure from 0.375 to 0.246 lb/in^2 [0.026-0.017 bar]) is obtained by sutracting the product of lbs of moisture per lb of interior air and heat content per lb of moisture at the interior wet bulb temperature from the equivalent product for the exterior air. In this case it amounts to 0.01645 lb/lb x 1095 btu/lb - 0.0111 lb/lb x 1085 btu/lb = 5.95 btu (1.5 kgcal)/lb of air to be extracted.

Total heat to be extracted is thus 10.71 btu (2.7 kgcal)/lb for the air and vapour mixture. Quantity of dry air in the incoming stream of fresh air at 125 ft^3 (3.5 m^3)/min, $w = Pa.Q/R.T$ where Pa is dry air atmospheric pressure, Q is quantity of fresh air supplied, R is gas constant for air and T is the absolute temperature of incoming air. Thus w = 8.67 lb (3.94 kg)/min and total heat to be extracted = 5575 btu (1405 kgcal)/hour with moisture to be removed totalling 0.0464 lb (0.02 kg)/min. Total cooling load thus amounts to 10,575 btu (2665 kgcal)/h.

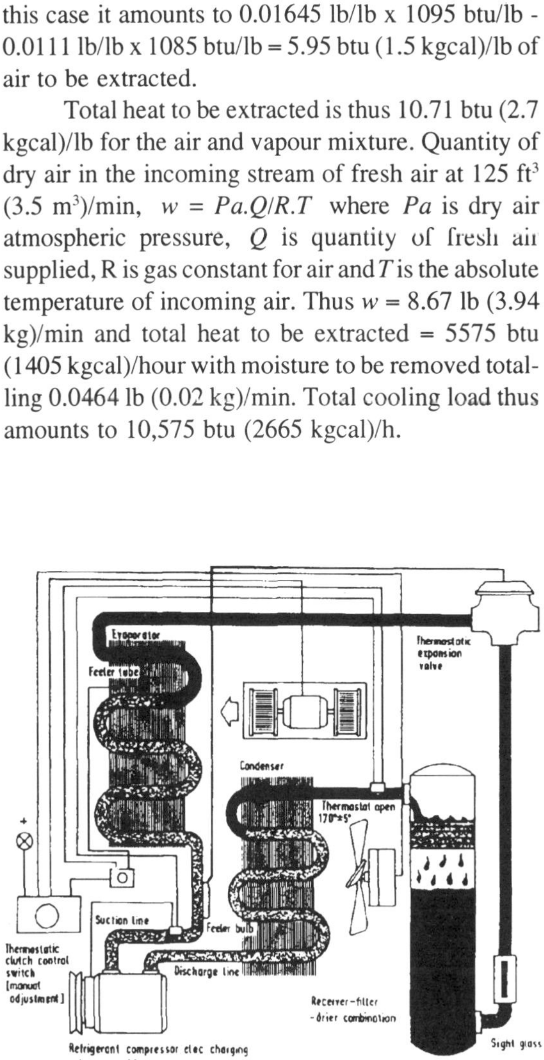

Fig 12: Kysor truck-cab unit

Cadillac was one of the first car models for which a full fresh-air air-conditioning system was developed, Fig 9, complete with automatic control. After all incoming air is cooled and dried by the evaporator a variable amount is diverted and re-heated sufficiently to provide the required duct air temperature where it is remixed with the remainder. Flow is then automatically proportioned between the outlets, recirculation being used only during initial cool down from high ambient temperatures. Blower speed is varied automatically to suit demand.

In the case of buses and coaches, the cooling loads are sufficiently high to warrant the size of industrial systems used for room air-conditioning. There is a more frequent requirement with buses than with cars for the unit to be run while the vehicle is at rest and therefore separate motor systems are often specified. The Leyland National bus with its familiar rear-end roof-pod represented an important step forward in bus air-conditioning which grew out on an interesting development aimed at a combined engine cooling/saloon heating system. In the event, the system realised, Fig 10, is a unitised heater which feeds air for conditioning from a built-in duct in the vehicle roof — which can be either cooled or heated as required by the operation. The saloon temperature controller has a transistorised amplifier which responds to three thermistors, sensing ambient, heater-outlet and saloon temperatures respectively. The amplified signal energises a servo-motor driving a water regulating valve. Each heater unit has an output of 60,000 btu (15120 kgcal)/hour/100 C temperature difference and supplies 750 ft^3 (21 m^3)/minute air-flow against a duct resistance of 0.75 inches water gauge (0.188 bar) — and providing 25 air changes per hour in the single-deck vehicle. The design allowed either tandem heater units or a side-by-side refrigeration unit to be installed according to operating climate. The original development which was not productionised culminated in a patent specification (GB1205941), Fig 11.

Truck cabs can be fitted with roof-top systems, of which the Kysor one shown in Fig 12 was a pioneer. An engine driven compressor absorbs 4.25 bhp when operating in a 100 F (37.8 C) ambient with 50% RH. The largest model had a thermal capacity of 16,000 btu (4032 kgcal)/h. Some of the latest systems under consideration are based on electric motor rather than engine driven arrangements.

Refrigeration efficiency would be increased by redesign of the refrigeration condenser unit, Fig 13, giving 75% higher co-efficient of heat transfer and 20% reduction in pressure drop over the conventional serpentine type. Fig 14 shows the trade off between improved efficiency and larger area required. These improvements have been shown in tests to result in substantial reductions in electrical powere demand.

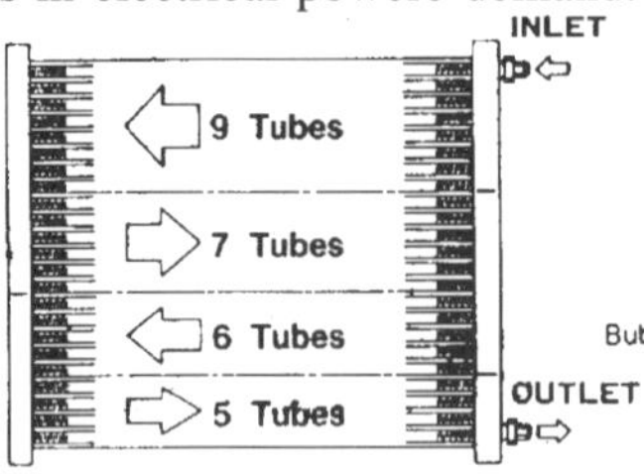

Fig 13: Optimising the condenser

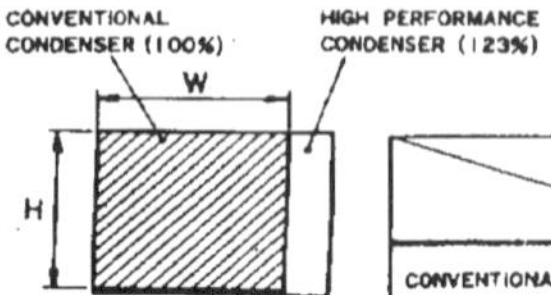

	W (mm)	H (mm)	T (mm)
CONVENTIONAL CONDENSER (SERPENTINE TYPE)	410	355	22
HIGH PERFORMANCE CONDENSER (PARALLEL FLOW TYPE)	500	360	16

Fig 14: Efficiency vs area

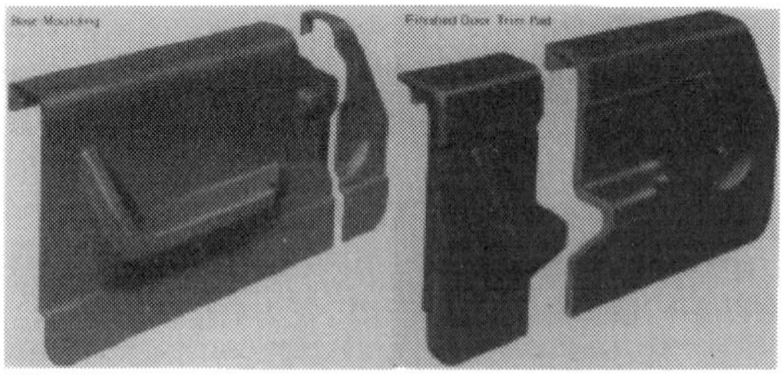

Fig 15: Wood-fibre based panel

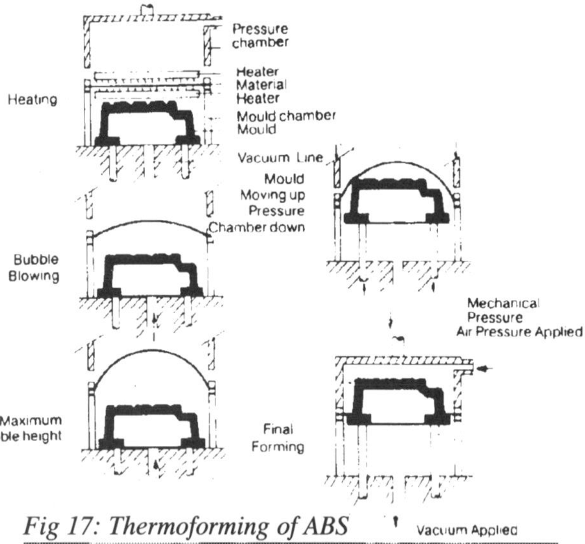

Fig 17: Thermoforming of ABS

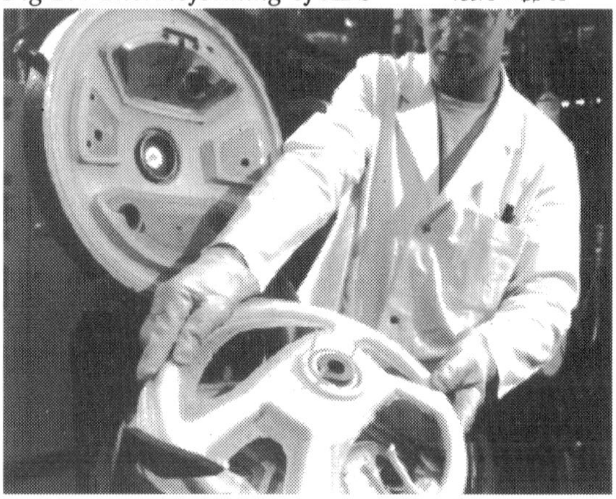

Fig 21: De-moulding a steering wheel cover

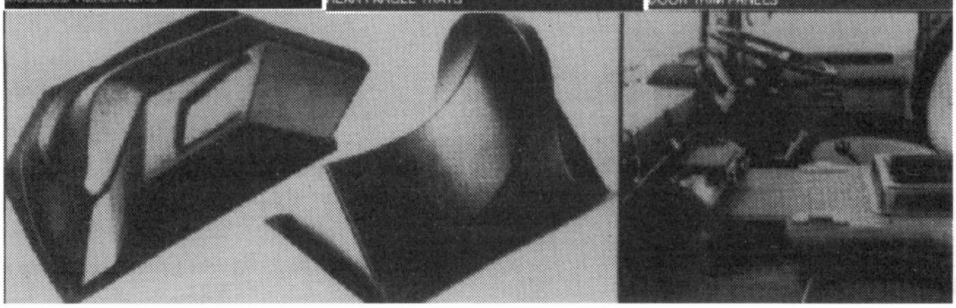

Fig 16: Polypropylene/wood-flour panels (top)

Fig 18: CV panels in Cycolac ABS (bottom)

Interior trim

The earlier section on timber products draws attention to the use of pressed woodfibre board as a trim base, Fig 15 showing a panel produced by British Moulded Fibre. Because of the relatively inexpensive tooling, the process is also attractive for the moulding on of fabrics and trimcloth to produce 'integrated' trim panels. Soft feel can be obtained with foam interlayers beneath the trim cloth which is usually vacuum-drawn on to the fibreboard.

An alternative trim-base uses wood-flour rather than woodfibre. This fine sawdust-like material is blended with polypropylene and moulded in a press with somewhat more expensive tooling but at least having potential for medium volumes. Trimcloth covering is also possible in a single closing of the mould, with such materials as vinyls, textile fabrics or carpet. Fig 16 shows a selection of items made with the process by John Cotton (Colne).

Materials used for seat upholstery, such as those described in the later section, are also used extensively in keying the door and side panels of car saloons and CV cabs to the texture of the seat. Particularly good leather simulations have been developed in PVC by such firms as Weston Hyde. Vynide is a solid PVC-coated fabric which was developed about 20 years ago. Perforated forms of the material make it attractive for headlinings. When used with knitted backcloths, allowing two-way stretch, quite sharply curved surfaces can be trimmed. Surfaces also can be embossed to obtain leather or other 'grain' effects. Flutes, too, can be moulded-in to obtain 'traditional' effects. Thermoforming plastics are another alternative to 'cloth' covered boards. Fig 17 shows a process developed by Borg Warner for thermoforming ABS. The same company has produced design rules relating thicknesses and draw depths available with the material. Well seasoned timber or composite tools can be used for very small runs — or to make prototype parts while metal powder filled heat-resistant epoxides can be used for medium runs. Draft angle should be 2-3 degrees and depth of draw should be less than half the finish width of the panel. Fig 18 shows Cycolac ABS parts and assemblies for CV cabs. Polyurethanes in various forms are also valuable as trim-base materials with their ability to exist in rigid, semi-rigid and flexible states. Hot and cold cure foams are both used, the former requiring aluminium moulds. The table in Fig 19 shows some of the properties which are typically obtainable with cold and hot-cure foams. Cold-cure foams can be produced in wood or GRP moulds but normally it is difficult to produce foams less dense than 45 kg/m^3. They do however have intrinsic flame-retardent qualities. Semi-rigid and self-skinning foams may be produced either by the so-called prepolymer or one-shot processes. In the former, excess toluene diisocyanate is reacted with polyol as a first step and the foam reaction is completed subsequently, with addition of water and catalysts during moulding. With the one shot system all the chemicals are mixed together at the time of moulding. Adjustment of the chemical formulation results in typical properties for a crash pad panel as shown in the table in Fig 20. Pads may be produced with a variety of cover materials such as vac-formed ABS or PVC; blow-moulded, blow-formed, rotationally-cast or slush-moulded PVC. The last two methods particularly suit arm-rest production. Flooring systems are also produced in PUR by spray-forming a PVC skin into an electroform,

Property (BS3379)	Typ hot cure	Typ cold cure	32 kg/m^2 cold cure
Density (kg/m^3)	24-45	45-70	32
Tensile strength (kN/m^2)	110-165	69-97	62-67
Tear strength (N/m) ASTM-D1564	350-610	157-200	157-200
% elongation at break	250-350	90-110	80-90
75% compression set	6-9	5-7	5-7
90% compression set	6-9	5-7	5-7
Hardness (12 in ind 40% comp),kg	18-35	15-45	20-24
Isoprene coating (% elong at break)		600	
Dynamic fatigue (% height loss)	4-12	1-4	1-2
Resistance (% hardness loss) (to ISO draft spec TC45/WGH-N84)	15-35	5-10	15-20

Fig 19: Properties available with cold- and hot- cure foams

Density kg/m^3	48-160
Tensile Strength kN/m^2 (BS3379)	138-414
Tear Strength N/m ASTM-D 1564	157-525
Compression Set 75% (BS3379)	25-35
Compression Set 90% (BS3379)	80-90
Elongation at Break % (BS3379)	50-90

Fig 20: Properties of semi-rigid polyether polyurethane foam

the skin, after curing, being filled with semi-rigid foam backed in the foaming operation by an 0.08 mm thick sheet of PVC. Door inner panels by contrast are usually made in the material either by vac- or blow-forming the covers in ABS. Self-skinning foams find applications in steering wheel trim and both acrylic and PUR colour coatings can be applied. An interesting development for in-mould coating systems for PUR panels is Trimbond by MacPherson Industrial Coatings. The paint formulation contains its own mould release agent and dries in a few seconds on the mould surface prior to injecting the PUR. After three minutes the coated panel can be withdrawn from the mould; a variety of surface textures are possible. A steering wheel cover by BP Chemicals is shown in demoulding in Fig 21. Trends towards softer touch trim finishes has led to the use of PVC compounds and the dry-powder slush-moulding process, Fig 22. Claims for the process include greater design freedom than that allowed by vac-forming as well as a softer skin and deeper grain. It is said to be less costly than wet slush moulding and a variation is called double slush moulding, where skin and PVC foam are formed integrally in one tool. The table in Fig 23 lists rival processes competing for trimmed instrument panel construction with advantages and disadvantages, due to US consultants Phillip Townsend Associates. In door panels, there is sign of a trend away from ABS towards polypropylene due to lower cost and better acoustic performance. Fig 24 due to Hoescht shows noise test results carried out on B-post trim panels for an executive car.

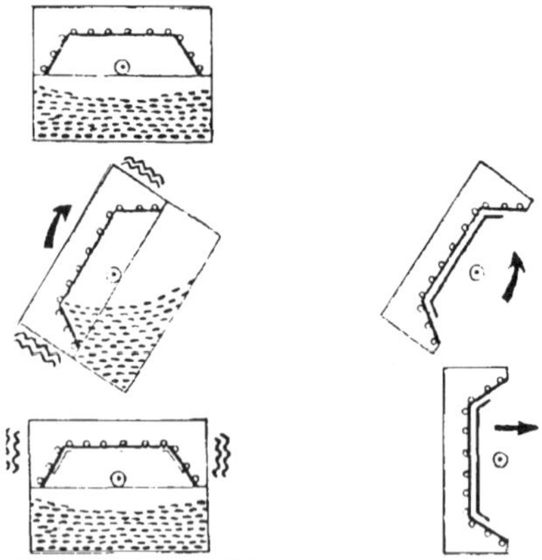

Fig 22: Dry slush-moulding

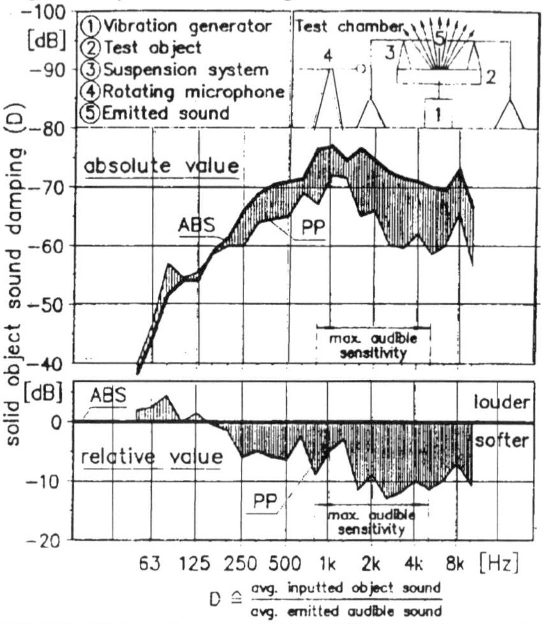

Fig 24: Noise characteristics of B-post trim panels

PROCESS:	DP-PVC	DP-PVC	Wet slush	Negative thermoform	Positive thermoform	Thermoform	Spray
PROCESS VARIANT:	Single slush	Double slush					
SKIN MATERIAL:	PVC	PVC	PVC	PVC/ABS alloy	PVC/ABS alloy	EPVC	PU
SKIN THICKNESS (MM):	1.5	1.5-2	1.8	1.1	1.1	1.1	1.1
GRAIN REPRODUCTION:	Excellent	Excellent	Fair	Fair	Excellent	Fair	Fair
FOAM:	PUF	PVC	PUF	PUF	PUF	PVC	PUF
FOAM PROCESS:	Injected	Integral	Injected	Injected	Injected	Integral	Injected
MOLD COST ($'000):	200	20	20	50	50	50	40
CYCLE TIME (SECONDS):	315	320	420	45	45	45	150
SOFTNESS:	Excellent	Excellent	Fair	Fair	Excellent	Fair	Fair
RELATIVE COST INDEX (UNITS/SKIN):	200	?	260	100	120[e]	100	157

Fig 23: Rival processes for instrument-panel trimming

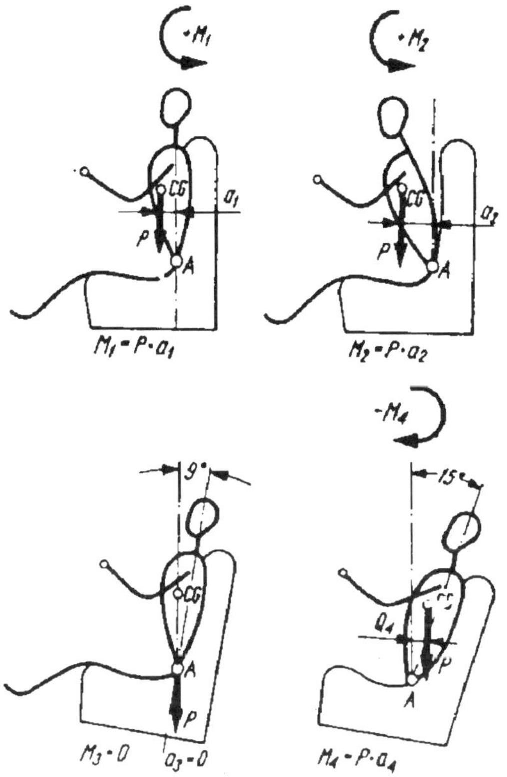

Fig 25: Results of misplaced back-rest

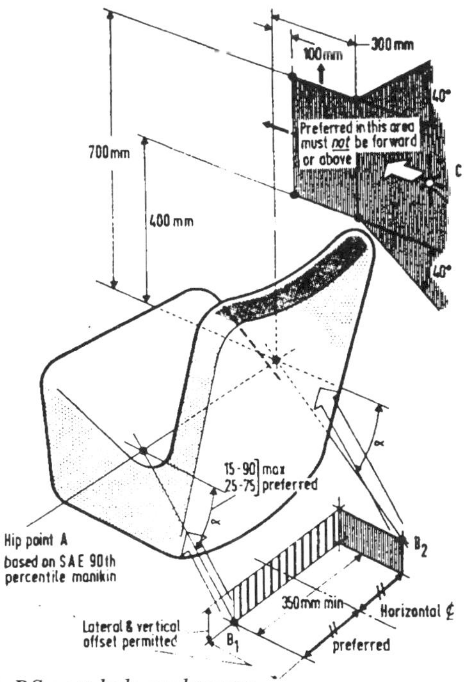

Fig 26: BS seat belt anchorage

Seating and seat trim

Seats and associated upholstery trim design can have dramatic effect on a potential customer both at the point of sale of the vehicle and at the end of a long journey. Leaving the selection of seat parameters until the 'afterthought' stage in the design of the complete vehicle can easily lead to failure in the market place for most categories of vehicle.

The earlier section on ergonomic layout of vehicle interiors made clear the importance of seat dimensioning, and provision of adjustment, to both the comfort of occupants and the adequate control of the vehicle. Fig 25 shows how even the wrong backrest inclination has the potential to cause severe discomfort over a long journey. Equally important is the choice of cushion springing and damping characteristics as well as the contouring of the seat and squab profiles and the choice of upholstery trimcloth.

The importance of seating to vehicle control leads the writer to recall the words of a latter-day automotive wit, commenting on the occupant's anatomical relationship with the primitive vehicle seats of days gone by. 'To seat two spheres upon one other and regard it as stable is foolish...' this is an extreme view of the driver's difficulty in maintaining a position of stability with respect to the controls but it nicely emphasises the risk of losing the essential 'seat of the pants' feedback of vehicle motions necessary to properly involve the driver in the 'control loop'. Medical scientists argue that muscular interaction with vehicle motions, sensed through the seat, is governed by anticipation of (experience) reaction and by a sense of balance. While the interaction becomes a reflex condition, considerably more energy is required with conventional seat systems, to maintain a relationship with the controls, than is necessary to actual control the vehicle.

Fig 26 shows the seat belt anchorage positions defined by British Standard BS AU 4B:1965 which are chosen to provide safety restraint in impact without undue discomfort in normal driving. In fact, much driving discomfort has been traced to vehicle manufacturers providing too great a static deflection for the seat springs in order to enhance showroom appeal. In general, it is argued, this can lead to undesirable overrun-bounce — an undamped motion in phase with the vehicle's wheel suspension frequency.

Suspension seats

For commercial and industrial vehicles, there is a strong case for suspension seats in which controlled levels of springing and damping can be introduced, Fig 27 showing one of the original Bostrom designs. In designing seat systems it is useful to observe measured comfort criteria such as those established by Janeway, illustrated in the earlier section on vehicle ride. A simplified approach is to regard the driver on his or her seat as a single-degree-of-freedom spring/mass/damper system with the familiar response characteristic shown in Fig 28. Maximum mechanical response of different parts of the body of a 170 lb (77 kg) man sitting on a rigid chair, and standing, are shown in Fig 29, due to Bostrom. This company has also measured vibration levels in a range of vehicle categories as shown by Fig 30. (A) represents on-highway trucks, wheeled earthmovers and farm tractors; (B) represents corresponding off-highway working and (C) is the same for non-suspended seats.

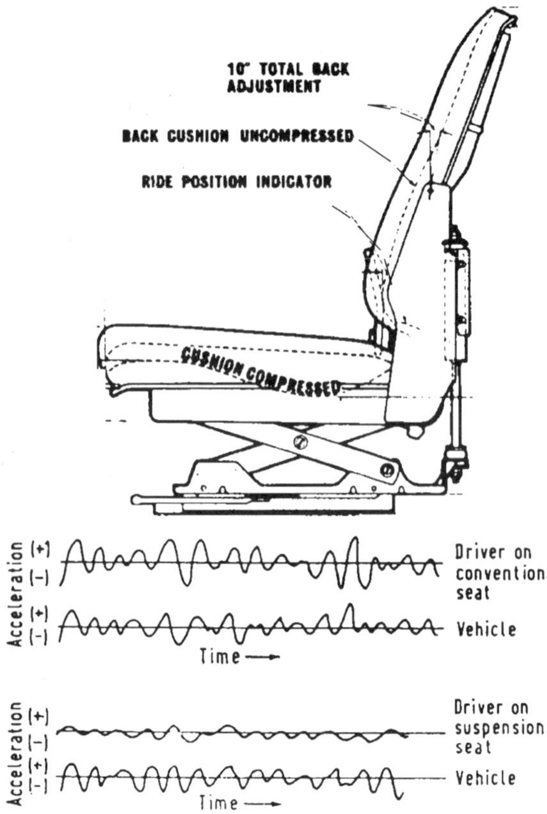

Fig 27: Bostrom suspension seat, with responses

Seat cushions

Cushion design should observe the anatomical necessity of providing firm support only for the bones immediately above the buttocks since the surrounding areas have little pressure capacity. Too deep a cushion deflection would overload the surrounding areas and cause discomfort — an acceptable maximum is put at 3 in (76 mm). Fig 31 shows belt-level response of a man subject to 1/4 in (6.35 mm) amplitude vibration when sitting on three typical cushions, without squabs, which unlike suspension seats cannot have their characteristics altered to isolate offending frequencies. Cushion design has been revolutionised by moulding techniques developed for low density flexible foams. While hot moulding has been established for over twenty years, the recent existence of cold moulding has helped lower costs due to lower energy requirements and cheaper tooling. Two-component dispensing machines are required, with manually opened GRP moulds, as the basic equipment. Materials used are polyethers expanded by isocyanates. Fig 32 shows systems available from ICI Polyurethanes. Hardness of foam can be varied by adjusting the isocyanate component in different parts of the mould — giving the possibility of moulding dual-hardness cushions to obtain a soft centre with stiffer side bolsters.

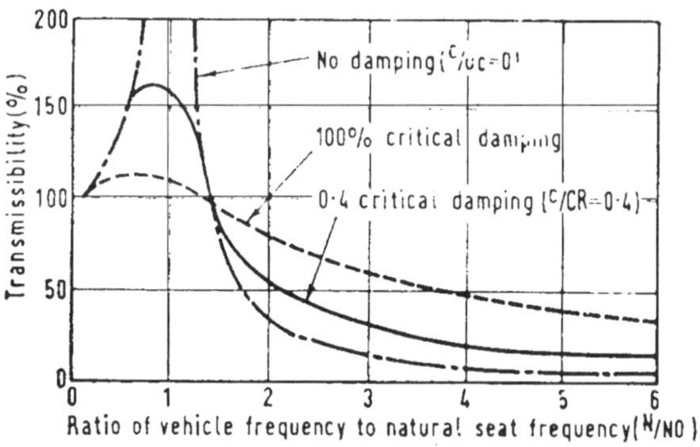

Fig 28: Response of single degree system

Upholstery

Techniques are also available for moulding the cushion on to the upholstery system when temperature tolerant coldmoulding is used. For industrial vehicles, composite moulding of slush-moulded PVC plastisol covers, filled with cold-moulded foam, is common. For other vehicles stretch-fabrics can be employed as the top layer of a multilayer laminate — over 23 mm thick peeled flexible foam, a PVC film and further inner layer of 1 mm thick foam. The so-called Skinform process involves such a laminate, being held under a controlled tension which is chosen according to the cushion design, drawn by vacuum into specially designed moulds. The coldcure cushion-foam is then moulded in-situ. The more traditional way of covering a moulded foam cushion pad is shown in Fig 33 for a car and in Fig 34 for a touring coach.

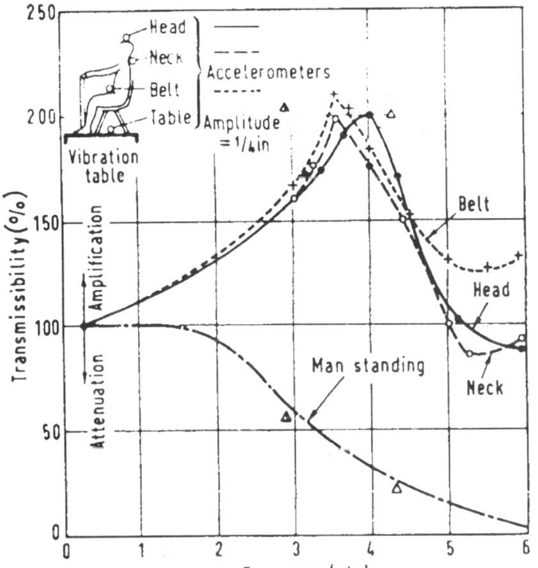

Fig 29: Different body area responses

Fabric moquettes in 85/15% wool/nylon are now available for coach seat covering, with backing materials which allow a thick pile effect to be created. For cars, nylon faced fabrics are typical, backed by foam and a layer of knitted cotton skrim. The exposed top surface is usually proofed against stain marking. PVC coated fabrics are also available which provide excellent simulation of leather — even to the

Formulation	Typical automotive system	Typical furniture system	
'Suprasec' VM 28 (pbw)	–	55.4	47.8
'Daltocel' C 971 (pbw)	–	100	–
'Suprasec' VM 25 (pbw)	63.3	–	–
'Daltocel' C 1565 (pbw)	100	–	–
'Daltocel' C 969 (pbw)	–	–	100
CFM-11 (pbw)	0 to 10	8	–
Isocyanate index	100	88	80
Processing conditions			
Polyol blend temperature (°C)	20 to 25	20 to 30	20 to 30
Isocyanate temperature (°C)	20 to 25	20 to 30	20 to 30
Mould temperature (°C)	40 to 45	ca 30	ca 30
Mould occupation time (minutes)	4 to 6	4 to 6	4 to 6
Post-cure	None	None	None
Foam properties			
Density (kg/m^3)	35 to 55	39	53
Compression set (50% compression 22 h at 70°C) - (% set)	7 to 10	4	7
Tensile strength (KPa)	145	60	80
Elongation at break (%)	140	110	105
Tear strength (N/m)	200 to 300	110	160
Indentation hardness at 40% (N) (ISO 2439)	150 to 300	165	200

Fig 32: Some MDI-based cold-cure moulding systems

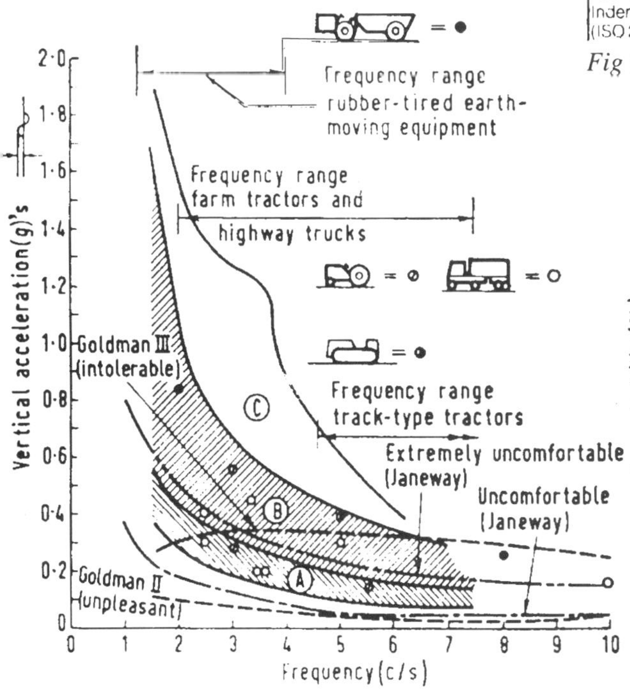

Fig 30: Different vehicle category responses

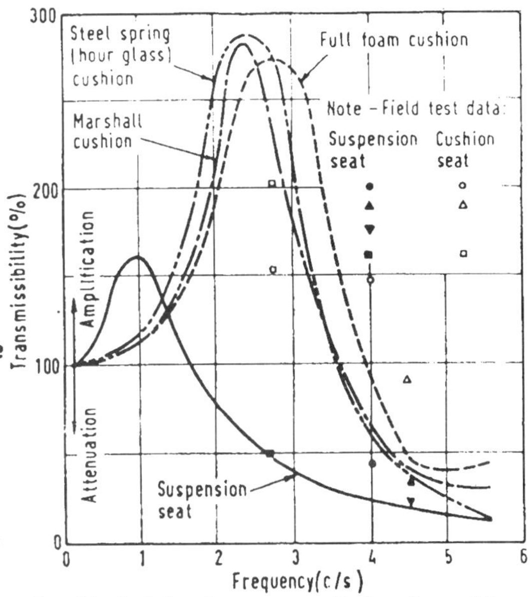

Fig 31: Belt-level response of vibrating subject

extent of fine-perforation for breathing. Foam backings and two-way stretch backing-cloths allow cooling flutes to be moulded-in and a variety of embossed surface effects are possible. Suppleness of the PVC can be varied by the amount of plasticizer added and the material can last up to ten years before any signs of hardening-off are detected.

Process advances in body trim

The need to reduce costs by consolidation of parts in such assemblies as instrument panels, door cases and seats is leading to one-shot moulding techniques. Environmental legislation is also leading to the use of common materials.

For door-rim panels a principal technique in use by major suppliers John Cotton is a trim base panel of moulded Woodstock. This material is bought by the company as an extruded blend of polypropylene and woodflour and then processed on a machine which permits forming, covering and trimming in one operation. Its properties are in the table of Fig 35. Coverings commonly in use are supported expanded vinyls (SEV), textile fabrics or carpeting. The company can also offer a wide range of alternative trim-base materials as seen in the table of Fig 36, which is an excellent survey of materials currently applied to vehicle trim.

In self-supporting moulded headliners the company uses an extensive range of substrates including resin-bonded and glass reinforced polyurethanes (500-610) and felts; FSL Dielark styrene (580-640), PPO foam and Eften corrugated board (975); figures in brackets being the weight in gms/metre². Coverings

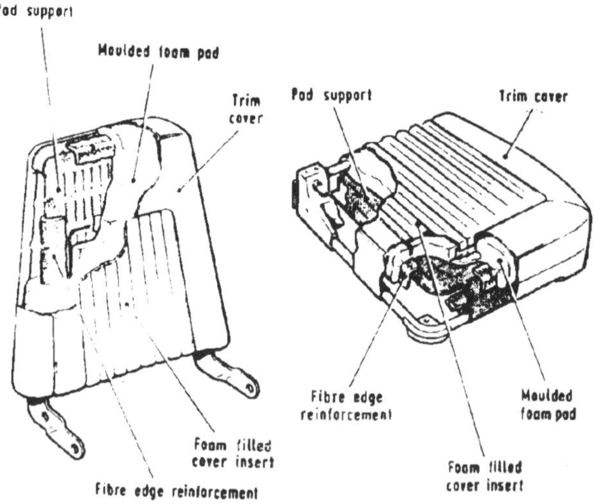

Fig 33: Car-seat upholstery

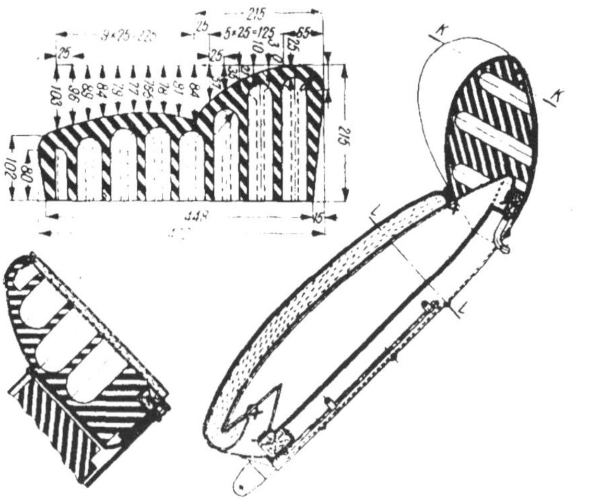

Fig 34: Coach-seat upholstery

Material	Fimit Lignatok	BMF	Mip Van Dresser	Northern Fibres	Lignatok	Cekazell	Rover 200 Woodstock	Woodstock Equivalant HMPP	Strapazzini Resine	Fimit Felt	Glass Mat Pu Rim Exd 113/1	Empe-worke Pelliform
Application	Fiat Tray	Montego	Maestro		Sierra	Parcel Tray	Rover 200	Matl. Data	Lancia	Fiat	Matl. Data	
Description	Wood Fibre Pehnolic Resin	Wood Fibre Aromatic Polyester	Wood Fibre Phenolic Resin	Wood Fibre Phenolic Resin	Wood Fibre Aromatic Polyester	Waste Fibre	Woodflour Filled Poly-propylene	40% Wood Filled Poly-propylene	20% Milled Glass Filled Resin	Felt Phenolic Resin	450 g/m² Glass Mat PU	100% Jute Thermoplastic Resin
Density (kg/m³)	1050	650	656	760	820	860	890	1000	720	1000	480	770
Thickness mm	2.8	2.2	3.5	3.3	2.4	4.5	2.4	3.0	3.8	3.0	3.5	2.5
Mass per unit area (kg/m²)	2.9	1.4	2.3	2.5	2.0	3.9	1.7	3.0	2.7	3.0	1.7	1.9
Flexural Strength (MN/m²)	56	20	24	34	46	34	>27	38	24 41	76	31	46
Flexural Modulus (MN/m²)	–	–	–	–	–	1700	–	4000	2500	–	1200	1500
Flammability	Self Ext 30 secs.	Burns Slowly	Self Ext 45 secs.	Self Ext 45 secs.	Burns Readily	Self Ext	Burns Slowly	Burns	Self Ext 30 seconds	Self Ext 30 seconds		Burns Slowly
Hardness (double thick-ness) (N)	810 (> 2000)	200 (360)	(970)	190 (1200)	550 (1020)	790 (1200)	(> 2000)	1350 (> 2000)	370 (> 690)	900 (> 2000)		500-1000
Wicking Length (mm)	0	0	45	35	55	–	0	0	0	30	0	–

Fig 36: Range of alternative trim systems

include non-woven fabrics of polyester, nylon or viscose in spun-bonded, point-bonded or needle-punched construction; polyester or nylon knitted fabrics, normally as a raised tricotte knit to give a velour effect; other options are woven fabrics, artificial suede, PVC washables and flock coatings. With the popular knitted fabrics, the material is flame-

Fig 35: Woodstock properties

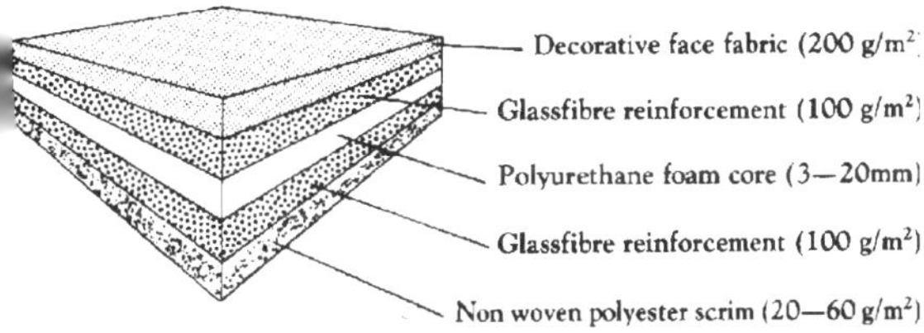

Fig 39: PUR headliner construction

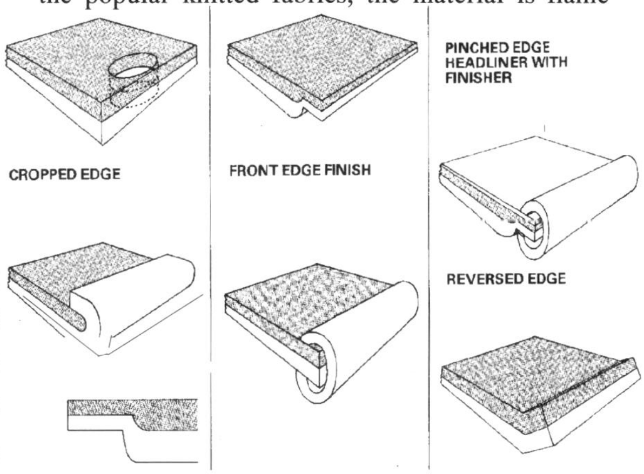

Fig 37: Edge conditions

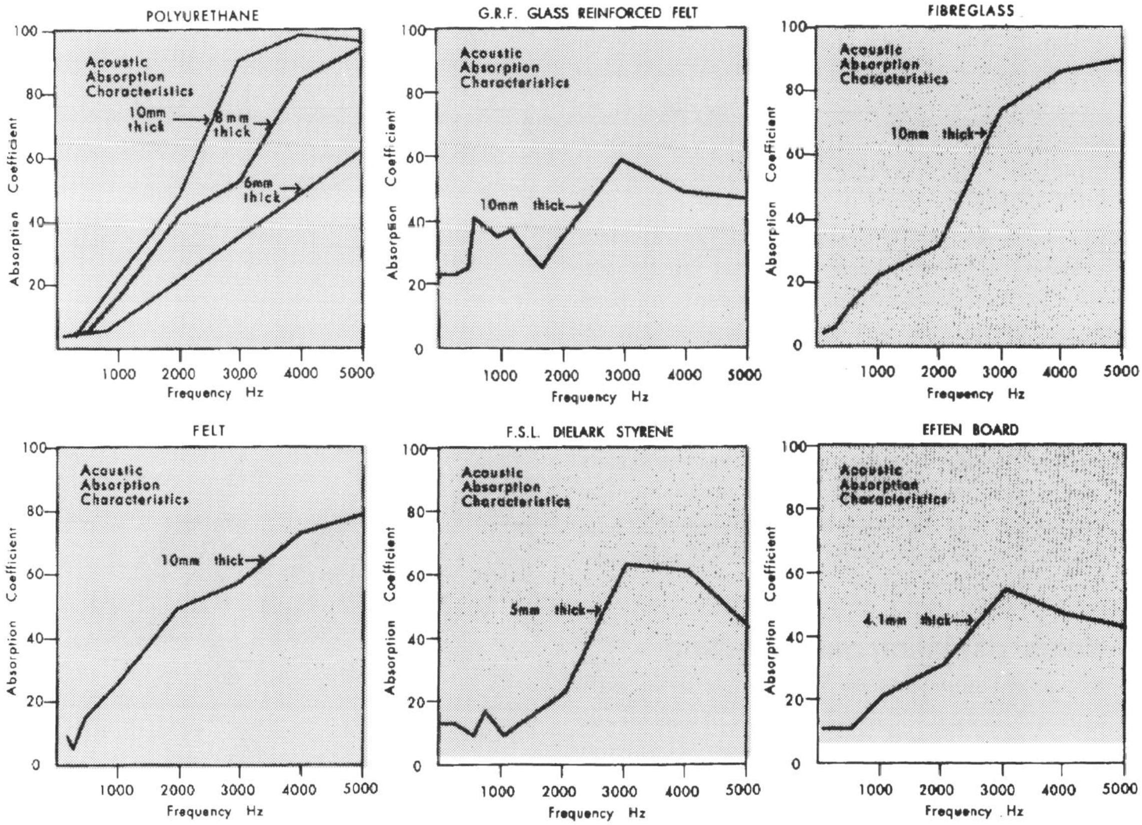

Fig 38: Acoustic performance

laminated or sinter-bonded on to the substrate foam of 1-4 mm thickness which gives a cloth of appealing surface texture and soft-touch feel. Knitted fabrics can also be drawn into most shapes and are readily dyed to colour key with other parts of the vehicle interior. A variety of edge conditions is offered, Fig 37; With the cropped edge, a finisher or an overlapping adjacent panel is normally necessary. The cloth-wrapped edge removes the need for a finisher but as the substrate has to be trimmed to size first two separate operations are involved. With the pinched edge, a compressed edge is designed into the moulding which is subsequently cropped to size, having first moulded the substrate and covering as an integral unit. Sometimes the so-called reversed edge is acceptable, as usually only the corner bend is visible. Acoustic performance of headliners made from the different substrates is seen in Fig 38.

Polyurethane headliners have been developed by the company which can be made in one piece for a car or cab roof. Construction is shown in Fig 39; the PUR core is based on an open-celled BASF polyester foam which is thermoformable and has unusually low density and good sound-absorbent characteristics. Hardness can be varied right through from flexible to rigid state. It has good dimensional stability up to 100C and has survived simulation testing for tropical conditions (200 hours at 40C and 90% relative humidity). It meets FMVSS flammability rating, and with all covering layers in place, weighs 1 kg/metre^2. Glass-reinforced felt has also recently been developed by the company, for headliners, and has strength characteristics which suit applications involving the suspending of sunroof frames, light fittings or speaker grilles. Traditional resin-bonded felts are enhanced by a glass reinforcement to provide twice the rigidity for a given density. Fixing clips and brackets can be integrally moulded and weight is in the order of 1300 gm/metre^2. Fig 40 shows typical weatherstripping employed at the junctions with closure panels.

Noise insulation and recycling

Recent developments by ICI involve a scientific approach being made to vehicle interior noise reduction by exact specification of trim material properties and dimensions. Heavy covering layers of carpeting and similar materials are mounted on flexible foams of known stiffness so that the spring/mass can be exploited in noise isolation, Fig 41. The company co-operated with German supplier GNM in the production of flexible foam backed acoustic mats for Audi and Mercedez Benz. Fig 42 shows the press used in form-fit bonding the material. PUR foam densities from ICI are now down to 48 kg/m^2 and mould cycle time is down from 5 to 2 minutes. So-called tactile

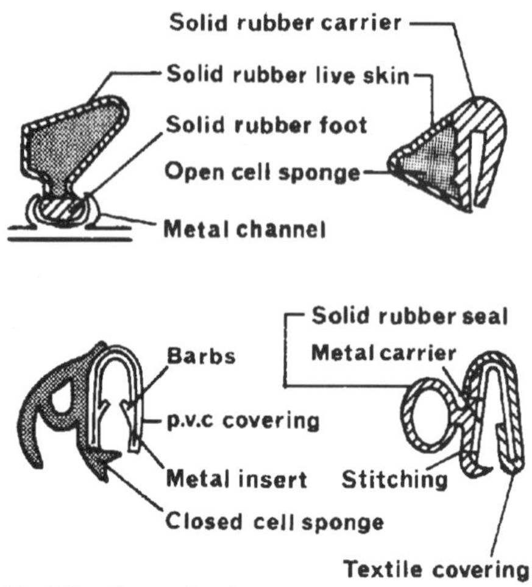

Fig 40: Weatherstripping

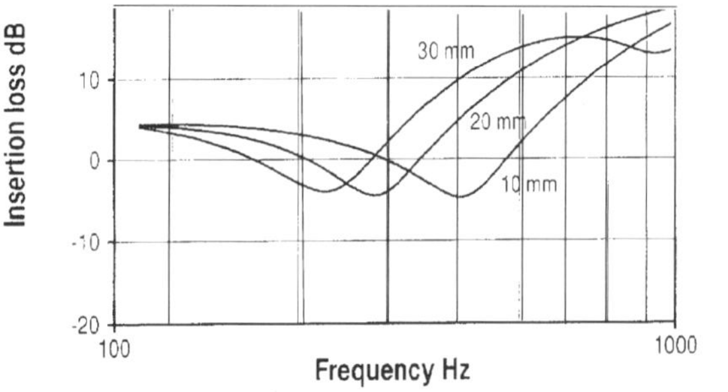

Fig 41: Using spring-mass system for noise isolation

Fig 42: Form-fit bonding press

foams have been developed for the formed carpet application which ensures the absence of voids between foam and floor-pan.

With the prospect of future legislation specifying fixed quantities of recyclate in plastics systems used in vehicle trim, there is an onus on designers to select suitable systems for future vehicles. The trend is to reduce the number of material types used in such items as instrument panels and door casings. Polypropylene is being favoured to replace a number of current mixed material systems. Also, despite warnings of problems with fogging and halogen emission in manufacture, PVC coverings have been developed which are now more environmentally acceptable. Soft feel versions are now available for covering skins of instrument panels and for substrates the trend is to talc-filled polypropylene. Eventually the plastics suppliers hope to be able to produce PP skins of equal acceptance in order to commonise the material usage. As a substrate the material claims to have an advantage over ABS in eliminating squeaks and rattles. Adhesion and cost problems have yet to be overcome with PP covering skins.

PP fabrics are available for upholstering of seats, however, and these are said to be strong, durable, light-fast and have good heat tolerance. Development work is still underway on improving abrasion resistance and proneness to electrostatic attraction of dust. Seating in single material systems is however leading some companies to develop polyester fabrics which `conform' with expanded polyester cushion foams. Du Pont recently launched Ecofil, a compression-mouldable fibre cluster material which

Fig 43: Double-decker test mock-up

can be formed into seat cushions and a 20% weight saving was claimed over expanded PUR foams. Polyester matrix composites are being mooted for the associated seat frames and Dacron polyester fabric for the upholstery. Woollen fabrics are being promoted for top-class vehicles and for some time it has been offered by Jaguar and a number of continental manufacturers. Claims for the material are good tensile strength, tear strength and abrasion resistance. Lack of static 'dust' problems, and the resilience which allows moulding around complex shapes are also claimed. Comfort is enhanced by low thermal conductivity and safety by its flame retardancy.

PSV access and comfort

Unique to the PSV category, of course, is the effect on vehicle design of passenger movement on and off the vehicle, and between saloons in the case of a double-decker. Ogle Design made an important study in this area, on behalf of London transport and Leyland bus, using the full-scale mock-up shown in Fig 43. With this equipment it was possible too infinitely adjust the staircase and exit positions with respect to the entrance and so test the validity of ticketing systems and entrance step configurations. Extensive user trials were carried out with test subjects from all quarters of the population. These lead to the favouring of designs which permitted dual stream boarding and optimum handrail layouts were determined as well as a design of angled split steps.

Fig 44 shows the effect of varying step to kerb distance; the split step reduces the exclusion of ageing and infirm sections of the population from the buses from a 26% to around 6%. The angled split step avoids the usual split-step problem of squeezing two steps into the space normally occupied by one and increases the tread area by 50%. Staircase layout trials, Fig 45, showed the ideal arrangement was one in which the bottom of the staircase is aligned opposite the front half of the exit doors so that dual-stream alighting is possible. Both benefit from being as close to the front of the vehicle as possible so as to improve driver-supervision and enhancing passenger flow by minimising the number of longitudinal seats forward of the exit. Staircase design evaluation brought requests for straight steps and absence of any protrusion of the bottom step into the lower saloon. The longer staircase shown in Fig 46 was shown in the user-trials to be considered an improvement over conventional layouts along with its developed handrail profile. Elliptical sections were preferred for the recommended handrail layout for the entrance — as shown in Fig 47.

Studies by Leyland Bus into air-conditioning for double deckers also showed the considerable extra capacity required of air-conditioning systems for occupant comfort. Passenger loadings as high as 102 seated and 30 standee passengers for such vehicles operating in Hong Kong amounted to a high cooling load as did high solar gain arising from large surface areas, a large proportion of which is glass — also the frequent door-openings involved in multi-stop duties. High idle speeds are required of the prime-moving engine to drive the refrigeration plant through the gearbox power-take-off. Because roof-mounted units are not possible, substantial saloon space is required for both plant and associated distributive ducting. Saloon temperature range for this particular application was 22-28 C for which a cooling capacity requirement in the upper saloon was 27kW and 18.5 kW in the lower saloon — with corresponding evaporator airflows of 3000 cubic metres per hour for each saloon. There was a dedicated compressor, condenser, evaporator and control system for each saloon. An engine idle speed of 700 rpm was require to drive the two compressors which together absorbed 30 bhp. In the associated blower system a 240 amp alternator is required to provide 24 volts DC to all the system motors.

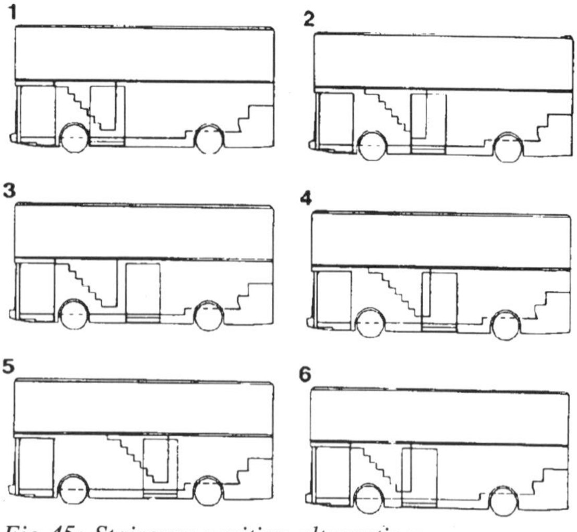

Fig 45: Staircase position alternatives

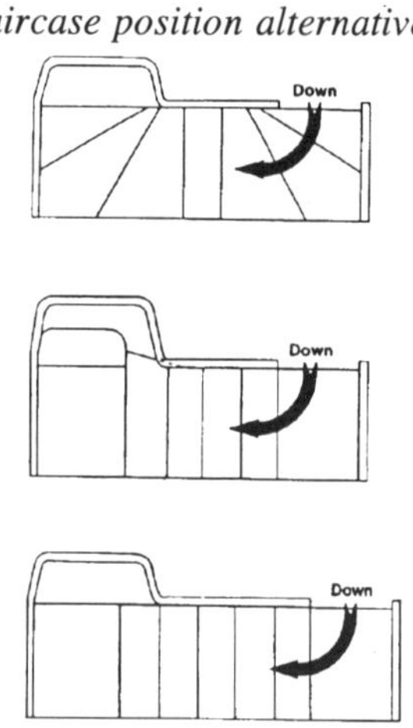

Fig 46: Staircase types: top, angled; centre, MCW Mark 1; bottom, Ogle's design

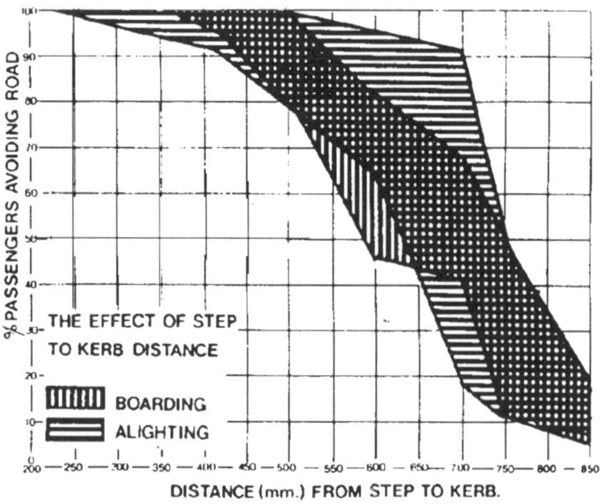

Fig 44: Effect of entrance dimensions

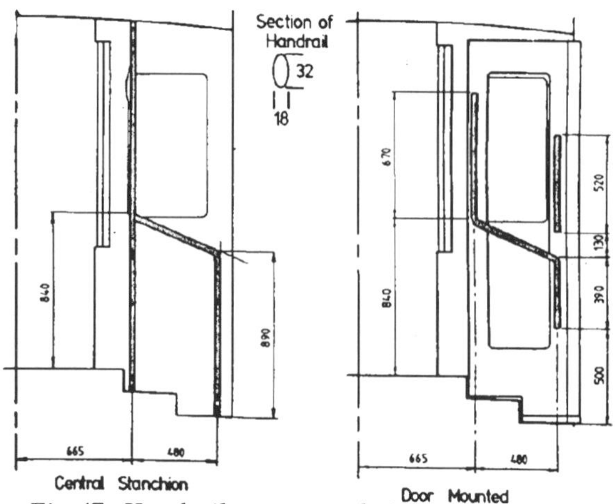

Fig 47: Handrail recommendations

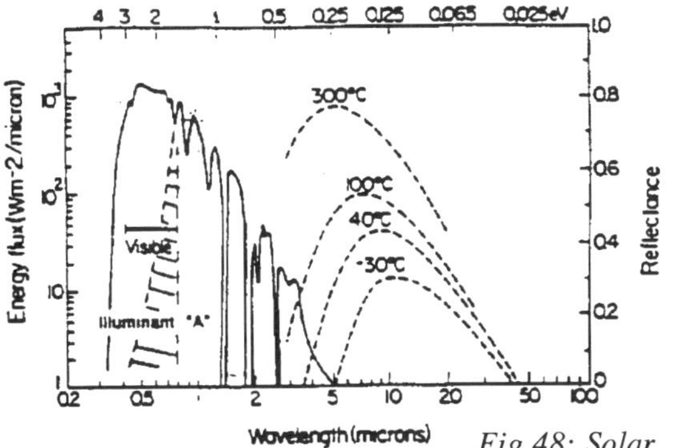

Fig 48: Solar load spectrum

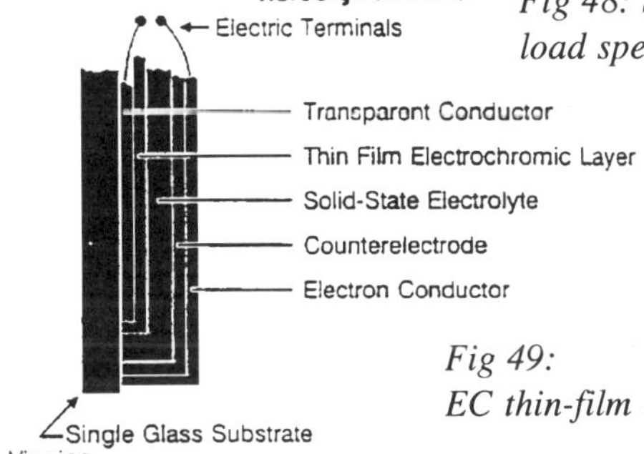

Fig 49: EC thin-film cell

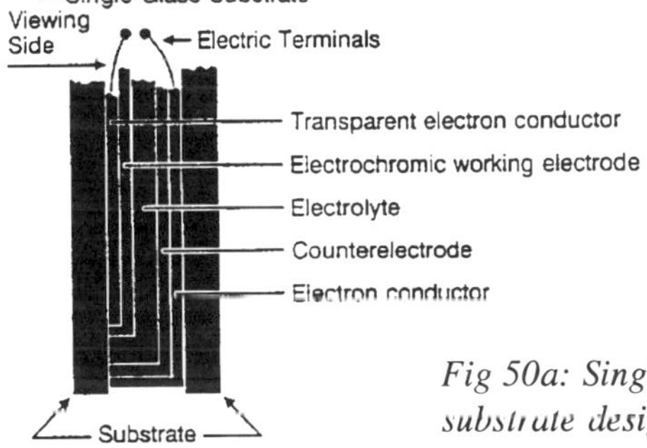

Fig 50a: Single-substrate design

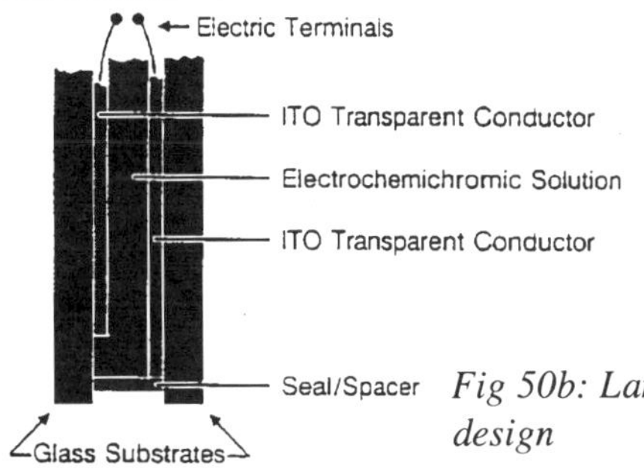

Fig 50b: Laminate design

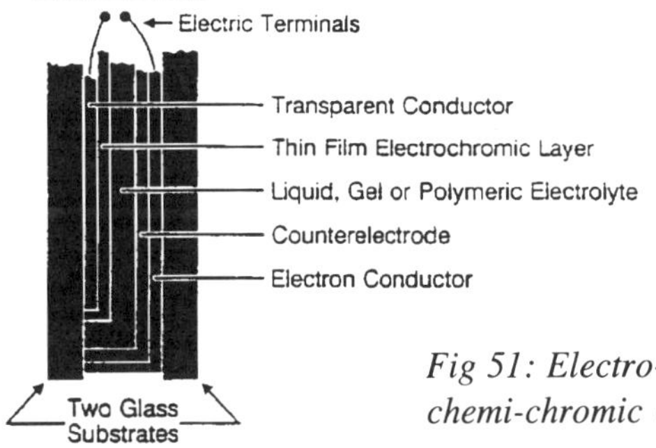

Fig 51: Electro-chemi-chromic cell

Smart-glass windows

Solid-state techniques for tinting of glass are leading to smart window systems which have the potential to reduce considerably vehicle interior heat build-up by solar radiation.

Donnelly Corporation have pioneered several systems for the control of glass transmissibility. Chromogenic technologies include Liquid Crystal and Electrochromic, they report. A variety of designs include laminated constructions in which two glass sheets enclose the electrochromically active medium and solid-state devices made by deposition of thin films into a single-layer glass substrate. Systems must, of course, meet safety legislation for auto-glass governing luminous transmission and shatter protection. The projected increase in glass usage on vehicles, for styling and aerodynamic considerations is, of course, stimulating the use of smart glass systems for controlling the solar heat loading inside the vehicle as well as making possible special stylistic effects.

Fig 48 shows the solar load spectrum incident on a vehicle from which it can be noted that if all infra-red radiation could be effectively filtered, half of the solar radiation could be rejected without any loss in visibility. A number of existing tinted glass systems are partially effective in this respect. Donnelly feel that more effective glazing systems could result in the downsizing of air-conditioners by 50% or more. Smart windows would have the objective of transmitting the full solar spectrum in cold winter months while being able to reflect most of the infra-red in summer.

Electrochromism is a reversible colour change in a material induced by an applied electric field. Devices may be EC thin-film or chemi-chromic, the latter involving solutions of the materials, subject to redox reactions, rather than films. Fig 49 shows the first category, comprising a working electrode that colours and a counter-electrode that is a sink and source for the ions used during colouration. These are separated by an electrolyte and there are conducting layers of a transparent material such as indium tin oxide which provide electrical contact to the electro-active layers.

When a solid-state electrolyte is used the device can be a stack of films deposited on to a single substrate, Fig 50a, and for liquid or gel systems a laminate design is usual, Fig 50b. Thin-film devices retain colour when voltage is removed, important in automotive glazing for long-term parking without power. The material also has the property of maintaining a degree of colouration according either to the magnitude or the duration of the applied voltage.

Construction for electro-chemichromic technology is shown in Fig 51. The effect of a small applied voltage through the transparent conductors causes the solution to colour. Donnelly point out that the gel variant has been manufactured by the Japanese Asahi Glass Company of the type shown in Fig 52. Spectral transmission and reflectance as a function of applied voltage is in the graph and table of Fig 53 a and b. Speed of response is dependent on window size and using 10 ohm/ft^2 (100 ohm/m^2) resistant ITO film, a 30 x 40 cm window takes about 120 seconds to dim to 30% transmission, whereas one a tenth of the area would take about 5 seconds.

In an electrochromic sun-roof developed for a Nissan concept car the construction shown in Fig 54 was used. Here a third auxiliary electrode was required because, initially, both of the electrochromic layers (tungsten oxide and prussian blue) are in the non-transparent, oxidized, state. Its purpose is to reduce the blue to a colourless white which forms a redox couple with the tungsten oxide. Panels measuring 40 x 40 cm have been demonstrated which dim from 60% to 5% visible transmission.

Encapsulated liquid crystal technology also has potential automotive application. Devices usually consist of LC droplets, of 1-10 microns, within a polymer matrix. These are further sandwiched between two conductive electrodes. In the off state the LC molecules are aligned so that their refractive index differs from that of the matrix. Such systems were illustrated in the earlier section on glazing.

	Transmission Characteristics	
	% of Solar Spectrum*	%. Weighted Illuminant A
Bleached	59.2	79.6
Colored @ 0.50 V	26.8	42.8
Colored @ 0.75 V	5.6	10.0

	Reflection Characteristics	
	% of Solar Spectrum*	%. Weighted Illuminant A
Bleached	15.4	12.7
Colored @ 0.50 V	26.8	42.8
Colored @ 0.75 V	5.6	10.0

* With reference to air mass 2 (AM2) between 300 nm to 2500 nm

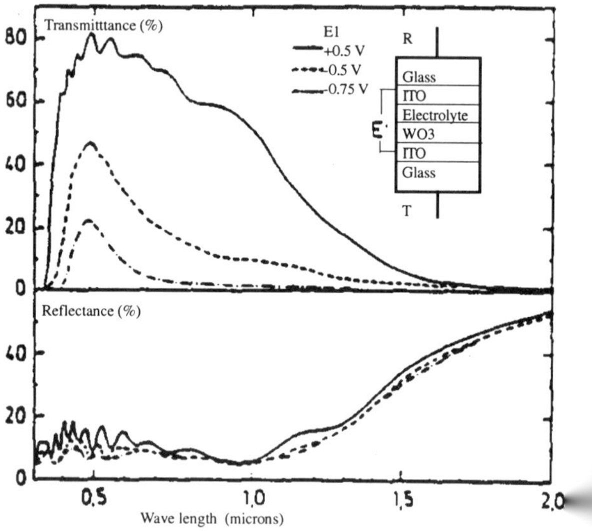

Fig 53: Performance characteristics

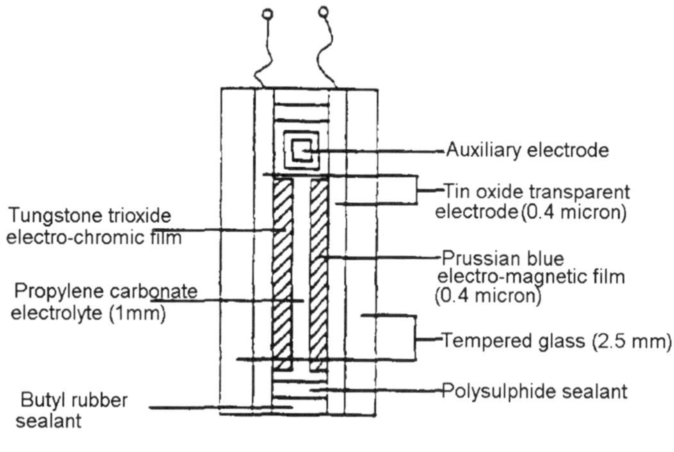

Fig 52: Cell with gel electrolyte

Refraction and Reflection.
1 Incident beam. 2 Refracted beam. 3 Reflected beam.

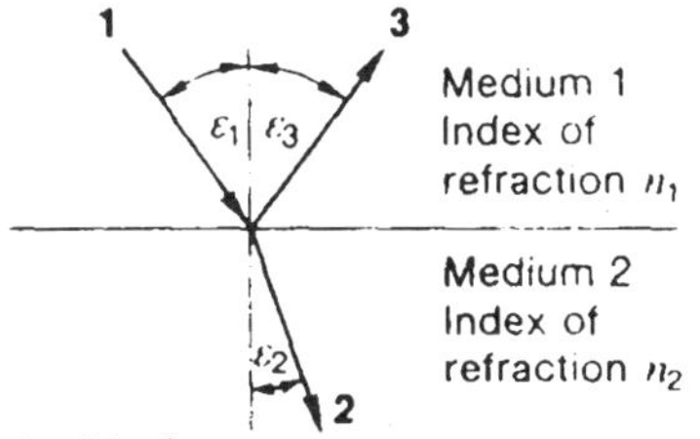

Fig 54: Optical indexes

Cylindrical lens element of a headlamp lens.

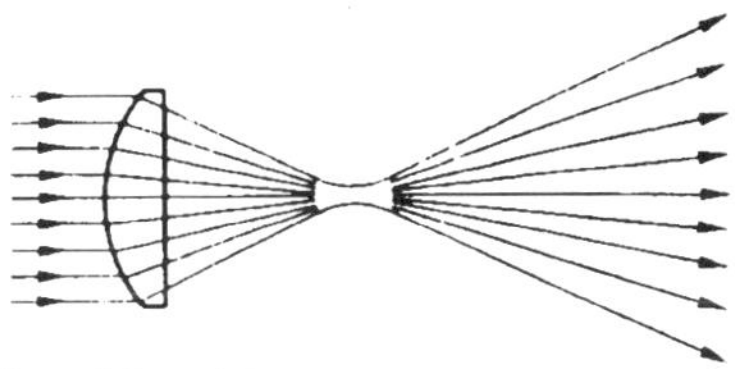

Fig 55: Conventional head-lamp

Parabolic reflector.
F Focal point, S Vertex, f Focal length

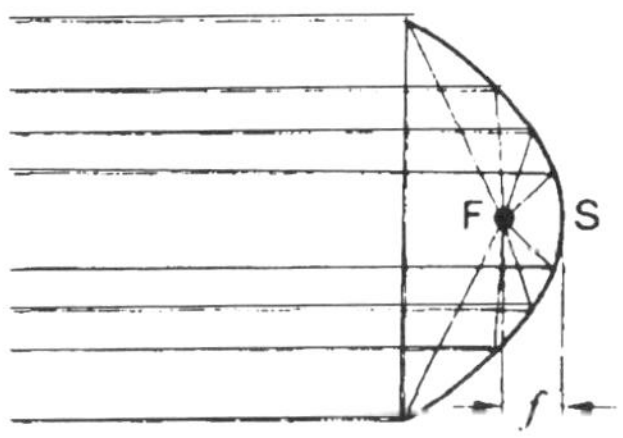

Fig 56: Parallel beam

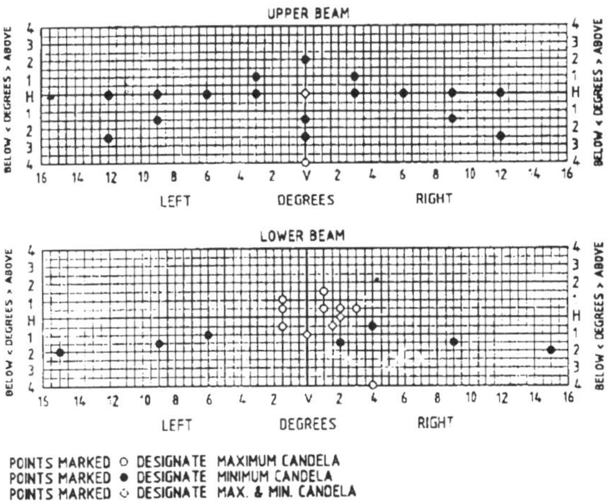

Fig 57: Beam assessment diagram

Lighting design/development

The specification of headlamp systems can be helped by an appreciation of the basic optics of lighting and an understanding of development trends past and and present. Bosch have done much to disseminate knowledge of the theory, developed here, as well as being prominent in developing improved systems.

The optics of lighting has to be understood for a proper appreciation of existing and developing systems. Simple geometric laws can be used to trace the path of 'light rays' which are refracted as they pass from one medium (air) to another (glass). The law of refraction gives the relationship

$$n_1 \sin e_1 = n_2 \sin e_2$$

where n_1 and n_2 are the indices of refraction, Fig 54. These are a function of the wavelength dispersion. The reflected ray is diverted from the surface at equal angles to the normal and the degree of reflection depends both on the angle of incidence and the refractive indices of the air and glass. Total reflection occurs if the angle of incidence is greater than or equal to the angle of total reflection. Degree of reflection varies from 70% for polished chrome to 89% for vacuum-deposited aluminium.

Conventional headlamp lenses comprise cylindrical elements, parallel rays being made to converge in a focal line, Fig 55. Other prismatic elements are also used to aid the process, to direct the rays into an acceptable beam. Incident parallel rays remain parallel after deflection by the prisms. From the 'point' light source, such as a bulb filament, rays emanate radially and are conventionally gathered by a parabolic reflector to be reflected as an approximately parallel beam. The point source is thus enlarged and its luminous intensity is a product of the light source luminance, the degree of reflection and the aperture area of the reflector, Fig 56.

As visible radiation sensitivity varies from person to person, a generalised function has been established to describe it. This is the spectral brightness response Φ. Luminous flux is the radiant power emitted by a source, assessed on the basis of Φ according to:

$$\Phi = K_m \int P_\lambda V\lambda . d\lambda$$

where K_m is known as the maximum luminous efficacy (683 lumens/watt). Values of F for a 2 degree field of vision are listed in European standards such as DIN 5031, Part 3. P_l is known as the spectral distribution of radiant energy.

Procedures for measuring photometric properties of automotive lamps are covered in SAE J1330. Other SAE standards cover headlamp beam switching, manual (J564) and automatic (J565). Performance requirements for headlamps is covered in J1383, with the system of beam assessment being shown as Fig 57. Conventional headlamp reflectors with focal length about one quarter of the diameter have provided a traditionally accepted solid angle. A reflector diameter D and source area A_s would cause an area illuminated at distance L of

$$A_s(L/f)^2 \text{ or } A_s(4L/D)^2$$

and the intensity of illumination is proportional to

$$f^2/L^2 \text{ or } D^2/16L^2$$

Eye sensitivity

Fig 58 shows the relative spectral sensitivity $V(v)$ of the light-adapted eye. Other definitions which affect the physiology of vision are luminous energy, being the spectral radiation energy assessed on the basis of V_l; the luminous intensity, being the ratio of luminous flux to the irradiated solid angle; the illuminance, being the ratio of incident luminous flux to the area of illuminated surface; luminance, being the ratio of luminous intensity to the apparent projected- area of the surface from which it is emitted; luminous emittance, being the ratio of the emitted luminous

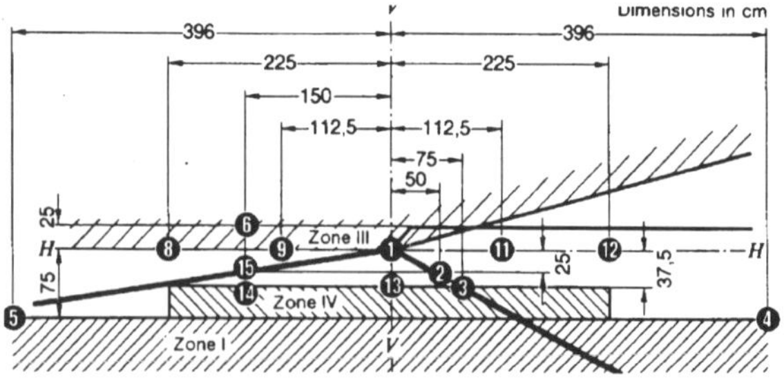

Lower beam			
Measurement points	Specified lx values	Measurement points	Specified lx values
1. E_{HV}	≤ 0.7	Max. value in Zone I	≤ 2 E_{50R}
2. E_{75R}	≥ 12	Max. value in Zone III	≤ 0.7
3. E_{50R}	≥ 12	Min. value in Zone IV	≥ 3
4. E_{25R}	≥ 2	13. E_{50V}	≥ 6
5. E_{25L}	≥ 2	14. E_{50L}	≤ 15
6. E_{B50L}	≤ 0.4	15. E_{75L}	≤ 12

Upper beam	
Measurement points	Specified lx values
7. E_{max}	> 48 < 16 × E_{75R} (< 240)
8. $E_{H-5.15°}$	≥ 6
9. $E_{H-2.55°}$	≥ 24
10. E_{HV}	≥ 0.8 E_{max}
11. $E_{H+2.55°}$	≥ 24
12. $E_{H+5.15°}$	≥ 6

Fig59: Measurement points for illuminance

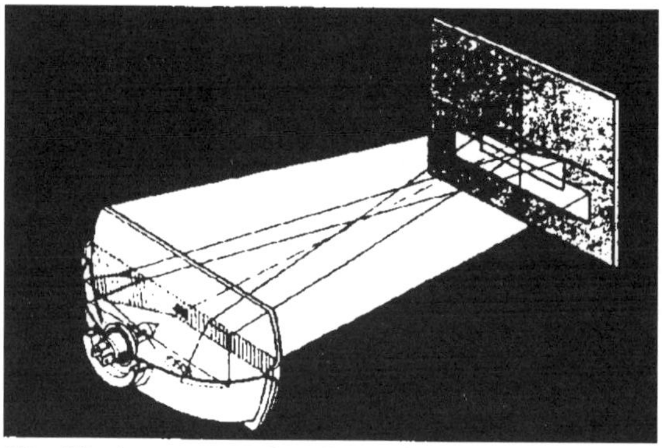

Fig 60: Asymmetric low-beam light

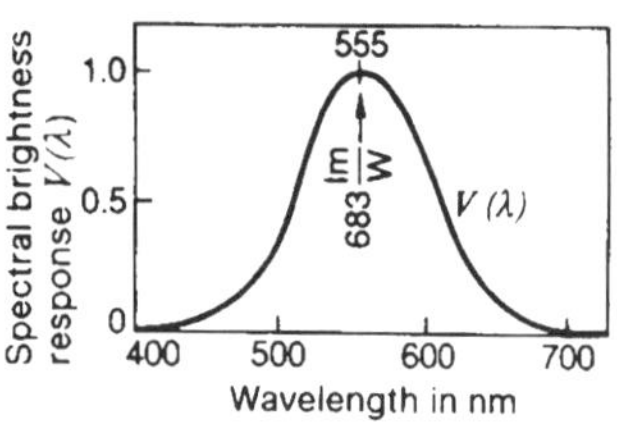

Fig 58: Spectral density of light-adapted eye

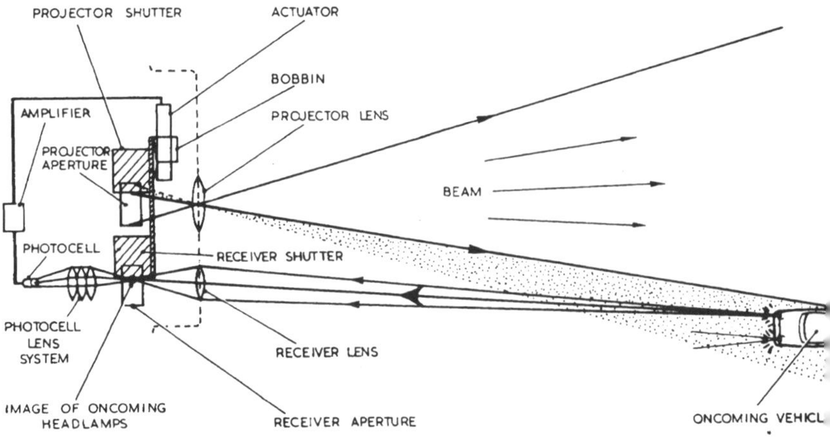

Fig 61: Lucas Autosensa

flux to the power absorbed and, finally, the solid angle which is the ratio of the irradiated portion of the surface of a sphere (concentric with the light source) to the square of the sphere's radius. While a number of these may not be used in design, they are of value when checking the specifications of proprietary lamps which may be expressed in different ways.

It is also useful to differentiate between direct glare, experienced in the line of sight, and indirect glare seen at the extremities of the angle of sight. Luminance values above about 20 cd/cm^2 produce absolute glare from a source which cannot be reduced in its effect by additional illumination of its background. Good lighting depends on three eye sensitivities: that for discrimination, that for shape, and that for speed of perception. In the case of glare, one European country stipulates that glare is effectively eliminated if each headlamp generates no more than 0.7 lx, at headlamp height, at a distance of 25 metres. Specified values are given in Fig 59 which apply to a screen at that distance with point (1) lying in the headlamp axis.

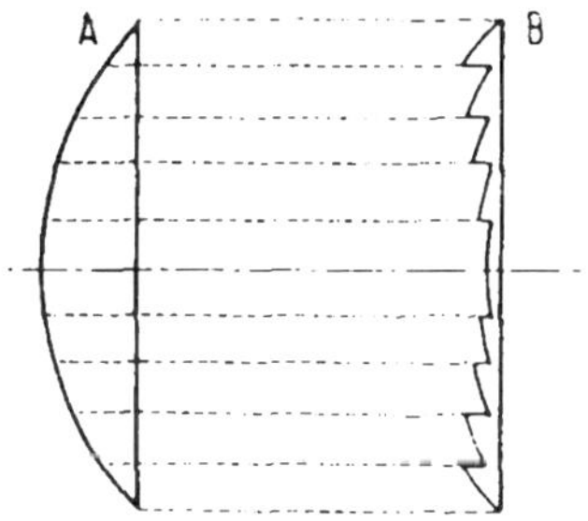

Fig 62: Fresnel lens

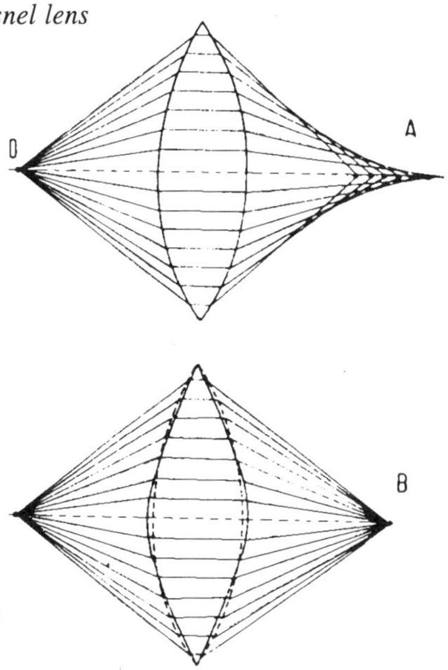

Fig 63: Overcoming aberration

Key parameters for conventional parabolic reflectors, Fig 56, are focal point F vertex S and focal length *f*. The range of the lamp is defined as that distance at which the luminous flux has a specified value, say 1 lux. Then a geometric range is considered to be the distance of the horizontal component of the light/dark cut-off on the road surface with 1% low beam inclination. Typically focal lengths lie between 15 and 40 mm for reflectors whose forward projected areas are referred to as the 'illuminated area'. The lamps 'effective luminous flux' is that portion of the filament's flux which is beamed by the reflector, via the refractive lens, to the road surface. This will be higher for short focal lengths because the reflector converts more of the flux into a light beam.

With increased traffic densities high-beam driving is more of a rarity and thus much development has gone into improving low-beam performance. Asymmetric low-beam lamps have extended visual range to the near-side while halogen bulbs have substantially raised luminous intensity. Low-beam is conventionally achieved by the appropriate filament being ahead of the focal point while a cap keeps the lower area of the reflector unlit; the cap's edge defines the cut-off, Fig 60. Departures from convention have included stepped reflectors with parabaloid sections of different focal lengths. Another type is the Homofocal reflector in which different quadrants of the reflector have the same focal point but have differing focal lengths. These quadrants aid near-field and lateral illumination but do not increase range and its economic production was made possible by plastics moulding.

Another ingenious development was the Lucas Autosensa, Fig 61, in which a receiver and projector was combined in one unit. An amplified signal from a photocell operated a linear actuator to move a shutter across the aperture which 'homed' on the image of an approaching vehicle; a second shutter moved in unison to cut-off the off-side edge of the beam.

The important contribution of the lens glass can be visualised by imagining the collapse of a conventional lens into a Fresnel type, Fig 62. It is possible to achieve such shapes in plastic that were hard to obtain in glass. Also it is possible to overcome the aberration effects of glass lenses with spherical curvature, Fig 63. The spherical lens cannot focus all incident rays at the same point, resulting in the aberration of the image, whereas a non spherical surface can be more easily moulded in plastic.

Recent lighting developments

Complex Surface headlamps are a relatively new departure, in which, instead of using a parabolic reflector, Valeo have one which is developed by computer analysis of some 10,000 surface points, Fig 64. Luminous flux is increased by the more efficient use of reflector surface and some 80% more light output is claimed, in low-beam configuration, than for a conventional lamp of equivalent reflector surface area. The latest versions require no inner lens and so can use a clear outer glass.

Both Valeo and Bosch are also developing gas-discharge lamps using electric arcs in lieu of tungsten filaments. A small quartz bulb contains zenon under high pressure as well as mercury and solid-state halogenures. After an initial arc is struck in the zenon, on turning on the lamp, a high current is

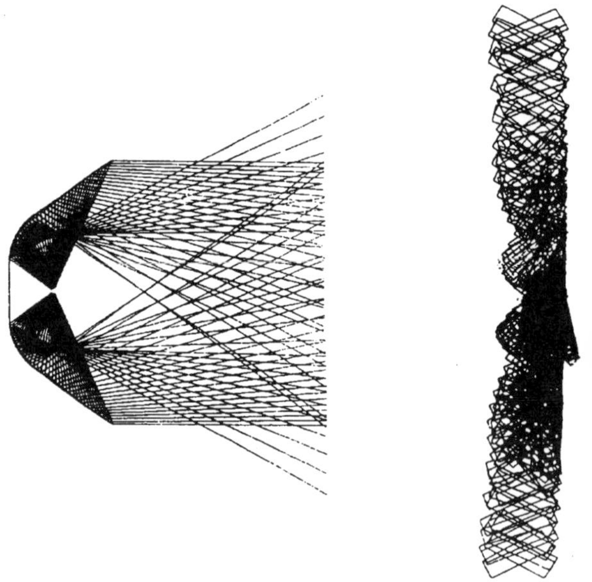

Fig 64: Complex surface system with beam pattern(below)

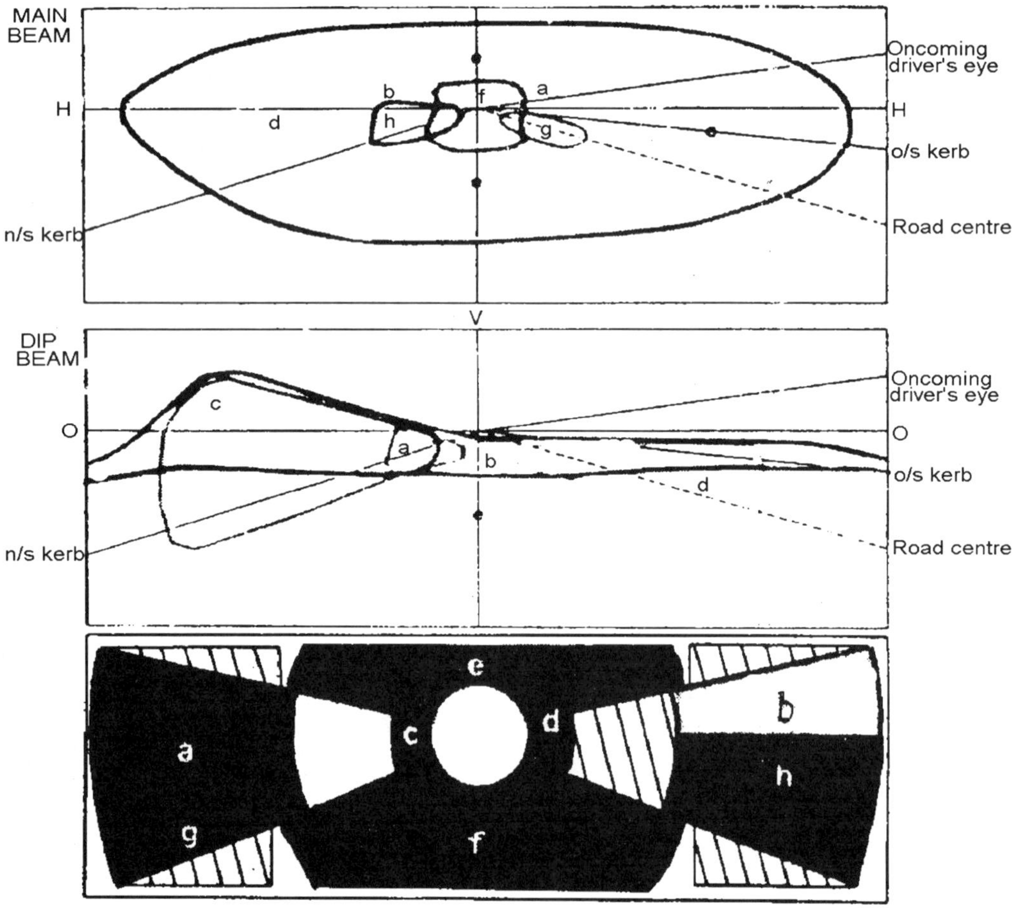

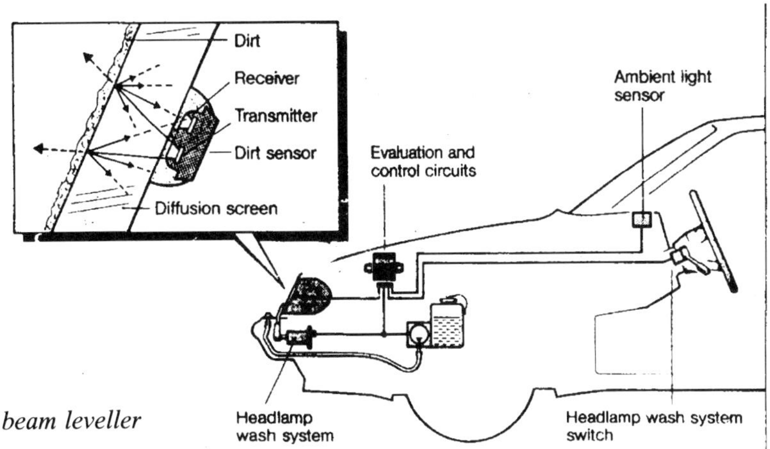

Fig 65: Automatic beam leveller

applied which evaporates the other constituents — increasing luminous flux. An initial over-voltage ignites the light quickly; thereafter a ballast device optimises power — reducing from the initial level of 75 W to a stable state of 35 W (initial strike voltage is 6-12 kV for the arc, made possible by a high step-up transformer and an AC supply). A doubling in field of vision is claimed over comparable lamps with a conventional halogen bulb of 55-65 W. The gas-discharge lamp is also 2.5 times brighter and gives a bluer light. Bulb life is also substantially enhanced with the prospects of lasting the whole vehicle life, for certain categories, by virtue of the lower power requirement and cooler running. About 2-3 seconds will be required to reach full luminosity for the automotive versions — compared with the one minute for street lamps of this type. First applications are expected on performance cars using four-headlamp systems. They will be used as low-beam lamps in pair with conventional high beam ones. The higher intensity beam would require an automatic levelling device.

The Bosch automatic beam leveller is another significant advance; it uses servo-motors to keep beam-aim within controlled limits. Another likely requirement, for which Bosch has also produced a solution, is a means for automatic washing after dirt build-up on the lens. A sensor for this purpose uses infra-red to gauge the equivalent amount of white light reflected back inside a lamp, from a dirty lens, Fig 65. The company's polyellipsoidal PES headlamp for low profile applications, in which an aperture of only 60 mm diameter provides normal levels of distribution using intense focusing of light, has also been upgraded. The PES Plus has had its cutoff mask modified and additional reflector elements incorporated so that 50% greater light intensity has been achieved for the 520 metre (near) field of vision, Fig 66.

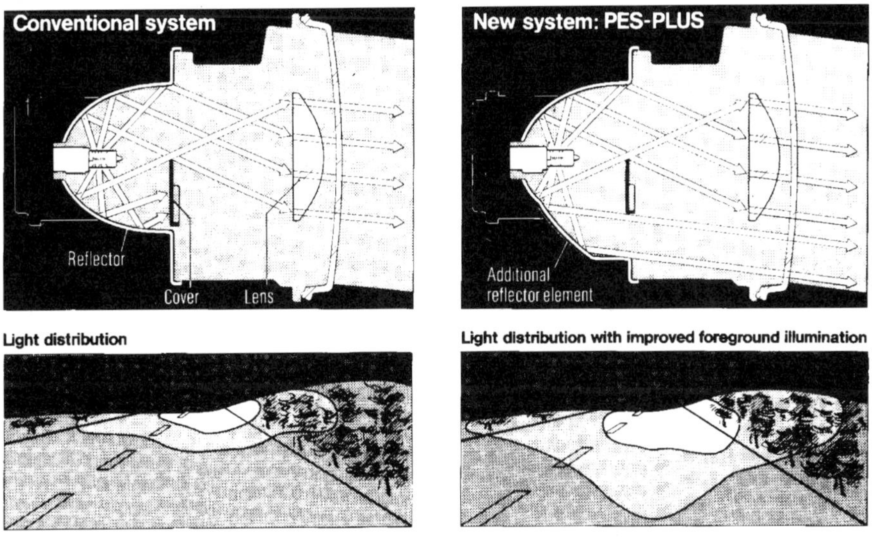

Fig 66: Conventional vs PES lamp

Electro magnetic compatibility

The successful use of vehicle on-board electronic controls depends upon the ability of individual systems to share the electro-magnetic spectrum of frequencies without degradation of performance. The systems also need protection from electro-magnetic interference (EMI) from road-side and other sources of radio-frequencies.

According to RFI Shielding Ltd[1], the beginning of 1992 saw the introduction of European Directive 89/336/EEC on EMC compliance for electronic goods sold on the market and a legal stimulus is added to the need to ensure good protection. There is also the safety influence of strong EM fields possibly causing the malfunction of anti-lock braking or engine-management systems. Most electronic devices involve rapid changes in voltage and/or current levels with the resultant development of electro-magnetic energy over a range of frequencies. This will be radiated in the space surrounding the circuit in question and be conducted into the wiring which communicates with the circuit.

Fig 67 shows the effect of providing a metal barrier against the incoming wave, the latter having two main component fields, one electrical and one magnetic — each at right-angles to one another and the direction of travel. A degree of reflection occurs at each surface of the barrier. The effectiveness of the screening is often quoted in decibels — such that the reduction in strength of the emergent field to one tenth of the incident value would be given an effectiveness of 20 dB; see table in Fig 68.

Absorption loss depends on the thickness of the barrier as shown in Fig 69 depending on the

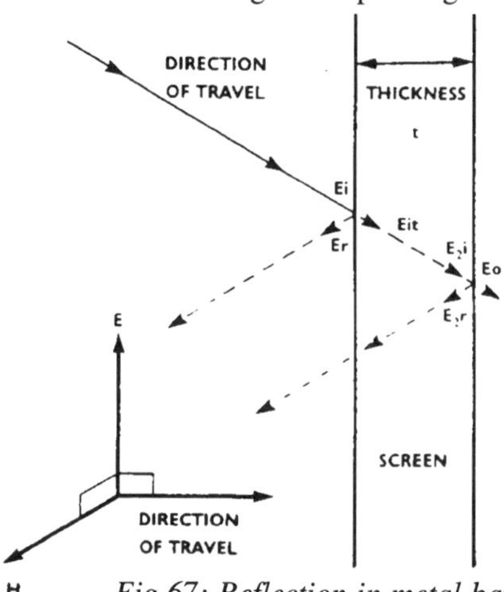

Fig 67: Reflection in metal barrier

S = SCREENING EFFECTIVENESS (dB)
A = Absorption loss (dB)
R = Reflection loss (dB)
B = Correction factor (dB) (for multiple reflections in thin screens)
S = A + R + B

dB	Percentage reduc
0	0
20	90
40	99
60	99.9
80	99.99
100	99.999, etc

Fig 68: Screen decibel values

A = $0.1315 \cdot t \sqrt{f \cdot \sigma \cdot \mu}$ (dB)
t = screen thickness (mm)
f = frequency (Hz)
σ = conductivity relative to copper
μ = permeability relative to copper
Note: For screen thickness (t) in inches replace the constant 0.1315 with 3.34

Material	σ	μ
Copper	1.00	1
Aluminium	0.61	1
Brass	0.61	1
Tin	0.15	1
Steel (SAE 1045)	0.10	1000
Monel	0.04	1
Stainless steel	0.02	500
Electroless nickel	0.02	1

Fig 69: Thickness/permeability effect

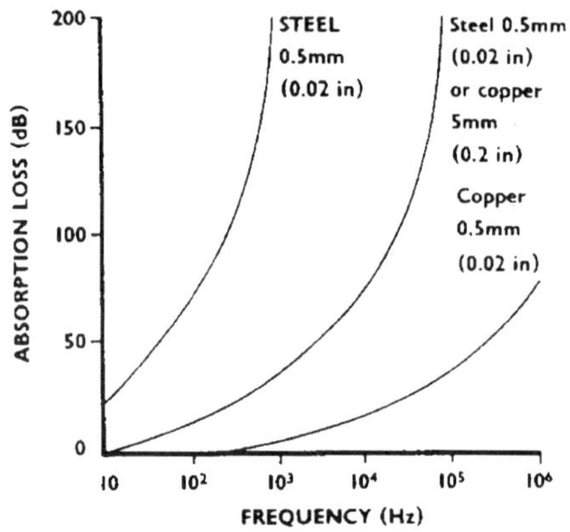

Fig 70: Frequency effect

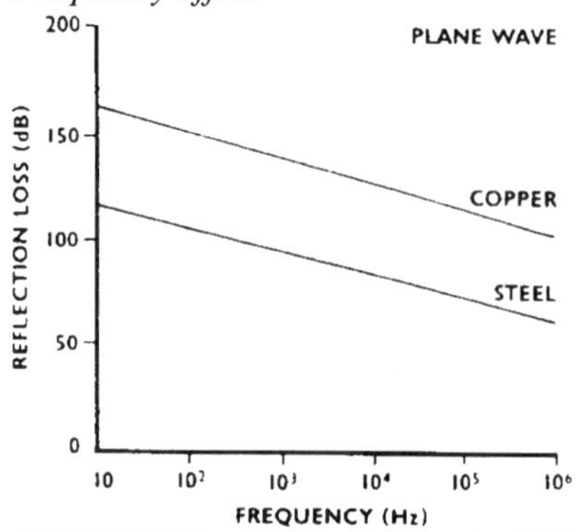

R (PLANE WAVE) = $168 - 10 \log(f) - 10 \log(\mu/\sigma)$ (dB)

Fig 71: Total effectiveness

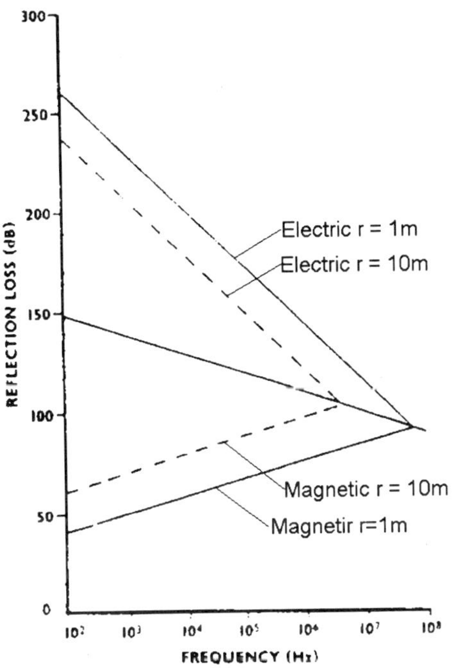

Fig 72: Straight antenna radiation (r is distance form source to screen (m)

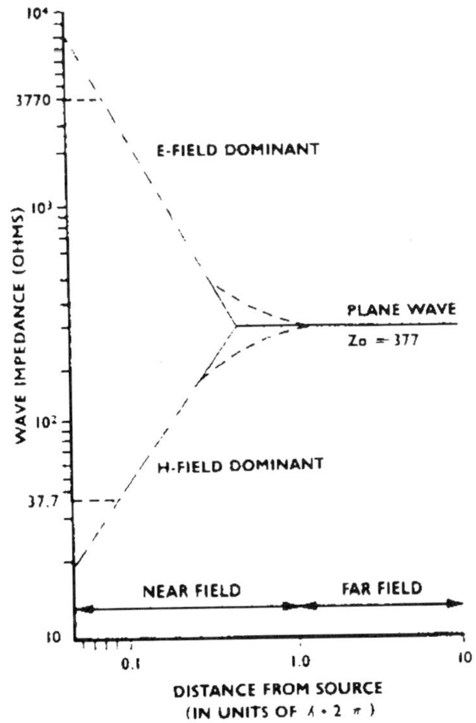

Fig 73: Screen impedance

conductivity and permeability of the materials from which it is made. Variation of absorption loss with frequency for two screening materials is shown in Fig 70. Reflection loss increases with ratio of impedance of the incident wave to the impedance of the screen material. Beyond a distance of one sixth of a wave-length from the source wave impedance is constant at about 377 ohms — while the impedance of the screen material is proportional to the square root of the frequency times the permeability divided by the conductivity.

Total effectiveness for a shield, allowing for absorption/reflection loss of plane waves, is seen in Fig 71. This refers to a copper 0.5 mm thick screen in the 'far-field' where the wave-front is plane.

Near and far fields

Below the one sixth wavelength distance from the source, 'near-field' conditions are said to exist. For a straight rod-antenna emitting radiation, wave impedance is high near the source, Fig 72. However, if the emitting antenna is a small wire loop, the field is predominantely magnetic and the wave impedance is low near the source, Fig 73.

For small screened devices, working in the near-field is the norm and formulae for calculation are shown in Fig 74 the correction factor B is normally only computed for the near-field case and then only if the absorption loss is less than 10 dB. Variation of reflection loss with distance and frequency, for a copper screen, is shown in Fig 75

REFLECTION LOSS IN THE NEAR FIELD

R (Electric) = 321.8 − 20.Log(r) − 30.Log(f) − 10.Log(μ/σ) (dB)

R (Magnetic) = 14.6 + 20.Log(r) + 10.Log(f) + 10.Log(σ/μ) (dB)

r = distance from source to screen (m)

f = frequency (Hz)

μ = permeability relative to copper

σ = conductivity relative to copper

CORRECTION FACTOR B

B = 20.Log(1-exp(−2.t/δ)) (dB)

t = screen thickness (mm)

δ = skin depth

= 0.102 ÷ $\sqrt{f.\mu.\sigma}$ (mm)

For (t/δ) = 0.1, B = −15dB

= 0.5, = −4dB

= 1.0, = −1dB

Fig 74: Calculation of losses

indicating the closer the screen is to the source, reflection loss for electrical fields is higher and the reverse applies to magnetic fields.

EMC materials and construction

At very low frequencies only high-permeability materials are effective for screening; cheaper materials are suitable for high-frequencies. Joints or holes in the screen will, of course, limit its effectiveness. By considering the hole in the screen as a wave-guide for the emitting radiation, its size can be optimised for screening purposes. If wavelength of the EM energy is too long relative to the gap dimension of the hole, little energy passes through and the 'cut-off' condition for the wave guide is said to to exist. Fig 76 shows formulae for cut-off frequencies in round and rectangular wave-guides as well as effectiveness indications at well below cut-off — a function of the length to diameter ratio.

An alternative way to obtain a good length to diameter ratio is to bond a small tube on the aperture for those cases where penetration of the enclosure is essential. These considerations are also used in the design of multi-hole perforated screens.

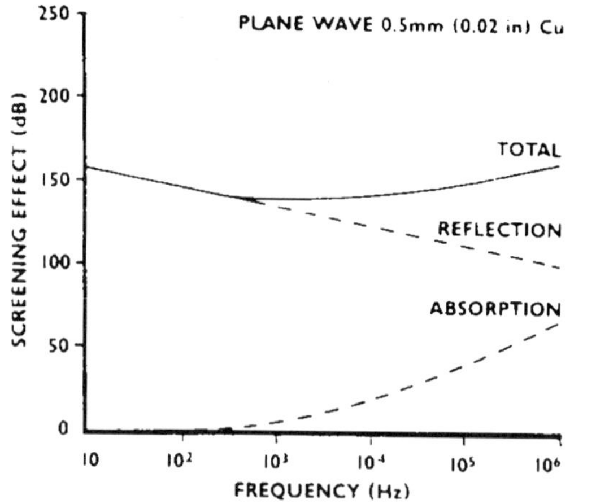

Fig 75: Loss variation with distance and frequency

WAVEGUIDE CUT-OFF FREQUENCY (fc)

In round guide, fc	= 175.26/d GHz (6.9/d in.)
d	= waveguide diameter (mm)
In rectangular guide, fc	= 149.86/a GHz (5.9/a in.)
a	= largest dimension of waveguide cross-section (mm)

SHIELDING EFFECTIVENESS (S) OF WAVEGUIDE

For operating frequencies well below cut-off

S (round)	= 32.t/d (dB)
S (rectangular)	= 27.2.t/a (dB)
t	= Screen thickness

Fig 76: Cut-off frequencies

PART 4:
SPECIALIST VEHICLE DESIGN

Chapter 11: Off-road vehicles

Off-road vehicle technology
Farm tractors
Light-weight ATVs
Farm support vehicles
All-terrain trucks
Earthworking vehicles
Military vehicle design

With the output of the global multinationals becoming increasingly standardised, opportunities are increasing for specialist vehicle categories to satisfy niche markets untouched by the more bland designs of car and commercial vehicle. With increased flexibility in manufacture there also exists an increasing manufacturing opportunity for specialist duty designs, particularly in commercial and industrial vehicle categories.

Off-road vehicle technology

Interaction between vehicles and the ground is a study that is very much younger than the development of the motor vehicle and only recently have analytical techniques been exploited in understanding the principles of soft-ground mobility. Now soil mechanics theory from civil engineering has combined with traction theory from mechanical engineering to provide a cross-disciplinary technology of very real value to the off-road vehicle designer.

Much soft ground mobility work has its roots in military vehicle studies and early concepts of 'flotation' have been questioned by AFV (armoured fighting vehicle) engineers. The idea of 'nominal ground pressure' based on dividing the projected area of wheel and/or track, 'touching' the ground, by vehicle weight is losing favour as a basis for mobility studies. Now the accepted theory is based on MMP (mean maximum pressure). This allows for that fact that, in a tracklaying vehicle for example, peaks of maximum pressure occur beneath each bogie wheel — and that weight distribution can vary between wheels.

For tracked vehicles, MMP is obtained from

$$1.26W/2mcb(pd)^{1/2}$$

where W = vehicle weight, m = number of axles, d = wheel diameter, b = track width, p = track link pitch, c = track link profile factor = area/pb. For wheeled vehicles on cohesive fine grained soil, the expression becomes

$$kW/2mb_0 85d_1 15(D/h)^{1/2}$$

where d and b become tyre diameter and breadth, D/h the tyre deflection on hard ground and the factor K taken from Fig 1. There are also modified forms of these expressions which apply to wheels on coarse-grained sand, for belt tracks with pneumatic tyres and for wheels of different size on different axles.

MMP gives an indication of vehicle sinkage into the soil and thus a lead to resistance to motion. It is also a guide to traction over differing terrains but it is suggested that only differences of 20 kPa between vehicles should be considered as 'significant'. For military vehicles standards have been set that make less than 230 kPa desirable and less than 280 kPa essential for heavy AFVs (armoured fighting vehicle) — with 140/230 corresponding figures for light AFVs and 280/700 for wheeled trucks according to gross weight. Desirable values for different terrains are given in Fig 2. Some values for familiar vehicles are as follows: Challenger tank 282, Swedish S tank 290, Alvis Stalwart 401, 1/2 tonne Land Rover 278 and Bedford TM 656.

Number of Axles	Proportion of Axles Driven						
	1	3/4	2/3	3/5	1/2	1/3	1/4
2	3·65	-	-	-	4·4	-	-
3	3·9	-	4·35	-	-	5·25	-
4	4·1	4·44	-	-	4·95	-	6·05
5	4·32	-	-	4·97	-	-	-
6	4·6	-	5·15	-	5·55	6·2	-

Fig 1: K-factor for MMP calculation

Condition	MMP levels for performance priority		
	Ideal	Satisfactory	Max acceptable
Temperate climate, fine wet soil	150	200	300
Tropical climate, fine wet soil	90	140	240
European bogs	5	10	15
Muskeg	30	50	60
Over-snow	10	25-30	40

Fig 2: Recommended MMP values

The resistance to motion from sinkage

$$R/W = 0.28(M/C)^{1.95}$$

where resistance R is expressed as a percentage; M is MMP and C is the soil cone index — a US originating measure of soil strength. This measure is based on defining soils in terms of their resistance to the penetration of a rod tipped with a right circular cone.

Design of new vehicles, of course, needs a greater knowledge of the soil mechanics than merely the cone index. The pioneer work of MG Bekker in off-the-road locomotion led eventually to his design of the moon buggy — the first wheeled vehicle in space. His early work led him to show that a vehicle track or tread under vertical load W developed a tractive force H caused by shear resistance of the soil between track-plates or tread lugs and the main body of the ground beneath.

For 'plastic' ground such as wet clay or snow H remains practically constant irrespective of W — explained by the cohesion of the soil particles, de

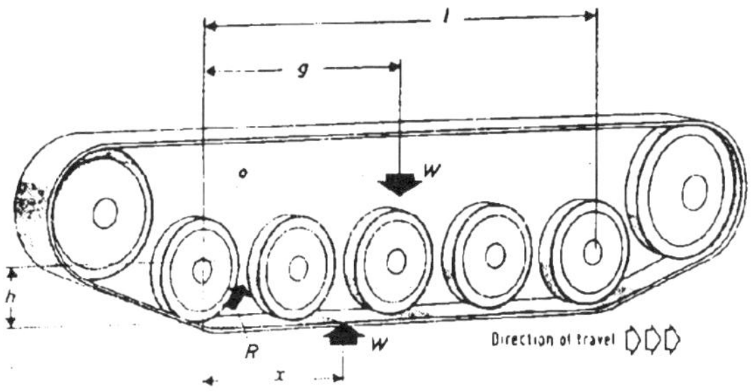

Fig 3: Dynamic load transfer

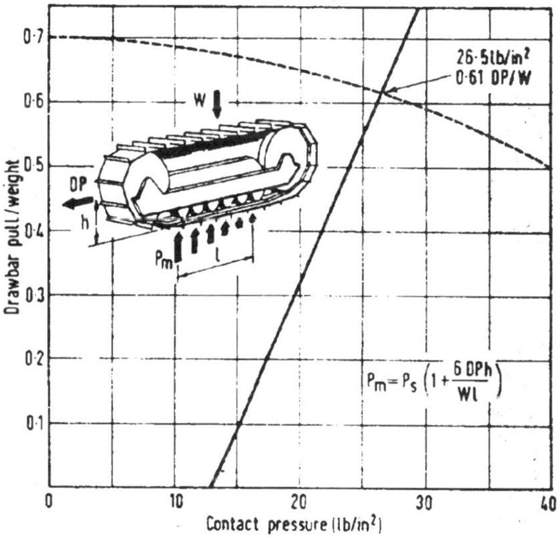

Fig 4: Wheel vs track performance

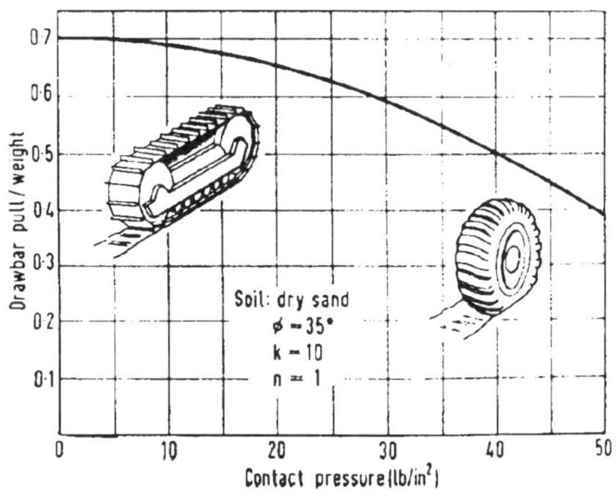

ig 5: Tracklayer performance effects

fined by a factor *c*. With dry sand, or powdered snow, the tractive force is developed by friction between the grains; this involves a horizontal force equal to the weight factored by the tangent of the friction angle, *F*. In practice, most soils contain an element of both conditions, so tractive force becomes

$$= Ac + W \tan F$$

Bekker further refined this expression in the first English language book on the subject, published in 1960, based on experiments with model vehicles in a 'soil bin' — a tool as important to soil mechanics as the windtunnel is to aerodynamics.

As a tracked vehicle advances in soft ground, vertical load distribution changes such that the pressure on the rear of the track can be increased by 15 to 20%. There is also a dynamic load transfer effect accentuating the rearward shift. Fig 3 shows the forces involved which result in a centre-of-pressure position defined by

$$x = g - Rh/W.$$

The relative disadvantage facing the wheel rather than the track is shown in Fig 4 in which the ratio of tractive force to weight has been plotted against ground pressure. It is based on the assumption of 35 degree soil friction angle and capability of supporting 10 lb/in^2 (69 kN/m^2) at one inch (25 mm) sinkage (compaction factor $k = 10$) also that sinkage is directly proportional to load. The expression for comparing tracks with wheels is

$$H/W = \tan F - bp^2/Wk$$

assuming a common width/weight ratio $b/W=1/800$. This would apply to a 20 in (508 mm) wide track loaded to 16000 lb (7.25 tonnes) and a 26.5 in (635 mm) section tyre carrying 22000 lb (10 tonnes). Comparing the track having a nominal contact pressure of 8 lb/in^2 (55 kN/m^2) with the tyre imposing a maximum soil pressure of 47 lb/in^2 (324 kN/m^2) (inflation pressure plus carcass stiffness), wheel pull is only 60% of that of the track.

In fact the tracklayer is not at such an advantage when the non-uniform ground pressure distribution is taken into account. Extra drawbar resistance due to soil compaction results in a moment around the rear overturning axis, Fig 5, equal to *DPh* which displaces the centre of pressure as a fraction of *W*. Maximum pressure on the rear of the track then becomes

$$P_m = P_s[1 + (6DPh/Wl)]$$

for static pressure P_s and ground contact length l. The figure demonstrates the effect for a crawler tractor having static pressure of 13 lb/in² and 0.28 ratio of drawbar height to track contact length. The dotted curve is that from Fig 4 superimposed. The intercept shows that the effect of load distribution is to increase maximum soil pressure to 26.5 lb/in² with a reduction of pull to 61% of tractor weight.

For a ribbed wheel, tests have shown tractive force on yielding soil to follow the characteristic shown in Fig 6; close examination of a smooth wheel on yielding soil reveals a particular contact area. The available thrust depends on the amount of soil distortion which is a function of distance to the point of maximum slip at which the soil parts company with the wheel after distorting progressively from the point of initial contact, Fig 7.

Performance-enhancing measures

For minimum sinkage in swampy conditions, contact pressure must be reduced as far as possible — and with some tracklayers this can be down to 1 lb/in² (6.9 kN/m²), against 20 lb/in² (138 kN/m²) for a conventional cross-country vehicle. Large wheel diameters usefully increase contact area but the apparent advantage of twin-formation truck tyres can be lost on soft ground by the bulldozing effect which results in build-up of mud in front of the tyres. Rolling resistance due to sinkage rises with increase in tyre inflation pressure but rolling resistance due to carcass deflection increases — so a compromise has to be found. An important factor in tyre performance is the self-clearing ability of the tread, requiring deep close-pitch ribs of soft rubber.

Changes in soil pressure distribution on off-road vehicles also arise from external forces from equipment like farm implements and earthmoving tools. Fig 8 shows the forces acting on various earthworking vehicles. At (a), centre of pressure is displaced from the static position by

$$[r(H_1 + H_2) - p(r - h)]/W$$

and rear wheel load is increased to

$$W[1 - (g - x)/l]$$

For the tracklayer at (b), respective expressions are

$$[Hr + p(r - h)]/W \text{ — and } 2pa[2 - 3(g - x)/l]$$

for increase in pressure under rear of track. The forces applying to a conventional bulldozer are shown in (c), in which *V* normally equals 15-20% *P*, while at

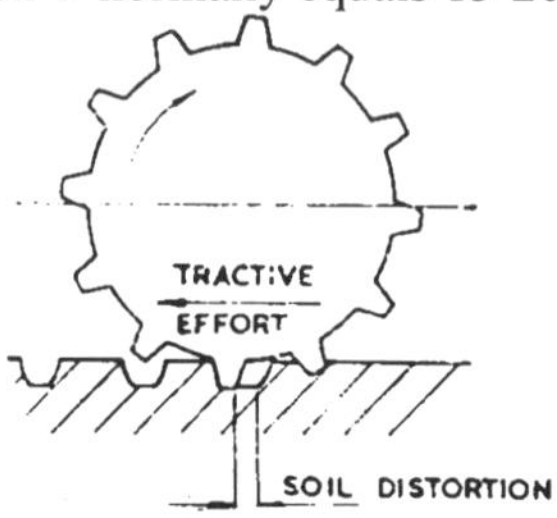

Fig 6: Ribbed wheel on soil

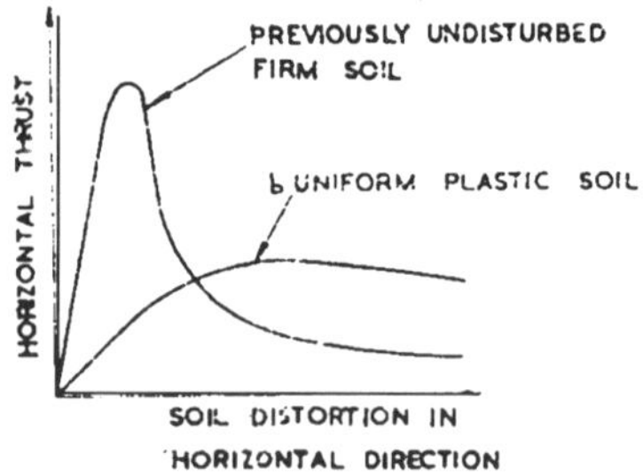

Fig 7: Thrust developed along tyre-soil contact patch

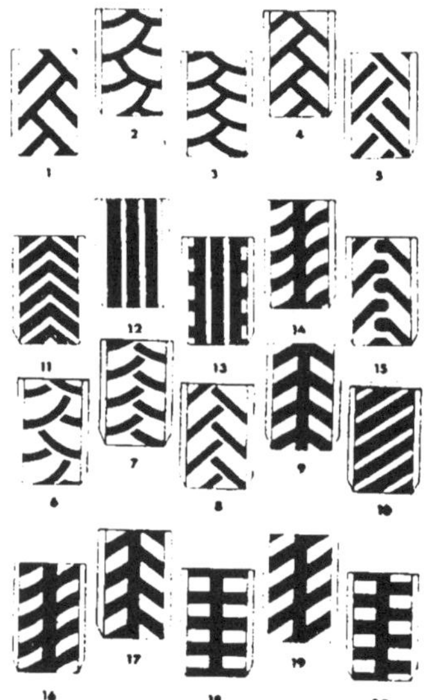

Fig 9: Vac-u-lug mouldable tread patterns

(d) is a means of obtaining the thrust resultant at ground level — involving uniform load distribution on traction surfaces. The good load distribution for a self-propelled tractor-scraper is shown at (e) and at (f), distribution on a tractor-hitched tractor scraper with a favourable linkage arrangement.

Fig 9 shows tread patterns that could be moulded in place by the Vac-u-Lug organisation. Numbers in the figure mean: 1. flatter tread profile, to provide increased ground contact; 2. more rounded profile, to prevent damage by roadway ruts; 3. higher than standard lug, for increased grip; 4. fitting of intermediate lugs, to provide extra casing protection; 5. lugs at increased angle, to improve lug-to-lug load transfer; 6. sidewall fenders, to protect cord body; 7,8. field tyres with open centres, for improved clearance; 9-17. special earthmover treads, for protection against chipping damage and 10. for hard rock conditions; 11. for general earthmoving; 12,13. circumferential ribs for trailing wheels; 14,16,19. for hard surfaces where little traction is required; 15. general purpose earthworking; 18, 20. non-directional to suit timber haulage work.

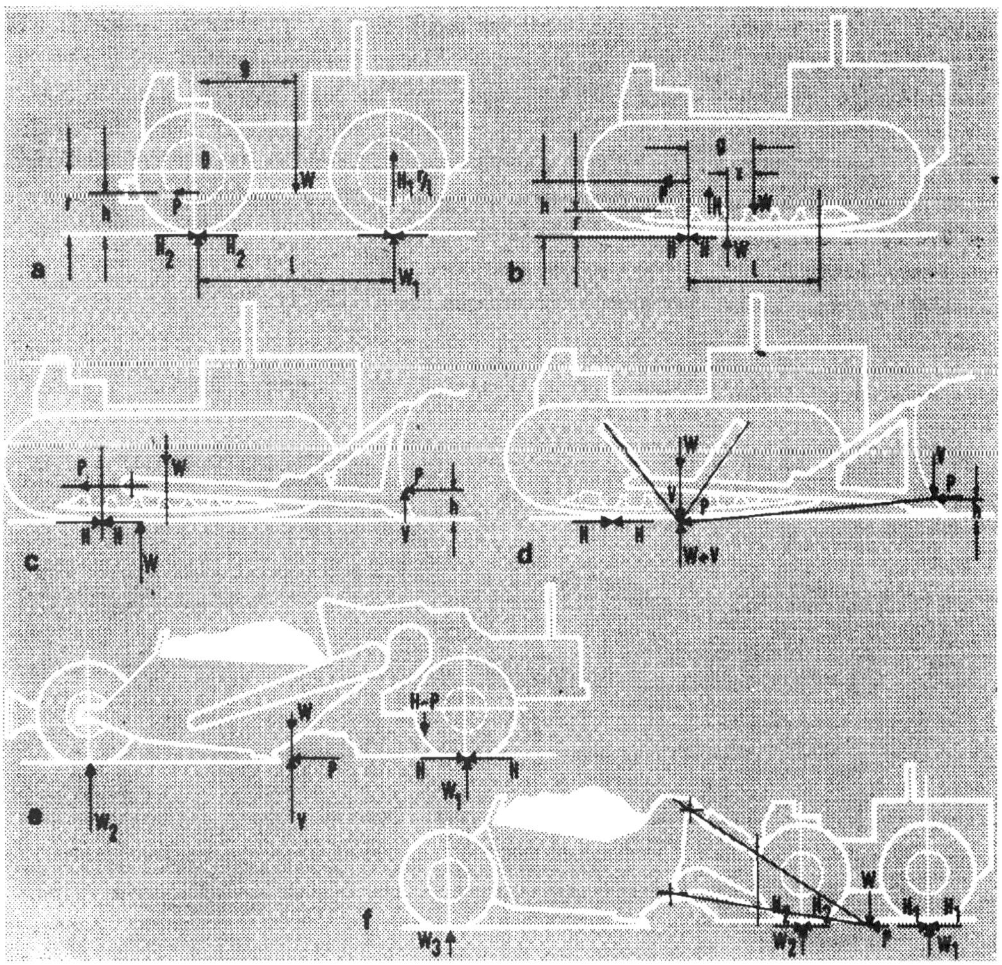

ig 8: Forces acting on earthworking vehicles

Farm tractor technology

AFRC have also developed techniques for predicting tractive performance of wheeled vehicles on soft terrains, with particular reference to farm tractors. The organisation have also been prominent in re-examining the traditional role of the tractor with respect to its tillage and general transport duties.

Because in slower gears the co-efficient of traction is so high that power available is limited by wheel slip according to the formula

$$[W_1 C_{tmax}(1 - e^{-ki}) - W_2 C_{rr}]v(1 - i)$$

where W_1/W_2 are weights on driving/undriven wheels; v is forward speed without wheelslip; i is wheelslip fraction; C_t is traction co-efficient; C_{rr} is rolling resistance co-efficient; k a constant and e the exponential co-efficient. The formula can be used to determine wheelslip corresponding to maximum drawbar power.

Work reported at Newcastle University in the 1970s usefully approximated the effect of tyre size on flotation and traction based on the simple geometric relations of Fig 10. Maximum tyre deflection of about 25 % section height is considered acceptable. Load carried by the tyre is

$$pb^{3/2}d^{1/2}$$

where p is inflation pressure. Maximum thrust is then obtained by multiplying the maximum soil shear strength by the contact area. The effect of soil strength was discussed in the previous section and the work yielded the useful charts of Fig 11, showing general relation between running gear type, soil conditions, vehicle duty and speed — also the relation between overall vehicle form as determined by number and type of ground-drive elements and operating conditions.

Tillage vs haulage

The introduction by JCB of its High Mobility Vehicle opposite has seriously challenged the traditional role of the farm tractor as conventionally designed. The HMV can haul a 24 tonne GTW, with payload of 14 tonnes, at road speeds of over 40 mph. The design philosophy was based on the growing realisation that more than 70% of a farm tractor's life is spent in transportation — yet the tillage function which ac-

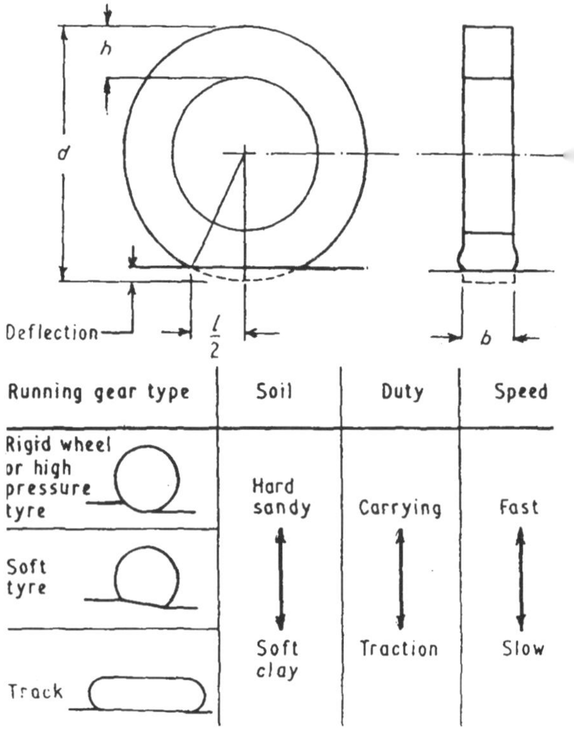

Fig 10: Flotation parameters

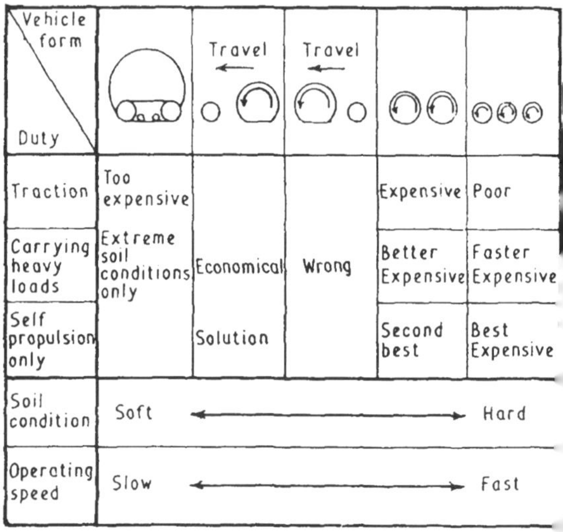

Fig 11: Wheel configuration recommendations

Fig 12: JCB HMV

Fig 13: Trantor truck/tractor

counts for most of the design specification is a much smaller proportion. The HMV, Fig 12, is well able to meet the requirements of deep draft ploughing but in addition its soft independent wheel suspension and power train provides high over ground performance for a variety of farm duties.

The vehicle has air-operated disc brakes on all four wheels which are equal size all round. There is a heavy duty three point linkage capable of handling the heaviest, fully mounted five furrow reversible ploughs and a front linkage with 2.5 tonne capacity is specified. There is also 2.5 tonne rear deck load capacity and self-levelling rear suspension is fitted. The cab interior is well ahead of current farm tractor practice and can have up to two passenger seats. Ride comfort is aided by front coil-sprung suspension along with the rear hydro-pneumatic system. Its Perkins engine of up to 145 is capable of running on red diesel as a farm tractor and the vehicle is also homologated to meet all CV construction and use regulations.

It mirrors the earlier efforts of the Trantor half-truck/half-tractor with which a much smaller company predated the HMV. Design of the Trantor is worthy of examination to understand the change in conventional power-farming philosophy. It was based on research in farm tractor usage which showed the results in Fig 14. Seen from the rear, the vehicle had a light truck like load deck but which could be fully enclosed by glazed hatches to afford a warm, spacious interior. Its optional four wheel drive system relied on hydraulic wheel motors at the front.

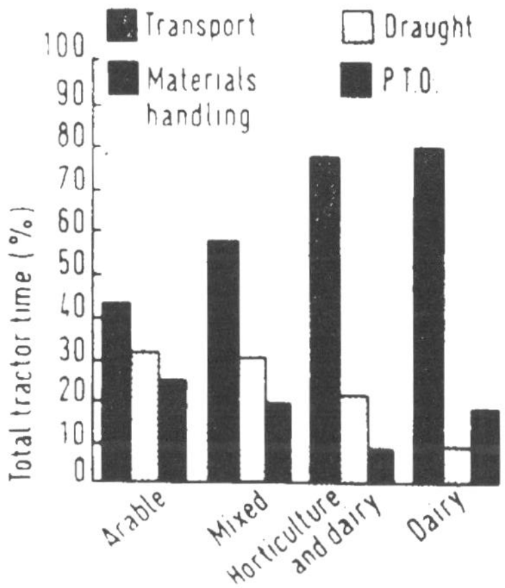

Fig 14: Tractor duty break-down

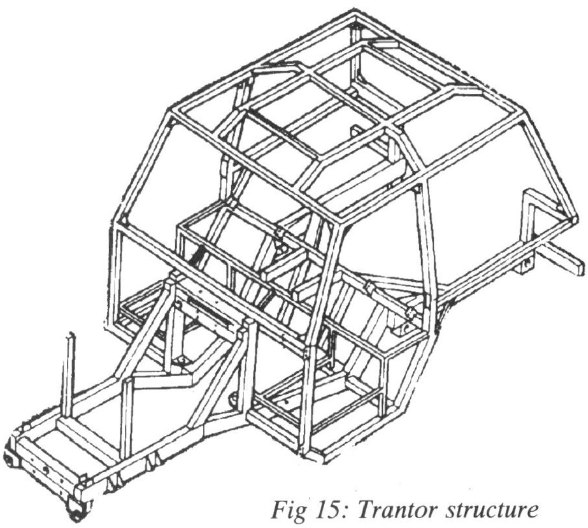

Fig 15: Trantor structure

It was one of the first farm tractors to depart totally from the 'skid-unit' construction of the power-train — having a chassis frame and truck-type drivetrain layout. In fact chassis and body structure were integrated as a space frame, Fig 15, to which a self-levelling mechanism for the implements, Fig 16, was attached. A two-way hydraulic-cylinder, A, had horizontally-opposed pistons separated by a valve block unit. Initially fluid pressure urged the foremost ram rearwards against the transverse leaf spring buffer. After a given displacement of the front ram the valves caused it to 'go solid' causing the swinging arm supporting the towing hook to raise the trailed implement. During towing, any trailer reaction to bump motion of the axle caused the rearmost ram to tension the buffer spring. The reverse happened on rebound and the pumping effect of vertical axle motion served to keep the hitch in the correct position.

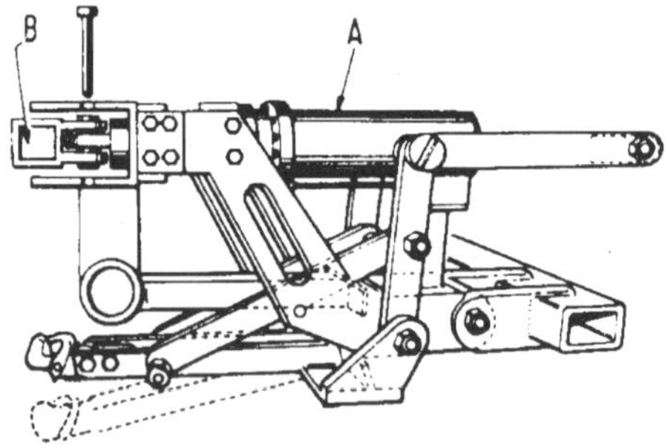

Fig 16: Self-levelling hitch

The full move towards the farm/truck tractor is more seen in the AFRC Engineering Farm Transport Vehicle, Fig 17, which uses a demountable body system to achieve complete versatility. The vehicle can thus be forage harvester, manure spreader and seed-drill as well as a transporter of grain. The active suspension system used on the vehicle is said to allow a performance which result in 30% higher work rates than on conventional farm tractors.

Fig 17: AFRC farm transport vehicle

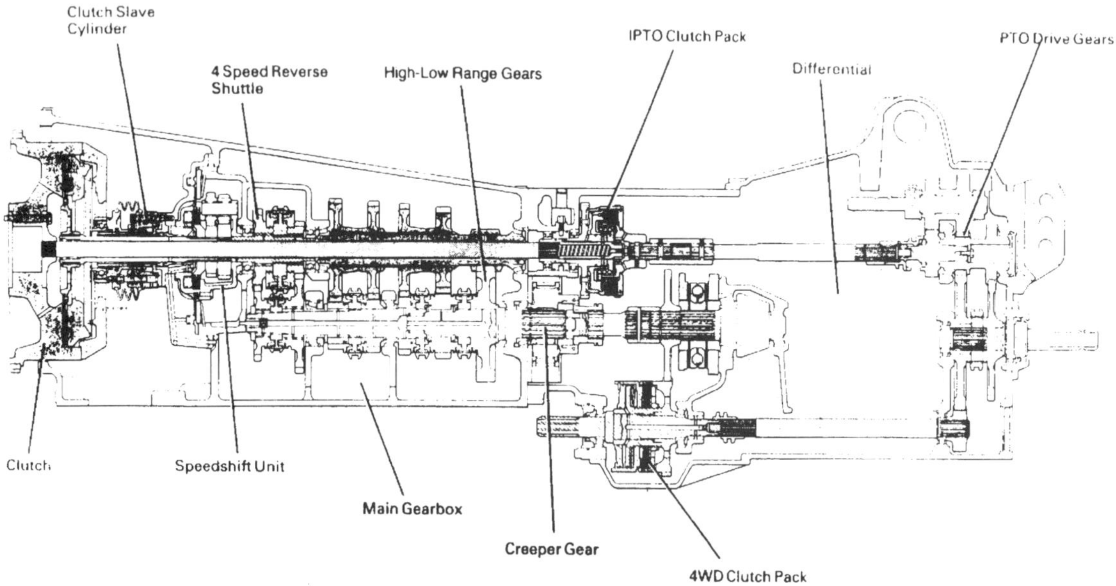

Fig 18: MF transaxle

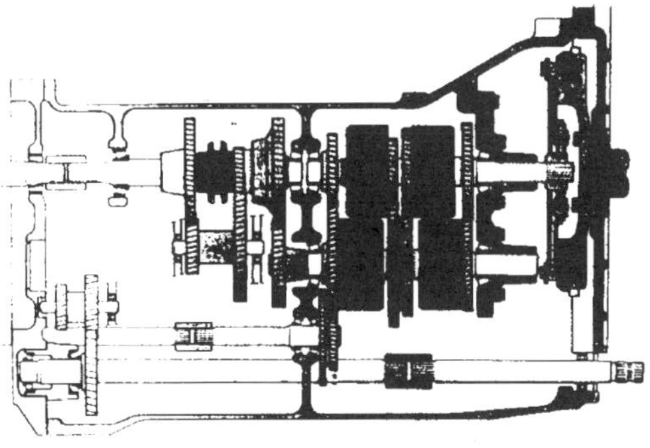

Fig 19: Case IH Power Shift transmission

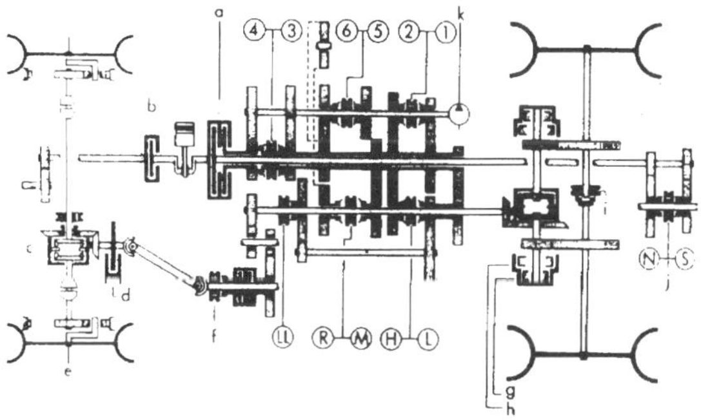

Fig 20: Deutz-Fahr DX 4 drivetrain

Fig 21: AFRC tracklayer suspension

Tractor transmission systems

More conventional machines typified by the Massey Ferguson 3000 series now have highly developed transmission systems and rear transaxle. Shown in Fig 18, the former comprises five major castings making up its case. It is an eight speed transmission with a reverse shuttle arrangement fitted forward of the main gearbox. When either of the two forward speeds is selected, the input shaft drives the mainshaft. Gear shift between the two is powered hydraulically as the lever passes across the neutral gate. When either of the two reverse speeds is selected drive is transmitted by a reverse idler shaft. This provides an overall of 16 forward and 16 reverse speeds. A speed shift unit can also be fitted to double this to 32. Being of epicyclic design it can be operated under load without interrupting drive. A creeper gear is also available which provides a further 4:1 reduction over the final drive.

The Case IH Maxxum is another comparatively recent introduction featuring a significant transmission development. It is a 16 x 12 speed Power Shift with 8 x 8 creep speed option, Fig 19. Electronically controlled hydraulic valves are used to energise the clutch packs which achieve pedal-less power shifting. Another interesting transmission variant is that specified for heavy draft work on the Deutz-Fahr DX 4. It has 16 forward and 6 reverse speeds — with a further 6 speeds coming from a creep gear. Fig 20 shows the complete drivetrain.

Another possible development in tractor traction could come from work on rubber tracks being carried out by AFRC Engineering. The company have devised an experimental system with equivalent rating to a 16.9 R 34 radial ply driving wheel tyre. In eight different operational field trials the track system exhibited an average of 20% more pull but its rolling resistance was more than twice as high. This resulted in an overall 6 % higher tractive efficiency of the tyre. If a positively 'geared' track drive was used, a firm advantage in favour of the track would be found, say AFRC, with the added benefit of much lower soil compaction. A possible suspension system is seen in Fig 21.

Light-weight ATVs

While weight distribution between axles, and distance above ground is important to many vehicle categories, it takes on paramount importance in lightweight ATVs whose gradient-ability, stability on side slopes and tractive performance are all critically dependent on this factor.

Fig 22 plots wheel adhesion and stability on an incline for a front drive vehicle driving forwards. An idealised loading condition can also be drawn, Fig 23, so that weight distribution can be predicted as accurately as possible at the design stage. Whereas the first section of the chapter outlined the main factors affecting mobility over relatively soft ground conditions, from the theories expounded it is clear that increasing tyre size to raise contact area, and an increase in the number of wheels, similarly, is the way to improve flotation and mobility. Rolling resistance decreases with tyre size, Fig 24, but reduces with inflation pressure, Fig 25, so that there is a limit to the decrease in tyre pressures that can be made to improve flotation by reducing ground pressure. These factors have no doubt influenced the multi-wheeled `doughnut-tyre' ATV configurations which are becoming commonplace. Wheel diameters cannot be raised too much because the high rotational inertias involved would detract from the performance of such vehicles.

By raising tyre section width, contact area increases almost in proportion; a significant reduction in the depth of rut results over soft ground with clear benefits to rolling resistance. Fig 26 shows the effect of shape, of a typical tyre profile, on performance for different values of *H*/*B* ratio. On decreasing *H*/*B*, the rigidity of the tyre continually increases, width of contact also increases and the contact area reaches its maximum at *H*/*B* = 0.5. Calculated values of performance indices for the wheel for different values of *H*/*B* are shown at (b) in Fig 27: values are 1=0.25; 2=0.5; 3=0.75; 4=1; 5=wet soil; 6=wet grass; 7=snow; 8=loam. At Fig 28(a) is seen the corresponding changing shape of the profile and at (b) the relationship with the contact parameters.

While Fig 29 shows the types of tyre available to achieve different *H*/*B* ratios, Fig 26 shows the key dimensions involved in tyre deformation. Sizes of the contact area for normal and minimum values of inflation pressure are also shown. Up to 35 percent can be achieved with certain configurations, limited

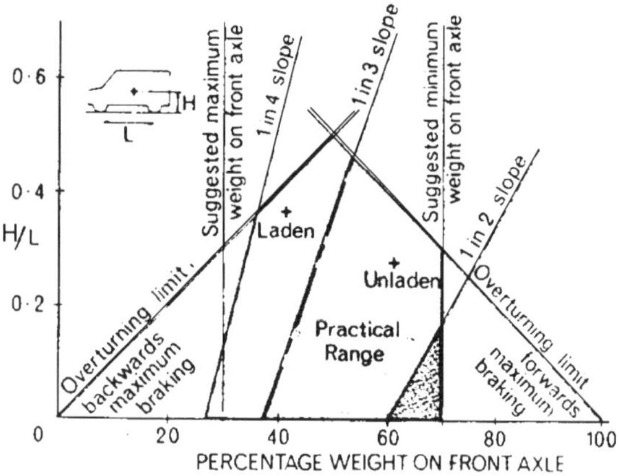

Fig 22: Traction and stability factors

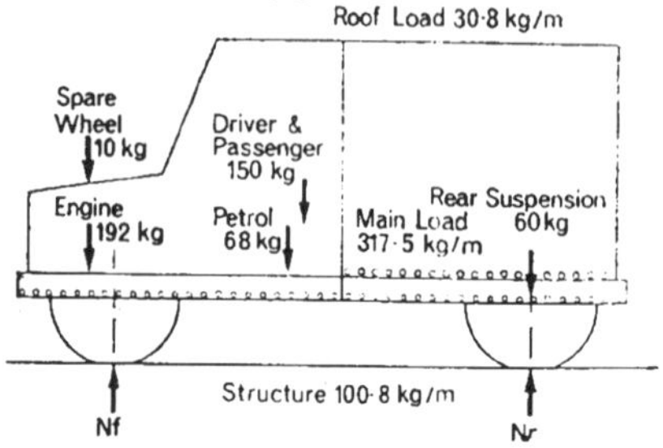

Fig 23: Idealised loading condition

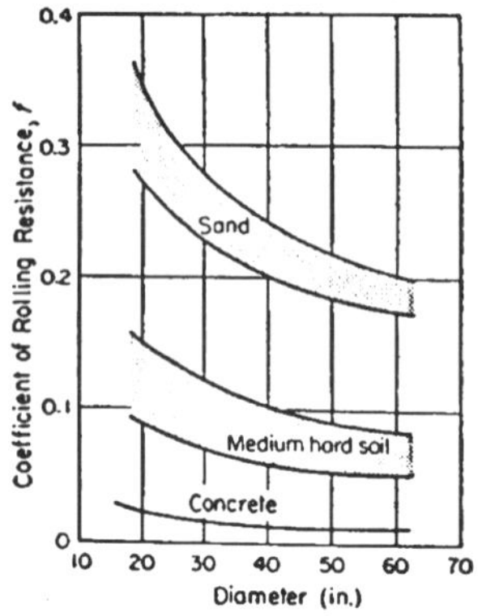

Fig 24: Rolling-resistance vs tyre size

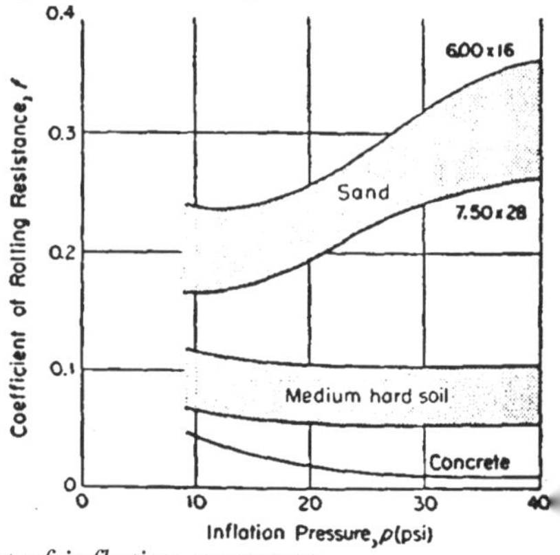

Fig 25: Effect of inflation pressure

eventually by loss of lateral stability. The arched tyres have short sidewalls with considerable bending rigidity and at large vertical deflections, the crown of the tyre deflects like a membrane redistribution contact pressure distribution over the contact patch to give very low sinkage on soft ground.

It is possible to quantify the effect of number of axles in relation to vehicle traction by assuming a rectangular tyre/ground contact patch (with $l/b = v$) and even pressure p distribution across it. Width of the contact surface b relates to the weight of the vehicle G and number of axles n in:

$$b = (G/2vnp)$$

On the basis of the maximum number of passages N the vehicle can make over the same track an expression, for sandy loams, has been calculated by Russian researchers:

$$N = 2\, t_r v/[(\pi aG/2Eh) + (1+p/X)(2Gnv/p)]^{1/2}$$

where h is rut depth; X and E are soil parameters. Fig 27a shows the effect of inflation pressure on traction performance. Traction is most sensitive to the internal cohesion value of soils on soft ground — but also, of course, to surface roughness on hard ground.

The drive configuration on multi-axle systems

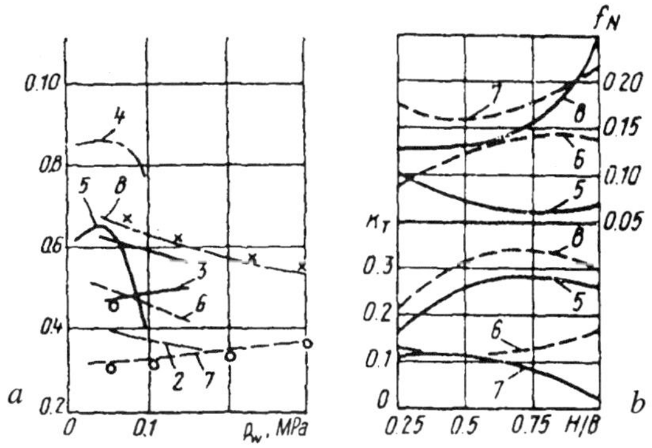

Fig 27: Inflation effect and performance (full-line: 1140-700; dashed line 1200-18; chained-dot 1200 x 1200 x 500; 1,5,6 sand; 3,4,7 sandy loam)

	toroidal with adjustable pressure	wide-sectioned	arched	pneumatic
H/B	1	0.7	0.37	0.32
D/B	3	2.5	1.6	1.0
B_F/B	0.75	0.8	1.0	1.0
h_{tr}/H	0.1	0.12	0.3	0.2
f_{tire} rolling resistance	0.015	0.012	0.020	0.028

Fig 26: Tyre parameters

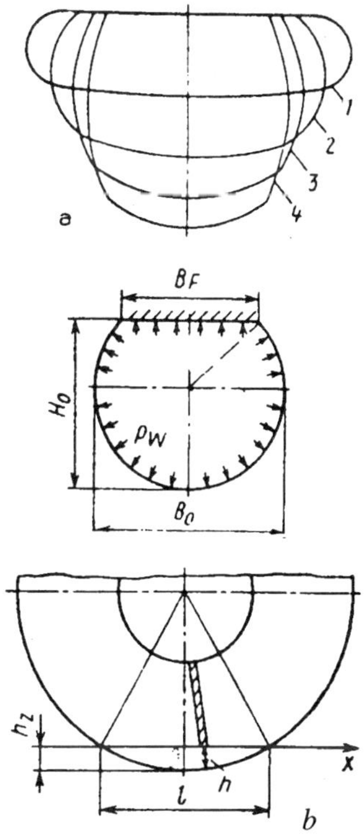

Fig 28: Parameters of tyre deformation

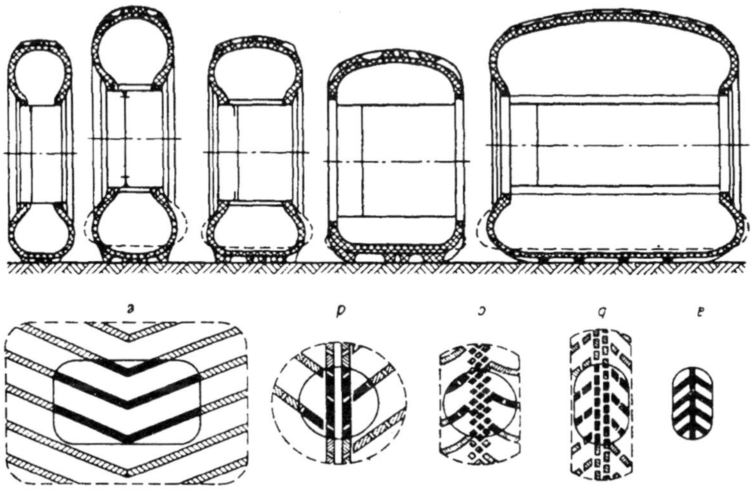

Fig 29: Tyre profile effect

is also important, multi-axle vehicles with all-wheel drive being prone to quite considerable transmission losses which soon show the competitive advantage of rubber-tracked ATVs such as the Snow-trac vehicle of Fig 30. An interesting alternative is the articulating chassis layout for a multi-axled vehicle, exemplified by the USA-built Gamma-Goat, Fig 31, with its front section having two and its rear a single axle. Up to 30 degrees cross articulation, about a horizontal axis, is possible and 40 degrees about a transverse axis. This is compared in the figure with the Swiss Metrak which is only articulated (up to 30 degrees) about a transverse axis and steers by angular displacement of both front and rear axles. A locking system is incorporated to lock up the articulating chassis when crossing ditches.

Layouts of some other well known ATVs and light on-off road vehicles are seen in Fig 32. Chassis of the Suzuki LJ80 4WD is seen at (a) to have a highly cranked frame for achieving the necessary axle bump clearances. Steep angle drives to the axles, particularly at the front, also characterise the configuration as does the very high-mounted engine. Another classic layout is that of the Citroen 2CV at (b) which became the base for the Mehari on/off roader. Here the chassis was underslung and the leading/trailing arm wheel suspension, front/rear, facilitated the use of torsion bar springing and a coupled no-pich suspension.

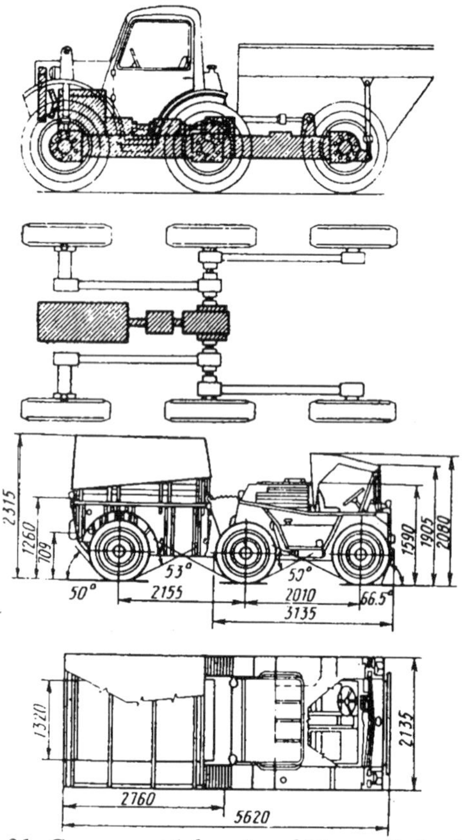

Fig 31: Gama-goat (above) and Metrac (below)

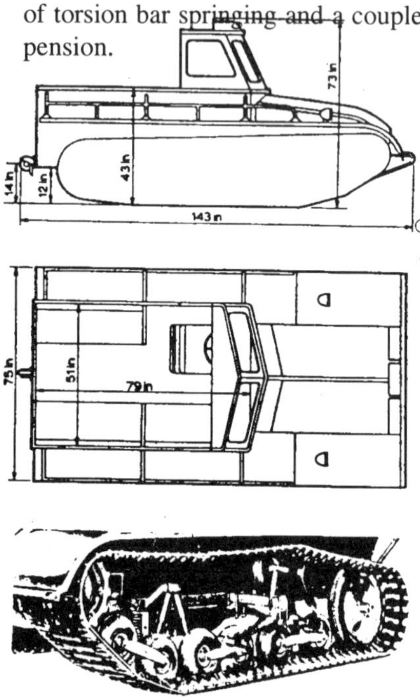

Fig 30 : Snow-trac

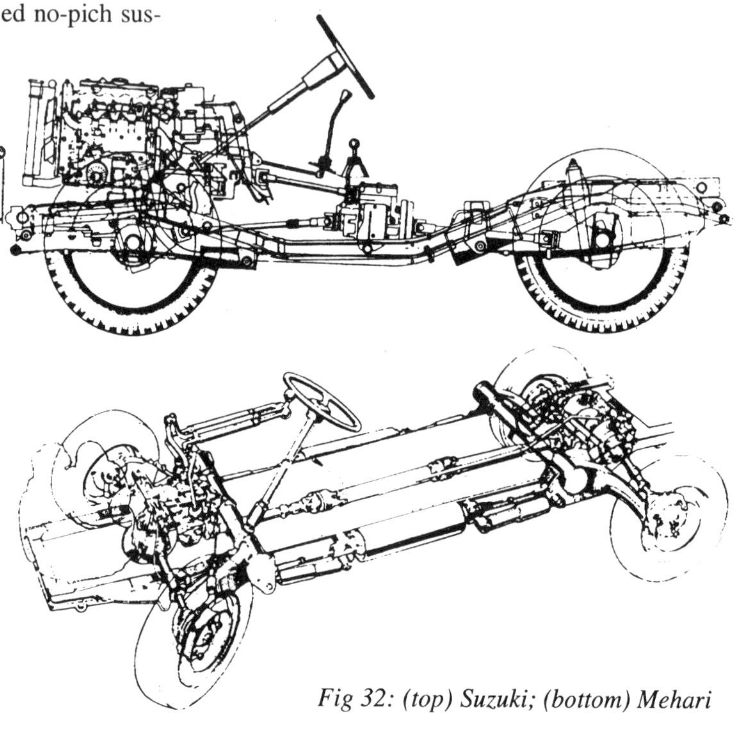
Fig 32: (top) Suzuki; (bottom) Mehari

Farm support vehicles

Here, the long-term development at AFRC Engineering of the so-called Farm Transport Vehicle, introduced in the penultimate section, is examined in greater detail. The company has extended this concept into a multi-purpose farm-worker having demountable bodies adapted to a wide range of duties.

As well as optimising the design of the base unit for maximum drawbar pull, in the case of the tillage-tractor concept, this vehicle is also designed for high-speed cross-country operation to optimise efficiency in other farm support roles. Bodies can be interchanged in a confined space, thanks to a patented dismount mechanism, and an articulating chassis enhances manoeuvrability. Pivoting axles provide 2- or 4- wheel steering and a 6 metre swept turning radius is achievable.

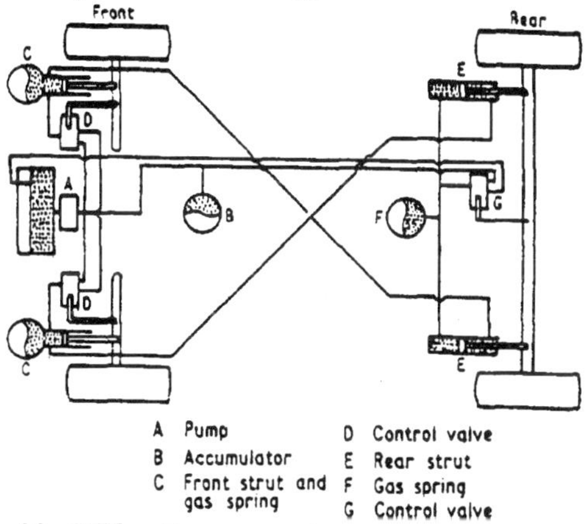

Fig 33: FTV active suspension

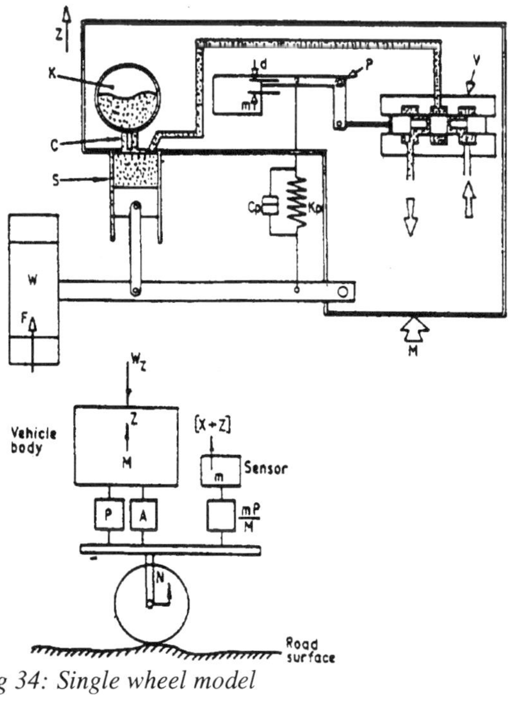

Fig 34: Single wheel model

Both front and rear power-take-off is available with the machine and implements can be carried either over the vehicle or on an A-frame at the front. In its handling mode, a lift mast rises at the rear of the vehicle; this has 3 tonnes capacity and allows the vehicle to function like a fork-truck. At half-tilt the mast is used for such jobs as skip loading. The vehicle is powered by a 75 kW Perkins engine, drive from which passes through a 5-ratio change-speed gearbox and through a transfer gearbox which divides torque between front and rear axles. The oleo-pneumatic active suspension system is also interconnected front and rear, Fig 33, a single wheel schematic being shown in Fig 34. Active and passive elements of the suspension are in parallel, the force exerted by the active one, on the body, being a function of the relative displacement of the sensor mass. This is attached to a pendulum coupled to the hydraulic spool valve and held horizontal by the spring/damper connection to the suspension arm.

While acting as a dynamic levelling valve in steady-state conditions, it has to discriminate between weight changes due to control inputs and those due to ground profile. The dynamic motion of the control mass models the required dynamic behaviour of the body, any error between the two initiating height correction. The pendulum has two effective degrees of freedom, one vertical and one rotational. For matching suspension motion the vertical component models suspension natural frequency and damping; the angular natural frequency can be regulated by modifying wheel to pendulum-arm input ratio, being related to speed of system response. In trials, the suspension has permitted 30% higher work rates, in general, the interchangeable bodies enabling the machine to be used as a forage harvester, manure spreader, grain transporter, sugar-beat handler and seed drill.

Another radical approach to farm-support from AFRC is the use of mobile gantries for a variety of duties. Both wheeled and tracked forms have been developed, first of the former having a 45 kW air-cooled engine driving through two wheels via hydraulic pump and motor transmission. High speed, low torque, motors drive the wheels through 28:1 reduction units. Wheels are at a track width of 11.5 metres, supported by an intervening square-section space frame, akin to a Warren truss structure. The remaining, trailer, wheels are castor-pivoted and swing into the chassis to narrow the clearance of the vehicle when it travels in the opposite, lengthwise, mode between fields — via gateways. For these purposes the driver sits in a rotating cab and has switch-control between sideways and lengthwise steering modes. An alternative, crawler-driven, gantry has been developed for vegetable production, Fig 35. Benefits include less damage to the crop and easier travel over it in wet-soil conditions, as well as more efficient crop density and harvesting. The machine carries a large work-platform with overhead weather protection so that crops can even be packaged on-board, to save double handling. Designed to span a 9 metre wide bed, all operations from seeding to chopping-up of plant debris have been successfully performed with the machine. In the case of the wheeled machine, used for grain cultivation, yield of wheat was increased from 5.72 to 6.77 tonnes/hectare by the benefit of having an untrafficked plot. There were advantages, too, in the reduced energy requirement in shallow-ploughing after harvesting in wet weather, due to the absence of soil-compaction. Spraying costs could also be reduced by the more accurate spraying of smaller quantities of liquid. Energy requirements in MJ/ha for spring cultivation and drilling, following autumn ploughing, are shown in the table of Fig 36. A gantry adapted for horticultural work has been used to raise palletised plots and lift them out of the beds into adjoining pathways. Another has been adapted for grain crops and has a 7 tonne/hour combine harvester, with a 2 metre grain-stripping rotor, which can be winched laterally across the machine.

Treatment	Energy use calculated from fuel measurements, MJ/ha
CD	47.5
LD	39.8
ZD	35.5
CS	178.1
LS	181.6
ZS	51.0

C=Conventional, L=LGP, Z=Zero traffic
D=Direct drill, S=Shallow cultivated

Fig 36: Energy requirements

Fig 37: Unimog truck/tractor

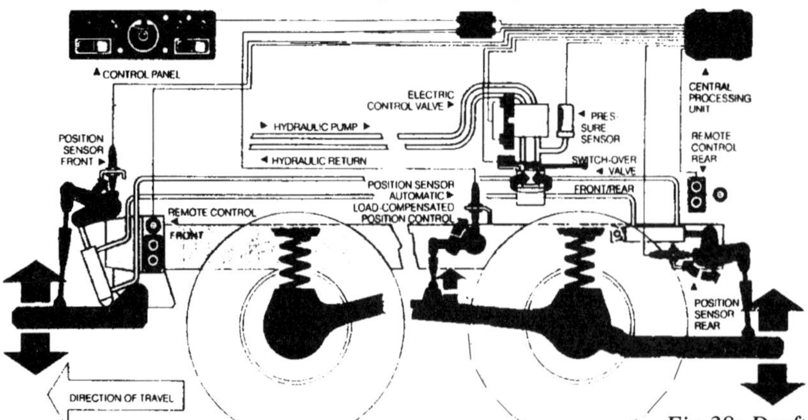

Fig 38: Draft control system

Truck/tractor vs tractor/truck

The Mercedes Benz Unimog approaches the concept of tillage from the different direction of a cross-country truck. The latest U1600 AG model, Fig 37, now has a draft control system, Fig 38, that gives the vehicle a dual role. The vehicle can travel at 64 km/h laden, on-highway and has a 16-speed synchromesh transmission and four wheel drive with diff-locks on both axles. Equal tyre sizes are used, front and rear, on the coil-sprung portal axles which have geared hub-reduction. The Terramatic hitch system allows for both front and rear mounted implements. Truly constant draft-depth control is claimed for the electro-hydraulic system — (with automatic load compensating position-control for spreading and spraying) which has automatic re-setting of linkage height. The chassis frame has tubular cross-members welded into channel sidemembers and in standard trim 60% of the vehicle weight is borne by the front axle. The portal axles give high ground clearance and are hydraulically damped front and rear. Mechanical PTOs drive through to front and rear of the vehicle and can be selected together or independently at either 540 or 1000 rpm.

JCB's HMV, or Fastrac, seems to be neither truck nor tractor derived and the company's claim to have designed 'from a clean sheet' appears to have substance. It has a 72 km/h speed capability, can pull payloads up to 14 tonnes and bear a direct imposed load of 2.5 tonnes — yet has the draft capacity of an equivalent conventional tractor. As well as a coil-sprung front axle it has a self-levelling gas-sprung rear suspension. Its interesting chassis is effectively an A-frame with single tubular crossmember between Z-section longitudinals, Fig 39.

Recently, models have been added which provide 65 km/h top-speed both on and off-road without need to change wheels and tyres. This follows the fitting of Goodyear 480/70R30 radial tyres which raise the ground clearance by 4 to 17.7 in (100-450 mm). The rear 3-point hitch is now a full category III design but will also accept category II implements and lift capacity has been raised to 6.5 tonnes at the rear and 2.9 tonnes at the front. There is also a flow-divider specified which allows controlled flow to a spreader motor regardless of the demands from the 3-point linkages; it can be set at between 5 and 55 litres/min.

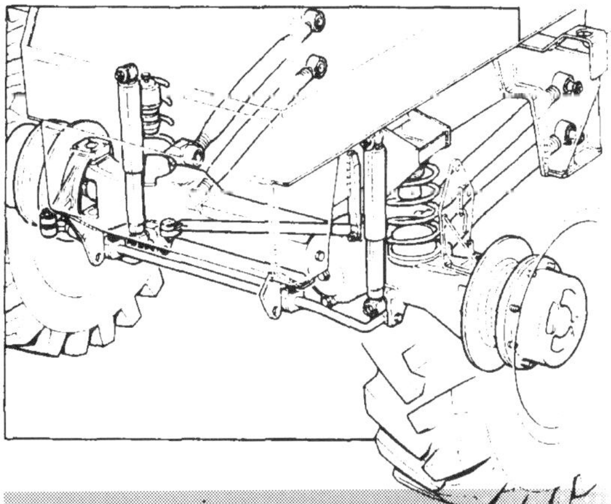

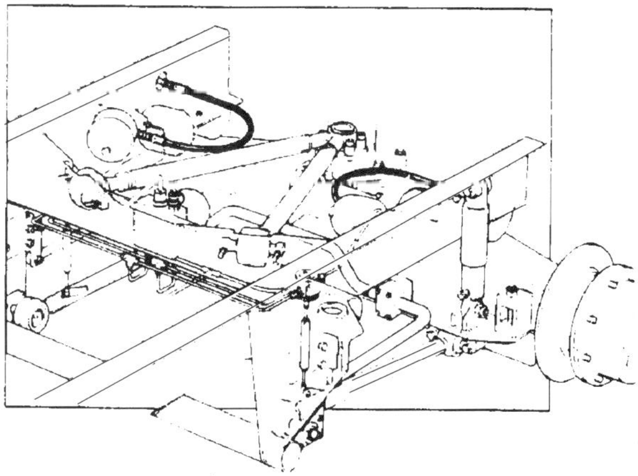

Fig 35: Tracked gantry

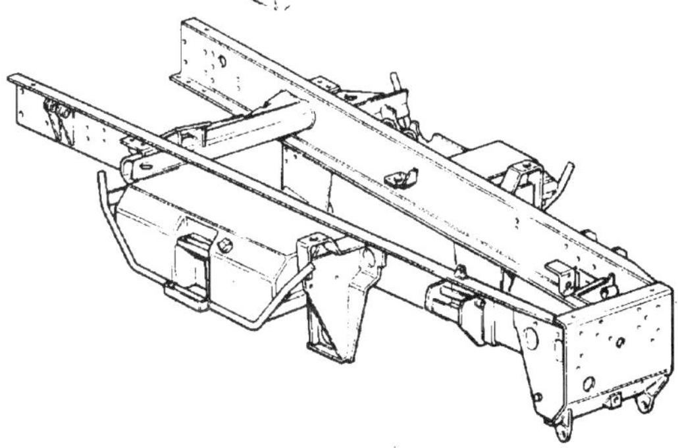

Fig 39: Fastrac suspension and chassis

All-terrain trucks

The unfolding technology of soil mechanics is increasingly making the prediction of terrain vehicle performance a more reliable proposition. Expressions for the different resistances to motion are provided for wheeled vehicles in this section — which also examines performance over obstacles and some of the different mechanical configurations for all-terrain trucks.

For wheeled vehicles, a useful work by Wong[1] has devoted its last two chapters to reviewing the analytical techniques for wheeled vehicle performance in off-road conditions. Recent experimental work has led to a better understanding of the complex flows of soft ground under a vehicle wheel. Fig 40 shows the flow pattern in dry sand in a longitudinal plane beneath the wheel for the driven (a), towed (b), locked (c) and 100% slipping wheel (d). Wheel-ground interaction gives rise to normal and shear stress at the interface which varies considerably with operating conditions. Fig 41 shows the distribution of these stresses measured for a towed rigid wheel in dry sand.

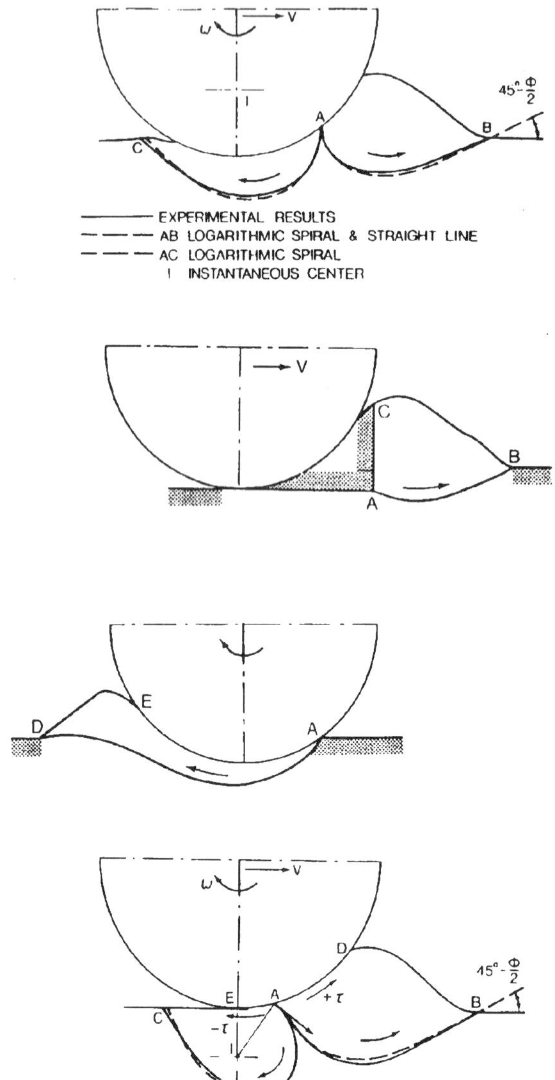

Fig 40: Soil-flow patterns for different wheel conditions

The classic semi-empirical work of Bekker[2] is based on the assumption that the deformation of the tyre will be insignificant compared with that of the terrain. The sinkage of the wheel can be determined from the parameters shown in Fig 42 such that:

$$z_e = [p_g/(k_c/b + k_\phi)]^{1/n}$$

where b is the smaller dimension of the projected contact patch of the wheel; P_g is average ground pressure and n, k_c, kf are pressure-sinkage parameters obtained from soil-mechanics testing, Fig 43. Corresponding 'compaction resistance' to motion caused by sinkage is obtained from:

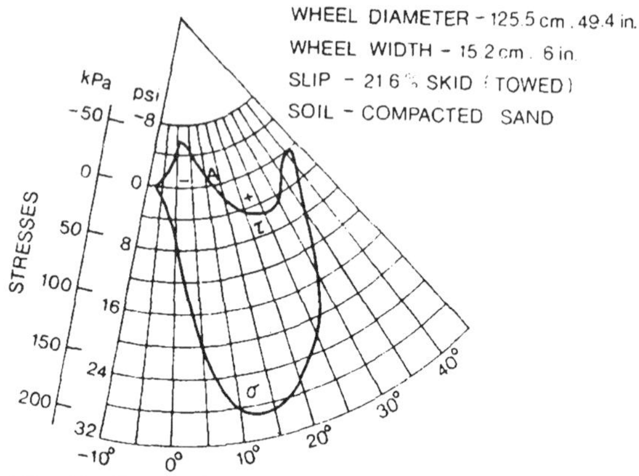

Fig 41: Stress distribution

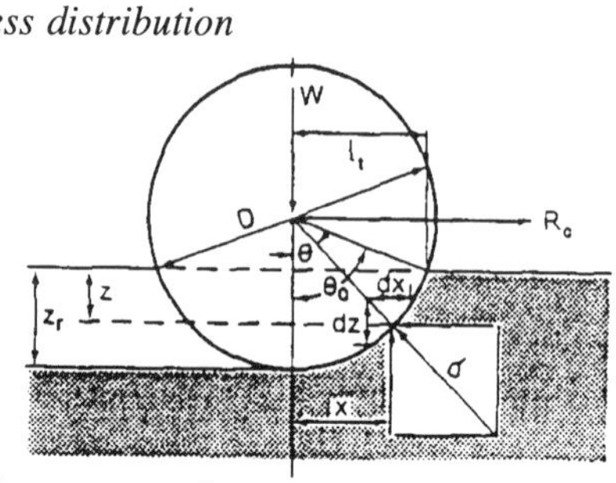

Fig 42: Sinkage parameters

$$R_c = b_{tr}[z_e^{n+1}/(n+1)][(k_c/b) + k_\phi]$$

where b_{tr} is the tread width of the tyre.

A further resistance to motion is caused by the bulldozing effect, Fig 44. The soil in front of the tyre is assumed to be in general shear failure and the resistance given by:

Terrain Type	Constants for Bekker's Equation		
	n	k_c (kN/m^{n+1})	k_ϕ (kN/m^{n+2})
LETE Sand	0.705	6.94	505.8
	0.611	1.16	475.0
	0.804	3.93	599.5
	0.728		1348
	0.578	9.08	2166
	0.781	47.8	6076
	0.806	155.9	4526
Upland Sandy Loam	1.10	74.6	2080
	0.97	65.5	1418
	1.00	5.7	2293
	0.74	26.8	1522
	1.74	259.0	1643
	0.85	3.3	2529
	0.72	59.1	1856
	0.77	58.4	2761
	1.09	24.9	3573
	0.70	70.6	1426
	0.75	55.7	2464
Rubicon Sandy Loam	0.66	6.9	752
	0.65	10.5	880
North Gower Clayey Loam	0.73	41.6	2471
	0.85	6.8	1134
Grenville Loam	1.01	0.06	5880
	1.02	66.0	4486

Fig 43: Parameter values from tests

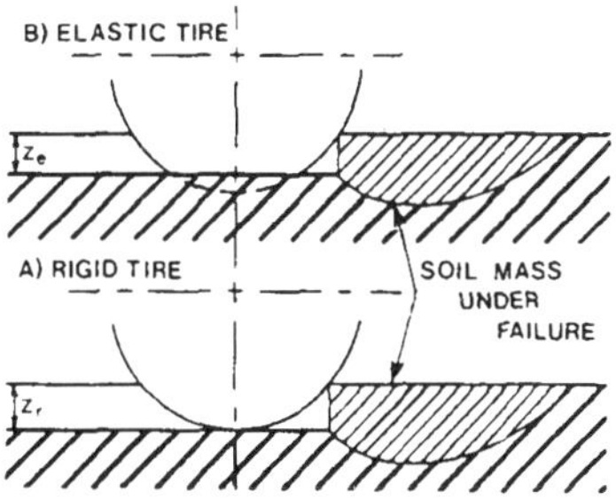

Fig 44: Bulldozing effect

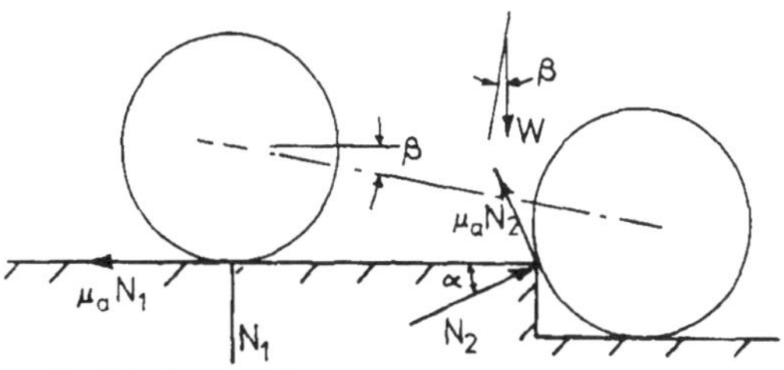

Fig 45: Ditch-crossing parameters

$$R_b = b_{tr}(czK_{pc} + 0.5z^2\gamma K_{p/\gamma})$$

where c is the cohesion of the terrain, z is tyre sinkage and g is the terrain density. Factors K_{pc} and K_{pg} are obtainable from soil analysis of bearing capacity and internal shearing resistance. The co-efficient of resistance to horizontal terrain deformation in fact varies between 0.0813 and 0.2454 depending on soil conditions, according to Bekker[3]. The third motion resistance is due to the flexing of the tyre carcase and a further empirical expression has been devised by Bekker and co-workers to account for this effect.

Earlier work by Bekker[4] defined the mobility of an off-road vehicle in terms of its ability to surmount obstacles in its path. The crossing of the

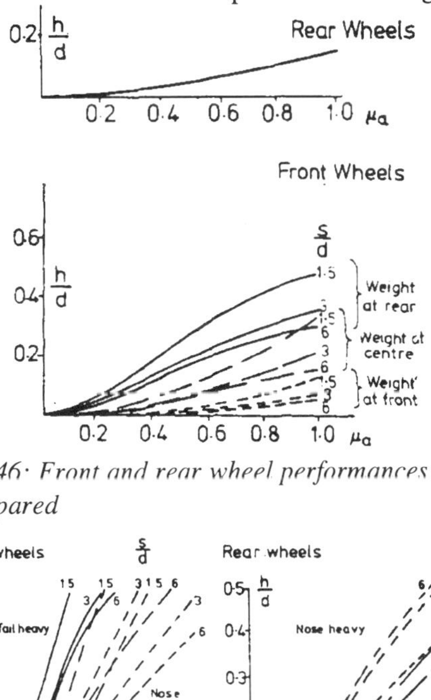

Fig 46: Front and rear wheel performances compared

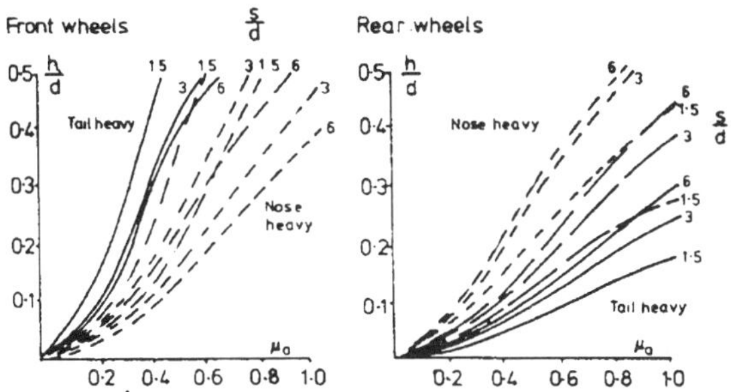

Fig 47: 4 x 4 performance

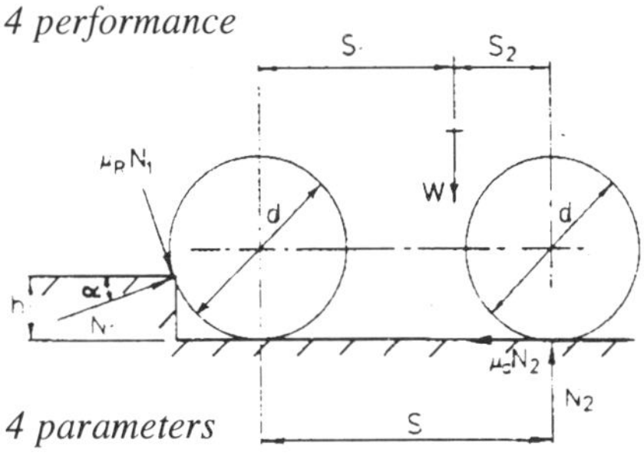

Fig 48: 4 x 4 parameters

ditch in Fig 45 was one example; here basic parameters are related by:

$$s/d = 2[(b/d)-(b/d)^2]^{1/2}$$

and the front wheel performance of a 4 x 2 vehicle can be defined by:

$$b/d = \{1-1/[1+(\mu_a S_1/S)^2/(1-S_1/S-d/2S)^2]^{1/2}\}/2$$

where m_a is adhesion co-efficient, the results being plotted in Fig 46 to compare with rear wheel performance. It is seen that the front wheels will surmount higher obstacles as the weight is shifted rearwards and the wheelbase ratio is reduced — whereas the rear wheel analysis reveals that it is the bump obstacle and not the plateau which limits performance, given by:

$$b/d = [1 - (1-\mu_a^2)^{-1/2}]/2$$

purely a function of adhesion co-efficient. In the case of 4 x 4 vehicles, more complex expressions result and these are plotted in Fig 47 based on the parameters of Fig 48.

All-terrain truck configurations

The Irish Timoney company have made some of the most interesting proposals for off-road heavy-vehicle operation — much of them destined for military and crash-tender application. The company have also produced cargo trucks in 4 x 4, 6 x 6 and 8 x 8 configurations, Fig 49. These exploit the well-known independent suspension system shown as Fig 50 in alternative forms with coil springs attached to either upper or lower wishbones. At the lighter end of the weight spectrum, the Spanish URO F14, Fig 51, uses a much more conventional construction but relies on very high degrees of chassis twist to obtain high levels of wheel articulation, typical for the smaller truck.

Another interesting approach is the central backbone layout of Tatra, Fig 52, which encloses the main transmission shafting and is convertible to a wide variety of wheel/drive configurations as shown by Fig 53. The articulated dump-truck concept of DJB Engineering is also of particular note, Fig 54.

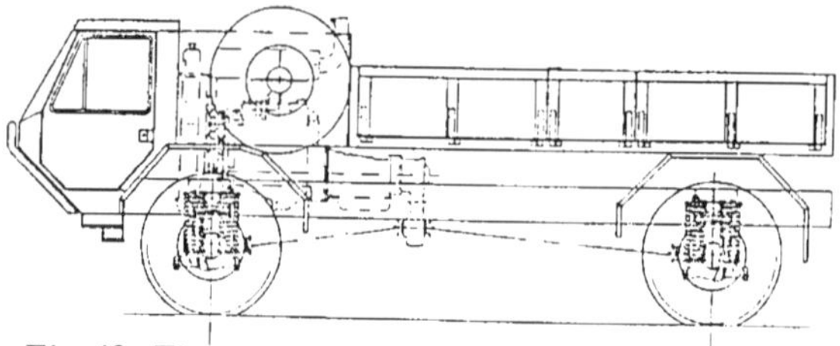

Fig 49: Timoney terrain truck

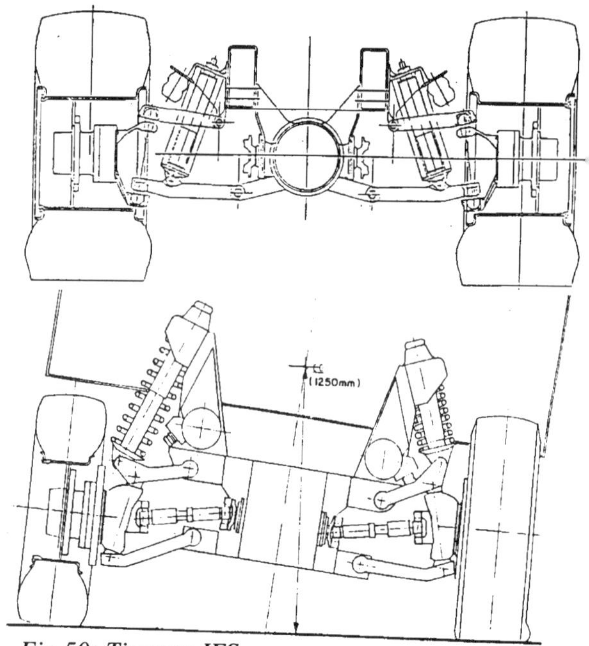

Fig 50: Timoney IFS

Fig 51: URO F14 truck

Fig 54: DJB 550 articulated dump-truck

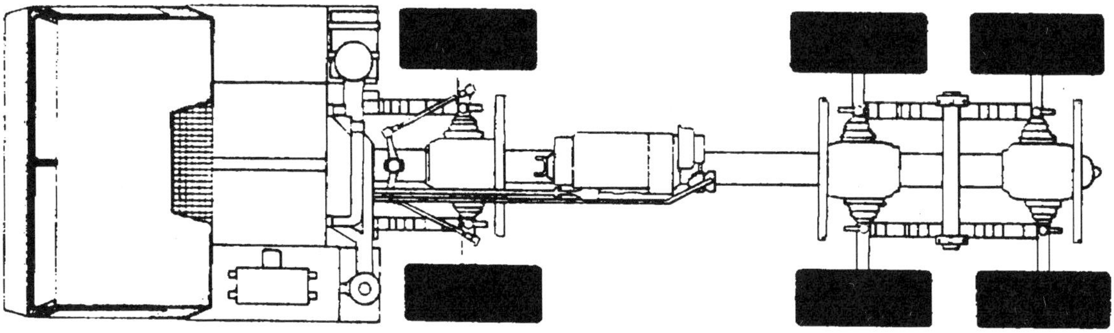

Fig 52: Tatra spine-chassised truck

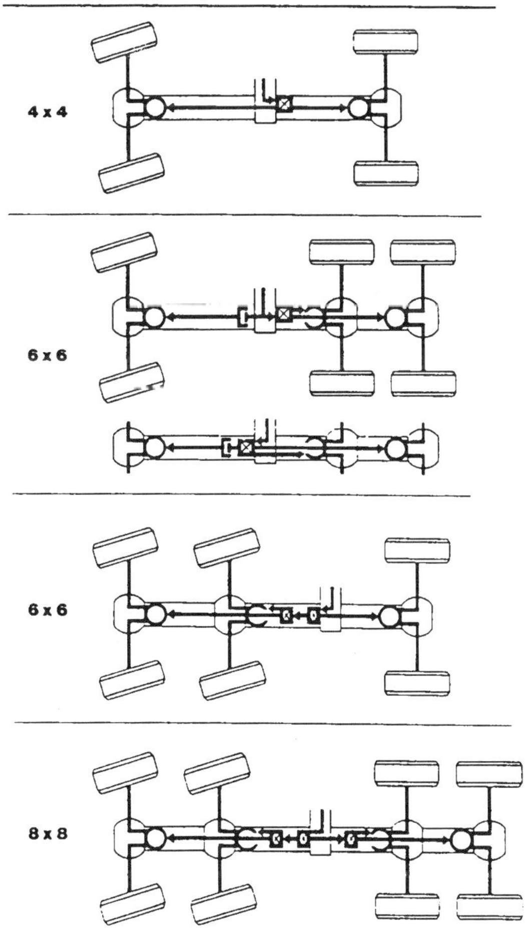

Fig 53: Alternative drive configurations

Earthworking vehicles

Because the high loads imposed on earthworkers considerably exceed those required for load-carrying and self-propulsion, additional criteria are required for performance prediction. Some techniques for estimating earthmoving forces, and relating them to traction requirements, are discussed here.

In the introductory section on off-road vehicle technology, factors such as the resistance to motion due to sinkage in soft ground were discussed, ending with an initial look at the effects of external forces such as bulldozer and scraper blade loads on vehicle performance. Nearly all earth-working vehicles are faced with the problem of overcoming additional motion resistances to those simply due to the sinkage of the prime-moving wheels and tracks. Excavators and other vehicles having long booms reaching out to their digging shovels have the additional problem of reacting highly offset loads.

The cutting of a shallow layer of soil by a wide inclined blade, such as a dozer or scraper, is a process of wedging a triangular chip of soil upwards between the inclined surface formed by the blade, at the back, and a sloping plane formed within the soil, at the front. In strong cohesive soils the slice of soil cut will roll in front of the mouldboard to form a 'swiss-roll'. In more friable soils a large loose heap will push forward in front of the blade. Researchers at Newcastle University have developed a hypothetical failure pattern for the friable soil case as in Fig 55.

Force P is the force requireed to cut the wedge and slide it upwards against the surcharge force *Q;* force *F* is that required to slide the front wedge along the uncut soil surface. A theoretical analysis, backed by tests on a Caterpillar D8 tractor, resulted in typical load values as seen in Fig 56. In heavier soil the force required to make a 6 inch (152 mm) cut was 4500 lbf (1 tonne) and the extra force required to overcome down force from the surcharge was 20,000 lbf (9 tonnes), while the force required to slide the triangular heap forward was 15,000 lbf (6.8 tonnes). Using such derived figures an efficiency for the machine can be predicted. This is based on the 'necessary' work which the machine is required to do — lifting the soil from half the depth of the cut, below surface level, to the heap centre-of-gravity as well as slicing the layer from the soil below. For this example eficiency would be (4500 + 1200)/40,000 or 15%.

Later machines have efficiencies as high as 30%, brought about by such measures as arranging for the dozer cutting edge to simply cut and lift the soil without any collected soil on top of the cut so as to reduce basic cutting force to around 5000 lbf (2.27 tonnes). Configuring the blade so part of the spoil weight contributes to the traction of the vehicle is another form of improvement. Means of reducing the frictional force of sliding the soil along the ground would enhance efficiency even further, of course.

Earthworker structural considerations

The Alcan company carried out pioneer work on large, 35 tonne capacity, dumper bodies in the loading of large rocks — to prove the suitability of aluminum alloy in this arduous operation. Observation of loading operations confirmed that maximum stresses occurred at the midline section of the bodies where the first load from the shovel was normally dropped. Loading tests were then carried out with strain-gauged bodies, positioning being shown in Fig 57. The purpose of each gauge was as :

1., 4. Maximum tensile stress due to bending in the floor cross bearer at mid-span.
2. To check stress concentration at the edge of the cut-out in the longitudinal stiffener at that joint.
3. (L&T) Longitudinal and transverse stresses in the floor plate at the point of highest impact.
5. Maximum stress in the side stiffener flange
6. Tensile flange stresses in horizontal wall-stiffeners at mid-span
7. Bending stresses at mid-length of the top-rail and compressive stress due to bucket-impact.
8. (L&T) Tensile stresses in the floor plate at the front of the body.
9. Tensile stresses in the continuous cross-member.
10. Magnitude of loading on 3-columnn support pad.
11. Longitudinal compressive stress at the bend in the floor-plate during body hoisting.

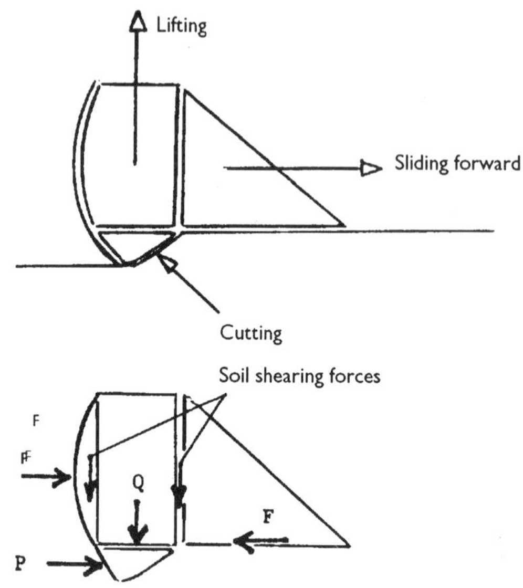

Fig 55: Dozer blade forces

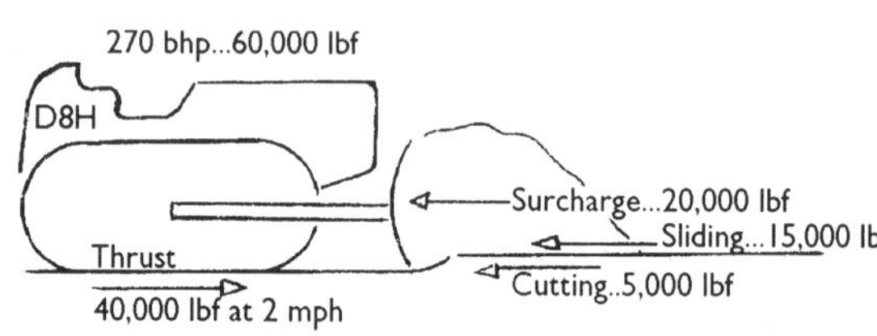

Fig 56: Loads on a D8 tractor

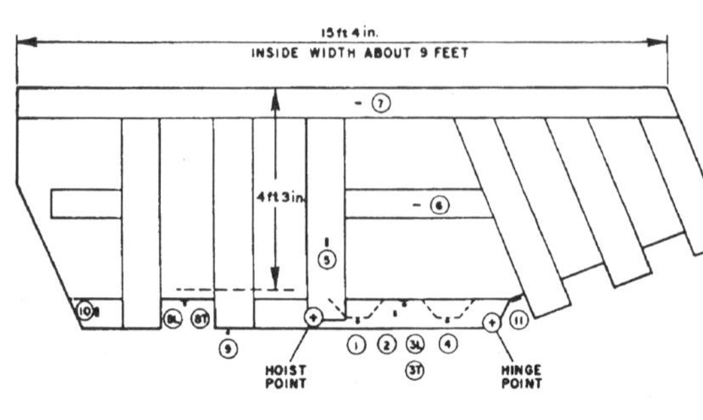

Fig 57: Strain-gauged dump-truck body

All parts of the body were subjected to falling rocks and strains were also measured whilst dumping individual rocks, in loading of the body to capacity; during terrain travelling under full-load and during dumping operations. The side of the body was also impacted by the shovel around gauge 7. Fig 58 shows the resulting maximum stress at each gauge location. While in general stress levels were low, those in 4 and 9 show high stiffening members can be stressed during severe loading. The high figure in 8 showed the effect of a heavy rock falling some 12 feet, its pointed edge causing a dished area in the floor 1/2 in deep x 10 in (12.7 x 154) diameter.

Tracklayer performance

The load distribution in motion of tracklayers differs from the static case and there is an influence from track-layout and suspension design which needs to be considered. Fig 3, at the start of this chapter, shows the change in relationship betweeen centre-of-pressure and track length as a vehicle moves forward in soft ground. Up to 20% increase in pressure can occur at the rear of the track. The effect on dimension y of increased load and consequent rear sinkage is as follows:

$$p_m = 2p_a\{2\text{-}3x/\{1+y[2[((3x/(1+y))\text{-}1]]^{1/n}\}\}$$

where p_a is the average track pressure without sinkage, l the nominal length of track contact, x the distance from the overturning axis to the dynamic centre of pressure, z the average sinkage and y (extension of track contact) is $z/\tan\,\alpha$, α being the angle of approach of track to ground.

Dynamic load transfer also affects rearward displacement of centre-of-pressure, due to external resistance forces acting about the rear overturning axis. With a tractor, the combined effect of drawbar resistance and rolling resistance can factor up the rear track ground pressure several times.

The assumption of even ground pressure along the length of the track is also challenged by the finite number of rolling wheels which can be arranged between the driving wheels. The effect of sinkage due to track deflection between the wheels, Fig 59, varies with different soils. Amount of deflection is normally decided as a compromise between too tight a track causing power loss and too loose a track being liable to throwing. It will generally not exceed 0.06 of dimension q; generally track tension induced by soil pressure on the free length does not involve

Maximum recorded stress in 1000 lb/in²

Test condition	Gauge number																									
	1		2		3L		3T		4		5		6		7		8L		8T		9		10		11	
	T	C	T	C	T	C	T	C	T	C	T	C	T	C	T	C	T	C	T	C	T	C	T	C	T	C
Static stress, full load of about 30 tons		1	<¼		<¼			<¼	½		½		1¼		1½			¼		<¼	<¼			<¼	¾	
Truck moving with full load, off highway	¼	¼	0	<¼	0	¼	0	0	¼	¼							<¼	¼	¼	¼	<¼	½				
Dumping full load	1	3	1	1	½	¼	½	¼	1	½	½	1	1	1	1	1	<¼	¼	1	½	2	1½	0	0	0	2
Side of body hit by 5 yd³ shovel	0	0							0	0	5	0	0	1	0	3				1	1	0	0	0	0	0
Tipping one boulder about 5 tons into empty body †	4	2	5	1	4	1	5	½	2	2							2	2	2	2	14	1				
Dropping one boulder (about 5 tons) about 9 ft into various parts of empty body	13½	4½	8	2½	1	9½	3	3	27	2	3	6¾	15	24*	5	7¾	28	2	31	2	36	6¼	1	3	1½	9½

* This figure is considered as faulty owing to a damaged lead wire.

† Boulders too large to pass through the shovel were picked up on top of the shovel and jostled off over the teeth of the shovel.

T Tension.
C Compression.

Fig 58: Recorded stresses at gauge locations

disproportionate friction loss, confirmed by:

$$T = bp_a q^2/8d$$

where T is track tension and d the deflection of the track between the wheels.

Implement resistance is another key factor affecting load distribution, on both tracked and wheeled machines. For a 4WD tractor, the forces involved are shown in Fig 60. The tractive effort H developed at tyre radius r is equal to the horizontal reaction of the soil and the driving torque reaction, partly reacted by the drawbar pull P; this results in an opposing torque about the axis of the rear axle to displace the centre-of-pressure from its static position by:

$$x=[(H_1r+H_2r)-P(r-h)]/W$$

increasing the rear wheel load to:

$$W_2=W[1-(g-x)/l]$$

Note that maximum displacement of load takes place not at maximum drawbar pull but when peak rolling resistance occurs. An even greater shift occurs in the crawler tractor, Fig 61, where:

$$x = [Hr+P(h-r)]/W$$

Work on bulldozers at the Aachen Technische Hochschule helped to establish the generally accepted optimum cutting angle for the blade of 30 degrees. It also found that maxmum output was obtained with a blade of involute form inclined at 15 degrees, resulting in the force distribution of Fig 62 in which the magnitude of V depends on the angle of the blade, depth of cut and soil friction properties; usually it is said to be 15-20% of the horizontal force. The centre-of-pressure height h is from 20-25% of blade height, according to depth of cut and speed. Fig 63 shows a proposal by consulting engineer L F Little[6], aimed at giving uniform distribution of load on traction surfaces. A positive downward force is induced by increasing rearward inclination of the blade and the dozer yoke is mounted on a linkage giving an instantaneous centre at ground level and mid-track. The objective is a thrust which eliminates rearward load transfer and increases the uniformly

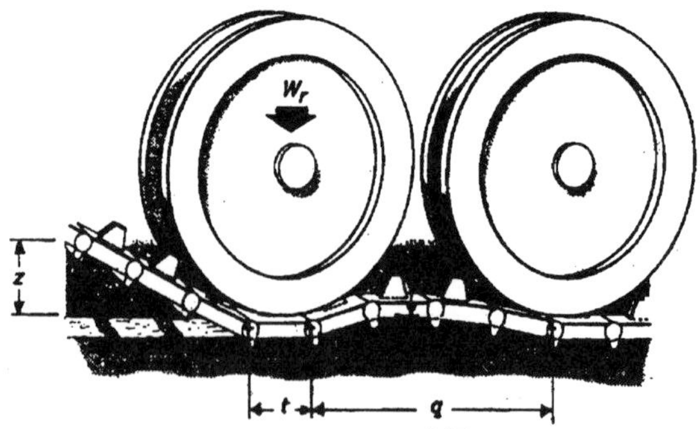

Fig 59: Track deflection and sinkage

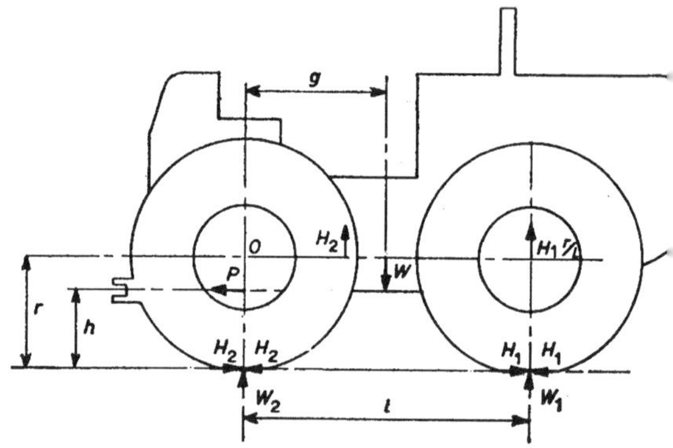

Fig 60: Forces on 4WD tractor

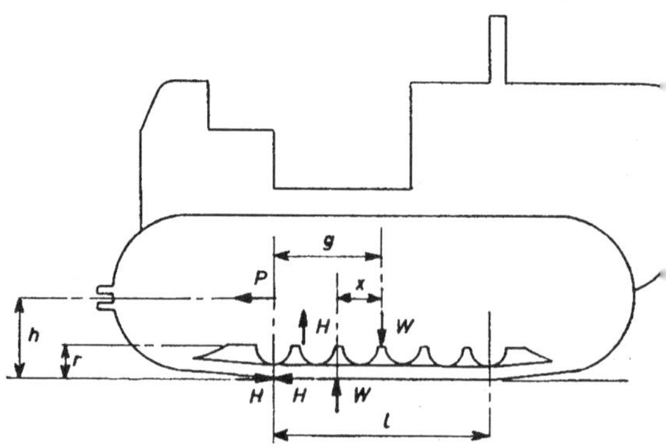

Fig 61: Forces on crawler tractor

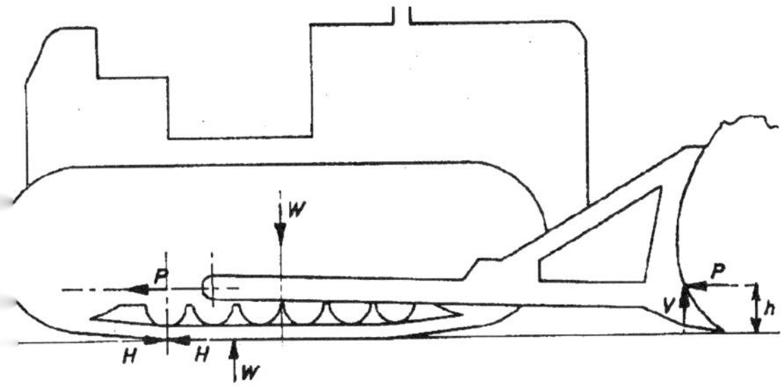

Fig 62: Operational forces on bulldozer

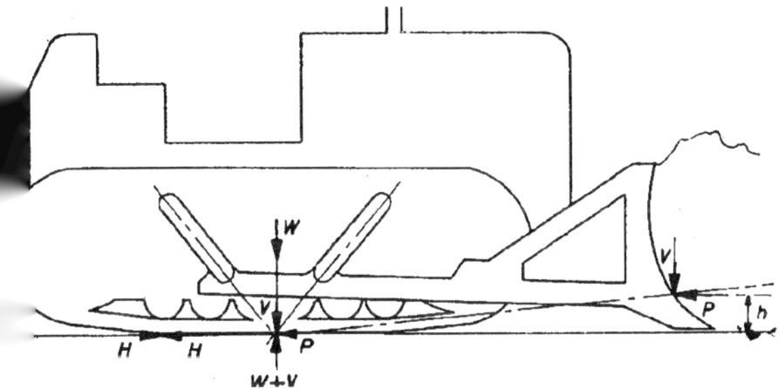

Fig 63: Self-propelled scraper

distributed load on the traction surfaces.

The success in recent decades of self-propelled scrapers is attributed by Little to their good load-distribution characteristics, Fig 64. Force P due to soil resistance at the cutting blade, plus rolling resistance is equal and opposite to tractive effort H. The direction of V is usually upward as more than the initial cut is collected and the bowl entry is arranged close to the laden centre-of-gravity so that it has little effect on weight distribution. The Richard motor scraper of Fig 65 uses a 4WD tractor coupled by a space linkage whose geometry provides a virtual hitch point close to ground level between the tractor drive axles and also maintains a constant weight distribution ideally suited for traction.

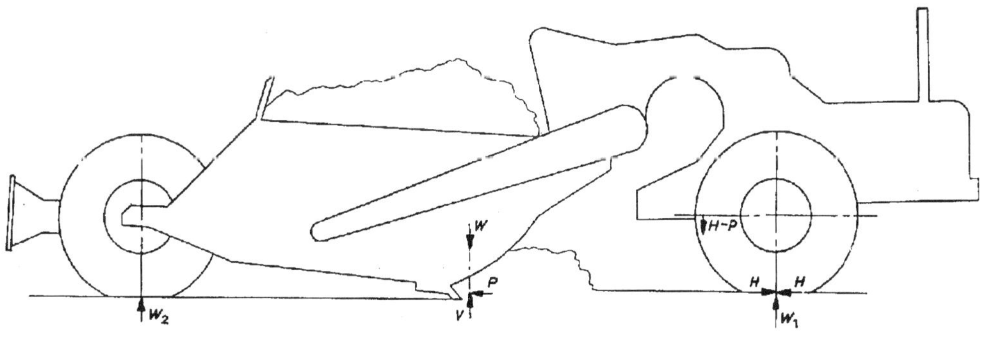

Fig 64: Little dozer-mounting system

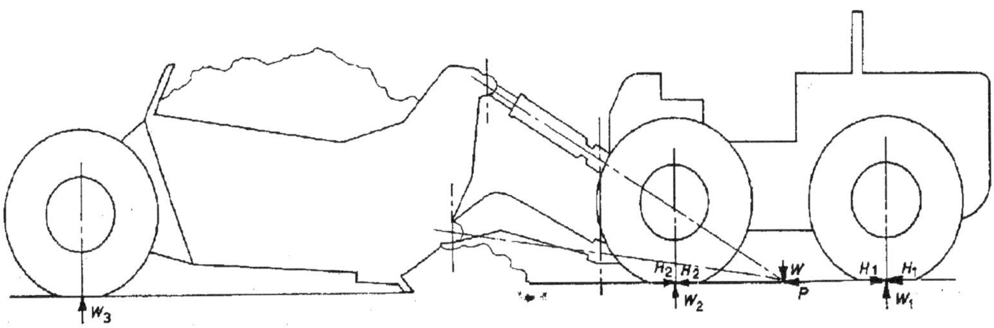

Fig 65: Richard motor scraper

Military vehicle design

The design of military vehicles is inevitably clouded in secrecy but the publication, over a number of years, of technical papers by the major research establishments, and builders, throws some light on the subject

Military vehicles are characterised by relatively high-speed cross-country operation and pioneer work by Cleare[7] has provided useful empirical data based on medium-weight high-speed military vehicles. This author argues that motion resistance is affected by the influence of track pressure on both the internal resistance of track and suspension and on the actual propelling power of the track, in turn affected by front or rear driving sprocket arrangement. Fig 66 records track tension effect for an 12 tonne vehicle with a rubber-bushed track and five 24 inch dual road wheels per side. The value recorded with track removed shows 63% of power loss to be due to track alone and the remainder due to suspension internal resistance.

The effect of road wheel diameter is shown in Fig 67. Generally the effects of wheel loading, tyre-section and rubber compound are small in relation to diameter effect; note that 32 in (0.812 m) is shown to be the limiting useful diameter for this particular vehicle configuration. The resistance on level paved

Table 1

Type of suspension: Drive sprocket	Top run of track	Weight (lb)	P factors (hp/ton/mile/h): Derived from towing	Propelling vehicle on level	Exerting drawbar pull	Climbing gradients
Front	unsupported	23 300	0·24	0·24	0·3	0·4
Front	supported	38 000	0·32	0·33	0·39	0·4
Rear	supported	47 500	—	—	—	0·21
Rear	supported	82 500	0·22	0·22	0·22	—
Rear	supported	102 700	—	0·26	0·24	0·25

Table 2

Terrain P factors (hp/ton/mile/h)

Concrete	0·24	Sand (dry)	0·6
Hard-packed gravel	0·30	Loam	0·58
Hard turf	0·38	Mud	1·0

Table 3

Gear	Inner	Outer	
5	0·941 V	1·059 V	Actual sprocket speeds
4	0·921 V	1·079 V	
	0·875 V	1·125 V	
2	0·800 V	1·200 V	
1	0·493 V	1·507 V	

Fig 71: P factors in Figs 66-70 (top); for other terrain (centre); ratios for 74-75 (bottom)

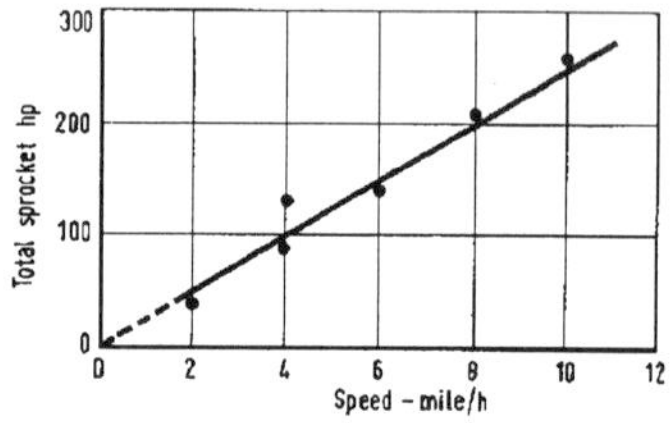

Fig 72: Sinkage effect of medium-weight vehicle

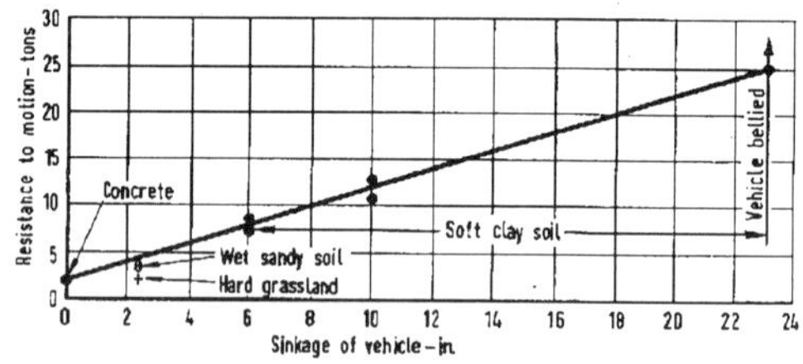

Fig 73: Sinkage effect of heavy vehicle

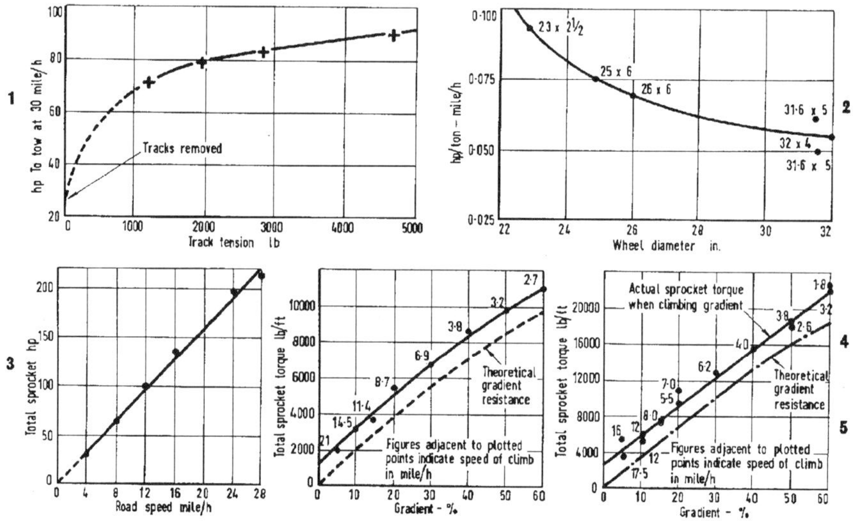

Figs 66 to 70: Cleare's empirical data (figures labelled to left and right of group)

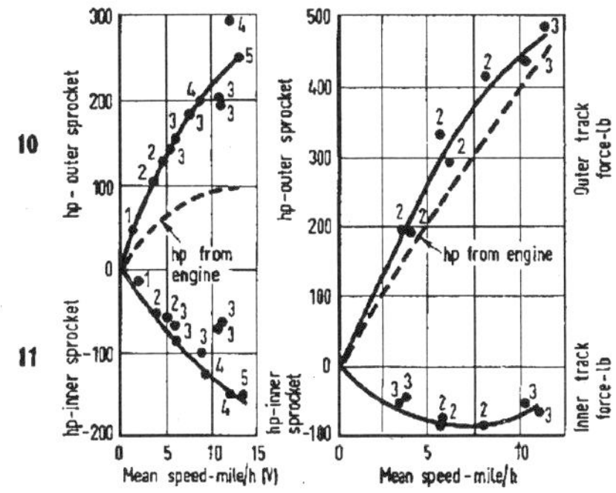

Figs 74 (left) and 75 (right): Resistance due to steering losses

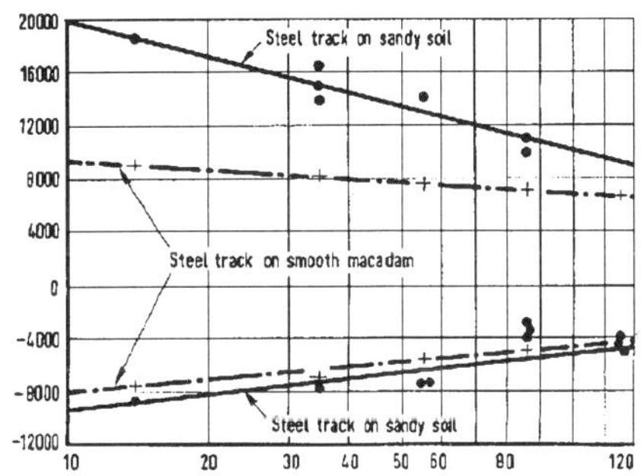

Fig 76: Resistance due to track slewing

surfaces is shown in Fig 68 based on data taken from strain gauge dynamometers fitted to the sprocket drive. By expressing results in non-dimensional form, propulsion resistance for various classes of vehicle can be compared by a `P' factor of bhp/ton/mile/hour. Figs 69 and 70 record gradient resistance which include additions to the resolved component of force parallel to the slope, from internal and terrain resistances. The two figures refer to vehicles of 10.4 and 16.95 tons (10.23 and 16.68 tonnes) respectively on normal road surfaces, sprocket radii of 0.797 and 0.944 ft (0.24 and 0.29 m) applying respectively while inset numbers are climbing speeds in mph. `P' factors for some representative vehicles are provided in the table of Fig 71.

A heavier vehicle of 37 tons ((37.6 tonnes) was used to obtain the resistance due to soil-compaction, hull drag and bulldozing effect in soft terrain shown in Fig 72, a secondary effect being due to track slip. The graph applies to a vehicle in 9% moisturised sand while the table in Fig 71 (centre) gives `P' factors for other terrains. A yet heavier vehicle of 50 tons (50.8 tonnes) is used to show the motion resistance due to sinkage in Fig 73. The effect of resistance due to steering losses is shown as Figs 74 and 75, using sprocket torquemeters to determine intertrack power transference during turns with a regenerative steering system. Here a 30 ton (30.5 tonne) vehicle is used as the subject, having a 147 in (3.8 m) track base and

1 Lubricating nipple
2 Breather
3 Torsion bar (left hand)
4 Torsion bar (right hand)
5 Torsion bar anchor lock
6 Axle arm
7 Bearing housing
8 Seal
9 Bearing
10 Bearing
11 Seal
12 Tab washer
13 Ring
14 Retaining ring
15 Rebound pad
16 PTFE washer
17 Taper nut
18 Split ring
19 Bump stop
20 Retaining ring
21 Stub axle nut
22 Socket headed screw
23 Stub axle
24 Serrated plates
25 Inner seal seat
26 Seal
27 Wheel hub (Stn 5)
28 Washer
29 Hub cover
30 Nut
31 Filler plug
32 Bearing
33 Gasket
34 Bearing
35 Outer seal seat
36 Axle arm
37 Seal seat (rear)
38 Seal seat (front)
39 Cover plate
40 Wheel nuts
41 Hub cover nuts
42 Hub cover
43 Intermediate road wheels

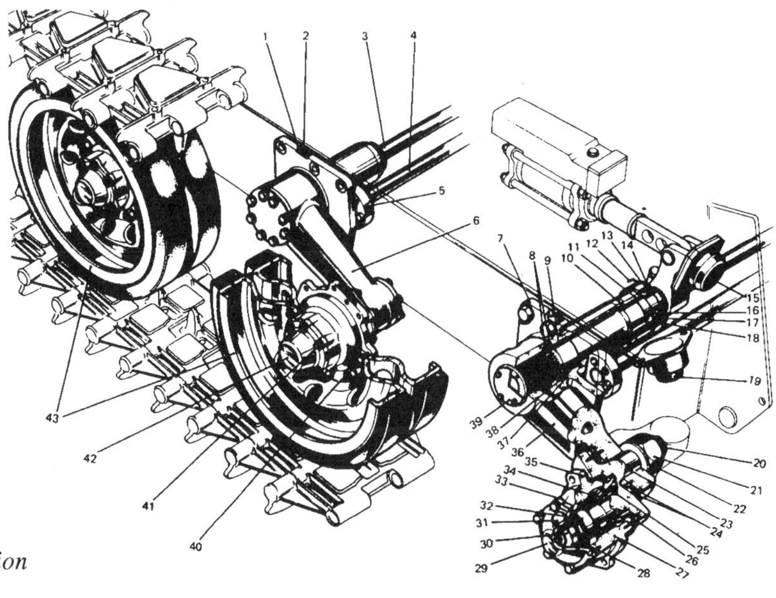

Fig 77: Scorpion track suspension

100.25 in (2.54 m) centres, turning on both tarmac and mud surfaces. Inset numbers refer to gear ratios, which control the minimum turn radii; see the table in Fig 71 (bottom). Note the reduced power generated on the mud surface, where vehicle sinkage was 4 in (102 mm). Finally, Fig 76 shows resistance arising from reduced track forces in slewing, which increases with increasing turn radius. The graphs show tangential sprocket force (track pull) for various turn radii and at zero-radius (neutral turn), track forces are equal and opposite.

Track suspensions have traditionally used torsion-bar systems such as the staggered arrangement for the Scorpion AFV shown in Fig 77. The layout suits parallel sided hulls as very little protrusion is required. Top rollers are not required in this particular configuration because the track cannot foul other suspension components — as these are always protected by the wheels. The disadvantage of the layout is lack of space for outside-mounted dampers. The figure shows how internal mounting is used to solve this problem; the damper, Fig 78, also provides a lock-out facility for earthmoving and this, rather than maximum damping pressure, governs its overall dimension. In the Combat Engineer Tractor version of the vehicle, Fig 79, forward speeds of 55 kph are typical and in this role the idler wheel and track adjuster is acting as a sprung wheel. In the earthmover mode the idler wheel is locked out, it being required to crush rubble and allow spoil to fall freely away so as to minimise the chance of track shedding. Good cross-country performance requires long wheel travel and high energy absorption and the torsion bar spring is limited in its effectiveness by the vehicle width; however, performance can be enhanced by the addition of torsion tubes, Fig 80.

An MVEE (now RARDE) design, Fig 81 increased the effective torsion bar length by 50% using a concentric bar and tube system. Some consider Hydrogas springing to be a better solution as these have the advantage of being mountable outside

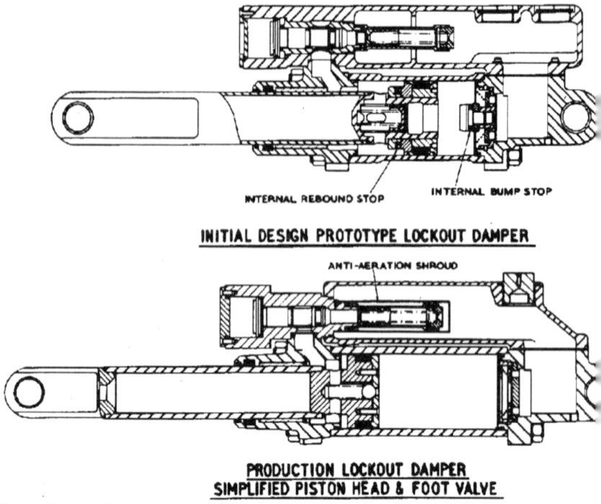

Fig 78: Scorpion damper

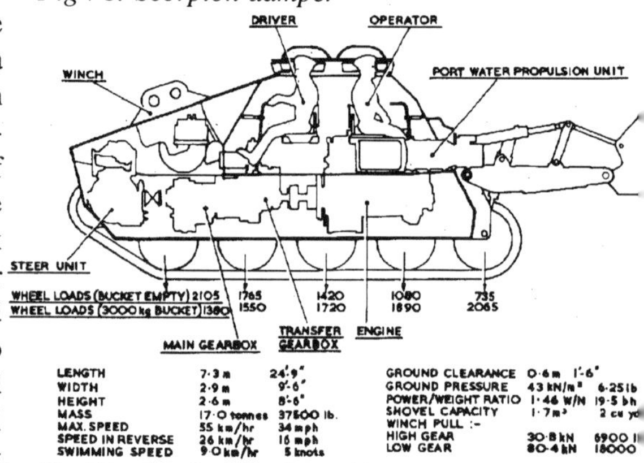

Fig 79: Combat Engineer Tractor

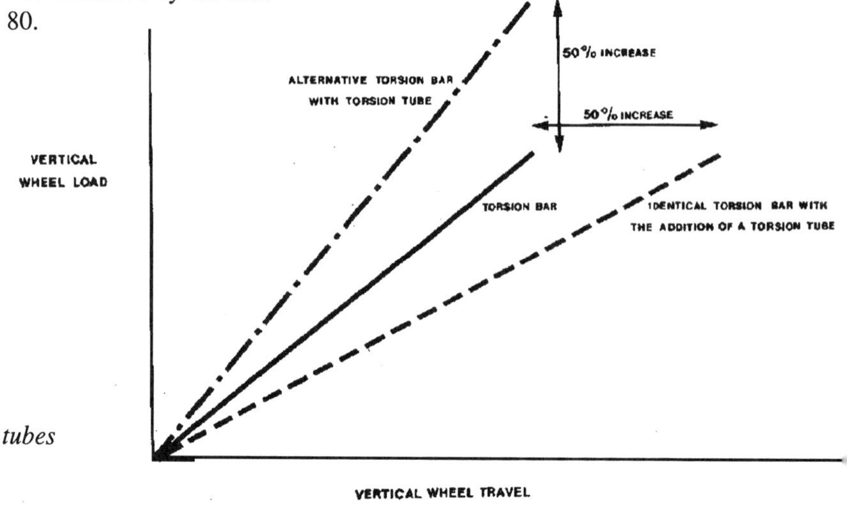

Fig 80: Effect of torsion tubes

the hull. They are claimed to be lighter than rival systems of equivalent damping performance. Fig 82 shows the system fitted to the Challenger tank; gas pressure at static wheel load (normal ride height) is 12.4 kN/m^2 rising to 54 kN/m^2 at full bump. Suspension weight on the Challenger, excluding hubs, is 5.5% of vehicle weight.

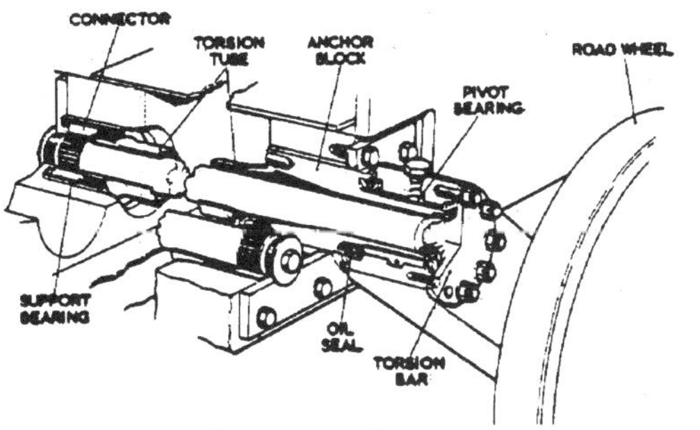

Fig 81:Torsion tube suspension

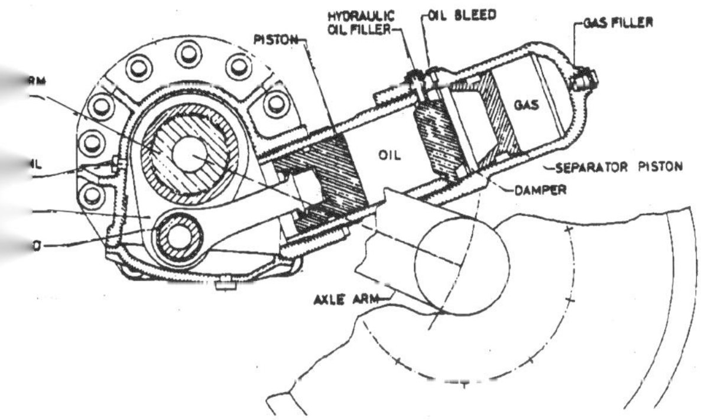

Fig 82: Hydrogas on Challenger

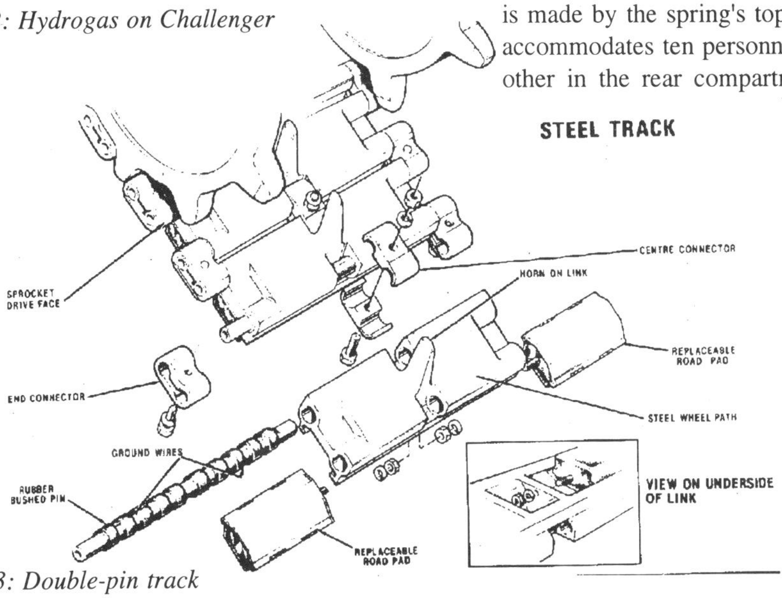

Fig 83: Double-pin track

The form of track favoured by western countries for their military vehicles is shown in Fig 83 and known as the double-pin rubber-bushed system, the cast steel body having the bushed pins inserted into the body lugs. The links are connected by two steel end connectors and one centre connector, the two end ones transmitting the sprocket drive. A double-pin track is considered necessary for a main battle tank since a single pin track cannot have sufficient rubber in the track bushes. This involves greater weight unless an aluminium alloy body is used; if so, it is necessary for the track horn to be included on the centre connector as an al-alloy horn could not withstand the abrasive wear.

Wheeled-vehicle/tracked-mobility

The provision of high vertical wheel travel by using advanced independent suspension systems has been exploited by the Irish company, Timoney Technology who have created high-mobility wheeled military vehicles exemplified by the 6x6 Armoured Personnel Carrier, Fig 84. Its wheel hubs are also fitted with a tyre inflate/deflate system which allows mobility to be geared to terrain requirements. This vehicle is also fitted with steerable water propulsion propellers to allow amphibious operation without special preparation.

The suspension allows vertical wheel travel of 400 mm and as the spring-damper units act on the lower wishbone minimum intrusion into crew space is made by the spring's top anchorage. The vehicle accommodates ten personnel, seated facing one another in the rear compartment of the body; three further personnel can sit forward-facing in the centre compartment. The driver sits slightly ahead of the front wheel centreline to one side of the 224 kW Cummins engine which occupies the equivalent space on the other side of the vehicle front-compartment. The three axles use similar drive modules, Fig 85; in the front axle the second output (through drive) is blanked off and in the rear axle the water propulsion system drive is installed.

A four-axled vehicle, the MWG ATV (Fig 86) is also being used to challenge tracklaying military vehicles. This is a 5.5 tonne GVW vehicle with carrying capacity of 2 tonnes at the same time as towing a 2 tonne trailer. The 4-litre Perkins Phaser of 110 bhp drives through a 3-speed automatic gearbox into two Land Rover transfer gearboxes, with high/low ratios and manually-engaged differential locks. The third and fourth, driving, axles also have cross-differential locks. Suspension is by long-travel coil springs and tyres are 305/75 x 16s with self-cleaning treads and are inflated to 7.7 lb/in^2 (53 kN/m^2). Maximum speed is 70 mph, gradient-ability 45 degrees and maximum wading depth one metre.

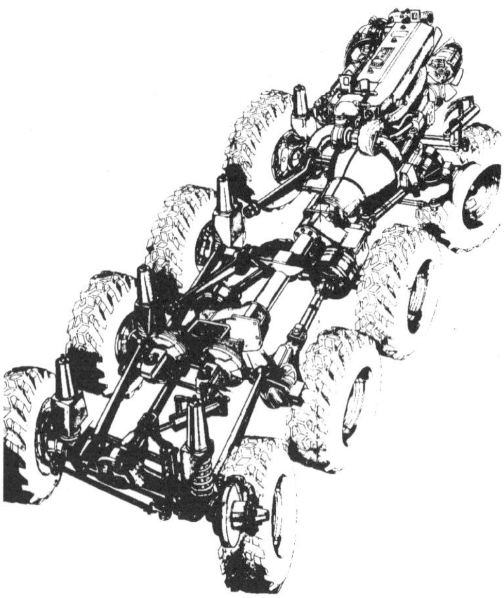

Fig 86: MWG ATV

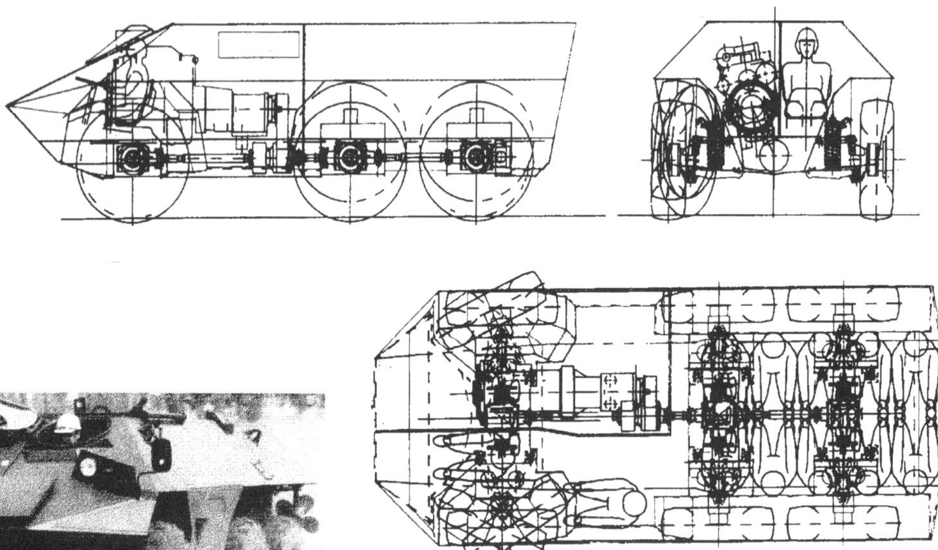

Fig 85: Timoney personnel carrier drive layout

Fig 84: Timoney personnel carrier

Chapter 12: Light special-vehicles

Specialist road cars

The increasing disenchantment shown by the general public for the stereotyped designs from the volume producers is in turn leading to improved prospects for specialist builders who produce individualised designs.

Chapter 8 dealt with the influence of the mechanical package on body configuration. Equally, of course, the chosen specialist road car function will affect the design of mechanical systems. The earliest road vehicles, with modestly powered engines, usually had their power units tucked beneath the rear seat. As powers increased, power-units moved forward to the position once occupied by the horse — in the days before the horseless carriage. Here it has remained despite the almost universal adoption of rear-engine/rear-drive by racing cars. The breakthrough for specialist road cars came with the so-called mid engine position pioneered by the Lamborghini Miura, Fig 1, in which the dictates of mechanical efficiency and body aerodynamic function began to coincide. The layout allowed the freedom to lower the front end while accommodating a

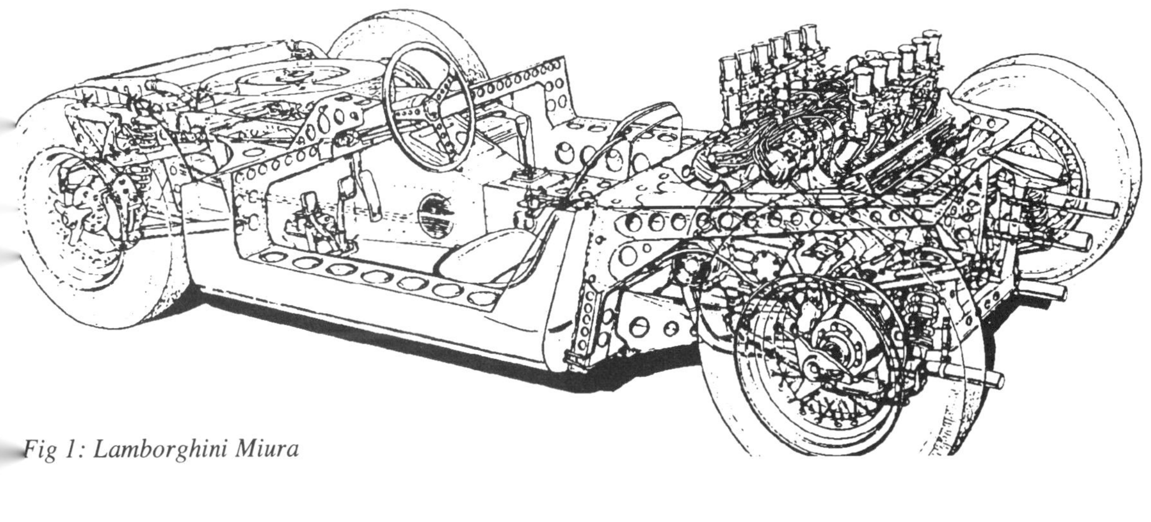

Fig 1: Lamborghini Miura

large transverse engine towards the rear where the body upswept into a 'wedge' profile.

The use of integral structures on many of the latest specialist sports cars also allows body function and user comfort to dictate the vehicle design — without having to compromise wheel position, say, to suit a particular chassis layout. As noted in the section on vehicle directional control in Chapter 15, an early stage in design for good vehicle handling is the determination of roll centre heights given by the wheel suspension systems, front and rear. These are one of the first things to be entered on a general arrangement drawing of the complete vehicle, having sketched in the packaging of occupants and major mechanical units on the basis of preliminary performance-prediction and ergonomic-dimensional calculations. The suspension linkages can next be laid out and the vertical wheel travel determined — wheel clearance envelopes being a major constraint on the overall vehicle package. The undertray line can next be drawn in to determine ground clearance at full bump motion of the suspension.

The specialist car builder can at this stage

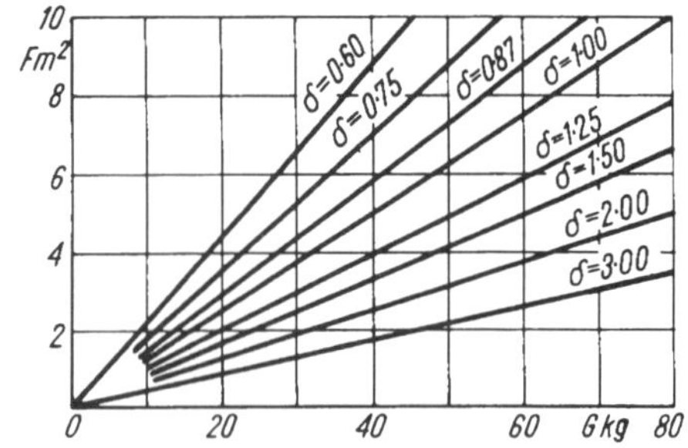

Alignment chart for the weight of steel sheets of different thicknesses. For small areas, 0–3 m², a larger scale graph is used

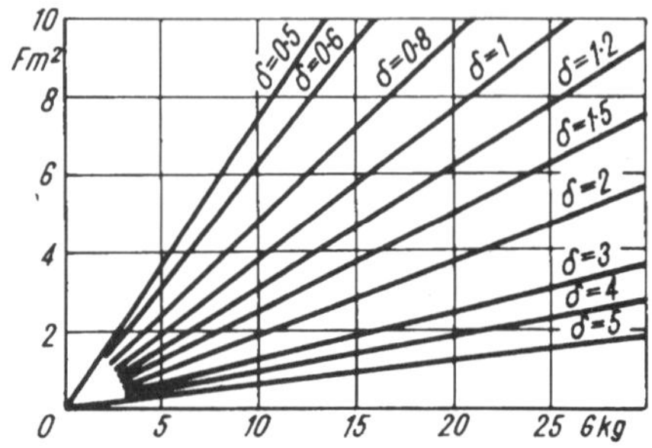

Alignment chart for the weight of aluminium sheets of different thicknesses

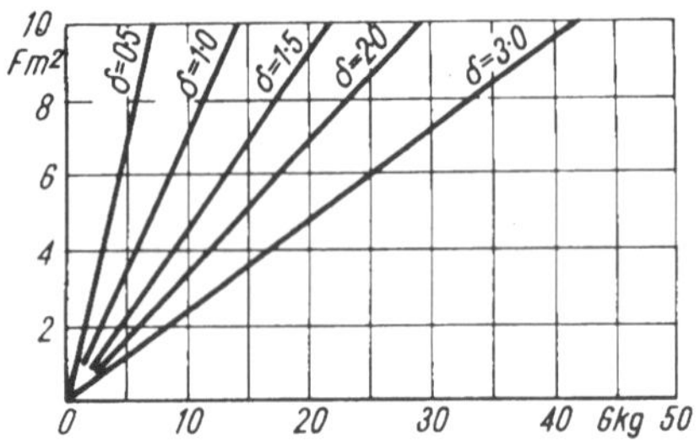

Alignment chart for the weight of trim material of different thicknesses

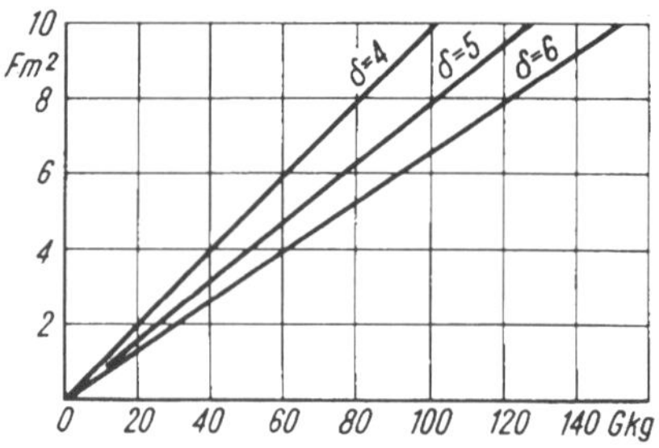

Alignment chart for the weight of glass sheets of different thicknesses

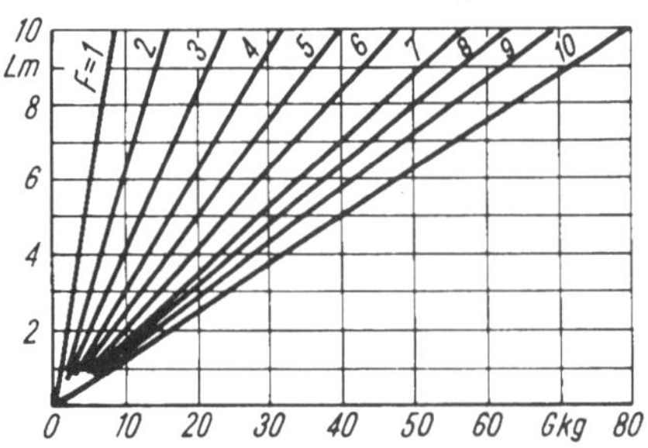

Alignment chart for the weight of various sections

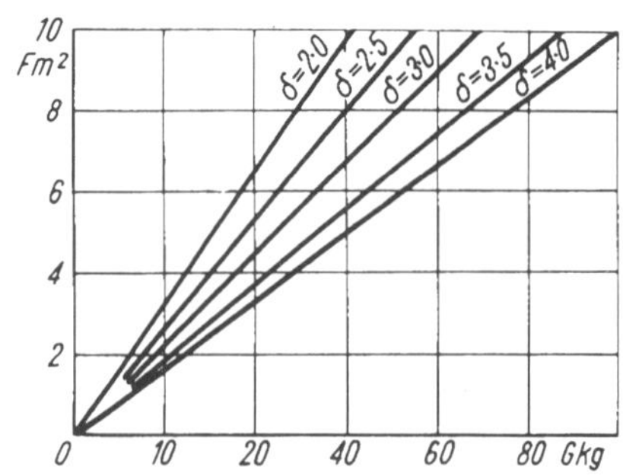

Alignment chart for the weight of rubber mats of different thicknesses

Fig 2. Alignment charts

benefit considerably from ergonomic techniques for positioning instruments and controls (Chapter 8). This helps to position the major bulkheads of the vehicle structure at seat-back and dash-panel locations. A most important adjunct to this is the accurate prediction of weights for each element of the vehicle structure, power-train and running-units. For this purpose, alignment charts as shown in Fig 2 can provide a valuable guide to the use of different materials in sheet and rail form. This will lead to an early appreciation of front-to-rear weight distribution upon which ride and handling performance is dependent — as well as the selection of tyre size and construction. Fig 3 shows a tabular approach to gravimetric analysis and the use of moment equations to determine axle loadings.

Trendsetting specialist cars

Aside from the Miura design mentioned above a number of recent introductions have marked substantial new directions in specialised road cars. Jaguar's XJ220, Fig 4, is a road car which is almost on the boundary between road and track design — being originally conceived as a vehicle which could compete successfully in Group B racing, had regulations allowed. Crucial to the design was the means used to achieve 4-wheel drive — a quill shaft between cylinder banks of the vee engine and linking the engine via torque tube to the front drive, Fig 5. The car also had to achieve very high speeds without generating aerodynamic lift and without resort to 'inelegant' wings and spoilers. A 48-valve variant of the Jaguar V12 engine was developed to power future variants of the

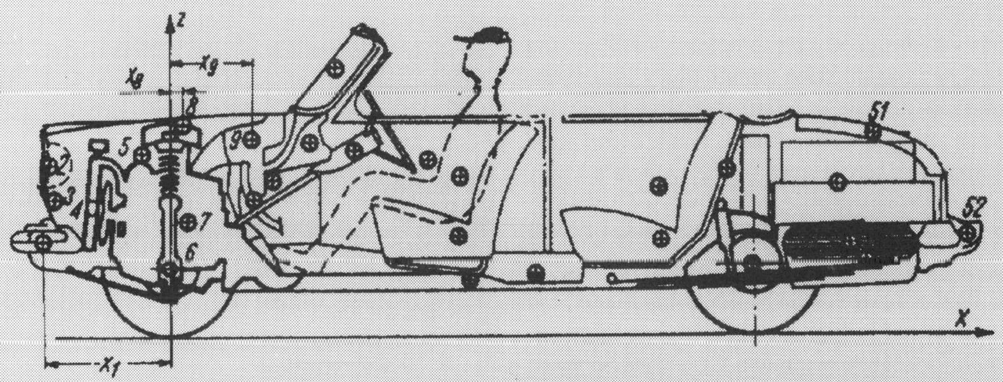

1—front bumper; 2—headlamps; 3—radiator grille; 4—radiator with water; 5—front mudguard; 6—front suspension complete; 7—engine, etc., up to 52—rear bumper

Lp.	Element	Q (kg)	X (m)	ΔM (kg.m) −m	ΔM (kg.m) +m	Correction	Remarks
1	Front bumper	4	−0·62	2·5			
2	Headlamps	6	−0·58	3·5			
3	Grille	5	−0·57	3·0			
4	Radiator with water	15	−0·50	7·5			
51	Boot lid	9	+3·10		27·8		
52	Rear bumper	4	+3·50		14·0		

$\Sigma Q = G_c$ $\quad \Sigma \Delta M = M$

Fig 3: Results of gravimetric analysis

vehicle giving an eventual potential of 700 bhp. This required purpose-designed rear transaxle and front final drive units to be produced in conjunction with FF Developments. Viscous control units are used in the centre, rear and, optionally, front differentials. The centre differential is incorporated within the rear transaxle, forward and concentric with the gearbox output shaft. Some 70% of the torque passes further forward to the rear final drive pinion while the remainder passes to a transfer gear set into a shaft passing above the rear differential and into the vee of the engine. The mid-engined layout has the gearbox overhanging the rear final drive. By designing the front final drive with a negative hypoid offset, and by slightly inclining the drive-line downwards towards the front, roughly equal heights are achieved for front and rear final drive shafts. So, unlike many 4-WD vehicles, both power unit and drive to the front lie on the centre line of the car. This allows well forward positioning of occupant's feet within the wheel arches. The body was designed with completely flat floor and fully enclosed transmission tunnel, Fig 6. Box member bulkheads join the large sill sections and bonded aluminium alloy was employed. Double curvature glass forms a structural part of the vehicle's upper skin.

Fig 4: Jaguar XJ220

Fig 5: Drive layout of XJ 220

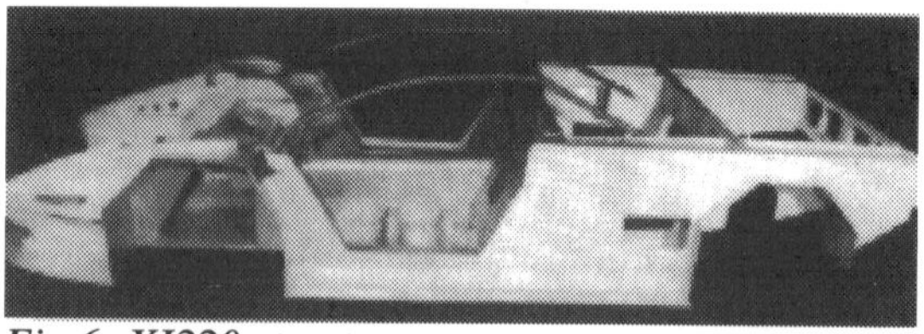

Fig 6: XJ220 structure

Moving down the power scale, Volvo's LCP 2000 concept car, Fig 7, also broke new ground in design. This is an extension of the sports estate type vehicle pioneered by Ogle for their Reliant Scimitar. LCP stood for Light Component Project and stressed

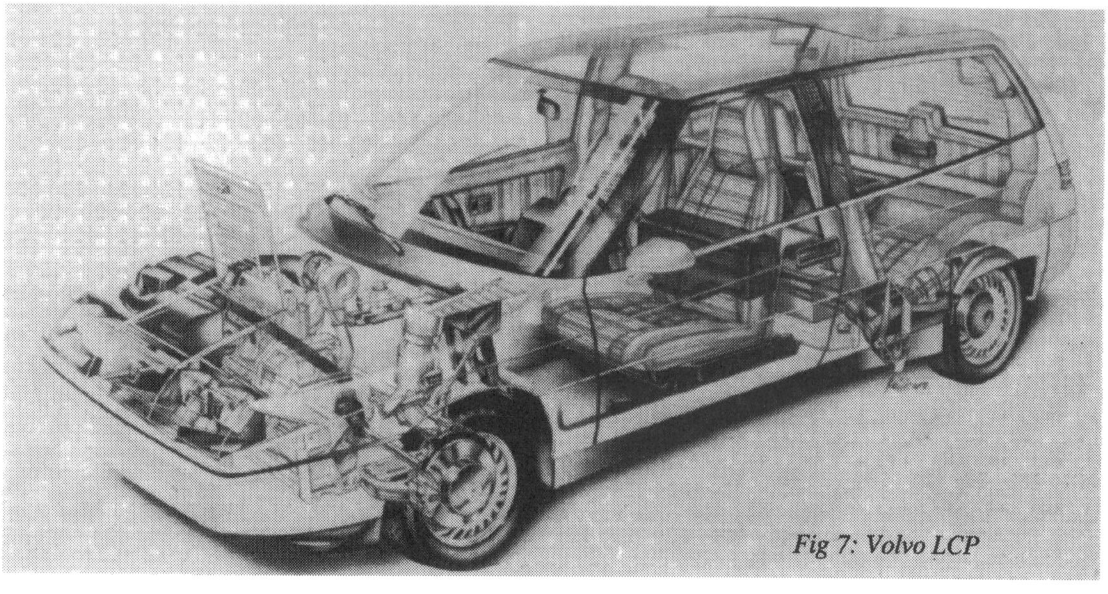

Fig 7: Volvo LCP

the overwhelming importance of light weight in achieving low fuel consumption and exhaust emissions. Dry weight of 645 kg was achieved against an industry standard of 830 kg for a similar specification vehicle — resulting in a diesel fuel consumption of 100 mpg from a 3-cylinder 1300 cc engine designed by Ricardo for multi-fuel capability. The vehicle base structure was made from aluminium alloy using the weld bonding technique to reduce spot welds from 4000 to 500. Skin panels were in thermoplastics including side windows in clear polycarbonate. Most of the major power train castings were in magnesium alloy and carbon fibre was used to reinforce the slender door arches and other parts of the plastic skin. A key departure was the use of a 'platform' structure, to which other major assemblies of the vehicle were attached as modular units.

In the 'mini' car class, Fiat's experimental CNR project car, the X1/75, broke new ground for that end of the specialist road car spectrum. The metal base structure is of interest for its stiff-jointed space frame construction used to attach removable plastic panels, Fig 8. Because of the wide load variation on the rear axles of mini-cars, according to passenger load, the designers chose a rigid axle with levelling compensation using air springs. The powertrain comprised a 2-cylinder engine coupled to a CVT gearbox arranged for front-wheel-drive. Key dimensions for the vehicle were 3,43 m long x 1.525 m wide x 1.41 m high with 2.27 m wheelbase.

SPVs for the sick/disabled

With some 10% of the population having a permanent mobility disability, and a substantial proportion at any one time suffering temporary illness or accident injury, this special vehicle category is of substantial importance. The specification and conversion of vehicles in the medium-sized car to panel-van size category is examined here (the design of individual invalid carriages [wheel chairs] is considered in a later section).

The Institute of Consumer Ergonomics at Loughborough University of Technology have published a guide called 'The ins and outs of car choice'

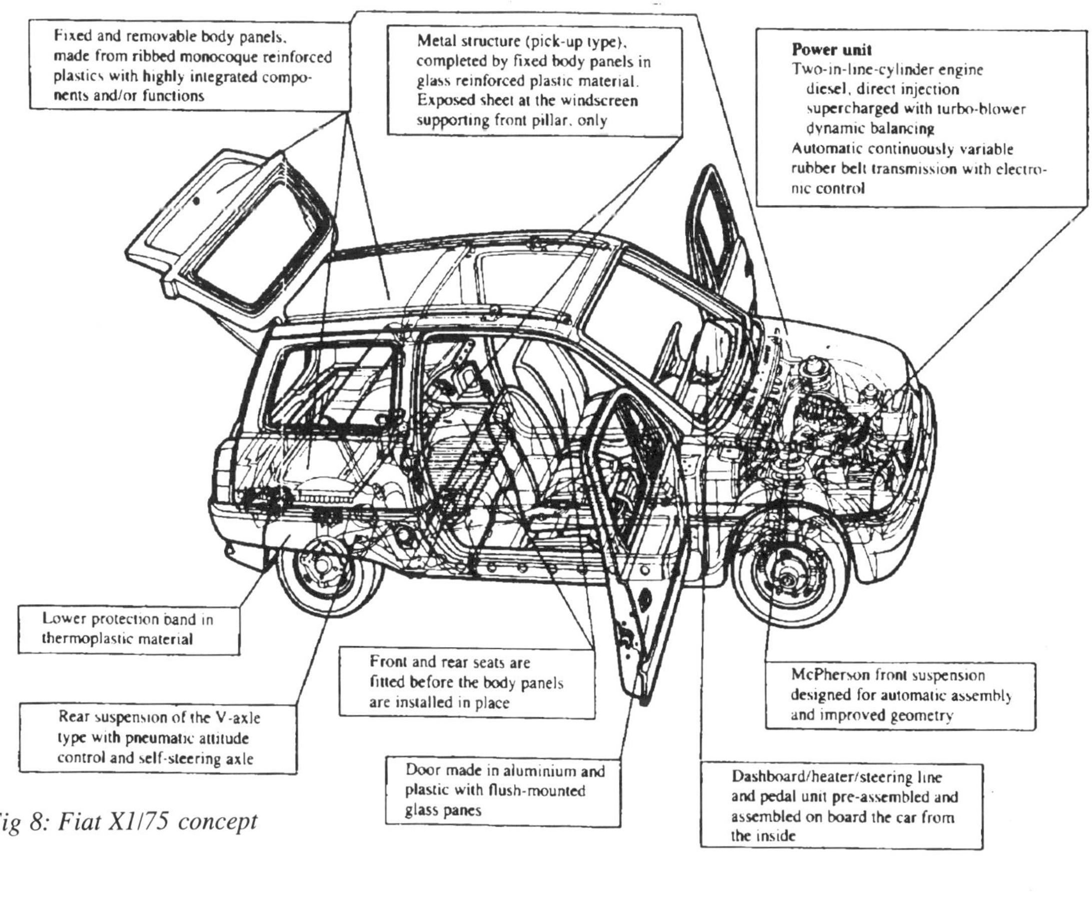

Fig 8: Fiat X1/75 concept

which found a number of key entry/exit problems with conventional cars. Problem areas for elderly and arthritically handicapped people centre around: low seat height; low door frame height; high door sill height; wide door sills and deep floor-wells — also narrow door opening apertures. Other problems are door handles, supports, seat adjustment, seat belts, centre console, handcontrol positioning and height of the boot sill. Fig 9 shows the supports commonly grasped in entry and exit with the percentages applying to each. These were observed in carefully controlled trials on high volume produced cars which led to dimensional recommendations as follows: minimum sill height above ground 270 mm; maximum footwell depth 60 mm; recommended seat height 500-550 mm above ground, 250 mm above sill; minimum A-post to seat front distance 440 mm; optimum door opening angle 70 degrees; minimum cant rail height from ground 1370 mm; maximum distance from seat edge to sill outer edge 160 mm; minimum distance from A-post to B-post 980 mm.

Servo-controls for disabled drivers

The expansion of on-vehicle electronics for conventional cars has also some promise for the disabled in the availability of improved servo-controls. Workers at Birmingham University have produced a system adaptable to a wide range of disabilities. A prototype vehicle has been built with a two-axis joystick. The search for low cost electric actuators led the researchers to use a high performance printed-armature DC motor to drive throttle mechanism and automatic gearbox linkages directly. For controlling the brake circuits a high pressure hydraulic system is regulated by a simple modulated solenoid valve. The most difficult part of the project, conducted in the Low Cost Automation Centre, is the steering servo-control for which response has to vary according to conditions and the rate of response has to be adaptive (these variables also need to be programmable). A small gearbox fitted on to the standard steering column of the adapted Ford Escort operate the servomechanisms.

Proprietary control systems are also available from Feeny & Johnson Ltd who have historically been in the business of vacuum servo foot brake assistors. The company's hand control conversion system for cars and commercial vehicles is shown in Fig 10. Throttle and brake levers have handles which form concentric arcs within easy reach of a hand on the steering wheel. The company were also involved in a project led by Reading University and the Transport and Road Research Laboratory in which only electric connection is provided between hand control and steering mechanism. A gearbox on the steering column contains three feedback potentiometers driven by anti-backlash gears. Two electric motors turn the column through worm gears and a 100 mm control handle is moved through a 40 degree arc to control them. On moving the handle, the potentiometers send signals to the motor. A photo pulse generator behind the vehicle's speedometer provides change of gain to the control handle as the speed increases.

Car and taxicab conversions

An alternative to modifying the controls of a conventional car is to provide access into the car for a wheelchair which is then clamped into the driving

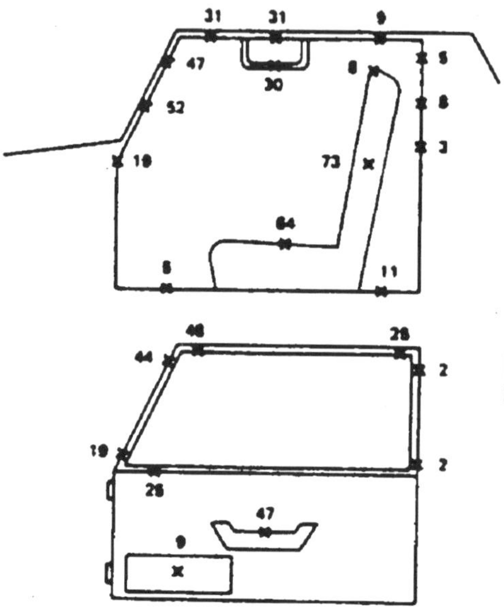

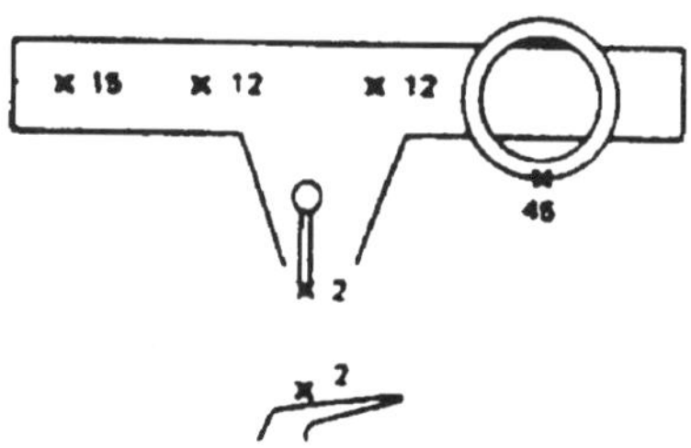

Fig 9: Study of supports grasped in entry/exit

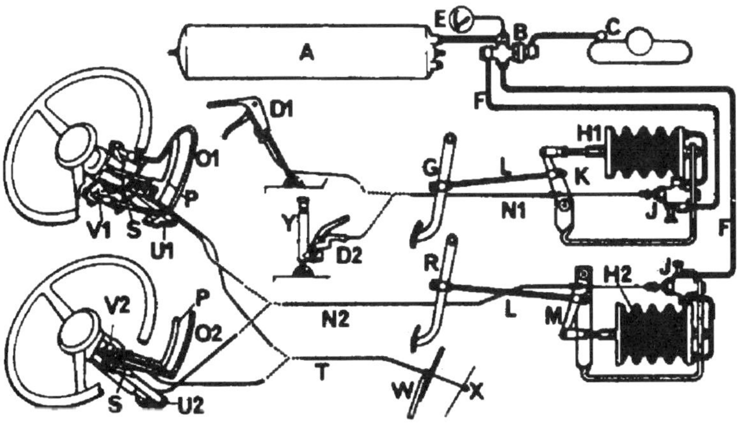

A Vacuum reservoir B Non-return valve C Induction fitting D1 Clutch handle std D2 Clutch handle tee E Vacuum gauge F Vacuum hose 5 in. I.D. G Clutch pedal H1 No. 1 Power unit H2 No. 2 Power unit J Control valves K Clutch relay lever L Rods C/W sliding links M Brake relay lever N1 Clutch control cable N2 Brake control cable O1 Brake handle Std O2 Brake handle slimline P Accel. control handle R Brake pedal S Accel. Pre-set lever T Accel. control cable U1 Column clamp block Std U2 Column clamp block "U" pin V1 Pivot claw V2 Pivot block slimline W Accel pedal X Accel cable retainer Y Clutch handle sleeve

Fig 10:Servo-assisted hand-actuated main brake

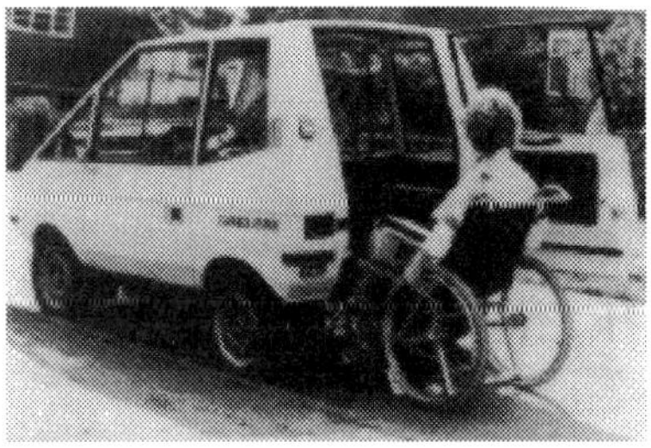

Fig 11: Elswick Envoy

position so that the user never has to leave his/her seat. This approach was used on the Elswick Envoy shown in Fig 11; for severe disability cases; this could be accompanied by a push-pull mechanism as shown in Fig 12 which achieves braking and acceleration from one handle. A variety of alternative devices are shown in Fig 13 from the catalogue of Petri + Lehr GmbH.

A number of efforts have been made to modify the standard London taxicab to make the vehicle more acceptable to the wheelchair-bound. Already the regulations governing basic layout fit the vehicle well for entry and exit by the less severely disabled. The Carbodies Ltd London Cab (Fig 14) now has many updated features which make conversion attractive. The vehicle features five speed all-syncromesh overdrive gearbox and has four wheel hydraulic full servo braking; key dimensions shown demonstrate the potential for conversion.

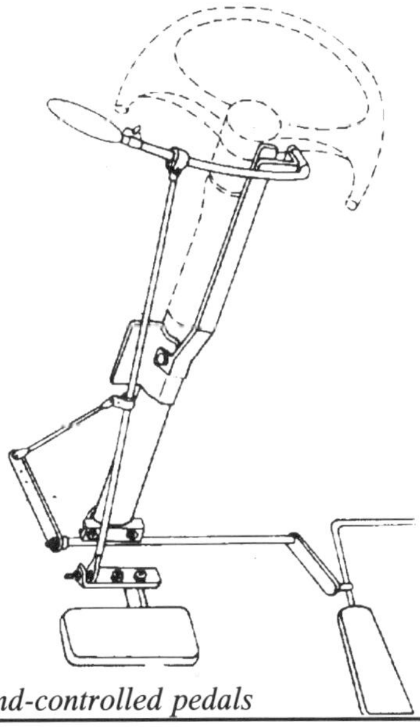

Fig 12: Hand-controlled pedals

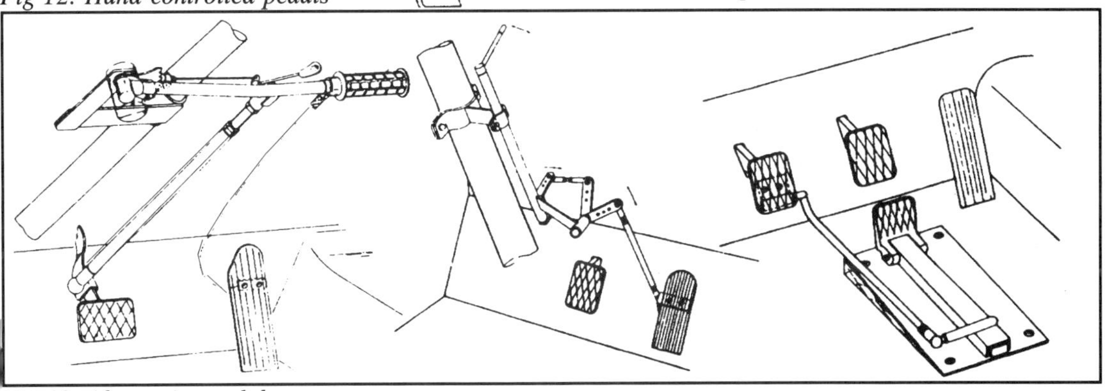

Fig 13: Alternative pedal movers

Ambulance design

Vehicles for the temporarily disabled have been the subject of considerable controversy in their history. Opinion differs in the medical profession as to whether ambulances should provide a heavier 'mobile hospital' type vehicle or a lighter minimally equipped one that would speed the injured between the scene of the accident, or serious illness, to the hospital. In earlier years this contrast could be seen in France's converted cars compared with Britain's converted trucks. More recently there has been a drawing together as designers strive for the optimum arrangement. Fig 15 shows a typical contemporary ambulance which has a Taurus body on an Iveco Ford 35.8 chassis. The 'sitting-case' model illustrated has twin 'one-third' doors at the rear and up to ten people can be carried. Fig 16 shows the inside of a car-derived van conversion, for a stretcher case, by KFB Extramobile of West Germany. Both of these contrast strongly with an early 1970s design proposal shown in Fig 17. This was by Dr Bothwell of the Birmingham University Accident Research Unit who believed that as first-aiders are not accident surgeons, high speed easy-access ambulances were needed to cut down the journey time from accident location to operating table. Key features of the Bothwell design were a front mounted underfloor flat engine, pitch-free front/

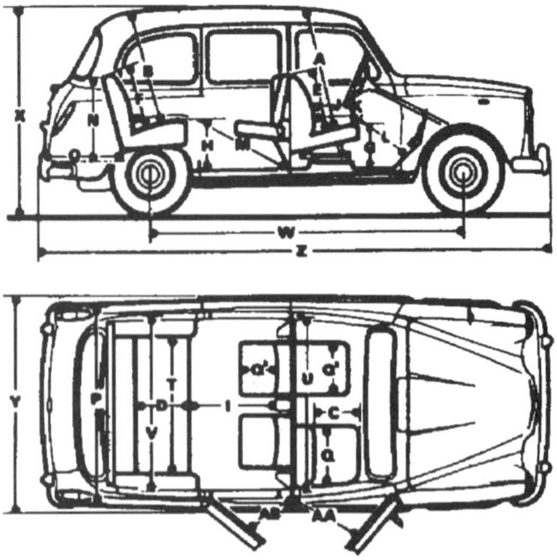

*A (Max.) 36½" (0.93 m.)	*A (Min.) 32" (0.81 m.)	*B 38" (0.97 m.)	*C 16" (0.41 m.)	*D 17½" (0.44 m.)	*E 18" (0.46 m.)
*F 20" (0.51 m.)	*G (Max.) 17" (0.43 m.)	*G (Min.) 13½" (0.34 m.)	*H 14" (0.36 m.)	*I 31¼" (0.79 m.)	*J (Max.) 15½" (0.39 m.)
*J (Min.) 10" (0.25 m.)	*K (Max.) 8½" (0.22 m.)	*K (Min.) 4½" (0.11 m.)	*L (Max.) 40" (1.02 m.)	*L (Min.) 35" (0.89 m.)	*M 57" (1.45 m.)
N 26½" (0.67 m.)	O 25½" (0.65 m.)	P 65" (1.65 m.)	*Q 20" (0.51 m.)	*†Q2 17" (0.43 m.)	*†Q3 13" (0.33 m.)
*T 45½" (1.15 m.)	U 55" (1.40 m.)	V 55½" (1.41 m.)	W 110⅝" (2.81 m.)	X 69½" (1.77 m.)	Y 68½" (1.74 m.)
Z 180½" (4.58 m.)	AA 25½" (0.65 m.)	AB 24" (0.61 m.)		Track – front 56" (1.42 m.) Track – rear 56" (1.42 m.) Turning circle 25 ft. (7.62 m.)	

† These measurements apply to all tip-up seats
* Approximate measurements

Fig 14: London cab and its key dimensions

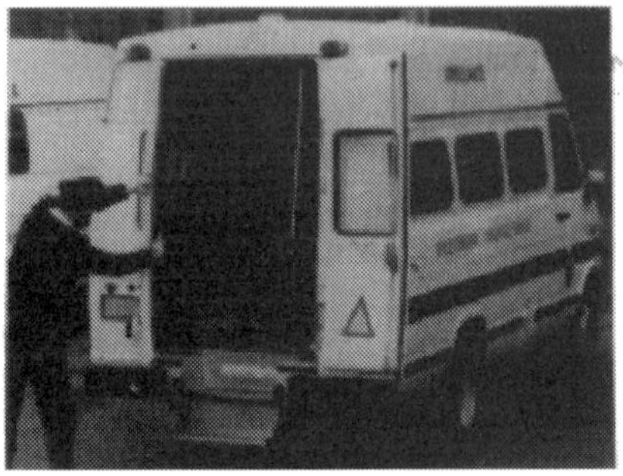

Fig 15: CV-based ambulance

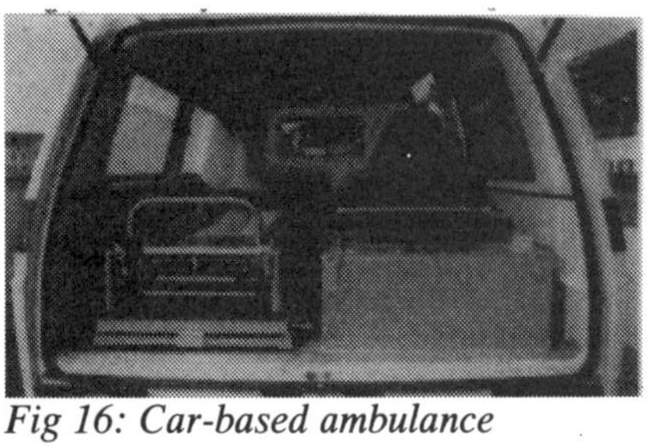

Fig 16: Car-based ambulance

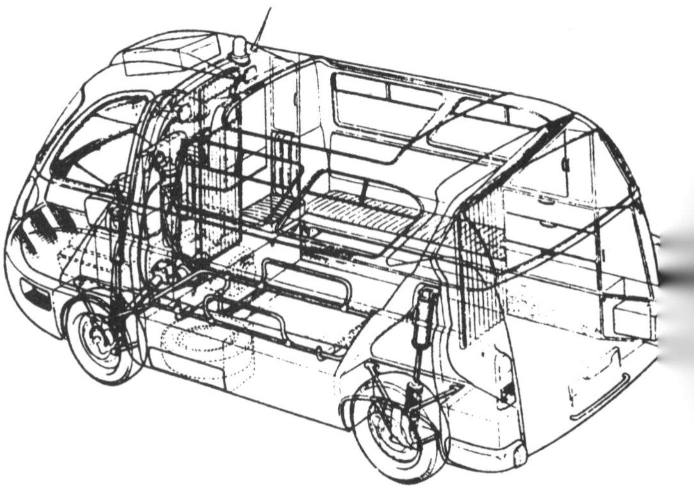

Fig 17: Bothwell concept ambulance

rear coupled suspension with automatic levelling and kneel facility; front wheel drive and wishbone rear wheel location to maximise rear-end access; monocoque chassisless construction for optimum interior space utilisation; up-and-over rear door in conjunction with concertina heat-retention curtains.

Racing car engineering

Changes in the method of financing racing cars in recent years, with the now widespread adoption of advertising sponsorship, has also had a marked effect on design in the respect of monies spent on product engineering to obtain competitive advantage. Nonetheless there are still developments which have their roots in the days when manufacturers financed their own cars — but more widespread is the use of advanced aerospace techniques which bring performance at almost any price.

The first move towards the current rear-engine layout in racing cars can be traced to the success of pioneering Cooper and Lotus cars in the early 1960s. Minimisation of power loss and reduced frontal area were the principal stimuli — coupled with low centre of gravity. In 1963, the Lotus 25 also pioneered the use of monocoque construction and the near reclining posture of the driver, from which the Lotus 33 in Fig 18 was derived with its 'monocoque-chassis' structure, shown inset, and wide-based double-wishbone IRS with familiar long trailing radius-rods, Fig 19. The Coventry Climax engine became an integral part of the structure as a rearward cantilever supporting rear suspension, gearbox and drive. The width of the engine governed the width of the car. The 1960s also saw the emergence of aerofoils, the negative lift device appearing on Formula One cars of 1968. However, 1969 saw the establishment of regulations which prohibited them being mounted on the unsprung elements of the car and restricted their size. But it

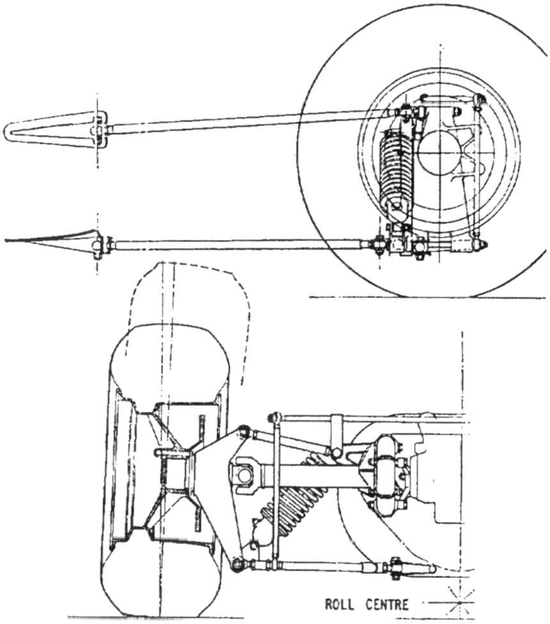

Fig 19: Lotus 33 suspension

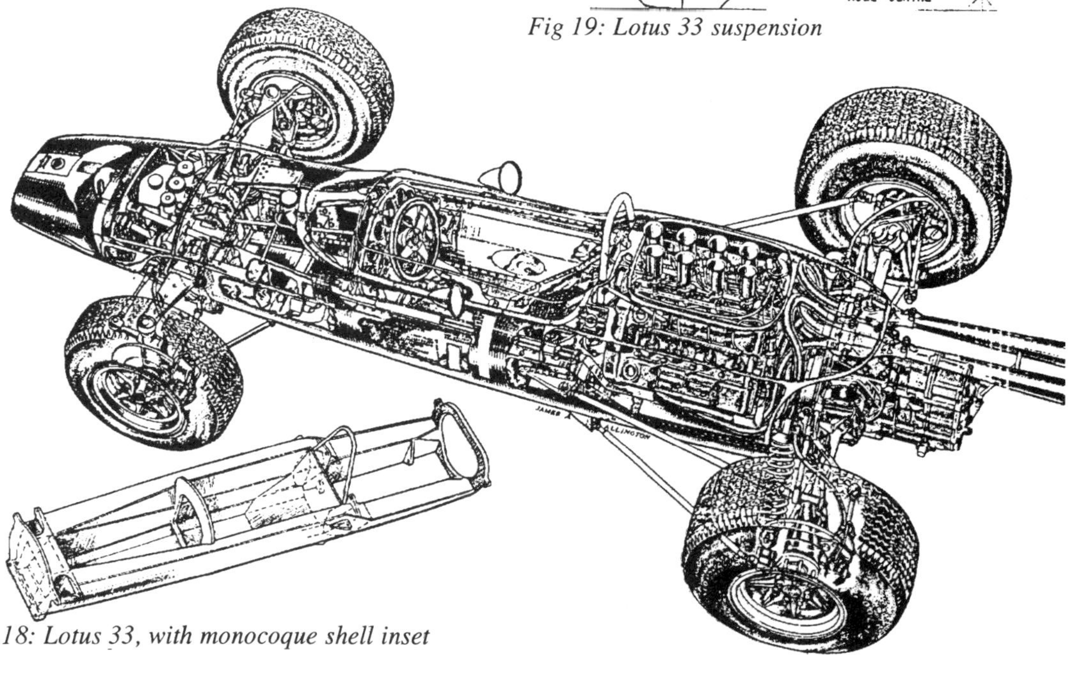
Fig 18: Lotus 33, with monocoque shell inset

was not until 1977 that Lotus introduced the first ground-effect car, aided by skirts, in which the full plan area of the car contributed to the development of negative lift — raising it to twice the weight of the car — at 250 km/h.

The main structure of the car amounts to a an inverted 4 square metre plan area wing, Fig 20, with Fig 21 showing the pitching and heaving characteristics. The aerodynamic loading was a spur towards the very stiff monocoque structures of composite construction which are commonplace today. Aerodynamics were also an important factor behind the development of active suspension systems in the need to maintain the critical ground clearance on ground-effect cars. Another development was towards two separate suspension systems, one supporting the 'chassis' structure, the other supporting the body, Fig 22. The aerodynamic part maintained a constant attitude relative to the ground, while the inner part moved separately with respect to the body.

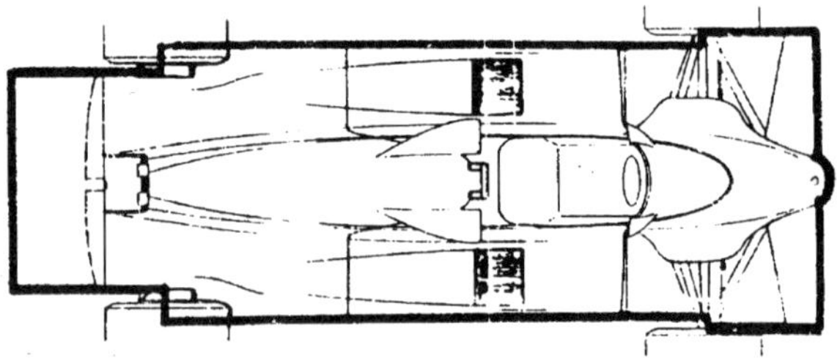

Fig 20: Wing plan area

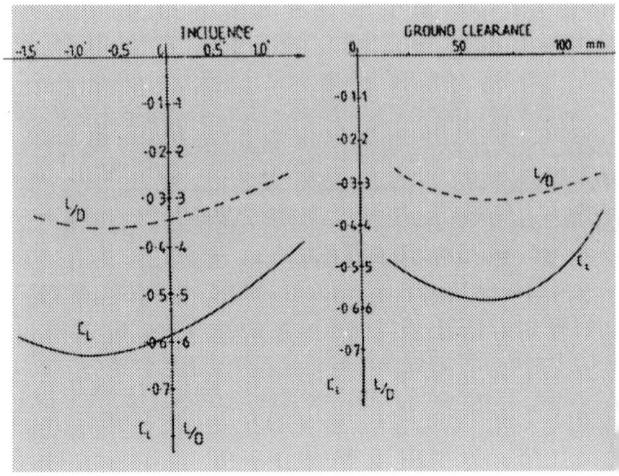

Fig 21: Pitching and heaving characteristics (C_l = lift co-efficient; L/D = lift/drag)

Structural trends in racing cars

Important contributors to high-strength composite laminates for modern racing car structures are carbon and para-aramid fibres. Sandwich panels of Nomex honeycomb core materials are also a key ingredient. This material is introduced in the section on GRP and is typically bonded to skins of high strength laminates by an adhesive film such as the Ciba Geigy Fibredux. For curved surfaces, the core is heat-formed at 170-200 C. Key property of this high performance sandwich panel, in this application, is very high resistance to buckling for very low weight, Fig 23. Typical edging methods, joints, corners and facings are shown inset.

Builders favour carbon-fibre reinforced laminate skins for ultimate strength and para-aramids such as DuPont's Kevlar for toughness. The latter is available in a variety of cloth weaves in weights ranging from 60 to 460 grammes per square metre. Typically an aramid/epoxy-resin laminate of 1.33 specific gravity has an elastic modulus of 42 N/mm^2 x 10^3 and tensile strength of 760 N/mm^2. A carbon-fibre/epoxy-resin laminate using a material such as Courtauld's Grafil has modulus up to 190 N/mm^2 x 10^3 and tensile strength of up to 1900 N/mm^2 with individual fibre strengths up to 1.9 N/mm^2 x 10^3. Weave of the cloth is important to laminated strength:

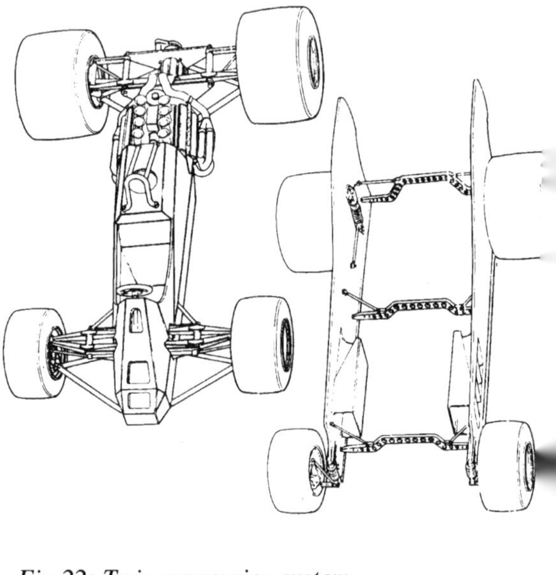

Fig 22: Twin suspension system

plain weave, alternate warp and weft, is stable and easily handled while satin weave (weft passing over three yarns and under one) is suitable for draping over deep-form surfaces. For high strength applications, twill weave, in which groups of warp and weft yarns are woven alternately, is particularly suitable.

Sandwich laminates are formed either in a vacuum bag or autoclave, an intermediate perforated membrane and blotting paper being used to absorb surplus resin squeezed out during the compression process. For aerodynamic wings, foam cores rather than honeycomb cores are often used. In the case of the Tyrrell 011 Formula One car, shown in Fig 24 as a finite-element mesh of its structure, the main body has two full length side panels forming the outer skin and joined by a Nomex honeycomb core to a Kevlar reinforced inner. Bulkheads down the body are machined from aluminium alloy. Weight of the car is 585 kg, with oil and water, for an overall length of 6.65 metres. Another example is the Williams Grand Prix car whose main structure is shown in Fig 25. This follows the now classic construction of a central semi-monocoque accommodating driver and front suspension; the rear-attached engine supports the gearbox casing on which, in turn, is mounted the rear suspension. The purpose of the bulkheads, Fig 26, is to feed attachment loads (and those due to driver and fuel inertia) into the main structure. Roll-over protection structures in front of and above the driver must withstand 7.5 g downward, 5.5 g rearward and 1.5 g sideways without collapsing. By regulation, for this formula, the car must not exceed 500 kg.

At Reynard, innovative techniques have been used to produce substantial batches within an economic framework — annual production now running at some 240 cars. A recently installed autoclave, Fig 27, operates at 7 bar and can be programmed to a temperature cycle which optimises cure of the laminate monocoques. As can be seen, the part is first

Fig 23: Composite sandwich panel

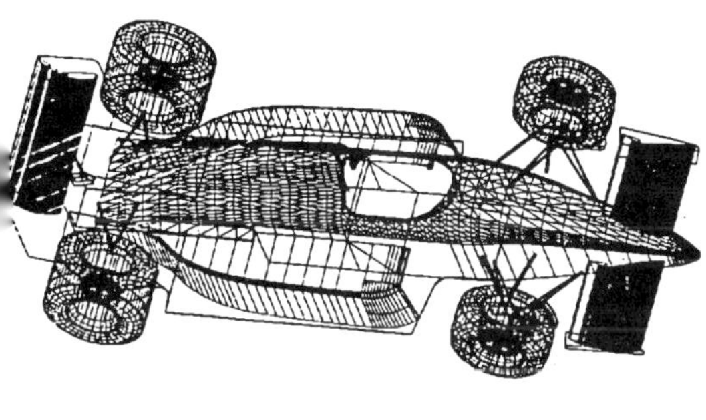

Fig 24: FE mesh for Tyrell F1 car

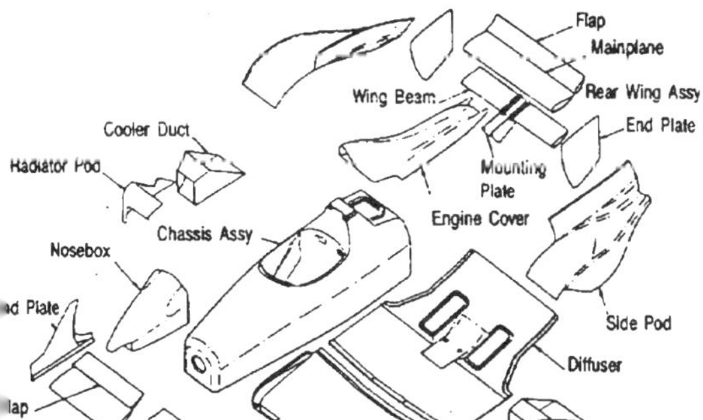

ig 25: Williams structure

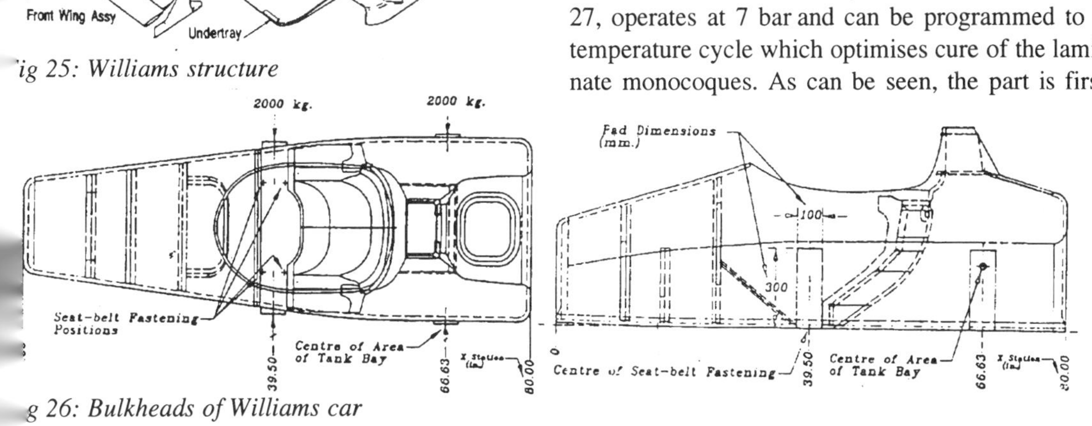

g 26: Bulkheads of Williams car

enclosed in a vacuum bag to ensure no air permeates the laminate. Layup, Fig 28, is carried out in clean-room conditions at 21 C and 50% relative humidity, materials having been previously stored in a cold room at 18 C for a maximum of six months. Cyanamid 77 carbon fibre is the currently favoured material for the body shells which must withstand a downforce in excess of 750 kg under track conditions. The latest cars have asymmetric nose cones which have U-shaped honeycomb reinforcements on their lower, extended, regions to withstand 740 kg test pendulum load impacting at 20 metres/second. Recent additional legislation also requires that a 2000 kg sideforce, applied over a 20 in^2 (130 cm^2) area, must also be withstood.

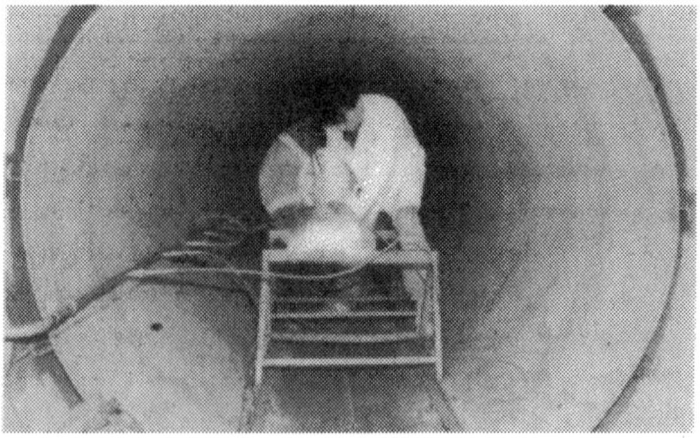

Fig 27: Autoclave at Reynard

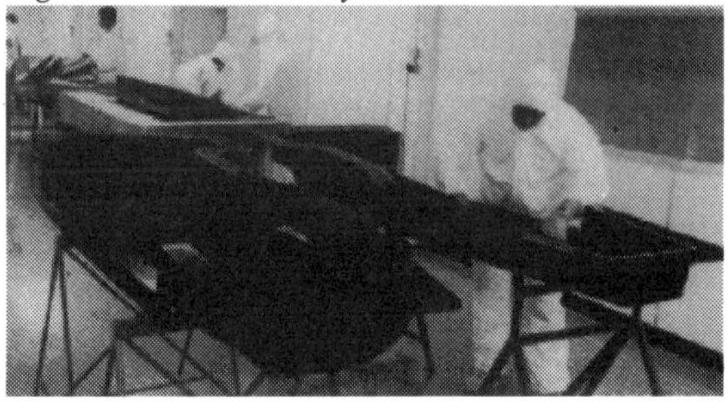

Fig 28: Lay-up of the laminate

Race-car mechanical developments

An important milestone was reached in the 1960s, for establishment of future racing car design trends, by Len Terry's work for BRM. The designer had a background in Indianapolis car design which no doubt affected the Formula One contender for the 1968 racing season, shown here alongside a contemporary Lola, with which it bears considerable resemblance in profile and driver posture, Fig 29. The structure was a full monocoque with spring-damper units inside it and actuated by rocker type top wishbones. At the rear, for the first time, a bottom trailing link of the parallel arm type was used to overcome the disadvantages of previous designs of the period — with rearward inclined wishbone axes to minimise bump steer. The wishbone became a 4-bar linkage which ensured the wheel plane remained parallel on bump, Fig 30.

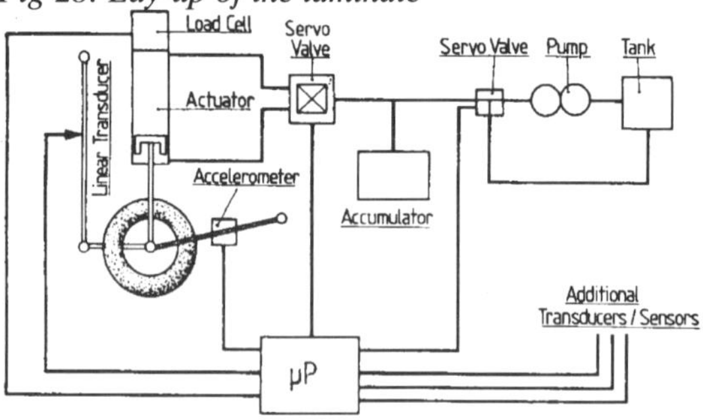

Fig 31: Control for Lotus active suspension

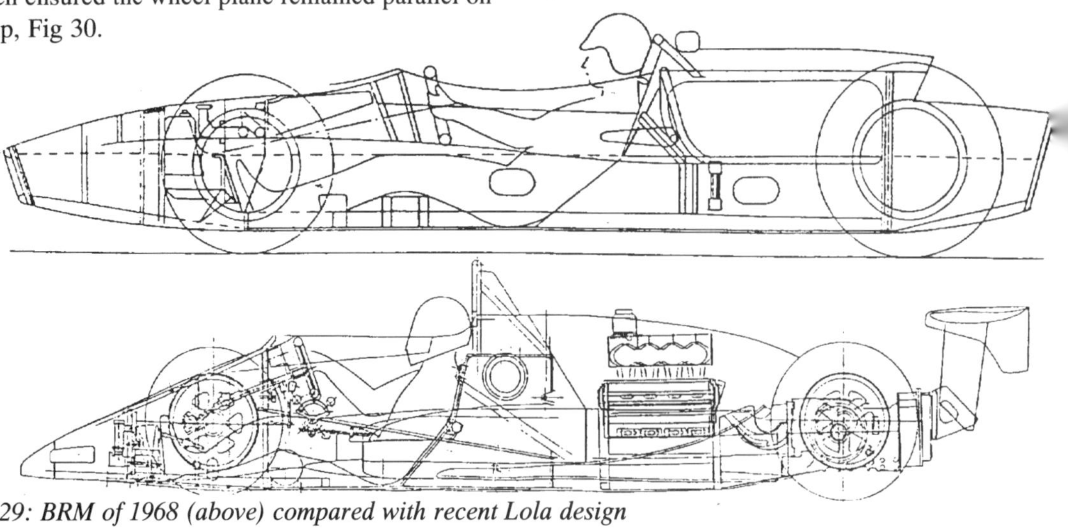

Fig 29: BRM of 1968 (above) compared with recent Lola design

Following the banning of the twin chassis Lotus 88F1 car, referred to earlier, came the need for another major change in suspension design, this time to maintain a constant ride height for skirted ground effect bodywork of recent years. This led to the Lotus active suspension system using double-acting hydraulic actuators in place of conventional dampers. Fig 31 is a schematic of the control system, the heart of which are the Moog servo valves.

The body-to-ground clearance-space beneath a car can be likened to a half-venturi having an expansion ratio which increases without limit until the duct stalls and causes a sudden drop in suction load as the distance to ground changes. At speeds in the order of 240 km/h, with the previously mentioned high downthrusts, any cornering can also cause massive additions in load on the suspension. Cornering accelerations in excess of 3.5g are possible and drivers can get transient loading of 4g/second. Stiffnesses are thus inevitably high to tolerate 3 to 1 changes in vertical loading.

Fig 32 Spydersport chassis

Fig 33 Spydersport race-car rolling-chassis/structure

Design of sports-racers

The design and construction of sports-racers is very much geared to economic factors and varies widely between budget kit-cars and expensive supercars. The latter category increasingly draw on the techniques of the volume sector with respect to CADCAM and the use of sophisticated shell structures while the former relies on workshop ingenuity

Specialist sports racers are often based on the fitment of proprietary running units into tubular space-frames on a try-and-test basis. Space frames, too, can be purchased as ready built units from such suppliers as Spydersport, a company which established itself with substitute frames for such cars as Lotus Europa and Elan, Fig 32, for which it also manufactured proprietary suspension assemblies and designed structures which would accommodate proprietary transmissions. Now the company has extended into rolling sports-racer chassis assemblies, Fig 33, either of pure space-frame or semi-monocoque construction. These provide well-engineered bases for the small-series builder who need only concentrate on body design and construction.

Other small-series builders have taken the root of devising replica designs based on notable past successes such as Ford's GT40 and GT70 projects. The GTD 40 is a case in point, Fig 34a, where the best in the earlier technology is retained and updated to be fully competitive with contemporary designs. The original GT 40 car, Fig 34b, was built for Ford by JW Automotive which was an early pioneer of monocoque construction for this class of vehicle. The later GT 70, Fig 35, was to have a simpler structure based on a perimeter rail, Fig 36. This design was due to Len Bailey who went on to create the Panther Solo, the structure of which, Fig 37, was a composite sandwich construction of woven glass-fibre and carbon-fibre reinforced epoxy resin skins with an aluminium-alloy honeycomb core. The car was also of interest for its four-wheel-drive configuration.

For the open sports-car a greater structural challenge is set for the designer. For the current Scimitar car, SDRC Engineering Services were sought to carry out a full computer-aided concept-design in a determined effort to overcome the low-frequency dynamic behaviour problems which have dogged many open-car structures. The steel underframe was designed around a central armature to react torsional loads — with a peripheral beam

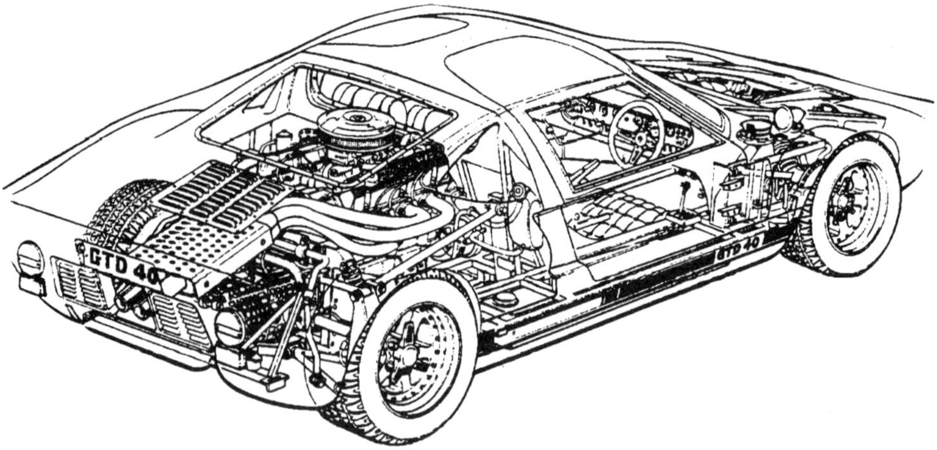

Fig 34a: GTD40

structure to carry the running units plus the vehicle occupants and luggage. The structure was notionally divided into shell and beam regions, the beam section properties being factored to allow for joint inefficiencies.

The resultant analytical model obtained is shown alongside the system responses later in the section. The central tunnel was subjected to a finite element analysis using shell elements while the remainder used beam elements. So as to retain 'designer-feel' during the mathematical analysis, a pre-processor with an advanced graphics capability was used to allow the model to be presented to the designers at regular review sessions. When the model was felt to truly represent the structure it was subject to analysis using SUPERB FE code. Test loads were applied to derive overall torsion and bending stiffnesses which were compared with those of known vehicle designs. Areas of high distortion were then located on the basis of elemental strain energy densities.

Such areas were artificially 'stiffened' and the program re-run to evaluate the effect and that of strain re-distribution through the remainder of the structure. Results of this preliminary analysis are shown in the table in Fig 38, from which an unacceptably low torsional stiffness resulted. The weak points found are annotated in the computer responses which showed that the peripheral frame structure was failing to strain the central tunnel sufficiently.

This led to the design modifications shown in later in the analysis which proved adequate with the exception of the front longitudinals which could not be sufficiently stiffened — hence a redesign of the

Fig 34b: Ford GT40

Fig 35 Ford GT70

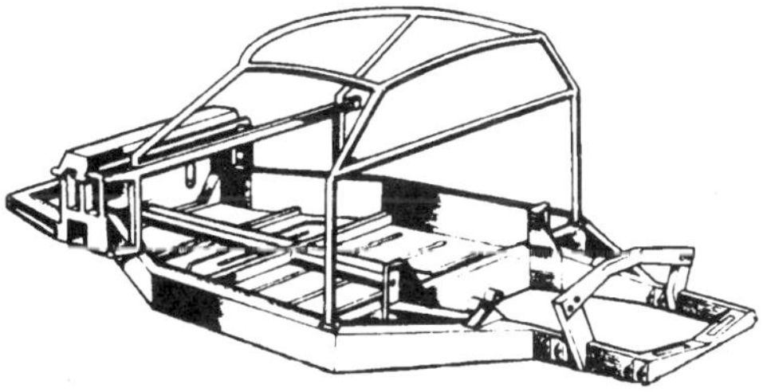

Fig 36 GT70: structure

Fig 37: Panther Solo structure

front-end. To achieve greater vertical bend stiffness between front-suspension reaction point and the tunnel, a lower longitudinal was added, together with a shear-panel with vertical edge stiffeners front and rear. Their addition to the model is also seen in the computer simulation and the table in Fig 39 shows that the predicted improvement to torsional rigidity closely aligned with the measured value on the first prototype.

The computer analysis commenced with a description of the original design, Fig 40, followed by an annotation of the weak points, Fig 41, a description of the necessary rectifications, Fig42, and the model redesign, Fig 43. For dynamic analysis it was necessary to add to the dydnamic model, Fig 44, the mass of vehicle components, Fig 45. This was used to study vibrational behaviour of the structure and its associated connecting masses so as to minimise the coupling paths between known dynamic forces and the vehicle occupants. An examination of the vibrational mode shapes led the designers to conclude that the critical fundamental torsional mode occurred at 18Hz — high enough to avoid interaction with the wheel-hop modes and low enough to avoid excitation by the engine at idle.

MODEL CONFIGURATION	TORSIONAL STIFFNESS Nmm/Degree (lbft/degree)	BENDING STIFFNESS N/mm (lbf/in)
Initial Design	2.1×10^6 (1.6×10^3)	6.1×10^3 (35.0×10^3)
Initial Design plus Rigidised Cpts	5.0×10^6 (3.7×10^3)	
Minor Modifications	2.8×10^6 (2.1×10^3)	
Final Design	4.2×10^6 (3.1×10^3)	

Fig 38: Stiffness predictons

Initial Design	2.1×10^6 (1.6×10^3)	2.0×10^6 (1.5×10^3)
Minor Modification	2.8×10^6 (2.1×10^3)	
Final Concept lst Prototype)	4.2×10^6 (3.1×10^3)	

Fig 39: Torsional stiffness correlation

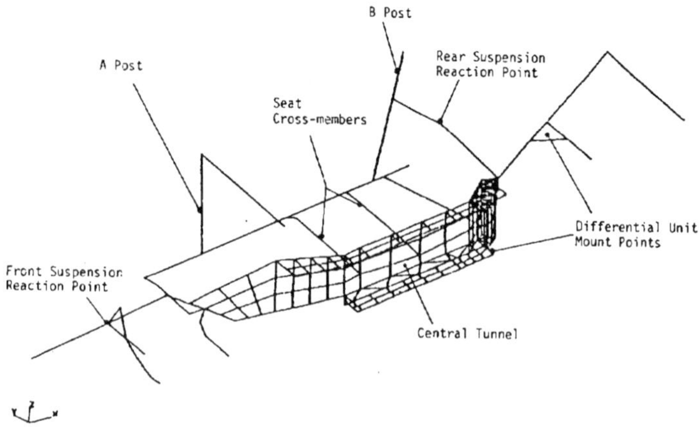

Fig 40: Analytical model

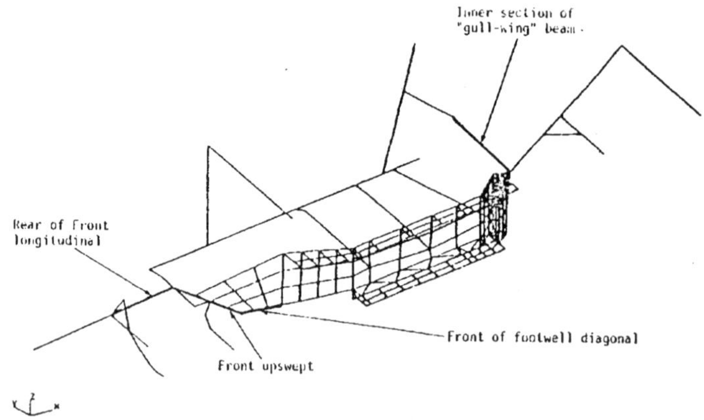

Fig 41: Weak beams made rigid for study

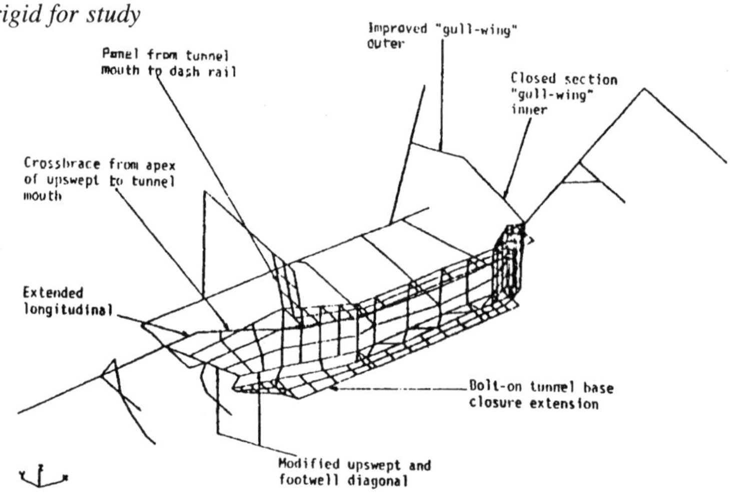

Fig 42: Design modifications in first revision

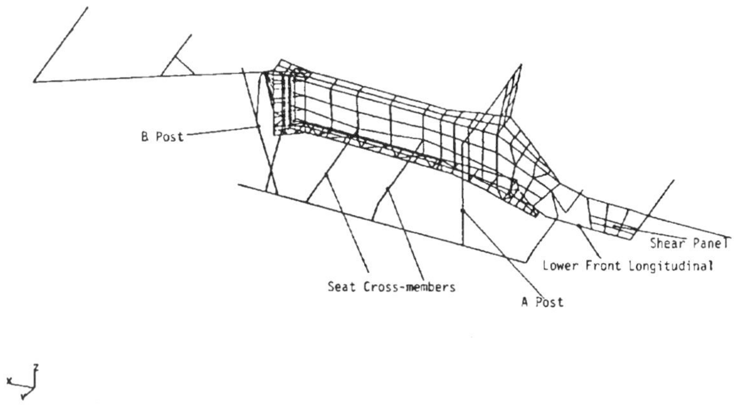

Fig 43: Sub-structure redesign

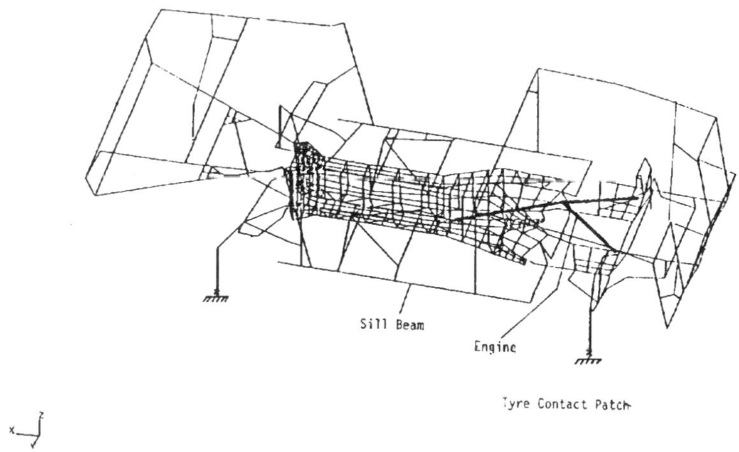

Fig 44: Full dynamic model

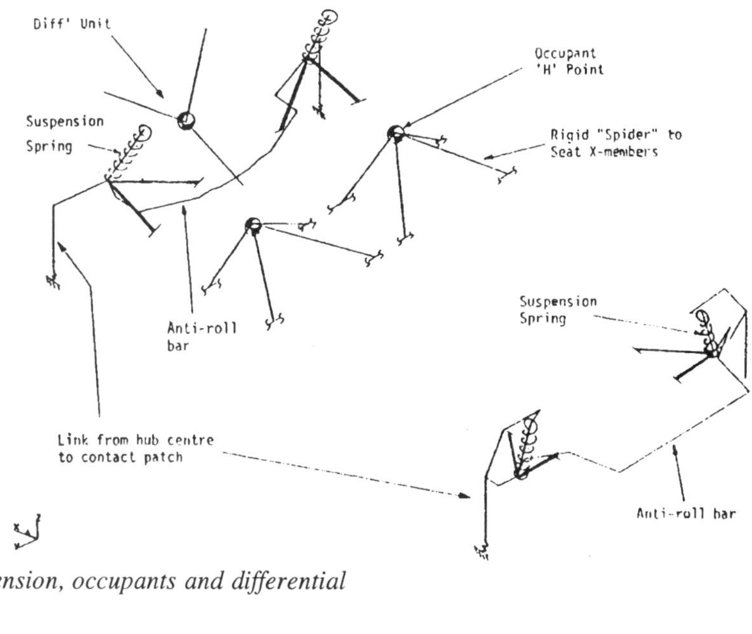

Fig 45: Dynamic model: suspension, occupants and differential

Designing a supercar

The ultimate automobile has been the challenge of specialist car designers over the years and many see the McLaren F1 as fulfilling that dream. The A-Z stages in that fulfilment are discussed here alongside the latest development stages in a rival contender in the GT stakes, the Lamborgini Diablo and that of two supercars in the luxury class, the Mercedes S-class and Bentley Continental R. The similarities and contrasts in approach are of equal interest to some of the techniques used in trying to obtain perfection.

One of the early sketches for the McLaren F1, Fig 46, showed the first steps towards packaging evolving around the idea of a central driving position. Another sketch set out the basic requirements for weight distribution and low polar moment of inertia, Fig 47. Weight target was 1000 kg and width had to be kept within 1.8 metres. A stable centre of pressure was deemed another essential for avoiding high speed instability problems. The vehicle was packaged around a five litre V-12 engine tuned to output some 100 bhp/litre — about half that of a racing engine — and fitted with a transverse gearbox to keep the main masses within the wheelbase. BMW Motorsport suggested going to six litres to guarantee the 550 bhp target but McLaren insisted it had to be no more than 600 mm in length. An offset final drive gear alongside the flywheel made the powertrain particularly compact.

Aesthetic design of the car was undertaken by Peter Stevens and interior packaging by Barry Lett who used a seating buck to establish the body height of 1140 mm. No styling sketches were undertaken until the package had been finalised so that no unfortunate compromises could result. At the stage when cabin area and windscreen rake were known, as well as height and width, wind tunnel of model variants commenced. Key parameters were rear glass slope, tail height, nose shape and wing design. A modest 0.35 Cd was required with the centre of pressure almost coincident with the centre-of-gravity; 160 lb of downforce at 150 mph was needed and very little change to the centre of downforce as pitch or attitude changed. Pop-up headlights were quickly discarded as causing high increase in downforce ahead of the lights as the centre of pressure would be moved ahead of the car at high speeds.

Main constraints on the style were an air-intake

Fig 46: F1 elevation

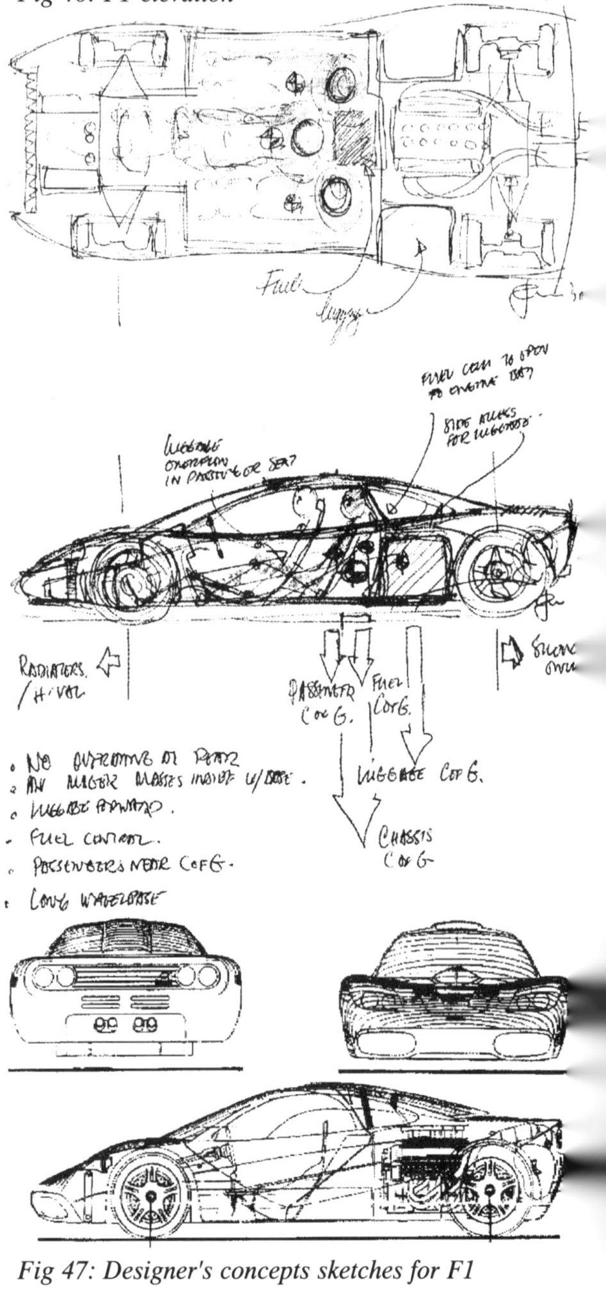

Fig 47: Designer's concepts sketches for F1

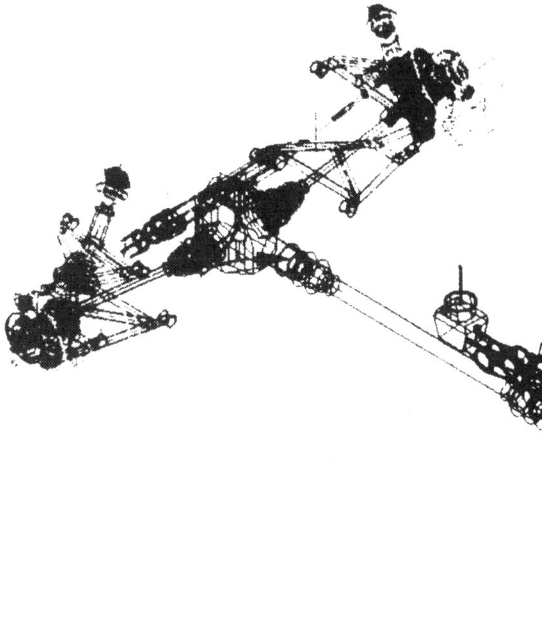

on the roof, for clean air and easy water separation, committing the designers to a spine frame. Twin radiators required a high nose line. While the door arrangement, scissor style, pioneered by the Countach/ Diablo was seriously considered, this was not possible for reasons of access and egress, and something akin to that used by the Porsche 962 was adopted.

Fig 48: Diablo with its mechanical, package

Suspension design was helped by McLaren's insistence that weight distribution (42/58, front/rear) would change by less than one per cent from a full to empty fuel load. Longitudinal weight transfer in cornering was to be minimised, too, by equalising the distances between suspension roll centre and body mass centroid at front and rear. Since roll centres had to be low, to avoid jacking effects, the engine too had to be as low as possible in the body, a situation helped by dry-sump lubrication.

Progressive rate springing was avoided on the grounds that in excess it can suddenly increase weight transfer when a wheel hits a mid-corner bump, making handling unpredictable. Adaptive damping and ride height control were also avoided, this time on grounds of weight. Finally wheel travel front and rear was set at 3.5 inches bump and 3.1 rebound while target unladen bounce frequencies were 1.43 Hz front and 1.80 rear. Such relatively soft springing had to be carefully matched to available downforce since high speed handling would otherwise have been dangerously unpredictable.

The double-wishbone suspension had longitudinal compliance built in to allow rearward wheel movement, on striking an obstacle, to improve ride quality. Compliance varied between front and rear wheel stations; at the front, priority was to prevent castor wind-off under braking (limited on the car to about one degree per one *g* of braking deceleration), which otherwise compromises stability. The company adopted the technique of ground-plane shear centre in which the wishbones were mounted on subframes having suspension bushes arranged to constrain arcuate motion so that it centred on the tyre ground contact patch.

At the rear, the inclined shear axis technique was adopted which accounts for traction reactive forces acting through the wheel hub. Engine and suspension mounts were designed as an integrated system aimed at minimising toe-in or toe-out under traction and braking. These were limited to 0.08 and 0.04 degrees per g of acceleration/deceleration respectively. Front castor angle of 4.6 and king-pin inclination of 8 degrees were accompanied by a larger than usual wheel offset of 25 mm.

Evolution of a super-roadster

The beginnings of the Lamborghini Diablo date back to 1971 when the company's Countach, conceived by Bertone, was launched. Marcello Gandini was to interpret Bertone's concepts into a classic Italian supercar design. The Diablo was the heir to this classic, to be unveiled in June 1986 shortly before the Chrysler acquisition of the company. Subsequent changes resulted in the production version with power of 492 bhp, top speed of 325 kph and 0-100 acceleration in 4.1 seconds, Fig 48. 1991 saw the announcement of the Roadster version whose evolution from the basic car was a triumph of CADCAM.

The Roadster's evolvement owed much to Control Data's Integrated Computer Aided Engineering (ICEM) system since the essential surface information of the base Diablo was already held on this system. The package was able to create surface design faster and more accurately than had previously been possible. With the associated SURF software basic surface data from concept renderings could be made to produce a computer-design suitable for producing machinable surfaces in just one step, without the time-consuming iterative steps of earlier systems. Surface smoothness on generated curves is said to be within 0.25 mm of measured inputs from the styling physical model. Once the data is organised by the package, surfaces can be directly manipulated by control-points, lifting-points and boundary conditions, Fig 49.

The closed version of Diablo has a skeleton and roof of steel, front and rear ends in aluminium alloy and closure panels in carbon-fibre composites. For the Roadster a significant number of these panels were changed. The drag co-efficient of the base car was a relatively high at 0.31 but the design is purposely arranged to exert maximum downforce. The space frame chassis is made of tubes with cross-shaped sections, welded to a centre tunnel, to which the front and rear wishbone suspensions are mounted.

The car's V-12 engine has a separate electronic controller for each bank of cylinders. Combustion chambers are trapezoidal with narrow valve angle and central spark plug; piston crowns are lenticular with squish lips.

Camshaft grinding is carried out on a CNC machine with which cam profiles can be computer-generated by the same software that generates lift and acceleration diagrams for the cams.

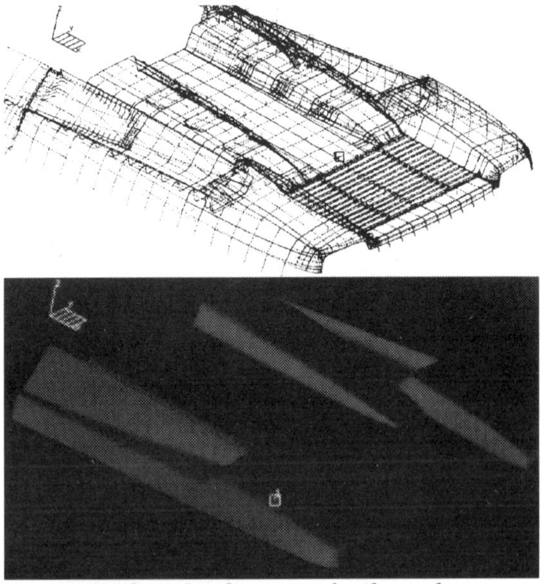

Fig 49: Surface development of tail panel

Fig 50: Mercedes S-class

Fig 51: Bentley Continental R

Fig 52: Mercedes V12 of over 400 bhp

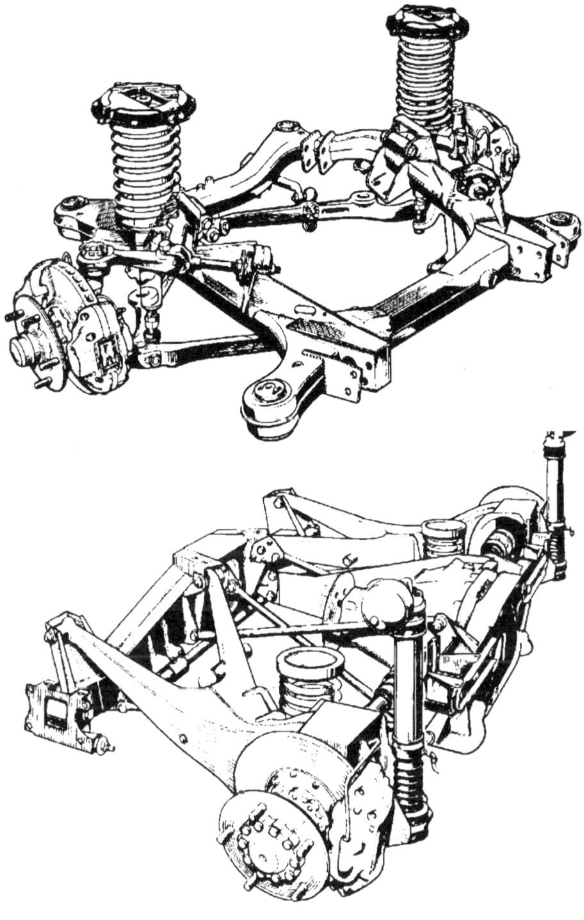

Fig 53 a: Bentley front and rear suspension

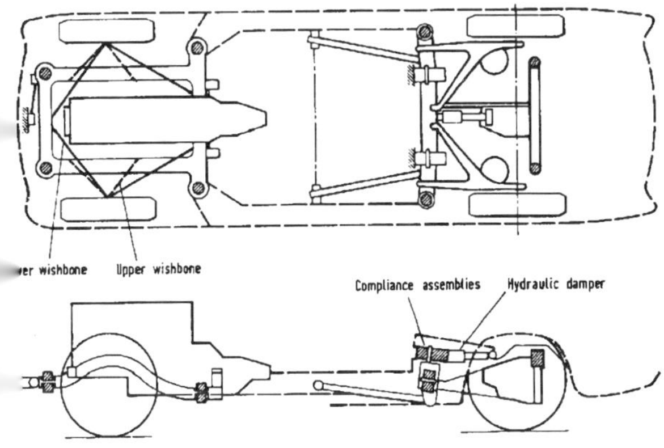

Fig 53b: Wide anchorage pivots for the R-R wishbones

Luxury closed-coupes compared

The latest Mercedes Benz S-class coupe variant, the S420, completes a trio started by the S500 and S600 derivatives of the main-line saloons, Fig 50. The latter are smaller than equivalent R-R models but the V-12 engine introduced in 1991 develops 400 bhp at 5200 rev/min from 6 litres making it one of the world's most powerful passenger car engines, Fig 52. This uses variable intake camshaft timing and Bosch LH Jetronic fuel injection. As in BMW's V-12 the two cylinder banks are independently managed with each cylinder having its own program. The S models use a front subframe system, for engine and suspension, and the front suspension is unusual in the aluminium alloy upper arms pivoting around an oblique axis such that those of the left and right hand arms converge towards the front. The five-link rear suspension is effectively a double wishbone system but incorporates an extra link for toe-in/out correction. This is also subframe-mounted in common with the final drive.The Bentley Continental R, Fig 51, is the first uniquely styled version of the marque since 1952 — with distinctive coupe body. Integral bumpers, air-dam and rear skirt emphasise the low side profile of the vehicle which now has reduced frontal area and flush glass to aid aerodynamics. The interior transformation is marked by a through centre console which extends through to the rear seats which are individual contoured units. Powered by the Turbo R engine coupled to a 4-speed auto-transmission, top speed is now 145 mph and 0-60 mph acceleration a creditable 6.6 seconds for this 2.42 tonne kerb weight vehicle. There is a two-piece prop-shaft and final-drive ratio is 3.07:1; wheels too are new and sized at 16 inch diameter and 7.5 inch rim width. The 6.75 litre 90 degree V-8 has a compression ratio of just 8:1 and is fuelled by the Bosch K-Motronic system. Mid-range power of the engine has been increased by 20 per cent despite reductions in fuel consumption and exhaust emissions. Larger inlet ports and exhaust valves have been introduced to improve volumetric and combustion efficiency, to effect the 18% fuel consumption reduction, as well as multi-point air injection and transient boost control for the turbo-charger. Mechanical layout of R-R models dates back to the company's switch to integral construction in 1965 commencing with the Shadow model and is particularly notable for very widely spaced anchorages for the front suspension wishbones, Fig 53.

System design for kit-cars

In stark contrast to the purpose-built supercar, the kit-car provides a valuable contender in that niche of the market which requires individuality of style at a readily affordable price. By applying a systematic approach to design, best use can be made of the systems which come from the donor vehicle(s) and the use of GRP body construction techniques can achieve high-style economical bodies.

The ever increasing popularity of GRP as a body material, which allows widespread choice of shape, has led to the materials-supplying companies providing a valuable back-up of design and construction data. New materials entering the market are also helping to achieve better engineered bodies. Combinations of woven roving and chopped-strand material are now available, for example. Such bi-directional materials reduce crimp tendency and allow the full tensile properties of the reinforcement to be realised. It also allows more consistent resin to glass ratio over the surface of the body and the moulded panel's strength as a laminate is enhanced.

Fig 54 compares bi-directional woven roving with conventional woven roving. Fig 55 compares the physical properties of different reinforcements and Fig 56 is a nomograph giving glass-weight per square metre against laminate thickness. The two examples illustrated concern a glass weight of 450 grms/metre^2 (1.5 oz/sq ft) and laminate thickness of 1.5 mm, giving glass content of 21.5%; the second shows that, for a glass content requirement of 30%, a 1 mm laminate corresponds to 425 grms/m^2.

Design safety factors range from 2 for static to short-term loads, 5 for repetitive loads, and 10 for impact loads. Joining mouldings together is crucial to overall performance of the overall body shell — thickening of the laminate being involved, Fig 57; sometimes mouldings can be butted together at a

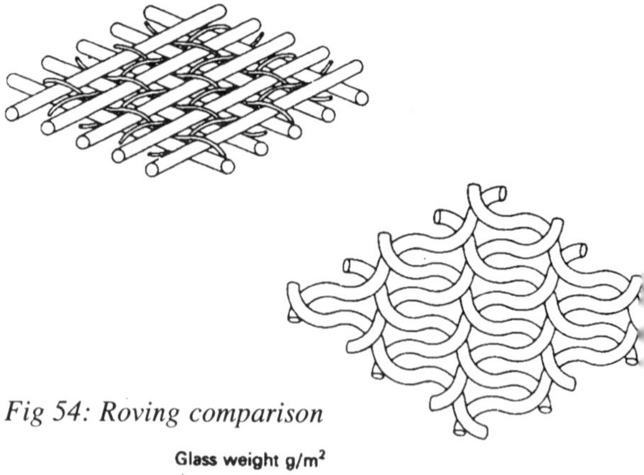

Fig 54: Roving comparison

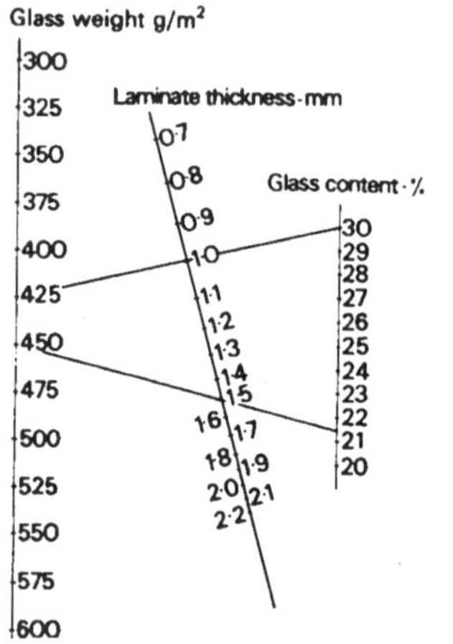

Fig 56: Nomograph for glass weight

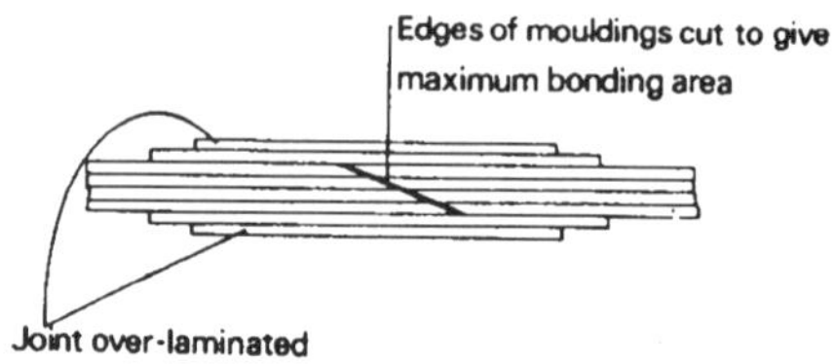

Fig 57: Joint design

Properties	*Unit*	*Chopped strand mat*	*Woven rovings*	*Satin weave cloth*	*Continuous rovings*
Glass content	% weight	30	45	55	70
	% volume	18	29	38	54
Specific gravity		1.4	1.6	1.7	1.9
Tensile strength	MPa	100	250	300	800
Tensile modulus	GPa	8	15	15	40
Compressive strength	MPa	150	150	250	350
Bend strength	MPa	150	250	400	1000
Modulus in bend	GPa	7	15	15	40
Impact strength, Izod, unnotched	kJ/m^2	75	125	150	250
Coefficient of linear expansion	x10^{-6}/°C	30	15	12	10
Thermal conductivity	W/m K	0.20	0.24	0.28	0.29

Fig 55: Reinforcement properties

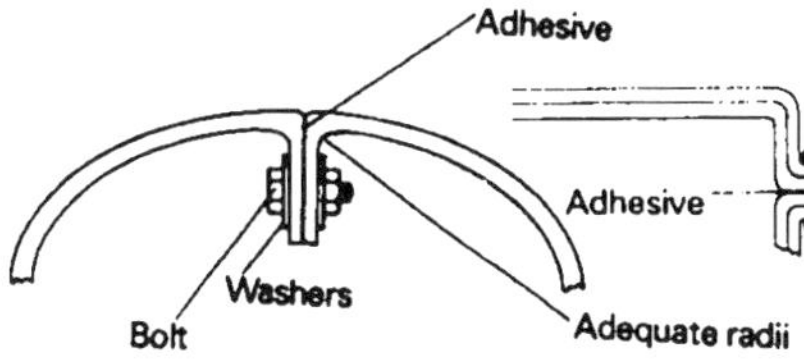

Fig 58: Bolting on a style line

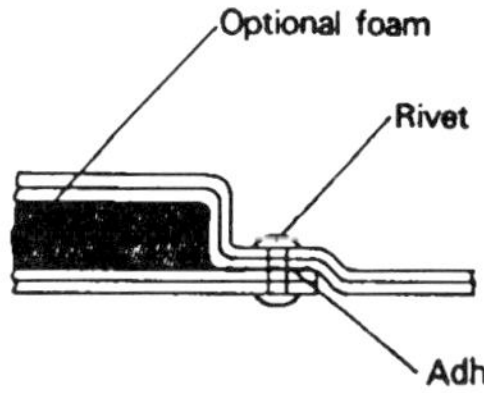

Fig 59: Foamed core floor joint

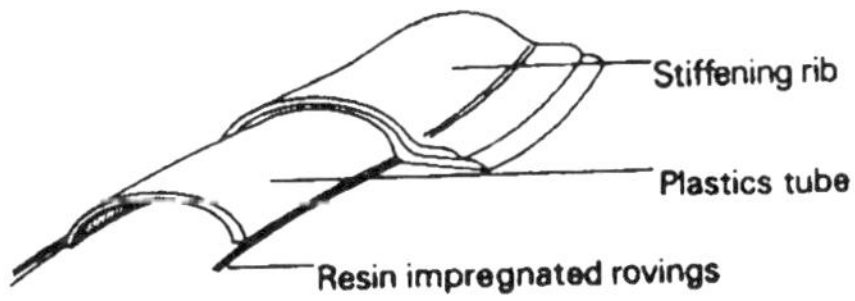

Fig 60: Stiffening ribs

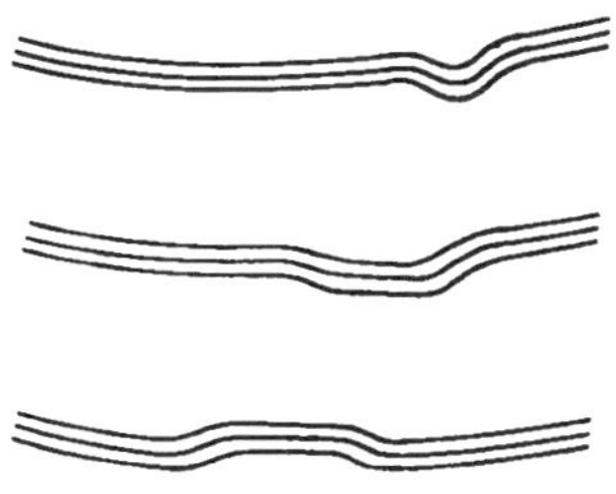

Fig 61: Rigidity from style lines

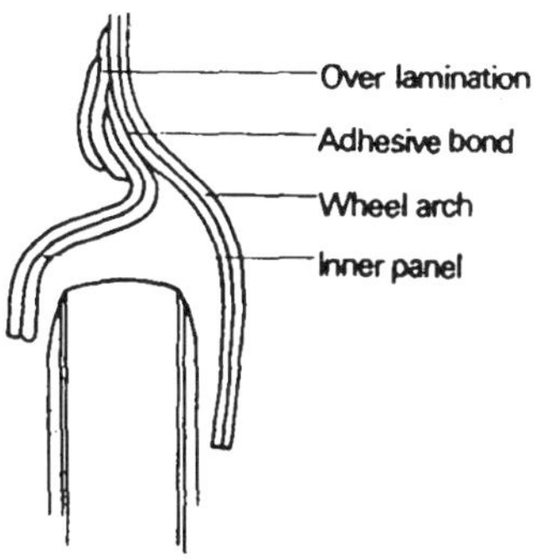

Fig 62: Double-skinned wheel-arch

style line of the vehicle, Fig 58, where they can have returned edges and be joined by adhesives or bolts. When mouldings join at right angles, a snap-on channel can be used to cover the join. In floor sections requiring high rigidity, double skinning may be involved, with a foam core, in which case the joint of Fig 59 is suitable.

When moulding-in stiffening ribs, Fig 60, the latter should be bedded down against the moulding, in the mould just prior to gelling. The over-laminating operation should ensure there is adequate area of the laminate in contact with the panel at each side of the stiffener. General technique in three-dimensional design is to build in maximum contour and where necessary supplement the stiffness with double-skin construction. Exterior style lines also serve to give rigidity, Fig 61. Double-skinning of wheel arches, as in Fig 62, can also be used to enhance rigidity.

Initial styling can be performed on a Plasticine model to a scale of typically 1:12 and a plaster mould cast from it — which is split to allow the undercuts to be withdrawn. After smoothing, this mould can be waxed, re-assembled and a GRP moulding laid up using surface tissue followed by layers of the lightest chopped-strand mat. After curing, the GRP model is removed and lightly sanded — then scribed-off at equal intervals using the front-axle centre-line as datum. The model is then cut into sections at each station interval and the outlines traced on to squared paper for scaling up to obtain the full sized plywood templates, Fig 63. Fine-mesh wire netting between templates serves as an armature for subsequent plastering, to form a master pattern. When hard, the plaster surface is given several coats of shellac and a smooth coat of PVA release agent.

To produce a monocoque structure, two major moulds would be made, one for floor and bulkheads, the other for the outer shell. A scale drawing taken from the master model would then be used to construct full-size mock-ups of the floor and shell, Fig 64. A 3 x 2 in (75 x 50 mm) timber frame with contiboard fill panels has been suggested. The body-shell 'plug' uses formers over bulkheads and hardboard is used to support the clay. Female templates made from the master model are then offfered up to the clay to achieve the design shape. A mould can then be taken from it for subsequent body production. Panels may incorporate foam infills such as that shown in Fig 65 for a door-sill.

Tractive performance

Some simple calculations and graphical constructions help to rationalise the choice of mechanical systems. Tractive effort P in lbs = $\varepsilon mt/R$ where ε is overall mechanical efficiency, m the overall gear ratio, t the engine torque (lb-in) and R the tyre-rolling radius (inches). Thus $P = k_1 t$ where $k_1 = mt/R$. The road speed of the vehicle can be expressed as

$$V = (N/m)(2\pi R/12)(60/5280) = k_2 N$$

where $k_2 = Rm/6723$; N is engine rpm.

An engine torque speed curve would then represent a tractive-effort x road-speed curve — if appropriately scaled — and T-E vs road-speed curves can readily be plotted for each gear ratio as in Fig 66. The motion resistance forces to the vehicle can be calculated as set out in the later chapter on performance prediction (approximated as in Fig 67) — and the gear ratios selected as set out in the chapter on manual gearbox design.

Curve AB is the top gear tractive effort and CD etc are those for the lower gears. JK is the total resistance curve and others for several gradients are shown. Where AB intersects JK is the point L where maximum level-road top speed is attainable. Any intercept such as MN gives net tractive force available for acceleration. PQRS is the tractive effort that would be available if the gearbox ratio could be continuously adjusted as in a CVT. Time required for a given change of speed can be obtained from

$$T_2 - t_1 = (W/g)\int_{v2}^{v1} [1/f(v)]dv$$

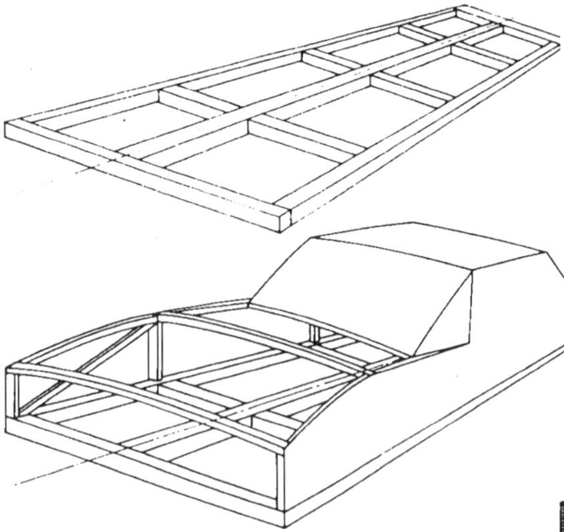

Fig 64: Floor and shell mock-ups

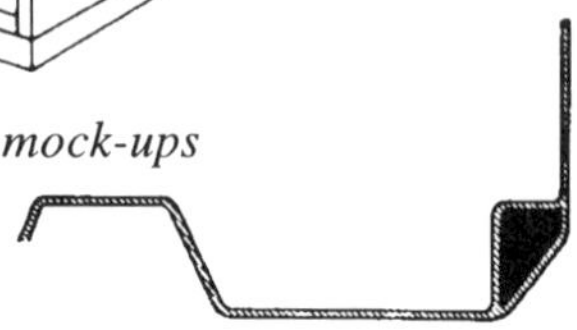

Fig 65: Foam infilling

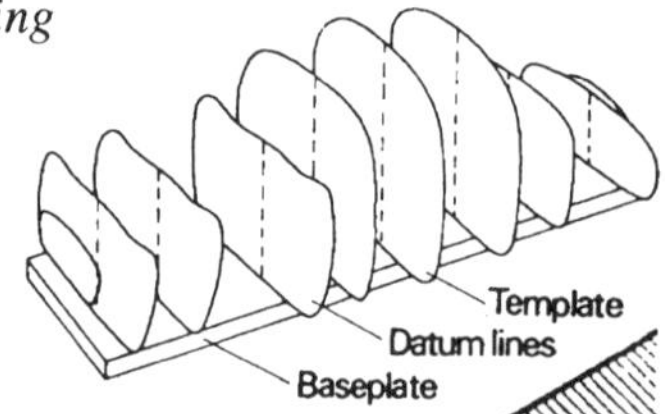

Fig 63: Body section templates

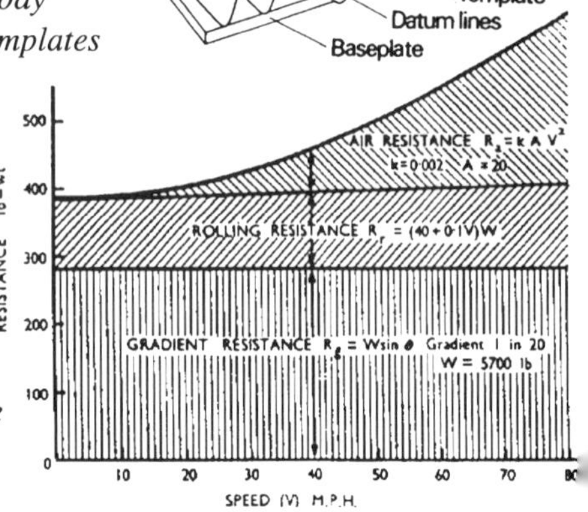

Fig 67: Motion resistance forces

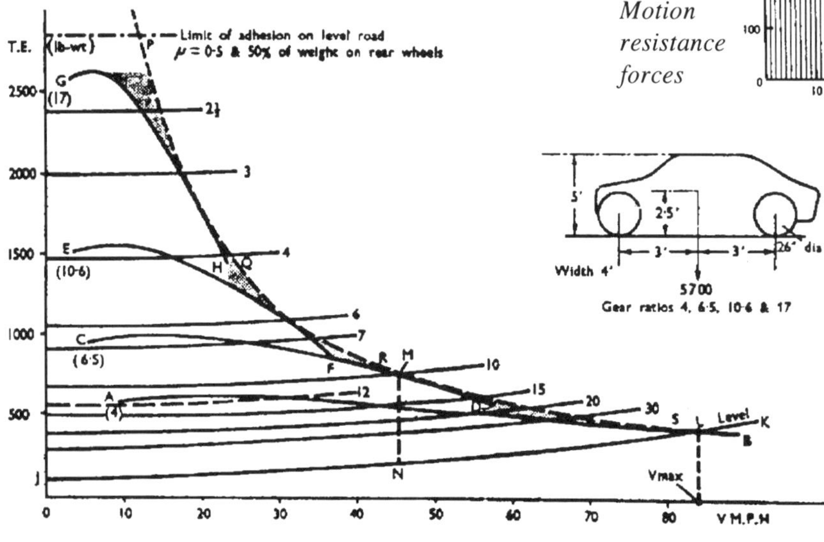

Fig 66: Tractive effort curves

where *f(v)* is the function giving tractive effort with respect to speed. This can be obtained from a graph made by plotting the intercepts corresponding to MN against the corresponding speed as in Fig 68a, curve AB. The integration can be performed graphically by drawing the curve of *1/f(v)* on a speed basis as in CD, the ordinate EF being made equal to 1/EH to a convenient scale. The shaded area between speed limits of interest can lead to a corresponding time interval — to provide the time/speed graph of Fig 68b with ordinate JL made equal to the corresponding area in Fig 68a. Thus:

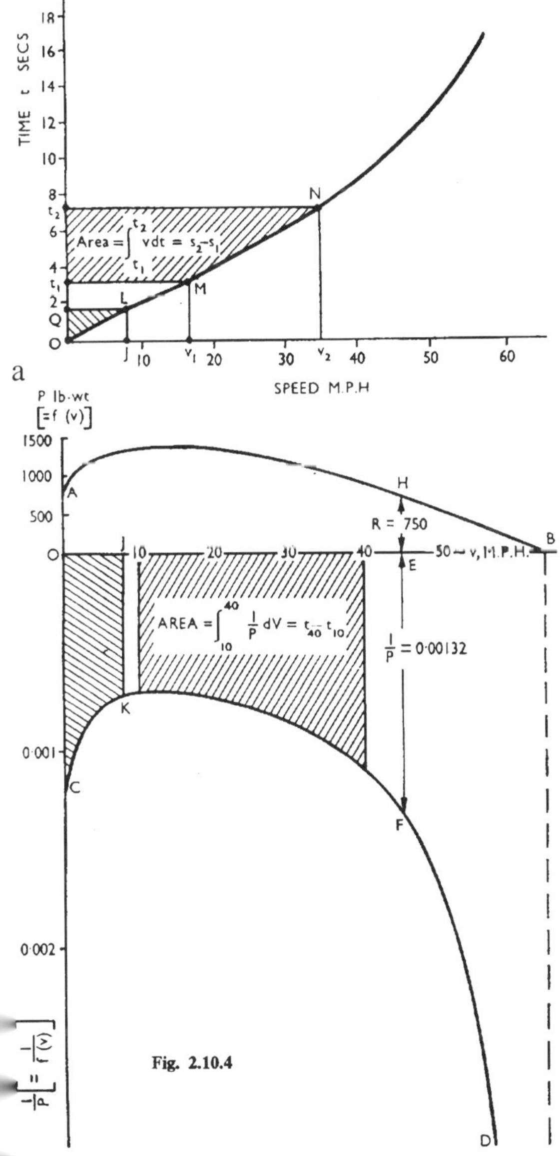

Fig 68: Speed/time curves

$$v=(ds/dt)\int_{v2}^{v1} v\,dt = \int_{v2}^{v1} ds = s_2 - s_1$$

This integral is the area between the time-speed graph and the time axis. The shaded area then gives the time taken t_2 - t_1 to accelerate between v_1 and v_2.

Vehicle systems design

The systems of the vehicle can be further examined using some of the methods set out in other chapters. Steering geometry and basic evaluation of handling on constant speed/constant radius bends is covered in Chapter 17. Some basic calculations can also be made on the structure with the sections on open-platform and closed shell structures Chapter 2. Chapter 9 deals with the elementals of the electrical system. The ergonomics of the vehicle interior dimensions and controls can be set out with the help of the section in Chapter 8 while Chapter 16 extends the performance prediction calculations given above to take in such effects as vehicle weight distribution and the inertia of the rotating masses.

Acrodynamic basics are given in Chapter 8 and in Chapter 1 is an introduction to framework analysis for those builders using a space-frame structure. A section in Chapter 18 covers braking fundamentals and in Chapter 17 the fundamentals of designing for ride comfort. The transmission gearbox design section of Chapter 19 includes a section on the rationalising of gear-ratio selection in terms of stepping intervals while factors affecting choice of engine are covered in Chapter 16 which also has a section on cooling of the engine. The section on driveline calculation (Chapter 18) covers design considerations of clutch, drive-shafts and universal joints while lighting appears in Chapter 10 and engine mounting in Chapter 16.

Case study in kit-car design

A feel for how the series-production specialist approaches the design of a new model can be gauged by case-study. The celebrated designer Frank Costin recently advanced the design of a single-seater commuter car whose road-performance targets were more than a match for many kit-cars. Looking for a 90 mph top speed and 0-60 mph acceleration in less than 10 seconds he chose the Reliant 850 cc engine supplied by Beans Engineering to power the vehicle shown as a concept layout in Fig 69. That company also suply the 'matching' Reliant gearbox and drive-axle.

His other targets were 50 mph fuel consumption and 0.8*g* cornering force. Engine, gearbox and axle weigh 128, 31 and 72 lbs (58.06, 14.06 and 32.66 kg) respectively. Price of the structure is kept down by the use of standard rectangular steel tube of constant section throughout the frame.

Maximisation of interior space dictated the use of radius rods in fore-aft wheel location and lateral control by panhard rods. At this stage of the design an unladen weight could be reliably forecast and the graphs in Fig 70 were used in performance prediction — resulting in 0-60 mph acceleration time of 9.4 seconds with the engine tuned to 55-60 bhp.

Body cost is minimised by limiting the unit to five principal parts. Low-cost pre-preg GRP sheets are used, involving a vacuum-bag lay-up comprising two pre-preg laminates followed by Nomex honeycomb — finally enclosed by a single prepreg laminate. This proved to be less costly than a 3-4 oz hand lay-up but very much stiffer as a torsional shell.

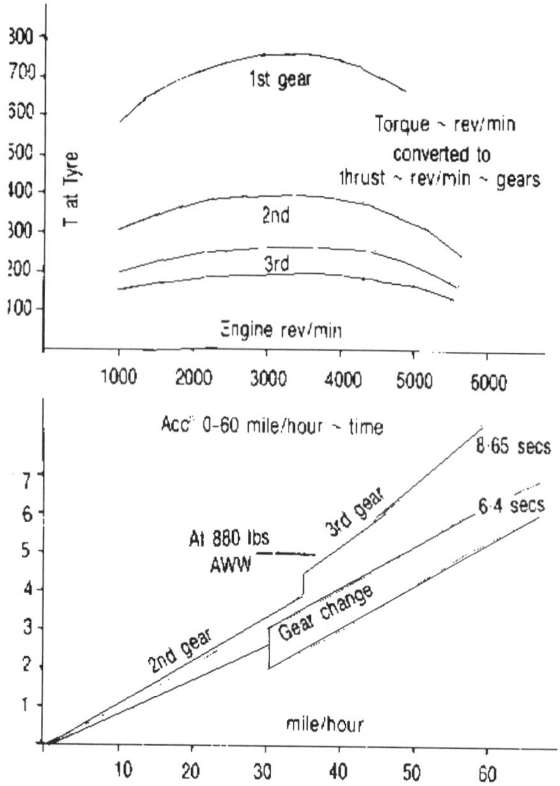

Fig 70: Performance predictions

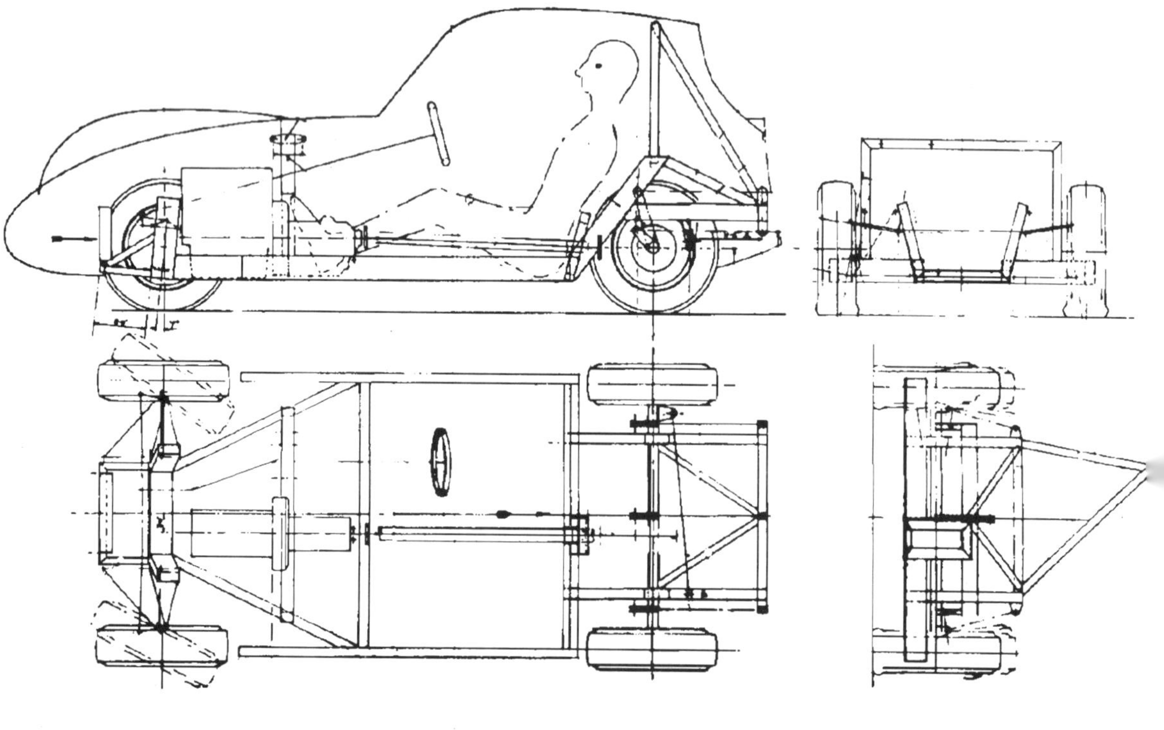

Fig 69: Layout of Costin 2-seater

Three-wheeled vehicles

Steepening oil prices and the desire for maximum manoeuvrabilty could bring a renewed interest in the 3wheeler configuration for both commuter and budget sportscars as long as they are not marketed as any more than motor tricycles.

While the earliest three-wheelers had their configuration derived from carriages such as the pony-and-trap, those to come may yet result from the need to minimise fuel consumption and maximise manoeuvrability. The world's first vehicle to move under its own power, the French steam-powered Cugnot of 1769, was a tricycle and despite the market failure of the Sinclair-C5, its low rolling-resistance configuration is one of its aspects that its many detractors paid tribute to. By the time Carl Benz made the first motor car in 1885, his Motorwagen probably derived its layout from pedal-driven tricycles of the time, motor engineering as a whole developing from cycle engineering at the end of the last century. The Benz's single wheel at the front configuration was particularly suited to front entry into the mid-wheel-base bench seat, an ergonomic factor that could affect the design of a 2-seater C5-type vehicle which has often been mooted.

It has been argued that it was mainly misinformed motoring-press comment that changed the majority of light 'cycle-cars' from three to four wheel configuration, between 1905 and 1910; few could foresee the coming of balloon tyres with well-based rims which would remove the danger of a single wheel 'burst' not uncommon with beaded-edge tyres. Morgan, of course, persisted and success in competition from 1910 onwards brought market success even up until the early post World-war 2 years, backed by a tax concession which was later to benefit Reliant in attracting custom from former riders of motor cycle combinations. About the same time, economy 'bubble' cars such as those of Heinkel and Isetta rose in popularity. High oil prices forecast for 2000 could again bring popularity to such vehicles.

Fig 71: Sinclair C5

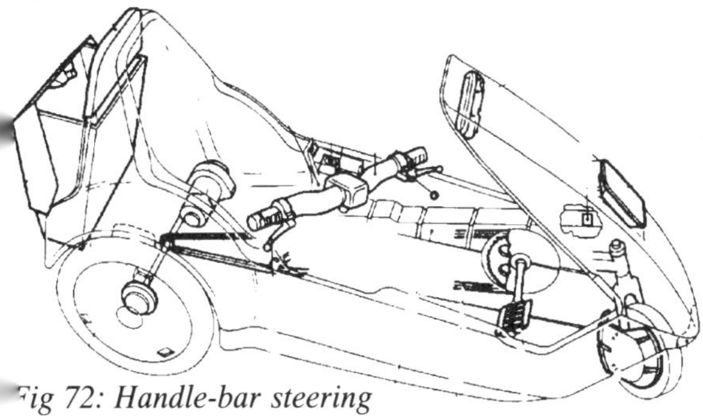

Fig 72: Handle-bar steering

A closer look at C5 technology reveals how a very low cost commuter tricycle could be obtained given the lessons learned in hindsight about the vulnerability of very low profile, poor weather protection and inadequate battery capacity. The vehicle's greatest detraction seem to be an over exuberant marketing campaign that allowed the motoring press to interpret the vehicle as a 'car' rather than cycle derivative. A more acceptable reception might have followed from the selling of the vehicle through cycling and motor-cycling outlets and indeed configuring the vehicle around the higher silhouette of a conventional pedal cycle with power-assistance to a much smaller initial market but one with greater long-term growth potential.

The prone riding position, Fig 71, had ergonomic advantages for applying pedal assistance as did the handlebar steering for accurate handling control, without the expense of conventional gear and linkage control, Fig 72. Conventional size cycle wheels of 26-30 inch diameter would have raised the silhouette substantially above that obtained with the 12.5 - 16 inch ones used, albeit at the cost of greater overall length. While the thermoplastic body moulding was at the time the largest single-piece moulding produced, current technology could no doubt achieve the larger size needed to enshroud a 28 inch wheel machine. The Lotus-designed backbone chassis (Fig 73) was heralded as a neat solution to a low-cost rigid base structure — with flexural compliance built into its rear arms providing a degree of suspension travel at the rear.

Performance-wise, an important feature was the achievement of less than one percent, the 45 kg unladen weight, rolling resistance from the use of narrow section tyres, a technique since exploited by a number of solar-powered vehicles. The aerodynamic drag figure of *CdA = 0.23* metre^2 would inevitably rise with the higher body profile, of course. The epicyclic reduction gear was an extension of cycle practice, the two-stage reduction, with 13:1 overall ratio, being achieved with glass reinforced nylon gear teeth. While a high castor angle was used on the front steered wheel, a modest castor trail of 11 mm was used. Front calliper and rear drum brakes also lent on cycle technology. Transfer drive to the rear axle was via tooth belt and the permanent magnet motor was rated at 250 W and worked in conjunction with a custom-chip centred electronic controller which, in conjunction with an LED display encouraged the rider to guard against motor overheating or undue battery drain. By far the most important lesson of the C5's abortive launch was that the occupant should have been seen as a 'rider' and not a 'driver' (Fig 74), the latter category expecting no compromise from the 'room on wheels' concept with which the contemporary car is perceived. It is for this reason above most that the interest in 3-wheeler car-derivatives from such companies as Morgan, originally, and more recently, Reliant, is on the wane. The famous two-speed Super Sports Morgans which won racing acclaim sixty years ago could no longer expect to generate the size of market for road derivatives which the early vehicles enjoyed. There is a significant minority market, however, which the nostalgia for such cars has kept alive. This is well catered for by such kit-cars as the Lomax 223, Fig 75. Here the Citroen 2CV is cleverly converted as the donor chassis, with an about-face application of a single rear wheel suspension being adapted for the steerable front wheel. The normally fan cooled engine is 'stripped bare to the wind' in the Lomax application, in the manner of the exposed JAP engines of the 2-speed Morgans. A somewhat similar theme marked the design of Reliant's Bond Bug, Fig 76, but some say this suffered from lack of the classic configuration of the 2-speed Morgan. The body shell was an appealing departure but relatively high cost and somewhat conservative mechanical specification probably limited the appeal.

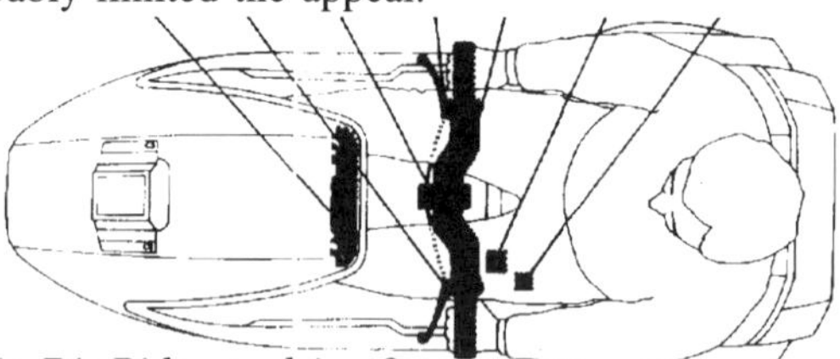

Fig 74: Rider or driver?

Fig 75: Lomax 223 and its modified 2CV front end

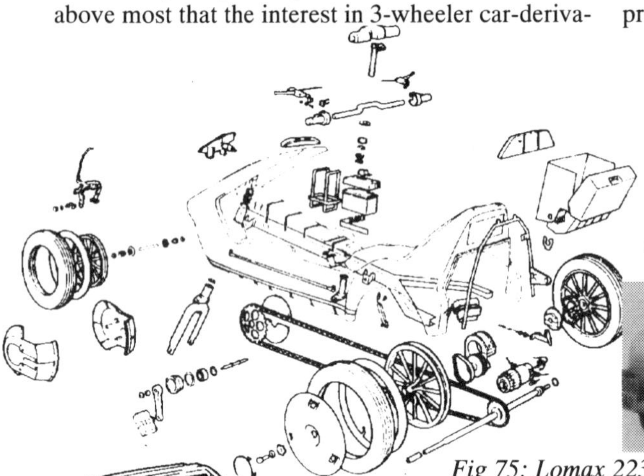

Fig 73: Chassis and running gear

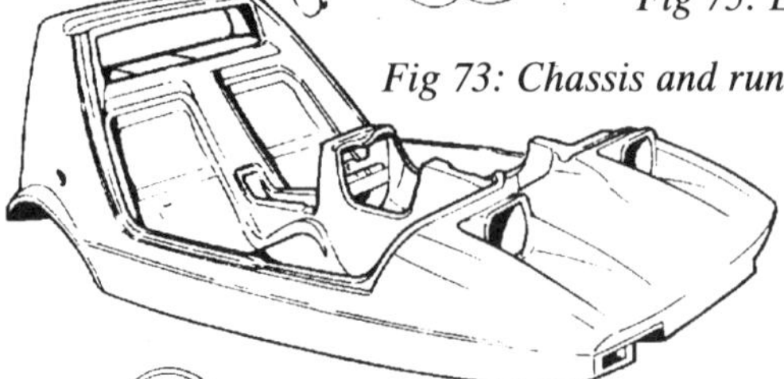

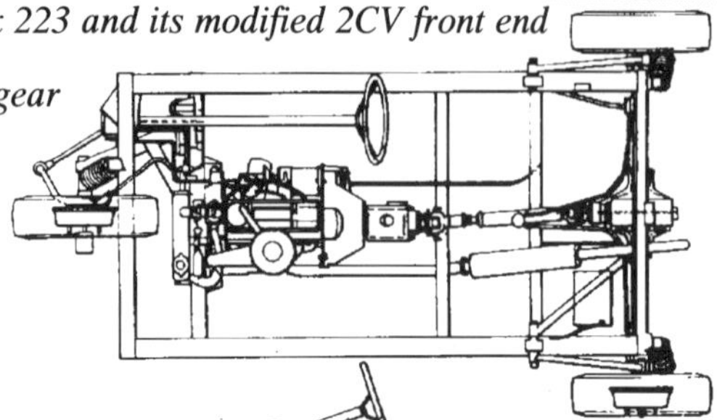

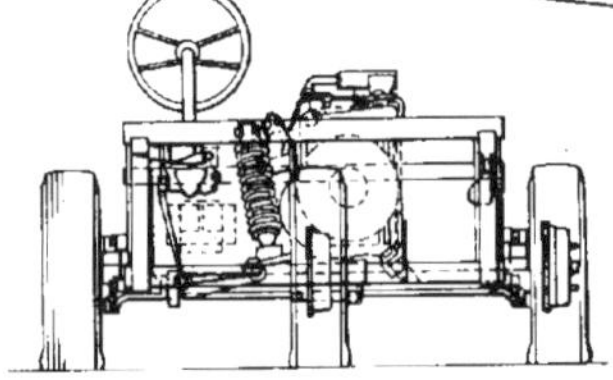

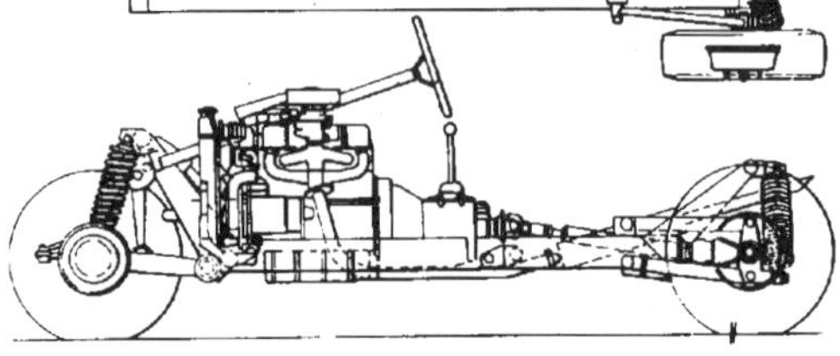

Fig 76: Bond Bug

Motor-cycle engineering

The design of single-track vehicles has a number of distinct approaches relating to the stability of the machine. This section examines the basic force systems involved as well as observing some developing trends in this specialist market.

While the chief virtue of motor-cycles was once perceived as economy, the effect of sport has turned motor-cycling into a pleasure pursuit. The bar-graph in Fig 77 is based on a 1991 Gallup poll and shows that while over two-thirds of motorists seldom drive for pleasure, three quarters of motor-cyclists claim to ride purely for pleasure. There is also an upsurge in urban use for both commuting and courier work. A key characteristic of single-track vehicles is that cornering is performed with a roll-angle adapted to lateral acceleration. The motor-cycle tyre has its maximum of circumferential force, without simultaneous lateral force, at a very small slip-angle — around 0.06-0.08 radians — and lateral forces are mainly developed by roll-angles. Knowledge of power transfer between tyre and road is thus a prime issue, Fig 78. For a given lateral acceleration, wheel slip is far from constant and depends upon curve radius, friction potential, driving speed and tyre condition. Acceleration and braking torques also have a substantial effect. In tests carried out by Darmstadt Technische Hochschule, the roll angle force was found to be caused by the tendency of the cambered tyre to rotate around the point of intersection of the steering axis with the road. Because the tyre is forced on to a much larger radius, however, it is necessary from a balance standpoint to keep tyre slip angle as low as possible when cornering. Tests on a 130/90 V17 tyre held at a constant inflation pressure of 2.5 bar, showed in Fig 79 that dynamic tyre radius depends considerably on roll-angle. Friction maximum is obtained at about 0.8% slip, Fig 80, which shows the typical steep rise at the lower slip range when riding without both slip- and roll- angles. Steep descent of the curve after reaching 100% braking results from the change from rolling friction to sliding friction. Above 100% slip there is another change to rolling friction.

Corresponding data for a roll angle of 45 degrees show the change in characteristics. Friction

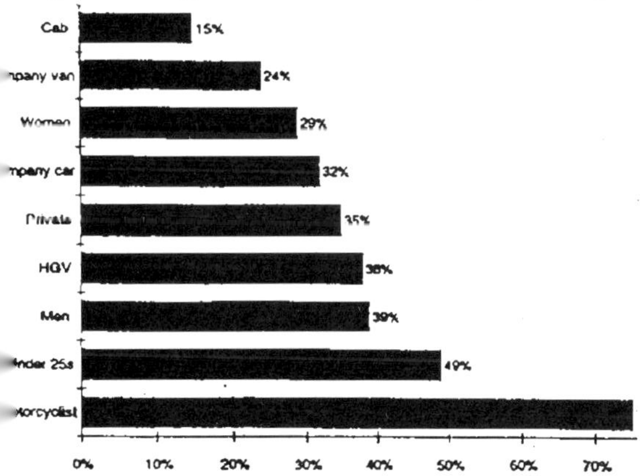

Fig 77: 'Driving for pleasure' survey

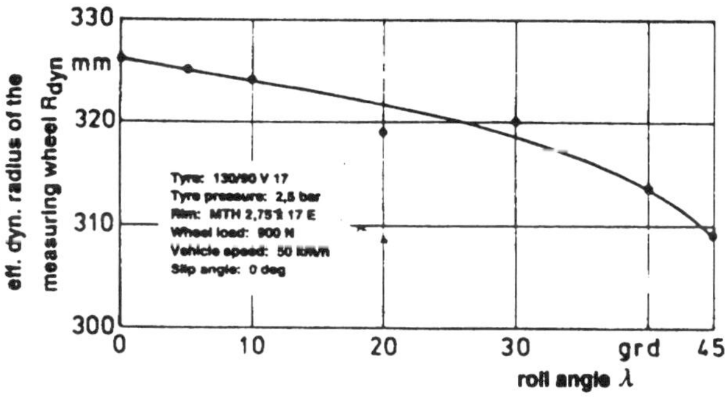

Fig 79: Dynamic tyre radius

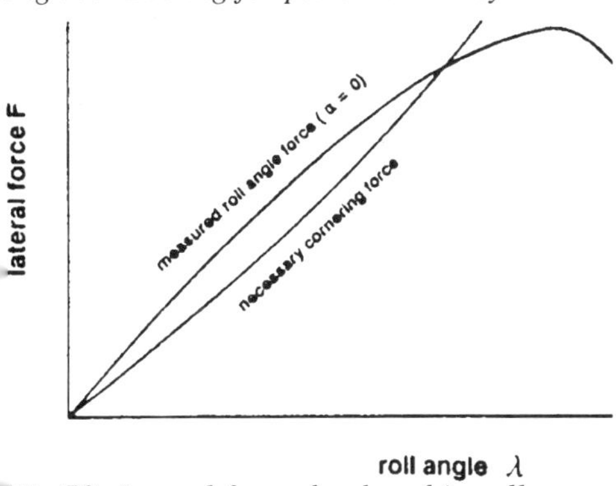

Fig 78: Lateral force developed in roll

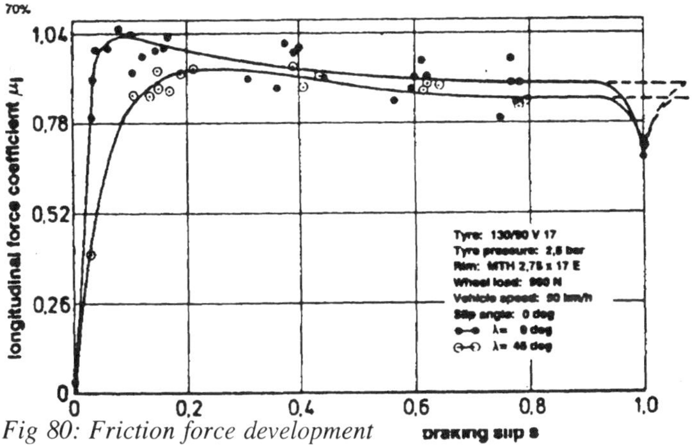

Fig 80: Friction force development

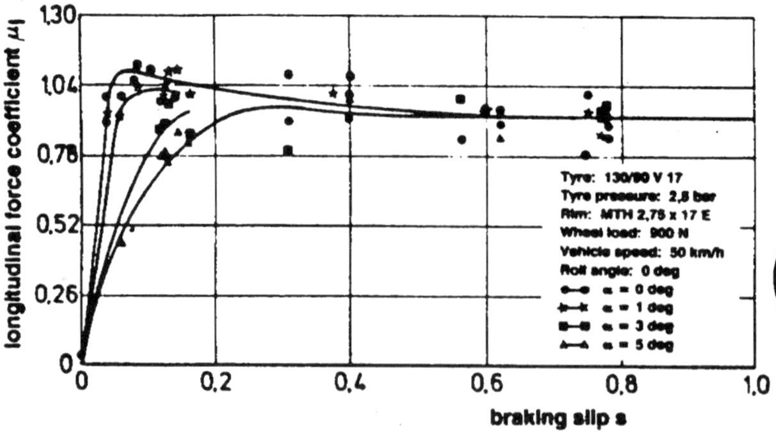

Fig 81: Slip and roll observed simultaneously

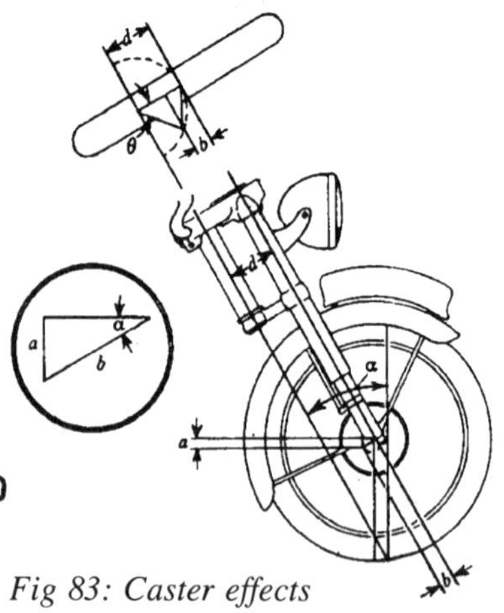

Fig 83: Caster effects

maximum is not as distinct and is reached only between 20-30% braking slip. At higher slip ranges friction levels are only slightly lower than at a roll angle of 0 degrees. When observing slip- and roll-angles at the same time, Fig 81, the roll-angle influence on transferable force predominates in the lower and negative slip-angle range, decreasing with increasing slip angle.

The steering and stability of single-track vehicles was examined in a classic study of the same title by R A Wilson-Jones[1] of the former Enfield company. It was clear that the idea that stability of the rider in 'dynamic' balance on the cycle did not just arise from the gyroscopic effects of the wheels. In fact this only occurs at high speeds and lower down the speed range it is the tendency as the machine tends to fall. That is from the wheels steered ahead position — and the gyroscopic action of the front wheel merely turns the steering in the direction of the fall. This same front-wheel gyroscopic effect also provides self-centering torque for the wheel when it is steered from the straight-ahead when the machine is at speed. The way in which the resultant of the weight of the machine plus centrifugal force acts in the plane of the wheel is clear in view (c) of Fig 82. When the machine corners at slower speeds in the near upright condition of (d) the point of action of AB is behind the steering axis by the extent of the caster trail which provides the self-centring action. These self-centring effects can, of course, be dangerously delayed by excessive friction in the steering head. The effects can also be tuned by fitment of steering dampers, altering the inertia of the steered assembly and by altering gyroscopic torque by changing wheel weight. Steering head angle is another important consideration which can aid self-centring due to the effect of the vehicle weight being lifted as the steering is turned. This is a comparatively small effect, raising the machine by: $d \sin\alpha(1-\cos\theta)$. A second effect is that it tends to increase the angle of lean of the front wheel when the steering is turned and at high steer angles, a point at the side of the tyre contacts the ground, thus lowering the steering head. More important is that the head angle produces the necessary caster trail without negating the initial effect of lifting the machine.

CG height of machine and driver has two opposing effects. As the trail causes the contact point of tyre and road to move left when the steering is

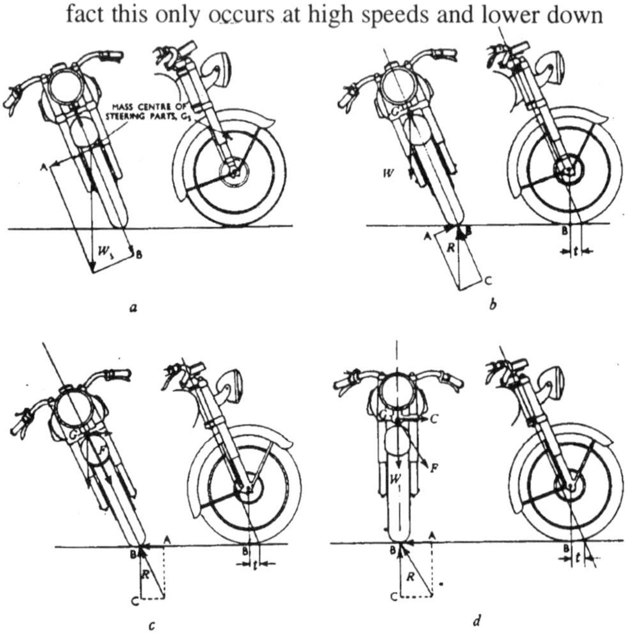

Fig 82: Resolving weight and centrifugal force

turned left, and vice versa, the front end of the line joining these points is deflected to one side of the machine, on steering, Fig 83, and a high CG machine is required to lean over a greater distance. However, due to the width and curvature of the tyre tread, the line of contact with the ground is also displaced to one side as the machine leans and the effects oppose one another Typifying recent trends, the new departure for BMW (who now feature ABS on some machines, Fig 84) of the F 650 Euro bike developed in conjunction

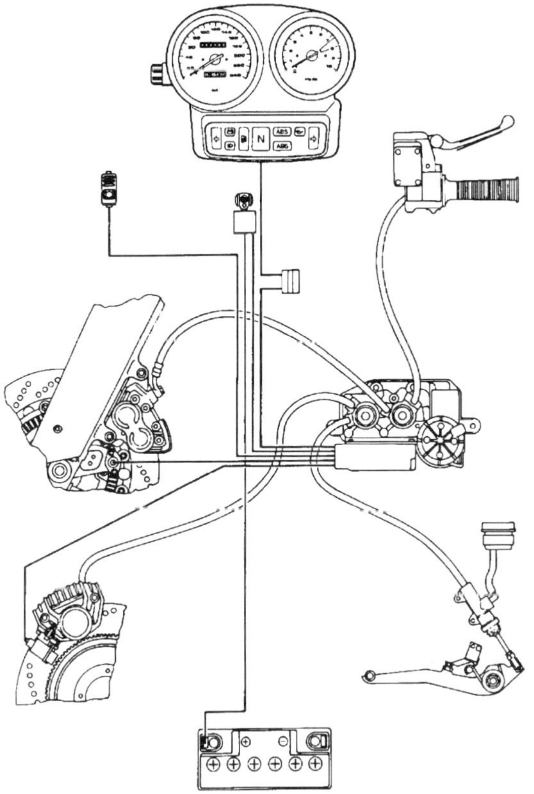

Fig 84: ABS II by BMW

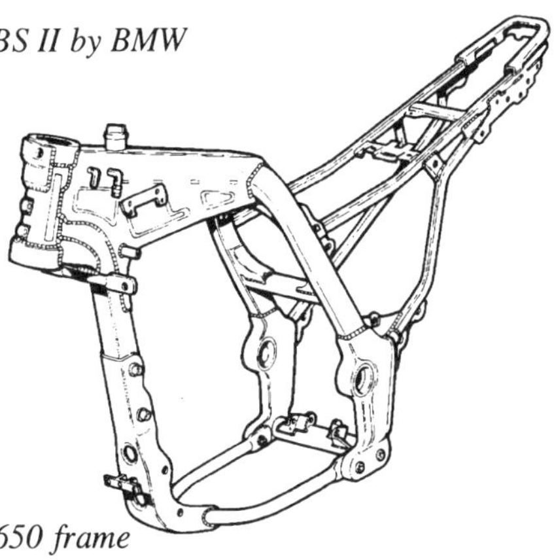

'ig 85: F650 frame

with Aprilia and Bombardier-Rotax, uses the latter company's 650 cc engine. Designed to combine the attributes of fun-bike and endurance machine, the F650 has engine outputs in either 34 or 48 bhp levels, a weight of just 417 lb (190 kg) and modest seat height of 32 in (813 mm). The engine is a liquid-cooled four valve unit developing up to 57 Nm torque. The gearbox is integrated within the aluminium alloy engine housing which is split vertically in two halves. The covers for the engine housing, having clutch at rear-left and alternator rear-right, are also made in aluminium alloy while the chain pinion cover is in plastic. Cylinder liners are coated in low-friction nickel-silicon while the cast piston is in 'box' configuration, cooled on its underside by splash oil.

To minimise vibration on this large single-cylinder engine, a gear-driven compensation shaft is provided which runs at crankshaft speed. The engine has dry-sump lubrication and obtains its oil from a 1.5 litre reservoir incorporated in the upper part of the frame. Fuel is supplied through two Mikuni constant-vacuum, automatic choke, carburettors of 33 mm choke size while air is fed to them via a 6.5 litre silencer/filter. The mapped capacitor-ignition system uses two sparking plugs in order to shorten flame travel and minimise pre-ignition, as well as improving part-load combustion. The stainless steel exhaust system has a 7.5 litre expander volume. Transmission of power is via a multi-plate clutch and dog-type five-speed gearbox with gears lubricated by splash oil. An interesting aspect is the recyclability, by registered BMW dealers, of the optional exhaust catalyser to preserve the precious metal. Otherwise the machine is serviced on a 10,000 km interval basis.

The frame of the machine, Fig 85, has a pressed steel section single loop, which also incorporates square-section tubing, further reinforcement being by bolted-on round-section tubing. While the front wheels run in conventional telescopic forks, the rear wheel is located by a delta shaped boxed swinging-arm. Front and rear spring struts are by the fork maker Showa and are adjustable for spring pre-tension and damper-rate. Progressive rate is obtained at the rear by a lever mounting system for the rear strut. Overall wheel travels are 170 mm front and 165 rear. The rear wheel is driven by an O-ring chain, with permanent lubrication between the rollers, running at 1:2.93 transmission ratio. Rubber cushion supports are built in to the rear drive sprocket.

Developing-country vehicles

Application of an 'appropriate-technology' philosophy is necessary in design of vehicles for operation in very poor countries where conventional manufacturing facilities are simply not available. A combination of the intelligent use of systems from donor vehicles, or bin-picking from major volume builders' overseas plants, with the choice of the most rudimentary engineering solutions, is the usual approach.

The Intermediate Technology Development Group have been prominent in promoting the view that "appropriate" technology should be applied to third world transport needs. Low cost motorised vehicles are particularly seen as the solution to poor areas where an intermediate range vehicle, between that of human porterage and conventional motor vehicles, is appropriate. Opportunities exist, it is argued, for transfer of low-cost vehicle technologies from regions of use to those of non-use.

The Group focus on utilitarian load-carriers for goods or several passengers. In rural India, for example, it is reported that 3-wheeled vehicles operating on level ground with diesel engines of 4.5 kW carry loads up to 1000 kg and those with engines of 8.25kW up to 2000 kg. Travel speeds are 20-25 km/h. Four wheeled vehicles having 6-8.25 kW diesel engines have payload capacity of 1500-2000 kg. In farming areas, using single-axled tractors powered by 4.5-8.25 kW diesels, trailers of 750-1500 kg capacity are also used for transport tasks. There are also 4-wheeler conversions of motor-cycles of the type shown in Fig 86, in use in such Far-Eastern locations as the Phillipines, and 3-wheelers made from chassis extensions of motor-cycle frames, Fig 87.

In the latter case the centre of the rear load platform will be just ahead of the rear axle. Weight bias of the load should not exceed 60% in front of the axle, however. Stability will be interdependent on wheel size and fore-aft CG position, the Group suggest. To provide a simple differential drive unit for the rear axle, former car axles are often converted to chain drive in India. A sprocket is attached in place of the crown-wheel and the rear of the housing cut away to accommodate the drive chain and alternative lubrication methods provided! The axle is reversed to allow the pinion drive flange to face rearwards. Where used-car drive axles are unavailable, a simplified free and fixed hub beam axle of the type shown in Fig 88 may be fabricated. Heavy gauge tube of some 50 mm diameter forms the beam which is machined at either end to receive bearings. A final-drive sprocket is bolted to a boss on the tube.

For purpose-built three-wheelers the Group favours a single driven wheel at the front to overcome the differential problem, despite inferior traction and load-carrying. The engine is recommended to pivot

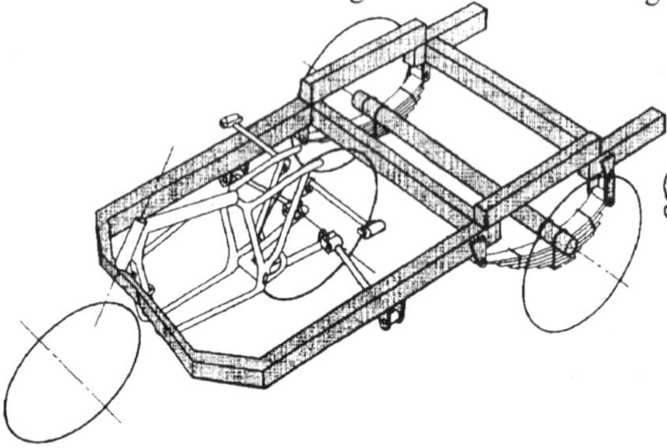

Fig 86: Four-wheeler conversion

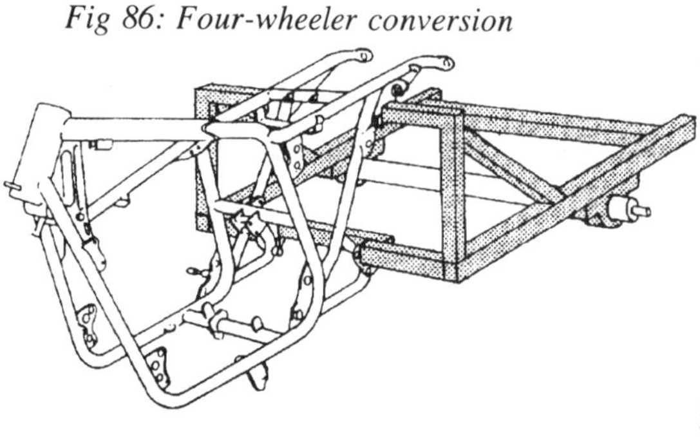

Fig 87: Three-wheeler conversion

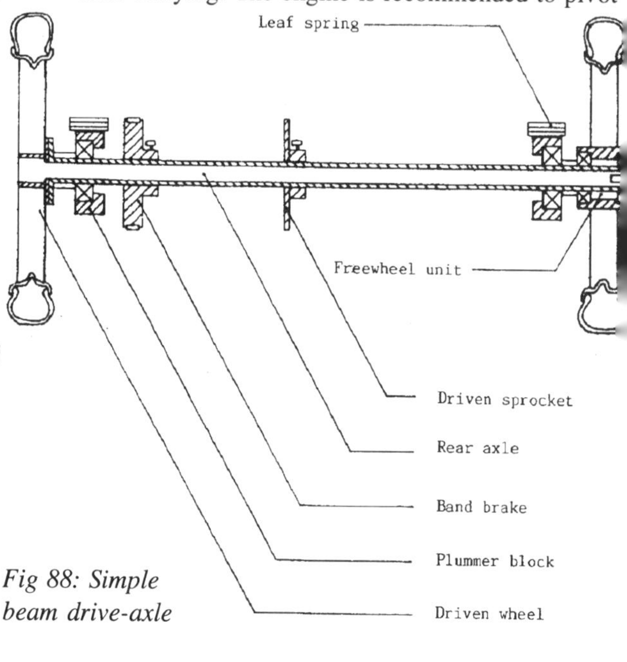

Fig 88: Simple beam drive-axle

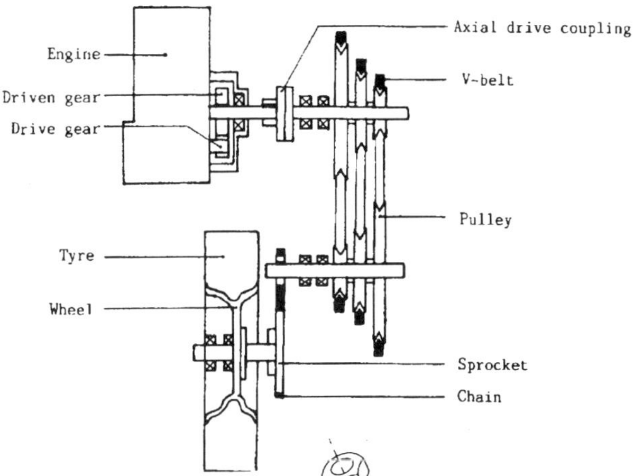

Fig 89: Trag change-speed mechanism

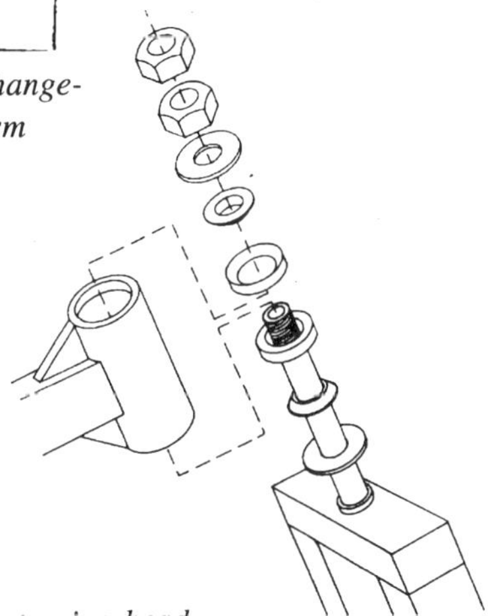

Fig 90: Simple steering head

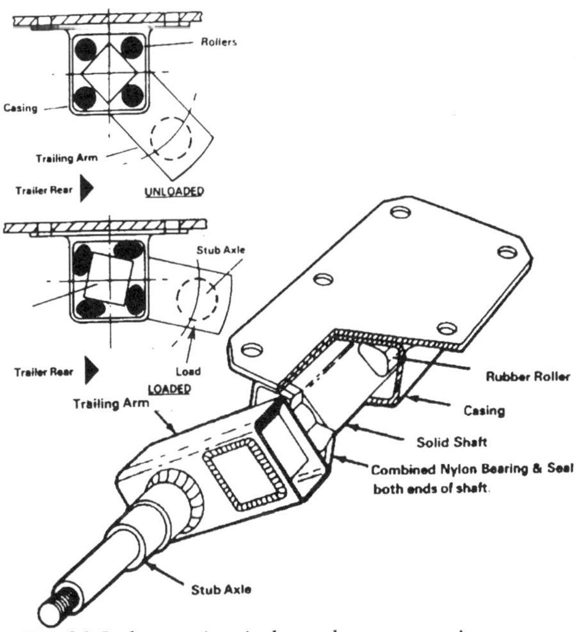

Fig 91:Indespension independent suspension

with the front steered wheel, to simplify the drive-line, and handle-bar control is used. A tensioned-belt clutch may be used as a simpler solution than the conventional plate clutch. Here a twin-pulley vee-belt drive is arranged with the belt set deliberately slack. A spring-loaded idler is used to take up the slack in the belt and the drive disengaged by releasing idler pressure and vice versa. Power engagement is less smooth than with a friction clutch but belt flexibility will take up some of the shock. Where the engine and clutch are mounted independently on the chassis, a flexible coupling should be used to take up some of the inevitable misalignment. Where a speed reduction is required to the final drive, a chain-drive with different sprocket sizes may be the easiest way to achieve this and gross misalignments can be thus better tolerated.

When a change-speed mechanism is required this may be most easily achieved with a stepped-pulley belt-drive system. Fig 89 shows the arrangement designed for the Trag vehicle. Here the engine has a factory-equipped step-down drive of 6:1 while the variable speed portion of the transmission is made up of an upper and lower set of three pulleys (of 1:2, 1:1 and 2:1 ratios) with equal length v-belts positioned over them. Three corresponding levers are positioned on the handle-bars; when one is pulled, the spring-loaded idler-wheel on the desired belt pushes the belt tight and causes it to grip the pulley. The arrangement is reported to have given good service in many parts of the world.

The main modification required to the chassis frame of a three-wheeled vehicle is the attachment of a steering head, a steel tube supporting the front wheel in bearings, Fig 90. Diagonal members are also often needed to brace the rectangular frame. Engine mounting for a stationary-type unit usually involves standing the feet on a suitable base-plate or framed-platform of sufficient rigidity. Practice has proved the suitability of 38 x 8 mm angle sections for this purpose, with engines up to 6 kW. Slotted holes in the fore-aft direction on the platform are a convenient way of adjusting the engine to the drive system before tightening down the mounting bolts.

Suspension systems for Third-World vehicles are usually adaptations of factory-made systems but proprietary caravan suspensions can be a useful standby. The Indespension system of Fig 91 has a square-section member inserted into a hollow axle at an interior face angle of 45 degrees and a rubber roller located in the resulting triangular section openings. Rotation of the radius arm causes compression and shear of the rubbers to create the suspension springing.

The front suspension must accommodate steering of the wheels whereas the rear must be capable of providing continuous drive to the wheels. For the three-wheeler, leading-link forks from a suitable motor-cycle are likely to be the easiest solution. For the rear wheels an effective but somewhat complex solution is to use chain drive with driven wheels mounted independently on trailing arms, in the manner of the Bajaj layout shown in Fig 92. The drive chains run from sprockets on an intermediate shaft mounted parallel to the pivoting axle of the trailing arms; the chains lie parallel to the trailing arms with the driven sprockets affixed to the rear wheels. A differential unit is incorporated within the intermediate shaft.

Stability is substantially affected by the castor trail on the front forks of three-wheelers. Two-wheelers require more trail because they are inherently less stable, of course, than three-wheelers so there is likely to be a reduction required in castor trail when mounting motor-cycle forks to three-wheelers. However, the amount of castor angle required is likely to be less, due to the heavier front-end weight; but a greater angle is required if high-speed operation on poor surfaces is the aim of the design.

Where the function of the vehicle is extended to agricultural as well as transport operations a more studied approach to the design is necessary, as exemplified by the Tinkabi tractor which has operated successfully in Swaziland. Here the requirement was to replace the use of oxen with a vehicle having a platform to carry a minimum 500 kg while having under-clearance to perform row-crop work. The resulting dimensions are shown in Fig 93. A payload of 1000 kg was achieved, 70% of the unballasted weight being over the rear wheels. Oxen-style implements are clamped to the rear drawbar which is hand-raised by a spring-assisted lever. Pumps, circular-saws and sprayers can also be drawbar-mounted and belt-driven from a pulley attached to the engine flywheel.

The tractor is powered by a Kirloskar 2-cylinder, air-cooled, direct-injection diesel which is hand-started and produces 12kW at 2000 rpm. Power is transmitted to the rear wheels by a hydrostatic transmission with variable-displacement pump and hy-

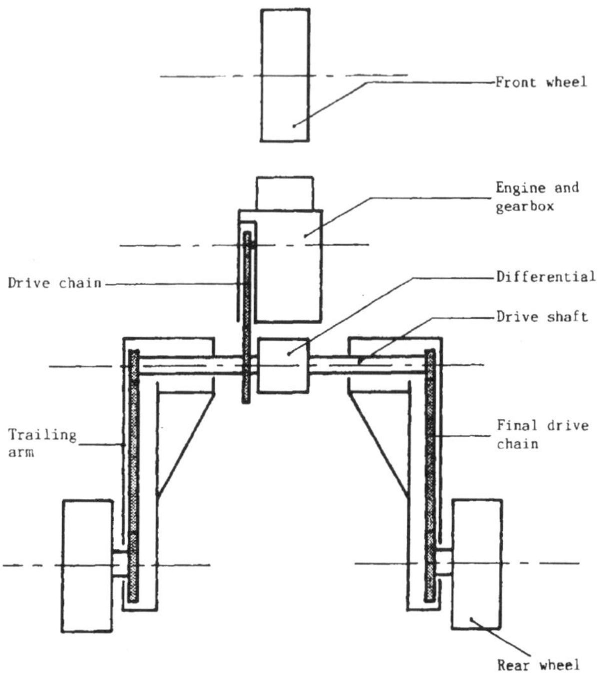

Fig 92: Bajaj rear wheel drive

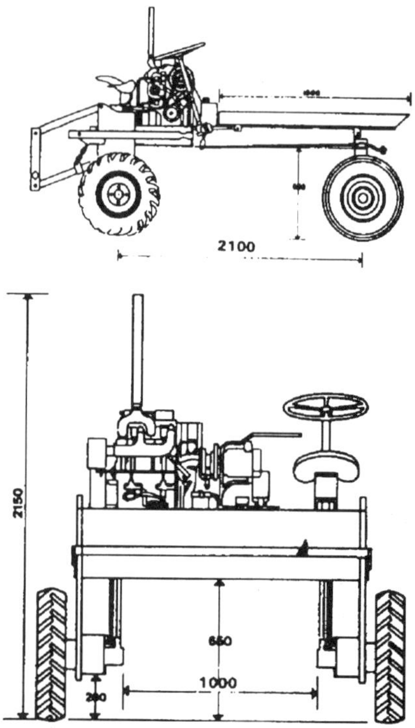

Fig 93: Tinkabi tractor

	Cost per unit	
	$US CIF Swaziland	$US FOB ex works
Diesel engine	756	630
Hydraulic pump	258	241
Hydraulic motors	515	471
Steering unit	63	55
350 kg mild steel	122	
Tyres 2 × 14 in	82	
Rims 2 × 15 in	64	
Oil		
(50 litres SAE 20)	29	
Bearings/fasteners	30	
Sundry items	20	
	1939	
Implement materials	120	
Fasteners/sundries	20	
Total per set	140	

Fig 94: Tinkabi costings

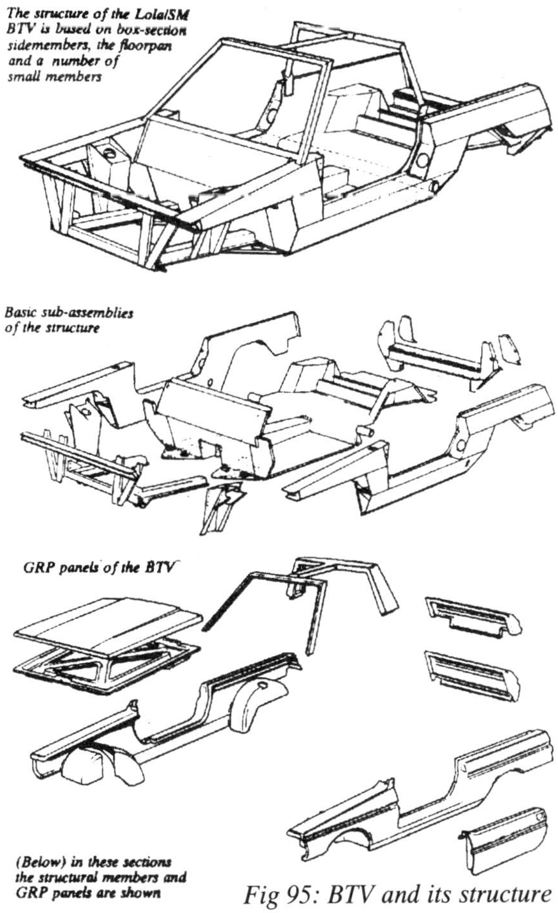

Fig 95: BTV and its structure

draulic motors on the wheel mountings. The pump is gear-driven from the engine crankshaft at 10% step-up speed and has a displacement of 75 litres/min at 2000 rpm. The motors are bi-directional and drive the wheels through 4:1 epicyclic reduction gears. Each has a maximum power output of 11 kW. A single lever control actuates the swashplate mechanism of the pump to control speed direction and braking of the tractor. Maximum drawbar pull of 818 kg is obtainable and speed is infinitely variable up to a maximum of 10 kph. The tractor will pull a plough, capable of turning a single furrow 300 mm wide by 200 mm deep, at a speed of 6.4 kph. Fig 94 shows tables of component and production costs at the time of the design of the machine.

A yet more sophisticated vehicle known as the BTV (Basic Transport Vehicle), Fig 95, was the subject of a study between Lola Cars and Specialised Mouldings Ltd, with Chrysler UK. By applying structural expertise to the base-frame of the vehicle, a more efficient Third-World workhorse results which also has pleasing body-style, with the aid of GRP mouldings. Key structural elements are floor, box-section sidemembers extending from front to rear, dash-bulkhead and two crossmembers, all of which can be made on a press-brake from guillotined panels. The floor-pan comprises two sub-assemblies, and a cross-tube, while the dash is a simple folded panel with flanges turned over at the edges and contoured to fit the sidemembers. The rear floor is basically an angle section, consisting of heel-board and floor, cut out from the wheel-arches. These are reinforced by Z-section transverse members above the rear axle. A tube beneath the floor reacts the loads from the rear-spring front-hangers. Forward of the floor there is a pair of box-sectioned longitudinals comprising top-hat sections and closing plates. At mid-span of these the towers for the front suspension struts are fixed with their bases just forward of the triangular extensions from the dash-panel.

Chapter 13: Heavy special-vehicles

Specialist goods vehicles
Heavy haulage vehicles
Road tankers
Pressurised tankers
Bulk powder vehicles
Open top bulkers/tippers
Calculating tipper stability
Bulker charging/discharging

Specialist goods vehicles

The need to meet the requirements of a wide range of special vehicle operations has led most builders of goods vehicles to establish SVO departments which are effectively special vehicle units within the volume chassis-cab builders. From them, strict regulatons are issued to CV body-builders for producing specialist variants; a selection of the rules are seen here.

Because the chassis-cab is essentially a compromise design to meet a multiplicity of operations, considerable engineering effort has to be devoted to adaptation for special purpose. With volume-build constraints on the number of available wheelbase and chassis length options, the most frequent adaptation is change in frame length. Some manufacturers, while forbidding the welding of attachments to chassis frames, do allow welding to be used in the cut-and-shut process of frame shortening — albeit under carefully controlled conditions. Requirements normally include use of dry welding-rods and the use of visual inspection methods after fabrication. The weld hump has usually to be ground flat and any strike marks or spatter ground off to avoid stress raisers. When holes are closer than normally allowed following the 'shut', selective ones must be filled by bevelling the hole to two-thirds metal-thickness then fitting a copper plate to the inside of the hole — and

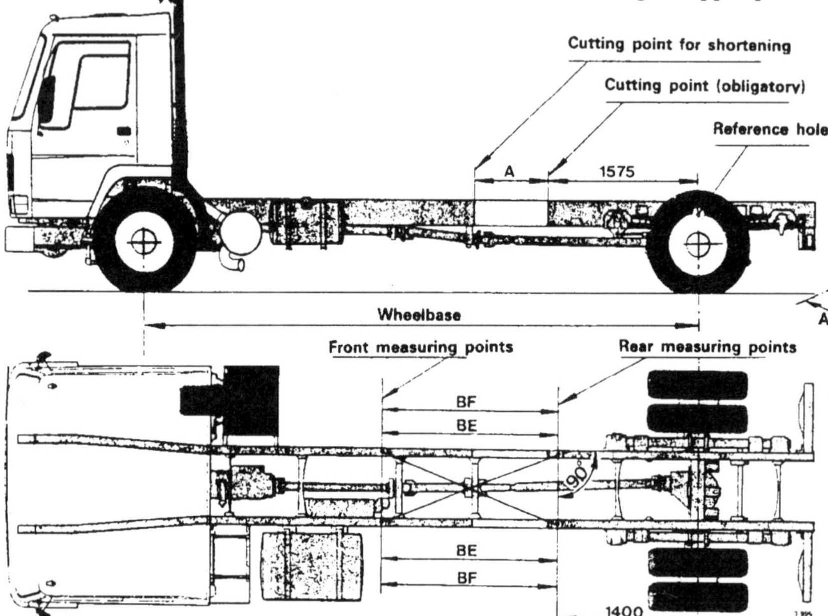

A = Shortening or extending section
B = Distance between measuring points before wheelbase alteration
BE = Distance between measuring points after wheelbase alteration

Fig 1: Recommendations for chassis cutting

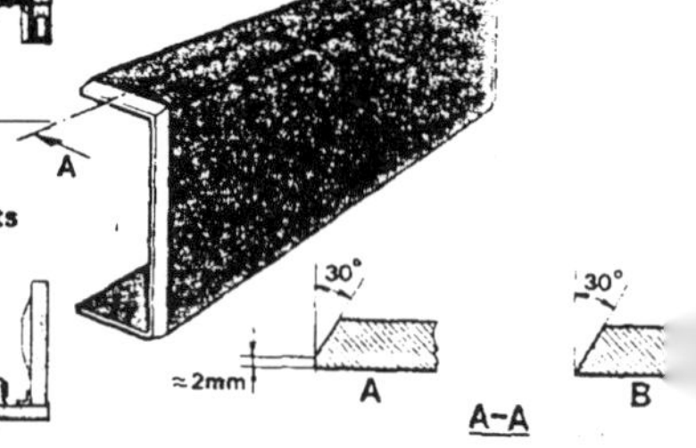

A = metal arc welding
B = gas-metal arc welding (MAG)

Fig 2: Bevelling of the joint to be welded

when the hole is larger than 20 mm diameter there may be a requirement for using a bevelled washer to block the hole prior to plug welding, which must be finished from the inside. Generally there is the requirement to strengthen the frame joint with a reinforcing profile to cater for any loss of temper in the steel as it is welded. This will typically extend 400 mm either side of the joint and in the case of a wheelbase extension a single reinforcing profile should cover both joints. Where the lengthening operation increases the allowable distance between crossmembers, additional members must be fitted. Where the crossmember position coincides with the reinforcing plate inside the chassis sidemember channel profile, the part should be cut at its centre span, reduced by the appropriate length, and re-welded.

Chassis-makers will normally recommend cutting points, those applying to the Volvo FL7/10 being shown, for example, in Fig 1. That company's procedure is to stand the truck on a flat floor and remove prop-shaft, rake-lines and electrics in the vicinity of the joint; then to mark out the position of the rear measuring points, making sure they are perpendicular to the web of the channel section; the distance BF to the front measuring point is measured so that the checking measurement BE, after alteration, is at least

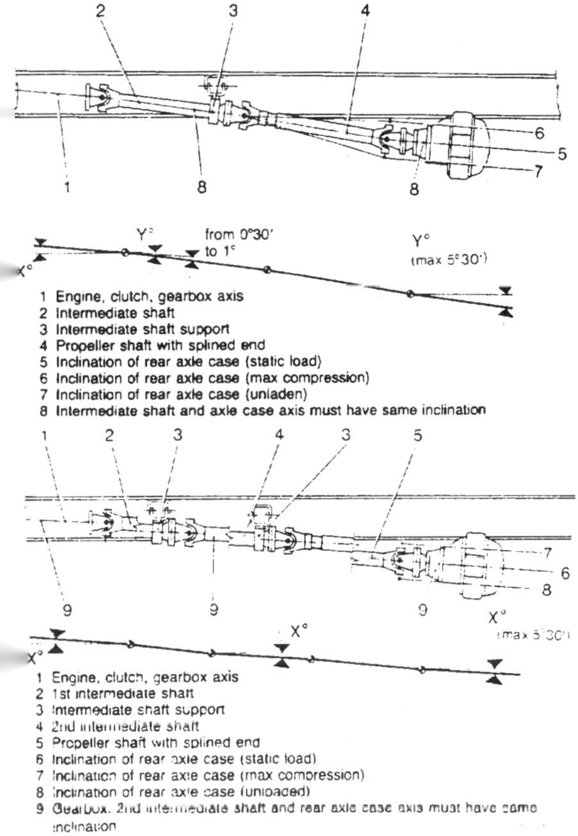

Fig 3: Single and double intermediate shafts

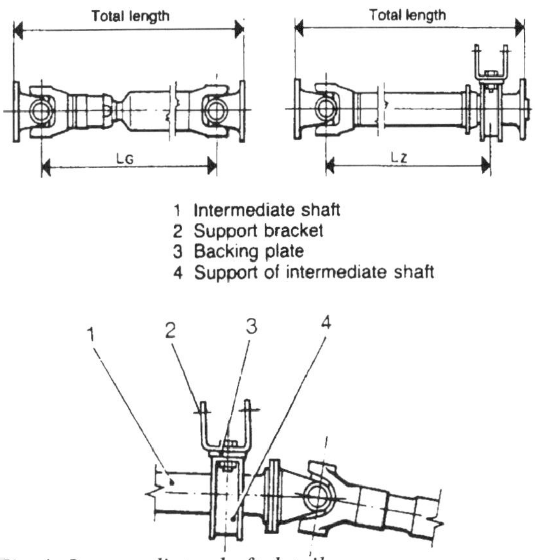

Fig 4: Intermediate shaft detail

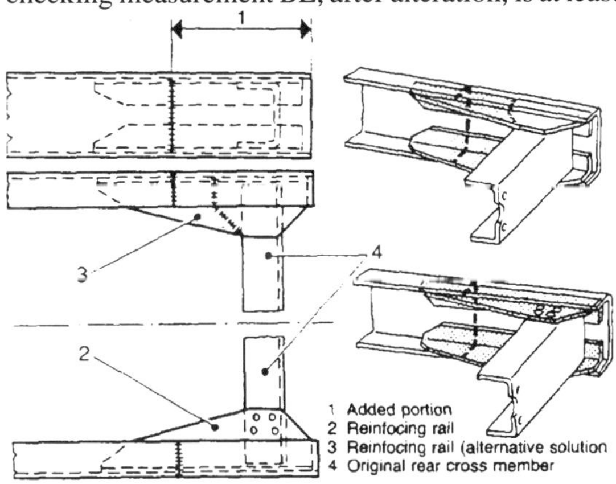

Fig 5: Welded crossmembers

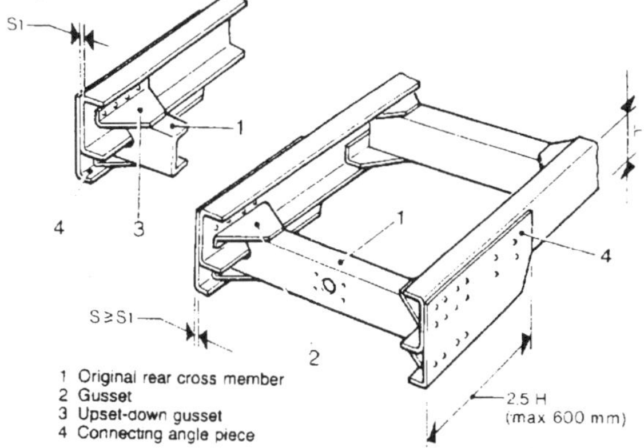

Fig 6: Riveted crossmembers

1200 mm. Trestles are then placed under the chassis to keep the frame truly horizontal; the member is then cut to a true right angle and its weld surface bevelled, together with that of the extension profile, as in Fig 2. The frame is then repositioned to obtain the correct BE measurement on both sides and that diagonals are equal to within 2 mm. This measurement procedure is repeated after tack welding — before the full weld is applied.

HGV drive-line

Prop-shaft dimensioning is critical on the new-length vehicle and manufacturers will define strict dimensional limits between flanges and between bearing supports, according to maximum rotational speed of the installed shaft — Fig 3, from Iveco. It is a company rule that individual shafts must be approximately the same length in multi-segment systems, and the difference in length between a driven and splined shaft must not exceed 600 mm (400 mm for similar duty shafts), Fig 4a. A margin of 25 mm minimum should be allowed for travel when the splined shaft is totally retracted. When extended the sliding sleeve should cover the splined stub for a length at least twice the spline diameter. Difference inclination between axle pinion axis and that of the gearbox output should not exceed limits of 0.5 - 1.0 degree. This can be adjusted by wedging the spring support pad on the rear axle — or by adjusting the torque rods of an air-suspension — but maximum axle pinion tilt is 5.5 degrees. When the vehicle is laden, these inclinations must also be checked: according to whether the rear-axle flange is lower or higher than that of the gearbox, inclination of the axle pinion and driven shaft must greater or less than, respectively, the inclination of the gearbox output axis. For substantial wheelbase lengthening an intermediate shaft, Fig 4b, may be necessary — requiring the same inclination as these axes.

Towing provision for heavy vehicles

Increase in frame overhang, rather than wheelbase, brings particular recommendations with respect to rear crossmember. In the case of Iveco, views in Fig 5 show ways of arranging a bolted crossmember and Fig 6 a riveted one. Fig 5a covers length increases of not more than 300-400 mm — the reinforcing L-gussets being the same thickness as the original plate. Connections are made with bolts, to class 8.8, having the same diameter as the original rivets; an alternative welded connection is also shown. For length increases over 350 mm, the configuration in Fig 5b is rcommended. A lowered rear crossmember may be required for certain towing applications and alternative configurations in Fig 7 are suggested by Iveco. Thickness of the outer reinforcing angles must be not less than that of the sidemembers and they must extend over an area at least 2.5 times the height of the sidemember and bolt sizes must be such as to account for the greater moment that is transmitted by the downward displacement of the towing pintle.

The towing of trailers with rigid drawbars imposes additional stress on the towing vehicle chassis for which the rear crossmember must be further strengthened. The alternative proposal by Iveco in Fig 7, uses a C-section reinforcement — the company insisting that the connection between sidemember also be reinforced. This shows the reinforcement on the inside of the crossmember, joining it to the

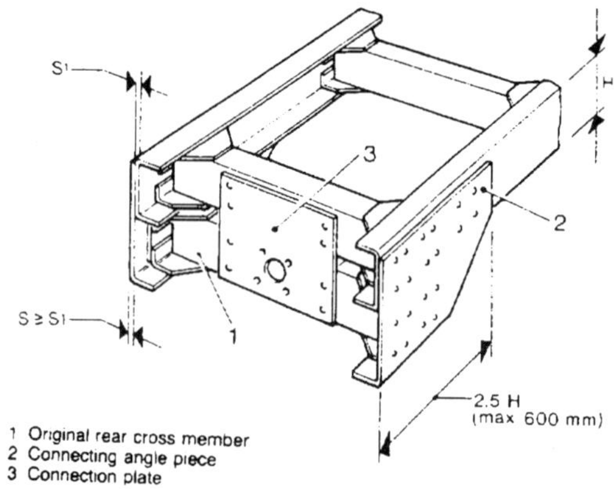

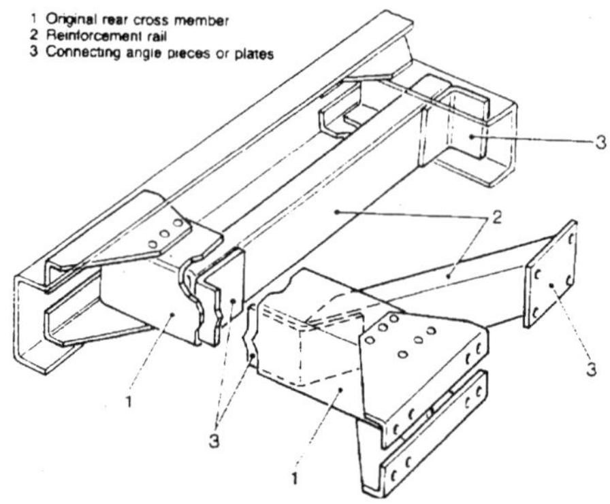

Fig 7: Alternative reinforcement schemes for drawbar mounting

vertical web of the sidemember — or to the following crossmember of the frame. The alternative is to mount a box-section beneath the crossmember — anchored at its ends to the same web and joined at its centre to the crossmember.

Additional axles for HGVs

The fitting of auxiliary axles requires modification to chassis frames such as to upgrade their loading capacity according to the additional carrying capacity of the axles. Fig 8 shows some possible alternatives, each of which must extend over the full span of the sidemember. Body mounting points should be used as a means of connecting the reinforcement to the sidemember in these Iveco proposals and a shear resistant connection should be used between midspan and the rear of the frame. Figs 9 a and b compares installations forward and aft of the drive axle, the former requiring reduction and the latter, extension.

Power-take-offs (PTOs) for HGVs

Chapter 19 deals with gearbox PTOs but other types and other loactions on the chassis are available. Fig 10, due to Leyland DAF, shows some of the preferred ones for different SPV adaptations. The company recommend that for a Z or W drives, the configuration should satisfy the criteria shown as Figs 11 a and b. Angles described by the drive-shafts should be equal and not exceed 8 degrees (6 for front-end PTOs): inset is shown how to obtain proper alignment for ensuring constant-velocity operation of the driven equipment. The sliding joint should provide at least 8 mm, forward, and 6 mm, rearward, of movement. In PTOs from the front of the engine the company gives recommendations as in Fig 12 for vee-belt drive from a pulley co-axial with the crankshaft for two typical model series. For flywheel PTOs, Fig 13 applies.

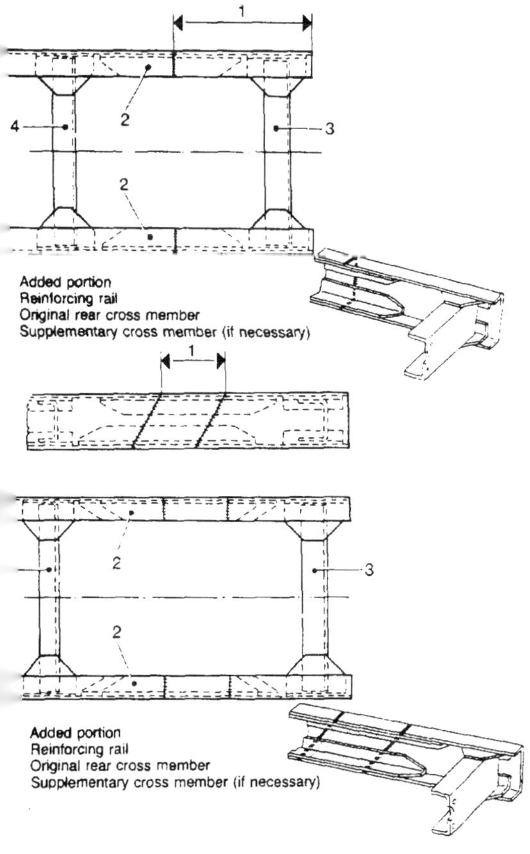

Fig 8: Frame bracing for extra axles

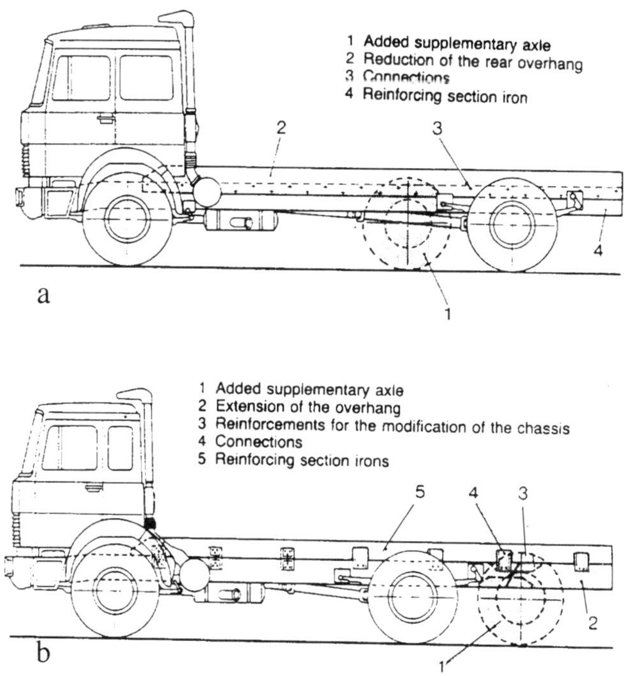

Fig 9: Fore and aft added axles

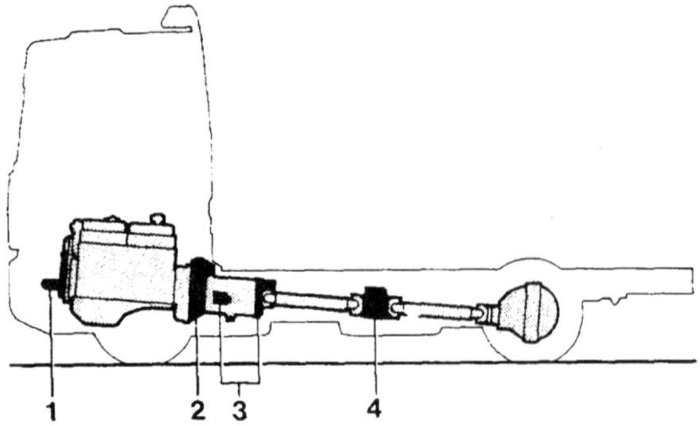

Application	P.T.O. 1	2	3	4
Refrigerated transport	x			
Air-conditioning	x			
Fibre transport			x	x
Tankers			x	
Bulk transport			x	
Aircraft refuellers		x		
Refuse collectors	x		x	
Road sweepers	x		x	
Gully emptiers	x	x	x	
Sewer cleaners		x	x	x
Tippers			x	
Demountables			x	
Concrete mixers		x		
Concrete pumps		x	x	x
Vehicle loading cranes			x	
Recovery vehicles			x	
Winches			x	
Hydraulic platforms			x	
Turntable vehicles			x	
Fire engines	x	x	x	

Fig 10: Preferred PTO locations

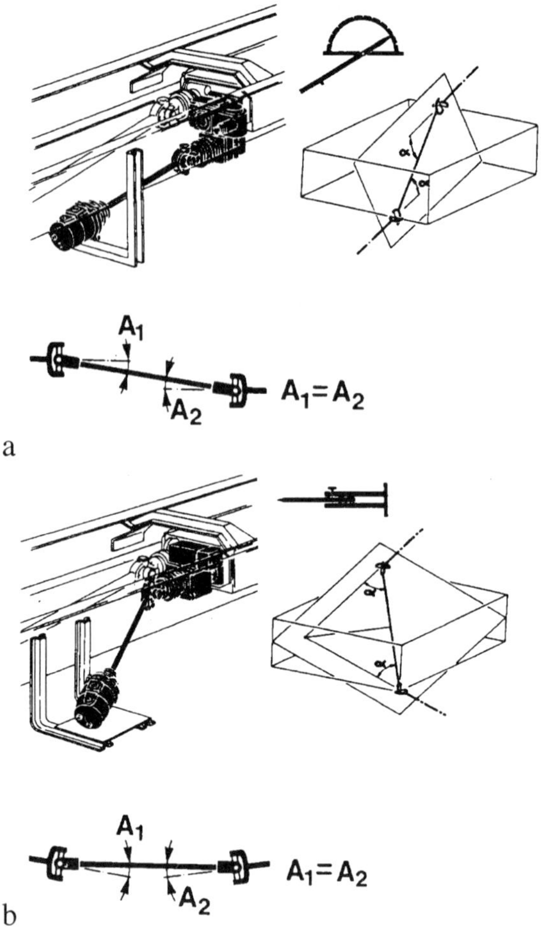

Fig 11a and b: Z and W drives

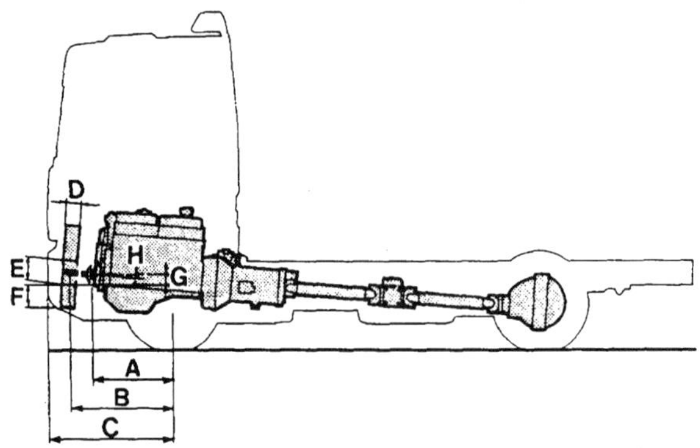

Fig 12: Engine-front PTO

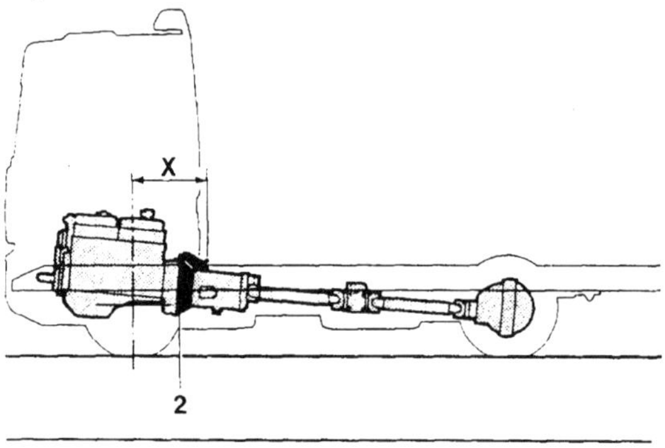

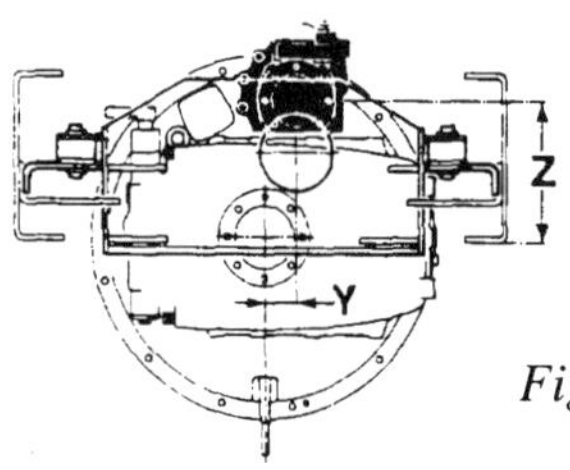

Fig 13: Engine-rear PTO

Heavy haulage vehicles

This vehicle category is required to operate outside the normal limits of the UK Construction and Use regulations and is subject to Special Types Orders covering the carriage of abnormally large and/or heavy indivisible loads. An understanding of the legal requirements is crucial to the design of such vehicles, which are usually tractor/trailer combinations. Trailer frame design for abnormal loads has some commonality with bridge design as seen in the following analysis.

The UK Special Types orders sub-divide heavy vehicles into three groups according to gross vehicle weights — 46, 80 and 150 tonnes.The lightest category vehicles have to comply with C & U regulations with respect to axle weights and spacings, as well as braking requirements, between 38 and 46 tonne GVW; see Fig 14. It also allows vehicles to be fitted with four closely spaced axles — beyond the normal remit of C & U. In the next category, upweight, axle weight increases up to 12.5 tonnes are permissible with the provision of 1.35 metres minimum axle spacing. Down to a 1.1 metre spacing maximum axle weight is just 12 tonnes. In all cases, vehicle tractor-trailer combinations must have at least five axles and and each 'bogie' is restricted to a total imposed load of 50 tonnes.

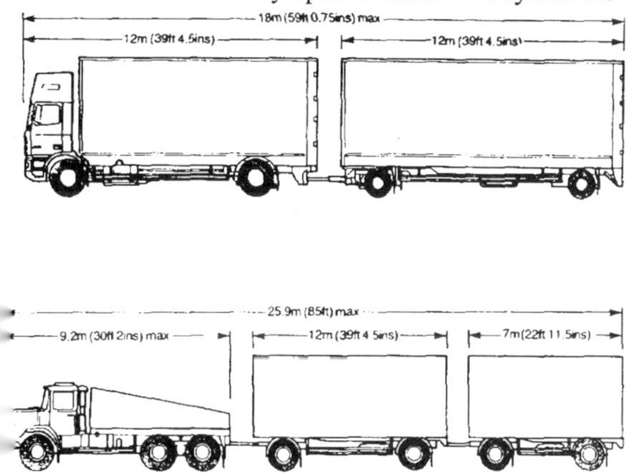

Fig 14: Length limits

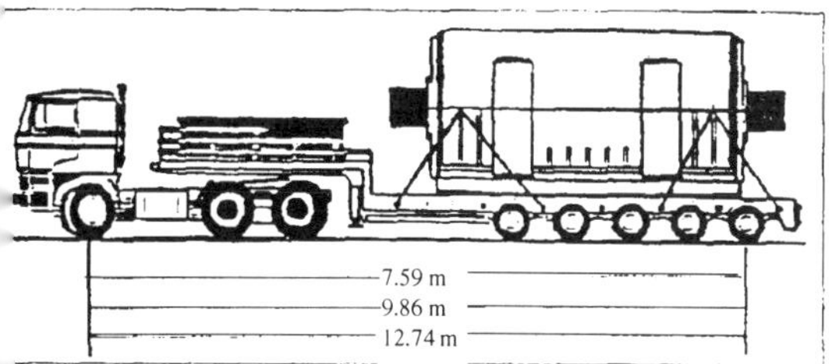

ig 15: Wheelbase requirements for 100, 125 and 00 tonnes GVW

	Motorways mph	Dual carriageways mph	Other roads mph
Category 1	60	50	40
Category 2	40	35	30
Category 3	30	25	20
Vehicles carrying loads exceeding 4.3m but not exceeding 6.1m	30	25	20

Fig 16: Speed limits

There are also minimum wheelbase requirements as set out in Fig 15. In the heaviest category, the same axle spacing dimension limits are imposed but the respective axle weights rise to 16.5 and 15 tonnes. The vehicle combination is required to have at least six axles and bogie limits of 100 tonnes apply, reducing to 90 tonnes where inter-axle spacing falls below 1.35 metres. Full details are given in section 44 of the UK Road Traffic Act and general Order SI 1198/79 of the Motor Vehicles (Authorization of Special Types) and careful reading is required before committing expensive special vehicles for this purpose to metal. Overall widths must not exceed 2.9 metres unless specially required to support extra wide loads in safety and then an absolute maximum of 6.1 metres applies — when a police escort will usually be required for its movement. Overall lengths must not exceed 27.4 metres unless by special permission of the Transport Secretary.

Differential speed limits are imposed for the three categories, Fig 16, to regulate the possible incidence of road damage. Special dispensation is given for the movement of engineering plant giving GVW increase to 152.4 tonnes provided individual wheel loads do not exceed 11.43 tonnes. Overall axle spreads for the three main categories are shown in Fig 17.

Since the imposition of these special types orders in 1988 weight restrictions applying to existing heavy haulage trailers have led to the development of multi-wheeler tractors such as Leyland DAF's FTD which uses a Ginaf second steering axle added in front of the rear bogie of a three-axle tractive unit. This 8 x 4 vehicle entitled the FTD 95.380, Fig 18, can operate at up to 200 tonnes GTW. The hydro-pneumatic suspension for the tri-axle bogie uses the spring/reservoir units shown in Fig 19. Steering for the first and second axles is interconnected while the second third and fourth axles share a four-spring fully-compensating suspension.

The vehicle has an imposed load capacity of

47.5 % as a tractor. Extra traction is obtainable on the third and fourth, drive, axles by using the interconnected oleo system to reduce the load on the second, trailing, axle by some 25%. In unladen running the axle can be lifted clear of the ground. The bogie suspension as a whole is self-levelling and proximity switches sense the piston positions in the six oleo springs.

The Unipower Contractor 6 x 4, Fig 20, is an on/off road tractor with gross train weight capability of up to 244 tonnes under the earlier regulations (on-road). This would be as a general service slow-speed drawbar combination, or as a special purpose one, up to 488 tonne GTW, in double formation road trains. The vehicle is also familiar as a high-speed tank-transporter in articulated form as illustrated, with 152.5 tonne capability on-road. Special 6 x 6 versions were designed for off-road duty. Each axle of the tandem through drive bogie has a torque capacity of 1.35 kNm and an inter-axle differential is provided. Gear ratio reductions allow a tractive effort potential of 26.75 tonne, gradient-ability up to 1 in 7.26 and road speeds up to 60 mph.

More recent designs include Volvo's NL 10/12 normal control tractor and the Faun Goliath, Fig 21. whose speeds, gradabilities and tractive efforts are listed in Fig 22. Trailers for heavy haulage are also built in a very specific way and are usually affected in a major way by the means used to on- and off-load indivisible payloads. Low-loading semi-trailers may be of the detachable back-axle variety or of the type using a detachable goose-neck, Fig23. An interesting variation was the folding goose-neck by Dyson in Fig 24; for very heavy loads multi-wheeled trailers are often built up as shown in Fig 25. Here two

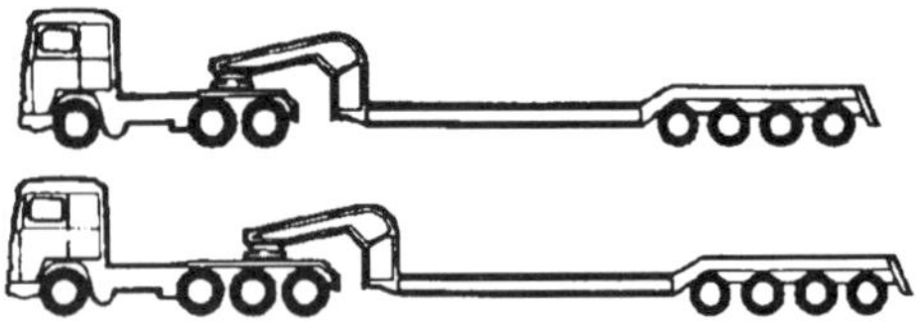

Relevant axle spacing in metres*	Weight (kg)
at least 6.5	40,000
at least 7.0	42,000
at least 7.5	44,000
at least 8.0	46,000

*Meaning distance between rearmost axle of semi-trailer and rearmost axle of tractive unit.

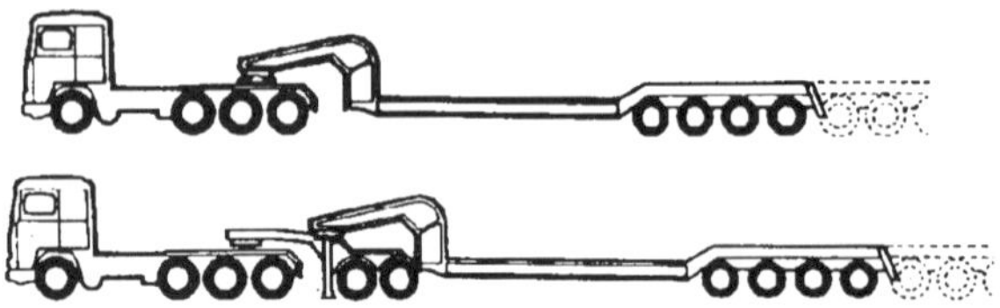

Distance between foremost and rearmost axles (metres)*	Weight (kg)
5.07	38,000
5.33	40,000
6.0	45,000
6.67	50,000
7.33	55,000
8.0	60,000
8.67	65,000
9.33	70,000
10.0	75,000
10.67	80,000

Distance between foremost and rearmost axles (metres)	Weight (kg)
5.77	80,000
6.23	85,000
6.68	90,000
7.14	95,000
7.59	100,000
8.05	105,000
8.50	110,000
8.95	115,000
9.41	120,000
9.86	125,000
10.32	130,000
10.77	135,000
11.23	140,000
11.68	145,000
12.14	150,000

Fig 17: Axle spreads

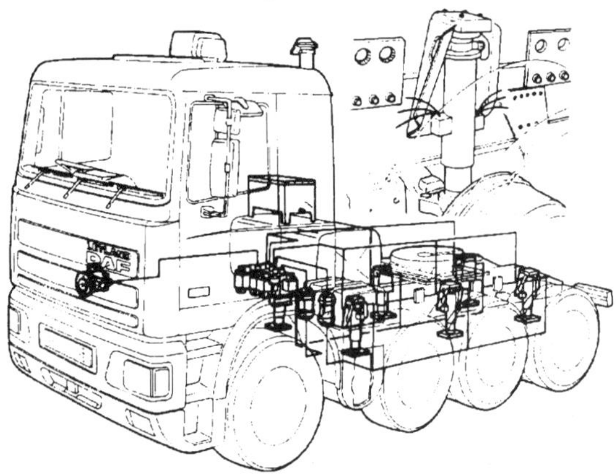

Fig 18: DAF FTD 95.380

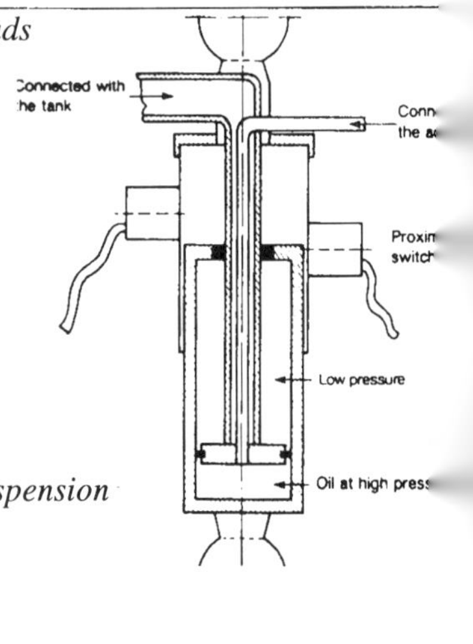

Fig 19: Oil/air suspension

twelve-wheeled bogies are attached to a common load carrying frame; in this German design wheels are evenly spaced rather than divided into bogies. It is possible by altering the direction of the steering motions to couple up to four of these units together in order to transport such items as heavy transformers while still retaining the wheel tracking facility. A 'standard' range of step-frame semi-trailers up to 60 tonne payload capacity are shown in Fig 26.

The design of heavy haulage trailer chassis inevitably leans heavily on beam theory developed for girder bridges. When the span or the loading are too great to permit the employment of rolled steel beams, singly or in combination, plate girders are used. Flanges are built up of plates or flats and attached to a web plate by means of flange angles. The web plate is stiffened by a series of vertical members, Fig 27. Whereas the most economical form is that with parallel flanges in which the area of the flanges is everywhere proportional to the bending moment, sometimes plate girders of varying web depth are feasible, Fig 28.

Occasionally more than one web can be used, in the form of a box girder, Fig 29. Built up beams of this type are recommended to have mid-span depth not less than 1/20 th of span. Ratio of breadth to

Gearbox step	Speeds in km/h at max. engine r.p.m.	Speeds in km/h at 70% converter range	Speeds in km/h at max. engine torque	Gradabilities in (%) at a total train weight of 100 t	200 t	250 t	300 t	Tractive effort in N
Off-Road								
1st gear				41.9	18.9	14.7	12.0	393 786***
	6.0	2.1	4.3	30.6	14.0	10.9	8.8	301 634**
				19.5	8.9	6.8	5.4	202 632*
2nd gear				26.4	12.1	9.3	7.5	264 723***
	8.9	3.2	6.4	19.5	8.9	6.8	5.4	202 771**
				12.5	5.5	4.1	3.1	136 210*
3rd gear				19.0	8.6	6.6	5.2	197 878***
	11.9	4.2	8.5	14.1	6.2	4.7	3.7	151 566**
				8.9	3.7	2.7	2.0	101 802*
4th gear				12.1	5.3	3.9	3.0	132 903***
	17.8	6.3	12.7	8.9	3.7	2.7	2.0	101 789**
				5.6	2.0	1.3	0.8	68 338*
5th gear				8.6	3.5	2.5	1.8	98 447***
	24.0	8.5	17.1	6.2	2.3	1.6	1.1	75 386**
				3.7	1.1	0.6	0.2	50 565*
6th gear				5.2	1.9	1.2	0.7	65 959***
	35.8	12.7	25.6	3.6	1.1	0.6	0.2	50 472**
				1.9	0.2	0.0	0.0	33 731*
rev. gear				57.6	24.9	19.4	15.8	504 046***
	4.7	1.7	3.3	40.9	18.5	14.4	11.7	386 093**
				25.8	11.8	9.1	7.3	259 373*
On-Road								
1st gear				22.5	10.5	7.9	8.3	230 284***
	10.3	3.6	7.3	16.7	7.5	5.7	4.5	176 391**
				10.6	4.5	3.3	2.5	118 484*
2nd gear				14.4	6.4	4.8	3.8	154 809***
	15.3	5.4	10.9	10.6	4.5	3.3	2.5	118 572**
				6.6	2.6	1.7	1.2	79 625*
3rd gear				10.4	4.4	3.2	2.4	115 718***
	20.4	7.2	14.6	7.6	3.0	2.1	1.5	88 622**
				4.6	1.5	0.9	0.5	59 479*
4th gear				6.4	2.5	1.7	1.1	7[illegible] 721***
	30.4	10.8	21.7	4.6	1.5	0.9	0.5	59 496**
				2.6	0.5	0.1	0.0	39 843*
5th gear				4.4	1.4	0.8	0.5	57 571***
	41.0	14.6	29.3	3.0	0.7	0.3	0.0	44 030**
				1.5	0.0	0.0	0.0	29 349*
6th gear				2.4	0.5	0.1	0.0	38 573***
	61.2	21.7	43.7	1.5	0.0	0.0	0.0	29 394**
				0.5	0.0	0.0	0.0	19 233*
rev. gear				29.8	13.6	10.6	8.5	294 764***
	8.0	2.8	5.7	22.0	10.1	7.7	6.2	225 783**
				14.1	6.2	4.7	3.7	151 672*

Fig 22: Goliath performance based on 0.015 rolling resistance co-efficient, 0.7 static friction co-efficient, 12-764 axle reduction ratio and 1.71:1 transfer box reduction

Fig 21: Faun Goliath

Fig 20: Unipower 224 tonne tractor

unsupported length should not exceed 1/40. Where the ratio l/B exceeds 20, the compressive stress should not exceed:

9.4 - 0.13l/B

Joints in web plates are made with butted joints, usually with butted joints having double cover plates, Fig 30. Such joints must be of sufficient strength to carry the shear force at the joint as well as bending moment and a web stiffener may be used to form a joint cover. As well as stiffening the web against buckling by the shear force, the stiffeners must also prevent the girder racking sideways and distribute any concentrated load over the depth of the web plate. Girder bridge makers recommend safe shearing force per unit length of web depth as:

5.5t - d/20

where *t* is web thickness and *d* is the stiffener spacing.

The curve of the goose-neck can be accounted for in design analysis by using the Unit Load thereom which was described in Chapter 1. In the example shown in Fig 31, the downward deflection of the point B due to the bending deformation of AB and BC is found using the analysis alongside:

$$\delta_B = \frac{1}{EI}\int M_0 m_1 \, ds \text{ then:}$$

	M_0	m_1
AB	$R_{0A}.s \quad \frac{ws^2}{2}$	$R_{1a}.s$
BC	$R_{0C}.r\sin\theta$	$R_{1C}.r\sin\theta$

(θ is measured from OC and s from A) and

$$\delta_B = \frac{1}{EI}\left[\int_O^l \left(R_{0a}.s - \frac{ws^2}{2}\right)\left(R_{1a}.s\right) ds + 2\int_0^{\frac{\pi}{2}} R_{1C}.R_{0C}.r^2.\sin^2\theta.r.d\theta\right]$$

$$R_{1a} = \frac{1}{3};\ R_{1C} = \frac{2}{3};\ R_{0C} = \frac{20 \times 4 \times 4/2}{2+4} = 26{\cdot}7;$$

$$R_{0a} = 80 \quad 26{\cdot}7 = 53{\cdot}3\ kN$$

Given the 2nd moment of area of the sidemember

$$= 0{\cdot}25 \times 10^{-4}\ m^4 \text{ and } E_{steel} = 210\ GN/m^2;$$

$$\delta_B = \frac{10^3 \times 10^3}{210 \times 10^9 \times 0{\cdot}25 \times 10^{-4}}\left[\frac{64}{3}\left(\frac{53{\cdot}3}{3} - \frac{80}{8}\right) + 26{\cdot}7 \times \frac{2}{3} \times 8 \times \frac{1}{2} \times \frac{\pi}{2}\right]$$

$$\therefore \delta_B = 53 \text{ mm}$$

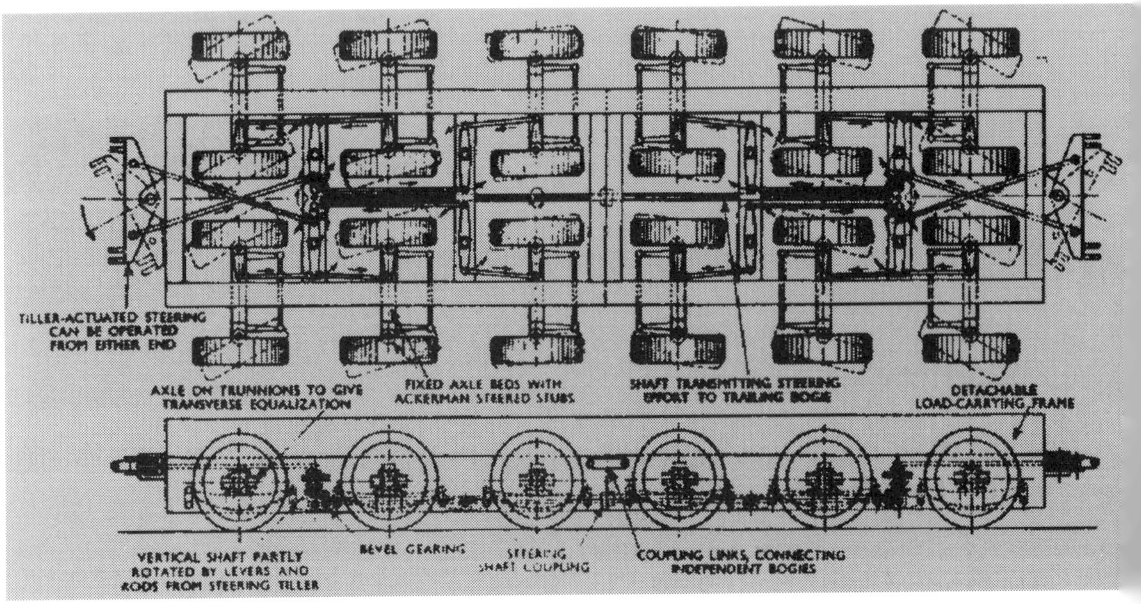

Fig 25: Multi-wheel trailer

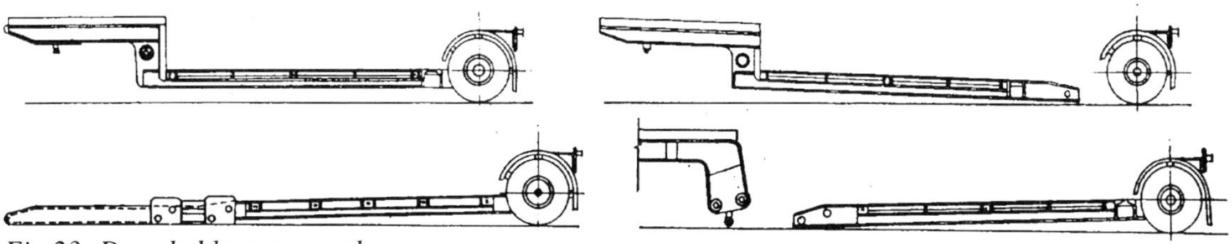

Fig 23: Detachable goose-neck

A NECK LENGTH	B PLATFORM LENGTH	C UNLADEN HEIGHT	D OVERALL LENGTH	E SWING CLEARANCE	NOMINAL PAYLOAD KG	KING PIN LOAD KG	BOGIE LOAD KG	F FIFTH WHEEL HT	W OVERALL WIDTH	TYRES	SUSPENSION
2.9 m	7 m 9 m	1.14 m	9.9 m 11.9 m	1.98 m	15,000	10,000	10,000	1.32 m	2.5 m	10.00 x 20-16 (4 off)	Single TST1/7
2.9 m	7 m 9 m	1.14 m	9.9 m 11.9 m	1.98 m	25,000	10,000	20,000	1.32 m	2.5 m	10.00 x 20-16 (8 off)	Tandem TST2/7
2.9 m	8 m 10 m	1.22 m	10.9 m 12.9 m	1.98 m	30,000	15,000	22,000	1.32 m	2.5 m	10.00 x 20-16 (8 off)	Tandem TST2/8
3.5 m	8 m 10 m	1.30 m	11.5 m 13.5 m	2.58 m	40,000	20,000	30,000	1.38 m	2.9 m	11.00 x 20-16 (8 off)	Tandem DN7
3.5 m	8 m 10 m 12 m	1.32 m	11.5 m 13.5 m 15.5 m	2.58 m	45,000	20,000	35,000	1.38 m	2.9 m	12.00 x 20-16 (8 off)	Tandem DN8
3.5 m	8 m 10 m 12 m	1.25 m	11.5 m 13.5 m 15.5 m	2.58 m	50,000	25,000	39,000	1.38 m	2.9 m	10.00 x 20-16 (12 off)	Tri-Axle TST3/8HD
3.5 m	8 m 10 m 12 m	1.30 m	11.5 m 13.5 m 15.5 m	2.58 m	60,000	30,000	45,000	1.38 m	3.0 m	11.00 x 20-16 (12 off)	Tri-Axle TST3/8HD

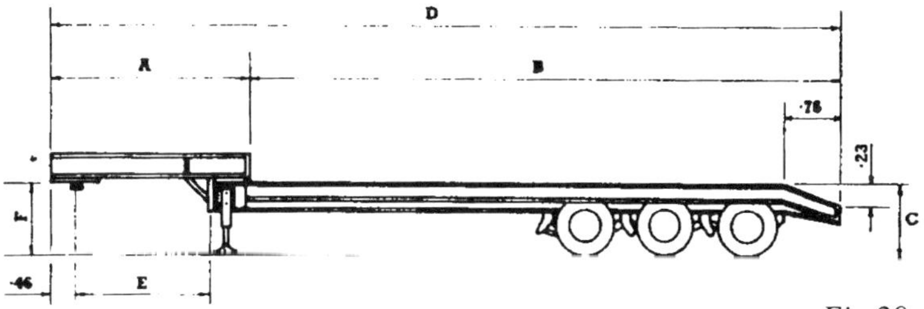

Fig 26: Typical trailer range for heavy haulage

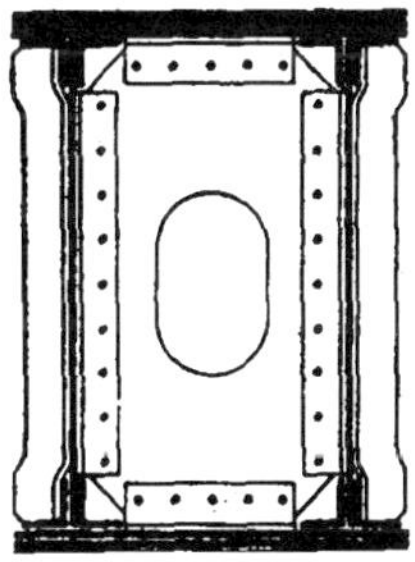

Fig 29: Box girder

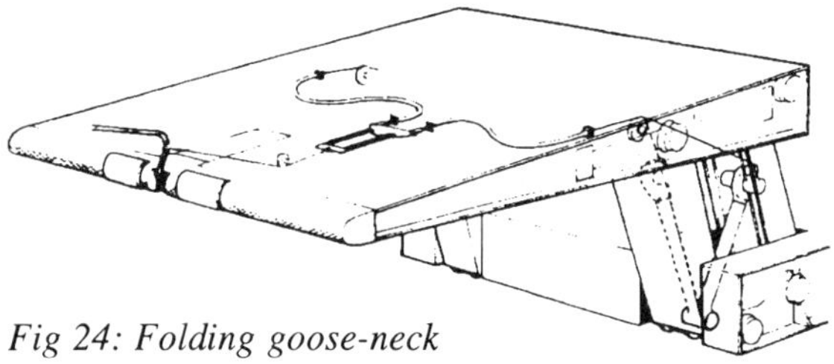

Fig 24: Folding goose-neck

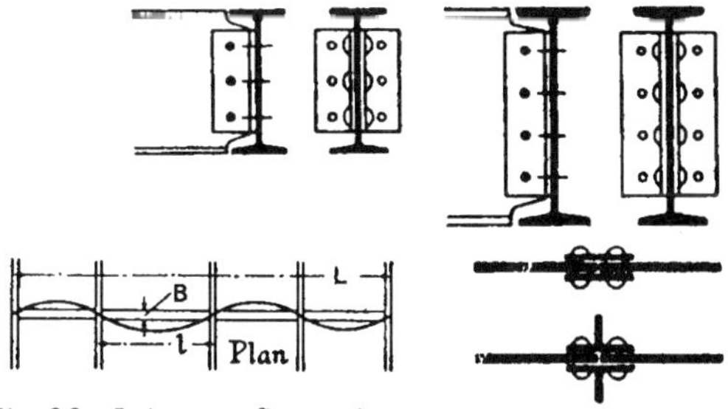

Fig 30: Joint configurations

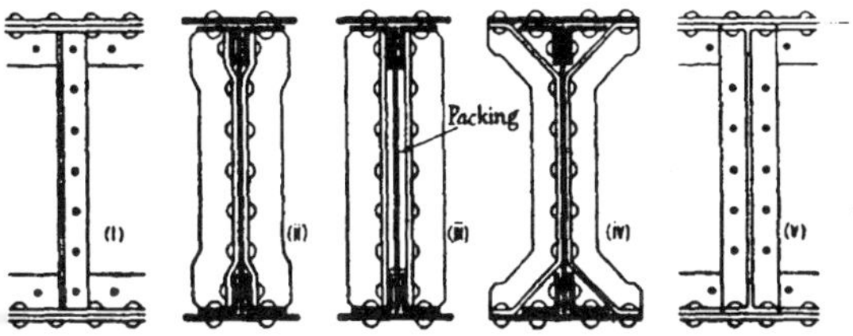

Fig 27: Webstiffeners

Fig 28: Tapered beam span

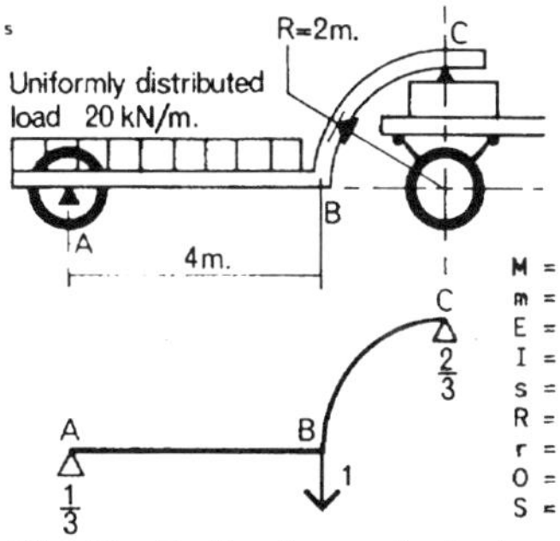

M = Moment due to real loads
m = Moment due to unit load
E = Elastic modulus
I = Inertia (2nd) moment
s = distance along member
R = Reaction force
r = radius
O = angle subtended by arc
S = deflection

Fig 31: Deflection calculation parameters

Road tankers

As one of the first truly chassisless vehicles, the monocoque road tanker was also one of the first vehicle categories to exploit the use of structural theory in design. According to the commodity being carried, the design can also be considerably influenced by Construction and Use regulations as well as the chemical and physical properties of the product.

The original 5000 litre/compartment limit for road tankers was one of the restrictions which forced the use of bulkheads along the length of a tanker, in the case of road-going vehicles. For airfield-only operation, a later regulation allowed refuelling tankers to have single compartment shells as large as 20,000 litres and, of course, very different design constraints apply to that particular category. In practice, however, these large airfield tankers do normally have baffles every 1.5 metres or so which help stabilise the shell structures.

In an effort to improve the operating safety of road tankers, the West German authorities recently constructed the TOPAS experimental vehicle, Fig 32, which could well affect future designs. This 39,000 litre unit has a centre of gravity some 300 mm lower than conventional designs and uses 70% aspect-ratio tyres at the maximum legal width dimension of 2.5 metres. The Haller-made tank has built-in roll-over bars and foam-filled double-skin front and rear bulkheads. Like most modern petrol tankers it is built for bottom loading and vapour recovery avoiding the risk of fuel spillage in roll-over.

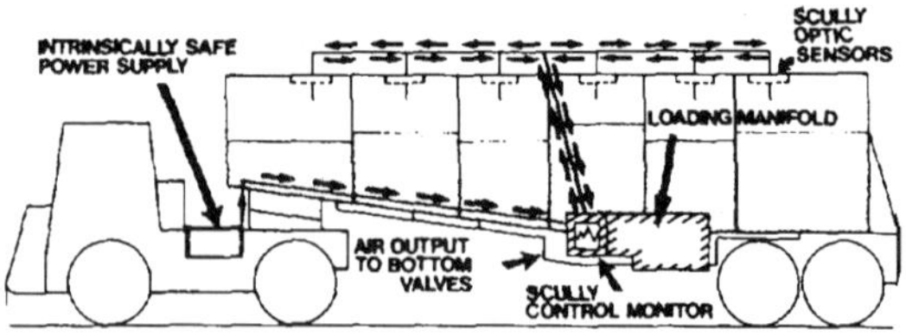

Fig 33: Skully overfill prevention system

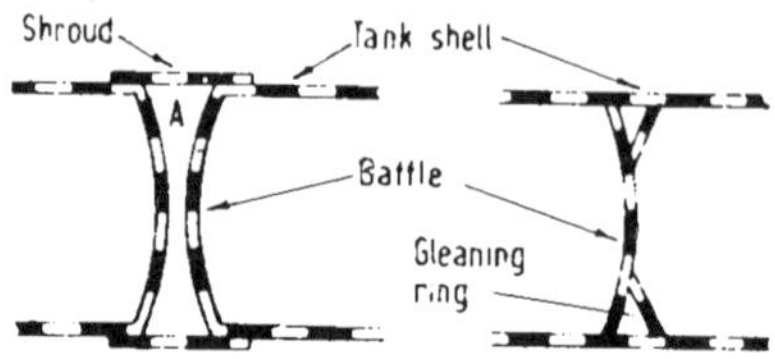

Fig 34: Small tanks joined by shrouds

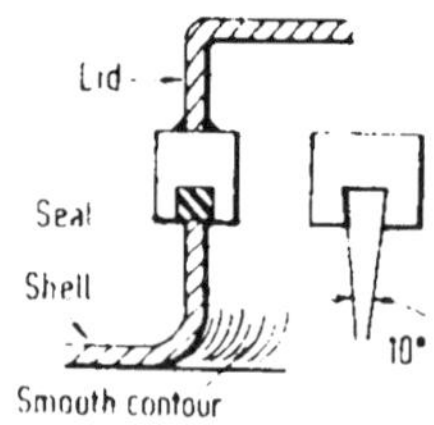

Fig 35: Man-lid details

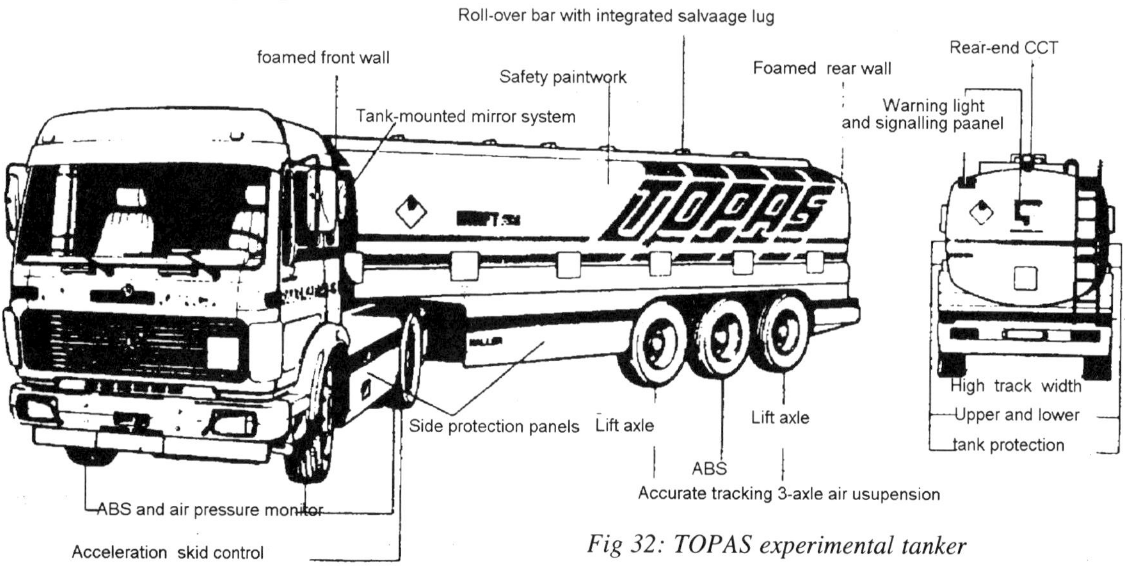

Fig 32: TOPAS experimental tanker

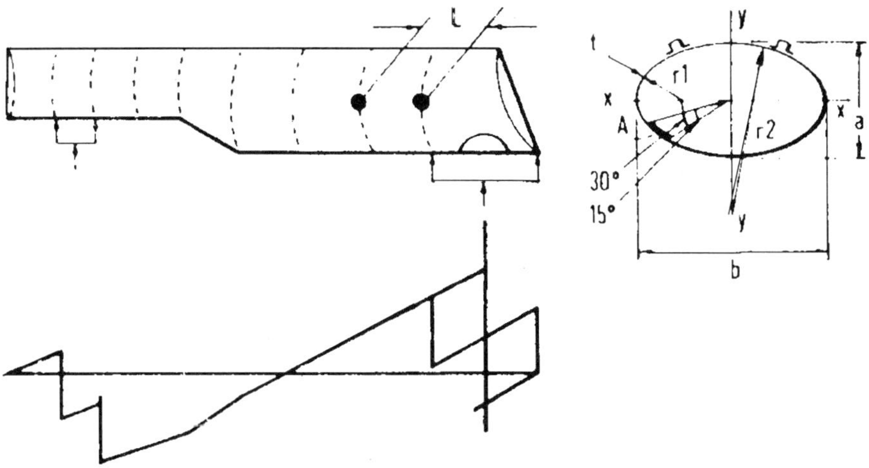

Fig 36: Tank dimensions and shear force distribution

Under the Dangerous Substances (Conveyance by Road in Road Tankers and Tank Containers) Regulations 1981, minimum tank specifications were prescribed for different commodities. Health and Safety Regulations broadly followed those applying to HM Goverment Petrol Regulations. A vessel was required to have a minimum wall thickness of 5 mm and to withstand a design pressure of 2.7 bar. In conjunction with the Institute of Petroleum, the HSE also supported a design code calling for centre of gravity of the load to be lower than 90% of the wheel track.

With tanks being filled under pressure, from beneath, through the same hoses used for discharge, an overfill protection device is often used. One such is the Scully Dynacheck, Fig 33, which uses electronic pulsing detected by optic sensors which can shut off the air valves on the delivery hoses. Some tanks are required to resist vacuum which puts different structural demands on the shell. The 3500 gallon KiloWhale by Whale Tankers is an example, used to lift waste through heights of up to 10 metres. Others, for special commodities like liquid sugar, use gravity discharge involving special shaped tanks to ease flow.

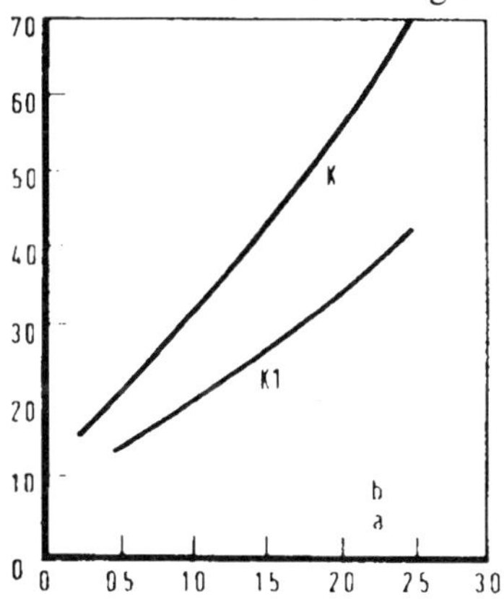

Fig 37: Shell shear stress factors

Shell construction techniques

Aluminium alloy of NS6 grade has been commonly used for conventional tankers of chassisless construction. The aluminium companies have been a valuable source for structural design procedures used in tanker shell design. Useful codes have also been drawn up by the major oil and chemical companies with large tanker fleets. One such draws the distinction between conventional and pressure discharge tanks; in the latter case, individual small tanks may be joined up as in Fig 34, by shrouds, to make up the overall shell. Pressure entry points must be made to the enclosed spaces (A) as well as the compartments. An alternative approach is to use dished bulkheads braced by gleaning rings (also shown). Man-lids of 20 in (508 mm) diameter are typically provided, arranged as in Fig 35 to avoid contamination traps in the neck-ring to shell joint. Where ring stiffeners rather than bulkheads are employed, on elliptical section tanks the arc of the bottom surface is recommended for the suspension subframe mounting.

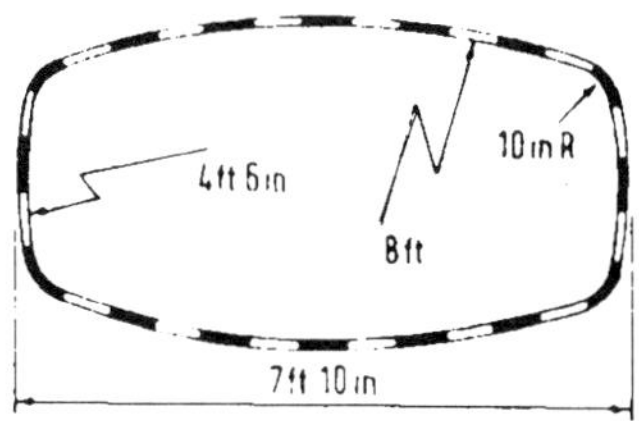

Fig 38: Max-section

Simple stressing of road tankers

A stressing procedure given in an ALCOA publication[1] gives the critical shear stress at which buckling might take place in the side walls of an elliptical tank under transverse shear as

$$0.313EB(r_1/t)^{3/2}$$

where E is elastic modulus and

$$B = 1.2[(2r_1/t)^{1/2}/(L/2r_1)]^{1/2}$$

The lower diagram Fig 36 shows the vertical shearing force V distribution along the length of the tanker. This gives rise to a corresponding shear stress $VQ/2tI$ for section moment of inertia I about the neutral axis and Q about the point on the section farthest from it. These values can be found from Fig 37 using $Q = K_1 a_2 t/4$ and $I = Ka_3 t/8$, the safety factor derived on the vertical axis, relating maximum stress to critical stress.

Higher volume capacity of the tank is obtained by going from an elliptical to a so-called 'max-section' cross-section as shown in Fig 38. For the dimensions given in this example, a radius to thickness ratio of under 500 is recommended to avoid panel 'panting'. Circumferential welds around the shell should be arranged to avoid the pitfalls shown in Fig 39 and tank ends should be flanged so as to lap into the main shell by 3/4 to 1 in (19-25.5mm), keeping the weld away from the flange radius.

A worked example for the stressing procedure is provided by the stainless steel tank design shown in Fig 40, devised some years ago by International Nickel. The tank has weights and dimensions as shown in Fig 41. Rear shell section is straight-

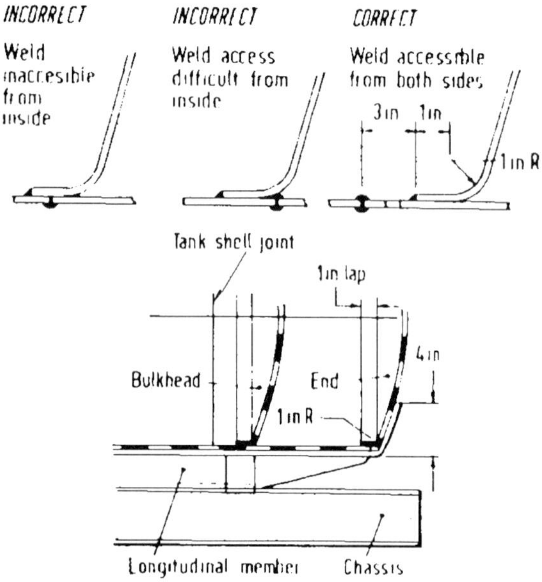

Fig 39: Shell welding recommendations

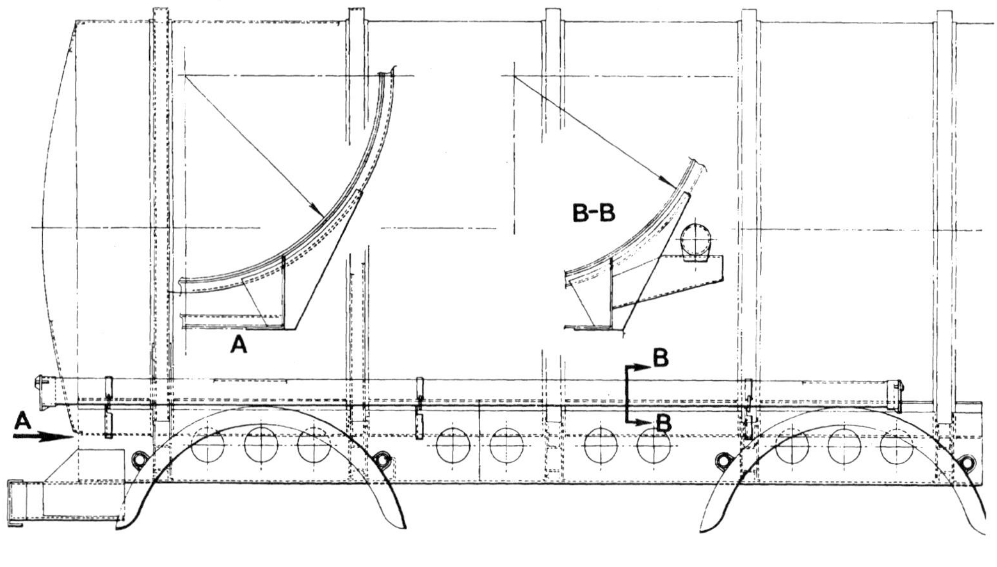

Fig 40: International Nickel stainless steel tanker construction

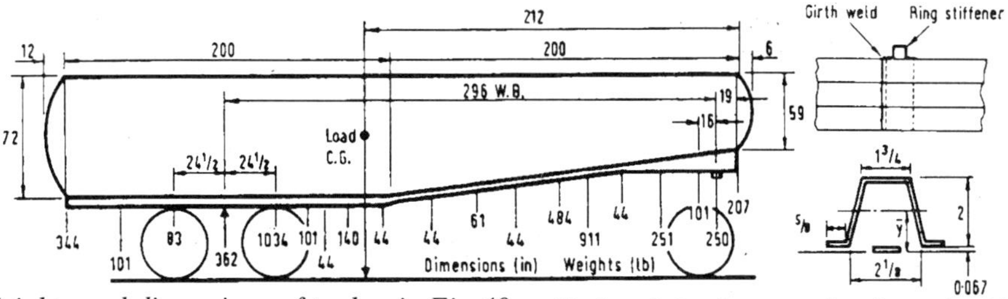

Fig 41: Weights and dimensions of tanker in Fig 40

sided and the front section is conical in shape. Ring stiffeners are 12 gauge top-hat section welded to the shell. One of them is directly adjacent to the girth weld where the two sections of the tank are joined. Load reactions are first determined at A and B as R_a = *16,200* lb (7.3 tonnes) and R_b = *30,515* lb (13.75 tonnes), Fig 42 shows the laden weight distribution over the rear suspension and front king-pin reaction points — also the shear force diagram. Critical shear stress at the point of maximum shear is 9660 lb/in² (6.66 MN/m²), from the expression above, while the

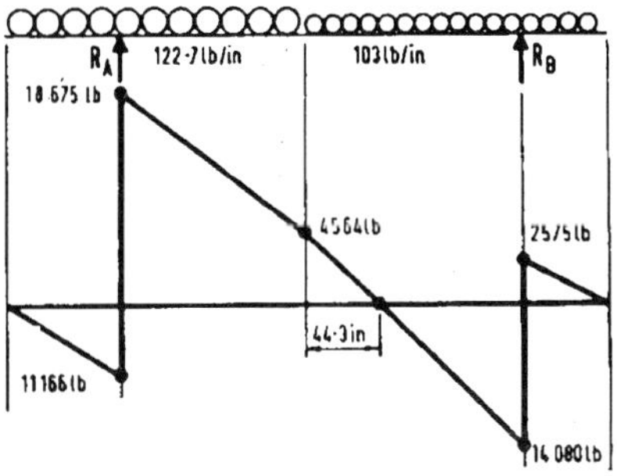

Fig 42: Loading and shear force diagrams

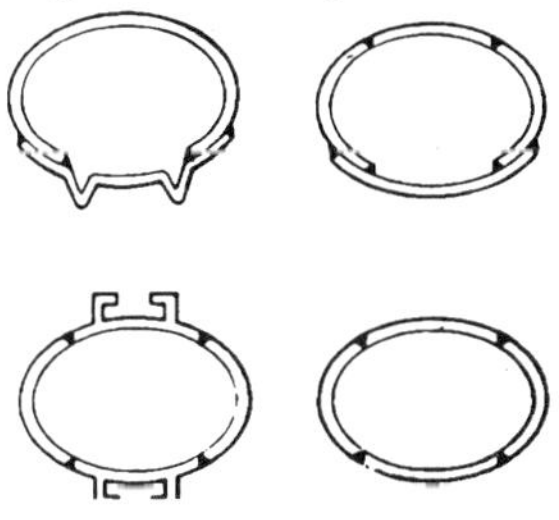

Fig 43: Alternative welding configurations

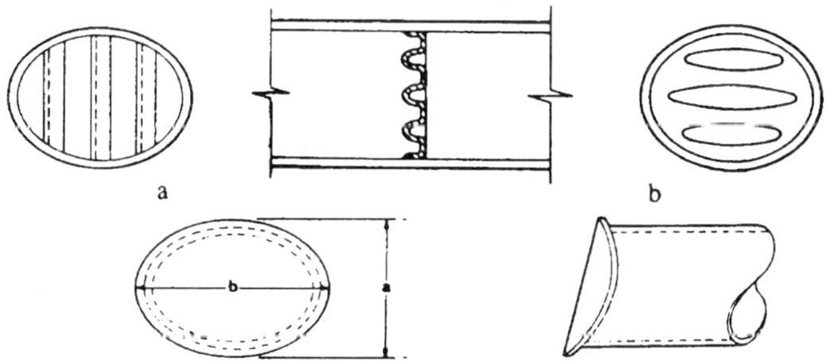

Fig 44: Bulkhead and end alternatives

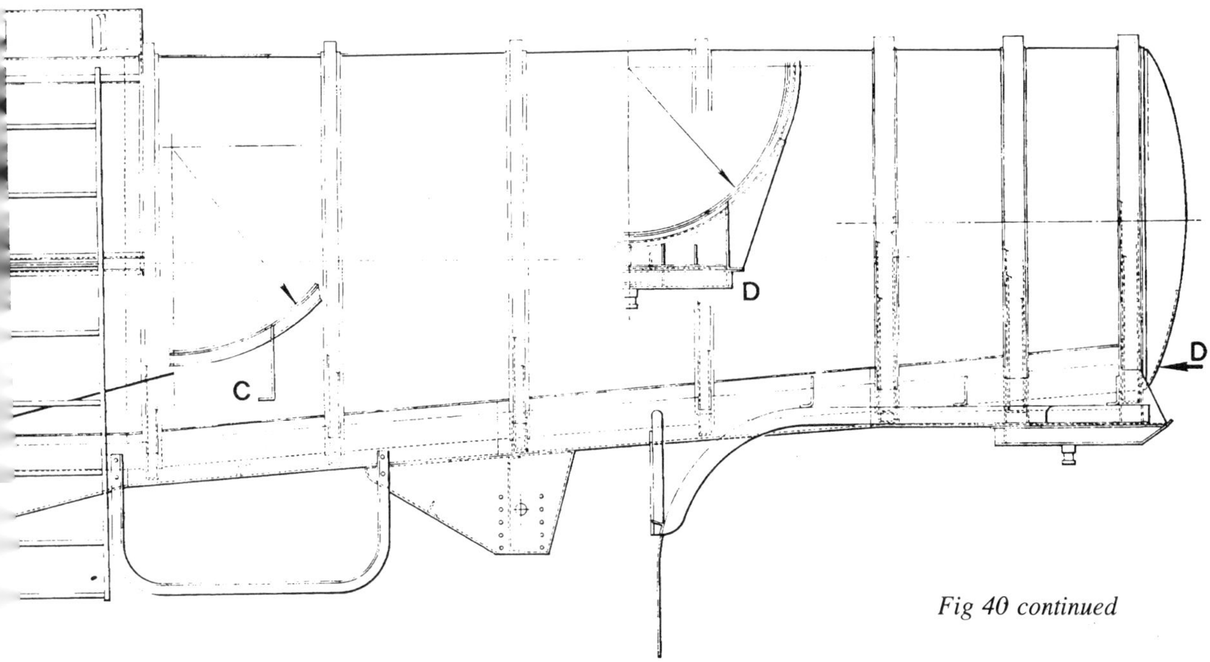

Fig 40 continued

actual shear stress, similarly derived, was 4800 (3.33) — to show a safety factor of approximately 2.

Detail design of road tankers

Further details of tank construction recommendations are given in an Alcan Industries publication**. It points out that the shell of an aluminium alloy tank is usually composed of sheets, curved to the desired radius and welded together. Sheets are butt-welded to form a complete cylinder and the bulkheads inserted and welded in place. The lower sheets can be wrapped around the bulkheads with the upper sheet and then lap-welded. Longitudinal extrusions often form the upper and lower parts of the tank; see Fig 43 for these alternatives. Instead of simple dishes, bulkheads may be stiffened or swaged flat sheet as shown in Fig 44a and have end-heads the same or simple curved sheet welded to the shell as shown in Fig 44b.

In designing supports, it should be noted that the weight of the cargo is carried first by circumferential tension in the shell. These tension stresses create shear stresses in the shell which are maximum at the supports where load feeds down from the shell walls via the bulkheads. The support reaction can be provided by a saddle or by total exterior rings, Figs 45 a and b. The collapse load for the ring shown, not welded to tank, is

$$2t[c + 2(Rt)^{1/2}]\sigma_y$$

where t is shell thickness, c ring width, $R = b^2/2a$ and s_y the alloy yield strength. In calculating transfer of load downward to the vehicle suspension, stress calculations should be based on the thickness of fillet weld either side of the saddle webs. Do's and dont's of alternative bulkhead and support arrangements are shown in Fig 46.

Torsional loading of tankers

Because of the very large diameters of tanks relative to the twisting moments applied to them, by road wheels striking assymmetrical bumps at speed, torsional stresses are comparatively low. However they can combine with the shear stresses computed above to lower the safety factor from the critical stress. An earlier section of Chapter 1 gave an expression for shear stress arising in twisting a relatively thick walled cylinder. Thin-walled closed tubes, by contrast, resist torsion by the area of the cross-section bounded by the mean thickness of the wall so that shear stress $\tau = T/2A$ for torque T and bounded section area A while the twist/unit length is $(T/4GA^2)\Sigma ds/t$ for shear modulus G, wall thickness t.

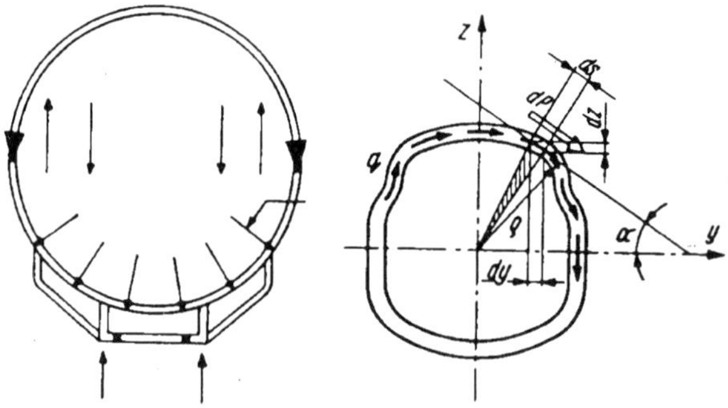

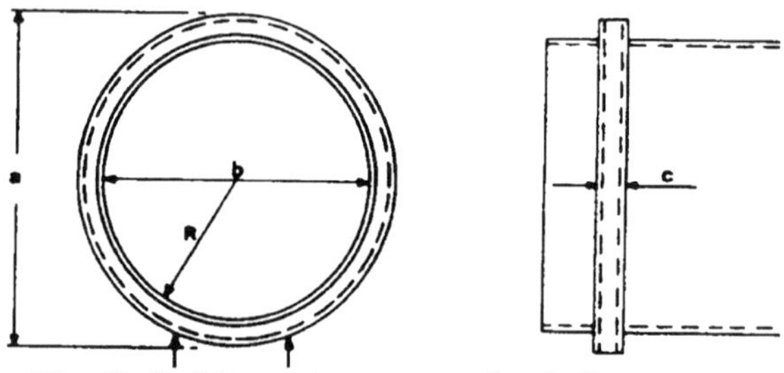

Fig 45: Saddle or ring support for shell

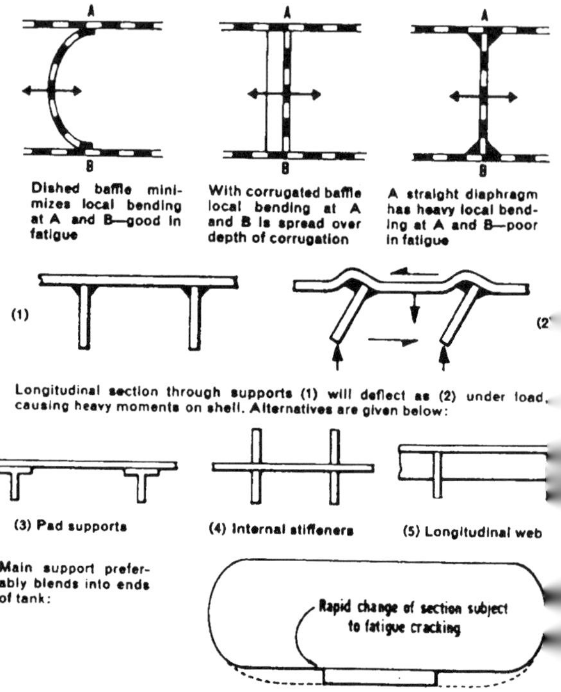

Fig 46: Do's and dont's of bulkhead welding

Pressurised tankers

The design of low-pressure bulk powder tankers is generally carried out to BS 5500. When higher pressures are required for special purpose, use can be made of analytical work catalogued by ESDU (Engineering Science Data Unit) whose pressure vessel studies have now been written into computer programs. In all cases an understanding of the basic theories which underline these practices is useful in appreciating the design limitations.

An introduction to semi-empirical stressing methods (simple Engineers theory of bending factored by values obtained from the results of experimental measurement) for road tankers was given in the last section. Pressure tank design is also discussed in the next section, on bulk powder vehicles, based on the methods of BS 1500, Part 3. However, a better understanding of the factors involved in pressurised road tankers for bulk liquids comes from the basic theory of thin shells subject to internal pressure, Fig 47.

The force system shown arises from a half full tank of liquid, in this example. For internal pressure p and tank length l, $P = pld$ and the transverse skin stress $\sigma_1 = pd/2t$. Pressure on the ends of the tank also produce a complementary longitudinal stress of $\sigma_2 = pd/4t$. In a cylindrical shell with hemispherical ends of a different thickness, Fig 48, the hoop strain in the main tank will be $(pd/4t_1E)(2 - \nu)$ where E is elastic modulus and n is Poisson's ratio; t is shear stress. The corresponding value for the dished ends has t_2 in place of t_1. These two strains should be equal, for compatibility, at the junction of tank and ends. so that:

$$t_2/t_1 = (1 - \nu)/(2 - \nu)$$

This ratio is 7/17 where ν is typically 0.3; for equality of stress between tank and ends the ratio should be reduced to 0.5. In the case of a rigid-vehicle tanker for which torsional loading due to road irregularities is combined with pressure, the shear stress on transverse planes needs to be added to direct stresses to obtain a principal stress:

$$(1/2)(\sigma_1 + \sigma_2) +/- [(\sigma_1 - \sigma_2)^2 + 4\tau^2]$$

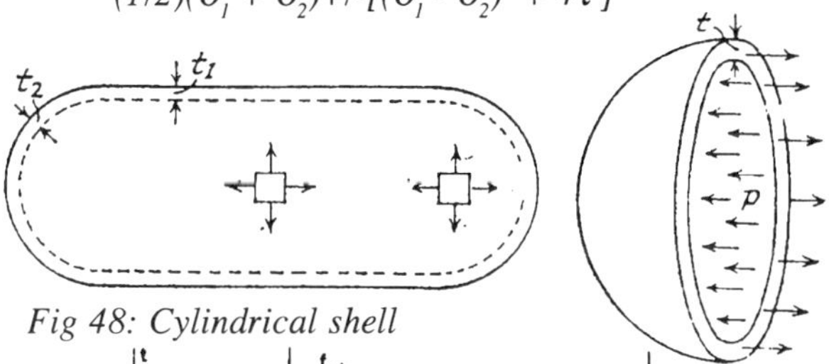

Fig 48: Cylindrical shell

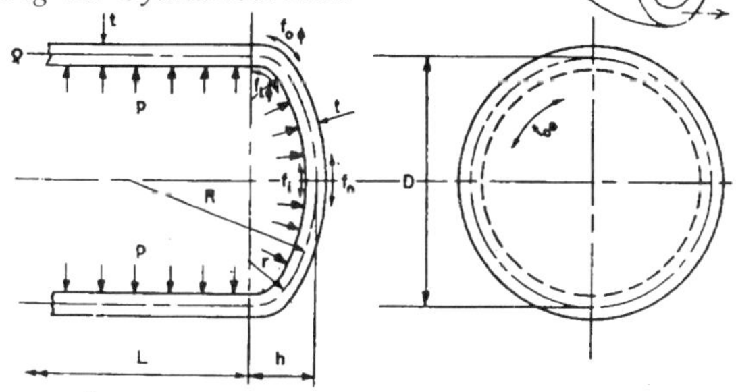

Fig 49: Dimensional notation: Dished-end cylinder (stress σ notated f)

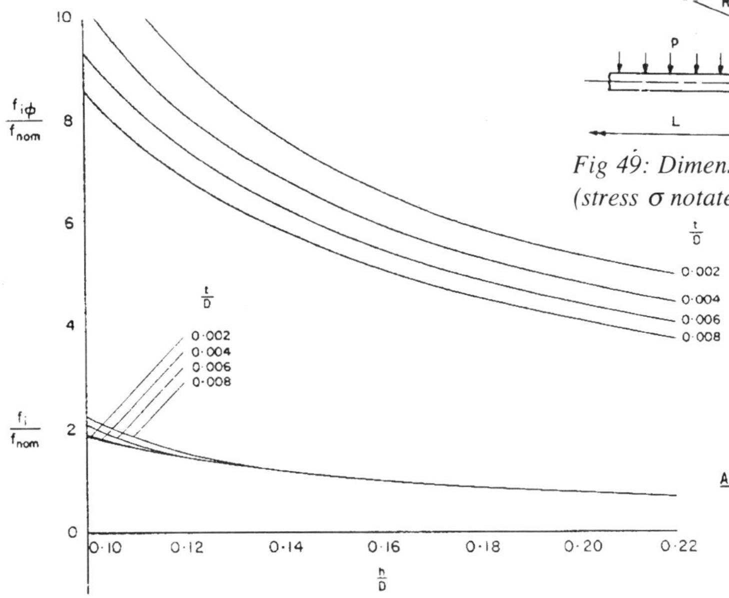

Fig 50: Stress characteristics: dished-end tank

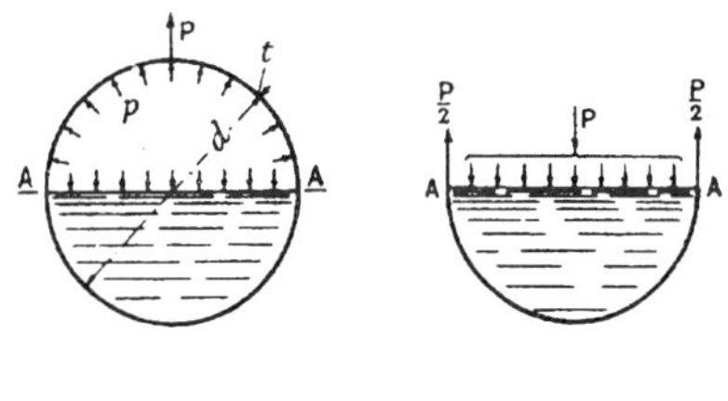

Fig 47: Shell under internal pressure

More exact analyses of pressure vessels of many different shapes and loading conditions can be found in 'Roark's classic work[3]. Many bulk powder tankers use conical ends where tipping is used to assist discharge. To obtain even stresses in tank and ends for the same material thickness usage the ends are made with an angle of 60 degrees to the axis of the cylinder. When the cone is truncated the ideal diameter of the flat portion, from the standpoint of pressure retention is

$$[16t^3\sigma/3p]^{1/2}$$

The computerisation of pressure vessel design has been eased by the recent release by the Engineering Science Data Unit of PV Designer-BS, a program geared to BS 5500. Its calculation procedures deal with cylindrical shells, stiffening rings, heads and bases; flanges, nozzle openings, flat plates and cone-cylinder reinforcing.

ESDU also publish three useful conventional data sheets of value in pressure vessel design. Cylindrical tanks with flat heads are dealt with in 66010. Stresses and deflections in the flat ends are dealt with in 65002/3 and torispherical ends in 67017. Fig 49 shows the force system and dimensional notation for the cylindrical tank and Fig 50 is typical of the curves derived from thin-shell theory. The shell stresses are nominal ones at the junction of tank and end resulting from a combination of bending stresses due to the effect of the end on the tank as well as direct stresses due to pressure thrust of the end and circumferential membrane stresses. Fig 51 shows the load case and notation for the torispherical dished end and Fig 52 a sample curve. Usually the maximum tensile stress occurs in or near the knuckle.

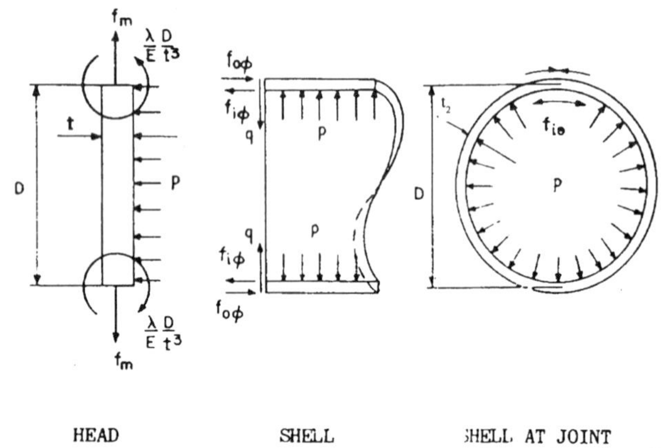

Fig 51: Dimensional notation: torispherical end tank (stress σ notated f)

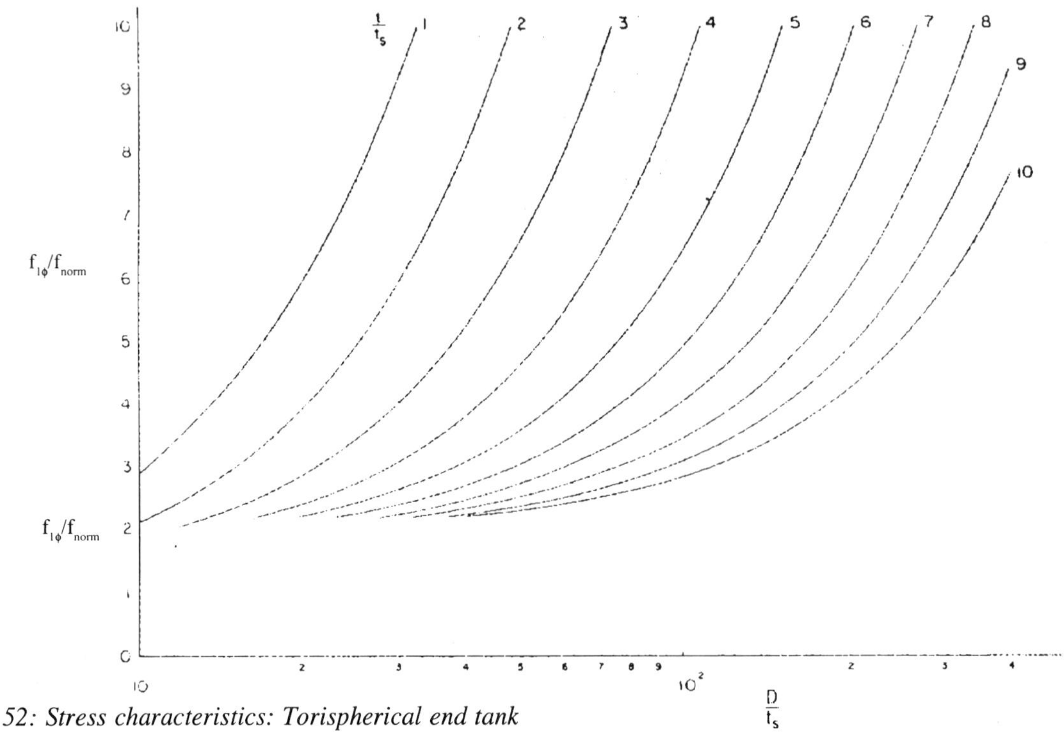

Fig 52: Stress characteristics: Torispherical end tank

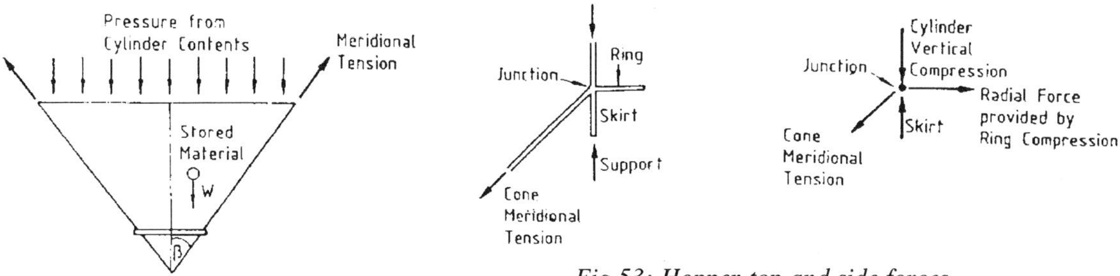

Fig 53: Hopper-top and side forces

Vertical cylinders for bulkers

A number of road-going low-pressure vessels for powders and viscous fluids are conical-bottomed vertical cylinders using pressure-assisted gravity discharge. To some extent stressing methods used for elevated silos can be used in these cases. The walls of such vessels are subject to both normal pressures and the frictional shears from the retention and discharge of the product. The former cause hoop stresses and the latter vertical compression stresses. Modes of failure include bursting, buckling under vertical or circumferential compressive stress or buckling under membrane shear stress. Work at Edinburgh University has found that the tension at the hopper top, Fig 53, may be from global equilibrium (a), the vertical component of this tension being resisted by the support (b) and the radial component by a ring. The ring thus carries a high circumferential compressive stress and if it is not provided by addition the hopper, cylinder and skirt will deform to form an effective ring. Failure modes of the ring may be either plastic collapse or buckling.

Metalair's tipping and non-tipping bulkers are each built to BS 5500; the company's tank-container design of up to 50 cubic metres capacity which can withstand pressures up to 2 bar is seen in Fig 54.

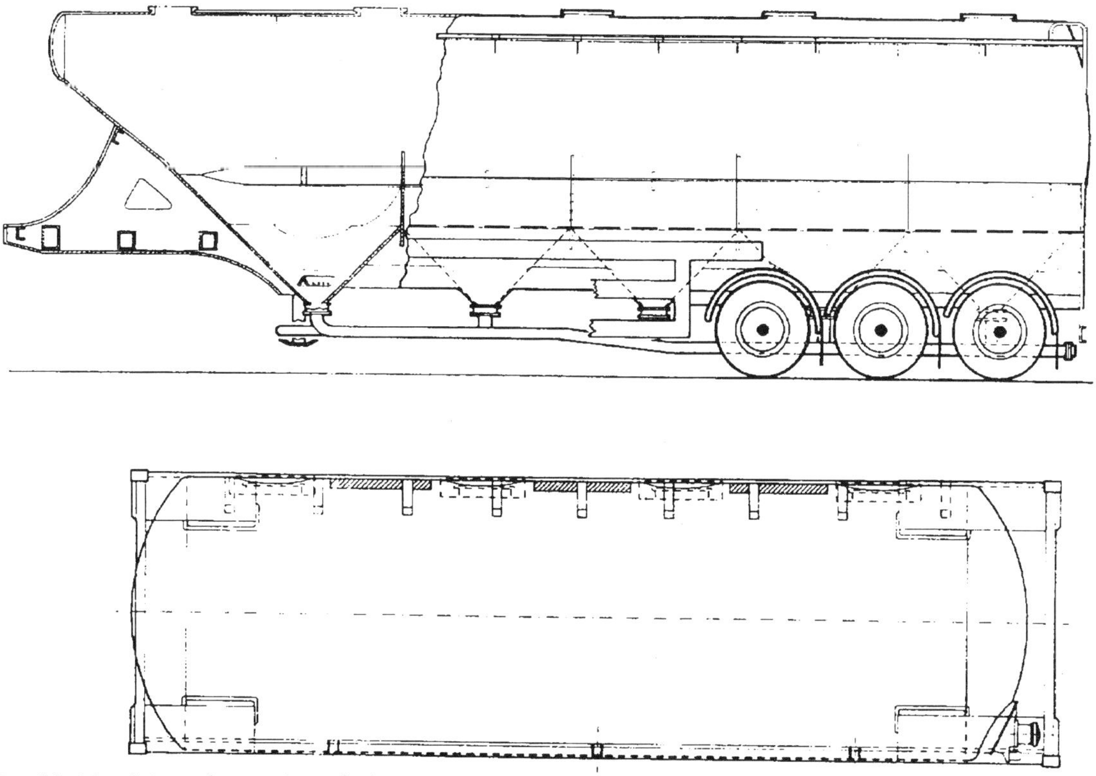

Fig 54: Metalair tank-container design

Bulk powder vehicles

The transport of bulk powders by so-called bulker vehicles involves careful design and selection of body type. If the material being carried is only required to be dumped at ground level then obviously a high-sided tipper could well suffice. If access to the dumping point is restricted then the use of a screw- or extending flat belt- conveyor would also be valuable — and a non-tipping body might suffice. For below-ground material 'dumps', or those accessible by overhead tracks, then gravity-discharge through the bottom of a hopper body might prove to be the solution.

Where both the horizontal and vertical distance between the point of vehicle access and the material 'dump', or store, is particularly large then pneumatic conveying is probably going to be required. This could involve a special-purpose pressurised body. Two common options are a fully pressurised body or a non-pressurised body having a rotary valve feeding a small pressurised compartment. The valve acts like a revolving door — it is a mechanical seal allowing material to be fed from an area of atmospheric pressure into one that is connected to a pipe used to convey the material at some 2/3 bar above atmospheric.

The economics of the operation often depend on the most intensive utilisation of the vehicle. Thus a quick turn-around of the vehicle, offered by a high pressure body operating at say 40 tonne/hour discharge rate may be preferred to one that might take twice the time — by a low-pressure, rotary valve, method. The relative potential payloads of the body types must also be considered and even the possibility of employing highly specialised rectangular section bulker tanks with the prospect of carrying 'return-loads' on top. Another factor is the weight of the thicker walls required for a high pressure bulker — which might adversely affect the maximum payload of the vehicle. In some cases a site air-pressure supply might be available for obtaining pneumatic discharge, avoiding the need for a vehicle-mounted compressor.

The ability for a vehicle to suck up material from bulk bay storage can also be an advantage, where overhead hopper feeds do not exist. Maintenance costs can also be crucial to overall economics and an operator may decide to have different demountable bodies for different materials — that

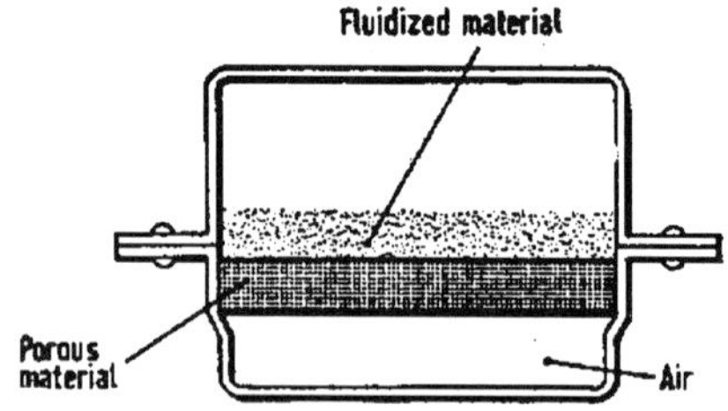

Fig 55: Fluidised bed

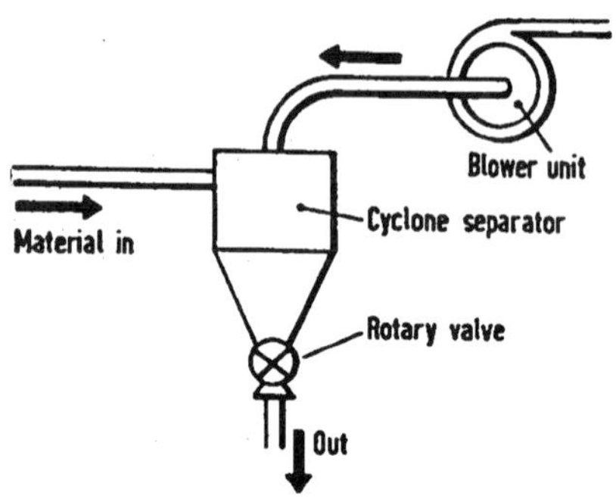

Fig 56: Suction handling system

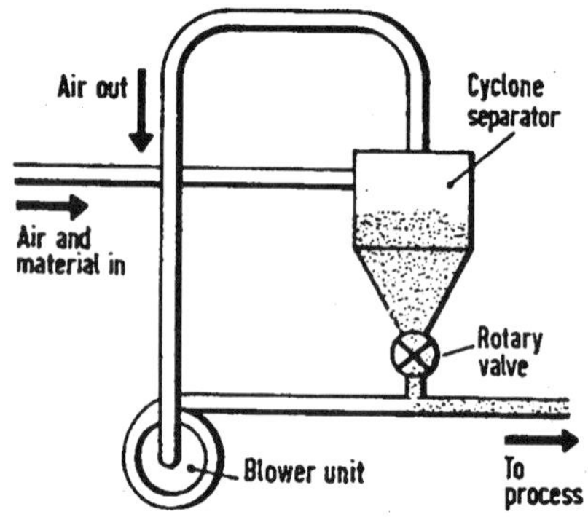

Fig 57: Combined suction and blowing system

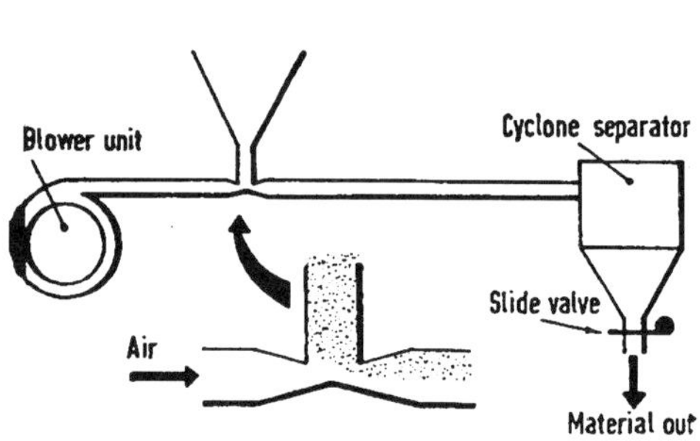

Fig 58: Low-pressure blowing system

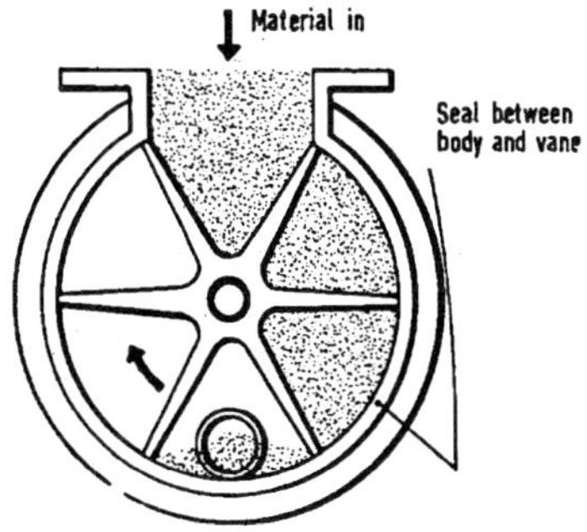

Fig 59: Medium pressure blowing slide-valve

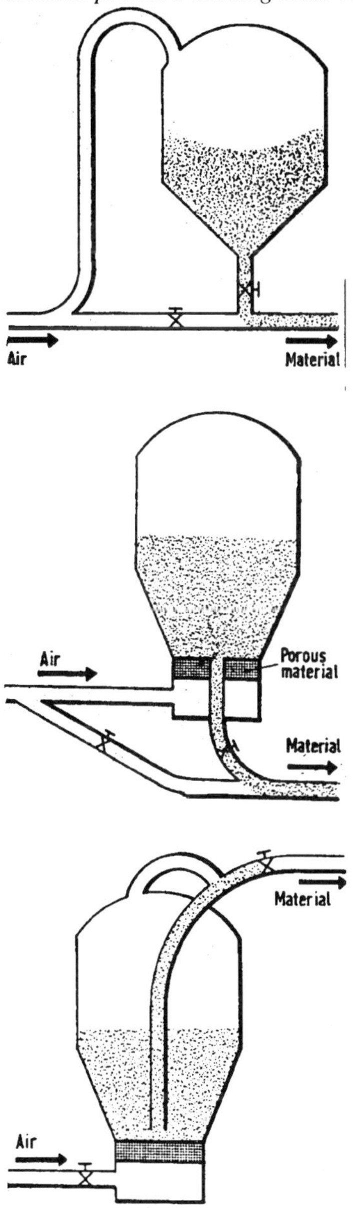

Fig 60: Fluidisation systems

might otherwise cause undue wear of the rotary valve, for example. Air conveying itself can vary in kind — between that in which the material is moved due to the energy contained within the air stream and a fluidised gravity flow arrangement. In the latter case the bulk-powder is converted effectively to a 'liquid' which will flow down a slight incline to convey horizontally without added kinetic energy. Fig 55 shows a fluidisation bed in which a porous membrane is subject to an air pressure beneath it, to disperse the powder above it.

Figs 56 - 60 show alternative blowing systems. A variety of factors might affect their selection not least of which are the properties of the individual materials to be carried. Abrasiveness affects any moving parts, for example; particle size governs the quantity of compressed-air required — while variation in size could lead to particle segregation problems with some equipment. Friability could affect the generation of unwanted dust — which can become explosive with certain air mixtures. Corrosive and heat-sensitive materials need special consideration and the materials' proneness to compaction can cause difficulties in fluidisation.

Fig 56 shows a suction system, limitation of which is that the inlet to outlet pressure differential is limited to less than one atmosphere — so that large bore pipes and big quantities of air are necessary. However, there is little difficulty entraining the particles in the air stream and it is easy to manoeuvre the inlet nozzle so that manual trimming of the product in the receipt-container is minimised. A combined suction and blowing system as shown in Fig 57 may be the best solution. Careful control is needed, however, over separating air from dust — any not trapped in the cyclone could pass through the blower.

Low-pressure blowing, Fig 58, can have problems of entraining particles in the airstream, leading to the requirement for a venturi constriction for assisting direct flow of the material. Throughputs are thus usually restricted to 3 tonnes/hour — however, only a limited power requirement can lead to a moderate cost installation. Medium pressure blowing, Fig 59, involves the use of a mechanised seal to feed the material into the compressed air pipe — while high pressure blowing involves pressurising the vehicle container tank itself and the material entraining in the airstream by gravity/fluidisation. Greater efficiency is obtained because less energy is required to entrain the powder and more material can be moved per kilo of air employed. Fig 60 shows different means of fluidising the powder solids.

Hopper design

For gravity-discharge vehicles, care is needed with hopper design to obviate against 'bridging' or 'rat-holing' disturbing the free-flow of bulk solids. Work by AW Jenike[4] is valuable in helping to solve such problems. He showed that 'doming' occurring in gravity feed of the product will collapse for slot-width less than or equal to σ_c/ρ (unconfined yield-strength/bulk density) or $\sigma_c/2\rho$ where the slot is not narrow but tends more towards a square or circular hole of diameter equivalent to the width dimension. σ_c strength depends on the hopper shape according to the flow factor *ff*, a high value of which signifies arching of the product. See lower curve of Fig 61; when the *ff* curve intersects the strength curve, this indicates critical consolidation pressure.

Stresses at the surface of the material on a 'dome' or 'pipe' divided into this pressure yield the critical flow factor. Plots show material characteristics from which average effective angle of friction is obtained. This can then be used in conjunction with Fig 62 for a particular hopper-side angle (30 deg in the example shown) to determine *ff* for constructing $x = ff$ lines on Fig 61. Intersection of these lines with the material property curves then gives the consolidating pressure P and σ_c.

A more accurate friction angle can next be obtained by inserting this pressure value as the upper curve in Fig 61 and a new value used in Fig 62 to obtain a more correct *ff* — leading to the more accurate blobbed curve in Fig 61. This in turn leads to a new consolidating pressure and a value of bulk density obtained from the figure. The minimum outlet size to prevent 'piping' is obtained in a similar manner but using Fig 63 and the expression $4(\sigma_c/\rho)A$ for controlling diagonal (or diameter) where A is the piping factor.

In the case of conical hoppers, Jenike suggests the maximum hopper angles shown in Fig 64 applying to hoppers having smooth transition radii to the vertical walls of the vessel. It is also assumed there are no impact conditions that might cause consolidation of the material during filling. A shelf should be provided in the path of the filling trajectory to avoid 'direct hits'. For a container with a long rectangular

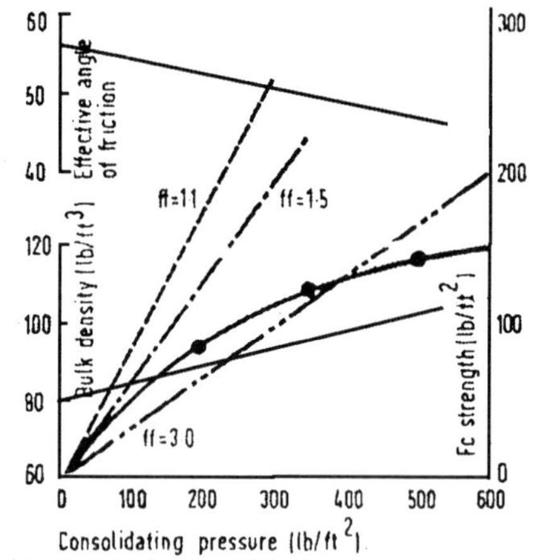

Fig 61: Critical consolidating pressure determination

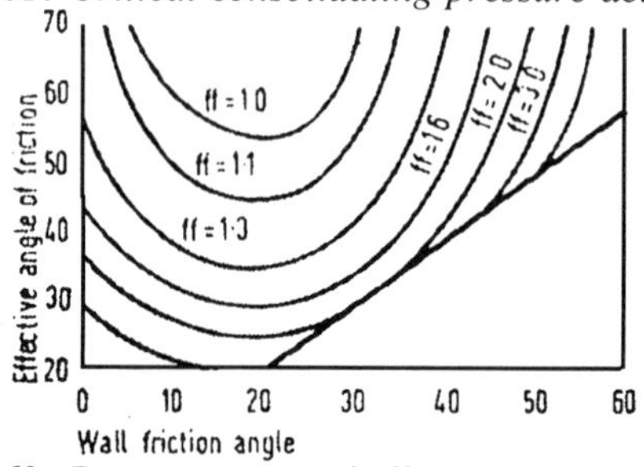

Fig 62: Determination of effective angle of friction

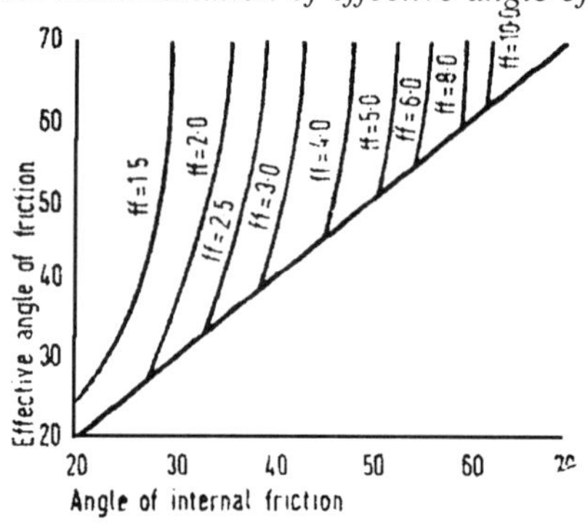

Fig 63: Determining outlet size to prevent piping

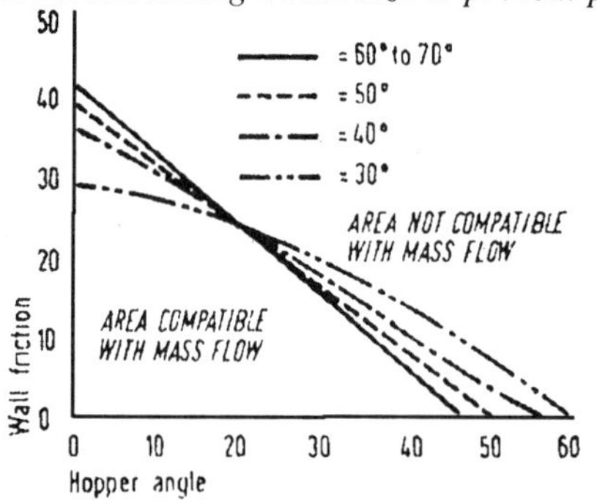

Fig 64: Minimum angles for conical hoppers

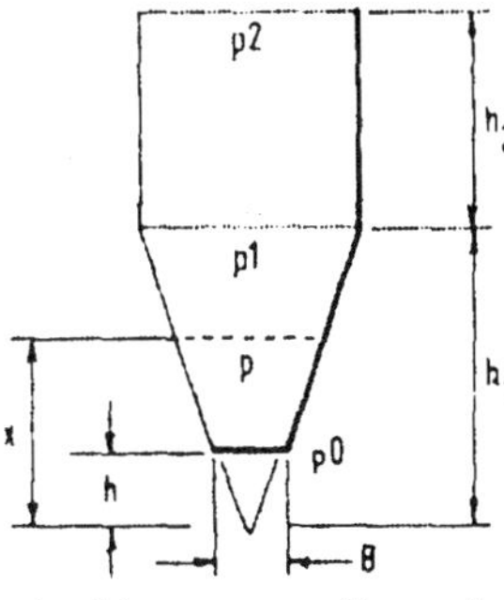

Fig 65: Typical hopper configuration

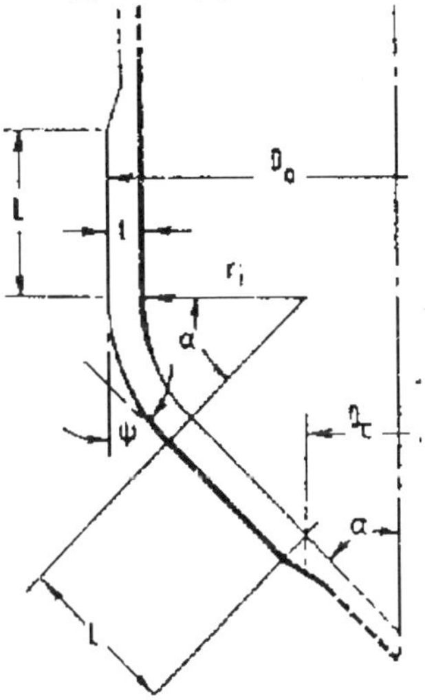

Fig 69: Analysis of tank top corner

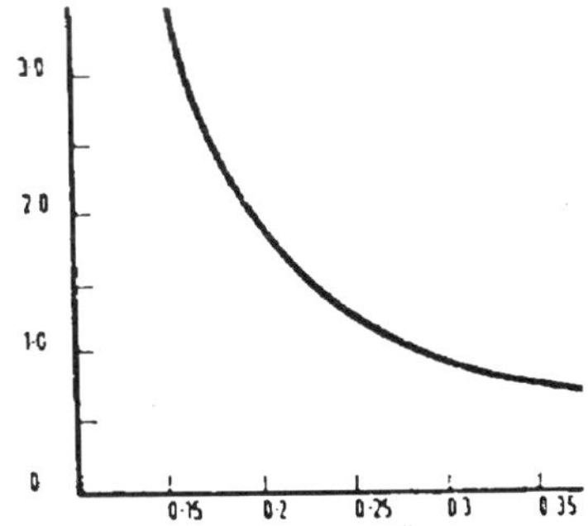

Fig 66: Tank-end design factor, k (x-axis) against h_0/d_0 (y-axis)

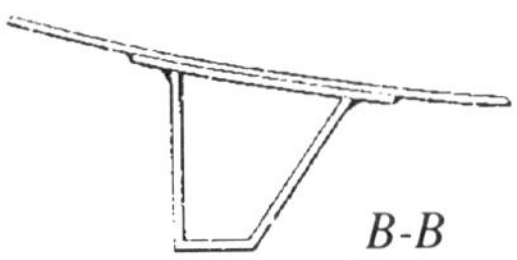

B-B

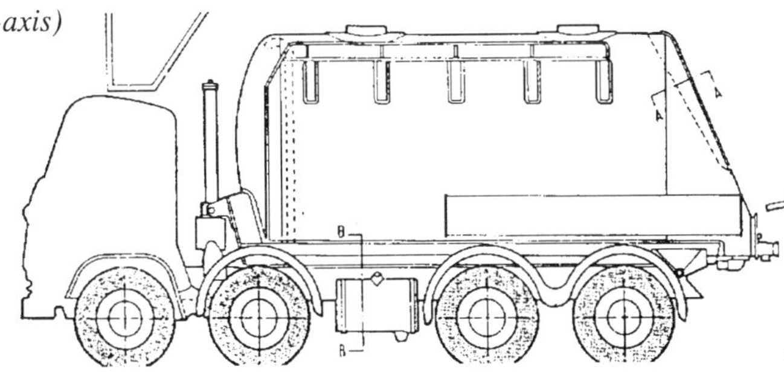

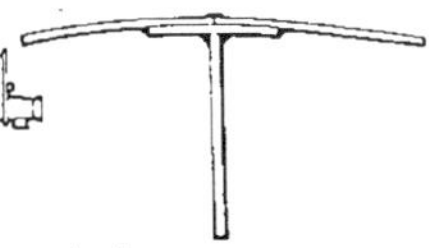

A-A

Fig 67: Inclined-axis conical bulker

Fig 68 (left-right): Key sections

plan section, plane-flow conditions are said to exist. In a configuration such as that in Fig 65, top-pressurisation causes a pressure gradient to build up from atmospheric p_0 at the bottom to p_2 at the top. The difference between them equals

$$agh[1 + h(h_2 - h_1)/h_1^2]$$

and in the design of the hopper, critical flow factor is divided by $(1 + a)$. For all values of h within the hopper, a must be sufficiently large for $ff/(1 + a)$ to be less than or equal to Ps_c. The required rate of pressurising air flow is given by

$$(A/R)ag(h/h_1)^2$$

where A is the section area of the vessel and R is the resistance to air flow of the material.

Pressure tank design

Aluminium construction for pressure tank design is covered in British Standard BS 1500 Part 3 for static units — but generous safety factors (2-5) have to be allowed for in vehicular duty. While tanks rarely actually travel under pressure, wracking and surging loads have to be designed for. With cylindrical tanks, shell thickness is determined from

$$t = (f.D_i)/(2PJ - f)$$

where f is design stress, D_i is internal diameter and P is design pressure (joint factor J). The expression assumes support conditions along the length of the tank.

With gravity-discharge bodies, the first consideration is to ensure the hopper sides are steeper than the angle of repose of the material, after which dimensions can be fine-tuned, as above, to avoid compaction effects. Undue bulging of the hopper walls can be avoided by designing within deflection limits prescribed by established plate-deflection (under-pressure) formulae, the pressure on the plate being given by $k\rho h$ where

$$k = (1 - \sin\gamma)/(1 + \sin\gamma)$$

and where g is the angle of repose, r the density of the material and h the depth of the plate below its surface.

Design of pressure-tank ends is dealt with in BS 1966; these ideally should be hemispherical — and thickness is given by

$$fD_i/(4PJ - F)$$

Should the semi-ellipsoid design be needed, to save on vehicle length, thickness is obtained from $fD_0k_1/2PJ$ where k_1 is obtained from Fig 66.

Conical ends of symmetrical pattern are covered in BS 1500; however, as these are almost universally of the inclined axis type (Figs 67/ 68) the worst-case condition of the top corner of the tank (Fig 69) needs to be considered. Thickness at the junction is given by

$$[(fD_0)/(2PJ - f)][1/\cos a]$$

Detail design of both gravity and pressure discharge tanks is important and typical layouts are compared in Figs 68 and 70. The use of doubler panels to avoid stress concentration at critical areas of pressure tanks is particularly important.

It is also helpful to consider the valley angles between the hopper walls when evaluating flow

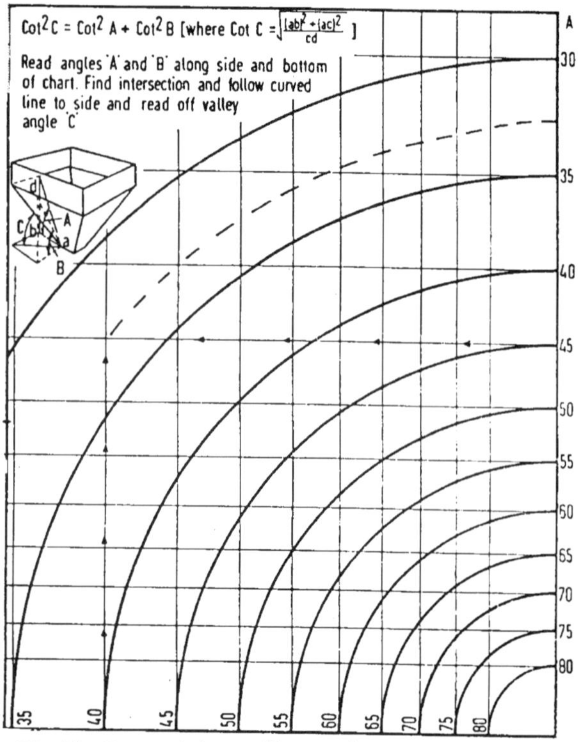

Fig 71: Determining valley angles

Fig 72a: Bulk tipper

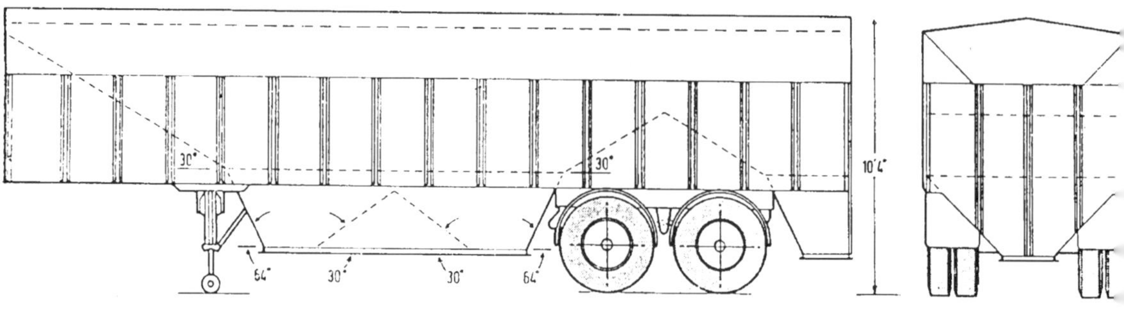

Fig 70: Gravity-discharge bulker

Breadth of plate, in.	1/16	1/8	3/16	1/4	5/16	3/8
6	3.32	13.3	30.0	53.2	83.4	120
7	2.45	9.80	22.0	39.2	61.2	88.0
8	1.88	7.50	16.9	30.0	46.8	67.6
9	1.49	5.93	13.3	23.72	37.0	53.2
10	1.20	4.80	10.8	19.20	30.0	43.2
11	0.99	3.96	9.93	15.84	24.8	39.72
12	0.84	3.34	7.50	13.36	20.8	30.00
13	0.71	2.84	6.40	11.36	17.7	25.60
14	0.61	2.45	5.51	9.80	15.3	22.04
15	0.53	2.13	5.07	8.52	13.3	20.28
16	0.47	1.88	4.22	7.50	11.7	16.88
17	0.41	1.66	3.74	6.64	10.4	14.96
18	0.37	1.48	3.33	5.72	9.26	13.32
19	0.33	1.33	2.99	5.32	8.30	11.96
20	0.30	1.20	2.70	4.80	7.50	10.80
21	0.27	1.09	2.45	4.36	6.80	9.80
22	0.25	0.99	2.23	3.96	6.20	8.92
23		0.908	2.04	3.632	5.67	8.16
24		0.834	1.87	3.472	5.21	7.50
25		0.769	1.73	3.076	4.80	6.92
26		0.71	1.60	2.84	4.44	6.40
27		0.66	1.47	2.63	4.11	5.92
28		0.61	1.38	2.45	3.83	5.52
29		0.57	1.28	2.28	3.56	5.12
30		0.53	1.20	2.13	3.33	4.80
31		0.50	1.12	2.00	3.12	4.48
32		0.47	1.06	1.87	2.93	4.22

Correction factors for use with the above Table for other than square plates. Pressures given above should be multiplied by the appropriate factor.

L/B	1	1.1	1.2	1.3	1.4	1.5	1.6	1.7
Factor	1	0.895	0.83	0.783	0.752	0.73	0.713	0.7

L/B	1.8	1.9	2.0	2.25	2.5	2.75	3.0
Factor	0.69	0.682	0.677	0.665	0.66	0.656	0.65

Fig 73: Pressure in lb/in² to give stress of 10,000 lb/in² in flat square mild steel plates with fixed edges

Breadth of plate, in.	1/16	1/8	3/16	1/4	5/16	3/8
6	**20.1**	**80.4**	**273**	**643**		
7	**10.9**	**43.4**	**147**	**347**		
8	**7.0**	**28.0**	**86.5**	**204**		
9	**4.0**	**16.0**	**54.0**	**127**		
10	**2.7**	**10.8**	**35.4**	**83.3**		
11	**1.78**	**7.11**	**24.2**	**56.9**		
12	**1.26**	**5.03**	**17.1**	**40.2**	**78.8**	
13	**0.91**	**3.65**	**12.4**	**29.2**	**57.3**	
14	**0.68**	**2.71**	**9.2**	**21.7**	**42.5**	**73.6**
15	0.51	2.05	**7.0**	**16.4**	**32.3**	**56.0**
16	0.40	1.61	**5.4**	**12.7**	**24.9**	**43.2**
17	0.31	1.25	**4.25**	**10.0**	**19.6**	**34.0**
18	0.25	1.00	**3.38**	**7.95**	**15.6**	**27.0**
19	0.20	0.80	2.72	**6.39**	**12.6**	**21.76**
20	0.163	0.65	2.21	**5.21**	**10.2**	**17.68**
21	0.135	0.54	1.82	4.29	**8.41**	**14.56**
22	0.123	0.49	1.51	3.56	**6.98**	**12.08**
23	0.099	0.39	1.27	2.98	**5.85**	**10.16**
24		0.31	1.07	2.51	4.94	**8.56**
25		0.269	0.91	2.13	4.19	**7.28**
26		0.228	0.77	1.82	3.58	6.20
27		0.196	0.66	1.57	3.08	5.33
28			0.58	1.36	2.66	4.61
29			0.50	1.18	2.31	4.00
30			0.44	1.03	2.02	3.50
31			0.38	0.90	1.77	3.07
32			0.34	0.80	1.56	2.70

Pressures shown in bold figures result in a stress of more than 10,000 lbf/in² and should not be used for this reason in normal tanks.

Correction factors for use with the above Table for other than square plates. Pressures given above should be multiplied by the appropriate factor.

L/B	1	1.1	1.2	1.3	1.4	1.5	1.6	1.7
Factor	1	0.885	0.812	0.756	0.722	0.695	0.676	0.664

L/B	1.8	1.9	2.0	2.25	2.5	2.75	3.0
Factor	0.65	0.645	0.638	0.637	0.621	0.617	0.614

Fig 74: Pressure in lbf/in² to give central deflection of 1/32 inch of square steel plate

inside a gravity-discharge vessel. Valley angles are always less than the limiting angle of repose that causes product to hang up in the corners. Fig 71, from a DATA design guide[5], enables valley angles to be read off when the hopper side angles are known.

An expression for the deflection of a uniform plate constrained at its edges is given by Rees[6]. Depending on the values predicted, the option of using tie bars could be considered with the possibility, in turn, of reducing wall thickness.

Open-top bulk vehicles

The last two sections covered closed bulk vehicles for carriage of solids and liquids respectively. Here, open bulk-carriers are considered with particular reference to their design to withstand structural loading. These vehicles are, of course, characterised by more arduous operating conditions, particularly during loading and unloading, and involve considerably more rugged construction — especially in the case of off-road operation.

Open-top bulktippers (Fig 72a), for on-road, and dump-trucks (Fig 72b), for off-road operation, have been constructed in steel and aluminium alloys and to a limited extent in GRP. The ability to withstand considerable drop loads in the case of dump-trucks, which might be loaded by tractor shovels pitching in massive rock boulders to their bodies, is matched by the need to withstand large side-wall pressures, for the case of low-viscosity payloads carried in high-sided vehicles. By far the most important input for the vehicle designer is operating history of earlier designs. Only when a very radical departure in material usage, or operating condition takes place will any recourse be made to other than comparative techniques in structural design.

The prospective use of GRP focuses again on the use of predictive techniques to calculate the internal loads in the structural elements. Finite-element techniques of dividing the structure into small inter-connecting parts, prior to the analysis, would be the normal way to prove the final design before commitment to hardware. At the earlier concept stage, a return to fundamental engineering principles is a good way to assess the effects of loading extremes. How much stress, for example, could be induced from a boulder falling on the floor of a dump-truck. Individual rocks, each weighing 3 tonnes are not unknown in the spoil being removed during blasting to provide the core of an 'earth' dam.

Gravitational acceleration will result in the boulder striking the floor and giving up its kinetic energy into the structure to an extent which depends on the local elasticity of the floor panels as well as the overall compliance of the vehicle suspension. The impact velocity can be found by equating the loss of potential energy in falling height h with the gain in kinetic energy as $2h^{1/2}$. The transfer of energy to floor also depends on the respective co-efficients of restitution of the two impacting materials and this effect will typically factor the relative velocity ratio before and after impact by about a half. Thus in the worst case of the floor having virtually no structural compliance there is a gain in strain energy across the area of impact far in excess of that caused by gradually resting the rock on the surface. A factor of fifty times increase in resultant stress would not be untypical.

Most materials carried in open tippers have an angle of internal friction which governs the angle which the body has to be tipped in order for the material to flow from it. Depending on the degree of moisture present, the angle reduces until a virtually liquid-like behaviour applies to wet soils — which can apply substantial forces to body sidewalls as well as to floors. How much bulging of the wall plates can be tolerated is debatable. Some say bulging to the extent of up to one hundredth of the supported plate-width (up to a 1/32nd of an inch limit) is a 'visually' acceptable maximum. A formula due to Grashoff is used to determine the deflection of a flat rectangular panel subject to lateral pressure as

$$l^4pb^4/(l^4 + b^4)32Et^3$$

where length l, breadth b and thickness t are in inches while pressure p and elastic modulus E are in lb/in^2.

The table in Fig 73 due to C W Rees provides a quick reference for steel plates up to their visually tolerable bulge limit. This sort of approach can be used to check the effect of rail-pitch for giving edge support to the side panels. The table in Fig 74 can be used to further check whether panel stresses are likely to be kept within safe working limits. Lateral pressure is, of course, the product of 'head' (height to the surface) times density for a liquid material (with zero cohesion). This results in a triangular distribution of pressure intensity toward the bottom of the tank — but for design purposes it would be reasonably safe to

Fig 72b: Dump truck

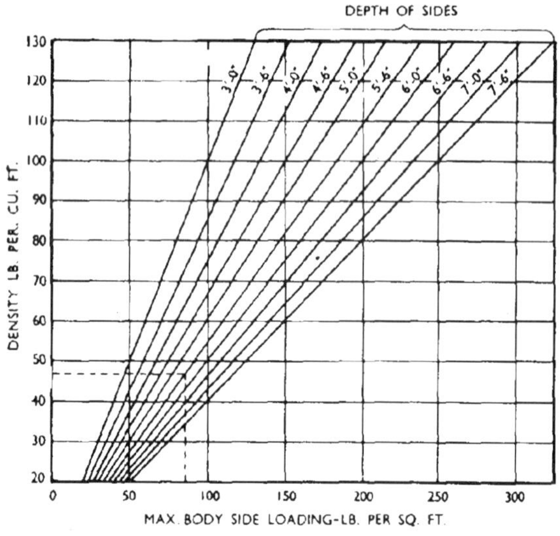

Fig 75: Pressure on body sides

Height of side (ft.)	4	5	6	7	8
Maximum body side loading (lb./sq. ft.)	67	83	100	117	133
Maximum pillar spacing (in.)					
2¼" Top-hat section 12873	117	60	33½	22	15
1¼" Top-hat section 12674	48½	25	14	9	6

Fig 76: Spacing of side pillars for product density of 50 lb/ft^3

Side Panels (BA.25 WP)*	Side Pillar Spacing					
	12"	15"	18"	21"	24"	30"
SHEET						
12 s.w.g. (0·104")	460	300	205	150	115	75
14 s.w.g. (0·080")	275	175	120	90	70	45
16 s.w.g. (0·064")	175	115	80	55	45	
18 s.w.g. (0·048")	100	65	45			

Fig 77: Side panelling — recommended maximum loading (lb/ft^2)

assume uniform distribution at the maximum level. Fig 75 shows the pressure at the base of the body sides for a range of depths and material densities, assuming a 30 degree internal friction angle (of repose) and the table of Fig 76 the proposed spacing of ribs in the case of an aluminium-alloy body, with recommended panel loadings in the table of Fig 77. The stress can be also calculated in absolute terms as:

$$b^2pl^4/2t^2(l^4 + b^4)$$

The design of ribs to support the side plates will be affected both by the need for support against the above pressure and by a robustness necessary to survive the hazards of operation on construction sites and similar conditions. The ribs must also stabilise the sidewalls against shear deformation in the overall bending and twisting of the body as a stiffened box tube. Top and bottom rails of the body must also withstand direct stresses due to overall bending and torsion.

One approach to assessing the pressure retention capacity of stiffening ribs is to assume each rib supports a strip of plating extending half way to the next stiffener on either side. Maximum bending stress in the stiffener is thus

$$Wly/12I \text{ lbf.in}$$

where W is total distributed load (lb) and l is stiffener length; y is neutral-axis to extreme-fibre distance and I the stiffener section second moment of area. The deflection of the stiffener is given by

$$Wl^3/384EI$$

Standard bending and torsion formuli given in earlier sections can then be used to calculate overall body structural loads while either the Moment Distribution or Unit Load methods, discussed in earlier chapters, can be used to calculate the effect of the stiff joints between the 'rib' members on the magnitude and distribution of internal forces. Fig 78 shows the initial bending stress calculation phase for a 20 tonne capacity tipping semitrailer.

Recommendations for floor loadings and dimensions have been provided by British Aluminium[7]; The table in Fig 79 reproduced here is based on the criterion of a maximum deflection of 1/200th the supported span while the table in Fig 80 is to withstand a working stress of 10,000 lb/in^2 — both using BA25WP grade aluminium-alloy sheet. Choice of crossbearer design depends on floor-loading, crossbearer-spacing and method of body attachment. The table in Fig 81 gives some recommended maximum distributed loads for a selection of crossbearer sections. To ensure that the body sidewalls contribute to the longitudinal strength of a tipping body, the crossbearer must also be adequately jointed to the side-pillar ribs as well as these being supported by substantial top-rails and corner posts, Fig 82.

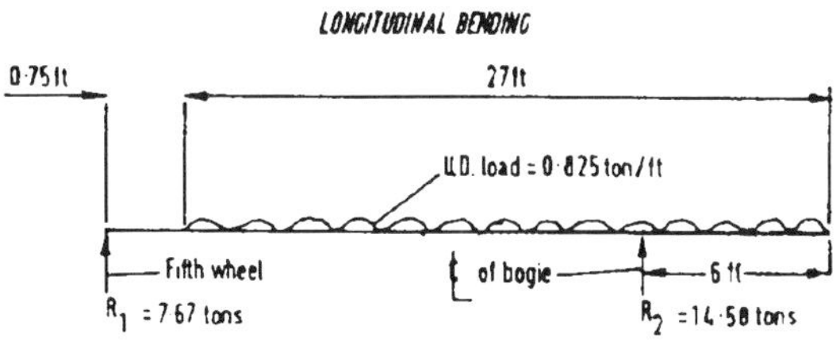

Max. bending moment = 41·5 ton-ft
Second moment of area = 21 520 in^4
Neutral axis 43·3 in below top rail
Max. static stress = 2250 lb/in^2 (compression)
Dynamic load factor = 3
Safety factor = 5
Fully factored stress = 15 × 2250 = 33 750 lb/in^2
Flexural strength of material = 65 000 lb/in^2

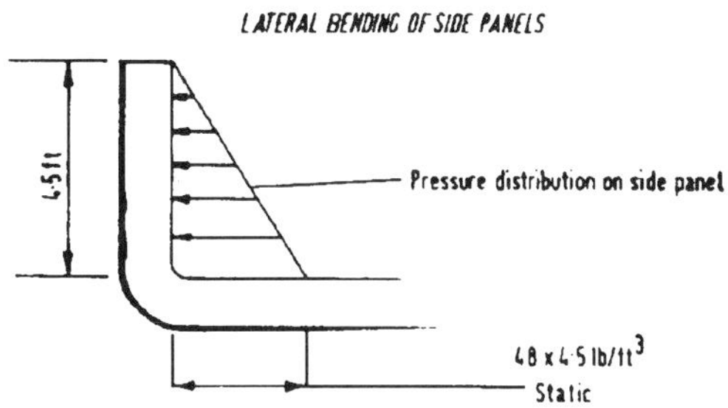

Max. bending per stanchion = 1312 lb-ft
Dynamic factor = 3
Second moment of area = 8·34 in^4
Neutral axis = 2·82 in from outside profile
Max. static stress = 16 000 lb/in^2 (compression)
Safety factor = 4
Fully factored stress = 64 000 lb/in^2

Fig 78: Bending stress calculation

Cross-bearer Spacing	Floor Loading (lb./ft²) Uniformly Distributed					
	50	75	100	150	200	250
15″	(0·128″)	(0·128″)	(0·128″)	(0·128″)	(0·144″)	(0·160″)
18″	(0·104″)	(0·128″)	(0·144″)	(0·160″)	(0·188″)	(0·188″)
21″	(0·128″)	(0·144″)	(0·160″)	(0·188″)	(0·219″)	(0·219″)
24″	(0·144″)	(0·160″)	(0·188″)	(0·219″)	(0·250″)	(0·250″)
27″	(0·160″)	(0·188″)	(0·219″)	(0·250″)	(0·281″)	(0·281″)

Fig 79: Recommended flooring thickness based on deflection

Cross-bearer Spacing	Floor Loading (lb. ft.²) Uniformly Distributed					
	50	75	100	150	200	250
15″	10 s.w.g.	10 s.w.g.	10 s.w.g.	10 s.w.g.	10 s.w.g.	8 s.w.g.
18″	10 s.w.g.	10 s.w.g.	10 s.w.g.	8 s.w.g.	8 s.w.g.	3/16″
21″	10 s.w.g.	10 s.w.g.	10 s.w.g.	8 s.w.g.	3/16″	7/32″
24″	10 s.w.g.	10 s.w.g.	8 s.w.g.	3/16″	7/32″	7/32″
27″	10 s.w.g.	8 s.w.g.	8 s.w.g.	7/32″	1/4″	1/4″

Fig 80: Recommended flooring thickness based on stress

Section	Weight (lb. per ft.)	Side Overhang	Cross-bearer Spacing				
			15″	18″	21″	24″	27″
	1·52	26″	256	213	183	160	142
		28″	192	160	137	120	107
		30″	147	122	105	92	81
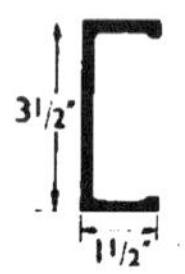	1·42	26″	314	262	224	196	174
		28″	235	196	167	147	131
		30″	181	151	129	113	100
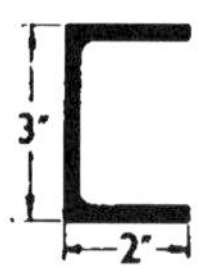	1·81	26″	319	266	228	199	177
		28″	239	199	170	149	133
		30″	184	153	131	115	102
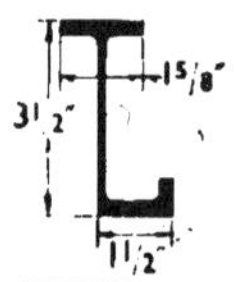	1·60	26″	380	317	271	237	211
		28″	285	237	203	178	158
		30″	219	182	156	137	121
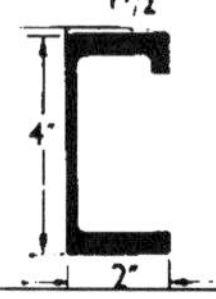	2·38	26″	774	644	552	483	430
		28″	581	485	415	363	323
		30″	446	372	319	279	248

Fig 81: Recommended maximum distributed load (lb/ft²)

In the case of all-welded bodies, the handbook suggests the corresponding joints shown in Fig 83. This is for a typical body type, for an 8.5 yd^3 (6.5 m^3) capacity, fabricated entirely from 1/4 and 3/16 in (6.35 and 4.76 mm) BA28 plate using MIG welding. Unless warranted by volume, stiffeners would be formed on a press-brake. Folding in of the lower side panels by 45 degrees permits strengthening of the connection between side-pillars and crossbearers. In this case, the body internal width tapers by two inches (50 mm) increase from bottom to top. Subframe longitudinals are of zed section 1/4in (6.35 mm) plate slotted for allowing crossbearers to pass through.

When a hopper-discharge body is employed, it is suggested that hopper panelling of 10 or 12 swg BA25WP plate is used rising to 3/16th in (4.76 mm) in the case of particularly abrasive materials. The hopper sides normally slope 5-15 degrees more than the internal friction angle of the material. Fig 84 is one design approach.

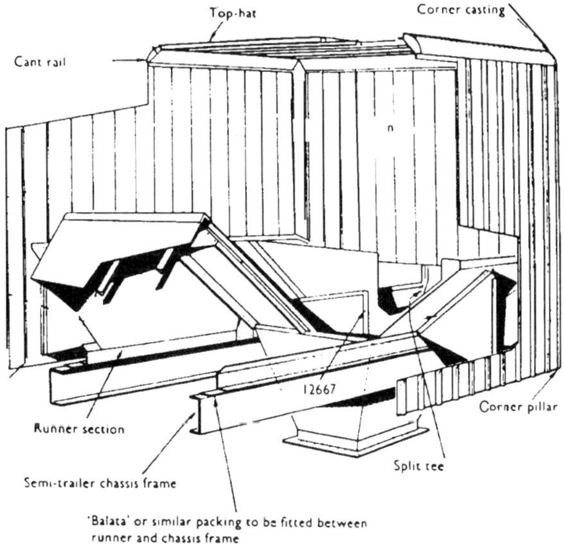

Fig 84: Hopper-discharge body

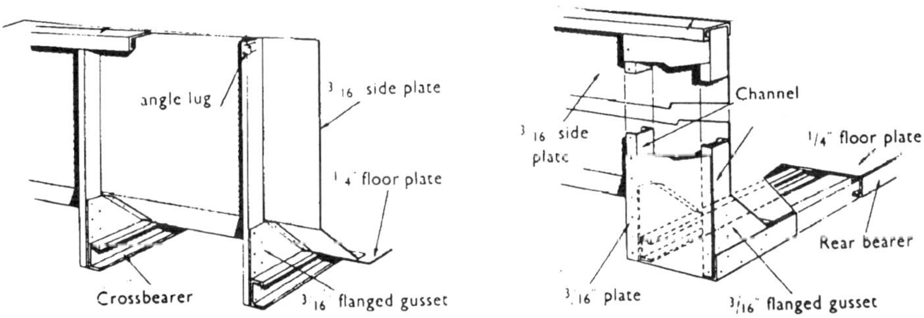

Fig 82: Side-pillar to crossbearer joint

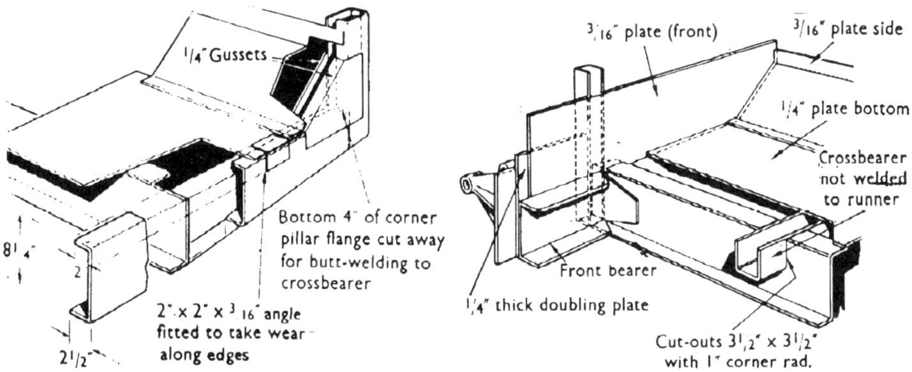

Fig 83: Welded-body joints

Calculating tipper stability

The IRTE tipper stability code has focused new attention on designing for the best possible resistance to overturning when long bodied tippers discharge on uneven ground. The problem is also being examined by university researchers who are demonstrating that a rigorous analysis can throw useful light on the problem. Manufacturers are also responding with such techniques as gusseted box section crossmembers, Fig 85.

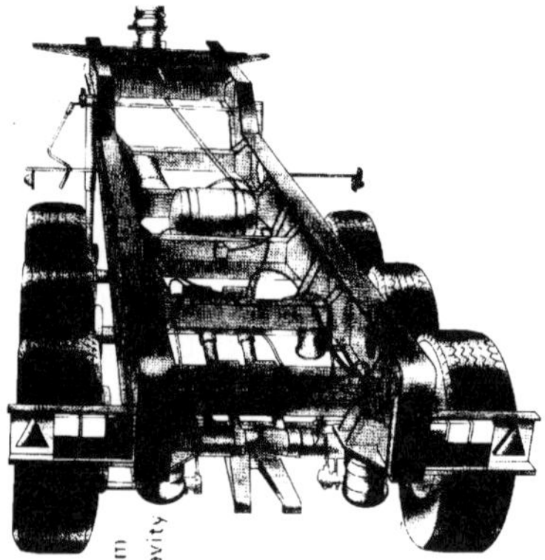

Fig 85: Tipper trailer chassis with gusseted crossmembers

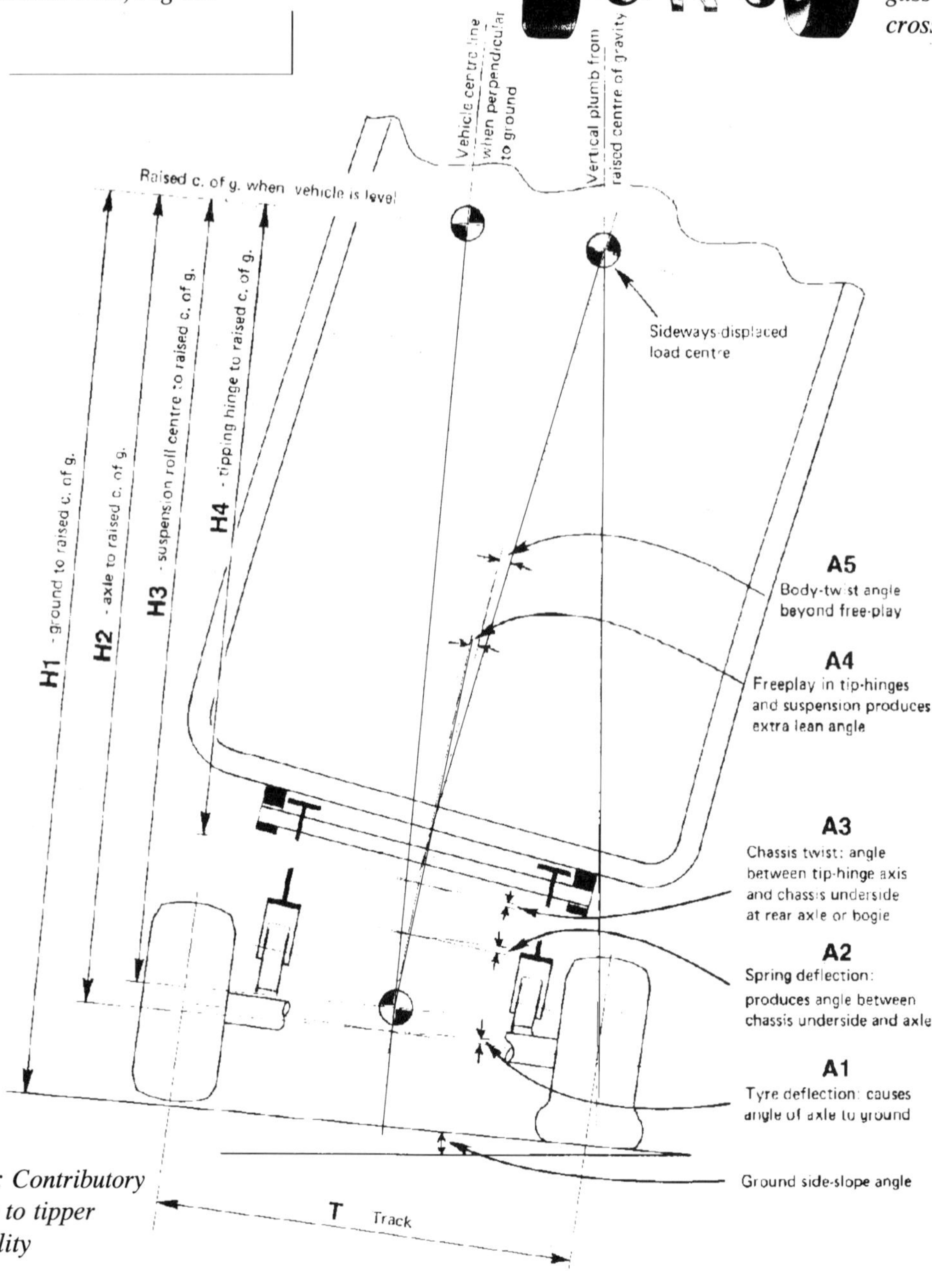

Fig 86: Contributory factors to tipper instability

Improved stability for tipper trucks is the objective of the code-of-practice recently published by the Institute of Road Transport Engineers. The institute drew on several independent expert consultants including Dr George Tramal of the Bristol Faculty of Engineering and former Crane Fruehauf engineering chief Norman Simmons who now operates as NCS Engineering Consultancy Services. The last company was responsible for the diagram of contributory factors shown as Fig 86.

This is drawn up on the basis of the commonest type of accident, falling over sideways when discharging load. While the situation can be worsened by such factors as soft ground, sticking payload or a side-wind gust, tolerance of such adversities can be built in to the design of the vehicle and observance of the basic parameters illustrated gives the tipper the best chance for stability under most conditions.

Two categories of minimum stability are recommended by the code: Category B covers the ability to discharge on hard level surfaces and Category A covers the unlevel off-road situation. The criterion of stability is the sideslope angle which the vehicle can withstand without wheel lift when the body is fully raised and fully loaded — which can be both measured and calculated. Because of the sticking load situation it is necessary for the vehicle to have its body fully raised with the load in place and then withstand uneven discharge.

For Category B operating conditions the minimum acceptable sideslope stability for any tipping vehicle must be 5 degrees. This increases to 7 degrees for Category A operation and for both cases it is recommended that the main safety points in operating the vehicle should be marked on it. For tipper manufacturers to have a reliable basis for assessment of stability the chassis maker should supply, on request, the torsional resistance of the chassis frame from hinge point to rear suspension centre-line; the roll-stiffness of the suspension at the point of wheel lift on one side; the effective spring-base dimension; roll-centre height for the suspension and mounting guidelines for both bodies and tipping-gear.

General guidelines in design are for the centre-of-gravity to be as low as possible; high suspension roll-stiffness at the point of overturning; torsional stiff chassis between hinge and suspension; minimum free-play in suspension and hinges (which must be as widely spaced as possible) also close as possible distance between hinge and suspension. Add-on features such as sideways restraint frames between chassis and body can also be beneficial.

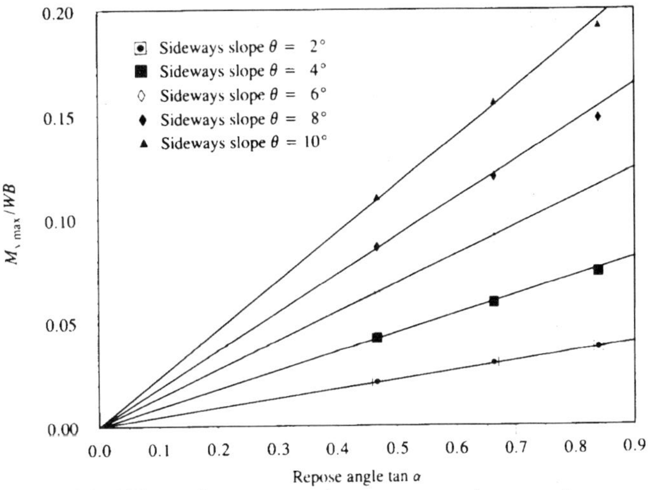

Fig 90: Effect of repose angle on overturning couple

Researchers at Nottingham University[8] have also been involved in tipper stability. This has led to method for calculating roll-over couples, for granular materials with different repose angles, when tipping from different sideslopes. Figs 87/88/89, overleaf, give a representation of the tipping body and payload which is assumed to be free-flowing and create no friction with the body walls. The researchers have examined the shape of the sectional plane IJKA through the payload and used co-ordinate geometry in three planes to calculate its shape. A computer program has also been written to relate a shape factor with payload volume, mass and centre-of-gravity co-ordinates. So far the work has been restricted to analysis of the body and payload only but further work is envisaged that will include factors pertaining to a large articulated vehicle.

For the body and payload maximum value of roll-over couple due to a free-flowing payload has been found as

$$0.023q \tan a/WB$$

for the corresponding tipping angle: α - *12* where W is weight of full payload and B is body width; θ is sideways slope of body and α is angle of repose. Fig 90 shows the effect of repose angle on overturning couple.

The various factors that go into the build-up of overbalancing angle, which equates, in a cumulative formula:

$$\sin^{-1}\left\{\frac{(T \div 2)-(H2 \times \sin A1)-(H3 \times \sin A2)-(H3 \times \sin A3)-(H4 \times \sin A4)-(H4 \times \sin A5)}{H1}\right\}$$

This will not give a precise prediction of the overturning angle but will normally be good enough for comparison purposes. Typical values are given below for a tri-axle trailer.

TYRES

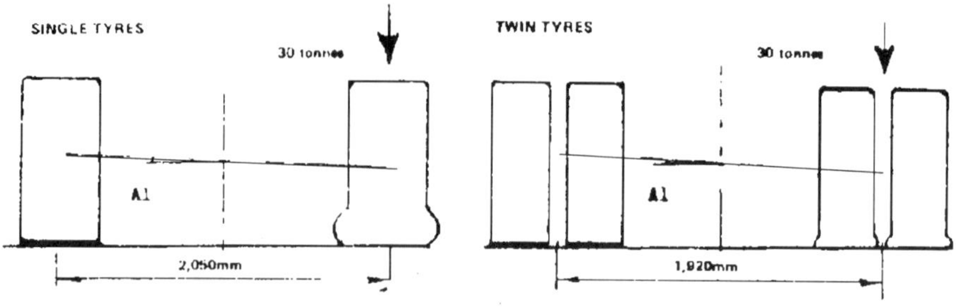

Deflection = 120mm Angle A1 = 3.35° Deflection = 95mm Angle A1 = 2.8°

SPRINGS

Deflection = −17mm 67mm

Angle A2 = 4°

Deflection = −47mm 73mm

Angle A2 = 7°

Air suspension A2 is less than 2°

FREEPLAY

1,200mm SPRING CENTRES	1,000mm SPRING CENTRES
For 30mm play, Angle A3 = 1.43°	For 30mm play, Angle A3 = 1.7°
For zero play, Angle A3 = 0	For zero play, Angle A3 = 0

BODY STIFFNESS

Assume Angle A5 = 2°

CHASSIS STIFFNESS

For a torsionally flexible chassis, Angle A4 approximately 3° (few, or channel or I-section cross-members).

For a torsionally stiff chassis, Angle A4 approximately 1° (tubular or box cross-members).

It is possible to get significant variations in these values. These figures are by way of example only.

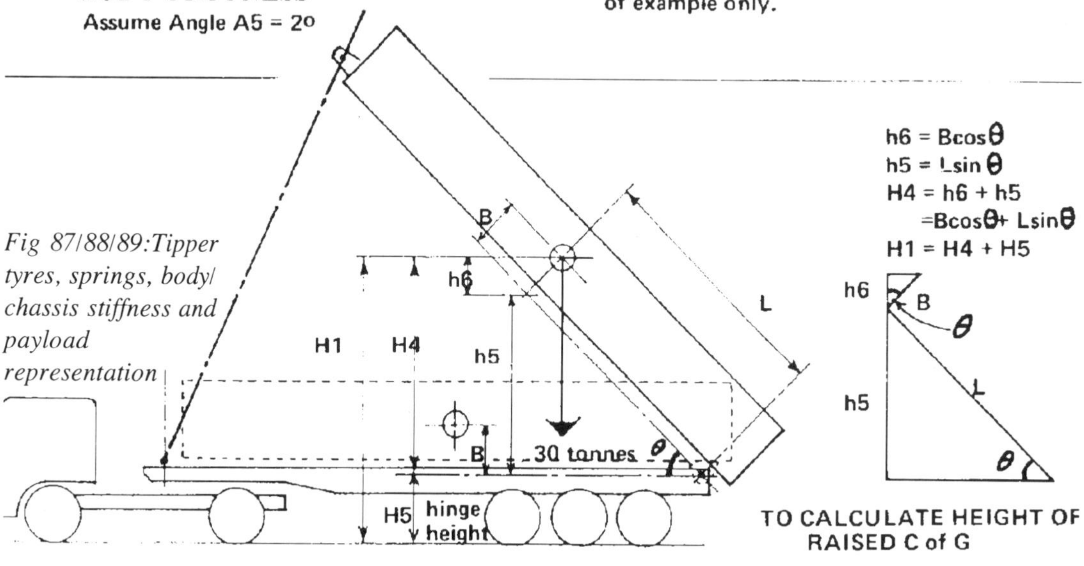

Fig 87/88/89:Tipper tyres, springs, body/chassis stiffness and payload representation

Powder bulker charge/discharge

The successful design of bulk powder vehicles depends as much on the equipment for loading and unloading the vehicle as on the container for carrying the material in transit.

According to Dr N Mainwaring of Simon Macawber, the physical properties of the bulk solid material are critical to the successful understanding of pneumatic conveying equipment. He sees conveying equipment as split into two groups, positive and negative pressure systems, depending respectively on blowing and suction — the latter being restricted to relatively low product throughputs and short conveying distances. Positive pressure systems can then be sub-divided into dilute and dense phase flows.

While the former are more versatile in terms of the wide range of products that can be conveyed, more recently, dense phase flow systems have increased in use due to their ability to reduce power consumption, minimise particle degradation and lessen the incidence of erosive wear on pipe bends. Development effort is directed towards making dense phase systems suit a wider range of products.

In defining the two main systems the criterion of amount of product continuously held in suspension is increasingly used, dense phase systems involving a significant fall-out of product over the conveying distance. Several modes of flow are involved depending on the type of product and the method of feeding air, and product, into the pipeline, Fig 90. Reducing air velocity from the dilute flow regime at (a) leads to the situation at (b) where a small strand of material slides along the bottom of the pipeline. Further reduction in velocity lads to 'moving bed' flow (c), typically occurring at air velocities of 8-10 metres/sec.

Often the moving bed part of the flow involves a pulsating stop-start regime. By reducing velocity to 3- 6 metres/sec, the flow pattern of (d) occurs, when the material in the line briefly fills the entire cross-section and any further velocity reduction would lead to a blocked pipe. Type (d) flow seems to suit material exhibiting good air retention, the author argues, such as cement, pulverised coal and fly-ash. Products with very poor air-retention can, when air velocity is reduced from the flow (a) situation, cause a plug of material to form which will then move at high speed around the pipeline and possibly cause excessive vibration.

If, instead of reducing air velocity from the flow (a) situation, flow commences with a very low air velocity, which is subsequently increased, then

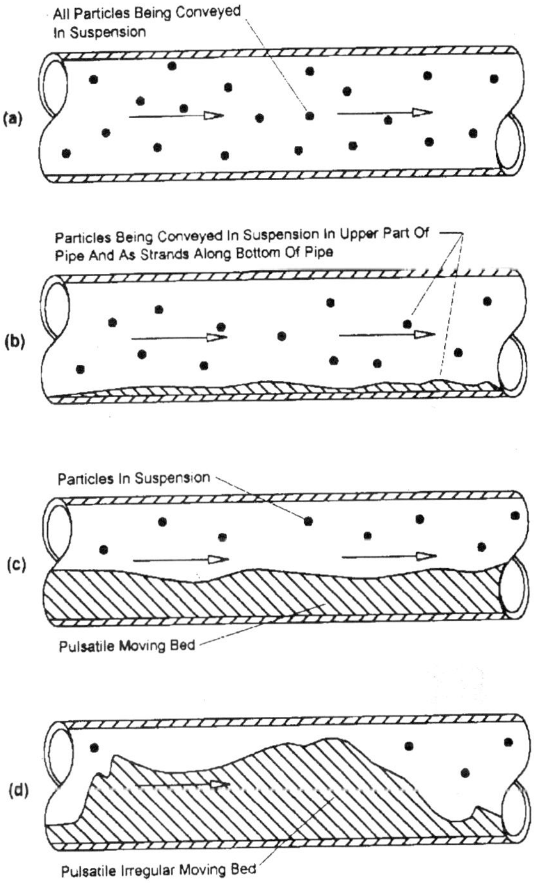

Fig 90: Dilute flow systems

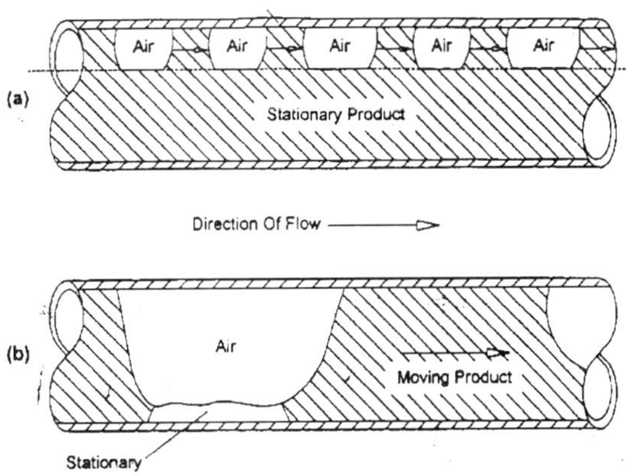

Fig 91: Dense flow systems

different flow modes are experienced. Initially a shearing type flow is experienced as shown in Fig 91(a), then slugs progress along the pipe above the shear line. Slugs progress along the pipe by dropping material off the back of the slug and picking it up from the front of the next one, maintaining constant depth to the stationary bed. Increasing air velocity results in transition to a full bore plug flow, (b). Moving plugs occupy the complete cross-section and again material is swept up at the front and dropped off at the rear. Air velocities are in the order of 0.5-2.0 metres/sec and plugs are separated by an air-space, with a thin stationary layer still existing. This sort of flow is suited to coarse permeable products. With such materials there is an unstable transition between dense and dilute phase flows when excessive vibration is encountered.

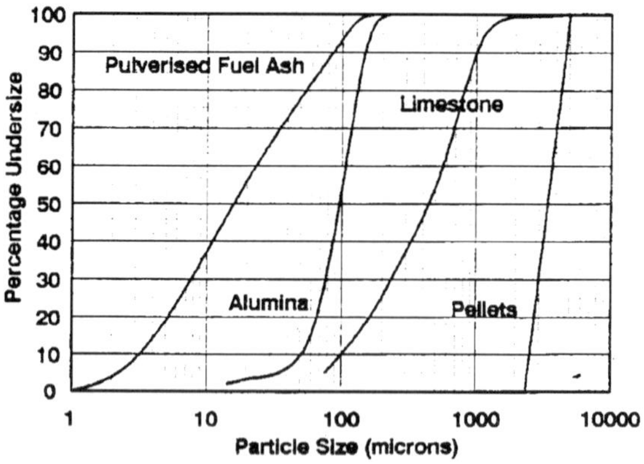

Fig 92: Particle size analysis

A material's suitability to be conveyed in a particular manner is most influenced by particle size distribution, shape distribution, permeability and air retention. The first can be readily measured and generally materials with wide particle size distribution are more problematic than, say, cement. With narrow size distribution materials there is also a direct particle size effect. Those of 150-200 microns are more difficult to convey than those of typically 3000 microns, where a dense plug mode is used. Fig 92 is a size analysis provided by the author. Particle shape distribution is usually qualitatively assessed and its effect associated with size distribution. Per-

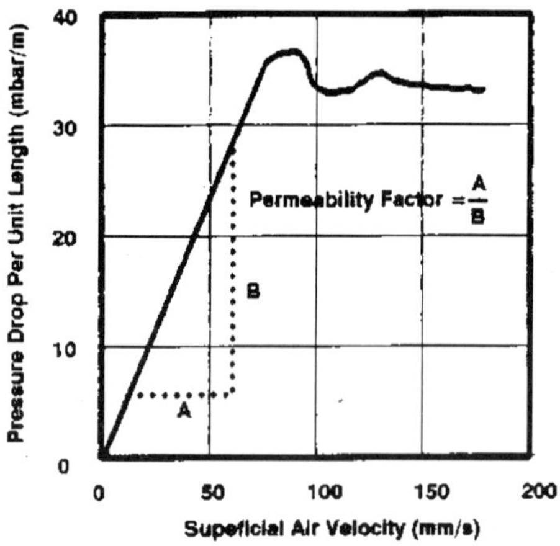

Fig 93: Permeability

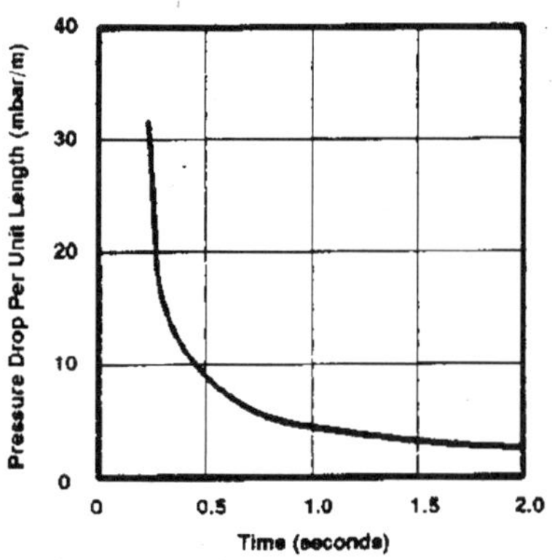

Fig 94: Pressure drop

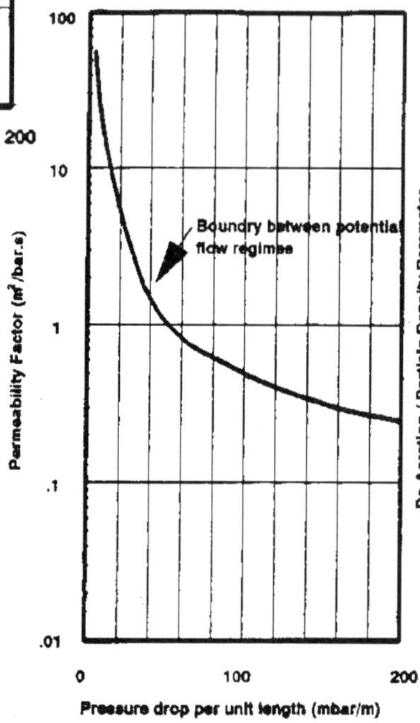

Fig 95: Permeability factor

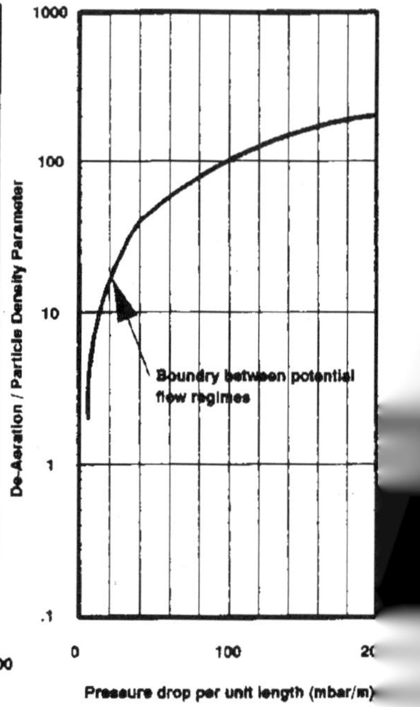

Fig 96: Air-retention factor

meability can be measured with equipment described in the paper; it is expressed as the relationship between superficial velocity of gas passing through a continuous bed, of such a product, and the pressure drop that results in the direction of the flow, Fig 93. Pressure drop is directionally proportional to air velocity over a certain range, corresponding to flow through a 'packed' bed. Another test described in the paper can be used to determine air-retention and a result shown in Fig 94 is obtained. If material is sufficiently permeable as expressed by being above the line in Fig 95 it is likely to adopt the dense phase mode whereas if it is not permeable, but sufficiently air-retentive expressed by being above the line in Fig 96, the same effect is obtained.

Powder conveying systems

Most commonly used dilute phase systems will utilise a rotary valve as a feeding device to control flow of material into the conveying line and to provide a pressure barrier to the source of the material. Generally a pressure differential of 1 bar is the rule. For dense phase conveying, several alternative systems are available; generally these involve a pressure vessel into which a batch of bulk solid is admitted. The material inlet valve is then closed and are air continuously fed in. When blowtank pressure reaches a sufficient level the product will pass into the conveying line resulting in the blowtank contents gradually discharging.

Use of two instead of one blowtank can lead to continuous rather than batch flow. By proportioning air supply between the blowtank and a supplementary air inlet position, control of feed rate can be obtained. Figs 97 and 98 show two possible arrangements; use of two air input positions results in a particularly versatile system with which, by proportioning the air, it can convey in both dilute and dense phase modes.

Hopper shapes for bulkers

Where gravity dischage bulkers are being considered, the advantages and disadvantages of the principal discharge patterns of material flowing in hoppers, silos, bins and bunkers have useful design indications for bulk powder vehicles. Prof AR Reed[9] of the University of Greenwich has provided useful data on the shapes of such facilities that lead to these flow patterns.

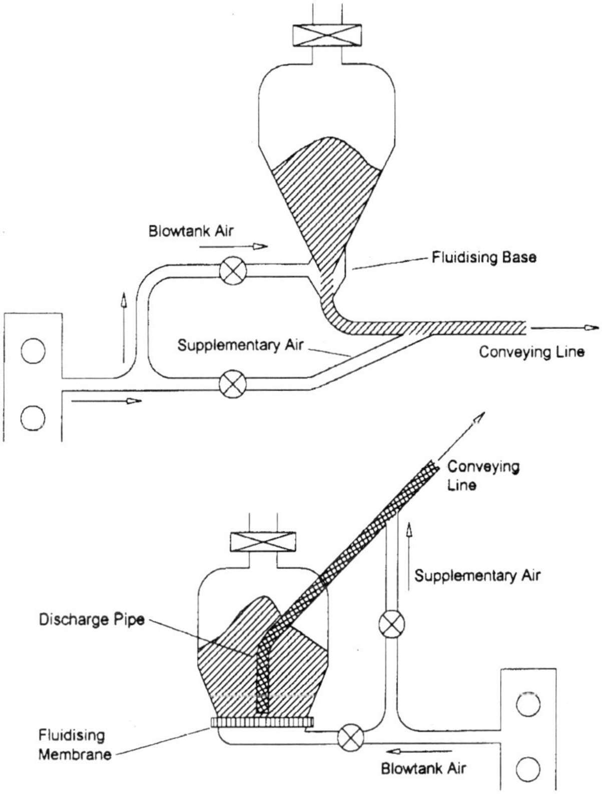

Figs 97 and 98: Alternative systems for proportioning flow

In general, he maintains, for free flowing products where particle segregation is of no concern, a funnel/core flow mode of discharge is an acceptable and economic approach. Determination of which flow type to adopt requires undertaking a simple bench scale test to determine the frictional characteristics between samples of the product to be carried and the material from which the internal surfaces of the converging section of the vehicle discharge hoppers are to be constructed. This test will yield an 'angle of wall friction', which, by consulting graphs of the form shown in Fig 99, will, depending on the basic shape of the converging section under consideration, suggest the angle of the converging section that is necessary to bring about the required discharge pattern.

The two basic flow/discharge patterns, mass flow and funnel flow (sometimes called core flow), are seen in Fig 100. With funnel or core flow the material sloughs off the top surface and discharges through a vertical channel which forms inside the bin. This channel extends from the outlet upward through the material, with the material outside of this channel being stationary. With free flowing products, that material in the stationary region will then flow into the discharging channel once the level of the surface moves down to permit it to do so. Funnel flow is often referred to as a 'first in, last out' mode of flow, since with this pattern of flow much of the material that is first to fill the container, particularly around the periphery, will be the last to flow out.

Funnel flow usually occurs in flat bottom bins or facilities where the angle of the converging section, measured from the horizontal, is relatively shallow. For obvious reasons, such a configuration serves to maximise the stored volume for a given headroom. Prior to developing an understanding of how to design for mass flow, a common, but very often unsatisfactory approach to hopper design was to place an unconstrained sample of product on a piece of material from which the hopper was to be constructed and then by tilting the plate noting the angle at which the produce flowed off the plate. Since the majority of materials will flow in such a situation at angles less than 60° to the horizontal, this figure together with 45° in the case of very free flowing materials, were chosen by many designers of silos as 'convenient' angles of manufacture of the converging section. Subsequent work has show that even with the most free flowing materials it is likely that 60°, and certainly 45°, will result in a funnel flow discharge pattern. However, the fact that many silos are operating perfectly satisfactorily under these conditions demonstrates that this is an acceptable discharge pattern with many products.

With mass flow, the bulk solid is in motion at substantially every point in the bin whenever any material is allowed to flow through the outlet. As such, the flowing channel of material coincides with the walls of the bin, which is brought about when the walls of the converging section are sufficiently steep and smooth. This flow pattern is often referred to as 'first in, first out', in that the first material to be charged into the bin will be first discharged, thereby preserving the loading sequence.

Just how steep the angle of converging section should be to bring about this mode of flow will depend on the frictional characteristics exhibited between the stored product and the material from which the internal surfaces of the converging section are to be constructed. However, to put the geometry of a bin operating in a mass flow pattern of discharge in perspective with funnel flow, the former will necessitate the angle of the converging section typically being in excess of 70° (measured from the horizontal) for a conical configuration of hopper. Obviously, such a geometry can result in a significant reduction in stored volume for a given headroom when compared with operating in a funnel flow mode on discharge. Other requirements to achieve a mass

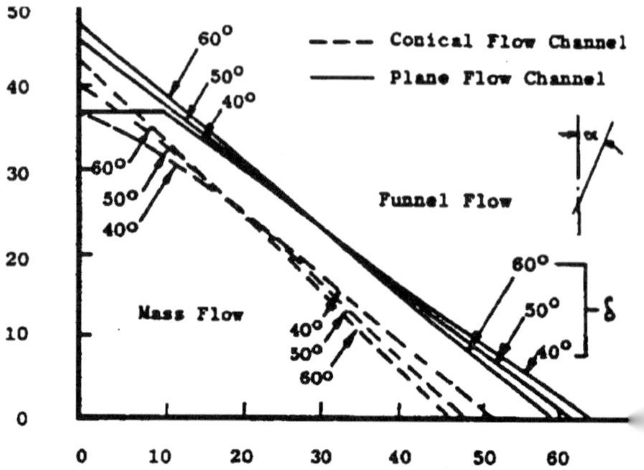

Fig 99: Limits for mass flow in conical and plane flow channels, shown in wall friction angle ϕ (y-axis) against hopper half-angle α (y-axis)

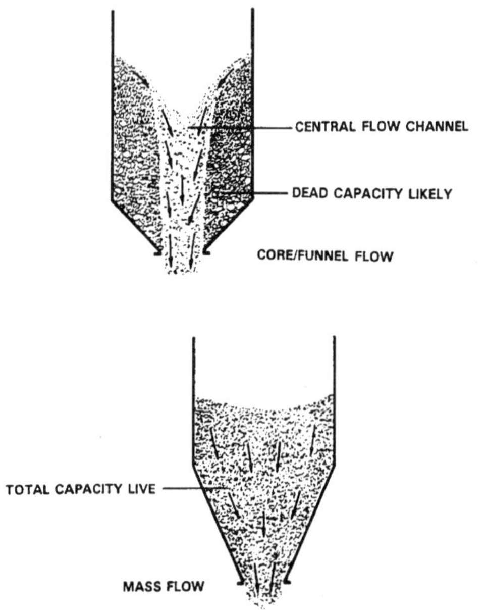

Fig 100: Basic patterns of discharge from storage containers

flow pattern of discharge are that the container has no abrupt transitions, the converging section has no inflowing valleys and the feeder is active across the whole area of the outlet.

Clearly, the 'first in, last out' characteristics of funnel flow can, with some products, lead to potential product contamination and or reduction in live capacity if the bin is continually topped up before being permitted to empty completely, Fig 101. With some products, the extended periods of storage of material in the dead zone, can, for a variety of reasons, lead to the material setting hard or 'caking', such that when material in this region is given the opportunity to discharge, it will not do so under the action of gravity alone. The main reason for this effect is often a difference in temperature between the stored product and the walls of the bin, which, in turn, leads to the migration of moisture between the particles comprising the bulk. Sugar and salt are examples of materials that can exhibit this behaviour.

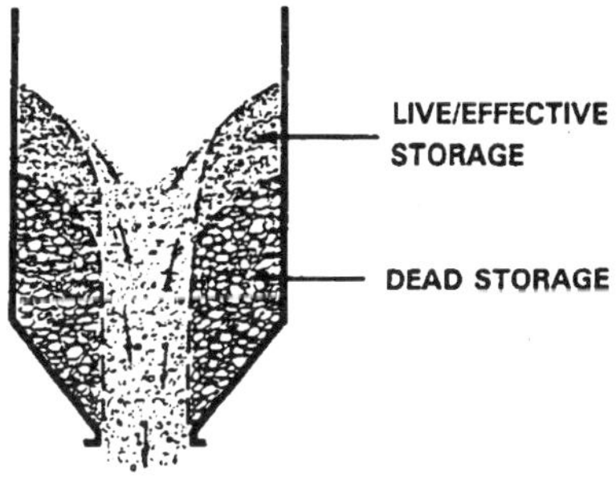

Fig 101: Reduction in the effective storage capacity of a funnel flow hopper as a result of 'topping-up'

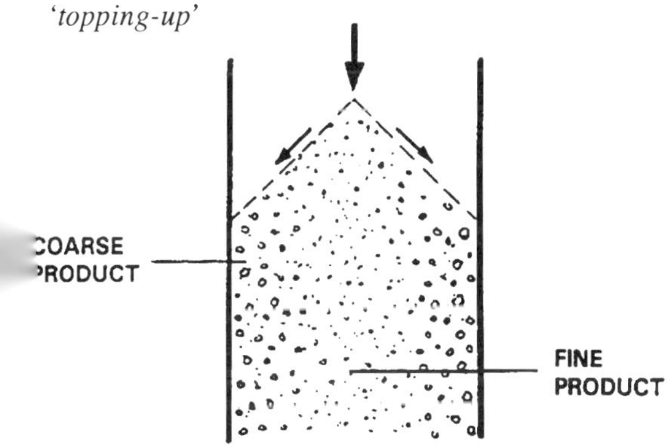

Fig 102: Particle segregation on charging a container

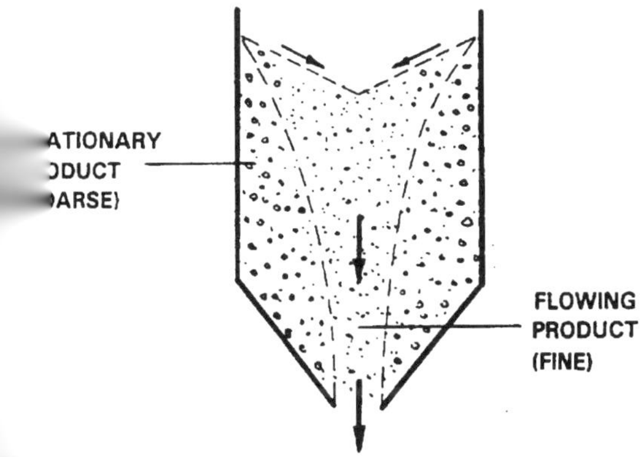

Fig 103: Particle segregation on discharge in funnel flow

Materials that are essentially free flowing, and have a wide particle size distribution, can be readily segregated in terms of particle size as a result of bulk handling in funnel flow mode. On filling the vessel, larger, and hence heavier particles will roll towards the periphery of the bin, with the finer elements remaining in a central core, Fig 102. As a consequence, a funnel mode will, in the first instance, discharge material that is rich in the finer elements, Fig 103. The proportion of fines will gradually decrease through the discharge cycle, until, at the end of the cycle, the material will be rich in the coarser elements. A similar phenomenon occurs with materials where the constituent particles are all of a similar size, but vary in density. Clearly, these segregation effects can result in quality control problems, and may have a significant impact on the operation of downstream processes. Whilst it is important to realise that this basic problem results from the segregation occurring on filling, which in itself is largely independent of hopper shape, the funnel flow mode of discharge resulting from the hopper shape serves to accentuate this effect.

Another potential drawback associated with funnel flow containers is that of product 'flooding' or 'flushing'. This is a particular problem with material of small particle size; i.e. typically less than 100-150 microns. It is known that such materials when aerated can take on liquid-like characteristics and flow uncontrollably; an effect commonly know as 'flooding' or 'flushing'. Such an effect can occur when such materials are discharged from containers that operate in a funnel flow mode. This is a particular problem when there is no feeder at the bottom to regulate or control discharge. The effect is caused by the material flowing off the top surface of the stored product into the flowing channel. Depending on the required discharge rate of product from the container, this might result in the material within the flowing channel acquiring a significant downward velocity. As the particles move through the interstitial air in the flow channel, this effect might be sufficient to aerate the product. Clearly, this situation can lead to wide variations in product density, which can result in difficulties in proportioning such materials for subsequent operations.

In general, many of the advantages and disadvantages associated with a facility operating in a funnel flow mode of discharge are reversed with mass flow. For example, it is clear that the steep angled converging section that is often necessary to bring about mass flow, will, when compared with the squatter proportions of a funnel flow container, reduce the potential storage capacity corresponding to a given headroom. Given that mass flow is essentially a 'first in, first out' mode of operation, with no dead regions in the stored product, it is clear that the time for which material will be in storage will be minimised. This, in turn, will minimise or eliminate problems such as product spoilage/degradation or caking, which are often time dependent. The 'first in, first out' pattern associated with mass flow means that the loading order is preserved and segregated material is re-mixed in the converging section. Therefore, mass flow is commonly recommended for products that readily segregate on handling. In a somewhat similar vein, the more controlled discharge associated with mass flow serves to lessen or eliminate potential flooding or flushing, thereby resulting in material of a more uniform density flowing through the outlet. This then makes it easier to proportion or meter the material into subsequent processes.

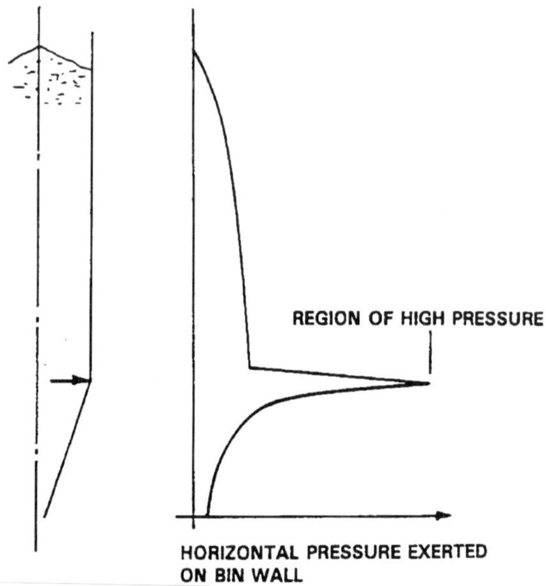

Fig 104: High radial pressure in region of transition from vertical to converging section of a mass flow container

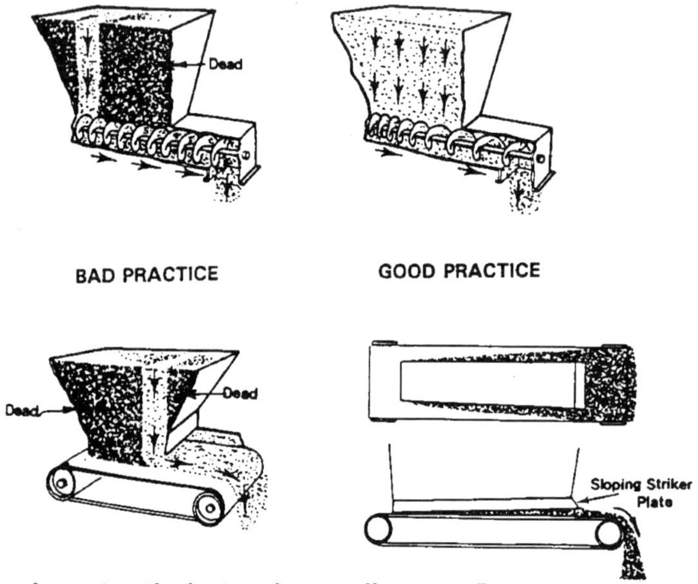

Fig 105: Importance of container/feeder interface to effect mass flow

Structural design of hoppers

Although not necessarily a drawback, the necessity for significant structural support for a mass flow container in the region of the transition from the vertical to converging section is seen by some as a disadvantage. The necessity for such support stems from the fact that, when discharging in a mass flow mode where a significant vertical surcharge is present, the pressure exerted is a horizontal direction on the walls of the bin in the region of the transition from the vertical and converging sections can, compared with the pressures present in funnel flow, be considerable, Fig 104. Therefore, to account for this high pressure considerable reinforcing might be required in this region so as to ensure the integrity of the resulting structure. Failure to recognise this important point has led to the structural failure of some containers.

Detailed design of hoppers

It is important to recognise that attention to detail is essential if mass flow is to be achieved. As well as ensuring smooth transitions, consideration should also be given to the wear of the walls of the container that will occur as a result of the abrading/sliding action when handling abrasive materials. This is likely to be more noticeable in the lower regions of the converging section where the velocity of the discharging particles will be at their greatest. With regard to the feeder being active across the whole area of outlet, failure to observe this requirement may destroy the required flow pattern in an otherwise properly designed mass flow container, Fig 105.

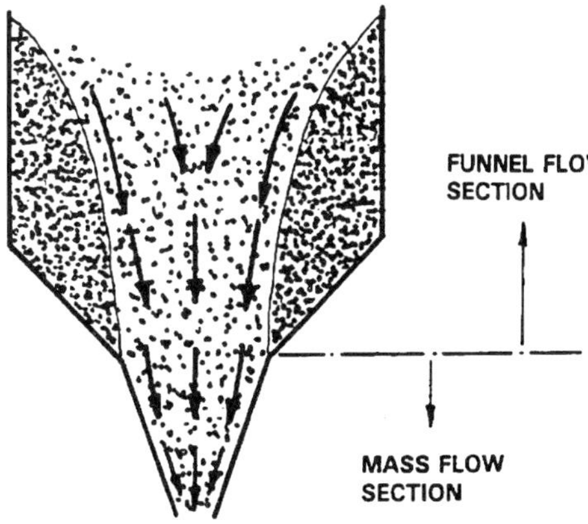

Fig 106: Composite flow

It is possible to blend some of the advantages of both funnel flow and mass flow into a single facility, by way of a composite/expanded flow hopper, Fig 106. This is achieved by way of a funnel flow container interfacing with a mass flow hopper section. The major advantage of such an approach is that if affords a potentially significant increase in stored volume when compared with a pure mass flow approach, whilst, at the same time, with cohesive products, it achieves reliable flow through an outlet that is smaller than would be necessary for a container operating solely in a funnel flow mode. Additionally, the density of feedstock flowing through the outlet will be more uniform than might otherwise be the case with a bin operating in a funnel flow mode. Although the application of such an approach is usually limited, when carefully applied it can be a very effective means of combining these advantages.

Chapter 14: Buses, RCVs and vans

PSV design
PSV systems
Light van design
Delivery vans
CV body kits
Refrigerated vehicles
Refuse vehicles
Handling trucks
Trailer technology

Buses and coaches

Design of buses and coaches is largely influenced now by the requirement to provide a better passenger environment for new-prospect users of PSVs and touring coaches. The influence of weight reduction and greater safety requirement is also affecting structural configurations.

A number of attempts to produce buses and coaches on a high volume basis have faltered because of the less than enthusiastic reception by operators for 'standardised' designs. The preference by fleet engineers for vehicles designed for local operating conditions, and other specifics of fleet use, ensures a strong future for buses and coaches in the specialist vehicle sector. This lack of volume base also ensures a continuance of separate chassis structures being made alongside integral designs — particularly so with the requirement for separate sizes of body to suit seating capacities for mini-, midi- and full size service buses as well as meeting the separate requirements of double-deckers and touring coaches.

The de-regulation of UK bus services brought about by the 1985 Transport Act has also affected the demands on design, as has the relatively low UK total industry volume. The effect of the single European market is beginning to counteract this particular constraint and running-gear suppliers are increasingly able to tailor mechanical units for the PSV market. There is a realisation at last that buses need no longer be aimed at the bottom of the market, as continental designs have shown how good design can attract passengers right across the social spectrum. Much needed improvements in the service infrastructure will also yield more passengers and these can be matched by well designed vehicle systems in the areas of entry/exits, seat layouts and ticket issuing systems. Coach comfort, too, is set to be revolutionised by continental practice.

The traditional 'light-weight' sector of the coach market has also virtually disappeared with the 'heavies' (premium coaches) occupying some 90% of the market. Roll-over protection has also affected design with respect to the large 'picture' windows that were once in vogue; these designs are also less favoured for touring coaches operating in semi-tropical conditions. High floors are likely to become more widespread with the requirement for better passenger visibility and increased underfloor luggage accommodation. In the sub-sector of minicoaches, an increased luxury specification is the norm with the trend away from van conversions as purpose built minibuses are put out to hire at off-peak periods.

Typical understructure construction naturally depends on whether design is chassised or integral. In the former, steel sidemembers of 6 to 8 mm are usually involved while in the latter, 3 to 6 mm would be the norm, at usually not much more than 400 MN/m^2 — to prevent difficulty in forming, drilling and punching operations. For integral underframes of thinner gauge, a zinc or similar coating is usually applied, such coated steels being deep-drawable up to 5 mm thickness. An 18 micron thickness of zinc would require a 50 micron thick layer over it to provide long life protection. Where aluminium alloy underframes are involved, typical alloys are BS 1474 HE 15 and 30 from the 6000 series while for body parts the NS3 alloy from the 3000 series would be typical.

Consultants Booz, Allen and Hamilton, have carried out some key specificational comparisons for recent service buses in which they have highlighted the trend towards making buses more accessible to the aged and disabled. This factor is likely to be one of the main driving influences on future bus design. Some valuable studies by the Transport and Road Research Laboratory have systemised this design approach as shown in Fig 1.

Detail recommendations such as those in Fig 2 have made made on the design of handrail systems (in this case for a two step entrance involving rises of 311 and 216 mm) as well as those for levels of braking and acceleration tolerable to infirm standee passengers. This has led to derivative of acceleration values, 'jerk', being defined: 1.5 is prescribed as a maximum tolerable level. The greatest incidence of accidents was found to be during entry/exit situations and the researchers designed the retractable step system in Fig 3 to help alleviate this problem. The mechanism was designed at Cambridge University and the step retracts into the side of the vehicle without affecting ground clearance. A sensitive strip on the edge of the step prevents injury during extension and sensitive matting on top prevents the step rising when anything comes in contact.

The Department of Transport have now issued a recommended specification, in TRRL Report 137, for buses used on local services to improve accessibility for passengers with mobility disorders. A preferred interior arrangement is shown in Fig 4 and designers are asked to improve on PSV (Conditions of Fitness, Equipment, Use and Certification) Regulations 1981. Headroom should be more than 1.8 metres in the gangway and stanchions/handrails should have circular sections of between 30 and 35 mm diameter, with 50 mm clearance to adjacent surfaces. They should be non-slip in surface texture and coloured to contrast sharply with surroundings. The design of centre handrails for a dual stream entrance/exit should follow the layout in Figs 5a,b and step dimensions in Fig 6. Seat cushion height at the front edge should be between 430 and 460 mm and cushions should not slope more than 30 mm over their depth. Headroom above the cushion top surface should be over 1 metre.

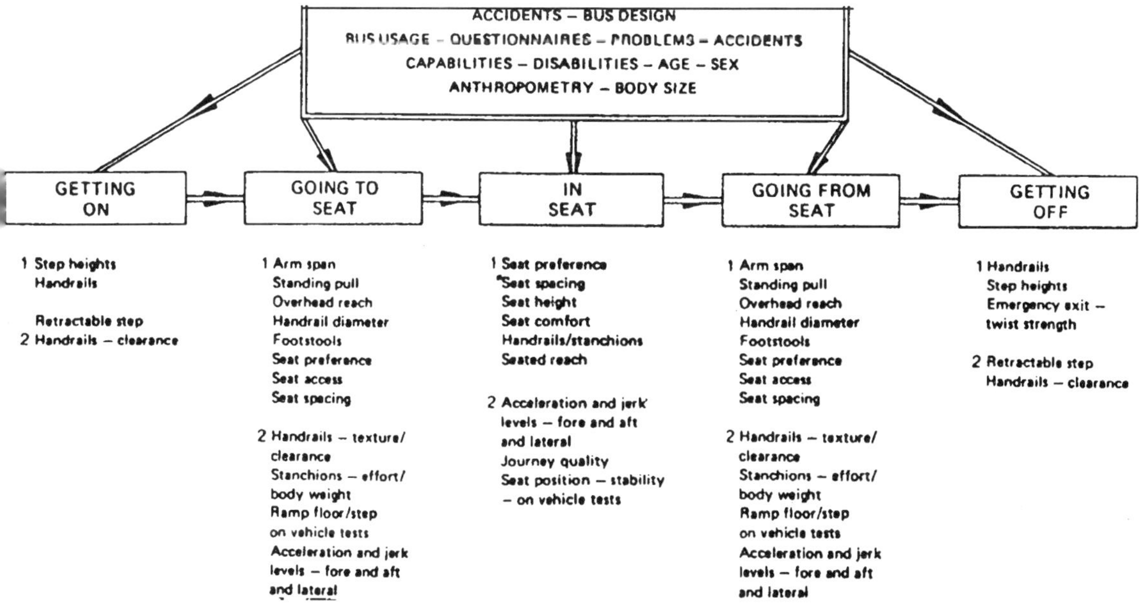

Fig 1: Systemised design approach (TRL)

Drive layout for PSVs

Interior layout for full-size service buses can be optimised when a grouped rear engine/drive system leaves an uninterrupted central section. Transmission makers ZF have recently made alternative proposals for engine/drive grouping dependent on their available range of gearboxes, drive-axles and angle drives, Fig 7. A particularly low floor arrangement is seen at (h) for which the engine is seen at (1), the gearbox at (2), rear doors at (3) and angle/spur gear drive at (5,6). The 80 degree angle drive has a 217 mm offset spur gear drive mounted directly to the Ecomat automatic transmission. A drop-axle, type AV130, is featured and the clearance requirements for the other units are provided by this supplier.

These layouts by no means exhaust the possible drive configurations for buses and coaches as evidenced by Fig 8 which shows some recent designs in current operation. At (a) Leyland National integral single-decker with fore-aft parallel drive and 'north-south' engine; (b) Leyland Olympian with shallow angle drive, 'east-west' engine and independent front suspension; (c) Mercedes Benz 0405 city bus, again with north-south engine and drive; (d) mid-engined Leyland Tiger coach chassis; (e) MCW Metrobus again with modest angle drive; (f) Volvo's early Ailsa bus with front engine and rear drive; (g) the integral structured Saviem SC-10 and (h) the mid to rear engined Dennis Javelin coach with arched chassis. The midi-bus is exemplified by the Dennis Dart, Fig 9, which is of semi-integral construction with a stainless steel frame providing space for 39 seated and 15 standing passengers in its 9 metre overall length.

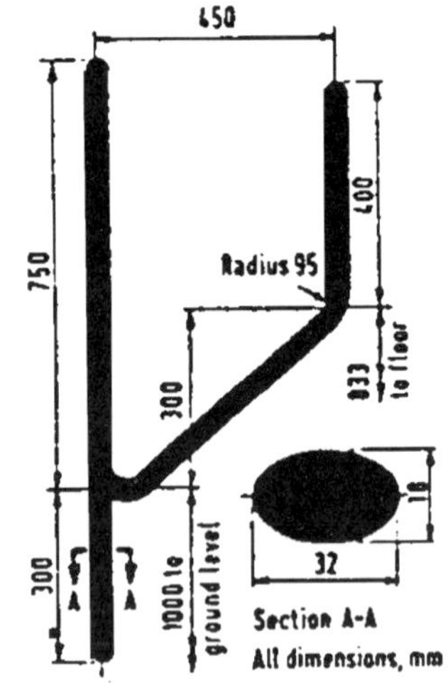

Fig 2: Hand-rail recommendations

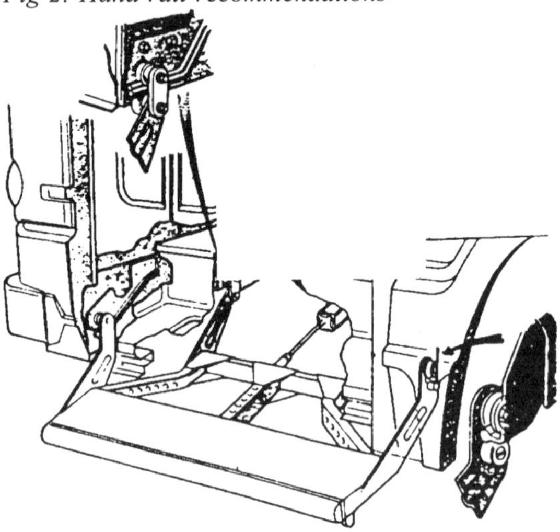

Fig 3: Retractable step mechanism

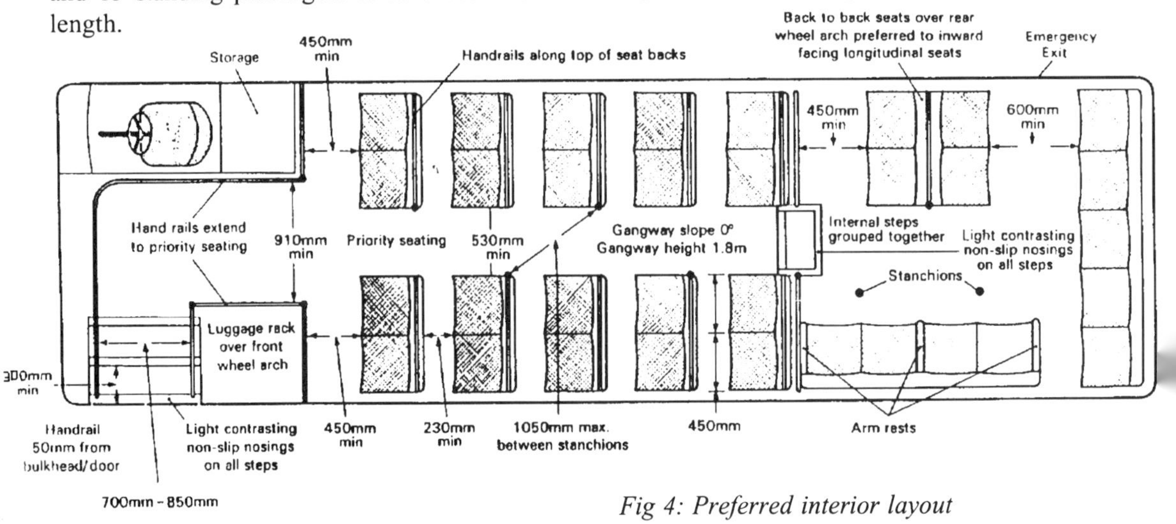

Fig 4: Preferred interior layout

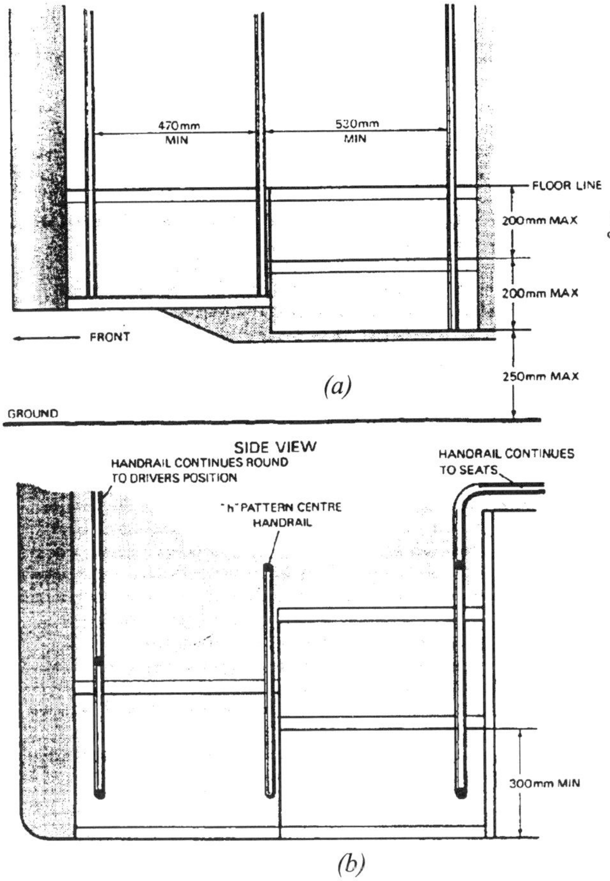

Figs 5: Dual stream access layout

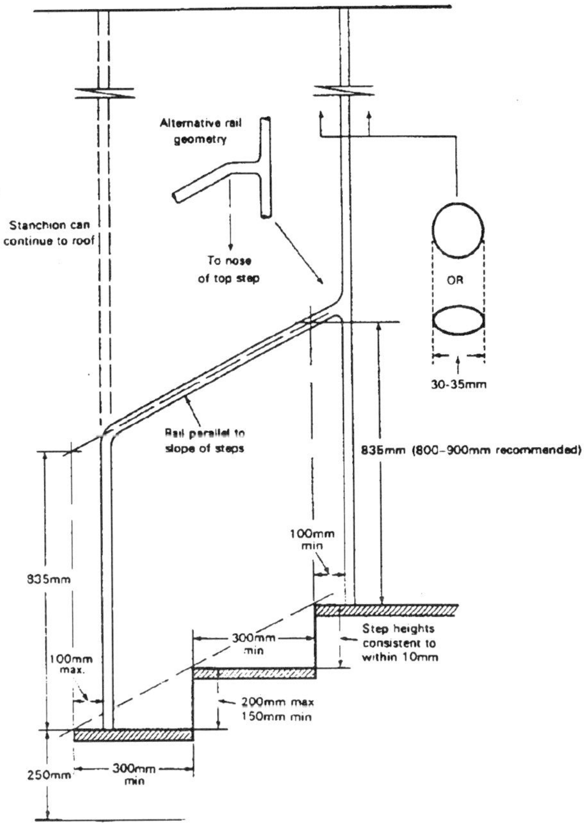

Fig 6: Preferred step dimensions

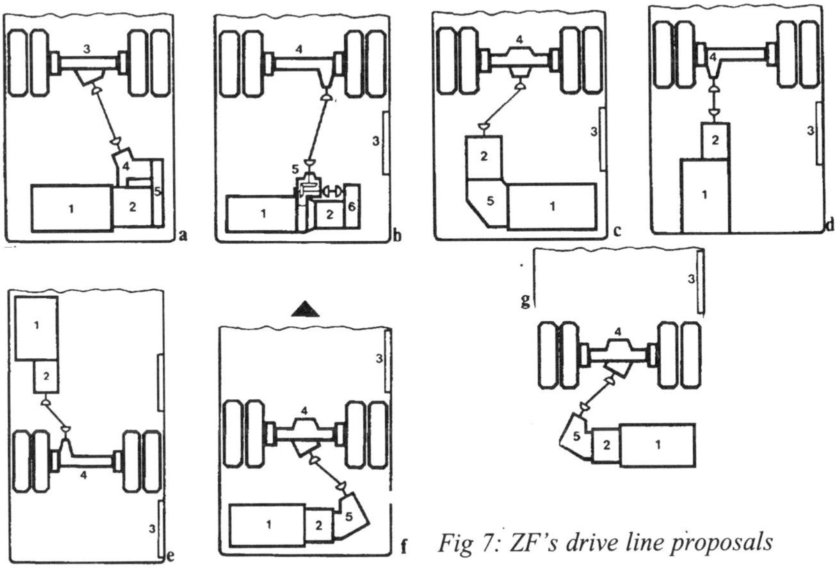

Fig 7: ZF's drive line proposals

Structural design of PSVs

The construction of semi-integral buses and coaches requires a new level of co-operation between chassis and body builder. DAF Bus have gone as far as formalising a written agreement between the two parties. This has led the chassis-maker into an in-depth programme of structural design involving finite element predictive techniques which allow com-

a b e

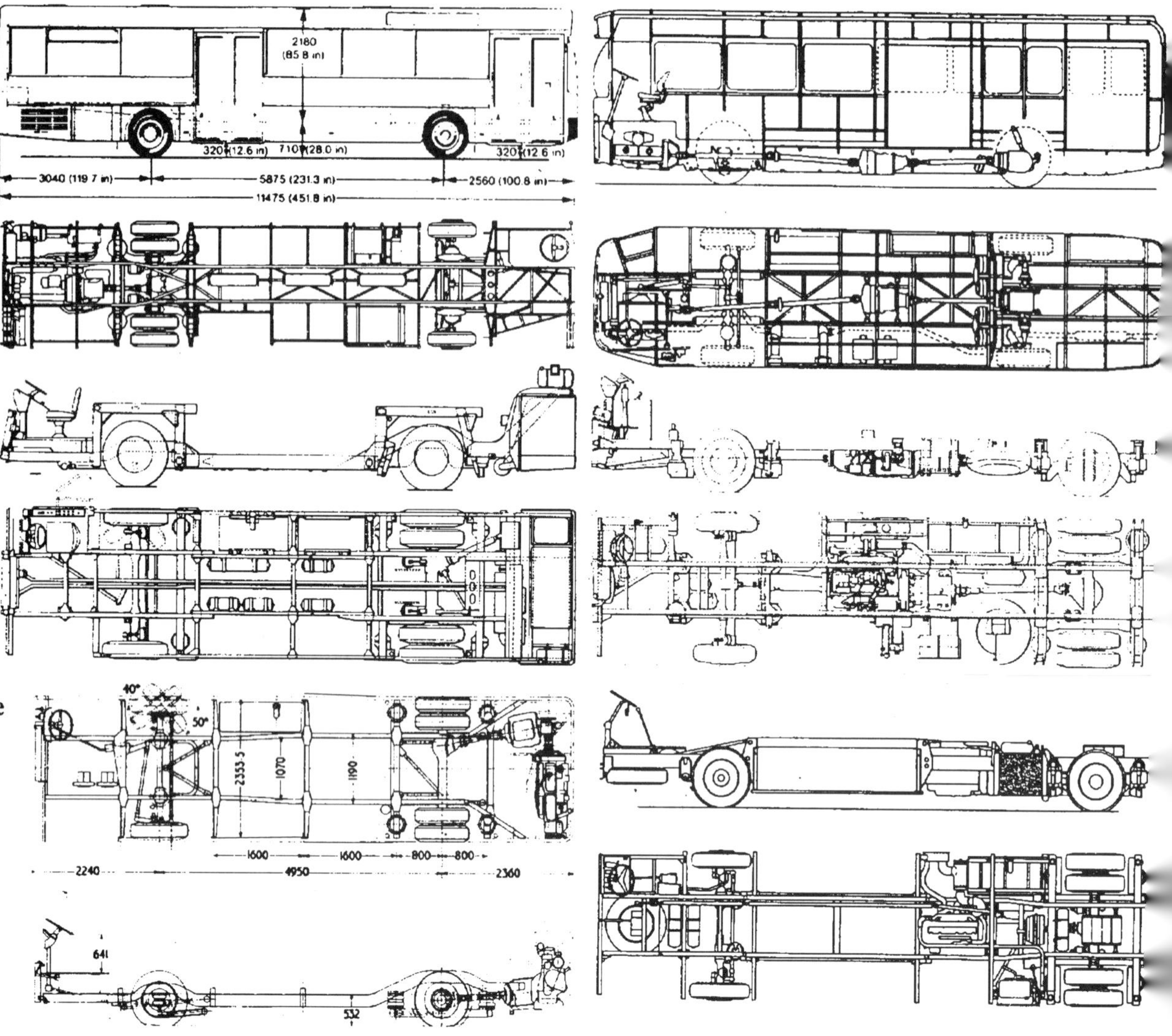

f

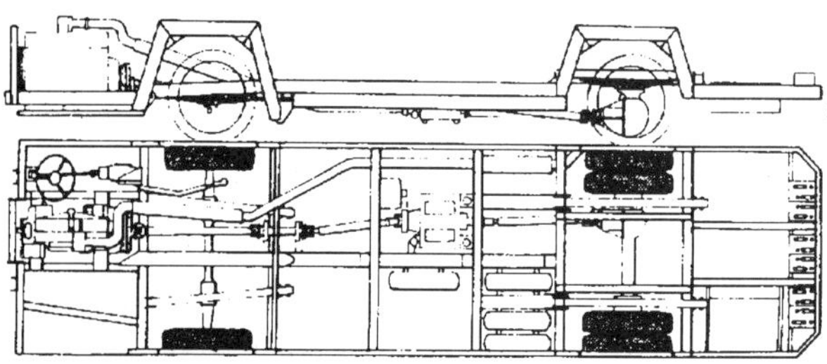

Fig 8: Recent PSV configurations

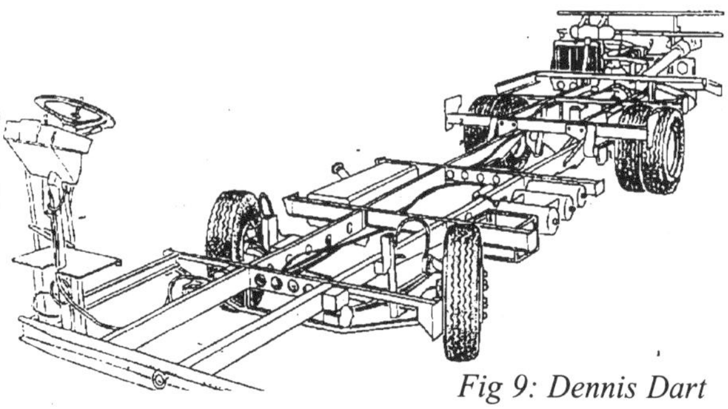

Fig 9: Dennis Dart

puter simulation of even the vibration of the structure, Fig 10. The difference in vibration mode shape for different sides of the bus is of particular interest — arising from the effect of door cut-outs in the structure. Fig 11 shows the good conformity between measured and predicted distortion mode of the side frame. To model the stiffness of the doorways, Fig 12, vertical and rotational spring representation were used. Fig 13 shows the magnitude of the bending moment in the roof panel for different frequencies corresponding to speeds driven over a rough road. Fig 14 shows two sideframe structural alternatives and the stress patterns superimposed. In the case of the maximum stress above the rear door, this is reduced from 21.2 to 8.4 N/mm^2 by fitting bonded rather than rubber-mounted windows.

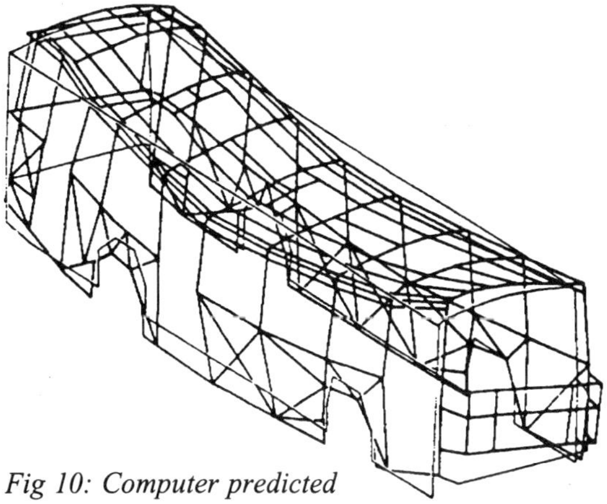

Fig 10: Computer predicted structural vibration

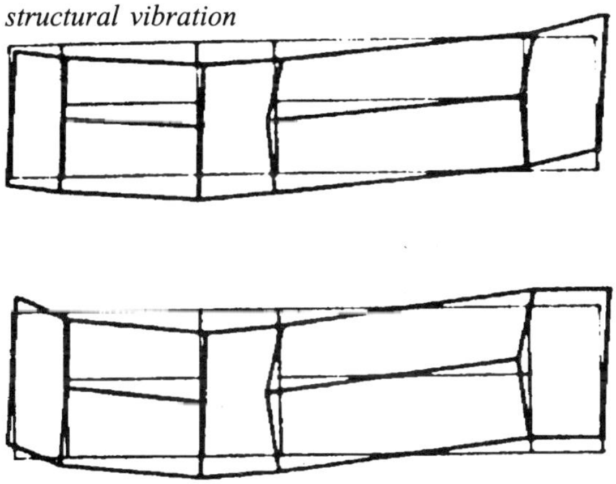

Fig 11: Measured and calculated sideframe deflections

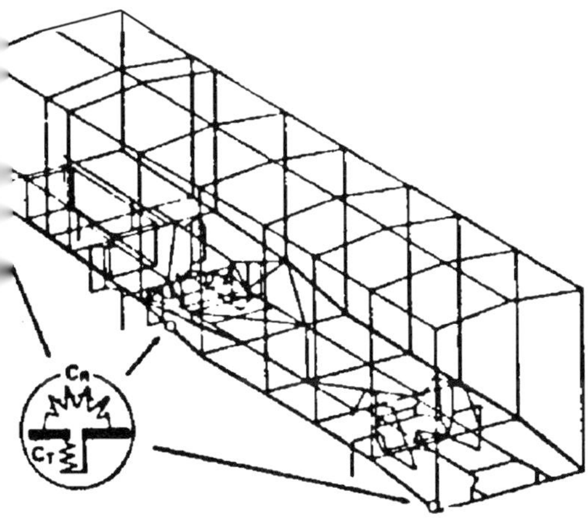

Fig 12: Modelling doorway stiffness

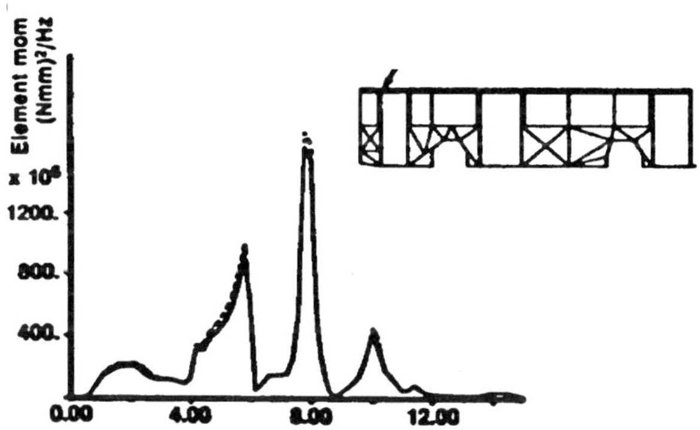

Fig 13: Roof-panel bending moment with respect to frequency in Hz

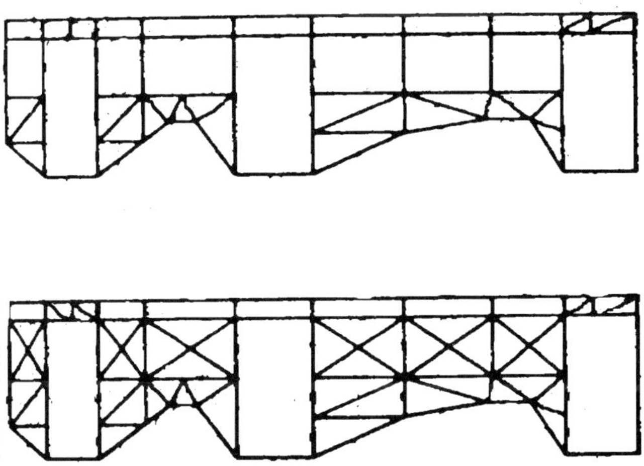

Fig 14: Alternative sideframe structures: above — rubber-mounted windows; below — bonded ones

An extension of the Simple Structural Surface method, discussed in Chapter 2, is also available for the design analysis of sideframes. The stiffness of the luggage compartment of a high-floor coach can be seen to effect in the SSS model of Fig 15. The structure is shown here under torsional loading — the load couple from the front suspension being in equilibrium with the reaction R_f in the sidewall and K_{26} in the front panel, similarly with R_r and $K_{88\text{-}89}$ in the rear wall. High bending moments occur on the window posts in this mode.

A number of bus body builders, including Optare, use the Alusuisse System M 5438 for body construction in aluminium alloy. This has standards for welded and bolted joints, Fig 16. Specially extruded sections, Fig 17, are available in widths up to 0.8 metre. The adaption of this system by Optare to the DAF SB220 chassis resulted in an unladen weight of 970 kg.

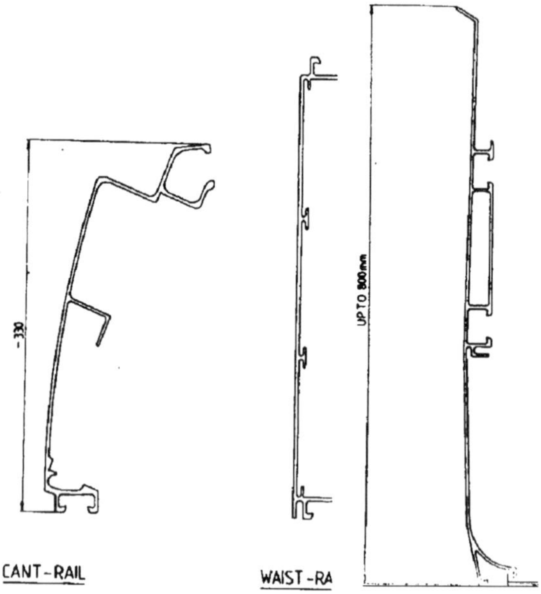

Fig 17: Special Alusuisse sections

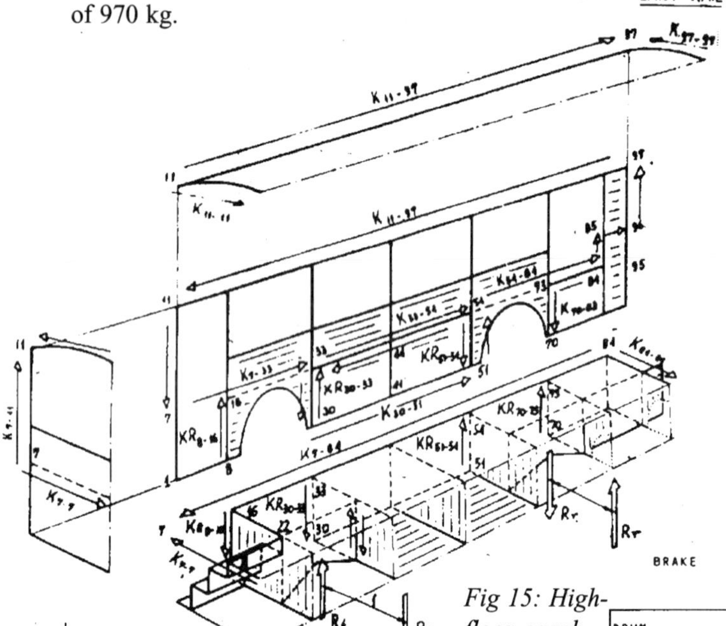

Fig 15: High-floor coach idealisation

Fig 18: Valeo twin-disc brake

BRAKE	SURFACE (cm2)					ROTOR MASS
	Rotor swept(a)	Linings (b)	Net swept(c)	Total rotor(d)	Net rotor(e)	(Kg.)(f)
DRUM Ø 414 x 150	1950	960	990	~4635	~ 3675	45
DRUM Ø 410 x 220	2833	1728	1105	~8239	~ 6511	72
SINGLE DISC Ø 392 x 48	1625	374	1251	~4450	~ 4076	28,5
VALEO TWO-DISC (Ø417 x 30)x2	3306	500	2806	~7916	~ 7416	38
SINGLE DISC (WHEEL 24,5") Ø 440 x 50	2136	500	1636	~ 5767	~ 5267	40

Fig 19: Brake thermal performance compared

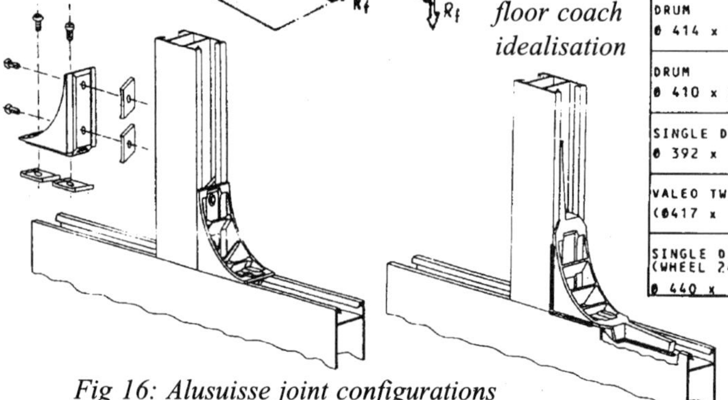

Fig 16: Alusuisse joint configurations

Bus mechanical systems

Braking of large buses also requires specific systems — to meet the EEC directive requiring deceleration of 5 metres/ second^2 for buses weighing more than 5 tonnes or with more than 8 passengers. Loss of efficiency up to 20% of this minimum value is allowed with hot brakes. A number of car categories only involve decelerations of 7-8 metres/ second^2 — for comparison. The Valeo company developed their twin-disc brake for such duties, Fig 18, which was designed for homologation on both front and rear axles having load capacities from 6-13 tonnes. Compared with the disc brake installation of a car the problem is one of dissipating heat within a wheel diameter less than twice that of a car but with braking loads up to ten times higher.

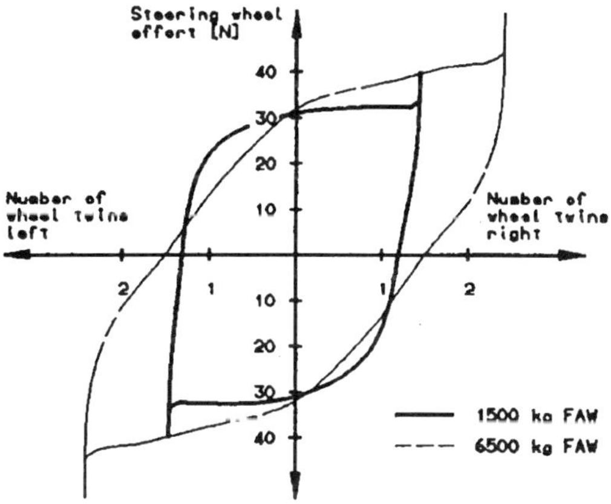

Fig 20: Steering efforts for small and large buses

The company tabulated the characteristics of existing brake systems against their twin-disc design in terms of thermal performance, Fig 19. They gauged thermal performance in terms of the area of swept surface not facing the linings (c). Also considered was the total useful surface for heat exchange from the rotor (e) and the ratio of useful rotor surface to rotor mass (e/f).

Buses also put special demands on steering systems and steering gear makers ZF have shown in Fig 20 the difference in measured effort between a small PSV with power steering from lock-to-lock at manoeuvring speed (full lines) and with the dotted lines a double deck bus illustrating the problem of sensing road feed-back — light and heavy vehicles requiring similar manual work being done in straight ahead driving. As vehicle speeds increase beyond manoeuvring speed, the forces needed to steer the vehicle generally reduce resulting in a yet smaller range of efforts within which the driver can evaluate the road contact. Fig 21 shows road feed-back when driving the light vehicle at low and high speeds straight ahead. This of course is opposite to the ideal situation and leads to consideration of steering-gears which can enhanced in this way.

Newly developed systems even avoid the need to continually adjust gears for reducing backlash and the avoidance of excess energy-transfer into the hydraulic power system. This is achieved by using higher operating pressures and reducing flow rate as well as using suction-control with a radial piston pump, Fig 22.

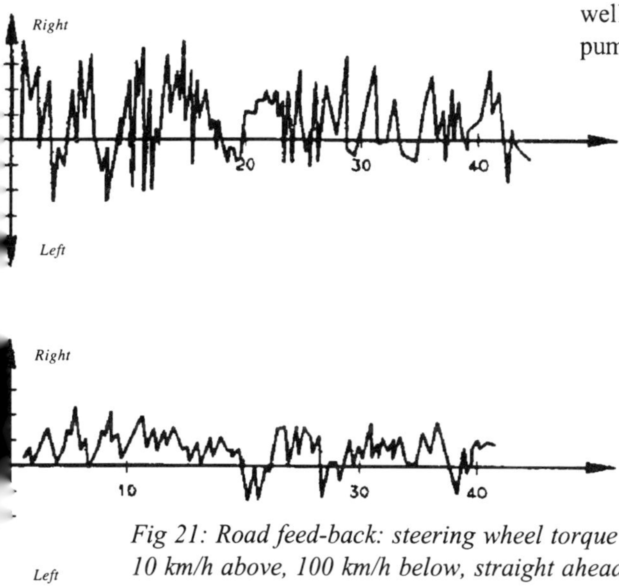

Fig 21: Road feed-back: steering wheel torque (Nm), 10 km/h above, 100 km/h below, straight ahead

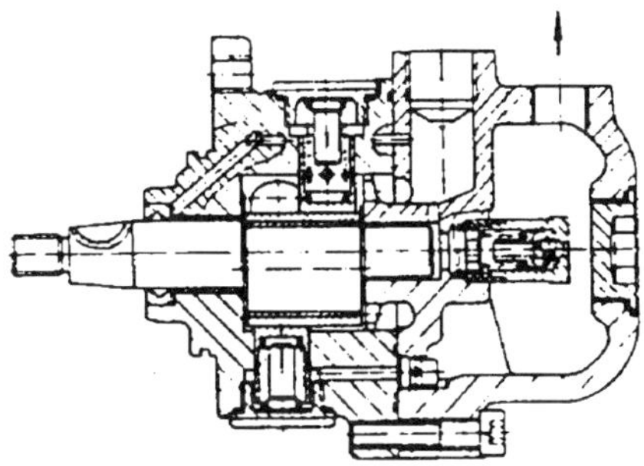

Fig 22: Radial piston pump

Design peculiarities of PSVs, according to ZF, centre round installation conditions involving long and tortuous input linkages — which can sometimes cause self-centring problems. There are also suspension design effects, both factors contributing to high forces on the steering systems which must be reacted by the vehicle structure without undue compliance. Fig 23 shows a bus gear of the integral power type involving recirculating ball and nut drive. It features hydraulic reaction, with cut-off and limiter valves — the latter being activated only at full lock to reduce loads on the vehicle structure. A vane-type pump, Fig 24, would typically run at 140 bar. An ideal valve characteristic would be as in Fig 25.

The Servotronic variant, Fig 26, first developed for cars helps to solve the feed-back problem by providing electronic sensed and modulated speed-proportional steering. The driver feels safer, and is safer, when the ratio of input to output torque increases with speed, Fig 27 showing the degree of change to the steering characteristic possible.

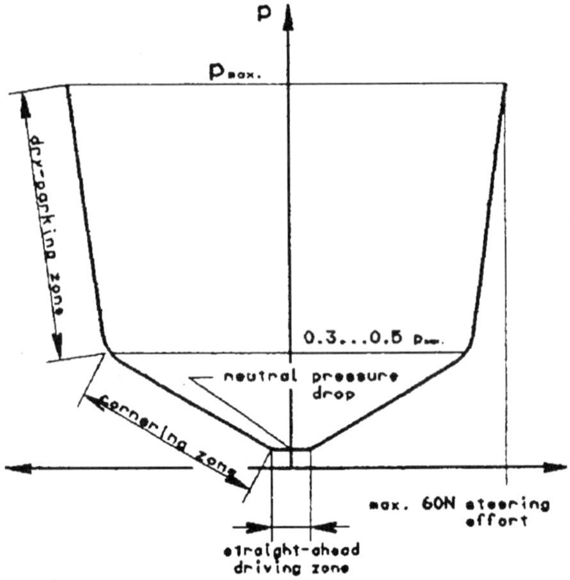

Fig 25: Ideal valve characteristic

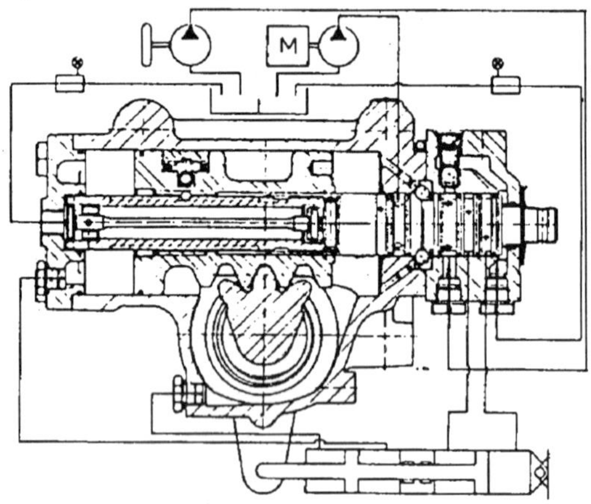

Fig 23: Integal power steering gear

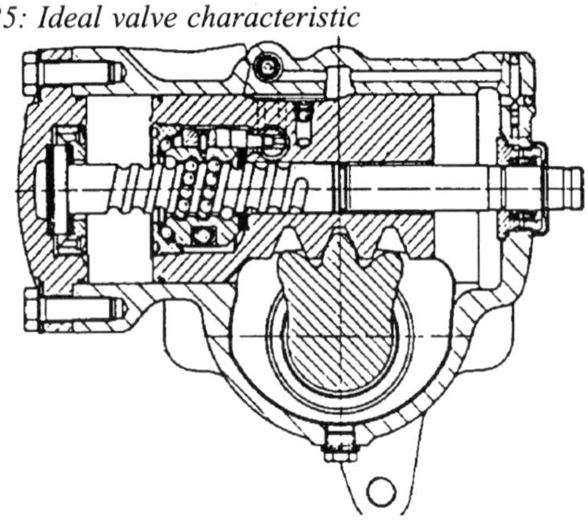

Fig 26: Sevotronic steering gear

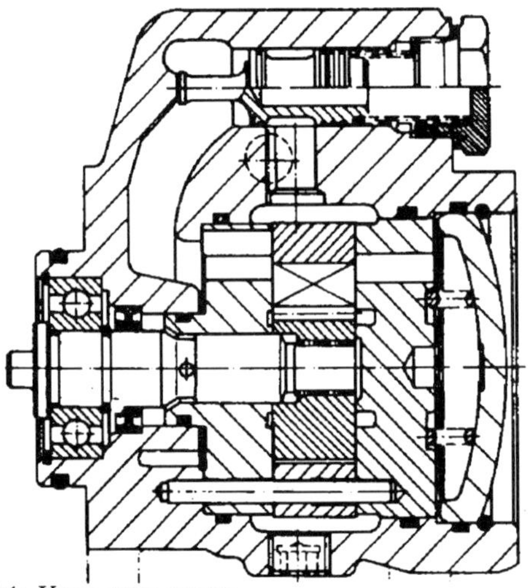

Fig 24: Vane-type pump

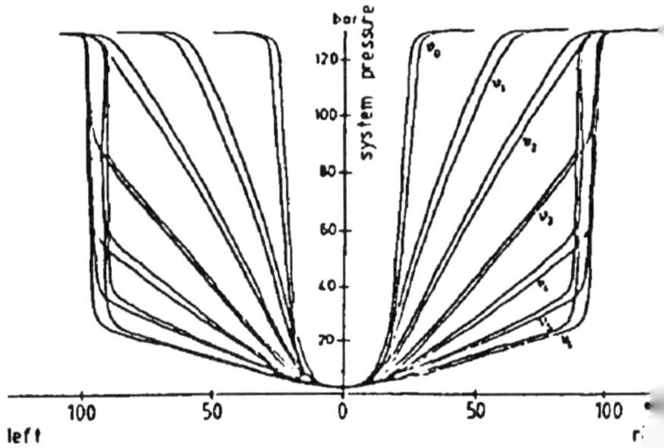

Fig 27: Modification to steering characteristic

Fig 28: Reliant Fox

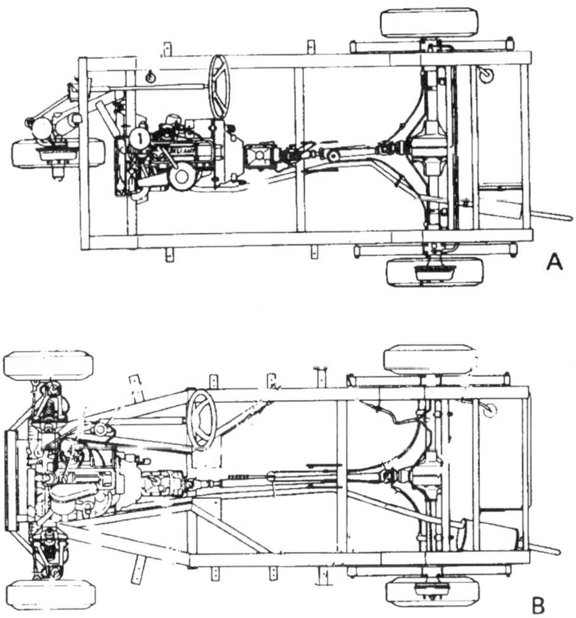

Fig 29: Robin and Kitten chassis

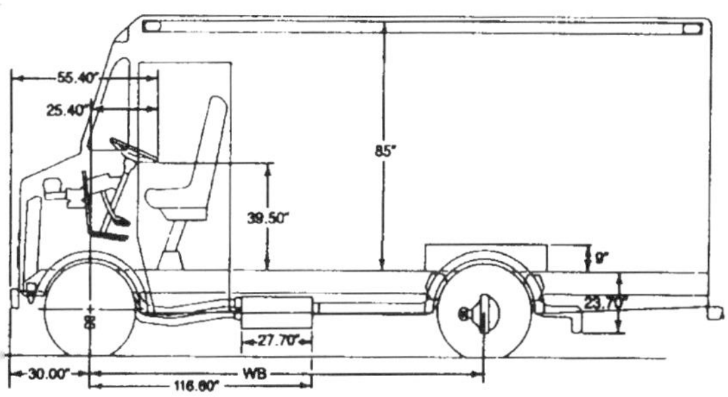

Fig 30: International Metro van

Light van design

The light van sector is affected considerably by the volume vehicle builders as much of the market is served by volume-car-derived vans at lower gross vehicle weights. At higher GVWs the volume sector are, of course, heavily into panel vans — but right across the weight range there are many important niches to be filled by the specialists.

The mini-sized volume car is often not available as a van derivative and volume panel vans can be restrictive in terms of optimum access to both cab and load space; opportunities also exist for specialist bodybuilders in the weight bracket between top-of-the-range panel vans and the smallest chassis-mounted custom box vans. The demise of the Rover Minivan in the UK presented the opportunity for the Reliant Fox, Fig 28, a vehicle which offers outstanding economy by virtue of its very low unladen weight and modest engine capacity. With optional final drive gearing, the vehicle is in fact more economical than almost any volume-produced small car. Ready conversion from closed van to open pick-up body is an important aspect of the design. The base for the vehicle was the chassis of the earlier 4-wheeled Kitten saloon, itself having much in common with the company's 3-wheeler chassis, Fig 29. Extensive use of aluminium alloy in the power-train and GRP for the bodywork helped to achieve the below 406.5 kg (8 cwt) target of the early vehicle to meet the 3-wheeler taxation requirements.

Beyond the top end of the panel van market, an interesting example is the novel integral van made by Navistar in the USA — which overlaps with the large delivery van sector covered in the next section. The International Metro II version, Fig 30 is available in body lengths of 14 to 24 ft (4.3-7.3 m) and has inside dimensions 85 in high x 93 in wide (2.16x2.36 m). This also has a separate chassis frame but the integral cab/body is factory-produced. The cab is 'walk-through' and the body can be accessed from it. Side walls are of GRP skins bonded to a tough core and are said to be considerably more scratch resistant than those of conventional aluminium alloy box vans.

Car-derived vans

The 7-10 cwt (356-508 kg) weight-capacity car-derived van, with volume-capacities typically 60 to 100 ft^3 (1.67-2.83 m^3), are subject to legislation detailed in Figs 31 and 32 which affects the overall design. Countries such as Sweden have additional legislation to be met. Vision, package layout, petrol tank location, rear end configuration, lamps and bumpers are among the items affected. Obviously dimensions are related to those of the base car and roof heights, too, are governed by the need for an aesthetically, and aerodynamically, smooth blend between car and expanded body surface. Suspension system is crucial to the inter wheel arch clearance and appropriate re-design can bring substantial benefits over the base car system — probably sacrificing ride comfort for improved cargo space. Rear access to the space can be improved, too, by appropriate choice of closure between half-lift/half-drop gates, lift tail gate, single/double rear doors or even roller shutter.

Generally, front wings and doors are carried over from the base car to the derived van. New doors could solve the frequent problem found on volume-built vans of the door tumblehome adversely affecting the shape and so the volume capacity, of the van body. Fig 33 shows five suggested ways of assembling the body side to an existing front-end. The most economic solution is that of (d) whereby the existing body side-panel is trimmed along the window line and thus only a new upper panel is required. At (e), the A-post and front cant rail is built up from small panels and a rear side panel assembled to the front-end just forward of the body centre-pillar.

Panel vans

Design techniques used in notable volume panel vans can often be exploited by the specialist builder. A look at some key points is thus worthwhile. The Leyland EA was an early panel van which successfully exploited the tilted engine concept to obtain good walk-through ability on a forward control layout. The low-slung engine-gearbox necessitated a down swept boxed crossmember, Fig 34, at the front, with 40 degree inclined rubber bobbin engine mounts. The tail of the gearbox was carried on twin conical mounts. State-of-the-art is represented by such vans as the Ford Transit. The layout and structure of the base vehicle, Fig 35, contains some important design features. The MacPherson strut IFS is arranged to clear the footwells and a semi forward control gives a nice compromise between available cube, access and driving comfort. Cab floor is a continuation of that of the load space so that good front access to the body is available.

Multipurpose vans

One of the greatest recent excitements in small van design is the multi-convertible Concept '92 vehicle of Hawtal Whiting, Fig 36. This vehicle also doubles as a fun-car and people-carrier. The chassis is a 3-

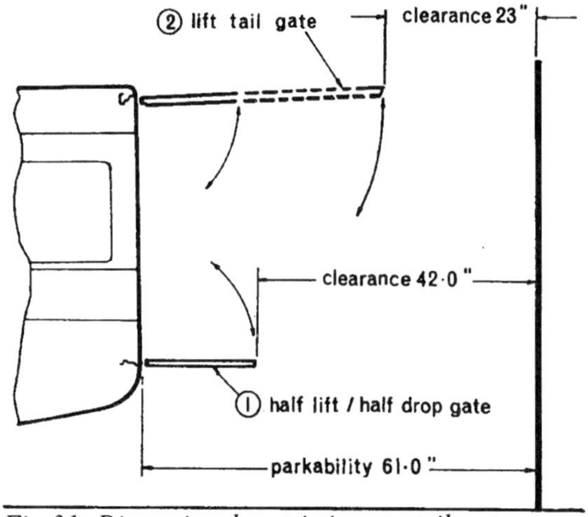

Fig 31: Dimensional restrictions to tail-gate

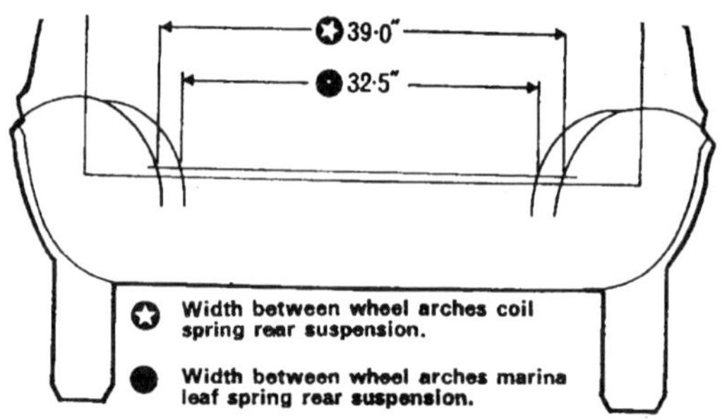

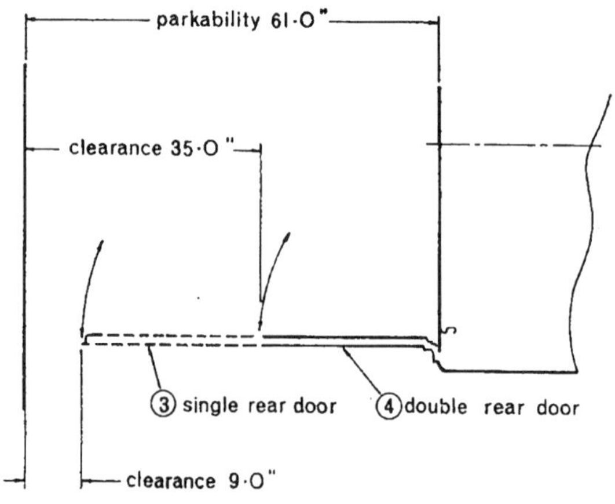

Fig 32: Dimensional restrictions to rear-doors

dimensional semi- monocoque which is common base for the variants upon which partially-structural and non-structural cladding can be attached. The optional petrol/electric/lpg power unit fits into a self-contained structural sub-frame which also reacts the suspension loads. The assembly becomes a removable pod for remote servicing of the engine as required. The 3 metre length is the same as the Rover

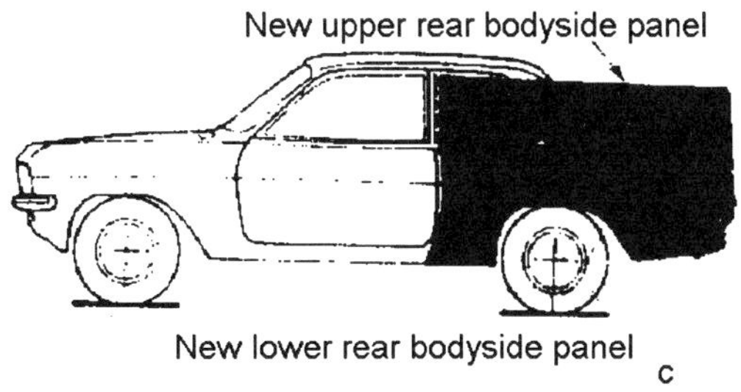

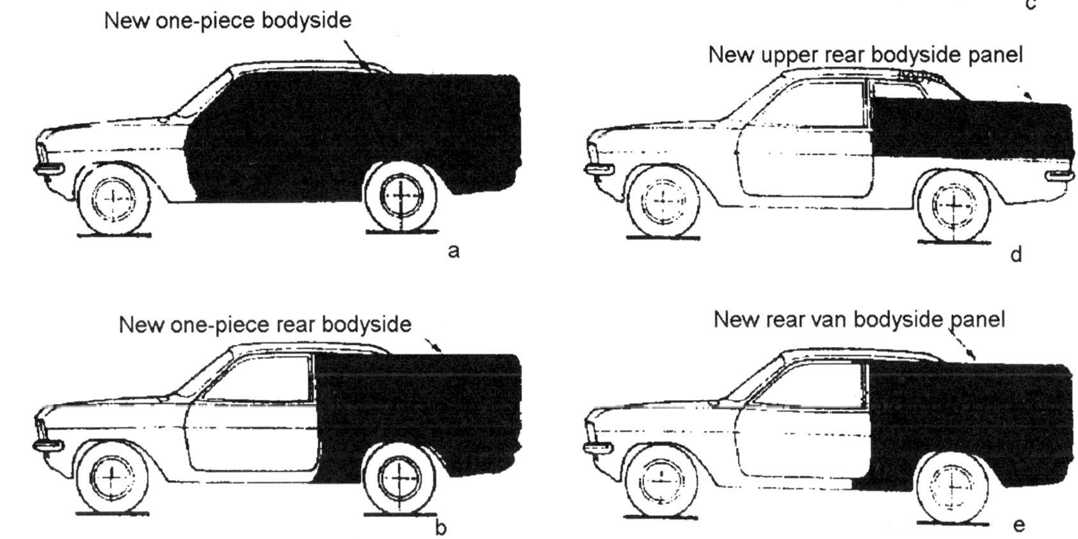

Fig 33: Body integration alternatives

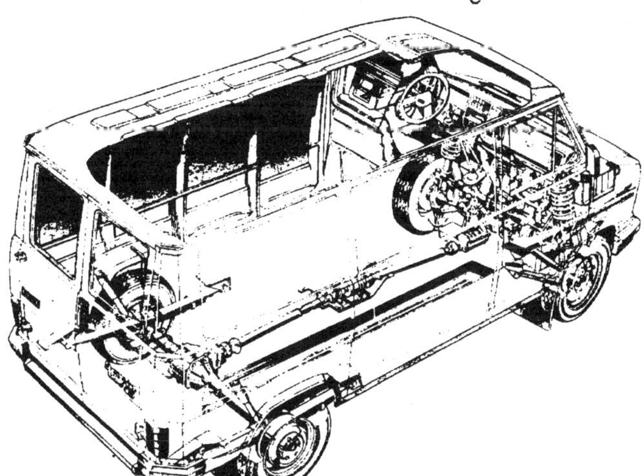

Fig 35: Transit mechanical package

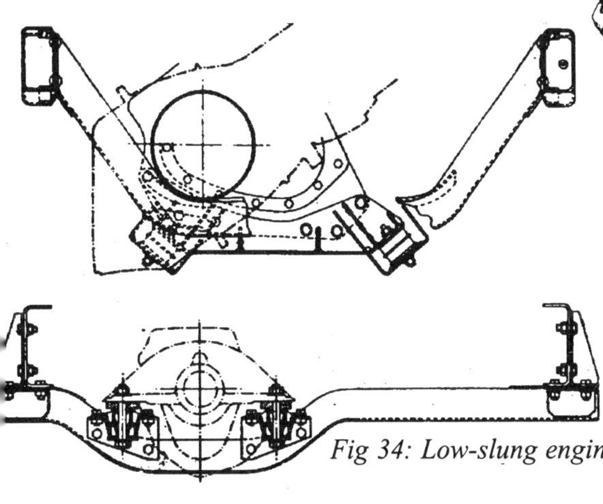

Fig 34: Low-slung engine suspension

Mini but considerably greater interior space has been achieved. The vehicle is taller than a Mini but not ungainly so in view of the relatively wide track. A three-across bench seat has therefore resulted and wide doors allow the ingress of wheel chairs. The two front seats are located slightly inboard to clear the wheel arches and can rotate through 45 degrees to improve access to the rear seat.

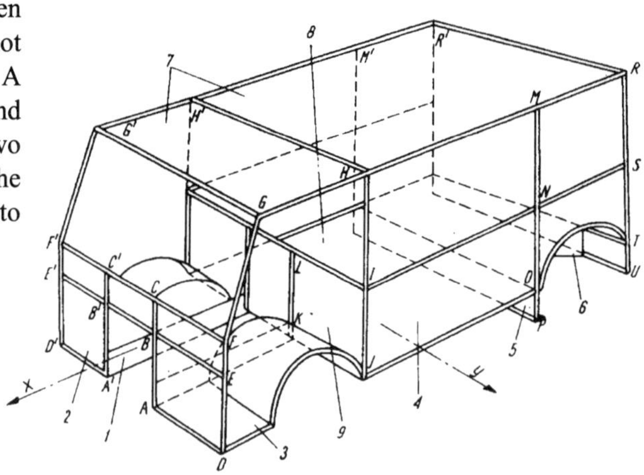

Fig 38: Integral-van proposal

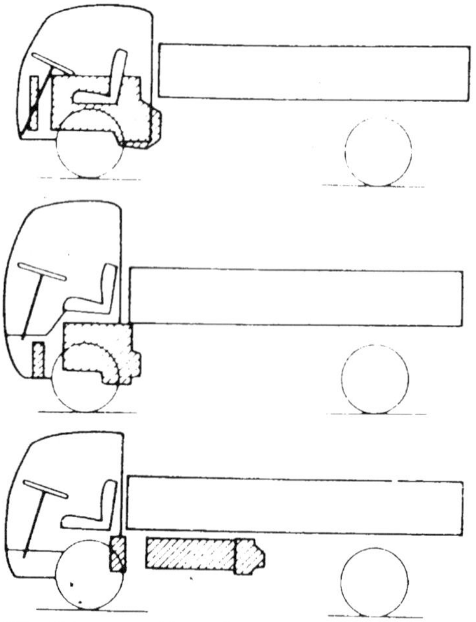

Fig 37: CV engine/drive layouts

Fig 39: Albion Claymore chassis

Fig 36: Hawtal Whiting Concept 92 with its body variants

Optimising the delivery vehicle

This section considers the multistop delivery vehicle of 3.5 tonnes GVW and over with reference to the improvements in fitness-for-purpose which could arise if conceived as a specialist vehicle in its own right.

A considerable challenge exists for engineers to design and build a high volume van which is also suitable for multi-stop delivery work. Panel vans between around 0.75 and 1.5 tonne carrying capacity are generally well arranged for this work but above this payload, the multi-purpose chassis-cab unit of 3.5 tonne GVW upwards tends to be the base for the larger multi-stop delivery vehicle. While cab design for light trucks has improved in respect of accessibility and general convenience, the requirements of the delivery vehicle operator can easily be compromised in favour of the tipper operator, or whomsoever, within the truck chassis's many functions.

The general-purpose cab has inevitably to be designed around the chassis frame and drive-train layout favoured for general duty. With the basic box-tube shape of a van body, it would be possible, as in the case of the light panel van, to use a more flexible integral-construction approach which avoided a high chassis front-end preventing a walk-through layout of the cab floor. There could be similar benefits in the adoption of a mid-engine layout, Fig 37, to ease the cab floor space problems. An integral structure proposal, due to Pawlowski, for a large van is shown in Fig 38. Engine support beams (1) are provided between the wheel-arches in this case. Other structural elements are front-wall (2), cab-floor and wheel-arch (3), side-wall (4), crossmember (5), rear-spring top beam (6), roof (7), floor-plate (8) and bulkhead (9).

Fig 40:C itroen H-van

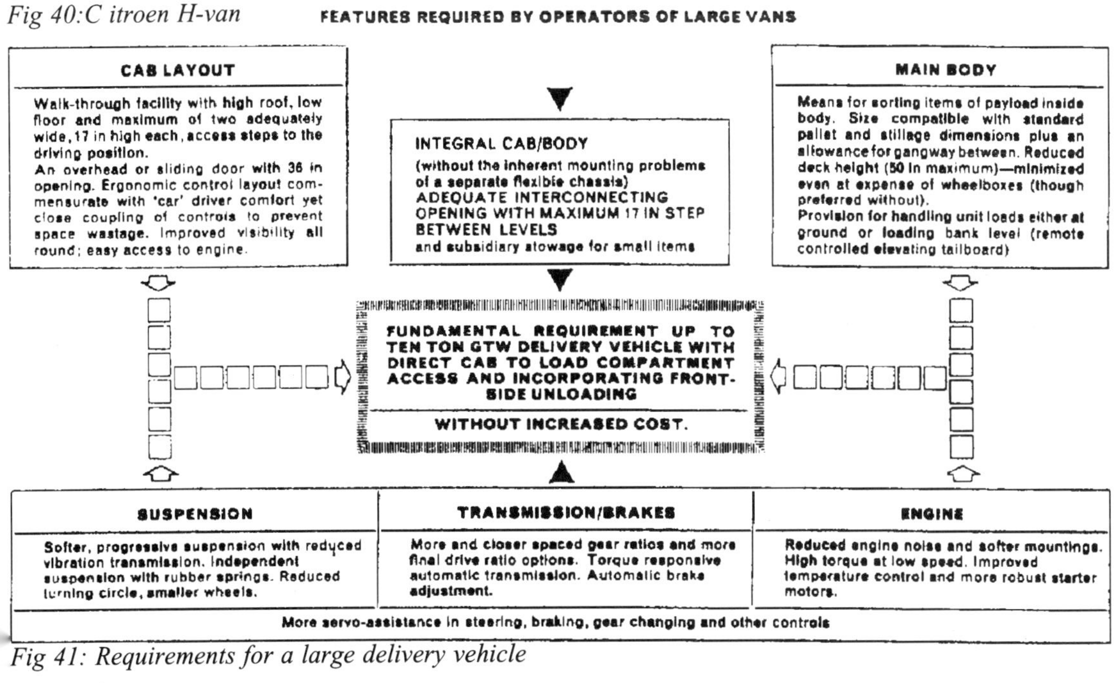

Fig 41: Requirements for a large delivery vehicle

The structure is considered integral if the structural surfaces which form the windscreen frame (F-G-G'-F'), the cab doors (D-G-H-J) and the rear doors (R-U-U'-R') are able to resist shear deformation.

A radical proposal for delivery van layout was made in Chapter 8, in terms of general layout. This will be examined here in greater detail. The concept of a purpose-built large delivery van is not new, of course, as seen in the Albion Claymore of Fig 39, an early 1960s model from a successful medium size company prior to absorption into the ill-fated BLMC. Even this vehicle was compromised by attempts to make it a 'Jack of all trades' but its flat-four mid engine and low drop-front frame were clear pointers in the direction of construction for purpose and a good access was obtained into an otherwise conventional cab. At the same period, the innovative Citroen company was in production with its H van, Fig 40, which showed how front-wheel-drive could also benefit van layout, particularly in respect of the rear cargo area whose low floor was additionally accessed from a mid-forward side door. The H-frame structure showed the way, disposing of the conventional chassis frame in favour of a base from which cargo and cab enclosures could be cantilevered.

The various requirements of fleet operators for large van specifications are summarised in the table of Fig 41, major aspect of which is the cab to cargo area walk-through facility which is absent in many large vans. Fig 42 is a more detailed picture of the above mentioned new-concept integral van, previously illustrated in diagrammatic form. The Albion Claymore flat-four engine envelope is again used in this proposal with drive taken back to a transfer box and then forward to a proprietary steer-drive axle mounted on rubber torsion springs — but with A-frame location from the transfer box. The H-frame in this case cantilevers a sandwich panelled box tube for

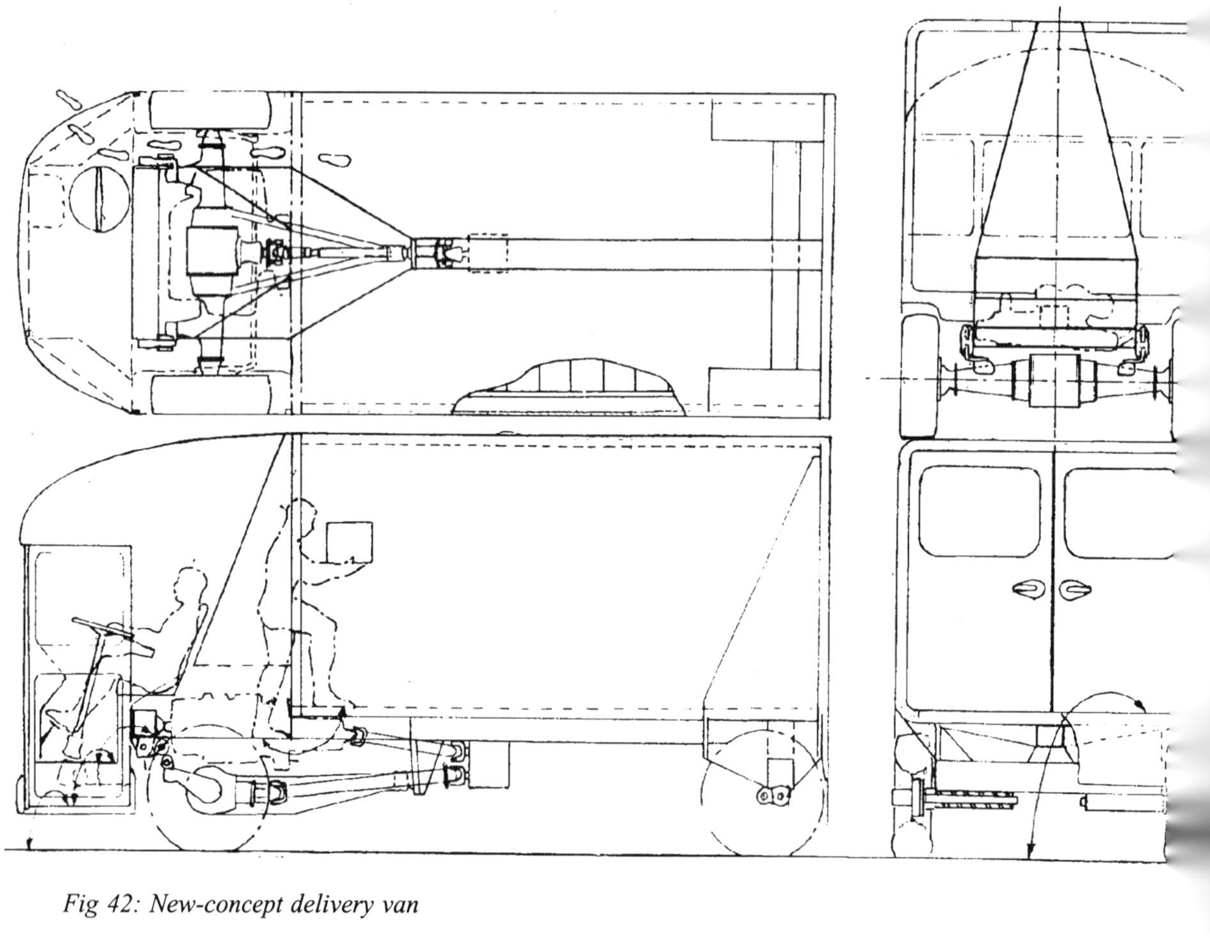

Fig 42: New-concept delivery van

the cargo area, the floor panel of which acts like a simple truss to transfer floor loading out to the sidewalls. A ring-frame at the rear is tapered forward to support the torsion rubber rear-suspension and provide support for the rear door hinges.

The foam-plastic core material of the sandwich panel would provide structural stability for the skin panels as well as giving good thermal insulation for the payload. The box tube is of such a symmetry as to readily provide optional lengths of vehicle while being relatively simple to analyse as a beam, for strength and rigidity under different load conditions. The structure would be virtually rattle-free compared with conventional separate chassis/body arrangements used for integral box-van conversions. The cab ahead of engine position, and well-aft placement of both axles combine with the use of single-formation wide tyres to make front-wheel drive a practical proposition. Better braking will also result from more regular front rear weight distribution whether laden or unladen. The compaction of mechanical components with FWD also means a general space saving on the vehicle. Modest step heights result from ground-level through cab-floor to that of the cargo area.

Integral bodies on large vans date back to such vehicles as the Bedford-Hawson Easy Access van, Fig 43, used extensively for parcels delivery work. This vehicle would have benefitted considerably from the more recent initiative by Bruce on the Scottish & Newcastle Brewery delivery-vehicle, Fig 44, whose imaginative front drive conversion allowed a usefully low rear deck height. The Leyland TX450 concept distribution vehicle (Fig 45) was also a very important step forward. While not precisely aimed at the local-delivery market, it is nonetheless probably one of the most sophisticated integral van designs yet built.

Fig 43: Hawson Easy-Access van

Fig 44: Bruce delivery vehicle chassis

Fig 45: Leyland TX450 concept

Box-vans based on kit-bodies

CV kit bodies are an attraction to operators for their ability to be assembled, and altered to suit later requirements by add-on modules, by skilled fitters at the operator's own premises, and because of the perceived quality and proven design of the kits.

Because kit-bodies have usually been the subject of joint engineering expertise by body-builder and supplier of panels and sections there is good expectancy of optimum exploitation of component properties. Body-builders, too, stand to benefit from purpose-designed systems which give some confidence of structural integrity. The past two decades have seen a considerable growth in CV kit-body and systems producers with the widening appreciation of the versatility of aluminium alloy extrusions and their adaptability into relative simple-to-assemble structures. Many options become availabe to builders from systems suppliers of made-to-measure bodywork such as Alusuisse; a typical kit from this company is seen in Fig 46.

Dawnpress Ltd supply their Eurokit system, which is available in aluminium alloy or GRP forms as either dry-freight (Glasply) or insulated (Glasfoam) bodywork for refrigerated operation. The former are plywood cored sandwich panels with polyester resin bonded GRP skins, lamination taking place under

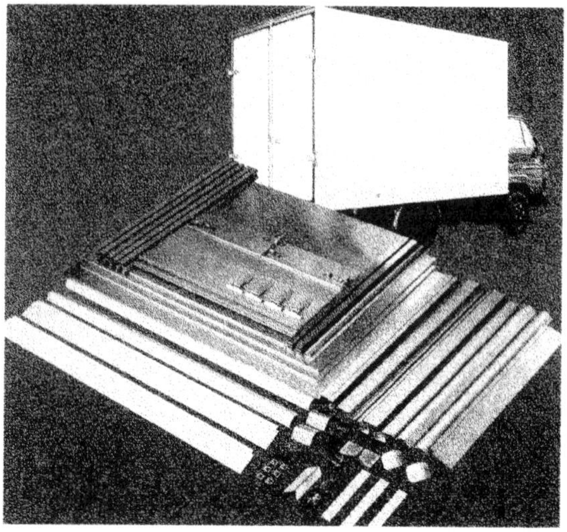

Fig 46: Alusuisse body kit

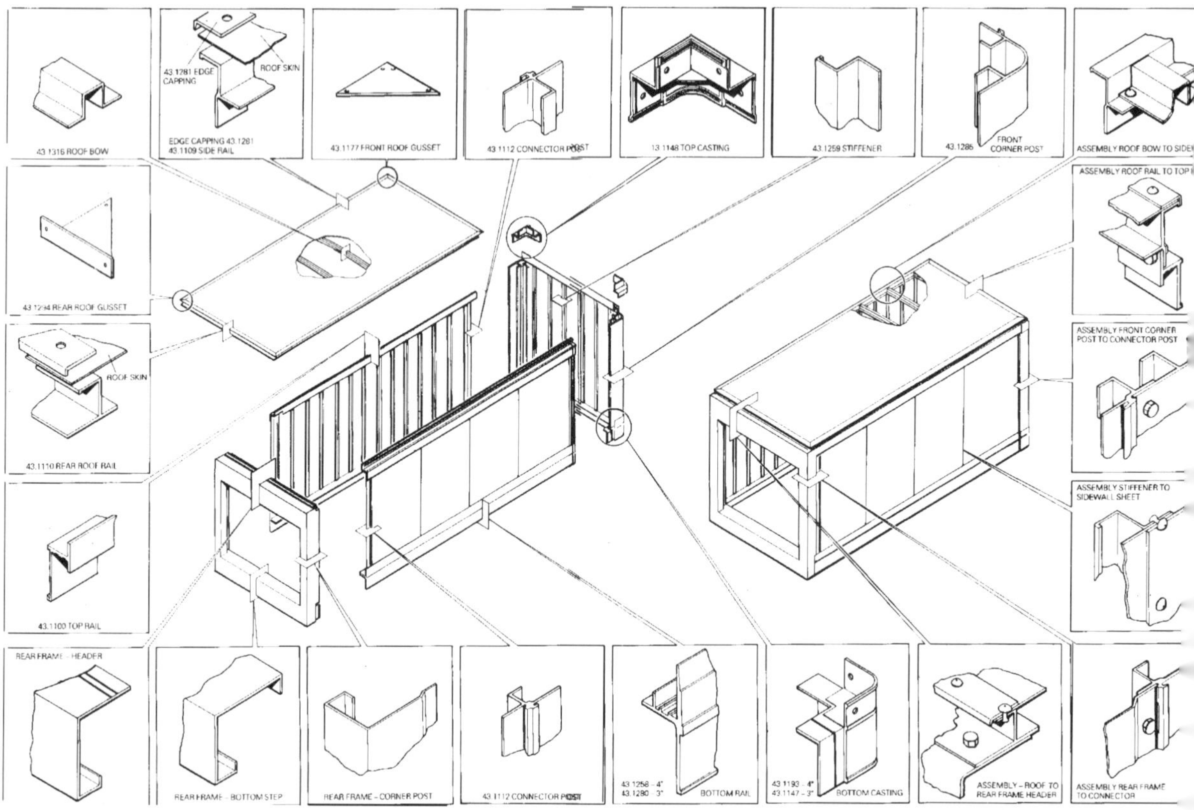

Fig 47: Eurokit by Dawnpress

vacuum following wet lay-up, curing taking place at 120 C. Seamless panels in lengths up to 13.42 metres can be supplied, with widths up to 3.66 mm, and thicknesses ranging from 6 to 24 mm in 7 steps. Steel inserts are provided within the core to form anchorages for cargo restraint systems. Glasfoam panels come in the same lengths and widths but with thicknesses up to 150 mm and containing the slabstock polyurethane foam core which is interspersed with GRP vertical webs at 100 mm centres to give high beam strength. Variants include 2LF construction where duplex layers of foam are separated by a GRP membrane and the webs are spaced at 600 mm. For the floor-panel variant timber supports are provided at 600 mm spacing between the ply-backed GRP skins. Fig 47 shows the Aluminium Eurokit in which section details of the members and joints are shown. Fig 48 shows how Glasfoam floor and sidewall panels are assembled.

Aluvan recently announced the Klemmbau kit which has the distinction of offcring a smooth cxtcrior walled body. This is achieved by joining panels with a unique clinching system, Fig 49. The clinching member also acts as a stiffener member which in the latest version is just 25 mm in section depth. These are at a pitch of 406 mm on the S-type body, to give the construction shown in Fig 50. Options in the kit include single and double side doors, a selection of rear frames, transparent roof-panels, sliding side-windows and lashing rails. Reinforced models are also available which have stiffeners spaced at 305 mm pitch.

A standard procedure has been established for assembly, comprising the following steps: first set up the front-end frame vertically and apply sealant along the length of the connection profile post. Then

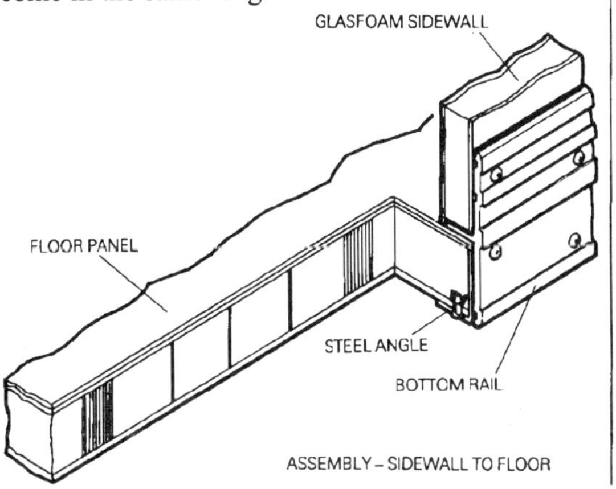

Fig 48: Glasfoam corner arrangement

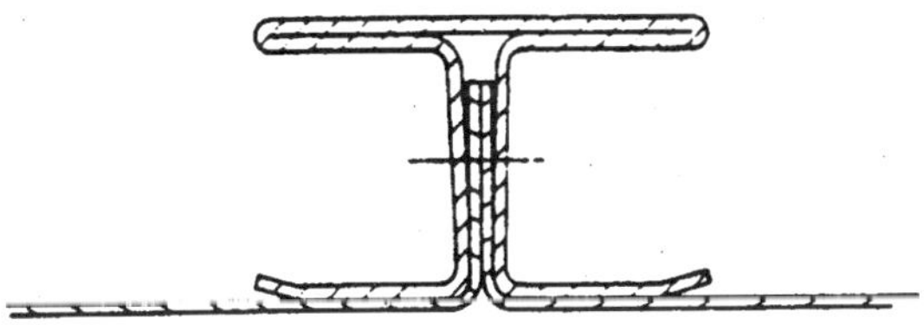

Fig 49: Aluvan panel clinching system

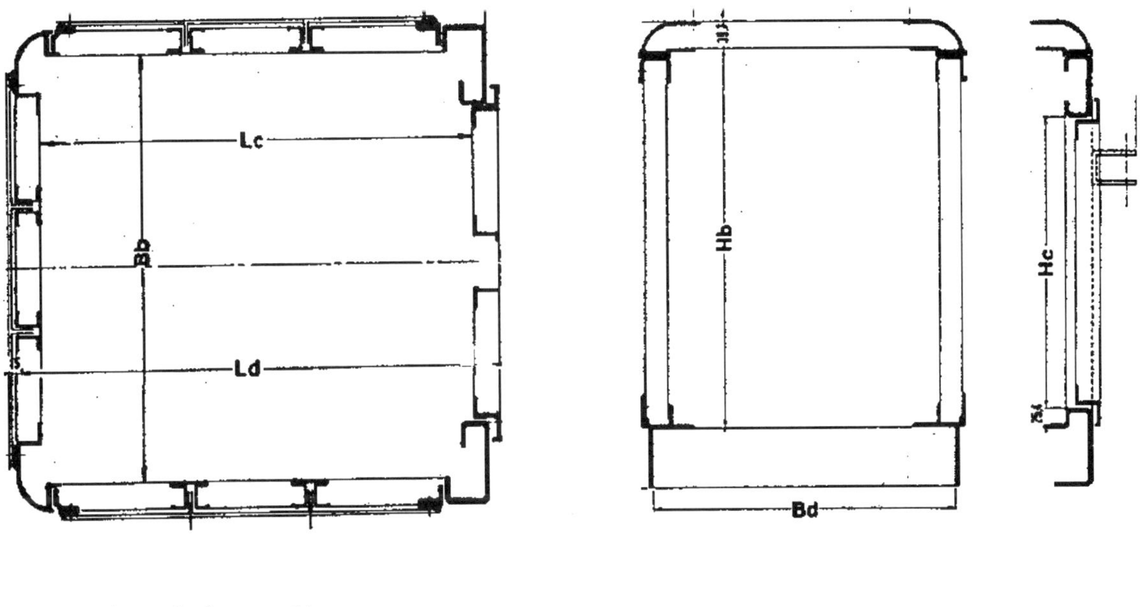

Fig 50: Aluvan body assembly

offer up a side-wall panel to it using positioning pins through a selection of the mating holes and a small number of securing bolts through some others. This is repeated for the second side-wall and the rear-frame is also applied with sealant and offered up to the sub-assembly. The resulting structure is then accurately levelled and made square and sealant applied to the upper edge prior to mounting the roof panel. Bolting should start from the front, one side at a time and an alignment check made prior to final bolting all round. Finally the corner post castings are finish-drilled to align with the body mounts on the vehicle chassis.

Tailor-made bodies from proprietory profiles

Body-builders wanting greater flexibility than is available with kits can still take advantage of extruded sections which have been specifically designed for CV bodies. The wide variety of sections available is somewhat daunting but the Aluminium Federation co-ordinates the activities of the various extruders in the UK, for example, through its member organisation, the Shapemakers. For a drop-side truck body two interesting sections are shown in Fig 51; such sections usually employ 6063 alloy in T6 temper. The first fixing method shown, for joining the sections, is a permanent one and requires only end-stiffeners. Continous hinges can also be built into the lower section.

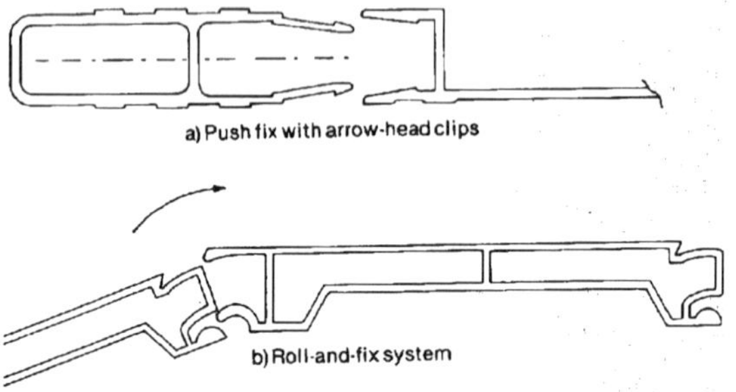

Fig 51: Sections for drop-siders

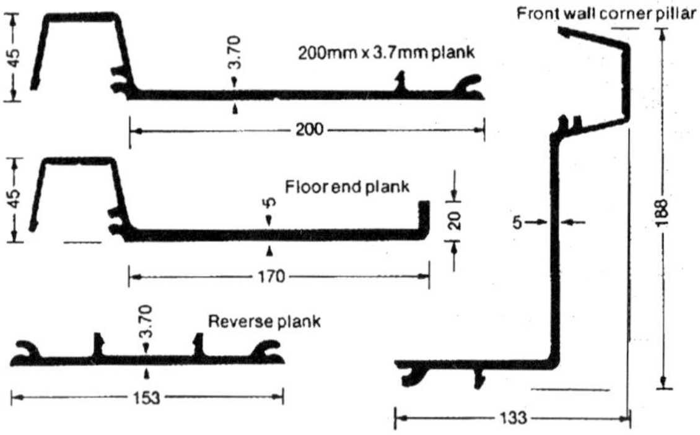

Fig 52: Sections for tippers

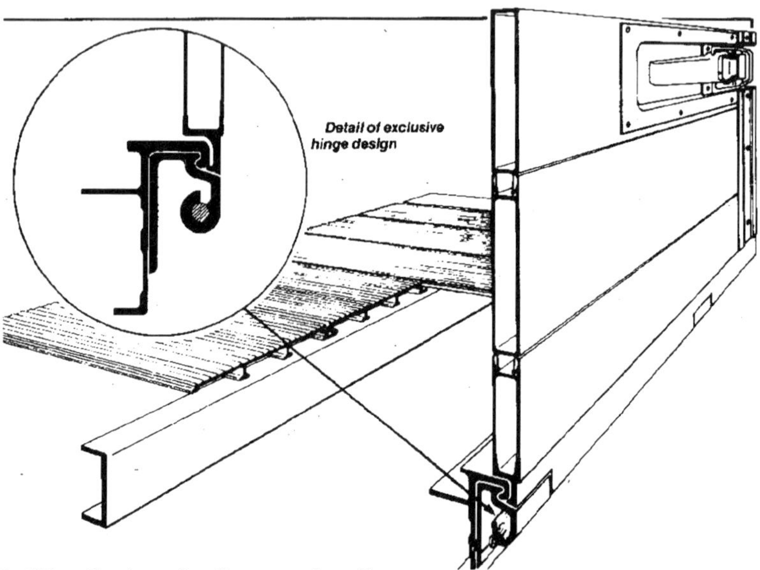

Fig 53a: Sections for floors and walls

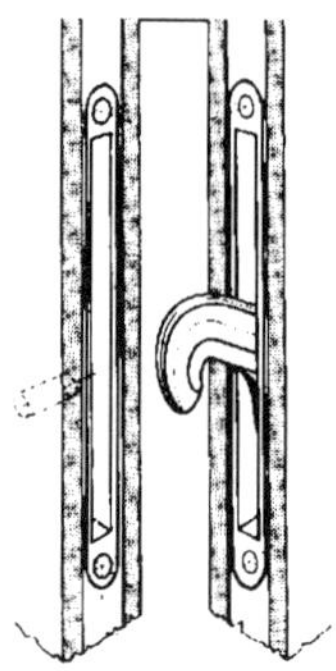

Fig 54: Isofix system

For sidewalls and end-walls, a range of sections up to 250 mm width is typically available in 6082 alloy at T6 temper, particularly suiting heavy-duty applications such as tippers, Fig 52. These are clipped together by a roll-and-fit system, with sufficiient holding power to allow jigless assembly before welding, the channelled portion of which doubles as a panel stiffener. Fig 53 shows a typical set of extrusions for a floor and wall assembly from Hydro Aluminium which illustrates a dropside hinge design. Another major supplier of profiles is Reynolds Transport Systems. Box-van body systems are based on snap-lock extruded "planks" to form either single or double-walled bodies. For insulated bodywork, the company have developed the Isofix system with which, again, panel members are coupled together without the need for pop-rivets or coupling strips, Fig 54. A synthetic hook-fastener avoids any thermal bridging at the panel intersections. For insulated bodies, Plastifont panels are used, while for fully-refrigerated bodies, Isofont panels, Fig 55, are available; for special-purpose light-weight applications, Carbofont is a third alternative.

The Isofix system involves a sandwich of 0.7 mm galvanised steel sheet over a PUR core of thickness tailored to thermal requirements. The bottoms of the panels have a hardwood edge and sizes of panels range up to 1200 mm wide x 2700 mm high, each containing 4-5 hook-fasteners on the vertical edges, which a secured by a socket-head wrench after PUR adhesive has been applied.

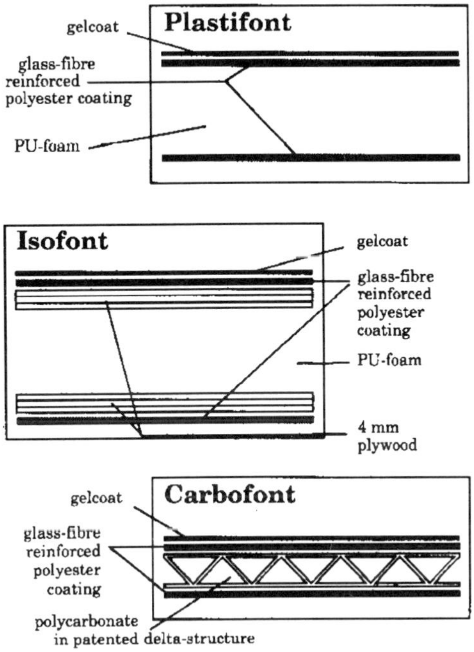

Fig 55: Insulated panel alternatives

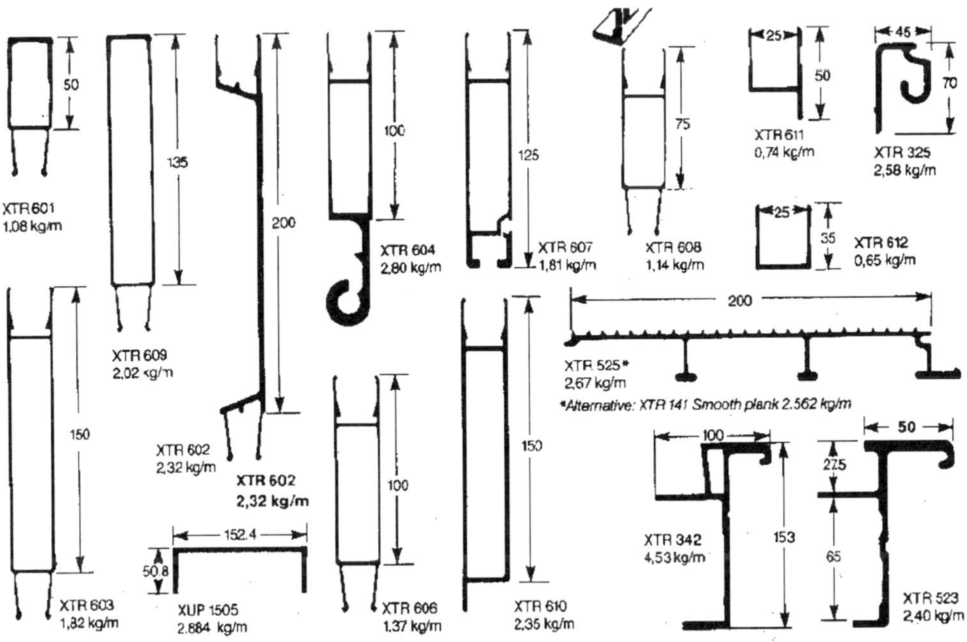

Fig 53: Sections for floors and walls

Refrigerated vehicles

There is intense competition between refrigerated vehicle builders to produce a thermally efficient, lightweight vehicle with maximum internal cube. This section looks at some of the techniques of analysis and construction used to wring out that final margin of performance over the competition.

The basic principles of heat transfer are a prerequisite to an understanding of body thermal efficiency — the ability to hold low temperatures over a complete journey with minimum expenditure of energy. Heat flows by conduction through a material; by radiation from one surface to another, via the intervening space, or by natural/forced convection in which the heat is carried by moving air (fluid) currents from one surface to another. As a vehicle moves along an open road the body will be subject to varying degrees of solar radiation which warm its outer surface. In general, heat will flow by conduction through the thickness of the wall and forced air from the refrigeration plant helps to induce convection currents on to its cooling coils or plates. Conversely when a door is open to atmosphere into the cool chamber cold air will 'fall' out of the body by natural convection and be replaced by warm air from the atmosphere.

Heat flow, being analogous to electrical current, follows the line of least resistance from the atmospheric ambient of, say, +15 C to the body interior which may be held down as low as 18 C for such products as frozen food and ice cream. A principal concern of body design and build is thus to avoid heat paths between outer and inner surfaces; high conductivity metal bridges between inner and outer skins are obviously something to avoid. So too is it necessary to seal against moisture ingress as water also has high conductivity relative to insulating materials. Sandwich panels are, of course, widely used as these have no structural pillars or rails penetrating the insulant. Fig 56 shows Concargo's Permacold sandwich panel body kit which includes pre-engineered door assembly and interlinking corner rebates.

A typical foam plastic insulant will have thermal conductivity of 0.025 kcal m/hr m^2 C. This will vary with density and structural strength and from material to material. Expanded PVC is one of the strongest materials related to weight and thermal insulation. Neoprene is an adhesive which is suitable for bonding expanded-PVC cores to metal skins and Fig 57 shows the kind of vacuum table which can be used to fabricate the panels from slabstock insulant and metal sheets — the latter should be degreased with a suitable solvent. Another common technique is foamed in-situ expanded polyurethane.

Insulated vehicle practice

For large refrigerated articulated trailers, Schmitz Cargobull have moulding presses installed for producing 40 ft (12.2 m) long sandwich panels using expanded-PVC insulant bonded panel-to-panel by polyurethane gap-filling adhesive on the edges. Expanded polyurethane is used as a roof -panel insulant to protect against the higher skin temperatures. PVC foams of 40 and 55 kg/m^3 density are used according to the structural requirements of the panel. Normally the exterior skin is prepainted white aluminium alloy, in 4 ft (1.22 m) wide sheets positioned vertically, but 10% of vans go out with electro-galvanised steel skins. Each is bonded to the cores by epoxy resin. Exhausting by 10 lb/in^2 (69 kN/m^2) in the vacuum bag is equivalent to a side thrust of 250 tonnes. Resulting side-panels are 45, 60 and 80 mm thick.

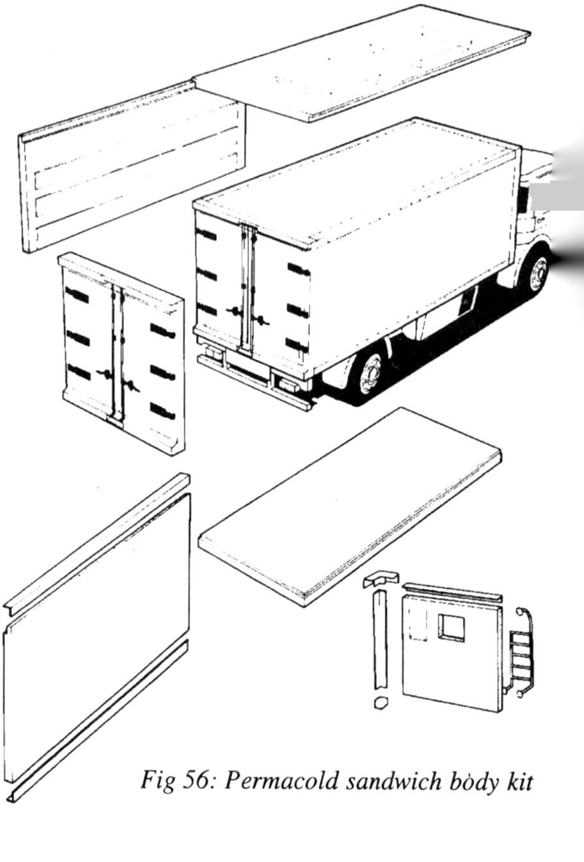

Fig 56: Permacold sandwich body kit

The floor sandwich comprises a 1 mm thick galvanised under-skin, adjacent to a 15 mm thick plywood layer. The rubbing plate above the kingpin is just the 1 mm thick under-skin but the plywood inner panel is overlayed with aluminium alloy above the kingpin frame, which is embedded in the insulant. In final body assembly the precision moulded panels are self-jigging and no mechanical fasteners are used to supplement the twin-pack adhesive. Corners are, however, capped with aluminium alloy angle sections over a vapour-sealing PVC tape. The one-piece stainless steel rear frame is fabricated from channel sections which cap the rear edges of all four main body panels. Doors have no lap joint so that either can be opened independently.

Interior body joints are also capped with angles, and sealed with mastic, except the bottom joints which are sealed when a layer of polyester resin is floated over the whole floor — which is impregnated with carborundum grit and topped by a sealing screed. Having an integrated kingpin frame and rubbing plate results in a 4.5 in (114 mm) gain in interior height to 2.5 metres — on a 48 in (1.21 m) high fifth-wheel coupling turntable and within the 4 metres continental overall height restriction. With interior length of 12.2 metres, and interior width of 2.45

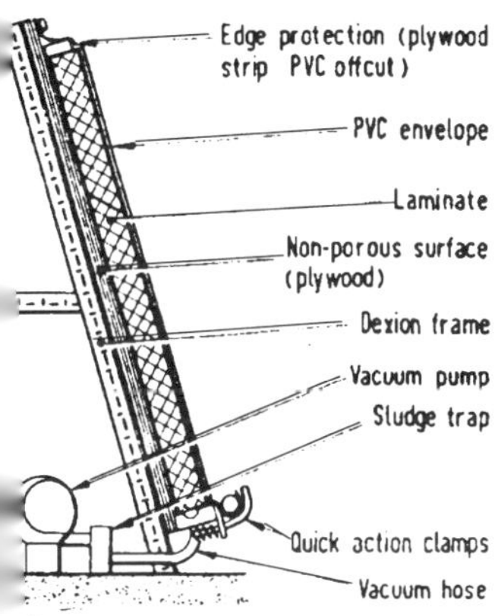

Fig 57: Vacuum table for assembling sandwich panels

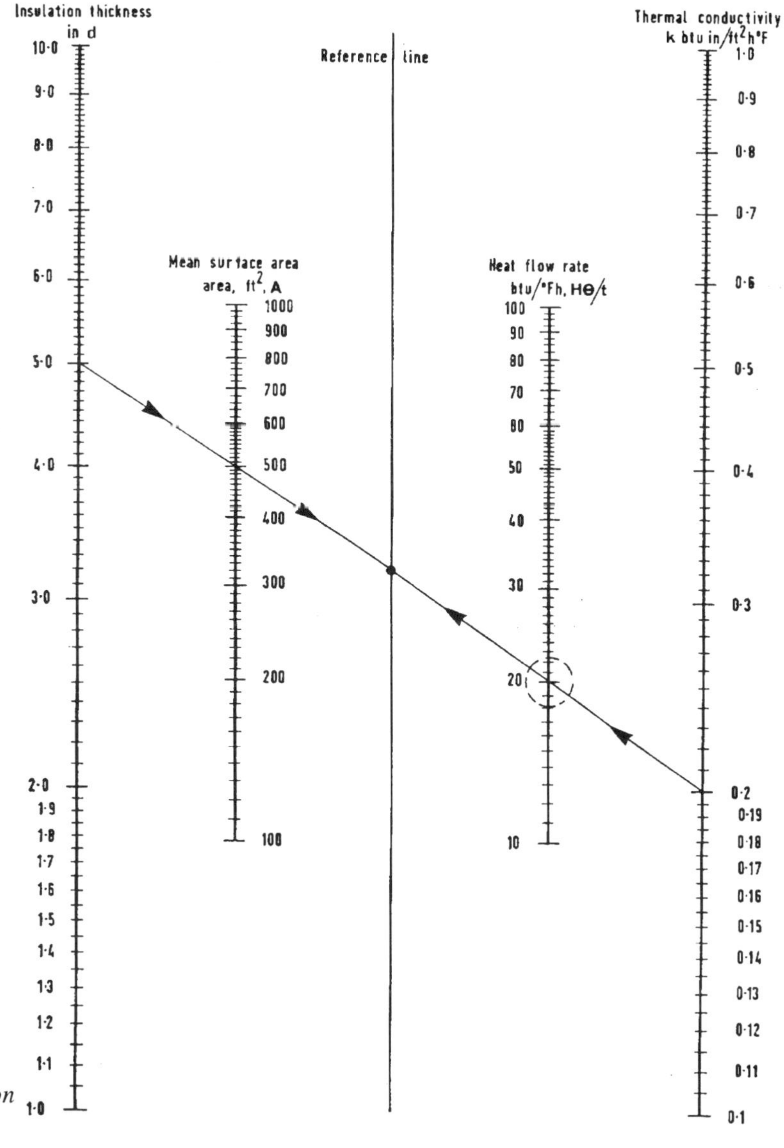

Fig 58: Insulated panel selection nomograph

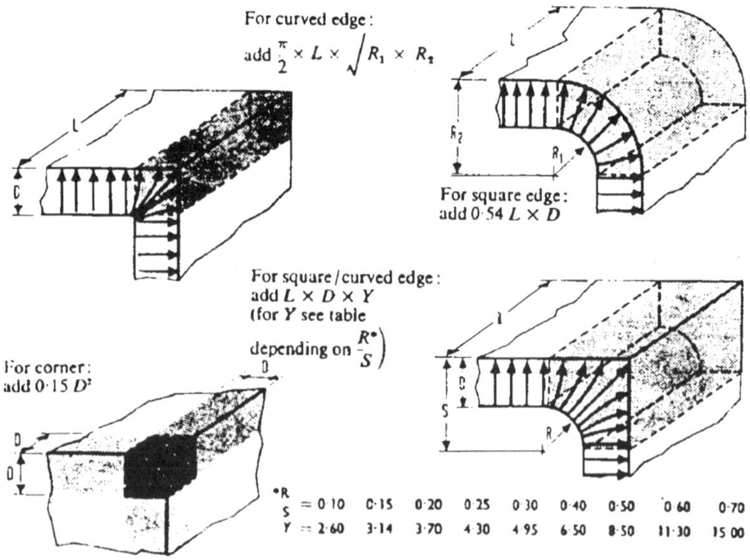

Fig 59: Shape factors for edges and corners

metres, internal cube is just below 75 cubic metres. Unladen weight of the integral trailer on tandem axles is below 14000 lb (6.35 tonnes) — for operation at 38 tonne GTW.

Analysis of insulated bodies

First assessment of insulation thickness required to achieve a given thermal performance by a sandwich panel can be made using the nomograph in Fig 58

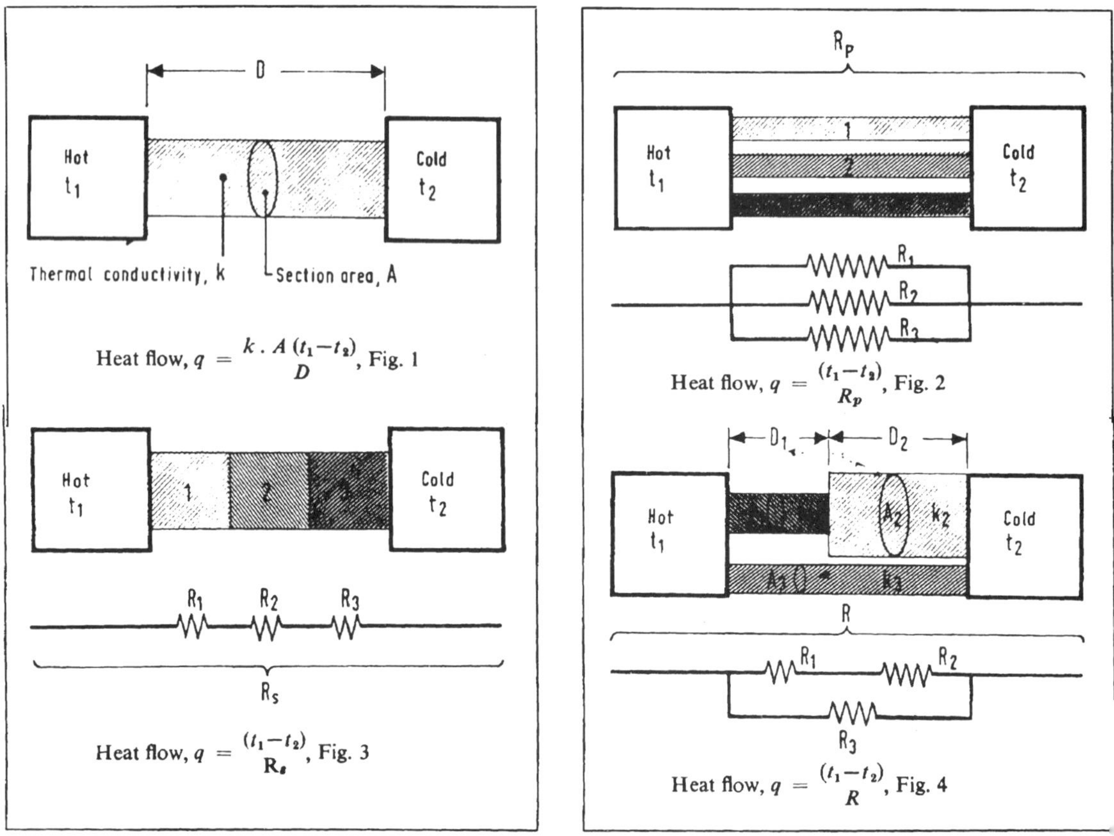

Fig 60: Electrical analogues for heat flow

which quotes thermal conductivity in Imperial units. To estimate the thickness d of insulation to give minimum heat leakage H through panel surface area A with insulant conductivity k, the equation

$$H/\theta t = kA/d$$

applies which is represented by the nomograph shown — in which a worked example shows intersects drawn for $d = 5$ in (127 mm), $A = 500$ ft^2 (46.45 m^2) and $k = 0.2$ BTU in/ft^{2}h F.

In calculating the heat flow through a complete box body, additional surface area to allow for edges and corners must be allowed for in calculation. These 'shape factors' are suggested in Fig 59 and are an estimate of the two and three dimensional flows which occur at these locations. Fig 60 shows how the electrical analogy to heat flow can be put to effect in calculation — the heat paths being considered as electrical resistances. The basic conduction equation can only be applied in case (a) whereas in other cases it is useful to draw up a thermal circuit to calculate an approximate heat flow for series, parallel and combined networks. Note, of course, that $R_s = R_1 + R_2 + R_3$ whereas $R_p = (1/R_1) + (1/R_2) + (1/R_3)$. Also the actual, and not 'specific', thermal resistances must be used in calculation so that areas and lengths of heat paths are accounted for. Thus in case (d), $R_1 = [D_1/k_1A_1](R2, R3, etc.)$ and $(1/R) = [1/(R_1 + R_2)] + 1/R_3$. Only complete accuracy can be obtained where the two and three dimensional heat flow effects are accounted for. These can be predicted by finite element techniques but it may be preferable to use physical models of the shapes in question and test them in electrical flow analogue experiments.

In the case, say, of a high conductivity rib extending into an insulated wall, the parallel lines of heat flow are disturbed around the locality of the rib as shown in Fig 61. Pioneer work by the Ship Owners Cargo Research Organisation obtained shape factors for such sections and the table of Fig 62 shows extra thickness of wall, X as a fraction of plain wall thickness 1 for rib depth m and spacing n. Other shapes shown in Fig 63 were found to have shape factors obtainable from the graph in Fig 64. The heat flow through door frames can be estimated by such techniques and it is a comparatively simple matter to

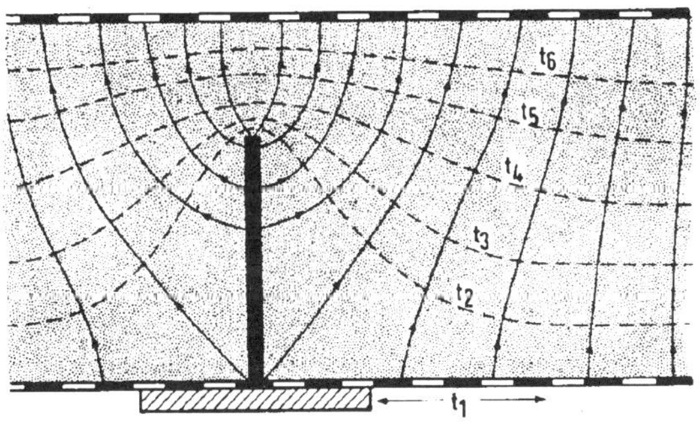

Fig 61: Heat flow from rib

$\frac{n}{l}$	0·333	0·5	1·0	2·0	3·0	∞
For $\frac{m}{l} = \frac{3}{4}$	0·507	0·573	0·609	0·611	0·612	0·611
For $\frac{m}{l} = \frac{1}{2}$	0·1834	0·207	0·2201	0·2205	0·2208	0·2206

Fig 62: Shape factors for rib

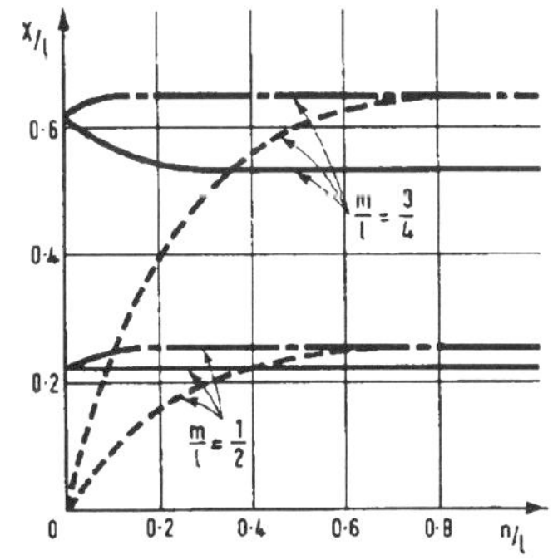

Fig 64: Shape factors for other sections

Values of X/l for various shapes

m/l	$n/l = 0$ ≡ Rib	$n/l = \infty$ Buttress	$n/l = \infty$ T-rib
0·1	0·0079	0·0122	0·0115
0·25	0·050	0·065	0·060
0·5	0·221	0·250	0·221
0·75	0·611	0·650	0·537
0·9	1·181	1·222	0·929
a	b	c	d

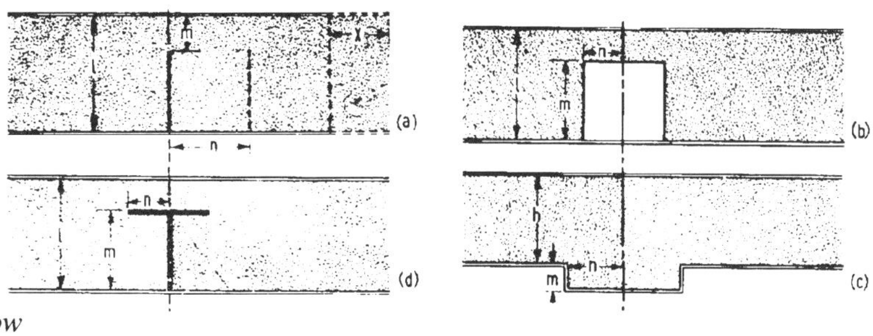

Fig 63: Shapes tested for heat flow

have computer programs written to simulate thermal performance of the vehicle prior to build, as a summation of all the local heat flows. As an illustration, the door-frame and seal section in Fig 65 is accompanied by its electrical resistance analogue for which the elemental resistances can be evaluated by the methods just outlined.

'Insulated' rather than 'refrigerated' vehicles, for the carriage of perishable rather than frozen goods, are sometimes termed 'half-isothermal' vans and may involve limited thermal bridges between inner and outer skins. Such construction is typified by frame pillars in expanded-PVC enclosed by expanded polyurethane insulation as in Fig 66. The importance of the vapour film is vital as it is possible for a large box van body to accumulate as much as 1 tonne of water in a year's operation — which can markedly affect thermal efficiency.

On-board refrigeration plant

Isothermal vans are characterised by on-board refrigeration systems which may be eutectic plates, mechanical units and/or liquid-nitrogen/solid-carbon-dioxide evaporative cooling. Eutectic plates are filled with refrigerant which is only circulated, through a land-based refrigeration plant, when the vehicle is 'in dock'. The plates act as storage coolers in transit and though somewhat heavy when made in metal, new plastic tubular versions are proving to be acceptable to operators.

Mechanical systems, of course, give maximum mobility to the operator but their weight, too, is not insignificant. A large reefer can involve a plant weighing 0.75 tonne which is usually mounted on the leading face of the trailer front panel. The mounting will usually involve thermal bridges in the frontpanel reinforcement frame which must be allowed for in the thermal efficiency calculation. Direct evaporation systems such as those for liquid nitrogen involve a container to which a series of small bore copper pipes are connected and installed usually at the upper extremities of the van, running from front to rear. A series of jets at predetermined centres spray the liquid over the cargo during initial temperature 'pulldown'. This can take only three minutes, from initial ambient temperature, as the liquid boils at 196 C. Low weight of the equipment is an attraction — being in the order of 50 kg for a 4.8 metre long box body. Fig 67 shows an early system from BOC alongside an early ICI

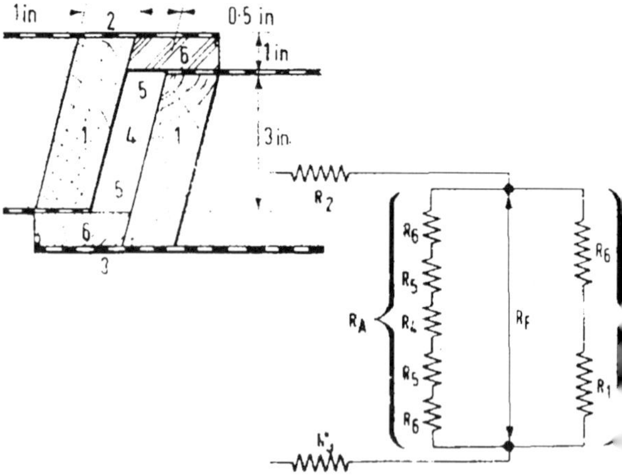

Fig 65: Door frame/seal with its electrical analogue

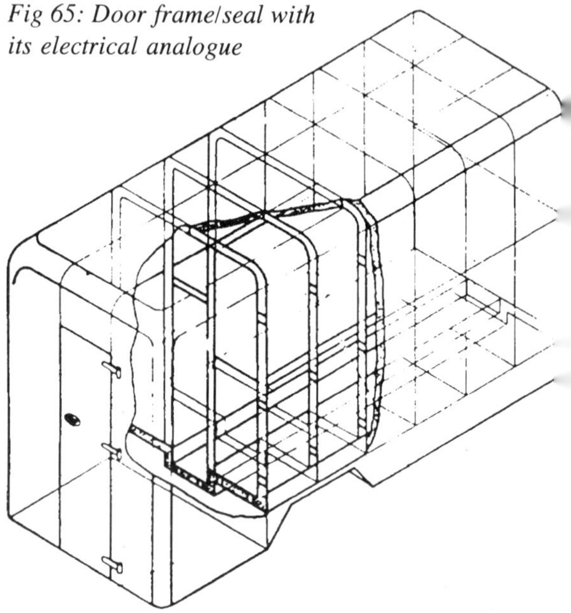

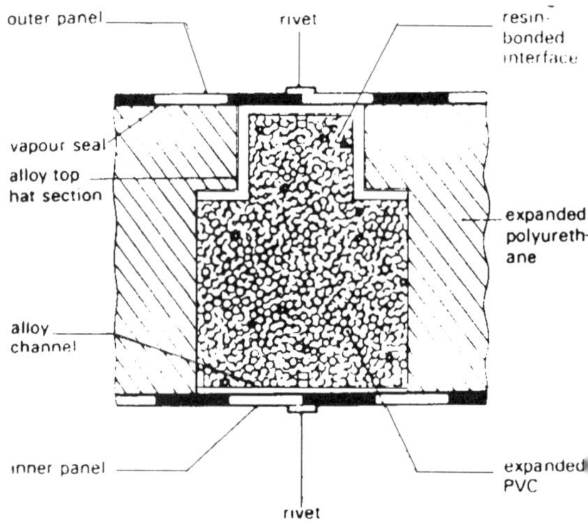

Fig 66: Bridging of inner and outer skins

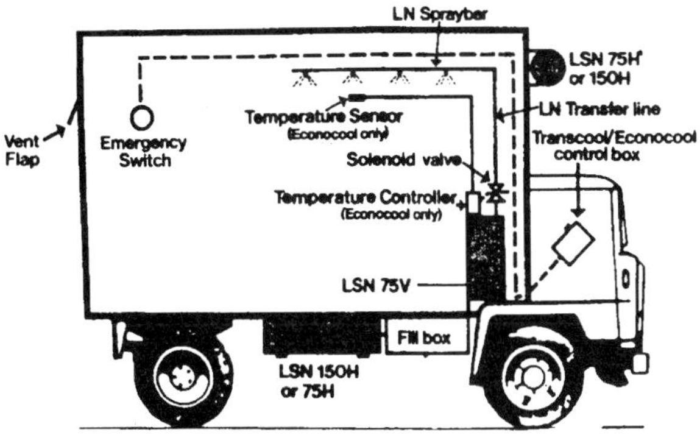

Fig 67: BOC liquid nitogen system

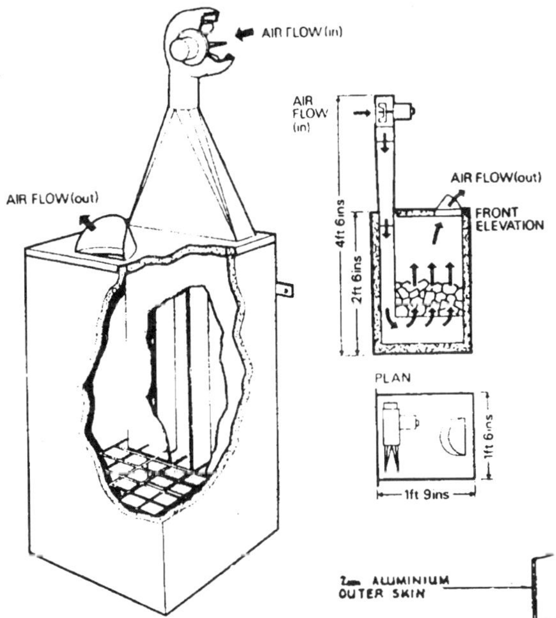

Fig 68: ICI dry ice system

evaporator system for solid carbondioxide (dryice), Fig 68. The latter occupied 18 x 21 in (457-533 mm) of floor space and was 30 in (762 mm) high. It weighed 62 lb (28 kg) empty and contained 120 lb (54.5 kg) of dry ice. Cooling capacity up to 20800 BTU/hr for short periods could be obtained with continuous running at 2000-6000 BTU/hr.

Performance of refrigeration systems depends upon a well sealed body with minimum cold air fall-out. Leakage rates of nine cubic metres per hour have been measured on reefers with poor rear door sealing. An interesting door sealing development by Wilsdon is shown in Fig 69 in which the thermal and weather-proofing duties of the seal are combined. Door jambs are moulded integrally with the wall panels and a close tolerance is held on shut clearance by jigging techniques. The sealing gasket works in compression against the wall-panel thus overcoming conventional shear drag.

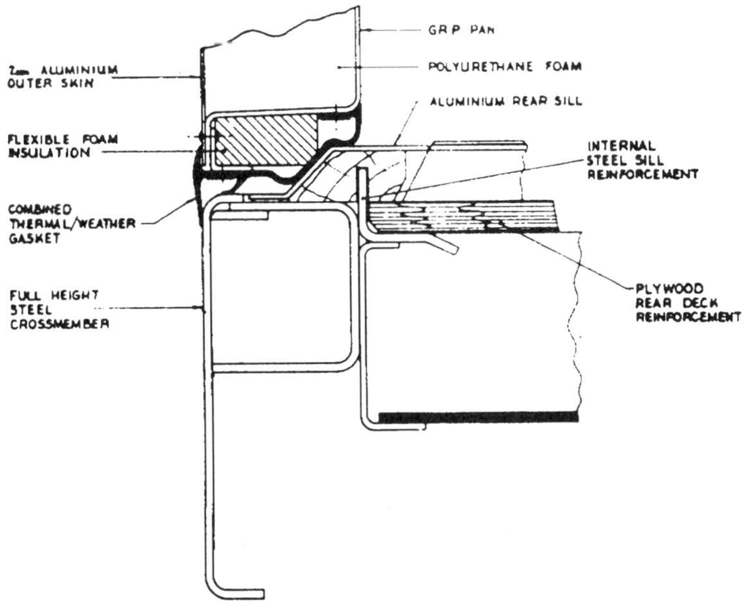

Fig 69: Wilsdon door seal

Refuse/cleansing vehicles

Refuse vehicles encompass a variety of technologies and call for considerable creative skill on behalf of design engineers who must successfully integrate them. Some new and unusual approaches to integration are reviewed here alongside consideration of two prevailing technologies, power-take-offs and hydrostatics.

The move from converted truck chassis to small purpose-built sweeper vehicles probably began in the 1970s with an unusual adaptation of the Reliant three-wheeler chassis for sweeper duty in Melford Engineering's Precinct model, Fig 70. Hydrostatic drive was used on the front steered wheel so that the vehicle could drive forward into a blind 90 degree corner and sweep the ground surface in one pass. A total steering angle of 200 deg was achieved enabling the two rotating brushes at the front corner of the machine to be traversed out of a corner in such a way that the only area unswept was in the very corner beyond the 28 in (711 mm) diameter of the brush.

The Schmidt SK152 model, Fig 71, has also advanced the state of the art; based on the Design Council Award winning SK150, it has special steering geometry on the oscillating front axle, and hydrostatic power steering, which allows this 4-wheeler machine also to sweep out a 90 deg internal corner. A patented water circulation system allows water injected into the suction nozzle and tube to be recycled and ensures a dust-free environment during sweeping. An electro-hydraulic anti-stall system operates between the vehicle transmission and the sweeper drive.

The Hestair Eagle Griffon was an interesting development in truck-mounted sweepers, Fig 72, whose considerable suction capacity comes from a diesel auxiliary engine driven centrifugal fan (1). The water tank (2) and 5 cubic metres refuse compartment (3) are constructed in curved high tensile steel panels to withstand the suction vacuum. Shown at (4) is the air-discharge silencer alongside brush-gear (5) and subframe (6); at (7) is the automatic rear

Fig 70: Melford Precinct sweeper

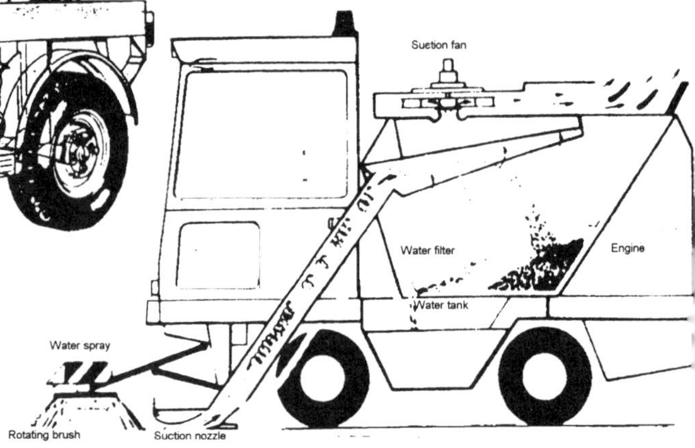

Fig 71: Water circulation system

Fig 72: Eagle Griffon

Fig 73: Laird Rotopress

Fig 74: Dempster Dumpmaster

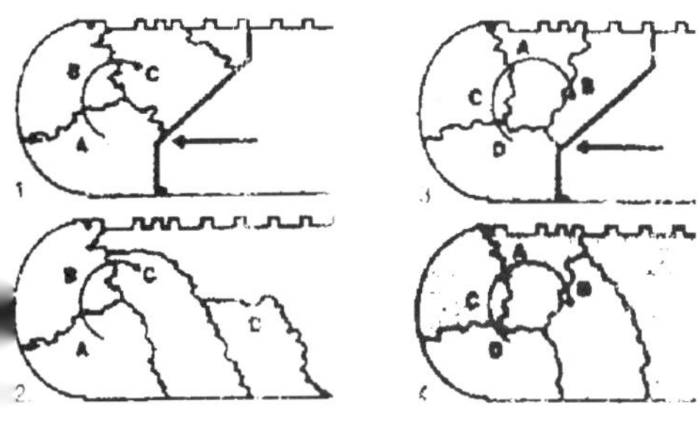

Fig 75: Curved tailgate for compaction and
Fig 76 (below): for ejection

door and at (8) the full-width filter screen. Twin inboard fuel tanks (9) are featured also deflectors (10) to improve refuse distribution.

An interesting departure in refuse collection vehicles is the Laird Rotopress (Fig 73) which favours a screw packer rather than the hydraulic one typified by traditional RCVs. On the Dempster Dumpmaster, Fig 74, another alternative is the curved tailgate which is designed to rotate the refuse as the compaction cycle proceeds, Fig s75/76. At (1) the packer head drives refuse (C) up and over; (2) as the packer head retracts, compacted refuse (C) in the upper body falls into empty head space; (3) new load (D) is compactcd against (C) and packer hcad drivcs refuse (C) up and around again; (4) refuse (B) falls into empty space left by the packer head. Gully emptier and jetter vehicle design has been advanced by the Peabody Holmes Vactor Combi, Fig 77, whose operating system is shown at Fig 78. The air flow system is driven by an independent diesel engine and the variable rate jetting pump is hydraulically driven from the engine PTO.

Fig 77: Vector Combi

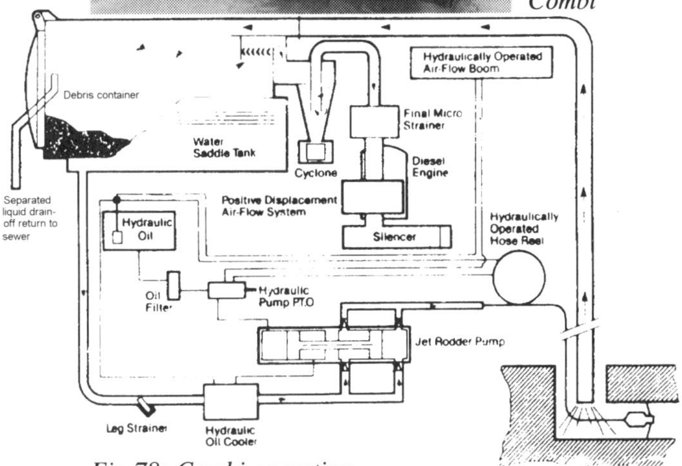

Fig 78: Combi operation

Power-take-off selection

A distinctive characteristic of refuse vehicles is the reliance put on engine and transmission power-take-offs to drive such items as water pumps, suction fans, waste packers and sweeping brushes. Openings provided in the top, side or bottom of transmission gearboxes are used to access constantly meshing gears for PTO drives to either hydraulic pumps or drive shafts. Typical sidemounting configurations, due to Edbro, are shown in Fig 79 employing the SAE standard four or six bolt facings; 8-bolt facings also exist for heavy duty drives. As well as these layshaft gear drives, other take-off points are front of engine crankshaft; sandwich drives between engine and gearbox and split propshaft drives. Rear mounted layshaft PTOs are another possibility usually involving a splined dog engagement; often step-up gearing is provided in this type of drive, Fig 80. Various adapter plates are also available, Fig 81, for such functions as directional reversing and even replacing a gearbox side plate with another containing PTO opening. Torque throughputs are typically 140 lb ft (190 Nm) for 6-bolt facings while 8-bolt facings can have up to full engine powers transmitted through them. In the case of automatic transmissions, having no layshaft, the drive gear is usually large, being mounted on the outer gear of the planetary train: thus requiring reduction gearing. PTO speed available varies in relation to the load imposed by the driven device with this type. The split driveshaft type of PTO, between gearbox and rear axle, Fig 82, might involve a belt drive in certain cases.

Selection procedure involves putting the questions: how many speeds are required and in which direction of rotation; what shaft powers are needed corresponding to these speeds; what type of transmission gearbox is fitted and is it capable of the required PTO powers? If, for example a 25 bhp blower is required to run at 1000 rpm for optimum efficiency from a PTO which only allows 500 rpm, then step-up gearing will be required which will double the torque throughput on the PTO. Generally the higher the pitch line speed of the transmission gearbox PTO gear, the higher the potential power take off capacity. Speed of the pick-off gear with relation to engine speed is also needed: generally pitch line velocities should not fall below 400 and 650 ft (122 and 198 m)/min for 15 an 25 bhp PTOs respectively. Gaskets should be chosen carefully for the PTO face to obtain

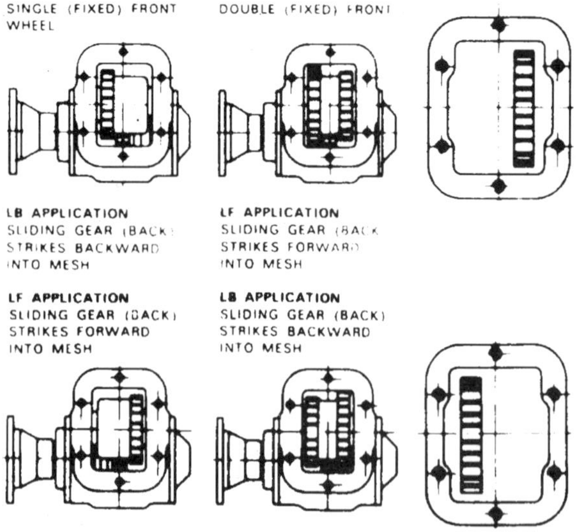

Fig 79: Edbro PTO configurations

Fig 80: Step-up gearing

Fig 81: Adaptor plates

Fig 82: Split drive-shaft PTO

optimum 0.008 to 0.012 in (0.2-0.3 mm) of gear backlash with fastening bolts tightened to 20 lb ft (27 Nm). PTO control systems vary from direct-lever arrangements to flexible cable remote controls — or even air shifting in and out of gear. With air shift, it is usual to spring-load the PTO gears out of mesh as a fail-safe measure in the event of losing air supply. Electrical solenoid controls are also available, with microprocessor switching.

Use of elementary fluid-mechanics theory is valuable in assessing the power required for the various functions of refuse vehicles. Emptying gullies and cesstanks involves forces and pressures obtained from the basic laws of hydrostatics. Liquids are usually assumed to be incompressible in the analysis and the weight of water is taken as 62.4 lb/ft^3 (1 tonne/m^3). Pressure of water is normal to the wall retaining it so it is useful to remember, with a cylindrical vessel, that total thrust force is calculated on the projected area rather than the total area of the vessel wall. In calculating the force applied by a simple plunger pump required to raise the level of water in a vertical cylindrical tank the mechanical advantage is simply the ratio to tank area to plunger area. Pumping power can thus be calculated on this basis. The pressure head of a liquid is another useful concept in calculation — based on a liquid being subjected to pressure due to its own weight. Pressure equals head (or depth) times density. In Fig 83, intensity of pressure due to weight in a vertical direction is also reflected by an equal horizontal pressure as shown by the triangular pressure profile. This profile affects the position of the centre of pressure on a submerged liquid-retaining hatch.

For low speed flow in pipes, the frictional forces can be neglected and streamline-flow conditions are assumed. For the general flow condition of Fig 84, velocity at (aa) times section area equals velocity at (bb) times its section area. To find the work done in overcoming liquid pressure in a tank, the set-up in Fig 85 can be considered. If water is forced into the tank at velocity v against the internal pressure of p, work done is WH where W is weight of water injected per second and H the equivalent static head in feet of water. If water then flows through the same pipe under the action of the weight of water in the tank $WH = Wv^2/2g$ which is the kinetic energy of the water ejected; the 'static head' then becomes the 'velocity head'. In pumps with a long suction pipe and a low delivery head, pressure due to inertia of the column of water in the suction pipe will be large compared with the pressure on the outside of the delivery valve. Fig 86 shows a work diagram for a pump, the vertical ordinate being plunger pressure either side of a double acting pump. Line (f.e) represents atmospheric pressure while (d.c) is cylinder pressure during the suction stroke and (a.b) cylinder pressure during delivery. Corresponding works done on the suction and delivery strokes are given by dcef and abef provided the pump is operated at low speed.

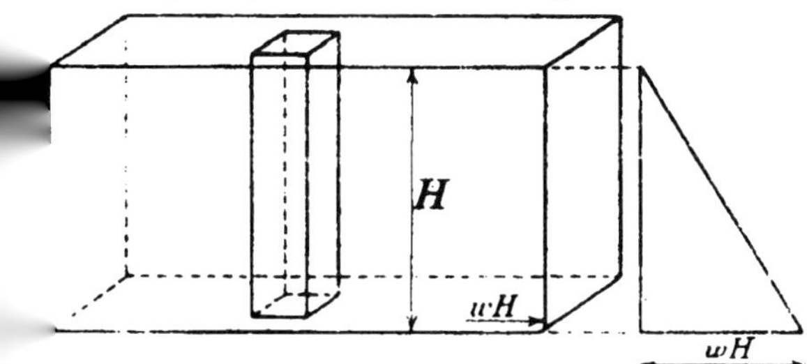

Fig 83: Pressure profile

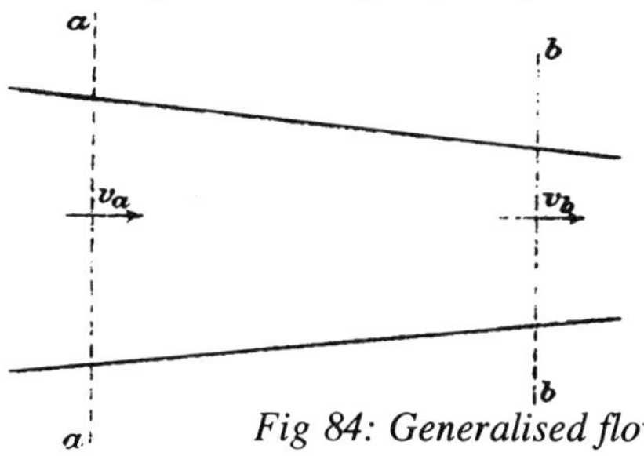

Fig 84: Generalised flow conditions

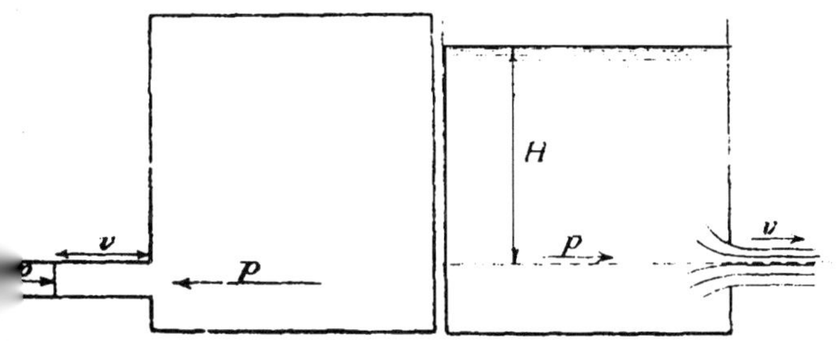

Fig 85: Work done in overcoming pressure

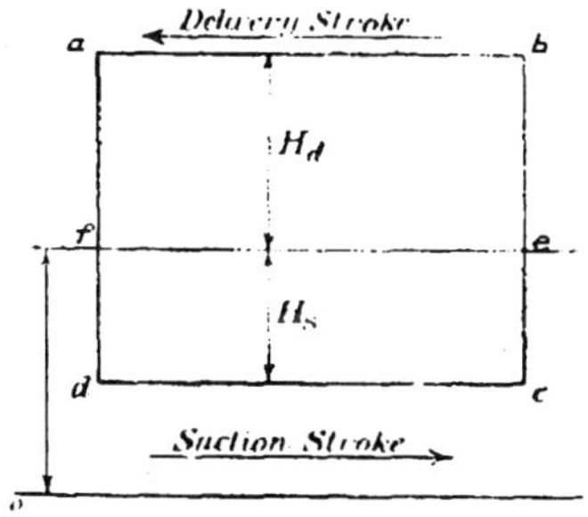

Fig 86: Pump work diagram

Handling trucks

Design of the industrial fork-lift truck is governed to a major extent by the compactness of the aisles in which it has to work and manoeuvre. Ergonomics of the driver's work area must not be compromised, however, otherwise work cycle efficiency will be reduced by the effects of driver fatigue. The stability of the truck is also a crucial factor for a truck which carries its payload outside its wheelbase—as well as the drive motor, batteries and controller.

Work at Birmingham University's Department of Engineering Production has shown that operator fatigue can be considerably reduced in fork-trucks by careful attention to seat design and control positioning—which can dictate the basic design layout of the machine. The optimum seat dimensions recommended by ergonomists, Fig 87, were found by the Birmingham team to have often been exceeded in commercially available trucks. Height limits for seats have been reportedly exceeded by as much as 3.5 inches, resulting in fatigueing under-thigh pressures for drivers in the lower percentiles of stature, Fig 88. Over-long seat squabs can cause drivers to adopt an unnecessarily slumped position and too high a back causes restriction of the shoulders which require freedom for both steering and reversing.

A recommended space envelope around the median plane of the body is often intruded into by the steering wheel — again causing discomfort. The need for the driver to have his lower leg vertical, as shown, is also important to ease frequent entry and egress. Common difficulties are forward projections from beneath the seat, arising from the designer's belief that a leg posture akin to a car driving position would be possible. Another common form of discomfort is overheating of the lower leg caused by proximity to a poorly insulated engine compartment. Finally it is necessary for the safety of the driver for the seat reference point (SRP) to be at least 20 in (508 mm) from the truck side so that no part of the driver's body can project over the edge of the truck.

Hand reach limits, Figs 89/90, worked out by the Birmingham team follow the approach made by classical researchers such as Rebiffe, and described in Chapter 8. On many commercial designs hand brake and horn were found to be out of reach as, on occasions, were gear and mast control levers. A special problem arises with fork-trucks in that full single handed rotation of the steering wheel is often

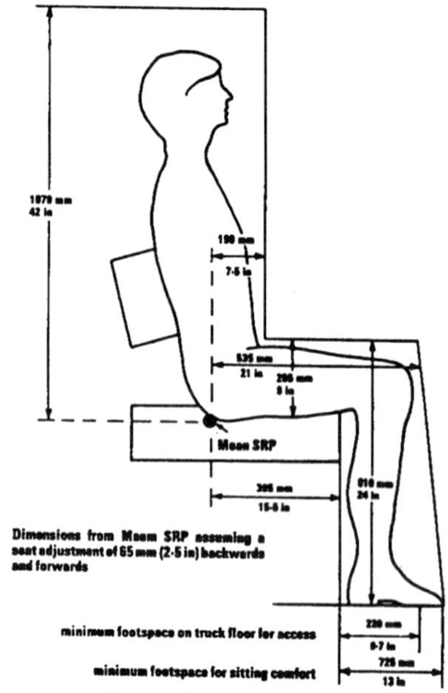

Fig 87: Optimum seat dimensions

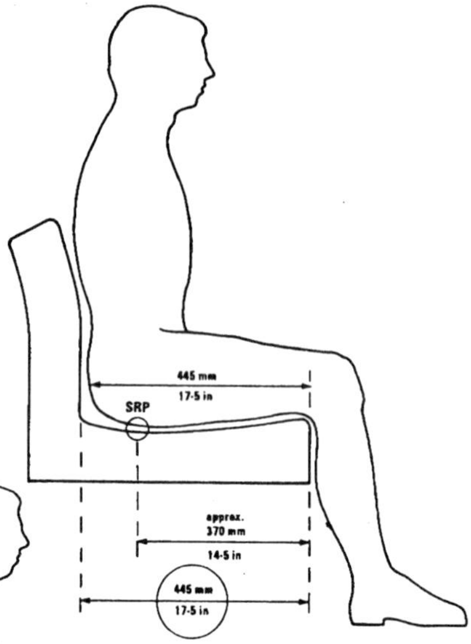

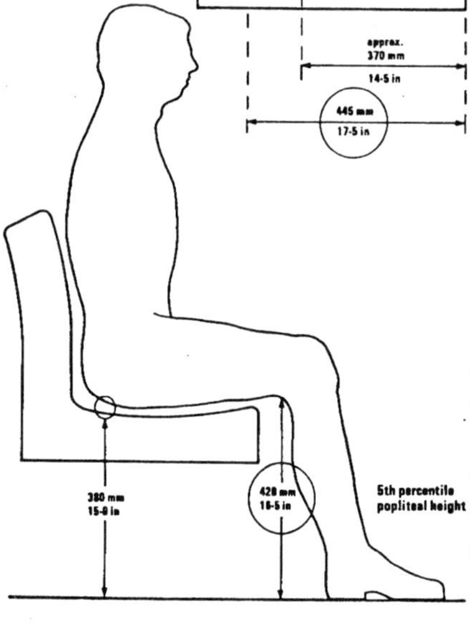

Fig 88: Seat height limits:max. squab lengt[h] (above), min[imum] seat height (below)

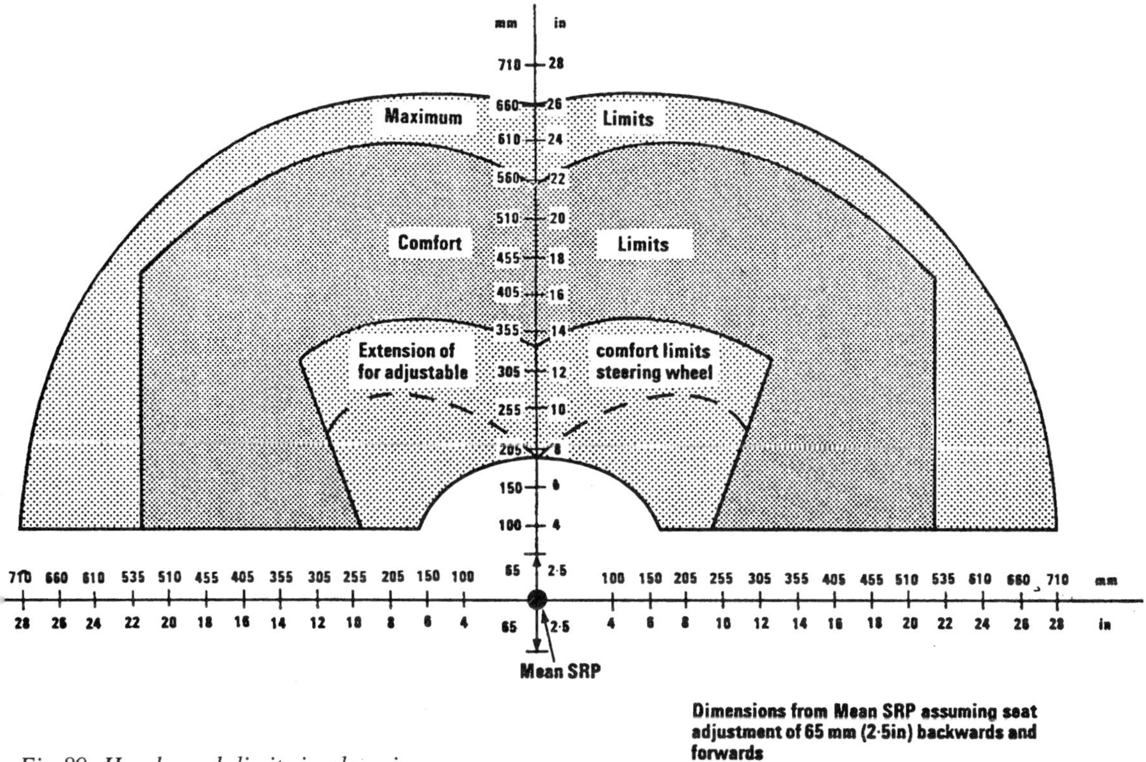

Fig 89: Hand-reach limits in plan view

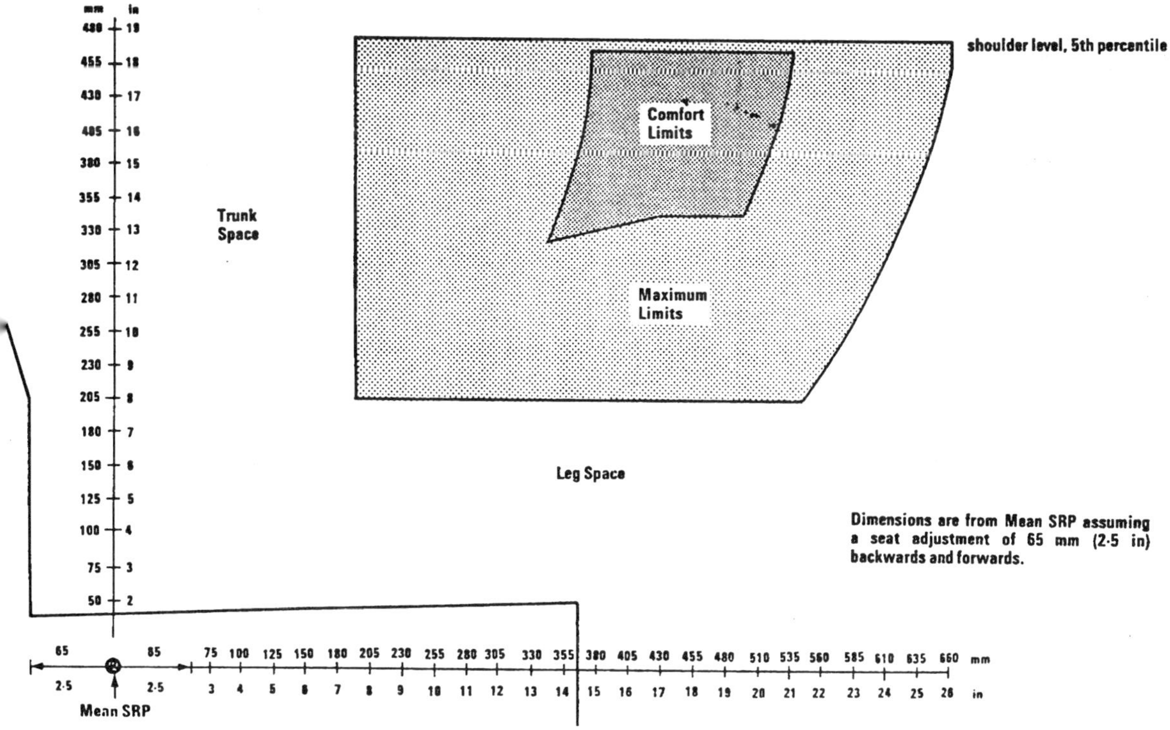

Fig 90: Hand-reach limits in side view for non-adjustable steering wheel

required which can cause problems if an unusually large diameter wheel has been used to bring it within the normal reach envelope for both hands. A well sited and adjustable-position wheel of smaller diameter, coupled with power-assisted steering, is often the answer. Pedal location limits, defined by foot reach envelopes, are unique to the upright driving position required of fork-trucks.

Stability factors of fork-lifts

The other principal constraint on fork-truck layout is that due to static stability of a vehicle whose inherent design involves carriage of its payload outside the wheelbase. This problem has been examined by the US Towmotor Corporation for the worst case of a truck decelerating down a slope, Fig 91. By equating resolved forces parallel to the slope and arranging terms so that a stability influencing factor can be defined as the reciprocal of moments due static level + those due to slope + vehicle acceleration + load acceleration. As this is a cumbersome equation to manipulate and if no slope criteria are specified, then level ground stability is conveniently determined as a first step, as shown in Fig 92. Here the stability factor is $L_f W / L_q Q$.

The constant relationship between the load Q and moment arm L_q is shown by the hyperbolic curves in the figure which apply to an example of truck empty weight of 12,200 lb (5.5 tonnes) with cg location at L_f = 46 and H = 30 in (1.17 and 0.76 m). Nominal load capacity is 11,000 lb (5 tonnes) at L_q = 40 in (1.03 m) and the curves represent a stability margin range of zero, 25 % and 50 %.

For the nominal capacity of this truck example there is a stability margin of 27%. The effect of gradient θ can be found by substituting in:

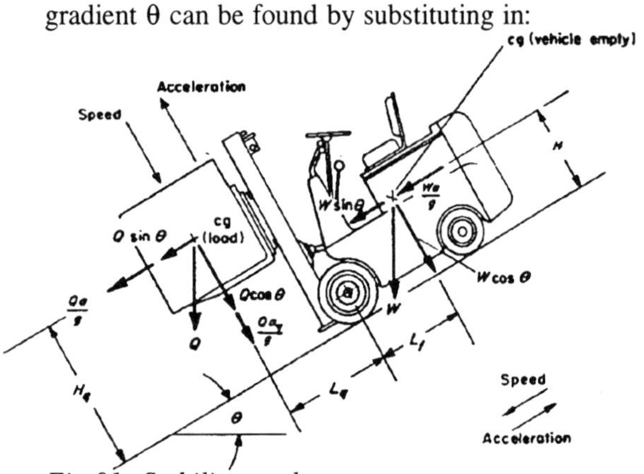

Fig 91: Stability on slope

$$\tan\theta_{max} = (L_f W - L_q Q)/(H_q Q + HW)$$

for the example truck in which cg height of the raised and lowered loads is taken as 100 and 40 in (2.54 and 1.02 m), for which maximum grades work out as 8.2 and 15%. The same expression, transposed, can be used to calculate the maximum translatory deceleration as 0.082 g and 0.15g in the respective cases.

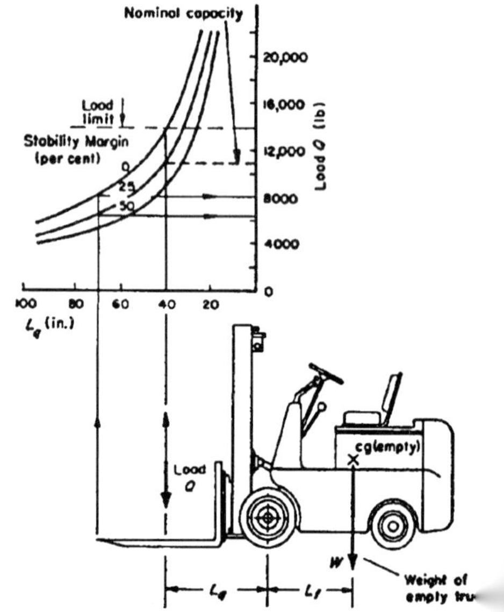

Fig 92: Level-ground stability

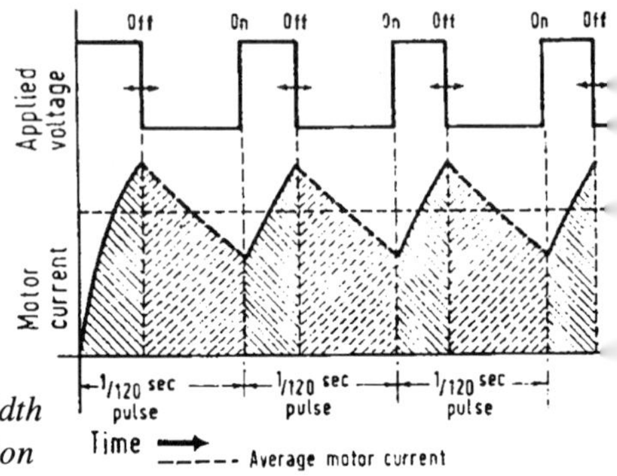

Fig 93: Pulse-width modulation

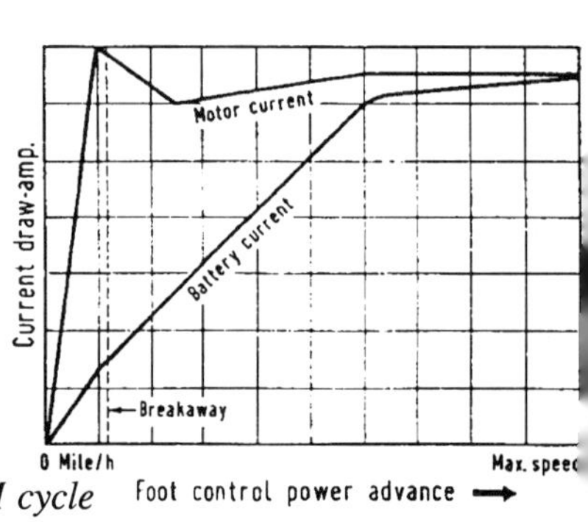

Fig 94: PWM cycle

Electric drive of handling trucks

The wide use of electric drive for inside factory and warehouse operations has made the fork truck one of the major EV categories. DC motors are the rule for these applications. Very high starting currents are required for moving off under load. Around 800 amps is common and silicon controlled rectifiers have been used for a number of years to regulate current feed to the traction motors. Sometimes a single electric motor powers the hydraulic systems of both steering and lift/tilt functions. By means of SCR pulse control, lift speed can be controlled electrically rather than hydraulically. Typically such a motor will have three speeds: say, 500 rpm for steering, 1000 rpm for tilt and slow lift then 3000 rpm for rapid lift.

The principal of pulse width modulation is shown in Fig 93. Typically a complete switching cycle will occupy 1/120 th of a second, applied voltage being either full battery output, say 36 V, and then zero. During the 'on' pulse, motor current rises on a curve governed by inductance or electrical inertia of the windings. At switch-off the current does not stop immediately, but continues flowing through a by-pass 'one-way' diode, dying away on another curve until the transistor switch closes for the next pulse, causing the resultant sawtooth waveform. Varying pulse length alters relative rise-and-fall times, longer pulses permitting higher peaks. Average value can be raised by lifting the whole saw-tooth up or down the graph by moving the trailing edge of the switching pulse sideways within each cycle. Resulting current characteristics are shown in Fig 94. A typical electronic circuit is described in Chapter 15.

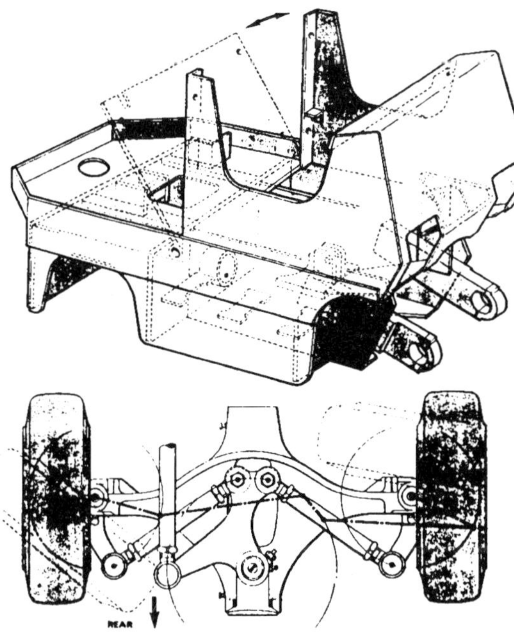

Fig 95: Yale KBE frame and rear-steering

Researchers at Imperial College produced a useful method for calculating the range of a multistop EV, applicable to categories including forktrucks, in terms of the drive motions only, not the work motions. To some extent the latter are easier to calculate separately by computing potential and kinetic energy changes in lifting tilting and swivelling the load for different duty cycles.

By first converting the energy density of the battery typically 25 watt-hours/kg for its weight at about a 5-hour rate of discharge, a figure of torque per kg battery weight can be computed (since 1 Wh/sec = 1 Nm), division of which by the tractive resistance would give a distance travelled D if the percentage of vehicle weight taken up by the battery is 100.

Fraction of battery energy used in moving itself S km against T N/kg tractive resistance is $f_s=ST/D$. If this energy were used entirely in giving up kinetic energy to the battery, the speed obtained would be given by v in $(v^2/2)=D$. Another value would be obtained by factoring down this energy by a percentage representing the starting energy losses. Fraction of battery energy f_u used in accelerating the truck to speed v is thus $(v/u)^2$. If then the battery occupies a fraction f_b of the vehicle weight the energy fraction above becomes $(v/u)^2/f_b$. And if the load on the battery be such that it is a fraction B of that due to a five hour rate the expression becomes $(v/u)^2/f_bB$ (B in fact is unity for those vehicles with intermittent duties which take several hours to discharge the battery). When motor and transmission efficiencies are also accounted for, the two expressions become:

$$f_s=ST/Df_bBE \quad \text{and} \quad f_u=(v/u)^2/f_bBE$$

The duty of the vehicle will be specified as P starts per km to a speed of v km/h and for such a duty the total fraction f of battery capacity required per km will be $f=(Pf_u+f_s)$ where f_s corresponds to 1 km. Total range is thus:

$$(1/f)=f_bBE/[P(v/u)^2 + T/D]$$

Development history of fork trucks

The past three decades have seen the development of fork trucks into their present state of the art. The 1960s saw the integral monocoque-type frame, exemplified by the Yale KBE , all-welded from plate. As a 4-wheeler this had a rear wheel steering system representative of then current practice, shown as Fig 95. A 1970s introduction was the Coventry Climax R-15 reach truck with full free lift mast operation permitting unusually good forward vision. The model, Fig 96, was noted for its ergonomic attention to layout of the work-motion levers — involving a staggered fan formation. In diesel engine designs, the Still R-70 is representative of current practice, Fig 97. Another interesting variation is the articulated chassis truck, exemplified by the Translift Bendi, Fig 98.

Fig 97: Still R 70

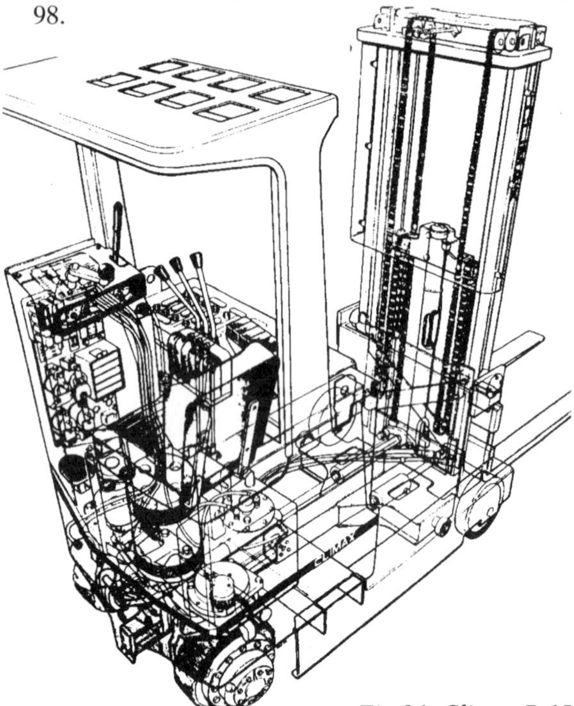

Fig 96: Climax R 15

Fig 98: Translift Bendi

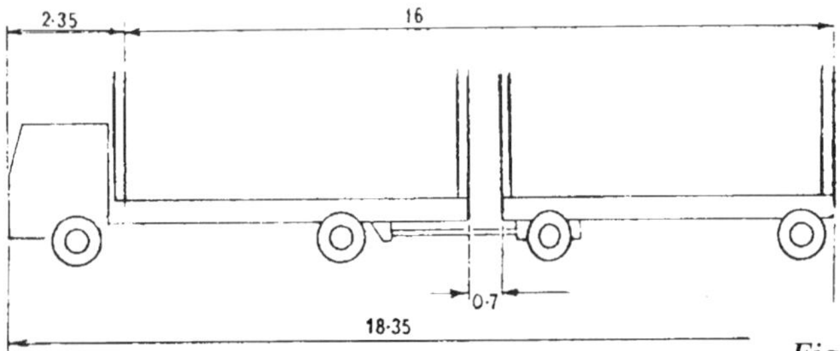

Fig 99: EC proposed dimensions for drawbar units

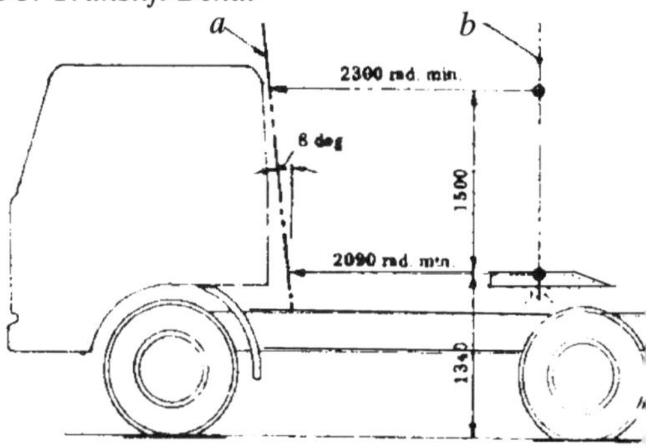

Fig 100: Clearance diagram for artics: a is conical clearance plane and b is fifth wheel centre

Trailer technology

To exploit the use of goods-carrying drawbar and articulated truck/trailer combinations, a careful dimensional analysis has to be carried out. This affects the manoeuvrabilty of the rig as well as the available cargo volume which can be carried. Later chapters will also examine how the stability of trailers affects their design and also the requirements of their interdependent braking systems.

The design of trailers and semi-trailers is much affected by legal dimensional limits and these will be subject to frequent alteration if the trend of the past few decades continues. Because such vehicles inevitably form part of maximum weight articulated- and drawbar- combinations they are for ever in the legislator's eye. The European Commission have made dimensional proposals for drawbar combinations as shown in Fig 99. There is also a proposal that the combinations are capable of steering within a curved corridor of inner and outer radii of 5.3 and 12.5 metres respectively. A further proposal is to raise overall length to 19 metres when using sleeper cabs which allows two bodies, truck and trailer, of just over 8 metres each. Some truck makers such as Mercedes Benz have produced ultra-short cab variants (of 1.47 metres) which allow two 8 metre bodies within 18 metres. Mercedes is also one of the manufacture's to introduce six-wheeled articulated tractive units and it is therefore instructive to note their coupling dimensional recommendations. A kingpin position 915 mm ahead of the drive axle is recommended, 115 mm beyond the IRTE advisory limit for avoiding fouls. Such dimensions are crucial for articulated vehicles and the disparity between traditional UK and continental legal dimension requirements means that many existing UK vehicles will not be able to benefit from EC proposals. An increase in length limit for such combinations has been proposed to 16.5 metres, but the long nosed trailers being introduced to exploit this concession are finding that available UK tractive units are too short. Specially lengthening conversions are therefore in demand. Fig 100 shows dimensions that control coupling position in relation to cab when the trailer has a front corner swing of 2.04 metres. The tractor rear axle position will, of course, be governed by the moment arm about the trailer axle required to react the forward imposed load of the trailer on the tractor. Some cabs are designed with rear wall tapered for accommodating vertical articulation as the combination traverses a hole in the road surface.

Three axled tractors are, of course, less critical of coupling position with respect to any possible overload of the tractor rear axle by the trailer imposed load. Generally an outer axle spread of 3.8 metres is involved and it is only necessary to ensure the coupling is at least 2.3 metres from the back of the cab to allow for the trailer swing 1.5 metres above the level of the coupling turntable. Existing UK regulations state that up to 5.09 tonnes vertical load can be applied to an axle only 1.05 metres from a 10.17 tonne driving axle. Another approach for a tractor chassis convertor is to use 1.2 metre spacing and to use added axle capacity of 4.76 tonnes to achieve

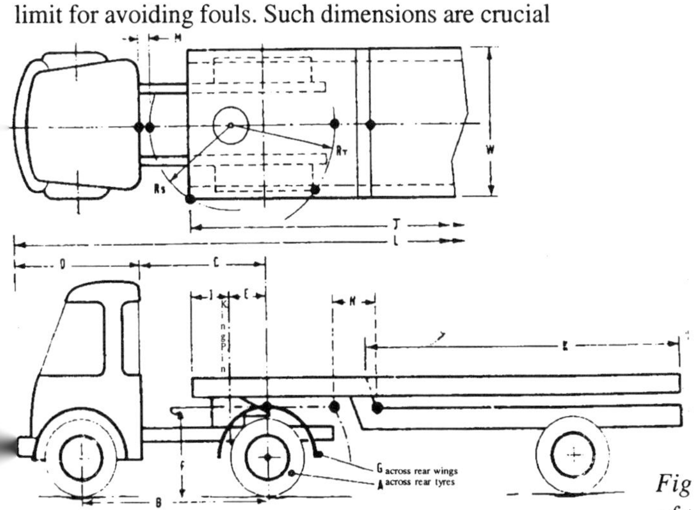

Fig 101: Dimensions for calculating path

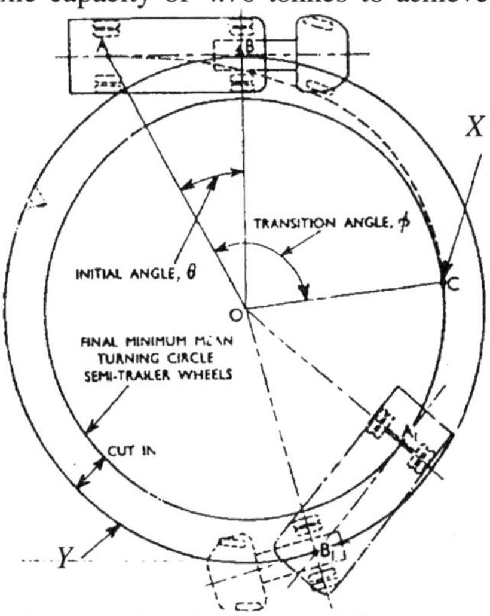

Fig 102: Articulated vehicle path: X is start of turn; Y is tractor wheel turning circle.

drive axle capacity of 10.5 tonnes.

While the above applies to particular legislative dimensional limitations and proposals, a more generalised approach to calculating layouts and clearances uses the diagram of Fig 101. By choosing a tractor trailer clearance dimension of 6 and 9 inches, for two and three axled tractors respectively, then, subject to appropriate choice of dimensions, E, I and C, the sum of the cab length, D, and the clearance dimension governs the maximum feasible length of a straight-frame trailer. Stepped-frame trailers need also to consider the clearance of the frame crank with the tractor rear axle swing. Prior to the intoduction of 'long-nose' trailers, ease of coupling with the standard SAE king-pin and turntable demanded maximum value of 42 in (1.07 m) which fixed the value of R in the relation

$[I^2 + (W/2)^2]^{1/2}$ making the trailer length:

$L - \{ [I^2 + (W/2)^2] + M + D\} + I$ given that:

$C - E$ is not less than $M + [I^2 + (W/2)^2]^{1/2}$

In the case of the cranked-frame trailer a length can also be defined rearward of the tractor rear-axle swing governed by $D + C$ and width G or A according to wheel position E. Assuming a trailer rear deck at equivalent weight to the tractor rear mudwings, gives an available deck length:

$$K = L - (D + C - E) - [E^2 + (G/2)^2)]^{1/2} - N$$

where $R_t = [(E^2 + (G/2)^2]^{1/2}$.

The course followed by an articulated vehicle can also be calulated using the diagram in Fig 102 due to Dyson Trailers. The relationship between the transition angle and the initial angle, defined by the vehicle dimensions and its minimum turning radius, has first to be calculated. A similar course diagram for drawbar trailers in Fig 103. In each case the cut-in may be obtained from:

$$(L/\tan\theta) - [(L/\tan\theta)^2 - L^2]^{1/2}$$

where $q = \tan^{-1}(L/R)$ where R is tractor turning radius and L is the distance between king-pin and trailer axle. Appropriate choice of dimensions can in theory be made to reduce cut-in to the point where the trailer virtually 'tracks' the tractor. Avoidance of tyre scrub

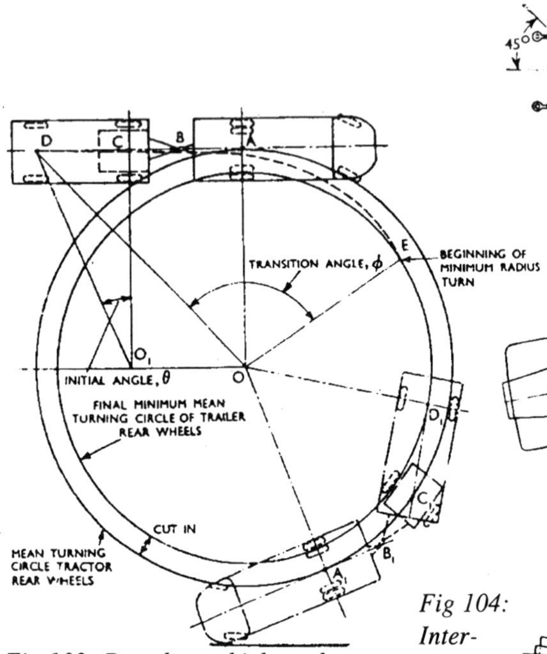

Fig 103: Drawbar vehicle path

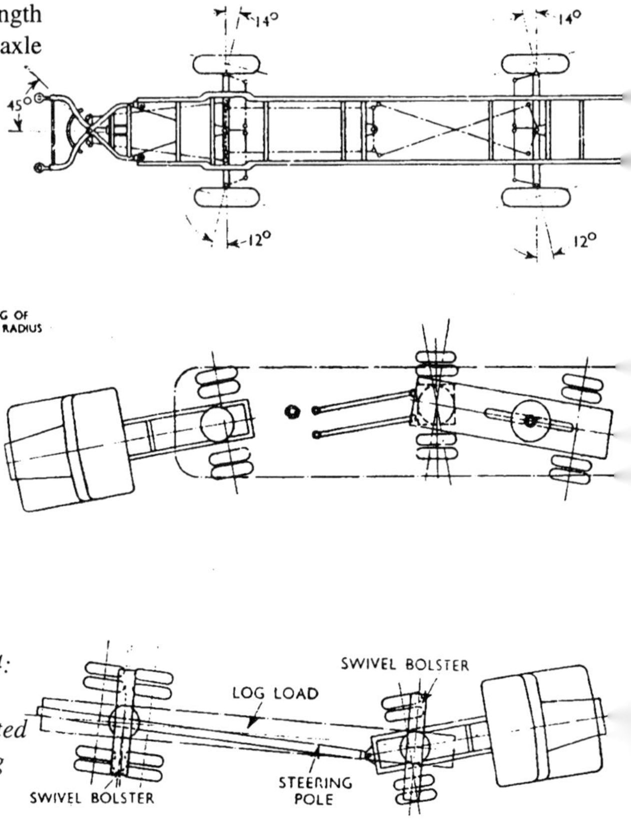

Fig 104: Inter-connected steering axles

is rated by many as important as reducing cut-in and this has led to trailers with interconnected steering axles as a means of reducing cut-in, Fig 104. The logging trailer is a particularly interesting example in demonstrating how the trailer changes from 'drawbar' to 'artic' as the log-load is applied. Another alternative configuration, as far as the drawbar trailer is concerned, is the use of either mechanically telescopic or powered sliding drawbars. EC steering corridor proposals, interpreted by DAF, are shown for the latter in Fig 105 in comparison with those for a conventional A-frame.

Extending-couplings help to allow drawbar combinations run at close coupling distances without violating the UK limit requiring maximum overall length in the straight ahead position of 18 metres. The extending coupling lengthens automatically when the combination starts to make a turn so as to reduce the standard coupling gap of 1.3-1.4 metres down to about 0.5 metres with a turntable trailer or 150 mm with a centre-axle unit. Ray Smith's Decker Supavol is shown in Figs 106. It is air-operated and as well as operating on turns it also responds to the bodies closing in at the top on vertical articulation. The body is moved along the chassis on needle-roller bearings, the hydraulic cylinder acting as a damper and slowing down the body's return on sudden straightening.

For articulated combinatioons, IRTE has made the coupling code of Fig 107. The king-pin dimensions for articulated combination coupling gears are now standardised to ISO 337-1981 as shown in Fig 108 with dimensional characteristics for mounting and interchangeability. In the UK, the standard is available as British Standard BSAU 1c:1989. This mates with the fifth-wheel coupling turntable typified by the arrangement shown in Fig 109. On full trailers, by comparison, a turntable supports the front steering axles, Fig 110, except in cases where

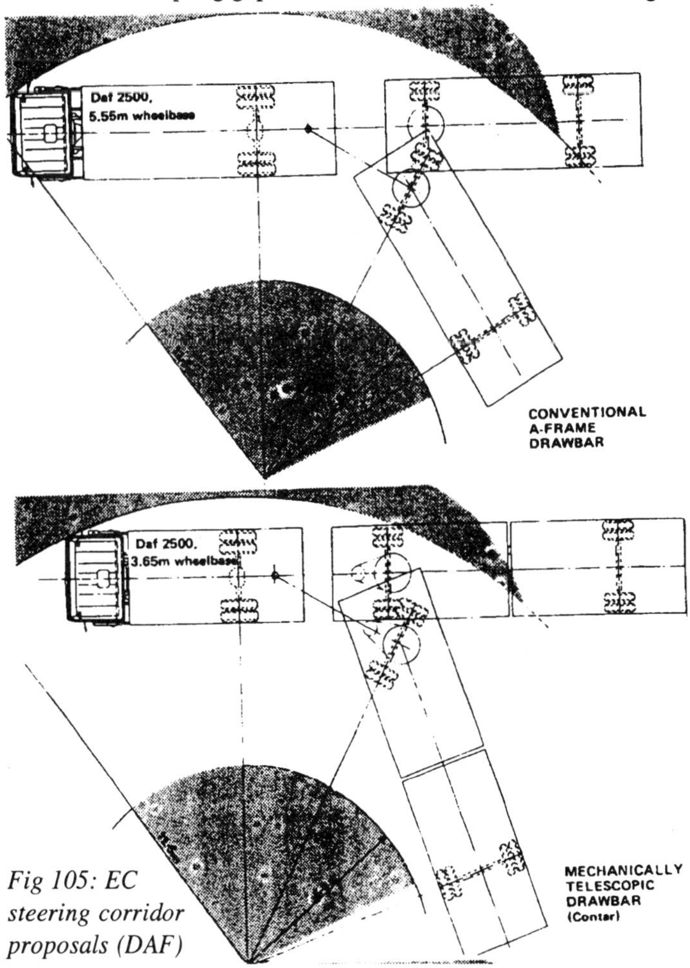

Fig 105: EC steering corridor proposals (DAF)

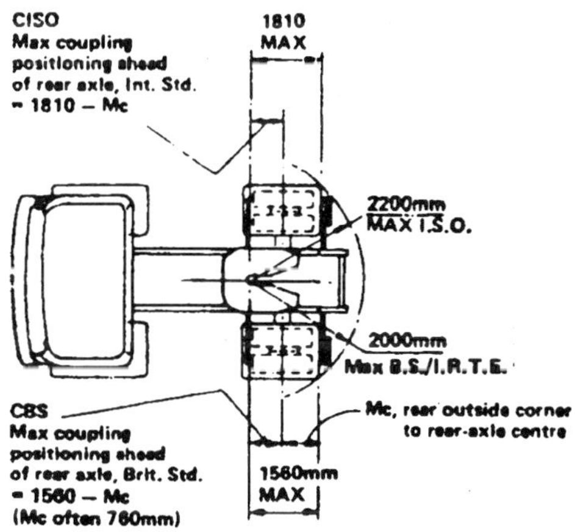

Fig 107: IRTE coupling code dimensions

Ackermann steering is provided for the front wheels. Usually the turntable is located by locking roller-rings rather than king-pins. As well as considering the turning corridors for the wheels of trailers and semi-trailers, it is also important to understand the sweep of the tractor and trailer body on a curve so that unduly high outswing can be avoided. This can be be predicted by graphical analysis. Fig 111 is an example produced by the former Scammell company using three steerable axles on the trailer allowing it to prescribe a path exactly related to that of the tractor coupling. Relative market preference for artics and drawbars is still in the hands of the legislators. By 1999 it is proposed that four-axled combinations will have a further lift to 36 tonnes with five-axled ones rising to 40 tonnes. By that date, however, 5-axled artics will be allowed to match the same GTW — and to 44 tonnes with ISO containers.

A further requirement of artic vehicle couplings is to provide for vertical articulation of the combination in negotiating road-humps and dips, usually by providing trunnion-mounting of the turntable. Eight degrees angular tilting either side of the level provision is the agreed standard. Also the fifth wheel has to be mounted high enough on the tractor to prevent the trailer contacting the tractor rear mudwings at full inclination. This clearance problem also affects how far forward the coupling can be mounted on the tractor chassis. Overall maximum legal dimensions affect this calculation and currently the EC ruling is a maximum width of 2.5 metres, height of 4.0 metres, combination length of 15.5 metres and GTW of 38 tonnes. The IRTE code of practice overcomes the limitations of ISO 1726 in allowing for the particular constraints of the UK over international legislation. This fixes a target height of 1.24 metres for coupling turntables. Fig 112 gives IRTE recommendations on maximum forward positioning of the coupling on the tractor chassis.

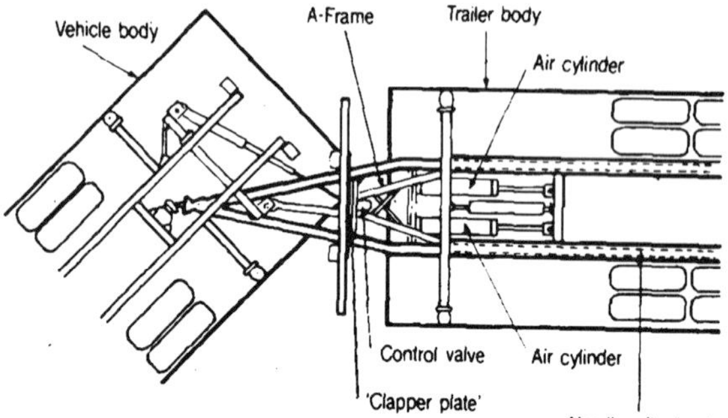

Figs 106: Ray Smith Decker-Supavol

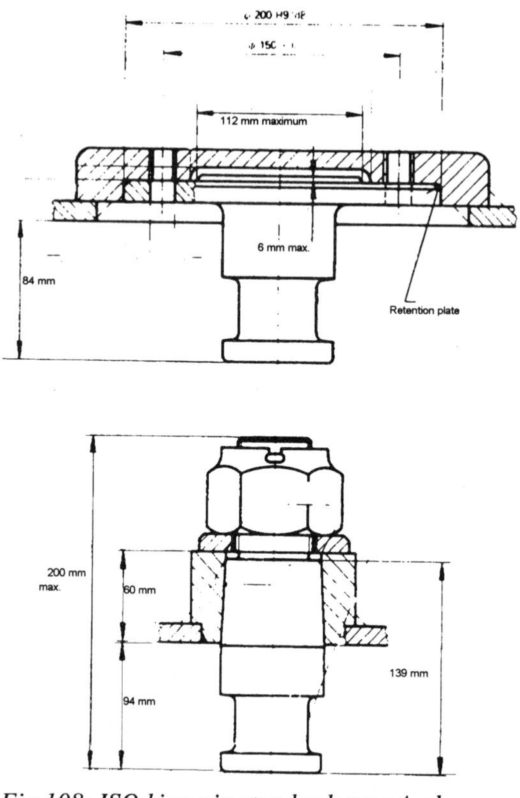

Fig 108: ISO king-pin standard: type A above , type B below

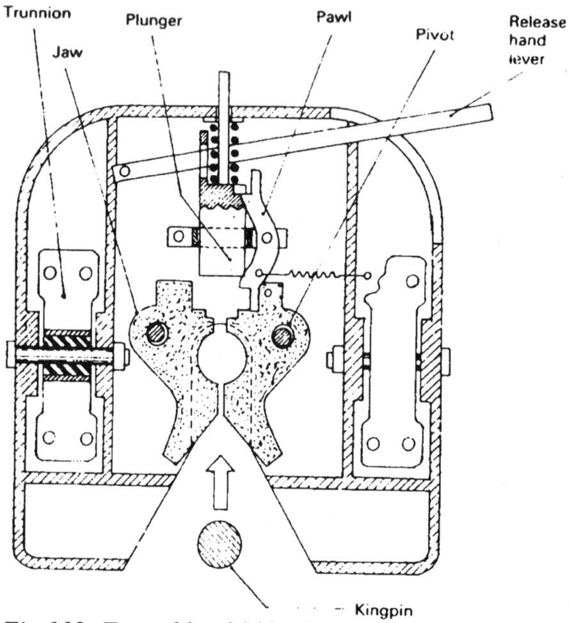

Fig 109: Turntable of fifth wheel coupling

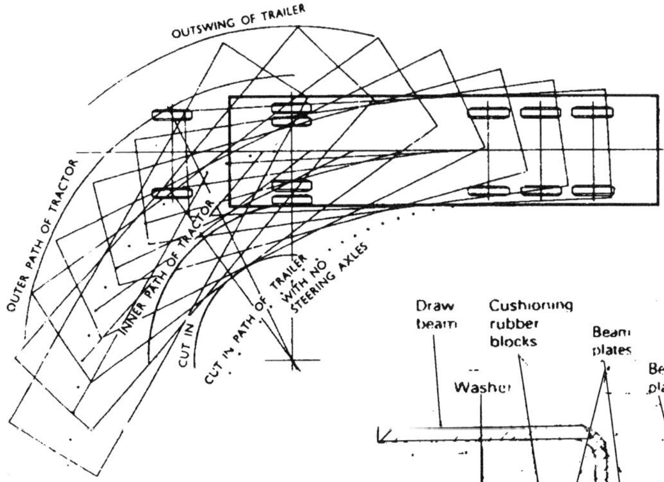

Fig 111: Determining trailer sweep

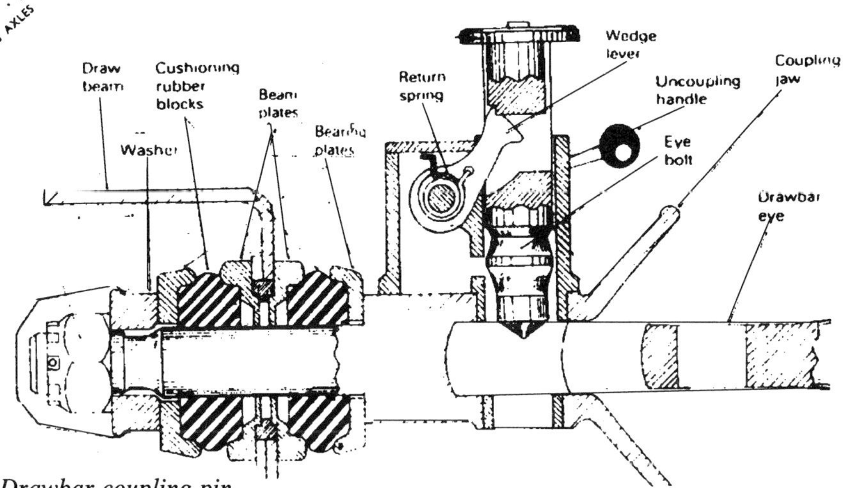

Fig 110: Drawbar coupling pin

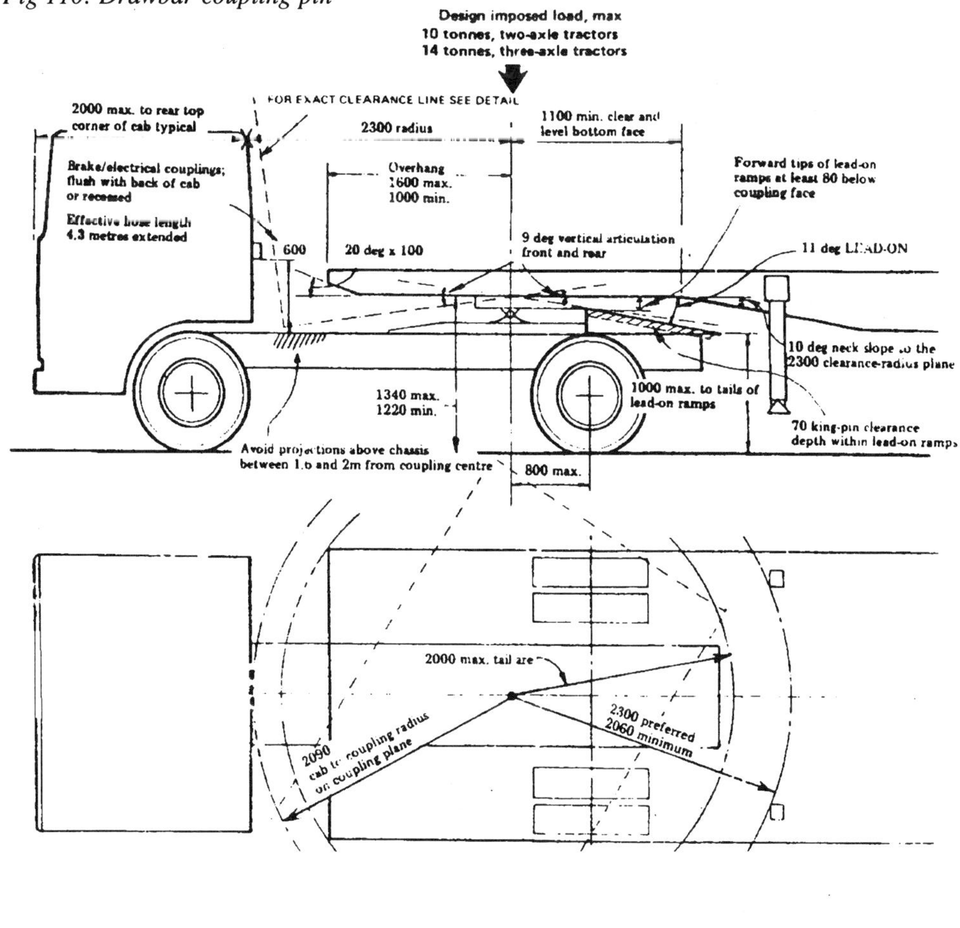

Fig 112: Coupling position recommendations

Chapter 15: Electric vehicles

Electric-drive
Electric storage
Electric car technology
Hybrid-drive vehicles
Electric trucks
Solar vehicles
Production electric vehicles
Hybrid electric buses
Practical hybrid systems

Battery-electric vehicles

For passenger cars, electric vehicle technology seems always to be waiting for that 'breakthrough around the corner' in battery development; however, for industrial vehicles, many operating environments rule that EV technology is the only solution. This feature examines the possibilities of exploiting the advanced systems considered for prototype electric cars — to see whether they have potential for the appropriate commercial vehicle.

Among the operating environments which have attracted commercial/industrial EVs are early-morning urban milk delivery, in-factory materials handling and airport cargo/baggage handling — and aircraft ground-towing. Future possibilities lie in noise and exhaust-fume sensitive city goods deliveries and public passenger-transport — also the possibility of individual passenger transport in future new-town residential and/or industrial areas incorporating suitable purpose-built carriageways. Draconian measures already introduced to remove individual transport from congested city centres could be rescinded in favour of sympathetic route planning and vehicle development which would enhance prospects for special purpose electric cars.

In the USA early calculations showed that most vans and LCVs travel less than 60 miles/day and do not require high top speeds or acceleration levels. Some years ago 31 UK-built GM Griffon vans (based on Lucas Chloride Bedford ECVs described below), after two years of operation, covered over 480000 miles at an operating cost of just $ 0.07/mile over that required for an equivalent petrol van (including first-cost depreciation).

Electric cars

Another early key US initiative was the Ford EV project EXT11 which has been exploring the use of an alternating current drive motor in the Aerostar Minivan, Fig 1. A sodium sulphur battery was employed and a single-shaft propulsion system. A bat-

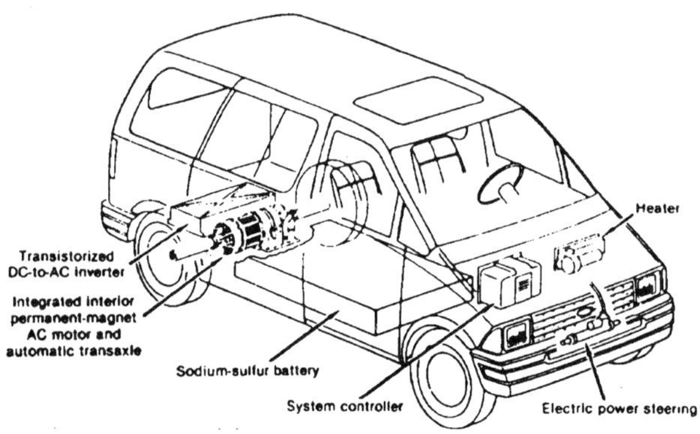

Fig 1: AC-driven Ford Aerostar conversion

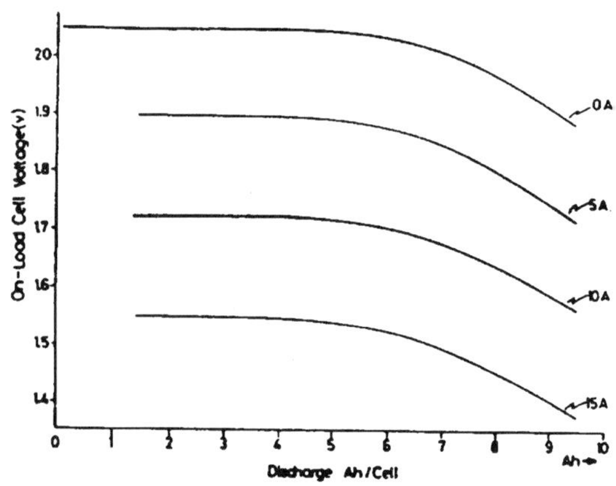

Fig 2: Discharge performance of s-s cells

tery with the following specification was involved: nominal voltage 200V; minimum voltage at 60 kW 135V; 50 kW capacity on FUDS cycling; 60 kW maximum power (20 seconds) and 35 kW continuous power (40 minutes). Much of the technology has since been carried over to the Ford Ecostar, described later.

Dimensions of the battery were 1520 x 1065 x 460 mm and it was based on the use of small (10 Ah)

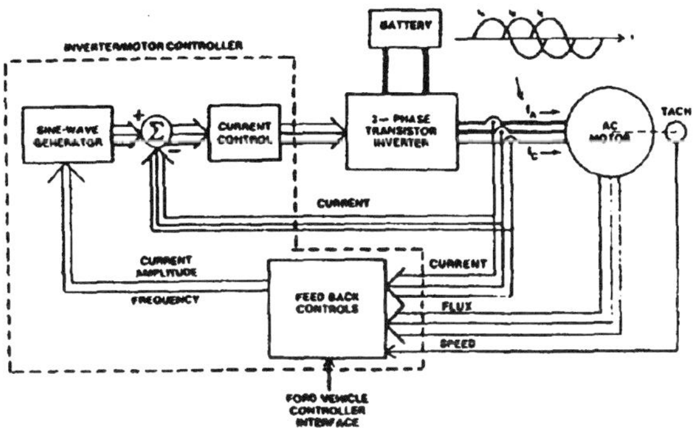

Fig 3: AC drive

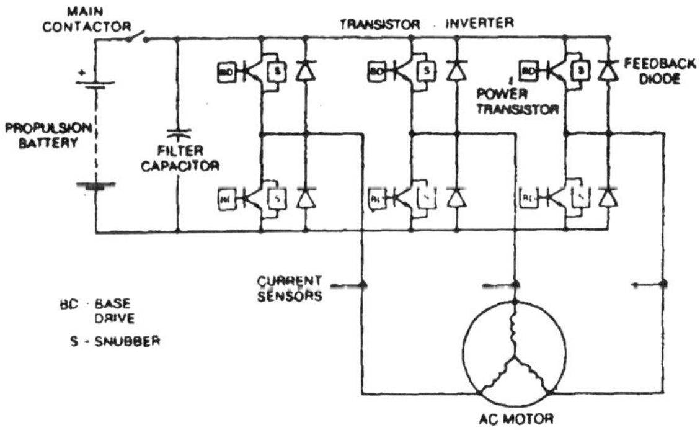

Fig 4: Transistorised inverter

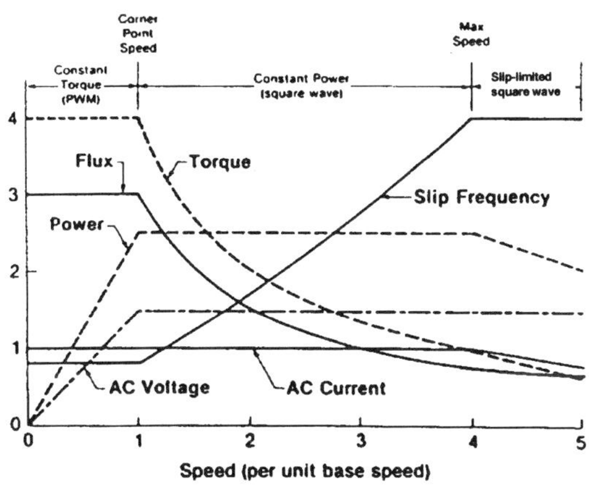

Fig 5: AC control strategy

cells connected in 4-cell series strings with parallel 8V banks arranged to provide the required capacity. A voltage of 200 required 96 cells in series (each had a voltage of 2.076) and a parallel arrangement of 30 cells gave the 300 Ah capacity; overall 2880 cells are used and their discharge performance is shown in Fig 2. Battery internal resistance was 30 m.ohms with an appropriate thermal management system under development. The cells rested on a heater plate which incorporates a 710 W element used particularly in the initial warm-up. An associated air cooling system can dissipate 6 kW. Temperature must be maintained above 300 C to achieve an internal resistance value that will allow sensible current flow.

The US General Electric Company was also a partner in the programme, specifically on the AC drive system, Fig 3. This comprised a 50 bhp induction motor within a two-speed transaxle, a liquid-cooled transistor inverter being used to convert the 200 battery volts into variable-voltage 3-phase AC, Fig 4. The induction motor was a 2-pole design with just under 13 cm stack length — within a stator diameter just under 23 cm. Calculated stall torque was104 Nm and the transition between constant torque and constant power operation occurred at 3800 rpm. Design speed was 9000 rpm and absolute maximum, 12000 rpm. Full power was developed at a line current of 244.5 A. Control strategy relating inverter to motor is shown in Fig 5.

Ford produced the transaxle on an in house basis having coaxial motor and drive, Fig 6. The motor rotor is hollow to allow one of the axle shafts to pass through it from the bevel-gear differential. The installation in a test saloon car achieved 30% gradientability, 0-50 mph acceleration in less than 20 seconds, 60 mph top speed and 0.25 kWh/mile energy consumption. Gear ratios were 15.52 and 10.15:1 and rearward speed obtained by electrically reversing the motor and in 'neutral' the motor was electrically disconnected.

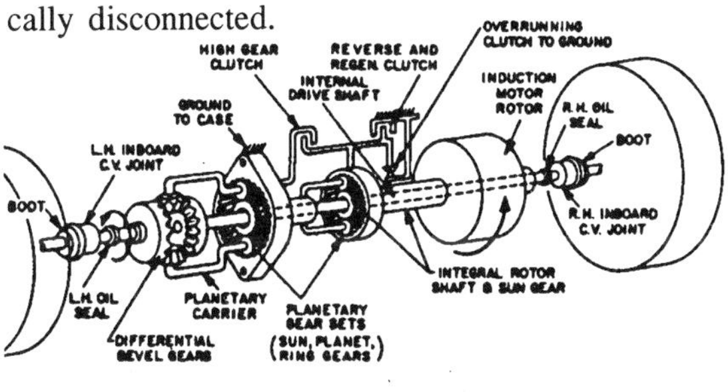

Fig 6: Transaxle drive with co-axial motor

Electric vans

The Lucas Chloride converted Bedford CF van, marketed as the Griffon in the USA, Fig 7, has two-pedal control and a simple selector for forward/reverse. Most of the vehicle's braking is regenerative and batteries are of the tubular cell lead-acid type. Thirty-six monobloc units of 6 volts are used — connected in series to give a 216 V, 188 Ah pack. The rear-mounted traction motor drives the wheels through a primary reduction unit coupled to a conventional rear axle, via a propshaft. Measured performance of the monobloc is shown in Figs 8/9. An energy density of 34 Wh/kg is involved, at the 5 hour rate, and a 4 year service life is claimed.

The motor used is a separately-excited type in order to allow the electronics maximum flexibility in determining the power curve. It weighs 15 kg and has a controlled output of 40 kW; working speed is 6100 rpm corresponding to a vehicle speed of about 60 mph. The motor control system uses an electronic by-pass to leave the main thyristor uncommutated during field control. The latter uses power transistors which handle up to 25 A.

Electric buses

Fiat was an early exponent of the hybrid diesel/electric bus. In 'parallel' configuration, the latter involves part of the traction energy being converted into electrical energy and then again into mechanical energy, part passing to the wheels via a mechanical transmission. In 'series' configuration all the diesel energy is converted into electrical and then into mechanical energy. The traction subsystem consists of a DC compound motor with separate field control and an electronic controller. Motor and characteristics are seen in Figs 10/11; the unit features interpoles and compensating windings in order to increase efficiency in both the driving and generating (as well as braking) modes. Power conditioning of the motor is operated by varying the armature current by means of a thyristor chopper — from zero up to the base speed — thereafter by varying the field current with an independent chopper. The motor is direct-connected to the rear axle without intermediate gearing.

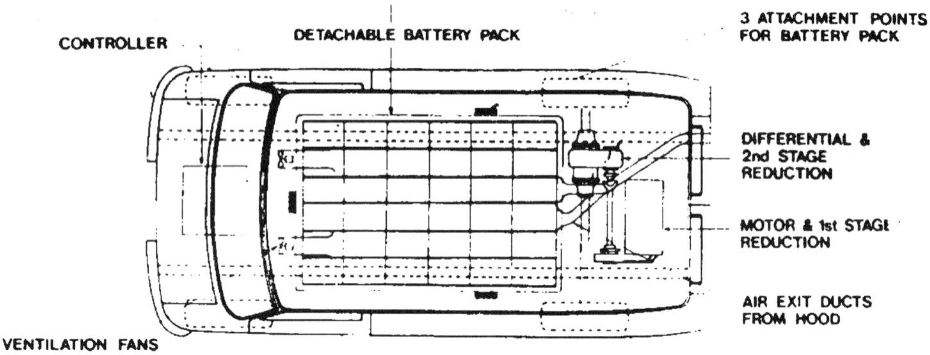

Fig 7: GM Griffon

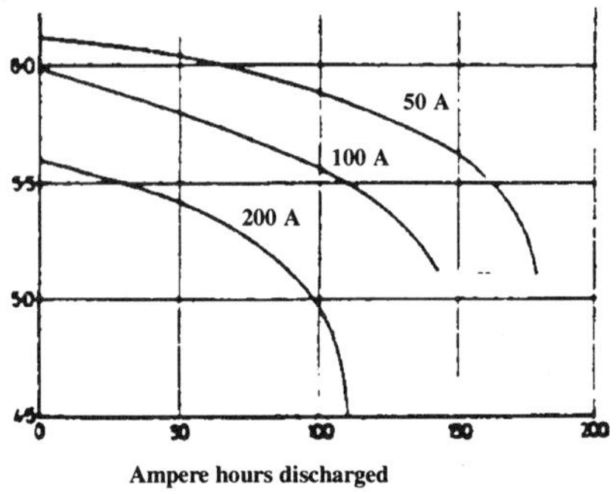

Fig 8: Terminal volts per 6V module

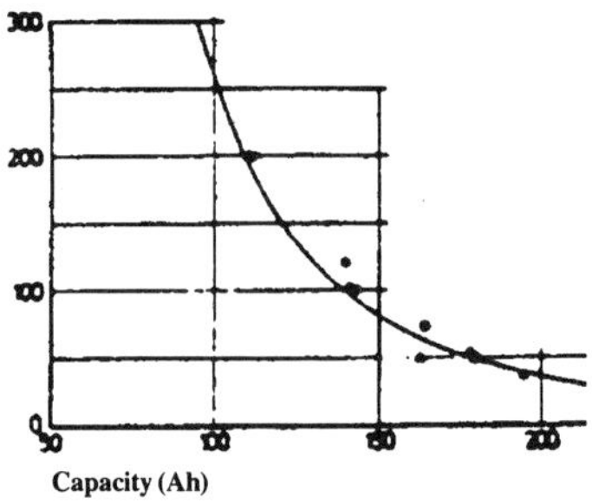

Fig 9: Discharge current in amps

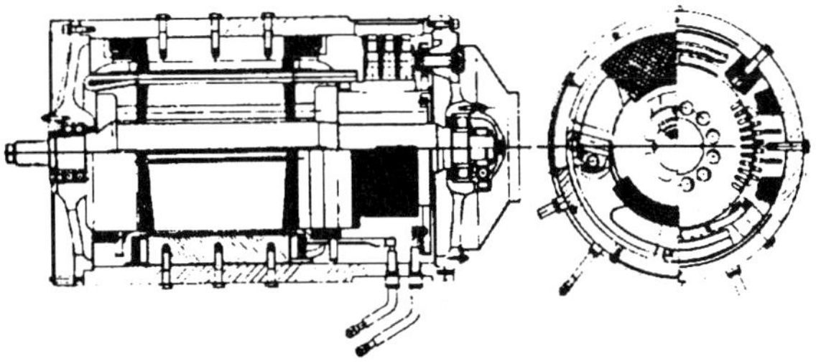

NOMINAL VOLTAGE 600 V
CONTINUOUS POWER 90 KW
MAXIMUM POWER (3') 180 KW
FIELD CONTROL RATIO AT CONTINUOS POWER: 1150 ÷ 3200 r.p.m.
EFFICIENCY AT CONTINUOS POWER IN FIELD CONTROL: 92 ÷ 90 %

Fig 10/11: Fiat hybrid bus motor and characteristics

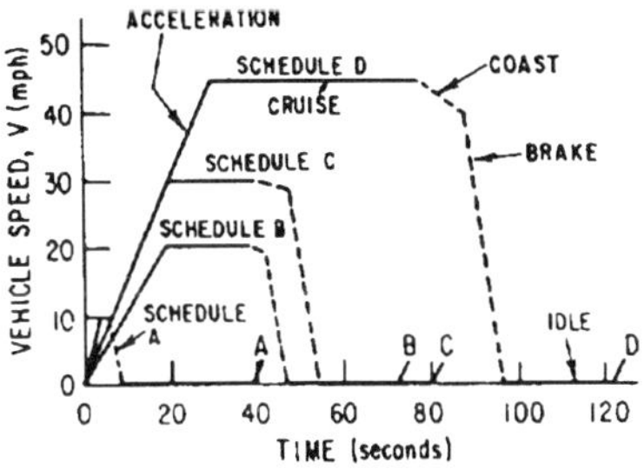

Fig 12: SAE EV drive cycle

Parameter	SAE J227a Schedule A	B	C	D
Maximum speed, mph	10 ± 1	20 ± 1	30 ± 1	45 ± 1
Acceleration time	4 ± 1	19 ± 1	18 ± 2	28 ± 2
Cruise time	0	19 ± 1	20 ± 1	50 ± 2
Coast time	2 ± 1	4 ± 1	8 ± 1	10 ± 1
Brake time	3 ± 1	5 ± 1	9 ± 1	9 ± 1
Idle time	30 ± 2	25 ± 2	25 ± 2	25 ± 2
Total time, s	39 ± 2	72 ± 2	80 ± 2	122 ± 2

Fig 13: Driving schedule

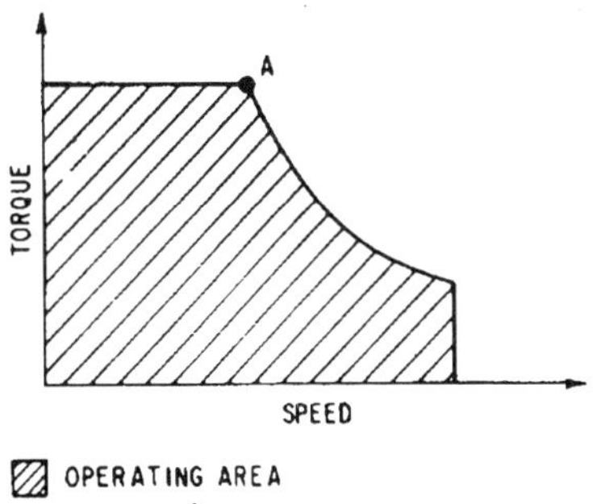

Fig 14: Optimum operation

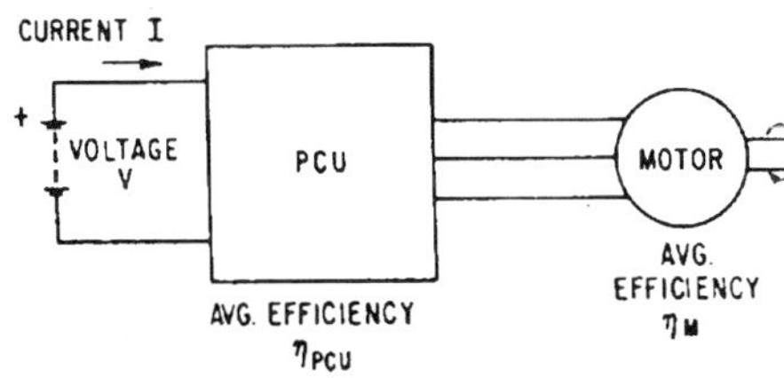

Fig 15: Component efficiencies

The diesel generating set has a power rating of 56 kW and maximum power of 78 kW. Generator has a nominal voltage of 600 V and speed range of 1800-2700 rpm. The storage battery comprises fifty 12V cells and has nominal capacity of 135Ah at 5 hours discharge rate. Battery weight at 1930 kg corresponds to 12% of the GVW. In control, the traction subsystem logic determines the electric power flow between generator/storage subsystem and traction motor — acting through the inputs of the throttle and brake pedals: affecting the armature and field power conditioners. In the traction mode, the power conditioning is operated on the basis of an armature current feedback, by imposing a given current demand through the throttle pedal. Its release provides the first level of regenerative braking, progressively increased by the brake pedal — but always on the basis of an armature current feedback. The generating subsystem logic determines the splitting of the electric power between diesel generator and battery, automatically. This is according to a maximum efficiency criterion — inputs being electrical power required for traction and state of battery charge. When the power required exceeds the maximum deliverable, the fuel pump rack is moved to a maximum setting and extra battery power supplied automatically.

A design approach to the selection of EV propulsion systems has been suggested by the US General Electric Co. This is based on the SAE J227a EV driving schedule shown in Fig 12 and the table of Fig 13. Minimisation of motor and controller ratings can be achieved by adopting the characteristic shown in Fig 14. Point A defines the peak power requirements; knowing battery voltage and component efficiencies (Fig 15), it then allows peak battery current to be calculated. The table of Fig 16 shows motor parameters, for alternative types, in the case of a 20 bhp EV requirement.

	DC Series	DC Separately excited	AC Induction	AC Permanent magnet synchronous
Wt (1b)	255	217	99	~85
Dimensions	21" × 14" diam	17" × 12" diam	12" × 12" diam	7" × 18" diam
Max. rpm	5000	5000	15 000	15 000
Max. efficiency	0.85	0.90	0.92	0.95
Approx. cost ratio	0.9	1.0	0.5	0.5

Fig 16: Motor parameters

Electric storage systems

The comparison of different types of electric storage cell is engaging most EV designers who seek maximum energy density with minimum cost and weight as well as quick charging and long life characteristics. The big-three US auto makers recently announced a consortium approach to evaluating new battery technologies, so great is the concern to find the optimum solution. Some of the more serious contenders are examined here.

The lead acid battery is still very much a serious contender and is attractive for its comparatively low cost and an existing infrastructure for charging, servicing and recyclable disposal. A number of special high energy versions have been devised such as that shown in Fig 17 due to researchers at the University of Idaho. This battery module has three cells each having a stack of double-lugged plates separated by micro-porous glass mats. High specific power is obtained by using narrow plates with dual current collecting lugs and a 1:4 height to width aspect ratio. Grid resistance is thus reduced by shortening conductor lengths and specific energy improved by plates that are thinner than conventional ones. They have higher active mass utilisation at discharge rates appropriate to EV use. At an operating temperature of 110F specific energy was 35.4 W-hr/kg and specific power 200 W/kg. Over 600 discharge cycles were performed in tests without any serious deterioration in performance. Fig 18 lists the main parameters of the battery.

The US company Unique Mobility Inc have compared advanced lead-acid batteries with other proposed systems. In carrying out trials on an advanced EV-conversion of a Chrysler Minivan the company obtained the comparisons shown in Fig 19. The graphs also show the extent to which specific energy content of batteries is reduced as specific power output is increased. Trojan and Chloride 3ET205 are commercial wet acid batteries while the Sonnenshein DF80 and JCI 12V100 are gelled electrolyte maintenance-free units which involve an energy density penalty. The Eagle pitcher battery is a nickel iron one taking energy density up to 50 Wh/kg at the 3 hour rate. The Beta and Delta units are sodium sulphur batteries offering nominal energy density of 110 Wh/kg. The company listed the characteristics of the batteries as in Fig 20.

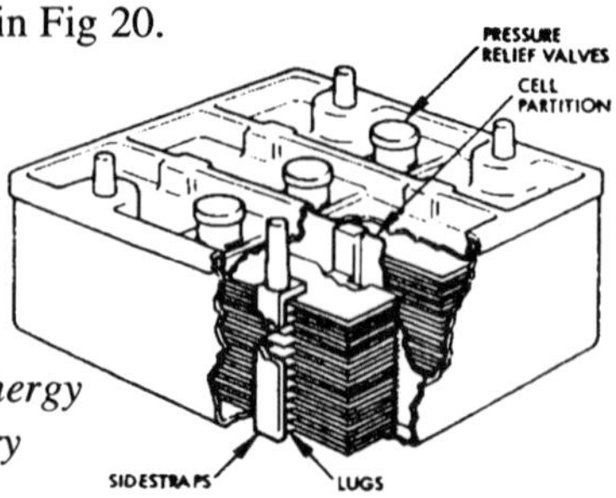

Fig 17: High-energy lead-acid battery

Fig 18: Parameters of H-E battery

35 Wh/kg		2hr rate
200 W/kg		20% state of charge
600 cycles at		80% DOD
Charge/discharge efficiency		85%
Battery volts		240 nominal
Weight		600 kg
Number of cells		40
Cell voltage		6 V
Cell weight		15 kg
Cell dimensions		10x7x6 inches

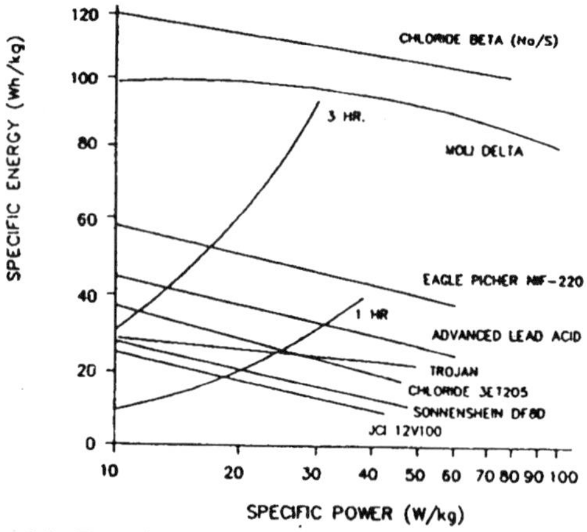

Fig 19: Energy-storage comparisons

Fig 20 Battery characteristics

Battery Type	Module Weight (lb)	Average Module Voltage (volts)	C/3 AMP Hour	Nominal Vehicle Capacity KWH^^	Nominal Energy Density Wh/Kg	Calculated EV Energy Density C-cycle Wh/Kg	D-cycle Wh/Kg	35 mph Wh/Kg	55mph Wh/Kg
GC 12V 100	68.0	11.1	72	16.0	25.9	22.1	18.0	24.9	17.0
DF8D	141.1	11.4	150	16.5	26.7	23.7	19.7	26.5	18.6
MET205	70.5	6.0	187	21.7	35.2	31.2	26.9	34.1	25.8
Ni-Fe	57.0	6.3	225	33.8	54.8	54.5	50.8	57.0	49.8
NaS	1102.0	220.0	250	67.9	110.0	115.6	111.6	118.4	110.6
Adv Pb acid	141.1	11.4	225	24.7	40.1	42.9	38.7	45.9	37.5

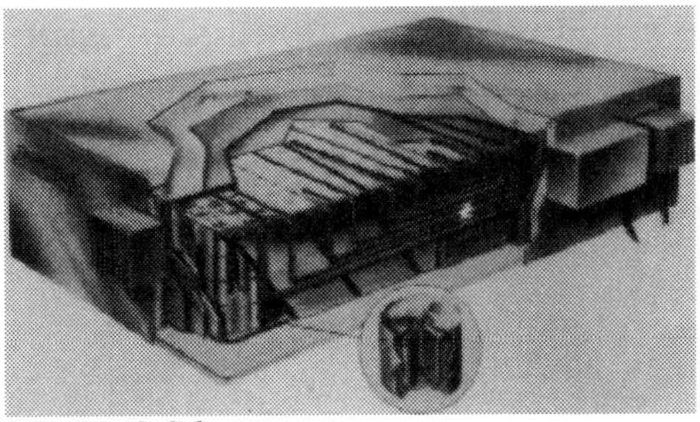

Fig 22: S-S battery

	Lead acid	Sodium-Sulphur		
Battery energy (kWh)	40	40	60	85
Range (miles) with 1t payload	52	70	105	150
Maximum payload (tonnes)	1.0	1.9	1.8	1.7
Battery weight (kg)	1250	330	424	580
Power available (kW-2h rate)	19	19	27	39

Fig 23: Projected performances

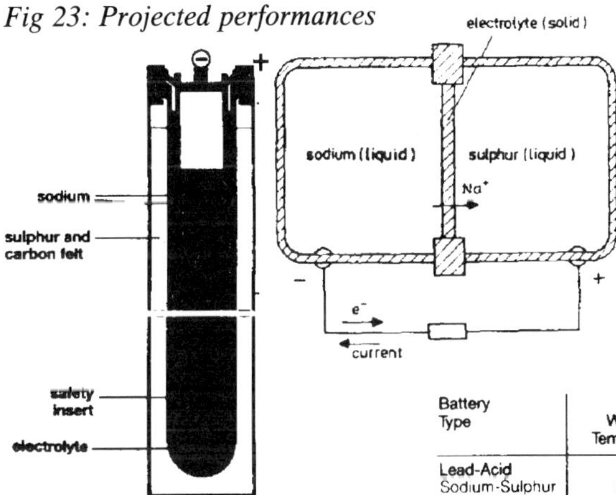

Fig 24: ABB s-s cell

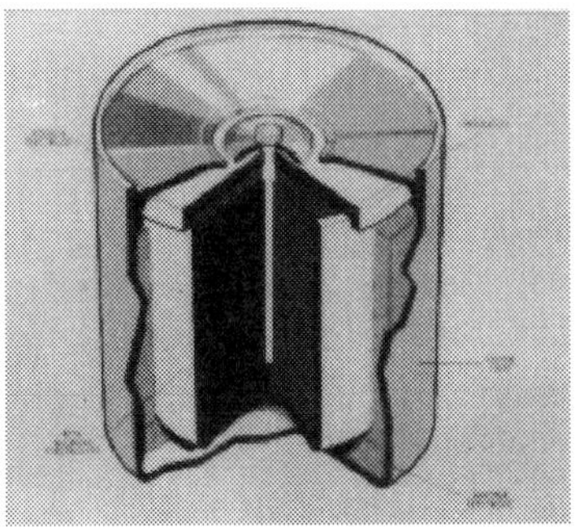

Fig 21: Sodium-sulphur cell

Sodium sulphur batteries have been produced to meet the considerable demands (250,000 cells per year in pilot trials) from vehicle manufacturers for test/evaluation purposes. The cathode of the cell is liquid sodium immersed in which is a current collector of beta-alumina. This is surrounded by a sulphur anode in contact with the outer case. The cells are inside a battery box containing a heater to maintain them at their operating temperature of 300-350 C. This is electrically powered and contained within the charge circuit. When discharging, internal resistance produces sufficient heat for the electrode but some 24 hours is required to reach running temperature from cold. In a typical EV application 100 cells would be connected in series to obtain 100 V and give a battery of 300 Ah, 60 kWh. In use batteries would typically be charged nightly to bring them up to voltage after daily discharge and to keep the electrode molten. The cell is shown in Fig 21 and a typical battery installation in Fig 22 with expected vehicle performance, compared with lead-acid, shown as Fig 23. The Chloride cells are based on an electropheritic process while those from the German company Asea Brown Boveri are made by isostatic pressing.

The ABB cell is seen at Fig 24 and its claimed characteristics compared with competitive systems shown as Fig 25. The electronic current flowing

Battery Type	Max. Working Temperature	Energy Density Wh/kg	Power Density W/kg	Self-Discharge	Efficiency %
Lead-Acid	60 C	30	100	Yes	70
Sodium-Sulphur	300 C	120	185	No	90
Nickel-Cadmium	50 C	40	90	Yes	70
Zinc-Borium	50 C	70	100	Yes	75

Fig 25: Comparative performances

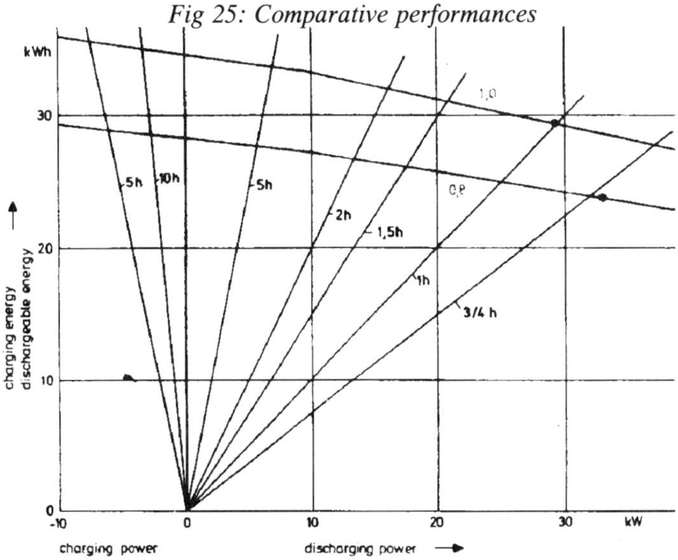

Fig 26: Discharge characteristics

through the external load resistor during discharge corresponds to a flow of sodium ions through the electrolyte from the sodium side to the sulphur side. Voltage is from 1.78 to 2.08 according to the degree of discharge involved, Fig 26. A cell with a capacity of 45 AH has a diameter of 35 mm and length of 230 mm. Its internal resistance is 7 m.ohms and 384 cells of this type can be installed in a battery of 0.25 litre volume. An example produced by ABB has external dimensions 1.42 x 0.485 x 0.36 metres. The cells account for 55 percent of the total weight of 265 kg. By connecting the cells in four parallel strings of 96, the battery has an open circuit voltage of 170-200 and a capacity of 180 Ah.

The electrical energy which can be drawn from the battery is shown in Fig 27 as a function of the (constant) discharge power. With a complete discharge in two hours, energy content is 32 kWh, corresponding to a density of 120 Wh/kg. Associated discharge efficiency is 92 percent. Complete discharge at constant power is possible in a minimum of one hour, and an 80 percent discharge in less than three quarters of an hour. Fig 28 shows that the battery can cope with a load of up to two-thirds of the no-load voltage for a few minutes. This corresponds to a rating of about 50 kW or 188 W/kg. The portion of the heat loss not removed by the cooling system which is incorporated into the battery is stored in the heated-up cells — and covers losses up to 30 hours. Additional heat must be supplied for longer standstill periods either from the electric mains or from the battery itself. Effective, vacuum-type, thermal insulation maintains the power loss at just 80 W so that when fully charged it can maintain its temperature for 16 days. In order to maintain the battery in a state of readiness, the battery must be held above a minimum temperature and it takes about 4-10 hours to heat up the battery from cold — but a limit of 30 freeze-thaw cycles is prescribed. Life expectancy of the battery, otherwise, is ten years and 1000 full discharge cycles, corresponding to an EV road distance of 200,000 km.

The aluminium air battery, or fuel cell, is seen as another EV contender. Alcan's Research & Development centre in Canada has produced a unit with 80 W/kg specific energy output. A cell weighing 100 kg has been made with 36.5 kWh capacity and occupies 103 litres, to provide energy density of some 360 Wh/kg. The company have experimented with an interesting 'hybrid' battery system in which a vehicle has

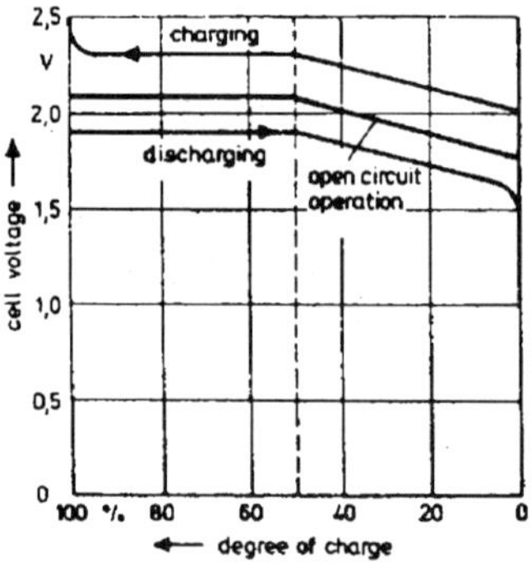

Fig 27: Energy capability

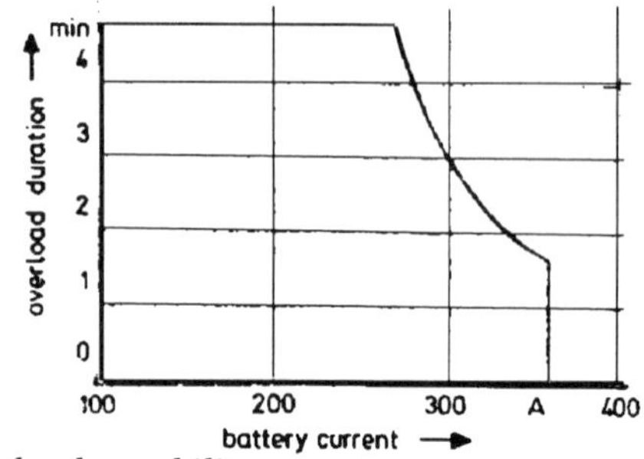

Fig 28: Overload capability

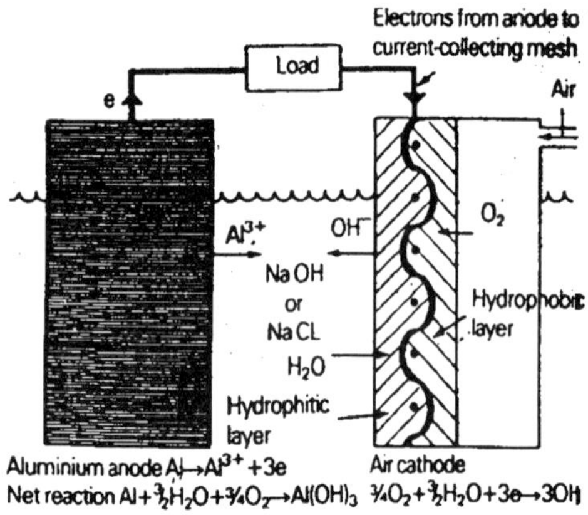

Fig 29: Aluminium-air cell

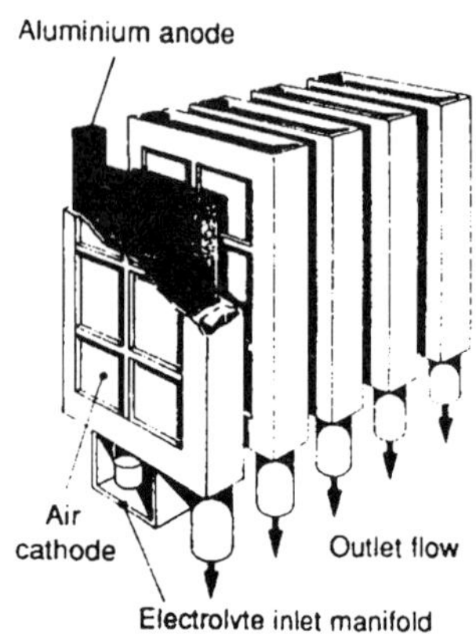

Fig 30: Alupower cell

been fitted with 35 Ah lead acid batteries in addition to two al-air batteries having an output of 15 mA/cm squared — with 250 W output per pack and each weighing 22 kg. This raised the range by a factor of three of the original vehicle equipped only with 24V 100AH lead-acid batteries. A nickel grid acts as a current collecting mesh within the air cathode, Fig 29. The aluminium anode is replaced as it becomes consumed, a typical light EV requiring 660 kg of aluminium to travel 15,000 km per year, the company estimate. It is possible, however, to re-cycle the aluminium which is dissolved on the salt water electrolyte. The company's joint venture subsidiary with Chloride, Alupower, has produced a less experimental al-air battery with 250 Wh/kg energy density, its full performance being delivered within 30 minutes of activation. Along with an aluminium cathode the Alupower battery has cathodes comprising two active carbon layers, a nickel grid substrate and a hydrophobic porous PTFE backing layer. Fig 30 shows a cell design for the communications industry. It contains an electrolyte feed tube through which electrolyte is pumped upwards and overflows a weir at the top and gravitating down through a drain tube into the reservoir.

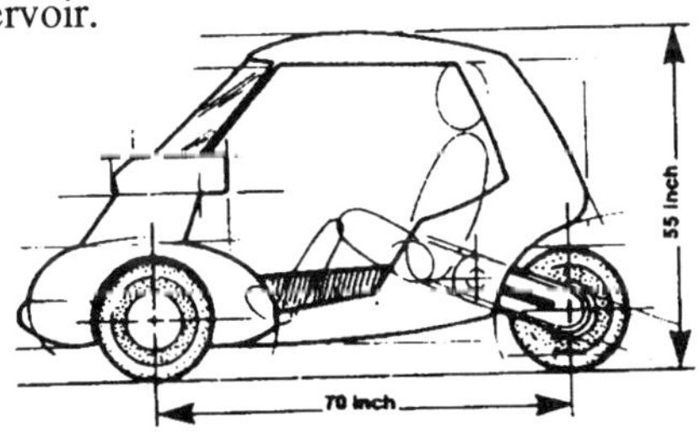

Fig 31: Sinclair C10 proposal

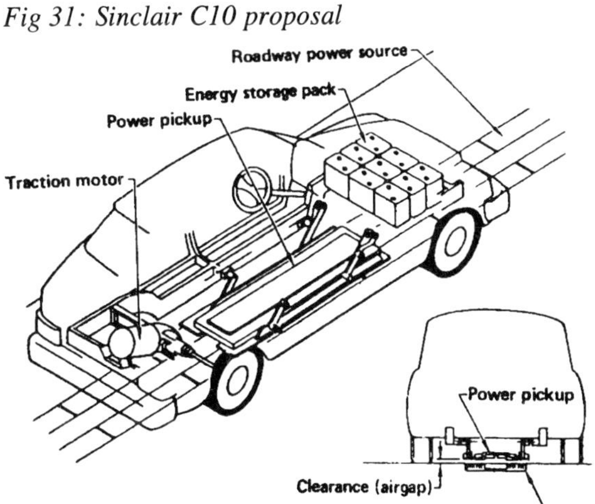

Fig 32: Road-induced electricity

Electric car technology

An introduction to electric vehicle technology was provided in the first section of this chapter. Here the particular case of battery cars is examined in the light of changing economic, environmental and political circumstances which have brought a new impetus to the study of drive systems for this category.

According to Sir Clive Sinclair whose abortive efforts to market an electric tricycle have led him to concentrate on economical bicycle conversions, peak efficiencies of 90% are available with EVs for converting electricity into tractive energy — and that attainable electrical generating efficiencies of over 50% meant a 45% fuel conversion efficiency could be compared with 30% for the petrol engine. His C10 proposal shown in Fig 31 means his faith in the future electric car is by no means exhausted. Of course the petrol-car thrives on the very high energy density of petrol — and thus its relatively small on-car storage volume for a reasonable vehicle range. This last factor is likely to keep EVs to relatively specialist sectors such as city cars in the short term.

There are other initiatives, too, such as the desire to make motorway driving under very high density peak traffic conditions less dangerous and less tiring. This is generating fresh interest in reserved lanes for vehicle guidance systems. Where these additionally provide roadway-induced powering, Fig 32, as described by researchers from the Lawrence Livermore National Laboratory*, a case for a car to suit relatively long-distant commuters can be made out.

The success of trials on GM's Impact electric car, so far, have pointed to the very considerable importance of light weight, good aerodynamics and low rolling resistance but the electrical breakthrough has come in the electronics technology of the DC/AC converter. Ford, too, have had very promising prototype results from their Ecostar 1 car-derived van, using a transistorised DC to AC inverter.

An interesting variant on the AC-motored theme of these projects is the use of a two-speed transaxle gearbox which reduces the otherwise required weight of the high speed motor and its associated inverter. A system developed by Eaton Corporation is shown in Fig 33 and has a 4-kW battery charger incorporated into the inverter. A three-phase induction motor operates at 12,500 rpm — the speed being unconstrained by slip-ring commutator systems. A block diagram of the arrangement is in Fig 34 and based on an induction motor with 18.6 kW one-hour rating — and base speed of 5640 rpm on a 192 V battery pack. The pulse width modulated inverter employs 100 A transistors and its bridge pack is seen as Fig 35, the key element of which is the transformer, T501, with windings P1, P2 and S. When in inverter mode, P2 is disconnected by relay 500 and P1 connects to the main battery. In the battery charging mode, P2 is connected to either 220 or 110 V AC lines and P1 is disconnected. P2 combines with S and transistor SB1 to form an ac-dc charger at relatively low cost. Fig 36 shows the controller drive system functions in association with the inverter.

In an AC induction motor, current is applied to the stator windings and then induced into the windings of the rotor. Motor torque is developed by the interaction of rotor currents with the magnetic field in the air-gap between rotor and stator. When the rotor is overdriven by coasting of the vehicle, say, it acts as a generator. Three phase winding of the stator armature suits motors of EV size; the rotor windings comprise conducting 'bars' short-circuited at either end to form a 'cage'. Rotation speed of the magnetic field in the air gap is known as the synchronous speed which is a function of the supply-current frequency and the number of stator poles. The running speed is related to synchronous speed by the 'slip'.

If two alternators were connected in parallel, and one was driven externally, the second would take current from the first and run as a 'synchronous motor' at a speed depending on the ratio of each machine's number of poles. While it is a high-efficiency machine which runs at constant speed for all normal loads, it requires constant current for the rotor poles; it is not self-starting and will stop if

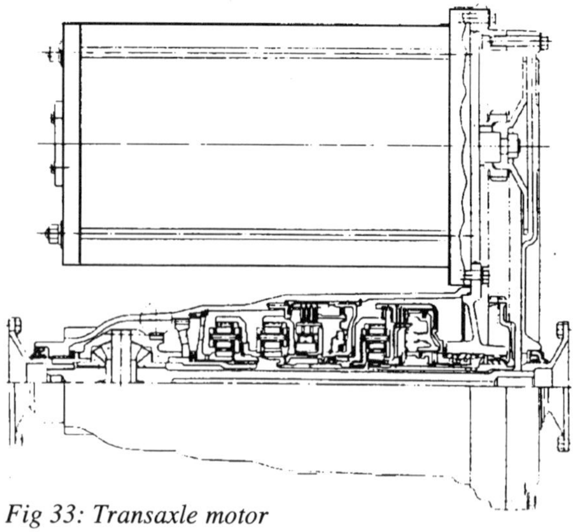

Fig 33: Transaxle motor

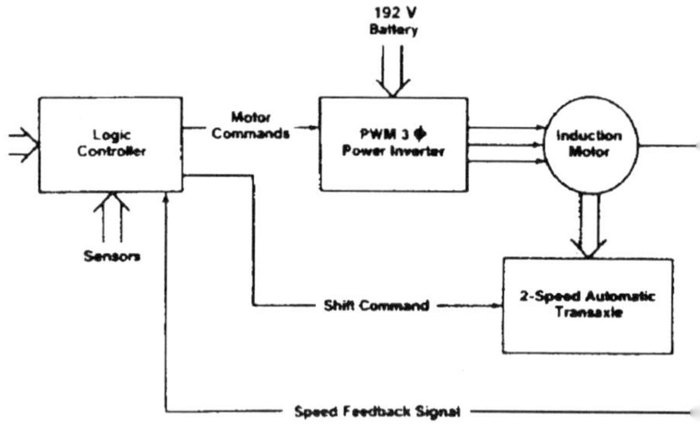

Fig 34: Controller arrangement

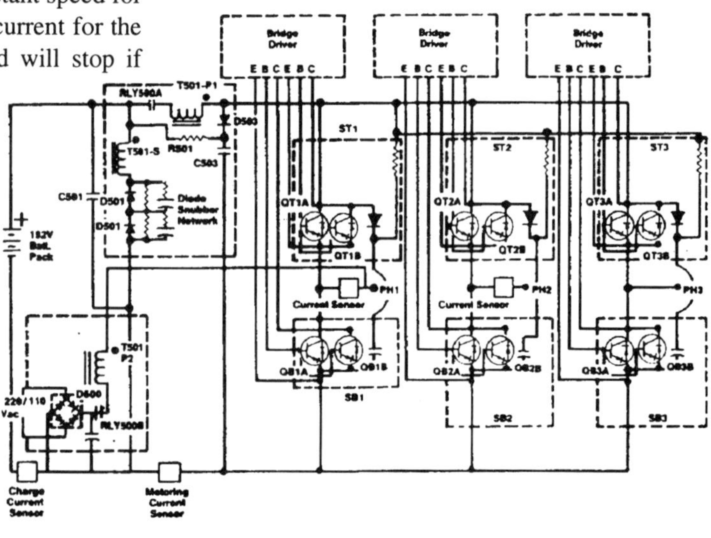

Fig 35: Transistor bridge circuit

overloaded enough for the rotor to slip too far behind the rotating stator-field. Normally, the synchronous motor is similar in construction to an induction motor but has no short-circuited rotor — which may be of the DC-excited, permanent-magnet or reluctance type.

Fig 37 shows a stator winding for a 2-pole 3-phase induction motor in diagrammatic form. If supply current frequency is f_s, then stator field speed is f_s/p for number of poles p. Rotor current frequency $f_r = sf_p$ where s is the slip. Power supplied to a 3-phase motor can be expressed as $EI(3)^{1/2}f_s n$ where n is efficiency. When a synchronous motor has no exciting voltage on the rotor it is termed a 'reluctance' motor which has very simple construction and, when used with power-transistors, can be applied as a variable speed drive. Axial air-gap versions are possible as in Fig 38; such an electronically commutated motor can operate with a DC source by periodic reversal of the rotor polarities.

The general form of control, with DC link inverter, for induction motors is shown diagrammatically in Fig 39. The thyristors of the inverter are generally switched so as to route current through the stator winding as though it were connected to a 3-phase AC. Frequency can be varied by timing of pulses to the thyristors and is typically 5-100 Hz to give a speed range of almost twice synchronous speed. According to engineers from Chloride EV Systems Division (SAE Paper 900578, 1990), devices for building such an inverter are not yet available for current levels associated with EV traction. This situation may have already changed in the nine months or so since these words were spoken, however.

The more widespread adoption of electric cars may be spurred by the development of improved solar cells. Work at Clark University in the USA involves a so-called photo-electrochemical cell in which the semi-conductor is in contact with an electrolyte. Around 17 % operating efficiency has been achieved in converting light- to electric-energy using calcium selenide and a water-based electrolyte. Unlike solid-state photo-voltaic cells, these devices can also store solar energy.

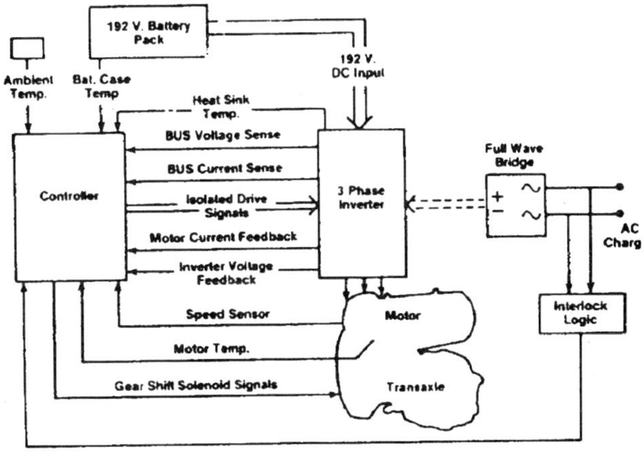

Fig 36: Controller drive system

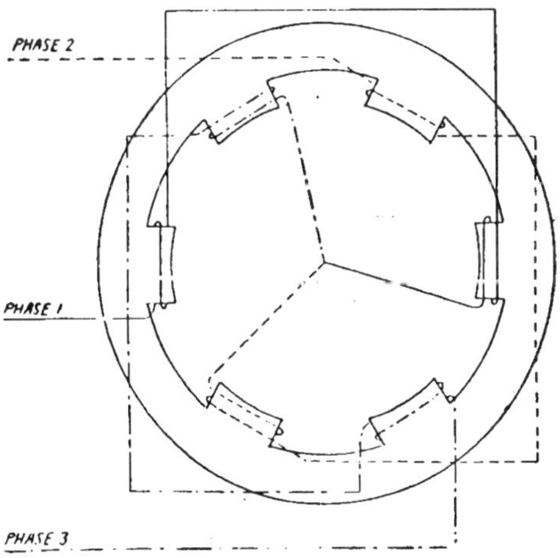

Fig 37: Three phase induction motor stator

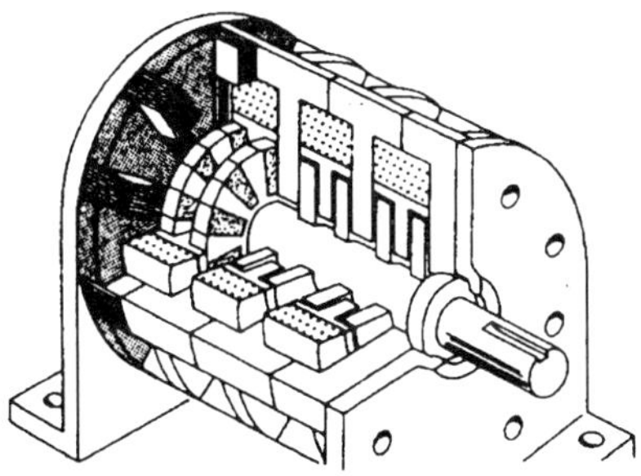

Fig 38: Air-gap type reluctance motor

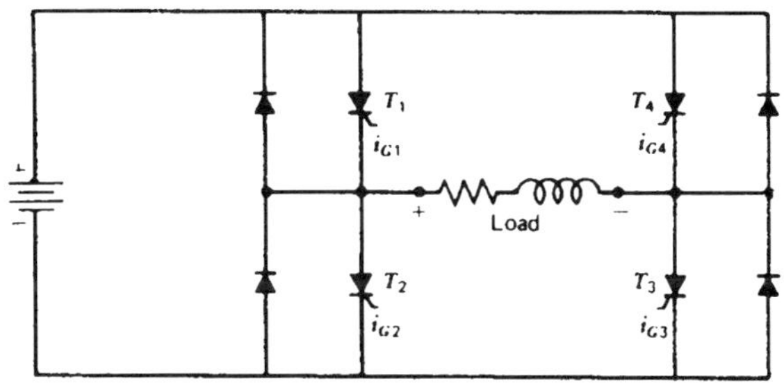

Fig 39: Induction motor controller

Hybrid drive vehicles

Of the various battery-electric technologies competing for commercial exploitation in vehicles, hybrid drive is seen by many as the most realistic contender. However, careful systems analysis is required to optimise the various parts of the system and different systems are seen to suit different vehicle categories.

While the IC engine is currently superior in terms of available performance and range, it has disadvantages such as drop in efficiency under conditions of part load operation, and increasing emissions prior to optimum speed operation, which have led to the design of vehicles in which the IC engine is specified only for optimum-power operation. Difference between power required and that generated is usually made up by an electric drive and the IC engine is run only at its peak efficiency of operation. Usually either one, series, or two alternative parallel drive arrangements have been tried — as seen in Fig 40. Letters M and G show whether the operation is engine or generator based. The disadvantage of the series arrangement is the need to convert energy several different times even though the IC engine can be operated at peak efficiency. The latter advantage is lost with the parallel arrangements but in the engine-parallel one, starting and gearing systems required of the engine are included in the electric drive branch and speed variation of both engine and electric motor is limited. The transmission-parallel one has better transmission efficiency than the previ-

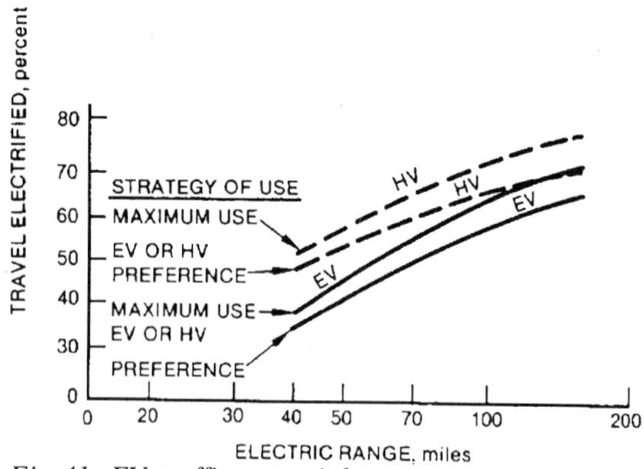

Fig 41: EV traffic potential

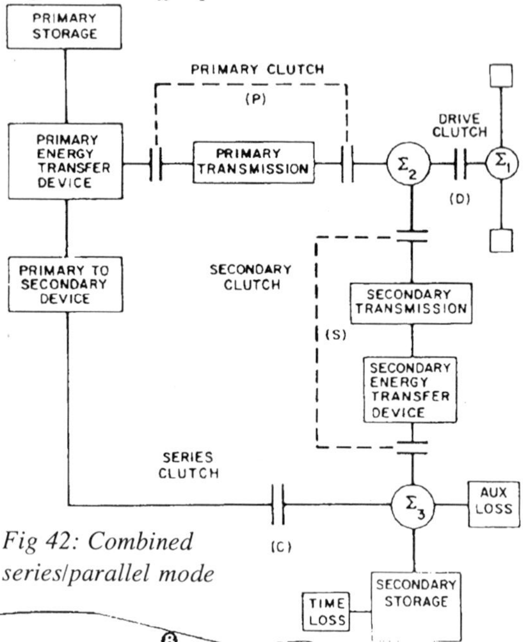

Fig 42: Combined series/parallel mode

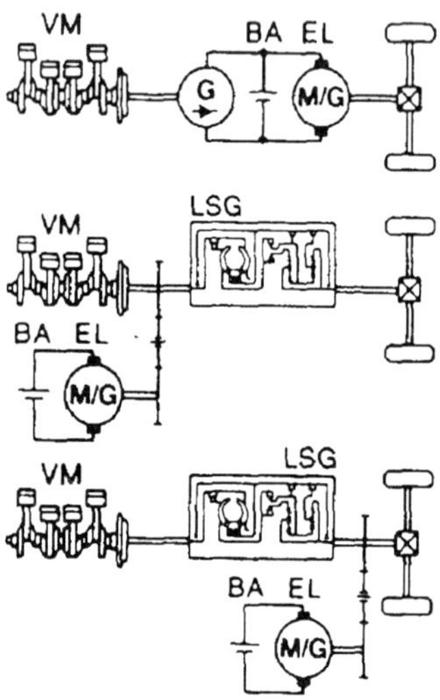

Fig 40: Series and parallel configurations

Fig 43: Daimler-Benz hybrid city-bus: 1 motor, 2 air-compresser, 3 motor fan, 4diesel generator, 5 battery fan, 6 electronic controller, 7 batteries and 8 battery cooler

ous arrangement but speed control of the motor is required from zero through to maximum.

Studies carried out at the general Research Corporation in California, where legislation on zero emission vehicles is imminent, have shown that the 160 km range electric car could electrify some 80% of urban travel based on the average range requirements of city households, Fig 41. It is unlikely however that a driver would take trips such that the full range of electric cars could be totally used before switching to the IC engine car for the remainder of the day's travel. This does not arise with a hybrid car whose entire electric range could be utilised before switching and it has been estimated that with similar electric range such a vehicle would cover 96 of urban travel requirements. In two or more car households, the second (and more) cars could meet 100% of urban demand, if of the hybrid drive type.

Because of the system complexities of hybrid drive vehicles, computer techniques have been developed to optimise the operating strategies. Ford researchers[2], as well as studying series and parallel systems, have also examined the combined series/parallel one of Fig 42. The complexity of the analysis is shown by the fact that in that one system, having four clutches, there are sixteen possible configurations depending on state of engagement. They also differentiated between types with and without wall-plug re-energisation of the batteries between trips.

Mercedes were early protagonists of hybrid buses, Fig 43. The modified OE 305 shown can carry 100 passengers and has a maximum speed of 43 mph. The diesel is rated at 100 bhp while the electric motor can develop up to 200 bhp. Operating voltage is 360 for the five 275 amp-hour batteries which weigh 3.5 tonnes. Electric range is 30-45 miles compared with total range of 190 miles for the 19 tonne GVW vehicle.

The Lucas Chloride hybrid, Fig 44, has engine (3) driving through the motor (1) but midships positioning of the batteries (4) with on-board charger (5) at the rear. Clutches are shown at (6) while (7) and (8) are alternator and control unit. This used Reliant's 848 cc engine developing 30 kW alongside a 50 kW Lucas CAV traction motor. The 216 V battery set had capacity of 100 amp-hour on a 5-hour rate. Maximum speed in electric drive of 120 km/h rises to 137 km/h in combined mode, Fig 45.

A later example with automatic control of drive selection is the Volkswagen/Bosch hybrid car in which the standard VW car diesel runs alongside a 6 kW asynchronous motor generator located in the bell housing, increasing the length of the engine/gearbox unit by only 58.5mm — and imposing a weight penalty of 29 kg. Alternator and starter motor are eliminated while the batteries are below the boot floor and provide an electric range of 30 miles in a Golf saloon. An inverter and thyristor power controller are also involved in the system. The car has an electronic clutch controlled by microswitch on the gearshift knob and the decision to drive by IC or electric is also made electronically. Fuel consumption in combined mode is just 113 mpg on the ECE urban cycle — with additional input of 16.3 kWh of electrical energy is required from the mains during the cycle. Another configuration, for buses, has been developed as a proprietarypropulsion system by Magnet-Motor GmbH. It has been likened to a diesel electric locomotive, the key element of which is the

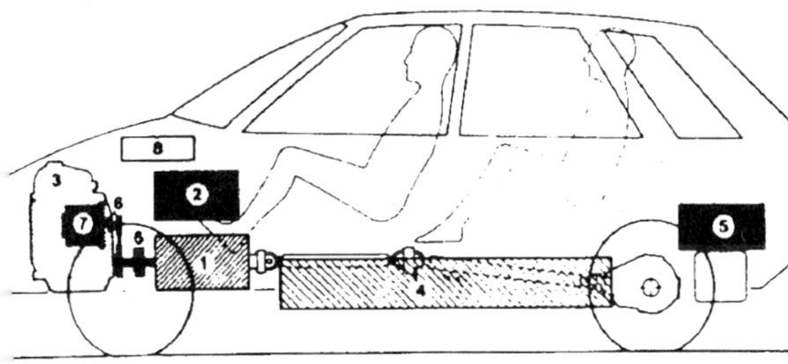

Fig 44: Lucas Chloride hybrid car

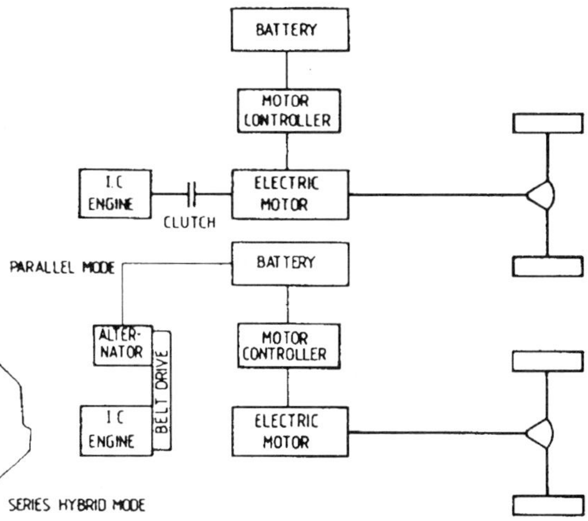

Fig 45: Bi-mode drive system

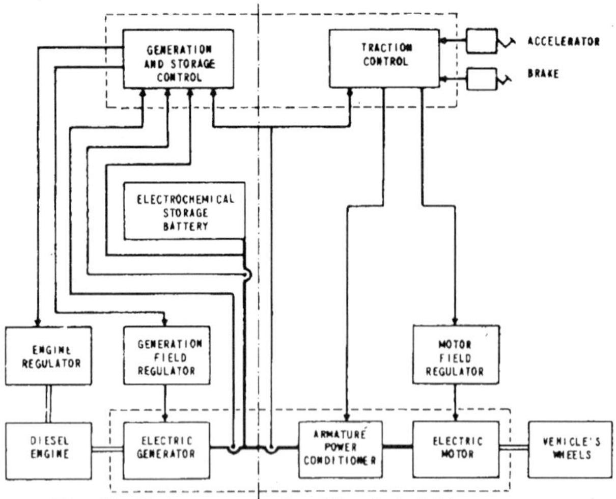

Fig 46: Drive system for Fiat hybrid bus

Magneto-dynamic storage unit using a multipole permanent magnet motor. Its rotor is a steel cylinder with magnets on its periphery. The stator comprises a large number of electro-magnets and the motor is said to produce 2.5 times more torque than a conventional electric motor, and has no slip-rings. It rotates on a vertical axis to avoid gyroscopic effects on the bus — which is also powered by a BMW turbo-diesel of 90 bhp. The generator is attached directly to the engine which runs at its optimum speed.

The Alupower electrical storage sytem described in an earlier section of this chapter, Figs 29 and 30, can also be regarded as part of a hybrid drive system . It has been adapted to golf-cart type vehicles with additional lead-acid batteries. This is thus a hybrid which which uses both batteries and fuel cells as the alternative drives.

The aluminium air cell works on the electrochemical reaction between aluminium and oxygen in air, an alkaline electrolyte being pumped between aluminium anode and cathode. Electricity is produced as the aluminium oxidises and aluminium hydroxide is formed as a by-product, precipitated out and collected in a sump. Air is blown through the cell to provide the oxygen. The aluminium which is consumed is recharged by replacing the anode plates, at which time the electrolyte is also replenished. A condenser continuously removes oxygen depleted air during operation. An energy density of 350 Wh/kg is claimed but power density is low — hence the need for additional lead-acid batteries in the hybrid vehicle. A total of 18 Chloride 3ET-205 6-volt batteries are combined with 48 fuel cells to drive an 80 bhp dc motor in a Chrysler minivan used to test the concept.

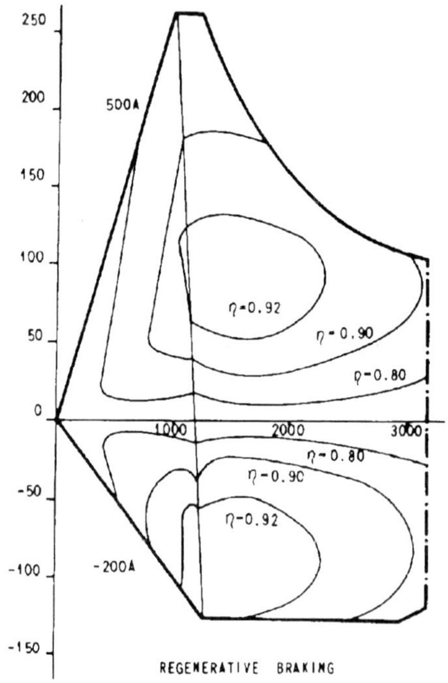

Fig 47: Motor characteristics: power in kW (y-axis) vs speeed in r/min

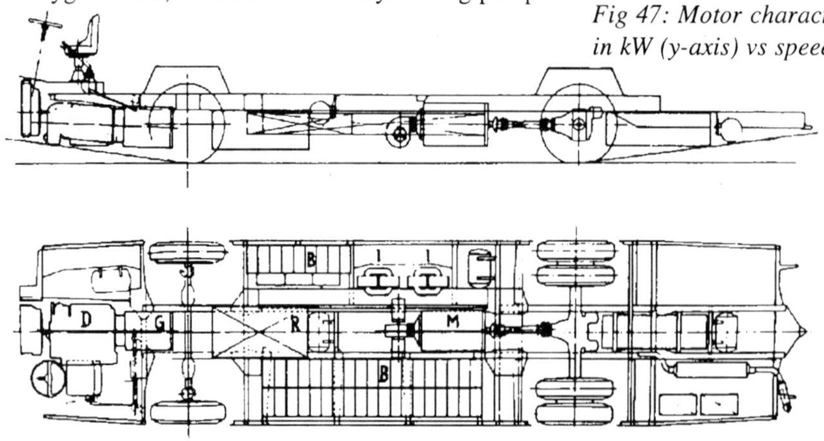

Fig 48: Layout of pioneer Fiat hybrid bus

Hybrid drive buses

In a series hybrid configuration, part of the traction energy is converted into electrical energy, and then into mechanical energy, and part flows to the wheels directly via a mechanical transmission. It is argued that this configuration can potentially offer higher overall efficiency. In a series layout, all the IC engine energy is converted into electrical energy and then into mechanical energy. Such a configuration could offer advantages where the electric motor is designed as a very high efficiency unit regardless of the load upon it. Futhermore, the ability to use pure electrical transmission allows for flexibility, in system management, in optimising engine operating conditions and reducing noise output. It also allows greater freedom in mechanical packaging, including electric motors direct-driving the wheels, and for the incorporation of future fuel-cell technology.

This was the thinking behind the choice of a series system by Fiat in their pioneering hybrid bus introduced earlier in this chapter, a flow diagram for the system being shown in Fig 46. The DC compound motor used with separately excited field, Fig 11 in this chapter, had interpoles and compensating windings to expand the load range of maximum efficiency operation both as a motor and generator. Armature current was varied, by means of a thyristor chopper, from zero to base speed, above which field current was varied by an independent chopper. The reversible chopper allowed regenerative braking, down to zero speed. The motor had characteristics as seen in Fig 47 and was direct-coupled to the rear-axle driving head. Its nominal voltage was 600V, continuous power 90 kW, maximum power 180 kW and field control ratio at continuous power 1150/3200 rpm. Efficiency at continuous power in field control was 92/90 %. The 600V generating unit comprised diesel gen-set with power rating of 56 kW and maximum power 78 kW. Battery storage comprised fifty 12V lead-acid cells of 135 Ah capacity at a 5 hour discharge rate, their weight of 1930 kg corresponding to 12% of the GVW. Vehicle layout was as in Fig 48; in tests the vehicle recorded diesel consumption of 32.3 kg/100 km compared with 37.8 for a conventional vehicle, with battery state of charge found to be the same at the beginning and end of the tests. Range in purely electric drive was 30 km of city driving from 100% to 20% state of charge of the batteries.

During the design stages Fiat examined a typical town route between two termini as seen in Fig 49. It was also established that the maximum acceptable acceleration for standing passengers was 1.5 m/sec^2 on level ground and 0.27 at a gradient. The subtended area in the power diagram gave energy required

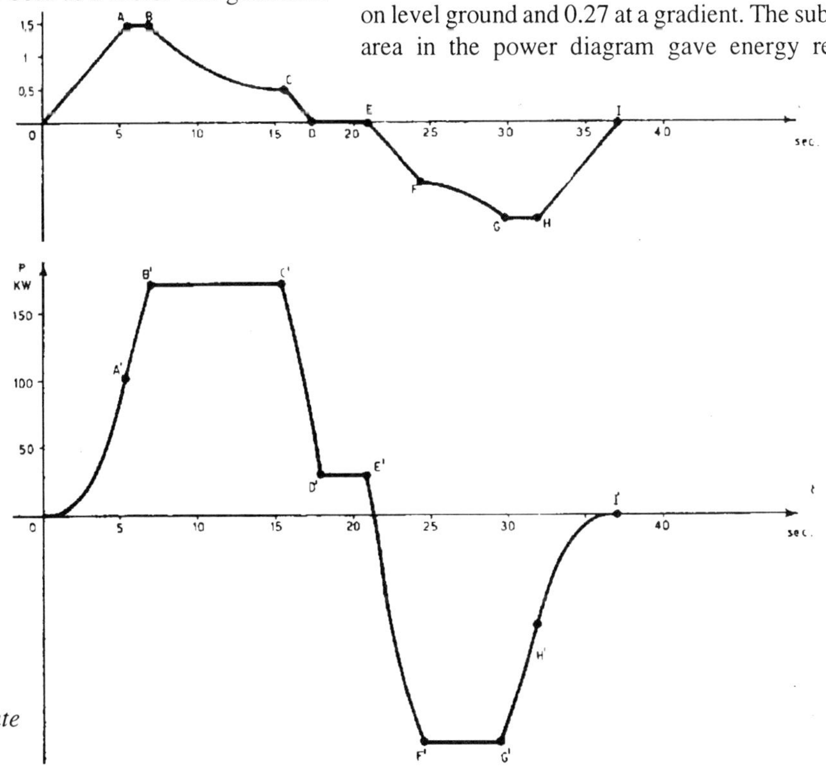

Fig 49: Typical route

between stops, the negative portion representing energy flowing back to the batteries having taken the various system efficiencies into account.

For E_s the total energy required at the wheels, E_m the engine energy and E_r the mains electrical energy, then

$$E_s = (\eta_g E_m + E_r)\eta_b \eta_t \eta_o$$

where efficiencies subscripted g, b, t and o refer to generator, battery, motors/controllers and transmission respectively, with their product the overall efficiency h. Then for total duration of daily service T_s, terminus turn-around time T_c and number of daily runs between termini N, then power required from the engine is:

$$P_m=(E_s-\eta E_r)/[\eta_g \eta\{T_s-(N-1)T_c\}]$$

The total daily energy is calculated for a typical route of length L_p, divided into N_t segments of length L_t. The heaviest cycle is shown in Fig 50.

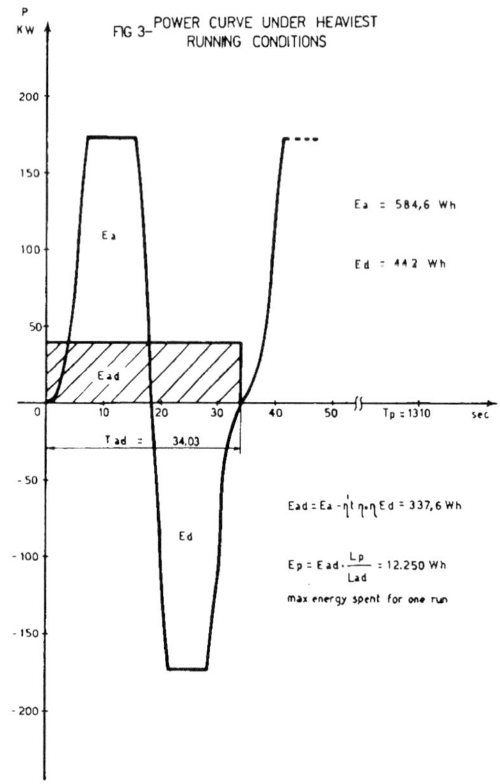

Fig 50: Heaviest duty cycle

CNG-electric hybrid

Smaller buses have been built with pure electric and alternative forms of hybrid drive. An interesting project by Unique Mobility in North America put a CNG-electric hybrid system into a 25 ft (7.62 m), 24 passenger, vehicle. Here the compactness and locational flexibility of the hybrid drive elements meant that considerable gains could be obtained in packaging the vehicle occupants. Using high power-density permanent-magnet motors driving the rear wheels allowed a particularly low floor of 12 in (305 mm) from the ground.

The CNG-engine generator provided steady state power and was augmented by storage batteries to supply the power required above that base level while recharging of the batteries would take place when the power requirement fell below the base level. CNG tanks were roof-mounted while batteries were positioned over the rear wheel wells and inside the engine compartment. The 11 tonne GVW bus is seen in Fig 51.

The 90 bhp gas engine drove the generator through a flywheel-positioned step-up planetary gear set and an engine management system allowed engine speed and power to vary with load conditions. The rate at which speed was increased was minimised

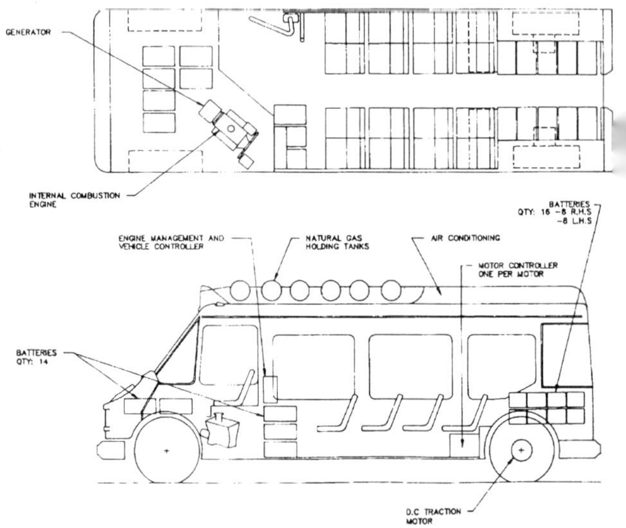

Fig 51: Unique Mobility midibus

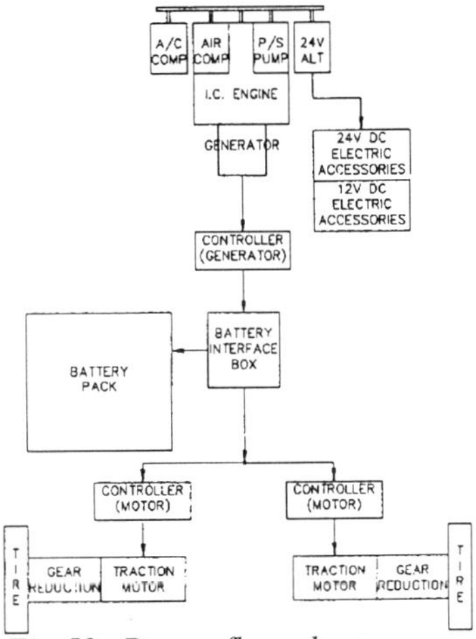

Fig 52: Power flow charts

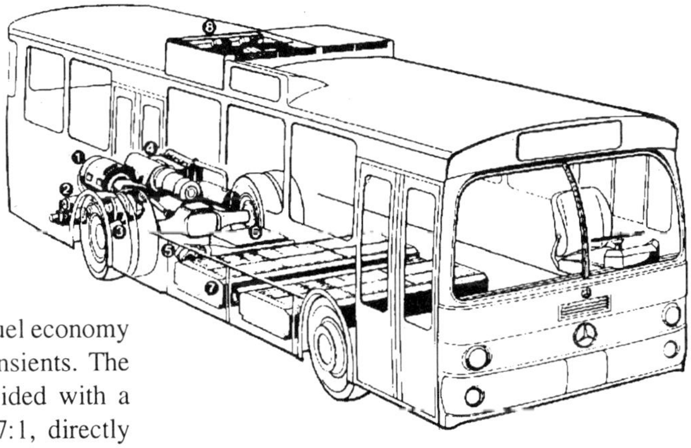

Fig 53: Daimler Benz OE305 conversion

by the controller in order to avoid poor fuel economy and high emissions associated with transients. The two 70 kW traction motors were provided with a single planetary reduction gear of 2.77:1, directly coupled to the drive wheels through a secondary set of 5.2:1 included in the wheel hub to give an attainable speed of 55 mph. Rear suspension was an independent trailing arm system with traction motors direct-mounted to the arms, so as to maximise floor area between the wheels. Motor differential speeds for cornering are electronically controlled with reference to steering wheel angle and road wheel speed.

Fig 52 shows the power flow charts for different modes of operation. In the first, on IC engine power only, a speed of 37 mph was achieved. On IC engine and battery power, higher speeds were possible and a reserve was available for gradients and acceleration; in the final mode of regenerative braking with the IC engine operating, the engine provided power only to the accessories and that from braking was fed into the storage batteries. The latter were used primarily for supplying accelerative power and were 12 V units with 160 Ah capacity. Two series strings of 15 batteries were connected in parallel to yield 180 V and 320 Ah total capacity.

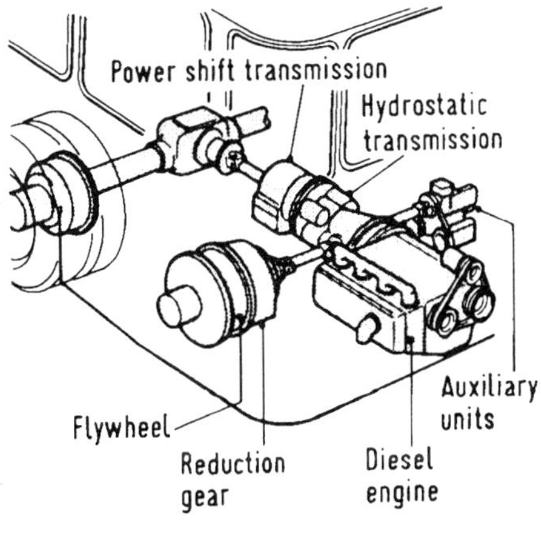

Fig 54: Flywheel drive hybrid

Another pioneering series of hybrid buses has been the Daimler Benz OE 305 city bus conversion, some twenty of which were evaluated in trials in German cities in the early 1980s. Electric drive in the city centre and diesel drive in the suburbs was the mode of operation. Seen in Fig 53, the set-up was electric motor (1), air-compressor and power-steering pump (2), motor-fan (3), diesel engine and generator (4), battery-fan (5), electronic controller (6), traction batteries (7) and battery cooling unit (8). Range was 30-45 miles on batteries alone and 190 miles as a hybrid diesel combination. The 100 passenger vehicle had a maximum speed of 43 mph and the 19 tonne GVW vehicle had batteries weighing 3.5 tonnes. The motor could develop up to 200 bhp while the diesel engine was rated at 100 bhp. The five 275 Ah batteries operated at 360 V.

At the same period a Daimler Benz 305 was also converted to flywheel hybrid operation in a study which involved M.A.N. and Berlin University, too. The team calculated the flywheel storage energy requirements of a city bus to be 750 kW for absorbing the kinetic energy of the vehicle at top speed and a 220 lb flywheel was chosen, with a 1500 kW total energy to allow a reserve, which suffered 2kW power loss at 12,000 rpm. For a typical urban operating cycle, characteristics were plotted, from the starting point of the bus stationary with diesel engine at idle and flywheel charged up from previous operation, for the configuration shown in Fig 54.

During acceleration the mechanical gear stage is automatically shifted as the vari-drive hydraulic transmission changes speed. This results in the almost constant slope (full-line) in the curves shown in Fig 55 compared with a conventional bus (dotted-line). Initially the flywheel alone is used, then the diesel is brought in during deceleration, as seen in the bottom half of the figure. Flywheel power P_s is 260 kW immediately before the constant-power cruising phase, during which the diesel drives. During deceleration the flywheel is recharged and its power is 200 kW corresponding diesel power being shown by P_d. Flywheel energy content and losses are seen as Fig 56; the inertia losses are replaced by diesel energy. The diesel losses are seen in Fig 57 together with the transmission losses between engine and drive wheels, compared with those for a conventional vehicle.

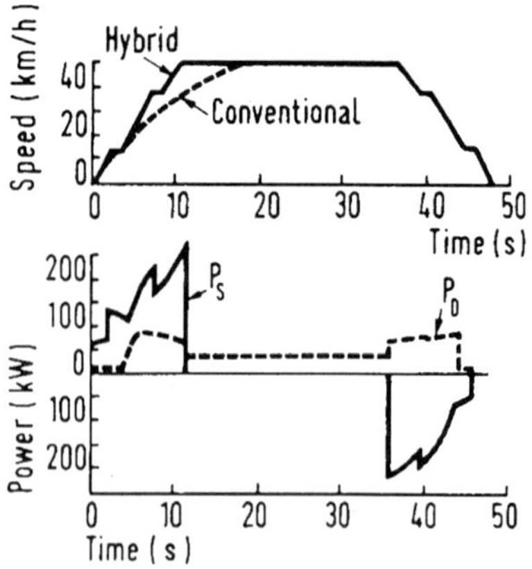

Fig 55: Characteristics of flywheel hybrid

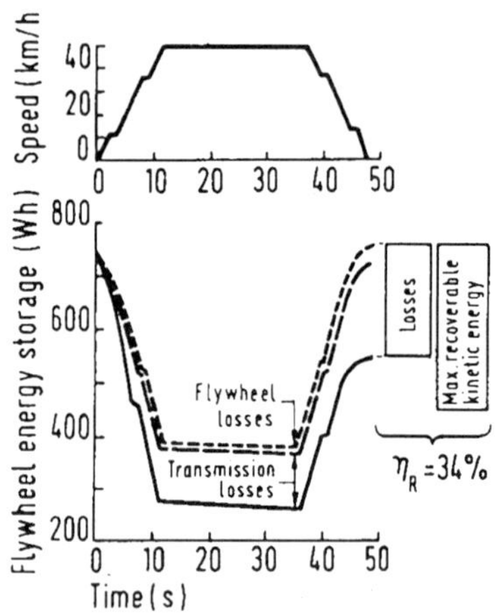

Fig 56: Flywheel losses

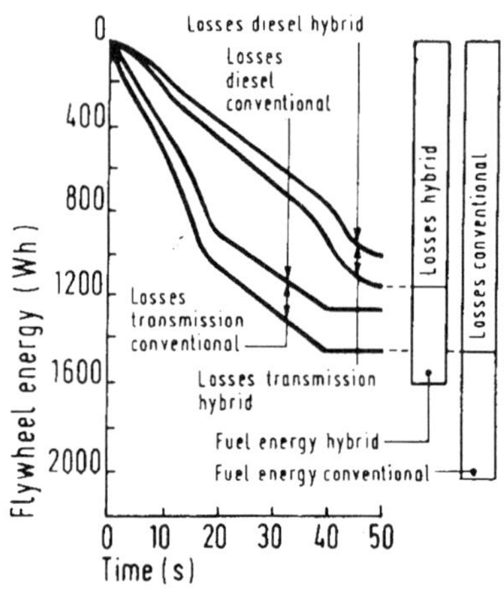

Fig 57: Diesel losses

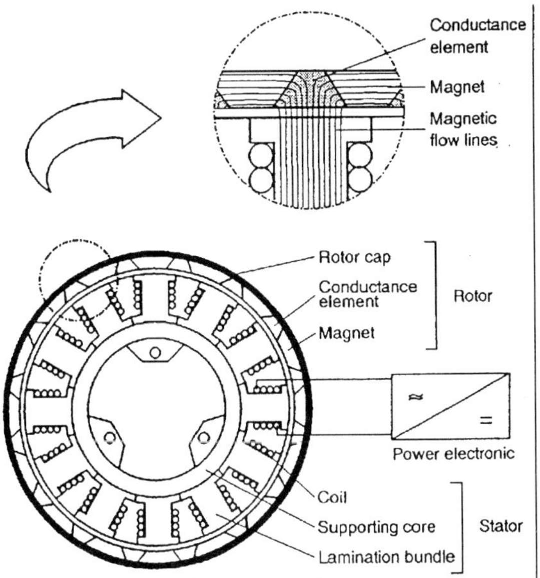

Fig 58: Trapezoidal iron conductors

Performance	appr. 100 kW
Max. input torque	appr. 190 Nm
Maximum speed	appr. 6000 rpm
Generator volume	appr. 13 l
Power electronic volume	appr. 20 l
Generator mass	≤ 25 kg
Power electronic mass	≤ 13 kg
Efficiency at rated performance	≥ 0.93
Joint cooling method	liquid cooling

Fig 59: Motor performance

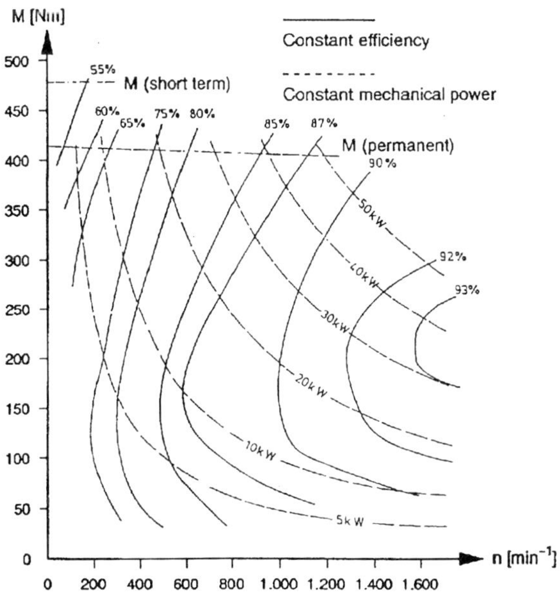

Coolant temperature: 50°C

Fig 60: Motor curves

Hybrid drive, a staged approach

Work carried out at Fichtel & Sachs and Magnet Motor GmbH aimed at optimising the design and installation of electric machines for EVs, and the better integration of electronic systems in the drive train, is further paving the way to achieving viable hybrid vehicles.

A new method of construction for permanent magnet motors exploits the relationship that specific motor torque, on the basis of weight and bulk, is proportional to the product of magnetic field in the air gap, radius of the air gap squared and the axial length of the motor. The requirement for maximum air-gap focuses on the construction of the outside of the rotor. On the inside tangentially magnetised permanent magnets are fitted to the circumference. Shown inset in Fig 58, trapezoidal iron conductors are seen between the rare-earth element magnets which are also trapezoidal in section. These collect magnetic field and divert it to the stator. The laminations of the stator are arranged radially and wound in individual coils which are connected individually or in groups, series or parallel, to the single phase power electronic DC/AC converters.

The latter are supplied with power from a DC link circuit and commutate the coil current at the amplitude required by the rotor angle as detected by remote sensors. The convertors only supply the section of the motor assigned to them and thus work independently. They are made as 4 quadrant controllers and have IGBT switching. These are described as Multiple Electronic Permanent (MEP) magnet motors and are made by the Magnet Motor Co. With liquid cooling, the specific performance of these MEP machines is significantly increased over conventional EV motors as seen by the table of Fig 59 and the corresponding characteristic curves of Fig 60. Both companies are jointly engaged in further development for volume production with target performances seen in the tables of Figs 61 and 62.

A test vehicle based on an Audi 100 Quattro of 100 kW has been constructed and the drive configuration chosen is seen in Fig 63. The four MEP motors have a nominal performance of 25 kW each and are direct connected to each road wheel without the need for mechanical reduction. The tandem configuration of the motors has the advantages of an electrical torque apportioning differential, the motor is part of the vehicle sprung mass and relatively long drive shafts can be used. The generator is also of MEP construction and direct-flange-connected to the IC engine. Power electronics are used to provide power through a DC link circuit to the four motors.

A drive-by-wire arrangement is involved in the IC-engine throttle control, an ECU matching engine speed to generator output. Fig 64 shows the speed-torque map of the engine set to a low SFC value. The ECU also controls the commutation of the MEP machines. The drive configuration allows the engine to operate at constant speed and the wheel speed reductions are also controlled by software in the ECU and thus different driving programs can be instituted. Handling control is also affected by the software in that different torque distribution to the road wheel can be programmed. The MEP generator also acts as a starter motor for the IC engine.

Test results and simulated performance are seen in Fig 65. The simulation has also been used to test the effect of further developments such as the removal of the IC engine flywheel, redesign of the Audi platform to better exploit electric propulsion and the achievement of the target MEP machine perfomances previously listed. These estimated performances are seen in Fig 66. A 10 % fuel consumption is achieved for road performance equivalent to the standard car and work is under way to design ULEV versions incorporating on-board storage batteries.

Maximum torque (short term)	482 Nm
(permanent)	410 Nm
Maximum mechanical performance	49 kW
Mass, including power electronic	30 kg
Maximum efficiency	0.935
Operating voltage	600 V DC

Performance	appr.	2 x 55 kW
Max. output torque	appr.	2 x 550 Nm
Maximum speed	appr.	1800 rpm
Tandem motor volume	appr.	17 l
Power electronic volume	appr.	20 l
Tandem motor mass	≤	40 kg
Power electronic mass	≤	13 kg
Efficiency at rated performance	≥	0.93
Joint cooling method		liquid cooling

Figs 61/62: Target performances

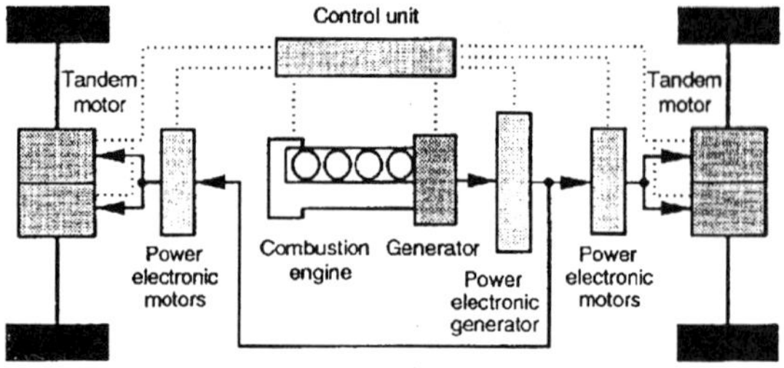

Fig 63: Drive configuration

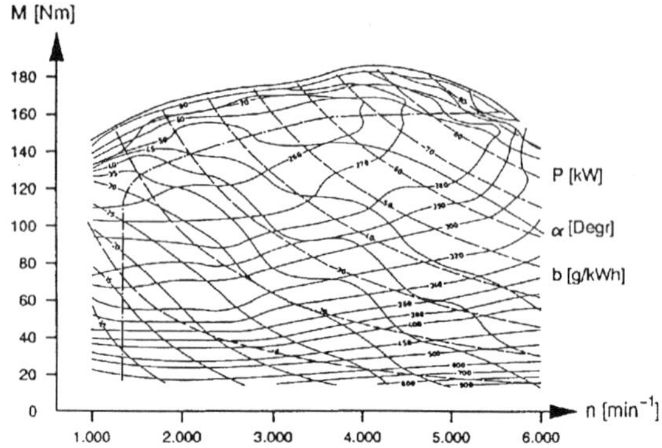

Fig 64: Speed/torque map

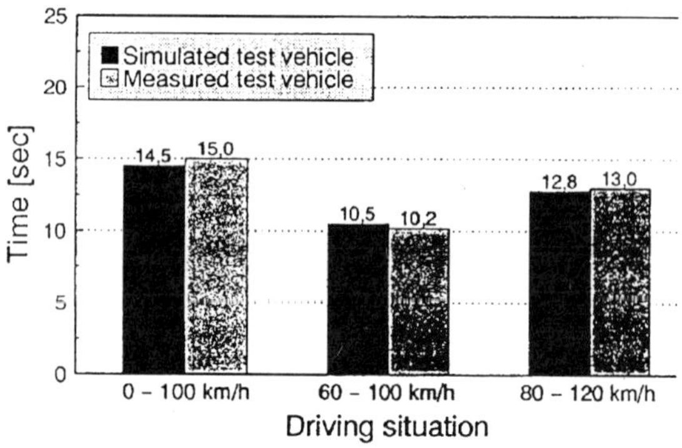

Fig 65: Simulated performance test

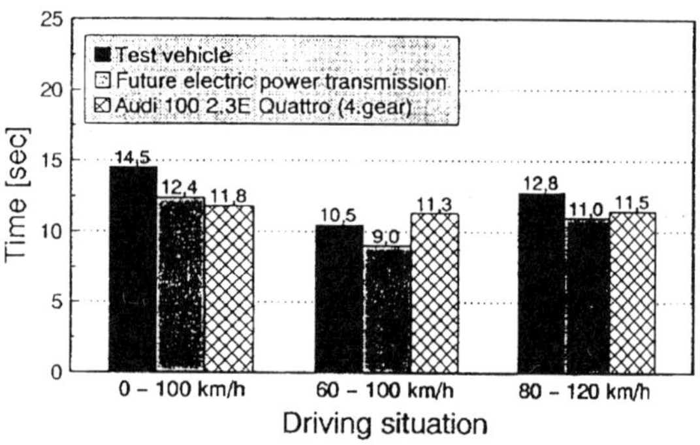

Fig 66: Estimated future performance

Designing electric trucks

For the next weight category up from the electric van and milk-float, electric trucks present a special challenge to designers in trading off battery weight against gross vehicle weight. Use of state-of-the-art controllers and motors such as those described here in some detail can result in effective vehicle designs.

In its manual of good practice for battery electric vehicles the Electric Vehicle Association lays down some useful ground rules for conceptual design of road-going electric trucks. Exploiting the obvious benefits of EV-technology is the first consideration. Thus an ultra-low floor walk-through cab is a real possibility when batteries and motors can be mounted remote from it. Lack of fuelling requirement, ease of start-up and get-away — also driving simplicity of two-pedal control without gearshifting — all these factors lend themelves to operations, such as busy city-centre deliveries where a substantial part of the driver's time is spent in off-loading and order-taking. Any aspect of vehicle design which minimizes the driving task thus maximizes his or her other workload duties. Successful builders of electric trucks are thus, say the EVA, are specialist in assembling bought-in systems and components. Required expertise is in tailoring a motor/battery/speed-controller package to a given application. Principles to be followed in this process are keeping top-speeds and motor power as low as possible consistent with fulfilling the task; using controllers which prevent any unnecessary acceleration once the vehicle has reached running speed in stop-start work; using generously rated motors rather than overspecifying battery capacity — the latter because large batteries cost more, displace payload and waste energy providing tractive force required for their extra weight.

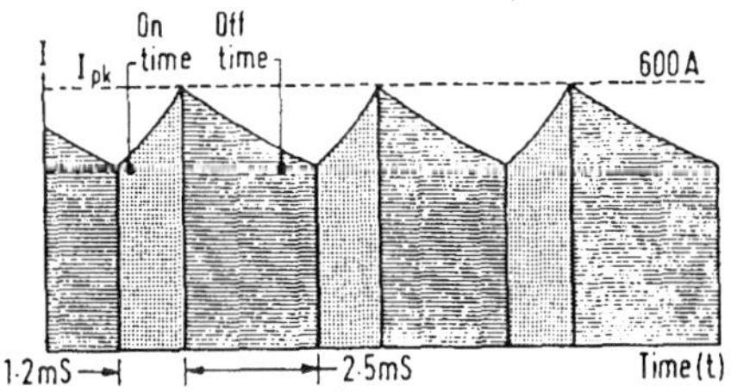

Motor consumption at current limit; ripple is 170 A while average motor current of 500 A compares with battery current of 165 A.

Fig 67: Thyristor controller pulses

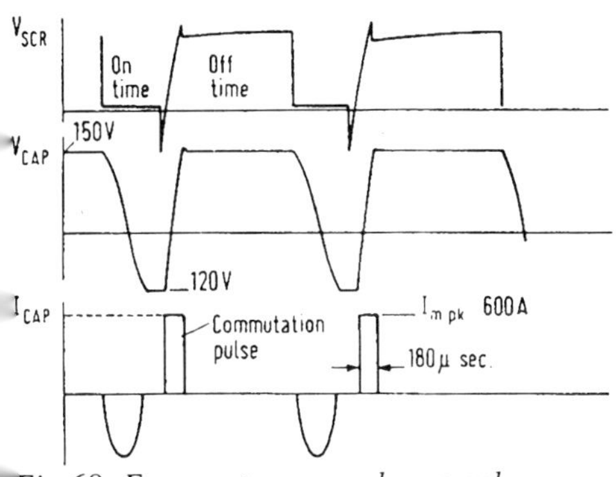

Fig 68: Energy storage and reversal

Whereas over a decade ago the rule-of-thumb applied that one tonne of batteries plus one of vehicle and one of payload could be transported at 20 mph over a 20 mile range including 200 stop-starts per charge, now 40/40 to 50/50 mph/miles range was feasible by careful design. Belted radial tyres could now be run at 50% above normal inflation pressures. High voltage series motors of 72 V and above have working efficiencies of 85-90% over a 3:1 speed range and electronic controllers cut peak acceleration currents and avoid resistive losses — to extend ranges by 15-20% over resistance controllers. A further 10% range increase is possible if high-frequency controllers (15,000 Hz) are used —also much lower internal-resistance batteries reduce voltage fall-off and the use of DC-DC converters means that cells do not have to work so deeply to power auxiliaries, the EVA explain. Percentage of GVW allocated to the battery is a key parameter which should be kept below 35%. Lightweight lead-acid batteries giving 15-25% greater specific capacity than the BS 2550 standard traction cell should be considered, as they also have better discharge characteristics, but there is usually a trade-off in equivalent percentage life reductions. Separately-excited motors and regenerative braking give further bonuses.

Thyristor control has been an important advance and units such as the controller by Sevcon are suitable for 24-72 V working with a current limit rating of 500 A for 5 minutes. Continuous rating at any frequency would be 160 A and a bank of eight 50 mF, 160 V commutation capacitors are employed. Overall dimensions are 430 x 301 x 147 mm and weight is 14 kg. Like other units of this type, it supplies the motor with fixed width pulses at a variable repitition frequency to provide stepless control over the average voltage supplied to the motor. Fig 67 shows the the typical pulse width of 12 ms, giving a frequency range from 10 to 750 Hz and a good motor current form factor with minimal heating effect commensurate with some iron losses. Ratio of on-time to the total period would be 30-50 % while the commutation circuit which switches the main conducting thyristor operates every 2.5-3.5 milliseconds. Fig 68 shows energy being stored and reversed each cycle by the main commutation capacitor.

A typical DC chopper circuit for thyristor drive is shown in Fig 69. Thyristor T1 is the 'on' one and T2 th 'off'; inductive load of a series motor is shown as Rl and Ll while D1 is a so-called 'free-wheel' diode. When capacitor C is charged to a voltage *Vb* in a direction opposing the battery voltage, Ti and T2 are off and load current *Il* is flowing through Rl, Ll and D. A pulse is applied to the gate of T1 to turn it on; D1 becomes reverse-biassed and load current *Il* begins to flow through Rl, Ll and T1. This causes a short circuit, effectively, across CL and D2, creating a tuned circuit. Its resonance drives a current through T2, d2, L and c, sinusoidally rising to a maximum then decaying to zero, at which point C is effectively charged to *V2* in the reverse direction.

When the current attempts the second half of its cycle, D2 prevents any reversal and the reverse charge on c is caught by D2, the 'catching' diode. At time *t*.on a pulse applied to the gate of T2 turns it on, applying reverse voltage on C across T1 — which rapidly turns off and diverts load current through Rl, Ll, C, L and T2. C is thus charged at a linear rate by *Il* passing through it. When capacitor voltage again approaches *Vb*, opposing battery voltage, load current begins to flow through Rl, Ll and D1 again, current through T2 dropping below the holding value so turning off T2 — completing the cycle ready for reinitiation by pulsing the gate of T2 again. The circuit thus represents a rapid on/off switch which enables effective DC output voltage to be variable according to:

$$[t_{on}/(t_{on} + t_{off})]Vb$$

and output from the device is a train of rectangular voltage pulses. The steady current related to such a train is obtained by connecting D1 across the motor load. Output voltage can thus be controlled by altering $t._{off}/t.$on while holding $t._{on} + t._{off}$ constant — a constant frequency variable mark/space chopper — or altering $1/(t._{on} + t._{off})$ while holding $t._{on}$ constant — a variable frequency constant mark chopper.

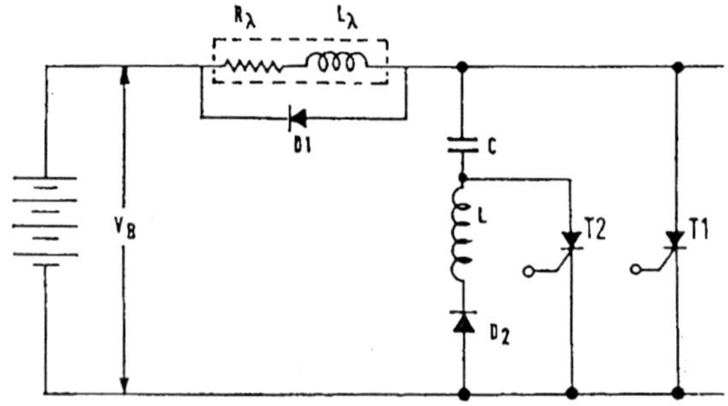

Fig 69: DC chopper circuit

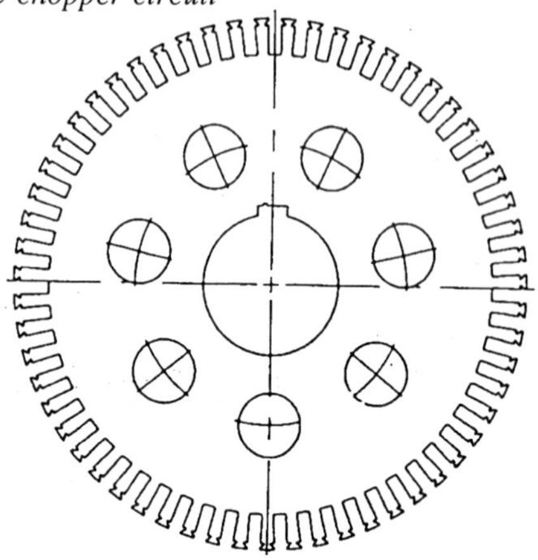

Fig 70: Armature laminations

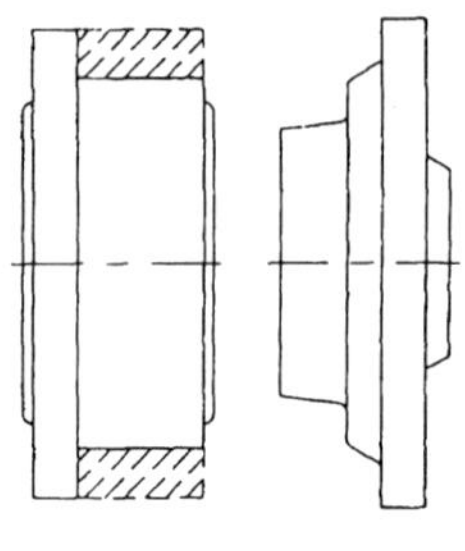

Fig 71: Faceplate adaptor

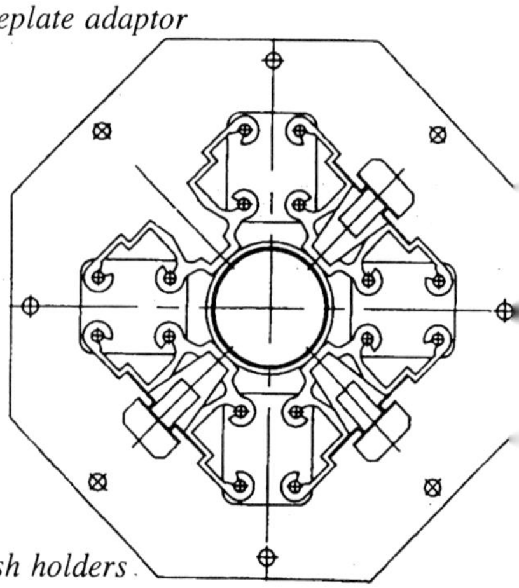

Fig 72: Brush holders

Electric motor considerations

According to EV motor makers Nelco, requirements for traction motors can be summarised as light weight, wide speed range, high efficiency, maximum torque and long life. The company have now developed their diagonal frame Nexus II motor, for more general electric truck operation. In this motor, active iron and copper represent 50 and 30 % respectively of the motor weight. Holes in the armature lamination, Fig 70, have resulted in some weight reduction and the use of a faceplate commutator has also helped keep weight down, Fig 71 — with only 30 % of the copper required for a barrel-type commutator — because the riser forms part of the brush contact face. With use of aluminium alloy for the non-active parts, such as brush holders (Fig 72) of the motor, weight of the 132L motor is held to 80 kg, a power to weight ratio of 450 watts/kg. Tolerance of high accelerations comes from perfection of the faceplate commutator to retain brush track surface stability. Usually the constraint on high power at high speeds, particularly when field strengths are reduced, is commutation ability, the company maintain.

The patented segmented frame of the Nexus, Fig 73, makes the provision of interpoles quite an easy option — to optimise commutation at all current loadings, so reducing brush heating losses and compensating for interpole coil resistance losses. As output torque is a function of armature current, flux and the number of conductors, all these must be maximised. Short time high-current densities, over the constant torque portion of the performance envelope, are possible given adequate cooling. Cost is held down by such measures as use of a segmented yoke/pole assembly, Fig 74; extruded brush holders are also used, Fig 75. Fig 76 shows rating and efficiency curves for the N180L machine.

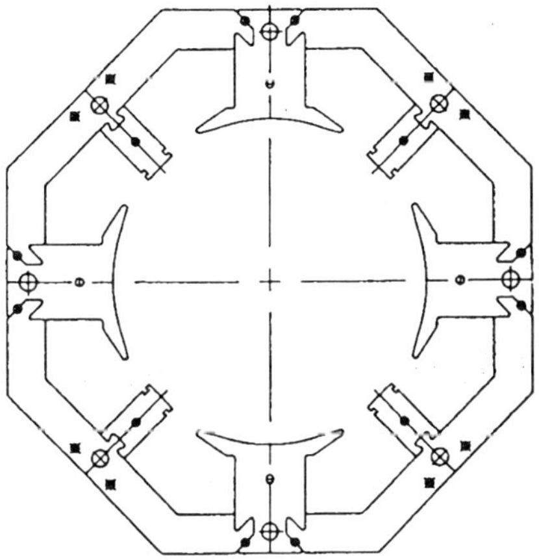

Fig 73: Segmented frame

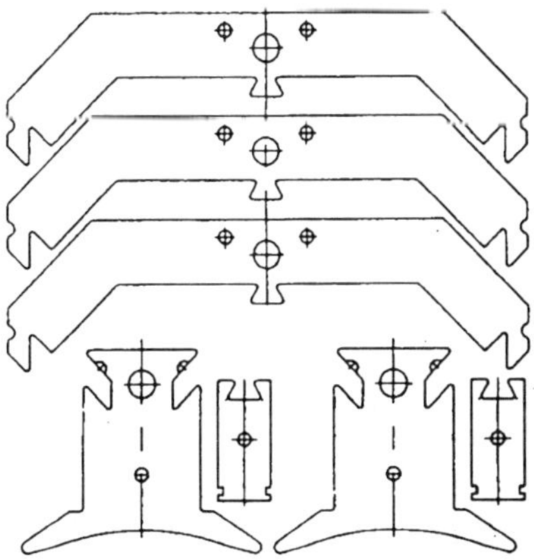

Fig 74: Segmented yoke/pole assembly

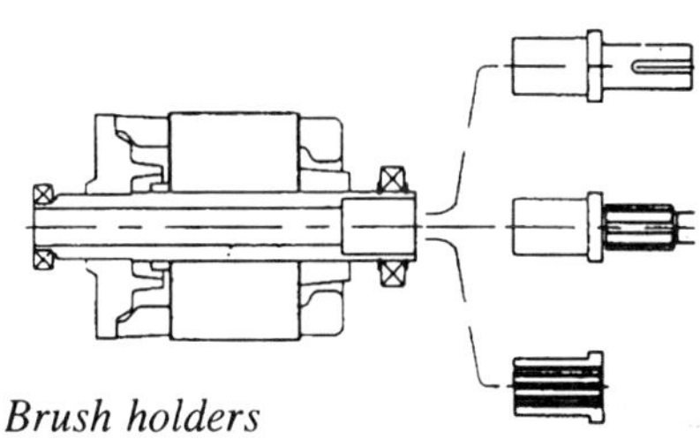

Fig 75: Brush holders

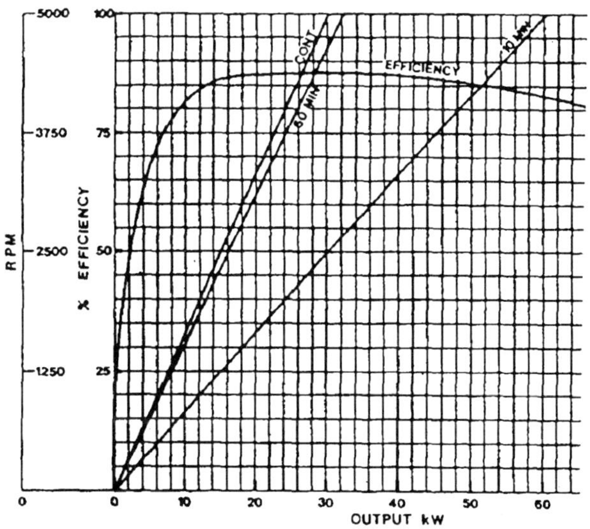

Fig 76: N180L motor characteristics

Solar vehicles

The most ardent critics of electric vehicles usually find it difficult to find fault with solar-powered vehicles; on environmental grounds, the argument about transferring pollution from roadway to power-station is no longer valid. Until higher efficiencies are achieved in photo-voltaic cells, however, and their cost comes down, EV power will usually be supplemented by chemical storage batteries, themselves ideally recharged in solar-powered electricity generating stations — for operation in appropriate climates.

The trans-Australia runs achieved in the Sun-Racer challenge series have demonstrated the possibilities for such vehicles but it has also shown that mere conversions of petrol-driven cars, as in the case of many battery-electrics, is right out of the question. The style of driving required also precludes conventional urban traffic operation but prospects are good for those more sparsely populated regions of the Third-World where a transition from pedestrian to wheeled traffic, without the mixed blessings of the internal combustion engine, could be a reality.

A recent report suggests battery power density versus state of charge as in Fig 77 sourced from the US Argonne Laboratories. Super quick charging characteristics, sourced from Nissan, were reproduced in Fig 78. Performance and cost status of EV batteries under development is shown in Fig 79, sourced from the mid-1992 Australian Electromotive Exposition.

Key vehicle-system developments, to make up for available energy shortfall compared with current vehicles, are transmission and motors; weight saving; rolling and air drag reduction; reducing parasitic losses like heating/ air-conditioning; reclaiming lost energy by regenerative braking. Of the four basic motor types, DC series and separately-excited; AC asynchronous, DC permanent magnet, brushless, the majority of EVs still use the simpler DC approach, with chopper control of tractive-performance. Now there is a move to AC drives with reduced size and more efficiency. Recent efforts to reduce rolling resistance include the use of 0.0005 rolling resistance co-efficient tyres by Nissan, on its FEV vehicle; another interesting development is the use by the US Washington State Ecology Department of two tyres on a common rim, the inner one of hard rubber providing low rolling resistance and the outer one of soft rubber only making ground contact under vehicle-roll conditions, to enhance stability. For reducing parasitic losses the report suggests that heat-pump type air-conditioning may become the norm on EVs. Regenerative energy savings can arise from such systems as the Brembo Kelsey Hayes system, for braking, with which between 25-30% of kinetic energy becomes recoverable.

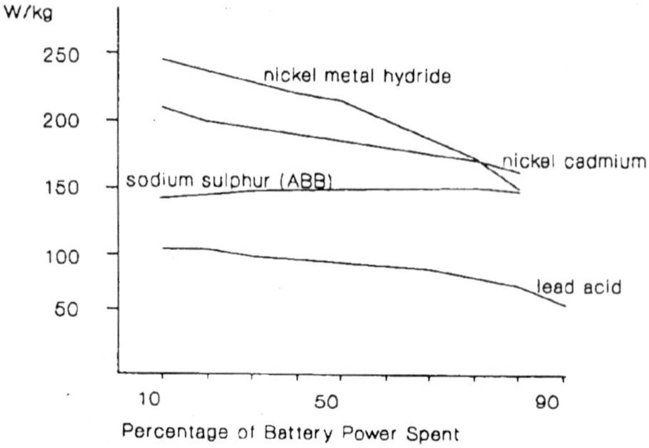

Fig 77: Power density comparison

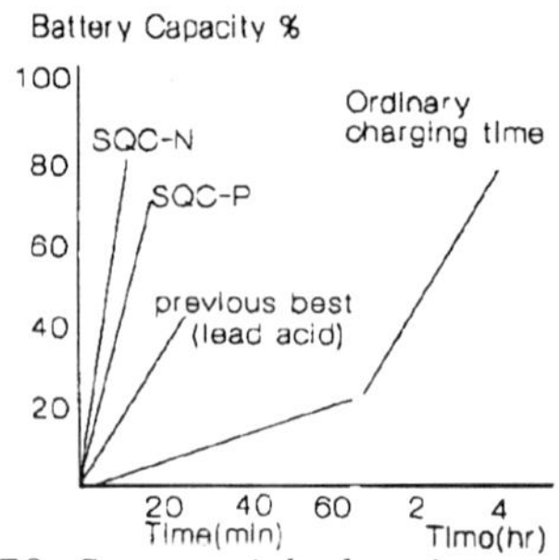

Fig 78: Super quick charging

	Energy Wh/kg	Peak Power W/kg	Energy Efficiency %	Cycle Life	Self discharge % per 48 hr	Cost US$/kW
Lead acid	30-50	70-130	>80	50-1500	0.6	70-130
Zinc bromine	70-80	35-100	75	200-2000	40	150-300
Nickel hydrogen	60-70	300	70	2000 +	6	700-200
Nickel iron	40-60	80-150	60	500-2000	3	200-140
Nickel zinc	55-85	170-260	70	500	1.2	200-500
Aluminium air	200-300	160	<50	?	?	?
Iron air	100	90	60	500 +	?	50
Zinc air	120-200	30-80	60	800 +	?	110
Sodium sulphur	100-140	130-180	85	800 +	0	110
Sodium nickel chloride	90-130	60-130	?	1000-2000	?	?
Lithium iron sulphide	60-130	100-200	80	1000 +	?	110
Lithium composite	85-130	?	80-85	100	0.03	100

Fig 79: Performance cost comparisons

Successful exploitation of solar vehicles requires a yet more radical approach to EV design in energy-saving generally, and weight saving in particular. Such a design inevitably marginalises its operation unless from the outset customers realise that the vehicles are designed for special duty.

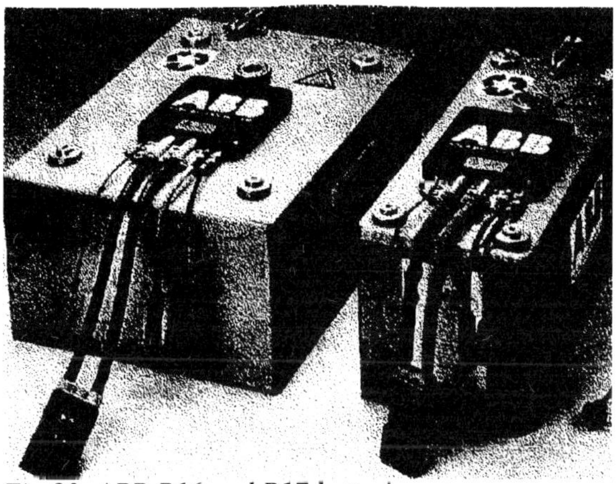
Fig 80: ABB B16 and B17 batteries

Table 1: Technical data of the B16 high-energy battery

Nominal energy content		9.6		kWh
No. of cells (A08)		120		
Dimensions (L x B x H)		730 x 312 x 315		mm
Height of connection zone		25		mm
Weight		103		kg
Heat losses		70		W
Parallel strings	4	5	10	
Nominal voltage	60	48	24	V
Open-circuit voltage, fully charged	62.4	49.9	25	V
Minimum discharge voltage	56.1	44.9	22.4	V
Internal resistance (310 °C)	60	38	9.6	mΩ
Maximum current (< 3 min)	250	300	600	A
Continuous discharge current	160	200	400	A
Continuous charging current	80	100	200	A
Capacity	160	200	400	Ah

Table 2: Technical data of the B17 high-energy battery

Nominal energy content		19.2		kWh
No. of cells (A08)		240		
Dimensions (L x B x H)		730 x 540 x 315		mm
Height of connection zone		25		mm
Weight		175		kg
Heat losses		90		W
Parallel strings	4	5	10	
Nominal voltage	120	96	48	V
Open-curcuit voltage, fully charged	124.8	99.9	50	V
Minimum discharge voltage	112.2	89.9	44.9	V
Internal resistance (310 °C)	120	76.8	19.2	mΩ
Maximum current (< 3 min)	250	300	600	A
Continuous discharge current	160	200	400	A
Continuous charging current	80	100	200	A
Capacity	160	200	400	Ah

Fig 81: ABB battery specifications

Battery selection

A solar-cell-assisted vehicle would need supplementary power storage capacity — particularly if the ideal is to be met of operating in an area where solar-power generation is also available to charge such a back-up system. For such a battery the selection criteria are much the same as for the 'normal' EV, save perhaps for the even lighter weight requirement — on a vehicle already carrying a large solar cell. There would perhaps be a slightly lower concern for deep-discharge performance, however, if the solar cell is providing complementary power.

General requirements, Nelco suggest, are for very low internal resistance; excellent gas recombination; totally sealed construction; physical robustness; long life; room- temperature operation; reasonable deep-discharge rate and compact construction. Nelco cite the Hawker-Siddeley lead-acid foil battery used on aircraft as a possibility, being one which retains far more charge from regeneration than conventional designs and can be subjected to rapid recharge rates. To do this, cells must be kept at uniform even temperature, requiring some liquid cooling in certain conditions and heating, by dumped waste heat from the motor in others.

According to ABB Hochenenergiebatterie GmbH, their sodium sulphur battery now offers energy-to-weight ratio in excess of 100 Wh/kg. The company's new-generation A08 cells are higher density than their A04 units and liquid rather than air-cooling has been adopted. Each cell has 40 Ah capacity, has outside-diameter of 38 mm and is 225 mm long. Up to 480 can be arrayed vertically side-by-side as a battery. Heat retained in the liquid coolant is used to optimise the operating temperature of the cells. So far the production battery options are the B16 and B17 shown in Fig 80, of 120 and 240 cells; seven other customised variants are available. Fig 81 tabulates their specification and Fig 82 shows the discharge characteristics of a B17 battery with nominal output of 96V.

Optimum utilisation of battery energy within an EV depends, says ABB, on control by a battery management system which must detect and process signals from the battery; monitor battery condition; transmit data to the processor within the drive controller, Fig 83, and regulate battery temperature.

Solar cell technology

The technology of solar cells has been given a recent boost by the Swiss Federal Institute of Technology who claim to have outperformed nature in the efficiency of conversion of sunlight to electricity even under diffuse light conditions. The cell has a rough surface of titanium dioxide semiconductor material and is 8% efficient in full sunlight rising to 12% in diffuse daylight. For more conventional cells, such as those making a Lucas solar panel, these are available in modules of five connected in series to give maximum output of 1.3 watts (0.6 A at 2.2 V). Some ten modules might be used in a solar panel giving 13 watts output in summer conditions. Power vs voltage and current vs voltage are shown in Fig 84 for so-called 'standard' and 'typical' operating conditions. 100 mW/cm^2 solar intensity, 0 C cell temperature at sea level defines the standard conditions against 80 mW/cm^2 and 25 C which represents 'typical' conditions at which power output per cell drops to 1 W. Temperature co-efficients for modules are 0.45% change in power output per 1 C rise in temperature, relative to 0 C; cell temperatures will be 20 C above ambient at 100 mW/cm^2 incident light intensity. Variation of solar energy at 52 degrees north latitude, assuming a clear atmosphere, is shown in Fig 85. On this basis the smallest one person car with speed of 15 mph and weight of 300 lb with driver would require 250 W or 50 ft^2 (4.65 m^2) of 5% efficient solar panel — falling to 12.5 ft^2 (1.18 m^2) with the latest technology cells. A 100 Wh sealed nickel cadmium battery would be fitted to the vehicle for charging by the solar panel while parked.

The future, of course, lies with the further development of advanced cell systems such as those by United Solar Systems in the USA. Their approach is to deposit six layers of amorphous silicon (two identical n-i-p cells) on to rolls of stainless steel sheet. The 4 ft^2 (0.37 m^2) panels are currently 6.2% efficient and made up of layers over an aluminium/zinc oxide back reflector. The push to yet higher efficiencies comes from the layer cake construction of different band-gap energy cells, each cell absorbing a different part of the solar spectrum. Researchers recently obtained 10% efficiency in a 12in^2 (0.09 m^2) module.

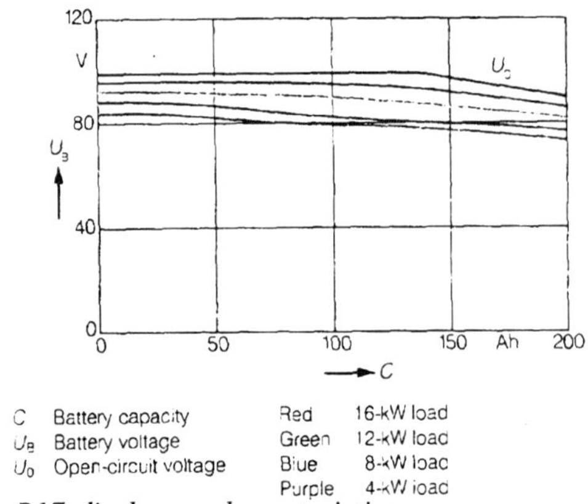

Fig 82: B17 discharge characteristics

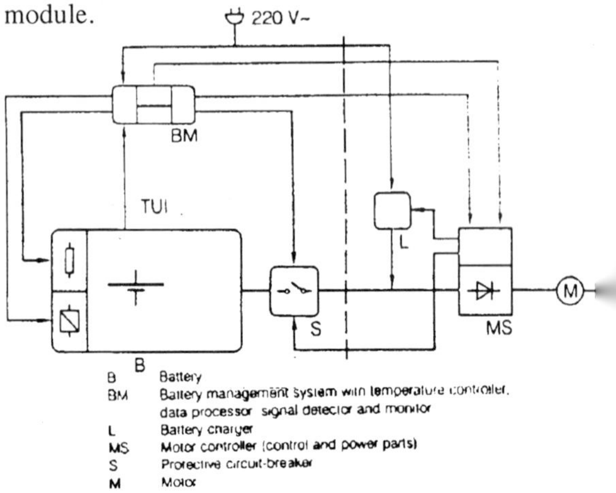

Fig 83: Battery management system

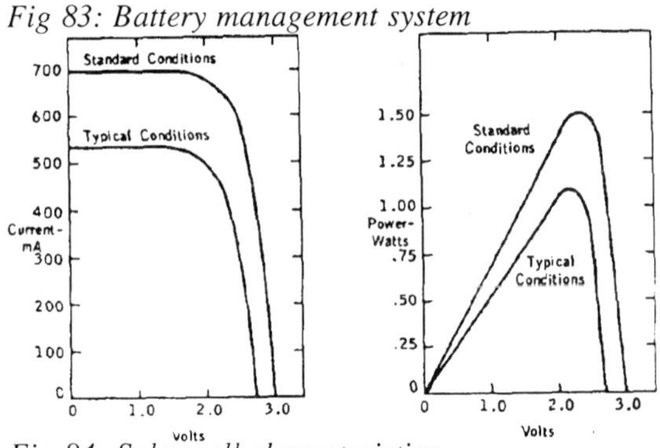

Fig 84: Solar cell characteristics

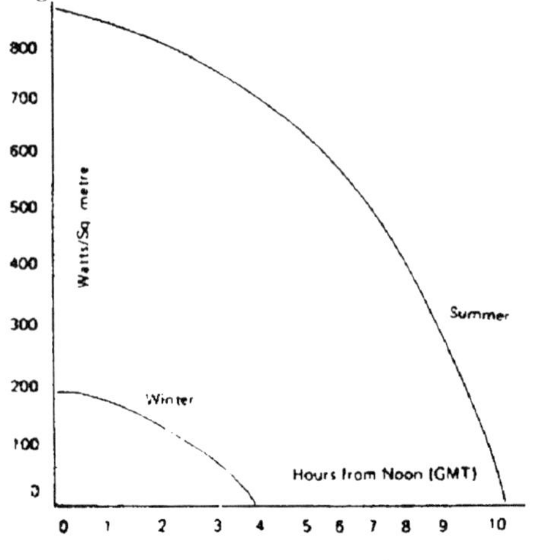

Fig 85: Solar energy variatio — left curve, winter; right curve, summe

The way forward in EVs?

A director of battery makers Varta, Dr Reinhard Gereth maintains that "those who want to wait for some miracle battery will only delay the introduction of the electric car instead of promoting it". He recommends that what is now feasible should be implemented. The company's nickel hydride system is claimed to double the range available with lead-acid batteries and offer a possible route. It can be recharged up to 75% of capacity in a few minutes and fully recharged in an hour . The negative electrode is a hydrogen energy-storage alloy while nickel hydroxide is the positive electrode. An optimum design would have weight around 300 kg, and capacity of 15 kWh, with life of 2000 discharge cycles. For buses the company have devised a mobile charging station, in co-operation with Neoplan, which will allow round-the clock operation of fleets. This removes the need for fixed sites and allows battery charging and changing to be carried out by the bus driver in a few minutes. The mobile station is based on a demountable container which can be unloaded by a conventional truck. Trials have shown that a bus covering a daily total distance of 75 miles on a three-mile-long route needs to stop at the station after eight journeys. Discharged batteries are changed semi-automatically on roller-belt arms, by a hand-held console.

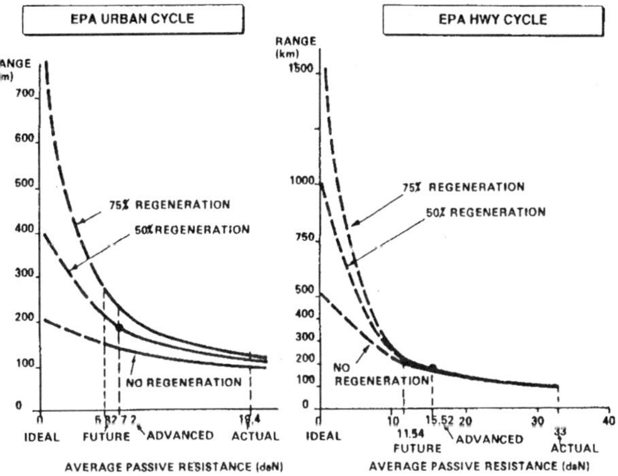

Fig 87: Drive cycle representation by Turin Polytechnic

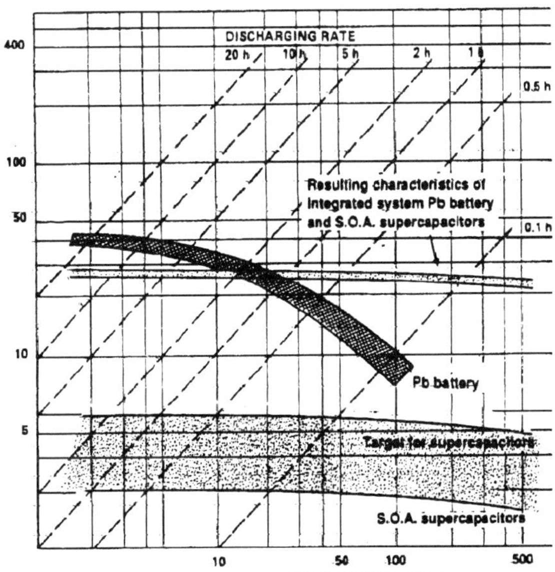

Fig 88: Effect of supercapacitors

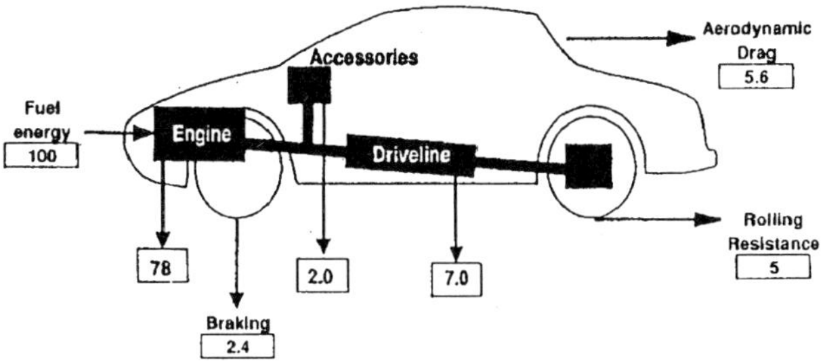

Fig 86: Energy breakdown for urban and highway cycles

Although arguments still rage about power-station emissions, according to Siemens, solar technology is a probable solution for Third World tropical countries. Solar modules are available from the company to supply 12 V 100 Ah batteries from a 50 W solar module. The company recently installed a system on the Cape Verde Islands with a collective power output of 550 kW at each of five island sites. Even in Bavaria, the village of Flanitzhutte, which has an average 1700 hours annual sunshine period, has severed its links with the national grid with the installation of 840 solar modules, with a total area of 360 square metres, to provide peak power of 40 kW. Maintenance-free batteries provide a cushion on grey days. Work at Turin Polytechnic is showing the importance of advance aerodynamic solutions in making alternative energy sources viable. This cites a study by GM who estimated the breakdown of energy consumption for combined city and highway driving cycles as in Fig 86. The researchers have represented the driving cycle in the form shown in Fig 87 and note that for typical cars, of around 0.3 drag coefficient and 0.015 rolling resistance coefficient, that the periods in which braking energy is dissipated are a not inconsiderable portion of the total time, 20% for highway and 12% for urban.

By examining the equivalent drive cycles for `advanced' and `future' design cars with drag coefficients and rolling resistances of 0.19, 0.13 and 0.0048, 0.004 respectively. The passive energy plots of Fig 87 show the considerable improvements possible. By adopting such means and adding supercapacitor accumulators to an EV's traction battery system, improved specific energies shown in Fig 88 are feasible. Supercapacitors are described as power-peaking devices to increase power acceptance and delivery at engine braking and acceleration respectively. The units are linked electronically to the storage batteries.

University of the West of England researchers report magneto-dynamic storage devices to help resolve energy supply and demand problems. A magnet motor has been designed for an electric bus which comprises a permanent-magnet alternator with rotating excitation, around which a composite flywheel is mounted. Power flow in the alternator is controlled and reversible by means of a static converter. High power density and storage efficiency, combined with almost unlimited cycle life, makes it an attractive proposition for urban traffic. The researchers showed Fig 89 as a way for obtaining the combined power and energy density and this could be applied to, say, the combination of a fuel cell and supercapacitor and suggests that equivalent performance and range to a conventional vehicle (without the additional weight) was in sight, but with negligible emissions.

The US Jet Propulsion Laboratory were early into the methodology for relating battery performance requirements to projected vehicle performance, as in Fig 90. Part 1a is the result of determining battery performance requirements (for a particular range) as a function of battery mass fraction based on vehicle performance and weight characteristics; this curve forms the basis for 1d; the volumetric characteristics of vehicle and battery are accounted for in the development of 1b which yields the maximum mass fraction possible as a function of battery mass density. The battery capabilities, on the discharge curve of 1c are finally considered. For meaningful comparison, the 30 second specific power capability (50% state-of-charge, 2/3 open circuit voltage assumed) and the specific energy capability in vehicular use (urban and highway cycles combined) must be determined. A particular vehicle design implies a point designating the battery capability on 1d.

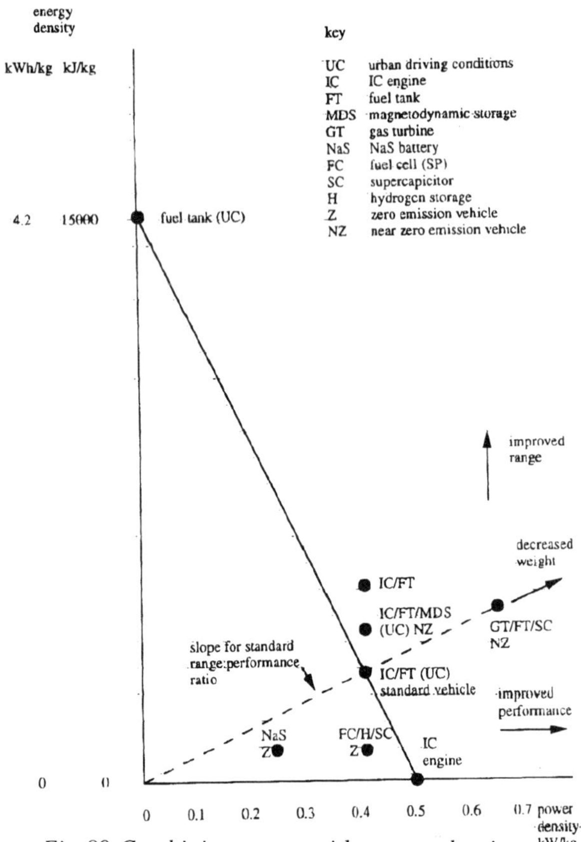

Fig 89 Combining power with energy density

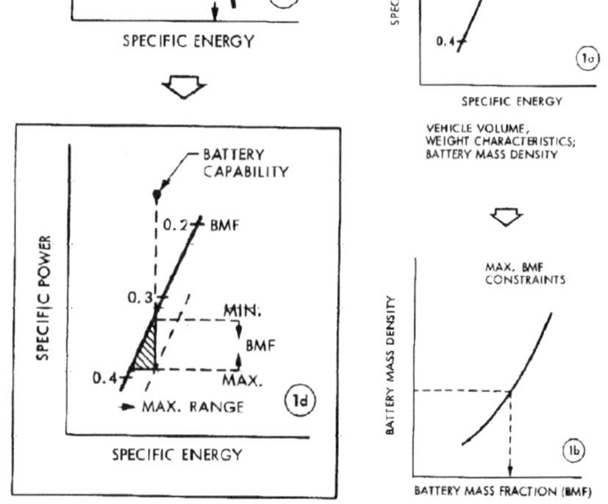

Fig 90: Relating battery perform-ance to vehicle performanc

Production contenders for electric/hybrid drives

Siemens estimate the potential world market for electric cars to reach half a million units by year 2001and this has prompted the company to set up an EV Drives Division at Wurzburg; these are already have twenty different current projects with vehicle makers. The company are concentrating their efforts on three drive concepts: DC motors with parallel-switched excitation; 3-phase AC motors with induced field (asynchronous motors) and those with permanent magnet field (PM-synchronous motors). There are also a number of smaller European contenders including an interesting proposal for a substitute hybrid power pack.

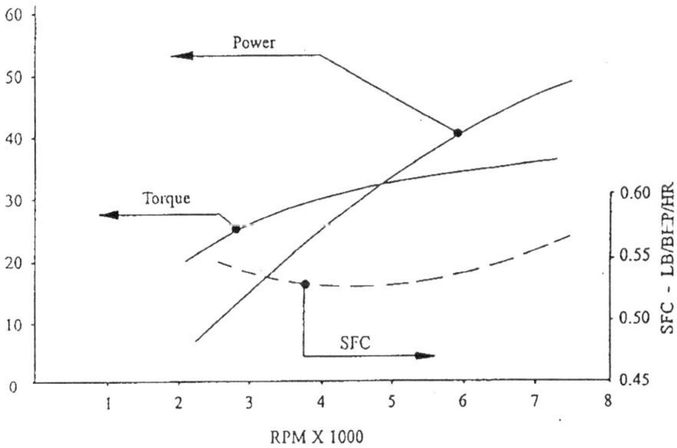

Fig 91: Rotary engine power characteristics

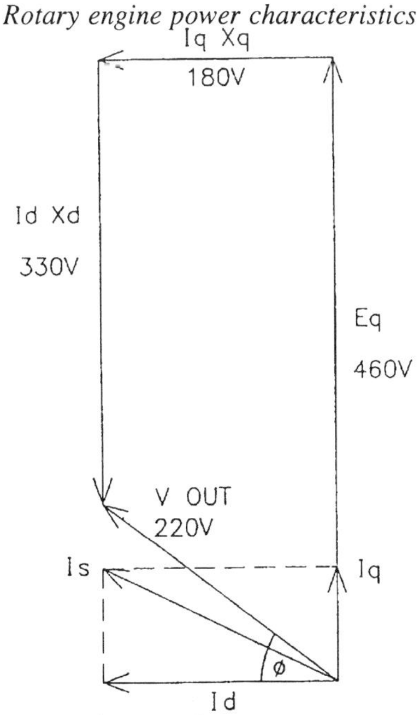

Fig 92: Vector diagram for motor

The capital cost of building a completely new EV is avoided by making a tailored hybrid-drive unit that can be readily substituted for the engine and gearbox of a conventional vehicle. Accordingly, Nelco have developed a parallel hybrid configuration for a mini-sized car which would provide a range of 500 miles and 80 mile/h continuous cruising speed capability. A mini-car installation would use 18 30 AH 12 V aerospace batteries, weighing 162 kg and the potential of 1100 cycles, when discharged to a depth of 80%, is possible. A pack measuring 1200 x 1200 x 40 mm could be under-floor mounted but a thermal management system would be required to maintain the cells at 30-40 C. Proposed IC-engine for the hybrid set-up is a 22.4 kW Wankel of 22.4 kW, chosen for its quick warm-up and level torque speed characteristic, Fig 91. The corresponding Nelco electric motor would be a liquid-cooled PM brushless DC unit weighing 45 kg.

Batteries alone would supply the transient power, at 44.7 kW, to give a total power of around 67 kW, some 60 kW being available without gear-changing, from 2500 to 7000 rev/min. Starting torque for the Wankel is only 8 Nm, allowing a vehicle to be bump-started without difficulty. The motor would be vector-controlled and have a trapezoidal-type drive with vane switches and an interrupter disc on the encoder for field-current timing.

Above 1500 rpm motor voltage equals DC bus voltage and the motor operates in free-field mode. Below that speed it runs as a current drive source. In the first case armature reaction from the reactive component of the current reduces back-EMF of the PM field (Fig 92) and the motor operates with a square-wave below 1 Hz, with a leading power factor resulting in no switching losses. The result is high efficiency over the cruise range from 1600 to 6000 rpm. Speed control is by pulse-width modulation at low speeds and at mid speeds inverter efficiency is further increased by switching at motor frequency. Vector control means that the stator current is resolved into two components, the reactive one controlling field and the real one controlling power. The torque-speed requirement for a typical mini-car is thus met Nelco believe, Fig 93; In the first region voltage rises with speed while in the second it is held constant by the introduction of a circulating current. The latter produces 220 V across the inductance of the machine winding as well as armature reaction to weaken the motor field, the combined field giving both lower air gap flux and lower iron losses.

The use of two control systems, current source and voltage source, for the drive optimises battery energy by operating motor and drive at optimum efficiency in the critical cruise mode. Current source control permits the use of trapezoidal wave control, chosen by the company for control-system economy. A current loop compares torque with throttle pedal demand, any error driving the PWM unit so as to gate the six switches in sequence and ensure smooth torque control is available right up to standstill. In the voltage source region motor torque per given current reduces as speed increases and so a DC link current is a better measure of feedback. At high speeds, therefore, the motor is energised with square waves at the motor frequency which are shifted in phase to control acceleration and braking. Fig 94 shows the drive system package which weighs 100 kg .

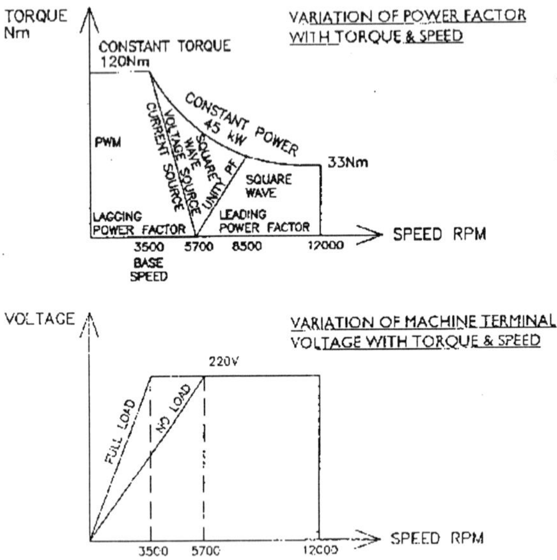

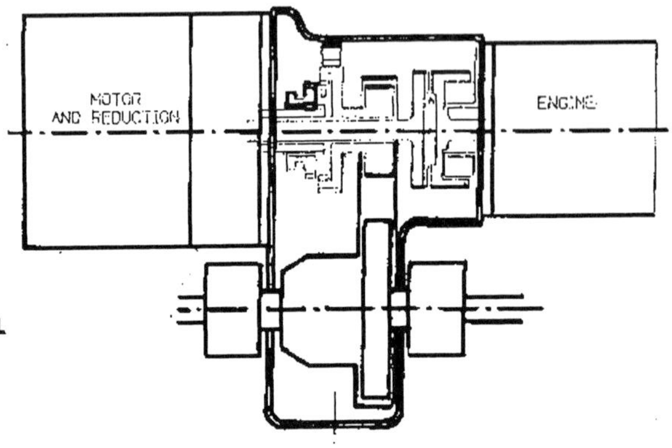

Fig 94: Nelco hybrid drive system package

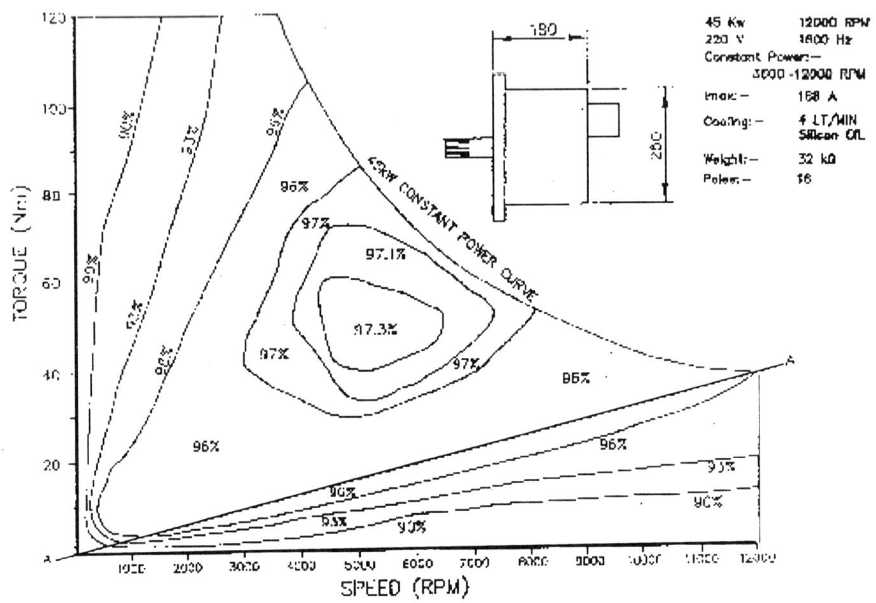

Fig 93: Matching motor output to vehicle performance requirement

PART 5:
VEHICLE SYSTEMS

Chapter 16: The power unit

Performance-calculation
Engine design
Engine air-intake
Engine fuelling systems
Exhaust system analysis
Gasoline emission control
Diesel particulate control
Engine mounting and balancing
Heating and cooling
IC engine noise control
Heat exchanger calculation

A number of specialist texts, including one edited by the author[1], deal with power unit technology in greater detail than is done in this chapter. The purpose here is to try and provide the vehicle engineer with guidance in the selection of engine fitment and acquaint him with some of the key development areas, driven mostly by the trade-off between fuel economy and emission control.

Performance prediction

If a clear picture of the performance requirements of the vehicle are available before the design commences, performance prediction techniques could lead to precise unit requirements in terms of purchase specification. An approach due to J J Taborek[1] provided a systematic use of known dynamic and kinematic relationships. This early work provides the systems approach, described here, to calculate performance from a variety of sources of motion resistance data.

Tractive force distribution

A key relationship in performance considerations is that tractive force P is equal to load on the driving axle W times tyre-to-ground adhesion co-efficient m. While W for a static vehicle can easily be found by a moment calculation, the dynamic axle reactions of a moving vehicle are more difficult to work out. The dynamic weight transfer:

$$dW = (H/L)(T_d/r - fW) \text{ and } P = (T_e RE)/r$$

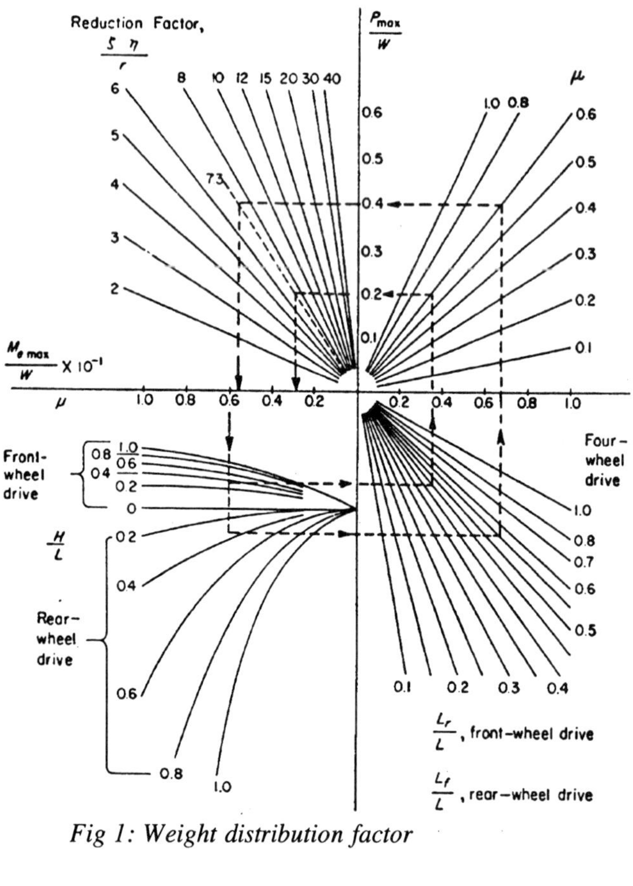

Fig 1: Weight distribution factor

where T_d is torque at drive-axle, T_e torque at engine, R total reduction ratio, E transmission efficiency and r rolling radius. H is height of vehicle centre of gravity (CG) above ground, L the wheelbase and f the symbol for rolling resistance coefficient.

With front and rear wheel drive, dynamic front axle load $= W(L_r + fH)/(L + \mu H)$ where L_r is the distance from CG to rear axle and μ is adhesion coefficient; this leads to the maximum transferable tractive force as

$$P_{f,max} = \mu W[(L_r + fH)/(L + \mu H)]$$

the square-bracketed term being the weight distribution factor $dW/W = w$ so that

$$W_{df} = W_{wf} P_{max} = \mu W_{wf} \text{ and } P_{r.max} = \mu W dr$$

With four-wheel-drive, $P_{4.max} = \mu W$ and $w_4 = 1$; otherwise the chart in Fig 1 can be used to determine the relative distributions. The weight distributions derived above are dimensionless, being derived from ratios, and the chart also contains only dimensionless factors, with the weight of the vehicle being eliminated by dividing the tractive force equation by W, giving $P_{max}/W = \mu w$.

The dotted line shows an example for a vehicle with $h/L = 0.35$ and $L_f/L = 0.45$ defining its CG position, for which it is required to find maximum transferrable tractive force and the maximum engine torque for a low gear reduction factor of $RE/r = 7.3$ — given also μ = 0.6 for the adhesion coefficient.

Starting in the lower left quadrant, the intersection of the m = 0.6 line cuts curves for front and rear wheel drive at $h/L = 0.35$. Horizontal projections from those points similarly lead to the P_{max}/W scale. Values thus obtained give the first solution while projecting into the top left quadrant gives the second. For a vehicle weight of 4000 kg, $P_{max} = 1620$ kg and $T_{e.max} = 224$ kNm (front drive) or 840 kg and 112 kNm (rear drive).

Gradient vs speed approximation

This data can be used in a chart, such as that shown in Fig 2, to obtain available vehicle speeds on different gradients. The chart plots maximum values of tractive force at the wheel, as functions of speed, in different gears. Also on the chart, gradient slopes are plotted against vehicle weight. The 0% line is drawn at a slope designating a rolling resistance of 35.4 lb/ton (80 kg/tonne) chosen for example. Propelling force P lines are obtained by graphical subtraction of wind resistance from maximum tractive force in this method — due to Goldschmidt and Hadar of the Israel Institute of Metals. Dotted lines illustrate worked examples for vehicles weighing 1910 and 1875 lb respectively. The first is shown to be able to climb a slope of 10% at 36 mph reducing to 9% at the same speed if that speed is maintained. The second climbs an 8% slope at 43.5 and 38.5 mph in third and fourth gears, respectively.

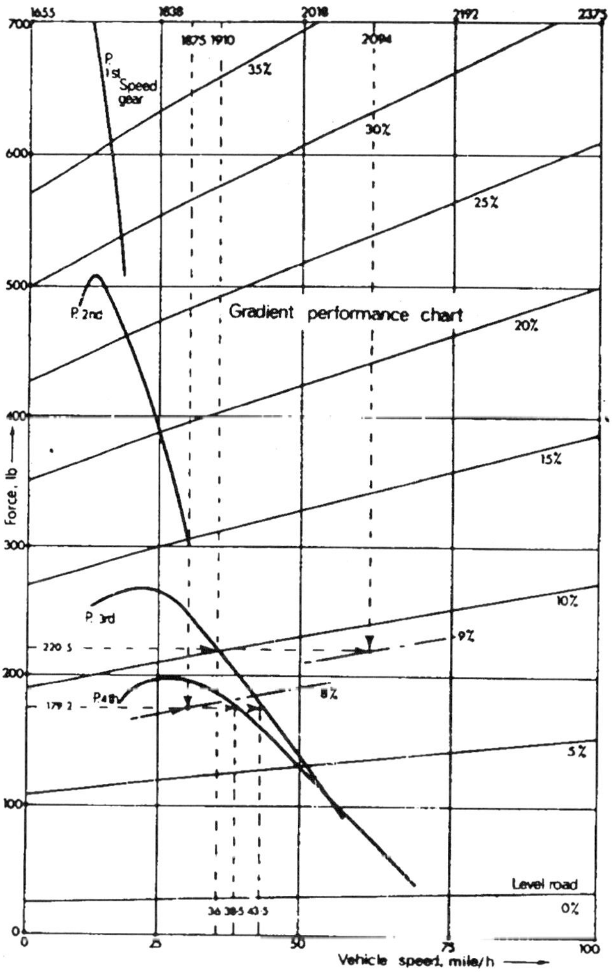

Fig 2: Tractive effort

Transmission inertia resistance

While translatory-mass inertia calculation is relatively easy, that of rotary-mass is more difficult. It is necessary to sum the rotating inertia of the separate transmission components by relating relative torque to the driving axle as $T_i = \Sigma IaR^2$ — for moment of inertia I and angular acceleration a. Since wheel circumferential speed is equal to vehicle translatory velocity, an 'equivalent mass' can be considered as concentrated at the rolling radius — having the same effect on the inertia of translatory motion as the summation of individual rotary inertias.

Thus effective inertia mass $M_i = M_e r^2 \alpha d = \Sigma IR^2$ where equivalent mass $M_e = \Sigma IR^2/r^2$, the effective mass being $M_l = M + M_e = MY$ where Y is termed the rotary mass factor, a valuable tool of the method.

To find its value, the rotary masses are divided into parts rotating with wheels — and those rotating with engine. The latter gain in importance in the lower ranges since M_e is proportional to R^2. Average values quoted in the literature for a fully laden commercial vehicle are fourth-gear, 1.09, then 1.2, 1.6 through to 2.5 in first. The value of Y is obtained from

$$Y = 1 + M_e/M = 1 + [(\Sigma I_w/Mr^2) + (\Sigma IR^2/Tr^2)]$$

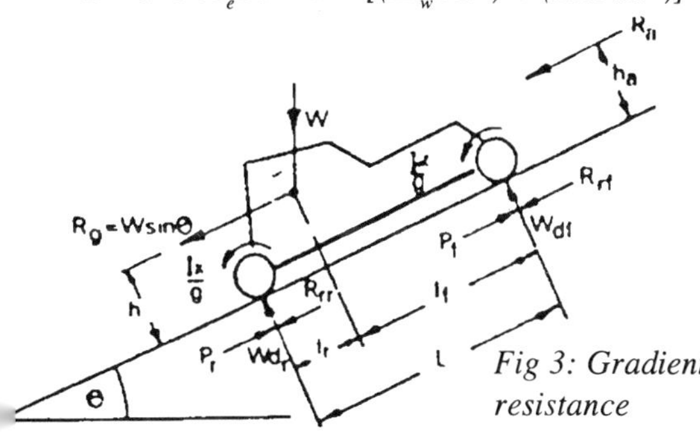

Fig 3: Gradient resistance

where ΣI_w is wheels inertia and I_e is that for 'engine speed' parts and typically $Y = 1 + (0.4 + 0.0025R^2)$.

Performance calculation

Rolling resistance R_r can be expressed as a dimensionless co-efficient f (= 0.013 for radial tyres) while transmission resistance is obtained from a summation of the power consumed at various stages along the powertrain. Overall efficiency E is a product of typically 0.99 for clutch, 0.98 for gearbox (direct) and 0.95 (indirect); final drive is assumed to be 0.98, giving E = 0.90 in direct drive and 0.85 in lower gears. Air resistance is obtained from:

$$R_a = C_d A \rho V^2/2g$$

for vehicle speed V, gravity acceleration g, projected frontal area A and air-density r (R_a rising to 0.8 for a bluff box van). Gradient resistance is the component of gravitational force parallel to vehicle attitude, Fig 3. Since
$W_d m = fW + W tan\, q_{max}$, maximum gradient $= Wm - f$

Accelerative performance

Selection of gear ratio will depend upon chosen operation for the vehicle. Ratio for maximum acceleration can be estimated from

$$G_{max} = (Rr/T_e) \pm [(Rr^2/T_e) + (Wr^2/g + I_w/r)/I_e]^{1/2}$$

where motion resistance $R = R_r + R_a$ and I_e and I_w are engine and wheel inertias. Engine speed relates to vehicle speed as

$$V = 2\pi N_e\, r.3600/12.60 E_s G.5280 = N_e r/168 G E_s$$

and tyre slip efficiency E_s can be assumed equal to 0.965 for this calculation.

Thus N_e and T_e for each gear reduction may be tabulated against road speed by means of the engine torque/speed curve. Motion resistance forces are subtracted from P to give free tractive force P_f then transferable tractive force is calculated. To obtain acceleration, $\alpha = T_f G/YW$ is used and values can be plotted in a curve such as Fig 4, a time/speed curve which yields times for accelerating up to (and between) given speeds.

Engine characteristics

A useful method for estimating the engine torque T_e *vs* speed curve is due to G G Lucas of Loughborough University of Technology. This starts with knowledge only of the maximum power and torque (at speeds) of the engine. The approximation uses the fact that the power/torque *vs* speed characteristics are zero at their peak values. The technique is to use the power formula for an engine

$$P_e = 2\pi T_e N/33{,}000$$

for torque in lb ft and engine speed N in rpm — to calculate the torque curve slope at maximum power and the power curve slope at maximum torque. From there an approximate power curve can be derived. A computer program could be written to perform this procedure of successive approximations. It could also be possible for the method to be accurate enough for engines employing inlet or exhaust tuning provided the appropriate resonant speeds are known.

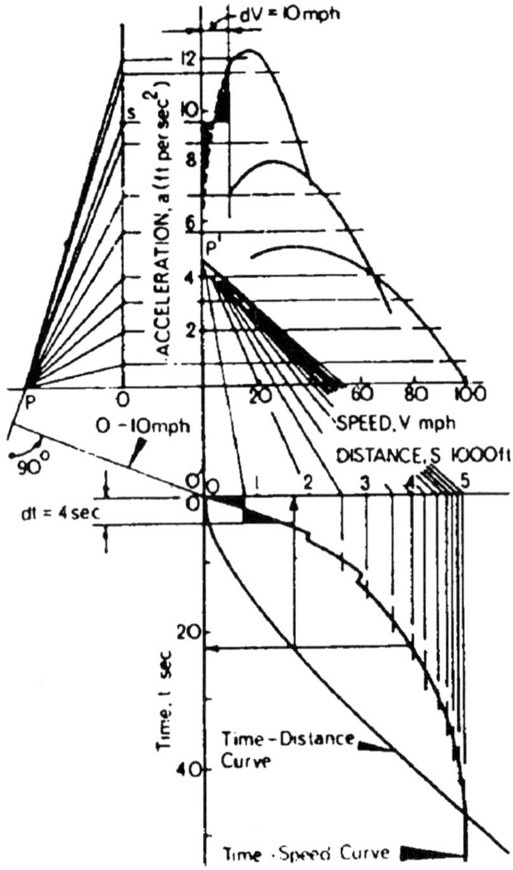

Fig 4: Accelerative performance

By partially differentiating the above expression, relating power and torque, with respect to power and torque in turn, slope at maximum power for the torque curve is

$$5250\, P_{e.max}/N^2$$

conversely, slope at maximum torque, for the power curve, is

$$T_{e.max}/5250.$$

A bench test carried out by Lucas yielded the dotted curve shown in Fig 5, against the full lines obtained by the approximation, for a 997 cc engine, and showed the validity for performance calculation purposes.

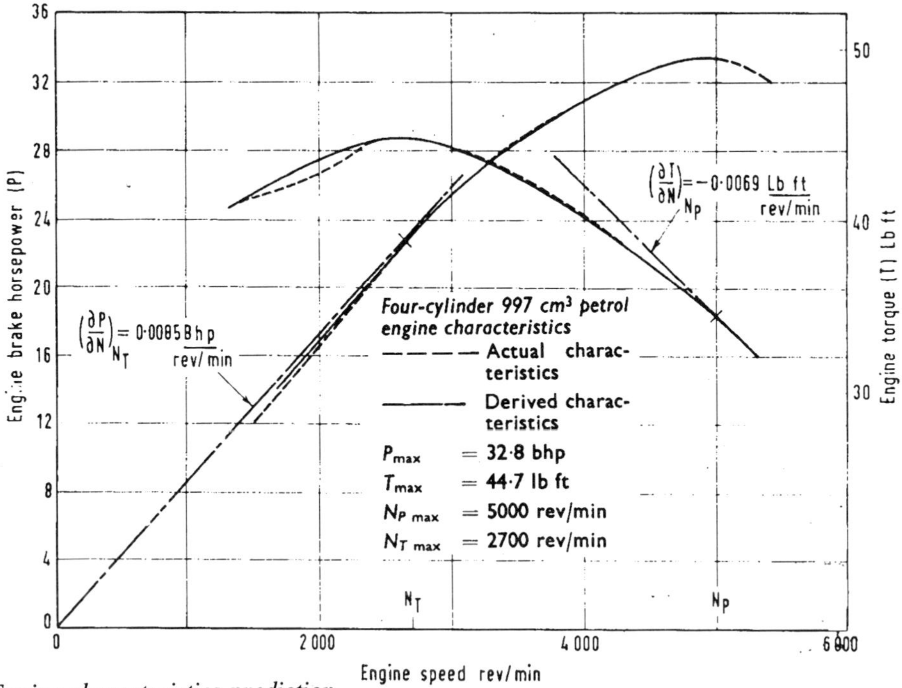

Fig 5: Engine-characteristics prediction

Engine efficiency factors

While the previous section dealt with vehicle performance prediction, and related an engine's characteristics with the performance of the vehicle to which it is fitted, here the engine design parameters which affect performance characteristics are discussed and consideration is given to engine specification for a particular vehicle duty.

Energy available from fuel is expressed as calorific value which ranges from 44 for petrol, through 43 for diesel down to 41 for 'heavy' (high viscosity) fuel oils, units being megajoules per kilogram. Engine thermal efficiency governs the rate at which this heat potential is converted into work and in percentage terms the diesel/petrol ratio is 40/32. Engine power depends on how much air it takes in and how efficiently it apportions air and fuel; pouring in a large excess of fuel does not, of course, increase the power. The cubic capacity (cylinder displacement) of the engine is only the amount of air which can be drawn in at low speed with unrestricted entry. Power output is the product of torque and speed, the torque being proportional to cylinder displacement and the extent of throttle opening or fuel addition. Highest torques are generally developed at medium speeds because at low speeds there is excess heat loss and at high speeds the short cycle time restricts the mass of air taken in and the fuel is adjusted to suit, unless supercharging is employed. The merit of the piston engine is its ability to vary speed and torque independently, without undue loss of efficiency.

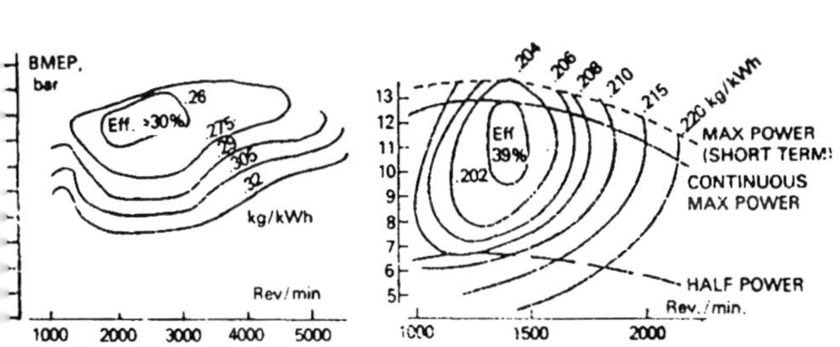

Fig 6: Engine efficiency contour map

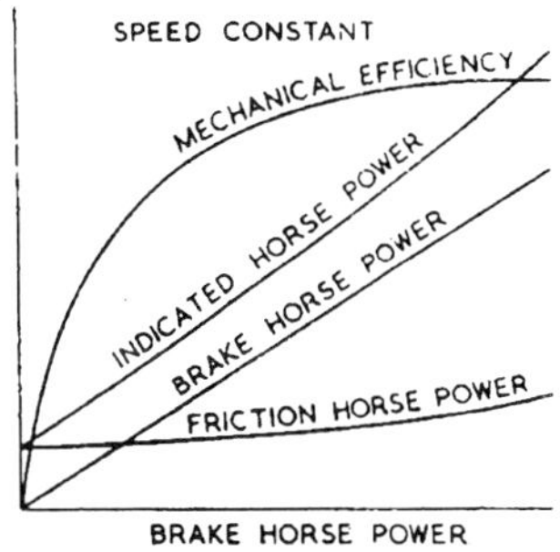

Fig 7: Factors affecting mechanical efficiency

Efficiency is conveniently presented as a contour map, Fig 5, the power being expressed by brake mean effective pressure (BMEP). An engine's efficiency can be obtained by instruments which record its 'indicator diagram' at different speeds. This is a plot of pressure against volume during the full cycle (compression, combustion, power and exhaust) of the engine. The indicated MEP is obtained by dividing the area of the plot by its length along the volume axis to obtain the mean height. The BMEP is this value corrected for friction in, expressed by mechanical efficiency of, the engine. As power equals torque times angular velocity, BMEP

$$= 120.2\pi T / (100.60.V) \text{ bars}$$

for a 4-stroke engine, and half that value for a 2-stroke, where T is torque in Nm, V is total displacement volume in litres and power is $2pnT/60$ watts for engine speed n rpm.

The contours of the map show brake specific fuel consumption (kg/kWh) which is readily converted into efficiency via the calorific value of the fuel. In the left hand plot (a) for a petrol engine, BSFC of 0.26 represents 29.5% thermal efficiency and in the diesel plot (b) 39% . The higher BMEPs of the diesel are often due to turbocharging. Note that in both plots the efficiency falls at high speeds because oil drag is proportional to the square of speed and power — also the finite time for combustion limits efficiency at high speeds as burning finishes too late to contribute fully to work. At low BMEP, under part load, friction losses share a relatively larger share of energy. Good efficiency is obtained by optimising air/fuel ratio which can be calculated knowing the chemical composition of the fuel and by ensuring an even distribution of charge between the cylinders. In diesel engines, the fuel has less time for evaporation than in petrol engines so considerably more excess

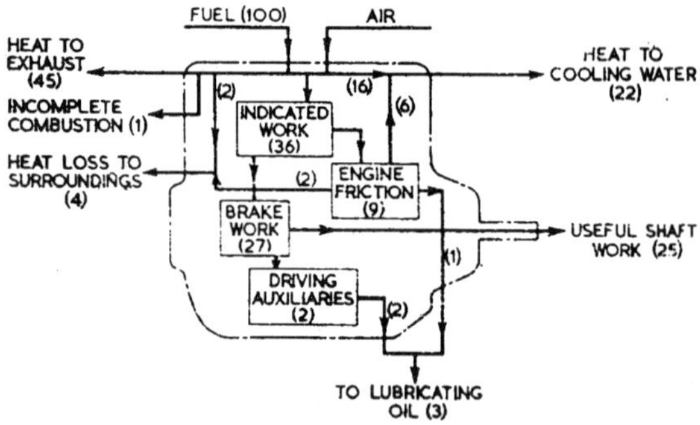

Fig 8: Overall energy balance

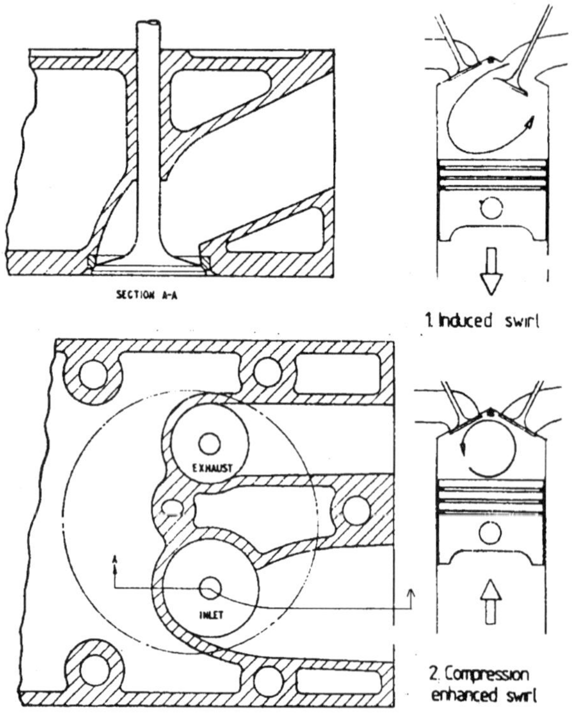

Fig 9: Tangential and barrel swirl ports

Type of Engine	Exhaust Opens b.t.d.c. deg	Exhaust Closes a.t.d.c. deg	Inlet Opens b.t.d.c. deg	Inlet Closes a.b.d.c. deg
Diesel: Low speed	40	8	8	35
High speed	50	12	12	42
Petrol: Family cars	45	10	10	45
Higher performance	55	15	15	50
Sports	60	22	28	68
Racing	80	60	60	80

Fig 11: Diesel/petrol engine valve timing compared

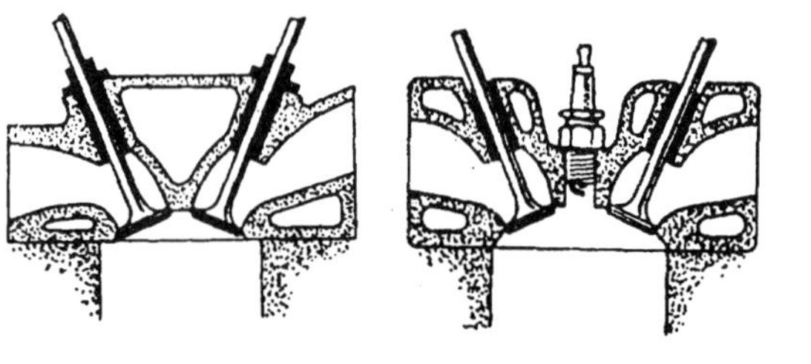

Fig 10: Chambers with and without squish

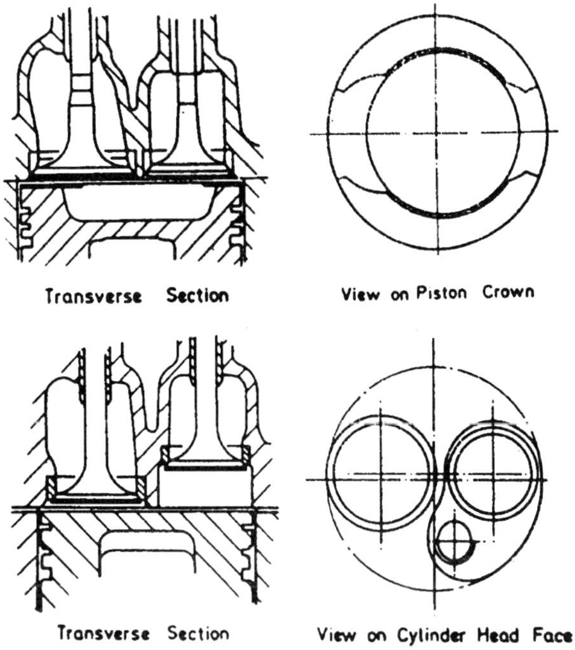

Fig 12: Ricardo high-compression chambers

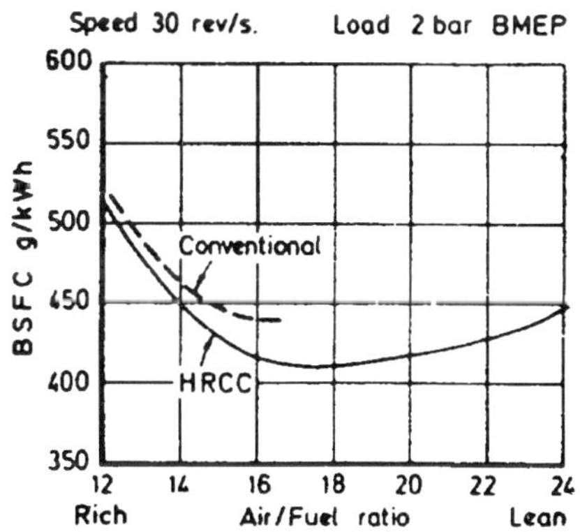

Fig 13: High-compression and conventional chambers compared

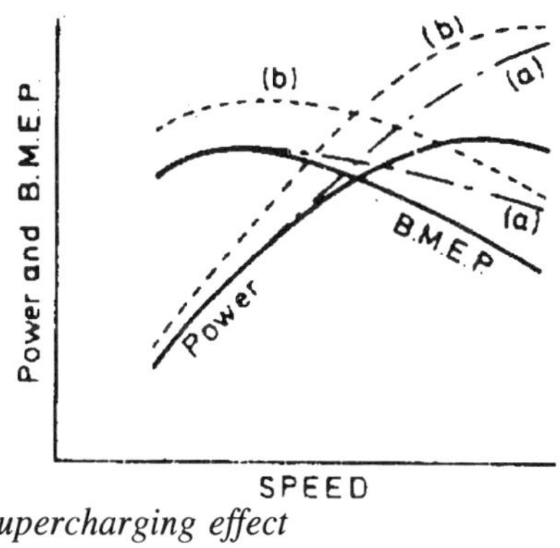

Fig 14: Supercharging effect

air, over the chemically correct mixture is required. The greater efficiency of the diesel comes from the greater compression ratio (and thus expansion ratio) of the engine which means that less pressure is wasted at the end of the power stroke. Two-strokes are more efficient than four strokes because the working parts are always fully employed on high-pressure duties of compression and expansion.

Fig 7 shows the graphical relation between factors affecting mechanical efficiency, at constant speed, while Fig 8 shows a typical overall energy balance in an internal combustion (IC) engine showing the various areas where potential improvements could be made.

Volumetric efficiency can be improved either by supercharging or improving the breathing of the naturally-aspirated engine as the indicated power is directly proportional to the weight of air burnt in a given time. To assist the almost instantaneous burning process various devices are used to increase air motion and induce air turbulence in the combustion chamber. Swirl ports and masked valves are two of these and chamber shapes which produce 'squish' are another. Fig 9 shows a tangential swirl port and a barrelswirl port compared while Fig 10 illustrates chambers with and without squish area between piston and cylinder head.

Increasing the amount, rather than acceleration and direction, of air can be achieved by increasing valve opening area and lift, for which cylinder head and camshaft conversions often exist in the case of petrol engines. There is generally a direct correlation between volumetric efficiency and mean charge velocity through the inlet valve throat. The ratio of this area with the piston area will enable this velocity to be obtained from mean piston speed for a four stroke engine. In the case of a two-stroke, the air is not completely drawn in by the piston as there has to be air blown in to the cylinder at higher than normal pressure.during the scavenge period in order to blow out residual exhaust gases.

For diesel engines it has been found empirically for most engines that maximum power and torque correspond respectively to fixed inlet valve air speeds of 240 and 140 ft/s (73-43 m/s). To obtain the same flow area through the annulus between valve seat and head, the valve lift must be at least 25% of valve throat diameter. Valve timing is also critical and typical values are shown in Fig 11 for various types of vehicle duty. Finally, cam form design is the other variable which without too much difficulty can be made to change the acceleration of air charge into the cylinder.

Air cycle efficiency is also increased with increase in compression ratio but in petrol engines the limit is something like 13:1 set by fuel limitations and heat loss through the chamber walls. Compact chambers are the key to high compression ratio tolerance, exemplified by the Ricardo designs of Fig 12 for in-head and in-piston chambers. Much leaner air-fuel ratios can be tolerated and part-load economy can be considerably improved. Less ignition advance is required and improved air utilisation gives scope for increased torque potential. Normally, poor efficiency at part loads comes from operating the engine in a less desirable condition, Fig 13. This situation can also be improved by optimising the gearing and final drive ratio of the vehicle. Optimum conditions can be obtained with automatic transmission systems which offer both 'sport' and 'economy' modes of operation.

Supercharging and turbocharging can also be used to improve part load efficiency on both petrol and diesel engines. The improvement is smaller on a diesel because of the need to lower compression ratio for avoiding detonation at full load. Fig 14 shows the effect of petrol-engine supercharging at different rates shown by chained and dotted lines against the naturally aspirated case (full lines). At (a) the rate is barely sufficient to maintain volumetric efficiency at the maximum value throughout the speed range. At (b) torque increase would cause unacceptable increase in mechanical loading and peak pressures would require reduction by either compression ratio decrease or less ignition advance.

Diesel engine charging is affected by different criteria. Because mixing of fuel with air has to take place within a period of just under 40 degrees of crank rotation, it is possible to use only 80% of the air supplied and generally larger cylinder dimensions than a petrol engine, for a given power output, are the result. At maximum speed and power, the engine's thermal loading is generally at its maximum without any increase brought about by supercharging. At lower speeds, however, the margin is greater and a blower can be provided with low-speed characteristics so as to give torque which rises with decrease in speed to obviate the need for frequent gear changes.

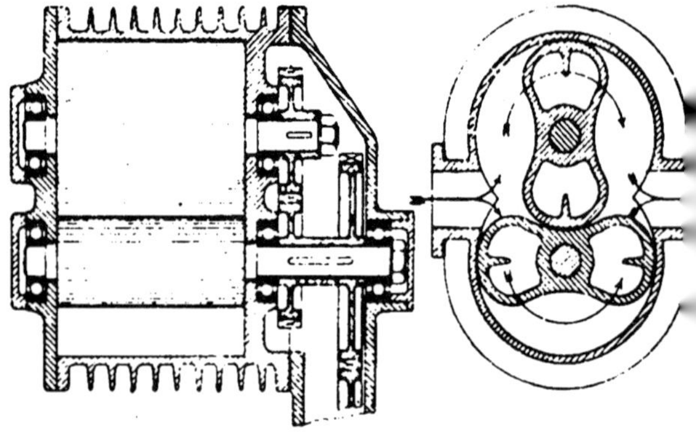

Fig 15: Roots-type supercharger

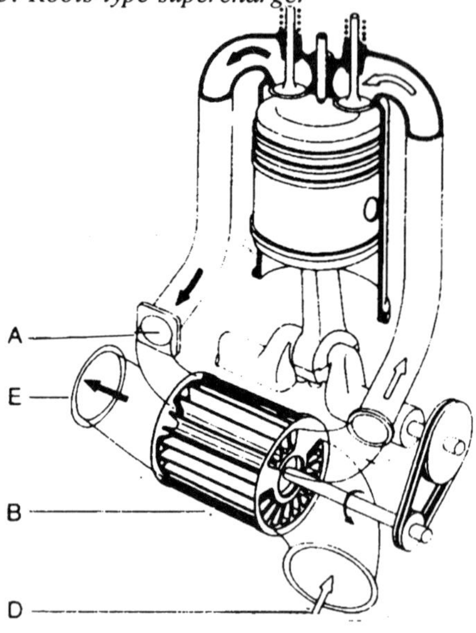

Fig 16: Comprex supercharger

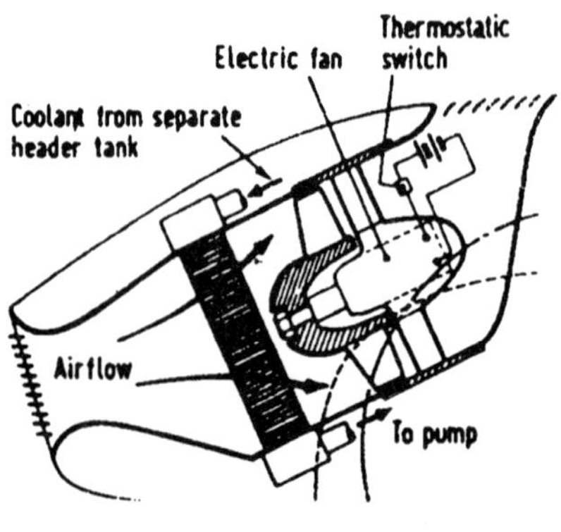

Fig 17: Ducted cooling system

This capacity to 'lug' at low speeds is a major attribute of the turbocharged diesel. Also the denser air of the supercharge tends to reduce the delay period and gives better control of the rate of pressure rise thus reducing knock tendency. Consequently, to double the normal torque requires only 30% increase in maximum pressure.

Fig 15 shows a Roots type blower, of the positive-displacement type which has a delivery rising slowly with speed, compared with a Centric vane type which gives better delivery at low speeds. Another quite separate supercharging principle is provided by the Comprex blower of Fig 16. Here the energy of the exhaust gas is transferred directly in the form of pressure pulses to the inlet air/fuel mixture rather than via turbine and rotary compressor as in the case of the exhaust turbocharger. Energy exchange takes place within a drum B, which absorbs little power as no gas compression is involved. During each drum revolution, one open end of each rotor cell passes exhaust port A into an instantaneously closed cell. Further rotation causes the pressure pulse to enter the inlet port as the opposite end of the cell opens. Further rotation connects the cell to the exhaust pipe E and yet further creates a suction pulse in the inlet tract D.

Engine auxiliary systems

Design of exhaust and cooling systems can each affect engine performance as well as affecting noise output and aerodynamic drag. The above mentioned requirement for maximum charge induction per stroke leads to considerable overlap in the opening of inlet and exhaust valves which can make the engine sensitive to exhaust back pressure. In fact careful design of the exhaust can be used to harness the pressure pulses to enhance exhaust gas scavenging. At the same time advanced silencer designs such as that shown for the Government-sponsored Quiet Heavy Vehicle project in Fig 17 can dramatically reduce noise output. Used on the 238 kW Rolls Royce diesel, it produced a back pressure of just 39 mm of mercury while recording a drive-by noise level of just 70 dB(A) at 7.7 metres.

While the design of the coolant radiator is a function of the amount of waste heat to be dissipated, additionally the differential between coolant and air temperature must be kept as high as possible to effect an efficient heat transfer. Radiator size is also governed by volume of air which flows through the core and the power of the fan chosen. Considerable improvements in efficiency can be obtained by ducted systems such as that shown in Fig 18.

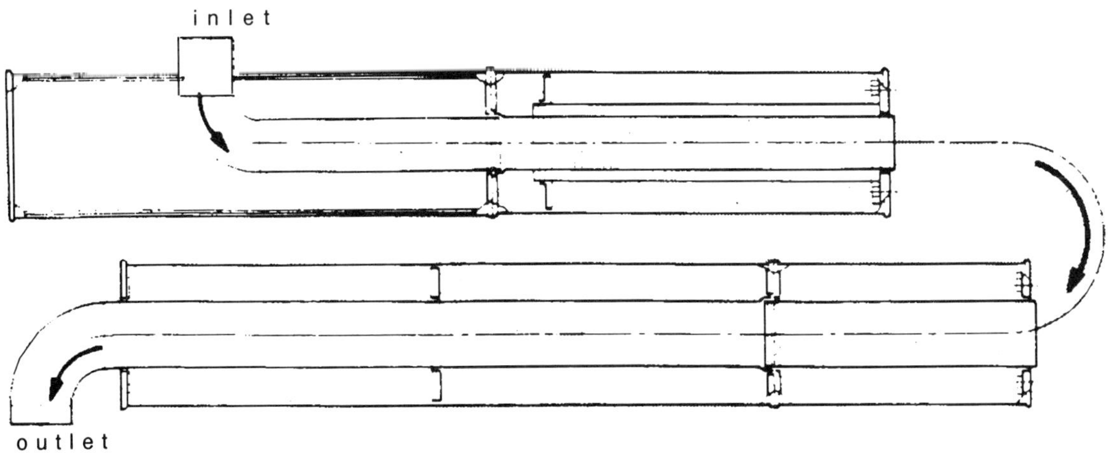

Fig 18: Advanced silencer design

Engine fuel/air intake systems

A more detailed look is taken here at engine intake parameters in relation to the improved breathing of the engine required for enhanced power and improved efficiency. The objective, generally, is to provide largest possible valve area combined with a suitable valve lift and opening period — at the same time providing a cam and valve train which will operate up to the highest possible engine speed.

There is approximate co-relation between mean air velocity through the valve throat and volumetric efficiency, Fig 19. For a naturally-aspirated 4-stroke engine the air velocity is calculated from the mean piston velocity and bore area as $v=LnD^2/360d^2$ where L is stroke in inches, n is engine speed in rev/min, D is cylinder bore and d the valve throat diameter.

Traditional engine designs gave figures of 140 and 240 ft/s (43 and 73 m/s) for maximum inlet valve velocity at maximum torque and maximum power respectively, according to classic investigation work by Ricardo[2]. The use of multi-valve heads and hemispherical combustion chambers is now changing this limitation which was governed by the limiting valve port to bore diameter of about 80% for more conventional combustion chambers. The use of venturi ports to accelerate air-flow at the valve throat is another common measure which allows large valve throats without penalising water jacket space. Valve lift is conventionally set at 25% of throat diameter to make the area of the opening annulus equal to the throat area but a higher value than this is actually desirable to maximise air throughput under dynamic conditions. The valve head should also be accelerated as fast as possible, by the cam profile, to obtain the greatest throughput of air, it is argued.

Flow rigs are used to measure the actual air flow through the valves in order to measure the effect of the valve discharge co-efficient on the calculated air-flow. Fig 20 shows a typical curve for the variation of discharge co-efficient (measured/calculated airflow) which shows the small improvement obtained by lifting the valve more that 25% of throat diameter (26% was traditionally considered to be the optimum figure).

Valve timing

The timing of valve opening is also crucial; the exhaust gas must be opened as late as possible to maximise the expansion ratio so as to lower the gas pressure and temperature against which the valve must be lifted. A compromise is reached between having a reasonable opening period up to top-dead-centre and encroachment on the suction stroke. The inlet valve is opened shortly before TDC to give it a worthwhile opening at the point when the piston starts to move down on the suction stroke. Its closing is a compromise between retaining reasonable opening area at bottom-dead-centre, for minimising suction losses, and the need to convert as much kinetic energy in the air moving through the inlet tract into pressure energy by the time the valve closes.

Cam profile design is, of course, a key factor. Fig 21 shows typical lift acceleration curves for a parabolic acceleration cam but a number of other

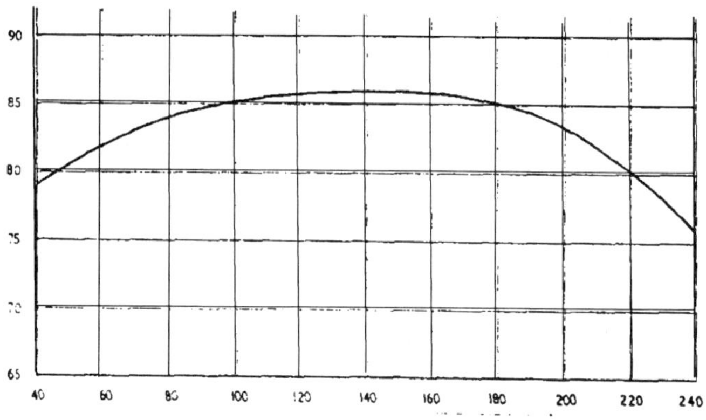

Fig 19: Basic engine-breathing relationship: volumetric efficiency at ITP (y-axis) vs mean gas velocity through inlet valve throat in ft/sec (x-axis)

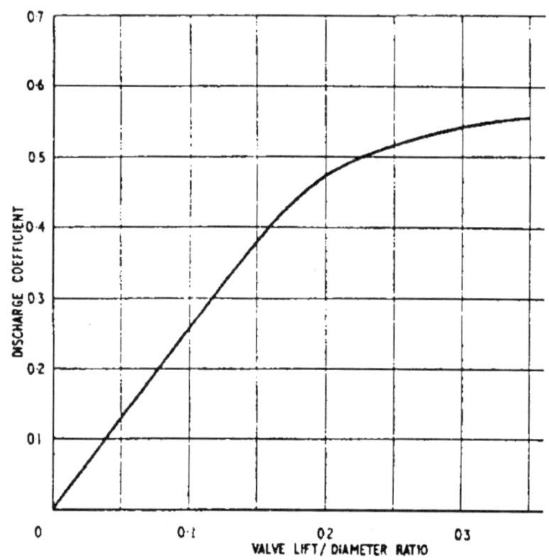

Fig 20: Effect of lifting valve more than 25% of thr

acceleration laws are also used to avoid instantaneous changes of acceleration and maintain the nose-to-flank acceleration ratios between the normal limits of 2 and 3. Polynomial cams are particularly favoured for their amenability to computer calculation of lift, the lift curve being of the form:

$$y = C_0 = C_p.x^p + C_q.x^q + C_r.x^r + \text{etc.}$$

Successive differentiation of the velocity, acceleration and jerk give rise to equations for the solution of the co-efficients C_n by assuming values for the power indices. Fig 22 shows some powers which may be used and the resulting shape variations for the acceleration curves.

Configuring the manifold

Inlet manifold design is a strong weapon in enhancing engine efficiency and tuned systems have now been perfected to work even on volume-produced power-units. While the use of large tract diameters for enhancing volumetric efficiency is desirable, undue size leads to poor drivability at low speeds because of the reduced velocity of air/fuel mixture flow at low engine speeds. A reasonably high velocity is required to assist atomisation of the fuel spray. A 'safe' flow rate of about 70 metres/second has been suggested but higher speeds can be tolerated if sufficient manifold heating is available. D-section shapes are often used, the flat of the D forming the floor of the passage on to which liquid fuel can drain down from the curved walls into a shallow pool of large surface area, which improves rate of subsequent evaporation.

The square or rectangular sections are preferred in those configurations with a tendency for swirling of the mixture around the axis of the tract which would tend, by centrifugal action, to deposit fuel on the walls. Circular sections are favoured near the valve throats and are generally more efficient in presenting the miniumum surface area and therefore resistance to flow, for a given air-fuel volume throughput. Without the action of the valves, pressure in the manifold would vary sinusoidally according to the movement of the engine piston within its cylinder. Valve opening and closing superimposed on this, in fact has the effect of initiating pressure pulses which can profoundly affect power output. For the single-cylinder case, gas velocity rises from zero to a mamimum just beyond mid-point of the induction stroke and the inertia of the moving gas column is maintained even after the piston has passed BDC. While on a single cylinder engine the valve timing

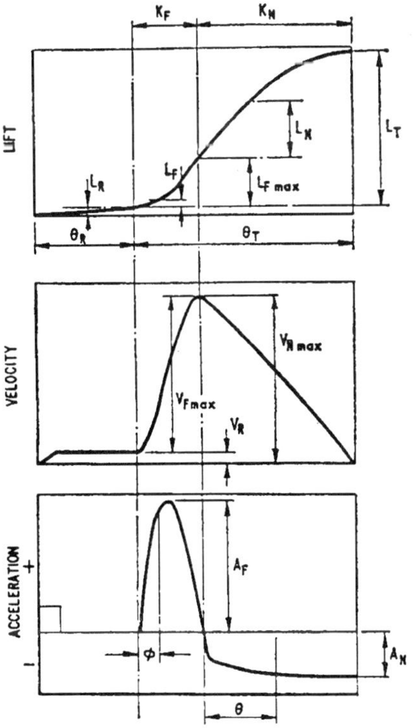

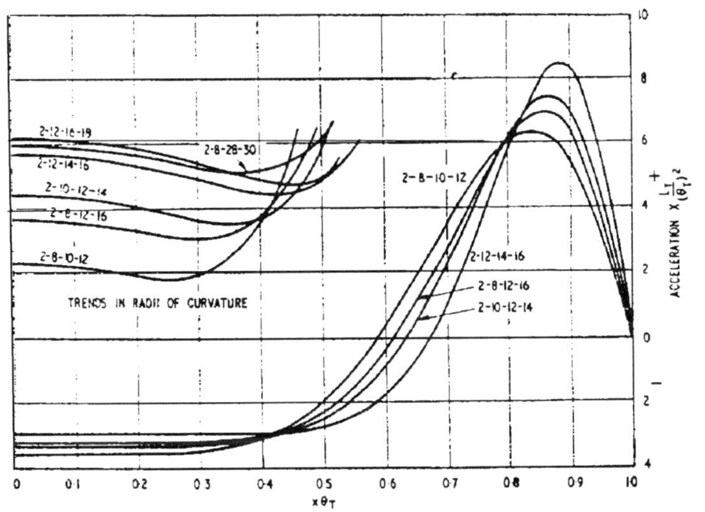

Fig 21: Lift diagram for parabolic cams *Fig 22: Powers for polynomial cams*

gives a 480 degree closed period within the 720 degrees crank rotation in a 4-stroke cycle, this closed period, in which the mixture tends to stagnate, is 120 degrees for a twin-cylinder engine with crank throws set at 360 degrees. There is thus no overlap in inlet valve opening periods and inlet ports can be siamesed.

On four cylinder engines, adverse effects of valve opening periods can be avoided by using a twin-tract manifold, according to Garrett[3], the engine becoming effectively a pair of twins. To maximise performance, however, a better arrangement is a twin-barrelled carburettor, one between each pair of cylinders, Fig 23. The restricted balance orifice between the two sets of two branch pipes equalises the pressures in the two sections of the manifold.

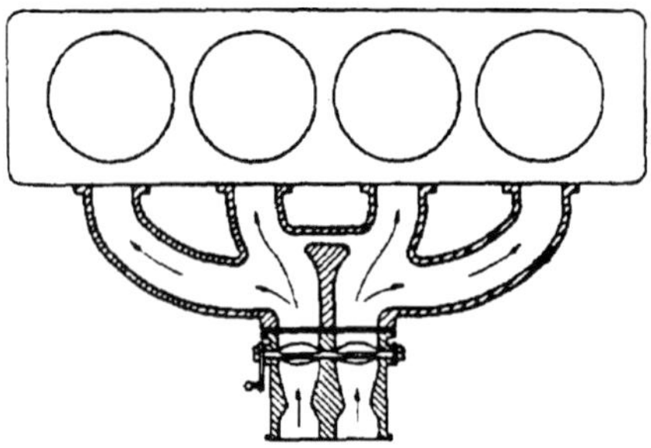

Fig 23: One barrel per cylinder pair

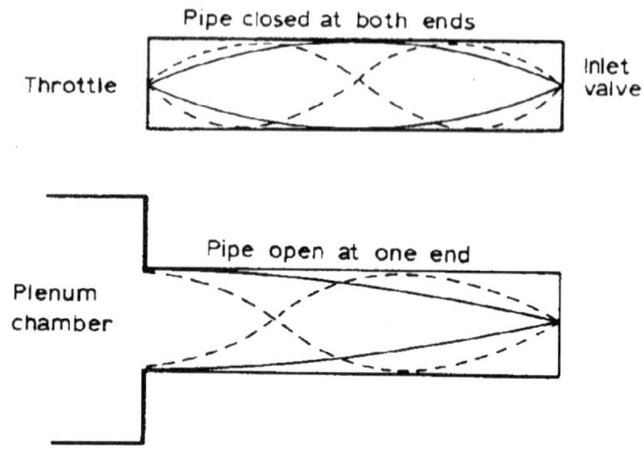

Fig 24: Ram tuning

Tuned manifolds

Manifold design for injection engines is similar to that for carburettors but the 'riser pipes' may be replaced by horizontal connection to the body of the throttle valve. Vauxhall's Dual Ram induction system was a successful volume production exploitation of tuned manifolds to an injection engine. Ram tuning is based on the resonance of the columns of air in the induction tracts, Fig 24, and is effected by selecting lengths of pipe such that they resonate at a fundamental frequency equal to that at which the valves open. The resulting pressure pulse drives the charge into the cylinder. The Dual ram arrangement made the system effective on six cylinder engines for the first time. The 6-cylinder GM engine with Dual Ram is tuned as a conventional six at high speeds but at lower speeds, in manifolding terms, becomes two 3-cylinder units. The torque curve is effectively a composite, Fig 25. Two induction pipes are taken from the throttle valve to a plenum chamber half way to the inlet ports. The plenum is in two parts, each maintaining the continuity of a set of four pipes. Flow from each of the two pipes from the throttle passes into the corresponding group of three induction tracts. As each inlet valve opens in turn, the air column throughout the whole length between it and the 90 degree pipe bend resonates. But the flap valve can destroy the continuity of the four pipe system and double its volume. This makes it large enough, Fig 26, to form an effective discontinuity between the two pipes from the throttle and the six tracts — so only air columns in the latter can resonate, at a correspondingly higher frequency.

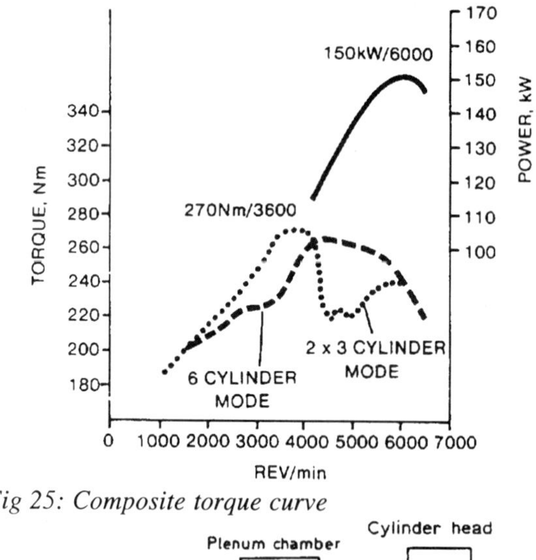

Fig 25: Composite torque curve

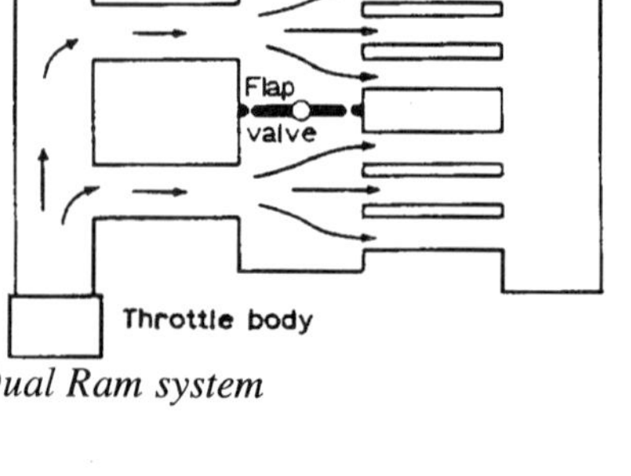

Fig 26: Dual Ram system

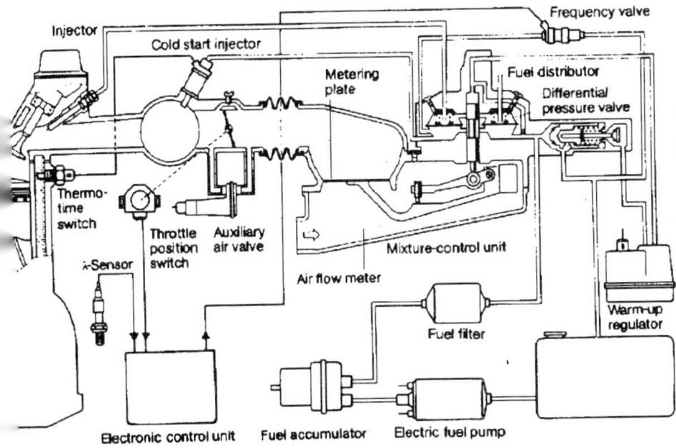

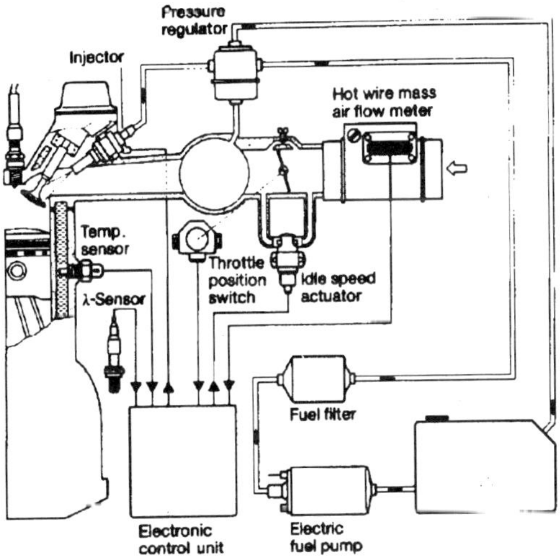

Engine fuelling

While carburettors have to carry out a number of functions simultaneously by mechanical means (air-flow measurement by the venturi, metering of the fuel by several orifices and fuel-mixture preparation), fuel injection works by separate fuel and air metering; a number of petrol and diesel systems to suit different vehicles are discussed.

Modern usage of petrol injection systems dates from the indirect-injection system used on Connaught racing cars in 1953, though earlier developments by Mercedes Benz resulted in its first appearance in a post-war production car, the 300 SL, in 1954. This was a Bosch direct-injection system based on a cam and plunger type pump. The Lucas shuttle valve system followed in 1958 and was fitted to the Maserati V8 engine but the main commercial breakthrough came in 1968, with the Bosch D-Jetronic, to be followed a few years later (Fig 27) by the addition of the Lambda sensor for detecting oxygen in the exhaust gases and effecting a closed-loop control system, key to successful emissions control.

Most current systems involve fuel being delivered, by an electric pump, at the appropriate pressure for the injectors; the phasing and opening period of the injectors governs the required air-fuel ratio, independent of any mechanical drive from the engine. The associated electronic control unit uses sensors to

Fig 27a: The K-Jetronic system from Bosch is a mechanically controlled continuous injection system with lambda control. Deflection of the sensor plate which senses air-flow also controls the fuel distributor. Fuel reaches the injectors at about 4 bar and they inject continuously, oscillating at about 1500 Hz. The KE version, shown above, introduced electronic control, used to monitor the electro-hydraulic actuator mounted to the fuel distributor. The L-Jetronic (below) incorporates air-flow metering and has injectors which fire twice per engine cycle. The LH Jetronic has a hot wire air mass meter in lieu of vane type flow meter. A derived form is the Motronic shown in the lower picture which is an integrated system which also controls spark timing. The single-point version is known as the Mono-Jetronic. Latest derivative is the Motronic M5, with OBD2.

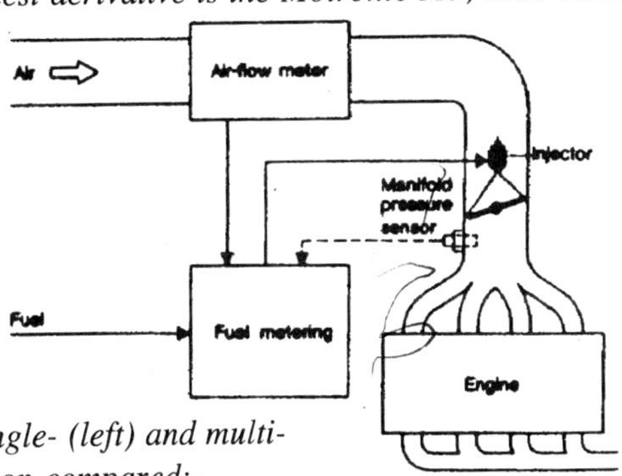

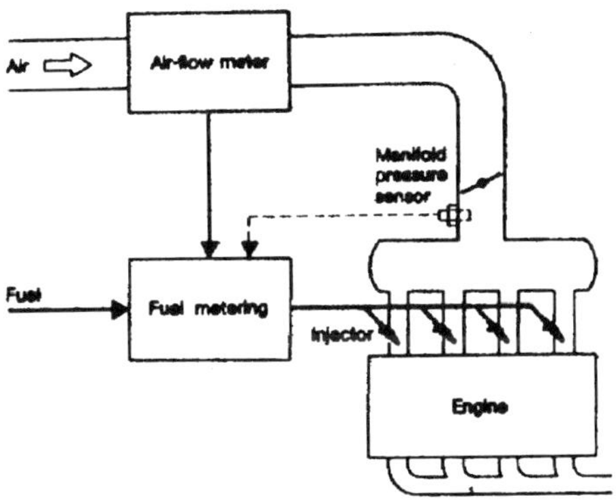

Fig 27b: Single- (left) and multi-point injection compared:-

monitor appropriate engine parameters. Injection is often into the manifold, at a relatively low-pressure area, before the inlet valve is opened, giving time for evaporation and mixing with the intake air. If injection is into the ports within the cylinder it is directed on to the hot inlet valves which assists evaporation and lessens the chance of condensation on the walls of the manifold.

Continuous, timed or sequential injection systems are chosen according to cost and performance trade-offs. With the former, the rich mixture upstream of the inlet valve is swept into the cylinder where swirl and turbulence convert it to an ignitable mixture. Most systems run at constant injection pressure but vary the volume of fuel being delivered, for controlling A/F ratio, by incorporating a flow control valve. In timed systems control of A/F ratio is by regulating the duration of flow through the injector and usually all injectors open and close simultaneously. With sequential injection the start of injection is phased, as well as the duration being controlled, relative to the opening of the inlet valve, for each individual cylinder.

The two latter systems have the advantage of being more effectively controlled by the ECU, to avoid pinking, in conjunction with a knock sensor linked up with the ignition timing system. It is also relatively easy to enrich the mixture for starting, acceleration and peak power demand.

Choice of single (throttle-body) or multi point injection systems is also usually made on performance/cost trade-off. SPI systems obviously share the A/F distribution problems, cylinder to cylinder, with single-carburettor engines. Throttle body heating is also often required to assist evaporation and avoid icing.

Recently there has been interest in direct in-cylinder injection systems and Ricardo is among the companies who have announced developments. This company has two variants under examination in four-valve head engines. The first is as in Fig 28, with injector located symmetrically between the inlet valves, and uses a central spark plug; the second has the injector in the end of the head, between inlet and exhaust valves. The best of these has produced HC and NO_x emission levels equivalent to the MPI engine on which it was based, at stoichiometric A/F ratio as seen in Fig 29.

Lean running stability is extended to better

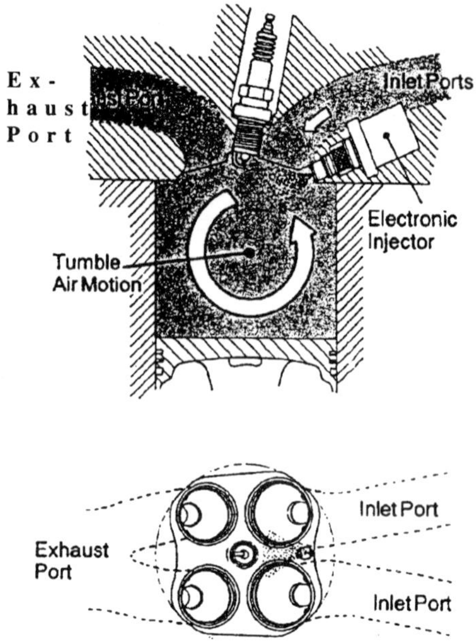

Fig 28: Ricardo DI petrol system

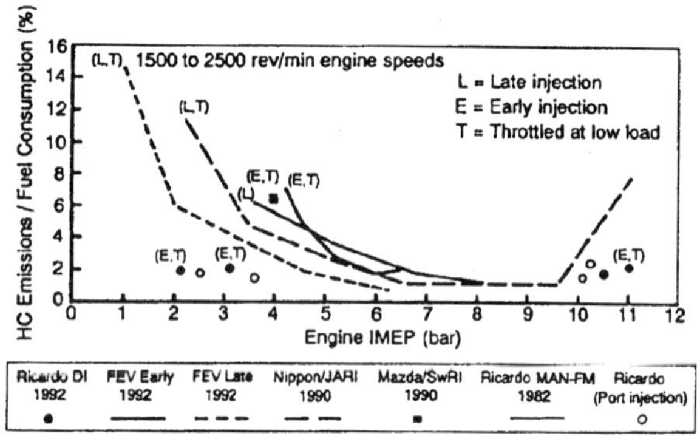

Fig 29: Emission results

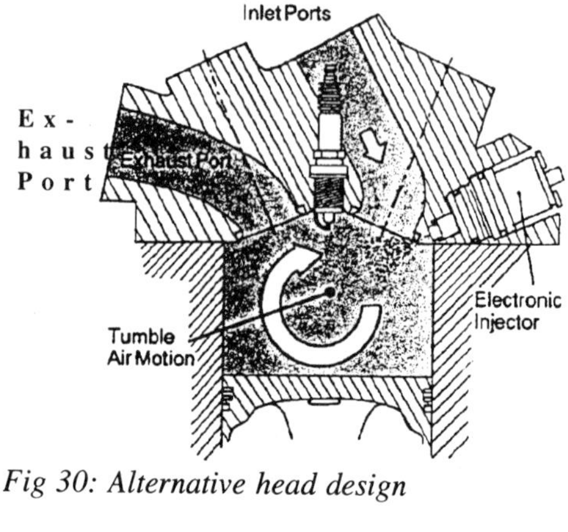

Fig 30: Alternative head design

than 21:1 and the DI engine was able to run, in tests, 8-10% leaner than the MPI engine, in simulated cold-start conditions. The DI engine also showed improved torque output over the MPI base-unit as its volumetric efficiency was superior by 4-5%. For optimum running, however, variable tumble-swirl ratio is required between full and part load. This could be provided by a variable geometry inlet system, say Ricardo. An alternative head design, Fig 30, has the potential for better A/F matching over the load and speed ranges.

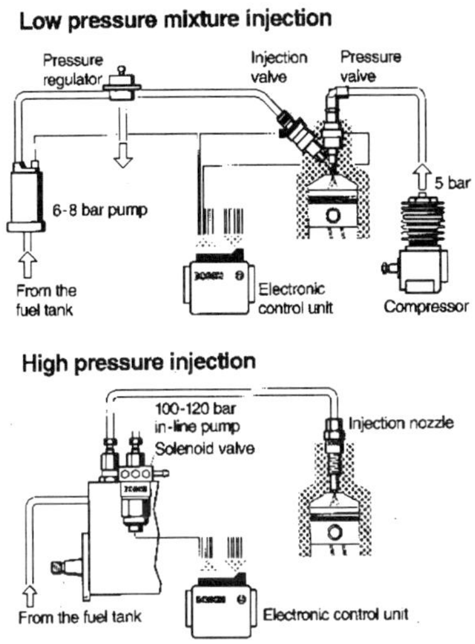

Fig 31: Bosch DI system

At Bosch, work has also been undertaken on direct petrol injection and two options have been examined, Fig 31. These are high-pressure fuel injection at 100-120 bar and low pressure mixture injection at 5-7 bar, involving an injection valve with air shrouding (right). Both require fuel droplets down to 10 microns in size and charge stratification in the region of the spark plug. The company have also introduced a throttle valve with integrated hot-film air-mass flow meter, Fig 32, to ascertain the precise amount of air injected in order to better control the exact quantity of fuel injected. This design protects the meter from engine back-fire and provides a particularly compact layout.

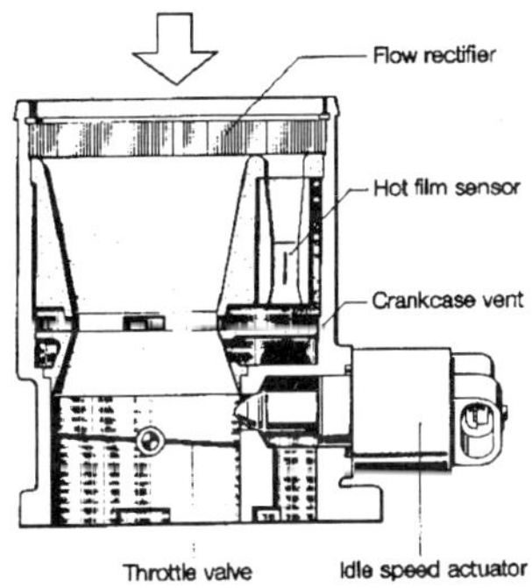

Fig 32: Air mass flow meter

A key refinement to fuel injection systems has been the Lucas disc injector, Fig 33. Its low-mass armature and open-orifice valve seat make it highly resistant to plugging and it is considerably quieter than current injectors. Precise control of fuel delivery depends on a linear relationship between injector pulse width and delivery, since the latter is the product of pulse width and static flow rate. Most injectors suffer from the time lag between beginning of the pulse and start of flow; this is minimised in the Lucas design by the inertia of the armature, compared with others, seen in Fig 34.

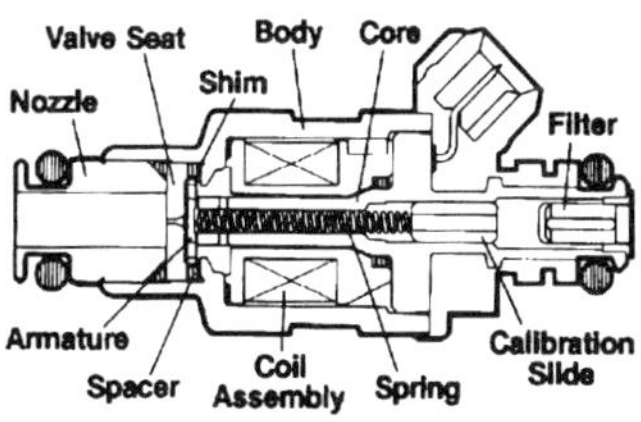

Fig 33: Lucas disc injector

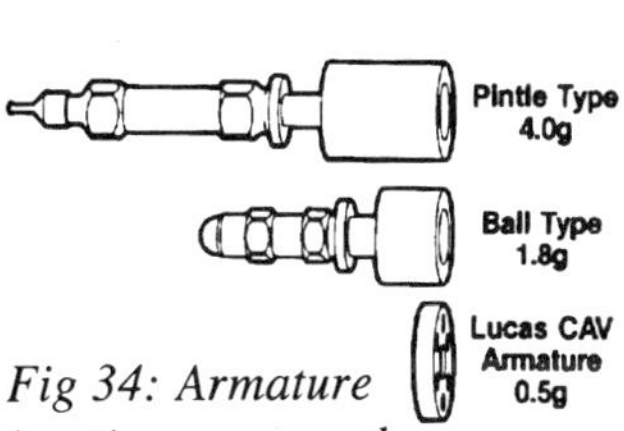

Fig 34: Armature inertias compared

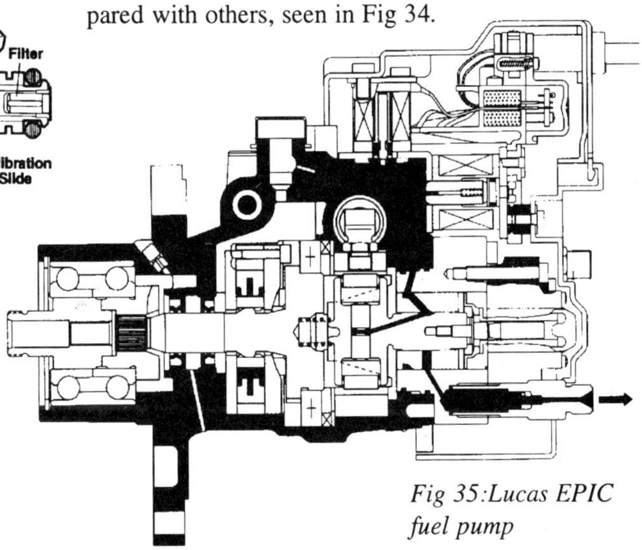
Fig 35:Lucas EPIC fuel pump

Diesel systems

Lucas is also prominent in diesel fuel injection systems and has recently had its electronically-controlled design, EPIC (Fig 35), accepted by such builders as Ford, Volvo and Mercedes on vans, trucks and cars respectively. Fuel injection has of course been the universal approach to diesel fuelling, because of the fuel's relatively low volatility and its ability to lubricate mechanical pumps and injectors used in the equipment. As a result of the recent interest in reducing noxious exhaust emissions from diesels to a minimum, electronically controlled systems are in favour. The EPIC system enables engines to meet Euro 96 emissions regulations and there is a 3% potential improvement in fuel economy by closer control of injection during warm starts and deceleration. Electronic mapping employed is said to result in a 20% reduction in NO_x emission, over a mechanical rotary pump, when the system is employed to control an incorporated EGR arrangement with inlet air pressure sensor.

Mapping allows programming of the EPIC pump with a model of the optimum fuel quantity and timing settings across a range of operating conditions such that sensor signals of the actual instantaneous demands of the engine can be compared with the 'ideal' conditions and continuous corrections made accordingly. The injection pump has been designed to maximise the advantages of electronic control and has half the moving parts of an equivalent mechanical design. Fuel volumes are metered by axial move-

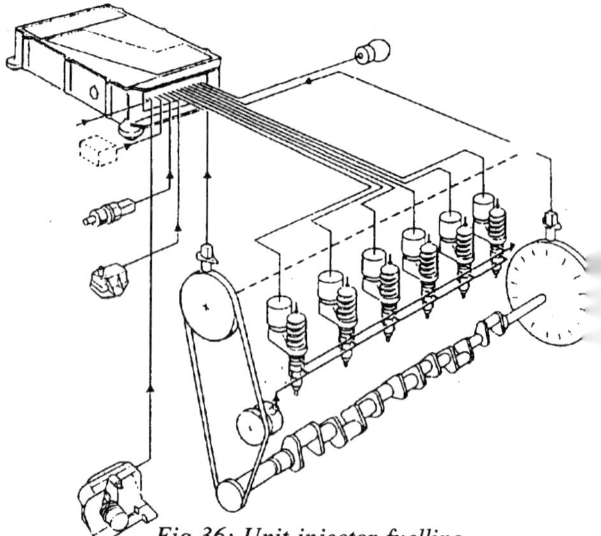

Fig 36: Unit-injector fuelling

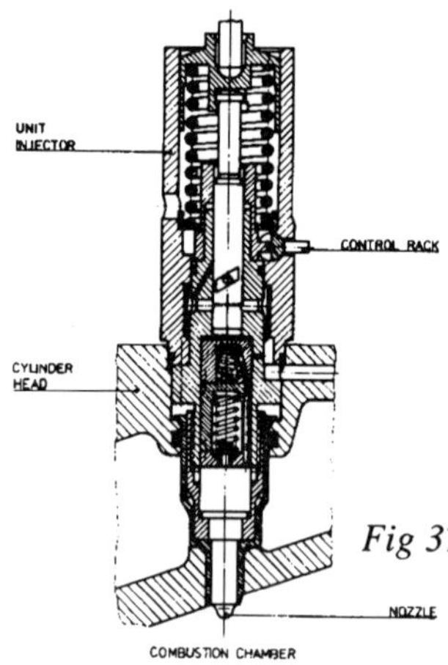

Fig 37: Unit injector

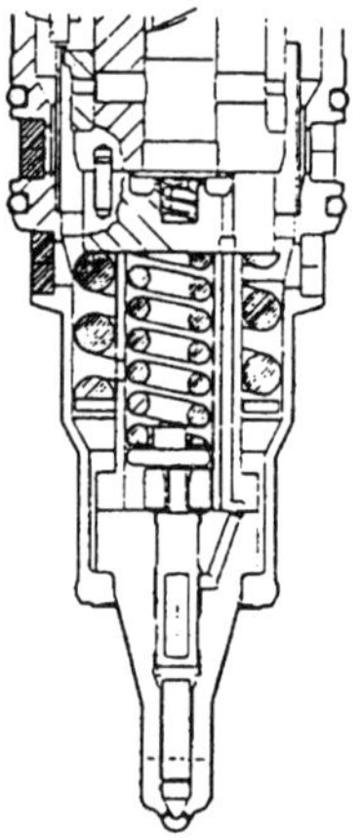

Fig 38: Two-spring nozzle

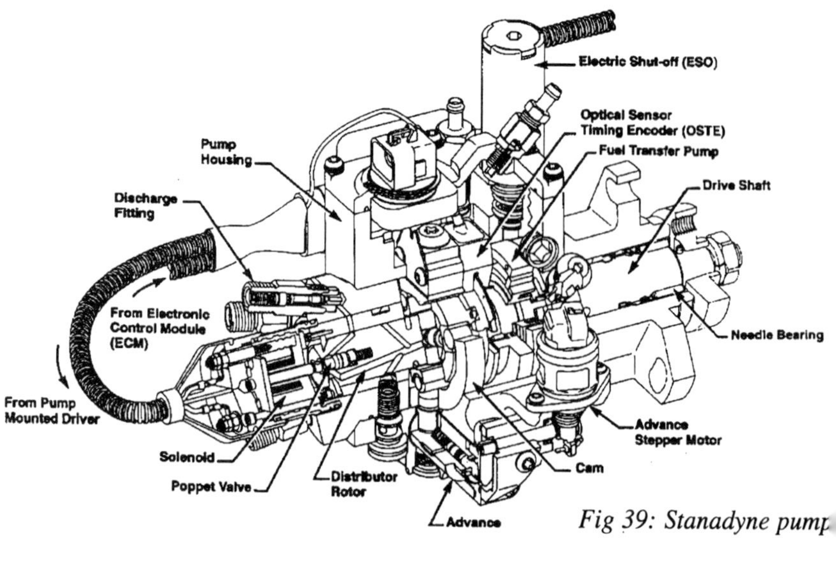

Fig 39: Stanadyne pump

ment of the rotor, altering the available stroke of the pumping plungers. Rotor movement is controlled by two electrical actuators, and monitored by an inductive sensor which completes a closed-loop circuit of information back to the ECU. Timing is controlled by moving the cam ring in the hydraulic head, altering the point at which pumping commences. Again its position is determined by the ECU and an electrical actuator which moves the ring with hydraulic pressure.

While the EPIC uses relatively standard injectors fed at high pressure, Lucas have developed the alternative approach of unit-injector fuelling, Fig 36, in which injectors are fed at low pressure and have individual pressure intensifiers built in to them. These can produce the very high injection pressures (1500 bar) deemed to be necessary to meet emissions levels required of heavy-duty diesels. Recently the system has been further developed to increase the spill valve area and decrease the cam rate, Fig 37. This has resulted in lower initial and faster final injection rates, meaning faster A/F mixing, which allows the longer injection period to match total burn period and hence control peak temperature for effecting reduced NO_x. Fuel feed to the injector is within the cylinder head itself and spilled fuel from the injector is taken to a second gallery in the head. Pumping energy is provided by an overhead camshaft

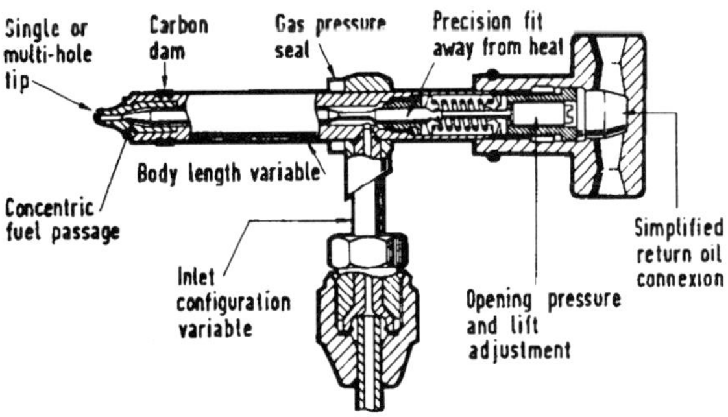

Fig 40: Roosa Master Pencil injector

which actuates the pumping plunger. The injector comprises pumping plunger, nozzle and spring assembly together with an electronically controlled spill valve.

As the plunger descends it closes off the fuel feed and fuel continues to be displaced by the plunger past the spill valve and into the spill gallery. The ECU signals the spill valve to close and injection commences; the valve is de-energised after a fixed time and pressure collapses to terminate injection. A single actuator thus controls start of injection and fuel quantity. Low internal volume allows for much higher injection pressures and higher rates of spill are possible without cavitation and secondary injection problems. For meeting future emissions requirements, investigations are being carried out on a two-spring nozzle design, Fig 38, with improved injection characteristics, which has been used on the experimental Steyr M1 engine.

Electronic control has also been introduced by Stanadyne on to more conventional diesel fuel pumps. In place of the mechanical governor, quantity of fuel is controlled by a solenoid-actuated spill valve of the poppet type and a stepper motor is used to actuate the cam ring mechanism. The spill valve is housed coaxially in a counterbore at the end of the distributor rotor such that the volume of fuel subjected to injection pressures is very small and higher pressures can be adopted without risk of compressibility effects. Cam rollers and tappets are housed in the large diameter drive shaft so that the distributor rotor is isolated from the driving loads, avoiding problems of torsional deflections. Electronic control strategy is angle-based rather than time-based which is said to improve performance in transient conditions, Fig 39.

The company were also pioneers of the Roosa Master pencil injector, Fig 40, which has recently been adopted by Ford for the advantages it gives in reducing space requirements in diesel cylinder heads. Body diameter is less than 3/8th of an inch (9.5 mm) and simple internal construction is involved. Lift of the valve is controlled by a simple stop with screw-adjustment, the stop running inside the main spring-pressure adjuster such that one locking nut fixes both adjustments.

Exhaust system analysis

A variety of parameters affect the design of the exhust system even beyond considerations of pollution control. Silencing and scavenging are among the two key factors examined here.

While the prime requirements of the exhaust system are to remove engine emissions to the safest exit point from the vehicle while minimising environmental noise and pollution, there are secondary requirements of providing adequate clearance between the system and other components: to enable low-cost manufacture for a replacement item and often to suit the system to a variety of engines in a vehicle range; to provide an induction hot-spot facility while minimising under-bonnet temperature and finally to provide adequate scavenging at low-speeds while minimising power loss at high speeds.

Because of high valve overlap required to obtain the fullest possible burning from the air/fuel charge, the modern engine is sensitive to back-pressure. It is important to avoid inter-cylinder interferences, Fig 41, possible with short-stub manifolds. Morrison[4] has noted from work on a single cylinder research engine the variations in back-pressure during an engine cycle, Fig 42 recording this alongside the preferred 'ideal' variation. Two straight-through pipes are compared; the shorter (bold curve) had a lower mean back-pressure while the longer (dashed curve) had a particular length which brought about a negative pressure at the end of the exhaust stroke. The basic kinetic theory of scavenging is seen as a pressure "slug", Fig 43, thc gas column here moving at an average speed of 60-90 metres/sec. The sudden impulse created by valve opening to cylinder pressure causes superimposed waves moving at around 400-500 metres/sec.

The authors suggest that a high-velocity head of gas is partially converted to a static head as it enters the pipe. Cambel and Jennings[5] were early exponents of using gas-dynamics theory for engine exhaust analysis, allowing as it does for the behaviour of compressible fluids. Key relationships are:

$$[P/P_0]^{((n-1)/n)} = T/T_o$$

$$\rho = \rho_0 [1 - \alpha_m (T - T_o)]$$

where n is compression index, α_m is expansion co-

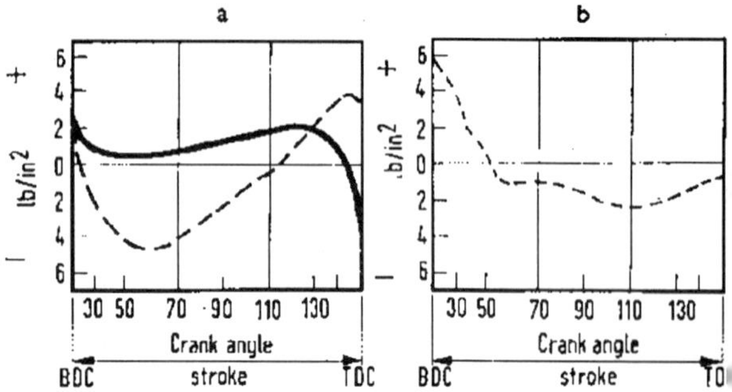

Fig 42: Back-pressure variation

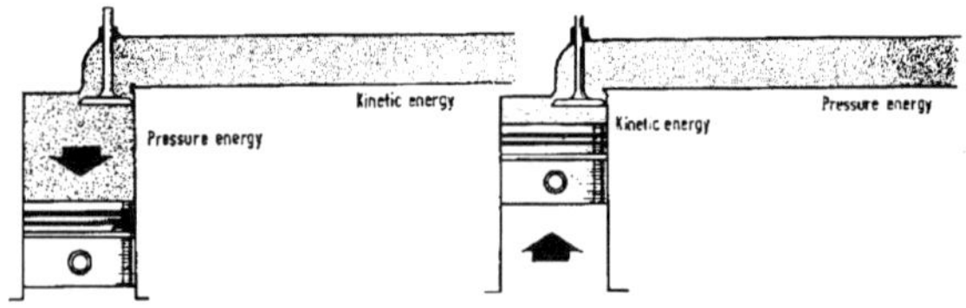

Fig 43: Pressure slug

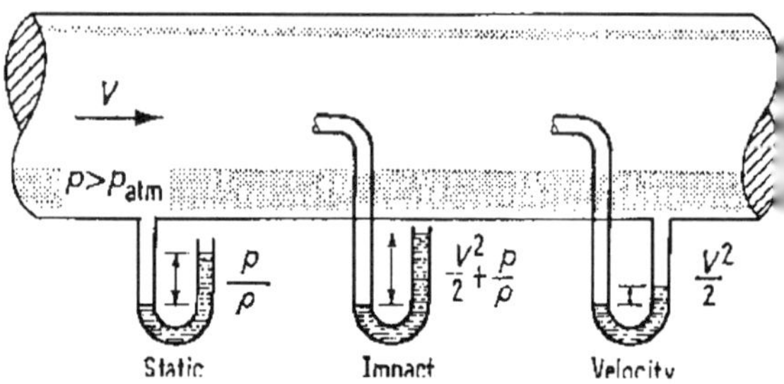

Fig 44: Forms of pressure measurement

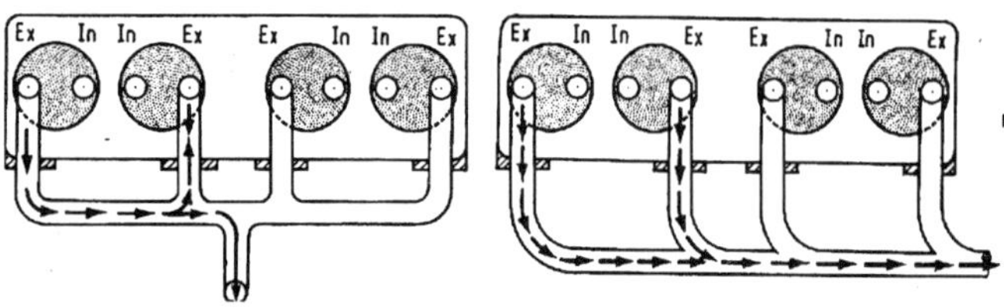

Fig 41: Inter-cylinder interference

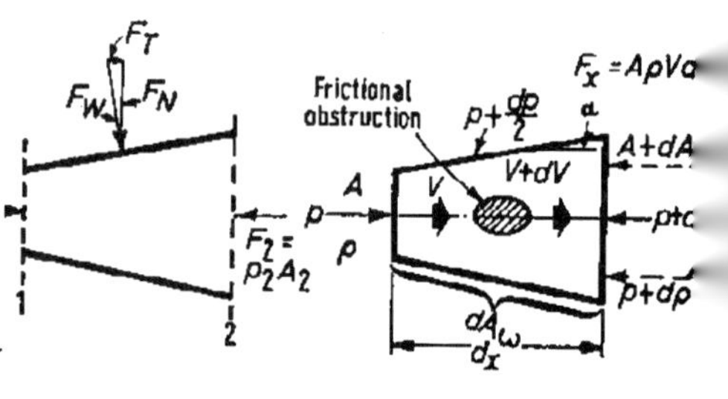

Fig 45: Exhaust duct element

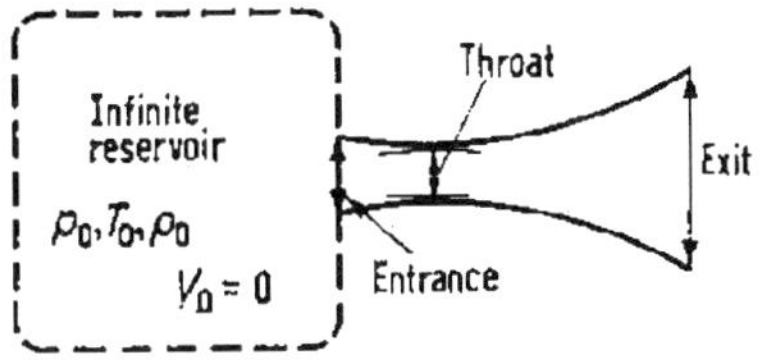

Fig 46: Generalised flow at throat

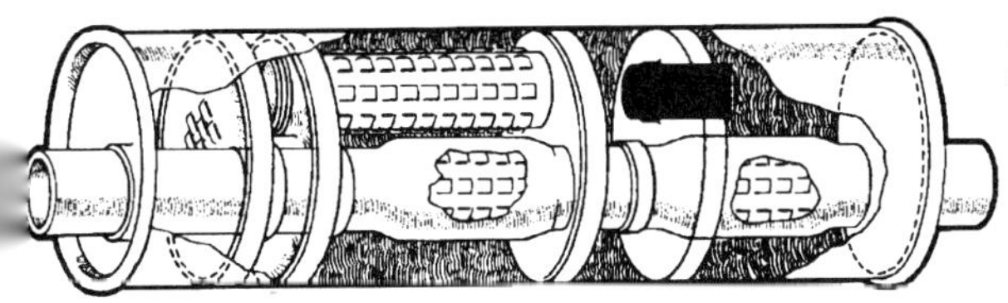

Fig 47: Helmholtz resonator

efficient, ρ is density, P is pressure and T is temperature. The three forms of pressure measurement are shown as Fig 44. At gas velocities above 120 metres/sec pressure change associated with the generation of velocity is high enough to vary the density and thus the onset of compressible flow for which the second expression above is inappropriate. By combining expressions of the basic laws of conservation of mass, momentum, energy and fluid-state, for an element of an exhaust duct, Fig 45, the pressure and kinetic-energy relationship can be derived:

$$(V_1^2/2) + [\gamma/(\gamma - 1)](P_{1}/\rho_1) - V_2^2/2 + [\gamma/(\gamma \;\; 1)(P_2/\gamma_2)]$$

where V = volume and γ is the gas constant. Since acoustic velocity is equivalent to the square root of the rate of change of pressure with density, a critical Mach number at the throat of a generalised flow system, Fig 46, can be expressed as:

$$[2/\gamma \; -1)]^{1/2}[(P_0/P_{cr})^{(\gamma-1)/\gamma} - 1]^{1/2}$$

Silencing techniques

The general approach to silencer development is one of allowing the exhaust gases to expand into chambers of large cross section area. At the speeds of shock waves mentioned above, and corresponding gases leaving the combustion chamber, there is an appreciable distance differential between the noise created by the valve opening and the slug of gas with which it was originally associated. Expansion of the low frequency wave front, successively, into volumes adding up to one third of the engine swept-volume is the conventional practice. Noise 'roughness' following in the wake of the wave front is reduced by the use of louvred tubes within the closed portions of the silencer which act as Helmholtz generators for attenuating high frequencies.

The combination of a large cross-section area with an adequate volume is important since the predominant frequencies are relatively low but have to occupy a broad band of the spectrum. While the main silencer can deal with the shock waves, there is usually a peak 'exhaust boom' frequency to be dealt with. In designing a suitable Helmholtz resonator for this, a chamber with a small diameter tube in the wall is calibrated so that resonant frequencies created by that chamber are out of phase with the peak frequency which has to be isolated. Fig 47 shows a silencer with

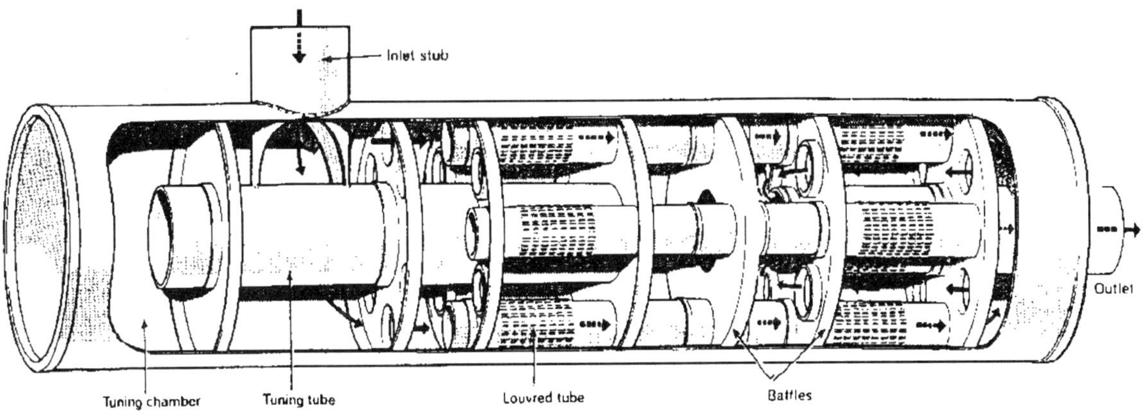

Fig 48: Truck-type silencer

chamber based on a Helmholtz resonator. Fig 48 shows a truck-type silencer, combining expansion chambers and Helmholtz resonators. Expansion chambers are arranged to cause a reversal in flow so as to create frequencies which are identical to, but out-of-phase with, the original pulse — to partially nullify its effect.

Scavenging techniques

All pipes exhibit a natural peak pressure at a given speed and generally an increase in length will result in an increase in pressure until a critical length is reached. This critical length decreases as engine speed increases. Peak pressure also varies with pipe diameter, pressures generally being in proportion to pipe diameter. It is usual to choose a pipe diameter which keeps peak pressure down to around just under half an atmosphere.

Use can be made of so-called interference pipes in the design of exhaust manifolds to improve scavenging, Fig 49. The principle of interference can be explained by a simplifying diagram such as Fig 50. The blind end A represents the closed exhaust valve on one cylinder as another is opened at B. A pulse entering the pipe spreads in both directions BA and BC, The pulse is reflected at A and follows the pulse that is already some way along BC. This extends the period of the positive pulse and afterwards adds to the negative period, helping to damp vibrations.

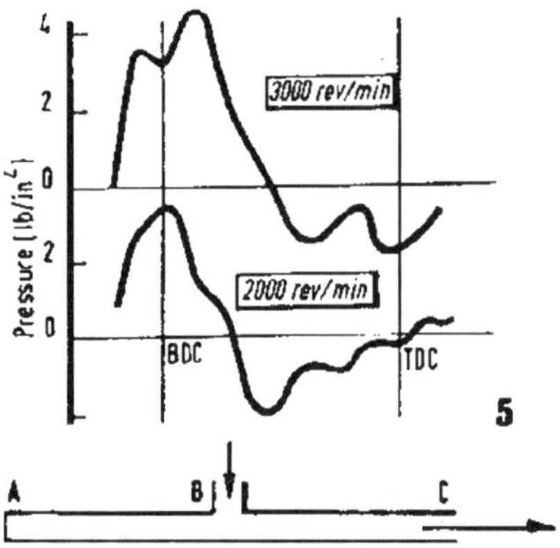

Fig 50: Effect of interference

Gasoline engine emission control

The rapidly developing environmental legislation scene is driving the major engine design consultancies and systems suppliers into perfecting reliable devices for controlling exhaust pollutants throughout the life of the vehicle. Existing production and future developing systems are considered here.

A three-way catalyst with closed-loop lambda control is becoming the norm for the stricter emission requirements facing the European market. Fig 51 shows the emission characteristics for the three principal toxic gases as a function of the excess-air factor lambda, with and without a catalyst. At the stoichiometric point where lambda = 1 there is just sufficient air for complete combustion; above 1 there is an excess-air condition. At about lambda = 1.25 all three pollutants have been reduced by the three-way catalyst. If lambda = 1 could be maintained for all operating conditions of the engine a one-way catalyst

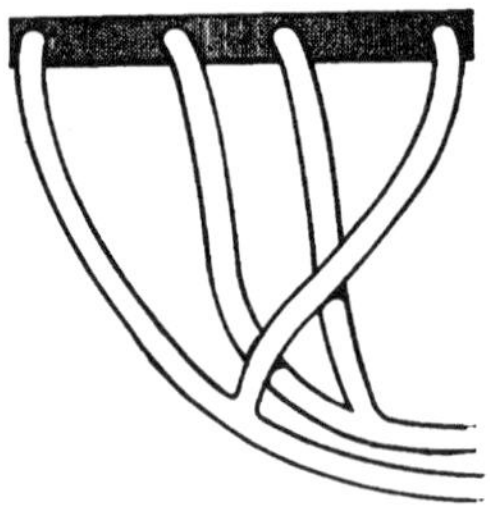

Fig 49: Interference pipes

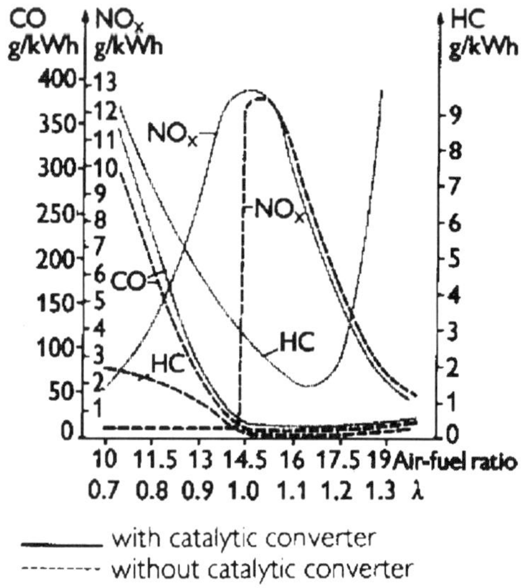

Fig 51: Emission gases with and without 3-way catalyst

would be adequate, just to deal with NOx — but no fuel-management system on the market could reach this level of performance.

The accuracy of control provided by closed loop lambda sensor systems, as described in the earlier section on fuelling systems , is very considerable and ensures the reliable working of the three-way catalyst. The sensor is essentially an oxygen concentration cell with a solid-state electrolyte that conducts oxygen ions at high temperatures. For quick warm-up the sensor has had to be close to the engine which could endanger the optimum continuous operating temperature of the device and its durability. This led to the heated sensor, Fig 52, which has a 100,000 mile service life and can even tolerate splash-water when mounted some way down the exhaust system. It reaches optimum temperature about 25 seconds after cold start.

The effect of engine operation on pollutant emission levels is considerable. During warm-up from a cold start CO* is particularly high as it is, even after warm-up, during engine idle. HC* is also high at idle and during over-run deceleration. NOx* rises as engine load rises such as in maintaining constant vehicle speed up a gradient; at idle however NOx is low.

Basic principle of the catalytic converter is seen in Fig 53. The honeycomb structure presents about 400 cells/in^2 (62/cm^2) to the exhaust flow and the ratio of catalyst volume to engine swept volume is about 1.1 to 1.23:1. A washcoat over the extruded steel or ceramic substrate acts as a key for the precious-metal catalyst and base-metal promoter of catalytic reaction. Generally the gas passes through the catalyst in about 1/10th of a second during which the pollutants are reduced to 10% of their initial concentration. While 3-way gasoline engine ones control HC, CO and NOx, diesel engine catalysts control HC, CO and smoke. The figure shows the addition of a third strategy to the basic 3-way con-

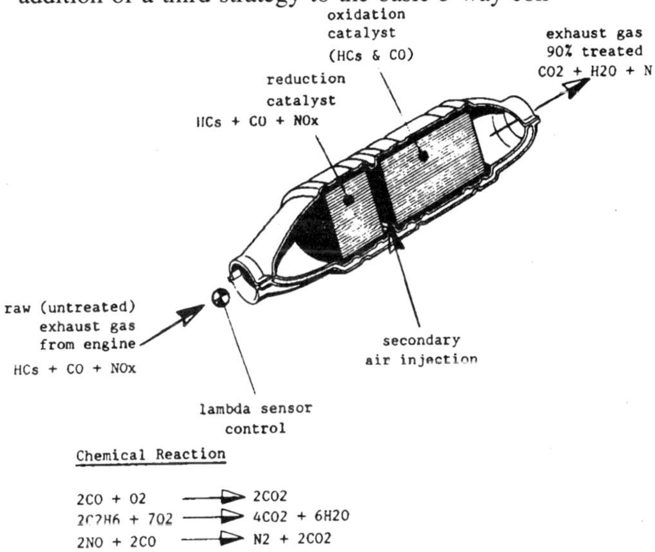

Fig 53: Principle of catalyst

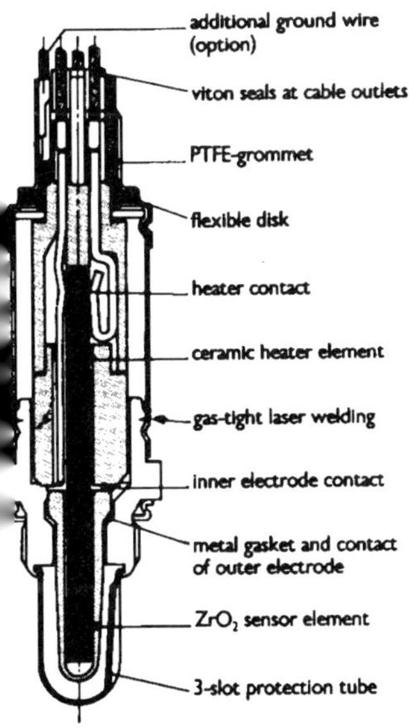

Fig 52: Heated oxygen sensor

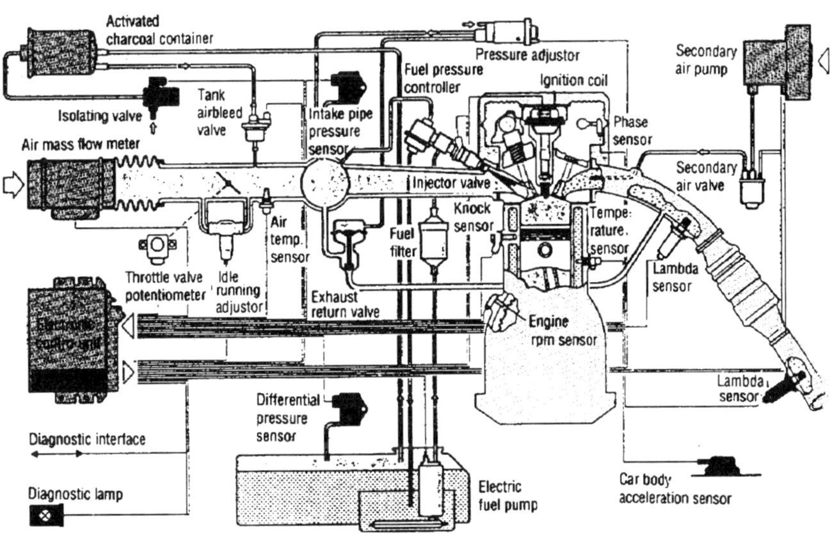

CO = carbon monoxide; HC = hydrocarbons; NOx = oxides of nitrogen

verter, the back-up of an oxidation catalyst to control residual CO and HCs. This requires additional air-injection while the main catalyst, up-stream, requires a relatively rich A/F ratio which can compromise fuel economy.

One of the highest standards of petrol engine exhaust pollution control has been achieved by Saab with their Trionic engine management system, for which the claim is made that it can actually clean urban air. It can achieve 99.37% reduction in CO and 99.6% in HC. The 32-bit controller continuously monitors and adjusts engine timing, fuel metering, fuel injection and turbo boost pressure. It is also adaptive in the sense of compensating for changes in engine behaviour throughout service life. It individually controls the direct-injection valves on each engine cylinder, computing the correct parameters before each opening. The engine also incorporates the Saab capacitive direct-ignition system, with separate ignition coils for each cylinder involving continuous measurement of the ionisation current in each cylinder. This is the means of providing individual combustion control for each cylinder to prevent detonation, pre-ignition and misfiring, without the need for a separate knock-sensor.

The current (1995) gasoline-engine emission level maxima of 2.72 gm/km of CO and 0.97 for HC and NOx is set to reduce imminently to 2.2 and 0.5 respectively in and is driving developments by the major engine design consultancies. This might involve measures beyond 3-way catalysts just described. There are also changes being discussed that will alter test cycles to more closely represent cold-start conditions. On top of this is the requirement in the USA for OBD2* which specifies an on-board diagnostic system which will monitor the system and signal any faults to the driver via a dashboard display. There is also the demand for less CO_2 emission by those concerned with greenhouse effect.

When the HC emissions are plotted against time from start-up, Fig 54, during an FTP 75 drive-cycle three quarters of the emission occurred in the first 200 seconds of the cycle. Bosch have introduced an idling speed actuator for their Motronic 5 fuel injection system with OBD2 which measures air consumption very accurately at these low speeds. To circumvent the problem of slow catalyst warm-up the company have devised a method of air injection into the exhaust just downstream of the valve to

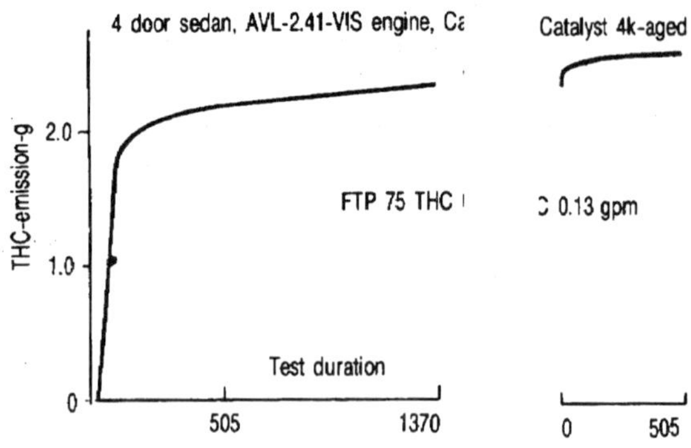

Fig 54: Emission related to time after cold-start

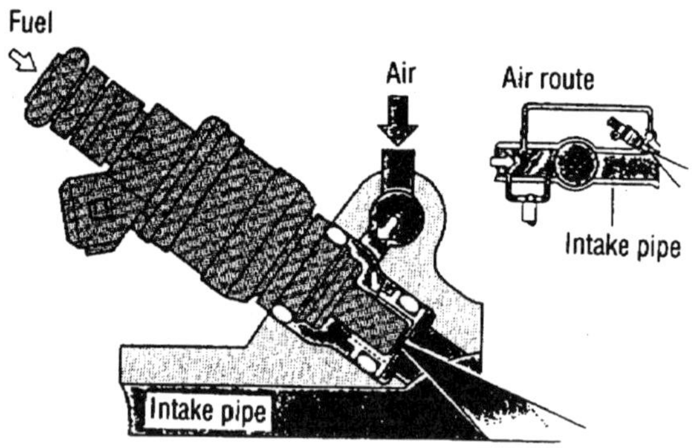

Fig 55: Air-injection at the injector nozzle

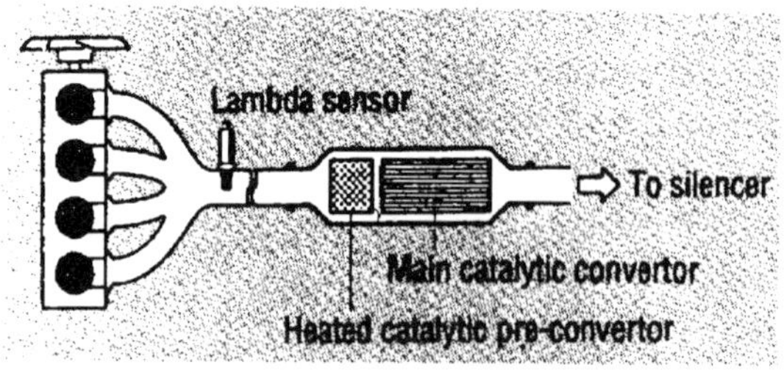

Fig 56: Catalytic pre-convertor

**OBD2 = phase 2 of on-board diagnostic regulations*

reduce HC. Another technique is to increase idle speed and retard the ignition to speed up converter heating. A further technique is to introduce high-speed air at the fuel-injector nozzle to improve dispersion of the sprayed fuel, Fig 55. This reduces droplet size below 50 microns for idle. Another measure is the use of heated catalytic pre-converters which allow catalytic action to commence immediately after cold-start, Fig 56. There is also the proposal to tighten the control of the lambda sensor over the life of the vehicle for 0.5 to 0.3% deviation. This is achieved by a dual-sensor system one before and one after the converter.

For monitoring system performance as required by OBD2, a key parameter is ignition since any misfire increases HC emission; normal technique is to use a speed sensor to detect any perturbations of crankshaft speed.

Future emission levels

According to AVL, in the US the Emissions Legislation up to the year 2000 is almost finalised, Fig 57. Due to the higher NOx-limit for diesel engines, this engine can survive until the year 2000 even in bigger passenger cars. With southern California suffering from the worst ozone problem it has embarked on an even more restrictive set of standards by defining 'Low Emissions Vehicles', Fig 58. These Standards will be phased in from 1994 onwards, starting with TLEV (Transition low emission vehicle)

Examining the formation of emissions during FTP 75, Fig 59, the importance of the first 50 - 100 sec after start can be seen. Eighty to ninety percent of total FTP 75 emissions are produced during this phase before light off of the catalyst. Significant differences between the emission behaviour of TLEV, LEV and ULEV (ultra low emission vehicle) vehicles can be seen only during the first 50 sec of FTP 75 test. When warmed up the differences in actual emissions are much smaller than recorded limits for TLEV, LEV and ULEV.

For low vehicle emission both fast light off of the catalyst and reduction of engine-out emissions have to be obtained. Whereas improved transient tuning of the engine by advanced engine management systems and more sophisticated development routines can improve the emission behaviour significantly compared to current production engines, Fig 60, obtaining 'Low Emission Vehicle' standards requires additional measures:

- * improvement of injector application for mini mum wall wetting
- * improvement of combustion system for lower HC-feed-gas
- * sheet steel insulated (3 layer) exhaust mani fold

Retarded spark advance during warm up reduces HC-feed-gas emission and raises exhaust gas temperatures by up to 250 C. This strategy is especially applied at low engine load, idle and engine overrun, whereas at higher engine load retarded spark timing would result in worse driveability. The temperature raise can be increased further by secondary air injection which is necessary for CO- and HC-conversion with rich operation of the engine. The

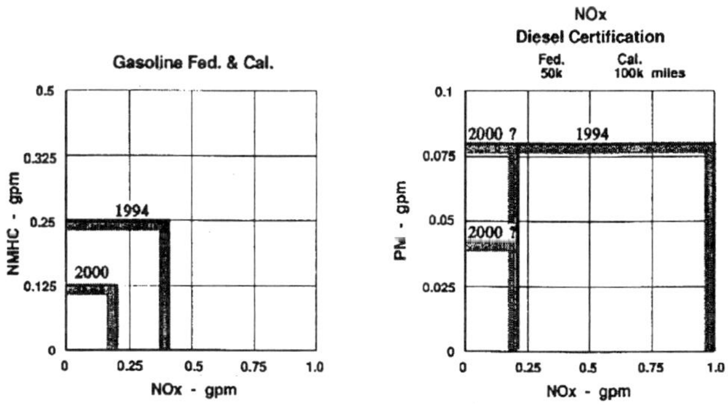

Fig 57: 1995 emission requirements

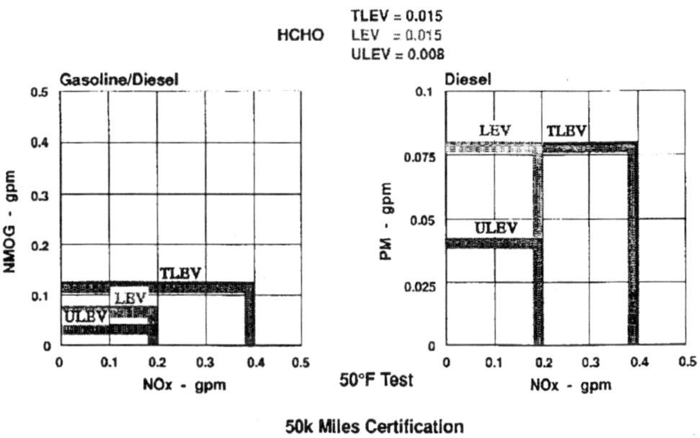

Fig 58: Projected (2000) emission limits

resulting deterioration of fuel economy varies depending on engine type. If spark retard has to be applied excessively, fuel economy in FTP 75 can be reduced by up to 3%.

Electrically heated catalysts are very often considered to be the most effective tool for obtaining Low Emission Vehicle targets. This technology does not require a modification of the engine itself but of the electrical system (increased battery capacity >100 Ah, alternator etc.) and has to be combined with secondary air. Heating strategies both with and without pre-cranking heating (15 - 20s) have to be considered. The high heating power required (up to 4 KW) not only requires high power cables and control systems but also reduces fuel economy by typically 2 - 3% (FTP 75).

Bypass start catalysts offer a similar fast temperature raise in the main catalyst as electrically heated catalysts but do not require electrical heating power. The start catalyst is close-coupled in a controlled bypass and only operated when the temperature in the main catalyst is too low for sufficient emission conversion. Thus the start catalyst wash coat/loading can be formulated for optimum light off characteristics, and not be affected by durability requirements.

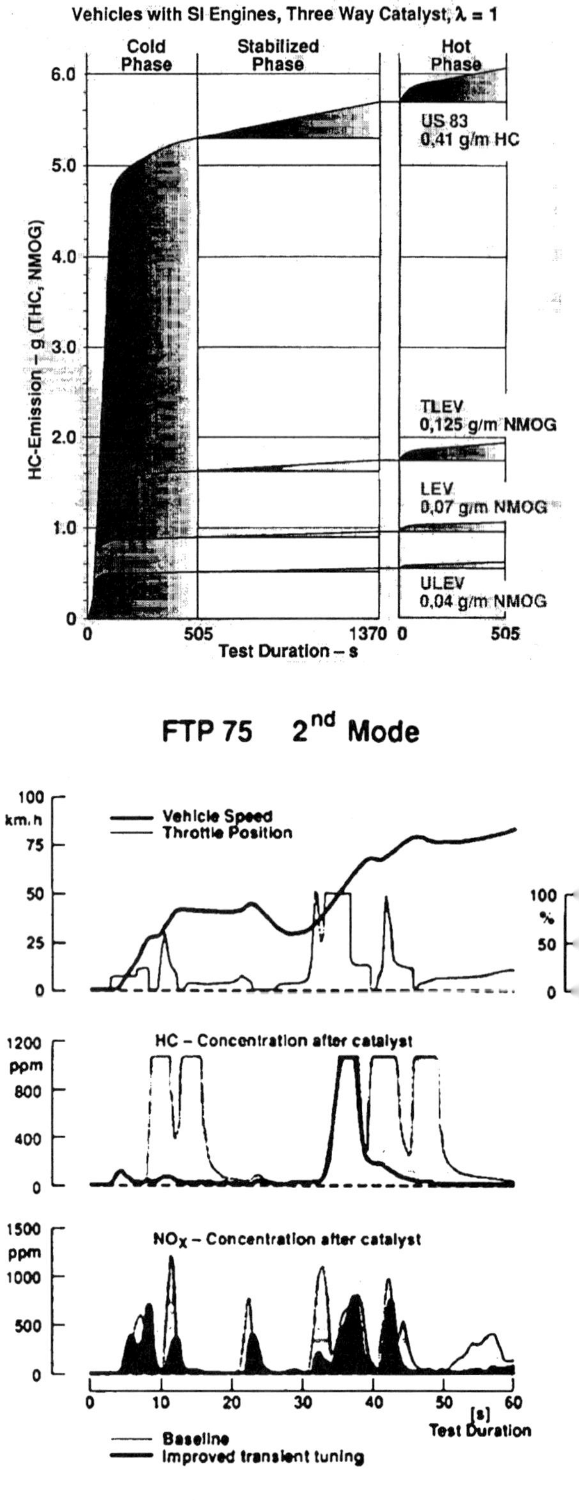

Fig 59: Emission formation during FTP 75 cycle

Diesel particulate filtration

Diesel cars and the majority of commercial/industrial vehicles will be significantly affected by the needs of the intensifying emissions regulations. As well as CO, NOx and HC emissions, the filtering of particulates has to be considered

Solid carbon, or soot, is produced in the combustion chamber as diesel fuel is burnt as a confined turbulent diffusion-flame. Some of this carbon is also burnt during this combustion process, due to recirculation, and that remaining is termed the "nett soot release", some of which may also burn in the initial stages of expansion while gas temperatures are high enough. The remainder will appear in the exhaust and its order of magnitude is some 0.5% of full-load fuel input, being equivalent to 4 units of density on a Bosch smoke meter.

At a given injection timing an increase in the rate of fuel injection reduces soot formation. Reduc-

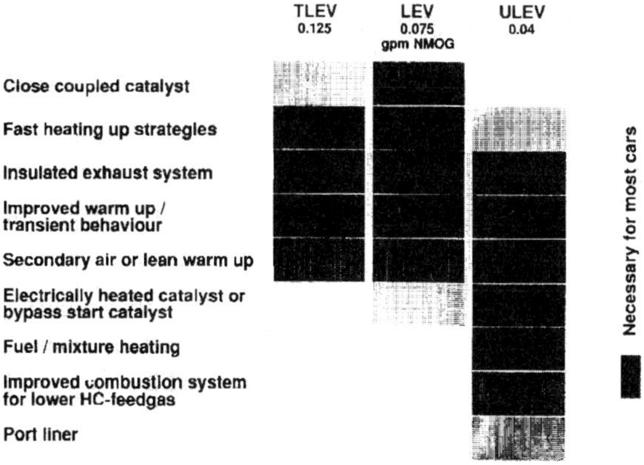

Fig 60: Additional measures for LEVs

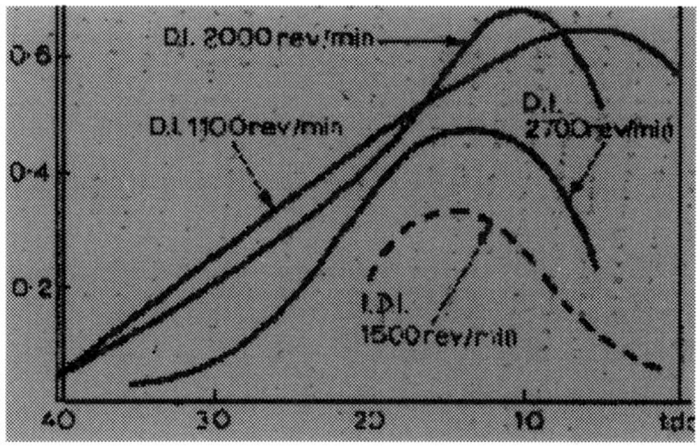

Fig 61: Factors affecting soot formation: smoke in exhaust (x-axis) vs timing in degrees BTDC

tion in the initial rate of injection for a given period of injection, or a reduction of the total injection period, both alter the amount of fuel entering the chamber after ignition — thus altering the relative importance of diffusion burning. Injection timing also affects soot formation to a considerable extent as seen by Fig 61.

Particulate size is typically 0.2 microns so if a filter has a collection efficiency of, say, 70% then for an emission rate of about 0.2 gm/km the filter will collect about 14 gm every 100 km. Generally it is necessary to regenerate the filter by burning off the particulates when it contains more than 20 gm, or at 140 km intervals. Normally ceramic honeycomb and wire mesh diesel particulate filters (DPFs) have been favoured.

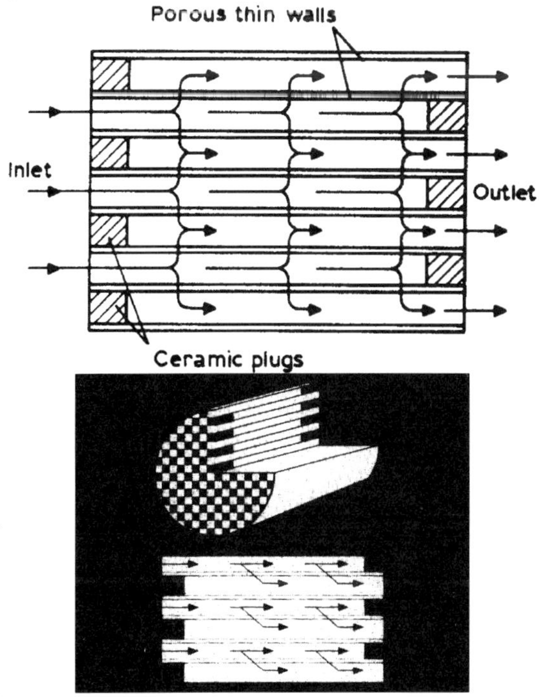

Fig 62: Oxidation trap construction

The burning of carbon requires a certain temperature and oxygen content. Means of achieving the required temperature include mounting the DPF as close as possible to the exhaust ports, throttling the intake manifold and the use of a burner. While the latter method allows good regeneration at low speeds and loads, it needs a precisely controlled air supply for stable combustion at different exhaust gas flow rates.

Besides soot, another group of diesel particulates is the subject of attention by environmentalists — polynuclear aromatic hydrocarbons (PAH), in the soluble organic fraction (SOF), which have been identified to exhibit carcinogenic activity in animals. These will appear condensed on to the soot particles as the exhaust gas is cooled. Although catalytic reactions will reduce the gas phase concentration of hydrocarbons, PAH reactions might also occur in the solid particle state.

The US Environmental Protection Agency set emission regulations for heavy-duty diesels in 1990 which prescribed 0.1 gm/bhp/h, from 1994, with engines being required to meet the standard for 290,000 miles. Unfortunately, tuning for the low NOx emissions of 5 gm/bhp/h from 1994 causes an adverse trade-off with particulates and has led to the use of traps and after-burning devices. These are also known as trap-oxidisers, as oxidation of the trapped particulate matter is needed to regenerate the filter, in order to restore acceptable exhaust back-pressure on the engine. While there is sufficient exhaust oxygen usually present to burn the trapped particulates it would have typically to be heated, so as to light off the particulates at some 550-600 C. This requires either an electric or diesel-fired heater upstream of the trap. To keep supplemental energy for regeneration to a minimum the exhaust flow is diverted around the trap during the process and combustion air for the particulats provided by electric motor. At the current state of system development, however, reliability of bypass valves and heaters has proved problematic in some cases.

The alternative approach to the particulate trap is therefore an oxidation catalyst doped on to a ceramic monolith. Here the catalyst lowers the temperature of ignition for the particulates down to around 325-400 C, which falls within the range of operational exhaust temperatures. Such catalytic particulate traps thus become passive devices, regenerating on their own whenever exhaust temperature exceeds soot ignition point. By regenerating under full exhaust flow, better heat transfer results and decreasing thermal stresses on the monolith is an aid to reliability. Total reliability has not been reached, however, due to poisoning of the catalyst or missizing of the filter. Latest thinking is to use the catalyst in the trap for oxidising the gaseous CO and HC and using heat released from the exothermal reaction to raise the temperature of the particulates so as to increase their rate of oxidation.

Construction of an oxidation trap is typified by a wall flow filter comprising a cellular monolith, similar to that used on petrol engines for gas catalysing but modified by blocking alternate cells, in chessboard fashion, on the entrance face. The exit face is similarly blocked, but one step removed so that exhaust gas is forced through the porous walls to exit through an adjacent cell, thus offering high filter area for a given volume. Filter efficiency is then controlled by altering wall thickness and porosity. Large area filters, for heavy-duty diesels, have a number of ceramic segments cemented together, Fig 62.

Two different active regeneration systems are shown in Fig 63; in the first, the technique of throttling the air to one or more of the engine-cylinders is seen, for increasing HC and CO concentration, while in the second a diesel-oil burner is shown. The simpler self-regenerating systems are in Fig 64 which includes means for positioning the trap closer to the engine by substituting it for the exhaust manifold.

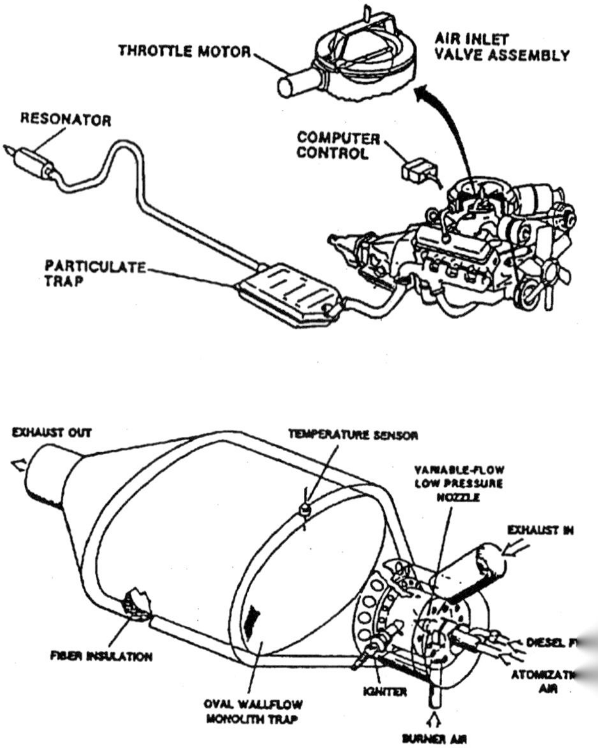

Fig 63: Active regeneration systems

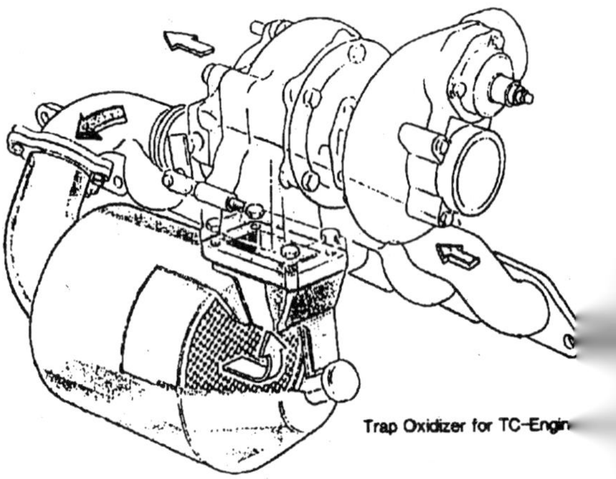

Fig 64: Self-regeneration system

Reducing emissions during transient operation

Turn of the century emissions proposals to be embodied in such legislation as Euro 3, and its USA counterpart, is encouraging development teams to study engine transient behaviour, throughout the vehicle drive cycles, prescribed by the codes.

According to Ricardo engineers[6], control strategies currently built into engine management systems that meet steady-state requirements of current legislation will need to be refined to give proper emission control in transient operation. Otherwise, the effect of sudden throttle opening and closing will lead to unacceptable excursions from optimum air/fuel ratio. These could lead to so-called catalyst break-through, due to such factors as delayed fuelling and wetting of manifold walls.

Engine load sensors will be required to be compensated for the effects of in-cycle pressure or flow fluctuations and to correct for the time lag between engine load change and its subsequent recognition by the management system prior to signalling fuelling requirements. Delays in the physical process of fuel injection will also need compensation and rate of injection will probably be based on the projected airflow into the engine at the next induction stroke. An allowance must also be made for fuel which condenses on the cold walls of engine ports just after start-up and then evaporates as the engine warms up. Baseline fuel pulse width in a conventional engine management system would require transient compensation based on such parameters as rate of change of throttle position and manifold pressure, extremely difficult in practice to calibrate.

The authors see new control strategies based on models, Fig 65, of the air and fuel flow processes in the engine, running in real time within the EMS. This "observer" system is modified by the authors to include a "predictor" for A/F control. Its requirements are seen in Fig 66 and a variety of measurements are take to compute fuel pulse width; the time delays in these are compensated for in the strategy. In tests on a 3 litre, 200 bhp, V-6 engined vehicle NOx was reduced from 0.044 g/mile to 0.012 and HC from 0.031 to 0.011 g/mile.

Another approach to reducing transient emissions is to use an afterburner for addressing the cold-start problem. The success of such devices has been shown by Ma *et al*[7] to depend on hydrogen concentration, to mix with the fuel-rich exhaust mixture, in the afterburner. Instantaneous ignition occurs immediately after a cold start if the critical hydrogen threshold is exceeded. The afterburner serves the dual purpose of removing emissions by burning them and releasing heat for bringing the downstream catalytic converter rapidly to light-off. Fig 67 shows a typical set-up, with which considerable excess air is introduced to the engine immediately after start-up.

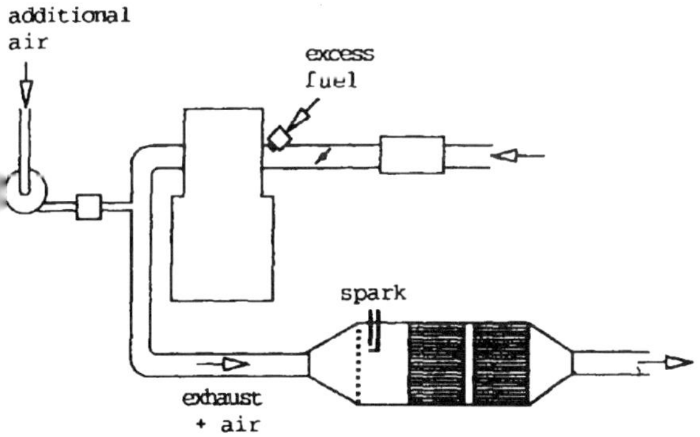

Fig 67: Afterburner for exhaust-gas ignition

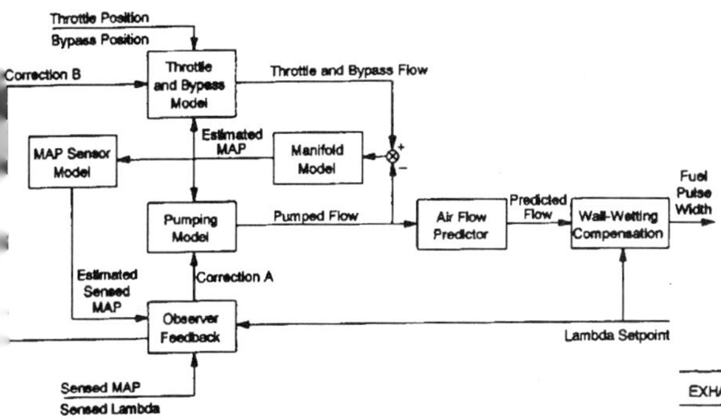

Fig 65: Model-based control strategy

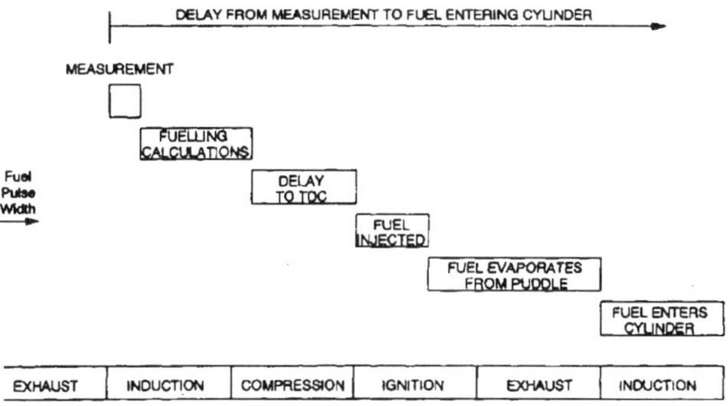

Fig 66: Delays in the fuel delivery system

Hydrogen is released in the secondary `water gas reaction' where CO and water vapour combine to give off H_2 and CO_2. Concentration of hydrogen will depend on H/C ratio of the fuel and concentration of CO produced by rich combustion at start-up.

The authors carried out experiments to study the ignitable limit of this rich mixture by taking samples from the afterburner with the engine running at different levels of excess fuel and with additional air introduced. The samples were ignited in a cold combustion bomb. For samples taken as the mixture was progressively richened from 11:1 ignition became increasingly successful. But such conditions do not exist within the afterburner, where turbulence and pressure-pulsations of the exhaust gases had the effect of delaying immediate ignition to the 8:1 enrichened condition. Fig 68 shows plots of H_2 and O_2 concentration in the afterburner for different A/F mixtures supplied to the engine alone and to the engine and exhaust system combined. Under the ideal conditions of the combustion bomb, minimum requirements for ignitability of the exhaust/air mixture are that H_2 concentration must exceed 3% by volume at the same time as the O_2 concentration exceeds 6% by volume, for example at point C. Operating point A is where thermal oxidation of CO and HC is achieved by secondary air injection. Point B is where the afterburner is fired up by ignition of CO and HC when these gases are hot and reactive. Neither of these points will support cold ignition just after start-up. Nor even will point C, due to the less than ideal conditions in the afterburner and point D defines where ignition occurs in practice at some 6% O_2 concentration. Within 5 seconds of ignition the front face of the catalytic converter reaches red-heat.

Other Ricardo researchers have proposed a revised combustion system for meeting ULEV emission standards. Proposed modifications to the FTP test cycle in the USA to add a `fourth bag' corresponds with the European plan for including starting and idling emissions within the test cycle sampling. Among the company's strategies for compliance are: a new approach to combustion that will increase tolerance of exhaust gas recirculation devices (EGR); direct in-cylinder gasoline injection to reduce emissions within the 30 second period after cold-start; a total-system approach to engine fuelling, exhaust after-treatment and engine management.

EGR tolerance is improved by a technique of stratifying the recirculating gases in the cylinder. The combustion chamber design involve a 4-valve pent-roof system with combustion control by vortex stratification (CCVS), Fig 69 showing the four stages of the combustion process. At idle and part-load a stoichiometric mixture is supplied via inlet-valve 2 and a strong circumferential swirling motion is induced. At low EGR levels through inlet valve 1 there will be insufficient flow to induce tumble swirl. As

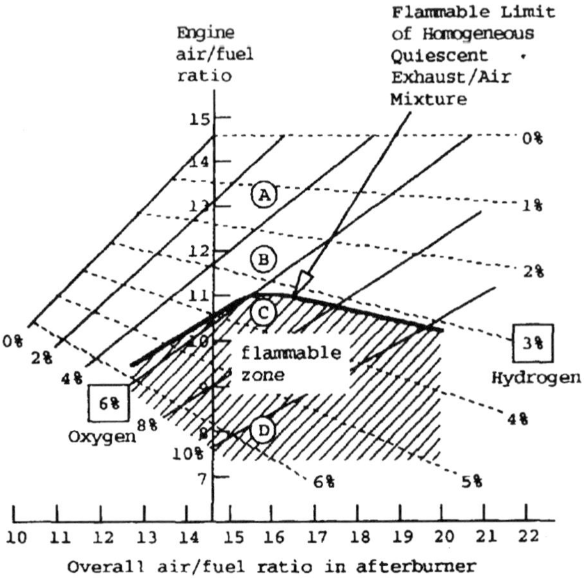

Fig 68: Flammability of exhaust/air mixture

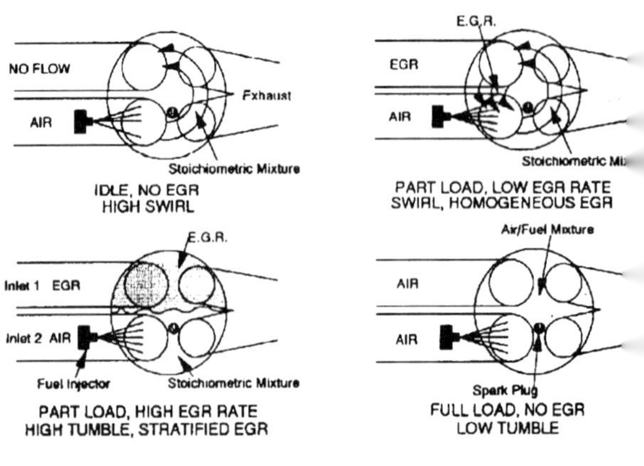

Fig 69: Four stages in the combustion process

EGR rates increase the charge naturally stratifies and greater EGR tolerance is obtained, the offset spark-plug location towards the fuelled port further helping to improve combustion. Towards full-load a point is reached where insufficient charge air can be supplied through one port and air is introduced at both inlet valves. Compared with a conventional 4-valve head, the CCVS uses a relatively high tumble to circumferential swirl ratio for inducing stratification.

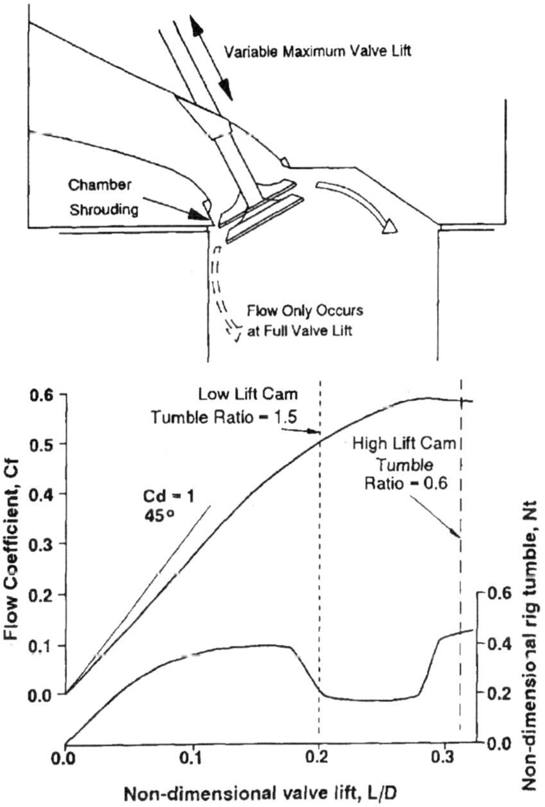

Fig 70: Configuration for inducing tumble swirl

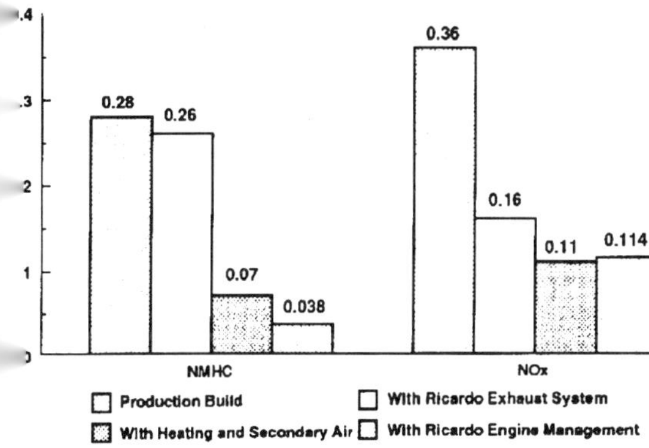

Fig 71: Incremental reductions in emissions

The mechanism for inducing variable tumble at part load is shown in Fig 70. This involves a cam switching device to give two stages of inlet valve timing and lift, shrouding on the liner side of the valve ensuring flow is directed across the combustion chamber to produce strong tumble at low lifts. At high loads full lift is used, with port/valve-angle geometry arranged to provide low tumble as seen in the graphs. The system also has a twin inlet-manifold and an electronically controlled EGR valve. A potential 90% reduction in NO_x emission is possible with the added benefit of lowered fuel consumption.

The second approach of direct gasoline injection was described earlier in this section while the third, total-system approach combines an integrated design of inlet-system, fuel-injection, engine, exhaust after treatment system, transmission characteristics and engine management. The company have applied this to a 3-litre saloon weighing 3500 lb (1.6 tonnes) and using a multi-point fuel-injected V-6 developing 200 bhp. Both engine banks exhausted into an electrically heated pre-catalyst and downstream main catalyst converters. Low thermal inertia manifolding was specified, secondary air injection applied and the model-based engine management system incorporated with advanced control sensors. The incremental improvements in emissions reduction are in Fig 71.

Anti-pollution ystem developments

The reluctance of vehicle manufacturers to adopt electrically-heated catalyst systems may be overcome by a new system developed by Corning R&D . It involves an extruded metal core based on an alloy of iron, chromium and aluminium with good oxidation resistance. It is formed by extrusion from powdered metal followed by sintering to a very low porosity. Desired electrical resistance, for heating, is achieved by slitting the core monolith — while the container is insulated by a high-temperature resistant ceramic fibre beneath the surface of the stainless steel can.

An FTP test carried out on the system, after ageing on a dynamometer, involved a 2.2 litre engined car with port-injection. To find the effect on cold-start emissions, bag 1 of the FTP cycle was split into two, the first representing 0-220 seconds and the second, 220 to 505 seconds. At 220 seconds the main converter temperature was found to be relatively stable and high enough to promote rapid chemical reaction. Prior to that point the temperature rose from ambient to 600 C with a corresponding increase in chemical activity. The contribution to the total FTP emissions from the remaining three bags were 0.009, 0.001, 0.276 and 0.065 g/mile respectively for total HC, non-methane HC, CO and NO_x. Electrical energy was supplied in the tests from both 12 and 24 V battery systems, using pre-heating or post-heating. Pre-heat is based on time power is supplied before the start of the cycle (initial cranking of the engine) while post-heat is energy supplied during cranking. Air-injection was applied at a rate of 141.6 litre/min and the electrically heated catalyst was `close-coupled', 33 in (0.84 m), from the engine.

Effect of EHC mass was found to be crucial, in the tests, for achieving the 400 C necessary for low NMHC emissions. Mass was changed by altering both frontal area and length of the EHC. Fig 72 shows the effect of mass on temperature change per watt hour and the superiority of Corning's new system (2) over the existing one (1). NMHC (non methane hydrocarbon) emissions with post-heat were found to be about twice those of the pre-heat strategy (0.018 g/mile) at energy levels between 11-15 Wh.

For the main catalyst thermal energy comes from both the EHC (entrapped hydrocarbon) and exothermic reaction of HC and CO. Tests showed that energy for activating the EHC is all that is required to initiate the chemical reaction, after which sufficient thermal energy is generated by the reaction itself.

Diesel particulate emissions can be effectively handled by the continuously regenerating trap (CRT) recently announced by Johnson Matthey, Fig 73, given appropriate fuel quality. The CRT combines oxidation catalyst and filter technology to remove particulates as well as reducing levels of CO, HC and diesel odour. Unlike previous particulate filters, the unit is able to trap solids on a continuous basis at the vehicle's normal operating temperature. The principle of such devices in experimental form, were discussed in the previous section and the first commercial development was an important breakthrough. It must operate with fuels containing less than 100 ppm of sulphur. Conversion of the principal pollutants is said to be 90%. Higher percentages still have been reported in trials on ten vehicles operated by

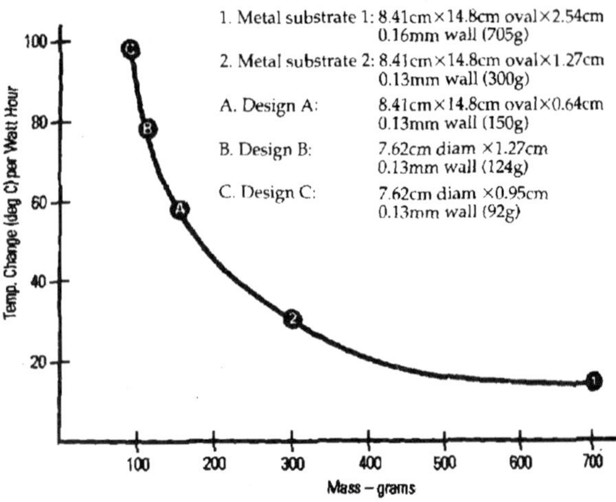

Fig 72: Effect of mass on temperature change

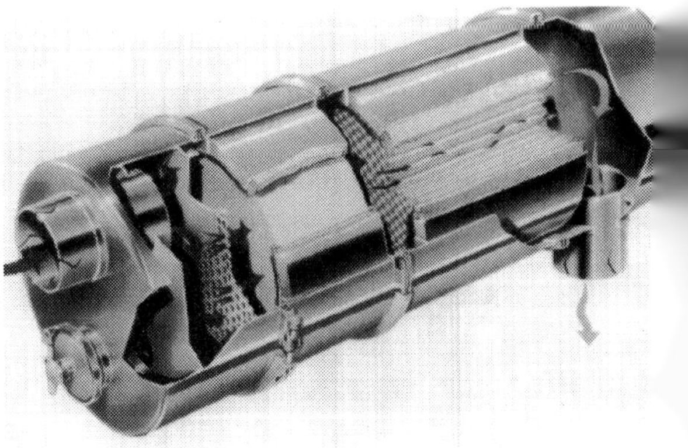

Fig 73: Johnson Matthey's Continuously Regenerating Trap

Swebus in Gothenburg.

According to Johnson Matthey a key factor of diesel exhaust is that it is far more oxidising than that of petrol exhaust. In their development work the company found that carbon in the form of soot was being oxidised not by activated oxygen but by NO_2. Higher combustion pressures and leaner A/F ratios with diesels leads to greater NO_x production. This is mostly NO which is converted to NO_2 over a catalyst using excess oxygen present in the exhaust. Its subsequent reaction with carbon is responsible for the observed temperature reduction at which a catalysed exhaust filter can be regenerated. It was thus found possible to generate the necessary NO_2 upstream of the reactor containing a non-catalysed trap. Platinum on alumina was found to be the best catalyst for the oxidation reaction and since the process of soot trapping and destruction are continuous at temperatures above 275 C, the CRT was successfully commercialised.

The substrate of the filter is a ceramic monolith with 100 gas channels/in^2 (15.5/cm^2), alternate ones being blocked at one end to obtain reverse flow in the `wall-flow' filter configuration. The catalyst substrate has a density of 400 cells/in^2 (62/cm^2) and has a platinum loading of 50 g/ft^2 (555/m^2). The device also produces the degree of sound attenuation of an ordinary silencer, which can be dispensed with.

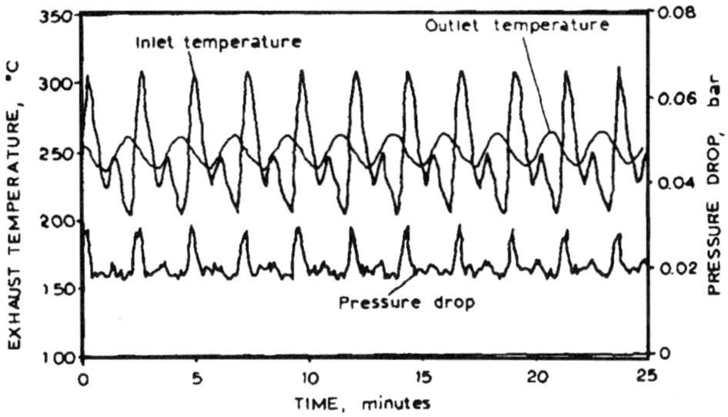

Fig 74: Endurance cycle

Tests on a 6-litre 6-cylinder diesel were carried out to cycles shown in the table of Fig 74 to give the conversion efficiencies indicated. Endurance testing was also performed with an ageing cycle that would mimic continuous operation in cool conditions, Fig 75. Pressure drop across the filter was measured over a period of 1000 hours during which no increase was said to have been observed.

Test cycle	Sulphur content of fuel, per cent	Conversion, per cent		
		TPM	HC	CO
ECE R49 (13 mode)	0.005	80	96	99
U.S. HDT, catalyst I	0.05	87	98	99
U.S. HDT, catalyst II	0.05	92	94	99
Braunschweig (bus)	0.005	92	96	98

Fig 75: Conversion efficiencies in recognised cycles

Engine-cooling/vehicle-heating

Engine cooling and vehicle interior heating systems are mutually dependent. It is thus useful to consider calculation procedures for both systems at the same time and indeed arrive at a simultaneous design solution if this is possible.

Heat flow from an internal combustion engine is directed either into the outside atmosphere, in the case of conventional cooling and exhaust systems, or to the vehicle cab/saloon interior in the case of the vehicle heating system. As ecological pressures continue to make engines run more efficiently, a greater quantity of fuel energy is converted to work and there is less heat rejected to exhaust and coolant. The move to so-called adiabatic engines, designed to increase efficiency by running at very high temperatures means that 'heat to coolant' may be reduced to very low levels. Careful design of the heat exchange system is thus required to obtain proper levels of interior heating as well as maintaining the engine at the correct running temperature.

Flow from the engine cylinder is less difficult to predict than might be supposed from the cyclic nature of gas temperature from one cylinder firing to the next. In fact thermal inertia is such that heat flow to coolant is largely independent of the fluctuations. The main difficulty in applying heat conduction formulae across the walls of the combustion chamber lies in the complex shapes involved. Shape factors akin to those, introduced in Chapter 14, on refrigerated vehicles have to be used; these might involve building the kind of electrical analogues, mentioned, to simulate the heat paths.

Heat flow from the combustion gases to the cylinder walls, and from the cylinder walls to the coolant is principally convective and it is usual to use empirical formulae based on the work of early experimenters to calculate heat flows. The Eichelberg formula states that for cylinder pressure P, absolute temperature T and mean piston speed V, heat transfer co-efficient is

$$2.1V^{1/3}(PT)^{1/2}$$

On the coolant side of the cylinder walls, heat flow is optimised by reducing any tendency for local boiling to occur at the cylinder wall surface by appropriate choice of water pump flow rate. Experiments have also shown that gross heat flow from an

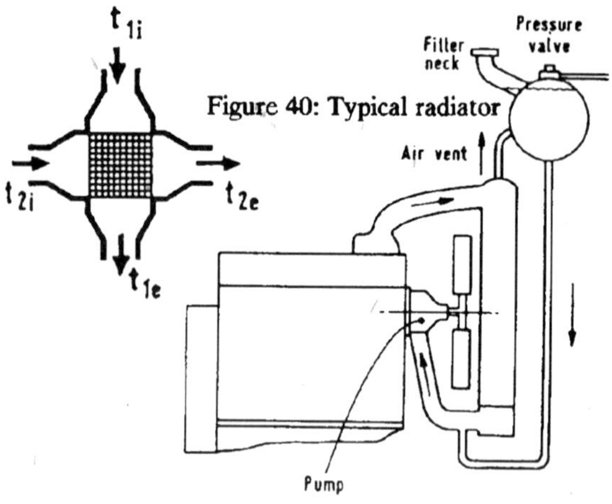

Fig 76: One-pass cross-flow system

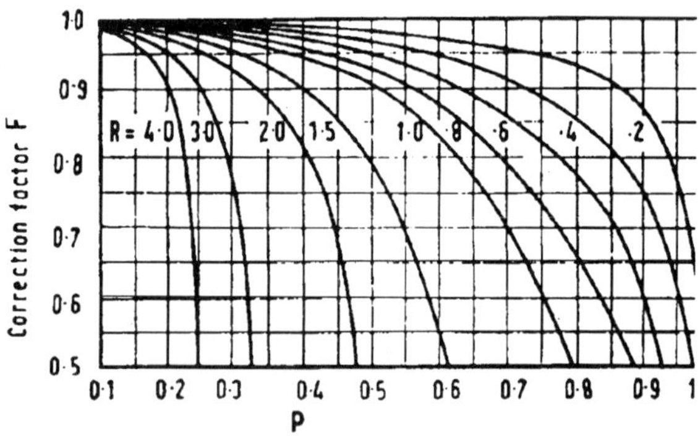

Fig 77: Correction factors

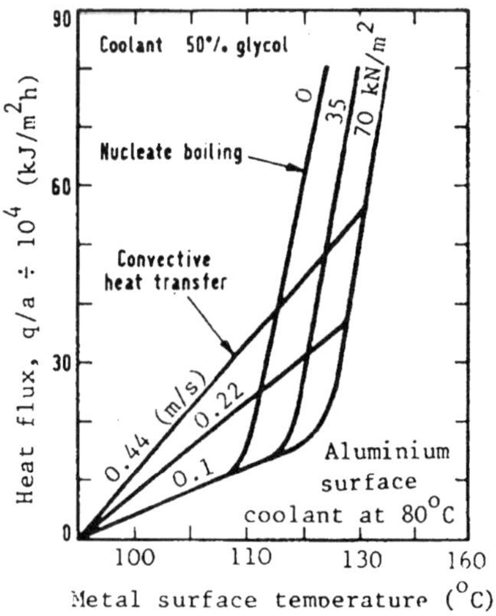

Fig 78: Curves for cylinder-wall heat transfer

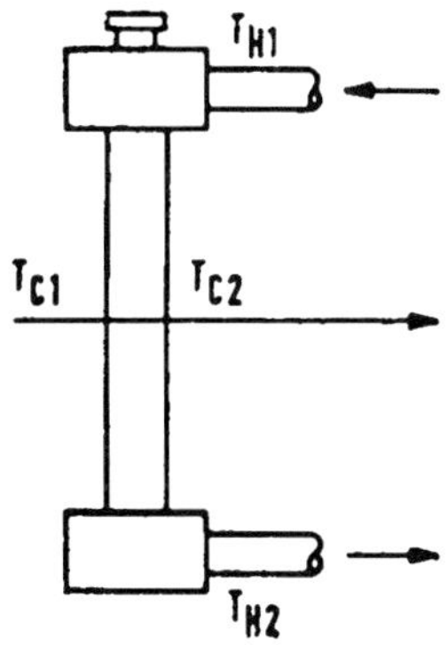

Fig 79: Radiator notation

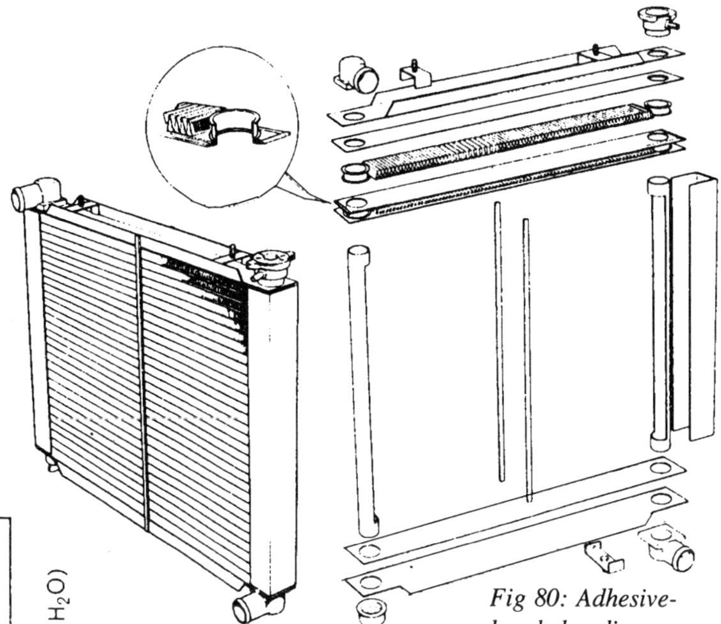

Fig 80: Adhesive-bonded radiator

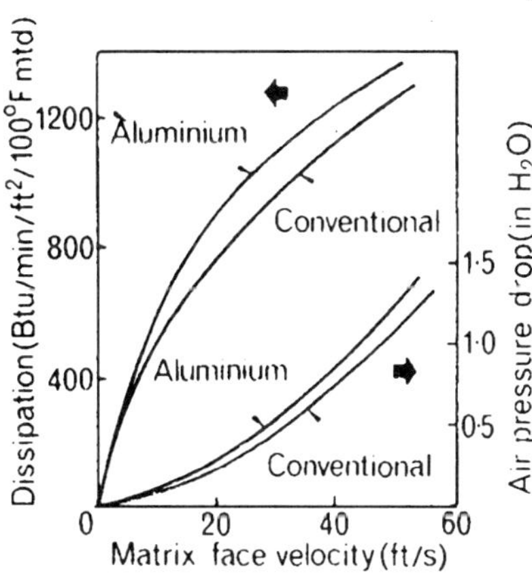

Fig 81: Curves for triangular-finned radiator

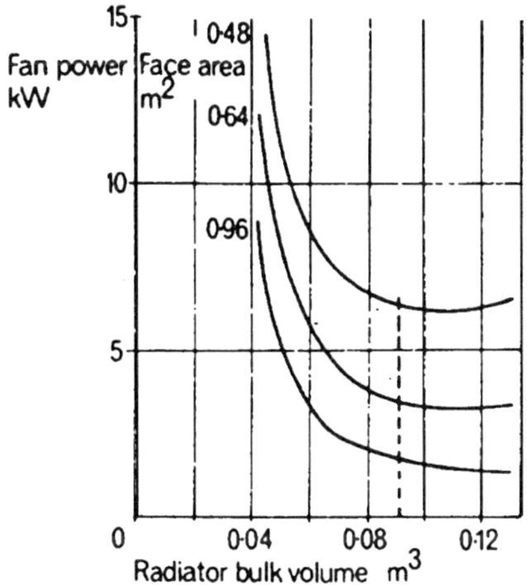

Fig 82: Fan performance curves

engine can be equated to fuel supply rate raised to the power 0.64-0.68. At the coolant heat exchanger, historically called a 'radiator' for some reason, the transfer of heat is usefully considered diagrammatically as in Fig 76 — represented as a one-pass, cross-flow system with both fluids unmixed in heat-transfer parlance.

For this system, mean temperature difference is

$$(A - B)/\log_e(A/B)$$

where A is the larger and B is the smaller temperature difference of $(t_{1e} - t_{2e})$ or $(t_{1e} - t_{2i})$ in which t_{1i} is inlet hot, t_{2i} inlet cold, t_{1e} outlet hot and t_{2e} outlet cold. Where the configuration differs from pure contra-flow, correction factors must be applied according to Fig 77. Again, empirical formulae based on the results of experiments are required to determine heat transfer co-efficients applying to the different elements of the heat exchanger because of the complex shape factors involved.

Conventionally, it has been the aim to control the temperature of the water coolant at around 85 C — corresponding to limiting water jacket temperatures of around 260 C for cast iron and 200 C for aluminium. The importance in keeping up coolant temperature is reflected in the claim that a reduction in temperature from 88 to 49 C has reportedly caused a 3% increase in fuel consumption.

Typically radiators have been designed to allow a drop in coolant temperature, through them, of 6 C and it is important to remember that the higher the coolant temperature, the smaller the radiator. The importance of maintaining the high coolant flow requirement discussed previously is seen in the relationship between heat transfer and metal surface temperature in the hot regions of the engine (such as around the exhaust valve) as shown in the characteristic curves of Fig 78. Any increase in the boiling point obtained by system pressurisation is reflected in a proportional increase in metal surface temperature while increasing coolant velocity for given heat flow suppresses the onset of nucleate boiling by lowering surface temperatures.

In design of a typical radiator system as shown in Fig 79, diagrammatically, heat exchanger effectiveness is considered as the ratio of obtainable heat flow to maximum heat flow given by:

$$CH(TH_1 - TH_2)/C_{min}(TH_1 - TC_1)$$
$$= CC(TC_2 - TC_1)/C_{min}(TH_1 - TC_1)$$

where C_{min} is the smaller of *CH* and $_{CO}$. This is because, in a cross-flow radiator, $C_{min} = CC$ as the radiator cannot absorb any more heat from the coolant than the radiator-airflow is capable of accepting.

Calculation of heat flow through the radiator can be made on the assumption of ignoring thermal resistance due to scaling deposits (although factors for this can be found in tables published by the Tubular Heat Exchanger Manufacturers Association). Formula for overall thermal resistance is:

$$(1/h0_A) = 1/h_a(A_p + nfA_s) + r_a/(A_p + A_s) + 1/A_m$$
$$k/t + rw/A_w + 1/h_wA_w$$

where r_a, r_w are air and water side fouling factors respectively; h_a, h_w are air and water side heat transfer coefficients (HTC); A_p, A_s are air side primary and secondary heat transfer areas; A_m, A_w are mean tube and water side primary heat transfer areas; *k, t* are tube material thermal conductivity and thickness while *nf* is air side secondary surface fin efficiency. The value of $h0_A$ obtained from this is then equal to the heat dissipation/mean temperature difference.

The most economical radiator matrix is one

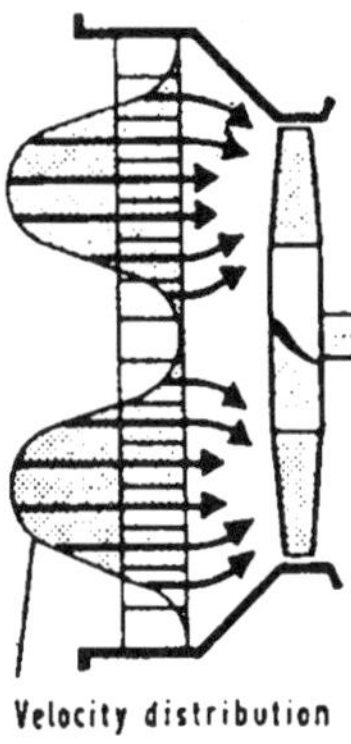

Fig 83: Ducted fan

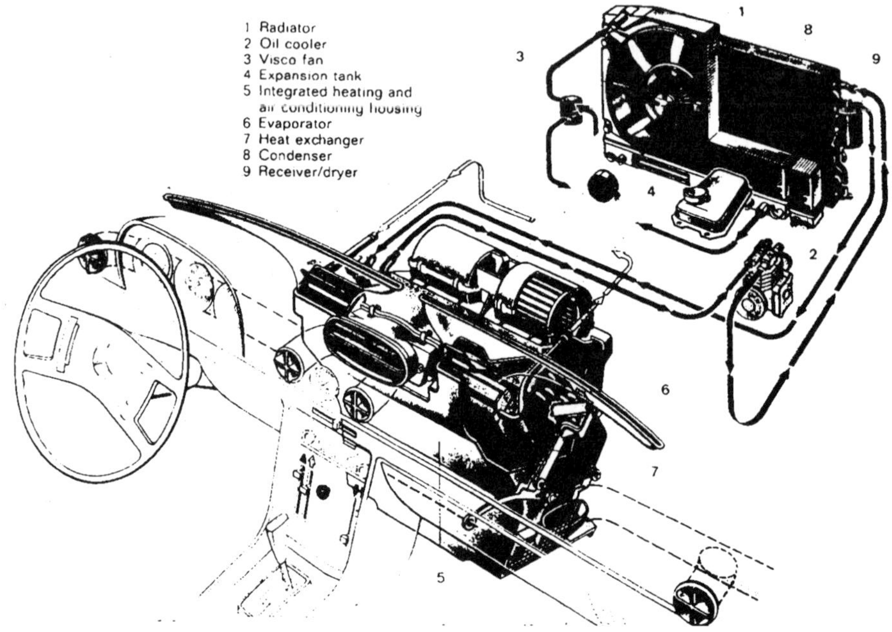

Fig 84: Combined vehicle-heating/engine-cooling

which gives highest airside HTC subject to a configuration which optimises fin efficiency without compromising fouling resistance. Fig 80 shows an adhesive-bonded aluminium matrix made by Covrad. Triangular finning is used rather than the earlier zigzag type array. Characteristic curves, compared with the conventional copper radiator, are shown in Fig 81. Radiator bulk volume affects fan performance also and Fig 82 shows how it is necessary to use the largest practicable face area and an adequate total bulk volume. The curves in the diagram assume a fan efficiency of 50% and a required heat dissipation of 210 kW with 50 C air/water temperature differential. The corresponding specific dissipation of 2300 kW/m^3, at the 0.9m^3 bulk volume selected from the graph is considered a reasonable design target.

A typical axial impeller fan used in a CV application would have a pressure coefficient of around 0.07 while a 'high solidity' axial fan, having 20 blades within a 710 mm diameter, say, would have pressure coefficient around 0.15 at 3000 rpm. Ducted systems such as that shown in Fig 83 will of course raise overall efficiency. Integrated engine-cooling and interior-heating/ air-conditioning systems, such as the one produced by Behr in Fig 84 are gaining in acceptance. Centrifugal rather than axial fans are the rule for interior heaters. Heater performance calculation is, of course, dependent on ambient temperature. The relationship between the temperature differential which can be maintained between it and saloon interior temperature for a car installation can be calculated from

$$H(t_{wi}-t_a)+nh_p[(1/0.7R_t)+dT/2)]$$
$$/(CQ-CR+K)[(1/0.7R_t)+dT/2)+HR/Q]$$

where t_a is ambient temperature, t_{wi} water inlet temperature, Q the total air flow and R the recirculation air flow through the heater. R_t is the ratio of Q to the standard airflow at which the heater has been bench tested and H the specific output of the heater under the same test conditions. The water temperature drop across the heater per degree temperature difference between mean water and air temperatures is dT; K is mean heat loss coefficient of the vehicle body, n the number of occupants and p the radiated heat per occupant. Constant C depends upon the units employed (= 0.0182 for Imperial units). Fig 85 shows variation of heat input required with temperature differential and fresh air supply in Imperial units.

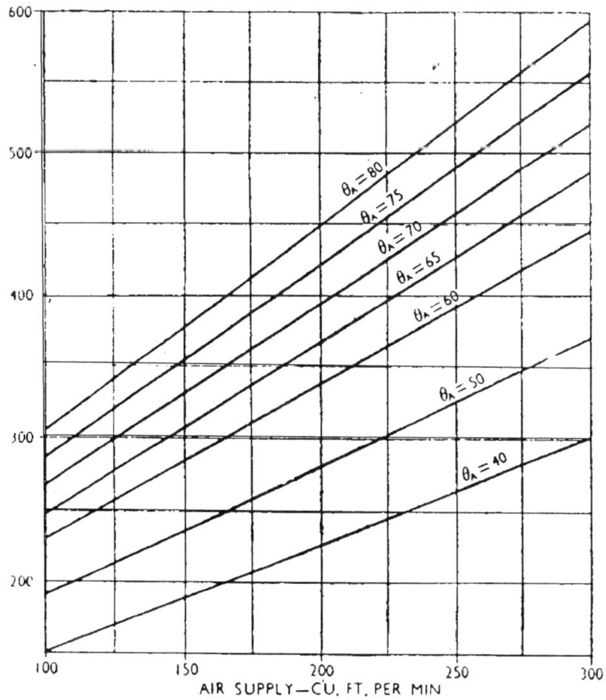

Fig 85: Heat input (btu/min) for temperature differential

IC engine noise control

Chapter 4 provided the basic definitions of noise intensity and loudness as well as showing some of the principal transmission paths. As the engine is the major noise source in the vehicle, this section shows how the above concepts can be used to understand one serious problem prior to effecting a solution for that source.

As well as being a primary noise source the engine is also a secondary one in being the main exciter of the body structure, in the case of integral vehicles such as saloon cars and vans. As noise legislation has intensified, all vehicle makers have had to effect major redesigns but most volume builders will have arrived at solutions very close to the legal limits of acceptable noise intensity. There is thus an opportunity for the specialist sector to meet the undoubted demand by a substantial minority of the overall market for vehicles which substantially outperform the basic legal requirements.

The heightened environmental consciousness among buyers of both private/leisure and commercial/industrial vehicles means that demand for alternative vehicles such as battery-electrics is based not so much on their intrinsic performance but on the knowledge that noise and noxious emission levels are so much below legal limits.

The important work carried out by the Transport and Road Research Laboratory on the Government-sponsored Quiet Heavy Vehicle (QHV) project has shown that EV levels can almost be reached even by hard working commercial vehicles. The engine noise source can be tackled in several ways — fan design, combustion control, improved structural and cover design. These treatments are separate from the use of noise enclosures for the complete engine/transmission unit which will be considered later. There is even scope for substantial noise control by arranging for a comparatively generous capacity engine to be run at relatively low speed, Fig 86.

Some of the most rewarding techniques for engine noise control at source are smoothing of the cylinder pressure development during combustion; cushioning of impacts between piston and cylinder wall; reducing impacts between timing gears and localising the impacts and high frequency components of torque reaction from engine ancillaries. Beyond this 'source' treatment, there is considerable scope for reduced transmission of vibration between

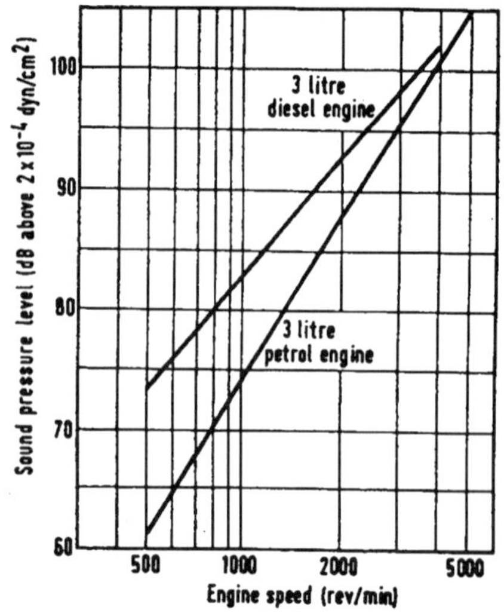

Fig 86: Effect of engine speed

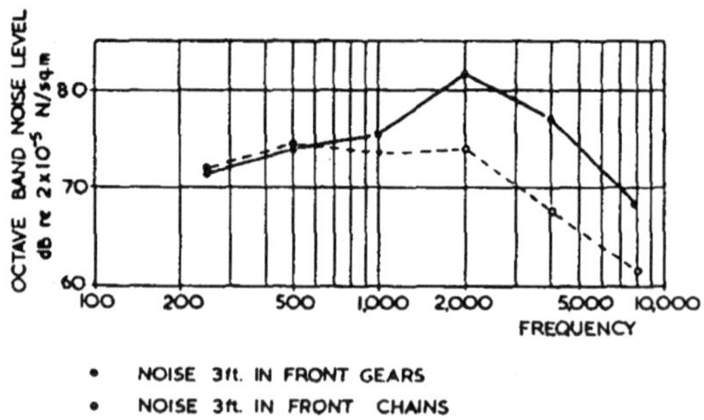

Fig 87: Chain vs belt drive

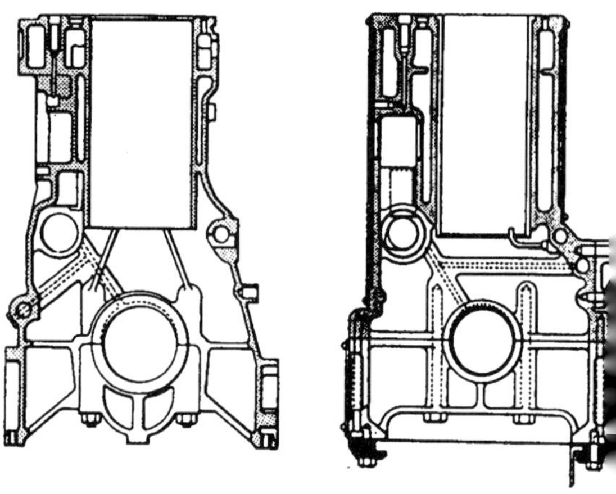

Fig 88: QHV1 engine structure

the sources and the radiating surfaces as well as controlling the vibration of those surfaces. The latter may be tackled by de-tuning the major resonances of

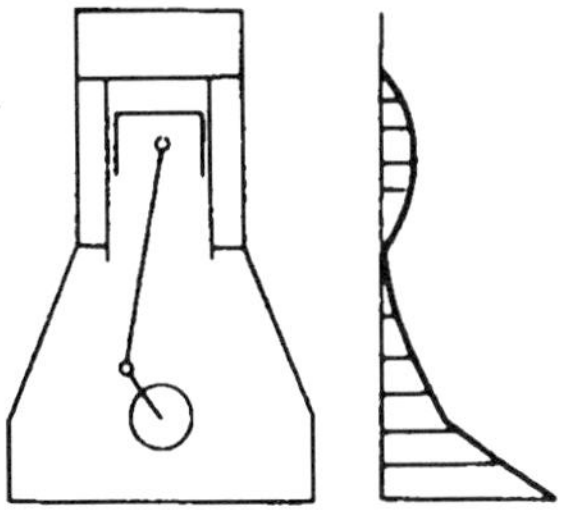

Fig 89: Crankcase wall flapping

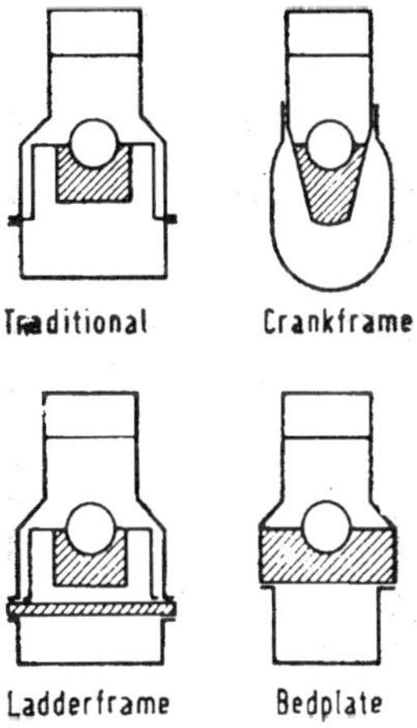

Fig 90: Engine structure types

these surfaces; by isolating non load-carrying surfaces from the engine structure; by damping the vibrations in flexible covers and by redesign of the engine structure to reduce vibration level at points of attachment of noise-radiating covers.

The actual cause of a perceived noise can of course be difficult to trace subjectively and many observers have, for example, confused 'combustion knock' with a mechanical noise source. Fig 87 shows the reduction in mechanical noise obtained on a particular engine found eventually to be suffering from timing gear rattle; by switching to chain drive of the valve train, quite substantial noise reductions were made. In the first Quiet Heavy Vehicle programme, one of the most important contributions to noise reduction was in redesign of the basic engine structure, Fig 88. The Institution of Sound and Vibration Research (ISVR) made important changes to the engine block/crankcase of the Rolls Royce Eagle diesel. The bottom deck of the cylinder block was moved down to the lower end of the cylinder, forming a rigid section to reduce wall flexing caused by crankcase distortion. Note also the noise-shielding cover panels fitted to the sidewalls of the engine. A gudgeon pin offset of 2 mm towards the thrust side was used to reduce impact noise and timing gears were changed from straight-cut to helical.

The flapping vibration typical of crankcase

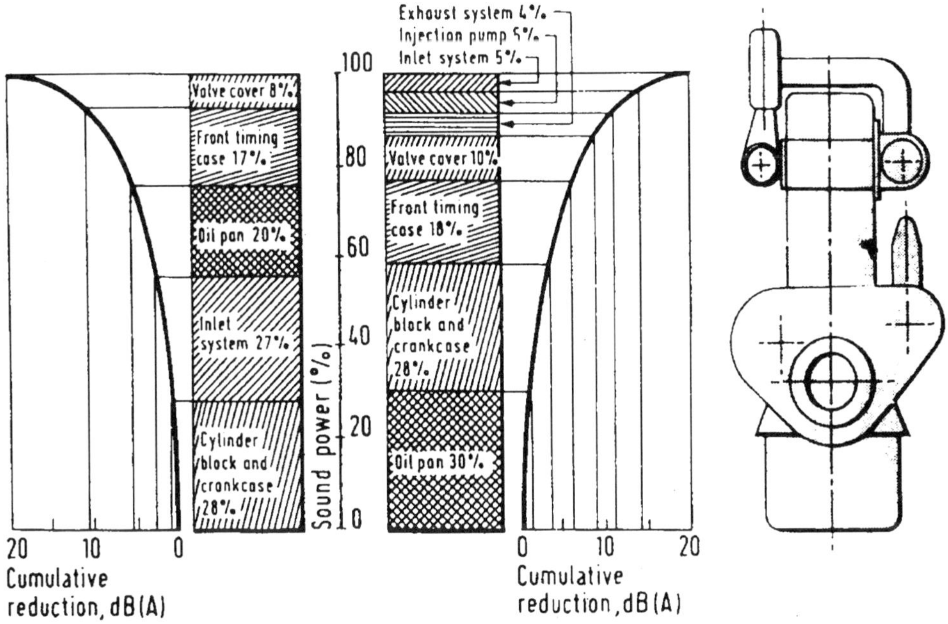

Fig 91: Crankcase resonances

walls unmodified by the above kind of approach is shown in Fig 89. In the programme of work leading up to QHV the ISVR went through the stages shown in Fig 90. The referred effects on other parts of the engine caused by the resonance of the traditional engine crankcase are shown in Fig 91. The main problem with the structure of the crankcase is that it is weak torsionally and vibrations tend to be greatest at its base — with the sump amplifying the effect.

SDRC Engineering Services have been prominent in the establishment of structured design procedures for low-noise engines. In the case of four-cylinder engines, the company point out that several components make up the total stuctural noise of an engine. Even harmonics of the engine rotational speed distinguish a 4-cylinder and 6-cylinder engine. If the amplitude of second order vibration exceeds 90 dB (linear), the vehicle is likely to exhibit a boom vibration when a shell body is fitted. Fig 92 is a characteristic signature. Half-order vibrations manifest themselves as rumble giving the engine noise a 'hard' quality. Rumble involves harmonic patterns and not merely a single, boom, frequency. It is pointed out that rumble is likely to be cause for concern if at least three adjacent half orders have amplitudes within 5 dB of each other; or if adjacent harmonics are in phases; if rumble harmonics are within 10 dB of the second order amplitude there is likely to be a deterioration in perceived quality. Fig 93 shows a signature containing even, odd and half-order harmonics. Having identified a repeatable noise problem, SDRC's approach is to make recordings of the vehicle interior noise. The company has found most petrol-engine noise problems can be traced to harmonic detail below 1000 Hz. Next stage is to use the technique of 'route tracking' to identify the path of the noise. SDRC argue that below 500 Hz most noise is transmitted mechanically and the first step in determining its path through the engine mountings is to determine the three orthogonal forces in each elastomeric mount. This is done by fitting accelerometers to them during test runs and the signals processed to obtain displacements; these are then related to the load/deflection characteristics of the mounts. The next step is to determine the vehicle structure's sensitivity to these mount forces. This is done by exciting the body with an electro-magnetic shaker and a microphone within the structure used to obtain a Frequency Response Function relating force to noise computed.

Fig 94 shows a resulting noise transfer function obtained for a specialist car by SDRC. This shows a good sensitivity is in the order of 50 to 55 dB per N applied force whereas a poor example can show in excess of 70 dB per N. Forces and noise transfer functions can then be multiplied together to calculate the contribution of individual routes. Fig 95 shows contribution of engine and suspension routes to an example second order boom problem. In this case only the frequency content at the engine 2nd order (firing frequency) is used. Final stage in the trouble-shooting process is to evaluate potential modifications to the vehicle by comparing recordings in the original and modified conditions.

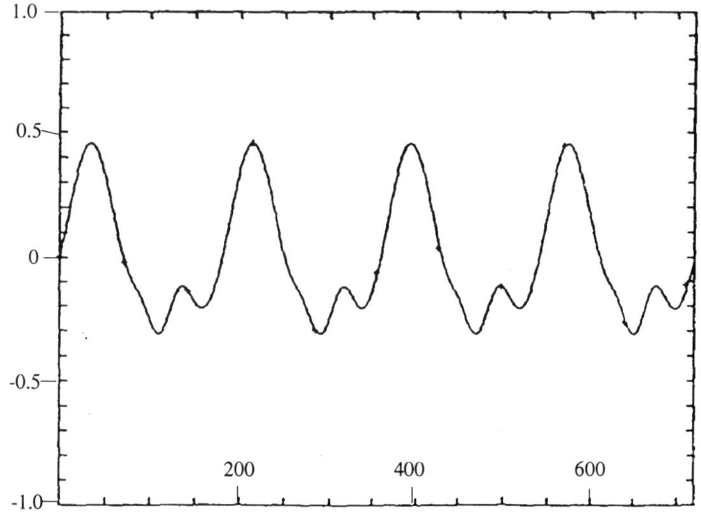

Fig 92: Vibration harmonics in terms of time response for 4-cylinder firing cycle, even orders only, at 3000 r/min: amplitude (y-axis) vs degrees of crank rotation (x-axis)

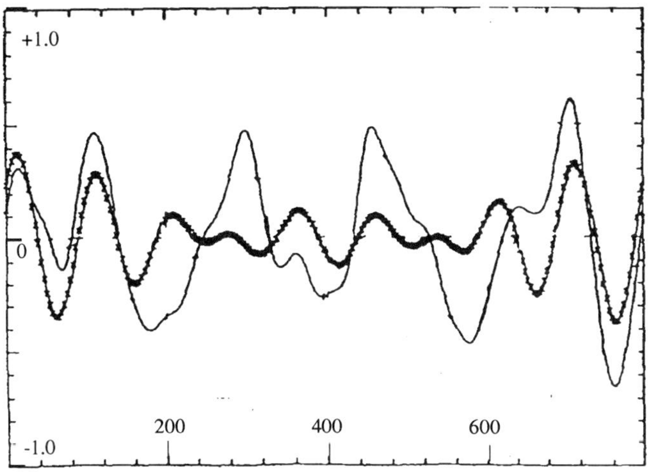

Fig 93: Characteristic `signature': for rumbly engine otherwise as in Fig 92, amplitude vs crank rotation

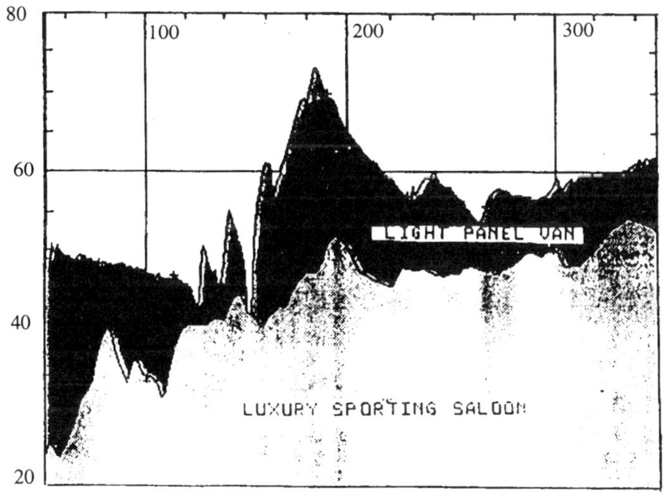

Fig 94: Transfer function: linear frequency in Hz on the x-axis

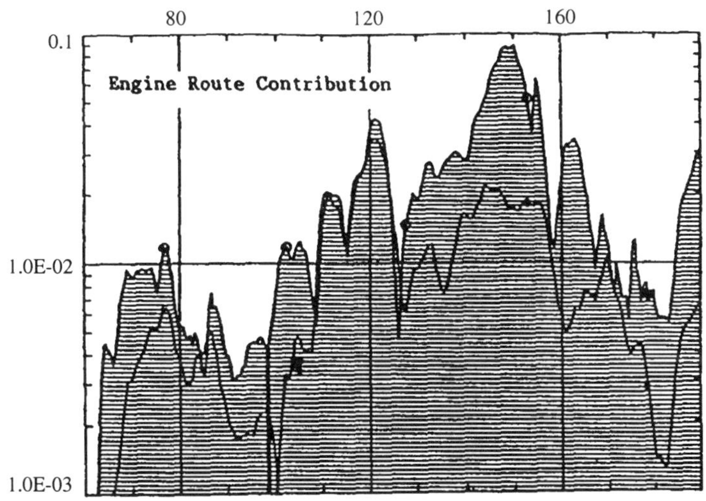

Fig 95: Engine contribution to boom: log PA (y-axis) vs linear frequency in Hz (x-axis)

Engine mounting

Designing effective engine mountings, which isolate the vehicle structure from out-of-balance forces and couples, requires a compromise between such factors and the isolation of high frequency noise. Both traditional techniques and those involving recently introduced hydraulically damped mounts are discussed.

Serious consideration of engine mounting problems probably started with the widespread adoption of 4-cylinder engines in vehicles using integral, chassisless construction in the early 1950s. Now, of course, nearly all vehicles are deemed to require flexibly mounted engines. Much of the art in mounting design is controlling the motion of the suspended engine/gearbox unit with respect to its principal axes.

Early research by Metalastik Ltd[8] found that the ideal suspension involved mountings arranged symmetrically about all axes, Fig 96, resulting in three each rotational and translational modes of vibration around and along the principal axes. More typically the suspension system is as in Fig 97 — with mountings only symmetrical with reference to the plane containing the longitudinal principal axis. While there are still six modes of vibration, resonant vibration will be about axes intercepting the plane of symmetry at A1, A2 and A3. Perpendicular to this plane vibration can also occur about B1B1, B2B2 and B3B3. To keep resonant speeds as low as possible, the objective is to use a suspension which decouples these vibrations as far as possible. Uncoupled frequencies f are calculated from :

$$f = \{\text{stiffness/mass}\}^{1/2}$$

with the assumption that the structure supporting the mounts is infinitely stiff and has infinite mass.

While normal six-cylinder engines have only harmonics of the crankshaft's torsional frequency, four cylinder 4-stroke engines, with cranks at 180 degrees, also produce secondary out-of-balance forces. The secondary force $F.a$ is seen in Fig 98 to act at right angles to the crankshaft centre-line, halfway between cylinders 2 and 3 — with a frequency of two vibrations per rev. Most troublesome is the large OY component which requires sufficient vertical deflection of the mountings to insulate it.

With the two firing impulses per rev of the four cylinder engine, major torque harmonics are second order, and the multiples fourth, sixth, eighth and so on. The major second order one has mounting component $T \cos\theta$ about OZ. Care is needed in choosing the natural frequency about the OZ axis in order to avoid resonance with the 1/2, 1 or 1 1/2 order at idling speed.

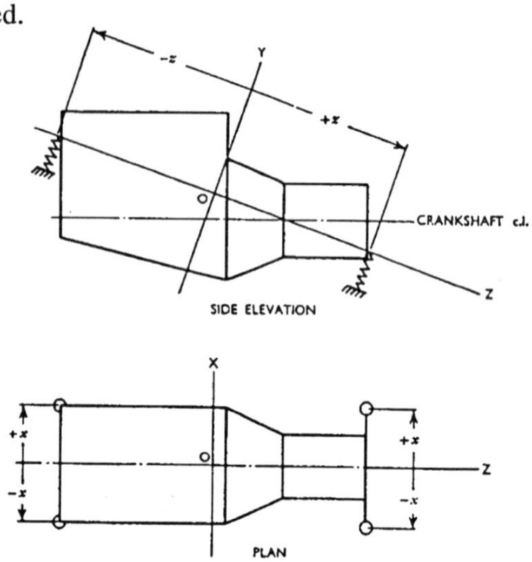

Fig 96: Symmetric mounting

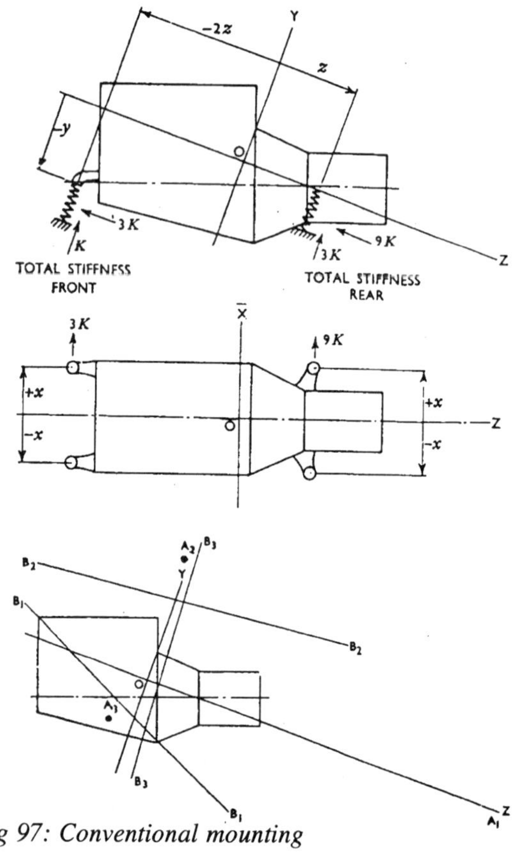

Fig 97: Conventional mounting

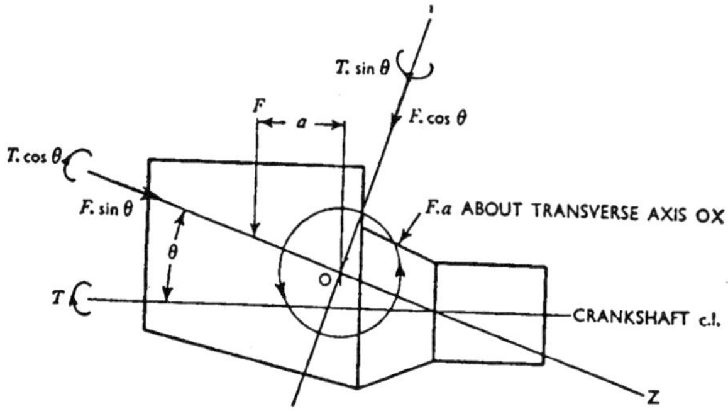

Fig 98: Balance forces on 4-cylinder engine

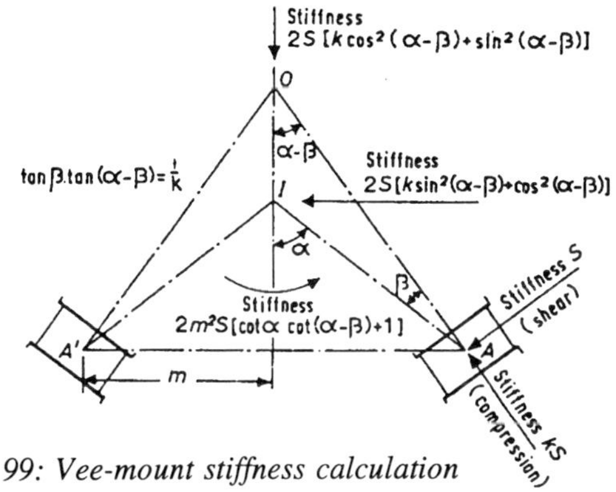

Fig 99: Vee-mount stiffness calculation

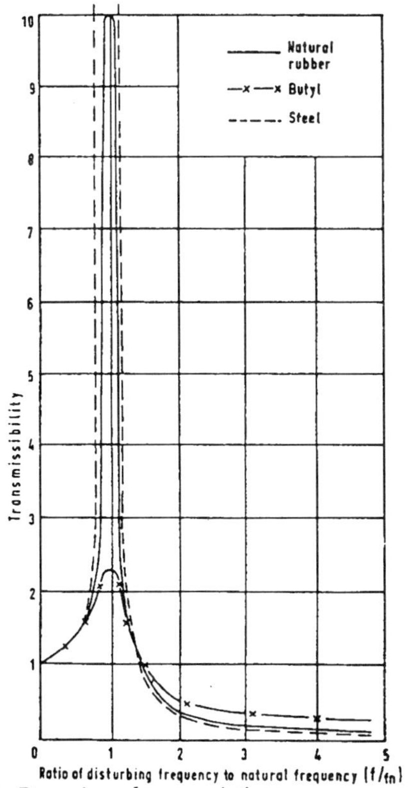

Fig 100: Damping characteristics

Certain configurations of engine such as three cylinders in line and V-4 or V-6 layouts can give problems if their makers have not compensated for relatively large out-of-balance primary or secondary forces/torques which will occur, by provision of such devices as contra-rotating Lanchester-type balancers. It must always be remembered that these forces are proportional to the square of speed, whereas gas torque harmonics stay practically constant with change in engine speed.

A centre of percussion position is sometimes used with the front mountings for reducing the tendency to pitch on rough terrain — and to relieve the rear mountings of shock loads. Another useful hint is to ensure at the design stage that the vertical spring frequency of the engine on its mountings is not too close to the wheel hop frequency of the front wheels on their suspension springs. Sometimes it is possible to tune the engine frequency to the vertical bending frequency of the vehicle structure so that the engine acts as a harmonic damper for body boom. A first approximation to the frequency of the vertical engine motions on the mountings can be obtained from:

$$f = (60/2\pi)(K.g/W)$$

Thus a static deflection of 0.3 in (7.6 mm) would give f = 350 c/min. Basic formulae for calculating stiffnesses of vee-type mountings are shown in Fig 99. Butyl rubbers are now favoured for such mountings, as they have higher internal damping than natural rubbers if appropriately compounded, Fig 100. It is important to observe the temperature effect on the damping characteristic in order to obtain equal performance regardless of weather conditions.

Increasingly, engine mountings are analysed as part of the the total vibrating system of the vehicle. This is because the secondary effects of such factors as beaming vibration of the chassis frame, covered in Chapter 4, are crucial to the obtaining of an accurate performance prediction. Such a study requires an idea of the magnitude of the engine forces involved. The so-called engine 'secondary' forces, for the vertically-mounted 4-cylinder in-line engine, have a vertical direction governed by the piston reciprocation axis. For an engine speed range of 750 to 6000 rpm, forcing frequency would be 25-200 Hz and the acceleration of the power-unit in the direction of the force is proportional, like the force, to the square of engine speed — giving a constant displacement amplitude for most of the frequency range.

Besides these out of balance inertia forces, firing forces are important, too, particularly in the case of diesel engines at idle. Since uneven firing causes forcing at half and first orders, in addition to second order, forcing frequencies for an idle speed of 75 rpm would be 6.25, 12.5 and 25 Hz. These very different input frequencies set the major challenge to the engine mount designer: this is in isolating unbalance forces between 25-200 Hz while giving the engine/gearbox unit a suspended natural frequency of some 6-12 Hz. Generally it is not, in fact, necessary to keep other than the vertical mode below 10-20 Hz. The vertical mode is held typically at 8-10 Hz, with highly damped mounts to control any engine shake excited by wheel-hop frequencies.

Shake involves amplitudes above 0.3 mm whereas higher frequency,30-200 Hz, noise transmission involves amplitudes less than 0.1 mm. The conflict between the high damping required for shake isolation and the low damping for dealing with noise has led to the development of hydraulic mounts. Fig 101 shows the Avon Hydramount alongside characteristics of conventionally compounded engine mounts at Fig 102. Both types typically use a conical rubber element but the Hydramount has a partition attached to its base, below which is a lower chamber bounded by an elastomeric bellows containing damping fluid. Orifice damping forces are generated as this fluid flows, to and fro, from upper to lower chamber. Much higher damping forces are generated at high amplitudes by means of an active control system whereby an inlet manifold vacuum is used to control the 'air-spring' element of the Hydramount. Fig 103 shows damping characteristics; *above:* the low-frequency characteristic shows that by switching off the vacuum, stiffness can be substantially reduced even at high amplitudes; top curves are non-active and bottom curves active, with full, dash and chain lines representing vacuum levels of 0, 0.5 and 0.3. *Below:* the high frequency characteristic demonstrates that very low stiffness levels can be achieved at particular frequencies; here full, chain, chain-dot and dashed lines represent standard-mount, 0.3, 0.1 and 0 bar vacuum respectively.

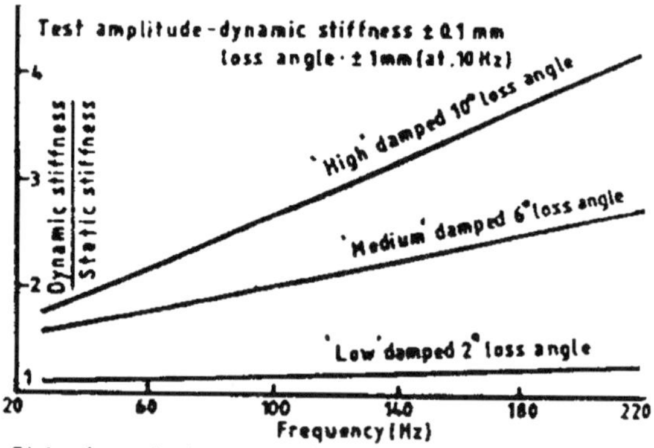

Right Avon hydramount: A top element; B air vent; C diaphragm; D bellow; E air space; F damping orifice; G snubber plates

Fig 102: Comparative performance

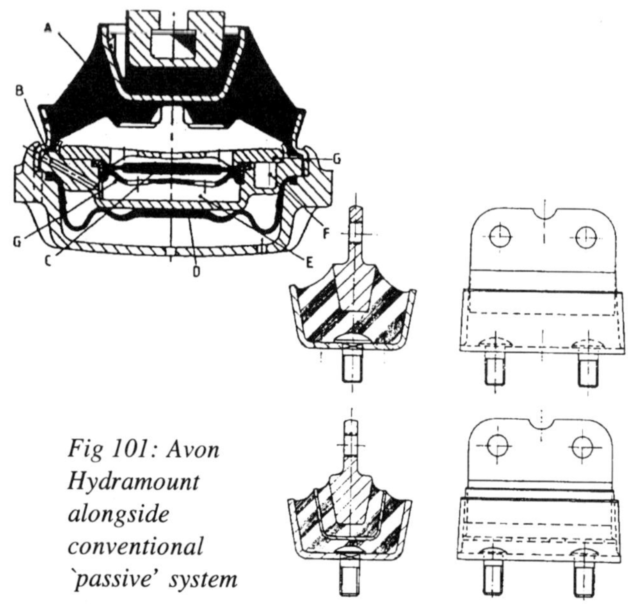

Fig 101: Avon Hydramount alongside conventional 'passive' system

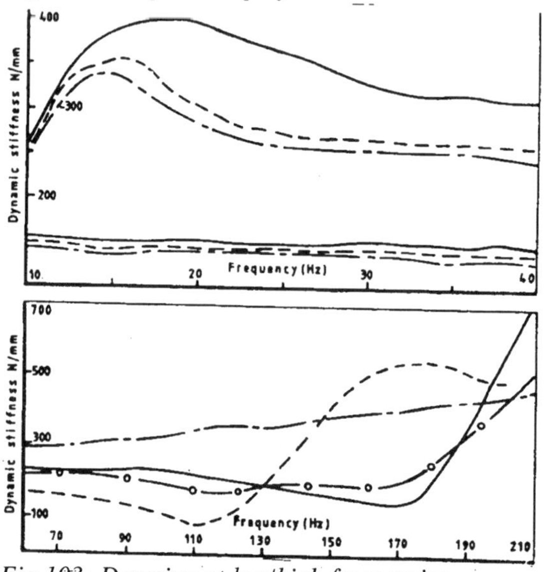

Fig 103: Damping at low/high frequencies

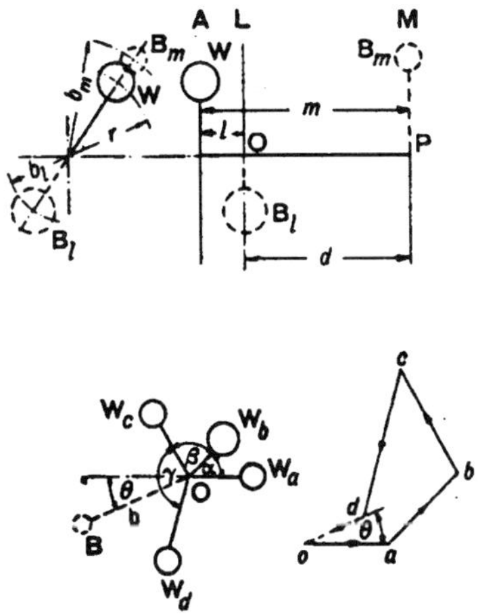

Fig 104: Masses in separate planes

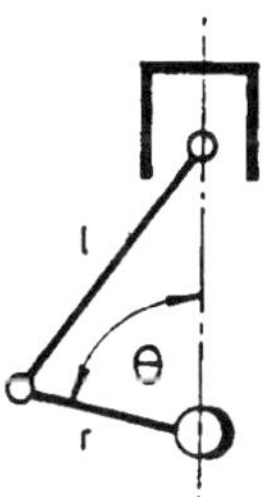

Fig 105:
Reciprocating masses

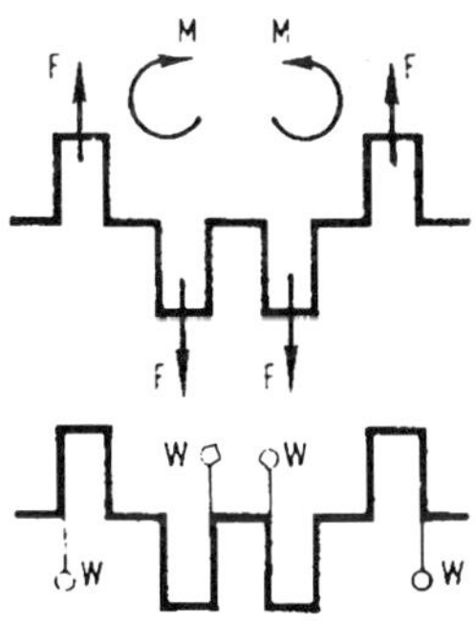

Fig 106: Symmetrical crankshaft

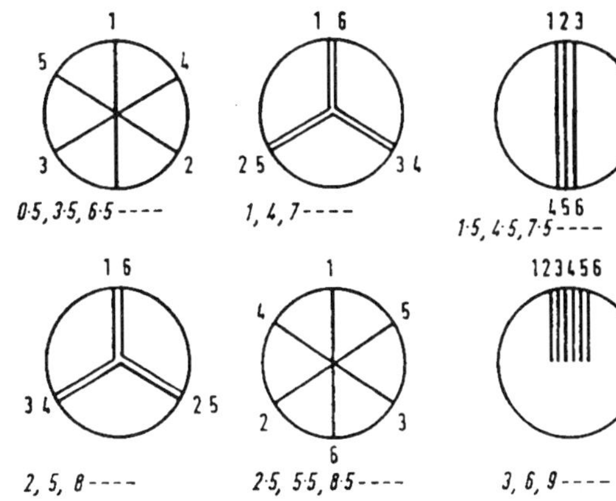

Fig 107: Forward and reverse cranks

Engine balancing

A more fundamental understanding of engine balance is required for achieving the best possible behaviour of the engine prior to designing the mounting system. This will be necessary where a high level of vehicle refinement is required with respect to NVH (Chap. 4) or where vibration problems would prevent the very highest specific power being obtained.

In order to balance a rotating mass which is offset from the axis of rotation, the normal course would be to set up an equivalent mass in the same plane of rotation but in a radially opposite position. The balance weight can be smaller if its radius to the axis is increased proportionally; in order to produce equivalent centripetal force, it is the product of mass and radius which matters. Because generally it is difficult to mount the balancing mass in the same rotational plane as the out-of-balance mass, special measures have to be taken to balance the secondary effects.

With the balancing mass in a separate plane, the lines of action of the two forces are different and cause a couple, tending to rock the shaft in its bearings. Two separate balancing masses must therefore be used in a way that the three masses produce no resultant dynamic force or couple. By taking moments of the force system shown in Fig 104 for the generalised case, values of the balancing masses can be calculated.

When several masses are eccentrically mounted on a shaft, a polygon of forces can be drawn to determine the resultant out-of-balance force, assuming the masses are in the same plane of rotation. Where the masses are at different intervals along the shaft, the moment method can again be used to find, for each of the two planes in which the balancing masses are to be positioned, the element of mass necessary to balance each of the disturbing masses. The data is conveniently set out as a table with one row for each disturbing mass.

For reciprocating masses such as pistons and conn-rod the inertia force exerted by these components can be calculated from the geometry of the system, Fig 105, as

$$(W/g)w^2r[\cos\theta+(\cos 2\theta/n)] = k_1\cos 4\theta+k_2\cos 6\theta+etc$$

where W is weight of reciprocating parts (piston + conn-rod/2), $r.a$ is crank radius and q the crank angle;

n is conn-rod length/*r*.

Third and subsequent terms of the equation are small enough to be ignored while the first and second define the 'primary' and 'secondary' balance forces. The first term is the force produced by a reciprocating mass in pure simple harmonic motion (as for example in a scotch yoke engine). The second term is the approximate correction for the obliquity of the conn-rod and is $1/n$ that of the primary but has twice the frequency. A reciprocating mass cannot be balanced by a rotating mass; if a weight is added to the crank so as to revolve in phase with the reciprocating mass it will cancel the component in the line of the stroke but produce an equal component at 90 degrees to that line. This is customarily put to advantage in 90 degree vee engines, with one bank of pistons providing secondary balance for the other.

As well as inertia forces, there are couples created by the reciprocating masses. Those produced by rotating masses are the primary ones and those by the reciprocating masses, the secondary ones. If one half of the crankshaft is the mirror image of the other the couples are in opposite sense and cancel one another out, Fig 106. As shown, external bob-weights can be added to offset these couples and so reduce bearing loads if the crankshaft is not stiff enough to react the resulting torque imposed. The method of 'forward and reverse cranks', based on rotating vectors is used to predict the effects of multiple orders, Fig 107. The order is the number of vibrations per revolution of the crank and in a 4-stroke engine there are half- as well as whole- orders. Different crank arrangements excite different orders and the analysis of crankshaft vibration requires determination of the receptive orders. In the example of Fig 108, for a 6-cylinder engine crankshaft having natural frequency of 1800 cycles/min, the 2000 and 3000 rpm bands show excitation of the ninth and sixth order harmonics respectively. Main sources of engine unbalance are build-up of assembly tolerance, variation of piston and conn-rod weights. Other factors include trapped oil in the galleries of these parts and piston-pin offset. Build-up of deposits on the reciprocating parts can also affect the equation as can any offset of the conn-rod centre-of-gravity. Pin-offset effect can be calculated using the representation of Fig 109 as $-(A/L)\sin\omega t$ representing a third term for the equation given earlier to replace the small-order terms after the terms for the primary and secondary inertia forces.

In the case of a 90 degree vee engine, the primary reciprocating unbalance force and the 'pin-offset' force rotate with the throw and remain at a constant magnitude and so can be compensated by counterweights. The resultant of these forces leads the crank-throw direction by 90 degrees as a rotating vector and has magnitude $(W/2g)R\omega^2(A/L)$. Resultant rotating couples are also produced by the primary unbalance and pin-offset forces; these are also phased at 90 degrees and can be compensated by balance weights.

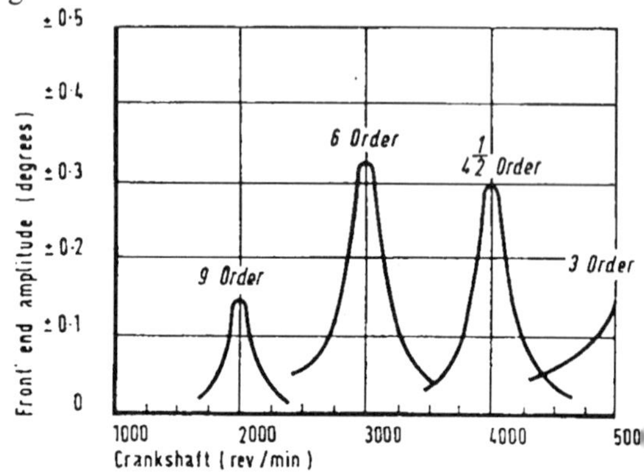

Fig 108: Six-cylinder engine example

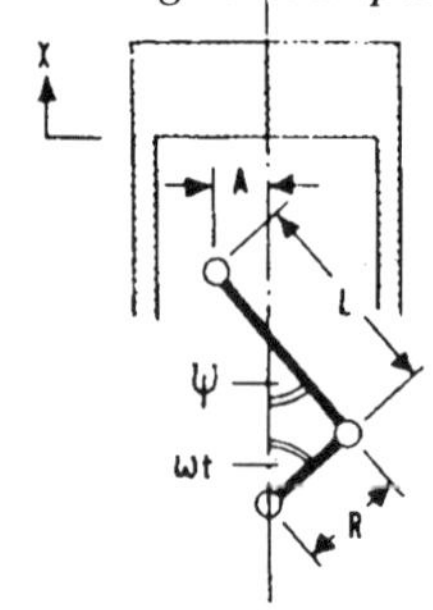

A = positive gudgeon pin offset; L = connecting rod length; R = crank radius; X = piston displacement; W = weight of reciprocating components

Fig 109: Allowing for pin-offset

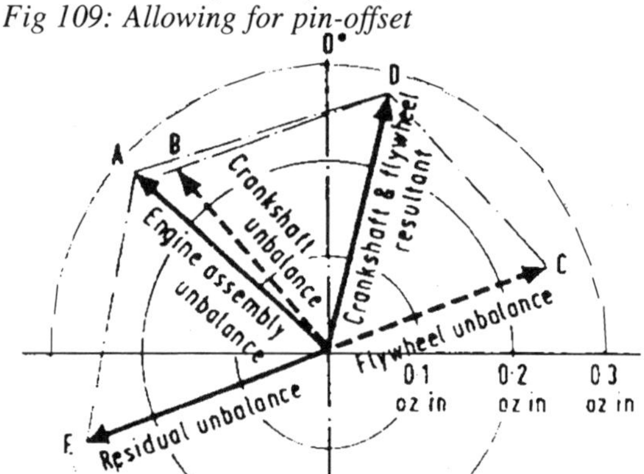

Fig 110: Determining residual unbalance

In a 90 deg-vee with crankshaft having equidistant crank-pins complete balance is achieved by appropriate counterweights. Fig 110 shows one manufacturer's vectorial representation of the procedure necessary to determine residual unbalance compensation for the flywheel end of the engine. Total engine unbalance measured on test-rigs is plotted as vector A. Separate balancing machine tests for crankshaft and flywheel imbalance are used to obtain vectors B and C — which are summed at D. Vectorial difference between A and D gives the required compensation E.

A general approach to vertical in-line engines can be made using the first equation simplified and converted for imperial units as:

Vertical forces:

Primary, 1 cycle/rev:

$F_v = 28.4(W_r + W_i)RK(N/1000)^2$

Secondary, 2 cycles/rev:

$F_v = 28.4\ W_iRK(N/1000)^2$

Transverse forces:

$F_h = 28.4\ W_rK(N/1000)^2$

— no secondary transverse force

Vertical couple:

Primary

$C_v = 28.4(W_r + W_i)RLK(N/1000)^2$

Secondary

$C_v = 28.4\ W_iRLK(N/1000)^2$

Transverse couple:

$C_h = 28.4\ W_rRLK(N/1000)^2$

— no secondary transverse couple

where W_r,W_i are weights of rotating, reciprocating parts referred to crank-radius (lb); R is crank radius, L is cylinder pitch (inches); N is engine rpm and K takes values from the table in Fig 111.

In the case of vee-engines, overall balance can be determined by combining the imbalance of each row with that of each bank. The factors given in that part of the table are qualified by bracketed terms which apply when the balance weights are increased to include the reciprocating parts for one cylinder which is added to the rotating parts *per crank*.

Crank diagram	Firing order	Firing interval Crankshaft degrees	Mass balancing characteristics: External forces		External couples		Internal couples	
			Primary	Secondary	Primary	Secondary	Primary	Secondary
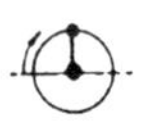	—	720°	1·00	$\frac{1\cdot00}{n}$	Nil	Nil Referred to engine centre line	Nil	Nil
	1–2	Uneven 180° and 540°	Nil	$\frac{2\cdot00}{n}$	1·00	Nil* referred to engine centre line	Nil	Nil
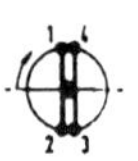	1–3–4–2	180°	Nil	$\frac{4\cdot00^*}{n}$	Nil	Nil referred to engine centre line	1·00	$\frac{2\cdot00}{n}$
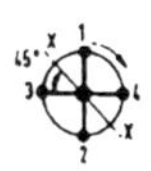	1–3–4–2	Uneven 90°–180° and 270°	Nil	Nil	1·41	$\frac{4\cdot00^*}{n}$	0·71	Nil
	1–5–3–6–2–4	120°	Nil	Nil	Nil	Nil	1·73	$\frac{1\cdot73}{n}$
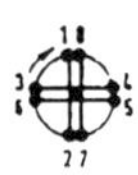	1–6–2–5–8–3–7–4	90°	Nil	Nil	Nil	Nil	1·41	$\frac{4\cdot00}{n}$
	1–5–2–6–4–8–3–7	90°	Nil	Nil	Nil	$\frac{16\cdot00}{n}$	Nil	Nil

Fig 111: K-factors for different crankshafts

Heat exchanger calculation

The fundamentals of heat exchanger calculation are well set out by Piskola[9] but the refinements necessary in examining forced convection effects require the application of dimensional number analysis suggested here.

The most common type of heat exchanger is the shell-and-tube; if one of the fluids involved condenses or evaporates then the system is known as a condenser or evaporator, familiar to air-conditioning systems and so on. The tube bundle of a typical unit is circular in cross-section and involves many closely-spaced tubes; the hollow shell encloses the tube bundle and baffles may be used to cause flow back and forth along the tube bundle to exploit the higher heat transfer effects of turbulent flow.

Heat transfer rate is defined by the expression:

$$Q = UA\Delta t$$

where U is overall heat transfer coefficient; A is transfer area and Dt the temperature difference between the respective fluids flowing in the tubes and shell. This is applied to the engine coolant "radiator" in the earlier partwork section on engine-cooling/vehicle heating.

In more generalised heat-exchangers for recovering engine exhaust heat, for example, a closer study of the fluid flow patterns may be necessary. Where the Reynold's number (Tube diameter D [ft] x mass velocity$G$ [lb/hour ft^2] / viscosityμ [centipoises x 2.42, lb/ft hour]) is less than 2100 flow is streamline and above this value turbulent flow exists. Film heat transfer coefficients at the boundary between fluid and adjacent wall are found from separate expressions for the two tube-side flow regimes

$$h_i D/k = 1.86[(DG_i/\mu)(c\mu/k)(D/L)]^{1/3}(\mu/\mu_w)^{0.14}$$

$$h_i D/k = 0.027(DG_i/\mu)^{0.8}(c\mu/k)^{1/3}(\mu/\mu_w)^{0.14}$$

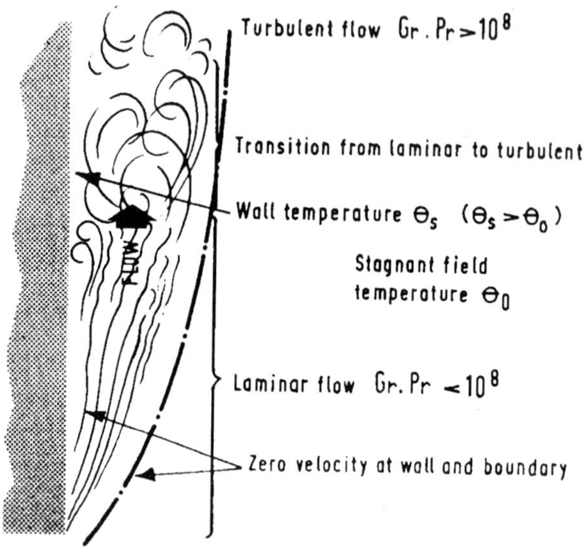

Fig 114: Boundary layer growth in natural convection

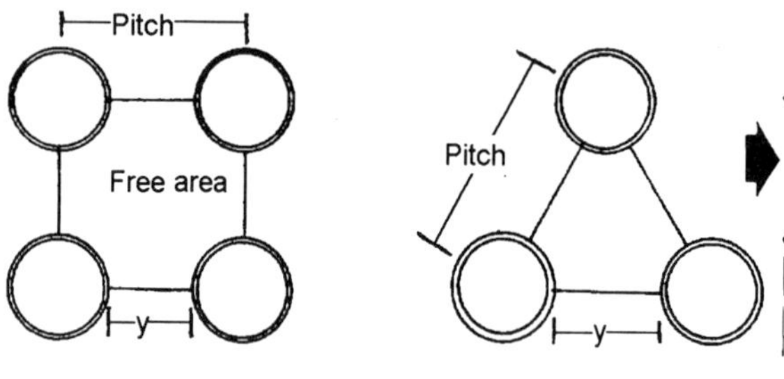

Fig 112: Tube configurations

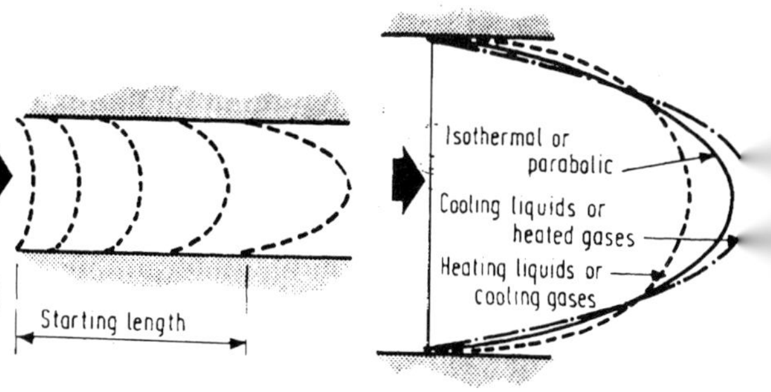

Fig 115: Velocity and temperature profiles

Grashof No. $Gr = \frac{L^3\beta g\Delta\theta}{\nu^2} = \frac{\text{buoyancy force}}{\text{viscous force}}$

Nusselt No. $Nu = \frac{hD}{k} = \frac{\text{heat transfer}}{\text{heat conducted}}$

Prandtl No. $P_2 = \frac{C_p\mu}{k} = \frac{\text{kinematic viscosity}}{\text{thermal diffusivity}}$

Reynolds No. $Re = \frac{\nu L}{\nu}$ or $\frac{GD}{\mu} = \frac{\text{inertia force}}{\text{viscous force}}$

Stanton No. $St = \frac{h}{GC_p} = \frac{\text{heat transfer}}{\text{heat transport}}$

Fig 113: Dimensionless constants

for streamline and turbulent flow respectively and for the shell side

$$h_0 D_e/k = 0.36(D_e G_s/\mu)^{0.55}(c\mu/k)^{1/3}(\mu/\mu_w)^{0.14}$$

Equivalent diameter D_e = 4 x free area / wetted perimeter, according to tube layout, Fig 112. The shell-side mass velocity G_s is the weight of fluid flowing in lb/hour divided by cross flow area *a* in ft² where

$$a = D_s y/(\text{tube pitch} \times 144)$$

Because the temperature difference between tube and shell side fluids does not remain constant, a mean temperature difference is calculated from

$$(\Delta t_2 - \Delta t_1)/2.3 \log_{10}(\Delta t_2/\Delta t_1)$$

where Dt_1 is the outlet temperature of fluid one minus the inlet temperature of fluid 2 and Dt_2 is the inlet temperature of fluid 1 minus the outlet temperature of fluid 2.

In studying surface effects in more detail,

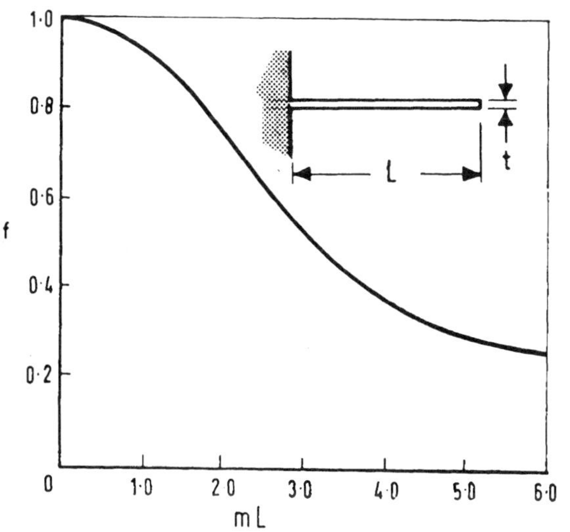

Fig 116: Extended surface effect

dimensional analysis techniques are often used based on the constants defined by Reynolds and other classic researchers in fluid dynamics shown in Fig 113. In natural convection flow is due to buoyancy forces as seen in Fig 114. In streamline forced convection the build-up of flow is shown in Fig 115 over a so-called starting length before full streamline flow applies. The right hand view shows temperature profiles in a tube which vary according to heating or cooling.

Turbulent forced convection is the likely condition in practical designs and would exist for Reynolds numbers above 5000-10 000. An expression for flow in a circular tube is

$$Nu = 0.023Re^{0.8}Pr^n$$

where n is 0.3 for cooling and 0.4 for heating of the fluid. In the case of finned heat exchangers, the extended surface effect can be studied by considering the expression for flow from an isothermal flat plate in turbulent forced convection as

$$= 0.037Re^{-0.2}$$

for Reynolds numbers above 500,000. Fins are a special case for which both interrupted flow and non-isothermal surface conditions apply. Fin effectiveness depends on the end conditions existing and is given by

$$\tanh mL/mL$$

for long fins of large surface to end area ratio, Fig 116, where $m = (2h/kt)^{1/2}$, *h* being the surface convective heat transfer coefficient and *k* the thermal conductivity of the fin material.

Nomographs are given in Figs 117,118 and 119 for mass flow in a tube, Reynolds number for circular tubes and flow from an extended surface.

Recent work by Magneti Marelli has shown the advantages of using internally spiral-finned tubes and test results have been reported on tube configurations shown in Fig 120. Spiral fins have a substantial effect at low Reynolds numbers where the combination of increasing the tube's heat transfer surface and the turbulence effect due to the spiral fin. Fig 121 records comparative performance with smooth and straight-finned tubes.

To use the chart, mark values of A and V in the left-hand columns, join up and extend to intersect the reference line. Then, for any value of ρ, extend line joining the reference point and this value to intersect the mass flow column.

A = cross section area of conduit, ft²
V = flow velocity, ft/s
ρ = fluid density, lb/ft³
$w = \rho . A . V$ = mass flow, lb/h

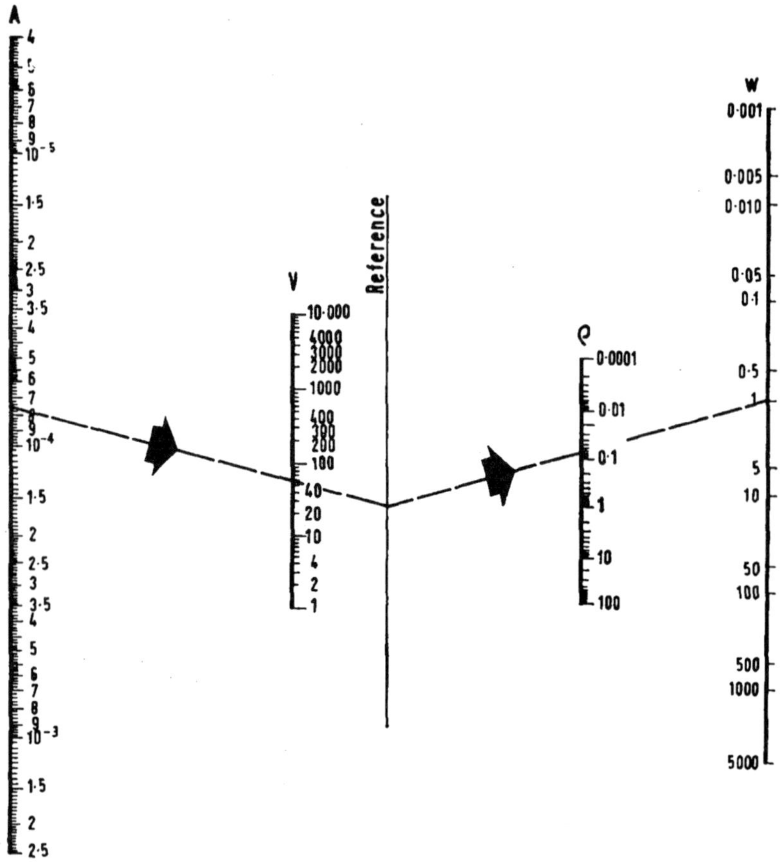

Fig 117: Mass flow in a tube

w = mass flow, lb/h
D = diameter, ft
μ = coefficient of dynamic (or absolute) viscosity, lb (mass)/h-ft
These units of viscosity are equivalent to $\frac{\text{pdl.h}}{\text{ft}^2}$

One centipoise ≡ 2·419 lb (mass)/h-ft
(Poise = grammes/sec.cm = dyn.s/cm²)
Viscosity of water, for example, as a saturated liquid at 70°F
= 2·372 lb (mass)/h-ft
= 0·98 centipoise

$$\text{Re} = \frac{4}{\lambda}\left(\frac{w}{D\mu}\right) = \text{Reynolds Number (dimensionless)}$$

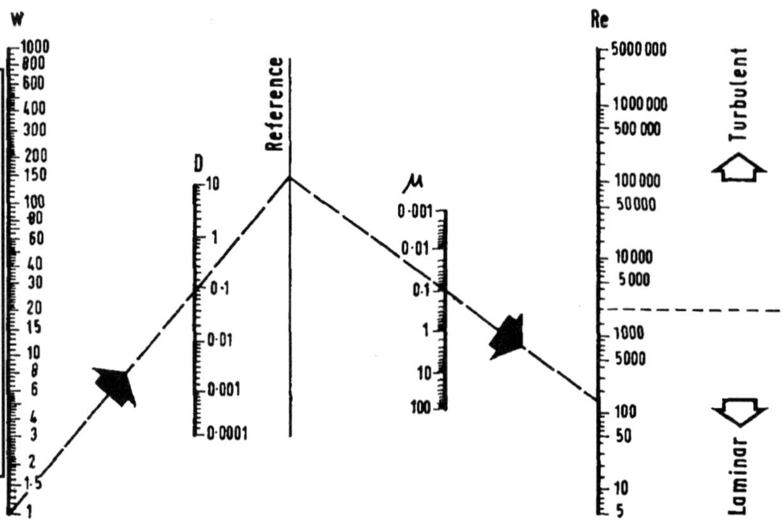

Fig 118: Reynolds number for circular tube

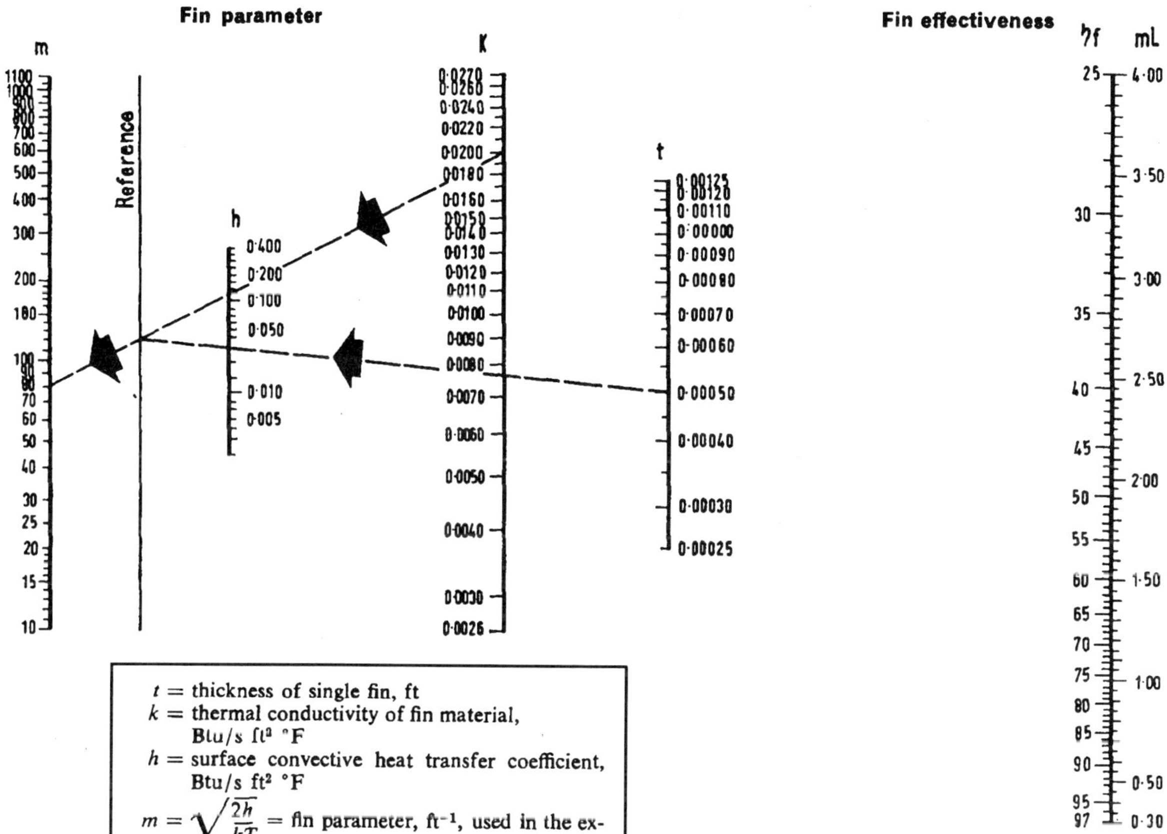

t = thickness of single fin, ft
k = thermal conductivity of fin material, Btu/s ft^2 °F
h = surface convective heat transfer coefficient, Btu/s ft^2 °F
$m = \sqrt{\dfrac{2h}{kT}}$ = fin parameter, ft^{-1}, used in the expression: fin effectiveness $\eta_f = \dfrac{\tanh mL}{mL}$
where L is the fin length.

'ig 119: Flow from extended surface

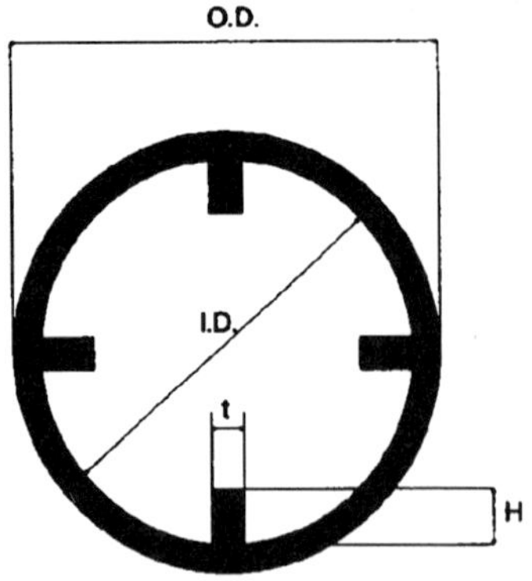

Sample	O.D. [mm]	I.D. [mm]	t [mm]	H [mm]	fin number	spiral pitch [mm]
FINNED A	8.0	7.3	0.5	1.0	4	180 220 280
FINNED B	8.0	7.0	0.75	1.5	4	130 180 280

Fig 120: Spiral fin configuration

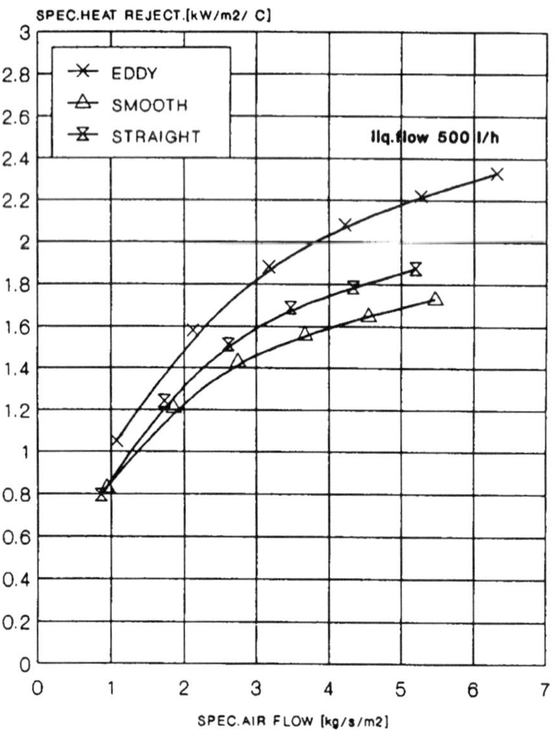

Fig 121: Comparative performance

Chapter 17: Ride and handling

Vehicle ride
Steering geometry
Steering/suspension linkages
Directional control
Laminated suspension springs
Suspension coil-spring design
Auxiliary suspension systems
Active suspension
Full active suspension
Air suspension application
Handling analysis in racing cars

Ride parameters and comfort criteria

Vehicle suspension design must start with a consideration of what ride comfort levels are desirable for passengers or what vibration levels are tolerable for the range of cargoes to be carried. The predicted ride motions of the vehicle can then be matched to these, at the design stage, if the fundamental rules of mechanics are followed.

Ride comfort may be affected by ground irregularities, aerodynamic forces and vibrations of the power train. Technique in design is to control the resulting vehicle vibration so that the passenger's sensation of discomfort does not exceed a certain level. Fig 1 shows a set of comfort criteria suggested by Janeway[1]. This defines the acceptable amplitude (total distance moved during one oscillation) of vibration in terms of its frequency (number of oscillations per second or Hz) and, obviously, as the frequency increases, the acceptable amplitude decreases. The criteria change also for low, medium and high frequency bands.

In the 1 to 6 Hz band, the criterion is a maximum constant value of 'jerk' (rate of change of acceleration given by the product of amplitude and the cube of frequency). The limiting value is 12.6

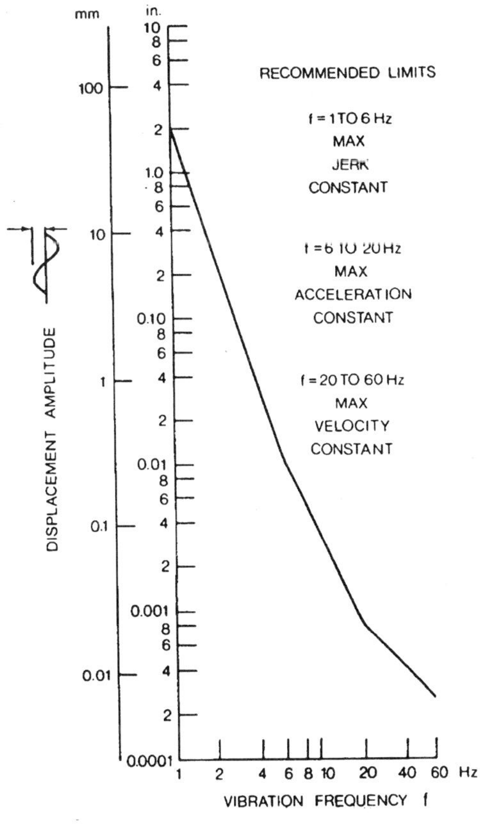

Fig 1: Comfort criteria

metres/second[3]. So at 1Hz, which is close to the normal ride frequency of softly sprung cars (60 cycles per minute), the recommended amplitude limit is 50mm. In the 6 to 20 Hz band a peak acceleration value of 0.33 m/s^2 is the criterion while in the 20 to 60 Hz range, a peak velocity of 2.7 mm/s is the criterion.

For commercial and industrial vehicles where available spring travel, in relation to sprung weight, may be limited, standards governing the 'tolerance' limits for human body vibration are often used, ISO 2631 (1974). Fig 2 shows the decreased efficiency boundaries for vertical and transverse vibration defined in terms of acceleration (as a dynamic average or root mean square RMS value). The boundary lowers as the daily exposure time increases.

Establishing suspension parameters

Much of the modern approach to vehicle ride and comfort analysis goes back to the pioneer work of Lanchester. The comfort of passengers, and safe carriage of cargo, he observed, were affected both by vertical motions, reflecting the negotiation of road obstacles, and by secondary lateral motions related to vehicle centre of gravity (CG) height, as the vehicle rocked under the effect of individual wheel obstacles, as in Fig 3. He also observed the relationship of these motions with the height above the ground of the lateral location of the springs, later to be defined as roll centre height. He preferred this to be as near as possible to ground level in order to minimise transverse oscillation effects but pointed out the conflict of increasing roll-centre to CG distance causing decreased roll stability.

The second compromise faced by vehicle engineers, he considered, was that relating the fore-aft rocking motions with direct vertical motions. Either central mass concentration, as far as possible, for the sprung mass of the vehicle having high pitching frequency or the opposite situation in which mass was concentrated at the extremities (with high freqency and low inertia) was preferred by Lanchester to the intermediate situation. He demonstrated the importance of pitching frequency to wheelbase relationship by considering the pendulum analogy in Fig 4. Negotiating a sudden change in road level is like a horizontal displacement of the pendulum fulcrum. In the configuration shown, sudden displacement of the front wheels and sudden return gives insufficient time for the suspended body, in 'midswing', to recover. If the rear wheel disturbance is a complete period later, the front wheel disturbance is nullified; for a half-period interval between disturbances, then they are amplified.

The three predominant ride motions experienced by the vehicle passenger — bounce, pitch and roll — are usually interrelated by the coupling effect of the wheel suspensions. Lanchester's simplified ride model of Fig 5 can be used to show it is possible

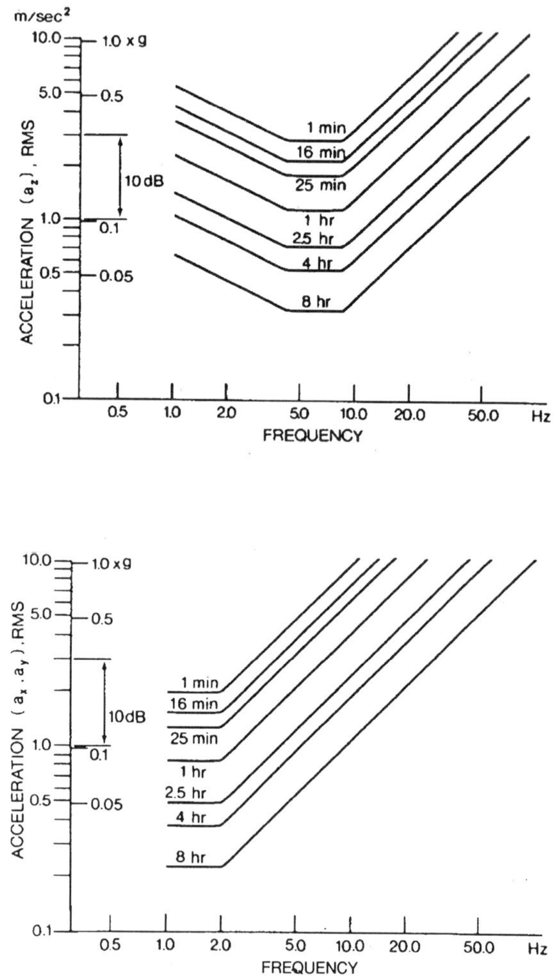

Fig 2: Decreased efficiency boundaries

to decouple the pitch and bounce modes (Fig 6) by making the length of a suspended symmetrical mass equal twice the radius of gyration R of the mass M in the pitch-oscillation mode. The product MR^2 is of course the moment of inertia of the body (sprung mass) in the pitch mode. In this situation the front and rear of the vehicle would behave as separate vibrating systems. In practice, the wheelbase would be longer than the idealised springbase required to achieve this effect so on most vehicles the pitch frequency is too high on all but the most soft suspensions. In roll, the actual springbase is usually, by contrast, too narrow in respect to the transverse moment of inertia and an anti roll bar is needed to provide a righting moment in proportion to the square of the springbase.

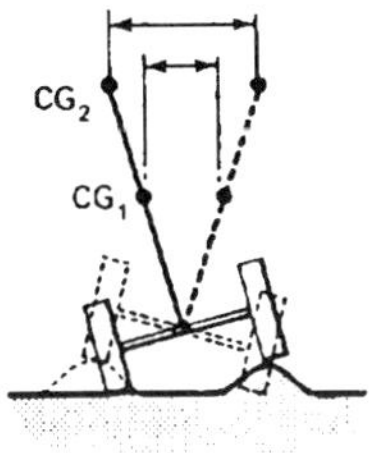

Fig 3: Secondary lateral motion

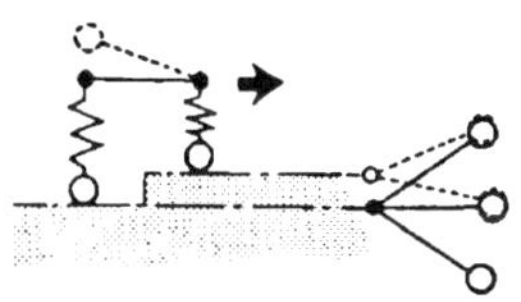

Fig 4: Pendulum analogy for pitch

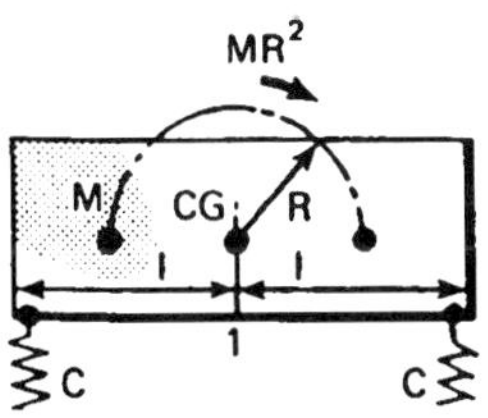

Fig 5: Simplified ride model

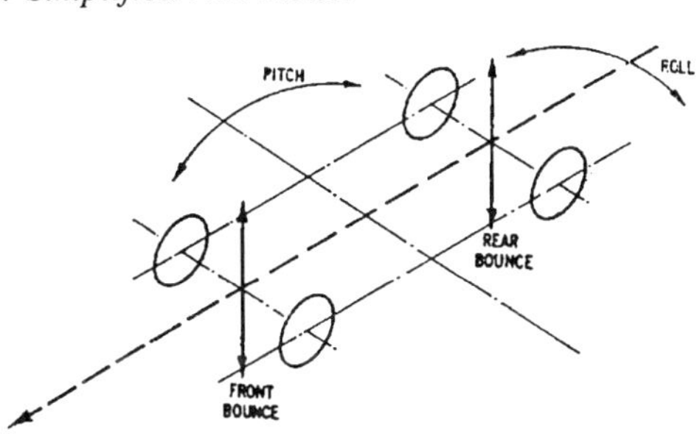

Fig 6: Ride oscillation modes

Idealisation of the ground surface as a regular sine wave (Fig 7) can be used to obtain approximate values for the suspension parameters. For wave length S and depth x, if the vehicle moves at speed V, the time of a half oscillation induced by the wave profile is S/V and tyre/ground contact is maintained when

$$F(\{w_f/W_f\}/2)^{1/2} \text{ and } x = d(1 + w_f/W_f)$$

governs the maximum depth of the wave at which wheel ground contact is maintained for static spring deflection d. While spring rate should be as low as possible for comfortable ride, to maintain ground contact, it is also necessary to reduce unsprung mass as much as possible. The force on the sprung mass is of course the product of that mass with its acceleration A_f which must equate with the equivalent force and acceleration a_f on the unsprung mass to give $A_f = (w_f/W_f)a_f$ showing the force reduction, and hence displacement reduction effect.

Considering again the pitch motion model of Fig 8 but substituting l_f and l_r to define the centre of gravity in respect to the wheels (instead of just l), allows an expression for pitching frequency to be obtained similar to that for bounce from the basic relationship of simple harmonic motion $F = [d/A]^{1/2}$ as $[(d/A)K^2/l_f l_r]^{1/2}$

Controlling ride motions

Other requirements for suspensions include the need to keep suspension rebound movement as low as possible in order to avoid the additive effect of successive waves and to avoid sharp changes in body motion at full bump. While front and rear suspension frequencies should be nearly equal, to minimise free pitching, front should be slightly less than rear to compensate for phase differences in road waves.

Damping force in the suspension system is usually arranged to be proportional to the linear vertical velocity of the wheels relative to the body. However, to meet the compromise between controlling body resonance (free vibration on the road springs) on the one hand, and avoiding excitation of the axle on the other, velocity rate is reduced at the ends of the strokes and the rebound rate is increased over the bump rate. A second resonance, wheel hop, has to be avoided: that of the axle freely vibrating on the tyre spring (usually some ten times stiffer than the suspension spring); the body in this condition is 'seismic' (does not respond to wheel motions) but the suspension spring contributes an extra 10% to the rate of the system. The vibration is sometimes excited at motorway speeds when the wheel out-of-balance forces resonate with the wheel-hop frequency. If there is static friction in the suspension/damping system, then 'lock-up' can occur at small wheel displacements giving rise to the phenomenon of 'boulevard-jerk', a name derived from the steps between successive cast slabs of concrete roadway causing the body to vibrate on the tyresprings only at about 4 Hz.

Spring rate of the suspension must also cater for the variation in laden to unladen running of the vehicle. With linear rate springs, sprung mass frequency can be obtained from $F = 187.8/d^{1/2}$ cycles/minute where static deflection d is in inches. Static deflection must of course be large for low frequency suspensions but as the static load varies for different payload conditions, mean spring deflection will vary. A wheel supporting sprung mass weighing 500 lb (227 kg) with a spring rate of 100 lb/in (179 kg/cm) has a vertical freedom of +/- 2 in (50 mm) from its mean position; increasing the sprung mass by 100 lb (45.3 kg) will compress the spring by a further 1 inch so the available wheel travel then becomes 1 inch towards bump and 3 inches towards rebound. With loading in excess of 200 lb, no bump movement is available at all. For change in static wheel deflection y, linear spring rate is P/y and natural frequency is thus proportional to $(P/W_f)^{1/2}$ for payload P. This leads to the requirement for progressive, non-linear, springs on heavy vehicles, having high laden/unladen weight ratios, to maintain the sprung mass frequency constant as the load changes. Alternatively a levelling system can be used whereby a servomechanism acts on the body to maintain a constant ride height independent of payload. A lower limit on spring rate must also be considered — in relation to spring travel available to negotiate significant ground obstacles without vertical body motion. In the example spring rate quoted above, maximum spring load to provide vertical acceleration is 2 in (5 cm) x 100 lb/in (180) kg/cm) or 200 lb (90 kg) so maximum acceleration

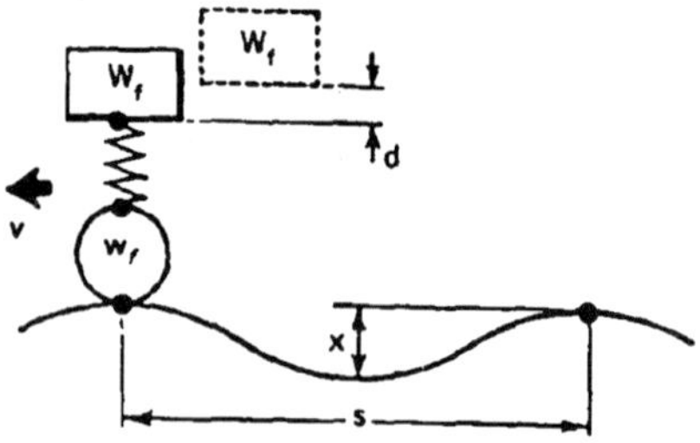

Fig 7: Sinusoidal road inputs

Main suspension		0.5-2.5
	bounce	lower range
	pitch	upper range
	roll	1.5-2.0
Boulevard jerk		3.0-5.0
Wheel hop		7.0-16.0
Engine on mounts		7-18
Body structural		13-40
Engine internals		10-200
Drive line beaming		8-300
Road noise		40-400

Fig 8: Main ride frequency ranges, Hz

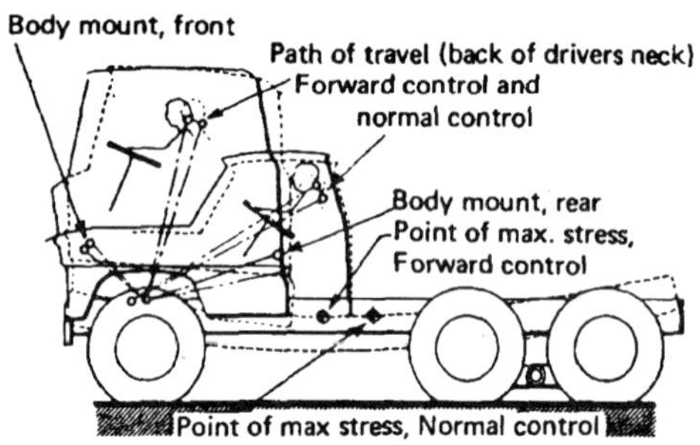

Fig 9: Beaming vibration of truck chassis

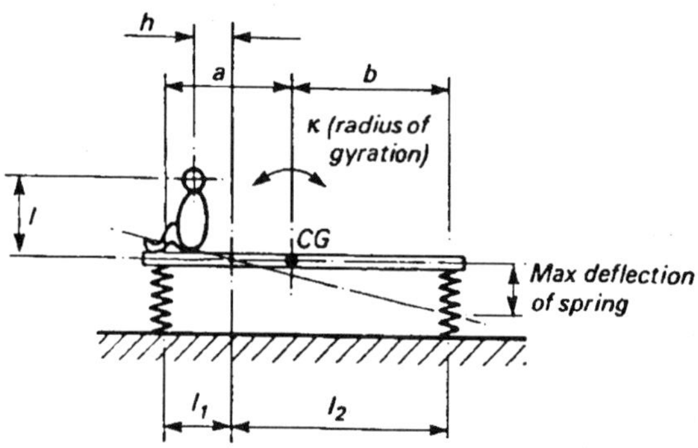

Fig 10: Ride and driver model

due to the spring will be 200/500 or 0.4 *g*. Damper forces are only available in the transient motions so at the ends of the suspension stroke, bump and rebound stops must be provided. The spring rate at the wheel, considered in these calculations, will of course be different from the rate at the spring by the installation ratio of the suspension linkage.

The table of Fig 8 shows predominant ride frequencies for a wide range of special vehicles from light cars to heavy off-road vehicles. Component vibrations on the power train of the vehicle can cause discomfort if improperly controlled. In saloon cars, some engines transmit troublesome amplitudes (around +/0.002 in) to the body structure — involving a 'booming' above 100 Hz. Other parts of the driveline can impart 'road-roar' above this frequency and 'rumble','thump' and 'harshness' below it. Body 'shake' at 525 Hz is sometimes differentiated from harshness, at 25-200 Hz. Vehicle parameters affecting the structure's response include: input impedence of suspension links, anchorages and sub-frames; body input impedence governed by local reinforcements at the anchorages and transmissibility of the whole body — also vibration modes of the body cavity and sound pressure patterns due to standing waves. In trucks and similar chassised vehicles, beaming modes of the chassis are prevalent, Fig 9. As well as stiffening the frame longitudinals, alteration of the system moment of inertia by judicious positioning of body mounts can sometimes result in cures.

Designing to meet ride criteria

A knowledge of ground surface necessary in design generally comes from past survey work. In the case of road vehicles, past studies have shown good roads to involve amplitudes below 0.005 metres, medium quality roads less than 0.013 m and others, above that value. Surface undulations with amplitudes from 0.019 to 0.025 are known historically to cause discomfort at normal speeds on poorly suspended vehicles. Irregularities have been found to occur at a mean pitch of around 4 m while the average distance travelled per wheel revolution is quoted as 2 m. By contrast, test 'washboard' surfaces have ripple height of 0.025 m at 0.75 mm pitch while longwave pitching tracks — each said to simulate worst cases — have sinusoidal waves 12 m long of 0.1 m amplitude. Effect of wheel radius is also significant in relation to the road disturbance. A drop in wheel radius from 0.37 to 0.25 m is said to give an increase in wheel vertical velocity over the first 20% of lift, over a 0.025 m step, of about 15%. In the simple driver and ride model of Fig 10, an expression for the value of jerk at the driver's head, useful in testing the Janeway comfort criteria, is $8\pi 3SF_p{}^3h/l2$. Using values of 5.04 in (128 mm) for static deflection, 2.19 Hz for pitch frequency, h = 50 in (1.27 m) and l_2 = 80 in (2.03 m), horizontal jerk is 7400 in (188 m)/sec^3. An additional complexity arises with articulated vehicles, as idealised in Fig 11. The relative motions of tractor and trailer means the imposition of further derived movements at the driver's head due to trailer vibration. By simulating the behaviour of this model on a computer, MIRA researchers have obtained the vibration modes and values for a typical outfit as shown in Fig 12, from which horizontal and vertical motions at the driver's head could be derived.

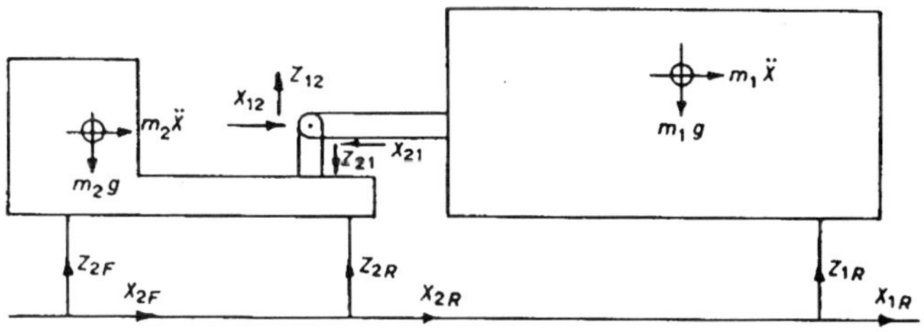

Fig 11: Articulated vehicle model

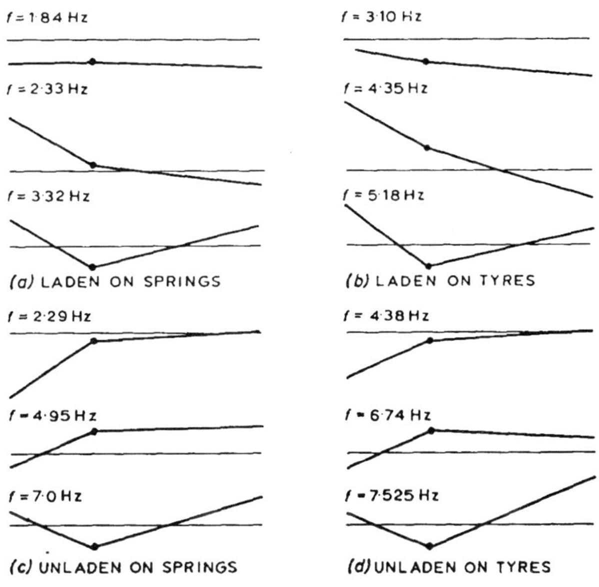

Fig 12: Ride motions for articulated vehicle

Steering geometry and vehicle handling

There are conflicting requirements in trying to set out the steering geometry for a new vehicle design. A very tight turning circle may be required for low speed manoeuvring while at high speeds strong directional control will be probably required around the zero steer-angle position. On cross-country vehicles large vertical wheel, and spring displacements will be required to ensure ground adhesion and ride comfort. Because conventionally steered and sprung vehicles are free to pitch, roll and yaw — as well as move in vertical bounce on the suspension springs, plus some lateral and horizontal movement against compliance which may be introduced to soften a variety of vibration effects — all these can make the handling response as complex even, or more so, than an aircraft in flight.

Avoidance of tyre scrub in slow speed vehicle manoevering, as opposed to high speed handling control, involves careful linkage design to counter any difficulties with the relatively large steer angles. The initial wheel angles will probably be selected to meet the 'true' steering condition and then tuned to meet the requirements of handling stability at full road speeds. The initial values suggested below apply to commercial vehicles — though reading across to cars is possible given the necessary design tuning to allow for front or four wheel drive, soft-ride independent suspension and the special problems of handling control mentioned above.

Selecting initial steering geometry

Caster angle promotes good straight line running and provides self-centering action for the steered wheels. Typical CV values are in the range 1.5 to 2.5 degrees although even negative caster angle could be specified if the suspension is of the type that induces high caster effect during cornering deflection of the springs.

Swivel-pin inclination, alongside caster angle, defines the ground-plane 'offset'. CV angles of 4.5 to 6.5 degrees are typically employed — which can reach 7 to 9 degrees when drive-steer axles are employed. Offset, in governing the moment arm for ground-plane forces around the svivel-pin axis, determines the heaviness of the steering in slow-speed parking and similar manoevres. Values are typically 60-70 mm with a limit set by the additional steering effort to 'raise' the vehicle at high lock angles.

Camber angle introduces a preload into the steering linkage by the tendency of the tyres to roll as cones — as their rotatioal axes are tilted. Current CV practice is to design within the range 1 to 1.5 degrees. Adjustment is also, of course, provided in the steering linkage for setting the wheels to toe-in for counteracting deflections in the system caused by rolling resistance — thus maintaining parallel running in forward motion of the vehicle.

Avoiding steer fight

Steering feedback is minimised by choosing a linkage geometry which avoids any 'fight' between the horizontal steering displacements and the vertical suspension movements. In the leaf-sprung beam axle axle arrangement of Fig 13, as near as possible coincidence between the swing centre of the spring and pivot point of the drag link is obviously desirable in the fore-aft upright plane. In the transverse plane, where for example double wishbone IFS is employed with rack and pinion steering, it is clearly also

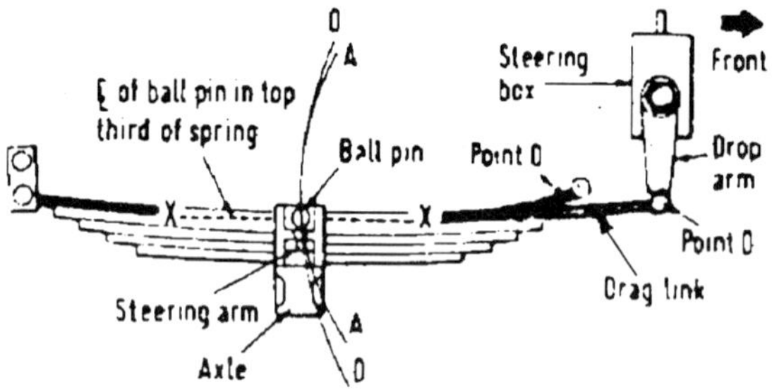

Fig 13: Leaf-sprung beam axle

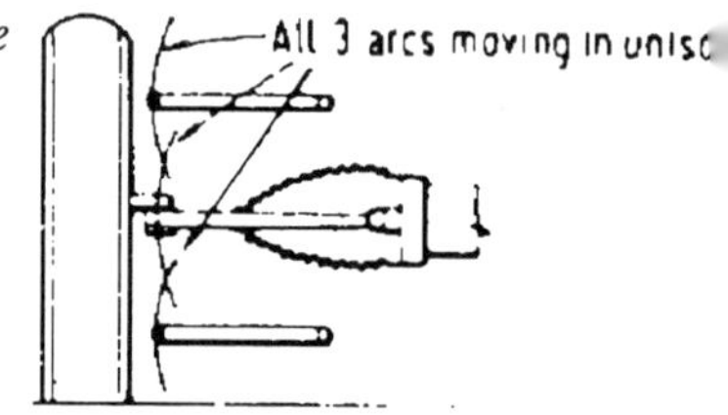

Fig 14: Wishbone suspension with rack-and-pinion

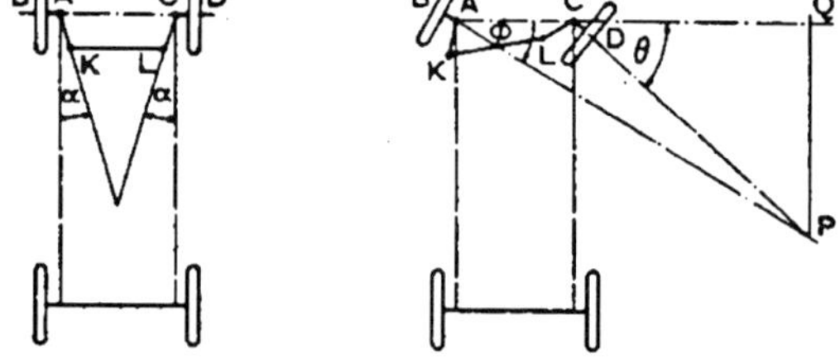

Fig 15: Ackermann steering

desirable to obtain sympathetic motions of wishbones and steering arms as shown in Fig 14.

Steer angle relationship for 'true' steering is traditionally achieved by one or other variant of the Ackermann mechanism. This is a four-bar linkage, in the ground plane, typically defined by the unequal length swivel pin centre distance and the track rod, on the one hand, and the equal length horizontal projections of the steering arms, on the other, Fig 15. The value off for a given value ofq depends upon the ratio AK/AC and angle a. Relative values can be obtained graphically or by trigonometry.

Results show 'true' steering, in which P lies on the axis of the rear wheels, only occurs for one value of θ. This value is about 24 degrees for a vehicle in which AC is typically 0.4 times the wheelbase. Generally the errors from true steering are small except for very large steer-angles — when speeds are inevitably low. Experience has shown that for a beam axle layout, with track rod located behind the axle, angle θ should be chosen to make QP equal two-thirds of the wheelbase.

Fig 16a shows a graphical construction for such a layout. Radii aa' and bb' are arcs of the steering arm ball centres and from A, lines are drawn which intersect aa' and EY (defined by inclination $\tan^{-1} T/2W$), indicating angles ϕ in 5 degree increments. Arcs HH' and JJ' define steering arm ball loci, H and J being established from $\theta = \tan^{-1} 3S / 4W$.

On arc HH', movements of steering arm in 5 degree increments are plotted as well as the corresponding movements on JJ'. These are transferred to bb' and lines from B are drawn to intersect common radials from A. Through the intersections the steering error curve EX can be drawn. Fig 16b shows a similar linkage diagram for double-wishbone independent front suspension (IFS) with rack and pinion steering in which F and E are steering-arm centre and ball in the straight ahead position.

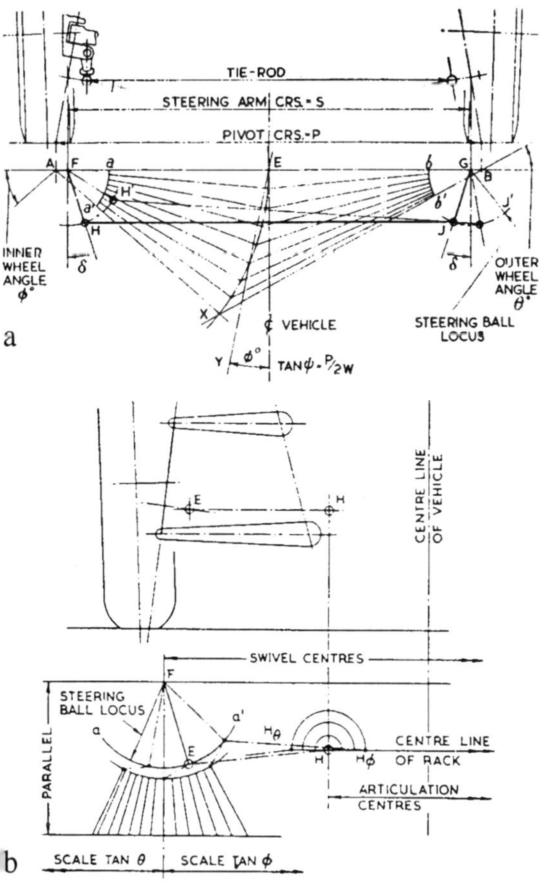

Fig 16: Graphical construction for steering geometry: a — beam axle; b — IFS

Handling requirements

To fix some initial steering geometry values in design, it is useful to consider the motions of the vehicle as though it moves in a horizontal plane only and, in cornering, that only fixed radius bends are encountered — at constant vehicle speed. This allows a 'steady-state' handling analysis to be carried out — with the simplifying assumptions yielding reasonably quick values for the geometric variables. The basic handling relationship for this condition is that the steer angle = heading angle + rear slip angle - front slip angle.

The slip (or drift) angle is best visualised by imagining the tyre-ground contact patch to be a row of springs which represent the tyre's cornering stiffness. There is a proportional (linear) relationship between cornering force and slip angle (between path and heading) until slip occurs (at B in Fig 17) — then slip angle increases rapidly and aligning torque *AT* decreases. The effect of radial load *RL* increase is to increase *L* and therefore increase the cornering stiffness. In Fig 17, slip angle θ can be recognised from steer angle ϕ. When, at comparatively low speeds, the vehicle steers about a point O, then steer angles are related by $\cot\theta_o - \cot\theta_i = T/W$.

As speed increases there is a build-up of slip angles and drift forces which correspond to those required to resist centripetal force of the vehicle as

$F_f = Fb/W$ and $F_r = Fa/W$

Cornering force (drift force) is also required to overcome thrust due to camber change in roll and differential induced drag caused by outward weight transfer. The latter leads to a torque in the ground plane which must be reacted by increased front- and decreased rear- side forces from the tyres. Also to be reacted are the combined self-centering torques due to tyre self-aligning, caster trail and king-pin inclination — also due to the product of tyre forces and king-pin offset as well as the effect of steered axle braking, in the opposite direction.

Tyre and wheel parameter interrelation

From Fig 18 can be seen how camber angle *C* causes camber thrust F_c in the ground plane for radial force *RL* (up to the nominal rated load *RLo* of the tyre) as $F_c = RL \tan C$. In a relatively high speed turn, an overturing couple causes a large 'roll-camber' angle — creating a powerful drift force in the same direction of the force causing the overturning couple. About 1 - 1.5 deg camber change at 0.5 g lateral acceleration gives a thrust of around 5% of the centrifugal thrust.

Induced drag *ID* as well as rolling drag *RD* exists under a heading to path slip-angle as $ID = R \tan\theta$, rolling drag being related to rolling resistance in $f = 0.01(1 + V/100)$ approximately (for vehicle speed *V*). Self aligning torque is due to the combination of caster trail *CT* and pneumatic trail which may be approximated as 1/6 of the contact patch and as CT = RRtanϕ, $AT = CF[(L/6) + RR\tan\phi]$ and king pin offset causes a torque created by the tyre/ground friction co-efficient μ in

$$T_s = \mu RL(OS^2 + k^2)^{1/2}$$

where radius of gyration *k* can be approximated from k^2 = (tread width)^2/8. In Fig 19, it can be seen that camber angle can also change with steer angle, firstly due to king-pin inclination and secondly due to caster, changing from *C'* to *C"* according to

$$\sin C' = \sin C + \cos(\gamma + C)(1 - \cos\phi \sin\theta)$$
$$\sin C'' = (\sin C + \sin\theta \sin\gamma)$$

values of which angles can then be summed.

Roll effects

Roll modifies inter-axle weight distribution according to roll-axis position and the relative front to rear roll stiffness of the vehicle suspension systems. With the wishbone IFS and beam-axle IRS layout shown in Fig 20, for lateral accelaration α, torque about the roll axis is $\alpha WH + WH \tan\varepsilon$ and for small roll angles respective axle torques are $(SR/SF + SR)(\alpha + \varepsilon)WH$ at the front and $(SR/SF + SR)(\alpha + \varepsilon)WH + FRh$ at the rear.

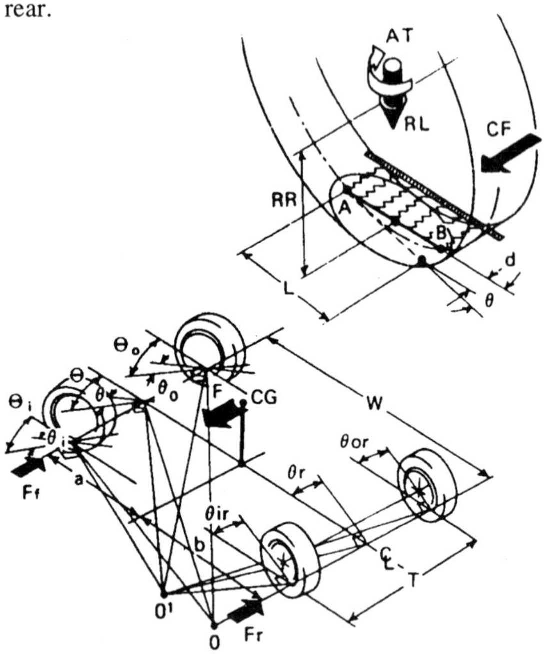

Fig 17: Tyre scrub simulation (top) and steer/slip angle (below)

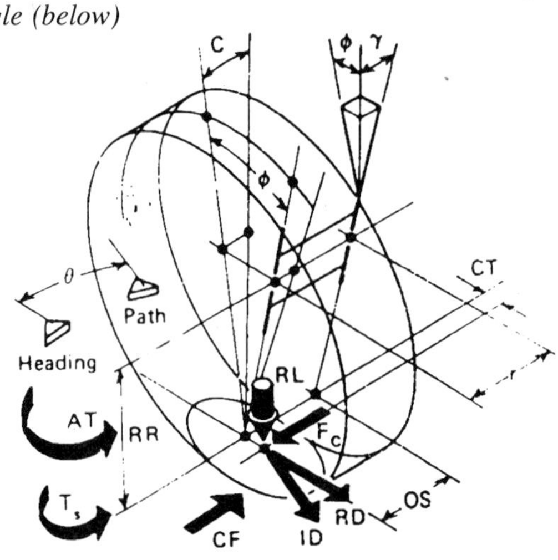

Fig 18: Camber, caster and swivel-pin inclination

Roll stiffness *S* relates spring and tyre tilt with spring-base and track, rear roll-steer angles being related to roll angle and effective spring length (represented by a link) as

$$tan\theta = x/z = (x/\{T/2\})/(z/\{T/2\}) = \varphi/\varepsilon$$

for a spring effective link (Fig 13) tilting downwards during suspension travel.

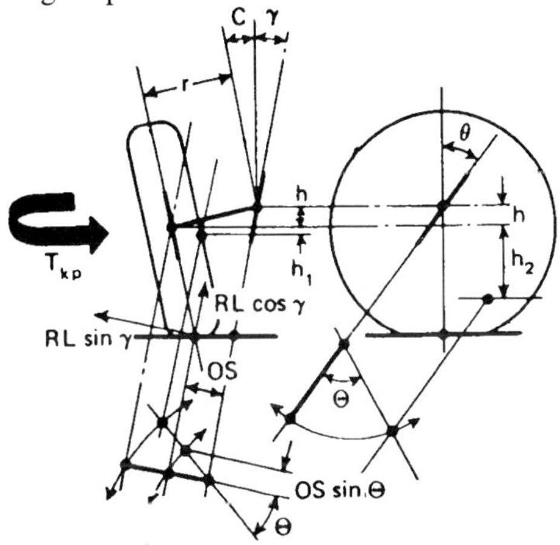

Fig 19: Camber change with steer angle

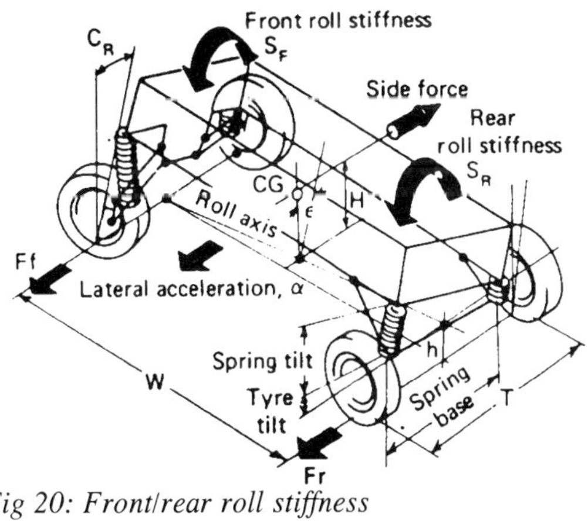

Fig 20: Front/rear roll stiffness

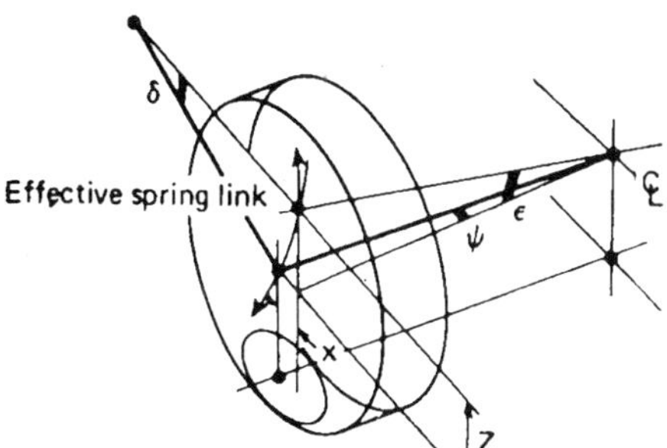

Fig 21: Roll geometry

Handling for the whole vehicle

If the centre of gravity of the vehicle is positioned symmetrically on the vehicle longitudinal centreline — at distances a from the front and *b* from the rear axle along the wheelbase *L* — then for cornering force CF on the whole vehicle

$$SR + CF/[1 + (b/L)] \text{ and } SR = CF/[1 + (a/b)]$$

also $CF = (W/g) - (V^2/r)$ relates vehicle speed with cornering radius *r* and the force can be expressed as a percentage *n* of vehicle weight *W*.

Plotting handling curves The degree of understeer, represented by the difference between front and rear slip angles, can be plotted against *n* to give the response curve. It is then usual to determine the effect of cambering the tyres, say, to produce extra side thrust during roll (by adjustiing roll-stiffness and suspension geometry), creating differential dynamic load transfer by varying roll-centre height through suspension geometry manipulation. Alternatively anti-roll bars with differential front/rear stiffness can be used to obtain a similar effect. The degree of load transfer depends on CG height and magnitude of sprung mass while extent of front/rear redistribution could be limited by conflicting spring rate requirements for ride comfort.

The logical procedure in constructing the curves would be to write down static weight distribution and CG position then determine by geometry the front and rear roll centre positions for the suspensions. Then express the tilting moment in terms of *x*,*n*,*g* and equate it against the suspension restoring torques based on spring rate, spring-base and roll-angle. The resulting expression can be solved to obtain roll-angle and a table set up with equal increments of lateral acceleration corresponding to side loads derived from force = mass x acceleration. This table would show the minimum to maximum load-transfer condition. Finally, from side-force/slip-angle/radial-load curves available from tyre manufacturers, mean (between inner and outer) slip angles can be obtained to tabulate for front and rear of the vehicle in similar increments. These can be plotted against lateral acceleration as the reponse curve for the vehicle — yielding degree of understeer for percentage-g turn.

Steering and suspension linkages

A dual role is played by steering and suspension linkages in providing both a path for road loads into the vehicle structure and ensuring that road wheels are kept in the correct relationship with the vehicle for every attitude and load condition of the vehicle. Considerable ingenuity by automotive engineers has resulted in progressively more standardised wheel-location geometries which are crucial to fixing the initial scantlings of a new-concept vehicle.

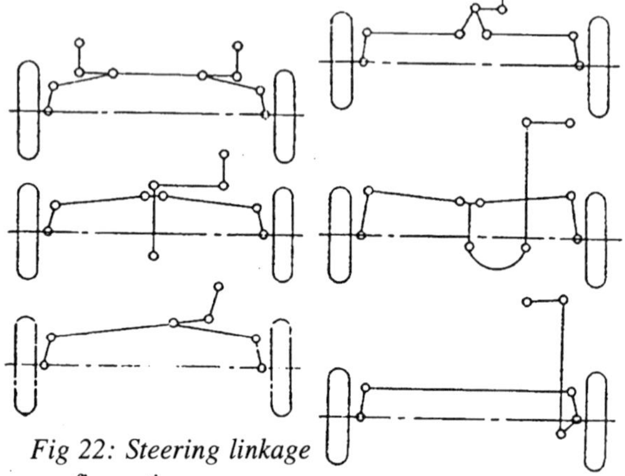

Fig 22: Steering linkage configurations

Steering linkages have varied widely in configuration, Fig 22, but there is a tendency towards the simpler layouts associated with rack-and-pinion steering at least for lighter vehicles. Apart from freak impact loads involving curbing at speed, highest steering linkage loads occur as the steering wheel is turned when the vehicle is stationary. An approximation to the torque required to turn a stationary tyre is

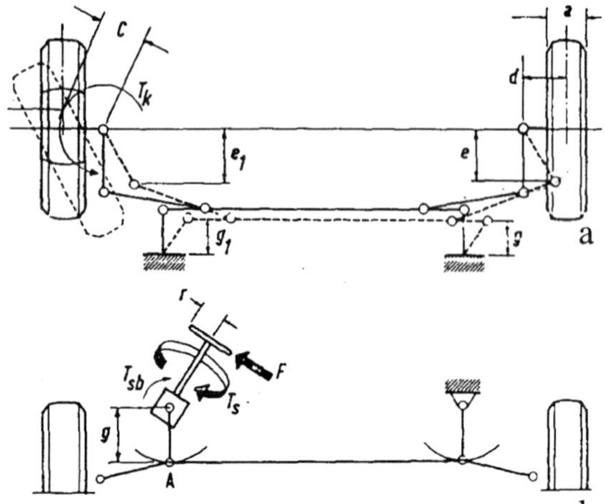

Fig 23: Steering rod load parameters

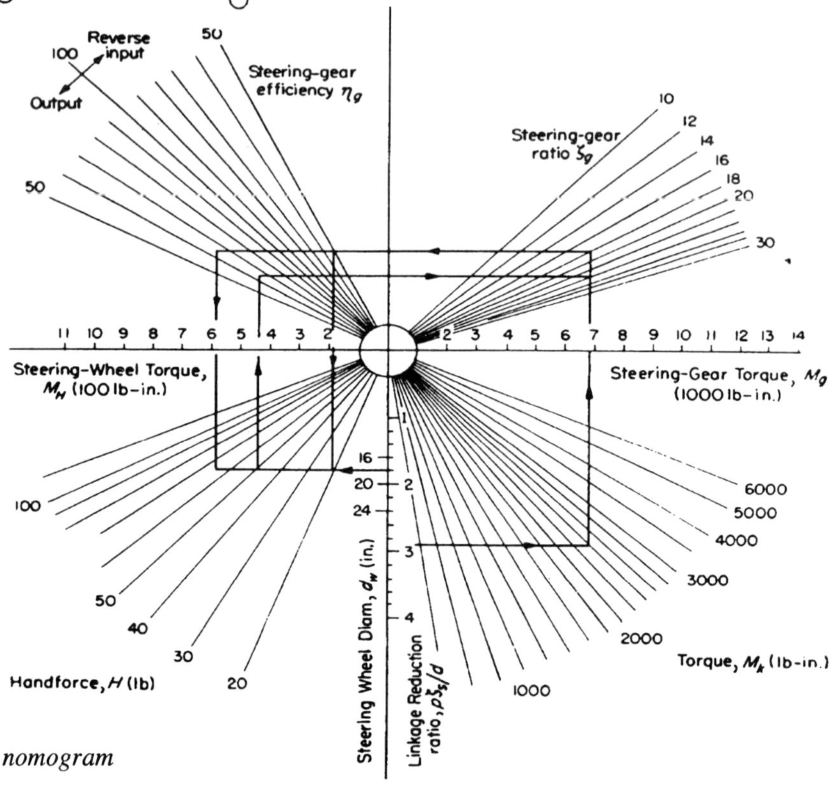

Fig 25: Force calculation nomogram

$(\mu W^{3/2})/3P^{1/2}$

for wheel load W, tyre-ground friction m and inflation pressure P. When the tyre is sunk into soft terrain, the value of torque can rise up to twice the hard-ground value for which m is often assumed to be unity. The previous section examined the effect of steering geometry on self -aligning torque as well as showing how linkages could be configured to obtain 'true' steering. Another accepted approximation for the maximum steering torque for a moving vehicle is one third the static value. A number of forces in play include centrifugal, dynamic 'weight', braking and obstacle impacts with possible added components duc to cambcr, castcr and slip angles. Steering rod loads measured by engineers at AP Group, on a moving car, are shown in Fig 23a to vary considerably with vehicle operation. Empirical formulae have been devised to relate these figures so that in Fig 23b a hand force F at the steering wheel gives torque at the steering box output shaft of rFS_b, representing a force at joint (A) of rFS_b/g where S_b is steering-box ratio. For the commercial vehicle linkage of Fig 24 a similar expression can be derived which leads to the calculation chart of Fig 25 due to Taborek[2]. In the case of power-steering, the following formula gives the required cylinder force to be

$$MkS_l/bcos\beta$$

where S_l is the steering linkage ratio. Fig 26 can then be used to select the necessary power-steering pump capacity corresponding to the idling speed of the vehicle engine.

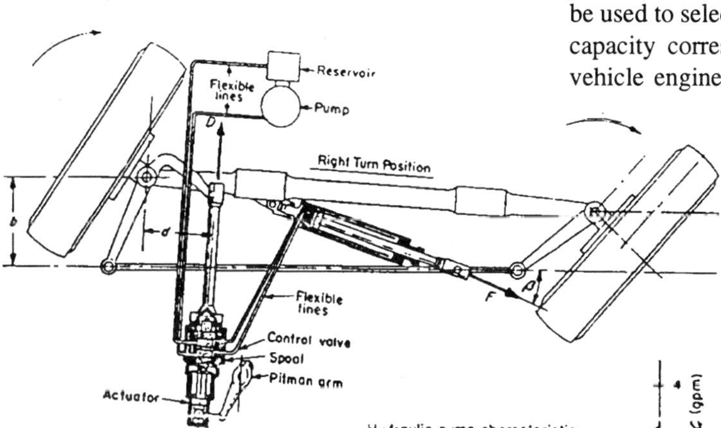

Fig 24: CV steering linkage

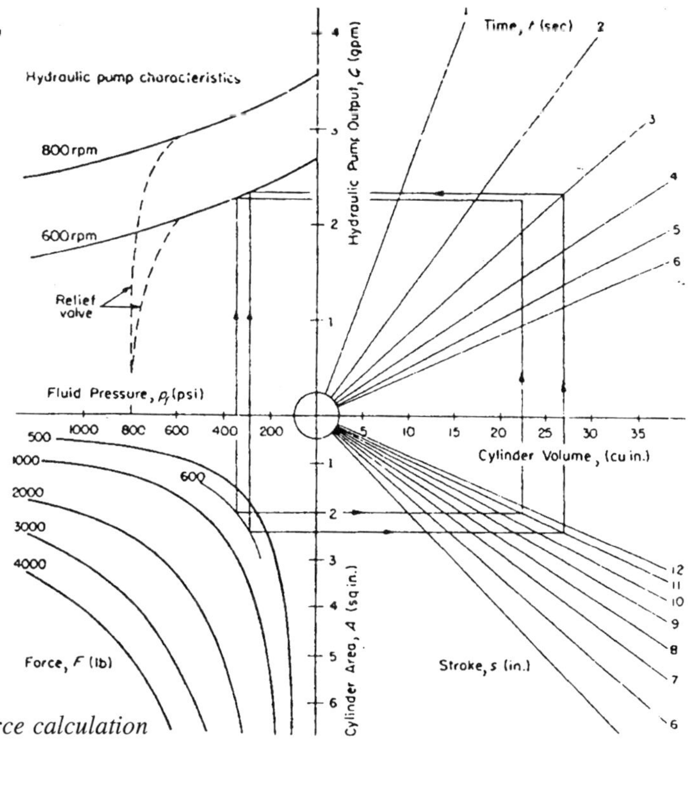

Fig 26: Power-steering force calculation

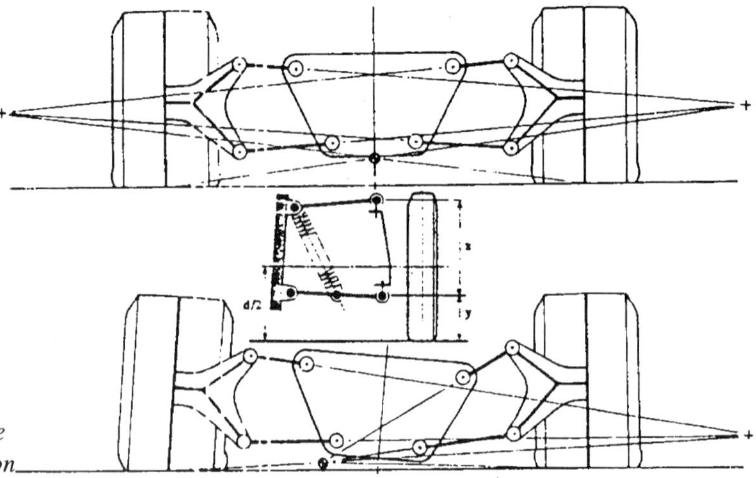

Fig 27: Wishbone linkage suspension

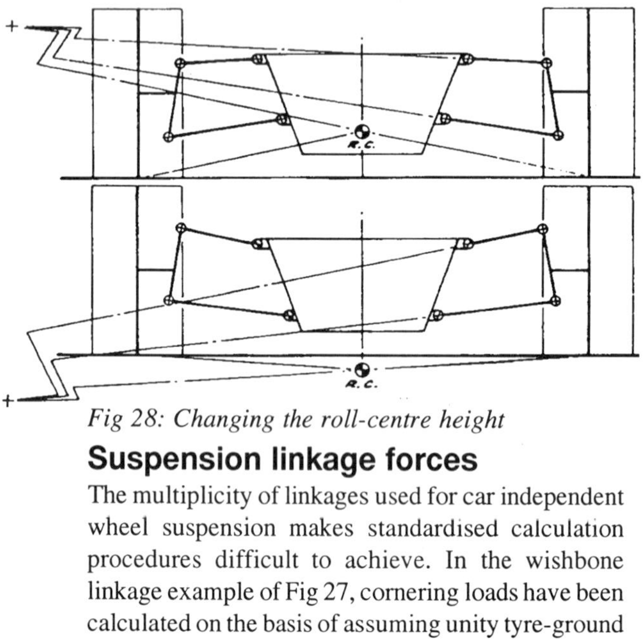

Fig 28: Changing the roll-centre height

Suspension linkage forces

The multiplicity of linkages used for car independent wheel suspension makes standardised calculation procedures difficult to achieve. In the wishbone linkage example of Fig 27, cornering loads have been calculated on the basis of assuming unity tyre-ground friction coefficient; one quarter the gross vehicle weight on the wheel, W_v; a one g braking load biassed 0.65 to the front and a $3g$ bump load. This gives groundplane force:

$$W_h = [(3W_f/2)^2 + (0.65GVW/2)^2]^{1/2}$$

Taking moments about the king-post upper and lower ball joints to obtain side loads, the combined bending and direct stress in the ball pin is

$$(W_h x/y)b.32/\pi d_2^3 + W_v 4/\pi d^2$$

where d is pin ball diameter.

Suspension linkage motions

It is necessary, of course, to consider the dynamic as well as the static force inputs through the steering system and therefore valuable to consider the vehicle behaviour under a sustained swerving manoeuvre. At a speed of 70 mph, a steering input frequency of 1 Hz is equivalent to a swerve wavelength of about 100 ft (30 m). Roll frequencies of about 2 Hz would cause the tyre to be disturbed by the rolling of the body at about 50 ft (15 m) wave lengths. In the limit of adhesion, the side force on the vehicle is μW and the maximum tilting moment of the body can be found from the position of the vehicle centre of gravity with relation to the roll axis. There will be a restoring moment from the springs and stabilisers and equating the moments will yield the tilt angle of the body and the lateral load transfer wheel to wheel. Suspension linkage geometry governs the height of the roll-axis about which the sprung part of the structure will tilt under the action of centrifugal force. Front and rear suspensions each involve a roll-centre through which the axis passes. Co-incidence of roll-axis with centre-of-gravity height, which would rule out tilt, is not really a practicality except in very low slung racing cars. If the front roll-centre is lower than the rear, the tendency is for the body front to tilt more than the rear so that the front springs have to resist a higher roll moment. However, because the body is almost rigid in torsion, the nett effect is that the rear springs have to apply greater roll resistance than the weight imposed on them would normally require them to produce and there is more than usual weight transfer from inner to outer tyre. Tilt of the roll axis can thus influence an oversteer/understeer condition.

Choice of front suspension linkage, on all but the heavier beam-axled vehicles, has narrowed down to MacPherson strut *vs* double-wishbone IFS. There are several variations of the latter and there is a discernable trend to one fully-fledged, lower, wishbone and an upper one merely comprising a swinging arm. Fig 28 shows how linkage geometry can be altered to effect a change in height of the roll centre. There is usually a space limitation which constrains an inordinately high roll centre, causing banking-in of the body and undue negative camber of the wheel, with a resulting severe reduction of clearance between them; an undesirable jacking effect also occurs as well as tyre scrub caused by track change, when negotiating symmetrical (both-wheel) bumps. Care must be taken to avoid undesirable shift of the roll-centre in the body-tilted condition. Usually it is arranged for the roll-centre to move slightly upwards and towards the centre of the corner, laterally.

The MacPherson strut, now widely used on volume cars, has a related layout in the Chapman strut suspension — but this uses the drive shaft as one of the links of the suspension. The appeal of both layouts is the small change in wheel camber angle that can be realised — while having widely spaced anchorage points which reduce loading of the body structure. Roll centre height is calculated as shown in Fig 29. The construction shows how it is easy to alter roll-centre height by changing the angle of the lower

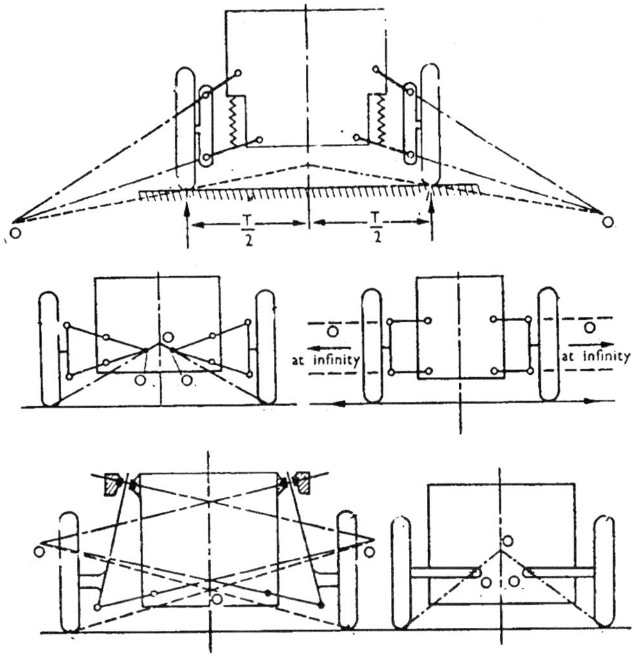

Fig 29: Construction for roll-centre height

link. As a disadvantage, however, under full tilt condition of the body, roll- centre height falls and moves away from the more heavily loaded outboard tyre — thus encouraging the use of an anti-roll bar.

Angling the lower wishbone in either layout can be used to produce an anti-dive and anti-squat geometry, to stabilise the vehicle's geometry on braking and acceleration. Rear suspensions on light as well as heavy vehicles, may employ a live axle — either to ease the drive arrangement, in the case of rear-drive, or for lightness in the case of front- drive. As shown in Fig 30, roll-centre height corresponds with the axle centre-line. The increasing trend on light vehicles is towards IRS, however. Fig 31 shows the swing-axle type which has an even higher roll-centre but suffers the drawback of jacking effect which weighs against its simplicity. The main trend is to semi-trailing arm IRS in which the wheel is articulated on the drive shaft but has a very substantial lower arm on an oblique hinge line. The fulcrum angle has to be a compromise between avoiding changes in wheel angle and the amount of plunge tolerable to the drive shaft. The system again has the benefits of widely spaced anchorage points. Alternatives to it also include MacPherson and double-wishbone rear suspensions.

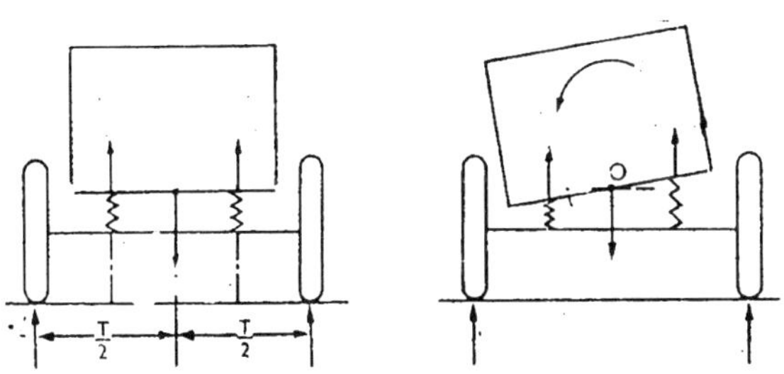

Fig 30: Roll-centre for beam axle

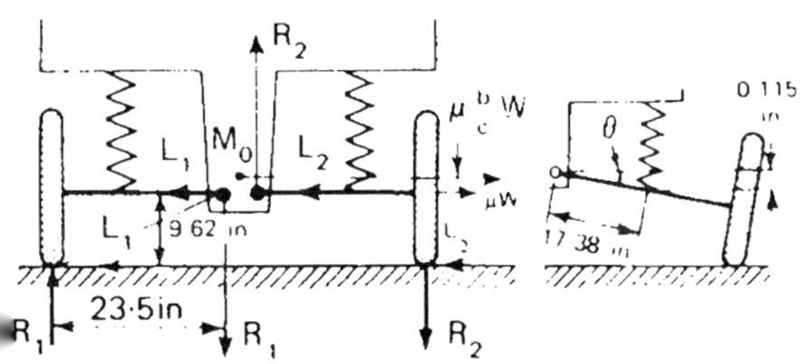

Fig 31: Swing-axle suspension calculation

Directional control

Good vehicle handling comes from proper attention to the design of the vehicle as a whole rather than just attending to suspension parameters. Here design for good directional stability is considered in terms of both steady-state and transient handling behaviour.

Ideally a vehicle should follow the driver's steering command and not deviate from the course set, whether this be in straight ahead driving or cornering. In fact most vehicles are designed to understeer slightly — such that they corner less sharply than the driver expects. This generally improves directional stability as the driver can get back on course without undue 'over-correction' of the steering wheel. It is also desirable that the vehicle does not switch too suddenly from an understeer to an oversteer condition with increase in vehicle speed.

The earlier section on steering geometry introduced the relationship between wheel geometry, established for near-stationary steering of the vehicle, and the constant speed dynamic handling of the vehicle. Fig 32 shows a desirable handling response curve for constant speed,'steady-state', cornering in which understeer builds up progressively at higher cornering accelerations. Up to about 0.25 g sideways acceleration, side loading is typically caused by cross-winds, road-camber or driving through shallow curvature bends. Above 0.25 g, heavier cornering loads from significant steering inputs occur usually causing visible roll of the vehicle.

As there is little dynamic weight transfer between wheels in the first phase up to 0.25 g, the only

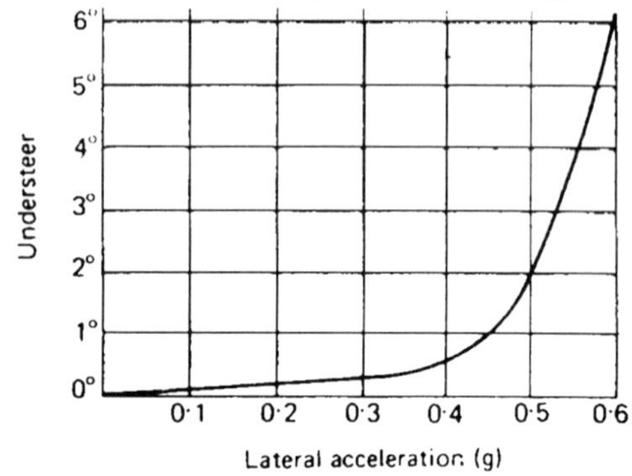

Fig 32: 'Desirable' steady-state handling curve

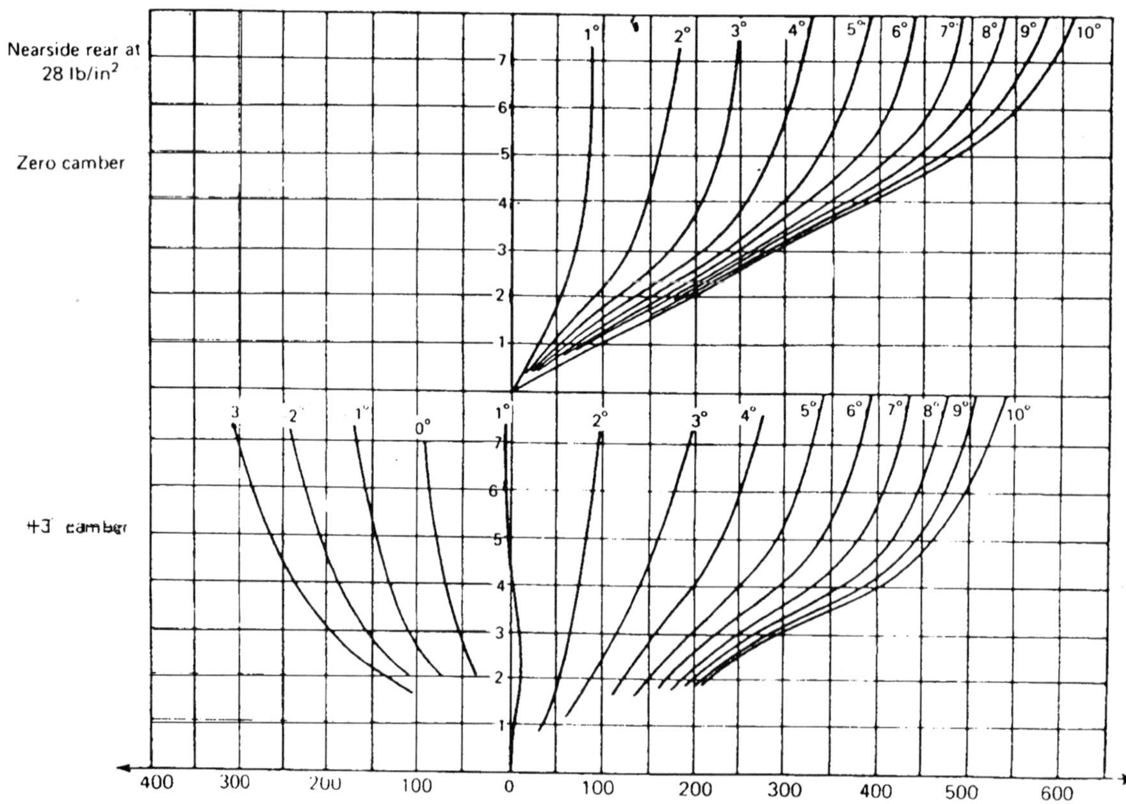

Figs 33/ 34: Side force generated, left/right, at 0 and + 3 deg cambers

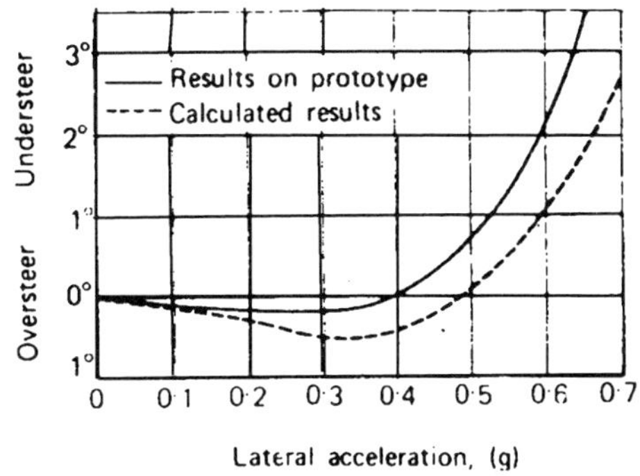

Fig 35: Response curves for small rear-engined saloon

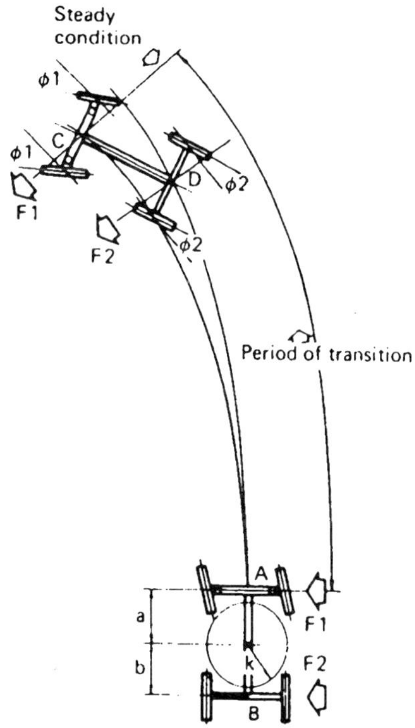

Fig 36: Cornering stability

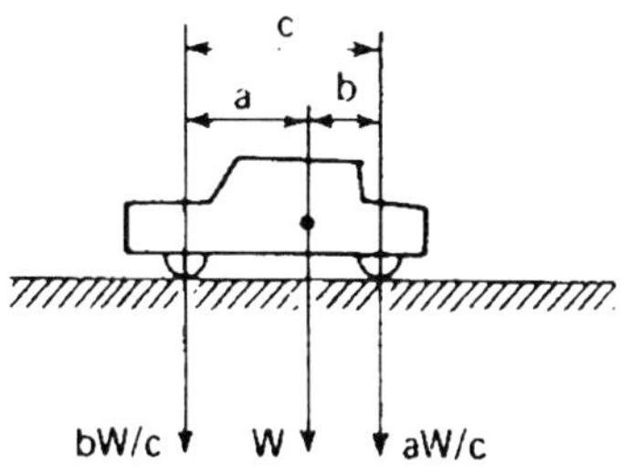

Fig 37: Moment diagram

changes which can significantly affect vehicle stability are those due to changes in weight distribution, tyre characteristics and wheel geometry. In phase two, changes to spring rate, roll axis and stiffness are effective.

Key to the prediction of steady-state handling behaviour at the design stage of the vehicle is an accurate assessment of the sprung mass of the vehicle and the position of its centre of gravity. Roll axis centres for the front and rear wheel suspension system need also to be predicted by the methods given in the section on suspension geometry. Distance between sprung mass CG and roll axis gives the moment arm for calculating roll couples on the vehicle as a product with the unsprung mass cornering acceleration and road adhesion co-efficient. The restoring moments presented by the front and rear suspension systems can then be calculated from the suspension geometry and spring stiffnesses for equating with the roll couple. By this means, the front and rear wheel transverse dynamic load transfer can be calculated.

Fig 31 of the previous section shows an example configuration for a swing-axle suspension vehicle showing the vehicle front-end loads applying in the cornering situation where c is wheelbase and b distance from spring mass CG to the front wheel axis. Fig 31 inset illustrates the well known jacking effect of the swing axle suspension in which unbalanced couples either side of the vehicle can result in a jacking force $(R2 - R1)$. This of course will indirectly result in wheel camber change and hence in the potential cornering power of the tyres.

Figs 33/34 show values of front and rear cornering acceleration for wheel cambers used in the design of a small rear-engined vehicle based on tyre characteristics measured on a mobile tyre test rig. These gave the response curve shown in Fig 35 for this case, dotted, alongside the values measured on the prototype vehicle shown in full line.

When a vehicle is not under the steady state cornering conditions of constant radius bends and constant forward speed then cornering force will vary throughout the bend. For the vehicle path shown in Fig 36, for example, then Newton's laws of motion give

$$[r + dr]GL = rGL + 2frd\theta$$

In the case of vehicle acceleration f being

constant, $GLdr = 2frd\theta$ and there is a Coriolis component of acceleration, $= 2V_r\omega$, where V_r is the radial velocity which can be shown to be four times the vehicle acceleration. Its effect can be determined by considering Fig 37 and taking moments about the rear wheel contact point to show

$$W_R = W[1 - (1/WB)(LR - 4fx/g)]$$

Thus for a cornering radius of 5.3 m and $\theta = 48$ degrees then for road width $RW = 9$ m and 1g maximum lateral acceleration, then $f = 0.0946g$ or $4f = 0.378g$ with a 60:40 weight distribution and CG height $x = 0.3$ m on a 2.3 m wheelbase vehicle. This would result in a 27.2 kg front to rear weight transfer.

Transient handling

In passing from one steady state cornering condition to another the vehicle passes through a transient condition where the behaviour may not be quite as expected, Fig 38, by applying the methods of statics. This can be explained, for example, by front slip angle, and hence lateral acceleration, building up more quickly than at the rear.

It is thus better to examine the dynamics of the vehicle as a whole using the notation of Fig 39. Then it can be shown that if the polar moment of inertia of the car about a vertical axis through its CG is such that K^2/ab is unity (where K is radius of gyration), the vehicle can be represented by masses W_f and W_r equal to loads carried by front and rear wheels placed at the axle centres. When this value is less than unity, the two masses lie within the wheelbase and the angular acceleration will have the effect of inducing an inward force at the rear wheels. In the less usual case of the value being above unity, the masses lie outside the wheelbase and an outward force is created at the rear wheels — resulting in an oversteer condition. This relationship explains why high polar moment vehicles thus designed to improve ride by decoupling the pitch and bounce modes may, with conventional suspension systems, be compromised by poor handling. Fig 40 shows that during the transient stage of entering or leaving a bend, tyre cornering force values will tend to lag behind what would be expected in steady conditions. Roll-steer effects can also give under or over-steering tendencies as seen in Fig 41.

One approach to the analysis is to consider the

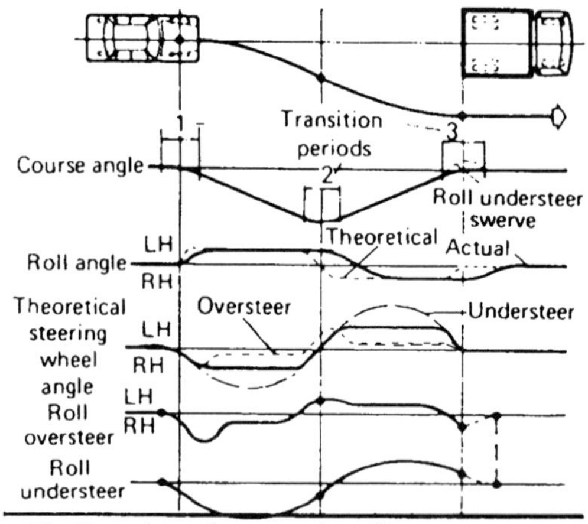

Fig 38: Transition between handling states

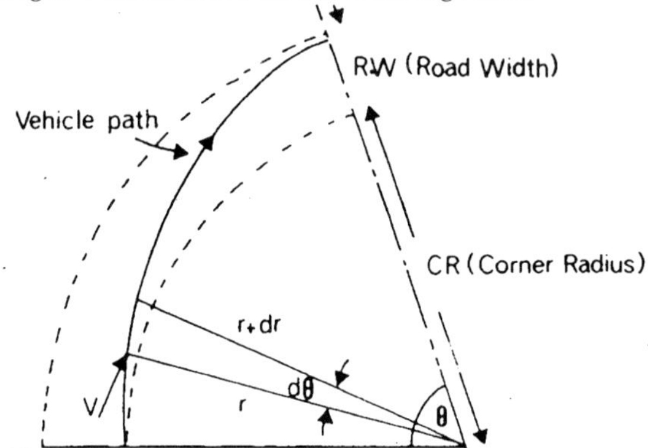

Fig 39: Transient handling notation

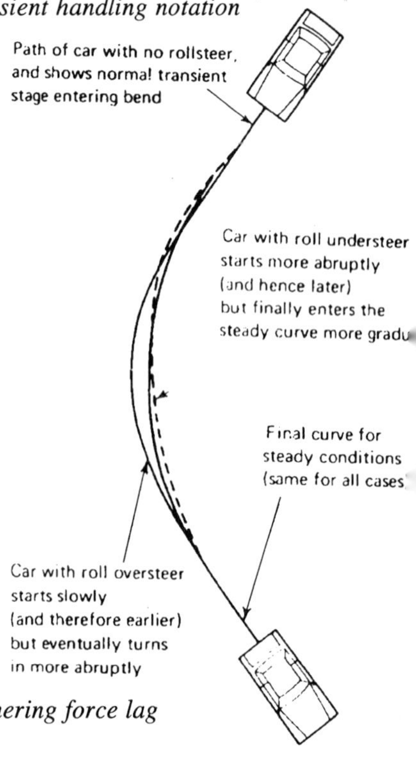

Fig 40: Cornering force lag

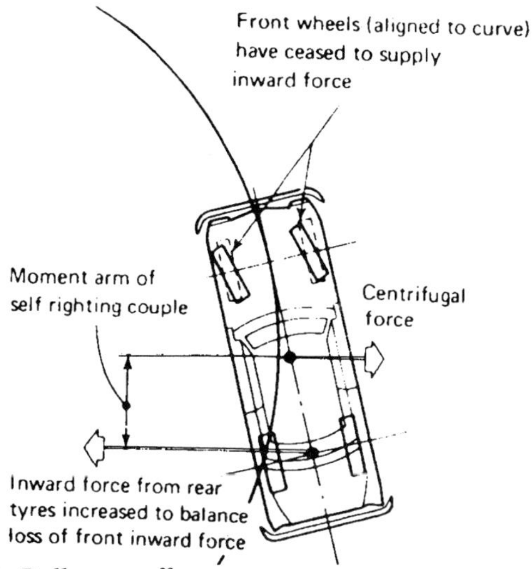

Fig 41: Roll-steer effects

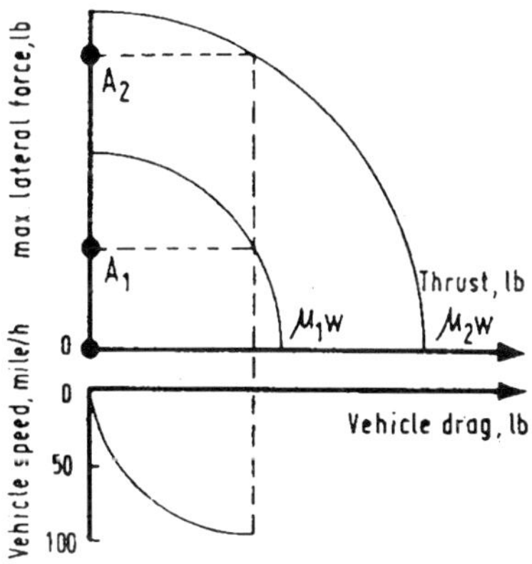

Fig 42: Friction circle

vehicle behaving like an enlarged springless Go-Kart by using a two-degree-of-freedom model. Skidding of a wheel occurs when it passes through a circle of maximum adhesive force as represented by the diagram in Fig 42. The residue of force available after driving and braking is able to steer the vehicle. The drag curve shown implies that tyre thrust available at 100 mph is *OY* while maximum lateral force available is *OA1* and *OA2* for front and rear axle respectively as a proportion of μW. Fig 43 shows a representation of the springless vehicle with front and rear 'axles' modelled as single tyres. The vehicle is side-slipping and yawing; in this condition equations of velocity perpendicular to the heading of the vehicle can be constructed to show values of the side forces, front and rear, in terms of the tyre cornering force co-efficients obtainable from tyre performance characteristics. The front wheel steer angle can be shown equal to

$$[(l_1+l_2)/R]+[M/(l_1+l_2)][(l_1/C2)-(l_2/C1)]U^2/R$$

where U is forward speed, M vehicle mass, C cornering-co-efficient and R the bend radius. The following worked example is calculated for a vehicle which has l_1 = 4.25 ft (1.3 m), l_2 = 3.25 ft (10 m) and M = 150 slugs. To find steer angle corresponding to 0.5g lateral acceleration on a 100 ft (30.5 m) radius bend tyre, co-efficients for 5.20-13 rear and 6.40-13 in front tyres are obtained from the slopes below 2 degree slip-angle of the characteristic curves in Fig 44. Steer angle in radians is thus (7.5/100)+(150/7.5)[(4.25/28600)-(3.25/48700)]0.5 = 5.8 degrees.

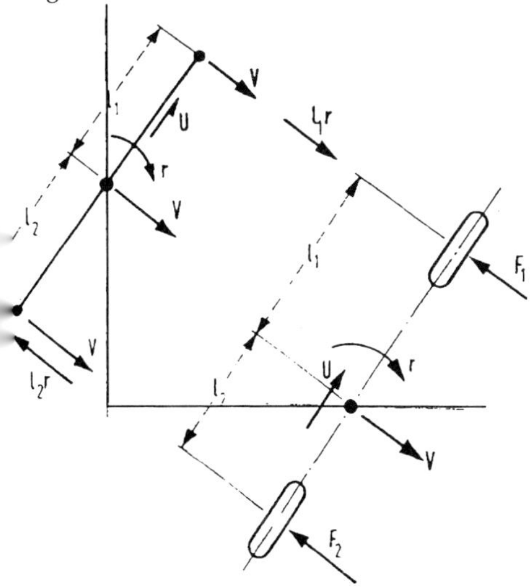

Fig 43: Springless vehicle

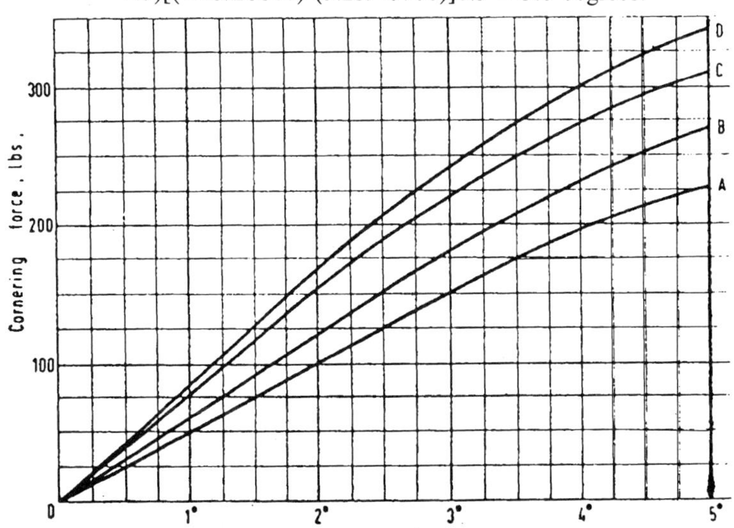

Fig 44: Tyre characteristic

Suspension spring design

Of the different suspension spring media, steel is still the most prominent and the laminated leafspring variant important on many commercial, industrial and specialist vehicles. The techniques for maximising the efficiencies of such springs and modifying their installed performance are discussed here.

Laminated (leaf) springs are probably the longest established category of suspension spring and are covered by SMMT Standard 42 with respect to leaf widths and thicknesses, rolling tolerances and preferred materials. Being locaters of axles as well as supplying the suspension springing medium, they are relatively efficient in terms of specific energy — which can be expressed as $\sigma_{max}^2/1.7\,E^2$ for a theoretical diamond plan-shape lamina from which the spring leaves would be made given perfect utilisation of material and frictionless contact between the leaves. The figure would also vary with leaf-tip shaping and overall leaf taper — values of 300-450 being typical in practice. This figure can be increased by grooving the tension faces of the leaves in order to shift the

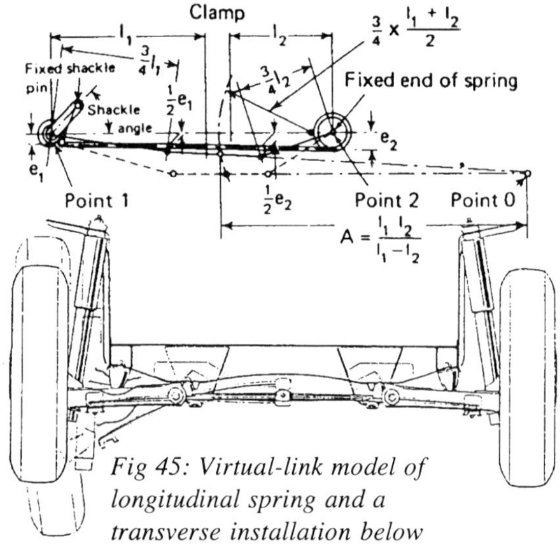

Fig 45: Virtual-link model of longitudinal spring and a transverse installation below

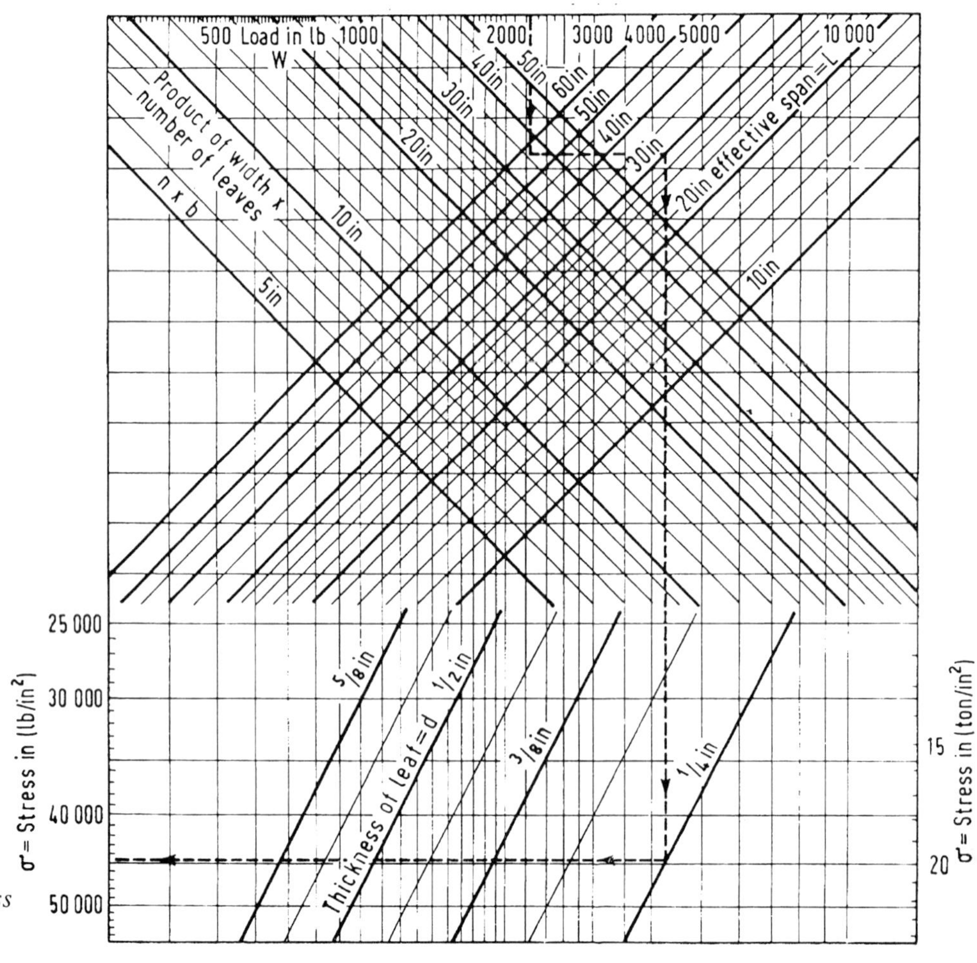

Fig 46: Stress nomograph

neutral axis of bending away from it and thus reduce likelihood of the more common tension-face failure.

Variable load/deflection rate is achieved on laminated springs by introducing extra 'helper' leaves which as well as adding stiffness after a set amount of the full stroke, due to their own deflection, can also affect the bending curvature of the main leaves. There can also be substantial increase of the nominal rate of a spring due to twisting by the axle as the vehicle rolls in corners. Shackle geometry can also affect rate by the imposition of end-loads which add to the flexure of the spring due to loads normal to the leaf surface. The latter effects can be examined by considering the spring as a system of virtual links as in Fig 45. Quarter-elliptic springs have a virtual centre of rotation of the spring-eye which is found to be one-quarter of the length from the clamp edge. Displacement of this centre from the main leaf centreline is said to be in the opposite direction to the spring-eye turn-around by a distance equal to the spring eye radius. This leads to the positioning of the links in the figure which can be used for drawing layout of the spring in order to plot the articulation of the axle. Stresses and deflections can be calculated from nomographs as shown in Figs 46 and 47 which apply to symmetrical semi-elliptics (uninstalled) with leaves stepped by an equal length l such that, for an effective spring length (not including axle-seat clamp), L, and number of leaves, n in $L = 2n.l$ with equal thickness leaves having depth, d, and breadth, b; tips

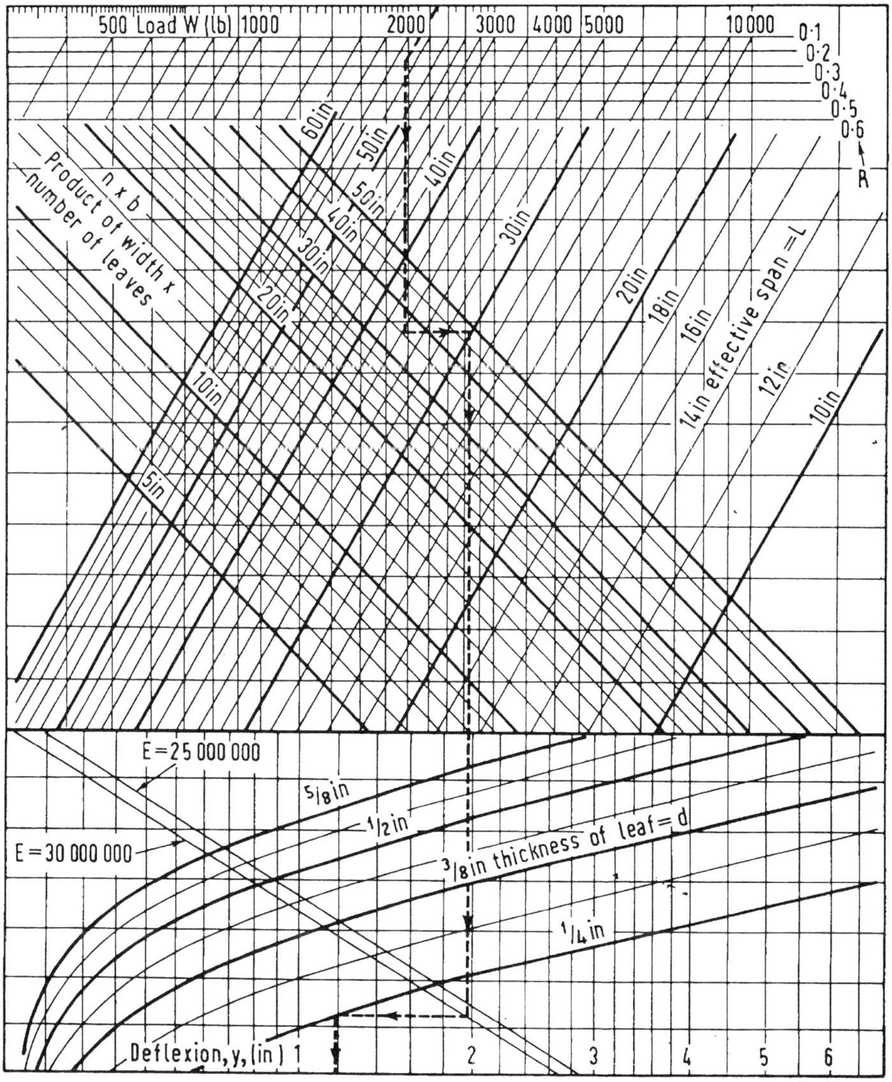

Fig 47: Deflection nomograph

are assumed to taper in such a way as to give a constant stress throughout of $(3/2)WL/nbd^2$ and a central deflection of

$$[2/(2 + R)].[3/8].WL^3/nbd^3E$$

where $R = L - 2l$ and is equal to the number of main leaves divided by the total number of leaves; E is, of course, the elastic modulus. The dotted line applies to an example in which $W = 1$ ton (1.02 t), $L = 30$ in (0.76 m), $b = 3$ in (76 mm), $d = 1/4$ in (63.5 mm) and $n = 12$, for three main leaves. In the case of asymmetric springs, where L is made up of $L1$ and $L2$ then

substitute $[16L1^2.L2^2/(L1 + L2)]^{1/3}$

for L when deriving spring rate from the deflection formula. If different stepping and leaf thicknesses are employed to reduce the top leaf stress below those of the others — in giving it an extra margin for axle location loads — then deflection is given by

$$(W/3E)[(l_1^3/I_1) + (l_2^3 - l_1^3)/I_2 + (l_3^3 - l_2^3)/I_3 + etc.]$$

where I_1, I_2, I_3 etc are the inertias of main leaf (or pair), main + 2nd, main + 2nd + 3rd, etc and l_1, l_2, l_3, etc are the lengths of first, second, third and subsequent steps. Since it is generally impossible to taper off all the leaves to produce uniform stress throughout, the expression for deflection becomes

$$(C/16).WL^3/16nbd^3.E \quad \text{where } C \text{ is}$$

$$[N/(N-1)][(1/2) - 1/(N - 1) + \log_e N/(N - 1)^2]$$

and N is minimum as a proportion of maximum width of the developed diamond-shape spring plan, Fig 48. Recent developments of leaf-springs, and their installation configuration, have centred around

- reducing inherent friction
- increasing working stresses
- providing compliant anchorages
- exploiting vehicle roll-control capability.

Longer and wider springs with fewer leaves have resulted in lighter and more durable installations with less friction. In cars, transverse spring

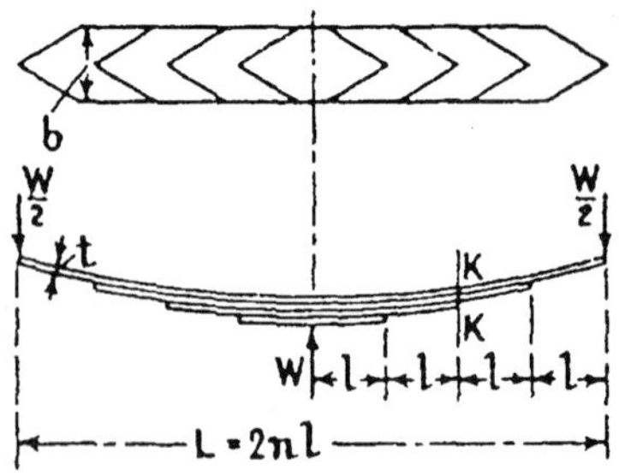

Fig 48: Elements of diamond-shaped spring

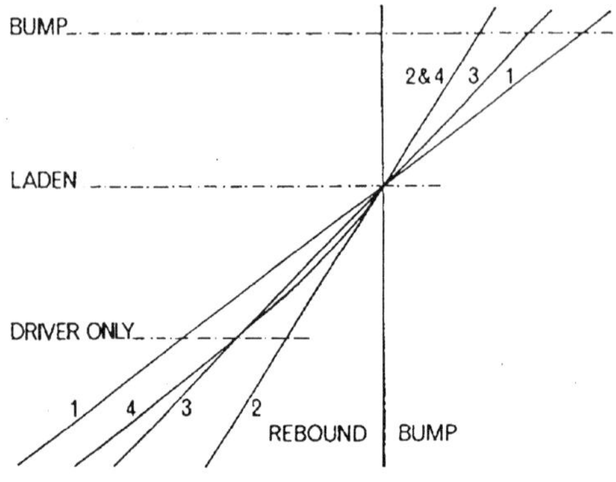

Fig 49: Conventional and up-rated springs compar

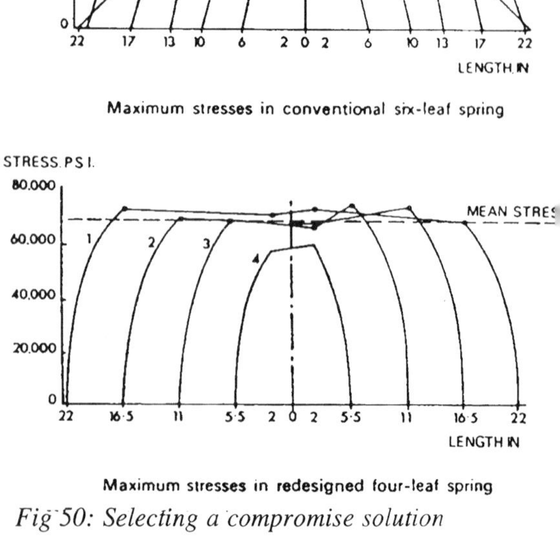

Fig 50: Selecting a compromise solution

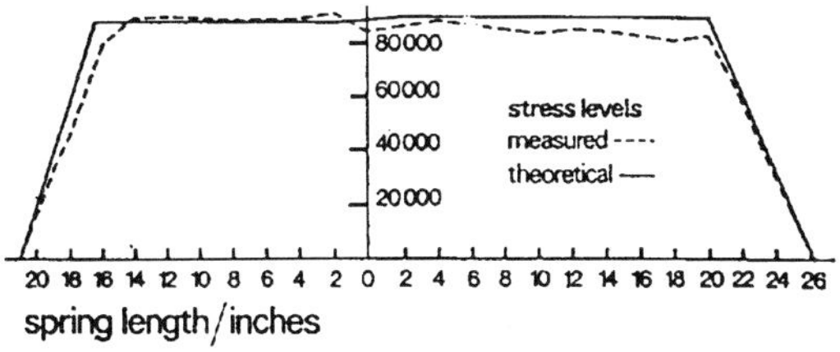

Fig 51: Taperlite stress levels

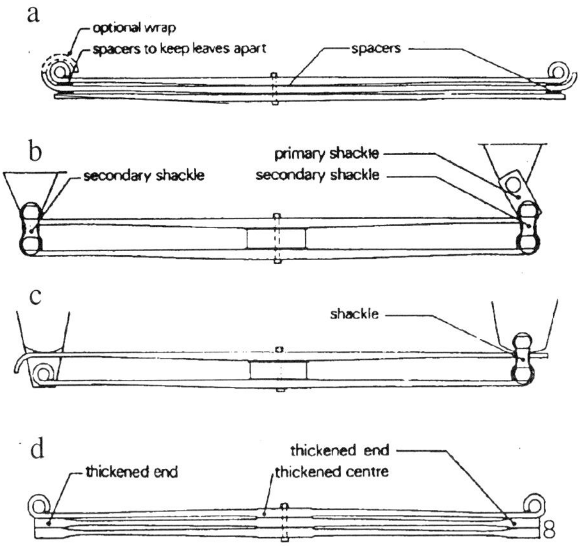

Fig 52: Taperlite configurations (a top to d bottom)

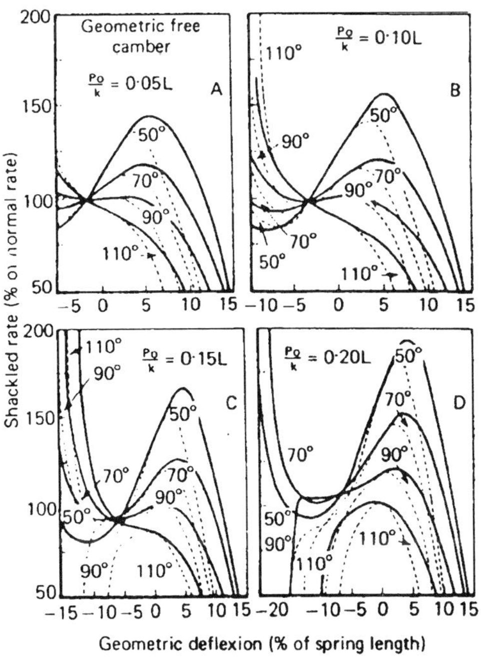

Fig 53: Camber and shackle angle change

layouts are now being used in conjunction with both front and rear independent wheel suspensions. Front end systems usually involve providing equal co-planar torques under normal loading, with the spring anchored at either its centre or two points equi-distant from it. Unequal torques occur under roll conditions thus providing a stabilising effect without necessarily needing the addition of a separate anti-roll bar. Fig 49 compares a linear low-rate spring with a high-rate one. The first would be too soft under laden conditions so would require a long progressive-rate bump-stop. The second would give a somewhat harsh ride unladen. Fig 50 shows how compromise linear or progressive-rate ones could be chosen as a compromise. The Taperlite spring from Tinsley Bridge has a parabolic leaf contour. Fig 51 shows expected stress levels calculated and measured for this spring. A typical steel used is SAE 5160 and for a panel-van application, a typical leaf profile of 0.5 in 12.7 mm) thick x 2.3 in (58.4 mm) wide would be used. A single leaf spring in this case would weigh 13.25 lb (6 kg) against a 3-leaf equivalent 'conventional' spring of 22.25 lb (10 kg). Various combinations of such springs, Fig 52, can be used in heavier vehicles including those using cam supports to achieve variable rate. Shot-peening and pre-setting is used to induce residual stresses which can be used to modify the rate characteristics . Stress distribution in a non-prestressed leaf is seen at (a); that after plastic deformation at (b); residual stress distribution in a pre-stressed leaf at (c) with a resulting operational stress distribution at (d). Net result is a design surface stress lower than the apparent one and a more efficent steel usage.

Spring installation effects

Since there is a change in chord length as the laminated spring deflects, some sort of shackle or slipper-end support is necessary. Change in shackle angle results in alteration of the direction of the resultant 'vertical' load. Horizontal load components increase the spring rate when in compression — and vice-versa in tension. Up to 50% rate changes are possible by changes in both shackle angle and camber of the spring, Fig 53. The 'geometric deflection' is measured along the line of load application and is used in the calculation of camber. It is zero when the shackle link is flat and positive for increasing loads. It is equal to the '*opening*' + *(1/2) x* '*shackle drop*' + *(t + e)/2*.

The geometric free camber is used to define load — which is zero when the linkage is flat and independent of shackle movement. It is equal to the load required to flatten the shackle linkage divided by nominal rate. Shackle effects are more pronounced with shorter shackles, most being designed to be within 5-10 % of spring-length. It should also be remembered that similar end-loadings on the spring to those of the 'shackle-effect' can be induced by vehicle braking and acceleration — according to the configuration of the installation. Graphs in Fig 54 cover both compression and tension effects.

In symmetrical springs, vertical applied forces cause no rotation of the spring seat; conversely, a 'wind-up' torque causes no vertical motion of the seat, Fig 55. Thus $K = P/F$ and for torque T causing rotation, $tan.V = F/(L/2)$ so that wind-up stiffness is given by $K.L2/4$ and for the asymmetric case, an equivalent cantilever, length A, is considered to deflect fc such that $V = fc/A$ and $A/L = [y/(z - y) + 1/(y + 1)]$ for circular-arc deformation of the spring(Fig 56) — uniform stress case $z = Y3$ and

$$A = YL/(Y2 - 1) = l_1.l_2/(l_1 - l_2)$$

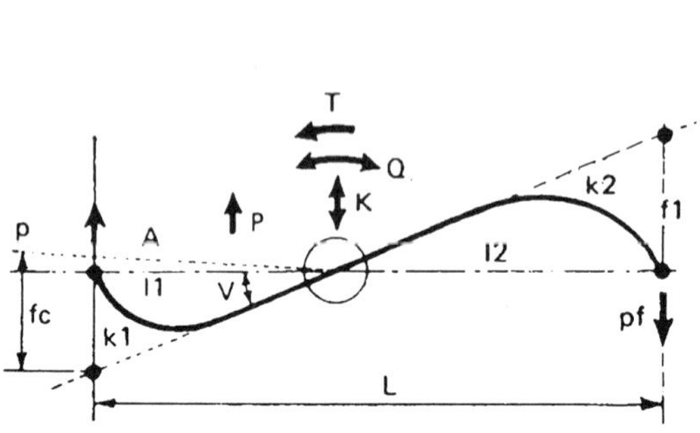

Fig 55: Spring wind-up

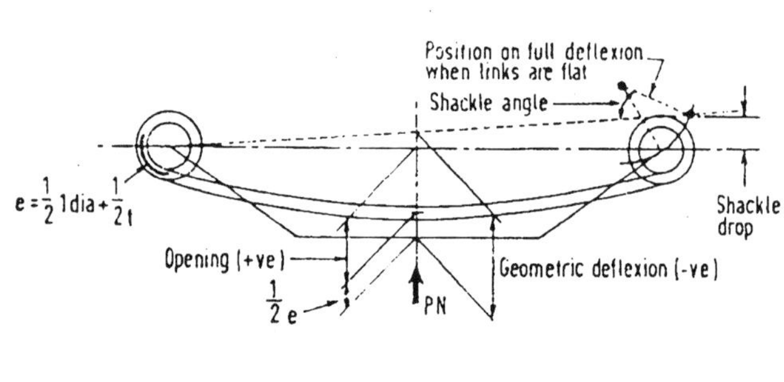

Fig 56: Deflection model for symmetrical spring

Roll-stiffness effect is determined by considering total vertical spring stiffness (under wind-up restraint), $K = k_1 + k_2$ only if $z = y$, in the symmetrical spring case. For an asymmetric spring, K is less than $k_1 + k_2$ and additional roll-stiffness is available. The ratio $(k_1 + k_2)/K$ is shown in Fig 57. The effect is added for the two springs, either side of the axle, since the axle constrains the spring seats parallel to one another. Increase in roll stiffness of 20-40 % is available with this effect.

Suspension coil springs

The coiled helical spring has increasingly replaced the laminated leaf spring in the suspension of light vehicles. This has followed the demand for softer ride, higher spring-deflections and the perfection of suspension linkages able to maintain wheel alignment even under high spring deflections.

Techniques have been developed for building in variable rate characteristics into coil springs, the basic one being the conical helical spring described in Chapter 6. A number of other configurations are

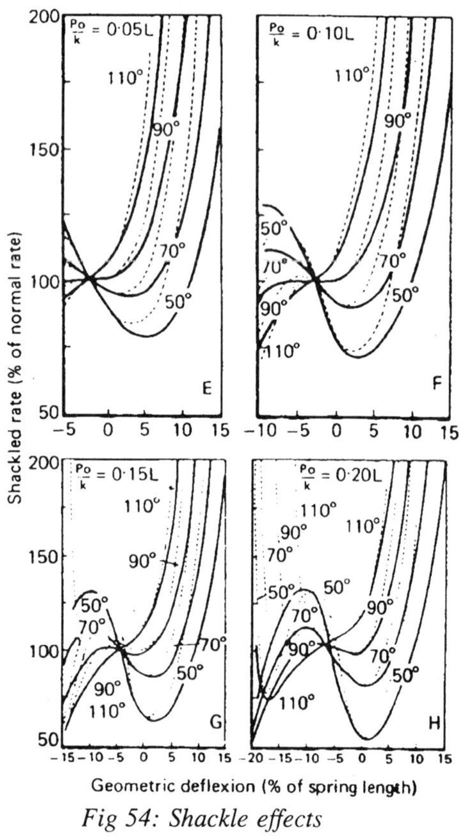

Fig 54: Shackle effects

possible, such as double-conical, and variable pitch of the coils can also be utilised.

The successful analysis of such systems requires an understanding of the basic solid mechanics applying to an element of the spring such as the half-coil shown in Fig 58 of an open-coiled spring. Essentially the steel rod used to wind the spring is subject to both torsion and bending and a higher resilience is possible compared with a leaf spring — which can lead to lighter weight provided a suitable wheel location linkage can be designed.

A spring is said to be close-coiled when the helix angle is small enough for bending and shear effects to be neglected. Applying the formula for torsion to this case and noting that length of an equivalent straight shaft in torsion is given by

$$l = pDn$$

where D is the helix diameter and n the number of coils, spring stiffness can then be derived as

$$k = Gd^4/8D^3n$$

where d is rod diameter, and G the shear modulus. In the open-coil case there will be axial load W on the spring plus a torque, plus a reaction torque T due to the restraint at the seats either end of the spring. The effect at point O are couples $WD/2$ about OH and T about OV giving a combined twisting couple of

$$(WD/2)\cos a = T \sin a$$

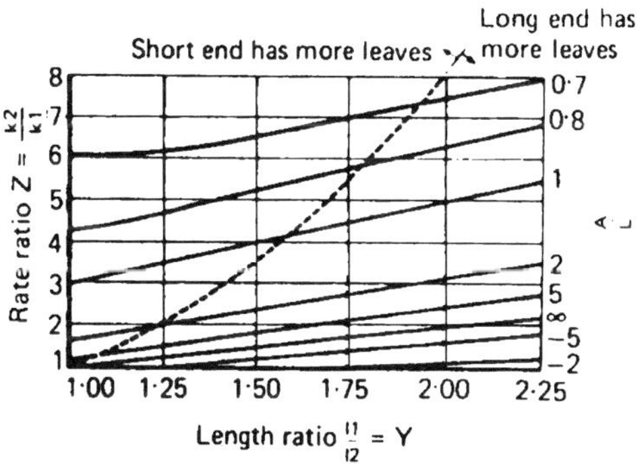

Fig 57: Length and stiffness ratio curves

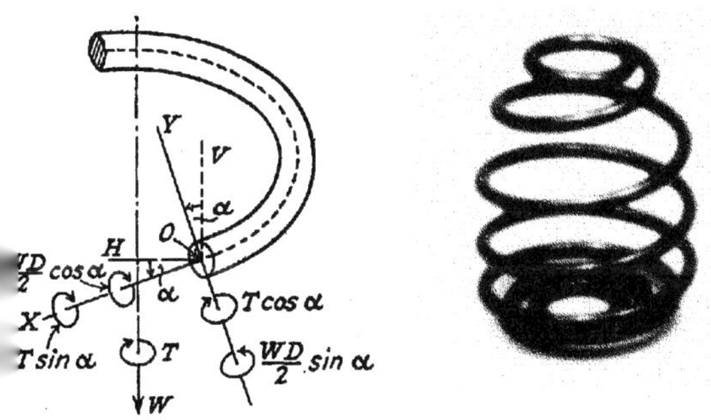

Fig 58: Half coil of typical suspension spring

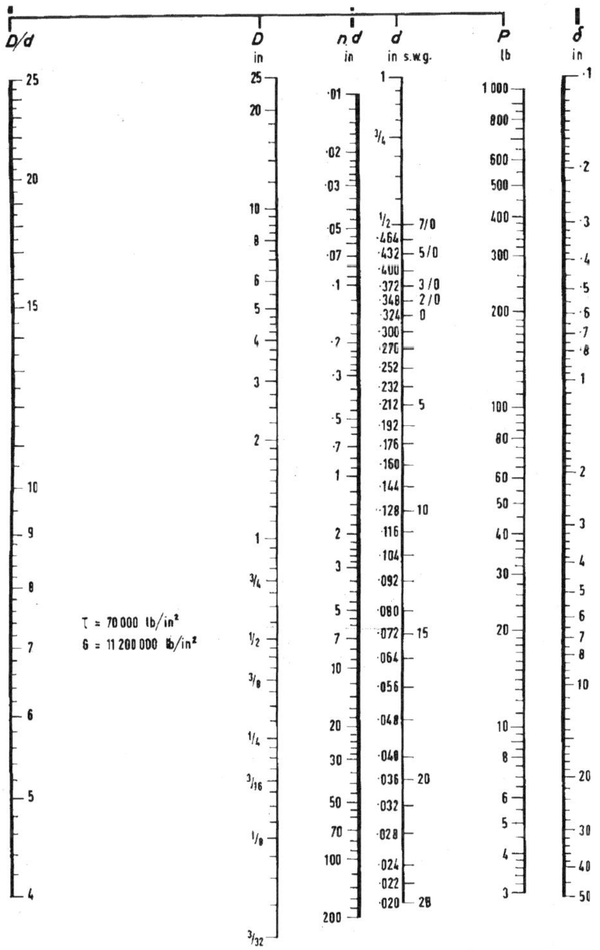

Fig 59: Spring design nomograph

and a combined bending couple

$$T\cos a - (WD/2)\sin a.$$

These couples can be substituted for *Mt* and *Mb* in the expression for total strain energy (resilience) given by

$$Mt.2l/2GJ + Mb.2l/2EI$$

where *J* and *I* are polar and normal moments of inertia for the rod section.

The shear stress in a close-coiled spring of circular section wire can be found from the expression:

$$\tau = 8WD/\pi d^3$$

and the deflection:

$$\delta = nD2\tau/Gd.$$

Fig 59 is a nomograph for these two expressions which is useful in the initial sizing of the spring for design purposes. Maximum allowable shear stress is taken as 70,000 lbf/in^2 (483 MN/m^2) and *G* for steel as 11.2 x10^6 lbf/in^2 (77 GN/m^2). A *D*/*d* index of 9 would be typical. A solution is obtained generally by selecting a point on the *D*/*d* line and tracing two lines from it to the point of known full-load deflection and the point of maximum load — to lead absolute values of *D* and *d*. The fully compressed

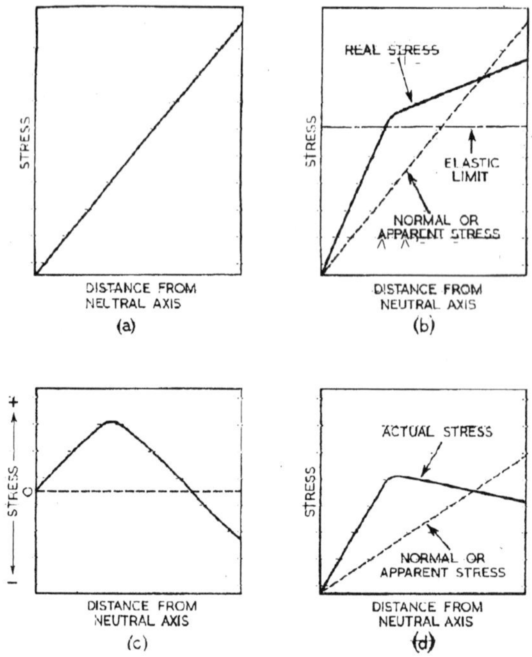

Fig 60: Effect of scragging

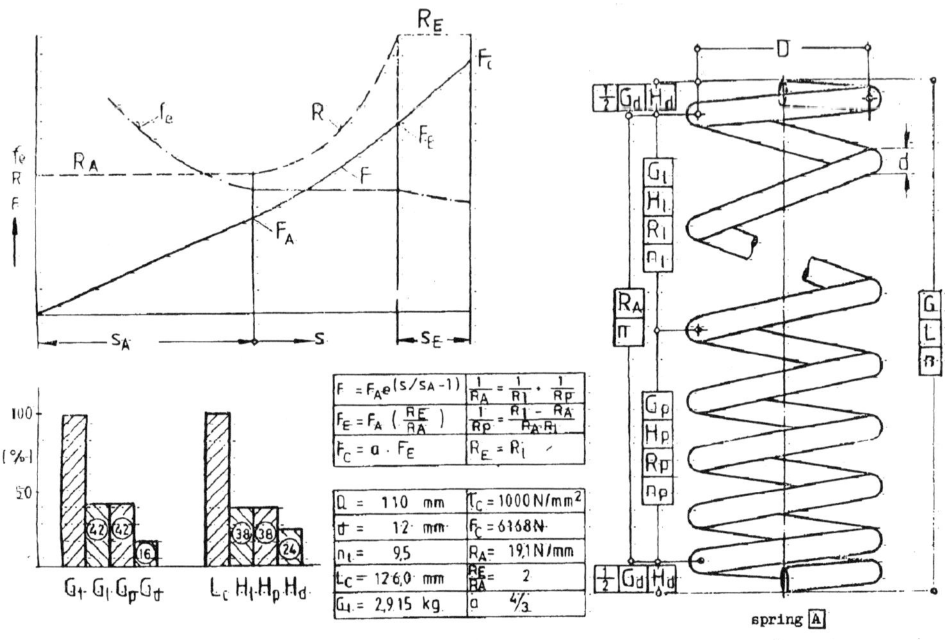

Fig 61: Analysis of progressive rate spring

length of the spring is given, of course, by *nd*. Where spring diameter and deflection under full load are known, graphical methods can also be used to obtain the full dimensions of the spring[3].

The Wahl correction factor should also be taken into account — based on the fact that parts of the outer surface of the spring are more highly stressed than those on the inside of the rod:

$$\{(4r-1)/(4r-4)] + (0.615/r)$$

where $r = D/d$. By specifying a scragged spring, where the spring has been compressed beyond the elastic limit, the stress at any section is no longer proportional to its distance from the neutral axis, the permanent deflection causing a non-linear stress distribution, Fig 60.

Variable coils and tapered rod

A progressive rate spring of variable coil pitch, Fig 61, usually comprises three regions, the end-coils which function as locators without participating in the spring energy, *Gd* and *Hd*; the coils which become inactivated as the compression proceeds until the final rate *Re* is reached,

Gp and *Hp*; coils which are active over the whole compression stroke, *G1* and *H1*. The bar-chart in the figure shows typical percentages for these ranges. Researchers at spring-makers Hoesch[2] recommend plotting for a given spring the calculated stress distribution at Forces *Fa*, *Fb*, and *Fc* from the previous figure as a function of the bar volume, Fig 62. This means that those portions of the rod available for weight reduction are immediately revealed — the size of the hatched area representing what is theoretically achievable. Influential factors for weight

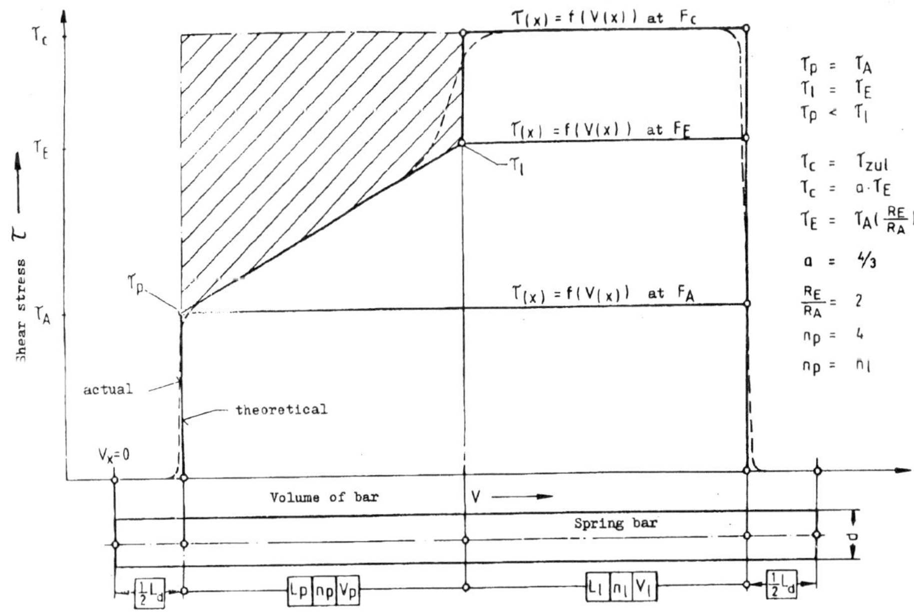

Fig 62: Plot for optimising spring configuration

Weight- and height portions	As a function of	May be influenced by
G_1 H_1	R_E	τ_{zul} τ_{zul}; D or n_1
G_p H_p	R_E/R_A	$\tau(x)/\tau_{zul}$ $\tau(x)/\tau_{zul}$; D or n_p
G_d H_d	R_A	τ_p; τ_{zul}; D; end configuration

Fig 63: Factors for weight reduction

reduction for this example can also be seen in the table of Fig 63. Effect of end coil configuration is seen in Fig 64; 1 and 3 give better results than 2 but higher production costs are involved in the former solutions and 3 requires, like 2, a specially shaped spring-seat. Studies on rod-section profile and variable coil-diameter led the researchers to conclude a preference for large coil diameters in the area of active coils, small ones at the dead end-coils and a variable rod diameter throughout the length of the spring.

Buckling of coil springs

Depending on the mounting conditions applying to a suspension spring installation, the installed length needs to be considered with respect to any tendency to distort sideways under load. A slender helical spring will become laterally unstable, and buckle in the same way as a long strut, when the axial compressive load exceeds a certain critical value. With a helical spring, rigidity in shear and change in length due to loading also have to be taken into consideration. The spring will buckle if its relative deflexion *d/Lo* equals or exceeds the critical value:

$$(\delta/L_o)_{cr} = A\{1 + [1 - B(C/L_o D)^2]^{1/2}\}$$

where, for circular section spring steel, $A = 0.811$ and $B = 6.89$

The values of C for various methods of end fixing are in Fig 65. Results obtained from the above equation are shown in the table of Fig 66 and Fig 67 where $C/(L_o/D)$ is plotted against $(\delta/Lo)_{cr}$. The curve for any type of spring divides the plane of the diagram into two areas, one representing stability and the other instability. To determine whether a spring with a degree of slenderness L_o/D and end-fixing factor C is stable for a relative deflexion δ/L_o, draw the horizontal straight line through the point $C/(L_o/D)$ on the vertical axis and the vertical straight line through the point δ/Lo on the horizontal axis. If these two lines intersect in the area of stability with respect to the curve, the spring is stable and, if it is physically possible, the relative deflexion could be increased to almost $(\delta/Lo)_{cr}$ without instability occurring. If $C/(Lo/D)$ is larger than $(1/B)^{1/2}$, the horizontal line and the curve will not intersect, and the spring will not buckle, however great the deflexion.

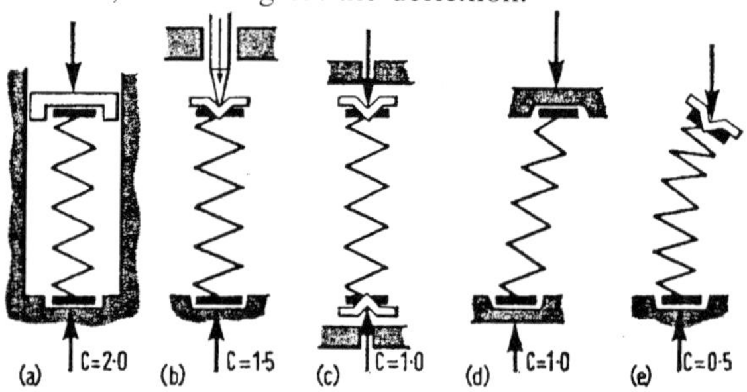

Fig 65: End-fixing factor C for various end conditions: (a) Both ends fixed, laterally guided (b) One end pivoted, laterally guided (c) Both ends pivoted, laterally guided (d) Both ends fixed, laterally guided (e) One end pivoted, laterally unguided

$(\delta/L_o)_{cr}$	0	0·05	0·1	0·2	0·3	0·4	0·5	0·6	0·7	0·8	0·9	1·0
$\frac{C}{L_o/D}$	0	0·133	0·183	0·251	0·296	0·329	0·352	0·368	0·378	0·382	0·380	0·372

Fig 66: Critical relative deflection for buckling

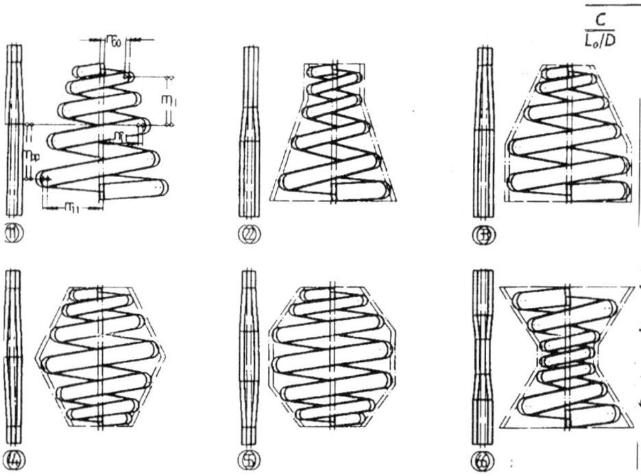

Fig 64: Effect of end-coil configuration: 1 truncated cone; 2 funnel; 3 bell; 4 double cone; 5 barrel; 6 hourglass

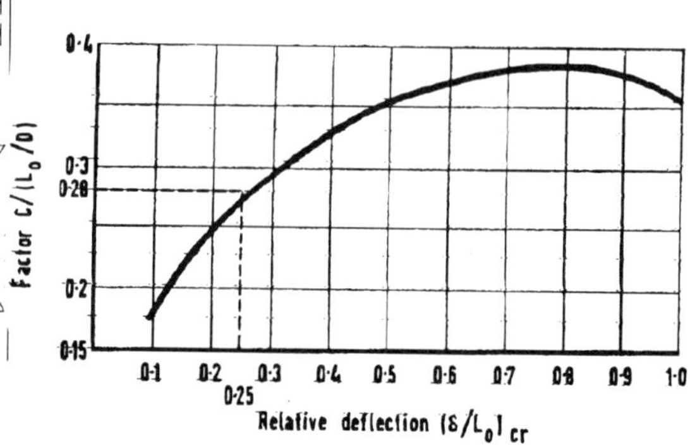

Fig 67: Spring buckling curve

The values of C as given in Fig 65 hold good for perfectly straight springs with strictly plane and parallel faces. However, even with such springs, the line of action of the load may not quite coincide with the geometrical axis of the spring. Although the values of C are on the small side, since they are calculated on the total free length instead of the effective length, it is strongly recommended that they be reduced about a further 20%. Danger of eccentric loading is considerably reduced if the total number of coils is an odd multiple of 1/2.

Fig 67 is strictly correct only for Poisson's ratio $\nu = 0.304$, corresponding to conventional spring steel; it may become necessary to calculate new curves if a spring material has a considerably different Poisson's ratio. In such a case it is useful to remember that the apex of any curve has the abscissa $(\delta/L_o)_{cr} = A$ and the ordinate $C/(L_o/D) = (1/B)^{1/2}$. Although it will suffice for all practical purposes to calculate only the rising parts of the curves (negative sign in equation), it should be noted that points of equal ordinate on the increasing and the decreasing slopes—positive sign in equation —are equidistant from the vertical through the apex. Consequently, the part of the curve with decreasing slope can be drawn without further calculation.

Notation

A, B numerical factors depending on the shape of the material cross-section
C factor depending on the method of end fixing
Lrl total free length of spring
D mean coil diameter
δ deflexion of spring
$L\delta/D$ degree of slenderness
δ/Lo relative deflexion
$(\delta/Lo)cr$ critical relative deflexion
G/E ratio of shear modulus to Young's modulus
ν Poisson's ratio

Auxiliary suspension springs

The difficulty in obtaining progressive spring rates with laminated leaf springs led to the use of so-called helper leaves in commercial vehicle installations in which special pads on the chassis frame contact shorter-than-normal, and higher curvature, leaves to provide a stepped rate; more recent alternatives are discussed here

While helper leaves can be tapered, and even comprise laminated sets, to reduce the severity of the stepping, rubber snubber springs are increasingly used to create a more progressive build-up in rate. The Aeon hollow-rubber spring is a good example of such a system; it incorporates a patented anti-tilt profile to maintain a vertical disposition, without support, to its full deflection, which normally represents one-third of its free-height. The unit can be used singly or in sets to obtain a wide variety of rate characteristics. The units are also available with one to four convolutions on each element to cater for different load ratings as shown in Fig 68.

There is also a measure of hysteritic damping available as seen in a typical rate characteristic, Fig 69 and it is possible to produce specific combined rate effects by using different rate characteristics in parallel. The difference in height between the two units may represent the travel corresponding to laden and unladen conditions of a vehicle, where the taller of the two-springs is effectively the 'sole-suspension' for empty-running operation. As indicated in Fig 70, further Aeon units can be added outboard of the main springs to obtain anti-roll stiffness adjust-

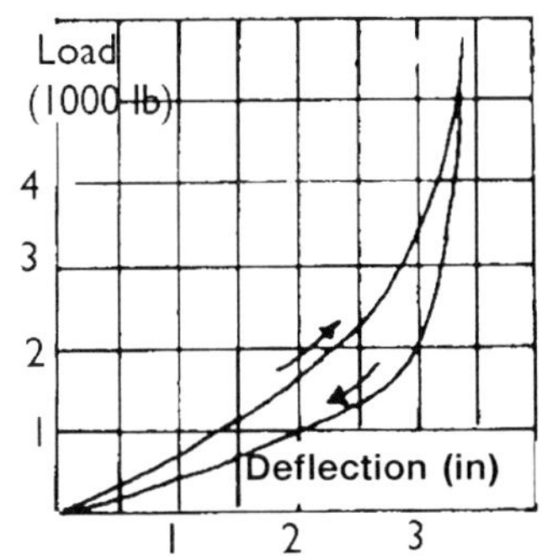

Fig 69: Hysteritic damping

	ONE-			TWO-CONVOLUTION													THREE-			FOUR-		
Height (in)	2·12	3·56	4·75	3·12	3·12	3·75	3·94	4·75	4·75	4·75	5·12	5·12	5·19	5·19	3·75	3·75	2·38	9·25	9·25	4.63	4·63	4·63
Outside dia. (in)	2·12	4·25	5·56	2 50	2·50	3·38	2·38	3·38	3·38	3·38	3·94	3·94	4·31	4·31	3·75	3·75	1·38	5·31	5·31	2·63	2·63	2·63
Bump load (1000 lb)	0·55	5·0	10·0	1 0	0·7	1·5	0·87	2·0	2·2	2·6	4·5	6·0	6·8	8·0	5·4	5·0	0·45	13·4	13·4	0·72	1·0	0·9
Full deflexion (in)	1·25	2·0	3·0	2·0	2·12	2·5	2·38	3·38	3·0	2·75	3·44	3·38	3·31	3·0	2·62	2·88	1·38	5·50	5·50	2·62	2·50	3·1

Fig 68: Aeon specifications

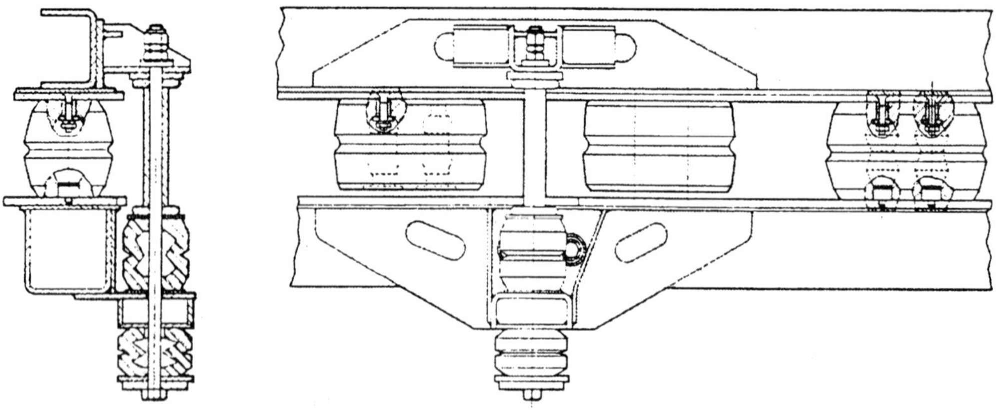

Fig 70: Aeon springs in combination

able to specific operation by means of the adjustable lock-nut.

Recently, Kemira Polymers have developed a grade of polyurethane elastomer known as Autothane which is said to have particularly advantageous properties for use auxiliary springing. The material has a closed-cell structure which provides a useful flexing characteristic and again, by building up convolutions, different progressive rates can be provided for different mixes of the elastomer.

Configuration of a spring is shown as Fig 71 while Fig 72 shows how reasonable rate conformity is maintained over considerable temperature differences. The diameter of a typical spring-aid increases only slightly when compressed: 10% at 40% compression and 35% at 80% compression. Density can be varied between 350 and 650 kg/m^2 to give a yet wider span of performance characteristics.

Spring rate modifiers

Where the helper spring only affects the main spring rate to a small degree the term rate modifier is appropriate, exemplified by the anti-roll bar bushing assembly shown in Fig 73. Here sliding rubber surfaces are used, rather than the trapped rubber of a suspension anchorage, to allow complete torsional freedom without compromising the flexing properties of the rubber. The specially compounded layer of hard rubber acts as a bearing and sealing surface while the more resilient backing provides the required flexibility. A fully bonded anchorage can also be used as a rate-modifier of the type seen in Fig 74.

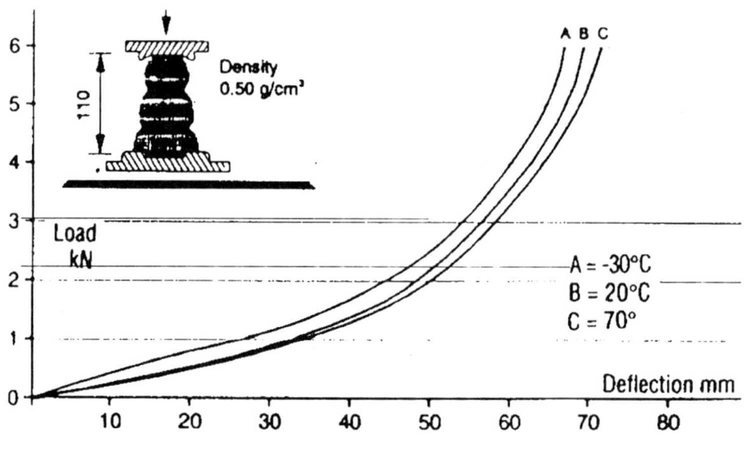

Fig 71: Autothane spring onfiguration

Fig 72: Autothane characteristics

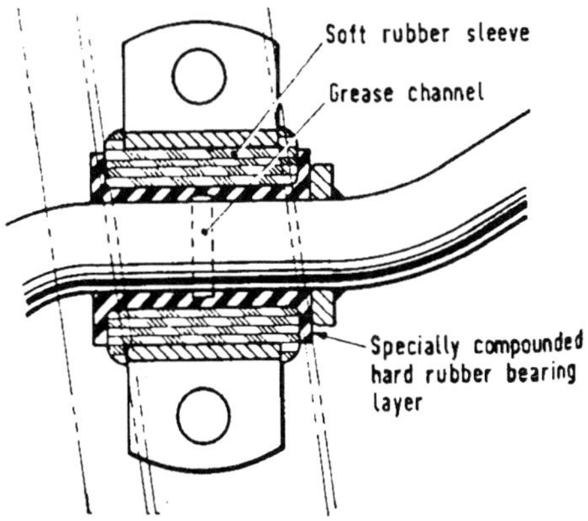

Fig 73: Anti-roll helper

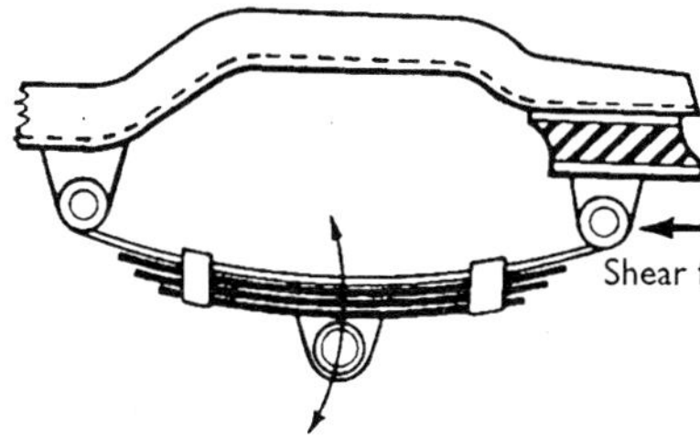

Fig 74: Fully-bonded anchorage

This uses a rubber sandwich mounting to dispense with the conventional spring shackle. Near the flat, normal-ride, configuration of the top-leaf, small vertical motions of the axle will be assisted by the shear force in the mounting, tending to buckle the spring and lower its vertical rate in this region.

A more sophisticated form of this device is the Contrasonic shackle (Fig 75), for leaf springs by Metalastik. Careful selection of the mounting angle and the right amount of rubber meets a three-fold requirement: fore-and-aft and torsional flexibility; good sideways control and sound insulation. On one upper-medium size car installation (Fig 76), the system gave vertical deflection of 3/8 in (95 mm) from no-load to full static load condition, the deflections of conventional shackles being in the order of 0.02-0.04 in (0.5-1 mm). Shear deflection was 1.375 in (35 mm) from fully laden to full-rebound condition and torsional deflection was 28 degrees. The units can also be configured and fitted to provide an S-shaped variable-rate spring characteristic which depends on the installed camber of the leaf spring, Fig 77.

Tunesi recently launched the Flexstab spring reinforcement blade made from a 60% unidirectional glass and 40% epoxy resin GRP mix. On mounting above the top link of a laminated leaf spring suspension, the main spring can work at close to its normal ride height operation even at substantial overload. The auxiliary leaves are clamped to the top leaf, Fig 78, by U-bolts and very little weight penalty is involved.

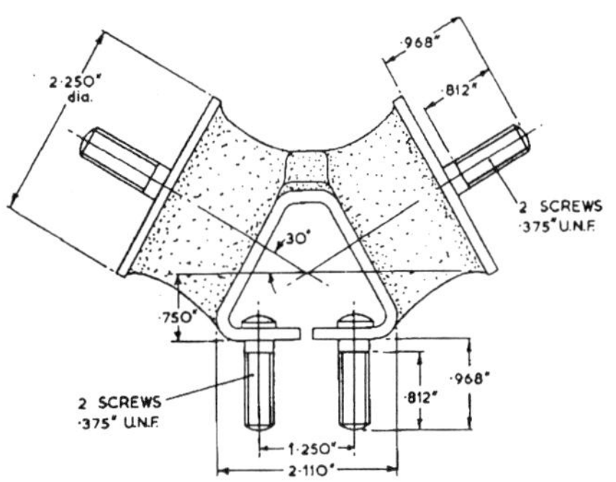

Fig 75: Contrasonic shackle

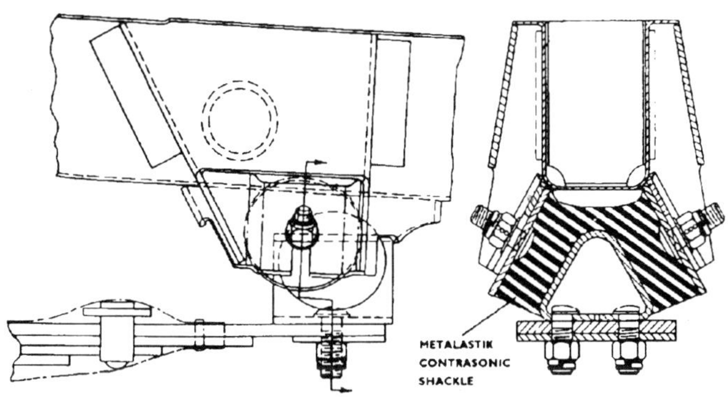

Fig 76: Upper-medium car installation of the Contrasonic shackle

Active suspension systems

Because of the complexity and high-energy requirements of the first active suspension systems, developed for racing cars, a number of intermediate systems have been developed for other vehicle categories.

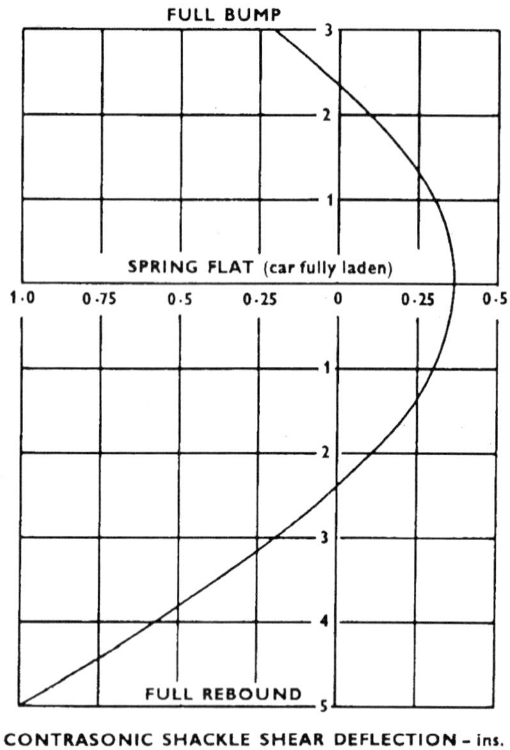

Fig 77: Variable spring rate in terms of axle deflection in inches (y-axis)

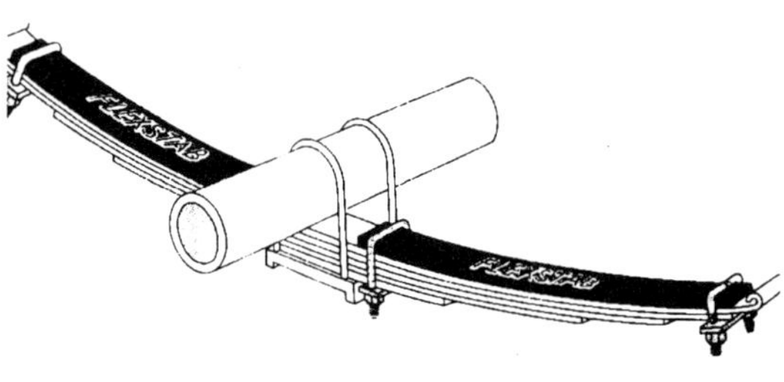

Fig 78: Flexstab helper blades

In fully active suspension a "spring-rate" and "damper characteristic" do not exist in the conventional sense, the spring and damper being replaced by an actuator tuned to maximise ride motions. The vehicle becomes analogous to a skier whose knee angle and leg muscles react to minimise body disturbance, absorbing movement of the skies over ground irregularities by control of the reaction forces. Normally a hydraulic actuator adjusts the forces, in the vehicle case, and a control valve implements the decisions of a computer.

While optimising ride the system is still compromised by the different requirements of the suspension for vehicle handling, considerable engine power is needed to carry out corrective control of body motions. Weight and cost are also likely to increase with fully active systems. Intermediate systems between full-active and full-passive have thus been developed.

Limits on passive systems have been set by lower limits of spring softness, causing sickness at ride frequencies below 0.75 Hz and attitude changes resulting from variations in load. Also the damping control of the unsprung mass can only be obtained by the larger sprung mass. What is required is instantaneous variation in spring rate and/or damping force as the wheel follows the contour of the road disturbance. But the need to maintain contact between tyre and road sets a limit on the amount of rebound damping that can be used.

Use of softer springs, with standing height control, is one of the first steps towards activity in suspensions but, as now generally applied, its speed of operation has to be restricted to avoid ill-timed intervention. Also, reductions in energy transfer to the sprung mass can be achieved by active control of damping so as to reduce it when disadvantageous and increase it when beneficial — at the same time providing a fast enough response for the standing height control to cause undue pitch or roll under longitudinal and lateral accelerations/deccelerations. Active control of damping requires a small electrical power input while height control by hydro-pneumatic cylinders requires moderate input to the oil pressure pump if response is to be quick enough.

The Armstrong Active Suspension Control system is an example of intermediate active suspension. It offers two levels of damping with electro-mechanical switching between the two. The twin-tube damper, Fig 79, actually has a third tube to enclose a gas spring which can act on its own or in conjunction with a lower than usual main spring rate. This spring achieves height control usually only at the rear of the vehicle. Sensors in the spring-damper units provide data, to the controller, on suspension movement while others feed in data on speed, steering, braking, acceleration and door-opening, Fig 80.

On encountering a surface undulation causing the springs to deflect beyond a set limit the computer will signal the dampers to set hard, so as to arrest the body movement arising from the input force. If it is an isolated undulation the dampers resume the low setting and vice versa. As speed increases, the limit on allowable body movement is reduced commensurate with payload. Sudden steering movement will also signal the dampers to set hard, again dependent on speed and load, Fig 81. As spring load and rate is automatically adjusted to the load being carried, a near constant ride frequency results, Fig 82.

Slow-active and semi-active systems

Alternatives to switched damper systems are those working on the so-called `slow-active' principle. Here some form of actuator is provided in series with the `passive' suspension spring. Because the actuator bandwidth need only be up to 3-4 Hz the systems are also sometimes called `narrow bandwidth'. This also means the systems can be cheaper and less complex than fully-active ones.

An example is Citroen's hydro-pneumatic system in which a proportional flow control valve controls the volume of hydraulic fluid, Fig 83. As hydraulic flow requirements are always at high pressures, around 50 bar, a method of reducing power is required so that the actuator need only provide dynamic fluctuations. Fig 84 shows a possible solution in which implementation is theoretically possible using only pneumatic components.

In semi-active systems energy dissipation is controlled in the most effective way but there is no mechanism for putting energy into the suspension. Control over energy dissipation is achieved by a damper which is capable of tracking a force demand signal, independent of the velocity across it. One method is to use a motor-controlled variable orifice.

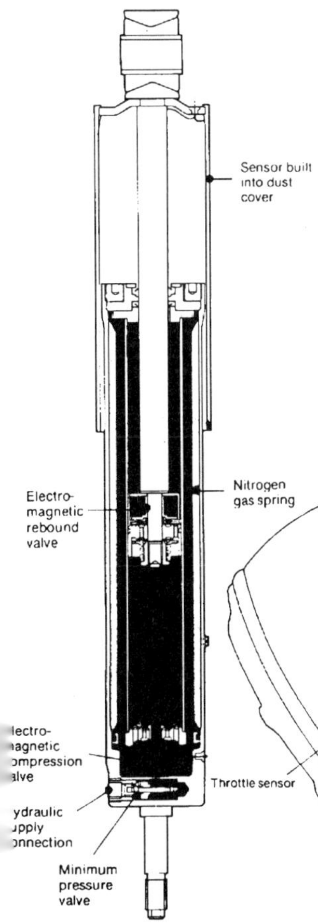

Fig 79: Twin-tube damper

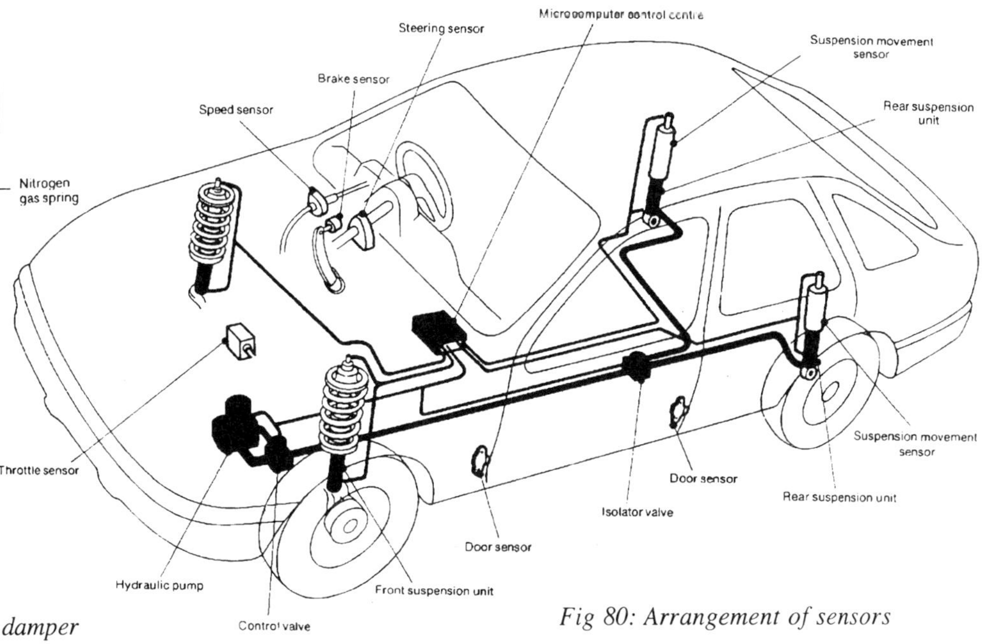

Fig 80: Arrangement of sensors

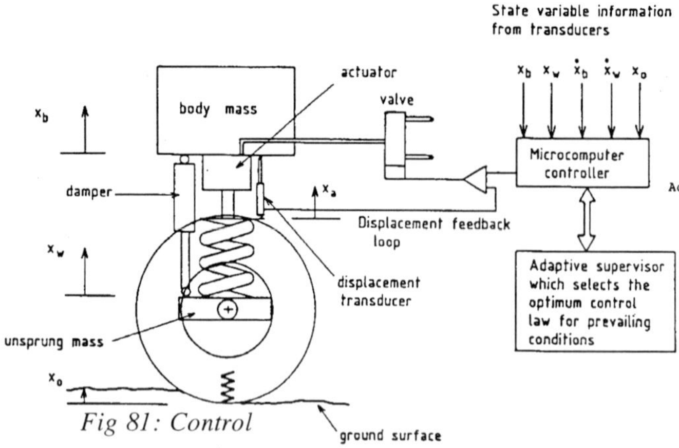

Fig 81: Control arrangement

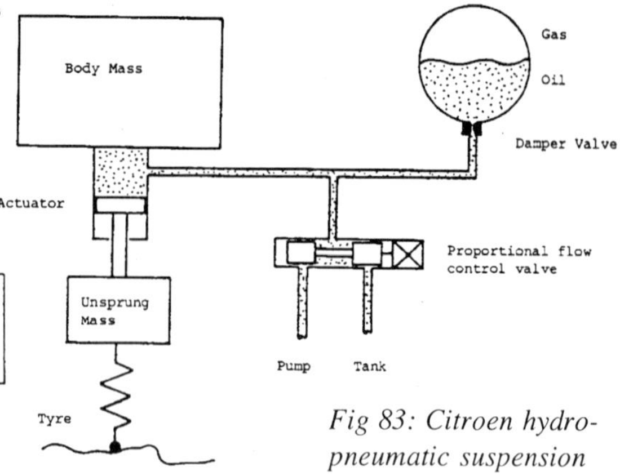

Fig 83: Citroen hydro-pneumatic suspension

Recently-released low-bandwidth suspension

The Anthony Best Dynamics suspension is based on oleo-pneumatic suspension units incorporating a flexible diaphragm that rolls between a tapered piston and the tapered skirt, Fig 85. The tapered piston is connected to the lower suspension arm via a low friction rolling joint. At its upper end, the piston supports the inner bead ring of the rolling diaphragm, and the outer bead ring is held by the pressed steel casing. The diaphragm forms the oil seal at the base of the displacer chamber, at the top of which is the damper pack. Above the damper is the gas spring, an egg-shaped chamber which contains oil below a flexible separator with nitrogen above it.

As the road wheel moves upwards on a bump, oil is displaced by the rolling diaphragm through the damper valve and into the gas spring. Spring rate is a combination of four factors. One is a hydraulic, due to the compression of the gas; another is due to the taper whereby pressure of the oil acting on the changing area of the diaphragm causes increasing force as the piston moves relative to the casing. There is also a parasitic rate element due to deflection of the rubber bushes and an element due to drop-angle involving change in leverage as the wheel moves relative to the vehicle body. Their relative contribution is shown in Fig 86.

Damping is achieved with disc-type valves to obtain the characteristic shown in Fig 87. The design of the gas spring is controlled by the performance requirements seen in Fig 88. Considering the separator at mid-load condition (d), as the temperature decreases, and gas contracts, oil will have to be pumped in to maintain vehicle ride height (e). As vertical load increases oil will also have to be pumped

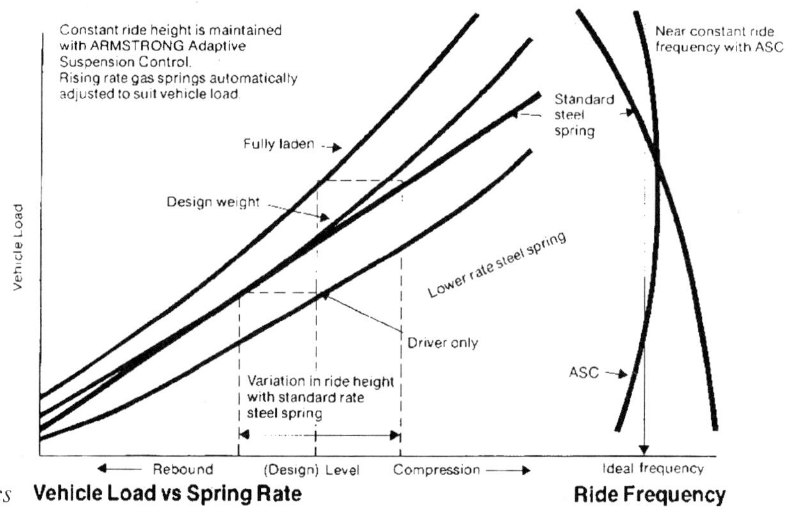

Fig 82: Rate characteristics

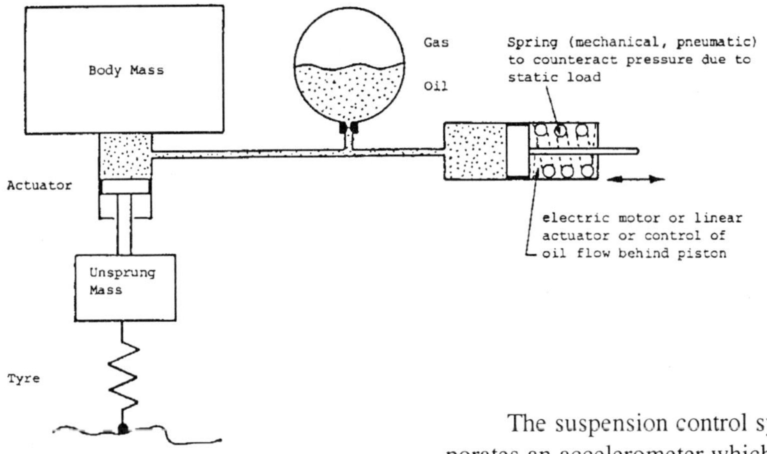

Fig 84: Method of reducing power required

in to maintain trim (f). If a road-bump is encountered the wheel moves to full-bump (g). Load decrease or temperature increase cause the separator to move in the opposite direction and (a) and (g) define maximum volume and pressure for the spring.

As static pressure at normal-ride height is a function of vehicle weight, leverage and unit area, the main parameters can be co-related as in Fig 88, dashed lines. Also shown are the required gas spring volumes. For one design installation the chosen leverage was 3.5 front and 3.4 rear. The front installation, Fig 89, shows the unit inclined outwards, with the gas turnover flange resting on the sub-frame. The rear installation comprised a moving unit with piston directly connected to the lower suspension arm and the turnover resting on a conical cup that pivots on a roller joint at its upper end, a technique for minimising bending moment on the lower arm, Fig 90.

The suspension control system, Fig 91, incorporates an accelerometer which measures body vertical acceleration and a displacement transducer measuring distance of the body from the suspension. Signals from these are fed to the controller which governs the quantity of oil supplied by an engine-driven swash plate axial piston pump to maintain system pressure, with peak flows being met by the accumulator. Output from the controller is fed to the three-way flow-control valve at each wheel station.

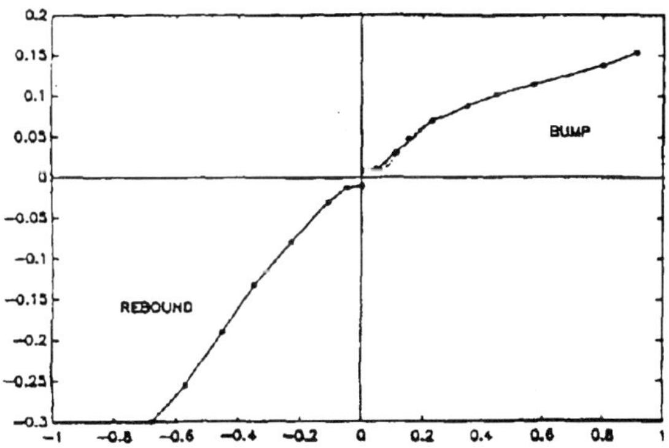

Fig 87: ABD suspension characteristic — body acceleration (g) in terms of wheel velocity (m/s)

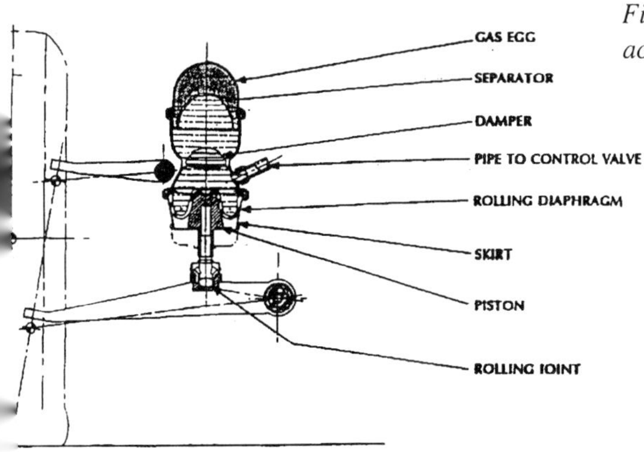

Fig 85: ABD active suspension

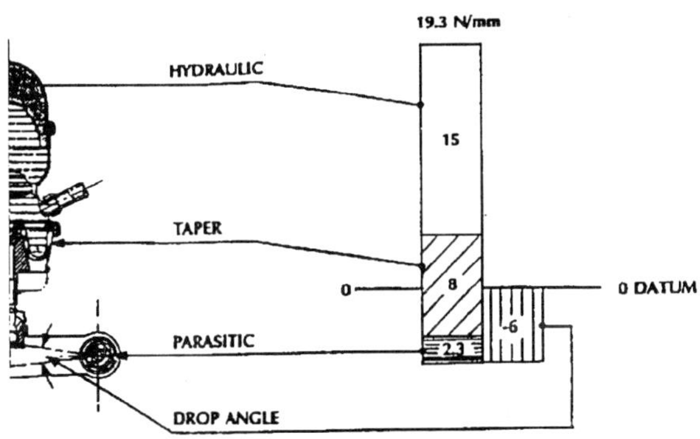

Fig 86: Relative contribution of factors

Full active suspension

The previous section dealt with a number of 'intermediate' systems such as slow-active, semi-active and low-bandwidth. Here the full active versions are considered and the more general case argued for the improvement over passive systems.

An early example of active suspension was the proprietory Automotive Products system shown in Figs 92 and 93. According to researchers at Bolton Institute of Technology, such systems were satisfactory to one specific vehicle but could not adequately be adapted to different vehicles without redesign of the system. It is argued that a specific advantage of active supensions is the ability to apply mathematical design procedures for which passive systems are not amenable. Rather than analysing pre-defined configurations, as in the passive case, with active systems a synthetic approach to design is possible which allows the desired dynamic characteristics to be assigned before the suspension is constructed so that the most effective configuration can be realised.

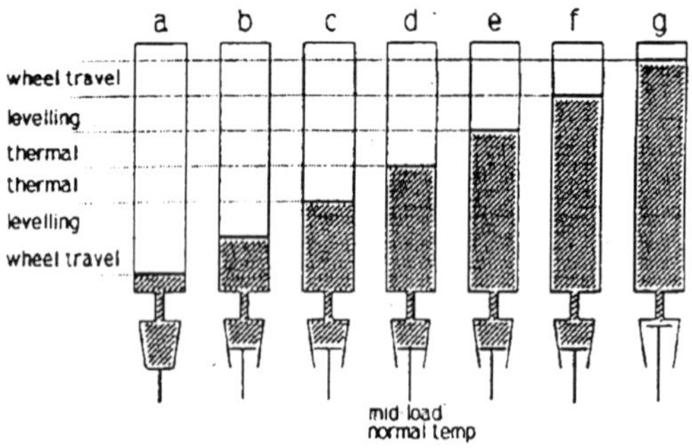

Fig 88: Performance requirements

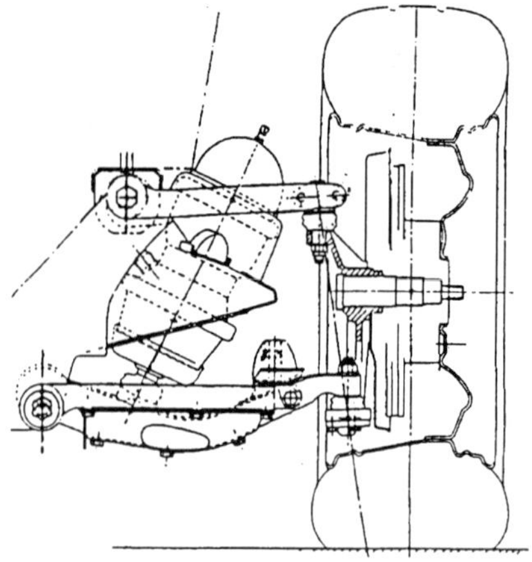
Fig 89: Front installation

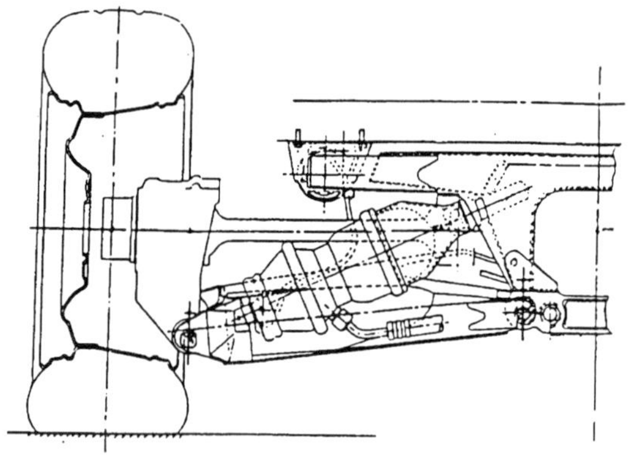
Fig 90: Rear installation

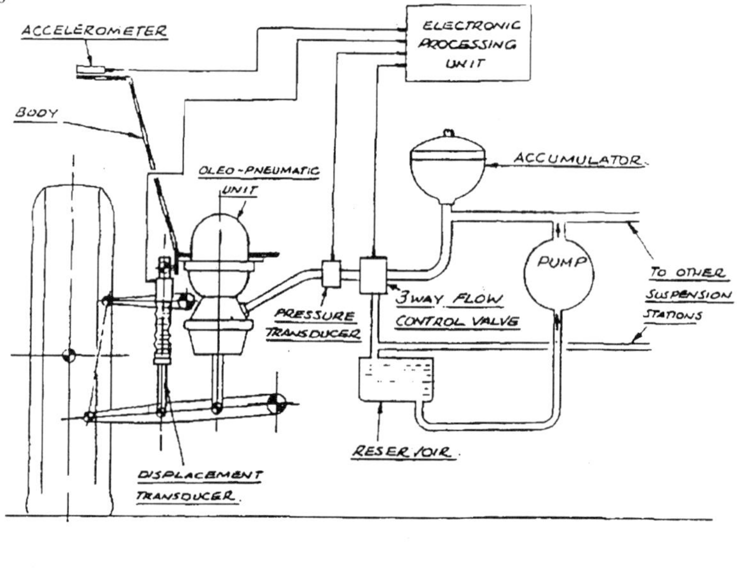

Fig 91: Control system

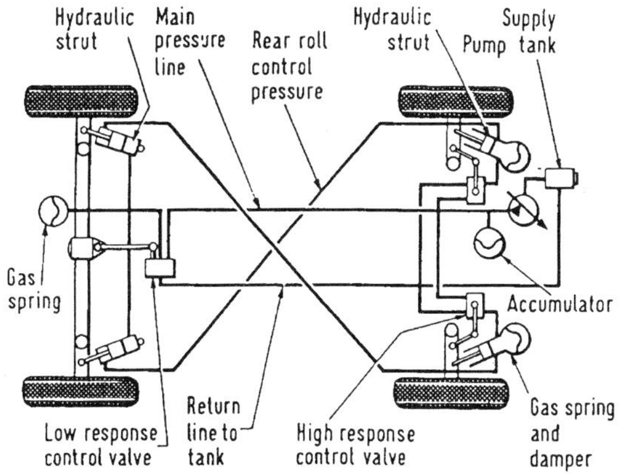

Fig 92: AP active system: physical layout

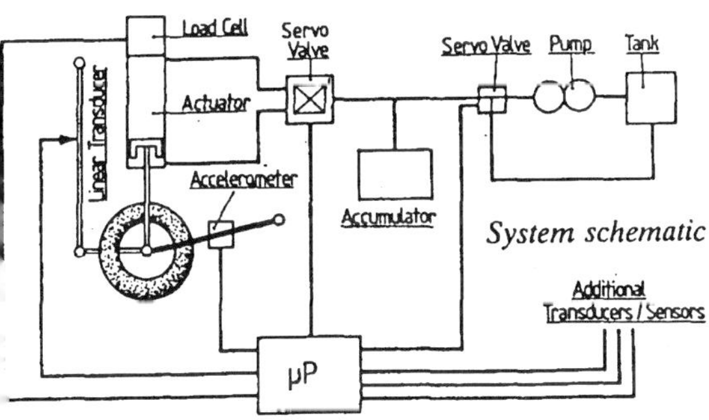

Fig 93: Control layout of AP system

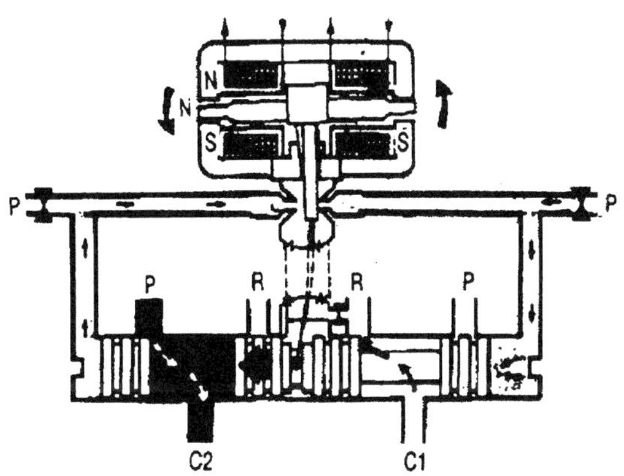

Fig 94: Operation of Moog servo valve

This circumvents the compromise of having to select components which might satisfy the degree of isolation required but result in undue static and dynamic wheel displacements. By applying closed-loop control theory to active suspensions, a performance index can be minimised for suspension behaviour with respect to such variables as acceleration, velocity and displacement. The technique involves minimising a function of one or more of these variables to obtain optimum characteristics. However, it is argued that there is still difficulty in deciding the representative form of performance index and in choosing the correct variables.

An alternative approach is to use modal theory in which an open-loop suspension is initially defined and operated in such a way that its control parameters are replaced by assigned ones such that the progressively modified system possesses a prescribed dynamic response. A control law can thus be formulated which indicates the required variables to be measured for feedback purposes and the magnitudes and polarities of the feedback gains.

In fully-active suspensions, an important component is the servo-hydraulic valve. The one made by Moog for the Lotus system is the so-called E050 series. Its function is to convert electrical signals from the controlling computer into hydraulic flow to the suspension struts. High levels of amplification are required to convert 64mW of electrical power to 25kW of hydraulic power, and very fast levels of response are exemplified by the need to typically enable 95% change in flow in 5ms. High sensitivity is a further factor, with a step change of just 5% being typically required to alter valve output flow. Fig 94 shows the way electrical control of the spool valve takes place. Electrical input moves the flapper valve to shift the spool as required for directing flow appropriately into ports C1 and C2.

Racing car active suspension

In the Lotus system the layout of the hydraulics is as shown in Fig 95. The transducers used for control include four on each actuator for displacement, load and wheel hub travel; one each for centre-of-gravity longitudinal acceleration, lateral acceleration; one for hydraulic supply pressure and one for flow rate. The control system, Fig 96, involved a micro-computer to regulate hydraulic pressure. Another feature was included to control car attitude dynamically. A car can thus be programmed not to respond to centrifugal force and even to dive under both braking and acceleration if required. Prototype testing led to an hydraulic pump capacity of 16 litres per minute being required, matched to actuators having a cross section area of 645 mm^2. A Dowty gear pump was chosen to run at a maximum of 4000 rpm and pressure control was achieved by a Dowty 4651 electro-hydraulic servo valve.

Development dated back to the need for the ride of a ground effect racer to be improved after the banning of the twin-chassis Lotus 88 F1 car. The first system ran in a Turbo Elite road car in 1982 and was designed to overcome the limitations of conventional suspensions with which there was an inherent inability to match inertias and effective stiffnesses. These constraints are removed by the provision of double-acting hydraulic actuators. By measuring instantaneous load and displacement, and by suitable processing and control, the actuators become synthetic springs whose characteristics can be controlled by a microprocessor. In the system schematic of Fig 97, for one wheel, the basic mechanical/hydraulic set-up is shown.

With the racing car version no fewer than 40 parameters can be controlled from the cockpit, using a hand-held computer. Principal variables are modal frequencies and damping; attitude rsponse to dynamic loading and control of cornering stability by altering front/rear roll stiffness ratio.

Off-road vehicle active suspension

One of the most interesting applications of active suspension has been to the NIAE-developed Farm Transport Vehicle. The analytical model used in its development, Fig 98, has ten degrees of freedom, body vertical roll and pitch, left and right front whe vertical , rear axle vertical, and roll, and three angle of the control pendulum. Response of the A suspension controller simulated on the comput model was as seen in Fig 99 for a step input of 80 kg static loading increase from 0-3 seconds of forwar travel, compared with passive and unsuspende configurations. The semi-active suspension reduce

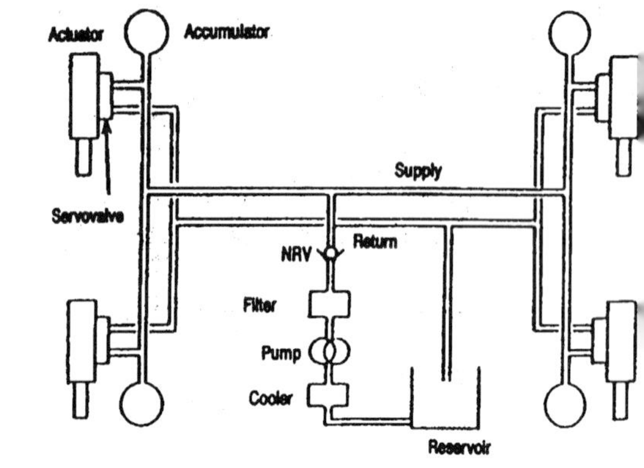

Fig 95: Lotus system: hydraulic layout

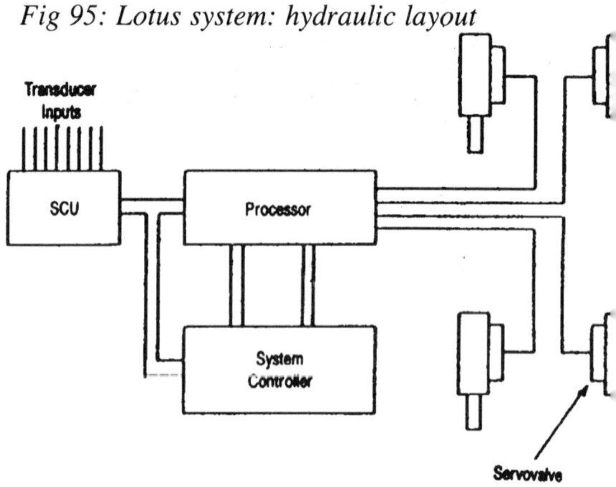

Fig 96: Lotus system: control arrangement

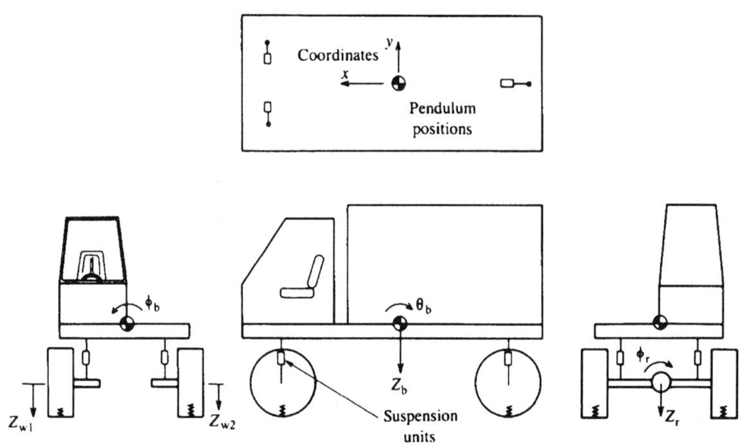

Fig 98: NIAE system model

the static deflection so that only the tyre deflection remains and it has fully responded to the input within 0.5 seconds. Similar response curves can be simulated which show the effect of pitch and roll disturbances.

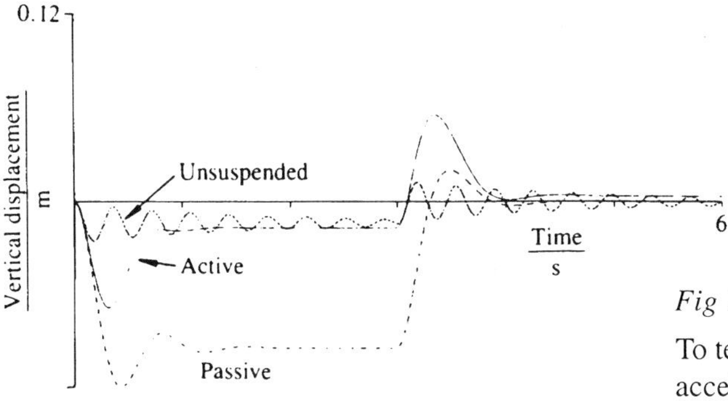

Fig 99: Response of NIAE model

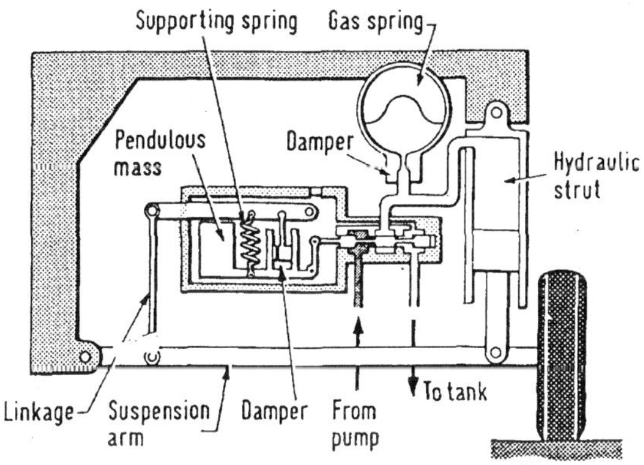

Fig 97: Mechanical/ hydraulic layout

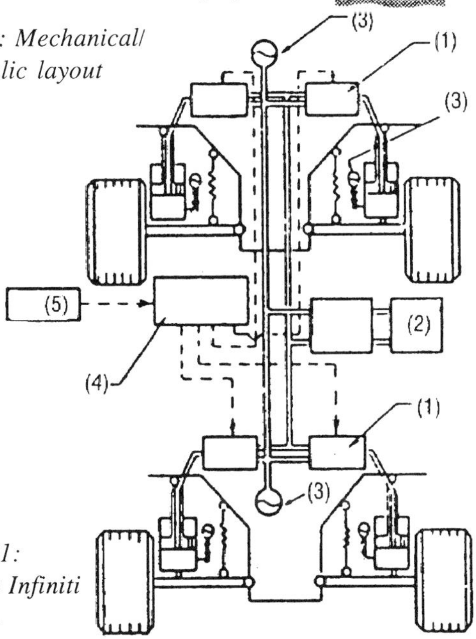

Fig 101: Nissan Infiniti system

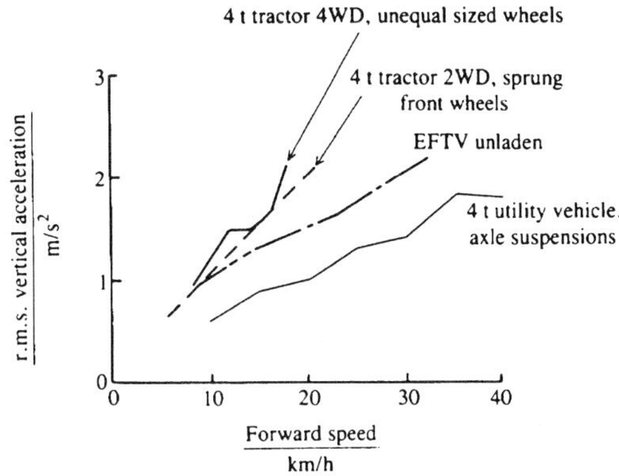

Fig 100: Response of NIAE system

To test the actual vehicle responses, six strain-gauge accelerometers were fitted to a tee-piece fastened under the driver's seat and connected to signal conditioning and recording equipment. Results for traversing an off-road track are shown in Fig 100.

Luxury car active suspension

The use of so-called 'sky-hook' dampers is illustrated in the Nissan Infiniti Q45 system, which is said to improve vehicle stability at high speed, Fig 101. There is also an arrangement for frequency-sensitive pitch and roll control. Three vertical accelerometers, a longitudinal one and three lateral ones, signal the controller to regulate the pressure in the hydraulic cylinders at each wheel station. (1) is the feedback-type pressure control valve, (2) the radial plunger pump, (3) the free-piston accumulator, (4) the controller and (5) an accelerometer. The skyhook damper generates a force to contol vertical body motions, proportional to body velocity in free space, determined by integration of the accelerometer signals. Cornering roll and pitching is suppressed by a counter movement created by differential pressure in the inner and outer wheel cylinders. Since handling is affected by the cornering force developed by individual tyres, related to vertical load on each wheel, fluctuations are monitored by the controller as a function of body vertical acceleration. The active system demeonstrates substantially reduced tyre load variation, and hence yaw-rate change, over a passive system. The hydraulic control valve acts as the damping restrictor for the fluid and the damping constant can be varied according to road input frequency. The system can also use roll control to vary longitudinal variation in anti-roll moment according to vehicle condition.

Active suspension with preview

While the previous sections have examined full and part active suspensions, here the added preview facility has been considered, based on reports of work carried out at Cranfield University and Renault.

It is possible to simulate the feedback laws required for an active suspension with preview according to researchers at Cranfield University[6]. Full bandwidth systems, it is argued, offer improvements in ride and handling but may suffer from other problems such as high power consumption and noise, vibration and harshness (NVH) problems. A good active system mainly gains over a conventional passive one by damping the body resonance much more effectively. It cannot do much better than the passive system higher up the frequency range. Slow-active systems can also offer significant improvements in the control of body resonances, with wheel-hop resonances being dealt with by a suitable choice of damper rate. Slow-active systems incorporate an actuator of limited bandwidth (around 5 Hz) in series with a passive spring, the combination being in parallel with a passive damper. This type of system offers advantages over full bandwidth active systems in the form of reduced power consumption and lower performance components (and hence lower capital cost) due to the lower frequency operation required, and they have been preferred for production application. In fact, the series actuator and spring are realised commercially by gas springs and electro-hydraulic elements.

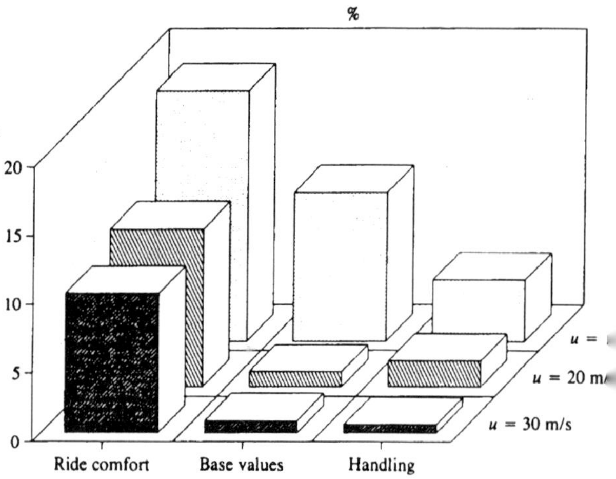

Fig 103: Improvements in cost function through wheelbase preview

The use of preview information with a full bandwidth system can give an improvement in vehicle suspension system performance. Preview can be of two types: (a) wheelbase preview, where knowledge of the front wheel states is used to improve performance at the rear of the vehicle, and (b) look-ahead preview, where a sensor of some description is used to measure the road profile ahead of the front

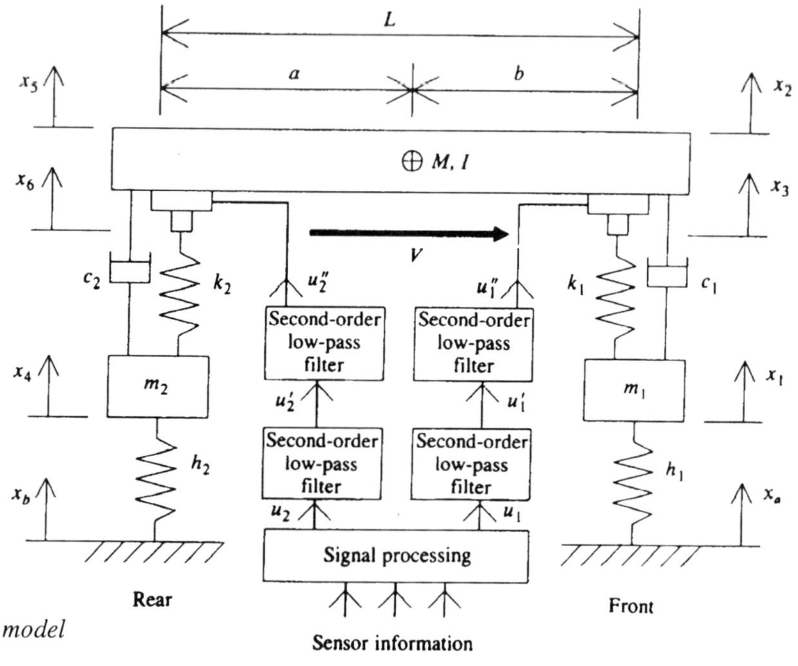

Fig 102: Half-car computer model

wheels. Both types of preview can give substantial improvements in performance, depending to some extent on the control scheme adopted. Discrete optimal control theory has been used to establish ultimate possibilities while the results of other work has shown that restriction to state feedback control only reduces performance capabilities a little (in the case of unlimited bandwidth systems). No corresponding information has yet been published for limited bandwidth systems.

A recent paper presented by the Cranfield team contains the results of an investigation into the performance potential of a limited bandwidth system with wheelbase preview. Control is restricted by assuming only state feedback and good control laws for the preview case are found using the Simplex method. The non-preview (uncorrelated input) case, calculated using linear quadratic regulator (LQR) theory, is used as a starting point for these calculations. The amount of preview is determined by the vehicle speed and its wheelbase. The effect of changing the amount of preview by varying the vehicle speed is investigated, as is the effect of placing the emphasis in the control scheme on either ride comfort or vehicle handling. Frequency response results are presented for a passive vehicle and for a slow-active system (with and without preview) for different vehicle speeds.

The half-car computer model is shown as Fig 102, road input being an integrated delta-function (that is, a unit step). A sub-optimal control scheme makes use of wheelbase preview, accounting for the fact that rear wheel input is a time-delayed version of front wheel input. The limited bandwidth nature of each actuator is represented in the model by including two similar second-order low-pass filters in series with a perfect actuator, restricting it to dealing with lower frequency body modes.

Active suspension simulation

Improvements in the performance index deriving from wheelbase preview are shown in Fig 103. These results show that at all vehicle speeds there is a significant advantage to be gained in ride comfort by taking into account the preview information available for the slow-active model, but little to be gained in handling performance. The improvement is greater at lower vehicle speeds, since the lower the vehicle speed, the longer the preview time. This result could have been expected because of the limited bandwidth of the actuator. Body resonances which occur in saloon cars at frequencies around 1-1.5 Hz can be dealt with by an actuator with a bandwidth of about 5 Hz, but wheel-hop resonances, which typically occur at around 10 Hz, are out of the range of effective control of the actuator.

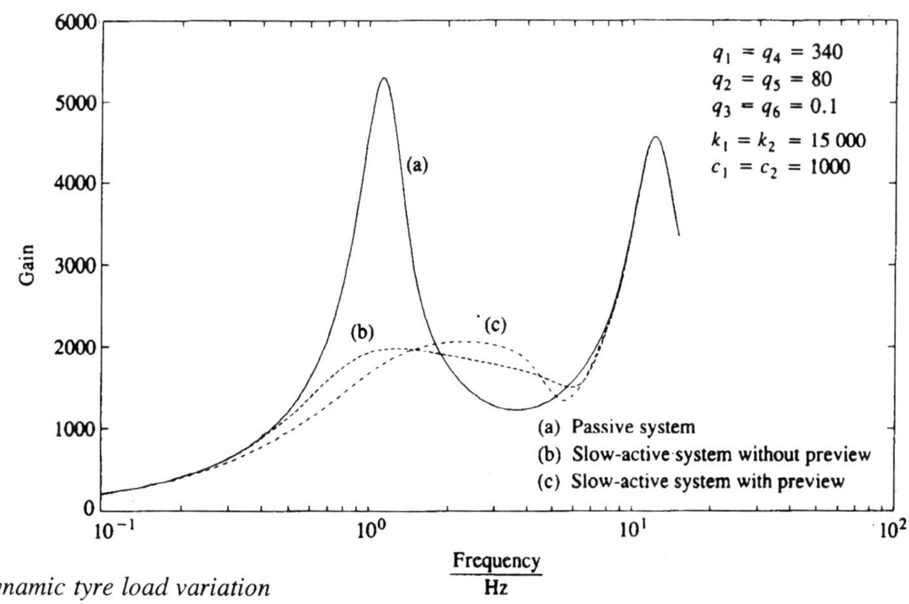

Fig 104: Front dynamic tyre load variation

Several simulations were carried out at vehicle speeds between 30 and 5 m/s, corresponding to different amounts of preview. The improvements in the cost function from the uncorrelated (non-preview) case to the correlated (wheelbase preview) case increased as the speed was reduced to around 10 m/s, but there was no further improvement available on reducing the speed below 10 m/s (0.26 s of preview). This shows that the point of diminishing returns lies at around a quarter of a second of preview, a figure close to those obtained for full bandwidth systems. Fig 104 shows a typical frequency response function, for a vehicle speed of 30 m/s. The graphs indicate that there is little difference between the three cases at higher frequencies (around wheel hop resonance) because of the limited bandwidth.

The dynamic tyre load variation is little altered at the front of the vehicle, since the preview information is only able to exert a significant influence on the performance at the rear. Rear dynamic tyre load variation in the preview case is improved over the non-preview case up to about 5 Hz, while the rear suspension working space response deteriorates up to about 2 Hz. At lower vehicle speeds (around 10 m/s) the rear suspension working space is improved between about 0.5 and 3 Hz, but is still worse below 0.5 Hz. Bounce acceleration is generally improved up to 5 or 6 Hz, but the improvement is less marked at lower vehicle speeds (below 15 m/s).

The pitch angle response is worse below about 2 Hz, although at lower speeds it is improved, except at very low frequencies below about 0.5 Hz. The fact that the improvements in the bounce and pitch angle responses are dependent on the vehicle speed is related to the fact that at different vehicle speeds the

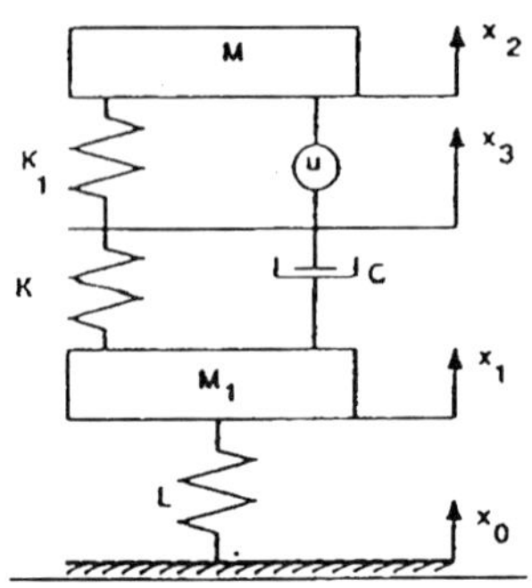

Fig 105: Model of CADHY

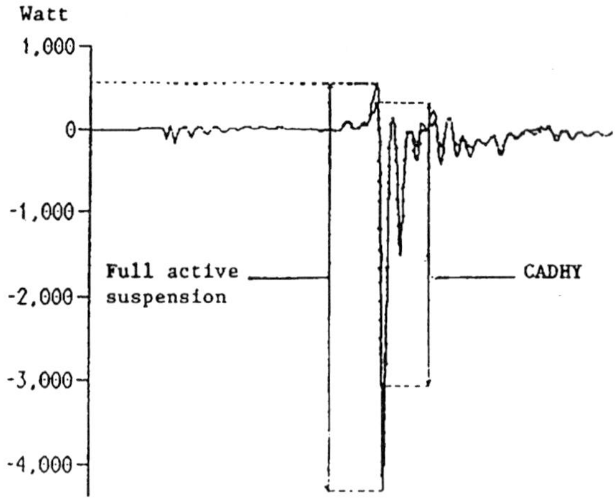

Fig 106: Comparison with full-active system

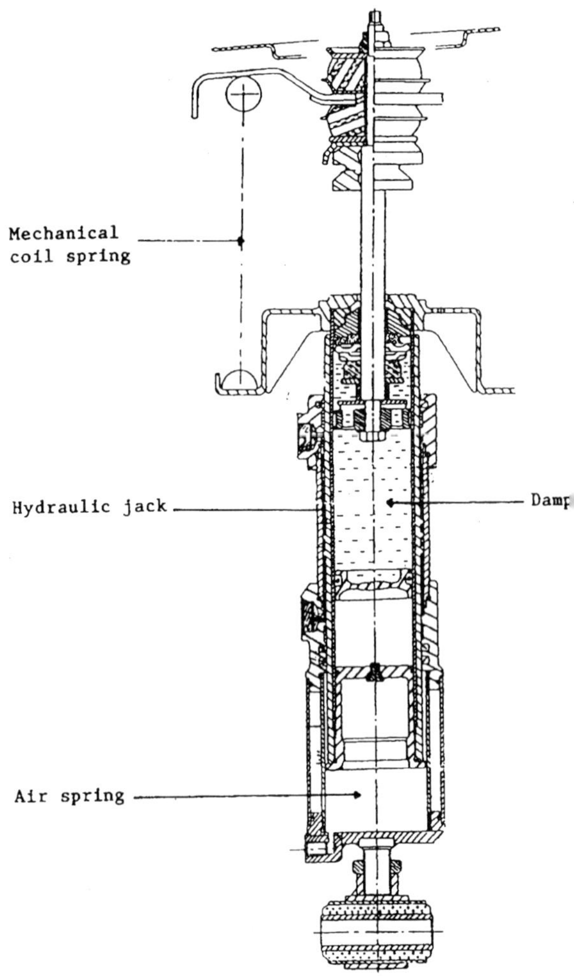

Fig 107: Physical set-up

phase angle between the front and rear inputs varies. Hence, if there is a whole number of wavelengths between the wheels, the inputs are in phase and the bounce acceleration response will be worse than if the inputs are out of phase, in which case the pitch angle response will be large. The results included in the paper are of some practical significance, the authors argue, since slow-active suspensions are now seen to be commercially viable and any performance improvements available from the use of wheelbase preview can be realised at very low cost through additional signal processing.

Dynamic ride-attitude corrector

According to C Berdah of Renault[7], the power demanded of fully active suspension systems is leading to the adoption of alternative active suspension systems. In the CADHY system a hydraulic jack is put in series with the damper in an otherwise conventional passive spring/damper suspension system. This has been modelled by the author's company, as seen in Fig 105 in which *L* is the stiffness of the tyre. The force *u* is of the form:

$$u(t) = g1(x_2 - x_3) + g2(x_3 - x_1) + g3V1 + g4V2$$

where *V1* and *V2* are the velocity of the unsprung and sprung masses respectively. Simulations have shown equal responses to a conventional active suspension system but with lower peaks of instantaneous power requirement, as seen in Fig 106.

A form of preview positional control has been developed for CADHY involving an ultrasonic sensor ahead of the front wheels of the vehicle and a control law derived based on the road profile. The effectiveness of this has been tested on pave-type and sinusoidal profiles. These have been verified by experimental tests with the physical set-up shown in Fig 107. The physical tests showed that comfort was only improved at frequencies above 10Hz, Fig 108. This was attributed to the active displacement of the hydraulic jack being insufficient. In the sinusoidal excitation it was noted that energy consumption decreased with increasing frequency. At 1 Hz the power requirement was 300 W per wheel station.

To cover losses and inefficiencies a pump of 1.8 kW is therefore considered to be necessary. In the case of the pave track, power consumption requirement rose to 2.7 kW on the same basis, only one quarter of the requirement of a fully active system. The tests indicated several improvements on the original physical set-up. These included using an air-spring in place of the coil spring; and an increased projected area of the jack piston.

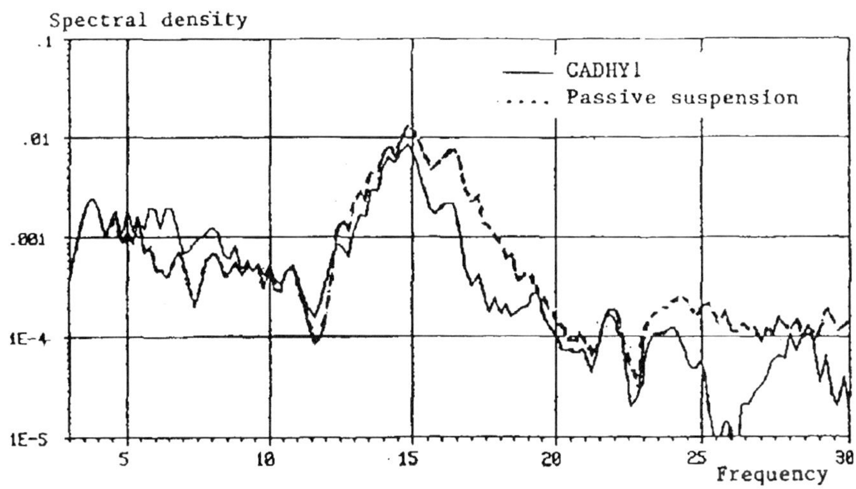

Fig 108: Vertical acceleration of the body

Extending the application of air-suspension

While the extension of air-suspension applications into commercial vehicles has continued, with the need for greater road-friendliness, on trucks, and greater passenger comfort, on buses, the take up on cars has been comparatively slow. Here the fundamentals of air-springing are reconsidered alongside some of the linkage vibration isolation problems which have to be solved.

Typical CV air-suspension applications such as those for drive and trailing axle suspensions shown in Fig 108 do not readily "read-over" to passenger cars whose light shell bodies are sensitive to road induced vibrations transmitted through the wheel hub retention links. Packaging problems also arise unless the air-springs can be built into more familiar suspension linkage configurations for cars, as proposed in Fig 109. There has been progress on some US car models over the years and proprietary systems are available from such suppliers as Delco Products (now Delphi) which are readily adaptable for specialist car fitment. The company's AIR-2000 system has been designed to optimise the space envelope for the air-spring, Fig 110. The system can be set up with almost constant natural frequency whatever the vehicle load and comes complete with air compressor, three height sensors, and air-line connections.

Air bellows elements are generally classified as 'rolling-lobe diaphragm' or 'convolution' types, or combined systems, the first-mentioned sometimes having profiled 'pedestals'. To lower the rate, the volume is often increased by the use of integral or separate surge tanks. All types develop their load reaction by air pressure changes arising from volume variations during stroking of the spring. The bellows is subjected to changes in effective projected area, which also affects the developed reaction load and therefore the spring rate. In the case of the convolution bellows, the effective area (enclosed by the locus of the centre of the lobe radius) is determined by layout drawing for various positions of the spring stroke. For the rolling-lobe bellows it has been shown by theoretical analysis that the effective area could be obtained from a formula giving the diameter of the effective circular area

$$d_e = d_0 - 2r - 2t$$

This value has been modified by replacing the

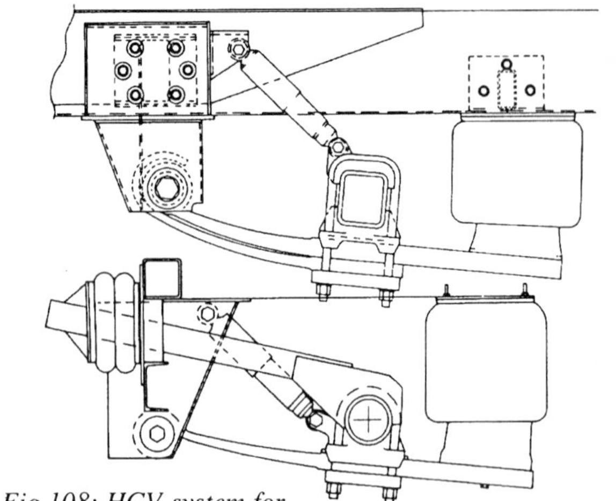

Fig 108: HGV system for drive (above) and trailing (below) axles

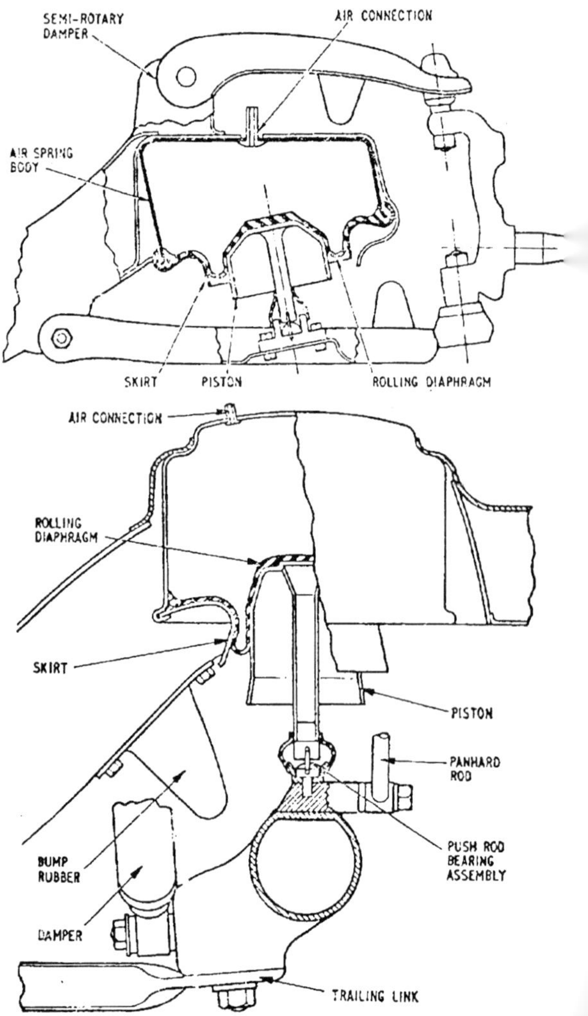

Fig 109: Possible front (above) and rear (below) a

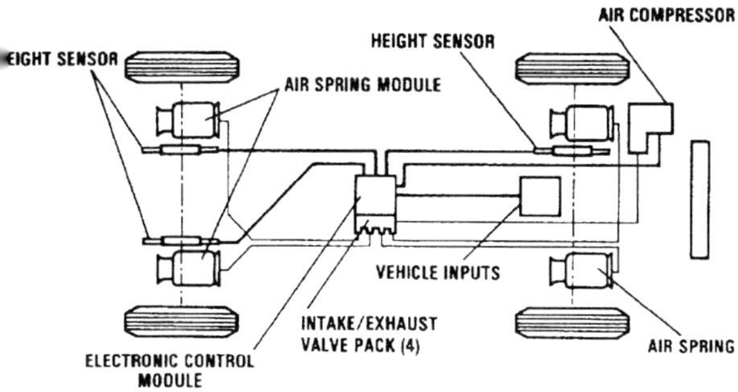

Fig 110: AIR-2000 air system for cars

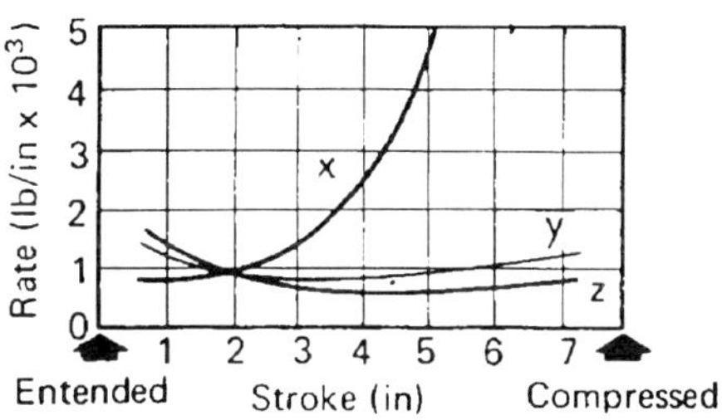

Fig 111: Effect of surge tank: x without; y 1770 in^3, z 3540 in^3 auxiliary volume

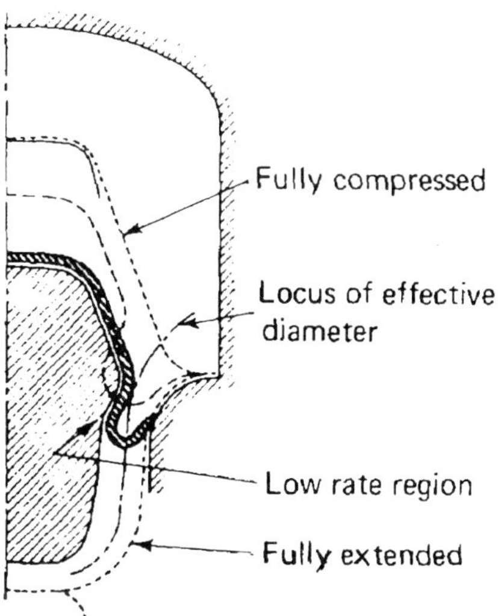

Fig 112: Mechanical control of effective area

vertical-sided piston by a profiled piston or "pedestal", as mentioned earlier.

In general, spring load is the product *pa*, and the bellows rate has been quoted as

$$d(pa)/dx = p(da/dx) + a(dp/dv)(dv/dx)$$

(assuming isothermal conditions) which helps to define the differences between the two basic types. For pre determination of the effect of a given surge tank, the auxiliary volume must be added to the bellows volume, and the gas law $p_1v_1^n = p_2v_2^n$ applied. Effect of surge tank volume on a typical air spring for 5000 lb (2.27 tonnes) load at 71 lb/in^2 (0.5 MN/m^2) is shown in Fig 111. Notation for equations is: *p*, pressure; *v*, volume; *a*, effective area; *d*, outside diameter of bellows; *r*, inside radius of lobe; *t*, bellows wall. thickness; p_1 v_1 pressure and volume at normal ride levelled position; p_2 v_2 pressure and volume at any position during spring stroke; *n*, index of compression/expansion (normally determined experimentally — of the order of 1·2 to 1·4).

Air spring installation effects

Former Dunlop engineering chief Eric Gough argued that while some of the benefits of air suspension are being realized in CVs, full potential can only be achieved in cars through a proper understanding of the problems associated with the transmission of vibrations through the suspension, and impulsive forces generated at the tyre/ground contact in the longitudinal and lateral as well as the vertical direction; the last of which has hitherto tended to be the only one considered relevant to ride comfort problems, he asserted.

Factors in the adoption of air suspension on heavy vehicles have been the relative simplicity of automatic adjustment of spring rate with load, and control of the mean position of the suspension. These systems can be designed to provide a reasonable ride over a wide range of loading conditions, including the lightly laden state which is particularly difficult with conventional laminated springs. These features, and the related one of the possibility of maintaining constant loading-platform height over a wide range of load conditions, have helped the progress of air suspension in this field. Further developments are likely as vehicle designs become lighter in unladen weight.

Air spring variable-rate effects

The relative compactness of the air spring has permitted springing elements having a wide enough effective spacing to give adequately high roll stiffness, even with low main ride or vertical stiffness. Obviously, application of air springing to heavy vehicles already having compressed air systems avoids one of the complications and objections to its use on private cars. Another of the main features of air springs is that of obtaining a non-linear load/deflexion response tailored to suit a specific design criterion. This non-linearity depends on two factors, change of air pressure with deflexion and change of area with deflexion. Considering the spring-rate equation, above, the rate of change of air pressure with deflection can be modified by change of the total volume of air being compressed while rate of change of area with deflection is controllable by modifying the diaphragm shape during deflection, Fig 112. Change of air pressure during dynamic motion follows an adiabatic law, whereas slow movements result in isothermal changes; this portion of the spring rate is then higher dynamically than statically.

The lateral stiffness of an air spring unit is generally low. For this reason link-work is required to locate and guide the motions of the wheels in relation to the chassis. The link-work must transmit (from a road wheel to the main chassis of the vehicle) the tractive effort, braking drag and brake torque reaction, together with the cornering forces that serve to guide the vehicle. It must, at the same time, maintain a suitable geometrical relationship between all wheels, to achieve vehicle control, stability of motion and adequate road holding.

Vibration transmission in air springs

Inevitably, this link-work, including the dampers, provide paths for the transmission of vibrations generated by fluctuations in tyre/ground reactions caused by road surface and tyre. Other vibratory forces — such as those caused by unbalance and, on smooth roads, the effects associated with tyre non-uniformity, or tyre/wheel eccentricity — are transmitted through the links. Road roar on rough tarmac surfaces, harshness and shake on pave and similar roads, impulses from 'cats' eyes' and so on, all pass through the link-work to the main structure of the vehicle; refer to Figs 113 and 114.

It follows that the change from, say, a helical steel spring to an air spring cannot automatically eliminate unwanted vibrations that disturb the ride. In the case of a simple double-convolution air bellows, the load carried by the bellows structure itself is very small, as can be inferred from the load/deflexion relationships over a pressure range as in Fig 115. It would therefore be expected that the amplitude of high-frequency vibrations would be less than if transmitted along a steel helical spring. As this is not the major path for vibration transmission, and link-work of some form must be employed, the change of spring unit may well result in there being no detectable change in the character of ride disturbance transmission.

Optimum values of both longitudinal stiffness and damping will depend in part on the tyres fitted: cross-ply tyres tend to generate vibrations that cause road roar, while radial-ply tyres can generate vibrations at a lower frequency, and the suspension should preferably attenuate those frequency bands of maximum energy. The choice of the optimum will also depend on the transmissibility characteristics over a wide range of frequencies of vibration, and the human reaction to such vibrations.

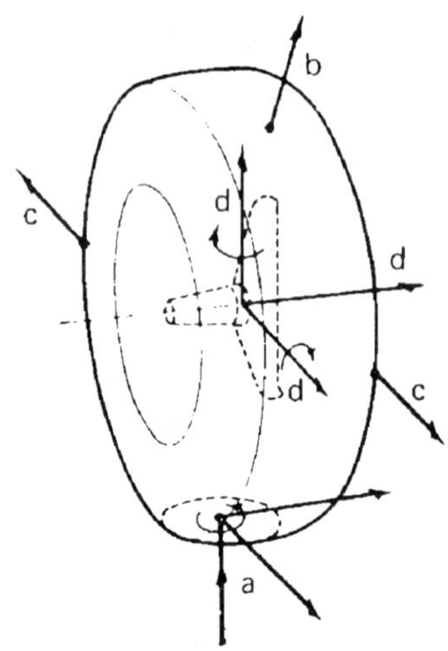

a Road/tyre generated force fluctuations
b Unbalance force – rotating vector
c Unbalance couple – rotating vectors
d Fluctuating reactions on stub axle due to summation of systems a, b and c

Fig 113: Fluctuating forces from wheel and tyre to stub axle

Practical air suspension systems

Leaf springs have long been recognized as practical means of locating axles, and to some extent they minimize the transmission of vibrations, at least over some frequency ranges. Link-work for air springs can incorporate such members; several HGV suspensions do feature them, so arranged as to give adequate roll stiffness, while the air springs give a softer ride.

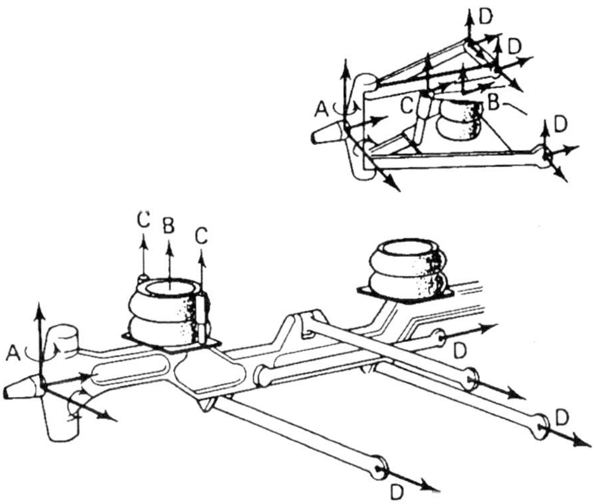

A Fluctuating reaction at stub axle
B Steady load and low-frequency force fluctuations due to spring deflexions
C High-frequency force fluctuations transmitted through dampers
D Reactions transmitted to chassis by linkwork locating axle

Fig 114: Transmission of fluctuating forces in two simple suspension layouts

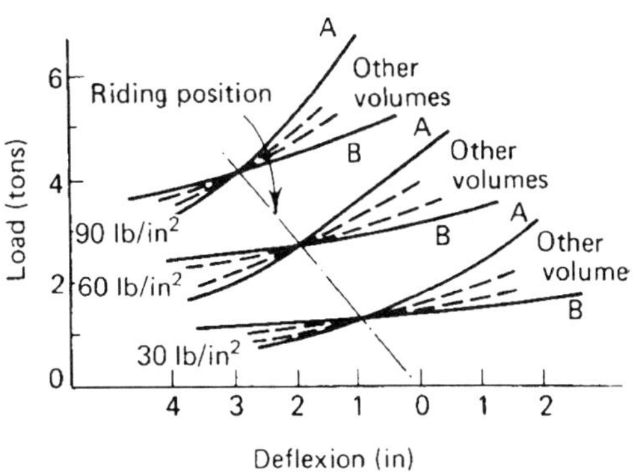

A No extra volume
B Infinite extra volume
dotted: other volumes

Fig 115: Effect of extra volume on rate

Because the load carried directly by the textile-reinforced structure of the double-convolution bellows is very small, the main load carried by the air. This means that the inherent damping in an air spring is small, and this in turn necessitates the use of dampers on all vehicles, even in the ranges where they have not previously been fitted. Need for dampers will be emphasized as tyres of lower rolling resistance and associated lower damping losses are developed.

There are possibilities in the two-chamber air suspension with orifice damping of the main ride motions of thc chassis, although some additional means of providing damping of the motion of the unsprung mass at its main mode would be needed. If a successful form of inertia damper could be evolved to cope with these axle frequencies, then one direct path of vibration transmission from axle to vehicle body would be eliminated.

Handling analysis for racing cars

William and Douglas Milliken, best known for their historic handling work at the Cornell Air Laboratory, have authored an important contribution to the analyis of handling in racing cars[8]. The major controlling input to the road vehicle is, of course, the tyre lateral force development and CAL were pioneers in studying the tyre affect on control as they would the control surfaces of an aeroplane

Eric Gough at Dunlop in the UK had already established that the force computed from the area under a curve of lateral force (stress) distribution along the tyre-to-ground footprint, and perpendicular to the wheel plane, equalled the lateral force measured at the axle. Side force was thus the lateral tyre force arising from distortion of the footprint plus any frictional sliding forces in the aft portion of the print.

Fig 116 shows data taken from a Gough type tyre test rig on a shaved racing car tyre at high (15 degrees) slip angle. For a complete slip angle range the characteristic of Fig 117 is given by the authors for a P215/60 R15 Goodyear Eagle GT-S tyre shaved for racing. This defines the elastic, transitional and frictional operating ranges. In dry conditions race tyres reach their peak side force at slip angles in the range 3-7 degrees whereas in the wet the peak will in general be lower and the fall off after the peak more rapid. At and beyond the peak most of the print is sliding and side force is the result of friction between tyre and road surface.

In fact the normal concept of dry friction does not fully apply and in comparing tyre characteristics at different radial loads it is noted that peak side force 'friction' coefficient actually varies, being higher for lighter loads and decreasing for heavier loads, the phenomenon of tyre load sensitivity. This is a key factor in racing and the coefficient can be increased with the use of 'stickier' rubber compounds. The value could reach 1.8 for a GP tyre at light load and its variation can have major effect on the balance of a car near its operating limit of adhesion.

In traction mode, Fig 118, the driving torque produces a forward reaction, the tractive force of which moves the footprint forward relative the axle. This compresses the tread elements ahead of the print and deforms the tyre circumferentially. The authors compare the compressed length C_l with the uncompressed length C_2 to show the compressed tread elements adhere to the ground on entering the print, the shear stress between tyre and ground being in the forward direction so that tread elements are bent forward. As more elements enter the print and move aft under increasing local load, shear stress builds up linearly as seen in curve 2. Without, traction longitudinal shear forces also exist because rolling radius is forced to change as tread elements enter the print. For constant angular velocity there is thus a change in circumferential velocity resulting in stress reversals seen in curve 1. The net effect is seen in curve 3. Longitudinal shear force falls off as the rear of the print becomes unloaded, with sliding taking place and the elements that were bent tending to straighten out, longitudinal velocity in the print relative to the velocity of the undeflected tyre being seen in the bottom graph. Sliding in the print thus results in greater revs per mile in traction than in free rolling.

Braking effects, Fig 119, are comparable with the torque in the reverse direction. A slip ratio can be defined, as the difference in angular velocities of the driven (or braked) wheel and the free-rolling one, of which traction and braking forces are a function. The forces rise rapidly to a given slip ratio and then fall off. Up to the peak the elastic properties of the tyre are crucial and after it such factors as tread compound and road texture are paramount. Although braking is initiated on the straight prior to a corner, the authors explain, it is generally carried on into the turn; similarly tractive effort may begin after the turn apex and continue on the straight.

Work due to Sakai has resulted in a single

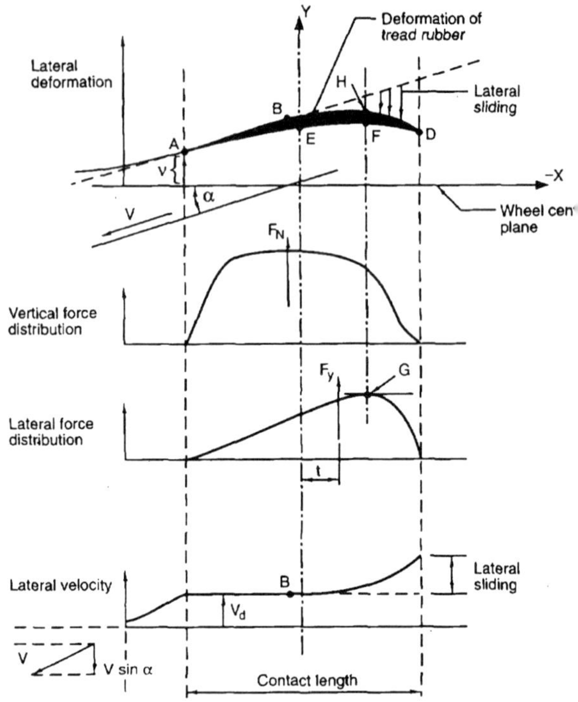

Fig 116: Tyre footprint characteristics

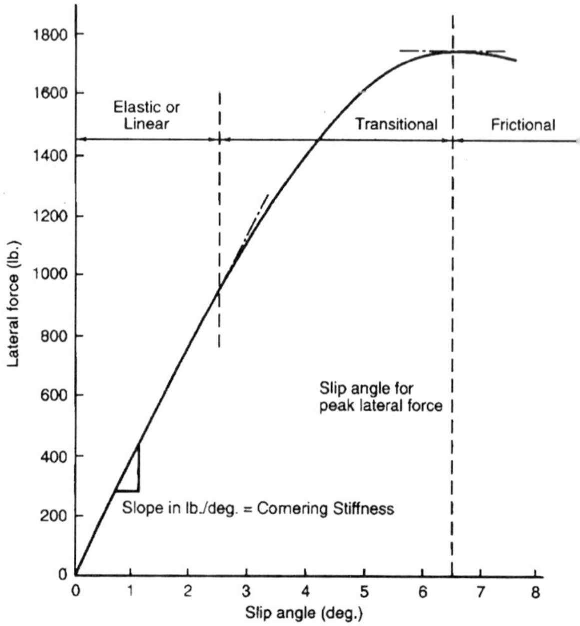

Fig 117: Side force vs slip angle for racing tyre

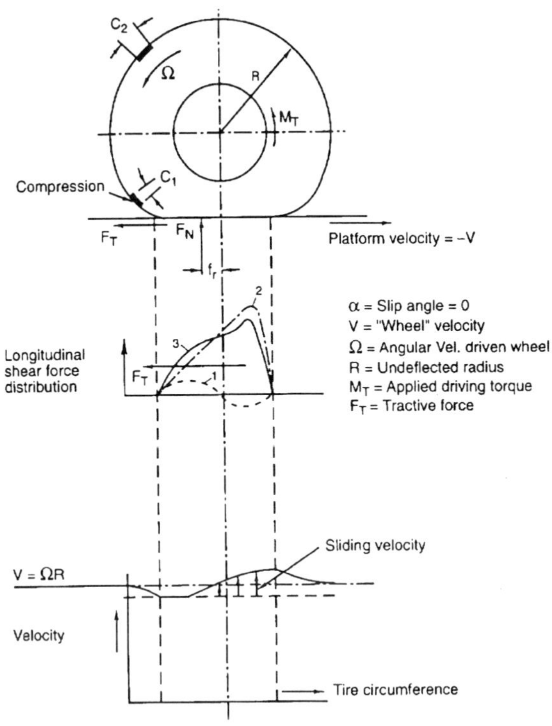

Distribution of forces and sliding velocity, over the contact length of a tire under the action of a driving torque M_T.

Fig 118: Tyre print characteristics in traction

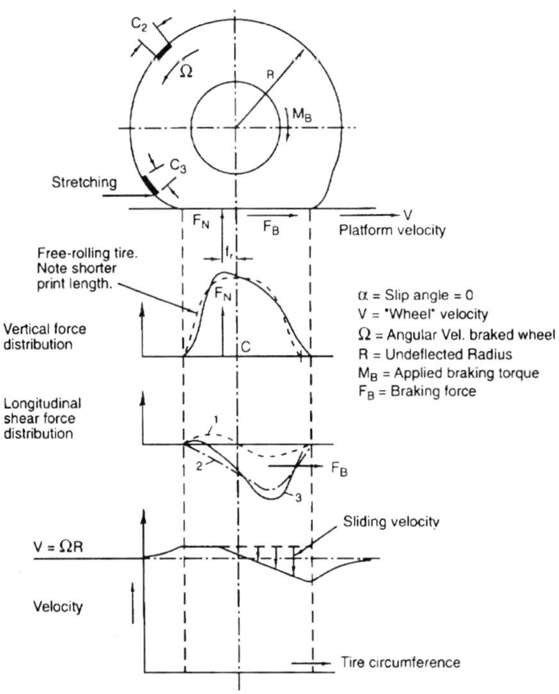

Distribution of forces and sliding velocity over the contact length of a tire under the action of a braking torque M_B.

Fig 119: Tyre print characteristics in braking

diagram combining the traction and braking composed of so-called friction ellipses and the friction circle. The latter represents the force-producing limit of the tyre for a given set of operating conditions and for a single tyre/wheel is conceptually similar to a friction circle diagram for the whole vehicle, as discussed in the last issue of this column. Fig 120 is a plot of side forces vs longitudinal forces for a series of slip angles and slip ratios. While this is for a passenger- rather than racing- car tyre, the authors clearly think such data would be valuable to race car engineers if tyre makers carried out the appropriate tests. At the origin the tyre is free-rolling with longitudinal forces only due to rolling resistance. Moving along the x-axis side forces and corresponding slip angles are noted. On the y-axis -ve is up for traction and +ve down for braking. At point A, a typical figure represents combined traction/braking with lateral force of 800 lb lateral and 500 longitudinal, generated by 4 degrees of slip angle and +0.036 of slip ratio. The boundary of the diagram corresponds to a friction circle for which the maximum force that the tyre can generate is represented; for some tyres the circle could become an ellipse.

In racing the tyre is operated at or near the friction circle and at point B, for example, the tyre is giving pure lateral force, at its best slip angle, of 1100 lb at 13 degrees slip angle. If 600 lb of tractive force is applied the available lateral force drops to 920 lb at point C. The combined force, vector sum, is still 1100 lb. Note that the slip angle curves hook back as they approach the vertical axis, indicating, the authors explain, that the peak of resultant force vs resultant slip velocity has been reached.

The g-g handling diagram

A more complete way of quantifying the capability envelope of the car is in the g-g diagram which is a further elaboration of the friction circle concept. The latter is limited in not accounting for tread pattern, print shape according to camber and so on. If the load sensitivity was zero and a constant friction coefficient applied at all wheels then friction circle radius would be determined by the product of friction coefficient and vehicle weight. In fact the vehicle cannot reach this theoretical friction circle for these reasons.

The power limitations of the vehicle over the gear-speed range cannot produce the slip ratio for

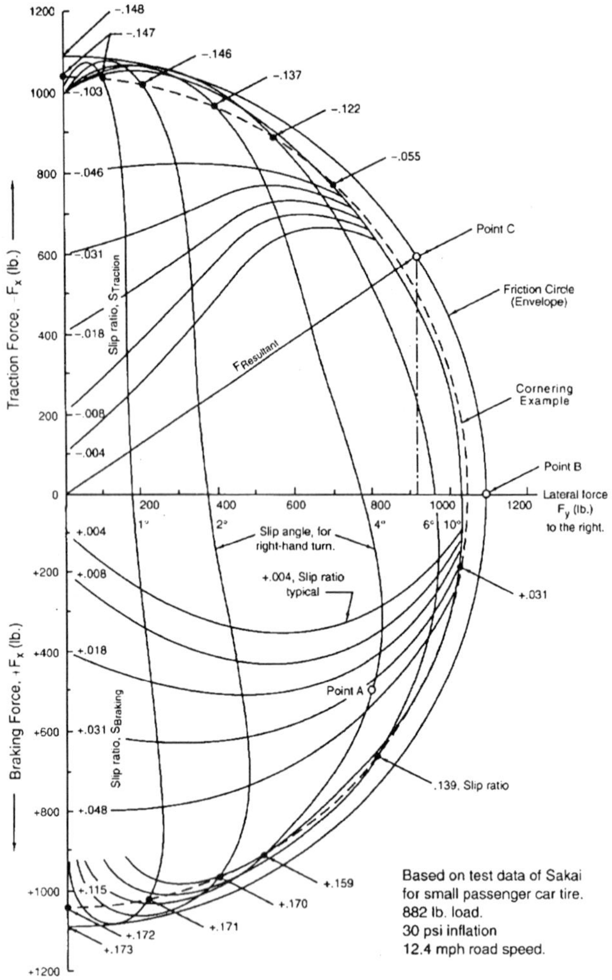

Fig 120: Friction circle for RH corner

maximum resultant force for part of the diagram. Lateral and longitudinal tyre forces are not linearly related to wheel load transfers owing to tyre load sensitivity. Suspension geometry changes and anchorage compliances can change wheel orientations and thus tyre forces. Operating conditions are such that each wheel does not reach its performance limit at the same time due to understeer/oversteer effects (racing cars are, however, often designed to achieve neutral steer near the limit). Lastly, brake balance may inhibit the use of friction limit by both axles simultaneously.

Computer modelling, by for example, the MRA Moment Method (MMM) allows simulation of the friction circles at each of the four wheels and the corresponding point on the vehicle friction circle may be computed for a particular operating condition. Fig 121 gives the example of a large rear engined car cornering at 0.76 g under reduced throttle and moderate braking, with the vehicle in equilibrium on a RH turn. As well as plotting friction circle for each wheel it tabulates slip and camber-inclination angles and the resultant horizontal force vector is shown at each wheel. In this case the inside rear wheel is utilising all the available force while the other wheels are below their limits. The vehicle resultant force vector can be obtained by adding longitudinal and lateral components of the wheel force vectors. The radii of the wheel friction circles is summed to find a point A on the vehicle friction circle and the solid ellipse through this point corre-

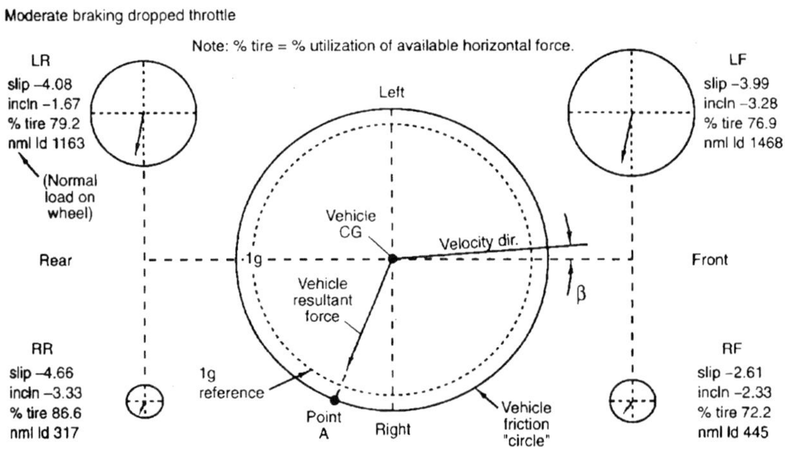

delsw (δ)	beta (β)	A_x	A_Y	C_N	N	V	r	R	phi (φ)
deg.	deg.	g	g	-	ft. - lb.	mph	deg. / sec.	ft.	deg.
13.06	-3.761	-0.318	0.757	0.0	0.0	60.00	15.63	323.4	-3.100

Fig 121: Rear wheel drive car braking and decelerating

sponds only to this operating condition. These are the steps in constructing the g-g diagram, the final one of which is to convert to g units so that the resultant forces are normalised by dividing by wheel weight; the 1*g* reference circle is shown dotted. Classic work by RS Rice has demonstrated that the g-g presentation can also be used to show how much vehicle capability is utilized by the driver.

Application of the g-g diagram to racing requires knowledge of the theoretical limits of the vehicle (given a perfect driver who fully utilises them). Experimental determination is made difficult by aerodynamic downforce and drag which increase with the square of speed. Peter Wright of Team Lotus was a pioneer in this respect and produced the g-g diagram of Fig 122 from limited experimental and theoretical information which accounted for the increase in limits available with downforce.

The authors attribute David Segal of MRA as being the first to calculate the g-g boundary for a real car. Figs 123 and 124 show his diagrams relating to a car with rear drive and 54% front weight distribution against another with front drive and 64% front weight distribution. These were at constant 60 mph and included aerodynamic forces applying to these high-performance sports saloons. Each figure is based on an MMM simulation. The diagrams essentially comprise horizontal lines of constant drive (or brake) torque plotted against longitudinal and lateral acceleration. Each solution point represents momentary steady-state cornering while accelerating/braking at a (momentary) constant speed.

The $T_D = 0$ curve represents a free-wheeling car cornering at various levels of lateral acceleration and is offset from zero longitudinal g by aero-drag and rolling resistance; it tends towards increased negative longitudinal acceleration because of increase in sideslip drag at higher lateral g. The zero longitudinal acceleration axis represents road load — enough driving torque to overcome all drag and maintain

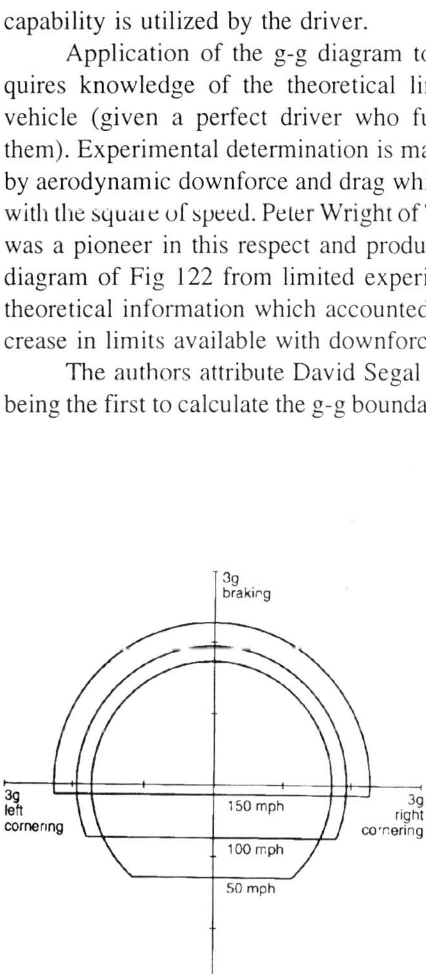

Fig 122: Vehicle g-g estimate

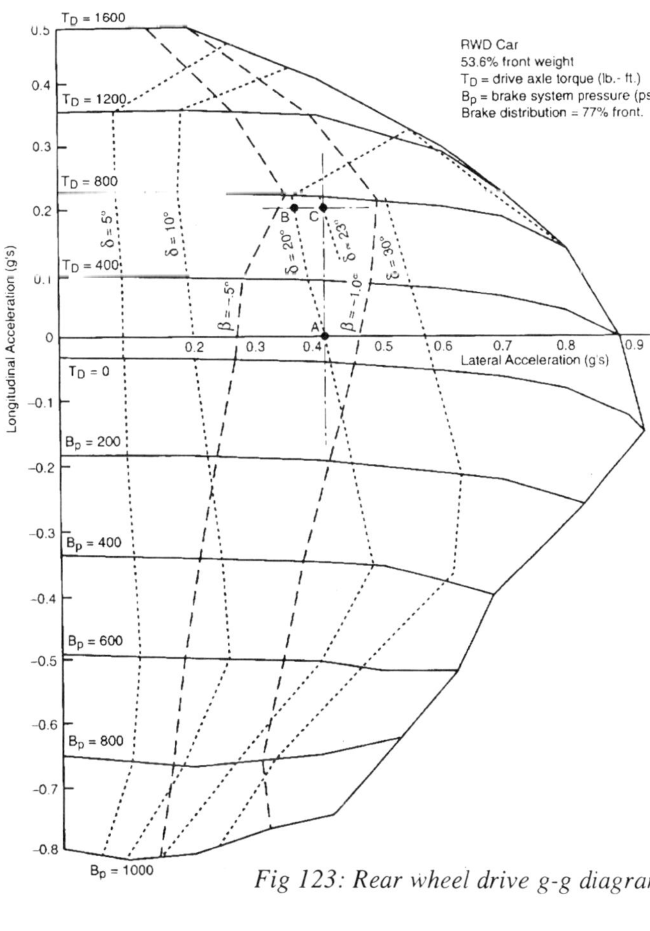

Fig 123: Rear wheel drive g-g diagram

constant speed cornering. The RH boundary of the diagram represents maximum trimmed lateral acceleration at different levels of longitudinal acceleration. Steer angle δ and vehicle slip angle β lines provide numeric measure of longitudinal trim change sensitivity of the car, effect on lateral as longitudinal is changed. For the RWD car cornering at road load at 0.41g with 20 degree steer angle (point A) suddenly accelerating to +0.2g with fixed steering, the lateral acceleration is reduced to 0.36 g (point B); this implies increase in path radius and to maintain the same lateral trim (point C) the steering must be increased to some 23 degrees.

The authors consider that it appears practical to map out other variables such as load transfers and slip angles on these diagrams. Future work is aimed at automating the production of g-g diagrams from MMM vehicle and tyre data.

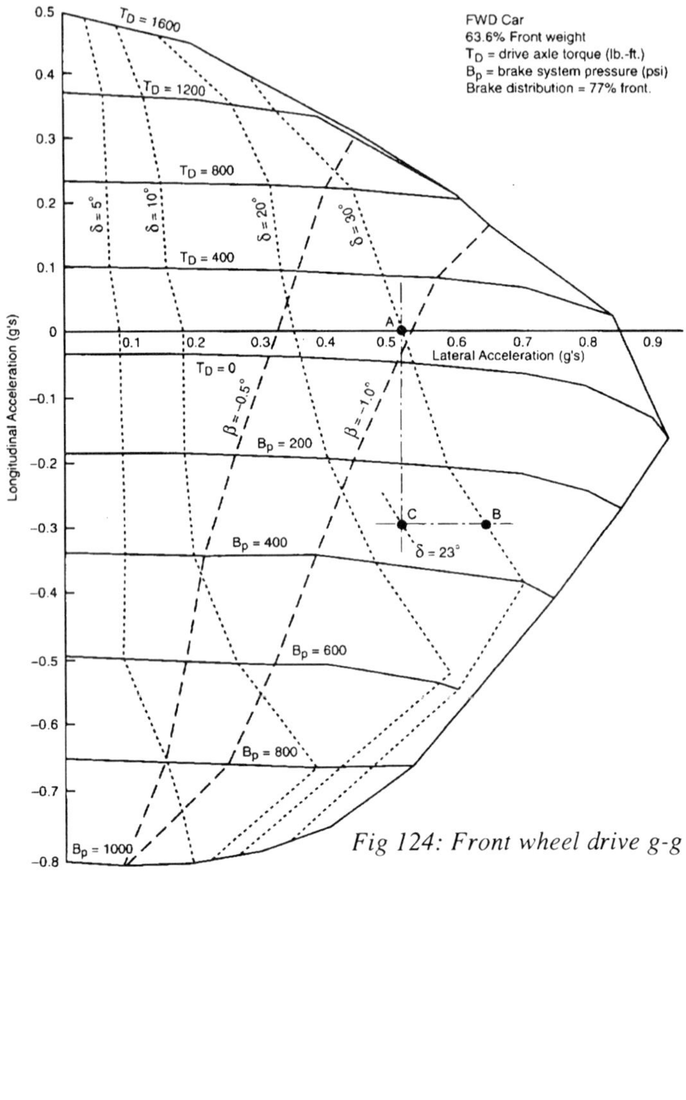

Fig 124: Front wheel drive g-g diagram

Chapter 18: Running gear, drive-line and brakes

Traction vs stability
Wheels & hubs
Braking systems
Anti-lock braking systems
Brake piping
Driveline calculations
Drive axles & drive-line
Driveline vibration
Light-weight drivelines

Traction vs stability

A fundamental trade-off between longitudinal tractive force (or decelerating force in the braking situation) and lateral cornering force, developed at the tyre-ground contact patch is centtral to the the balance between traction and stability of the wheeled vehicle.

Road camber or lateral wind forces on the vehicle will rob the tyre frictional force available for traction (or braking) so a knowledge of the magnitude of the interacting forces is necessary in performance prediction calculations.

Simplification of the vehicle to a rigid rectangukar frame carrying a wheel at each corner which cannot be steered helps to visualise the force system involved. Lateral forces developed can be considered proportional to slip-angle for the simulation and an axis system can be drawn in the ground plane as shown as Fig 1. An analysis due to Rocard[1], dating from the 1950s used this simplification to obtain an expression for the stability of the vehicle from which the oscillatry motion of the idealised vehicle in swerving from its straight-ahead path could be examined. These are described by Steeds[2] who rearranges them to define a critical speed for the vehicle at which the vehicle changes from a stable to an unstable state:

$$=\{[2K_1K_3(a+b)^2]/[M(K_1a - K_3b)]\}^{1/2}$$

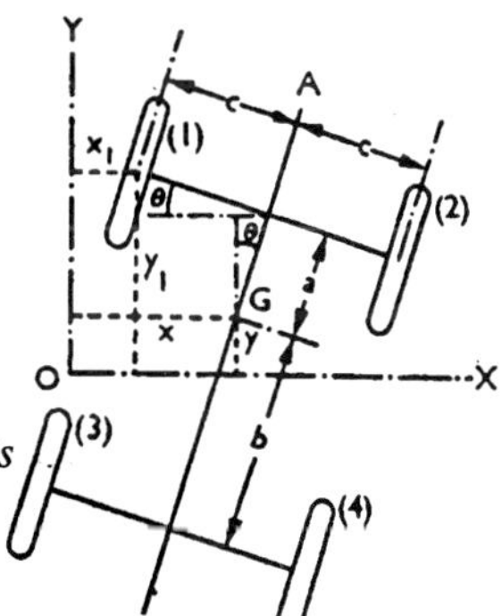

Fig 1: Ground-plane axis system

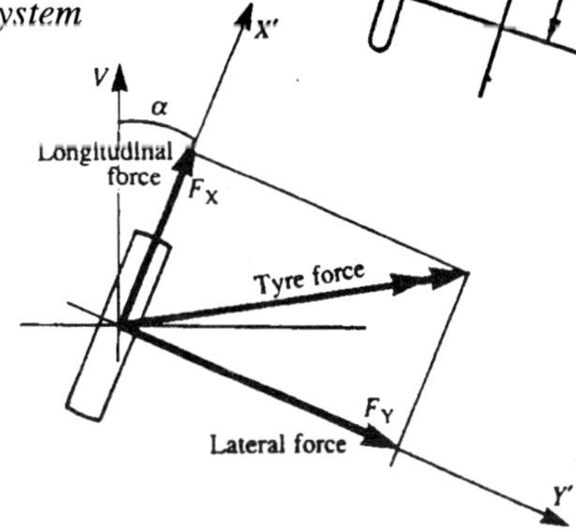

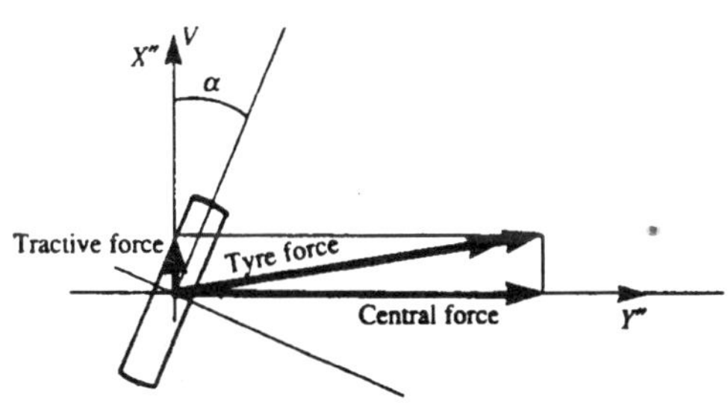

Fig 2: Lateral and central forces

where $K_{1,3}$ are cornering force co-efficients for the different wheel stations and M is the vehicle mass acting through G.

From this it can be seen that if K_3b is greater than K_1a the motion is stable at all speeds. Thus the choice of steering force characteristics for front and rear tyres in relation to the vehicle centre of gravity is fundamental to stability problems. As tractive forces (or braking ones) upset the balance of available side forces, these must be considered in design. The effects of suspension springs and body roll effects on these relationships has been obtained from a more complex model of the vehicle by such pioneer researchers as Segel[3].

When the wheels are steered there is also a difference between the lateral force developed in relation to the longitudinal axis of the vehicle (path) and that with repect to the plane of the tyre (heading), Fig 2. The force along Y' is the lateral force comprising cornering force due to slip angle plus camber force due to camber angle.

The force perpendicular to vehicle motion (path) is now referred to as the central force, acting with tractive force, and two axis systems are used to describe separately the wheel and the vehicle. The tractive force determines whether the vehicle rounds a corner at constant speed (steady-state handling) or not and is effective in developing tangential acceleration.

The mechanism of friction development between tyre and ground is examined by Dixon[4] who has noted the dependence on sliding speed and temperature and the difficulty in quantifying the effects of temperature change. He shows the dynamic friction co-efficient for different vehicle-category tyres (in the locked-wheel braking case) as Fig 3. The lower values for commercial vehicles are due to the hard tread compounds used for low wear and to high contact pressures. The speed dependence of friction is explained by their being one 'static' co-efficient applicable to the non-sliding part of the tyre footprint and a lower 'dynamic' one applying to the sliding part. At high slip angles the sliding part will have reduced co-efficient so that that the cornering force peaks at moderate slip-angle and then tails off — but its decline will be less dramatic than that of locked wheel braking as the relative sliding speed is lower.

Combined longitudinal and lateral forces are of practical importance particularly in accident avoidance manoeuvres when stability effects are crucial. These are functions of lateral and longitudinal slip and one is usually at the expense of the other, Fig 4.

Disturbing forces and stability

A criterion of stability, handed on in earlier times by the aircraft industry, is 'stability-margin' between the centre-of-gravity of the vehicle and the 'neutral steer line'. This such that at any point along the line lateral force can be applied without tendency of the vehicle to rotate about a vertical axis — stability being achieved when the CG is ahead of the neutral steer line. There is also the concept of steering stability in which a stable vehicle has its front slip angle greater than the rear so that in general, centrifu-

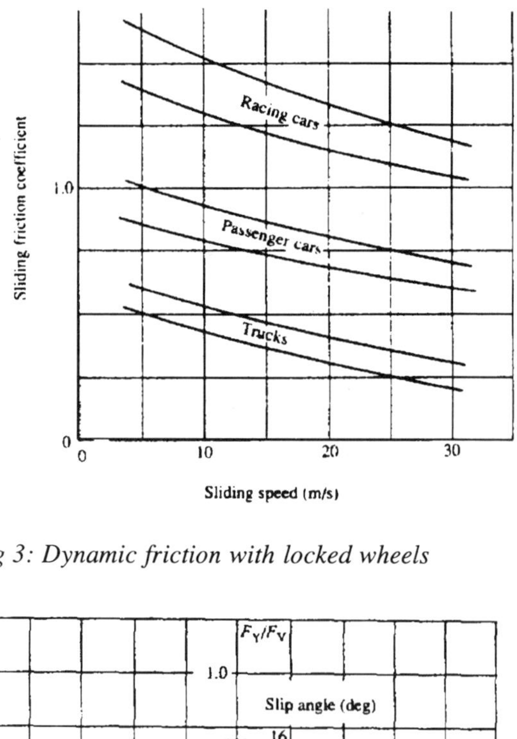

Fig 3: Dynamic friction with locked wheels

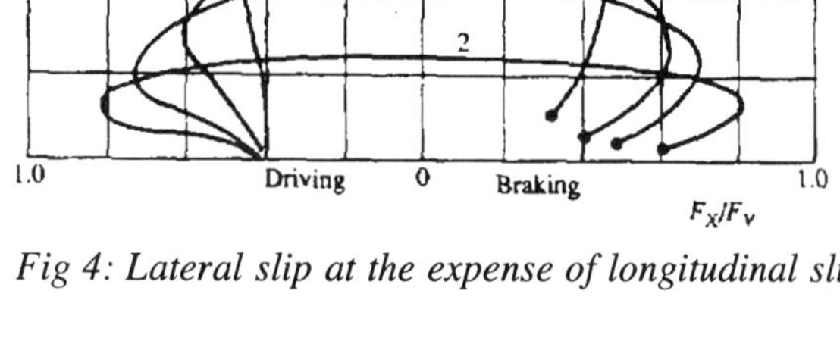

Fig 4: Lateral slip at the expense of longitudinal slip

gal force opposes disturbance and the vehicle tends to run straight — the condition of 'understeer'.

There is also the concept of aerodynamic stability with lift and pitching moments transferring load front front to rear tyres. A simplified consideration of aerodynamic stability assumes the centre of pressure at the same height as the centre of gravity and side wind force produces only yawing moment. Longitudinal forces, side thrust and yawing moment may be combined vertically to give a single aerodynamic force acting horizontally through the centre of pressure as in Fig 5. A side-thjrust co-efficient can thus be defined:

$$= C_L cos\gamma + C_D sin\gamma$$

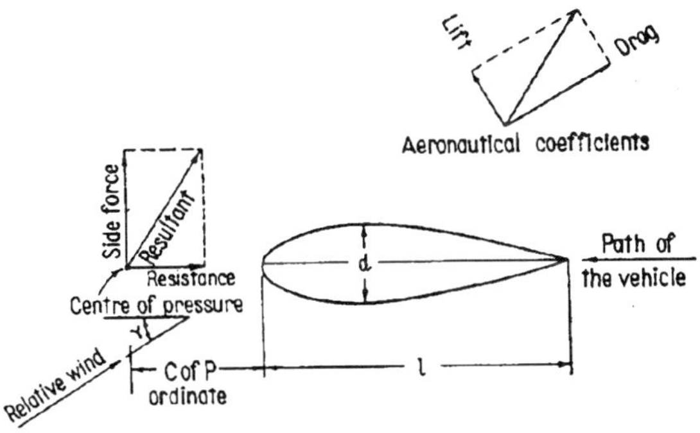

Fig 5: Resolved aerodynamic forces

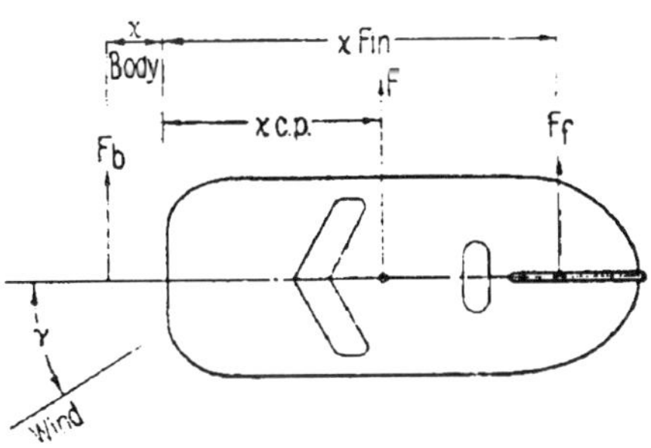

Fig 6: Forces on streamline body

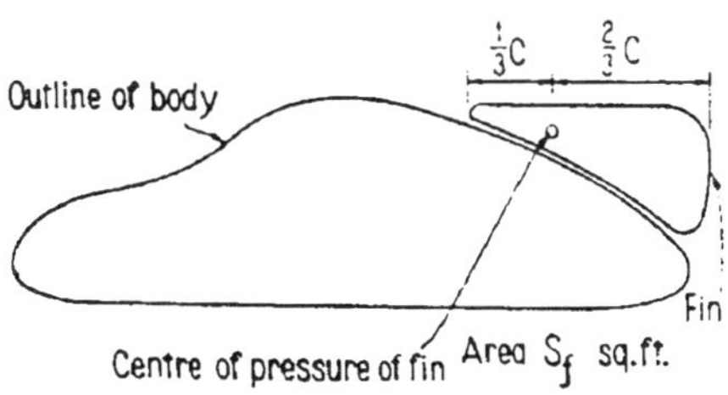

Fig 7: Effect of rear fins

where g is the angle of attack. Fig 6 shows the results of measurements on a streamlined body with fineness ratio $l/d = 6$ such that the yawing couple:

$$= (1/2)\rho V^2 A l C_Y$$

For such a long streamlined body the centre of pressure lies well ahead of the nose and a small side force can produce considerable disturbance in direction. The dotted lines show that for a rectangular sectioned squat body the side force co-efficient is considerably higher. C_S values limited to 1.0 apply to ground vehicles — involving considerable side force. So for ultra-light streamlined vehicles it is important it is important to restrain the centre of pressure in a rearward position. This is achieved by rear fins, Fig 7; that shown having its centre of pressure about 1/3 of its length back from the leading edge. The point at which the resultant side force takes effect can be found by taking moments. The cross-wind response of vehicles is examined in detail by Howell[5].

Graphical method for stability

Stability on a curve under vehicle inertia forces is considered by Taborek[6] who suggests that traction and braking forces acting on the steered wheels should be balanced by other motion-affecting forces ideally acting through the vehicle cenre of gravity. All these produce side force components which add to or subtract from the centrifugal force on the vehicle. Distribution of side forces to the wheels is the key factor in directional stability — safe manoevrability being determined by the wheel which first starts to slide. He commends vectorial analysis using graphical techniques to determine the distribution. In Fig 8 representing accelerated motion through a curve example forces are centrifugal C = 340 lbf, total rolling resistance $R_r = W_f = 80$ lb (356 N) and an acceleration $a = 3$ ft/s^2 (0.9 m/s^2). Thus the inertia force $R_i = (W/g)a = 375$ lb (1.67 kN). Vectors are drawn in the figure for rear-wheel drive and front-wheel steering and the same with front wheel drive. Compared with the case of the free-wheeling vehicle without traction, the external forces all produce side components that increase the side force reactions on the rear-drive vehicle. On the front-drive vehicle, however, traction forces considerably decrease side thrust.

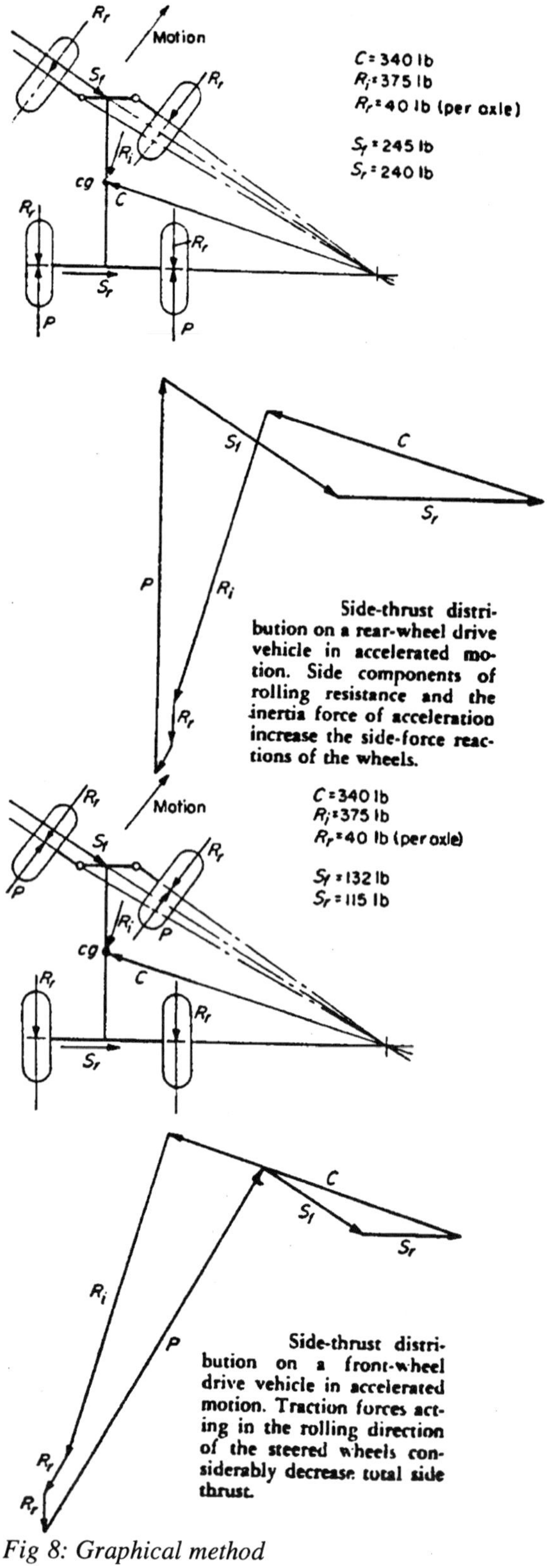

Fig 8: Graphical method

Roadwheels and hubs

Wheels and hubs used by lighter, volume-built, vehicles generally involve pressed and roll-formed steel road-wheels supported on hubs by either taper-roller or angular-contact ball bearings. In the case of very high volume built cars there are also moves to integrate other wheel suspension components into the hub structure and even to integrate parts of the brake into the wheel. The design of both wheel and hub also vary widely in design, according to vehicle duty, as can be seen from the comparison of a racing car wheeel/hub assembly with that of a commercial vehicle as shown in Fig 9.

Tyre design and specification, to be considered in a later section, govern the rim dimension and it is helpful to be armed with a copy of British Standard BS AU 50 Part 2 if a new wheel is being designed for a particular chosen tyre size. In the case of commer-

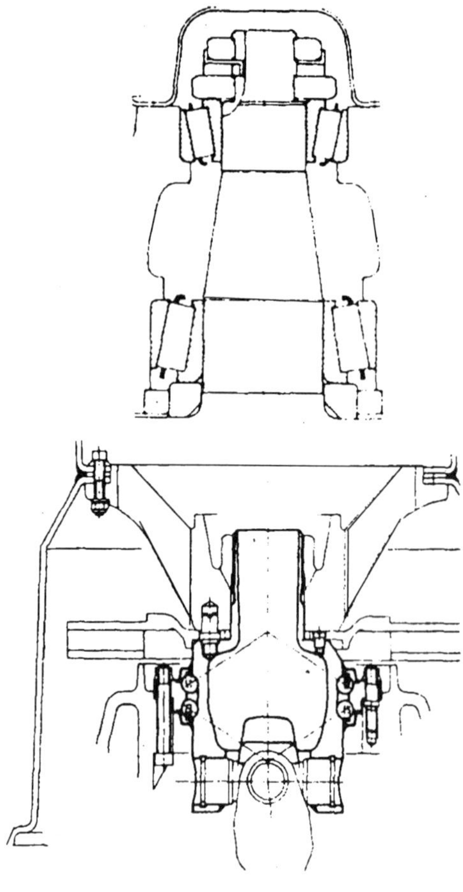

Fig 9: CV (above) and race-car (below) hubs compared

cial vehicles, a variety of standardised rim design alternatives are available, as in Fig 10, according to the type of construction chosen. A key factor in the choice of wheel construction is the fixing of the wheel to the hub — governed by the in-service loading, amongst other considerations. The fixings are also crucial to the accurate centring of the wheel and most are of the conical seat stud type as shown in Fig 11. BS AU 50 gives a strong warning against mixing spherical and conical seats and studs on any installation. Many CV wheels use spigot centring for which case plain stud holes and flat nuts would be used.

Car wheels are almost always designed with drop centre rims of the type shown in Fig 12. While this shows the common double-hump configuration, for wheel diameter *D*, centre-hole diameter *N*, stud hole diameter *L* and rim width *M*, other variants are used which can provide more room for brakes and make possible the use of run-flat tyres. Fig 13 shows a combination of Dunlop Denloc and Michelin TR rims which are of this type; Goodyear's 'asymmetric' rim is a further variant. Also approaching production acceptance is the Continental Conti system (Fig 14) said to allow complete run-flat ability for the unorthodox tyre contour — which could lead to the total discontinuance of the 'spare wheel'.

Wheel design also varies with material usage and fabrication process. For commercial vehicles, the Trilex system with cast 'spoked' wheels comprises rim and spider. The removable rims are connected to the spider by means of clips and screws. Another development is the Monoflex 15 degree tapered bead seat shown in Fig 15 with and without rim offset. For the more conventional CV wheel, a guide to traditional availability of sizes for the three main CV categories is shown in the table of Fig 16.

For those contemplating the design of special road wheels, it is useful to consider the forces on an elemental length of rim (Fig 17) when trying to determine rim stresses. The stress derived from standard solid mechanics theory is given by

$$6\,P.l/h_1^2 + (P/h_1)$$

Work due to Svenson[6] considered the imposition of stresses from both horizontal and vertical forces at the tyre/ground contact patch, leading to an expression for dynamic factoring of the statically derived stress values as

$$1 + 2.6C/P$$

where *C* is the spring rate of the tyre.

The complex shape of the wheel disc of the conventional pressed variety makes it difficult to

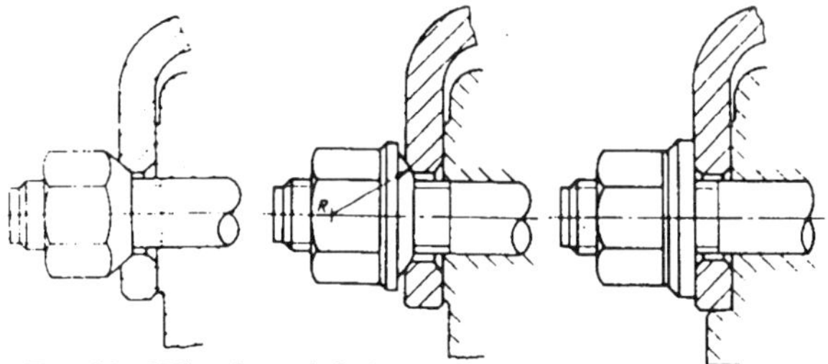

Fig 11: Wheel stud fixings

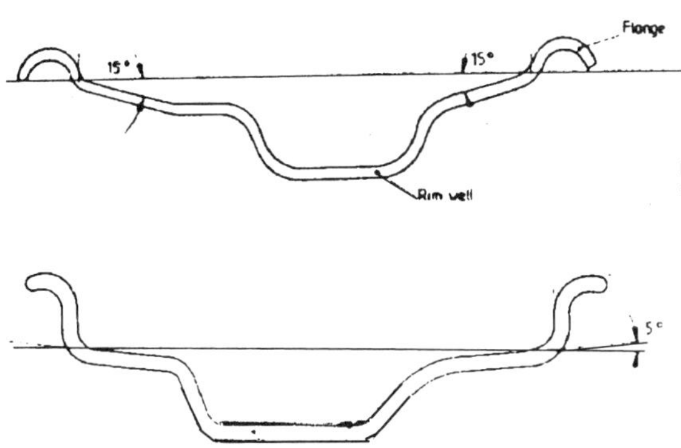

Fig 12: Drop-centre rim for car

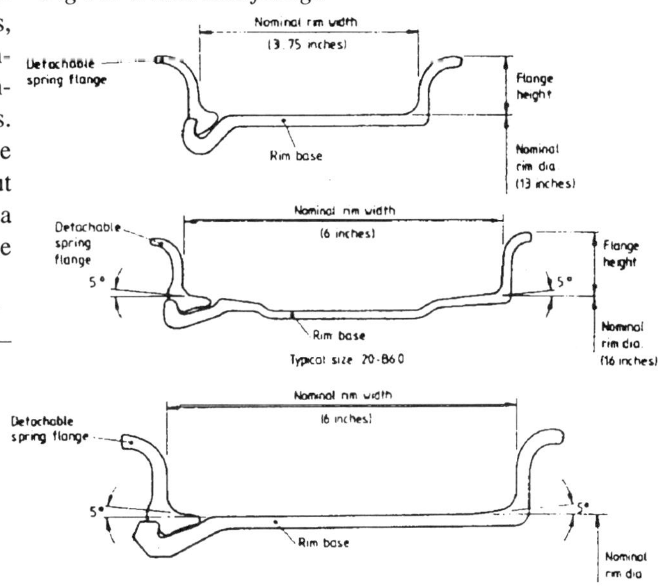

Fig 10: CV rim configurations

calculate the stress from simple theory. Empirical work carried out at MIRA has shown that bending moments imposed on wheel discs are increased not only by lateral weight transfer but also by fore-and-aft weight transfer, body roll and tyre distortion, as illustrated in Fig 18. Analysis shows shows torque on stub axle to be :

$M_A = -V_s e_s$ *static and*
$M_A = V_c e_c$ *in cornering*
while torque on wheel is
$M_W = V_s cos g_1 d + V_s sin g_1 r_{sl}$ *static and*
$M_W = V_c cos g_2 d + V_c sin g_2 r_{cl} + H r_c$

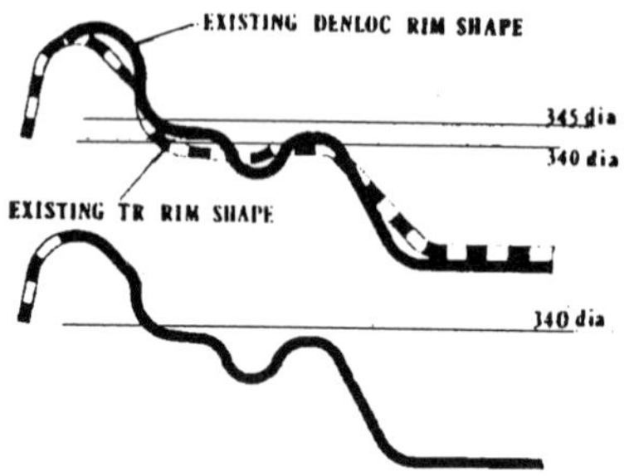

Fig 13: Run-flat rim proposals

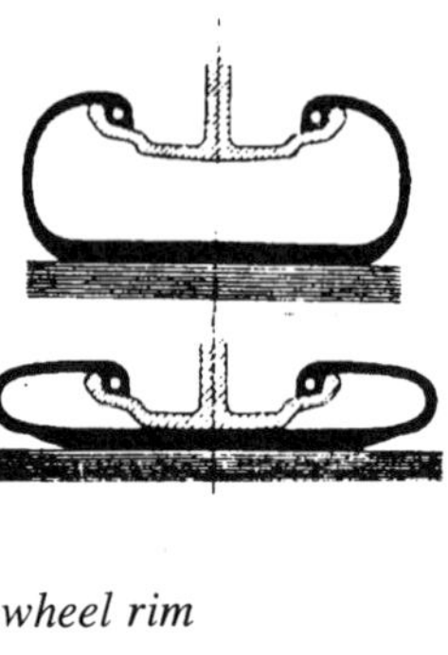

Fig 14: Conti wheel rim

in cornering. Work at Michelin has shown the potential for substantial weight savings, from those conventionally employed, by careful design of wheel discs. One path of study has resulted in the so-called Roue Globale design in which the brake drum is integral with the disc, as in Fig 19.

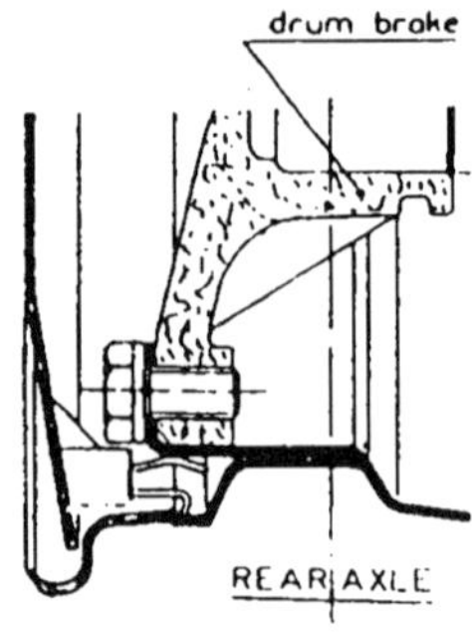

Fig 15: Monoflex wheel fixing

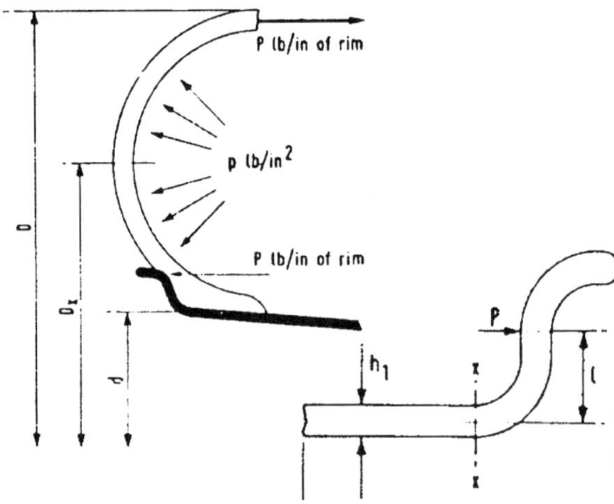

Fig 17: Element of rim section and tyre dimensions

Load, cwt	No. of stud holes	Stud dia	Pitch circle dia (mm)	Bore, mm	Nave thick-ness (in)	Nave dia (in)
17½	6	18 mm	205	161	5/16	10
22½	8	18 mm	275	221	5/16	12·75
22½–34	8	⅞ in BSF	275	221	5/16	12·75
27½	8	18 mm	275	221	3/8	12·75
27½–34	8	⅞ in BSF	275	221	3/8	12·75
31½	8	18 mm	275	221	13/32	12·75
31½–34	8	⅞ in BSF	275	221	13/32	12·75
47½–74	10	⅞ in BSF	335	221	13/32	15·0

Fig 16: Traditional CV wheel sizes

Hub and bearing configurations

Fig 20 compares traditional roller and ball bearing configurations for front hubs, which in the case of volume-built vehicles are tending to give way to the close-coupled bearing arrangements shown as Fig 21; these can be assembled more quickly and promise maintenance-free service for the life of the vehicle. Double row angular-contact (DRAC) bearings in Generation Two form can have integral flanges included (Fig 22), for wheel and brake disc attachment. A further development employs a detachable inner raceway to ease the assembly of balls into an otherwise 'solid' inner and outer race. This approach is used in another form on racing-car hubs — as shown for the Lotus F1 arrangements of about a decade ago seen in Fig 23. A particularly interesting development is the SKF 'hub-unit' bearing concept used on the Fiat Panda, Fig 24, in which both inner and outer races are integral with the wheel-flange and suspension system.

A particular advantage of close-coupled taper-roller systems is the avoidance of the bearing

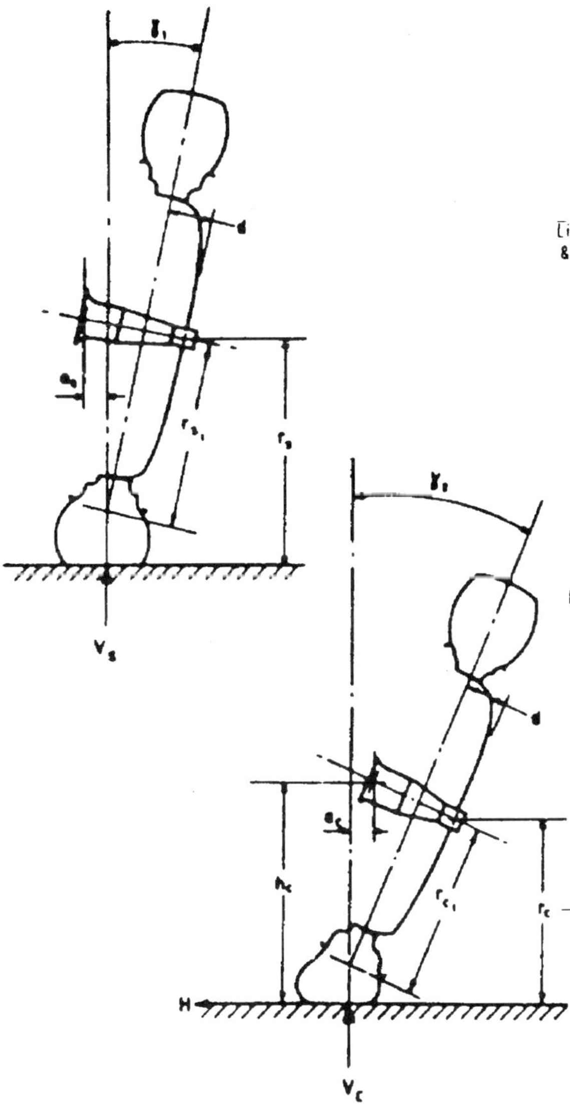

Fig 18: Wheel disc analysis (static, top; cornering, bottom)

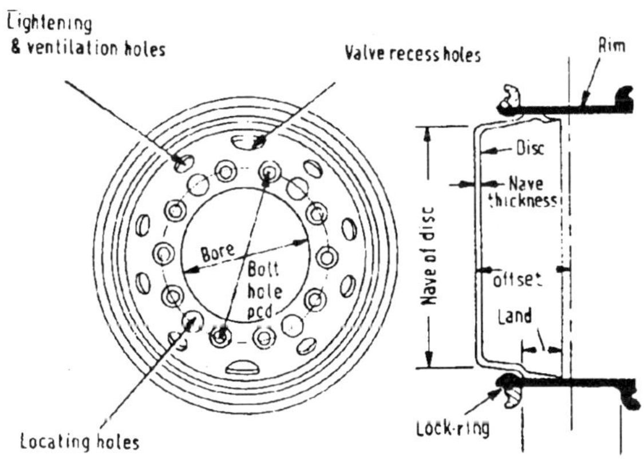

Fig 19: Drum and wheel integrated

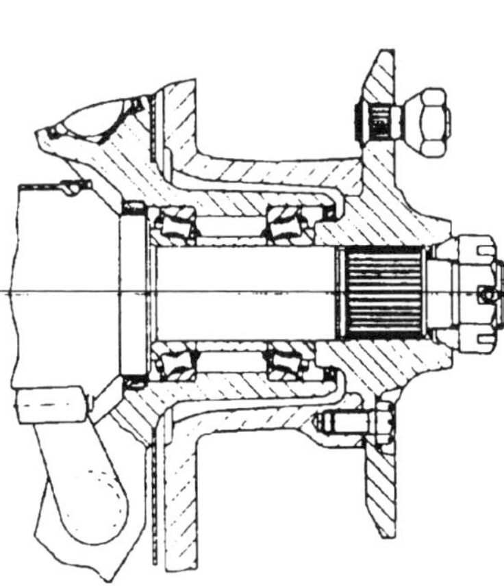

Fig 20: Front hub bearing arrangement

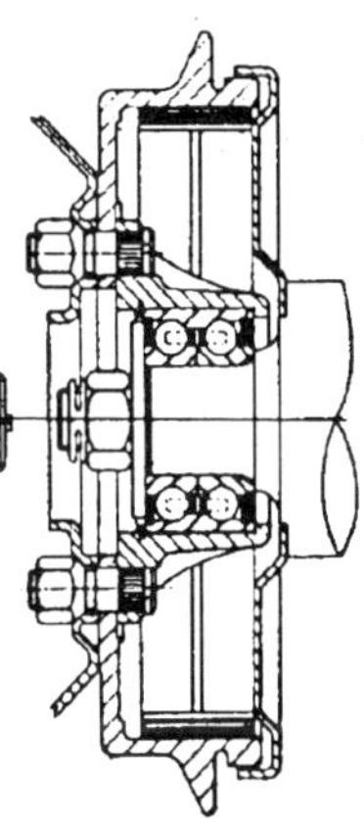

Fig 21: Close-coupled hub bearing

spacer of traditional designs which can lead to tolerance-stack problems. The so-called Timken Set-Right system requires no adjustment in service and the TDOUP packaged bearing concept of Fig 25 uses only one outer double-track race. The assembly is supplied complete with seals and pre-packed with grease.

Evaluation of new hub designs is often on the basis of programmed laboratory testing based on simulated service histories. Data for these has been created by extensive road test measurements made by MIRA. Typical bending moments on the stub axles of medium cars have been measured at 800 Nm, with accompanying axial and radial loads of 5 and 8 kN, in 0.6 g cornering. Relationship between cornering force and bearing loads depends on position of resultant force and axle weight, those in Fig 26 due to RHP showing bearing loads on a 2-litre car hub at various axle weights for a range of lateral accelerations.

RHP say the ISO formula for ball bearing fatigue life is based on the inverse cube of the 'equivalent' load. From that expression, blocks of fatigue wear can be calculated and plotted as a distribution such as the one shown from which it will be noted how rapidly equivalent loads increase with positive g cornering. The area under the graph represents total life fatigue wear. If a bearing shows an equivalent load of 1 at 0 g (straight-ahead) and 7, say, at 0.8 g, then load ratio of 7:1 becomes a fatigue:wear ratio of $(7)^3$:1 or 343:1, meaning one mile at 0.8 g causes as much bearing fatigue as 343 miles of straight -ahead driving.

Safety factors are fixed by choosing conditions of operation and expected driver behaviour. Given that the datum is the average driver on a variety of roads with vehicle lightly laden, then factoring up is undertaken towards the extreme cases as follows:

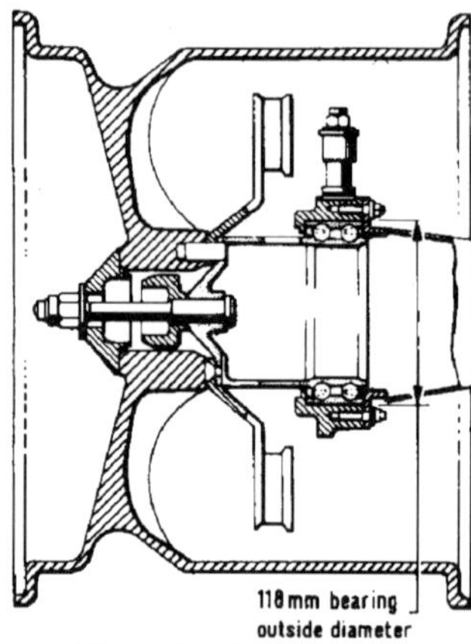

Fig 23: Lotus F1 racing hub bearing

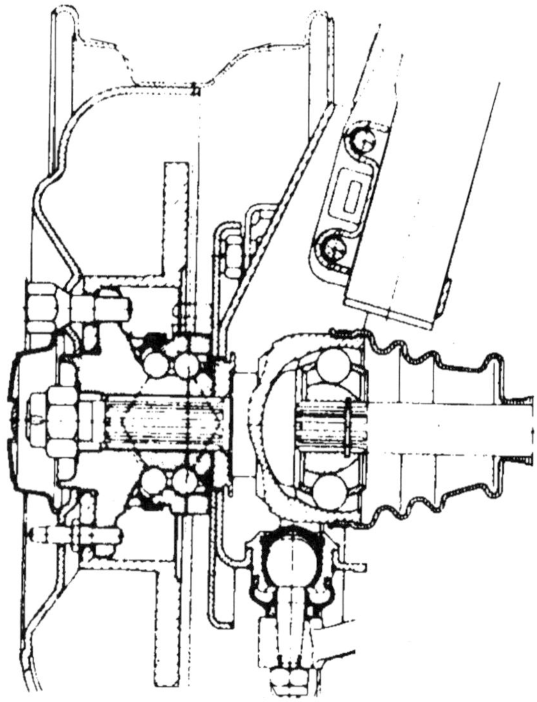

Fig 24: Hub integrated with bearing

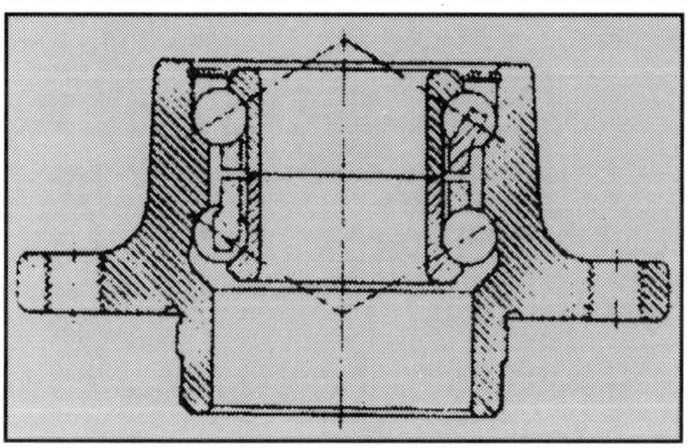

Fig 22: Double-row angular-contact bearing

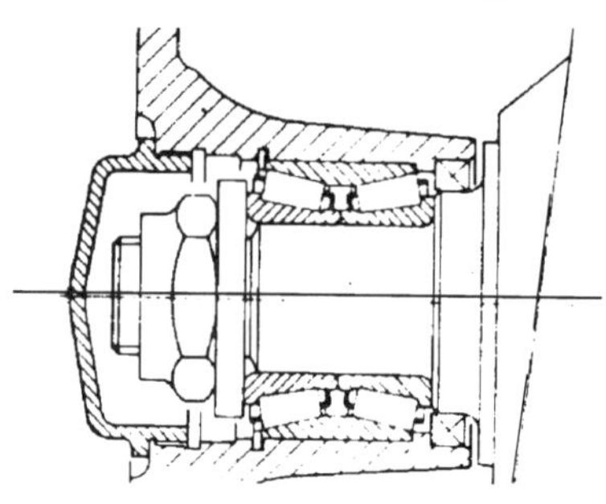

Fig 25: Timken TDOUP system

consider only the fast driver; assume whole life spent on traffic-free rural roads; assume vehicle is fully laden up to 0.2 *g* and midladen between 0.2 and 0.5 *g*; assume 0.8 *g* occurs only occasionally when vehicle is lightly laden. Timken Company engineers[7] have claimed that better fatigue life of the latest taper roller bearings meant that a reduced hub package size could be obtained from a given load carrying ballbearing hub. The table in Fig 27 shows a life load cycle said to be typical for a modern car. While existing taper-roller and ballbearing designs showed similar performance, size for size, Timken claim that latest taper-roller designs, with profiled roller and track sections arranged to uniformly distribute stress along the contact length, resulted in enhanced life for the taper-roller layout.

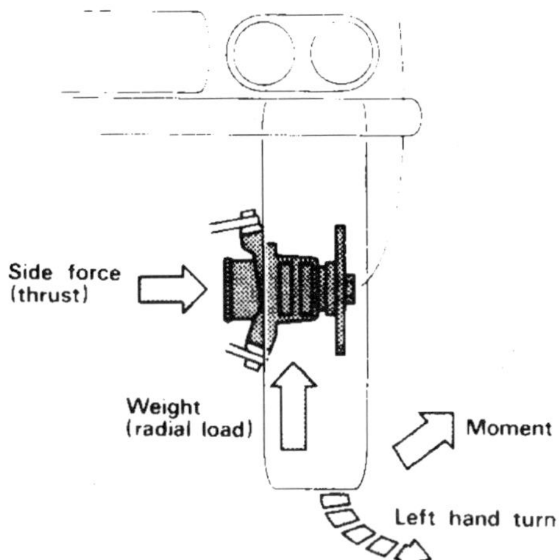

Condition	Load (N)	Load (N)	Speed RPM	% Time
1	5340	2890 (I)*	207	1.58
2	7115	3650 (I)	207	0.47
3	7115	6230 (I)	114	0.08
4	8895	2225 (I)	207	1.11
5	8895	0	1037	39.14
6	16,015	-1780 (O)	207	0.58
7	8005	-2445 (O)	207	0.36
8	4450	-2670 (O)	207	0.36
9	6670	-2445 (O)	1037	1.08
10	8895	0	1037	39.14
11	8895	2225 (I)	1037	8.69
12	6670	3335 (I)	1037	0.42
13	6005	2670 (I)	1037	6.97

* (I) denotes inboard thrust and (O) denotes outboard thrust

Fig 27: Bearing loads on a car hub: second column, radial; third column, thrust

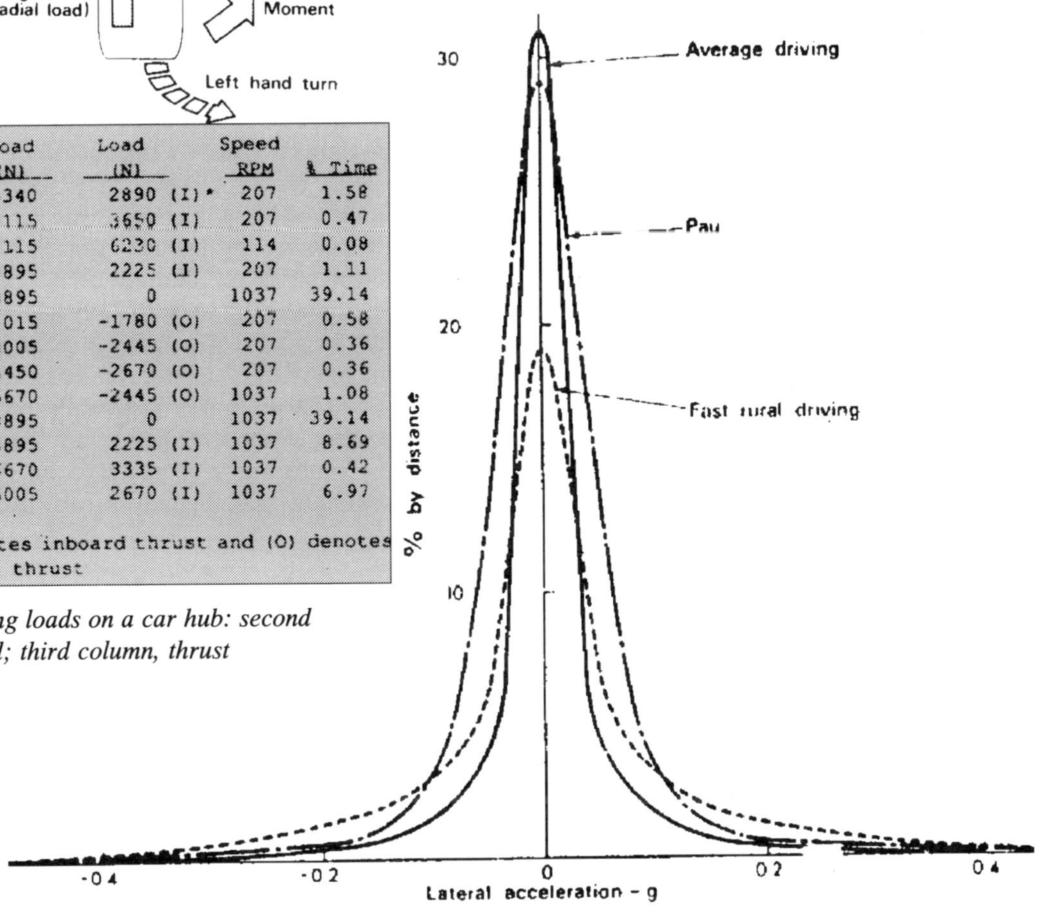

Fig 26: Bearing load distribution

Braking systems

Braking calculations carried out at the early stages of design help to establish the size of foundation brake and any special cooling requirements that have to be built in. Conversion of kinetic and potential energy of the vehicle into heat can involve considerable cooling loads in the case of high speed vehicles so that design for maximum ventilation is a key factor. The analysis is conveniently carried out at the same time as the prediction of accelerative and gradient performance of the vehicle, as covered in Chapter 16.

The fundamental formulae derived from Newton's laws of motion relate the basic parameters so that braking time t and braking distance s can be found from

$$t = (2s/a)^{1/2} \text{ and } s = at^2/2$$

and the energy to be dissipated in braking from speed v down slope of height h is

$$(mv^2/2) + mh$$

where m is vehicle mass and a is acceleration/deceleration. Since the kinetic energy is proportional to the square of vehicle speed, importance of cooling to dissipate the heat energy from braking high speed vehicles is clear. The efficiency of a brake system is the ratio of the applied retarding force to the magnitude of the gravitational force on the vehicle mass. Limiting deceleration is

$$3600\ P/\ mv + g\sin\theta$$

where P is braking power and g gravitational acceleration (SI units).

Practical stopping distance exceeds the theoretical one by that taken to cover driver reaction time and system response time. Also maximum deceleration is conventionally assumed to have been reached when half the brake pressure build-up time has been attained. These are the three so-called 'lost times'. Depending on the driver, brake reaction time is considered to vary between 0.3 and 1.7 seconds.

Minimum braking performance for vehicles is given in European Directive 71/320/EEC (equivalent to UN ECE Regulation 13) while maximum braking performance is limited by the available tyre/ground adhesion. The appropriate Directive must be read in its latest published form to obtain the current requirements but, as a guide, in recent years most vehicles have been required to reach 0.5 g and 0.25 g deceleration with service and parking brakes respectively. This might typically involve designing for 0.6 and 0.3 g and for the vehicle to hold on a 0.19 gradient, in the case of cars. For CVs, the situation is more complex but a requirement to decelerate at least 4.4 m/s^2 (0.45 g) from 50 to 40 km/h for a maximum pedal effort of 70 kg is another general guide, depending on vehicle weight category. Obviously the vehicle brakes must meet these performance criteria as well as being tolerant of changes in payload and road surface conditions without skidding or loss of directional control due to imbalance of braking forces from one side of the vehicle to another. Brakes must also be tolerant of extremes in severity of use and for the life of the vehicle, given proper maintenance.

Drum brakes historically provided varying degrees of self-servo (wrapping) effect, according to the configuration of the shoe linkages, while the coming of disc brakes on cars, and their current

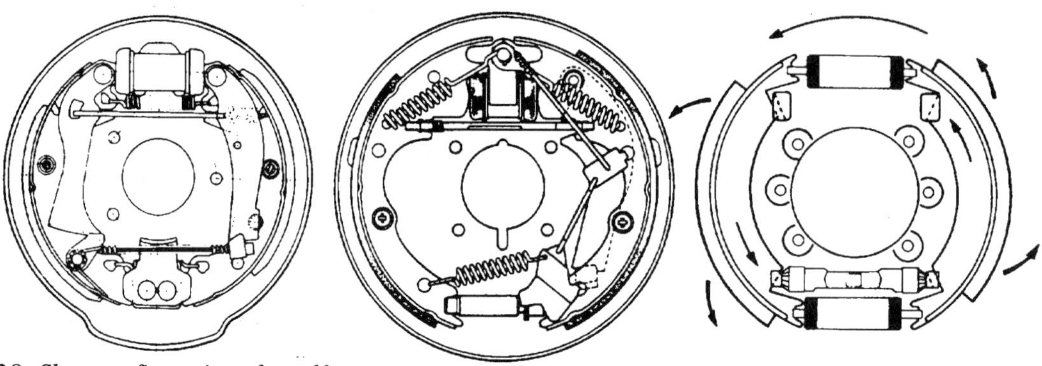

Fig 28: Shoe configurations for self-servo

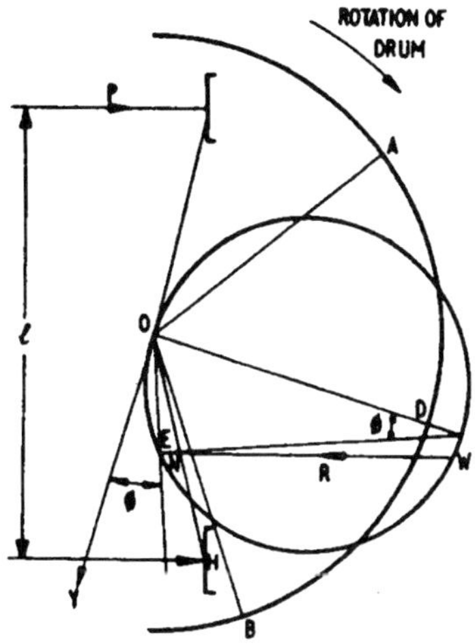

Fig 29: Construction for self-servo

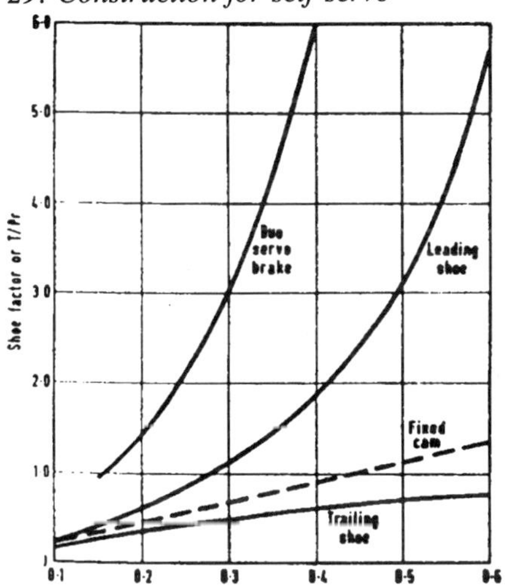

Fig 30: Shoe-factor variation

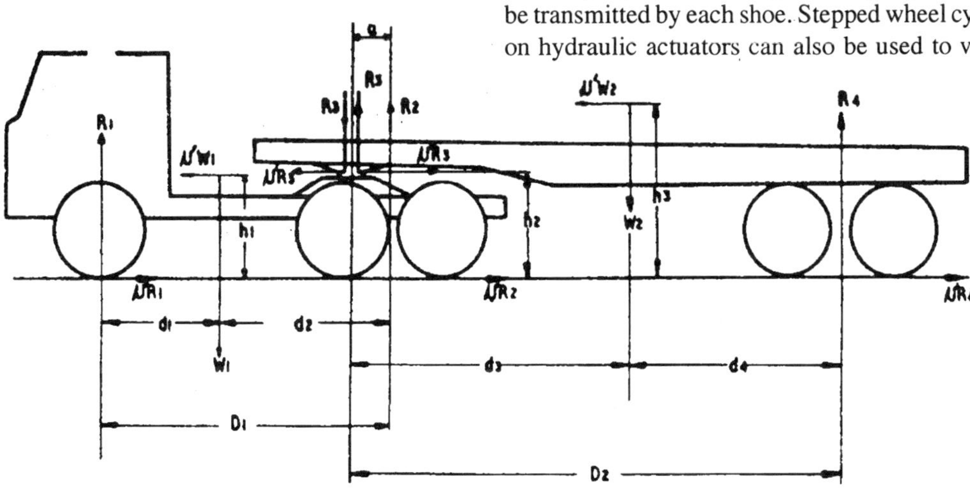

Fig 31: Dynamic weight transfer for articulated vehicle

beginnings on commercial vehicles, means that much less self-servo effect — and therefore a much better chance of achieving even braking force on either side of the vehicle — should result. Heavier vehicles still need the self-wrapping performance of drum brakes to achieve the required decelerations without exceeding reasonable input loads to the foundation brakes. Twin-trailing shoe layouts can be used in drum brakes to reduce self-servo effect to the minimum but on lighter, and particularly, faster vehicles better heat dissipation can be achieved from discs as well as greater ability to avoid brake-fade — thermal expansion of the rubbing surface away from the friction material.

Fig 28 shows some of the various shoe configurations for achieving different levels of self-servo effect; these can include (a) leading/trailing, (b) duo-servo, (c) twin-leading and twin-trailing layouts. Geometric constructions are available for determining the degree of self-servo (or shoe-factor), that applying to a leading shoe with parallel abutment actuation being shown in Fig 29. The technique is to determine point E on the basic circle which lies on the line of action of R. Since forces on the shoe are parallel, EW represents the direction of the resultant reaction of the drum on the shoe and W is the centre of pressure. Shoe factor of a leading shoe is then given by $S = l.\ ON / r.\ HN$ where N is the point of intersection of R with OH.

Two types of expander mechanism are common, one which will apply equal actuating forces to the tips of each shoe and those which displace the shoes by equal amounts, permitting equal torque to be transmitted by each shoe. Stepped wheel cylinders on hydraulic actuators can also be used to vary the force on each shoe and therefore to nullify the difference in self-servo effect between leading and trailing shoe, say. Varying cam forms on mechanical actuators can also be used to obtain different characteristics. The S-cam brake, for example, uses an involute profile to give a constant lift as the cam rotates. Fig 30 shows the variation of shoe factor for some different shoe layouts and actuators. Note that for a friction coefficient of about 0.4, the torque output of a leading shoe is about three times that of a trailing shoe of the fixed pivot type. The torque developed by the camshaft is the product of the servo actuator fluid pressure and its cross section area times the mechanical advantage of the lever arm. This torque is then divided by the effective cam radius to obtain the load applied to the tip of the shoe. A typical value for shoe centre lift, to achieve 0.6 *g* stop at 5.86 bar servo pressure, is 2 mm corresponding to a stroke of 57 mm for servo-actuator areas of around 130 cm^2.

Brakes should provide sufficient torque to lock all wheels simultaneously on a surface with adhesion coefficient of 0.6, to achieve the 0.6*g* retardation. Given vehicle centre-of-gravity position and key dimensions, the dynamic load on each axle can be determined for a 0.6 *g* stop, taking load transfer into account. Fig 31 shows the difficult example of an articulated vehicle. Given that $W1 = 7.5$, $W2 = 30.5$ tonnes, static weight distribution $R'1 = 6$, $R'2 = 16$, $R'3 = 16$ tonnes; $D1 = 154$, $D2 = 282$, $d1 = 80$, $d2 = 74$, $d3 = 148$, $d4 = 134$ and $a = 0.64$ *m* — and that heights $h1$, $h2$ and $h3$ are 0.97, 1.12 and 1.78 m — dynamic weight distributions are

$$R1 = W1(\mu h1 + d2)/D1 + W2(\mu h2 + a)(d4 + \mu h3)/D1(D2 + \mu h2) = 10.5$$
$$R2 = W1 + W2 - R1 - R4 = 14.4$$
$$R3 = W2([d4 + \mu h3]/[D2 + \mu h2])$$
$$R4 = W2([D2 - d4 + \mu h2 \mu h3]/[D2 + \mu h2]) = 13.1 \text{ tonnes}$$

Respective braking forces thus become $F1$, $F2$ and $F4 = 6.3$, 8.64 and 7.86 tonnes and for a tyre rolling radius of 0.5 m, torque to be developed by a front brake is 16 kNm. Given a fixed cam brake of 0.91 shoe factor, required shoe tip load is 9 tonnes for drum diameter 0.42 m and overall brake efficiency 0.9, corresponding to a 5.86 bar air pressure on a 185 cm^2 chamber area and 8.55:1 lever-arm ratio — using this calculation method due to TP Newcomb[8] of Loughborough University. Width of lining selected is a function of heat dissipation capacity and should be chosen so all drums throughout the vehicle run at much the same temperature. Newcomb suggests using 17.75 cm wide for the trailer linings and 15.25 cm for the front brakes which are more effectively cooled but further formulae can be derived to obtain a more precise picture of the heat dissipation requirements.

The kinetic and potential energy converted into thermal energy can be made by using the 'J' constant for the mechanical equivalent of heat. It is suggested that disc brakes, Fig 32 can operate at 50% higher temperatures than drums with their inherently better cooling rates. With repeated brake applications the disc temperature eventually reaches an equilibrium value at which the heat loss between applications equals heat input during an application. This limiting temperature is given by $\theta_0 + d\theta/bt_0$ where $d\theta$ is the increase in temperature after one stop, t_0 the interval between applications and b a cooling constant. As an example, a car weighing 2800 lb repeatedly braked from 40 to 20 mile/h in neutral at a frequency of twice

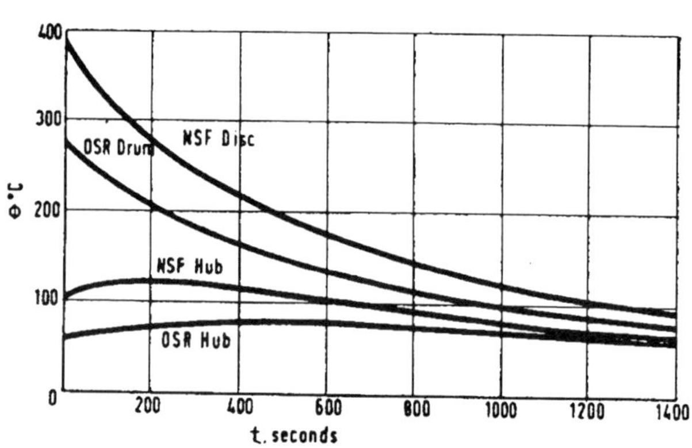

Fig 33: Brake thermal performance

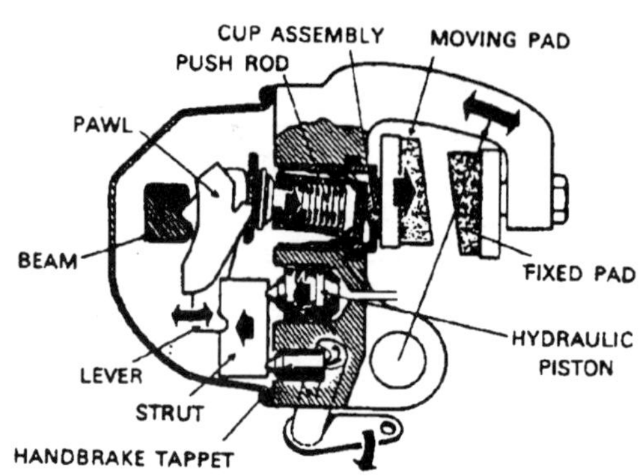

Fig 32: Disc brake constituents

per minute would generate 1.05 x 2800 x 882 / 64.4 x 778 x 3 Btu's at 0.6g deceleration; $b = 3 \times 10^{-3}$/s and temperature increase for one stop = 43.2/15.28; theoretical thermal performance, compared with measured values is shown in Fig 33.

Braking performance limits set by available friction at the tyre/ground contact patch are affected by the dynamic weight distribution during deceleration of the vehicle. Maximum transferable braking force for the front and rear axles respectively is

$$\mu W(Lr + fH)/(L - \mu H)$$
$$\mu W(Lf - fH)/(L + fW)$$

The condition of maximum brake performance, with braking front and rear, is when distribution of braking forces to the axles is in proportion to their respective dynamic weights, expressed approximately as front-to-rear braking force $Ff/Fr = (Lrg + bH)/(Lfg - bH)$ where Lf, Lr are distances to front and rear axles from vehicle centre-of-gravity and H is CG height. Usually an average value of road adhesion of 0.35 g is assumed corresponding to deceleration b of 3.4 m/sec^2. This can rise to 6 m/sec^2 on dry concrete (compared with over 152 m/sec^2 typical of head-on collisions. Parking brake holding power is calculated without recourse to dynamic weight transfers. Minimum friction co-efficient to hold the vehicle on a grade (Fig 34) of $G\%$ is

$$G - 100f = (100Lf/L) +/- (HG/L)$$

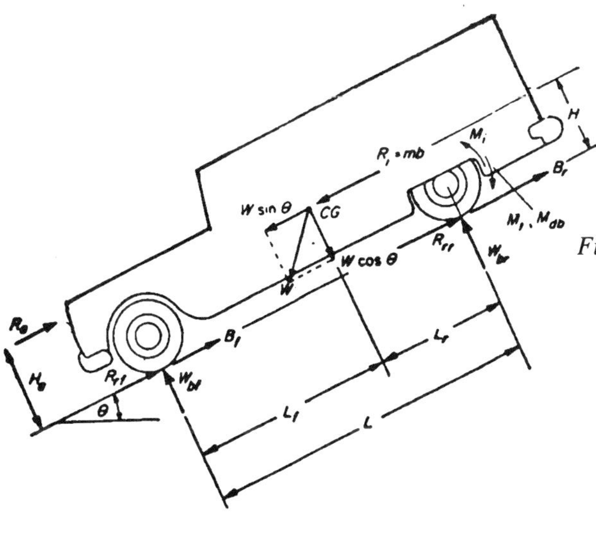

Fig 34: Braking on a gradient

Anti-lock braking

Increasingly becoming a requirement on many car and commercial vehicle categories, ABS is seen as a key active safety fitment. Choice of system depends on cost and drive configuration of the vehicle as well as on the availability or not of other on board electronic systems.

Anti-lock braking can be seen not only a means of preventing skid situations but also of improving brake performance on those vehicles which cannot warrant elaborate servo-assisted systems. Higher brake factors can be employed on comparatively simple drum brakes if the incidence of brake locking can be avoided. By continuously matching brake pressures to the demands of the ever changing tyre/road adhesion co-efficients optimum braking level can be achieved for a given level of foundation-brake sophistication.

There is also potential already to adapt ABS to obtain anti wheelspin capability and in prospect are systems which include such functions as brake-performance monitoring, aquaplane (temporary-lock) control, hill climb and hold aids, yaw-moment control, and multi-wheel drive torque distribution control. Integration with handbrake-by-wire and parking-steer systems has also been mooted as well as communication with other electronically managed

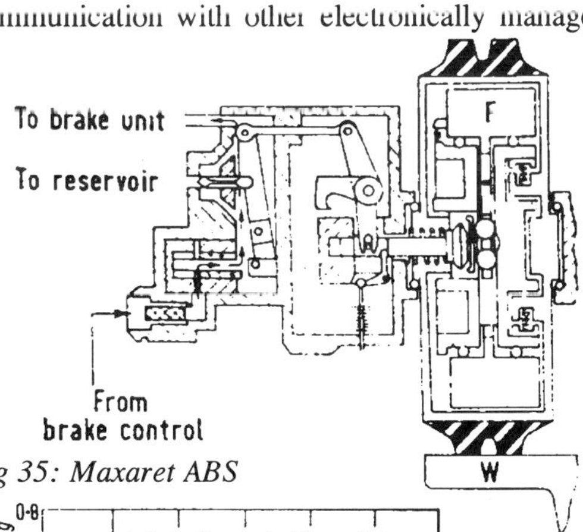

Fig 35: Maxaret ABS

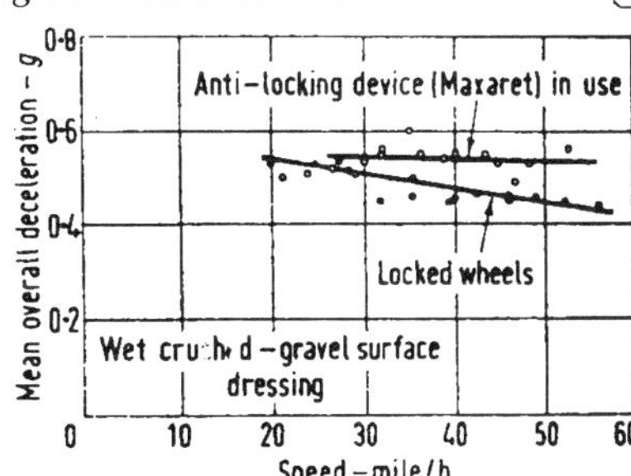

Fig 36: Braking improvement with ABS

vehicle systems.

A number of legal requirements on braking systems must also be honoured by the ABS. Since it is a requirement that front brakes lock before the rear in an emergency, priority to the front wheels should be given even though the system may be designed to prevent lock on all wheels. In the event of brake servo failure, an adequate system must still remain, to provide 0.3 *g* braking, without the driver exerting more than 500 n pedal pressure.

Commercially available systems involve two basic hydraulic concepts: 'return flow' and 'dynamic flow-in'. The former reduces pressure in the brake cylinders by return of drained brake fluid to the master-cylinder — against pedal force. By contrast, dynamic flow-in implies draining of the brake fluid volume required to reduce the brake pressure — into the non pressurised reservoir; it draws the fluid volume required for subsequent pressure restoration from an auxiliary energy source, feeding it to the brakes at an adapted pressure. Methods of achieving these concepts differ between system suppliers. Early automotive systems were derived from aircraft units such as the Dunlop Maxaret. Initial applications were to articulated trailers, as a means of curing the jack-knife problem. They included a device sensitive to rapid changes of rotational speed and so detected the violent checking of either wheel at the onset of brake-lock. A flywheel which was incorporated overran the axle shaft rotation and was caused by a helix to move axially against a spring-loaded valve. This operated, through a second relay valve, air-circuits which released the brakes. Fig 35 shows a derivative of the system developed for the Ferguson racing car. Fig 36 compares the behaviour of tyres braked on different surfaces with and without ABS and Fig 37 shows higher peak brake force co-efficents which are available when the wheels do not lock.

Second generation systems were able to overcome the disadvantage of having just one braking threshold level which was inherent in the mechanical and electro-hydraulic systems — by taking account of road conditions. Frictional co-efficient between tyre and road effectively varies according to whether

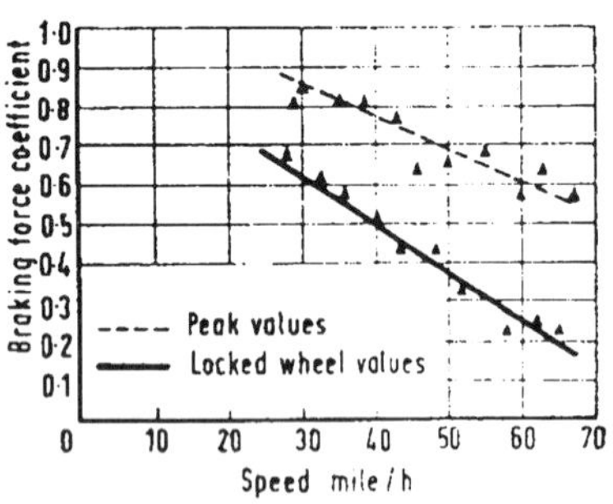

Fig 37: Achieving higher brake force co-efficient

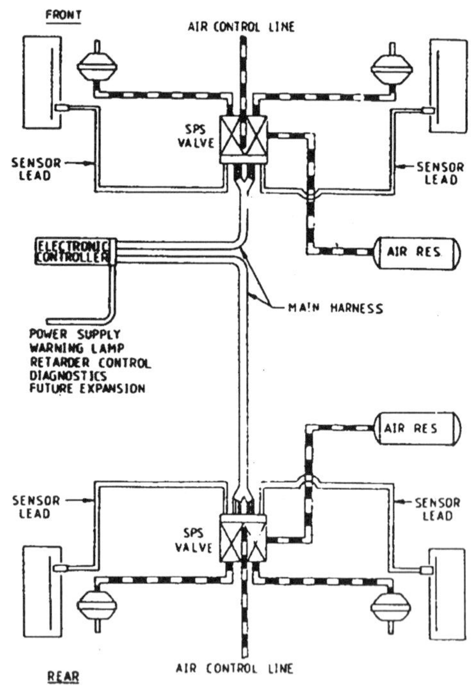

Fig 41: Skidchek BGX system

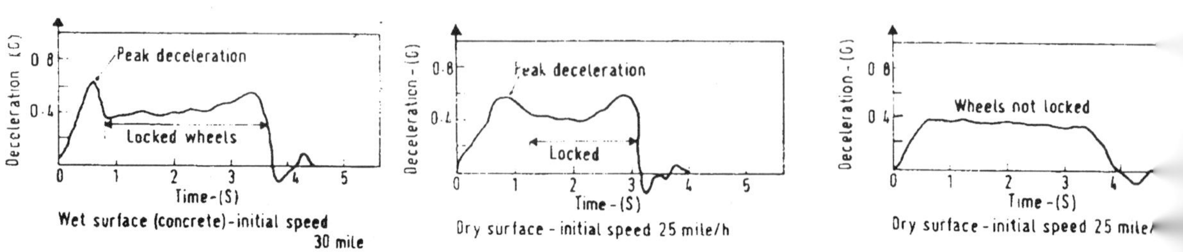

Fig 38/39/40: Friction co-efficient variation

the wheel is turning freely or locked, in the extremes, as seen in Fig 38/39/40. The shape of this curve also varies with ground surface, tread-pattern, inflation-pressure and temperature. By changing choice of locking parameter to the acceleration of the wheel — derived from the speed of a toothed disc — better susceptibility to these other factors is obtained.

As well as the wheel sensor, the other key components are the control-unit, to determine the point of brake release or re-application, and a modulator to actuate fluid pressure in braking as required by prevailing conditions. The controller requires specific logic for reselection — finding whether or not the danger of locking has been averted. It must also perform the prediction function — verifying whether locking conditions are satisfied or otherwise.

Anti-skid brakes on many truck-trailer combinations are now mandatory. Three types of system are listed by the EC brake directives. Category 3 is an individual axle system which might fit to a tractive drive axle; Category 2 is a complete vehicle system with the wheels on all axles sensed and must cope with split-mhu surfaces, either side of the vehicle; Category 1 has to use the maximum grip available at each wheel on a surface whose grip can vary at each wheel station. For trailers, the EC standards are a little less stringent but already some categories of semi-trailer in the UK require ABS fitment. In the UK only one axle need be sensed on tandem and triaxle trailers and systems should use a sensor on each wheel — while reducing air pressure on each when one wheel senses a low-grip surface.

The Lucas Skidchek BGX systems for CVs is seen in Fig 41 as a four-channel system, to each wheel station. It has the facility that on the front axle, although having two independent valve control elements, can be controlled so that on split-mhu surfaces, actions of the control elements are related. The gradual build up of unequal forces on the front axle allows the driver to anticipate further changes and make steering corrections. The company also have a low cost system for light vehicles known as the SCS and a 2, 3 or 4-channel modular electronic control system which can incorporate traction control.

The SCS has only four components, two modulator units, Fig 42, and two apportioning valves. The inertial sensors are built into the modulators and are sited in the brake lines between master-cylinder the point at which the diagonal brake circuits divide to front and rear axles. They are driven by toothed belts from the drive shafts either side of the transaxle. No electrical input is required. Fig 43 shows the operating characteristics.

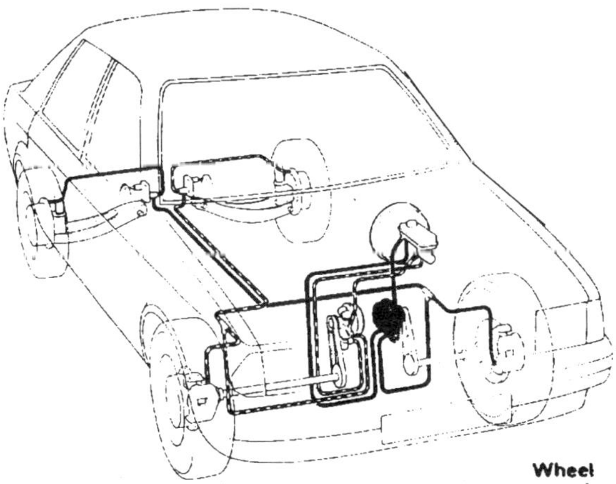

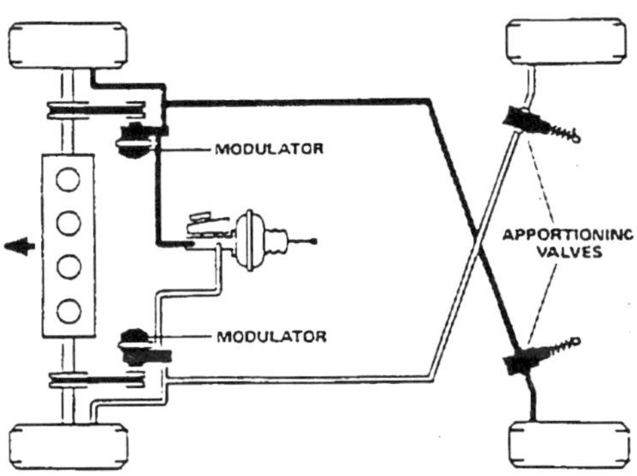

Fig 42: Lucas SCS schematic

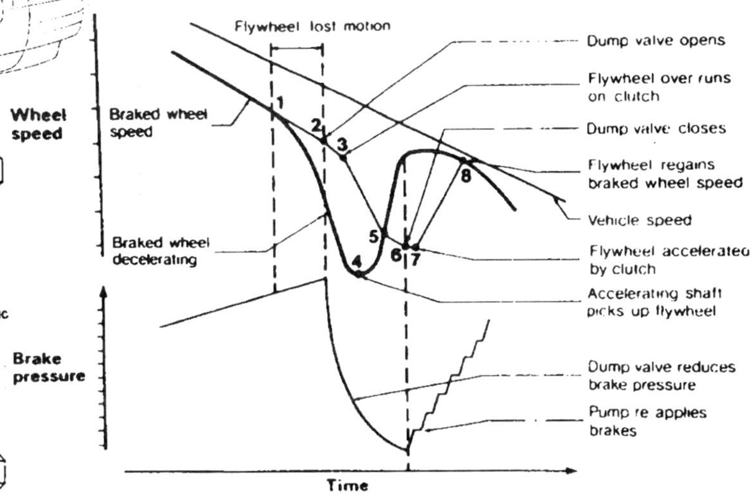

Fig 43: SCS operating characteristics

The Lucas electronic systems are challenged by those of Bosch and Teves, Figs 44/45. The latter's mark 4 system is illustrated complete with traction-control. Brake fluid is directed to the appropriate wheel cylinder via a motor-powered 2-circuit valve pump. In a traction control situation, fluid flows from pump or energy reservoir to the wheel cylinder at a pressure level determined by the electronic controller. Start slip control up to 30 kph comes from two additional valves in the modulator while traction control for the entire speed range comes from an additional high power and pressure energy reservoir. Engine intervention (with electro-magnetic butterfly valve adjustment) is necessary to modulate engine torque for both drive modes, in the case of traction control.

Type of pipe	Surface finish	Friction factor
Copper	70a	0.035
Rubber	38a	0.033
Nylon	2.5a	0.030

Fig 46: Roughnesses compared

Braking pipework

In the move towards vehicle weight reduction, nylon pipes are being increasingly used to replace steel and copper assemblies — also offering ease of fitment, relative freedom from fatigue failure and absence of corrosion. Nylon's resistance to creep under temperature makes it one of the preferred contenders.

Hydraulic actuation and brake servo system air lines are potential applications and self-swage fittings are readily available. Alternatives are drive-in fittings and nut-and-olive assemblies but both these can cause distortion of the pipe. Where bends are unusually tight, a degree of pre-forming is advisable. Surface smoothness offered by the material is particularly beneficial in those applications where low pressure drop is critical; Fig 46 shows the roughness of copper and steel systems against nylon and the

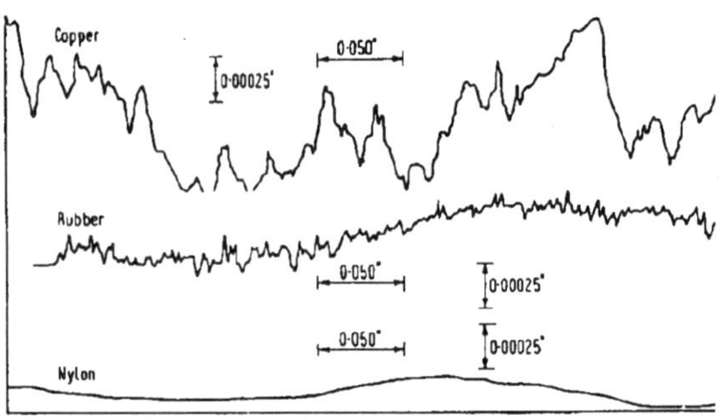

Fig 48: Surface finish applying to Fig 47

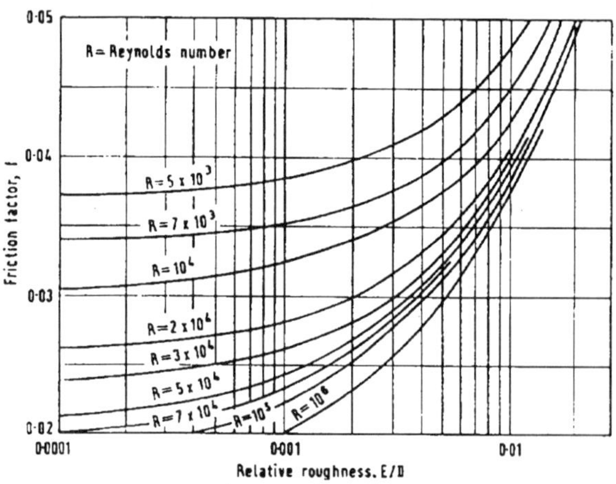

Fig 47: Friction factor effect in pipe

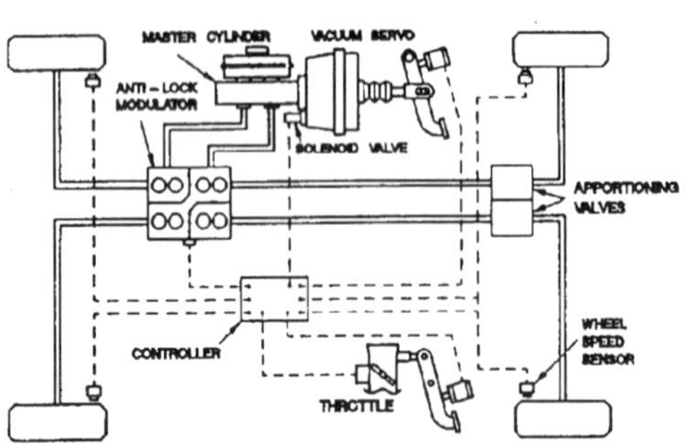

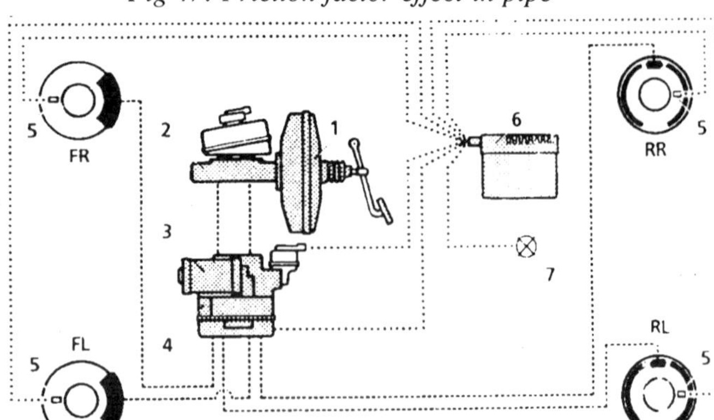

1) Vacuum booster
2) ABS tandem master cylinder
3) ABS hydraulic unit
4) TCS valve module
5) Sensors
6) Electronic ABS anti-lock system controller with integrated TCS logic
7) Warning lamp fittings
Brake, ABS anti-lock system, TCS operation

Figs 44/ 45: Bosch (left) and Teves (right) systems

Reynolds number plots in Fig 47 show the effect on friction factor in the pipe. For a 6.5 mm bore pipe having a surface finish as seen in Fig 48; with a constant Reynolds Number of 104, friction factors are as seen in the figure.

In hydraulic lines, circuit pressures are usually too high for unreinforced nylon tubing. In air braking applications, the coiled 'suzies' between artic tractors and trailers are one of the best known with the material's ability to be pre-coiled so that a tension spring effect is produced. SMMT data sheet NM 7 (1-9) provides useful background in relating properties to applications.

Resistance to fluid flow in pipes

The resistance in terms of the pressure drop needed to cause a unit volume to flow in unit time is given by $E = k\mu l/d^2$ where m is viscosity, d the diameter, and k a constant dependent on units. For pressure drop in absolute units the kinematic viscosity is required in equivalent units. The nomograph in Fig 49 enables the resistance factor to be found. Multiplying this by both the fluid specific gravity s and length of the pipe in inches will give pressure drop in lb/in² for each cubic inch of flow per second. The formula is valid for values of Reynolds number, Re, below 2000.

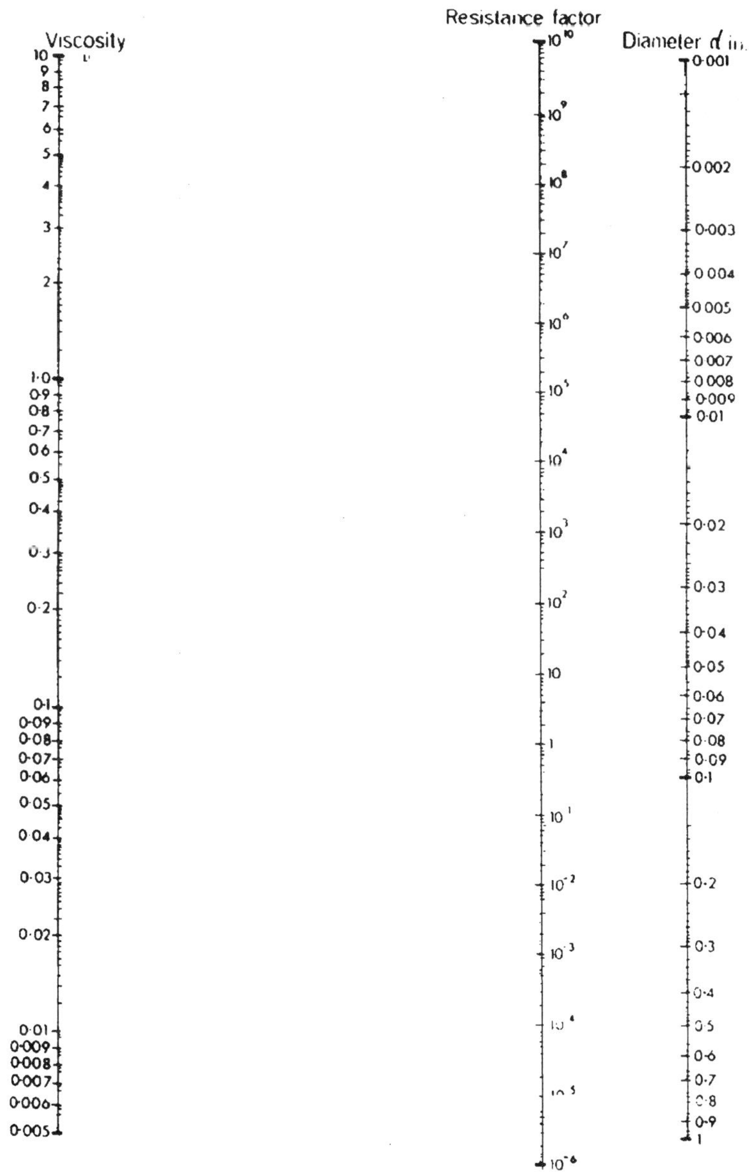

Fig 49: Nomograph for flow resistance

Drive-line calculation

Clutches, driveshafts and universal joints have become highly developed components which need to be specified with care in order to optimise driveline efficiency. Here some calculation procedures and design considerations are discussed which will help rationalise the choice of the drive-line assembly.

Two methods of calculating torque capacity of clutches are used — one based on uniform wear and one on uniform pressure. Both give closely tallying results, given accurate values of friction co-efficient and clamping load — each of which can vary according to operating conditions. The clamping load variation which is speed-dependent can be calculated. The friction value is more difficult as it can vary with temperature. This is significant in the face temperatures reached by different vehicles in accelerating from rest on a gradient, say.

During such an engagement, clutch slip occurs until the power source and driven member are rotating at the same speed. The clutch must be designed to absorb heat developed in this process without detriment to the friction value. The following formulae are based on the engine delivering constant torque at constant speed during the take-up — and the clutch being instantaneously engaged:

$$\text{Acceleration} = [(T.G.n/R) - (r + \theta)]q/W$$
$$\text{Slip time} \quad t = N.R/g$$
$$\text{Heat generated} = T.N.t/2P$$

per unit of pressure plate weight where T is engine torque, N the speed, R tyre rolling radius, G overall gear ratio and W the gross vehicle weight. Rolling resistance is r, gradient resistance q, transmission efficiency n and pressure plate weight, P.

The technique is to draw a gradeability curve, of the type produced by Borg and Beck in Fig 50. This shows that the maximum gradient on which the

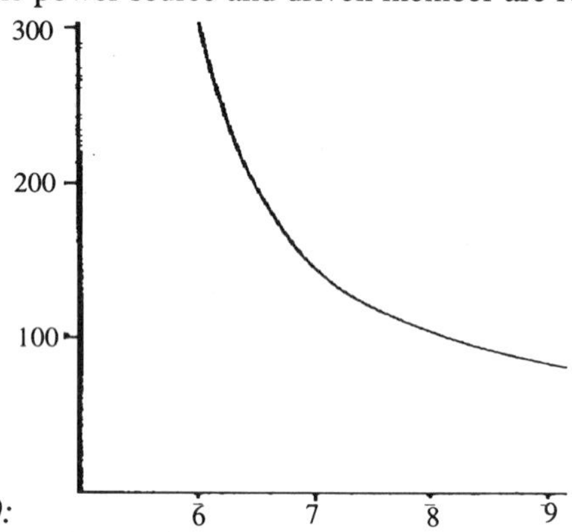

Fig 50: Gradeability curve: total heat generated in BTUs x 10^{-5} (y-axis) against gradient;1 (x-axis)

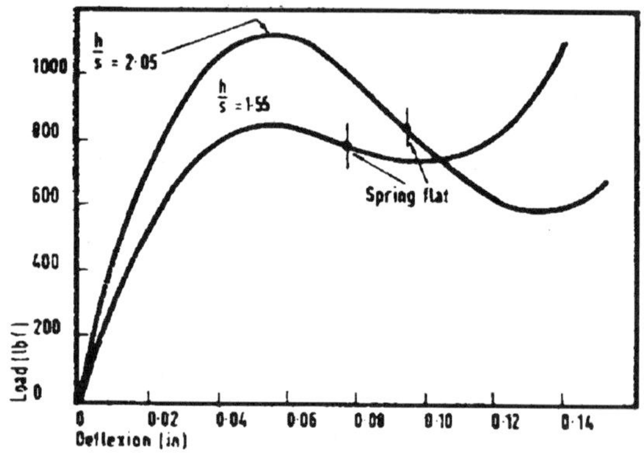

Fig 52: Spring release curves

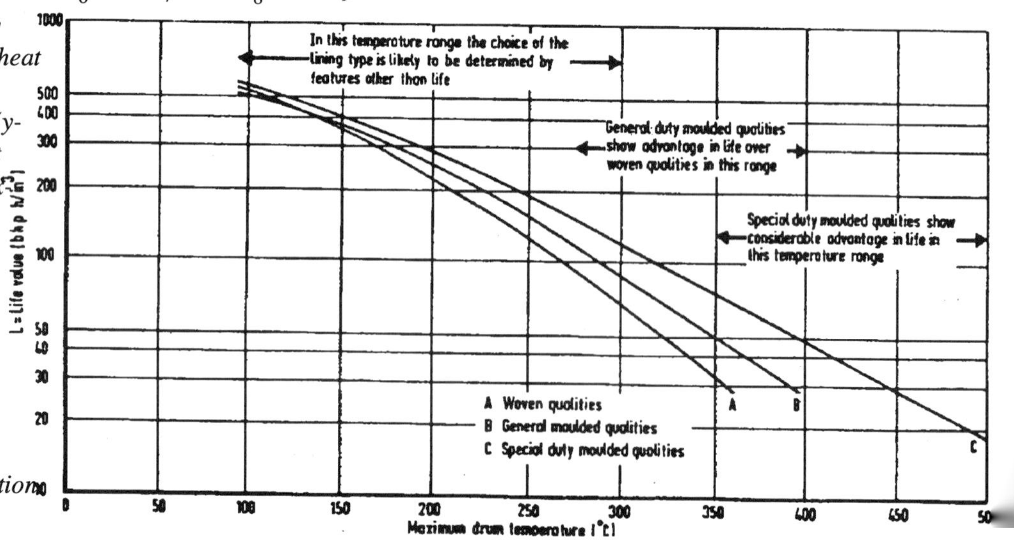

Fig 51: Friction material life

vehicle is expected to start should not be too far around the curve — otherwise slight variations in tractive effort can make large differences in the work done by the clutch. Although these are unlikely to result in clutch slip they could detrimentally affect its life.

The alternative approach of assessing clutch life by consideration of surface temperature is based on a formula for one-dimensional heat flow with the assumption of uniform thermal flux at the surface. Temperature will be affected by rise during a period of continuous slipping, rise during a single engagement, bulk temperature and localised surface temperatures. In continuous slipping the whole pressure plate can heat up and after initial surface temperature builds up; temperature rises at a rate inversely proportional to pressure-plate thickness. For single engagement, by contrast, the plate can be considered infinitely thick as only the surface vicinity will be heated.

Surface temperatures generated are of a transient nature and heat will subsequently raise the bulk temperature of the pressure plate by conduction from the surface. There will be cooling from the plate between applications — and for closely repeated or continuous ones the bulk temperature will rise to a limiting value. This can be calculated using Newton's law of cooling and assuming fixed intervals between successive applications for design purposes. Life values for woven and moulded friction materials are typified by those given for Ferodo linings, some years ago, in Fig 51.

Work done per application times frequency divided by the product of material rating and slipping time will provide the area of lining required. Thickness required comes from dividing the life factor from the figure into the energy dissipated per application expressed as the product of material rating and slipping time. Main considerations in diaphragm spring selection are the required clamp-load, pedal 'feel', lever ratio and the available release travel.

Release curve characteristics can be modified for a given spring diameter by changes to height h and thickness s, Fig 52. These curves are obtained from an empirical modification to the handbook formulae for Belleville springs. It is usual to check the stress occurring at points C,D,E and F in Fig 53a to obtain plots such as those in Fig 53b. Localised stresses up to 200,000 lb/in^2 (1.38 MN/m^2) are tolerable due to 'support' of adjacent material. The determination of pressure plate weight is a function of its ability to act as a heat sink for any temperature rise to a proximity of 1000 C possible face temperatures. In terms of power dissipated, 2 bhp per inch of lining face area has been suggested converted to 20 bhp per lb of

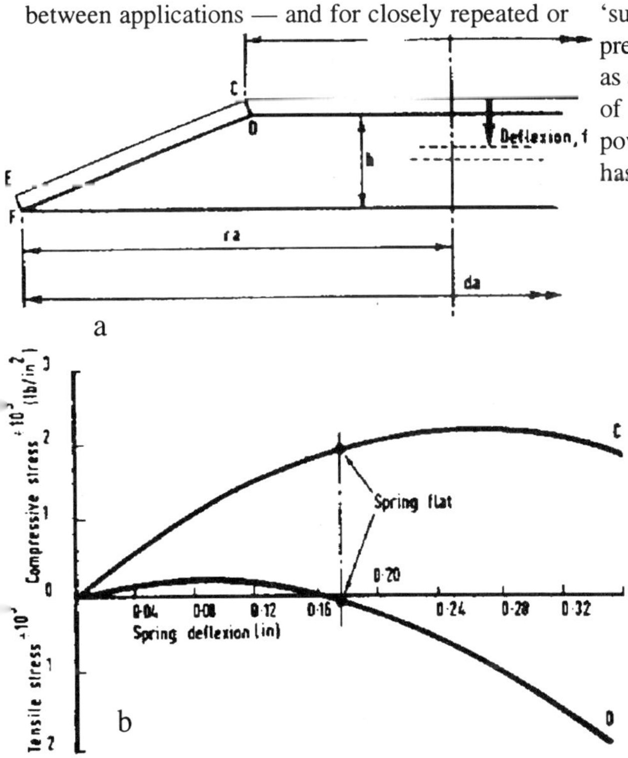

Fig 53: Disc spring stresses

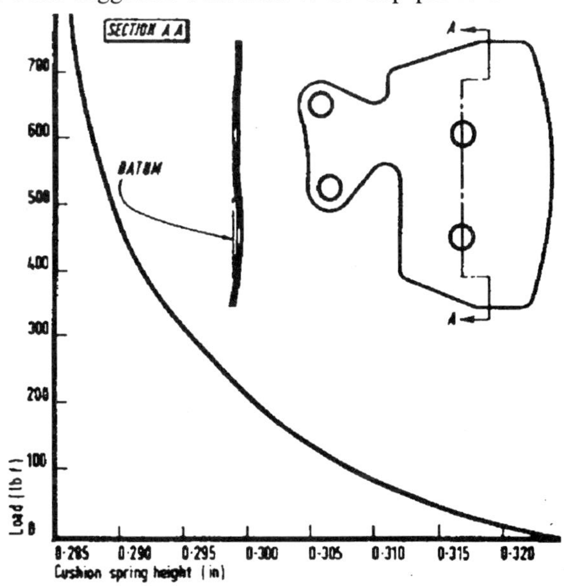

Fig 54: Cushion curve

spring as a guide for design. Also suggested in design are slip torque to engine torque ratios of about 1.35 and clamping load per facing area of about 45 lb/in^2.

Driven plate design is governed by needs for smooth engagement, progressive wind-up and torsional damping — commensurate with a life factor appropriate to operating conditions. Fig 52 shows variation in clamp load which must be observed for new and worn driven plates, respectively. Axial cushioning between friction facings is usually provided to both smooth the engagement and lower the release loads. Fig 54 shows a typical cushion curve for the spring segment shown inset and Fig 55 the modification to the release curve. Further springs tangential to the lining, Fig 56, are provided to reduce tendency for torsional vibration in conjunction with a rubber damping ring. The wind-up of these springs needs to be calculated and if required, delayed pick-up is provided on some springs to produce an overall characteristic such as that of Fig 57.

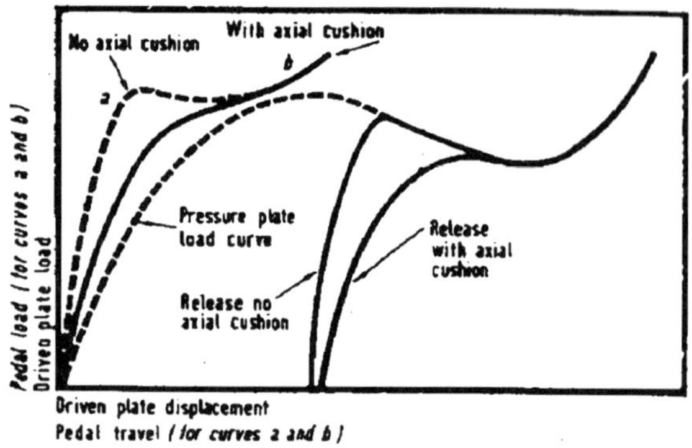

Fig 55: Release curve modification

To find the pressure which must be exerted by the spring, the total axial thrust can be expressed as

$$p.\pi(r_1^2 - r_2^2)$$

and the friction moment from each surface as

$$\mu p\pi(r_1^2 - r_2^2).(r_1 + r_2)$$

An idea of the different performance requirements for cars and commercial vehicles is seen in Fig 58, due to transmission consultant C G Giles[9], which compares tractive effort for the two categories in the moving from rest phase. This shows that the car has to be travelling between 5-10 mph before the lower gears are operational and the 5-tonne truck must be moving between 2-3 mph. To start the vehicle from rest, the starting resistance must be overcome, for the worst gradient. The engine is typically run up to medium speed and the clutch partially engaged so as to slip for about two seconds until positive connection was achieved. The engine flywheel also provides a reserve of thrust; most clutches are designed to carry at least twice engine torque before slipping and for steep hills the engine can be speeded to nearly its maximum before releasing the clutch reasonably abruptly to obtain torque transfer. A curve in the figure also shows how even greater starting torque can be obtained with a torque converter.

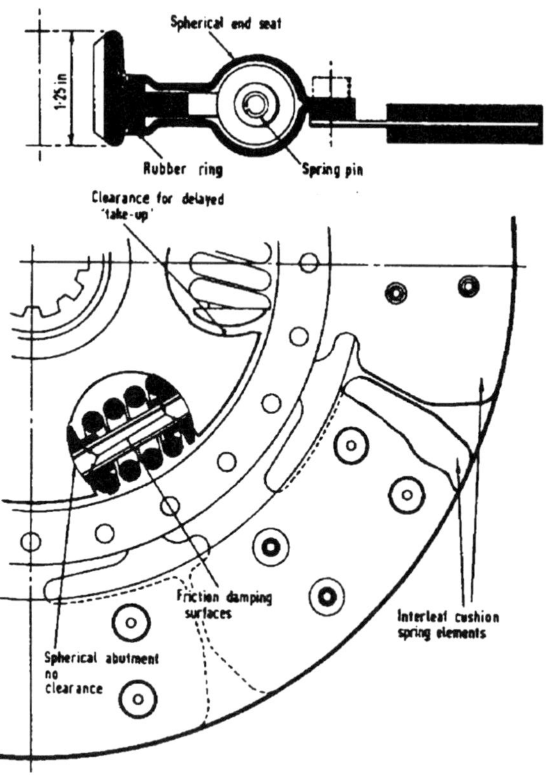

Fig 56: Anti-chatter springs

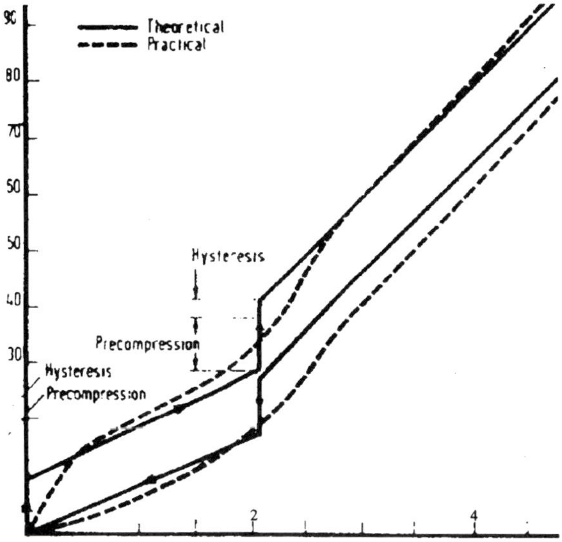

Fig 57: Desired characteristic: torque in lb ft (y-axis) vs degrees of wind-up

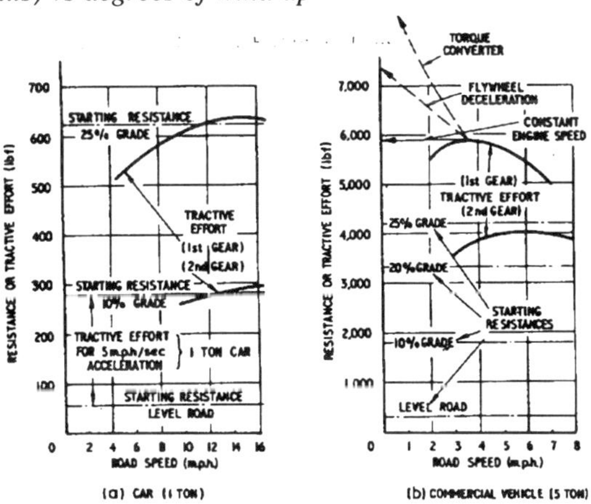

Fig 58: Car and truck requirements

Drive shafting and universal joints

Propeller shafts for cars and commercial vehicles are compared in Fig 59. Constant velocity operation through a propeller shaft can, of course, be obtained if the axes of the input and output shafts, beyond the universal joints at either end, are truly parallel. There are, however , torque and velocity variations in the elements of conventional Hookes joints which give rise to secondary rocking couples; these affect the support bearings of the input and output shafts. Joint angles should thus be as small as possible to minimise these effects.

In the case of shafts which drive direct from the gearbox to the wheels, it is particularly difficult to achieve this and use of constant-velocity universally-jointed half-shafts, Fig 60, have become the rule. These CV joints incorporate plunge action to avoid the need for using splines. In general the permissible load in geometrically similar designs is proportional to the cube of their linear dimensions. Reduction in the allowable working stress, for obtaining a given life expectancy, by an angle factor is shown in Fig 61. The recommended limit of speed at any angle is also shown by the family of curves covering the various joint sizes.

GKN engineers have devised techniques for calculating load/life relationships for CV joints, thc homokinetic behaviour of which depends on the torque-transmitting balls being guided in such a manner that they always lie in the plane bisecting the angle between driving and driven axes, Fig 62. Two

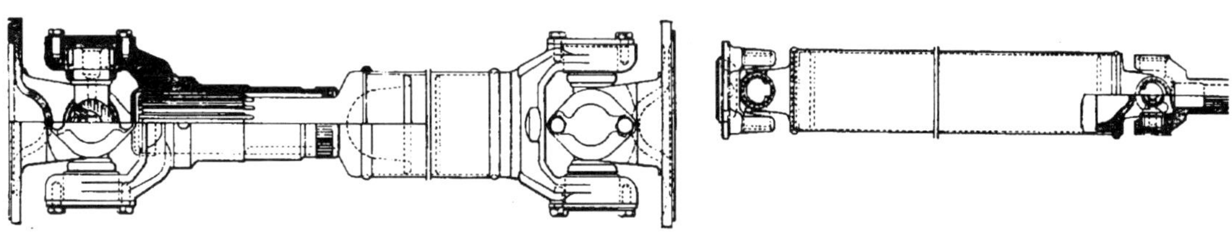

Fig 59: Truck and car prop-shafts

Fig 60: Constant-velocity jointed shaft

loading conditions are normally considered: momentary loading by sudden application of driving torque against stationary components — and fatigue loading caused by continuous variation in torque and speed when in motion. The first is the one normally employed in design and is obtained by factoring static torque capacity by 1.2 to 1.8. Torque transmitted

$$T = Pa.n.R.\sin.p$$

for pressure angle angle *p*, number of balls *n* at pitch circle radius *R*. Allowable stress *Pa* is derived from maximum permissible stress of around 240 tonf/in². Joint life is calculated from

$$L = 400[1000T_{100}^{\ 3}.A_x^{\ 3}/(T_x^{\ 3} + 0.094^3)N_x]$$

where N_x is working speed, T_x working torque, T_{100} joint application torque specified by the manufacturer; A_x is the corresponding angular factor.

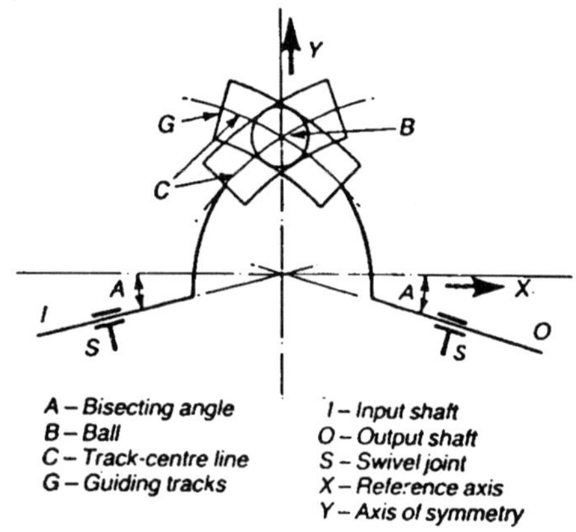

Fig 62: Load/life calculation

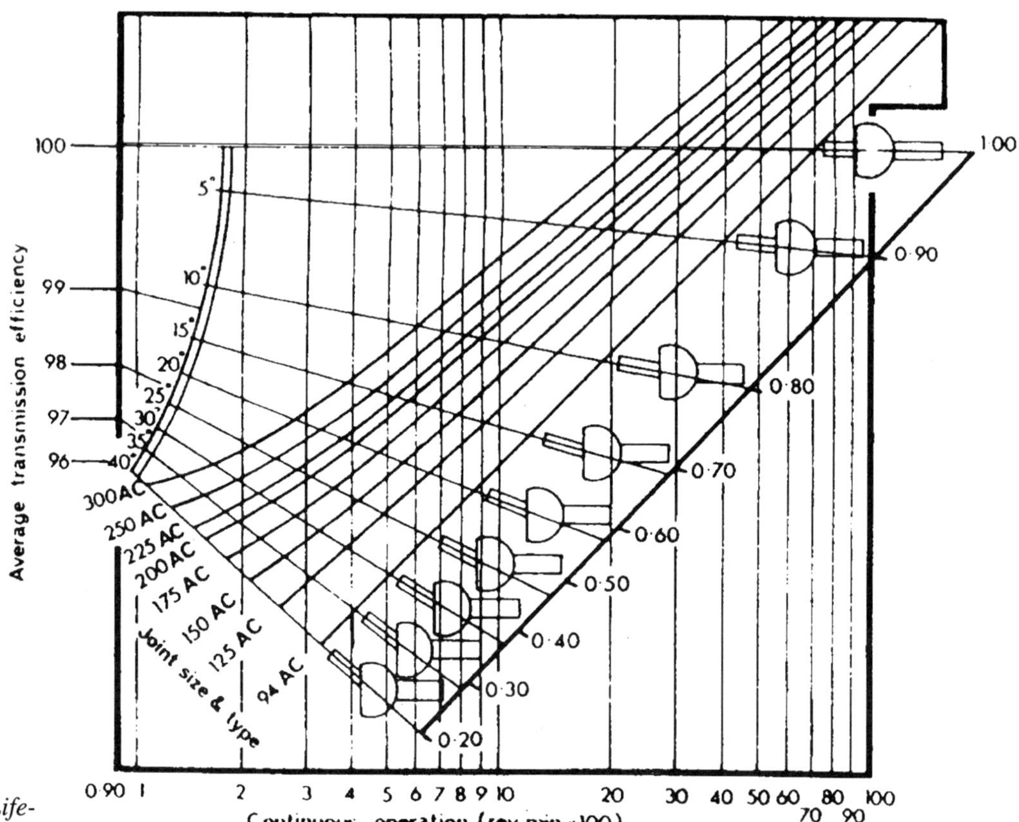

Fig 61: Life-expectancy chart

Drive axles

Usually comprising drive shafting and final drive, the drive axle includes differential gears, bearings and support casing, each of which need to be optimised to the particular drive configuration — layout of which was discussed in the section on mechanical system packaging in Chapter 8. The familiar front-engine rear drive case is examined in this section but many of the engineering principles read across to other configurations.

The section in Chapter 16 on performance prediction, gave an indication of the total external loading on a drivetrain at both the road wheel and power-unit ends of the system. From the road wheel hub, the connecting drive shaft is generally designed to fail near to the differential gears of the drive axle — usually at the differential side-gear spline location in the case of a heavier vehicle.

As part of a design procedure proposed by R Abinett[10], based on classical analyses by Heldt[11] and others, a typical quoted vehicle example might be a 2.4 tonne GVW vehicle powered by an engine of 27.76 lb ft (37.5 Nm) torque at 2800 rpm. Selecting a rolling radius value of 13.3 in (0.34 m) and a double reduction axle of overall ratio 21:1, first gear ratio might be assumed to be 4.3:1 for this example calculation. The resulting torque on the differential developed by the maximum engine torque would be 27.76 x 21 x 4.3 = 2500 lb ft. Meanwhile the torque on the differential arising from the maximum possible tractive effort, assuming a wheel to ground adhesion co-efficient of 1.0, is 2.36 x 2240 x 13.3/12 = 3465 lb ft (4.7 kNm). Thus for a straight sided 10-spline axle shaft, stress calculations can be carried out selecting the following example dimensions: major diameter 1.375 in (35 mm), minor diameter 1.22 (30.5 mm)and spline width = 0.215 (5.4 mm), depth = 0.0775 (1.96 mm) and engaged length = 1.625 in (41 mm).

Using the notation of Fig 63, considering first bending stress, then tangential load = $T_s/0.6Rsn$ where

$$R_s = [(1.375 + 1.22)/4] = 0.65$$

while n = 10 and 0.6 is a load mal-distribution factor. This would show a value of F_s of 5330 lb (2.42 tonnes) per axle shaft and a bending stress

$$6CF_s h_s/L_s w_s^2 = 66{,}100 \text{ lb/in}^2 \text{ (456 MN/m}^2\text{)}$$

(6 x 2 x 5330 x 0.0775/1.625 x 0.2152) for length L, width w and height h of spline (subscript s). For shear stress, a stress concentration factor of 1.1 is chosen instead of 2 in the bending stress case above and

$$q_s = 1.1 \times 5350/0.215 \times 1.625$$
$$= 16{,}750 \text{ lb/in}^2 \text{ (117 MN/m}^2\text{)}$$

Principal stress formulae can then be used to compute tensile and shear stress maxima of 70,050 and 37,000 lb/in² (483 and 255 MN/m²) respectively. This would provide safety factors of 2.0 (tensile) and 2.36 (shear) for an axle shaft steel such as EN25Y.

Final drive gears

To demonstrate the calculation for a bevelled crown wheel of the final drive axle, a larger vehicle of 6270 lb (2.85 tonnes) GVW is considered as an example — using a spiral-bevel pinion gear. Here an engine torque of 46 lb ft (317 Nm) at 1700 rpm is considered with overall double-reduction axle ratio of 15:1 and

Xb = speed factor	(s) is 'spline'
Yp = strength factor	T = torque
f = facewidth	R = radius
c = cone distance	n = number of
Sb = bending stress factor	F = force (tangential)
y = pinion pitch angle	C = concentration factor (stress)
pn = normal pitch	h = height (depth)
nt = number of teeth	l = length (engagement)
x = spiral angle	w = width
yc = crown wheel pitch angle	q = shear stress
Pd = pitch circle diameter	
Z = bending modulus	(e) is 'engine'; (p) is 'pinion'
Tc = crown wheel torque	M = miles (per gear)
a = acceleration	G = gear ratio
Ra = tyre rolling radius	Ra = reduction ratio (all gears)
Ter= reduced engine torque	Wg = weight (gross) of vehicle
Tgn= torque in gear (n)	gf = gradeability factor

Fig 63: Notation for calculation

4.75:1 final drive ratio (gearbox ratios being 4.3, 2.07 and 1:1). Tyre rolling resistance is this time 16.4 in (0.41 m) and the percentage time spent in each gear assumed to be 30% in first, 30% in second and 40% in third. The equivalent percentage road mileages are first calculated as 11.36, 23.55 and 65.09 for 1st, 2nd and 3rd gears. So for a total operating mileage of 60,000, say, equivalent miles per gear are 6,810, 14,100 and 39,100. The product MG_nT_e. then gives corresponding crown wheel torques of 35,700, 17,250 and 8,300 lb in, equivalent to pinion torques of 7520, 3640 and 1750 lb in (854, 393 195 Nm), dividing by the final drive ratio. Revolutions per mile of the tyres calculates to 610 and pinion revs/mile to 2900 giving total number of pinion revolutions as 174 x 10^6, divided as 19.7, 41 and 113.3 x 10^6 for the respective gear ratios.

Bevel pinion speeds corresponding to the maximum torque speed of the engine are next calculated as 125, 260 and 536 rpm respectively and the total equivalent revolutions of the bevel gear in 1st gear obtained from

$$100{,}000[197+(260/125)(2.07/4.3)^7 \times 410 + (536/125)(1/4.3)^7 \times 1133]$$

which equals 20.286 x 10^6 for a 100,000 mile vehicle life corresponding to a life in hours of 2700 at 125 rpm and torque of 520 lb in (59 Nm); from this, the life factor can be obtained as

$$[26{,}200/(200 + 2700)]^{1/7} = 1.37$$

and BS 436 can also be used to obtain gradeability factor as

$$= 100T_eR_a/W_gr$$

for maximum governed engine torque in first gear considered and life factor as

$$= (15 + g_f)/30$$

These formulae give $g_f = 33$ and $X_b = 1.60$ in this example. Permissible tangential pinion tooth load is

=f X_b (p_n/\pi)J_b S_b Y_p [(c - f)/1.1c]

from BS 545, Yp = 0.55 and the bending stress factor from the same source = 50,000 so that in this case

$$(fp_n/\pi)1.60 \times 1.37 \times 50{,}000 \times 0.55[(c - f)/1.1c]$$

Pinion pitch angle = $\tan^{-1}(8/38)$ so that $\sin y$ = 0.2062 and the actual tangential tooth load due to engine torque in first gear is

$$= 2T_e/2c\sin y = 11{,}350/f$$

for $c = 3.2f$ which is substituted in

$$p_n = (2\pi c/nt)\sin y \cos x$$

to give $f = p_n/0.412$, $pF = 12{,}000fp_n = 29{,}200p_n^2$ and $pFa = 4670/p_n$ which can be solved to give $p_n = 0.542$. The formula $\sin yc = Pd/Z_cS_o$ by cross-substitution, the minimum value of Pd can be obtained as

$$(2 \times 3.2 \times 0.542/0.412)\sin 90 - 0.2062 = 8.25.$$

Remaining gear data is then diametral pitch = 38/9.5 = 4, giving $p_n = \pi/dp \cos x = 0.643$ and face width = $p_n/0.412 = 1.56$.

Differential gears

Stressing the differential gears is often carried out on the basis of empirical formulae — typical of which gives pitch diameter of a 4-pinion unit as $0.24T_c^{1/3}$ where crown wheel torque capacity

$$= (18{,}000\, Y_dN/dp^3)(N^2 + nd^2)^{1/2}$$

For a differential unit of 3500 lb ft (4.75 kNm),typical values are $dp = 5.64$, $N = 20$ and $nd = 14$ with $Y = 0.1$ (Fig 64) which would give crown wheel torque capacity 4900 lb ft (6.6 kNm) and gear tooth thickness at the PCD of 0.278 in (7mm). Pinion and gear root thickness can be balanced to obtain equal strength characteristics.

With maximum pitch diameters of 14/5.64 and 20/5.64 in this example, cone distance

$$= [(3.543/2)^2 + (2.482/2)^2]^{1/2} = 2.162$$

For a 20 deg pressure angle, face width is taken as 0.475 x cone distance = 1.035 in (26 mm). The construction shown in Fig 65 can then be used to find

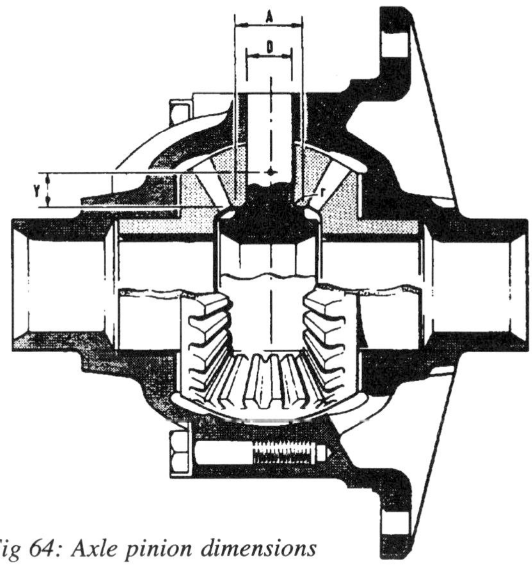

Fig 64: Axle pinion dimensions

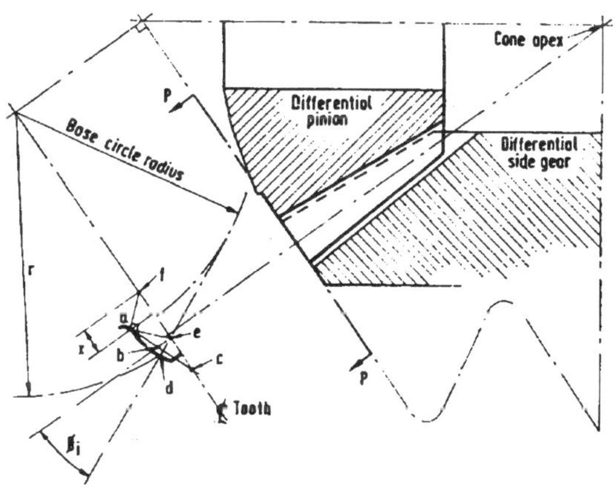

Fig 65: Construction to find gear load

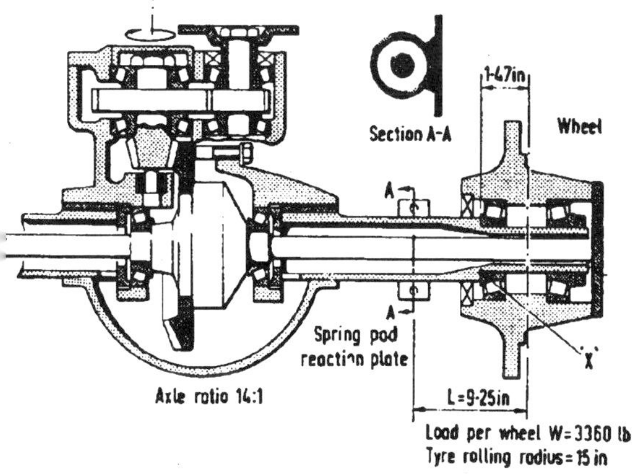

Fig 66: Axle tube calculation

whether the gears will take the design load and a modified value of T_c becomes (0.085/0.1) x 4900 = 4160 lb ft (5.6kNm). However, based on a differential torque of 3500 lb ft, pinion and wheel radii to top point of single tooth contact is $r = 1.593$ and $R = 3.17$ inch at the heel; torque per tooth is obtained from (4 x 2 meshes = 3500/8) x 12 = 5250 lb in (590 Nm) and normal load per tooth

$$= (5250/1.593)/1.035\backslash\cos 23$$
$$= 3460 \text{ lb/in (390 Nm)}$$

of face width and static root stress at the design-critical fillet is thus 3460/0.85 = 40,700 lbf/in^2 (282 MN/m^2) giving a safety factor of 2.46 against the steel tensile strength.

Axle casing

Axle tube loading can be calculated for a typical set-up shown in Fig 66, applying in this case to a 4.5 GVW electric vehicle developing 254 lb ft (345 Nm) at 725 rev/min having bending moment across A-A of 3360 x 9.25 = 31,000 lb in (3.5 kNm). For maximum tractive or braking force, the combined bending moment is $(1 + a^2W)^{1/2}$, developing tensile stress (lb/in^2) within the tube of

$$(W/2Z)\backslash\{ [L(1 + a^2)^{1/2}] + [a^2R^2 + (1 + a^2)L^2]^{1/2}\}$$

Maximum torque, W_gR, also puts a shearing stress of

$$(W_g/2Z)[a^2r^2 + (1 + a^2)L^2]^{1/2}$$

Taking transmission efficiency to be 0.9, maximum tractive force is $T_e \times 12 \times Ra \times 0.9/r$ where required Z = 1.9546 in^3 (32.5 cm^3). With this maximum tractive force being equivalent to 25.3% of the GVW, in this example, then stresses of f_t = 17,000 and f_s = 8800 lb/in^2 (117 and 62 kN/m^2) are obtained with respective safety factors of 3.53 and 3.95.

Axle bearings

Drive axle bearing loads can be calculated similarly, aided by a lower power/weight-ratio factor which governs the assumed percentage of maximum engine torque. For the ratio of engine maximum bhp/gvw in

GEAR	MILEAGE IN GEAR					
	Passenger Cars			Commercial Vehicles		Tractors
	3 speed	4 speed	5 speed	4 speed	5 speed	
TOP	0·90	0·87	0·75	0·79	0·75	0·34
4	—	—	0·14	—	0·14	—
3	—	0·10	0·08	0·14	0·07	—
2	0·90	0·025	0·025	0·065	0·035	0·60
1	0·01	0·025	0·025	0·065	0·005	0·06

Fig 67: Proportion of mileage in different gears

cwt equal to a factor of between 2 and 3 then 60% of maximum engine torque is assumed, changing to 50 and 70 for values above and below this range respectively. The data shown in the table of Fig 67 has been used by one bearing manufacturer to fix the proportion of total mileage spent in different gears which has led to the following formula for 'equivalent continuous torque' input to the final drive (assuming 0.9 overall transmission efficiency) as

$$= T_e[0.9(T_{g1}{}^3 f_{m1} + T_{g2}{}^3 f_{m2} + ...) + F_{m(top)}]^{1/3}$$

Cars usually use hypoid gears while heavier vehicles range between hypoid, spiral bevel or double-reduction involving helical gears for the secondary set. Hypoids are favoured for low drive lines — which arise from the pinion offset and quietness of running compared with spiral bevel — also the inherently larger and stronger pinion of the hypoid is preferred. Disadvantages of the hypoid are end thrust on the pinion and the requirement for extreme-pressure lubricants following from the higher sliding velocity between pinion and crownwheel teeth, Fig 68. In the spiral bevel, tangential velocities of pinion and crown wheel are equal to AB, whereas in the hypoid, for the same crown-wheel velocity CD (=AB), pinion velocity is CE and relative velocity is DE. Fig 69 shows the combination of this effect plus that of sliding of the flanks of both teeth. To limit sliding velocity, a maximum offset of 0.2 x crown wheel diameter is recommended for cars and 0.1 for CVs. Requirements for mounting rigidity of the pinion have been laid down by principal bevel gear-cutting machine makers, Gleason. Referring to a particular example in Fig 70, the general rule is that the pinion and crown wheel must not lift or depress more than 0.003 in (0.07 mm) and must not move more than this value, axially, in both directions. Also the crown wheel must not recede from the pinion more than 0.10 in (0.25 mm) at meshpoint.

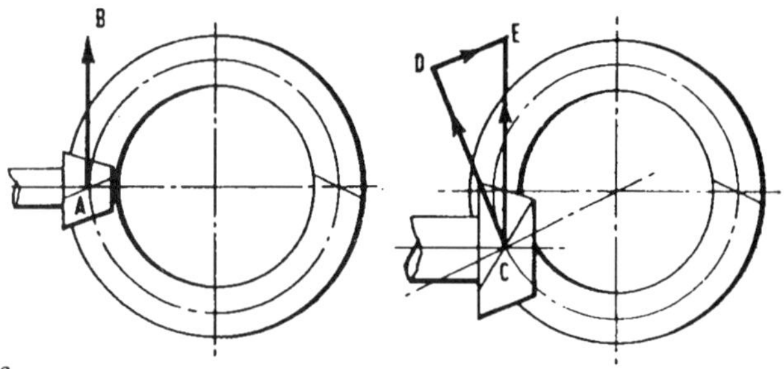

Fig 68: Pinion undercutting (left) and pinion offset (right)

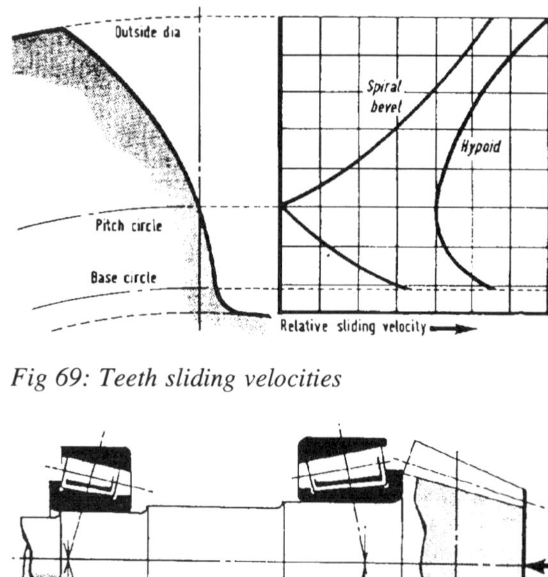

Fig 69: Teeth sliding velocities

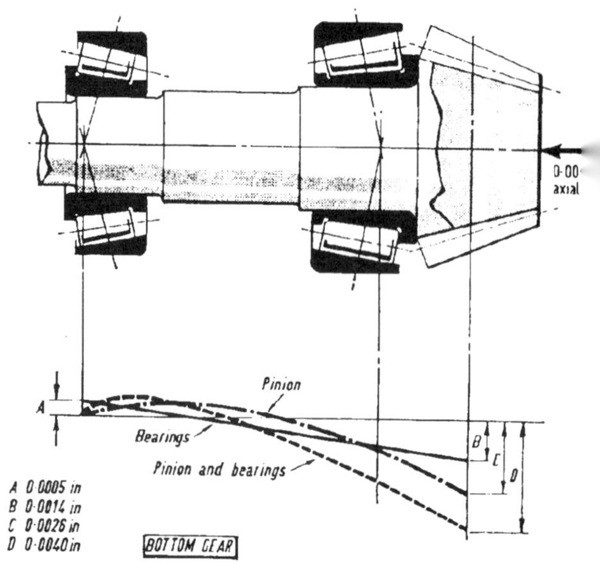

Fig 70: Mounting-rigidity requirement

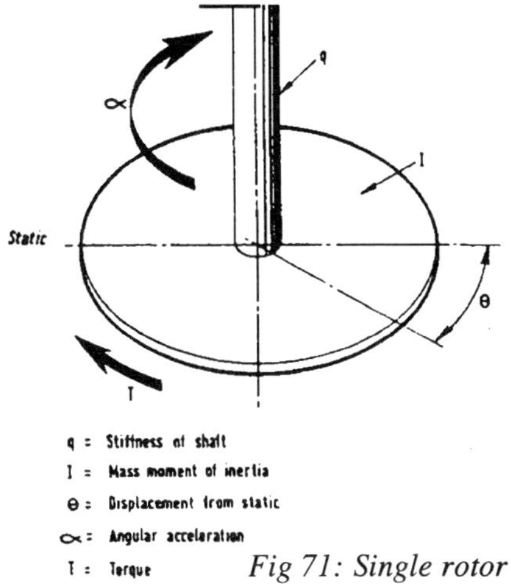

Fig 71: Single rotor system

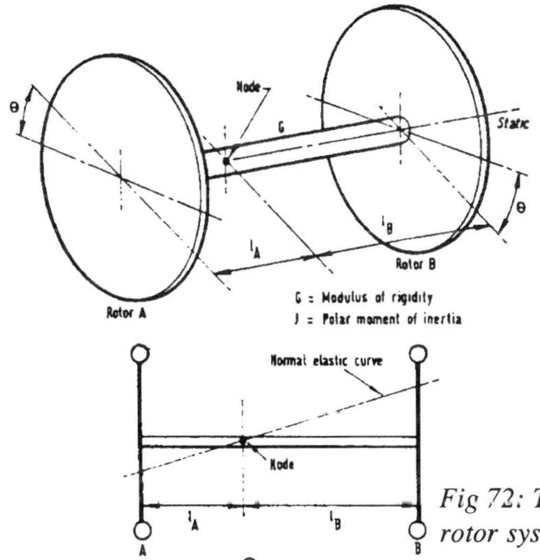

Fig 72: Two-rotor system

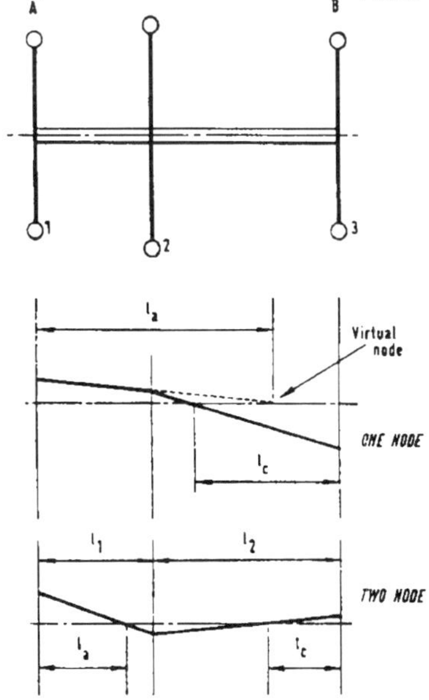

Fig 73: Three-rotor system

Drive line vibration

A basic understanding of the behaviour in torsion and bending of driveshaft assemblies is a prerequisite of solving driveline vibration problems. Substantial improvements can be made by appropriate tuning of the vibrating masses, reducing out-of-balance masses and introducing flexibility, if required, into the system.

For torsional vibration of drive shafts, the simplest analysis assumes simple harmonic motion of inertial masses on the shaft due to the elastic deflection of the shaft itself behaving as a torsion bar. For the mass concentrated as a single rotor at one end and the mass at the other, Fig 71:

$$q/I=\alpha/\theta$$

and the natural frequency:

$$N=(60/2\pi)(q/I)^{1/2}$$

For a two rotor system, Fig 72:

$$N=(60/2\pi)(GJ/I_a l_a)^{1/2}=(60/2\pi)(GJ/I_b l_b)^{1/2}$$

Since $I_a l_a = I_b l_b$, a node exists between the two rotors which are moving in opposite directions and an elastic curve can be drawn representing angular displacement at any point along the length.

For the three rotor system of Fig 73, there are two possible modes of vibration: where the outer rotor C opposes rotors A and B giving one node — and where the middle rotor B opposes outer rotors A and C, producing two nodes. Here the natural frequency is:

$$(1/2\pi)[(GJ/I_b)\{1/(l_1 - l_a)\}+\{1/(l_2 - l_c)\}]^{1/2}$$

For multi-rotor systems, Fig 74, the Holzer tabular method is used to simplify the analysis, the values shown in Fig 75 being calculated and summed.

Drive-line excitation sources

Engine torque fluctuation is the prime excitation-source of drive-line vibration and is a function of the number of engine cylinders. The next in importance is generally unbalance in reciprocating and rotating components; this aspect was discussed in the previous section. Another common source is universal joints having variable velocity ratios and two sources of lesser importance are pulsating loads caused by transfer gearing in the drive-line or by road irregularities.

The general approach to analysis is that of modelling the dynamic system taking into account both twist and bending of the shafts. Fig 76 shows a basic bending model for a prop-shaft represented by a simply-supported uniform beam. This has flexural rigidity expressed as the product of elastic modulus and section second moment of area *EI*. For weight per unit length \mu the model has a series of natural frequencies according to the vibration modes shown, given by:

$$(\pi^2 r^2/L^2)(EIg/\mu)^{1/2}$$

where $r = 1, 2, 3$ etc. This formula, applied to a steel tubular shaft of 2 in (50 mm) diameter and 1/16th in (1.6 mm) thick, for a length of 60 in (1.52 m), gives natural frequencies of 3650, 14,600 and 32,850 cycles/minute.

These values assume weightless UJ elements and rigid end-constraint for the shaft. More likely is the flexible end-constraint shown in Fig 77 with its corresponding vibration modes. For analysis, a stiffness parameter is introduced, $\rho = EI/kL^3$; as this value increases the frequencies decrease as shown in Fig 78. When the supports are very flexible, the frequencies of course correspond to the bounce and pitch frequencies of the beam as a rigid body, Fig 79, a third mode corresponding to the flexural vibration of a beam with free ends. For shafts of non-uniform cross-section an analysis similar to that applied for the flexural vibration of chassis frames can be used (*see Chapter 4*).

Alleviating drive-line vibration

Having ensured that as many of the excitation sources are minimized by attending to out-of-balance, and

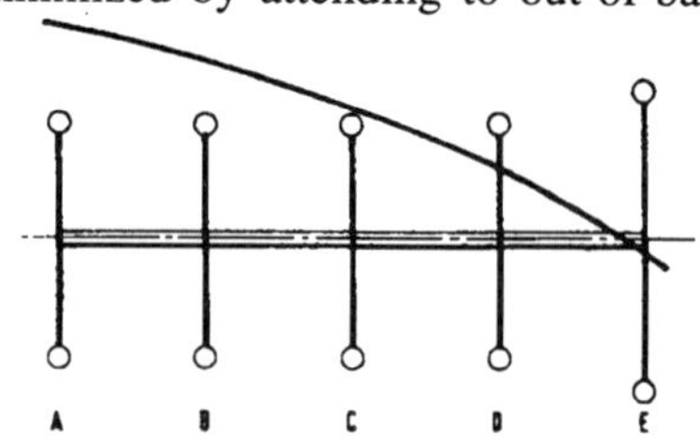

Fig 74: Multi-rotor system

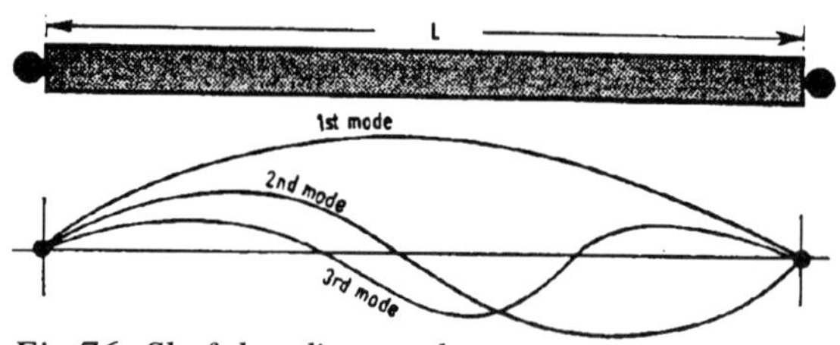

Fig 76: Shaft bending mode

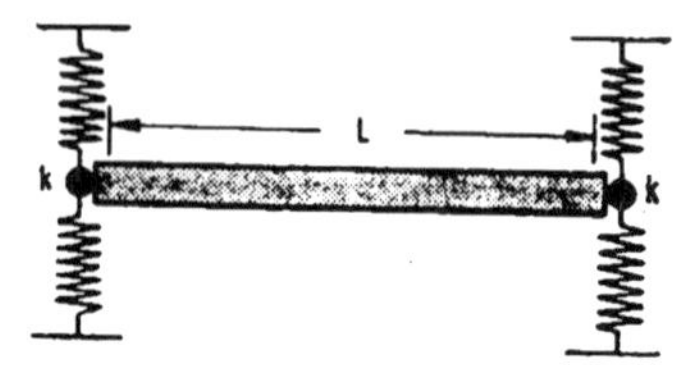

Fig 77: Flexible end-constraints

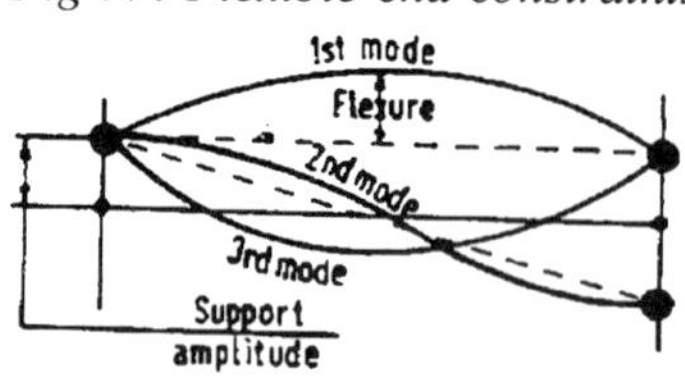

Fig 79: Vibration modes

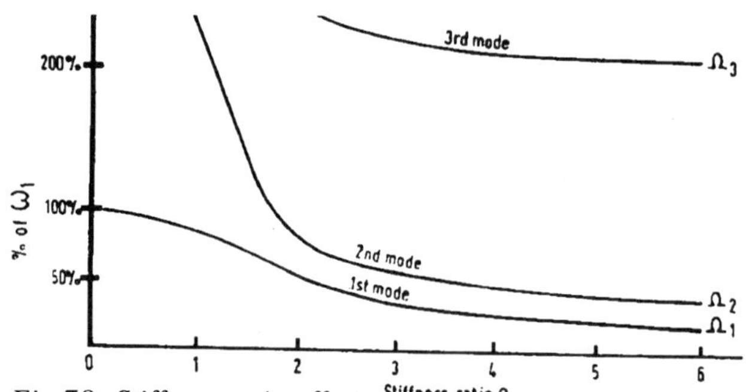

Fig 78: Stiffness ratio effect

Mass	Inertia	Inertia torque/rad	Deflection at J_n	Inertia torque	Shaft stiffness	Shaft twist	Sum of inertia torques 1 to n
n	J_n	$J_n\omega^2$	θ_n	$J_n\omega^2\theta_n$	q_n	$\frac{1}{q_n}$	$\Sigma_1 Jn\, \omega^2\theta_n$

Fig 75: Holzer tabulation

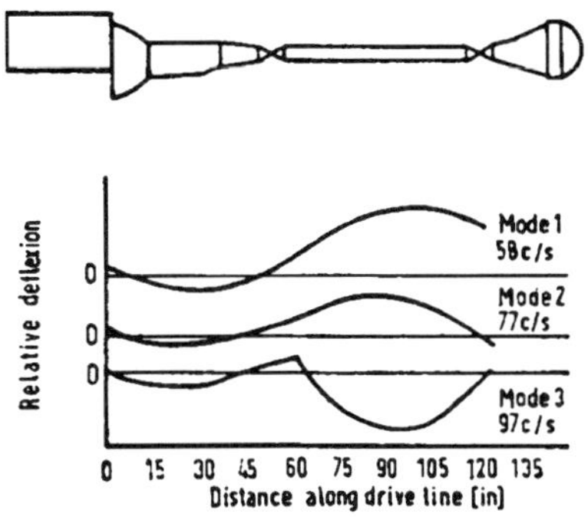

Fig 80: Typical vibration characteristics

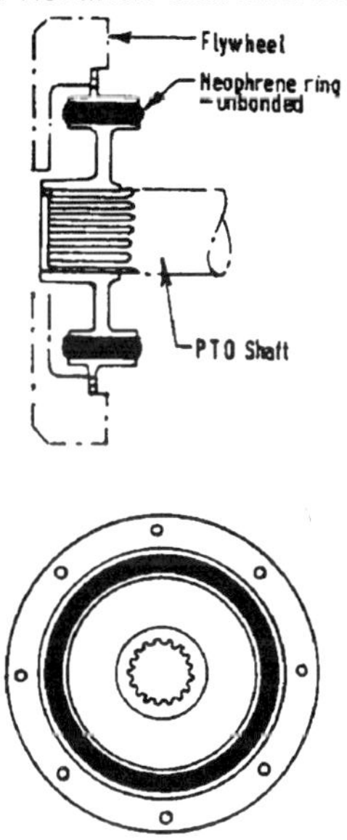

Fig 81: Flexible coupling

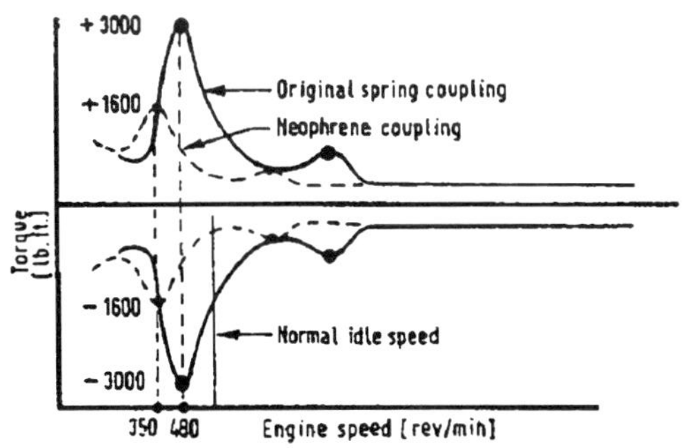

Fig 82: Modified characteristics

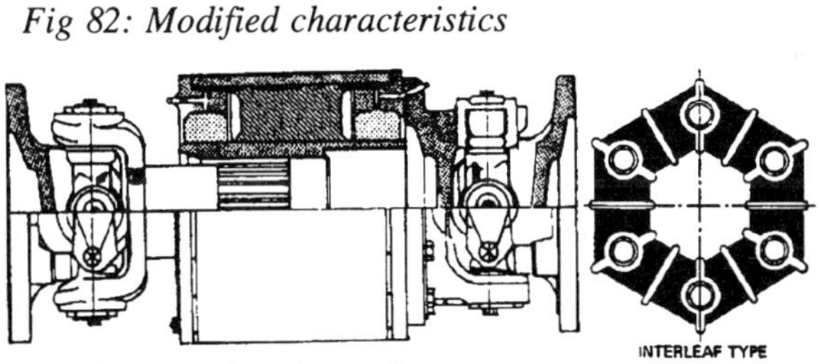

Fig 83: Metalastik couplings

the use of constant-velocity universal joints, the introduction of a flexible coupling into the system may be the appropriate way of reducing the resonant critical speed. An example is the drive-line with torsional vibration characteristics shown in Fig 80; by adding the coupling shown in Fig 81 a characteristic shown in Fig 82 was achieved by researchers at Newcastle-upon-Tyne University. The last figure shows both lowering of resonant speed and reduction in maximum torque amplitude. A further flexible isolator is seen in Fig 83 where a pair of universal joints is connected by a rubber torsional spring.

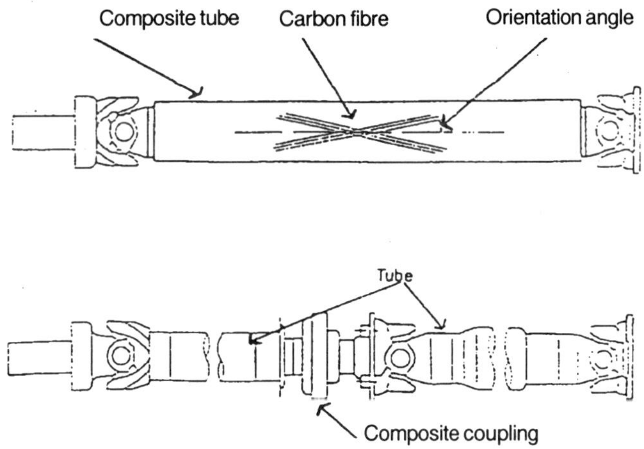

Fig 84: Steel (below) and composite (above) shafts

Light-weight drivelines

A concerted development programme by GKN has led to substantial weight reductions in most elements of the drive-line for front, rear and all-wheel drive systems. Reduction in weight has also had a knock-on effect in obtaining increased whirling speeds and therefore the ability to have less prop-shaft UJs without the risk of vibration problems.

With the group's new generation of half-shafts for both front and rear-end application, there are eleven standardized sizes available from GKN plants. Lower weight is accompanied by higher resistance to strain, reduced vibration and easier assembly/repairability. In the case of prop-shafts, improved acoustics have been achieved by decoupling of torsional and axial vibration as well as less run-out resulting in reduced wear. Transmission efficiency of universal joints has also been increased and systems developed for providing controlled collapse of the shaft under front-end crash loading of the vehicle. A choice of steel, aluminium alloy and high-strength composite systems allows tailoring shafts, and their couplings, for particular operational applications.

Steel shafts have been lightened by the use of high-tensile materials along with the appropriate design modifications and revised joining methods. Aluminium alloy shafts can be up to 50% lighter than conventional steel shafts provided appropriate design modifications are made. Metal matrix composite solutions, involving aluminium with ceramic inclusions, are also possible. Metal/composite combination shafts have also been successfully exploited on applications where three-shaft assemblies have been replaced by two-shaft systems. After a first generation of composite shafts, Fig 84, has proven the effectiveness of resin systems reinforced by glass and carbon-fibre in achieving up to 70% weight reductions over traditional shaft assemblies, the group has now announced a second generation. This involves end-fittings, as well as the shafts, in polymer construction and has resulted in overall weight savings of up to 75%.

In second-generation composite shaft systems, one approach has been to replace the Hookes-type universal joint with a composite disk UJ where only small angularity is required. In such cases a typical steel shaft system weighing 10 kg can be replaced by a first-generation assembly weighing 5 kg and a second generation one of only 2.7 kg, Fig 85. Other advantages of second-generation shafts include increased torque capacity, Fig 86, 40% increase in static break value and a new resin system which provides 15% increase in fracture value at temperatures of 120C. Techniques of introducing damping into the fibre compound help acoustic performance.

In terms of assistance given by the prop-shaft to passenger protection in frontal impact, a key factor is the possible removal of the centre-UJ/bearing where the shaft would potentially bend out-of-line on impact. By the shaft contributing to impact reaction, more protection can be provided to the foot-well of the vehicle. By use of a drop-weight rig the group is able to produce crash-optimised cardan shafts in a variety of material combinations and geometrical configurations, Fig 87. In the case of composite shafts, radially aligned reinforcing fibres can be used at the ends so that the end-pieces are pushed into the main tube on initial impact such that they lie up against the joint shoulders. Thereafter the tube is split open on one side or both sides with energy being reduced both by splitting and friction. With aluminium-alloy shafts, an axially weak point in the

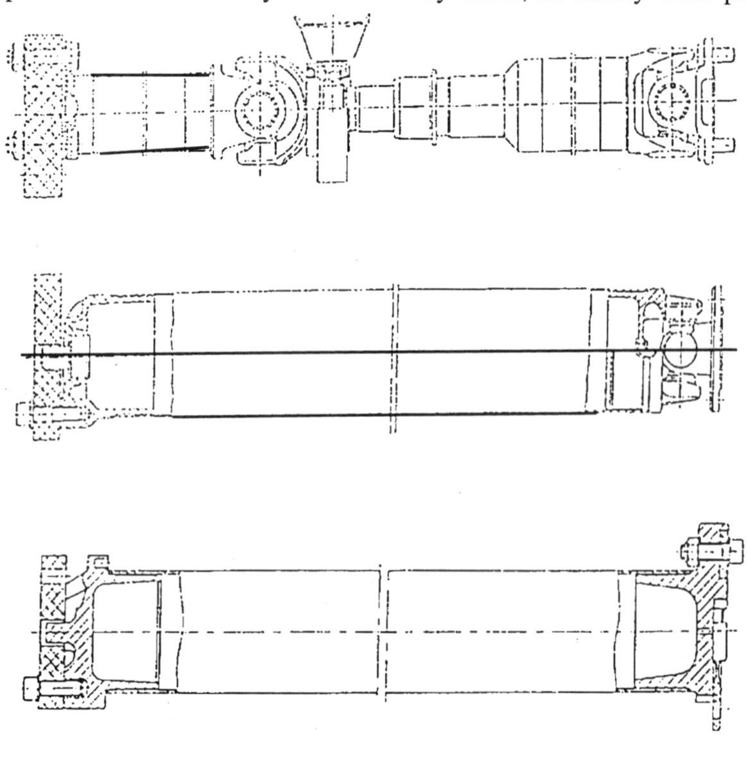

Fig 85: Traditional system (top), first generation (centre) and second generation (bottom) composite systems compared

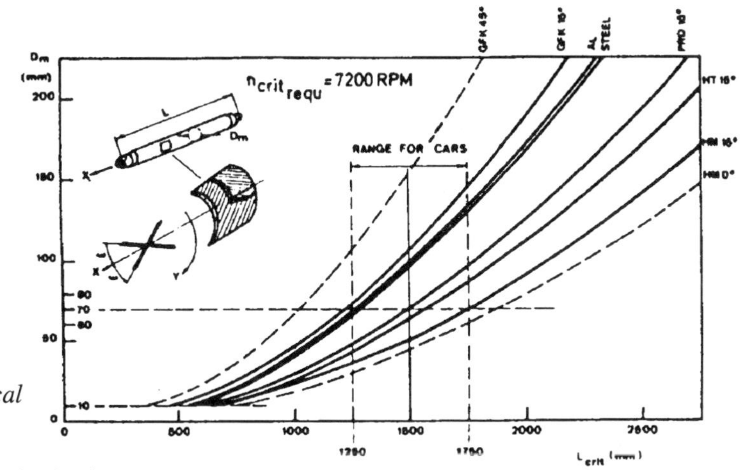

Fig 89: Shaft diameter vs critical speed

centre of the tube is the focal point; by using two different tube diameters either side of it, and carefully designing the transition area different impact characteristics can be obtained, by the smaller tube sliding inside the larger one. Special press fit connections of the end pieces can also be provided.

Figs 88 and 89 show the critical shaft length and minimum required diameter plotted for different shaft materials, and fibre reinforcements, at a critical rotating speed of 7200 rpm. Fibre-reinforced types have strengths related to fibre orientation as shown in the second figure. As the shear strength of a shaft with 0 degree fibre orientation is comparatively low, Fig 89 is based on a realistic orientation of 15 degrees. Usually multi-layer composites are involved and the elastic behaviour of any defined combination of layers is generally derived from the properties of one undidirectional layer, defined by the modulus parallel and perpendicular to the fibres, one of the two Poissons ratios and the in-plane shear modulus.

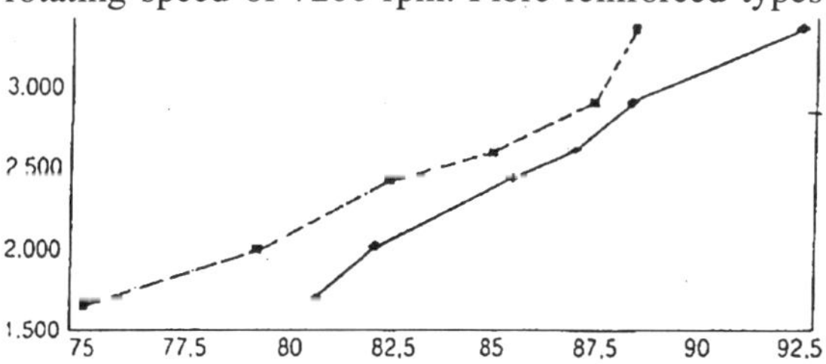

Fig 86: Torque capacity in Nm (x-axis) of new-generation shaft against external diameter in mm (y-axis): full line= current series; dashed line=new generation

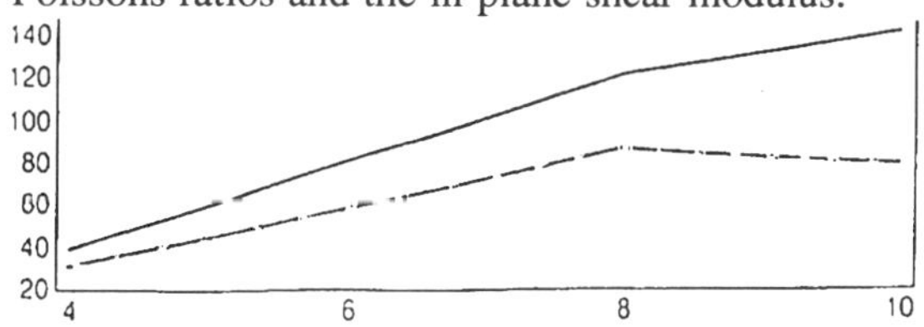

Fig 87: Axial force generation percentage (x-axis) against fibre orientation angle in degrees (y-axis) at 685 Nm toque and 340 rpm speed with conventional joint (full-line) and low-friction joint (broken line)

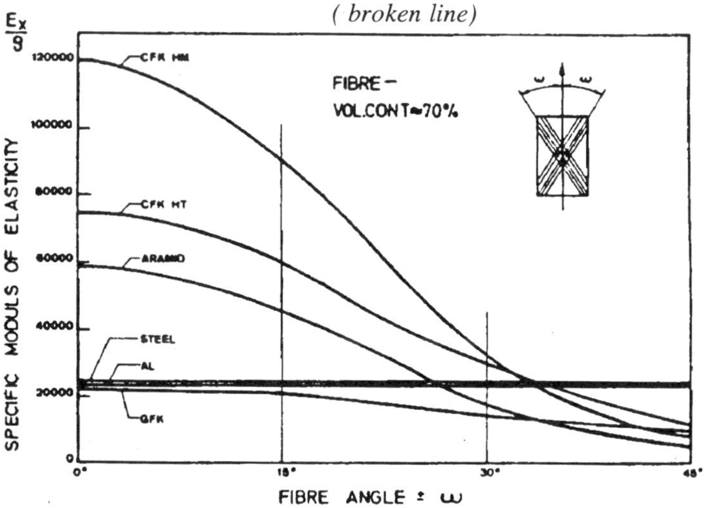

Fig 88: Modulus vs fibre-angle

Chapter 19: The drive-train

Transmission gearboxes

Automotive gearboxes will often use different gear tooth profiles from those used in static machinery that is not subject to frequent speed-changing and severe changes in duty. Thus special considerations are needed in predicting the performance of gear teeth as well as in designing the overall gearbox system.

Before examining 'corrected' tooth geometries for automotive use it is useful to examine those of constant-meshing, steady load and speed, gears to establish the principles of operation and the conventions of geometric measurement. Proper meshing requires centre distance, between meshing gear axes, to be such that maximum penetration of any tooth into the mating tooth-space of the opposite gear to be equal to the standard 'working depth' of the British Standard basic rack used to generate the teeth, Fig 1.

Centre distance is thus half the sum of the blank diameters less the working depth ($= 2p_n/\pi$) and below the teeth a radial distance of $0.6p_n$ is required as a minimum to give proper tooth support. The flank angle φ_n, normal pitch p_n and helix angle σ for the teeth (Fig 2) all relate to the cutter rack, in the geometric convention, as they are otherwise meaningless without reference to the diameter at which they apply. Axial pitch is $p_n cosec\sigma$, the distance measured parallel to the axis of the gear, between corresponding flanks of adjacent teeth. The other basic dimension for an involute-rack tooth is the normal base pitch p_{no} between adjacent teeth flanks ($= p_n cos\varphi$). For number of teeth t, diameter of reference cylinder is

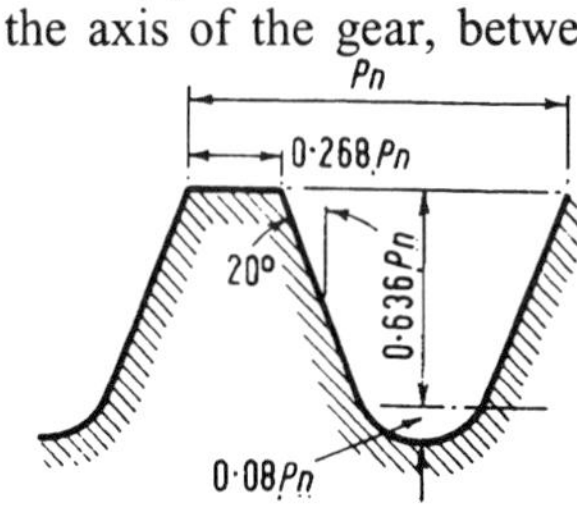

Fig 1: British Standard basic rack

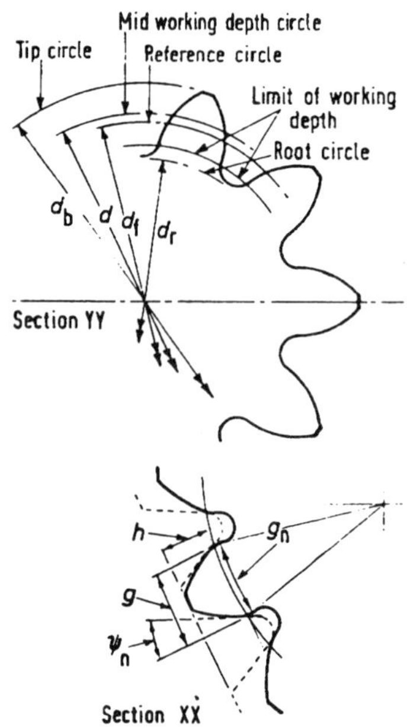

Fig 2: Gear tooth parameters

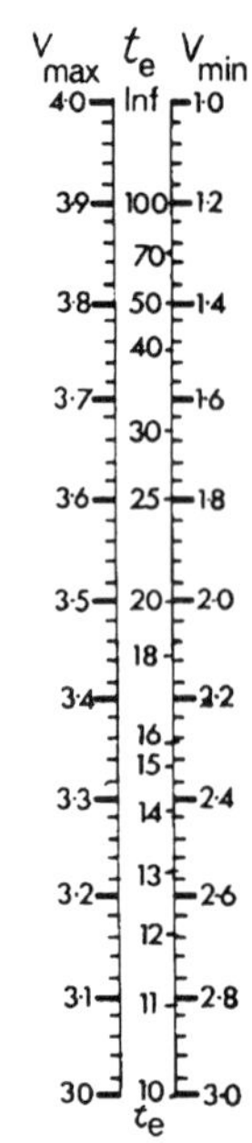

Fig 3: V-values

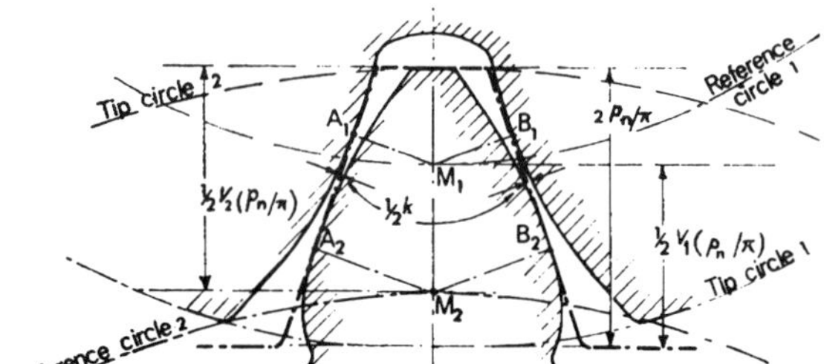

Fig 4: Rack and helical gear engagement

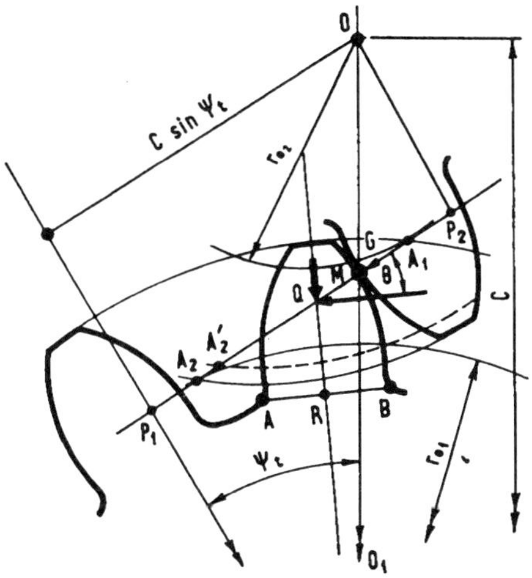

Fig 5: Dimensioning procedure for helical gears

R_g = speed of pinion ÷ speed of wheel. C = centre distance (axis distance)

Shaft position	Range of gear ratio R_g	Tooth speed (ft/min)	Type of gear	Suggested t_1	Minimum t_1
Parallel	1 to 10	Up to 1000	Spur or helical	For case-hardened steel:	
	,,	Up to 4000	Helical or ground spur	$0{\cdot}55 \times (17+30/R_g)$	$10 \cos^3\sigma$
	,,	Over 1000	Helical		

Fig 6: BS load rating procedure

$$t(p_n/\pi)\sec\sigma = (t/\pi)p_a p_n/[(p_a - p_n)(p_a + p_n)]^{1/2}$$

When $p_n = p_{no}$, the reference diameter is the base diameter; the blank diameter exceeds the reference diameter by $V(p_n/\pi)$ where V takes values from Fig 3, upper limit being set by need to avoid tooth crest widths less than around $0.1p_n$. Effective number of teeth

$$t_e = t.\sec^3\sigma\sec\varphi$$

Blank diameter can be found from

$$(p_n/\pi)(t\sec\sigma + V)$$

and when a pinion and wheel are meshing at the correct depth, centre distance is

$$(1/2)(p_n/\pi)[(t_1 + t_2)\sec\sigma + V1 + V2 - 4]$$

Chordal thickness at the reference cylinder is g and normal arc thickness g_n, Fig 4 showing a normal plane section of a rack in engagement with a helical gear. For a flank angle of 20 deg, and teeth cut to give

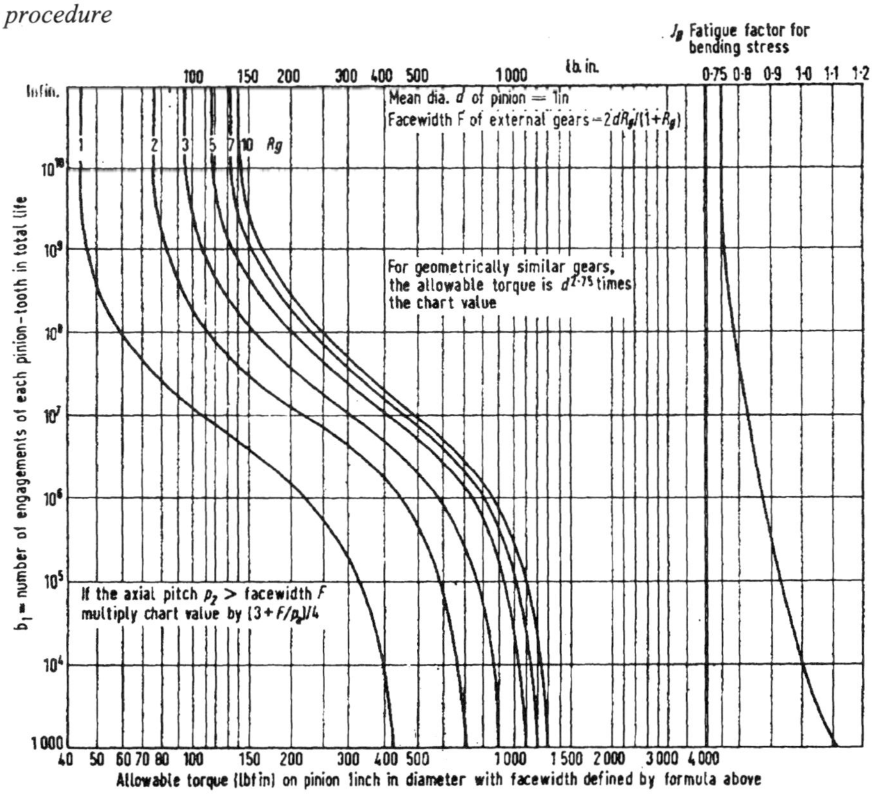

Fig 7: Dynamic loading calculation chart

backlash b in gears with basic backlash (see below) of k, g_n equals

$$(0.843 + 0.364V)(p_n/\pi) + 0.53\ (k - b)$$

When gear teeth are cut to exactly full depth at the designed centre distance they are said to have 'basic' backlash, depending on number of teeth and value V. The procedure for dimensioning helical gears for specified applications is shown in Fig 5.

Gear load capacity

Tuplin[1] has provided a method of refining the British Standard load rating procedure for gear teeth starting with the Lewis formula, see Fig 6, which states that tooth line loading intensity for unit face width $G_{jt} = S_b(p_n/p)\ Y$ for maximum bending strength S where Y equals

$$\pi(AB/p_n)\cos\varphi/[6\cos\theta(QR/AB) - \sin\theta]$$

Y_{min} being the strength factor.

As well as bending strength, ability to resist intense local direct, Hertzian, stresses has to be considered, the maximum tolerable surface stress being given by

$$0.59[F_0G_{jt}\{(1/r_1)+(1/r_2)\}\ /\{(1/E_1)+(1/E_2)\}]^{1/2}$$

and the associated shear stress is about three times this value, failure by one or the other being a function of crucial stress S_c which equals FG_{jt}/R_r where R_r is relative radius of curvature which at any point, radius r from the gear centre, on an involute flank, and equals $(r_2 - r_0^2)^{1/2}$. The worst point for 'contact loading' is nearest to the pinion gear root whereas the worst for bending is, of course, near the tip. Total intensity of loading for 20 deg flank BS teeth becomes

$$Z_sS_cd_1R_g/(1 + R_g)$$

where Z_s is zone factor (0.015 for spur gears). In the case of helical gears, total intensity of loading is

$$Z_hS_cd_1R_g/(1 + R_g)$$

where zone factor $Z_h = 0.24$. Speed and life factors can also be used in calculation as well as the effect of dynamic loading of gear teeth where appropriate. A useful chart for design purposes, due to Tuplin, is shown as Fig 7.

Gearbox design approach

The design of manual shift transmission gearboxes depends considerably on performance prediction calculations for a new-concept vehicle as described in Chapter 16. Fig 8 provides a typical tractive force/speed curve from which gear ratios can be optimised. Conflicting operational requirements of fuel economy and accelerative performance can make the choice difficult. Thus the construction of engine utilisation curves, such as that in Fig 9 due to C G Giles[2], includes engine fuel consumption loops for different throttle openings in a way that the reduced power

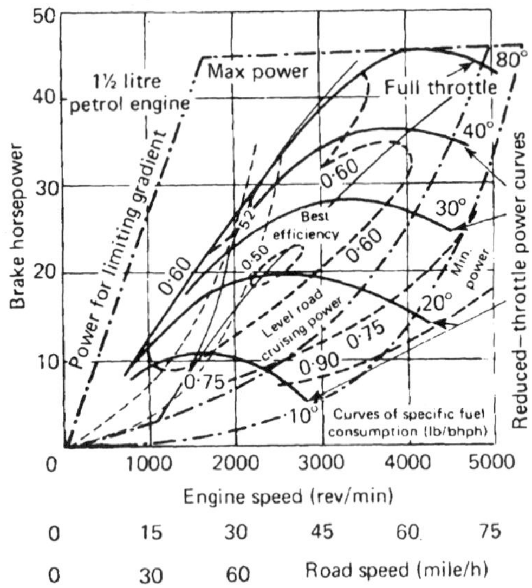

Fig 8: Tractive force vs road speed curves

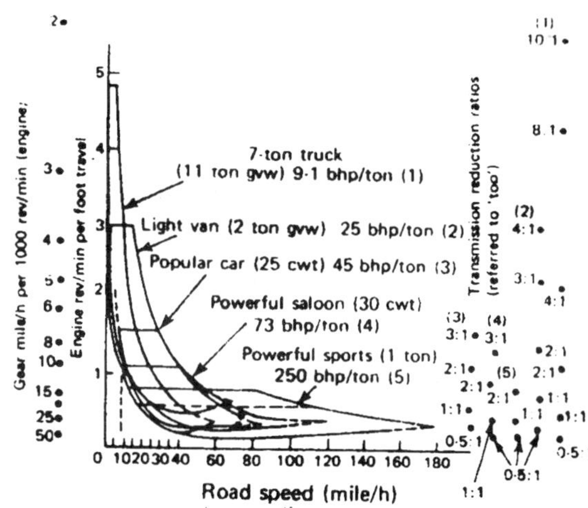

Fig 9: Engine utilisation curves

demand in cruising can be optimised. The utilisation curve can be used to create power requirement curves.

For example, a minimum power requirement of around half the power required for level road cruising could be used. A requirement envelope is then enclosed by limiting gradient or road adhesion. Suggested procedure is (1) to calculate power absorbed in overcoming level-road resistances so as to find maximum level road speed for a given power; (b) choose minimum power criterion and find corresponding engine-speed increments from that curve up to maximum speed; (c) write down engine speed for distance travelled per maximum engine output, for fractions of the maximum engine speed, up to the point where power required = power available; (d) similarly write down engine speed per distance travelled relationship for minimum power requirement; (e) obtain the lowest gear by relating maximum gradient, or limiting traction, to available engine torque.

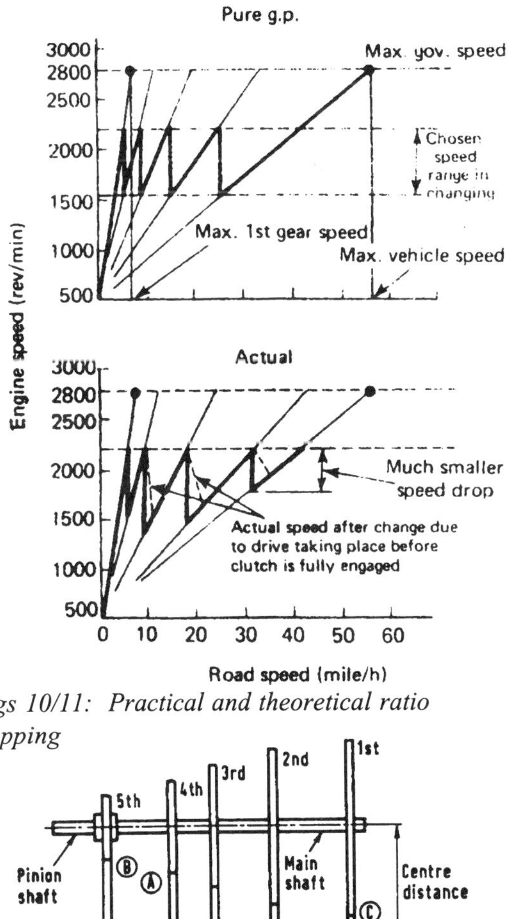

Figs 10/11: Practical and theoretical ratio stepping

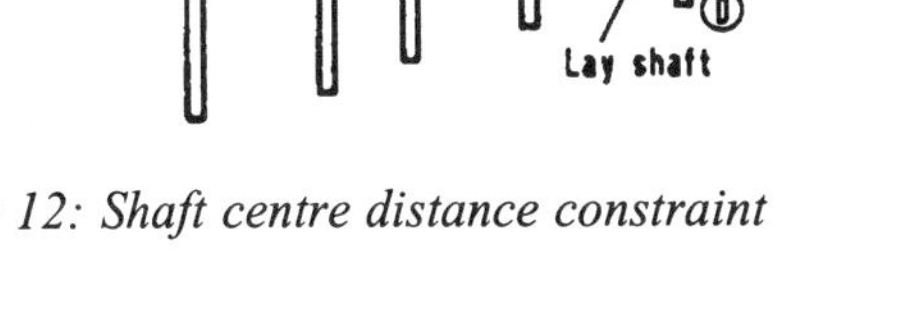

Fig 12: Shaft centre distance constraint

Envelopes shown apply to different category vehicles and show at any given speed which ratios could usefully be employed. Choice of steps between ratios is often the subject of debate for different vehicle categories. In general, torque/speed curves in each ratio must overlap the maximum torque point on adjacent curves in order to make even down-changes. Ratio spacing should otherwise be such as not to unduly strain the engine by either too high or low an engine speed at the time of engagement. A geometric progression of ratios is sometimes favoured for industrial and goods vehicles but in practice a closer step is often provided between the upper two ratios. This is dictated by traffic conditions resulting in an actual arrangement, compared with the theoretical one, as shown in Fig11.

Lowest gear ratio can also be affected by geometrical considerations of the gearbox itself, shaft centre distance between layshaft and mainshaft being a key dimension in this respect. Since the situation could arise that the higher intermediate gear pairs (A), in Fig 12, could not cope with a large stepup speed, the first-reduction constant-mesh pair (B) is limited in degree of step-down. So, for first-gear, the major part of the speed reduction is in the first gear pair (C) and the tooth strength of the smallest layshaft pinion (D) becomes the design criterion. Centre distance also affects choice of tooth pitch — too fine a pitch necessitating a large centre distance for keeping tooth stresses within acceptable limits and too cause a pitch giving rise to roughness. Normal diametral pitches and corresponding shaft centre distances have been suggested as 12 in cars (centre distance $0.5 + 0.5T^{1/3}$) of under 200 bhp and 10 in those above 200 bhp. For CVs (centre distance $0.8T^{1/3}$), DP of 8 is suggested below 200 bhp; of 7 for 200-250 bhp and 6 for over 250 bhp. Total number of teeth in any gear pair can be reduced by increasing circular pitch and helix angle for helical gears in turn reducing side thrust.

In reducing helix angle, recommendations have also been made for overlap ratio, given by

$F\sin\sigma/p$, to be not less than 1 for cars (0.75 for CVs). Starting with as small as possible layshaft first speed pinion, it is then possible to calculate exact numbers of teeth, and pitch circles by direct proportioning. Resulting overall gear ratios obtained can then be substituted in $F = 2Pm_eR_{go}/S.1000d$, where S is obtained from Table 1, and, in checking the overlap ratio, the helix angle is obtained from $\cos^{-1}[P(T + t)/2\pi 2C$.

Automotive teeth proportions

British Standard stressing procedures result in excessively large tooth proportions for comparatively lightly laden car gearbox gears used relatively little in the intermediate gears. Because comparatively high short-term stresses are involved in car gearboxes, it is argued that bending fatigue rather than surface deterioration should be the criterion of failure. As interchangeability of gears was not universally necessary, standardised proportions were not necessary. Thus corrected forms which allowed simultaneous theoretical failure of both wheels in a pair was seen as the most economical route. A drawing layout as in Fig 13 is thus often carried out to obtain corrected forms — with bending stresses arranged so relative cycles to failure are approximately in reverse ratio to the number of teeth, typical fatigue curves for teeth being obtained by experiment. A typical target for life requirements at maximum input torque (mainshaft cycles/10^6) is in Table 2.

In the illustrated example, involving a 38 tooth gear wheel at 79,540 lb/in^2 (55 MN/m^2) bending stress, estimated pinion life was 4.5 x 10^6 cycles and wheel life 0.2 x 10^6 close to the 2:1 gear ratio. Suggested limit to correction has been stated as increase in root width up to 1/8 in (3.2 mm) each side in excess of the mating gear facewidth. Other hints

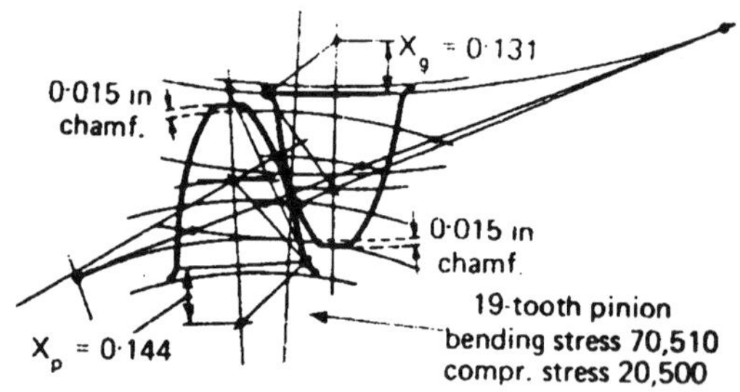

Fig 13: Drawing layout for tooth proportioning

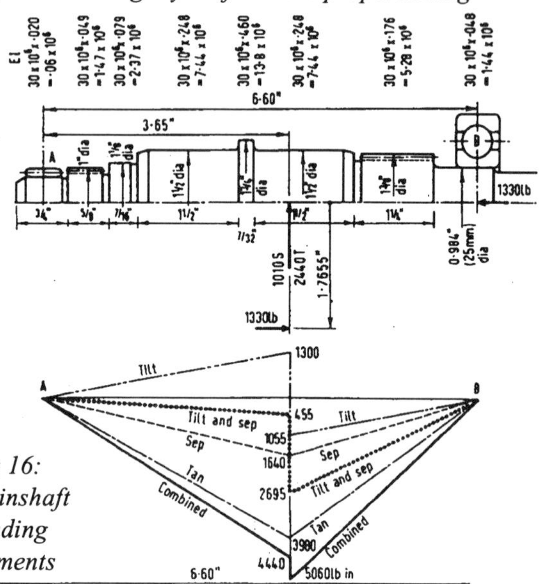

Fig 16: Mainshaft bending moments

Table 1: Values of S

Gear pair		Car	CV
Layshaft driving		18-21	11-13
Intermediate	4th	21-24	13-15
	3rd	21-24	15-17
	2nd	21-24	20-22
Low (bottom)		28-32	26-30

Table 2: Life target

Gear pair	LCV	Car
1st	0.75	0.5
2nd	3.5	2.0
3rd	7.0	3.0

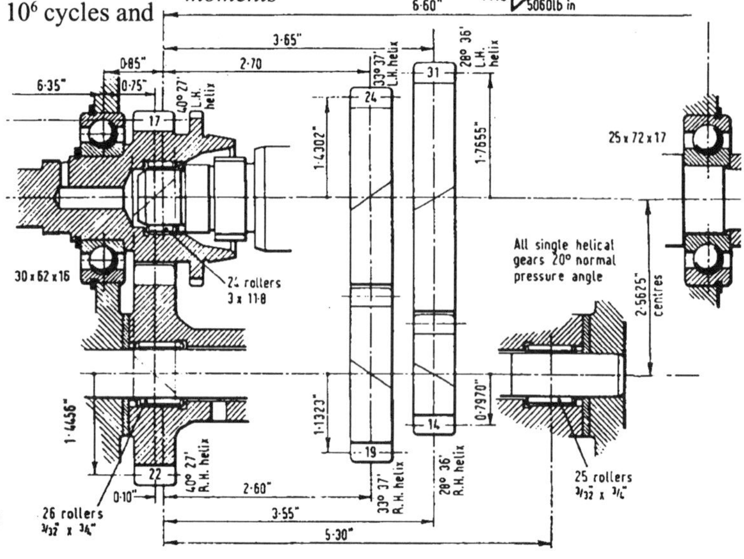

Fig 15: Typical mainshaft intallation

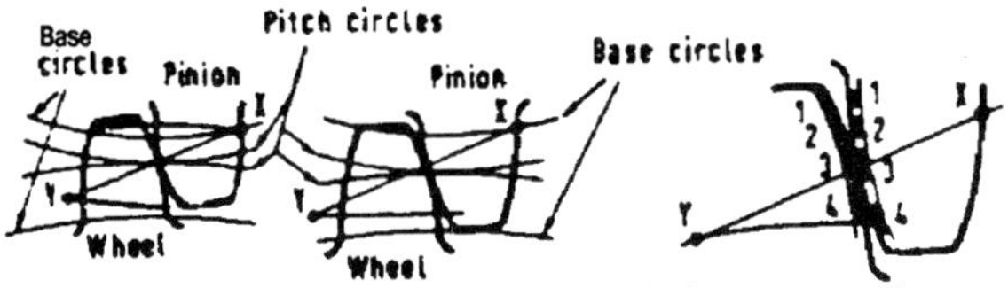

Fig 14: Pinion undercutting — left, corrected teeth; right, uncorrected teeth

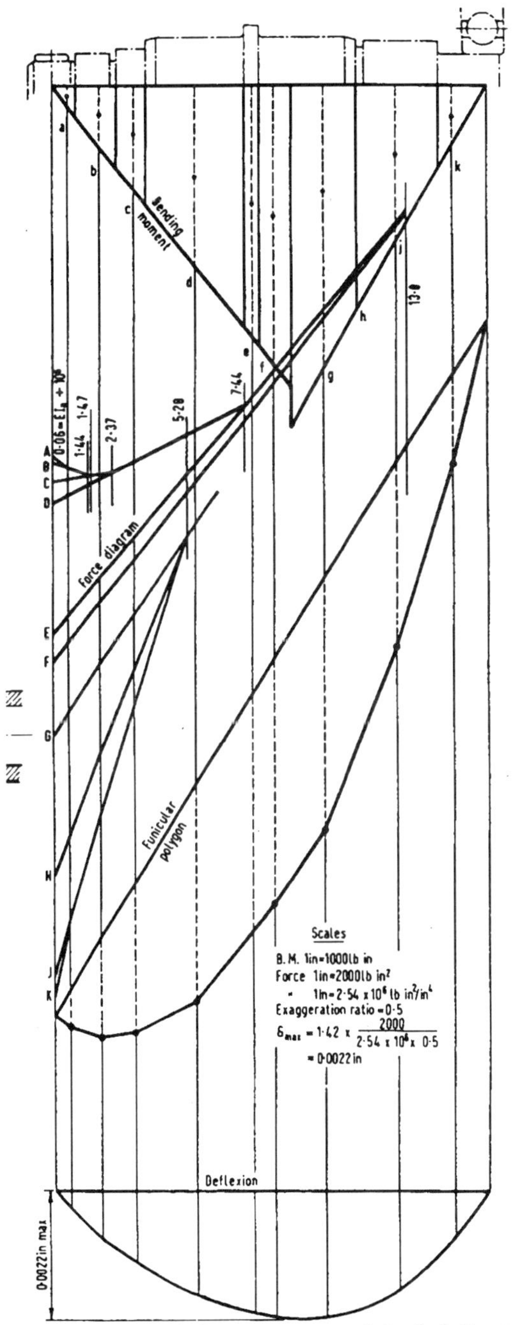

Fig 17: Graphical determination of shaft deflection

given by early workers for tooth correction include: modified helix angle to offset detrimental effects of shaft deflections; slight lateral offset of mating helical gears to avoid tooth corner fracture; provision of protective tooth tip chamfers; reducing noise by limiting helical overlap to below 1, pitch circle runout within 0.0002 in (0.005 mm) and tooth spacing errors to within 0.0005 in (0.013mm). Otherwise a traditional approach to tooth correction has been to increase addendum and decrease dedendum in the pinion and vice versa on the wheel. Fig 14 shows undercutting of the pinion for this purpose and the removal of the redundant portion of the wheel tooth. Continuity of tooth engagement must be observed with contact line XY being greater than base pitch. True rolling must take place at the pitch circle, with limited sliding permissible at the tips.

Transmission gearbox bearings

The SKF company established design procedures for these installations in the 1960s. It was then recommended that shaft deflection within a gearbox should be limited to 0.001 cm per cm of span. For the installation shown in Fig 15, main-shaft deflection in first gear is calculated from a standard beam deflection formula to be about 0.002 in (0.05 mm) — ignoring the couple due to the end thrust. If this couple is accounted for the bending moment diagram will be as seen in Fig 16 and a graphical method (Fig 17) can be used to predict the deflections.

The deflection of the ball bearing is relatively small and is equal to:

$(0.002\cos\alpha)[Q^2/D_w]^{1/3}$ which is modified to

$0.0006Q^{0.9}/[\cos\alpha l_a^{0.8}]$

for roller bearings. This is for contact angle α and load on the heaviest-loaded rolling-element Q — and element diameter D_w. In the example shown the bearing defections are about 1/10 th of the shaft central defections due to bending.

The general approach to gearbox bearing load calculation is to start with the maximum engine torque figure and work out from it the loads in the various gears — then to ascertain what percentage of the maximum torque can be considered normal, using the power/weight ratio of the vehicle as a guide.

After deciding on the proportion of mileage in

each gear, aggregate mileage is calculated using the tyre diameter, axle and gearbox ratios. Finally the Hertzian stress is checked. Those in the top row in Fig 18 have been used to relate the ratio r = maximum engine power (bhp)/laden weight (cwt) with *Ti*, the normal input torque as a fraction of maximium engine torque.

Next stage is to determine the design life of the vehicle and the proportion of mileage spent in the different gears, the lower half of Fig 18 showing figures suggested by the company at the time. Prior to graphical analysis for determining bearing loads it is recommended first to resolve all forces into three mutually perpendicular planes. A worked example describes the procedure.

Maximum engine power of 86 bhp and torque of 120 lb ft (163 Nm) is assumed for a vehicle weighing 30 cwt (1.52 tonnes) with 8:33 axle ratio and 12.5 in (0.32 m) tyre rolling radius. Capacities and thrust factors are taken from a contemporary SKF catalogue. From the gearwheel teeth numbers in Fig15 layshaft torque is calculated as 120 x 12 x 23/17 = 1950 lb in (220 Nm). Tangential forces, end-thrusts and separating forces for the gears are computed as in Fig 19, the force system for the latter being shown in Fig 20. The corresponding loads on the bearings are shown in Fig 21.

Only bearings E and B are subject to end thrust, the 'equivalent radial load' being obtained on bearing E in 2nd gear, for example, from $P = XF_r + F_a$ where subscripts a and r define axial and radial directions for loads. Here the bearing is Type 6206 with a catalogue static capacity, C_0 of 2200. Thrust is 1150 lb (520 kg) in both 1st and 2nd gears so $F_a/C_0 = 1150/2200 = 0.52$; radial load on 2nd gear is 750 lb (340 kg) so $F_a/F_r = 1.5$. This gives an X-value of 0.31 and a Y of 1.2, such that in 2nd gear, $P = 0.31 \times 750 + 1.2 \times 1150 = 1610$ lb (730 kg) and for the other gears as in Fig 22. The total revolution mileage count is seen in Fig 23 and the equivalent continuous load for bearing E would be 515 lb (233 kg) using the formula given above — leading to the life in revolutions, converting to 90,000 miles.

Needle roller bearings are also strong contenders for gearbox work and suppliers such as Ina Bearing argue that very considerable space savings are available compared with other rolling-element bearings. Fig 24 shows specific load carrying capacity for needle rollers (top) compared with cylindrical

T	> 3	2–3	< 2	tractors
Ti	0·5	0·6	0·7	0·8

Gear	Proportion of total milage					
	Passenger cars			Commercial vehicles		Tractors
	3-speed	4-speed	5-speed	4-speed	5-speed	3-speed
Top	0·90	0·87	0·75	0·79	0·75	0·34
4th	—	—	0·14	—	0·14	—
3rd	—	0·10	0·08	0·14	0·07	—
2nd	0·09	0·025	0·025	0·065	0·035	0·60
1st	0·01	0·005	0·005	0·005	0·005	0·06

Fig 18: Mileage in gears

Tangential forces

Primary gear: 1950/1·4450 = 1350 lb

Second gear: 1950/1·1323 = 1715 lb

First gear: 1950/0·7970 = 2440 lb

End thrusts

Primary: 1350 × tan 40°27′ = 1350 × 0·853 = 1150 lb→

Second: 1715 × tan 38°37′ = 1715 × 0·665 = 1140 lb←

First: 2240 × tan 28°36′ = 2440 × 0·545 = 1330 lb←

It will be seen that, on the layshaft, thrusts almost cancel out in 2nd gear and do not attain particularly high values in 1st gear, a fact that makes use of plain thrust washers quite practicable.

Separating forces

Primary: $\dfrac{1350 \times \tan 20^\circ}{\cos 40^\circ 27'} = \dfrac{1350 \times 0.364}{0{\cdot}761}$ 645 lb

Second: $\dfrac{1715 \times \tan 20^\circ}{\cos 33^\circ 37'} = \dfrac{1715 \times 0{\cdot}364}{0{\cdot}833} = 750$ lb

First: $\dfrac{2440 \times \tan 20}{\cos 28^\circ 36'} = \dfrac{2440 \times 0{\cdot}364}{0{\cdot}878} = 1010$ lb

Fig 19: Computation of forces

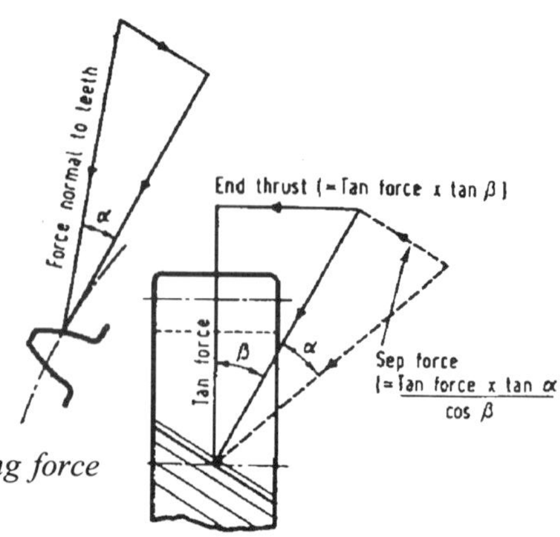

Fig 20: Separating force diagram

Loads on bearings due to tangential forces

Loads in plane of paper downwards marked ⊕

" " " " " upwards " ○

due 2nd $\begin{cases} A = 1715 \times (6{\cdot}6 - 2{\cdot}7)/6{\cdot}6 = 1010 \text{ lb } \oplus \\ B = 1715 - 1010 = 705 \text{ lb } \oplus \end{cases}$

" 1st $\begin{cases} A = 2440 \times (6{\cdot}6 - 3{\cdot}65)/6{\cdot}6 = 1090 \text{ lb } \oplus \\ B = 2440 - 1090 = 1350 \text{ lb } \oplus \end{cases}$

due primary $\begin{cases} C = 1350 \times (5{\cdot}3 + 0{\cdot}1)/5{\cdot}3 = 1375 \text{ lb } \oplus \\ D = 1375 - 1250 = 25 \text{ lb } \circ \end{cases}$

" 2nd $\begin{cases} C = 1715 \times (5{\cdot}3 - 2{\cdot}6)/5{\cdot}3 = 875 \text{ lb } \\ D = 1715 - 875 = 840 \text{ lb } \end{cases}$

" 1st $\begin{cases} C = 2440 \times (5{\cdot}3 - 3{\cdot}55)/5{\cdot}3 = 810 \text{ lb } \circ \\ D = 2440 - 810 = 1630 \text{ lb } \circ \end{cases}$

and similarly for E and F.

Loads on bearings due to separating forces

due 2nd $\begin{cases} A = 750 \times (6{\cdot}6 - 2{\cdot}7)/6{\cdot}6 = 445 \text{ lb } \uparrow \\ B = 750 - 445 = 305 \text{ lb } \uparrow \end{cases}$

" 1st $A = 450 \text{ lb } \uparrow$ $B = 560 \text{ lb } \uparrow$

and similarly for C, D, E and F.

Loads due to tilting effect on end thrusts

due 2nd $\begin{cases} A \\ B \end{cases} = 1140 \times \dfrac{1{\cdot}4302}{6{\cdot}6} = 250 \text{ lb } \begin{matrix} \downarrow \\ \uparrow \end{matrix}$

Resultant radial loads

on A 2nd gear $= \sqrt{[1010^2 + (450 - 250)^2]} = 1030 \text{ lb}$

on A 1st gear $= \sqrt{[1090^2 + (450 - 325)^2]} = 1100 \text{ lb}$

on B 2nd gear $= \sqrt{[(750^2 + 305 + 250)^2]} = 905 \text{ lb}$

on B 1st gear $= \sqrt{[1350^2 + 560 + 325)^2]} = 1620 \text{ lb}$

on C 2nd gear $= \sqrt{[(1375 - 875)^2 + (655 + 380 - 315 + 245)^2]} = 1080 \text{ lb}$

on C 1st gear $= \sqrt{[(1375 - 810)^2 + (655 + 335 - 315 + 180)^2]} = 1030 \text{ lb}$

and similarly for D, E and F.

Fig 21: Bearing loads

Location	Type number	C	Gear	F_r	F_a	X	Y	Max. torque P	Normal torque P_r*	Equiv. cont. load P_e	$\frac{C}{P_e}$	$\left(\frac{C}{P_e}\right)^a$	Milage
B	6305NR	3600	3	—	—	—	—	—	—	470	7·6	440	130 000
			2	905	1140	0·31	1·2	1645	990				
			1	1620	1330			1950	1170				
E	6206NR	3350	3	—	—	—	—	—	—	515	7·0	343	90 000
			2	750	1150	0·31	1·2	1610	970				
			1	745				1610	965				

Fig 22: Thrust tabulation

roller and ball bearings (bottom). In the trans-axle unit of Fig 25 the needle bearing unit shown enlarged in Fig 26 is only subject to radial loads.

Ina Bearing has recommended a similar approach to that above for calculating gearbox bearing requirements and for this transaxle case they suggested the approach shown in Figs 27/28, and Table 3, based on the assessment of mean torques and time spent for each gear engagement. In considering the effect of shaft deflection on bearing life it is important to factor the results by a value given by the bearing manufacturer.

Total revolutions

The milage is arrived at thus:
Mainshaft revolutions per mile
$=(5280 \times 12)/(3{\cdot}142 \times 25) \times 33/8 = 3320.$

Layshaft revolutions per mile
$3320 \times 31/14 = 7350$ in 1st
$3320 \times 24/19 = 4280$ in 2nd
$3320 \times 17/22 = 2570$ in top

Primary shaft
$7350 \times 22/17 = 9500$ in 1st
$4280 \times 22/17 = 5550$ in 2nd
$2570 \times 22/17 = 3320$ in top

Mainshaft front
$9500 - 3320 = 6180$ in 1st
$5550 - 3320 = 2230$ in 2nd
$3320 - 3320 = 0$ in top

Fig 23: Total mileage count

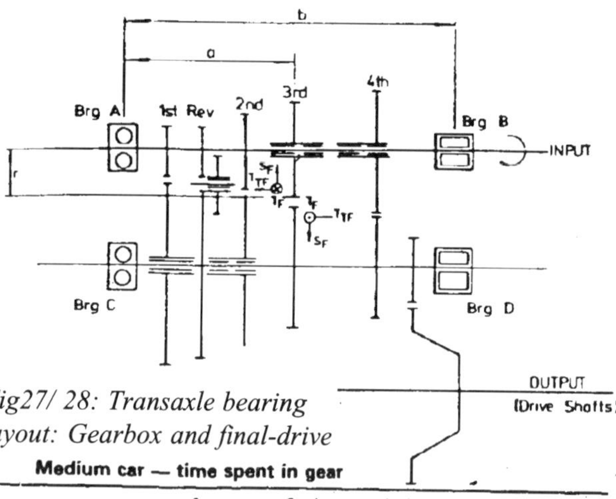

Fig27/ 28: Transaxle bearing layout: Gearbox and final-drive

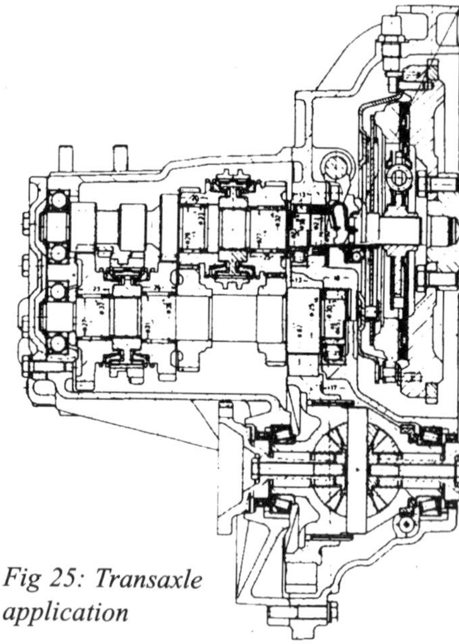

Fig 25: Transaxle application

Medium car — time spent in gear

	1st	2nd	3rd	4th	5th
Time (%)	1	5	12	25	57

$$\text{Weighted average life} = \frac{100\,000}{\frac{q_1}{L_1} + \frac{q_2}{L_2} + \dots + \frac{q_n}{L_n}} \text{ Hours}$$

where q_1, q_2 = percentage time in each gear and L_1, L_2 = calculated life in that gear ($\div 10^3$ hrs)

$$L_h = \left(\frac{1}{0{\cdot}70}\right)^P \times \text{ full torque life (hours)}$$

Percentage torque estimated at 70 may also vary for each individual gear.

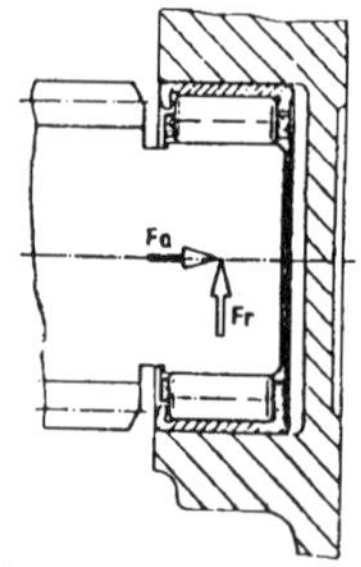

Fig 26: Bearing close-up

Gear tooth forces

Tangential: $T_F + T/r$
Separating: $S_F + T_F \operatorname{Tan} \propto / \cos \beta$
Thrust: $T_{TF} + T_F \tan \beta$
(T_{TF} direction is determined from the helix direction for the driving member)

Bearing loads

$$P_A = \left[\left(T_F \frac{(b-a)}{b}\right)^2 + \left(S_F \frac{(b-a)}{b} + T_{TF} \frac{r}{b}\right)^2\right]^{\frac{1}{2}}$$

$$P_B = \left[\left(T_F \frac{a}{b}\right)^2 + \left(S_F \frac{a}{b} - T_{TF} \frac{r}{b}\right)^2\right]^{\frac{1}{2}}$$

Bearing life

$$L_h = \frac{16\,666}{n} \left(\frac{C}{P}\right)^P \text{ hours where:}$$

T = torque (Nm)
r = gear Pitch Circle Radius (m)
a & b = distance (m)
∞ = gear pressure angle (deg)
β = gear Helix angle (deg)
n = bearing speed (rev/min)
C = bearing dynamic basic load rating (N)
P = bearing load (N)
P = life exponent (10/3 for needle and roller bearings and 3 for ball bearings)

Table 3: Transaxle bearing calculation

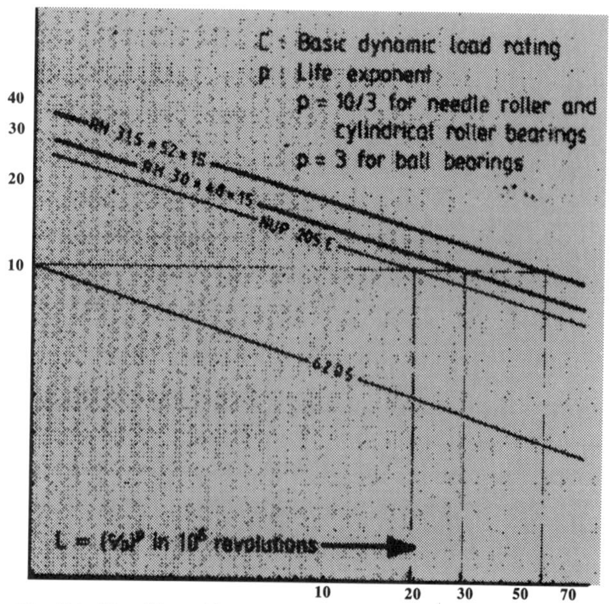

Fig 24: Needle-roller load capacity — load P in kN

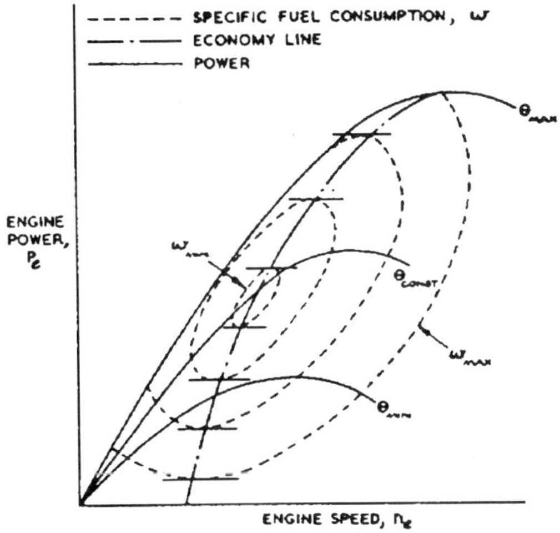

Fig 29: Economy range of IC engines

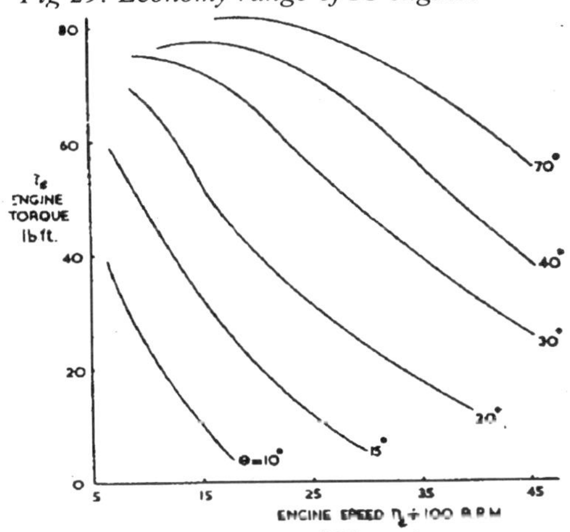

Fig 30: Torque range of IC engines

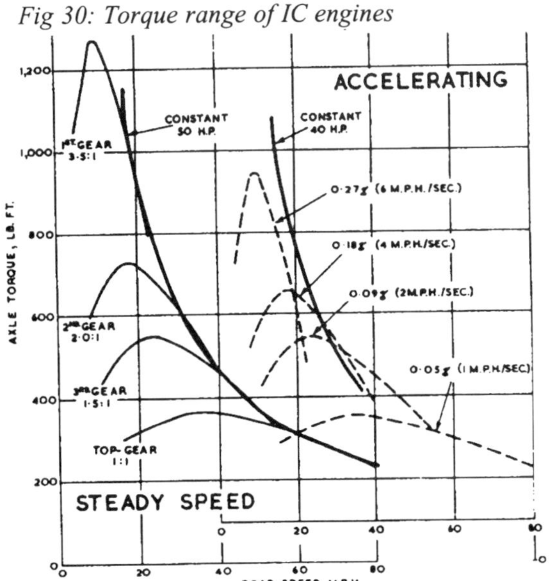

Fig 31: Transmission characteristics for 1.5 litre car

Automatic transmissions

The choice of automatic transmission system is influenced by vehicle category and the need to properly match engine characteristics to the vehicle weight distribution and performance requirements. Here the special attributes of hydrokinetic systems are examined alongside an introduction to matching performance characteristics with those of the vehicle.

On the multilane freeways of the Americas there has been a significant trend to automatics historically, as an exception to world car trends, but this has tended to be in relatively high-powered vehicles. Only the gradual perfection of continuously variable transmission (CVT) has caused any inroads into the lower-powered car market. The factor which is really likely to change this situation is the rapidly increasing price of fuel — aided by sophisticated electronics which allow transmission management systems to track the most efficient operating regime of the engine.

The great speed and load flexibility of petrol and diesel engines means otherwise that a surprisingly wide torque range can be obtained in one gear merely by throttle control. This is clear in Figs 29 and 30 based on historical studies at MIRA which show also the regions of specific fuel consumption overlaid. The largely linear relationship between torque and speed for different throttle angles is useful in designing optimum drive speed ratios to correspond with them.

When considering speed ratio change under power, it is important to account for the considerable rotary inertias in the vehicle which result in very high energy transfers between engine and drive train. Approximations used in initial design calculations have assumed 0.7 engine rotary inertia acting at the flywheel and vehicle inertia equal to gross weight x (tyre rolling radius)^2/ (overall transmission ratio)^2. These assumptions were used in the characteristics calculated for a 1.5 litre engined car, Fig 31, due to J G Giles[3]. As ratio is lowered to gain increased axle torque, engine inertia absorbs a greater proportion of power output during vehicle acceleration. Fig 32 shows some estimates of rotary inertias for different engines, obtained by Giles. Maximum acceleration is obtained when engine and vehicle inertias are equal but even then only half engine power is available for acceleration, owing to flywheel inertia.

Engine inertia is also important in deceleration of vehicles with automatic transmission. This is because the inability to declutch during severe braking reduces the retardation by the ratio vehicle-inertia / total-inertia and can cause problems at lower speed-ratios. Energy changes in gear shifts are shown in Fig 33 for a medium sized car. In the case of a much heavier vehicle such as an HGV, engine inertias may be ten times as great, because speed change of the flywheel may be only half that of a car and energy change may be only twice that of the car. Where a slipping clutch or fluid coupling is employed between the two, typically only half work expended during engagement is usefully converted. For a given throttle setting, engine speed change can bring about engine torque change. In Fig 34, fixed-throttle characteristics are drawn to a road speed base for third and fourth speeds of a 4-speed light vehicle; the chain-line region where ratio change does not effect axle torque is useful for selection of ratio-change points corresponding to throttle settings.

Fluid converter/couplings are valuable for smoothing automatic gearshifts and most are derived from the hydrokinetic type patented by Fottinger in 1905 in which a driveshaft-mounted impeller imparts kinetic energy to a fluid which is transferred to the driven turbine member. Power and torque capacity is proportional to (impeller speed)^2 and (impeller o/diameter)^5. In its simplest form, the fluid coupling, Fig 35, comprises impeller (1) comprising flat radial vanes integral with the outside casing (3). Similar radial blades (4) are formed in the turbine. A guide

Engine size		No. of cyl.	Inertia lb ft sec² units	Inertia lb in² units
Petrol engines	1 l.	4 cyl.	0·10–0·15	450–700
	1½ l.	4 cyl.	0·15–0·20	700–900
	2½ l.	6 cyl.	0·20–0·30	900–1400
	5–6 l.	V-8	0·35–0·50	1600–2400
Diesel engines	1½ l.	4	0·2–0·25	900–1300
	2–4 l.	4	0·25–0·60	1000–3000
	5–6 l.	6	0·35–1·0	1600–5000
	8–10 l.	6	1·2–2·0	6000–10,000

Fig 32: Rotary inertias of typical engines

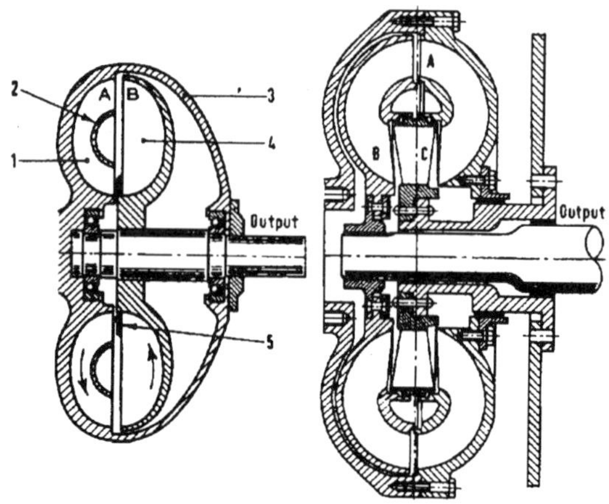

Fig 35: Fluid coupling (left) and torque converter

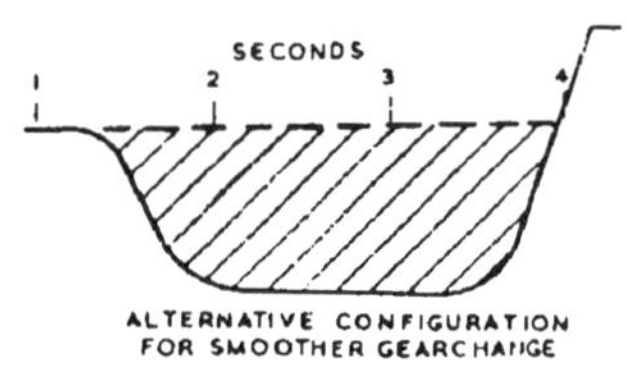

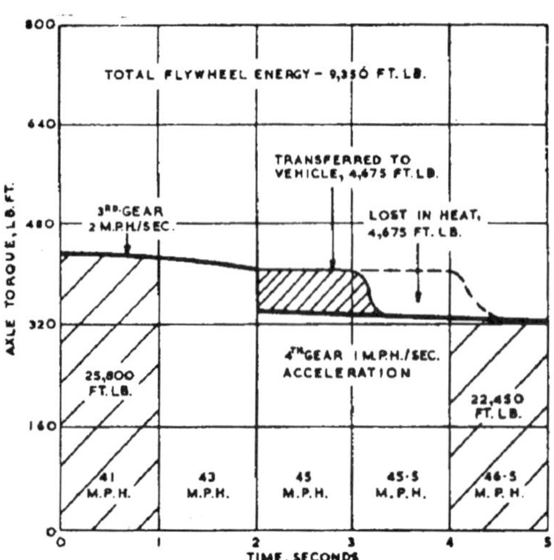

Fig 33: Energy changes in gear shifting

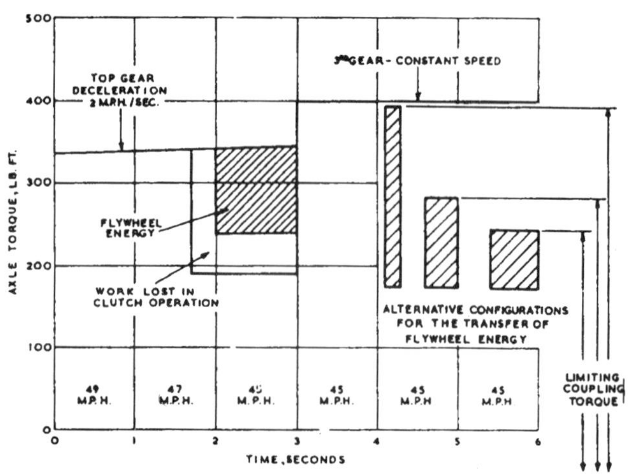

annulus (2) assists in establishing fluid circulation between the two and an antidrag baffle (5) may be incorporated. On rotation of the coupling, fluid centrifuges outward from the impeller to create a vortex like circulation between it and the turbine. Because of a 24% slip in speed between the two, centrifugal pressure at A is always slightly higher than at B to ensure circulation in the direction of the arrows. Fluid velocity increase from the inside to the outside of the coupling creates the kinetic energy which is given up to the turbine.

The torque converter differs from the coupling in itself having the capability of limited speed ratio change (with associated torque conversion). Here an additional reaction member is introduced between impeller and turbine. This comprises a set of circular blades inclined to the axis of rotation and held still in a circular frame. Turbine torque depends on fluid speed change from A to B; impeller torque depends on corresponding change from C to A. As change in fluid speed can now be effected in the reaction zone between B and C a torque on the reaction member results. This prevents the very high pressures (and flows) that would otherwise build up in a pure fluid coupling under heavy load and high slip. The blade sections in Fig 36 show how high the high flows in such conditions are converted into 'jets' of fluid driving the impeller forward and thus adding to the driving torque.

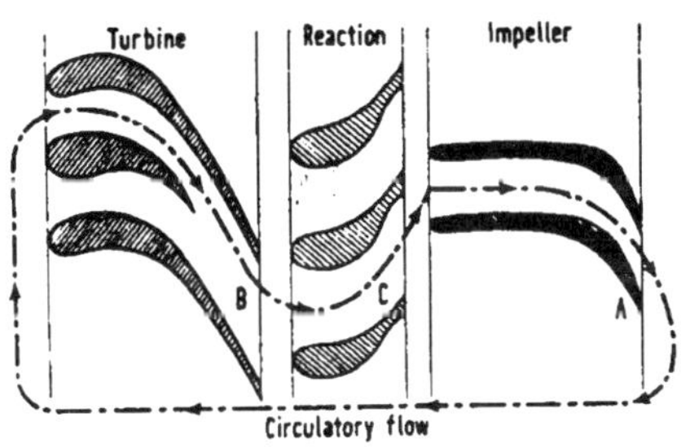

Fig 36: Blade sections

Some systems include adjustable reaction vanes which result in a relatively wide change in speed ratio for the coupling but usually the converters are used in conjunction with epicyclic gear trains with ratio changing effected by band brakes.

Typical performance characteristics for a torque converter are shown in Fig 37 for a small-vehicle system with constant input torque of 50 lb ft (67.8 Nm). The right hand diagram repeats the left hand one but converts it to ratio form. Efficiency in this case is the product of torque ratio and speed ratio and the end points of the reactor and output torque plots define the stall torque ratio and coupling point efficiency respectively. A lockout clutch can be used to prevent unduly high rise in engine speed degrading converter performance to provide a mechanical drive in top gear.

The success of different configurations of automatic transmission seems to be governed by the speciality of the vehicle application. The hydrostatic type has had most success in off-highway farm, construction and handling truck applications. The torque-converter/epicyclic type seem to have had most sales on executive/luxury cars and municipal vehicles. The design and configuration of the associated epicyclic gearboxes will be considered in a later section as will the continuously variable mechanical transmissions now finding favour in small cars and vans.

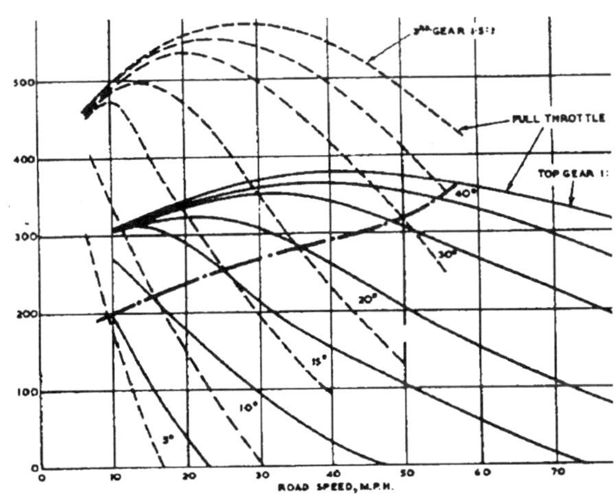

Fig 34: Selecting ratio-change points (torque, y-axis, in lb ft)

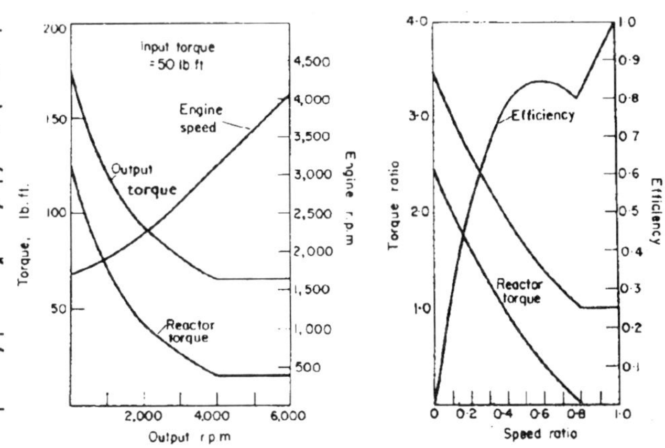

Fig 37: Converter characteristics

Epicyclic gears

Both semi-automatic and fully automatic transmissions, using fluid-couplings and torque-converters respectively, have conventionally used epicyclic gears for changing speed-ratios. Fig 38 shows a power flow diagram for a traditional automatic system which serves to show the operation of band brakes to effect speed-changing. The smooth transition between one speed-ratio and the next, and the ease of automating the actuation of the band brakes and/or clutches, are among the main advantages of epicyclics for automatic shifting. Automated spur gear automatics are now of course challenging epicyclics, for certain vehicle categories, with improving control technology. CVTs based on variable radius driving elements are yet another challenge but these, too, are tending to be combined with epicyclics.

Final drive differential units are the greatest users of epicyclics and these are variations on the basic planetary set shown in Fig 39 where A is the sun gear, B the ring gear and C the planets-carrier. Because, unlike countershaft gear trains, axes of rotation of the gear wheels are not all fixed, it is difficult to visualise behaviour of the gearset. The key factor is that relative motion between a pair of mating gears is the same whether the axes move or not — a principle which has led to a helpful technique for understanding how speed ratios are developed.

In Fig 40, for example, at (a) B and C are the pitch circles of two wheels contacting at pitch point P. As in this example carrier arm A for gear C is free to rotate about B. After moving through an angle θ this arm moves about the point on C which coincided with the pitch point, to T, such that RT = arc RP and QP becomes ST. As the angles:

$$\text{UST} = \text{USR} + \text{RST} = \theta + \phi \text{ and } \phi/\theta = T_b/T_c$$

where T is the number of teeth on the respective gears.

$$\text{Angle UST} = \theta(1 + T_b/T_c)$$

C thus turns in the same direction as the arm and $(1 + T_b/T_c)$ as fast as the arm. This technique can then be extended to evaluate the most complex gear trains. However, as the complexity of the calculation increases with the complexity of the train, a tabular method is often used both to ease the calculation and obtain a better appreciation of the relative motions

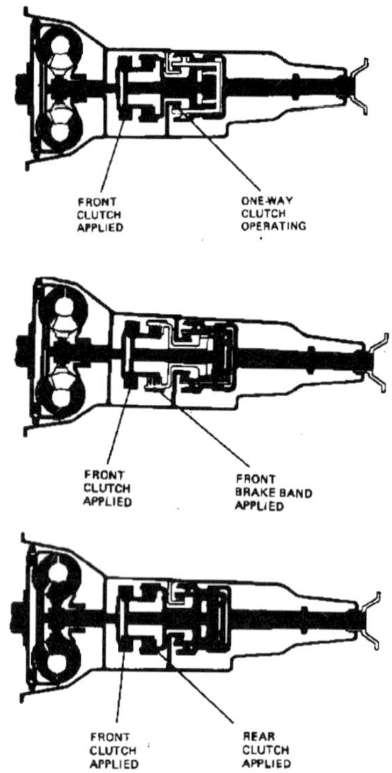

Fig 38: Power flow for Borg-Warner automatic

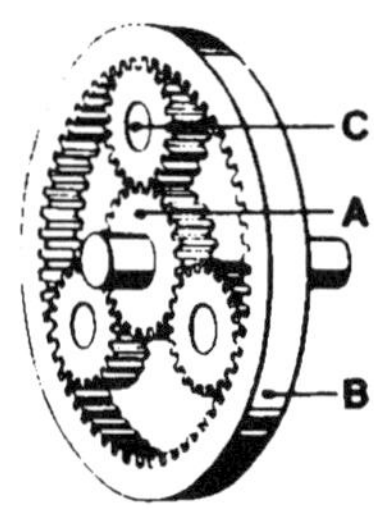

Fig 39: Planetary gear set

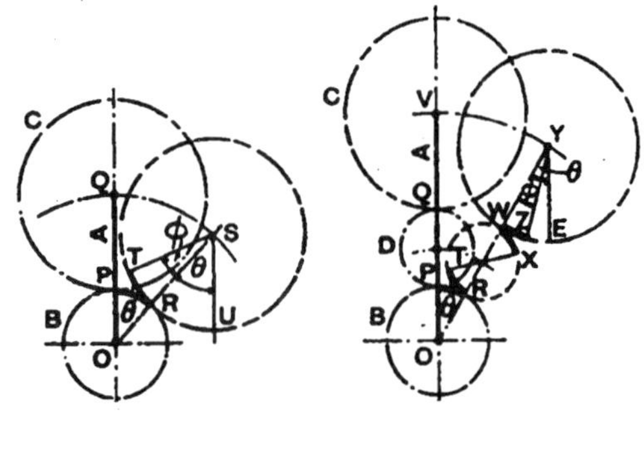

Fig 40: Plotting relative motion

within the system. Again this method is best understood by example, using the simple train of Fig 39 for which the ring gear is assumed to have 59 teeth, the planet wheels 20 and the sun gear 19 teeth. The technique is then to hold the three gear elements stationary in turn — and tabulate the motions, related to the number of teeth, of the remaining elements.

The ring gear, or annulus, is given a theoretical 1000 rpm input first with the planet-carrier held stationary; speeds and directions of other members are tabulated as shown in Fig 41, column 1, indicating reverse rotation by a minus sign. Next the sunwheel is made the input by multiplying column 1 values by the speed ratio, just derived, between it and the annulus, to give column 2 values.

Next stage is to make the sunwheel stationary by adding an equal quantity of the opposite sign to its speed value to achieve the column 3 situation. The carrier is first considered as input in this stage, a sign reversal will make all rotations clockwise as in column 3. The annulus is next the input by multiplying these values by its speed ratio with the carrier to give column 5 results.

The final stage is to bring the annulus to rest, as before by applying reverse input to it and the other elements, to give column 6 values. Then reverse these directions to show a forward drive result as in column 7. First, make the sunwheel the input, then follow by making the carrier the input by factoring by the speed ratio of the pair once more to obtain the column 8 results, which describe the overall performance of the train.

			1		2		3	4		·5		6	7		8
			Annulus Input Carrier Locked		Sunwheel Input Carrier Locked		Sunwheel Locked	Carrier Input Sunwheel Locked		Annulus Input Sunwheel Locked		Annulus Locked	Sunwheel Input Annulus Locked		Carrier Input Annulus Locked
A	59	+ 1000	+ 1000		− 323		− 1323	+ 1323	$\times \frac{1000}{1323}$	+ 1000	− 1000	0	0		0
P	20	+ 1000 × 59/20	+ 2900		− 936		− 1936	+ 1936		+ 1460		+ 460	− 460		− 1936
S	19	− 1000 × 59/19	− 3100	$\times \frac{1000}{3100}$	+ 1000	− 1000	0	0		0		− 1000	+ 1000		+ 4100
C	0	0	0		0		− 1000	+ 1000		+ 756		− 244	+ 244	$\times \frac{1000}{244}$	+ 1000

Fig 41: Tabular analysis for basic gear set

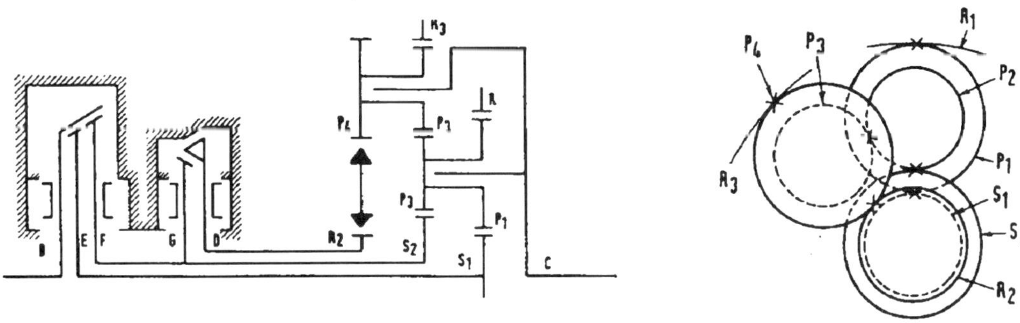

Fig 42: Compound epicylic train

						1st		2nd				3rd			4th				REV
27	S_1	+1000		+1310	$\times \frac{1000}{1310}$	+1000		+1800		+1900		+1440			+1000		+ 807	$\times \frac{1000}{807}$	+1000
30	P_1	− 900		− 590		− 450		− 812		0		0			+1000		−1093		− 1354
87	R_1	− 310	+310	0		0		0		+ 590		+ 448			+1000		− 503		− 623
39	S_2	+ 416		+ 726		+ 554	$\times \frac{1000}{726}$	+1000		+1316	$\times \frac{1000}{1316}$	+1000		LOCK	+1000		+ 223		+ 276
18	P_2	− 900		− 590		− 450		− 812		0		0		LOCK	+1000		−1093		− 1354
18	P_3	+ 900		+1210		+ 924		+1666		+1800		+1367		LOCK	+1000		+ 707		+ 876
33	P_4	+ 900		+1210		+ 924		+1666		+1800		+1367		LOCK	+1000		+ 707		+ 876
33	R_2	− 900		− 590		− 450		− 812	+900	0		0	+1000	LOCK	+1000		−1093		− 1354
84	R_3	+ 193		+ 503		+ 384		+ 692		+1093		+ 832			+1000	− 193	0		0
C				+ 310		+ 237		+ 427		+ 900		+ 684			+1000		− 193		− 239

Fig 43: Tabular analysis for compound train

For the sort of compound train seen in a gearbox, Fig 42, the tabulation in Fig 43would result. Here sunwheel S1 is the input for reverse and first with ring-gears R3 and R1, the reaction members respectively. In second, sunwheel S2 reacts again with R1 and in third it reacts against R2. For fourth gear S2 is still the input but it is locked with R2 by clutch GD, D being simultaneously released from the casing, allowing the train to rotate as a whole.

Advantages of such compound systems include the ability to effect changes in ratio under power without interrupting the drive; compactness is another benefit as well as the sharing of tooth loads between two or three contact points, helped by carrier rotation reducing pitch line velocity; there is also a greater range of ratios from a given number of elements than in many countershaft systems. In assembling such trains a useful rule to remember is that for A and S the numbers of teeth on annulus and sunwheel respectively, and n the number of planets, $(A + S)/n$ must be a whole number; otherwise certain combinations of tooth numbers cannot be assembled because of the need of equal spacing on the planets. For large ratio steps it is also desirable to use either one-way free-wheel clutches or self-energising bandbrakes; for smaller steps, cone or single plate clutches are acceptable.

An epicyclic gearbox made in very large numbers for buses was the Leyland Pneumocyclic. This four-speed unit, Fig 44, obtained first, second and third indirect ratios by arresting the epicyclic reaction members by means of power-operated bandbrakes. For compact front-drive cars such as the Mini an interesting alternative, Fig 45, was the bevel-geared epicyclic transmission. This exploited the particular feature of epicyclics that torque reactions transmitted to the casing, in forward drive, are relative to the numerical gear reactions minus on and the main consideration was thus to provide longitudinal stiffness of the assembly.

Spiral bevel gears were employed and just

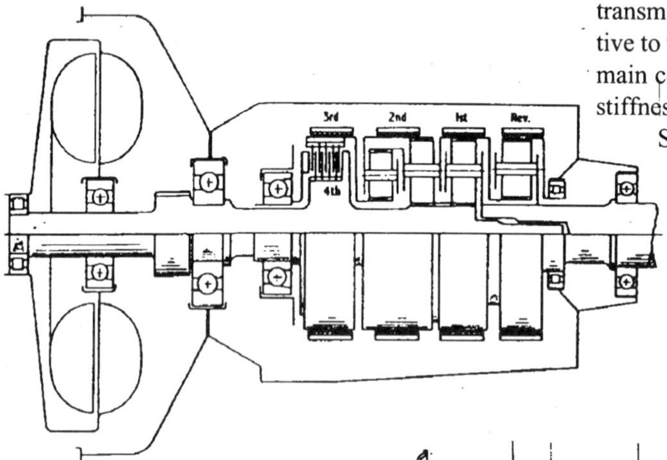

Fig 44: Pneumocyclic transmission

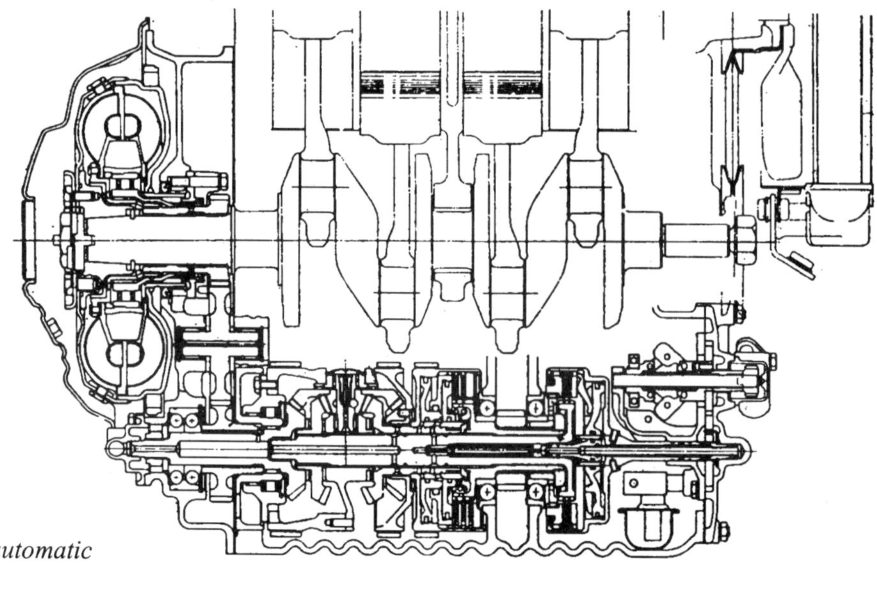

Fig 45: Mini transaxle automatic

eight in number gave four well-spaced ratios. The compactness of the assembly allowed very short mainshafts allowing relatively low bending stresses to be produced by the band brakes. Diagrammatically the four-speed arrangement was as Fig 45. This was developed from an earlier two-speed unit shown, in which the bevels are so arranged that the input gear with its meshing planetaries form the outer gear cluster, involving adjacent input and output wheels such that the output shaft was taken through the centre of the gear unit. To obtain a three speed unit a way was found of holding the reverse gear stationary and allowing the gear unit to planet about this newly found sun wheel. In adding a fourth speed, a further sun wheel was created about which the gear cluster could planet. From the ratio graph in Fig 46 it can be seen that the contra rotational speeds within the gear train are relatively low.

Fig 47 shows the diversity of epicyclic trains used on established gearbox designs. At (i), the SRM has two-stage turbine, a, normal drive band, b, low-drive band, c, direct-drive clutch, d, and gearset, e; at (ii) the ZF Hydromedia high-gear plate clutch, a, reversed free-wheel for overrun braking, b, reaction member free-wheel, c, reverse cone clutch, d, low-gear plate clutch, e, and low-gear free-wheel, f;

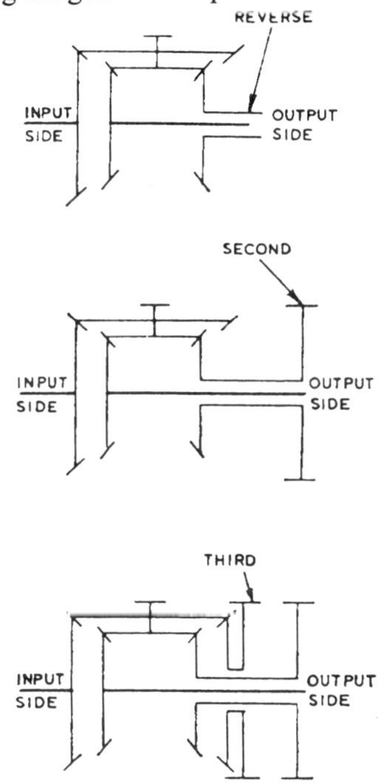

Fig 46: Ratio stepping arrangements

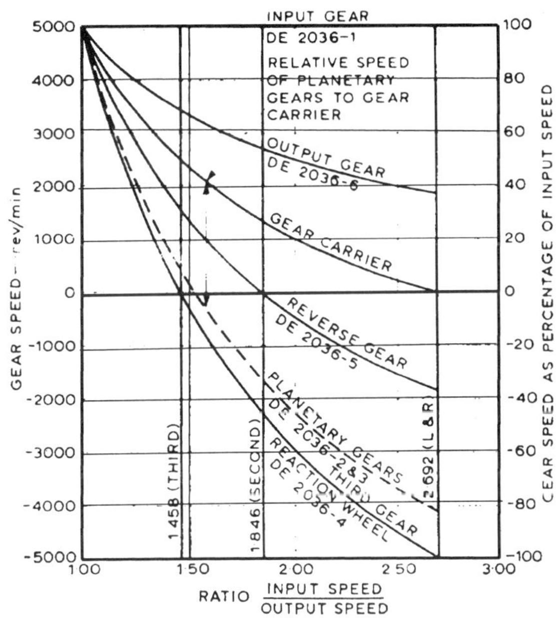

Fig 47: Gear ratio characteristics

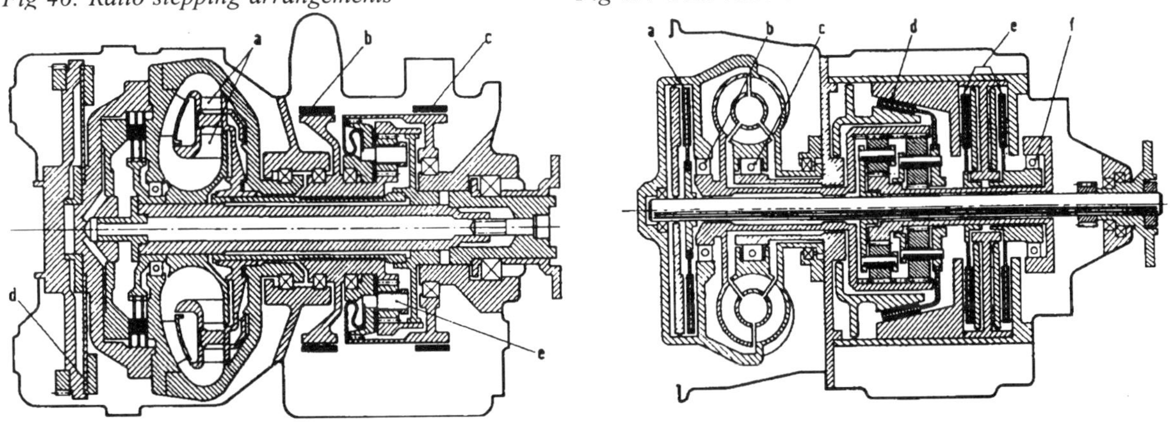

Fig 48: Epicyclic configurations

Automatic clutches

The relatively high cost of automatic transmission has prompted consideration of means to automate existing manual change-speed gearboxes. Automatic clutches are a substantial step in that direction and need only a satisfactory means of either servo-assistance or automation of the gear-shift to achieve either most of the benefits of an automatic change or at least the two-pedal control of the vehicle.

To improve upon the early attempts at such systems, which were a market failure, a proposal by ERA Ltd in the late 1950s sought to use a clutch basically similar in design to a manually-operated one and effect its actuation by means of a hydraulic cylinder, using automated control. The cylinder replaced the conventional clutch springs and an engine-driven pump generated hydraulic pressure with which to load the clutch. To disengage the clutch, a hydraulic valve was operated by a solenoid — switched by an electrical contact on the gear-shift lever, Fig 49. Fig 50 shows a centrifugal oil governor which served as the pump.

The pump was kept full of oil which was made to rotate at a speed as close as possible to that of the rotor. This was achieved by continuously feeding oil from the engine pump to the axis of the governor, allowing it to overflow freely back to the sump so as to avoid pressure build-up; the governor housing was incorporated in the engine block and the rotor mounted on the crankshaft.

To avoid too rapid re-engagement of the automatic clutch a softening device was proposed, sensitive to engine overrun. Oil pressure was thus reduced to a value at which the clutch was able just to transmit the maximum braking torque that can be developed by the engine. By this means, after rapid down-change with no throttle opening, the clutch would slip and prevent any unpleasant jerks or locking of the driving wheels. The overrun condition could be sensed by an inertia-switch or engine torque-reaction rock could be used to operate a micro-switch. That would energise a solenoid-operated valve and connect the hydraulic cylinder to a relief valve set to the required operating pressure.

Fig 51 shows the complete system, in which higher than average clamping loads could be used for the clutch since these were not required to be released by a clutch pedal. Clutch plate inertia could thus be

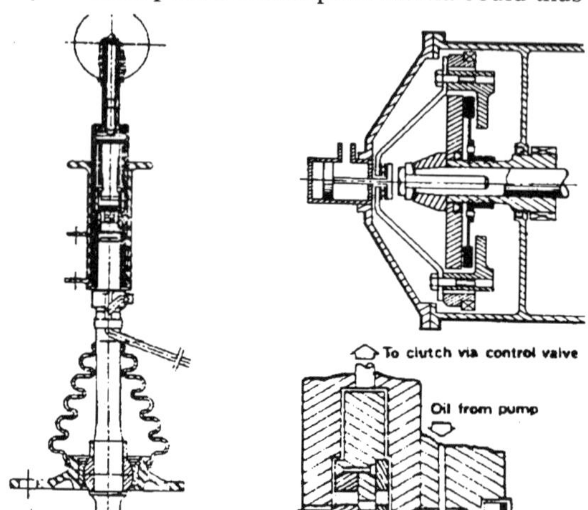

Fig 49: Hydraulic clutch operation and gear-lever actuator

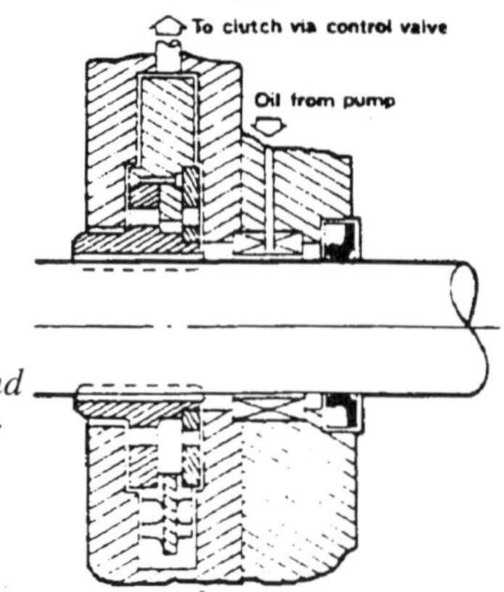

Fig 50: Centrifugal oil-governor

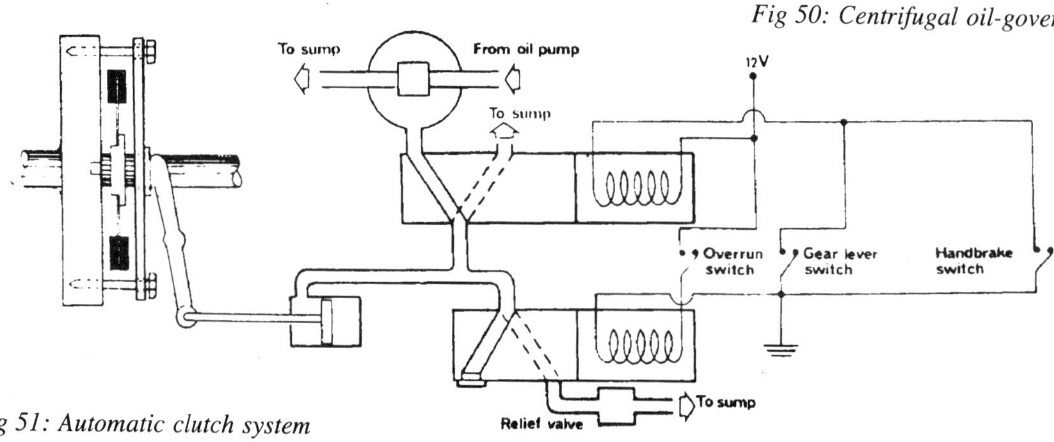

Fig 51: Automatic clutch system

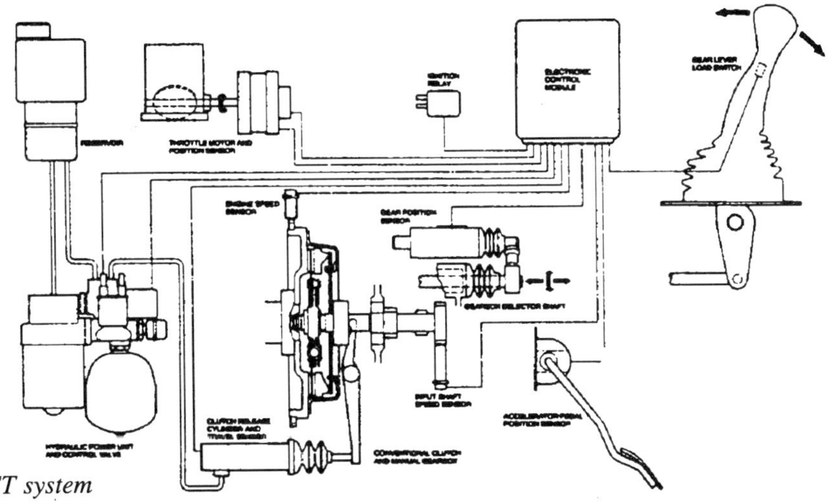

Fig 52: ACT system

reduced in order to ease the task of the synchromesh mechanism. An anti-creep switch is shown which prevented oil pressure being applied to the cylinder when the handbrake was applied. Such a system was considered superior to the automated clutch systems which were marketed with only limited success in the early post-war period and it paved the way towards the successful modern systems which have benefited so much from the improvement in control electronics.

Auto-clutch systems

One of the early post-war systems was the Automotive Products Manumatic. That company have revived the 2-pedal control concept in a more refined form recently as the ACT system; this again used hydraulic operation, favoured because of the relatively high force requirement. The later AP system incorporates drive-by-wire throttle control which can control engine speed during gearshifts. Speed sensing for the throttle control is from the clutch driven plate and the throttle pedal does not have to be 'eased' when changing up. The clutch is automatically disengaged below a pre-determined engine

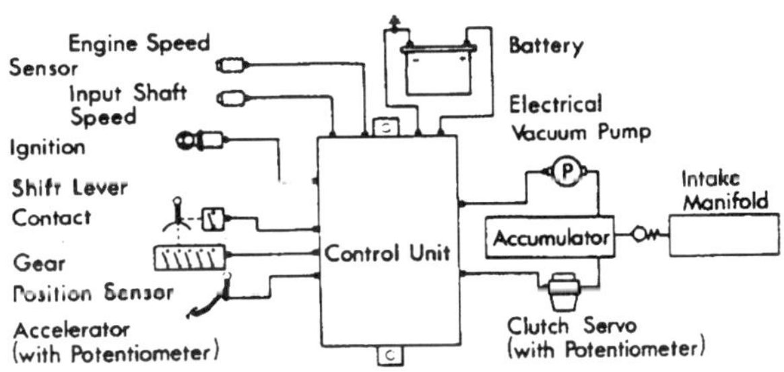

Fig 54: Servo-pneumatic actuator

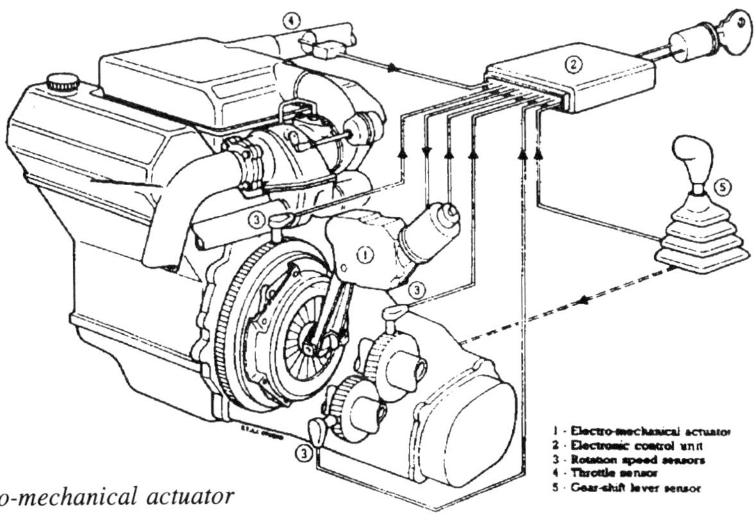

Fig 53: Electro-mechanical actuator

speed and the release bearing is given ample capacity for a prolonged 'hold-out'.

An electronic controller protects against excessive slipping should the gear selector have been left in a higher speed prior to taking off from rest. The electronics also provides compensation against clutch-lining wear and protects against post cold-start fast-idle which would otherwise cause premature engagement. It will also take account of the higher engine speeds required for clutch engagement on a hill-start. Fig 52 shows the ACT system layout.

Electrical actuation is favoured by Valeo for its automatic clutch. A computer operates the clutch though the electro-mechanical actuator, Fig 53. Sensors provide input data to the computer — on engine speed, gearbox input shaft speed, gear-shift lever movements and throttle valve position. The system is fitted to the Lancia Delta Integrale and the Ferrari Mondiale T and faster acceleration times are said to follow from increased speed of gear-shifting on both of these applications.

The Fichtel & Sachs EKS automatic clutch system uses an electric motor driven screw-jack actuator which operates the clutch via a master and slave hydraulic cylinder. For heavy-vehicle applications air-servo assistance can be provided. ZF have automated an F& S clutch using wholly pneumatic actuation while another supplier, Kongsberg Automotive, use hydraulic actuation based on a proportional pressure control valve. This way there is no need for position feedback from the clutch arm. An auxiliary spring in the actuator helps to simulate the typical S-characterstic of the manual clutch diaphragm spring.

Yet another actuation system has been used by VW who also extended the automatic clutch arrangement for both engine cut-off arrangements and hybrid drive configurations. Servo-pneumatic operation has been used, fed by engine manifold vacuum, Fig 54 showing the various sensor inputs. Different conditions of engagement and disengagement, the latter being required, of course, when transmission input speed is less than engine idle speed. Another quite different characteristic is required when driving away from standstill; here engine speed is always greater than transmission input speed. Between these two extremes engine speed can be less than, equal to or greater than the transmission speed, as indicated in Fig 55.

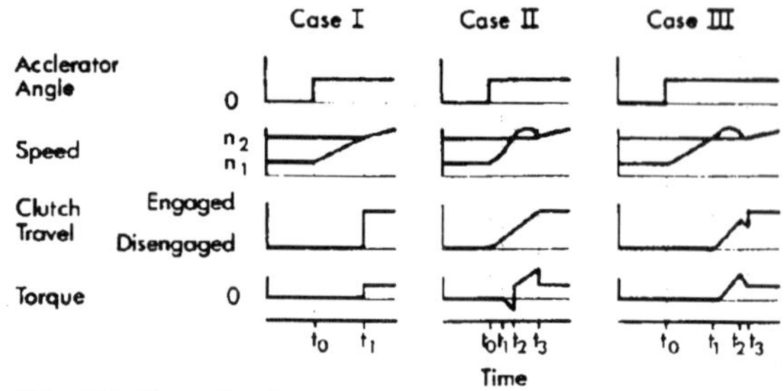

Fig 55: Speed relationship cases

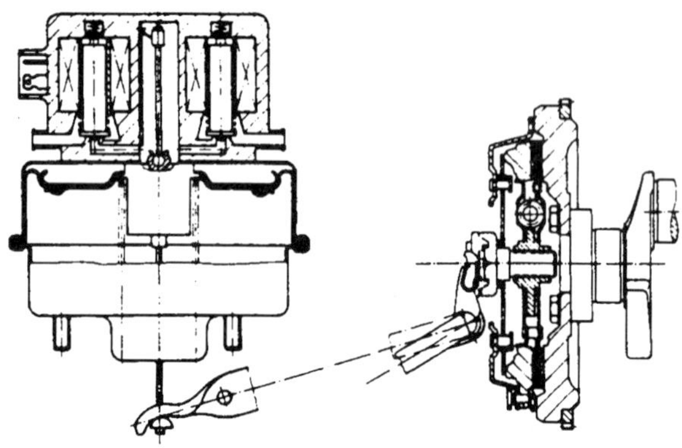

Fig 56: Servo-unit components

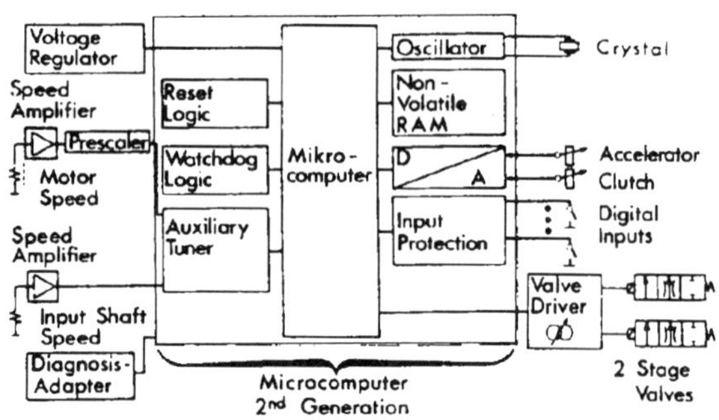

Fig 57: Electronic control

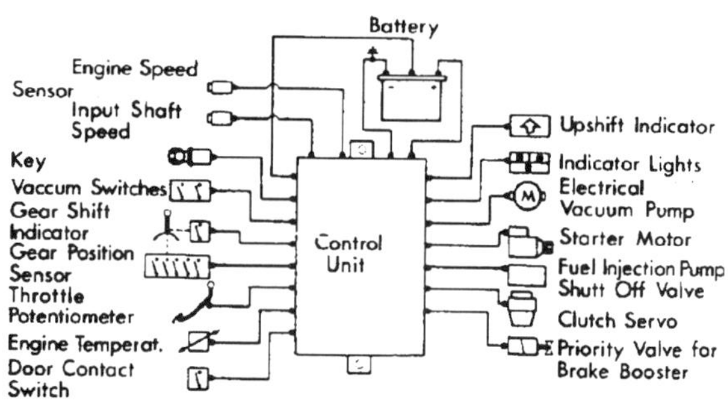

Fig 58: Incorporating stop-start

VW chose a combination of cases II and III for their solution. Digital valves were used in the clutch travel circuit, which are either fully open or closed, and careful consideration had to be given to the following factors:

- dead time between electrical actuation and mechanical reaction at the servo
- delay time between switch-off of the actuator and the end of the servo action
- change in travel as a function of the accumu lator pressure
- dependence on valve temperature, voltage and operating time
- minimum time between two pulses without manual intervention

Two-stage valves were developed, small ones being used for fine positioning and large ones for rapid changes in travel, Fig 56 showing the two components of the servo unit — position cylinder and valve block. The vacuum side of the flexible diaphragm is connected to the potentiometer wiper, to measure the effective position of the clutch release lever. Fig 57 is a block diagram for the electronic controls.

As well as the straight-forward single automatic clutch and series gearbox arrangement, alternative drives which have been used by VW include a stop-start system, Fig 58, in which the engine is switched off in the idling phase during vehicle standstill and overrun conditions. Another configuration, Fig 59, allows hybrid drive to be used. Here the flywheel can be replaced by an electric motor so that the vehicle can be electrically powered when one clutch is closed or by IC engine when both are closed. The high inertia of the electric motor rotor allows unusually high starting speeds for the IC engine to be realised.

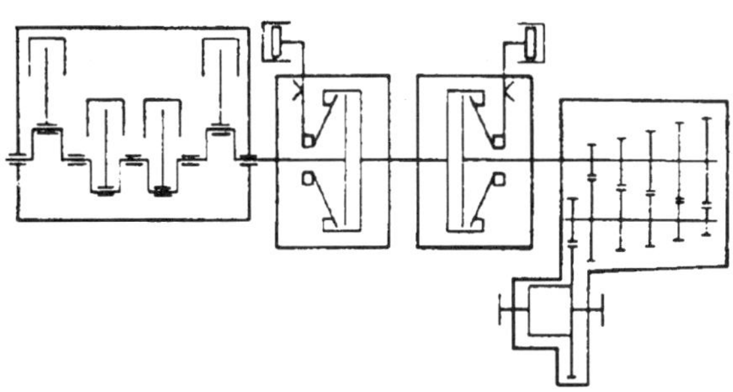

Fig 59: Incorporating dual, hybrid, drive

Continuously variable transmission

The development of CVTs has been dominated by two distinct types, originally for different vehicle categories. Recent trends have seen a convergence of the vehicle applications so the choice of fitment is likely to be made on the different characteristics of each system. Here the development history is traced on a chronological basis in order to provide a guide to relative strengths and weaknesses.

CVTs were originally conceived for manufacturing machines and prior to the advent of digital control electronics had limited prospects for automotive exploitation. Such infinitely variable speed drives convert a constant input speed into output speeds steplessly variable within certain limits. An early contender was the Perbury gear, similar in principal to the Hayes unit used in Austin cars between 1934 and 1938 — but improved to ensure automatic equalisation of load between discs and rollers. Traction between the rolling surfaces are now supported by viscous drag in the interposed oil film. This type of transmission, involving metal rollers, contrasts with the variable pitch pulley and belt drive type, the latest development of which is by Van Doorne, Fig 60.

The Perbury gear, Fig 61, transmitted about 100 bhp at 92 percent efficiency in its original form (1960s), with an overall speed ratio of 4.7:1. The input drive carries two discs, which drive the double-

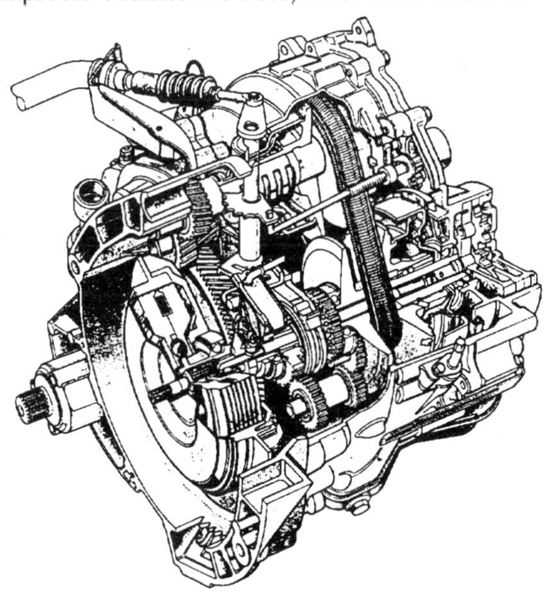

Fig 60: Van Doorne CVT

sided output disc through two sets of tilting rollers. Axial loading is effected hydraulically. The Van Doorne gear overtook the Perbury in terms of post-war application success and became sole fitment to the DAF 33 car. The so-called Variomatic unit was infinitely variable between 16.4 and 3.9:1. In the 1970s the transmission was improved for the DAF 66 car, Fig 62, with such features as a roller-actuated centrifugal plate clutch. On the pulleys, centrifugal weights were introduced to compress the vee-belt drive causing it to ride up and give a higher ratio with increasing speed. This effect is countered by rolling resistance at the road wheels tending to increase belt tension and hence give a lower ratio. The resulting balance could be overridden by the electro- pneumatic system, sensing throttle depression and pressurising the drive pulley system accordingly.

In 1976, Van Doorne introduced the Transmatic transmission which heralded the steel drive belt having separate tension and thrust members; this considerably increased torque capacity. Axial force applied to the pulley sheaves tensioned the belt; then, the combination of lateral and radial force on the blocks was sufficient to transmit drive from pulley to belt. Maximum speed ratio was 6:1 — limited by the size of the pulleys involved — and input torque capacity 90 lb ft (122 Nm). To give a ratio spread of either 12:1 or 16: in the car, a fully automatic range-change was incorporated. Transmission efficiency of about 90% was claimed and gear ratios varied from 2.31 to 0.58:1. Unit weight complete with clutch was 60 kg, Fig 63

By 1983, the Van Doorne transmission had been developed for commercial vehicles and a unit was installed in a Leyland National bus. Fig 64 shows that an epicyclic gear train was incorporated and microprocessor control was introduced. Six instead of four rollers were used, bearing on discs mounted on a two-piece layshaft. Parallel to it a mainshaft drives from the engine to the planet carrier of the epicyclic gears while the layshaft drives the sun gear. One clutch takes drive from annulus gear to the output shaft and another connects to it via the variator and sun gear. The control system operates the 'steering' of the variator rollers electro-hydraulically.

Fig 65 shows how the rollers of the successor to the Perbury transmission rotate in carriages which tilt about trunnions at their ends, these being on Y-shaped rockers, the legs of which extend inwards into

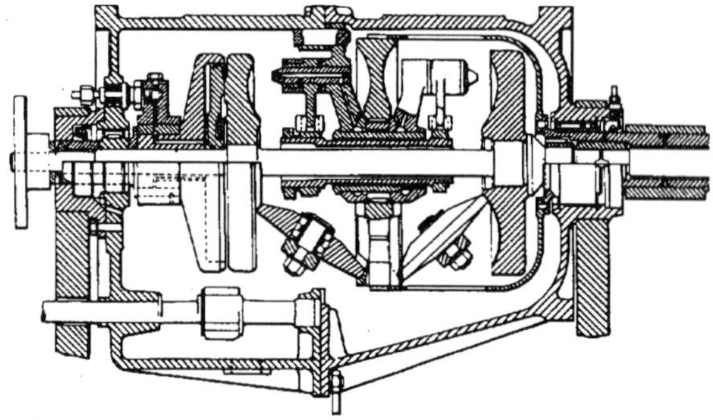

Fig 61: Perbury gearbox

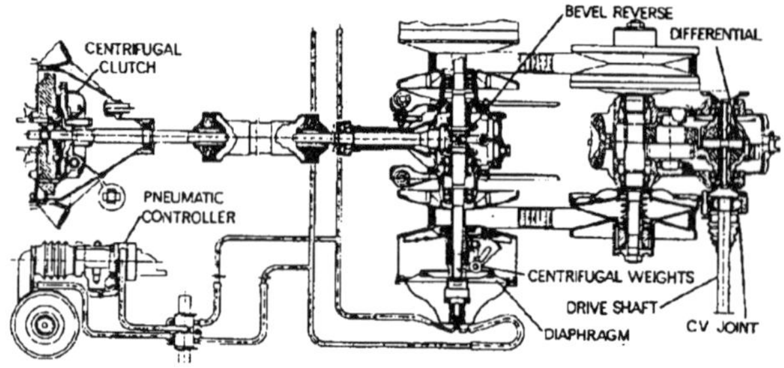

Fig 62: DAF Variomatic

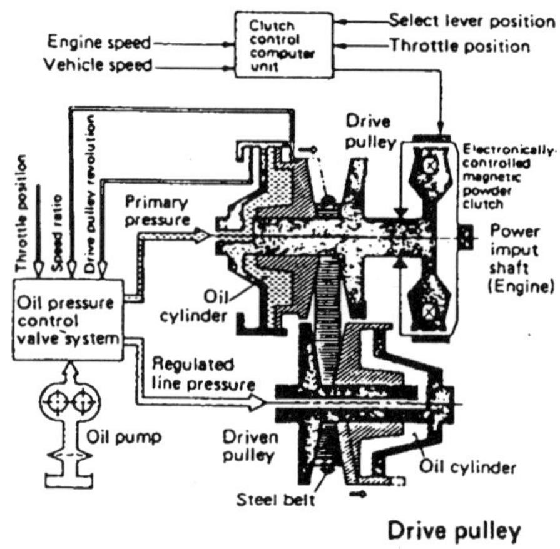

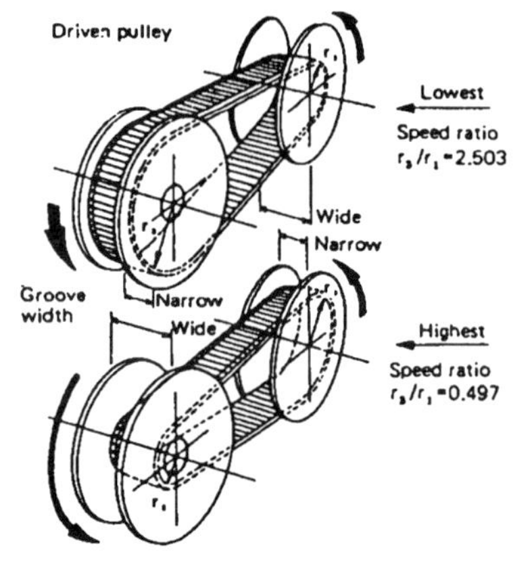

Fig 63: Variator mechanism

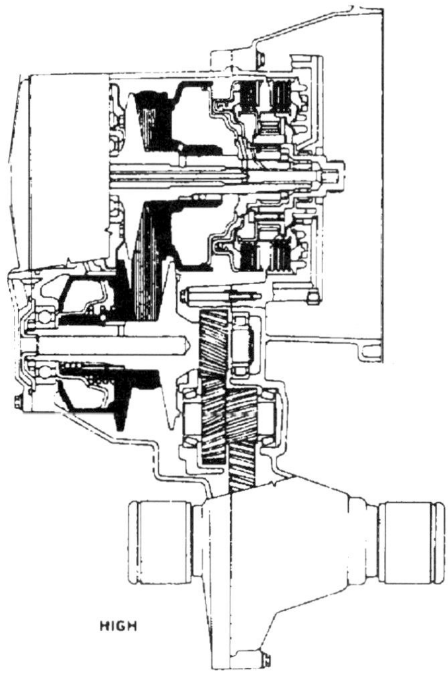

Fig 64: Transmatic unit for Leyland National

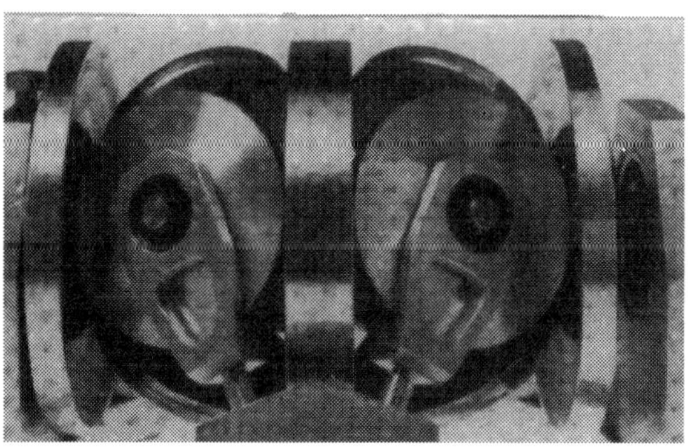

Fig 65: Roller arrangement

slots on a control sleeve which is rotated a few degrees in either direction to steer the rollers. Torroidal tracks in the surface of the discs ensure the rollers stay in continuous contact as they are steered. Because the variator can only run in one direction, the epicylic gear is used to obtain reverse and neutral, Fig 66 showing the torque/speed characteristic. These modes are obtained by using the variator to increase or decrease sun-gear speed relative to that of the carrier — thus driving the annulus backwards or forwards, or, in the intermediate condition, to hold it stationary.

In 'low' regime, the annulus is clutched and the sun declutched; at 0.43:1, both sun and annulus rotate at the same speed and direction — the synchronous speed of the variator. Here the sun is clutched and the annulus declutched. Thereon the transmission is in 'high', further increase in variator ratio taking overall ratio up to 1.8:1 overdrive. In this mode the variator carries all the torque while the epicyclic gears rotate unladen. The electronic control can, however, work the clutches to hold the transmission in 'low' when engine braking is required. Since the variator works best at relatively low torques and high speeds it is geared up from the mainshaft speed. In this vehicle installation, constant speed running of the engine was largely achieved and speed changes of the vehicle achieved by the transmission only. However, acceleration levels demanding significant increase in power involved momentarily relieving the engine of transmission load for it to operate at a higher speed, then gradual recoupling to a new steady-state condition — achieved by the electronic controls.

1984 saw the development of the Van Doorne CVT for adaptation to the Fiat Uno and the Ford Fiesta cars, also with the integration of epicyclic

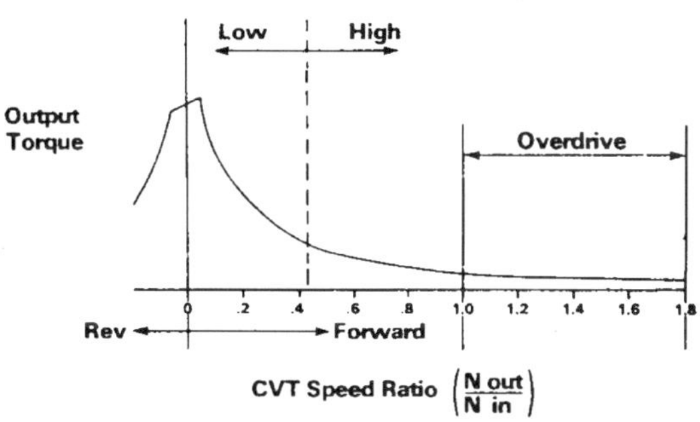

Fig 66: Torque/speed characteristic

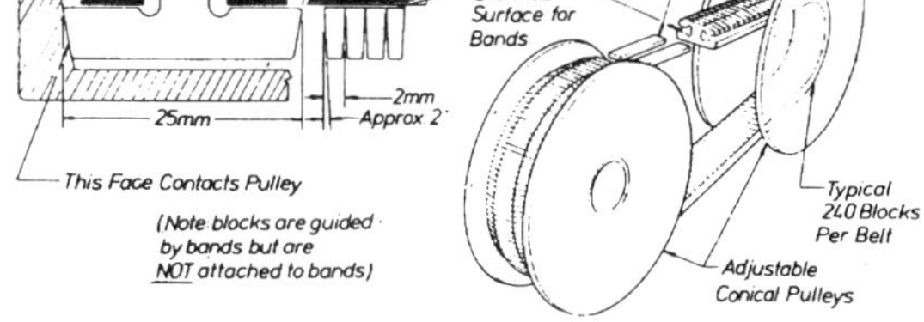

Fig 67: Belt clamping

trains. The centrifugal clutch was replaced by a hydraulically operated rubber- plated one. Control of transmission ratio is effected by varying the primary pulley width by means of a hydraulic cylinder while applying an axial force to the secondary pulley by means of a spring and second hydraulic cylinder — to prevent pulley slip but at the same time avoid excessive clamping loads that would waste energy, Fig 67.

The British technology group's development of the Perbury gear has resulted in the Torotrak transmission, Fig 68, which has involved changing the mechanical control of the variator rollers to a hydraulic one. Also while there are still two torroids, there are only three disc members, the central, output, one being double-sided. Hydraulic actuation helps ensure the rollers work exactly in unison so as to avoid losses due to internal power recirculation, the forked carriages now terminating in hydraulic pistons working in cylinders housed in the transmission casing. Quicker responding ratio changes are also said to result. Contact reaction balancing the pistons' control forces are created by the shearing of the oil film through the rollers, precession angle which creates speed differences across the contacts. Thus when either the control force or the working load changes, the carriages move to steer the rollers to a new precession angle. Control is therefore effected by torque and not ratio.

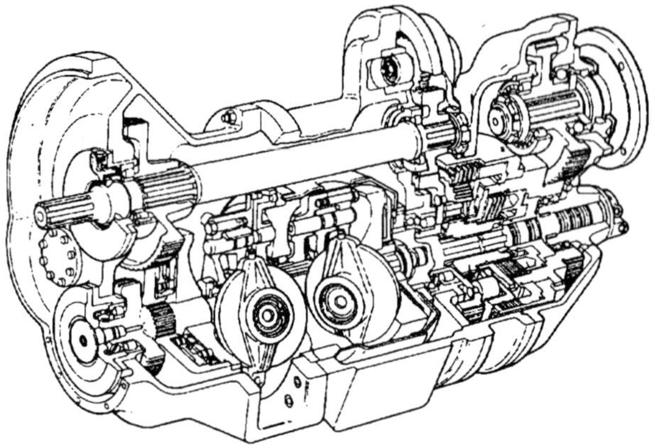

Fig 68: Torotrak transmission

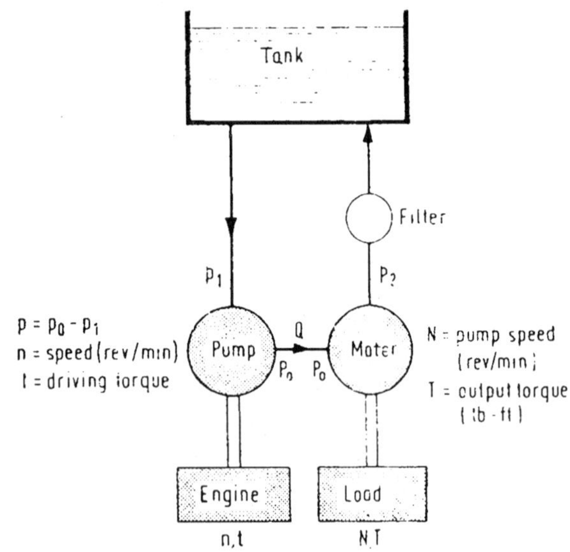

Fig 69: Idealised gravity-fed system

TYPE	DELIVERY Fixed	DELIVERY Variable	MAX. FLOW, gal/min	MAX. INPUT h.p.	MAXIMUM WORKING PRESSURE, lb/in²	WORKING SPEED RANGE, rev/min	MAXIMUM OVERALL EFFICIENCY, %	SPECIFIC WEIGHT lb/h.p.
Gear	×		270	284	2500	500–3000	90	1–7
Vane	×		90	110	2000	600–2000	88	1–17
Radial piston	×	×	120	222	6000	up to 2000	not known	1·5–22
Axial piston	×	×	3000	5250	5000	up to 4000	96	0·5–12
In-line plunger	×	×	23	—	7000	1440	not known	not known
Ball piston	×	×	—	45	—	1500	not known	not known

TYPE	DISPLACEMENT Fixed	DISPLACEMENT Variable	MAX. FLOW gal/min	MAX. OUTPUT h.p.	MAXIMUM WORKING PRESSURE, lb/in²	WORKING SPEED RANGE, rev/min	MAXIMUM OVERALL EFFICIENCY, %	SPECIFIC WEIGHT lb/h.p.
Gear	×		270	284	2500	500–3500	85	1–7
Vane	×		77	100	2000	50–2000	88	2–6
Radial piston	×	×	95	150	3600	0–750	97·5	0·7–18
Axial piston	×	×	3000	5250	4000	30–4000	96	0·7–9
Ball piston	×	×	—	45	—	1500	not known	not known

Fig 70: Types of pump/motor

Hydrostatic drives

A reasonably wide spectrum of special vehicle types has enjoyed the benefits of hydrostatic transmission. By careful design to overcome the problems of inefficiency, compared with geared transmissions, this application spread could further widen to include road vehicles.

The advantages of hydrostatic drives are well known to makers of farm, construction and industrial vehicles whose work motions, and power take off requirements favour the ready availability of infinitely variable transmission in forward and reverse modes. Although having slightly lower efficiency than gear transmissions, the ability to place the constituent pump and motor units almost anywhere in the vehicle, alongside their stepless speed changing, makes them attractive to specialist vehicles. There is the added advantage of in-built braking as the system is hydraulically locked when the pump is at zero flow.

It is important to differentiate between hydrostatic and hydrodynamic drives. While the latter derive power from the kinetic energy of the working fluid, the former only involve the principle of pressure in a liquid column actuating a hydraulic unit attached to the load. There is positive displacement of fluid between continuously running pump and motor combinations. Pump throughput, for given speed and capacity, is almost independent of pressure; motor speed, too, is independent of pressure but depends on capacity and flow — while pump pressure depends on the motor load. Fig 69 shows an idealised circuit, gravity fed from a header tank. Power pumped into the fluid is

pQ/1428 bhp and as mechanical power output is

n.t/5252, and the power absorbed by the motor is

$(p_0 - p_2)Q/1428$ bhp and the motor output

= *NT/5252* bhp.

Fig 70 shows some of the common types of pump and motor units with their different operating characteristics. Variable capacity can be achieved with many of these systems using a tiltable swashplate, Fig 71. The most common arrangement is for a variable displacement pump to be used in conjunction with a fixed-displacement motor — with which output speed ranges from zero to any specific value depending on ratio of motor to pump capacities. Motor output torque is dependent on the pressure difference across it. Fig 72 shows the theoretical performance characteristic for this system compared with an actual one in Fig 73. HP_{in}, P_t and T_t are the theoretical, 100% efficiency, values while HP_{out}, P_a

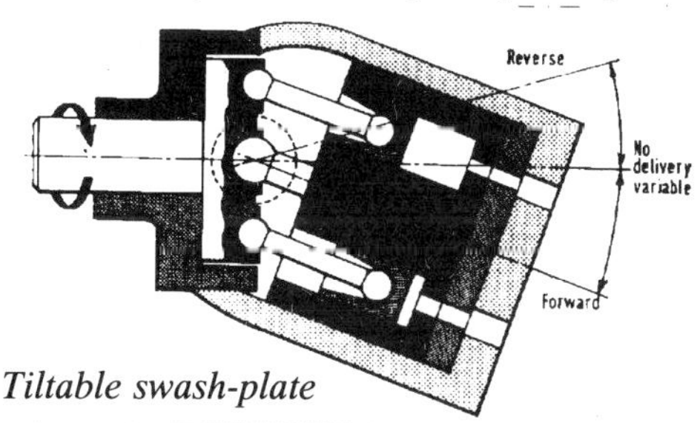

Fig 71: Tiltable swash-plate

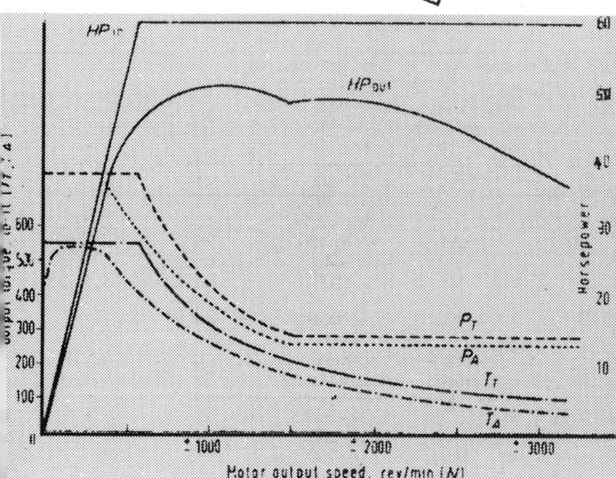

Fig 72: Theoretical performance

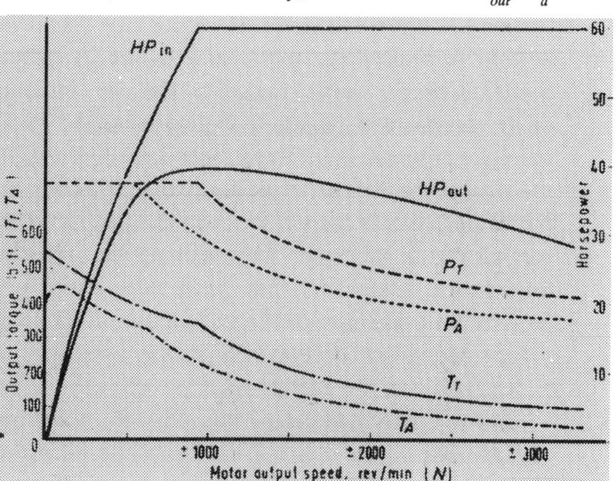

Fig 73: Actual performance

and T_a show the attainable output values for actual units. The plots show efficiency of the transmission at maximum input power and the need to avoid reducing the capacity of a variable motor too much.

In variable capacity pumps and motors, the effort required to change capacity under load is often considerable, thus a main pump usually incorporates a servo-valve for moving a two-way ram which tilts the swash-plate using boost pressure. A typical closed circuit sytem is shown in Fig 74 in which leakage from pump and motor bodies returns to the tank and a boost pump continuously replenishes the circuit, excess being bled off through a relief valve.

Hydrostatic transmissions

Vehicle hydrostatic transmissions are usually of the constant power type. In variable displacement pump/motor systems, the pump displacement is increased from zero to maximum, first, then the motor displacement is reduced from maximum to minimum. The correct sequence may be controlled by a cam, hydraulic servo or linkage as shown in Fig 75. Here the Dowty Taurodyne transmission has a linkage, the arms of which are hollowed out to double as fluid feeds. Its characteristics are shown in Fig 76.

Hydrostatic transmissions are readily adapable to the steering of off-road vehicles, particularly tracklayers (Fig 77). Here parallel motors for the main drive have differential ability; auxiliary variable units are set to give required speed ratio at the driving sprockets, power being transferred from one track to the other through the hydraulic link. Hydostatic steering can also be achieved by a double differential system as shown in Fig 78. Here the cross-shaft is driven by hydrostatic pump and motor in either direction over a limited range of speeds, continuously variable with respect to engine speed. Most of the drive power is transmitted mechanically through the change speed gears so the relative inefficiency of the hydrostatics is not too serious. The pump/motor can also lock the cross-shaft making the vehicle intrinsically straight-running when steer is not required. If a brake is added to the annulus shaft then pivot-steering can be obtained by the initiating in neutral mode of the change-speed gearbox.

While most hydrostatic transmissions based on variable displacement pumps have transmission ratio controlled by the driver through a single lever control, it is also possible to control the ratio automatically. Care must be taken to avoid wheel slip at take-off when ratio is nearly zero and torque consequently very high. Use is made of a constant power pump which sets itself to the high-displacement mode when

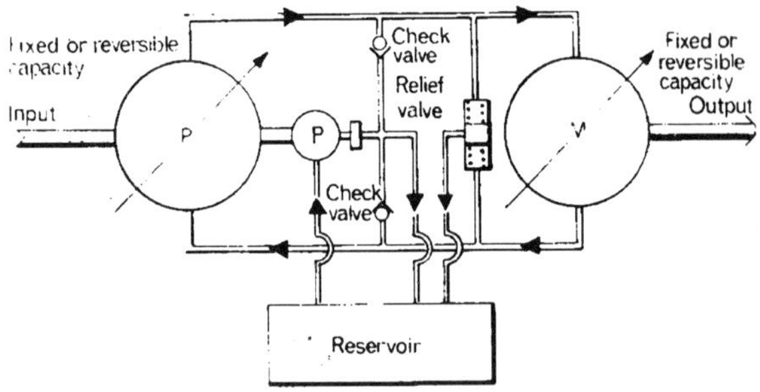

Fig 74: Closed-circuit system with servo valve

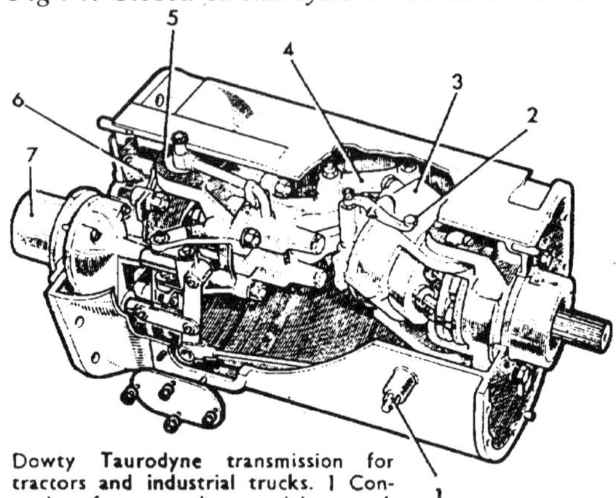

Dowty Taurodyne transmission for tractors and industrial trucks. 1 Connexion for manual or pedal control, 2 unloading valve, 3 variable motor, 4 computing link, 5 variable pump, 6 make-up pump, 7 servo

Fig 75: Link-controlled transmission

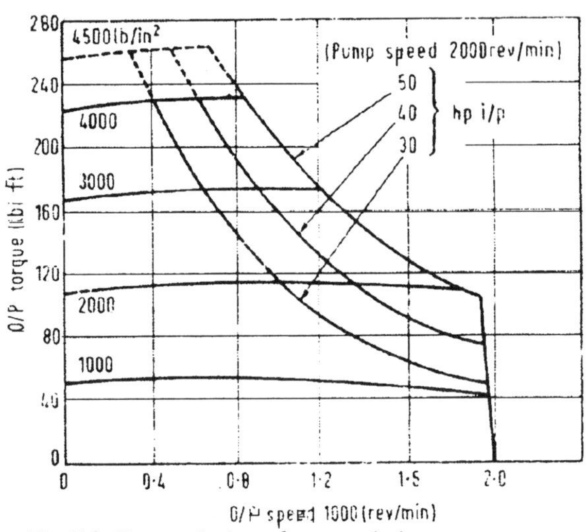

Fig 76: Transmission characteristic

the demand pressure is low — gradually 'changing down' as demand pressure rises above a pre-set level. The limitation of such a system, using a fixed displacement motor would be lack of facility for engine start and warm-up; inability of the vehicle to 'coast'; no way of preventing engine stall when the vehicle stops and no low-gear hold facility for descending hills.

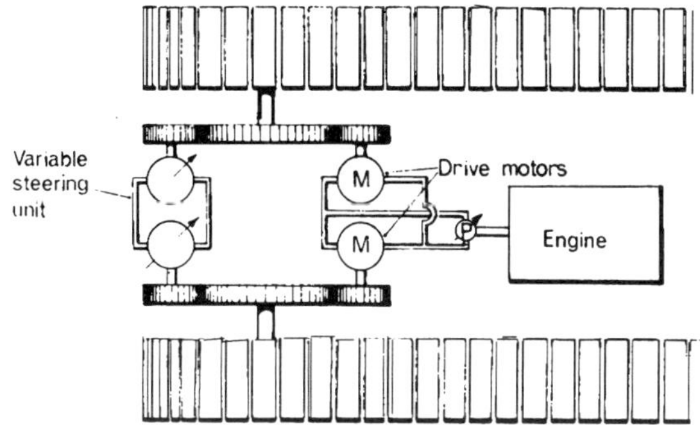

Fig 77: Hydrostatic drive/steer transmission

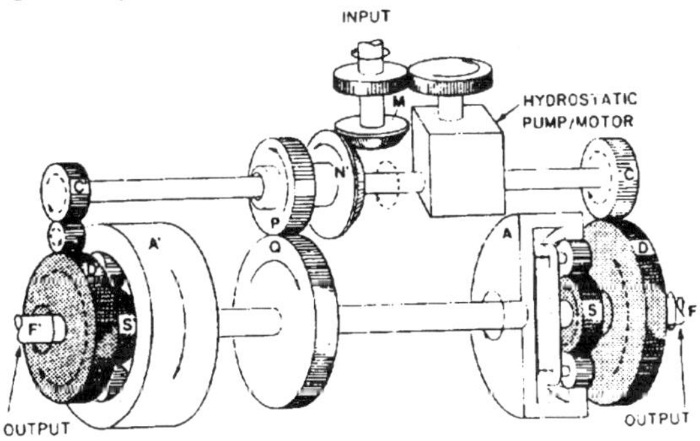

Fig 78: Double differential steering

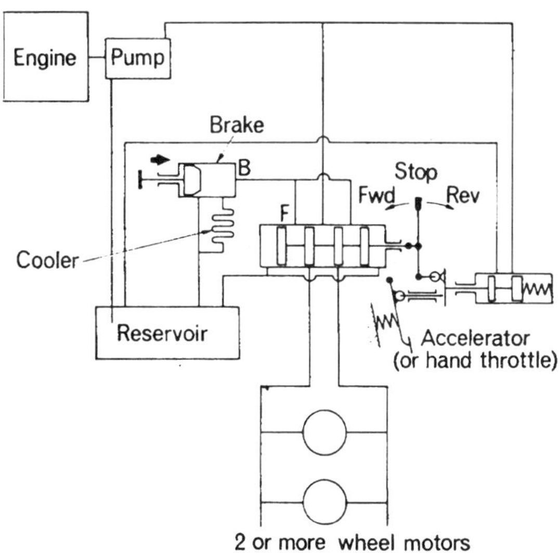

Fig 79: Hydrostatic transmission proposal

A neat elaboration, attributable to researchers at Sheffield University, is shown in Fig 79 in which (I) is an idling valve. The latter is held open either by a cam on the forward-reverse selector which operates in the 'stop' position. Another important contributor to the exploitation of hydrostatic drives in vehicles has been the National Engineering Laboratory who have provided formulae to help the first approximation of pump and motor size as in Fig 80. The organisation co-operated with several hydraulic equipment manufacturers in the development of ball piston wheel motors to facilitate projected road vehicle applications. In the early scheme of Fig 81 wheel motors each developing 18 bhp were installed in a transverse engine front drive car and fluid flow to the hubs was arranged through hollow wheel suspension wishbones via special swivel ball joints. For this application dual fixed-capacity pumps were employed, also of the ball-piston type, — with valving to combine the pumps and motors to give four fixed speed ratios.

MOTOR DISPLACEMENT

Decide i) torque required M (lbf in)
ii) working pressure p (lbf/in²)

Then motor displacement per revolution V (in³/rev)

$$= \frac{2\pi M}{V}$$

MAXIMUM FLOW REQUIRED TO MOTORS

Decide i) motor displacement V (in³/rev)
ii) maximum motor speed n (rev/min)

Then flow required $Q = \frac{Vn}{60}$ in³/s

PUMP DISPLACEMENT TO GIVE REQUIRED MOTOR SPEED

Decide i) maximum pump speed n (rev/min)
ii) maximum flow required Q (in³/s)

Then pump displacement V(in³/rev) $= \frac{60Q}{n}$

POWER ABSORBED BY PUMP

Decide i) pump flow required Q (in³/s)
ii) maximum working pressure required at pump flow given in (i) p (lbf/in²)

Power absorbed P (horsepower) $= \frac{pQ}{6600}$

Fig 80: NEL calculation procedure

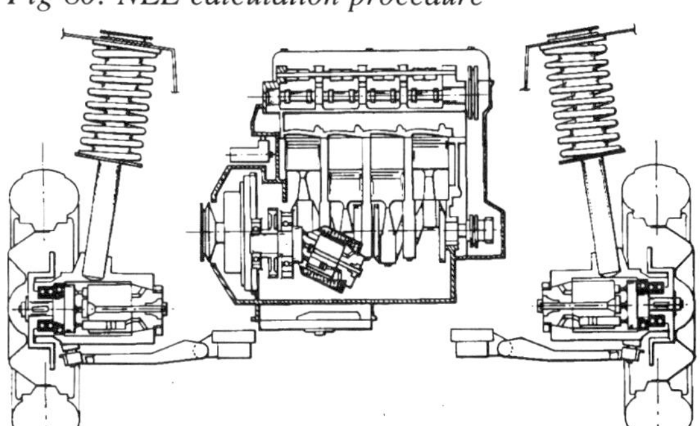

Fig 81: NEL ball-piston wheel motors

All-wheel-drive

As well as the need for greater traction available with four-wheel drive systems, there is also the demand from some users,particularly in the private-car sector, for traction to have a reduced effect on the handling of the vehicle. Permanently engaged and selectable systems are vying for the attention of designers in most vehicle sectors.

Historically the off-road sector was the first to require four-wheel-drive — owing to the demand to have wheeled vehicles approach the performance of tracklayers — as discussed in Chapter 11. Now the interest in on/off road vehicles and high-performance road-cars derived from successful 4WD track racers has sharpened demand. Much was done to promote the systems by the launch of the Ferguson Formula in the 1960s when the possibility of controlling inter-axle torque split was commercially exploited, Fig 82.

The four basic essentials of the Ferguson Formula were:

- two additional differentials, one for the sec on driven axle and an inter-axle one in the drive shaft to the second axle
- free-wheels to control overrun and underrun within selected limits
- acceleration/deceleration sensing device to control braking and traction.

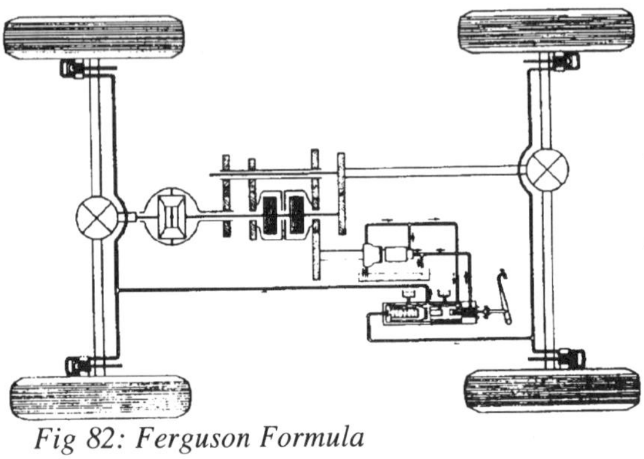

Fig 82: Ferguson Formula

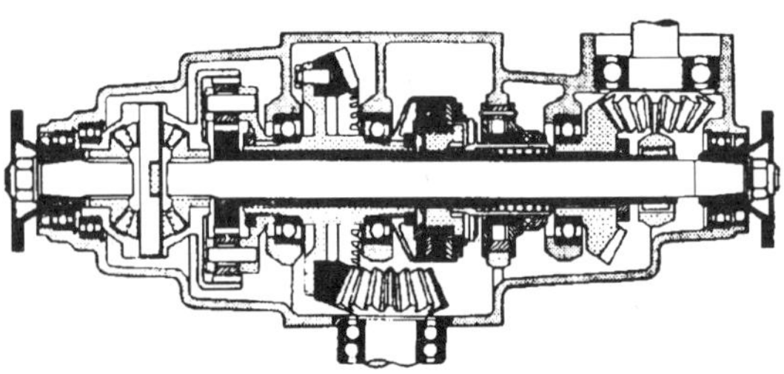

Fig 83: Integral sensor/diff-unit

The means for incorporating the sensing device and differential gear into one unit was later patented, Fig 83. An early application was to an American police car based on a Plymouth Fury which had a second drive to the front axle passing beneath and to one side of the front-mounted fore-aft engine. Fig 84

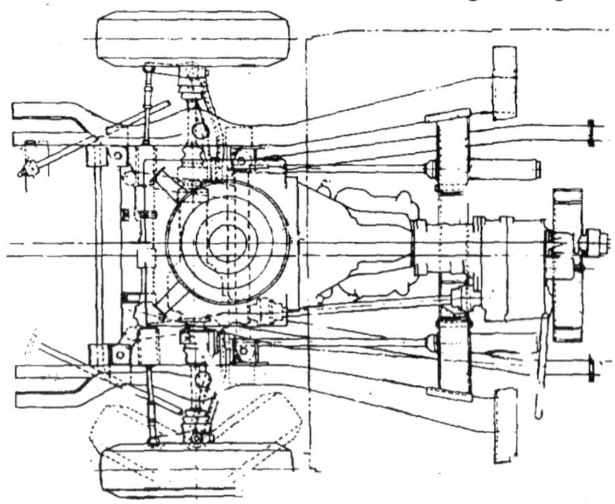

Fig 84: Plymouth Fury installation

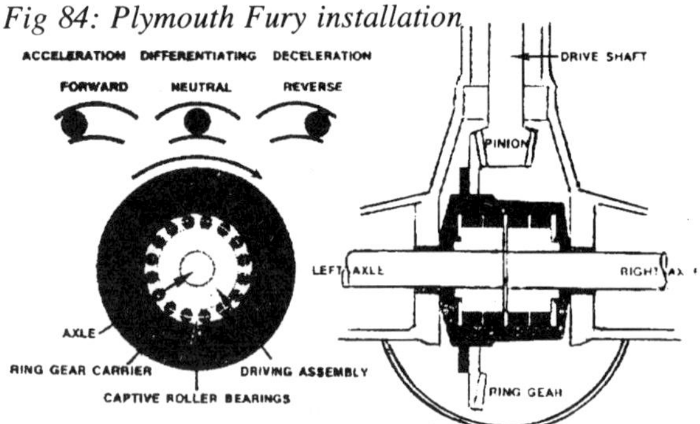

Fig 85: Sprag-clutch differential

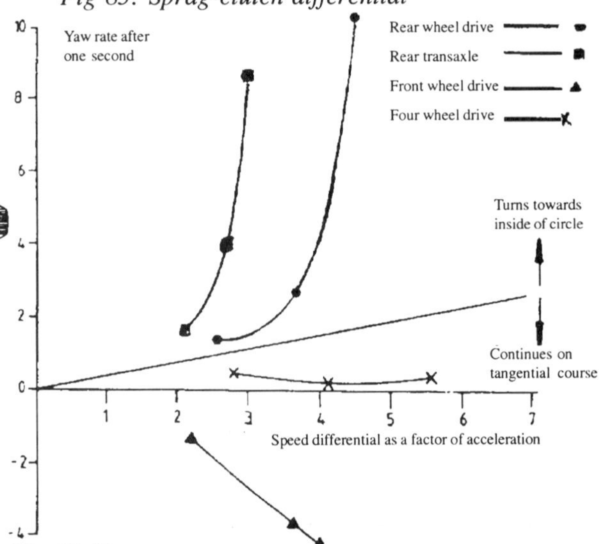

Fig 86: Yaw response

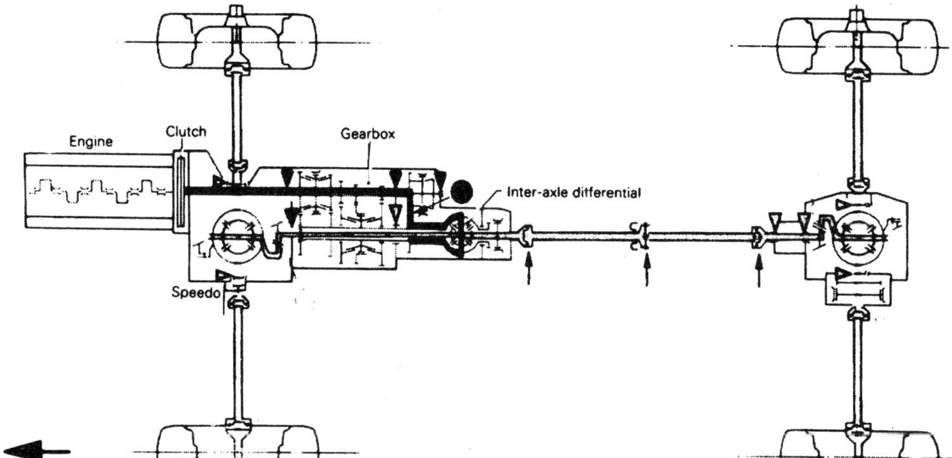

Fig 87: Quattro configuration

shows the installation which involved a 310 lb (141 kg) weight penalty to the 4120 lb (1880 kg) car.

Another device which was to use free-wheels to replace one of the differential gears was the Hy-Torq unit which was introduced soon after in the USA. Effectively, two 2-way sprag clutches replace the differential gear, Fig 85, to provide the effect of a free-wheel in either direction. When drive is applied in either direction, the cage shifts the sprags so that they bear on the driving face of the recesses in the outer drive member. In cornering, the whole of the driving torque is applied to the inside wheel and the heavily loaded outer wheel is able to generate greater cornering force.

The Land Rover field car is one of the chief protagonists of four-wheel-drive; latest versions have permanently engaged 4WD in a transfer gearbox of over 270 Nm capacity incorporating an inter axle differential lock. Front and rear output shafts are co-axial and at 230 mm centre distance from the input shaft to which it is angled at 25 degrees. Low and high output ratios are provided. The two-stage reduction unit involves a 3-shaft 6-gear design, range change being effected by dogs on the output gears. An alternative selective two or four wheel drive system is also available in which the front drive can be isolated by free-wheel hubs.

Audi were the trend setters in road-car systems with their permanent 4WD Quattro models. In justifying these systems, Audi engineers have pointed out that an unduly large amount of tractive force exerted at one wheel can result in resulting loss of cornering force potential within the total grip force available to the tyre. Fig 86 shows test results for the Quattro and three comparison vehicles cornering at a radius of 45 metres with a constant steering wheel angular input;

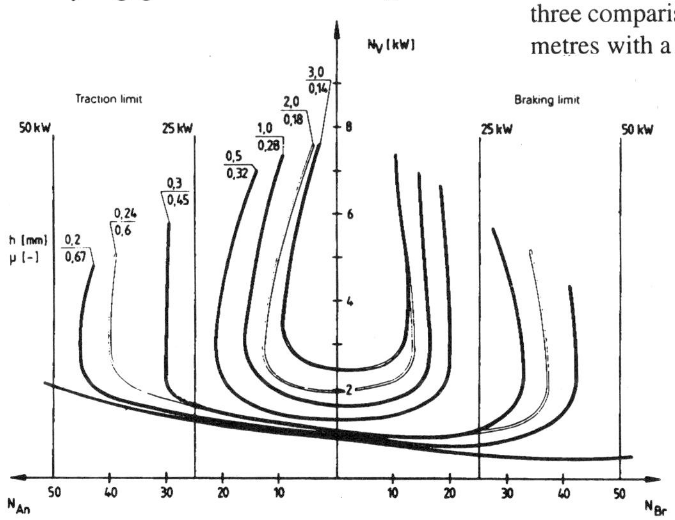

Fig 88: Power losses in traction

the curves show the extent to which the rotation of the vehicles in yaw changes one second after the start of acceleration — as a function of acceleration-rate. Where the change in yaw rate corresponds most closely to the straight reference line means the vehicle maintains its direction under acceleration.

Fig 87 shows the mechanical set up used in the Quattros with one output shaft tubular and integrated with the output gears. The rear end of this shaft is splined to couple it with the inter axle differential. Both it and the rear axle differential can be locked to improve traction in difficult conditions. Fig 88 shows the power losses, on the vertical axes, versus the tractive power, horizontal axes as a function of road water film thickness at 60 km/h. The μ values (adhesion co-efficients) apply to the 175/170 HR14 tyres on the vehicle. Recently the company has added Torsen differential units to its cars, Fig 89. The inter-axle unit provides a measure of torque sensitivity such that the 50:50 split has given way to one that is more responsive to the instantaneous weight and adhesion conditions applying to each axle. now torque can be split in ratios up to 25:75, Fig 90.

Both electromagnetic and viscous couplings are also used in the distribution of torque but do so in relation to differential speed. The latter comprise a housing, hub and some plates alternatively attached to the housing and the hub and having a viscous fluid between them, Fig 91. In the event of a speed difference between the housing and the hub the viscous fluid is sheared to generate a locking torque which can be finely tuned to the differential. In Fig 92, the Dana transfer case uses electromagnetically controlled clutches to provide the torque split which can be up to 100 % in either direction. The control of the clutches is carried out electronically with microprocessors monitoring signals from points throughout the drivetrain including transmission, steering, throttle-position and wheel speed.

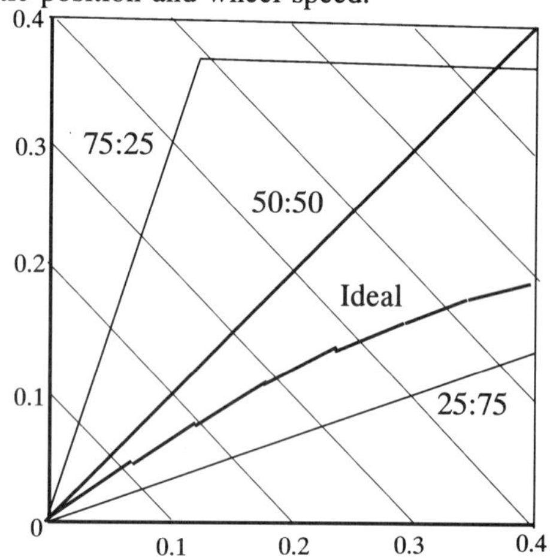

Fig 90: Torque split: Front tractive force per vehicle weight (y-axis) against rear tractive force per vehicle weight (x-axis)

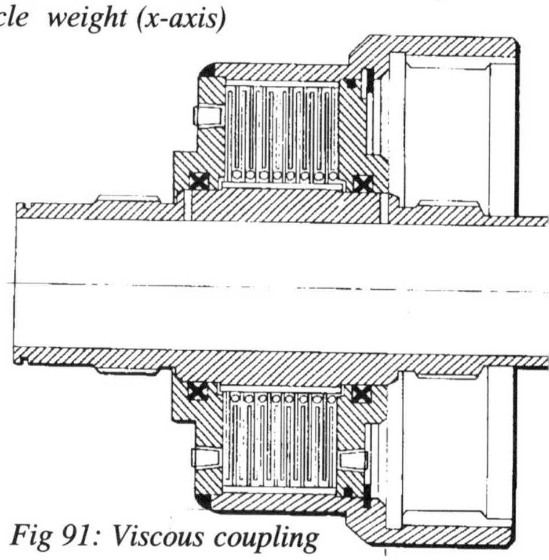

Fig 91: Viscous coupling

Fig 89: Torsen differential

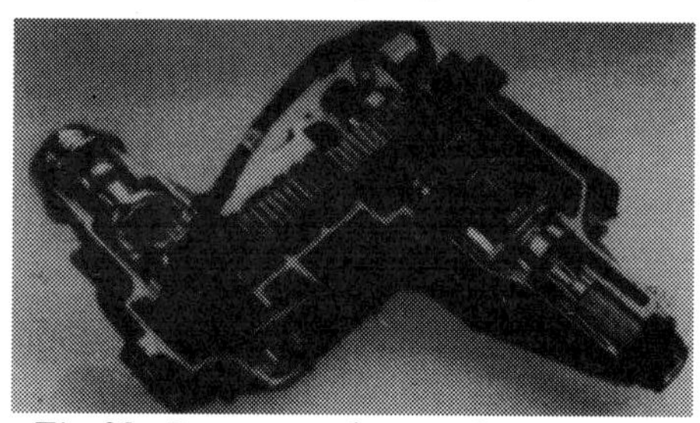

Fig 92: Dana transfer gearbox

Viscous couplings

Viscodrive have done much to develop the versatile viscous coupling concept. It has found several uses including limited slip differentials and four-wheel drive inter-axle torque apportioning systems. Different alternatives are described here alongside a calculation technique developed by the company.

The viscous coupling is essentially a sealed drum-shaped housing, inside of which are two independent packs of perforated and slotted steel plates, Fig 93. One pack is splined to a free-wheeling input shaft while the second, attached to the drum, is driven by a viscous silicone oil and an output shaft splined to the drum transmits the drive onwards. There is no disengaging device, as on a conventional clutch, power being transmitted according to the shear resistance of the fluid. Thus, with small differences in rotational speed, the plates allow slight slip between input and output but stiffen as the rotational speed difference, and shearing force, increases.

An air-space is included above the fluid to allow it to expand with the frictional heat which builds up in shearing, without subjecting the drum to excess pressure. As transmitted torque increases, molecular changes in the fluid effectively 'lock-up' the coupling, with consequent lowering of the temperature. Torque is the transmitted by Coulomb friction between the plates and the coupling moves into so-called 'hump' mode of induced torque amplification, Fig 94. Where external pressure is applied, torque is controlled by a control-algorithm, as 'externally induced toque amplification'. When pressure is built up internally by supplying energy ('self-induced torque amplification') the torque adapts to suit prevailing operating conditions and a multiple of the initial viscous torque may be transmitted.

The coupling is generally used either as a differential brake, to prevent slip between front and rear axles in permanent four-wheel drive vehicles (such as the Ford Sierra 4x4), or as a self-regulating transmission unit on its own (as in the Volkswagen Syncho or Opel Vectra/Vauxhall Cavalier 4 x 4). Visco-control couplings are also used as integral self-locking systems, for the differential, in more powerful rear-wheel drive vehicles, Fig 95.

Use of Visco-control couplings in front-wheel drive (FWD) vehicles is much less common, so far, because the device sometimes requires changes in the front suspension and steering geometry for optimum

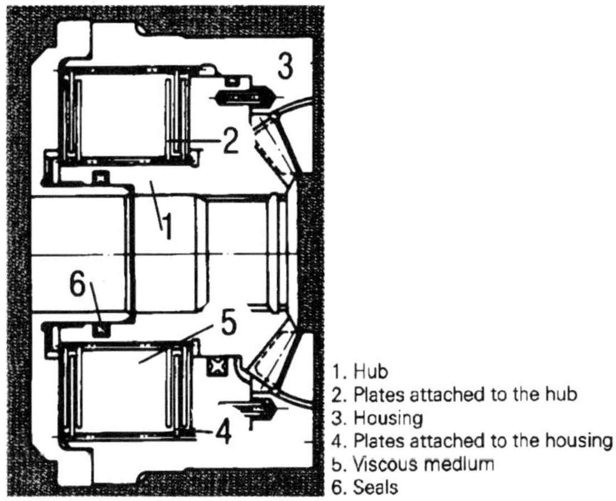

Fig 93: Viscous coupling elements

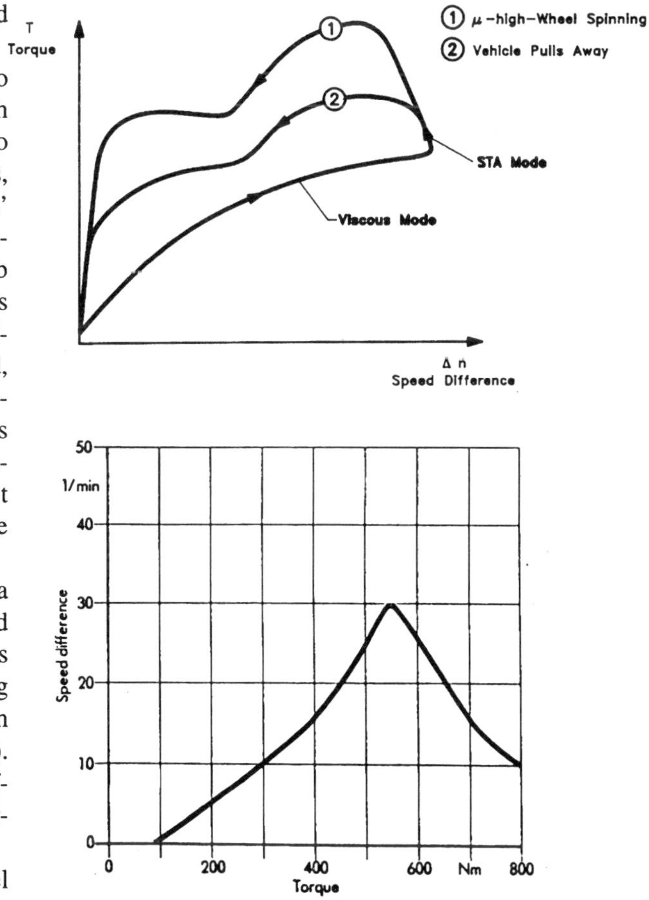

Fig 94: Coupling torque/speed representation (below) and hump characteristic (above)

benefits to be obtained without disturbing side effects. When these factors are taken into account, however, besides traction, appreciable improvements can be made to the handling and safety, particularly of FWD cars with high power outputs.

Two different types of Viscodrive units have been designed by GKN for axle drive applications. The so-called 'shaft-to-shaft' type, Fig 96, has its external and internal discs directly linked to their respective output shafts, which in a transverse FWD transmission are the half shafts to each front wheel. Visco-control couplings can be integrated within both spherical spur gear and epicyclic differentials. In the spherical spur gear type it simply bridges the two differential elements. In the epicyclic type, the bevel wheel driving one half shaft is also connected to the internal Viscodrive discs, while the bevel gear carrier driving the other half shaft is connected to the external discs. An alternative and simpler Viscodrive unit is known as the 'shaft-to-housing' design, Fig 97. Here one of the half shafts is connected to the internal disc carrier of the visco-control coupling, while the external disc carrier is connected to the differential gear assembly. This design can usually be produced more cost effectively, but because of the gearing effect it requires a 'nominal torque' about three times that of the 'shaft-to-shaft' type. This in turn means that the dimensions of the coupling must be increased, sometimes to the point where it can no longer be packaged within the differential housing. To provide for this eventuality, GKN Viscodrive has developed an external unit, as a sub-system.

Many front-wheel drive cars with transverse engines and offset transmissions now use an intermediate drive shaft to equalise the half-shaft lengths and prevent torque-steer effects in high powered models. The company's external Viscodrive assembly is easily installed on this intermediate shaft in such a way that the internal discs are integrated with the shaft and the external discs are carried by the coupling housing, which is connected to the differential assembly.

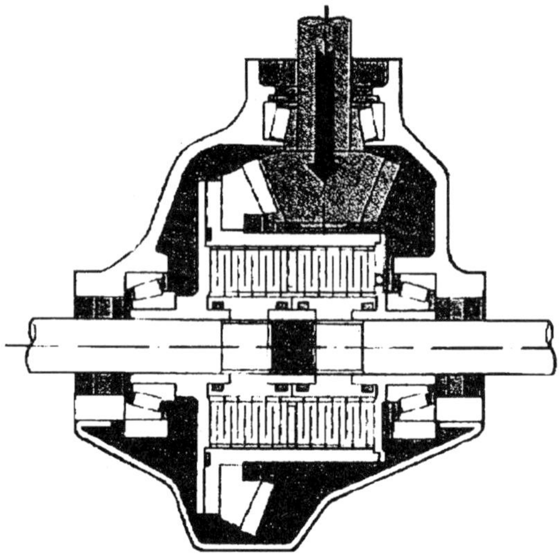

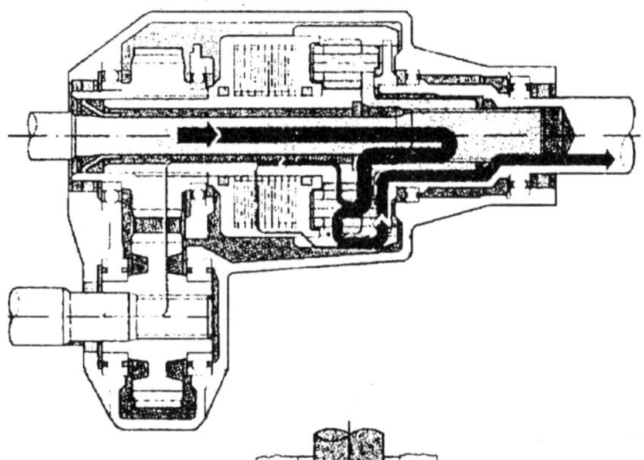

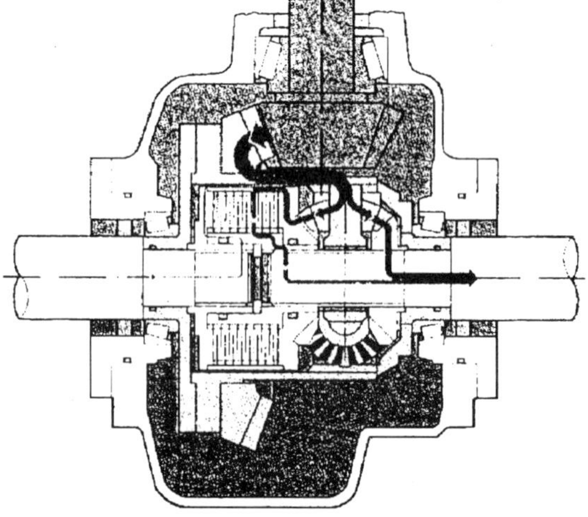

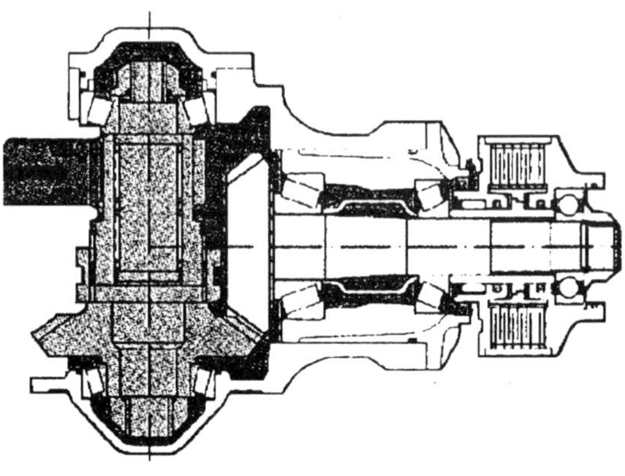

Fig 95: Viscous coupling applications by GKN

External Viscodrive units suitable for use with front-wheel-drive transmissions are easily packaged within the clearances provided for the intermediate shaft. They also require only minor modifications to the transmission and differential housing design. This feature makes external Viscodrive FWD-sub systems installable as optional equipment.

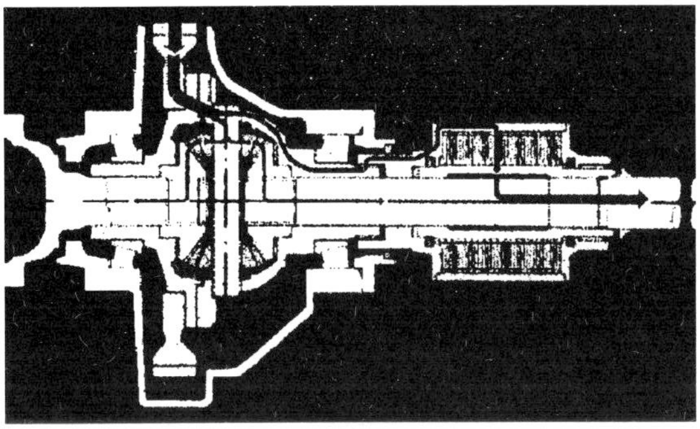

Fig 96: Shaft-to-shaft coupling

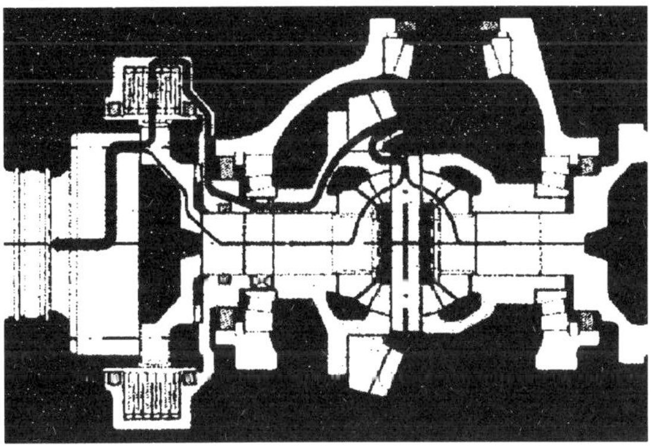

Fig 97: Shaft-to-housing coupling

= number of working faces
= radius
= gap width

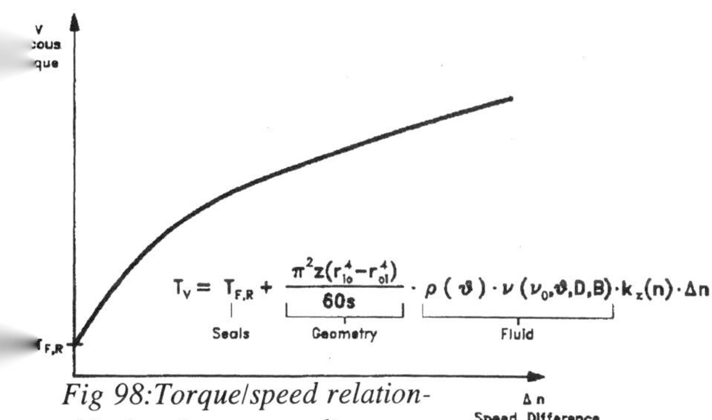

Fig 98: Torque/speed relationship for viscous coupling

Front-wheel drive viscous couplings

Conventional differentials act as torque distributors, supplying identical tractive efforts to each front wheel. They allow the wheels to turn at different speed in corners, but still drive evenly. This characteristic becomes a handicap when the road surface has different adhesion qualities for the left and right front wheels (the so-called split-m condition). In this case the driving torque is restricted to that of the wheel with the lower of the two friction coefficients. With a self-locking Viscodrive differential, the higher of two coefficients can be utilised instead, improving traction and increasing driver control while retaining stability.

The self-locking Viscodrive is said to be both fast- acting and progressive in its control. It responds much more rapidly than a driver can reduce power (the usual reaction to spinning wheels) and claims to be superior in operation to most forms of electronic traction control, because it is much closer to the action. Although the immediate effect of unequal driving forces is to turn the vehicle towards the side with the lower mu value, the principle of slip-sensitive fluid friction in the couplings modulates the turning moment (or swerving couples) so it can be more easily controlled than with a torque-sensing device.

Although self-locking differentials are primarily intended to improve traction, the visco types also produce beneficial effects on the stability of straight line running, especially in powerful FWD cars. Disturbances such as side gusts or road irregularities tend to change the steered angle of the front wheels very slightly, inducing small differences in the wheel speeds. With a Viscodrive differential this asymmetric distribution of forces generates a turning moment that automatically brings the vehicle back on line. This self-stabilising effect is particularly noticeable at high speed.

A secondary benefit of using visco-control couplings on front-wheel drive vehicles is that the dynamic changes in stability caused by lifting off the power in mid-corner are reduced. Since the driving forces are more equally distributed between the left and right hand sides of the vehicle, the turning moment, when driving torque is suddenly reversed to engine braking, is much less severe.

Powerful FWD cars with free differentials also tend to pick up the inside front wheel when accelerating hard in a low gear through a turn (such as when joining a line of moving traffic from a side street). When a torque-sensing self-locking differential is used to improve traction, strong torque-steer effects can be introduced as the wheel speeds are equalised. The progressive characteristics of the viscous coupling are much less harsh, making it far easier to maintain full control in these extreme conditions.

Use of self-locking differentials with front-wheel drive can generate different driving forces in the front wheels which influence the feel of the steering control. The relative importance of these influences depends on many different factors, such as the geometry of the front suspension, the type of self-locking device and the characteristics of the power assistance, if fitted. Locking differentials that rely on torque-sensing for their effect are not suitable for comfortable front-wheel drive because they operate too abruptly. Visco-control units are much more progressive and can be more easily handled by the average driver. If they are developed with the suspension from the early stages of a new vehicle concept they can provide considerable dynamic benefits, with minimum influence on the steering behaviour.

Viscous coupling calculation

The characteristic of a coupling can be calculated by a method due to Dipl Ing Taureg of Viscodrive[4], based on Newton's formula: $t = rnD$ where t is shear stress, r is fluid density and n is kinematic viscosity. The relationship of torque with speed difference Dn is seen in Fig 98, where z is the number of working faces, r the radius and s the gap width. Viscosity varies with both temperature and input/output energy. It also varies with shear rate in the gap which rises proportionally with radius and speed difference increase. The filling degree of the coupling is given by $D= prDn/30s$ which depends also on fluid temperature but the effect is modified by the temperature effect on viscosity. It also depends on gas bubble volume. Factor k_z takes into account the small increase in viscous torque dependent on the absolute speed; the characteristic is seen in Fig 99.

These formulae remain valid as long as filling degree is below 100% or contains gas in physically undisolved form. With energy input, however, the temperature and pressure rise and the gas is dissolved

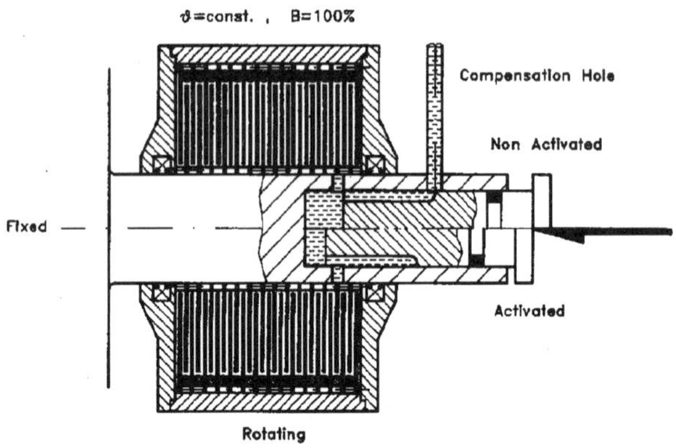

Fig 99: Externally induced torque amplification coupling

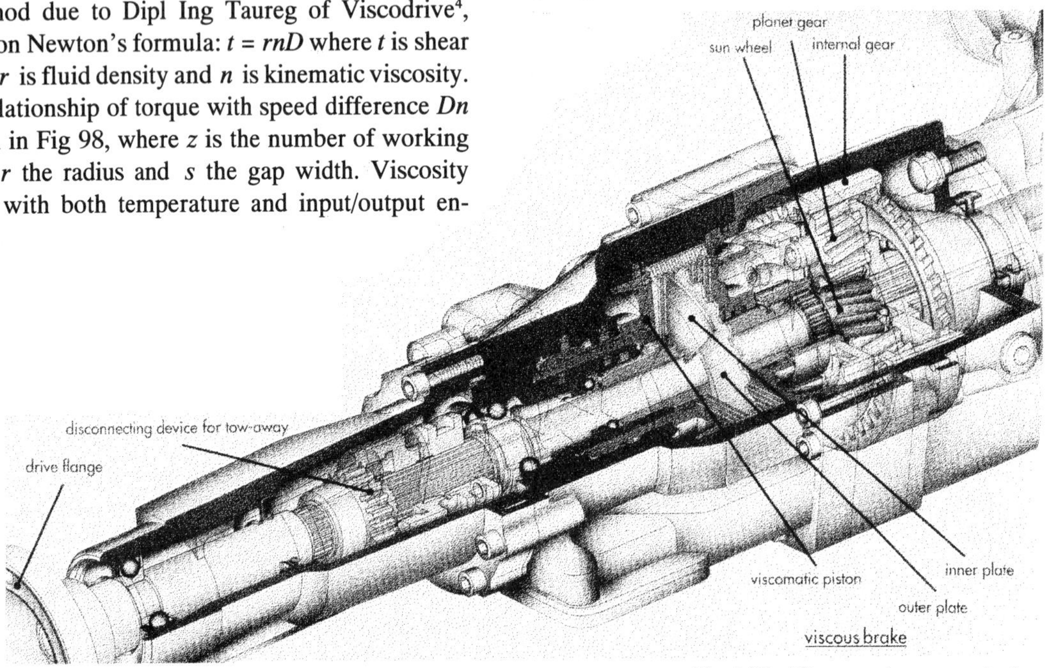

Fig 100: Viscomatic transmission

so that the coupling becomes 100% fluid-filled, the hump phenomenon.

Fig 98 shows torque and speed difference values for a vehicle with viscous axle diff-lock when pulling away. When the hump appears the speed difference falls again as torque increases in this example of self-induced torque amplification. To illustrate externally induced torque amplification, Fig 99 shows an experimental coupling, 100% filled with low-viscosity fluid, having expansion compensation and external pressure control.

Plate perforation performs four separate functions: absorption of most of the undisolved air in the viscous mode; asymmetric control of the inner plates in the gap via the profiled perforation edges during the activation phase; expulsion of the fluid from the gap during the ETA or STA mode; compensation of asymmetrical thermal stresses in both these modes.

Another prominent supplier of transmission systems involving viscous couplings is Steyr Daimler Puch. The Viscomatic, Fig 100, is an all-wheel-drive system which claims to be the first to offer a variable torque-speed characteristic. It gives infinitely variable torque distribution, front-to-rear, between rigid-connection and complete-separation of axles. It is a combination of planetary gear-set and controllable viscous brake, one set of plates being attached to the housing and the other to the sun wheel. By modulating the brake torque for the sun gear, torque transfer is completely controlled. A hydraulically actuated piston reduces distance between plates to increase torque; at the same time the action of the piston entering the housing reduces bubble air volume to increase filling ratio and thus torque. The unit has been used on the Alfa Romeo Proteo concept car with electronic control arranged so that active influence is maintained on the mode of driving. Control strategy is shown in Fig 101.

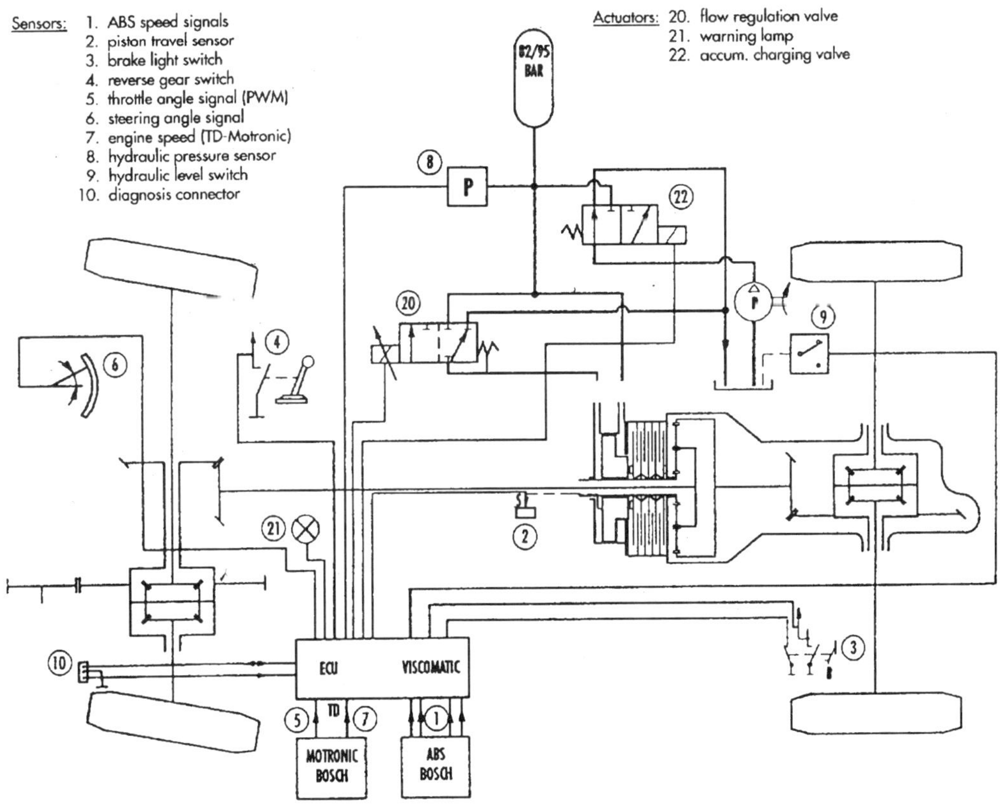

Fig 101: Control strategy for Viscomatic transmission

References

Chapter 1

1. Campbell, C, The sports car, its design and performance, London, Chapman and Hall, 1959
2. Platt, M, Structure of automobile, Proc IMechE, AD no. 13, 1959/60
3. Short, M, Possibilities of aircraft structures in ground vehicles, SAE Transactions, 1945.
4. Peery, D, Aircraft structures, New-York, McGraw-Hill, 1950, 566 pp.
5. Garrett, K, Automobile dynamic loads, *Automobile Engineer*, 1953, vol. 43, pp 60—64; Structural design. Part 1: Analytical method for chassisless vehicle design. *Automobile Engineer*. 1953, vol. 43, pp. 103—111; Structural design Part 2: The front end structure. *Automobile Engineer*, 1953, vol. 43, pp. 152—157.
6. Costin and Phipps, Racing and sports car chassis design, Batsford, 1962.
7. British Standard Code of Practice CP118:1969, The structural use of aluminium (also BS8118, 1991)
8. Cooke, C, Torsional stiffness analysis of car bodies. Proc IMechE, 1965—66, Vol 180, part 2A, no. 1
9. Sharman, P, Optimum stiffness-weight design of peripheral and ladder frames, Proc IMechE, Vol 182, Part 2A, No 3, 167-68
10. Automotive steel design manual, American Iron and Steel Institute, 2000 Town Centre, Suite 1900, Southfield, MI 48075-1138, USA.

Structural design basics

Giles, C, Body construction and design, Iliffe, Automotive technology series vol 6, 1971

Framework analysis and stiff-jointed frames

Goodwin, E, The unit load method of structural analysis, Draughtsmen's and Allied Technicians Association booklet, session 1966/7

Costin, Competition car chassis design, *Automotive Engineer*, Dec 1983

Bolland, G, The static indeterminacy of a shell structure, *Automotive Design Engineering*, Sep 1966

Column and sheet stability

Buckling of plates, Engineering Science Data Unit ESDU data sheets 71005/71014/71015

Williams, Structures Report No 2, Royal Aircraft Establishment, Farnborough, UK

Timoshenko and Gere, Theory of elastic stability, McGraw Hill

Nichols, Stability of honeycomb sandwich panels in shear, College of Aeronautics thesis, Cranfield, UK, 1955

Timoshenko, Theory of plates and shells, McGraw-Hill Reports 1505 and 1583, US Forest Products Laboratory

Barwell, R&M 2467, Aeronautical Research Council, UK

Williams, D, Sandwich construction — a practical approach for the use of designers. RAE Report no. 2, Structures, Report and Memoranda 2466, 1948, Aeronautical Research Council, UK

Williams, Leggett and Hopkins, Report and Memoranda 1987, Aeronautical Research Council, UK

Williams D, An introduction to the theory of aircraft structures, Arnold, 1960

Faupel, Engineering Design, Wiley, 1964

Pippard and Baker, The analysis of engineering structures, Cambridge University Press

Timoshenko and Gere, Theory of Elastic Stability, McGraw Hill, 1961

Hargreaves, T, Introduction to plate buckling, *Engineering Materials and Design*, 25 Aug 1972

Lutz, L, Structural design using the automotive steel design manual, SAE paper 870463

Chapter 2

1. Pawlowski, J, Vehicle body engineering, Business Books, 1969
2. Sharman, P, Torsion of chassis frames, *Automotive Engineer*, Oct/Nov 1976
3. Seitler, W (Saurer), Utility vehicle frame developments, *Technische Rundschau*, No 16/62
4. Cooke, C, Vehicle structural analysis, paper of IMechE seminar, 15 May 1968
5. Sherman D, Stresses and deformations in truck side-rail attachments, SAE paper 690175
6. Tidbury et al, Torsional stiffness of commercial vehicle chassis frames, FISITA conference 1968
7. Tidbury, G, Stress Analysis of Vehicle Structures, ASAE Note 1, College of Aeronautics, Cranfield, 1965
8. Ed. Donald, I, Thin-walled structures, Elsevier, 1990.
9. Beerman, H, The analysis of commercial vehicle structures, Mechanical Engineering Publications, 1989
10. Guide to stability design criteria for metal structures' published by Wiley (1988) for the Structural Stability Research Council
11. Marshall, P, Design of welded tubular connections, Elsevier (1992)
12. Lowe, W, et al, Impact behaviour of small scale model motor coaches, *Journal of Automotive Engineering*, January 1972.
13. Young, W, Roark's formulas for stress and strain, McGraw Hill
14. Timoshenko and MacCulloch, Elements of strength of materials, Van Norstrand
15. Peterson, R, Stress concentration design factors, Wiley

Chassis analysis

Sharman, P, Optimum stiffness-weight design of peripheral and ladder frames, Proc IMechE, vol 182, part 2a, no. 3, 1967-8

Dean-Averns, R, Automotive Chassis Design, Iliffe.

Erz K. Uber die durch Unebenheiten der Fahrbahn hervorgerufene Verdrehung von Strassenfahrzeugen. Tell I, *Automobiltechnische Zeitschrift (ATZ)*, 1957, N. 4, SS. 89—96. Tell 2, ATZ, 1957, N. 6. SS. 163—170. Tell 3, *ATZ*, 1957, N. 11, SS. 345—347. Tell 4, *ATZ*, 1957, N. 12, SS. 371—375.

Fenton, J, Truck chassis design, *Automotive Engineer*, Jun/Jly 1976

Carver, G, Application of variable depth side-rail to heavy truck frames. SAE Transactions, 1969, vol. 78, Paper 690174.

McNitt, L, Truck frame side-rail buckling stresses. SAE Transactions, 1969, vol. 78, Paper 690176.

Megson, T, Increasing semi-trailer torsional stiffness could enhance operational safety, *Automotive Design Engineering*, May 1970

Sidelko, W, An objective approach to highway truck frame design, SAE Transactions, vol. 75, Paper SP 276, 1966

Body shell analysis

Bruhn, E, Analysis and design of flight vehicle structures, USA, Tri-State Offset Co., 1965.

Peach, R, Design to prevent buckling in rectangular plates, *Machine Design*, 1963, vol 35, no. 30

Fagan, M, Finite element analysis, Longman, 1992

Ali et al, The application of finite element techniques to the analysis of an automobile structure, Proc IMechE, 1970-71

Myers, R, Body stress analysis, SAE paper 660005

Kirioka, K, An analysis of body structures, SAE paper 979A, 1965

Swallow, W, Unification of body and chassis-frame, Proc IAE, 1938-39

Argyris and Kelsey, Energy theorems and structural analysis, Butterworth

Lucas, A, The Argyris method of structural analysis using matrix algebra, *Automotive Design Engineering*, 1963, vol 2, no. 6

Wang, C, Matrix methods of structural analysis, International Textbook Co

Amos, R, Structural design analysis of commercial vehicle cabs, IMechE conference paper C134/84

Fenton, J, Shell beams and the Argyris method, *Automotive Design Engineering*, March 1963

Norville and Mills, An applied finite element displacement analysis of a simple integral construction vehicle body shell, Intl J Mech Sci, Pergamon, 1972, Vol 14, pp 809-821

hapter 3

I. Barley, G W, Mills B, A study of impact behaviour through the use of geometrical similar models, IMechE conference on Body Engineering, 1970, Paper N3

2. Baracos and Rhodes, Ford CC9 frame, SAE paper 690004

3. Miura and Kawamura, Analysis of deformation mechanism in head-on collision, SAE paper 680484

4. Emmerson, W, The application of computer simulation in vehicle safety, TRRL conference Towards safer road vehicles, 1972

5. Finch, P, Vehicle frontal barrier impacts, TRRL conference Towards safer road vehicles, 1972

6. Grime and Jones, Car collisions. — The movements of cars and their occupants in accidents. Proc IMechE, (AD) vol 184, part 2A, no. 5, 1969/70

7. Murray, N, When it comes to the crunch, ISBN 981-02-2096-0, World Scientific Publishing

8. Duffin, A, ISATA, 1993

Design for crashworthiness

United Nations World Papers 29/372, 373 463, 477.

Searle and Brabin, The Invertube, Motor Industry Research Association Bul. 1970, no. 2

Johnson, W, Impact strength of materials, Edward Arnold, 1972

Grime, G, Handbook of road safety research, Butterworth

Ed Murthy and Brebbia, Structural design and crashworthiness of automobiles, Springer-Verlag, 1987

Vehicle and road design for safety, IMechE conference report, 1968

Yamaguchi et al, Efficient energy absorption of automobile side rails, Tech Session 3, Experimental safety vehicles, TRRL hosted Oxford conference report, US Dept of Transport, NHTSA, 1985

Ed Dorgham, M, Vehicle safety, Inderscience Enterprises, 1986

Kipp, W, Impact without rebound: an advanced tool for laboratory crash simulation, SAE paper 700406

Cornell Aeronautical Laboratory report YB-2684-V-7 on engine underrun deflectors for crash impact.

Hardy and Suthurst, Simulations to assess the influence of car lateral impact characteristics on occupant kinematics, Proc IMechE, Vol 199, D2, 1985

Kamal, SAE paper 700414 on dynamic impact coefficients

Progress towards safer passenger cars in the UK, IMechE conference report, Jun 1980

Parsons, D, Predicting crash performance, IBCAM Boditek conference, 1987

Haman, J, Proc FISITA 1972, paper on `target/bullet' impacts

Occupant and cargo restraints
Rodger, M, Occupant protection in side impact, TRRL report, 1972
Wolfe, J, Use of models in automobile impact research, TRRL 1972
Coombs, G, Safety frames for tractors, *Farm Machine Design Engineering*, Dec 1971
Dept of Transport Code of Practice, Safety of loads on vehicles, HMSO, 1984
Pan et al, The use of a computer simulation program to predict occupant motion and injury levels in a coach front impact, Cranfield Impact Centre report

Chapter 4
1. Industrial Products Catalogue, Mezeler / Lord Gimetall
2. Autotech 1993 Paper C462/17/141
2. Autotech 1993 Paper C462/31/043
4. Autotech 1993 Paper C462/17/022

Noise and vibration basics
Priede and Jha, Low frequency noise in cars, *The Journal of Automotive Engineering*, Jul 1970
Guignard, J, Human response to intense low-frequency noise and vibration, Proc I Mech E vol. 182, part 1, no. 3,1967/8
Jha S, Noise inside and out, *Automotive Design Engineering*, 1972, vol 11
Gibbs and Richards, Stress, vibration and noise analysis in vehicles, Applied Science Publishers, 1975
Brughmans et al, The application of FEM-EMA correlation and validation techniques on a body-in-white, paper 3 in IMechE Vehicle NVH and refinement conference report C487, 1994
Burton and Southall, Noise, vibration and harshness in the automotive industry, paper 3, session 3R, Design Engineering Show conference, 1982.
Miki et al, Elastic vibration analysis of passenger car bodies, SAE paper 690272
Halvorsen and Brown, Impulse technique for structural frequency response testing, *Sound and Vibration*, Nov 1977

Designing against fatigue; rubber as an isolator
Newman and Coats, Investigation of fatigue strength of welded thin gauge construction box-tube, British Welding Research Association Report FE 23, no. 5, 1956
Newman R. P. Fatigue of spot-welded box beams. British Welding Research Association Reports FE 23, 1958, no. 6, 7.
Murakami, Y, The rainflow method in fatigue, Butterworth Heinemann, 1992
Designing for finite life, *Automotive Engineer*, Feb/Mar 1983
Metalastik, Body mounting, *Automobile Engineer*, March 1958
Prebble, C, Recommended methods of body mounting, Vauxhall Motors, 1963
Boden, E, Some aspects of vehicle ride, IMechE conference on Vehicle ride, 1962

Chapter 5
1. Tidbury G, The structural design of bus bodies, Paper 8 at IMechE conference on Body Engineering, 1970,
2. Phillips, L, Improving racing car bodies, *Composites* 1(1)50
3. Hollaway, L, Glass reinforced plastics in construction, Surrey University Press
4. Parry-Williams, D, Panther Solo, the first production composite vehicle structure, paper 15 in IMechE conference C387
5. Potts, H, Application of advanced composite materials in racing cars, Paper 29, British Plastics Federation conference, October 1990.
6. O'Rourke, Paper 4, I Mech E seminar on Materials, fabricating a novel approach, 1993
7. Pitchford, N, Adhesive bonding for aluminium structured vehicles, IMechE seminar on Materials, fabricating a novel approach
8. SAE Paper 922112

The design, construction and operation of public service vehicles, IMechE conference report C130, 1977
Pawlowski, J, Van idealisation for analysis, *Auto-Technika Motoryzacyjna*, no. 1, 1988
Hundy and Broadstock, The use of aluminium for commercial vehicle structures , a feasibility study, Proc IMechE, Vol 192, no. 13, 1978
Grieg, J, Concept structure development through design and analysis: an automotive case study using Hydroform tubes, IMechE conference paper C462/23/105
Pressel and Strifler, Analytical comparison of integral and chassis design on buses, IMechE conference paper C144/77

Chapter 6
1. Houldcroft, P, Which Process?, Abington,1990
2. Gibson/Smith, Basic welding, Macmillan, 1993
3. Hicks, J, A guide to designing welds, Abington, 1989
4. Pearce, R, Sheet Metal Forming, Adam Hilger Publishing
5. Institute of Materials conference Moving forward with steel automobiles, 1993

Hartley, J, A worldwide review of manufacturing trends, Economist Intelligence Unit, 1992
Roberts, D, Designing vehicles with an economic framework. *Automotive Design Engineering*, vol 5, 1966
Timings, R, Manufacturing Technology, Longmans, 1993

Fastening and welding
Sawhill and Sawdon, A new mechanical joining technique for steel compared with spot-welding, SAE paper 830128
Jubb, J, Design for spot welding, *Automotive Design Engineering*, 1966
Snap-fits and press-fits, *DuPont review Engineering Design*, E-96323
Handbook for disc springs, Adolf Schnorr Kg, Maichingen bei Stuttgart, Germany
Structural Adhesives in engineering, IMechE conference report 6, 1986
Ed Sadek, M, Industrial applications of adhesive bonding
James, P, Isostatic pressing technology, Applied Science Publishers, 1983
Hicks, J, Welded joint design, BSP Professional Books, 1987
Redford and Chell, Design for assembly, McGraw Hill, 1994

Metal cutting and forming
Cole A, Paper at conference of Institute of Sheet Metal Engineering, Design for tooling, *Sheet Metal Industries*, 1968, p 812.
Johannisson, T, Flex-forming in the automotive industry, *ABB Review*, 5/90
Bull et al, The application of a knowledge-based CAD system to press tool design, SAE paper 900797
Grainger and Morse, Design for steel pressings, *Automotive Design Engineering*, March 1971
Cole, A, Design for economical press tooling, *Automotive Design Engineering*, December 1968
Wright and Neate, Cold extrusion, *Automotive Design Engineering*, Feb 1965
McRobert and Watson, Design optimization of forgings, Drop Forging Congress, Cologne, June 1953
Pearce, R, Superplasicity and automotive engineering, *Automotive Design Engineering*, May 1971

Anti-corrosion treatment and painting
Ed Baboian, R, Automotive corrosion by de-icing salts, US National Association of Corrosion Engineers, 1981
Fenton, J, Car body structure design for corrosion prevention, paper 3 at *ADE*/Hatfield Polytechnic conference on Nature and prevention of vehicle corrosion, 1992
Corrosion of motor vehicles, IMechE conference report, 1974
McArthur, H, Motor vehicle corrosion prediction and prevention on vehicles, Expert Books, 1990
Child, P, Vehicle fine finishing, BSP Professional Books, 1987
Loop, F, Cathodic automotive electro-deposition, SAE paper 780189
Johnson, B, Development of an automotive paint film concept, SAE paper 890352

Chapter 7
1. West, GH, Manufacturing in Plastics, PRI
2. UK DTI pamphlet on reinforced thermoplastics available from the Design Council, 28 Haymarket, London SW1Y 4SU
3. Designers Guide to Rotational Moulding, T&D Rotomoulding, Victoria Street, Pontycymmer, Bridgend, CF32 8LR, UK
4. Design Optimization of Forgings, GKN Technology Ltd, Birmingham New Road, Wolverhampton, UK.

Hartley, J, The materials revolution in the motor industry, Economist Intelligence Unit, 1993
Ed Young and Shane, Materials and processes, Marcel Dekker, 1985

Designing in plastics/composites
Gutman, H, New concept bumper in plastic, SITEV conference, 1990
Jutte and McCluskey, Development of SC for automotive body panel use, SITEV conference, 1983
Thermoplastic matrix composites, Profile series, Materials Information Service, DTI
Wootton, A, Development of a novel RP suspension system, Soc of the Plas Ind/Composites Inst conference report, 1989
West, G, Engineering Design in Plastics, Plastics and Rubber Institute
Chant P, Reservoir moulding technique: a new process for glass-reinforced resins. Reinforced Plastic Conference, 1972, paper 15, British Plastics Federation Reinforced Plastic Group
Owen M, How fatigue affects glass and carbon-fibre composites. *OEM design*, 1972, vol. 1, no 4
Wood, R, Automotive engineering plastics, Pentech Press, 1991
Design in composite materials, IMechE conference report 2, 1989
Maxwell, J, Plastics in the automobile, Woodhead Publishing, 1994
Plastics and rubbers, engineering design and applications, Mechanical engineering publications, 1989
Macosko, C, Reaction injection moulding, Hanser Publications, 1989
Limits of plastics uses in vehicle construction, VDI-Verlag, 1981
Plastics on the road, PRI/IMechE conference report, 1988
Sabol et al, A design procedure for thermoplastic bumpers, SAE paper 870109
Data for design in Propathene polyurethane, Tech Service Note PP110, ICI

Steel and nonferrous metals
Engineering steels, Profile series, Materials Information Service, DTI
Nardini and Seeds, Structural design considerations for bonded aluminium structured vehicles, SAE paper 890716
Ruden et al, Design and development of a magnesium/ aluminium door frame, SAE paper 930413
User manual for 3CR12 steel, Cromweld Steels
Cowie, G, The AISI automotive steel design manual, SAE paper 870462
Using copper alloys, Profile series, Materials Information Service, DTI
Commercial vehicle body-building in aluminium, British Aluminium Company, 1965.
Mazurek, Y, Chrysler XLV truck cab has been designed for low-volume production overseas, SAE paper 650023.
Bouge, L, Semi-trailer van-design consideration, SAE paper 660077.
Reidelbach, W, Zur I-'rage der Gestaltfestigkeit von Karosserie-Oberbauten. *Automobiltechnische Zeitschrift*, 1964, N.3, SS.63-66.
Schreiner et al, Heat treatment processes for the reduction of wear in vehicle components, Mechanical Engineering Publications, 1991
Schiek, H, The design of aluminium alloy castings, *Engineers' Digest*, March 1966

Chapter 8
1. Flurscheim, C, Industrial design in engineering, The Design Council, London, 1983.
3. Thiel, P, Visual awareness and design, University of Washington Press, 1983
2,4. Whitford, F, Bauhaus, Chapters 10, 12, Thames and Hudson, 1988
5. *Illuminating Engineering*, USA, Vol 57, number 12,
6. Costin F, The aerodynamics of the modern car, *Automotive Engineer*, vol 1, no 1, 1975
7. Muller, G, TECHNA, a concept-car to challenge automotive engineers, SAE paper 690267.
8. Murrell H, Control and instruments, design procedure, *Automotive Design Engineering*, vol 2, no. 14, 1963
9. White, R, A rating method for assessing vehicle drag coefficients, Motor Industry Research Association Report, 1967
10. SAE Paper 890600

Clausen, H, The case for the engineering designer, *Automotive Design Engineering*, Jun 1965
Fores, M, Technik: the missing concept, Higher Education review, Spring 1979

CADCAM and surface development
Curved surfaces in engineering, CADCentre conference report, IPC Science & Technology, 1972
Effective CADCAM, IMechE conference report 9, 1987
Ward, P, Surface modelling, a part of the computer integrated manufacturing cycle, Proc ISATA, 85062
Ranger, T, Surface reappoximation for automotive body panels, *Automotive Engineer*, Apr/May 1987, pp 25-27
McClelland, M, Computers and the design process, *Materials and Design*, Vol 5, Jun/Jly 1984
Karp, H, CAE applications in the automotive industry, SAE paper 850446
Ostrowsky, O, Engineering drawing with CAD applications, Edward Arnold, 1989.

Ergonomics and packaging
Ergonomics and safety in motor car design, IMechE symposium report, 1966
Woodson, W, Human engineering guide for equipment designers, Berkeley Univ of Cal Press, 1954
Rebiffe et al, Efficient vehicle packaging with front-wheel drive, SAE paper 780131
Everden, I, Designing for comfort and convenience, *Automotive Design Engineering*, Jun 1963
Tilbury, Body interior dimensioning, *Automotive Design Engineering*, Apr 1964
Design of commercial vehicle cabs in relation to driver environment, IMechE conference report, 1973
Davey, J, Looking after the lorry driver, *Design*, April 1965

Aerodynamics
Cooper, K, The effect of front-edge rounding and rear edge shaping on the aerodynamic drag of bluff vehicles in ground proximity, SAE paper 850288
Sibor-Rylski, A, Aerodynamic characteristics of the wedge shaped racing car, *Automotive Design Engineering*, July 1969
Scibor-Rylski, A, Road vehicle aerodynamics, Pentech Press, 1984
Ed Dorgham, M, Impact of aerodynamics on vehicle design, Inderscience Enterprises, 1993

Chapter 9
1. Jones, T, 'Commercial Vehicle Electronics' 1986 SAE Ray Buckendale lecture,

Molian, S, Some key problems in the theory of plane mechanisms, Borough Polytechnic Mechanisms conference, 1969.
Ed Giles, J, Vehicle Equipment, Chapter 3 on Mechanisms, Iliffe, 1969
MacDonald, K, Automotive air conditioning, Howard Sams, 1978
Washino, S, Automobile electronics, Gordon and Breach Scientific, 1989
Ribbens and Mansour, Understanding automotive electronics, Texas Instruments Understanding Series

Bromley, P, Advanced automotive electronics, IFS Publications, 1989
Miller, G, Introduction to microprocessors, SAE paper 790234
Phail, F, Controller Area Network, an in-vehicle network solution, SAE paper 880588
Electronics for mechanical engineers, *OEM Design*, Oct 1971-Oct 1972
Elementary AC Theory, *Automotive Design Engineering*, Jun 1967
Embrey et al, A non multiplexed microprocessor based driver interface switching system for automotive application, AB Electronic Products
Hillyer, M, Using motors in automotive multiplex systems, SAE paper 860394

Chapter 10
1. Materials Design Manual, RFI Shielding, Warner Drive, Springwood Industrial Estate, Rayne Road, Braintree, CM7 7YW, UK

Monk and Morgan, The manufacture of a new generation of headlamp reflectors, Proc IMechE, Vol 199, B3, 1985.
Lindae, G, Improvements of low-beam pattern by use of polyellipsoidal (PES) headlamps, SAE Paper 850228
Steffens, E, Electrical conductive plastic compounds for EMI shielding, Proc SITEV, 1990
Watts, A, Automotive seating with dual-hardness foams, Automotive Engineer, Oct/Nov 1984
Bedford, A, Development of a headlining, Proc IBCAM Boditek 1987
Lee, W, Effect of loading geometry on stress-strain behaviour of plastic foams, SAE Paper 710040
Stikeleather et al, A study of vehicle vibration spectra as related to seating dynamics, SAE Paper 720001
Garry, K, A review of commercial vehicle aerodynamic drag reduction techniques, Proc IMechE, Vol 199, D3, 1985
Lynam, N, Smart windows for automobiles, SAE Paper 900419

Chapter 11
1. Wong, J, Terramechanics and off-road vehicles, Elsevier, 1989
1. Wong and Preston-Thomas, A comparison between a conventional method and an improved one for predicting tracked vehicle performance, Proc Conf of Int Soc of Terrain Vehicle Systems, Vol 1, 361-380, 1985
2. Bekker, M, National Research Council of Canada Report 24146, 1985
3. Bekker, M, Off-the-road locomotion, University of Michigan Press, 1960
4. Bekker, M, Theory of Land Locomotion, University of Michigan Press, 1956
6. Little, L, Modern soil theory and tracked vehicles, *Automotive Design Engineering*, Dec 1962
7. Cleare, Factors affecting the performance of high speed tracklayers, Proc IMechE, 1964

Esmonde, R, Aluminium for earthmoving and rock hauling equipment, paper 11 of the Aluminium in road vehicles conference report, IMechE, 1963
Off-highway vehicles, tractors and equipment, IMechE conference report, 1975
Ageikin, I, Off-the-road mobility of automobiles, AA Balkema, 1987
Maclaurin, B, Progress in British tracked vehicle suspension systems, SAE paper 830442
Dunn, M, Soil mechanics and the wheeled vehicle, *Automotive Design Engineering*, May 1963
Elder, J, Suspensions and tracks, paper 3 of the High technology in defence symposium, Defence Manufacturers Assn of GB, 1981
Liston, Correlation between predicted and actual off-road vehicle performance, SAE paper 670170
Mobile construction plant design conference report, Design Engineering Publications, 1971
Crolla et al, Active suspension control for an off-road vehicle, Proc IMechE, Vol 201, D1, 1987
Parmee and Cernes, High mobility vehicle design, *Journal of Automotive Engineering*, May, July 1972, Feb 1973

Chapter 12
Fothergill et al, Computer-aided concept design of a sports car chassis system, IMechE conference on Vehicle Structures, no 7, 1984
The design and manufacture of low-cost motorised vehicles, Intermediate Technology Publications, 1988
Wilson-Jones, R, Steering and stability of single-track vehicles, Proc IMechE, 1952
Spicer, J, Car aperture design for easy access, vehicles for the disabled session, Autotech 1987
Dunn, M, The Bentley Turbo R, Proc IMechE, vol 201, no. 95, 1987
Steeds, W, Mechanics of road vehicles, Iliffe, 1960

Chapter 13
1. ALCOA publication, Notes on the design of aluminium monocoque semi-trailer transport tanks
2. ALCAN publication, Semi-trailer road tanker design and details
3. Formulas for stress & strain, McGraw Hill
4. Experimental Station Bulletins 108 and 123, University of Utah

5. Draughtsman and Allied Technicians Association Information Sheet Number Nine
6. *Engineering Designer*, December 1966
7. British Aluminium publication, Commercial vehicle body-building in aluminium
8. Proc IMechE, Vol 207, No D4, 1993.
9. IMechE symposium on Hoppers and Silos (S308 Bulk Materials Handling Committee)

Commercial vehicles, engineering and operation, IMechE conference report, 1967
Ed Montgomerie, G, Vehicle engineering handbook, Freight Transport Association, 1990
Bulk materials handling towards the year 2000, IMechE conference report, no 12, 1991
Jenike, A, Gravity flow of bulk solids, Storage and flow of solids, bulletins 108 and 123 of University of Utah Engineering Experimental Station
Spencer et al, New tools from research make bins, hoppers flow, *Materials Handling Engineering*, June 1965
Rees, C, Design of tanks and cylindrical fluid containing vessels, *The Engineering Designer*, Dec 1966
D'Arcy-Smith, F, Bulk handling of powders in the chemical industry, Institute of Materials Handling conference, 1961
Rotter, J, Strength criteria for steel silo structures, IMechE conference paper C418/035

Chapter 14

Recommended specification for buses used to operate local services, Disabled Persons Transport Advisory Committee (DIPTAC), DoT, 1988
Dickson-Simpson, J, Era of articulation, Transport Press services, 1968
Ed Hoelze et al, Utility vehicle design handbook, SAE, 1991
Brooks et al, Passenger problems on moving buses, TRRL/Leyland Vehicles report
Bus 92, The expanding role of buses towards the 21st century, Proc IMechE, no. 2, 1992
Evaluation of unit load handling equipment, IMechE conference report, no. 1, 1973
Webster and McGowan, Planning for the bus, TRRL conference Moving people in cities, 1973
McHugh and Tidbury, The design of touring coach bodywork for safety, IMechE conference paper C200/73
Fenton, J, New concept for delivery vehicles, *Automotive Design Engineering*, Aug 1968
Hales, F, The lateral stability of a simplified articulated vehicle, IMechE symposium on Control of vehicles during braking and cornering, 1963
Coupling articulated vehicles, IRTE Code of Practice, 1986

Chapter 15

1 SAE Paper 830350
2 SAE Paper 850225

EVA manual, Electric Vehicle Association of GB Ltd
Unnewehr and Nasar, Electric vehicle technology, Wiley, 1982
Ed Dorgham, M, Electric and hybrid vehicles, Inderscience Enterprises, 1982
Electric vehicle technology, MIRA seminar repot, 1992
Battery electric and hybrid vehicles, IMechE seminar report, 1992
Ed Lovering, D, Fuel cells, Elsevier, 1989
Copus, A, DC traction motors for electric vehicles, Electric vehicles for Europe, EVA conference report, 1991
Hodkinson, R, Systems approach to realising an effective hybrid EV drive, *Automotive Engineer*, Aug/Sep 1993

Chapter 16

1 *Machine Design*, May to December 1957
2 Ed Giles, C, Engine Design, Chap 3, Valve gear and breathing, Iliffe, 1968
3 Garrett,K, Automotive fuels and fuel systems, Vol 1, Pentech Press, 1991
4 Morrison, Proc IAE, Vol 28,1933
5 Cambel and Jennings, Gas Dynamics, McGraw-Hill Books
6. Beaumont et al, Air fuel ratio control technology for ultra low emission vehicles, Proc ISATA, 93ENO32.
7. Ma, Collings, Hands, Reduction of exhaust emissions during engine cold start by means of an afterburner operating in the hydrogen ignitable regime, Proc EAEC, 1993.
9. AESD booklet, An introduction to heat exchangers

Engine design

Ed Fenton, J, Gasoline engine analysis, Mechanical Engineering Publications, 1986
Smith, P, Exhaust and intake systems, Foulis, 1962
Ed Nelson, P, Transportation noise reference book, Butterworth, 1987
Polak, P, Theory and practice in engineering thermodynamics, Macmillan, 1983
Thomson, W, Fundamentals of automotive engine balance, Mechanical Engineering Publications, 1978
Vehicle thermal management systems, IMechE conference report, 1995
Ed Evans, R, Automotive engine alternatives, Plenum Press, 1986
Taylor, C, The internal combustion engine in theory and practice, MIT Press, 1985
Poole et al, Cosworth MBA engine, SAE paper 920849

Emission and noise control
Worldwide engine emission standards and how to meet them, IMechE conference report, 1993
Tyler, J, The TRRL Quiet Heavy Vehicle project, Proc IMechE, Vol 193, no. 23, 1979
Wunche and Wojik, The new passenger car diesel engine designed for low emission, Graz Engine Symposium, 1993

Engine balancing and mounting
McLernon and Phelps, Balancing the Chevrolet V8 engine, *GM Engineering Journal*, Part 3, 1962
Timpner, F, Design considerations in engine mounting, SAE paper 966B, 1965
Horovitz, M, Suspension of internal-combustion engines in vehicles, Proc IMechE, no. 1, 1957/8
Crankshaft torsional vibration and balance, *Automotive Design Engineering*, Data Sheet 55

Chapter 17
1. SAE Ride and Vibration Data Manual (J6a)
2. Taborek, J, Mechanics of vehicles, Penton Publishing
3. Graphical spring design, The Machinery Publishing Co
4. Vondracek and Nadorff, SAE Paper 820162
5. Number 2 of a spring design series published by the Spring Manufacturers' Research Association
6. Pilbeam & Sharp, Active suspension, Proc IMechE, Vol 207,D4.
7. C Berdah, Proc EAEC, 1993, Pages 159-168
8 Race car vehicle dynamics, Society of Automotive Engineers, 1995

Vehicle ride
Janeway, R, Ride and vibration, HS-6 data sheets, SAE, 1950
Grylls, S, Traction vs stability in passenger cars, *Journal of Automotive Engineering*, May 1972
Lanchester, F, Motor car suspension and independent springing, Proc IAE, 1935-6
Campbell, The sports car, its design and performance, Chapman and Hall, 1954
Cain, Vibration of rail and road vehicles, Pitman, 1949
Ed Tidbury, G, Advances in automobile engineering: vehicle ride, Pergamon, 1962
Le Fevre, The Highway Truck Ride Problem, SAE paper SP-260, 1965
Sylvester, Vibration reduction in motor cars, Society of Environmental Engineers symposium, Volume 4, 1966
Goldwyn, An Introduction to the simulation and measurement of large-amplitude non-linear wheel bounce, ASAE, Cranfield, thesis 1962
Smith, Frame beaming, fifth-wheel location, special body mounting and loading problems, SAE paper SP 260, 1965
Ellis, Vehicle simulation, Society of Environmental Engineers symposium, Volume 4, 1966
Janeway, A better truck ride for driver and cargo — problems and practical solutions, SAE paper SP-154, 1958.
Ungar, Mechanical Design and Systems Handbook, Section 6, McGrawHill, 1965.
Barson et al, Some aspects of tyre and vehicle vibration testing, IMechE (AD) symposium, Instrumentation and test techniques for motor vehicles, 1967.
Mitschke, Influence of road and vehicle dimensions on the amplitude of body motions and dynamic wheel loads, SAE Transactions, 1961.
Ashley and Mills, The laboratory investigation of vehicle vibration, IMechE (AD)/ASAE symposium, Instrumentation and test techniques for motor vehicles, 1967.
Barnacle, Mechanics of automobiles, Pergamon

Suspension geometry and linkages
Ashton, B, The rubber road friendly suspension, Proc IMechE, Tech Note, Vol 206, 1992
Gough and Shearer, Front suspension and tyre wear, Proc IMechE, no. 6, 1955/6
Hartley, J, Suspension design, *Automotive Design Engineering*, May 1973
Bastow, D, Independent rear suspension, Proc IMechE, 1951/2
Fenton, J, Steering geometry, *Automotive Design Engineering*, Feb 1965
Townsend and Eames, Forces in steering linkages, *Automotive Design Engineering*, Jul 1965

Steering-geometry/directional control
IMechE conference report, no. 9, 1988
Ellis, J, Vehicle handling dynamics, Mechanical Engineering Publications, 1994
Ed Giles, G, Steering, suspension and tyres, Iliffe, 1968
Bastow and Howard, Car suspension and handling, Pentech Press, 1987
French, T, Tyre technology, Adam Hilger, 1988
Subtleties of steering, Automotive Products Co, Leamington
Evans, Properties of tyres affecting riding, steering and handling, *SAE Journal*, Feb 1935
Olley, M, Road manners of the modern car, Proc IAE, 1937-8
Bastow, D, Steering problems and layout, Proc IAE, 1937-8
Stonex, Car control factors and their measurement, *SAE Journall*, no. 3, 1941
Bergman, W, The basic nature of vehicle understeer-oversteer, SAE publication 957B
Bundorf et al, Vehicle handling response to aerodynamic

inputs, SAE publication 716B
Hodkin, D, The status of vehicle handling in Great Britain, SAE publication 957A
White, H, Analytical approach to suspension behaviour, the Hillman Imp, *Automotive Design Engineering*, Jun 1963
Milliken et al, Research in automobile stability and control and in tyre performance, Proc IMechE, no. 7, 1956/7
Lind Walker, G, Directional stability, *Automobile Engineer*, Aug 1950
Joy, T, Tyre characteristics as applicable to vehicle stability problems, Proc IMechE, no.6, 1953/4
Ball, A, Tyre behaviour in steering, *SAE Journal*, Aug 1939
Wilson, J, Tyres and the suspension system, IRI conference on Rubber in transport engineering, 1970
Hodkin, R, Vehicle Handling, Chapter 6 in Vol 1 Automotive Technology Series, Iliffe.

Spring design, active and air suspension
Graphical spring design, Yellow Back Series no. 32, Machinery Publishing
The buckling and transverse loading of helical compression springs, Spring Manufacturer's Research Association, 1964
Hodkin, R, A new development in air suspension, Proc FISITA, 1966
Cebon, A, Interaction between heavy vehicles and roads, SAE paper SP-951, 1993
Wright and Williams, The application of active suspension to high performance road vehicles, IMechE conference paper C239/84
Sainsbury, J, Air suspension for road vehicles, Proc IMechE, no. 3, 1957/8
Engineering design with natural rubber, Nat Rubber Producers Res Assn

Chapter 18
1. Rocard, Y, L'instbilite en Mechanique, Masson et Cie, Paris (1954)
2. Steeds, Mechanics of Road Vehicles, Iliffe, London (1960)
3. Dixon, Tyres, Suspension and Handling, Cambridge University Press (1991)
4. Howell, Crosswind response of cars, Proc. SITEV, 1988
5. Taborek, Mechanics of vehicles, Machine Design, Penton, 1957.
6. ATZ (Number 8, 1967),
7. Paper 890869, 1989 SAE congress
8. Newcomb and Spurr, Design series on vehicle braking, *Automotive Design Engineering*, Nov 1968, Feb 1969, Mar 1969
9. Giles, C, Factors influencing the friction between tyre and road under wet conditions, Proc IMechE, Symposium on Control of Vehicles, 1967
10. Abinett, R, Design calculations for drive axles, Automotive Design Engineering, *Jun 1970*
11. Heldt, P, Torque convertors or transmissions, Chilton, 1955

Wheels and hubs
Adams, G, Automotive wheel design, *Automotive Design Engineering*, Feb 1967
Davisson, J, Design and application of commercial type tyres, SAE paper SP-344, 1969
Automobile wheels and tyres, IMechE conference report, no. 12, 1983
Svenson, The working stresses of wheel rims, *ATZ*, no.8, 1967
Automobile hub design, *Automotive Design Engineering*, May 1972

Braking systems and ABS
Traction control and anti-wheel-spin systems for road vehicles, IMechE conference report, no. 6, 1988
Disc brakes for commercial vehicles, IMechE conference report, no. 10, 1988
Braking of road vehicles, IMechE conference report, no. 2, 1993
Wroten, C, Pneumatic line losses, *Machine Design*, Dec 11 1969
Resistance to fluid flow through pipes, *Automotive Design Engineering*, Nov 1964
Clapson, B, Ultra low profile tyres, *Journal of Automotive Engineering*, Jun 1970

Driveline and drive axles
Schmelz et al, Universal joints and driveshafts, Springer Verlag, 1991
Neale et al, Couplings and shaft alignment, Mechanical Engineering Publications, 1991
Driveline 84, IMechE conference report, 1984
Heldt, P, Automotive chassis, Chilton, 1955
Ruegg, C, Carbon and aramid reinforced plastics in the manufacture of automotive propeller shafts, VDI

Chapter 19
1 Tuplin, Gear load capacity, Pitman, 1961
2 Giles, J, Gears and transmissions, Iliffe, 1969
3 Giles, J, Transmission ratio requirements of road vehicles, MIRA report 1957-54.
4 Taureg, H, Calculation of torque in the viscous mode, SITEV conference report, 1990

Macielinski, J, The design and selection of universal joints, *Journal of Automotive Engineering*, May-June 1970
Wood, M, Hooke's coupling analysis, *Automotive Design Engineering*, Jun-Jul 1970

Davies, T, Connecting skew shafts, *Automotive Design Engineering*, May 1965
Knaggs, M, Final drive bearings, *Automotive Design Engineering*, May 1967
Ramachandra Rao, G, Clutch engagement characteristics, *Automotive Design Engineering*, Jun 1968
Burton, R, Diaphragm-spring clutch design, *Automotive Design Engineering*, Aug 1967

Transmission gearboxes and bearings

Knaggs, M, Ball & roller bearings for gearboxes, *Automotive Design Engineering*, Dec 1965
Jacobson, M, Gear design, lessons from failures, *Automotive Design Engineering*, Aug 1969-Jan 1970
Stott, T, Problems in the design and development of an economical automobile gearbox, Proc IMechE, no. 1, 1953/4
Ed Carter, N, Advances in automobile engineering: automatic transmissions, Pergamon, 1964
Stokes, A, Manual gearbox design, Butterworth Heinemann, 1992
Abinett, R, Important design criteria for a 5-speed automotive gearbox, *The Engineering Designer*, Nov, Dec 1963
Radcliffe, T, Gearing and gearboxes, Automobile Engineers Reference Book, Newnes
Machine cut helical and spur gears', BS 436: 1940
Dean-Averns, R, Automobile Chassis Design, Iliffe
Buckingham, E, Dynamic loads on gear teeth, ASME Special Research Report, 1931.
Abbott, W, Machine drawing and design, Blackie
Tuplin, W, Dynamic loads on gear teeth, *Machine Design*, October 1953
Shipley, E, Dudley's Gear Handbook, chapter 14, McGraw Hill
Tuplin, W, Mechanical Design and Systems Handbook, section 32, McGraw Hill
Mitchell and Wildig, Synchromesh mechanisms, *Automotive Design Engineering*, Feb 1966
Riddle, E, Manual gearboxes for trucks, session 1, paper 5, IMechE Conference on driveline engineering, 1970

Automatic gearboxes and hydrostatic drives

Saward, J, Survey of infinitely variable speed transmissions, *Machine Design Engineering*, Aug-Dec, 1963
Lamburn and Whateley, Epicyclic gear trains speed ratio determination, *Automotive Design Engineering*, Dec 1963
Automatic and semiautomatic gearboxes for heavy commercial vehicles, IMechE conference report, no. 4, 1978
Hydrostatic transmissions for vehicle applications, IMechE conference report, no. 8, 1981
Giles, J, Shunt hydrostatic transmissions, IMechE (AD) conference on Drive line engineering report, 1970
Thoma, H, High-pressure hydraulic power transmission, Proc IMechE, 1957

Index